Lecture Notes in Computer Science 3580

Commenced Publication in 1973
Founding and Former Series Editors:
Gerhard Goos, Juris Hartmanis, and Jan van Leeuwen

Editorial Board

David Hutchison
  Lancaster University, UK
Takeo Kanade
  Carnegie Mellon University, Pittsburgh, PA, USA
Josef Kittler
  University of Surrey, Guildford, UK
Jon M. Kleinberg
  Cornell University, Ithaca, NY, USA
Friedemann Mattern
  ETH Zurich, Switzerland
John C. Mitchell
  Stanford University, CA, USA
Moni Naor
  Weizmann Institute of Science, Rehovot, Israel
Oscar Nierstrasz
  University of Bern, Switzerland
C. Pandu Rangan
  Indian Institute of Technology, Madras, India
Bernhard Steffen
  University of Dortmund, Germany
Madhu Sudan
  Massachusetts Institute of Technology, MA, USA
Demetri Terzopoulos
  New York University, NY, USA
Doug Tygar
  University of California, Berkeley, CA, USA
Moshe Y. Vardi
  Rice University, Houston, TX, USA
Gerhard Weikum
  Max-Planck Institute of Computer Science, Saarbruecken, Germany

Luís Caires   Giuseppe F. Italiano
Luís Monteiro   Catuscia Palamidessi
Moti Yung (Eds.)

Automata, Languages and Programming

32nd International Colloquium, ICALP 2005
Lisbon, Portugal, July 11-15, 2005
Proceedings

Volume Editors

Luís Caires
Universidade Nova de Lisboa, Departamento de Informatica
2829-516 Caparica, Portugal
E-mail: Luis.Caires@di.fct.unl.pt

Giuseppe F. Italiano
Universitá di Roma "Tor Vergata"
Dipartimento di Informatica, Sistemi e Produzione
Via del Politecnico 1, 00133 Roma, Italy
E-mail: italiano@disp.uniroma2.it

Luís Monteiro
Universidade Nova de Lisboa, Departamento de Informatica
2829-516 Caparica, Portugal
E-mail: lm@di.fct.unl.pt

Catuscia Palamidessi
INRIA Futurs and LIX, École Polytechnique
rue de Saclay, 91128 Palaiseau, France
E-mail: catuscia@lix.polytechnique.fr

Moti Yung
RSA Laboratories and Columbia University
Computer Science Department
1214 Amsterdam Av., New York, NY 10027, USA
E-mail: moti@cs.columbia.edu

Library of Congress Control Number: 2005928673

CR Subject Classification (1998): F, D, C.2-3, G.1-2, I.3, E.1-2

ISSN   0302-9743
ISBN-10   3-540-27580-0 Springer Berlin Heidelberg New York
ISBN-13   978-3-540-27580-0 Springer Berlin Heidelberg New York

This work is subject to copyright. All rights are reserved, whether the whole or part of the material is concerned, specifically the rights of translation, reprinting, re-use of illustrations, recitation, broadcasting, reproduction on microfilms or in any other way, and storage in data banks. Duplication of this publication or parts thereof is permitted only under the provisions of the German Copyright Law of September 9, 1965, in its current version, and permission for use must always be obtained from Springer. Violations are liable to prosecution under the German Copyright Law.

Springer is a part of Springer Science+Business Media

springeronline.com

© Springer-Verlag Berlin Heidelberg 2005
Printed in Germany

Typesetting: Camera-ready by author, data conversion by Scientific Publishing Services, Chennai, India
Printed on acid-free paper   SPIN: 11523468   06/3142   5 4 3 2 1 0

Preface

The 32nd International Colloquium on Automata, Languages and Programming (ICALP 2005) was held in Lisbon, Portugal from July 11 to July 15, 2005. These proceedings contain all contributed papers presented at ICALP 2005, together with the papers by the invited speakers Giuseppe Castagna (ENS), Leonid Libkin (Toronto), John C. Mitchell (Stanford), Burkhard Monien (Paderborn), and Leslie Valiant (Harvard). The program had an additional invited lecture by Adi Shamir (Weizmann Institute) which does not appear in these proceedings.

ICALP is a series of annual conferences of the European Association for Theoretical Computer Science (EATCS). The first ICALP took place in 1972. This year, the ICALP program consisted of the established track A (focusing on algorithms, automata, complexity and games) and track B (focusing on logic, semantics and theory of programming), and innovated on the structure of its traditional scientific program with the inauguration of a new track C (focusing on security and cryptography foundation).

In response to a call for papers, the Program Committee received 407 submissions, 258 for track A, 75 for track B and 74 for track C. This is the highest number of submitted papers in the history of the ICALP conferences. The Program Committees selected 113 papers for inclusion in the scientific program. In particular, the Program Committee for track A selected 65 papers, the Program Committee for track B selected 24 papers, and the Program Committee for track C selected 24 papers. All the work of the Program Committees was done electronically.

ICALP 2005 was held in conjunction with the Annual ACM International Symposium on Principles and Practice of Declarative Programming (PPDP 2005). Additionally, the following workshops were held as satellite events of ICALP 2005: the 2nd Workshop on Automated Reasoning for Security Protocol Analysis (ARSPA), the 1st International Workshop on Verification of COncurrent Systems with dynaMIC Allocated Heaps (COSMICAH), the 1st International Workshop on New Developments in Computational Models (DCM), the 4th International Workshop on Parallel and Distributed Methods in Verification (PDMC), the 4th International Workshop on Proof Theory, Computation, Complexity (PCC), the Workshop on Structures and Deduction — The Quest for the Essence of Proofs (DS), the 2nd Workshop on Structural Operational Semantics (SOS), and the Workshop on Semigroups and Automata (WSA).

We wish to thank all authors who submitted papers for consideration, the Program Committees for their hard work, as well as the external reviewers who assisted the Program Committees in the evaluation process.

We thank the sponsors and the Gulbenkian Foundation of Lisbon for hosting ICALP 2005. We are also grateful to the Department of Informatics of the

Faculty of Sciences and Technology, New University of Lisbon, in particular the administrative office and the technical support service.

Last but not least, we would like to thank Andrei Voronkov for providing the conference management software EasyChair. It was of great help in handling the submissions and the electronic PC meeting.

Luís Caires
Giuseppe F. Italiano
Luís Monteiro
Catuscia Palamidessi
Moti Yung

Organization

Program Committee

Track A

Lars Arge, Duke University, USA
Giorgio Ausiello, University of Rome "La Sapienza", Italy
Surender Baswana, Max-Planck-Institut für Informatik, Saarbrücken, Germany
Hans Bodlaender, University of Utrecht, The Netherlands
Véronique Bruyère, University of Mons-Hainaut, Belgium
Adam Buchsbaum, AT&T Labs Research, USA
Josep Diaz, Universitat Politècnica de Catalunya, Spain
David Eppstein, University of Irvine, USA
Andrew Goldberg, Microsoft, USA
Monika Henzinger, Google and ETH Lausanne, Switzerland
Giuseppe F. Italiano, University of Rome "Tor Vergata", Italy (Chair)
Marios Mavronicolas, University of Cyprus, Cyprus
Peter Bro Miltersen, University of Aarhus, Denmark
Mike Paterson, University of Warwick, UK
Dominique Perrin, Université de Marne la Vallée, France
Seth Pettie, Max-Planck-Institut für Informatik, Saarbrücken, Germany
Yuval Rabani, Technion, Israel
Antonio Restivo, University of Palermo, Italy
José Rolim, University of Geneva, Switzerland
Dorothea Wagner, University of Karlsruhe, Germany
Tandy Warnow, University of Texas at Austin, USA
Christos Zaroliagis, CTI and University of Patras, Greece

Track B

Kenichi Asai, Ochanomizu University, Japan
Jos Baeten, Eindhoven University of Technology, The Netherlands
Peter Buneman, University of Edinburgh, UK
Zoltan Esik, University of Szeged, Hungary and Rovira University, Spain
Javier Esparza, University of Stuttgart, Germany
Marcelo Fiore, Christ's College and University of Cambridge, UK
Manuel Hermenegildo, Universidad Politècnica de Madrid, Spain
Delia Kesner, Université Paris VII, France
Kim Guldstrand Larsen, University of Aalborg, Denmark
Gopalan Nadathur, University of Minnesota, USA

Uwe Nestmann, EPFL, Switzerland
Catuscia Palamidessi, INRIA, France (Chair)
Amr Sabry, University of Indiana, USA
Davide Sangiorgi, Università di Bologna, Italy
Roberto Segala, Università di Verona, Italy
Harald Søndergaard, University of Melbourne, Australia

Track C

David Basin, ETH Zurich, Switzerland
Christian Cachin, IBM Research, Switzerland
Alfredo De Santis, Università di Salerno, Italy
Cynthia Dwork, Microsoft Research, USA
Matt Franklin, U.C. Davis, USA
Michael Goodrich, U.C. Irvine, USA
Andrew D. Gordon, Microsoft Research, UK
Roberto Gorrieri, Università di Bologna, Italy
Yuval Ishai, Technion, Israel
Phil MacKenzie, DoCoMo Labs, USA
Tatsuaki Okamoto, NTT Labs, Japan
David Pointcheval, ENS Paris, France
Tal Rabin, IBM Research, USA
Omer Reingold, Weizmann Institute, Israel
Adi Rosen, Technion, Israel
Amit Sahai, UCLA, USA
Andre Scedrov, University of Pennsylvania, USA
Igor Shparlinski, Macquarie University, Australia
Nigel Smart, University of Bristol, UK
Moti Yung, Columbia University and RSA Laboratories, USA (Chair)

Organizing Committee

Luís Caires, Conference Co-chair
Luís Monteiro, Conference Co-chair
António Ravara, Workshops Co-chair
Vasco Vasconcelos, Workshops Co-chair
Margarida Mamede
João Costa Seco
José Pacheco

List of External Referees

Track A

Karen Aardal
Scott Aaronson
Saurabh Aggarwal
Marjan van den Akker
Cyril Allauzen
Jean-Paul Allouche
Luca Allulli
Carme Alvarez
Andris Ambainis
Marcella Anselmo
Sanjeev Arora
Albert Atserias
Vincenzo Auletta
Jose Balcazar
Jeremy Barbay
Amotz Bar-Noy
Tugkan Batu
Michael Baur
Marie-Pierre Béal
Luca Becchetti
Philip Bille
Yvonne Bleischwitz
Maria J. Blesa
Avrim Blum
Luc Boasson
Vincenzo Bonifaci
Paola Bonizzoni
Vasco Brattka
Gerth Stølting Brodal
Peter Buergisser
Harry Buhrman
Luciana S. Buriol
Costas Busch
Cristian S. Calude
Massimiliano Caramia
Jean Cardinal
Olivier Carton
Patrick Cegielski
Julien Cervelle
J.-M. Champarnaud
Sunil Chandran

Moses Charikar
Hubie Chen
Joseph Cheriyan
Janka Chlebikova
Bogdan Chlebus
Christian Choffrut
George Christodoulou
Serafino Cicerone
Julien Clément
Andrea Clementi
Eric de La Clergerie
Bruno Codenotti
Edith Cohen
Anne Condon
Pier Francesco Cortese
Stefano Crespi-Reghizzi
Peter Damaschke
Fabrizio d'Amore
Camil Demetrescu
Kedar Dhamdhere
Christoph Dorr
Petros Drineas
Christoph Durr
Stephan Eidenbenz
Amr Elmasry
Thomas Erlebach
Alex Fabrikant
Rolf Fagerberg
Jacques Farré
Lene Favrholdt
Rainer Feldmann
Stephen A. Fenner
Antonio Fernandez
Henning Fernau
Paolo Ferragina
Jiri Fiala
Irene Finocchi
Fedor Fomin
Lance Fortnow
Dimitris Fotakis
Paolo G. Franciosa

Gudmund Frandsen
Alan Frieze
Andrea Frosini
Marco Gaertler
Martin Gairing
Emden Gansner
Naveen Garg
William Ian Gasarch
Leszek Gasieniec
Georgiadis Georgios
Kostis Georgiou
Arkadeb Ghosal
Dora Giammarresi
Raffaele Giancarlo
Aristides Gionis
Ashish Goel
Paul Golberg
Robert Görke
Fabrizio Grandoni
Serge Grigorieff
Alexander Grigoriev
Joachim Gudmundsson
Rachid Guerraoui
Dan Gusfield
Gus Gutoski
M. Hajiaghayi
Magnus M. Halldorsson
Kristoffer Hansen
Sariel Har-Peled
Ramesh Hariharan
Herman Haverkort
Illya V. Hicks
Mika Hirvensalo
John Hitchcock
Martin Holzer
Han Hoogeveen
Peter Hoyer
Juraj Hromkovic
Cor Hurkens
Lucian Ilie
Costas Iliopoulos

Piotr Indyk
Garud Iyengar
Kamal Jain
Petr Jančar
Klaus Jansen
Mark Jerrum
David Johnson
Adrian Johnstone
Marcin Jurdzinski
Erich Kaltofen
Juhani Karhumäki
Anna Karlin
Marek Karpinski
Claire Kenyon
Richard Kenyon
Iordanis Kerenidis
Leonid Khachiyan
Rohit Khandekar
Pekka Kilpelainen
Lefteris Kirousis
Ralf Klasing
Rolf Klein
Bettina Klinz
Adam Klivans
Pascal Koiran
Jochen Konemann
Spyros Kontogiannis
Guy Kortsarz
Arie Koster
Manolis Koubarakis
Elias Koutsoupias
Daniel Kral
Evangelos Kranakis
Dieter Kratsch
Michael Krivelevich
Ravi Kumar
Viraj Kumar
Dietrich Kuske
Shay Kutten
Gregory Lafitte
Jens Lagergren
Sophie Laplante
Michel Latteux
Luigi Laura
Van Bang Le

Thierry Lecroq
Stefano Leonardi
Pierre Leone
Xiang-Yang Li
Paolo Liberatore
Christian Liebchen
Michael Loizos
Thomas Luecking
George Lueker
Alejandro Maas
Marina Madonia
Malik Magdon-Ismail
Frederic Magniez
Mohammad Mahdian
Christos Makris
Sebastian Maneth
Alberto Spaccamela
Maurice Margenstern
Vangelis Markakis
Chip Martel
Giancarlo Mauri
Jacques Mazoyer
Pierre McKenzie
Frank McSherry
Steffen Mecke
Dieter van Melkebeek
Carlo Mereghetti
Wolfgang Merkle
Ramgopal Mettu
Ulrich Meyer
Dimitrios Michail
Christian Michaux
Filippo Mignosi
Vahab Mirrokni
Michael Mitzenmacher
Shuichi Miyazaki
Kousha MoaveniNejad
Mehryar Mohri
Burkhard Monien
Cris Moore
Shlomo Moran
Burkhard Morgenstern
Kenichi Morita
Gabriel Moruz
Thomas Moscibroda

Philippe Moser
Anca Muscholl
Umberto Nanni
Konstantinos Nedas
Mark-Jan Nederhof
Jaroslav Nesetril
Frank Neven
Sotiris Nikoletseas
John Noga
Rasmus Pagh
Jakob Illeborg Pagter
Rina Panigrahy
Anindya Patthak
Christian N.S. Pedersen
David Peleg
Sriram Pemmaraju
Giovanni Pighizzini
Jean-Eric Pin
Giuseppe Pirillo
Nadia Pisanti
Andrzej Proskurowski
J. Radhakrishnan
Harald Raecke
Mathieu Raffinot
Srinivasa Rao
David Rappaport
Jean-François Raskin
S.S. Ravi
John Reif
Jan Reimann
Omer Reingold
Eric Rémila
Christophe Reutenauer
Michel Rigo
Adi Rosen
Martin Rotteler
Tim Roughgarden
Gilles Roussel
Alexander Russell
Jacques Sakarovitch
Peter Sanders
Pierluigi San Pietro
Miklos Santha
Martin Sauerhoff
Guido Schaefer

Thomas Schank
Christian Schindelhauer
Torsten Schlieder
Anita Schöbel
Sylvain Schmitz
Étienne Schramm
Frank Schulz
Elizabeth Scott
Luc Segoufin
Helmut Seidl
Pranab Sen
Géraud Sénizergues
Maria Serna
Rocco Servedio
Jeffrey Shallit
Micha Sharir
Peter Shor
Riccardo Silvestri
Alistair Sinclair
Spiros Skiadopoulos
Martin Skutella
Roberto Solis-Oba
Robert Spalek
Klaus Ambos Spies

Paul Spirakis
Venkatesh Srinivasan
Ludwig Staiger
Yannis Stamatiou
Cliff Stein
David Steurer
Leen Stougie
Howard Straubing
Martin Strauss
K.S. Sudeep
Peng Sun
Maxim Sviridenko
Mario Szegedy
Claude Tadonki
Kunal Talwar
Gerard Tel
Dimitrios Thilikos
Wolfgang Thomas
Karsten Tiemann
Luca Trevisan
Panayiotis Tsaparas
Kostas Tsichlas
Marc Uetz
Ugo Vaccaro

Kasturi Varadarajan
Vijay V. Vazirani
S. Venkatasubramanian
Adrian Vetta
Eric Vigoda
Emanuele Viola
Rakesh V. Vohra
Heribert Vollmer
Nicolai Vorbjov
Osamu Watanabe
Pascal Weil
Klaus Wich
Peter Widmayer
Jef Wijsen
Gerhard Woeginger
Alexander Wolff
Deng Xiaotie
Hiroaki Yamamoto
Mihalis Yannakakis
Norbert Zeh
Li Zhang
Wieslaw Zielonka
Uri Zwick

Track B

Elvira Albert
Thorsten Altenkirch
Rajeev Alur
Sergio Antoy
André Arnold
Benjamin Aziz
Brian Babcock
James Bailey
Vincent Balat
José Balcázar
Michael Baldamus
Jiri Barnat
Gerd Behrmann
Martin Berger
Jan Bergstra
Luca Bianco
Lars Birkedal
Frédéric Blanqui

Benedikt Bollig
Johannes Borgström
Dragan Bošnački
Debora Botturi
Ahmed Bouajjani
Patricia Bouyer
Julian Bradfield
Mario Bravetti
Franck van Breugel
Sébastien Briais
Geoffrey Brown
Glenn Bruns
Antonio Bucciarelli
Francisco Bueno
Nadia Busi
Luís Caires
Cristiano Calcagno
Manuel Campagnolo

Manuel Carro
D. Caucal
Witold Charatonik
Krishnendu Chatterjee
Chiyan Chen
James Cheney
Tom Chothia
Horatiu Cirstea
Rance Cleaveland
John Cochran
Thomas Colcombet
Andrea Corradini
Flavio Corradini
Alin Deutsch
Silvano Dal-Zilio
Vincent Danos
Alexandre David
Anuj Dawar

Soeren Debois
Yuxin Deng
M. Dezani-Ciancaglini
Volker Diekert
Rachid Echahed
Norm Ferns
Thomas Hildebrandt
Matthew Flatt
Cédric Fournet
Michael Franssen
Fabio Gadducci
Jacques Garrigue
Floris Geerts
Blaise Genest
Dan R. Ghica
Rob van Glabbeek
Patrice Godefroid
Jan Friso Groote
Sudipto Guha
Vesa Halava
James Harland
Russ Harmer
Tobias Heindel
Holger Hermanns
Thomas Hildebrandt
Kees Huizing
Hans Hüttel
Atsushi Igarashi
Jacob Illum Rasmussen
Anna Ingólfsdóttir
Radha Jagadeesan
Achim Jung
Marcin Jurdziński
Yukiyoshi Kameyama
Deepak Kapur
Claude Kirchner
Christoph Koch
Simon Kramer
Antonín Kučera
Werner Kuich
Ruurd Kuiper
K. Narayan Kumar
Orna Kupferman
Marcos Kurban
Martin Kutrib

Barbara König
Salvatore La Torre
Daniel Leivant
Stéphane Lengrand
Michael Leuschel
Leonid Libkin
Didier Lime
Jim Lipton
Kamal Lodaya
Markus Lohrey
Pedro López
Etienne Lozes
Michael Luttenberger
Bas Luttik
Angelika Mader
A. Maggiolo Schettini
Istvan Majzik
Luc Maranget
Julio Mariño
Hidehiko Masuhara
Sjouke Mauw
Guy McCusker
Paul-André Melliès
Michael Mendler
Massimo Merro
Dale Miller
Kevin Millikin
Alexandre Miquel
Alberto Momigliano
Madhavan Mukund
Anca Muscholl
Anders Møller
Francesco Zappa Nardelli
Damian Niwinski
Dirk Nowotka
Jan Obdržálek
Martin Otto
Matthew Parkinson
Justin Pearson
Simon Peyton Jones
Frank Pfenning
Iain Phillips
Sophie Pinchinat
G. Michele Pinna
François Pottier

Marc Pouzet
John Power
Germán Puebla
Jean-François Raskin
Anders Ravn
Henrik Reif Andersen
Didier Rémy
Eike Ritter
Francesca Rossi
Wojciech Rytter
Jean-Paul Sansonnet
Vijay Saraswat
Stefan Schwoon
Géraud Senizergues
Natalia Sidorova
Petr Sosik
Jeremy Sproston
Jiri Srba
Graham Steel
Martin Steffen
Colin Stirling
Oldřich Stražovský
Martin Strecker
Thomas Streicher
Martin Sulzmann
Stephanie Swerich
Paulo Tabuada
Vanessa Teague
P.S. Thiagarajan
Hayo Thielecke
Marc Tommasi
Lorenzo Tortora de Falco
Frank D. Valencia
Dirk Van Gucht
Daniele Varacca
Helmut Veith
Bob Veroff
Alicia Villanueva
Erik de Vink
Walter Vogler
Marc Voorhoeve
Jérôme Vouillon
Roel de Vrijer
David S. Warren
Hiroshi Watanabe

Stephanie Weirich
Joe Wells
J. Winkowski
Anthony Wirth
James Worrell

Eric Van Wyk
Hongwei Xi
Alexander Yakhnis
Mihalis Yannakakis
Dachuan Yu

Hans Zantema
Marc Zeitoun
Wieslaw Zielonka
Pascal Zimmer

Track C

Martín Abadi
Michel Abdalla
Alessandro Acquisti
Saurabh Agarwal
Alessandro Aldini
Giuseppe Ateniese
Michael Backes
Zuzana Beerliova
Kamel Bentahar
Carlo Blundo
Marcello Bonsangue
Xavier Boyen
Marzia Buscemi
Jan Camenisch
Marco Carbone
Dario Catalano
Qi Cheng
Jung Hee Cheon
Mika Cohen
Hubert Comon-Lundh
Scott Contini
Nicolas Courtois
Silvia Crafa
Paolo D'Arco
Stephanie Delaune
Giovanni Di Crescenzo
Pierpaolo Degano
Christophe Doche
Seiji Doi
Paul Hankes Drielsma
Claudiu Duma
Orr Dunkelman
Antonio Durante
Sandro Etalle
Pooya Farshim
Serge Fehr
Sebastian Fischmeister

Riccardo Focardi
Pierre-Alain Fouque
Cédric Fournet
Jessica Fridrich
Martin Gagne
Steven Galbraith
Pierrick Gaudry
Rosario Gennaro
Craig Gentry
Rob Granger
Claudio Guidi
Shai Halevi
Amir Herzberg
Omer Horvitz
Markus Jakobsson
Marc Joye
Bruce Kapron
Hartmut Klauck
Ralf Kuesters
Sebastien Kunz-Jacques
Eyal Kushilevitz
Peeter Laud
Kristin Lauter
Peter Leadbitter
Shiyong Lu
Ben Lynn
Anna Lysyanskaya
Matteo Maffei
Toshiaki Makita
John Malone-Lee
Heiko Mantel
Barbara Masucci
Alexander May
Willi Meier
Phong Nguyen
Jesper Buus Nielsen
Kobbi Nissim

Dan Page
Enes Pasalic
Rafael Pass
Kenny Paterson
Manas Patra
Erez Petrank
Duong Hieu Phan
Krzysztof Pietrzak
Benny Pinkas
Alexander Pretschner
Zulfikar Ramzan
Oded Regev
Leonid Reyzin
Mike Roe
Alon Rosen
Sabina Rossi
Michael Scott
Andrei Serjantov
Ronen Shaltiel
Vitaly Shmatikov
Christoph Sprenger
Martijn Stam
Pante Stanica
Ron Steinfeld
Jacques Stern
Koutarou Suzuki
Tamir Tassa
Yael Tauman-Kalai
Luca Trevisan
A. Troina
Luca Vigano
Ivan Visconti
Bogdan Warinschi
Brent Waters
Diego Zamboni

Sponsors

Fundação para a Ciência e Tecnologia, Ministério da Ciência e Ensino Superior
Centro de Informática e Tecnologias da Informação/FCT/UNL
Centro de Lógica e Computação/IST/UTL

Table of Contents

Invited Lectures

Holographic Circuits
    Leslie G. Valiant 1

Probabilistic Polynomial-Time Semantics for a Protocol Security Logic
    Anupam Datta, Ante Derek, John C. Mitchell, Vitaly Shmatikov,
    Mathieu Turuani 16

A Gentle Introduction to Semantic Subtyping
    Giuseppe Castagna, Alain Frisch 30

Logics for Unranked Trees: An Overview
    Leonid Libkin 35

Nash Equilibria, the Price of Anarchy and the Fully Mixed Nash
Equilibrium Conjecture
    Martin Gairing, Thomas Lücking, Burkhard Monien,
    Karsten Tiemann 51

Data Structures I

The Tree Inclusion Problem: In Optimal Space and Faster
    Philip Bille, Inge Li Gørtz 66

Union-Find with Constant Time Deletions
    Stephen Alstrup, Inge Li Gørtz, Theis Rauhe, Mikkel Thorup,
    Uri Zwick 78

Optimal In-place Sorting of Vectors and Records
    Gianni Franceschini, Roberto Grossi 90

Towards Optimal Multiple Selection
    Kanela Kaligosi, Kurt Mehlhorn, J. Ian Munro, Peter Sanders 103

Cryptography and Complexity

Simple Extractors via Constructions of Cryptographic Pseudo-random
Generators
    Marius Zimand 115

Bounds on the Efficiency of "Black-Box" Commitment Schemes
Omer Horvitz, Jonathan Katz 128

On Round-Efficient Argument Systems
Hoeteck Wee 140

Computational Bounds on Hierarchical Data Processing with
Applications to Information Security
Roberto Tamassia, Nikos Triandopoulos 153

Data Structures II

Balanced Allocation and Dictionaries with Tightly Packed Constant
Size Bins
Martin Dietzfelbinger, Christoph Weidling 166

Worst Case Optimal Union-Intersection Expression Evaluation
*Ehsan Chiniforooshan, Arash Farzan,
Mehdi Mirzazadeh* 179

Measure and Conquer: Domination – A Case Study
Fedor V. Fomin, Fabrizio Grandoni, Dieter Kratsch 191

Cryptography and Distributed Systems

Optimistic Asynchronous Atomic Broadcast
Klaus Kursawe, Victor Shoup 204

Asynchronous Perfectly Secure Communication over One-Time Pads
Giovanni Di Crescenzo, Aggelos Kiayias 216

Single-Prover Concurrent Zero Knowledge in Almost Constant Rounds
Giuseppe Persiano, Ivan Visconti 228

Graph Algorithms I

LCA Queries in Directed Acyclic Graphs
Miroslaw Kowaluk, Andrzej Lingas 241

Replacement Paths and k Simple Shortest Paths in Unweighted
Directed Graphs
Liam Roditty, Uri Zwick 249

Deterministic Constructions of Approximate Distance Oracles and
Spanners
   Liam Roditty, Mikkel Thorup, Uri Zwick 261

An $\tilde{O}(m^2n)$ Randomized Algorithm to Compute a Minimum Cycle
Basis of a Directed Graph
   Telikepalli Kavitha 273

Security Mechanisms

Basing Cryptographic Protocols on Tamper-Evident Seals
   Tal Moran, Moni Naor 285

Hybrid Trapdoor Commitments and Their Applications
   Dario Catalano, Ivan Visconti 298

On Steganographic Chosen Covertext Security
   Nicholas Hopper 311

Classification of Boolean Functions of 6 Variables or Less with Respect
to Some Cryptographic Properties
   An Braeken, Yuri Borissov, Svetla Nikova, Bart Preneel 324

Graph Algorithms II

Label-Guided Graph Exploration by a Finite Automaton
   Reuven Cohen, Pierre Fraigniaud, David Ilcinkas, Amos Korman,
   David Peleg 335

On the Wake-Up Problem in Radio Networks
   Bogdan S. Chlebus, Leszek Gąsieniec, Dariusz R. Kowalski,
   Tomasz Radzik 347

Distance Constrained Labelings of Graphs of Bounded Treewidth
   Jiří Fiala, Petr A. Golovach, Jan Kratochvíl 360

Optimal Branch-Decomposition of Planar Graphs in $O(n^3)$ Time
   Qian-Ping Gu, Hisao Tamaki 373

Automata and Formal Languages I

NFAs With and Without ε-Transitions
   Juraj Hromkovič, Georg Schnitger 385

On the Equivalence of ℤ-Automata
  Marie-Pierre Béal, Sylvain Lombardy, Jacques Sakarovitch 397

A Tight Linear Bound on the Neighborhood of Inverse Cellular
Automata
  Eugen Czeizler, Jarkko Kari 410

Groupoids That Recognize Only Regular Languages
  Martin Beaudry, François Lemieux, Denis Thérien 421

Signature and Message Authentication

Append-Only Signatures
  Eike Kiltz, Anton Mityagin, Saurabh Panjwani, Barath Raghavan 434

Hierarchical Group Signatures
  Mårten Trolin, Douglas Wikström 446

Designated Verifier Signature Schemes: Attacks, New Security Notions
and a New Construction
  Helger Lipmaa, Guilin Wang, Feng Bao 459

Single-Key AIL-MACs from Any FIL-MAC
  Ueli Maurer, Johan Sjödin 472

Algorithmic Game Theory

The Efficiency and Fairness of a Fixed Budget Resource Allocation Game
  Li Zhang 485

Braess's Paradox, Fibonacci Numbers, and Exponential
Inapproximability
  Henry Lin, Tim Roughgarden, Éva Tardos, Asher Walkover 497

Automata and Logic

Weighted Automata and Weighted Logics
  Manfred Droste, Paul Gastin 513

Restricted Two-Variable FO+MOD Sentences, Circuits and
Communication Complexity
  Pascal Tesson, Denis Thérien 526

Computational Algebra

Suitable Curves for Genus-4 HCC over Prime Fields: Point Counting Formulae for Hyperelliptic Curves of Type $y^2 = x^{2k+1} + ax$
Mitsuhiro Haneda, Mitsuru Kawazoe, Tetsuya Takahashi 539

Solvability of a System of Bivariate Polynomial Equations over a Finite Field
Neeraj Kayal 551

Cache-Oblivious Algorithms and Algorithmic Engineering

Cache-Oblivious Planar Shortest Paths
Hema Jampala, Norbert Zeh 563

Cache-Aware and Cache-Oblivious Adaptive Sorting
Gerth Stølting Brodal, Rolf Fagerberg, Gabriel Moruz 576

Simulated Annealing Beats Metropolis in Combinatorial Optimization
Ingo Wegener 589

On-line Algorithms

Online Interval Coloring and Variants
Leah Epstein, Meital Levy 602

Dynamic Bin Packing of Unit Fractions Items
Wun-Tat Chan, Tak-Wah Lam, Prudence W.H. Wong 614

Reordering Buffer Management for Non-uniform Cost Models
Matthias Englert, Matthias Westermann 627

Security Protocols Logic

Combining Intruder Theories
Yannick Chevalier, Michaël Rusinowitch 639

Computationally Sound Implementations of Equational Theories Against Passive Adversaries
Mathieu Baudet, Véronique Cortier, Steve Kremer 652

Password-Based Encryption Analyzed
Martín Abadi, Bogdan Warinschi 664

Random Graphs

On the Cover Time of Random Geometric Graphs
  Chen Avin, Gunes Ercal 677

On the Existence of Hamiltonian Cycles in Random Intersection Graphs
  Charilaos Efthymiou, Paul G. Spirakis 690

Optimal Cover Time for a Graph-Based Coupon Collector Process
  Nedialko B. Dimitrov, C. Greg Plaxton 702

Stability and Similarity of Link Analysis Ranking Algorithms
  Debora Donato, Stefano Leonardi, Panayiotis Tsaparas 717

Concurrency I

Up-to Techniques for Weak Bisimulation
  Damien Pous 730

Petri Algebras
  Eric Badouel, Jules Chenou, Goulven Guillou 742

A Finite Basis for Failure Semantics
  Wan Fokkink, Sumit Nain 755

Spatial Logics for Bigraphs
  Giovanni Conforti, Damiano Macedonio, Vladimiro Sassone 766

Encryption and related Primitives

Completely Non-malleable Schemes
  Marc Fischlin 779

Boneh-Franklin Identity Based Encryption Revisited
  David Galindo 791

Single-Database Private Information Retrieval with Constant Communication Rate
  Craig Gentry, Zulfikar Ramzan 803

Concurrent Zero Knowledge in the Public-Key Model
  Giovanni Di Crescenzo, Ivan Visconti 816

Approximation Algorithms I

A Faster Combinatorial Approximation Algorithm for Scheduling
Unrelated Parallel Machines
  Martin Gairing, Burkhard Monien, Andreas Woclaw 828

Polynomial Time Preemptive Sum-Multicoloring on Paths
  Annamária Kovács 840

The Generalized Deadlock Resolution Problem
  Kamal Jain, MohammadTaghi Hajiaghayi, Kunal Talwar 853

Facility Location in Sublinear Time
  Mihai Bădoiu, Artur Czumaj, Piotr Indyk, Christian Sohler 866

Games

The Complexity of Stochastic Rabin and Streett Games
  Krishnendu Chatterjee, Luca de Alfaro, Thomas A. Henzinger 878

Recursive Markov Decision Processes and Recursive Stochastic Games
  Kousha Etessami, Mihalis Yannakakis 891

Decidability in Syntactic Control of Interference
  James Laird 904

Idealized Algol with Ground Recursion, and DPDA Equivalence
  Andrzej Murawski, Chin-Hao Luke Ong, Igur Walukiewicz 917

Approximation Algorithms II

From Primal-Dual to Cost Shares and Back: A Stronger LP Relaxation
for the Steiner Forest Problem
  Jochen Könemann, Stefano Leonardi, Guido Schäfer,
  Stefan van Zwam 930

How Well Can Primal-Dual and Local-Ratio Algorithms Perform?
  Allan Borodin, David Cashman, Avner Magen 943

Approximating Max kCSP – Outperforming a Random Assignment
with Almost a Linear Factor
  Gustav Hast 956

Lower Bounds

On Dynamic Bit-Probe Complexity
Corina E. Pătraşcu, Mihai Pătraşcu 969

Time-Space Lower Bounds for the Polynomial-Time Hierarchy on Randomized Machines
Scott Diehl, Dieter van Melkebeek 982

Lower Bounds for Circuits with Few Modular and Symmetric Gates
Arkadev Chattopadhyay, Kristoffer Arnsfelt Hansen 994

Probability

Discrete Random Variables over Domains
Michael W. Mislove 1006

An Accessible Approach to Behavioural Pseudometrics
Franck van Breugel, Claudio Hermida, Michael Makkai, James Worrell 1018

Noisy Turing Machines
Eugene Asarin, Pieter Collins 1031

Approximation Algorithms III

A Better Approximation Ratio for the Vertex Cover Problem
George Karakostas 1043

Stochastic Steiner Trees Without a Root
Anupam Gupta, Martin Pál 1051

Approximation Algorithms for the Max-coloring Problem
Sriram V. Pemmaraju, Rajiv Raman 1064

Automata and Formal Languages II

Tight Lower Bounds for Query Processing on Streaming and External Memory Data
Martin Grohe, Christoph Koch, Nicole Schweikardt 1076

Decidability and Complexity Results for Timed Automata via Channel Machines
Parosh Aziz Abdulla, Johann Deneux, Joël Ouaknine, James Worrell 1089

Congruences for Visibly Pushdown Languages
Rajeev Alur, Viraj Kumar, P. Madhusudan, Mahesh Viswanathan ... 1102

Approximation Algorithms IV

Approximation Algorithms for Euclidean Group TSP
Khaled Elbassioni, Aleksei V. Fishkin, Nabil H. Mustafa, René Sitters 1115

Influential Nodes in a Diffusion Model for Social Networks
David Kempe, Jon Kleinberg, Éva Tardos 1127

An Optimal Bound for the MST Algorithm to Compute Energy Efficient Broadcast Trees in Wireless Networks
Christoph Ambühl 1139

New Approaches for Virtual Private Network Design
Friedrich Eisenbrand, Fabrizio Grandoni, Gianpaolo Oriolo, Martin Skutella 1151

Algebraic Computation and Communication Complexity

Hadamard Tensors and Lower Bounds on Multiparty Communication Complexity
Jeff Ford, Anna Gál 1163

Lower Bounds for Lovász-Schrijver Systems and Beyond Follow from Multiparty Communication Complexity
Paul Beame, Toniann Pitassi, Nathan Segerlind 1176

On the l-Ary GCD-Algorithm in Rings of Integers
Douglas Wikström 1189

Concurrency II

A Fully Abstract Encoding of the π-Calculus with Data Terms
Michael Baldamus, Joachim Parrow, Björn Victor 1202

Orthogonal Extensions in Structural Operational Semantics
  Mohammad Reza Mousavi, Michel A. Reniers 1214

Basic Observables for a Calculus for Global Computing
  Rocco De Nicola, Daniele Gorla, Rosario Pugliese 1226

Compositional Verification of Asynchronous Processes via Constraint Solving
  Giorgio Delzanno, Maurizio Gabbrielli 1239

String Matching and Computational Biology

Optimal Spaced Seeds for Faster Approximate String Matching
  Martin Farach-Colton, Gad M. Landau, S. Cenk Sahinalp, Dekel Tsur 1251

Fast Neighbor Joining
  Isaac Elias, Jens Lagergren 1263

Randomized Fast Design of Short DNA Words
  Ming-Yang Kao, Manan Sanghi, Robert Schweller 1275

Quantum Complexity

A Quantum Lower Bound for the Query Complexity of Simon's Problem
  Pascal Koiran, Vincent Nesme, Natacha Portier 1287

All Quantum Adversary Methods Are Equivalent
  Robert Špalek, Mario Szegedy 1299

Quantum Complexity of Testing Group Commutativity
  Frédéric Magniez, Ashwin Nayak 1312

Analysis and Verification

Semantic-Based Code Obfuscation by Abstract Interpretation
  Mila Dalla Preda, Roberto Giacobazzi 1325

About Hoare Logics for Higher-Order Store
  Bernhard Reus, Thomas Streicher 1337

The Polyranking Principle
  Aaron R. Bradley, Zohar Manna, Henny B. Sipma 1349

Geometry and Load Balancing

Approximate Guarding of Monotone and Rectilinear Polygons
  Bengt J. Nilsson 1362

Linear Time Algorithms for Clustering Problems in Any Dimensions
  Amit Kumar, Yogish Sabharwal, Sandeep Sen 1374

Dynamic Diffusion Load Balancing
  Petra Berenbrink, Tom Friedetzky, Russell Martin 1386

Concrete Complexity and Codes

On the Power of Random Bases in Fourier Sampling: Hidden Subgroup Problem in the Heisenberg Group
  Jaikumar Radhakrishnan, Martin Rötteler, Pranab Sen 1399

On the Hardness of Embeddings Between Two Finite Metrics
  Matthew Cary, Atri Rudra, Ashish Sabharwal 1412

Improved Lower Bounds for Locally Decodable Codes and Private Information Retrieval
  Stephanie Wehner, Ronald de Wolf 1424

Model Theory and Model Checking

Preservation Under Extensions on Well-Behaved Finite Structures
  Albert Atserias, Anuj Dawar, Martin Grohe 1437

Unsafe Grammars and Panic Automata
  Teodor Knapik, Damian Niwiński, Paweł Urzyczyn, Igor Walukiewicz 1450

Signaling P Systems and Verification Problems
  Cheng Li, Zhe Dang, Oscar H. Ibarra, Hsu-Chun Yen 1462

Author Index 1475

Holographic Circuits

Leslie G. Valiant*

Division of Engineering and Applied Sciences,
Harvard University, Cambridge,
MA 02138, USA

Abstract. Holographic circuits are defined here to be circuits in which information is represented as linear superpositions. Holographic circuits when suitably formulated can be emulated on classical computers in polynomial time. The questions we investigate are those of characterizing the complexity classes of computations that can be expressed by holographic circuits.

1 Introduction

A holographic reduction [V04] between two computational problems is a reduction that preserves the sum of the solutions without preserving any correspondences among the individual solutions. A polynomial time holographic algorithm is an algorithm derived by applying a polynomial time holographic reduction to a problem having a classical polynomial time algorithm. Such holographic algorithms have been derived for several counting problems for which no polynomial time algorithms had been known before. This previous work used the Fisher-Kasteleyn-Temperley (FKT) algorithm for counting perfect matchings in planar graphs as the starting classical algorithm, which itself may be viewed as a classical reduction to the problem of computing the Pfaffian of an antisymmetric matrix.

This present paper develops holographic computation in two directions. First, the object of evaluation is now a computation or circuit, rather than a combinatorial problem. Second, it uses the Pfaffian itself as the starting point, and bypasses the FKT construction and the constraint of planarity.

2 Universal Operations Sets for Complexity Classes

Holographic circuits offer a new approach to computation in the following sense. Suppose we view a conventional computation as an acyclic circuit embedded in a plane. Then one possible listing of the atomic constituents of computation is:

---

* This research was supported in part by grants NSF-CCR-03-10882, NSF-CCR-98-77049, NSF-CCF-04-27129 and by the National Security Agency (NSA) anad Advanced Research and Development Activity (ARDA) under Army Research Office (ARO) contract DAAD19-01-1-0506.

(i) Inputs of constants 0 or 1,
(ii) Boolean gates AND, OR, and NOT,
(iii) Fanout or replication,
(iv) Crossovers of wires in the planar embedding,

For nondeterministic computation we would add:

(v) Nondeterministic generation of 0 or 1 as inputs, and
(vi) Summation of circuit values over all generated input vectors.

In the holographic framework the sum (vi) over all possible computations of a circuit will be, by design, computable in polynomial time. This becomes meaningful if (v), or some similar nondeterministic operation, is available, since that offers polynomial time simulation of nondeterministic computations. In classical computation, of course, (i)–(iv) are trivial, while nondeterministic summation (vi) appears highly problematic. In the holographic formulation, it is the other way around: nondeterministic summation (vi) is easy but (i)–(iv), or (v), may be problematic. As we shall show, we can realize various subsets of (i)–(v) consistently. If all five could be so realized then $P^{\#P} = P$ would follow. By consistently we mean here that they can be realized with a *common basis*, as explained in the following sections.

We note that there are numerous other possible listings of the atomic constituents of computation besides the one above. For complexity classes X, Y we define such a *universal operations set for X with respect to Y*, (or a *UOS for X with respect to Y*) to be a list of operations such that if they can be realized in a common basis then $X = Y$ is implied. In this paper we shall restrict attention to and assume the case that $Y = NC2$. It is clearly interesting then to investigate such UOS's for $X = NP$, $X = \oplus P$, and $X = \#P$, among others.

The present paper is therefore oriented towards the simulation of computational circuits as opposed to the solution of particular combinatorial problems. We have to emphasize that known algebraic relations, the matchgate indentities [V02a, V02b], present obstacles to some of the more obvious attempts at encoding general classes of computations, such as NP or BQP, in this manner at least when using the smallest bases. However, the question of whether these obstacles can be overcome, perhaps by means of larger bases, remains open.

3 Matchgates and Matchcircuits

We start with some standard graph-theoretic notions and their relation to the Pfaffian of a matrix [BR91, M00]. We then go on to define the notions of PfaffianSum, matchgates, character matrices, and matchcircuits, closely following the treatment in [V02a, V02b].

A weighted undirected graph, or simply a *graph*, G is a triple (V, E, W) where V is a set of *vertices* each represented by a distinct positive integer, E is a set of *edges* or unordered pairs (i, j) of the vertices $i, j \in V$, and W is the set of *weights*, each weight $w(i, j)$ corresponding to the edge $(i, j) \in E$.

An $n \times n$ matrix B is *skew-symmetric* if for all i, j $(1 \leq i, j \leq n)$ $B(i,j) = -B(j,i)$. The *matrix of the graph* $G = (V, E, W)$ where $V = \{1, 2, \ldots, n\}$ is the $n \times n$ matrix $M(G)$ where the (i, j) entry $M(G)(i, j)$ is defined to equal:

(i) $w(i, j)$ if $i < j$,
(ii) $-w(i, j)$ if $i > j$, and
(iii) 0 otherwise.

In the more general case that $V = \{k_1, k_2, \ldots, k_n\}$ where $k_1 < k_2 < \ldots < k_n$, weight $w(k_i, k_j)$ replaces $w(i, j)$ in (i) and (ii) in this definition. For brevity we shall abbreviate $M(G)$ by G whenever it is clear that a matrix is intended.

The Pfaffian of an $n \times n$ skew-symmetric matrix B is defined to be zero if n is odd, one if $n = 0$, and if n is even with $n = 2k$ and $k > 0$ then it is defined as:

$$\mathrm{Pf}(B) = \sum_\pi \epsilon_\pi w(i_1, i_2) w(i_3, i_4) \ldots w(i_{2k-1}, i_{2k})$$

where

(i) $\pi = [i_1, i_2, i_3, \ldots, i_{2k}]$ is a permutation on $[1, 2, \ldots, n]$,
(ii) summation is over all such permutations π where further
$i_1 < i_2, i_3 < i_4, \ldots, i_{2k-1} < i_{2k}$, and
$i_1 < i_3 < i_5 < \ldots < i_{2k-1}$, and
(iii) $\epsilon_\pi \in \{-1, 1\}$ is the sign of the permutation π, i.e., it is -1 or $+1$ according to whether the number of transpositions or swaps of pairs of distinct elements i_j, i_k, needed to reorder π to the identity permutation is odd or even. (An equivalent definition in this context is that it is the sign or parity of the number of overlapping pairs, where a pair of edges (i_{2r-1}, i_{2r}), (i_{2s-1}, i_{2s}) is *overlapping* iff $i_{2r-1} < i_{2s-1} < i_{2r} < i_{2s}$ or $i_{2s-1} < i_{2r-1} < i_{2s} < i_{2r}$. Note that it is implicit here that $i_{2r-1} < i_{2r}$ and $i_{2s-1} < i_{2s}$.)

A *matching* $E^* \subseteq E$ of G is a set of edges such that if (i, j), (r, s) are distinct edges in E^* then i, j, r, s are all distinct vertices. In a graph with an even number $2k$ of nodes a matching E^* is *perfect* if it contains k edges. (Then every $i \in V$ is an endpoint of, or is *saturated* by, some edge in E^*.)

We shall use the following graph-theoretic interpretation of the Pfaffian. If B is the matrix of the graph G then there is a one-to-one correspondence between monomials in the Pfaffian of B and the perfect matchings in G. The monomial $w(i_1, i_2)\, w(i_3, i_4) \ldots w(i_{2k-1}, i_{2k})$ in $\mathrm{Pf}(G)$ corresponds to the perfect matching $\{(i_1, i_2), (i_3, i_4), \ldots, (i_{2k-1}, i_{2k})\}$ in G. The coefficient ϵ_π of this monomial will be the parity of the numbers of overlapping pairs of edges, in the sense defined above.

For an $n \times n$ matrix B and any set $A = \{i_1, \cdots, i_r\} \subseteq \{1, \cdots, n\}$ we denote by $B[A]$ the $(n-r) \times (n-r)$ matrix obtained by deleting from B all the rows and columns indexed by A. The following is from [V02a]:

Definition. The *PfaffianSum* of an $n \times n$ skew-symmetric matrix B is the polynomial over indeterminates $\lambda_1 \ldots \lambda_n$

$$\mathrm{PfS}(B) = \sum_A \left(\prod_{i \in A} \lambda_i\right) \mathrm{Pf}(B[A]).$$

Summation here is over the various principal minors obtained from B by deleting some subset A of the indices. In this paper we shall only need the instances in which each λ_i is fixed to be 0 or 1. The i for which $\lambda_i = 0$ can be thought of as the *unomittable* indices, and those with $\lambda_i = 1$ as the *omittable* indices. Then for this *(0,1)-case* the PfaffianSum is simply the sum of the $\mathrm{Pf}(B[A])$ over those A that contain only omittable indices.

We shall simulate each gate of a computation by what we call a matchgate. A *matchgate* Γ is a quadruple (G, X, Y, T) where G is a graph (V, E, W), $X \subseteq V$ is a set of *input* vertices, $Y \subseteq V$ is a set of *output* vertices, and $T \subseteq V$ is a set of *omittable* vertices such that (i) X, Y and T are all disjoint, and (ii) $\forall i \in T$ if $j \in X$ then $j < i$ and if $j \in Y$ then $j > i$. A matchgate is an (m, n)-*matchgate* if it has m input nodes and n output nodes. It is an input gate if $n = 0$ and an *output* gate if $m = 0$. It is an *even* gate if it has no omittable nodes.

The matchings we consider will be those that saturate all the unomittable nodes, i.e. $V - T$, and also some, possibly empty, subset of T. Whenever we refer to the PfaffianSum of a matchgate fragment, such as G' in the following paragraph, we shall assume the substitutions $\lambda_i = 1$ if $i \in T$, and $\lambda_i = 0$ otherwise.

We call $X \cup Y$ the *external* nodes. For $Z \subseteq X \cup Y$ we consider the matchings of a larger graph that contains G as a subgraph such that the Z nodes are matched by edges external to G, and the remaining elements of $X \cup Y$ by edges of G. We define the *character* $\chi(\Gamma, Z)$ of Γ with respect to $Z \subseteq X \cup Y$ to be the product

$$\mu(\Gamma, Z)\mathrm{PfS}(G')$$

where: (a) $G' = (V - Z, E', W')$ where further E' is the restriction of E to edges with both endpoints in $V - Z$, and W' is the corresponding restriction of W, and (b) the *modifier* $\mu(\Gamma, Z) \in \{-1, 1\}$ counts the parity of the number of overlaps between matched edges in E' and matched external edges. We consider there to exist one matched external edge from each node in $X \cap Z$ and from each node in $Y \cap Z$. The other endpoint of each of the former is some node of lower index than any in V, and of each of the latter is some node of index higher than any in V.

The character of a matchgate, therefore, characterizes the contribution of G to the PfaffianSum of a larger graph, and takes into account overlaps between its internal edges and the external edges that link its external nodes to the rest of the graph. The significance of condition (ii) in the definition of matchgates is that it guarantees that the modifier $\mu(\Gamma, Z)$ is always well defined: for any fixed Z the external edges that arise are uniquely defined, but it has to be guaranteed that the parity of the overlap of any one such external edge with *every* matching of E' that saturates all the unomittable nodes is the same. Condition (ii) ensures this by not allowing an omittable node in the gate to be numbered intermediate between the endpoints of an external edge. (That case might produce different overlap parity for the given external edge and the various internal matchings depending on whether the omittable node was in the matching.) To verify this, note that if for $i \in X \cap Z$ there are r nodes $j < i$ where $j \in V - Z$, then the parity of the overlap of the external edge from i with the internal edges is the parity of r.

We define the *character* $\chi(\Gamma)$ of Γ as the $2^{|X \cup Y|}$ values of $\chi(\Gamma, Z)$ for the various $2^{|X \cup Y|}$ possible choices of Z. In particular, we think of the character as a $2^{|X|} \times 2^{|Y|}$ matrix where the rows represent the subsets of the inputs X, and the columns the subsets of the outputs Y. Matchgates with $|X| = |Y| = k$ can then be regarded as matrix transformations defined by a square character matrix. For example $k = 1$ corresponds to one-bit 2×2 matrix transformations, and $k = 2$ corresponds to two-bit 4×4 transformations. In all cases we need to specify a correspondence between subsets of X and the rows of the matrix, and another correspondence between subsets of Y and the columns of the matrix. In this paper we shall specify these correspondences as necessary. In general, here as in [Va02a] and [Va02b] we assume what we call a *normal ordering* in which both the rows and columns are in increasing order when the subsets of externally matched nodes Z are represented in binary in the sense of the following example. Suppose $X = \{1, 2, 3\}$ and $Y = \{7, 8, 9\}$. Then the rows will be ordered 1,2,...,8, as Z ranges over $\emptyset$, $\{1\}$, $\{2\}$, $\{1,2\}$, $\{3\}$, $\{1,3\}$, $\{2,3\}$, $\{1,2,3\}$. The columns are in the same order if 1 is identified with 9, 2 with 8, and 3 with 7. (N.B. The definition for this in [Va02b] was inaccurate.)

We say that matchgate $\Gamma = (G, X, Y, T)$ with $G = (V, E, W)$ has *normal numbering* if the numbering of V is consecutive and X, Y have minimal and maximal numbers respectively. Formally, $V = \{1, 2, \cdots, |V|\}$ and $\forall i \in X, \forall j \in Y$ and $\forall k \notin X \cup Y$ it is the case that $i < k < j$.

We shall construct matchcircuits by composing normally numbered matchgates in the manner of Figure 1. This is the same as in [V02a] except that we are making the input and output gates more general. The purpose of the construction is to ensure that the PfaffianSum properties of the circuit conform to the composition of the corresponding properties of the individual matchgates, without interfering sign effects. Then for any input vector generated by the in-

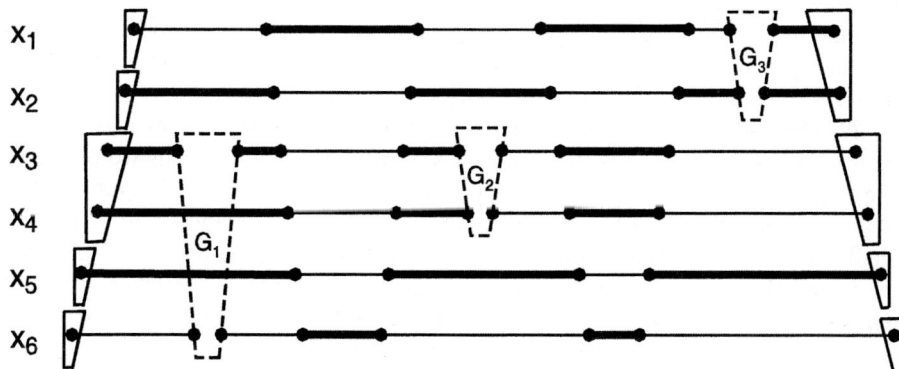

Fig. 1. An example of a matchcircuit composed of five input gates on the left, four output gates on the right, and three intermediate gates G_1, G_2, G_3. The internal nodes and edges in the gates are not shown. The nodes are numbered in increasing order from left to right. The bold edges show an example of the edges in a perfect matching of the circuit, excluding any matched edges internal to the individual gates

put gates the remaining circuit will compute a *value* equal to the composition of the functions computed by the gates. The proof there of the Main Theorem supports the following more general statement.

Main Theorem. *Consider a matchcircuit Γ composed of gates as shown in Figure 1. Suppose that every gate is:*

(i) *a gate with diagonal character matrix,*
(ii) *an even gate applied to consecutive bits $x_i, x_{i+1}, \cdots x_{i+j}$ for some $j > 0$,*
(iii) *an arbitrary gate applied to bits $x_1, \cdots, x_j$ for some $j > 1$, or*
(iv) *input or output gates on consecutive bits $x_i, x_{i+1}, \cdots, x_{i+j}$ for some $j > 0$.*

Then the PfaffianSum of Γ equals the sum over all the input vectors generated by the input gates of the value of the circuit defined by the remaining gates.

4 Bases

A *basis* **b** of size k is a set of distinct nonzero vectors of length 2^k with entries from a field F. In this paper the set will always consist of two vectors corresponding to the Boolean 1 and 0, and denoted by p and n respectively. Each of the 2^k components of p or n is interpreted as corresponding to an element of $\{0,1\}^k$, where the 1's represent the subset of the input or output nodes in a matchgate that are to be matched by external edges. We denote the size k of **b** by $|$ **b** $|$, and a basis of size k is also called a *k-basis*. If q, r are two vectors of length l_1, l_2 we denote by $s = q \otimes r$ their tensor product, which is of length $l_1 l_2$ and in which $s_{il_2 + j} = q_i r_j$ for $0 \leq i < l_1$ and $0 \leq j < l_2$.

The *standard basis* is the one with $k = 1, p = (0,1), n = (1,0)$. An *even basis* is one in which every component p_i and n_i is zero if i corresponds to an odd number of nodes being matched externally. Note that an *even 2-basis* has nonzero components in p, n, only for the vector components representing 00 and 11. A special case of an even 2-basis is the *two-rail basis*, in which the only nonzero component in p is for 11, and the only nonzero component in n is for 00.

We shall seek to construct matchgates that perform various operations over various bases. For example, we say that a 2-input 1-output matchgate performs the AND operation over basis **b** if for the four combinations of inputs $n \otimes n$, $n \otimes p$, $p \otimes n$, and $p \otimes p$, the outputs are n, n, n, and p, respectively. Also, we shall use the name of an operation, such as AND, to represent the set of bases for which there exists a matchgate to realize that operation. Further, we shall denote by X^* (e.g., AND* if the operation is AND) the set of bases for which there is an even matchgate that realizes operation X.

5 Basis Classes

We shall seek to use matchgates to simulate Boolean circuits. Thus we shall endeavor to simulate such basic functions as "and", "swap", "fanout" and "cnot" gates. In the definitions below we shall assume that $\mathbf{b} = (n, p)$ is a basis of size k, and hence that n, p, are of length 2^k.

We shall now define some basis classes that each correspond to something that can be viewed as a primitive constituent of computation. Interspersed with the definitions will be statements of some of their simpler properties.

(i) **Generating Constants**

Defn. GEN0: Set of $\mathbf{b}$ such that there is a $(0, |\mathbf{b}|)$-matchgate with $\chi = n$.
Defn. GEN1: Set of $\mathbf{b}$ such that there is a $(0, |\mathbf{b}|)$-matchgate with $\chi = p$.
Defn. GEN01: Set of $\mathbf{b}$ such that there is a $(0, |\mathbf{b}|)$-matchgate with $\chi = n$ and another with $\chi = p$. (Clearly GEN01 = GEN0 ∩ GEN1.)
Defn. GEN0+1: Set of $\mathbf{b}$ such that there is a $(0, |\mathbf{b}|)$-matchgate with $\chi = n + p$.

Proposition 1. *GEN0, GEN1, GEN01, GEN0+1 each contain all the 1-bases and 2-bases for any field.*

Proof. This can be deduced from Proposition 1 in [V02a], or proved more directly as follows. For 2-bases consider the following matchgate where node 1 is omittable, and the output nodes are $\{3, 4\}$.

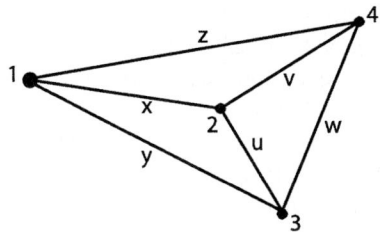

Suppose we wish to generate the 1×4 character matrix (a_1, a_2, a_3, a_4) where ordering $(\emptyset, \{4\}, \{3\}, \{3,4\})$ is implied on the matrix columns. Then, by inspection,

$$a_1 = zu - vy + xw,$$
$$a_2 = u,$$
$$a_3 = -v,$$
$$a_4 = x.$$

For example, by the definition of the character matrix, a_3 is the Pfaffian of the graph with node set $\{3\}$ omitted, adjusted by the multiplicative modifier $\mu = -1$, since this is the overlap between the internal and external edges (i.e. between the internal edge (2,4) and the external edge from 3.)

Clearly we can set a_2, a_3, a_4 to arbitrary values by setting u, v, x appropriately. If at least one of a_2, a_3, a_4 is nonzero then we can also set a_1 arbitrarily by setting z, y, w appropriately. In the special case that $a_2 = a_3 = a_4 = 0$, we modify the matchgate so that node 1 is not omittable and $x = 0$. This sets

$a_2 = a_3 = a_4 = 0$. Also $a_1 = zu - vy$ can then be set arbitrarily. Finally, for 1-bases consider the output nodes to be just $\{4\}$. □

(ii) Swapping

Defn. SWAP: Set of bases **b** such that there is a $(2|\mathbf{b}|, 2|\mathbf{b}|)$-matchgate where $(p \otimes p)\chi = p \otimes p, (n \otimes p)\chi = p \otimes n, (p \otimes n)\chi = n \otimes p, (n \otimes n)\chi = n \otimes n$.

Proposition 2. *Every even basis belongs to SWAP* for any field.*

Proof. Suppose the basis is a k-basis. We imagine $2k$ parallel wires. The task is to interchange the contents of the top k wires with the contents of the bottom k wires. To do this we use the matrix

$$\begin{pmatrix} 1 & 0 & 0 & 0 \\ 0 & 0 & 1 & 0 \\ 0 & -1 & 0 & 0 \\ 0 & 0 & 0 & 1 \end{pmatrix}$$

as a "pseudoswap" for pairs of bits on adjacent wires. To implement a SWAP gate between two sets of k wires we have the two sets of k wires cross and place a pseudoswap gate at all the k^2 intersections. Since we have an even basis, in every situation that contributes to the Pfaffian, an even number of the wires in each group of k will have a 1 bit on it. Hence 1 values will need to be "interchanged" with 0 values an even number of times. Hence an even number of the pseudoswap gates will have their -1 entry invoked, and hence the overall contribution will always be +1. Finally note that this is realizable by a matchgate over the standard basis by virtue of Proposition 4 in [V02a], and in particular by the following matchgate with input ordering $(\emptyset,\{1\},\{2\},\{1,2\})$ and output ordering $(\emptyset,\{4\},\{3\},\{3,4\})$. □

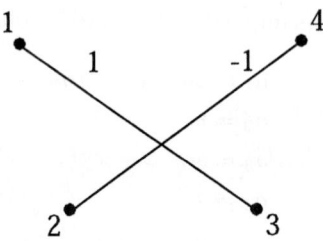

Proposition 3. *Every basis over any field of characteristic 2 belongs to SWAP*.*

Proof. Immediate. □

Proposition 4. *There is no independent 1-basis for SWAP for the complex numbers.*

Proof. This can be deduced from the definition of SWAP by solving the set of equations that define SWAP, together with the five matchgate identities [V02a], using a polynomial solver such as Singular [GPS01]. □

(iii) **Fanout**

The replication of influence, or fanout, is a very basic constituent of computation.

Defn. FANOUT: Set of **b** such that there is a $(|\mathbf{b}|, 2|\mathbf{b}|)$-matchgate with $n\chi = n \otimes n$ and $p\chi = p \otimes p$.

Perhaps surprisingly, there does exist a 1-basis for FANOUT.

Proposition 5. *For any field with characteristic different from 2 there exists a 1-basis for FANOUT*.*

Proof. Consider the one bit basis $n, p = (1, -1), (1, 1)$. It can be verified that

$$\begin{pmatrix} 1 & -1 \\ 1 & 1 \end{pmatrix} \begin{pmatrix} 1 & 0 & 0 & 1 \\ 0 & 1 & 1 & 0 \end{pmatrix} = \begin{pmatrix} 1 & -1 & -1 & 1 \\ 1 & 1 & 1 & 1 \end{pmatrix}$$

and that the second matrix is realized by the matchgate with $V = \{1, 2, 3, 4\}$, with input $\{1\}$, outputs $\{3, 4\}$, and $V = \{(1, 2), (2, 3), (2, 4), (3, 4)\}$ where all the edges have weight one except for (2,4) which has weight -1. □

Proposition 6. *There is no independent 1-basis of characteristic 2 for FANOUT.*

Proof. From the matchgate identities. □

Propositions 4 and 6 show that for 1-bases SWAP ∩ FANOUT = ∅ for any field. We therefore go on to consider notions of replication that are no stronger, and possibly more restricted than fanout. These are GENEQ and GENOPP which output pairs of equal or opposite values, but take no inputs.

Defn. GENEQ: Set of **b** such that there is a $(0, 2|\mathbf{b}|)$-matchgate with $\chi = p \otimes p + n \otimes n$.

Defn. GENOPP: Set of **b** such that there is a $(0, 2|\mathbf{b}|)$-matchgate with $\chi = p \otimes n + n \otimes p$.

Proposition 7. *There is no independent even 2-basis for GENEQ* for any field.*

Proof. Consider an even 2-basis $(a, 0, 0, b), (c, 0, 0, d)$. Then any (0,4)-matchgate for GENEQ has to generate the (1×16)-character matrix

$$(a^2 + c^2, 0, 0, ab + cd, 0, 0, 0, 0, 0, 0, 0, 0, ab + cd, 0, 0, b^2 + d^2).$$

But from this matchgate one can easily construct a (2,2)-matchgate with
(4 × 4) character matrix

$$\begin{pmatrix} a^2 + c^2 & 0 & 0 & ab + cd \\ 0 & 0 & 0 & 0 \\ 0 & 0 & 0 & 0 \\ ab + cd & 0 & 0 & b^2 + d^2 \end{pmatrix}$$

The construction requires the creation of 2 new input nodes each connected by chains of 2 edges to the first two output nodes, which are no longer considered as output nodes.

But the first matchgate identity then implies that $ad = bc$, which contradicts $(a, 0, 0, b), (c, 0, 0, d)$ being idependent. □

Proposition 8. $GEN0+1 \cap FANOUT \subseteq GENEQ$ *for any field.*

Proof. If we append, in the manner of Figure 3 in [V02a], the matchgate for FANOUT to one for GEN0+1 we obtain one for GENEQ. □

Proposition 9. *For any field GENEQ and GENOPP contain all 1-bases.*

Proof. Immediate from Proposition 1 applied to 2-bases. □

Proposition 10. *There is no independent even 2-basis for GENOPP* for any field.*

Proof. Similar to the proof of Proposition 7. □

(iv) **Boolean operations**

Defn. NOT: Set of **b** such that there is a $(|\mathbf{b}|, |\mathbf{b}|)$-matchgate with $p = n\chi$ and $n = p\chi$.

Defn. AND: Set of **b** such that there is a $(2|\mathbf{b}|, |\mathbf{b}|)$-matchgate where $(p \otimes p)\chi = p$, and $(n \otimes n)\chi = (n \otimes p)\chi = (p \otimes n)\chi = n$.

Proposition 11. *The standard basis does not belong to AND*, nor does any independent 1-basis over a field of characteristic 2.*

Proposition 12. *There is a 1-basis for AND* over any field with characteristic different from 2.*

Proof. $k = 1 : n, p = (1, 1), (2, 0),$

$$\chi = \begin{pmatrix} \frac{1}{2} & 0 \\ 0 & \frac{1}{2} \\ 0 & \frac{1}{2} \\ \frac{1}{2} & 0 \end{pmatrix}$$

The matchgate that realizes this is (V, E) where $V = \{1, 2, 3, 4\}$, the inputs are $\{1, 2\}$, the output is $\{4\}, E = \{(1, 2), (1, 3), (2, 3), (3, 4)\}$ with weights $1, -1/2, 1/2,$
$1/2$, respectively. □

Defn. BOOL: Set of **b** such that there are AND, OR and NOT matchgates for **b**.

Defn. MON: Set of **b** such that there are AND and OR matchgates for **b**.

Various relations are immediate. For example BOOL $\supseteq$ NAND $\cap$ GEN1 since NOT can be obtained from NAND using GEN1, and OR can be obtained from AND and NOT. Note that neither swap nor replication is needed in these classical Boolean simulations.

Proposition 13. *The 2-rail basis belongs to BOOL for any field.*

Proof. We can realize AND using $k = 2 : n, p = (1,0,0,0), (0,0,0,1)$. The matchgate that realizes this is (V, E) where $V = \{1, 2, 3, 4, 5, 6\}$, inputs are $\{1, 2, 3, 4\}$, outputs are $\{5, 6\}$, $E = \{(1, 6), (2, 5), (2, 3), (3, 6), (4, 5)\}$ all with weight 1. Similarly NOT and hence OR are also realizable. $\square$

(v) Universality in the sense of quantum computation

Defn. ONEBITUNIV: Set of **b** such that for all complex numbers $\alpha, \beta, \gamma, \delta$, there is a $(|\mathbf{b}|,|\mathbf{b}|)$-matchgate with $n\chi = \alpha n + \beta p$, and $p\chi = \gamma n + \delta p$.

Proposition 14. *The two rail basis belongs to ONEBITUNIV for any field.*

Proof. For realizing the mapping

$$\begin{pmatrix} \alpha & \beta \\ \gamma & \delta \end{pmatrix}$$

in the two rail basis use the matchgate that realizes

$$\begin{pmatrix} \alpha & 0 & 0 & \beta \\ 0 & \alpha & \beta & 0 \\ 0 & \gamma & \delta & 0 \\ \gamma & 0 & 0 & \delta \end{pmatrix}.$$

If $\delta = 0$ then by inverting the input or output we first move a nonzero entry to the lower right position in the matrix. $\square$

Defn: CNOT: Set of **b** such that there is a $(2|\mathbf{b}|, 2|\mathbf{b}|)$-matchgate where $(n \otimes n)\chi = n \otimes n, (n \otimes p)\chi = n \otimes p, (p \otimes n)\chi = p \otimes p, (p \otimes p)\chi = p \otimes n$.

Defn. ISWAP: Set of **b** such that there is a $(2|\mathbf{b}|, 2|\mathbf{b}|)$-matchgate where $(n \otimes n)\chi = n \otimes n, (n \otimes p)\chi = ip \otimes n, (p \otimes n)\chi = in \otimes p, (p \otimes p)\chi = p \otimes p$, where i is the square root of -1.

Defn. UNIV: Set of **b** over the complex numbers such that a universal set of gates, in the sense of quantum computation, can be realized for **b**.

Note that CNOT [BBC+95,CN01] together with the set of all one-bit gates form a universal set. The same holds for ISWAP [EWD+01] together with all one-bitgates.

Proposition 15. *There is no independent 1-basis for CNOT for the complex numbers.*

Proposition 16. *The standard basis can realize ISWAP for the complex numbers.*

Proof. ISWAP over the standard basis requires a character matrix:
$$\begin{pmatrix} 1 & 0 & 0 & 0 \\ 0 & 0 & i & 0 \\ 0 & i & 0 & 0 \\ 0 & 0 & 0 & 1 \end{pmatrix}$$
This can be realized by the matchgate (V, E) where $V = \{1, 2, 3, 4\}$, inputs are $\{1, 2\}$, outputs are $\{3, 4\}, E = \{(1, 3), (2, 4)\}$ with weights $-i, -i$ respectively. □

(vi) Translations from and to the standard basis

Defn. ENCODE: Set of **b** such that there is a $(1, |\mathbf{b}|)$-matchgate where the first row of χ is n and the second is p.

Proposition 17. *There is no even 2-basis for ENCODE.*

Proof. The matchgate identities preclude this. □

Defn. DECODE: Set of **b** such that there is a $(|\mathbf{b}|, 1)$-matchgate with $n\chi = (1, 0)$ and $p\chi = (0, 1)$.

Proposition 18. *ENCODE, DECODE each contain all the 1-bases.*

Proof. Similar to proof of Proposition 1. □

6 Some Universal Operations Sets

There appear to be an enormous variety of interesting universal operations sets with respect to NC2. Here we shall give examples for each of #P, ⊕P, QBP, and P.

Our first example shows that one does not need the general FANOUT operation for universality over #P or ⊕P. The ability to produce complementary pairs of bits is sufficient.

Proposition 19. *(i){GENOPP, SWAP*, MON} over any field of infinite characteristic is a universal operations set for #P, as is also {GENEQ, SWAP*, MON}.*
(ii){GENOPP, MON} over GF[2] is a universal operations set for ⊕P.

Proof. We need to establish that fanout, of the limited form of producing pairs of opposite bits (GENOPP), is enough to give #P- and ⊕P-completeness, as is also GENEQ for #P-completeness. For this we shall need that the counting and parity problems of read-twice Boolean formulae (i.e. where each variable occurs twice) are #P- and ⊕P-complete. For read-twice CNF formulae #P-completeness was proved in the monotone case [BD97], and this will be sufficient for the GENEQ result. Also, for more general formulae consisting of ∧ and ∨ gates #P- and ⊕P-completeness can be proved if each variable occurs just once positively and once negated [Va05]. This will give the GENOPP results. Let us therefore consider formulae F obtained in one of these two ways.

We now construct a matchcircuit to evaluate such a formula F. The circuit has no inputs and starts with the GENOPP/GENEQ gates on the left. There will be k output nodes corresponding to the k bits of the output if the basis has size k. Wires crossing are implemented by SWAP* gates, which exist over GF[2] by virtue of Proposition 3. Finally the gates will be assembled in the manner of Figure 1.

We view each GENOPP/GENEQ gate as assigning a value $p \otimes n$ or $n \otimes p$ / $p \otimes p$ or $n \otimes n$ in the first instance. For each combination of assignments to the, say m, GENOPP/GENEQ gates the Boolean gates will simulate the execution of the formula being encoded, so that the final output will have n or p according to whether the formula would have the Boolean value 0 or 1.

In the actual matchcircuit all 2^m assignments are computed simultaneously and added, so that if K of the assignments give value 1 and $2^m - K$ give value 0 then the PfaffianSum of the matchcircuit will be $Kp + (2^m - K)n$. Pick any component, say i, in which vectors n and p differ. There must be one since they are distinct. Suppose that $p_i - n_i = x$. Now delete the ith output node so as to insist that the corresponding component has value 1, and make the remaining output nodes omittable. Then the PfaffianSum of the matchcircuit must equal $2^m n_i + Kx$. The number K of solutions can be obtained then immediately. (Note that the Pfaffian can be computed with the correct sign in polynomial time (e.g. [GM94, L97]).

This completes the proof except that for the case of infinite characteristic we need to explain why even gates are not necessary for MON or GENOPP/GENEQ. The main reason is that since we assume that we have even gates for SWAP*, we can ensure that all other operations are done on consecutive bits $x_1, x_2, \cdots, x_k$ so that the conditions of the Main Theorem are satisfied. We note that the GENOPP/GENEQ gates have no inputs and can be placed at the left of the circuit so that each one generates a consecutive sequence of bits (e.g. for 3-bases the second such gate could generate $x_7, x_8, \cdots, x_{12}$.) □

Next we observe that existing results in quantum computation are sufficient to provide the following universal operations set for BQP.

Proposition 20. *The set {GEN01, UNIV*, SWAP*} over the complex numbers is a universal operations set for BQP. If GEN01 is replaced by GEN0+1 then it is a UOS for #P.* We note that UNIV is a broad class. For example, it has been shown that any entangling 2-bit transformation in conjunction with all 1-bit transformations is in UNIV [BD+02, BB02]. A transformation is *entangling* if it maps some tensor product of two one-bit vectors to something that cannot be expressed as a tensor product of two one-bit vectors.

We can also formulate a question that corresponds to P=?NC2.

Proposition 21. *The sets { GEN01, FANOUT, BOOL, SWAP* } and { GEN01, FANOUT*, BOOL*} over any field are universal operations sets for P. If GEN01 is replaced by GEN0+1 then over fields of characteristic zero they are UOS's for #P, and over characteristic two, for $\oplus P$.*

7 Interpretation

As described in the introduction, we can regard deterministic computation as the composition of four components: Boolean operations, some form of fanout, swap, and constant inputs. Nondeterministic computation is the composition of these with the components of nondeterministic constant generation, and exponential summation. In conventional models of computation the first four components are trivial, and the last two appear to be problematic.

In our formulation each of the four components is achievable separately in some basis, and also in many combinations together. Proposition 19 identifies the three components MON, GENOPP, and SWAP as being sufficient for universality over #P. Our results show that common bases exist for some interesting pairs of operations. The basis $\{(1,1),(2,0)\}$ is common to AND and GENOPP, the 2-rail basis$\{(1,0,0,0),(0,0,0,1)\}$ is common to BOOL and SWAP, and the standard basis $\{(1,0),(0,1)\}$ over GF[2] is common to SWAP and GENOPP. Also, Propositions 20 and 21 give some distinct decompositions in terms of UNIV, and in terms of BOOL and FANOUT, respectively. Numerous further such decompositions can be described.

Our approach therefore gives a new methodology for searching for exotic algorithms for problems that are currently believed to be intractable. It is possible that such searches will fail but in that case there is some hope that our model will serve as a useful "restricted" model of computation for which lower bounds can be proved. If one fixes a field and the appropriate parameters then the question of whether any of the set intersections considered is nonempty, is equivalent to determining whether a fixed set of polynomial equations has a common solution. Small instances of such problems can be solved mechanically by computer algebra systems. We have shown that there is considerable richness in the properties of nonstandard bases for some fundamental computational operations, and therefore expect that exploring these further will be of interest whatever the ultimate outcome.

References

[BBC+95] A. Barenco, et al., Elementary gates for quantum computation, *Phys. Rev.* A52 (1995) 3457.

[BD97] R. Bubley and M. Dyer, Graph orientations with no sink and an approximation for a hard case of #SAT, In *Proceedings of the Eighth Annual ACM-SIAM Symposium on Discrete Algorithms*, pages 248-257, New Orleans, Louisiana a, 5-7 January 1997

[BD+02]   M.J. Bremmer, C.M. Dawson, J.L. Dodd, A. Gilchrist, A.W. Harrow, D. Mortimer, M.A. Nielsen, and T.J. Osborne, A Practical scheme for quantum computation with any two-qubit entangling gate, *Phys. Rev. Lett.*, **89**, 247902 (2002).

[BR92]   R.A. Brualdi and H.J. Ryser, Combinatorial Matrix Theory, Cambridge University Press, Cambridge, 1991.

[BB02]   J.L. Brylinski and R. Brylinski, Universal quantum gates, in *Mathematics of Quantum Computation*, Chapman & Hall/CRC Press, Boca Raton, Florida, 2002 (edited by R. Brylinski and G. Chen).

[CN01]   M. A. Chuang and I. L. Nielsen, *Quantum Computation and Quantum Information*. Cambridge University Press, 2001.

[EWD+01]   P. Echternach, *et al.* Universal quantum gates for single Cooper pair box bases quantum computing, *Quantum Information and Computation*, (2001) 143-150. (Also quant-ph/0112025.)

[GM94]   G. Gallbiati and F. Maftioli, *Discrete Applied Math.* v51 (1994) 269-275.

[GPS01]   G.-M. Greuel, G. Pfister, and H. Schönemann, Singular 2.0: A Computer Algebra System for Polynomial Computations. Centre for Computer Algebra, University of Kaiserslautern, (2001).

[HS90]   H.B. Hunt III and R.E. Stearns. The complexity of very simple boolean formulas with applications. *SIAM J. Comput.* 19:1 (1990) 44-70.

[L97]   P.D. Lax, *Linear Algebra*, Wiley, New York, 1997.

[M00]   K. Murota, *Matrices and Matroids for Systems Analysis*, Springer-Verlag, Berlin, 2000.

[P95]   C.H. Papadimitriou. *Computational Complexity*, Addison Wesley, 1995.

[V02a]   L. G. Valiant, Quantum circuits that can be simulated classically in polynomial time. *SIAM J. on Computing,* 31:4 (2002) 1229-1254.

[V02b]   L. G. Valiant. Expressiveness of matchgates, *Theoretical Computer Science,* 289:1(2002) 457-471.

[V04]   L.G. Valiant. Holographic algorithms, *Proc. 45th Annual IEEE Symp. on Foundations of Computer Science* (2004) IEEE Press, 306-315.

[V05]   L.G. Valiant, Completeness for parity problems, manuscript, 2005.

Probabilistic Polynomial-Time Semantics for a Protocol Security Logic[*]

Anupam Datta[1], Ante Derek[1], John C. Mitchell[1],
Vitaly Shmatikov[2], and Mathieu Turuani[3]

[1] Dept. Computer Science, Stanford University, Stanford, CA
[2] Dept. Computer Science, University of Texas, Austin, TX
[3] LORIA-INRIA Nancy, France

Abstract. We describe a cryptographically sound formal logic for proving protocol security properties without explicitly reasoning about probability, asymptotic complexity, or the actions of a malicious attacker. The approach rests on a new probabilistic, polynomial-time semantics for an existing protocol security logic, replacing an earlier semantics that uses nondeterministic symbolic evaluation. While the basic form of the protocol logic remains unchanged from previous work, there are some interesting technical problems involving the difference between efficiently recognizing and efficiently producing a value, and involving a reinterpretation of standard logical connectives that seems necessary to support certain forms of reasoning.

1 Introduction

Security analysis of network protocols is a successful scientific area with two important but historically independent foundations, one based on logic and symbolic computation, and one based on computational complexity theory. The symbolic approach, which uses a highly idealized representation of cryptographic primitives, has been a successful basis for formal logics and automated tools. Conversely, the computational approach yields more insight into the strength and vulnerabilities of protocols, but it is more difficult to apply and it involves explicit reasoning about probability and computational complexity. The purpose of this paper is to suggest that formal reasoning, based on an abstract treatment of cryptographic primitives, can be used to reason about probabilistic polynomial-time protocols in the face of probabilistic polynomial-time attacks.

---

[*] This work was partially supported by NSF CyberTrust Grant 0430594, Collaborative research: High-fidelity methods for security protocols, by the DoD University Research Initiative (URI) program administered by the Office of Naval Research under Grant N00014-01-1-0795, by OSD/ONR CIP/SW URI through ONR Grant N00014-04-1-0725, by NSF CCR-0121403, Computational Logic Tools for Research and Education, and by the NSF Cybertrust grant to the PORTIA project. M. Turuani's activities at Stanford were also funded by a postdoctoral grant from INRIA.

We do this by proposing a new semantics for a variant of an existing logic. The new semantics brings forward some interesting distinctions that were not available in the coarser symbolic model, and also raises some apparently fundamental issues about the inherent logic of asymptotic probabilistic properties.

The *Protocol Composition Logic* [2, 7, 8, 10] uses a modal operator similar to Floyd-Hoare logic. Intuitively, the formula $\psi\,[P]_X\,\varphi$ means that if ψ is true at some point in the execution of a protocol (in the presence of a malicious attacker), then φ will be true after agent X performs the sequence P of actions. The pre- and post-conditions may describe actions taken by various principals and characterize the information that is available to or hidden from them. The semantics we explore in this paper recasts the methods of [15] in a logical setting, and reflects accepted modelling approaches used in the field of cryptography, particularly [5, 17].

Our central organizing idea is to interpret formulas as operators on probability distributions on traces. Informally, representing a probability distribution by a set of equi-probable traces (each tagged by the random sequence used to produce it), the meaning of a formula φ on a set T of traces is the subset $T' \subseteq T$ in which φ holds. This interpretation yields a probability: the probability that φ holds is the ratio $|T'|/|T|$. Conjunction and disjunction are simply intersection and union. There are several possible interpretations for implication, and it is not clear at this point which will prove most fruitful in the long run. In the present paper, we interpret $\varphi \implies \psi$ as the union of $\neg\varphi$ and the composition of ψ with φ; the latter is also the conditional probability of ψ given φ. This interpretation supports a soundness proof for a sizable fragment of the protocol logic, and resembles the probabilistic interpretation of implication in [16]. Since the logic does not mention probability explicitly, we consider a formula "true" if it holds with asymptotically overwhelming probability.

In previous work [2, 7, 8, 10] over a symbolic semantic model, the atomic formula $\mathsf{Has}(X,m)$ means that m is in the set of values "derivable," by a simple fixed algorithm, from information visible to X. The simple fixed algorithm is central to what is called the Dolev-Yao model, after [9] and much subsequent work by others. In replacing the symbolic semantics with a computational semantics based on probabilistic polynomial time, we replace the predicate Has with two predicates, $\mathsf{Possess}$ and Indist. Intuitively, $\mathsf{Possess}(X,m)$ means that there is an algorithm that computes the value of m with high probability from information available to X, while $\mathsf{Indist}(X,m)$ means that X cannot feasibly distinguish m from a random value chosen according to the same distribution. However, certain technical problems discussed in Section 7 lead us to work with slightly simplified semantics of these predicates that capture our intuition most strongly when the possessing principal is assumed honest (in the sense of following the protocol) and the predicate Indist only appears with positive polarity. Fortunately, these syntactic conditions are met in many formulas expressing authentication and secrecy properties.

Several groups of researchers have either formulated connections between symbolic logic and feasible probabilistic computation, or developed relationships between symbolic and computational models. In particular, Abadi and Rogaway [1] propose a logical characterization of indistinguishability by passive eavesdroppers that has been studied by a number of others, and Kapron and Impagliazzo suggest a formal logic for reasoning about probabilistic polynomial-time indistinguishability [13]. Some semantic connections between symbolic and computational models have been developed by a team at IBM Zurich, *e.g.*, [3], with other connections explored in a series of related papers by Micciancio, Warinschi, and collaborators [15, 18, 6]. Herzog [11, 12] shows that if a protocol attack exists in a Dolev-Yao model, there is an attack in a computational model. More recent related work also appears in [14, 6].

Section 2 presents the syntax for defining roles of a protocol, while the syntax of the logic appears in Section 3. Some axioms and proof rules are described in Section 4, followed by a short proof example in Section 5. Section 6 presents the probabilistic polynomial-time execution and attacker model. The semantics of the logic are given in Section 7, and concluding remarks in Section 8.

2 Protocol Syntax

We use a simple "protocol programming language" based on [10, 7, 8] to represent a protocol by a set of roles, such as "Initiator", "Responder" or "Server", each specifying a sequence of actions to be executed by a honest participant. The syntax of terms and actions is given in Table 1.

Names, sessions and threads: We use $\hat{X}, \hat{Y}, \ldots$ as *names* for protocol participants. Since a particular participant might be involved in more than one session at a time, we will give unique names to sessions and use $(\hat{X}, s)$ to designate a particular *thread* being executed by $\hat{X}$. All threads of a participant $\hat{X}$ share the same asymmetric key denoted by X. As a notational convenience, we will sometimes write $\tilde{X}$ for an arbitrary thread of $\hat{X}$.

Terms, actions, and action lists: Terms name messages and their parts, such as nonces, keys, variables and pairs. For technical reasons, we distinguish *basic terms* from *terms* that may contain encryption. To account for probabilistic encryption, encrypted terms explicitly identify the randomness used for encryption.

Table 1. Syntax of protocol terms and actions

Terms:		Actions:
$N ::= \hat{X}$	(name)	a ::=
$K ::= X$	(key)	$\mid$ new T, n
$S ::= s$	(session)	$\mid$ $V :=$ enc T, t, K
$n ::= r$	(nonce)	$\mid$ $V :=$ dec T, t, K
$T ::= (N, S)$	(thread)	$\mid$ match $T, t/t$
$V ::= x$	(term variable)	$\mid$ send T, t
$t_B ::= V \mid K \mid T \mid N \mid n \mid \langle t_B, t_B \rangle$	(basic term)	$\mid$ receive T, V
$t ::= t_B \mid \{t\}_K^n \mid \langle t, t \rangle$	(term)	

Specifically, $\{t\}_K^n$ indicates the encryption of t with key K using randomness n generated for the purpose of encryption. We write $m \subseteq m'$ when m is a subterm of $m' \in t$.

Actions include nonce generation, encryption, decryption, pattern matching, and communication steps (sending and receiving). An *ActionList* consists of a sequence of actions that contain only basic terms. This means that encryption cannot be performed implicitly; explicit enc actions, written as assignment, must be used instead. We assume that each variable will be assigned at most once, at its first occurrence. For any $s \in ActionList$, we write $s_{|X}$ to denote the subsequence of s containing only actions of a participant (or a thread) X.

Strands, roles, protocols and execution: A *strand* is an *ActionList*, containing actions of only one thread. Typically we will use notation $[ActionList]_{\tilde{X}}$ to denote a strand executed by thread $\tilde{X}$ and drop the thread identifier from the actions themselves. A *role* is a strand together with a basic term representing the initial knowledge of the thread. A *protocol* is a finite set of *Roles*, together with a basic term representing the initial intruder knowledge.

An *execution strand* is a pair $ExecStrand ::= InitialState(\mathcal{I}); ActionList$ where $\mathcal{I}$ is a data structure representing the initial state of the protocol, as produced by the initialization phase from Section 6. In particular, this includes the list of agents and threads, the public/private keys and honesty/dishonesty tokens of each agent, and the roles played by each thread.

3 Logic Syntax

The syntax of formulas is given in Table 2. Protocol proofs will usually use modal formulas of the form $\psi[P]_{\tilde{X}}\varphi$, as explained intuitively in the introduction of the paper. Most formulas have the same intuitive meaning in the computational semantics as in the symbolic model [7, 8], except for predicates Possess and Indist. We summarize the meaning of formulas informally below, with precise semantics in the next section.

Action Predicates:
a ::= $\mathsf{Send}(T, t) \mid \mathsf{Receive}(T, t) \mid \mathsf{New}(T, n)$

Formulas:
φ ::= a $\mid t = t \mid \mathsf{Start}(T) \mid \mathsf{Possess}(T, t) \mid \mathsf{Indist}(T, t) \mid \mathsf{Fresh}(T, t) \mid \mathsf{Honest}(N) \mid$
$\mathsf{Start}(T) \mid \mathsf{Contains}(t, t) \mid \mathsf{ContainsOut}(t, t, t) \mid \mathsf{DecryptsHonest}(T, t) \mid$
$\mathsf{Source}(T, t, t) \mid \varphi \wedge \varphi \mid \varphi \vee \varphi \mid \exists Var. \varphi \mid \forall Var. \varphi \mid \neg \varphi \mid \varphi \supset \varphi \mid \varphi \Rightarrow \varphi$

Modal formulas:
Ψ ::= $\varphi \, [Strand]_T \, \varphi$

Table 2. Syntax of the logic

For every protocol action, there is a corresponding action predicate which asserts that the action has occurred in the run. For example, $\mathsf{Send}(\tilde{X}, t)$ holds in a run where the thread $\tilde{X}$ has sent the term t. $\mathsf{Fresh}(\tilde{X}, t)$ means that the value of t generated by $\tilde{X}$ is "fresh" in the sense that no one else has seen any messages containing t, while $\mathsf{Honest}(\hat{X})$ means that $\hat{X}$ is acting honestly, i.e., the actions of every thread of $\hat{X}$ precisely follows some role of the protocol. The Source predicate is used to reason about the source of a piece of information, such as a nonce. Intuitively, the formula $\mathsf{Source}(\tilde{Y}, u, \{m\}^r_X)$ means that the only way for a thread $\tilde{X}$ different from $\tilde{Y}$ to know u is to learn u from the term $\{m\}^r_X$, possibly by some indirect path.

The predicate Fresh is definable by $\mathsf{Fresh}(\tilde{X}, v) \equiv \mathsf{New}(\tilde{X}, v) \wedge \neg(\exists u. \mathsf{Send}(\tilde{X}, u) \wedge \mathsf{Contains}(u, v))$ and classical implication is definable by $A \supset B \equiv \neg A \vee B$.

In the symbolic model [7, 8], the predicate Has states that a principal can "derive" a message or its contents from the information gathered during protocol execution. We use $\mathsf{Possess}(\tilde{X}, t)$ to state that it is possible to derive t by Dolev-Yao rules from $\tilde{X}$'s view of the run and $\mathsf{Indist}(\tilde{X}, t)$ to state that no probabilistic polynomial-time algorithm, given $\tilde{X}$'s view of the run, can distinguish t from a random value from the same distribution. Typically, we use Possess to say that some honest party obtained some secret, and Indist to say that the attacker does not have any partial information about a secret.

4 Proof System

The proof system used in this paper is based on the proof system developed in [7, 8, 2]. Some example axioms and rules are given in Table 3; the full presentation is deferred to the extended version of this paper. These axioms express reasoning principles that can be justified using complexity-theoretic reductions, information-theoretic arguments, and asymptotic calculations. However, the advantage of the proof system is that its justification using cryptographic-style arguments is a one-time mathematical effort; protocol proofs can be carried out symbolically using the proof system without explicitly reasoning about probability and complexity. Another advantage of the axiomatic approach is that different axioms and rules rest on different cryptographic assumptions. Therefore, by examining the axioms and rules used in a specific proof, we can identify specific properties of the cryptographic primitives that are needed to guarantee protocol correctness. This provides useful information in protocol design because primitives that provide weaker properties often have more efficient constructions.

Axioms: Axioms **AN2** and **AN3** capture some of the properties of nonce generation. Informally, **AN2** states that if a thread $\tilde{X}$ generates a fresh nonce x and does not perform any additional actions, then x is indistinguishable from a random value for all other threads. The soundness of this axiom is established by a simple information-theoretic argument. The informal interpretation of axiom **S1** (also called the "Source" axiom) is that, unless a ciphertext is decrypted, a thread which does not possess the decryption key cannot extract any par-

Axioms:

$\mathbf{AN2}: \top[\text{ new } x]_{\tilde{X}} \tilde{Y} \neq \tilde{X} \Rightarrow \mathsf{Indist}(\tilde{Y}, x)$

$\mathbf{AN3}: \top[\text{ new } x]_{\tilde{X}} \mathsf{Fresh}(\tilde{X}, x)$

$\mathbf{S1}: \mathsf{Source}(\tilde{Y}, u, \{m\}_X^r) \wedge \neg \mathsf{DecryptsHonest}(\hat{X}, \{m\}_X^r) \wedge \hat{Z} \neq \hat{X} \wedge \hat{Z} \neq \hat{Y} \wedge$
$\mathsf{Honest}(\hat{X}) \wedge \mathsf{Honest}(\hat{Y}) \Rightarrow \mathsf{Indist}(\tilde{Z}, u)$

Proof rules:

$$\dfrac{\theta[P]_X\varphi \quad \theta' \supset \theta \quad \varphi \supset \varphi'}{\theta'[P]_X\varphi'} \; \mathbf{G3} \qquad \dfrac{\theta[P_1]_X\varphi \quad \varphi[P_2]_X\psi}{\theta[P_1 P_2]_X\psi} \; \mathbf{SEQ}$$

$$\dfrac{\varphi \quad \varphi \Rightarrow \psi}{\psi} \; \mathbf{MP} \qquad \dfrac{\varphi}{\forall x.\varphi} \; \mathbf{GEN}$$

Table 3. Fragment of the proof system

tial information about the plaintext. The soundness of this axiom is proved by a complexity-theoretic reduction. Specifically, we show that if an attacker can break this property, then there is another attacker that can break the underlying IND-CCA2 secure encryption scheme [4].

Inference rules: Inference rules include generic rules from modal logics (e.g. **G3**), sequencing rule **SEQ** used for reasoning about sequential composition of protocol actions and a rule (called the honesty rule) for proving protocol invariants using induction. These rules are analogous to proof rules from our earlier work [7, 8].

First-order axioms and rules: We use two implications: a conditional implication $\Rightarrow$, discussed and defined precisely in section 7, and a classical implication $\supset$ with $A \supset B \equiv \neg A \vee B$. While standard classical tautologies hold for classical implication, some familiar propositional or first-order tautologies may not hold when written using $\Rightarrow$ instead of $\supset$. However, modus ponens and the generalization rule above are sound. The soundness of modus ponens relies on the simple asymptotic fact that the sum of two negligible functions is a negligible function. In future work, we hope to develop a more complete proof system for the first-order fragment of this logic.

5 Example

In this section, we present a simple protocol and state a secrecy property that can be proved using the proof system. The interested reader is referred to [10, 7, 8] for further explanation and examples. The two protocol roles are:

$$\mathbf{Init} \equiv [\text{ new } x; \; y := \mathsf{enc}\langle x, \tilde{X}\rangle, Y; \; \mathsf{send} \; \hat{X}, \hat{Y}, y]_{\tilde{X}}$$
$$\mathbf{Resp} \equiv [\text{ receive } z; \; \mathsf{match} \; z/\langle \hat{X}, \hat{Y}, z'\rangle; \; z'' := \mathsf{dec} \; z', Y]_{\tilde{Y}}$$

The initiator generates a new nonce and sends it encrypted to the responder. The responder receives the message and recovers the nonce by decrypting the ciphertext. We can prove that if $\tilde{X}$ completes the protocol with $\tilde{Y}$, then x will be a shared secret between them, provided both agents are honest. Formally,

$$\mathsf{Start}(\tilde{X})[\mathbf{Init}]_{\tilde{X}} \mathsf{Honest}(\hat{X}) \wedge \mathsf{Honest}(\hat{Y}) \wedge (\tilde{Z} \neq \tilde{X}) \wedge (\tilde{Z} \neq \tilde{Y}) \Rightarrow \mathsf{Indist}(\tilde{Z}, x)$$

Since the meaning of $\mathsf{Indist}(\tilde{Z}, x)$ (formally defined in Section 7) is that $\tilde{Z}$ cannot distinguish the secret nonce x from a randomly chosen nonce, this formula expresses a standard form of secrecy used in the cryptographic literature.

The axiomatic proof uses **AN2**, a variant of **S1**, and modus ponens **MP**. The proof idea is that at the point the initiator produces the nonce x, by **AN2**, it is indistinguishable from random to everyone else other than $\tilde{X}$ and $\tilde{Y}$. It continues to remain indistinguishable since it appears on the network under encryption with a public key whose corresponding private key is not available to the attacker. This part of the reasoning is codified by an axiom that is similar to **S1** and relies on the fact that the encryption scheme used is IND-CCA2 secure. Modus ponens is used in the general first-order reasoning involved in the proof.

6 Protocol Execution

Given a protocol, adversary, and value of the security parameter, we define a set of protocol traces, each associated with the random bits that produce this sequence of actions and additional randomness for algorithms used in the semantics of formulas about the run. The definition proceeds in two phases. In the initialization phase, we assign a set of roles to each principal, identify a subset which is honest, and provide all entities with private-public key pairs and random bits. In the execution phase, the adversary executes the protocol by interacting with honest principals, as in the accepted cryptographic model of [5].

Initialization: We fix the protocol Q, adversary A, security parameter η, and some randomness R of size polynomially bounded in η. Each principal and each thread (*i.e.*, an instance of a protocol role executed by the principal) is assigned a unique bitstring identifier. We choose a sufficiently large polynomial number of bitstrings $i \in I \subseteq \{0,1\}^\eta$ to represent the names of principals and threads. Randomness R is split into r_i for each honest $i \in I$ (referred to as "coin tosses of honest party i") and R_A (referred to as "adversarial randomness").

The adversary designates some of the principals as *honest* and the rest of the principals as *dishonest*. Intuitively, honest principles will follow one or more roles of the protocol faithfully. The adversary chooses a set of threads, and to each thread it assigns a strand (a program to be executed by that thread), under the restriction that all threads of honest principals are assigned roles of protocol Q.

The key generation algorithm $\mathcal{K}$ of a public-key encryption scheme $(\mathcal{K}, \mathcal{E}, \mathcal{D})$ is run on $\mathbf{1}^\eta$ for each participant a using randomness r_a, and producing a public-private key pair (pk_a, sk_a). The public key pk_a is given to all participants and to the adversary A; the private key is given to all threads belonging to this principal and to the adversary if the principal is dishonest.

Generating Computational Traces: Following [5], we view an agent i trying to communicate with agent j in protocol session s as a (stateful) oracle $\Pi_{i,j}^s$. The state of each oracle is defined by a mapping λ from atomic symbols to bitstrings (with variables and nonces renamed to be unique for each role) and a counter c. Each oracle proceeds to execute a step of the protocol as defined by actions in the corresponding role's action list, when activated by the adversary.

We omit the details of communication between the adversary and the oracles, and focus on computational interpretation of symbolic protocol actions. Let a_c be the current action in the *ActionList* defining some role of participant i in session s, i.e., $Thread = (i', s')$ where $i = \lambda(i'), s = \lambda(s')$.

If $a_c = (\text{ new } (i', s'), v)$, then update λ so that $\lambda(v) = NonceGen(R_i)$, where $NonceGen$ is a nonce generation function(e.g., $NonceGen$ simply extracts a fresh piece of R_i). If $a_c = (v := \text{enc } (i', s'), j, u)$, then update λ so that $\lambda(v) = \mathcal{E}(\lambda(u), pk_j, R_i)$ where $\mathcal{E}(\lambda(u), pk_j, R_i)$ is the result of executing the public-key encryption algorithm on plaintext $\lambda(u)$ with public key pk_j and fresh randomness extracted from R_i. For brevity, we omit computational interpretation of decryption and matching (pairing, unpairing, and equality-test) actions. Sending a variable $\text{send } (i', s'), v$ is executed by sending $\lambda(v)$ to the adversary, and receiving $\text{receive } (i', s'), v$ is executed by updating λ so that $\lambda(v) = m$ where m is the bitstring sent by the adversary.

At any time during the protocol execution, the adversary A may record any internal, private message on a special *knowledge tape*. This tape is not read by any participant of the protocol. However, its content will be made available to the test algorithms used to decide if a given security formula containing Indist(...) is valid or not. Let K be $[(i_1, m_1), .., (i_n, m_n)]$ the list of messages m_k written by A on the knowledge tape, indexed by the number of actions i_k already executed when m_k was written (position in the protocol execution). This index will be useful to remember a previous state of the knowledge tape.

At the end of the protocol execution, the adversary A outputs a pair of integers (p_1, p_2) on an *output tape*. When the security formula is a modal formula $\theta[P]_X \varphi$, these two integers represent two positions in the protocol execution where the adversary claims that the formula is violated, i.e. that θ is true in p_1 but φ is false in p_2, with P between p_1 and p_2. Let O be this pair (p_1, p_2) of integers written on the output tape.

The symbolic trace of the protocol is the execution strand $e \in ExecStrand$ which lists, in the order of execution, all honest participant actions and the dishonest participant's send and receive actions. This strand contains two parts: $InitialState(\mathcal{I})$ stores the initialization data, and the rest is an ordered list of all exchanged messages and honest participants' internal actions.

Definition 1. *(Computational Traces) Given a protocol Q, an adversary A, a security parameter η, and a sequence of random bits $R \in \{0,1\}^{p(\eta)}$ used by the honest principals and the adversary, a run of the protocol is the tuple $\langle e, \lambda, O, K, R \rangle$ where e is the symbolic execution strand, $\lambda : Term(e) \to \{0,1\}^{p(\eta)}$ maps the symbolic terms in e to bitstrings, O is the pair of integers written on* the output tape, and K is the indexed list of messages written on the knowledge tape. Finally, $p(x)$ is a polynomial in x.

A computational trace is a run with two additional elements: $R_T \in \{0,1\}^{p(\eta)}$, a sequence of random bits used for testing indistinguishability, and $\sigma : FVar(\varphi) \to \{0,1\}^{p(\eta)}$, a substitution that maps free variables in a formula to bitstrings. The set of computational traces is

$$T_Q(A, \eta) = \{\langle e, \lambda, O, K, R, R_T, \sigma \rangle \mid R, R_T \text{ chosen uniformly}\}.$$

Definition 2. *(Participant's View) Given a protocol Q, an adversary A, a security parameter η, a participant $\hat{X}$ and a trace $t = \langle e, \lambda, O, K, R, R_T, \sigma \rangle \in T_Q(A, \eta)$, $View_t(\tilde{X})$ represents $\tilde{X}$'s view of the trace. It is defined precisely as follows:*

If $\hat{X}$ is honest, then $View_t(\tilde{X})$ is the initial knowledge of $\tilde{X}$, a representation of $e_{|\tilde{X}}$ and $\lambda(x)$ for any variable x in $e_{|\tilde{X}}$. If $\hat{X}$ is dishonest, then $View_t(\tilde{X})$ is the union of the knowledge of all dishonest participants $\tilde{X}'$ after the trace t (where $View_t(\tilde{X}')$ is defined as above for honest participants) plus K, the messages written on the knowledge tape by the adversary.

The following three definitions are used in the semantics of the predicate Indist(). Informally, based on some trace knowledge K, the distinguisher D tries to determine which of two bitstrings is the value of a symbolic term. One of the bitstrings will be the computational value of the term in the current run, while the other will be a random bitstring of the same structure, chosen in a specific way. The order of the two bitstrings presented to the distinguisher is determined by an LR Oracle using a random selector bit.

Definition 3. *(LR Oracle) The LR Oracle [4] is used to determine the order in which two bitstrings are presented depending on the value of the selector bit, i.e. $LR(s_0, s_1, b) = \langle s_b, s_{1-b} \rangle$.*

Definition 4. *(Distinguishing test input) Let u be a symbolic term and σ be a substitution that maps variables of u to bitstrings. We construct another bitstring $f(u, \sigma, r)$, whose symbolic representation is the same as u. Here, r is a sequence of bits chosen uniformly at random. The function f is defined by induction over the structure of the term u.*

- Nonce $u : f(u, \sigma, r) = r$
- Name/Key $u : f(u, \sigma, r) = \sigma(u)$
- Pair $u = \langle u_1, u_2 \rangle : f(\langle u_1, u_2 \rangle, \sigma, r_1; r_2) = \langle f(u_1, \sigma, r_1), f(u_2, \sigma, r_2) \rangle$
- Encryption $u = \{v\}_K^n : f(\{v\}_K^n, \sigma, r_1; r_2) = \mathcal{E}(f(v, \sigma, r_1), \sigma(K), r_2)$

Definition 5. *(Distinguisher) A distinguisher D is a polynomial time algorithm which takes as input a tuple $\langle K, t, \langle s_0, s_1 \rangle, R, \eta \rangle$, consisting of knowledge K, symbolic term t, two bitstrings s_0 and s_1, randomness R and the security parameter η, and outputs a bit b'.*

The next definition is used while defining semantics of modal formulas. Given a set T of traces and a strand P of actions executed by a thread $\hat{X}$, the set T_P includes only those traces from T which contain P. $Pre(T_P)$ is obtained from T_P by taking the initial segment of each trace upto the point where P starts. The precondition of a modal formula is evaluated over this set. $Post(T_P)$ is similarly defined; the only difference is now the trace is cut at the point that P ends. The postcondition of a modal formula is evaluated over this set. The begin and end positions are determined by the component O in the trace.

Definition 6. *(Splitting computational traces)* Let T be a set of computational traces and $t = \langle e, \lambda, O, K, R, R_T, \sigma \rangle \in T$. $O = \langle p_1, p_2 \rangle$, $e = InitialState(\mathcal{I}); s$, and $s = s_1; s_2; s_3$ with p_1, p_2 the start and end positions of s_2 in s. Given a strand P executed by participant $\hat{X}$, we denote by T_P the set of traces in T for which there exists a substitution σ' which extends σ to variables in P such that $\sigma'(P) = \lambda(s_{2|\hat{X}})$. The complement of this set is denoted by $T_{\neg P}$ and contains all traces which do not have any occurrence of the strand P. We define the set of traces $Pre(T_P) = \{t[s \leftarrow s_1, K \leftarrow K_{\leq p_1}, \sigma \leftarrow \sigma'] \mid t \in T_P\}$, where $K_{\leq p}$ is the restriction of the knowledge tape K to messages written before the position p. We define the set of traces $Post(T_P) = \{t[s \leftarrow s_1; s_2, K \leftarrow K_{\leq p_2}, \sigma \leftarrow \sigma'] \mid t \in T_P\}$.

7 Computational Semantics

The semantics of a formula φ on a set T of computational traces is a subset $T' \subseteq T$ that respects φ in some specific way. For many predicates and connectives, the semantics is essentially straightforward. For example, an action predicate such as Send selects a set of traces in which a send occurs. However, the semantics of predicates Indist and Possess is inherently more complex.

Intuitively, an agent possesses the value of an expression (such as another agent's nonce or key) if the agent can compute this value from information it has seen, with high probability. If an agent is honest, and therefore follows the rules of the protocol, then it suffices to use a simple, symbolic algorithm for computing values from information seen in the run of a protocol. For dishonest agents, we would prefer in principle to allow any probabilistic polynomial-time algorithm. However, quantifying over such algorithms, in a way that respects the difference between positive and negative occurrences of the predicate in a formula, appears to introduce some technical complications. Therefore, in the interest of outlining a relatively simple form of computational semantics, we will use a fixed algorithm. This gives a useful semantics for formulas where Possess($\hat{X}, u$) is used under the hypothesis that $\hat{X}$ is honest. We leave adequate treatment of the general case for future work.

Intuitively, an agent has partial information about the value of some expression if the agent can distinguish that value, when presented, from a random value generated according to the same distribution. More specifically, an agent has partial information about a nonce u if, when presented with two bitstrings of the appropriate length, one the value of u and the other chosen randomly, the agent has a good chance of telling which is which. As with Possess, there are technical issues associated with positive and negative occurrences of the predicate. For positive occurrences of Indist, we should say that *no* probabilistic polynomial-time algorithm has more than a negligible chance, where as for ¬Indist(...) we want to say that *there exists* a probabilistic polynomial-time distinguisher. In order to present a reasonably understandable semantics, and establish a useful basis for further exploration of computational semantics of symbolic security logics, we give an interpretation that appears accurate for formulas that have only positive occurrences of Indist and could be somewhat anomalous for formulas that contain negative occurrences. This seems adequate for reasoning about many secrecy properties, since these are expressed by saying that at the end of any run of the protocol, a value used in the run is indistinguishable from random.

Conditional implication $\theta \Rightarrow \varphi$ is interpreted using the negation of θ and the conditional probability of φ given θ. This non-classical interpretation of implication seems to be essential for relating provable formulas to cryptographic-style reductions involving conditional probabilities. In particular, the soundness proof for the "source" axiom **S1**, not proved in this conference paper, uses the conditional aspect of this implication in a fundamental way. On the other hand, $\Rightarrow$ coincides with $\supset$ in formulas where Indist does not appear on the right hand size of the implication.

We inductively define the semantics $[\![\varphi]\!](T, D, \epsilon)$ of a formula φ on the set T of traces, with distinguisher D and tolerance ϵ. The distinguisher and tolerance are not used in any of the clauses except for Indist, where they are used to determine whether the distinguisher has more than a negligible chance of distinguishing the given value from a random value. In definition 7 below, the tolerance is set to a negligible function of the security parameter and $T = T_Q(A, \eta)$ is the set of traces of a protocol Q with adversary A.

- $[\![\mathsf{Send}(\tilde{X}, u)]\!](T, D, \epsilon)$ is the collection of all $\langle e, \lambda, O, K, R, R_T, \sigma \rangle \in T$ such that some action in the symbolic execution strand e has the form send $\tilde{Y}, v$ with $\lambda(\tilde{Y}) = \sigma(\tilde{X})$ and $\lambda(v) = \sigma(u)$. Recall that σ maps formula variables to bitstrings and represents the environment in which the formula is evaluated.
- $[\![\mathsf{a}(\cdot, \cdot)]\!](T, D, \epsilon)$ for other action predicates a is similar to $\mathsf{Send}(\tilde{X}, u)$.
- $[\![\mathsf{Honest}(\hat{X})]\!](T, D, \epsilon)$ is the collection of all $\langle e, \lambda, O, K, R, R_T, \sigma \rangle \in T$ where $e = InitialState(\mathcal{I}); s$ and $\sigma(X)$ is designated *honest* in the initial configuration $\mathcal{I}$. Since we are only dealing with static corruptions in this paper, the resulting set is either the whole set T or the empty set ϕ depending on whether a principal is honest or not.
- $[\![\mathsf{Start}(\tilde{X})]\!](T, D, \epsilon)$ includes all traces $\langle e, \lambda, O, K, R, R_T, \sigma \rangle \in T$ where $e = InitialState(\mathcal{I}); s$ and $\lambda(s)_{|\sigma(\tilde{X})} = \epsilon$. Intuitively, this set contains traces in which $\tilde{X}$ has executed no actions.
- $[\![\mathsf{Contains}(u, v)]\!](T, D, \epsilon)$ includes all traces $\langle e, \lambda, O, K, R, R_T, \sigma \rangle \in T$ such that there exists a series of decryptions with $\{\lambda(k) \mid k \in Key\}$ and projections (π_1, π_2) constructing $\sigma(v)$ from $\sigma(u)$. This definition guarantees that the result is the whole set T if v is a symbolic subterm of u.

- $[\![\,\mathsf{ContainsOut}(u,v,t)\,]\!]\,(T,D,\epsilon)$ includes all traces $\langle e,\lambda,O,K,R,R_T,\sigma\rangle \in T$ such that there exists a series of projections (π_1,π_2) and decryptions with $\{\lambda(k)\mid k\in Key\}$, where $\sigma(t)$ is never decomposed, creating $\sigma(v)$ from $\sigma(u)$. This definition ensures that the result is the whole set T if v is a symbolic subterm of u but is not a subterm of t.
- $[\![\,\theta\wedge\varphi\,]\!]\,(T,D,\epsilon) = [\![\,\theta\,]\!]\,(T,D,\epsilon)\cap[\![\,\varphi\,]\!]\,(T,D,\epsilon)$.
- $[\![\,\theta\vee\varphi\,]\!]\,(T,D,\epsilon) = [\![\,\theta\,]\!]\,(T,D,\epsilon)\cup[\![\,\varphi\,]\!]\,(T,D,\epsilon)$.
- $[\![\,\neg\varphi\,]\!]\,(T,D,\epsilon) = T\setminus [\![\,\varphi\,]\!]\,(T,D,\epsilon)$.
- $[\![\,\exists x.\varphi\,]\!]\,(T,D,\epsilon) = \bigcup_\beta ([\![\,\varphi\,]\!]\,(T[x\leftarrow \beta],D,\epsilon)[x\leftarrow\sigma(x)])$
  with $T[x\leftarrow\beta] = \{t[\sigma[x\leftarrow\beta]] \mid t = \langle e,\lambda,O,K,R,R_T,\sigma\rangle \in T\}$, and β any bitstring of polynomial size.
- $[\![\,\theta\Rightarrow\varphi\,]\!]\,(T,D,\epsilon) = [\![\,\neg\theta\,]\!]\,(T,D,\epsilon)\cup[\![\,\varphi\,]\!]\,(T',D,\epsilon)$, where $T' = [\![\,\theta\,]\!]\,(T,D,\epsilon)$. Note that the semantics of φ is taken over the set T' given by the semantics of θ, as discussed earlier in this section.
- $[\![\,u = v\,]\!]\,(T,D,\epsilon)$ includes all traces $\langle e,\lambda,O,K,R,R_T,\sigma\rangle \in T$ such that $\sigma(u) = \sigma(v)$.
- $[\![\,\mathsf{DecryptsHonest}(\tilde{Y},\{u\}_X^r)\,]\!]\,(T,D,\epsilon) = [\![\,\varphi\,]\!]\,(T,D,\epsilon)$ with $\varphi = \mathsf{Honest}(\hat{X})\wedge \exists v.\, v := \mathsf{dec}\,\tilde{Y},\{u\}_X^r$.
- $[\![\,\mathsf{Source}(\tilde{Y},u,\{m\}_X^r)\,]\!]\,(T,D,\epsilon) = [\![\,\exists v.\forall w.\,\varphi\,]\!]\,(T,D,\epsilon)$ with :

$$\varphi = \mathsf{New}(\tilde{Y},u)\wedge \mathsf{Contains}(m,u)$$
$$\wedge\,\mathsf{Contains}(v,\{m\}_X^r)\wedge \mathsf{Send}(\tilde{Y},v)$$
$$\wedge\,\neg\mathsf{ContainsOut}(v,u,\{m\}_X^r)$$
$$\wedge\,(v\neq w\wedge \mathsf{Contains}(w,u))\Rightarrow \neg\mathsf{Send}(\tilde{Y},w)$$

- $[\![\,\mathsf{Possess}(\tilde{X},u)\,]\!]\,(T,D,\epsilon)$ includes all traces $t=\langle e,\lambda,O,K,R,R_T,\sigma\rangle \in T$ such that $\sigma(u)$ can be built from $View_t(\sigma(\tilde{X}))$ with the Dolev-Yao deduction rules.
- $[\![\,\mathsf{Indist}(\tilde{X},u)\,]\!]\,(T,\epsilon,D) = T$ if

$$\frac{|\{D(View_t(\sigma(\tilde{X})),u,LR(\sigma(u),f(u,\sigma,r),b),R_D,\eta) = b \mid t\in T\}|}{|T|} \leq \frac{1}{2}+\epsilon$$

and the empty set ϕ otherwise. Here, the random sequence $b;r;R_D = R_T$, the testing randomness for the trace t.

- $[\![\,\theta[P]_{\tilde{X}}\varphi\,]\!]\,(T,D,\epsilon) = T_{\neg P}\cup[\![\,\neg\theta\,]\!]\,(Pre(T_P),D,\epsilon)\cup[\![\,\varphi\,]\!]\,(Post(T_P),D,\epsilon)$ with $T_{\neg P}, Pre(T_P)$, and $Post(T_P)$ as given by Definition 6.

Definition 7. *A protocol Q satisfies a formula φ, written $Q\models \varphi$, if $\forall A$ providing an active protocol adversary, $\forall D$ providing a probabilistic-polynomial-time distinguisher, $\forall \nu$ giving a negligible function, $\exists N, \forall \eta\geq N$,*

$$|[\![\,\varphi\,]\!]\,(T,D,\nu(\eta))| \,/\, |T| \geq 1-\nu(\eta)$$

where $[\![\varphi]\!]\,(T,D,\nu(\eta))$ is the subset of T given by the semantics of φ and $T = T_Q(A,\eta)$ is the set of computational traces of protocol Q generated using adversary A and security parameter η, according to Definition 1.

Theorem 1. *(Soundness)* $\quad \forall Q,\, \forall \varphi, \quad Q \vdash \varphi \Rightarrow Q \models \varphi$

8 Conclusion and Future Work

We propose a computational semantics for a variant of the Protocol Composition Logic presented in [2,7,8,10]. The associated soundness theorem implies that it is possible to reason symbolically, and at a high level, about probabilistic polynomial-time security properties. Although omitted from this conference paper, the soundness proof uses a combination of information-theoretic arguments, calculations about negligible functions, and cryptographic-style reductions involving encryption. While the semantics given here has some imperfections, such as an interpretation of indistinguishability that only seems appropriate for formulas where Indist appears with positive polarity, the general approach seems promising. We look forward to future efforts to lift certain restrictions on the logic, explore the semantics and axiomatization of logical connectives over probabilistic polynomial-time interpretations, and extend the approach suggested here to additional cryptographic primitives, such as signatures and hash functions. One interesting research direction might be to develop a version of this semantics based on information-theoretic security, since that may provide some useful insight into problems we encountered in developing the semantics.

Acknowledgments: Thanks to Bogdan Warinschi, Andre Scedrov, and Dan Boneh for many insightful comments and suggestions.

References

1. M. Abadi and P. Rogaway. Reconciling two views of cryptography (the computational soundness of formal encryption). *Journal of Cryptology*, 15(2):103–127, 2002.
2. M. Backes, A. Datta, A. Derek, J. C. Mitchell, and M. Turuani. Compositional analysis of contract signing protocols. In *Proceedings of 18th IEEE Computer Security Foundations Workshop*. IEEE, 2005. To appear.
3. M. Backes, B. Pfitzmann, and M. Waidner. A universally composable cryptographic library. Cryptology ePrint Archive, Report 2003/015, 2003.
4. M. Bellare, A. Boldyreva, and S. Micali. Public-key encryption in a multi-user setting: Security proofs and improvements. In *Advances in Cryptology - EUROCRYPT 2000, Proceedings*, pages 259–274, 2000.
5. M. Bellare and P. Rogaway. Entity authentication and key distribution. In *Proceedings of the 13th Annual International Cryptology Conference on Advances in Cryptology (CRYPTO '93)*, pages 232–249. Springer-Verlag, 1994.
6. V. Cortier and B. Warinschi. Computationally sound, automated proofs for security protocols. In *Proceedings of 14th European Symposium on Programming (ESOP'05)*, Lecture Notes in Computer Science, pages 157–171. Springer-Verlag, 2005.

7. A. Datta, A. Derek, J. C. Mitchell, and D. Pavlovic. A derivation system for security protocols and its logical formalization. In *Proceedings of 16th IEEE Computer Security Foundations Workshop*, pages 109–125. IEEE, 2003.
8. A. Datta, A. Derek, J. C. Mitchell, and D. Pavlovic. A derivation system and compositional logic for security protocols. *Journal of Computer Security*, 2005. To appear.
9. D. Dolev and A. Yao. On the security of public-key protocols. *IEEE Transactions on Information Theory*, 2(29):198–208, 1983.
10. N. Durgin, J. C. Mitchell, and D. Pavlovic. A compositional logic for proving security properties of protocols. *Journal of Computer Security*, 11:677–721, 2003.
11. J. Herzog. The Diffie-Hellman key-agreement scheme in the strand-space model. In *Proceedings of 16th IEEE Computer Security Foundations Workshop*, pages 234–247, 2003.
12. J. Herzog. *Computational Soundness for Standard Assumptions of Formal Cryptography*. PhD thesis, MIT, 2004.
13. R. Impagliazzo and B. M. Kapron. Logics for reasoning about cryptographic constructions. In *Proceedings of the 44th Annual IEEE Symposium on Foundations of Computer Science (FOCS '03)*, pages 372–383. IEEE, 2003.
14. R. Janvier, L. Mazare, and Y. Lakhnech. Completing the picture: Soundness of formal encryption in the presence of active adversaries. In *Proceedings of 14th European Symposium on Programming (ESOP'05)*, Lecture Notes in Computer Science, pages 172–185. Springer-Verlag, 2005.
15. D. Micciancio and B. Warinschi. Soundness of formal encryption in the presence of active adversaries. In *Theory of Cryptography Conference - Proceedings of TCC 2004*, volume 2951 of *Lecture Notes in Computer Science*, pages 133–151. Springer-Verlag, 2004.
16. N. J. Nilsson. Probabilistic logic. *Artificial Intelligence*, 28(1):71–87, 1986.
17. V. Shoup. On formal models for secure key exchange (version 4). Technical Report RZ 3120, IBM Research, 1999.
18. B. Warinschi. A computational analysis of the Needham-Schroeder(-Lowe) protocol. In *Proceedings of 16th Computer Science Foundation Workshop*, pages 248–262. ACM Press, 2003.

A Gentle Introduction to Semantic Subtyping[*]

Giuseppe Castagna[1] and Alain Frisch[2]

[1]CNRS, École Normale Supérieure de Paris, France
[2]INRIA, Rocquencourt, France

Abstract. Subtyping relations are usually defined either syntactically by a formal system or semantically by an interpretation of types into an untyped denotational model. This work shows how to define a subtyping relation semantically in the presence of boolean connectives, functional types and dynamic dispatch on types, without the complexity of denotational models, and how to derive a complete subtyping algorithm. The presentation is voluntarily kept informal and discursive and the technical details are reduced to a minimum since we rather insist on the motivations, the intuition, and the guidelines to apply the approach.

1 Introduction

Many recent type systems rely on a subtyping relation. Its definition generally depends on the type algebra, and on its intended use. We can distinguish two main approaches for defining subtyping: the *syntactic* approach and the *semantic* one. The syntactic approach—by far the more used—consists in defining the subtyping relation by axiomatising it in a formal system (a set of inductive or coinductive rules); in the semantic approach (for instance, [1, 4]), instead, one starts with a model of the language and an interpretation of types as subsets of the model, then defines the subtyping relation as the inclusion of denoted sets, and, finally, when the relation is decidable, derives a subtyping algorithm from the semantic definition.

The semantic approach has several advantages but it is also more constraining. Finding an interpretation in which types can be interpreted as subsets of a model may be a hard task. A solution to this problem was given by Haruo Hosoya and Benjamin Pierce [9, 8, 7] with the work on XDuce. The key idea is that in order to define the subtyping relation semantically one does not need to start from a model of the whole language: a model of the types suffices. In particular Hosoya and Pierce take as model of types the set of values of the language. Their notion of model cannot capture functional values. On the one hand, the resulting type system is poor since it lacks function types. On the other hand, it manages to integrate union, product and recursive types and still keep the presentation of the subtyping relation and of the whole type system quite simple.

In [6, 5], together with Véronique Benzaken, we extended the work on XDuce and reframed it in a more general setting: we show a technique to define semantic sub-

---

[*] Joint ICALP/PPDP 2005 keynote talk, short overview. The full article is included in the proceedings of PPDP '05 [3].

typing in the presence of a rich type system including function types, but also arbitrary boolean combinations (union, intersection, and negation types) and in the presence of lately bound overloaded functions and type-based pattern matching. The aim of [6, 5] was to provide a theoretical foundation on the top of which to build the language CDuce [2], an XML-oriented transformation language. This motivation needed a rather heavy technical development that concealed a side—but important—contribution of the work, namely a generic and uniform technique (or rather, a cookbook of techniques) to define semantic subtyping when straightforward set-theoretic interpretation does not work, in particular for arrow types. Here we concentrate on this second aspect of the work: we get rid of many features (e.g. patterns and pattern matching, full-fledged overloading, pattern variable type inference,...), skip many technical details, and focus on the basic intuition to gradually introduce our approach. This results in a presentation along which we explain the reader how to take her/his favourite set of type constructors (e.g. arrows, but also records, lists, pointers, channels, etc.) and add to it a complete set of boolean combinators: union, intersection and negation types.

Our hope is that this work will provide the reader with enough intuition and a detailed roadmap to decide whether it is possible/interesting to endow her/his favourite language with a set-theoretically defined subtyping relation.

2 Overview of the Approach

Our objective is to define give a semantic definition of a subtyping relation by interpreting types as sets and subtyping as set inclusion. We thus need to define a set-theoretic model of types. This model is usually defined by starting from a model of the terms of the object language. For a language with function types, we have to interpret the duality of functions as terms and as functions on terms. This yields the need to solve complicated recursive domain equations that hardly combines with a set-theoretic interpretation of types, whence the introduction of restrictions in the definition of semantic subtyping (e.g. no function types, no negation types, etc ...).

Note however that in order to define semantic subtyping all we need is a set-theoretic *model of types*. The construction works even if we do not have a model of terms. The definition of a semantic subtyping relation needs neither an interpretation for applications (that is, an applicative model) nor the solution of complicated domain equations.

The first key idea to generalise semantic subtyping is, then, to dissociate the *model of types* from the *model of terms* and define the former independently from the latter. In other words, the interpretation of types must not forcedly be based on an interpretation of terms. More formally, what we do is first to define an interpretation function for types $[\![\ ]\!] : \textit{Types} \to \mathcal{P}(\mathcal{D})$, where $\mathcal{P}$ denotes the powerset, and $\mathcal{D}$ is some domain (which may not be expressive enough to interpret terms). Then, we define the subtyping relation as follows:

$$s \leq t \stackrel{\text{def}}{\iff} [\![s]\!] \subseteq [\![t]\!].$$

The second key idea is that we do not need $[\![\ ]\!]$ to state what types *mean*, but just to describe how types *are related*. And to this end, all that matters is when $[\![\ ]\!]$ is equal to zero. Indeed,

$s \leq t \iff [\![s]\!] \subseteq [\![t]\!] \iff [\![s]\!] \cap \overline{[\![t]\!]} = \emptyset \iff [\![s]\!] \cap [\![\neg t]\!] = \emptyset \iff [\![s \wedge \neg t]\!] = \emptyset$

where the overbar denotes complementation in $\mathcal{D}$. Thus, we can concentrate our efforts just on the $[\![\ ]\!]$-counterimage of the empty set and disregard the behaviour of $[\![\ ]\!]$ on the remaining types.

Types as Sets of Values. Nevertheless, to ensure type safety, the meaning of types has to be somewhat correlated with the language. A classical solution is to interpret types as sets of *values*, that is, as the results of *well-typed* computations in the language. More formally, the values of a typed language are all the terms that are well-typed, closed, and in normal form. Interpreting types a sets of values is much easier than interpreting them as sets of terms: since a closed application usually denotes a redex, then by restricting to the sole values we avoid the need to interpret application and, therefore, also the need to solve complicated domain equations. This is the solution adopted by XDuce, where values are XML documents and types are sets of documents.

But if we consider a language with arrow types, that is a language with higher order functions, then the applications come back again: arrow types must be interpreted as sets of function values, that is, as sets of well-typed closed lambda abstractions, and applications may occur in the body of these abstractions. Here is where XDuce stops and it is the reason why it does not include arrow types.

A Circularity to Break. Introducing arrow types is problematic because it slips applications back again in the interpretation of types. However this does not mean that we need a semantic interpretation for application, it just implies that we must define how application is typed. In particular, since functional values are *well-typed* lambda abstractions, then to interpret functional types we must be able to type the applications that occur in the body of lambda abstractions. Now this is not an easy task in our context: in the absence of higher order functions the set of values of type constructors such as products or records can be inductively defined from basic types without resorting to any typing relation (this is why the Hosoya Pierce approach works smoothly). With the arrow type constructor, instead, this can be done only by using a typing relation which yields to a circularity: in order to define the subtyping relation we need an interpretation of the types of the language; for this we have to define which are the values of an arrow type; this needs that we define the typing relation for applications, which in turns needs the definition of the subtyping relation.

Thus, if we want to define the semantic subtyping of arrow types we must find a way the avoid this circularity. The simplest way to avoid it is to break it. We already said that to define set-theoretic subtyping we *must* have a model of types; it is also clear that the typing relation *must* use subtyping; on the contrary it is not strictly necessary for our model to be based on the interpretation of values, this is just convenient since it ties the types with the language the types are intended for. This is therefore the weakest link and we can break it. So the idea is to start from a model (of the types) defined independently (but not too much) from the language the types are intended for (and therefore independently from its values), and then from that define the rest: subtyping, typing, set of values.

For a given type algebra and a language, the approach can be summarised as follows:

1. First, we define a notion of *set-theoretic model* of the type algebra. In such a model, set-theoretic connectives in the type algebra (that is union, intersection, and negation) must be interpreted in a set-theoretic way. Also, the notion of model must capture the essence of the type constructors, and in particular of the function types. For instance, in [6,5], we show that an extensional interpretation of functions as binary relations is suitable for a language with overloaded functions and dynamic type dispatch.
2. There might be several models, each of them induces a specific subtyping relation on the type algebra. We only need to prove that there exists at least one model, and to pick one, which we call the *bootstrap model*. In [6,5], we exhibit a model with an universal property: it induces the largest possible subtyping relation.
3. Now we have a (hopefully) suitable subtyping relation available, we can focus again on the language itself and consider its typing rules (which depend on the subtyping relation). The rich set-theoretic structure of models helps us to carry out the meta-theoretic study of the system, and to prove easily classical properties such as the transitivity of subtyping, subsumption elimination, and so on.
4. The typing judgement for the language produces a new natural set-theoretic interpretation of types: a type t denotes the set of values of type t. This interpretation induces a new subtyping relation, which might be different from the one we used to define the type system. However, if the definition of the models and the typing rules have been carefully chosen, then we can expect the two subtyping relations to coincide. This is the case in [6,5].

The last step is the most critical one. It consists in "closing the circle": even if we have used a detour to create a subtyping relation, in the end we obtain a consistent system where types are interpreted as sets of values and subtyping is exactly set-inclusion of types. The rest of the story is standard: we can define the operational semantics for the language (we consider a language whose semantics is driven by types) and prove a type preservation result by a classical syntactic argument. This proof requires a lot of intermediate properties about the subtyping relation, which can be obtained easily using the definition of model. Also, we can use set-theoretic reasoning to derive from this definition an actual algorithm to compute the subtyping relation. As for the subtyping relation, this algorithm obviously depends on the bootstrap model we have chosen.

In the full version [3] we show the details of this construction and hint at how it can be applied to other type constructors such as product, lazy, reference, and channel types.

References

1. A. Aiken and E. L. Wimmers. Type inclusion constraints and type inference. In *Proc. of the 7th ACM Conference on Functional Programming and Computer Architecture*, 1993.
2. V. Benzaken, G. Castagna, and A. Frisch. CDuce: an XML-friendly general purpose language. In *ICFP '03, 8th ACM International Conference on Functional Programming*, 2003.
3. G. Castagna and A. Frisch. A gentle introduction to semantic subtyping. In Proc. of *PPDP '05, the 7th ACM Symposium on Principles and Practice of Declarative Programming*, 2005.

4. F. Damm. Subtyping with union types, intersection types and recursive types II. Research Report 816, IRISA, 1994.
5. Alain Frisch. *Théorie, conception et réalisation d'un langage de programmation fonctionnel adapté à XML*. PhD thesis, Université Paris 7, December 2004.
6. A. Frisch, G. Castagna, and V. Benzaken. Semantic Subtyping. In Proc. of *LICS '02, the 7th Annual IEEE Symposium on Logic in Computer Science*, pages 137–146, 2002.
7. H. Hosoya and B. Pierce. XDuce: A typed XML processing language. *ACM Transactions on Internet Technology*, 3(2):117–148, 2003.
8. H. Hosoya. *Regular Expression Types for XML*. PhD thesis, University of Tokyo, 2001.
9. H. Hosoya and B. C. Pierce. Regular expression pattern matching for XML. In Proc. of *POPL '01, the 25th Annual ACM Symposium on Principles of Programming Languages*, 2001.

Logics for Unranked Trees: An Overview

Leonid Libkin*

University of Toronto
libkin@cs.toronto.edu

Abstract. Labeled unranked trees are used as a model of XML documents, and logical languages for them have been studied actively over the past several years. Such logics have different purposes: some are better suited for extracting data, some for expressing navigational properties, and some make it easy to relate complex properties of trees to the existence of tree automata for those properties. Furthermore, logics differ significantly in their model-checking properties, their automata models, and their behavior on ordered and unordered trees. In this paper we present a survey of logics for unranked trees.

1 Introduction

Trees arise everywhere in computer science, and there are numerous formalisms in the literature for describing and manipulating trees. Some of these formalisms are declarative and based on logical specifications: for example, first-order logic, or monadic second-order logic, or various temporal or fixed-point logics over trees. Others are procedural formalisms such as various flavors of tree automata, or tree transducers, or tree grammars. All these formalisms have found numerous applications in verification, program analysis, logic programming, constraint programming, linguistics, and databases.

Until recently, most logical formalisms for trees dealt with *ranked* trees [18, 58]: in such trees, all nodes have the same fixed number of children (or, a bit more generally, the number of children of a node is determined by the label of that node). Over the past several years, however, the focus has shifted towards *unranked* trees, in which there are no restrictions on the number of children a node can have. Although unranked trees have been considered in the 60s and 70s, and are related to feature trees over an infinite set of features that have been investigated by computational linguists, their systematic study was initiated by the development of XML (eXtensible Markup Language). XML is a data format which has become the lingua franca for information exchange on the world wide web. XML data is typically modeled as labeled unranked trees [42].

This connection has led to a renewed interest in logical and procedural formalisms for unranked trees: one uses logical formalisms for expressing declarative queries, and procedural formalisms for evaluating them. Logics over unranked trees appeared in large numbers over the past 7–8 years, and they come in many flavors in shapes. Common to them is a close connection to automata models, and quite often to temporal and modal logics, especially when one describes properties of paths through a document.

---

* Complete version of this survey can be found at www.cs.toronto.edu/˜libkin/publ.html.

Let us now review some of the parameters according to which logics for unranked trees can be classified.

The yardstick logic. Most formalisms are "inspired" by either *first-order logic* (FO), or *monadic second-order logic* (MSO) that extends FO by quantification over sets. Query languages and schema formalisms for XML tend to use MSO as the yardstick: for example, XML DTDs are (almost) equivalent to MSO sentences, and various language for extraction of data from XML documents have the power of MSO unary queries. On the other hand, navigational aspects of XML, in particular, logics capturing various fragments of XPath, are usually related to FO and its fragments.

Arity of queries. Most commonly one considers Boolean or unary queries. Boolean queries evaluate to *true* or *false*. Checking if an XML document conforms to a schema specification is represented by a Boolean query. Unary queries correspond to formulae in one free variable, and thus produce a set of nodes. E.g., extracting sets of nodes, or evaluating XPath expressions relative to the root naturally give rise to unary queries.

Complexity of model-checking. The model-checking problem asks whether a tree T satisfies a logical sentence φ. If φ is an MSO sentence φ, it can be evaluated in linear time in the size of T, by converting to a tree automaton. But there is a price to pay: in terms of the size of φ, the complexity becomes non-elementary. This type of trade-offs is one of the central issues in dealing with logics over trees.

Ordered vs. unordered trees. In unranked XML trees, children of the same node are ordered by a *sibling ordering*. If such an order is present, we speak of ordered unranked trees. In many cases, however, this ordering is irrelevant, and some models, such as feature trees, do not impose any ordering on siblings. There is considerable difference between the expressiveness of logics and automata models depending on the availability of sibling ordering. The presence of ordering also affects the yardstick logic, since without order often counting is needed to match the power of automata models [19].

The paper is organized as follows. After we give basic definitions in Section 2, we review logics for ordered trees in Section 3. We start with MSO-related logics, including syntactic restrictions of MSO, a datalog-based logic, and the μ-calculus. We then turn to FO-related logics, present analogs of LTL and CTL* that have been studied for expressing navigational properties, and also look at conjunctive queries over trees. In Section 4 we turn to trees that lack the sibling ordering, and show that in many logics some form of counting needs to be added to compensate for the missing ordering. In Section 5 we look at the model-theoretic approach in the spirit of automatic structures.

2 Trees, Logics, and Automata

Tree domains, trees, and operations on trees. Nodes in unranked trees are elements of $\mathbb{N}^*$ – that is, finite strings whose letters are natural numbers. A string $s = n_0 n_1 \ldots$ defines a path from the root to a give node: one goes to the n_0th child of the root, then to the n_1th child, etc. We write $s_1 \cdot s_2$ for the concatenation of strings s_1 and s_2.

We need some basic binary relations on $\mathbb{N}^*$ – the *child* and *next-sibling* relations:

$$s \prec_{\text{ch}} s' \Leftrightarrow s' = s \cdot i \text{ for some } i \in \mathbb{N};$$
$$s \prec_{\text{ns}} s' \Leftrightarrow s = s_0 \cdot i \text{ and } s' = s_0 \cdot (i+1) \text{ for some } s_0 \in \mathbb{N}^* \text{ and } i \in \mathbb{N}.$$

We also use the *first child relation*: $s \prec_{\text{fc}} s \cdot 0$. We shall use * to denote the reflexive-transitive closure of a relation. Thus, $\prec_{\text{ch}}^*$ is the *descendant* relation (including self), and $\prec_{\text{ns}}^*$ is a linear ordering on siblings.

Definition 1 (Tree domain). *A tree domain D is a finite prefix-closed subset of $\mathbb{N}^*$ (i.e., if $s \in D$ and s' is a prefix of s, then $s' \in D$) such that $s \cdot i \in D$ implies $s \cdot j \in D$ for all $j < i$.*

Let Σ be a finite alphabet.

Definition 2 (Σ-trees). *An ordered unranked Σ-labeled tree T is a structure*

$$T = \langle D, \prec_{\text{ch}}^*, \prec_{\text{ns}}^*, (P_a)_{a \in \Sigma} \rangle,$$

where D is a tree domain, $\prec_{\text{ch}}^*$ and $\prec_{\text{ns}}^*$ are the descendant relation and the sibling ordering, and P_a's are interpreted as disjoint sets whose union is the entire domain D. An unordered unranked tree is defined as a structure $\langle D, \prec_{\text{ch}}^*, (P_a)_{a \in \Sigma} \rangle$.

Thus, a tree consists of a tree domain together with a labeling on its nodes: if $s \in P_a$, then the label of s is a. In this case we write $\lambda_T(s) = a$.

First-order and Monadic Second-Order Logic. We only consider relational vocabularies: finite lists $(R_1, \ldots, R_m)$ of relation symbols, each R_i with an associated arity n_i. Over trees, relation symbols are binary (e.g., $\prec_{\text{ch}}, \prec_{\text{ns}}, \prec_{\text{ch}}^*$) or unary ($P_a$'s for $a \in \Sigma$).

Formulae of *first-order* logic (FO) are built from atomic formulae $x = x'$, and $R(\bar{x})$, where x, x' are variables, and $\bar{x}$ is a tuple of variables, using the Boolean connectives $\vee, \wedge, \neg$ and quantifiers $\exists$ and $\forall$. If a formula φ has free variables $\bar{x}$, we shall write $\varphi(\bar{x})$.

Formulae of *monadic second-order* logic (MSO) in addition allow quantification over sets. We shall normally denote sets of nodes by upper case letters. Thus, MSO formulae have the usual first-order quantifiers $\exists x \varphi$ and $\forall x \varphi$ as well as second-order quantifiers $\exists X \varphi$ and $\forall X \varphi$, and new atomic formulae $X(x)$, where X is a second-order variable and x is a first-order variable. An MSO formula may have both free first-order and second-order variables. If it only has free first-order variables, then it defines a relation on the universe of the structure.

Note that relations $\prec_{\text{ch}}$ and $\prec_{\text{ns}}$ are definable, in FO, from $\prec_{\text{ch}}^*$ and $\prec_{\text{ns}}^*$. In MSO one can define $\prec_{\text{ch}}^*$ from $\prec_{\text{ch}}$; however, it is well-known that in FO this is *not* possible. This is why we chose $\prec_{\text{ch}}^*$ and $\prec_{\text{ns}}^*$, rather than $\prec_{\text{ch}}$ and $\prec_{\text{ns}}$, as our basic relations.

Definition 3 (Definability in logic). *Given a logic $\mathcal{L}$, we say that a set of trees $\mathcal{T}$ is definable in $\mathcal{L}$ if there is a sentence φ of $\mathcal{L}$ such that $T \in \mathcal{T}$ iff $T \models \varphi$. We say that a unary query Q (that selects nodes from trees) is definable in $\mathcal{L}$ if there is a formula $\psi(x)$ of $\mathcal{L}$ such that $s \in Q(T)$ iff $T \models \psi(s)$, for every tree T and a node s in T.*

<u>Unranked Tree Automata.</u> An *nondeterministic unranked tree automaton*, NUTA [56, 9], over Σ-labeled trees is a triple $\mathcal{A} = (Q, F, \delta)$ where Q is a finite set of states, $F \subseteq Q$ is the set of final states, and δ is a mapping $Q \times \Sigma \to 2^{Q^*}$ such that $\delta(q, a)$ is a regular language over Q (normally represented by a regular expression over Q). A run of $\mathcal{A}$ on a tree T with domain D is a function $\rho_{\mathcal{A}} : D \to Q$ such that, if s is a node with n children, and it is labeled a, then the string $\rho_{\mathcal{A}}(s \cdot 0) \cdots \rho_{\mathcal{A}}(s \cdot (n-1))$ is in $\delta(\rho_{\mathcal{A}}(s), a)$.

In particular, if s is a leaf labeled a, then $\rho_{\mathcal{A}}(s) = q$ implies that $\varepsilon \in \delta(q, a)$. A run is *accepting* if $\rho_{\mathcal{A}}(\varepsilon) \in F$, that is, the root is in an accepting state. A tree T is *accepted* by $\mathcal{A}$ if there exists an accepting run. We let $L(\mathcal{A})$ denote the set of all trees accepted by $\mathcal{A}$. Such sets of trees will be called *regular*.

Binary trees and translations. A *binary tree domain* is a prefix-closed subset D of $\{0, 1\}^*$ such that if $s \cdot i \in D$, then $s \cdot (1 - i) \in D$ (that is, a node is either a leaf, or both its children are in D). A (binary) *nondeterministic tree automaton*, NTA, is a quadruple $\mathcal{A}_b = (Q, q_0, F, \delta)$ where Q and F are as before, q_0 is the initial state, and δ is a function $Q \times Q \times \Sigma \to 2^Q$. A run $\rho_{\mathcal{A}_b}$ on a binary tree T with domain D is a function from D to Q such that if s is a leaf labeled a, then $\rho_{\mathcal{A}_b}(s) \in \delta(q_0, q_0, a)$, and if $s \cdot 0, s \cdot 1$ belong to D, and s is labeled a, then $\rho_{\mathcal{A}_b}(s) \in \delta(\rho_{\mathcal{A}_b}(s \cdot 0), \rho_{\mathcal{A}_b}(s \cdot 1), a)$. A run is accepting if $\rho_{\mathcal{A}_b}(\varepsilon) \in F$, and $L(\mathcal{A}_b)$ is the set of all binary trees for which there exists an accepting run. Such sets are called regular.

There is a well-known regularity-preserving translation between ranked and unranked trees. It was used in [49] to show decidability of SωS (but here we shall apply it only to finite tree domains). The idea of the translation is that the first successor in the binary tree corresponds to the first child, and the second successor to the next sibling. More precisely, we define a mapping $\mathcal{R} : \mathbb{N}^* \to \{0, 1\}^*$ such that $\mathcal{R}(\varepsilon) = \varepsilon$, and if $\mathcal{R}(s) = s'$, where $s = s_0 \cdot i$, then $\mathcal{R}(s \cdot 0) = s' \cdot 0$ and $\mathcal{R}(s_0 \cdot (i+1)) = s' \cdot 1$. If D is an unranked tree domain, we let $\mathcal{R}(D)$ be $\{\mathcal{R}(s) \mid s \in D\}$ together with $\mathcal{R}(s) \cdot 1$ if s is a non-leaf last child, and $\mathcal{R}(s) \cdot 0$ if s a leaf, other than the last sibling (these additions ensure that $\mathcal{R}(D)$ is a binary tree domain). We define $\mathcal{R}(T)$ to be a tree with domain $\mathcal{R}(D)$, where $\mathcal{R}(s)$ has the same label as s, and the added nodes are labeled by a symbol $\bot \notin \Sigma$. The following is a folklore result.

Lemma 1. *For every NUTA $\mathcal{A}$, there is an NTA $\mathcal{A}_b$ such that $L(\mathcal{A}_b) = \{\mathcal{R}(T) \mid T \in L(\mathcal{A})\}$, and for every NTA $\mathcal{A}_b$ there is an NUTA $\mathcal{A}$ such that the above holds.*

3 Ordered Trees

In this section we only deal with ordered unranked trees. We first survey MSO-based logics, and then move to FO-based ones.

3.1 MSO and Its Relatives

As we mentioned already, MSO is often used as a yardstick logic for trees, because of its close connection to regular languages. The following result belonged to folklore, and was explicitly stated in [41].

Theorem 1. *A set of unranked trees is regular iff it is definable in MSO.*

When one considers binary trees, this result says that regular sets of binary trees are precisely those MSO-definable, and if we look at strings, which may be viewed as trees without branching, we obtain that regular languages are precisely those MSO-definable. Of course these are well-known results by Büchi [10], and Thatcher, Wright [57].

There is also a close connection between automata, MSO, and a formalism for describing XML schemas, called DTDs (which are essentially extended context-free grammars). A DTD d over an alphabet Σ is a collection of rules $a \to e_a$, where $a \in \Sigma$ and e_a is a regular expression over Σ. We shall assume there is at most one such rule for each $a \in \Sigma$. A Σ-labeled tree T satisfies d, if for each node s of T with n children, and $\lambda_T(s) = a$, the string $\lambda_T(s \cdot 0) \cdots \lambda_T(s \cdot (n-1))$ is in the language denoted by e_a.

Each DTD is easily definable by an unranked tree automaton: in fact its states just correspond to labels of nodes. This, however, is too restrictive to capture full definability in MSO, but a slight extension of DTDs does precisely that. An *extended DTD* over Σ is a triple (Σ', d', g) where $\Sigma' \supseteq \Sigma$, with g being a mapping $g : \Sigma' \mapsto \Sigma$, and d' is a DTD over Σ'. A Σ-labeled tree T satisfies (Σ', d', g) if there is a Σ'-labeled tree T' that satisfies d' such that $T = g(T')$. The following was established in [56].

Proposition 1. *A set of unranked trees is MSO definable iff it is the set of all trees satisfying some extended DTD (Σ', d', g).*

Theorem 1 talks about MSO sentences, but it can be extended to unary MSO queries using the concept of *query automata* [44]. A (nondeterministic) *query automaton* over unranked Σ-labeled trees is a quadruple $\mathcal{QA} = (Q, F, \delta, S)$ where $\mathcal{A} = (Q, F, \delta)$ is an UNTA, and S is a subset of $Q \times \Sigma$. Such a query automaton defines a unary query $\mathcal{Q}_{\mathcal{QA}}$ that selects nodes s in T such that $(\rho_{\mathcal{A}}(s), \lambda_T(s)) \in S$ for some accepting run $\rho_{\mathcal{A}}$.

Theorem 2. (see [44, 41, 24]) *A unary query $\mathcal{Q}$ on unranked trees is MSO-definable iff it is of the form $\mathcal{Q}_{\mathcal{QA}}$ for some query automaton.*

One can also define the semantics universally ($(\rho_{\mathcal{A}}(s), \lambda_T(s)) \in S$ for all accepting runs) and the result still holds. Query automata have a deterministic counterpart; however, in the deterministic version, two passes over the tree are required; see [44].

Theorems 1 and 2 are constructive. In particular, every MSO sentence φ can be effectively transformed into an automaton $\mathcal{A}_\varphi$ that accepts a tree T iff $T \models \varphi$. Since tree automata can be determinized, this gives us a $O(\|T\|)$ algorithm to check whether $T \models \varphi$, if φ is fixed[1]. However, it is well-known that the size of $\mathcal{A}_\varphi$ (even for string automata) cannot be bounded by an elementary function in $\|\varphi\|$ [55]. An even stronger result of [23] says that there could be no algorithm for checking whether $T \models \varphi$ that runs in time $O(f(\|\varphi\|) \cdot \|T\|)$, where f is an elementary function, unless PTIME=NP.

Nonetheless, these results do not rule out the existence of a logic $\mathcal{L}$ that has the same power as MSO and yet permits faster model-checking algorithms. Even looking at a simpler case of FO on strings, where results of [23] also rule out $O(f(\|\varphi\|) \cdot |s|)$ algorithms for checking if a string s satisfies φ, with f being an elementary function, the logic LTL (linear-time temporal logic) has the same expressiveness as FO [33] and admits model-checking algorithm with running time $2^{O(\|\varphi\|)} \cdot |s|$.

Logic ETL. The first logic for unranked trees that has the power of MSO and model-checking complexity matching that of LTL appeared in [43] and was called ETL (*efficient tree logic*). It was obtained by putting syntactic restrictions on MSO formulae, and at the same time adding new constructors for formulae, which are not present in MSO, but are MSO-definable.

---

[1] We use the notation $\|T\|, \|\varphi\|$ to denote the sizes of natural encodings of trees and formulae.

The atomic formulae of ETL are the same as for MSO, except that we are allowed to use both $\prec_{ch}$ and $\prec_{ch}^*$ and are *not* allowed to use the next-sibling relation $\prec_{ns}^*$. ETL is closed under Boolean combinations (which are required to be in DNF), *guarded quantification*, and *path formulae*. The rules for guarded quantification are:

- if $\varphi(x, y, X)$ is an ETL formula, then $\exists y\ (x \prec_{ch} y \wedge \varphi)$ and $\exists y\ (x \prec_{ch}^* y \wedge \varphi)$ are ETL formulae;
- if $\varphi(x, X)$ is an ETL formula, then $\exists X\ (x \prec_{ch}^* X \wedge \varphi)$ is an ETL formula. Here $x \prec_{ch}^* X$ means that X only contains descendants of x. In this case φ cannot contain vertical path formulae (defined below).

Path formulae are defined as follows:

- if e is a regular expression over ETL formulae $\psi(u, v)$, then $e^{\downarrow}(x, y)$ is a (vertical path) ETL formula. The semantics is as follows: $T \models e^{\downarrow}(s, s')$ if there is a child-relation path $s = s_0, s_1, \ldots, s_n = s'$ in T and a sequence of ETL formulae $\psi_i(u, v)$, $i \leq n - 1$, such that $T \models \psi_i(s_i, s_{i+1})$ for each $i \leq n - 1$, and the sequence $\psi_0 \ldots \psi_{n-1}$ matches e.
- if e is a regular expression over ETL formulae $\psi(u, \bar{X})$, then $e^{\rightarrow}(x, \bar{X})$ is a (horizontal path) ETL formula. Then $T \models e^{\rightarrow}(s, \bar{X})$ if children $s \cdot i, i \leq k$ of s can be labeled with ETL formulae $\psi_i(u, \bar{X})$ such that $T \models \psi_i(s \cdot i, \bar{X})$ for all i, and the sequence $\psi_0 \ldots \psi_k$ matches e.

Theorem 3. (see [43]) *With respect to Boolean and unary queries,* ETL *and* MSO *are equally expressive. Furthermore, each* ETL *formula φ can be evaluated on a tree T in time $2^{O(\|\varphi\|)} \cdot \|T\|$.*

Monadic datalog. Another approach to obtaining the full power of MSO while keeping the complexity low is based on database query language *datalog* (cf. [1]). A datalog program is a sequence of rules $H{:}{-}P_1, \ldots, P_k$ where H and all P_i's are atomic formulae. The predicate H is called the head of the rule, and every variable that appears in H is required to appear in one of the P_i's. Given a datalog program $\mathcal{P}$, predicates which appear as a head of some rule are called intensional, and other predicates are called extensional. If all intensional predicates are monadic (of the form $H(x)$), then $\mathcal{P}$ is a *monadic* datalog program. The semantics is a standard fixed-point semantics, see, e.g., [1]. An intensional unary predicate of a program $\mathcal{P}$ defines a unary query.

For extensional predicates, we shall need *Leaf, LastChild*, and *Root*. Given a tree domain D, they are interpreted as $\textit{Leaf} = \{s \in D \mid \neg \exists s' \in D : s \prec_{ch} s'\}$, $\textit{LastChild} = \{s \cdot i \in D \mid s \cdot (i+1) \notin D\}$ and $\textit{Root} = \{\varepsilon\}$.

Theorem 4. (see [25]) *A unary query over unranked trees is definable in* MSO *iff it is definable in monadic datalog over extensional predicates $\prec_{fc}, \prec_{ns}$, Leaf, LastChild, Root, and $P_a, a \in \Sigma$. Furthermore, each monadic datalog query $(\mathcal{P}, H)$ can be evaluated on a tree T in time $O(\|\mathcal{P}\| \cdot \|T\|)$.*

μ-calculus. Yet another way of getting a logic equivalent to MSO is suggested by a close connection between MSO and the modal μ-calculus L_μ on ranked trees, which can easily be extended to the unranked case by using the connection between ranked

and unranked trees. It was shown in [22, 47] that every property of infinite binary trees definable in MSO is also be definable in L_μ. To deal with unranked trees, we shall define L_μ over Σ-labeled structures that have several binary relations $E_1, \ldots, E_m$, cf. [2]. Formulae of $L_\mu[E_1, \ldots, E_m]$ are given by

$$\varphi := a\ (a \in \Sigma) \mid X \mid \varphi \vee \varphi \mid \neg \varphi \mid \Diamond(E_i)\varphi \mid \mu X\ \varphi(X),$$

where in $\mu X\ \varphi(X)$, the variable X must occur positively in φ. Given a tree T with domain D, $s \in D$, and a valuation v for free variables (each $v(X)$ is a subset of D), we define the semantics (omitting the rules for letters $a \in \Sigma$ and Boolean connectives) by
- $(T, v, s) \models X$ iff $s \in v(X)$.
- $(T, v, s) \models \Diamond(E_r)\varphi$ iff $(T, v, s') \models \varphi$ for some s' with $(s, s') \in E_r$.
- $(T, v, s) \models \mu X\ \varphi(X)$ iff s is in the least fixed point of the operator defined by φ.

An L_μ formula φ without free variables naturally defines a unary query on trees ($\{s \mid (T, s) \models \varphi\}$) and a Boolean query on trees (by checking if $(T, \varepsilon) \models \varphi$).

Using the translation into ranked trees, it is easy to show (see [3]):

Proposition 2. *The class of Boolean* MSO *queries on unranked trees is precisely the class of Boolean queries defined by* $L_\mu[\prec_{\text{fc}}, \prec_{\text{ns}}]$.

It is also possible to characterize unary MSO queries over unranked trees in terms of the *full* μ-calculus L_μ^{full} (cf. [59]) which adds backward modalities $\Diamond(E_i^-)\varphi$ with the semantics $(T, s) \models \Diamond(E_i^-)\varphi$ iff $(T, s') \models \varphi$ for some s' such that $(s', s) \in E_i$.

Proposition 3. (see [3]) *The class of unary* MSO *queries on unranked trees is precisely the class of queries defined by* $L_\mu^{\text{full}}[\prec_{\text{ch}}, \prec_{\text{ns}}]$.

3.2 FO and Its Relatives

While much is known about FO on both finite and infinite strings, it has not been as extensively studied for trees until recently. Recall that over strings – which we can view as trees with only unary branching – FO defines precisely the star-free languages (cf. [58]), and over both finite and infinite strings FO has exactly the power of LTL [33].

In contrast, the natural analog of star-free expressions over binary trees captures not FO but MSO [48]. One well-known equivalent logical description of FO on binary trees is Hafer-Thomas's theorem [31] stating that over finite binary trees, FO = CTL* (CTL* is a branching time temporal logic widely used in verification, cf. [16], and it will be defined shortly). Actually, the result of [31] shows that CTL* is equivalent to MSO with second-order quantification over paths only, but over finite trees this fragment of MSO is equivalent to FO.

The interest in logics over unranked trees whose power is equal to or subsumed by that of FO stems from the fact that navigational features of XPath can be described in FO. XPath [17] is a W3C standard for describing paths in XML documents. Thus, it is very natural to look for connections between XPath, FO on trees, and temporal logics, which are designed to talk about properties of paths.

Logics introduced in the context of studying XPath, and more generally, navigational properties of XML documents, can be roughly subdivided into two groups. Firstly, one may try to establish analogs of Kamp's theorem (stating that FO = LTL over strings) for trees. Secondly, one can try extended Hafer-Thomas's theorem (the equivalence FO = CTL*) from binary to unranked trees.

XPath and Temporal Logics. First, recall the syntax of LTL over alphabet Σ:

$$\varphi, \varphi' := a, a \in \Sigma \mid \varphi \vee \varphi' \mid \neg \varphi \mid \mathbf{X}\varphi \mid \mathbf{X}^-\varphi \mid \varphi \mathbf{U}\varphi' \mid \varphi \mathbf{S}\varphi'.$$

Formulae of LTL are interpreted over finite or infinite strings over Σ. Given a string $s = a_0 a_1 \ldots$, the semantics is as follows: $(s,i) \models a$ iff $a_i = a$, $(s,i) \models \mathbf{X}\varphi$ ("next" φ) iff $(s, i+1) \models \varphi$; $(s,i) \models \mathbf{X}^-\varphi$ iff $(s, i-1) \models \varphi$; $(s,i) \models \varphi \mathbf{U}\varphi'$ (φ "until" φ') if there exists $j \geq i$ such that $(s,j) \models \varphi'$ and $(s,k) \models \varphi$ for all $i \leq k < j$, and the semantics of the dual $\varphi \mathbf{S}\varphi$ (φ "since" φ') is that there exists $j \leq i$ such that $(s,j) \models \varphi'$ and $(s,k) \models \varphi$ for all $j < k \leq i$. (Note: it is possible to avoid $\mathbf{X}$ and $\mathbf{X}^-$ by defining a strict semantics for $\mathbf{U}$ and $\mathbf{S}$, without requiring φ to be true in (s,i)).

A logic TL^{tree} (*tree temporal logic*) is a minor extension of LTL:

$$\varphi, \varphi' := a, a \in \Sigma \mid \varphi \vee \varphi' \mid \neg \varphi \mid \mathbf{X}_*\varphi \mid \mathbf{X}^-_*\varphi \mid \varphi \mathbf{U}_*\varphi' \mid \varphi \mathbf{S}_*\varphi',$$

where $*$ is either 'ch' (child) or 'ns' (next sibling). We define the semantics with respect to a tree and a node in a tree: $(T,s) \models a$ iff $\lambda_T(s) = a$; $(T,s) \models \mathbf{X}_{\text{ch}}\varphi$ if $(T, s \cdot i) \models \varphi$ for some i; $(T,s) \models \mathbf{X}^-_{\text{ch}}\varphi$ if $(T, s') \models \varphi$ for the node s' such that $s' \prec_{\text{ch}} s$; $(T,s) \models \varphi \mathbf{U}_{\text{ch}} \varphi'$ if there is a node s' such that $s \prec^*_{\text{ch}} s'$, $(T, s') \models \varphi'$, and for all $s'' \neq s'$ satisfying $s \prec^*_{\text{ch}} s'' \prec^*_{\text{ch}} s'$ we have $(T, s'') \models \varphi$. The semantics of $\mathbf{S}_{\text{ch}}$ is defined by reversing the order in the semantics of $\mathbf{U}_{\text{ch}}$, and the semantics of $\mathbf{X}_{\text{ns}}, \mathbf{X}^-_{\text{ns}}, \mathbf{U}_{\text{ns}}$, and $\mathbf{S}_{\text{ns}}$ is the same by replacing the child relation with the next sibling relation.

As L_μ, the logic TL^{tree} naturally defines unary and Boolean queries on trees.

Theorem 5. (see [38]) *A unary or Boolean query over unranked trees is definable in FO iff it is definable in TL^{tree}.*

In both CTL* and XPath formalisms there are two kinds of formulae: those evaluated in nodes of trees, and those evaluated on paths in trees (these are state and path formulae of CTL* and filter and location path expressions of XPath).

We now look at XPath-inspired logics, and present them using a slight modification of the syntax that keeps all the main XPath constructions and yet makes the connection with temporal logics more visible. The language CXPath [38] (*Conditional XPath*) is defined to have *node formulae* α and *path formulae* β given by:

$$\alpha, \alpha' := a, a \in \Sigma \mid \neg \alpha \mid \alpha \vee \alpha' \mid \mathbf{E}\beta$$
$$\beta, \beta' := ?\alpha \mid \text{step} \mid (\text{step}/?\alpha)^+ \mid \beta/\beta' \mid \beta \vee \beta'$$

where **step** is one of the following: $\prec_{\text{ch}}, \prec^-_{\text{ch}}, \prec_{\text{ns}}$, or $\prec^-_{\text{ns}}$. Intuitively $\mathbf{E}\beta$ states the existence of a path starting in a given node and satisfying β, $?\alpha$ tests if α is true in the initial node of a path, and / is the composition of paths.

Formally, given a tree T, we evaluate each node formula in a node s, and each path formula in a pair of nodes (s, s'). The main semantic rules are:

- $(T,s) \models \mathbf{E}\beta$ iff there is s' such that $(T, s, s') \models \beta$;
- $(T, s, s') \models ?\alpha$ iff $s = s'$ and $(T, s) \models \alpha$;

- $(T, s, s') \models \mathsf{step}$ iff $(s, s') \in \mathsf{step}$;
- $(T, s, s') \models \beta/\beta'$ iff for some s'' we have $(T, s, s'') \models \beta$ and $(T, s'', s') \models \beta'$;
- $(T, s, s') \models (\mathsf{step}/?\alpha)^+$ if there exists a sequence of nodes $s = s_0, s_1, \ldots, s_k = s'$, $k > 0$, such that each (s_i, s_{i+1}) is in step, and $(T, s_{i+1}) \models \alpha$ for each $i < k$.

The language Core_XPath [26] is obtained by only allowing step^+ as opposed to $(\mathsf{step}/?\alpha)^+$ in the definition of path formulae. Notice that since $\mathsf{step}^+ = (\mathsf{step}/?\mathit{true})$, where $\mathit{true} = \bigvee_{a \in \Sigma} a$, we have Core_XPath $\subseteq$ CXPath.

Core_XPath corresponds to XPath as defined by W3C [17], while CXPath represents an addition to XPath proposed by [38]. Node formulae of either CXPath or Core_XPath naturally define unary queries on trees. These can be characterized as follows.

Theorem 6. *a)* (see [38]) *The node formulae of* CXPath *have precisely the power of* FO *unary queries.*

b) (see [39]) *The node formulae of* Core_XPath *have precisely the power of unary* FO^2 *queries (that is,* FO *with two variables) in the vocabulary* $\prec_{\mathrm{ch}}, \prec_{\mathrm{ch}}^*, \prec_{\mathrm{ns}}, \prec_{\mathrm{ns}}^*$.

A CTL^*-*Like Logic.* CTL^* is a branching time temporal logic used in verification of reactive systems. Here we define it with past connectives, using the syntax close to that of [35]. In CTL^*, one also has node (normally called state) formulae and path formulae, but path formulae are evaluated on paths, not on arbitrary pairs of nodes.

We define CTL^*_{past} node formulae α, and child and sibling path formulae β_*, for $*$ being 'ch' or 'ns', as follows:

$$\alpha, \alpha' := a \ (a \in \Sigma) \mid \neg \alpha \mid \alpha \vee \alpha' \mid \mathbf{E}\beta_{\mathrm{ch}} \mid \mathbf{E}\beta_{\mathrm{ns}}$$

$$\beta_*, \beta'_* := \alpha \mid \neg \beta_* \mid \beta_* \vee \beta'_* \mid \mathbf{X}_*\beta_* \mid \mathbf{X}_*^-\beta_* \mid \beta_* \mathbf{U}_* \beta'_* \mid \beta_* \mathbf{S}_* \beta'_*$$

The semantics is standard and omitted here. The following can be seen as an analog of the equivalence FO = CTL^* for finite binary trees [31].

Theorem 7. (see [3]) *A unary or Boolean query over unranked trees is definable in* FO *iff it is definable in* CTL^*_{past}.

Conjunctive Queries Over Unranked Trees. Conjunctive queries are a very important class of database queries: they correspond to the $\exists, \wedge$-fragment of FO. These are the same queries that can be expressed by selection, projection, and join in relational algebra, and thus they form the core of database queries. The complexity of evaluating a conjunctive query φ over a database $\mathcal{D}$ is in NP, in terms of both the size of φ and the size of $\mathcal{D}$. In fact, the problem is NP-hard, and there has been a large body of work on classifying tractable cases (see, e.g., [28, 30]).

In the case of unranked trees, conjunctive queries are formulae of the form $\varphi(\bar{x}) = \exists \bar{y} \ R_1 \wedge \ldots \wedge R_k$, where each R_i is either $P_a(z)$ or $z \prec z'$, where z, z' are variables among $\bar{x}, \bar{y}$, and $\prec$ is one of $\prec_{\mathrm{ch}}, \prec_{\mathrm{ch}}^*, \prec_{\mathrm{ns}}$, or $\prec_{\mathrm{ns}}^*$. We write $\mathrm{CQ}(\prec_1, \ldots, \prec_m)$ to denote the class of conjunctive queries over unranked trees in which only unary predicates P_a and binary predicates among $\prec_i$ can be used.

Theorem 8. (see [27]) *The maximal tractable classes of queries* $\mathrm{CQ}(\prec_1, \ldots, \prec_m)$, *where all* $\prec_i$'s *are among* $\{\prec_{\mathrm{ch}}, \prec_{\mathrm{ch}}^*, \prec_{\mathrm{ns}}, \prec_{\mathrm{ns}}^*\}$, *are* $\mathrm{CQ}(\prec_{\mathrm{ch}}, \prec_{\mathrm{ns}}, \prec_{\mathrm{ns}}^*)$ *and* $\mathrm{CQ}(\prec_{\mathrm{ch}}^*)$; *all others are NP-hard.*

4 Unordered Trees

In unordered trees, nodes can still have arbitrarily many children, but the sibling ordering $\prec_{ns}$ is no longer available. Logics considered for unordered unranked trees typically introduce some form of *counting*, see [3, 19, 20, 21, 40, 46, 51, 53, 54].

An explanation for this comes from a modified notion of automata for unordered unranked trees. A *counting nondeterministic unranked tree automaton* is a tuple $\mathcal{A}_c = (Q, F, \delta)$, where Q is a set of states, and $F \subseteq Q$ is a set of final states. Let V_Q be the set of variables $\{v_q^k \mid q \in Q, k > 0\}$. Then the transition function δ maps each pair $(q, a) \in Q \times \Sigma$ into a Boolean function over V_Q. A run of $\mathcal{A}$ on an unordered tree T with domain D is a mapping $\rho_{\mathcal{A}_c} : D \to Q$ such that if $\rho_{\mathcal{A}_c}(s) = q$ for a node s labeled a, then the value of $\delta(q, a)$ is 1, where each variable $v_{q_i}^k$ is set to 1 if s has at least k children s' with $\rho_{\mathcal{A}_c}(s') = q_i$, and to 0 otherwise. A run is accepting if $\rho_{\mathcal{A}_c}(\varepsilon) \in F$, and the set of unordered trees accepted by $\mathcal{A}_c$ is denoted by $L_u(\mathcal{A}_c)$.

A *counting query automaton* $\mathcal{QA}_c$ is defined as (Q, F, δ, S) where $S \subseteq Q \times \Sigma$; it selects nodes s in a run ρ where $(\rho_{\mathcal{A}_c}(s), \lambda_T(s)) \in S$. The following appears not to have been stated explicitly, although it follows easily from results in [41, 44, 53].

Theorem 9. *a) A set of unordered unranked trees is MSO-definable iff it is of the form $L_u(\mathcal{A}_c)$ for a counting nondeterministic unranked tree automaton $\mathcal{A}_c$.*

b) A unary query over unordered unranked trees is MSO-definable iff it is definable by a counting query automaton $\mathcal{QA}_c$.

MSO *and* FO *Over Unordered Trees.* Define the *counting μ-calculus* C_μ (cf. [32]) as an extension of L_μ with formulae $\Diamond^{\geq k}(E)\varphi$. The semantics of $(T, s) \models \Diamond^{\geq k}(E)\varphi$ is as follows: there exist distinct elements $s_1, \ldots, s_k$ such that $(s, s_i) \in E$ and $(T, s_i) \models \varphi$ for every $1 \leq i \leq k$. The next result follows from [60], as was noticed in [32]:

Theorem 10. *Over unordered unranked trees, MSO and $C_\mu[\prec_{ch}]$ have precisely the same power with respect to Boolean queries.*

For first-order logic, counting extensions of both the temporal logic TL^{tree} and CTL^* give us analogs of Kamp's and Hafer-Thomas's theorems. Define $\text{TL}^{\text{tree}}_{\text{count}}$ as a version of TL^{tree} in which only modalities for the child relation are used, but in addition we have formulae $\mathbf{X}_{\text{ch}}^k \varphi$, with the semantics that $(T, s) \models \mathbf{X}_{\text{ch}}^k \varphi$ iff there are at least k children s' of s such that $(T, s') \models \varphi$.

We also extend CTL^* to a logic $\text{CTL}^*_{\text{count}}$ in which we have new state formulae $\mathbf{EX}_{\text{ch}}^k \alpha$, where α is a state formula, with the same semantics as above.

Theorem 11. (see [40, 51]) *Over unordered unranked trees, the classes of Boolean queries expressed in FO, $\text{TL}^{\text{tree}}_{\text{count}}$, and $\text{CTL}^*_{\text{count}}$ over binary relation $\prec_{ch}$, are the same.*

For unary queries, the equivalence FO = $\text{TL}^{\text{tree}}_{\text{count}}$ still holds [51], and FO can be shown to be equivalent to an extension of CTL^* with both counting and the past [3].

Extensions and More Powerful Counting. Consider now a scenario in which we deal with unordered trees, but in our formulae we can refer to some arbitrary ordering on siblings: after all, in any encoding of a tree, siblings will come in some order. Of course we do not want any particular order to affect the truth value, so we want our formulae, even if they use an ordering, to be independent of a particular ordering that was used.

This is the standard setting of *order-invariance*, an important concept in finite model theory, cf. [36]. We say that an MSO sentence φ over vocabulary including $\prec_{ch}^*$ and $\prec_{ns}^*$ is $\prec_{ns}$-*invariant* if for every unordered tree T and every two expansions $T^{\prec_{ns}^1}$ and $T^{\prec_{ns}^2}$ with sibling-orderings $\prec_{ns}^1$ and $\prec_{ns}^2$ we have $T^{\prec_{ns}^1} \models \varphi \Leftrightarrow T^{\prec_{ns}^2} \models \varphi$. A $\prec_{ns}$-invariant sentence defines a Boolean query on unordered trees.

We now define MSO_{mod} [19] as an extension of MSO with *modulo quantifiers*: for each set variable X, and $k > 1$, we have set new formulae $Q_k(X)$ which are true iff the cardinality of X is congruent to 0 modulo k.

Theorem 12. (see [20]) *Over unordered unranked trees, $\prec_{ns}$-invariant Boolean queries are precisely the Boolean queries definable in* MSO_{mod}.

Further extensions in terms of arithmetic power have been considered [53, 54]. Recall that Presburger arithmetic refers to the FO theory of the structure $\langle \mathbb{N}, + \rangle$. Define *Presburger* MSO, or PMSO, as an extension of MSO over unordered trees with the following rule: if $\varphi(\bar{x}, y, \bar{X})$ is a PMSO formula and $\alpha(\bar{v})$ a Presburger arithmetic formula with $|\bar{X}| = |\bar{v}| = n$, then $[\varphi/\alpha](\bar{x}, y, \bar{X})$ is a PMSO formula. Given valuation $\bar{s}, s_0, \bar{S}$ for free variables, with $\bar{S} = (S_1, \ldots, S_n)$, let m_i be the cardinality of $\{s' \mid s_0 \prec_{ch} s' \text{ and } s' \in S_i\}$. Then $[\varphi/\alpha](\bar{s}, s_0, \bar{S})$ is true iff $\alpha(m_1, \ldots, m_n)$ is true.

It is easy to see that $\text{MSO} \subsetneq \text{MSO}_{\text{mod}} \subsetneq \text{PMSO}$ over unordered trees. Still, PMSO is captured by a decidable automaton model.

Define Presburger unordered tree automata just as counting automata except that δ maps pairs from $Q \times \Sigma$ into Presburger formulae over v_q, for $q \in Q$. We interpret v_q as the number of children in state q, and a transition is enabled if the corresponding Presburger formula is true in this interpretation.

Theorem 13. (see [53]) *Presburger unordered tree automata and* PMSO *are equivalent. Furthermore, both emptiness and universality are decidable for Presburger unordered tree automata.*

Further extensions with counting have been considered for fixed-point logics [54] and the μ-calculus with modulo-quantifiers [3].

Edge-Labeled Unordered Trees. There are several areas where edge-labeled trees play a prominent and role, and traditionally logical formalisms have been designed for such data. For example, there are feature logics, used extensively in computational linguistics [15], or spatial logics used for describing networks and mobile agents [14]: in both cases one deals with unordered edge-labeled trees.

In the setting of feature trees, one has an infinite set of features $\mathcal{F}$, and in an unordered unranked tree every edge is labeled by an element $f \in \mathcal{F}$ such that each node s has at most one outgoing edge labeled f for each $f \in \mathcal{F}$. Furthermore, nodes may be labeled by elements of some alphabet Σ, as before. It is thus natural to model feature trees as structures $\langle D, (E_f)_{f \in \mathcal{F}}, (P_a)_{a \in \Sigma} \rangle$ such that the union of all E_f's forms the child relation of a tree, and no node has two outgoing E_f-edges. In the context of computational linguistics, one commonly used [5] logic for feature trees is the propositional modal logic that, in the context of feature structures (not necessarily trees), is also often supplemented with path-equivalence [50], as well as regular expressions [34].

Ambient logics are modal logics for trees that have been proposed in the context of mobile computation [14] and later adapted for tree-represented data [12, 13]. One views trees as edge-labeled and defines them by the grammar

$$T, T' := \Lambda \mid T|T' \mid a[T], \; a \in \Sigma,$$

with the equivalences that $|$ is commutative and associative, and that $T|\Lambda \equiv T$. Here Λ is the empty tree, $|$ is the parallel composition, and $a[T]$ adds an a-labeled edge on top of T. If we extend $\equiv$ to a congruence in the natural way, then every tree is equivalent to one of the form $a_1[T_1]|\ldots|a_m[T_m]$, which is viewed as a tree whose root has m outgoing edges labeled $a_1, \ldots, a_m$, with subtrees rooted at its children being $T_1, \ldots, T_m$.

There were several similar logics proposed in [11, 12, 13, 14, 21]. Here we consider the logic from [11] whose formulae are given by

$$\varphi, \varphi' := \bot \mid \Lambda \mid \varphi \wedge \varphi' \mid \neg \varphi \mid \varphi|\varphi' \mid \varphi \triangleright \varphi' \mid a[\varphi] \mid \varphi@a, \quad a \in \Sigma.$$

The semantics is as follows: $\bot$ is *false*; Λ is only true in a tree equivalent to Λ, $T \models \varphi_1|\varphi_2$ iff $T \equiv T_1|T_2$ with $T_i \models \varphi_i, i = 1, 2$; $T \models \varphi \triangleright \varphi'$ if for every $T' \models \varphi$ we have $T|T' \models \varphi'$; $T \models a[\varphi]$ iff $T \equiv a[T']$ with $T' \models \varphi$, and $T \models \varphi@a$ iff $a[T] \models \varphi$.

The study of ambient logics for trees took a different path compared to other logics seen in this survey; in particular, the focus was on type systems for tree languages and thus on proof systems for logics, rather than model-checking, its complexity, automata models, and comparison with other logics.

However, the ambient logic above does not take us outside of the MSO expressiveness: this can be seen by going from edge-labeled trees to node-labeled ones. The translation is simple: the label of each edge (x, y) becomes the label of y. The root will have a special label *Root* that cannot occur as a label of any other node. The only modification in the logic is that now we have formulae Λ_a for $a \in \Sigma$, which are true in a singleton-tree labeled a. The resulting logic is easily translated into MSO. For example, $\varphi|\varphi'$ states that the children of the root can be partitioned into two sets, X and X', such that the subtree that contains all the X-children satisfies φ and the subtree that contains all the X'-children satisfies φ'. For $\varphi \triangleright \varphi'$, one can consider $\neg(\varphi \triangleright \varphi')$ saying that there exists a tree T' such that $T' \models \varphi$ and $T|T' \models \neg \varphi'$, and use nondeterministic counting automata to guess this tree T'.

5 Automatic Structures

In this section we look at a different kind of logics for unranked trees, using the standard approach of model theory. Let $\text{TREE}(\Sigma)$ be the set of all Σ-labeled unranked trees. We consider structures of the form $\mathfrak{M} = \langle \text{TREE}(\Sigma), \Omega \rangle$ where Ω is a set of relation, constant, and function symbols.

Let $\text{Def}_n(\mathfrak{M})$ be the family of n-*dimensional definable sets* over $\mathfrak{M}$: that is, sets of the form $\{\bar{T} \in \text{TREE}(\Sigma)^n \mid \mathfrak{M} \models \varphi(\bar{T})\}$, where $\varphi(x_1, \ldots, x_n)$ is an FO formula in the vocabulary Ω. We shall be looking at structures $\mathfrak{M}$ so that definable sets would be relations definable in MSO or other logics. In particular, such relations will be given by automata, and thus structures $\mathfrak{M}$ of this kind are called *automatic structures*.

Following known automatic structures for strings [4, 6], we introduce several predicates on trees: the extension predicate, node tests, and domain equality. For two trees T_1 and T_2 with domains D_1 and D_2, we say that T_2 is an *extension* of T_1, written $T_1 \preceq T_2$, if $D_1 \subseteq D_2$, and the labeling function of T_2 agrees with the labeling function of T_1 on D_1. It will actually be more convenient to work with two extension relations: extension on the right $\preceq_\rightarrow$ and extension down $\preceq_\downarrow$. For $T_1 \preceq_\rightarrow T_2$, we require that every $s \in D_2 - D_1$ be of the form $s' \cdot i$ when $s' \cdot j \in D_1$ for some $j < i$. For $T_1 \preceq_\downarrow T_2$, we require that every $s \in D_2 - D_1$ have a prefix s' which is a leaf of T_1. Define L_a to be true in a tree T if the rightmost node is labeled a. Finally, $T_1 \approx_{\text{dom}} T_2$ iff $D_1 = D_2$.

Now we have the following structures:

$$\mathfrak{T}_{\text{univ}} = \langle \text{TREE}(\Sigma), \preceq_\rightarrow, \preceq_\downarrow, (L_a)_{a \in \Sigma}, \approx_{\text{dom}} \rangle$$
$$\mathfrak{T} = \langle \text{TREE}(\Sigma), \preceq_\rightarrow, \preceq_\downarrow, (L_a)_{a \in \Sigma} \rangle$$

Theorem 14. (see [37]) *a) For every $n \geq 1$, $\text{Def}_n(\mathfrak{T}_{\text{univ}})$ is precisely the class of regular n-ary relations over $\text{TREE}(\Sigma)$.*

b) $\text{Def}_1(\mathfrak{T}) = \text{Def}_1(\mathfrak{T}_{\text{univ}})$ is the class of regular unranked tree languages, but for every $n > 1$, $\text{Def}_n(\mathfrak{T}) \subsetneq \text{Def}_n(\mathfrak{T}_{\text{univ}})$.

Working with $\mathfrak{T}_{\text{univ}}$ makes it easy to write rather complicated properties of tree languages, and then Theorem 14 implies that those languages are regular. For example, if $X \subseteq \text{TREE}(\Sigma)$ is regular, then the set of trees T such that all their extensions can be extended on the right to a tree in X is regular. Indeed, this is easy to write in FO over $\mathfrak{T}_{\text{univ}}$, if we have a membership test for X, which is definable by Theorem 14. Also, conversions from formulae to automata are effective for both $\mathfrak{T}$ and $\mathfrak{T}_{\text{univ}}$, which implies decidability of their theories.

Other logics over unranked trees can be naturally represented over these structures: for example, Boolean FO queries are precisely sets of trees definable over $\mathfrak{T}$ if quantification is restricted to single branches [37].

A Different View of Unranked Trees. We conclude by presenting a different view of unranked trees and a different structure for them that makes it easy to talk about about their extensions in which new children may be inserted between existing ones. For example, if we have a tree T with domain $D = \{\varepsilon, 0, 1\}$, and we want to add more children of the root, they would have to be added on the right, e.g, we may have an extension with domain $\{\varepsilon, 0, 1, 2, 3\}$. But what if we want to add a child on the left of 0, and two children between 1 and 2? Intuitively, we need a new tree domain $\{\varepsilon, -1, 0, \frac{1}{3}, \frac{2}{3}, 1\}$ then. We now capture this situation and present a different automatic structure that makes it easy to derive that certain relations on trees are regular.

A *rational unranked tree domain* is a finite prefix-closed subset of $\mathbb{Q}^*$. Relation $\prec_{\text{ch}}^*$ is defined for rational domains just as before, and relation $\prec_{\text{ns}}^*$ is now given by $s \cdot r \prec_{\text{ns}}^* s \cdot r'$ iff $r \leq r'$. Then an unranked tree T over a rational unranked tree domain is, as before, a structure $T = \langle D, \prec_{\text{ch}}^*, \prec_{\text{ns}}^*, (P_a)_{a \in \Sigma} \rangle$.

Let $\text{TREE}_\mathbb{Q}(\Sigma)$ be the set of all unranked trees with rational unranked tree domains. Note that different elements of $\text{TREE}_\mathbb{Q}(\Sigma)$ may be isomorphic as trees; we denote this isomorphism relation by $\cong$.

Define the extension relation $\preceq$ over trees in $\text{TREE}_\mathbb{Q}(\Sigma)$ as before. A *branch* is a tree $T \in \text{TREE}_\mathbb{Q}(\Sigma)$ such that the set $\{T' \mid T' \preceq T\}$ is linearly ordered by $\preceq$. It follows from the definition of rational unranked tree domains that the domain of a branch consists of all the prefixes of some string $s \in \mathbb{Q}^*$. Let $L_a(T)$ be true iff T is a branch whose leaf is labeled a, and let $T_1 <_{\text{lex}} T_2$ be true iff T_1 and T_2 are branches with leaves s_1 and s_2, and $s_1 <_{\text{lex}} s_2$. We then define the structure

$$\mathfrak{T}^\mathbb{Q}_{\text{univ}} = \langle \text{TREE}_\mathbb{Q}(\Sigma), \preceq, <_{\text{lex}}, \approx_{\text{dom}}, (L_a)_{a \in \Sigma} \rangle.$$

Proposition 4. *The structure $\mathfrak{T}^\mathbb{Q}_{\text{univ}}$ is interpretable in $\mathfrak{T}_{\text{univ}}$. Furthermore, there is a definable subset of the image of $\text{TREE}_\mathbb{Q}(\Sigma)$ that contains exactly one representative of each $\cong$-equivalence class.*

That is, under the mapping $\iota : \text{TREE}_\mathbb{Q}(\Sigma)/\cong \longrightarrow \text{TREE}(\Sigma)$, definable sets over $\mathfrak{T}^\mathbb{Q}_{\text{univ}}$ become precisely the regular tree languages. Hence, expressing properties of unranked trees in first-order logic over $\mathfrak{T}^\mathbb{Q}_{\text{univ}}$ allows us to conclude easily that certain tree languages are regular, and thus MSO-definable.

6 Other Directions and Conclusions

We present very briefly some directions for future work (for more detailed discussion, see the full version).

Among problems that need to be addressed are the following: (a) How does one compare different logics over unranked trees? One way is in terms of their succinctness [29]. (b) Connection between ambient logics and other logics presented there is not yet adequately understood. (c) We do not know much about logics over string representations of trees (which occur naturally, for example, in streaming XML applications [52]). (d) Nor do we know much about handling data values which are present in XML trees. Some early results were reported in [45, 8], complemented recently by a nice decidability result that works on strings with data values [7].

Acknowledgments. I am grateful to Cristiana Chitic, Christoph Koch, Maarten Marx, Frank Neven, Joachim Niehren, Gerald Penn, Thomas Schwentick, and Luc Segoufin for their comments.

References

1. S. Abiteboul, R. Hull, V. Vianu. *Foundations of Databases*, Addison Wesley, 1995.
2. A. Arnold, D. Niwinski. *Rudiments of μ-calculus*. Elsevier, 2001.
3. P. Barceló, L. Libkin. Temporal logics over unranked trees. In *LICS'05*.
4. M. Benedikt, L. Libkin, T. Schwentick, L. Segoufin. Definable relations and first-order query languages over strings. *J. ACM*, 50 (2003), 694–751.

5. P. Blackburn. Structures, languages and translations: the structural approach to feature logic. In *Constraints, Language and Computation*, AP, 1994, pages 1–27.
6. A. Blumensath and E. Grädel. Automatic structures. In *LICS'00*, pages 51–62.
7. M. Bojanczyk, C. David, A. Muscholl, T. Schwentick, L. Segoufin. Two-variable logic on words with data. 2005.
8. P. Bouyer, A. Petit, D. Thérien. An algebraic characterization of data and timed languages. In *CONCUR 2001*, pages 248–261.
9. A. Brüggemann-Klein, M. Murata, and D. Wood. Regular tree and regular hedge languages over unranked alphabets: Version 1, 2001. HKUST Tech. Report.
10. J.R. Büchi. Weak second-order arithmetic and finite automata. *Zeit. Math. Logik Grundl. Math.* 6 (1960), 66–92.
11. C. Calcagno, L. Cardelli, A. Gordon. Deciding validity in a spatial logic for trees. *J. Funct. Progr.*, to appear.
12. L. Cardelli. Describing semistructured data. *SIGMOD Record* 30 (2001), 80–85.
13. L. Cardelli, G. Ghelli. A query language based on the ambient logic. In *ESOP 2001*, pages 1–22.
14. L. Cardelli, A. Gordon. Anytime, anywhere: Modal logics for mobile ambients. In *POPL 2000*, pages 365–377.
15. B. Carpenter. *The Logic of Typed Feature Structures*. Cambridge, 1992.
16. E. Clarke, O. Grumberg, and D. Peled. *Model Checking*. The MIT Press, 1999.
17. J. Clark and S. DeRose. XML Path Language (XPath). W3C Recommendation, Nov. 1999. www.w3.org/TR/xpath.
18. H. Comon et al. *Tree Automata: Techniques and Applications*. Available at www.grappa.univ-lille3.fr/tata. October 2002.
19. B. Courcelle. The monadic second-order logic of graphs I: Recognizable sets of finite graphs. *Inf.&Comput.* 85 (1990), 12–75.
20. B. Courcelle. The monadic second-order logic of graphs V: On closing the gap between definability and recognizability. *TCS* 80 (1991), 153–202.
21. S. Dal-Zilio, D. Lugiez, C. Meyssonnier. A logic you can count on. In *POPL 2004*, pages 135–146.
22. E. A. Emerson, C. Jutla. Tree automata, mu-calculus and determinacy. In *FOCS 1991*, pages 368–377.
23. M. Frick, M. Grohe. The complexity of first-order and monadic second-order logic revisited. In *LICS 2002*, 215–224.
24. M. Frick, M. Grohe, C. Koch. Query evaluation on compressed trees. In *LICS 2003*, pages 188-197.
25. G. Gottlob, C. Koch. Monadic datalog and the expressive power of languages for web information extraction. *J. ACM* 51 (2004), 74–113.
26. G. Gottlob, C. Koch, R. Pichler, and L. Segoufin. The complexity of XPath query evaluation and XML typing. *J. ACM*, 2005, to appear.
27. G. Gottlob, C. Koch, K. Schulz. Conjunctive queries over trees. In *PODS 2004*, pages 189–200.
28. G. Gottlob, N. Leone, and F. Scarcello. The complexity of acyclic conjunctive queries. *J. ACM*, 48 (2001), 431–498.
29. M. Grohe, N. Schweikardt. Comparing the succinctness of monadic query languages over finite trees. In *CSL 2003*, pages 226–240.
30. M. Grohe, T. Schwentick, and L. Segoufin. When is the evaluation of conjunctive queries tractable? In *STOC 2001*, pages 657–666.
31. T. Hafer, W. Thomas. Computation tree logic CTL* and path quantifiers in the monadic theory of the binary tree. *ICALP 1987*, pages 269–279.

32. D. Janin, G. Lenzi. Relating levels of the mu-calculus hierarchy and levels of the monadic hierarchy. In *LICS 2001*, pages 347–356.
33. H.W. Kamp. *Tense Logic and the Theory of Linear Order*. PhD Thesis, UCLA, 1968.
34. B. Keller. *Feature Logics, Infinitary Descriptions and Grammar*. CSLI Press, 1993.
35. O. Kupferman, A. Pnueli. Once and for all. In *LICS'95*, pages 25–35.
36. L. Libkin. *Elements of Finite Model Theory*. Springer, 2004.
37. L. Libkin, F. Neven. Logical definability and query languages over unranked trees. In *LICS 2003*, pages 178–187.
38. M. Marx. Conditional XPath, the first order complete XPath dialect. In *PODS 2004*, pages 13–22.
39. M. Marx and M. de Rijke. Semantic characterizations of XPath. In *TDM Workshop on XML Databases and Information Retrieval*, 2004.
40. F. Moller, A. Rabinovich. Counting on CTL*: on the expressive power of monadic path logic. *Information and Computation*, 184 (2003), 147-159.
41. F. Neven. *Design and Analysis of Query Languages for Structured Documents*. PhD Thesis, U. Limburg, 1999.
42. F. Neven. Automata, logic, and XML. In *CSL 2002*, pages 2–26.
43. F. Neven, Th. Schwentick. Expressive and efficient pattern languages for tree-structured data. In *PODS 2000*, pages 145–156. Corrigendum at http://www.mathematik.uni-marburg.de/˜tick/
44. F. Neven, Th. Schwentick. Query automata over finite trees. *Theor. Comput. Sci.* 275 (2002), 633–674.
45. F. Neven, Th. Schwentick, V. Vianu. Towards regular languages over infinite alphabets. In *MFCS 2001*, pages 560–572.
46. J. Niehren, A. Podelski. Feature automata and recognizable sets of feature trees. *TAPSOFT 1993*, pages 356–375.
47. D. Niwinski. Fixed points vs. infinite generation. In *LICS 1988*, pages 402–409.
48. A. Potthoff, W. Thomas. Regular tree languages without unary symbols are star-free. In *FCT 1993*, pages 396–405.
49. M. Rabin. Decidability of second-order theories and automata on infinite trees. *Trans. AMS* 141 (1969), 1–35.
50. W. C. Rounds, R. Kasper. A logical semantics for feature structures. In *24th Annual Meeting of the Assoc. for Computational Linguistics*, 1986, pages 257–266.
51. B.-H. Schlingloff. Expressive completeness of temporal logic of trees. *Journal of Applied Non-Classical Logics* 2 (1992), 157–180.
52. L. Segoufin, V. Vianu. Validating streaming XML documents. In *PODS 2002*, pages 53–64.
53. H. Seidl, Th. Schwentick, A. Muscholl. Numerical document queries. In *PODS 2003*, 155–166.
54. H. Seidl, Th. Schwentick, A. Muscholl, P. Habermehl. Counting in trees for free. In *ICALP 2004*, pages 1136–1149.
55. L. Stockmeyer and A. Meyer. Cosmological lower bound on the circuit complexity of a small problem in logic. *Journal of the ACM*, 49 (2002), 753–784.
56. J.W. Thatcher. Characterizing derivation trees of context-free grammars through a generalization of finite automata theory. *JCSS* 1 (1967), 317–322.
57. J.W. Thatcher and J.B. Wright. Generalized finite automata theory with an application to a decision problem of second-order logic. *Mathematical Systems Theory*, 2(1):57–81, 1968.
58. W. Thomas. Languages, automata, and logic. In *Handbook of Formal Languages, Vol. 3*, Springer-Verlag, 1997.
59. M. Y. Vardi. Reasoning about the past with two-way automata. In *ICALP 1998*, pages 628–641.
60. I. Walukiewicz. Monadic second-order logic on tree-like structures. *TCS* 275 (2002), 311–346.

Nash Equilibria, the Price of Anarchy and the Fully Mixed Nash Equilibrium Conjecture*

Martin Gairing, Thomas Lücking,
Burkhard Monien, and Karsten Tiemann**

Department of Computer Science, Electrical Engineering and Mathematics,
University of Paderborn, Fürstenallee 11, 33102 Paderborn, Germany
{gairing, luck, bm, tiemann}@uni-paderborn.de

1 Introduction

Motivation-Framework. Apparently, it is in human's nature to act selfishly. Game Theory, founded by von Neumann and Morgenstern [39, 40], provides us with *strategic games*, an important mathematical model to describe and analyze such a selfish behavior and its resulting conflicts. In a strategic game, each of a finite set of players aims for an optimal value of its *private objective function* by choosing either a *pure* strategy (a single strategy) or a *mixed* strategy (a probability distribution over all pure strategies) from its *strategy set*. Strategic games in which the strategy sets are finite are called *finite strategic games*. Each player chooses its strategy once and for all, and all players' choices are made *non-cooperatively* and *simultaneously* (that is, when choosing a strategy each player is not informed of the strategies chosen by any other player). One of the basic assumption in strategic games is that the players act *rational*, that is, consistently in pursuit of their private objective function. For a concise introduction to contemporary Game Theory we recommend [25].

One of the most widely used solution concepts for strategic games is the concept of *Nash equilibrium*. It represents a stable state in which no player wishes to leave unilaterally its own strategy in order to improve the value of its private objective function. A Nash equilibrium is called *pure* if all players choose a pure strategy, otherwise *mixed*. Many algorithms have been developed to compute a Nash equilibrium (see [27] for an overview). Though the celebrated results of Nash [30, 31] ensure the existence of a mixed Nash equilibrium, the complexity to compute such a Nash equilibrium is widely unknown. Papadimitriou [32] advocates it to be *"the most important concrete open question on the boundary of $\mathcal{P}$ today"*.

Rosenthal [33] introduced a special class of strategic games, now widely known as *congestion games*. Here, the strategy set of each player is a subset of the power set of given *resources*. The players share a private objective function, defined as the sum (over their chosen resources) of functions in the *number* of players sharing this resource. In his seminal work, Rosenthal [33] showed with help of a *potential function* that con-

---
* This work has been partially supported by the DFG-SFB 376 and by the European Union within the 6th Framework Programme under contract 001907 (DELIS).
** International Graduate School of Dynamic Intelligent Systems.

gestion games (in sharp contrast to general strategic games) always admit at least one pure Nash equilibrium. Later, Milchtaich [28] considered two extensions of congestion games, namely *weighted congestion games* in which the players have *weights* and thus different influence on the congestion of the resources, and congestion games with *player-specific payoff-functions* in which the players do not share a private objective function.

Another class of (weighted) congestion games are (weighted) *network congestion games* [8, 11] in which the strategy sets correspond to paths in a network. Koutsoupias and Papadimitriou [21] considered a very simple member of this class, now known as *KP-model*. The network consists of a single *source* and a single *destination* which are connected by parallel *links*. Associated with each link is a *capacity* representing the rate at which the link processes *load*, that is, the total weight of players assigned to this link. Thus, the latency functions are linear. Each of the players selfishly routes from the source to the destination by choosing a probability distribution over the links. The private objective function of a player is defined as its expected latency.

Koutsoupias and Papadimitriou [21] were not only interested in the computational complexity of Nash equilibria but also in the degradation of the social welfare of the system due to the selfish behavior of the players. In order to measure this social welfare, they introduced a global objective function, usually coined as *social cost*, which is defined as the expected maximum latency on a link, where the expectation is taken over all random choices of the players. The *price of anarchy*, also called *coordination ratio*, measures the extent to which non-cooperation approximates cooperation. It is defined as the worst-case ratio between the value of social cost in a Nash equilibrium and that of some social optimum. So, the price of anarchy represents a rendezvous of Nash equilibrium, a concept fundamental to Game Theory, with approximation, an ubiquitous concept in Theoretical Computer Science today (see, e.g., [38]).

Mavronicolas and Spirakis [26] introduced the notion of a *fully mixed* Nash equilibrium in which each player chooses every link with positive probability. Gairing *et al.* [15] conjectured that, in case of its existence, the *fully mixed* Nash equilibrium is the worst Nash equilibrium with respect to social cost. This so-called *Fully Mixed Nash Equilibrium Conjecture* is simultaneously intuitive and natural. To support intuition, observe that the fully mixed Nash equilibrium favors collisions between different players (since each player assigns its item with positive probability to every link). This increased probability of collisions should favor an increase to social cost. To support significance, note that the Fully Mixed Nash Equilibrium Conjecture identifies the worst-case Nash equilibrium of *all* instances. We stress that, in sharp contrast, the price of anarchy only determines the worst-case Nash equilibrium of *worst-case* instances.

Recently, the KP-model was extended to *restricted strategy sets* [2, 13] where the strategy set of each player is a *subset* of the links. In addition, the KP-model was extended to general latency functions and studied with respect to different definitions of social cost [1, 14]. Inspired by the arisen interest in the price of anarchy, the much older *Wardrop-model* [3, 6, 41] was re-investigated [35, 36]. In this weighted network congestion game, weight can be split into arbitrary pieces. The social welfare of the system is defined as the sum of the edge latencies. An equilibrium in the Wardrop-model can be interpreted as a Nash equilibrium in a game with infinitely many players, each carrying an infinitesimal amount of weight. Finally, the price of anarchy found its way into congestion games [4, 11].

In this paper, we give a thorough survey on the most exciting results on finite (weighted) congestion games and the special classes mentioned above. In particular, we review the findings on the existence and computational complexity of pure Nash equilibria. Furthermore, we discuss results on the price of anarchy. Last but not least, we survey known facts on fully mixed Nash equilibria.

Overview. The rest of this paper is organized as follows. After a formal definition of (weighted) congestion games in Section 2, we turn our attention to the existence and computational complexity of pure Nash equilibria in Section 3. In Section 4, we consider the price of anarchy before we investigate fully mixed Nash equilibria in Section 5. We conclude, in Section 6, with some open problems.

2 Definitions and Notations

For all integers $k \geq 0$, we denote $[k] = \{1, \ldots, k\}$.
A *weighted congestion game* Γ is a tuple

$$\Gamma = \left(n, E, (w_i)_{i \in [n]}, (S_i)_{i \in [n]}, (f_e)_{e \in E}\right).$$

Here, n is the number of *players* and E is the finite set of *resources*. For every player $i \in [n]$, w_i is the *weight* and $S_i \subseteq 2^E$ is the *strategy set* of player i. Denote $W = \sum_{i \in [n]} w_i$ and $S = S_1 \times \ldots \times S_n$. For every resource $e \in E$, the *latency function* $f_e : \mathbb{R}^+ \mapsto \mathbb{R}^+$ describes the *latency* on resource e.

In a *congestion game*, the weights of all players are equal. Thus, the private cost of a player only depends on the *number* of players choosing the same resources. A congestion game is *symmetric* if the players share a strategy set.

2.1 Strategies and Assignments

A *pure strategy* for player $i \in [n]$ is some specific $s_i \in S_i$ whereas a *mixed strategy* $P_i = (p(i, s_i))_{s_i \in S_i}$ is a probability distribution over S_i, where $p(i, s_i)$ denotes the probability that player i chooses the pure strategy s_i.

A *pure assignment* is an n-tuple $\mathbf{L} = (s_1, \ldots, s_n) \in S$ whereas a *mixed assignment* $\mathbf{P} = (P_1, \ldots, P_n)$ is represented by an n-tuple of mixed strategies. A mixed assignment is *fully mixed* if $p(i, s_i) > 0$ for all $i \in [n]$ and $s_i \in S_i$.

2.2 Private Cost

Fix any pure assignment $\mathbf{L}$, and denote by $l_e(\mathbf{L}) = \sum_{i \in [n], s_i \ni e} w_i$ the *load* on resource $e \in E$. The *private cost* of player $i \in [n]$ is defined by

$$\mathsf{PC}_i(\mathbf{L}) = \sum_{e \in s_i} f_e(l_e(\mathbf{L})).$$

For a mixed assignment $\mathbf{P}$, the *private cost* of player $i \in [n]$ is

$$\mathsf{PC}_i(\mathbf{P}) = \sum_{\mathbf{L} \in S} p(\mathbf{L}) \cdot \mathsf{PC}_i(\mathbf{L}).$$

2.3 Social Cost

Associated with a weighted congestion game Γ and a mixed assignment $\mathbf{P}$ is the *social cost* as a measure of social welfare. We consider the following three definitions of social cost:

- Sum of private costs $\quad\quad\quad \mathsf{SC}_{\mathsf{SUM}}(\mathbf{P}) = \sum_{i\in[n]} \mathsf{PC}_i(\mathbf{P})$
- Maximum of private costs $\quad \mathsf{SC}_{\mathsf{MAX}}(\mathbf{P}) = \max_{i\in[n]} \mathsf{PC}_i(\mathbf{P})$
- Expected maximum latency $\quad \mathsf{SC}_{\infty}(\mathbf{P}) = \sum_{\mathbf{L}\in S} p(\mathbf{L}) \cdot \max_{i\in[n]} \mathsf{PC}_i(\mathbf{L})$

Let $* \in \{\mathsf{SUM}, \mathsf{MAX}, \infty\}$. The *optimum* associated with a weighted congestion game is defined by $\mathsf{OPT}_* = \min_{\mathbf{P}} \mathsf{SC}_*(\mathbf{P})$.

2.4 Nash Equilibria and Price of Anarchy

We are interested in a special class of (mixed) assignments called Nash equilibria [30, 31] that we describe here. Given a weighted congestion game and an associated mixed assignment $\mathbf{P}$, a player $i \in [n]$ is *satisfied* if it can not improve its private cost by unilaterally changing its strategy. Otherwise, player i is *unsatisfied*. The mixed assignment $\mathbf{P}$ is a *Nash equilibrium* if and only if all players $i \in [n]$ are satisfied. Depending on the type of assignment, we differ between *pure*, *mixed* and *fully mixed* Nash equilibria.

The *mixed price of anarchy*, also called *coordination ratio* and denoted PoA_{mixed}, is the maximum value, over all instances Γ and Nash equilibria $\mathbf{P}$, of the ratio $\frac{\mathsf{SC}_*(\mathbf{P})}{\mathsf{OPT}_*}$. If we restrict to pure Nash equilibria, then we speak of the *pure price of anarchy* and denote it by PoA_{pure}.

2.5 Selfish Steps

Fix any pure assignment $\mathbf{L}$. In a *selfish step*, exactly one unsatisfied player is allowed to change its pure strategy such that its private cost decreases. A selfish step is *greedy* if the player chooses its best strategy. Clearly, selfish steps define a neighborhood of pure assignments that can be reached from $\mathbf{L}$. The assignment $\mathbf{L}$ has an empty neighborhood if and only if $\mathbf{L}$ is a Nash equilibrium. Thus, a pure Nash equilibrium corresponds to a local optimum. This stresses the close relationship of selfish steps on the one hand and local search processes on the other hand.

2.6 Special Weighted Congestion Games

Weighted Network Congestion Games. In a *weighted network congestion game* the strategies of a player correspond to paths from a source to a destination in a network. Thus, this class of games can be interpreted as *routing games*. If the players share the same source and destination, then we have a weighted *single-commodity* network congestion game, otherwise a weighted *multi-commodity* network congestion game. The underlying network of a weighted single-commodity network congestion game is called *l-layered* if all paths from source to destination have length l.

KP-Model. Koutsoupias and Papadimitriou [21] considered a special weighted network congestion game, now widely known as the *KP-model*. In this model, each of the n players is allowed to use exactly one of m resources (here called *links*), that is, $S_i = [m]$ for all $i \in [n]$. The players are called *identical* if all weights are equal, otherwise *arbitrary*. Associated with each link $j \in [m]$ is a *capacity* c_j representing the rate at which link j processes *load*. Clearly, the latency on link j is $f_j(l_j) = \frac{l_j}{c_j}$, showing that the latency functions are linear. If $c_1, \ldots, c_m$ are equal, then the resources are *identical*, otherwise *related*. Denote $C = \sum_{j \in [m]} c_j$. In order to measure the social welfare of the system, Koutsoupias and Papadimitriou [21] considered the expected maximum latency.

A natural goal is to identify a Nash equilibrium with worst social cost for a given instance. For the model of related links, Gairing *et al.* [15] conjectured that, in case of its existence, the fully mixed Nash equilibrium is the worst Nash equilibrium with respect to social cost.

Fully Mixed Nash Equilibrium Conjecture ([15]). *Consider the model of arbitrary players and related links. Then, for any instance such that a fully mixed Nash equilibrium $\mathbf{F}$ exists, and for any associated Nash equilibrium $\mathbf{P}$, $\mathsf{SC}_\infty(\mathbf{P}) \leq \mathsf{SC}_\infty(\mathbf{F})$.*

Routing Games on Parallel Links. We also consider variants of the KP-model to which we refer as *routing games on parallel links*. In particular, we investigate *restricted strategy sets* in which the players are only allowed to choose from a subset of links, that is, $S_i \subseteq [m]$ for all $i \in [n]$.

2.7 Exact Potential Games

A function $\Phi : (S_1 \times \ldots \times S_n) \mapsto \mathbb{R}$ is an *exact potential function* for a game Γ if for every pure strategy profile $\mathbf{L} = (s_1, \ldots, s_n)$, for every player $i \in [n]$ and for every strategy $s'_i \in S_i$, $\mathsf{PC}_i(\mathbf{L}') - \mathsf{PC}_i(\mathbf{L}) = \Phi(\mathbf{L}') - \Phi(\mathbf{L})$, where $\mathbf{L}' = (s_1, \ldots, s_{i-1}, s'_i, s_{i+1}, \ldots, s_n)$. In this case, Γ is an *exact potential game*. Since all exact potential games admit a pure Nash equilibrium (see e.g. [29]) these games are of interest in this paper.

3 Existence and Computation of Pure Nash Equilibria

Even though Nash was able to show that every finite game possesses a mixed Nash equilibrium, the question which class of games admits a pure Nash equilibrium remains open. In the case of its existence, it is of interest whether it is possible to compute a pure Nash equilibrium in polynomial time. In this section, we give some positive and some negative answers to both questions concerning the existence and the polynomial time computation. We start in Section 3.1 with routing games on parallel links and continue in Sections 3.2 and 3.3 with congestion games and weighted congestion games.

3.1 Routing Games on Parallel Links

We begin our survey with results on the KP-model. Afterwards we focus on games with restricted strategy sets.

KP-model. We first turn our attention to the problem of computing a pure Nash equilibrium. Basically, two different approaches can be found in the literature.

The first approach is to directly compute a pure Nash equilibrium. Fotakis *et al.* [10] showed that the LPT algorithm, first explored by Graham [16], yields some pure Nash equilibrium. Clearly, this holds for parallel links with *arbitrary non-decreasing* latency functions. For related links, the social cost of the Nash equilibrium computed by LPT approximates the social cost of an optimal assignment by a factor between 1.52 and 1.67 [12].

The second approach is to convert a given pure assignment into a Nash equilibrium without increasing the social cost. This conversion process is called *nashification*. Since selfish steps do not increase the social cost and any sequence of selfish steps eventually reaches a pure Nash equilibrium, selfish steps seem to be suitable for nashification. However, we have to use them carefully since the number of selfish steps may be exponential in the number of players before reaching a pure Nash equilibrium.

Theorem 1 ([7]). *Consider the model of arbitrary players and identical links. Then, there exists an instance and associated pure assignment for which the maximum length of a sequence of greedy selfish steps is at least*

$$\frac{\left(\frac{n}{m-1}\right)^{m-1}}{2(m-1)!}.$$

Though there exist sequences of greedy selfish steps of exponential length, it is possible to use selfish steps to compute a Nash equilibrium in polynomial time if the links are identical. In particular, always moving an unsatisfied player with maximum weight to its best link requires at most n greedy selfish steps [15]. For related links, it is unknown whether selfish steps can be used to implement nashification in polynomial time. Feldmann *et al.* [9] chose a different approach not only based on selfish steps. Their algorithm relies on the following crucial observation.

Lemma 1 ([9]). *Consider the model of arbitrary players and related links. Then, for any pure assignment, a greedy selfish step of an unsatisfied player $i_1 \in [n]$ with weight w_{i_1} from a link $j_1 \in [m]$ to a link $j_2 \in [m]$ with $c_{j_1} \leq c_{j_2}$ makes no satisfied player $i_2 \in [n]$ with weight $w_{i_2} \geq w_{i_1}$ unsatisfied.*

The algorithm of Feldmann *et al.* [9] works in two phases. In the first phase, it fills up links with small capacities with players with small weight as close to $\mathsf{SC_{MAX}}(\mathbf{L})$ as possible (but without exceeding $\mathsf{SC_{MAX}}(\mathbf{L})$), and it collects all these users in a set $\mathcal{U}$. In the second phase, the algorithm performs greedy selfish steps for unsatisfied players in $\mathcal{U}$ in non-increasing order of the weights. Lemma 1 allows to show that this procedure results in a pure Nash equilibrium. Implementing the algorithm in a proper way, we get:

Theorem 2 ([9]). *Consider the model of arbitrary players and related links. Then, for any pure assignment $\mathbf{L}$, a pure Nash equilibrium $\mathbf{L}'$ with $\mathsf{SC}_\infty(\mathbf{L}') \leq \mathsf{SC}_\infty(\mathbf{L})$ can be computed using $O(m^2 n)$ time.*

Thus, we can apply the PTAS of Hochbaum and Shmoys [17] for scheduling *jobs* on related *machines* and then convert the computed assignment into a pure Nash equilibrium in polynomial time, and we get:

Corollary 1. *There is a PTAS for computing a best pure Nash equilibrium.*

Restricted Strategy Sets. Gairing et al. [13] considered a variant of the routing game on parallel links where there exists at least one player $i \in [n]$ with $S_i \subsetneq [m]$. So, the strategy sets of the players are *restricted*.

Gairing et al. [13] combined ideas from blocking flows and the generic PREFLOW-PUSH algorithm to derive a nashification algorithm for games with restricted strategy sets on identical links.

Theorem 3 ([13]). *Consider the model of arbitrary players with restricted strategy sets and identical links. Then, for any pure assignment* $\mathbf{L}$, *a pure Nash equilibrium* $\mathbf{L}'$ *with* $\mathsf{SC}_\infty(\mathbf{L}') \leq \mathsf{SC}_\infty(\mathbf{L})$ *can be computed from* $\mathbf{L}$ *using* $O(rmA(\log W + m^2))$ *time, where* r *is the number of distinct weights and* $A = \sum_{i \in [n]} |S_i|$.

Lenstra et al. [23] showed that an optimum assignment can be approximated within a factor of 2. It is worth mentioning that the nashification algorithm of Gairing et al. [13] improves this result since, for *any* given assignment $\mathbf{L}$, it computes a pure Nash equilibrium $\mathbf{L}'$ with $\mathsf{SC}_\infty(\mathbf{L}') \leq (2 - \frac{1}{w_1}) \cdot \mathsf{OPT}_\infty$. Note that we can not hope to approximate an optimum assignment with factor less than $\frac{3}{2}$ unless $\mathcal{P} = \mathcal{NP}$ [23].

3.2 Congestion Games

In his seminal paper, Rosenthal [33] proved that $\Phi(\mathbf{L}) = \sum_{e \in E} \sum_{j=1}^{l_e(\mathbf{L})} f_e(j)$ is an exact potential function for congestion games. An immediate consequence follows:

Theorem 4 ([33]). *Every congestion game possesses a pure Nash equilibrium.*

Rosenthal's argumentation implies that every congestion game is an exact potential game. A result by Monderer and Shapley [29] shows that every exact potential game is closely related to a congestion game.

Theorem 5 ([29]). *Every finite exact potential game is isomorphic to a congestion game.*

Since every congestion game Γ possesses a pure Nash equilibrium the natural question arises whether it is possible to compute a pure Nash equilibrium for Γ in polynomial time. It is easy to see that this computational problem is in *PLS*. The class PLS (polynomial-time *local search*) introduced in [19] consists of local search problems for which local optimality can be verified in polynomial time. Many local search problems were shown to be complete for this class (see e.g. [19, 22, 37]), including graph partitioning, weighted satisfiability and traveling salesman problems. For none of these PLS-complete problems an algorithm is known that is able to compute a local optimum in polynomial time.

Using a sophisticated *PLS-reduction* Fabrikant et al. [8] proved that the computation of a pure Nash equilibrium for symmetric congestion games and asymmetric network congestion games is PLS-complete (see Figure 1). However, they showed that it is possible to calculate a pure Nash equilibrium for a symmetric network congestion game in polynomial time by using a min-cost flow algorithm.

	Symmetric	Asymmetric
Congestion Games	PLS-complete	PLS-complete
Network Congestion Games	Polynomial time	PLS-complete

Fig. 1. Complexity of computing pure Nash equilibria in congestion games [8]

We now switch to the class of *congestion games with player-specific payoff-functions* introduced by Milchtaich [28]. Here, a player always selects exactly one resource, that is, $S_1 = \ldots = S_n = E$. Furthermore, the private cost of a player $i \in [n]$ on a resource $e \in E$ is described by a load dependent non-increasing latency function $f_e^i : \mathbb{R}^+ \mapsto \mathbb{R}^+$ that may be different from the latency function f_e^j for another player $j \neq i$. Milchtaich [28] considered these games with respect to pure Nash equilibria and sequences of selfish step. He showed:

Theorem 6 ([28]). *Every congestion game with player-specific payoff-function possesses a pure Nash equilibrium.*

Theorem 7 ([28]). *There exists a finite congestion game with player-specific payoff-function that admits a cycle of selfish steps, that is, a sequence of selfish steps starting and ending in the same assignment.*

It follows from the last theorem that games with player-specific payoff-functions do not admit an exact potential function.

3.3 Weighted Congestion Games

In this section we deal with weighted congestion games where the players may have different weights. Fotakis *et al.* [11] showed that there are such games that possess no pure Nash equilibrium. Moreover, they were able to proof that there is a subclass of games for which the existence of pure Nash equilibria is guaranteed.

Theorem 8 ([11]). *There exist instances of weighted single-commodity network congestion games for which there is no pure Nash equilibrium.*

Theorem 9 ([11]). *For any weighted multi-commodity network congestion game with linear latency functions, at least one pure Nash equilibrium exists.*

4 Price of Anarchy

The *mixed price of anarchy*, also known as *coordination ratio*, has been defined in the seminal work by Koutsoupias and Papadimitriou [21] as a measure of the extent to which non-cooperation approximates cooperation. Recall that it is defined as the worst-case ratio between the value of social cost in a Nash equilibrium and that of a social optimum. We present results on the pure and the mixed price of anarchy for routing games on parallel links in Section 4.1, for congestion games in Section 4.2, and for weighted congestion games in Section 4.3.

4.1 Routing Games on Parallel Links

We start with results on the KP-model. We then focus on the extension of this model to restricted strategy sets. Finally, we investigate routing games on parallel links with social cost defined as the sum of the private costs of the players.

KP-model. In the KP-model, latency functions are linear, social cost is defined as the expected maximum latency and the players may choose any link. For the case of identical links the pure price of anarchy is upper bounded by a constant. This does not hold for related links or mixed Nash equilibria. The bounds for mixed Nash equilibria are shown by first bounding the maximum expected load on a link and then applying a Hoeffding inequality [18]. All bounds are summarized in Figure 2.

	Pure Price of Anarchy		Mixed Price of Anarchy	
Identical Links	$2 - \frac{2}{m+1}$	[15]	$\Theta\left(\frac{\log m}{\log \log m}\right)$	[5, 20]
Related Links	$\Theta\left(\frac{\log m}{\log \log m}\right)$	[5]	$\Theta\left(\frac{\log m}{\log \log \log m}\right)$	[5]

Fig. 2. Pure and mixed price of anarchy for the KP-model

Restricted Strategy Sets. In case of restricted strategy sets, even for identical links, the pure price of anarchy cannot be bounded by a constant. This also holds if the weights are identical. Figure 3 shows bounds on the pure price of anarchy. Note, that the bound for identical players and related links is only tight if $n = m$. Awerbuch et al. [2] further extended their result to mixed Nash equilibria.

Theorem 10 ([2]). *Consider the model of arbitrary players with restricted strategy sets and identical links. Then,*

$$PoA_{mixed} = \Theta\left(\frac{\log m}{\log \log \log m}\right).$$

	Identical Players		Arbitrary Players	
Identical Links	$\Theta\left(\frac{\log m}{\log \log m}\right)$	[2, 13]	$\Theta\left(\frac{\log m}{\log \log m}\right)$	[2, 13]
Related Links	$O\left(\frac{\log n}{\log \log n}\right)$	[13]	$m - 1 \leq PoA_{pure} \leq m$	[13]

Fig. 3. Pure price of anarchy for the KP-model with restricted strategy sets

Social Cost as Sum of Private Costs. Gairing et al. [14] considered another routing game on parallel links. In contrast to the KP-model, social cost is defined as the sum of the private costs of the players. This good natured definition of social cost makes the analysis significantly simpler and allows the investigation of general non-decreasing non-constant latency functions. For identical players, Gairing et al. [14] carried over an upper bound on the pure price of anarchy from the Wardrop-model [35] to the discrete setting.

Proposition 1 ([14]). *Consider the model of identical players and arbitrary links with non-decreasing and non-constant latency functions. If $x f_j(x) \leq \alpha \sum_{t=1}^{x} f_j(t)$ for all $x \in [n]$ and $j \in [m]$, then for any pure Nash equilibrium $\mathbf{L}$, $\mathsf{SC}_{\mathsf{SUM}}(\mathbf{L}) \leq \alpha \cdot \mathsf{OPT}_{\mathsf{SUM}}$.*

Corollary 2 ([14]). *Consider the model of identical players and arbitrary links. If the latency functions are polynomials with non-negative coefficients and maximum degree d, then the pure price of anarchy is bounded by $d + 1$.*

In case that all links have the same latency function $f(x) = x^d$, one can show the following bound on the mixed price of anarchy. B_k is the k'th Bell number and counts the number of ways that a set of k elements can be partitioned into non-empty subsets.

Theorem 11 ([14]). *Consider the model of identical players and identical links with latency function $f(x) = x^d$, $d \in \mathbb{N}$. Then,*

$$\sup_{\mathbf{w},\mathbf{P}} \frac{\mathsf{SC}_{\mathsf{SUM}}(\mathbf{P})}{\mathsf{OPT}_{\mathsf{SUM}}} = B_{d+1} .$$

4.2 Congestion Games

Recently, the pure price of anarchy found its way into congestion games [1,4]. We restrict to results of Christodoulou and Koutsoupias [4] since only the abstract of the paper of Awerbuch et al. [1] was available (note that the latter paper also considers *weighted* congestion games). For congestion games with linear latency functions, Figure 4 summarizes results (both upper and lower bounds) on the pure price of anarchy. For the case of symmetric congestion games and social cost as the maximum of the private costs there is still a gap between the upper and the lower bound.

Christodoulou and Koutsoupias [4] also considered polynomial latency functions of degree d with non-negative coefficients. Figure 5 shows the corresponding bounds.

Both linear and polynomial latency functions were also considered in the Wardrop-model. Recall that in this model the social welfare of the system is defined as the sum

	$\mathsf{SC}_{\mathsf{SUM}}$	$\mathsf{SC}_{\mathsf{MAX}}$
Symmetric	$\frac{5n-2}{2n+1}$	$\frac{5n-2}{2n+1} \leq \mathsf{PoA}_{pure} \leq \frac{5}{2}$
Asymmetric	$\frac{5}{2}$	$\Theta(\sqrt{n})$

Fig. 4. Pure price of anarchy for congestion games with linear latency functions [4]

	SC$_{SUM}$	SC$_{MAX}$
Symmetric	$d^{\Theta(d)}$	$d^{\Theta(d)}$
Asymmetric	$d^{\Theta(d)}$	$\Omega(n^{d/(d+1)}), O(n)$

Fig. 5. Pure price of anarchy for congestion games with polynomial latency functions [4]

of the edge latencies. The pure price of anarchy for linear latency functions is $\frac{4}{3}$ [35] whereas the pure price of anarchy for polynomial latency functions of degree d turned out to be $\Theta(\frac{d}{\log d})$ [34].

4.3 Weighted Congestion Games

The mixed price of anarchy was also studied in weighted congestion games. Fotakis *et al.* [11] considered l-layered networks with identical edges each having the same linear latency function and social cost defined as the expected maximum latency.

Theorem 12 ([11]). *For weighted l-layered network congestion games with latency function $f_e(x) = x$ for all $e \in E$, the mixed price of anarchy for social cost as expected maximum latency is $O\left(\frac{\log m}{\log \log m}\right)$.*

This result is particularly interesting in comparison with the corresponding bound for the parallel link network (see Figure 2). It shows that under all l-layered networks the parallel link network has worst mixed price of anarchy.

5 Fully Mixed Nash Equilibria for Routing Games on Parallel Links

In routing games on parallel links, a *fully mixed* Nash equilibrium is a special Nash equilibrium, where each player chooses each link with strictly positive probability. Such a Nash equilibrium does not always exist. In this section, we give a characterization of instances with a fully mixed Nash equilibrium, we show its uniqueness and we study the Fully Mixed Nash Equilibrium Conjecture. We do this for two different routing games on parallel links. We would like to point out that there exist routing games on parallel links for which the Fully Mixed Nash Equilibrium Conjecture was disproved [24].

KP-model. Mavronicolas and Spirakis [26] were the first to consider fully mixed Nash equilibria. They showed for the KP-model, that if a fully mixed Nash equilibrium exists, it is unique and can be easily computed.

Theorem 13 ([26]). *Consider the model of arbitrary players and related links. Then, there exists a fully mixed Nash equilibrium* **F** *if and only if*

$$f_{ij} = \left(1 - \frac{mc_j}{C}\right) \cdot \left(1 - \frac{W}{(n-1)w_i}\right) + \frac{c_j}{C} \in (0,1)$$

for all $i \in [n]$ and $j \in [m]$. If $\mathbf{F}$ exists, then $\mathbf{F}$ is unique and $\mathbf{F} = (f_{ij})_{i \in [n], j \in [m]}$.

In particular, this implies that for the case of identical links the fully mixed Nash equilibrium uniquely exists and has probabilities $f_{ij} = \frac{1}{m}, \forall i \in [n], j \in [m]$.

In [15], the Fully Mixed Nash Equilibrium Conjecture was first explicitly stated for the KP-model, where social cost is defined as the expected maximum latency. Here, the ultimate settlement of this conjecture would reveal an interesting complexity-theoretic contrast between the worst-case pure and the worst-case mixed Nash equilibria. On the one hand, if the conjecture is valid, then the identification of the worst-case mixed Nash equilibrium is immediate in the cases where the fully mixed Nash equilibrium exists. On the other hand, Gairing *et al.* [15] showed that the worst-case pure Nash equilibrium is not $(2 - \frac{2}{m+1} - \varepsilon)$-approximable even on identical links.

Theorem 14 ([15]). *Consider the model of arbitrary players and identical links. If, for any ε with $0 < \varepsilon \leq 1 - \frac{2}{m+1}$, the worst-case pure Nash equilibrium is $(2 - \frac{2}{m+1} - \varepsilon)$-approximable, then $\mathcal{P} = \mathcal{NP}$.*

This result also unfolds an interesting contrast between best and worst-case pure Nash equilibria. For any $\varepsilon > 0$, a pure Nash equilibrium $\mathbf{L}$ with $SC_\infty(\mathbf{L}) \leq (1 + \varepsilon) \cdot OPT_\infty$ can be computed in polynomial time whereas the computation of a pure Nash equilibrium $\mathbf{L}'$ with $SC_\infty(\mathbf{L}') \geq (1 + \varepsilon) \cdot OPT_\infty$ is $\mathcal{NP}$-hard.

So far, the Fully Mixed Nash Equilibrium Conjecture has been proved only for some special cases, namely, two players on identical links [15], two identical players on related links and identical players on two identical links [24]. Furthermore, it was shown up to a factor of 49.02 in case of identical players and related links [10] and up to a factor of $2h(1+\varepsilon)$ for arbitrary players on identical links, if $n = m$ sufficient large [15], where h is the factor between the maximum and the average weight of the players.

On the other hand, Gairing *et al.* [15] proved that the private costs of all players in a Nash equilibrium are upper bounded by their private costs in the fully mixed Nash equilibrium. This directly implies:

Theorem 15 ([15]). *Consider the model of arbitrary players and related links. If the fully mixed Nash equilibrium $\mathbf{F}$ exists, then, for any mixed Nash equilibrium $\mathbf{P}$, we have $SC_{SUM}(\mathbf{P}) \leq SC_{SUM}(\mathbf{F})$ and $SC_{MAX}(\mathbf{P}) \leq SC_{MAX}(\mathbf{F})$.*

Social Cost as Sum of Private Costs. Fully mixed Nash equilibria were also considered for identical players and general non-decreasing and non-constant latency functions with respect to social cost defined as the sum of the private costs [14]. In order to characterize instances where the fully mixed Nash equilibrium exists, Gairing *et al.* [14] introduced two classes of links, namely *dead links* and *special links*. They showed that in any Nash equilibrium, none of the players is assigned to a dead link. Moreover, there exists at most one player who is assigned to any of the special links. Availing these results, they could give the following thorough characterization.

Theorem 16 ([14]). *Consider the model of identical players and links with non-decreasing and non-constant latency functions. Then, there exists a fully mixed Nash equilibrium* $\mathbf{F}$ *if and only if there are no special and no dead links. If* $\mathbf{F}$ *exists then* $\mathbf{F}$ *is unique.*

For every instance define the *generalized fully mixed Nash equilibrium* as the fully mixed Nash equilibrium for the instance where the links are restricted to non-special and non-dead links. If latency functions are non-decreasing, non-constant and *convex*, then one can show, that the private cost of each player in a Nash equilibrium is upper bounded by its private cost in the generalized fully mixed Nash equilibrium. This directly implies:

Theorem 17 ([14]). *Consider the model of identical players and links with non-decreasing, non-constant and convex latency functions. Then, for any Nash equilibrium* $\mathbf{P}$ *and generalized fully mixed Nash equilibrium* $\mathbf{F}$, $SC_{SUM}(\mathbf{P}) \leq SC_{SUM}(\mathbf{F})$ *and* $SC_{MAX}(\mathbf{P}) \leq SC_{MAX}(\mathbf{F})$.

6 Open Problems

The flourishing interest in weighted congestion games resulted in a multitude of results and methods, but raised even more questions remaining tantalizingly open. We only state some of them:

- Although the results of Nash [30, 31] guarantee the existence of a Nash equilibrium in strategic games, the computational complexity of computing a Nash equilibrium is open even if only two players are involved.
- Which classes of symmetric weighted network congestions games possess a pure Nash equilibrium? For which classes is it possible to compute such a pure Nash equilibrium in polynomial time?
- It is impossible to approximate a worst-case pure Nash equilibrium within a factor better than $2 - \frac{2}{m+1}$ in the KP-model with identical links [15]. To which extent is it possible to approximate a worst-case pure Nash equilibrium in the KP-model with related links or in more general settings?
- Most of the known bounds on the price of anarchy for network congestion games were shown with respect to social cost defined as sum or maximum of the private costs of the players [1, 4]. What is the price of anarchy if social cost is defined as expected maximum latency?
- For the KP-model, Gairing *et al.* [15] showed that the private costs of all players in a Nash equilibrium are bounded from above by their private costs in the fully mixed Nash equilibrium. For which classes of network congestion games does this property still hold?
- If the players are identical and the links are related, then the Fully Mixed Nash Equilibrium Conjecture holds up to a factor of $2h(1 + \varepsilon)$, where h is the factor between the maximum and the average weight of the players [15]. Does there exist an approximation factor independent of h?

References

1. B. Awerbuch, Y. Azar, and A. Epstein. The Price of Routing Unsplittable Flow. In *Proceedings of the 37th Annual ACM Symposium on Theory of Computing (STOC'05)*, 2005.
2. B. Awerbuch, Y. Azar, Y. Richter, and D. Tsur. Tradeoffs in Worst-Case Equilibria. In *Proceedings of the 1st International Workshop on Approximation and Online Algorithms (WAOA'03)*, LNCS 2909, pages 41–52, 2003.
3. M. Beckmann, C. B. McGuire, and C. B. Winsten. *Studies in the Economics of Transportation*. Yale University Press, 1956.
4. G. Christodoulou and E. Koutsoupias. The Price of Anarchy of Finite Congestion Games. In *Proceedings of the 37th Annual ACM Symposium on Theory of Computing (STOC'05)*, 2005.
5. A. Czumaj and B. Vöcking. Tight Bounds for Worst-Case Equilibria. In *Proceedings of the 13th Annual ACM-SIAM Symposium on Discrete Algorithms (SODA'02)*, pages 413–420, 2002. Also accepted to *Journal of Algorithms* as Special Issue of SODA'02.
6. S. C. Dafermos and F. T. Sparrow. The Traffic Assignment Problem for a General Network. *Journal of Research of the National Bureau of Standards, Series B*, 73(2):91–118, 1969.
7. E. Even-Dar, A. Kesselmann, and Y. Mansour. Convergence Time to Nash Equilibria. In *Proceedings of the 30th International Colloquium on Automata, Languages, and Programming (ICALP'03)*, LNCS 2719, pages 502–513, 2003.
8. A. Fabrikant, C. H. Papadimitriou, and K. Talwar. The Complexity of Pure Nash Equilibria. In *Proceedings of the 36th Annual ACM Symposium on Theory of Computing (STOC'04)*, pages 604–612, 2004.
9. R. Feldmann, M. Gairing, T. Lücking, B. Monien, and M. Rode. Nashification and the Coordination Ratio for a Selfish Routing Game. In *Proceedings of the 30th International Colloquium on Automata, Languages, and Programming (ICALP'03)*, LNCS 2719, pages 514–526, 2003.
10. D. Fotakis, S. Kontogiannis, E. Koutsoupias, M. Mavronicolas, and P. Spirakis. The Structure and Complexity of Nash Equilibria for a Selfish Routing Game. In *Proceedings of the 29th International Colloquium on Automata, Languages, and Programming (ICALP'02)*, LNCS 2380, pages 123–134, 2002.
11. D. Fotakis, S. Kontogiannis, and P. Spirakis. Selfish Unsplittable Flows. Accepted to *Theoretical Computer Science*.
12. D. K. Friesen. Tighter Bounds for LPT Scheduling on Uniform Processors. *SIAM Journal on Computing*, 16(3):554–560, 1987.
13. M. Gairing, T. Lücking, M. Mavronicolas, and B. Monien. Computing Nash Equilibria for Scheduling on Restricted Parallel Links. In *Proceedings of the 36th Annual ACM Symposium on Theory of Computing (STOC'04)*, pages 613–622, 2004.
14. M. Gairing, T. Lücking, M. Mavronicolas, B. Monien, and M. Rode. Nash Equilibria in Discrete Routing Games with Convex Latency Functions. In *Proceedings of the 31st International Colloquium on Automata, Languages, and Programming (ICALP'04)*, LNCS 3142, pages 645–657, 2004.
15. M. Gairing, T. Lücking, M. Mavronicolas, B. Monien, and P. Spirakis. Extreme Nash Equilibria. In *Proceedings of the 8th Italian Conference on Theoretical Computer Science (ICTCS'03)*, LNCS 2841, pages 1–20, 2003. Also accepted to *Theoretical Computer Science*, Special Issue on *Game Theory Meets Theoretical Computer Science*.
16. R. L. Graham. Bounds on Multiprocessing Timing Anomalies. *SIAM Journal of Applied Mathematics*, 17(2):416–429, 1969.
17. D. S. Hochbaum and D. B. Shmoys. A Polynomial Approximation Scheme for Scheduling on Uniform Processors: Using the Dual Approximation Approach. *SIAM Journal on Computing*, 17(3):539–551, 1988.

18. W. Hoeffding. Probability Inequalities for Sums of Bounded Random Variables. *American Statistical Association Journal*, 58(301):12–30, 1963.
19. D. S. Johnson, C. H. Papadimitriou, and M. Yannakakis. How Easy is Local Search? *Journal of Computer and System Sciences*, 37(1):79–100, 1988.
20. E. Koutsoupias, M. Mavronicolas, and P. Spirakis. Approximate Equilibria and Ball Fusion. *Theory of Computing Systems*, 36(6):683–693, 2003.
21. E. Koutsoupias and C. H. Papadimitriou. Worst-Case Equilibria. In *Proceedings of the 16th International Symposium on Theoretical Aspects of Computer Science (STACS'99)*, LNCS 1563, pages 404–413, 1999.
22. M. W. Krentel. On Finding and Verifying Locally Optimal Solutions. *SIAM Journal of Computing*, 19(4):742–729, 1990.
23. J. K. Lenstra, D. B. Shmoys, and É. Tardos. Approximation Algorithms for Scheduling Unrelated Parallel Machines. *Mathematical Programming*, 46:259–271, 1990.
24. T. Lücking, M. Mavronicolas, B. Monien, M. Rode, P. Spirakis, and I. Vrto. Which is the Worst-Case Nash Equilibrium? In *Proceedings of the 28th International Symposium on Mathematical Foundations of Computer Science (MFCS'03)*, LNCS 2747, pages 551–561, 2003.
25. A. Mas-Colell, M. D. Whinston, and J. R. Green. *Microeconomic Theory*. Oxford University Press, 1995.
26. M. Mavronicolas and P. Spirakis. The Price of Selfish Routing. In *Proceedings of the 33rd Annual ACM Symposium on Theory of Computing (STOC'01)*, pages 510–519, 2001.
27. R. D. McKelvey and A. McLennan. Computation of Equilibria in Finite Games. In *Handbook of Computational Economics*, 1996.
28. I. Milchtaich. Congestion Games with Player-Specific Payoff Functions. *Games and Economic Behavior*, 13(1):111–124, 1996.
29. D. Monderer and L. S. Shapley. Potential Games. *Games and Economic Behavior*, 14(1):124–143, 1996.
30. J. F. Nash. Equilibrium Points in n-Person Games. *Proceedings of the National Academy of Sciences of the United States of America*, 36:48–49, 1950.
31. J. F. Nash. Non-Cooperative Games. *Annals of Mathematics*, 54(2):286–295, 1951.
32. C. H. Papadimitriou. Algorithms, Games, and the Internet. In *Proceedings of the 33rd Annual ACM Symposium on Theory of Computing (STOC'01)*, pages 749–753, 2001.
33. R. W. Rosenthal. A Class of Games Possessing Pure-Strategy Nash Equilibria. *International Journal of Game Theory*, 2:65–67, 1973.
34. T. Roughgarden. The Price of Anarchy is Independent of the Network Topology. *Journal of Computer and System Sciences*, 67(2):341–364, 2003.
35. T. Roughgarden and É. Tardos. How Bad Is Selfish Routing? *Journal of the ACM*, 49(2):236–259, 2002.
36. T. Roughgarden and É. Tardos. Bounding the Inefficiency of Equilibria in Nonatomic Congestion Games. *Games and Economic Behaviour*, 47(2):389–403, 2004.
37. A. A. Schäffer and M. Yannakakis. Simple Local Search Problems that are Hard to Solve. *SIAM Journal of Computing*, 20(1):56–87, 1991.
38. V. Vazirani. *Approximation Algorithms*. Springer Verlag, 2001.
39. J. von Neumann. Zur Theorie der Gesellschaftsspiele. *Mathematische Annalen*, 100:295–320, 1928.
40. J. von Neumann and O. Morgenstern. *Theory of Games and Economic Behavior*. Princeton University Press, 1944.
41. J. G. Wardrop. Some Theoretical Aspects of Road Traffic Research. In *Proceedings of the Institute of Civil Engineers, Pt. II, Vol. 1*, pages 325–378, 1956.

The Tree Inclusion Problem: In Optimal Space and Faster

Philip Bille* and Inge Li Gørtz

The IT University of Copenhagen,
Department of Theoretical Computer Science,
Rued Langgaards Vej 7, 2300 Copenhagen S, Denmark
{beetle, inge}@itu.dk

Abstract. Given two rooted, ordered, and labeled trees P and T the tree inclusion problem is to determine if P can be obtained from T by deleting nodes in T. This problem has recently been recognized as an important query primitive in XML databases. Kilpeläinen and Mannila (SIAM J. of Comp. 1995) presented the first polynomial time algorithm using quadratic time and space. Since then several improved results have been obtained for special cases when P and T have a small number of leaves or small depth. However, in the worst case these algorithms still use quadratic time and space. In this paper we present a new approach to the problem which leads to a new algorithm which uses optimal linear space and has subquadratic running time. Our algorithm improves all previous time and space bounds. Most importantly, the space is improved by a linear factor. This will make it possible to query larger XML databases and speed up the query time since more of the computation can be kept in main memory.

1 Introduction

Let T be a rooted tree. We say that T is *labeled* if each node is a assigned a symbol from an alphabet Σ and we say that T is *ordered* if a left-to-right order among siblings in T is given. All trees in this paper are rooted, ordered, and labeled. A tree P is *included* in T, denoted $P \sqsubseteq T$, if P can be obtained from T by deleting nodes of T. Deleting a node v in T means making the children of v children of the parent of v and then removing v. The children are inserted in the place of v in the left-to-right order among the siblings of v. The *tree inclusion problem* is to determine if P can be included in T and if so report all subtrees of T that include P.

Recently, the problem has been recognized as an important query primitive for XML data and has received considerable attention, see e.g., [15, 16, 18, 17]. The key idea is that an XML document can be viewed as an ordered, labeled

---

* This work is part of the DSSCV project supported by the IST Programme of the European Union (IST-2001-35443).

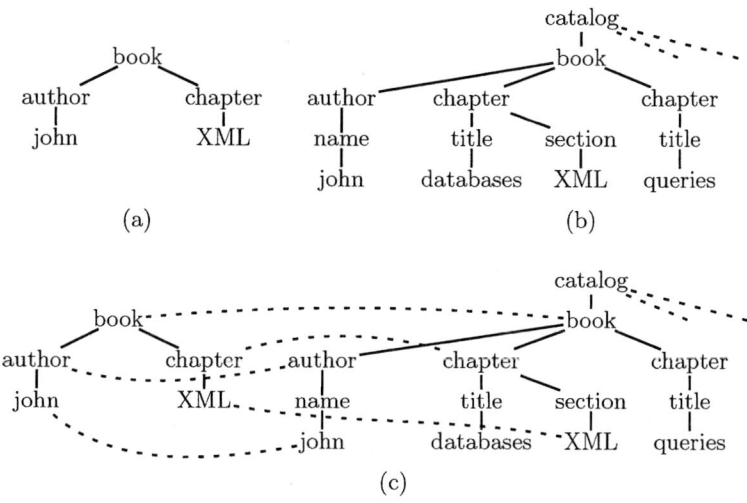

Fig. 1. Can tree (a) be included in tree (b)? Yes. The embedding is given in (c)

tree and queries on this tree correspond to a tree inclusion problem. As an example consider Fig. 1. Suppose that we want to maintain a catalog of books for a bookstore. A fragment of the tree, denoted D, corresponding to the catalog is shown in (b). In addition to supporting full-text queries, such as find all documents containing the word "John", we can also utilize the tree structure of the catalog to ask more specific queries, such as "find all books written by John with a chapter that has something to do with XML". We can model this query by constructing the tree, denoted Q, shown in (a) and solve the tree inclusion problem: is $Q \sqsubseteq D$? The answer is yes and a possible way to include Q in D is indicated by the dashed lines in (c). If we delete all the nodes in D not touched by dashed lines the trees Q and D become isomorphic. Such a mapping of the nodes from Q to D given by the dashed lines is called an *embedding* (formally defined in Sec. 3).

The tree inclusion problem was initially introduced by Knuth [11–exercise 2.3.2-22] who gave a sufficient condition for testing inclusion. Motivated by applications in structured databases [9, 12] Kilpeläinen and Mannila [10] presented the first polynomial time algorithm using $O(n_P n_T)$ time and space, where n_P and n_T is the number of nodes in a tree P and T, respectively. During the last decade several improvements of the original algorithm of [10] have been suggested [8, 1, 14, 4]. The previously best known bound is due to Chen [4] who presented an algorithm using $O(l_P n_T)$ time and $O(l_P \min\{d_T, l_T\})$ space. Here, l_S and d_S denotes the number of leaves of and the maximum depth of a tree S, respectively. This algorithm is based on an algorithm of Kilpeläinen [8]. Note that the time and space is still $\Theta(n_P n_T)$ for worst-case input trees.

In this paper we improve all of the previously known time and space bounds. Combining the three algorithms presented in this paper we have:

Theorem 1. *For trees T and P the tree inclusion problem can be solved in $O(\min(\frac{n_P n_T}{\log n_T}, l_P n_T, n_P l_T \log \log n_T))$ time using optimal $O(n_T + n_P)$ space.*

Hence, for worst-case input this improves the previous time and space bounds by a logarithmic and linear factor, respectively. When P has a small number of leaves the running time of our algorithm matches the previously best known time bound of [4] while maintaining linear space. In the context of XML databases the most important feature of our algorithms is the space usage. This will make it possible to query larger trees and speed up the query time since more of the computation can be kept in main memory.

Techniques. Most of the previous algorithms, including the best one [4], are essentially based on a simple dynamic programming approach from the original algorithm of [10]. The main idea behind this algorithm is following: Let $v \in V(P)$ and $w \in V(T)$ be nodes with children $v_1, \ldots, v_i$ and $w_1, \ldots, w_j$, respectively. To decide if $P(v)$ can be included $T(w)$ we try to find a sequence of numbers $1 \leq x_1 < x_2 < \cdots < x_i \leq j$ such that $P(v_k)$ can be included in $T(w_{x_k})$ for all $k, 1 \leq k \leq i$. If we have already determined whether or not $P(v_s) \sqsubseteq T(w_t)$, for all s and t, $1 \leq s \leq i$, $1 \leq t \leq j$, we can efficiently find such a sequence by scanning the children of v from left to right. Hence, applying this approach in a bottom-up fashion we can determine, if $P(v) \sqsubseteq T(w)$, for all pairs $(v, w) \in V(P) \times V(T)$.

In this paper we take a significantly different approach. The main idea is to construct a data structure on T supporting a small number of procedures, called the *set procedures*, on subsets of nodes of T. We show that any such data structure implies an algorithm for the tree inclusion problem. We consider various implementations of this data structure which all use linear space. The first simple implementation gives an algorithm with $O(l_P n_T)$ running time. As it turns out, the running time depends on a well-studied problem known as the *tree color problem*. We show a general connection between data structures for the tree color problem and the tree inclusion problem. Plugging in a data structure of Dietz [5] we obtain an algorithm with $O(n_P l_T \log \log n_T)$ running time.

Based on the simple algorithms above we show how to improve the worst-case running time of the set procedures by a logarithmic factor. The general idea used to achieve this is to divide T into small trees or forests, called *micro trees* or *clusters* of logarithmic size which overlap with other micro trees in at most 2 nodes. Each micro tree is represented by a constant number of nodes in a *macro tree*. The nodes in the macro tree are then connected according to the overlap of the micro trees they represent. We can efficiently preprocess the micro trees and the macro tree such that the set procedures use constant time for each micro tree. Hence, the worst-case running time is improved by a logarithmic factor to $O(\frac{n_P n_T}{\log n_T})$.

Our results rely on a standard RAM model of computation with word size $\Theta(\log n)$. We use a standard instruction set such as bitwise boolean operations, shifts, and addition. Most of the proofs are omitted due to lack of space. They can be found in the full version of the paper [3].

2 Notation and Definitions

In this section we define the notation and definitions we will use throughout the paper. For a graph G we denote the set of nodes and edges by $V(G)$ and $E(G)$, respectively. Let T be a rooted tree. The root of T is denoted by $\text{root}(T)$. The *size* of T, denoted by n_T, is $|V(T)|$. The *depth* of a node $v \in V(T)$, $\text{depth}(v)$, is the number of edges on the path from v to $\text{root}(T)$ and the depth of T, denoted d_T, is the maximum depth of any node in T. The set of children of a node v is denoted $\text{child}(v)$. A node with no children is a leaf and otherwise an internal node. The set of leaves of T is denoted $L(T)$ and we define $l_T = |L(T)|$. We say that T is *labeled* if each node v is a assigned a symbol, denoted $\text{label}(v)$, from an alphabet Σ and we say that T is *ordered* if a left-to-right order among siblings in T is given. All trees in this paper are rooted, ordered, and labeled.

Let $T(v)$ denote the subtree of T rooted at a node $v \in V(T)$. If $w \in V(T(v))$ then v is an ancestor of w, denoted $v \preceq w$, and if $w \in V(T(v)) \backslash \{v\}$ then v is a proper ancestor of w, denoted $v \prec w$. If v is a (proper) ancestor of w then w is a (proper) descendant of v. A node z is a common ancestor of v and w if it is an ancestor of both v and w. The nearest common ancestor of v and w, $\text{nca}(v,w)$, is the common ancestor of v and w of largest depth. The *first ancestor of w labeled α*, denoted $\text{fl}(w,\alpha)$, is the node v such that $v \preceq w$, $\text{label}(v) = \alpha$, and no node on the path between v and w is labeled α. If no such node exists then $\text{fl}(w,\alpha) = \bot$, where $\bot \notin V(T)$ is a special *null node*.

For any set of pairs U, let $U|_1$ and $U|_2$ denote the *projection* of U to the first and second coordinate, that is, if $(u_1, u_2) \in U$ then $u_1 \in U|_1$ and $u_2 \in U|_2$.

Lists. A *list*, X, is a finite sequence of objects $X = [v_1, \ldots, v_k]$. The *length* of the list, denoted $|X|$, is the number of objects in X. The ith element of X, $X[i]$, $1 \leq i \leq |X|$ is the object v_i and $v \in X$ iff $v = X[j]$ for some $1 \leq j \leq |X|$. For any two lists $X = [v_1, \ldots, v_k]$ and $Y = [w_1, \ldots, w_k]$, the list obtained by *appending* Y to X is the list $X \circ Y = [v_1, \ldots, v_k, w_1, \ldots, w_k]$. We extend this notation such that for any object u, $X \circ u$ denotes the list $X \circ [u]$. For simplicity in the notation we will sometimes write $[v_i \mid 1 \leq i \leq k]$ to denote the list $[v_1, \ldots, v_k]$. A *pair list* is a list of pairs of object $Y = [(v_1, w_1), \ldots, (v_k, w_k)]$. Here the first and second element in the pair is denoted by $Y[i]_1 = v_i$ and $Y[i]_2 = w_i$. The projection of pair lists is defined by $Y|_1 = [v_1, \ldots, v_k]$ and $Y|_2 = [w_1, \ldots, w_k]$.

Orderings. Let T be a tree with root v and let $v_1, \ldots, v_k$ be the children of v from left-to-right. The *preorder traversal* of T is obtained by visiting v and then recursively visiting $T(v_i)$, $1 \leq i \leq k$, in order. Similarly, the *postorder traversal* is obtained by first visiting $T(v_i)$, $1 \leq i \leq k$, and then v. The *preorder number* and *postorder number* of a node $w \in T(v)$, denoted by $\text{pre}(w)$ and $\text{post}(w)$, is the number of nodes preceding w in the preorder and postorder traversal of T, respectively. The nodes to the *left* of w in T is the set of nodes $u \in V(T)$ such that $\text{pre}(u) < \text{pre}(w)$ and $\text{post}(u) < \text{post}(w)$. If u is to the left of w, denoted by $u \triangleleft w$, then w is to the *right* of u. If $u \triangleleft w$, $u \preceq w$, or $w \prec u$ we write $u \trianglelefteq w$. The null node $\bot$ is not in the ordering, i.e., $\bot \ntriangleleft v$ for all nodes v.

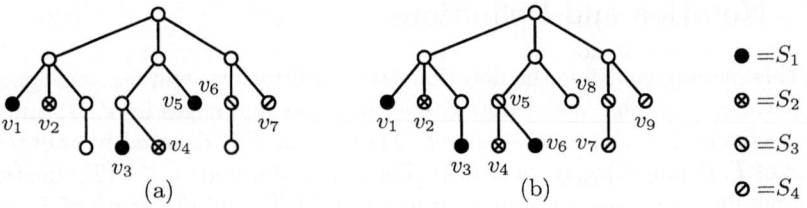

Fig. 2. In (a) we have $\text{mop}(S_1, S_2, S_1, S_3, S_4) = \{(v_3, v_7)\}$ and in (b) we have $\text{mop}(S_1, S_2, S_1, S_3, S_4) = \{(v_1, v_7), (v_3, v_9)\}$

Deep Sets. A set of nodes $V \subseteq V(T)$ is *deep* iff no node in V is a proper ancestor of another node in V.

Minimum Ordered Pair. For deep sets of nodes $V_1, \ldots, V_k$ let $\Phi(V_1, \ldots, V_k) \subseteq (V_1 \times \cdots \times V_k)$, be the set such that $(v_1, \ldots, v_k) \in \Phi(V_1, \ldots, V_k)$ iff $v_1 \triangleleft \cdots \triangleleft v_k$. If $(v_1, \ldots, v_k) \in \Phi(V_1, \ldots, V_k)$ and there is no $(v'_1, \ldots, v'_k) \in \Phi(V_1, \ldots, V_k)$, where either $v_1 \triangleleft v'_1 \triangleleft v'_k \trianglelefteq v_k$ or $v_1 \trianglelefteq v'_1 \triangleleft v'_k \triangleleft v_k$ then the pair (v_1, v_k) is a *minimum ordered pair*. The set of minimum ordered pairs for $V_1, \ldots, V_k$ is denoted by $\text{mop}(V_1, \ldots, V_k)$. Fig. 2 illustrates mop on a small example. The following lemma shows that we can compute $\text{mop}(V_1, \ldots, V_k)$ iteratively by first computing $\text{mop}(V_1, V_2)$ and then $\text{mop}(\text{mop}(V_1, V_2)|_2, V_3)$ and so on.

Lemma 1. *For any deep sets of nodes* $V_1, \ldots, V_k$: $(v_1, v_k) \in \text{mop}(V_1, \ldots, V_k)$ *iff there exists a* v_{k-1} *such that* $(v_1, v_{k-1}) \in \text{mop}(V_1, \ldots, V_{k-1})$ *and* $(v_{k-1}, v_k) \in \text{mop}(\text{mop}(V_1, \ldots, V_{k-1})|_2, V_k)$.

3 Computing Deep Embeddings

In this section we present a general framework for answering tree inclusion queries. As in [10] we solve the equivalent *tree embedding problem*. Let P and T be rooted labeled trees. An *embedding* of P in T is an injective function $f : V(P) \to V(T)$ such that for all nodes $v, u \in V(P)$,

(i) $\text{label}(v) = \text{label}(f(v))$. (label preservation condition)
(ii) $v \prec u$ iff $f(v) \prec f(u)$. (ancestor condition)
(iii) $v \triangleleft u$ iff $f(v) \triangleleft f(u)$. (order condition)

Lemma 2 ([10]). *For any trees P and T, $P \sqsubseteq T$ iff there exists an embedding of P in T.*

An example of an embedding is given in Fig. 1(c). We say that the embedding f is *deep* if there is no embedding g such that $f(\text{root}(P)) \prec g(\text{root}(P))$. The *deep occurrences* of P in T, denoted $\text{emb}(P, T)$ is the set of nodes,

$$\text{emb}(P, T) = \{f(\text{root}(P)) \mid f \text{ is a deep embedding of } P \text{ in } T\}.$$

Note that emb(P,T) must be a deep set in T. Furthermore, by definition the set of ancestors of nodes in emb(P,T) is the set of subtrees $T(u)$ such that $P \sqsubseteq T(u)$. Hence, to solve the tree inclusion problem it is sufficient to compute emb(P,T) and then, using additional $O(n_T)$ time, report all ancestors (if any) of this set.

The key idea in our algorithm for computing deep embeddings is to construct a data structure that allows a fast implementation of the following procedures, called the *set procedures*. For all $V \subseteq V(T)$, $U \subseteq V(T) \times V(T)$, $\alpha \in \Sigma$ define:

PARENT$_T(V)$. Return the set $R := \{\text{parent}(v) \mid v \in V\}$.
NCA$_T(U)$. Return the set $R := \{\text{nca}(u_1, u_2) \mid (u_1, u_2) \in U\}$.
DEEP$_T(V)$. Return the set $R := \{v \in V \mid \nexists w \in V \text{ such that } v \prec w\}$.
MOP$_T(U, V)$. Return the set of pairs R such that for any pair $(u_1, u_2) \in U$, $(u_1, v) \in R$ iff $(u_2, v) \in \text{mop}(U|_2, V)$.
FL$_T(V, \alpha)$. Return the set $R := \{\text{fl}(v, \alpha) \mid v \in V\}$.

With the set procedures we can compute deep embeddings. The following procedure EMB$_T(v)$, $v \in V(P)$, recursively computes the set of deep occurrences of $P(v)$ in T. Fig. 3 illustrates how EMB works on a small example.

EMB$_T(v)$ Let $v_1, \ldots, v_k$ be the sequence of children of v ordered from left to right. There are three cases:
1. $k = 0$ (v is a leaf). Set $R := \text{DEEP}_T(\text{FL}_T(L(T), \text{label}(v)))$.
2. $k = 1$. Recursively compute $R_1 := \text{EMB}_T(v_1)$.
   Set $R := \text{DEEP}_T(\text{FL}_T(\text{DEEP}_T(\text{PARENT}_T(R_1)), \text{label}(v)))$.
3. $k > 1$. Compute $R_1 := \text{EMB}_T(v_1)$ and $U_1 := \{(r, r) \mid r \in R_1\}$. For i, $1 \leq i \leq k$, compute $R_i := \text{EMB}_T(v_i)$ and $U_i := \text{MOP}_T(U_{i-1}, R_i)$.
   Finally, compute $R := \text{DEEP}_T(\text{FL}_T(\text{DEEP}_T(\text{NCA}_T(U_k)), \text{label}(v)))$.

If $R = \emptyset$ stop and report that there is no deep embedding of $P(v)$ in T. Otherwise return R.

Lemma 3. *For any two trees T and P, EMB$_T(v)$ computes the set of deep occurrences of $P(v)$ in T.*

Proof. By induction on the size of the subtree $P(v)$. If v is a leaf we immediately have emb(v, T) = DEEP$_T$(FL$_T(L(T), \text{label}(v)))$. Suppose that v is an internal node with children $v_1, \ldots, v_k$, $k \geq 1$. We show that emb($P(v), T$) = EMB$_T(v)$.

If $k = 1$, $w \in$ EMB$_T(v)$ implies label(w) = label(v) and there is a node $w_1 \in$ EMB$_T(v_1)$ such that fl(parent(w_1), label(v)) = w, i.e., no node on the path between w_1 and w is labeled label(v). By induction EMB$_T(v_1)$ = emb($P(v_1), T$) and thus w is the root of an embedding of $P(v)$ in T. Since EMB$_T(v)$ is the deep set of all such nodes we have $w \in$ emb($P(v), T$). Conversely, if $w \in$ emb($P(v), T$) then label(w) = label(v), there is a node $w_1 \in$ emb($P(v_1), T$) such that $w \prec w_1$, and no node on the path between w and w_1 is labeled label(v), that is, fl(w_1, label(v)) = w. Hence, $w \in$ EMB$_T(v)$.

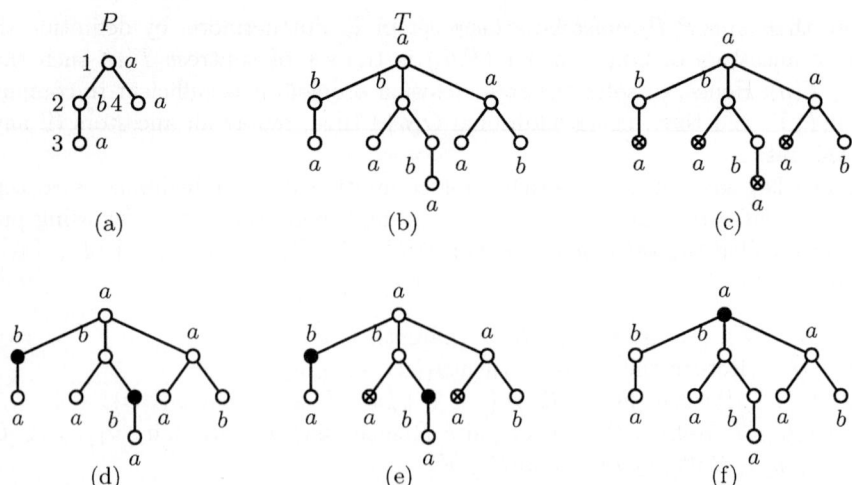

Fig. 3. Computing the deep occurrences of P into T depicted in (a) and (b) respectively. The nodes in P are numbered 1–4 for easy reference. (c) Case 1 of EMB. Since 3 and 4 are leaves and label(3) = label(4) we have $\text{EMB}_T(3) = \text{EMB}_T(4)$. (d) Case 2 of EMB. Note that the middle child of the root(T) is not in the set since it is not a deep occurrence. (e) Case 3 of EMB: The two minimal ordered pairs of the sets of (d) and (c). (f) nca of the pairs in (e) both give the root node of T which is the only (deep) occurrence of P

Before considering case 3 we show that $U_j = \text{mop}(\text{EMB}_T(v_1), \ldots, \text{EMB}_T(v_j))$ by induction on j, $2 \leq j \leq k$. For $j = 2$ it follows from the definition of MOP_T that $U_2 = \text{mop}(\text{EMB}_T(v_1), \text{EMB}_T(v_2))$. Hence, assume that $j > 2$. We have $U_j = \text{MOP}_T(U_{j-1}, \text{EMB}_T(v_j)) = \text{MOP}_T(\text{mop}(\text{EMB}_T(v_1), \ldots, \text{EMB}_T(v_{j-1})), R_j)$. By definition of MOP_T, U_j is the set of pairs such that for any pair $(r_1, r_{j-1}) \in \text{mop}(\text{EMB}_T(v_1), \ldots, \text{EMB}_T(v_{j-1}))$, we have $(r_1, r_j) \in U_j$ if and only if $(r_{j-1}, r_j) \in \text{mop}(\text{mop}(\text{EMB}_T(v_1), \ldots, \text{EMB}_T(v_{j-1}))|_2, R_j)$. It now follows from Lemma 1 that $(r_1, r_j) \in U_j$ iff $(r_1, r_j) \in \text{mop}(\text{EMB}_T(v_1), \ldots, \text{EMB}_T(v_j))$.

Consider case 3. If $k > 1$, $w \in \text{EMB}_T(v)$ implies label(w) = label(v) and there are nodes $(w_1, w_k) \in \text{mop}(\text{emb}(P(v_1), T), \ldots, \text{emb}(P(v_k), T))$ such that $w = \text{fl}(\text{nca}(w_1, w_k), \text{label}(v))$. Clearly, w is the root of an embedding of $P(v)$ in T. Assume for contradiction that w is not a deep embedding, i.e., $w \prec u$ for some node $u \in \text{emb}(P(v), T)$. Since $w = \text{fl}(\text{nca}(w_1, w_k), \text{label}(v))$ there must be nodes $u_1 \triangleleft \cdots \triangleleft u_k$, such that $u_i \in \text{emb}(P(v_i), T)$, $1 \leq i \leq k$, and $u = \text{fl}(\text{nca}(u_1, u_k), \text{label}(v))$. However, this contradicts the fact that $(w_1, w_k) \in \text{mop}(\text{emb}(P(v_1), T), \ldots, \text{emb}(P(v_k), T))$. If $w \in \text{emb}(P(v), T)$ a similar argument implies that $w \in \text{EMB}_T(v)$. □

When the tree T is clear from the context we may not write the subscript T in the procedure names. Note that since the $\text{EMB}_T(v)$ is a deep set we can assume that PARENT, FL, NCA, and MOP take deep sets as input.

4 A Simple Tree Inclusion Algorithm

In this section we a present a simple implementation of the set procedures which leads to an efficient tree inclusion algorithm. Subsequently, we modify one of the procedures to obtain a family of tree inclusion algorithms where the complexities depend on the solution to a well-studied problem known as the *tree color problem*.

Preprocessing. To compute deep embeddings efficiently we require a data structure for T which allows us, for any $v, w \in V(T)$, to compute $\text{nca}_T(v, w)$ and determine if $v \prec w$ or $v \lhd w$. In linear time we can compute $\text{pre}(v)$ and $\text{post}(v)$ for all nodes $v \in V(T)$, and with these it is straightforward to test the two conditions. Using a data structure by Harel and Tarjan [7] we can answer nearest common ancestor queries in $O(1)$ time using $O(n_T)$ space and preprocessing time. Hence, our data structure uses linear preprocessing time and space.

Implementation of the Set Procedures. To answer tree inclusion queries we give an efficient implementation of the set procedures. The idea is to represent the node sets in a left-to-right order. For this purpose we introduce some helpful notation. A *node list*, X, is a list of nodes. If $v_i \lhd v_{i+1}$, $1 \leq i < |X|$ then X is *ordered* and if $v_i \unlhd v_{i+1}$, $1 \leq i < |X|$ then X is *semiordered*. A *node pair list*, Y, is a list of pairs of nodes. We say that Y is ordered if $Y|_1$ and $Y|_2$ are ordered, and semiordered if $Y|_1$ and $Y|_2$ are semiordered.

The set procedures are implemented using node lists and node pair lists. All lists used in the procedures are either ordered or semiordered. As noted in Sec. 3 we may assume that the input to all of the procedures, except DEEP, represent a deep set, that is, the corresponding node list or node pair list is ordered. We assume that the input list given to DEEP is semiordered. Hence, the output of all the other set procedures must be semiordered.

PARENT$_T(X)$. Return the list $Z := [\text{parent}(X[i]) \mid 1 \leq i \leq |X|]$.
NCA(Y). Return the list $Z := [\text{nca}(Y[i]) \mid 1 \leq i \leq |Y|]$.
DEEP$_T(X)$. Initially, set $v := X[1]$ and $Z := []$. For each i, $2 \leq i \leq k$, compare v and $X[i]$: If $v \lhd X[i]$ set $Z := Z \circ v$ and $v := X[i]$. If $v \prec X[i]$, set $v := X[i]$ and otherwise $(X[i] \prec v)$ do nothing. Finally, set $Z := Z \circ v$ and return Z.
MOP$_T(X, Y)$. Initially, set $Z := []$. Find the minimum j such that $X[1]_2 \lhd Y[j]$ and set $x := X[1]_1$, $y := Y[j]$, and $h := j$. If no such j exists, stop. As long as $h \leq |Y|$ do the following: For each i, $2 \leq i \leq |X|$, do: Set $h := h+1$ until $X[i]_2 \lhd Y[h]$. Compare $Y[h]$ and y: If $y = Y[h]$ set $x := X[i]_1$. If $y \lhd Y[h]$ set $Z := Z \circ (x, y)$, $x := X[i]_1$, and $y := Y[h]$. Finally, set $Z := Z \circ (x, y)$ and return Z.
FL$_T(X, \alpha)$. Initially, set $Y := X$, $Z := []$, and $S := []$. Repeat until $Y := []$: For $i = 1, \ldots, |Y|$ if label$(Y[i]) = \alpha$ set $Z :=$ INSERT$(Y[i], Z)$ and otherwise set $S := S \circ \text{parent}(Y[i])$. Set $S := \text{DEEP}_T(S)$, $Y := \text{DEEP}^*_T(S, Z)$, $S := []$. Return Z.

Procedure FL calls two auxiliary procedures: INSERT(v, Z) takes an ordered list Z and insert the node v such that the resulting list is ordered, and DEEP$^*(S, Z)$

takes two ordered lists and returns the ordered list representing the set DEEP($S \cup Z$) $\cap$ S, i.e., DEEP$^*(S, Z) = [s \in S | \nexists z \in Z : s \prec z]$. Below we describe the implementation of FL in more detail.

We use one doubly linked list to represent all the lists Y, S, and Z. For each element in Y we have pointers Pred and Succ pointing to the predecessor and successor in the list, respectively. We also have at each element a pointer Next pointing to the next element in Y. In the beginning Next = Succ for all elements, since all elements in the list are in Y. When going through Y in one iteration we simple follow the Next pointers. When FL calls INSERT($Y[i], Z$) we set Next(Pred($Y[i]$)) to Next($Y[i]$). That is, all nodes in the list not in Y, i.e., nodes not having a Next pointer pointing to them, are in Z. We do not explicitly maintain S. Instead we just save PARENT($Y[i]$) at the position in the list instead of $Y[i]$. Now DEEP(S) can be performed following the Next pointers and removing elements from the doubly linked list accordingly to procedure DEEP. It remains to show how to calculate DEEP$^*(S, Z)$. This can be done by running through S following the Next pointers. At each node s compare Pred(s) and Succ(s) with s. If one of them is a descendant of s then remove s from the doubly linked list.

Using this linked list implementation DEEP$^*(S, Z)$ takes time $O(|S|)$, whereas using DEEP to calculate this would have used time $O(|S| + |Z|)$.

Complexity of the Algorithm. For the running time of the node list implementation observe that, given the data structure described above, all set procedures, except FL, perform a single pass over the input using constant time at each step. Hence we have,

Lemma 4. *For any tree T there is a data structure using $O(n_T)$ space and preprocessing which supports each of the procedures* PARENT, DEEP, MOP, *and* NCA *in linear time (in the size of their input).*

The running time of a single call to FL might take time $O(n_T)$. Instead we will divide the calls to FL into groups and analyze the total time used on such a group of calls. The intuition behind the division is that for a path in P the calls made to FL by EMB is done bottom up on disjoint lists of node in T.

Lemma 5. *For disjoint ordered node lists $V_1, \ldots, V_k$ and labels $\alpha_1, \ldots, \alpha_k$, such that any node in V_{i+1} is an ancestor of some node in* DEEP(FL$_T(V_i, \alpha_i)$), $2 \le i < k$, *all of* FL$_T(V_1, \alpha_1), \ldots,$ FL$_T(V_k, \alpha_k)$ *can be computed in $O(n_T)$ time.*

The proof is omitted due to lack of space. The basic idea in the proof is to show that any node in T can be in Y at most twice during all calls to FL.

Using the node list implementation of the set procedures we get:

Theorem 2. *For trees P and T the tree inclusion problem can be solved in $O(l_P n_T)$ time and $O(n_P + n_T)$ space.*

Proof. By Lemma 4 we can preprocess T in $O(n_T)$ time and space. Let $g(n)$ denote the time used by FL on a list of length n. Consider the time used by EMB$_T$(root(P)). We bound the contribution for each node $v \in V(P)$. From

Lemma 4 it follows that if v is a leaf the cost of v is at most $O(g(l_T))$. Hence, by Lemma 5, the total cost of all leaves is $O(l_P g(l_T)) = O(l_P n_T)$. If v has a single child w the cost is $O(g(|\text{EMB}_T(w)|))$. If v has more than one child the cost of MOP, NCA, and DEEP is bounded by $\sum_{w \in \text{child}(v)} O(|\text{EMB}_T(w)|)$. Furthermore, since the length of the output of MOP (and thus NCA) is at most $z = \min_{w \in \text{child}(v)} |\text{EMB}_T(w)|$ the cost of FL is $O(g(z))$. Hence, the total cost for internal nodes is,

$$\sum_{v \in V(P) \setminus L(P)} O(g(\min_{w \in \text{child}(v)} |\text{EMB}_T(w)|) + \sum_{w \in \text{child}(v)} |\text{EMB}_T(w)|) \leq \sum_{v \in V(P)} O(g(|\text{EMB}_T(v)|)).$$

Next we bound the sum $\sum_{v \in V(P)} O(g(|\text{EMB}_T(v)|))$. For any $w \in \text{child}(v)$ we have that $\text{EMB}_T(w)$ and $\text{EMB}_T(v)$ are disjoint ordered lists. Furthermore we have that any node in $\text{EMB}_T(v)$ must be an ancestor of some node in $\text{DEEP}_T(\text{FL}_T(\text{EMB}_T(w), \text{label}(v)))$. Hence, by Lemma 5, for any leaf to root path $\delta = v_1, \ldots, v_k$ in P, we have that $\sum_{u \in \delta} g(|\text{EMB}_T(u)|) \leq O(n_T)$. Let Δ denote the set of all root to leaf paths in P. It follows that, $\sum_{v \in V(T)} g(|\text{EMB}_T(v)|) \leq \sum_{p \in \Delta} \sum_{u \in p} g(|\text{EMB}_T(u)|) \leq O(l_P n_T)$.

Since this time dominates the time spent at the leaves the time bound follows. Next consider the space used by $\text{EMB}_T(\text{root}(P))$. The preprocessing of described above uses only $O(n_T)$ space. Furthermore, by induction on the size of the subtree $P(v)$ it follows immediately that at each step in the algorithm at most $O(\max_{v \in V(P)} |\text{EMB}_T(v)|)$ space is needed. Since $\text{EMB}_T(v)$ a deep embedding, it follows that $|\text{EMB}_T(v)| \leq l_T$. □

An Alternative Algorithm. In this section we present an alternative algorithm. Since the time complexity of the algorithm in the previous section is dominated by the time used by FL, we present an implementation of this procedure which leads to a different complexity. Define a *firstlabel data structure* as a data structure supporting queries of the form $\text{fl}(v, \alpha)$, $v \in V(T)$, $\alpha \in \Sigma$. Maintaining such a data structure is known as the *tree color problem*, see e.g., [5, 13]. With such a data structure available we can compute $\text{FL}(X, \alpha)$ as the list $[\text{fl}(X[i], \alpha) \mid 1 \leq i \leq |X|]$.

Theorem 3. *Let P and T be trees. Given a firstlabel data structure using $s(n_T)$ space, $p(n_T)$ preprocessing time, and $q(n_T)$ time for queries, the tree inclusion problem can be solved in $O(p(n_T) + n_P l_T \cdot q(n_T))$ time and $O(n_P + s(n_T) + n_T)$ space.*

Proof. Constructing the firstlabel data structures uses $O(s(n_T))$ space and time $O(p(n_T))$. As in the proof of Thm. 2 we have that the total time used by $\text{EMB}_T(\text{root}(P))$ is bounded by $\sum_{v \in V(P)} g(|\text{EMB}_T(v)|)$, where $g(n)$ is the time used by FL on a list of length n. Since $\text{EMB}_T(v)$ is a deep embedding and each fl takes $q(n_T)$ we have, $\sum_{v \in V(P)} g(|\text{EMB}_T(v)|) \leq \sum_{v \in V(P)} g(l_T) = n_P l_T \cdot q(n_T)$. □

Several firstlabel data structures are available, for instance, if we want to maintain linear space, we can use a data structure by Dietz [5] that supports firstlabel queries in $O(\log \log n_T)$ time using $O(n_T)$ space and $O(n_T)$ expected preprocessing time. Plugging in this data structure we obtain,

Corollary 1. *For trees P and T the tree inclusion problem can be solved in $O(n_P l_T \log \log n_T)$ time and $O(n_P + n_T)$ space.*

Since the preprocessing time $p(n)$ of the firstlabel data structure is expected the running time of the tree inclusion algorithm is also expected. However, the expectation is due to a dictionary using perfect hashing and we can therefore use the deterministic dictionary of [6] with $O(n_T \log n_T)$ worst-case preprocessing time instead. This does not affect the overall complexity of the algorithm.

5 A Faster Tree Inclusion Algorithm

In this section we present a new tree inclusion algorithm which has a worst-case subquadratic running time. Due to lack of space we will only give a rough sketch of the algorithm. A full description of the algorithm can be found in the full version of the paper [3].

The first step is to divide T into small connected subgraphs, called *micro trees* or *clusters*. Using a technique from [2] we can construct in linear time a *cluster partition* of T, consisting of $O(n_T/\log n_T)$ clusters each of size $O(\log n_T)$, with the property that any cluster shares at most two nodes with any other cluster. Each micro tree is represented by a constant number of nodes in a *macro tree*. The nodes in the macro tree are then connected according to the overlap of the micro trees they represent. Note that the total number of nodes in the macro tree is $O(n_T/\log n_T)$.

In linear time of the tree T we preprocess all the micro trees and the macro tree such that the set procedures use constant time for each micro tree. Using a compact node representation we can then implement all the set procedures in $O(n_T/\log n_T)$ time.

Lemma 6. *For any tree T there is a data structure using $O(n_T)$ space and $O(n_T)$ expected preprocessing time which supports all of the set procedures in $O(n_T/\log n_T)$ time.*

The proof of the lemma and all details in the implementation of the set procedures can be found in the full version of the paper. We can now compute the deep occurrences of P in T using the procedure EMB of Sec. 3 and Lemma 6. Since each node $v \in V(P)$ contributes at most a constant number of calls to set procedures it follows that,

Theorem 4. *For trees P and T the tree inclusion problem can be solved in $O(\frac{n_P n_T}{\log n_T})$ time and $O(n_P + n_T)$ space.*

Combining the results in Theorems 2, 4 and Corollary 1 we have the main result of Theorem 1.

Acknowledgments. We thank the reviewers for the many insightful comments.

References

1. L. Alonso and R. Schott. On the tree inclusion problem. In *Proc. of Math. Foundations of Computer Science*, pages 211–221, 1993.
2. S. Alstrup, J. Holm, K. de Lichtenberg, and M. Thorup. Minimizing diameters of dynamic trees. In *Proc. of Intl. Coll. on Automata, Languages and Programming (ICALP)*, pages 270–280, 1997.
3. P. Bille and I. Gørtz. The tree inclusion problem: In optimal space and faster. Technical Report TR-2005-54, IT University of Copenhagen, January 2005.
4. W. Chen. More efficient algorithm for ordered tree inclusion. *J. Algorithms*, 26:370–385, 1998.
5. P. F. Dietz. Fully persistent arrays. In *Proc. of Workshop on Algorithms and Data Structures (WADS)*, pages 67–74, 1989.
6. T. Hagerup, P. B. Miltersen, and R. Pagh. Deterministic dictionaries. *J. Algorithms*, 41(1):69–85, 2001.
7. D. Harel and R. E. Tarjan. Fast algorithms for finding nearest common ancestors. *SIAM J. Comput.*, 13(2):338–355, 1984.
8. P. Kilpeläinen. *Tree Matching Problems with Applications to Structured Text Databases*. PhD thesis, University of Helsinki, Department of Computer Science, 1992.
9. P. Kilpeläinen and H. Mannila. Retrieval from hierarchical texts by partial patterns. In *Proc. of Conf. on Research and Development in Information Retrieval*, pages 214–222, 1993.
10. P. Kilpeläinen and H. Mannila. Ordered and unordered tree inclusion. *SIAM J. Comp.*, 24:340–356, 1995.
11. D. E. Knuth. *The Art of Computer Programming, Volume 1*. Addison Wesley, 1969.
12. H. Mannila and K. J. Räihä. On query languages for the p-string data model. *Information Modelling and Knowledge Bases*, pages 469–482, 1990.
13. S. Muthukrishnan and M. Müller. Time and space efficient method-lookup for object-oriented programs. In *Proc. of Symp. on Discrete Algorithms*, pages 42–51, 1996.
14. T. Richter. A new algorithm for the ordered tree inclusion problem. In *Proc. of Symp. on Combinatorial Pattern Matching (CPM)*, pages 150–166, 1997.
15. T. Schlieder and H. Meuss. Querying and ranking XML documents. *J. Am. Soc. Inf. Sci. Technol.*, 53(6):489–503, 2002.
16. T. Schlieder and F. Naumann. Approximate tree embedding for querying XML data. In *Proc. of Workshop On XML and Information Retrieval*, 2000.
17. H. Yang, L. Lee, and W. Hsu. Finding hot query patterns over an xquery stream. *The VLDB Journal*, 13(4):318–332, 2004.
18. L. H. Yang, M. L. Lee, and W. Hsu. Efficient mining of XML query patterns for caching. In *Proc. of Conference on Very Large Databases (VLDB)*, pages 69–80, 2003.

Union-Find with Constant Time Deletions

Stephen Alstrup[1], Inge Li Gørtz[1], Theis Rauhe[1],
Mikkel Thorup[2], and Uri Zwick[3]

[1] Department of Theoretical Computer Science,
IT University of Copenhagen, Denmark
{stephen, inge, theis}@itu.dk
[2] AT&T Research Labs, USA
mthorup@research.att.com
[3] School of Computer Science, Tel Aviv University, Israel
zwick@cs.tau.ac.il

Abstract. A union-find data structure maintains a collection of disjoint sets under *makeset*, *union* and *find* operations. Kaplan, Shafrir and Tarjan [SODA 2002] designed data structures for an extension of the union-find problem in which elements of the sets maintained may be deleted. The cost of a *delete* operation in their implementations is the same as the cost of a *find* operation. They left open the question whether *delete* operations can be implemented more efficiently than *find* operations. We resolve this open problem by presenting a relatively simple modification of the classical union-find data structure that supports *delete*, as well as *makeset* and *union*, operations in *constant* time, while still supporting *find* operations in $O(\log n)$ worst-case time and $O(\alpha(n))$ amortized time, where n is the number of elements in the set returned by the *find* operation, and $\alpha(n)$ is a functional inverse of Ackermann's function.

1 Introduction

A union-find data structure maintains a collection of disjoint sets under the operations *makeset*, *union* and *find*. A *makeset* operation generates a singleton set. A *union* operation takes two sets and unites them, destroying the two original sets. A *find* operation takes an element and returns a reference to the set currently containing it. The union-find problem is one of the most fundamental data structure problems. It has many applications in a wide range of areas. For an extensive list of such applications, and for a wealth of information on the problem and many of its variants, see the survey of Galil and Italiano [7].

An extremely simple union-find data structure (attributed by Aho *et al.* [1] to McIlroy and Morris), which employs two simple heuristics, *union by rank* and *path compression*, was shown by Tarjan [12] (see also Tarjan and van Leeuwen [13]) to be amazingly efficient. It performs a sequence of M *find* operations and N *makeset* and *union* operations in $O(N+M\,\alpha(M,N))$ total time. Here $\alpha(\cdot,\cdot)$ is an extremely slowly growing functional inverse of Ackermann's function. In other words, the amortized cost of each *makeset* and *union* operation is $O(1)$, while the amortized cost of each *find* operation is $O(\alpha(M+N, N))$, only marginally more than a constant. Fredman and Saks [6] obtained a matching lower bound in the cell probe model of computation, showing that this simple data structure is essentially optimal in the amortized setting.

The union by rank heuristics on its own implies that *find* operations take $O(\log n)$ worst-case time. Here n is the number of elements in the set returned by the *find* operation. All other operations take constant worst-case time. It is possible to trade a slower *union* for a faster *find*. Smid [11], building on a result of Blum [4], gives for any k a data structure that supports *union* operations in $O(k)$ time and *find* operations in $O(\log_k n)$ time. When $k = \log n/\log\log n$, both the *union* and *find* operation take $O(\log n/\log\log n)$ time. Fredman and Saks [6] (see also Ben-Amram and Galil [3]) again show that this tradeoff is optimal, establishing an interesting gap between the amortized and worst-case complexities of the union-find problem. Alstrup et al. [2] present union-find algorithms with simultaneously optimal amortized and worst-case bounds.

Local Amortized Bounds. As noted by Kaplan et al. [8], the standard amortized bounds for *find* are global in terms of the total number N of elements ever created whereas the worst-case bounds are local in terms of the number n of elements in the current set we are finding. Obviously n may be much smaller than N. To state more local amortized bounds, we need a non-standard parameterization of the inverse Ackermann function. For integers $k \geq 0$ and $j \geq 1$, define an Ackermann function $A_k(j)$ as follows

$$A_k(j) = \begin{cases} j+1 & \text{if } k = 0, \\ A_{k-1}^{(j+1)}(j) & \text{if } k \geq 1. \end{cases}$$

Here $f^{(i)}(x)$ is the function f iterated i times on x. Now, define the inverse of the function $\bar{\alpha}(j, i)$, for integer $i, j \geq 0$, as

$$\bar{\alpha}(j, i) = \min\{k \geq 2 \mid A_k(j) > i\}.$$

(For a technical reason, $\bar{\alpha}(j, i)$ is defined to be at least 2 for every $i, j \geq 0$.) Relating to the standard definition of α, we have $\alpha(M, N) = \Theta(\bar{\alpha}(\lceil M/N \rceil, N))$. Kaplan et al. [8] present a refined analysis of the classical union-find data structure showing that the amortized cost of *find*(x) operation is only $O(\bar{\alpha}(\lceil (M+N)/N \rceil, n))$. Kaplan et al. [8] state their results equivalently in terms of a three parameter function that we will not define here. To get a purely local amortized cost for *find*, we note that $\bar{\alpha}(\lceil (M+N)/N \rceil, n) \leq \bar{\alpha}(1, n) = O(\alpha(n, n)) = O(\alpha(n))$.

Union-Find with Deletions. In the traditional version of the union-find problem elements are created using *makeset* operations. Once created, however, elements are never destroyed. Kaplan et al. [8] consider a very natural extension of the union-find problem in which elements may be *deleted*. We refer to this problem as the *union-find with deletions* problem, or *union-find-delete* for short.

Using relatively straightforward ideas (see, e.g., [8]) it is possible to design a union-find-delete data structure that uses only $O(N^*)$ space, handles *makeset*, *union* and *delete* operations in $O(1)$ worst-case time, and *find* operations in $O(\log N^*)$ worst-case time and $O(\alpha(N^*))$ amortized time, where N^* is the current number of elements in the whole data structure. The challenge in the design of union-find-delete data structures is to have the time of a *find*(x) operation depend on n, the size of the set currently containing x, and not on N^*, the total number of elements currently contained in all the sets.

Using an incremental background rebuilding technique for each set, Kaplan et al. [8] describe a way of converting any data structure for the classical union-find problem into a union-find-delete data structure. The time bounds for *makeset*, *find* and *union* operations change by only a constant factor, while the time needed for a *delete*(x) operation is the same as the time needed for a *find*(x) operation followed by a *union* operation with a singleton set. As a *union* operation is usually much cheaper than a *find* operation, Kaplan et al. [8] thus show that in both the amortized and the worst-case settings, a *delete* operation is not more expensive than a *find* operation. Combined with their refined amortized analysis of the classical union-find data structure, this provides, in particular, a union-find-delete data structure that implements *makeset* and *union* operations in $O(1)$ time, and *find*(x) and *delete*(x) operations in $O(\alpha(n))$ amortized time and $O(\log n)$ worst-case time. They leave open, however, the question whether *delete* operations can be implemented *faster* than *find* operations.

Our Results. We solve the major open problem raised by Kaplan et al. [8] and show that *delete* operations can be performed in *constant* worst-case time, while still keeping the $O(\bar{\alpha}(\lceil (M+N)/N \rceil, n)) = O(\alpha(n))$ amortized cost and the $O(\log n)$ worst-case cost of *find* operations, and the constant worst-case cost of *makeset* and *union* operations. We recall here that N is the total number of elements ever created, M is the total number of *find* operations performed, and n is the number of elements in the set returned by the *find* operation. The data structure that we present uses linear space and is a relatively simple modification of the classical union-find data structure. It is at least as simple as the data structures presented by Kaplan et al. [8].

As a by-product we also obtain a very concise potential-based proof of the $O(\bar{\alpha}(\lceil (M+N)/N \rceil, n))$ bound, first obtained by Kaplan et al. [8], on the amortized cost of a *find* operation in the classical setting. We believe that our potential-based analysis is much simpler than the one given by Kaplan et al. [8].

Our Techniques. Our new union-find-delete data structure, like most other union-find data structures, maintains the elements of each set in a rooted tree. As elements can now be deleted, not all the nodes in these trees will contain active elements. Nodes that contain elements are said to be *occupied*, while nodes that do not contain elements are said to be *vacant*. When an element is deleted, the node containing it becomes vacant. If proper measures are not taken, then a tree representing a set may contain too many vacant nodes. As a result, the space needed to store the tree, and the time needed to process a *find* operation may become too large. The new data structure uses a simple collection of local operations to *tidy up* a tree after each delete operation. This ensures that at most half of the nodes in a tree are vacant. More importantly, the algorithm employs local constant-time *shortcut* operations in which the grandparent, or a more distant ancestor, of a node becomes its new parent. These operations, which may be viewed as a local constant-time variant of the path compression technique, keep the trees relatively shallow to allow fast *find* operations.

As with the simple standard union-find, the analysis is the most non-trivial part. The analysis of the new data structure uses two different potential functions. The first potential function is used to bound the *worst-case* cost of *find* operations. Both potential functions are needed to bound the *amortized* cost of *find* operations. The second potential function on its own can be used to obtain a simple derivation of the refined amortized bounds of Kaplan et al. [8] for union-find without deletions.

We end this section with a short discussion of the different techniques used to analyze union-find data structures. The first tight amortized analysis of the classical union-find data structure, by Tarjan [12] and Tarjan and van Leeuwen [13], uses *collections of partitions* and the so-called *accounting method*. The refined analysis of Kaplan et al. [8] is directly based on this method. A much more concise analysis of the union-find data structure based on potential functions can be found in Kozen [9] and Chapter 21 of Cormen et al. [5]. The amortized analysis of our new union-find-delete data structure is based on small but crucial modifications of the potential function used in this analysis. As a by product we get, as mentioned above, a simple proof of the amortized bounds of Kaplan et al. [8]. Seidel and Sharir [10] presented recently an intriguing top-down amortized analysis of the union-find data structure. Our analysis is no less concise, though perhaps less intuitive, and has the additional advantage of bounding the cost of an amortized operation in terms of the size of the set returned by the operation.

2 Preliminaries

The Union-Find and Union-Find-Delete Problems. A classical union-find data structure supports the following operations:

- *make-set(x)*: Create a singleton set containing x.
- *union(A,B)*: Combine the sets A and B into a new set, destroying A and B.
- *find(x)*: Return an identifier of the set containing x.

The only requirement from the identifier, or name, returned by a *find* operation is that if two elements x and y are currently contained in the same set, then the calls *find(x)* and *find(y)* return the same identifier. Kaplan et al. [8] studied data structures that also support *delete* operations:

- *delete(x)*: Delete x from the set containing it.

A *delete* operation should not change the identifier attached to the set from which the element was deleted. It is important to note that a *delete* operation does not receive a reference to the set currently containing x. It only receives the element x itself. As mentioned, Kaplan et al. [8] essentially showed that *delete* operations are not more expensive than *find* operations.

Standard Worst-Case Bounds for Union-Find. We briefly review here the simple standard union-find data structure that supports *makeset* and *union* operations in constant time and *find* operations in $O(\log n)$ time, as it forms the basis of our new data structure for the union-find-delete problem.

The elements of each set A are maintained in a rooted tree $T = T_A$. The identifier of the set A is the root of T. Fixing some terminology, the *height* of a node $v \in T$, denoted by $h(v)$, is defined to be 0, if v is a leaf, and $\max\{h(w) \mid w \text{ is a child of } v\} + 1$, otherwise. Let $root(T)$ denote the root of T. The height of a tree is the height of its root. For a node $v \in T$ let $p(v)$ denote the *parent* of v. A node $x \in T$ is an *ancestor* of a node $y \in T$ if x is on the path from y to the root of T—both y and the root included. A node $x \in T$ is a *descendant* of a node $y \in T$ if y is an ancestor of x.

Each node v has an assigned integer rank $rank(v)$. An important invariant is that for the parent of a node always has a strictly higher rank than the node itself. The rank of a tree is defined to be the rank of the root of the tree.

We implement the operations as follows.

find(x): Follow parent pointers from x all the way to the root. Return the root as the identifier of the set.

make-set(x): Create a new node x. Let $p(x) \leftarrow x$, $rank(x) \leftarrow 0$.

union(A,B): Recall that A and B are root nodes. Assume w.l.o.g. that $rank(A) \geq rank(B)$. Make B a child of A. If $rank(A)=rank(B)$, increase $rank(A)$ by one.

Analysis. Trivially, *makeset* and *union* operations take constant time. Since ranks are strictly increasing when following parent pointers, the time of a *find* operation applied to an element in a set A is proportional to $rank(A)$. We prove, by induction, that $rank(A) \leq \log_2 |A|$, or equivalently, that

$$|A| \geq 2^{rank(A)}. \tag{1}$$

When A is just created with *make-set(x)*, it has rank 0 and $2^0 = 1$ elements. If C is the set created by *union(A,B)*, then $|C| = |A| + |B|$. If C has the same rank as A, or the same rank as B, we are trivially done. Otherwise, we have $rank(A) = rank(B) = k$ and $rank(C) = k + 1$, and then $|C| = |A| + |B| \geq 2^k + 2^k = 2^{k+1}$. This completes the standard analysis of union-find with worst-case bounds.

3 Augmenting Worst-Case Union-Find with Deletions

Each set in the data structure is again maintained in a rooted tree. In the standard union-find data structure, reviewed in Section 2, the nodes of each tree were identified with the elements of the set. In the new data structure, elements are attached to nodes, not identified with them. Some nodes in a tree are *occupied*, i.e., have an element attached to them, while others are *vacant*, i.e., have no element attached to them. An element can then be deleted by simply removing it from the node it was attached to. This node then becomes vacant. The name of a set is taken to be its root node. As the name of a set is a node, and not an element, names do not change as a result of *delete* operations.

An obvious problem with this approach is that if we never remove vacant nodes from the trees, we may end up consuming non-linear space. To avoid this, we require our union-find trees to be *tidy*:

Definition 1. *A tree is said to be* tidy *if it satisfies the following properties:*

- *Every vacant non-root node has at least two children,*
- *Every leaf is occupied and has rank 0.*

It is easy to tidy up a tree. First, we remove vacant leaves. When a node becomes a leaf, its rank is reduced to 0. Next, if a vacant non-root node v has a single child w, we make the parent of v the parent of w and remove v. We call this *bypassing* v. The following Lemma is now obvious.

Lemma 1. *At most half of the nodes in a tidy tree may be vacant.*

Tidy trees thus use linear space. However, tidyness on it own does not yield a sublinear time bound on *find* operations. (Note, for example, that a path of occupied nodes is tidy.) Our next goal would be to make sure that the depth of a tree is logarithmic in the number of occupied nodes contained in it. Ideally, we would want all trees to be *reduced*:

Definition 2. *A tree is said to be* reduced *if it is either*

- *A tree composed of a single occupied node of rank 0, or*
- *A tree of height 1 with a root of rank 1 and occupied leaves of rank 0.*

Naturally, we will not manage to keep our trees reduced at all times. Reduced trees form, however, the base case for our analysis.

Keeping the Trees Shallow During Deletions. This section contains our main technical contribution. We show how to implement deletions so that for any set A,

$$|A| \geq (2/3)(6/5)^{rank(A)} . \tag{2}$$

Consequently, $rank(A) \leq \log_{6/5}(3|A|/2) - O(\log|A|+1)$. As the rank of a tree is always an upper bound on its height, we thus need to follow at most $O(\log|A|+1)$ parent pointers to get from any element of A to the root identifier.

The key idea is to associate the following value with each node v:

Definition 3. *The value* $val(v)$ *of a node* v *is defined as*

$$val(v) = \begin{cases} (5/3)^{rank(p(v))} & \text{if } v \text{ is occupied,} \\ (1/2)(5/3)^{rank(p(v))} & \text{if } v \text{ is vacant.} \end{cases}$$

Here, if v is a root, $p(v) = v$. The value of a set A is defined as the sum the values of all nodes in the tree T_A representing A: $VAL(A) = \sum_{v \in T_A} val(v)$.

The value 5/3 is chosen to satisfy Equation 2 and Lemma 2, 4, and 9 below. In fact, we could have chosen any constant value in $[(1+\sqrt{5})/2, 2)$. We are going to implement deletions in such a way that

$$VAL(A) \geq 2^{rank(A)}. \tag{3}$$

Since the tree representing a set A contains exactly $|A|$ occupied nodes, each of value at most $(5/3)^{rank(A)}$, and at most $|A|$ vacant nodes in T_A, each of value at most $(5/3)^{rank(A)}/2$, it will follow that

$$|A| \geq \frac{2^{rank(A)}}{(3/2)(5/3)^{rank(A)}} = (2/3)(6/5)^{rank(A)},$$

so (3) will imply (2).

The essential operation used to keep trees shallow is to *shortcut* from a node v, giving v a parent higher up over v in the tree. For example, path compression shortcuts from all nodes in a search path directly to the root. Since ranks are strictly increasing up through the tree, shortcutting from v increases the value of v by a factor of at least 5/3. This suggests that we can make up for the loss of a deleted node by a constant number of shortcuts from nearby nodes of similar rank. Before proceeding, let us check that reduced trees satisfy (3).

Lemma 2. *If the tree representing a set A is reduced then $VAL(A) \geq 2^{rank(A)}$.*

Proof. If A is of height 0, then $VAL(A) = (5/3)^0 = 1$ and $2^{rank(A)} = 1$. If A is of height 1, then $VAL(A) \geq (5/3)^1 + (1/2)(5/3)^1 = 5/2$ while $2^{rank(A)} = 2$. □

Let us for a moment assume that we have an implementation of *delete* that preserves, i.e., does not decrease, value, and let us check that the other operations preserve (3). A *makeset* operation creates a reduced tree, so (3) is satisfied by Lemma 2. Also, when we set $C := union(A, B)$, we get $VAL(C) \geq VAL(A) + VAL(B)$, and hence (3) follows just like (1).

Paying for a Deletion via Local Rebuilding. We now show how we can implement a delete operation in constant time, either without decreasing value of the set from which the element is deleted, or ending up with a reduced tree representing the set. Suppose we delete an element of A attached to a node u. As u becomes vacant, we immediately loose half its value. Before u was vacant the tree was tidy, but now we may have to tidy the tree. If u is not a leaf, the only required tidying up is to bypass u if it has a single child. If instead u was a leaf, we first delete u. If $p(u)$ is now a leaf, its rank is reduced to zero, but that in itself does not affect any value. If $p(u)$ is vacant and now has only one child, we bypass $p(u)$. This completes the tidying up.

Lemma 3. *Let v be the parent of the highest node affected by a* delete, *including tidying up. If $rank(v) = k$, then the maximal loss of value is at most $(9/10)(5/3)^k$.*

Proof. It is easy to see that the worst-case is when $v = p(p(u))$, where u is a deleted leaf and $p(u)$ is bypassed. Now u lost at most $(5/3)^{k-1}$ and $p(u)$ lost $(5/3)^k/2$, while the other child of $p(u)$ gained at least $((5/3)^k - (5/3)^{k-1})/2$ from the bypass. Adding up, the total loss is $(9/10)(5/3)^k$. □

Below we show how to regain the loss from a *delete* using a pointer to v from Lemma 3. To find nearby nodes to shortcut from, we maintain two doubly linked lists for each node v; namely $C(v)$ containing the children of v, and $G(v)$ containing the children of v that themselves have children. Thus, to find a grandchild of v, we take a child of a child in $G(v)$. Both lists are easily maintained as children are added and deleted: if a child u is added to v, it is added to $C(v)$. If u is the first child of v, we add v to $G(p(v))$. Finally, we add u to $G(v)$ if $C(u)$ is non-empty. Deleting a child is symmetric. Using these lists, we first prove

Lemma 4. *In a tidy tree, if node x has rank k and grandchildren, we can gain $\Omega((5/3)^k)$ value in $O(1)$ time.*

Proof. Using $G(x)$, find a child y of x that have children. If y is occupied, we can take any child z of y and shortcut to x. This increases the value of z by at least $(1/2)((5/3)^k - (5/3)^{k-1}) = (1/5)(5/3)^k$. We note that y may have rank much lower than $k-1$, but that would only increase our gain. If z is the last child of y, we remove y from $G(x)$. If, on the other hand, y is vacant, we have two cases. First note that since the tree is tidy, $|C(y)| \geq 2$. If $|C(y)| > 2$, we can just take any child z of y and shortcut to x as above. Otherwise $C(y) = \{z, z'\}$. If both z and z' are occupied, we shortcut both z and z' to x and remove y. This gives a gain of at least $2((5/3)^k - (5/3)^{k-1}) - (1/2)(5/3)^k = (3/10)(5/3)^k$. Otherwise, one of them, say z is vacant. Tidyness implies that z has at least two children. If more than two, any one of them can be shortcut to x gaining at least $(1/2)((5/3)^k - (5/3)^{k-2}) = (8/25)(5/3)^k$. If exactly two, then one of them is shortcut to y and the other to x while z is removed. The gain in value is at least $(1/2)((5/3)^k + 2(5/3)^{k-2}) = (7/50)(5/3)^k$. We note that all the above shortcuts preserves tidyness. □

The following lemma shows how we—using Lemma 4—can regain the value lost due to a deletion.

Lemma 5. *In a tidy tree with a pointer to a node v of rank k, we can increase the value by $t \cdot (5/3)^k$ or get to a reduced tree in $O(t)$ time.*

Proof. The proof is constructive. We set $x = v$ and repeat the following until either we have gained enough value, or reach the base case of a reduced tree:

1. While $G(x)$ is non-empty and there is more value to be gained, apply Lemma 4.
2. If x is not the root, set $x = p(x)$.

In case 1, we gain $\Omega((5/3)^k)$ per constant time iteration due to Lemma 4. We cannot get to case 2 twice in a row without getting to case 1, since $p(x) \in G(p(p(x)))$. Thus, in $O(t)$ time, we either gain $t \cdot (5/3)^k$ in value, or we end with x the root but with no grand children, that is, a tree of height at most 1.

If we are in the base case with a tree of height 0 or 1, we set $rank(x)$ to 0 or 1, respectively. □

Combining Lemmas 2, 3, and 5 with $t = O(1)$, we implement a deletion in constant time so that either we have no loss, meaning that (3) is preserved, or obtaining a reduced tree that satisfies (3) directly. Thus we have proved

Theorem 1. *In union-find with deletion we can implement each* makeset, union, *and* delete *in constant time, and each* find *in $O(\log n)$ time.*

4   Faster Amortized Bounds

We will now show that we can get much faster amortized bounds for *find*, yet preserve the previous worst-case bounds. All we have to do is to use *path compression* followed by tidying up operations. Path compression of a path from node $v \in T$ to node $u \in T$ makes every node on the path a child of u. When we perform a *find* from a node v, we compress the path to the root. Our analysis is new and much cleaner analysis than was previously known even without deletes.

Before going further, we note that path compression consists of shortcuts that increase value of the previous section, so intuitively, the path compression can only help the deletions. Below, we first present our new analysis without the deletions, and then we observe that deletions are only helpful.

Analysis. We assign a potential $\phi(x)$ to each node x in the forest. To define the potential we need some extra functions. Define $Q = \lceil \frac{M+N}{N} \rceil$ and $\alpha'(n) = \bar{\alpha}(Q, n)$. Note that $Q \geq 2$ whenever $M > 0$. Our goal is to prove that the amortized cost of *find* is $O(\alpha'(n))$ where n is the cardinality of the set found. We also define $rank'(v) = rank(v) + Q$.

Definition 4. *For a non-root node x we define*

$$level(x) = \max\{k \geq 0 \mid A_k(rank'(x)) \leq rank'(p(x))\},$$

$$index(x) = \max\{i \geq 1 \mid A^{(i)}_{level(x)}(rank'(x)) \leq rank'(p(x))\}.$$

We have

$$0 \leq level(x) < \bar{\alpha}(rank'(x), rank'(p(x))) \leq \alpha'(rank'(p(x))), \quad (4)$$

$$1 \leq index(x) \leq rank'(x). \quad (5)$$

Definition 5. *The potential $\phi(x)$ of a node x is defined as*

$$\phi(x) = \begin{cases} \alpha'(rank'(x)) \cdot (rank'(x) + 1) & \text{if } x \text{ root,} \\ (\alpha'(rank'(x)) - level(x)) \cdot rank'(x) - index(x) + 1 \\ \qquad \text{if } x \text{ not root and } \alpha'(rank'(x)) = \alpha'(rank'(p(x))), \\ 0 & \text{otherwise.} \end{cases}$$

The potential $\Phi(x)$ of a set A is defined as the sum of the potentials of the nodes in the tree T_A representing the set A: $\Phi(A) = \sum_{x \in T_A} \phi(x)$.

At first sight the potential function looks very similar to the standard one from [5], but there are important differences. Using $\alpha(rank(x))$ instead of $\alpha(N)$ we get a potential function that is more locally sensitive. To get this change to work, we use the trick that the potential of a node is only positive if $\alpha'(rank'(x)) = \alpha'(rank'(p(x)))$.

From (4) and (5) it immediately follows that the potential of a node x with $\alpha'(rank'(x)) = \alpha'(rank'(p(x)))$ is strictly positive. We also note that the only potentials that can increase are those of roots. All other nodes keep their ranks while the ranks of their parents increase and that can only decrease the potential.

We will now analyze the change in potential due to the operations.

Lemma 6. *The cost of* makeset *is amortized as a constant per* makeset *plus a constant per* find.

Proof. When we create a new set A with rank 0, it gets potential $\bar{\alpha}(Q,Q)(Q+1) = 2(Q+1) = O((M+N)/N)$. Over N makeset operations, this adds up to a total increase of $O(M+N)$. □

Lemma 7. *The operation* union(A, B) *does not increase the potential.*

Proof. Suppose we make A the parent of B. If the rank of A is not increased, there is no node that increases potential, so assume that $rank'(A)$ is increased from k to $k+1$. Then k was also the rank of B. If $\alpha'(k+1) > \alpha'(k)$, then B gets zero potential along with any previous child of A. The potential of B is reduced by $\alpha'(k) \cdot (k+1)$. On the other hand, the potential of A is increased by $(\alpha'(k)+1) \cdot (k+2) - \alpha'(k) \cdot (k+1) = \alpha'(k) + k + 2$, which is less than $\alpha'(k) \cdot (k+1)$, as $k \geq 2$ and $\alpha'(k) \geq 2$. (Here we use the fact that $\bar{\alpha}(j,i) \geq 2$, for every $i, j \geq 0$.)

Finally, if $\alpha'(k+1) = \alpha'(k)$, then the potential of A increases by $\alpha'(k)$ while the potential of B decreases by at least $\alpha'(k)$, since B was a root with potential $\alpha'(k) \cdot (k+1)$ and now becomes a child with potential at most $\alpha'(k) \cdot k$. □

Lemma 8. *A path compression of length ℓ from a node v up to some node u decreases the potential by at least $\ell - (2 \cdot \alpha'(rank'(u)) + 1)$. In particular, the amortized cost is at most $O(\alpha'(rank'(u)))$.*

Proof. The potential of the root does not change due to the path compression. We will show that at least $\max\{0, \ell - (2 \cdot \alpha'(rank'(u)) + 2)\}$ nodes have their potential decreased by at least one.

There are at most $\alpha'(rank'(u))$ nodes x on the path that had $\alpha'(rank'(x)) < \alpha'(rank'(p(x)))$ before the operation. The potentials of these nodes do not change.

If node x had $\alpha'(rank'(x)) = \alpha'(rank'(p(x))) < \alpha'(rank'(u))$, then its potential drops to 0, and the decrease in x's potential is therefore at least one.

It remains to account for the nodes x with $\alpha'(rank'(x)) = \alpha'(rank(u))$. Let x be a node on the path such that x is followed somewhere on the path by a node $y \neq u$ with $level(y) = level(x) = k$. There can be at most $\alpha'(rank'(u)) + 1$ nodes on the path that do not satisfy these constraints: The last node before u, u, and the last node on the path for each level, since level$(y) < \alpha'(\textit{rank}'(u))$. Let x be a node that satisfies the conditions. We show that the potential of x decreases by at least one. Before the path compression we have $\textit{rank}'(p(y)) \geq A_k(\textit{rank}'(y)) \geq A_k(\textit{rank}'(p(x))) \geq A_k(A_k^{(\text{index}(x))}(\textit{rank}'(x))) = A_k^{(\text{index}(x)+1)}(\textit{rank}'(x))$. After the path compression we have $\textit{rank}'(p(x)) = \textit{rank}'(p(y))$ and thus $\textit{rank}'((p(x)) \geq A_k^{(\text{index}(x)+1)}(\textit{rank}'(x))$, since $\textit{rank}'(x)$ does not change and $\textit{rank}'(p(y))$ does not decrease. This means that either index(x) or level(x) must increase by at least one. Thus $\phi(x)$ decreases by at least one. $\square$

We conclude that the amortized cost of *find* in a set A is

$$O(\alpha'(\textit{rank}'(A))) = O(\bar{\alpha}(Q, \textit{rank}(A) + Q + c)) = O(\bar{\alpha}(Q, \textit{rank}(A))).$$

The last step follows because $\bar{\alpha}$ is defined to be at least 2. Recall that $Q = \lceil \frac{M+N}{N} \rceil$ and that $\textit{rank}(A) \leq \log_2 |A|$, so without deletions, this is the desired bound.

Deletion and Path Compression. We now combine the path compression and amortized analysis with deletions. The potential used in the amortization is identical for vacant and occupied nodes. It is clear that deletions and tidying up can only decrease this potential, so they have no extra amortized cost. Likewise, a path compression can only increase value as it only performs shortcuts. However, after a path compression, there may be some cleaning to do if some vacant nodes go down to 0 or 1 children. We start the path compression from a tidy tree where each vacant node has at least two children, and the compression takes at most one child from each node on the path. Hence the only relevant tidying up is to bypass some of the nodes on the path. The tidying up takes time proportional to the length of the path, so the cost of a *find* is unchanged.

The tidying up does decreases value, but the loss turns out less than the gain from the compression.

Lemma 9. *Path compression followed by tidying up operations does not decrease the value of a tree.*

Proof. The path compression involves nodes $v_0, ..., v_\ell$ starting in some occupied node v_0 and ending in the root which has some rank k. After the compression, all nodes $v_0, .., v_{\ell-1}$ are children of the root v_ℓ. If node v_i is not bypassed when tidying up, its value gain is at least $((5/3)^{rank(v_\ell)} - (5/3)^{rank(v_{i+1})})/2$. If v_i is bypassed, then $0 < i < \ell$, and v_i is vacant, so the loss is $(5/3)^{rank(v_{i+1})}/2$. However, then v_i has a child w_i which gains at least $((5/3)^{rank(v_\ell)} - (5/3)^{rank(v_i)})/2$, so the total change is

$$((5/3)^{rank(v_\ell)} - (5/3)^{rank(v_{i+1})} - (5/3)^{rank(v_i)})/2$$

Since ranks are strictly increasing along a path, this change is positive for all but $i = \ell - 1$. On the other hand, the first node v_0 is always occupied, and has a gain of at least $(5/3)^{rank(v_\ell)} - (5/3)^{rank(v_1)}$, where $1 \leq \ell - 1$. We can use the value gained by v_0 to pay for the value lost by bypassing both v_1 and v_{l-1}. There are two cases.

If both v_{l-1} and v_l is bypassed we must have $l \geq 4$. Combining the changes in potential for the nodes v_0, v_1, and v_{l-1} we get,$(5/3)^{rank(l)} - (5/3)^{rank(v_1)} - (1/2)(5/3)^{rank(v_2)} - (1/2)(5/3)^{rank(v_{l-1})} > 0$.

If v_1 is not bypassed, we get that the total gain for v_0 and v_{l-1} is at least, $(5/3)^{rank(v_l)} - (5/3)^{rank(v_1)} - (1/2)(5/3)^{rank(v_l)}$, which is always positive. Thus the overall change in value is positive, or zero if the path has length 0 or 1 and no compression happens. □

Since our values and hence (3) are preserved, for any set A, we get $rank(A) = O(\log |A|)$. Thus our amortized cost of a *find* operation is $O(\bar{\alpha}(Q, O(\log |A|)) = \Theta(\bar{\alpha}(\lceil \frac{M+N}{N} \rceil, |A|))$. Summing up, we have proved

Theorem 2. *If we do a total of M find operations on a total of N makeset operations, then the operation times can be amortized as follows. We pay only a constant for each makeset, union, and delete, and for a find on an element in a set A, we pay $O(\bar{\alpha}(\lceil \frac{M+N}{N} \rceil, |A|))$. Meanwhile, the worst-case bounds of Theorem 1 are preserved.*

References

1. A. V. Aho, J. E. Hopcroft, and J. D. Ullman. *The Design and Analysis of Computer Algorithms*. Addison-Wesley, Reading, 1974.
2. S. Alstrup, A. M. Ben-Amram, and T. Rauhe. Worst-case and amortised optimality in union-find. In *Proceedings of the Thirty-First Annual ACM Symposium on Theory of Computing (STOC'99)*, pages 499–506, May 1999.
3. A. M. Ben-Amram and Z. Galil. A generalization of a lower bound technique due to Fredman and Saks. *Algorithmica*, 30(1):34–66, 2001.
4. N. Blum. On the single-operation worst-case time complexity of the disjoint set union problem. *SIAM J. Comput.*, 15(4):1021–1024, 1986.
5. T. H. Cormen, C. E. Leiserson, R. L. Rivest, and C. Stein. *Introduction to Algorithms*. MIT Press, 2nd edition, 2001.
6. M. Fredman and M. Saks. The cell probe complexity of dynamic data structures. In *Proceedings of the 21st Annual Symposium on Theory of Computing (STOC '89)*, pages 345–354, New York, May 1989. ACM Association for Computing Machinery.
7. Z. Galil and G. F. Italiano. Data structures and algorithms for disjoint set union problems. *ACM Computing Surveys*, 23(3):319, Sept. 1991.
8. H. Kaplan, N. Shafrir, and R. E. Tarjan. Union-find with deletions. In *Proc. of the 13th ACM-SIAM Symp. On Discrete Mathematics (SODA)*, pages 19–28, 2002.
9. D. L. Kozen. *The Design and Analysis of Algorithms*. Springer, Berlin, 1992.
10. R. Seidel and M. Sharir. Top-down analysis of path compression. *SIAM J. Comput.*, 34(3):515–525, 2005.
11. M. Smid. A data structure for the union-find problem having good single-operation complexity. *ALCOM: Algorithms Review, Newsletter of the ESPRIT II Basic Research Actions Program Project no. 3075 (ALCOM)*, 1, 1990.
12. R. E. Tarjan. Efficiency of a good but not linear disjoint set union algorithm. *Journal of the ACM*, 22:215–225, 1975.
13. R. E. Tarjan and J. van Leeuwen. Worst-case analysis of set union algorithms. *Journal of the ACM*, 31(2):245–281, Apr. 1984.

Optimal In-place Sorting of Vectors and Records

Gianni Franceschini and Roberto Grossi

Dipartimento di Informatica, Università di Pisa,
Largo Bruno Pontecorvo 3, 56127 Pisa, Italy
{francesc, grossi}@di.unipi.it

Abstract. We study the problem of determining the complexity of optimal comparison-based in-place sorting when the key length, k, is not a constant. We present the first algorithm for lexicographically sorting n keys in $O(nk + n \log n)$ time using $O(1)$ auxiliary data locations, which is *simultaneously* optimal in time *and* space.

1 Introduction

We study the computational complexity of the classical problem of comparison-based sorting by considering the case in which the keys are of non-constant length, k. We aim at minimizing simultaneously the time and space bounds under the assumption that the keys are vectors $x \in \Sigma^k$ of k scalar components over a totally ordered, possibly unbounded set Σ. Denoting the ith scalar component of vector x by $x(i)$ for $1 \leq i \leq k$, we indicate the vector's *chunks* by $x(i, j)$, which are the contiguous portions of x consisting of $x(i)$, $x(i+1)$, ..., $x(j)$, where $1 \leq i \leq j \leq k$. The lexicographic (or alphabetic) order, $x \leq y$, is defined in terms of the scalar components: either $x(1) < y(1)$ or recursively $x(2, k) \leq y(2, k)$ for $x(1) = y(1)$. The model easily extends to k-field records in $\Sigma_1 \times \Sigma_2 \times \cdots \times \Sigma_k$, but we prefer to keep the notation simple.

We are given a set $\mathcal{V} \subseteq \Sigma^k$ of n vectors stored in n *vectorial locations*, one vector of $\mathcal{V}$ per location. We permit two kinds of operations on the vector locations: (1) exchange any two vectors in $O(k)$ time; (2) access the ith scalar component of any two vectors for comparison purposes in $O(1)$ time. Hence, determining the lexicographic order of any two vectors from scratch takes $O(k)$ time. We are also given a number of *auxiliary locations*, each location storing one integer of $O(\log n)$ bits. We employ the standard repertoire of RAM instructions on the auxiliary locations, with $O(1)$ time per operation.

The model resulting from the above rules naturally extends the comparison model to keys of non-constant length. (We obtain the comparison model by fixing $k = 1$.) We are interested in exploring algorithms using the minimal number of auxiliary locations, referring to the model using just $O(1)$ auxiliary locations as the *in-place model for vectors*. This model is useful for studying, in an abstract way, the complexity of in-place sorting and searching for a variety of keys: k-length strings, k-field records, k-dimensional points, k-digit numbers, etc.

One significant example is how to perform in-place searching on a set $\mathcal{V}$ of n vectors. With sophisticated techniques for proving upper and lower bounds on the complexity of searching $\mathscr{V}$ in lexicographic order, Andersson, Hagerup, Håstad and Petersson have proved in [1] that it requires

$$\Theta\left(\frac{k \log \log n}{\log \log(4 + \frac{k \log \log n}{\log n})} + k + \log n\right)$$

time. This bound is worse than $\Theta(k + \log n)$, obtained by searching $\mathscr{V}$ *plus* $O(n)$ auxiliary locations (e.g., Manber and Myers [18]). Using permutations other than those resulting from sorting is a way to reach optimality: Franceschini and Grossi [10] have shown that for any set $\mathscr{V}$ of n vectors in lexicographic order, there exists a permutation of them allowing for $\Theta(k + \log n)$ search time using $O(1)$ auxiliary data locations.

In-place sorting is an even more intriguing example in this scenario. Any optimal in-place sorting algorithm for constant-sized keys can be turned into an $O(nk \log n)$-time in-place algorithm for vectors, losing optimality in this way. The lower bound of $\Omega(nk + n \log n)$ time easily derives from decision trees [14]. If the number of comparison is to be minimized, the best up-to-date result for in-place sorting is $n \log n + O(nk \log^* n)$ scalar comparisons and $n \log n + O(nk)$ vector exchanges by Munro and Raman [20]. Since each vector exchange takes $O(k)$ time, the time complexity sums up to $O(nk^2 + nk \log n)$. For the same reason, the multikey Quicksort analyzed by Bentley and Sedgewick [4] yields a non-optimal algorithm of cost $O(nk \log n)$ when adapted to run in the in-place model for vectors, since it requires $O(n \log n)$ vector exchanges. The original version of the algorithm takes $O(nk + n \log n)$ time since it can exploit $O(n)$ auxiliary locations to store the pointers to the vectors. It exchanges the pointers rather than the vectors, following the *address table sorting* suggested in Knuth [14–p.74]. Recently, Franceschini and Geffert [9] have devised an optimal in-place algorithm for constant-sized keys with $O(n)$ data moves. Subsequent results by Franceschini [7, 8] have shown how to achieve cache-obliviousness or stableness for in-place sorting. However, the $O(k)$-time cost of each vector comparison makes these methods non-optimal in our setting. The bit encoding for vectors in Franceschini and Grossi [10] cannot help either, as it assumes that vectors are initially sorted while this is actually the major goal in this paper.

The above discussion highlights the fact that the known algorithms, to the best of our knowledge, are unable to simultaneously achieve time optimality and space optimality for sorting vectors (in place). Our main result is that of obtaining the first optimal bounds for sorting an arbitrary set of n vectors in place, taking $\Theta(nk + n \log n)$ time and using $O(1)$ auxiliary locations. An implication of our result is that we can provide optimal in-place preprocessing for efficient in-place searching [1, 10, 11, 12, 15] when the vectors are initially arranged in any arbitrary order, with a preprocessing cost of $O(nk + n \log n)$ time. Another implication is that sorting bulky records can be done optimally in place by exchanging them directly without using the $O(n)$ auxiliary locations required by Knuth's address table sorting.

2 High-Level Description

We present our in-place sorting algorithm for vectors in a top-down fashion. We describe how to reduce the original problem to a sequence of simpler sorting problems to solve. In our description, we identify the n input vectors in $\mathscr{V}$ with their vectorial locations. At the beginning of the computation, $\mathscr{V}[i]$ represents the ith vectorial location, for $1 \leq i \leq n$, and contains the ith input vector. At the end of the computation, $\mathscr{V}[i]$ contains the vector of rank i after the sorting (ties for equal vectors are broken arbitrarily). During the intermediate steps, we solve several instances of a *general vector sorting problem*, denoted $GVSP\{m,p,h\}$ (see Figure 1). Given n vectors in $\mathscr{V}$, we refer to $GVSP\{m,p,h\}$ as the problem of sorting a subset of m contiguous vectors in $\mathscr{V}$, using

- $O(1)$ auxiliary locations,
- p vectors as *placeholders* taken from a contiguous subsequence of p locations in $\mathscr{V}$,
- h *heavy* bits suitably encoded by h pairs of vectors taken from two contiguous subsequences, each consisting of h locations in $\mathscr{V}$,

under the requirement that $m + p + 2h \leq n$ and that the four subsequences of h, p, m, and h vector locations, respectively, are pairwise disjoint as shown in Figure 1. Placeholders and heavy bits are defined in the rest of the section.

The general notation of $GVSP\{m,p,h\}$ is useful for expressing the various sorting instances that we get by reducing our initial problem, $GVSP\{n,0,0\}$, to simpler problems (with suitable values of m, p and h). Some basic instances occur just a constant number of times in the reduction and are easy to solve.

Lemma 1. *Any instance of $GVSP\{O(n/\log n), 0, 0\}$ takes $O(nk)$ time.*

Proof. We employ the in-place mergesort of [21] and pay a slowdown of $O(k)$ in the time complexity, since we run it on $O(n/\log n)$ vectors, each of length k. The cost is $O(k \times (n/\log n) \log(n/\log n)) = O(nk)$ time.

We now present the high-level structure of our reduction. In the following, for any two vectors x and y, we denote the length of their longest common prefix by $lcp(x,y) = \max(\{0\} \cup \{1 \leq \ell \leq k : x(1,\ell) = y(1,\ell)\})$.

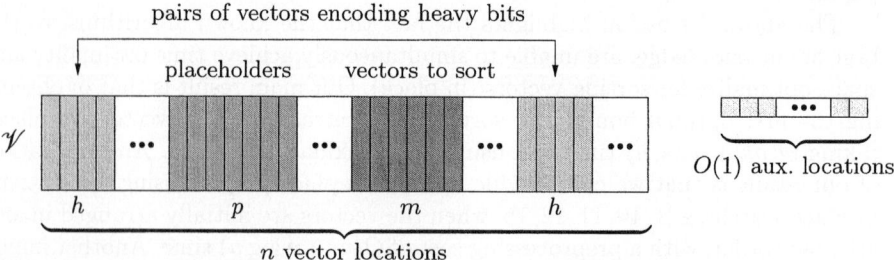

Fig. 1. An instance of $GVSP\{m,p,h\}$. Each of the n vector locations in $\mathscr{V}$ contains one vector. Each of the $O(1)$ auxiliary locations contains one integer of $O(\log n)$ bits

Heavy Bits (Section 3). To begin with, we reduce an instance of $GVSP\{n,0,0\}$ to a suitable instance of $GVSP\{n-o(n), 0, O(n/\log^2 n)\}$ plus a constant number of instances of $GVSP\{O(n/\log n), 0, 0\}$. We partition in place the sequence $\mathscr{V}$ into contiguous subsequences $\mathscr{L}$, $\mathscr{M}$, and $\mathscr{R}$, such that for each $x \in \mathscr{L}$, $y \in \mathscr{M}$ and $z \in \mathscr{R}$, we have $x \le y \le z$. Moreover, the number of vectors in $\mathscr{L}$ equals that of $\mathscr{R}$, namely, $|\mathscr{L}| = |\mathscr{R}| = O(n/\log n)$. Assuming that $\max \mathscr{L} \ne \min \mathscr{R}$ (otherwise sorting is trivial), we consider the pairs $P = \{\langle \mathscr{L}[i], \mathscr{R}[i]\rangle,$ for $1 \le i \le |\mathscr{L}|\}$. Note that for every pair $\langle x, y \rangle \in P$, vectors x and y are distinct ($x < y$) and their first mismatching scalar component is at position $lcp(x,y)+1$. Based on this observation we identify a subset H of the pairs in P satisfying two constraints:

1. $|H| = \Omega(n/\log^2 n)$.
2. There exists an interval $[l,r] \subseteq [1,k]$ of size $\max\{1, k/\log n\}$, such that $lcp(x,y) + 1 \in [l,r]$ for every pair $\langle x, y \rangle \in H$.

Under constraints 1–2, we can use the vectors in H for implicitly representing $O(n/\log^2 n)$ bits, called *heavy bits*, so that decoding one heavy bit requires $O(1 + k/\log n)$ time while encoding it takes $O(k)$ time. Let us see why do we need these bits. When designing an optimal in-place algorithm, the constraint on using just $O(1)$ auxiliary locations, namely, $O(\log n)$ extra bits of information, is rather stringent. Fortunately, permutations of the keys encode themselves further bits of informations. Potentially, we are plenty of $\log h!$ bits by permuting h distinct keys. Based on this idea, bit stealing [19] is a basic technique for implicitly encoding up to h bits of information by pairwise permuting h pairs of keys. In its original design, the technique encodes a bit with each distinct pair of keys x and y, such that $x < y$. The bit is of value 0 if x occurs *before* y in the permutation; it's of value 1 if x occurs *after* y in the permutation. The main drawback of this technique in our setting is that we need $O(k)$ time for encoding and decoding one bit since x and y are vectors. As we shall see, we will require an amortized number of $O(1)$ encoded bits and $O(\log n)$ decoded bits per vector, so that we have to decrease the cost of decoding to $O(1 + k/\log n)$ to stay within the claimed bounds.

At this stage, sorting $\mathscr{V}$ reduces to sorting $\mathscr{M}$ as an instance of $GVSP\{n - o(n), 0, O(n/\log^2 n)\}$. After that, it also reduces to sorting $\mathscr{L}$ and $\mathscr{R}$ as instances of $GVSP\{O(n/\log n), 0, 0\}$ solved by Lemma 1.

Buffering and Session Sorting (Sect. 4). We solve $GVSP\{n - o(n), 0, O(n/\log^2 n)\}$ reducing to $O(\log n)$ instances of $GVSP\{O(n/\log n), O(n/\log n), O(n/\log^2 n)\}$ and to $O(1)$ instances of $GVSP\{O(n/\log n), 0, 0\}$ solved by Lemma 1. We logically divide $\mathscr{M}$ into contiguous subsequences $\mathscr{M}_1, \ldots, \mathscr{M}_{s-1}, \mathscr{M}_s$, where $|\mathscr{M}_2| = \cdots = |\mathscr{M}_{s-1}| \le |\mathscr{M}_s| = O(n/\log n)$ and $s = O(\log n)$. Moreover, $|\mathscr{M}_1| = O(n/\log n)$ has a sufficiently large multiplicative constant, so that $\mathscr{M}_1$ can host enough vectors playing the role of placeholders. With reference to Figure 1, we sort the $m = O(n/\log n)$ vectors in each individual $\mathscr{M}_i$, $i \ne 1$, using the $p = O(n/\log n)$ placeholders in $\mathscr{M}_1$ and the $h = O(n/\log^2 n)$ heavy bits encoded by the pairs in $H \subseteq \mathscr{L} \times \mathscr{R}$.

Let us first give some motivation for using the placeholders while sorting. Having just n vector locations, we cannot rely on a temporary area of vector locations for efficiently permuting the vectors with a few moves. We therefore exploit a virtual form of temporary area using the internal buffering technique of Kronrod [16]. We designate the vectors in $\mathcal{M}_1$ as placeholders for "free memory" since we do not care to sort them at this stage. Hence, they can be scrambled up without interplaying with the sorting process that is running on a given $\mathcal{M}_i$, $i \neq 1$. When we need to move a vector of $\mathcal{M}_i$ to the temporary area, we simulate this fact by exchanging the vector with a suitable placeholder of $\mathcal{M}_1$. At the same time, we should guarantee that this exchange is somehow reversible, allowing us to put the placeholders back to the "free memory" in $\mathcal{M}_1$ without perturbing the sorting obtained for $\mathcal{M}_i$, $i \neq 1$.

Assuming to have obtained each of $\mathcal{M}_2, \ldots, \mathcal{M}_{s-1}, \mathcal{M}_s$ in lexicographic order, we still have to merge them using the heavy bits in H and the placeholders in $\mathcal{M}_1$. It turns out that this task is non-trivial to be performed. Just to have a rough idea, let us imagine to run the 2-way in-place mergesort for $O(\log s) = O(\log \log n)$ passes on them. This would definitively give a non-optimal time cost for the vectors since the number of vector exchanges would be $\omega(n)$, losing optimality in this way. We introduce a useful generalization of the technique in [7, 16], thus obtaining what we call *session sorting*. Let us assume that the vectors are distinct (we shall disregard this assumption in Section 4).

The main goal of session sorting is that of rearranging all the vectors in $\mathcal{M}_2, \ldots, \mathcal{M}_{s-1}, \mathcal{M}_s$, so that they are not too far from their final destination. If any such vector has rank r among all the other vectors in $\mathcal{M}_2 \mathcal{M}_3 \cdots \mathcal{M}_{s-1} \mathcal{M}_s$, and occupies a position $g > r$ after session sorting, we guarantee that $g - r \leq |\mathcal{M}_i|$. (Note that we do not claim anything regarding the case $g \leq r$.) Using this strong property, we show that the sequence of 2-way in-place operations for merging $\mathcal{M}_i$ and $\mathcal{M}_{i+1}$ for $i = 2, 3, \ldots, s-1$ (in this order) yields the sorted sequence. (We remark that this is not generally true if we do not apply session sorting.) As a result, the entire sequence $\mathcal{M}_2 \mathcal{M}_3 \cdots \mathcal{M}_{s-1} \mathcal{M}_s$ is in lexicographic order with a linear number of moves.

What remains to do is sorting $\mathcal{L}$, $\mathcal{M}_1$, and $\mathcal{R}$ individually as instances of $GVSP\{O(n/\log n), 0, 0\}$ by Lemma 1. Merging them in place with the rest of sorted vectors is a standard task giving $\mathcal{V}$ in sorted order. Hence, we are left with an instance of $GVSP\{O(n/\log n), O(n/\log n), O(n/\log^2 n)\}$, which corresponds to sorting a given $\mathcal{M}_i$, $i \neq 1$, using the placeholders initially hosted in $\mathcal{M}_1$ and the heavy bits encoded by the pairs in H.

Sorting Each $\mathcal{M}_i$ Individually (Section 5). We describe this stage in general terms. For a given $i \neq 1$, let $\mathcal{M}' = \mathcal{M}_i$ and $\mathcal{M}_B = \mathcal{M}_1$, for the instance of $GVSP\{|\mathcal{M}'|, |\mathcal{M}_B|, |H|\}$ that we are going to solve with the heavy bits in H (see Figure 1). Using $\mathcal{M}_B$ as a "free memory" area, we simulate the sorting of the $m' = |\mathcal{M}'|$ vectors by inserting them into a suitable structure that is incrementally built inside $\mathcal{M}_B$. Each insertion of a vector $x \in \mathcal{M}'$ into the internal structure of $\mathcal{M}_B$ exchanges x with a placeholder. After each such exchange we permute some of the vectors inside $\mathcal{M}_B$, so as to dynamically maintain a set of

$O(m'/\log^2 m')$ pivot vectors in the internal structure. The pivots have buckets associated inside $\mathcal{M}_B$ for the purpose of distributing the non-pivot vectors inserted up to that point, like in distribution sort. Each bucket contain $\Theta(\log^2 m')$ vectors that are kept *unsorted* to minimize the number of vector exchanges needed to maintain the internal structure of $\mathcal{M}_B$.

The pivots inside $\mathcal{M}_B$ are kept searchable by a suitable blend of the techniques in [10, 13, 18], requiring to decode $O(\log n)$ heavy bits per inserted vector (which is fine since decoding takes $O(1 + k/\log n)$ time). In particular, we logically divide each vector x into a concatenation of $O(\log m') = O(\log n)$ equally sized chunks. We only store the *lcp* information for the chunks considered as "meta-characters," thus obtaining an approximation of the *lcp* information for the vectors. After that the distribution completes by inserting all the vectors of $\mathcal{M}'$ into the internal structure of $\mathcal{M}_B$, we sort the buckets individually by using a constant number of recursive iterations of session sorting whose parameters are suitably adapted to the buckets' size. The base case of the recursion consists in solving $GVSP\{O(\sqrt{\log m'}), O(\sqrt{\log m'}), 0\}$, for which we design an optimal ad-hoc algorithm. After completing the individual sorting of the buckets, which still reside in $\mathcal{M}_B$, we exchange them with the placeholders that were temporarily moved to $\mathcal{M}'$. We place back the sorted buckets and their pivots to $\mathcal{M}'$ according to their relative order, which means that the m' vectors in $\mathcal{M}'$ are in lexicographic order. Since this stage is quite full of technicalities, we give more details in the full paper.

Known Tools. We use a few optimal algorithmic tools for atomic keys: in-place stable mergesort and in-place merge [21]; in-place selection for order statistics [17]. We apply these algorithms to vectors in a straightforward way by paying a slowdown of $O(k)$ per elementary step in their time complexity. We also use Hirschberg's linear scanning method [11] for searching in place a set of n vectors of length k, with the simple bound of $O(k + n)$ time. We go through the convention that the last lowercase letters—$\ldots, x, y, w, z$—denote vectors and the middle ones—$\ldots, i, j, k, l, \ldots$—are auxiliary indices or parameters.

3 Heavy Bits

We detail how to reduce the problem of sorting in place n vectors—an instance of $GVSP\{n, 0, 0\}$—to an instance of $GVSP\{n - o(n), 0, O(n/\log^2 n)\}$ plus a constant number of instances of $GVSP\{O(n/\log n), 0, 0\}$. (The notation for $GVSP\{m, p, h\}$ is defined in Section 2 and illustrated in Figure 1.) We recall that we partition in place the sequence $\mathcal{V}$ into $\mathcal{L}$, $\mathcal{M}$, and $\mathcal{R}$, where $|\mathcal{L}| = |\mathcal{R}| = p = O(n/\log n)$. We obtain this partition by performing order statistics in place [17] so as to identify the pth and the $(n - p + 1)$st elements of $\mathcal{V}$ in $O(nk)$ time. In the rest of the paper we assume that $w_L \neq w_R$; otherwise, $\mathcal{M}$ is made up of all equal vectors and sorting is trivially solved by applying Lemma 1 to $\mathcal{L}$ and $\mathcal{R}$.

Let us consider the set of pairs of vectors thus obtained, $P = \{\langle \mathscr{L}[i], \mathscr{R}[i]\rangle : 1 \leq i \leq p\} \subseteq \mathscr{L} \times \mathscr{R}$. Let us conceptually divide each of these vectors into chunks of $k/\ell = O(1 + k/\log n)$ scalar components, where $\ell = \min\{k, \log n\}$. We index these chunks from 1 to ℓ, in the order of their appearance inside the vector. We assign an integer label j to each pair $\langle \mathscr{L}[i], \mathscr{R}[i]\rangle$, where $1 \leq j \leq \ell$ and $1 \leq i \leq p$. Since $\mathscr{L}[i] < \mathscr{R}[i]$ by construction, label j is the index of the chunk containing the first mismatching position for $\mathscr{L}[i]$ and $\mathscr{R}[i]$; that is, it satisfies $(j-1)\,k/\ell \leq lcp(\mathscr{L}[i], \mathscr{R}[i]) < j\,k/\ell$. By the pigeon principle, there must exist a value of j for which at least $p/\ell = \Omega(n/\log^2 n)$ pairs in P are labeled j. We can identify that value by running at most ℓ in-place scans of $\mathscr{L}$ and $\mathscr{R}$, with an overall cost of $O(\ell \times pk) = O(nk)$ time. With a further scan of $\mathscr{L}$ and $\mathscr{R}$, we single out $h = \Theta(p/\ell) = \Theta(n/\log^2 n)$ pairs in P that have label j, moving them in place at the beginning of $\mathscr{L}$ and $\mathscr{R}$, respectively. Consequently, we identify these vectors in the first h locations in $\mathscr{L}$ and $\mathscr{R}$ by a set of pairs, denoted H:

- $H \subseteq P$ and $|H| = h = \Theta(n/\log^2 n)$;
- $H = \{\langle \mathscr{L}[i], \mathscr{R}[i]\rangle : 1 \leq i \leq h\}$ after the preprocessing;
- there exists $j \in [1, \ell]$ such that $(j-1)\,k/\ell \leq lcp(x, y) < j\,k/\ell$ for every pair $\langle x, y\rangle \in H$.

We steal bits in H using the knowledge of j as follows. For $1 \leq i \leq h$, we encode the ith bit of value 1 by exchanging $\mathscr{L}[i]$ and $\mathscr{R}[i]$ in $O(k)$ time; namely, $\mathscr{L}[i]$ occupies now position i inside $\mathscr{R}$ and $\mathscr{R}[i]$ does it inside $\mathscr{L}$. If the bit is 0, we leave them at their position (no exchange). In order to decode the ith bit, we only compare their jth chunk to find their mismatching position in the interval $[(j-1)\,k/\ell+1, j\,k/\ell]$. In this way, we can establish whether or not the two vectors have been exchanged during encoding (and so we decode either 0 or 1). Decoding performs at most k/ℓ scalar comparisons and thus takes $O(1 + k/\log n)$ time. The non-constant cost of bit stealing motivates our choice of referring to these bits as *heavy*.

Lemma 2. *We can encode $h = \Theta(n/\log^2 n)$ heavy bits by the pairwise permutation of vectors in $H \subseteq \mathscr{L} \times \mathscr{R}$. Encoding one bit requires $O(k)$ time while decoding it requires $O(1 + k/\log n)$ time. Preprocessing requires $O(nk)$ time using $O(1)$ auxiliary locations.*

We keep $\mathscr{L}$ and $\mathscr{R}$ unsorted for encoding bits until the end of the algorithm. At that point, we can in-place sort $\mathscr{L}$ and $\mathscr{R}$ by Lemma 1, in $O(nk)$ time. Consequently we are left with the problem of sorting $\mathscr{M}$.

Lemma 3. *There exists an $O(nk)$-time reduction from $GVSP\{n, 0, 0\}$ to $GVSP\{n - o(n), 0, O(n/\log^2 n)\}$, using $O(1)$ auxiliary locations.*

4 Buffering and Session Sorting

In this section, we detail how to sort the vectors in $\mathscr{M}$, which is an instance of $GVSP\{n - o(n), 0, O(n/\log^2 n)\}$. We logically divide $\mathscr{M}$ into contiguous subsequences $\mathscr{M}_1, \ldots, \mathscr{M}_{s-1}, \mathscr{M}_s$, called *blocks*, where $|\mathscr{M}_2| = \cdots = |\mathscr{M}_{s-1}| \leq$

$|\mathcal{M}_s| = O(n/\log n)$ and $s = O(\log n)$. In the following, we assume without loss of generality that $|\mathcal{M}_s| = |\mathcal{M}_{s-1}|$ (if not, we treat $\mathcal{M}_s$ differently, applying Lemma 1 to it). We remark that only a constant number of blocks can be sorted with the bounds of Lemma 1. Hence we should proceed otherwise. We designate the $O(n/\log n)$ vectors in $\mathcal{M}_1$, for a sufficiently large multiplicative constant, to act as placeholders [16]. In this way we obtain $O(\log n)$ instances of $GVSP\{O(n/\log n), O(n/\log n), O(n/\log^2 n)\}$, plus a constant number of instances of $GVSP\{O(n/\log n), 0, 0\}$ solved by Lemma 1.

We are still missing a crucial part of the reduction performed at this stage, namely, how to obtain *all* the vectors in $\mathcal{M}_2 \mathcal{M}_3 \cdots \mathcal{M}_{s-1} \mathcal{M}_s$ in lexicographic order. We introduce the *right-bounded* permutations, since they rearrange the vectors so that each vector cannot occupy a position beyond a bounded distance to the right of its final position in the sorted sequence. As we will prove, the net effect of the right-bounded permutation is that we can simulate the in-place merging scheme by the following scheme: IN-PLACE-MERGE($\mathcal{M}_2, \mathcal{M}_3$); IN-PLACE-MERGE($\mathcal{M}_3, \mathcal{M}_4$); ...; IN-PLACE-MERGE($\mathcal{M}_{s-1}, \mathcal{M}_s$). We describe this permutation in general terms as it is of independent interest.

4.1 Right-Bounded Permutations

We are given three positive integers m, p, q, such that q divides p and p divides m, satisfying

$$\left(\frac{m}{p} - 1\right) \times (q - 1) \leq p. \tag{1}$$

Given a sequence $\mathcal{B}$ of m vectors, we logically divide it into m/q sub-blocks of q vectors each, denoted $\mathcal{S}_1, \ldots, \mathcal{S}_{m/q}$. The sub-blocks are grouped into blocks of p/q sub-blocks each, thus logically dividing $\mathcal{B}$ into m/p blocks of p vectors each, denoted $\mathcal{B}_1, \ldots, \mathcal{B}_{m/p}$. A right-bounded permutation is the arrangement of the vectors in $\mathcal{B}$ resulting from steps P1–P2, with steps P3–P4 yielding the sequence in lexicographic order:

P1. For $j = 1, \ldots, m/p$, sort each block $\mathcal{B}_j$ individually.
P2. Sort *stably* the m/q sub-blocks $\mathcal{S}_1, \ldots, \mathcal{S}_{m/q}$ according to their first vector (i.e., comparisons are driven by the minimum vector in each sub-block, and the rest of the vectors are considered as "satellite data").
P3. For $j = 1, \ldots, m/p$, sort each block $\mathcal{B}_j$ individually (note that the content of the blocks changed!).
P4. For $j = 1, \ldots, m/p - 1$, merge the vectors contained in blocks $\mathcal{B}_j$ and $\mathcal{B}_{j+1}$.

Lemma 4. *For each vector $\mathcal{B}[i]$, $1 \leq i \leq m$, let g_i be the number of vectors $\mathcal{B}[j] > \mathcal{B}[i]$ such that $1 \leq j < i$ right after steps P1–P2. Then*

$$g_i \leq \left(\frac{m}{p} - 1\right) \times (q - 1). \tag{2}$$

Proof. Let us consider the arrangement of the vectors in $\mathscr{B}$ right after steps P1–P2. In order to prove equation (2), we need to consider the intermediate arrangement of the vectors in $\mathscr{B}$ *after* step P1 and *before* step P2. Recall that we logically divide $\mathscr{B}$ into blocks and sub-blocks, indexing the blocks from 1 to m/p. We assign a unique type to each block based on its index, namely, block $\mathscr{B}_t$ is assigned *type t*, where $1 \leq t \leq m/p$, since it is the tth block in $\mathscr{B}$. For the intermediate arrangement above, we say that a vector has type t if it belongs to $\mathscr{B}_t$ (recall that $\mathscr{B}_t$ is sorted). We can assign type t to the sub-blocks of each $\mathscr{B}_t$ in the same manner, since each sub-block contains vectors of the same type t by construction. Hence the type of a sub-block is well defined. We refer to the first vector of each sub-block, which is also the minimum in it, as the *header* of the sub-block.

Let us now resume the arrangement of the vectors in $\mathscr{B}$ *right after* steps P1–P2. Consider a generic vector $\mathscr{B}[i]$ belonging to a sub-block, say $\mathscr{S}'$ of type t', and let g_i be defined as above. We give an upper bound to g_i so as equation (2) holds. Specifically, we count the maximum number of vectors contributing to g_i. Let us discuss them by their type. By the stability of the sorting process in step P2, we know that the vectors of type t' have maintained the relative order they had in the intermediate arrangement (after step P1 and before step P2) and so they cannot contribute to g_i.

Let x be the header of the sub-block $\mathscr{S}'$ containing $\mathscr{B}[i]$. Let us evaluate the contribution to g_i for the vectors of type $t'' \neq t'$. Consider all sub-blocks of type t'': we claim that at most one of them, say $\mathscr{S}''$, can contain vectors contributing to g_i. Precisely, $\mathscr{S}''$ is the sub-block of type t'' having the largest header less than or equal to x. Let $y \leq x$ be the header of $\mathscr{S}''$ and z be one of such contributing vectors in $\mathscr{S}''$. Sub-block $\mathscr{S}''$ is laid out before $\mathscr{S}'$ by construction but $z > \mathscr{B}[i]$ by definition of g_i. Note that there can be at most $q-1$ such vectors z in $\mathscr{S}''$. For any other sub-block of type t'', we show that its vectors cannot contribute to g_i. Since the block of type t'' is sorted after step P1, there are two possibilities for its sub-blocks $\mathscr{S}''' \neq \mathscr{S}''$: (a) $\mathscr{S}'''$ contains all vectors that are less than or equal to $y \leq x$ (i.e., $\mathscr{S}'''$ is laid out before $\mathscr{S}''$); they do not contribute to g_i by transitivity since $x \leq \mathscr{B}[i]$. (b) $\mathscr{S}'''$ contains all vectors that are greater than or equal to $z > \mathscr{B}[i] \geq x$ (i.e., $\mathscr{S}'''$ is laid out after $\mathscr{S}''$); they do not contribute because the header of $\mathscr{S}'''$ is strictly larger than x by transitivity and so $\mathscr{S}'''$ is laid out after $\mathscr{S}'$. Summing up, the total contribution to g_i for the vectors of type $t'' \neq t'$ is at most $q-1$ (a subset of the vectors in $\mathscr{S}''$). Since there are $\frac{m}{p} - 1$ different types other than t', we obtain the upper bound for equation (2).

Theorem 1. *After steps P1–P4, the sequence $\mathscr{B}$ is sorted.*

Proof. We proceed by induction on the length of prefixes of blocks in $\mathscr{B}$. The base case is obvious, as we know that $\mathscr{B}_1$ is sorted by step P3. Let us assume that the jth prefix of blocks $\mathscr{B}_1 \mathscr{B}_2 \cdots \mathscr{B}_j$ is sorted by induction, for $j \geq 1$. After step P3, the upper bound in equation (2) still holds for any vector v in block $\mathscr{B}_{j+1}$ (modulo the inner permutation due to the sorting of $\mathscr{B}_{j+1}$). Indeed, the number of vectors $z > v$ that are laid out before v cannot increase; those inside $\mathscr{B}_{j+1}$ disappear after sorting it and so the upper bound in equation (2) is anyway valid. By equation (1), we derive that p, the size of each block, is larger than the upper bound of equation (2). As a result, the number of vectors $z > v$ that belong to the jth prefix of blocks cannot exceed p. Hence, they should be contained in the last locations of block $\mathscr{B}_j$ since $p = |\mathscr{B}_j|$ and $\mathscr{B}_1 \mathscr{B}_2 \cdots \mathscr{B}_j$ is sorted by induction. This allows us to conclude that after merging $\mathscr{B}_j$ and $\mathscr{B}_{j+1}$, the $(j+1)$st prefix of blocks $\mathscr{B}_1 \mathscr{B}_2 \cdots \mathscr{B}_{j+1}$ is sorted, thus proving the statement of the theorem.

4.2 Session Sorting

We apply the steps stated in Theorem 1 to sorting the vectors in $\mathscr{M}$ into sessions. We choose $\mathscr{M}_1$ of size $O(n/\log n)$ for the placeholders. We then fix $q = \log^3 n$, $p = qn/\log^4 n = \Theta(n/\log n)$, and we pick m as the largest multiple of p such that $m \leq |\mathscr{M}| - |\mathscr{M}_1|$. These values satisfy equation (1). We therefore obtain the logical division of $\mathscr{M}$ into blocks $\mathscr{M}_1, \ldots, \mathscr{M}_{s-1}, \mathscr{M}_s$, as expected. We comment on how to apply steps P1–P4 to $\mathscr{M}_2, \ldots, \mathscr{M}_s$ (assuming w.l.o.g. that $|\mathscr{M}_s| = |\mathscr{M}_{s-1}|$).

In steps P1 and P3, we have to solve a number of $m/p = O(\log n)$ instances of $GVSP\{O(n/\log n), O(n/\log n), O(n/\log^2 n)\}$ (see Section 5).

In step P2, we have just $m/q = O(n/\log^3 n)$ vectors to sort, which are the minimum in each sub-block. We refer to them as *headers* and to the rest of the vectors as *satellite data* (with $q - 1$ vectors each). We associate a unique implicit index in the range from 1 to m/q with the satellite data in each sub-block. We employ the heavy bits in H so as to form a sequence of m/q integers $h_1, h_2, \ldots, h_{m/q}$ of $\log n$ bits each, employed to encode a permutation of these indexes. Note that we have direct access to any h_j, $1 \leq j \leq m/q$, in $O(k \log n)$ time for encoding it and $O(k + \log n)$ time for decoding it by Lemma 2.

At the beginning of step P2, we set $h_j = j$ and exchange the jth header with the jth placeholder in $\mathscr{M}_1$, for $1 \leq j \leq m/q$. We then apply the in-place *stable* mergesort on the headers thus collected in $\mathscr{M}_1$. Each comparison cost is $O(k)$ time while each exchange requires $O(k \log n)$ time. Indeed, when exchanging two headers inside $\mathscr{M}_1$, say at position j' and j'', we have also to swap the values of $h_{j'}$ and $h_{j''}$, involving their decoding and encoding in H. Note that the satellite data is not exchanged but $h_{j'}$ and $h_{j''}$ are correctly updated to maintain the association of the headers with their satellite data in the sub-blocks. At the end of the mergesort, we exchange the jth header in $\mathscr{M}_1$ with the placeholder temporarily hosted in the h_jth sub-block. The total cost is $O((m/q)\log(m/q) \times (k \log n)) = O((n/\log^3 n)\log n \times (k \log n)) = o(nk)$.

We now have to permute the sub-blocks according to the values of $h_1, \ldots, h_{m/q}$ encoded in H. Specifically, the h_jth sub-block must occupy the jth position among the sub-blocks to reflect the stable sorting of their headers. We employ an additional sequence of integers $r_1, r_2, \ldots, r_{m/q}$ encoded in H, initializing $r_i = j$ if and only if $h_j = i$. We proceed incrementally for $j = 1, 2, \ldots, m/q - 1$ (in this order), preserving the invariant that we have correctly placed the first $j - 1$ sub-blocks, with $h_1, \ldots, h_{m/q}$ and $r_1, r_2, \ldots, r_{m/q}$ suitably updated to reflect the fact that one permutation is the inverse of the other (in particular, $h_{j'} = r_{j'} = j'$ for $1 \leq j' < j$, so the invariant is meaningful for the rest of the indexes). Note that some of the sub-blocks may have exchanged in order to place the first $j-1$ sub-blocks. Hence, when we refer to the jth and the h_jth sub-blocks, they are taken from the current arrangement of sub-blocks. If $j = h_j$, the current sub-block is already correctly placed and the invariant is trivially preserved. Otherwise, we exchange the jth and the h_jth sub-blocks by pairwise exchanging their ith vectors for $i = 1, 2, \ldots, q$. In order to preserve the invariant, we simultaneously swap the values of h_j and h_{r_j} and the values of r_j and r_{h_j}, respectively, re-encoding them in H. Since the exchange of sub-blocks requires the pairwise exchange of q vectors plus the encoding and decoding of $O(1)$ values among $h_1, \ldots, h_{m/q}$ and $r_1, r_2, \ldots, r_{m/q}$, the cost is $O(qk + k\log n)$. When $j = m/q - 1$, the last two sub-blocks are placed correctly and we have performed a total of $O(m/q)$ such exchanges. The final cost is $O(m/q \times (qk + k\log n)) = O(n/\log^3 n \times k\log^3 n) = O(nk)$. Hence, the total cost of step P2 is $O(nk)$.

Finally, in step P4, we use the in-place merging with comparison cost $O(k)$. As a result, we obtain a total cost of $O(m/p \times pk) = O(nk)$ for step P4 (and $\mathcal{M}$ is sorted).

Lemma 5. *There is an $O(nk)$-time reduction from $GVSP\{n - o(n), 0, O(n/\log^2 n)\}$ to a number of $O(\log n)$ instances of $GVSP\{O(n/\log n), O(n/\log n), O(n/\log^2 n)\}$, using $O(1)$ auxiliary locations.*

5 Sorting Each Block Individually

We have to solve an instance of $GVSP\{O(n/\log n), O(n/\log n), O(n/\log^2 n)\}$ (see Figure 1). We reformulate it as $GVSP\{m', O(m'), O(m'/\log m')\}$, where $m' = O(n/\log n)$ vectors in $\mathcal{M}'$ should be sorted using a sufficiently large number $O(m')$ of placeholders in $\mathcal{M}_B$. We need to encode $O(1)$ sequences of integers of $O(\log m') = O(\log n)$ heavy bits each in $H \subseteq \mathcal{L} \times \mathcal{R}$, totalizing $O(m'/\log m')$ heavy bits. We sort $\mathcal{M}'$ by repeatedly inserting its vectors in an internal structure maintained inside $\mathcal{M}_B$ to mimic a distribution sort into buckets of $O(\log^2 m')$ vectors each. Each bucket is sorted by applying a constant number of recursive calls to session sorting (Section 4.2). The base case is an instance of $GVSP\{O(\sqrt{\log m'}), O(\sqrt{\log m'}), 0\}$. We first rank the vectors by linking them in a sorted list without moving the vectors (we mimic the insertion sort in a list without moving vectors). The list pointers of $O(\log \log m')$ bits each, however, are not encoded with heavy bits in this case. Since we sort one bucket at a time and have $O(\sqrt{\log m'})$ such pointers, we can keep the $O(\sqrt{\log m'} \log \log m') = o(\log n)$ bits for all the pointers in one auxiliary location. We can access any such pointer in constant time, and we can append a new pointer to them within the same complexity by using RAM operations. We apply Hirschberg's linear scanning to add a new vector to the sorted list and mimic insertion sort. Hence, the cost per vector is $O(k + \sqrt{\log m'})$. After setting up the linked list that reflects the sorted order, we permute the vectors using the temporary buffer of $O(\sqrt{\log m'})$ placeholders. Thus the time complexity of $GVSP\{O(\sqrt{\log m'}), O(\sqrt{\log m'}), 0\}$ is bounded by $O(k\sqrt{\log n} + \log n)$. We summarize the resulting bounds, leaving several technical details to the full paper.

Lemma 6. *An instance of $GVSP\{O(n/\log n), O(n/\log n), O(n/\log^2 n)\}$ takes $O(n + nk/\log n)$ time using $O(1)$ auxiliary locations.*

Theorem 2. *An arbitrary set of n vectors of length k can be sorted in place optimally, taking $O(nk + n\log n)$ time and using $O(1)$ auxiliary locations.*

References

1. A. Andersson, T. Hagerup, J. Håstad, and O. Petersson. Tight bounds for searching a sorted array of strings. *SIAM Journal on Computing*, 30(5):1552–1578, 2001.
2. L. Arge, P. Ferragina, R. Grossi, and J.S. Vitter. On sorting strings in external memory. ACM STOC '97, 540–548, 1997.
3. M.A. Bender, E.D. Demaine, and M. Farach-Colton. Cache-oblivious B-trees. IEEE FOCS '00, 399–409, 2000.
4. J.L. Bentley and R. Sedgewick. Fast algorithms for sorting and searching strings. ACM-SIAM SODA '97, 360–369, 1997.
5. G.S. Brodal, R. Fagerberg, and R. Jacob. Cache oblivious search trees via binary trees of small height. ACM-SIAM SODA '02, 39–48. 2002.
6. T.H. Cormen, C.E. Leiserson, R.L. Rivest, and C. Stein. *Introduction to Algorithms.* MIT Press, 2001.
7. G. Franceschini. Proximity mergesort: Optimal in-place sorting in the cache-oblivious model. ACM-SIAM SODA '04, 284–292, 2004.
8. G. Franceschini. Sorting stably, in-place, with $O(n\log n)$ comparisons and $O(n)$ moves. STACS '05, to appear, 2005.
9. G. Franceschini and V. Geffert. An In-Place Sorting with $O(n\log n)$ Comparisons and $O(n)$ Moves. IEEE FOCS '03, 242–250, 2003.
10. G. Franceschini and R. Grossi. No Sorting? better Searching! IEEE FOCS '04, 491–498, 2004.
11. D.S. Hirschberg. A lower worst-case complexity for searching a dictionary. Proc. 16th Allerton Conference on Comm., Control, and Computing, 50–53, 1978.
12. D.S. Hirschberg. On the complexity of searching a set of vectors. *SIAM J. Computing*, 9(1):126–129, 1980.
13. A. Itai, A.G. Konheim, and M. Rodeh. A sparse table implementation of priority queues. ICALP '81, 417–431, 1981.
14. D.E. Knuth. *The Art of Computer Programming III: Sorting and Searching.* Addison–Wesley, 1998.
15. S.R. Kosaraju. On a multidimensional search problem. ACM STOC '79, 67–73, 1979.
16. M.A. Kronrod. Optimal ordering algorithm without operational field. *Soviet Math. Dokl.*, 10:744–746, 1969.
17. T.W. Lai and D. Wood. Implicit selection. SWAT '88, 14–23, 1988.

18. U. Manber and G. Myers. Suffix arrays: A new method for on-line string searches. *SIAM Journal on Computing*, 22(5):935–948, 1993.
19. J.I. Munro. An implicit data structure supporting insertion, deletion, and search in $O(\log^2 n)$ time. *Journal of Computer and System Sciences*, 33(1):66–74, 1986.
20. J.I. Munro and V. Raman. Sorting multisets and vectors in-place. WADS '91, 473–480, 1991.
21. J. Salowe and W. Steiger. Simplified stable merging tasks. *Journal of Algorithms*, 8(4):557–571, 1987.
22. D.E. Willard. Maintaining dense sequential files in a dynamic environment. ACM STOC '82, 114–121, 1982.

Towards Optimal Multiple Selection

Kanela Kaligosi[1], Kurt Mehlhorn[1], J. Ian Munro[2], and Peter Sanders[3]

[1] Max Planck Institut für Informatik, Saarbrücken, Germany
{kaligosi, mehlhorn}@mpi-sb.mpg.de
[2] University of Waterloo, Ontario, Canada
imunro@uwaterloo.ca
[3] Universität Karlsruhe, Germany
sanders@ira.uka.de

Abstract. The multiple selection problem asks for the elements of rank $r_1, r_2, \ldots, r_k$ from a linearly ordered set of n elements. Let B denote the information theoretic lower bound on the number of element comparisons needed for multiple selection. We first show that a variant of multiple quickselect — a well known, simple, and practical generalization of quicksort — solves this problem with $B + \mathcal{O}(n)$ expected comparisons. We then develop a deterministic divide-and-conquer algorithm that solves the problem in $\mathcal{O}(B)$ time and $B + o(B) + \mathcal{O}(n)$ element comparisons.

1 Introduction

We consider the problem of determining the elements of rank $r_1, r_2, \ldots, r_k$ in an ordered set S with n elements using as few comparisons as possible. The problem has been studied extensively, primarily at its extreme ends, when one either requires one or two order statistics (such as the median or the maximum and minimum) or when all ranks are to be determined, and so the set is to be sorted. The more general problem can be very helpful in statistical analyses, providing a clean summary of the data by giving elements of a chosen set of ranks. Multiple selection is also an ingredient of *sample sort* [1].
Let $\Delta_j := r_j - r_{j-1}$ where $r_0 = 0$ and $r_{k+1} = n$ and let[1]

$$B = n \lg n - \sum_{j=1}^{k+1} \Delta_j \lg \Delta_j - \mathcal{O}(n) = \sum_{j=1}^{k+1} \Delta_j \lg \frac{n}{\Delta_j} - \mathcal{O}(n).$$

A comparison-based multi-selection algorithm identifies the Δ_1 smallest elements, the Δ_2 next smallest elements, ..., and hence an additional $\sum_j \Delta_j \lg \Delta_j + \mathcal{O}(n)$ comparisons suffice to sort the entire set. Thus B is a lower bound for the number of comparisons needed by any multiselection algorithm, even in the expected case [5]. Note that this lower bound takes both the input size n and the structure of the required ranks R into account.

---
[1] Throughout this paper, $\lg x$ stands for $\log_2 x$.

1.1 Background and Ancient History

The problem of sorting is one of the oldest and most heavily studied in computing. "Interesting" sorting algorithms date back to the 1940's, or earlier. The information theoretic lower bound of $\lg(n!) = n \lg n - n \lg e + \mathcal{O}(\lg n)$ comparisons is missed by only about $n(\lg e - 1)$ comparisons by mergesort, when n is a power of 2. Despite such an easy approach to optimality, even Ford-Johnson [9] merge insertion and other very complex methods still leave a $\Theta(n)$ gap (with an admittedly small constant).

It seems that Charles Dodgeson (Lewis Carroll) was the first to point to the unfairness of declaring the runner-up (the player losing to the last game to the winner in a single knockout competition) to be the "second best". Hoare [8] proposed a simple $\mathcal{O}(n)$ expected case (median) selection algorithm as an addendum to his original work on Quicksort. Close to a decade later Blum, Floyd, Pratt, Rivest and Tarjan [2] provided the first median finding technique with $\mathcal{O}(n)$ worst case behavior. The result was certainly a major surprise at the time, although it is now often presented as a basic example of divide and conquer. At essentially the same time, two of those authors, Floyd and Rivest [7], gave a probabilistic method for selection using $n + \min(r_1, n - r_1) + o(n)$ comparisons in the expected case. This bound is optimal, at least up to $o(n)$ terms, as shown in [4]. We note that this lower bound exceeds that of the information theoretic argument, above, for selecting the median by $n/2$. The difference is even greater for selecting other ranks.

The contrast between the simple randomized techniques and the complex worst case approach is striking. The development of the $3n + o(n)$ worst case method of Schönhage et al. [14] and the $(2.94 \cdots)n$ method of Dor and Zwick [6] further amplifies

Multiple selection problems have received much less attention. Chambers [3] proposed multiple quickselect — a natural generalization of quicksort: Pick a *pivot* element m as an estimate for the median of S and partition S into two sets $S_\leq = \{s \in S : s \leq m\}$ and $S_> = \{s \in S : s > m\}$. Output m as the element of rank $r_m := |S_\leq|$ if $r_m \in R$. Recurse on $(S_\leq, \{r \in R : r < r_m\})$ and $(S_>, \{r - r_m \in R : r > r_m\})$. The recursion can be broken when $R = \emptyset$.

Pohl [12] gave a slick deterministic algorithm for finding the maximum and minimum of a set in the optimal number of $\lceil 3n/2 \rceil - 1$ comparisons. Cunto and Munro [4] provided optimal and near optimal randomized techniques for finding elements of two or three ranks. The "full blown" deterministic multiselection problem that we address appears to have first been considered by Dobkin and Munro [5], who suggest a method based on median finding that solves the problem using $3B + \mathcal{O}(n)$ comparisons.

A quarter century after Chamber's paper, there has been renewed interest in the analysis of multiple quickselect. Prodinger [13] has given an exact formula for the expected number of comparisons performed by multiple quickselect with random pivot which can be written as $2B/\lg(e) + \mathcal{O}(n)$. Panholzer [11] has derived very accurate bounds for the expected number of comparisons for the case that the pivot is the median of $2t+1$ random sample elements and where R

is a *random* k element subset of $\{1,\ldots,n\}$. These bounds show that $n \lg k + \mathcal{O}(n)$ expected comparisons are sufficient if $t = \Omega(\lg n)$. Note that for general R the lower bound B might be much smaller. For example, consider the set of ranks $R := \{\lceil n/i \rceil : i \leq n\}$. Observe further that this set has $k = \Theta(\sqrt{n})$ elements, i.e., $n \lg k = \Theta(n \lg n)$. On the other hand, $B \approx n \sum_i \frac{\ln(i(i+1))}{i(i+1)} = \Theta(n)$, i.e., there is a logarithmic factor gap between Panholzer's upper bound and the lower bound B. It is a natural question whether multiple quicksort is still a near optimal algorithm for inputs with nonuniformly spread ranks in R. For example, it was initially not clear to us whether one should choose m such that S is approximately halved, such that R is approximately halved, or some compromise between these two strategies.

1.2 The Rest of the Paper

Section 2 introduces our basic proof technique by analyzing an idealized variant of multiple quickselect that is given the exact median for free. This algorithm needs at most $n \lg n - \sum_{j=1}^{k+1} \Delta_j \lg \Delta_j + r_k - r_1 + 2n$ comparisons. In Section 3 we show that multiple quickselect with pivot selection based on random sampling can closely approximate the idealized algorithm with respect to the expected number of comparisons. The method is reasonably straightforward and so usable in practice: one uses about n comparisons to split S into elements above and below a pivot value which, with high probability, is quite close to the median. Some care is required in sampling, but as long as one does not insist on minimizing lower order terms too vigorously, this aspect of the analysis is fairly straightforward. Handling the details of the splits is a bit trickier.

Section 4 deals with the deterministic algorithm. As in the deterministic median algorithm of Blum et al. [2], we represent sets as collections of sorted sequences. Each sequence has length about $\ell = \lg(B/n)$. The number of sequences is about n/ℓ. We determine the median of the median of the sequences, call it m, and partition each sequence at m using binary search. This has cost $\mathcal{O}((n/\ell) \lg \ell)$ which is sublinear, if $B = \omega(n)$. After partitioning, we have two collections of sorted sequences on which we recurse. In each collection, we have short sequences (length less than ℓ) and long sequences (length between ℓ and 2ℓ). As long as we have two short sequences of the same length, we merge them using an optimal number of comparisons. The key insight is that the amortized cost of merging is $nH(\alpha)$ where $n\alpha$ is the rank of the pivot element and $H(\alpha) = -\alpha \lg \alpha - (1-\alpha) \lg(1-\alpha)$ is the binary entropy function. A simple induction shows that the total cost of merging is $B + o(B) + \mathcal{O}(n)$ and that the total cost of partitioning is $o(B) + \mathcal{O}(n)$.

2 An Idealized Algorithm

Consider multiple quickselect where we pick the exact median of the current subproblem as a pivot. Let $T(n, R)$ denote the number of comparisons needed in the partitioning steps, i.e., we disregard the cost needed for obtaining the pivot.

Theorem 1. $T(n, R) \leq n \lg n - \sum_{j=1}^{k+1} \Delta_j \lg \Delta_j + r_k - r_1 + 2n.$

Proof: The proof is by induction. For $n = 1$ the claim $0 = T(n, R) \leq 0 - 0 + 1 - 1 + 2$ is true. Now let r_i denote the largest element of R with $r_i \leq n/2$. For the induction step we distinguish two cases and substitute the induction hypothesis for the subproblems in each case.

Case $0 < i < k$:

$$T(n, R) = n - 1 + T\left(\frac{n}{2}, \{r_1, \ldots, r_i\}\right) + T\left(\frac{n}{2}, \{r_{i+1} - \frac{n}{2}, \ldots, r_k - \frac{n}{2}\}\right)$$

$$\leq n - 1 + n \lg \frac{n}{2} - \sum_{j=1}^{k+1} \Delta_j \lg \Delta_j + r_i - r_1 + r_k - r_{i+1} + 2n$$

$$+ \Delta_{i+1} \lg \Delta_{i+1} - \left(\frac{n}{2} - r_i\right) \lg \left(\frac{n}{2} - r_i\right) - \left(r_{i+1} - \frac{n}{2}\right) \lg \left(r_{i+1} - \frac{n}{2}\right)$$

$$\leq n \lg n - \sum_{j=1}^{k+1} \Delta_j \lg \Delta_j - r_k + r_1 + 2n$$

For the last estimate we have used $n - 1 + n \lg(n/2) = n \lg n - 1 \leq n \lg n$ and Lemma 1 below with $x = n/2 - r_i$ and $y = r_{i+1} - n/2$.
Case $i = k$: (the case $i = 0$ is symmetric)

$$T(n, R) = n - 1 + T\left(\frac{n}{2}, R\right)$$

$$\leq n - 1 + \frac{n}{2} \lg \frac{n}{2} - \sum_{j=1}^{k+1} \Delta_j \lg \Delta_j + r_k - r_1 + 2\frac{n}{2}$$

$$+ (n - r_k) \lg(n - r_k) - \left(\frac{n}{2} - r_k\right) \lg \left(\frac{n}{2} - r_k\right)$$

Now we use $n - 1 + 2n/2 \leq 2n$ and, by Lemma 1,
$(n - r_k) \lg(n - r_k) - \left(\frac{n}{2} - r_k\right) \lg \left(\frac{n}{2} - r_k\right) \leq \frac{n}{2} \lg \frac{n}{2} + n - r_k$.

$$T(n, R) \leq n \lg \frac{n}{2} - \sum_{j=1}^{k+1} \Delta_j \lg \Delta_j + r_k - r_1 + 2n + n - r_k \ .$$

Finally, observe that $n \lg \frac{n}{2} + n - r_k = n \lg n - r_k \leq n \lg n$. □

Lemma 1. For $x, y \geq 0$ we have $(x+y) \lg(x+y) - x \lg x - y \lg y - (x+y) \leq 0$.

Proof: Let $s = x + y$ and $x = \gamma s$, with $0 \leq \gamma \leq 1$. Then
$s(\lg s - \gamma \lg(\gamma s) - (1 - \gamma) \lg((1 - \gamma)s) - 1)$
$= s(-\gamma \lg(\gamma) - (1 - \gamma) \lg(1 - \gamma) - 1) = s(H(\gamma) - 1) \leq 0$. □

3 The Randomized Algorithm

We now analyze a variant of quickselect where the pivot m is the median of $n^{3/4}$ sample elements chosen uniformly at random from S with replacement. Let $T(n, R)$ denote the expected number of comparisons needed to execute this of multiple quickselect.

Theorem 2. $T(n, R) \leq n \lg n - \sum_{j=1}^{k+1} \Delta_j \lg \Delta_j + r_k - r_1 + 2n + \mathcal{O}\left(k^{1/4} n^{3/4}\right).$

The remainder of this section outlines a proof for the above theorem. Most probabilistic aspects of the problem are packaged in the following lemma:

Lemma 2. *The rank of the pivot m is outside the range $[n/2 - n^{3/4}, n/2 + n^{3/4}]$ with probability at most $1/n$.*

The proof is in the full paper. It is based on Chernoff bounds and very similar to the proof used in the analysis of the Floyd Rivest algorithm [7] in [10].

Using Lemma 2 we can conclude that

$$T(n, R) \leq n - 1 + \mathcal{O}\left(n^{3/4}\right) + \frac{1}{n} T(n, R)$$
$$+ \left(1 - \frac{1}{n}\right) \max_{|\frac{n}{2} - n'| \leq n^{3/4}} (T(n', R') + T(n'', R''))$$

where $R' = \{r \in R : r < n'\}$, $R'' = \{r - n' : r \in R, r > n'\}$ and $n'' = n - n'$. The term $n - 1$ stems from the comparisons of elements in $S \setminus \{m\}$ with m. The term $\mathcal{O}(n^{3/4})$ accounts for the expected cost of finding m in the sample using a (randomized) linear time selection algorithm. The next term is the probability $1/n$ that m is an inaccurate estimate of the median times a conservative estimate of the cost in this case — partitioning can never make the subproblems more difficult than the original problem. Subtracting $T(n, R)/n$ and dividing by $1 - 1/n$ yields

$$T(n, R) \leq n + \mathcal{O}\left(n^{3/4}\right) + \max_{|\frac{n}{2} - n'| \leq n^{3/4}} (T(n', R') + T(n'', R'')) \qquad (1)$$

We now use induction over n using the induction hypothesis

$$T(n, R) \leq n \lg n - \sum_{j=1}^{k+1} \Delta_j \lg \Delta_j + r_k - r_1 + 2n + (a k^{1/4} - b) n^{3/4}$$

where a and b are constants whose value will be implied by constraints showing up in the proof. First note that for any constant n we can make this work by choosing a value for a that is sufficiently much larger than b.

For the induction step we will use the following two lemmata which encapsulate a number of calculations that we defer to the full paper.

Lemma 3. $f(\delta) := (1+\delta)\lg(1+\delta) + (1-\delta)\lg(1-\delta) \le 2\delta^2$ for $\delta \in [-1, 1]$.

Lemma 4. For $0 \le i \le k$ and $-1 \le \delta \le 1$ we have
$$i^{1/4}(1+\delta)^{3/4} + (k-i)^{1/4}(1-\delta)^{3/4} \le 2(k/2)^{1/4}$$

To simplify the max term in Eq. (1), consider a fixed value $0 \le \delta \le 2n^{-1/4}$ such that $n' = (1+\delta)\frac{n}{2}$ (the case $\delta < 0$ is symmetric). Let r_i denote the largest element of R with $r_i \le n'$. We get three cases:

Case $0 < i < k$: We use the abbreviation $I(x) := x \lg x$.

$$T(n', R') + T(n'', R'') \le I((1+\delta)\frac{n}{2}) + I((1-\delta)\frac{n}{2}) - \sum_{j=1}^{k+1} I(\Delta_j)$$
$$+ I(\Delta_{i+1}) - I(n' - r_i) - I(r_{i+1} - n')$$
$$+ r_i - r_1 + 2n' + (ai^{1/4} - b)n'^{3/4}$$
$$+ r_k - r_{i+1} + 2n'' + (a(k-i)^{1/4} - b)n''^{3/4}$$
$$= n \lg n - n + \frac{n}{2}(I(1+\delta) + I(1-\delta)) - \sum_{j=1}^{k+1} I(\Delta_j)$$
$$+ I(\Delta_{i+1}) - I(n' - r_i) - I(r_{i+1} - n') - (r_{i+1} - r_i)$$
$$+ 2^{-3/4}a(i^{1/4}(1+\delta)^{3/4} + (k-i)^{1/4}(1-\delta)^{3/4})n^{3/4}$$
$$- 2^{-3/4}((1+\delta)^{3/4} + (1-\delta)^{3/4})bn^{3/4} + r_k - r_1 + 2n$$

Now we can apply a number of estimates:

$I(1+\delta) + I(1-\delta) \le 2\delta^2 \le 2(2n^{-1/4})^2 = \mathcal{O}(n^{-1/2})$ (using Lemma 3). As in the idealized case, Lemma 1 can be used to eliminate the $I(x)$ terms not in $\sum_{j=1}^{k+1} I(\Delta_j)$. Now it follows (using Lemma 4) that
$i^{1/4}(1+\delta)^{3/4} + (k-i)^{1/4}(1-\delta)^{3/4} \le 2(k/2)^{1/4} = 2^{3/4}k^{1/4}$. Finally, $-(1+\delta)^{3/4} - (1-\delta)^{3/4} \le -2 + \mathcal{O}(n^{-1/2})$ using calculus. So

$$\le n \lg n - n - \sum_{j=1}^{k+1} I(\Delta_j) + \mathcal{O}(\sqrt{n}) + ak^{1/4}n^{3/4} - 2^{1/4}bn^{3/4} + r_k - r_1 + 2n.$$

Substituting this back into Eq. (1) we obtain

$$T(n, R) \le n \lg n - \sum_{j=1}^{k+1} I(\Delta_j) + \mathcal{O}(n^{3/4}) + ak^{1/4}n^{3/4} - 2^{1/4}bn^{3/4} + r_k - r_1 + 2n .$$

Since $2^{1/4} > 1$ we can use the term $b(2^{1/4} - 1)n^{3/4}$ to cancel the $\mathcal{O}(n^{3/4})$ term by choosing a sufficiently large value for b.

Case $i = k$: This case is a slight generalization of the reasoning from the idealized algorithm and from the case $0 < i < k$. Compared to the idealized case, we get various $\mathcal{O}(n^{3/4})$ terms that we need to cancel. For this purpose we use the term $ak^{1/4}((1+\delta)\frac{n}{2})^{3/4} = 2^{-3/4}ak^{1/4}n^{3/4} + \mathcal{O}(k^{1/4}n^{1/2})$. Since $2^{-3/4} < 1$ we can choose a sufficiently large to cancel all the unwanted lower order terms.

Case $i = 0$: This case is similar to the case $i = k$ but "easier" because the non-trivial recursion is on a smaller input. We therefore omit the detailed proof. ∎

4 The Deterministic Algorithm

In this section we present a deterministic algorithm for the multiple selection problem. It performs $B + o(B) + \mathcal{O}(n)$ comparisons.

The algorithms sets $\ell = \max(1, \lceil \lg(B/n) \rceil)$ and is based on the recursive procedure *multiselect*. The input to *multiselect* is a collection C of non-empty sorted sequences and a non-empty set $R = \{r_1, \ldots, r_k\}$ of ranks. Each sequence in the collection has length at most 2ℓ. The procedure returns the elements of rank $r_1, r_2, \ldots, r_k$ in the input set. We solve the multiselection problem by calling *multiselect* with a collection of n sequences of length one each and our set R of desired ranks.

We call a sequence *short* if it has length less than ℓ. A collection C satisfies the length invariant if it contains no two short sequences of the same length.

Procedure *multiselect* works as follows: If the total number of elements in the sequences of C is less than $4\ell^2$, then solve the problem by sorting all the elements and if R is empty, simply return. Otherwise, we first establish the length invariant. As long as there are two short sequences of the same length, we merge them into a single sequence. Recall that two sorted sequences of length s can be merged using $2s - 1$ comparisons. After the merge step, we have a collection D of q sequences satisfying the length invariant. We determine the median of the medians of the sequences in D, call it m, and split every sequence in D at m, i.e., into the elements smaller than or equal to m and into the elements larger than m. This gives us collections C' and C'' of size p' and p'', respectively. We have $p' \leq q$, $p'' \leq q$ and $p' + p'' \geq q$. We also split the ranks and make one or two recursive calls.

This ends the description of the algorithm. We turn to the analysis. For a sequence c we use $|c|$ to denote its length. For a collection C of sorted sequences, let $I(C) = \sum_{c \in C} |c| \lg |c|$ be the *information content* of the collection. Multiselect generates a recursion tree T. We use v to denote an arbitrary node of the recursion tree and introduce the following notation: $I_v = I(C)$ is the information content of the collection entering node v, $p_v = |C|$ is the number of sequences in the collection and $n_v = \sum_{c \in C} |c|$ is the total length of the sequences in the collection. $J_v = I(D)$ is the information content of the collection after the merging step and q_v is the number of sequences in the collection. $I'_v = I(C')$ and $I''_v = I(C'')$ are the information content of the sequences resulting from the partitioning step and p'_v and p''_v are the number of sequences in these collections, respectively. Also, the total number of elements in C' is $\alpha_v n_v$. Finally, we use C_v^m for the number of comparisons spent on merging in node v and C_v^p for the number of comparisons spent on computing the median and partitioning in node v. The total number of comparisons in the calls with $n_v < 4\ell^2$ is $\mathcal{O}(n \lg \ell)$, since the calls are applied to disjoint sets of elements.

Lemma 5. $C_v^m \leq J_v - I_v + q_v - p_v$.

Proof: It takes at most $2s - 1$ comparisons to merge two sequence of length s. Also $2s \lg(2s) - 2(s \lg s) + 1 - 2 = 2s - 1$. □

Lemma 6. $J_v \leq I'_v + I''_v + n_v H(\alpha_v)$.

Proof: Let $D = \{d_1, \ldots, d_q\}$ be the collection after the merging step and let $n = n_v$ and $\alpha = \alpha_v$. Each d_i is split into two sequences d'_i and d''_i and C' is formed by the non-empty sequences in $\{d'_1, \ldots, d'_q\}$. C'' is defined analogously. Define α_i as $\alpha_i = d'_i/d_i$. We have (using $0 \lg 0 = 0$)

$$J_v - I'_v - I''_v = \sum_{1 \leq i \leq q} (d_i \lg d_i - d'_i \lg d'_i - d''_i \lg d''_i)$$

$$= \sum_{1 \leq i \leq q} d_i (\lg d_i - \alpha_i \lg(\alpha_i d_i) - (1 - \alpha_i) \lg((1 - \alpha_i) d_i))$$

$$= \sum_{1 \leq i \leq q} d_i H(\alpha_i).$$

Next observe that $\sum_i d_i/n = 1$, $\sum_i \alpha_i d_i / n = \sum_i d'_i/n = \alpha$. Thus $\sum_i (d_i/n) H(\alpha_i)$ is a convex combination of the values of H at the arguments α_i. Since H is convex this is bounded by the value of H at the corresponding combination of the arguments, i.e., by the value at $\sum_i (d_i/n) \alpha_i$. Thus $\sum_i (d_i/n) H(\alpha_i) \leq H(\sum_i (d_i/n) \alpha_i) = H(\alpha)$. □

Lemma 7. $C_v^m \leq n_v H(\alpha_v) + I'_v + I''_v - I_v + p'_v + p''_v - p_v$.

Proof: This follows immediately from the two lemmas above and the inequality $q_v \leq p'_v + p''_v$. □

The next Lemma shows that the splitting process is reasonably balanced. A similar Lemma is used in the analysis of the deterministic median algorithm of Blum et al.

Lemma 8. If $n_v \geq 4\ell^2$ then $1/16 \leq \alpha_v \leq 15/16$ and $H(\alpha_v) \geq 1/4$.

Proof: We count the number of elements that are smaller than or equal to the median of the medians m after partitioning.

After the merge step, we have a collection of q_v sequences, each of length at most 2ℓ. Therefore, $q_v \geq \lceil n_v/2\ell \rceil$. Recall that by the invariant of the algorithm there are no two sequences of the same length less than ℓ. Or, in other words, there can be at most one sequence for each of the lengths $1, \ldots, \ell-1$. Consider the sequences whose median is less than or equal to m. Their number is $\lceil q_v/2 \rceil \geq n_v/4\ell$. In the worst case this collection contains one sequence from each of the lengths $1, \ldots, \ell-1$. Thus, the number of non-short sequences is at least $n_v/4\ell - (\ell-1)$ and each of them has at least $\lceil \ell/2 \rceil$ elements that are less than or equal to m. Thus, in total they have at least $(n_v/4\ell - (\ell-1))\lceil \ell/2 \rceil \geq n_v/8 - \ell(\ell-1)/2$ elements. Moreover, the number of elements contained in the short sequences that are smaller than or equal to m are $\sum_{i=1}^{\ell-1} \lceil i/2 \rceil \geq 1/2 \sum_{i=1}^{\ell-1} i = \ell(\ell-1)/4$. Therefore, over all $\alpha_v n_v \geq n_v/8 - \ell(\ell-1)/4$. Similarly, for the number of elements greater than m we can show that $(1-\alpha_v)n_v \geq n_v/8 - \ell(\ell-1)/4$ and the claim follows. □

Consider the recursion tree T generated by the algorithm. This is a binary tree. The leaves correspond to calls with either $n_v < 4\ell^2$ or empty R_v. Interior nodes have $n_v > 4\ell^2$ and non-empty R_v. The total cost of merging is

$$\sum_{v \in T} C_v^m \leq \sum_{v \in T} (n_v H(\alpha_v) + I_v' + I_v'' - I_v + p_v' + p_v'' - p_v).$$

We first bound $\sum_{v \in T} n_v H(\alpha_v)$.

Lemma 9. *Let* $T(n, R) = \sum_{v \in T} n_v H(\alpha_v)$. *Then*

$$T(n, R) \leq \sum_{j=1}^{k+1} \Delta_j \lg \frac{n}{\Delta_j} + r_k - r_1 + 16n = n \lg n - \sum_{j=1}^{k+1} \Delta_j \lg \Delta_j + r_k - r_1 + 16n$$

whenever R is non-empty[2].

Proof: We use induction. If $n < 4\ell^2$ and R is non-empty, $T(n, R) = 0$ and the right hand side is at least zero. If R is non-empty, we split into subproblems (n', R') and (n'', R''), with $n' = \alpha n$, $n'' = (1-\alpha)n$, $R' = \{r : r \in R, r < n'\}$ and $R'' = \{r - n' : r \in R, r > n'\}$. Then $nH(\alpha) = n \lg n - n' \lg n' - n'' \lg n''$. Let $k' = |R'|$. By symmetry, we may assume $k' > 0$. Let $\Delta_1', \ldots, \Delta_{k'+1}'$ and $\Delta_1'', \ldots, \Delta_{k''+1}''$ be the sequence of Δ's for the two subproblems. Then $\Delta_1', \ldots, \Delta_{k'}', \Delta_{k'+1}' + \Delta_1'', \Delta_2'', \ldots, \Delta_{k''+1}''$ is the sequence of Δ's in the original problem. We need to distinguish cases.

Assume first $k'' > 0$. Then we can apply the induction hypothesis to both subproblems and obtain (writing x for $\Delta_{k'+1}'$ and y for Δ_1'')

---

[2] If R is empty, $k = 0$ and $r_k = 0$ and $r_1 = n$. This will not make sense in the inductive proof.

$$T(n,R) \le nH(\alpha) + T(n',R') + T(n'',R'')$$

$$\le n\lg n - n'\lg n' - n''\lg n'' + n'\lg n' - \sum_{j=1}^{k'+1} \Delta'_j \lg \Delta'_j + r'_k - r_1 + 16n'$$

$$+ n''\lg n'' - \sum_{j=1}^{k''+1} \Delta''_j \lg \Delta''_j + r_k - r_{k'+1} + 16n''$$

$$= n\lg n - \sum_{j=1}^{k+1} \Delta_j \lg \Delta_j + r_k - r_1 + 16n$$

$$+ (x+y)\lg(x+y) - x\lg x - y\lg y - (r_{k'+1} - r_{k'})$$

and we have established the induction step using Lemma 1 and $x+y = r_{k'+1} - r_{k'}$.

Assume next $k'' = 0$. We apply the induction hypothesis to the first subproblem and obtain

$$T(n,R) \le nH(\alpha) + T(n',R')$$

$$\le n\lg n - n'\lg n' - n''\lg n'' + n'\lg n' - \sum_{1 \le j \le k+1} \Delta'_j \lg \Delta'_j + r_k - r_1 + 16n'$$

$$= n\lg n - \sum_{1 \le j \le k+1} \Delta_j \lg \Delta_j + r_k - r_1 + 16n$$

$$+ (\Delta'_{k+1} + n'')\lg(\Delta'_{k+1} + n'') - \Delta'_{k+1} \lg \Delta'_{k+1} - n''\lg n'' - 16n''$$

We have $\Delta'_{k+1} + n'' \le n \le 16n''$ by Lemma 8. An application of Lemma 1 completes the induction step. □

Lemma 10. $\sum_{v \in T}(I'_v + I''_v - I_v + p'_v + p''_v - p_v) \le n\lg(2\ell)$.

Proof: The sum telescopes to

$$-I_{\text{root}} - p_{\text{root}} + \sum_{\substack{v \in T \text{ and there is} \\ \text{no recursive call for } C'}} (I'_v + p'_v) + \sum_{\substack{v \in T \text{ and there is} \\ \text{no recursive call for } C''}} (I''_v + p''_v).$$

The collection entering the root consists of sequences of length 1 and hence $I_{\text{root}} = 0$ and $p_{\text{root}} = n$. The collections for which there is no recursive call are disjoint and hence their total length is n. Also no sequence is longer than 2ℓ and every sequence is non-empty. Thus their contribution is bounded by $n\lg(2\ell) + n$. □

We next turn to the cost of computing the median of the medians and partitioning our sequences at the median of medians. We let β be a constant such that the median of z elements can be computed with βz comparisons.

Lemma 11. *The total cost C^p of computing the median of the medians and partitioning is bounded by*

$$\frac{4(\lg 2\ell + \beta)}{\ell} \sum_{j=1}^{k+1} \Delta_j \lg \frac{n}{\Delta_j} + \frac{4(\lg 2\ell + \beta)}{\ell}(r_k - r_1) + \mathcal{O}(n) + \mathcal{O}(\beta \ell n) + \mathcal{O}(n\ell \lg \ell).$$

Proof: We split the cost for computing medians and partitioning into two parts: the cost arising from non-short sequences and the cost arising from short sequences.

The first three terms refer to the cost of finding the median of medians and partitioning arising from the non-short sequences. The median of each sequence is known since they are sorted. At each node v the number of non-short sequences is at most $\lfloor n_v/\ell \rfloor \leq n_v/\ell$ and hence the share of the non-short sequences is $\beta n_v/\ell$ comparisons in node v. In order to partition at m we perform binary search on each sequence with cost at most $\lg 2\ell$, thus, the cost over all sequences is at most $n_v \lg 2\ell/\ell$. Therefore, the comparisons spent in node v for non-short sequences is $(\lg 2\ell + \beta)n_v/\ell$. Next observe that $1 \leq 4H(\alpha_v)$ for any vertex v and hence $(\lg 2\ell + \beta)n_v/\ell \leq 4(\lg 2\ell + \beta)H(\alpha_v)n_v/\ell$. The bound now follows from Lemma 9.

The fourth term refers to the cost contributed by the short sequences for computing the median of medians. Since we have at most $\ell - 1$ short sequences in each node, they contribute $\beta(\ell - 1)$ to the number of comparisons at each node of the recursion tree. Hence, they contribute an $\mathcal{O}(\beta \ell n)$ over all.

The last term refers to the cost of partitioning the short sequences at m. At each node this cost is at most $\sum_{i=1}^{\ell-1} \lg i \leq \ell \lg \ell$ and the bound follows. □

Lemma 12. *The total cost for the base case of the recursion is bounded by βn.*

Proof: The collections for which there is no recursive call are disjoint and the cost of selecting the median of s elements is βs. □

Theorem 3. *The deterministic multi-selection algorithm uses $B + o(B) + \mathcal{O}(n)$ comparisons.*

Proof: Summing the bounds in Lemmas 9, 10, 11, and 12 we obtain the following upper bound:

$$B + \mathcal{O}(n) + \mathcal{O}(n \lg \ell) + \mathcal{O}((1 + \lg \ell)/\ell) \cdot B + \mathcal{O}(n) + \mathcal{O}(n\ell \lg \ell)$$
$$= B + \mathcal{O}(n) + \mathcal{O}\left(\frac{1 + \lg \lg(B/n)}{\lg(B/n)} + \frac{\lg(B/n) \lg \lg(B/n)}{B/n}\right) \cdot B.$$

This is $B + o(B) + \mathcal{O}(n)$. □

5 Conclusion

We have shown that multiple quickselect performs an optimal number of comparisons up to a linear term. There are several ways to improve this linear term which is particularly interesting when the lower bound is linear itself. For example, when $r_k < n/2$ (or $r_1 > n/2$) it pays to choose the pivot such that its rank is just a bit larger than r_k (or a bit smaller than r_1). It can then be shown that the linear term is reduced from $2n$ to $3n/2$. Also note that for $k = 1$ the algorithm then becomes a special case of the optimal algorithm by Floyd and Rivest [7]. Likewise, the lower bound could be further refined. However, since there is even a remaining linear size gap between upper and lower bound for sorting, completely closing the gap remains a distant possibility.

There is a long standing constant factor gap between the upper bounds for the best deterministic selection algorithms and the lower bound. Our deterministic algorithm shows that this gap becomes a lower order term for multiple selection.

Multiple quickselect is obviously highly practical. It might be quite complicated to implement our *deterministic* algorithm efficiently but at the end it might work quite well: Its cost is dominated by merging operations that are known to be quite fast. Moreover, since the algorithm executes batches of many small binary merging operations, one might execute them in parallel using multithreading or instruction parallelism. This approach might mitigate the impact of data dependencies and branch mispredictions.

References

1. G. E. Blelloch, C. E. Leiserson, B. M. Maggs, C. G. Plaxton, S. J. Smith, and M. Zagha. A comparison of sorting algorithms for the connection machine CM-2. In *3rd ACM Symposium on Parallel Algorithms and Architectures*, pages 3–16, 1991.
2. M. Blum, R. W. Floyd, V. R. Pratt, R. L. Rivest, and R. E. Tarjan. Time bounds for selection. *Journal of Computer and System Sciences*, 7(4):448–461, 1973.
3. J. Chambers. Partial sorting (algorithm 410). *Communications of the ACM*, 14:357–358, 1971.
4. W. Cunto and J. I. Munro. Average case selection. *J. ACM*, 36(2):270–279, 1989.
5. D. P. Dobkin and J. I. Munro. Optimal time minimal space selection algorithms. *Journal of the ACM*, 28(3):454–461, 1981.
6. D. Dor and U. Zwick. Selecting the median. In *SODA: ACM-SIAM Symposium on Discrete Algorithms*, 1995.
7. R. W. Floyd and R. L. Rivest. Expected time bounds for selection. *Commun. ACM*, 18(3):165–172, 1975.
8. C.A.R. Hoare. Find (algorithm 65). *Communications of the ACM*, 4(7):321–322, 1961.
9. L. R. Ford Jr. and S. B. Johnson. A tournament problem. *AMM*, 66(5):387–389, 1959.
10. R. Motwani and P. Raghavan. *Randomized Algorithms*. Cambridge University Press, 1995.
11. A. Panholzer. Analysis of multiple quickselect variants. *Theor. Comput. Sci.*, 302(1-3):45–91, 2003.
12. I. Pohl. A sorting problem and its complexity. *Commun. ACM*, 15(6):462–464, 1972.
13. H. Prodinger. Multiple quickselect - Hoare's find algorithm for several elements. *Information Processing Letters*, 56:123–129, 1995.
14. A. Schönhage, M. Paterson, and N. Pippenger. Finding the median. *J. Comput. Syst. Sci.*, 13:184–199, 1976.

Simple Extractors via Constructions of Cryptographic Pseudo-random Generators

Marius Zimand

Department of Computer and Information Sciences,
Towson University, Baltimore
http://triton.towson.edu/~mzimand

Abstract. Trevisan has shown that constructions of pseudo-random generators from hard functions (the Nisan-Wigderson approach) also produce extractors. We show that constructions of pseudo-random generators from one-way permutations (the Blum-Micali-Yao approach) can be used for building extractors as well. Using this new technique we build extractors that do not use designs and polynomial-based error-correcting codes and that are very simple and efficient. For example, one extractor produces each output bit separately in $O(\log^2 n)$ time. These extractors work for weak sources with min entropy λn, for arbitrary constant $\lambda > 0$, have seed length $O(\log^2 n)$, and their output length is $\approx n^{\lambda/3}$.

1 Introduction

This paper puts forward a new framework for constructing extractors based on a new connection between extractors and pseudo-random generators. A pseudo-random generator takes as input a short random string called the seed and outputs a long string that cannot be distinguished from a truly random string by any test that is computable by circuits of bounded size. An extractor has two inputs: (a) The first one comes from an imperfect (i.e., with biased bits and correlations among bits) distribution on binary strings of some length and it is called the weakly-random string; (b) the second one is a short random seed. The output is a long string that cannot be distinguished from a truly random string by *any* test. One difference between pseudo-random generators and extractors is the number of inputs (one versus two). From a technical point of view this difference is minor because the known constructions of pseudo-random generators implicitly do use an extra input which is a function that in some sense is computationally hard. The fundamental difference is in the randomness requirement for the output. Thus, while the output of a pseudo-random generator looks random in a complexity-theoretic way, the output of an extractor is random (or very close to random) in an absolute information-theoretic way. Consequently pseudo-random generators and extractors appear to belong to two very different worlds, and, for many years, the developments in the construction of pseudo-random generators and extractors went along distinct research lines.

Trevisan [Tre01] has made a breakthrough contribution in this area by observing that the (apparently superficial) similarity between extractors and pseudo-random generators extends to some of the methods to build the two kind of objects. For the reasons mentioned above, Trevisan's result has been extremely surprising. It has also been an isolated example of a transfer from the complexity theory standard arsenal of techniques to the information theoretical area. In this paper we extend Trevisan's observation and establish that, as far as construction methods are concerned, there is a truly close relationship between pseudo-random generators and extractors. Specifically, we show that the other major route (than the one followed by Trevisan) that leads to pseudo-random generators (of a somewhat different kind) can also be used to construct extractors. Some explanations are in order at this point. There are two known approaches for constructing pseudo-random generators. One approach uses as a building block a hard function f and, in one typical setting of parameters, for any given $k \in \mathbf{N}$, builds a pseudo-random generator g with outputs of length n that is secure against adversary tests computable in time n^k. The running time to compute $g(x)$ is $n^{k'}$, for some $k' > k$. This kind of pseudo-random generators can be used for derandomizing BPP computations. They cannot be used in cryptography, because in this setting, it is unwise to assume that the adversary is endowed with less computational power (n^k) than the legitimate users $(n^{k'})$. Henceforth we will call this type of pseudo-random generator a "derandomization pseudo-random generator" (also known as a Nisan-Wigderson pseudo-random generator). The second approach uses as a building block a hard object of a more sophisticated type, namely a one-way function (the hardness of such a function f consists in the difficulty to invert it, but f must satisfy an additional property, namely, it should be easy to calculate $f(x)$ given x). It is known that given a one-way function, one can construct a pseudo-random generator [HILL99]. An easier construction produces a pseudo-random generator from any one-way length-preserving permutation. This second approach has the disadvantage that is using as a building block a more demanding type of object. The advantage of the method is that a pseudo-random generator g constructed in this way can be used in cryptography because $g(x)$ can be calculated in time significantly shorter than the time an adversary must spend to distinguish $g(x)$ from a truly random string. Henceforth we will call this type of pseudo-random generator a "crypto pseudo-random generator" (also known as a Blum-Micali-Yao pseudo-random generator).

Trevisan has shown that the known methods for constructing derandomization pseudo-random generators also produce extractors. More precisely, he has shown that the constructions of pseudo-random generators from hard functions given by Nisan and Wigderson [NW94] and Impagliazzo and Wigderson [IW97] can be used almost directly to produce extractors. His method has been extended in a number of papers to build extractors with increasingly better parameters (see the survey paper by Shaltiel [Sha02]). In the paper [Tre99], the conference version of [Tre01], Trevisan has suggested that the methods to construct crypto pseudo-random generator cannot be used to build extractors. We show that in fact they can, at least for a combination of parameters that, even though not optimal, is not trivial. Moreover, we show that the extractors constructed in this way are very simple and efficient. The first extractor built in this paper follows almost directly the classical construction of a pseudo-random generator from a one-way permutation. It runs in $O(n \log n)$ time (in the standard RAM model) and is very simple. The following is a complete description of it. The input consists of the weakly-random string X, of length $n = \tilde{n}2^{\tilde{n}}$ for some integer $\tilde{n}$, and of the seed $((x_1, \ldots, x_\ell), r)$, with $|x_i| = \tilde{n}$, $\ell = O(\tilde{n})$, and $|r| = \ell\tilde{n}$. We view X as a function $X : \{0,1\}^{\tilde{n}} \to \{0,1\}^{\tilde{n}}$, and, using the standard procedure, we transform X into a circular permutation $R : \{0,1\}^{\tilde{n}} \to \{0,1\}^{\tilde{n}}$. For $i = 0$ to $m - 1 = n^{\Omega(1)}$, we calculate b_i as the inner product modulo 2 of r and $(R^i(x_1) \ldots R^i(x_\ell))$. The output is $b_0 \ldots b_{m-1}$. A type of efficiency which has received a lot of attention recently is that of sublinear time. It may be the case that in some applications we only need the i-th bit from the sequence of random bits that are extracted from the weakly-random string. We would like to obtain this bit in time polynomial in the length of the index i, which typically means polylog time in the input length (under the assumption that each input bit can be accessed in one time unit). By analogy with the case of list-decodable codes, we call an extractor with this property, a *bitwise locally computable extractor*.[1] The second extractor that we build is of this type. It relies on the same basic method used in the construction of the first extractor, combined with the idea of taking consecutive inputs of the hard function as in the extractor of Ta-Shma, Zuckerman and Safra [TSZS01]. This second extractor is even simpler and its complete description is as follows. The input consists of the weakly-random string X of length $n = \tilde{n} \cdot 2^{\tilde{n}}$, for some natural number $\tilde{n}$, and of the seed $((x_1, \ldots, x_\ell), r)$, with $|x_i| = \tilde{n}$, for all i, $\ell = O(\tilde{n})$, and $|r| = \ell\tilde{n}$. We view X as the truth-table of a function $X : \{0,1\}^{\tilde{n}} \to \{0,1\}^{\tilde{n}}$. For $i = 0$ to $m - 1 = n^{\Omega(1)}$, we calculate b_i as the inner product modulo 2 of r and $(X(x_1 + i), \ldots, X(x_\ell + i))$, where the addition is done modulo $2^{\tilde{n}}$. The output is $b_0 \ldots b_{m-1}$.

The parameters of the extractors constructed in this paper are not optimal. Both extractors that have been described above work for weak sources having min-entropy λn, for arbitrary constant $\lambda > 0$, use a random seed of length $O(\log^2 n)$, and the output length is approximately $n^{\lambda/3}$. A variant of the second extractor has seed length $O(\log n)$ (here, for simplicity, we assume that the extractor's error parameter ϵ is a constant), but the output length reduces to $2^{O(\sqrt{\log n})}$.

Lu's extractor [Lu04] coupled with the constructions of designs from the paper of Hartman and Raz [HR03] can be seen to be also a bitwise locally computable extractor with parameters similar to those of our second extractor (note that the designs in [HR03] appear to imply extractors with seed length $\Omega(\log^2 n)$). Lu's extractor is using expander graphs and the designs from [HR03]

---

[1] The simpler name *locally computable extractor* is already taken by a different kind of efficient extractors, namely by extractors computable in space linear in the output length, see [Vad04], [Lu04].

need somewhat unwieldy algebraic objects. It seems to us that the extractors presented in this paper are simpler than all the extractors from the literature.[2] At the highest level of abstraction, our extractors follow the "reconstruction paradigm" (see [Sha02]) typical to Trevisan's extractor and to its improvements [RRV99, TSZS01, SU01]. The major differences are that our extractors avoid (1) the use of designs (in this respect they are similar to the extractors in [TSZS01] and [SU01]), and, perhaps more strikingly, (2) the encoding of the weakly-random string with an error-correcting code having a good list-decoding property. Our extractors can be implemented very easily and are thus suitable for practical applications. For example, they can be utilized to generate one-time pad keys in cryptosystems based on the bounded-storage model (see the papers of Lu [Lu04] and Vadhan [Vad04]), or for constructions of error-correcting codes using the scheme in [TSZ01] (the extractors built in this paper are actually strong extractors—for definition see, for example [Sha02]—as required by this scheme). They may also have theoretical applications in situations where the kind of efficiency achieved by our extractors is essential.

1.1 Definitions

Notations: $x \odot y$ denotes the concatenation of the strings x and y, $|x|$ denotes the length of the string x, and $\|A\|$ denotes the cardinality of the set A. For two binary strings x and r of the same length, $b(x, r)$ denotes the inner product of x and r viewed as vectors over the field GF(2). Let $n \in \mathbf{N}$. Let X_n, Y_n be two distributions on Σ^n. The statistical distance between X_n and Y_n is denoted $\Delta_{\text{stat}}(X_n, Y_n)$ and is defined by $\Delta_{\text{stat}}(X_n, Y_n) = \max_{A \subseteq \{0,1\}^n} |\text{Prob}(X_n \in A) - \text{Prob}(Y_n \in A)|$. The computational distance between X_n and Y_n relative to size S is denoted $\Delta_{\text{comp},S}(X_n, Y_n)$ and is defined by $\Delta_{\text{comp},S}(X_n, Y_n) = \max |\text{Prob}(C(X_n) = 1) - \text{Prob}(C(Y_n) = 1)|$, where the maximum is taken over all circuits C of size $\leq S$. Abusing notation, we identify a circuit C with the set of strings x for which $C(x) = 1$. Thus, $x \in C$ is equivalent to $C(x) = 1$. The min-entropy of a random variable taking values in $\{0,1\}^n$ is given by $\min \left\{ \log \frac{1}{\text{Prob}(X=a)} \;\middle|\; a \in \{0,1\}^n, \text{Prob}(X = a) \neq 0 \right\}$. For each $n \in \mathbf{N}$, let U_n denote the uniform distribution over $\{0,1\}^n$.

Definition 1.1. (Extractor) *The values n, k, d, m are integer parameters, and $\epsilon > 0$ is a real number parameter. A function $E \colon \{0,1\}^n \times \{0,1\}^d \to \{0,1\}^m$ is a (k, ϵ)-extractor if for every distribution X on $\{0,1\}^n$ with min-entropy at least k, the distribution $E(X, U_d)$ is ϵ-close to the uniform distribution U_m in the statistical sense, i.e., $\Delta_{\text{stat}}(E(X, U_d), U_m) \leq \epsilon$.*

We fix parameters n, d, m and ϵ and a function $E \colon \{0,1\}^n \times \{0,1\}^d \to \{0,1\}^m$. Let us consider an arbitrary set $W \subseteq \{0,1\}^m$ and a string $x \in \{0,1\}^n$. We say

---

[2] We note that Dziembowski and Maurer [DM04] give a similarly simple construction of an object that is related to extractors.

that x hits W ϵ-correctly via E if $\left| \frac{\|\{E(x,y) \mid y \in \{0,1\}^d\} \cap W\|}{\|\{0,1\}^d\|} - \frac{\|W\|}{\|\{0,1\}^m\|} \right| \leq \epsilon$. The folowing lemma has appeared more or less explicitly in the literature (see, for example, [Sha02]).

Lemma 1.2. *Let $E\colon \{0,1\}^n \times \{0,1\}^d \to \{0,1\}^m$ and $\epsilon > 0$. Suppose that for every $W \subseteq \{0,1\}^m$, the number of $x \in \{0,1\}^n$ that do not hit W ϵ-correctly via E is at most 2^t, for some t. Then E is a $(t + \log(1/\epsilon), 2\epsilon)$-extractor.*

1.2 Overview and Comparison with Trevisan's Approach

Trevisan's method is based on the constructions of pseudo-random generators from hard functions given in [NW94] and in [IW97]. These constructions use a function f as a block-box and construct from it a function g_f that stretches the input (i.e., $|g_f(x)| \gg |x|$) and which has the following property. If there exists a circuit D that distinguishes $g_f(x)$, when x is randomly chosen in the domain of g_f, from the uniform distribution, then there is a small circuit A, which uses D as a subroutine, such that A calculates f (or an approximation of f, depending on whether we are using the method in [IW97] or the one in [NW94]). Therefore if f is a hard function, there can be no circuit D as above of small size and thus g_f is a pseudo-random generator. Trevisan has observed that (1) the truth-table of f can be viewed as a string produced by a weak source that can serve as an extra input of the pseudo-random generator, and (2) the circuit A invoking D can be considered as a special type of a circuit that is endowed with D-gates. By a standard counting argument, it can be shown that, for any circuit D, regardless of its size, the set of functions that can be calculated by small circuits with D-gates is small. A circuit D can be viewed statically as a statistical test (more exactly, the statistical test associated to the circuit D is the set of strings accepted by D). In the new terminology, the fact that D distinguishes the distribution of $g_f(x)$ from the uniform distribution with ϵ bias can be restated as "f does not hit D ϵ-correctly via g." The main property mentioned above can be restated as saying that the set of functions f that do not hit D ϵ-correctly is included in the set of functions computable by small circuits with D-gates. Since the latter set is small, the former set is small as well, and thus, by Lemma 1.2, the construction yields an extractor. In a nutshell, Trevisan's method replaces hard functions (a complexity-theoretic concept) with random functions (an information-theoretic concept) and takes advantage of the fact that a random function is hard and thus the construction carries over in the new setting.

We would like to follow a similar approach for the construction of crypto pseudo-random generators from one-way permutations. Those constructions do use a one-way permutation R as a black box to construct a pseudo-random generator g_R, and thus a truth-table of R can be considered as an extra input of the pseudo-random generator. Also, the proof is a reduction that shows that if a circuit D distinguishes $g_R(x)$ from the uniform distribution, then there is a small circuit A, invoking the circuit D, that inverts R on a large fraction of inputs. To close the proof in a similar way to Trevisan's approach, we would need to argue that the vast majority of permutations are one-way. It seems that we hit a major obstacle because, unlike the case of hard functions, it is not currently known if even a single one-way function exists (and we are seeking an unconditional proof for the extractors that we build). We go around this obstacle by allowing algorithms to have oracle access to the function they compute. Thus, in the above analysis, the circuit A, in addition to invoking the circuit D, will also have oracle access to the permutation R. In this setting all permutations are easy to compute because, obviously, there is a trivial constant-time algorithm that, for any permutation $R : \{0,1\}^n \to \{0,1\}^n$, given the possibility to query R, calculates $R(x)$. We need to argue that only few permutations R are invertible by algorithms that can query R in a bounded fashion. More precisely we need to estimate the size of the set of permutations $R : \{0,1\}^n \to \{0,1\}^n$ that can be inverted on a set of T elements in $\{0,1\}^n$ by circuits that can pose Q queries to R. This problem has been considered by Impagliazzo [Imp96] and by Gennaro and Trevisan [GT00]. Their techniques seem to work for the case $T \cdot Q < 2^n$ and lead to extractors that work only for sources with high min-entropy.[3]

We obtain better parameters by restricting the type of one-way permutations and the type of circuits that attempt to invert them. A closer look at the standard construction of Blum-Micali-Yao pseudo-random generators reveals that the circuit A with D-gates manages to determine x using only the values $R(x), R^2(x), \ldots, R^m(x)$ (where m is the generator's output length). It is thus enough to consider only circuits that use this pattern of queries to the permutation R. Intuitively, for a random permutation R, the value of x should be almost independent of the values of $R(x), R^2(x), \ldots, R^m(x)$, and thus, a circuit A restricted as above cannot invert but a very small fraction of permutations. If we take R to be a random circular permutation, the above intuition can be easily turned into a proof based on a Kolmogorov-complexity counting argument. A circular permutation $R : \{0,1\}^n \to \{0,1\}^n$ is fully specified by the sequence $(R(1), R^2(1), \ldots, R^{N-1}(1))$, where $N = 2^n$. If a circuit A restricted as above inverts $R(x)$ for all x, then the permutation R is determined by the last m values in the above sequence, namely $R^{N-m}(1), R^{N-(m-1)}(1), \ldots, R^{N-1}(1)$. Indeed, given the above values, the circuit A can determine $R^{N-m-1}(1)$, which is $R^{-1}(R^{N-m}(1))$, and then $R^{N-m-2}(1)$, and so on till $R(1)$ is determined. Therefore such a permutation R, given the circuit A, can be described concisely using only $m \cdot n$ bits (for specifying, as discussed, the last m elements in the above sequence). In fact, in our case, the circuit A does not invert $R(x)$ for all $x \in \{0,1\}^n$, and, therefore, the values of R at the points where the inversion fails have to be included in the description. A further complication is that even for the successful cases, the circuit A only list-inverts $R(x)$, which means that A on input $R(x)$ produces a relatively short list of elements, one of which is x. Thus, one also has to include in the description of R the rank of x in the

---

[3] On the other hand, these extractors have the interesting property that their output looks random even to statistical tests that have some type of access to the weakly-random string. These results will be reported in a separate paper.

list produced by A. The quantitative analysis of the standard construction of a crypto pseudo-random generator shows that if the permutation R does not hit D ϵ-correctly, then the circuit A with D-gates is only able to produce for an ϵ/m fraction of $R(x), x \in \{0,1\}^n$, a list with m^2/ϵ^2 elements one of which is x. For interesting values of m (the pseudo generator's output length), the ϵ/m fraction is too small and needs to be amplified to a value of the form $(1-\delta)$, for a small constant δ. This can be done by employing another technique that is well-known in the context of one-way functions. Namely, we use Yao's method of converting a weak one-way function into a strong one-way function by taking the direct product. In other words, we start with a circular permutation R, define (the direct product) $\overline{R}(x_1, \ldots, x_\ell) = R(x_1) \odot \ldots \odot R(x_\ell)$ (where $\odot$ denotes concatenation), for some appropriate value of ℓ, and use $\overline{R}$ in the definition of the extractor (instead of R in our tentative plan sketched above). It can be shown that, for $\ell = O((1/\delta)\log(1/\gamma))$, if a circuit A list-inverts $(y_1, \ldots, y_\ell)$, with list size $T = m^2/\epsilon^2$, for a $\gamma = \epsilon/m$ fraction of ℓ-tuples $(y_1, \ldots, y_\ell) \in (\{0,1\}^n)^\ell$, then there is a probabilistic algorithm A' that list-inverts $R(x)$ with list size $O(n \cdot T \cdot (1/\delta) \cdot (1/\gamma) \cdot \log(1/\gamma))$ for a $(1-\delta)$ fraction of $x \in \{0,1\}^n$. By fixing the random bits and the queries that depend on these random bits, we can obtain a brief description of R as in our first tentative plan. It follows that only few permutations R can hit D ϵ-incorrectly and, therefore, by Lemma 1.2, we have almost obtained an extractor (we also need to convert an arbitrary function $X : \{0,1\}^n \to \{0,1\}^n$ into a circular permutation $R : \{0,1\}^n \to \{0,1\}^n$, which is an easy task).

The second extractor starts from this idea and the observation that, for the sake of building an extractor, we can work with a function $\overline{X}$ (i.e., not necessarily a permutation) and consider consecutive values $\overline{X}(\overline{x}), \overline{X}(\overline{x}+1), \ldots, \overline{X}(\overline{x}+m)$, as in the extractor of Ta-Shma, Zuckerman, and Safra [TSZS01]. That extractor (as well as all the extractors using the "reconstruction paradigm") takes $\overline{X}$ to be the encoding of an arbitrary function X with a good list-decoding property and some other special algebraic properties. This is necessary, among other things, for the same type of amplification as in our discussion above. We use instead a direct-product construction that is much simpler to implement (however, the cost is a longer seed length).

Because of the space constraints, most of the proofs are omitted. A full version of the paper is available [Zim05].

2 An Extractor from a Crypto Pseudo-random Generator

Restricted permutations, restricted circuits

The space from where we randomly choose permutations consists of permutations of a special form. First we consider the set CIRC of all circular permutations $R : \{0,1\}^n \to \{0,1\}^n$. Next, for some parameter $\ell \in \mathbf{N}$, we take the ℓ-direct product of CIRC. This means that for any $R \in$ CIRC, we define

$\overline{R}_\ell : \{0,1\}^{\ell n} \to \{0,1\}^{\ell n}$ by $\overline{R}_\ell(x_1 \odot x_2 \odot \ldots \odot x_\ell) = R(x_1) \odot R(x_2) \odot \ldots \odot R(x_\ell)$. We let PERM_ℓ be the set $\{\overline{R}_\ell \mid R \in \text{CIRC}\}$. We will drop the subscript ℓ when its value is clear from the context or when it is not relevant in the discussion.

We want to argue that no circuit that queries $\overline{R}$ in a restricted way can invert a "large" fraction of $\overline{R}(\overline{x})$ except for a "small" fraction of permutations $\overline{R}$ in PERM. In order to obtain adequate values for "large" and "small" we will impose the following restriction on the pattern of queries that the circuit can make.

Definition 2.1. *An oracle circuit C on inputs of length at least $\ell \cdot n$ is L-restricted if on any input x and for all oracles $\overline{R} \in \text{PERM}_\ell$, C only queries $x_{\text{first}}, \overline{R}(x_{\text{first}}), \overline{R}^2(x_{\text{first}}), \ldots, \overline{R}^{L-1}(x_{\text{first}})$, where x_{first} is the string consisting of the first $\ell \cdot n$ bits of x.*

We will allow the circuits to attempt to invert $\overline{R}$ in a weaker form: On input $\overline{R}(\overline{x})$, $C^{\overline{R}}$ outputs a small list of strings one of which (in case C succeeds) is $\overline{x}$. When this event happens, we say that $C^{\overline{R}}$ *list-inverts* $\overline{x}$. We are interested in estimating the number of permutations $\overline{R} \in \text{PERM}$ so that $C^{\overline{R}}$ list-inverts $\overline{R}(\overline{x})$ for a large fraction of $\overline{x}$.

Definition 2.2. *Let C be an oracle circuit. A permutation $\overline{R}$ is (γ, T)-good for C if for at least a γ fraction of $\overline{x} \in \{0,1\}^{\ell n}$, $C^{\overline{R}}$ on input $\overline{R}(\overline{x})$ outputs a list of T elements that contains $\overline{x}$.*

The next lemma shows that a permutation that is (γ, T)-good for a restricted circuit C admits a short description conditioned by C being given.

Lemma 2.3. *Let $\gamma > 0$, $n \in \mathbf{N}, L \in \mathbf{N}$, and $T \in \mathbf{N}$. Let $N = 2^n$. Let $\delta > 0$ and let $\ell = \lceil \frac{3}{\delta} \cdot \log\left(\frac{2}{\gamma}\right)\rceil$. Assume $\delta \geq 2e^{-n}$ and $\ell < L + 1$. Let C be an L-restricted circuit, having inputs of length ℓn, and let $\overline{R} \in \text{PERM}_\ell$ be a permutation that is (γ, T)-good for C. Then, given C and ℓ, $\overline{R}$ can be described using a number of bits that is bounded by $2\delta N n + L n + N \log n + (\log 6) N + N \log(1/\delta) + N \log\log(2/\gamma) + N \log(1/\gamma) + N \log T + 18 n^2 \cdot L \cdot \frac{1}{\gamma} \cdot \left(\frac{1}{\delta}\right)^2 \left(\log \frac{2}{\gamma}\right)^2$.*

The above lemma allows us to estimate the number of permutations that are (γ, T)-good for some L-restricted circuit C. We state the result for a particular combination of parameters that will be of interest in our application.

Lemma 2.4. *Let $n \in \mathbf{N}, m \in \mathbf{N}, \epsilon > 0, \delta > 0$. Let $N = 2^n$. Consider $\gamma = \epsilon/m$ and $T = m^2 \cdot (1/\epsilon^2)$. Let $\ell = \lceil (3/\delta) \log(2/\gamma) \rceil$. Assume that $\delta = O(1)$ and $m^2 \cdot (1/\epsilon) = o(N/n^4)$. Let C be an m-restricted circuit, with inputs of length ℓn. Then the number of permutations $\overline{R}$ in PERM_ℓ that are (γ, T)-good for C is bounded by 2^h, where $h = 3\delta \cdot N \cdot n + 3N \log m + 3N \log(1/\epsilon)$.*

Analysis of the construction of pseudo-random generators from one-way permutations

We recall the classic construction (Blum and Micali [BM84] and Yao [Yao82]) of a pseudo-random generator from a one-way permutation.

The function $G_{\overline{R}}(\overline{x}, r)$ is defined by the following algorithm.
```
Input: R̄ a permutation of {0,1}^ℓn, x̄ ∈ {0,1}^ℓn, r ∈ {0,1}^ℓn.
For i = 0 to m − 1, b_i = b(r, R̄^i(x̄)).
Output b_0 ⊙ b_1 ⊙ ... ⊙ b_{m−1}.
```

The following lemma (whose proof follows closely the classical proof – see for example [Zim04]– and, in addition, analyzes the pattern of queries) establishes the properties of the above function in an information-theoretic context.

Lemma 2.5. *Let C_4 be a circuit. Then there are $2^{m+1} - 4$ circuits $C_{1,1}, \ldots, C_{1,2^{m+1}-4}$ such that*

(1) *If $\overline{R}$ is a permutation with $|\text{Prob}_{\overline{x},r}(G_{\overline{R}}(\overline{x}, r) \in C_4) - \text{Prob}(U_m \in C_4)| > \epsilon$, (i.e., $\overline{R}$ does not hit C_4 ϵ-correctly via G), then there is some circuit $C_{1,i}$ such that for at least a fraction $\frac{\epsilon}{m}$ of $\overline{x}$, $C_{1,i}^{\overline{R}}$ on input $\overline{R}(\overline{x})$ outputs a list of $m^2 \cdot \left(\frac{1}{\epsilon}\right)^2$ strings that contains $\overline{x}$ (i.e., $\overline{R}$ is $(\epsilon/m, m^2/\epsilon^2)$-good for $C_{1,i}$).*
(2) *All the circuits $C_{1,i}$ are $(m-2)$-restricted.*

The extractor

We first build a special type of extractor in which the weakly-random string is the truth-table of a permutation in PERM.

The following parameters will be used throughout this section. Let $\epsilon > 0, \delta > 0$, and $n, m \in \mathbb{N}$ be parameters. Let $N = 2^n$. Let $\ell = \lceil (3/\delta) \log(2m \cdot (1/\epsilon)) \rceil$. We consider the set of permutations PERM_ℓ. We assume that $\delta = O(1)$ and $m^2 \cdot (1/\epsilon) = o(N/n^4)$. Let $G : \text{PERM}_\ell \times (\{0,1\}^{\ell n} \times \{0,1\}^{\ell n}) \to \{0,1\}^m$ be the function defined by the following algorithm (the same as the algorithm for $G_{\overline{R}}$ from the previous section).

```
Parameters: ℓ ∈ N, m ∈ N.
Input: R̄ ∈ PERM_ℓ, (x̄, r) ∈ {0,1}^ℓn × {0,1}^ℓn.
For i = 0 to m − 1, b_i = b(r, R̄^i(x̄)).
Output b_0 ⊙ b_1 ⊙ ... ⊙ b_{m−1}.
```

The following lemma, in view of Lemma 1.2, shows that G is an extractor for the special case of weakly-random strings that are truth-tables of permutations in PERM_ℓ.

Lemma 2.6. *Let C_4 be a test for strings of length m (i.e., $C_4 \subseteq \{0,1\}^m$). Let $\text{GOOD}(C_4) = \{\overline{R} \in \text{PERM}_\ell \mid \overline{R} \text{ does not hit } C_4 \text{ } \epsilon\text{-correctly via } G\}$. Then $\|\text{GOOD}(C_4)\| < 2^{m+h+1}$, where $h = 3\delta Nn + 3N \log m + 3N \log(1/\epsilon)$.*

Proof. Let $C_{1,1}, \ldots, C_{1,2^{m+1}-4}$ be the $2^{m+1} - 4$ circuits implied by Lemma 2.5 to exist (corresponding to the test C_4). Let $\overline{R}$ be in $\text{GOOD}(C_4)$. Then Lemma 2.5 shows that there is a circuit $C_{1,i}$ from the above list having the following property: For at least a fraction $\gamma = \epsilon/m$ of strings $\overline{x} \in \{0,1\}^{\ell n}$, $C_{1,i}^{\overline{R}}$ on input $\overline{R}(\overline{x})$ returns a list having $T = m^2 \cdot (1/\epsilon^2)$ strings, one of which is $\overline{x}$. Thus, $\overline{R}$ is (γ, T)-good for $C_{1,i}$ (recall Definition 2.2). It follows that the set of permutations $\overline{R} \in \text{PERM}_\ell$ that do not hit C_4 ϵ-correctly via G is included in

$\bigcup_1^{2^{m+1}-4}\{\overline{R} \in \text{PERM}_\ell \mid \overline{R} \text{ is } (\gamma,T)\text{-good for } C_{1,i}\}$. Lemma 2.4 shows, that, for each $i \in \{1,\ldots,2^{m+1}-4\}$, $\|\{\overline{R} \in \text{PERM}_\ell \mid \overline{R} \text{ is } (\gamma,T)\text{-good for} C_{1,i}\}\| \leq 2^h$. □

In order to obtain a standard extractor (rather than the special type given by Lemma 2.6), the only thing that remains to be done is to transform a random binary string X into a permutation $R \in \text{CIRC}$, which determines $\overline{R} \in \text{PERM}_\ell$ that is used in the function G given above. Note that a permutation $R \in \text{CIRC}$ is specified by $(R(1), R^2(1), \ldots, R^{N-1}(1))$, which is an arbitrary permutation of the set $\{2,3,\ldots,N\}$. Consequently, we need to generate permutations of the set $\{1,2,\ldots,N-1\}$ (which can be viewed as permutations of $\{2,3,\ldots,N\}$ in the obvious way). We can use the standard procedure that transforms a function mapping $[N-1]$ to $[N-1]$ into a permutation of the same type. To avoid some minor truncation nuisances, we actually use a function $X : [N] \to [N]$.

> Input: $X : [N] \to [N]$.
>   Loop 1: for $i = 1$ to $N-1$, $R(i) = i$.
>   Loop 2: for $i = 1$ to $N-1$, $Y(i) = 1 + (X(i) \mod i)$.
>   Loop 3: for $i = 1$ to $N-1$, swap $R(i)$ with $R(Y(i))$.
> Output: permutation $R : [N-1] \to [N-1]$.

It is easy to see that the number of functions $X : [N] \to [N]$ that map via the above procedure into a given permutation $R : [N-1] \to [N-1]$ is bounded by 2^{2N}.

We can now present the (standard) extractor. We choose the parameters as follows. Fix $n \in \mathbb{N}$ and let $N = 2^n$ and $\overline{N} = n \cdot 2^n$. Let $\lambda \in (0,1)$ be a constant. Let $\alpha > 0, \beta > 0$ be constants such that $\alpha < \lambda/3, \beta < (\lambda - 3\alpha)/4$. Let $\epsilon \geq N^{-\beta}$ and $m \leq N^\alpha$. Take $\delta = (\lambda - 4\beta - 3\alpha)/4$ and $\ell = \lceil (3/\delta)\log(2m \cdot (1/\epsilon)) \rceil$.

Parameters: $n \in \mathbb{N}, \overline{N} \in \mathbb{N}, \lambda > 0, \epsilon > 0, \ell \in \mathbb{N}, m \in \mathbb{N}$, satisfying the above requirements.

Inputs: The weakly-random string $X \in \{0,1\}^{\overline{N}}$, viewed as the truth-table of a function $X : [N] \to [N]$; the seed $y = (\overline{x}, r) \in \{0,1\}^{\ell n} \times \{0,1\}^{\ell n}$.

Step 1. Transform X into a permutation $\overline{R}_X \in \text{PERM}_\ell$. The transformation is performed by the above procedure which yields a permutation $R \in \text{CIRC}$, and, next, $\overline{R}_X$ is the ℓ-direct product of R.

Step 2. Run G on input $\overline{R}_X$, $(\overline{x}, r)$.

We have defined a function $E : \{0,1\}^{\overline{N}} \times \{0,1\}^{2\ell n} \to \{0,1\}^m$. Note that the seed length $2\ell n$ is $O(\log^2 \overline{N})$ and the output length m is $\overline{N}^\alpha$, for an arbitrary $\alpha < \lambda/3$.

Theorem 2.7. *The function E is a $(\lambda \overline{N}, 2\epsilon)$-extractor.*

Proof. Let C_4 be a subset of $\{0,1\}^m$. Taking into account Lemma 1.2, it is enough to show that the number of strings $X \in \{0,1\}^{\overline{N}}$ that do not hit C_4 ϵ-correctly via E is at most $2^{\lambda \overline{N} - \log(1/\epsilon)}$. Let $X \in \{0,1\}^{\overline{N}}$ be a string that does not hit C_4 ϵ-correctly via E. By the definition of E, it follows that $\overline{R}_X$ does not hit C_4 ϵ-correctly via G. By Lemma 2.6, there are at most 2^{m+h+1} permutations $\overline{R} \in \text{PERM}_\ell$ that do not hit C_4 ϵ-correctly via G, where $h = 3\delta Nn + 3N\log m + 3N\log(1/\epsilon)$. Since the number of functions $X : [N] \to [N]$ that map into a given permutation $\overline{R} \in \text{PERM}_\ell$ is at most 2^{2N}, it follows that $\|\{X \in \{0,1\}^{\overline{N}} \mid X \text{ does not hit } C_4 \text{ }\epsilon\text{-correctly}\}\| < 2^{2N} \cdot 2^{m+h+1} < 2^{\lambda \overline{N} - \log(1/\epsilon)}$, where the last inequality follows from the choice of parameters. □

3 A Bitwise Locally-Computable Extractor

We present a bitwise locally-computable extractor: Each bit of the output string can be calculated separately in $O(\log^2 \overline{N})$, where $\overline{N}$ is the length of the weakly-random string. The proof uses the same plan as for the extractor in Section 2, except that the weakly-random string X is viewed as the truth-table of an arbitrary function (not necessarily a permutation) and the "consecutive"' values that are used in the extractor are $\overline{X}(\overline{x}), \overline{X}(\overline{x}+1), \ldots, \overline{X}(\overline{x}+m-1)$ (instead of $\overline{R}(\overline{x}), \overline{R}^2(\overline{x}), \ldots, \overline{R}^{m-1}(\overline{x})$ used in Section 2).

The parameter $n \in \mathbf{N}$ will be considered fixed throughout this section. We denote $N = 2^n$ and $\overline{N} = n \cdot N$. The parameter $m \in \mathbf{N}$ will be specified later (it will be a subunitary power of N). The weakly-random string X has length $\overline{N}$, and is viewed as the truth-table of a function $X : \{0,1\}^n \to \{0,1\}^n$. For some $\ell \in \mathbf{N}$ that will be specified later we define $\overline{X} : \{0,1\}^{\ell n} \to \{0,1\}^{\ell n}$ by $\overline{X}(x_1 \odot \ldots \odot x_\ell) = X(x_1) \odot \ldots \odot X(x_\ell)$, i.e., $\overline{X}$ is the ℓ-direct product of X. We also denote $\overline{x} = x_1 \odot \ldots \odot x_\ell$. The seed of the extractor will be $(\overline{x}, r) \in \{0,1\}^{\ell n} \times \{0,1\}^{\ell n}$. We define $\overline{x}+1 = (x_1+1) \odot \ldots \odot (x_\ell+1)$ (where the addition is done modulo 2^n) and inductively, for any $k \in \mathbf{N}$, $\overline{x}+k+1 = (\overline{x}+k)+1$. The extractor is defined by

$$E(X, (\overline{x}, r)) = b(\overline{X}(\overline{x}), r) \odot b(\overline{X}(\overline{x}+1), r) \odot \ldots \odot b(\overline{X}(\overline{x}+m-1), r). \quad (1)$$

It can be shown that, for any constants $\lambda > 0$ and $\alpha < \lambda/3$, for $m \leq N^\alpha$ and $\epsilon \geq N^{(-(\lambda-3\alpha)/4)}$, for some $\ell = O(n)$, E is a $(\lambda \overline{N}, 2\epsilon)$-extractor. The proof shows that if some $X \in \{0,1\}^{\overline{N}}$ does not hit ϵ-correctly some test $D \subseteq \{0,1\}^m$ via E, then for "many" $x \in \{0,1\}^n$, $X(x)$ can derived from $X(x-m+1), \ldots, X(x-1)$ and from the value of X in a few additional points that do not depend on x. It is then shown that less than $2^{\lambda \overline{N} - \log(1/\epsilon)}$ elements $X \in \{0,1\}^{\overline{N}}$ can have such a property. The conclusion follows from Lemma 1.2.

The construction scheme of the extractor given in Equation (1) allows some flexibility in the choice of parameters. We can obtain a quite simple extractor that has seed length $O(\log(\overline{N}))$, is capable to extract from sources with min-entropy $\lambda \overline{N}$, for arbitrary constant $\lambda > 0$, and has output length $\approx 2^{(1/3)\sqrt{\log(\overline{N})}}$. This extractor has a good seed length, however the output length is much smaller than the min-entropy of the source.

Acknowledgments

I am grateful to Luca Trevisan for his insightful comments on an earlier draft of this work.

References

[BM84]   Manuel Blum and Silvio Micali. How to generate cryptographically strong sequences of pseudo-random bits. *SIAM Journal on Computing*, 13(4):850–864, November 1984.

[DM04]   Stefan Dziembowski and Ueli Maurer. Optimal randomizer efficiency in the bounded-storage model. *Journal of Cryptology*, 17(1):5–26, January 2004. (Conference version appeared in Proc. of STOC'02.).

[GT00]   R. Gennaro and L. Trevisan. Lower bounds on the efficiency of generic cryptographic constructions. In *Proceedings of the 41st IEEE Symposium on Foundations of Computer Science*, 2000.

[HILL99] J. Håstad, R. Impagliazzo, L. Levin, and M. Luby. Construction of a pseudo-random generator from any one-way function. *SIAM Journal on Computing*, 28(4), 1999.

[HR03]   T. Hartman and R. Raz. On the distribution of the number of roots of polynomials and explicit logspace extractors. *Random Structures & Algorithms*, 23(3):235–263, Oct. 2003.

[Imp96]   R. Impagliazzo. Very strong one-way functions and pseudo-random generators exist relative to a random oracle. (manuscript), January 1996.

[IW97]   Russell Impagliazzo and Avi Wigderson. P = BPP if E requires exponential circuits: Derandomizing the XOR lemma. In *Proceedings of the 29th Annual ACM Symposium on the Theory of Computing (STOC '97)*, pages 220–229, New York, May 1997. Association for Computing Machinery.

[Lu04]   C.J. Lu. Encryption against storage-bounded adversaries from on-line strong extractors. *Journal of Cryptology*, 17(1):27–42, January 2004.

[NW94]   N. Nisan and A. Wigderson. Hardness vs. randomness. *Journal of Computer and System Sciences*, 49:149–167, 1994.

[RRV99]   R. Raz, O. Reingold, and S. Vadhan. Extracting all the randomness and reducing the error in trevisan's extractor. In *Proceedings of the 30th ACM Symposium on Theory of Computing*, pages 149–158. ACM Press, May 1999.

[Sha02]   R. Shaltiel. Recent developments in explicit constructions of extractors. *Bulletin EATCS*, 77:67–95, June 2002.

[SU01]   R. Shaltiel and C. Umans. Simple extractors for all min-entropies and a new pseudo-random generator. In *Proceedings of the 42nd IEEE Symposium on Foundations of Computer Science*, 2001.

[Tre99]   L. Trevisan. Constructions of near-optimal extractors using pseudo-random generators. In *Proceedings of the 30th ACM Symposium on Theory of Computing*, pages 141–148. ACM Press, May 1999.

[Tre01]   L. Trevisan. Extractors and pseudorandom generators. *Journal of the ACM*, 48(4):860–879, 2001.

[TSZ01]   A. Ta-Shma and D. Zuckerman. Extractor codes. In *Proceedings of the 33rd ACM Symposium on Theory of Computing*, pages 193–199, 2001.

[TSZS01] A. Ta-Shma, D. Zuckerman, and S. Safra. Extractors from Reed-Muller codes. In *Proceedings of the 42nd IEEE Symposium on Foundations of Computer Science*, 2001.

[Vad04]   S. Vadhan. On constructing locally computable extractors and cryptosystems in the bounded-storage model. *J. of Cryptology*, 17(1):43–77, January 2004.
[Yao82]   A. Yao. Theory and application of trapdoor functions. In *Proceedings of the 23rd IEEE Symposium on Foundations of Computer Science*, pages 80–91, 1982.
[Zim04]   Marius Zimand. *Computational Complexity: A Quantitative Perspective*. North-Holland Mathematics Studies. vol. 196. Elsevier, 2004.
[Zim05]   Marius Zimand. Simple extractors via constructions of cryptographic pseudo-random generators. Technical Report 0501075, Computing Research Repository, January 2005. Available at http://arxiv.org/abs/cs.CC/0501075.

Bounds on the Efficiency of "Black-Box" Commitment Schemes

Omer Horvitz* and Jonathan Katz**

Department of Computer Science,
University of Maryland, College Park, MD 20742
{horvitz, jkatz}@cs.umd.edu

Abstract. Constructions of cryptographic primitives based on general assumptions (e.g., the existence of one-way functions) tend to be less efficient than constructions based on specific (e.g., number-theoretic) assumptions. This has prompted a recent line of research aimed at investigating the best possible efficiency of (black-box) constructions based on general assumptions. Here, we present bounds on the efficiency of statistically-binding commitment schemes constructed using black-box access to one-way permutations; our bounds are tight for the case of *perfectly*-binding schemes. We present the bounds in an extension of the Impagliazzo-Rudich model; that is, we show that any construction beating our bounds would imply the unconditional existence of a one-way function (from which a commitment scheme could be constructed "from scratch"). Our analysis is the first in the area to pertain directly to an information-theoretic component of the security notion.

1 Introduction

A central goal of modern cryptography has been to characterize the minimal assumptions needed to construct cryptographic tools and protocols. For example, we now know that one-way functions are sufficient for constructing pseudorandom generators (PRGs) [2, 18, 8, 9], universal one-way hash functions (UOWHFs) and digital signature schemes [14, 16], private-key encryption schemes [6] and commitment schemes [13]. In each of these cases, one-way functions are also known to be necessary [10, 16], making the characterization exact. While important from a theoretical point of view, the above constructions have had limited practical impact due to their inefficiency. In practice, more efficient constructions based on stronger assumptions tend to be used. Moreover, for all of the examples listed above, no constructions from general assumptions that improve on the efficiency of the original solutions are known.

The tension between general and efficient constructions has motivated a recent line of research, aimed at understanding the minimal assumptions needed for *efficient* constructions of various primitives. Put another way, we ask what is

---

* Supported by U.S. Army Research Office award DAAD19-01-1-0494.
** Supported by NSF CAREER award #0447075.

the best possible efficiency that can be achieved (for a particular primitive of interest) if we limit ourselves to constructions based on the weakest cryptographic assumptions. Initial work by Kim, Simon, and Tetali [12] demonstrated (non-tight) bounds on the efficiency of constructing universal one-way hash functions from one-way permutations. Extending their results, Gennaro, Gertner, Katz, and Trevisan [4] showed that known constructions of UOWHFs based on one-way permutations are in fact *optimal*; they also show efficiency bounds for the case of PRGs, private-key encryption schemes, and digital signatures based on one-way permutations, as well as for the case of public-key encryption schemes based on trapdoor permutations.

It is noted that in all the above bounds, analysis pertains directly to a *computational* aspect of the security notion at hand: for PRGs, it is the indistinguishability of the output from random; for UOWHFs – the computational hardness of finding a collision with a given element for a random member of the family; for encryption – the computational indistinguishability of encryptions of distinct messages; and for signatures – the computational hardness of generating a forgery. For commitment schemes, appealing to computational formulations of both the hiding and binding properties resists similar analysis. Taking either of the properties as *information-theoretic* introduces new challenges. Indeed, finding lower bounds on the efficiency of commitments from one-way permutations is left as an open problem in [4].

Our results. Here, we prove lower bounds on the efficiency of (black-box) constructions of statistically-binding, interactive commitment schemes based on one-way permutations. A commitment scheme for m-bit messages is an interactive protocol between a sender $\mathcal{S}$ and a receiver $\mathcal{R}$, where $\mathcal{S}$ takes as input a message $M \in \{0,1\}^m$. Call the view of $\mathcal{R}$ during an interaction with $\mathcal{S}$ the *commitment*. Informally, a statistically-binding commitment scheme satisfies two properties (defined more formally in Section 2.2): (1) *Hiding*: the distribution over commitments to M_0 is computationally indistinguishable from the distribution over commitments to M_1, for any M_0, M_1 and a possibly malicious receiver; and (2) (statistical) *Binding*: the probability (over coin tosses of the honest receiver $\mathcal{R}$) that there exist different M, M' and coins s, s' for $\mathcal{S}$ such that the corresponding commitments to M, M' are identical is at most ε_b. When $\varepsilon_b = 0$ we say the scheme is *perfectly binding*. A permutation $\pi : \{0,1\}^n \to \{0,1\}^n$ is *one-way with security S* if any circuit of size $\leq S$ inverts π on a fraction $\leq 1/S$ of its inputs.

We show that any black-box construction[1] of a statistically-binding commitment scheme based on one-way permutations must have the sender query its oracle $\Omega((m - \log(1 + 2^m \cdot \varepsilon_b))/\log S)$ times. For perfectly-binding schemes, the bound translates to $\Omega(m/\log S)$; for non perfectly-binding schemes, the bound becomes $\Omega(\log \varepsilon_b^{-1}/\log S)$ when $m > \log \varepsilon_b^{-1}$, and is essentially $\Omega(m/\log S)$ otherwise. We prove our bounds in the model of Gennaro, et al. [4] which, in turn,

---

[1] Using the terminology of [17] (see also [4]), our bounds hold for *weak* black-box constructions and thus rule out stronger types of black-box constructions as well.

extends the model of Impagliazzo and Rudich [11, 15] (see [4] for a detailed comparison). In particular, we show that the existence of a construction beating our efficiency bound would imply the unconditional existence of a one-way function (and hence $P \neq NP$).

For the case of perfectly-binding schemes constructed from one-way permutations, our bound matches the efficiency achieved by the construction of Blum [1] as extended by Goldreich and Levin [8] (see also [5]). For statistically-binding schemes, our bound matches the efficiency achieved by the construction of Naor [13] for single-bit messages (setting $\varepsilon_b = 2^{-S}$). Closing the gap for the case of longer messages remains an interesting open problem.

2 Preliminaries

2.1 One-Way Functions and Permutations

Let A^f denote a circuit A with oracle access to the function f. We say that a function $f : \{0,1\}^n \to \{0,1\}^n$ is (S, ε)-*one-way* if for every circuit A of size $\leq S$ we have

$$\Pr_x[A^f(f(x)) \in f^{-1}(f(x))] \leq \varepsilon.$$

To reduce the number of parameters, we will call a function S-*hard* if it is $(S, 1/S)$-one way.

Let Π_t denote the set of all permutations over $\{0,1\}^t$. Then:

Theorem 1 ([4]). *For sufficiently large t, a random $\pi \in \Pi_t$ is $2^{t/5}$-hard with probability at least $1 - 2^{-2^{t/2}}$.*

Let $a \| b$ denote the concatenation of strings a and b. For $t < n$, let $\Pi_{t,n}$ denote the subset of Π_n such that $\pi \in \Pi_{t,n}$ iff $\pi(a\|b) = \hat{\pi}(a)\|b$ for some $\hat{\pi} \in \Pi_t$. A corollary to Theorem 1 is that if $t = 5 \log S_p$, then for $n > t$, a randomly chosen $\pi \in \Pi_{t,n}$ is S_p-hard with high probability.

Corollary 1 ([4]). *For sufficiently large t and $n > t$, a random $\pi \in \Pi_{t,n}$ is $2^{t/5}$-hard with probability at least $1 - 2^{-2^{t/2}}$.*

2.2 Commitment Schemes

A *commitment scheme* for m-bit messages is defined by a pair of probabilistic, interactive algorithms $(\mathcal{S}, \mathcal{R})$. The inputs to $\mathcal{S}$, the *sender*, are a message $M \in \{0,1\}^m$ and randomness s, while the input to $\mathcal{R}$, the *receiver*, is randomness r. Let $\langle \mathcal{S}(M;s), \mathcal{R}(r) \rangle$ denote the receiver's view of an interaction with the sender on the specified inputs; this view simply consists of the receiver's randomness and the messages it receives from the sender during the interaction (when the receiver makes queries to an oracle, the view additionally includes the answers it receives from the oracle). We call $\langle \mathcal{S}(M;s), \mathcal{R}(r) \rangle$ a *commitment*. A *decommitment* consists of a message and sender randomness. We say that a commitment $C = \langle \mathcal{S}(M; s), \mathcal{R}(r) \rangle$ can be *decommitted* to a message M' if there exists a string s' such that $\langle \mathcal{S}(M'; s'), \mathcal{R}(r) \rangle = C$. For a message M, let

$$\langle \mathcal{S}(M), \mathcal{R} \rangle \stackrel{\text{def}}{=} \left\{ s, r \stackrel{R}{\leftarrow} \{0,1\}^* : \langle \mathcal{S}(M; s), \mathcal{R}(r) \rangle \right\}.$$

We say that two distributions $\mathcal{X}, \mathcal{Y}$ are (S, ε)-*indistinguishable*, and write $\mathcal{X} \stackrel{(S,\varepsilon)}{\approx} \mathcal{Y}$, if for every circuit A of size $\leq S$, we have

$$\left| \Pr_{x \in \mathcal{X}}[A(x) = 0] - \Pr_{x \in \mathcal{Y}}[A(x) = 0] \right| \leq \varepsilon.$$

Definition 1. *Let $(\mathcal{S}, \mathcal{R})$ be a commitment scheme for m-bit messages. We say that $(\mathcal{S}, \mathcal{R})$ is (S_h, ε_h)-hiding if for every circuit $\mathcal{R}^*$ of size $\leq S_h$, for all $M_0, M_1 \in \{0,1\}^m$, we have*

$$\langle \mathcal{S}(M_0), \mathcal{R}^* \rangle \stackrel{(S_h, \varepsilon_h)}{\approx} \langle \mathcal{S}(M_1), \mathcal{R}^* \rangle.$$

We say that $(\mathcal{S}, \mathcal{R})$ is ε_b-binding if

$$\Pr_r \left[\begin{array}{c} \exists M, M' \in \{0,1\}^m, s, s' \text{ s.t.} \\ \langle \mathcal{S}(M; s), \mathcal{R}(r) \rangle = \langle \mathcal{S}(M'; s'), \mathcal{R}(r) \rangle \end{array} \right] \leq \varepsilon_b.$$

We say that $(\mathcal{S}, \mathcal{R})$ is ε_b-binding for an honest sender if for all $M \in \{0,1\}^m$, we have

$$\Pr_{s,r} \left[\begin{array}{c} \exists M' \neq M, s' \text{ s.t.} \\ \langle \mathcal{S}(M; s), \mathcal{R}(r) \rangle = \langle \mathcal{S}(M'; s'), \mathcal{R}(r) \rangle \end{array} \right] \leq \varepsilon_b.$$

If $\varepsilon_b = 0$ we say the scheme is perfectly-binding. *Finally, we say that $(\mathcal{S}, \mathcal{R})$ is $(S_h, \varepsilon_h, \varepsilon_b)$-secure (resp., secure for an honest sender) if $(\mathcal{S}, \mathcal{R})$ is (S_h, ε_h)-hiding and ε_b-binding (resp., binding for an honest sender).*

A construction *of a commitment scheme for m-bit messages (based on one-way permutations) is a pair of oracle algorithms $(\mathcal{S}^{(\cdot)}, \mathcal{R}^{(\cdot)})$ such that, for all $\pi \in \Pi_n$, $(\mathcal{S}^\pi, \mathcal{R}^\pi)$ is a commitment scheme for m-bit messages. We say that $(\mathcal{S}^{(\cdot)}, \mathcal{R}^{(\cdot)})$ is $(S_p, S_h, \varepsilon_h, \varepsilon_b)$-secure (resp., secure for an honest sender) if for every $\pi \in \Pi_n$ that is S_p-hard, $(\mathcal{S}^\pi, \mathcal{R}^\pi)$ is (S_h, ε_h)-hiding and for every $\pi \in \Pi_n$ (even those not S_p-hard), $(\mathcal{S}^\pi, \mathcal{R}^\pi)$ is ε_b-binding (resp., binding for an honest sender).* ◇

We note that, in the terminology of [17], our definitions of security for commitment constructions fall in the category of *weak black-box* constructions. This is evident in that in the definition of the hiding property, the distinguishing algorithm is *not* given oracle access to π. As we prove lower bounds, our results apply to stronger notions of black-box constructions as well, and in particular to *semi-* and *fully black-box* constructions (see [17] for further details).

2.3 Pairwise-Independent Hashing

Let H be a family of functions mapping m-bit strings to m'-bit strings. We assume that the following can be done in time polynomial in m: (1) selecting a function $h \in H$ uniformly at random; (2) given $h \in H$ and $x \in \{0,1\}^m$, evaluating $h(x)$; and (3) given h^*, deciding whether $h^* \in H$ or not. We say that H is a *pairwise-independent hash family* (following [3]) if for any distinct $x_1, x_2 \in \{0,1\}^m$ and any $y_1, y_2 \in \{0,1\}^{m'}$ we have:

$$\Pr_{h \in H}[h(x_1) = y_1 \wedge h(x_2) = y_2] = 2^{-2m'}.$$

3 The Lower Bound

Let $(\mathcal{S}^{(\cdot)}, \mathcal{R}^{(\cdot)})$ be an $(S_p, S_h, \varepsilon_h, \varepsilon_b)$-secure construction of a commitment scheme for m-bit messages (based on one-way permutations). We prove that unless $\mathcal{S}$ queries its oracle at least $\Omega((m - \log(1 + 2^m \cdot \varepsilon_b))/\log S_p)$ times, there exists (constructively) a commitment scheme $(\bar{\mathcal{S}}, \bar{\mathcal{R}})$ secure for an honest sender which *does not* require access to any oracle (i.e., the scheme is secure unconditionally). Strengthening a result of Impagliazzo and Luby [10] (cf. Lemma 1), this implies the unconditional existence of a one-way function.

We describe here the intuition behind our proof, assuming for ease of exposition that $(\mathcal{S}, \mathcal{R})$ is a non-interactive, perfectly-binding scheme. As in [4], our starting point is that a random $\pi \in \Pi_{t,n}$ (for $t = \Theta(\log S_p)$) is S_p-hard with all but negligible probability (cf. Corollary 1). So, consider a scheme in which the sender $\mathcal{S}'$ simulates a random $\pi \in \Pi_{t,n}$ for $\mathcal{S}$, using independent random coins[2] y. Note that decommitment here consists of both s and y.

It is quite straightforward to show that the resulting scheme still satisfies hiding. Binding, however, is another matter. Note first that we need only show binding for an *honest* sender. That is, we may assume that s and y have been chosen honestly (i.e., uniformly at random) by $\mathcal{S}'$. Letting P denote the pairs of t-bit prefixes of oracle queries made by $\mathcal{S}'$ during the computation and t-bit prefixes of corresponding answers, the question then is whether there exist s', y' (with associated query/answer prefixes P') that form a decommitment of C to a different message. We first observe, informally, that having $P' = P$ will not help: this is because the original scheme is binding for any choice of π. In fact, this further implies that for any choice of P' there exists at most one message to which C can be decommitted. Said another way, this means that C can be decommitted to at most

$$2^{2t|P|} = 2^{2tq}$$

different messages (where q is the number of oracle queries made by the original sender). Although this clearly violates binding, it does limit the space of possible

---

[2] This can easily be done by selecting random t-bit answer-prefixes to new t-bit query-prefixes, as needed.

messages *somewhat* as long as $2^{2tq} < m$. We now show how to "bootstrap" this to achieve "full" binding with noticeable probability.

The idea is to modify the scheme as follows: the sender $\bar{\mathcal{S}}$ now chooses a function h uniformly at random from a pairwise-independent hash family mapping m-bit strings to m-bit strings. On message M, $\bar{\mathcal{S}}$ now runs $\mathcal{S}'$ *twice* (simulating a random π as before), first on input M and then on input $h(M)$. The commitment also includes h. Hiding for the derived scheme follows easily, as before. As for binding, denote the two sub-commitments of $\bar{\mathcal{S}}$ by C_1 and C_2. We have said already that for each of these commitments there are sets S_1, S_2, each of size at most $2^{2t|P|} < 2^m$, such that C_1 might potentially be decommitted to any message from S_1 and similarly for C_2, S_2. But now the commitment $C_1 \| C_2 \| h$ can only be decommitted as a different message if there exists an $M' \in S_1$ (with $M' \neq M$) for which $h(M') \in S_2$. The crux of our proof is to show that, with constant probability over choice of h (and for certain values of $|P|$), such an M' does not exist. This immediately implies that binding holds with constant probability.

3.1 A Technical Lemma

We begin by showing that the existence of a commitment scheme secure for honest senders implies the existence of a one-way function. Although the result can be derived from [10], we give a simple and more direct proof here.

Lemma 1. *Let $(\mathcal{S}, \mathcal{R})$ be a commitment scheme for m-bit messages which is $(S_h, \varepsilon_h, \varepsilon_b)$-secure for an honest sender. Let $S_\mathcal{S}, S_\mathcal{R}$ be the sizes of the circuits computing $\mathcal{S}, \mathcal{R}$, respectively. Then for any $\ell \in \mathbb{N}$, there exists an $(S_h - \ell(S_\mathcal{S} + S_\mathcal{R}), \ell\varepsilon_h + 2\varepsilon_b^\ell)$-one-way function.*

Proof. Via a standard hybrid argument, running ℓ independent interactions between $\mathcal{S}$ and $\mathcal{R}$ on the same, fixed message yields an $(S_h, \ell\varepsilon_h, \varepsilon_b^\ell)$-secure commitment scheme for m-bit messages. Let $(\mathcal{S}_\ell, \mathcal{R}_\ell)$ denote this scheme.

Let $f(M, s, r) \stackrel{\text{def}}{=} \langle \mathcal{S}_\ell(M; s), \mathcal{R}_\ell(r) \rangle$. We claim that f is $(S_h - \ell(S_\mathcal{S} + S_\mathcal{R}), \ell\varepsilon_h + 2\varepsilon_b^\ell)$-one-way. Assume the contrary. Then there exists a circuit B of size at most $S_h - \ell(S_\mathcal{S} + S_\mathcal{R})$ such that

$$\text{Succ}_{B,f}^{\text{owf}} \stackrel{\text{def}}{=} \Pr_{M,s,r}[B(f(M,s,r)) \in f^{-1}(f(M,s,r))] > \ell\varepsilon_h + 2\varepsilon_b^\ell.$$

We use B to construct a circuit A that violates the hiding property of $(\mathcal{S}_\ell, \mathcal{R}_\ell)$. On input (M_0, M_1, C), A computes $(M', s', r') \leftarrow B(C)$, and checks whether $f(M', s', r') \stackrel{?}{=} C$ and whether $M' \stackrel{?}{=} M_0$. If both hold, A outputs 0; otherwise, it outputs 1. Note that $|A| = |B| + \ell(S_\mathcal{S} + S_\mathcal{R}) \leq S_h$.

Let $\text{Bad} \stackrel{\text{def}}{=} \{(M, s, r) | \exists M' \neq M, s' : \langle \mathcal{S}_\ell(M; s), \mathcal{R}_\ell(r) \rangle = \langle \mathcal{S}_\ell(M'; s'), \mathcal{R}_\ell(r) \rangle\}$. In what follows, note that if $(M', s', r') \in f^{-1}(f(M, s, r))$ then $r' = r$, as r is included in the receiver's view. We have:

$$\Pr_{\substack{M_0,M_1 \\ C\in\langle\mathcal{S}_\ell(M_0),\mathcal{R}_\ell\rangle}} [A(M_0,M_1,C)=0]$$

$$= \Pr_{\substack{M_0,M_1 \\ s,r}} \left[\begin{array}{c} (M',s',r') \leftarrow B(f(M_0,s,r)): \\ (M',s',r') \in f^{-1}(f(M_0,s,r)) \wedge M'=M_0 \end{array} \right]$$

$$\geq \Pr_{\substack{M_0,M_1 \\ s,r}} \left[\begin{array}{c} (M',s',r') \leftarrow B(f(M_0,s,r)): \\ (M',s',r') \in f^{-1}(f(M_0,s,r)) \wedge (M_0,s,r)\notin \mathsf{Bad} \end{array} \right]$$

$$= \Pr_{\substack{M_0,M_1 \\ s,r}} \left[\begin{array}{c} (M',s',r') \leftarrow B(f(M_0,s,r)): \\ (M',s',r') \in f^{-1}(f(M_0,s,r)) \end{array} \right]$$

$$- \Pr_{\substack{M_0,M_1 \\ s,r}} \left[\begin{array}{c} (M',s',r') \leftarrow B(f(M_0,s,r)): \\ (M',s',r') \in f^{-1}(f(M_0,s,r)) \wedge (M_0,s,r)\in \mathsf{Bad} \end{array} \right]$$

$$\geq \Pr_{\substack{M_0,M_1 \\ s,r}} \left[\begin{array}{c} (M',s',r') \leftarrow B(f(M_0,s,r)): \\ (M',s',r') \in f^{-1}(f(M_0,s,r)) \end{array} \right] - \Pr_{\substack{M_0,M_1 \\ s,r}} [(M_0,s,r)\in \mathsf{Bad}]$$

$$\geq \mathsf{Succ}^{\mathsf{owf}}_{B,f} - \varepsilon_b^\ell \;>\; \ell\varepsilon_h + \varepsilon_b^\ell.$$

Furthermore, we have:

$$\Pr_{\substack{M_0,M_1 \\ C\in\langle\mathcal{S}_\ell(M_1),\mathcal{R}_\ell\rangle}} [A(M_0,M_1,C)=0]$$

$$= \Pr_{\substack{M_0,M_1 \\ s,r}} \left[\begin{array}{c} (M',s',r') \leftarrow B(f(M_1,s,r)): \\ (M',s',r') \in f^{-1}(f(M_1,s,r)) \wedge M'=M_0 \end{array} \right]$$

$$\leq \Pr_{\substack{M_0,M_1 \\ s,r}} \left[\begin{array}{c} (M',s',r') \leftarrow B(f(M_1,s,r)): \\ (M',s',r') \in f^{-1}(f(M_1,s,r)) \wedge (M_1,s,r)\in \mathsf{Bad} \end{array} \right]$$

$$\leq \Pr_{\substack{M_0,M_1 \\ s,r}} [(M_1,s,r)\in \mathsf{Bad}] \leq \varepsilon_b^\ell.$$

Putting everything together, we have:

$$\left| \Pr_{\substack{M_0,M_1 \\ C\in\langle\mathcal{S}_\ell(M_0),\mathcal{R}_\ell\rangle}} [A(M_0,M_1,C)=0] - \Pr_{\substack{M_0,M_1 \\ C\in\langle\mathcal{S}_\ell(M_1),\mathcal{R}_\ell\rangle}} [A(M_0,M_1,C)=0] \right| > \ell\varepsilon_h.$$

But this implies that there exist two messages M_0, M_1 for which A can distinguish $\langle\mathcal{S}_\ell(M_0),\mathcal{R}_\ell\rangle$ from $\langle\mathcal{S}_\ell(M_1),\mathcal{R}_\ell\rangle$ with probability $> \ell\varepsilon_h$, contradicting the hiding property of $(\mathcal{S}_\ell,\mathcal{R}_\ell)$. ∎

3.2 Main Result

Theorem 2. *Let $(\mathcal{S}^{(\cdot)}, \mathcal{R}^{(\cdot)})$ be an $(S_p, S_h, \varepsilon_h, \varepsilon_b)$-secure construction of a commitment scheme for m-bit messages that expects an oracle $\pi \in \Pi_n$. Let $t = 5\log S_p$. If $\mathcal{S}$ makes $q_S \leq (m-2-\log(1+2^{m+1}\cdot\varepsilon_b))/4t$ queries to its oracle, and $\varepsilon_h \leq 1/8 - 2^{-S_p}$, then there exists a commitment scheme for m-bit messages which is $(S_h, 1/4, 1/4)$-secure for an honest sender (without access to any oracle).*

Applying Lemma 1, this implies the existence of a one-way function (without access to any oracle).

Proof. We construct non-interactive commitment scheme $(\bar{\mathcal{S}}, \bar{\mathcal{R}})$ for m-bit messages. The construction makes use of a procedure $\mathcal{SIM}$ that simulates a random permutation in $\Pi_{t,n}$, as follows [4]: $\mathcal{SIM}$ takes as input a list L. On a query $x\|x'$, where $|x| = t$, $\mathcal{SIM}$ checks whether there exists a y such that $(x, y) \in L$. If so, $\mathcal{SIM}$ returns $y\|x'$. Otherwise, it picks uniformly at random a $y \in \{0,1\}^t \setminus \{\hat{y} \mid \exists \hat{x} : (\hat{x}, \hat{y}) \in L\}$, adds (x, y) to L, and returns $y\|x'$. As usual, we let $\mathcal{SIM}(L; y)$ denote an execution of $\mathcal{SIM}$ on input L and randomness y. We let $\mathcal{SIM}_y$ denote $\mathcal{SIM}(\emptyset; y)$.

Let H be a pairwise-independent family of functions from m-bit strings to m-bit strings. Define $\bar{\mathcal{S}}$ as follows. On input a message $M \in \{0,1\}^m$, $\bar{\mathcal{S}}$ chooses uniformly at random $h \in H$ and values $s_1, r_1, y_1, s_2, r_2, y_2$. It then computes $C_1 = \langle \mathcal{S}^{\mathcal{SIM}_{y_1}}(M; s_1), \mathcal{R}^{\mathcal{SIM}_{y_1}}(r_1) \rangle$ and $C_2 = \langle \mathcal{S}^{\mathcal{SIM}_{y_2}}(h(M); s_2), \mathcal{R}^{\mathcal{SIM}_{y_2}}(r_2) \rangle$. The resulting commitment is $C_1\|C_2\|h$.[3] Decommitment consists of all the random coins used by $\bar{\mathcal{S}}$. We claim that $(\bar{\mathcal{S}}, \bar{\mathcal{R}})$ is $(S_h, 1/4, 1/4)$-secure for an honest sender. This follows from the following two lemmata.

Lemma 2. $(\bar{\mathcal{S}}, \bar{\mathcal{R}})$ *is* $(S_h, 1/4)$-*hiding.*

Proof (of lemma). The hiding property of $(\mathcal{S}^{(\cdot)}, \mathcal{R}^{(\cdot)})$ guarantees that for any $\pi \in \Pi_n$ that is S_p-hard, for any circuit B of size $\leq S_h$, and for any $M_0, M_1 \in \{0,1\}^m$, we have

$$\left| \Pr_{C \in \langle \mathcal{S}^\pi(M_0), \mathcal{R}^\pi \rangle}[B(C) = 0] - \Pr_{C \in \langle \mathcal{S}^\pi(M_1), \mathcal{R}^\pi \rangle}[B(C) = 0] \right| \leq \varepsilon_h.$$

A straightforward hybrid argument shows that for any $\pi_1, \pi_2 \in \Pi_n$ that are S_p-hard, for any circuit B of size $\leq S_h$, and for any $M_0, M_1 \in \{0,1\}^m$, we have

$$\left| \Pr_{\substack{h \in H \\ C_1 \in \langle \mathcal{S}^{\pi_1}(M_0), \mathcal{R}^{\pi_1} \rangle \\ C_2 \in \langle \mathcal{S}^{\pi_2}(h(M_0)), \mathcal{R}^{\pi_2} \rangle}}[B(C_1\|C_2\|h) = 0] - \Pr_{\substack{h \in H \\ C_1 \in \langle \mathcal{S}^{\pi_1}(M_1), \mathcal{R}^{\pi_1} \rangle \\ C_2 \in \langle \mathcal{S}^{\pi_2}(h(M_1)), \mathcal{R}^{\pi_2} \rangle}}[B(C_1\|C_2\|h) = 0] \right|$$

$$\leq 2\varepsilon_h.$$

Corollary 1 shows that a random $\pi \in \Pi_{t,n}$ is S_p-hard except with probability at most $2^{-S_p^{5/2}} \leq 2^{-S_p}$. Using the union bound and a simple averaging argument, we get that for any circuit B of size $\leq S_h$ and for any $M_0, M_1 \in \{0,1\}^m$,

---

[3] The permutations simulated by $\mathcal{SIM}$ in the computations of C_1, C_2 may be different. The theorem can be strengthened (improving the bounds on ε_h) by having $\mathcal{SIM}$ provide a consistent simulation for both computations. We forgo this for simplicity.

$$\left| \Pr_{\substack{\pi_1,\pi_2 \in \Pi_{t,n} \\ h \in H \\ C_1 \in \langle \mathcal{S}^{\pi_1}(M_0), \mathcal{R}^{\pi_1}\rangle \\ C_2 \in \langle \mathcal{S}^{\pi_2}(h(M_0)), \mathcal{R}^{\pi_2}\rangle}}[B(C_1\|C_2\|h) = 0] - \Pr_{\substack{\pi_1,\pi_2 \in \Pi_{t,n} \\ h \in H \\ C_1 \in \langle \mathcal{S}^{\pi_1}(M_1), \mathcal{R}^{\pi_1}\rangle \\ C_2 \in \langle \mathcal{S}^{\pi_2}(h(M_1)), \mathcal{R}^{\pi_2}\rangle}}[B(C_1\|C_2\|h) = 0] \right|$$

$$\leq 2\varepsilon_h + 2^{1-S_p}.$$

Since $\mathcal{SIM}$ perfectly simulates a random $\pi \in \Pi_{t,n}$, we have

$$\left| \Pr_{\substack{y_1,y_2 \\ h \in H \\ C_1 \in \langle \mathcal{S}^{\mathcal{SIM}_{y_1}}(M_0), \mathcal{R}^{\mathcal{SIM}_{y_1}}\rangle \\ C_2 \in \langle \mathcal{S}^{\mathcal{SIM}_{y_2}}(h(M_0)), \mathcal{R}^{\mathcal{SIM}_{y_2}}\rangle}}[B(C_1\|C_2\|h) = 0] \right.$$

$$\left. - \Pr_{\substack{y_1,y_2 \\ h \in H \\ C_1 \in \langle \mathcal{S}^{\mathcal{SIM}_{y_1}}(M_1), \mathcal{R}^{\mathcal{SIM}_{y_1}}\rangle \\ C_2 \in \langle \mathcal{S}^{\mathcal{SIM}_{y_2}}(h(M_1)), \mathcal{R}^{\mathcal{SIM}_{y_2}}\rangle}}[B(C_1\|C_2\|h) = 0] \right| \leq 2\varepsilon_h + 2^{1-S_p}.$$

But that precisely means that

$$\left| \Pr_{C \in \langle \bar{\mathcal{S}}(M_0), \mathcal{R}^*\rangle}[B(C) = 0] - \Pr_{C \in \langle \bar{\mathcal{S}}(M_1), \mathcal{R}^*\rangle}[B(C) = 0] \right| \leq 2\varepsilon_h + 2^{1-S_p} \leq 1/4$$

for any $\mathcal{R}^*$ and any B of size $\leq S_h$, where the last inequality uses the assumption that $\varepsilon_h \leq 1/8 - 2^{-S_p}$. The hiding property therefore holds as claimed. □

Lemma 3. $(\bar{\mathcal{S}}, \bar{\mathcal{R}})$ *is 1/4-binding for an honest sender.*

Proof (of lemma). (See also Fig. 1.) Fix an arbitrary $M \in \{0,1\}^m$. Define

$$C(M, s, r, y) \stackrel{\text{def}}{=} \langle \mathcal{S}^{\mathcal{SIM}_y}(M; s), \mathcal{R}^{\mathcal{SIM}_y}(r)\rangle.$$

Since $(\bar{\mathcal{S}}, \bar{\mathcal{R}})$ is non-interactive, we are interested in the following probability:

$$\text{NoBind} \stackrel{\text{def}}{=} \Pr_{\substack{h \in H \\ s_1, r_1, y_1 \\ s_2, r_2, y_2}}\left[\begin{array}{l} \exists M' \neq M, s_1', y_1', s_2', y_2' \text{ s.t.} \\ C(M', s_1', r_1, y_1') = C(M, s_1, r_1, y_1), \\ C(h(M'), s_2', r_2, y_2') = C(h(M), s_2, r_2, y_2) \end{array}\right]$$

(here, we use the fact that r_1, r_2 and h are explicit in the receiver's view).

Let Perm_t^q denote the set of injective functions from t-bit strings to t-bit strings over domains of size q. Let $\text{queries}_\mathcal{S}(M, s, r, y) \in \text{Perm}_t^{q_S}$ denote the t-bit prefixes of the queries/answers made by $\mathcal{S}$ to its oracle during the execution of $C(M, s, r, y)$ (i.e., an oracle query $a\|a'$ by $\mathcal{S}$ is answered with $b\|a'$ in $C(M, s, r, y)$ iff $(a, b) \in \text{queries}_\mathcal{S}(M, s, r, y)$). Letting $q_\mathcal{R}$ denote the number of queries $\mathcal{R}$ makes to its oracle, we define $\text{queries}_\mathcal{R}(M, s, r, y) \in \text{Perm}_t^{q_\mathcal{R}}$ similarly.

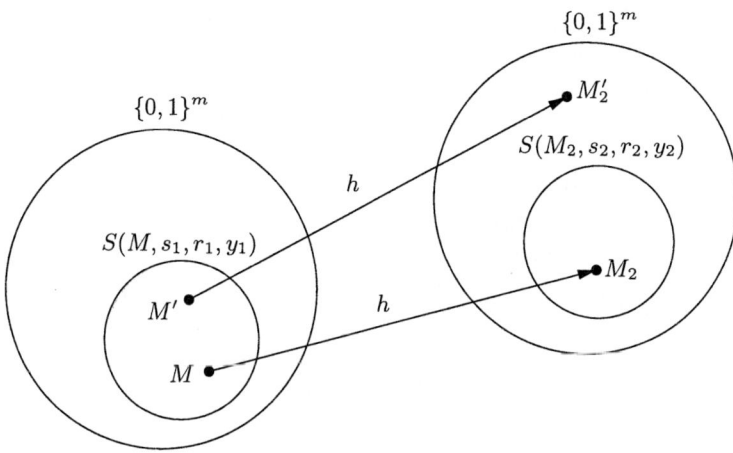

Fig. 1. $\bar{\mathcal{S}}$ uses the given scheme to generate commitments C_1 to M and C_2 to $M_2 = h(M)$ (simulating random permutations for $(\mathcal{S}, \mathcal{R})$, as needed). C_1 (resp., C_2) may be decommitted to any of the messages in the set $S(M, s_1, r_1, y_1)$ (resp., $S(h(M), s_2, r_2, y_2)$). But for a small number of sender oracle-queries, the sizes of these sets are small, and the probability that a random h maps a message from one to the other can be shown to be less than a constant

Say $C(M, s, r, y)$ is *good for* $P \in \text{Perm}_t^{qs}$ if there do not exist distinct values M', M'' along with s', s'', y', y'' such that

- $C(M', s', r, y') = C(M'', s'', r, y'') = C(M, s, r, y)$; and
- $\text{queries}_{\mathcal{S}}(M', s', r, y') = \text{queries}_{\mathcal{S}}(M'', s'', r, y'') = P$.

Say $C(M, s, r, y)$ is *good* if it is good for all $P \in \text{Perm}_t^{qs}$. Now, if $C(M, s, r, y)$ is not good for some P, and $M', M'', s', s'', y', y''$ are the witnesses, then it is also the case that

$$\text{queries}_{\mathcal{R}}(M', s', r, y') = \text{queries}_{\mathcal{R}}(M'', s'', r, y'') = \text{queries}_{\mathcal{R}}(M, s, r, y),$$

as queries/answers of the receiver to its oracle are explicit in the receiver's view. Furthermore, $P \cup \text{queries}_{\mathcal{R}}(M, s, r, y)$ may be extended to a permutation in $\Pi_{t,n}$. Therefore, for any M, s, y and P, we have

$$\Pr_r[C(M, s, r, y) \text{ is not good for } P]$$

$$\leq \Pr_r \left[\begin{array}{c} \exists \pi \in \Pi_{t,n} \text{ extending } P \cup \text{queries}_{\mathcal{R}}(M, s, r, y) \\ \exists M', M'', s', s'' \text{ s.t.} \\ \langle \mathcal{S}^\pi(M'; s'), \mathcal{R}^\pi(r) \rangle = \langle \mathcal{S}^\pi(M''; s''), \mathcal{R}^\pi(r) \rangle \end{array} \right] \leq \varepsilon_b,$$

following the binding property of $(\mathcal{S}, \mathcal{R})$. Applying the union bound over all $\binom{2^t}{qs} \cdot \prod_{i=0}^{qs-1}(2^t - i) \leq 2^{2tqs}$ permutations in Perm_t^{qs}, we obtain, for any M, s, y:

$$\Pr_r[C(M, s, r, y) \text{ is not good}] \leq 2^{2tqs} \varepsilon_b.$$

Assume $C \stackrel{\text{def}}{=} C(M, s, r, y)$ is *good*. Then there are at most 2^{2tqs} possible messages to which C may be decommitted. This is because for any $P \in \text{Perm}_t^{qs}$, if $C(M', s', r, y') = C(M'', s'', r, y'') = C$ and $\text{queries}_S(M', s', r, y') = \text{queries}_S(M'', s'', r, y'') = P$, then it must be the case that $M' = M''$. Let $S(M, s, r, y)$ denote this set.

Viewed this way, we have:

$$\text{NoBind} \leq \Pr_{\substack{h \in H \\ s_1, r_1, y_1 \\ s_2, r_2, y_2}} \left[\begin{array}{c} \exists M' \neq M \text{ s.t.} \\ M' \in S(M, s_1, r_1, y_1) \\ h(M') \in S(h(M), s_2, r_2, y_2) \end{array} \middle| \begin{array}{c} C(M, s_1, r_1, y_1), \\ C(h(M), s_2, r_2, y_2) \\ \text{are good} \end{array} \right] + 2^{2tqs+1} \cdot \varepsilon_b,$$

where the right term represents the probability that either of $C(M, s_1, r_1, y_1)$ or $C(h(M), s_2, r_2, y_2)$ is not good. Continuing with the left term, we have:

LeftTerm

$$= \Pr_{\substack{h \in H \\ s_1, r_1, y_1 \\ s_2, r_2, y_2}} \left[\begin{array}{c} \exists M' \in S(M, s_1, r_1, y_1) \setminus M \\ \exists M'_2 \in S(h(M), s_2, r_2, y_2) \text{ s.t.} \\ h(M') = M'_2 \end{array} \middle| \begin{array}{c} C(M, s_1, r_1, y_1), \\ C(h(M), s_2, r_2, y_2) \\ \text{are good} \end{array} \right]$$

$$= \sum_{M_2} \Pr_{\substack{h \in H \\ s_1, r_1, y_1 \\ s_2, r_2, y_2}} \left[\begin{array}{c} \exists M' \in S(M, s_1, r_1, y_1) \setminus M \\ \exists M'_2 \in S(M_2, s_2, r_2, y_2) \text{ s.t.} \\ h(M') = M'_2, h(M) = M_2 \end{array} \middle| \begin{array}{c} C(M, s_1, r_1, y_1), \\ C(M_2, s_2, r_2, y_2) \\ \text{are good} \end{array} \right]$$

$$\leq \sum_{\substack{M_2 \\ \hat{s}_1, \hat{r}_1, \hat{y}_1 \text{ s.t.} \\ C(M, \hat{s}_1, \hat{r}_1, \hat{y}_1) \text{ good,} \\ \hat{s}_2, \hat{r}_2, \hat{y}_2 \text{ s.t.} \\ C(M_2, \hat{s}_2, \hat{r}_2, \hat{y}_2) \text{ good,} \\ M' \in S(M, \hat{s}_1, \hat{r}_1, \hat{y}_1) \\ M'_2 \in S(M_2, \hat{s}_2, \hat{r}_2, \hat{y}_2)}} \Pr_{\substack{h \in H \\ s_1, r_1, y_1 \\ s_2, r_2, y_2}} \left[\begin{array}{c} s_1 = \hat{s}_1, r_1 = \hat{r}_1, y_1 = \hat{y}_1 \\ s_2 = \hat{s}_2, r_2 = \hat{r}_2, y_2 = \hat{y}_2 \\ h(M') = M'_2, h(M) = M_2 \end{array} \middle| \begin{array}{c} C(M, s_1, r_1, y_1), \\ C(M_2, s_2, r_2, y_2) \\ \text{are good} \end{array} \right]$$

$$= 2^{-2m} \cdot \sum_{\substack{M_2 \\ \hat{s}_1, \hat{r}_1, \hat{y}_1 \text{ s.t.} \\ C(M, \hat{s}_1, \hat{r}_1, \hat{y}_1) \text{ good,} \\ \hat{s}_2, \hat{r}_2, \hat{y}_2 \text{ s.t.} \\ C(M_2, \hat{s}_2, \hat{r}_2, \hat{y}_2) \text{ good,} \\ M' \in S(M, \hat{s}_1, \hat{r}_1, \hat{y}_1) \\ M'_2 \in S(M_2, \hat{s}_2, \hat{r}_2, \hat{y}_2)}} \Pr_{\substack{s_1, r_1, y_1 \\ s_2, r_2, y_2}} \left[\begin{array}{c} s_1 = \hat{s}_1, r_1 = \hat{r}_1, y_1 = \hat{y}_1 \\ s_2 = \hat{s}_2, r_2 = \hat{r}_2, y_2 = \hat{y}_2 \end{array} \middle| \begin{array}{c} C(M, s_1, r_1, y_1), \\ C(M_2, s_2, r_2, y_2) \\ \text{are good} \end{array} \right],$$

using the pairwise independence of H. Since $|S(M, s, r, y)| \leq 2^{2tqs}$, we obtain

$$\text{LeftTerm} \leq 2^{-2m} \cdot 2^m \cdot 2^{4tqs} \leq 2^{4tqs-m}.$$

Putting it together, we have

$$\text{NoBind} \leq 2^{4tqs-m} + 2^{2tqs+1} \cdot \varepsilon_b \leq 2^{4tqs} \cdot (2^{-m} + 2\varepsilon_b) \leq 1/4,$$

where the last inequality is due to the assumption that $q_S \leq (m - 2 - \log(1 + 2^{m+1} \cdot \varepsilon_b))/4t$. □

This completes the proof of the theorem. ∎

Acknowledgments. We thank Virgil Gligor and Chiu-Yuen Koo.

References

1. M. Blum. Coin Flipping by Phone. In *24th IEEE Computer Conference (CompCon)*, pp. 133–137, 1983. (See also *SIGACT News*, vol. 15(1), 1983.)
2. M. Blum and S. Micali. How to Generate Cryptographically Strong Sequences of Pseudo-Random Bits. In *SIAM J. Computing*, vol. 13(4), pp. 850–864, 1984.
3. J. Carter and M. Wegman. Universal Classes of Hash Functions. In *Journal of Computer and System Sciences*, vol. 18, pp. 143–154, 1979.
4. R. Gennaro, Y. Gertner, J. Katz, and L. Trevisan. Bounds on the Efficiency of Generic Cryptographic Constructions. *SIAM J. Computing*, to appear.
5. O. Goldreich. Foundations of Cryptography, vol. 1: Basic Tools. Cambridge University Press, 2001.
6. O. Goldreich, S. Goldwasser, and S. Micali. On the Cryptographic Applications of Random Functions. In *Advances in Cryptology — CRYPTO 84*, LNCS vol. 263, Springer-Verlag, pp. 276–288, 1985.
7. O. Goldreich, S. Goldwasser, and S. Micali. How to Construct Random Functions. In *J. ACM*, vol. 33(4), pp. 792–807, 1986.
8. O. Goldreich and L. Levin. Hard-Core Predicates for any One-Way Function. In *21st ACM Symposium on Theory of Computing (STOC)*, ACM, pp. 25–32, 1989.
9. J. Håstad, R. Impagliazzo, L. Levin, and M. Luby. A Pseudorandom Generator From any One-Way Function. In *SIAM J. Computing*, vol. 28(4), pp. 1364–1396, 1999.
10. R. Impagliazzo and S. Luby. One-Way Functions are Essential for Complexity-Based Cryptography. In *30th IEEE Symposium on Foundations of Computer Science (FOCS)*, IEEE, pp. 230–235, 1989.
11. R. Impagliazzo and S. Rudich. Limits on the Provable Consequences of One-Way Permutations. In *21st ACM Symposium on Theory of Computing (STOC)*, ACM, pp. 44–61,1989.
12. J.H. Kim, D.R. Simon, and P. Tetali. Limits on the Efficiency of One-Way Permutation-Based Hash Functions. In *40th IEEE Symposium on Foundations of Computer Science (FOCS)*, IEEE, pp. 535–542, 1999.
13. M. Naor. Bit Commitment Using Pseudorandomness. In *J. Cryptology*, vol. 4(2), pp. 151–158, 1991.
14. M. Naor and M. Yung. Universal One-Way Hash Functions and their Cryptographic Applications. In *21st ACM Symposium on Theory of Computing (STOC)*, ACM, pp. 33–43, 1989.
15. S. Rudich. Limits on the Provable Consequences of One-Way Functions. Ph.D. thesis, University of California at Berkeley, 1988.
16. J. Rompel. One-Way Functions are Necessary and Sufficient for Secure Signatures. In *22nd ACM Symposium on Theory of Computing (STOC)*, ACM, pp. 387–394, 1990.
17. O. Reingold, L. Trevisan, and S. Vadhan. Notions of Reducibility Between Cryptographic Primitives. In *1st Theory of Cryptography Conference*, LNCS vol. 2951, Springer-Verlag, pp. 1–20, 2004.
18. A.C.-C. Yao. Theory and Application of Trapdoor Functions. In *23rd IEEE Symposium on Foundations of Computer Science (FOCS)*, IEEE, pp. 80–91, 1982.
19. A.C.-C. Yao. How to Generate and Exchange secrets. In *27th IEEE Symposium on Foundations of Computer Science (FOCS)*, IEEE, pp. 162–167, 1986.

On Round-Efficient Argument Systems

Hoeteck Wee[*]

Computer Science Division, UC Berkeley
hoeteck@cs.berkeley.edu

Abstract. We consider the problem of constructing round-efficient public-coin argument systems, that is, interactive proof systems that are only computationally sound with a constant number of rounds. We focus on argument systems for $\mathsf{NTime}(T(n))$ where either the communication complexity or the verifier's running time is subpolynomial in $T(n)$, such as Kilian's argument system for NP [Kil92] and universal arguments [BG02, Mic00]. We begin with the observation that under standard complexity assumptions, such argument systems require at least 2 rounds. Next, we relate the existence of non-trivial 2-round argument systems to that of hard-on-average search problems in NP and that of efficient public-coin zero-knowledge arguments for NP. Finally, we show that the Fiat-Shamir paradigm [FS86] and Babai-Moran round reduction [BM88] fails to preserve computational soundness for some 3-round and 4-round argument systems.

1 Introduction

1.1 Background and Motivation

Argument systems are like interactive proof systems, except we only require computational soundness, namely that it is *computationally infeasible* (and not impossible) for a prover to convince the verifier to accept inputs not in the language. The relaxation in the soundness requirement was used to obtain protocols for NP that are perfect zero-knowledge [BCC88], or constant-round with low communication complexity [Kil92], and in both cases, seems to also be necessary [For89, GH98].

In this paper, we focus on the study of round-efficient argument systems for $\mathsf{NTime}(T(n))$ that do not necessarily satisfy any notion of secrecy, such as witness indistinguishability (WI), or zero-knowledge (although we do indulge in the occasional digression). We will however require that either the communication complexity or the verifier's running time be subpolynomial in $T(n)$ which is necessary in some applications, and to rule out the trivial one-round proof system. Argument systems of the latter type with bounded verifier's running time are a crucial component in the use of non-black-box techniques in cryptography [CGH98, Bar01, Bar04, GK03].

---

[*] Work supported by US-Israel BSF Grant 2002246.

The study of round-efficient argument systems was initiated by Kilian [Kil92], who constructed a 4-round public-coin argument system for NP with polylogarithmic communication complexity based on a probabilistically checkable proof (PCP) system for NP. Micali [Mic00] introduced *CS Proofs*, an argument system for NEXP satisfying a relatively efficient prover condition and wherein the verifier runs in polynomial time (much less than the time needed to verify an NEXP relation). In addition, he provided a non-interactive construction in the random oracle model, which is essentially derived from scaling up and then applying the Fiat-Shamir transformation to Kilian's argument system. Barak and Goldreich [BG02] adapted Kilian's construction to obtain *universal arguments (of knowledge)*, which is a single argument system for any language in NP, and in addition, satisfies a weak proof-of-knowledge property. We stress that in a universal argument, the communication complexity and the verifier's running time is bounded by an a-priori fixed polynomial in the input length, whereas the length of the witness may be any arbitrary polynomial in the length of the input. Both of the constructions in [Kil92] and in [BG02] rely on the existence of collision-resistant function ensembles.

In this work, we initiate a systematic study of round-efficient argument systems.

- What is the minimal round complexity of argument systems with bounded communication complexity or verifier's running time?
- What are the minimal assumptions and cryptographic primitives needed for the existence of such argument systems? Are collision-resistant function ensembles really necessary? What kind of security parameters do we require from these primitives (possibly as a function of communication complexity)?
- How useful is improving the round efficiency of argument systems for the construction of round-efficient cryptographic protocols?
- Is there an efficient function ensemble with which we could securely realize the Fiat-Shamir transformation for the 4-round argument systems in [Kil92] and [BG02] (as conjectured by Micali in [Mic00])? More generally, is there some generic round reduction technique that preserves computational soundness?

We provide partial answers for all of these questions in this paper.

1.2 Our Results

We begin with the observations (possibly known in "folklore") that under standard complexity assumptions, the argument systems we are interested in require at least 2 rounds, and anything provable with such an argument system can be proven in 4 rounds. Refer to Sec 3 for the precise statements.

Necessity of Hardness Assumptions. We show that under standard complexity assumptions, the existence of 2-round argument systems for NP with subpolynomial communication complexity implies the existence of hard-on-average search problems in NP, that is, there is samplable distribution over

CSAT instances (circuit satisfiability, namely given a circuit, decide whether the circuit has a satisfying assignment) with the property that most instances (say a constant fraction) are satisfiable, but any nonuniform polynomial-time algorithm on input a random instance from the distribution succeeds in finding a satisfying assignment for that instance with only negligible probability. Note that the existence of hard-on-average search problems in NP is possibly weaker than that of one-way functions and collision-resistant function ensembles.

Zero-Knowledge and 2-Round Argument Systems. We note that the existence of a 2-round public-coin universal argument of knowledge secure against subexponential-sized circuits yields a 4-round public-coin zero-knowledge argument for NP with negligible soundness error; this follows readily from the work of Barak et al. [Bar01, BLV04]. Such an argument system is almost round-optimal, as there is no 2-round zero-knowledge argument system for languages outside of BPP [GO94]. We also relate the existence of non-interactive zero-knowledge arguments to that 2-round witness-indistinguishable arguments for NP where the length of the common reference string, messages and proofs are subpolynomial in the input length. This follows readily from a similar characterization in [DN00].

Insecurity of the Fiat-Shamir Transformation. We observe that the constructions of Goldwasser and Kalai [GK03] demonstrating the insecurity of the Fiat-Shamir transformation as applied to identification schemes also yield a 4-round argument system such that the instantiation of the Fiat-Shamir transformation with any efficiently computable function results in a 2-round protocol that is no longer computationally sound. Note that Barak's zero-knowledge argument system [Bar01] already yields a 6-round argument system for which the Fiat-Shamir transformation is insecure [DNRS03]. We also prove that there exists a 4-round universal argument of knowledge for which the Fiat-Shamir transformation fails to preserve the weak proof-of-knowledge property.

Insecurity of Babai-Moran Round Reduction. Babai and Moran [BM88] used a round reduction procedure to prove that any language having a constant-round public-coin interactive proof system also has a 2-round public-coin proof system. In particular, the round reduction procedure preserves soundness of proof systems. Here, we construct 3-round and 4-round argument systems for which the round reduction procedure fails to preserve computational soundness.

A Note on Presentation: We state our results for argument systems with either bounded communication complexity or bounded verifier's running time, depending on which of the two leads to a cleaner statement. In most cases, an analogous statement can be deduced for the other set-up. Note that a subpolynomial bound on verifier's running time must necessarily imply a subpolynomial bound on the communication complexity.

1.3 Additional Related Work

Dwork et al. [DLN+04] investigated the possibility of constructing 2-round argument systems for NP with poly-logarithmic communication complexity based on a suggestion of Aiello, Bhatt, Ostrovsky and Rajagopalan, namely, to compose a PCP system for NP with computational private information retrieval scheme; their results are mostly negative. Goldreich and Håstad [GH98] proved that NP does not have constant-round public-coin proof systems with subpolynomial communication complexity, unless NP has probabilistic subexponential time algorithms. Barak et al. [BLV04] proved that the Fiat-Shamir transformation is in fact secure for proof systems under a non-standard but very plausible and concrete assumption.

2 Definitions and Setup

Due to space limitations, we refer the reader to [Gol01] to definitions of interactive protocols, zero-knowledge and witness-indistinguishability.

2.1 Interactive Proofs and Argument Systems

For a relation $R \subseteq \{0,1\}^* \times \{0,1\}^*$, the *language associated with R* is $L_R = \{x : \exists y \, (x,y) \in R\}$.

Definition 1 (interactive proof system). *An interactive protocol (P,V) is an* interactive proof system *for a language L if there is a relation R such that $L = L_R$, and functions $c, s : \mathbb{N} \to [0,1]$ such that $1 - c(n) > s(n) + 1/poly(n)$ and the following holds:*

- *(efficiency): the length of all the messages are polynomially-bounded, and V is computable in probabilistic polynomial time.*
- *(completeness): If $(x,w) \in R$, then V accepts in $(P(w), V)(x)$ with probability at least $1 - c(|x|)$,*
- *(soundness): If $x \notin L$, then for every P^*, V accepts in $(P^*, V)(x)$ with probability at most $s(|x|)$.*

We call $c(\cdot)$ the *completeness error* and $s(\cdot)$ the *soundness error*. We say that (P,V) has *negligible error* if both c and s are negligible. We say that it has *perfect completeness* if $c = 0$. P is an *efficient prover* if $P(w)$ is computable by a probabilistic polynomial-time algorithm when $w \in R_x$. The *communication complexity* of the proof system is the total length of all the messages exchanged by both parties. For a public-coin protocol (P,V), view$(V(x))$ is the set of accepting transcripts on common input x. We also use $\mathsf{AM}_{c,s}(m(n))$ to denote constant-round public-coin interactive proof systems with completeness error c, soundness s and communication complexity bounded by $m(n)$.

Definition 2 (argument system). *An* argument system *(P,V) is defined in the same way as an interactive proof system, with the following modification:*

- The soundness condition is replaced with computational soundness: For every nonuniform PPT P^* and for all sufficiently long $x \notin L$, the verifier V accepts in $(P^*, V)(x)$ with probability at most $s(|x|)$.

2.2 Universal Arguments

We begin with the universal language L_U: the tuple (M, x, t) (where t is specified in binary) is in L_U is M is a non-deterministic Turing machine that accepts x within t steps. We use R_U to denote the associated relation.

Definition 3 (universal argument). *A universal argument for* $\mathsf{NTime}(T(n))$ *is an argument system* (P, V) *for* $L_U \cap \mathsf{NTime}(T(n))$ *that satisfies the following properties:*

- *(completeness by a relatively-efficient prover) For every* $((M, x, t), w) \in R_u$ *with* $(M, x, t) \in \mathsf{NTime}(T(n))$,

$$\Pr[V \text{ accepts } (P(w), V)(M, x, t)] = 1$$

  Furthermore, there exists a polynomial p such that the total time spent by $P(w)$, on common input (M, x, t), is at most $p(T_M(x, w)) \leq p(t)$.
- *(computational soundness) For every nonuniform PPT P^*, there exists a negligible function $\epsilon(n)$ such that for every n and every $(M, x, t) \in \{0, 1\}^n \setminus L_U$, the verifier V accepts in $(P^*, V)(M, x, t)$ with probability at most $\epsilon(n)$.*

In addition, we call (P, V) a *universal argument of knowledge* if it satisfies the weak proof-of-knowledge property [BG02]. Informally, this means that there is an efficient oracle machine (the knowledge extractor) that given oracle access to a cheating prover that convinces the verifier with inverse polynomial probability, outputs an implicit description of a witness. Both the running time and the success probability of the knowledge extractor are allowed to depend on the success probability of the cheating verifier.

Theorem 1 ([BG02]). *The existence of (standard) collision-resistant function ensembles implies the existence of a 4-round public-coin universal argument of knowledge $(P_{\mathsf{ua}}, V_{\mathsf{ua}})$ for $\mathsf{NTime}(n^{\log n})$. In addition, if the collision-resistant function ensemble is secure against circuits of size 2^{n^ϵ} for some $\epsilon > 0$, then $(P_{\mathsf{ua}}, V_{\mathsf{ua}})$ is a universal argument of knowledge against circuits of size $2^{O(n^\epsilon)}$.*

3 Simple Bounds on Round Complexity

The results in this section are probably known in "folklore". As pointed out in [BP04], non-interactive (one-round) arguments are equivalent to non-interactive (one-round) proof systems, since if there exists a prover message that can convince the verifier of a false statement, the non-uniform prover that has this message "hard-wired into it". This essentially rules out non-interactive argument systems for NP with subpolynomial communication complexity.

Proposition 1. *Unless* NP $\subseteq$ BPTime($2^{n^{o(1)}}$), *non-interactive argument systems with subpolynomial communication complexity for* NP *do not exist.*

In the context of efficient-prover argument systems, we have a collapse to 4 rounds (as pointed out to us by Salil Vadhan).

Proposition 2. *Suppose there exists collision-resistant function ensembles secure against 2^{n^ϵ}-sized circuits for some $\epsilon > 0$ and a language in* E *with $2^{\Omega(n)}$ circuit complexity. Then, any language L with an efficient-prover argument system has a 4-round, public-coin, efficient-prover argument system with subpolynomial (in fact, poly-logarithmic) communication complexity.*

This follows from the observation in [BLV04] that any language with an efficient-prover argument system is contained in MA, which collapses to NP under the given derandomization assumption. The proposition then follows from Kilian's protocol [Kil92].

4 Necessity of Hardness Assumptions

We present hardness assumptions that are necessary for 2-round argument systems for NP with subpolynomial communication complexity. Under complexity assumptions, such a protocol cannot be a proof system [GH98]. Hence, there exists infinitely many NO instances that are merely "computationally sound", from which we may construct hard-on-average search problems in NP.

Note that we may assume the 2-round argument system has negligible soundness error, which can be achieved with $\omega(\log n)$ parallel repetitions [BIN97]. Parallel repetition blows up the communication complexity by a $\omega(\log n)$ multiplicative factor, but preserves prover's complexity, perfect completeness and public-coin property.

Lemma 1. *Suppose a promise problem $\Pi = (\Pi_Y, \Pi_N)$ has a 2-round public-coin argument system (P, V) with communication complexity $m(n)$, perfect completeness and negligible soundness error. Then, there exists a subset $I \subset \Pi_N$ such that:*

- *Ignoring inputs in I, Π has a $\mathsf{AM}_{1,1/2}(m(n))$ proof system. Formally, $(\Pi_Y, \Pi_N \setminus I) \in \mathsf{AM}_{1,1/2}(m(n))$.*
- *When $x \in I$, the predicate $V(x, \cdot, \cdot)$ induces a hard-on-average search instances in* NP. *That is, for every $x \in I$:*

$$\Pr_r[\exists\, y : V(x, r, y) = 1] \geq 1/2,$$

but for every n, every $x \in I \cap \{0,1\}^n$ and every nonuniform PPT A, there exists a negligible function $\epsilon(n)$ such that ,

$$\Pr_r[V(x, r, A(r)) = 1] < \epsilon(n)$$

Remark 1. Note that we may boost the probability of generating a satisfying assignment for the hard-on-average search instance to $1 - 1/\operatorname{poly}(n)$ while maintaining the same hardness parameters by taking the OR of $O(\log n)$ independent copies of $V(x, \cdot, \cdot)$.

Theorem 2. *Suppose* NP *has a 2-round public-coin argument system* (P, V) *with communication complexity* $n^{o(1)}$, *perfect completeness and negligible soundness error. Then, at least one of the following is true:*

- NP $\subseteq$ AM$_{1,1/2}(n^{o(1)})$
- *There exists an infinite set I such that for all $x \in I$, the predicate $V(x, \cdot, \cdot)$ induces a hard-on-average search instance in* NP *(as formalized in Lemma 1). This yields an auxiliary-input samplable distribution over search instances in* NP *that is infinitely-often hard on average.*

Remark 2. The first statement is unlikely to be true as it would imply that NP $\subseteq$ BPTime($2^{n^{o(1)}}$) [GH98]. On the other hand, the latter is possibly weaker than the existence of (auxiliary input, i.o.) one-way functions. However, it does imply that there is no probabilistic polynomial-time algorithm for the circuit satisfiability problem where the number of variables is bounded by $n^{o(1)}$.

Remark 3. Salil Vadhan pointed out that if there exists a hard-on-average decision problem in NP where the instances and witnesses have length bounded by $m(n)$, then every language has a 2-round argument system with communication complexity $m(n)$. However, the argument system does not satisfy the efficient prover constraint, though the constraint is (trivially) satisfied if we consider the empty language. This shows that the conclusion in Theorem 2 is essentially the strongest we can hope for without making additional assumptions about the argument system, for instance, that it has an efficient prover, that it is WI, or that it is an argument of knowledge.

5 Zero-Knowledge and 2-Round Argument Systems

Barak et. al [BLV04] constructed a 2-round argument for NP that is zero-knowledge against cheating verifiers of bounded non-uniformity assuming the existence of a 2-round universal argument secure against 2^{n^ϵ}-sized circuits. We observe that if we strengthen the soundness requirement on the universal argument to an argument of knowledge, it follows readily from [Bar01, BLV04] that there exists a 4-round zero-knowledge argument for NP. The idea is to convert the universal argument of knowledge into a WI universal argument of knowledge (with a subexponential-time knowledge extractor) without any overhead in the number of rounds. To accomplish this, we encrypt the messages of the universal argument using a weak commitment scheme and prove correctness using a WI proof for NP [DN00].

Theorem 3 ([Bar01, BLV04]). *Suppose there exist 2-round public-coin universal argument of knowledge for $\mathsf{NTime}(f(n))$ for some super-polynomial $f : \mathbb{N} \to \mathbb{N}$, enhanced trapdoor permutations and collision-resistant function ensembles secure against 2^{n^ϵ}-sized circuits for some constant $\epsilon > 0$. Then, there exists a 4-round public-coin (auxiliary-input) zero-knowledge argument system for NP, with perfect completeness, negligible soundness error, an efficient prover and a simulator that runs in strict polynomial time.*

Another open problem is whether there exists non-interactive zero-knowledge (NIZK) arguments or 2-round WI arguments for NP with subpolynomial communication complexity and randomness [FLS99, KP98, DLN+04]. We do not know how to construct either primitive starting from an argument system for NP with subpolynomial communication complexity, but it follows from the characterization of zaps (a 2-round public-coin WI proof system for NP) in [DN00] that they are almost equivalent:

Theorem 4 ([FLS99, DN00]). *Suppose there exist one-way functions secure against 2^{n^ϵ}-sized circuits for some constant $\epsilon > 0$. Then, the following statements are equivalent:*

– *There exists a 2-round public-coin efficient-prover honest-verifier WI argument for NP with subpolynomial communication complexity.*
– *There exists an efficient-prover NIZK argument for NP where the length of the common reference string and the proof are subpolynomial in the length of the input.*

Theorem 4 is weaker than the characterization of zaps in [DN00] in that we can only deduce the existence the existence of honest-verifier WI (but not cheating-verifier WI) arguments for NP from NIZK. This is because the construction of zaps from NIZK protocols requires that the underlying NIZK protocol be a proof system in order to preserve soundness. On the other hand, we observe that honest-verifier WI is sufficient for the construction of a NIZK argument for NP.

6 Insecurity of the Fiat-Shamir Transformation

Goldwasser and Kalai [GK03] proved the existence of a (secure) 3-round public-coin identification scheme for which any instantiation of the Fiat-Shamir transformation with an efficiently computable function ensemble yields an insecure signature scheme. As both the identification scheme and the signature scheme are defined in the public-key model, there is a fairly natural interpretation of the construction as obtaining a 2-round argument system from a 4-round argument system via the Fiat-Shamir transformation. The main (albeit minor) technical difference is in handling auxiliary inputs inherent to argument systems, as the set-up in [GK03] is inherently uniform (there, the variable is the security parameter, and messages to be signed are thought of as having constant size[1]). We also feel that viewing the constructions of [GK03] in the context of argument systems yields a clearer and simpler presentation of their constructions and results. The following result has been independently observed by the authors of [GK03] (but was not explicitly mentioned in [GK03]):

Theorem 5 ([GK03]). *Suppose there exists (standard) collision-resistant function ensembles. Then, there exists a 4-round public-coin argument system with negligible soundness error, but for which the instantiation of the Fiat-Shamir transformation with any efficiently function ensemble yields a 2-round protocol that is not computationally sound (that is, it has a polynomial-sized cheating prover that succeeds with non-negligible probability).*

Remark 4. The cheating prover in the proof of Theorem 5 succeeds with only a non-negligible probability. It is therefore conceivable while the Fiat-Shamir paradigm does not in general preserve soundness of 4-round argument systems, the Fiat-Shamir paradigm along with parallel repetition does preserve soundness of 4-round argument systems (since parallel repetition does reduce the soundness error for 2-round argument systems to a negligible quantity [BIN97]).

We also observe that the Fiat-Shamir transformation fails to preserve the weak proof-of-knowledge property. The proof goes via a case analysis similar to that in [GK03] (except a lot simpler). Suppose the statement holds for (P_{ua}, V_{ua}); then we are done. Otherwise, we have a 2-round public-coin universal argument of knowledge which combined with Barak's non-uniform generation protocol [Bar01] yields the desired argument system.

Theorem 6. *Suppose there exists (standard) collision-resistant function ensembles. Then, there exists a 4-round public-coin universal argument of knowledge, but for which the instantiation of the Fiat-Shamir transformation with any efficiently function ensemble yields a 2-round protocol that does not satisfy the weak proof-of-knowledge property.*

7 Insecurity of Babai-Moran Round Reduction

We start by describing Babai-Moran round reduction. For a public-coin proof system $\Pi = (P, V)$ of at most 4 rounds, this procedure has a simple description and comprises two steps, for some parameter $k = poly(n)$. First, the residual protocol after the prover's first message is repeated k times in parallel and the new verifier accepts if all k repetitions are accepting. Next, second, the order of the prover's first message and the verifier's next message are reversed. We denote the new protocol by $\Pi^{rr(k)}$. For protocols with 3 or 4 rounds, the resulting protocol has 2 rounds.

---

[1] Alternatively, we may consider the forger as forging a family of uniformly computable messages of length polynomial in the security parameter, infinitely often.

Intuitively, Babai-Moran round reduction fails to preserve computational soundness for the following reasons:

- Parallel repetition fails to reduce soundness error at an exponential rate beyond $1/\operatorname{poly}(n)$ if we require a black-box proof of security [BIN97].
- A cheating prover can gain significant advantage upon round-switching, wherein the verifier reveals his coin tosses before the prover sends his next message.

We exploit the former reasoning in our construction of the 3-round argument system, as the latter does not seem to apply in this case (made precise in Prop 3) as the first message of a 3-round argument system is "unconditionally sound". For the 4-round argument system, we exploit the latter reasoning in an essential manner so as to obtain a result that holds even with a non-black-box proof of security.

Theorem 7 (Babai-Moran round reduction).

(i) Suppose there exists collision-resistant function ensembles secure against $n^{\log n}$-sized circuits. Then, there exists a 4-round public-coin argument system with negligible soundness error for which Babai-Moran round reduction yields a 2-round argument system that is not computationally sound.

(ii) There exists a 3-round (relativized) public-coin argument system with negligible soundness error for which Babai-Moran round reduction yields a 2-round argument system that is not computationally sound if limited to a black-box proof of security.

In both constructions, the cheating prover succeeds with probability $1 - \operatorname{neg}(n)$. This means that even upon applying parallel repetition to the resulting 2-round argument systems, we would not obtain a computationally sound protocol.

Both constructions are for the empty language $L_\emptyset$. The 4-round protocol, specified in Fig 1, is a straight-forward simplification of the argument system in [Kil92]. For 3-round argument systems, we only rule out the case with a black-box proof of security. In this setting, it suffices to construct a relativized protocol, wherein all parties (provers, cheating provers, verifier) have oracle access to a permutation π, as shown in Fig 2. It helps to think of π as a one-way permutation, although we will require a stronger property that we only know how to prove in a relativized setting:

Lemma 2 ([GT00]). *For all sufficiently large n, there exists a permutation π on $\{0,1\}^n$ such that for all oracle circuits A of size $n^{\log n}$,*

$$\Pr[\sigma \leftarrow \{0,1\}^n;\ A^{\pi, I_\sigma}(\sigma) - y;\ \pi(y) - \sigma] < \frac{1}{n^{\log n}}$$

where I_σ is an oracle that on input $\sigma' \neq \sigma$ returns $\pi^{-1}(\sigma')$, and $\perp$ otherwise.

We note that overcoming the limitation to black-box proof of security for 3-round argument systems will require resolving a well-known open problem:

Common input: 1^n

1. (V1) verifier sends a random h from $\mathcal{H}$ (collision-resistant function ensemble).
2. (P1) prover sends a Merkle-tree commitment to B, where B is an array of $n^{\log n}$ blocks of 0^n.
3. (V2) verifier sends β at random from 1 to $n^{\log n}$ and γ at random from $\{0,1\}^n$.
4. (P2) prover decommits to $B[\beta]$.

Verification: verifier accepts if $B[\beta]$ decommits to γ.

---

1. (V1) verifier sends a random h from $\mathcal{H}$, and $\beta_1, \ldots, \beta_k$ at random from 1 to $n^{\log n}$ and $\gamma_1, \ldots, \gamma_k$ at random from $\{0,1\}^n$.
2. (P1) prover sends a Merkle-tree commitment to B, which is an array of $n^{\log n}$ blocks of 0^n, and decommits to $B[\beta_1], \ldots, B[\beta_k]$.

Verification: verifier accepts if $B[\beta_i]$ decommits to γ_i for all $i = 1, \ldots, k$.

Fig. 1. 4-round protocol Π_1 and 2-round protocol $\Pi_1^{rr(k)}$ for the empty language $L_\emptyset$

Common input: 1^n, oracle access to π (a permutation on $\{0,1\}^n$)

1. (P1) prover sends $z \in \{0,1\}^n$.
2. (V1) verifier sends a random σ in $\{0,1\}^n$.
3. (P2) prover sends $y \in \{0,1\}^n$.

Verification: verifier accepts iff $\pi(y) = z \oplus \sigma$.

---

1. (V1) verifier sends random $\sigma_1, \ldots, \sigma_k$ in $\{0,1\}^n$.
2. (P1) prover sends $z, y_1, \ldots, y_k \in \{0,1\}^n$.

Verification: verifier accepts iff $\pi(y_i) = z \oplus \sigma_i$, for all $i = 1, \ldots, k$.

Fig. 2. 3-round relativized protocol Π_2 and 2-round protocol $\Pi_1^{rr(k)}$ for $L_\emptyset$

Proposition 3. *Suppose parallel repetition on 2-round argument systems can reduce the computational soundness error exponentially fast to $2^{-\text{poly}(n)}$, then Babai-Moran round reduction yields a collapse of 3-round argument systems to 2-round argument systems.*

8 Conclusion

We hope that the collection of observations, connections and results presented in this paper (one that is perhaps better regarded as a survey) clarifies our understanding of round-efficient argument systems and motivates further work in this area, and perhaps a resolution of the main open problem – determining the exact round complexity of non-trivial argument systems.

Acknowledgments

I am very grateful to Yael Tauman Kalai, Luca Trevisan and Salil Vadhan for their encouragement and insightful discussions on the subject; Salil also gave me very valuable feedback on earlier versions of this paper. I thank Kobbi Nissim for bringing [DLN+04] to my attention, and the anonymous referees for helpful suggestions on the write-up.

References

[Bar01] Boaz Barak. How to go beyond the black-box simulation barrier. In *Proc. 42nd FOCS*, 2001.

[Bar04] Boaz Barak. *Non-Black-Box Techniques in Cryptography*. Ph.D., Weizmann Institute of Science, January 2004.

[BCC88] Gilles Brassard, David Chaum, and Claude Crépeau. Minimum disclosure proofs of knowledge. *JCSS*, 37(2):156–189, 1988.

[BG02] Boaz Barak and Oded Goldreich. Universal arguments and their applications. In *Proc. CCC '02*, 2002.

[BIN97] Mihir Bellare, Russell Impagliazzo, and Moni Naor. Does parallel repetition lower the error in computationally sound protocols? In *Proc. 38th FOCS*, 1997.

[BLV04] Boaz Barak, Yehuda Lindell, and Salil Vadhan. Lower bounds for non-black-box zero knowledge. Cryptology ePrint Archive, Report 2004/226, 2004. Extended abstract in *Proc. 44th FOCS*, 2003.

[BM88] László Babai and Shlomo Moran. Arthur-Merlin games: a randomized proof system, and a hierarchy of complexity class. *JCSS*, 36(2):254–276, 1988.

[BP04] Boaz Barak and Rafael Pass. On the possibility of one-message weak zero-knowledge. In *Proc. 1st TCC*, 2004.

[CGH98] Ran Canetti, Oded Goldreich, and Shai Halevi. The random oracle methodology, revisited. In *Proc. 30th STOC*, 1998.

[DLN+04] Cynthia Dwork, Michael Langberg, Moni Naor, Kobbi Nissim, and Omer Reingold. Succint proofs for NP and spooky interactions. manuscript, 2004.

[DN00] Cynthia Dwork and Moni Naor. Zaps and their applications. In *Proc. 41st FOCS*, 2000.

[DNRS03] Cynthia Dwork, Moni Naor, Omer Reingold, and Larry Stockmeyer. Magic functions. *JACM*, 50(6):852–921, 2003.

[FLS99] Uriel Feige, Dror Lapidot, and Adi Shamir. Multiple noninteractive zero knowledge proofs under general assumptions. *SICOMP*, 29(1):1–28, 1999.

[For89] Lance Fortnow. The complexity of perfect zero-knowledge. *Advances in Computing Research*, 5:429–442, 1989.

[FS86] Amos Fiat and Adi Shamir. How to prove to yourself: practical solutions to identification and signature problems. In *Proc. Crypto '86*, 1986.

[GH98] Oded Goldreich and Johan Håstad. On the complexity of interactive proofs with bounded communication. *IPL*, 67(4):205–214, 1998.

[GK03] Shafi Goldwasser and Yael Tauman Kalai. On the (in)security of the Fiat-Shamir paradigm. In *Proc. 44th FOCS*, 2003.

[GO94] Oded Goldreich and Yair Oren. Definitions and properties of zero-knowledge proof systems. *J. Cryptology*, 7(1):1–32, 1994.
[Gol01] Oded Goldreich. *Foundations of Cryptography: Basic Tools*. Cambridge University Press, 2001.
[GT00] Rosario Gennaro and Luca Trevisan. Lower bounds on efficiency of generic cryptographic constructions. In *Proc. 41st FOCS*, 2000.
[Kil92] Joe Kilian. A note on efficient zero-knowledge proofs and arguments. In *Proc. 24th STOC*, 1992.
[KP98] Joe Kilian and Erez Petrank. An efficient noninteractive zero-knowledge proof system for NP with general assumptions. *J. Cryptology*, 11(1):1–27, 1998.
[Mic00] Silvio Micali. Computationally sound proofs. *SICOMP*, 30(4):1253–1298, 2000.

Computational Bounds on Hierarchical Data Processing with Applications to Information Security*

(Extended Abstract)

Roberto Tamassia and Nikos Triandopoulos

Department of Computer Science, Brown University
{rt, nikos}@cs.brown.edu

Abstract. We introduce *hierarchical data processing (HDP)* problems, a class of computations over a collection of values associated with a set of n elements, based on a directed acyclic graph (DAG). We present an $\Omega(\log n)$ lower bound on various computational cost measures for HDP problems and we develop an efficient randomized DAG scheme for HDP problems. We apply our results to *data authentication through cryptographic hashing* and *multicast key distribution using key-graphs*. We show that both problems involve HDP and prove logarithmic lower bounds on their computational and communication costs. Using our new DAG scheme, we present a new efficient authenticated dictionary and a new skip-list version with expected search complexity $1.25 \log_2 n + O(1)$.

1 Introduction

In this paper, we present a unified analysis and design of algorithms and data structures for two important, and seemingly unrelated, information security problems: the authentication of membership queries in the presence of data replication at untrusted directories and the distribution of cryptographic keys by the controller of a dynamic multicast group. For both problems, we provide logarithmic lower bounds on various time and space cost measures, develop new efficient data structures and give an accurate analysis of their performance, taking into account constant factors in the leading asymptotic term.

Our unified approach is based on the definition of the class of *hierarchical data processing (HDP)* problems, where a directed acyclic graph (DAG) describes the computation of a collection of output values from an input set of n elements. We define structural cost measures for DAGs that express computational costs in an HDP problem and prove $\Omega(\log n)$ lower bounds for them using a reduction from the problem of searching by comparisons in an ordered set. We also design a new

---

* This work was supported in part by NSF grants CCF–0311510, IIS–0324846 and CNS–0303577.

efficient randomized DAG scheme for HDP problems, based on a variation of the skip-list. Our results for the two information security problems are obtained by showing that they can be modeled as an HDP problem and by appropriately applying to their domain the general lower bounds and our new DAG scheme. This extended abstract omits the details of our work. A full version is available in [14]. Our contributions are summarized as follows.

Hierarchical Data Processing. In Section 2, we introduce the class of hierarchical data processing (HDP) problems and initiate their study. This class models computations on a dynamic set of elements that share the following characteristics. Associated with the elements is a structured collection of values, organized according to a DAG. Update operations change elements and require adjusting the values. Also, queries on elements are issued, where typically the answer to a query is a subset of the associated values. The computational cost of an update or query operation depends on certain structural properties of the underlying DAG. In general, HDP provides an abstract model for problems where computations are performed sequentially and hierarchically according to a DAG and their complexity depends on the structure of the DAG.

We define various structural cost measures for DAGs and we relate them to the space and time complexity of queries and updates in an HDP problem. Using a reduction from the problem of searching by comparisons in an ordered set, we prove an $\Omega(\log n)$ lower bound on the space and time complexity of query and update operations in an HDP problem of size n. We also show that with respect to HDP problems, trees are optimal DAG structures compared with general DAGs. In Section 3, we design and analyze a new randomized DAG scheme for HDP problems, called *multi-way skip-list DAG scheme*, which is based on a variation of the skip-list data structure [12]. Our DAG scheme has better cost measures than previous DAG schemes based on the skip-list data structure.

Data Authentication. An *authenticated data structure (ADS)* is a distributed model of data authentication, where a *directory* answers queries on a data structure on behalf of a trusted *source* and provides to the *user* a cryptographic proof of the validity of the answer. In the important class of *hash-based authenticated data structures*, a digest of the data set is computed by hierarchically applying a cryptographic hash function over the data set and is digitally signed by the source.

Early work on ADSs has focused on hash-based authenticated dictionaries, where Merkle trees are used to authenticate answers to membership queries incurring logarithmic authentication cost. General techniques for hash-based query authentication are presented in [6,7]. Beyond dictionaries, hash-based ADSs have been developed for various type of queries, including connectivity queries in graphs and range queries in two-dimensional point sets [3,6,7]. An alternative approach to the design of an authenticated dictionary, based on the *RSA accumulator* (e.g.,[1]) is presented in [4]. Related to ADSs is also work on zero-knowledge sets and consistency proofs [8,11]. All existing hash-based authenticated dictionaries have logarithmic authentication cost. Naor and Nissim [10] posed as an open problem the question of whether one can achieve sublogarithmic authentication cost for dictionaries. We answer this question negatively for hash-based ADSs.

In Section 4, we present a detailed analysis of the cost of ADSs, focusing on dictionaries. We show how a hash-based dictionary ADS can be modeled as an HDP problem. We consider a general authentication technique where hashing is hierarchically performed over the data set according to a DAG and multiple digests can be digitally signed by the source. Applying our HDP framework to this domain, we prove the first nontrivial lower bound on the authentication cost for dictionaries: any hash-based authenticated dictionary of size n where the source signs k digests of the data set has $\Omega(\log \frac{n}{k})$ update cost, verification cost, and communication cost in the worst case. We also present a new hash-based dictionary ADS based on our new DAG scheme.

Multicast Key Distribution. *Multicast key distribution* refers to a model for realizing secrecy in multicast communications among a dynamic group of n users. Users share secret keys and a common *group-key* with which they encrypt multicast messages using secret-key encryption. When updates in the multicast group occur, in order to preserve (forward/backward) security, the group-key needs to be securely updated. A *group controller* is responsible for distributing an initial set of keys to the users and for updating the keys accordingly. Costs associated with this problem include the number of messages transmitted after an update and the time spent for key encryptions by the controller. We focus on the widely studied *key-graph* scheme (see, e.g., [15]), where possession of keys by users and key updates are modeled through a DAG. Known constructions based on key-trees achieve $O(\log n)$ communication cost for a group of size n. In [2], the first lower bounds are given for a special class of key distribution protocols. In [13], an amortized logarithmic lower bound is presented on the number of messages sent after an update. A similar amortized logarithmic lower bound is shown in [9] for a more general class of key distribution protocols, where one can additionally employ a pseudorandom generator to extract (in a one-way fashion) two new keys from one key and where multiple nested key encryptions can be used.

In Section 5, we show that the multicast key distribution problem using key-graphs is an HDP problem and we apply our results to this domain. We perform the first study of general key-graphs and show that trees are optimal structures. Our new DAG scheme for multicast key distribution achieves costs closer to the theoretical optimal. We also prove the first worst-case logarithmic lower bound on both the communication cost and the computational cost of a single update operation. We prove that for any instance of the problem of size n, there exists an update whose communication cost is at least $\lfloor \log_2 n \rfloor$. All previous lower bounds are amortized, i.e., they refer to a sequence of updates such that the average cost of an update in the sequence is logarithmic.

Skip-Lists. The skip-list [12] is an efficient randomized data structure for dictionaries. A search in a skip-list with n elements takes $1.5 \log_2 n + O(1)$ expected comparisons. As an application of our improved DAG scheme to search structures, in Section 6, we present a new version of the skip-list such that the expected number of comparisons in a search is $1.25 \log_2 n + O(1)$.

2 Hierarchical Data Processing and Its Theoretical Limits

In this section, we define structural cost measures for subgraphs of a DAG and prove lower bounds on them. These cost measures are related to the computational complexity of operations in HDP problems, as will be shown in the next sections.

DAG Scheme for HDP. We define here some graph notation. Let $G = (V, E)$ be a directed acyclic graph. For each node v of G, $indeg(v)$ denotes the in-degree of v and $outdeg(v)$ denotes the out-degree of v. We denote with $V_{so} \subset V$ the set of *source* nodes of G, i.e., nodes v such that $indeg(v) = 0$ and with $V_{si} \subset V$ the set of *sink* nodes of G, i.e., nodes v such that $outdeg(v) = 0$. A subgraph H of G is said to be *weakly connected* if it is connected when one ignores edge directions. For any node v in a DAG G, we denote with G_v the subgraph of G consisting of the nodes that can be reached from v through directed paths.

Definition 1. *A DAG scheme Γ is a quadruple (G, S, n, k), where $G = (V, E)$ is a directed acyclic graph without parallel edges, $S \subset V$ is a set of special nodes and n and k are integers such that: (i) $|V_{so}| = n$; (ii) $|V|$ is bounded by a polynomial in n; and (iii) $|S| = k$, $S \supset V_{si}$ and $S \cap V_{so} = \emptyset$.*

We define three structural cost measures for a subgraph of a DAG and based on them, we define three cost measures for a DAG scheme Γ.

Definition 2. *Let $H = (V_H, E_H)$ be a weakly connected subgraph of a DAG G. With respect to G: (i) The node size $size(H)$ of H is the number of nodes in H, i.e., $size(H) = |V_H|$; (ii) the degree size $indeg(H)$ of H is the sum of the in-degrees (with respect to G) of the nodes of H, i.e., $indeg(H) = \sum_{v \in H} indeg(v)$; (iii) the combined size $comb(H)$ of H is the sum of its node and degree sizes, i.e., $comb(H) = size(H) + indeg(H)$; (iv) the boundary size $bnd(H)$ of H is the number of edges of G that enter nodes of H but are not in H.*

Definition 3. *Given a DAG scheme $\Gamma = (G, S, n, k)$, let s be a source node of G. Let P_s^t denote the set of directed paths connecting node s to node t in G. The associated path π_s of s is a directed path in G_s that starts at s, ends at a node of S and has the minimum combined size among all such paths, i.e., $comb(\pi_s) = \min_{u \in S, p \in P_s^u} comb(p)$. We define the following cost measures for Γ: (i) the update cost $\mathcal{U}(\Gamma)$ of Γ is $\mathcal{U}(\Gamma) = \max_{s \in V_{so}} comb(G_s)$; (ii) the query cost $\mathcal{Q}(\Gamma)$ of Γ is $\mathcal{Q}(\Gamma) = \max_{s \in V_{so}} comb(\pi_s)$ (i.e., $\max_s \min_{u \in S, p \in P_s^u} comb(p)$); and the sibling cost $\mathcal{S}(\Gamma)$ of Γ is $\mathcal{S}(\Gamma) = \max_{s \in V_{so}} bnd(\pi_s)$.*

Observe that if $\Gamma = (G, S, n, k)$ is a DAG scheme, H is a subgraph of G and p is any directed path in G, then we have that (with respect to G): (i) $comb(H) = \sum_{v \in H}(1 + indeg(v))$ and $bnd(p) = 1 + indeg(p) - size(p)$; (ii) $comb(H) > indeg(H) \geq size(H)$ and $indeg(H) \geq bnd(H)$; (iii) $\mathcal{U}(\Gamma) \geq \mathcal{Q}(\Gamma) > \mathcal{S}(\Gamma)$.

Our motivation for introducing DAG schemes is that they model an abstract class of computational problem where a DAG G holds a collection of n input *elements* (stored at source nodes) and a collection of output *values* (stored at non-source nodes) that are computed using the DAG. Query operations on elements return a collection of values. Update operations modify the DAG G and the input elements, causing corresponding changes to the set of values. Computations are performed sequentially and hierarchically, according to the hierarchy induced by the underlying DAG G. The computational cost (time, space, or communication complexity) of query and update operations can be expressed as the combined, degree or boundary size of a subgraph (usually G_s or π_s, for a source node s of G), where every node v in the subgraph contributes an amount proportional to $indeg(v)$ to the cost. Generally, any computational cost measure for a problem in this class is completely characterized by structural cost measures of subgraphs of DAG G. We refer to such problems (informally defined due to space limitations) as *hierarchical data processing (HDP) problems*.

We study the cost measures of general DAG schemes, derive results that explain the inherent computational limits that exist in any HDP problem and characterize optimal DAG scheme structures for these problems. We first show that the cost measures for a tree-based DAG scheme are related to the number of comparisons for searching in an ordered set, by drawing a direct analogy between the sibling cost of any tree-based DAG scheme and the number of comparisons performed in a search tree corresponding to the DAG scheme. This result forms the basis for a reduction from searching by comparisons to any computational procedure of an HDP problem, with cost that is expressed by the sibling cost of a tree-based DAG scheme. Using this reduction we get the following.

Theorem 1. *Any DAG scheme $\Delta = (T, S, n, 1)$ such that T is a directed tree has $\Omega(\log n)$ update, query and sibling costs.*

We show that among all possible DAGs, trees have optimal cost measures.

Theorem 2. *Let $\Gamma = (G, S, n, 1)$ be a DAG scheme. There exists a DAG scheme $\Delta = (T, S, n, 1)$ such that T is a directed tree and $\mathcal{U}(\Delta) \leq \mathcal{U}(\Gamma)$, $\mathcal{Q}(\Delta) \leq \mathcal{Q}(\Gamma)$, and $\mathcal{S}(\Delta) \leq \mathcal{S}(\Gamma)$.*

The above result also applies to a general DAG scheme (G, S, n, k) with k special nodes. In this case, there exists a forest of trees achieving better performance with respect to the cost measures being considered. The following theorem summarizes the results of this section with respect to the cost measures of any DAG scheme.

Theorem 3. *Any DAG scheme $\Gamma = (G, S, n, k)$ has $\Omega(\log \frac{n}{k})$ update, query and sibling costs.*

The above results establish a reduction from searching by comparisons to computations related to HDP problems (computations performed sequentially and hierarchically according to the hierarchy induced by the underlying DAG scheme) and give us lower bounds on the costs of these computations. Also, the optimality of tree-based DAG schemes over general graphs further characterizes the optimal schemes for HDP problems. The connection between HDP problems and

DAG schemes is illustrated in Sections 4 and 5, where we model two information security problems as HDP problems and translate the above results to their domain.

3 A New DAG Scheme Based on Skip-Lists

In view of the logarithmic lower bounds and the optimality of tree structures for DAG schemes, we present a new DAG scheme that is based on skip-lists. Our *multi-way skip-list DAG scheme* $\Delta = (T, S, n, 1)$ is based on the skip-list data structure and achieves cost measures close to the theoretical optimal.

Skip-Lists and Bridges. A *skip-list* [12] with *probability parameter* p is a set of lists $L_1, ..., L_h$, where L_1 stores the element of a totally ordered set $(X, \preceq)$ of size n (sorted according to $\preceq$) and, for each i, each of the elements of list L_i is independently chosen to be contained in L_{i+1} with probability p. Lists are viewed as *levels* and we consider all elements of the same value that are stored in different levels to form a *tower*. The *level* of a tower is the level of its top element. Each node of a tower has a forward pointer to the successor element in the corresponding list and a pointer to the element one level below it. A *header* tower that stores sentinel element $-\infty$ is included in the skip-list as the left-most tower of level one more than the maximum level of any other tower in the skip-list. A node of the skip-list is a *plateau* node if it is the top node of its tower. We introduce next the notion of a *bridge* and define related concepts.

Definition 4. *In a skip-list: (i) a* bridge *b is a maximal sequence of towers of the same level such that no higher tower lies between them and the plateau nodes of the towers are all reachable in a sequence using forward pointers; (ii) the* size $|b|$ *of bridge b is the number of towers in the bridge and the* bridge size *of a tower is the size of the bridge that the tower belongs to; (iii) a* child bridge *of b is a bridge that is contained under b and to which a tower of b connects through forward pointers; (iv) the* plateau towers *of a tower t are the towers whose plateau nodes can be reached by a node of t using one forward pointer.*

Directed Tree T. We now describe the multi-way skip-list DAG scheme $\Delta = (T, r, n, 1)$, where the directed tree T is defined with respect to a skip-list. By *list node* we refer to a node of the skip-list and by *DAG node* to a node of tree T. An edge (v, u) in T is directed towards node u. If $v, v_1, \ldots, v_l$ are nodes in T, then operation $New(v, v_1, \ldots, v_l)$ on existing DAG nodes creates in T new nodes $u_1, \ldots, u_l$ and new edges $(v_1, u_2), \ldots, (v_{l-1}, u_l), (v_l, v), (u_1, u_2), \ldots, (u_{l-1}, u_l)$ and (u_l, v), where DAG node u_1 becomes a new *source* node of T. The notion of a bridge is essential in skip-list DAG T. For each bridge b in the skip-list, a corresponding node $v(b)$ is created in T. We call $v(b)$ the DAG node of b. Node $v(b)$ is connected in T with the DAG nodes of all the child bridges of b. Thus, DAG T is defined in a recursive way with respect to a skip-list. First, all bridges in the skip-list are identified and the DAG node for the outer bridge (header tower) is created. Then, given that the DAG node $v(b)$ of a bridge b is created, using

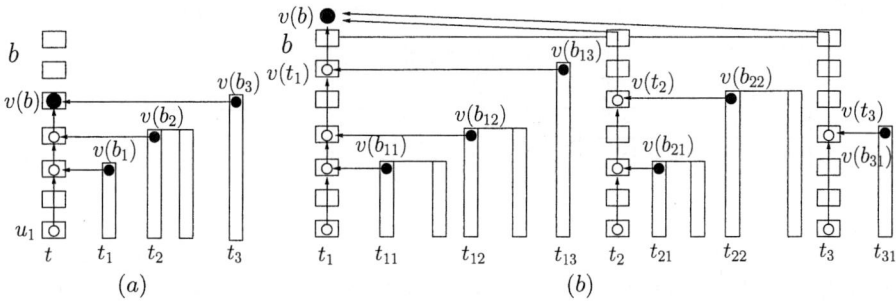

Fig. 1. Multi-way skip-list DAG Δ, where circle nodes are DAG nodes (bridge DAG nodes are solid) and square nodes are nodes of the skip-list. DAG node $v(b)$ of bridge b is recursively connected to the DAG nodes of the child bridges depending on the following two cases: **(a)** $|b| = 1$ and **(b)** $|b| > 1$

operation $New(\cdot)$, $v(b)$ is connected with paths in T to the newly created DAG nodes of the child bridges of b, as follows (see Figure 1).

If $|b| = 1$ (Figure 1(a)), let $t_1, ..., t_l$ (in increasing level) be the plateau towers of the tower t of b. If plateau tower t_i belongs to bridge b_i, we perform operation $New(v(b), v(b_1), \ldots, v(b_l))$ where $v(b)$ is the DAG node of b.

If $|b| > 1$ (Figure 1(b)), then let $k = |b|$ and let $t_1, \ldots, t_k$ be the towers of b. For each such tower t_i, we create a new DAG node $v(t_i)$, $1 \leq i \leq k$. For tower t_k, we consider its, say l, plateau towers $t_{k1}, \ldots, t_{kl}$ and perform operation $New(v(t_k), v(b_{k1}), \ldots, v(b_{kl}))$, where $b_{k1}, \ldots, b_{kl}$ are the child bridges of b that plateau towers $t_{k1}, ..., t_{kl}$ belong to. Moreover, for each tower t_i, $i < k$, of, say $l + 1$, plateau towers, we consider its l lowest plateau towers $t_{i1}, \ldots, t_{il}$, that is, for $i < k$, tower t_{l+1} is *omitted* from this sequence. Let $b_{i1}, \ldots, b_{il}$ be the child bridges of b that plateau towers $t_{i1}, ..., t_{il}$ belong in. For tower t_i, $i < k$, we perform operation $New(v(t_i), v(b_{i1}), \ldots, v(b_{il}))$. Finally, we add k new edges in T: edge $(v(t_i), v(b))$ for $1 \leq i \leq k$. By this construction, T is a directed tree, the root of T is the DAG node r of the header bridge, and T has exactly n leaves.

We can show that multi-way skip-list Δ achieves cost measures that are close to the theoretical optimal value of $\lfloor \log_2 n \rfloor + 1$.

Theorem 4. *With respect to a skip-list with probability parameter p, the multi-way skip-list DAG scheme $\Delta = (T, r, n, 1)$ has the following expected cost measures for any fixed source node s of T with corresponding source to root path π_s:*

1. $E[size(\pi_s)] \leq 2(1-p) \log_{\frac{1}{p}} n + O(1)$;
2. $E[indeg(\pi_s)] \leq (1-p)\frac{(1+p)^2}{p} \log_{\frac{1}{p}} n + O(1)$;
3. $E[bnd(\pi_s)] \leq \frac{(1-p)(1+p^2)}{p} \log_{\frac{1}{p}} n + O(1)$; and
4. $E[size(T)] \leq (1 + pq^2 + pq + \frac{p}{q(2-pq^2)})n$, where $q = 1 - p$.

In particular, our multi-way skip-list DAG scheme Δ has better expected cost measures than the DAG scheme of [5, 12], also based on the skip-list, which we

Table 1. Comparison of three tree DAG schemes in terms of their cost measures

	$E[size(\pi_s)]$	$E[indeg(\pi_s)]$	$E[bnd(\pi_s)]$	$E[size(T)]$
red-black tree	$\log_2 n$	$2\log_2 n$	$\log_2 n$	$2n$
standard skip-list	$1.5\log_2 n$	$3\log_2 n$	$1.5\log_2 n$	$2n$
multi-way skip-list	$\log_2 n$	$2.25\log_2 n$	$1.25\log_2 n$	$1.9n$

call *standard skip-list DAG scheme*. Table 1 summarizes the comparison results for probability parameter $p = 1/2$. The comparison between multi-way skip-list scheme and red-black tree scheme shows an interesting trade-off on the performance with respect to the cost measures we study.

4 Data Authentication Through Hashing

We apply our results of Sections 2 and 3 to *data authentication through cryptographic hashing*. We focus on *authenticated dictionaries*, ADSs that authenticate answers to membership queries. ADSs provide a model of computation, where an untrusted *directory* answers queries issued by a *user* on a data structure on behalf of a trusted *source* and provides a proof of the validity of the answer to the user. In this model, authentication is achieved by having the data source signing some *digest* of the data, where, for data authentication through hashing, a hash function is systematically used to produce this digest. On any query, along with the answer, the signed digest and some information that relates the answer to this digest are given to the user and these are used for the answer verification.

Authenticated Dictionary. Let X be a data set owned by the source that evolves through update operations insert and delete. Membership queries contains are issued about X. A (multivariate extension of a) cryptographic hash function h is used to produce a digest of set X which is signed by the source (see [6]). In our study, we actually consider a more general model where several digests are produced and signed by the source. These digests are computed through a *hashing scheme* over a DAG that has k *signature* nodes $t_1, \ldots, t_k$ and stores the elements of X at the source nodes. Each node u of G stores a *label* (hash value) $L(u)$ such that, if u is a source of G, then $L(u) = h(e_1, \ldots, e_p)$, where $e_1, \ldots, e_p$ are elements of X, else (u is not a source of G) $L(u) = h(L(w_1), \ldots, L(w_l), e_1, \ldots, e_q)$, where $(w_1, u), \ldots, (w_l, u)$ are edges of G, $e_1, \ldots, e_q$ are elements of X and p, q and l are some non negative integers. Without loss of generality, we focus our study on the *canonical* case where $p = 1$ and $q = 0$, noting that any general hashing scheme is *equivalent* to a canonical one. We view the labels $L(t_i)$ of the signature nodes t_i of G as the digests of X, computed via the above DAG G.

The authentication technique is based on the following general approach. The source and the directory store identical copies of the data structure for X and maintain the same hashing scheme on X. The source periodically signs the digests of X together with time-stamps and sends the signed time-stamped digests to the directory. When updates occur on X, they are sent to the directory together with the new signed time-stamped digests. When the user poses a query, the directory returns to the user some *answer authentication information*, which consists of: (i) one signed time-stamped digest of X, (ii) the answer to the query and (iii) a proof consisting of a small collection of labels from the hashing scheme (or of data elements if needed) that allows the recomputation of the digest. The user validates the answer by recomputing the digest, checking that it is equal to the signed one and verifying the signature of the digest. Security against forgery of proofs by the directory follows from the properties of the signature scheme and the hash function.

Authentication Overhead. Now we study the performance overhead of computations related to authentication in an authenticated dictionary based on a hashing scheme (the analysis is valid for any ADS). This *authentication overhead* consists of *time overhead* for the (i) maintenance of the hashing scheme after updates, (ii) generation of the answer authentication information in queries, and (iii) verification of the proof of the answer; *communication overhead*, defined as the size of the answer authentication information; *storage overhead*, given by the number of hash values used by the authentication scheme; and *signature overhead*, defined as the number of signing operations performed at the source (and thus the number of signatures sent by the source). The time for computing a hash function is a few orders of magnitude larger than the time for comparing two basic numerical types. Thus, the rehashing overhead dominates the update time and the practical performance of an ADS is characterized by the *authentication overhead*, which depends on the hash function h in use and the mechanism for realizing a multivariate hash function from h.

Cryptographic Hash Functions. A *collision-resistant hash function* $h(x)$ maps a bit string x of arbitrary length to a hash value of fixed length, such that collisions are hard to find. We refer to h simply as *hash function*. Generic constructions of hash functions are modeled by *iterative computations* based on a *compression function* $f(\cdot)$ that is applied iteratively on the input. For this class of hash functions, there exist constants c_1 and c_2 (which may depend on a security parameter) such that, given an input string x, the computation $h(x)$ takes time $T(x) = c_1|x| + c_2$. Moreover, without loss of generality, we extend h to a multivariate function using *string concatenation*. We define $h(x_1, ..., x_d) = h(x_1\|...\|x_d)$.

Cost of Data Authentication Through Hashing. Let G be any hashing scheme used to implement a hash-based authenticated dictionary for set X of size n, where k signature nodes store hash values signed by the source. Hashing scheme G along with the signature nodes can be viewed as a DAG scheme $\Gamma = (G, S, n, k)$, where special nodes are signature nodes and there are exactly n source nodes in G storing elements in X. Each cost parameter of the authentication overhead depends *linearly* on $size(H)$ and $indeg(H)$ for some subgraph H of G. The node size corresponds to the number of performed *hash operations* and the degree size to the total number of hash values that participate as operands in these hash operations. In particular, the following holds.

Lemma 1. *Let $\Gamma = (G, S, n, k)$ be any hashing scheme used to implement a hash-based authenticated dictionary for set X, where special nodes are signature nodes. Let s be a source node of G storing element $x \in X$, G_s be the subgraph of G that is reachable from s, and π_s the associated path of s. We have: (i) an update operation on element x has update time overhead that is lower bounded by $comb(G_s)$; (ii) a query operation on element x has verification time overhead that is lower bounded by $comb(\pi_s)$ and communication overhead that is lower bounded by $bnd(\pi_s)$; (iii) the storage overhead is $size(G)$. All involved computations are performed according to the hierarchy induced by G.*

Thus, hash-based authentication of membership queries is an HDP problem, where operations insert/delete are related to the update cost and operation contains is related to the query cost of the underlying DAG scheme. Our results on DAG schemes indicate that signing multiple hash values does not help and that tree structures are optimal.

Theorem 5. *In the data authentication model through hashing, any hashing scheme with k signature nodes that implements an authenticated dictionary of size n has (i) $\Omega(\log \frac{n}{k})$ worst-case update and verification time overheads; (ii) $\Omega(\log \frac{n}{k})$ worst-case communication overhead; and (iii) $\Omega(k)$ signature overhead.*

Theorem 6. *An authenticated dictionary of size n implemented with a multiway skip-list DAG scheme with probability parameter p achieves the following expected performance, where c_1 and c_2 are constants that depend on the hash function h in use and $q = 1 - p$:*

1. *the expected hashing overhead of an update or verification operation is at most $(1-p)\left(2c_2 + \frac{(1+p)^2}{p}c_1\right)\log_{\frac{1}{p}} n + O(1)$,*
2. *the expected communication cost is at most $(1-p)\left(\frac{1+p^2}{p}\right)\log_{\frac{1}{p}} n + O(1)$ and*
3. *the expected storage overhead is at most $\left(1 + pq^2 + pq + \frac{p}{q(2-pq^2)}\right)n$.*

5 Multicast Key Distribution Using Key-Graphs

In this section, we apply our results of Section 2 and 3 to multicast key distribution.

Multicast Key Distribution. The problem refers to secure communication in multicast groups. A *group* consists of a set of n *users* and a *group key controller*. Secret-key cryptography is used to transmit *encrypted* multicast messages among users of the group. Messages are encrypted using a *group key* available to all the current users of the group. The security problem arises when updates on the group are performed through users addition to or deletions from the group. The goal is to achieve *confidentiality*, i.e., only group members can decrypt messages, and *forward (backward) secrecy*, i.e., users deleted from (added to) the group can not decrypt messages transmitted in the future (past). In this model, the group controller is responsible for distributing secret keys to the users and updating them appropriately after user updates. The idea is that a set of keys, known to the controller, is distributed to the users, so that a key is possessed by multiple users and a user possesses multiple keys. In particular, any user is required to have, except from the group key, a *secret key* that no other user knows. On user updates, subsets of these keys need to be updated appropriately by the controller.

Key-Graphs. *Key-graphs* (e.g., [15]) provide a framework to implement this idea. A key-graph models the possession of keys by users and the computations (encryptions at the controller, decryptions at the users) and message transmissions that need to be performed after an update. A key-graph is a single-sink DAG G, known by the group controller and the users, that facilitates key updates. The source nodes of G correspond to users and store their individual secret keys. The non-source nodes of G correspond to keys that are shared among subsets of users. The user associated with a source node s of G possesses *all* and *only* the keys that correspond to the subgraph G_s of G (i.e., the nodes reachable from s). On any update of user s, the keys in G_s have to change to achieve forward and backward secrecy. The group controller distributes each new key by sending it encrypted with a previously distributed key.

Cost parameters. The cost parameters of a key distribution scheme using key-graphs are defined as follows: (i) the *encryption cost* is the computational cost at the controller for encrypting all new keys and thus producing the messages for transmission; (ii) the *decryption cost* is the computational cost at a user for decrypting received messages and updating her keys; (iii) the *communication cost* is the number of transmitted messages; and (iv) the *key-storage cost* is the total number of keys stored at the key controller and at the users.

We can view a key-graph G as DAG scheme $\Gamma = (G, S, n, 1)$, where S consists of the unique sink node of G, called *group* node, and the source nodes correspond to the users. Each cost parameter of the key distribution scheme depends linearly on $size(H)$ and $indeg(H)$, for some subgraph H of G. The node size corresponds to keys stored at users, decryptions and key generations and the degree size corresponds to the number of messages sent and the number of encryptions performed during the update.

Definition 5. *Let $\Gamma = (G, S, n, 1)$ be a key-graph scheme and let v be a node in G. The support $Sup(v)$ of v is the set of users that possess the key stored at v, i.e., $s \in Sup(v)$ if and only if $v \in G_s$. Let $U = \{u_1, ..., u_k\}$ be a set of nodes of G and let $T \subseteq V_{so}$ be a set of source nodes of G. We say that set U spans set T if $\bigcup_{1 \leq i \leq k} Sup(u_i) = T$. A node v is said to be safe if v is a source node or if $indeg(v) > 1$ and any node set that spans $Sup(v)$ and does not include v has size at least $indeg(v)$. Key-graph scheme Γ is said to be reduced, if all nodes in G are safe.*

Lemma 2. *Let $\Gamma = (G, S, n, 1)$ be a reduced key-graph scheme used for the multicast key distribution problem. Then we have: (i) an update operation on a user that corresponds to a source node s has communication cost at least $indeg(G_s)$ and encryption cost at least $comb(G_s)$; and (ii) the key-storage cost is $size(G)$.*

All involved encryptions and decryptions are performed according to the hierarchy induced by G.

Multicast key distribution using reduced key-graphs is an HDP problem, where the cost of an update in the group is related to the update cost of the underlying DAG scheme. By studying more carefully reduced key-graph schemes and using Theorems 1 and 2, we can prove the main result of the section.

Theorem 7. *For a multicast key distribution problem of size n using key-graphs, in the worst case, an update operation in the group requires at least $\lfloor \log_2 n \rfloor$ communication cost and $\Omega(\log n)$ encryption cost. Also, key-tree structures are optimal over general key-graphs.*

6 A New Skip-List Version

From the relation of the sibling cost of any tree-based DAG scheme and the number of comparisons performed in a search tree corresponding to the DAG scheme and Theorem 4, we get a new version of the skip-list data structure with an expected number of comparisons closer to the theoretical optimal $\lfloor \log_2 n \rfloor + 1$, up to an additive constant term, than the one of the standard skip-list. The new skip-list version can be viewed as a *multi-way* extension of the skip-list data structure. We omit the details due to space limitations.

Theorem 8. *There is a multi-way version of a skip-list for set X of size n and probability parameter p, where the expected number of comparisons performed while searching in X for any fixed element is at most $\frac{(1-p)(1+p^2)}{p \log_2 \frac{1}{p}} \log_2 n + O(1)$, or $1.25 \log_2 n + O(1)$ for $p = \frac{1}{2}$.*

References

[1] J. Camenisch and A. Lysyanskaya. Dynamic accumulators and application to efficient revocation of anonymous credentials. In *Proc. CRYPTO*, 2002.

[2] R. Canetti, T. Malkin, and K. Nissim. Efficient communication - storage tradeoffs for multicast encryption. In *Proc. EUROCRYPT*, 1999.

[3] P. Devanbu, M. Gertz, C. Martel, and S. G. Stubblebine. Authentic data publication over the Internet. *Journal of Computer Security*, 11(3), 2003.

[4] M. T. Goodrich, R. Tamassia, and J. Hasic. An efficient dynamic and distributed cryptographic accumulator. In *Proc. ISC*, 2002.

[5] M. T. Goodrich, R. Tamassia, and A. Schwerin. Implementation of an authenticated dictionary with skip lists and commutative hashing. In *Proc. DISCEX*, 2001.

[6] M. T. Goodrich, R. Tamassia, N. Triandopoulos, and R. Cohen. Authenticated data structures for graph and geometric searching. In *Proc. RSA-CT*, 2003.

[7] C. Martel, G. Nuckolls, P. Devanbu, M. Gertz, A. Kwong, and S. G. Stubblebine. A general model for authenticated data structures. *Algorithmica*, 39(1), 2004.

[8] S. Micali, M. Rabin, and J. Kilian. Zero-Knowledge sets. In *Proc. FOCS*, 2003.

[9] D. Micciancio and S. Panjwani. Optimal communication complexity of generic multicast key distribution. In *Proc. EUROCRYPT*, 2004.

[10] M. Naor and K. Nissim. Certificate revocation and certificate update. In *Proc. USENIX Security*, 1998.

[11] R. Ostrovsky, C. Rackoff, and A. Smith. Efficient consistency proofs for generalized queries on a committed database. In *Proc. ICALP*, 2004.

[12] W. Pugh. Skip lists: a probabilistic alternative to balanced trees. *CACM*, 33(6), 1990.

[13] J. Snoeyink, S. Suri, and G. Varghese. A lower bound for multicast key distribution. In *Proc. INFOCOMM*, 2001.

[14] R. Tamassia and N. Triandopoulos. Computational bounds on hierarchical data processing with applications to information security. Manuscript, Brown University, 2005. Full version http://www.cs.brown.edu/cgc/stms/papers/cbhdp.pdf.

[15] C. K. Wong, M. Gouda, and S. S. Lam. Secure group communications using key graphs. *IEEE/ACM Transactions on Networking*, 8(1), 2000.

Balanced Allocation and Dictionaries with Tightly Packed Constant Size Bins

(Extended Abstract)

Martin Dietzfelbinger[1,*] and Christoph Weidling[2,**]

[1] Technische Universität Ilmenau, 98684 Ilmenau, Germany
martin.dietzfelbinger@tu-ilmenau.de
[2] sd&m AG, 63065 Offenbach am Main, Germany
christoph.weidling@sdm.de

Abstract. We study an aspect of the balanced allocation paradigm (also known as the "two-choices paradigm"). Assume there are n balls and $m = (1+\varepsilon)n/d$ bins of capacity d each, for a fixed $d \geq 1$. To each of the balls two possible bins are assigned at random. We show that $\varepsilon > (2/e)^{d-1}$ is sufficient to guarantee that with high probability each ball can be put into one of the two bins assigned to it, without any bin overflowing. Further, it takes constant time on average for changing the arrangement to accommodate a new ball, if $\varepsilon > \gamma \cdot \beta^d$, for some constants $\gamma > 0$, $\beta < 1$. The problem may also be described in data structure language. Generalizing cuckoo hashing (Pagh and Rodler, 2001), we consider a hash table with m positions, each representing a bucket of capacity $d \geq 1$. Key x may be stored in bucket $h_1(x)$ or $h_2(x)$, for two fully random hash functions h_1 and h_2. For arbitrary $\varepsilon > 0$, we obtain an implementation of a dynamic dictionary that accommodates n keys in $m = (1+\varepsilon)n/d$ buckets of size $d = O(\log(1/\varepsilon))$. For a lookup operation only two hash functions have to be evaluated and two contiguous segments of d memory cells have to be inspected. The expected time for inserting a new key is constant.

1 Introduction: Bounded Balanced Allocation, d-Orientability, and Blocked Cuckoo Hashing

In this paper, we study a data allocation problem that can be described in different terminologies. In the **"balanced allocation paradigm"** (also known as the "two-choices paradigm") we have n balls and m bins. To each ball two bins are assigned at random. Each ball is to be placed into one of the two bins assigned to it; the aim is to keep the maximum load in the bins small. Much work has been devoted to analyzing the online version of this experiment, where the balls arrive one after the other and are put into a bin upon arrival, and this

---

[*] Corresponding author.
[**] Affiliated with the Technische Universität Ilmenau while this work was done.

placement can never be changed later. Not that much is known about the offline version, where all balls with their two choices are given as input. However, it has been known for quite a while that with high probability[1] for $n = O(m)$ constant bin size is enough to place all balls. Other authors have considered "dynamic" situations in which the positions of balls may change in the course of a random process.

In this paper, we ask how well we can utilize the space in the bins if we fix the capacity of the bins to some number $d \geq 1$. That means, given n, we wish to keep the space overhead $\varepsilon = \frac{dm-n}{n}$ as small as possible and still be able to place the balls w.h.p. We show that $\varepsilon > (2/e)^{d-1}$ is sufficient to guarantee this. We also consider a dynamic version of the problem. Assume that n balls are placed and a new ball arrives, with two randomly assigned bins. We show that the expected time needed to rearrange the balls so that the new ball can be accommodated as well is constant, as long as $\varepsilon > \gamma \cdot \beta^d$, for some constants $\gamma > 0$, $\beta < 1$. This implies that the expected time to place n balls is $O(n)$. No estimates for the density that can be achieved asymptotically for fixed bin size d larger than some small constant have been known before.

The allocation problem with bounded bin size d is equivalent to a version of the **d-orientability problem** [12] for random graphs. We say an undirected graph $G = (V, E)$ of m nodes and n edges is *d-orientable* if the edges can be directed in such a way that every node has outdegree not larger than d. Given a constant d, we ask for upper and lower bounds on the edge density $\frac{n}{m}$ so that a graph with n randomly placed edges (including loops and multiple edges) is d-orientable w.h.p. Also, we ask how long it takes to adapt a given edge orientation when a new random edge arrives.

Yet another formulation of the same problem can be given in **data structure** language. We wish to implement dynamic dictionaries so that constant lookup time is guaranteed. Dynamic dictionaries store keys from a universe U (possibly together with satellite data) and support the operations *insert*, *delete*, and *lookup*. Pagh and Rodler's **"cuckoo hashing"** method [17] assumes that each one of n keys is assigned to two locations $h_1(x)$ and $h_2(x)$ in a hash table of size m, and can be stored in one of the two locations. Each location has capacity 1. We generalize this approach by considering buckets of capacity d, for some arbitrary constant $d \geq 1$. Our construction results in the following. Assuming that fully random hash functions are available, we obtain an implementation of a dynamic dictionary that for given $\varepsilon > 0$ stores n keys in space $(1+\varepsilon)n$ in such a way that a lookup for x requires evaluating two hash functions and probing two contiguous blocks of d memory cells. The expected cost of inserting a new key is $(1/\varepsilon)^{O(\log \log(1/\varepsilon))}$. This compares favorably with the performance of "*d-ary cuckoo hashing*", a different generalization of cuckoo hashing by Fotakis, Pagh, Sanders, and Spirakis [9]. There $d = O(\log(1/\varepsilon))$ independent hash functions are used to achieve a similar space utilization. The access procedure of our scheme is more local and hence more suited for cache architectures. Experiments (see [7])

---

[1] abbreviated: "w.h.p.", meaning "with probability $1 - \frac{1}{\text{poly}(n)}$".

support the hope that the new scheme is competitive with d-ary cuckoo hashing [17] as far as space utilization is concerned, and allows faster accesses.

Remark 1. For reference, we describe the connection between the hashing formulation and the d-orientability formulation from [2,12,19] in detail. Given is a set S of n keys and two random hash functions h_1, h_2. We consider a random (multi)graph $G_u = (V, E_u)$ with labeled edges. The node set is $V = [m] = \{0, \ldots, m-1\}$, the set of labeled edges is $E_u = E_u(S, h_1, h_2) = \{\{h_1(x), h_2(x)\} \mid x \in S\}$. We say that G_u is d-orientable if the n edges in E_u can be directed in such a way that each node has outdegree at most d. Assigning such directions to edges is equivalent to storing the keys in a table with m blocks with maximal load d, as follows: the edge $\{y, y'\} = \{h_1(x), h_2(x)\}$ is directed from y to y' if and only if x is stored in block y. Below, any directed version of G_u with outdegree bounded by d will be called $G = (V, E)$.

Terminology. In this paper, the algorithms and the analysis are described in the language of a hash table with two functions. This also makes it possible to describe a new workaround for the assumption that fully random hash functions are available for free, see Section 2.3. However, all results readily translate into the terminology of the balanced allocation paradigm or the graph orientation paradigm.

1.1 Background and Related Work

Early contributions to the fixed sized bin allocation problem were made in connection with the balanced allocation paradigm. (For a survey of this area, see [14].) In the seminal paper [1] by Azar, Broder, Karlin, and Upfal it was noted [10] (also see [2]) that if a set of $n \leq 1.67m$ balls is allocated to m cells, with two choices per ball, then with high probability the keys can be placed so that no bin holds more than two balls. This immediately extends to a scheme for storing n balls in m bins with a maximum load of $d = 2 \cdot \lceil n/(1.67m) \rceil \leq 2 + 1.2\frac{n}{m}$, which for m, n large corresponds to a space overhead of $\varepsilon = 0.2$ in our notation.

Simultaneously, observations concerning the existence of such placements were made in papers on the simulation of parallel random access machines on distributed memory machines by redundantly storing data (e.g., [3,13]). In particular, it was shown there that a maximum load of $O(1 + n/m)$ is achievable with high probability, even if for allocating the bins hash functions from classes described in [22] are used.

Sanders et al. [20,21] studied the static allocation problem with fixed bin sizes as the combinatorial abstraction of a scheme called "Randomized Duplicate Allocation (RDA)", used for storing data blocks on disks. In [21] it was shown that with high probability a bin size of $d = 1 + \lceil n/m \rceil$ is sufficient (this would correspond to a bound $d \geq \frac{1}{\varepsilon}$ in our notation). In [20] the question was asked how close n/m might be to $d = \lceil n/m \rceil$ so that still block size d is sufficient. No asymptotic relation between n/m and d was derived. In the analysis of the static case we start with the condition (2) on d and ε that has been noted already in

[20]. The transformation of this condition into the general relation $d > 1 + \frac{\ln(1/\varepsilon)}{1-\ln 2}$ by means of calculus has escaped other researchers until this date.

In [2] the online version of the case of heavily loaded bins (i.e., $\frac{n}{m} \to \infty$) was studied. In [4] it was demonstrated that perfect balance ($n = dm$, with no slack at all) is impossible w.h.p. if $d < \gamma_1 \ln n$ for a suitable constant γ_1, while perfect balance is possible w.h.p. if $d > \gamma_2 \ln n$ for some larger constant γ_2. Further, a randomized rebalancing procedure was described and analyzed, with running time polynomial in n.

On the data structures side, the paper by Pagh and Rodler [17] showed that $m = (2+\varepsilon)n$ cells are enough if each cell may have load 1 and a key x may be stored in one of the locations given by two hash functions. It is easy to see that space below $2n$ is not sufficient for cuckoo hashing in the simple form. As a remedy for this situation, Fotakis et al. [9] suggested "d-ary cuckoo hashing". Their scheme amounts to the balanced allocation problem with bin size 1 and d random targets for each ball. They show that with n balls and $m = (1+\varepsilon)n$ bins it is sufficient to have $d = O(\log(1/\varepsilon))$ for statically placing the balls and inserting new balls in expected constant time, w.h.p. This leads to an implementation of a dictionary for n keys in space $(1+\varepsilon)n$, where a lookup requires evaluating $d = O(\log(1/\varepsilon))$ hash values and probing d random locations in the worst case. Inserting a key takes expected constant time. Our result leads to a comparable space utilization, but has the advantage that only two hash functions have to be evaluated and two contiguous blocks of d memory cells must be probed in a search, which has advantages in architectures with caches. The basic structure of our analysis for the dynamic case is the same as in [9]; however, quite a few extra technical obstacles have to be overcome.

Recently, Panigrahy [18] studied the dynamic version of the allocation problem (in the formulation for dynamic hash tables with insertions) for two choices and bin size $d = 2$. He established, by analyzing related branching processes, that inserting keys is possible in expected constant time as long as $n \leq 1.67m$.

For lack of space, not all details of the analysis are given in this extended abstract. They may be found in the technical report version [7].

2 The Results

2.1 The Static Case

A set S of n keys from the universe U is to be stored. We use an array $T[0..m-1]$ consisting of $m = n(1+\varepsilon)/d$ blocks (subarrays) of d cells each. Inside each block we store up to d keys sequentially. Given two hash functions $h_1, h_2 \colon U \to [m]$, we say that h_1, h_2 are *suitable* for S and d if it is possible to store each key x from S in one of the blocks $h_1(x), h_2(x)$ without any block receiving more than d keys. If the keys from S are stored according to h_1, h_2, a *lookup procedure* is obvious, which involves evaluating two hash values and searching two blocks. Our first result is the following theorem, whose proof is outlined in Section 3.

Theorem 1. *Let $\varepsilon > 0$ be arbitrary. Assume that $d \geq 1 + \frac{\ln(1/\varepsilon)}{1-\ln 2}$. Let n be sufficiently large, let $S \subseteq U$ be an arbitrary set of n keys, and let T be a table with m blocks of size d each, where $dm \geq (1+\varepsilon)n$. Further assume that $h_1, h_2\colon U \to [m]$ are fully random hash functions. Then with probability $1 - O(1/m^{d-1})$ the functions h_1, h_2 are suitable for S and d.*

2.2 Updates: The Cuckoo Insertion Procedure

Assume n keys are stored in the table T according to h_1, h_2, with blocks of size d. Inserting a new key x can best be described in terms of the directed graph G from Remark 1. In G, find a directed path $y_0, y_1, \ldots, y_\ell$ with $y_0 \in \{h_1(x), h_2(x)\}$ and y_ℓ a node that is "free", i.e., has outdegree smaller than d. (This means that block y_ℓ contains an empty cell.) The edges that form the path correspond to keys $x_1, \ldots, x_\ell$ such that x_i is stored in y_{i-1}, but may be stored in y_i. After moving x_i from y_{i-1} to y_i, for $1 \leq i \leq \ell$ (this corresponds to flipping the edges on the path), node (block) y_0 is free, and hence we can store x there. We call a path $y_0, y_1, \ldots, y_\ell$ as described an "augmenting path" for G and x.

It is very easy to see that if h_1, h_2 are suitable for $S \cup \{x\}$, then there is an augmenting path. So the problem is to find an augmenting path fast. As proposed in [9], a simple approach for this is *breadth-first-search* (*BFS*) in G, starting from $\{h_1(x), h_2(x)\}$. The time for this is proportional to the number of edges probed before a free node is found. Since the nodes in the part of G that is searched have outdegree d, this number is not larger than $2(d + d^2 + \cdots + d^\ell) < 4d^\ell$, where ℓ is the length of a *shortest* path from $\{h_1(x), h_2(x)\}$ to a free node. Thus we will have to analyze (the distribution of) the distance between $\{h_1(x), h_2(x)\}$ and the set of free nodes. The proof of the following theorem is outlined in Section 4.

Theorem 2. *Let $\varepsilon > 0$ be arbitrary. Assume that $d \geq 90.1 \cdot \ln(1/\varepsilon)$. Let n be sufficiently large, let S an arbitrary set of n keys, let $x \in U - S$, and let T be a table with m blocks of size d each, where $dm \geq (1+\varepsilon)n$. Assume that $h_1, h_2\colon U \to [m]$ are fully random hash functions, and that the keys from S have been stored in T by an algorithm that is ignorant of $h_1(x), h_2(x)$. Then the expected time to insert x by the BFS procedure is $(1/\varepsilon)^{O(\log d)}$.*

The constants in the bound are certainly not optimal. In particular, the horrendous factor 90.1 is only an artefact of our proof. Numerical estimates suggest that the approach we use could be tuned to get by with $d \geq 4\ln(1/\varepsilon)$, for ε very small. — In this extended abstract, we do not address the issue of extra space (up to $O(n)$) needed by the BFS, for the following reason: If the technique described in Section 2.3 below is used, no more than $n^{2/3}$ keys have to be handled at any time, so the scratch space problem vanishes.

An alternative approach to insertion (also suggested in [9]) is to search an augmenting path by a certain kind of random walk in G, as follows: Assume x is to be inserted. Repeat the following, starting with $z := x$:

Calculate $h_1(z)$ and $h_2(z)$. If one of the two blocks $h_1(z)$ or $h_2(z)$ is not full, store z in one such block, and stop. Ties are broken arbitrarily. If both blocks are full, randomly choose one of the keys stored in these blocks, call it z', kick z' out from its block (this is the "cuckoo step") and insert z in its place. Let $z := z'$, and start again.

Of course, some rules for stopping the loop have to be incorporated. The implementations used in our experiments [7] are based on this random walk idea, not on the BFS procedure. It is an intriguing open problem to provide an analysis of the random walk insertion procedure. (The same question is open for d-ary cuckoo hashing [9].)

2.3 Sharing Fully Random Hash Functions

For the analysis to carry through, we assume that the hash functions h_1, h_2 behave fully randomly on S. If in the course of inserting a key it turns out that h_1, h_2 are not suitable, we might want to rehash the whole set, using new hash functions h_1, h_2. Although in the balanced allocation literature the full randomness assumption is routinely used, this is not the case in the hashing literature. Earlier work on hashing (e.g., [3, 9, 13, 17]) has, very carefully, pointed out ways of working around this problem, for example by using functions from high-performance universal classes like in [22]. (This would not be sufficient for d-ary cuckoo-hashing, though.) In [8, 16] it was demonstrated that full randomness can be simulated by universal hashing at the cost of $O(n)$ words of extra space. However, using such a construction would be unsatisfactory in our context, since we aim at getting by with εn extra space.

We propose the following workaround, which might be helpful also in other contexts. Let $\varepsilon > 0$ and n be given. Using high-performance hash classes [6, 22] we may choose a function $h \colon U \to [n^{1/3}]$ so that with probability $1 - n^{-c}$ (for some constant c) the set S is split into $n^{1/3}$ pieces $S_i = \{x \in S \mid h(x) = i\}$ of size $\leq (1 + \frac{\varepsilon}{2})n^{2/3}$. (There is no need that S is known or the pieces are listed.) For each of the pieces S_i we run cuckoo hashing with blocks of size d in a separate table T_i of size $(1 + \varepsilon)n^{2/3}$. It is an easy exercise, using polynomial hash functions and techniques described in [6], to provide a pair h_1, h_2 of hash functions that with high probability behaves fully randomly on a single S_i, if we are allowed to use space $n^{5/6}$ for storing h_1, h_2. Since the data structures for the pieces S_i do not interact, we may use the *same* hash function pair h_1, h_2 for all pieces S_i, $i = 0, \ldots, n^{1/3} - 1$. The overall space is $O(n^{5/6}) = o(n)$. The algorithms described in the present paper then has to be applied to each of the pieces separately. (For details see [7].)

3 Analysis for the Static Case: Proof of Theorem 1

In [20, 21] it is shown that a set S of n keys can be stored in $m = (1 + \varepsilon)n/d$ blocks of size d if and only if for every $X \subseteq S$ it holds that $|\Gamma(X)| \geq \frac{1}{d}|X|$, where $\Gamma(X) = \{h_1(x), h_2(x) \mid x \in X\}$. (This may be seen directly: It is immediate that S can be stored if and only if the bipartite graph $(S, [dm], E)$, $E = \{(x, d \cdot h_i(x) + j) \mid x \in S, i \in \{1, 2\}, 0 \leq j < d\}$ has a matching that covers all nodes in S. Hall's marriage theorem [5, p. 31] implies that such a matching exists if and only if $|\Gamma(X)| \geq \frac{1}{d}|X|$ for all $X \subseteq S$.) Thus, to prove Theorem 1 it is sufficient to establish an upper bound on the probability F of the event that there is some $X \subseteq S$ such that $|\Gamma(X)| < \frac{1}{d}|X|$.

Lemma 1. *If* $\varepsilon \leq 0.25$ *and* $d > 1 + \frac{\ln(1/\varepsilon)}{1-\ln 2}$, *then* $F = O(m^{1-d})$.

Proof (Outline). For $1 \leq j \leq m/(1+\varepsilon)$, let $F(j)$ be the probability that there is a set Y of j blocks such that some set $X \subseteq S$, $|X| = dj$, satisfies $\Gamma(X) \subseteq Y$. Clearly, $F \leq \sum_{1 \leq j \leq m/(1+\varepsilon)} F(j)$. Using the Chernoff-Hoeffding bound (13) and the binomial bound (11), we get

$$F(j) \leq \binom{m}{j} \left(\frac{n(j/m)^2}{jd}\right)^{jd} \left(\frac{n - n(j/m)^2}{n - jd}\right)^{n-jd}$$

$$\leq (1+\varepsilon)^{-jd} m^{\frac{m(1+\varepsilon-d)}{1+\varepsilon}} j^{j(d-1)} (m-j)^{j-m} \left(\frac{m^2 - j^2}{m - j(1+\varepsilon)}\right)^{d\frac{m-j(1+\varepsilon)}{1+\varepsilon}}, \quad (1)$$

which was already observed in [20, 21]. We examine the expression on the right-hand side of (1), which we abbreviate by $f(j, \varepsilon)$, in different ranges of j. For $j = 1$ we find $f(1, \varepsilon) = O(m^{1-d})$. — For j, $2 \leq j < e^{-4}m$, we prove that $f(j, \varepsilon)$ is a decreasing function of ε; thus, we can concentrate on the case $\varepsilon = 0$. The sequence $f(j, 0), j = 2, \ldots, \lfloor e^{-4}m \rfloor$ turns out to be geometrically decreasing; hence we get $\sum_{2 \leq j \leq e^{-4}m} F(j) = O(m^{2-2d})$. — The most involved calculation concerns the range $e^{-4}m \leq j \leq (1-2\varepsilon)m$. Here we read off from (1) by substituting $j = \alpha m$, that $f(j, \varepsilon) = O(c^m)$ for a constant $c < 1$, if

$$d > \frac{\alpha \ln \alpha + (1-\alpha) \ln(1-\alpha)}{\alpha(\ln \alpha - \ln(1+\varepsilon)) + \frac{(1-\alpha(1+\varepsilon))(\ln(1-\alpha^2) - \ln(1-\alpha(1+\varepsilon)))}{1+\varepsilon}}, \quad (2)$$

for $e^{-4} \leq \alpha \leq 1 - 2\varepsilon$. We prove that the right-hand side of (2) is bounded by

$$g(\alpha) := \frac{\alpha \ln \alpha + (1-\alpha) \ln(1-\alpha)}{\alpha \ln \alpha + (1-\alpha) \ln(1+\alpha)} \quad (3)$$

By calculus we show that g is an increasing function and that $g(1 - 2\varepsilon) < 1 - \frac{\ln \varepsilon}{1-\ln 2}$. — For the range $j > (1-2\varepsilon)m$, we observe that for each $\varepsilon \leq 0.25$ the right-hand side of (2) is decreasing in α, if $1 - 2\varepsilon \leq \alpha \leq 1/(1+\varepsilon)$. □

4 The Expected Insertion Time: Proof of Theorem 2

We want to examine the expected time of the BFS algorithm for inserting a new key x in T, as described in Section 2.2 (also recall Remark 1). Just before x is inserted, a set S of n keys is stored in T. We assume that the directed graph

G determined by this placement is independent of the hash values $h_1(x), h_2(x)$ — this is the case if x was never inserted before. We start a BFS in G from $\{h_1(x), h_2(x)\}$ with the aim of finding a shortest path to a node in $Y_0 := \{y \in V \mid y \text{ is free in } G\}$. The time for the BFS in G is $O(\min\{n, d^{\ell+1}\})$, where ℓ is the length of such a shortest path. Our aim is to see that the expectation of the number of edges we have to inspect before a free node is found is $O(1)$. For this, we analyze the distribution of the number of nodes at different distances to Y_0. (The analysis runs along the lines of that of [9], but we are dealing with quite a different graph.) — Recursively define, for $k \geq 1$:

$$X_k := \{x \in S \mid h_1(x) \in Y_{k-1} \text{ or } h_2(x) \in Y_{k-1}\} \text{ and}$$
$$Y_k := \{y \in [m] \mid \exists x \in X_k \colon x \text{ is stored in } y\}.$$

We say that the keys of X_k "hit" Y_{k-1}. It is easy to see that Y_k is the set of nodes from which Y_0 can be reached in at most k steps in G. By the definitions, $Y_{k-1} \subseteq Y_k$, for $k = 1, 2, \ldots$. Inserting a new key x has cost roughly $O(d^{\ell+1})$, if ℓ is minimal with $h_1(x) \in Y_\ell$ or $h_2(x) \in Y_\ell$. The proof strategy is as follows. We show that with high probability a large constant fraction of the nodes are in $Y_{k^*+\ell^*}$ for suitable constants k^*, ℓ^*. (Lemmas 2, 3, 4, and 5). Beyond that, the complements $[m] - Y_k$ shrink at a rate of roughly $d^{-2/3}$ (Lemmas 6 and 7). If a new key x arrives, with constant probability one of the values $h_1(x), h_2(x)$ will hit $Y_{k^*+\ell^*}$ (causing constant cost). The probability that *both* $h_1(x)$ and $h_2(x)$ hit nodes at a distance $k^* + \ell^* + j$ from Y_0 shrinks geometrically with growing j, at a rate of $(d^{-2/3})^2 = d^{-4/3}$. This probability is small enough to compensate the high cost of $O(d^{k^*+\ell^*+j})$ for reaching Y_0 should this happen. Overall, the expected insertion cost remains a constant (Lemma 8).

Lemma 2. *Let $\varepsilon \leq 0.1$ and $d \geq 90.1 \cdot \ln(\frac{1}{\varepsilon})$. Then there is a constant $\beta < 1$ such that with probability $1 - O(m\beta^m)$ each set of blocks Y that satisfies $\frac{\varepsilon}{1+\varepsilon}m \leq r = |Y| \leq \frac{5}{13}m$ is hit by at least $\frac{4}{3}rd$ keys from S.*

Proof (Outline). A set of blocks Y, $|Y| = r$, is hit by $\frac{4}{3}rd$ keys with h_1 or h_2 if at most $n - \frac{4}{3}rd$ keys *avoid* Y with both hash functions h_1 and h_2. Let $F(r)$ denote the probability that there is a set Y of size r such that there are more than $n - \frac{4}{3}rd$ keys that avoid Y with both hash functions. Employing (13):

$$F(r) \leq \binom{m}{r} \left(\frac{n\left(1-\frac{r}{m}\right)^2}{n-\frac{4}{3}rd}\right)^{n-\frac{4}{3}rd} \left(\frac{n - n\left(1-\frac{r}{m}\right)^2}{\frac{4}{3}rd}\right)^{\frac{4}{3}rd}. \quad (4)$$

By $n = dm/(1+\varepsilon)$, the binomial bound (11), and the substitution $r = \alpha m$ we observe that $F(r) < \beta^m$ for a suitable $0 < \beta < 1$, provided that

$$d > \frac{\alpha \ln \alpha + (1-\alpha)\ln(1-\alpha)}{\left(\frac{1}{1+\varepsilon} - \frac{4}{3}\alpha\right)\left(2\ln(1-\alpha) - \ln\left(1 - \frac{4(1+\varepsilon)\alpha}{3}\right)\right) + \frac{4\alpha}{3}\ln\left(\frac{3(2-\alpha)}{4(1+\varepsilon)}\right)} \quad (5)$$

for all α between $\frac{\varepsilon}{1+\varepsilon}$ and $\frac{5}{13}$. The right-hand side of (5) can be shown to be bounded by $90.1\ln(1/\varepsilon)$, for $0 \le \varepsilon \le 0.1$ and $\frac{\varepsilon}{1+\varepsilon} \le \alpha \le \frac{5}{13}$. □

Lemma 3. *If the digraph G induced by S being stored meets the conclusion of Lemma 2, then there is some $k^* \le 2 + \log_{\frac{4}{3}}\left(\frac{5(1+\varepsilon)}{13\varepsilon}\right)$ such that $|Y_{k^*}| > \frac{m}{2}$.*

Proof. Assume $k \ge 1$ and $|Y_{k-1}| \le \frac{5}{13}m$. Consider the set X_k of keys that hit Y_{k-1}. By the assumption, $|X_k| \ge \frac{4}{3}d|Y_{k-1}|$. By definition, all keys $x \in X_k$ are stored in blocks in Y_k. Only $d|Y_{k-1}|$ of them can be stored in Y_{k-1}, so at least $\frac{1}{3}d|Y_{k-1}|$ of them must be stored in $Y_k - Y_{k-1}$, which implies that $|Y_k - Y_{k-1}| \ge \frac{1}{3}|Y_{k-1}|$; hence $|Y_k| \ge \frac{4}{3}|Y_{k-1}|$.

Now let k' be minimal with $|Y_{k'}| > \frac{5}{13}m$. By the above, it is easy to see that either $|Y_{k'}| > \frac{1}{2}m$ (then we let $k^* = k'$) or $|Y_{k'+1}| \ge \frac{4}{3} \cdot \frac{5}{13}m > \frac{1}{2}m$ (then we let $k^* = k' + 1$). (This holds even if $5m/13 < |Y_{k'}| \le m/2$, since then we apply the conclusion of Lemma 2 to a subset of $Y_{k'}$ of size $\lfloor 5m/13 \rfloor$.)

Because there are exactly εn free cells in the table, we have $|Y_0| \ge \frac{\varepsilon n}{d} = \frac{\varepsilon}{1+\varepsilon}m$. Thus, $|Y_k| \ge \frac{\varepsilon}{1+\varepsilon}m\left(\frac{4}{3}\right)^k$ for $0 \le k < k'$, whence we get $k^* \le 1 + k' \le 2 + \left\lfloor \log_{\frac{4}{3}}\left(\frac{5(1+\varepsilon)}{13\varepsilon}\right) \right\rfloor$. □

Lemma 4. *Let $d \ge 8$. Then there is some $\beta < 1$ such that with probability $1 - O(\beta^m)$ we have that each set Y of blocks with $\frac{m}{2} \le |Y| \le m - \frac{4m}{e^4 d^3}$ is hit by at least $n - \frac{9}{10}d(m - |Y|)$ keys.*

Proof (Outline). Let $\overline{Y} = [m] - Y$ and $r = |\overline{Y}|$. By (13) and (11) we find that the probability that there is a set Y with $r = m - |Y|$ in $[\frac{4m}{e^4 d^3}, \frac{m}{2}]$ such that more than $\frac{9}{10}dr$ keys hit $\overline{Y}$ with both hash functions is bounded by

$$\frac{m^m}{r^r(m-r)^{m-r}}\left(\frac{10r}{9(1+\varepsilon)m}\right)^{\frac{9}{10}rd}\left(\frac{10(m^2 - r^2)}{(10m - 9r(1+\varepsilon))m}\right)^{\frac{dm}{1+\varepsilon} - \frac{9}{10}rd}. \quad (6)$$

We denote the expression in (6) by $f(r, \varepsilon)$. It is not hard to see that $f(r, \varepsilon)$ is decreasing in ε, thus we can concentrate on the case $\varepsilon = 0$. After replacing ε by 0 in (6), and substituting $\alpha = r/m$, an expression for a function $g(\alpha, d)^m$ results. For each fixed α the function $g(\alpha, d)$ is decreasing in d. A look at a Maple plot reveals that $g(\alpha, 8)$ does not exceed $\max\{g(\frac{4}{e^4 d^3}, 8), g(0.5, 8)\} = g(\frac{4}{e^4 d^3}, 8) < 1$. □

Lemma 5. *If G meets the conclusions of Lemmas 2 and 4, and k^* satisfies $|Y_{k^*}| \ge \frac{m}{2}$, then there is some ℓ^*, $\ell^* = O(\log d)$ such that $m - |Y_{k^*+\ell^*}| \le \frac{4m}{e^4 d^3}$.*

Proof. The induction proof, which uses Lemma 4, is omitted. □

The following lemma states a standard expansion property of bipartite random graphs, see [15, p. 109].

Lemma 6. *Let $d \geq 20$, and $\gamma = \frac{4}{e^4 d^3}$. With probability $1 - O(m^{-d/2})$ we have that each set $X \subseteq S$ of keys with $d \leq |X| \leq \gamma dm$ hits more than $\Delta |X|$ different blocks, where $\Delta = \frac{1}{d^{1/3}} + \frac{1}{d}$.*

Proof. The probability that there is a set X of j keys, $d \leq j \leq \gamma dm$, that hits no more than Δj blocks can be bounded by

$$\sum_{d \leq j \leq \gamma dm} \binom{n}{j} \binom{m}{\lfloor \Delta j \rfloor} \left(\frac{\lfloor \Delta j \rfloor}{m}\right)^{2j} \leq \sum_{d \leq j \leq \gamma dm} \left(\frac{en}{j}\right)^j \left(\frac{em}{\Delta j}\right)^{\Delta j} \left(\frac{\Delta j}{m}\right)^{2j}. \quad (7)$$

Straightforward simplifications, using that $d \geq 20$ implies that $\Delta < \frac{1}{2}$, lead to the bound $\sum_{d \leq j \leq \gamma dm} \left((e/2)^{3/2} \cdot d \cdot \sqrt{j/m}\right)^j$ for the right hand side in (7). For $j = d, d+1, \ldots, \lfloor \gamma dm \rfloor$ the terms in this sum are geometrically decreasing by a factor smaller than $\frac{1}{2}$, hence the sum is bounded by $O(m^{-d/2})$. □

We conclude that for $k \geq k^* + \ell^*$ the complements of the sets Y_k shrink fast.

Lemma 7. *Assume that $|[m] - Y_{k^*+\ell^*}| \leq \gamma m$ and that the hash functions h_1, h_2 meet the conclusion of Lemma 6. Then the cardinalities $a_j = |[m] - Y_{k^*+\ell^*+j}|$, $j = 0, 1, 2, \ldots$, satisfy $a_j \leq d^{-2/3} \cdot a_{j-1}$ for $j = 1, 2, 3, \ldots$. Hence,*

$$|[m] - Y_{k^*+\ell^*+j}| \leq \gamma d^{-2j/3} m, \text{ for } j = 0, 1, 2, \ldots.$$

(In particular, there is some L with $Y_{k^*+\ell^*+L} \neq [m] = Y_{k^*+\ell^*+L+1}$.)

Proof. Fix $j \geq 1$. If $a_j = 0$, there is nothing to prove; thus assume $a_j \geq 1$. Then by the definitions, all a_j nodes in $[m] - Y_{k^*+\ell^*+j}$ are full. That means that the set E_j of edges in G with tails in $[m] - Y_{k^*+\ell^*+j}$ has cardinality da_j. By the assumption that the conclusion of Lemma 6 is satisfied the edges in E_j touch at least $\Delta da_j = (d^{2/3} + 1)a_j$ nodes overall. Only a_j of these nodes are in $[m] - Y_{k^*+\ell^*+j}$. By the definition of the sets Y_k, no edge in G can run from a node in $[m] - Y_{k^*+\ell^*+j}$ to a node in $Y_{k^*+\ell^*+(j-1)}$. Hence the heads of the edges in E_j hit at least $(d^{2/3}+1)a_j - a_j = d^{2/3} a_j$ distinct nodes in $[m] - Y_{k^*+\ell^*+(j-1)}$, which implies that $a_{j-1} \geq d^{2/3} a_j$. □

Lemma 8. *Assume $Y_0, Y_1, \ldots, Y_{k^*}, \ldots, Y_{k^*+\ell^*}, \ldots, Y_{k^*+\ell^*+L}$ are fixed and fulfill the expansion properties from Lemmas 3, 5, and 7. Assume further that $h_1(x), h_2(x)$ are random values in $[m]$. Then the expected number of edges probed in the BFS insertion procedure for x is $(1/\varepsilon)^{O(\log d)}$.*

Proof. Let N_x be the number of edges of G probed when x is inserted. Let $\sigma_k = \sum_{0 < \kappa \leq k} d^\kappa < 2d^k$, for $k \geq 0$, and $k_x = \min\{k \mid h_1(x) \in Y_k \text{ or } h_2(x) \in Y_k\}$. Then the number of edges of G probed when inserting x is not larger than $2\sigma_{k_x}$. Thus, it is sufficient to estimate $\mathbf{E}(\sigma_{k_x})$. We have

$$\mathbf{E}(\sigma_{k_x}) = \sum_{q \geq 1} \mathbf{Prob}(\sigma_{k_x} \geq q) = \sum_{k \geq 1} \mathbf{Prob}(k_x \geq k) \cdot d^k. \quad (8)$$

The last sum in (8) is estimated in two pieces. We have

$$\sum_{1 \leq k \leq k^* + \ell^*} \mathbf{Prob}(k_x \geq k) \cdot d^k \leq (k^* + \ell^*) d^{k^* + \ell^*}. \tag{9}$$

For the rest of the sum in (8), we notice that by Lemma 7

$$\sum_{k^* + \ell^* < k \leq k^* + \ell^* + L} \mathbf{Prob}(k_x \geq k) \cdot d^k = d^{k^* + \ell^*} \cdot \sum_{1 \leq j \leq L} \mathbf{Prob}(k_x \geq k^* + \ell^* + j) \cdot d^j$$

$$\leq d^{k^* + \ell^*} \cdot \sum_{1 \leq j \leq L} \mathbf{Prob}(h_1(x), h_2(x) \in [m] - Y_{k^* + \ell^* + (j-1)}) \cdot d^j$$

$$\leq d^{k^* + \ell^*} \cdot \sum_{1 \leq j \leq L} (d^{-2(j-1)/3})^2 \cdot d^j < 2 d^{k^* + \ell^* + 1}. \tag{10}$$

The sum of the parts in (9) and (10) is bounded by $(k^* + \ell^* + 2) d^{k^* + \ell^* + 1} = O(d)^{O(\log(1/\varepsilon))} = (1/\varepsilon)^{O(\log d)}$. This shows that the expected number of edges probed in inserting x is bounded by $(1/\varepsilon)^{O(\log d)}$. □

To prove Theorem 2 we note that with probability $1 - O(m^{-d/2})$ the graph G satisfies the conclusions of Lemmas 2, 4, and 6. If this is the case, then the expansion properties of Lemmas 3, 5, and 7 hold, and we may apply Lemma 8 to obtain the claimed bound on the expected insertion time. If G does not have the expansion properties from Lemmas 3, 5, and 7, and Y_0 is reachable from $\{h_1(x), h_2(x)\}$, the BFS will find an augmenting path in time $O(n)$ — this gives a contribution of $O(m^{1-d/2})$ to the expected insertion time. In case Y_0 is not reachable from $\{h_1(x), h_2(x)\}$, the functions h_1, h_2 are not suitable for $S \cup \{x\}$, and we must perform a total rehashing for all these keys. By Theorem 1 this happens with probability $O(m^{1-d})$. It is easily seen that even if we simply insert the keys by the BFS procedure, and rehash again if necessary, the expected time for rebuilding the table is $O(n)$. Hence, this last case contributes $O(m^{2-d})$ to the expected insertion cost.

5 Conclusion

We obtained new results for a natural data allocation problem arising in different contexts: balanced allocation with two choices, edge orientation in random graphs, dynamic dictionaries with worst case constant access time. It is an intriguing open problem to analyze at least one variant of the random-walk insertion procedure from Section 2.2, if possible establishing a bound of $O(\ln(1/\varepsilon))$ on the expected number of blocks probed, while maintaining the bound $d = O(\log(1/\varepsilon))$.

Acknowledgement. Remarks from several anonymous referees, which helped in improving the exposition, are gratefully acknowledged.

References

1. Y. Azar, A. Z. Broder, A. R. Karlin, and E. Upfal. Balanced allocations. *SIAM J. Comput.*, 29:180–200, 2000.
2. P. Berenbrink, A. Czumaj, A. Steger, and B. Vöcking. Balanced allocations: The heavily loaded case. In *32nd STOC*, pp. 745–754. ACM, 2000.
3. A. Czumaj, F. Meyer auf der Heide, and V. Stemann. Contention resolution in hashing based shared memory simulations. *SIAM J. Comput.*, 29:1703–1739, 2000.
4. A. Czumaj, Ch. Riley, and Ch. Scheideler. Perfectly balanced allocation. In *RANDOM-APPROX*, LNCS 2764, pp. 240–251. Springer, 2003.
5. R. Diestel. *Graph Theory*. Springer, New York, 1997.
6. M. Dietzfelbinger and F. Meyer auf der Heide. Dynamic hashing in real time. In Buchmann, J., et al., editor, *Informatik · Festschrift zum 60. Geburtstag von Günter Hotz*, pp. 95–119. B. G. Teubner, 1992.
7. M. Dietzfelbinger and C. Weidling. Balanced allocation and dictionaries with tightly packed constant size bins. Technical Report, http://www.tu-ilmenau.de/fakia/md-papers.html, 2005.
8. M. Dietzfelbinger and P. Woelfel. Almost random graphs with simple hash functions. In *35th STOC*, pp. 629–638. ACM, 2003.
9. D. Fotakis, R. Pagh, P. Sanders, and P. Spirakis. Space efficient hash tables with worst case constant access time. *Theory of Computing Systems*, 38:229–248, 2005.
10. A. Frieze. Personal communication in [1]. 1990.
11. T. Hagerup and Ch. Rüb. A guided tour of Chernoff bounds. *Inf. Process. Lett.*, 33:305–308, 1990.
12. R. Karp. Random graphs, random walks, differential equations and the probabilistic analysis of algorithms. In *15th STACS*, LNCS 1373, pp. 1–2. Springer, 1998.
13. R. Karp, M. Luby, and F. Meyer auf der Heide. Efficient PRAM simulations on a distributed memory machine. *Algorithmica*, 16:517–542, 1996.
14. M. Mitzenmacher, A. W. Richa, and R. Sitaraman. The power of two random choices: A survey of techniques and results, vol. 1, pp. 255–312. In Rajasekaran et al., editor, *Handbook of Randomized Computing*. Kluwer Academic Press, 2001.
15. R. Motwani and P. Raghavan. *Randomized Algorithms*. Cambridge University Press, 1995.
16. A. Östlin and R. Pagh. Uniform hashing in constant time and linear space. In *35th STOC*, pp. 622–628. ACM, 2003.
17. R. Pagh and F. F. Rodler. Cuckoo hashing. *J. Algorithms*, 51:122–144, 2004.
18. R. Panigrahy. Efficient hashing with lookups in two memory accesses. In *16th SODA*. ACM-SIAM, 2005.
19. P. Sanders. Fast priority queues for cached memory. In *1st Workshop ALENEX*, LNCS 1619, pp. 312–327. Springer, 1999.
20. P. Sanders. Reconciling simplicity and realism in parallel disk models. In *12th SODA*, pp. 67–76. ACM-SIAM, 2001.
21. P. Sanders, S. Egner, and J. Korst. Fast concurrent access to parallel disks. In *11th SODA*, pp. 849–858. ACM-SIAM, 2000.
22. A. Siegel. On universal classes of fast high performance hash functions, their time-space tradeoff, and their applications. In *30th FOCS*, pp. 20–25. IEEE, 1989.

A Some Inequalities

We will be using the following upper bounds for binomial coefficients:

$$\binom{n}{k} \leq \frac{n^n}{k^k(n-k)^{n-k}} = \left(\frac{1}{\mu^\mu(1-\mu)^{1-\mu}}\right)^n, \text{ for } 0 \leq k \leq n, \quad (11)$$

where $\mu = k/n$, and

$$\binom{n}{k} \leq \left(\frac{e \cdot n}{k}\right)^k. \quad (12)$$

Further, a standard version of the Chernoff-Hoeffding bounds is used repeatedly: If $X_1, \ldots, X_n$ are independent 0-1-valued random variables and $X = X_1 + \cdots + X_n$, then for $\mathbf{E}(X) \leq a \leq n$ we have

$$\mathbf{Prob}(X \geq a) \leq \left(\frac{\mathbf{E}(X)}{a}\right)^a \left(\frac{n - \mathbf{E}(X)}{n - a}\right)^{n-a}. \quad (13)$$

(For a proof, see e.g. [11].)

Worst Case Optimal Union-Intersection Expression Evaluation

Ehsan Chiniforooshan, Arash Farzan, and Mehdi Mirzazadeh

School of Computer Science, University of Waterloo
{echinifo, afarzan, mmirzaza}@cs.uwaterloo.ca

Abstract. We consider the problem of evaluating an expression consisting of unions and intersections of some sorted sets. Given the expression and the sizes of the sets, we are interested in the worst-case complexity of evaluating the expression in terms of the sizes of the sets. We assume no set is repeated in the expression. We show a lower bound on this problem and present an algorithm that matches the lower bound asymptotically.

1 Introduction

In this paper, we study the problem of evaluating a set expression consisting of a number of union and intersection operators. Sets are known to be sorted and we also assume that no set is repeated more than once in the input. While the worst case complexity in terms of the size of the whole combined input is straightforward, we measure the running time of algorithms as a function of the sizes of the input sets; we are interested in a worst-case optimal algorithm.

The problem arises in the context of evaluating search queries in text database systems; search engines maintain a set $S(w)$ for each word w consisting of all documents that contain w [1,7,11]. Thus, answering to a query such as "Database OR Search AND Engine", requires evaluation of the expression $S(\mathsf{Database})\,\cup\,(S(\mathsf{Search})\,\cap\,S(\mathsf{Engine}))$. Note that the queries and their corresponding expressions can become very complicated if the queries are automatically generated [6].

Different variations of the problem have been studied before. The simplest case which is finding intersection or union of two sets is equivalent to the problem of merging two ordered sets of sizes m and n, which was studied by Hwang and Lin [5]. They present an algorithm that matches the information theoretic lower bound $\lceil \log \binom{m+n}{n} \rceil$. They choose sorted arrays as the format of the input and a list of cross references (pointers) between arrays as the output format. Later Brown and Tarjan [2,3] and Pugh [9] showed how data structures such as AVL-tree, B tree, or skip-list can be used as the format of the input and output.

Later, Demaine, López-Ortiz, and Munro [4], studied a somewhat more general case in which the number of input sets can be more than two. The expressions they looked at, though, consist of just one type of operand: they are either all unions or all intersections. Their algorithm is adaptive; they do not focus on the worst-case complexity of the problem. They define the *difficulty* of every possible input I as a function $D(I)$, which measures how complicated a proof for the input I is; they focus on minimizing the maximum value of $\frac{f(I)}{D(I)}$ among all inputs I of size n, where $f(I)$ is the running time of the algorithm on I. The adaptive problem was generalized by Mirzazadeh [8] to general expressions consisting of both union and intersection operators.

Neither of the algorithms that we mentioned work optimally in the worst case in terms of the sizes of the input sets. In this paper, we consider the worst case complexity as mentioned. We present a lower bound and then an algorithm that matches the lower bound.

The rest of this paper is organized as follows: In Section 2 we give some definitions and preliminary observations. In Section 3, Theorem 4, we present our lower bound and finally, in Section 4, the optimal algorithm is explained.

2 Definitions and Preliminaries

We study the problem of evaluating a set expression when the inputs are ordered sets and the output is required to be an ordered set as well. We formally define an *input* as a pair (T, Γ), where T and Γ are defined as follows. T is an *union-intersection tree* representing the expression: every internal node v is assigned a union or an intersection operator $\pi(v)$ and each leaf v of T represents an input set and is assigned an integer size(v). We call T the *signature* of the input I. Also, Γ is an *assignment function* that assigns an ordered set of size size(v) to each leaf v. For an internal node v with k children $u_1, \ldots, u_k$, we denote the union or intersection of $\Gamma(u_1), \Gamma(u_2), \ldots,$ and $\Gamma(u_k)$, depending on the operator assigned to v, by $\Gamma(v)$. By the *result* of an input (T, Γ) we mean the set $\Gamma(\text{Root}(T))$. We denote the set of nodes of a tree T and the set of leaves of T by V_T and leaves(T), respectively. Without loss of generality, we assume that every internal node has at least two children, and that the operator assigned to every internal node other than root differs from the operator assigned to its parent.

In this paper we focus on the *comparison-based algorithms* which are those that, for any input $I = (T, \Gamma)$, use only comparisons in the input sets to compute the result. In our model, the algorithm has oracle access to Γ, which means that the algorithm reads the signature of the input and can later submit queries of the form (x, y) to the oracle, where x and y are members of the input sets. Then, the oracle informs the algorithm of the relative values of x and y, that is, the algorithm is told whether x is less than, equal to, or greater than y according to Γ. In such situations we say x and y are *touched* by the algorithm. We show the interaction between the algorithm $\mathcal{A}$ and the oracle O on the input (T, Γ) by $\langle \mathcal{A}, O(\Gamma) \rangle(T) = (q_1, r_1, \ldots, q_k, r_k, R)$ where q_i is the ith query of the algorithm, r_i is the response of O to the ith query, and $R = \Gamma(\text{Root}(T))$ is the result. We expect the algorithm to specify ranges of input sets that appear in the result, rather than to write all elements of the result. This allows us to generate the output in sub-linear time if possible. More precisely, we define the output format below. We use $S[i]$ to denote the ith element of a sequence S.

Definition 1. *Consider an input $I = (T, \Gamma)$ and a set S. A cross reference representation of S is a sequence of triples $(v_1, b_1, b'_1), \ldots, (v_n, b_n, b'_n)$ where v_i is a leaf of T and $1 \leq b_i \leq b'_i \leq \text{size}(v_i)$, for every $1 \leq i \leq n$, $\Gamma(v_j)[b'_j] < \Gamma(v_{j+1})[b_{j+1}]$, for every $1 \leq j < n$, and $S = \cup_{i=1}^{n} \cup_{j=b_i}^{b'_i} \{\Gamma(v_i)[j]\}$.*

A leaf v of an expression tree T is a *shallow* leaf if v is a child of Root(T) and $\pi(\text{Root}(T)) = \cup$.

We define $\binom{s}{s_1,\ldots,s_n}$, when $s \leq \sum_{i=1}^{n} s_i$ as the number of ways to select sets X_1, $\ldots$, X_n of sizes $s_1, \ldots, s_n$, respectively, such that X_i's are subsets of a given set X of size s and $\cup_{i=1}^{n} X_i = X$. Note that this definitions matches definition of the well known notation $\binom{s}{s_1,\ldots,s_n}$ when $\sum_{i=1}^{n} s_i = s$. Also for a union-intersection tree, we define functions ψ^* and ψ over the set of nodes of T as follows: for a leaf v we define $\psi(v) = \text{size}(v)$. If v is an internal node and $u_1, \ldots, u_k$ is the list of children of v, we define $\psi(v) = \min_{i=1}^{k} \psi(u_i)$ when v is an intersection node, and $\psi(v) = \sum_{i=1}^{k} \psi(u_i)$, otherwise. In fact $\psi(v)$ is the maximum potential size of $\Gamma(v)$. Also, for every node v we define $\psi^*(v) = \min \psi(u)$, where the minimum is taken over all ancestors u of v, including v itself. Note that the values of ψ and ψ^* for all nodes of an expression tree T can be evaluated in time $O(|V_T|)$.

Observation 1. *Suppose v is an internal node with k children $u_1, \ldots, u_k$. $\sum_{i=1}^{k} \psi^*(u_i) \geq \psi^*(u)$ if u is a union node and $\sum_{i=1}^{k} \psi^*(u_i) \geq 2\psi^*(u)$, otherwise.*

We present an algorithm such that for every signature T, the maximum running time of the algorithm, over all possible inputs with the signature T, is minimum.

3 Lower Bounds

In this section, fixing an arbitrary union-intersection tree T, we present a lower bound on the maximum number of comparisons performed by any algorithm when it is run on inputs with the signature T. For this purpose, we design an adversary $\mathcal{B}$ that for any given algorithm $\mathcal{A}$ and any signature T, as the algorithm $\mathcal{A}$ proceeds and compares members of the input, $\mathcal{B}$ fixes relative values of more members and responds to $\mathcal{A}$. In this way, an assignment function Γ is constructed gradually and we make sure that there is at least one Γ such that the responses of $\mathcal{B}$ to $\mathcal{A}$ are consistent with Γ. For two members x and y, if some certain conditions (which will be defined later in this section) hold, we say x and y are *similar*. We suppose that when a query (x, y) is submitted, in addition to relative values of x and y, $\mathcal{A}$ is informed of whether x and y are similar or not. It is clear that any lower bound for algorithms working in this new model is a lower bound for algorithms working in the comparison model as well.

Considering a set O_T of size $\psi^*(\text{Root}(T))$, $\mathcal{B}$ responds queries such that O_T becomes the result of the final input. We spread the elements of O_T down to the nodes of T such that every vertex v is *labeled* by a subset of O_T of size $\psi^*(v)$ in such a way that for every union (intersection) vertex v, the union (the intersection) of labels of its children is the label of v. As a simple observation, suppose that S is the set of leaves having a certain member s of O_T in their "labels". Then, if s appears in the set associated with every leaf in S, s will appear in the result. Responses of $\mathcal{B}$ will be such that rather than appearing in sets associated with all leaves in S, s will appear in the sets associated with leaves in just a subset of S, yet still s will appear in the result of the input. This subset will be determined based on the behavior of $\mathcal{A}$.

Now we define the above labeling more formally. For convenience, rather than using real numbers, we will use triples of integers for representing members of our sets. Triples are compared to each other according to their lexicographic order. We define the set O_T as $\{(1,0,0),(2,0,0),\ldots,(m,0,0)\}$ where $m = \psi^*(\text{Root}(T))$. Given a triple $x = (i,j,k)$, we call i, j, and k the first, the second, and the third coordinates of x, respectively.

Definition 2. *Given a signature T, $\Lambda : V_T \mapsto 2^{O_T}$ is a proof labeling for T if it has the following properties: First, $\Lambda(\text{Root}(T)) = O_T$. Second, for every vertex $v \in V_T$, $|\Lambda(v)| = \psi^*(v)$. Third, if $v \in V_T - \text{leaves}(T)$ and $u_1, \ldots, u_k$ are children of v, $\cup_{i=1}^{k} \Lambda(u_i) = \Lambda(v)$ if v is a union node; otherwise, $\Lambda(u_i) = \Lambda(v)$, for every i, $1 \leq i \leq k$.*

$\mathcal{B}$ chooses a proof labeling Λ arbitrarily from all possible labellings. Then, $\mathcal{B}$ divides the sequence of members of every leaf v of T into $\psi^*(v)$ regions of sizes $\lfloor \frac{\text{size}(v)}{\psi^*(v)} \rfloor$ or $\lceil \frac{\text{size}(v)}{\psi^*(v)} \rceil$. For a leaf v and integers i and a, if the ith biggest member of $\Lambda(v)$ is $(a,0,0)$, then the ith region of v is called an a-region. For any a and any a-region $\mathcal{R}$, $\mathcal{B}$ sets the first coordinates of all members of $\mathcal{R}$ to a at the beginning. Thus, given a member x of an a-region and a member y of a b-region such that $a \neq b$, whenever a query (x,y) is submitted, $\mathcal{B}$ can answer the query without knowledge on the second and the third coordinates.

For any region $\mathcal{R}$, the second coordinate of exactly one element of $\mathcal{R}$, which is called the *key member of $\mathcal{R}$*, will be zero. The strategy is to determine the second coordinates of triples of a region $\mathcal{R}$ in such a way that $\mathcal{A}$ does not touch the key member of $\mathcal{R}$ before touching $\log |\mathcal{R}|$ members of $\mathcal{R}$ where $|\mathcal{R}|$ denotes the length of $\mathcal{R}$. The second coordinates of members of a region are all distinct and the third coordinates of non-key members are zero. Moreover, the third coordinates of the key members are determined in such a way that $\mathcal{A}$ needs to touch all key members (actually we will prove a stronger fact).

Next, we explain the strategy of determining second coordinates of the triples. For every region $\mathcal{R}$ we consider a variable S storing a subsequence in $\mathcal{R}$, initially $\mathcal{R}$. At any point, the following condition will hold: The second coordinate of every member in $\mathcal{R} \setminus S$ is already fixed, the second coordinate of every member of $\mathcal{R}$ placed before members of S is at most $-|S|$, and the second coordinate of every member of $\mathcal{R}$ placed after members of S is at least $|S|$. Now whenever a member s of S is touched, if s is the only member of S, $\mathcal{B}$ sets its second coordinate to zero. Otherwise, depending on whether s is in the first half or in the second half of S, $\mathcal{B}$ considers members of S placed after or before S, fixes second coordinates of these members as explained in Figure 1, and deletes them from S. Then, by touching each member of $\mathcal{R}$ the length of S is divided by at most two. Since the value of 0 is not assigned to the second coordinate of any member unless $|\mathcal{S}| = 1$, $\log |\mathcal{R}|$ members of $\mathcal{R}$ have already been touched at the time the key member of $\mathcal{R}$ is being touched. Whenever a member is touched in which the second coordinate is not determined before, before attempting to answer the query $\mathcal{B}$, determines the second coordinate according to the method we described here. Therefore, we have the following theorem.

Theorem 1. *If all key members of $L \subseteq \text{leaves}(T)$ are touched by $\mathcal{A}$ in $\langle \mathcal{A}, \mathcal{B} \rangle(T)$, $\mathcal{A}$ has submitted at least $\sum_{v \in L} \psi^*(v) \cdot \lg \lceil \frac{\text{size}(v)}{\psi^*(v)} + 1 \rceil$ queries.*

From now on, when we talk about the strategy of $\mathcal{B}$ for responding a given query (x, y), we assume second coordinates of x and y are determined.

---

if $|\mathcal{S}| = 1$ then
    – set the second coordinate of s equal to zero;
    – set $\mathcal{S}$ equal to the empty sequence;
else
    suppose s is the ith member of $\mathcal{R}$;
    if $i < |\mathcal{S}| - i + 1$ then
      – assign values $-(|\mathcal{S}| - 1)$, $-(|\mathcal{S}| - 2)$, ..., $-(|\mathcal{S}| - i)$ to the second coordinates of the first i members of $\mathcal{S}$;
      – Remove the first i members of $\mathcal{S}$ from it;
    else
      – assign values $i - 1$, i, ..., $|\mathcal{S}| - 1$ to the second coordinates of the last $|\mathcal{S}| - i + 1$ members of $\mathcal{S}$;
      – Remove the last $|\mathcal{S}| - i + 1$ members of $\mathcal{S}$ from it;

Fig. 1. How to determine the second coordinates of members

Given a query (x, y) if x and y are from two a-regions, for some a, and the second coordinate of one of x or y is non-zero (that is at most one of x and y is a key member), $\mathcal{B}$ has enough information to answer the query. We define two members x and y to be *similar* if x and y are key members of two a-regions, for some a. As noted before, $\mathcal{A}$ will be informed if x and y are similar. In Subsection 3.1 we show that $\mathcal{B}$ can respond to queries on similar members in such a way that at the end for each member x, $\mathcal{A}$ knows which members are similar to x.

Theorem 2. *For any signature T, and any deterministic comparison-based algorithm $\mathcal{A}$, if after an interaction $\langle A, B \rangle(T) = (q_1, r_1, \ldots, q_k, r_k, R)$, $\mathcal{A}$ knows all sets of similar members, then $k \geq \frac{1}{2}(l_1 + \log_6 l_2)$ where l_1 and l_2 are defined below, L is the set of non-shallow leaves of T, and $u_1, \ldots, u_k$ are children of v in the expressions.*

$$l_1 = \sum_{v \in L} \psi^*(v) \cdot \lg(\tfrac{\text{size}(v)}{\psi^*(v)} + 1)$$

$$l_2 = \prod_{v : \pi(v) = \cup} \binom{\psi^*(v)}{\psi^*(u_1), \ldots, \psi^*(u_k)}$$

Proof. For any proof labeling Λ that $\mathcal{B}$ chooses, any key member of a non-shallow leaf is similar to at least another member. Therefore, $\mathcal{A}$ has touched all key members of non-shallow leaves of T in $\langle \mathcal{A}, \mathcal{B} \rangle(T)$. Thus, by Theorem 1, we have $k \geq l_1$.

Also, for any member of the result R, $\mathcal{A}$ is aware of all similar members to that member to that member. This means that $\mathcal{A}$ has enough information to figure out what proof labeling $\mathcal{B}$ has chosen and thus Λ can be expressed as a function of the sequence of responses of $\mathcal{B}$, $(r_1, \ldots, r_k)$. Hence, $(r_1, \ldots, r_k)$ is different for different Λ's. So, since r_i can have only six different values and the number of possible proof labellings is l_2, we have $k \geq \log_6 l_2$. Putting this upper bound for k together with the previous upper bound we conclude that $k \geq \frac{1}{2}(l_1 + \log_6 l_2)$. □

3.1 The Game

In this part we fix an a and focus on answering queries on key members of a-regions (which are similar). Let's focus on the subtree of T consisting of leaves of T that contain a-regions and their ancestors, and ignore the rest of T. We have a two player game here between $\mathcal{A}$ and $\mathcal{B}$, in which $\mathcal{A}$ submits queries between key members of a-regions until it knows weather a key member of an a-region appears in the result; $\mathcal{B}$ aims to prevent the game from finishing before $\mathcal{A}$ has enough information for calculating the set of members similar to a key member of an a-regions. The configuration of the game at some specific point is a tuple $(\mathcal{T}, \mathcal{G})$ where $\mathcal{T}$ is a union-intersection tree in which the root is not a union node or has only one child and $\mathcal{G}$ is a graph defined on leaves of $\mathcal{T}$ storing the history of queries submitted. Each leaf is representing only one member (the key member of its a-region, in our application) and thus we can view a query as a pair of leaves of $\mathcal{T}$. Moreover, each edge of $\mathcal{G}$ is labeled with one of $<, =,$ or $>$, demonstrating the respond to that query. $\mathcal{A}$ is supposed to not submit a query twice during the game. $v \in \mathsf{leaves}(\mathcal{T})$ is $\mathcal{G}$-*identical* to $u \in \mathsf{leaves}(\mathcal{T})$ if there is a path from v to u in $\mathcal{G}$ such that labels of all edges in the path are $=$.

The game finishes when the root in $(\mathcal{T}, \mathcal{G})$ has a *witness* where a witness of a node v is a subset of nodes recursively defined as follows. The only witness of a leaf v is $\{v\}$. A set S is a witness of a union node v if $v \in S$ and $S \setminus \{v\}$ is a witness of a child of v. For an intersection node v, S is a witness of v if $v \in S$ and every child u of v has a witness W_u such that $S = \{v\} \cup \bigcup_u W_u$ and every two leaves in S are $\mathcal{G}$-identical. When a game finishes, $\mathcal{B}$ wins if G is connected and the root has a witness; otherwise $\mathcal{A}$ is the winner. We prove that $\mathcal{B}$ has a winning strategy.

Moreover, let $\mathsf{component}(u)$ be the set of all leaves of $\mathcal{T}$ that are connected to u including u itself. By "$(\mathcal{T}, \mathcal{G})$ can be reduced to $(\mathcal{T}', \mathcal{G}')$" we mean that if $\mathcal{B}$ has a wining strategy in $(\mathcal{T}', \mathcal{G}')$, then $\mathcal{B}$ has a winning strategy in $(\mathcal{T}, \mathcal{G})$. In addition, we say that a vertex of $\mathcal{T}$ is *unfinished* if either it is an internal node or if it is the only version of its connected component in G. We define four operations which can be used to reduce a game instance to a smaller one.

Throwing away: A leaf $u \in \text{leaves}(T)$ can be thrown away if u is a child of a union node and is unfinished. By this action, we remove u from T and G. The resulted game is denoted by $\text{throw}_u(T, G)$.

Joining: A leaf $u \in \text{leaves}(T)$ can be joined to its sibling $v \in \text{leaves}(T)$ if u is a child of an intersection node and v is $\mathcal{G}$-identical to u. By this action, we remove u from T and G, and we denote the resulted game by $\text{join}_{u \to v}(T, G)$.

Contracting: An internal node $x \in V_T - (\text{leaves}(T) \cup \{\text{Root}(T)\})$ that is a child of $p \in V_T$ can be contracted if it has only one child, say $q \in V_T$. By this action, we remove x from T and make q a child of p. Otherwise, we remove both x and q from T and make the children of q children of p. We denote the resulted game by $\text{contract}_x(T, G)$.

Dispersing: An internal node $x \in V_T - (\text{leaves}(T) \cup \{\text{Root}(T)\})$ can be dispersed if all its children $u_1, \ldots, u_k$ are leaves, $|\text{component}(u_i)| > 1$ for $1 \leq i \leq k$, and there is at least one child u_j of x that is not $\mathcal{G}$-identical to any other leaf. Moreover, no vertex can be contracted or joined to another vertex. By this action we remove $u_1, \ldots, u_k$ from T and G, and also, we remove x from T. The resulted game is denoted by $\text{disperse}_x(T, G)$.

The proof of the next lemma is omitted due to the lack of space.

Lemma 1. *Consider a game $(\mathcal{T}, \mathcal{G})$ in which $\mathcal{G}$ has an edge between every two leaves that are $\mathcal{G}$-identical to each other. Then, if one of the above four operations can be applied on $(\mathcal{T}, \mathcal{G})$ to obtain $(\mathcal{T}', \mathcal{G}')$, $(\mathcal{T}, \mathcal{G})$ can be reduced to $(\mathcal{T}', \mathcal{G}')$.*

Theorem 3. *Suppose that (T, G) is a game. Then, if the following three conditions hold $\mathcal{B}$ has a winning strategy.*

1. *Labels of all edges of G are $=$.*
2. *If x is an internal node of T with label $\cup$, then all its children are unfinished.*
3. *If x is an internal node of T with label $\cap$, then at least one of its children is unfinished.*

Proof. Suppose we have a game $(\mathcal{T}, \mathcal{G})$ such that the number of vertices of $\mathcal{T}$ plus the number of connected components of $\mathcal{G}$ is n. We can assume that $\mathcal{T}$ does not have any internal node that can be contracted and also no leaf can be joined to another leaf in $\mathcal{T}$. Otherwise, we can reduce $(\mathcal{T}, \mathcal{G})$ to a smaller game by either contraction or joining and because the smaller game will satisfy all three properties, $\mathcal{B}$ has a winning strategy. In other words, $\mathcal{T}$ does not have any internal node that has only one child. Also, no two children of an internal node can be in the same connected component of $\mathcal{G}$. Furthermore, without loss of generality we can assume that $\mathcal{G}$ has an edge between every two leaves that are $\mathcal{G}$-identical to each other; otherwise, $(\mathcal{T}, \mathcal{G})$ can be reduced to a game that has this property by adding more edges. Because of properties 2 and 3, $(\mathcal{T}, \mathcal{G})$ has no witness, and thus, $\mathcal{A}$ has not won yet. Therefore, $\mathcal{A}$ must submit a query (u, v) where $u, v \in \text{leaves}(\mathcal{T})$.

The strategy that $\mathcal{B}$ should follow to answer $\mathcal{A}$ is summarized in Table 1. In the table, if a leaf u is labeled with "F", "N", or "L" it means that u is a finished

Table 1. The strategy of $\mathcal{B}$ for answering

$F(u)$	$F(v)$	u	v	Response	Justification
∪	×	N	×	<>	$(\mathcal{T},\mathcal{G})$ is reduced to the smaller game $\text{throw}_u(\mathcal{T},\mathcal{G})$.
∩	∩	F	F	=	The number of components of $\mathcal{G}$ is reduced.
∩	∩	N	F	=	The number of components of $\mathcal{G}$ is reduced.
∩	∩	N	N	=	The number of components of $\mathcal{G}$ is reduced.
∩	∩	L	×	<>	$(\mathcal{T},\mathcal{G})$ is reduced to the smaller game $\text{disperse}_{F(u)}(\mathcal{T},\mathcal{G})$.

leaf, an unfinished leaf that has another unfinished sibling, or an unfinished leaf that does not have any other unfinished sibling, respectively. Note that those cases which can be obtained from one of the cases in Table 3.1 by switching u and v are omitted due to the symmetry. □

Having proved Theorem 3, we know the adversary can respond to queries on key members of a-regions such that the algorithm knows all members similar to the key member of each a-region, for every $1 \leq a \leq m$, and also that $(a, 0, 0)$ is in the result of the root. Thus, Theorem 2 yields the next theorem.

Theorem 4. *For any signature T and any deterministic comparison-based algorithm $\mathcal{A}$, there is an input with the signature G such that $\mathcal{A}$ submits $\Omega(l_1 + \lg l_2)$ queries where l_1 and l_2 are defined as in Theorem 2.*

4 The Worst-Case Optimal Algorithm

In this section, we present our algorithm that matches the lower bound in Section 3. First we study two special cases separately; these special problems come in handy in solving the general problem.

The first special case has the form $X_1 \cup X_2 \cup \ldots \cup X_k$. This problem in case when $k = 2$ has been studied before [5]; to compute $A \cup B$, tight lower and upper bounds of $\Theta\left(\lg\binom{|A|+|B|}{|A|}\right)$ exist. So suppose $k > 2$. Defining $s_i = |X_i|$, for $1 \leq i \leq k$ and $s = \sum_{i=1}^{k} s_i$, w. l. o. g., we suppose $s_1 = \max\{s_1, s_2, \ldots, s_k\}$. The way we compute this union of k sets is as follows: we temporarily ignore the first set (i.e. X_1) and compute the union of $\bigcup_{i=2}^{k} X_i$ in time $O((s - s_1) \lg(k - 1))$ by using the well-known k-way merging algorithm [10]. Then, we take the union of X_1 with the result of the merge. The union can be found in time $O\left(\lg\binom{s}{s_1}\right)$, as described by Hwang and Lin [5], but the result is in the cross reference format. It can be proved that $(s - s_1) \lg(k - 1) \in O\left(\sum_{i=2}^{k} \lg\binom{s}{s_i}\right)$ which is in $O\left(\lg\binom{s}{s_1,\ldots,s_k}\right)$ according to the following lemma (the proof is omitted).

Lemma 2. *If $s \leq \sum_{i=1}^{n} s_i$, then $\lg\binom{s}{s_1,\ldots,s_n} \geq \frac{1}{2} \sum_{i=1}^{n} \lg\binom{s}{s_i}$.* □

Lemma 3. *A cross reference representation of the union of sets $X_1, X_2, \ldots, X_k$ can be computed in time $O\left(\lg\binom{s}{s_1,\ldots,s_k}\right)$ where $s_i = |X_i|$ and $s = \sum_{i=1}^{k} s_i$.* □

The complete proof of Lemma 3 is omitted. In the end, we expand the ranges of the output to have the union in the sorted list format again. The time this takes is proportional to the size of the output, which is at most $O(\sum_{i=1}^{k}|X_k|)$. This yields the following theorem:

Theorem 5. *A sorted array representation of the union of sets $X_1, X_2, \ldots, X_n$ can be computed in time $O\left(s + \lg \binom{s}{s_1,\ldots,s_k}\right)$ where $s_i = |X_i|$ and $s = \sum_{i=1}^{k}|X_i|$.* □

The second special case has the form $Y \cap (X_1 \cup X_2 \cup \ldots \cup X_k)$, given that $|Y| \geq |X_i|$ for each i. This problem for the case when $k=1$ (i.e. computing $Y \cap X$) has been studied before [5] and tight lower and upper bounds of $\Theta\left(|X| \lg \frac{|X|+|Y|}{|X|}\right)$ already exist. To solve the problem for $k > 1$, we first create a boolean array $\mathcal{B}$ of size $|Y|$, so that each element y in Y has an associated element in the array ($\mathcal{B}[y]$). We initialize all the elements in it to false; we, then, compute the intersection of each X_i with Y separately: $Y_i = Y \cap X_i$. According to the result just mentioned, this takes $O\left(\sum_{i=1}^{k}|X_i| \lg \frac{|X_i|+|Y|}{|X_i|}\right)$ time. When Y_i's are all computed, we consider them one by one and for each Y_i, for all $y \in Y_i$, we set $\mathcal{B}[y] = true$. Then we scan array $\mathcal{B}$ and return, as output, each element b iff $\mathcal{B}[b]$ is true. It is clear that scanning through all Y_i's will take $\sum_{i=1}^{k}|Y \cap X_i|$ which is less than the time consumed for all Y_i's. Also creating $\mathcal{B}$ in the beginning and scanning it in the end takes time $O(|Y|)$. Therefore:

Theorem 6. *The result set of $Y \cap (X_1 \cup X_2 \cup \ldots \cup X_k)$, given that $|Y| \geq |X_i|$, for every $1 \leq i \leq k$, can be computed in $O\left(|Y| + \sum_{i=1}^{k}|X_i| \lg \frac{|X_i|+|Y|}{|X_i|}\right)$ time.* □

We now turn to the general case and describe the algorithm. We generalize the problem a little and define two types of problems: in the first type, we are interested in computing $\Gamma(v) \cap U$, for a given "universal set" U. In the second type, we are solely asked to compute $\Gamma(v)$. The procedures COMPUTE (v, U) and COMPUTE(v) (Figure 2) are designed to solve these two types of problems. The intuition behind the universal set U in COMPUTE (v, U) is the following: consider an intersection node v with its children $u_1, \ldots, u_k$. Suppose we somehow have processed the subtree rooted at u_i for some i, and have obtained $\Gamma(u_i)$. It makes perfect sense to pass $\Gamma(u_i)$ as a universal set to subtrees rooted at children of v other than u_i so that in reporting back elements, they confine themselves to the universal set and ignore those that do not appear in the universal set. As for COMPUTE(v) it turns out that, for some nodes v, the size of the possible result of a node is less than any universal set we can possibly provide with in advance. In these cases we do not pass any universal set as it will not save any computation time.

Next, we investigate the correctness and the running time of the algorithm. The following theorem can be proved easily by considering different calls to procedures COMPUTE(v, U) and COMPUTE(v) in Algorithms 1 and 2.

Theorem 7. *At each invocation of the procedure COMPUTE(v, U) in Algorithm 1, the precondition $|U| \leq \psi^*(v)$ holds and the procedure computes $\Gamma(v) \cap U$.*

```
Procedure COMPUTE(v, U);                    Procedure COMPUTE(v);
1  // precondition: |U| ≤ ψ*(v).            1  // precondition: ψ(v) = ψ*(v).
   begin                                       begin
      switch type of node v do                    switch type of node v do
2        case Leaf: return Γ(v) ∩ U ;       2       case Leaf: return Γ(v) ;
         case Union:                                case Union:
            foreach uᵢ child of v do                   foreach uᵢ child of v do
               if ψ(uᵢ) < |U| then          3             Xᵢ ← Compute(uᵢ)
3                 Xᵢ ← Compute(uᵢ)          4          return X₁ ∪ X₂ ∪ ... ∪ Xₖ ;
               else                                 case Intersection:
4                 Xᵢ ← Compute(uᵢ, U)                  j ← minindex(ψ(uᵢ)) ;
5           return U ∩ (X₁ ∪ X₂ ∪ ... ∪ Xₖ) 5          X ← Compute(uⱼ) ;
         case Intersection:                         foreach uᵢ child of v do
            X ← U ;                                    if i ≠ j then
            foreach uᵢ child of v do        6             X ← Compute(Uᵢ, X)
6              X ← Compute(Uᵢ, X)                   return X
            return X
   end                                         end
```

Algorithm 1 Computing the intersection of U with the result set of the subtree rooted at v (i.e. $\Gamma(v) \cap U$).

Algorithm 2 Computing the result set of the subtree rooted at v (i.e. $\Gamma(v)$).

Fig. 2. The general algorithm

Also, at each invocation of that procedure COMPUTE(v) in Algorithm 2, the precondition $\psi(v) = \psi^*(v)$ holds and the procedure computes $\Gamma(v)$. □

Next, we analyze the running times of the procedures by measuring the time we spend at each node v of the tree, not taking into account the time we spend in recursive calls. The total running time of the algorithm will be, of course, the sum of such processing times in nodes of the tree. It is easy to see that no computation is involved in intersection nodes. Here, we analyze two other types of nodes (i.e. leaf and union) separately:

Processing Time in Union Nodes: Line 5 is the only one in Algorithm 1 on which we spend some computing time. Also, in Algorithm 2, only line 4 involved computation. These two are exactly the special cases we studied in the beginning of this section. We can prove the following lemma.

Lemma 4. *Processing time in a union node v takes time of* $O\left(\sum_{i=1}^{k} \psi^*(u_i) + \lg \binom{\psi^*(v)}{\psi^*(u_1),\ldots,\psi^*(u_k)}\right)$ *where $u_1, \ldots, u_k$ are children of v.* □

We make a slight change in the algorithm to save time: in the case when the root of the whole tree is of type union, we use the algorithm in Lemma 3 to compute the union in the root node as opposed to the algorithm in Theorem 5. That is, we do not expand the ranges in the result and we keep it in the cross reference format. Then, in the case when v is the root and is a union node, we can get a better result than Lemma 4.

Lemma 5. *If the root is a union node, processing time in the root takes time of* $O\left(\lg\left(\frac{\psi^*(root)}{\psi^*(u_1),\ldots,\psi^*(u_k)}\right)\right)$ *where* $u_1,\ldots,u_k$ *are children of the root.* □

Here we claim that the term $\sum_{i=1}^{k}\psi^*(u_i)$ in Lemma 4 is negligible when it is summed over all union nodes. In the sum, ψ^* of all the children of union nodes are added together, which means the sum is over all the intersection nodes and leaves. Now we argue that if S is the set of all intersection nodes of T, we have $\sum_{v\in S}\psi^*(v) \le \sum_{v\in L}\psi^*(v)$ where L is the set of non-shallow leaves. This can be proved by summing up together the inequalities of Observation 1 for all nodes of the tree.

Theorem 8. *Processing in union nodes and leaves takes time of* $O\left(t + \sum_{v\in L}\psi^*(v) + \sum_{\substack{union\\nodes\ v}} \lg\left(\frac{\psi^*(v)}{\psi^*(u_1),\psi^*(u_2),\ldots,\psi^*(u_k)}\right)\right)$ *where* L *is the set of non-shallow leaves and t is the time we spend in non-shallow leaves.* □

Processing Time in Leaf Nodes: If v is a leaf, in line 2 in Algorithm 1, we compute the intersection of $\Gamma(v)$ and U. As a precondition, we know that $|U| < \psi^*(v)$ and also by definition that $\psi^*(v) \le \psi(v) = \mathsf{size}(v)$, so $|U| < \mathsf{size}(v)$. In the first special case, we showed how to compute the intersection in time $O(|U|\lg\frac{|U|+\mathsf{size}(v)}{|U|})$. Since $|U| < \psi^*(v) \le \mathsf{size}(v)$, the processing time is in $O\left(\psi^*(v)\lg\frac{\psi^*(v)+\mathsf{size}(v)}{\psi^*(v)}\right)$.

In line 2 of Algorithm 2, we simply return $\Gamma(v)$ which, by precondition, has size $\psi^*(v)$. In case v is a shallow leaf by the argument mentioned in Theorem 8, we use a slightly different method to take the union at the root, and therefore we do not spend any time in the shallow leaves (we do spend, however, some time in the root for computing the union, which has been accounted for in Theorem 8.) Thus the following theorem holds:

Theorem 9. *In Algorithms 1 and 2, the time spent in each non-shallow leaf is* $O\left(\psi^*(v)\lg(\frac{\mathsf{size}(v)}{\psi^*(v)}+1)\right)$ *and we spend no time in shallow leaves.* □

We conclude from Theorems 4, 9, and 8 that our algorithm is optimum.

5 Conclusion and Extensions

We studied the problem of evaluating an expression of sorted sets with union and intersection operands. Complexity of algorithms were measured in terms of the sizes of the input sets. We proved lower bounds on the worst case complexity of algorithms that can solve this problem, and later presented an algorithm that asymptotically matches the lower bound.

An immediate extension to this work is changing its format of input/output to a more appropriate format. Our assumption for the format of the input is lists of elements of sets, and the format of the output is a list of cross-references which specify the ranges of the elements. However, with a little effort, the format of both the input and the output can be changed to *balanced search trees*. More specifically, we choose B-trees. Adapting the lower bounds is straightforward; Theorem 4 will still hold without any modification. As for the upper bound, it is sufficient to show that we can handle the two special cases in the beginning of Section 4 with the same time complexity, since the general algorithm only uses these two for computation. These two special cases can be dealt with in the same way Demaine et al. [4] handled B-tree representations of their input sets. It is easy to see that the extra work for assembling and disassembling the B-trees in their scheme does not affect our bounds in Theorems 5 and 6.

As a future work, one can consider expressions that can have operands of type complement besides those of type union and intersection.

Acknowledgments

The authors would like to thank Alex López-Ortiz, Peyman Afshani, Reza Dorri, Narges Simjour, and anonymous referees for their useful comments.

References

1. S. Brin and L. Page. The anatomy of a large-scale hypertextual web search engine. In *Proceedings of the seventh international conference on World Wide Web 7*, pages 107–117. Elsevier Science Publishers B. V., 1998.
2. M.R. Brown and R.E. Tarjan. A fast merging algorithm. *J. ACM*, 26(2):211–226, 1979.
3. M.R. Brown and R.E. Tarjan. Design and analysis of a data structure for representing sorted lists. *SIAM Journal of Computing*, 9(3):594–614, Aug. 1980.
4. E.D. Demaine, A. Lopez-Ortiz, and J.I. Munro. Adaptive set intersections, unions, and differences. In *Proc. of Eleventh ACM-SIAM Symposium on Discrete Algorithms, SODA*, pages 743–752, 2000.
5. F.K. Hwang and S. Lin. A simple algorithm for merging two disjoint linearly ordered sets. *SIAM Journal on Computing*, 1(1):31–39, 1972.
6. G. Lee, M. Park, and H. Won. Using syntactic information in handling natural language quries for extended boolean retrieval model. In *Proceedings of the 4th international workshop on information retrieval with Asian languages*, 1999.
7. Mauldin, M.I.Lycos Inc., and PA Pittsburgh. Lycos: design choices in an internet search service. *IEEE Expert*, 12(1):8–11, 1997.
8. M. Mirzazadeh. Adaptive comparison-based algorithms for evaluating set queries. Master's thesis, School of Computer Science, University of Waterloo, 2004.
9. W. Pugh. A skip list cookbook. Technical Report CS-TR-2286.1, University of Maryland, 1990.
10. Ronald L. Rivest and Charles E. Leiserson. *Introduction to Algorithms*. McGraw-Hill, Inc., 1990.
11. I.H. Witten, T.C. Bell, and A. Moffat. *Managing Gigabytes: Compressing and Indexing Documents and Images*. John Wiley & Sons, Inc., 1994.

Measure and Conquer:
Domination – A Case Study

Fedor V. Fomin[1,*], Fabrizio Grandoni[2,**], and Dieter Kratsch[3]

[1] Department of Informatics,
University of Bergen, N-5020 Bergen, Norway
fomin@ii.uib.no
[2] Dipartimento di Informatica, Università di Roma "La Sapienza",
Via Salaria 113, 00198 Roma, Italy
grandoni@di.uniroma1.it
[3] LITA, Université de Metz, 57045 Metz Cedex 01, France
kratsch@sciences.univ-metz.fr

Abstract. Davis-Putnam-style exponential-time backtracking algorithms are the most common algorithms used for finding exact solutions of NP-hard problems. The analysis of such recursive algorithms is based on the bounded search tree technique: a measure of the size of the subproblems is defined; this measure is used to lower bound the progress made by the algorithm at each branching step.

For the last 30 years the research on exact algorithms has been mainly focused on the design of more and more sophisticated algorithms. However, measures used in the analysis of backtracking algorithms are usually very simple. In this paper we stress that a more careful choice of the measure can lead to significantly better worst case time analysis.

As an example, we consider the minimum dominating set problem. The currently fastest algorithm for this problem has running time $O(2^{0.850n})$ on n-nodes graphs. By measuring the progress of the (same) algorithm in a different way, we refine the time bound to $O(2^{0.598n})$. A good choice of the measure can provide such a (surprisingly big) improvement; this suggests that the running time of many other exponential-time recursive algorithms is largely overestimated because of a "bad" choice of the measure.

Keywords: Algorithms and data structures, exponential-time exact algorithm, NP hard problem, dominating set.

1 Introduction

The interest in exact and fast exponential-time algorithms solving hard problems dates back to the sixties and seventies [13, 25]. The last decade has led to much re-

---

* Supported by Norges forskningsråd project 160778/V30.
** Supported by Web-Minds project of the Italian Ministry of University and Research, under the FIRB program.

search in fast exponential-time algorithms. Examples of recently developed exponential algorithms are algorithms for Maximum Independent Set [15, 23], (Maximum) Satisfiability [4, 14, 17, 19, 24, 26], Coloring [2, 3, 6], Treewidth [8], and many others (see the recent survey written by Woeginger [27] for an overview).

Most of the currently fastest exact algorithms for NP-hard problems are recursive algorithms. In order to bound the total number of subproblems generated by such algorithms, the *bounded search tree* technique is often used: one defines a suitable *measure* of the size of the subproblems. This measure is used to lower bound the "progress" made by the algorithm at each branching step.

Though the algorithms considered may be rather complicated, the measures used in their analysis are usually very simple. In this paper we remark that a more careful choice of the measure can lead to much tighter time bounds.

In order to show that, we consider one of the best known NP-hard problems: the *minimum dominating set* problem. The currently fastest exact algorithm for this problem is a recursive algorithm of running time $\mathcal{O}^*(2^{0.850n})$ on n-nodes graphs [10, 11][1]. Here we present a refined analysis, based on a different measure of the size of the subproblems generated and show that the same algorithm has indeed running time $\mathcal{O}^*(2^{0.598n})$. This surprisingly big improvement suggests the possibility that the running times of many other exponential-time recursive algorithms (including possibly the one presented here) are largely overestimated because of a "bad" choice of the measure in their analysis. Despite the importance of the problem, only few works address this issue [2, 7].

Since the current tools do not seem to be strong enough to support an analysis of exponential-time recursive algorithms providing tight running time upper bounds, it is natural to ask for lower bounds (notice that we are concerned with lower bounds on the complexity of a particular algorithm and not with lower bounds on the complexity of an algorithmic problem). A lower bound may give an idea of how far the analysis is from being tight. There are several results known on lower exponential bounds for different branching algorithms for SAT (see e.g. [1, 18]) but we are not aware of lower bounds for existing exponential-time recursive graph algorithms. One of the reasons to this could be that for most of the graph problems the construction of good lower bounds is often difficult even for very simple algorithms. In this paper we prove a $\Omega(2^{0.333n})$ lower bound on the time complexity of our minimum dominating set algorithm. The large gap between the upper bound and the lower bound suggests the possibility that the analysis of the algorithm can be further refined (possibly by measuring the size of the subproblems in a smarter way).

Previous results on dominating set. The *minimum dominating set* problem (MDS) is a classic NP-hard graph optimization problem which fits into the broader class of *domination* and *covering* problems on which hundreds of papers have been written; see e.g. the survey [12] by Haynes et al. The dominating set problem

---

[1] Throughout this paper we use a modified big-Oh notation that suppresses all polynomially bounded factors. For functions f and g we write $f(n) = \mathcal{O}^*(g(n))$ if $f(n) = \mathcal{O}(g(n)poly(n))$, where $poly(n)$ is a polynomial.

is also one of the basic problems in parameterized complexity [5]; it is W[2]-complete and thus it is unlikely that the problem is fixed parameter tractable. What are the best time complexities for the dominating set problem in n-node graphs $G = (V, E)$ that we can possibly hope for? It has been observed in [9] that (unless some very unexpected things happen in computational complexity theory) there is no sub-exponential time (i.e. of running time $c^{o(n)}$ for some constant c) algorithm solving dominating set problem. There is the trivial $O^*(2^n)$ algorithm that simply searches through all the 2^n subsets of V. Hence, we can only hope for time complexities of the form $O^*(2^{cn})$, with some small value $c < 1$. Although MDS is a natural and very interesting problem concerning the design and analysis of exponential-time algorithms, no exact algorithm for MDS faster than the trivial one had been known until very recently. In 2004 three different sets of authors seemingly independently published algorithms breaking the trivial "2^n-barrier". The algorithm of Fomin et al. [9] uses a deep graph-theoretic result due to Reed [21], providing an upper bound on the domination number of graphs of minimum degree three. The most time consuming part of their algorithm is an enumeration of all subsets of nodes of cardinality at most $3n/8$, thus the overall running time is $\mathcal{O}^*(2^{0.955n})$. The algorithm of Randerath and Schiermeyer [20] uses a very nice and cute idea (including matching techniques) to restrict the search space. The most time consuming part of their algorithm enumerates all subsets of nodes of cardinality at most $n/3$, thus the overall running time is $\mathcal{O}^*(2^{0.919n})$. Finally, the fastest algorithm known prior to our work is due to Grandoni [10, 11], who described a $\mathcal{O}^*(2^{0.850n})$ algorithm for MDS.

Our Results. We show that MDS can be solved in $\mathcal{O}^*(2^{0.610n})$ time using polynomial space. The running time of our algorithm can be reduced at the cost of exponential space to $\mathcal{O}^*(2^{0.598n})$ which is a significant improvement of all known results on MDS. To solve the problem we represent MDS as a set cover problem which allows us to use a search tree based algorithm. This idea was first used in [10, 11]. To obtain running time $\mathcal{O}^*(2^{0.610n})$ we do not add more and more sophisticated rules to existing algorithms which is a usual practice to improve on the exponential base. Instead we give a simple and easy to implement algorithm and observe how the careful choice of the measure changes the algorithm analysis dramatically. Our refined analysis leads to a multivariate recurrence. For a general treatment of this type of recurrences we refer to Eppstein's paper [7]. Since the analysis of our search tree based algorithms is so depended on the choice of the measure, it is natural to ask for (exponential) lower bounds on the running time of the algorithm. We prove that our algorithm requires $\Omega(2^{0.333n})$ steps.

2 Definitions and Basic Algorithm

Let $G = (V, E)$ be an n-node undirected, simple graph without loops. The open *neighborhood* of a node v is denoted by $N(v) = \{u \in V : uv \in E\}$, and the closed neighborhood of v is denoted by $N[v] = N(v) \cup \{v\}$. A set $A \subseteq E$ of edges of

$G = (V, E)$ is an *edge cover*, if every node of G is incident to an edge of A; the edge set A is a *matching* if no node of G is incident to two edges of A.

The minimum dominating set problem. Let $G = (V, E)$ be a graph. A set $D \subseteq V$ is called a *dominating set* for G if every node of G is either in D, or adjacent to some node in D. The *domination number* $\gamma(G)$ of a graph G is the minimum cardinality of a dominating set of G. The *Minimum Dominating Set* problem (MDS) asks to determine $\gamma(G)$.

The minimum set cover problem. In the *Minimum Set Cover* problem (MSC) we are given a universe $\mathcal{U}$ of elements and a collection $\mathcal{S}$ of (non-empty) subsets of $\mathcal{U}$. The aim is to determine the minimum cardinality of a subset $\mathcal{S}' \subseteq \mathcal{S}$ which *covers* $\mathcal{U}$, that is such that

$$\cup_{S \in \mathcal{S}'} S = \mathcal{U}.$$

The *frequency* of $u \in \mathcal{U}$ is the number of subsets $S \in \mathcal{S}$ in which u is contained. For the sake of simplicity, we always assume in this paper that $\mathcal{S}$ covers $\mathcal{U}$:

$$\mathcal{U} = \mathcal{U}(\mathcal{S}) \triangleq \cup_{S \in \mathcal{S}} S.$$

With this assumption, an instance of MSC is univocally specified by $\mathcal{S}$.

We recall that, if all the subsets of $\mathcal{S}$ are of cardinality two, MSC can be solved in polynomial time via the following standard reduction to maximum matching. Consider the graph $\tilde{G}$ which has a node u for each $u \in \mathcal{U}$, and an edge uv for each subset $S = \{u, v\}$ in $\mathcal{S}$. Thus we have to compute a minimum edge cover of $\tilde{G}$. To compute a minimum edge cover of $\tilde{G}$ we compute a maximum matching M in $\tilde{G}$. Then, for each unmatched node u, we add to M an arbitrary edge incident to u (if no such edge exists, there is no set cover at all). The subsets corresponding to M form a minimum set cover.

MDS can be naturally reduced to MSC by imposing $\mathcal{U} = V$ and $\mathcal{S} = \{N[v] | v \in V\}$. Note that $N[v]$ is the set of nodes dominated by v, thus D is a dominating set of G if and only if $\{N[v] | v \in D\}$ is a set cover of $\{N[v] | v \in V\}$. Thus every minimum set cover of $\{N[v] | v \in V\}$ corresponds to a minimum dominating set of G.

At first view such a transformation from one NP-hard problem to another seems to be completely useless: The only known exact algorithms for MSC are brute force $\mathcal{O}^*(2^{|\mathcal{S}|})$ and $\mathcal{O}^*(2^{|\mathcal{U}|})$ [9] dynamic programming algorithms. Both algorithms result in an $\mathcal{O}^*(2^n)$ algorithm for MDS and it seems that such an approach is not interesting. Not at all! On second thought the transformation from MDS to MSC becomes very helpful. It enables the use of a search tree based algorithm to solve MSC, and thus also MDS.

Basic algorithm. We consider a simple recursive algorithm msc for solving MSC. The algorithm is a slight modification of the algorithm from [11] and it makes use of the following observation.

Lemma 1. *For a given MSC instance $\mathcal{S}$:*

1. *If there are two distinct sets S and R in $\mathcal{S}$, $S \subseteq R$, then there is a minimum set cover which does not contain S.*

```
1    int msc(S) {
2        if(|S| = 0) return 0;
3        if(∃S, R ∈ S : S ⊆ R) return msc(S\{S});
4        if(∃u ∈ U(S)∃ a unique S ∈ S : u ∈ S) return 1+msc(del(S, S));
5        take S ∈ S of maximum cardinality;
6        if(|S| = 2) return poly-msc(S)
7        return min{msc(S\{S}), 1+msc(del(S, S))};
8    }
```

Fig. 1. A recursive algorithm for minimum set cover

2. *If there is an element u of $\mathcal{U}$ which belongs to a unique $S \in \mathcal{S}$, then S belongs to every set cover.*

Note that each subset of cardinality one satisfies exactly one of the properties in Lemma 1.

A basic version of msc is described in Figure 1. If $|\mathcal{S}| = 0$ (line 2), $msc(\mathcal{S}) = 0$. Otherwise (lines 3 and 4), the algorithm *tries* to reduce the size of the problem without branching, by applying one of the Properties 1 and 2 of Lemma 1. Specifically, if there are two sets S and R, $S \subseteq R$, we have $msc(\mathcal{S}) = msc(\mathcal{S}\backslash S)$. If there is an element u which is contained in a unique set S, we have $msc(\mathcal{S}) = 1 + msc(del(S, \mathcal{S}))$, where $del(S, \mathcal{S}) = \{Z | Z = R\backslash S \neq \emptyset, R \in \mathcal{S}\}$ is the instance of MSC which is obtained from $\mathcal{S}$ by removing the elements of S from the subsets in $\mathcal{S}$, and by eventually removing the empty sets obtained.

If none of the two properties above applies, the algorithm takes (line 5) a set $S \in \mathcal{S}$ of maximum cardinality. If $|S| = 2$ (line 6), the algorithm directly solves the problem with the polynomial time algorithm poly-msc based on the reduction to maximum matching. Otherwise (line 7), it branches on the two subproblems $\mathcal{S}_{IN} = del(S, \mathcal{S})$ (the case where S belongs to the minimum set cover) and $\mathcal{S}_{OUT} = \mathcal{S}\backslash S$ (corresponding to the case S is not in the minimum set cover). Thus

$$msc(\mathcal{S}) = \min\{msc(\mathcal{S}\backslash\{S\}), 1 + msc(del(S, \mathcal{S}))\}.$$

Notice that with simple modifications, the algorithm can also provide one minimum set cover (besides its cardinality).

To emphasize the importance of the measure we sketch the analysis of the algorithm with a simple measure (taken from [11]). Let us choose the following measure $k(\mathcal{S}')$ of the size of a MSC instance $\mathcal{S}'$,

$$k(\mathcal{S}') = |\mathcal{S}'| + |\mathcal{U}(\mathcal{S}')|.$$

Let $\ell(k)$ be the number of leaves in the search tree generated by the algorithm to solve a problem of size $k = k(\mathcal{S})$. If one of the conditions of lines 3 and 4 is satisfied, $\ell(k) \leq \ell(k-1)$. Let S be the set selected in line 5. If $|S| = 2$, the algorithm directly solves the problem in polynomial time ($\ell(k) = 1$). Otherwise ($|S| \geq 3$), the algorithm branches on the two subproblems $\mathcal{S}_{OUT} = \mathcal{S}\backslash\{S\}$

and $\mathcal{S}_{IN} = del(\mathcal{S}, S)$. The size of $\mathcal{S}_{OUT}$ is $k - 1$ (one set removed from $\mathcal{S}$). The size of $\mathcal{S}_{IN}$ is at most $k - 4$ (one set removed from $\mathcal{S}$ and at least three elements removed from $\mathcal{U}$). This brings us to $\ell(k) \leq \ell(k-1) + \ell(k-4)$. We conclude that $\ell(k) \leq \alpha^k$, where $\alpha = 1.3802\ldots < 1.3803$ is the (unique) positive root of the polynomial $(x^4 - x^3 - 1)$. It turns out that the total number of subproblems solved is within a polynomial factor from $\ell(k)$. Moreover, solving each subproblem takes polynomial time. Thus the complexity of the algorithm is $\mathcal{O}^*(\ell(k)) = \mathcal{O}^*(\alpha^k) = \mathcal{O}^*(1.3803^{|\mathcal{S}|+|\mathcal{U}|}) = \mathcal{O}^*(2^{0.465(|\mathcal{S}|+|\mathcal{U}|)})$.

In next section we will show how to refine the running time analysis to $\mathcal{O}^*(2^{0.305(|\mathcal{S}|+|\mathcal{U}|)})$ via a more careful choice of the measure $k(\mathcal{S}')$ (without modifying the algorithm!).

3 Refined Analysis

In this section we show that algorithm msc has time complexity $\mathcal{O}^*(2^{0.305(|\mathcal{S}|+|\mathcal{U}|)})$.

Our result is based on the following observation. Removing a large set has a different impact on the "progress" of the algorithm than removing a small one. In fact, when we remove a large set, we decrease the frequency of many elements. Decreasing elements frequency pays of on long term, since the elements of frequency one can be filtered out (without branching). A dual argument holds for the elements. Removing an element of high frequency is somehow preferable to removing an element of small frequency. In fact, when we remove an element occurring in many sets, we decrease the cardinality of all such sets by one. This is good on long term, since sets of cardinality one can be filtered out. Both phenomena are not taken into account in the measure used in [10]. With that measure, by removing one set (element), we decrease the size of the problem by one, no matter which is the cardinality (frequency) of the set (element) considered.

This suggests the idea to give a different "weight" to sets of different cardinality and to elements of different frequency. In particular, let n_i denote the number of subsets $S \in \mathcal{S}$ of cardinality i. Let moreover m_j denote the number of elements $u \in \mathcal{U}$ of frequency j. We will use the following measure $k = k(\mathcal{S})$ of the size of $\mathcal{S}$:

$$k(\mathcal{S}) = \sum_{i \geq 1} w_i\, n_i + \sum_{j \geq 1} v_j\, m_j,$$

where the weights $w_i, v_j \in (0, 1]$ will be fixed in the following. Note that $k \leq |\mathcal{S}| + |\mathcal{U}|$. The quantities

$$\Delta w_i = \begin{cases} w_i - w_{i-1} & \text{if } i \geq 3, \\ w_2 & \text{if } i = 2, \end{cases} \quad \text{and} \quad \Delta v_i = \begin{cases} v_i - v_{i-1} & \text{if } i \geq 3, \\ v_2 & \text{if } i = 2, \end{cases}$$

turn out to be useful in the analysis. Intuitively, Δw_i (Δv_i) is the reduction of the size of the problem corresponding to the reduction of the cardinality of a set (of the frequency of an element) from i to $i - 1$. Note that this holds also in the case $i = 2$. In fact, in that case the size of the problem first increases by $1 - w_2$ $(1 - v_2)$, but the new set of cardinality one (the new element of frequency one) introduced is removed before the next branching, with a reduction of the size by one. Thus one has an overall reduction by $1 - (1 - w_2) = w_2 (1 - (1 - v_2)) = v_2)$.

Theorem 1. *Algorithm* msc *solves* MSC *in time* $\mathcal{O}^*(2^{0.305(|\mathcal{U}|+|\mathcal{S}|)})$.

Proof. The correctness of the algorithm is trivial. In order to simplify the running time analysis, we will make the following assumptions:

- $w_1 = v_1 = 1$ and $w_i = v_i = 1$ for $i \geq 6$;
- $0 \leq \Delta w_i \leq \Delta w_{i-1}$ for $i \geq 2$.

Note that this implies $w_i \geq w_{i-1}$ for every $i \geq 3$ (excluding sets of cardinality one, larger sets have larger weights). Moreover, $\Delta w_i = \Delta v_i = 0$ for $i \geq 7$.

Let $P_h(k)$ be the number of subproblems of size h, $0 \leq h \leq k$, solved by msc to solve a problem of size k. Clearly, $P_k(k) = 1$. Consider the case $h < k$ (which implies $|\mathcal{S}| \neq 0$). If one of the condition of lines 3 and 4 holds, one set S is removed from $\mathcal{S}$. Thus the reduction of the size of the problem is at least w_2 (corresponding to the case $|S| = 2$) and $P_h(k) \leq P_h(k - w_2)$. Otherwise, let S be the subset selected in line 5. If $|S| = 2$, no subproblem is generated ($P_h(k) = 0$). Otherwise ($|S| \geq 3$), msc generates two subproblems $\mathcal{S}_{IN} = del(S, \mathcal{S})$ and $\mathcal{S}_{OUT} = \mathcal{S} \setminus S$.

Consider the subproblem $\mathcal{S}_{OUT}$. The size of $\mathcal{S}_{OUT}$ decreases by $w_{|S|}$ because of the removal of S. Let r_i be the number of elements of S of frequency i. Note that there cannot be elements of frequency 1. Consider an element $u \in S$ of frequency $i \geq 2$. When we remove S, the frequency of u decreases by one. As a consequence, the size of the subproblem decreases by Δv_i. Thus the overall reduction of the size of $\mathcal{S}_{OUT}$ due to the reduction of the frequencies is at least

$$\sum_{i \geq 2} r_i \Delta v_i = \sum_{i=2}^{6} r_i \Delta v_i.$$

Suppose that there is an element $u \in S$ of frequency 2. Let $R \neq S$ be the other set containing u. When we remove S, we have to include R in the set cover. Thus we reduce the size of the problem by at least w_2 (corresponding to the case $|R| = 2$). Also $R \setminus S$ is not empty (otherwise condition of line 3 of the algorithm is met) and thus all elements of $R \setminus S$ are removed when we include R in the set cover. This reduces the size by at least v_2 (corresponding to the case that the frequency of z is 2). Therefore the overall reduction of the size of $\mathcal{S}_{OUT}$ due to the removal of the sets R is at least $r_2 w_2 + \delta(r_2) v_2$, where $\delta(r_2) = 0$ for $r_2 = 0$, and $\delta(r_2) = 1$ otherwise.

Consider now the subproblem $\mathcal{S}_{IN}$. The size of $\mathcal{S}_{IN}$ decreases by $w_{|S|}$ because of the removal of S. Consider an element $u \in S$ of frequency i ($i \geq 2$). The size of $\mathcal{S}_{IN}$ further decreases by v_i because of the removal of u. Thus the overall reduction due to the removal of the elements u of S is

$$\sum_{i \geq 2} r_i v_i = \sum_{i=2}^{6} r_i v_i + r_{\geq 7},$$

where $r_{\geq i}$ is the number of elements of S of frequency at least i. Let R be a set sharing an element u with S. Note that $|R| \leq |S|$. By removing u, the cardinality of R is reduced by one. This implies a reduction of the size of S_{IN} by $\Delta w_{|R|} \geq \Delta w_{|S|}$. Thus the overall reduction of S_{IN} due to the reduction of the cardinalities of the sets R is at least:

$$\Delta w_{|S|} \sum_{i \geq 2} (i-1) r_i \geq \Delta w_{|S|} \left(\sum_{i=2}^{6} (i-1) r_i + 6 \cdot r_{\geq 7} \right).$$

Note that this quantity is 0 for $|S| \geq 7$. Putting all together, for all the possible values of $|S| \geq 3$ and of the r_i such that

$$\sum_{i=2}^{6} r_i + r_{\geq 7} = |S|,$$

we have the following set of recursions

$$P_h(k) \leq P_h(k - \Delta k_{OUT}) + P_h(k - \Delta k_{IN}),$$

where

- $\Delta k_{OUT} \triangleq w_{|S|} + \sum_{i=2}^{6} r_i \Delta v_i + r_2 w_2 + \delta(r_2) v_2$,
- $\Delta k_{IN} \triangleq w_{|S|} + \sum_{i=2}^{6} r_i v_i + r_{\geq 7} + \Delta w_{|S|} \left(\sum_{i=2}^{6} (i-1) r_i + 6 \cdot r_{\geq 7} \right)$.

Since $\Delta w_{|S|} = 0$ for $|S| \geq 7$, we have that each recurrence with $|S| \geq 8$ is "dominated" by some recurrence with $|S| = 7$. For this reason, we restrict our attention only to the cases $3 \leq |S| \leq 7$. Thus we consider a large but finite number of recurrences. For every fixed 8-tuple $(w_2, w_3, w_4, w_5, v_2, v_3, v_4, v_5)$ the number $P_h(k)$ is upper bounded by α^{k-h}, where α is the largest number from the set of real roots of the set of equations

$$\alpha^k = \alpha^{k-\Delta k_{OUT}} + \alpha^{k-\Delta k_{IN}}$$

corresponding to different combinations of values $|S|$ and r_i. Thus the estimation of $P_h(k)$ boils up to choosing the weights minimizing α. This optimization problem is interesting in its own and we refer to Eppstein's work [7] on quasi-convex programming for general treatment of such problems.

We numerically obtained the following values of the weights:

$$w_i = \begin{cases} 0.3774 & \text{if } i = 2, \\ 0.7548 & \text{if } i = 3, \\ 0.9095 & \text{if } i = 4, \\ 0.9764 & \text{if } i = 5, \end{cases} \quad \text{and} \quad v_i = \begin{cases} 0.3996 & \text{if } i = 2, \\ 0.7677 & \text{if } i = 3, \\ 0.9300 & \text{if } i = 4, \\ 0.9856 & \text{if } i = 5, \end{cases}$$

which yields $\alpha \leq 1.2352\ldots < 1.2353$. In Table 1 the values of $|S|$ and r_i of the eight worst case recurrences are listed.

Table 1. The eight worst case recurrences

$\|S\|$	$(r_2, r_3, r_4, r_5, r_6, r_{\geq 7})$
6	$(0, 0, 0, 0, 0, 6)$
5	$(0, 0, 0, 0, 4, 1)$
5	$(0, 0, 0, 0, 5, 0)$
4	$(0, 0, 0, 0, 4, 0)$
4	$(0, 0, 0, 4, 0, 0)$
3	$(0, 0, 3, 0, 0, 0)$
3	$(0, 3, 0, 0, 0, 0)$
3	$(3, 0, 0, 0, 0, 0)$

Let K denote the set of the possible sizes of the subproblems solved. Note that $|K|$ is polynomially bounded. The total number $P(k)$ of subproblems solved satisfies:

$$P(k) \leq \sum_{h \in K} P_h(k) \leq \sum_{h \in K} \alpha^{k-h} \leq |K|\alpha^k.$$

The cost of solving a problem of size $h \leq k$, excluding the cost of solving the corresponding subproblems (if any), is a polynomial $poly(k)$ of k. Thus the time complexity of the algorithm is

$$\mathcal{O}^*(poly(k)|K|\alpha^k) = \mathcal{O}^*(1.2353^{|\mathcal{U}|+|\mathcal{S}|}) = \mathcal{O}^*(2^{0.305(|\mathcal{U}|+|\mathcal{S}|)}). \quad \square$$

As already observed, MDS can be reduced to MSC by imposing $\mathcal{U} = V$ and $\mathcal{S} = \{N[v]\|\ v \in V\}$. The size of the MSC instance obtained is at most $2n$. By simply combining this reduction with algorithm msc one obtains:

Corollary 1. *There is a $\mathcal{O}^*(2^{0.305(2n)}) = \mathcal{O}^*(2^{0.610n})$ algorithm for MDS.*

3.1 An Exponential Space Algorithm

The time complexity of msc can be reduced at the cost of an exponential space complexity via the *memorization* technique by Robson [22]. The general idea is the following: The algorithm keeps the solutions of all the subproblems solved. If the same subproblem turns up more than once, the algorithm is not to run a second time, but the already computed result is looked up. Note that the corresponding data structure can be implemented in such a way that the *query time* is logarithmic in the number of solutions stored [22].

Theorem 2. *Algorithm msc, modified as above, solves MSC in $\mathcal{O}^*(2^{0.299(|\mathcal{S}|+|\mathcal{U}|)})$ time.*

Corollary 2. *There is an algorithm which solves MDS in time $\mathcal{O}^*(2^{0.299(2n)}) = \mathcal{O}^*(2^{0.598n})$.*

Due to space restrictions, the proof of Theorem 2 is omitted here.

4 An Exponential Lower Bound

By carefully measuring the size of the subproblems, we obtained a much tighter running time bound. However the bound achieved might still be only a pessimistic estimation of the worst case running time of the algorithm. Therefore it is natural to ask for lower bounds: A lower bound may give an idea of how far is the bound computed from the real worst case running time.

Let us consider the $\mathcal{O}^*(2^{0.610n})$ polynomial-space MDS algorithm mds based on the reduction to MSC and (the polynomial-space version of) algorithm msc.

Theorem 3. *The worst case running time of* mds *is* $\Omega(2^{n/3}) = \Omega(2^{0.333n})$.

Proof. Consider the following input graph G_n ($n \geq 1$): the node set of G_n is $\{a_i, b_i, c_i : 1 \leq i \leq n\}$. The edge set of G_n consists of two types of edges: for each $i = 1, 2 \ldots, n$, the vertices a_i, b_i and c_i induce a triangle T_i; and for each $i = 1, 2, ..., n - 1$: $\{a_i, a_{i+1}\}$, $\{b_i, b_{i+1}\}$ and $\{c_i, c_{i+1}\}$ are edges.

Each node of the search tree corresponds to a subproblem of the MSC problem with input $(\mathcal{U}; \mathcal{S} = \{S_v : v \in V\})$ where $S_v = N[v]$.

We give a selection rule for the choice of the vertices v (respectively sets S_v) to be chosen for the branching. Clearly the goal is to choose them such that the number of nodes in the search tree obtained by the execution of algorithm msc on graph G_n is as large as possible.

In each round i, $i \in \{2, 3, \ldots, n - 1\}$, we start with a pair $P = \{x_i, y_i\}$ of nodes (belonging to triangle T_i), where $\{x, y\} \subset \{a, b, c\}$. Initially $P = \{a_2, b_2\}$. Our choice makes sure that for each branching node x the cardinality of its set S_x is five in the current subproblem $\mathcal{S}$, and that no other rules of the algorithm will apply to a branching node than the one of line 5. Consequently, by line 7 of msc either the set S_v is taken into the set cover ($\mathcal{S} := del(\mathcal{S}, S_v)$), or S_v is removed ($\mathcal{S} := \mathcal{S} \setminus S_v$).

For each pair $P = \{x_i, y_i\}$ of nodes we branch in the following 3 ways
1) take S_{x_i}
2) remove S_{x_i}, and then take S_{y_i}
3) remove S_{x_i}, and then remove S_{y_i}

The following new pairs of nodes correspond to each of the three branches:
1) $P_1 = \{a_{i+2}, b_{i+2}, c_{i+2}\} \setminus x_{i+2}$
2) $P_2 = \{a_{i+2}, b_{i+2}, c_{i+2}\} \setminus y_{i+2}$
3) $P_3 = \{x_{i+1}, y_{i+1}\}$

On each pair P_j we recursively repeat the process. Thus of the three branches of T_i two are proceeded on T_{i+2} and one is proceeded on T_{i+1}.

Let $T(k)$ be the number of leaves in the search tree when all triangles up to T_k have been used for branching. Thus $T(k) = 2 \cdot T(k-2) + T(k-1)$, and hence $T(k) \geq 2^{k-2}$. Consequently the worst case number of leaves in the search tree of msc for a graph on n vertices is at least $2^{n/3 - 2}$. □

The lower bound above can be easily improved by considering disconnected graphs formed by several (disconnected) copies of a carefully chosen small subgraph. We did not consider such lower bounds, since algorithm mds can be easily modified in order to invalidate them (it is sufficient to solve each disconnected subproblem separately, and then combine the partial solutions).

We may notice that there is a large gap between the $\mathcal{O}^*(2^{0.610n})$ upper bound and the $\Omega(2^{0.333n})$ lower bound. This could suggest the possibility that the analysis of algorithm mds can be further refined (possibly via a further refined measure of the size of the MSC instances).

5 Conclusions

We investigated the impact of different measures of the size of the problem in the analysis of exponential-time recursive algorithms. In particular, we considered the minimum dominating set problem. We showed how a more careful choice of the measure leads to a much tighter running time bound on the fastest know algorithm for the problem. Specifically, we reduced the time bound from $\mathcal{O}^*(2^{0.850n})$ to $\mathcal{O}^*(2^{0.598n})$ (without modifying the algorithm).

The impressive reduction of the running time achieved for minimum dominating set, suggests the possibility that the time complexity of many other exponential-time exact algorithms is largely overestimated because of a bad choice of the measure. Indeed, this could be the case also for our refined analysis of minimum dominating set. This possibility is somehow supported by the large gap between the $\mathcal{O}^*(2^{0.598n})$ upper bound and the $\Omega(2^{0.333n})$ lower bound we managed to prove.

Another natural problem to play with measure is Independent Set. The best running time $\mathcal{O}^*(2^{n/4})$ for this problem was claimed by Robson [23]. Though Robson's algorithm is extremely technical and complicated, the measure used in its analysis is very simple (the number of nodes in the graph). Can we refine the analysis of this algorithm via a different choice of the measure? Moreover, how fast really are simple algorithms for Independent Set?

References

1. M. Alekhnovich, E.A. Hirsch, and D. Itsykon. Exponential lower bounds for the running time of DPLL algorithms on satisfiable formulas. Proceedings of the *31st International Colloquium on Automata, Languages and Programming (ICALP 2004)*, Springer LNCS vol. 3142, 2004, pp. 84–96.
2. R. Beigel and D. Eppstein. 3-coloring in time $O(1.3446^n)$: a no-MIS algorithm. Proceedings of the *36th IEEE Symposium on Foundations of Computer Science (FOCS 1995)*, pp. 444–452.
3. J. M. Byskov. Enumerating maximal independent sets with applications to graph colouring. *Operations Research Letters*, 32:547–556, 2004.
4. E. Dantsin, A. Goerdt, E. A. Hirsch, R. Kannan, J. Kleinberg, C. Papadimitriou, P. Raghavan, and U. Schöning. A deterministic $(2 - 2/(k+1))^n$ algorithm for k-SAT based on local search. *Theoretical Computer Science*, 289(1):69–83, 2002.

5. R. G. Downey and M. R. Fellows. *Parameterized complexity*. Springer-Verlag, New York, 1999.
6. D. Eppstein. Improved algorithms for 3-coloring, 3-edge-coloring, and constraint satisfaction. Proceedings of the *12th ACM-SIAM Symposium on Discrete Algorithms (SODA 2001)*, pp. 329–337.
7. D. Eppstein. Quasiconvex analysis of backtracking algorithms. Proceedings of the *15th ACM-SIAM Symposium on Discrete Algorithms (SODA 2004)*, pp. 781–790.
8. F. V. Fomin, D. Kratsch, and I. Todinca. Exact algorithms for treewidth and minimum fill-in. Proceedings of the *31st International Colloquium on Automata, Languages and Programming (ICALP 2004)*, Springer LNCS vol. 3142, 2004, pp. 568–580.
9. F. V. Fomin, D. Kratsch, and G. J. Woeginger. Exact (exponential) algorithms for the dominating set problem. Proceedings of the 30th Workshop on Graph Theoretic Concepts in Computer Science (WG 2004), Springer LNCS vol. 3353, 2004, pp. 245–256.
10. F. Grandoni. *Exact Algorithms for Hard Graph Problems*. PhD thesis, Università di Roma "Tor Vergata", Roma, Italy, Mar. 2004.
11. F. Grandoni. A note on the complexity of minimum dominating set. *Journal of Discrete Algorithms*. To appear.
12. T. W. Haynes, S. T. Hedetniemi, and P. J. Slater. *Fundamentals of domination in graphs*. Marcel Dekker Inc., New York, 1998.
13. M. Held and R.M. Karp. A dynamic programming approach to sequencing problems. *Journal of SIAM*, pages 196–210, 1962.
14. K. Iwama and S. Tamaki. Improved upper bounds for 3-SAT. Proceedings of the *15th ACM-SIAM Symposium on Discrete Algorithms (SODA 2004)*, p.328.
15. T. Jian. An $O(2^{0.304n})$ algorithm for solving maximum independent set problem. *IEEE Transactions on Computers*, 35(9):847–851, 1986.
16. E.L. Lawler. A note on the complexity of the chromatic number problem. *Information Processing Letters* 5(3):66–67, 1976.
17. R. Niedermeier and P. Rossmanith. New upper bounds for maximum satisfiability. *Journal of Algorithms*, 36(1):63–88, 2000.
18. P. Pudlak and R. Impagliazzio. A lower bound for DLL algorithms for k-SAT. A lower bound for DLL algorithms for k-SAT (preliminary version). Proceedings of the *11th ACM-SIAM Symposium on Discrete Algorithms (SODA 2000)*, pp. 128–136
19. R. Paturi, P. Pudlak, M. E. Saks, and F. Zane. An improved exponential-time algorithm for k-SAT. Proceedings of the *39th IEEE Symposium on Foundations of Computer Science (FOCS 1998)*, pp. 628–637.
20. B. Randerath and I. Schiermeyer. Exact algorithms for MINIMUM DOMINATING SET. Technical Report, zaik-469, Zentrum für Angewandte Informatik Köln, April 2004.
21. B. Reed. Paths, stars and the number three. *Combinatorics, Probability and Computing* 5:277–295, 1996.
22. J. M. Robson. Algorithms for maximum independent sets. *Journal of Algorithms*, 7(3):425–440, 1986.
23. J. M. Robson. Finding a maximum independent set in time $O(2^{n/4})$. Technical Report 1251-01, LaBRI, Université Bordeaux I, 2001.

24. U. Schoning. A Probabilistic Algorithm for k-SAT and Constraint Satisfaction Problems. Proceedings of the *40th IEEE Symposium on Foundations of Computer Science (FOCS 1999)*, pp. 410-414.
25. R. Tarjan and A. Trojanowski. Finding a maximum independent set. *SIAM Journal on Computing*, 6(3):537–546, 1977.
26. R. Williams. A new algorithm for optimal constraint satisfaction and its implications. Proceedings of the *31st International Colloquium on Automata, Languages and Programming (ICALP 2004)*, Springer LNCS vol. 3142, 2004, pp. 1227–1237.
27. G. J. Woeginger. Exact algorithms for NP-hard problems: A survey. *Combinatorial Optimization – Eureka, You Shrink*, Springer LNCS vol. 2570, 2003, pp. 185–207.

Optimistic Asynchronous Atomic Broadcast

Klaus Kursawe[1] and Victor Shoup[2]

[1] KU Leuven
[2] New York University

Abstract. This paper presents a new protocol for atomic broadcast in an *asynchronous* network with a maximal number of *Byzantine* failures. It guarantees both *safety* and *liveness* without making any timing assumptions. Under normal circumstances, the protocol runs in an extremely efficient "optimistic mode," while in rare circumstances the protocol may briefly switch to a less efficient "pessimistic mode."

1 Introduction

Atomic broadcast is a fundamental building block in fault tolerant distributed computing. By ordering broadcast requests in such a way that they are delivered in the same order to all honest recipients, a synchronization mechanism is provided that deals with many of the most problematic aspects of asynchronous networks. We present a new protocol for atomic broadcast in an *asynchronous* network with a maximal number of *Byzantine* failures. It guarantees both *safety* and *liveness* without making any timing assumptions or using any type of "failure detector," and under normal circumstances is just as efficient as a simple "Bracha broadcast."

The FLP "impossibility" result [F+85] implies that there is no deterministic protocol for Byzantine agreement (and hence, for atomic broadcast) that guarantees both safety and liveness. However, there are *randomized* protocols that terminate quickly with very high probability.

A protocol for asynchronous Byzantine agreement may be used as a building block for atomic broadcast. Canetti and Rabin's protocol [CR93] runs in polynomial time, but is in fact, highly impractical. The protocol of Cachin *et al.* [C+00] is a practical, polynomial-time protocol that makes use of public-key cryptographic primitives that can be proven correct in the "random oracle" model [BR93], assuming a computationally bounded adversary; this protocol relies on a trusted dealer during system set-up, but after this, an arbitrary number of instances of the protocol can be executed. Building on [C+00], the paper [C+01] presents a fairly practical protocol for atomic broadcast. However, this protocol still uses a lot of fairly expensive, public-key operations, and may not be fast enough for some applications.

Our protocol is inspired by the innovative work of Castro and Liskov [CL99b, CL99a, C00]. Like their protocol, our protocol works in two phases: an *optimistic phase* and a *pessimistic phase*. The optimistic phase is very "lightweight" —

each request is processed using nothing more than a "Bracha broadcast" [B84] — in particular, no public-key cryptography is used (only message authentication codes, which are very cheap, are used). As long as the network is reasonably behaved, the protocol remains in the optimistic phase — even if some number of parties, barring a designated leader, are corrupted. If there are unexpected network delays, or the leader is corrupted, several parties may "time out," shifting the protocol into the pessimistic phase. The pessimistic phase is somewhat more expensive than the optimistic phase — both in terms of communication and computational complexity. Nevertheless, it is still reasonably practical, although certainly not as efficient as the optimistic phase. The pessimistic phase cleans up any potential "mess" left by the current leader, re-synchronizing the protocol, after which the optimistic phase starts again with a new leader.

The optimistic phase of our protocol is essentially the same as that of Castro and Liskov. While [CL99b] relies extensively on expensive public-key cryptographic operations, the optimized versions in [CL99a, C00] do not use on public-key cryptography in the optimistic phase. Therefore, we expect that in practice, our protocol is just as efficient as theirs. However, our pessimistic phase is quite different. In the Castro/Liskov protocol, the new leader is responsible for re-synchronizing the protocol; this re-synchronization may fail, either because the new leader is corrupt or because of unexpected network delays, in which case yet another leader must take on the task of re-synchronization. In contrast, in our protocol, this re-synchronization is done using a distributed computation, based on randomized Byzantine agreement, and is guaranteed to succeed, regardless of the behavior of the corrupted parties and regardless of any network delays.

Castro and Liskov's protocol is completely deterministic, and hence is subject to the FLP impossibility result. Indeed, although their protocol guarantees safety, it does not guarantee liveness, unless one makes additional *timing assumptions*. Our protocol guarantees both safety and liveness without making any timing assumptions *at all*, while being just as efficient in practice as the Castro/Liskov protocol. The trade-off, of course, is that our protocol is randomized and relies on more cryptographic assumptions; however, there is no practical downside to this, in terms of either security or performance, and so it seems to be a trade-off worth making.

Our work builds on the work of [C+01] in two ways: we use the same definitional framework as [C+01], and we make novel use of a protocol in [C+01] for *multivalued Byzantine agreement*.

Other Related Work. There is a rich literature on ordering broadcast channels, including several implementations and a broad theoretical basis. However, most work in the literature is done in the crash-failure model; much less work has been done in the Byzantine failure model. Rampart [R94] and SecureRing [K+98] directly transfer crash-failure protocols into the Byzantine setting by using a modified failure detector along with digital signatures. The disadvantage of this approach is that it is relatively expensive, as a large number of public-key cryptographic operations need to be performed. Furthermore, there are attacks on the failure detector [A+95] that can violate the safety of these protocols. Doudou *et al.* [D+00] take a similar approach to that of Castro and Liskov. However, their protocol is described in a more abstract and modular way, leading to a protocol that is somewhat less complex and easier to analyze.

2 System Model and Problem Statement

Our formal system model and definitions of security are the same as in [C+00, C+01], which models attacks by *computationally bounded* adversaries. We refer the reader to [C+01] for complete details. We give only a brief summary here. We assume a network of n parties $P_1, \ldots, P_n$, t of which are corrupted and fully controlled by an adversary. We shall assume that $t < n/3$. We also assume a trusted dealer that is needed only at system set-up time. Informally, the adversary also has full control over the network; the adversary may insert, duplicate, and reorder messages at will.

More formally, at the beginning of the attack, the trusted dealer is run, initializing the internal state of the honest parties; the initial state information for the corrupted parties is given to the adversary. The attack then proceeds in steps. In each step of the attack, the adversary delivers a single message to an honest party, upon receipt of which the party updates its internal state and generates one or more response messages. These response messages indicate their origin and intended destination; however, the adversary is free to do with these messages what he wishes: to deliver them when he wishes, in any order that he wishes; he may also deliver them more than once, or not all. We do assume, however, that the adversary may not modify messages or "fake" their origin. This assumption is reasonable, since this property can be effectively enforced quite cheaply using message authentication codes.

We assume that the adversary's corruptions are *static*: the set of corrupted parties is chosen once and for all at the very beginning of the attack. However, it should be straightforward to prove that our protocol is secure in an *adaptive* corruption model, assuming all underlying cryptographic primitives are secure in this model.

Because we want to use cryptographic techniques, it does not make sense to consider "infinite runs" of protocols, but rather, we only consider attacks that terminate after some bounded amount of steps. The number of steps in the adversary's attack, as well as the computational complexity of the adversary, are assumed to be bounded by a polynomial in some security parameter.

Our protocols are defined such that they are only guaranteed to make progress to the extent to which the adversary actually delivers messages. To ensure that such a protocol behaves well in practice, an implementation would have to resend messages until receiving (secure) acknowledgments for them. We do not discuss any of these implementation details any further in this paper.

In our formal model, there is no notion of time. However, in making the transition from the optimistic phase to the pessimistic phase of our protocol, we need a way to test if an unexpectedly large amount of time has passed since some progress has been made by the protocol. That is, we need a "time out" mechanism. This is a bit difficult to represent in a formal model in which there is no notion of time. Nevertheless, we can effectively implement such a "time out" as follows: to start a timer, a party simply sends a message to itself, and when this message is delivered to that party, the clock "times out." By representing time outs in this way, we effectively give the adversary complete control of our "clock."

We define the *message complexity* of a protocol as the number of messages generated by all honest parties. This is a random variable that depends on the adversary and the value of the security parameter, and is denoted $MC(ID)$, where ID identifies a particular protocol instance. The term *probabilistically uniformly bounded* is a technical term that we borrow from [C+01]. Let X be a random variable associated with a run of a protocol that depends on the adversary and the value of the security parameter. Informally, "X is probabilistically uniformly bounded" means that X is distributed "very tightly" around a quantity Y, where Y is bounded by a polynomial in the security parameter that is independent of the adversary. See [C+01] for the formal definition.

Our definition of atomic broadcast comes directly from [C+01], with just some minor notational changes. As we define it, an atomic broadcast primitive offers one or several broadcast channels, each specified by some channel identifier ID. Before a party can use a channel, it must be explicitly *opened*. Formally speaking, this is done by the adversary. At any point, the adversary may deliver the message $(ID, \text{in}, \text{a-broadcast}, m)$ to some honest party, where m is an arbitrary bit string (of bounded size); we say the party *a-broadcasts the request* m at this point. At any point, an honest party may generate an output message $(ID, \text{out}, \text{a-broadcast}, m)$, which is given to the adversary; we say the party *a-delivers the request* m at this point. We adopt the following terminological convention: a "request" is something that is *a-broadcast* or *a-delivered*, while a "message" is something that is sent or delivered in the implementation of the protocol.

To give higher level protocols the option to block the atomic broadcast protocol, the delivering party waits for an acknowledgment after every *a-delivery* of a request. That is, the number of *a-delivered* requests is equal to either the number of acknowledgments or the number of acknowledgments plus one. This is necessary so that higher-level protocols may satisfy a property analogous to the *efficiency* property (see Definition 1 below). Without this ability to synchronize protocol layers, a low-level atomic broadcast protocol could generate an arbitrary amount of network traffic without a higher-level protocol ever doing anything useful.

At any point in time, for any honest party P_i, we define $\mathcal{B}^{(i)}$ to be the set of requests that P_i has *a-broadcast*, and we define $\mathcal{D}^{(i)}$ to be the set of requests that P_i has *a-delivered*. We say that one request in $\mathcal{B}^{(i)}$ is *older* than another if P_i *a-broadcast* the first request before it *a-broadcast* the second request. At any point in time, we also define $\mathcal{D}^* = \cup_{\text{honest } P_i} \mathcal{D}^{(i)}$.

In discussing the values of the sets $\mathcal{B}^{(i)}$, $\mathcal{D}^{(i)}$, or $\mathcal{D}^*$ at particular points in time, we consider the sequence of events $E_1, \ldots, E_k$ during the adversary's attack, where each event but the last is either an *a-broadcast* or *a-delivery* by an honest party, and the last event is a special "end of attack" event. The phrase "at time τ," for $1 \leq \tau \leq k$, refers to the point in time just *before* event E_τ occurs.

Definition 1 (Atomic Broadcast). *A protocol for* atomic broadcast *satisfies the following conditions, for all channels ID and all adversaries, with all but negligible probability.*

Agreement: *If some honest party has a-delivered m on channel ID, then all honest parties a-deliver m on channel ID, provided the adversary opens channel ID for all honest parties, delivers all associated messages, and generates acknowledgments for every party that has not yet a-delivered m on channel ID.*

Total Order: *Suppose one honest party has a-delivered $m_1, \ldots, m_s$ on channel ID, and another honest party has a-delivered $m'_1, \ldots, m'_{s'}$ on channel ID with $s \leq s'$. Then $m_l = m'_l$ for $1 \leq l \leq s$.*

Validity: *There are at most t honest parties P_j with $\mathcal{B}^{(j)} \backslash \mathcal{D}^* \neq \emptyset$, provided the adversary opens channel ID for all honest parties, delivers all associated messages, and generates all acknowledgments.*

Fairness: *There exist a quantity Δ, which is bounded by a fixed polynomial in the security parameter (independent of the adversary), such that the following holds. Suppose that at some time τ_1, there is a set $\mathcal{S}$ of $t+1$ honest parties, such that for all $P_j \in \mathcal{S}$, the set $\mathcal{B}^{(j)} \backslash \mathcal{D}^*$ is non-empty. Suppose that there is a later point in time τ_2 such that the size of $\mathcal{D}^*$ increases by more than Δ between times τ_1 and τ_2. Then there is some $P_j \in \mathcal{S}$, such that the oldest request in $\mathcal{B}^{(j)} \backslash \mathcal{D}^*$ at time τ_1 is in $\mathcal{D}^*$ at τ_2.*

Efficiency: *At any point in time, the quantity $MC(ID)/(|\mathcal{D}^*|+1)$ is probabilistically uniformly bounded.*

Integrity: *Every honest party a-delivers a request m at most once on channel ID. Moreover, if all parties follow the protocol, then m was previously a-broadcast by some party on channel ID.*

3 Protocol Conventions and Notations

At each step of an attack, the adversary delivers a message to an honest party, and activates the honest party: the party performs some computations, updates its internal state, generates messages, and then returns control to the adversary. Messages delivered to a party are appended to the rear of an *incoming message queue*. When activated, the party may examine this queue, and remove any messages it wishes.

A party consists of one or more threads of execution. When a party is activated, each thread is in a *wait state*, waiting for one of the corresponding conditions to be satisfied. If the condition upon which any thread is waiting is satsified, the corresponding thread is activated (if several threads could be activated, one is chosen arbitrarily), and this thread runs until it reaches another wait state. This process continues until all threads are in wait states whose conditions are not satisfied, and then control returns to the adversary.

Our protocol syntax is rather self explaining, with one exception. A *wait* condition can either *receive* messages or *detect* them. In the former case, the messages are deleted from the queue (and thus do not trigger any further conditions) while in the latter, they remain in the queue. We also define an abstract timeout mechanism, which allows a process to start or stop the timer or wait for a timeout. In our model, this is implemented by the party simply sending a message to itself. As the adversary has full control over message delivery, this gives him full control over the timer, too.

4 Our New Protocol for Atomic Broadcast

The protocol operates in epochs, each epoch $e = 0, 1, 2$, etc., consisting of an optimistic and a pessimistic phase. In the optimistic phase, a designated leader orders incoming requests by assigning sequence numbers to them and initiating a Bracha broadcast [B84]; the optimistic phase guarantees the *agreement* and *total order* properties, but not the *validity* or *fairness* properties; however, the protocol can effectively determine if *validity* or *fairness* are potentially threatened, and if so, switch to the pessimistic phase, which cleans up any "mess" left by the current leader; then the optimistic phase starts again with a new leader.

4.1 Overview and Optimistic Phase

In the optimistic phase of epoch e, when a party *a-broadcasts* a request m, it *initiates* the request by sending a message of the form $(ID, \texttt{initiate}, e, m)$ to the leader for epoch e. When the leader receives such a message, it *0-binds* a sequence number s to m by sending a message of the form $(ID, \texttt{0-bind}, e, m, s)$ to all parties. Sequence numbers start at zero in each epoch. Upon receiving a *0-binding* of s to m, an honest party *1-binds* s to m by sending a message of the form $(ID, \texttt{1-bind}, e, m, s)$ to all parties. Upon receiving $n - t$ such *1-bindings* of s to m, an honest party *2-binds* s to m by sending a message of the form $(ID, \texttt{2-bind}, e, m, s)$ to all parties. A party also *2-binds* s to m if it receives $t + 1$ *2-bindings* of s to m — this has the effect of "amplifying" *2-bindings*, which is used to ensure *agreement*. Upon receiving $n - t$ such *2-bindings* of s to m, an honest party *a-delivers* m, provided all messages with lower sequence numbers were already delivered, enough acknowledgments have been received, and m was not already *a-delivered*.

A party only sends or reacts to *0-*, *1-*, or *2-bindings* for sequence numbers s in a "sliding window" $\{w, \ldots, w + \mathit{WinSize} - 1\}$, where w is the number of requests already *a-delivered* in this epoch, and $\mathit{WinSize}$ is a fixed system parameter. Keeping the "action" bounded in this way is necessary to ensure *efficiency* and *fairness*.

The number of requests that any party *initiates* but has not yet *a-delivered* is bounded by a parameter $\mathit{BufSize}$: a party will not *initiate* any more requests once this bound is reached. We denote by $\mathcal{I}$ the set of requests that have been *initiated* but not *a-delivered*, and we call this the *initiation queue*. If sufficient time passes without anything leaving the initiation queue, the party "times out" and *complains* to all other parties. These *complaints* are "amplified" analogously to the *2-bindings*. Upon receiving $n-t$ *complaints*, a party enters the pessimistic phase of the protocol. This strategy will ensure *validity*. Keeping the size of $\mathcal{I}$ bounded is necessary to ensure *efficiency* and *fairness*.

Also to ensure *fairness*, a party keeps track of the "age" of the requests in its initiation queue, and if it appears that the oldest request is being ignored, i.e., many other requests are being *a-delivered*, but not this one, then the party simply refuses to generate *1-bindings* until the problem clears up. If $t+1$ parties block in this way, they effectively prevent the remaining parties from making any progress in the optimistic phase, and thus, the pessimistic phase will be entered, where the fairness problem will ultimately be resolved.

We say that an honest party P_i *commits* s to m in epoch e, if m is the sth request (counting from 0) that it *a-delivered* in this epoch, optimistically or pessimistically.

Now the details. The state variables for party P_i are as follows.

Epoch number e: The current epoch number, initially zero.
Delivered set $\mathcal{D}$: All requests that have been *a-delivered* by P_i. It is required to ensure that requests are not *a-delivered* more than once; in practice, however, other mechanisms may be employed for this purpose. Initially, $\mathcal{D}$ is empty.
Initiation queue $\mathcal{I}$: The queue of requests that P_i *initiated* but not yet *a-delivered*. Its size is bounded by *BufSize*. Initially, $\mathcal{I}$ is empty.
Window pointer w: w is the number of requests that have been *a-delivered* in this epoch. Initially, $w = 0$. The optimistic phase of the protocol only reacts to messages pertaining to requests whose sequence number lies in the "sliding window" $\{w, \ldots, w + \textit{WinSize} - 1\}$. Here, *WinSize* is a fixed system parameter.
Echo index sets $BIND_1$ **and** $BIND_2$: The sets of sequence numbers which P_i has *1-bound* or *2-bound*, respectively. Initially empty.
Acknowledgment count $acnt$: Counts the number of acknowledgments received for *a-delivered* requests. Initially zero.
Complaint flag *complained*: Set if P_i has issued a complaint. Initially *false*.
Initiation time $it(m)$: For each $m \in \mathcal{I}$, $it(m)$ is equal to the value of w at the point in time when m was added to $\mathcal{I}$. Reset to zero across epoch boundaries. These variables are used in combination with a fixed parameter *Thresh* to ensure *fairness*.
Leader index l: The index of the leader in the current epoch; we simply set $l = (e \bmod n) + 1$. Initially, $l = 1$.
Scheduled request set $\mathcal{SR}$: Only maintained by the current leader. It contains the set of messages which have been assigned sequence numbers in this epoch. Initially, it is empty.
Next available sequence number $scnt$: Only maintained by the leader. Value of the next available sequence number. Initially, it is zero.

/* Initiate m. */
upon receiving a message $(ID, \text{in}, \text{a-broadcast}, m)$ for some m such that $m \notin \mathcal{I} \cup \mathcal{D}$ and $|\mathcal{I}| < \textit{BufSize}$ (note that we take the *oldest* such message first):
    Send the message $(ID, \text{initiate}, e, m)$ to the leader.
    Add m to $\mathcal{I}$; set $it(m) \leftarrow w$.

/* 0-bind $\textit{scnt}$ to m. */
upon receiving a message $(ID, \text{initiate}, e, m)$ for some m, such that $i = l$ and $w \leq \textit{scnt} < w + \textit{WinSize}$ and $m \notin \mathcal{D} \cup \mathcal{SR}$:
    Send the message $(ID, \text{0-bind}, e, m, \textit{scnt})$ to all parties.
    Increment $\textit{scnt}$ and add m to $\mathcal{SR}$.

/* 1-bind s to m. */
upon receiving a message $(ID, \text{0-bind}, e, m, s)$ from the current leader for some m, s such that $w \leq s < w + \textit{WinSize}$ and $s \notin \textit{BIND}_1$ and $((\mathcal{I} = \emptyset)$ or $(w \leq \min\{it(m) : m \in \mathcal{I}\} + \textit{Thresh}))$:
    Send the message $(ID, \text{1-bind}, e, m, s)$ to all parties; add s to $\textit{BIND}_1$.

/* 2-bind s to m. */
upon receiving $n - t$ messages of the form $(ID, \text{1-bind}, e, m, s)$ from distinct parties that agree on s and m, such that $w \leq s < w + \textit{WinSize}$ and $s \notin \textit{BIND}_2$:
    Send the message $(ID, \text{2-bind}, e, m, s)$ to all parties; add s to $\textit{BIND}_2$.

/* Amplify a 2-binding of s to m. */
upon detecting $t + 1$ messages of the form $(ID, \text{2-bind}, e, m, s)$ from distinct parties that agree on s and m, such that $w \leq s < w + \textit{WinSize}$ and $s \notin \textit{BIND}_2$:
    Send the message $(ID, \text{2-bind}, e, m, s)$ to all parties; add s to $\textit{BIND}_2$.

/* Commit s to m. */
upon receiving $n - t$ messages of the form $(ID, \text{2-bind}, e, m, s)$ from distinct parties that agree on s and m, such that $s = w$ and $\textit{acnt} \geq |\mathcal{D}|$ and $m \notin \mathcal{D}$ and $s \in \textit{BIND}_2$:
    Output $(ID, \text{out}, \text{a-deliver}, m)$; increment w; add m to $\mathcal{D}$, and remove it from $\mathcal{I}$ (if present); **stop timer**.

/* Start timer. */
upon (timer not running) and (not *complained*) and $(\mathcal{I} \neq \emptyset)$ and $(\textit{acnt} \geq |\mathcal{D}|)$:
    start timer.

/* Complain. */
upon timeout:
    if not *complained*: send the message $(ID, \text{complain}, e)$ to all parties; set *complained* $\leftarrow$ *true*.

/* Amplify complaint. */
upon detecting $t + 1$ messages $(ID, \text{complain}, e)$ from distinct parties, such that not *complained*:
    Send the message $(ID, \text{complain}, e)$ to all parties; set *complained* $\leftarrow$ *true*; **stop timer**.

/* Go pessimistic. */
upon receiving $n - t$ messages $(ID, \text{complain}, e)$ from distinct parties, such that *complained*:
    Execute the procedure *Recover* below.

Fig. 1. The optimistic phase

The protocol for party P_i consists of two threads. The first is a trivial thread that simply counts acknowledgments for *a-delivered* requests; it consists of an infinite loop whose body is as follows:
    wait until receiving an acknowledgment; increment *acnt*
The main thread is an infinite loop whose body is as follows:
    case *MainSwitch* **end case**
where the *MainSwitch* is a sequence of **upon** clauses described in Figure 1.

4.2 Fully Asynchronous Recovery

The recovery protocol is invoked if the optimistic phase appears to not work properly; this happens if either the leader is faulty or the network is too slow. Its job is to synchronize the parties by *a-delivering* all broadcasts that any honest party *may* have already *a-delivered*, and to guarantee the efficiency of the overall protocol by assuring that *some* messages are *a-delivered*. Finally, it hands over to a new leader to restart a new optimistic phase.

Validated Multivalued Byzantine Agreement. Our recovery-protocol builds on top of validated multivalued Byzantine agreement (i.e., the agreement is not restricted to a binary value), as defined and implemented in [C+01]. The final agreement value must be legal according to some validation function, which guarantees that it is some "useful" value. The definition of the validation function is clear from the context as the exact form of a valid proposal is defined in the protocol description. In the atomic broadcast protocol, we use the phrase "*propose X_i for multivalued Byzantine agreement on X*" to denote the invocation of a validated multivalued Byzantine agreement protocol, where X_i is P_i's initial proposal, and X the resulting agreement value.

Overview of the Recovery Procedure. We distinguish between three types of requests: (i) requests for which it can be guaranteed that they have been *a-delivered* by an honest party; (ii) requests that potentially got *a-delivered* by an honest party; (iii) requests for which it can be guaranteed that they have not been *a-delivered* by an honest party. For the first two kinds of requests, an order of delivery might already be defined, and has to be preserved. The other requests have not been *a-delivered* at all, so the recovery protocol has complete freedom on how to order them. They can not be left to the next leader, however, as an adversary can always force this leader to be thrown out as well. To guarantee efficiency, the recovery procedure has to ensure that *some* request is *a-delivered* in every epoch. This is precisely the property that Castro and Liskov's protocol fails to achieve: in their protocol, without imposing additional timing assumptions, the adversary can cause the honest parties to generate an arbitrary amount of messages before a single request is *a-delivered*. According to the three types of requests, the recovery protocol consists of three parts.

Part 1: Requests whose sequence number is determined. A "watermark" $\hat{s}_e$ is jointly computed, which has the property that at least one honest party optimistically committed the sequence number $\hat{s}_e$, and no honest party optimistically committed a sequence number higher than $\hat{s}_e + 2 \cdot \textit{WinSize}$. After computing the watermark, all parties "catch up" to the watermark, i.e., commit all sequence numbers up to $\hat{s}_e$, by simply waiting for $t + 1$ consistent *2-bindings* for each sequence number up to the watermark. The work performed in this part constant, and especially independent of the number of unfinished requests.

Part 2: Requests whose sequence number may be determined. Here, we deal with the requests that might or might not have been *a-delivered* by some honest party in the optimistic phase of this epoch. We have to ensure that if some honest party has optimistically *a-delivered* a request, then all honest parties *a-deliver* this request as well. The sequence numbers of requests with this property lie in the interval $\hat{s}_e + 1 \ldots \hat{s}_e + 2 \cdot \textit{WinSize}$. Each party makes a proposal that indicates what action should be taken for all sequence numbers in this critical interval. Again, multivalued Byzantine agreement will be used to determine which of possibly several valid proposals should be accepted. While this part is relatively expensive, we can guarantee an upper bound of the number of requests processed here, which is determined by the window-size parameter.

Part 3: Undetermined Requests. This part is the one that guarantees that some messages are *a-delivered* in this epoch. We use a multivalued Byzantine agreement protocol to agree on a certain set of additional requests that should be *a-delivered* this epoch. We need to do this to ensure fairness and efficiency.

Terminology of the Recovery Procedure. For any party P_i, and any message α, we denote by $\{\alpha\}_i$ a signed version of the message, i.e., α concatenated with a valid signature under P_i's public key on α, along with P_i's identity.

For any $s \geq -1$, a *strong consistent set* Σ for s is a set of $t + 1$ correctly signed messages from distinct parties, each of the form $\{(ID, \texttt{s-2-bind}, e, s')\}_j$ for some j and $s' \geq s$.

A *valid watermark proposal* $\mathcal{M}$ is a set of $n - t$ correctly signed messaged from distinct parties, each of the form $\{(ID, \texttt{watermark}, e, \Sigma_j, s_j)\}_j$ for some j, where Σ_j is a strong consistent set of signatures for s_j. The maximum value s_j appearing in these watermark messages is called the *maximum sequence number* of $\mathcal{M}$.

For any $s \geq 0$, a *weak consistent set* Σ' for s is a set of $n - t$ correctly signed messages from distinct parties — each of the form $\{(ID, \texttt{w-2-bind}, e, s, m_j)\}_j$ for some j — such that either all $m_j = \bot$ (indicating no *2-binding* for s), or there exists a request m and all m_j are either m or $\bot$. In the former case, we say Σ' *defines* $\bot$, and in the latter case, we say Σ' *defines* m.

A *valid recover proposal* $\mathcal{P}$ is a set of $n - t$ correctly signed messages from distinct parties each of the form $\{(ID, \texttt{recover request}, c, \mathcal{Q}_j)\}_j$ for some j, where $\mathcal{Q}_j$ is a set of at most *BufSize* requests.

The protocol for the pessimistic phase is presented in Figure 2.

A proof of security of the complete atomic broadcast protocol, as well as a number of other details, can be found in the full version of this paper [KS01].

/* Part 1: Recover requests with a determined sequence-number */
Send a the signed message $\{(ID, \texttt{s-2-bind}, e, \max(BIND_2 \cup \{-1\}))\}_i$ to all parties.
wait until receiving a strong consistent set Σ_i for $w-1$.
Send the signed message $\{(ID, \texttt{watermark}, e, \Sigma_i, w-1)\}_i$ to all parties.
wait until receiving a valid watermark proposal $\mathcal{M}_i$.
Propose $\mathcal{M}_i$ for multivalued Byzantine agreement on a valid watermark proposal $\mathcal{M}$.
Set $\hat{s}_e \leftarrow \tilde{s} - \textit{WinSize}$, where $\tilde{s}$ is the maximum sequence number of $\mathcal{M}$.
while $w \leq \hat{s}_e$ **do**:
 wait until receiving $t+1$ messages of the form $(ID, \texttt{2-bind}, e, m, w)$ from distinct parties that agree on m, such that $acnt \geq |\mathcal{D}|$.
 Output $(ID, \texttt{out}, \texttt{a-deliver}, m)$; increment w.
 Add m to $\mathcal{D}$, and remove it from $\mathcal{I}$ (if present).

/* Part 2: Recover requests with a potentially determined sequence-number */
For $s \leftarrow \hat{s}_e + 1$ to $\hat{s}_e + (2 \cdot \textit{WinSize})$ **do**:
 If P_i sent the message $(ID, \texttt{2-bind}, e, m)$ for some m, set $\tilde{m} \leftarrow m$; otherwise, set $\tilde{m} \leftarrow \bot$.
 Send the signed message $(ID, \texttt{w-2-bind}, e, s, \tilde{m})$ to all parties.
 wait until receiving a weak consistent set Σ'_i for s.
 Propose Σ'_i for multivalued Byzantine agreement on a weak consistent set Σ' for s.
 Let Σ' define m.
 If $(s \geq w$ and $m \in \mathcal{D})$ or $m = \bot$, exit the for loop and go to Part 3.
 If $m \notin \mathcal{D}$: **wait until** $acnt \geq |\mathcal{D}|$; output $(ID, \texttt{out}, \texttt{a-deliver}, m)$; increment w; add m to $\mathcal{D}$, and remove it from $\mathcal{I}$ (if present).

/* Part 3: Recover undetermined Requests */
If $w = 0$ and $\mathcal{I} \neq \emptyset$: send the message $(ID, \texttt{recover-help}, e, \mathcal{I})$ to all parties.
If $w = 0$ and $\mathcal{I} = \emptyset$: **wait until receiving** a message $(ID, \texttt{recover-help}, e, \mathcal{Q})$, such that $\mathcal{Q}$ is a non-empty set of at most $\textit{BufSize}$ requests, and $\mathcal{Q} \cap \mathcal{D} = \emptyset$.
If $w \neq 0$ or $\mathcal{I} \neq \emptyset$: set $\mathcal{Q} \leftarrow \mathcal{I}$.
Send the signed message $\{(ID, \texttt{recover-request}, e, \mathcal{Q})\}_i$ to all parties.
wait until receiving a valid recover proposal $\mathcal{P}_i$.
Propose $\mathcal{P}_i$ for multivalued Byzantine agreement on a valid recover proposal $\mathcal{P}$.
Sequence through the request set of $\mathcal{P}$ in some deterministic order, and for each such request $m \notin \mathcal{D}$, do the following:
 wait until $acnt \geq |\mathcal{D}|$; output $(ID, \texttt{out}, \texttt{a-deliver}, m)$; increment w; add m to $\mathcal{D}$, and remove it from $\mathcal{I}$ (if present).

/* Start New Epoch */
Set $e \leftarrow e+1$; $l \leftarrow (e \bmod n) + 1$; $w \leftarrow scnt \leftarrow 0$; $\mathcal{SR} \leftarrow BIND_1 \leftarrow BIND_2 \leftarrow \emptyset$; $\textit{complained} \leftarrow \textit{false}$.
For each $m \in \mathcal{I}$: send the message $(ID, \texttt{initiate}, e, m)$ to the leader; set $it(m) \leftarrow 0$.

Fig. 2. The pessimistic phase

References

[A+95] E. Anceaume, B. Charron-Bost, P. Minet, and S. Toueg. On the formal specification of group membership services. Tech. Rep. TR95-1534, Cornell University, Computer Science Department, 1995.

[BR93] M. Bellare and P. Rogaway. Random oracles are practical: A paradigm for designing efficient protocols. In *1st ACM Conf. on Computer and Communications Security*, pp. 62–73, 1993.

[B84] G. Bracha. An asynchronous $[(n-1)/3]$-resilient consensus protocol. In *Proc. of the 3rd Ann. ACM Symp. on Principles of Distributed Computing*, pp. 154–162, 1984.

[C00] M. Castro. *Practical Byzantine Fault Tolerance*. PhD thesis, Massachusetts Institute of Technology, 2000.

[C+01] C. Cachin, K. Kursawe, F. Petzold, and V. Shoup. Secure and efficient asynchronous broadcast protocols. In *Advances in Cryptology—Crypto 2001*, pp. 524–541, 2001.

[C+00] C. Cachin, K. Kursawe, and V. Shoup. Random Oracles in Constantinople: Practical Asynchronous Byzantine Agreement using Cryptography. In *ACM SIGACT-SIGOPS Symp. on Principles of Distributed Computing*, pp. 123–132, 2000.

[CL99a] M. Castro and B. Liskov. Authenticated byzantine fault tolerance without public-key cryptography. Tech. Memo MIT/LCS/TM-589, MIT Laboratory for Computer Science, 1999.

[CL99b] M. Castro and B. Liskov. Practical Byzantine fault tolerance. In *Proc. 3rd Symp. Operating Systems Design and Implementation*, 1999.

[CR93] R. Canetti and T. Rabin. Fast asynchronous Byzantine agreement with optimal resilience. In *Proc. 25th Ann. ACM Symp. on Theory of Computing*, pp. 42–51, 1993.

[D+00] Doudou, Guerraoui, and Garbinato. Abstractions for devising byzantine-resilient state machine replication. In *SRDS: 19th Symp. on Reliable Distributed Systems*, 2000.

[F+85] M. J. Fischer, N. A. Lynch, and M. S. Paterson. Impossibility of distributed consensus with one faulty process. *J. ACM*, 32(2):374–382, 1985.

[K+98] K. P. Kihlstrom, L. E. Moser, and P. M. Melliar-Smith. The SecureRing protocols for securing group communication. In *31st Hawaii International Conference on System Sciences*, pp. 317–326, 1998.

[KS01] K. Kursawe, V. Shoup. Optimistic asynchronous atomic broadcast. Cryptology ePrint Archive, Report 2001/022, http://eprint.iacr.org, 2001.

[R94] M. K. Reiter. Secure agreement protocols: Reliable and atomic group multicast in Rampart. In *Proc. of the 2nd ACM Conference on Computer and Communication Security*, pp. 68–80, 1994.

Asynchronous Perfectly Secure Communication over One-Time Pads[*]

Giovanni Di Crescenzo[1] and Aggelos Kiayias[2,**]

[1] Telcordia, Piscataway, NJ, USA
giovanni@research.telcordia.com
[2] CSE Dept., University of Connecticut, Storrs, CT, USA
aggelos@cse.uconn.edu

Abstract. The "One-Time Pad" is a fundamental cryptographic protocol as it represents the ideal in secure unidirectional communication (i.e., in cases where there is a designated sender and a designated receiver) both in terms of security (in the presence of eavesdroppers) as well as in terms of computational efficiency. Surprisingly, no modeling and investigation of this protocol has been done in important practical settings, as distributed and asynchronous ones. In this work we introduce an asynchronous model for multidirectional and multi-player One-Time Pad asynchronous communication protocols. In this model the random pad is shared by all players, and there is no designated sender and receiver; in fact any participating player can act as a receiver at any given time, players communicate in a totally asynchronous fashion and may arbitrarily go off-line.

We define the problem of designing One-Time Pad asynchronous communication protocols, where the goal is that of maximizing the amount of the shared pad used before new randomness needs to be generated, with the constraint of mantaining the security property under reasonable adversarial assumptions on the relative behavior of the players and the network. We present lower bounds and protocol solutions for this problem that significantly improve over the obvious scenario where parties use an equal fraction of the pad. Our constructions are *non-interactive* in the sense that they require no additional synchronizing communication beyond the (usual) information that accompanies each ciphertext.

1 Introduction

The "One-Time Pad" is a well-known private-key encryption scheme, originally invented by Vernam [14] in 1918. Assume Alice and Bob agree on a random key K (a random "pad"); then, they can communicate securely (that is, without the eavesdropper Eve obtaining any information about their message) as follows: On input message m, Alice computes the ciphertext $c = m \oplus [K]$ where $[K]$

---

[*] Copyright Telcordia.
[**] Research partly supported by NSF CAREER Award CNS-0447808.

denotes a substring of K of appropriate length and sends it to Bob. Given c, Bob can recover message m, by decrypting c as $m = c \oplus [K]$. Here, $\oplus$ is the "exclusive OR" operator, and K is at least as large as m. What makes one-time pad encryption remarkable is the following two facts (i) as shown by Shannon, in [13], it holds that any provably-secure (in the information-theoretic sense) encryption scheme must satisfy $|K| \geq |m|$ and more specifically one-time pad is optimally secure in the information theoretic sense. (ii) the encryption and decryption operations are essentially optimal in terms of time-complexity (being a mere exclusive-or operation). While the length of the key is inappropriate for most practical cryptographic applications, one-time pads are widely utilized as atomic components of more elaborate encryption systems by employing pseudo-random generators to generate arbitrarily long sequences of pseudo-random bits given only a short shared random seed (see e.g. [12,7]). In this case the resulting pseudorandom sequence is used as a pad. The employment of such pseudorandom generators allows the transmission of messages longer than the shared key and security would rely solely on the unpredictability of the pseudo-random generator (since one-time pad over a truly random pad is information theoretically secure).

Now consider the setting where Alice and Bob both wish to play the sender or receiver role engaging in a *conversation* without a predetermined schedule and in an asynchronous communication setting. In the obvious solution, both players divide the pad K in two halves and each one uses a separate portion. Nevertheless, it would be much preferable to allow players to dynamically balance the pad portion they use depending on their need as this is determined by the progress of their interaction. In this sense the protocol allowing $|K|/2$ maximum total communication by each player individually loses a lot of the one-time pad efficiency in terms of utilizing their private storage and computation power (we stress that expanding the shared pad is an expensive operation and players would wish to utilize the jointly generated pad as efficiently as possible). In the bidirectional setting, a solution with more efficient pad utilization would allow the segment that can be expended by Alice in some conversations to be expended by Bob in others. Note that in the asynchronous setting we consider we cannot rely on participants being online to execute synchronization steps (and in this case even deterministic consensus decision is unattainable, [6]), while other synchronization techniques employing e.g. randomization [2], are clearly quite expensive in terms of communication for our setting.

The Problem. The problem we consider in this paper is then, as follows: Is it possible for $m \geq 2$ players, sharing a one-time pad of a certain size, to communicate securely (that is, without re-using portions of the one-time pad) and efficiently (that is, by using as much pad material as possible before re-generating new randomness)? In particular we will seek answers to the above problem where (i) the communication model is *multidirectional* and totally *asynchronous*; (ii) players do not use additional messages to synchronize themselves; (iii) players might arbitrarily go off-line or crash. Note that in order to satisfy these constraints we will opt for *non-interactive* protocol constructions: when a player wishes to transmit a message he should make a decision based only on his local data on which pad to use next. The adversarial action that we will consider deals with the relative behavior of the players and the possible delays/swaps that are introduced during message transmission.

Our results. To describe the class of protocols solving the above problem we introduce the notion of one-time pad asynchronous communication (OTPAC) protocols. Our formal definition of secure OTPAC protocols considers an adversary arbitrarily delaying messages and introduces an undelivery parameter d meaning that, at any time, and for each sender, at most d messages were already sent by other players but have not been delivered to him yet. Note that this adversarial behavior refers to the relative behavior of the participants rather than properties of the network that is assumed to be totally asynchronous. The adversarial behavior is formally defined by specifying some natural property of a family of communication patterns (described as graphs) among the players. As a consequence, protocols for secure communication in this model can be shown to be secure according to this family of communication patterns.

In order to compare protocols in terms of the allowed flexibility in expending the shared pad, we define a measure called the "efficiency ratio" that is associated with each OTPAC protocol. Informally, the ratio is a metric for the maximum number of random bits from the one-time pad that can be used by any player before the pad needs to be refreshed. Under this metric, the protocol of the unidirectional or the synchronous setting exhibits ratio 1 (thus, optimal), while the protocol of m players equally splitting the pad has ratio $1/m$.

We present concrete constructions of secure OTPAC protocols in the asynchronous setting whose efficiency ratio is larger than $1/m$ and is expressed as a function of the undelivery parameter d, the number of players m and the number of pads n. Our general approach for OTPAC protocol construction ensures improved expendability of the pad by having players maintaining various types of "logical clocks" [11]. We completely characterize the two party case, by exhibiting a protocol achieving efficiency ratio $1 - d/n$ for $d \leq \lfloor n/2 \rfloor$, and a lower bound that matches this protocol when $d \leq \lfloor n/2 \rfloor$ and the protocol of 2 players equally splitting the pad otherwise. In the m-party case, we exhibit a protocol with efficiency ratio $1 - 1/\mathsf{poly}(n)$ for typical parameter values, and a lower

Protocol	Number of players	Efficiency Ratio	Proven Secure for	Other Constraints
$\mathcal{A}_1$	m	$\frac{1}{m}$	unconditional	
$\mathcal{A}_2$	m	1	undelivery 0	
$\mathcal{A}_3$	2	$1 - \frac{d}{n}$	undelivery d	$d < n/2$
$\mathcal{A}_4$	m	$1 - \frac{d}{n}(L-2)^{\log m - 1} - \frac{\log m - 2}{L}$	undelivery d	$d < nL^{1 - \log m}$

Fig. 1. Comparison of 4 Protocols. Note that in all cases we assume $n \gg m$. $\mathcal{A}_1$ refers to the protocol where players use disjoint pad segments, where $\mathcal{A}_2$ refers to the protocol where players employ a jointly maintained counter (players are synchronized). These two protocols are listed for the sake of comparison. Our main constructions are protocols $\mathcal{A}_3, \mathcal{A}_4$. L is a parameter of $\mathcal{A}_4$

bound of $1 - d/n$ for $d \leq \lfloor n(m-1)/m \rfloor$ and $1/m$ otherwise, the latter matching the protocol of m players equally splitting the pad. Our upper bound results are overviewed in figure 1. We note that we know of no previous work in the literature that considered one-time pads in the asynchronous communication setting. The problem of designing OTPAC protocols relates to the Do-All (DA) problem (see, e.g., [5, 4, 10]) and Write-All (WA) problem (see, e.g., [9, 1, 3, 8]) from the Distributed Computing literature. In fact, from this literature's viewpoint, designing OTPAC protocols can be thought of as solving a "Do/Write-Once" kind of problem; this interrelationship will be further analyzed in upcoming work.

2 One-Time Pad Asynchronous Communication

In this section we present definitions for the communication model and protocols, the security requirements as well as an efficiency metric, which we instantiate through examples of basic protocols.

OTPAC Protocols. In our setting, m players $A_1, \ldots, A_m$ wish to communicate securely taking advantage of a shared random pad. Players are assumed to communicate through a network that supports broadcast transmission and each player can be a sender or a receiver in a certain communication. There is no global clock that the players can use for synchronization and message delays are unbounded; all messages are eventually delivered but their order is not necessarily preserved by the network (such assumptions are typical in the Distributed Computing literature). Each player A_i is initialized by executing an Initialization procedure, and subsequently sequences of two events may happen to A_i: Send and Receive, according to the requests of sending messages that the specific application environment demands from the players in the protocol. Each player A_j is assumed to have access to the same shared private *random pad* p that can be written as the concatenation of n words $p_1, \ldots, p_n$, of k bits each, k being the length of one message-block. The communication is driven by a sequence of *send requests* for the players (that can be thought to be adversarially generated) and are ordered according to some (unknown to the players) global notion of time; the send requests form the *schedule* of the protocol denoted by $s = ((s_1, t_1), \ldots, (s_v, t_v))$, where $t_1 \leq t_2 \leq \cdots \leq t_v$ correspond to the time of the send requests and each $s_\ell = (j, u)$ stands for the u th request of the j-th player. To achieve the highest possible generality we will assume that each execution of the protocol takes as input a pad p and follows an arbitrary schedule s. Let us denote by $[n] = \{1, \ldots, n\}$. We can now proceed with a formal definition.

Definition 1. Let m, n be positive integers, let Head $= [m] \times [n]$, and let Conf denote a set of possible configurations; at local time l we denote the configuration of player A_j as $\Pi_j[t] \in$ Conf. We define a *One-Time-Pad Asynchronous Communication* (OTPAC) protocol as a triple of polynomial-size circuits $\langle I, S, R \rangle$, with the following functionality:

$I : [m] \to$ Conf $S : [m] \times$ Conf $\to \{\bot\} \cup ($Conf$\times [n])$ $R : [m] \times$ Conf $\times$ Head $\to$ Conf

The operation of an OTPAC protocol is as follows: initially every player A_j executes $I(j) \to \Pi_j[0]$ to set its initial configuration. Then the sequence of send requests in the schedule s is sequentially scanned. Given a send request at time t, player A_j generates a Send event, by executing $S(\Pi_j[t]) \to \langle \Pi_j[t+1], i \rangle$ and resulting in the transmission of the ciphertext $\langle j, i, p_i \oplus M \rangle$, where $M \in \{0,1\}^k$ is the message that player A_j wishes to transmit and $\langle j,i \rangle \in [m] \times [n]$ is the ciphertext header (note that frequently we will write $S(\cdot)$ instead of $S(j, \cdot)$ if the sending player is clear from the context; similarly for $R(\cdot)$). If $S(\Pi_j[t]) = \bot$ player A_j ceases to send messages. Every ciphertext has length $\log m + \log n + k$. A Receive event is triggered by the arrival of a ciphertext $\langle j', i', C \rangle$ at player A_j in time t. Player A_j decrypts the message by computing $p_{i'} \oplus C$ and updates its configuration to $\Pi_j[t+1] \leftarrow R(\Pi_j[t], j', i')$. We assume that the S, R circuits satisfy the property $S(\Pi_j[t]) = \bot \Rightarrow S(R(\Pi_j[t], j, i)) = \bot$ for any $\langle j, i \rangle \in$ Head. An OTPAC protocol stops when all parties A_j enter in a configuration $\Pi_j[t]$ such that $S(\Pi_j[t]) = \bot$. A particular execution of an OTPAC protocol (namely, an execution of algorithms I, S, R, for fixed random tapes of all participants) can be characterized as in the following definition.

Definition 2. A feasible *asynchronous communication pattern* (ACP) for an m-player OTPAC protocol $\langle I, S, R \rangle$ on input an n-word pad $\boldsymbol{p}$ and a schedule $\boldsymbol{s}$ is a finite directed acyclic graph $\mathcal{P}$ so that the set of its nodes is a subset of $\mathbb{N} \times \mathbb{N}$ and is partitioned into m subsets. Every node $\langle j, t \rangle$ is labeled by an element of Conf, denoted by $\Pi_j[t]$. $\mathcal{P}$ includes the nodes $\langle 1, 0 \rangle, \ldots, \langle m, 0 \rangle$ labeled by $I(1), \ldots, I(m)$ respectively, as well as the nodes $\langle 1, t_{1,\max} \rangle, \ldots, \langle m, t_{m,\max} \rangle$ such that for any $t > t_{j,\max}$, $\langle j, t \rangle \notin \mathcal{P}$. For any $j, t < t_{j,\max}$ there is an edge from $\langle j, t \rangle$ to $\langle j, t' \rangle$ for some t' with $t < t'$ and such that for all $t^* \in \{t+1, \ldots, t'-1\}$, $\langle j, t^* \rangle \notin \mathcal{P}$; the node $\langle j, t' \rangle$ is the *subsequent* node of $\langle j, t \rangle$. The nodes of $\mathcal{P}$ and their labels fall into either one of the following three categories:

- *Send Nodes.* It is a node $\langle j, t \rangle$ with $S(\Pi_j[t]) \neq \bot$ that has m outgoing edges to $\langle 1, t_1 \rangle, \ldots, \langle m, t_m \rangle$ s.t. $t_j = t+1$ and at most one incoming edge. The label of $\langle j, t+1 \rangle$ is equal to Π where $\langle \Pi, i \rangle \leftarrow S(\Pi_j[t])$.
- *Receive Nodes.* It is a node $\langle j, t \rangle$ with one incoming edge coming from some Send-node $\langle j', t' \rangle$ $j' \neq j$. The label of $\langle j, t+1 \rangle$ is equal to $R(\Pi_j[t], j', i')$ where i' is such that $\langle \Pi', i' \rangle \leftarrow S(\Pi_{j'}[t'])$.
- *Idle Nodes.* It is a node $\langle j, t \rangle$ that is neither a Send or Receive node. If $t > 0$, the node has one incoming and one outgoing edge to its subsequent node $\langle j, t' \rangle$ (recall $t' > t$); if $\langle j, 0 \rangle$ is idle then it has no incoming edge. The label of $\langle j, t' \rangle$ is equal to $\Pi_j[t]$.

Any OTPAC protocol $\mathcal{A} := \langle I, S, C \rangle$ for m players using a random pad $\boldsymbol{p}$ of n k-bit words and a schedule $\boldsymbol{s}$ defines a family of feasible ACP's denoted by $\mathcal{F}_{\mathcal{A},\boldsymbol{s}}^{m,n}$. (By $\mathcal{F}_{\mathcal{A}}^{m,n}$ we denote the union, over all schedules $\boldsymbol{s}$, of the family of feasible ACP's $\mathcal{F}_{\mathcal{A},\boldsymbol{s}}^{m,n}$.)

Security and efficiency of OTPAC protocols. We define the security of OTPAC protocols by extending the security notion of the one-time pad from the unidirectional synchronous setting (namely, that the same portion of the pad should not be used twice) to the asynchronous setting considered here.

Let $\mathcal{A}$ be an OTPAC protocol, and $\mathcal{F}_\mathcal{A}^{m,n}$ be the set of feasible ACP's. While players $A_1, \ldots, A_m$ are attempting to communicate, the adversary is capable of controlling the network so to both choose a particular schedule of send requests and swap or delay messages being sent. The goal of the adversary is to force the re-use of some portion of the pad so that the "one-time" property is lost and some information can be derived from the transcript. It follows that the only mechanism available to players to protect the security of their communication is to reach the $\bot$ output of their Send-event state update function S, before the adversary is capable of violating the one-time property. So given an OTPAC protocol $\mathcal{A}$, the adversary selects a schedule s that is input to $\mathcal{A}$, and furthermore selects a pattern $\mathcal{P} \in \mathcal{F}_\mathcal{A}^{m,n}$; the protocol is executed following the pattern $\mathcal{P}$; if $\mathcal{P}$ contains two Send events that use the same pad segment, then the adversary wins. In conclusion, security of an OTPAC protocol would be argued for a subfamily of $\mathcal{F}_\mathcal{A}^{m,n}$ for which no pad is used twice. Formally,

Definition 3. The Security Property. Let $\mathcal{A} := \langle I, S, C \rangle$ be an OTPAC protocol, let $\mathcal{F}_{\mathcal{A},s}^{m,n}$ the be family of feasible ACP's for $\mathcal{A}$ on input schedule s and let $\mathcal{F}_{\mathcal{A},s}^{m,n} = \cup_s \mathcal{F}_{\mathcal{A},s}^{m,n}$. We say that $\mathcal{A}$ is *insecure on* $\mathcal{P} \in \mathcal{F}_{\mathcal{A},s}^{m,n}$ if $\mathcal{P}$ includes two Send nodes $\langle j, t \rangle$ and $\langle j', t' \rangle$ so that if $\langle \Pi, i \rangle \leftarrow S(\Pi_j[t])$ and $\langle \Pi', i' \rangle \leftarrow S(\Pi_{j'}[t'])$ it holds that $i = i'$. Let $\mathcal{G} = \cup_s \mathcal{G}_s$ be a family of ACP's, where for each s, $\mathcal{G}_s \subseteq \mathcal{F}_{\mathcal{A},s}^{m,n}$. We say that $\mathcal{A}$ is *secure on family* $\mathcal{G}$, if there exists no schedule s and no pattern $\mathcal{P} \in \mathcal{G}_s$ such that $\mathcal{A}$ is insecure on $\mathcal{P}$.

Naturally, for any given $\mathcal{A}$ it would be the most desirable that it is secure for any $\mathcal{P} \in \mathcal{F}_\mathcal{A}^{m,n}$. Considering only such protocols though is very restrictive. Instead, as it is standard in cryptography as well as in distributed systems theory, we will opt to restrict the behavior of the adversary in meaningful ways and consider protocols that can be proven secure in such adversarial setting.

Adversarial Setting: Undelivery. A communication pattern $\mathcal{P} \in \mathcal{F}_{\mathcal{A},s}^{m,n}$ admits various total orderings $\prec$ between the vertices of $\mathcal{P}$ (topological sortings of the graph $\mathcal{P}$). One such sorting $\prec$ corresponds to the actual "real global time" (unknown to the participants). Such total orderings can also be called "runs" of an asynchronous communication pattern $\mathcal{P}$, following the standard terminology in distributed systems. Note that, as it is common in distributed systems theory, we assume that no two events happen at the same time, i.e. the total ordering $\prec$ is strict. A Send event $\langle j, t \rangle$ is said to have d *undelivered messages* w.r.t. a topological sorting $\prec$ of $\mathcal{P}$, if there exist d Send events $\langle j_1, t_1 \rangle, \ldots, \langle j_d, t_d \rangle$, $j \notin \{j_1, \ldots, j_d\}$ for which it holds $\langle j_\ell, t_\ell \rangle \prec \langle j, t \rangle$ for $\ell = 1, \ldots, d$, and if $\langle j, s_1 \rangle, \ldots, \langle j, s_d \rangle$ are the corresponding Receive events at player A_j it holds that $s_\ell > t$ for $\ell = 1, \ldots, d$. For any schedule s, let $\mathcal{U}_{\mathcal{A},s,d}^{m,n}$ denote the subfamily of all ACP's $\mathcal{P} \in \mathcal{F}_{\mathcal{A},s}^{m,n}$ that have no Send events with d undelivered messages w.r.t. any possible topological sorting of $\mathcal{P}$. Also, let $\mathcal{U}_{\mathcal{A},d}^{m,n} = \cup_s \mathcal{U}_{\mathcal{A},s,d}^{m,n}$.

Note that undelivery does not describe any fixed property of the network. More specifically, if an ACP contains a Send event that has d undelivered messages this does not mean that the network delays messages by a fixed upper time-bound that is proportional to d. Instead, the adversary controlled the message delivery of the network and the random coin tosses of the players' random processes in such a way that one of the players initiated a Send-event before having heard of d Send events of other players that happened concurrently. We remark that if it was possible for the players to synchronize themselves (e.g., by using a global clock) it would be easy to restrict the adversary to produce only ACP's that belong to $\mathcal{U}_{\mathcal{A},1}^{m,n}$, i.e., if a player sends a message the other players are silent until they receive it. Clearly in such a scenario designing an OTPAC protocol is a trivial matter. Tackling the asynchronous case where the parameter $d > 1$ is the motivation for our investigations.

Efficiency of OTPAC protocols. We introduce a natural metric of the efficiency of an OTPAC protocol with respect to usage of the shared pad vector.

We start with some definitions. The *message-count* of an ACP $\mathcal{P} \in \mathcal{F}_{\mathcal{A}}^{m,n}$, denoted by $\mathsf{mc}(\mathcal{P})$, is a tuple $\langle N_1, \ldots, N_m \rangle$ denoting the number of Send events for each of the players $A_1, \ldots, A_m$. Let $\mathcal{G} = \cup_s \mathcal{G}_s$ be a family of ACP's, where for each s, $\mathcal{G}_s \subseteq \mathcal{F}_{\mathcal{A},s}^{m,n}$. Next we define the *order* of a family $\mathcal{G} \subseteq \mathcal{F}_{\mathcal{A}}^{m,n}$ as $N = \min_s \min_{\mathcal{P} \in \mathcal{G}_s} \sum_{i=1}^n N_i$. Intuitively, the order of $\mathcal{G}$ is the maximum number of messages that all players can send using protocol $\mathcal{A}$ when, on input schedule s, the communication pattern is drawn from $\mathcal{G}_s$. Using this notion, we define our main metric for OTPAC protocols, which we call the "efficiency ratio," standing for the maximum percentage of the pad all players are allowed to use.

Definition 4. The Efficiency Ratio. Let $\mathcal{A}$ be an OTPAC protocol with corresponding ACP family $\mathcal{F}_{\mathcal{A}}^{m,n}$ and let $\mathcal{G} = \cup_s \mathcal{G}_s$ be an ACP family such that for each s, $\mathcal{G}_s \subseteq \mathcal{F}_{\mathcal{A},s}^{m,n}$. The *efficiency ratio of algorithm $\mathcal{A}$ on family $\mathcal{G}$* is defined as $\alpha := \frac{N}{n}$ where N is the order of family $\mathcal{G}$.

Although it does not appear explicitly in the above definition, we stress that this metric only makes sense for secure executions, as insecure executions can always have efficiency ratio 1 or even greater. To understand the above notion it is helpful to recall the two simple protocols described in figure 1: $\mathcal{A}_1$ will be the OTPAC protocol where all players use a disjoint segment in the pad and $\mathcal{A}_2$ will be the protocol that all players use the whole pad at the same time while maintaining a joint counter. It is straightforward to show that $\mathcal{A}_1$ has efficiency ratio $1/m$ and can be proven secure for all feasible ACP's $\mathcal{F}_{\mathcal{A}_1}^{m,n}$, whereas $\mathcal{A}_2$ has efficiency ratio 1, but can be proven secure only for $\mathcal{U}_{\mathcal{A}_2,1}^{m,n}$, the family of ACP's that are effectively synchronous (no Send-event has an undelivered message). The protocols $\mathcal{A}_1, \mathcal{A}_2$ represent the two different ends of the spectrum: $\mathcal{A}_1$ allows arbitrary adversarial action but has very small efficiency ratio; on the other hand $\mathcal{A}_2$ allows the highest efficiency ratio nevertheless it can only be proven secure if we restrict the adversary so much that players communicate in an effectively synchronous fashion. This motivates our investigations in the rest of the paper, where we construct OTPAC protocols that trade adversarial action for efficiency.

3 OTPAC Protocol Lower Bounds

In this section we prove lower bounds on the efficiency ratio of OTPAC protocols, as a function of the number n of elements in the common pad, the number m of players, and the undelivery parameter d. We start with OTPAC protocols among 2 players, and then extend the analysis to any number m of players.

The Two-Player Case. Let $\mathcal{A}$ be an OTPAC protocol among 2 players. Recall that we consider adversaries that can choose the request schedule s input to protocol $\mathcal{A}$ and can arbitrarily delay messages, so to force any Send event to have up to d undelivered messages, where d is an unrestricted parameter (in particular, note that $d \geq n$ is equivalent to assuming $d = \infty$). We obtain the following:

Theorem 1. Any OTPAC protocol $\mathcal{A}$ secure on $\mathcal{U}_{\mathcal{A},d}^{2,n}$ can achieve efficiency ratio α such that:
- $\alpha \leq 1 - d/n$ if $d \leq \lfloor n/2 \rfloor$,
- $\alpha \leq 1/2$ if $d > \lfloor n/2 \rfloor$.

The proof of Theorem 1 starts by assuming, towards contradiction, that there exists an OTPAC protocol $\mathcal{A}$ that can achieve efficiency ratio larger than the claimed bound on the ACP family $\mathcal{U}_{\mathcal{A},d}^{2,n}$; finally, it reaches contradiction by showing that $\mathcal{A}$ is not secure on this family. An intuition on how the contradiction is reached goes as follows. First, we show the existence of a schedule s_1 on which the efficiency ratio larger than the claimed bound is achieved by portions of the random pad only used by party P_1. Then we show the existence of a schedule s_2 with the analogue property for P_2. Finally, we show the existence of a schedule s and of an adversary, such that in an execution of $\mathcal{A}$ on input s the following holds: at any time the adversary keeps up to d undelivered messages; the view of P_1 is as if $s = s_1$; and the view of P_2 is as if $s = s_2$. As a consequence, both parties use a large number of portions of the random pad, and at least one portion is used by both parties, thus implying that the protocol is not secure.

The m-Player Case. We now consider an OTPAC protocol $\mathcal{A}$ among m players, for $m > 2$. We note that in this case we count one undelivered message for each player that has not received the message yet (that is, if there are 2 receivers that have not received the same message, then the undelivery parameter satisfies $d = 2$). The lower bound on the efficiency ratio of $\mathcal{A}$ is then obtained as an appropriate generalization of the bound in the 2 player case.

Theorem 2. Let $m \leq n$. Any OTPAC protocol $\mathcal{A}$ secure on $\mathcal{U}_{\mathcal{A},d}^{m,n}$ can achieve efficiency ratio at most $1 - d/n$ if $d \leq \lfloor n(m-1)/m \rfloor$ or at most $1/m$ otherwise.

The above theorems imply that protocol $\mathcal{A}_1$ (defined in figure 1) has *optimal* efficiency ratio for values of $d \geq n/2$. Note that this range of values strictly includes the range $d \geq n$, that, in turn, is equivalent to assuming $d = \infty$; that is, when no bound at all can be assumed on the undelivery parameter.

4 OTPAC Protocol Constructions: Head-on Collision Runs

In this section we present two protocol constructions that exhibit more refined trade-offs between security and efficiency. To increase the efficiency ratio we allow players to use common pad segments while at the same time maintaining safe distance between the segments that are actively used during a conversation. The manner with which players will be expending pad portions will resemble a "head-on collision run" over the pad vector. We will start with the simpler two-player case and then discuss the m-player setting.

4.1 The Two Player Setting

Informal Description of the $\mathcal{A}_3$ protocol. The two players initialize counters, each one at the far end of the pad vector. Players expend pads in a "head-on collision run" to each other while maintaining a "safety distance" from each other's pad segment. The safety factor is determined by a parameter d. A formal description of the protocol will appear in the full version; a graphical representation is in figure 2. In the lemma below we characterize the ACP's for which the security of $\mathcal{A}_3$ fails.

Fig. 2. A head-on collision run: protocol $\mathcal{A}_3$

Lemma 1. *Let $\mathcal{P} \in \mathcal{F}_{\mathcal{A}_3}^{m,n}$ for which $\mathcal{A}_3$ is insecure. Then, $\mathcal{P}$ has* Send-*event with at least $d+1$ undelivered messages.*

Theorem 3. *The OTPAC protocol $\mathcal{A}_3$ is secure for $\mathcal{U}_{\mathcal{A}_3,d}^{2,n} \subseteq \mathcal{F}_{\mathcal{A}_3}^{2,n}$ The efficiency ratio of $\mathcal{A}_3$ over $\mathcal{U}_{\mathcal{A}_3,d}^{2,n}$ equals $1 - d/n$.*

Because of Theorem 1, we obtain that protocol $\mathcal{A}_3$ has *optimal* efficiency ratio over family $\mathcal{U}_{\mathcal{A}_3,d+1}^{2,n} \subseteq \mathcal{F}_{\mathcal{A}_3}^{2,n}$ for $d \leq \lfloor n/2 \rfloor$.

4.2 The Multi-player Setting

In this section we will generalize the OTPAC protocol $\mathcal{A}_3$ to the multi-player setting. The protocol $\mathcal{A}_4$ is a multilayer recursive generalization of protocol $\mathcal{A}_3$. Refer to figure 3 for a graphical depiction of the four player setting.

Below we describe the protocol $\mathcal{A}_4$ and we assume for simplicity that it involves $m = 2^\delta$ players. Note that to avoid cluttering of notation and due to lack of space we will only give an informal (but sufficiently detailed) description

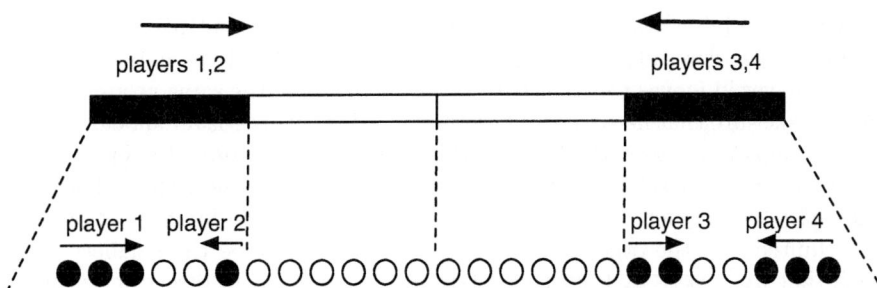

Fig. 3. Multilayered head-on collision runs: protocol $\mathcal{A}_4$

of the algorithms $\langle I, R, S \rangle$. We give some notation first: every player is identified by a δ-bitstring id. For a string a we denote by $[a]_j$ the symbol in the j-th location. It holds that $a = [a]_1 \ldots [a]_\delta$ if a is a δ-long string. If b is a bit, we denote by $\overline{b}$ its complement. The neighbor set (res. native set) of degree j for a player id is a set of bitstrings a so that (i) $[a]_\ell = [\mathrm{id}]_\ell$ for $\ell = 1, \ldots, j-1$ (ii) $[a]_j = \overline{[\mathrm{id}]_j}$, (respectively : $[a]_j = [\mathrm{id}]_j$), and (iii) $[a]_\ell \in \{0, 1\}$ for all $\ell = j+1, \ldots, \delta$. The pad vector will be divided into a number of sub-divisions using a parameter L (thought to be a small function in n). The sub-division of degree ℓ for $\ell = 1, \ldots, \delta - 1$ will be comprised of L^ℓ components. Finally in the sub-division of degree δ we will have actual pad segments of length $s := n/L^{\delta-1}$.

The main idea of the protocol is that at the δ-subdivision we will have individual players implementing the two-player protocol in head-on collision run fashion. Then, at the $\delta - 1$ subdivision we will have groups of 2 players operating in head-on collision fashion, and so on.

A player id maintains a position variable for himself $p_{\mathrm{id}} \in \{1, \ldots, L\}^{\delta-1} \times \{1, \ldots, s\}$. Note that a position maps to a specific pad inside the n-long pad-vector by specifying one of the $L^{\delta-1}$ sub-divisions of degree $\delta - 1$ as well as a pad location inside this sub-division. Every player assumes an initial position in the pad vector. In particular player id assumes the initial position $p_{\mathrm{id}}^{\mathrm{init}}$ with $[p_{\mathrm{id}}^{\mathrm{init}}]_\ell = 1([\mathrm{id}]_\ell =_? 0) + L([\mathrm{id}]_\ell =_? 1)$ for $\ell = 1, \ldots, \delta - 1$ and $[p_{\mathrm{id}}^{\mathrm{init}}]_\delta = 1([\mathrm{id}]_\ell =_? 0) + s([\mathrm{id}]_\ell =_? 1)$. This $p_{\mathrm{id}} = p_{\mathrm{id}}^{\mathrm{init}}$ will be the first pad used by player id if he wishes to send a message. The header of a ciphertext will be of the form $\langle \mathrm{id}, p_{\mathrm{id}} \rangle$.

Every player will maintain information about the movements of his neighboring subsets of users. In particular, player id will maintain δ counters $v_1, \ldots, v_\delta$ for the neighboring subsets of users. These counters are initialized as follows: $v_\ell = 1(\overline{[\mathrm{id}]_\ell} =_? 0) + L(\overline{[\mathrm{id}]_\ell} =_? 1)$ and $v_\delta = 1(\overline{[\mathrm{id}]_\delta} =_? 0) + s(\overline{[\mathrm{id}]_\delta} =_? 1)$ If a message is received that has header $\langle \mathrm{id}^*, p^* \rangle$, player id will compute the function $\alpha := \alpha(\mathrm{id}, \mathrm{id}^*) = j \subset \{1, \ldots, \delta\}$ such that $[\mathrm{id}]_\ell - [\mathrm{id}^*]_\ell$ for all $\ell < j$; this α is called the nativity degree between two players. Conceptually, for two players id and id* the nativity degree specifies the highest level at which the two players are facing each other in a head-on collision run (either by themselves, or as members of larger groups).

Upon receiving a message with header $\langle \text{id}^*, p^* \rangle$, player id takes the following actions: (i) it computes the nativity degree α. (ii) for all the sub-divisions for which player id moves together with id^* as members of the same group, player id must make sure that he advances together with his group. Player id belongs to the same group of players with id^* for all the subdivisions of degree $\ell \in \{1, \ldots, \alpha-1\}$ (note: if $\alpha = 1$ this set is empty). If id is not "trailing" behind then it should hold that $[p_{\text{id}}]_\ell = [p^*]_\ell$ for all $\ell = 1, \ldots, \alpha - 1$. In this case no further action would be required. On the other hand, let $j \in \{1, \ldots, \alpha - 1\}$ be the smallest integer such that $[p_{\text{id}}]_j \neq [p^*]_j$. Player id determines his movement direction within the group at the subdivision of degree j, which is "to the right" if $[\text{id}]_j = 0$ or "to the left" if $[\text{id}]_j = 1$. Then, in case the direction is to the right, the player performs the operation: if $[p_{\text{id}}]_j < [p^*]_j$ set $[p_{\text{id}}]_j = [p^*]_j$. In case the direction is to the left, the player does instead: if $[p_{\text{id}}]_j > [p^*]_j$ set $[p_{\text{id}}]_j = [p^*]_j$. If one of the above two modifications takes place and the value $[p_{\text{id}}]_j$ is updated, then id resets the remaining values of p_{id} as follows $[p_{\text{id}}]_\ell = [p_{\text{id}}^{\text{init}}]_\ell$ for $\ell = j+1, \ldots, \delta$. The above update allows to a player that is trailing behind on a certain level to "catch up" with his native players; it follows that a player may change his position in the pad vector because of the movement of his native players. (iii) Regarding the neighboring players, one neighboring counter needs to be updated: v_α; to update counter v_α player id determines the movement direction of the neighboring set of players at the α-level (to the right if $[\text{id}^*]_\alpha = 0$, or to the left if $[\text{id}^*]_\alpha = 1$) and performs the update: if $v_\alpha < [p^*]_\alpha$ and $[\text{id}^*]_\alpha = 0$ then $v_\alpha = [p^*]_\alpha$; if $v_\alpha > [p^*]_\alpha$ and $[\text{id}^*]_\alpha = 1$ then $v_\alpha = [p^*]_\alpha$. Note that player id will ignore any information provided by p^* and refers to a level greater to the nativity degree.

Having described the semantics of the counters we proceed now to see how a player id having just used the pad at position p_{id}, advances to the next pad that he can use. The player needs to determine the status of every level $\ell \in \{1, \ldots, \delta\}$ among two possible states: stay and jump. The state computation for level $\ell = 1, \ldots, \delta - 1$ is as follows: the player checks whether $|[p_{\text{id}}]_\ell - v_\ell| \geq 2$ and in this case assigns to level ℓ the state stay; for the δ-th level the state stay is assigned if $|[p_{\text{id}}]_\delta - v_\delta| > d$ where d is a given fixed parameter. If all levels are assigned state stay then player id will use the adjacent pad, i.e., if $p_{\text{id}} = l_1 l_2 \ldots l_{\delta-1} l_\delta$ the next pad will be at $p_{\text{id}}^{\text{new}} = l_1 l_2 \ldots l_{\delta-1}(l_\delta + 1)$ if $[\text{id}]_\delta = 0$ (movement to the right) or $p_{\text{id}}^{\text{new}} = l_1 l_2 \ldots l_{\delta-1}(l_\delta - 1)$ if $[\text{id}]_\delta = 1$ (movement to the left). If all levels are assigned the state jump then player id terminates. Now let j be the highest level that is assigned a stay state (and it holds that the $(j+1)$-th state is at the jump state). Then, player id will assume the position $p_{\text{id}}^{\text{new}}$ defined as (i) $[p_{\text{id}}^{\text{new}}]_\ell = [p_{\text{id}}]_\ell$ for $\ell = 1, \ldots, j$, (ii) $[p_{\text{id}}^{\text{new}}]_{j+1} = [p_{\text{id}}^{\text{init}}]_{j+1} + 1$ if $[\text{id}]_{j+1} = 0$, or $[p_{\text{id}}^{\text{new}}]_{j+1} = [p_{\text{id}}^{\text{init}}]_{j+1} - 1$ if $[\text{id}]_{j+1} = 1$; finally (iii) $[p_{\text{id}}^{\text{new}}]_\ell = [p_{\text{id}}^{\text{init}}]_\ell$ for $\ell = j+2, \ldots, \delta$ (recall that $p_{\text{id}}^{\text{init}}$ is the initial position of player id). In addition to the above, player id must update all its neighbor counters v_ℓ for the levels $\ell = j+1, \ldots, \delta$ to their initial position, as in the beginning of the protocol, $v_\ell = 1(\overline{[\text{id}]}_\ell =_? 0) + L(\overline{[\text{id}]}_\ell =_? 1)$ and $v_\delta = 1(\overline{[\text{id}]}_\delta =_? 0) + s(\overline{[\text{id}]}_\delta =_? 1)$.

A formal description of the protocol will appear in the full version.

Lemma 2. Assume parameters $L, d, \delta = \log m$ such that $n/L^{\delta-1} > d$. Let $\mathcal{P} \in \mathcal{F}_{\mathcal{A}_4}^{m,n}$ be an an asynchronous communication pattern for which protocol $\mathcal{A}_4$ is insecure. Then, $\mathcal{P}$ has Send-event with at least $d+1$ undelivered messages.

Theorem 4. The OTPAC protocol $\mathcal{A}_4$ is secure for $\mathcal{U}_{\mathcal{A}_4,d}^{m,n} \subseteq \mathcal{F}_{\mathcal{A}_4}^{2,n}$ The efficiency ratio of $\mathcal{A}_4$ over $\mathcal{U}_{\mathcal{A}_4,d}^{m,n}$ equals $1 - 2(\delta-2)/L - (L-2)^{\delta-1}d/n$.

We note that in the typical case $m = O(1), d = O(n^\epsilon), \epsilon < 1$, it is possible to select the parameter L so that the efficiency ratio becomes $1 - 1/\text{poly}(n)$, i.e., arbitrarily close to 1 for sufficiently large n.

Acknowledgement. The authors thank O. Kornievskaia and A. Shvartsman for helpful discussions.

References

1. R. Anderson and H. Woll, *Algorithms for the Certified Write-All Problem*, in SIAM Journal on Computing, vol. 139(1), 1997, p. 1–16.
2. James Aspnes, *Randomized protocols for asynchronous consensus*, Invited survey paper for Distributed Computing, PODC 20th anniversary issue. 2002.
3. J. Buss, P. Kanellakis, P. Ragde, and A. Shvartsman, *Parallel Algorithms for Process Failures and Delays*, in Journal of Algorithms, vol. 20(196), 1997, p. 45–86.
4. R. De Prisco, A. Mayer and M. Yung, *Time-Optimal Message-Efficient Work-Performance in the Presence of Faults*, in Proc. of PODC 94.
5. C. Dwork, J. Halpern and O. Waarts, *Performing Work Efficiently in the Presence of Faults*, in SIAM Journal on Computing, vol. 27, 1998, p. 1457–1491.
6. Michael J. Fischer, Nancy A. Lynch and Mike Paterson, *Impossibility of Distributed Consensus with One Faulty Process*, JACM 32(2): 374-382 (1985)
7. Oded Goldreich, *The Foundations of Cryptography - Vol. 1*, Cambridge University Press, 2001.
8. J. Groote, W. Hesselink, S. Mauw, and R. Vermeulen, *An Algorithm for the Asynchronous Write-All Problem based on Process Collision*, in Distributed Computing, vol. 14(2), 2001, p. 75–81.
9. P. Kanellakis and A. Shvartsman, *Efficient Parallel Algorithms Can Be Made Robust*, in Distributed Computing, vol. 5(4), 1992, p. 201–217.
10. D. Kowalski and A. Shvartsman, *Performing Work with Asynchronous Processors: Message-Delay-Sensitive Bounds*, in Proc. of PODC 2003, p. 265–274
11. Leslie Lamport, *Time, Clocks, and the Ordering of Events in a Distributed System* CACM 21(7): 558-565 (1978)
12. Michael Luby, *Pseudorandomness and Cryptographic Applications*, Princeton University Press, 1996.
13. C. E. Shannon, *A Mathematical Theory of Communication*, The Bell System Technical Journal 27, 1948, 379-423, 623-656.
14. G. Vernam, *Secret Signaling Systems*, US Patent, 1919.

Single-Prover Concurrent Zero Knowledge in Almost Constant Rounds*

Giuseppe Persiano and Ivan Visconti

Dipartimento di Informatica ed Appl.,
Università di Salerno, Italy
{giuper, visconti}@dia.unisa.it

Abstract. In this paper we study the round complexity of *concurrent zero-knowledge* arguments and show that, for any function $\beta(n) = \omega(1)$, there exists an *unbounded* concurrent zero-knowledge argument system with $\beta(n)$ rounds. Our result assumes that the *same* prover is engaged in several concurrent sessions and that the prover has a counter whose value is shared across concurrent executions of the argument. Previous constructions for concurrent zero knowledge required a (almost) logarithmic number of rounds [Prabhakaran et al. - FOCS 2002] in the plain model or seemingly stronger set-up assumptions.

Moreover, we construct two $\beta(n)$-round unbounded concurrent zero-knowledge arguments that are mutually concurrent simulation sound for any $\beta(n) = \omega(1)$. Here we assume that each party has access to a counter and that the two protocols are used by the *same* two parties to play several concurrent sessions of the two protocols.

1 Introduction

Since its introduction, the concept of a zero-knowledge proof system and the simulation paradigm have been widely used to prove the security of many protocols. The notion of *concurrent zero knowledge* [2] formalizes security in a scenario in which several verifiers access concurrently a prover and maliciously coordinate their actions so to extract information from the prover. In [3] it has been showed that in the black-box model $\tilde{\Omega}(\log n)$ round are necessary for concurrent zero knowledge for non-trivial languages. The first concurrent zero-knowledge proof system for $\mathcal{NP}$ has been given by [4] that showed that $O(n^\epsilon)$ are sufficient for any $\epsilon > 0$. Poly-logarithmic round-complexity was achieved in [5] and, finally, in [1] it is shown that $\tilde{O}(\log n)$ rounds are sufficient. The proof systems presented in [4,5,1] are black-box zero knowledge and the round-complexity of the proof system of [1] is almost optimal in view of the lower bound proved in [3]. Thus unlike the stand-alone case, black-box concurrent zero knowledge cannot be achieved in a constant number of rounds.

---

* Work supported by Network of Excellence ECRYPT under contract IST-2002-507932.

Different models have been presented in which round-efficient black-box concurrent zero knowledge is possible. In [2, 6, 7] constant-round concurrent zero-knowledge proof systems have been presented by relaxing the asynchrony of the model or the zero-knowledge property. In [8, 9], constant-round concurrent zero-knowledge proof systems have been presented assuming the existence of a common reference string or a shared random string (i.e., a trusted third party) while in [10] a constant-round concurrent zero-knowledge with concurrent soundness argument system is shown by assuming that there exists a public repository that contains the public keys of the verifiers. Furthermore, Pass [12] gave a constant-round concurrent zero-knowledge argument with a super-polynomial-time simulator. In [13], Barak presented a *non-black-box* constant-round *bounded-concurrent* zero-knowledge argument system. The construction of [13] assumes that the maximum number of concurrent sessions is known in advance.

Simulation-sound zero knowledge. Simulation-sound zero knowledge has been introduced in [14] for the purpose of constructing cryptosystems secure against adaptive chosen-ciphertext attacks. This concept is related to the concept of non-malleability introduced in [15]. Indeed, both notions deal with an adversary (called the *man-in-the-middle*) that simultaneously participates to many executions of two proof systems and acts as a prover in the former and as a verifier in the latter. The adversary has complete control over the scheduling of the messages in the executions of the protocols. Informally, two zero-knowledge proof systems are said mutually concurrent *simulation sound* if the information that the man-in-the-middle adversary collects as a verifier from concurrent sessions played with a *simulated* prover of the former proof system does not help him to prove a *false* statement in the latter proof system and vice versa. Here the man-in-the-middle can choose to see *simulated* proofs of true and false statements.

Simulation-sound zero knowledge plays an important role for proving the security of protocols. Indeed, when the simulation paradigm is used to prove the security of a protocol, the simulator could, in some cases, need to simulate the proof of a false statement. Here simulation soundness is crucial since the adversary could gain knowledge from such a proof in order to prove a false statement in another protocol.

Our results. In this paper we show that, for any function β such that $\beta = \omega(1)$, there exists a $\beta(n)$-round concurrent zero-knowledge argument system for any language in $\mathcal{NP}$. Our argument system assumes that: 1) the prover is equipped with a counter that counts the total number of bits he has sent so far in all sessions; 2) the argument system remains zero knowledge provided that one *single* prover is engaged by the adversarial verifier in any polynomial number of (unbounded) concurrent sessions.

We stress that our set-up assumptions seem no-stronger than the ones made in [2, 6, 7, 8, 10]. Comparing our result with the bounded-concurrent result of [13], we stress that our construction does not assume knowledge of an *a-priori* bound on the number of concurrent sessions. On the other hand, our construction requires a super-constant number of rounds. Finally, we remark that the concurrent zero-knowledge property of our construction is obtained by means of an efficient simulator.

Relying upon our construction of concurrent zero knowledge we present two $\beta(n)$-round argument systems that are concurrent zero knowledge and mutually concurrent simulation sound for any $\beta = \omega(1)$. Here we require set-up assumptions inherited from the concurrent zero-knowledge argument on which they are based: 1) each party is equipped with a counter that is used to keep track of the total length of some of the messages sent; 2) the two protocols are played by the *same* two players with roles inverted.

Comparison with known assumptions. In the past, round-efficient concurrent zero knowledge has been achieved by making assumptions that decrease the power of an adversarial concurrent verifier. For example timing assumptions (used in [2, 6, 7]) limit the power of adversarially scheduling the communication between prover and verifier (by imposing a bound on the delays experienced by the messages), bounded concurrency (used in [13]) upper bounds the number of concurrent sessions. When concurrent zero knowledge is obtained in presence of a common reference string [8, 9], the adversarial concurrent verifier is assumed not to corrupt the party that generated a common reference string and that this party securely preserves (or erases) any secret information about the string.

In contrast, our assumptions are quite different in nature. Let us discuss the case of concurrent zero knowledge (the same discussion can be used for the two arguments that are mutually concurrent simulation-sound and concurrent zero-knowledge). First of all, we claim that the need of a counter for a prover is a very weak assumption. Indeed, we only require that the adversarial concurrent verifier cannot modify the value of the counter of the prover. Our results continue to hold if the adversary is allowed to read the value of the counter.

Our second requirement forbids an adversarial concurrent verifier the execution of concurrent sessions with different provers. In comparison with the assumptions made in the literature, in our case the adversarial verifier can still open any polynomial number of concurrent sessions with complete control on the scheduling of the messages. There is no bound on the delay of the messages and no trusted party is assumed. Instead, the zero-knowledge property concerns an attack of an adversarial verifier that tries to gain knowledge from a *given* prover. It is interesting that such a restriction allows the design of round-efficient and concurrent-secure protocols.

Constant-round arguments. Our construction leaves as an open problem the construction (even under our set-up assumptions) of a constant-round concurrent zero-knowledge argument for all $\mathcal{NP}$ and of constant-round mutually concurrent simulation-sound and concurrent zero-knowledge arguments. Indeed for the case of zero knowledge one would like to prove the following informal statement.

Statement 1 (Open problem.). *There exist a constant c and an argument system (P, V) such that for all constants k and all adversaries $V^\star$ that open at most n^k concurrent sessions, (P, V) takes at most c rounds and the view of $V^\star$ can be simulated in polynomial time.*

Instead we prove the following weaker statement which can be seen as an intermediate result towards the proof of Statement 1.

Statement 2. *There exists an argument system (P,V) such that for all constants k and all adversaries $V^\star$ that open at most n^k concurrent sessions, there exists a constant c such that (P,V) takes at most c rounds and the view of $V^\star$ can be simulated in polynomial time.*

Indeed, in our construction the number of rounds depends on the degree of concurrency. However, we show that a super-polynomial number of sessions are needed in order to force a super-constant number of rounds. Statement 2 should be contrasted with the following statement proved in [13] (corresponding to bounded-concurrent zero knowledge).

Statement 3 ([13]). *There exists a constant c such that for all constants k there exists an argument system (P,V) such that for all adversaries $V^\star$ that open at most n^k of concurrent sessions, (P,V) takes at most c rounds and the view of $V^\star$ can be simulated in polynomial time.*

2 Concurrent Zero Knowledge

We now define the zero knowledge requirement that is of interest for this paper: concurrent zero knowledge.

Definition 1. *Let $\langle P,V \rangle$ be an interactive or argument system for a language L. We say that a probabilistic polynomial-time adversarial verifier $V^\star$ is a concurrent adversary if it concurrently runs a polynomial number of interaction with a prover P, without any restrictions over the scheduling of the messages in the different interactions with P. Moreover we say that the transcript of such a concurrent interaction consists of the common inputs and the sequence of prover and verifier messages exchanged during the interaction. We refer to $\text{view}_{V^\star}^P(x)$ as the random variable describing the content of the random tape of $V^\star$ and the transcript of the concurrent interactions between P and $V^\star$.*

Definition 2. *Let $\langle P,V \rangle$ be an interactive argument system for a language L. We say that $\langle P,V \rangle$ is concurrent zero knowledge if, for each probabilistic polynomial-time concurrent adversary $V^\star$ there exists a probabilistic polynomial-time algorithm $S_{V^\star}$ such that the ensembles $\{\text{view}_{V^\star}^P(x)\}_{x \in L}$ and $\{S_{V^\star}(x)\}_{x \in L}$ are computationally indistinguishable.*

The above definition (and our construction of a concurrent zero-knowledge argument system) considers the case of a single prover that potentially runs several sessions with the adversarial verifier. Such a definition can be extended to a multi-prover setting in which the adversarial verifier concurrently runs a polynomial number of sessions with possibly different provers. In the standard model when a concurrent adversary is considered, the multi-prover setting and the single-prover setting coincide since a single-prover in each interaction is oblivious of the existence of each other interaction. Instead in our settings, we rely on the fact that the prover knows an upper bound on the length of the view of the verifier he is interacting with and this is possible if there is only one prover.

We now present our construction of a concurrent zero-knowledge argument system for any language in $\mathcal{NP}$. We start by describing the model in Section 2.1, we give a high-level description of the protocol in Section 2.2, we describe its main components in Section 2.3 and 2.4, and, finally, the protocol and the proofs of its properties are presented in Section 2.5.

2.1 The Model

Our (unbounded) concurrent zero-knowledge argument system requires a model richer than the standard plain model for concurrent zero knowledge used by the previous unbounded-concurrent black-box zero-knowledge argument systems of [4, 1, 5] and the bounded-concurrent non-black-box zero-knowledge argument system of [13]. Indeed, our argument system is based on the following two set-up assumptions.

1. The prover is equipped with a counter that counts the total number of bits that he has sent in all the sessions. The counter can not be modified by the adversarial verifier.
2. The argument system is zero knowledge provided that the *same* prover is engaged by the adversarial verifier in any polynomial number of concurrent sessions.

The first assumption requires that the prover is stateful as the counter is shared by all concurrent sessions. Moreover, the adversarial verifier cannot modify this value. The fact that the adversarial verifier can run many concurrent sessions against only one prover is a consequence of the first assumption since in general a stateful prover behaves differently from another stateful prover when their states are different. Obviously, this does not constitute an issue in the plain model for concurrent zero knowledge where the prover is stateless.

2.2 A High Level Description

Our protocol follows the FLS paradigm [16] that has been used for the bounded-concurrent zero-knowledge argument system of Barak [13]. An FLS-type protocol is composed of two subprotocols: a *preamble* subprotocol and a *witness-indistinguishable argument* subprotocol. In an FLS-type protocol, the goal of the preamble subprotocol is to allow prover and verifier that are interacting on input a statement "$x \in L$", to create an augmented statement "$(x \in L) \vee (\tau \in \Lambda)$" to be proved later in the witness-indistinguishable argument subprotocol. In [13], the auxiliary statement "$\tau \in \Lambda$" informally stands for "com is the commitment of $h(A)$ where A is a program that on input a sufficiently short string tr outputs r in time bounded by a slightly super-polynomial function". The simulator of [13] sets A equal to the description of the adversary and tr to its view. To guarantee soundness it is enough that r is sufficiently longer than tr.

Since we want *unbounded-concurrent* zero knowledge, we cannot give a bound on the length of the string tr. Instead, our preamble consists of several iterations (corresponding to increasing values for the length of tr) in which the prover asks the verifier for increasingly long strings r. The prover stops when he receives a string r whose length is at least twice the length of the transcript of all the concurrent sessions[1]. The prover maintains a counter to count the length of the transcript of the concurrent sessions. Notice that since the number of rounds of the preamble is not known at the beginning of the protocol, the randomness owned by the verifier could be not sufficient to complete the protocol. Therefore, we assume that the verifier has at least an n-bit random string that he stretches round by round of the preamble by means of a pseudorandom generator.

A non-black-box simulator S, interacting with a concurrent adversary $V^\star$ and given access to the code of $V^\star$, commits to its code and randomness by setting $A = V^\star$. Then in the argument subprotocol S proves that there exists a string tr (the transcript of the interaction between the simulator and $V^\star$) for which the machine whose code has been committed (that is $V^\star$) would give as output r when receiving tr as input.

The statement is obviously true and S has a witness for it (i.e., the decommitment, the description of the adversary and the sufficiently short transcript). S finishes by running the prover's program for the witness indistinguishable argument. Notice that the simulator runs the same number of sub-preambles played by the prover.

For the round complexity of the protocol, it can be seen that for each constant k there exists a constant c_k such that, if the adversary starts at most n^k concurrent sessions, the protocols will take at most c_k rounds. Thus, for any function $\beta(n) = \omega(1)$, we can conclude that, for any polynomial number of concurrent sessions, our protocol takes (for sufficiently large n) at most $\beta(n)$ rounds.

2.3 The Preamble Subprotocol

Auxiliary inputs. On input x of length n, prover P has a witness w for $x \in L$ and a counter c that is shared by all concurrent executions of Γ.

The preamble step by step. Let $\mathcal{H} = \{h_\alpha\}$ be an ensemble of collision-resistant hash functions secure against $n^{O(\log n)}$-time adversaries. Each function of $h_\alpha \in \mathcal{H}$ maps any string to a string of length $3|\alpha|$. V randomly picks a hash function h by randomly picking an n-bit string α and sends (a description of) $h = h_\alpha$ to P. Moreover V picks a random seed s for the pseudorandom generator $\mathcal{G}$ and will use the output of $\mathcal{G}$ on input s as a random tape.

---

[1] The transcript of a session that we consider here consists of all the messages sent by the prover.

Then the prover uses a statistically binding commitment scheme $(\mathcal{G}, \mathsf{Com}, \mathsf{Dec})$ to compute $(\mathsf{com}, \mathsf{dec}) = \mathsf{Com}(h(0^n))$, sends com to the verifier and increments c by $|\mathsf{com}|$.

Now the prover and the verifier repeat the following iteration, starting with $i = 1$, until the preamble is declared completed:

1. at the i-th iteration, V randomly picks a $2n^i$-bit string r_i and sends it to the prover;
2. if $c \leq n^i$ then the preamble is declared completed and P sends the bit "1" to V to mark the end of the preamble; otherwise, P sends the bit "0" to ask V to perform iteration $i = i + 1$. In both cases, P increments the counter c by 1.

2.4 The Argument Subprotocol

The argument subprotocol consists in the execution of a witness-indistinguishable universal argument for proving statement "$(x \in L) \vee (\tau \in \Lambda)$." We stress that during the execution of the argument subprotocol the prover increments c to keep track of the number of bits he has sent. Let us now describe the language Λ and the string τ.

We say that $(h, \mathsf{com}, r) \in \Lambda$ if $(\mathsf{com}, \mathsf{dec}) = \mathsf{Com}(h(A))$ and there exists tr of length $|\mathsf{tr}| < |r|/2$ such that A, on input tr, outputs r in at most $n^{\log \log n / 2}$ steps.

The triple $\tau = (h, \mathsf{com}, r)$ used by the prover in the argument subprotocol consists of the (description of the) collision-resistant hash function h picked by the verifier, the commitment com sent by the prover and the last string r sent by the verifier during the preamble protocol. Obviously, the prover runs the witness indistinguishable universal argument subprotocol for $\mathtt{Ntime}(n^{\log \log n})$ for proving statement "$(x \in L) \vee (\tau \in \Lambda)$" using w such that $(x, w) \in R_L$ as witness. The simulator instead uses his knowledge of the code of the verifier to compute $(\mathsf{com}, \mathsf{dec}) = \mathsf{Com}(h(V^\star))$ in the preamble. Therefore, if $V^\star$ in the i-th iteration of the loop of the preamble subprotocol, on input a transcript tr of the messages sent by the simulator such that $|\mathsf{tr}| \leq n^i$, outputs a $2n^i$-bit message r, the triple $(\mathsf{dec}, V^\star, \mathsf{tr})$ is a witness for $(h, \mathsf{com}, r) \in \Lambda$ and therefore for "$(x \in L) \vee (\tau \in \Lambda)$". Thus the simulator runs the code of the prover of the witness-indistinguishable universal argument subprotocol using $(\mathsf{dec}, V^\star, \mathsf{tr})$ as witness.

2.5 The Concurrent Zero-Knowledge Argument System

Our concurrent zero-knowledge argument system for all languages in $\mathcal{NP}$ combines the preamble subprotocol from Section 2.3 and the witness-indistinguishable universal argument for $\mathtt{Ntime}(n^{\log \log n})$ from Section 2.4. By using the techniques introduced in [17] our construction only needs the existence of collision-resistant hash function ensembles that are secure with respect to polynomial-time algorithms.

Theorem 1. *Assuming the existence of collision-resistant hash function ensembles that are secure against polynomial-time algorithms, then there exists (constructively) a $\beta(n)$-round concurrent zero-knowledge argument system for $\mathcal{NP}$ for any $\beta(n) = \omega(1)$.*

The proof of Theorem 1, is omitted from this extended abstract.

3 Concurrent Simulation Soundness

In this section we show that the concurrent zero-knowledge argument system presented in Section 2 can be used to construct two mutually concurrent simulation-sound and concurrent zero-knowledge argument systems. More precisely, in this section we show that, under set-up assumptions similar to those used to construct the concurrent zero-knowledge argument of Section 2, there exists (constructively) a pair of $\beta(n)$-round concurrent zero-knowledge argument systems that are mutually concurrent simulation sound for any $\beta = \omega(1)$.

Man-in-the-middle adversary. The strong notion of *simulation-sound zero knowledge* deals with an adversary $\mathcal{A}$ that mounts a man-in-the-middle attack at two arguments $SSP_0 = \langle P_0, V_0 \rangle$ and $SSP_1 = \langle P_1, V_1 \rangle$. The adversary $\mathcal{A}$ acts as a verifier in an instance of protocol SSP_0 and as a prover in a (concurrently played) instance of protocol SSP_1. $\mathcal{A}$ has complete control of the communication channel and can decide the scheduling of the messages. Informally, SSP_0 is *simulation sound with respect to* SSP_1 if the "simulated" proof of a (possibly false) statement seen by $\mathcal{A}$ as a verifier of SSP_0 does not help him to prove a false statement in SSP_1. If $SSP_0 = SSP_1$ then we say that SSP_0 is *self simulation sound*. If SSP_0 is simulation sound with respect to SSP_1 and SSP_1 is simulation sound with respect to SSP_0 then we say that SSP_0 and SSP_1 are *mutually simulation sound*.

A *concurrent man-in-the-middle* adversary $\mathcal{A}$ is allowed to play several concurrent instances of SSP_0 and SSP_1 (instead of just one for each protocol). In this case, if the "simulated" proofs of both true and false statements in SSP_0 do not help $\mathcal{A}$ to prove a false statement in SSP_1 we say that SSP_0 is *concurrently simulation sound* with respect to SSP_1. We will consider this stronger notion of simulation soundness.

We denote by $\{\text{out}_{\mathcal{A},V_1}^{S_0,\mathcal{A}}(x_1,\ldots,x_{\texttt{poly}(n)}, x'_1, \ldots, x'_{\texttt{poly}(n)})\}$ the distribution of the output of V_1 after a concurrent man-in-the-middle attack of $\mathcal{A}$. We assume that in SSP_0, S_0 simulates the proofs for both true and false statements $(x_1,\ldots,x_{\texttt{poly}(n)})$ and in SSP_1 $\mathcal{A}$ tries to prove statements $(x'_1,\ldots,x'_{\texttt{poly}(n)})$. The output of V_1 in such an experiment is therefore a vector of bits $(b_1,\ldots,b_{\texttt{poly}(n)})$ where $b_i = 1$ means that V_1 accepted the proof for x'_i while $b_i = 0$ means that V_1 rejected the proof for x'_i, for $i = 1,\ldots,\texttt{poly}(n)$.

We now give a formal definition of concurrent simulation soundness that we use in our construction.

Definition 3. Let $SSP_0 = \langle P_0, V_0 \rangle$ and $SSP_1 = \langle P_1, V_1 \rangle$ be two argument systems for a language L. We say that SSP_0 is concurrently simulation-sound with respect to SSP_1 if, for any concurrent man-in-the-middle adversary $\mathcal{A}$, there exists a probabilistic polynomial-time algorithm $S_{\mathcal{A}}$ such the probability that the i-th bit of $\{\text{out}_{\mathcal{A}, V_1}^{S_{\mathcal{A}}, \mathcal{A}}(x_1, \ldots, x_{\text{poly}(n)}, x'_1, \ldots, x'_{\text{poly}(n)})\}$ is 1 and $x'_i \notin L$ for $i \in \{1, \ldots, \text{poly}(n)\}$ is negligible.

3.1 The Additional Assumptions of Our Model

We show in the next section a pair of mutually (unbounded) concurrent simulation-sound and concurrent zero-knowledge arguments. For our constructions we will need set-up assumptions very similar to the ones used for concurrent zero knowledge. As for the case of concurrent zero knowledge, we do not assume the existence of any trusted third party nor of trusted sources of shared randomness. Our protocol is based on the following two set-up assumptions.

1. Each party is equipped with a counter that gives at each step the total number of bits he has sent so far in all sessions in which he acts as a prover in both protocols and as a verifier in only one of the two protocols.
2. SSP_0 and SSP_1 are played by the *same* two players with roles inverted.

As in the previous construction, the first assumption makes the parties stateful since they need to propagate the value of the counter across concurrent sessions. The fact that we need to restrict the adversary to mount an attack against only one player (although this single player is allowed to play both as a prover and as a verifier in several concurrent sessions) is a consequence of our first set-up assumption. Indeed for stateless parties, it does not make a difference whether the prover and verifier that are interacting with the man-in-the-middle adversary are the same or not.

We stress that in this model, the argument system of Section 2 is also concurrent non-malleable (we stress that non-malleability requires that proofs of *true* statements do not help the adversary for proving a different statement) and concurrent zero-knowledge. Indeed, the fact that we only have to deal with two parties implies that the simulator controls both the prover and the verifier played by the honest parties (in particular the simulator has access to the randomness used by these algorithms) which makes things much easier. This approach does not work for obtaining simulation soundness; in this case the adversary can request to see the (simulated) proofs of polynomially many true and false statements which makes the design of concurrent simulation-sound argument systems more difficult. Let us now concentrate on simulation soundness.

3.2 The Mutually Concurrent Simulation-Sound Argument Systems

In this section we describe the two argument systems that are mutually concurrent simulation-sound and concurrent zero knowledge, that is both are concurrent zero knowledge and each one is concurrently simulation sound with respect to the other one.

For our construction, we use some techniques introduced in [11], and therefore we need cryptographic primitives that are secure with respect to super-polynomial-time adversaries.

Using an ensemble of collision resistant hash functions that is secure against $T(n)^{O(1)}$-adversaries, the universal argument presented in [17, 13] is sound with respect to adversaries running in time $T(n)$. By plugging such a strengthened universal argument in the concurrent zero-knowledge argument system of Section 2, and assuming that even in the preamble such stronger hash functions are used, we have that the resulting concurrent zero-knowledge argument systems is sound against adversaries running in time $T(n)$.

In order to obtain two mutually concurrent simulation-sound and concurrent zero-knowledge argument systems, we use the following approach. We show a concurrent zero-knowledge argument system SSP_1 that has also a straight-line simulator that by running in time $n^{O(\log n)}$ inverts a one-way permutation (here we use the recent techniques of [12, 11]). We then show a concurrent zero-knowledge argument system SSP_0 that is sound with respect to $n^{O(\log^2 n)}$ adversaries. Informally, concurrent simulation soundness of SSP_1 with respect to SSP_0 is proved in the following way. An adversarial prover $P_0^\star$ for SSP_0 that proves a false statement while concurrently interacting in any polynomial number of sessions as a prover of SSP_0 and as a verifier of SSP_1 can be used to break the stand-alone soundness of SSP_0. We use here the existence of an $n^{O(\log n)}$-time straight-line simulator for SSP_1, since it can be easily extended to play also the role of verifier in SSP_0[2], therefore a relay strategy for the session in which $P_0^\star$ proves a false statement can be use to break in time $n^{O(\log n)}$ the stand-alone soundness of SSP_0 that is assumed to work against $n^{O(\log^2 n)}$ adversaries.

For proving the concurrent simulation soundness of SSP_0 with respect to SSP_1 we use a different technique since the previous approach can not work in both directions at the same time. The idea is that for proving the simulation soundness of SSP_0 with respect to SSP_1, it is necessary to consider an adversary $P_1^\star$ that plays both the role of prover in concurrent sessions of SSP_1 and the role of verifier in concurrent sessions of SSP_0. The zero-knowledge simulator of SSP_0 suffices here for simulation soundness but it has to consider the view of the adversary $P_1^\star$ that also includes the messages that he receives in SSP_1. In case $P_1^\star$ has proved a false statement for SSP_1, we get a contradiction with respect to the stand-alone soundness of SSP_1. This can be achieved by performing a relay strategy with a real verifier V_1. The original parameter (i.e., the counter) of the concurrent zero-knowledge argument system of Section 2 includes in SSP_0 the messages of the sessions of SSP_0 received by $P_1^\star$ when playing the role of verifier and the messages of SSP_1 received by $P_1^\star$ when playing the role of prover. Since the same does not hold for SSP_1 (where the prover only counts the bits sent as

---

[2] This additional work does not damage the simulation of SSP_1 since such simulation is not based on the knowledge of the code of the adversary and its input. Instead, the simulation is based on the power of breaking a primitive that is assumed to be hard for $n^{o(\log n)}$-time algorithms.

a prover of SSP_1), the communication complexity of SSP_1 is short enough to allow the desired round complexity in SSP_0.

We now give the details of the two protocols while the formal proofs are omitted from this extended abstract.

The first protocol. The first of the two mutually concurrent simulation-sound and concurrent zero-knowledge arguments is referred to as SSP_0 and is the concurrent zero-knowledge protocol of Section 2 with the following modification. First of all, SSP_0 is the strengthened concurrent zero-knowledge argument system (as discussed above) that is sound against $n^{O(\log^2 n)}$-adversaries (i.e., we set $T(n) = n^{O(\log^2 n)}$). For this reason, we assume the existence of an ensemble of hash functions that are collision resistant with respect to algorithms running in time $n^{O(\log^2 n)}$. Moreover, the counter is incremented also by $|m|$ when a message m is sent as a verifier in some concurrent execution of SSP_1. Obviously, when only concurrent zero knowledge is considered, this last case never occurs. During the universal argument phase the prover proves the statement "$x \in L \vee \tau = (h, \text{com}, r) \in \Lambda$" where h is the collision resistant hash function selected by the verifier, com is the commitment sent by the prover and r is the last message sent by the verifier during the preamble. It is easy to see that the modified protocol is still concurrent zero knowledge and has the same round complexity as the one discussed in Section 2. For proving (stand-alone) soundness, notice that by the soundness and the weak proof of knowledge properties of the witness indistinguishable universal argument, an extractor algorithm obtains in time $n^{O(\log \log n)}$ the witness used by the adversarial prover for proving a false statement. Therefore, in this case he obtains a triple $(\text{dec}, \mathcal{M}, \text{tr})$ such that $\mathcal{M}(\text{tr}) = r$ where $|\text{tr}| \leq |r|/2$, and $\mathcal{M}$ outputs r in at most $n^{\log \log n/2}$ steps.

The second protocol. The second of the two mutually concurrent simulation-sound arguments is referred to as SSP_1 and is the concurrent zero-knowledge protocol of Section 2 with the following modifications. First of all, SSP_1 is the strengthened concurrent zero-knowledge argument system (as discussed above) that is witness indistinguishable with respect to $n^{O(\log n)}$-adversaries (i.e., we set $T(n) = n^{O(\log n)}$). Then we assume the existence of a one-way permutation f such that f is hard to invert with respect to algorithms running in time $n^{O(\log \log n)}$ but can be inverted in time $n^{O(\log n)}$. In the first round the verifier still chooses an hash function. In addition, V picks a random string u in the domain of f and sends $v = f(u)$ to the prover. The language Λ is the same used for the concurrent zero-knowledge argument system of Section 2: $\tau = (h, \text{com}, r) \in \Lambda$ if com is a commitment of $h(A)$ and there exists a string tr such that $|\text{tr}| \leq |r|/2$, and $\mathcal{M}(\text{tr})$ outputs r in at most $n^{\log \log n/2}$ steps.

During the witness indistinguishable universal argument of knowledge the prover proves knowledge of a witness y for either $(x, y) \in L$ or $\tau = (h, \text{com}, r) \in \Lambda$ or such that $f(y) = v$.

It is easy to see that the modified protocol is still concurrent zero knowledge and has the same round complexity as the original one. Only the (stand-alone) soundness property is affected by this update. However, notice that the only difference with respect to soundness of the argument system of Theorem 1 is that the extractor of the universal argument could also extract a witness y such that $f(y) = v$. However, since the extractor runs in time $n^{O(\log \log n)}$, we have that the one-way permutation is inverted in time $n^{o(\log n)}$ that contradicts the assumed hardness of f.

Protocol SSP_1 admits also a quasi-polynomial-time simulator that by running in time $n^{O(\log n)}$ inverts any polynomial number of one-way permutations and therefore can simulate in a straight-line fashion any polynomial number of session of SSP_1 without using knowledge of the description of the adversarial verifier. Notice that such a simulator still has to run the algorithm of the prover during the preamble, in order to maintain the same round complexity of the real prover.

References

1. Prabhakaran, M., Rosen, A., Sahai, A.: Concurrent Zero-Knowledge with Logarithmic Round Complexity. In Proc. of FOCS '02, IEEE Computer Society Press 366–375
2. Dwork, C., Naor, M., Sahai, A.: Concurrent Zero-Knowledge. In Proc. of STOC '98, ACM (1998) 409–418
3. Canetti, R., Kilian, J., Petrank, E., Rosen, A.: Black-Box Concurrent Zero-Knowledge Requires $\omega(\log n)$ Rounds. In Proc. of STOC '01, ACM (2001) 570–579
4. Richardson, R., Kilian, J.: On the Concurrent Composition of Zero-Knowledge Proofs. Proceeding of Eurocrypt '99. Vol. 1592 of LNCS, Springer-Verlag (1999) 415–431
5. Kilian, J., Petrank, E.: Concurrent and Resettable Zero-Knowledge in Poly-Logarithmic Rounds. In Proc. of STOC '01, ACM (2001) 560–569
6. Dwork, C., Sahai, A.: Concurrent Zero-Knowledge: Reducing the Need for Timing Constraints. In Proc. of Crypto '98. Vol. 1462 of LNCS, Springer-Verlag (1998) 442–457
7. Goldreich, O.: Concurrent Zero-Knowledge with Timing, Revisited. In Proc. of STOC '02, ACM (2002) 332–340
8. Damgard, I.: Efficient Concurrent Zero-Knowledge in the Auxiliary String Model. In Proc. of Eurocrypt '00. Vol. 1807 of LNCS, Springer-Verlag (2000) 418–430
9. Blum, M., De Santis, A., Micali, S., Persiano, G.: Non-Interactive Zero-Knowledge. SIAM J. on Computing **20** (1991) 1084–1118
10. Di Crescenzo, G., Persiano, G., Visconti, I.: Constant-Round Resettable Zero Knowledge with Concurrent Soundness in the Bare Public-Key Model. In Proc. of Crypto '04. Vol. 3152 of LNCS, Springer-Verlag (2004) 237–253
11. Pass, R., Rosen, A.: Bounded-Concurrent Secure Two-Party Computation in a Constant Number of Rounds. In Proc. of FOCS '03, IEEE Computer Society Press (2003)
12. Pass, R.: Simulation in Quasi-Polynomial Time and Its Applications to Protocol Composition. In Proc. of Eurocrypt '03. Vol. 2045 of LNCS, Springer-Verlag (2003) 160–176
13. Barak, B.: How to Go Beyond the Black-Box Simulation Barrier. In Proc. of FOCS '01, IEEE Computer Society Press (2001) 106–115

14. Sahai, A.: Non-Malleable Non-Interactive Zero Knowledge and Adaptive Chosen-Ciphertext Security. In Proc. of FOCS '99, IEEE Computer Society Press (1999) 543–553
15. Dolev, D., Dwork, C., Naor, M.: Non-Malleable Cryptography. SIAM J. on Computing **30** (2000) 391–437
16. Feige, U., Lapidot, D., Shamir, A.: Multiple Non-Interactive Zero Knowledge Proofs Under General Assumptions. SIAM J. on Computing **29** (1999) 1–28
17. Barak, B., Goldreich, O.: Universal Arguments and Their Applications. In: IEEE Conference on Computational Complexity (CCC '02), IEEE Computer Society Press (2002)

LCA Queries in Directed Acyclic Graphs

Miroslaw Kowaluk[1,*] and Andrzej Lingas[2,**]

[1] Institute of Informatics, Warsaw University, Warsaw
kowaluk@mimuw.edu.pl
[2] Department of Computer Science, Lund University, 22100 Lund
Fax +46 46 13 10 21
Andrzej.Lingas@cs.lth.se

Abstract. We present two methods for finding a lowest common ancestor (LCA) for each pair of vertices of a directed acyclic graph (dag) on n vertices and m edges.

The first method is surprisingly natural and solves the all-pairs LCA problem for the input dag on n vertices and m edges in time $O(nm)$. As a corollary, we obtain an $O(n^2)$-time algorithm for finding genealogical distances considerably improving the previously known $O(n^{2.575})$ time-bound for this problem.

The second method relies on a novel reduction of the all-pairs LCA problem to the problem of finding maximum witnesses for Boolean matrix product. We solve the latter problem and hence also the all-pairs LCA problem in time $O(n^{2+\frac{1}{4-\omega}})$, where $\omega = 2.376$ is the exponent of the fastest known matrix multiplication algorithm. This improves the previously known $O(n^{\frac{\omega+3}{2}})$ time-bound for the general all-pairs LCA problem in dags.

1 Introduction

The problem of finding a *lowest common ancestor* (LCA) in a tree, or more generally, in a *directed acyclic graph* (dag) is one of the basic algorithmic problems. An LCA of vertices u and v in a dag is an ancestor of both u and v which has no descendant that is an ancestor of u and v, see Fig. 1 for example. We consider the problem of preprocessing a dag such that LCA queries can be answered quickly for any pair of vertices. It has a variety of important applications, e.g., in object inheritance in programming languages, analysis of genealogical data and lattice operations for complex systems (see [2] for details and further references).

For trees, linear-time preprocessing is sufficient to answer LCA queries in constant time [7]. For general dags, after an $O(n^{\frac{\omega+3}{2}})$-time preprocessing, LCA queries can be answered in constant time [2] (where n is the number of vertices and $\omega = 2.376$ is the exponent of the fastest known matrix multiplication algo-

---

* Research supported by KBN grant 4T11C04425.
** Research supported in part by VR grant 621-2002-4049.

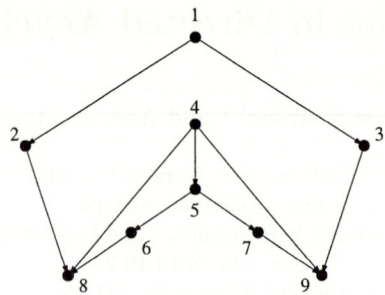

Fig. 1. The LCA of 8 and 9 are 1 and 5

rithm). A lower bound $\Omega(n^w)$ by reduction of the transitive closure problem to all-pairs LCA in dags is also given in [2].

We present two methods of efficiently preprocessing a directed graph on n vertices and m edges in order to answer an LCA query for any pair of vertices in constant time, subsuming the previously known best results from [2].

The first method is surprisingly natural and solves the all-pairs LCA problem for the input dag on n vertices and m edges in time $O(nm)$. For sparse dags, this method is optimal and substantially faster than the known $O(n^{\frac{w+3}{2}})$-time general method from [2]. As a corollary, we obtain an $O(n^2)$-time algorithm for finding genealogical distances considerably improving the previously known $O(n^{2.575})$ time-bound for this problem [2].

The second method efficiently reduces the all-pairs LCA problem to the problem of finding maximum (index) witnesses for Boolean matrix product. We solve the latter problem and hence also the all-pairs LCA problem in time $O(n^{2+\frac{1}{4-\omega}})$. Since $2+\frac{1}{4-\omega} \approx 2.616$ and $\frac{w+3}{2} \approx 2.688$, our result subsumes the previously known $O(n^{\frac{w+3}{2}})$ time-bound for the general all-pairs LCA problem in dags [2].

The first and second methods are respectively described in Sections 2 and 3 whereas Section 4 presents the algorithm for finding genealogical distances. Our paper concludes with final remarks.

2 Optimal Method for Sparse Dags

First, we shall describe preprocessing for answering queries about existence of a common ancestor for arbitrary pair of vertices in constant time.

For the input dag, we shall denote by n and m its number of vertices and edges, respectively. Also for a vertex v in the dag, $indeg(v)$ and $outdeg(v)$ stand respectively for the in-degree and out-degree of v. If $outdeg(v) = 0$ then v is called a *terminal vertex* and if $indeg(v) = 0$ then v is called a *source vertex*.

We may assume without loss of generality that the input dag is connected since otherwise we can decompose it into connected components and solve the problem for each component separately. For technical reasons, we shall also assume that every vertex is its own ancestor.

The following lemma immediately follows from the definition of a dag.

Lemma 1. *If two vertices have a common ancestor then there is a source vertex that is their common ancestor.*

In the first stage of the preprocessing, for each vertex of the input dag we form a table containing its descendants. In other words, we create the transitive closure of the dag which obviously can be done in time $O(nm)$. For the sake of Section 4, we describe this stage in more details below.

We initialize the tables in time $O(n^2)$ and start from the terminal vertices, filling their tables with single vertices in time $O(n)$. Next we iterate the following step: remove the vertices of out-degree 0 with incident edges and fill the tables for the new vertices v of out-degree 0 by merging the information from the tables associated with the removed direct descendants of v, and taking into account the set of direct descendants of v. We also add v to its table. For each vertex v such an operation takes time $O(n) \times outdeg(v)$. Thus, for the whole graph it takes $O(nm)$ time.

Lemma 2. *The tables of descendants for all vertices can be formed in time $O(nm)$.*

In the second stage of the preprocessing, we determine for each vertex v the set of vertices which have a common ancestor with v. We proceed similarly as in the first stage of preprocessing starting from source vertices instead of the terminal ones. For the source vertices s, the sets are already computed, they are just the sets of descendants of s. Next, we iterate the following step: remove the vertices of in-degree 0 with incident edges and fill the tables for the new vertices v of in-degree 0 by merging the information from the tables associated with the removed direct ancestors of v. For each vertex v such an operation takes time $O(n) \times indeg(v)$. Thus, for the whole graph it takes $O(nm)$ time.

By the *height* of a vertex v in a dag, we shall mean the length of the longest path from a source vertex to v in the dag.

Note that the set of vertices having a common ancestor with a vertex v is the union of the sets of vertices having common ancestors with the ancestors of v (recall that v is also an ancestor of itself). Hence, we obtain the following lemma by induction on the height of v.

Lemma 3. *For all vertices v, the tables of vertices having a common ancestor with v can be computed in time $O(nm)$.*

In order to answer LCA queries we need to refine the preprocessing slightly. During the second descending phase of the preprocessing we additionally enumerate the vertices in their visiting order. Since an ancestor is always visited before its descendant, we obtain the following lemma.

Lemma 4. *A vertex of a higher number cannot be an ancestor of a vertex of a lower number.*

For all vertices v, in the table keeping vertices w having a common ancestor with v, we keep also the maximum of the numbers assigned to the common ancestors of v and w. To achieve this, when we merge the information from the tables of direct ancestors of v, we pick the maximum number of a common ancestor of a direct ancestor of v and w. Clearly, the refinement can be accomplished within the same asymptotic time $O(mn)$. By induction, we obtain the following lemma.

Lemma 5. *For all vertices v, the tables of vertices w having a common ancestor with v with a pointer to a lowest common ancestor of v and w can be computed in time $O(nm)$.*

Hence, we obtain immediately the following theorem.

Theorem 1. *A dag on n vertices and m edges can be preprocessed for constant-time LCA queries in time $O(nm)$.*

If $m = O(n)$ then the preprocessing is optimal.

Corollary 1. *The all-pairs LCA problem for a dag on n vertices and m edges can be solved in time $O(n(n+m))$.*

3 $O(n^{2+\frac{1}{4-\omega}})$-Time Method for General Dags

If an entry $C[i, j]$ of the Boolean product of two Boolean matrices A and B is equal to 1 then any index k such that $A[i, k]$ and $B[k, j]$ are equal to 1 is a *witness* for $C[i, j]$. If k is the largest possible witness for $C[i, j]$ then it is called the *maximum witness* for $C[i, j]$.

In [3], Galil and Margalit presented an $O(n^{\omega+\epsilon})$-time method for the problem of computing witnesses for all positive entries of the Boolean product of two $n \times n$ Boolean matrices. Their method (too involved to describe shortly) can be viewed as a sequence of algorithms for a generalization of the problem. The first algorithm corresponds to the straightforward cubic method testing all the n witness possibilities for each positive entry of the product. The consecutive algorithms partition the input into blocks. Next, they use the fast algorithm for Boolean matrix product to compute the product of the blocks pairwise, and use the resulting products to partition the problem into subproblems. In the subproblems, for a row of the first input matrix and a column of the second input matrix, only an unique index fragment induced by the block partition and containing a witness is considered. The subproblems are solved recursively by permuting rows and columns and using the previous algorithms from the sequence.

Only the first two algorithms in the sequence of algorithms constructed by their recursive method do not rely on row and column permutations. Therefore, the method does not seem adaptable to produce the maximum witnesses without altering its asymptotic time.

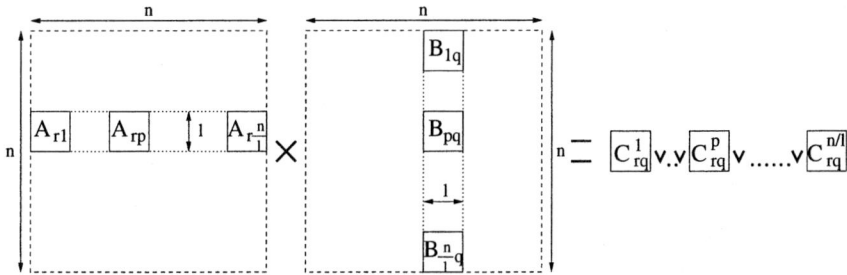

Fig. 2. The relationship between $A'_{rp}s$, $B'_{pq}s$ and $C'_{rq}s$

Our method for maximum witnesses of the Boolean product C of two $n \times n$ Boolean matrices A and B can be viewed as a modification of the second of the algorithms for witnesses of C in the aforementioned sequence of algorithms from [3].

Let l be a positive integer smaller than n. Partition the matrices A and B into $l \times l$ sub-matrices A_{rq}, B_{rq}, where $1 \leq r, q \leq n/l$, such that for $1 \leq r \leq n/l$, the sub-matrices A_{rq}, $1 \leq q \leq n/l$, cover the rows $(r-1)l + 1$ through rl of A whereas for $1 \leq q \leq n/l$, the sub-matrices B_{rq}, $1 \leq r \leq n/l$, cover the columns $(q-1)l + 1$ through ql of B.

For $1 \leq r, q \leq n/l$, $p = 1, ..., n/l$, compute the Boolean product C^p_{rq} of A_{rp} and B_{pq} using the fast algorithm. The following remark is straightforward.

Remark. Suppose that the (i, j) entry of the product matrix C is positive and $(r-1)l < i \leq rl$ and $(q-1)l < j \leq ql$. Let p' be the maximum value of p such that the entry of C^p_{rq} which is the dot product of the row of A_{rp} corresponding to the i-th row of A and the column of B_{pq} corresponding to the j-th column of B is 1. The maximum witness of the (i,j) entry of the Boolean product of A and B belongs to the interval $[(p'-1)l + 1, \, p'l]$.

By this remark, after computing all the products C^p_{rq}, $1 \leq p, r, q \leq n/l$, we need $O(l)$ time per positive entry of C to find the maximum witness. Thus, the total time taken by our method for maximum witnesses is $O((\frac{n}{l})^3 l^\omega + n^2 l)$.

By solving the equation $(\frac{n}{l})^3 l^\omega = n^2 l$, we conclude that for $l = n^{\frac{1}{4-\omega}}$ our method achieves minimum worst-case time complexity at $O(n^{2+\frac{1}{4-\omega}})$. Hence, we obtain the following theorem.

Theorem 2. *The maximum witnesses for all positive entries of the Boolean product of two $n \times n$ Boolean matrices can be computed in time $O(n^{2+\frac{1}{4-\omega}})$.*

The following obvious lemma leads to an efficient reduction of the problem of all pairs LCA in a dag to that of determining maximum witnesses of the Boolean product of two Boolean matrices.

Lemma 6. *Let G be a dag and let G^* be its transitive closure. For vertices u, v in G, let w be its common ancestor of highest rank among all common ancestors* of u and v in the ordering resulting from a topological sort of G^*. The vertex w is a lowest common ancestor of u and v.

Our algorithm for all pairs LCA in a dag is as follows.

Algorithm 1

1. Compute the transitive closure of the input dag G.
2. Topologically sort the vertices of G and number them by their ranks in the resulting sorting order.
3. Form two Boolean $n \times n$ matrices A and B such that for $i, k \in \{1, ..., n\}$ the k-th coordinate of the i-th row of A and the i-th column of B is set to 1 if the k-th vertex is an ancestor of the i-th vertex, or $k = i$, otherwise these two coordinates are set to 0.
4. Find maximum witnesses for the Boolean product C of A and B and for each non-zero entry $C[i,j]$ output the vertex whose number is the index of maximum witness of $C[i,j]$ as the lowest common ancestor of the i-th and j-th vertices.

The correctness of the algorithm follows from Lemma 6. Step 1 can be implemented in time $O(n^\omega)$. Steps 2 and 3 take $O(n^2)$ time. Finally, Step 4 requires $O(n^{2+\frac{1}{4-\omega}})$ time by Theorem 2. Hence, we obtain our second main result.

Theorem 3. *For a dag on n vertices, we can determine for each pair of vertices having a common ancestor their lowest common ancestor in time $O(n^{2+\frac{1}{4-\omega}})$.*

4 Shortest Genealogical Distances

The authors of [2] discuss the so called *pedigree graphs* which are sparse dags used to model human ancestor relations. Since each human has at most two parents, a pedigree graph has maximum in-degree bounded by two. For the fundamental applications of pedigree graphs in the identification of genes associated with genetic diseases the reader is referred to [4,6]. In these applications, computing the so called *shortest ancestral distance* between a pair of vertices in a pedigree graph is important [2]. The shortest ancestral distance between two vertices u and v in a dag is defined as the length of a shortest path between u and v which passes through a common ancestor of u and v (observe that the common ancestor is not necessarily the lowest one [1]). Bender et al. showed that the all-pairs shortest ancestral distances can be computed in time $O(n^{2.575})$ [2]. In this section, we show that the all-pairs shortest ancestral distances can be optimally computed for sparse dags, in particular, pedigree graphs.

---

[1] One can also consider the so called shortest ancestral lca distance where the common ancestor is required to be lowest [2].

We can modify our first method to obtain an $O(mn)$-time algorithm to compute the all-pairs shortest ancestral distances as follows. In the ascending phase, for each vertex v, and for each descendent u of v, we additionally compute the shortest directed distance between u and v. This can be easily accomplished within the same asymptotic time $O(mn)$. At the beginning of the descending phase, the previously computed shortest directed distances yield the shortest ancestral distances between sources and their descendents. While descending the shortest ancestral distances between the parents of the current vertex v and each other vertex u are increased by one. Next, the minimum of them and the shortest directed distance between v and u (it can be infinite) is taken as the shortest ancestral distance between v and u. In this way for all pairs of vertices v and u the shortest ancestral distance is computed.

Similarly, the so modified descending phase can be also implemented in time $O(mn)$. We conclude with the following theorem.

Theorem 4. *For a dag on n vertices and m edges, the all-pairs shortest ancestral distances can be computed in time $O(nm)$.*

Corollary 2. *For a pedigree graph on n vertices, the all-pairs shortest ancestral distances can be computed in time $O(n^2)$.*

5 Final Remarks

The problems of finding LCA are classical and central in the area of algorithms and data structures [2, 5, 7]. In spite of the long history of studies devoted to LCA problems, we have succeeded to design two quite natural methods for finding LCA in dags considerably subsuming the previously known best results [2].

The problem of finding maximum witnesses of Boolean matrix product seems to be of interest in its own rights. At first glance it seems that the recursive $O(n^{\omega+\epsilon})$-time method of Galil and Margalit [3] could be adapted to produce the maximum witnesses by considering the fragments containing maximum witnesses in the subproblems without substantially altering its asymptotic time. However, the aforementioned method may permute rows or columns in recursive steps which may disturb the search for maximum witnesses. Thus, the problem of whether or not our $O(n^{2+\frac{1}{4-\omega}})$-time method is optimal is open.

It is also an interesting question whether or not the instances of the problem of finding maximum witnesses of Boolean matrix product occurring in our reduction from the LCA problem in dags are computationally easier than the general ones.

Acknowledgments

The authors are grateful to Pavel Sumazin for inspiration and to Leszek Gąsieniec for some discussions.

References

1. N. Alon and M. Naor. Derandomization, Witnesses for Boolean Matrix Multiplication and Construction of Perfect hash functions. Algorithmica 16, pp. 434-449, 1996.
2. M.A. Bender, G. Pemmasani, S. Skiena and P. Sumazin. Finding Least Common Ancestors in Directed Acyclic Graphs. Proc. the 12th Annual ACM-SIAM Symposium on Discrete Algorithms (SODA), pp. 845-853, 2001.
3. Z. Galil and O. Margalit. Witnesses for Boolean Matrix Multiplication and Shortest Paths. Journal of Complexity, pp. 417-426, 1993.
4. R.W. Cottingham Jr., R.M. Idury, and A.A. Shäffer. Genetic linkage computations. American Journal of Human Genetics, 53, pp. 252-263, 1993.
5. M. Nykänen and E. Ukkonen. Finding lowest common ancestors in arbitrarily directed trees. Inf. Process. Lett., 50(6), pp. 307-310, 1994.
6. A.A. Shäffer, S.K. Gupta, K. Shriram, and R.W. Cottingham Jr. Avoiding recomputation in linkage analysis. Human Heredity, 44, pp. 225-237, 1994.
7. R.E. Tarjan. Applications of path compression on balanced trees. Journal of the ACM 26(4), pp. 690-715, 1979.

Replacement Paths and k Simple Shortest Paths in Unweighted Directed Graphs

Liam Roditty and Uri Zwick

School of Computer Science,
Tel Aviv University, Tel Aviv 69978, Israel

Abstract. Let $G = (V, E)$ be a *directed* graph and let P be a shortest path from s to t in G. In the *replacement paths* problem we are required to find, for every edge e on P, a shortest path from s to t in G that avoids e. We present the first non-trivial algorithm for computing replacement paths in unweighted directed graphs (and in graphs with small integer weights). Our algorithm is Monte-Carlo and its running time is $\tilde{O}(m\sqrt{n})$. Using the improved algorithm for the replacement paths problem we get an improved algorithm for finding the k *simple* shortest paths between two given vertices.

1 Introduction

Let $G = (V, E)$ be a graph, let $s, t \in V$ be two vertices in G, and let P be a shortest path from s to t in G. In certain scenarios, edges in the graph G may occasionally fail, and we are thus interested in finding, for every edge e on the path P, the shortest path from s to t in G that avoids e. This problem is refereed to as the *replacement paths* problem.

The replacement paths problem for *undirected* graphs is a well studied problem. An $O(m + n \log n)$ time algorithm for the problem was given by Malik *et al.* [13]. A similar algorithm was independently discovered, much later, by Hershberger and Suri [7]. Hershberger and Suri [7] claimed that their algorithm also works for directed graphs, but this claim turned out to be false (see Hershberger and Suri [8]). Nardelli *et al.* [14] gave an $O(m\alpha(m,n))$ time algorithm for the undirected version of the problem using the linear time single source shortest paths algorithm of Thorup [18].

All the results mentioned above for the replacement paths problem work only for undirected graphs. This situation is partially explained by an $\Omega(m\sqrt{n})$ lower bound for the replacement paths problem for directed graphs in the *path-comparison* model of Karger *et al.* [10] given by Hershberger *et al.* [9].

The replacement paths problem in directed graphs can be trivially solved in $O(|P|(m + n \log n)) = O(mn + n^2 \log n)$ time by removing each edge on P from the graph and finding a shortest path from s to t. No faster algorithm for the problem was previously known.

The replacement paths problem in directed graphs is strongly motivated by the following applications:

The fastest algorithm to compute a set of k simple shortest paths in a directed graph uses in each iteration a replacement paths algorithm. This algorithm which was given independently by Yen [20] and Lawler [12], has a running time of $O(kn(m + n\log n))$. An $o(mn)$ algorithm for the replacement paths problem implies immediately on $o(mn)$ algorithm for the k simple shortest paths problem.

The second motivation for studying replacement paths is the *Vickrey pricing* of edges. Suppose we like to find the shortest path from s to t in a directed graph G in which edges are owned by selfish agents. As noted by Nisan and Ronen [15], a mechanism that offers to pay $d_{G|e=\infty}(s,t) - d_{G|e=0}(s,t)$ to the owner of edge e, for any edge e on the shortest path from s to t, and zero otherwise, forces the edge owners to reveal their true cost. This kind of pricing is called *Vickrey pricing*. Computing the first quantity for every edge in the graph is equivalent to computing the replacement paths between s and t. (For further details see Hershberger and Suri [7] and Demetrescu et al. [4]).

We present here the first non-trivial algorithm for the replacement paths problem in directed graphs. It improves immediately the running time of the two applications mentioned above. Our algorithm is randomized and its running time is $\tilde{O}(m\sqrt{n})$ time. This seemingly matches the lower bound of Hershberger et al. [9]. Unfortunately, our algorithm works only for unweighted directed graphs, or directed graphs with small integer weights, while the lower bound of [9] is for generally weighted directed graphs.

One of the ingredients used in our algorithm for the replacement paths problem is a simple sampling technique used before to develop parallel algorithms (Ullman and Yannakakis [19]), static algorithms (Zwick [21]) and dynamic algorithms (Henzinger and King [6], Baswana et al. [1,2] Roditty and Zwick [17,16]) for paths problems. This technique on its own, however, does not supply an improved algorithm for the replacement paths problems and other ideas are needed.

Demetrescu and Thorup [3] considered the more general problem of finding, for *every* pair of vertices $u,v \in V$ and every edge $e \in E$, a shortest path from u to v that avoids e. They devise a data structure of size $O(n^2 \log n)$ capable of answering each such query in $O(\log n)$ time. The preprocessing time needed for constructing this data structure is, however, $\tilde{O}(mn^2)$. The preprocessing time can be reduced to $\tilde{O}(mn^{1.5})$ at the price of increasing the size of the data structure to $O(n^{2.5})$. (For a recent improvement, see Demetrescu et al. [4].)

We also consider two variants of the replacement paths problem. Assume again that $G = (V, E)$ is a directed graph and that P is a shortest path from s to t in G. In the *restricted replacement paths* problem, we are required to find, for every edge $e = (u,v)$ on the path P, a shortest path from u to t in G that avoids e. This corresponds to a scenario in which the failure of the edge $e = (u,v)$ is only detected at u (see Figure 1(b) for example). In the *edge replacement paths* problem we are required to find, for every edge $e = (u,v)$ on the path P, a shortest path from u to v in G that avoids e, (see Figure 1(c) for example). Our $\tilde{O}(mn^{1/2})$ time algorithm for the replacement paths problem can be adapted to solve these two versions of the problem.

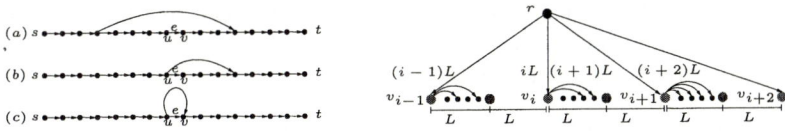

Fig. 1. Three detours and the auxiliary graph used to find short detours

We next turn our attention to the *k shortest paths problem*. Given a graph $G = (V, E)$, two vertices $s, t \in V$ and an integer k, we are required to find the k shortest paths from s to t in G. Eppstein [5], gave an $O(m + n \log n + k)$ time algorithm for the directed version of the problem. However, the paths returned by Eppstein's algorithm are not necessarily simple. i.e., they may visit certain vertices more than once. In the *k simple shortest paths problem*, the paths returned should all be simple. Katoh *et al.* [11] gave an $O(k(m+n \log n))$ time algorithm for solving the k simple shortest paths problem for undirected graphs. Yen [20] and Lawler [12] gave an $O(kn(m+n \log n))$ time algorithms for solving the problem for directed graphs. It is interesting to note that, as for the replacement paths problem, the directed version of the problem seems to be much harder than the undirected version. Using our $\tilde{O}(m\sqrt{n})$ time algorithm for the replacement paths problem we obtain a randomized $\tilde{O}(km\sqrt{n})$ time algorithm for the k simple shortest paths problem for unweighted directed graphs and for directed graphs with small integer weights.

We also show that computing the k simple shortest paths can be reduced to $O(k)$ computations of a *second* simple shortest path between s and t, each time in a different subgraph of G. Thus, to obtain an $o(kmn)$ time algorithm for the k simple shortest paths problem it is enough to obtain an $o(mn)$ time algorithm for the second shortest path problem.

The rest of this extended abstract is organized as follows. In the next section we describe our replacement paths algorithm and its adaption to the different variants of the replacement paths problem mentioned above. In Section 3 we show that the k simple shortest paths between two given vertices can be found by at most $2k$ invocations of an algorithm for finding the second shortest path between a given pair of vertices. As the second simple shortest path can be trivially found by solving the replacement paths problem, we obtain an $\tilde{O}(km\sqrt{n})$ time algorithm for the k simple shortest paths problem for unweighted graphs. We end in Section 4 with some concluding remarks and open problems.

2 Replacements Paths

In this section we describe an algorithm for solving the replacement paths problem for unweighted directed graphs and directed graphs with small integer weights. Let $G = (V, E)$ be a directed graph and let s and t be two vertices in the graph. Let $P(s,t) = \langle u_0, u_1, \ldots, u_\ell \rangle$ be a shortest path from $s = u_0$ to $t = u_\ell$. Let $P^E(s,t) = \langle (u_0, u_1), (u_1, u_2), \ldots, (u_{l-1}, u_\ell) \rangle$ be the set of edges of this path. The objective of a replacement path algorithm is to find for every edge $e \in P^E(s,t)$ a shortest path $P'(s,t)$ in the graph $G_e = (V, E \setminus \{e\})$.

We start by defining detours:

Definition 1 (Detours). *Let $P(s,t)$ be a simple path from s to t. A simple path $D(u,v)$ is a detour of $P(s,t)$ if $D(u,v) \cap P(s,t) = \{u,v\}$ and u precedes v on $P(s,t)$.*

Three detours of a shortest path from s to t are depicted in Figure 1. A shortest path from s to t that avoids the edge (u_i, u_{i+1}) of the shortest path $P(s,t) = \langle u_0, u_1, \ldots, u_\ell \rangle$ is composed of an initial portion $\langle u_0, u_1, \ldots, u_j \rangle$ of P, where $0 \leq j \leq i$, a detour $D(u_j, u_{j'})$, where $i+1 \leq j' \leq \ell$, and then the final portion $\langle u_{j'}, \ldots, u_\ell \rangle$ of P.

Let L be a parameter to be chosen later. A detour is said to be *short* if its length is at most L. Otherwise, it is said to be *long*. (Note that we are considering here only the length of the detour, not the total length of the resulting path from s to t. For example, the detour in Figure 1(a) is longer than the one in Figure 1(b), but the resulting paths may have the same length.)

We find separately the best short detours and the best long detours. The short detours are found in Section 2.1 in $\tilde{O}(mL)$ time. The long ones are found in Section 2.2 in $\tilde{O}(mn/L)$ time. Setting $L = \sqrt{n}$, we get that the running time of both algorithms is $\tilde{O}(m\sqrt{n})$. By choosing for every edge the best short or long detour, we obtain all the optimal replacement paths.

2.1 Finding Short Detours

We now describe an $\tilde{O}(mL)$ time algorithm for finding the best detours of length at most L. We can easily find the best detours that start in a given vertex u on the shortest path $P(s,t)$ by running the BFS algorithm from u in the graph $G - P^E$. However, doing so from each vertex on P may require n BFS computation which is too time consuming. The main observation made in this Section is that if $v_0, v_1, \ldots, v_k$ are vertices on $P(s,t)$ that are at a distance of at least $2L$ apart from each other, then the best short detours from all these vertices can be found by one run of Dijkstra's algorithm on a suitably modified graph. Thus, $O(L)$ runs suffice to find all short detours.

More specifically, to find the best detours from the vertices $u_0, u_{2L}, \ldots, u_{2kL}$, were $k = \lfloor \frac{\ell}{2L} \rfloor$, we consider the graph $G - P^E$ to which we add a new source vertex r and an edge (r, u_{2iL}) of weight iL, for every $0 \leq i \leq k$. The weight of all the edges of $E - P^E$ is set to 1. We denote the weight function of the auxiliary graph with wt. Note that the weight assigned to the edge (r, u_{2iL}) is iL, and not $2iL$ as might have been expected. Also, note that even though we are interested in detours that are of length at most L, the distance between every two selected vertices should be at least $2L$. The reason to that will become clear in the proof of Theorem 1. The resulting auxiliary graph, which we denote by G^A, is depicted in Figure 1.

We claim that by running Dijkstra's algorithm, from r, on G^A we find all the best short detours that start in one of the selected vertices. We then run this

```
algorithm ShortDet(P, b, L)              algorithm ShortRepPath(P, L)

⟨u₀, u₁, ..., u_ℓ⟩ ← P                    ⟨u₀, u₁, ..., u_ℓ⟩ ← P
V' ← V ∪ {r}                              for b ← 0 to 2L − 1
E' ← E \ P^E                                 ShortDet(P, b, L)
for each e ∈ E' do wt(e) ← 1              Q ← φ
for i ← 0 to ⌊(ℓ − b)/(2L)⌋               for i ← 0 to ℓ − 1
   E' ← E' ∪ {(r, u_{2iL+b})}                for j ← i + 1 to min{i + L, ℓ}
   wt(r, u_{2iL+b}) ← iL                        Insert(Q, (i, j), RD[i, j − i] + i + ℓ − j)
δ' ← Dijkstra(r, (V', E', wt))            for j ← max{i − L, 0} to i − 1
                                              Delete(Q, (j, i))
for i ← 0 to ⌊(ℓ − b)/(2L)⌋               (a, b) ← findmin(Q)
   g ← 2iL + b                            len ← RD[a, b − a] + a + ℓ − b
   for j ← 1 to L                         RP[i] ← ⟨len, a, b⟩
      if δ'(r, u_{g+j}) ≤ (i + 1)L then
         RD[g, j] ← δ'(r, u_{g+j}) − iL
      else
         RD[g, j] ← ∞
```

Fig. 2. The algorithm for finding short detours and short replacement paths

algorithm $2L - 1$ more times to find short detours emanating from the other vertices of $P(s, t)$. In the i-th run we find the short detours emanating from one of the vertices $u_i, u_{2L+i}, \ldots, u_{2kL+i}$.

We let $\delta(u, v)$ denote the distance from u to v in the graph G. We let $\delta^-(u, v)$ denote the distance from u to v in the graph $G - P^E$, i.e., the graph G with the edges of the path P removed. (The minus sign is supposed to remind us that the edges of P are removed from the graph.) We let $\delta^A(u, v)$ denote the distance from u to v in the auxiliary graph G^A. We now claim:

Theorem 1. *If $\delta^A(r, u_{2iL+j}) \leq (i+1)L$, where $0 \leq i \leq k$ and $1 \leq j \leq L$, then $\delta^-(u_{2iL}, u_{2iL+j}) = \delta^A(r, u_{2iL+j}) - iL$. Otherwise, $\delta^-(u_{2iL}, u_{2iL+j}) > L$.*

Proof. For brevity, let $v_i = u_{2iL}$ and $v_{ij} = u_{2iL+j}$. Assume at first that $\delta^A(r, v_{ij}) \leq (i+1)L$. Consider a shortest path from r to v_{ij} in G^A. Let (r, v_q) be the first edge on the path. If $q < i$, then we have

$$\delta^A(r, v_{ij}) = qL + \delta^-(v_q, v_{ij}) \geq qL + 2(i - q)L + j$$
$$= (2i - q)L + j \geq (i + 1)L + j > (i + 1)L,$$

a contradiction. Note that if the distance between any v_i and v_{i+1} was L instead of $2L$ then for $q < i$ we do have $\delta^A(r, v_{ij}) \leq (i+1)L$.

Similarly, if $q > i$ then we again have $\delta^A(r, v_{ij}) = qL + \delta^-(v_q, v_{ij}) > (i+1)L$. Thus, we must have $q = i$ and $\delta^A(r, v_{ij}) = iL + \delta^-(v_i, v_{ij})$, as required.

On the other hand, if $\delta^-(v_i, v_{ij}) \leq L$, then clearly

$$\delta^A(r, v_{ij}) \leq wt(r, v_i) + \delta^-(v_i, v_{ij}) \leq iL + L = (i+1)L,$$

as required. □

A description of the resulting algorithm, which we call **ShortDet** is given in Figure 2. By running the algorithm with the parameter b ranging from 0 to $2L-1$ we find, for every vertex on the path, the best short detours starting at it. This information is gathered in the table RD.

The entry $RD[i,j]$ gives us the length of the shortest detour starting at u_i and ending at u_{i+j}, if that length is at most L. To find the shortest path from s to t that avoids the edge (u_i, u_{i+1}) and uses a short detour, we need to find indices $i - L \leq a \leq i$ and $i < b \leq i + L$ for which the expression

$$\delta(s, u_a) + \delta^-(u_a, u_b) + \delta(u_b, t) = a + RD[a, b-a] + (\ell - b)$$

is minimized. An algorithm, called **ShortRepPath**, for finding such replacement paths is given in Figure 2. Algorithm **ShortRepPath** uses a priority queue Q. When looking for the shortest replacement path for the edge (u_i, u_{i+1}), the priority queue Q contains all pairs (a, b) such that $i - L \leq a \leq i$ and $i < b \leq i + L$. The key associated with a pair (a, b) is naturally $a + RD[a, b-a] + (\ell - b)$. In the start of the iteration corresponding to the edge (u_i, u_{i+1}), we insert the pairs (i, j), for $i + 1 \leq j \leq i + L$ into Q, and remove from it the pairs (j, i), for $i - L \leq j \leq i$. A *findmin* operation on Q then returns the minimal pair (a, b). It is easy to see that the complexity of this process is only $\tilde{O}(nL)$. Thus, the total running time of the algorithm is $\tilde{O}(mL)$, as required. We have thus proved:

Theorem 2. *Algorithm* **ShortRepPath** *finds all the shortest replacement paths that use short detours. Its running time is $\tilde{O}(mL)$.*

2.2 Finding Long Detours

To find long detours, i.e., detours that are of length at least L, we use the following simple sampling lemma. (To the best of our knowledge, it was not used before in the context of finding replacement paths).

Lemma 1. *Let $D_1, D_2, \ldots, D_q \subseteq V$ such that $|D_i| \geq L$ for $1 \leq i \leq q$ and $|V| = n$. If $R \subseteq V$ is a random subset obtained by selecting each vertex, independently, with probability $(c \ln n)/L$, for some constant c, then with probability of at least $1 - q \cdot n^{-c}$ we have $D_i \cap R \neq \phi$ for every $1 \leq i \leq q$.*

For every pair of vertices u and v on the path P for which the shortest detour from u to v is of length at least L, let $D(u, v)$ be such a shortest detour. By the lemma, if R is a random set as above, then with a probability of at least $1 - n^{2-c}$ we have $D(u, v) \cap R \neq \phi$, for every such pair u and v. The choice of the random set R is the only randomization used by our algorithm.

Our algorithm for finding the best replacement paths that use long detours starts by calling $sample(V,(4\ln n)/L)$ which selects a random set R in which each vertex $v \in V$ is placed, independently, with probability $(4\ln n)/L$. The expected size of R is clearly $\tilde{O}(\frac{n}{L})$. We assume, throughout the section, that $D(u,v) \cap R \neq \phi$, whenever $|D(u,v)| \geq L$.

For every sampled vertex $r \in R$, the algorithm maintains two priority queues $Q_{in}[r]$ and $Q_{out}[r]$ containing indices of vertices on P. When looking for a replacement path for the edge (u_i, u_{i+1}) we have $Q_{in}[r] = \{0, 1, \ldots, i\}$ and $Q_{out}[r] = \{i+1, \ldots, \ell\}$. The key associated with an element $j \in Q_{in}[r]$ is $j + \delta^-(u_j, r)$. The key associated with an element $j \in Q_{out}[r]$ is $\delta^-(r, u_j) + (\ell - j)$.

Recall that $\delta^-(u,v)$ is the distance from u to v in $G - P^E$. The algorithm computes $\delta^-(r,v)$ and $\delta^-(v,r)$, for every $r \in R$ and $v \in V$, by running two BFS's from r, for each $r \in R$, one in $G - P^E$ and one in the graph obtained from $G - P^E$ by reversing all the edges. (Only one of these BFS's is explicitly mentioned in **LongRepPath**.) The total running time of computing these distances is $\tilde{O}(mn/L)$.

To find the shortest replacement path for the edge (u_i, u_{i+1}) that passes through a given vertex $r \in R$, the algorithm needs to find an index $0 \leq a \leq i$ which minimizes the expression $a + \delta^-(u_a, r)$, and an index $i < b \leq \ell$ which minimizes the expression $\delta^-(r, u_b) + (\ell - b)$. The minimizing index a is found by a $findmin$ operation on $Q_{in}[r]$ and the minimizing index b is found by a $findmin$ operation on $Q_{out}[r]$.

It is not difficult to check that the total running time of the algorithm is $\tilde{O}(mn/L)$, as required. We have thus proved:

Theorem 3. *Algorithm* **LongRepPath** *finds, with very high probability, all the shortest replacement paths that use long detours. Its running time is $\tilde{O}(mn/L)$.*

2.3 The Replacement Paths Algorithm and Its Variants

The algorithms **ShortRepPath** and **LongRepPath** find the best short and long replacement paths available to bypass every edge on a given shortest path. By passing on their output and picking the minimal path found for every edge we obtain the solution for the replacement paths problem as promised.

There are another two natural variants of replacement paths that can be solved by our short and long detours detection.

Let $G = (V, E)$ be a directed graph and let P be a shortest path from s to t in G. In the *restricted replacement paths* problem, we are required to find, for every edge $e = (u, v)$ on the path P, a shortest path from u to t in G that avoids e. This corresponds to a scenario in which the failure of the edge $e = (u, v)$ is only detected at u. In the *edge replacement paths* problem we are required to find, for every edge $e = (u, v)$ on the path P, a shortest path from u to v in G that avoids e.

To solve the above two problems the main idea of short and long detours remains unchanged. The only change is in the set from which we choose the best detour to be used to bypass a given edge. This set is now updated according to the structural restrictions given by the problem definition.

For the restricted replacement paths after finding the short detours by the algorithm **ShortDet** we maintain the heap subject to the following constraint: If we currently searching for a restricted replacement path to bypass the edge (u_i, u_{i+1}) the heap contains only detours that emanate from u_i.

In a similar manner, when searching for a restricted replacement path composed from a long detour to bypass the edge (u_i, u_{i+1}) we use only detours that emanate from u_i. Thus, we only need one heap for this process. A pseudo-code of the algorithm will be given in the full version of this paper. We claim:

Theorem 4. *Algorithm* **ResRepPath**, *with* $L = \sqrt{n}$, *finds, with very high probability, all the restricted shortest replacement paths. Its running time is* $\tilde{O}(m\sqrt{n})$.

Using similar constraints on the set from which paths are picked we can adapt our ideas to solve also the edge replacement paths problem.

3 The k Simple Shortest Paths Problem

The k simple shortest paths problem (also known as the k shortest loopless paths) is a fundamental graph theoretic problem. Let $G = (V, E)$ be a directed graph and let s and t be two vertices of the graph. Let k be an integer. The target is to find the k simple shortest paths from s to t. This version of the problem is considered to be much harder than the general version in which non-simple paths (i.e. paths that may contain a loop) are allowed to be among the k shortest paths. The k shortest non-simple paths can be computed in time of $O(m + n \log n + k)$ using an algorithm of Eppstein [5]. In cases that a shortest paths tree can be computed in $O(m + n)$, Eppstein's algorithm has a running time of $O(m + n + k)$. However, the running time of the restricted problem is much worse. The best algorithm is due to Yen [20] and Lawler [12]. It has a running time of $O(kn(m + n \log n))$.

In this section we show that for unweighted directed graphs (and for graphs with small integer weights) the running time of $O(kn(m + n \log n))$ can be significantly improved using our new replacement paths algorithm. We obtain a randomized algorithm with running time of $\tilde{O}(km\sqrt{n} \log n)$.

We also reduce the problem of computing k simple shortest paths to $O(k)$ computations of a second shortest path each time in a different subgraph of G. This reduction works in weighted graphs. Both Yen [20] and Lawler [12] use $O(k)$ computations of replacement paths. Our reduction implies that we can focus our efforts in improving the second shortest path algorithm, which may turn out to be an easier problem than the replacement paths problem. We only deal in this section with simple paths thus we refer to a simple path simply by saying a path.

The algorithm for computing k shortest paths works as follow. It maintains a priority queue Q of paths from s to t. The key attached with each path is its length. The algorithm preforms k iterations. The priority queue is initialized

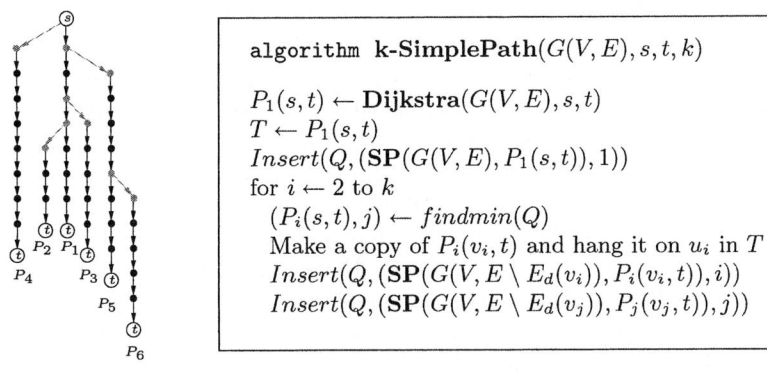

Fig. 3. A deviations tree and our k-simple shortest paths algorithm

with a second shortest path of a shortest path from s to t. In the i-th iteration the algorithm picks from Q the path with the minimal length and remove it. Let $P_i(s,t) = \langle s, u_1, \ldots u_{\ell-1}, t \rangle$ be the i-th path picked by the algorithm. This path is added to the output as the i-th shortest path. To describe the output structure we need the following definition:

Definition 2 (Deviation edge and Deviations tree). *For $k=1$ the deviations tree is simply a copy of a shortest path from s to t. Suppose that the tree already exists for $i-1$ paths. Let $P_i(s,t) = \langle s, u_1, \ldots, u_l \rangle$ be the i-th shortest path to be output. Let $P_i(s, u_j)$ be the longest subpath of $P_i(s,t)$ that was already part of the output and thus part of the tree. We make a copy only from $P_i(u_{j+1}, t)$ and hang it on the copy of u_j in the tree. We say that the edge (u_j, u_{j+1}) is the deviation edge of $P_i(s,t)$.*

Note that by this definition a vertex may have more than one occurrence in the deviations tree. However, there are at most k copies of each vertex in the deviations tree, thus, the size of the tree is $O(kn)$. An example of a deviations tree is given in Figure 3. The deviation edges are in light color.

The main challenge is to quickly obtain the paths to be added to Q in each iteration. Suppose we have extracted the i-th shortest path $P_i(s,t)$ from Q. After having $P_i(s,t)$ in the deviations tree we need to find the new paths to be added to Q. Let (u_i, v_i) be the deviation edge of $P_i(s,t)$. In Yen's algorithm the path $P_i(u_i, t) = \langle w_1, w_2, \ldots, w_l \rangle$, where $w_1 = u_i$ and $w_l = t$, is scanned. For each vertex $w_j \in P_i(u_i, t)$, the algorithm finds a shortest path $P'(w_j, t)$ from w_j to t which does not use the edge (w_j, w_{j+1}). In the special case of w_1 the path $P'(w_1, t)$ is obtained when all the edges emanate from the copy of w_1 in the tree are forbidden to use. Each such a path is concatenated to $P_i(s, w_j)$ and added to Q. This is essentially a restricted replacement paths problem for the path $P_i(u_i, t)$. Thus, we can claim the following:

Theorem 5. *The algorithm of Yen combined with our restricted shortest paths algorithm computes k simple shortest paths in $\tilde{O}(km\sqrt{n})$ time.*

However, this process generates many paths and most of them are not needed. Many of these paths can be ruled out without being actually computed by making just two computations of second shortest paths in the i-th iteration.

Recall that $P_i(s,t)$ is the i-th shortest path extracted from Q and (u_i, v_i) is its deviation edge. Let $E_d(v_i)$ be the set of deviation edges emanate from $P_i(v_i, t)$. We compute a second shortest path for the path $P_i(v_i, t)$ in the graph $G(V, E \setminus E_d(v_i))$, concatenate it to $P_i(s, v_i)$ and add it to Q. This path is associated to $P_i(s,t)$. Note that in the first computation of a second shortest path for $P_i(v_i,t)$ the set $E_d(v_i)$ is empty by its definition. However, we are not done yet. The extracted path $P_i(s,t)$ is associated to some other path $P_j(s,t)$, where $j < i$. Since we have extracted the path associated to $P_j(s,t)$ we need to find a new path, other than $P_i(s,t)$, to associate to $P_j(s,t)$. Let (u_j, v_j) be the deviation edge of $P_j(s,t)$. We compute a second shortest path for the path $P_j(v_j,t)$ in the graph $G(V, E \setminus E_d(v_j))$. By its definition the set $E_d(v_j)$ contains in this stage the deviation edge of $P_i(s,t)$, thus, the resulting second shortest path will be other than $P_i(s,t)$. We concatenate this path to $P_j(s,v_j)$ and add it to Q. The algorithm is given in Figure 3. We assume that the deviation edge of the path $P_i(s,t)$ if exists is (u_i, v_i). For the path $P_1(s,t)$ we treat s as the head of a deviation edge. We use the algorithm of $Dijkstra$ to compute a shortest path from s to t and the algorithm SP to compute a second shortest path given a shortest path.

Next, we justify why the extracted path is only associated to one other path or more precisely why a path cannot be added to Q as a second shortest path of two different paths.

Lemma 2. *In any stage all paths in Q are distinct.*

Proof. To the purpose of the proof only we divide the paths of the graph into disjoint sets. Each set is associated with a path which already was picked from Q. The set C_i is associated with the path $P_i(s,t)$. We prove that these sets exist and any second shortest path in Q is associated to a different set. This implies that all paths in Q are distinct.

We set C_1 to be all the paths in the graph. After finding a second shortest path for the first time we divide C_1 into two sets. Let (u_2, v_2) be the deviation edge of $P_2(s,t)$ then C_2 is set to be all the paths from C_1 that have the prefix $P_2(s, v_2)$ and C_1 is set to $C_1 \setminus C_2$. Obviously, C_1 and C_2 are disjoint. Now computing a second shortest path for $P_2(v_2, t)$ and concatenating it with $P_2(s, v_2)$ results in a path from C_2 and computing a second shortest path for the path $P_1(s,t)$ in $G(V, E \setminus E_d(s))$ results in a second shortest path in C_1. (Note that $E_d(s)$ contains the edge (u_2, v_2).) Thus, the paths added to Q are from two disjoint sets. We prove by induction that in general it also holds.

Suppose that right before the i-th extraction we have $i-1$ disjoint sets, such that for any $j \leq i-1$ the set C_j is associated to $P_j(s,t)$ and a path composed from the concatenation of $P_j(s, v_j)$ and a second shortest path of $P_j(v_j, t)$ in the graph $G(V, E \setminus E_d(v_j))$ is the path in Q from C_j. We show that right before the $i+1$-th extraction this invariant still holds. Let $P_i(s,t)$ be the path picked from Q in the i-th extraction. By the induction hypothesis we know that there is a set

C_j, disjoint from all the others, such that $P_i(s,t) \in C_j$ and $P_j(s,t)$ is associated with C_j. Let (u_i, v_i) be the deviation edge of $P_i(s,t)$. We divide C_j into two sets as follow, C_i will have the paths from C_j with the prefix $P_i(s, v_i)$ and C_j is set to $C_j \setminus C_i$. The resulted sets are disjoint. Now computing a second shortest path for $P_i(v_i, t)$ and concatenating it with $P_i(s, v_i)$ results in a path from C_i and computing a second shortest path for the path $P_j(v_j, t)$ in $G(V, E \setminus E_d(v_j))$ and concatenating it with $P_j(s, v_j)$ results in a second shortest path in C_j. In this process the two paths added to Q are from different disjoint sets and the above invariant still holds. □

It follows that once a path is out of Q only two second shortest path computations have to be done. We now claim the correctness of the algorithm.

Lemma 3. *The algorithm computes k simple shortest paths.*

Proof. The proof is by induction. For $i = 1$ the claim trivially holds. Suppose that the $i - 1$ first paths are found by our algorithm. We prove that the i-th path is found also. Let W be the weight of the i-th shortest path. Let $P_i(s,t)$ be an i-th shortest path. Let (u_i, v_i) be its deviation edge and let $P_j(s,t)$ be the path $P_i(s,t)$ deviates from. We will show that the path $P_i(s,t)$ or other path of the same length must be associated to $P_j(s,t)$.

Let (u', v') be the closest deviation edge to (u_i, v_i) on the path $P_j(s, u_i)$, if exists, or let $v' = s$ otherwise. Consider the last time a second shortest path computation was done for the path $P_j(v', t)$ before the i-th extraction. The weight of the path P obtained in this computation is at most W since all the edges of $P_i(u_i, t)$ are eligible to use. Suppose that the length of P is strictly less than W then by the induction hypothesis the path P is extracted before the i-th extraction. By our algorithm when P is extracted we recompute a second shortest path for $P_j(v', t)$, the path that P was associated to. However, the second shortest path computation that have added P was the last computation done for $P_j(v', t)$ before the i-th extraction, a contradiction. □

The following Theorem stems from Lemma 2 and Lemma 3.

Theorem 6. *The algorithm described above computes correctly the k simple shortest paths of a directed graph by $O(k)$ computation of second shortest paths.*

4 Concluding Remarks and Open Problems

We presented a randomized $O(m\sqrt{n})$ time algorithm for the replacement paths problem in unweighted graphs and in graphs with small integer weights. Many problems are still open, however. In particular, is it possible to obtain an $o(mn)$ time algorithm for the replacement paths problem in weighted directed graphs? Is it possible to obtain an $o(mn)$ time algorithm for the second simple shortest path in weighted directed graphs. A positive answer to one of these questions will yield an $o(kmn)$ time algorithm for finding the k simple shortest paths problem in weighted directed graphs.

References

1. S. Baswana, R. Hariharan, and S. Sen. Improved decremental algorithms for transitive closure and all-pairs shortest paths. In *Proc. of 34th STOC*, pages 117–123, 2002.
2. S. Baswana, R. Hariharan, and S. Sen. Maintaining all-pairs approximate shortest paths under deletion of edges. In *Proc. of 14th SODA*, pages 394–403, 2003.
3. C. Demetrescu and M. Thorup. Oracles for distances avoiding a link-failure. In *Proc. of 13th SODA*, pages 838–843, 2002.
4. C. Demetrescu, M. Thorup, R. Alam Chaudhury, and V. Ramachandran. Oracles for distances avoiding a link-failure.
5. D. Eppstein. Finding the k shortest paths. *SIAM Journal on Computing*, 28(2):652–673, 1998.
6. M. Henzinger and V. King. Fully dynamic biconnectivity and transitive closure. In *Proc. of 36th FOCS*, pages 664–672, 1995.
7. J. Hershberger and S. Suri. Vickrey prices and shortest paths: what is an edge worth? In *Proc. of 42nd FOCS*, pages 252–259, 2001.
8. J. Hershberger and S. Suri. Erratum to "vickrey pricing and shortest paths: What is an edge worth?". In *Proc. of 43rd FOCS*, page 809, 2002.
9. J. Hershberger, S. Suri, and A. Bhosle. On the difficulty of some shortest path problems. In *Proc. of the 20th STACS*, pages 343–354, 2003.
10. D.R. Karger, D. Koller, and S.J. Phillips. Finding the hidden path: time bounds for all-pairs shortest paths. *SIAM Journal on Computing*, 22:1199–1217, 1993.
11. N. Katoh, T. Ibaraki, and H. Mine. An efficient algorithm for K shortest simple paths. *Networks*, 12(4):411–427, 1982.
12. E.L. Lawler. A procedure for computing the K best solutions to discrete optimization problems and its application to the shortest path problem. *Management Science*, 18:401–405, 1971/72.
13. K. Malik, A. K. Mittal, and S. K. Gupta. The k most vital arcs in the shortest path problem. *Operations Research Letters*, 8(4):223–227, 1989.
14. E. Nardelli, G. Proietti, and P. Widmayer. A faster computation of the most vital edge of a shortest path. *Information Processing Letters*, 79(2):81–85, 2001.
15. N. Nisan and A. Ronen. Algorithmic mechanism design. *Games and Economic Behavior*, 35:166–196, 2001.
16. L. Roditty and U. Zwick. Dynamic approximate all-pairs shortest paths in undirected graphs. In *Proc. of 45th FOCS*, 2004. 499-508.
17. L. Roditty and U. Zwick. On dynamic shortest paths problems. In *Proc. of 12th ESA*, 2004. 580–591.
18. M. Thorup. Undirected single-source shortest paths with positive integer weights in linear time. *Journal of the ACM*, 46:362–394, 1999.
19. J.D. Ullman and M. Yannakakis. High-probability parallel transitive-closure algorithms. *SIAM Journal on Computing*, 20:100–125, 1991.
20. J.Y. Yen. Finding the K shortest loopless paths in a network. *Management Science*, 17:712–716, 1970/71.
21. U. Zwick. All-pairs shortest paths using bridging sets and rectangular matrix multiplication. *Journal of the ACM*, 49:289–317, 2002.

Deterministic Constructions of Approximate Distance Oracles and Spanners

Liam Roditty[1], Mikkel Thorup[2], and Uri Zwick[1]

[1] School of Computer Science, Tel Aviv University, Israel
[2] AT&T Research Labs, USA

Abstract. Thorup and Zwick showed that for any integer $k \geq 1$, it is possible to preprocess any positively weighted undirected graph $G = (V, E)$, with $|E| = m$ and $|V| = n$, in $\tilde{O}(kmn^{1/k})$ *expected* time and construct a data structure (a $(2k-1)$-*approximate distance oracle*) of size $O(kn^{1+1/k})$ capable of returning in $O(k)$ time an approximation $\hat{\delta}(u,v)$ of the distance $\delta(u,v)$ from u to v in G that satisfies $\delta(u,v) \leq \hat{\delta}(u,v) \leq (2k-1) \cdot \delta(u,v)$, for any two vertices $u, v \in V$. They also presented a much slower $\tilde{O}(kmn)$ time *deterministic* algorithm for constructing approximate distance oracle with the slightly larger size of $O(kn^{1+1/k} \log n)$. We present here a *deterministic* $\tilde{O}(kmn^{1/k})$ time algorithm for constructing oracles of size $O(kn^{1+1/k})$. Our deterministic algorithm is slower than the randomized one by only a logarithmic factor.

Using our derandomization technique we also obtain the first deterministic *linear* time algorithm for constructing optimal spanners of weighted graphs. We do that by derandomizing the $O(km)$ expected time algorithm of Baswana and Sen (ICALP'03) for constructing $(2k-1)$-spanners of size $O(kn^{1+1/k})$ of weighted undirected graphs without incurring *any* asymptotic loss in the running time or in the size of the spanners produced.

1 Introduction

Thorup and Zwick [16] showed that for any integer $k \geq 1$, any graph $G = (V, E)$, with $|V| = n$ and $|E| = m$, can be preprocessed in $\tilde{O}(kmn^{1/k})$ *expected* time, producing an approximate distance oracle of size $O(kn^{1+1/k})$ capable of returning, in $O(k)$ time, a stretch $2k-1$ approximation of $\delta(u,v)$, for any $u, v \in V$. As discussed in [16], the stretch-size tradeoff presented by this construction is believed to be optimal. The approximate distance oracles of [16] improve previous results of [4] and [7]. For other results dealing with approximate distances, see, [1],[6],[8],[10],[11] [12].

We present here two independent extensions of the result of [16]. The first extension deals with situations in which we are only interested in approximate distances from a specified set $S \subseteq V$ of sources. We show that both the construction time and the space requirements of the appropriate data structure can be reduced in this case. More specifically, we show that if $|S| = s$, then the expected preprocessing time can be reduced from $\tilde{O}(mn^{1/k})$ to $\tilde{O}(ms^{1/k})$ and the space required can be reduced from $O(kn^{1+1/k})$ to $O(kns^{1/k})$. This is significant when $s \ll n$. We call the obtained data structures *source-restricted approximate distance oracles*.

We then move on to solve a major open problem raised in [16], namely the development of *deterministic* algorithms for constructing approximate distance oracles that are almost as efficient as the randomized ones. The deterministic construction in [16] first computes exact APSP in $\tilde{O}(mn)$ time, and then uses the complete distance matrix to derandomize the randomized construction algorithm. In addition to being much slower, the space used by the constructed stretch $2k-1$ oracles is increased to $O(kn^{1+1/k}\log n)$. Our new derandomization loses only a logarithmic factor in running time and suffers no asymptotic loss in space. Thus we get a deterministic $\tilde{O}(mn^{1/k})$ time algorithm for constructing stretch $2k-1$ approximate distance oracles of size $O(kn^{1+1/k})$, solving the problem from [16]. For the source-restricted distance oracles with s sources, the deterministic construction time and space is $\tilde{O}(ms^{1/k})$ and $O(kns^{1/k})$, respectively.

The techniques we use to obtain the new deterministic algorithm can also be used to derandomize the expected linear time algorithm of Baswana and Sen [5] for constructing $(2k-1)$-spanners of size $O(kn^{1+1/k})$, retaining the linear running time and the $O(kn^{1+1/k})$ size of the spanners. Similarly, they can be used to improve the deterministic algorithm of Dor et al. [10] for the construction of 2-emulators (surplus 2 additive spanners) of unweighted graphs. The size of the emulators produced is reduced by a factor of $O(\sqrt{\log n})$ to the optimal $O(n^{3/2})$, with a similar improvement is obtained in the running time. Furthermore, our techniques can be used to improve the the algorithm of Baswana and Sen [6] for the construction of approximate distance oracles for unweighted graphs and make it run, deterministically, in $O(n^2)$ time, which is optimal in terms on n. (Due to lack of space we will not elaborate on this result here.)

The new deterministic algorithm uses two new ingredients that are of interest in their own right and may find additional applications. They are both simple and implementable. The first ingredient is an $\tilde{O}(qm)$ time algorithm that given a weighted directed graph $G = (V, E)$, a subset $U \subseteq V$ of sources, and an integer q, finds for every vertex $v \in V$ the set of the q vertices of U that are closest to v.

The second ingredient is a linear time deterministic algorithm for constructing *close dominating sets*. For a definition of this concept, see Section 4.

The rest of this extended abstract is organized as follows. In the next Section we present the construction of source-restricted approximate distance oracles. In Section 3 we present the algorithm for finding the nearest neighbors. In Section 4 we describe the linear time algorithm for constructing close dominating sets. In Section 5 we then present the main result of this paper, an efficient deterministic algorithm for constructing approximate distance oracles. Due to lack of space we cannot describe here the deterministic version of the linear time spanner construction algorithm of Baswana and Sen [5]. This algorithm will appear in the full version of the paper.

2 Source-Restricted Approximate Distance Oracles

We present here an extension of the approximate distance oracle construction of [16].

Theorem 1. *Let $G = (V, E)$ be an undirected graph with positive weights attached to its edges. Let $k \geq 1$ be an integer, and let $S \subseteq V$ be a specified set of sources. Then, it is*

```
algorithm prepro_k(G, S)

A_0 ← S ; A_k ← φ
for i ← 1 to k − 1
    A_i ← sample(A_{i−1}, |S|^{−1/k})
for every v ∈ V
    for i ← 0 to k − 1
        let δ(A_i, v) ← min{ δ(w, v) | w ∈ A_i }
        let p_i(v) ∈ A_i be such that δ(p_i(v), v) = δ(A_i, v)
    δ(A_k, v) ← ∞
    let B(v) ← ∪_{i=0}^{k−1} { w ∈ A_i − A_{i+1} | δ(w, v) < δ(A_{i+1}, v) }
    let H(v) ← hash(B(v))
```

Fig. 1. The randomized preprocessing algorithm

possible to preprocess G in $\tilde{O}(km|S|^{1/k})$ expected time, and produce a data structure of size $O(kn|S|^{1/k})$, such that for any $u \in S$ and $v \in V$ it is possible to produce, in $O(k)$ time, an estimate $\hat{\delta}(u,v)$ of the distance $\delta(u,v)$ from u to v in G that satisfies $\delta(u,v) \leq \hat{\delta}(u,v) \leq (2k-1) \cdot \delta(u,v)$.

Thorup and Zwick [16] prove Theorem 1 for the case $S = V$. The proof of Theorem 1 is obtained by slightly modifying the construction of [16]. For the sake of completeness, we give the full details. This also allows us to review the randomized construction of [16] before presenting a deterministic version of it later in this paper.

Proof. A high level description of the preprocessing algorithm is given in Figure 1. The algorithm starts by defining a hierarchy $A_0 \supseteq A_1 \supseteq A_2 \supseteq \cdots \supseteq A_k$ of subsets of S in the following way: We begin with $A_0 = S$. For every $1 \leq i < k$, we let A_i be random subset of A_{i-1} obtained by selecting each element of A_{i-1}, independently, with probability $|S|^{-1/k}$. Finally, we let $A_k = \phi$. The elements of A_i are referred to as *i-centers*. A similar hierarchy is used in [16]. There, however, we have $A_0 = V$, and A_i, for $1 \leq i < k$, is obtained by selecting each element of A_{i-1} with probability $n^{-1/k}$. Interestingly, this is the *only* change needed with respect to the construction of [16].

Next, the algorithm finds, for each vertex $v \in V$, and each $1 \leq i < k$, the distance $\delta(A_i, v) = \min\{\delta(w, v) \mid w \in A_i\}$ and an i-center $p_i(v) \in A_i$ that is closest to v. (We assume $A_{k-1} \neq \phi$.) For every vertex $v \in V$ it then defines the *bunch* $B(v)$ as follows:

$$B(v) = \cup_{i=0}^{k-1} B_i(v) \quad , \quad B_i(v) = \{w \in A_i \mid \delta(w,v) < \delta(A_{i+1}, v)\}.$$

Note that $B_i(v) \subseteq A_i - A_{i+1}$ as if $w \in A_{i+1}$ then $\delta(A_{i+1}, v) \leq \delta(w, v)$. We show later that the centers $p_i(v)$ can be found in $\tilde{O}(km)$ time and that the bunches $B(v)$, and the distances $\delta(w, v)$, for every $w \in B(v)$, can be found in $\tilde{O}(km|S|^{1/k})$ expected time.

Finally, for every vertex $v \in V$ the preprocessing algorithm constructs a hash table $H(v)$ of size $O(|B(v)|)$ that stores for each $w \in B(v)$ the distance $\delta(w, v)$. The hash table is constructed in $O(|B(v)|)$ expected time using the algorithm of Fredman *et al.*

[13]. (For a deterministic version, see Alon and Naor [2].) For every $w \in V$ we can then check, in $O(1)$ time, whether $w \in B(v)$ and if so obtain $\delta(w, v)$.

The total size of the data structure produced is $O(kn + \sum_{v \in V} |B(v)|)$. We next show that for every $v \in V$ we have $E[|B(v)|] \leq k|S|^{1/k}$, and thus the expected size of the whole data structure is $O(kn|S|^{1/k})$.

Lemma 1. *For every vertex $v \in V$ we have $E[|B(v)|] \leq k|S|^{1/k}$.*

Proof. We show that $E[|B_i(v)|] \leq |S|^{1/k}$, for $0 \leq i < k$. For $i = k - 1$ the claim is obvious as $B_{k-1}(v) \subseteq A_{k-1}$, and $E[|A_{k-1}|] = |S|^{1/k}$. Suppose, therefore, that $0 \leq i < k - 1$, and suppose that A_i was already chosen, while A_{i+1} is now about to be chosen. Let $w_1, w_2, \ldots, w_\ell$ be the vertices of A_i arranged in non-decreasing order of distance from v. If $w_j \in A_{i+1}$, then $B_i(v) \subseteq \{w_1, w_2, \ldots, w_{j-1}\}$. Thus $\Pr[w_j \in B_i(v)] \leq \Pr[w_1, w_2, \ldots, w_{j-1} \notin A_{i+1}]$. As each vertex of A_i is placed in A_{i+1}, independently, with probability $p = |S|^{-1/k}$, we get that $\Pr[w_j \in B_i(v)] \leq (1-p)^{j-1}$ and thus $E[|B_i(v)|] \leq \sum_{j \geq 1} (1-p)^{j-1} = p^{-1} = |S|^{1/k}$, as required. □

The algorithm used to answer approximate distance queries is given in Figure 2.

Lemma 2. *For every $u \in S$ and $v \in V$, algorithm $\mathbf{dist}_k(u, v)$ runs in $O(k)$ time and returns an approximate distance $\hat{\delta}(u, v)$ satisfying $\delta(u, v) \leq \hat{\delta}(u, v) \leq (2k-1)\delta(u, v)$.*

Proof. Let $\Delta = \delta(u, v)$. We begin by proving, by induction, that at the start of each iteration of the while loop we have $w \in A_i$ and $\delta(w, u) \leq i\Delta$. This clearly holds at the start of the first iteration, when $i = 0$, as $w = u \in S = A_0$ and $\delta(w, u) = 0$. (Here is were we use the assumption that $u \in S$.) Suppose, therefore that the claim holds at the start of some iteration, i.e., $w \in A_i$ and $\delta(w, u) \leq i\Delta$, and that the while condition, i.e., $w \notin B(v)$, is satisfied. Let $w' = p_{i+1}(v) \in A_{i+1}$. As $w \notin B(v)$, we get, by the definition of $B(v)$, that $\delta(w', v) = \delta(A_{i+1}, v) \leq \delta(w, v)$. We therefore have

$$\delta(w', v) \leq \delta(w, v) \leq \delta(w, u) + \delta(u, v) \leq i\Delta + \Delta = (i+1)\Delta .$$

Thus, by incrementing i, swapping u and v and letting $w \leftarrow w'$ we reestablish the invariant condition. (The algorithm performs these operations in a slightly different order.)

In each iteration of the while loop the algorithm performs only a constant number of operations. (To check whether $w \in B(v)$ it uses the hash table $H(v)$.) As $B(v) \supseteq A_{k-1}$ and $w \in A_i$, the algorithm performs at most $k - 1$ iterations and hence the running time is $O(k)$.

When the while loop terminates, we have $\delta(w, u) \leq i\Delta$, $w \in B(v)$ and $i \leq k - 1$. The algorithm then returns the estimate $\hat{\delta}(u, v) = \delta(w, u) + \delta(w, v)$ which satisfies

$$\begin{aligned}\delta(w, u) + \delta(w, v) &\leq \delta(w, u) + (\delta(w, u) + \delta(u, v)) \\ &= 2\delta(w, u) + \Delta \leq 2(k-1)\Delta + \Delta \leq (2k-1)\Delta ,\end{aligned}$$

as required. □

All that remain, therefore, is to explain how the preprocessing algorithm can be implemented to run in $\tilde{O}(km|S|^{1/k})$ time. Finding for each vertex $v \in V$ and every $0 \le i < k$ the vertex $p_i(v) \in A_i$ closest to v is fairly easy. For every $0 \le i < k$ we add a new source vertex s_i to the graph and connect it with zero weight edges to all the vertices of A_i. By running Dijkstra's algorithm (see [9]) we compute the distances from s_i to all other vertices and construct a shortest paths tree rooted at s_i. The distances thus computed are exactly $\delta(A_i, v)$, for every $v \in V$. Using the shortest paths tree it is easy to identify for every $v \in V$ a vertex $p_i(v) \in A_i$ for which $\delta(p_i(v), v) = \delta(A_i, v)$. The whole process requires only $\tilde{O}(km)$ time.

We next describe an $\tilde{O}(km|S|^{1/k})$ algorithm for constructing the bunches $B(v)$, for every $v \in V$. Instead of computing the bunches directly, we compute their 'duals'. For every i-center $w \in A_i - A_{i+1}$ we define the *cluster* $C(w)$ as follows:

$$C(w) = \{v \in V \mid \delta(w, v) < \delta(A_{i+1}, v)\} \quad , \quad \text{for } w \in A_i - A_{i+1}.$$

It is easy to see that $v \in C(w)$ if and only if $w \in B(v)$. We now claim:

Lemma 3. *If $v \in C(w)$, and u is on a shortest path from w to v in G, then $u \in C(w)$.*

Proof. Suppose that $w \in A_i - A_{i+1}$. If $u \notin C(w)$, then $\delta(A_{i+1}, u) \le \delta(w, u)$. But then $\delta(A_{i+1}, v) \le \delta(A_{i+1}, u) + \delta(u, v) \le \delta(w, u) + \delta(u, v) = \delta(w, v)$, contradicting the assumption that $v \in C(w)$. □

It follows that the cluster $C(w)$ can be constructed using the modified version of Dijkstra's algorithm given in Figure 3. For the straightforward correctness proof, the reader is referred to [16]. The running time of the algorithm, when Fibonacci heaps [14]

```
algorithm dist_k(u,v)
(Assumption. u ∈ S)
w ← u ; i ← 0
while w ∉ B(v)
    i ← i + 1
    (u, v) ← (v, u)
    w ← p_i(u)
return δ(w, u) + δ(w, v)
```

```
algorithm cluster(G, w, A)
d[w] ← 0 ; C ← φ
Q ← φ ; insert(Q, w, d[w])
while Q ≠ φ
    u ← extract-min(Q)
    C ← C ∪ {u}
    for every (u, v) ∈ E
        d ← d[u] + ℓ(u, v)
        if d < δ(A, v) then
            if v ∉ Q then
                d[v] ← d ; insert(Q, v, d[v])
            else if d < d[v] then
                d[v] ← d ; decrease-key(Q, v, d)
return C
```

Fig. 2. Answering a distance query

Fig. 3. Constructing a cluster

(see also [9]) are used to implement the priority queue Q, is $O(m_w + n_w \log n_w) = O(m_w + n_w \log n)$, where $n_w = |C(w)|$ and m_w is the total number of edges touching the vertices of $C(w)$. This is $O(\log n)$ time per vertex v in $C(w)$ and constant time per edge touching such a vertex v. However, $v \in C(w) \iff w \in B(v)$ and $E[|B(v)|] \le ks^{1/k}$, so the total expected running time needed for constructing all clusters is $O((m+n\log n)ks^{1/k})$, as required. The running time can be reduced to $O(kms^{1/k})$ using the techniques of Thorup [15]. This completes the proof of Theorem 1. □

3 A Deterministic Algorithm for Finding the q Nearest Centers

Let $G = (V, E)$ be a weighted directed graph and let $U \subseteq V$ be an arbitrary set of sources, or centers. We start with a formal definition of the set $U_q(v)$ of the q nearest centers from U of a vertex $v \in V$. We assume that all edge weights are positive. We also assume, without loss of generality, that $V = \{1, 2 \ldots, n\}$.

Definition 1 (Nearest centers from U). *Let $G = (V, E)$ be a directed graph with positive edge weights assigned to its edges. Let $U \subseteq V$ be an arbitrary set of sources, and let $1 \le q \le |U|$ be an integer. For every $v \in V$, the set $U_q(v)$ is defined to be the set of the q vertices of U that are closest to v. Ties are broken in favor of vertices with smaller indices. More precisely, for every $v \in V$ we have $U_q(v) \subseteq U$, $|U_q(v)| = q$ and if $w_1 \in U_q(v)$ while $w_2 \notin U_q(v)$ then either $\delta(w_1, v) < \delta(w_2, v)$ or $\delta(w_1, v) = \delta(w_2, v)$ and $w_1 < w_2$.*

The following lemma, which is reminiscent of Lemma 1, is easily verfied.

Lemma 4. *If $u \in U_q(v)$ and w lies on a shortest path from u to v in G, then $u \in U_q(w)$.*

We now claim:

Theorem 2. *Let $G = (V, E)$ be a directed graph with positive weights assigned to its edges. Let $U \subseteq V$ be an arbitrary set of sources, and let $1 \le q \le |U|$ be an integer. Then, the sets $U_q(v)$, for every $v \in V$, can be computed by performing q single-source shortest paths computations on graphs with at most $O(n)$ vertices and $O(m)$ edges.*

Proof. We begin by finding for every vertex $v \in V$ its nearest neighbor in U. This is easily done by adding a new source vertex s to the graph, connecting it with 0 length edges to all the vertices of U, and computing a tree of shortest paths in the resulting graph. This gives us $U_1(v)$, for every $v \in V$.

Suppose now that we have already computed $U_{i-1}(v)$, for every $v \in V$. We show that $U_i(v)$, for every $v \in V$ can be obtained by finding a tree of shortest paths in an auxiliary graph with $O(n)$ vertices and $O(m+n)$ edges. This auxiliary graph is constructed as follows:

1. Add to G a new source vertex s and *copies* of all vertices of U. If $u \in U$, we let $\bar{u}$ denote the copy of u. Add a 0 length edges from s to $\bar{u}$, for every $u \in U$.
2. For every edge $(v, w) \in E$:
   (a) If $U_{i-1}(v) = U_{i-1}(w)$, keep the edge (v, w).

(b) Otherwise, if $U_{i-1}(v) \neq U_{i-1}(w)$ and u is the first vertex in $U_{i-1}(v) - U_{i-1}(w)$, replace the edge (v,w) by an edge $(\bar{u}, w)$ of length $\delta(u,v) + \ell(v,w)$.

The auxiliary graph thus contains $n + |U| + 1$ vertices and $m + |U|$ edges. It is not difficult to check that u is the i-th nearest neighbor from U of all the vertices in the subtree of $\bar{u}$ in the tree of shortest paths from s in this auxiliary graph. The proof is fairly straightforward and is omitted due to lack of space. □

4 A Deterministic Construction of Close Dominating Sets

Instead of dealing directly with the close dominating sets from the introduction, it is convenient first to consider a simpler case phrased in terms of a matrix. In that context, we will talk about *early hitting sets*: Let M be an $n \times k$ matrix whose elements are taken from a finite set S of size $|S| = s$. We assume that the elements in each row of M are distinct. A set A is said to be a *hitting set* of M if and only if every row of M contains an element of A. A standard calculation shows that if each element of S is placed in A, independently, with probability $(c \ln n)/k$, for some $c > 1$, then with a probability of at least $1 - n^{1-c}$ the resulting set A is a hitting set of M. The expected size of A is then $(c s \ln n)/k$. We are interested in hitting sets of small size that hit the rows of M close to their beginnings, at least on average.

Definition 2 (Hitting sums). *Let M be an $n \times k$ matrix, let A be a set, and let $P \geq 0$ be a penalty. Let $hit(M_i, A)$ be the index of the first element of M_i, the i-th row of M, that belongs to A, or $k + P$, if no element of M_i belongs to A. Let $hit(M, A) = \sum_{i=1}^{n} hit(M_i, A)$ be the hitting sum of A with respect to M.*

Note that a set A need not be a hitting set of M for the hitting sum $hit(M, A)$ to be defined. A penalty of P, plus the length of the row, is paid, however, for each row that is not hit. Typically, the goal is to hit all rows avoiding all penalties. A set A with a small hitting sum $hit(M, A)$ is informally referred to as an *early* hitting set. The following simple probabilistic lemma proves the existence of small early hitting sets.

Lemma 5. *Let M be an $n \times k$ matrix whose elements are taken from a finite set S of size $|S| = s$ and let $P \geq 0$ be a penalty. Then, for every $0 < p < 1$ there exists a set $A \subseteq S$ for which $\frac{n}{p^2 s}|A| + hit(M, A) \leq 2n/p + (1-p)^k Pn$. In particular, if $pP \geq 3n$ and $pP(1-p)^k \leq 1$ than all rows are hit with $|A| < 3ps$ and $hit(M, A) < 3n/p$.*

Proof. Let A be a random subset of S obtained by selecting each element of S, independently, with probability p. It is easy to see that

$$E[\,|A|\,] = ps,$$

$$E[\,hit(M_i, A)\,] = \sum_{j=1}^{k}(1-p)^{j-1} + (1-p)^k P < p^{-1} + (1-p)^k P,$$

and thus $E[\frac{n}{p^2 s}|A| + hit(M, A)] \leq \frac{2n}{p} + (1-p)^k Pn$. This proves the existence of the required set.

Concerning the last statement, the condition $pP(1-p)^{-k} < 1$ implies that the right hand side is at most $3n/p$. By the first condition, this corresponds to at most a single penalty, but since we have other costs, we conclude that we pay no penalties. The bounds on $|A|$ and $hit(M, A)$ follow because each term on the left hand size is non-zero and strictly smaller than the right hand side. □

The main result of this section is a deterministic *linear time* algorithm for constructing early hitting sets that almost match the bounds of Lemma 5. Quite naturally, the algorithm is based on the *method of conditional expectations* (see, e.g., Alon and Spencer [3]). The challenge is to get a running time linear in the size of the matrix M.

Theorem 3. *Let M be an $n \times k$ matrix whose elements are taken from a finite set S of size $|S| = s$ and let $P \geq 0$ be a penalty. Let $0 < p < 1$. Then, there is a deterministic $O(nk)$ time algorithm that finds a set $A \subseteq S$ for which $\frac{n}{p^2 s}|A| + hit(M, A) \leq 3n/p + (1-p)^k Pn$. In particular, if $pP \geq 4n$ and $pP(1-p)^k \leq 1$ then all rows are hit with $|A| < 3ps$ and $hit(M, A) < 3n/p$.*

Proof. Let $A_0, A_1 \subseteq S$ be two disjoint sets. Define

$$hit(M \mid A_0, A_1) = E[\frac{n}{p^2 s}|A| + hit(M, A) \mid A_1 \subseteq A \subseteq A_0^c].$$

In other words, $hit(M|A_0, A_1)$ is the (conditional) expectation of the random variable $\frac{n}{p^2 s}|A| + hit(M, A)$ where the set A is chosen in the following way: Each element of A_1 is placed in A. Each element of A_0 is *not* placed in A. Each other element is placed in A, independently, with probability p.

Lemma 5 states that $hit(M, A) = hit(M \mid \phi, \phi) \leq \mu = 2n/p + (1-p)^k Pn$. Our goal it to deterministically find a set $A \subseteq S$ such that $hit(M \mid A^c, A) \leq \mu$. Suppose that we have already found two disjoint sets $A_0, A_1 \subseteq S$ such that $hit(M \mid A_0, A_1) \leq \mu$ and that $e \in S - (A_0 \cup A_1)$. We then have

$$hit(M \mid A_0, A_1) = p \cdot hit(M \mid A_0, A_1 \cup \{e\}) + (1-p) \cdot hit(M \mid A_0 \cup \{e\}, A_1).$$

Thus, at least one of the two conditional expectations appearing above is at least μ. We choose it and then consider another element that was not yet placed in either A_0 and A_1. Continuing in this way, we get two disjoint sets $A_0, A_1 \subseteq S$ such that $A_0 \cup A_1 = S$ and $hit(M \mid A_0, A_1) \leq \mu$, as required. This is precisely the method of conditional expectations.

The remaining question is the following: Given $hit(M \mid A_0, A_1)$ and an element $e \in S - (A_0 \cup A_1)$, how fast can we compute $hit(M \mid A_0, A_1 \cup \{e\})$ and $hit(M \mid A_0 \cup \{e\}, A_1)$? Let us focus on the computation of the conditional expectations $hit(M_i \mid A_0, A_1 \cup \{e\})$ and $hit(M_i \mid A_0 \cup \{e\}, A_1)$ corresponding to the i-th row of M.

Let $n_i = n_i(A_1)$ be the index of the first element in M_i that belongs to A_1. If none of the elements of M_i belongs to A_1, we let $n_i = \infty$. Let $n_{i,j} = n_{i,j}(A_0)$ be the number of elements among the first j elements of M_i that do *not* belong to A_0. (We let $n_{i,0} = 0$.) It is easy to see that

$$hit(M_i \mid A_0, A_1) = \sum_{j=1}^{\min\{n_i, k\}} (1-p)^{n_{i,j-1}} + \begin{cases} (1-p)^{n_{i,k}} P & \text{if } n_i = \infty \\ 0 & \text{otherwise} \end{cases}.$$

Maintaining the penalty term $(1-p)^{n_{i,k}}P$ is easy. To simplify the presentation we therefore ignore this term. (In other words we assume that $P=0$. The changes needed when $P>0$ are minimal.) Let

$$x_{i,j} = \begin{cases} (1-p)^{n_{i,j}-1} & \text{if } j \leq n_i, \\ 0 & \text{otherwise}. \end{cases}$$

With this notation, and with the assumption $P=0$, we clearly have $hit(M_i \mid A_0, A_1) = \sum_{j=1}^{k} x_{i,j}$. We now consider the changes that should be made to the $x_{i,j}$'s when an element e is added to A_0 or to A_1. If e does not appear in M_i, then no changes are required. Assume, therefore, that $M_{ir} = e$, i.e., the e is the r-th element in M_i. If $r > n_i$, then again no changes are required. Assume, therefore, that $r < n_i$.

If e is added to A_1, then the required operations are $n_i \leftarrow r$ and $x_{i,j} \leftarrow 0$, for $r < j \leq k$. If e is added to A_0, then the required operations are $n_{i,j} \leftarrow n_{i,j} - 1$, for $r \leq j \leq k$, and therefore $x_{i,j} \leftarrow x_{i,j}/(1-p)$, again for $r \leq j \leq k$. In both cases, the new conditional expectation is the new sum $\sum_{j=1}^{k} x_{i,j}$.

These operations can be implemented fairly efficiently using a data structure that maintains an array $x = [x_1, x_2, \ldots, x_q]$ of q real numbers under the following update operations: $init(x)$ – initialize the array x; $scale(i,j,a)$ – multiply the elements in the sub-array $[x_i, \ldots, x_j]$ by the constant a; sum – return the sum $\sum_{i=1}^{k} x_i$; and $undo$ – undo the last update operation. (The $undo$ operation is required for tentatively placing new elements in A_0 and then in A_1.) Using standard techniques it is not difficult to implement such a data structure that can be initialized in $O(k)$ time and that can support each update operation in $O(\log k)$ time. However, the description of such a data structure is not short, and the resulting algorithm would have an over all non-linear running time of $O(nk \log k)$. Luckily, there is a simpler to implement, and a more efficient, solution. Let us define the following variant of hitting sums:

Definition 3 (Dyadic hitting sums). *Let M be an $n \times k$ matrix, let A be a set, and let $P \geq 0$ be a penalty. Let $\overline{hit}(M_i, A) = 2^{\lceil \log_2 hit(M_i, A) \rceil}$ be the smallest power of 2 greater or equal to the index of the first element of M_i that belongs to A, or $q+P$, if no element of M_i belongs to A. Let $\overline{hit}(M, A) = \sum_{i=1}^{n} \overline{hit}(M_i, A)$ be the dyadic hitting sum of A with respect to M.*

Clearly $hit(M, A) \leq \overline{hit}(M, A) < 2 \cdot hit(M, A)$. Thus, as in the proof of Lemma 5, we get that $E[\frac{n}{p^2 s}|A| + \overline{hit}(M, A)] \leq 3n/p + (1-p)^k P n$. The conditional expectation $\overline{hit}(M \mid A_0, A_1)$ is defined in the obvious analogous way. Now define

$$\bar{k} = \lceil \log_2 k \rceil, \quad \bar{n}_i = \lceil \log_2 n_i \rceil, \quad \bar{n}_{i,j} = n_{i,2^j},$$

$$\bar{x}_{i,j} = \begin{cases} (1-p)^{\bar{n}_{i,j}-1} & \text{if } j \leq \bar{n}_i, \\ 0 & \text{otherwise}. \end{cases}, \quad \bar{y}_{i,r} = 1 + \sum_{j=1}^{r} \bar{x}_{i,j} 2^{j-1}$$

With these definitions we have $\overline{hit}(M_i \mid A_0, A_1) = \bar{y}_{i, \bar{k}-1}$.

Each update now trivially takes $O(\bar{k}) = O(\log k)$ worst-case time, even if we implement the updates naively. Furthermore, we argue that the *amortized* cost of each update is only $O(1)$!

procedure $init(i)$	procedure $update_0(i,r)$	procedure $update_1(i,r)$
$\bar{x}_{i,1} \leftarrow 1-p$	$\bar{r} \leftarrow \lceil \log_2 r \rceil$	$\bar{r} \leftarrow \lceil \log_2 r \rceil$
$\bar{y}_{i,1} \leftarrow 1+\bar{x}_{i,1}$	for $j \leftarrow \bar{r}$ to $\bar{k}-1$	for $j \leftarrow \bar{r}+1$ to $\bar{k}-1$
for $j \leftarrow 2$ to $\bar{k}-1$	$\bar{x}_{i,j} \leftarrow \bar{x}_{i,j}/(1-p)$	$\bar{x}_{i,j} \leftarrow 0$
$\bar{x}_{i,j} \leftarrow \bar{x}_{i,j-1}^2$	$\bar{y}_{i,j} \leftarrow \bar{y}_{i,j-1} + \bar{x}_{i,j} \cdot 2^{j-1}$	$\bar{y}_{i,j} \leftarrow \bar{y}_{i,j-1}$
$\bar{y}_{i,j} \leftarrow \bar{y}_{i,j-1} + \bar{x}_{i,j} \cdot 2^{j-1}$	return $\bar{y}_{i,\bar{k}-1}$	return $\bar{y}_{i,\bar{k}-1}$

Fig. 4. Updating the conditional expectations

A complete description of procedures used to initialize and update the conditional expectations is given in Figure 4. A call to $init(i)$ initializes $\bar{x}_{i,j} = (1-p)^{2^{j-1}}$ and $\bar{y}_{i,r} = 1 + \sum_{j=1}^{r} \bar{x}_{i,j} \cdot 2^{j-1}$, for $1 \leq r \leq \bar{k}-1$. Calls to $update_0(i,j)$ and $update_1(i,j)$, respectively, perform the necessary updates to the i-rows of the arrays $x[i,j]$ and $y[i,j]$ as a result of adding the element $e = M_{ir}$ to A_0, or to A_1, and return the new value of $\overline{hit}(M_i|A_0, A_1)$. The difference between the old and the new value of $\overline{hit}(M_i|A_0, A_1)$ should also be applied to the global sum $\overline{hit}(M \mid A_0, A_1) = \sum_{i=1}^{n} \overline{hit}(M_i|A_0, A_1)$. It is easy to implement an $undo(i)$ procedure that undoes the last update performed on the i-th row. We simply need to record the operations made and undo them in reverse order. To obtain $\overline{hit}(M, A_0, A_1)$, we simply sum $\overline{hit}(M_i|A_0, A_1)$ up, for $1 \leq i \leq n$. The correctness of the computation follows from the long discussion above.

All that remains is to analyze the complexity of the proposed algorithm. Each element $e \in S$ is considered once by the algorithm. For each appearance of $e = M_{ir}$ in M we need to call $update_0(i,r)$ and $update_1(i,r)$. The complexity of these calls is $O(\bar{k}-\bar{r}+1) = O(\lceil \log_2 k \rceil - \lceil \log_2 r \rceil + 1)$. For every $2^{\bar{k}-j} \leq r \leq 2^{\bar{k}-j+1}$, where $1 \leq j \leq \bar{k}$, the cost is $O(j)$. Thus, the total cost of handling all the elements of the i-th row is $O(k \sum_{j \geq 1} j 2^{-j}) = O(k)$. The total cost is therefore $O(kn)$, as required. The last statement of the theorem is derived like the last statement of Lemma 5. □

Theorem 4. *A close dominating set of any given size can be found in linear time.*

Proof. We now consider the closest dominating set problem from the introduction, modifying our early hitting set algorithm to solve this problem. The first change is to let each row M_i to have an individual length $k_i \leq k$. The total number of elements is then $m = \sum_{i=1}^{n} k_i$. We also make the change that there is only a penalty P for not hitting a full row M_i with $k_i = k$. It is straightforward to modify the previous early hitting set algorithm for these variable length rows. Essentially, we just replace k and $\bar{k}$ by k_i and $\bar{k}_i$, and drop the penalty for the partial rows. We then get a deterministic algorithm that in $O(m)$ time finds a hitting set A with the same properties as those stated in Theorem 3. In particular, if $pP \geq 4n$ and $pP(1-p)^k \leq 1$ then all full rows are hit with $|A| < 3ps$ and $hit(M, A) < 3n/p$.

We now need to transform our bipartite graph $G = (U, V, E)$ to the matrix form. The set S of elements that are placed in the matrix is simply the set U. The matrix constructed has a row for each vertex $v \in V$. Ideally, the row M_v would contain the neighboring centers u ordered according to the edge weights $\ell(u,v)$. The list should be truncated to only contain the $k = (s/h)(2 + \ln n)$ nearest centers. The lists with k centers are the full rows with a penalty P for not being hit. We use $p = h/(3s)$ and $P = 12ns/h$. Then $pP = 4n$ and

$$pP(1-p)^k < 4n\exp(-(h/s)(s/h(2+\ln n))) < 4/e^2 < 1.$$

Since the conditions are satisfied, we get a set A hitting all full rows with $|A| < 3ps = h$, and $hit(M, A) < 3n/p = 9ns/h$. This also means that A is a close dominating set.

Our only remaining problem is that we cannot sort neighboring centers according to distance. However, thanks to the dyadic solution, it suffices to apply a linear time selection algorithm (see, e.g., [9]). First, if a vertex v has more the k neighboring centers, we apply selection to find the k nearest centers. Next, for r decreasing from $\lfloor \log_2 k_i \rfloor$ down to 0, we identify the 2^r nearest centers. The total running time is linear, and this provides a sufficient sorting for the diadic hitting sum algorithm. □

5 A Deterministic Construction of Approximate Distance Oracles

In this section we present a *deterministic* algorithm for constructing (source-restricted) approximate distance oracles. The algorithm is slower than the randomized algorithm of Theorem 1 by only a logarithmic factor. Obtaining such a deterministic algorithm is one of the open problems mentioned in [16].

Theorem 5. *Let $G = (V, E)$ be an undirected graph with positive weights attached to its edges. Let $k \geq 1$ be an integer, and let $S \subseteq V$ be a specified set of sources. Then, it is possible to preprocess G, deterministically, in $\tilde{O}(km|S|^{1/k})$ time, and produce a data structure of size $O(kn|S|^{1/k})$, such that for any $u \in S$ and $v \in V$ it is possible to produce, in $O(k)$ time, an estimate $\hat{\delta}(u,v)$ of the distance $\delta(u,v)$ from u to v in G that satisfies $\delta(u,v) \leq \hat{\delta}(u,v) \leq (2k-1)\cdot\delta(u,v)$.*

```
algorithm detpre_k(G, S)
A_0 ← S ; A_k ← ∅
p ← ¼|S|^{-1/k} ; ℓ ← 3|S|^{1/k} ln n ; P ← n²
for i ← 1 to k − 1
    N_{i-1} ← near(G, A_{i-1}, ℓ)
    Create bipartite graph B from A_{i-1} to V with
    an edge (u, v) of length δ_G(u, v) if u ∈ N_{i-1}[v].
    A_i ← domset(B, s^{1-i/k})
for every v ∈ V
    B(v) ← A_{k-1}
    for i ← 0 to k − 2
        B(v) ← B(v)∪{ w ∈ N_i[v] | δ(w, v) < δ(A_{i+1}, v) }
```

Fig. 5. The deterministic preprocessing algorithm

Proof. The deterministic preprocessing algorithm is given in Figure 5. It is composed of $k-1$ iteration. The i-th iteration constructs the set A_i. We let $s = |S|$ and $\ell = \lceil s^{1/k}(2+\ln n)\rceil$, the iteration begins by finding for each vertex $v \in V$ the set $N_i[v]$ of the ℓ vertices of A_{i-1} that are nearest v, using algorithm **near** of Section 3. The running time of the algorithm is $\tilde{O}(ms^{1/k})$. Next we create a bipartite graph B from A_{i-1} to V with an edge (u,v) of length $\delta_G(u,v)$ if $u \in N_{i-1}[v]$. Using the algorithm of Theorem 4, which we here call **domset**, we now find a close dominating subset A_i of size $h_i = s^{1-i/k} = |A_{i-1}|/s^{1/k}$. Since each vertex have at least $s^{1/k}(2+\ln n) = |A_{i-1}|/h_i(2+\ln n)$ neighboring centers, we know that A_i hits all these neighborhoods. The result is that in the original graph G, the sum of the number of centers in from A_{i-1} nearer than then nearest center in A_i is at most $9|A_{i-1}|/h_i = O(ns^{1/k})$. It follows that the total size of the bunches returned by the algorithm is in $O(kns^{1/k})$, as required. □

References

1. D. Aingworth, C. Chekuri, P. Indyk, and R. Motwani. Fast estimation of diameter and shortest paths (without matrix multiplication). *SIAM Journal on Computing*, 28:1167–1181, 1999.
2. N. Alon and M. Naor. Derandomization, witnesses for Boolean matrix multiplication and construction of perfect hash functions. *Algorithmica*, 16:434–449, 1996.
3. N. Alon and J.H. Spencer. *The probabilistic method.* Wiley-Interscience, 2nd edition, 2000.
4. B. Awerbuch, B. Berger, L. Cowen, and D. Peleg. Near-linear time construction of sparse neighborhood covers. *SIAM Journal on Computing*, 28:263–277, 1999.
5. S. Baswana and S. Sen. A simple linear time algorithm for computing $(2k-1)$-spanner of $O(n^{1+1/k})$ size for weighted graphs. In *Proc. of 30th ICALP*, pages 384–296, 2003.
6. S. Baswana and S. Sen. Approximate distance oracles for unweighted graphs in $O(n^2 \log n)$ time. In *Proc. of 15th SODA*, pages 264–273, 2004.
7. E. Cohen. Fast algorithms for constructing t-spanners and paths with stretch t. *SIAM Journal on Computing*, 28:210–236, 1999.
8. E. Cohen and U. Zwick. All-pairs small-stretch paths. *Journal of Algorithms*, 38:335–353, 2001.
9. T.H. Cormen, C.E. Leiserson, R.L. Rivest, and C. Stein. *Introduction to algorithms.* The MIT Press, 2nd edition, 2001.
10. D. Dor, S. Halperin, and U. Zwick. All pairs almost shortest paths. *SIAM Journal on Computing*, 29:1740–1759, 2000.
11. M. Elkin. Computing almost shortest paths. In *Proc. of 20th PODC*, pages 53–62, 2001.
12. M.L. Elkin and D. Peleg. $(1+\epsilon,\beta)$-Spanner constructions for general graphs. *SIAM Journal on Computing*, 33(3):608–631, 2004.
13. M.L. Fredman, J. Komlós, and E. Szemerédi. Storing a sparse table with $O(1)$ worst case access time. *Journal of the ACM*, 31:538–544, 1984.
14. M.L. Fredman and R.E. Tarjan. Fibonacci heaps and their uses in improved network optimization algorithms. *Journal of the ACM*, 34:596–615, 1987.
15. M. Thorup. Undirected single-source shortest paths with positive integer weights in linear time. *Journal of the ACM*, 46:362–394, 1999.
16. M. Thorup and U. Zwick. Approximate distance oracles. *Journal of the ACM*, 52(1):1–24, 2005.

An $\tilde{O}(m^2n)$ Randomized Algorithm to Compute a Minimum Cycle Basis of a Directed Graph

Telikepalli Kavitha[*]

Indian Institute of Science, Bangalore, India
kavitha@csa.iisc.ernet.in

Abstract. We consider the problem of computing a minimum cycle basis in a directed graph G. The input to this problem is a directed graph whose arcs have positive weights. In this problem a $\{-1, 0, 1\}$ incidence vector is associated with each cycle and the vector space over $\mathbb{Q}$ generated by these vectors is the cycle space of G. A set of cycles is called a cycle basis of G if it forms a basis for its cycle space. A cycle basis where the sum of weights of the cycles is minimum is called a minimum cycle basis of G. The current fastest algorithm for computing a minimum cycle basis in a directed graph with m arcs and n vertices runs in $\tilde{O}(m^{\omega+1}n)$ time (where $\omega < 2.376$ is the exponent of matrix multiplication). If one allows randomization, then an $\tilde{O}(m^3n)$ algorithm is known for this problem. In this paper we present a simple $\tilde{O}(m^2n)$ randomized algorithm for this problem.

The problem of computing a minimum cycle basis in an *undirected* graph has been well-studied. In this problem a $\{0, 1\}$ incidence vector is associated with each cycle and the vector space over $\mathbb{F}_2$ generated by these vectors is the cycle space of the graph. The fastest known algorithm for computing a minimum cycle basis in an undirected graph runs in $O(m^2n + mn^2 \log n)$ time and our randomized algorithm for directed graphs almost matches this running time.

1 Introduction

Let $G = (V, A)$ be a directed graph with m arcs and n vertices. A *cycle* C in G consists of *forward arcs* C^+ and *backward arcs* C^- such that $C = C^+ \dot\cup C^-$ and reorienting all arcs in C^- results in a closed path. Associated with each cycle is a $\{-1, 0, 1\}$ vector, indexed on the arc set A. This vector, also called C, is defined as follows. For each arc $a \in A$

$$C(a) = \begin{cases} 1 & \text{if } a \text{ is a forward arc of } C \\ -1 & \text{if } a \text{ is a backward arc of } C \\ 0 & \text{if } a \notin C \end{cases}$$

---

[*] This research was partially supported by a "Max Planck-India Fellowship" provided by the Max Planck Society.

The *cycle space* of G is the vector space over $\mathbb{Q}$ that is generated by the incidence vectors of cycles in G. When G is connected, the cycle space has dimension $d = m - n + 1$. A *cycle basis* of G is a basis of the cycle space of G.

The arcs of G have positive weights assigned to them. A cycle basis where the sum of the weights of the cycles is minimum is called a *minimum cycle basis* of G. In this paper we consider the problem of computing a minimum cycle basis in a given digraph.

Background. The problem of computing a minimum cycle basis in a graph is well-studied. Apart from its interest as a natural question, it is motivated by its use as a preprocessing step in several algorithms. That is, a cycle basis is used as an input for a later algorithm, and using a minimum cycle basis instead of any arbitrary cycle basis usually reduces the amount of work that has to be done by this later algorithm. Such algorithms include algorithms for diverse applications like structural engineering [4], cycle analysis of electrical networks [5], and chemical ring perception [7]. And in many cases the network graphs of interest are directed graphs.

Undirected graphs. In an undirected graph $U = (N, E)$, with each cycle C we associate a $\{0,1\}$ incidence vector x, indexed on E, where $x_e = 1$ if e is an edge of C, $x_e = 0$ otherwise. The vector space over $\mathbb{F}_2$ generated by these vectors is called the *cycle space* of U. A minimum cycle basis of U is a set of linearly independent (over $\mathbb{F}_2$) cycles that span the cycle space of U and whose sum of weights is minimum.

For a directed graph G, we obtain the underlying undirected graph of G by removing the directions from the arcs. A set of cycles $C_1, ..., C_d$ of G projects onto an undirected cycle basis, if by removing the orientations of the arcs in the cycles, we obtain a cycle basis for the underlying undirected graph. If $\mathcal{C} = \{C_1, ..., C_d\}$ is a set of cycles in a directed graph G that projects onto an undirected cycle basis, then $\mathcal{C}$ is a cycle basis of G. But the the converse is not true. Similarly, a minimum cycle basis of a digraph need not project onto a cycle basis of the underlying undirected graph. Such examples are given in [13, 14]. In particular, [14] contains an example of a directed graph, which is a generalized Petersen graph, whose *unique* minimum cycle basis does not project onto a cycle basis of the underlying undirected graph. The books by Deo [6] and Bollobás [3] have an in-depth coverage of the subject of cycle bases.

Previous Results. Algorithms for computing a minimum cycle basis in an undirected graph have been well-studied [2, 5, 8, 9, 11] and the current fastest algorithm for this problem runs in $O(m^2 n + mn^2 \log n)$ time [11], where m is the number of edges and n is the number of vertices. The first polynomial time algorithm for computing a minimum cycle basis in a directed graph had a running time of $\tilde{O}(m^4 n)$ [10]. Liebchen and Rizzi [14] gave an $\tilde{O}(m^{\omega+1} n)$ algorithm for this problem, where $\omega < 2.376$ is the exponent of matrix multiplication. This is the current fastest deterministic algorithm known for this problem. But faster randomized algorithms are known. Kavitha and Mehlhorn [10] gave an $\tilde{O}(m^3n)$ Monte Carlo algorithm for this problem.

New Results. In this paper we present a simple $O(m^2 n \log n)$ randomized algorithm to compute a minimum cycle basis in a directed graph G with m arcs and n vertices. This algorithm always returns a cycle basis and we show that with high probability this cycle basis is a minimum cycle basis. We obtain this algorithm through an effective use of randomization, which enables us to work entirely over the finite field $\mathbb{F}_p$ for a randomly chosen prime p instead of working over $\mathbb{Q}$.

We recall here that the $\tilde{O}(m^3 n)$ randomized algorithm given in [10] works by sampling $\log^2 m$ random primes independently in each iteration of the algorithm. Then the algorithm either uses at least $\log m$ *suitable* primes from this sample to compute cycles in that iteration or it quits if there is no large enough subset of suitable primes in this sample.

In Section 2.1 we first present a deterministic algorithm, whose correctness is simple to show. But the numbers used in this algorithm grow very large and so this algorithm is not interesting from an implementation point of view. However, the analysis of this algorithm leads to our efficient randomized algorithm. Section 2.2 contains our randomized algorithm and its analysis.

A key step in our algorithm is a subroutine to compute a shortest cycle whose inner product with a given vector is non-zero modulo p. Such a subroutine was also used in [10] and we first review that method in Section 3.1. However that subroutine is not good enough for us since using it would make the running time of our algorithm $\tilde{O}(m^4 n)$. Here we modify Dijkstra's algorithm for this particular problem and improve this subroutine. Section 3.2 contains this implementation. This leads to an $O(m^3 + m^2 n \log n)$ randomized algorithm for computing a minimum cycle basis in a directed graph.

We can improve the running time even further. As mentioned earlier, the current fastest algorithm for computing a minimum cycle basis in an undirected graph has a running time of $O(m^2 n + mn^2 \log n)$. This running time is achieved through the use of fast matrix multiplication to speed up certain operations on vectors. In Section 4 we use the same technique to get rid of the m^3 term in our running time and we thus get an $O(m^2 n \log n)$ algorithm.

2 An $\tilde{O}(m^3)$ Randomized Algorithm

Our algorithm is broadly based on the approach used in [5, 2, 11, 10] for computing a minimum cycle basis. We are given a digraph $G = (V, A)$, where $|V| = n$ and $|A| = m$. There is no loss of generality in assuming that the underlying undirected graph of G is connected. Then $d = m - n + 1$ is the dimension of the cycle space of G. The notation $\langle v_1, v_2 \rangle$ denotes the standard inner product or dot product of the vectors v_1 and v_2. First we will assume that we have ordered the arcs in the arc set A so that the arcs $a_{d+1}, ..., a_m$ form the edges of a spanning tree T of the underlying undirected graph. This means that in the incidence vector representation of cycles, the first d coordinates correspond to arcs outside the tree T and the last $n-1$ coordinates are the arcs of T.

Before we present our randomized algorithm, let us first consider the following deterministic algorithm.

2.1 Deterministic-MCB

1. Initialize the vectors $S_1, \ldots, S_d$ of $\mathbb{Q}^m$ to the first d vectors $e_1, \ldots, e_d$ of the standard basis of $\mathbb{Q}^m$.
2. For $i = 1$ to d do
   - compute a shortest cycle C_i such that $\langle C_i, S_i \rangle \neq 0$.
   - for $j = i+1$ to d do

$$\text{update } S_j \text{ as: } S_j = S_j - S_i \frac{\langle C_i, S_j \rangle}{\langle C_i, S_i \rangle}$$

It can be easily shown that the set $\{C_1, \ldots, C_d\}$ is a minimum cycle basis. We can also show the following lemma using induction.

Lemma 1. *For $1 \leq i \leq d-1$ the above algorithm maintains the following invariant: at the end of the i-th iteration the vectors $S_{i+1}, \ldots, S_d$ are orthogonal to the cycles $C_1, \ldots, C_i$.*

Let us now understand the structure of the vectors S_j in Deterministic-MCB. The vector S_j gets updated in each iteration till iteration j. Call the version of S_j at the beginning of iteration i as S_j^i. And S_j^j is finally used in iteration j to compute the cycle C_j. (Let us denote the final version S_j^j by S_j itself.) S_j^i has the form $(r_1, r_2, \ldots, r_{i-1}, 0, \ldots, 0, 1, 0, \ldots, 0)$, where $r_1, \ldots, r_{i-1}$ are some rational numbers and the 1 occurs in the j-th coordinate. Since S_j^i is orthogonal to $C_1, \ldots, C_{i-1}$, we have $C_k \cdot (r_1, \ldots, r_{i-1}, 0, \ldots, 1, 0, \ldots)^T = 0$ for $k = 1, \ldots, i-1$.

Let the incidence vector of C_k be $(c_{k1}, \ldots, c_{km})$ and let $\tilde{C}_k$ be the restriction of this vector to its first $i-1$ coordinates. Then $(r_1, \ldots, r_{i-1})$ is a solution to

$$\tilde{C}_k \cdot (x_1, \ldots, x_{i-1})^T = -c_{kj} \text{ for } k = 1, \ldots, i-1. \tag{1}$$

We will show that this set of equations has a unique solution. Suppose the linear combination

$$\sum_{j=1}^{i-1} \alpha_j \tilde{C}_j = \mathbf{0} \tag{2}$$

and not all α_j are 0. Then consider the largest t such that $\alpha_t \neq 0$ and take the inner product of both sides of Equation (2) with $\tilde{S}_t$, where $\tilde{S}_t$ is the restriction of the vector S_t to its first $i-1$ coordinates.

Then the left hand side is $\sum_{k=1}^{t} \alpha_k \langle \tilde{C}_k, \tilde{S}_t \rangle = \sum_{k=1}^{t} \alpha_k \langle C_k, S_t \rangle$ since $\tilde{S}_t$ has all the non-zero entries of S_t for each $1 \leq t \leq i-1$. This is equal to $\alpha_t \langle C_t, S_t \rangle$ since $\langle C_k, S_t \rangle = 0$ for $k < t$. Since $\langle C_t, S_t \rangle \neq 0$ and the right hand side is 0 we get $\alpha_t = 0$ - a contradiction. Hence each α_k has to be 0 for $1 \leq k \leq i-1$. So the $\tilde{C}_k$'s are linearly independent. So we can conclude the following lemma.

Lemma 2. *For any i, the $(i-1) \times (i-1)$ matrix $\mathcal{M}_i$ whose k-th row is the vector $\tilde{C}_k$ for $1 \leq k \leq i-1$ is nonsingular.*

Thus Equation (1) has a unique solution, which is $(r_1, \ldots, r_{i-1})$. By Cramer's rule, each r_l is of the form $r_l = y_l/k_i$, where k_i is the determinant of $\mathcal{M}_i$ and y_l is the determinant of the matrix obtained by replacing the l-th column of $\mathcal{M}_i$ by the vector on the right hand side of Equation (1). So multiplying S_j^i with k_i gives us an integral vector $N_j^i = (y_1, \ldots, y_{i-1}, 0, \ldots, k_i, 0, \ldots)$. Since k_i is the determinant of an $(i-1) \times (i-1)$ matrix whose entries are $-1, 0, 1$, it follows from Hadamard's inequality that $|k_i| \leq (i-1)^{\frac{i-1}{2}}$. Similarly, the absolute value of each y_l is bounded from above by $(i-1)^{\frac{i-1}{2}}$. So we have $\|N_j^i\|_1 \leq i(i-1)^{\frac{i-1}{2}} \leq d^{\frac{d+1}{2}}$ since $i \leq d$. Let us denote each N_j^j by N_j, respectively.

Definition 1. *Call a prime p good if for each $i = 1, \ldots, d$: $\langle C_i, N_i \rangle \neq 0 (\bmod p)$. Call a prime p bad if it is not good.*

Lemma 3. *Let P be a set of d^2 primes, each of which is at least d^2. Then at least $3/4$-th of the set P is good.*

Proof. For any i, $\langle C_i, S_i \rangle \neq 0$ is equivalent to $\langle C_i, N_i \rangle \neq 0$ since $N_i = k_i S_i$ and $k_i = \det(\mathcal{M}_i) \neq 0$ by Lemma 2. So for each $1 \leq i \leq d$, it holds that $\langle C_i, N_i \rangle \neq 0$. Since C_i is a $\{-1, 0, 1\}$ vector we also get that $|\langle C_i, N_i \rangle| \leq \|N_i\|_1$. So $|\langle C_i, N_i \rangle| \leq d^{\frac{d+1}{2}}$.

Since $N_1 = S_1 = (1, 0, \ldots, 0)$, the number $\langle C_1, N_1 \rangle$ is always ± 1. So no prime can divide it. For $i \geq 2$, we will use $0 \neq |\langle C_i, N_i \rangle| \leq d^{\frac{d+1}{2}}$. Since each prime in P is at least d^2, at most $(d+1)/4$ elements in P can be divisors of $\langle C_i, N_i \rangle$. So the number of primes in P that can divide at least one of $\langle C_2, N_2 \rangle, \langle C_3, N_3 \rangle, \ldots, \langle C_d, N_d \rangle$ is at most $(d-1)(d+1)/4$. Hence the fraction of bad primes in P is at most $(d-1)(d+1)/4d^2 < 1/4$. □

Now we present the algorithm Randomized-MCB. This is similar to the algorithm Deterministic-MCB. But here we work over the field $\mathbb{F}_p$ for a randomly chosen prime p from the set P, instead of working over $\mathbb{Q}$.

2.2 Randomized-MCB

1. Compute a set P of d^2 primes $p_0, p_1, \ldots$ where each $p_j \geq d^2$. Choose a prime p uniformly at random from this set.
2. Initialize the vectors $X_1, \ldots, X_d$ of $\mathbb{F}_p^m$ to the first d vectors of the standard basis $e_1, \ldots, e_d$.
3. For $i = 1$ to d do
   - compute a shortest cycle B_i such that $\langle B_i, X_i \rangle \neq 0 (\bmod p)$.
   - for $j = i+1$ to d do

     update X_j (over the finite field $\mathbb{F}_p$) as: $X_j = X_j - X_i \dfrac{\langle B_i, X_j \rangle}{\langle B_i, X_i \rangle}$

     (An analogue of Lemma 1 shows that $X_{i+1}, \ldots, X_d$ are now orthogonal to $B_1, \ldots, B_i$ over $\mathbb{F}_p$.)

We will now show that $\{B_1, \ldots, B_d\}$ is always a cycle basis.

Lemma 4. *The cycles $B_1, ..., B_d$ are linearly independent.*

Proof. We know that $\langle B_j, X_i \rangle = 0 \pmod{p}$ for all $j < i$. It is now easy to see that B_i is linearly independent of $B_1, ..., B_{i-1}$ over $\mathbb{F}_p$. X_i is a witness of this linear independence since $\langle B_j, X_i \rangle = 0 \pmod{p}$ for each $j < i$, so the inner product of X_i with any linear combination of $B_1, ..., B_{i-1}$ has to be zero modulo p but $\langle B_i, X_i \rangle \neq 0 \pmod{p}$. Hence the whole set $\{B_1, ..., B_d\}$ is linearly independent over $\mathbb{F}_p$, which means that it is linearly independent over $\mathbb{Q}$. □

We will next show that the set $\{B_1, ..., B_d\}$ is a minimum cycle basis with probability at least 3/4. In the rest of this section, we will prove the following theorem.

Theorem 1. *When p is good, Randomized-MCB computes a minimum cycle basis.*

We will assume that the algorithm Deterministic-MCB breaks ties for the shortest cycle in the same way as Randomized-MCB breaks them. That is, both the algorithms use the same rule to determine the *shorter* cycle between two cycles of equal weight. Then we can show the following.

Lemma 5. *When p is good, $B_i = C_i$, for each $1 \leq i \leq d$.*

We will show this by induction. The vector $X_1 = S_1 = (1, 0, \ldots, 0)$. The inner product of any cycle with $(1, 0, ..., 0)$ is ± 1 or 0. The inner product will be ± 1 if and only if the cycle contains the arc a_1. Also, looking at the inner product modulo p does not change a 1 or a -1 to 0. So B_1 is a shortest cycle that contains the arc a_1. C_1 is also a shortest cycle that contains the arc a_1 and so by our assumption that both these algorithms break ties identically, we have that $B_1 = C_1$.

Let us now assume that $B_j = C_j$ for $j \leq i-1$. Recall that N_i is a vector in $\mathbb{Z}^m$ of the form $(y_1, ..., y_{i-1}, k_i, 0, ..., 0)$, where $(y_1, ..., y_{i-1})$ is the unique solution to

$$\begin{pmatrix} \tilde{C}_1 \\ \vdots \\ \tilde{C}_{i-1} \end{pmatrix} x = \begin{pmatrix} -k_i c_{1i} \\ \vdots \\ -k_i c_{(i-1)i} \end{pmatrix}. \qquad (3)$$

Recall that $\tilde{C}_j$ is the incidence vector of cycle C_j restricted to its first $i-1$ coordinates and $k_i = \det(\mathcal{M}_i)$, where $\mathcal{M}_i$ is the $(i-1) \times (i-1)$ matrix above whose rows are $\tilde{C}$'s. Note that $(y_1, ..., y_{i-1}) \bmod p$ is a solution to this set of equations in $\mathbb{F}_p$.

X_i is a vector in $\mathbb{F}_p^m$ of the form $(t_1, ..., t_{i-1}, 1, 0, ...)$ for some $t_1, ..., t_{i-1}$ in $\mathbb{F}_p$ and X_i is orthogonal to $B_1, ..., B_{i-1}$ in $\mathbb{F}_p$. Since $B_j = C_j$ for $j \leq i-1$, this means that X_i is orthogonal to $C_1, ..., C_{i-1}$ in $\mathbb{F}_p$. So in $\mathbb{F}_p$, $(t_1, ..., t_{i-1})$ is a solution to

$$\begin{pmatrix} \tilde{C}_1 \\ \vdots \\ \tilde{C}_{i-1} \end{pmatrix} x = \begin{pmatrix} -c_{1i} \\ \vdots \\ -c_{(i-1)i} \end{pmatrix}.$$

So $k_i(t_1, \ldots, t_{i-1}) \bmod p$ is a solution to Equation (3) in $\mathbb{F}_p$. We would like to prove that Equation (3) has a unique solution in $\mathbb{F}_p$.

Lemma 6. *If $\langle C_j, N_j \rangle \neq 0 (\bmod p)$ for $1 \leq j \leq i-1$, then $k_i \neq 0 (\bmod p)$.*

Let us assume the above Lemma and complete the argument. Then we will prove Lemma 6. Since $\det(\mathcal{M}_i) = k_i \neq 0 (\bmod p)$, Equation (3) should have a unique solution in $\mathbb{F}_p$. So $k_i(t_1, \ldots, t_{i-1}) \bmod p = (y_1, \ldots, y_{i-1}) \bmod p$. In other words, $k_i(t_1, \ldots, t_{i-1}, 1, 0, \ldots, 0) \bmod p = (y_1, \ldots, y_{i-1}, k_i, 0, \ldots, 0) \bmod p$. That is,
$$k_i \cdot X_i \ (\bmod p) = N_i \ (\bmod p).$$

So N_i is just a scalar multiple of X_i when these vectors are viewed as elements of $\mathbb{F}_p^m$. Hence for any cycle D, $\langle D, N_i \rangle \neq 0 (\bmod p)$ if and only if $\langle D, X_i \rangle \neq 0 (\bmod p)$. Since p is a good prime, $\langle C_i, N_i \rangle \neq 0 (\bmod p)$. So $\langle C_i, X_i \rangle \neq 0 (\bmod p)$. This proves that C_i is a candidate for a shortest cycle whose inner product with X_i is non-zero modulo p. And every cycle that has non-zero inner product with X_i modulo p is also a candidate cycle of Deterministic-MCB in its i-th iteration. Since C_i was the shortest among all these candidate cycles for Deterministic-MCB, we get that C_i also has to be the shortest cycle for Randomized-MCB in its i-th iteration. That is, $B_i = C_i$. This proves the induction step.

Proof of Lemma 6. We know that $\langle C_k, N_l \rangle = 0$ for $k < l$, so when we multiply the $(i-1) \times m$ matrix whose rows are C's with the $m \times (i-1)$ matrix whose columns are N's we get:

$$\begin{pmatrix} C_1 \\ \vdots \\ C_{i-1} \end{pmatrix} \cdot (N_1^T \ldots N_{i-1}^T) = \begin{pmatrix} \langle C_1, N_1 \rangle & 0 & 0 & \cdots & 0 \\ * & \langle C_2, N_2 \rangle & 0 & \cdots & 0 \\ * & * & \langle C_3, N_3 \rangle & \cdots & 0 \\ \vdots & \vdots & \vdots & \vdots & \vdots \\ * & * & * & \cdots & \langle C_{i-1}, N_{i-1} \rangle \end{pmatrix}$$

which is a lower triangular $(i-1) \times (i-1)$ matrix. Since each N_j has only 0's after its j-th coordinate, we can restrict the matrix of N's to its first $i-1$ rows and the matrix of C's to its first $i-1$ columns and we still have:

$$\begin{pmatrix} \tilde{C}_1 \\ \vdots \\ \tilde{C}_{i-1} \end{pmatrix} \cdot (\tilde{N}_1^T \ldots \tilde{N}_{i-1}^T) = \begin{pmatrix} \langle C_1, N_1 \rangle & 0 & 0 & \cdots & 0 \\ * & \langle C_2, N_2 \rangle & 0 & \cdots & 0 \\ * & * & \langle C_3, N_3 \rangle & \cdots & 0 \\ \vdots & \vdots & \vdots & \vdots & \vdots \\ * & * & * & \cdots & \langle C_{i-1}, N_{i-1} \rangle \end{pmatrix}$$

where $\tilde{N}_j$ is the restriction of N_j to its first $i-1$ coordinates. Now all the matrices are square matrices. The determinant of the matrix of $\tilde{C}$'s is k_i and the determinant of the matrix of $\tilde{N}_j$'s is an integer. So k_i divides the determinant on the right hand side, which is $\langle C_1, N_1 \rangle \cdots \langle C_{i-1}, N_{i-1} \rangle$. Since none of $\langle C_j, N_j \rangle$

is 0 modulo p for $1 \leq j \leq i-1$, the prime p does not divide this product. So p cannot divide k_i. Hence $k_i \neq 0 \pmod{p}$. □

This also completes the proof of Lemma 5 which says that when p is good, $B_i = C_i$, for $1 \leq i \leq d$. This immediately implies Theorem 1 since $\{C_1, \ldots, C_d\}$ is a minimum cycle basis. So the cycle basis computed by Randomized-MCB is a minimum cycle basis with probability at least 3/4 (from Lemma 3).

3   Running Time of Randomized-MCB

The value of $\pi(r)$, the number of primes less than r, is given by $r/6 \log r \leq \pi(r) \leq 8r/\log r$ [1]. So the elements in P can be bounded by $100d^2 \log d$. Using sieving, we can compute the set of primes in the first $100d^2 \log d$ numbers in $O(d^2 \log^2 d)$ time. So the set P can be determined in $O(d^2 \log^2 d)$ time.

3.1   Computing B_i

Now we consider the problem of computing a shortest cycle in G whose inner product with X_i is non-zero modulo p. Let us first review the technique in [10] and then we describe our improved algorithm for this problem. Using the digraph $G = (V, A)$ and the vector X_i, an undirected graph $U_{i,p}$ can be constructed. The graph $U_{i,p}$ can be visualized as a graph with p layers. Call these layers as layer 0, ..., layer $(p-1)$. Each layer has a copy of every vertex $v \in V$. Let v_j be the copy of vertex v in layer j. The edge set of $U_{i,p}$ also consists of p copies of each arc $a \in A$. The edges corresponding to arc $a = (u, v)$ are (u_j, v_k) where $k = (j + X_i(a))$ modulo p for each $j = 0, 1, \ldots, p-1$. For example, let $a = (u, v), X_i(a) = 3$ and $p = 5$. Then $U_{i,p}$ has 5 copies of a which are $(u_0, v_3), (u_1, v_4), (u_2, v_0), (u_3, v_1), (u_4, v_2)$. Each edge (u_j, v_k) in $U_{i,p}$ inherits the weight of its corresponding arc (u, v) of G.

The above construction gives us a well-defined map from the vertex set of $U_{i,p}$ to the vertex set of G and from the edge set of $U_{i,p}$ to the arc set of G. We can extend this map to paths of $U_{i,p}$. Any path in $U_{i,p}$ maps to a *chain*[1] in G by mapping the vertices and edges in $U_{i,p}$ to their images in G. We say that path $(e_0, \ldots, e_r)$ in the graph $U_{i,p}$ has *repeated* edges if e_i and e_j for some $i \neq j$, map to the same arc of G.

The following properties of $U_{i,p}$ capture the essence of this graph.

- any (v_0, v_ℓ) path in $U_{i,p}$ maps to a closed chain in G.
- a (v_0, v_ℓ) path in $U_{i,p}$ with no repeated edges maps to a cycle in G.
- the inner product of such a cycle with X_i is ℓ (in $\mathbb{F}_p$).

The following lemma from [10] is what we need. Its proof follows easily from the above properties.

---

[1] a chain is an alternating sequence of vertices and arcs $(x_0, a_1, x_1, a_2, \ldots, a_r, x_r)$ such that either $a_k = (x_{k-1}, x_k)$ or $a_k = (x_k, x_{k-1})$.

Lemma 7. *Let $q = \min_v \min_{\ell \neq 0}$ shortest (v_0, v_ℓ) path in the graph $U_{i,p}$. Then q corresponds to a shortest cycle in G whose inner product with X_i is non-zero modulo p.*

So B_i can be computed by running Dijkstra's algorithm from v_0 for each v in V and taking the minimum over v, of these shortest $(v_0, v_\ell), \ell \neq 0$ paths. Since $U_{i,p}$ has pn nodes and pm edges, Dijkstra's algorithm takes $O(pm + pn \log n)$ time for each v_0. Hence the total time taken to compute the cycle B_i is $O(n \cdot (pm + pn \log n))$.

Now we will show that we can modify Dijkstra's algorithm for this application so that we take $O(m \log n)$ time for each v_0 instead of $O(pm + pn \log n)$ time.

3.2 Improved Implementation of Computing a Shortest $(v_0, v_\ell), \ell \neq 0$ Path

We will not build the graph $U_{i,p}$ explicitly. Whenever we are at a vertex u_j, we know its neighborhood as follows. If there is an arc a between u and a vertex w in G, then in $U_{i,p}$, w_k is a neighbor of u_j where $k = (j + X_i(a)) \mod p$ if $a = (u, w)$ (directed from u to w), and $k = (j - X_i(a)) \mod p$ if $a = (w, u)$ (directed from w to u).

The key observation here is that to compute $\min_{\ell \neq 0}$ shortest (v_0, v_ℓ) path, it is enough to look at those intermediate vertices which are the closest or second closest of their "type" to v_0. That is, if u_j is a vertex in $\min_{\ell \neq 0}$ shortest (v_0, v_ℓ) path, then u_j is closest or second closest to v_0 among all of $\{u_0, u_1, ..., u_{p-1}\}$. So while running Dijkstra's algorithm to determine $\min_{\ell \neq 0}$ shortest (v_0, v_ℓ) path, we only explore neighborhoods of such vertices in the priority queue.

More formally, we will have $\text{dist}[u] = \infty$ for each vertex u in $U_{i,p}$ and the priority queue Q contains all the nodes of $U_{i,p}$, keyed by their dist values. Then we start computing single-source shortest paths for each vertex in layer 0. Call one such vertex as v_0. This procedure runs as follows:

- set $\text{dist}[v_0] = 0$.
- Maintain an array marked for the n vertices of G and initially $\text{marked}[u] = 0$ for each $u \in V$.
- Repeat
  - Extract the vertex x with the smallest dist value from Q.
    If x is v_ℓ for some $\ell \neq 0$, then store $\text{dist}[v_\ell]$ and the path computed to v_ℓ and quit the Repeat loop.
    Else let $x = u_k$.
    - if $\text{marked}[u] < 2$, then increment $\text{marked}[u]$ and for each neighbor w do

    $$\text{dist}[w] = \min(\text{dist}[w], \text{dist}[x] + \text{weight}(x, w))$$

    and update predecessor of w to x if necessary.
    - else do nothing.
- For each vertex whose distance was made finite in our loop, set its dist back to ∞ and insert back to Q the deleted vertices. (so that we can now run this procedure for another vertex w_0 of layer 0)

Remark. There is always a (v_0, v_ℓ) path for some $\ell \neq 0$ in the graph $U_{i,p}$ for each $v \in G$. This is because X_i on its last $n-1$ coordinates (which are the arcs of the spanning tree T) is 0 and $X_i \neq \mathbf{0}$. So each layer of the graph $U_{i,p}$ is connected and there is at least one edge from layer 0 to some non-zero layer.

Running Time of the Above Algorithm. We look at neighborhoods of only those vertices which are of the type u_j such that u_j is the first or second among all the vertices in $\{u_0, ..., u_{p-1}\}$ to be extracted from the priority queue. For such vertices we make the dist value of their neighbors to be finite. The total number of vertices whose distance is ever made finite in our loop is bounded by $\sum_{u \in G} \deg(u_j)$ for all u_j which are closest or second closest to v_0 among the "u" vertices. Since $\deg(u_j) = \deg(u)$, we get the bound of $2\sum_u \deg(u) = O(m)$. Let us implement the priority queue as a binary heap so that each of the operations needed above can be implemented in $O(\log(pn)) = O(\log n)$ amount of time. In the Repeat loop we charge the cost of extracting a vertex x to x's predecessor in shortest-path(v_0, u). So for each vertex u_j which is closest or second closest among "u" vertices to v_0, we do $O(\deg(u) \cdot \log n)$ amount of work. For the other vertices we do no work. We take $O(pn)$ time to build the binary heap. But we do this just once in the entire algorithm, at the beginning of our first iteration. Thereafter, we simply reset to infinity the dist value of only those vertices which were made finite while running our procedure for the previous vertex. Also, we insert the deleted vertices back into the heap. This takes $O(m \log n)$ work.

In iteration i, once we compute $\min_{\ell \neq 0}$shortest(v_0, v_ℓ) path for all $v \in V$, we have determined B_i. This takes time $O(n \cdot m \log n)$ given the priority queue Q containing all the vertices with their dist values. So the total amount of time to compute all the cycles $B_1, \ldots, B_d$ given the vectors $X_1, \ldots, X_d$ is $O(pn + d(n \cdot m \log n))$ which is $O(m^2 n \log n)$. All we need to show now is the following lemma. The proof will be given in the full version of the paper.

Lemma 8. *In order to compute* $\min_{\ell \neq 0}$ *shortest* (v_0, v_ℓ) *path in* $U_{i,p}$, *it is enough to look at vertices which are of the form: closest or second closest of their "type" to* v_0.

The Overall Running Time of Randomized-MCB. Under the assumption that arithmetic on $O(\log m)$ bits takes unit time, it follows that addition, subtraction and multiplication in $\mathbb{F}_p$ can be implemented in unit time since p is $O(d^2 \log d)$. However we also need to implement division efficiently since the update step of X_j involves division. Once p is chosen, we will compute the multiplicative inverses of all elements in $\mathbb{Z}_p^*$ by the extended Euclid's gcd algorithm by solving $ax = 1 \pmod{p}$ for each $a \in \mathbb{Z}_p^*$. This takes time $O(\log p)$ for each element and hence $O(p \log p)$ for all the elements. Thereafter, division in $\mathbb{F}_p$ gets implemented as multiplication with the inverse of the divisor.

We need to account for the time taken to update the vectors $X_{i+1}, ..., X_d$ in iteration i. Adding a scalar multiple of X_i to a vector X_j takes $\Theta(i)$ time. So the time taken by the update step in iteration i to update $d - i$ vectors is $\Theta(i(d-i))$. So the entire time taken by the update steps of all the iterations is

$\Theta(d^3)$. So the total time taken by the algorithm Randomized-MCB is $O(m^3 + m^2 n \log n)$.

4 Faster Implementation

Instead of spending $\Theta(m^3)$ time for the update step, using the technique in [11] we can implement the update step in $O(m^\omega)$ time, where $\omega < 2.376$ is the exponent of matrix multiplication. This then gives us an $O(m^2 n \log n)$ randomized algorithm for computing a minimum cycle basis. The algorithm FAST-Randomized-MCB is described below.

- Compute a set P of d^2 primes $p_0, p_1, \ldots$ where each $p_j \geq d^2$. Choose a prime p uniformly at random from this set.
- Call the procedure $extend\_cycle\_basis(\{\}, \{e_1, \ldots, e_d\}, d)$, where $e_1, \ldots, e_d$ are the first d vectors of the standard basis.

The procedure $extend\_cycle\_basis$ takes as input a partial cycle basis, say, $\{D_1, \ldots, D_i\}$ (denoted by $\mathcal{D}$), a parameter k, and k vectors $v_{i+1}, \ldots, v_{i+k}$ of $\mathbb{F}_p^m$ which are orthogonal to $\{D_1, \ldots, D_i\}$ over $\mathbb{F}_p$ and computes k new cycles $D_{i+1}, \ldots, D_{i+k}$ to extend the cycle basis. The role of $v_{i+1}, \ldots, v_{i+k}$ is identical to the role played by the vectors $X_{i+1}, X_{i+2}, \ldots$ at the beginning of iteration $i+1$ in the algorithm Randomized-MCB (Section 2.2). Just as the vectors X_j got updated in Randomized-MCB, the vectors v_j get updated during the course of $extend\_cycle\_basis$. But the difference is that we will update many v_j's with respect to many cycles in one bulk update step called $update$. We describe below the recursive procedure $extend\_cycle\_basis$. Note that all the arithmetic that we do here is over the field $\mathbb{F}_p$. (For clarity we will sometimes use the notation v_j^ℓ to denote the version of v_j that is orthogonal to the cycles $D_1, \ldots, D_{\ell-1}$.)

The procedure $extend\_cycle\_basis(\mathcal{D}, \{v_{i+1}, \ldots, v_{i+k}\}, k)$:

- if $k = 1$, compute a shortest cycle D_{i+1} such that $\langle D_{i+1}, v_{i+1} \rangle \neq 0 \pmod p$.
- if $k > 1$, use recursion. Let $t = \lfloor k/2 \rfloor$.
  1. call $extend\_cycle\_basis(\mathcal{D}, \{v_{i+1} \ldots, v_{i+t}\}, t)$ to extend the current cycle basis by t elements. That is, the cycles $D_{i+1}, \ldots, D_{i+t}$ are computed in a recursive manner.
  2. call $update(\{v_{i+1}, \ldots, v_{i+t}\}, \{v_{i+t+1}, \ldots, v_{i+k}\})$. This updates $\{v_{i+t+1}^{i+1}, \ldots, v_{i+k}^{i+1}\}$ en masse into the desired versions $\{v_{i+t+1}^{i+t+1}, \ldots, v_{i+k}^{i+t+1}\}$ that are orthogonal to $D_1, \ldots, D_{i+t}$.
  3. call $extend\_cycle\_basis(\mathcal{D} \cup \{D_{i+1}, \ldots, D_{i+t}\}, \{v_{i+t+1}, \ldots, v_{i+k}\}, k-t)$ to extend the current cycle basis by $k - t$ cycles. That is, the cycles $D_{i+t+1}, \ldots, D_{i+k}$ will be computed recursively.

The key subroutine $update$ can be implemented efficiently using fast matrix multiplication. Then the running time of FAST-Randomized-MCB becomes

$O(m^\omega + m^2 n \log n)$. We can assume that G is a simple graph, so $m \leq n^2$. Then $m^\omega < m^2 n$ and so the running time is $O(m^2 n \log n)$. The running time analysis and the analysis that the cycle basis computed here is the same as the cycle basis computed by the algorithm Randomized-MCB follow directly from similar analysis given in [11]. They will be presented in the full version of the paper. We conclude with the following theorem.

Theorem 2. *A minimum cycle basis of a directed graph, with positive weights on its arcs, can be computed with high probability in time $O(m^2 n \log n)$.*

Acknowledgments. I am grateful to Kurt Mehlhorn for useful discussions and his help in improving the presentation of the paper. I also wish to thank Jaikumar Radhakrishnan for his helpful comments and the referee for bringing the example in [14] to my attention.

References

1. T. M. Apostol. *Introduction to Analytic Number Theory*. Springer-Verlag, 1997.
2. F. Berger, P. Gritzmann, and S. de Vries. Minimum Cycle Bases for Network Graphs. *Algorithmica*, 40(1): 51-62, 2004.
3. B. Bollobás. *Modern Graph Theory*, volume 184 of *Graduate Texts in Mathematics*, Springer, Berlin, 1998.
4. A. C. Cassell and J. C. Henderson and K. Ramachandran. Cycle bases of minimal measure for the structural analysis of skeletal structures by the flexibility method *Proc. Royal Society of London Series A*, 350: 61-70, 1976.
5. J.C. de Pina. *Applications of Shortest Path Methods*. PhD thesis, University of Amsterdam, Netherlands, 1995.
6. N. Deo. *Graph Theory with Applications to Engineering and Computer Science*. Prentice-Hall Series in Automatic Computation. Prentice-Hall, Englewood Cliffs, 1982.
7. P. M. Gleiss. *Short cycles: minimum cycle bases of graphs from chemistry and biochemistry*. PhD thesis, Universität Wien, 2001.
8. Alexander Golynski and Joseph D. Horton. A polynomial time algorithm to find the minimum cycle basis of a regular matroid. In *8th Scandinavian Workshop on Algorithm Theory*, 2002.
9. J. D. Horton. A polynomial-time algorithm to find a shortest cycle basis of a graph. *SIAM Journal of Computing*, 16:359–366, 1987.
10. T. Kavitha and K. Mehlhorn. A Polynomial Time Algorithm for Minimum Cycle Basis in Directed Graphs *In Proc. of STACS*, LNCS 3404: 654-665, 2005.
11. T. Kavitha, K. Mehlhorn, D. Michail, and K. Paluch. A faster algorithm for Minimum Cycle Basis of graphs. *In Proc. of ICALP*, LNCS 3142: 846-857, 2004.
12. Christian Liebchen. Finding Short Integral Cycle Bases for Cyclic Timetabling. *In Proc. of ESA*, LNCS 2832: 715-726, 2003.
13. C. Liebchen and L. Peeters. On Cyclic Timetabling and Cycles in Graphs. Technical Report 761/2002, TU Berlin.
14. C. Liebchen and R. Rizzi. A Greedy Approach to compute a Minimum Cycle Basis of a Directed Graph. Technical Report 2004/31, TU Berlin.

Basing Cryptographic Protocols on Tamper-Evident Seals*

Tal Moran and Moni Naor**

Department of Computer Science and Applied Mathematics,
Weizmann Institute of Science, Rehovot, Israel

Abstract. In this paper we attempt to formally study two very intuitive physical models: sealed envelopes and locked boxes, often used as illustrations for common cryptographic operations. We relax the security properties usually required from locked boxes (such as in bit-commitment protocols) and require only that a broken lock or torn envelope be identifiable to the original sender. Unlike the completely impregnable locked box, this functionality may be achievable in real life, where containers having this property are called "tamper-evident seals". Another physical object with this property is the "scratch-off card", often used in lottery tickets. We show that scratch-off cards can be used to implement bit-commitment and coin flipping, but not oblivious transfer. Of particular interest, we give a strongly-fair coin flipping protocol with bias bounded by $O(1/r)$ (where r is the number of rounds), beating the best known bias in the standard model even with cryptographic assumptions.

1 Introduction

In this paper we consider the use of "tamper-evident seals" in cryptographic protocols. A tamper-evident seal is a primitive based on very intuitive physical models: the sealed envelope and the locked box. These are often used as illustrations for a number of basic cryptographic primitives: For example, encryption is often depicted as placing a message in a locked box (that cannot be opened without a key), while bit commitment is usually illustrated as a sealed envelope.

In a bit-commitment protocol one party, Alice, commits to a bit b to Bob in such a way that Bob cannot tell what b is. At a later time Alice can reveal b, and Bob can verify that this is indeed the bit to which she committed. The standard illustration used for a bit-commitment protocol is Alice putting b in a sealed envelope, which she gives to Bob. Bob cannot see through the envelope (so cannot learn b). When Alice reveals her bit, she lets Bob open the envelope so he can verify that she didn't cheat.

The problem with the above illustration is that a physical "sealed envelope" is insufficient for bit-commitment: Bob can always tear open the envelope before

---

* This work was partially supported by the Minerva Foundation.
** Incumbent of the Judith Kleeman Professorial Chair.

Alice officially allows him to do so. Even a locked box is unlikely to suffice; many protocols based on bit-commitment remain secure only if no adversary can *ever* open the box without a key. A more modest security guarantee seems to be more easily obtained: an adversary may be able to tear open the envelope but Alice will be able to recognize this when she sees the envelope again.

"Real" closures with this property are commonly known as "tamper evident seals". These are used widely, from containers for food and medicines to high-security government applications. Another common application that embodies these properties is the "scratch-off card", often used as a lottery ticket. This is usually a printed cardboard card which has some areas coated by an opaque layer (e.g., the possible prizes to be won are covered). The text under the opaque coating cannot be read without scratching off the coating, but it is immediately evident that this has been done.

In this paper we attempt to clarify what it means to use sealed envelopes in a cryptographic protocol. In particular, we study their applicability to coin flipping (CF), zero-knowledge protocols, bit commitment (BC) and oblivious transfer (OT), some of the most fundamental primitives in modern cryptography.

Our Results. In this paper we show that the sealed envelopes we consider can be used to implement standard cryptographic protocols. We construct protocols for "weakly fair" CF (in which the result is 0, 1 or *invalid*) and BC.

The existence of BC immediately implies the existence of a zero-knowledge proof system for any NP language [16]. In the full version, we also show a non-interactive zero knowledge proof system (with a preprocessing stage) for any NP language that can be implemented using pre-printed scratch-off cards.

A possibly practical application of our model is the "cryptographic randomized response technique" (CRRT), defined by Ambainis et al. [2]. "Randomized response" is a polling technique used when some of the answers to the poll may be stigmatizing (e.g., "do you use drugs?"). The respondent lies with some known probability, allowing statistical analysis of the results while letting the respondent disavow a stigmatizing response. In a CRRT, there is the additional requirement that a malicious respondent cannot bias the results more than by choosing a different answer. The techniques described by Ambainis et al. achieve this, but require "heavy" cryptographic machinery (such as OT), or quantum cryptography. In the full version of the paper we give a simple protocol for CRRT using scratch-off cards.

One of the most interesting results is a protocol for "strongly fair" CF (where the result for an honest player must be either 0 or 1 even if the other player quits before finishing the protocol) with bias bounded by $O(\frac{1}{r})$, where r is the number of rounds. In the standard model, even with cryptographic assumptions, the best known bias for a protocol with r rounds is $O(\frac{1}{\sqrt{r}})$ (due to Cleve [8]). For a large class of protocols (including any that rely on bit-commitment or weakly fair coin flipping as a black-box), an unpublished result of Cleve and Impagliazzo [9] shows this is the best possible bias.

An important contribution of this paper is the *formal* analysis for the models and protocols we construct. We show that the protocols are *Universally Com-*

posable in the sense of Canetti [7]. This allows us to use them securely as "black-boxes" in larger constructions.

On the negative side, we show that our protocol for strongly fair CF using scratch-off cards is optimal: it is impossible to do better than $O(\frac{1}{r})$ bias (this follows from a careful reading of the proof in [8]; the proof appears in the full version). We show that OT cannot be implemented using scratch-off cards without additional assumptions (note that we show the impossibility of *any* type of OT, not just universally composable realizations).

Seals in Different Flavors. In the full version of the paper, we consider some additional variants of tamper-evident seals. Roughly speaking, the difference arises from whether or not a party can distinguish their own containers from those of the other party without opening the seal, and whether or not an honest player can "break" the seal. Scratch-off cards correspond to the distinguishable case in which the honest party *can* break the seal. We show that in a distinguishable "weak-lock" model, where the honest party cannot break the seal, even BC cannot be implemented (the proof is similar to the impossibility of OT, while in the indistinguishable case it is possible to implement both BC and OT. Due to space constraints, this paper will concentrate solely on the "distinguishable envelope" (scratch-off card) model. Note that in the standard model of cryptography, where the parties exchange messages and there is no access to outside physical resources, we do not know how to implement any of these closures.

Related Work. To the best of our knowledge, this is the first attempt at using tamper evident seals for cryptographic protocols. Ross Anderson discusses "packaging and seals" in the context of security engineering [3], however the use of tamper-evidence does not extend to more complex protocols. Blaze gives some examples of the reverse side of the problem: cryptanalysis of physical security systems using techniques from computer science [4,5]. Using scratch-off cards in the lottery setting can be described as a very weak form of CF, however we do not believe this has ever been formally analyzed (or used in more complex protocols).

On the other hand, basing cryptographic protocols on physical models is a common practice. Perhaps the most striking example is the field of quantum cryptography. One of the inspirations for this work was the idea of "Quantum Bit Escrow" (QBE) [1], a primitive that is very similar to a tamper-evident seal and that can be implemented in a quantum setting. There are, however, significant differences between our definitions of tamper-evident seals and QBE. In particular, in QBE the adversary may "entangle" separate escrowed bits and "partially open" commitments. Thus, while unconditionally secure bit-commitment is impossible in the pure quantum setting [19, 18], it is possible in ours.

Much work has been done on basing BC and OT on the physical properties of communication channels, using the random noise in a communication channel as the basis for security. Both BC and OT were shown to be realizable in the *Binary Symmetric Channel* model [11, 10], in which random noise is added to the channel in both directions with some known, constant, probability. Later work shows that they can also be implemented, under certain conditions, in the weaker (but more convincing) *Unfair Noisy Channel* model [13, 12], where the error probability is not known exactly to the honest parties, and furthermore can be influenced by the adversary. Our construction for 1-2 OT uses some of the techniques and results from [13].

One of the motivations for this work was the attempt to construct cryptographic protocols that are implementable by humans without the aid of computers. This property is useful, for example, in situations where computers cannot be trusted to be running the protocol they claim, or where "transparency" to humans is a requirement (such as in voting protocols). Many other examples exist of using simple physical objects as a basis for cryptographic protocols that can be performed by humans, some are even folklore: Sarah Flannery [15] recounts a childhood riddle that uses a doubly-locked box to transfer a diamond between two parties, overcoming the corrupt postal system (which opens any unlocked boxes) despite the fact that the two parties have never met (and can only communicate through the mail). Fagin, Naor and Winkler [14] assembled a number of solutions to the problem of comparing secret information without revealing anything but the result of the comparison using a variety of different physical methods. Schneier devised a cipher [21] that can be implemented by a human using a pack of cards. In a lighter vein, Naor, Naor and Reingold [20] give a protocol that provides a "zero knowledge proof of knowledge" of the correct answer to the children's puzzle "Where's Waldo" using only a large newspaper and scissors. A common thread in these works is that they lack a formal specification of the model they use, and a formal proof of security.

Organization of the Paper. In Section 2, we give a formal definition for tamper-evident envelopes and the functionalities we attempt to realize using them. In Section 3 we discuss the capabilities of this model, showing that OT is impossible and giving protocols for BC and strongly-fair CF with bias $1/r$. Section 4 contains the discussion and some open problems.

2 The Model: Ideal Functionalities

Many two-party functionalities are easy to implement using a trusted third party that follows pre-agreed rules. In proving that a two-party protocol is secure, we often want to say that it behaves "as if it were performed using the trusted third party". A formalization of this idea is the "Universally Composable" model defined by Canetti [7]. In the UC model, the trusted third party is called the *ideal functionality*. The point of the model is that protocols that are secure in the UC have very strong security properties, such as security under composition and security that is retained when the protocol is used as a sub-protocol to replace an ideal functionality. This security guarantee allows us to simplify many of our proofs, by showing separately the security of their component subprotocols.

Note that our impossibility results are not specific to the UC model: the impossibility results for BC (in the full version of the paper), OT (Section 3) and the lower bound for strongly fair CF (also in the full version) hold even for the much weaker "standard" notions of these functionalities[1].

In this section we formally define tamper-evident envelopes in terms of their ideal functionalities. For completeness, we also give the definitions of the primitives we are trying to implement (CF, BC, and OT). We restrict ourselves to the two-party case, and to adversaries that decide at the beginning of the protocol whether to corrupt one of the parties or neither.

Distinguishable Envelopes. This functionality models an opaque envelope (or scratch-off card). Without opening the envelope it is impossible to determine its contents. Any party can decide to rip open the envelope (breaking the seal), but this will be evident to the envelope's creator if the envelope is returned.

In succeeding sections, we assume we are given a realization of this functionality and attempt to construct a protocol for a "target" functionality (these are described below)

Functionality $\mathcal{F}_{DE}$ contains an internal table that consists of tuples of the form $(id, value, creator, holder, state)$. The table represents the state and location of the tamper-evident envelopes, and contains one entry for each existing envelope, indexed by the envelope's id. We denote $value_{id}$, $creator_{id}$, $holder_{id}$ and $state_{id}$ the corresponding values in the table in row id (assuming the row exists). The table is initially empty. The functionality is as follows, running with parties $P_1, \ldots, P_n$ and adversary S:

Seal $(id, value)$. On receiving this command from party P_i, if this is the first message with id id store the tuple $(id, value, P_i, P_i, \text{sealed})$ in the table (if this is not the first message with id id, do nothing).

Send (id, P_j). On receiving this command from party P_i, the functionality checks if an entry for envelope id appears in the table and that $holder_{id} = P_i$. If so, it outputs (Receipt, id, P_i, P_j) to P_j and S and replaces the entry in the table with $(id, value_{id}, creator_{id}, P_j, state_{id})$. Otherwise, it does nothing.

Open id. On receiving this command from P_i, the functionality checks that an entry for envelope id appears in the table, that $holder_{id} = P_i$. If not, the message is ignored. Otherwise, it sends (Opened, $id, value_{id}, creator_{id}$) to P_i and S. It then replaces the entry in the table with
$(id, value_{id}, creator_{id}, holder_{id}, \text{broken})$.

Verify id. On receiving this command from P_i, the functionality checks that an entry for envelope id appears in the table and that $holder_{id} = P_i$. If not, the message is ignored. If so, it considers $state_{id}$. If $state_{id} = $ broken it sends (Verified, id, broken) to P_i and S. Otherwise, it sends (Verified, id, ok) to P_i and S.

A Note about Notation. In the interests of readability, we will often refer to parties "preparing", "verifying" and "opening" envelopes, instead of specifying that

---

[1] We *do* assume the "interface" supplied by the envelope primitives is the one defined by the ideal functionality (e.g., we do not allow putting one envelope inside another).

they send the corresponding messages (**Seal**, **Verify**, **Open**) to the functionality and wait for the appropriate response.

Weakly Fair Coin Flipping. This functionality models coin flipping in which the result of the coin flip can be 0, 1 or $\perp$. The result of the flip c should satisfy: $\Pr[c = 0] \leq \frac{1}{2}$ and $\Pr[c = 1] \leq \frac{1}{2}$. This is usually what is meant when talking about "coin flipping" (for instance, in Blum's protocol [6]). The $\perp$ result corresponds to the case where one of the parties deviated from (or prematurely aborted) the protocol. Under standard cryptographic assumptions (such as the existence of one-way functions), weakly fair coin flipping is possible. Conversely, in the standard model the existence of weakly fair coin flipping implies one-way functions [17].

Functionality $\mathcal{F}_{WCF}$ is as follows, with parties Alice and Bob (in this definition we only allow Bob to trigger an invalid output):

Value. The sender of this command is called the *initiator*. The other party is the *receiver*. When this command is received, the functionality chooses a uniform value $d \in \{0, 1\}$. If the receiver is corrupted, the functionality then outputs "approve d" to the receiver. In that case, the functionality ignores all input until it receives either a **Continue** command or a **Stop** command from the receiver. If the receiver is not corrupted, the functionality proceeds as if he had sent a **Continue** command.

Stop. When this command is received from a corrupt receiver (in response to an "approve d" message) the functionality outputs $\perp$ to all parties and halts.

Continue. When this command is received from the receiver (in response to an "approve d" message), the functionality outputs "coin is d" to all parties and halts.

Strongly Fair Coin Flipping with Bias p (Adapted from [7]). This functionality models a coin flip between two parties with a bounded bias. If both parties follow the protocol, they output an identical uniformly chosen bit. Even if one party does not follow the protocol[2], the other party outputs a random bit whose bias towards 0 or 1 is at most p.

Functionality $\mathcal{F}_{SCF}$ is as follows:

Value. When this command is received for the first time from any party, the functionality chooses a value $d \in \{0, 1\}$. It then outputs "accept d?" to the adversary. If the adversary responds with **no**, with probability p the functionality outputs "value $1 - d$" to all parties and with probability $1 - p$ outputs "value d" to all parties. If the adversary responds with **yes** (or does not respond), the functionality outputs "value d" to all parties.

---

[2] In "not following the protocol" we include halting (failing to respond). The UC model does not explicitly handle this behaviour, however we can treat it as a special "halt" command that can be sent by the adversary.

Bit Commitment (Adapted from [7]). Functionality $\mathcal{F}_{BC}$:

Commit b. The issuer of this command is called the sender, the other party is the receiver. On receiving this command the functionality records b and outputs "committed" to the receiver. It then ignores any other commands until it receives the **Open** command from the sender.

Open. On receiving this command from the sender, the functionality outputs "opened b" to the receiver.

2.1 Intermediate Functionalities

In order to simplify the presentation, in the following sections we will present protocols that realize functionalities that are slightly weaker than the ones we want. We then use standard amplification techniques to construct the "full" functionalities from their weak version. In this section we define these intermediate functionalities and give the amplification lemmas we use to construct the stronger versions of these primitives. These definitions are in the spirit of [13].

(p,q)-Weak Bit-Commitment. This functionality models bit commitment where a corrupt sender can cheat with probability q while a corrupt receiver can cheat with probability p. The result of failing to cheat is asymmetric. In the case of a corrupt receiver, an unsuccessful attempt to cheat causes the sender to be notified, while a corrupt sender risks nothing in attempting to cheat. Note that an (ϵ, ϵ)-WBC protocol is a regular bit-commitment protocol when ϵ is negligible. Formally, functionality $\mathcal{F}_{(p,q)-WBC}$ proceeds as follows:

Commit b. The issuer of this command is called the sender, the other party is the receiver. On receiving this command the functionality records b and outputs "committed" to the receiver. It then ignores any other commands until it receives an **Open** command from the sender, an **OpenFlip** command from a corrupt sender or a **Break** command from a corrupt receiver.

Open b. On receiving this command from the sender, the functionality checks that the sender previously sent a **Commit b** command. If so, or if the sender is corrupt and previously sent a **CanEquivocate** command whose response was "Can Equivocate", the functionality outputs "opened b" to the receiver. Otherwise the command is ignored.

CanEquivocate On receiving this command from a corrupt sender, choose a value r uniformly in $[0,1]$. If $r \geq q$ send "No Equivocation" to the sender and ignore further **CanEquivocate** messages. If $r < q$ send "Can Equivocate" to the sender.

Break On receiving this command from a corrupt receiver, choose a value r uniformly in $[0,1]$. If $r \geq p$ send "cheating receiver" to the sender and receiver and then halt. Otherwise, send b to the receiver.

We can amplify any (p,q)-WBC protocol when $p,q < 1$, meaning that the existence of such a protocol implies the existence of regular BC.

Theorem 1. *Let P be a (p,q)-WBC protocol and $p, q < 1$. Then there exists an (ϵ, ϵ)-WBC for any $\epsilon > 0$ using $O\left(\log^2\left(\frac{1}{\epsilon}\right)\right)$ invocations of P.*

The proof for this theorem is straightforward, and will be given in the full version.

(p, q)-Remotely Inspectable Seal (RIS). This functionality is used in our protocol for strongly fair CF. It is a strengthened version of a tamper-evident seal. With a tamper-evident seal, only its holder can interact with it. Thus, either the sender can check if it was opened, or the receiver can verify that the sealed contents were not changed, but not both at the same time. A remotely inspectable seal is one that can be tested "remotely" (without returning it to the sender). Unfortunately, we cannot realize the ideal version, and therefore relax it somewhat: we allow a corrupt receiver to learn the committed bit during the verification process, and only then decide (assuming he did not previously break the seal) whether or not the verification should succeed. Our definition is actually a further relaxation[3]: both sender and receiver may cheat with some probability. A corrupt sender can cause the result of the **Open** command to be a uniformly random value (instead of a specific value determined in the commit stage). The receiver will catch the sender with probability at least $1-q$. A corrupt receiver who opens the commitment before the verify stage will be caught with probability $1 - p$.

Formally, the functionality maintains an internal state variable $v = (v_b, v_s)$ consisting of the committed bit v_b and a "seal" flag v_s. It accepts the commands:

Commit b. The issuer of this command is called the sender, the other party is the receiver. b can be either 0, 1 or (if the sender is corrupt) $\bot$. If $b \in \{0, 1\}$ (the sender did not try to cheat), the functionality sets $v \leftarrow (b, \text{sealed})$. If $b = \bot$ (the sender tried to cheat) with probability q (the sender cheated successfully) $v \leftarrow (r, \text{sealed})$ (where r is randomly chosen from $\{0, 1\}$) and with probability $1 - q$ it sets $v \leftarrow (\bot, \text{sealed})$ In any case the functionality concludes by outputting "committed" to the receiver, and ignoring any subsequent **Commit** commands.

Open. This command is sent by the receiver. If $v_b \in \{0, 1\}$ the functionality outputs "opened b" to the receiver. Otherwise it outputs "invalid" to the receiver. If $v_s = $ sealed, with probability $1 - p$ the functionality sets $v_s \leftarrow $ open

Verify. If $v_s \neq $ sealed, the functionality outputs "invalid" to the sender. Otherwise (no opening was detected), the functionality outputs "verifying value b" to the adversary and waits for a response. If the adversary responds with **ok**, the functionality outputs "sealed" to the sender, otherwise it outputs "invalid" to the sender. After receiving this command from the sender (and responding appropriately), the functionality ignores any subsequent **Verify** commands.

---

[3] This second relaxation is only for convenience; we can remove it using amplification as noted in Theorem 2.

A (p,q)-Remotely Inspectable Seal can be amplified for any $p < 1$ and $q \leq 1$:

Theorem 2. *Let P be a (p,q)-RIS protocol $p < 1$ and $q \leq 1$. Then there exists an (ϵ, ϵ)-RIS for any $\epsilon > 0$ using $O\left(\log^2\left(\frac{1}{\epsilon}\right)\right)$ invocations of P*

Note that the amplification works even if $q = 1$: this is because the adversary doesn't have full control over the revealed bit but can only cause it to be a random bit. The proof of this theorem will appear in the full version.

3 Capabilities of the Distinguishable Envelope Model

Oblivious Transfer is Impossible. Let Alice be the sender and Bob the receiver. Consider Alice's bits a_0 and a_1, as well as Bob's input c, to be random variables taken from some arbitrary distribution. Alice's view of a protocol execution can also be considered a random variable $V_A = (a_0, a_1, r_A, N_1, \ldots, N_n)$, consisting of Alice's bits, random coins (r_A) and the sequence of messages that comprise the transcript as seen by Alice. In the same way we denote Bob's view with $V_B = (c, r_B, M_1, \ldots, M_n)$, consisting of Bob's input and random coins and the sequence of messages seen by Bob.

The essence of oblivious transfer (whether universally composable or not) is that Bob gains information about one of Alice's bits, but Alice does not know which one. We can describe the information Bob has about Alice's bits using Shannon entropy, a basic tool of information theory. The Shannon entropy of a random variable X, denoted $H(X)$ is a measure of the "uncertainty" that resides in X. When X has finite support: $H(X) = -\sum_x \Pr[X = x] \log \Pr[X = x]$.

Suppose Bob's view of a specific protocol transcript is v_B. What Bob learns about a_i ($i \in \{0, 1\}$) can be described by the conditional entropy of a_i given Bob's view of the protocol. We write this $H(a_i \mid V_B = v_B)$. If Bob knows a_i at the end of the protocol then $H(a_i \mid V_B = v_B) = 0$ since there is no uncertainty left about the value of a_i given Bob's view. If Bob has no information at all about a_i then $H(a_i \mid V_B = v_B) = 1$, since there are two equally likely values of a_i given Bob's view.

We show that in any protocol in the DE Model, Alice can calculate the amount of information Bob has about each of her bits:

Theorem 3. *For any protocol transcript where $V_A = v_A$ and $V_B = v_B$, both $H(a_0 \mid V_B = v_B)$ and $H(a_1 \mid V_B = v_B)$ are completely determined by v_A.*

In any OT protocol, Bob must have full information about a_c (c is Bob's "choice" bit): $H(a_c \mid V_B = v_B) = 0$, and no information at all about a_{1-c}: $H(a_c \mid V_B = v_B) = 1$. The theorem states that in this case Alice can determine c, violating the "obliviousness" requirement of the protocol.

Due to space considerations, we do not give the proof of Theorem 3 here. The basic idea is that for any injection f and any random variable Y, the event $Y = y$ is identical to the event $f(Y) = f(y)$. Therefore, for any two random variables X and Y, it holds that $H(X \mid Y = y) = H(X \mid f(Y) = f(y))$. We show an injection from V_B to V_A, r_B, c. Since r_B and c must be independent of (a_0, a_1), we can conclude that Alice can compute $H(a_0 \mid V_B = v_B)$ and $H(a_1 \mid V_B = v_B)$ for any transcript. A similar argument holds for Bob.

A $\left(\frac{3}{4}, \frac{1}{2}\right)$-Weak Bit Commitment Protocol. We will show a weak bit commitment protocol, and apply Theorem 1 to amplify the protocol and construct a "standard" bit commitment protocol.

To implement **Commit** b:

1. Bob prepares four sealed envelopes, two containing a 0 and two a 1 in random order. Bob sends the envelopes to Alice
2. Alice opens three envelopes (chosen randomly), and verifies that they are not all the same. Let r be the value in the remaining (sealed) envelope. Alice sends $d = b \oplus r$ to Bob.

To implement **Open**:

1. Alice sends b and the sealed envelope to Bob.
2. Bob verifies that it is sealed, then opens it. He checks that $d = b \oplus r$.

Intuitively, Bob can cheat only by sending three envelopes with the same bit in the commit phase. However, Alice will catch him with probability $\frac{1}{4}$. Alice can cheat by opening less than three envelopes (and guessing d). She will then be caught in the open phase with probability at least $\frac{1}{2}$. A proof of security will appear in the full version.

A Strongly Fair Coin Flipping Protocol with Bias $O(\frac{1}{r})$. The construction uses remotely inspectable seals (defined in Sec. 2.1), which we then show how to implement in the DE model. The idea is similar to the "standard" CF protocol using BC: Alice commits to a random bit a. Bob sends Alice a random bit b, after which Alice opens her commitment. The result is $a \oplus b$.

The reason that this is not a strongly fair CF protocol is that Alice learns the result of the toss before Bob, and can decide to quit before opening her commitment. Using RIS instead of BC solves this problem, because Bob can open the commitment without Alice's help.

Ideally, we would like to replace BC with RIS (and have Alice verify that Bob didn't break the seal before sending b). This seems to work; If Bob quits before verification, or if the verification fails, Alice can use a as her bit, because Bob had to have decided to quit before seeing a. If Bob quits after verification (and the verification passed), Alice can use $a \oplus b$, since Bob sent b before learning a. This idea fails, however. The reason is that RIS allows Bob to see the committed bit *during* verification. If he doesn't like it, he can cause the verification to fail.

We can overcome the problem with probability $1 - \frac{1}{r}$ by doing the verification in r rounds. The trick is that Alice secretly decides on a "threshold round": after this round a failure in verification won't matter. Bob doesn't know which is the threshold round (he can guess with probability at most $1/r$). If Bob decides to stop before the threshold round, either he did not attempt to illegally open a commitment (in which case his decision to stop cannot depend on the result of the coin flip), or he illegally opened all the remaining commitments (opening less than that gives no information about the result). In this case all subsequent verifications will fail, so he may as well have simply stopped at this round (note that the decision to open is made before knowing the result of the coin flip). Clearly, anything Bob does after the threshold round has no effect on the result. Only if he chooses to illegally open commitments during the threshold round can this have an effect on the outcome (since in this case, whether or not the verification fails determines whether Alice outputs a or $a \oplus b$). The full protocol is as follows:

1. Alice chooses r random bits $a_1, \ldots, a_r$ and commits to each bit using the RIS scheme (this is done in parallel). Denote $a = a_1 \oplus \cdots \oplus a_r$,
2. Bob chooses a random bit b. If Alice quits before finishing the commit stage, Bob outputs b. Otherwise, he sends b to Alice.
3. If Bob quits before sending b, Alice outputs a. Otherwise, Alice chooses a secret index $j \in \{1, \ldots, r\}$.
4. The protocol now proceeds in r rounds. Round i has the following form:
   (a) Alice verifies that Bob did not open the commitment for a_i.
   (b) Bob opens the commitment for a_i.
5. If the verification is successful up to round j (and Bob didn't quit up to round j), Alice outputs $a \oplus b$. Otherwise, Alice outputs a.
6. Bob always outputs $a \oplus b$ (If Alice quits before completing the verification rounds, Bob opens the commitments without verification).

Implementation of Remotely Inspectable Seals. We start by giving a $(\frac{1}{2}, 1)$-RIS protocol. We can then apply Theorem 2 to amplify it to (ϵ, ϵ)-RIS for some negligible ϵ. To implement **Commit** b:

1. Alice sends two envelopes to Bob, each containing the bit b.

To implement **Verify**:

1. Alice initiates a (weakly fair) coin flip with Bob.
2. Denote the result of the coin flip r. Bob returns envelope r to Alice.
3. Alice waits for the result of the coin flip and the envelope from Bob. If the result is $\bot$, or if Bob does not return an envelope, Alice outputs "opened". Otherwise, Alice verifies that Bob returned the correct envelope, and that it is still sealed. If both of these conditions are not satisfied, she outputs "sealed", otherwise she outputs "opened".

To implement **Open**:

1. Bob randomly chooses one of the envelopes in his possession (if he already returned one to Alice during the **Verify** stage then he opens the one he has left). He outputs the contents of the envelope.

4 Discussion and Open Problems

The protocols we describe in this paper can be performed by unaided humans, however they require too many "envelopes" to be practical for most uses. It would be useful to construct protocols that can be performed with a smaller number of envelopes or with a smaller number of rounds.

Another point worth mentioning is that the protocols we construct only require one of the parties to seal and verify envelopes. Thus, the binding property is only used in one direction, and the tamper-evidence and hiding properties in the other. This property is useful when we want to implement the protocols in a setting where one of the parties may be powerful enough to open the seal undetectably. (for instance, in the context of voting, where one of the parties could be "the government" while the other is a private citizen)

In both the weakly and strongly fair CF protocols, only the first round requires envelopes to be created, and their contents do not depend on communication with the other party. This allows the protocols to be implemented using scratch-off cards (which must be printed in advance). In particular, the weakly fair CF protocol can be implemented with a scratch-off card using only a small number of areas to be scratched (this protocol is given in the full version).

In the case of BC, our protocol requires the powerful party to be the receiver. It would be interesting to construct a BC protocol for which the powerful party is the sender (i.e., only the sender is required to to seal and verify envelopes).

References

1. D. Aharonov, A. Ta-Shma, U. V. Vazirani, and A. C. Yao. Quantum bit escrow. In *STOC '00*, pages 705–714, 2000.
2. A. Ambainis, M. Jakobsson, and H. Lipmaa. Cryptographic randomized response techniques. In *PKC '04*, volume 2947 of *LNCS*, pages 425–438, 2004.
3. R. J. Anderson. *Security Engineering: A Guide to Building Dependable Distributed Systems*. John Wiley & Sons, Inc., 2001.
4. M. Blaze. Cryptology and physical security: Rights amplification in master-keyed mechanical locks. *IEEE Security and Privacy*, March 2003.
5. M. Blaze. Safecracking for the computer scientist. *U. Penn CIS Department Technical Report*, December 2004. http://www.crypto.com/papers/safelocks.pdf.
6. M. Blum. Coin flipping over the telephone. In *Proceedings of IEEE COMPCON '82*, pages 133–137, 1982.
7. R. Canetti. Universally composable security: A new paradigm for cryptographic protocols. Cryptology ePrint Archive, Report 2000/067, 2000.
8. R. Cleve. Limits on the security of coin flips when half the processors are faulty. In *STOC '86*, pages 364–369, 1986.
9. R. Cleve and R. Impagliazzo. Martingales, collective coin flipping and discrete control processes. http://www.cpsc.ucalgary.ca/ cleve/pubs/martingales.ps, 1993.
10. C. Crépeau. Efficient cryptographic protocols based on noisy channels. In *Eurocrypt '97*, volume 1233 of *LNCS*, pages 306–317, 1997.
11. C. Crépeau and J. Kilian. Achieving oblivious transfer using weakened security assumptions. In *FOCS '88*, pages 42–52, 1988.

12. I. B. Damgård, S. Fehr, K. Morozov, and L. Salvail. Unfair noisy channels and oblivious transfer. In *TCC '04*, volume 2951 of *LNCS*, pages 355–373, 2004.
13. I. B. Damgård, J. Kilian, and L. Salvail. On the (im)possibility of basing oblivious transfer and bit commitment on weakened security assumptions. In *Eurocrypt '99*, volume 1592 of *LNCS*, pages 56–73, 1999.
14. R. Fagin, M. Naor, and P. Winkler. Comparing information without leaking it. *Commun. ACM*, 39(5):77–85, 1996.
15. S. Flannery and D. Flannery. *In Code: A Mathematical Journey*. Algonquin Books of Chapel Hill, 2002.
16. O. Goldreich, S. Micali, and A. Wigderson. Proofs that yield nothing but their validity or all languages in NP have zero-knowledge proof systems. *J. of the ACM*, 38(3):691–729, July 1991.
17. R. Impagliazzo and M. Luby. One-way functions are essential for complexity based cryptography. In *FOCS '89*, pages 230–235, 1989.
18. H.-K. Lo and H. F. Chau. Why quantum bit commitment and ideal quantum coin tossing are impossible. In *PhysComp '98*, pages 177–187, 1998.
19. D. Mayers. Unconditionally secure quantum bit commitment is impossible. *Phys. Rev. Lett.*, (78):3414–3417, 1997.
20. M. Naor, Y. Naor, and O. Reingold. Applied kid cryptography or how to convince your children you are not cheating, Mar. 1999. http://www.wisdom.weizmann.ac.il/ naor/PAPERS/waldo.ps.
21. B. Schneier. The solitaire encryption algorithm, 1999. http://www.schneier.com/solitaire.html.

Hybrid Trapdoor Commitments and Their Applications*

Dario Catalano[1] and Ivan Visconti[2,**]

[1] CNRS-Ecole Normale Supérieure,
Laboratoire d'Informatique 45 Rue d'Ulm,
75230 Paris Cedex 05 - France
dario.catalano@ens.fr

[2] Dip. di Informatica ed Appl. Università di Salerno Via S. Allende n. 2,
84081 Baronissi (SA) - Italy
visconti@dia.unisa.it

Abstract. We introduce the notion of hybrid trapdoor commitment schemes. Intuitively an hybrid trapdoor commitment scheme is a primitive which can be either an unconditionally binding commitment scheme or a trapdoor commitment scheme depending on the distribution of commitment parameters. Moreover, such two distributions are computationally indistinguishable. Hybrid trapdoor commitments are related but different with respect to *mixed* commitments (introduced by Damgård and Nielsen at Crypto 2002). In particular hybrid trapdoor commitments can either be polynomially trapdoor commitments or unconditionally binding commitments, while mixed commitment can be either trapdoor commitments or extractable commitments. In this paper we show that strong notions (e.g., simulation sound, multi-trapdoor) of hybrid trapdoor commitments admit constructions based on the sole assumption that one-way functions exist as well as efficient constructions based on standard number-theoretic assumptions. To further stress the difference between hybrid and mixed commitments, we remark here that mixed commitments seems to require stronger theoretical assumptions (and the known number-theoretic constructions are less efficient). The main application of our results is that we show how to construct concurrent and simulation-sound zero-knowledge proof (in contrast to the arguments recently presented in [1, 2, 3]) systems in the common reference string model. We crucially use hybrid commitment since we present general constructions based on the sole assumption that one-way functions exists and very efficient constructions based on number-theoretic assumptions.

---

* Extended abstract. The full version of this paper can be found at http://www.di.ens.fr/~catalano.
** Work partially done while Ivan Visconti was a post-doctoral fellow at the Département d'Informatique of the Ecole Normale Supérieure in Paris, France.

1 Introduction

Commitment schemes are arguably among the most important and useful primitives in cryptography. Intuitively a commitment scheme can be seen as the digital equivalent of a sealed envelope. If a party A wants to commit to some message m she just puts it into the sealed envelope, so that whenever A wants to reveal the message, she opens the envelope. Clearly, such a mechanism can be useful only if it meets some basic requirements. First of all the digital envelope should *hide* the message: no party other than A should be able to learn m from the commitment (this is often referred in the literature as the hiding property). Second, the digital envelope should be *binding*, meaning with this that A cannot change her mind about m, and by checking the opening of the commitment one can verify that the obtained value is actually the one A had in mind originally (this is often referred as the binding property). These two properties make commitments very useful in a wide range of cryptographic applications such as zero-knowledge protocols, multi-party computation, digital auctions and electronic commerce.

A *commitment scheme* is a primitive to generate and open commitments. More precisely a commitment scheme is a two-phase interactive protocol between two probabilistic polynomial time algorithms sender and receiver. In a first stage (called the *commitment* phase) sender commits to a bit b using some appropriate function Com which takes as input b and some auxiliary value r and produces as output a value y. The value y is sent to receiver as a commitment on b. In the second stage (called the *decommitment* phase) sender "convinces" receiver that y is actually a valid commitment on b by revealing b and the auxiliary input r (if receiver is not convinced, it just outputs some special string). The requirements that we make on a commitment scheme are the following ones. First, if both sender and receiver behave honestly, then at the end of the decommitment phase receiver is convinced that sender had committed to bit b with probability 1. This is often referred as the *correctness* requirement. Second a cheating receiver cannot guess b with probability significantly better than $1/2$. This is the so-called *hiding* property. Finally, a cheating sender should be able to open a commitment (i.e., to decommit) with both b and $1-b$ only with very small (i.e., negligible) probability (this is the *binding* property). Each of the last two properties (i.e., hiding and binding) can be satisfied unconditionally or relatively to a computational assumption. In our context (i.e., where only two parties are involved) this immediately implies that one cannot hope to build a commitment scheme where both the hiding and the binding properties hold unconditionally. Unconditionally binding commitment schemes have been constructed under the sole assumption that one-way functions exist [4] and in such a construction an initial message of the receiver is required. It is known how to construct non-interactive unconditionally binding commitment schemes by using any one-way permutation [5]. Constant-round (actually 2-round) unconditionally hiding commitment schemes have been constructed under the assumption that collections of claw-free functions exist [5].

Since commitment schemes are very useful primitives they are often used as building blocks to construct larger protocols. In this sense it is often the case that the two basic requirements described above turn out to be insufficient. For this reason commitment schemes with additional properties have been proposed.

A *trapdoor commitment scheme* (sometimes also called *chameleon* commitment), is a commitment scheme with associated a pair of public and private keys (the latter also called the *trapdoor*). Knowledge of the trapdoor allows the sender to open the commitment in more than one way (this is often referred as the *equivocality* property). On the other hand, without knowledge of the trapdoor, equivocality remains computationally infeasible. When the commitments computed by means of a trapdoor are distributed exactly as real commitments then the trapdoor commitment scheme is unconditionally hiding. Instead, the equivocality property allows only computationally binding trapdoor commitment schemes.

1.1 Our Contributions

In this paper we introduce the notion of hybrid trapdoor commitment schemes. Informally an hybrid trapdoor commitment scheme is a general commitment primitive that allows for two commitment parameters generation algorithms HGen and HTGen. If the commitment parameters are obtained as the output of HGen then the resulting scheme is an unconditionally binding commitment scheme, while if the parameters are generated by HTGen the produced scheme is actually a trapdoor commitment scheme. Moreover, as for mixed commitments, no polynomially bounded adversary, taking as input only the (public) commitment parameters, should be able to tell the difference between keys generated from HGen and keys produced by HTGen.

Comparison with Mixed Commitments. Notice that the notion of hybrid trapdoor commitment may look very similar to that of mixed commitment introduced in [6]. There is a crucial difference however. Depending on the way the parameters are generated a mixed commitment can be either an extractable commitment or a trapdoor commitment. In our case, on the other hand, we require only that the commitment is either unconditionally binding or a trapdoor commitment scheme. As mentioned before, mixed commitments have been introduced to construct universally composable commitments and indeed Damgård and Nielsen proved that it is possible to construct a universally composable commitment from a mixed commitment where the number of E-keys (over the total number of keys) is negligible and that the number of X-keys (over the total number of keys) is only negligibly less than 1. Interestingly, a recent result by Damgård and Groth [7] shows that universally composable commitments imply key exchange and, when implemented in the shared random string model, they imply oblivious transfer. Therefore it seems unlikely that universally composable commitments (in the sense of [6]) can be implemented from one-way functions only. In this paper, on the other hand, we show that hybrid trapdoor commitments can be constructed from any one-way function.

Efficient Implementations. To improve on efficiency we then turn our attention to specific number-theoretic constructions and in particular we propose a very efficient implementations that relies on the Decisional Diffie-Hellmann assumption (other constructions based on Paillier's [8] Decisional Composite Residuosity Assumption can be found in the full version of this paper).

Stronger Extensions. As a second major contribution of this paper, we study some stronger extensions of hybrid trapdoor commitments. In particular we show how to build hybrid simulation-sound trapdoor commitments (hybrid SSTC, for short) and hybrid multi-trapdoor commitments from the sole assumption that one-way functions exist. Note that for the case of multi-trapdoor commitments their equivalence to one-way functions was not known. In this paper we show that multi-trapdoor commitment schemes are actually equivalent to digital signatures which are secure with respect to *generic* chosen message attack. Informally in a generic chosen message attack the adversary can obtain signatures only on a list of messages chosen *before* the public key of the signer is published. This is clearly a weaker notion with respect to the standard one where the adversary is allowed to choose the messages adaptively. Since SSTC's are actually equivalent to standard secure signatures, from a practical point of view, our result further clarifies why the known (practical) implementations of multi-trapdoor commitments are more efficient than the corresponding implementations of SSTC.

Applications. We use the different variants of hybrid trapdoor commitments for achieving the following applications. Using hybrid trapdoor commitments we show how to construct 3-round concurrent zero-knowledge proof systems, in the common reference string model, for all $\mathcal{NP}$ languages. We give a construction based on the existence of any one-way function and an efficient construction that is based on the DDH assumption. These results improves the computational soundness achieved in a previous result by Damgård [1] in the sense that ours are actually zero-knowledge proofs rather than zero-knowledge arguments. Using either hybrid SSTC or hybrid multi-trapdoor commitments we show how to construct an unbounded simulation-sound zero-knowledge proof system in the common reference string model. This improves the recent results of [3, 2] where similar results were presented for unbounded simulation-sound zero-knowledge arguments (rather than proofs).

Proofs vs Arguments. A proof system has the following property: any adversarial prover (regardless of his computing power) has negligible probability of making a honest verifier accept a false statement. This strong notion of soundness differs from the corresponding notion of soundness of an argument system, where security for honest verifiers holds only against polynomial-time adversarial provers. The notions of argument and proof differ dramatically when zero knowledge is considered. For example, while it is known that any $\mathcal{NP}$ language has a perfect zero-knowledge argument [9], if an $\mathcal{NP}$-complete language has a perfect zero-knowledge proof then the polynomial hierarchy collapses to its second level [10, 11]. With respect to constant-round zero knowledge, the current state of knowledge gives us a constant-round (computational) zero-knowledge proof for $\mathcal{NP}$ under the assumption that collections of claw-free functions exist [5], while constant-round zero-knowledge arguments for $\mathcal{NP}$ are known to exist under the assumption that one-way functions exist [12].

As discussed above, in this paper we show that our new notion of commitment scheme can be used to obtain some strong variants of zero-knowledge proof systems improving the current state-of-the art in which only arguments have been shown.

2 Definitions

We now give some basic definitions that we will use in this paper. We use the notation $\{\beta_1, \ldots, \beta_k : \alpha\}$ to specify the probability distribution of α after the sequential executions of events $\beta_1, \ldots, \beta_k$. In general, we assume that an algorithm $\mathcal{A}$ has access to some random (auxiliary) input even though this is not explicitly specified. Moreover, if A is a probabilistic algorithm we denote with $A(x)$ the random variable describing the output of A on input x. We say that a function ν is *negligible* iff for all constants c there exists n_0 such that for all $n > n_0$ it holds that $0 < \nu(n) < \frac{1}{n^c}$. A binary relation R is polynomially bounded if it is decidable in polynomial time and there exists a polynomial p such that for all pairs $(x, y) \in R$ it holds that $|y| \leq p(|x|)$. We denote by $L_R = \{x | \exists y : (x, y) \in R\}$ the $\mathcal{NP}$-language associated with R. For an $\mathcal{NP}$-language L we denote by R_L the *witness relation* associate with L defined as $x \in L \Leftrightarrow \exists y : (x, y) \in R_L$. We now give definitions for several notions of commitment schemes. For readability we will use "for all x" to mean any possible string x of length polynomial in the security parameter. We start with the standard notion of commitment scheme with its two main variants (i.e., unconditionally binding and unconditionally hiding). Note that all definitions will use a commitment generator function that outputs the commitment parameters. Therefore, such commitments have a straightforward implementation in the common reference string model where a trusted third party generates a reference string that is later received as common input by all parties. In some cases the commitment parameters generated by the commitment generator function will be strings with uniform distribution; in such cases the corresponding commitments can be implemented in the shared random string model which is a set-up assumption weaker than the common reference string model. For the sole sake of simplicity, in the following definitions, we consider the case in which the commitment parameters are used for computing a single commitment. However all the definitions can be extended so that the same commitment parameters can be used for any polynomial number of commitments (and actually all our results hold in this stronger setting).

Definition 1. (Gen, Com, Ver) *is a commitment scheme if:*

- **efficiency:** Gen, Com *and* Ver *are polynomial-time algorithms;*
- **completeness:** *for all v it holds that*

$$\text{Prob}\left(\text{crs} \leftarrow \text{Gen}(1^k); (\text{com}, \text{dec}) \leftarrow \text{Com}(\text{crs}, v) : \text{Ver}(\text{crs}, \text{com}, \text{dec}, v) = 1\right) = 1;$$

- **binding:** *there is a negligible function ν such that for any polynomial-time algorithm* sender *it holds that*

$$\text{Prob}\left(\text{crs} \leftarrow \text{Gen}(1^k); (\text{com}, v_0, v_1, \text{dec}_0, \text{dec}_1) \leftarrow \text{sender}(\text{crs}) : \right.$$
$$\text{Ver}(\text{crs}, \text{com}, \text{dec}_0, v_0) = \text{Ver}(\text{crs}, \text{com}, \text{dec}_1, v_1) = 1) \leq \nu(k);$$

- **hiding:** *for all* crs *generated with non-zero probability by* $\text{Gen}(1^k)$, *for all* v_0, v_1 *where* $|v_0| = |v_1|$ *the probability distributions:*

$$\{(\text{com}_0, \text{dec}_0) \leftarrow \text{Com}(\text{crs}, v_0) : \text{com}_0\} \quad \text{and} \quad \{(\text{com}_1, \text{dec}_1) \leftarrow \text{Com}(\text{crs}, v_1) : \text{com}_1\}$$

are computationally indistinguishable.

If the binding property holds with respect to a computationally unbounded algorithm sender, *the commitment scheme is said* unconditionally binding; *if instead, the hiding property holds with respect to a computationally unbounded algorithm* receiver, *the commitment scheme is said* unconditionally hiding.

We now give the definition of a trapdoor commitment scheme.

Definition 2. (Gen, Com, TCom, TDec, Ver) *is a* trapdoor *commitment scheme (TCS, for short) if* $\text{Gen}(1^k)$ *outputs a pair* (crs, aux), Gen_{crs} *is the related algorithm that restricts the output of* Gen *to the first element* crs, $(\text{Gen}_{\text{crs}}, \text{Com}, \text{Ver})$ *is a commitment scheme and* TCom *and* TDec *are polynomial-time algorithms such that:*
- **trapdoorness:** *for all v the probability distributions:*

$$\{(\text{crs}, \text{aux}) \leftarrow \text{Gen}(1^k); (\text{com}, \text{dec}) \leftarrow \text{Com}(\text{crs}, v) : (\text{crs}, \text{com}, \text{dec}, v)\} \quad \text{and}$$

$$\{(\text{crs}, \text{aux}) \leftarrow \text{Gen}(1^k); (\text{com}', \text{aux}_{\text{com}'}) \leftarrow \text{TCom}(\text{crs}, \text{aux}); \text{dec}' \leftarrow \text{TDec}(\text{aux}_{\text{com}'}, v') :$$
$$(\text{crs}, \text{com}', \text{dec}', v')\}$$

are computationally indistinguishable.

The definitions of commitment and trapdoor commitment schemes presented above can be extended by adding one more input to algorithms Com, Ver and TCom that is, a label referred to as "tag". In this case, algorithm Ver has an additional constraint, it outputs 1 only if the same tag has been used as input by algorithms Com or TCom. In particular, we will use such a tag-based definition of commitment when we will consider the notion of simulation-sound trapdoor commitment.

The definitions of multi-trapdoor commitment schemes and simulation-sound trapdoor commitment schemes are deferred to the full version of this paper.
Now we are ready to introduce the notion of hybrid trapdoor commitment. As sketched in the introduction, such a notion consider the existence of two commitment generation functions whose outputs are computationally indistinguishable. Still the properties of the two resulting commitment schemes are very different. We start with the basic notion of hybrid trapdoor commitment scheme.

Definition 3. $(\mathsf{HGen}, \mathsf{HTGen}, \mathsf{HCom}, \mathsf{HTCom}, \mathsf{HTDec}, \mathsf{HVer})$ *is an hybrid* trapdoor *commitment scheme (HTCS, for short) if:*

- **binding:** $(\mathsf{HGen}, \mathsf{HCom}, \mathsf{HVer})$ *is an unconditionally binding commitment scheme;*
- **trapdoorness:** $(\mathsf{HTGen}, \mathsf{HCom}, \mathsf{HTCom}, \mathsf{HTDec}, \mathsf{HVer})$ *is a* trapdoor *commitment scheme.*
- **hybridness:** *let* HTGen' *be an algorithm that restricts the output* $(\mathrm{crs}, \mathrm{aux})$ *of* $\mathsf{HTGen}(1^k)$ *to* crs, *then the following probability distribution are computationally indistinguishable:* $\{\mathrm{crs}_0 \leftarrow \mathsf{HGen}(1^k) : \mathrm{crs}_0\}$ *and* $\{\mathrm{crs}_1 \leftarrow \mathsf{HTGen}'(1^k) : \mathrm{crs}_1\}$.

The notion given above can be extended to be a tag-based commitment scheme (as for the case of standard trapdoor commitment schemes).

We now define the notions of hybrid multi-trapdoor and hybrid simulation-sound trapdoor commitment schemes. We stress that for the latter we focus on tag-based commitments (obtained by adding a label - the tag - as input to the algorithms that compute and verify commitments).

Intuitively, since multi-trapdoor and simulation-sound trapdoor commitment schemes define families of trapdoor commitment schemes, for the *hybrid* variant of such primitives, we require that each scheme in the family is an *hybrid* trapdoor commitment scheme.

Definition 4. $(\mathsf{HGen}, \mathsf{HTGen}, \mathsf{HSel}, \mathsf{HTkg}, \mathsf{HCom}, \mathsf{HTCom}, \mathsf{HTDec}, \mathsf{HVer})$ *is an hybrid* multi-trapdoor *commitment scheme (HMTCS, for short) if:*

- **multi trapdorness:** $(\mathsf{HTGen}, \mathsf{HSel}, \mathsf{HTkg}, \mathsf{HCom}, \mathsf{HTCom}, \mathsf{HTDec}, \mathsf{HVer})$ *is a multi-trapdoor commitment scheme;*
- **hybridness:** *let* $\mathsf{HGen}'(1^k)$ *(resp.,* HTGen'*) be an algorithm that outputs* pk *(resp.,* $(\mathrm{pk}, \mathrm{tk})$*) if and only if* $(\mathrm{crs}, \mathrm{aux})$ *(resp.,* $\mathrm{crs}', \mathrm{aux}'$*) is the output of* $\mathsf{HGen}(1^k)$ *(resp.,* $\mathsf{HTGen}(1^k)$*) and* pk *(resp.,* $(\mathrm{pk}, \mathrm{tk})$*) is the output of* $\mathsf{HSel}(\mathrm{crs})$ *(resp.,* $\mathsf{HSel}(\mathrm{crs}')$ *and* $\mathsf{HTkg}(\mathrm{aux}', \mathsf{HSel}(\mathrm{crs}'))$*); then it holds that the following tuple of algorithms* $(\mathsf{HGen}', \mathsf{HTGen}', \mathsf{HCom}, \mathsf{HTCom}, \mathsf{HTDec}, \mathsf{HVer})$ *is an hybrid trapdoor commitment scheme.*

Definition 5. $(\mathsf{HGen}, \mathsf{HTGen}, \mathsf{HCom}, \mathsf{HTCom}, \mathsf{HTDec}, \mathsf{HVer})$ *is an hybrid* simulation-sound *trapdoor commitment scheme (HSSTCS, for short) if:*

- **simulation soundness:** $(\mathsf{HTGen}, \mathsf{HCom}, \mathsf{HTCom}, \mathsf{HTDec}, \mathsf{HVer})$ *is a simulation-sound trapdoor commitment scheme;*
- **hybridness:** $(\mathsf{HGen}, \mathsf{HTGen}, \mathsf{HCom}, \mathsf{HTCom}, \mathsf{HTDec}, \mathsf{HVer})$ *is an hybrid trapdoor commitment scheme.*

3 Hybrid Trapdoor Commitments: Constructions

We now show that hybrid trapdoor commitment schemes exist under standard assumptions. In particular, for each definition, we show both a construction based on general primitives and a practical construction based on number-theoretic assumptions. The constructions and therefore the proofs that we give under complexity-based assumptions are modular, thus we only briefly discuss the efficient implementations. We start with a construction for an hybrid trapdoor commitment scheme. The main idea of the proof is the following. The algorithm that generates the reference string uses Naor's commitment scheme (which is based on the existence of one-way functions) to write in the reference string a commitment com of the string 0^k. The commitment computed by the prover is the first message a of the Σ-protocol for proving that com is a commitment of 1^k. Since this is a false statement, there exists only one challenge m and a third message z such that (a, m, z) is an accepting transcript. Therefore a is an unconditionally binding commitment of message m. The sender can compute a commitment a of m by running on input m the simulator of the honest-verifier zero knowledge property, and obtains the pair (a, z) as output. The algorithm that generates the fake reference string, instead, computes com as a commitment of 1^k. In this case, for each valid first message a, and any possible challenge m, it is always possible to compute z such that (a, m, z) is an accepting transcript.

Theorem 1. *Under the assumption that one-way functions exist, there exists an hybrid trapdoor commitment scheme.*

Next, we – constructively – show how to construct an efficient scheme based on the decisional Diffie-Hellman assumption. More details can be found in the full version of this paper, here we only describe the basic idea underlying our construction. The common reference string contains a quadruple (g, h, g^{r_1}, h^{r_2}) which is either a Diffie-Hellman quadruple or a random one. Consider a Σ protocol ∇ to prove equality of two discrete logarithms. We use ∇ to prove that $g_1 = g^{r_1}$ and $h_1 = h^{r_2}$ have the same discrete logarithm with respect to bases g and h, respectively. If the quadruple is a random one the instance for the Σ-protocol is false. Consequently a commitment to a message m can be computed only using the simulator of the Σ protocol for obtaining an accepting transcript (a, m, z) where a is the commitment key and (m, z) is the decommitment key. On the other hand when the quadruple in the shared random string is a Diffie-Hellman one, then knowledge of $r_1 = r_2$ (the trapdoor) allows to send a commitment key a that can later be opened as any possible message m. This is because, by running the prover algorithm of the Σ-protocol on input $r_1 = r_2$ (as witness) and m (as challenge), it is always possible to find a z such that (a, m, z) is an accepting transcript. The efficiency of this commitment scheme directly follows from the efficiency of the considered Σ-protocol. This informal discussion leads to the following theorem.

Theorem 2. *Under the assumption that the Decisional Diffie-Hellman problem is hard, there exists an efficient hybrid trapdoor commitment scheme.*

A construction for hybrid multi-trapdoor (resp., simulation-sound trapdoor) commitment schemes may seem, at first, much harder to achieve. After all, multi-trapdoor commitments need more parameters (with respect to basic trapdoor ones) and, to have an hybrid version of them, we need to make sure that these parameters remain distributed in a way such that it should be hard to say which of the two commitment generation algorithms was used to produce them.

Informally, we solve this problem by composing a multi-trapdoor (resp, simulation-sound trapdoor) commitment scheme with an hybrid trapdoor commitment scheme as the one described so far. The composition is made by considering the concatenation of *both* commitment parameters. Moreover all the operations made by the committing and decommitting algorithms are performed twice, once for each *subscheme*. Intuitively, using this technique, when a multi-trapdoor (resp., simulation-sound trapdoor) commitment scheme is composed with the hybrid trapdoor commitment scheme instantiated as a trapdoor commitment scheme, the resulting scheme is still a multi-trapdoor (resp., simulation-sound trapdoor) commitment scheme. On the other hand, if the hybrid trapdoor commitment scheme is instantiated as an unconditionally binding commitment scheme, then then resulting scheme is unconditionally binding. By the indistinguishability of the commitment parameters of the two instantiations we obtain the desired result.

Theorem 3. *Under the assumption that multi-trapdoor commitment scheme exist there exists an hybrid multi-trapdoor commitment scheme.*

Note that, with the theorem above, we show how to construct hybrid multi-trapdoor commitments from the hypothesis that multi-trapdoor commitments exist. It is quite natural then to ask if is it possible to base the existence of multi-trapdoor commitments on some weaker' assumption. Here we give a positive answer to this question and in particular we show that multi-trapdoor commitments exist if and only if secure signature against *generic* chosen message attack exist. Notice that one-way functions are equivalent to secure signatures [13] in the sense of [14], which, in turn, imply secure signature against *generic* chosen message attack exist. This means that theorem 3 can be restated as follows.

Theorem 4. *Under the assumption that one-way functions exist, there exists an hybrid multi-trapdoor commitment scheme.*

As for the case of hybrid trapdoor commitments we give an efficient implementation.

Theorem 5. *Under the assumption that the strong RSA and DDH problems are hard, (or under the assumption that the strong Diffie Hellman [15] and DDH problems are hard), there exists an efficient hybrid multi-trapdoor commitment scheme.*

Similar results can be proved for the case of hybrid Simulation Sound trapdoor commitments.

Theorem 6. *Under the assumption that one-way functions exist, there exists an hybrid simulation-sound trapdoor commitment scheme.*

Theorem 7. *Under the assumption that the DSA signature scheme is secure and the DDH problem is hard, (or under the assumption that the Cramer-Shoup signature scheme [16] is secure and the DDH problem is hard), there exists an efficient hybrid simulation-sound trapdoor commitment scheme.*

4 Hybrid Trapdoor Commitments: Applications

In this Section we describe some important applications of our primitive. In particular we show that hybrid trapdoor commitments can be used to construct interactive protocols that achieve strong notions of zero knowledge [17]. More precisely we improve the concurrent zero-knowledge arguments of [1] and both the simulation-sound and the left-concurrent non-malleable zero-knowledge arguments of [2,3] by showing how to achieve zero-knowledge *proofs* (rather than *arguments*). Therefore the security of our constructions holds even against computationally unbounded provers. Moreover, our zero-knowledge proofs can be based on the same complexity-theoretic assumptions used in [1,2,3]. The efficient constructions also require the hardness of the DDH problem.

CONCURRENT ZERO-KNOWLEDGE PROOFS. In [1], 3-round concurrent zero-knowledge arguments in the common reference string model are presented. More precisely Damgård [1] presents a general protocol based on the existence of one-way functions only and an efficient implementation based on number-theoretic assumptions. In this section we improve on this result by showing the existence of 3-round concurrent zero-knowledge proof (in contrast to argument) systems in the common reference string model. The first construction needs the sole assumption that one-way functions exist, while the second, more efficient, construction relies on the decisional Diffie-Hellman assumption. Interestingly the first construction holds in the shared random string model as well. In our construction we consider unbounded black-box zero-knowledge proofs with a non-rewinding simulator (which, consequently, is also concurrent zero knowledge).

Theorem 8. *If one-way functions exist, there exists a 3-round concurrent zero-knowledge proof system in the common reference string model for any $\mathcal{NP}$ language.*

We remark here that, going through the details of the proof of Theorem 8 one can easily verify that reference string used in the proof of Theorem 8 is uniformly distributed. Thus we have the following corollary.

Corollary 1. *If one-way functions exist, there exists a 3-round concurrent zero-knowledge proof system in the shared random string model for any $\mathcal{NP}$ language.*

Theorem 9. *Given an $\mathcal{NP}$-language L that admits an efficient Σ-protocol, then under the DDH assumption there exists a 3-round concurrent zero-knowledge proof system in the common reference string model for L.*

SIMULATION-SOUND ZERO KNOWLEDGE. The notion of *simulation soundness* has been used for the design of many secure cryptographic primitives (see for instance [18]). Informally, a proof system is simulation sound if an adversary that plays the role of verifier when the proofs are simulated for both true and false instances, is not able to play as a prover another session of the protocol in which he convinces an honest verifier of a false statement. In [3], MacKenzie and Yang proposed 3-round unbounded simulation-sound zero-knowledge argument systems in the common reference string model, in particular their arguments use simulation-sound trapdoor commitment schemes, therefore they obtain efficient argument systems based on the security of DSA [19] or the Cramer-Shoup [16] signature schemes and argument systems based on the existence of one-way functions. The multi-trapdoor commitments presented in [2] allow for more efficient constructions of unbounded simulation-sound zero-knowledge argument systems. In this section we extend their results by showing the existence of 3-round unbounded simulation-sound zero-knowledge proof (in contrast to argument) systems in the common reference string model. We can achieve this result either by using hybrid simulation-sound trapdoor commitments instead of non-hybrid simulation-sound trapdoor commitments in the construction of [3] or by using hybrid multi-trapdoor commitments instead of non-hybrid multi-trapdoor commitments in the construction of [2]. For each of this two results we give a first construction that needs the sole assumption that one-way functions exist. Then we give a more efficient second construction that requires (on top of the assumptions described for the efficient constructions of [3] and [2]) the decisional Diffie-Hellman assumption.

Theorem 10. *If one-way functions exist, there exists a 3-round unbounded simulation-sound zero-knowledge proof system in the common reference string model for any $\mathcal{NP}$ language.*

Theorem 11. *Given an $\mathcal{NP}$-language L that admits an efficient Σ-protocol, then under the assumption that DSA is a secure signature scheme and that the DDH assumption hold (or that the Cramer-Shoup signature scheme is secure and that the DDH assumption holds) there exists an efficient 3-round unbounded simulation-sound zero-knowledge proof system in the common reference string model for L.*

We stress that the approach used to achieve simulation-sound zero-knowledge proofs from hybrid simulation-sound trapdoor commitments (using the construction of [3]) is quite general and can be used to build simulation-sound zero-knowledge proofs from hybrid multi-trapdoor commitments (using the construction of [2]). In the proof of Theorem 10 we show that, by replacing a (non-hybrid) simulation-sound trapdoor commitment scheme with an hybrid one, an unbounded simulation-sound zero-knowledge argument can be transformed into an unbounded simulation-sound zero-knowledge proof. This same approach can be used for the case of (non-hybrid) multi-trapdoor commitment schemes and their application to unbounded simulation-sound zero-knowledge arguments [2] (we stress that in [2] such a notion is referred to as left-concurrent non-malleable zero-knowledge arguments).

Theorem 12. *Given an $\mathcal{NP}$-language L that admits an efficient Σ-protocol, if collision-resistant hash functions exist then under the sRSA and the DDH assumption or the sDH and the DDH assumptions there exists an efficient 3-round unbounded simulation-sound zero-knowledge proof system in the common reference string model for L.*

Acknowledgments

The authors would like to thank Pino Persiano for many useful discussions on zero knowledge proofs in the shared random string model.

The work described in this paper has been supported in part by the European Commission through the IST Programme under Contract IST-2002-507932 ECRYPT.

References

1. Damgard, I.: Efficient Concurrent Zero-Knowledge in the Auxiliary String Model. In: Advances in Cryptology – Eurocrypt '00. Volume 1807 of LNCS, Springer-Verlag (2000) 418–430
2. Gennaro, R.: Multi-trapdoor Commitments and Their Applications to Proofs of Knowledge Secure Under Concurrent Man-in-the-Middle Attacks. In: Advances in Cryptology – Crypto '04. Volume 3152 of LNCS, Springer-Verlag (2004) 220–236
3. MacKenzie, P., Yang, K.: On Simulation-Sound Trapdoor Commitments. In: Advances in Cryptology – Eurocrypt '04. Volume 3027 of LNCS, Springer-Verlag (2004) 382–400
4. Naor, M.: Bit Commitment Using Pseudorandomness. Journal of Cryptology **4** (1991) 151–158
5. Goldreich, O., Kahan, A.: How to Construct Constant-Round Zero-Knowledge Proof Systems for NP. Journal of Cryptology **9** (1996) 167–190
6. Damgård, I., Nielsen, J.B.: Perfect Hiding and Perfect Binding Universally Composable Commitment Schemes with Constant Expansion Factor. In: Advances in Cryptology - Crypto '02. Volume 2442 of LNCS, Springer-Verlag (2002) 581–596
7. Damgård, I., Groth, J.: Non interactive and reusable non-malleable commitments. In: 35th ACM Symposium on Theory of Computing, ACM (2003) 426–437
8. Paillier, P.: Public-key cryptosystems based on composite degree residuosity classes. In Stern, J., ed.: EUROCRYPT '99, Volume 1592 of LNCS, Springer-Verlag (1999) 223–238
9. Brassard, J., Chaum, D., Crepéau, C.: Minimum Disclosure Proofs of Knowledge. Journal of Computer and System Science **37** (1988) 156–189
10. Fortnow, L.: The Complexity of Perfect Zero-Knowledge. In: 19th ACM Symposium on Theory of Computing (STOC '87). (1987) 204–209
11. Boppana, R., Hastad, J., Zachos, S.: Does co-NP Have Short Interactive Proofs? Inf. Process. Lett. **25** (1987) 127–132

12. Bellare, M., Jakobsson, M., Yung, M.: Round-optimal zero-knowledge arguments based on any one-way function. In Fumy, W., ed.: Advances in Cryptology – Eurocrypt '97. Volume 1223 of LNCS, Springer-Verlag (1997) 280–305
13. Rompel, J.: One-Way Functions are Necessary and Sufficient for Digital Signatures. In: 22nd ACM Symposium on Theory of Computing (STOC '90). (1990) 12–19
14. Goldwasser, S., Micali, S., Rivest, R.: A digital signature scheme secure against adaptive chosen message attacks. In: SIAM J. on Computing. Volume 17-(2). (1988) 281–308
15. Boneh, D., Boyen, X.: Short Signatures without Random Oracles. In: Advances in Cryptology – Eurocrypt '04. Volume 3027 of LNCS, Springer-Verlag (2004) 56–73
16. Cramer, R., Shoup, V.: Signature Schemes Based on the Strong RSA Assumption. In: 6th ACM Conference on Computer and Communications Security (CCS '99), ACM (1999)
17. Goldwasser, S., Micali, S., Rackoff, C.: The Knowledge Complexity of Interactive Proof-Systems. SIAM J. on Computing **18** (1989) 186–208
18. Sahai, A.: Non-Malleable Non-Interactive Zero Knowledge and Adaptive Chosen-Ciphertext Security. In: 40th Symposium on Foundations of Computer Science, (FOCS '99), IEEE Computer Society Press (1999) 543–553
19. NIST: Digital Signature Standard (DSS). FIPS PUB 186 (1998)

On Steganographic Chosen Covertext Security

Nicholas Hopper

University of Minnesota, 4-192 EECS,
200 Union St SE, Minneapolis MN 55455
hopper@cs.umn.edu

Abstract. At TCC 2005, Backes and Cachin proposed a new and very strong notion of security for public key steganography: secrecy against adaptive chosen covertext attack (SS-CCA); and posed the question of whether SS-CCA security was achievable for *any* covertext channel. We resolve this question in the affirmative: SS-CCA security is possible for any channel that admits a secure stegosystem against the standard and weaker "chosen hiddentext attack" in the standard model of computation. Our construction requires a public-key encryption scheme with ciphertexts that remain indistinguishable from random bits under adaptive chosen-ciphertext attack. We show that a scheme with this property can be constructed under the Decisional Diffie-Hellman assumption. This encryption scheme, which modifies a scheme proposed by Kurosawa and Desmedt, also resolves an open question posed by von Ahn and Hopper at Eurocrypt 2004.

1 Introduction

Suppose that Alice and Bob are prisoners, and that their prison warden has foolishly allowed them to send "harmless messages" between their cells, so long as he may listen to everything they say. *Steganography* is the study of techniques that allow Alice and Bob to hide arbitrary messages – *hiddentexts* – in their apparently harmless communications (normally, *covertexts*) so that the warden cannot detect the presence of these messages. The case where the prisoners share a secret key has been studied extensively in both information-theoretically [5] and computationally secure settings [13, 9]. Several recent papers have also addressed the case in which one or both of the prisoners has a public key [1, 3, 17]. *In this paper, we are only concerned with the* **bare public key scenario**, *considered in [3], in which only Bob publishes a public key, and any prisoner can send hidden information to Bob.*

A recent paper by Backes and Cachin [3] considers the scenario where the warden may also inject messages into the channel between Alice and Bob, and observe Bob's reaction to these messages. Roughly, [3] gives a formal model of this scenario and defines a strong sense of security against this adversary: a stegosystem is said to be *steganographically secure against adaptive chosen covertext attacks* (SS-CCA) if, even in this case, the warden cannot tell whether Alice's messages contain hiddentexts. Analogously to the standard cryptographic notion of a chosen ciphertext attack, this seems to be the most general type of attack possible on a system for steganography.

Backes and Cachin leave open the problem of constructing a stegosystem satisfying **SS-CCA**, and instead address a relaxed notion of security, against adaptive *replayable* chosen-covertext attacks (**SS-PDR-CCA**). Roughly, in this notion, the warden is still allowed to inject messages into the channel between Alice and Bob, *except* that he is now restricted from sending messages which are, in some sense, *replays* of previous messages sent by Alice. Intuitively, two covertexts are replays of each other with respect to a public key if they decode to the same hiddentext. Backes and Cachin construct public-key stegosystems which satisfy **SS-PDR-CCA** under a variety of assumptions.

While it is an important advancement to limit the adversary to replay attacks, these attacks still constitute a serious threat against steganography. Imagine that Alice sends Bob some message which prompts an "unusual" reaction; in a replay attack, the warden can construct an apparently harmless covertext which corresponds to the same hiddentext as Alice's message, and send it to Bob. If Bob has the same "unusual" reaction, in response to a *different message*, it suggests to the warden that Alice's covertext contained a hidden message.

In this paper, we show how the previously known schemes fail in defending against replay attacks, and modify them to demonstrate the feasibility of the **SS-CCA** security condition, for any *efficiently sampleable* channel. This is a stronger assumption on the channel than in many previous works on steganography [1, 18, 9, 3], which assume only oracle access to the channel distribution. However, [14] shows that any channel which admits a secure stegosystem at all (in the standard model of computation) must be efficiently sampleable. Thus this construction serves as a demonstration that the **SS-CCA** notion is *feasible*, even though our particular construction may not always be practical to implement.

Our construction relies on the existence of public-key encryption schemes which are pseudorandom against chosen-ciphertext attack, a nonstandard security notion for encryption schemes. We also show that such encryption schemes exist, *without need of the random oracle assumption*,[1] under the Decisional Diffie-Hellman assumption. The existence of an encryption scheme satisfying this notion was an open question posed by von Ahn and Hopper [1].

Related Work. In addition to the work of Backes and Cachin [3], which we build on, Le and Kurosawa [17] and von Ahn and Hopper [1] have both proposed notions of security against "chosen stegotext attack." The notion proposed in [17] seems to be equivalent to **SS-CCA**; however the construction proposed there requires that the receiver know the sender's public key in order to decode. Similarly, the **SS-CSA** notion of [1] explicitly includes the public key of the sender; it can be thought of as an "attacker-specific" notion of security. However, the security model of [1] is also intended to prevent forgery by the warden, which is not a concern in the present model.

---

[1] We note that several constructions in the random oracle model are known [4, 19].

Both of these schemes require the sender to publish a public key. While this may not be a concern for ordinary communication, it is undesirable for steganography. This is because the aim of the sender in steganography is to *avoid suspicion* – yet publishing a public key for a stegosystem may be inherently suspicious.

On the other hand, it is frequently the case, as [1] argue, that the receiver of steganography need not avoid suspicion. This could be the case when, for example, the receiver is a newspaper or government agency wishing to receive whistle-blowing reports. Or when the receiver is a human-rights organization that would like to receive reports from its volunteers in the field. Thus it is important to have a construction which is secure in the bare public key model.

Other recent papers on foundations of steganography have focused on the private key setting. Cachin [5] formulated a model for steganography in an information-theoretic setting. Hopper *et al* [13] gave the first rigorous formulation of steganography with computational security, and demonstrated the feasibility of the notion with provably secure constructions. They also proposed the model of communication which subsequent work has followed. Independently, Katzenbeisser and Petitcolas [15] proposed a similar security condition. Dedić *et al* [9] address bounds on communication rate for a generic stegosystem. Lysyanskaya and Meyerovich [18] consider the possibility of an imperfect covertext oracle.

Anderson and Petitcolas [2] first proposed the possibility of public-key steganography and gave a heuristic construction. Craver [8] proposed a notion of public-key steganography with heuristic security against removal of the hiddentext. von Ahn and Hopper [1] were the first to formulate rigorous security definitions for the public-key case and demonstrate that public-key steganography was feasible.

Notation. A function $\mu : \mathbb{N} \to [0,1]$ is said to be *negligible* if for every $c > 0$, for all sufficiently large n, $\mu(n) < 1/n^c$. We denote the length (in bits) of a string or integer s by $|s|$. The concatenation of string s_1 and string s_2 will be denoted by $s_1 \| s_2$. The assignment $a \|_l b = c$ means that a is the first l bits of c and b is the remaining $|c| - l$ bits of c. We assume the existence of efficient, unambiguous *pairing* and *un-pairing* operations, so (s_1, s_2) is not the same as $s_1 \| s_2$.

We let U_k denote the uniform distribution on k bit strings. If V denotes an event in some probability space, we denote its complement by $\overline{\mathsf{V}}$. If $\mathcal{D}$ is a probability distribution with finite support X, we define the *minimum entropy* of $\mathcal{D}$, by $H_\infty(\mathcal{D}) = \min_{x \in X} \{\log_2(1/\Pr_\mathcal{D}[x])\}$. For a probability distribution $\mathcal{D}$, we denote by $x \leftarrow \mathcal{D}$ the action of drawing a sample x according to $\mathcal{D}$. We denote the statistical difference between distributions $\mathcal{D}$ and $\mathcal{E}$, with finite support X, by $\|\mathcal{D} - \mathcal{E}\| = \frac{1}{2} \sum_{x \in X} |\Pr_\mathcal{D}[x] - \Pr_\mathcal{E}[x]|$.

2 Pseudorandomness Against Chosen-Ciphertext Attack

We will need to construct a public-key encryption scheme which satisfies a non-standard security notion: indistinguishability from random bits under chosen-ciphertext attack. A scheme satisfying this notion is also non-malleable [10] and has pseudoranom ciphertexts [1]; the existence of a scheme simultaneously satisfying these latter notions *without random oracles* was an open question posed by von Ahn and Hopper at Eurocrypt 2004 [1].

Let $\mathcal{E}$ be a public-key encryption scheme with message expansion function ℓ. We define a chosen-ciphertext attack against $\mathcal{E}$ as a game played by an oracle adversary A:

1. $A^{D_{SK}}(PK)$ outputs *challenge message* $m^* \in \{0,1\}^{l^*}$.
2. A is given a *challenge ciphertext* c^*, where either $c \leftarrow E_{PK}(m^*)$ or $c \leftarrow U_{\ell(l^*)}$.
3. A continues to query D_{SK} subject to the restriction that A may not query $D_{SK}(c^*)$. A outputs a bit.

We define A's CCA advantage against $\mathcal{E}$ by

$$\mathbf{Adv}^{\mathsf{cca}}_{\mathcal{E},A}(k) = \left| \Pr[A^{D_{SK}}(PK, E_{PK}(m^*)) = 1] - \Pr[A^{D_{SK}}(PK, U_\ell) = 1] \right| ,$$

where $m^* \leftarrow A^{D_{SK}}(PK)$ and $(PK, SK) \leftarrow G(1^k)$, and define the CCA insecurity of $\mathcal{E}$ by $\mathbf{InSec}^{\mathsf{cca}}_{\mathcal{E}}(t, q, \mu, l^*, k) = \max_{A \in \mathcal{A}(t,q,\mu,l^*)} \{\mathbf{Adv}^{\mathsf{cca}}_{\mathcal{E},A}(k)\}$, where $\mathcal{A}(t, q, \mu, l^*)$ denotes the set of adversaries running in time t, that make q queries of total length μ, and issue a challenge message m^* of length l^*. Then $\mathcal{E}$ is $(t, q, \mu, l^*, k, \epsilon)$-*indistinguishable from random bits under chosen ciphertext attack* if $\mathbf{InSec}^{\mathsf{cca}}_{\mathcal{E}}(t, q, \mu, l^*, k) \leq \epsilon$. $\mathcal{E}$ is called *indistinguishable from random bits under chosen ciphertext attack* (IND$-CCA) if for every probabilistic polynomial time (PPT) A, $\mathbf{Adv}^{\mathsf{cca}}_{A,\mathcal{E}}(k)$ is negligible in k.

We show a simple modification of an encryption scheme of Kurosawa and Desmedt [16] (which itself is a modification of the original Cramer-Shoup encryption scheme [7]) which satisfies IND$-CCA. The main modification to the scheme is to use a dense encoding of the DDH subgroup and rejection sampling to produce uniform k-bit strings.

Setup. We let p_k, Q_k be large primes such that $p = 2Q + 1$ and $2^{k+1} > Q > 2^k$. We let $g \in \mathbb{Z}_p^*$ have order Q, and define the maps $lr : \langle g \rangle \to \mathbb{Z}_Q$, $qr : \mathbb{Z}_Q \to \langle g \rangle$ such that $lr(v) = v$ if $v \leq Q$ and $lr(v) = -v \bmod p$ otherwise; and $qr(u) = u$ if u is a quadratic residue modulo p and $qr(u) = p - u$ otherwise. Notice that $qr \circ lr$ is the identity map on the quadratic residues and $lr \circ qr$ is the identity map on $\mathbb{Z}_Q$. We assume the Decisional Diffie Hellman (DDH) assumption: for any PPT A, $\mathbf{Adv}^{\mathsf{ddh}}_{A,g,p,Q}(k) = |\Pr_{x,y \leftarrow \mathbb{Z}_Q}[A(g^x, g^y, g^{xy}) = 1] - \Pr_{x,y,z \leftarrow \mathbb{Z}_Q}[A(g^x, g^y, g^z) = 1]|$ is negligible.

We assume the existence of a family of target collision-resistant hash functions $H : \{0,1\}^{2k} \to \mathbb{Z}_Q$,[2] A universal family of hash functions $\Lambda : \mathbb{Z}_Q \to \{0,1\}^{2k'}$, an IND$-CPA symmetric-key encryption scheme E, D with k'-bit keys,[3] and a

---

[2] So for any PPT A, $\mathbf{Adv}^{\mathsf{tcr}}_{A,H}(k) = \Pr_{h \leftarrow H}[h(A(h(x))) = x : x \leftarrow \mathbb{Z}_Q]$ is negligible.
[3] So for any PPT A, $\mathbf{Adv}^{\mathsf{cpa}}_{A,E}(k) = |\Pr_{K \leftarrow U_{k'}}[A^{E_K}(1^{k'}) = 1] - \Pr[A^{U_{|\cdot|}}(1^{k'}) = 1]|$ is negligible.

pseudorandom function family $F : \{0,1\}^{k'} \times \{0,1\}^* \to \{0,1\}^{\tau}$.[4] Note that the existence of all of these primitives is implied by the DDH assumption.

Key Generation. Choose random $g_1, g_2 \in \langle g \rangle$, and choose random $x_1, x_2, y_1, y_2 \in \mathbb{Z}_Q$. Compute the group elements $c = g_1^{x_1} g_2^{x_2}, d = g_1^{y_1} g_2^{y_2}$. Choose hash functions H, Λ. The public key is $(g_1, g_2, c, d, H, \Lambda)$ and the private key is (x_1, x_2, y_1, y_2).

Encryption. Given a message $m \in \{0,1\}^*$, repeat the following steps:

- Choose $r \leftarrow \mathbb{Z}_q$.
- Compute $u_1 = lr(g_1^r)$, $u_2 = lr(g_2^r)$

Until u_1, u_2 are both at most 2^k. Then compute $\alpha = H(u_1 \| u_2)$, $v = c^r d^{r\alpha}$, $(K, \kappa) = \Lambda(v)$, $e = E_K(m)$, $T = F_\kappa(e)$. The ciphertext is $u_1 \| u_2 \| e \| T$.

Decryption. To decrypt the ciphertext $u_1 \| u_2 \| e \| T$, first compute $\alpha = H(u_1 \| u_2)$ and compute $v = qr(u_1)^{x_1 + y_1 \alpha} qr(u_2)^{x_2 + y_2 \alpha}$, $K \| \kappa = \Lambda(v)$. Test whether $T = F_\kappa(e)$; if not output $\bot$, otherwise output $D_K(e)$.

Theorem 1. *If $k \geq 4k'$, then*

$$\mathbf{InSec}_{\mathcal{E}}^{\mathsf{cca}}(t, q, \mu, l, k) \leq 8\mathbf{InSec}_H^{\mathsf{tcr}}(t, k) + 12\mathbf{InSec}_{g,p,Q}^{\mathsf{ddh}}(t) + 4\mathbf{InSec}_E^{\mathsf{cpa}}(t, 1, l, k')$$
$$+ (16q + 4)\mathbf{InSec}_F^{\mathsf{prf}}(t, q, \mu, k') + 8q(2^{-\tau} + 2^{-k'+1}) + 2^{-k'+4}$$

The security proof appears in the full version and closely follows the security proof for Kurosawa and Desmedt's scheme given by Gennaro and Shoup [11].

3 Definitions

Channels. We follow previous work [13, 17, 1, 9] in modeling the communication between two parties by a *channel*. We define a channel $\mathcal{C}$ as a family of probability distributions on documents from a set D, indexed by sequences $h \in D^*$. A channel implicitly defines an indexed distribution on sequences of ℓ documents — given index h, draw $d_1 \leftarrow \mathcal{C}_h$, $d_2 \leftarrow \mathcal{C}_{(h,d_1)}$, ..., $d_\ell \leftarrow \mathcal{C}_{(h,d_1,\ldots,d_{\ell-1})}$. We call the index h the *history* and label this distribution on sequences by $\mathcal{C}_h^\ell$. A history $h = (d_1, d_2, \ldots, d_\ell)$ is called *legal* (denoted $h \in \mathcal{H}$) if for all i, $\Pr_{\mathcal{C}_{(d_1,\ldots,d_{i-1})}}[d_i] > 0$. A channel is *always informative* if for every legal history h, $H_\infty(\mathcal{C}_h^\ell) = \Omega(\ell)$.

We will require that a channel be *efficiently sampleable*: there is an efficiently computable algorithm channel such that $\mathsf{channel}(h, U_k)$ and $\mathcal{C}_h$ are computationally indistinguishable.[5] This is in contrast to the models of [13, 9, 1, 3], where the

---

[4] So for any PPT A, $\mathbf{Adv}_{A,F}^{\mathsf{prf}}(k) = |\Pr_{K \leftarrow U_{k'}}[A^{F_K}(1^{k'}) = 1] - \Pr_{f:\{0,1\}^* \to \{0,1\}^\tau}[A^f(1^{k'}) = 1]|$ is negligible.

[5] Some examples of widely used channels satisfying this notion include: scientific simulations, cryptography and security protocols, computer games, financial modeling, weather forecasts, etc.

channel is assumed to be accessible only via a probabilistic oracle. While results in that model are in some sense more general, we refer the reader to [14] for a proof that in the standard model of computation, sampleability is necessary for secure steganography.

Since it is widely believed that all natural processes can be computed in probabilistic polynomial time [12], we do not *in theory* rule out steganography for any realistic channels by requiring the channel to be sampleable. On the other hand, it is conceivable that there are channels which we can currently sample physically but not computationally, and thus in practice it is still an open problem to design a stegosystem which is SS-CCA secure for such channels.

Public-Key Stegosystem. A public-key stegosystem $\mathcal{S}$ is a triple of probabilistic algorithms:

- $\mathcal{S}$.Generate (abbreviated SG) takes as input a security parameter 1^k and generates a key pair $(\rho, \sigma) \in \mathcal{PK} \times \mathcal{SK}$.
- $\mathcal{S}$.Encode (abbreviated SE) takes as input a public key $\rho \in \mathcal{PK}$, a string $m \in \{0,1\}^*$ (the *hiddentext*), and a channel history h. $SE(\rho, m, h)$ returns a sequence of documents $s_1, s_2, \ldots, s_l$ (the *stegotext*) from the support of $\mathcal{C}_h^l$.
- $\mathcal{S}$.Decode (abbreviated SD) takes as input a secret key $\sigma \in \mathcal{SK}$, a sequence of documents $s_1, s_2, \ldots, s_l$, and a channel history h. $SD(\sigma, s, h)$ returns a hiddentext $m \in \{0,1\}^*$.

We require that a stegosystem is *correct:* for every polynomial $p(k)$ there exists a negligible $\nu(k)$ such that for every $m \in \{0,1\}^{p(k)}$, legal history h, and $(\rho, \sigma) \in [SG(1^k)]$, $\Pr[SD(\sigma, SE(\rho, m, h), h) = m] \geq 1 - \nu(k)$.

Chosen-Covertext Attack. In an adaptive chosen-covertext attack against a public-key stegosystem $\mathcal{S}$, a challenger draws a key pair $(\rho, \sigma) \leftarrow SG(1^k)$, and an adversary W is given PK and allowed oracle access to SD_σ. The attacker produces a *challenge hiddentext* m^* and history h^* and is given as a response a sequence of documents $s^* \in D^{\ell(|m^*|)}$. After this, the attacker continues to query SD with the restriction that he may not query $SD(s^*)$. (As always, W may depend on the channel distribution $\mathcal{C}$) At the conclusion of the attack, W must guess whether $s^* \leftarrow SE(\rho, m^*, h^*)$ or $s^* \leftarrow \mathcal{C}_{h^*}^{\ell^*}$. We define the (steganographic) *Chosen-Covertext Advantage* of W against $\mathcal{S}$ with respect to $\mathcal{C}$ by

$$\mathbf{Adv}^{\text{scca}}_{\mathcal{S},\mathcal{C},W}(k) = \left| \Pr[W^{SD_\sigma}(PK, SE(\rho, m^*, h^*)) = 1] - \Pr[W^{SD_\sigma}(\rho, \mathcal{C}_{h^*}^{l^*}) = 1] \right| ,$$

where $(m^*, h^*) \leftarrow W^{SD_\sigma}(\rho)$ and $(\rho, \sigma) \leftarrow SG(1^k)$. We define the sCCA insecurity of $\mathcal{S}$ with respect to $\mathcal{C}$ by

$$\mathbf{InSec}^{\text{scca}}_{\mathcal{S},\mathcal{C}}(t, q, \mu, l^*, k) = \max_{W \in \mathcal{W}(t, q, \mu, l^*)} \left\{ \mathbf{Adv}^{\text{scca}}_{\mathcal{S},\mathcal{C},W}(k) \right\} ,$$

where $\mathcal{W}(t, q, \mu, l^*)$ denotes the class of all W running in time t which make at most q oracle queries of μ bits and submit a challenge hiddentext of length at most l^*.

We say that $\mathcal{S}$ is $(t, q, \mu, l, k, \epsilon)$ secure against chosen-covertext attack with respect to $\mathcal{C}$ if $\mathbf{InSec}^{\text{scca}}_{\mathcal{S},\mathcal{C}}(t, q, \mu, l, k) \leq \epsilon$, and that $\mathcal{S}$ is secure against chosen-covertext attack with respect to $\mathcal{C}$ (SS-CCA) if $\mathbf{Adv}^{\text{scca}}_{\mathcal{S},\mathcal{C},\mathcal{W}}(k)$ is negligible for all PPT W.

4 Previous Constructions

Both previously known constructions of (bare) public-key steganography [1,3] have a common structure. Let $\mathcal{F}$ denote a strongly universal family of hash functions $f : D \to \{0,1\}$. Let $f \leftarrow \mathcal{F}$ be chosen as part of a public key, or fixed as a "common reference string." Then both constructions use the routine shown in figure 1 to hide uniformly chosen bits in $\mathcal{C}$.

Proposition 1. *Let $\mathcal{C}$ be always informative and $f \leftarrow \mathcal{F}$. Then for any $h \in \mathcal{H}$,*

$$\|(f, \mathsf{sample}(h, U_1, k)) - (f, \mathcal{C}_h)\| \leq 2^{-H_\infty(\mathcal{C}_h)/2} .$$

The proposition is a direct consequence of the leftover hash lemma. If the channel is always-informative, sample can operate on samples from $\mathcal{C}_h^k$ and induce only a negligible statistical difference in its output distribution. The basic construction of a stegosystem, HashRS, is shown in figure 2, where (G, E, D) is a public-key cryptosystem which has pseudorandom ciphertexts.

Informally, the scheme works by transforming the hiddentext into a uniform-looking ciphertext $c = E_{PK}(m)$. The ciphertext bits are then used one at a time (or w at a time, with sampling costs and statistical difference increased by a factor of 2^w) to select covertexts that hash (via f) to the bits of the ciphertext,

Procedure sample:
Input: target $c \in \{0,1\}$, history h, bound k
Let $j = 0$
repeat:
  sample $s \leftarrow \mathcal{C}_h$, increment j
until $f(s) = c$ OR $(j > k)$
Output: s

Fig. 1. Sampling routine

Procedure Encode:
Input: $m \in \{0,1\}^l$, h, PK
Draw $c_1 \cdots c_\ell \leftarrow E(PK, m)$
for $i = 1 \ldots \ell$ do
  set $s_i = \mathsf{sample}(c_i, (h, s_{1,\ldots,i-1}), k)$.
Output: $s_1, s_2, \ldots, s_\ell$

Procedure Decode:
Input: $s_1, s_2, \ldots, s_l$, SK
for $i = 1 \ldots l$ do
  set $c_i = f(s_i)$
set $c = c_1 || c_2 || \cdots || c_l$.
Output: $D(K, c)$

Fig. 2. HashRS Stegosystem

using sample. Since the ciphertext looks uniform, the covertexts thus selected will be indistinguishable from samples from $\mathcal{C}_h$. Decoding applies f to each covertext document to recover the ciphertext c, and then decrypts this ciphertext using SK to compute the hiddentext $m = D_{SK}(c)$.

The Backes-Cachin construction instantiates HashRS with a public-key encryption scheme which satisfies two properties. First, it must be PDR-CCA secure, as defined by Canetti et al [6]. Second, the encryption scheme should have *pseudorandom* ciphertexts: given the public key it was encrypted under, a ciphertext should be computationally indistinguishable from a random string of the same length. When instantiated with a public-key cryptosystem satisfying these properties, we call the resulting stegosystem $\mathcal{BC}$.

Intuitively, the SS-PDR-CCA security of the $\mathcal{BC}$ scheme arises from the fact that W is disallowed from submitting covertexts that decode to the same hiddentext. Thus an attack W against $\mathcal{BC}$ can easily be turned into a PDR-CCA attack A against the underlying encryption scheme. The main technical step is in simulating decryption queries: whenever W queries the decoding oracle on a covertext $s = s_1, \ldots, s_\ell$, the PDR-CCA attacker computes a ciphertext $c = c_1, \ldots, c_\ell$ by setting $c_i = f(s_i)$. If the ciphertext c is a replay of the challenge ciphertext c^*, then the stegotext s is also a replay, so A responds to W with $\bot$. Otherwise A queries his decryption oracle at c and returns the result to W.

This standard simulation technique also hints at a CCA attack against the $\mathcal{BC}$ stegosystem. We now formally describe the attack W. On input PK, W uniformly picks a challenge message $m^* \leftarrow U_{l^*}$. On receiving the challenge covertext s^*, W computes c^* by setting $c_i^* = f(s_i^*)$. W computes a "replay" covertext $s' \leftarrow \mathsf{sample}((h^*, s^*), c^*, k)$. Finally, W queries the decryption oracle on s'. If $SD_{SK}(s') = m^*$, W outputs 1 and otherwise W outputs 0. It is obvious that when $s^* \in SE(PK, m^*, h^*)$, then we will have that $SD(SK, s', h^*) = m^*$ except when encoding fails, since otherwise unique decryption requires that $D_{SK}(c^*) = D_{SK}(E_{PK}(m^*)) = m^*$. On the other hand, when $s^* \leftarrow \mathcal{C}_h^\ell$, then m^* and s^* are chosen independently of each other, so $\Pr[D_{SK}(c^*) = m^*] \leq 2^{-l^*}$.

Proposition 2. *For every l^*, there exists a negligible function $\nu(k)$ such that*

$$\mathbf{Adv}^{\mathsf{scca}}_{W,\mathcal{BC}}(k) \geq 1 - 2^{-l^*} - \nu(k)$$

Note that the "replay" covertext will be indistinguishable from a sample from the channel, so the decoder would have no reason not to decode it and act on any information contained in the hiddentext. Thus this attack is reasonable, in that it could be applied in a realistic scenario, rather than being merely an artifact of the model. Of course the adversary might further attempt to replay the exact stegotext; this latter attack is, however, impossible to defeat.

5 Our Construction

Intuitively, the reason the attack in the previous section succeeds is that even though the underlying *ciphertext* is non-malleable, there are many possible en-

Procedure DEncode:
Input: bits $c_1, \ldots, c_l$, history h, bound k, randomness $r_1, \ldots r_{lk} \in \{0,1\}^k$
Let $\iota = 0$; for $i = 1 \ldots l$ do
    Let $j = 0$; repeat:
        compute $s_i = \text{channel}((h, s_{1\ldots i-1}), r_\iota)$; increment j, ι
    until $f(s_i) = c_i$ OR $(j > k)$
Output: $s_1, s_2, \ldots, s_l$

Fig. 3. Deterministic Encode

Procedure Encode:
Input: $m \in \{0,1\}^\ell$, h, PK
Choose $r \leftarrow U_k$
Let $c = E_{PK}(r\|m)$
Let $\mathbf{r} = G(r)$
Output: DEncode$(c, h, k, \mathbf{r})$

Procedure Decode:
Input: $s_1, \ldots, s_l$, h, SK
Let $c = f(s_1)\|\cdots\|f(s_l)$
Let $r\|_k m = D_{SK}(c)$.
Set $\mathbf{r} = G(r)$.
If $s \neq \text{DEncode}(c, h, k, \mathbf{r})$ return $\perp$.
Output: m

Fig. 4. SCCA Stegosystem

codings of the ciphertext. This observation immediately suggests a possible improvement: design a sampling method such that each ciphertext corresponds to exactly one stegotext. Indeed, the construction of [17] seems to have this property, but this construction inherently requires a shared secret between the encoder and the decoder. Likewise, the "attacker-specific" construction of [1] seems to achieve a similar property, but validity of a stegotext is determined by the sender's public key. Our construction modifies this latter approach to remove this dependence on the sender, and also removes the reliance on the random oracle model from that construction.

We make use of the fact that we have an efficiently sampleable channel $\mathcal{C}$, and will make use of the "deterministic encoding" routine shown in figure 3. This algorithm works in a similar manner to the HashRS.Encode algorithm, with the exception that the randomness for sampling is an explicit argument. Thus for a given sequence of lk random inputs, this routine has exactly one possible encoding for any message $c \in \{0,1\}^l$. Thus if an l-bit, non-malleable, ciphertext can determine the lk bits of sampling randomness to be used in its encoding, we can prevent replay attacks. One way to do this is to apply a random oracle to the randomness used in producing the ciphertext; this approach was used by [1]. We instead use a pseudorandom generator to expand an k-bit seed into an lk-bit sequence and then include this seed in the plaintext. Proving the security of this approach requires some additional care, because now it is conceivable that the sampling algorithm could leak information about the plaintext.

We now formally describe our construction. We will assume that $\mathcal{E}$ is a public-key IND\$-CCA secure encryption scheme, and $(PK, SK) \leftarrow \mathcal{E}.G(1^k)$. Furthermore, we assume that for any l, $\Pr[D_{SK}(U_l) \neq \perp] \leq \nu(k)$ for some negligible ν. Thus, valid ciphertexts, which do not decrypt to $\perp$, have negligible density. For convenience, we assume that for all m, $|E_{PK}(m)| = \ell(|m|)$, for some polynomial ℓ. We will also assume that $G : \{0,1\}^k \to \{0,1\}^{k \times lk}$ is a pseudorandom generator. The final scheme SCCA is shown in Figure 4.

Theorem 2. *Let $f \leftarrow \mathcal{F}$ and let $\epsilon = \max_{h \in \mathcal{H}} \left\{ 2^{-H_\infty(C_h^k)/2} \right\} = 2^{-\Omega(k)}$. Then*

$$\mathbf{InSec}^{\mathsf{scca}}_{\mathsf{SCCA},\mathcal{C}}(t,q,\mu,l,k) \leq \mathbf{InSec}^{\mathsf{cca}}_{\mathcal{E}}(t',q,\mu,l,k) + \nu(k) + \ell(l+k)\epsilon + \mathbf{InSec}^{\mathsf{prg}}_G(t',k),$$

where $t' \leq t + O(lk)$.

Proof. Choose an arbitrary $W \in \mathcal{W}(t,q,\mu,l)$; let $(PK,SK) \leftarrow G(1^k)$ and let $(m^*, h^*) \leftarrow W^{SD_{SK}}(PK)$. We will bound $\mathbf{Adv}^{\mathsf{scca}}_{W,\mathsf{SCCA},\mathcal{C}}(k)$ by considering the following sequence of hybrid distributions:

- D_1: $C_{h^*}^{\ell(l+k)}$
- D_2: $\mathsf{DEncode}(U_{\ell(l+k)}, h^*, k, U_{k \times lk})$
- D_3: $\mathsf{DEncode}(U_{\ell(l+k)}, h^*, k, G(U_k))$
- D_4: $\mathsf{DEncode}(E_{PK}(r\|m^*), h^*, k, G(r))$, where $r \leftarrow U_k$

Clearly D_4 perfectly simulates the stegotext distribution, and likewise D_1 perfectly simulates the covertext distribution. For convenience, we will define the quantity $\mathbf{Adv}^i_W(k) = |\Pr[W^{SD}(PK, D_{i+1}) = 1] - \Pr[W^{SD}(PK, D_i) = 1]|$, and note that

$$\mathbf{Adv}^{\mathsf{scca}}_{W,\mathsf{SCCA},\mathcal{C}}(k) = |\Pr[W^{SD}(PK, D_4) = 1] - \Pr[W^{SD}(PK, D_1) = 1]|$$
$$\leq \mathbf{Adv}^1_W(k) + \mathbf{Adv}^2_W(k) + \mathbf{Adv}^3_W(k).$$

Thus we proceed to bound $\mathbf{Adv}^i_W(k)$ for $i \in \{1,2,3\}$.

Lemma 1. $\mathbf{Adv}^1_W(k) \leq \ell(l+k)\epsilon$

Proof. This follows because $\|f(C_h) - U_1\| \leq \epsilon$, and no (nonuniform) efficient process can increase statistical distance.

Lemma 2. $\mathbf{Adv}^2_W(k) \leq \mathbf{InSec}^{\mathsf{prg}}_G(t',k)$

Proof. We will construct a PRG adversary A for G such that $\mathbf{Adv}^{\mathsf{prg}}_{A,G}(k) = \mathbf{Adv}^2_W(k)$. A works as follows: first, A picks a key pair $(PK, SK) \leftarrow G(1^k)$ to use in responding to the queries W makes to SD. A is given as input a string $r \in \{0,1\}^{k \times lk}$ and asked to decide whether $r \leftarrow U_{k \times lk}$ or $r \leftarrow G(U_k)$. Then A can achieve advantage precisely $\mathbf{Adv}^2_W(k)$ by emulating W, responding to its decoding queries using SK, and responding to the challenge hiddentext (m^*, h^*) by drawing $c \leftarrow U_{\ell(l+k)}$ and giving the response $s = \mathsf{DEncode}(c, h, k, r)$. If $r \leftarrow U_{k \times lk}$, then $s \leftarrow D_1$, and if $r \leftarrow G(U_k)$, then $s \leftarrow D_2$. Thus A's advantage in distinguishing $G(U_k)$ and $U_{k \times lk}$ is exactly:

$$\mathbf{Adv}^{\mathsf{prg}}_{A,G}(k) = |\Pr[A(G(U_k)) = 1] - \Pr[A(U_{k \times lk}) = 1]|$$
$$= |\Pr[W^{SD}(D_2) = 1] - \Pr[W^{SD}(D_1) = 1]|$$
$$= \mathbf{Adv}^2_W(k)$$

Lemma 3. $\mathbf{Adv}_W^3(k) \leq \mathbf{InSec}_{\mathcal{E}}^{\mathsf{cca}}(t', q, \boldsymbol{\mu}, k) + \nu(k)$

Proof. We will construct an adversary A that plays the chosen-ciphertext attack game against $\mathcal{E}$ with advantage $\mathbf{Adv}_{A,\mathcal{E}}^{\mathsf{cca}}(k) \geq \mathbf{Adv}_W^3(k)$.

A starts by emulating W to get a challenge hiddentext, responding to decoding queries as follows: on query $(s_1, \ldots, s_l, h)$, A computes $c = f(s_1) \| \cdots \| f(s_l)$; A then uses its decryption oracle to compute $r \|_k m = D_{SK}(c)$. If $c \neq \perp$ and $s = \mathsf{DEncode}(c, h, k, G(r))$, A returns m, otherwise A returns $\perp$.

When W generates challenge (m^*, h^*), A chooses $r^* \leftarrow U_k$ and outputs the challenge $r^* \| m^*$. A is given the challenge ciphertext c^* and returns $s^* = \mathsf{DEncode}(c^*, h^*, k, G(r^*))$ to W.

A continues to emulate W, responding to queries as before, except that on decoding query $(s_1, \ldots, s_l, h)$, A first checks whether $f(s_1) \| \cdots \| f(s_l) = c^*$; if so, A returns $\perp$ rather than querying $D_{SK}(c^*)$.

In other words, A simulates running SCCA.Decode with its D_{SK} oracle, except that because A is playing the IND$-CCA game, he is not allowed to query D_{SK} on the challenge value c^*: thus a decoding query that has the same underlying ciphertext c^* must be dealt with specially.

Notice that when A is given an encryption of $r^* \| m^*$, he perfectly simulates D_4 to W, so that $\Pr[A^{D_{SK}}(PK, E_{PK}(r^* \| m^*) = 1] = \Pr[W^{SD}(PK, D_4) = 1]$. This is because when $c^* = E_K(r^* \| m^*)$ then the test $s = \mathsf{DEncode}(c, h, k, G(r))$ would fail anyways. Likewise, when A is given a random string, he perfectly simulates D_3 to W, *given that c^* is not a valid ciphertext*. Let us denote the event that c^* is a valid ciphertext by V, and the event that a sample from D_3 encodes a valid ciphertext by U; notice that by construction $\Pr[\mathsf{U}] = \Pr[\mathsf{V}]$. We then have that

$$\Pr[A^D(PK, U_\ell) = 1] = \Pr[A^D(PK, U_\ell) = 1 | \overline{\mathsf{V}}] \Pr[\overline{\mathsf{V}}] + \Pr[A^D(PK, U_\ell) = 1 | \mathsf{V}] \Pr[\mathsf{V}]$$
$$\leq \Pr[W^{SD}(PK, D_3) = 1 | \overline{\mathsf{U}}] \Pr[\overline{\mathsf{U}}] + \Pr[\mathsf{V}]$$
$$\leq \Pr[W^{SD}(PK, D_3) = 1] + \Pr[\mathsf{V}]$$
$$\leq \Pr[W^{SD}(PK, D_3) = 1] + \nu(k) ,$$

since $\Pr[\mathsf{V}] \leq \nu(k)$ by assumption on $\mathcal{E}$. Combining the cases, we find that

$$\mathbf{Adv}_{A,\mathcal{E}}^{\mathsf{cca}}(k) = \Pr[A^{D_{SK}}(PK, E_{PK}(r^* \| m^*)) = 1] - \Pr[A^{D_{SK}}(PK, U_\ell) = 1]$$
$$= \Pr[W^{SD}(PK, D_4) = 1] - \Pr[A^{D_{SK}}(PK, U_\ell) = 1]$$
$$\geq \Pr[W^{SD}(PK, D_4) = 1] - \Pr[W^{SD}(PK, D_3) = 1] - \nu(k)$$
$$= \mathbf{Adv}_W^3(k) - \nu(k)$$

Remark. As described, the stegosystem SCCA requires the decoder to know the algorithm channel used by the encoder to sample from $\mathcal{C}$. This can be avoided by changing the encoder to append a canonical encoding of this algorithm to the hiddentext before encrypting; the decoder then recovers this algorithm before running the final DEncode check. Since the length of the algorithm is constant, the security bounds for the resulting scheme are essentially unchanged.

6 Conclusion and Open Problems

We have argued for the importance of a SS-CCA-secure stegosystem in the bare public key model, and given the first construction which meets this criterion. This resolves an open question posed by Backes and Cachin [3]. Furthermore, our construction relies on a public-key cryptosystem which is pseudorandom against chosen-ciphertext attack in the standard model. The existence of a cryptosystem satisfying this notion was an open problem posed by von Ahn and Hopper [1]. Because replay attacks are a realistic possibility, this represents an important advance over previous work.

One interesting direction for future work is to investigate the relationship between efficiently sampleable channels and the probabilistic channel oracle notion of earlier work. Designing a SS-CCA stegosystem in this setting seems to be a challenging problem. Another important notion of security against active attacks is *robustness* — the property that an attacker is unable to "remove" the hiddentext from a message. Hopper *et al* [13] define a weak notion of robustness and give a robust construction in the private key case. To our knowledge, there is no provably secure construction satisfying this definition in the public-key case. It is interesting to note that SS-CCA and robustness are inherently contradictory, since robustness *requires* that a replay attack is possible. Thus it is also an interesting question whether some notion of robustness with decoding oracles can be achieved, even in the private key case.

References

1. L. von Ahn and N. Hopper. Public-Key Steganography. In: *Advances in Cryptology – Proceedings of Eurocrypt '04*, 2004.
2. R. J. Anderson and F. A. P. Petitcolas. On The Limits of Steganography. *IEEE Journal of Selected Areas in Communications*, 16(4), pages 474-481, 1998.
3. M. Backes and C. Cachin. Public-Key Steganography with Active Attacks. In: *Proc. Second Theory of Cryptography Conference (TCC)*, 2005.
4. M. Bellare and P. Rogaway. Random Oracles are Practical. In: *Proc. First ACM Conference on Computer and Communications Security (CCS 1993)*, 1993.
5. C. Cachin. An Information-theoretic model of steganography. In: *Information Hiding, 2nd International Workshop*, pages 306-318, 1998.
6. R. Canetti, H. Krawczyk, and J. Nielsen. Relaxing chosen-ciphertext security. In: *Advances in Cryptology – CRYPTO 2003*, 2003.
7. R. Cramer and V. Shoup. A practical public-key cryptosystem provably secure against adaptive chosen ciphertext attack. *Advances in Cryptology: CRYPTO 98*, Springer LNCS 1462, pages 13-27, 1998.
8. S. Craver. On Public-key Steganography in the Presence of an Active Warden. *Proceedings of Second International Information Hiding Workshop*, Springer LNCS 1525, pages 355-368, 1998.
9. N. Dedić, G. Itkis, L. Reyzin and S. Russell. Upper and lower bounds on black-box steganography. In: *Proc. Second Theory of Cryptography Conference (TCC)*, 2005.
10. D. Dolev and C. Dwork and M. Naor. Nonmalleable Cryptography. *SIAM J. Computing*, 30(2), pages 391–437, 2000.

11. R. Gennarro and V. Shoup. A Note on an Encryption Scheme of Kurosawa and Desmedt. *IACR e-print archive report 2004/194*, 2004.
12. O. Goldreich. *Foundations of Cryptography: volume 1 – Basic Tools*. Cambridge University Press, 2001.
13. N. J. Hopper, J. Langford, and L. Von Ahn. Provably Secure Steganography. In: *Advances in Cryptology – CRYPTO 2002*, Springer LNCS 2442, pages 77-92, 2002.
14. N.J. Hopper. Toward a theory of steganography. Ph.D. Thesis, Carnegie Mellon University, July 2004. Available online: http://reports-archive.adm.cs.cmu.edu/anon/2004/abstracts/04-157.html
15. S. Katzenbeisser and F. A. P. Petitcolas. Defining Security in Steganographic Systems. In: *Proceedings of the SPIE vol. 4675, Security and Watermarking of Multimedia Contents IV*, pp. 50-56, 2002.
16. K. Kurosawa and Y. Desmedt. A New Paradigm of Hybrid Encryption Scheme. In: *Advances in Cryptology – Proceedings of CRYPTO '04*, 2004.
17. T. V. Le and K. Kurosawa. Efficient public key steganography secure against adaptive chosen stegotext attacks. *IACR e-print archive report 2003/244*, 2003.
18. A. Lysyanskaya and M. Meyerovich. Steganography with imperfect sampling. At: *CRYPTO 2004 Rump Session*, August 2004.
19. B. Möller. A Public-Key Encryption Scheme with Pseudorandom Ciphertexts. In: *Computer Security – ESORICS 2004*, 2004.

Classification of Boolean Functions of 6 Variables or Less with Respect to Some Cryptographic Properties*

An Braeken[1], Yuri Borissov[2], Svetla Nikova[1], and Bart Preneel[1]

[1] Department Electrical Engineering - ESAT/SCD/COSIC,
Katholieke Universiteit Leuven, Kasteelpark Arenberg 10,
B-3001 Leuven, Belgium
{an.braeken, svetla.nikova, bart.preneel}@esat.kuleuven.ac.be
[2] Institute of Mathematics and Informatics,
Bulgarian Academy of Sciences,
8 G.Bonchev, 1113 Sofia, Bulgaria
yborisov@moi.math.bas.bg

Abstract. This paper presents an efficient approach to the classification of the affine equivalence classes of cosets of the first order Reed-Muller code with respect to cryptographic properties such as correlation-immunity, resiliency and propagation characteristics. First, we apply the method to completely classify with this respect all the 48 classes into which the general affine group $AGL(2,5)$ partitions the cosets of $RM(1,5)$. Second, after distinguishing the 34 affine equivalence classes of cosets of $RM(1,6)$ in $RM(3,6)$ we perform the same classification for these classes.

1 Introduction

Many constructions of Boolean functions with properties relevant to cryptography are recursive. The efficiency of the constructions relies heavily on the use of appropriate functions of small dimensions. Another important method for construction is the random and heuristic search approach. As equivalence classes are used to provide restricted input of such optimization algorithms, it is very important to identify which equivalence classes obtain functions with desired properties.

In this paper, we present an efficient approach (based on some group-theoretical considerations) for the classification of affine equivalence classes of cosets of the first order Reed-Muller code with respect to cryptographic properties such as correlation-immunity, resiliency, propagation characteristics and

---

* The work described in this paper has been supported in part by the European Commission through the IST Programme under Contract IST-2002-507932 ECRYPT and by Concerted Research Action GOA Ambiorix 2005/11 of the Flemish Government. An Braeken is research assistant of the FWO.

their combinations. We apply this method to perform a complete classification of all the 48 orbits of affine equivalent cosets of $RM(1,5)$ (classified by Berlekamp and Welch [1] according to weight distributions), with respect to the above mentioned cryptographic properties. Partial results for this case on the existence and their number have already been mentioned in [3, 13, 14, 16]. In this paper, we study this problem into more detail and show in which classes these functions appear and how to enumerate them. The method also allows us, if necessary, to generate all the Boolean functions of 5 variables that possess good cryptographic properties. Our approach can also be extended for Boolean functions of higher dimension. As an illustration we apply it to the cubic functions of 6 variables using a proper classification of the cosets of $RM(1,6)$ in $RM(3,6)$.

The paper is organized as follows. In Sect. 2, we present some general background on Boolean functions. In Sect. 3, we describe our approach which will be used in Sect. 4 for a complete classification of the affine equivalence classes of the Boolean functions of 5 variables. In Sect. 5, we first show how to derive the $RM(3,6)/RM(1,6)$ equivalence classes together with their sizes. Using this information we classify them according to the most important cryptographic properties.

2 Background on Boolean Functions

A Boolean function f is a mapping from $\mathbb{F}_2^n$ into $\mathbb{F}_2$. It can be represented by a *truth table*, which is a vector of length 2^n consisting of its function values $(f(\overline{0}), \ldots, f(\overline{1}))$. Another way of representing a Boolean function is by means of its *algebraic normal form* (ANF):

$$f(\overline{x}) = \bigoplus_{(a_1,\ldots,a_n) \in \mathbb{F}_2^n} h(a_1,\ldots,a_n)\, x_1^{a_1} \ldots x_n^{a_n},$$

where f and h are functions on $\mathbb{F}_2^n$. The *algebraic degree* of f, denoted by $\deg(f)$, is defined as the highest number of variables in the term $x_1^{a_1} \ldots x_n^{a_n}$ in the ANF of f.

Two Boolean functions f_1 and f_2 on $\mathbb{F}_2^n$ are called *equivalent* if and only if

$$f_1(\overline{x}) = f_2(\overline{x}A \oplus \overline{a}) \oplus \overline{x}\overline{B}^t \oplus b, \quad \forall x \in \mathbb{F}_2^n, \tag{1}$$

where A is a nonsingular binary $n \times n$-matrix, b is a binary constant, and $\overline{a}, \overline{B}$ are n-dimensional binary vectors. If $\overline{B}, b$ are zero, the functions f_1 and f_2 are said to be *affine equivalent*. A property is called affine invariant if it is invariant under affine equivalence.

The study of properties of Boolean functions is related to the study of *Reed-Muller codes*. The codewords of the r-th order Reed-Muller code of length 2^n, denoted by $RM(r,n)$, are the truth tables of Boolean functions with degree less or equal to r. The number of codewords is equal to $2^{\sum_{i=0}^{r} \binom{n}{i}}$ and the minimum number of positions in which any two codewords $\overline{u}, \overline{v}$ differ (denoted by $d(\overline{u},\overline{v})$)

is 2^{n-r}. The *Hamming weight* of a vector $\overline{v}$ is denoted by $wt(\overline{v})$ and equals the number of non-zero positions, i.e. $wt(\overline{v}) = d(\overline{v}, \overline{0})$.

In 1972, Berlekamp and Welch classified all 2^{26} cosets of $RM(1,5)$ into 48 equivalence classes under the action of the general affine group $AGL(2,5)$ [1]. Moreover for each equivalence class the weight distribution and the number of cosets in that class has been determined.

Before describing the cryptographic properties that are investigated in this paper, we first mention two important tools in the study of Boolean functions f on $\mathbb{F}_2^n$. The *Walsh transform* of f is a real-valued function over $\mathbb{F}_2^n$ that can be defined as

$$W_f(\overline{\omega}) = \sum_{\overline{x} \in \mathbb{F}_2^n} (-1)^{f(\overline{x}) \oplus \overline{x} \cdot \overline{\omega}} = 2^n - 2wt(f \oplus \overline{x} \cdot \overline{\omega}), \qquad (2)$$

where $\overline{x} \cdot \overline{\omega} = \overline{x}\overline{\omega}^t = x_1\omega_1 \oplus x_2\omega_2 \oplus \cdots \oplus x_n\omega_n$ is the *dot product* of $\overline{x}$ and $\overline{\omega}$. The *nonlinearity* N_f of the function f is defined as the minimum distance between f and any affine function which can be expressed as $N_f = 2^{n-1} - \frac{1}{2} \max_{\overline{\omega} \in \mathbb{F}_2^n} |W_f(\overline{\omega})|$.

The *autocorrelation function* of f is a real-valued function over $\mathbb{F}_2^n$ that can be defined as

$$r_f(\overline{\omega}) = \sum_{\overline{x} \in \mathbb{F}_2^n} (-1)^{f(\overline{x}) \oplus f(\overline{x} \oplus \overline{\omega})}. \qquad (3)$$

For two equivalent functions f_1 and f_2 such that $f_1(\overline{x}) = f_2(\overline{x}A \oplus \overline{a}) \oplus \overline{x}\overline{B}^t \oplus b$, it holds that [15]:

$$W_{f_1}(\overline{\omega}) = (-1)^{\overline{a}A^{-1}\overline{\omega}^t + \overline{a}A^{-1}\overline{B}^t + b} W_{f_2}(((\overline{\omega} \oplus \overline{B})(A^{-1})^t) \qquad (4)$$

$$r_{f_1}(\overline{\omega}) = (-1)^{\overline{\omega}\overline{B}^t} r_{f_2}(\overline{\omega}A). \qquad (5)$$

A Boolean function is said to be *correlation-immune* of order t, denoted by $CI(t)$, if the output of the function is statistically independent of the combination of any t of its inputs. If the function is also *balanced* (equal number of zeros and ones in the truth table), then it is said to be *resilient* of order t, denoted by $R(t)$. These definitions of correlation-immunity and resiliency can be expressed by spectral characterization as given by Xiao and Massey [8].

Definition 1. *[8] A function $f(\overline{x})$ is $CI(t)$ if and only if its Walsh transform W_f satisfies $W_f(\overline{\omega}) = 0$, for $1 \leq wt(\overline{\omega}) \leq t$. If also $W_f(\overline{0}) = 0$, the function is called t-resilient.*

A Boolean function is said to satisfy the *propagation characteristics* of degree p, denoted by $PC(p)$ if the function $f(\overline{x}) \oplus f(\overline{x} \oplus \overline{\omega})$ is balanced for $1 \leq wt(\overline{\omega}) \leq p$. If the function $f(\overline{x}) \oplus f(\overline{x} \oplus \overline{\omega})$ is also t-resilient, the function f is called a $PC(p)$ function of order t. Or, by using the autocorrelation and Walsh spectrum, the definition can also be expressed as follows:

Definition 2. *[14] A function $f(\overline{x})$ is $PC(p)$ if and only if its autocorrelation transform r_f satisfies $r_f(\overline{w}) = 0$, for $1 \leq wt(\overline{w}) \leq p$. If also $W_{f(\overline{x}) \oplus f(\overline{x} \oplus \overline{w})}(\overline{a}) = 0$ for all $\overline{a}$ with $0 \leq wt(\overline{a}) \leq t$, the function f is said to satisfy $PC(p)$ of order t.*

If $r_f(\overline{w}) = \pm 2^n$, the vector $\overline{w}$ is called a *linear structure* of the function f. It is easy to prove that the set of linear structures forms a linear space [6].

We now present some known results which will be used in the rest of the paper. First of all, we start with mentioning several trade-offs between the above described properties of a Boolean function.

Theorem 1. *(Siegenthaler's Inequality [17]) If a function f on $\mathbb{F}_2^n$ is $CI(t)$, then $\deg(f) \leq n - t$. If f is t-resilient and $t \leq n - 2$, then $\deg(f) \leq n - t - 1$.*

Theorem 2. *[14] If a function f on $\mathbb{F}_2^n$ satisfies $PC(p)$ of order t with $0 \leq t < n - 2$, then $\deg(f) \leq n - t - 1$ for all p. If $t = n - 2$ then the degree of f is equal to 2.*

Theorem 3. *[20] If a function f on $\mathbb{F}_2^n$ is t-resilient and satisfies $PC(p)$, then $p + t \leq n - 1$. If $p + t = n - 1$, then $p = n - 1$, n is odd and $t = 0$.*

Another important result is the following divisibility theorem proven by Carlet and Sarkar [4].

Theorem 4. *If a coset of the $RM(1, n)$ with representative Boolean function f of degree d contains $CI(t)$ (resp. t-resilient) functions, then the weights of the functions in $f + RM(1, n)$ are divisible by*

$$2^{t + \lfloor \frac{n-t-1}{d} \rfloor} \quad (\text{resp. } 2^{t+1+\lfloor \frac{n-t-2}{d} \rfloor}). \tag{6}$$

From this Theorem together with Dickson's theorem on the canonical representations of quadratic Boolean functions [11], we derive a classification of correlation-immune (resp. resilient) quadratic functions in any dimension.

Proposition 1. *If the coset of $RM(1, n)$ with representative $x_1 x_2 \oplus x_3 x_4 \oplus \cdots \oplus x_{2h-1} x_{2h} \oplus \varepsilon$ where ε is an affine function of x_{2h+1} through x_n and $h \leq \lfloor \frac{n}{2} \rfloor$ given by Dickson's theorem contains $CI(t)$ (resp. t-resilient) functions then*

$$h \leq n - t - \left\lfloor \frac{n-t-1}{2} \right\rfloor - 1 \ (\text{resp. } h \leq n - t - \lfloor \tfrac{n-t-2}{2} \rfloor - 2).$$

Proof. The weight of the function equals (depending on the parameter h) [11]:

weight	$2^{n-1} - 2^{n-h-1}$	2^{n-1}	$2^{n-1} - 2^{n-h-1}$
number	2^{2h}	$2^{n+1} - 2^{2h+1}$	2^{2h}

The statement of the proposition follows from the divisibility theorem of Carlet and Sarkar applied on the weights. □

Remark 1. Using Proposition 1 together with the bound $h \leq \lfloor \frac{n}{2} \rfloor$, we obtain that the order of resiliency for quadratic functions is less or equal to $\lceil \frac{n}{2} \rceil - 1$, which was also stated in [18].

3   General Outline of Our Method

In this section we describe our main approach for the classification of equivalence classes (also called orbits) of cosets of the first order Reed-Muller code $RM(1,n)$ with respect to cryptographic properties such as correlation-immunity, resiliency, propagation characteristics and their combinations. For the sake of simplicity we shall refer to such a property as a C-property. For a given function f we denote by ZC_f the set of vectors which are mapped to zero by the transform corresponding to the considered C-property (e.g. Walsh transform for correlation-immunity and resiliency, autocorrelation for propagation characteristics) and call it a *zero-set* of f with respect to this C-property. We also refer to any set of n linearly independent vectors in $\mathbb{F}_2^n$ as a basis.

Our method employs the idea behind the "change of basis" construction as previously used by Maitra and Pasalic [12], and Clark et al. [5].

Let $\mathcal{R}$ be a representative coset of a given orbit $\mathcal{O}$ under the action of the general affine group $AGL(2,n)$. $\mathcal{R}$ is partitioned into subsets consisting of affine equivalent functions. Denote by $\mathcal{T}$ the family of these subsets. Let us fix one $T \in \mathcal{T}$ and a function $f \in T$.

From equations (4) and (5) and the definition of the corresponding C-property, it follows that for any function with this property, affine equivalent to f, a basis in ZC_f with certain properties exists. Conversely, for any proper basis in ZC_f and a constant from $\mathbb{F}_2^n$ we can apply an invertible affine transformation to f (derived by the basis and the constant) such that its image $\tilde{f}$ possess the C-property. Therefore the number N_f of functions affine equivalent to f and satisfying a certain C-property can be determined by counting bases in ZC_f. Moreover it can be seen that this number does not depend on the specific choice of f from T, since for two different functions f_1 and f_2 from T there exists one-to-one correspondence between the sets of their proper bases in the zero-sets. It is important to note that in case of Walsh transform we use the fact that vector $\overline{B}$ defined in previous section is $\overline{0}$.

In the following theorem we prove the formula that gives the number $\mathcal{N}_C$ of functions with C-property in the orbit $\mathcal{O}$.

Theorem 5. *Let $\mathcal{R}$ be a representative coset of a given orbit $\mathcal{O}$ under the action of the general affine group $AGL(2,n)$. Then the number $\mathcal{N}_C$ of functions with C-property in this orbit can be computed by the formula:*

$$\mathcal{N}_C = K_\mathcal{O} \sum_{f \in \mathcal{R}} B_f, \qquad (7)$$

where B_f is the number of proper bases in ZC_f and $K_\mathcal{O} = \frac{n!|\mathcal{O}|}{|GL(2,n)|}$.

Proof. We will find the number of functions with C-property in the orbit $\mathcal{O}$ by counting bases in zero-sets ZC_f. But this way we count each function $|S(f)| = S_f$ times, where $S(f)$ is the stabilizer subgroup of function f in $AGL(2,n)$. Therefore taking into account considerations preceding the theorem, the number

N_T of functions equivalent to the functions from T and satisfying the C-property is equal to

$$N_T = N_f = \frac{2^n n! B_f}{S_f}, \tag{8}$$

where B_f is the number of proper bases in ZC_f. The factor $n!$ appears since any arrangement of a given basis represents different function. Let $|\mathcal{O}|$ be the number of cosets in the orbit $\mathcal{O}$. Then substituting $S_f = \frac{|AGL(2,n)|}{|\mathcal{O}||T|}$ in (8) we get

$$N_T = \frac{2^n n! |\mathcal{O}| B_f |T|}{|AGL(2,n)|} = K_\mathcal{O} B_f |T|, \tag{9}$$

where $K_\mathcal{O} = \frac{n!|\mathcal{O}|}{|GL(2,n)|}$ and $GL(2,n)$ is the general linear group.

Therefore the number of all functions with C-property belonging to the orbit $\mathcal{O}$ is:

$$\sum_{T \in \mathcal{T}} N_T = K_\mathcal{O} \sum_{T \in \mathcal{T}} B_f |T| = K_\mathcal{O} \sum_{f \in \mathcal{R}} B_f. \tag{10}$$

□

In order to avoid difficulties when determining affine equivalent functions in $\mathcal{R}$ we prefer to use the last expression of (10). Thus, to compute the number $\mathcal{N}_C$ of functions with C-property in the orbit $\mathcal{O}$ we shall apply the following formula

$$\mathcal{N}_C = K_\mathcal{O} \sum_{f \in \mathcal{R}} B_f. \tag{11}$$

4 Boolean Functions of Less Than 5 Variables

For the study of functions in n variables with $n \leq 4$, we refer to [3] and [14]. In [3, Sect. 4.2], a formula is derived for the number of $(n-3)$-resilient functions and the number of balanced quadratic functions of n variables. In [14, Table 1], the number of quadratic functions that satisfy $PC(l)$ of order k with $k + l \leq n$ are determined for $n \leq 7$. Consequently, taking into account the trade-offs mentioned in Sect. 2, to cover all classes only the class with representative $x_1 x_2 x_3 \oplus x_1 x_4$ with $n = 4$ should be considered in relation with its propagation characteristics. It can be easily computed by exhaustive search that its size is 26 880 and that it contains 2 816 $PC(1)$ functions.

We now count the number of functions satisfying correlation-immunity, resiliency, propagation characteristics and their combinations in each of the 48 affine equivalence classes of $RM(1,5)$ by using the method explained in Sect. 3. Note that only the cosets with even weight need to be considered. Numerical results can be found in tables 1 through 5. In the tables, the functions are represented by means of an abbreviated notation (only the digits of the variables) and the sum should be considered modulo 2. We refer to the extended version of the paper concerning details about the computation.

Table 1. The Number of functions satisfying 1-CI, 1-Resilient, 1-PC, 1-PC with resiliency properties

Representative	$\mathcal{N}_{CI(1)}$	$\mathcal{N}_{R(1)}$	$\mathcal{N}_{PC(1)}$	$\mathcal{N}_{PC(1)\cap Bal}$	$\mathcal{N}_{PC(1)\cap CI(1)}$	$\mathcal{N}_{PC(1)\cap R(1)}$
2345	512	0	0	0	0	0
2345+12	28 160	0	163 840	71 680	0	0
2345+23	1 790	0	0	0	0	0
2345+23+45	14 336	0	0	0	0	0
2345+12+34	1 146 880	0	0	0	0	0
2345+123	6 400	0	0	0	0	0
2345+123+12	76 800	0	0	0	0	0
2345+123+24	17 280	0	645 120	201 600	0	0
2345+123+14	385 400	0	737 280	253 440	640	0
2345+123+45	102 400	0	1 904 640	714 240	0	0
2345+123+12+34	230 400	0	0	0	0	0
2345+123+14+35	122 880	0	11 550 720	2 887 680	0	0
2345+123+12+45	7 680	0	0	0	0	0
2345+123+24+35	0	0	3 440 640	430 080	0	0
2345+123+145	138 240	0	276 480	77 760	0	0
2345+123+145+45	27 648	0	0	0	0	0
2345+123+145+24+45	414 720	0	1 966 080	614 400	4 160	0
2345+123+145+35+24	6 144	0	2 654 208	497 664	384	0
123	16 640	11 520	0	0	0	0
123+45	0	0	1 310 720	0	0	0
123+14	216 000	133 984	94 720	65 120	10 560	5 280
123+14+25	69 120	24 960	1 582 080	791 040	19 200	0
123+145	0	0	0	0	0	0
123+145+23	1 029 120	537 600	0	0	0	0
123+145+24	0	0	0	0	0	0
123+145+23+24+35	233 472	96 960	0	0	0	0
12	4 840	4 120	2 560	2 240	1 120	840
12+34	896	0	46 592	23 296	896	0

Table 2. The Number of 2-CI functions

Representative	$\mathcal{N}_{CI(2)}$	$\mathcal{N}_{CI(2)\cap PC(1)}$
123+145+23+24+35	384	0
12	640	120

Table 3. The Number of functions satisfying $PC(1)$ of order 1 and 2

Representative	$\mathcal{N}_{PC(1)\,of\,ord\,1}$	$\mathcal{N}_{PC(1)\,of\,ord\,2}$
123+45	5 120	0
123+14	30 720	0
12	2 240	960
12+34	13 952	704

Table 4. The Number of functions satisfying $PC(2)$

Representative	$\mathcal{N}_{PC(2)}$	$\mathcal{N}_{PC(2)\cap Bal}$	$\mathcal{N}_{PC(2)\cap CI(1)}$	$\mathcal{N}_{PC(2)\,of\,ord\,1}$	$\mathcal{N}_{PC(2)\,of\,ord\,2}$
2345+123+145+35+24	12 288	2 304	384	0	0
123+14+25	199 680	99 840	3 840	0	0
12+34	28 672	23 296	896	1 792	64

Table 5. The Number of functions satisfying $PC(3)$ and $PC(4)$

Representative	$\mathcal{N}_{PC(3)}$	$\mathcal{N}_{PC(4)}$	$\mathcal{N}_{PC(3)\cap Bal}$	$\mathcal{N}_{PC(4)\cap Bal}$	$\mathcal{N}_{PC(3)\,of\,ord\,1}$	$\mathcal{N}_{PC(4)\,of\,ord\,1}$
12+34	10 752	1 792	5 376	896	1 792	64

5 Boolean Functions of 6 Variables and Degree 3

In this section first we show how to find the 34 affine equivalence classes of $RM(3,6)/RM(1,6)$, together with the orders of their size. Then we count in each class the number of resilient and PC functions.

5.1 Classification of $RM(3,6)/RM(1,6)$

Table 1 in [9] presents the number of affine equivalence classes of $RM(s,6)$ in $RM(r,6)$ with $-1 \leq s < r \leq 6$. In $RM(3,6)/RM(1,6)$ there are 34 equivalence classes. In order to classify the affine equivalence classes in $RM(3,6)/RM(1,6)$, we use the 6 representatives $f_i \oplus RM(2,6)$ for $1 \leq i \leq 6$ of the equivalence classes of $RM(3,6)/RM(2,6)$ as given in [10]: $f_1 = 0, f_2 = 123, f_3 = 123 + 245, f_4 = 123 + 456, f_5 = 123 + 245 + 346, f_6 = 123 + 145 + 246 + 356 + 456$. For each representative, we run through all functions consisting only of quadratic terms and distinguish the affine inequivalent cosets of $RM(1,6)$ by using the frequency distribution of absolute values of the Walsh and autocorrelation distribution as affine invariants. These indicators suffice to distinguish all 34 affine equivalence classes.

In order to employ the approach described in Sect. 3 we also need to know the sizes of these orbits. They were computed during the classification phase by multiplying the final results by the sizes of the corresponding orbits in $RM(3,6)/RM(2,6)$ given in [10]. To check these results in the cases of f_2, f_4 and f_6 we obtained linear systems for unknown sizes by taking into account the weight distributions of the cosets of $RM(1,6)$ and the weight distribution of the corresponding representative of $RM(3,6)/RM(2,6)$ to which these cosets belong. Of course if $f_1 = 0$ one can use also [11, Theorem 1 and Theorem 2, p.436]. The results obtained in these two ways coincide. We refer to Table 6 for the sizes of the orbits.

Remark 2. The 150 357 affine equivalence classes were classified for the first time by Maiorana [7]. They also are mentioned on the webpage maintained by

Table 6. The number of resilient and PC functions in the classes of $RM(3,6)/RM(1,6)$

	Representative	$\mathcal{N}_{R(1)}$	$\mathcal{N}_{R(2)}$	$\mathcal{N}_{PC(1)}(\times 128)$	$\mathcal{N}_{PC(2)}(\times 128)$	Number of Cosets
f_1	12	51 800	14 840	121	0	651
	14+23	569 696	0	13 440	4 900	18 228
	16+25+34	0	0	13 888	13 888	13 888
f_2	0	532 480	44 800	0	0	1 395 × 8
	14	19 914 720	826 560	17 240	0	1 395 × 392
	24+15	49 257 600	268 800	1 249 440	52 080	1 395 × 2 352
	16+25+34	0	0	1 874 880	1 874 880	1 395 × 1 344
	45	0	0	929 280	0	1 395 × 3 584
	123+16+45	0	0	18 744 320	1 881 600	1 395 × 25 088
f_3	0	0	0	0	0	54 684 × 32
	13	416 604 160	5 174 400	0	0	54 684 × 320
	14	0	0	0	0	54 684 × 480
	16	0	0	21 396 480	0	54 684 × 7 680
	26	0	0	33 152	0	54 684 × 32
	26+13	264 627 040	1 411 200	4 659 200	47 040	54 684 × 320
	26+14	0	0	14 058 240	1 411 200	54 684 × 480
	13+15+26+34	0	0	10 499 328	10 499 328	54 684 × 192
	34+16	0	0	0	0	54 684 × 23 040
	34+13+15	1 89807·10^{10}	82 897 920	1 250 304	0	54 684 × 192
f_4	0	0	0	0	0	357 120 × 64
	14	0	0	2 486 400	0	357 120 × 3 136
	15+24	0	0	572 315 · 10^{10}	0	357 120 × 64
	34+25+16	0	0	505 258 · 10^{10}	1 290 240	357 120 × 64
f_5	0	0	0	0	0	468 720 × 448
	12+13	0	0	3 609 586	0	468 720 × 18
	15	0	0	60 211 200	0	468 720 × 14 336
	12+13+25	3 287 027 200	8 601 600	0	0	468 720 × 2 222
	14+25	0	0	75 018 240	0	468 720 × 1 344
	35+26+25+12	0	0	6 719 569 920	6 719 569 920	468 720 × 14 336
	25+15+16	0	0	1 434 240	0	468 720 × 64
f_6	0	0	0	1 326 080	0	166 656 × 3 584
	12+13	0	0	7 956 480	0	166 656 × 21 504
	23+15+14	0	0	37 079 040	0	166 656 × 7 680

Fuller: http://www.isrc.qut.edu.au/people/fuller/ together with the degree, nonlinearity, maximum value in autocorrelation spectrum and truth tables of Boolean functions of dimension 6. Here we describe another approach for finding the 34 affine equivalence classes of functions of degree 3. One reason for this is that our method requires the sizes of the orbits, which are not given by Fuller.

5.2 Cryptographic Properties

In order to count the number of functions that satisfy certain cryptographic properties, the same approach as used for $n = 5$ is applied on these 34 classes of $RM(3,6)/RM(1,6)$. In Table 6, we present the classes together with the numbers of functions in these classes that satisfy t-resiliency with $t \leq 2$ and propagation characteristics of degree less or equal to 2. The last columns represents the sizes of the orbits.

By the Siegenthaler's inequality, 3-resilient functions should have degree less or equal to 2. Only the class with representative x_1x_2 contains 3-resilient functions and there are in total 1 120 3-resilient functions of dimension 6 (see also [3]).

For functions satisfying PC of higher degree, we have the following results. Besides the bent functions which are $PC(6)$, only the class with representative $x_1x_4 \oplus x_2x_3$ contains $PC(3)$ functions with a total of 128×420, as also computed in [14].

6 Conclusions

In this paper, we present a complete classification of the set of Boolean functions of 5 variables with respect to the most important cryptographic properties. Our method can also be applied to Boolean functions of dimension 6. As an example, we compute the 34 affine equivalence classes of $RM(3,6)/RM(1,6)$ and determine the number of resilient and PC functions belonging to each class. Moreover, we show a practical way to find the affine equivalence classes of Boolean functions. This method can be extended to dimension 7.

References

1. E. Berlekamp, L. Welch, Weight Distribution of the Cosets of the (32,6) Reed-Muller Code, *IEEE Transactions on Information Theory*, Vol. 18, pp. 203-207, 1972.
2. E. Brier, P. Langevin, Classification of Boolean Cubic Forms of Nine Variables, *2003 IEEE Information Theory Workshop (ITW 2003)*, IEEE Press, pp. 179-182, 2003.
3. P. Camion, C. Carlet, P. Charpin, N. Sendrier, On Correlation-Immune Functions, *Crypto 1991*, LNCS 576, Springer-Verlag, pp. 86-100, 1992.
4. C. Carlet, P. Sarkar, Spectral Domain Analysis of Correlation Immune and Resilient Boolean Functions, *Finite Fields and Applications*, Vol. 8 (1), pp. 120-130, 2002.
5. J. Clark, J.L. Jacob, S. Stepney, S. Maitra, W. Millan, Evolving Boolean Functions Satisfying Multiple Criteria, *Indocrypt 2002*, LNCS 2551, Springer-Verlag, pp. 246-259, 2002.
6. J. H. Evertse, Linear Structures in Block Ciphers, *Eurocrypt 87*, LNCS 304, Springer-Verlag, pp. 249266.
7. J. Maiorana, A Classification of the Cosets of the Reed-Muller Code $R(1,6)$, *Mathematics of Computation*, vol. 57, No. 195, July 1991, pp. 403-414.
8. X. Guo-Zhen, J. Massey, A Spectral Characterization of Correlation-Immune Combining Functions, IEEE Transactions on Information Theory, Vol. 34 (3), pp. 569-571, 1988.
9. X. -D. Hou, $AGL(m,2)$ Acting on $RM(r,m)/RM(s,m)$, *Journal of Algebra*, Vol. 171, pp. 921-938, 1995.

10. X. -D. Hou, $GL(m,2)$ Acting on $R(r,m)/R(r-1,m)$, *Discrete Mathematics*, Vol. 149, pp. 99-122, 1996.
11. F. J. MacWilliams, N. J. A. Sloane, The Theory of Error- Correcting Codes, North-Holland Publishing Company, 1977.
12. S. Maitra, E. Pasalic, Further Constructions of Resilient Boolean Functions with Very High Nonlinearity, *IEEE Transactions on Information Theory*, Vol. 48 (7), pp. 1825-1834, 2002.
13. E. Pasalic, T. Johansson, S. Maitra, P. Sarkar, New Constructions of Resilient and Correlation Immune Boolean Functions Achieving Upper Bounds on Nonlinearity, *Workshop on Coding and Cryptography 2001*, pp. 425-435, 2001.
14. B. Preneel, W. Van Leekwijck, L. Van Linden, R. Govaerts, J. Vandewalle, Propagation Characteristics of Boolean Functions, *Eurocrypt 1990*, LNCS 473, Springer-Verlag, pp. 161-173, 1990.
15. B. Preneel, Analysis and design of cryptographic hash functions, PhD. Thesis, Katholieke Universiteit Leuven, 1993.
16. P. Stanica, S.H. Sung, Boolean Functions with Five Controllable Cryptographic Properties, *Designs, Codes and Cryptography*, Vol. 31, pp. 147-157, 2004.
17. T. Siegenthaler, Correlation-Immunity of Non-linear Combining Functions for Cryptographic Applications, *IEEE Transactions on Information Theory*, Vol. 30 (5), pp. 776-780, 1984.
18. Y. Tarannikov, P. Korolev, A. Botev, Autocorrelation Coefficients and Correlation Immunity of Boolean Functions, *Asiacrypt 2001*, LNCS 2248, Springer-Verlag, pp. 460-479, 2001.
19. Y. Zheng, X. M. Zhang, GAC - the Criterion for Global Avalanche Characteristics of Cryptographic Functions, *Journal for Universal Computer Science*, Vol. 1 (5), pp. 316-333, 1995.
20. Y. Zheng, X. M. Zhang, On Relationship Among Avalanche, Nonlinearity, and Propagation Criteria, *Asiacrypt 2000*, LNCS 1976, Springer-Verlag, pp. 470-483, 2000.

Label-Guided Graph Exploration by a Finite Automaton

Reuven Cohen[1,*], Pierre Fraigniaud[2,**], David Ilcinkas[2,**], Amos Korman[1], and David Peleg[1]

[1] Dept. of Computer Science, Weizmann Institute, Israel
{r.cohen, amos.korman, david.peleg}@weizmann.ac.il
[2] CNRS, LRI, Université Paris-Sud, France
{pierre, ilcinkas}@lri.fr

Abstract. A finite automaton, simply referred to as a *robot*, has to explore a graph, i.e., visit all the nodes of the graph. The robot has no a priori knowledge of the topology of the graph or of its size. It is known that, for any k-state robot, there exists a $(k+1)$-node graph of maximum degree 3 that the robot cannot explore. This paper considers the effects of allowing the system designer to add short labels to the graph nodes in a preprocessing stage, and using these labels to guide the exploration by the robot. We describe an exploration algorithm that given appropriate 2-bit labels (in fact, only 3-valued labels) allows a robot to explore all graphs. Furthermore, we describe a suitable labeling algorithm for generating the required labels, in linear time. We also show how to modify our labeling scheme so that a robot can explore all graphs of bounded degree, given appropriate 1-bit labels. In other words, although there is no robot able to explore all graphs of maximum degree 3, there is a robot $\mathcal{R}$, and a way to color in black or white the nodes of any bounded-degree graph G, so that $\mathcal{R}$ can explore the colored graph G. Finally, we give impossibility results regarding graph exploration by a robot with no internal memory (*i.e.*, a single state automaton).

1 Introduction

Let $\mathcal{R}$ be a finite automaton, simply referred to in this context as a *robot*, moving in an unknown graph $G = (V, E)$. The robot has no a priori information about the topology of G and its size. To allow the robot $\mathcal{R}$, visiting a node u, to distinguish between its edges, the $d = \deg(u)$ edges incident to u are associated to d distinct *port numbers* in $\{0, \ldots, d-1\}$, in a one-to-one manner. The port numbering is given as part of the input graph, and the robot has no a priori information about it. For convenience of terminology, we henceforth refer to

---

[*] Supported by the Pacific Theaters Foundation.
[**] Supported by the project "PairAPair" of the ACI Masses de Données, the project "Fragile" of the ACI Sécurité et Informatique, and by the project "Grand Large" of INRIA.

"the edge incident to port number l at node u" simply as "edge l of u". (Clearly, if this edge connects u to v, then it may also be referred to as "edge l' of v" for the appropriate l'.) The robot has a transition function f, and a finite number of states. If $\mathcal{R}$ enters a node of degree d through port i in state s, then it switches to state s' and exits the node through port i', where $(s', i') = f(s, i, d)$. The objective of the robot is to *explore* the graph, i.e., to visit all its nodes.

The first known algorithm designed for graph exploration was introduced by Shannon [8]. Since then, several papers have been dedicated to the feasibility of graph exploration by a finite automaton. Rabin [6] conjectured that no finite automaton with a finite number of pebbles can explore all graphs (a *pebble* is a marker that can be dropped at and removed from nodes). The first step towards a formal proof of Rabin's conjecture is generally attributed to Budach [2], for a robot without pebbles. Blum and Kozen [1] improved Budach's result by proving that a robot with three pebbles cannot perform exploration of all graphs. Kozen [5] proved that a robot with four pebbles cannot explore all graphs. Finally, Rollik [7] gave a complete proof of Rabin's conjecture, showing that no robot with a finite number of pebbles can explore all graphs. The result holds even when restricted to planar 3-regular graphs. Without pebbles, it was proved [4] that a robot needs $\Theta(D \log \Delta)$ bits of memory for exploring all graphs of diameter D and maximum degree Δ. On the other hand, if the class of input graphs is restricted to trees, then exploration is possible even by a robot with no memory (i.e., zero states), simply by DFS using the transition function $f(i, d) = i + 1 \mod d$ (see, e.g., [3]).

The ability of dropping and removing pebbles at nodes can be viewed alternatively as the ability of the robot to dynamically *label* the nodes. If the robot is given k pebbles, then, at any time of the exploration, $\sum_{u \in V} |l_u| \leq k$ where l_u is the label of node u and $|l_u|$ denotes the size of the label in unary. This paper considers the effects of allowing the system designer to assign labels to the nodes in a preprocessing stage, and using these labels to guide the exploration by the robot. The transition function f is augmented to utilize labels as follows. If $\mathcal{R}$ in state s enters a node of degree d, labeled by l, through port i, then it switches to state s' and exits the node through port i', where

$$(s', i') = f(s, i, d, l).$$

This model can be considered stronger than Rabin's pebble model since labels are given in a preprocessing stage, but it can also be considered weaker since, once assigned to nodes, the labels cannot be modified.

In this paper, we consider settings where it is expected that the graph will be visited by many exploring robots, and consequently, the system designer would like to preprocess the graph by leaving (preferably small) road-signs, or *labels*, that will aid the robots in their exploration task. As possible scenarios one may consider a network system where finite automata are used for traversing the system and distributing information in a sequential manner.

More formally, we address the design of *exploration labeling schemes*. Such schemes consist of a pair $(\mathcal{L}, \mathcal{R})$ such that, given any graph G with any port

Table 1. Summary of main results

Label size (#bits)	Robot's memory (#bits)	Time (#edge-traversals)
2	$O(1)$	$O(m)$
1	$O(\log \Delta)$	$O(\Delta^{O(1)} m)$

numbering, the algorithm $\mathcal{L}$ labels the nodes of G, and the robot $\mathcal{R}$ explores G with the help of the labeling produced by $\mathcal{L}$. In particular, we are interested in exploration labeling schemes for which: (1) the preprocessing time required to label the nodes is polynomial, (2) the labels are short, and (3) the exploration is completed after a small number of edge-traversals.

As a consequence of Rollik's result, any exploration labeling scheme must use at least *two* different labels. Our main result states that just *three* labels (e.g., three colors) are sufficient for enabling a robot to explore all graphs. Moreover, we show that our labeling scheme gives to the robot the power to stop once exploration is completed, although, in the general setting of graph exploration, the robot is not required to stop once the exploration has been completed, i.e., once all nodes have been visited. In fact, we show that exploration is completed in time $O(m)$, i.e., after $O(m)$ edge traversals, in any m-edge graph.

For the class of bounded degree graphs, we design an exploration scheme using even smaller labels. More precisely, we show that just *two* labels (i.e., 1-bit labels) are sufficient for enabling a robot to explore all bounded degree graphs. The robot is however required to have a memory of size $O(\log \Delta)$ to explore all graphs of maximum degree Δ. The completion time $O(\Delta^{O(1)} m)$ of the exploration is larger than the one of our previous 2-bit labeling scheme, nevertheless it remains polynomial.

All these results are summarized in Table 1. The two mentioned labeling schemes require polynomial preprocessing time.

We also prove several impossibility results for 1-state robots, i.e., robots that are oblivious. The behavior of 1-state robots depends solely on the input port number, and on the degree and label of the current node. In particular, we prove that for any $d > 4$ and for any 1-state robot using at most $\lfloor \log d \rfloor - 2$ colors, there exists a simple graph of maximum degree d that cannot be explored by the robot. This lower bound on the number of colors needed for exploration can be increased exponentially to $d/2 - 1$ by allowing loops.

2 A 2-Bit Exploration-Labeling Scheme

In this section, we describe an exploration-labeling scheme using only 2-bit (actually, 3-valued) labels. More precisely, we prove the following.

Theorem 1. *There exists a robot with the property that for any graph G, it is possible to color the nodes of G with three colors (or alternatively, assign each node a 2-bit label) so that using the labeling, the robot can explore the entire* graph G, starting from any given node and terminating after identifying that the entire graph has been traversed. Moreover, the total number of edge-traversals by the robot is $\leq 20m$.

To prove Theorem 1, we first describe the labeling scheme $\mathcal{L}$ and then the exploration algorithm. The node labeling is in fact very simple; it uses three labels, called colors, and denoted WHITE, BLACK, and RED. Let D be the diameter of the graph.

Labeling $\mathcal{L}$. Pick an arbitrary node r. Node r is called the *root* of the labeling $\mathcal{L}$. Nodes at distance d from r, $0 \leq d \leq D$, are labeled WHITE if $d \bmod 3 = 0$, BLACK if $d \bmod 3 = 1$, and RED if $d \bmod 3 = 2$.

The neighbor set $\mathcal{N}(u)$ of each node u can be partitioned into three disjoint sets: (1) the set $\mathsf{pred}(u)$ of neighbors closer to r than u; (2) the set $\mathsf{succ}(u)$ of neighbors farther from r than u; (3) the set $\mathsf{sibling}(u)$ of neighbors at the same distance from r as u. We also identify the following two special subsets of neighbors:

- $\mathsf{parent}(u)$ is the node $v \in \mathsf{pred}(u)$ such that the edge $\{u,v\}$ has the smallest port number at u among all edges leading to a node in $\mathsf{pred}(u)$.
- $\mathsf{child}(u)$ is the set of nodes $v \in \mathsf{succ}(u)$ such that $\mathsf{parent}(v) = u$.

For the root, set $\mathsf{parent}(r) = \emptyset$. The exploration algorithm is partially based on the following observations.

1. For the root r, $\mathsf{child}(r) = \mathsf{succ}(r) = \mathcal{N}(r)$.
2. For every node u with label $\mathcal{L}(u)$, and for every neighbor $v \in \mathcal{N}(u)$, the label $\mathcal{L}(v)$ uniquely determines whether v belongs to $\mathsf{pred}(u)$, $\mathsf{succ}(u)$ or $\mathsf{sibling}(u)$.
3. Once at node u, a robot can identify $\mathsf{parent}(u)$ by visiting its neighbors successively, starting with the neighbor connected to port 0, then port 1, and so on. Indeed, by observation 2, the nodes in $\mathsf{pred}(u)$ can be identified by their label. The order in which the robot visits the neighbors ensures that $\mathsf{parent}(u)$ is the first visited node in $\mathsf{pred}(u)$.

Remark. The difficulty of graph exploration by a robot with a finite memory is that the robot entering some node u by port p, and aiming at exiting u by the same port p after having performed some local exploration around u, has not enough memory to store the value of p.

Exploration algorithm. Our exploration algorithm uses a procedure called Check_Edge. This procedure is specified as follows. When Check_Edge(j) is initiated at some node u, the robot starts visiting the neighbors of u one by one, and eventually returns to u reporting one of three possible outcomes: "child", "parent", or "false". These values have the following interpretation:

(i) if "child" is returned, then edge j at u leads to a child of u;
(ii) if "parent" is returned, then edge j at u leads to the parent of u;

(iii) if "false" is returned, then edge j at u leads to a node in $\mathcal{N}(u)\setminus(\mathsf{parent}(u)\cup \mathsf{child}(u))$.

The implementation of Procedure Check_Edge will be described later. Meanwhile, let us describe how the algorithm makes use of this procedure to perform exploration.

Assume that the robot $\mathcal{R}$ is initially at the root r of the 3-coloring $\mathcal{L}$ of the nodes. $\mathcal{R}$ leaves r by port number 0, in state DOWN. Note that, by the above observations, the node at the other endpoint of edge 0 of r is a child of r.

Assume that $\mathcal{R}$ enters a node u via port number i, in state DOWN. Assume u is of degree d; all arithmetic operations in the following description are modulo d. $\mathcal{R}$ aims at identifying a child of u if one exists, or to backtrack along edge i of u if none exists. To do so it executes Procedure Check_Edge(j) for every port number $j = i+1, i+2, \ldots$ until the procedure eventually returns "child" or "parent" for some port number j. $\mathcal{R}$ then sets its state to DOWN in the former case and UP in the latter, and leaves u by port j.

Assume that $\mathcal{R}$ enters a node u via port number i, in state UP. Assume u is of degree d; all arithmetic operations in the following description are modulo d. $\mathcal{R}$ aims at identifying a child of u with port number $j \in \{i+1,\ldots,p-1\}$ if one exists (where p is the port number of the edge leading to $\mathsf{parent}(u)$), or to carry on moving up to the parent of u if there is no such child. To do so, $\mathcal{R}$ executes Procedure Check_Edge(j) for every port number $j = i+1, i+2, \ldots$ until the procedure eventually returns "child" or "parent" for some port number j. $\mathcal{R}$ then sets its state to DOWN in the former case and UP in the latter, and leaves u by port j.

If the robot does not start from the root r of the labeling $\mathcal{L}$, then it first goes to r by using Procedure Check_Edge to identify the parent of every intermediate node, and by identifying r as the only node with $\mathsf{pred}(r) = \emptyset$.

Moreover, the robot can stop after the exploration has been completed. More precisely, this can be done by introducing a slight modification of the robot behavior when it enters a node u of degree d via port number d in state UP. In this case, $\mathcal{R}$ first check whether u has a parent. If yes, then it acts as previously stated ($\mathcal{R}$ does not need to store d since d is the node degree). If not, the robot terminates the exploration.

Procedure Check_Edge. We now describe the actions of the robot $\mathcal{R}$ when Procedure Check_Edge(j) is initiated at a node u. The objective of $\mathcal{R}$ is to set the value of the variable edge to one of {parent, child, false}. We denote by v the other endpoint of the edge e with port number j at u. First, $\mathcal{R}$ moves to v in state "check_edge", carrying with it the color of node u. Let i be the port number of edge e at v. There are three cases to be considered.

(a) $v \in \mathsf{sibling}(u)$: Then $\mathcal{R}$ backtracks through port i and reports "edge = false".
(b) $v \in \mathsf{pred}(u)$: Then $\mathcal{R}$ aims at checking whether v is the parent of u, that is, whether u is a child of v. For that purpose, $\mathcal{R}$ moves back to u, and proceeds

as follows: $\mathcal{R}$ successively visits edges $j-1, j-2, \ldots$ of u until either the other endpoint of the edge belongs to $\mathsf{pred}(u)$, or all edges $j-1, j-2, \ldots, 0$ have been visited. $\mathcal{R}$ then sets "edge=false" in the former case and "edge=parent" in the latter. At this point, let k be the port number at u of the last edge visited by $\mathcal{R}$. Then $\mathcal{R}$ successively visit edge $k+1, k+2, \cdots$ until the other endpoint belongs to $\mathsf{pred}(u)$. Then it moves back to u and reports the value of edge.

(c) $v \in \mathsf{succ}(u)$: Then $\mathcal{R}$ aims at checking whether u is the parent of v. For that purpose, $\mathcal{R}$ proceeds in a way similar to Case (b), i.e., it successively visits edges $i-1, i-2, \ldots$ of v until either the other endpoint of the edge belongs to $\mathsf{pred}(v)$, or all edges $i-1, i-2, \ldots, 0$ have been visited. $\mathcal{R}$ then sets its variable edge to "false" in the former case and to "child" in the latter. At this point of the exploration, let k denotes the port number of the last edge incident to v that $\mathcal{R}$ visited. Then $\mathcal{R}$ successively visits edges $k+1, k+2, \ldots$ until the other endpoint w of the edge belongs to $\mathsf{pred}(v)$. Then it moves to w, and reports the value of edge.

This completes the description of our exploration procedure.

Proof of Theorem 1. Clearly, labeling all nodes by $\mathcal{L}$ can be done in time linear in m, the number of edges of the graph. Obviously, two bits are enough to encode the label of each node. More specifically, using two bits for a color that is present on at most one third of the nodes, and one bit for the two other colors, we obtain a labeling with average label size $4/3$. It remains to prove the correctness of the exploration algorithm.

It is easy to check that if Procedure Check_Edge satisfies its specifications, then the robot $\mathcal{R}$ essentially performs a DFS traversal of the graph using edges $\{u, v\}$ where $u = \mathsf{parent}(v)$ or $u \in \mathsf{child}(v)$. Thus, we focus on the correctness of Procedure Check_Edge(j) initiated at node u. Let v be other endpoint of the edge e with port number j at u, and let i be the port number of edge e at v. We check separately the three cases considered in the description of the procedure. By the previous observations, comparing the color of the current node with the color of u allows $\mathcal{R}$ to distinguish between these cases.

If $v \in \mathsf{sibling}(u)$, then v is neither a parent nor a child of u, and thus reporting "false" is correct. Indeed, $\mathcal{R}$ then backtracks to u via port i, as specified in Case (a).

If $v \in \mathsf{pred}(u)$, then $v = \mathsf{parent}(u)$ iff for every neighbor w_k connected to u by an edge with port number $k \in \{j-1, j-2, \ldots, 0\}$, $w_k \notin \mathsf{pred}(u)$. The robot does check this property in Case (b) of the description, by returning to u, and visiting all the w_k's. Hence, Procedure Check_Edge performs correctly in this case.

Finally, if $v \in \mathsf{succ}(u)$, then $v = \mathsf{child}(u)$ iff for every neighbor z_l connected to v by an edge with port number $l \in \{i-1, i-2, \ldots, 0\}$, $z_l \notin \mathsf{pred}(v)$. In case (c), the robot does check this property by visiting all the z_l's. At this point, it remains for $\mathcal{R}$ to return to u (obviously, the port number leading from v to u cannot be stored in the robot memory since it has only a constant number of states). Let k be the port number of the last edge incident to v that $\mathcal{R}$ visited before setting its variable edge to "false" or "child". We have $0 \leq k \leq i-1$, $z_l \notin \mathsf{pred}(v)$ for all $l \in \{k+1,\ldots,i-1\}$, and $u \in \mathsf{pred}(v)$. Thus u is identified as the first neighbor that is met when visiting all v's neighbors by successively traversing edges $k+1, k+2, \ldots$ of v. This is precisely what $\mathcal{R}$ does according to the description of the procedure in Case (c). Hence, Procedure Check_Edge performs correctly in this case.

Hence Procedure Check_Edge performs correctly in all cases and so does the global exploration algorithm. It remains to compute the number of edge traversals performed by the robot during the exploration (including the several calls to Check_Edge).

We use again the same notations as in the description and the proof of Procedure Check_Edge. Let us consider the Procedure Check_Edge(j) initiated at node u. Let v be other endpoint of the edge e with port number j at u, and let i be the port number of edge e at v. First observe that during the execution of the Procedure Check_Edge only edges incident to u and v are traversed. More precisely:

Case (a): $v \in \mathsf{sibling}(u)$. Then edge $e = \{u, v\}$ is traversed twice and no other edges are traversed during this execution of Procedure Check_Edge.

Case (b): $v \in \mathsf{pred}(u)$. Then $\mathcal{R}$ traverses only edges incident to u. Let k be the greatest port number of the edges leading to a node in $\mathsf{pred}(u)$ and satisfying $k < j$. If it does not exist, set $k = 0$. $\mathcal{R}$ explores twice each edge $j, j-1, \ldots, k+1$ of u, then twice edge k, and finally again twice edges $k+1, \ldots, j-1, j$. To summarize, edge k of u is explored twice, and edges $k+1, \ldots, j-1, j$ of u are explored four times.

Case (c): $v \in \mathsf{succ}(u)$. Then $\mathcal{R}$ traverses only edges incident to v. Let k be the greatest port number of the edges leading to a node in $\mathsf{pred}(v)$ and satisfying $k < i$. If it does not exist, set $k = 0$. $\mathcal{R}$ explores once edge j of u, twice each edge $i-1, i-2, \ldots, k+1$ of v, twice edge k, twice again edges $k+1, \ldots, i-2, i-1$, and finally once edge i of v (i.e., j of u). To summarize, edge i of u and edge k of v are explored twice and edges $k+1, \ldots, i-2, i-1$ of v are explored four times.

We bound now the number of times each edge e of the graph is traversed. Edge $e = \{u, v\}$ is labeled i at u and j at v. Let us consider different cases:

(1) $e = \{u, v\}$ with $v = \mathsf{parent}(u)$. The edge e is in the spanning tree, and thus is explored twice outside any execution of the Procedure Check_Edge. During Procedure Check_Edge(j) at v, edge e is explored twice. e is also explored four times during Check_Edge(i) at u, except if $i = 0$ where e is only explored twice during Check_Edge(i) at u. If there exists an edge $\{u', u\}$ labeled i' at u and i'' at u' such that $i' < i$ and $u' \in \mathsf{pred}(u)$, then edge e is explored twice during Procedure Check_Edge(i') at u and twice again during Procedure Check_Edge(i'') at u'. If there exists an edge $\{v', v\}$ labeled j' at v and j'' at v' such that $j' < j$ and $v' \in \mathsf{pred}(v)$, then edge e is explored four times during Procedure Check_Edge(j') at v and four times again during

Procedure Check_Edge(j'') at v'. To summarize, edge e is explored at most 20 times during a DFS.

(2) $e = \{u, v\}$ with $v \in \mathsf{pred}(u)$ but $v \neq \mathsf{parent}(u)$. During Procedure Check_Edge(j) at v, edge e is explored twice. e is also explored four times during Check_Edge(i) at u. If there exists an edge $\{u', u\}$ labeled i' at u and i'' at u' such that $i' < i$ and $u' \in \mathsf{pred}(u)$, then edge e is explored twice during Procedure Check_Edge(i') at u and twice again during Procedure Check_Edge(i'') at u'. If there exists an edge $\{v', v\}$ labeled j' at v and j'' at v' such that $j' < j$ and $v' \in \mathsf{pred}(v)$, then edge e is explored four times during Procedure Check_Edge(j') at v and four times again during Procedure Check_Edge(j'') at v'. To summarize, edge e is explored at most 18 times during a DFS.

(3) $e = \{u, v\}$ with $v \in \mathsf{sibling}(u)$. During Procedure Check_Edge(j) at v, edge e is explored twice. e is also explored twice during Check_Edge(i) at u. If there exists an edge $\{u', u\}$ labeled i' at u and i'' at u' such that $i' < i$ and $u' \in \mathsf{pred}(u)$, then edge e is explored four times during Procedure Check_Edge(i') at u and four times again during Procedure Check_Edge(i'') at u'. If there exists an edge $\{v', v\}$ labeled j' at v and j'' at v' such that $j' < j$ and $v' \in \mathsf{pred}(v)$, then edge e is explored four times during Procedure Check_Edge(j') at v and four times again during Procedure Check_Edge(j'') at v'. To summarize, edge e is explored at most 20 times during a DFS.

Therefore, our exploration algorithm completes exploration in time $\leq 20|E|$ where $|E|$ is the number of edges in the graph G. □

3 A 1-Bit Exploration-Labeling Scheme for Bounded Degree Graphs

In this section, we describe an exploration labeling scheme using only 1-bit labels. This scheme requires a robot with $O(\log \Delta)$ bits of memory for the exploration of graphs of maximum degree Δ. More precisely, we prove the following.

Theorem 2. *There exists a robot with the property that for any graph G of degree bounded by a constant Δ, it is possible to color the nodes of G with two colors (or alternatively, assign each node a 1-bit label) so that using the labeling, the robot can explore the entire graph G, starting from any given node and terminating after identifying that the entire graph has been traversed. The robot has $O(\log \Delta)$ bits of memory, and the total number of edge-traversals by the robot is $O(\Delta^{O(1)} m)$.*

To prove Theorem 2, we first describe a 1-bit labeling scheme $\mathcal{L}'$ for $G = (V, E)$, i.e., a coloring of each node in black or white. Then, we will show how to perform exploration using $\mathcal{L}'$.

Labeling $\mathcal{L}'$. As for $\mathcal{L}$, pick an arbitrary node $r \in V$, called the *root*. Nodes at distance d from r are labeled as a function of d mod 8. Partition the nodes into eight *classes* by letting

$$C_i = \{u \in V \mid \text{dist}_G(r, u) \bmod 8 = i\}$$

for $0 \leq i \leq 7$. Node u is colored white if $u \in C_0 \cup C_2 \cup C_3 \cup C_4$, and black otherwise. Let

$$\tilde{C}_1 = \{u \mid \text{dist}_G(r, u) = 1\}$$
$$\widehat{C} = \{r\} \cup \{u \in C_2 \mid \text{dist}_G(r, u) = 2 \text{ and } \mathcal{N}(u) = \tilde{C}_1\}.$$

Lemma 1. *There is a local search procedure enabling a robot of $O(\log \Delta)$ bits of memory to decide whether a node u belongs to $\widehat{C}$ and to $\tilde{C}_1$, and to identify the class C_i of every node $u \notin \widehat{C}$.*

Proof. Let **B** (resp., **W**) be the set of black (resp., white) nodes which have all their neighbors black (resp., white). One can easily check that the class C_1 and the classes $C_3, \ldots, C_7$ can be redefined as follows:

- $u \in C_6 \Leftrightarrow u \in \mathbf{B}$ and there is a node in **W** at distance ≤ 3 from u;
- $u \in C_7 \Leftrightarrow u \notin C_6$, u has a neighbor in C_6, and there is no node in **W** at distance ≤ 2 from u;
- $u \in C_1 \Leftrightarrow u$ is black, u has no neighbor in **B**, and u has a white neighbor v that has no neighbor in **W**.
- $u \in C_5 \Leftrightarrow u$ is black, and $u \notin C_1 \cup C_6 \cup C_7$;
- $u \in C_3 \Leftrightarrow u \in \mathbf{W}$, and there is a node in C_1 at distance ≤ 2 from u;
- $u \in C_4 \Leftrightarrow u$ has a neighbor in **W**, and there is no node in C_1 at distance ≤ 2 from u.

Based on the above characterizations, the classes C_1 and $C_3, \ldots, C_7$ can be easily identified by a robot of $O(\log \Delta)$ bits, via performing a local search. Moreover, the sets $\tilde{C}_1$ and $\widehat{C}$ can also be characterized as follows:

- $u \in \tilde{C}_1 \Leftrightarrow u \in C_1$ and u has no node in C_7 at distance ≤ 2;
- $u \in \widehat{C} \Leftrightarrow N(u) \subseteq \tilde{C}_1$ and every node v at distance ≤ 2 from u satisfies $|N(v) \cap \tilde{C}_1| \leq |N(u)|$.

Using this we can deduce:

- $u \in C_0 \setminus \widehat{C} \Leftrightarrow u \notin (\cup_{i=3}^{7} C_i) \cup C_1$ and u has a neighbor in C_7;
- $u \in C_2 \setminus \widehat{C} \Leftrightarrow u \notin \widehat{C}$, has a neighbor in C_1, but has no neighbor in C_7.

It follows that a robot of $O(\log \Delta)$ bits can identify the class of every node except for nodes in $\widehat{C}$. □

Proof of Theorem 2. The exploration algorithm for $\mathcal{L}'$ follows the same strategy as the exploration algorithm for $\mathcal{L}$. Indeed, for $u \in C_i$, we have

$$\begin{aligned}\mathsf{pred}(u) &= \mathcal{N}(u) \cap C_{i-1} \pmod{8} \\ \mathsf{succ}(u) &= \mathcal{N}(u) \cap C_{i+1} \pmod{8} \\ \mathsf{sibling}(u) &= \mathcal{N}(u) \cap C_i\end{aligned}$$

Therefore, due to Lemma 1, all instructions of the exploration algorithm using labeling $\mathcal{L}$ can be executed using labeling $\mathcal{L}'$, but for the cases not captured in Lemma 1, i.e., $\widehat{C}$.

To solve the problem of identifying the root, we notice that each of the nodes in $\widehat{C}$ can be used as a root, and all the others can be considered as leaves in C_2. Thus, when leaving the root, the robot should memorize the port P by which it should return to the root. When the robot arrives at a node $u \in \tilde{C}_1$ through a tree edge and is in the UP state, it leaves immediately through port P and deletes the contents of P, then it goes down through the next unexplored port if one is left. When the robot is in a node $u \in \tilde{C}_1$ and in the DOWN state, it will skip the port P.

If the exploration begins at the root, then the above is sufficient. To handle explorations beginning at an arbitrary node, it is necessary to identify the root. Since every node in $\widehat{C}$ can be used as a root, it suffices to find one node of $\widehat{C}$ by going up and then start the exploration from it as described above. □

4 Impossibility Results

Theorem 3. *For any $d > 4$, and for any 1-state robot using at most $d/2 - 1$ colors, there exists a graph (with loops) with maximum degree d and at most $d+1$ vertices that cannot be explored by the robot.*

Proof. Fix $d > 4$, and assume for contradiction that there exists a 1-state robot exploring all graphs of degree d colored with at most $d/2 - 1$ colors. Recall that when a 1-state robot enters a node v by port i, it will leave v by port j where j is depending only on i, d and the color c of v. Thus for fixed d, each color corresponds to a mapping from entry ports to exit ports, namely, a function from $\{0, 1, \cdots, d-1\}$ to $\{0, 1, \cdots, d-1\}$. Partition the functions corresponding to the colors of nodes of degree d into surjective functions $f_1, f_2, \cdots, f_t$ and non-surjective functions $g_1, g_2, \cdots, g_r$. We have $0 < t + r \leq d/2 - 1$. Let c_i be the color corresponding to f_i, and c_{t+i} be the color corresponding to g_i. For each g_i, choose p_i to be some port number not in the range of g_i. Let $p_0 \in \{0, 1, \cdots, d-1\} \setminus \{p_1, p_2, \cdots, p_r\}$ (it is possible because $d - r \geq 1$).

We will construct a family $\{G_0, G_1, \cdots, G_t\}$ of graphs such that, for every $k \in \{0, 1, \cdots, t\}$:

1. G_k has exactly one degree-d vertex v (possibly with loops);
2. the other vertices of G_k are degree-1 neighbors of v;
3. all edges are either loops incident to v, or edges leading from v to some degree-1 node;
4. edges labeled $p_1, p_2, \cdots, p_r$ at v (if any, i.e., if $r > 0$) are not loops (and thus lead to degree-1 nodes);
5. the edge labeled p_0 leads to some degree-1 node, denoted by u_0;

6. there exists a set $X_k \subseteq \{0, 1, \cdots, d-1\}$ such that $\{p_0, p_1, \cdots, p_r\} \subseteq X_k$ and $d - |X_k| > 2(t - k)$, and for which, in G_k, edges with port number not in X_k lead to degree-1 vertices.

We will prove the following property for any $k = 0, \cdots, t$:

Property P_k. In G_k, if the color of v is in $\{c_1, \cdots, c_k\}$, then the robot, starting at $u_0 \in V(G_k)$, cannot explore G_k. More precisely any vertex attached to v by a port $\notin X$ is not visited by the robot.

We prove P_k by induction on k. Let G_0 be the star composed of one degree-d vertex v and d leaf vertices. Let $X_0 = \{p_0, p_1, p_2, \cdots, p_r\}$. Recall that $t + r \leq d/2 - 1$. Thus, $t \leq d/2 - 1$ and hence $2t + r + 1 \leq d - 1$. Therefore, we have $d - |X_0| = d - (r+1) > 2t$. P_0 is trivially true.

Let $k > 0$, and let G_{k-1} and X_{k-1} be respectively a graph and a set satisfying the induction property for $k - 1$. Assume first that v is colored by color c_k and that the robot starts its traversal at u_0. If the robot never visits vertices attached to v by ports not in X_{k-1} then the graph G_{k-1} and the set X_{k-1} satisfy P_k. I.e., $G_k = G_{k-1}$ and $X_k = X_{k-1}$. Otherwise, let p be the first port not in X_{k-1} that is visited by the robot at v, when starting at u_0. For a port $i \in \{0, 1, \cdots, d-1\}$, set $\mathsf{twin}(i) = j$ if there exists a port j and a loop labeled by i and j in G_{k-1}; Set $\mathsf{twin}(i) = i$ otherwise. Define a sequence of ports $(i_l)_{l \geq 1}$ as follows. Let i_1 be the port in X_{k-1} such that $f_k(i_1) = p$. For all $l \geq 2$, let i_l be the port such that $f_k(i_l) = \mathsf{twin}(i_{l-1})$. This sequence is well defined because f_k is surjective.

Observe that there exists some l such that $i_l \notin X_{k-1}$. Indeed, suppose, for the purpose of contradiction, that $i_l \in X_{k-1}$ for all l. Since X_{k-1} is finite, there exists some $i_l = i_{l+m}$ for $m \geq 1$. Let i_l be the first port repeated twice in this process. If $l > 1$, then we have $f_k(i_l) = \mathsf{twin}(i_{l-1})$ and $f_k(i_{l+m}) = \mathsf{twin}(i_{l+m-1})$. Therefore $\mathsf{twin}(i_{l-1}) = \mathsf{twin}(i_{l+m-1})$, yielding $i_{l-1} = i_{l+m-1}$ by bijectivity of f_k, which contradicts the minimality of l. If $l = 1$, then we have $i_1 = i_{1+m}$, therefore $i_m = p$, contradicting $i_j \in X_{k-1}$ for all j.

From the above, let h be the smallest index such that $i_h \notin X$. Let $q = i_h$. If $q = p$, then set $G_k = G_{k-1}$ and $X_k = X_{k-1} \cup \{p\}$. If $q \neq p$, then connect ports p and q to create a loop, denote the new graph G_k and let $X_k = X_{k-1} \cup \{p, q\}$.

In G_k, if v is colored by color c_k, then by the choice of p, starting at u_0, the robot enters and exits v through ports in X_{k-1} until it eventually exits v through port p. After that, the robot goes back to v by port q. Port q was chosen so that it causes the robot to continue entering v on ports $i_{h-1}, i_{h-2}, \cdots i_1$, after which the robot exits v through port p, locking the robot in a cycle. Since the ports of v occurring in this cycle are all from X_k, the robot does not visit any of the ports outside X_k, as claimed. By induction, we have $d - |X_{k-1}| > 2(t - (k-1))$. By the construction of X_k from X_{k-1}, we have $|X_k| \leq |X_{k-1}| + 2$. Therefore $d - |X_k| > 2(t - k)$, which completes the correctness of G_k and X_k.

If the color of v in G_k is in $\{c_1, \cdots, c_{k-1}\}$ then the robot is doomed to fail in exploring G_k. Indeed since starting at u_0 in G_{k-1} the robot does not traverse any of the vertices corresponding to ports not in X_{k-1}, then in G_k too, the robot does not traverse any of the vertices corresponding to ports not in $X_k \supseteq X_{k-1}$, and thus fails to explore G_k because $d - |X_k| \geq 1$. This completes the proof of P_k and thus the induction.

In particular, G_t is not explored by the robot if the node v is colored with a color in $c_1, c_2, \cdots, c_t$. If v is colored c_{t+i} with $1 \leq i \leq r$, then assume that the robot starts the traversal at vertex u_0. Since the edge labeled p_i leads to a degree-1 vertex in G_t, this vertex will never be visited by the robot, by definition of p_i. Therefore the graph G_t cannot be explored by the robot. □

The theorem above makes use of graphs with loops. For graphs without loops we have the following theorem.

Theorem 4. *For any $d > 4$ and for any 1-state robot using at most $\lfloor \log d \rfloor - 2$ colors, there exists a graph of maximum degree d, without loops, that cannot be explored by the robot.*

5 Further Investigations

It was known that there is no 0-bit exploration-labeling scheme, even for bounded degree graphs. We proved that there is a 2-bit exploration-labeling scheme for arbitrary graphs, and that there is a 1-bit exploration-labeling scheme for bounded degree graphs. It remains open whether or not there exists a 1-bit exploration-labeling scheme for arbitrary graphs.

References

1. M. Blum and D. Kozen. On the power of the compass (or, why mazes are easier to search than graphs). In 19th Symposium on Foundations of Computer Science (FOCS), pages 132-142, 1978.
2. L. Budach. Automata and labyrinths. Math. Nachrichten, pages 195-282, 1978.
3. K. Diks, P. Fraigniaud, E. Kranakis, and A. Pelc. Tree Exploration with Little Memory. In 13th Annual ACM-SIAM Symp. on Discrete Algorithms (SODA), pages 588-597, 2002.
4. P. Fraigniaud, D. Ilcinkas, A. Pelc, G. Peer and D. Peleg. Graph Exploration by a Finite Automaton. In Proc. 29th Int. Symp. on Mathematical Foundations of Computer Science (MFCS), LNCS 3153, 451-462, 2004.
5. D. Kozen. Automata and planar graphs. In Fund. Computat. Theory (FCT), 243-254, 1979. Fundamentals of Computation Theory (FCT), pages 243-254, 1979.
6. M.O. Rabin, Maze threading automata. Seminar talk presented at the University of California at Berkeley, October 1967.
7. H.A. Rollik. Automaten in planaren Graphen. Acta Informatica 13:287-298, 1980 (also in LNCS 67, pages 266-275, 1979).
8. C. E. Shannon. Presentation of a maze-solving machine. In 8th Conf. of the Josiah Macy Jr. Found. (Cybernetics), pages 173-180, 1951.

On the Wake-Up Problem in Radio Networks

Bogdan S. Chlebus[1,*], Leszek Gąsieniec[2],
Dariusz R. Kowalski[2,3,**], and Tomasz Radzik[4]

[1] Department of Computer Science and Eng.,
UCDHSC, Denver, CO 80217, USA
[2] Department of Computer Science,
University of Liverpool, Liverpool L69 7ZF, UK
[3] Instytut Informatyki, Uniwersytet Warszawski,
Banacha 2, Warszawa, Poland
[4] Department of Computer Science,
King's College London, London WC2R 2LS, UK

Abstract. Radio networks model wireless communication when processing units communicate using one wave frequency. This is captured by the property that multiple messages arriving simultaneously to a node interfere with one another and none of them can be read reliably. We present improved solutions to the problem of waking up such a network. This requires activating all nodes in a scenario when some nodes start to be active spontaneously, while every sleeping node needs to be awaken by receiving successfully a message from a neighbor. Our contributions concern the existence and efficient construction of universal radio synchronizers, which are combinatorial structures introduced in [6] as building blocks of efficient wake-up algorithms. First we show by counting that there are (n,g)-universal synchronizers for $g(k) = \mathcal{O}(k \log k \log n)$. Next we show an explicit construction of (n,g)-universal-synchronizers for $g(k) = \mathcal{O}(k^2 \operatorname{polylog} n)$. By way of applications, we obtain an existential wake-up algorithm which works in time $\mathcal{O}(n \log^2 n)$ and an explicitly instantiated algorithm that works in time $\mathcal{O}(n \Delta \operatorname{polylog} n)$, where n is the number of nodes and Δ is the maximum in-degree in the network. Algorithms for leader-election and synchronization can be developed on top of wake-up ones, as shown in [7], such that they work in time slower by a factor of $\mathcal{O}(\log n)$ than the underlying wake-up ones.

1 Introduction

Radio networks model mobile wireless communication when processing units communicate using one wave frequency. We consider networks that are synchronous in the sense that there is a global time measured in rounds and local

---

[*] The work of this author is supported by the NSF Grant 0310503.
[**] Part of this work was done while the author was a postdoctoral fellow in Max-Planck-Institut für Informatik, Saarbrücken, Germany. The work of this author is supported by the KBN Grant 4T11C04425.

clocks at nodes are ticking at the same rate. Communication has the property that multiple messages arriving at the same round to a node interfere with one another and none can be reliably received. Radio networks are given as directed graphs with edges representing direct-transmission ranges among nodes. A message is said to be *heard* by a receiving node when it can be read correctly. Radio networks are characterized by the following properties:

(i) A node can transmit at most one message at a round.
(ii) All the out-neighbors of a transmitting node receive the message in the same round when it was transmitted.
(iii) A recipient of a transmission can hear it only when the message was the only one received at the round.

Distributed communication protocols may have nodes categorized into *active* and *passive*. The former know that the protocol is being performed and execute it according to the code, while the latter need to be activated to join the execution. A passive node becomes automatically active after hearing a message. In the *wake-up problem*, some node may become spontaneously active at independent rounds and each of them wants to activate the whole network. Observe that wake-up is a generalization of the *broadcast problem,* in which exactly one node is activated spontaneously with the goal to wake up the whole network.

A multi-hop radio network is modeled as a directed strongly-connected graph $G = (V, E)$, with $n = |V|$ nodes. We simply assume $V = [1..n] = [n]$. Size n is the only network parameter included in codes of protocols we consider. Let D be the maximum directed distance between a pair of nodes, measured in terms of lengths of directed paths, and Δ the maximum in-degree of a node.

The rounds when nodes wake up spontaneously are represented as a function $\omega : V \to \mathbb{N}$, where $\mathbb{N}$ is the set of natural numbers. This means that node v wakes up spontaneously at time $\omega(v) = \omega_v$, unless it was already awaken by hearing a message sent by its active neighbor. Function ω is often referred to as a *shift function*, because it defines shifts of local time at nodes when they wake up spontaneously.

An active node v executes a sequence of transmissions given by some *schedule*, which is a binary sequence $\mathcal{R}^v$ repeated cyclically. An occurrence of 0 means a pause and an occurrence of 1 a transmission. The positions in $\mathcal{R}^v$ correspond to consecutive rounds after wake-up, according to the local clock of v. A collection of schedules good for waking up a radio network is called a radio synchronizer. Next we recall precise definitions of such synchronizers.

Let n and m be positive integers. Let $\mathcal{R} = \{\mathcal{R}^v\}_{1 \le v \le n}$ be an $n \times m$ binary array. There are n rows and m columns in the array. Each sequence $\mathcal{R}^v = \mathcal{R}^v(1)\mathcal{R}^v(2)\ldots\mathcal{R}^v(m)$ of length m is a row in the array. The number m is called the *length* of $\mathcal{R}$; it is often given as a function of the number n of rows.

Consider a shift function ω for $\mathcal{R}$. We always assume that $0 \le \omega_v < m$ for $1 \le v \le n$. A row v is ω-*active* at position t, or just *active* if ω is understood, if $\omega_v < t \le \omega_v + m$. Shift function ω can be *applied* to array $\mathcal{R}$ by shifting row v to the right by ω_v columns and padding the empty entries with occurrences of 0. The resulting array $\mathcal{R}_\omega$ has at most $2m - 1$ columns. We often refer to the columns of $\mathcal{R}_\omega$ as *positions*. To simplify the exposition, we may assume that array $\mathcal{R}_\omega$ has exactly $2m-1$ positions and if position t satisfies either $1 \le t \le \omega_v$ or $\omega_v + m + 1 \le t \le 2m - 1$, then $\mathcal{R}_\omega[v, t] = 0$.

If L is a set of rows of $\mathcal{R}$, then let L_ω denote the sub-array of $\mathcal{R}_\omega$ restricted to rows in L. Given a set L of rows, a position t is *successful for L*, if there is exactly one occurrence of 1 at position t of all rows of L_ω. For a shift function ω and such a set L of rows, define $\omega_L = \min_{v \in L} \omega_v$. We assume throughout that $\omega_L = 0$ in a context when rows of L correspond to nodes that wake up spontaneously; this makes positions correspond to rounds counted from the moment when the first in-neighbor of a node became active.

Radio synchronizers:
Array $\mathcal{R}$ is an (n, k)-*synchronizer of length* m, where $1 \le k \le n$, if for any shift function ω and any set L of at most k rows, there is a successful position for L.

Let $g : \mathbb{N} \times \mathbb{N} \to \mathbb{N}$ be a non-decreasing function called the *delay function* in the context of universal synchronizers.

Universal radio synchronizers:
Array $\mathcal{R}$ is a (n, g)-*universal synchronizer* if, for any shift function ω and any set L of rows, with the property that $\omega_L = \min_{v \in L} \omega_v = 0$, there is a successful position for L between positions 1 and $g(n, |L|)$.

Notice that the length of a (n, g)-universal synchronizer can always be assumed to be equal to $g(n, n)$. Universal radio synchronizers were defined to capture the following quantitative requirement on a wake-up protocol: if the active nodes transmit according to a (n, g)-universal synchronizer, then a node with k in-neighbors always becomes active within $g(n, k)$ rounds from the time when its first in-neighbor becomes awake. Radio synchronizers were defined by Chrobak, Gąsieniec and Kowalski [7] in their solution to the problems of wake-up, leader election and synchronization of local clocks in multi-hop networks. The notion of radio synchronizers was already implicitly used by Gąsieniec, Pelc and Peleg [11] in their algorithms for waking up a multiple-access channel. Universal radio synchronizers were defined by Chlebus and Kowalski [6] as a strengthening of radio synchronizers with the goal to improve the time of wake-up.

A family of combinatorial structures or instantiations of an algorithm parametrized by size n is said to be *explicit*, if the n-th object can be obtained in time polynomial in n.

Our results. We show the existence and present explicit constructions of improved universal radio synchronizers. This is translated into a better performance of known protocols in multi-hop radio networks, that solve problems such as wake-up, leader election and synchronization of local clocks. The contributions are summarized as follows.

I. We show the existence of (n, g)-universal-synchronizers where $g(n, k)$ is of order $\mathcal{O}(k \log k \log n)$. The best previously known universal radio synchronizers were shown by Chlebus and Kowalski [6] to exist for $g(n, k)$ of order $\mathcal{O}(k \min\{k, \sqrt{n}\} \log n)$.

II. We construct explicit (n,g)-universal-synchronizers with $g(n,k)$ of order $\mathcal{O}(k^2\operatorname{polylog} n)$. This is the first explicit construction of universal synchronizers. This generalizes previously known related design of explicit (n,k)-synchronizers of length $\mathcal{O}(k^2\operatorname{polylog} n)$ given in [6].

III. Using our new universal synchronizers, we show how to instantiate protocols for wake-up, leader election and local-clocks synchronizations presented in [6, 7] to significantly improve their performance. The resulting existential wake-up solution works in time $\mathcal{O}(\min\{n, D\Delta\}\log\Delta\log n)$ and the explicit solution works in time $\mathcal{O}(\min\{n, D\Delta\}\Delta\operatorname{polylog} n)$. Ours is the first *explicit sub-quadratic* wake-up solution for $\Delta = o(n/\operatorname{polylog} n)$. The best previously known existential wake-up protocol has running time $\mathcal{O}(n^{3/2}\log n)$; it was developed in [6]. The best previously known quasi-explicit wake-up protocol for networks of diameter D operates in time $\mathcal{O}(Dn^{1+\varepsilon})$, for an arbitrary $\varepsilon > 0$; it can be constructed in a quasi-polynomial time $2^{\operatorname{polylog} n}$ following the construction given by Indyk [12]. The leader election and synchronization protocols have time performance slower than wake-up by an additional factor of $\mathcal{O}(\log n)$.

Previous work. Wake-up protocols were first considered for the multiple-access channel by Gąsieniec, Pelc and Peleg [11]. Such networks are also called single-hop radio networks and are modeled by complete directed graphs. Randomized wake-up protocols for the multiple-access channel were also studied by Jurdziński and Stachowiak [13], and explicit wake-up protocols were given by Indyk [12]. The general wake-up problem for multi-hop networks was first studied by Chrobak, Gąsieniec and Kowalski [7]. They introduced the notion of radio synchronizer and also developed leader-election and synchronization protocols. The fastest known wake-up protocol was given by Chlebus and Kowalski [6], who introduced the notion of universal radio synchronizers. Gąsieniec, Pelc and Peleg [11] showed that there are (n,n)-synchronizers of length $\mathcal{O}(n\log^2 n)$. Chrobak, Gąsieniec and Kowalski [7] showed that there are (n,k)-synchronizers of length $\mathcal{O}(k^2\log n)$. Chlebus and Kowalski [6] showed that for each n there is an (n,g)-universal synchronizer with delay $g(n,k)$ upper bounded by the function $\mathcal{O}(k\min\{k,\sqrt{n}\}\log n)$. A construction of (n,n)-synchronizers of length $\mathcal{O}(n^{1+\varepsilon})$, for any constant $\varepsilon > 0$, in a quasi-polynomial time $\mathcal{O}(2^{\operatorname{polylog} n})$ was given by Indyk [12]. Chlebus and Kowalski [6] described explicit (n,k)-synchronizers of length $\mathcal{O}(k^2\operatorname{polylog} n)$. Prior to this work, there was no known non-trivial explicit construction of *universal* synchronizers.

Clementi, Monti and Silvestri [8] showed that (n,k)-selective-families, as defined in Chlebus et al. [5], have to be of a size $\Omega(k\log(n/k))$. Radio synchronizers have the properties of selective families; more precisely, they have the properties of radio synchronizers for the special case where all rows are active at position 1. It follows that (n,k)-synchronizers have to be of length $\Omega(k\log(n/k))$. Indyk [12] gave an explicit construction of (n,k)-selectors of length $\mathcal{O}(k\operatorname{polylog} n)$; such selectors are more general structures than selective families.

Related work. The model of radio communication was introduced by Chlamtac and Kutten [3]. Chlamtac and Weinstein [4] gave a centralized explicit broadcast protocol working in time $\mathcal{O}(D\log^2(n/D))$. Clementi et. al [9] developed an explicit protocol for the same problem working in time $\mathcal{O}(D\log\Delta\log(n/D))$. Regarding distributed sub-quadratic broadcast, the first such a randomized protocol was given by Bar-Yehuda, Goldreich, and Itai [2], and the first deterministic one by Chlebus et al. [5]. The fastest known distributed deterministic existential broadcast protocol for directed networks works in time $\mathcal{O}(n\log^2 D)$; it was given by Czumaj and Rytter [10].

Document structure. The rest of the document is structured as follows. In Section 2 we show by the probabilistic method that short universal synchronizers exist. A polynomial-time construction of short universal synchronizers is given in Section 3. In Section 4 we present applications of universal synchronizers to the problems of wake-up, leader election and synchronization of local clocks in multi-hop radio networks. We conclude in Section 5. Proofs of statements labelled as Lemmas have been omitted.

2 Short Universal Synchronizers by Counting

In this section we show the existence of (n,g)-universal-synchronizers with a slowly growing delay function g. To avoid rounding, we assume that n is a power of 2. Otherwise prove the existence of (n',g)-universal-synchronizer, where $n' = 2^{\lceil\log n\rceil}$, and remove the last $n'-n$ rows to obtain an (n,g)-universal-synchronizer. The notation $\log x$ denotes the logarithm of x to the base 2. We often write $g(k)$ for $g(n,k)$ when n is understood.

Specification of random universal synchronizers. Now we determine the length m and the delay function g of a (n,g)-universal-synchronizer. Let $c \geq 4$ be a constant to be determined later.

Define $m(\ell) = c2^\ell \ell \log n$, for every $1 \leq \ell \leq \log n$. Let $m = \sum_{\ell \leq \log n} m(\ell)$. Define $g(n,k) = \sum_{\ell \leq \lceil \log k \rceil} m(\ell)$ for $k > 1$, and $g(1) = m(1)$. Note that $g(n,k) = \mathcal{O}(k\log k \log n)$.

Given a column j, for $1 \leq j \leq m$, there is a unique positive integer κ satisfying the inequalities $\sum_{\ell < \kappa} m(\ell) < j \leq \sum_{\ell \leq \kappa} m(\ell)$. This number κ is denoted by κ_j.

Consider a random $n \times m$ array $\mathcal{R} = [\mathcal{R}^v(j)]$, where the binary-valued random variables $\mathcal{R}^v(j)$ are all independent, for $1 \leq v \leq n$ and $1 \leq j \leq m$. Random variable $\mathcal{R}^v(j)$ is defined to be equal to 1 with probability $1/(\kappa_j 2^{\kappa_j})$.

Theorem 1. *The array $\mathcal{R}$ is an (n,g)-universal-synchronizer, for $g(n,k) = \mathcal{O}(k\log k \log n)$, with the probability of at least $1 - 1/n$, for a sufficiently large constant $c \geq 4$.*

The remaining part of this section is devoted to a proof of Theorem 1. Let us fix n, which allows to write $g(k)$ for $g(n,k)$. For a position t, let $A_\omega(t)$ be the set of rows ω-active at t; we write $A(t)$ when ω is clear from context. Let us partition shift functions into n classes such that ω *belongs to the i-th class* if i is the smallest positive integer for which the number of ω-active rows at position $g(i)$ is at most i; this is equivalent for i to be the smallest number such that the inequality $|A_\omega(g(i))| \le i$ holds. These classes are well defined because all rows are active at position m by the assumption that $\omega_v < m$, for $1 \le v \le n$, and because $m = g(n)$. For a shift function ω in the i-th class, we need to show that the universal synchronizer under consideration has the property that there is a successful position by the position $g(i)$. Therefore it is sufficient to consider only a partial function defined by the values of ω at rows v such that $\omega_v < g(i)$. This partial function is denoted as $\omega|^{g(i)}$.

Consider shift functions in the 1-st class. There is only one row v active on positions $[1..2c \log n]$, by the definition of this class. The probability that all these positions are unsuccessful is at most $2^{-2c \log n} \le e^{-c \log n}$. There are at most n possible rows v, hence the probability that there is a shift function in the 1-st class, where all positions $t \le g(1)$ are unsuccessful, is at most $e^{\ln n} \cdot e^{-c \log n} \le e^{-2 \ln n} = n^{-2}$, for $c \ge 3$.

Consider next the general case of shift functions in the k-th class, for $1 < k \le n$. By the definition, number k is the smallest positive integer such that there are at most k active rows at position $g(k)$. Let us fix an ω in this class, and let $K = K_\omega$ be the set of rows that are ω-active at position $g(k)$, that is, set K denotes $A(g(k)) = A_\omega(g(k))$.

For row $v \in K$ and position $t \le g(k)$, let $p_v(t)$ denote the probability that there is 1 in row v at position t, which means $\mathcal{R}^v(t - \omega_v) = 1$, with $p_v(t) = 0$ for every $t \le \omega_v$. Let $\mu_K(t)$ denote $\sum_{v \in K} p_v(t)$. We call a position $t \le g(k)$ balanced if the inequalities $1/(4\kappa_t) \le \mu_K(t) \le 4$ hold. Let κ denote $\kappa_{g(k)}$. Notice that $m(\kappa) = g(k) - g(k/2)$.

Lemma 1. *There are at least $m(\kappa)/2$ balanced positions between position $g(k/2)+1$ and position $g(k)$.*

Consider a balanced position $t \le g(k)$. The probability that t is successful is

$$\sum_{v \in K} p_v(t) \cdot \prod_{w \in K, w \ne v} (1 - p_w(t)) \ge \mu_K(t) \cdot 4^{-\mu_K(t)} \ge \frac{1}{4\kappa} 4^{-4} = \frac{1}{4^5 \kappa},$$

since $p_v(t) \le 1/2$ for every $v \in K$. Hence, by Lemma 1, the probability that all balanced positions are unsuccessful is at most

$$\left(1 - \frac{1}{4^5 \kappa}\right)^{m(\kappa)/2} \le e^{-(2c/4^6) 2^\kappa \log n} \le e^{-(2c/4^6) k \log n}.$$

The above considerations were for a fixed shift function ω from the k-th class. We do not need to know how ω precisely behaves for rows that are shifted by more than $g(k)$. The set K is uniquely defined by $\omega|^{g(k)}$, and the above probabilistic analysis is based on $\omega|^{g(k)}$. Such a partial function is determined by its domain of k elements and an assignment of values in the range from 0 to $g(k) - 1$ to them. It follows that there are at most $\binom{n}{k} \cdot (g(k))^k \le e^{k \ln(ne/k) + k \ln g(k)}$ different partial functions $\omega|^{g(k)}$ such that ω is in the k-th class.

The probability that there exists a shift function ω from the k-th class such that there is no successful position $t \le g(k)$ in $\mathcal{R}$ is at most

$$e^{k \ln(ne/k) + k \ln g(k)} \cdot e^{-(2c/4^6) k \log n} \le e^{-2k \ln n},$$

for a sufficiently large constant $c \geq 4$, since $\ln g(k) = \mathcal{O}(\log n)$. Summing up all the probabilities for all classes, the probability of the event that there is a shift function in some k-th class such that there is no successful position $t \leq g(k)$ in $\mathcal{R}$ is at most $\sum_{k=1}^{n} e^{-2k \ln n} \leq 1/n$, for $n > 1$. This completes the proof of Theorem 1.

3 Explicit Constructions

In this section we show how to construct (n, g)-universal-synchronizer with delay $g(n, k) = \mathcal{O}(k^2 \operatorname{polylog} n)$ in time polynomial in n. We recall the construction from [6], and then modify it to obtain synchronizers satisfying certain properties stronger than synchronization, to be used in the construction of universal synchronizers. It is enough to construct universal synchronizers for values of n being powers of 3 to guarantee the same overall asymptotic behavior. We want to construct a universal synchronizer in two steps. First is a concatenation of $(n, 3^x)$-synchronizers of length $m(3^x)$, for $x = 1, 2, \ldots, \log_3 n$, as constructed in [6], denoted by $\mathcal{Q}$. Second is replacing each occurrence of 0 in the obtained structure by a sequence of zeroes of a specific length $m_* = \operatorname{polylog} n$, and each occurrence of 1 in row v by the corresponding sequence $\mathcal{S}^v$ taken from a suitable synchronizer $\mathcal{S}$ of length m_*.

3.1 Explicit Synchronizers

We first present (n, k)-synchronizers denoted $\mathcal{F}(n, k)$, of length $f(k)$ that is of order $\mathcal{O}(k^2 \operatorname{polylog} n)$, based on the construction from [6]. Then we modify this construction and argue that the obtained synchronizers satisfy a stronger condition than the one in the definition of synchronizers.

Construction of (n, k)-synchronizers $\mathcal{F}(n, k)$. We briefly recall the construction from [6]. Assume that number n is sufficiently large and that k divides n. Let $\mathcal{P}$ be a set of k different primes between $k \log n$ and $3k \log n$. Set $\mathcal{P}$ is well defined by the Chebyshev Theorem, as given in [1]. Let p be in set $\mathcal{P}$. Denote by π_p a binary sequence of length $3k^2 \log n$ and such that $\pi_p(j) = 1$, when both p divides j and $0 < j \leq 3k^2 \log n$, and let $\pi_p(j) = 0$ otherwise. Let σ denote a sequence of zeroes of length $3k^2 \log n$. Number d is a positive integer, to be determined later, depending on numbers k and n and on applications. Let h be a function from $[n] \times [d]$ to $\mathcal{P}$. For $v \in [n]$, define schedule $\mathcal{S}^v$ to be a concatenation of schedules of the form $\langle \pi_{h(v,1)}, \sigma, \pi_{h(v,2)}, \sigma, \ldots, \pi_{h(v,d)}, \sigma, \sigma^{2d} \rangle$, where σ^{2d} denotes a sequence of zeros of length $6dk^2 \log n$.

Selective functions:
Function $h : [n] \times [d] \to \mathcal{P}$ is said to be (n, k, d)-*selective* when for every set $W \subseteq [n]$ of a size at most k, there is a number p in $\mathcal{P}$ such that set $h^{-1}(p) \cap (W \times [d])$ is a singleton.

Lemma 2 ([6]). *If function h is (n, k, d)-selective, then set $\mathcal{S} = \{\mathcal{S}^v\}_{v \in V}$ of schedules is an (n, k)-synchronizer of length $f(n, k) = \mathcal{O}(dk^2 \log n)$.*

The construction of Indyk [12] gives an explicit (n, k, d)-selective function for $d = $ polylog n. It relies on a polynomial-time construction of $(n, k', d', 1/2 - \varepsilon)$-dispersers developed by Ta-Shma, Umans and Zuckerman [16], for $d' = \mathcal{O}(\text{polylog } n)$ and for a constant $0 < \varepsilon < 1/2$. Using this explicit selective function and Lemma 2, the following fact can be shown:

Lemma 3 ([6]). *A family $\{\mathcal{F}(n, k)\}_{n \geq k}$ of (n, k)-synchronizers of length which is $\mathcal{O}(k^2 \text{polylog } n)$ can be constructed in time polynomial in n.*

Construction of (n, k)-synchronizers $\mathcal{F}^*(n, k)$. We use the general scheme of construction from [6] and modify it to obtain the construction of (n, k)-synchronizers satisfying the following property $\mathcal{M}(k)$. We often write $f(k)$ for $f(n, k)$ when n is understood.

Property $\mathcal{M}(k)$:
We say that an (n, k)-synchronizer of length $f(n, k)$ *satisfies property $\mathcal{M}(k)$* if for every set K of at most k rows shifted by at most $3f(n, k) - 1$ positions and such that at least one row is shifted by at least $f(n, k)$ and at most $2f(n, k) - 1$ positions, there is a successful position in the interval $[f(n, k) + 1..3f(n, k)]$.

The main idea is that if we take function h satisfying a certain stronger property, then the obtained structure is also (n, k)-synchronizer, of a somewhat larger length, but it satisfies property $\mathcal{M}(k)$. We need the following property to be satisfied by function h.

Strongly selective functions:
We say that function $h : [n] \times [d] \rightarrow \mathcal{P}$ is (n, k, d)-*strongly-selective* if for every set $W \subseteq [n]$ of a size at most k and every $w \in W$, there are numbers p in $\mathcal{P}$ and $d' \in [d]$ such that $\{(w, d')\} = h^{-1}(p) \cap (W \times [d])$.

Note that if h is (n, k, d)-strongly-selective, then it is also (n, k, d)-selective.

Lemma 4. *If function h is (n, k, d)-strongly-selective, then set $\mathcal{S} = \{\mathcal{S}^v\}_{v \in V}$ of schedules is an (n, k)-synchronizer of length $f(n, k) = \mathcal{O}(dk^2 \log n)$ and it satisfies property $\mathcal{M}(k)$.*

On the other hand, (n, k, d)-strongly-selective functions can be explicitly constructed for $d = \mathcal{O}(k \log^2 n)$, relying on a construction by Kautz and Singleton [14]. Define $\mathcal{F}^*(n, k)$ to be the structure with (n, k, d)-strongly-selective function h, as constructed explicitly in [14]. Combine an explicit construction of (n, k, d)-strongly-selective functions with $d = \mathcal{O}(k \log^2 n)$ [14] and Lemma 4 to obtain:

Theorem 2. *A family $\{\mathcal{F}^*(n, k)\}_{n \geq k}$ of (n, k)-synchronizers of length $f(n, k) = \mathcal{O}(k^3 \text{ polylog } n)$ and satisfying property $\mathcal{M}(k)$ can be constructed in time polynomial in n.*

3.2 Explicit Universal Synchronizers

Let $\mathcal{F}(n,k)$ be the synchronizer, defined in Section 3.1, of the corresponding length $\mathcal{O}(dk^2 \log n)$, for some d being polylog n. Let $\mathcal{S} = \mathcal{F}^*(n, 27d\log^2 n)$ be the explicit $(n, 27d\log^2 n)$-synchronizer of some length m_* which satisfies property $\mathcal{M}(27d\log^2 n)$, as described in Section 3.1.

To obtain universal synchronizer $\mathcal{R}^v$, we proceed in two steps:

First step: Concatenate rows of synchronizers $\mathcal{F}(n,k)$, for $k = 3^i$, from $i = 0$ through $i = \log_3 n$,
Denote by $\mathcal{Q}$ the obtained structure.
Second step: Replace each occurrence of 0 in row $\mathcal{Q}^v$ by the sequence of m_* zeros, and each occurrence of 1 in row $\mathcal{Q}^v$ by the sequence $\mathcal{S}^v$.

Structure $\mathcal{Q}$ is like a universal radio synchronizer but with a notion of a "successful position" replaced with the weaker notion of an "almost successful position," meant to be a position where 1 occurs at least once but at most $\mathcal{O}(\text{polylog } n)$ times. The reason why structure $\mathcal{Q}$ can only guarantee the existence of "almost successful positions" is that the $\log n$ concatenated synchronizers interfere with one another when the rows are arbitrarily shifted. Structure $\mathcal{S}$ guarantees a successful position in any set of $\mathcal{O}(\text{polylog } n)$ rows. Therefore the operation of expanding each 1 in $\mathcal{Q}$ with structure $\mathcal{S}$ turns each "almost successful position" into a sequence of consecutive positions with at least one of them being successful. A complete proof, that structure $\mathcal{R}$ is indeed a universal synchronizer, needs to account properly for the possibility that rows of $\mathcal{R}$ are shifted in such a way that the copies of structure $\mathcal{S}$ in different rows are not aligned.

Theorem 3. *$\mathcal{R}$ is an explicit (n, g)-universal-synchronizer with $g(n, k)$ of order $\mathcal{O}(k^2 \text{ polylog } n)$.*

In the remaining part of this section we prove Theorem 3. Let us fix n, which allows to write $f(k)$, $g(k)$ and $g'(k)$ rather than $f(n,k)$, $g(n,k)$ and $g'(n,k)$, respectively. Define function $g(n,k) = g(k) = f(3x) \cdot m_*$, where $x = \min\{3^{\lceil \log_3 k \rceil}, n/3\}$. Observe that estimate $g(k) = \mathcal{O}(k^2 \text{ polylog } n)$ holds, since $f(3x)$ is of order $\mathcal{O}(x^2 \text{ polylog } n)$ by Lemma 3, function $27d \log^2 n$ of n is polylog n and hence $m_* = \text{polylog } n$ by Theorem 2.

We first analyze structure $\mathcal{Q}$. Consider a shift function ω' on $\mathcal{Q}$, and let $g'(k) = f(3x)$. Let k be the first positive integer such that there are at most k active rows at position $g'(k)$. Recall that $A(t)$ stands for the set of active rows at position t. By the definition of universal synchronizers and how function g' is specified, we obtain that $|A(g'(k))| = k$. Let K denote set $A(g'(k))$.

By the definition of k we also have that $|A(g'(k/3))| > k/3$. Let L denote $A(g'(k/3))$. By the definition of $\mathcal{Q}$, all rows from L have their components taken from an (n,x)-synchronizer of length $f(x) \leq g'(k)$.

Lemma 5. *There is a prime number $p \in [x \log n, 3x \log n]$ for which there is exactly one row v such that $\mathcal{R}^v$ contains sequence π_p in its prefix by position $g'(k)$.*

Fix a prime number p, let t_* be the first position of sub-sequence π_p of $\mathcal{Q}^v$ and t^* be the last such a position, all as existing by Lemma 5. For every row $w \neq v$ in K, array $\mathcal{Q}^w[t_* .. t^*]$ contains at most $d \log_3 x$ different sub-sequences π_q, for primes $q \leq 3x \log n$, all these sequences different from π_p.

(p, q)-**Crowded positions:**
Position $1 < t \leq g'(k)$ is (p,q)-*crowded* if one of the following holds:
(a) 1 occurs at position t in both sequences π_p and π_q,
(b) 1 occurs at position t in π_p and at position $t-1$ in π_q,
(c) 1 occurs at position t in π_q and at position $t-1$ in π_p.

Lemma 6. *Any sequence π_q, for a prime $q \leq 3 \cdot 3x \log n$ and $q \neq p$, contributes at most nine (p,q)-crowded positions, if shifted arbitrarily.*

It follows from Lemma 6 that there are at most

$$9 \cdot k \cdot \log_3 x \cdot 3d \log n \leq k \cdot (27 d \log^2 n - 1)$$

(p,q)-crowded positions, for some primes $q \neq p$, while there are at least k occurrences of 1 in sequence π_p in row v. For some position t, there is 1 at position t in row v, and at most $27 d \log^2 n - 1$ occurrences of 1 in other rows at positions $t-1, t, t+1$ in total. Call such t a *candidate position*.

Now consider the whole structure $\mathcal{R}$, by position $g(k) = g'(k) m_*$, together with the shift function ω. Let ω' be the corresponding shift functions for the underlying structure $\mathcal{Q}$, determined by $\omega'_v = \lceil \frac{\omega_v}{m_*} \rceil$. It follows that in structure $\mathcal{Q}$, with such a shift function ω', there is a candidate position t'. Denote by $K' \subseteq K$ all rows $v \in K$ such that there is a 1 in row v at positions $t'-1, t', t'+1$. The inequality $|K'| \leq 27 d \log^2 n$ holds by the definition of t'. It follows that there is a sequence $\mathcal{S}^v$ of length m_* contained in row v in structure $\mathcal{R}$ between positions $(t'-1)m_* + 1$ and $(t'+1)m_*$, and at most $27 d \log^2 n - 1$ sequences $\mathcal{S}^w$, for different $w \neq v$, which intersect interval of positions $[(t'-2)m_* + 1, (t'+1)m_*]$. There is a successful position in row w in the interval $[(t'-1)m_* + 1, (t'+1)m_*]$, by property $\mathcal{M}(27 d \log^2 n)$ of $(n, 27 d \log^2 n)$-synchronizer $\mathcal{S}$. This completes the proof of Theorem 3.

4 Applications

We show that the problems of wake-up, leader election and synchronization can all be solved in deterministic time $\mathcal{O}(n \operatorname{polylog} n)$ by relying on universal synchronizers shown to exist, and in time $\mathcal{O}(n \Delta \operatorname{polylog} n)$ by construction of explicit universal synchronizers.

When developing leader-election and synchronization algorithms, one usually assumes that unique names are from some polynomial range, while the difference between the maximum and minimum-value names is $\mathcal{O}(n)$. To simplify the exposition, we assume that there are n nodes with unique names in an interval $[1..\mathcal{O}(n)]$. Recall that D denotes the diameter of the directed network, and Δ is its maximum in-degree.

For the sake of completeness, we recall briefly the wake-up protocol based on universal synchronizers, as proposed in [6]. It operates as follows. Take a (n,g)-universal synchronizer $\mathcal{S}$, where rows correspond to nodes. Each node v starts executing its sequence $\mathcal{S}^v \in \mathcal{S}$ immediately when it becomes active. This means that it performs a transmission in the i-th round exactly when the i-th bit in $\mathcal{S}^v$ is a 1. We refer to this protocol as CK.

We develop a general estimate on the complexity of the wake-up protocol CK, which we call β, that depends on parameters of the network and the delay function of the universal synchronizer that is a part of its code. For numbers n, D, Δ and a delay function g of a (n,g)-universal synchronizer, let $\beta(n, D, \Delta, g)$ denote the supremum of the function $\sum_{i=1}^{D} g(x_i)$, where integers $0 \le x_i \le \Delta$, for $1 \le i \le D$, satisfy an additional constraint $\sum_{i=1}^{D} x_i \le n$.

Lemma 7. *Protocol CK based on (n,g)-universal-synchronizer wakes up a radio network of n nodes with parameters D and Δ in $\beta(n, D, \Delta, g)$ rounds.*

Corollary 1. *Protocol CK can be instantiated such that it solves the wake-up problem in $\mathcal{O}(\min\{n, D\Delta\} \log \Delta \log n)$ rounds.*

Corollary 2. *Protocol CK can be explicitly instantiated such that it solves the wake-up problem within time bound $\mathcal{O}(\min\{n, D\Delta\} \Delta \operatorname{polylog} n)$.*

The estimate given in Corollary 1 is within a logarithmic factor away from $\Omega(\min\{n \log D, D\Delta \log(n/\Delta)\})$, which is a lower bound on broadcasting given in [8] that holds also for wake-up. A general protocol for leader election and synchronizing local clocks was given by Chrobak, Gąsieniec and Kowalski [7], we call it CGK. The protocol is built on top of a wake-up one.

Corollary 3. *Protocol CGK, based on suitably instantiated CK, can solve leader election and synchronization of local clocks in $\mathcal{O}(\min\{n, D\Delta\} \log \Delta \log^2 n)$ rounds.*

Corollary 4. *Protocol CGK, based on an explicitly instantiated CK, can solve leader election and synchronization of local clocks in $\mathcal{O}(\min\{n, D\Delta\} \Delta \operatorname{polylog} n)$ rounds.*

5 Conclusion

We obtain an existential distributed wake-up protocol for multi-hop ad-hoc radio networks working in time $\mathcal{O}(n \log^2 n)$ and an explicitly instantiated distributed protocol that works in time $\mathcal{O}(n \Delta \operatorname{polylog} n)$. Developing explicit sub-quadratic wake-up of radio networks remains an open problem.

On the technical level, our results concern universal synchronizers. We show the existence of (n,g)-universal synchronizers for $g(k)$ of order $\mathcal{O}(k \log k \log n)$. The best known lower bound is $\Omega(k \log(n/k))$, which is derived form a lower bound on selective families given in [8]. Selective families are apparently weaker structures than universal synchronizers. Improving these upper and/or lower bounds for universal synchronizers, in terms of function g, remains open. In particular, the following is an open problem: is there an explicit construction of (n, g)-universal-synchronizer for $g(k) = \mathcal{O}(k \operatorname{polylog} n)$?

If we have only an upper bound N on the size of the network and node labels are from the interval $[1..N]$, then we may use a (N, g)-universal synchronizer. Lemma 7 holds true, since its proof relies on the property that nodes are uniquely assigned to rows of the universal synchronizer. For this case we have $g(k) = k \log k \log N$, and so the corresponding modification on the estimates of β, and consequently on all the following results, is that we have expressions $\log N$ and polylog N instead of $\log n$ and polylog n, respectively.

It is a natural question to ask what is the inherent difference in time performance between broadcast and wake-up protocols. It turns out that such a difference is within a factor of $\mathcal{O}(\log n)$. The answer has already been known for randomized protocols, by the results in [7] for wake-up and [10, 15] for broadcast. This paper settles this for deterministic existential protocols, by obtaining a time $\mathcal{O}(n \log^2 n)$ wake-up, as efficient, in terms of n, as the broadcast in [10].

References

1. E. Bach, and J. Shallit, *"Algorithmic Number Theory,"* Vol. I, 1996, The MIT Press, Cambridge, Massachusetts.
2. R. Bar-Yehuda, O. Goldreich, and A. Itai, On the time complexity of broadcast in radio networks: an exponential gap between determinism and randomization, *Journal of Computer and System Sciences*, 45 (1992) 104 - 126.
3. I. Chlamtac, and S. Kutten, On broadcasting in radio networks - problem analysis and protocol design, *IEEE Transactions on Communications*, 33 (1985) 1240 - 1246.
4. I. Chlamtac, and O. Weinstein, The wave expansion approach to broadcasting in multihop radio networks, *IEEE Transactions on Communications*, 39 (1991) 426 - 433.
5. B.S. Chlebus, L. Gąsieniec, A. Gibbons, A. Pelc, and W. Rytter, Deterministic broadcasting in unknown radio networks, *Distributed Computing*, 15 (2002) 27 - 38.
6. B.S. Chlebus, and D.R. Kowalski, A better wake-up in radio networks, in *Proceedings, 23rd ACM Symposium on Principles of Distributed Computing (PODC)*, 2004, pp. 266 - 274.
7. M. Chrobak, L. Gąsieniec, and D.R. Kowalski, The wake-up problem in multi-hop radio networks, in *Proceedings, 15th ACM-SIAM Symposium on Discrete Algorithms (SODA)*, 2004, pp. 985 - 993.
8. A.E.F. Clementi, A. Monti, and R. Silvestri, Distributed broadcast in radio networks of unknown topology, *Theoretical Computer Science* 302 (2003) 337 - 364.
9. A.E.F. Clementi, P. Crescenzi, A. Monti, P. Penna, and R. Silvestri, On computing ad-hoc selective families, in *Proceedings, 5th International Workshop on Randomization and Approximation Techniques in Computer Science (RANDOM-APPROX)*, 2001, pp. 211 - 222.
10. A. Czumaj, and W. Rytter, Broadcasting algorithms in radio networks with unknown topology, in *Proceedings, 44th IEEE Symposium on Foundations of Computer Science (FOCS)*, 2003, pp. 492 - 501.

11. L. Gąsieniec, A. Pelc, and D. Peleg, The wakeup problem in synchronous broadcast systems, *SIAM Journal on Discrete Mathematics*, 14 (2001) 207 - 222.
12. P. Indyk, Explicit constructions of selectors and related combinatorial structures, with applications, in *Proceedings, 13th ACM-SIAM Symposium on Discrete Algorithms (SODA)*, 2002, pp. 697 - 704.
13. T. Jurdziński, and G. Stachowiak, Probabilistic algorithms for the wakeup problem in single-hop radio networks, in *Proceedings, 13th International Symposium on Algorithms and Computation (ISAAC)*, 2002, LNCS 2518, pp. 535 - 549.
14. W.H. Kautz, and R.R.C. Singleton, Nonrandom binary superimposed codes, *IEEE Transactions on Information Theory*, 10 (1964) 363 - 377.
15. D.R. Kowalski, and A. Pelc, Deterministic broadcasting time in radio networks of unknown topology, in *Proceedings, 22nd ACM Symposium on Principles of Distributed Computing (PODC)*, 2003, pp. 73 - 82.
16. A. Ta-Shma, C. Umans, and D. Zuckerman, Loss-less condensers, unbalanced expanders, and extractors, in *Proceedings, 33rd ACM Symposium on Theory of Computing (STOC)*, 2001, pp. 143 - 152.

Distance Constrained Labelings of Graphs of Bounded Treewidth*

Jiří Fiala[1], Petr A. Golovach[2], and Jan Kratochvíl[1]

[1] Inst. for Theoretical Computer Science** and Dept. of Applied Mathematics,
Faculty of Mathematics and Physics, Charles University, Prague, Czech Republic
{fiala, honza}@kam.mff.cuni.cz
[2] Matematicheskii Fakultet, Syktyvkar State University, Syktyvkar, Russia
golovach@ssu.komi.com

Abstract. We prove that the L(2,1)-LABELING problem is NP-complete for graphs of treewidth two, thus adding a natural and well studied problem to the short list of problems whose computational complexity separates treewidth one from treewidth two. We prove similar results for other variants of the distance constrained graph labeling problem.

1 Introduction

The notion of distance constrained graph labeling attracted a lot of attention in the past years both for its motivation by the practical frequency assignment problem, and for its interesting graph theoretic properties. The task of assigning frequencies to transmitters to avoid undesired interference of signals is modeled in several ways. The so called *channel assignment problem* assumes that a minimum allowed difference of channels is given for every two transmitters. Thus the input of this problem is a weighted graph whose vertices correspond to the transmitters, and the task is to assign nonnegative integers (channels) to the vertices so that for every edge, the difference of the assigned channels is at least the weight of the edge, and so that the largest channel used is minimized.

Another approach, and this one we follow in the present paper, is the *distance constrained graph labeling*. Here it is assumed that the distance of transmitters can be modeled by a graph, and that the distance of the transmitters influences possible interference in such a way that the closer two transmitters are, the farther apart their frequencies must be. Formally, an assignment of nonnegative integers to the vertices of a graph G is an $L(p_1, \ldots, p_k)$-labeling if for every two vertices at distance at most $i \leq k$, the difference of the integers (labels) assigned to them is at least p_i. Here $k \geq 1$ is the depth to which the distance constraints are applied, and integers $p_1 \geq p_2 \geq \ldots \geq p_k$ are parameters of the problem.

---

* Research supported in part by project KONTAKT 525 — DIMACS-DIMATIA-Rényi Cooperation in Discrete Mathematics.
** Supported by the Ministry of Education of the Czech Republic as project 1M0021620808.

Again, the goal is to minimize the maximum label used. The most studied of the distance constrained labelings is the case $k = 2, p_1 = 2, p_2 = 1$, i.e., the $L(2,1)$-*labeling*. In this case adjacent vertices must be assigned labels that differ by at least 2, while nonadjacent vertices with a common neighbor must be assigned distinct labels. The maximum label used is called the *span* of the labeling. The minimum span of an $L(2,1)$-labeling of a graph G will be denoted by $L_{(2,1)}(G)$.

The notion of $L(2,1)$-labeling was in fact first proposed by Roberts [21] and many nontrivial results were presented in a pioneer paper of Griggs and Yeh [15]. Let us mention their conjecture that $L_{(2,1)}(G) \leq \Delta^2(G)$ (where $\Delta(G)$ stands for the maximum vertex degree in G). This conjecture has been verified for various graph classes, but it is still open for general graphs (with $L_{(2,1)}(G) \leq \Delta(G)^2 + \Delta(G) - 1$ being the current record [16]). From the computational complexity point of view, Griggs and Yeh proved that determining $L_{(2,1)}(G)$ is an NP-hard problem, and this result was later strengthened by Fiala et al. [7] by showing that deciding $L_{(2,1)}(G) \leq k$ is NP-complete for every fixed $k \geq 4$. Griggs and Yeh also conjectured that it is NP-complete to compute the $L_{(2,1)}$ number of a tree, but this was somewhat surprisingly disproved by a dynamic programming polynomial time algorithm of Chang and Kuo [4].

The common expectation says that problems solvable in polynomial time for trees should also be polynomially solvable for graphs of bounded treewidth, though sometimes the extension to bounded treewidth is not straightforward (cf. e.g., the case of chromatic index [2]). (We informally recall that the treewidth is a graph invariant that describes how far is the graph from being a tree. For a formal definition the reader is referred to a survey [3] or to one of the original papers [1] introducing this invariant in terms of so called *partial k-trees*. For our purposes we only need the fact that graphs of treewidth at most two are exactly the graphs that do not contain a subdvision of K_4 as a subgraph, and connected graphs of treewidth one are exactly trees.) Only very few exceptions to this rule of thumb are known, and in fact very few problems are known to be hard for graphs of bounded treewidth. An example is, e.g., the MINIMUM BANDWIDTH problem (which is NP-hard already for trees [12]) or the closely related CHANNEL ASSIGNMENT problem which has been recently shown NP-complete for graphs of treewidth three [19]. The natural question of the complexity of $L(2,1)$-labelings for graphs of bounded treewidth has been posed many times and remained open since 1996. The main result of our paper settles it by showing that determining the $L_{(2,1)}$ number of graphs of treewidth two is NP-hard.

Before we formulate the result formally, we specify precisely what problem we deal with. The decision problem whether a given graph admits an $L(2,1)$-labeling of *fixed* span can be described in Monadic Second Order Logic (MSOL), and therefore is solvable in linear time for any class of graphs of bounded treewidth by a generic algorithm of Courcelle [5]. Thus we naturally assume that the span is a part of the input, and we consider the following problem.

$L(2,1)$-LABELING
Input: An integer λ and a graph G.
Question: Is $L_{(2,1)}(G) \leq \lambda$?

Theorem 1. *The* L(2,1)-LABELING *problem is* NP-*complete for graphs of treewidth at most two.*

So far we have only discussed the model in which interference of the frequencies (or channels) decreases linearly with their increasing difference. It is, however, plausible to consider also such models in which frequencies far apart may interfere (e.g., if one is a multiple of the other one). This means more complicated metrics in the frequency space. A concrete step in this direction is the cyclic metric introduced by van den Heuvel et al. [24]. In this metric, the graph of the channel space is the cycle of length λ. Similarly to the linear case, we talk about $C(2,1)$-labelings and denote by $C_{(2,1)}(G)$ the minimum span of a $C(2,1)$-labeling of G (note that in the cyclic metric, the span is the number of available channels, not the difference between the largest and smallest one). For general graphs, deciding if $C_{(2,1)}(G) \leq \lambda$ is NP-complete for every fixed $\lambda \geq 6$ [9]. For λ part of the input and graphs of bounded treewidth, we fully characterize the complexity of the C(2,1)-LABELING problem (which, given a graph G and an integer λ as input, asks if $C_{(2,1)}(G) \leq \lambda$):

Proposition 1. ([18,17]) *Let T be a tree with at least one edge, and $p \geq q$ nonnegative integers. Then $C_{(p,q)}(T) = q\Delta(T) + 2p - q$ where $\Delta(T)$ is the maximum degree of a vertex in T.*

Theorem 2. *The* C(2,1)-LABELING *problem is* NP-*complete for graphs of treewidth at most two.*

This theorem will be proven in the full version of the paper.

Fiala and Kratochvíl [9] defined the notion of $H(2,1)$-labeling as the utmost generalization in the case when the metric of the channel space can be described by a graph H, and showed that $H(2,1)$-labelings of a graph G are exactly locally injective homomorphisms from G into the complement of H. The complexity of the H(2,1)-LABELING problem for some parameter graphs H then follows from [8], but the complete characterization is not even in sight. On the other hand, if G has bounded treewidth, the H(2,1)-LABELING problem is solvable in polynomial time since for a fixed graph H, the existence of an $H(2,1)$-labeling of G can be expressed in MSOL.

It remains to study the case when both G and H are part of the input and we refer to it as the (2,1)-LABELING problem. Observe that the L(2,1)-LABELING problem is the restriction of (2,1)-LABELING to inputs such that H is a path. Hence it follows from Theorem 1 that (2,1)-LABELING is NP-complete for graphs of treewidth two. However, in this most general setting, we are able to prove dichotomy even with respect to pathwidth (for definition of pathwidth see [22, 23, 3], just recall that connected graphs of pathwidth one are exactly caterpillars):

Theorem 3. *For a tree T with m vertices and an arbitrary graph H with n vertices, one can decide in time $O(n^3 m^2)$ whether T allows an $H(2,1)$-labeling.*

Theorem 4. *The* (2,1)-LABELING *problem is* NP-*complete for graphs G of pathwidth at most two (the graph H may be arbitrary).*

The paper is organized as follows. In Section 2 we review technical definitions and notation and prove an auxiliary result on systems of distant representatives for symmetric sets. The main result, Theorem 1, is proved in Section 3. The technical proofs for the case of cyclic metric are omitted due to space restrictions. Theorems 3 and 4 are proved in Section 4. The last section contains concluding remarks and open questions.

2 Preliminaries

All graphs considered are finite and simple, i.e., with a finite vertex set and without loops or multiple edges. For a vertex u, the symbol $N(u)$ denotes the *open neighborhood* of u, i.e., the set of all vertices adjacent to u, and we denote by $\deg u = |N(u)|$ the *degree* of u.

A graph is called *series-parallel* if it can built from isolated edges with end-vertices called South and North poles by a sequence of series and parallel compositions (the former identifies the North pole of one component with the South pole of the other one, the latter unifies the North poles of the components into a common North pole, and likewise the South poles). It is well known that a graph has treewidth at most two if and only if all its 2-connected subgraphs are series-parallel.

The labels are always nonnegative integers, with 0 being the smallest label used. We use the notation $[x, y] = \{x, x+1, \ldots, y-1, y\}$ to denote intervals of consecutive integers. We say that a set S of integers is *symmetric within* an interval $[x, y]$ if $S \subseteq [x, y]$ and for every $i \in [x, y]$, $i \in S$ if and only if $y + x - i \in S$.

A system of distinct representatives for a set system $S_1, S_2, \ldots, S_n$ is a system of distinct elements $s_i \in S_i, i = 1, 2, \ldots, n$. The theory of SDR's is well developed, the necessary and sufficient condition for their existence is given by the well known Hall theorem, and an SDR can be found in polynomial time (e.g., by a bipartite matching algorithm). If the ground set $\bigcup_{i=1}^{n} S_i$ is equipped with a metric function, we can further impose conditions on the distance of the chosen representatives. We refer the reader to [11, 14] for a survey on the computational complexity of finding systems of distant representatives for sets in metric spaces and their applications in various graph labeling problems. Now we will use a special variant of this problem as an auxiliary tool:

SRL (Special representatives in the linear metric)
Input: An integer n and a collection of sets of integers $S_1, S_2, \ldots, S_m$ symmetric within the interval $[2, \lambda - 2]$, where $\lambda = 4n + 5$.
Question: Does there exist a collection of distinct integers
$s_1, s_2, \ldots, s_m, t_1, t_2, \ldots, t_n, u_1, u_2, \ldots, u_n$ such that
- $s_i \in S_i$ for every $i = 1, \ldots, m$,
- $t_i \in \{2i, \lambda - 2i - 1\}$ for every $i = 1, \ldots, n$,
- $u_i \in \{2i + 1, \lambda - 2i\}$ for every $i = 1, \ldots, n$,
- $|t_i - u_i| \geq 2$ for every $i = 1, 2, \ldots, n$?

Lemma 1. *The problem* SRL *is* NP-*complete.*

Proof. The proof is provided by a reduction from the following special variant of the 3-SAT problem (known NP-complete, cf. e.g. [6]).

2-3-SAT
Input: A Boolean formula Φ in conjunctive normal form, whose each clause consists of 2 or 3 literals and whose every variable has at most 2 positive and at most 2 negative occurrences.
Question: Is Φ satisfiable?

Let Φ have n variables $x_1, \ldots, x_n$ and m clauses $C_1, \ldots, C_m$. The number of variables n will be the n for the input of SRL. Recall that $\lambda = 4n + 5$. For every $j = 1, 2, \ldots, m$ the set S_j is constructed from the clause C_j as follows

$$S_j = \bigcup_{i: x_i \in C_j} \{2i, \lambda - 2i\} \cup \bigcup_{i: \neg x_i \in C_j} \{2i+1, \lambda - 2i - 1\}.$$

Thus every set S_j has 4 or 6 elements and is symmetric within $[0, \lambda]$.

Assume that Φ allows a satisfying assignment. If a variable x_i is assigned the value true, we set $t_i = 2i - 1$, $u_i = \lambda - 2i - 1$. Analogously for x_i negatively valued, we let $t_i = 2i$, $u_i = \lambda - 2i$. For each clause C_j we choose one satisfying literal. If C_j is satisfied by the literal x_i for some $i = 1, 2, \ldots, n$, we let $s_j = 2i$, if x_i is the first occurrence of x_i in Φ, and $s_j = \lambda - 2i$ for the second occurrence of x_i in Φ. In the case C_j is satisfied by $\neg x_i$ we choose $s_j = 2i + 1$ for the first occurrence of $\neg x_i$ and $s_j = \lambda - 2i - 1$ otherwise. Straightforwardly, the collection $s_1, \ldots, u_n$ satisfies all four properties from the definition of the SRL problem.

For the opposite direction suppose that $s_1, \ldots, u_n$ is a valid solution for the SRL problem. The crucial observation is that for every $i = 1, 2, \ldots, n$, there are only two possible choices for the values of t_i and u_i so that $|t_i - u_i| \geq 2$. Namely, either $t_i = 2i$ and $u_i = \lambda - 2i$ or alternatively $t_i = 2i + 1$ and $u_i = \lambda - 2i - 1$. In the first case we assign $x_i = $ false, and in the second one $x_i = $ true.

Then for each $j = 1, \ldots, m$, the value of u_j indicates the satisfying literal for the clause C_j: If $u_j = 2i$ or $\lambda - 2i$, then C_j is satisfied by the true assignment to the variable x_i. Alternatively, if $u_j = 2i + 1$ or $\lambda - 2i - i$ then the literal $\neg x_i$ satisfies C_j as the variable x_i is assigned false.

Since the size of the family $S_1, S_2, \ldots, S_m$ is polynomial in the size of Φ, 2-3-SAT $\propto$ SRL as claimed.

3 $L(2,1)$-Labeling of Graphs of Treewidth Two

This entire section is devoted to the proof of Theorem 1. We will utilize Lemma 1 and reduce from the SRL problem. Suppose we are given integers n and $\lambda = 4n + 5$, and m subsets $S_1, \ldots, S_m$ of $[2, \lambda - 2]$ which are all symmetric within this interval (we may further assume that all of them have size at most 6, but this is not important for our proof). Our aim is to construct a graph G' of treewidth two such that $L_{(2,1)}(G') \leq \lambda$ if and only if the given instance of SRL is feasible. The construction of G' is achieved in several steps.

3.1 Reduction to List Labeling

Construct the graph G on vertices $V_G = (v_0, v_1^s, \ldots, v_m^s, v_1^t, \ldots, v_n^t, v_1^u, \ldots, v_n^u)$ where v_0 is adjacent to all other vertices, and futhermore $(v_i^t, v_i^u) \in E_G$ for all $i = 1, \ldots, n$. (See Fig. 1.) To each vertex of $x \in V_G$ we assign a set of admissible labels as follows

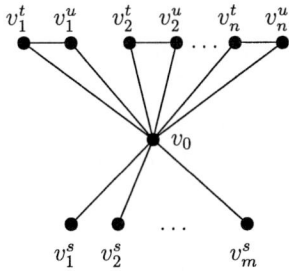

Fig. 1. The graph G

- $T(v_0) = \{0, \lambda\}$
- $T(v_i^s) = S_i$ for all $i = 1, \ldots, m$
- $T(v_i^t) = \{2i, \lambda - 2i - 1\}$ for all $i = 1, \ldots, n$
- $T(v_i^u) = \{2i + 1, \lambda - 2i\}$ for all $i = 1, \ldots, n$

and we call an $L(2,1)$-labeling c admissible if $c(x) \in T(x)$ for every $x \in V_G$. In any admissible $L(2,1)$-labeling, any pair of vertices must get distinct labels since G has diameter two. Moreover, as the vertices v_i^t and v_i^u are adjacent, they must be assigned labels that are at least two apart.

Hence $c(v_i^s) = s_i$, $c(v_i^t) = t_i$, and $c(v_i^u) = u_i$ is a one-to-one correspondence between admissible $L(2,1)$-labelings of G and systems of special representatives for $S_1, \ldots, S_m$ (the choice of $c(v_0) = 0$ or λ does not interfere with the labels of the remaining vertices). The graph G has clearly treewidth two. We will further design a collection of gadgets that will force the desired lists on the vertices of the graph G.

3.2 Labels of Neighbors of Vertices of Large Degrees

The following simple observation will be used repeatedly in our arguments. Let v be a vertex whose two neighbors w and w' have degree $\lambda - 1$, and let c be an $L(2,1)$-labeling of span λ. Denote $S = c(N(w) \setminus \{v\})$ the set of labels used on the neighbors of w other than v. Since w and w' have the maximum possible degree, they are assigned labels 0 and λ, and hence $c(v) \in [2, \lambda - 2] \setminus S$.

3.3 The Crucial Gadget

For every $i \in [1, \frac{\lambda-1}{2}]$, we construct the graph H_i with nonadjacent vertices z_i, z'_i of degree one inductively as follows.

1) H_0 is the cycle of length four and z_0, z'_0 are two nonadjacent vertices (of degree two).

2) To construct H_{i+1}, we take the graph H_i and

- insert the edge (z_i, z'_i),
- insert two new vertices z_{i+1}, z'_{i+1} and edges $(z_i, z_{i+1}), (z'_i, z'_{i+1})$,
- insert $\lambda - 5$ new common neighbors of z_i and z'_i.

(See Fig. 2 for an example.) Then H_i is a series-parallel graph whose number of vertices is polynomial in i and n (precisely, $|V_{H_i}| = i(\lambda - 3) + 4$). It has the following crucial property.

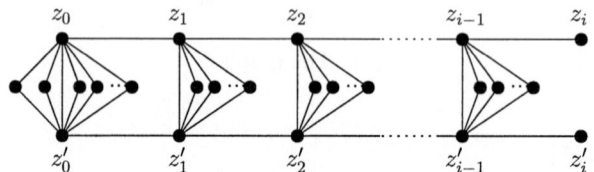

Fig. 2. Construction of the graph H_i

Lemma 2. *For every $i \geq 1$, in any $L(2,1)$-labeling of H_i of span λ, the vertices $z_{i-1}, z_i, z'_{i-1}, z'_i$ are assigned (in this order) labels $i-1, \lambda-i, \lambda-i+1, i$ or $\lambda-i+1, i, i-1, \lambda-i$.*

Proof. We prove the statement by induction on i. Let c be an $L(2,1)$-labeling of H_i of span λ.

1) For $i = 1$, observe that since z_0 and z'_0 have degree $\lambda - 1$, they must be assigned labels 0 and λ, or vice versa. Their $\lambda-3$ common neighbors are assigned distinct labels forming the interval $[2, \lambda-2]$ and hence $\{c(z_1), c(z'_1)\} = \{1, \lambda-1\}$.

2) By induction hypothesis, $\{c(z_{i-1}), c(z'_{i-1})\} = \{i-1, \lambda-i+1\}$ and $\{c(z_i), c(z'_i)\} = \{i, \lambda-i\}$. These two vertices have further $\lambda-5$ common neighbors that could be assigned only the labels forming the set $[0, i-2] \cup [i+2, \lambda-i-2] \cup [\lambda-i+2, \lambda]$. It is therefore easy to conclude that the two triples $(c(z_{i-1}), c(z'_i), c(z_{i+1}))$ and $(c(z'_{i-1}), c(z_i), c(z'_{i+1}))$ could be only the two consecutive triples $(i-1, i, i+1)$ and $(\lambda-i+1, \lambda-i, \lambda-i-1)$.

3.4 Forcing $T(v_0)$

Add $\lambda - 1 - 2n - m = 2n + 4 - m$ new neighbors to the vertex v_0. (We may assume $2n + 4 - m \geq 0$ since the SRL problem trivially has no system of distinct representatives if $2n + m > 2\lambda - 1$.) Then v_0 has degree $\lambda - 1$ and it can be assigned only labels 0 or λ by any $L(2,1)$-labeling of span λ.

3.5 Forcing $T(v_i^s)$

For each vertex $i \in [1, m]$, insert a new vertex x_i and make it adjacent to v_i^s. Further for each pair of labels l and $\lambda - l$ in the set $[2, \lambda - 2] \setminus S_i$, insert a new copy of the graph H_l and make x_i adjacent to the vertices z_l and z_l' of this new copy. Finally, add further new neighbors to the vertex x_i so that it has degree $\lambda - 1$ (see Fig. 3). It follows from the observation in 3.2 and Lemma 2 that the vertex v_i^s is now allowed to be assigned only a label from the set S_i as required.

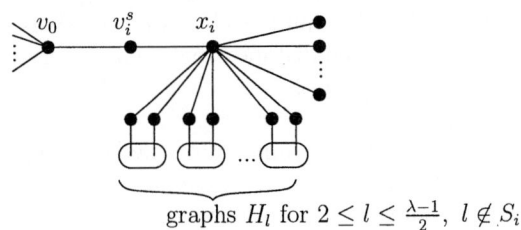

graphs H_l for $2 \leq l \leq \frac{\lambda-1}{2}$, $l \notin S_i$

Fig. 3. Forcing list S_i on the vertex v_i^s

3.6 Forcing $T(v_i^t)$ and $T(v_i^u)$

For each $i \in [1, n]$, insert vertices y_i, y_i' adjacent to v_i^t and v_i^u, respectively. Further take a copy of the graph H_{2i+1}, remove one common neighbor of z_{2i} and z_{2i}' and make y_i adjacent to z_{2i}, z_{2i+1} and y_i' to z_{2i}', z_{2i+1}' of this copy. For each label $l \in [2, \frac{\lambda-1}{2}] \setminus \{2i, 2i+1\}$, insert two new copies of the graph H_l (the second copy is denoted by H_l^*) and connect both vertices z_l, z_l^* to y_i and both $z_l', z_l^{*'}$ to y_i'. Finally, add three new neighbors to each vertex y_i, y_i' so that both have degree $\lambda - 1$ (see Fig. 4).

Suppose c is an $L(2, 1)$-labeling of span λ. Since both y_i, y_i' have degree $\lambda - 1$ and are at distance 2 from v_0 of the same degree, they are both assigned the same label, either 0 or λ. It also follows that in the copy H_{2i+1} the vertices z_{2i}, z_{2i}'

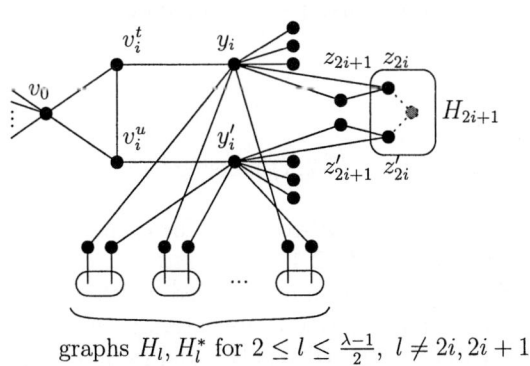

graphs H_l, H_l^* for $2 \leq l \leq \frac{\lambda-1}{2}$, $l \neq 2i, 2i+1$

Fig. 4. Forcing lists $\{2i, \lambda - 2i - 1\}$ and $\{2i+1, \lambda - 2i\}$ on the vertices v_i^t, v_i^u

behave as stated in Lemma 2, even if we removed one common neighbor (whose role was taken over by y_i and y'_i). Now according to observation in 3.2, the vertex v_i^t can be assigned only labels from $\{2i, 2i+1, \lambda-2i-1, \lambda-2i\} \setminus \{c(z_{2i}), c(z_{2i+1})\}$ and similarly for $c(v_i^u) \in \{2i, 2i+1, \lambda-2i-1, \lambda-2i\} \setminus \{c(z'_{2i}), c(z'_{2i+1})\}$. Since by Lemma 2 either $\{c(z_{2i}), c(z_{2i+1})\} = \{\lambda-2i, 2i+1\}$ or $\{2i, \lambda-2i-1\}$, and respectively, $\{c(z'_{2i}), c(z'_{2i+1})\} = \{2i, \lambda-2i-1\}$ or $\{\lambda-2i, 2i+1\}$, we get the desired admissible sets for both v_i^t and v_i^u (note here that the entire construction is symmetric with respect to vertices v_i^t and v_i^u).

By the above discussion, any $L(2,1)$-labeling of the resulting graph G' forces every vertex x of its subgraph G to be assigned labels from the list $T(x)$. During the construction of G' the distances between the original vertices of G were not changed, and hence any $L(2,1)$-labeling of G' restricted to G is an admissible $L(2,1)$-labeling for the lists $T(x), x \in V_G$.

The proof of the opposite implication (i.e., that any admissible $L(2,1)$-labeling of G can be extended to an $L(2,1)$-labeling of G') follows from the construction of all the gadgets and is straightforward.

Finally, observe that the size of G' is polynomial in the size of G (more precisely $|G'| = O(|G|^4)$, and also that all gadgets were constructed so that G' maintains treewidth two. This concludes the proof of Theorem 1.

4 (2,1)-Labelings of Graphs of Bounded Treewidth

Given graphs G and H, an $H(2,1)$-labeling of G is a mapping $f : V_G \longrightarrow V_H$ such that adjacent vertices of G are mapped onto distinct nonadjacent vertices of H (i.e., distance of the target vertices is at least 2, measured in the target graph H) and vertices with a common neighbor in G are mapped onto distinct vertices of H (i.e., the distance of the target vertices is at least 1) [9]. This definition generalizes both the $L(2,1)$-labelings (when H is a path whose length equals the span of the labeling) and the $C(2,1)$-labelings (when H is a cycle whose length again equals the span). The computational complexity of this problem for fixed parameter graphs H was studied and many particular results were proven in [8]. The case when the span is also part of the input corresponds to the following decision problem:

(2, 1)-LABELING
Input: Graphs G and H.
Question: Does G allow an $H(2,1)$-labeling?

Of course this problem is NP-complete for graphs G of treewidth two, since both L(2,1)-LABELING and C(2,1)-LABELING are its special cases. In this section we give a subtler separation of bounded width classes, namely in terms of pathwidth. Graphs of pathwidth one are caterpillars (trees obtained by pending any number of leaves to vertices of a path), and so the claim that (2, 1)-LABELING is solvable in polynomial time for graphs G of pathwidth one (and arbitrary H) follows from our Theorem 3.

Proof (of Theorem 3). The following algorithm is a straightforward extension of the algorithm for $L(2,1)$-labeling of trees of [4].

Given a tree T with m vertices, choose a leaf $r \in V_T$ and regard it as a root of T. For every edge $(u,v) \in E_T$ such that u is a child of v, denote by $T_{u,v}$ the subtree of T rooted in v and containing u and all its descendants. For every such edge and for every pair of vertices $x, y \in V_H$, we introduce a Boolean variable $\phi(u, v, x, y)$ which is true if and only if $T_{u,v}$ allows an $H(2,1)$-labeling f such that $f(u) = x$ and $f(v) = y$. Then T allows an $H(2,1)$-labeling if and only if $\phi(u, r, x, y) = \text{true}$ for some vertices $x, y \in V_H$ (and u being the only child of the root r). The function ϕ can be computed by the following dynamic programming algorithm:

1. Set the initial values $\phi(u, v, x, y) = \text{false}$ for all edges $(u,v) \in E_T$ and vertices $x, y \in V_H$.
2. If u is a leaf of T adjacent to its parent v, then set $\phi(u, v, x, y) = \text{true}$ for all distinct nonadjacent vertices $x, y \in V_H$.
3. Suppose that ϕ is already calculated for all edges of $T_{u,v}$ except (u,v). Denote by $v_1, v_2, \ldots, v_k$ the children of u. For all pairs of distinct nonadjacent vertices $x, y \in V_H$, construct the set system $\{M_1, M_2, \ldots, M_k\}$, where

$$M_i = \{z : z \in V_H, z \neq y \text{ and } \phi(v_i, u, z, x) = \text{true}\}$$

and set $\phi(u, v, x, y) = \text{true}$ if the set system $\{M_1, M_2, \ldots, M_k\}$ has a system of distinct representatives.

For the time analysis note that the recursive step requires, for each pair $x, y \in V_H$, time $O(nk)$ to construct the set system and time $O(k \cdot nk)$ for deciding if it has an SDR (e.g., by using the augmenting paths algorithm for a bipartite graph with at most nk edges and with k vertices in one bipartition class). Altogether the recursive step requires time $O(n^3k^2)$. If we denote by k_u the number of children of a nonleaf vertex $u \in V_T$, we have $\sum_{u \in V_T} k_u = m - 1$ (the number of edges of T), and hence the total running time is majorized by $O(\sum_{u \in V_T} n^3 k_u^2) = O(n^3 \sum_{u \in V_T} k_u^2) = O(n^3 (\sum_{u \in V_T} k_u)^2) = O(n^3 m^2)$.

A *2-path* is a graph constructed from a triangle (say Δ_0) by consecutive augmentation of triangles so that each Δ_i shares one edge with the previously augmented Δ_{i-1}, while the third vertex of Δ_i is a vertex newly added in this step. A graph has pathwidth at most two if and only if it is a subgraph of a 2-path. In particular, a fan of triangles obtained from a path by adding a vertex adjacent to all vertices of the path, has pathwidth two.

Proof (of Theorem 4). We reduce from HAMILTONIAN PATH which is well known to be NP-complete [13]. Given a graph H' with n vertices, let H be the disjoint union of the complement of H' and an isolated vertex x. Let G be obtained from a path of length $n-1$ on vertices $v_1, v_2, \ldots, v_n$ by adding a vertex w, which is adjacent to all v_i's. Then every $H(2,1)$-labeling f of G is an injective mapping from V_G to V_H (since G has diameter two), and f is necessarily bijective (since $|V_G| = |V_H|$). Without loss of generality $f(w) = x$, and hence $f(v_1), f(v_2), \ldots, f(v_n)$ is a Hamiltonian path in H', since $(v_i, v_{i+1}) \in E_G$ implies that $(f(v_i), f(v_{i+1})) \notin E_H$. The opposite implication is straightforward.

5 Concluding Remarks

1. We have fully characterized the computational complexity of (2,1)-distance constrained graph labelings in the case of linear and cyclic metrics in the channel space, with respect to the treewidth of the input graphs. Our results prove polynomial/NP-completeness dichotomy separating treewidth one from treewidth two, which is a rare phenomenon and has so far been known only for very few problems (namely the CUTWIDTH or MINIMUM LINEAR ARRANGEMENT which is polynomial for trees [25] while NP-hardness for graphs of treewidth two follows from [20]). With distance constrained labelings we have added a natural and important problem to this short list.

2. Let us remark that our main result is independent on the NP-completeness of the CHANNEL ASSIGNMENT problem, though both problems are related by the motivation in frequency assignment. The CHANNEL ASSIGNMENT is known NP-complete for graphs of treewidth three, but its complexity for treewidth two graphs is still open. The core of the NP-hardness of the two problems lies in different aspects of the problems and one does not straightforwardly follow from the other. On one hand, L(2,1)-LABELING relays to CHANNEL ASSIGNMENT by considering the second (distance) power of the input graph and assigning weights 2 to the original edges and 1 to the new ones. However, the graph constructed in this way will not have bounded treewidth. On the other hand, L(2,1)-LABELING involves only weights 2 and 1, while the NP-hardness of the CHANNEL ASSIGNMENT problem is based on large weights, the problem is not strongly NP-complete for graphs of bounded treewidth (it can be solved by dynamic programming algorithm in polynomial time if the weights are considered in unary encoding).

3. In the general (2,1)-LABELING problem, when both graphs come as parts of the input, we prove tight dichotomy with respect to pathwidth of the input (transmitters) graph. For both special metrics, $L(2,1)$ and $C(2,1)$, the complexity for graphs of bounded pathwidth is open.

4. To keep the paper well focused, we have stated most of the results for the simplest case of distance constraints (2,1). However, most of them can be extended to (p,q)- or at least $(p,1)$-labelings, see e.g. Proposition 1. It is known that L$(p,1)$-LABELING is polynomial for trees for every p (even the list and prelabeled versions), but when q does not divide p, the complexity of L(p,q)-LABELING for trees is open for all $q > 1$ (the list and prelabeled versions are known to be NP-complete [10]). To the contrary, C(p,q)-LABELING is polynomial for trees for all p,q as is shown by Proposition 1. Extension of our Theorem 3 to general $(p,1)$-LABELING of trees is trivial, since that follows by replacing H by its p-th distance power. An analog of Theorem 4 for general (p,q)-labelings can be proved by a more technical reduction.

References

1. ARNBORG, S., AND PROSKUROWSKI, A. Characterization and recognition of partial k-trees *Congr. Numerantium 47* (1985) 69-75.
2. BODLAENDER H. Polynomial algorithms for Graph Isomorphism and Chromatic Index on partial k-trees *J. Algorithms 11* (1990) 631-643.
3. BODLAENDER H. A tourist guide through treewidth *Acta Cybern. 11* (1993) 1-21.
4. CHANG, G. J., AND KUO, D. The $L(2,1)$-labeling problem on graphs. *SIAM Journal of Discrete Mathematics 9*, 2 (May 1996), 309–316.
5. COURCELLE, B. The monadic second-order logic of graphs. I: Recognizable sets of finite graphs. *Inf. Comput. 85*, 1 (1990), 12–75.
6. FELLOWS, M.R.; KRATOCHVÍL, J.; MIDDENDORF, M.; PFEIFFER, F. The complexity of induced minors and related problems *Algorithmica 13* (1995) 266-282.
7. FIALA, J., KLOKS, T., AND KRATOCHVÍL, J. Fixed-parameter complexity of λ-labelings *Discrete Applied Math. 113* (2001), 59-72.
8. FIALA, J., AND KRATOCHVÍL, J. Complexity of partial covers of graphs In: Algorithms and Computation (P.Eades and T.Takaoka, eds.), Proceedings ISAAC 2001, Christchurch, December 2001, Lecture Notes in Computer Science 2223, Springer Verlag, Berlin Heidelberg 2001, pp. 537-549
9. FIALA, J., AND KRATOCHVÍL, J. Partial covers of graphs *Discussiones Mathematicae Graph Theory 22* (2002) 89-99.
10. FIALA, J., KRATOCHVÍL, J., AND PROSKUROWSKI, A. Distance constrained labelings of precolored trees. In *Theoretical Computer Science, 7th ICTCS '01, Torino* (2001), no. 2202 in Lecture Notes in Computer Science, Springer Verlag, pp. 285–292.
11. FIALA, J., KRATOCHVÍL, J., AND PROSKUROWSKI, A. Systems of distant representatives. *Discrete Applied Mathematics 145*, 2 (2005), 306–316.
12. GAREY, M., GRAHAM, R., JOHNSON, D., AND KNUTH, D. Complexity results for bandwidth minimization. *SIAM J. Appl. Math. 34* (1978), 477–495.
13. GAREY, M. R., AND JOHNSON, D. S. *Computers and Intractability.* W. H. Freeman and Co., New York, 1979.
14. GOLOVACH, P. A. Systems of pairs of q-distant representatives and graph colorings. *Zap. nau. sem. POMI 293* (2002), 5–25. in Russian.
15. GRIGGS, J. R., AND YEH, R. K. Labelling graphs with a condition at distance 2. *SIAM Journal of Discrete Mathematics 5*, 4 (Nov 1992), 586–595.
16. KRÁL, D., AND SKREKOVSKI, R. A theorem about the channel assignment problem *SIAM J. Discrete Math. 16*, No.3 (2003) 426-437.
17. LEESE, R. A., AND NOBLE, S. D. Cyclic labellings with constraints at two distances. *Electr. J. Comb. 11*, 1 (2004).
18. LIU, D., AND ZHU, X. Circular Distance Two Labelings and Circular Chromatic Numbers. *Ars Combin. 69*, 4 (2003), 177–183.
19. MCDIARMID, C., AND REED, B. Channel assignment on graphs of bounded treewidth. *Discrete Math. 273*, 1-3 (2003), 183–192.

20. MONIEN, B., AND SUDBOROUGH, I. H. Min Cut is NP-Complete for Edge Weigthed Trees *Theor. Comput. Sci. 58*, No.1-3, (1988) 209-229.
21. ROBERTS, F.S. private communication to J. Griggs.
22. ROBERTSON, N., AND SEYMOUR, P. Graph minors. I. Excluding a forest. *J. Comb. Theory, Ser. B 35* (1983), 39–61.
23. PROSKUROWSKI, A. Separating subgraphs in k-trees: cables and caterpillars. *Discrete Math. 49* (1984), 275–285.
24. VAN DEN HEUVEL, J., LEESE, R. A., AND SHEPHERD, M. A. Graph labeling and radio channel assignment. *Journal of Graph Theory 29*, 4 (1998), 263–283.
25. YANNAKAKIS, M. A polynomial algoritm for the min-cut linear arrangements of trees *J. ACM 32* (1985) 950-988.

Optimal Branch-Decomposition of Planar Graphs in $O(n^3)$ Time

Qian-Ping Gu[1] and Hisao Tamaki[2]

[1] School of Computing Science,
Simon Fraser University, Burnaby BC Canada V5A 1S6
qgu@cs.sfu.ca
[2] School of Science and Technology, Meiji University,
Kawasaki, 214-8571 Japan
tamaki@cs.meiji.ac.jp

Abstract. We give an $O(n^3)$ time algorithm for constructing a minimum-width branch-decomposition of a given planar graph with n vertices. This is achieved through a refinement to the previously best known algorithm of Seymour and Thomas, which runs in $O(n^4)$ time.

1 Introduction

Let G be a graph. (Unless otherwise stated, all graphs in this paper are undirected). A *branch-decomposition* of G is a tree T such that the set of leaves of T is $E(G)$ and each internal node of T has degree 3. For each edge $\{p,q\}$ of T, let $L_T(p,q) \subseteq E(G)$ denote the set of leaves of T reachable in T from p without passing q. The *width* of edge $\{p,q\}$ in T with respect to G is the number of vertices of G incident with both an edge in $L_T(p,q)$ and an edge in $L_T(q,p)$. The width of T with respect to G is the maximum width of all the edges of T. Finally, the *branchwidth* of G is the minimum width of all the branch-decompositions of G.

Branchwidth is introduced by Robertson and Seymour [8] in relation to the more celebrated notion of treewidth [6,7]. They show that the branchwidth and treewidth of a graph G are very closely related, not only structurally but also in values: $\beta(G) \leq \omega(G) + 1 \leq \lfloor (3/2)\beta(G) \rfloor$, where $\beta(G)$ denotes the branchwidth and $\omega(G)$ the treewidth of G. A graph of small branchwidth or treewidth admits efficient dynamic programming algorithms for a vast class of problems [2,3] and therefore the problem of deciding the width and that of constructing the associated decomposition are of extreme importance. When the width of the graph is bounded by a constant, both of these tasks can be performed in linear time [4]. However, deciding whether the width of a given graph exceeds k is NP-complete if k is part of the input, both for treewidth [1] and for branchwidth [9]. Thus the tractability of treewidth and branchwidth are quite similar on general graphs. The situation is different on planar graphs: while no polynomial time algorithm is known for deciding the treewidth of a given planar graph, Seymour and Thomas [9] give a polynomial time decision algorithm for the branchwidth of planar graphs. Given a planar graph G with n vertices and a positive integer

k, their algorithm takes $O(n^2)$ time to decide whether G has a branchwidth not exceeding k. Through the relationship above between the two width values, this also gives a 1.5-approximation of the treewidth of planar graphs. Based on this result, Demaine et al. [5] develop approximation algorithms for the treewidth and branchwidth of graphs more general than planar graphs. For some other classes of graphs for which the treewidth can be computed in polynomial time, see Bodlaender's survey [3].

The decision algorithm of Seymour and Thomas for planar graphs works with a certain type of obstructions to the branch-decomposition and returns a positive answer when it detects the absence of such obstructions. Unfortunately, the evidence of the absence of obstructions that the algorithm constructs does not seem to provide a direct clue to the structure of a branch-decomposition within the specified width. Nonetheless, using the decision procedure as a subroutine, they give, in the same paper, an algorithm for constructing a minimum-width branch-decomposition of a given planar graph. This algorithm calls the decision procedure $O(n^2)$ times and runs in $O(n^4)$ time. We improve the running time to $O(n^3)$ in this paper.

Theorem 1. *Given a planar graph G with n vertices, a minimum-width branch-decomposition of G can be constructed in $O(n^3)$ time.*

We achieve this improvement within the framework of Seymour and Thomas through a non-trivial refinement that reduces the number of times the decision procedure is called to $O(n)$.

The rest of this paper is organized as follows. Section 2 gives basic material and summarizes the approach of Seymour and Thomas. Section 3 proves a key lemma that enables our savings in the use of the expensive decision procedure. Section 4 describes our refinement to the algorithm based on this lemma and proves that the number of calls to the decision procedure is $O(n)$ during the execution of the algorithm with this refinement. Section 5 sketches a rather straightforward analysis that the rest of the computation takes $O(n^3)$ time.

2 Preliminaries

In this paper, graphs are simple, i.e., without parallel edges or self-loops, unless they are explicitly referred to as multigraphs. We have already used some of the following standard graph notation. For a multigraph G, we denote by $V(G)$ the set of vertices of G and by $E(G)$ the set of edges of G. For each $X \subseteq V(G)$, we denote by $\delta_G(X)$ the set of edges incident with a vertex in X and a vertex in $V(G) \setminus X$. For disjoint $U, V \subseteq V(G)$, we denote by $E_G(U, V)$ the edge set $\delta_G(U) \cap \delta_G(V)$. We use symbols $\supset$ and $\subset$ for proper set inclusion.

Both the branchwidth and branch-decomposition algorithms of Seymour and Thomas for planar graphs use a reduction to the *carvingwidth/-decomposition* of a related multigraph.

A carving-decomposition of a multigraph G is a tree T such that the set of leaves of T is $V(G)$ (not $E(G)$ as in branch-decomposition) and the degree of each internal node of T is 3. The width of T with respect to G is defined to be the maximum of $\delta_G(L'_T(p,q))$ over all edges $\{p,q\}$ of T, where $L'_T(p,q) \subseteq V(G)$ is the set of leaves of T reachable in T from p without passing q. The *carvingwidth* of G is the minimum width of all the carving-decompositions of G.

Let G be a planar graph with a fixed embedding. The *medial graph* [9] $M(G)$ of G is the planar dual of the graph of incidences between the vertex set and the face set of G. Note that $M(G)$ in general is a multigraph but the number of edges of $M(G)$ is linear in the number of vertices of G. Seymour and Thomas [9] show that the carvingwidth of $M(G)$ is exactly twice the branchwidth of G. Moreover, an optimal carving-decomposition of $M(G)$ can be translated into an optimal branch-decomposition of G in linear time. The algorithms of Seymour and Thomas mentioned in the introduction indeed work on carvingwidth and decompositions. Similarly, Theorem 1 is an immediate corollary to the following theorem, which we prove in the sequel.

Theorem 2. *Given a planar multigraph G with n vertices and $O(n)$ edges, a minimum-width carving-decomposition of G can be constructed in $O(n^3)$ time.*

This is an improvement over the $O(n^4)$ result of Seymour and Thomas. We must remark, however, that their result for carving-decompositions is more general: it allows positive integer weights on edges and defines the width as the sum of these edge weights. Although this generalization does not seem to be relevant for our present application, namely the branch-decomposition, it does make a difference for another application of carving-decompositions, namely call routing. See [9] for details. Our algorithm runs for this general problem in $O(\min\{n^4, Wn^2\})$ time, where W is the total weight of the edges.

In the rest of this paper, G will always denote the planar multigraph that is the input to our algorithm, n the number of vertices of G, and k the carvingwidth of G. We also assume that G has at least 2 vertices and the number of edges of G is $O(n)$.

For the purpose of our prospective refinement, it turns out convenient to describe the carving decomposition algorithm of Seymour and Thomas [9] in a slightly different language.

We begin with viewing carving-decompositions as rooted binary trees. Let $\sigma(V)$ denote the set of all singleton subsets of a finite set V. A *binary merging* in V is defined inductively as follows.

1. $\sigma(V)$ is a binary merging in V.
2. If $\mathcal{M}$ is a binary merging in V and X, Y are two distinct maximal subsets of V in $\mathcal{M}$, then $\mathcal{M} \cup \{X \cup Y\}$ is a binary merging in V.

We say that a binary merging in V is *full* if V belongs to it. Each full binary merging $\mathcal{M}$ in $V(G)$ represents a carving decomposition $T_\mathcal{M}$ of G as follows. The node set of $T_\mathcal{M}$ is $\mathcal{M} \setminus \{V(G)\}$ and $T_\mathcal{M}$ has an edge between $X, Y \in \mathcal{M} \setminus \{V(G)\}$ if and only if X is a maximal proper subset of Y in $\mathcal{M}$, Y is a maximal proper subset of X in $\mathcal{M}$, or $X \cup Y = V(G)$. We may view $\mathcal{M}$ as representing a rooted version of tree $T_\mathcal{M}$, where the root is placed in the middle of the edge of the last type above. We define the width of a binary merging $\mathcal{M}$ in $V(G)$ to be the maximum of $|\delta_G(X)|$ over all $X \in \mathcal{M}$. When $\mathcal{M}$ is full, this definition of width coincides with the width of the carving decomposition $T_\mathcal{M}$. We say that a binary merging $\mathcal{M}'$ *extends* a binary merging $\mathcal{M}$ if $\mathcal{M}' \supseteq \mathcal{M}$. Let $\mathcal{M}$ be a binary merging in $V(G)$ with width k. We say that maximal elements X, Y of $\mathcal{M}$ are *mergeable* in $\mathcal{M}$ if there is a full binary merging of width k that extends $\mathcal{M} \cup \{X \cup Y\}$. We say disjoint subsets X, Y of $V(G)$ are *adjacent* if $E_G(X, Y) \neq \emptyset$.

We formulate the carving-decomposition algorithm of Seymour and Thomas in terms of binary mergings as follows. The algorithm grows a binary merging starting from $\sigma(V(G))$ towards a full one, keeping the invariant that the binary merging maintained is extendable to a full binary merging of width k. We assume that the input multigraph G is connected: if not, we can work on each connected component and trivially combine the results.

Algorithm ST
Input: A connected planar multigraph G with n vertices and $O(n)$ edges.
Output: A carving-decomposition in $V(G)$ of minimum width
(1) Decide the carvingwidth k of G.
(2) Set $\mathcal{M}^0 = \sigma(V(G))$, the set of all singleton subsets of $V(G)$. Set $i = 1$.
(3) Repeat the following while $i < n$.
(3.1) Scan adjacent pairs $\{X, Y\}$ of maximal elements of $\mathcal{M}^{i-1}$, testing if X and Y are mergeable, until a mergeable pair $\{X, Y\}$ is found, and set $X^i = X \cup Y$.
(3.2) Set $\mathcal{M}^i = \mathcal{M}^{i-1} \cup \{X^i\}$ and set i to $i + 1$.
(4) Return the carving-decomposition represented by $\mathcal{M}^{n-1}$.

The correctness proof of this algorithm consists in showing, by induction on i, that each $\mathcal{M}^i$ is a well-defined binary merging in $V(G)$ and is extendable to a full binary merging of width k. The base case is trivial and the induction step is a straightforward consequence of the following lemma.

Lemma 1. *Let $\mathcal{M}$ be a binary merging in $V(G)$ that is extendable to a full binary merging of the optimal width k. Then there are adjacent maximal elements X and Y of $\mathcal{M}$ that are mergeable.*

Proof. The following proof sketch is a translation of the original argument by Seymour and Thomas into our language. Let $G_\mathcal{M}$ denote the multigraph whose vertex set is the set of maximal elements of $\mathcal{M}$ and the edge set is $E(G)$, with each pair of vertices X and Y of this multigraph connected by all the edges in $E_G(X, Y)$. Since $\mathcal{M}$ is extendable to a full binary merging of width k, the carvingwidth of $G_\mathcal{M}$ is k or smaller. Seymour and Thomas [9] show that every connected planar multigraph H of carvingwidth k has a *bond-carving* of width k, where a bond-carving of H is a carving-decomposition T of H such that for every edge $\{p, q\}$ of T, the subgraph of H induced by $L'_T(p, q) \subseteq V(H)$ is connected. Let T be a bond-carving of $G_\mathcal{M}$ with width k or smaller. Then, T contains an internal node p_0 adjacent to two distinct leaves $X, Y \in V(G_\mathcal{M})$. Since T is a bond-carving, X and Y must be adjacent in $G_\mathcal{M}$. Let $\mathcal{M}' = \mathcal{M} \cup \{X \cup Y\}$. We may convert T into a carving-decomposition T' of $G_{\mathcal{M}'}$ by replacing p_0, together with the two incident leaves X and Y, by a leaf $X \cup Y$. Clearly the width of T' is k or smaller. Following the structure of T', we may extend $\mathcal{M}'$ into a full binary merging of width k. Therefore, the adjacent pair X and Y are mergeable in $\mathcal{M}$. □

To test if X and Y are mergeable in $\mathcal{M}$, we call the $O(n^2)$ time procedure of Seymour and Thomas to decide if $G_{\mathcal{M}'}$ has carving width not exceeding k, where $\mathcal{M}' = \mathcal{M} \cup \{X \cup Y\}$. We remark that their decision algorithm works for planar multigraphs, where the running time is quadratic in the number of edges. We call this mergeability test an *expensive test*. Since $O(n^2)$ expensive tests are performed during the execution of algorithm ST and the time for this task is clearly dominating, algorithm ST runs in $O(n^4)$ time.

3 Barrier Lemma

In this section, we prove a key lemma, called the barrier lemma, which enables our savings in the use of expensive tests.

Lemma 2. *Let $\mathcal{M}$ be a binary merging in $V(G)$ and X, Y distinct maximal elements of $\mathcal{M}$. Suppose that $X = X' \cup Z$ and $Y = Y' \cup W$ where $X', Y', Z, W \in \mathcal{M}$ and that $E_G(Z, W) = \emptyset$. If X and Y are mergeable in $\mathcal{M}$ and $|\delta_G(X' \cup Y')| \leq k$ then X' and Y' are mergeable in $\mathcal{M} \setminus \{X, Y\}$.*

Proof. Let $\mathcal{M}' = \mathcal{M} \setminus \{X, Y\}$ and $\mathcal{M}^*$ a full binary merging in $V(G)$ with width k that extends $\mathcal{M} \cup \{X \cup Y\}$, which exists because X and Y are mergeable in $\mathcal{M}$. Let $\mathcal{M}_1 = \mathcal{M}' \cup \{X' \cup Y', X' \cup Y' \cup Z\} \cup (\mathcal{M}^* \setminus \mathcal{M})$ and $\mathcal{M}_2 = \mathcal{M}' \cup \{X' \cup Y', X' \cup Y' \cup W\} \cup (\mathcal{M}^* \setminus \mathcal{M})$. Then, both $\mathcal{M}_1$ and $\mathcal{M}_2$ are full binary mergings in $V(G)$ that extend $\mathcal{M}' \cup \{X' \cup Y'\}$. See Figure 1. Therefore, in order to show that X' and Y' are mergeable in $\mathcal{M}'$, it suffices to show that either $\mathcal{M}_1$ or $\mathcal{M}_2$ has width k. Since $\mathcal{M}_1 \setminus \mathcal{M}^* = \{X' \cup Y', X' \cup Y' \cup Z\}$, $\mathcal{M}_2 \setminus \mathcal{M}^* = \{X' \cup Y', X' \cup Y' \cup W\}$, and $|\delta_G(X' \cup Y')| \leq k$ by assumption, it in turn suffices to show that either $|\delta_G(X' \cup Y' \cup Z)| \leq k$ or $|\delta_G(X' \cup Y' \cup W)| \leq k$. Let $A = X' \cup Y'$ and $B = V(G) \setminus (X \cup Y)$. We have

$$\delta_G(A \cup Z) = E_G(A, B) \cup E_G(Z, B) \cup E_G(A, W),$$

since $E_G(Z, W) = \emptyset$ by assumption, and hence

$$|\delta_G(A \cup Z)| = |E_G(A, B)| + |E_G(Z, B)| + |E_G(A, W)|,$$

since the edge sets in the right hand side are disjoint. Similarly, we have

$$|\delta_G(A \cup W)| = |E_G(A, B)| + |E_G(W, B)| + |E_G(A, Z)|$$
$$|\delta_G(A)| = |E_G(A, B)| + |E_G(A, Z)| + |E_G(A, W)|$$
$$|\delta_G(B)| = |E_G(A, B)| + |E_G(Z, B)| + |E_G(W, B)|.$$

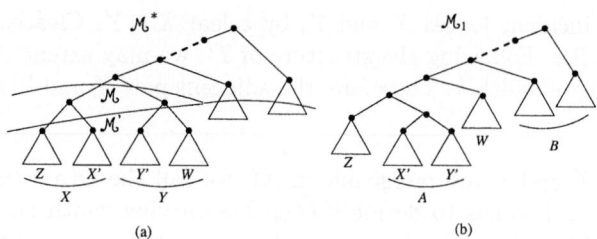

Fig. 1. Binary mergings in Lemma 2: (a) $\mathcal{M}, \mathcal{M}', \mathcal{M}^*$ and (b) $\mathcal{M}_1$

Therefore, we have

$$|\delta_G(A \cup Z)| + |\delta_G(A \cup W)|$$
$$= 2|E_G(A,B)| + |E_G(Z,B)| + |E_G(W,B)| + |E_G(A,Z)| + |E_G(A,W)|$$
$$= |\delta_G(A)| + |\delta_G(B)|.$$

Moreover, $|\delta_G(A)| \leq k$ is one of the assumptions in the lemma and we also have $|\delta_G(B)| = |\delta_G(X \cup Y)| \leq k$ since X and Y are mergeable in $\mathcal{M}$. Therefore, we have

$$|\delta_G(A \cup Z)| + |\delta_G(A \cup W)| \leq 2k$$

and hence either $|\delta_G(A \cup Z)| \leq k$ or $|\delta_G(A \cup W)| \leq k$ as desired. □

Let $\mathcal{M}$ be a binary merging in $V(G)$. A sequence $X_0 \supset X_1 \supset \ldots \supset X_m$ of elements of $\mathcal{M}$ is called a *chain* in $\mathcal{M}$ if each X_i for $1 \leq i \leq m$ is a maximal proper subset of X_{i-1} in $\mathcal{M}$. If we denote by C the chain above, then X_0 is called the *top* of C, denoted by $\top C$, and X_m is called the *bottom* of C, denoted by $\bot C$. A chain C is called a *barrier* if $|\delta_G(\top C \setminus \bot C)| > k$.

Lemma 3. *(Barrier lemma) Let $\mathcal{M}$ be a binary merging in $V(G)$ with width k or smaller and X, Y two maximal elements of $\mathcal{M}$. Suppose there is a binary merging $\mathcal{M}' \subseteq \mathcal{M}$ and maximal elements X' and Y' of $\mathcal{M}'$ with $X' \subseteq X$ and $Y' \subseteq Y$ such that the following conditions hold.*

1. *X and Y are mergeable in $\mathcal{M}$;*
2. *$|\delta(X' \cup Y')| \leq k$;*
3. *$E_G(X \setminus X', Y \setminus Y') = \emptyset$;*
4. *there is no barrier B in $\mathcal{M}$ with $\top B = X$ and $\bot B \supseteq X'$;*
5. *there is no barrier B in $\mathcal{M}$ with $\top B = Y$ and $\bot B \supseteq Y'$;*

Then, X' and Y' are mergeable in $\mathcal{M}'$.

Proof. Let $Z = X \setminus X'$ and $W = Y \setminus Y'$. Let $X_0 \supset X_1 \supset \ldots \supset X_h$ be the chain in $\mathcal{M}$ with top $X_0 = X$ and bottom $X_h = X'$. Similarly, let $Y_0 \supset Y_1 \supset \ldots \supset Y_j$ be the chain in $\mathcal{M}$ with top $Y_0 = Y$ and bottom $Y_j = Y'$. Let $\mathcal{M}_1$ be the binary merging in $V(G)$ obtained from $\mathcal{M}$ by removing X_i for $0 \leq i < h$ and Y_i for

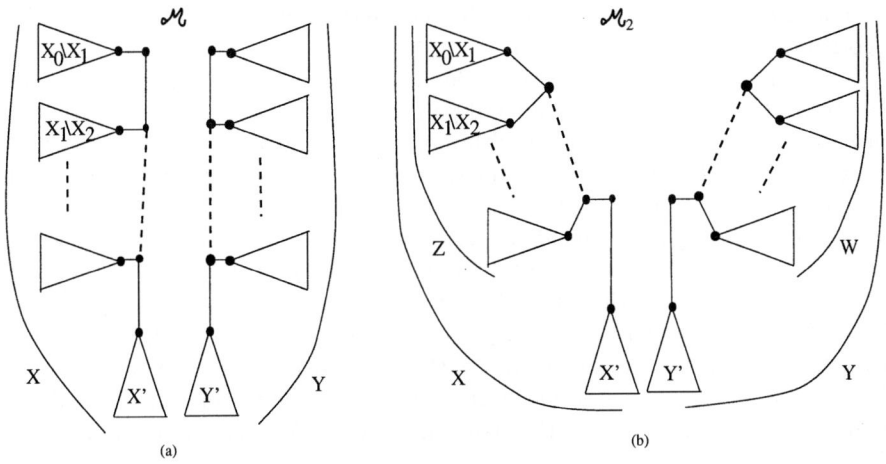

Fig. 2. Binary mergings in Lemma 3: (a) $\mathcal{M}$ is transformed into (b) $\mathcal{M}_2$

$0 \leq i < j$. Then, $\mathcal{M}_1$ extends $\mathcal{M}'$, $X_i \setminus X_{i+1}$ is a maximal element of $\mathcal{M}_1$ for $0 \leq i < h$, and $Y_i \setminus Y_{i+1}$ is a maximal element of $\mathcal{M}_1$ for $0 \leq i < j$. We extend $\mathcal{M}_1$ into $\mathcal{M}_2$ with X and Y as maximal elements, in which the two maximial subsets of X are X' and Z, and the two maximal subsets of Y are Y' and W, as follows. First, we successively extend $\mathcal{M}_1$ by adding $X_0 \setminus X_i$ for $i = 2, \ldots, h$, creating a chain with bottom $X_0 \setminus X_1$ and top $X_0 \setminus X_h = Z$. Similarly, we add a chain with bottom $Y_0 \setminus Y_1$ and top W. Finally, we add X and Y and call the result $\mathcal{M}_2$. See Figure 2. We have $\delta_G(X_0 \setminus X_i) \leq k$ for $i = 2, \ldots, h$, since if $\delta_G(X_0 \setminus X_i) > k$ then the chain in $\mathcal{M}$ with top X_0 and bottom X_i would be a barrier contradicting our assumption. Similarly, we have $\delta_G(Y_0 \setminus Y_i) \leq k$ for $i = 2, \ldots, j$. Since these are the only elements in $\mathcal{M}_2 \setminus \mathcal{M}$, the width of $\mathcal{M}_2$ does not exceed k. Since the set of maximal elements of $\mathcal{M}_2$ is equal to that of $\mathcal{M}$, the assumption that X and Y are mergeable in $\mathcal{M}$ implies that they are also mergeable in $\mathcal{M}_2$.

Thus, we have $X', Y', Z, W \in \mathcal{M}_2$, $X = X' \cup Z$ and $Y = Y' \cup W$ are maximal in $\mathcal{M}_2$, and X and Y are mergeable in $\mathcal{M}_2$. Moreover, by assumptions, we have $E_G(Z, W) = \emptyset$ and $|\delta_G(X' \cup Y')| \leq k$. Therefore we may apply Lemma 2 to conclude that X' and Y' are mergeable in $\mathcal{M}_2 \setminus \{X, Y\}$. Since $\mathcal{M}_2 \setminus \{X, Y\}$ extends $\mathcal{M}'$, X' and Y' are mergeable in $\mathcal{M}'$. □

4 Saving Expensive Tests

In this section, $\mathcal{M}^i$, $0 \leq i \leq n-1$, denotes the sequence of binary mergings in $V(G)$ as constructed by the execution of Algorithm ST. We assume Algorithm ST is executed with the method of mergeability testing described in this section. Even though the description of our testing method involves $\mathcal{M}^i$, the definitions are not circular, as the action taken by our method at the ith iteration step of algorithm ST depends only on $\mathcal{M}^{i-1}$.

The following notation is used in the description of our testing method. We denote by P^i the set of all unordered pairs on which the expensive test is performed since the start of the algorithm execution and prior to the addition of the element X^i. We denote by Q^i the set of pairs in P^i on which the test result was negative, i.e., "unmergeable".

Let X, Y be disjoint elements of $\mathcal{M}^{n-1}$ such that $E_G(X, Y) \neq \emptyset$. We denote by $\mu_Y(X)$ the minimal subset X' of X such that $E_G(X, Y) = E_G(X', Y)$. We also denote by $Q(X, Y)$ the set of distinct unordered pairs $\{X', Y'\} \in Q^{n-1}$ with $X \supseteq X' \supseteq \mu_Y(X)$ and $Y \supseteq Y' \supseteq \mu_X(Y)$. Note that $\mu_Y(X)$ and $Q(X, Y)$ can be determined from $\mathcal{M}^i$ and Q^i where i is the smallest index with $X, Y \in \mathcal{M}^i$ and therefore these sets are accessible by the algorithm when testing the mergeability of X and Y. Note also that the pairs in $Q(X, Y)$ are totally ordered by the element-wise set inclusion.

Parsimonious mergeability test
Input: maximal elements X and Y of $\mathcal{M}^{i-1}$.
Reports: if X and Y are mergeable in $\mathcal{M}^{i-1}$.
(1) If $|\delta(X \cup Y)| > k$ then report "NO".
(2) Otherwise, if $Q(X, Y) = \emptyset$ then do the expensive test and report the result.
(3) Otherwise, let $\{X', Y'\}$ be the maximal pair in $Q(X, Y)$. If there is no barrier B with $\top B = X$ and $\bot B \supseteq X'$ or $\top B = Y$ and $\bot B \supseteq Y'$, then report "NO". Otherwise, do the expensive test and report the result.

Lemma 4. *The testing procedure above reports a correct answer.*

Proof. We need only to show that, whenever the procedure skips the expensive test and reports "NO", X and Y are not mergeable. This is trivially the case, when the test returns in step (1). Consider the case where the procedure skips the expensive test in step (3). Let $\{X', Y'\}$ be as defined in the procedure and set $Z = X \setminus X'$ and $W = Y \setminus Y'$. The definition of $Q(X, Y)$ implies that $E_G(Z, W) = \emptyset$. Therefore, we may apply the barrier lemma with $\mathcal{M} = \mathcal{M}^{i-1}$ and $\mathcal{M}'$ being $\mathcal{M}^j$ with the smallest j that contains X' and Y' and conclude that X, Y are not mergeable in $\mathcal{M}^{i-1}$. □

In the remainder of this section, we concentrate on showing that, with this parsimonious mergeability testing, $O(n)$ expensive tests are performed during the execution of Algorithm ST.

We start with some properties of barriers. Let C be a chain in a binary merging $\mathcal{M}$ in $V(G)$. We define three disjoint edge sets:

$$\text{pass}(C) = E_G(\bot C, V(G) \setminus \top C)$$
$$\text{in}(C) = E_G(\bot C, \top C \setminus \bot C)$$
$$\text{out}(C) = E_G(\top C \setminus \bot C, V(G) \setminus \top C).$$

Proposition 1. *Assuming that the width of binary merging $\mathcal{M}$ does not exceed k, $|\operatorname{pass}(B)| < |\operatorname{in}(B)|$ for every barrier B in $\mathcal{M}$.*

Proof. Since the width of $\mathcal{M}$ does not exceed k, we have $|\delta_G(\top B)| \le k$. Since $\delta_G(\top B) = \operatorname{pass}(B) \cup \operatorname{out}(B)$ and these two sets are disjoint, we have $|\operatorname{pass}(B)| + |\operatorname{out}(B)| \le k$. Since B is a barrier and hence $|\operatorname{in}(B)| + |\operatorname{out}(B)| > k$, it follows that $|\operatorname{pass}(B)| < |\operatorname{in}(B)|$. □

We say that a chain C is *rich* if $|\operatorname{in}(C)| \ge k/4$.

Proposition 2. *Let B be a barrier in a binary merging in $V(G)$ of width k or smaller. If either $|\delta_G(\top B)| \le 3k/4$ or $|\delta_G(\bot B)| \ge k/2$ then B is rich.*

Proof. If $|\delta_G(\top B)| \le 3k/4$ then $|\operatorname{out}(B)| \le 3k/4$ and hence $|\operatorname{in}(B)| > k/4$ since B is a barrier. If $|\delta_G(\bot B)| \ge k/2$ then, since $\delta_G(\bot B) = \operatorname{in}(B) \cup \operatorname{pass}(B)$, it follows from the proposition above that $|\operatorname{in}(B)| \ge k/4$. □

The following proposition is an immediate consequence of the above.

Proposition 3. *Let B_1 and B_2 be two barriers with $\bot B_1 \supseteq \top B_2$. If neither B_1 nor B_2 is rich, then the chain C with $\top C = \bot B_1$ and $\bot C = \top B_2$ is rich.*

Recall that $\mathcal{M}^{n-1}$ denotes the full binary merging constructed by the algorithm, P^{n-1} the set of unordered pairs from $\mathcal{M}^{n-1}$ on which the expensive test is performed. Also recall related notation $\mu_X(Y)$ and $Q(X,Y)$.

Call a disjoint pair $\{X,Y\}$ with $X,Y \in \mathcal{M}^{n-1}$ *relevant* if $E_G(X,Y)$ is nonempty and there is some $i < n$ such that both X and Y are maximal in $\mathcal{M}^i$. Clearly, all pairs in P^{n-1} are relevant. We call a pair $\{X,Y\}$ from $\mathcal{M}^{n-1}$ *essential* if it is relevant and $\mu_Y(X) = X$ and $\mu_X(Y) = Y$. We first note that there are at most $O(n)$ essential pairs from $\mathcal{M}^{n-1}$. To see this, consider subsets of $E(G)$ of the form $E_G(X,Y)$ for an essential $\{X,Y\}$ from $\mathcal{M}^{n-1}$. Since those subsets are formed through successive binary merging, starting from singleton subsets, the number of such subsets is at most $2|E(G)|$. Since $E_G(X,Y) = E_G(\mu_Y(X), \mu_X(Y))$, by definition of $\mu_Y(X)$ and $\mu_X(Y)$, the number of such edge subsets is exactly the number of essential pairs from $\mathcal{M}^{n-1}$.

We are ready to bound the number of expensive tests. The conditions of the testing algorithm implies that exactly one expensive test is applied to each pair $X, Y \in P^{n-1}$. Therefore, it suffices to show that $|P^{n-1}| = O(n)$. Let $P' = \{\{X,Y\} \in P^{n-1} \mid |Q(X,Y)| \le 2\}$. Then, the cardinality of P' is at most three times the number of essential pairs from $\mathcal{M}^{n-1}$ and therefore is $O(n)$. To bound the cardinality of $P^{n-1} \setminus P'$, we need the following lemma. Recall that the pairs in $Q(X,Y)$ are totally ordered by element-wise set inclusion and note that, for $\{X,Y\} \in P^{n-1} \setminus P'$, we have $|Q(X,Y)| \ge 3$.

Lemma 5. *Let $(X,Y) \in P^{n-1} \setminus P'$ and let $\{X_1,Y_1\}$, $\{X_2,Y_2\}$, and $\{X_3,Y_3\}$ be the first, second, and third maximal elements of $Q(X,Y)$ respectively, with $X_1 \supseteq X_2 \supseteq X_3$ and $Y_1 \supseteq Y_2 \supseteq Y_3$. Then, there exists a rich chain C with either $X \supseteq \top C$ and $\bot C \supseteq X_3$ or $Y \supseteq \top C$ and $\bot C \supseteq Y_3$.*

Proof. The rule of our testing method demands that, in order for all the pairs $\{X_2, Y_2\}$, $\{X_1, Y_1\}$, and $\{X, Y\}$ to be included in P^{n-1}, there must be a barrier B_i with either $X_i \supseteq \top B_i$ and $\bot B_i \supseteq X_{i+1}$ or $Y_i \supseteq \top B_i$ and $\bot B_i \supseteq Y_{i+1}$ for $0 \leq i \leq 2$, where we set $X_0 = X$ and $Y_0 = Y$. We may assume without loss of generality that for at least two values of i we have $X_i \supseteq \top B_i$ and $\bot B_i \supseteq X_{i+1}$. If either of these two barriers are rich, then we are done. Otherwise, we obtain a desired rich chain between these two barriers applying Proposition 3. □

We say that a chain C is *minimal rich*, if it is rich and no proper subchain of C is rich. Let $\mathcal{R}^*$ denote the set of all minimal rich chains in $\mathcal{M}^{n-1}$. We say that a chain C supports a chain C' if $\top C \supset \bot C' \supseteq \bot C$. We say that a subset $\mathcal{R}$ of $\mathcal{R}^*$ supports $\mathcal{R}^*$ if for each $C \in \mathcal{R}^*$ there is some chain in $\mathcal{R}$ that supports C. We choose an arbitrary minimal subset of $\mathcal{R}^*$ that supports $\mathcal{R}^*$ and call it $\hat{\mathcal{R}}$.

Let $\boldsymbol{G}$ denote the directed version of G with directed edges (u, v) and (v, u) for each undirected edge $\{u, v\}$ of G. For a chain C, we define $\mathbf{in}(C) = \{(u, v) \in E(\boldsymbol{G}) \mid u \in \bot C, v \in \top C \setminus \bot C\}$. The directed edges in $\mathbf{in}(C)$ and undirected edges in $\mathbf{in}(C)$ are in a natural one-to-one correspondence.

Proposition 4. *For each directed edge e of $\boldsymbol{G}$, there are at most two chains $C \in \hat{\mathcal{R}}$ such that $e \in \mathbf{in}(C)$.*

Proof. Suppose to the contrary that some $(u, v) \in E(\boldsymbol{G})$ belongs to $\mathbf{in}(C)$ for three distinct chains C_1, C_2, C_3 in $\hat{\mathcal{R}}$. Since $\bot C_i$ contains u for $i = 1, 2, 3$, all these chains are subchains of the chain in $\mathcal{M}^{n-1}$ with top $V(G)$ and bottom $\{u\}$. Without loss of generality, we assume $\top C_1 \supset \top C_2 \supset \top C_3$. Note here that the tops of these chains are mutually distinct, due to the minimality of these chains with respect to richness. Let X be the minimal element of $\mathcal{M}^{n-1}$ that contains both u and v. Then, we must have $\top C_i \supseteq X \supset \bot C_i$ for $i = 1, 2, 3$. Therefore, we have

$$\top C_1 \supset \top C_2 \supset \top C_3 \supset \bot C_1 \supset \bot C_2 \supset \bot C_3$$

Let C be an arbitrary chain in $\mathcal{R}^*$ supported by C_2. We have $\top C_2 \supset \bot C \supseteq \bot C_2$ by the definition of supporting. If $\bot C \supseteq \bot C_1$ then C_1 supports C, since $\top C_1 \supset \top C_2 \supset \bot C$. Otherwise, i.e., if $\bot C_1 \supset \bot C$, we have $\top C_3 \supset \bot C$ and C_3 supports C, since $\bot C \supseteq \bot C_2 \supset \bot C_3$. Therefore, $\hat{\mathcal{R}} \setminus \{C_2\}$ supports $\mathcal{R}^*$, contradicting the minimality of $\hat{\mathcal{R}}$. □

Corollary 1. $|\hat{\mathcal{R}}| \leq 16|E(G)|/k$.

Proof. Each chain $C \in \hat{\mathcal{R}}$ is rich and hence $|\mathbf{in}(C)| \geq k/4$. Each of $2|E(G)|$ edges of $\boldsymbol{G}$ belongs to $\mathbf{in}(C)$ for at most two chains of $\hat{\mathcal{R}}$ and hence $|\hat{\mathcal{R}}| \cdot \frac{k}{4} \leq \sum_{C \in \hat{\mathcal{R}}} |\mathbf{in}(C)| \leq 4|E(G)|$, from which the inequality of the corollary follows. □

For each $C \in \hat{\mathcal{R}}$, we define a subset P_C of $P^{n-1} \setminus P'$ as follows. Let $\{X, Y\} \in P^{n-1} \setminus P'$ and $\{X_3, Y_3\}$ the third maximal pair in $Q(X, Y)$ with $X \supseteq X_3$ and $Y \supseteq Y_3$. Then, we put $\{X, Y\}$ in P_C if and only if either $X \supseteq \top C \supseteq X_3$ or $Y \supseteq \top C \supseteq Y_3$. Lemma 5 and the definition of $\hat{\mathcal{R}}$ imply that $P^{n-1} \setminus P' \subseteq \bigcup_{C \in \hat{\mathcal{R}}} P_C$.

We now bound the cardinality of each P_C. The definition of $Q(X,Y)$ implies that $E_G(X,Y) = E_G(X_3, Y)$ and hence that for each W such that $X \supseteq W \supseteq X_3$, $E_G(X,Y)$ is a subset of $\delta_G(W)$. The same holds for each W such that $Y \supseteq W \supseteq Y_3$. Therefore, for pair $\{X,Y\}$ to be in P_C, $E_G(X,Y)$ must contain some edge in $\delta_G(\mathsf{T}C)$.

Let $e = \{u,v\}$ be an arbitrary edge in $\delta_G(\mathsf{T}C)$ with $u \in \mathsf{T}C$ and $v \in V(G) \setminus \mathsf{T}C$. We want to know how many pairs X,Y with $e \in E_G(X,Y)$ can be in P_C. Let $\{X_i, Y_i\}$, $1 \le i \le a$, be the list of all pairs $\{X,Y\}$ in P^{n-1} such that $e \in E_G(X,Y)$. Assume without loss of generality that these pairs are ordered so that $X_1 \supseteq X_2 \supseteq \ldots \supseteq X_a \supseteq \{u\}$ and $Y_1 \supseteq Y_2 \supseteq \ldots \supseteq Y_a \supseteq \{v\}$. Suppose $\{X_i, Y_i\} \in P_C$ for some $1 \le i \le a$. Then, it must be that $|Q(X_i, Y_i)| \ge 3$ and the first, second and third maximal elements of $Q(X_i, Y_i)$ are $\{X_{i+1}, Y_{i+1}\}$, $\{X_{i+2}, Y_{i+2}\}$ and $\{X_{i+3}, Y_{i+3}\}$. Note that the expensive test on all of these three pairs must have failed, since if any of these pairs had been merged then the pair $\{X_i, Y_i\}$ would not appear later in the process. From the definition of P_C, we must have $X_i \supseteq \mathsf{T}C \supset X_{i+3}$ or $Y_i \supseteq \mathsf{T}C \supset Y_{i+3}$. The latter condition, however, cannot hold because $u \notin Y_i$. For a fixed C, at most three values of i can satisfy this condition. Therefore, for each C and each $e \in \delta_G(\mathsf{T}C)$, there are at most 3 pairs $\{X,Y\} \in P_C$ with $e \in E_G(X,Y)$. As $|\delta_G(\mathsf{T}C)| \le k$, we have $|P_C| \le 3k$.

Combining this with Corollary 1, we have $|P^{n-1} \setminus P'| \le \sum_{C \in \hat{\mathcal{R}}} |P_C| \le 48|E(G)|$ and therefore $|P^{n-1}| = O(n)$.

5 Total Running Time

In this section, we show that the total running time of Algorithm ST is $O(n^3)$, when the parsimonious mergeability test described in the previous section is used. We need only elementary data structures and the analysis is straightforward.

We represent the binary merging $\mathcal{M}^i$ constructed by the algorithm naturally by a binary forest. With each node of this forest we associate a member X of $\mathcal{M}^i$, in the form of a sorted list, and the edge set $\delta_G(X)$ also in the form a sorted list. As the merging of two members of $\mathcal{M}^i$ can be done in $O(n)$ time in this representation, the total time for maintaining this binary forest is $O(n^2)$. In addition, we maintain a graph G^i for each $0 \le i \le n-1$, where the vertices of G^i are the maximal elements of $\mathcal{M}^i$ and $X, Y \in V(G^i)$ are adjacent in G^i if and only if $E_G(X,Y) \ne \emptyset$. We maintain G^i in the standard edge list representation. Moreover, with each edge e of G^i, $0 < i \le n-1$, we associate the corresponding edge(s) in G^{i-1}: if the endvertices X and Y of e are both in $V(G^{i-1})$, then this association is to the edge between X and Y in G^{i-1}; otherwise, i.e., if $X = X_1 \cup X_2$ with $X_1, X_2 \in V(G^{i-1})$, say, then this association is to the edge between X_1 and Y, if present, and to the edge between X_2 and Y, if present. The representation of G^i can be constructed in $O(n)$ time, so the total maintenance cost of this part is $O(n^2)$. The set Q, the set of unordered pairs $\{X,Y\}$ on which the expensive test is performed with a negative result, is represented by marking the corresponding edge of G^i.

We now bound the cost of mergeability testing. We have already proved that the number of expensive tests is $O(n)$ and therefore the total running time for expensive tests is $O(n^3)$. Since the mergeability testing is performed $O(n^2)$ times, it suffices to show that other parts of the testing takes $O(n)$ time for each tested pair. The time for the degree test in step (1) is trivially $O(n)$. The set $Q(X, Y)$ in step (2) can be constructed as follows. We start from the edge e between X and Y in the current G^i and trace back its origin in G^{i-1}, G^{i-2}, ... This tracing stops when we reach G^0 or encounter an edge between an essential pair. Using the marks of the edges in the traced sequence, we construct $Q(X,Y)$ in $O(n)$ time. The maximal pair $\{X', Y'\} \in Q(X, Y)$ is the first element of $Q(X, Y)$ found in this process. Finally, the barrier computation can be done in $O(n)$ time as follows. Let $X_1 \supset X_2 \supset \ldots \supset X_h$ be the chain with top $X_1 = X$ and bottom $X_h = X'$. For $j = 2, \ldots, h$, we successively construct the edge set $\delta_G(X_1 \setminus X_j)$. Noting that $\delta_G(X_1 \setminus X_j)$ for $2 \leq j \leq h$ is the symmetric difference between $\delta_G(X_1 \setminus X_{j-1})$ and $\delta_G(X_{j-1} \setminus X_j)$, we represent the sets in this process with a single membership bit for each edge of G and a single counter for the cardinality of the set: when we process $\delta_G(X_{j-1} \setminus X_j)$, we flip the bit of each edge in this set and count up or down according to the result of this flip. Since the set $\delta_G(X_{j-1} \setminus X_j)$ for each j is available in the data structure and each edge enters the constructed set at most once and exits at most once, the total time for the construction is $O(n)$. If any of these constructed edge sets has cardinality greater than k, then we have detected a barrier.

This concludes the analysis of the running time and we have proved Theorem 2.

References

1. S. Arnborg, D.G. Corneil, and A. Proskurowski, Complexity of finding embeddings in a k-tree, SIAM Journal on Algebraic and Discrete Methods, 8: 277-284, 1987
2. S. Arnborg, J. Lagergren, and D. Seese, Easy problems for tree-decomposable graphs, Journal of Algorithms, 12: 308-340, 1991
3. H.L. Bodlaender, A tourist guide through treewidth, Acta Cybernetica, 11: 1-21, 1993.
4. H.L. Bodlaender, A Linear-Time Algorithm for Finding Tree-Decompositions of Small Treewidth, SIAM Journal on Computing, 25: 1305-1317, 1996.
5. E.D. Demaine, M.T. Hajiaghayi, N. Nishimura, P. Ragde, and D.M. Thilikos, Approximation algorithms for classes of graphs excluding single-crossing graphs as minors, Journal of Computer and System Sciences, 69-2: 166-195, 2004.
6. N. Robertson and P.D. Seymour, Graph minors I. Excluding a forest, Journal of Combinatorial Theory, Series B, 35: 39-61, 1983
7. N. Robertson and P.D. Seymour, Graph minors. II. Algorithmic aspects of tree-width, Journal of Algorithms, 7: 309-322, 1986
8. N. Robertson and P.D. Seymour, Graph minors. X. Obstructions to tree-decomposition, Journal of Combinatorial Theory, Series B, 52: 153-190, 1991
9. P.D. Seymour and R. Thomas, Call Routing and the Ratcatcher. Combinatorica, 14(2), 217-241, 1994

NFAs With and Without ε-Transitions*

Juraj Hromkovič[1] and Georg Schnitger[2]

[1] Department of Computer Science, ETH Zürich,
ETH Zentrum, RZ F2, CH-8092 Zürich, Switzerland
[2] Institut für Informatik, Johann Wolfgang Goethe-Universität,
Robert Mayer Straße 11–15, 60054 Frankfurt am Main, Germany

Abstract. The construction of an ε-free nondeterministic finite automaton (NFA) from a given NFA is a basic step in the development of compilers and computer systems. The standard conversion may increase the number of transitions quadratically and its optimality with respect to the number of transitions is a long standing open problem. We show that there exist regular languages L_n that can be recognized by NFAs with $O(n \log_2 n)$ transitions, but ε-free NFAs need $\Omega(n^2)$ transitions to accept L_n. Hence the standard conversion cannot be improved significantly. However L_n requires an alphabet of size n, but we also construct regular languages K_n over $\{0,1\}$ with NFAs of size $O(n \log_2 n)$, whereas ε-free NFAs require size $n \cdot 2^{c \cdot \sqrt{\log_2 n}}$ for every $c < 1/2$.

1 Introduction

One of the central tasks on the border between formal language theory and complexity theory is to describe infinite objects such as languages by finite formalisms such as automata, grammars, expressions etc., and to investigate the descriptional complexity and capabilities of these formalisms. Formalisms like expressions and finite automata have proven to be very useful in building compilers, and techniques converting one formalism into the other are used as basic tools in the design of computer systems [Tho68] such as UNIX (see [HMU01], p. 123). A typical application in lexicographical analysis starts with a regular expression that has to be converted into an ε-free nondeterministic finite automaton. Here, the descriptional complexity of an expression is its length and the descriptional complexity of a nondeterministic finite automaton (NFA) is the number of its transitions (edges).

Surprisingly ε-free NFA are capable of simulating regular expressions of length n with at most $O(n (\log_2 n)^2)$ transitions [HSW97] and hence in this case the loss of ε-transitions does only weakly deteriorate the conciseness of ε-free NFAs. In [Lif03] a lower bound of $\Omega(n(\log_2 n)^2 / \log_2 \log_2 n)$ is shown and as a consequence the upper bound of [HSW97] is almost tight.

In contrast to translating regular expressions into ε-free NFAs, no subquadratic transformation from NFAs into ε-free NFAs is known. Since any regular

---
* Supported in part by SNF grant 200021-107327/1 and DFG grant SCHN 503/2-2.

expression of length n can be recognized by unrestricted NFAs with $O(n)$ transitions, ε-free NFAs of size at least $\Omega(n(\log_2 n)^2/\log_2\log_2 n)$ are required for given NFAs of size n. We considerably improve this lower bound and show an almost quadratic increase in size.

Theorem 1. *Sequences $\{L_n\}_{n=1}^{\infty}, \{K_n\}_{n=1}^{\infty}$ of regular languages exist such that*

  (i) *L_n is defined over the alphabet $\{0,1\}^n$, K_n is defined over the alphabet $\{0,1\}$,*
 (ii) *L_n and K_n can be accepted by NFAs with $O(n \cdot 2^n)$ transitions,*
(iii) *every ε-free NFA accepting L_n has at least $\Omega(2^{2n})$ transitions and*
 (iv) *every ε-free NFA accepting K_n has at least $\Omega(2^{n+c\sqrt{n}})$ transitions for every constant $c < 1/2$.*

Hence, for $m = 2^n$, L_n and K_n can be accepted by NFAs with $O(m \log_2 m)$ transitions, but every ε-free NFA for L_n has at least $\Omega(m^2)$ transitions, whereas K_n requires ε-free NFAs of size $O(m \cdot 2^{c \cdot \sqrt{\log_2 m}})$ for every $c < 1/2$. Thus ε-free NFAs for L_n require almost quadratically more transitions than unrestricted NFAs for L_n and the obvious transformation from NFAs to ε-free NFAs cannot be improved considerably. The provable gap for binary alphabets is smaller, but still a far larger than poly-logarithmic gap is required. Moreover, for a large class of NFAs, we show in Lemma 4 that the lower bound of Theorem 1 (iv) is almost optimal.

We use methods from communication complexity. However communication arguments can be applied to ε-free NFAs and NFAs alike and hence we have to carefully mix communication arguments with graph-theoretical arguments.

The rest of this paper is organized as follows. In Section 2 we describe the languages L_n which are hard for ε-free NFAs. Lower bounds for large and small alphabets are presented in Sections 3 and 4 respectively. Conclusions and open problems are stated in Section 5.

2   Hard Languages for ε-Free NFAs

Our goal is to construct a sequence $\{L_n\}_{n=1}^{\infty}$ of regular languages that is considerably harder for ε-free NFAs than for NFAs. Since the basic component of our construction of L_n is the **Hadamard matrix** H_n, we start by describing its properties. H_n is a $1/-1$ matrix with 2^n rows and columns. H_n is defined recursively by setting $H_0[1,1] = 1$ and

$$H_{n+1} = \begin{bmatrix} H_n & H_n \\ H_n & -H_n \end{bmatrix}.$$

An inductive argument shows that the columns of H_n are pairwise orthogonal, and that every column (resp. row) of H_n, except for the first which consists of 1's only, has the same number of 1's and -1's.

H_n has the following crucial property: If a submatrix of H_n with row set X and column set Y covers only 1-entries, then

$$|X| \cdot |Y| \leq 2^n. \tag{1}$$

To show this fact replace any 1 (resp. -1) by a 0 (resp. 1) and call the resulting matrix $\overline{H}_n$. Another inductive argument verifies that $\overline{H}_n$ is the matrix of inner products modulo two: we may label rows and columns by vectors $u, v \in \mathbb{Z}_2^n$ such that $\overline{H}_n[u, v] = \langle u, v \rangle$ mod 2. If the sets $X, Y \subseteq \mathbb{Z}_2^n$ define a 0-chromatic submatrix of $\overline{H}_n$, then we may close X and Y under addition to obtain vector spaces $\overline{X}$ and $\overline{Y}$ (with $X \subseteq \overline{X}$ and $Y \subseteq \overline{Y}$) which still define a 0-chromatic submatrix. Thus $\overline{X}$ and $\overline{Y}$ are orthogonal vector spaces and $\dim(\overline{X}) + \dim(\overline{Y}) \leq n$ follows. Hence $|X| \cdot |Y| \leq 2^n$.

The language L_n is defined over the alphabet $\Sigma_n = \{0,1\}^n$ and we set

$$L_n = \{\varepsilon\} \cup \Sigma_n \cup \bigcup_{k \geq 2} L_{n,k}, \text{ where} \tag{2}$$

$$L_{n,k} = \{u_1 \cdots u_k \in \Sigma_n^k \mid H_n[u_i, u_{i+1}] = 1 \text{ for all } i < k\}. \tag{3}$$

Thus a sequence of letters of length n belongs to L_n iff the inner product of any two successive letters of the sequence is zero. (Observe that all words of length at most one belong to L_n.)

Proposition 1. *There is an NFA which accepts L_n with $O(n \cdot 2^n)$ transitions.*

Proof Sketch. One can construct a bounded degree switching network G_n of depth n that contains a path between one of the 2^n sources labelled by symbols from Σ_n to one of the 2^n sinks labelled by symbols from Σ_n if and only if the inner product between the corresponding labels is zero. We obtain an NFA for L_n, if we replace sources by "source transitions" of the same label, utilize all other transitions as ε-transitions and redirect edges, previously directed into a sink, into the starting point of the corresponding source transition instead. □

3 Large Alphabets

We first sketch the argument verifying that every ε-free NFA for L_n has at least $\Omega(2^{2n})$ transitions (edges). We state without proof that $L_{n,k}$ (see (3)) has surprisingly small ε-free NFA's for small values of k and, as a consequence, our argument has to concentrate on $L_{n,k}$ for large values of k.

Lemma 1. *$L_{n,k}$ is recognizable by ε-free NFA's of size at most $O(k \cdot 2^{2n - \frac{n}{k}})$.*

Let A_n be an NFA without ε-transitions which recognizes L_n and let k be an arbitrary integer. We first observe that A_n implicitly also recognizes $L_{n,k}$: we unravel the transition diagram of A_n into a layered directed acyclic graph $F_{n,k}$ with $k+1$ layers. Layer V_i (for $0 \leq i \leq k$) consists of all vertices (x, i) such that x is a vertex of the transition diagram of A_n and x is endpoint of a path of A_n which starts in q_0 and traverses i edges. Observe that layer V_0 consists only of the starting state q_0. We insert an edge $((x, i), (y, i+1))$ with label $a \in \{0, 1\}^n$ into $F_{n,k}$ iff (x, y) is an edge of A_n with label a.

Next we remove all vertices of $F_{n,k}$ which are not traversed by an accepting path starting in q_0 and ending in a vertex of the last layer V_k. Finally we replace all vertices of V_k by the new vertex q_f and insert an edge from $(u, k-1) \in V_{k-1}$ to q_f with label a, provided there was an a-transition from $(u, k-1)$ to an accepting vertex of V_k. Let $F_{n,k}$ be the resulting labelled graph.

Proposition 2. *(a) $F_{n,k}$ is a layered acyclic graph with $k+1$ layers.*
(b) $F_{n,k}$ has a path $q_0 \xrightarrow{*} q_f$ with label sequence $(u_1, \ldots, u_k)$ iff $u_1 \cdots u_k \in L_n$.
(c) For any i, the number of edges between vertices in V_i and V_{i+1} is bounded by the number of edges of the transition diagram of A_n.

Our goal is to show that there are successive layers V_i and V_{i+1} such that the number of edges between V_i and V_{i+1} has to be large, provided the total number $k+1$ of layers is sufficiently large. We utilize that the Hadamard matrix H_n has only monochromatic submatrices of size at most 2^n (see (1)). To apply this property we show that any inner vertex x of $F_{n,k}$ (i.e., any vertex different from q_0 and q_f) defines a monochromatic submatrix of H_n spanned by the rows with labels in $S(x)$ (resp. columns with labels in $T(x)$), where $S(x)$ and $T(x)$ are the sets of incoming and leaving labels of v respectively. As a consequence of (1) we obtain $|S(x)| \times |T(x)| \leq 2^n$, whereas in total at least 2^{2n-1} pairs of labels, namely all pairs (u, v) with $H_n[u, v] = 1$, have to be supported. Hence, if the sets $S(x)$ and $T(x)$ are all of identical size s respectively t, then $s \cdot t \leq 2^n$ and $\sum_{x \in V_1} |S(x)| \cdot |T(x)| = \sum_{x \in V_1} s \cdot t \geq 2^{2n-1}$ holds. Thus V_1 has to consist of at least 2^{n-1} vertices and it is optimal to choose $s = t = 2^{n/2}$. Thus at least $2^{3n/2}$ transitions are required.

We apply this procedure to vertices of the second layer by considering paths of length three starting in q_0. However this time we select for any label v a label $u(v)$ such that the pair $u(v)v$ is supported by the *least* number of paths of length two originating in q_0. Thus when considering all label sequences $u(v)vw$ we have in effect decreased the "fanin multiplicity" of label v for vertices of the second layer and this reduction can only be compensated for by increasing the "fanout multiplicity" of labels w for vertices of the second layer. (The construction in Lemma 1 also explicitly increases fanout multiplicity by copying edges.) We repeat this procedure n times to verify the required large number of transitions.

3.1 The Details

We start with the "submatrix property" of vertices and associate with an inner vertex x the set **fanin**(x) (resp. **fanout**(x)) of labels of incoming (resp. leaving) edges. In both sets we disregard multiple copies of the same label and hence we also introduce the *multisets* **fanin**$^*(x)$ and **fanout**$^*(x)$, where all labels of edges entering resp. leaving x are collected according to their true multiplicity.

We say that $(u, v) \in \{0, 1\}^n \times \{0, 1\}^n$ is a **Hadamard pair**, if $H_n[u, v] = 1$. Since any edge of $F_{n,k}$ is traversed by an "accepting path" of $F_{n,k}$, every pair $(u, v) \in$ fanin$(x) \times$ fanout(x) is a Hadamard pair: we obtain an accepting path traversing both the incoming and outgoing edge by splicing together two accepting paths traversing the incoming edge and the outgoing edge respectively.

Thus the submatrix fanin$(x) \times$ fanout(x) is a 1-chromatic submatrix of H_n and we say that the entries in fanin$(x) \times$ fanout(x) are **covered by** x. Therefore, as a consequence of (1), we have

$$|\text{fanin}(x)| \cdot |\text{fanout}(x)| \leq 2^n. \quad (4)$$

Observe that all ones of H_n are covered by submatrices and, since at least one half of all entries of a row of H_n are 1-entries, we obtain

$$\sum_{x \in V_i} |\text{fanin}(x) \cap K| \cdot |\text{fanout}(x)| \geq |K| \cdot 2^{n-1} \quad (5)$$

for an arbitrary subset K of labels. We say that vertex $x \in V_1$ covers the 1-entry (u, v) of H_n with multiplicity $m_x(u, v) = a \cdot b$, if u occurs in $\text{fanin}^*(x)$ with multiplicity a and v occurs in $\text{fanout}^*(x)$ with multiplicity b. Finally we say that (u, v) is covered with **multiplicity** $m(u, v) = \sum_{x \in V_1} m_x(u, v)$.

For a given set $K \subseteq \{0, 1\}^n$ of labels we define $\mathbf{L_{n,k}(K)}$ as the subset of L_n consisting of all words of length k beginning with a letter in K. Finally let $\mathbf{F_{n,k}(K)}$ be the subgraph of $F_{n,k}$ obtained by removing all edges between q_0 and V_1 with label outside of K. We say that a graph G recognizes $L_{n,k}(K)$, if G is a layered directed acyclic graph with $k+1$ layers $V_0, \ldots, V_k$. Moreover V_0 and V_k consist of exactly one state each and exactly the words from $L_{n,k}(K)$ appear as label sequences of paths starting in V_0 and ending in V_k. Observe that Proposition 2 implies that $F_{n,k}(K)$ recognizes $L_{n,k}(K)$.

Assume that graph G recognizes $L_{n,k}(K)$. We show that G has many edges incident with vertices in V_1, provided many 1-entries have to be covered often.

Lemma 2. *Let $K \subseteq \{0, 1\}^n$ be an arbitrary subset of at least 2^{n-1} labels. Assume that the graph G recognizes $L_{n,k}(K)$ and that*

$$\sum_{x \in V_1} |\text{fanin}^*(x)| = a_1 \cdot 2^n \quad \text{and} \quad \sum_{x \in V_1} |\text{fanout}^*(x)| = a_2 \cdot 2^n$$

holds for layer V_1 of G. If at least $|K| \cdot 2^{n-2}$ 1-entries from $K \times \{0,1\}^n$ are covered with multiplicity at least b, then $a_1 \cdot a_2 \geq b \cdot 2^{n-8}$.

Proof. We say that vertex $x \in V_1$ is *large*, if $|\text{fanin}^*(x)| \geq 2^4 \cdot a_1$, and *small* otherwise. Since $\sum_{x \in V_1} |\text{fanin}^*(x)| = a_1 \cdot 2^n$, there are at most 2^{n-4} large vertices. But according to (4) the submatrix $\text{fanin}(x) \times \text{fanout}(x)$ of a vertex $x \in V_1$ covers at most 2^n 1-entries and hence $\text{fanin}^*(x) \times \text{fanout}^*(x)$ also covers at most 2^n 1-entries, since no new 1-entries are introduced. Therefore at most $2^{n-4} \cdot 2^n = 2^{2n-4}$ 1-entries of H_n are covered by large vertices.

By assumption at least $|K| \cdot 2^{n-2} \geq 2^{2n-3}$ 1-entries from $K \times \{0,1\}^n$ are covered with multiplicity at least b. But then at least $2^{2n-3} - 2^{2n-4} = 2^{2n-4}$ 1-entries from $K \times \{0,1\}^n$ have to be covered with multiplicity at least b by small vertices only. We have

$$b \cdot 2^{2n-4} \leq \sum_{x \in V_1, x \text{ small}} |\text{fanin}^*(x)| \cdot |\text{fanout}^*(x)|$$

$$\leq \sum_{x \in V_1} 2^4 \cdot a_1 \cdot |\text{fanout}^*(x)| = 2^4 \cdot a_1 \cdot a_2 \cdot 2^n$$

and the claim follows. □

Observe as a first consequence of Lemma 2 that at least $\sqrt{b/2^8} \cdot 2^{3n/2}$ edges incident with vertices in V_1 are required, if the vertices of V_1 cover at least one fourth of all ones of H_n at least b times. This follows for $K = \{0,1\}^n$, since $\max\{a_1 \cdot 2^n, a_2 \cdot 2^n\}$, with $a_1 \cdot a_2 \geq b \cdot 2^{n-8}$ is minimized for $a_1 = a_2 = \sqrt{b/2^8} \cdot 2^{n/2}$. Thus at least $\Omega(2^{3n/2})$ edges are required to recognize $L_{n,2}(\{0,1\}^n)$. We improve this bound for $L_{n,k}(K)$ by repeatedly searching for inputs with low multiplicity.

Lemma 3. Let $K \subseteq \{0,1\}^n$ be an arbitrary subset of at least 2^{n-1} labels. Assume that the graph G recognizes $L_{n,k}(K)$ and that

$$\sum_{x \in V_1} |\text{fanin}^*(x)| = a_1 \cdot 2^n \quad \text{and} \quad \sum_{x \in V_1} |\text{fanout}^*(x)| = a_2 \cdot 2^n$$

holds for layer V_1 of G. Then there is a subset $K' \subseteq \{0,1\}^n$ of size 2^{n-1} and a graph G' which accepts exactly all words in $L_{n,k-1}(K')$. G' results from G by removing the first layer of G and introducing at most $2^9 \cdot a_1 \cdot a_2$ new edges from the starting state to the new first layer.

Proof. We first observe that less than $|K| \cdot 2^{n-2}$ pairs from $K \times \{0,1\}^n$ are covered with multiplicity at least $b = a_1 \cdot a_2/2^{n-9}$, since otherwise we get $a_1 \cdot a_2/2^{n-8} \geq b = a_1 \cdot a_2/2^{n-9}$ with Lemma 2.

For every label v select a label $u = u(v)$ such that the multiplicity $m(u,v)$ is minimal among all labels u with $H_n[u,v] = 1$. Observe that $m(u(v),v) \geq b$ holds for less than 2^{n-2} labels $v \in \{0,1\}^n$, since $m(u(v),v) \geq b$ implies that the $|K|$ pairs $K \times \{v\}$ are covered with multiplicity at least b. Thus we find a set K' of more than $2^n - 2^{n-2} \geq 2^{n-1}$ labels v with $m(u(v),v) \leq b$.

We say that edge e belongs to label v, if e leaves a vertex of layer V_1 and e is traversed by an accepting path for some input with prefix $u(v)v$. We say that e belongs to K' iff e belongs to a label of K'. A label $v \in K'$ appears on at most b accepting paths with prefix $u(v)v$ and hence at most $b \cdot |K'|$ edges belong to K'.

In order to construct G' from G we remove all vertices of layer V_1 and insert an edge (q_0, x) (with label $v \in K'$) from q_0 to a vertex $x \in V_2$ iff there is an accepting path from q_0 to x with prefix $u(v)v$. The number of layers decreases from $k+1$ to k after insertion of at most $b \cdot |K'| \leq b \cdot 2^n$ edges. The new graph G' recognizes $L_{n,k-1}(K')$. We are done, since $b \cdot 2^n = a_1 a_2/2^{n-9} \cdot 2^n = 2^9 \cdot a_1 a_2$. □

We can now conclude the argument. Let us assume that $F_{n,k}$ has $a_i \cdot 2^n$ edges ending in a vertex of V_i. We apply Lemma 3 and obtain a graph $F_{n,k}^{(1)}$ as well as a subset K_1 of at least 2^{n-1} labels such that $F_{n,k}^{(1)}$ accepts $L_{n,k-1}(K_1)$. We set

$$a_1^{(0)} = a_1 \quad \text{and} \quad a_1^{(i)} = \frac{2^9 \cdot a_1^{(i-1)} \cdot a_{i+1}}{2^n}$$

and instead of $a_1^{(0)} \cdot 2^n$ edges between layers V_0 and V_1 we now have at most $2^9 \cdot a_1^{(0)} \cdot a_2 = a_1^{(1)} \cdot 2^n$ edges between layer zero and the new layer one. We repeat this procedure i times and obtain a graph $F_{n,k}^{(i)}$ as well as a set K_i of 2^{n-1} labels such that $F_{n,k}^{(i)}$ recognizes $L_{n,k-i}(K_i)$ with at most $\frac{2^9 \cdot a_1^{(i-1)} \cdot a_{i+1}}{2^n} \cdot 2^n = a_1^{(i)} \cdot 2^n$ edges between layer zero and the new layer one. Hence we obtain

$$a_1^{(k-2)} = \frac{2^9 \cdot a_1^{(k-3)} \cdot a_{k-1}}{2^n} = \cdots = \frac{2^{9(k-2)} \cdot a_1 \cdots a_{k-1}}{2^{(k-2) \cdot n}}.$$

We apply Lemma 2 to $G = F_{n,k}^{(k-1)}$ (with $b = 1$) and obtain

$$\frac{2^{9(k-2)} \cdot a_1 \cdots a_k}{2^{(k-2) \cdot n}} = a_1^{(k-2)} \cdot a_k \geq 2^{n-8}.$$

Thus $a_1 \cdots a_k \geq 2^{(k-1) \cdot n}/(2^8 \cdot 2^{9(k-2)}) \geq 2^{(k-1) \cdot n}/2^{9(k-1)}$ and we have to solve the optimization problem

minimize $\max\{a_1 \cdot 2^n, \ldots, a_k \cdot 2^n\}$ subject to $a_1 \cdots a_k \geq 2^{(k-1) \cdot n}/2^{9(k-1)}$.

The optimal solution is obtained for $a_1 = \cdots = a_k$ and hence we obtain

$$\max\{a_1 \cdot 2^n, \ldots, a_k \cdot 2^n\} \geq 2^{n + \frac{k-1}{k} \cdot n}/2^{9(k-1)/k} \geq 2^{2n - \frac{n}{k}}/2^9.$$

Theorem 1 follows, if we set $k = n$. Observe that we have also shown that the construction in Lemma 1 is almost optimal. □

4 Small Alphabets

We apply our approach for large alphabets to small aphabets and first show an upper bound. In Section 4.1 we investigate a family of languages over $\{0,1\}$ which is hard for ε-free NFAs. The argument concludes in Section 4.2.

The regular language $L_{n,2}$ can be recognized by an NFA whose transition diagram consists of a layered acyclic graph with 2^n sources, 2^n sinks and $O(n)$ layers. Sources and sinks have to be expanded into edges labelled by letters from the alphabet $\{0,1\}^n$. What happens, if we work with the binary alphabet instead and expand sources and sinks into 0-1 labelled paths of length n? We call the corresponding class of NFAs "n-bipartite" NFAs.

Definition 1. *An NFA N is called n-bipartite iff its transition diagram G is a layered acyclic graph with 2^n sources and 2^n sinks. Each source (resp. sink) is starting point (resp. endpoint) of its own 0-1 labelled path of length n; the paths are node-disjoint. The endpoints of source paths and the starting points of sink paths are connected by n layers of 2^n vertices each. (The endpoints of source paths define layer one and the starting points of sink paths define layer n.) All edges between two layers are ε-transitions. The 2^n sources are the only initial states and the 2^n sinks are the only accepting states.*

The advantage of ε-transition decreases for the binary alphabet in comparison to large alphabets as the following observation shows.

Lemma 4. *If N is an n-bipartite NFA whose transition diagram has bounded fan-out, then there is an equivalent ε-free NFA with at most $2^{O(\sqrt{n})} \cdot 2^n$ transitions. (Observe that N has at most $O(n \cdot 2^n)$ transitions.)*

We omit the proof due to space limitations. Observe that we do not obtain asymptotically larger bounds, if arbitrary *sequences* of n-bit strings, as in the language L_n, have to be accepted: it suffices to provide transitions from a sink path back to the corresponding source path.

4.1 A Family of Hard Languages

We now introduce languages which turn out to be hard for ε-free NFAs.

Definition 2. *Let $f: \{0,1\}^n \to \{0,1\}^n$ be given. Then we define*

$$L_n(f) = \{u_1 u_2 \mid u_1, u_2 \in \{0,1\}^n,\ H_n[f(u_1), f(u_2)] = 1\}$$

as a language over the binary alphabet.

Observe that $L_n(f)$ can be recognized by an NFA with at most $O(n \cdot 2^n)$ transitions, if we proceed as in the proof of Proposition 1. Our goal is to show that $L_n(f)$ requires ε-free NFAs with $2^{n+\Omega(\sqrt{n})}$ transitions, provided f is chosen at random. We fix a function f and investigate an ε-free NFA $A_n(f)$ for $L_n(f)$. We unravel the transition diagram of $A_n(f)$ to obtain a layered acyclic graph $F_n(f)$ such that $F_n(f)$ has $2n+1$ layers and $O(n \cdot t(f))$ transitions, where $t(f)$ is the number of transitions of $A_n(f)$. Moreover $F_n(f)$ has a path with label sequence $u_1 u_2 \in \{0,1\}^{2n}$ iff $u_1 u_2 \in L_n(f)$ and all edges of $F_n(f)$ are traversed by paths of length $2n$.

Let V_i be the set of vertices of $F_n(f)$ in layer i. For a vertex $u \in \bigcup_{i=0}^{n} V_i$ we define fanin$'(u)$ as the set of label sequences of paths starting in the source of $F_n(f)$ and ending in u. The set fanout$'(u)$ consists of all label sequences of paths starting in u and ending in a vertex of V_n. We now make the crucial observation that, for f chosen at random, fanin$'(u)$ can only be large, if fanout$'(u)$ is small.

Proposition 3. *For sufficiently large n there is a function f such that*

(a) $|f^{-1}(y)| \leq 2 \cdot \sqrt{n}$ for all $y \in \{0,1\}^n$.
(b) *For any $l \leq n$, any $w \in \{0,1\}^l$ and any subspace V of $\mathbb{Z}_2^n$ with $\dim(V) \geq l$,*

$$|f(\{0,1\}^{n-l} w) \cap V| \leq 2n \cdot \frac{|V|}{2^l}.$$

(c) $|\text{fanin}'(u)| \leq 7 \cdot n$ or $|\text{fanout}'(u)| \leq 5 \cdot n$ for any vertex $u \in \bigcup_{i=0}^{n} V_i$, provided $F_n(f)$ has at most 2^{2n} vertices.

Proof. It suffices to show that any property (a), (b) or (c) does *not* hold with probability at most $\frac{1}{4}$, provided n is sufficiently large.

(a) We interpret a function f as the result of 2^n independent random trials x fixing the value $f(x)$. For any $x, y \in \{0,1\}^n$ we have $f(x) = y$ with probability 2^{-n} and hence the expected size of $f^{-1}(y)$ is one. We apply the Chernoff inequality and obtain

$$\text{prob}[|f^{-1}(y)| > 1 + \beta] \leq e^{-\beta^2/3}.$$

Hence size $2 \cdot \sqrt{n}$ is exceeded with probability bounded by $e^{-(2\sqrt{n}-1)^2/3} \leq 2^{-n-2}$ for large n and, with probability at most $\frac{1}{4}$, $|f^{-1}(y)| \geq 2 \cdot \sqrt{n}$ for some y.

(b) We fix l and $w \in \{0,1\}^l$ and again interpret f, restricted to $\text{Suffix}(w) = \{0,1\}^{n-l}w$, as the result of 2^{n-l} random trials. $f(y) \in V$ happens with probability $p = \frac{|V|}{2^n}$ and hence the expected size of $f(\text{Suffix}(w)) \cap V$ is $2^{n-l} \cdot p = \frac{|V|}{2^l}$. We again apply the Chernoff inequality and obtain

$$\text{prob}[|f(\text{Suffix}(w)) \cap V| > (1+\beta) \cdot \frac{|V|}{2^l}] \leq e^{-\beta^2 \cdot |V|/(3 \cdot 2^l)} \leq e^{-\beta^2/3}.$$

There are at most $n \cdot 2^n$ ways to fix l and w and there are at most 2^{n^2} subspaces V. We set $\beta = 2 \cdot n - 1$ and property (b) fails with probability at most $n \cdot 2^{n+n^2} \cdot e^{-(2n-1)^2/3} < 1/4$ for sufficiently large n.

(c) Assume that $|\text{fanin}'(u)| = s$ and $|\text{fanout}'(u)| = t$. Hence $s \cdot t$ label sequences $x^i y^j$ ($1 \leq i \leq s$, $1 \leq j \leq t$) have paths travelling through u and ending in vertices of V_n. The f-values of those label sequences, which end in the same vertex of V_n, induce a space of dimension at most $n - 1$. Thus there are vectors $z^1, \ldots, z^t \in \{0,1\}^n$ with the restriction $\langle f(x^i y^j), z^j \rangle = 0$. We fix x^i, y^j, z^j for all i, j and the $s \cdot t$ restrictions hold with probability at most $2^{-s \cdot t}$.

There are at most $2^{n \cdot s} \cdot 2^{2n \cdot t} = 2^{n(s+2t)}$ ways to fix the $s \cdot t$ vectors. But $2^{n(s+2t)} < 2^{s \cdot t - 2n}/4$, provided $s \geq 7 \cdot n$ and $t \geq 5 \cdot n$. (If $s \geq 2t$, then $n \cdot (s+2t) \leq n \cdot 2s \leq s \cdot t - 2n - 2$. If $s < 2t$, then $n \cdot (s+2t) < n \cdot 4t \leq s \cdot t - 2n - 2$.) Thus some vertex u with too large fanin' and fanout' sets exists with probability at most $\frac{1}{4}$. □

4.2 The Lower Bound

We fix a function f as guaranteed by Proposition 3 and have to show that the ε-free NFA $F_n(f)$ is large. Let $l = c \cdot \sqrt{n}$, where $c < 1/2$ is arbitrary, and define the set

$$X = \{u \mid u \in V_{n-l} \text{ and } |\text{fanin}'(u)| \geq 2^{\sqrt{n}}\}$$

of high fanin vertices. We say that string $s \in \{0,1\}^{2n}$ hits vertex u of $F_n(f)$ iff there is an accepting path with label sequence s which traverses u. If sufficiently many strings hit vertices in X, then we show in Lemma 5 that V_{n-l} has to contain many high fanin vertices and we are done. Thus we may assume that only "low fanin" vertices in V_{n-l} "count". But then Lemma 6 shows that the fanin' sets increase only gradually in size between layers V_{n-l} and V_n, if we disregard a few "exceptional" vertices. Since the situation is similar for "high fanout" vertices in V_{n+l}, the argument concludes with Lemma 7, which shows that the overall fanin *or* the overall fanout of vertices in V_n has to be large. We begin by showing that X has to be large, if X is hit by many strings.

Lemma 5. *Let $\alpha > 0$ be arbitrary. If at least $\alpha \cdot 2^{2n}$ strings hit vertices in X, then $|X| \geq \alpha \cdot \frac{2^{n+l}}{10n^2}$.*

Proof. Remember $l = c \cdot \sqrt{n}$. For $w \in \{0,1\}^l$ let $X(w)$ be the set of vertices in X which are hit by at least $2^{\sqrt{n}}$ strings from $\{0,1\}^{n-l} \cdot w \cdot \{0,1\}^n$. We pick an arbitrary vertex $u \in X(w)$ which is hit by t strings $\alpha_1 w \beta_1, \ldots, \alpha_t w \beta_t$. We define $V_u(w)$ as the vector space spanned by $f(\alpha_1 w), \ldots, f(\alpha_t w)$ and observe that $V_u(w)$ has dimension at least $\sqrt{n} \geq l$. Hence we may apply Proposition 3(b) and obtain that $t \leq |f(\{0,1\}^{n-l} w) \cap V_u(w)| \leq 2n \cdot \frac{|V_u(w)|}{2^l}$.

But then vertex u hits at most $2n \cdot \frac{|V_u(w)|}{2^l} \cdot \frac{2^n}{|V_u(w)|} = 2n \cdot 2^{n-l}$ strings from $\{0,1\}^{n-l} \cdot w \cdot \{0,1\}^n$, since all n-bit suffices belong to the orthogonal space $V_u(w)^\perp$. If $x(w)$ is the number of strings from $\{0,1\}^{n-l} \cdot w \cdot \{0,1\}^n$ hitting X, then $X(w)$ has to consist of at least $\frac{x(w)}{2n \cdot 2^{n-l}}$ vertices. We observe as a consequence of Proposition 3(c) that a vertex $u \in X$ satisfies $|\text{fanout}'(u)| \leq 5n$ and hence u appears in at most $5n$ sets $X(w)$. But $\sum_w x(w) \geq \alpha \cdot 2^{2n}$ by assumption and therefore

$$|X| \geq \sum_w \frac{|X_w|}{5n} \geq \sum_w \frac{1}{5n} \cdot \frac{x(w)}{2n \cdot 2^{n-l}} \geq \sum_w \frac{x_w}{10n^2 \cdot 2^{n-l}} \geq \alpha \frac{2^{2n}}{10n^2 \cdot 2^{n-l}} = \alpha \frac{2^{n+l}}{10n^2}.$$

The claim follows. □

The same result holds if we consider vertices u of depth $n + c \cdot \sqrt{n}$. In particular, if $l = c \cdot \sqrt{n}$, then define $\text{fanin}''(u)$ (resp. $\text{fanout}''(u)$) as the set of all l-bit (resp. $n - l$-bit) label sequences of paths from a vertex in V_n to u (resp. of paths from u to the sink of $F_n(f)$). We consider the set $X' = \{u \mid u \in V_{n+l}$ and $|\text{fanout}''(u)| \geq 2^{\sqrt{n}}\}$ and obtain $|X'| \geq \alpha \cdot \frac{2^{n+l}}{10n^2}$, provided at least $\alpha \cdot 2^{2n}$ strings hit vertices in X'. Since our goal is to show that ε-free NFAs have size $2^{n+\Omega(l)}$, we only have to deal with the situation that vertices in X as well as X' are hit by too few strings.

We remove X and X' and observe that only accepting paths for a set S of at most $2\alpha \cdot 2^{2n}$ different label sequences are lost. But now all vertices $u \in V_{n-c \cdot \sqrt{n}}$ and $v \in V_{n+c \cdot \sqrt{n}}$ have small fanin (i.e., $|\text{fanin}'(u)| \leq 2^{\sqrt{n}}$) and small fanout (i.e., $|\text{fanout}'(v)| \leq 2^{\sqrt{n}}$) respectively. If $F_n(f)$ has few edges, then we show next that fanins increase only gradually in the intermediate $c \cdot \sqrt{n}$ layers. The situation for fanouts is analogous. Let $t_0 = n - c \cdot \sqrt{n}$, $t_1 = n + c \cdot \sqrt{n}$ and set $\alpha = 1/16$.

Lemma 6. *Remove X and X'. If $|\text{fanin}'(u)| \leq 2^{\sqrt{n}}$ for all $u \in V_{t_0}$ and $|\text{fanout}''(u)| \leq 2^{\sqrt{n}}$ for all $u \in V_{t_1}$ and if $F_n(f)$ has at most $2^{n+\sqrt{n}}/(320 \cdot n^2)$ edges, then there is a subset $V_n^* \subseteq V_n$ of vertices v with $|\text{fanin}'(v)|, |\text{fanout}''(v)| \leq 2^{\sqrt{n}+c \cdot n}$ which is hit by at least $2^{2n}/4$ strings.*

Proof. For $t_0 \leq t < n$ let e_t be the number of edges between vertices in V_t and V_{t+1}. For the parameter β to be fixed later, we call a vertex $u \in V_{t+1}$ *fat*, if it receives more than $\beta \cdot e_t/2^n$ edges and if $|\text{fanin}'(u)| > 7n$. Observe that there is a total of at most $c\sqrt{n} \cdot \frac{2^n}{\beta}$ fat vertices in V_t for $t_0 \leq t < n$.

We claim that not too many strings hit fat vertices u. To see why, we first observe $|\text{fanout}'(u)| \leq 5n$. Hence there are at most $5n$ vector spaces $V_1 \times V_1^\perp, \ldots, V_{5n} \times V_{5n}^\perp$ such that $(f(x), f(y))$ belongs to one of the spaces for any string xy that hits u. Thus, with Proposition 3(a) at most $5n \cdot (2\sqrt{n})^2 \cdot 2^n = 20n^2 \cdot 2^n$ strings hit u and at most $(c\sqrt{n} \cdot \frac{2^n}{\beta}) \cdot (20n^2 \cdot 2^n) = 20c \cdot n^{2.5} \cdot \frac{2^{2n}}{\beta}$ strings hit fat vertices. Set $\beta = 320c \cdot n^{2.5}$.

We remove all fat vertices and destroy accepting paths for at most $2^{2n}/16$ strings. An inductive argument shows that $|\text{fanin}'(u)| \leq 2^{\sqrt{n}} \cdot \Pi_{\tau=t_0}^{t-1}(\beta e_\tau/2^n)$ for all vertices $u \in V_t$ for $t_0 \leq t < n$. Hence, since $e = \sum_{\tau=t_0}^{t} e_\tau \leq 2^{n+\sqrt{n}}/(320 \cdot n^2)$ by assumption, all fanin$'$ sets have size bounded by

$$2^{\sqrt{n}} \cdot (\beta \cdot \frac{e}{c\sqrt{n} \cdot 2^n})^{c \cdot \sqrt{n}} = 2^{\sqrt{n}} \cdot (\frac{320 \cdot n^2 \cdot e}{2^n})^{c \cdot \sqrt{n}} \leq 2^{\sqrt{n}} \cdot 2^{c \cdot n}.$$

We repeat this procedure, but now utilize the fanout restrictions. We remove (a new set of) fat vertices and lose additionally 2^{2n-4} strings. Counting also the strings lost by removing $X \cup X'$ and observing $\alpha = 1/16$, we have lost at most $2\alpha 2^{2n} + 2 \cdot 2^{2n-4} = 2^{2n-2}$ strings and at least 2^{2n-2} strings are still "alive". □

Hence, if $F_n(f)$ is not too large, then one half of all strings hit vertices in V_n with small fanin and small fanout. However we now show that the overall fanin or the overall fanout of V_n^* has to be large.

Lemma 7. *Remove X and X'. Then*

$$\sum_{v \in V_n^*} |\text{fanin}'(v)| \geq 2^{3n/2}/(16\sqrt{n}) \quad \text{or} \quad \sum_{v \in V_n^*} |\text{fanout}''(v)| \geq 2^{3n/2}/(16\sqrt{n}).$$

Proof. Remove all vertices in X and X'. According to Lemma 6 the vertices $v \in V_n^*$ define a cover of at least one half of all ones of H_n, if we associate the submatrix fanin$'(v) \times$ fanout$''(v)$ with v. We observe, with Proposition 3(a) and inequality (1), that vertex v is hit by at most $(2\sqrt{n})^2 \cdot 2^n = 4n \cdot 2^n$ different strings $u_1 u_2 \in L_n(f)$. Assume that the claim is false and we obtain

$$\sum_{v \in V_n^*} |\text{fanin}'(v)|, \sum_{v \in V_n^*} |\text{fanout}''(v)| \leq \frac{2^{3n/2}}{16\sqrt{n}} \quad \text{as well as} \quad (6)$$

$$|\text{fanin}'(v)| \cdot |\text{fanout}''(v)| \leq 4n \cdot 2^n. \quad (7)$$

When maximizing $\sum_{v \in V_n^*} |\text{fanin}'(v)| \cdot |\text{fanout}''(v)|$ subject to restrictions (6) and (7), it is optimal to choose $|\text{fanin}'(v)| = |\text{fanout}''(v)| = \sqrt{4n \cdot 2^n}$ for all $v \in V_n^*$. But then $|V_n^*| \leq \frac{2^{3n/2}}{16\sqrt{n}} \cdot \frac{1}{\sqrt{4n \cdot 2^n}}$ and hence $\sum_{v \in V_n^*} |\text{fanin}'(v)| \cdot |\text{fanout}''(v)| \leq \frac{2^{3n/2}}{16\sqrt{n}\sqrt{4n 2^n}} \cdot (4n 2^n) = 2^{2n-3}$, but H has at least 2^{2n-2} 1-entries. □

Proof of Theorem 1 (iv). If X or X' is hit by sufficiently many strings, then, according to Lemma 5, $F_n(f)$ consists of at least $\alpha \cdot \frac{2^{n+l}}{10n^2} = \Omega(\frac{2^{n+l}}{n^2})$ vertices, where $l = c \cdot \sqrt{n}$. Otherwise we have to differentiate two cases. Firstly $F_n(f)$ has at least $2^{n+\sqrt{n}}/(320 \cdot n^2) = \Omega(2^{n+\sqrt{n}}/n^2)$ edges. Secondly $F_n(f)$ has fewer edges and we may apply Lemma 6: we obtain a subset $V_n^* \subseteq V_n$ of vertices with individual fanin and fanout bounded by $2^{\sqrt{n}+c \cdot n}$ such that V_n^* is hit by at least $2^{2n}/4$ edges. But then Lemma 7 requires an overall fanin or an overall fanout of at least $\Omega(2^{3n/2}/\sqrt{n})$ and hence $\Omega(2^{3n/2}/(\sqrt{n} \cdot 2^{\sqrt{n}+c \cdot n})) = \Omega(2^{(3/2-c)n}/(\sqrt{n} \cdot 2^{\sqrt{n}}))$ vertices are required. For any constant $c < 1/2$ the graph $F_n(f)$ has size at least $\Omega(2^{n+c \cdot \sqrt{n}}/n^2)$ and we may drop the factor $1/n^2$ by increasing c slightly. □

5 Conclusions and Open Problems

We have shown an almost quadratic gap between the descriptional complexity of NFAs and ε-free NFAs. Our approach is based on communication complexity: "inner" vertices of an NFA recognizing $\{xy \mid x, y \in \{0,1\}^n, (x,y) \in L\}$ define a cover of the communication matrix of L. Additional graph-theoretic arguments are required for large as well as for small alphabets to obtain a separation of NFA and ε-free NFA.

Whereas Theorem 1 shows for large alphabets that the obvious transformation from NFAs to ε-free NFAs cannot be improved significantly, considerable improvements for small alphabets are not ruled out. Observe however that improvements of our lower bound require new ideas, since our analysis of "bipartite NFAs" is close to optimal according to Lemma 4.

Acknowledgement. Thanks to an anonymous referee for pointing out a mistake in an earlier version.

References

[HMU01] J.E. Hopcroft, R. Motwani, J.D. Ullman, "Introduction to Automata Theory, Languages and Computation", Addison-Wesley, 2001.

[HSW97] J. Hromkovič, S. Seibert, and T. Wilke, Translating regular expression into small ε-free nondeterministic automata, *Journal of Computer and System Sciences*, 62 (4), pp. 565-588, 2001.

[KN97] E. Kushilevitz and N. Nisan, Communication Complexity, *Cambridge University Press*, 1997.

[Lif03] Y. Lifshits, A lower bound on the size of ε-free NFA corresponding to a regular expression, *Inf. Process. Lett.* 85(6), pp. 293-299, 2003.

[Tho68] K. Thompson, Regular expression search, *Comm. ACM* 11, pp. 419-422, 1968.

On the Equivalence of $\mathbb{Z}$-Automata

Marie-Pierre Béal[1], Sylvain Lombardy[2], and Jacques Sakarovitch[3]

[1] Institut Gaspard-Monge, Université Marne-la-Vallée
[2] LIAFA, Université Paris 7
[3] LTCI, CNRS / Ecole Nationale Supérieure des Télécommunications. (UMR 5141)
beal@univ-mlv.fr, lombardy@liafa.jussieu.fr, sakarovitch@enst.fr

Abstract. We prove that two automata with multiplicity in $\mathbb{Z}$ are equivalent, *i.e.* define the same rational series, if and only if there is a sequence of $\mathbb{Z}$-coverings, co-$\mathbb{Z}$-coverings, and circulations of -1, which transforms one automaton into the other. Moreover, the construction of these transformations is effective.

This is obtained by combining two results: the first one relates coverings to conjugacy of automata, and is modeled after a theorem from symbolic dynamics; the second one is an adaptation of Schützenberger's reduction algorithm of representations in a field to representations in an Euclidean domain (and thus in $\mathbb{Z}$).

1 Introduction

Equivalence of $\mathbb{Z}$-automata is decidable with polynomial (cubic) complexity. This is not a new result: it is more than forty years old. We investigate it again in order to give more *structural information* on two equivalent $\mathbb{Z}$-automata. A first and simple example should make clear what we mean by that before we state the precise results we are aiming at.

An Example. Let us consider the two $\mathbb{Z}$-automata $\mathcal{A}_1$ and $\mathcal{B}_1$ of Figure 1. They are equivalent.[1] This can be proved by checking that series $|\mathcal{A}_1|$ and $|\mathcal{B}_1|$ have the same coefficients on every word of $\{a,b\}^*$ up to length 8 — which would be the algorithm derived from the *Equality Theorem* — or by verifying that the *reduced representation* of the series $|\mathcal{A}_1| - |\mathcal{B}_1|$ has dimension 0.

We aim here at the construction of the two $\mathbb{Z}$-automata $\mathcal{C}_1$ and $\mathcal{D}_1$ of Figure 2. They are equivalent as one is obtained from the other by multiplying by -1 the coefficients of both the incoming and outgoing transitions around the state 1. The automata $\mathcal{C}_1$ and $\mathcal{A}_1$ are equivalent as $\mathcal{A}_1$ is obtained from $\mathcal{C}_1$ by merging the states 1 and 3 on one hand, the states 2 and 4 on the other hand, as these merged states have *the same incoming transitions* and as the outgoing transitions are added. The automata $\mathcal{D}_1$ and $\mathcal{B}_1$ are equivalent as $\mathcal{B}_1$ is obtained from $\mathcal{D}_1$ by

---

[1] This equivalence expresses a "shuffle identity": $(a\,b)^* \between (-a\,b)^* = (-4a^2b^2)^*$ that was mentioned to us by M. Waldschmidt (personnal communication).

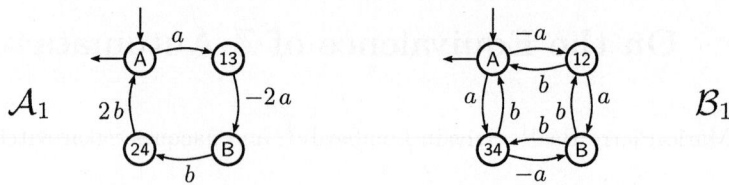

Fig. 1. Two equivalent $\mathbb{Z}$-automata

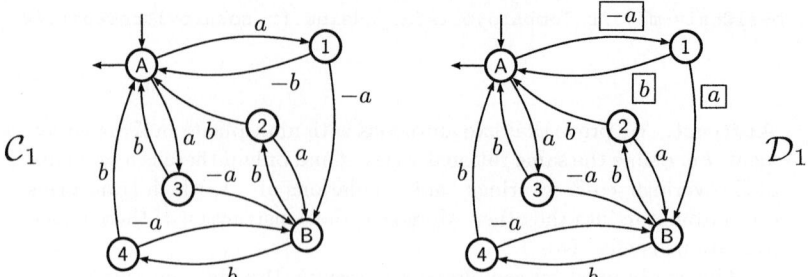

Fig. 2. Two other equivalent $\mathbb{Z}$-automata

merging the states 1 and 2 on one hand, the states 3 and 4 on the other hand, as these merged states have *the same outgoing transitions* and as the incoming transitions are added.

Equivalence of $\mathcal{A}_1$ and $\mathcal{B}_1$ boils down to the obvious three equivalences: $\mathcal{A}_1$ and $\mathcal{C}_1$, $\mathcal{C}_1$ and $\mathcal{D}_1$, and $\mathcal{D}_1$ and $\mathcal{B}_1$. The general case will not be that simple, but will nevertheless follow the same scheme.

The General Framework and Our Results. Finite automata with multiplicity in $\mathbb{Z}$ — with multiplicity in $\mathbb{Q}$ or in a commutative field $\mathbb{F}$ as well — were introduced as soon as 1961 by M.P. Schützenberger in a paper entitled: *On the definition of a family of automata* [17]. At that time, the emphasis was put on the investigation of new models of computation that allow the definition of *new families of languages* inside or outside Chomsky's hierarchy. The $\mathbb{Z}$-automata define one of the latter by way of the support of the series they realize. But Schützenberger's seminal paper also generalizes the theory of *rational formal power series* from one variable to several non commuting variables. Among other things, the existence of reduced $\mathbb{F}$-automata that realize a given $\mathbb{F}$-rational series, and the similarity between two equivalent reduced $\mathbb{F}$-automata is shown there. An algorithm is given — more or less explicitly — that computes a reduced $\mathbb{F}$-automaton. The decidability of the equivalence of $\mathbb{Z}$-automata is a direct consequence of the latter.

In [6] Eilenberg stated the *Equality Theorem* but disregarded the notion of reduced representation although the complexity of the corresponding equivalence algorithms blows up from polynomial to exponential without the latter. In other books on automata with multiplicity ([15, 10]) the equivalence of $\mathbb{Z}$-automata is reduced to the one of rational series over one variable. Schützenberger's reduction algorithm was made explicit in [5][2] and above all in [3]. The generalization to skew fields of coefficients is straightforward (as was noted for instance in [8]). The importance of the decidablity of the equivalence of automata with multiplicity in a (sub-semiring of a) skew field appears clearly with its role in the proof of the decidability of the equivalence of deterministic k-tape transducers [9].

Our contribution to this now well-established chapter of automata theory is based on two notions: *covering* and *conjugacy* of automata and develops in two directions: the generalization and reinterpretation of the *Finite Equivalence Theorem* of symbolic dynamics for $\mathbb{Z}$-automata on one hand and of the reduction algorithm for automata with multiplicity in an Euclidean domain on the other hand. The two results are combined in our last theorem (Theorem 4) that expresses the equivalence of two $\mathbb{Z}$-automata as a sequence of six coverings and a conjugacy of a special kind, all effectively computable.

Coverings, Conjugacy and the Conjugacy Theorems. A *finite* automaton $\mathcal{A}$ over an alphabet A with multiplicity in a semiring $\mathbb{K}$, or $\mathbb{K}$-automaton for short, can be written in a compact way as $\mathcal{A} = \langle I, E, T \rangle$ where E is a square matrix of finite dimension Q whose entries are linear combinations of letters (with coefficients in $\mathbb{K}$) and where I and T are two vectors — respectively row vector and column vector — with entries in $\mathbb{K}$ as well. We can view each entry $E_{p,q}$ as the label of a unique arc which goes from state p to state q in the graph whose set of vertices is Q (if $E_{p,q} = 0_\mathbb{K}$, we consider that there is *no* arc from p and q).

The *behaviour* of $\mathcal{A}$, denoted $|\mathcal{A}|$, is the series such that the coefficient of a word w is the coefficient of w in $I \cdot E^{|w|} \cdot T$. It is part of Kleene-Schützenberger Theorem that every $\mathbb{K}$-rational series is the behaviour of a $\mathbb{K}$-automaton of the form we have just defined. For missing definitions, we refer to [6, 3, 14].

$\mathbb{K}$-coverings and co-$\mathbb{K}$-coverings are generalizations of morphisms of classical (Boolean) automata to $\mathbb{K}$-automata; the precise definition will be given and discussed in Section 2 but typically a $\mathbb{K}$-*covering* is the map that sends the above automaton $\mathcal{D}_1$ onto $\mathcal{B}_1$ and a co-$\mathbb{K}$-*covering* is the map that sends $\mathcal{C}_1$ onto $\mathcal{A}_1$. The second notion, the conjugacy of ($\mathbb{K}$-)automata, comes from symbolic dynamics (we follow [12]) and can be described as follow.

Definition 1. *An automaton* $\mathcal{A} = \langle I, E, T \rangle$ *is conjugate to an automaton* $\mathcal{B} = \langle J, F, U \rangle$ *if there exists a matrix X with entries in $\mathbb{K}$ such that*

$$IX = J, \quad EX = XF, \quad \text{and} \quad T = XU.$$

The matrix X is the transfer matrix *of the conjugacy and we write* $\mathcal{A} \overset{X}{\Longrightarrow} \mathcal{B}$.

Obviously two conjugate automata are equivalent (*i.e.* have the same behaviour). Remark that in spite of the idea conveyed by the terminology, the conjugacy relation[3] *is not an equivalence* but a *preorder* relation. Suppose that

---
[2] If a paper written in French is not considered to be cryptic.
[3] A conjugacy of automata was called *a backward elementary equivalence* in [2]. The transfer matrix X is called a *simulation* from $\mathcal{A}$ to $\mathcal{B}$ in [4].

$\mathcal{A} \overset{X}{\Longrightarrow} \mathcal{B}$ holds; if $\mathcal{B} \overset{Y}{\Longrightarrow} \mathcal{C}$ then $\mathcal{A} \overset{XY}{\Longrightarrow} \mathcal{C}$, but if $\mathcal{C} \overset{Y}{\Longrightarrow} \mathcal{B}$ then $\mathcal{A}$ is not necessarily conjugate to $\mathcal{C}$, we write $\mathcal{A} \overset{X}{\Longrightarrow} \mathcal{B} \overset{Y}{\Longleftarrow} \mathcal{C}$ and we refer to this situation as "a chain of two (convergent) conjugacies".

We shall see that $\mathbb{K}$-coverings are realized by conjugacy with special transfer matrices. The following theorems express a kind of converse. (A matrix is *non degenerate* if it contains no zero row nor zero column.)

Theorem 1. *Let $\mathcal{A}$ and $\mathcal{B}$ be two $\mathbb{Z}$-automata. We have $\mathcal{A} \overset{X}{\Longrightarrow} \mathcal{B}$ with a nonnegative and nondegenerate transfer matrix X if and only if there exists a $\mathbb{Z}$-automaton $\mathcal{C}$ that is a co-$\mathbb{Z}$-covering of $\mathcal{A}$ and a $\mathbb{Z}$-covering of $\mathcal{B}$.*

One of the reasons that make Theorem 1 interesting in our opinion is the relationship it bears with the Finite Equivalence Theorem of symbolic dynamics and we develop this point at Section 2.3. In order to state the results under the most general form in the sequel, let us write $\mathbb{H}$ for a ring that is either $\mathbb{Z}$ or a division ring[4]. By the following, we free ourselves from the two hypotheses on the transfer matrix (we call *circulation matrix* a diagonal invertible matrix).

Theorem 2. *Let $\mathcal{A}$ and $\mathcal{B}$ be two $\mathbb{H}$-automata. We have $\mathcal{A} \overset{X}{\Longrightarrow} \mathcal{B}$ if and only if there exists two $\mathbb{H}$-automata $\mathcal{C}$ and $\mathcal{D}$ and a circulation matrix D such that $\mathcal{C}$ is a co-$\mathbb{H}$-covering of $\mathcal{A}$, $\mathcal{D}$ and a $\mathbb{H}$-covering of $\mathcal{B}$ and $\mathcal{C} \overset{D}{\Longrightarrow} \mathcal{D}$.*

This theorem describes precisely the situation in our first example: $\mathcal{A}_1$ is conjugate to $\mathcal{B}_1$ with the transfer matrix X_1 shown opposite.

$$X_1 = \begin{pmatrix} 1 & 0 & 0 & 0 \\ 0 & -1 & 1 & 0 \\ 0 & 1 & 1 & 0 \\ 0 & 0 & 0 & 1 \end{pmatrix}$$

Equivalence, Conjugacy and the Reduction Theorems. Our second contribution consists in establishing a kind of converse to the equivalence of conjugate $\mathbb{K}$-automata in two important cases: skew fields and Euclidean domains[5].

Theorem 3. *Let $\mathbb{L}$ be a skew field, or an Euclidean domain. If $\mathcal{A}$ and $\mathcal{B}$ are two equivalent $\mathbb{L}$-automata, then there exist two $\mathbb{L}$-automata $\mathcal{C}$ and $\mathcal{D}$ and three $\mathbb{L}$-matrices X, Y and Z such that:*

$$\mathcal{A} \overset{X}{\Longrightarrow} \mathcal{C} \overset{Y}{\Longleftarrow} \mathcal{D} \overset{Z}{\Longrightarrow} \mathcal{B}. \tag{1}$$

The alternative hypotheses correspond indeed to two different results, and two distinct proofs.

If $\mathbb{L}$ is a skew field, the proof is based on Schützenberger's algorithm that computes a reduced representation (*i.e.* an automaton with a minimal number of states) for a given $\mathbb{L}$-rational series. This algorithm may be interpreted as the effective computation of the transfer matrices of a chain of two (divergent

---

[4] *i.e.* a skew field. In Section 2.2, we take even more general hypotheses.
[5] An Euclidean domain is a principal *commutative* ring with no divisors of zero and where the gcd of any two elements is effectively computable.

$$\mathcal{A} \xleftarrow{\quad C \quad} \xleftarrow{\quad R'' \quad} \xleftarrow{\quad C' \quad} \mathcal{C} \xleftarrow{\quad D \quad} \mathcal{D} \xrightarrow{\quad R' \quad} \xrightarrow{\quad C'' \quad} \xrightarrow{\quad R \quad} \mathcal{B}$$

Fig. 3. The decomposition of the equivalence between $\mathbb{H}$-automata

or convergent) conjugacies. As the same algorithm implies that two minimal $\mathbb{L}$-automata are similar[6], Equation 1 holds, with the supplementary condition that the $\mathbb{L}$-automata $\mathcal{C}$ and $\mathcal{D}$ are minimal.

In the case where $\mathbb{L}$ is an Euclidean domain ($\mathbb{Z}$ for instance), we prove that the above reduction algorithm can be transformed in such a way that it still computes a reduced $\mathbb{L}$-representation. As far as we know, this is the first reduction algorithm for automata with multiplicity in such rings. Another step of proof is then necessary to establish Theorem 3 in this case, for minimal automata are not necessarily conjugate anymore.

Combining Theorem 2 and Theorem 3 together with some further properties of coverings yields the final result of this paper, illustrated in Figure 3 [as we shall see below, coverings and co-coverings are special cases of conjugacy, which we represent with simple arrows, solid for coverings, dashed for co-coverings; a dotted simple arrow represents conjugacy with a circulation matrix].

Theorem 4. *Two $\mathbb{H}$-automata $\mathcal{A}$ and $\mathcal{B}$ are equivalent if and only if there exist two $\mathbb{H}$-automata $\mathcal{C}$ and $\mathcal{D}$ such that there is a sequence of three $\mathbb{H}$-coverings and co-$\mathbb{H}$-coverings from $\mathcal{C}$ onto $\mathcal{A}$ on one hand and from $\mathcal{D}$ onto $\mathcal{B}$ on the other hand, and a conjugacy by a circulation matrix between $\mathcal{C}$ and $\mathcal{D}$.*

2 The Conjugacy Theorems

The two main ingredients in Theorem 2 are the notion of coverings and the property of *equisubtractivity* in a semiring, which albeit simple will be crucial for the proof.

2.1 $\mathbb{K}$-Coverings and Co-$\mathbb{K}$-coverings

The standard notion of morphisms of automata is not well-suited to automata with multiplicity in that it does not capture some similarities between these automata that we would like to be able to describe. Hence the definitions we take now. For the rest of the section, $\mathcal{A} = \langle I, E, T \rangle$ is a $\mathbb{K}$-automaton of dimension Q.

An equivalence φ on Q or, which is the same, a surjective map $\varphi \colon Q \to R$ is *Out-licit* (understood, *with respect to* $\mathcal{A}$) if for any two equivalent states p and p' modulo φ the *sum* of the labels of the transitions that go from p to *all the states of a whole class* modulo φ is equal to the *sum* of the labels of the transitions that go from p' to the same states *and* if any two entries of T indexed by equivalent states modulo φ are equal.[7] We denote by $[q]_\varphi$ the class of q modulo φ.

---

[6] *i.e.* conjugate with a transfer matrix which is invertible.
[7] This definition bears some resemblance with the one of block-stochastic matrix, as given in [10, Ex. 4.5]

Definition 2 ([14]). *A surjective map $\varphi\colon Q \to R$ is Out-licit with respect to $\mathcal{A}$ if the following holds:*

$$\forall p, p', q \in Q \quad p \equiv p' \mod \varphi \implies \begin{cases} \text{(i)} & \sum_{r \in [q]_\varphi} E_{p,r} = \sum_{s \in [q]_\varphi} E_{p',s} \\ \text{(ii)} & T_p = T_{p'} \end{cases} \quad (2)$$

If $\varphi\colon Q \to R$ is Out-licit, the $\mathbb{K}$-quotient of $\mathcal{A}$ by φ is the automaton $\mathcal{B} = \langle J, F, U \rangle$ of dimension R, defined by the following:

$$\forall (r,s) \in R^2, \quad \forall p \in \varphi^{-1}(r),$$
$$J_s = \sum_{q \in \varphi^{-1}(s)} I_q, \quad F_{r,s} = \sum_{q \in \varphi^{-1}(s)} E_{p,q}, \quad U_r = T_p. \quad (3)$$

The automaton $\mathcal{B}$ is called a $\mathbb{K}$-quotient of $\mathcal{A}$ and, conversely, $\mathcal{A}$ is called a $\mathbb{K}$-covering of $\mathcal{B}$. We write also $\varphi\colon \mathcal{A} \to \mathcal{B}$ and call it, by way of metonymy, a $\mathbb{K}$-covering from $\mathcal{A}$ onto $\mathcal{B}$.[8]

If $\varphi\colon Q \to R$ is a map, the above condition and construction may be elegantly described by means of the $Q \times R$-matrix H_φ naturally associated with φ: its (q,r) entry is 1 if $\varphi(q) = r$, 0 otherwise. Since φ is a map, each row of H_φ contains exactly one 1 and since φ is surjective, each column of H_φ contains at least one 1. We call H_φ the *amalgamation matrix*[9] associated with φ. The following expresses that a quotient is a conjugate.

Proposition 1. *There is a $\mathbb{K}$-covering φ from $\mathcal{A}$ onto $\mathcal{B}$ if and only if there exists an amalgamation matrix X such that $\mathcal{A} \xRightarrow{X} \mathcal{B}$ (and in this case $X = H_\varphi$).*

The notion of $\mathbb{K}$-quotient is *lateralized* in that it refers not to the transitions of the automaton but to the *outgoing* transitions from the states of the automaton. Somehow, it is the price we pay for extending the notion of morphism of automata. Therefore the *dual* notions of *In-licit* map, *co-$\mathbb{K}$-quotient* and *co-$\mathbb{K}$-covering* are defined in a natural way and we have:

Proposition 2. *There is a co-$\mathbb{K}$-covering ψ from $\mathcal{A}$ onto $\mathcal{B}$ if and only if there exists an amalgamation matrix X such that $\mathcal{A} \xRightarrow{{}^t X} \mathcal{B}$ (and in this case $X = H_\psi$).*

It follows that every $\mathbb{K}$-automaton is equivalent to any of its $\mathbb{K}$-quotients or co-$\mathbb{K}$-quotients. Clearly, if $\varphi : Q \to R$ and $\psi : R \to S$ are surjective maps, then $H_{\varphi\psi} = H_\varphi H_\psi$. Hence $\mathbb{K}$-coverings (resp. co-$\mathbb{K}$-coverings) are closed under composition.

---

[8] Definition 2 has probably been stated independently a number of times. We relied on [14] where both the definition and its matrix expression are given. It was used in full generality in [13]. If $\mathbb{K} = \mathbb{B}$, the Boolean semiring, a $\mathbb{B}$-quotient is a simulation in the sense of [1].

[9] This is the terminology proposed in [12, Def. 8.2.4].

2.2 Decomposition of Conjugacy

The proof of Theorem 2 involves indeed two properties.

We call *equisubtractive* a semiring is in which for all p, q, r and s such that $p+q = r+s$ there exist x, y, z and t such that $p = x+y$, $q = z+t$, $r = x+z$ and $s = y+t$. The semiring $\mathbb{N}$ and all rings are equisubtractive, and if $\mathbb{K}$ is equisubtractive, then so are $\mathbb{K}\langle A^* \rangle$ and $\mathbb{K}\langle\langle A^* \rangle\rangle$.

We say that a semiring has property (P) if *every element is a sum of units*. The ring $\mathbb{Z}$ and all fields have property (P). In any semiring with (P), every matrix X can be written as $X = CDR$ where C is an amalgamation, R a co-amalgamation and D a circulation matrix. In $\mathbb{Z}$, the dimension of D will be the sum of the absolute value of the entries of X.

Theorem 2 indeed holds for equisubtractive semiring $\mathbb{K}$ with (P). Its proof will be sketched with the following example.

Example 2. Let $\mathcal{A}_2 = \langle I_2, E_2, T_2 \rangle$ and $\mathcal{B}_2 = \langle J_2, F_2, U_2 \rangle$ be the two $\mathbb{Z}$-automata defined by:

$$I_2 = \begin{pmatrix} 1 & 1 & 0 \end{pmatrix}, \quad E_2 = \begin{pmatrix} b & 2a+b & a \\ b & a+b & a \\ 0 & a & 0 \end{pmatrix}, \quad T_2 = \begin{pmatrix} 0 \\ -1 \\ 1 \end{pmatrix},$$

and

$$J_2 = \begin{pmatrix} 2 & -1 \end{pmatrix}, \quad F_2 = \begin{pmatrix} 2a+2b & -a-b \\ a & -a \end{pmatrix}, \quad U_2 = \begin{pmatrix} 0 \\ 1 \end{pmatrix}.$$

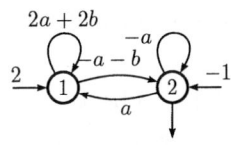

One can check that $\mathcal{A}_2 \stackrel{X_2}{\Longrightarrow} \mathcal{B}_2$, with $X_2 = \begin{pmatrix} 1 & 0 \\ 1 & -1 \\ 0 & 1 \end{pmatrix}$.

It then comes, $X_2 = C_2 D_2 R_2$ with

$$C_2 = \begin{pmatrix} 1 & 0 & 0 & 0 \\ 0 & 1 & 1 & 0 \\ 0 & 0 & 0 & 1 \end{pmatrix}, \quad D_2 = \begin{pmatrix} 1 & 0 & 0 & 0 \\ 0 & 1 & 0 & 0 \\ 0 & 0 & -1 & 0 \\ 0 & 0 & 0 & 1 \end{pmatrix}, \quad \text{and} \quad R_2 = \begin{pmatrix} 1 & 0 \\ 1 & 0 \\ 0 & 1 \\ 0 & 1 \end{pmatrix}.$$

The proof of Theorem 2 amounts then to computing $\mathcal{C}_2 = \langle K_2, G_2, V_2 \rangle$ and $\mathcal{D}_2 = \langle L_2, H_2, W_2 \rangle$ such that:

$$\mathcal{A}_2 \stackrel{C_2}{\Longrightarrow} \mathcal{C}_2 \stackrel{D_2}{\Longrightarrow} \mathcal{D}_2 \stackrel{R_2}{\Longrightarrow} \mathcal{B}_2 \ .$$

We set $K_2 = I_2 C_2$, $L_2 = K_2 D_2$, $W_2 = R_2 U_2$, $V_2 = D_2 W_2$ and we are left with the computation of $G_2 D_2 = D_2 H_2$, a 4×4 matrix. This matrix is composed of sub-matrices and the sum of the entries on every column and every row of each of these sub-matrices is given by the products $E_2 C_2 D_2$ and $D_2 R_2 F_2$. The sub-matrix decomposition and the "constraints" of our example are shown at Figure 4. The fact that $\mathcal{B}_2$ is conjugate to $\mathcal{A}_2$ ensures that

The dots figure the coefficients of the matrix, the cartouches figure their sums.

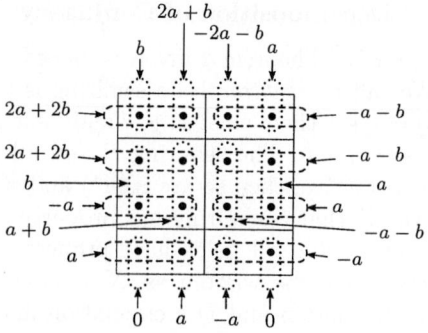

Fig. 4. Computation of $G_2 D_2 = D_2 H_2$ in Theorem 2

these constraints are consistent and the equisubtractivity of $\mathbb{Z}$ allows to compute a solution:

$$G_2 = \begin{pmatrix} b & 2a+b & 2a+b & a \\ b & 2a+b & a+b & 0 \\ 0 & -a & 0 & a \\ 0 & a & a & 0 \end{pmatrix}, \quad H_2 = \begin{pmatrix} b & 2a+b & -2a-b & a \\ b & 2a+b & -a-b & 0 \\ 0 & a & 0 & -a \\ 0 & a & -a & 0 \end{pmatrix}.$$

This completes Example 2, and with it the proof of Theorem 2.

Finally, we note that if X is a non negative and non degenerate matrix, D is the identity matrix and Theorem 1 follows.

2.3 The Link with Symbolic Dynamics

We claim that Theorem 1 is a generalization of (a part of) the Finite Equivalence Theorem, a standard result in symbolic dynamics (*cf.* [12, Theorem 8.3.7]).

Theorem 5 (FET). *Two irreducible sofic shifts are finitely equivalent if and only if they have the same entropy.*

This requires some definitions to be understood. Symbolic dynamics deals with sets of *bi-infinite words*, i.e. subsets of $A^{\mathbb{Z}}$, closed under the *shift operation* and called *shifts*. A *sofic shift* X is the set of labels of bi-infinite paths in a finite labelled graph G and is *irreducible* if such G can be chosen to be a strongly connected graph. The set of finite factors of words in a shift X is denoted $F(\mathsf{X})$, the *entropy* of X is defined by: $h(\mathsf{X}) = \lim_{n\to\infty} \frac{1}{n} \log_2 \operatorname{Card}(F(\mathsf{X}) \cap A^n)$ and is an effectively computable "dynamic invariant" of a sofic shift. A shift X is *of finite type* if it is defined by the condition that $F(\mathsf{X})$ does not contain a finite set of words (*i.e.* $A^* \setminus F(\mathsf{X})$ is a finitely generated ideal).

Let $\mathsf{X} \subseteq A^{\mathbb{Z}}$ and $\mathsf{W} \subseteq C^{\mathbb{Z}}$ be two sofic shifts. A map $\Phi\colon \mathsf{W} \to \mathsf{X}$ is called a *k-block map* if there exist a map $\overline{\Phi}\colon C^k \to A$ and two nonnegative integers m (for memory) and a (for anticipation) with $k = m + 1 + a$ such that $\Phi((c_n)_{n\in\mathbb{Z}}) = (a_n)_{n\in\mathbb{Z}}$ iff $\overline{\Phi}(c_{n-m} \ldots c_n \ldots c_{n+a}) = a_n$ for every $n \in \mathbb{Z}$. A block map is a

k-block map for some positive integer k. A block map[10] is *finite-to-one* if the cardinal of $\Phi^{-1}(\mathbf{x})$ is bounded (independently of $\mathbf{x}$).

Two sofic shifts X and Y are *finitely equivalent* if is there is a shift of finite type W together with finite-to-one and onto block maps such that $\Phi\colon \mathsf{W} \to \mathsf{X}$ and $\Psi\colon \mathsf{W} \to \mathsf{Y}$. All terms used in the FET are now defined. The connection with Theorem 1 requires one more definition, and a double encoding.

An *edge shift* X is a sofic shift with an underlying graph G whose edges all have distinct labels. It is thus completely described by the adjacency matrix X of G, a matrix with entries in $\mathbb{N}$.

Standard techniques in symbolic dynamics reduce the FET to the case of edge shifts X and Y whose adjacency matrices are denoted by X and Y respectively. The proof of sufficiency of the entropy condition relies then on two steps. The first step, known as Furstenberg's lemma, shows that, when X and Y have equal entropy, there is a nonnegative and nonnull matrix F such that $XF = FY$. The second step constructs the adjacency matrix W of an edge shift W, together with finite-to-one and onto block maps $\Phi\colon \mathsf{W} \to \mathsf{X}$ and $\Psi\colon \mathsf{W} \to \mathsf{Y}$.

Now, the edge shifts X and Y may be seen as $\mathbb{N}$-automata over a one letter alphabet. Theorem 1 then applies to these two automata and yields an automaton that corresponds to the edge shift W, the covering and co-covering corresponding to block maps Φ and Ψ. The computation of the matrix G in the proof of Theorem 2 corresponds to the original construction for the second step of the proof of the FET, known as "filling in the tableau" (see [12, Example 8.3.11]).

3 The Reduction Theorems

A $\mathbb{K}$-automaton $\mathcal{A} = \langle I, E, T \rangle$ can be seen as a triple (I, μ, T), where μ is the morphism from A^* into $\mathbb{K}^{n \times n}$ such that $E = \sum_{a \in A} \mu(a)a$. A $\mathbb{K}$-representation, or a $\mathbb{K}$-automaton, is *minimal* if it has a minimal dimension, or a minimal number of states, among all $\mathbb{K}$-representations, or all $\mathbb{K}$-automata, that realize the same series.

3.1 Reduction in a Skew Field

The computation of a minimal representation by Schützenberger's reduction algorithm [17, 3] has two symmetrical steps: a *left reduction* and a *right reduction*; it may be described within the framework of conjugacy of automata.

The left reduction of $\mathcal{A} = (I, \mu, T)$ consists in computing a matrix X whose rows form a basis of the vector space[11] $\langle I\mu(A^*)\rangle$. The matrix X uniquely defines

---

[10] Note that for $\Phi\colon \mathsf{W} \to \mathsf{X}$ being a (k-)block map does not depend on W and X whereas the property of being finite-to-one does, and this is the reason why we consider that the definition of Φ depend on W and X.

[11] Modules over a skew field are called vector spaces (*cf.* [11]). As $\mathbb{F}$ is non commutative, one should distinguish between *left* and *right* vector spaces. $I\mu(S)$ is the set of vectors $I\mu(w)$ for $w \in S$; $\langle U \rangle$ is the vector space generated by the set U of vectors.

the automaton $\mathcal{B}$ such that $\mathcal{B} \overset{X}{\Longrightarrow} \mathcal{A}$; the dimension of $\mathcal{B}$ is equal to the one of $\langle I\mu(A^*)\rangle$. Likewise, the right reduction of $\mathcal{A}$ consists in computing a matrix Y whose columns form a basis of the (right) vector space $\langle \mu(A^*)T \rangle$ and Y uniquely defines the automaton $\mathcal{C}$ such that $\mathcal{A} \overset{Y}{\Longrightarrow} \mathcal{C}$; the dimension of $\mathcal{C}$ is equal to the one of $\langle \mu(A^*)T \rangle$. The following property is the basis of the reduction algorithm:

Proposition 3. *Let $\mathbb{F}$ be a skew field. A left reduction followed by a right reduction applied to a $\mathbb{F}$-automaton $\mathcal{A}$ yields an equivalent minimal automaton.*

The computation of a left or right reduction (*i.e* the computation of bases of the appropriate subspaces) is made effective, *via* the *completion basis theorem* by the following lemma.

Lemma 1. *Let (I, μ, T) be a $\mathbb{F}$-representation and P a finite subset of A^* which contains the empty word 1_{A^*}. If, for every a in A and every w in P, $I\mu(wa)$ belongs to $\langle I\mu(P) \rangle$ then $\langle I\mu(P) \rangle = \langle I\mu(A^*) \rangle$ (and the symmetric for $\langle \mu(P)T \rangle$).*

The algorithm of left reduction consists in finding such a finite set P by considering words of A^* in the lexicographic order; P is prefix-closed and the set $I\mu(P)$ is a basis of $\langle I\mu(A^*) \rangle$. Likewise, the algorithm of right reduction yields a suffix-closed set of words. The reduction algorithm applied to two equivalent $\mathbb{F}$-automata (using the same order) yields the *same* minimal automaton and conversely we have the following.

Lemma 2 ([7]). *If $\mathbb{F}$ is a skew field, two minimal equivalent $\mathbb{F}$-automata are similar, i.e. conjugate with an invertible transfer matrix.*

If $\mathcal{A}$ and $\mathcal{B}$ are two equivalent $\mathbb{F}$-automata, then there exist two reduced automata $\mathcal{R}$ and $\mathcal{R}'$ such that $\mathcal{A} \overset{X}{\Longleftarrow} \overset{Y}{\Longrightarrow} \mathcal{R}$, and symmetrically $\mathcal{B} \overset{X'}{\Longrightarrow} \overset{Y'}{\Longleftarrow} \mathcal{R}'$. By Lemma 2, $\mathcal{R} \overset{Z}{\Longrightarrow} \mathcal{R}'$. Hence, $\mathcal{A} \overset{X}{\Longleftarrow} \overset{YZY'}{\Longrightarrow} \overset{X'}{\Longleftarrow} \mathcal{B}$, which proves Theorem 3.

3.2 Reduction in an Euclidean Domain

We now deal with automata with multiplicity in an Euclidean domain $\mathbb{K}$ instead of in a skew field $\mathbb{F}$. (In particular, $\mathbb{Z}$ is an Euclidean domain.) There is a *dimension theory* for the free modules[12] over $\mathbb{K}$ just as the one for the vector spaces over $\mathbb{F}$ — that is any two bases of a $\mathbb{K}$-module have the same cardinal. On the other hand the completion basis theorem does not hold anymore in $\mathbb{K}$-modules. We present here a reduction algorithm that overcomes this difficulty.

In fact the proof of Proposition 3 and Lemma 1 readily extends for Euclidean domain but does not yields an effective procedure anymore. The problem arises from the fact that two $\mathbb{K}$-modules can be strictly contained one in the other and still have the same dimension. Nevertheless, the following result implies the existence of an effective procedure for the reduction algorithm.

---

[12] All the modules we consider here are free and we just call them modules.

Proposition 4. *Let (I, μ, T) be a $\mathbb{K}$-representation. There exists a finite subset P of A^* such that $\langle I\mu(P)\rangle = \langle I\mu(A^*)\rangle$.*

In contrast with the case where the multiplicity is taken in a field, we have no a priori bound (given the dimension of (I, μ, T)) on the number of elements in the set P and the basis of $\langle I\mu(P)\rangle$ is not found in the set of vectors $\{I\mu(w) \mid w \in P\}$ but in the set of the linear combinations of them.

Example 3. Let $\mathcal{A}_3 = \langle I_3, \mu_3, T_3\rangle$ be the $\mathbb{Z}$-automaton defined by:

$$I_3 = \begin{pmatrix} 3 & 4 \end{pmatrix}, \; \mu_3(a) = \begin{pmatrix} -1 & 4 \\ 1 & -3 \end{pmatrix}, \; \mu_3(b) = \begin{pmatrix} -4 & 3 \\ 3 & -2 \end{pmatrix}, \; T_3 = \begin{pmatrix} 1 \\ 1 \end{pmatrix}.$$

Here $\langle I\mu(A^*)\rangle$ is $\mathbb{Z}^2$ and a finite set P such that $\langle I\mu(P)\rangle = \mathbb{Z}^2$ is for instance $\{\varepsilon, a, ab\}$. Neither $\{\varepsilon, a\}$ nor $\{\varepsilon, b\}$ — that are the only prefix-closed sets of cardinal 2 — corresponds to a basis.

The reduction algorithm yields then directly a chain of four conjugacies between two equivalent $\mathbb{K}$-automata. Lemma 2 does not hold anymore and reducing the length of the chain from four to three requires a new result from which Theorem 3 then follows easily.

Proposition 5. *Let $\mathbb{K}$ be an Euclidean domain. If $\mathcal{A}$ and $\mathcal{B}$ are equivalent $\mathbb{K}$-automata, with $\mathcal{B}$ right reduced, then there are a $\mathbb{K}$-automaton $\mathcal{C}$ and matrices X and Y such that: $\mathcal{A} \overset{X}{\Longleftarrow} \mathcal{C} \overset{Y}{\Longrightarrow} \mathcal{B}$.*

4 Final Result

We first establish a number of properties for coverings that can be seen as "backward Church-Rosser properties", that is properties that allow to complete a commutative diagram when the *lower part* of it is known.

Proposition 6. *Let $\mathbb{K}$ be an equisubtractive semiring and let $\mathcal{A}$, $\mathcal{B}$ and $\mathcal{C}$ be three $\mathbb{K}$-automata.*

(a) If $\mathcal{A}$ and $\mathcal{B}$ are $\mathbb{K}$-coverings of $\mathcal{C}$ (resp. co-$\mathbb{K}$-coverings of $\mathcal{C}$), there exists a $\mathbb{K}$-automaton $\mathcal{D}$ which is a $\mathbb{K}$-covering (resp. a co-$\mathbb{K}$-covering) of both $\mathcal{A}$ and $\mathcal{B}$.

(b) If $\mathcal{A}$ is a $\mathbb{K}$-covering of $\mathcal{C}$ and $\mathcal{B}$ is a co-$\mathbb{K}$-covering of $\mathcal{C}$, there exists a $\mathbb{K}$-automaton $\mathcal{D}$ which is both a co-$\mathbb{K}$-covering of $\mathcal{A}$ and a $\mathbb{K}$-covering of $\mathcal{B}$.

Lemma 3. *Let D be a circulation, C a co-amalgamation and R an amalgamation matrix such that DC and RD are defined. Then there exist circulation matrices D' and D'' such that $DC = CD'$ and $RD = D''R$.*

The proof of Theorem 4 now boils down to the diagram of Figure 5. Given equivalent $\mathbb{H}$-automata $\mathcal{A}$ and $\mathcal{B}$, Theorem 3 yields the bottom line of the diagram whereas Theorem 2 allows to build the diagrams 1, 2 and 3; Proposition 5 (a) gives 4 and 5, Lemma 3 6, 7 and 8; Proposition 5 (b) gives 9 and Lemma 3 again completes the full diagram with 10 and 11.

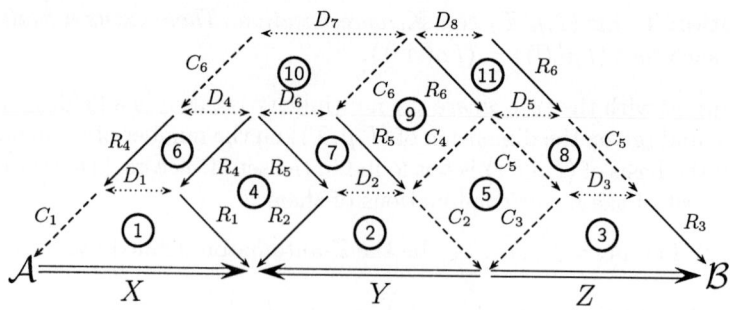

Fig. 5. Proof of Theorem 4

5 The Way for Further Developments

The gap that still remains between the above results and those formely known raise a number of questions. Let $\mathcal{A}$ and $\mathcal{B}$ be two equivalent $\mathbb{Z}$-automata. We know that they are not always conjugate one to the other and it is decidable (cf. [16]) in polynomial time (resp. in exponential time) if $\mathcal{A}$ is conjugate to $\mathcal{B}$ with a transfer matrix in $\mathbb{Z}$ (resp. with a transfer matrix in $\mathbb{N}$). Theorem 2 states the existence of a chain of three conjugacies between $\mathcal{A}$ and $\mathcal{B}$.

One then may ask whether *three* conjugacies are necessary (in general), and, if yes, whether it is decidable when *two* conjugacies suffice.

If moreover $\mathcal{A}$ and $\mathcal{B}$ are (equivalent) $\mathbb{N}$-automata, one may ask also whether the chain of conjugacies could be always realized with transfer matrices in $\mathbb{N}$ and, if not, whether it is decidable when this property holds.

By means of techniques different from the ones presented here, it can be shown that in both cases the stronger property holds: the answer to the first question is no and two conjugacies always suffice, the answer to the second question is yes and two equivalent $\mathbb{N}$-automata are joined by a chain of *four* conjugacies with transfer matrices in $\mathbb{N}$. This is the object of on-going work of the authors and will be presented in a forthcoming publication.[13]

Acknowledgements. The authors are grateful to a careful referee whose remarks helped them to make the definitions more precise and the presentation hopefully clearer and to Prof. W. Kuich who pointed to several references.

References

1. ARNOLD, A. *Finite Transitions Systems*. Prentice-Hall, 1994.
2. BÉAL, M.-P., AND PERRIN, D. On the generating sequences of regular languages on k symbols. *J. ACM* 50 (2003), 955–980.

---

[13] The above questions were quoted as open problems in the paper that was submitted to ICALP and reviewed by the PC. But since then, we have made progress and are able to answer them. They are thus not open problems anymore, and on the other hand we could not include the answers as results since they have not been refereed.

3. BERSTEL, J., AND REUTENAUER, CH. *Rational Series and their Languages.* Springer, 1988.
4. BLOOM, S., AND ESIK, Z. *Iteration Theories.* Springer, 1993.
5. CARDON, A., AND CROCHEMORE, M. Détermination de la représentation standard d'une série reconnaissable. *RAIRO Inform. Théor. 14* (1980), 371–379.
6. EILENBERG, S. *Automata, Languages, and Machines. Vol. A.* Academic Press, 1974.
7. FLIESS, M. Matrices de Hankel. *J. Math. Pures Appl. (9) 53* (1974), 197–222.
8. FLOURET, M., AND LAUGEROTTE, E. Noncommutative minimization algorithms. *Inform. Process. Lett. 64* (1997), 123–126.
9. HARJU, T., AND KARHUMÄKI, J. The equivalence problem of multitape finite automata. *Theoret. Comput. Sci. 78* (1991), 347–355.
10. KUICH, W., AND SALOMAA, A. *Semirings, Automata, Languages.* Springer, 1986.
11. LANG, S. *Algebra.* Addison Wesley, 1965.
12. LIND, D., AND MARCUS, B. *An Introduction to Symbolic Dynamics and Coding.* Cambridge University Press, 1995.
13. LOMBARDY, S. AND SAKAROVITCH, J. Derivatives of rational expressions with multiplicity. *Theoret. Comput. Sci. 332* (2005), 141–177.
14. SAKAROVITCH, J., *Eléments de théorie des automates*, Vuibert, 2003. English translation, Cambridge Universit Press, to appear.
15. SALOMAA, A., AND SOITTOLA, M. *Automata-theoretic Aspects of Formal Power Series.* Springer, 1978.
16. SCHRIJVER, A. *Theory of Linear and Integer Programming.* Wiley, 1986.
17. SCHÜTZENBERGER, M. P. On the definition of a family of automata. *Information and Control 4* (1961), 245–270.

A Tight Linear Bound on the Neighborhood of Inverse Cellular Automata

Eugen Czeizler[1] and Jarkko Kari[2,*]

[1] Department of Mathematics, FIN-20014 University of Turku,
Finland, and Turku Centre for Computer Science,
FIN-20520 Turku, Finland
eugenc@cs.utu.fi

[2] Department of Mathematics, FIN-20014 University of Turku,
Finland, and Department of Computer Science,
University of Iowa, Iowa City, IA 52242, USA
jjkari@cs.uiowa.edu

Abstract. Reversible cellular automata (RCA) are models of massively parallel computation that preserve information. They consist of an array of identical finite state machines that change their states synchronously according to a local update rule. By selecting the update rule properly the system has been made information preserving, which means that any computation process can be traced back step-by-step using an inverse automaton. We investigate the maximum range in the array that a cell may need to see in order to determine its previous state. We provide a tight upper bound on this inverse neighborhood size in the one-dimensional case: we prove that in a RCA with n states the inverse neighborhood is not wider than $n-1$, when the neighborhood in the forward direction consists of two consecutive cells. Examples are known where range $n-1$ is needed, so the bound is tight. If the forward neighborhood consists of m consecutive cells then the same technique provides the upper bound $n^{m-1} - 1$ for the inverse direction.

1 Introduction

Cellular automata (CA) are discrete dynamical systems consisting of a grid of identical finite state machines whose states are updated synchronously at discrete time steps according to a local update rule. Cellular automata possess several fundamental properties of the physical world: they are massively parallel, homogeneous and all interactions are local. It is therefore not surprising that physical and biological systems have been successfully simulated using cellular automata models. The physical nature of cellular automata may have even greater importance when applied in the opposite direction, that is, when using the physics to simulate cellular automata. Many cellular automata are computationally universal — including some extremely simple ones, as reported recently

---

* Research supported by the Academy of Finland grant 54102.

by S. Wolfram [12] — so the most powerful massively parallel computers in the future may be implementations of cellular automata based on some physical phenomena of microscopic scale. Energy efficiency of such an implementation requires that the simulated universal CA obeys the rules of physics, including reversibility and conservation laws. Non-reversibility always implies energy dissipation, usually in the form of heat.

A cellular automaton is called *reversible* if there is another cellular automaton — the *inverse CA* — that computes the inverse function. The inverse CA retraces the computation steps back in time. There are simple reversible cellular automata that are computationally universal [5]. Universality is even possible in the one-dimensional space [7], that is, when the cells are organized along a line. Reversible CA have been popular topics of study since the early years of CA research, and many interesting facts have been discovered.

It is well known that injectivity and reversibility of CA are equivalent concepts: if a CA function has an inverse (i.e. it is one-to-one) then this inverse is always a CA function [2, 9]. This means that in order to backtrack the computation, each cell only needs to know the states of a finite number of its neighbors. The question this article investigates is the extent of the neighborhood that may be needed. In two- and higher dimensional cellular automata this inverse neighborhood can be extremely large: there is namely no algorithm to determine if a given CA is reversible, which means that the extent of the inverse neighborhood cannot be bounded by any computable function of the number of states [4]. In the one-dimensional case the reversibility question is decidable, and a trivial quadratic upper bound $O(n^2)$ exists [1], where n is the number of states and the neighborhood in the forward direction has been fixed to two consecutive cells.

We improve this bound to linear $n - 1$ where n is the number of states. This bound is tight as examples of one-dimensional reversible CA are known whose inverse neighborhoods reach this bound [3]. If the neighborhood in the forward direction consists of m consecutive cells rather than two cells, then the same argument provides an upper bound $n^{m-1} - 1$ for the inverse neighborhood. This is not known to be tight: [3] only provides examples of cellular automata with $2n$ states whose inverse neighborhoods reach this size.

2 Definitions and Basic Properties

In this section we present precise definitions and some basic properties of reversible cellular automata, and Welch sets and indices. Our proofs are based on elementary linear algebra, so we also recall some linear algebra concepts.

2.1 Cellular Automata

Formally, a one-dimensional cellular automaton, CA for short, is a 3-tuple system

$$\mathcal{A} = (S, N, f),$$

where $S = \{1, 2, \ldots, n\}$ is a finite *state set*, N is a *neighborhood vector*

$$N = (x_1, \ldots, x_m) \in \mathbb{Z}^m$$

of m distinct integers, and f is a mapping from S^m to S representing the *local update rule* of the CA. The *cells* are laid on an infinite line and are indexed by $\mathbb{Z}$, the set of integers. The neighbors of a cell situated on position $x \in \mathbb{Z}$ are all the cells on positions $x + x_i, i = 1, \ldots, m$. The local update rule f determines the future state of a cell according to the states of its neighbors.

A *configuration* c of a CA $\mathcal{A}$ is a mapping

$$c : \mathbb{Z} \to S$$

which specifies the states of all the cells. We are denoting by $\mathcal{C}$ the set of all configurations. The *global transition function*

$$G : \mathcal{C} \to \mathcal{C}$$

describes the evolution of the CA and is obtained by a simultaneous application of the local update rule f on all cells:

$$G(c)(x) = f(c(x + x_1), \ldots, c(x + x_m)),$$

for all $x \in \mathbb{Z}$. It is common to identify a cellular automaton with its global transition function G, and talk about cellular automaton function G or, when there is no risk of confusion, simply cellular automaton G.

If the neighborhood vector is $(-r, \ldots, r)$ then the CA is called *radius-r* automaton. The special case $r = 1$ is the nearest neighbor neighborhood. In this work we mainly consider CA whose neighborhood is even smaller and consists of just two consecutive integers. If $N = (0, 1)$ we say that we have a *radius-$\frac{1}{2}$* CA. Figure 1 shows the trellis whose rows are consecutive configurations of a radius-$\frac{1}{2}$ cellular automaton, and the rows are shifted to make the neighborhood look symmetric. Note that any CA can be viewed as a radius-$\frac{1}{2}$ CA over a larger state set if we divide the configurations into sufficiently long blocks and use the blocks as "super cells". The partitioning may shift in time, but the computation is essentially the same.

Two CA are called equivalent if their global functions are identical. The following facts are easy to see: If two cellular automata are equivalent then there is a third equivalent CA whose neighborhood is the intersection of the neighborhoods of the first two CA. Hence, each CA function G has a minimal neighborhood, that is, a neighborhood that is contained in the neighborhoods of all CA that specify G. We call it the neighborhood of G. The interval from the smallest to the largest element of the minimal neighborhood is the neighborhood range for G. It is the smallest contiguous segment that can be used as the neighborhood to specify G.

A CA $\mathcal{A}$ with global function G is called *reversible*, for short RCA, if there exists another CA, called the *inverse automaton* of $\mathcal{A}$, whose global transition function is G^{-1}, the inverse of G. The minimal neighborhood of G^{-1} is called the

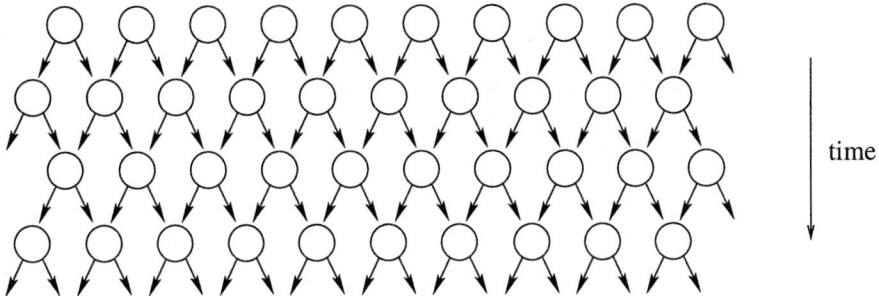

Fig. 1. Dependencies in a radius-$\frac{1}{2}$ cellular automaton

inverse neighborhood of $\mathcal{A}$ or G. Each cell can uniquely determine its previous state by looking only at the states contained in the inverse neighborhood.

A CA $\mathcal{A}$ is called *injective (surjective, bijective)* if its global transition rule $G : \mathcal{C} \to \mathcal{C}$ is an injective (surjective, bijective, respectively) function. It has been known since the early 60's that injective cellular automata are automatically also surjective [6, 8], while the converse is not necessarily true. It is also known that all bijective CA are reversible [2, 9]. We have

Property 1 ([2, 6, 8, 9]). *In cellular automata, reversibility, bijectivity and injectivity are equivalent. They imply surjectivity.*

2.2 Welch Sets and Indices

From now on we consider radius-$\frac{1}{2}$ RCA only. We frequently need to apply the CA on partial configuration where we only know the states on some contiguous interval. Since the exact location of the interval on the line is irrelevant, we specify such configurations as finite or infinite words. For the state set S we denote by S^* the set of all words over alphabet S, by S^k the set of words of length k, by S^ω the set of one-way infinite words that are infinite to the right, and by $^\omega S$ the set of words that are infinite to the left. CA $\mathcal{A} = (S, (0,1), f)$ specifies the functions $G : S^* \longrightarrow S^*$, $G : S^\omega \longrightarrow S^\omega$ and $G : {}^\omega S \longrightarrow {}^\omega S$ (all denoted by the same symbol G) defined by

- $G(a_1 a_2 \ldots a_k) = b_1 b_2 \ldots b_{k-1}$ where each $b_i = f(a_i, a_{i+1})$,
- $G(a_1 a_2 \ldots) = b_1 b_2 \ldots$ where each $b_i = f(a_i, a_{i+1})$,
- $G(\ldots a_2 a_1) = \ldots b_2 b_1$ where each $b_i = f(a_{i+1}, a_i)$.

For each $w \in S^\omega$ we set

$$L(w) = \{a \in S \mid G(au) = w \text{ for some } u \in S^\omega \}$$

and call it the *left Welch set* of w. It contains all the states that were possible one time step earlier at the leftmost cell that affects w, see Figure 2. Analogously, for any $w \in {}^\omega S$ we define the *right Welch set* as

$$R(w) = \{a \in S \mid G(ua) = w \text{ for some } u \in {}^\omega S \}.$$

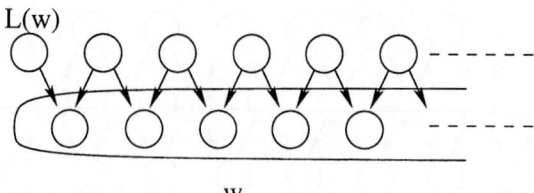

Fig. 2. The left Welch set $L(w)$ of the infinite word w consists of all possible states in the indicated cell

These sets were introduced already in [2], and have since been reinvented independently by many authors. The Welch sets have the following nice properties [2]:

Property 2. *Let $\mathcal{A} = (S, (0,1), f)$ be reversible and let $n = |S|$ be the number of states. Then for every $w \in S^\omega$ and $v \in {}^\omega S$ we have*

$$|L(w)| \cdot |R(v)| = n.$$

Consequently, the cardinalities $|L(w)|$ and $|R(v)|$ are independent of the choice of w and v.

We denote by n_L the size of left Welch sets and by n_R the size of the right Welch sets, and call them the *left* and the *right Welch index*. Then $n_L \cdot n_R = n$.

The following result is another useful property of the Welch sets [2]:

Property 3. *Let $\mathcal{A} = (S, (0,1), f)$ be reversible. Then for every $w \in S^\omega$ and $v \in {}^\omega S$ we have*

$$|L(w) \cap R(v)| = 1,$$

i.e. the intersection of any left Welch set with any right Welch set is a singleton.

The following proposition relates the Welch sets to the minimal inverse neighborhood of the CA:

Proposition 1. *Let $\mathcal{A} = (S, (0,1), f)$ be reversible. Then the inverse neighborhood of G is included in the interval*

$$\{-r, \ldots, l-1\}$$

if and only if

- $L(uw) = L(uv)$ for all $u \in S^l$ and $w, v \in S^\omega$, and
- $R(wu) = R(vu)$ for all $u \in S^r$ and $w, v \in {}^\omega S$.

Proof. First, notice that even though the given interval does not at first appear symmetric, it in fact contains l positions to the right of the cell and r positions to the left of the cell, if the cells are shifted to the right as in Figure 1.

($\Longrightarrow$) Suppose that $\{-r, \ldots, l-1\}$ contains the inverse neighborhood of G and let $u \in S^l$ and $w, v \in S^\omega$ be arbitrary. Let us prove that $L(uw) \subseteq L(uv)$. Then by symmetry we have $L(uw) = L(uv)$.

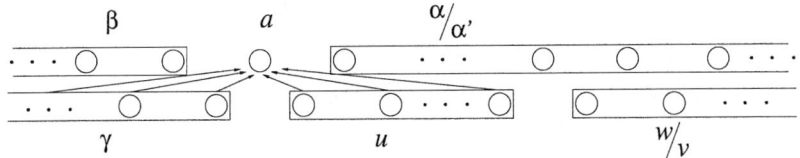

Fig. 3. Configurations $\gamma u w$ and $\gamma u v$ agree in all the positions up to $l-1$

If $a \in L(uw)$ then there exists some $\alpha \in S^\omega$ such that $G(a\alpha) = uw$. Pick an arbitrary $\beta \in {}^\omega S$ and let $\gamma = G(\beta a)$. Then $G(\beta a \alpha) = \gamma u w$, where letter a and the first letter of u are in the position 0 of the cellular array. Consider then the configuration $\gamma u v$, where the first letter of u is still in position 0, see Figure 3. It agrees with $\gamma u w$ in positions up to $l-1$. By applying the inverse cellular automata to $\gamma u v$ we must therefore get state a in position 0, which means that $a \in L(uv)$.

Analogously we get the claim concerning the right Welch sets.

($\Longleftarrow$) Suppose l and r are such that for all $v \in S^l$ and $\alpha_1, \alpha_2 \in S^\omega$ we have $L(v\alpha_1) = L(v\alpha_2)$, and that for all $u \in S^r$ and $\beta_1, \beta_2 \in {}^\omega S$ we have $R(\beta_1 u) = R(\beta_2 u)$. Then the inverse function G^{-1} is computed by the cellular automaton that uses the neighborhood $(-r, \ldots, l-1)$ and has the local update rule

$$g(uv) = R(\beta u) \cap L(v\alpha)$$

for $u \in S^r$, $v \in S^l$ and all $\alpha \in S^\omega$ and $\beta \in {}^\omega S$. The above intersection always contains a unique element, due to Property 3. □

2.3 Vector Interpretation of Sets

In our proofs we take advantage of dimension arguments on vector spaces. Any subset X of the state set $S = \{1, 2, \ldots, n\}$ is interpreted as the 0-1 vector $\vec{X}$ in $\mathbb{R}^n$ whose i'th coordinate is 1 if $i \in X$ and 0 if $i \notin X$. The single element sets $\{a\}$ then correspond to the unit coordinate vectors of $\mathbb{R}^n$ and they form a basis of the vector space $\mathbb{R}^n$. Notice that for any $X, Y \subseteq S$ the inner product $\vec{X} \cdot \vec{Y}$ is the cardinality of their intersection $X \cap Y$. The vectors $\vec{L}$ and $\vec{R}$ corresponding to left and right Welch sets L and R will be called left and right Welch vectors, respectively.

Let us denote by Θ the null space $\{(0, 0, \ldots 0)\}$ and by I the one-dimensional space generated by vector $(1, 1, \ldots, 1)$. For any $U \subseteq \mathbb{R}^n$ the subspace of $\mathbb{R}^n$ generated by U is denoted as $\langle U \rangle$.

Let $\mathcal{A} = (S, (0, 1), f)$ be reversible. For any $c \in S$ we define a linear function $h_c : \mathbb{R}^n \longrightarrow \mathbb{R}^n$ as follows. For every $b \in S$ we have $h_c(\vec{b}) = \vec{H}$ where $\vec{b}$ is the basis vector corresponding to b and $H = \{a \mid f(a, b) = c\}$. This uniquely specifies the linear function h_c. Vector $\vec{X}$, corresponding to a set $X \subseteq S$ of states, is mapped according to $h_c(\vec{X}) = \sum_{b \in X} h_c(\vec{b})$. Note that $h_c(\vec{X})$ is not always a 0-1 vector, so it does not necessarily represent a set. However, the next proposition states that if L is a left Welch set then $h_c(\vec{L})$ is a 0-1 vector representing a left Welch set:

Proposition 2. Let $\mathcal{A} = (S, (0, 1), f)$ be reversible, and let $c \in S$ be arbitrary. For every $w \in S^\omega$ we have $h_c(\vec{L}(w)) = \vec{L}(cw)$.

Proof. It is enough to show that (i) for every $a \in L(cw)$ there is a unique $b \in L(w)$ such that $f(a, b) = c$, and (ii) for any $a \notin L(cw)$ there is no $b \in L(w)$ such that $f(a, b) = c$. Parts (i) and (ii) imply then that the vector $h_c(\vec{L}(w))$ has 1 and 0 in coordinates i for all $i \in L(cw)$ and $i \notin L(cw)$, respectively.

Claim (ii) is trivial, as if there would exist $b \in L(w)$ such that $f(a, b) = c$ then $G(ab\alpha) = cw$ where $\alpha \in S^\omega$ is such that $G(b\alpha) = w$. This contradicts the assumption $a \notin L(cw)$.

Consider then claim (i). Since $a \in L(cw)$ there is some $b\alpha \in S^\omega$ such that $G(ab\alpha) = cw$. This b satisfies the condition in (i). If $b' \in L(w)$ is another state with the property $f(a, b') = c$ then $G(ab'\beta) = cw$ for some $\beta \in S^\omega$. But then $G(\gamma ab\alpha) = G(\gamma ab'\beta)$ for any $\gamma \in {}^\omega S$ which, by injectivity, implies that $b = b'$. □

Analogously, let us define linear functions $g_c(\vec{a}) = \vec{H}$ where $H = \{b \mid f(a, b) = c\}$. They naturally have the similar property concerning the right Welch sets:

Proposition 3. Let $\mathcal{A} = (S, (0, 1), f)$ be reversible, and let $c \in S$ be arbitrary. For every $w \in {}^\omega S$ we have $g_c(\vec{R}(w)) = \vec{R}(wc)$. □

3 The Inverse Neighborhood Range

In this section we prove that the size of the inverse neighborhood range of a radius-$\frac{1}{2}$ RCA $\mathcal{A} = (S, (0, 1), f)$ is less than or equal to $n - 1$, where n is the number of states. We do this by creating two decreasing chains of linear subspaces of $\mathbb{R}^n$ based on the Welch sets. The first elements of the chains are the subspaces

$$\mathcal{L}_0 = \langle \vec{L}(w) - \vec{L}(v) \mid w, v \in S^\omega \rangle, \text{ and}$$
$$\mathcal{R}_0 = \langle \vec{R}(w) - \vec{R}(v) \mid w, v \in {}^\omega S \rangle,$$

that is, the spaces generated by the differences between any two left Welch vectors and any two right Welch vectors, respectively. The goal is to prove the following theorem:

Theorem 1. Let $\mathcal{A} = (S, (0, 1), f)$ be reversible, and let $\mathcal{L}_0$ and $\mathcal{R}_0$ be the subspaces defined above. Then the inverse neighborhood range of G contains at most $\dim \mathcal{L}_0 + \dim \mathcal{R}_0$ elements. More precisely, the inverse neighborhood of G is included in the interval

$$\{-\dim \mathcal{R}_0, \ldots, \dim \mathcal{L}_0 - 1\}.$$

Proof. For every $k = 0, 1, 2, \ldots$ define the following subspaces of $\mathbb{R}^n$:

$$\mathcal{L}_k = \langle \vec{L}(uw) - \vec{L}(uv) \mid u \in S^k, w, v \in S^\omega \rangle, \text{ and}$$
$$\mathcal{R}_k = \langle \vec{R}(wu) - \vec{R}(vu) \mid u \in S^k, w, v \in {}^\omega S \rangle.$$

We make the following observations:

- There is l such that $\mathcal{L}_l = \Theta$, the null space,
- $\mathcal{L}_{k+1} \subseteq \mathcal{L}_k$ for every $k = 0, 1, 2, \ldots$, and
- if $\mathcal{L}_{k+1} = \mathcal{L}_k$ then $\mathcal{L}_j = \mathcal{L}_k$ for every $j \geq k$.

To prove the first fact, choose l and r such that the inverse neighborhood of G is included in the interval $\{-r, \ldots, l-1\}$. According to Proposition 1, $L(uw) = L(uv)$ for every $u \in S^l$ and $w, v \in S^\omega$. But then all generators of $\mathcal{L}_l$ are zero vectors, hence $\mathcal{L}_l = \Theta$.

The second fact is trivial since all the generators of $\mathcal{L}_{k+1}$ are among the generators of $\mathcal{L}_k$.

For the third fact, notice that $\vec{L}(cuw) - \vec{L}(cuv) = h_c(\vec{L}(uw) - \vec{L}(uv))$. This means that, for every $k = 0, 1, 2, \ldots$, the generators of $\mathcal{L}_{k+1}$ are obtained from the generators of $\mathcal{L}_k$ by applying the homomorphisms h_c with all $c \in S$. Consequently,

$$\mathcal{L}_{k+1} = \langle h_c(\vec{X}) \mid c \in S, \vec{X} \text{ is a generator of } \mathcal{L}_k \rangle$$
$$= \langle h_c(\vec{X}) \mid c \in S, \vec{X} \in \mathcal{L}_k \rangle.$$

In other words, $\mathcal{L}_{k+1}$ is determined by $\mathcal{L}_k$. It follows that if $\mathcal{L}_{k+1} = \mathcal{L}_k$ then $\mathcal{L}_{k+2} = \mathcal{L}_{k+1}$, and therefore $\mathcal{L}_j = \mathcal{L}_k$ for all $j \geq k$.

Our three facts imply that

$$\mathcal{L}_0 \supsetneq \mathcal{L}_1 \supsetneq \mathcal{L}_2 \supsetneq \ldots \supsetneq \mathcal{L}_l = \Theta$$

for some l. Since the dimension of the subspaces decreases at every step, we must have $l \leq \dim \mathcal{L}_0$.

The analogous reasoning can be done on the right Welch sets. We conclude that there are numbers $l \leq \dim \mathcal{L}_0$ and $r \leq \dim \mathcal{R}_0$ such that $\mathcal{L}_l = \mathcal{R}_r = \Theta$. Then l has the property that $L(uw) = L(uv)$ for every $u \in S^l$ and $w, v \in S^\omega$, and r has the property that $L(wu) = L(vu)$ for every $u \in S^r$ and $w, v \in {}^\omega S$. According to Proposition 1, the inverse neighborhood of G is included in the interval $\{-r, \ldots, l-1\}$ and hence also in the interval $\{-\dim \mathcal{R}_0, \ldots, \dim \mathcal{L}_0 - 1\}$. □

Upper bounds on the dimensions of the spaces $\mathcal{L}_0$ and $\mathcal{R}_0$ provide nice limits on the inverse neighborhood:

Corollary 1. *Let $\mathcal{A} = (S, (0, 1), f)$ be reversible. Then $\dim \mathcal{L}_0 + \dim \mathcal{R}_0 \leq n-1$ where n is the number of states. Hence the inverse neighborhood range of G has at most size $n - 1$*

Proof. This follows from the facts that vector spaces $\mathcal{L}_0$ and $\mathcal{R}_0$ are orthogonal to each other and also to the one-dimensional space I generated by the vector $(1, 1, \ldots, 1)$. Let $\vec{L}_1 - \vec{L}_2$ and $\vec{R}_1 - \vec{R}_2$ be two arbitrary generators of $\mathcal{L}_0$ and $\mathcal{R}_0$, respectively. Their inner product is

$$(\vec{L}_1 - \vec{L}_2) \cdot (\vec{R}_1 - \vec{R}_2) = \vec{L}_1 \cdot \vec{R}_1 - \vec{L}_1 \cdot \vec{R}_2 - \vec{L}_2 \cdot \vec{R}_1 + \vec{L}_2 \cdot \vec{R}_2 = 1 - 1 - 1 + 1 = 0,$$

where we have used Property 3 of the Welch sets. So spaces $\mathcal{L}_0$ and $\mathcal{R}_0$ are orthogonal to each other. With $(1,1,\ldots,1)$ we get the inner product

$$(\vec{L}_1 - \vec{L}_2) \cdot (1,1,\ldots,1) = n_L - n_L = 0,$$

where n_L is the left Welch index. Here we used Property 2 of the Welch sets. Analogously $\mathcal{R}_0$ is seen orthogonal to I.

Now we can reason as follows: Since the three spaces are orthogonal, we have

$$\dim \mathcal{L}_0 + \dim \mathcal{R}_0 + \dim I = \dim(\mathcal{L}_0 \oplus \mathcal{R}_0 \oplus I) \leq \dim \mathbb{R}^n = n,$$

so

$$\dim \mathcal{L}_0 + \dim \mathcal{R}_0 \leq n - 1.$$

□

We can also use our theorem to bound the inverse neighborhood from either side separately:

Corollary 2. *Let $\mathcal{A} = (S,(0,1),f)$ be reversible, and let n_L and n_R be its left and right Welch indices, respectively. Then $\dim \mathcal{L}_0 \leq n - n_L$ and $\dim \mathcal{R}_0 \leq n - n_R$ where n is the number of states. Hence the inverse neighborhood of G is contained in the interval*

$$\{n_R - n, \ldots, n - n_L - 1\}.$$

Proof. Consider the left Welch vectors $\vec{L}(u)$, $u \in S^\omega$. Each is a 0-1 vector with n_L ones. Every state belongs to some left Welch set, so each position has one in some of the vectors. Out of all this vectors, we can extract a set of linearly independent ones as follows. First, extract an arbitrarily vector. Then, for any state $a \in S$ such that the corresponding position is zero in all the vectors already selected, extract a left Welch vector having one in position a, and add it to the set of linearly independent vectors. Repeat the process until each position is covered by at least one selected vector. It is clear that the extracted vectors are linearly independent, and since each vector covers n_L positions there are at least $\frac{n}{n_L}$ vectors selected. Since $\frac{n}{n_L} = n_R$, it follows easily that there are at least n_R linearly independent left Welch vectors.

Next we use the following well known property: if $\vec{v}_1, \vec{v}_2, \ldots, \vec{v}_k$ are k linearly independent vectors, then $\vec{v}_2 - \vec{v}_1, \vec{v}_3 - \vec{v}_1, \ldots, \vec{v}_k - \vec{v}_1$ are $k-1$ linearly independent vectors. We apply this to the n_R linearly independent left Welch vectors and obtain the result that the generators of $\mathcal{L}_0$ contain at least $n_R - 1$ linearly independent vectors, so $\dim \mathcal{L}_0 \geq n_R - 1$. On the other hand we know by Corollary 1 that $\dim \mathcal{L}_0 + \dim \mathcal{R}_0 \leq n - 1$, so

$$\dim \mathcal{R}_0 \leq (n-1) - \dim \mathcal{L}_0 \leq (n-1) - (n_R - 1) = n - n_R.$$

Analogously we can prove that $\dim \mathcal{L}_0 \leq n - n_L$.

□

The previous corollaries were proved in [3] in the special case that one of the Welch indices is 1. This constraint simplifies the proofs very much. The techniques used in [3] were quite different. Examples were also provided in [3] of reversible CA with n states and Welch index 1 whose inverse neighborhood reached the size $n-1$. Hence the bound of Corollary 1 is tight. We do not know if there are such examples for other values of the Welch indices, and also we do not know if the bounds in Corollary 2 are tight.

4 Larger Neighborhoods

So far we have been concerned with radius-$\frac{1}{2}$ cellular automata. With larger forward neighborhoods larger inverse neighborhoods are possible. The notions of the Welch sets and indices can be generalized to such settings. Let m be the size of the neighborhood range in the forward direction, that is, m consecutive positions can be used as the forward neighborhood. Then the elements of the Welch sets are words of length $m-1$ over alphabet S, and the Welch indices n_L and n_R satisfy the relation $n_L \cdot n_R = n^{m-1}$. By a straightforward generalization of the proofs in the previous section we obtain the following results:

Theorem 2. *Let $\mathcal{A} = (S, (0, \ldots, m-1), f)$ be a reversible CA with n states and forward neighborhood range m. Then the inverse range has size at most $n^{m-1} - 1$. Moreover, the size of the left inverse neighborhood is less than or equal to $n^{m-1} - n_R$ while the size of the right inverse neighborhood is less than or equal to $n^{m-1} - n_L$, where n_L and n_R are the left and right Welch indices.*

The bound in Theorem 2 is not known to be tight. The best known examples are automata with $2n$ states whose inverse neighborhood have range n^{m-1} [3].

5 Final Remarks

We have shown that the inverse neighborhood of a one-dimensional reversible cellular automaton of size n is at most $n-1$ when the neighborhood in the forward direction consists of only two consecutive cells. We have also generalized this result for the case when the forward neighborhood is wider, i.e., if it contains m consecutive cells, then the size of the inverse neighborhood is bounded by $n^{m-1} - 1$. The proof uses several properties of the Welch sets, as well as some algebraic results concerning dimension of vector spaces.

The present paper gives rise to several open problems. E.g., we do not know if the generalized bound $n^{m-1} - 1$ for the size of the inverse neighborhood is tight. This is indeed the case if $m = 2$: see [3] for an example of a reversible cellular automaton with left and right Welch indexes equal to 1 and n respectively, and with inverse neighborhood size equal to $n-1$. Also, for any $n_L, n_R \in \mathbb{N}$, it remains open to find examples of reversible cellular automata with left and right Welch indexes equal to n_L and n_R respectively, such that the size of the inverse neighborhood is maximal, i.e., equal to $n-1 = n_L \cdot n_R - 1$.

There are quadratic time algorithms in the literature testing for surjectivity and injectivity of a given cellular automaton, see [1] and [10]. Although it is improbable that a linear algorithm exists, some improvements may be possible. For example, Lemma 3 from [1] can now be improved from quadratic to linear, although the time complexity of the injectivity algorithm, based on that result, does not change.

References

1. S. Amoroso and Y. Patt, Decision Procedures for Surjectivity and Injectivity of Parallel Maps for Tessellation Structures, Journal of Computer and System Sciences 6 (1972) 448–464.
2. G. Hedlund, Endomorphisms and automorphisms of shift dynamical systems, Mathematical Systems Theory 3 (1969) 320–375.
3. J. Kari, On the Inverse Neighborhood of Reversible Cellular Automata, in: Lindenmayer Systems, Impact in Theoretical Computer Science, Computer Graphics and Developmental Biology, G. Rosenberg, A. Salomaa, eds., 477–495, Springer-Verlag, Berlin-Heidenberg, 1989.
4. J. Kari, Reversibility and surjectivity problems of cellular automata, Journal of Computer and System Sciences 48 (1994) 149–182.
5. N. Margolus, Physics-like models of computation, Physica D 10 (1984) 81–95.
6. E.F. Moore, Machine Models of Self-reproduction, Proceedings of the Symposium in Applied Mathematics 14 (1962) 17–33.
7. K. Morita and M. Harao, Computation Universality of one-dimensional reversible (injective) cellular automata, IEICE Transactions E72 (1989) 758–762.
8. J. Myhill, The Converse to Moore's Garden-of-Eden Theorem, Proceedings of the American Mathematical Society 14 (1963) 685–686.
9. D. Richardson, Tessellations with Local Transformations, Journal of Computer and System Sciences 6 (1972) 373–388.
10. K. Sutner, De Bruijn graphs and linear cellular automata, Complex Systems 5 (1991) 19–31.
11. T. Toffoli and N. Margolus, Invertible cellular automata: a review, Physica D 45 (1990) 229–253.
12. S. Wolfram, A New Kind of Science, Wolfram Media, 2002.

Groupoids That Recognize Only Regular Languages

(Extended Abstract)

Martin Beaudry[1], François Lemieux[2], and Denis Thérien[3]

[1] Département de mathématiques et d'informatique,
Université de Sherbrooke, Sherbrooke (Qc) Canada, J1K 2R1
beaudry@dmi.usherb.ca

[2] Département d'informatique et de mathématique,
Université du Québec à Chicoutimi, Chicoutimi (Qc) Canada, G7H 2B1
flemieux@uqac.ca

[3] School of Computer Science, McGill University,
3480 rue University, Montréal (Qc) Canada, H3A 2A7
denis@cs.mcgill.ca

1 Introduction

Finite semigroups, i.e. finites sets equipped with a binary associative operation, have played a role in theoretical computer science for fifty years. They were first observed to be closely related to finite automata, hence, by the famous theorem of Kleene, to regular languages. It was later understood that this association is very deep and the theory of pseudo-varieties of Schützenberger and Eilenberg [5] became the accepted framework in which to discuss computations realized by finite-state machines. It is today fair to say that semigroups and automata are so tightly intertwined that it makes little sense to study one without the other.

In recent years, the usage of semigroups in the theory of computation has increased substantially; for example, it has proved useful in the study of shallow boolean circuits [1]. Word problems in finite semigroups provide instances of complete problems for the well-known circuit class $\mathbf{NC}^1$, i.e. circuits of logarithmic depth constructed with binary gates; moreover, several interesting subclasses of $\mathbf{NC}^1$ can also be characterized by word problems, by restricting appropriately the semigroups under consideration. Recently [12], new separation results have been proved within the realm of constant-depth circuits, largely based on an algebraic point of view. Other areas of theory of computation where semigroups have proved to be interesting include communication complexity [16] and computational learning theory [6].

When the axiom of associativity is dropped, we obtain algebraic structures known as groupoids. These also play a role in formal languages and there is an analog to Kleene's theorem relating groupoids to push-down automata and to context-free languages [10]. On the computational complexity side, these structures are now known to be of interest as well. Non-associativity can be seen as an algebraic form of non-determinism that has been used, for example, as an important ingredient in the separation of the monotone $\mathbf{NC}$ hierarchy [9]. On the other hand, word problems over groupoids provide instances of complete problems for the circuit class SAC^1, i.e. circuits of logarithmic depth and polynomial size constructed with binary AND gates and unbounded OR gates, and also for interesting subclasses like L and NL.

It is clear that groupoids have been much less studied than semigroups and moreover much of what has been done is of little help to understand computations. It seems important to develop a deeper knowledge of these algebras, in a form that is useful in theory of computation. The apparent absence of structure in the non-associative world should not be seen as an absolute barrier to comprehension. Meaningful results can be obtained; for example, the fact that loops (i.e. groupoids with an identity that obey the cancellation law) recognize exactly the regular open languages [3] provide a surprising link to the topology of the free monoid.

In this paper, we directly address what appears to us to be the most basic question. What differentiates a "weak" groupoid from a "powerful" one in a computing power perspective? For example, when can it be guaranteed that a groupoid recognizes only regular languages? We provide an algebraic sufficient condition on the groupoid for this property to hold. We also show that any regular language can be recognized by a groupoid that satisfies this condition. Finally, we concentrate on loops, which form a proper subclass of this family and have been studied at greater length. We show that the syntactic monoid of any regular open languages has interesting connection with the loops that recognize it. We introduce the notion of the *derived monoid* of a loop and establish the computational significance of this concept. All these results provide further evidence that non-associative structures can be analyzed in a way that sheds light on computational issues.

2 Preliminaries

A groupoid is a set closed under a binary operation. A semigroup S is a groupoid that satisfies the *associative law*: $(xy)z = x(yz)$ for all $x, y, z \in S$. A monoid is a semigroup in which there exists an identity element, i.e. an element $e \in S$ such that $ex = x = xe$ for all $x \in S$. A group is a semigroup G that satisfies the *cancellation laws*, i.e. for all $a, b \in G$ the equations $ax = b$ and $ya = b$ have exactly one solution. Observe that in the finite case, satisfaction of the cancellation and associative laws guarantees the presence of an identity. The *free* semigroup over an alphabet A is the set of all non empty words over A where the operation is the concatenation. It is denoted A^+. The free monoid A^* is obtained by adjoining an identity (the empty word) to A^+. We say that a semigroup S *divides* a semigroup T (written $S \prec T$) if S is a morphic image of a subsemigroup of T. All groupoids considered in this paper are finite, except for the free semigroup and the free monoid.

An important key to understand the structure of semigroups has been developed by Green in 1951 and consists of four equivalence relations denoted $\mathcal{R}, \mathcal{L}, \mathcal{H}$ and $\mathcal{J}$. Let S be a semigroup and let S^1 be the smallest monoid that contains S. The Green relations are defined as follows:

1. $a\mathcal{R}b$ iff $aS^1 = bS^1$
2. $a\mathcal{L}b$ iff $S^1a = S^1b$
3. $a\mathcal{H}b$ iff $a\mathcal{R}b$ and $a\mathcal{L}b$
4. $a\mathcal{J}b$ iff $S^1aS^1 = S^1bS^1$

The relation $\mathcal{J}$ is the coarsest among the four. Each $\mathcal{J}$-class (i.e. each equivalent class of the relation $\mathcal{J}$) can be visualized as a two-dimensional array where each row is an $\mathcal{R}$-class, each column is an $\mathcal{L}$-class and each cell is an $\mathcal{H}$-class.

An element $x \in S$ is an *idempotent* if $xx = x$. Idempotents play a fundamental role in semigroup theory. A $\mathcal{J}$-class that contains an idempotent is said to be *regular*. Any regular $\mathcal{J}$-class has the property that each row and each column contains at least one idempotent. An $\mathcal{H}$-class that contains an idempotent forms a subsemigroup which is a maximal group in S. A necessary and sufficient condition for a $\mathcal{J}$-class to be a subsemigroup is that each of its $\mathcal{H}$-classes contains an idempotent. We define **DS** to be the class of all semigroups for which each regular $\mathcal{J}$-class is a subsemigroup.

Some other classes of semigroups have to be mentioned here. A semigroup that contains no nontrivial group is said to be *aperiodic*. A semigroup is aperiodic if and only if each of its regular $\mathcal{H}$-classes is trivial (i.e contains only one element). The class of aperiodic semigroups contained in **DS** is denoted **DA**. A semigroup is called *orthodox* if the set of its idempotents forms a subsemigroup. The class **DO** consists of those semigroups in **DS** for which each regular $\mathcal{J}$-class is orthodox. A semigroup is $\mathcal{J}$-trivial if each of its $\mathcal{J}$-classes is trivial. We define similarly $\mathcal{R}$-trivial semigroups. We denote with **J** and **R** the classes of $\mathcal{J}$-trivial and $\mathcal{R}$-trivial semigroups, respectively. Finally, the class $\mathbf{J_1}$ contains all commutative semigroups in which all elements are idempotent. From these definitions we have the inclusions:

$$\mathbf{J_1} \subset \mathbf{J} \subset \mathbf{R} \subset \mathbf{DA} \subset \mathbf{DO} \subset \mathbf{DS}$$

Every semigroup S can be seen as a language recognizer in the following way. Given a finite alphabet A and a language $L \subseteq A^+$ we say that L is *recognized* by S if there exists a subset $F \subseteq S$ and a morphism $h : A^+ \to S$ such that $L = \{w \in A^+ \mid h(w) \in F\}$. A variant of Kleene's theorem asserts that a subset of A^+ is regular iff it can be recognized by a finite semigroup. The theorem also holds in the case of monoids recognizing subsets of A^*.

One fascinating aspect of this theory is the strong correspondence that exists between subclasses of regular languages satisfying certain combinatorial properties and subclasses of semigroups satisfying some algebraic properties. A classical example is the result of Schützenberger [13] stating that a language is recognized by a finite aperiodic monoid if and only if it is star-free.

The following more complex example is of particular interest for our work. Let $\mathcal{C}$ be a class of semigroups. The *polynomial closure* of $\mathcal{C}$ (denoted $\text{Pol}(\mathcal{C})$) is the set of languages which are finite unions of languages of the form $K = L_0 a_1 L_1 \cdots a_n L_n$, where each a_i is a letter and each L_i is a language recognized by some semigroup in $\mathcal{C}$. We say that the product defining K is unambiguous if every word w in K can have at most one factorization $w = w_0 a_1 w_1 \cdots a_n w_n$ with

w_i in L_i. The subset of $\mathrm{Pol}(\mathcal{C})$ obtained by restricting to unambiguous products is denoted $\mathrm{UPol}(\mathcal{C})$. Let $\mathbf{J_1} \vee \mathbf{G}$ be the class of all semigroups that divides a direct product of a semigroup in $\mathbf{J_1}$ and a group. Then, a theorem of Schützenberger [14] states that a language is recognized by a semigroup in $\mathbf{DO}$ if and only if it is in $\mathrm{UPol}(\mathbf{J_1} \vee \mathbf{G})$.

Associativity is not essential in the previous definition of recognition and it has been known for a long time [10, 7] that every groupoid can be seen as a pushdown automaton. Let us formalize this idea. Let G be a groupoid, $\mathcal{P}(G)$ its power set, and let G^+ denotes the free semigroup generated by the set of elements in G. Observe that evaluating a word in G^+ to an element in G depends on the evaluation tree since G is in general non-associative. We define the function $\mathrm{eval} : G^+ \to \mathcal{P}(G)$ by requesting that for each $w \in G^+$, $\mathrm{eval}(w)$ is equal to the set of elements that can be obtained by evaluating w with all possible evaluation trees. Given an alphabet A and a language $L \subseteq A^+$, we say that a groupoid G recognizes L if there exist a subset $F \subseteq G$ and a length-preserving morphism $h : A^+ \to G^+$ such that $L = \{w \in A^+ \mid \mathrm{eval}(h(w)) \cap F \neq \emptyset\}$. The analog of Kleene's theorem now states that a language is recognized by a finite groupoid if and only if it is context-free.

The structure of the groupoid being used as a recognizeer must necessarily determine the combinatorial properties of the languages being recognized. One would hope that this relationship can be made explicit for interesting cases. One positive example of this happening has to do with quasigroups. A *quasigroup* is a groupoid that satisfies the cancellation law. A quasigroup with identity is called a *loop*. In [3], it is proved that a language is recognized by a finite quasigroup (or loop) if and only if it is in $\mathrm{Pol}(\mathbf{G})$, where $\mathbf{G}$ is the class of all finite groups.

Loops and quasigroups are "weak" in the sense that they can only recognize regular languages. It is natural to ask what are the groupoids with this property. Our first result gives a sufficient condition for a groupoid to be "weak".

Let G be a groupoid. With any element $a \in G$ we define the mapping $R(a) : G \to G$ called the *right multiplication by a* which is defined by $xR(a) = xa$ and the mapping $L : G \to G$ called the *left multiplication by a* which is defined by $xL(a) = ax$. The *multiplication semigroup*[1] of G (denoted $\mathcal{M}(G)$) is the closure under composition of all left and right multiplications. When G is a quasigroup then $\mathcal{M}(G)$ is a group called the *multiplication group* of G.

Theorem 1. *Let G be a groupoid such that $\mathcal{M}(G) \in \mathbf{DO}$. Then, G recognizes only regular languages.*

Two particular cases of this theorem were already known : when $\mathcal{M}(G)$ belongs to $\mathbf{DA}$ (see [2]) and when G is a quasigroup (see [4]). We conjecture that this result is optimal in the sense that for any pseudovariety $\mathbf{V}$ not contained in $\mathbf{DO}$, there exists a groupoid whose multiplication semigroup is in $\mathbf{V}$ that can recognize a non-regular language.

---

[1] This terminology is standard in loop theory. A similar notion exists in universal algebra where this is called *translation semigroup*.

Now, what regular languages can be recognized by groupoids with multiplication semigroup in **DO**? The answer is given by our second result.

Theorem 2. *Any regular language is recognized by a groupoid whose multiplication semigroup is in **R** (which is a proper subclass of **DO**).*

In this context where "simple" groupoids are used to recognize "complicated" regular languages, it is not so clear what is the relation between the combinatorial structure of the language and the algebraic structure of the groupoid. As an example, consider the language PARITY over the alphabet $\{0,1\}$ that contains all words with an even number of 1. Any semigroup that recognizes this language must contain the cyclic group of order two. However, PARITY can be recognized by a groupoid of four elements ($00 = 0$, $01 = 10 = 1$, $11 = 22 = 2$, all other products gives 3) and neither this groupoid nor its multiplication semigroup contains any group. Our third result states that this situation does not arise with loops.

Recall that the syntactic congruence of a language $L \subseteq A^*$ is defined by $x \sim_L y$ whenever $uxv \in L \Leftrightarrow uyv \in L$ for all $u,v \in A^*$. The syntactic monoid of L is the quotient monoid $M(L) = A^*/\sim_L$. It can be shown that $M(L)$ divides any monoid that recognizes L.

Theorem 3. *Let $L \in A^*$ be recognized by a loop B and let $M(L)$ be its syntactic monoid. Then, any group that divides $M(L)$ also divides B.*

The proof of the previous theorem relies upon an algebraic tool which is introduced in this paper. For any groupoid G we define a congruence over G^+ by defining $x \sim y$ if and only if $\mathsf{eval}(sxt) = \mathsf{eval}(syt)$, for all $s,t \in G^*$. The *derived monoid* of G is the quotient monoid $D(G) = G^*/\sim$. It is easy to see that $D(G)$ is finite if and only if G recognizes only regular languages.

In the case of a loop B, we show that any language recognized by B is also recognized by $D(B)$. Moreover, a group divides B if and only if it divides $D(B)$. We show that there exists a relational morphism $\tau : D(B) \to \mathcal{M}(B)$ such that $D_1(B) = \tau^{-1}(1)$ is a $\mathcal{J}$-trivial submonoid of $D(B)$. Finally, denoting the wreath product by $\circ$, we prove the following noticeable identity:

Theorem 4. $D(B) \prec D_1(B) \circ \mathcal{M}(B)$

Theorems 1 to 4 are proved in Sections 3 to 6, respectively. Conclusion and open problems are given in Section 7.

3 Recognizing Only Regular Languages

Theorem 1 is proved in this section. Let G be a groupoid, $F \subseteq G$ and $L = \{w \in G^+ \mid \mathsf{eval}(w) \cap F \neq \emptyset\}$. Define $K_F = \{W \in \mathcal{M}(G)^+ \mid 1W \in F\}$.

Proposition 1. *If $K_F \in Pol(\mathcal{J}_1 \vee G)$ then L is regular.*

Proof. Let $w \in G^+$ be any word and consider some fixed evaluation tree of w. Each node n has a value $v(n) \in G$ that is the value of the subtree rooted at n. We associate to any path from a node n to a leaf a a word $p(n, a) \in \mathcal{M}(G)^+$ as follows. If n is a leaf then $p(a, a) = R(v(a))$. Otherwise, let n_l and n_r be, respectively, the left and right children of n. Then, $p(n, a)) = p(n_l, a)R(v(n_r))$ if n_l lies on the path, otherwise $p(n, a) = p(n_r, a)L(v(n_l))$ A simple inductive argument shows that $v(n) = 1p(n, a)$. In particular, for any leaf a we have $v(r) = 1p(r, a)$, where r is the root of the evaluation tree.

Let an *angle* along a path be a pair of consecutive edges with exactly one right edge and one left edge. In [2], it is shown that L is recognized by a PDA whose stack never needs to be higher than the maximal number of angles along any path in the tree. We will show that any evaluation tree is equivalent to a tree whose maximal number of angles is bounded by a constant. This will prove that L is regular.

Each angle along a path from a leaf a to the root r corresponds in $p(r, a)$ to two consecutive elements of the form $R(x)L(y)$ or $L(x)R(y)$. Let $p(r, a)$ contains a factor of the form $L(a)R(b_1) \cdots R(b_i)L(c_1) \cdots L(c_j)R(d)$ and replace this factor with $L(a)L(c_1) \cdots L(c_j)R(b_1) \cdots R(b_i)R(d)$. This keeps intact the ordering of the leaves and corresponds to a transformation of the evaluation tree that reduces the number of angles along the path from a to r without increasing that number on the other paths. However, nothing guarantees that the two trees evaluate to the same element.

In order to solve this problem, assume $w \in L$. Thus, $1p(r, a) \in F$ and $p(r, a) = x_0 a_1 x_1 \cdots a_n x_n$ where $a_i \in \mathcal{M}(G)$ and $x_i \in L_i$ which is recognized by some semigroup $M \in \mathbf{J_1} \vee \mathbf{G}$. Without loss of generality we can suppose that M is the same for all $0 \leq i \leq n$ and that $|M| = t$. If $p(r, a)$ contains more than $3nt$ angles then there must exists $0 \leq c \leq n$ such that x_c contains more than $3t$ angles. Let $x_c = u_0 B_1 u_1 B_2 u_2 \cdots B_t u_t$ be a factorization such that each B_m has the form $L(a)R(b_1) \cdots R(b_i)L(c_1) \cdots L(c_j)R(d)$. For each such B_m, define $B'_m = L(a)L(c_1) \cdots L(c_j)R(b_1) \cdots R(b_i)R(d)$. For each $0 \leq i \leq t$ define $y_i = u_0 B'_1 u_1 \cdots B'_i u_i B_{i+1} u_{i+1} \cdots B_t u_t$.

Let $\varphi : \mathcal{M}(G) \to M$ be the syntactic morphism of L_c. By the pigeon hole principle, there is $0 \leq i < j \leq t$ such that $\varphi(y_i) = \varphi(y_j)$. This implies that $\varphi(B_{i+1}u_{i+1} \cdots B_j u_j) = \varphi(B'_{i+1}u_{i+1} \cdots B'_j u_j)$ since $M \in \mathbf{J_1} \vee \mathbf{G}$. Hence, $\varphi(x_c) = \varphi(u_0 B_1 u_1 \cdots B_i u_i B'_{i+1} u_{i+1} \cdots B'_j u_j B_{j+1} u_{j+1} \cdots B_t u_t)$. This shows that if an evaluation tree contains a path with more than $3tn$ angles then we can find an equivalent tree that contains less angles along that path and no more angles along the other paths. Hence, if $w \in L$, there exists an evaluation tree with no more than $3tn$ angles along any path. We conclude that L is regular.

Theorem 1 is a simple corollary of the above proposition since, as it was mentioned in the preliminaries, monoids in **DO** recognize only languages in UPol($\mathbf{J_1} \vee \mathbf{G}$) which is a subclass of Pol($\mathbf{J_1} \vee \mathbf{G}$).

4 Recognizing all Regular Languages

In this section we prove that any regular language is recognized by a finite groupoid whose multiplication semigroup is in **R**. Let M be a finite monoid and $\{a_1, a_2, \ldots, a_k\}$ an enumeration of the elements of M. Let S_0 be the empty set and for $1 \leq i \leq k$ let $S_i = \{a_1, \ldots, a_i\}$. We write $v < w$ to mean that v is a strict non empty prefix of w. We define the following language:

$$D(a_i, a_j, t) = \{w \in M^+ \mid a_i w = a_j \text{ and } v < w \Rightarrow a_i v \notin S_t\}$$

We only have to show that Theorem 2 applies for languages of the form $D(a_i, a_j, 0) = \{w \in M^+ \mid a_i w = a_j\}$ which can be defined using the well known recurrence:

$$D(a_i, a_j, t) = D(a_i, a_j, t+1) \cup D(a_i, a_t, t+1) D(a_t, a_t, t+1)^* D(a_t, a_j, t+1)$$

if $0 \leq t < k$ and

$$D(a_i, a_j, k) = \{b \in M \mid a_i b = a_j\}$$

We define a groupoid G on the set $\{0\} \cup \{[a_i, a_j, t] \mid 1 \leq i, j \leq k, 0 \leq t \leq k\}$ with $[a_i, a_j, s][a_j, a_l, t] = [a_i, a_l, j-1]$ if $s \geq j-1$ and $t \geq j$ (these inequalities are crucial for the proof). In all other cases the product is defined to be 0. The absorbing element of G is $R(0) = L(0)$. Observe also that a necessary condition to evaluate $w = [a_{i_1}, a_{j_1}, t_1] \cdots [a_{i_n}, a_{j_n}, t_n]$ to a non zero element is that $a_{j_s} = a_{i_{s+1}}$ for all $1 \leq s < n$. Moreover, if $t_i = k$ for all i then there exists a way to evaluate w to $[a_{i_1}, a_{j_n}, m]$ where $m = \min\{i_2, \ldots, i_n\} - 1$. It is a simple exercise to prove:

Lemma 1. *The only idempotents in $\mathcal{M}(G)$ are $R(0)$ and the elements of the form $R([a_j, a_j, k])$, where $0 \leq j \leq k$. Moreover, if $W \in \mathcal{M}(G)^+$ can be evaluated to a nonzero element, then it contains at most $2k$ occurrences of non idempotent elements. Furthermore, two occurences of the same idempotent must be adjacent.*

Let $\mathcal{P}(G)$ be the power groupoid of G and let $H \subseteq \mathcal{P}(G)$ be the subgroupoid generated by the sets $U_b = \{[a_i, a_i b, k] \mid 1 \leq i \leq k\}$ for all $b \in M$. It is easy to see that, for any $1 \leq i, j \leq k$, $D(a_i, a_j, 0)$ is recognized by H using the morphism which maps each $b \in M$ to the element U_b of H. We want to show that the multiplication semigroup of H belongs to **R**.

There is an embedding from $\mathcal{M}(H)$ to $\mathcal{P}(\mathcal{M}(G))$. As in the next lemma, it will sometime be useful to see $\mathcal{M}(H)$ as a subsemigroup of $\mathcal{P}(\mathcal{M}(G))$. Due to lack of space, the proof of the following lemma is omitted.

Lemma 2. $X \in \mathcal{M}(H)$ *is an idempotent if and only if for all $A \in X$ there exists an idempotent $E \in X$ such that $A = AE$.*

Lemma 3. $\mathcal{M}(H)$ *is aperiodic.*

Proof. Let S be an element in $\mathcal{M}(H)$. We must show that there exists $m > 0$ such that $S^m = S^{m+1}$. By Lemma 1, for large m there exists a position i such that all elements $X_1X_2\cdots X_{m+1} \in S^{m+1}$ which evaluate to a nonzero element are such that X_i is an idempotent that occurs at least twice in X. Thus, we have $X = X_1\cdots X_{i-1}X_{i+1}\cdots X_{m+1} \in S^m$. This show that $S^{m+1} \subseteq S^m$. For the other direction, let S^m be an idempotent. By Lemma 2, for all $A \in S^m$ there exists an idempotent $E \in S^m$ such that $A = AE$. Now, if $E = R([a,a,k]) \in S^m$ then $E \in S$ as well. Hence, $A = AE \in S^{m+1}$ and $S^m \subseteq S^{m+1}$.

Lemma 4. *Let $X, Y \in \mathcal{M}(H)$ be two idempotents. Then, X and Y are $\mathcal{J}$-equivalent if and only if they contain the same idempotents in $\mathcal{M}(G)$. Moreover, if X and Y are $\mathcal{J}$-equivalent idempotents then they are $\mathcal{L}$-equivalent.*

Proof. Let $X, Y \in \mathcal{M}(H)$ be two $\mathcal{J}$-equivalent idempotents. Let $U, V \in \mathcal{M}(H^1)$ be such that $UXV = Y$. Then, for each $R([a,a,k]) \in Y$, there exist $u \in U$, $v \in V$ and $x \in X$ such that $uxv = R([a,a,k])$ and this is only possible if $x = R([a,a,k])$. This shows that $R([a,a,k]) \in X$ if and only if $R([a,a,k]) \in Y$.

Suppose now that X and Y are idempotents in $\mathcal{M}(H)$ containing the same idempotents in $\mathcal{M}(G)$. By Lemma 2 there exists an idempotent $e \in Y$ such that $y = ye$. By the definition of H, there is at most one $x \in X$ such that $yx \neq 0$. Since e also belongs to X then we have $YX = Y$. Similarly, we show that $X = XY$ proving that X and Y are $\mathcal{L}$-equivalent.

Proposition 2. $\mathcal{M}(H)$ *belongs to* **R**.

Proof. The proof of Lemma 4 shows that the product of two $\mathcal{J}$-equivalent idempotents is an idempotent in the same $\mathcal{L}$-class. Hence, each regular $\mathcal{J}$-class contains only one $\mathcal{L}$-class. Moreover, by lemma 3, $\mathcal{M}(H)$ is aperiodic. We conclude that $\mathcal{M}(H)$ is $\mathcal{R}$-trivial since it satisfies the equation $(xy)^\omega x = (xy)^\omega$.

5 Groups in Loops

This section is devoted to the proof of Theorem 3 which makes extensive use of the derived monoid of a loop defined earlier. It is easy to see that any language L recognized by a loop B is also recognized by $D(B) = B^+/\sim$. We denote by $[w]$ the element of $D(B)$ that contains $w \in B^+$. Before proving Theorem 3, we give some simple properties of the derived monoid. Let B be a loop with identity e. The set of elements $[w] \in D(B)$ such that $e \in \text{eval}(w)$ forms a submonoid of $D(B)$ that we denote $D_1(B)$.

Lemma 5. *If $[w]$ is an idempotent of $D(B)$ then $[w] \in D_1(B)$ and $\text{eval}(w)$ is a subloop of B.*

Proof. By the cancellation law, there exists $k > 0$ such that $e \in \text{eval}(a^k)$ for all $a \in B$. In particular we have $e \in \text{eval}(w^k)$ for all $w \in B^+$. Since $[w]$ is an idempotent then $[w^k] = [w]^k = [w]$, and so, $e \in \text{eval}(w)^k = \text{eval}(w)$.

Lemma 6. $D_1(B)$ *is $\mathcal{J}$-trivial.*

Proof. Let $X, Y \in D_1(B)$ such that $X \mathcal{J} Y$. Hence, there exists $A, B \in D_1(B)$ such that $AXB = Y$. This means that for all $s, t \in B^*$ we have $\text{eval}(sAXBt) = \text{eval}(sYt)$. Since both $\text{eval}(A)$ and $\text{eval}(B)$ contain the identity, then we must have $\text{eval}(sXt) \subseteq \text{eval}(sYt)$. Similarly, we show that $\text{eval}(sYt) \subseteq \text{eval}(sXt)$ for all $s, t \in B^*$. This proves that $X = Y$.

Proposition 3. *If H is a group that divides B, then H divides $D(B)$.*

Proof. We first show that if H is a loop that divides B then $D(H)$ divides $D(B)$. Let K be a subloop of B and let $\psi : K \to H$ be a morphism. Let $\varphi : D(K) \to D(H)$ be the morphism defined by $\varphi([w]) = [\psi(w)]$. Ideed, we have that $\varphi([u][v]) = \varphi([uv]) = [\psi(uv)] = [\psi(u)\psi(v)] = [\psi(u)][\psi(v)] = \varphi([u])\varphi([v])$, proving that φ is a morphism. The conclusion follows from the observation that $D(K) \prec D(B)$ and that if H is a group then $D(H) = H$.

The converse of this proposition needs a few lemmas. Let $H = \{[w_1], \ldots [w_n]\}$ be a subgroup of $D(B)$ where $[w_1]$ is the identity.

Lemma 7. *For all $[u], [v] \in H$ we have that $|\text{eval}(u)| = |\text{eval}(v)|$.*

Proof. Since H is a group, there exists $[w] \in H$ such that $[u][w] = [v]$. Hence, $\text{eval}(uw) = \text{eval}(v)$ and by the cancellation law $|\text{eval}(u)| \leq |\text{eval}(v)|$. Similarly, we show that $|\text{eval}(u)| \geq |\text{eval}(v)|$.

Lemma 8. *For all i, j, either $\text{eval}(w_i) = \text{eval}(w_j)$ or $\text{eval}(w_i) \cap \text{eval}(w_j) = \emptyset$.*

Proof. Let $a \in \text{eval}(w_i) \cap \text{eval}(w_j)$. Since $[w_1]$ is the identity, then $\text{eval}(w_i) = \text{eval}(w_1)\text{eval}(w_i) = \text{eval}(w_1)a = \text{eval}(w_1)\text{eval}(w_j) = \text{eval}(w_j)$, where the first and the last equality are given by Lemma 7.

Lemma 9. *For all $1 \leq i \leq n$, the set $N_i = \{[u] \in H : \text{eval}(u) = \text{eval}(w_i)\}$ is a singleton.*

Proof. We first show that for all i, j, we have $|N_i| = |N_j|$. Let $[u_i] \in N_i$ and $[u_j] \in N_j$. There exists N_k and $[u_k] \in N_k$ such that $[u_i][u_k] = [u_j]$. It follows from Lemma 7 that $\text{eval}(u_i)\text{eval}(u_k) = \text{eval}(u_j)$. This shows that $N_i N_k \subseteq N_j$ and that $|N_i| \leq |N_j|$. Similarly, we show that $|N_j| \leq |N_i|$. We conclude that $|N_i| = |N_j|$. Now, let N_1 contains the identity $[w_1] \in H$. Then, N_1 is a group that is a submonoid of $D_1(B)$, and since $D_1(B)$ is $\mathcal{J}$-trivial then N_1 must be trivial.

We are now ready to prove the converse of Proposition 3.

Proposition 4. *If H be a group that divides $D(B)$, then H divides B.*

Proof. It is known (see [5] Chap.3, prop. 4.5) that if a group G divides $D(B)$ then there exists a group H which is a submonoid of $D(B)$ and such that G divides H. Hence, we only have to consider the case where H is a group which is a submonoid of $D(B)$.

From the previous lemma, if $H = \{[w_1], [w_2], \ldots, [w_n]\}$ is a group that is a submonoid of $D(B)$ then $\mathsf{eval}(w_i) \cap \mathsf{eval}(w_j) \neq \emptyset$ implies that $i = j$. In other words, $\mathsf{eval}(w_1), \mathsf{eval}(w_2), \ldots, \mathsf{eval}(w_n)$ is a partition of $D = \{g \in B : \exists [w] \in H \text{ s.t. } g \in \mathsf{eval}(w)\}$ which is a subloop of B. Moreover, $\mathsf{eval}(w_1)$ is a normal subloop of D, and $D/\mathsf{eval}(w_1)$ is a group isomorphic to H.

Proof of Theorem 3: If L is recognized by B then it is also recognized by $D(B)$. This means that $M(L)$ divides $D(B)$. Hence, any group that divides $M(L)$ also divides B by the previous proposition.

6 Other Properties of the Derived Monoid

This section is devoted to the proof of Theorem 4. We first show the existence of a relational morphism from the derived monoid $D(B)$ to the multiplication group $\mathcal{M}(B)$. Recall that a relational morphism between two monoids N and M is a relation $\tau : M \to N$ satisfying: (1) $1 \in \tau(1)$; (2) $\tau(s) \neq \emptyset$; (3) $\tau(s)\tau(t) \subseteq \tau(st)$, for all $s, t \in M$.

Proposition 5. *Let B be a loop, $\mathcal{M}(B)$ its multiplication group, and $R(1) = L(1)$ the identity of $\mathcal{M}(B)$. There exists a relational morphism $\tau : D(B) \to \mathcal{M}(B)$ such that $D_1(B) = \tau^{-1}(R(1))$.*

Proof. We actually only need the subgroup $\mathcal{M}_R(B) \subseteq \mathcal{M}(B)$ that is generated by the right multiplications of B. Given $S \subseteq \mathcal{M}_R(B)$ and $a \in B$ let $aS = \{aW : W \in S\} \subseteq B$. Given a word $v \in B^+$ let $R(v) = \{R(a) : a \in \mathsf{eval}(v)\}$. We first define the relational morphism $\eta : B^* \to \mathcal{M}_R(B)$ as follows: $\eta(\epsilon) = \{R(1)\}$, $\eta(a) = \{R(a)\}$ for all $a \in G$, and $\eta(w) = \bigcup_{w=uv} \eta(u) R(v)$, where $u \in B^*, v \in B^+$. It should be clear that $\mathsf{eval}(w) = 1\eta(w)$ for all $w \in B^+$.

To prove that η is a relational morphism we need to show that $\eta(u)\eta(v) \subseteq \eta(uv)$ for all $u, v \in B^+$. We proceed by induction on the length of v. If $v = a \in B$ then $\eta(u)\eta(a) = \eta(u)R(a) \subseteq \eta(ua)$. Suppose now that v contains more than one letters and let $X \in \eta(v)$. Then, there exist $v_1 \in B^*, v_2 \in B^+$ such that $v = v_1 v_2$ and such that $X \in \eta(v_1) R(v_2)$. By the induction hypothesis we have $\eta(u)\eta(v_1) \subseteq \eta(uv_1)$. Putting everything together we have: $\eta(u)X \subseteq \eta(u)\eta(v_1)R(v_2) \subseteq \eta(uv_1)R(v_2) \subseteq \eta(uv_1 v_2) = \eta(uv)$. Let $h : B^* \to D(B)$ be the natural morphism. Then, $\tau = \eta h^{-1}$ is a relational morphism with the desired property.

To prove Theorem 4 we need some classical result from category theory. A *multigraph* C is a set of *objets* $\mathrm{Obj}(C)$ and, for each pair $a_1, a_2 \in \mathrm{Obj}(C)$, a set of *arrows* denoted $C(a_1, a_2)$. A *category* C is a multigraph with an associative composition rule that assign to every consecutive pair of arrows $s : a_1 \to a_2$

and $t : a_2 \to a_3$ an arrow $st : a_1 \to a_3$. Moreover, for each $a \in Obj(C)$ there is a unique *identity arrow* $1_a : a \to a$ satisfying $1_a t = t$ and $s1_a = s$ for all $s \in C(a', a)$ and $t \in C(a, a'')$ such that $a', a'' \in Obj(C)$.

Given a category C, a *subcategory* S is a sub-multigraph of C which is a category under the composition rule of C. If $S \subseteq Obj(C)$ then $C(S)$ denote the full subcategory whose set of objects is S. In particular, for each $a \in Obj(C)$, the subcategory $C(a)$ forms a monoid called the *local monoid* of C at a.

Let C and D be categories. A *relational morphism* $\tau : C \to D$ is an object function $\tau_1 : Obj(C) \to Obj(D)$ and, for each $a, b \in Obj(C)$, a relation $\tau_2 : C(a, b) \to D(\tau_1(a), \tau_1(b))$ such that for each pair of arrows s, t we have $\varphi_2(s)\varphi_2(t) \subseteq \varphi_2(st)$. Also, for all $a \in Obj(C)$ we have $1_{\varphi_1(a)} \in \varphi_2(1_a)$. Moreover, if $\tau_2 : C(a, b) \to D(\tau_1(a), \tau_1(b))$ is *injective* (i.e. $s \neq t$ implies $\tau_2(s) \cap \tau_2(t) = \emptyset$), then τ is called a *division*.

A category C *divides* a category D (denoted $C \prec D$) if there exists a division $C \to D$. If moreover $D \prec C$ then we say that C is *equivalent* to D and we write $C \sim D$.

Let $a, b \in Obj(C)$. We say a is a *retract* of b if there exists arrows $s \in C(a, b)$ and $t \in C(b, a)$ such that $st = 1_a$. The following three results are from [17]:

Lemma 10 (Tilson). *Let C be a category, and let $S \subseteq Obj(C)$. If every object of C is a retract of some object in S, then $C \sim C(S)$.*

Let $\tau : S \to T$ be a relational morphism of monoids. The *derived category* D_τ of τ is the category whose objects set is $Obj(D_\tau) = \tau(S)$ and $D_\tau(t, tt') = \{[t, (s, t'), tt'] : s \in \tau^{-1}(t)\}$. The composition rule is given by

$$[t, (s, t'), tt'][tt', (s', t''), tt't''] = [t, (ss', t't''), tt't'']$$

Lemma 11 (Tilson). *Let $\tau : S \to T$ be a relational morphism of monoids. Then, $D_\tau(1) = \tau^{-1}(1)$.*

Theorem 5 (Derived Category Theorem). *Let $\tau : S \to T$ be a relational morphism of monoids, and let M be a monoid satisfying $D_\tau \prec M$. Then, $S \prec M \circ T$.*

Proof of Theorem 4: Let $\tau : D(B) \to \mathcal{M}(B)$ be the relational morphism defined in Proposition 5. By Lemma 11 we have that $D_\tau(1)$ is equivalent to $\tau^{-1}(1) = D_1(B)$. Now, let $a \in Obj(D_\tau)$ and consider the arrows $s = [1, (1, a), a]$ and $t = [a, (1, a^{-1}), 1]$. Then, we have $st = [1, (1, a), a][a, (1, a^{-1}), 1] = [1, (1, 1), 1] = 1_1$ which is the identity arrow of the object 1. Hence, every object is a retract of 1 and, by Lemma 10, D_τ is equivalent to $D_\tau(1) \sim D_1(B)$. The conclusion follows from the Derived Category Theorem.

7 Conclusion

In this work we have investigated groupoids that can recognize only regular languages. We prove that if the multiplication semigroup of a groupoid G belongs to **DO** then G can only recognize regular languages. On the other hand, it is known from [2] that if the multiplication semigroup belongs to any pseudovariety which is not contained in **DS**, then G can recognize a language that is not regular. The case of **DS** is more obscure and its investigation is complicated by the fact that it is difficult to build non-trivial examples of groupoids whose multiplication semigroup is in **DS** but not in **DO**.

Groupoids who recognize only regular languages could be called "weak" since their Word Problems is always in $\mathbf{NC}^1$. However, there exist groupoids with this property that can recognize non-regular languages. For example, a groupoid that recognizes the non-regular language MAJORITY (set of words over $\{0,1\}$ that contain a majority of 1) is given in [9]. Moreover, all the Word Problems of this groupoid belong to $\mathbf{TC}^0$ a class that lies inside $\mathbf{NC}^1$. Knowledge of algebraic conditions that would force the Word problem of a groupoid to be in $\mathbf{TC}^0$ could be very useful for the important question of whether the inclusion $\mathbf{TC}^0 \subseteq \mathbf{NC}^1$ is strict or not.

Generally, given a groupoid G and a set $F \subseteq G$, it is undecidable to determine if the set $\{w \in G^+ \mid \mathsf{eval}(w) \cap F \neq \emptyset\}$ is regular. However, it is not clear if it is decidable to determine if a groupoid recognizes only regular languages. This is equivalent to the question of deciding if the derived semigroup of a groupoid is finite or not. An important related question is to decide whether a groupoid has its word problem in $\mathbf{NC}^1$ or not. Again, this could have interesting consequences on the study of small complexity classes.

Ackowledgment

This work has received support from FQRNT (Québec) and CRSNG (Canada).

References

1. D. Barrington and D. Thérien, *"Finite Monoids and the Fine Structure of NC^1"*, JACM **354**(1988)941–952
2. M. Beaudry, "Languages recognized by finite aperiodic groupoids", TCS **209**(1998)299–317
3. M. Beaudry, F. Lemieux and D. Thérien, *Finite loops recognize exactly the regular open languages,* Proc. 24th ICALP, LNCS 1256 (1997), 110-120.
4. H. Caussinus and F. Lemieux, *The complexity of computing over quasigroups,* Proc. 14th annual FST&TCS, 1994, pp.36-47.
5. S. Eilenberg, *Automata, Languages and Machines, vol. B,* Academic Press (1976)
6. R. Gavaldà and D. Thérien, *Learning expressions over monoids,* Proc. 18th Intl. Symposium on Theoretical Aspects of Computer Science (STACS'01). Springer-Verlag Lecture Notes in Computer Science 2010 (2001), 283–293.
7. F. Gécseg and M. Steinby, *Tree Automata,* Akadémiai Kiadó, Budapest, 1984.
8. S.C. Kleene, *Representations of events in nerve nets and finite automata,* Automata Studies, C.E. Shannon (ed.), vol. 3-41, Princeton, N.J. (1956) Priceton University Press.

9. F. Lemieux, *Finite groupoids and their applications to computational complexity*, Ph.D. Thesis, McGill University, May 1996.
10. J. Mezei and J.B. Wright, *Algebraic automata and context-free sets*, Inform. and Contr. **11** (1967) 3–29.
11. J.-E. Pin, *Varieties of Formal Languages*, Plenum Press, New York, 1986.
12. M. Koucky, P. Pudlak and D. Therien, *Bounded-depth circuits: separating wires from gates*, accepted in the 37th ACM Symposium on Theory of Computing (STOC 2005).
13. M.-P. Schützenberger *On finite monoids having only trivial subgroups*, Information and Control **8** (1965) 190–194.
14. M.-P. Schützenberger *Sur le produit de concaténation non ambigu*, Semigroup Forum **13** (1976) 45–75.
15. I. Simon *Piecewise Testable Events*, Proc. 2nd GI Conf., LNCS 33 Springer Verlag, (1975) 214–222.
16. P. Tesson and D. Thérien, *Complete Classifications for the Communication Complexity of Regular Languages*, Proceedings of the 20th International Symposium on Theoretical Aspects of Computer Science, (2003)
17. B. Tilson, *Categories as algebra: an essential ingredient in the theory of monoid*, J. Pure and Applied Algebra **48** (1987) 83–198.

Append-Only Signatures*

Eike Kiltz**, Anton Mityagin***, Saurabh Panjwani[†], and Barath Raghavan[‡]

Department of Computer Science and Engineering,
University of California, San Diego, USA
{ekiltz, amityagin, panjwani, barath}@cs.ucsd.edu

Abstract. We present a new primitive – Append-only Signatures (AOS) – with the property that any party given an AOS signature $\text{Sig}[M_1]$ on message M_1 can compute $\text{Sig}[M_1 \| M_2]$ for any message M_2, where $M_1 \| M_2$ is the concatenation of M_1 and M_2. We define the security of AOS, present concrete AOS schemes, and prove their security under standard assumptions. In addition, we find that despite its simple definition, AOS is equivalent to Hierarchical Identity-based Signatures (HIBS) through efficient and security-preserving reductions. Finally, we show direct applications of AOS to problems in network security. Our investigations indicate that AOS is both useful in practical applications and worthy of further study as a cryptographic primitive.

Keywords: Algebraic Signatures, Append-only Signatures, Hierarchical Identity-based Signatures.

1 Introduction

In many real-world applications, users and programs alike require notions of delegation to model the flow of information. It is often required that delegation from one party to another enables the delegatee to "append" to the information it received but to do nothing more. For example, in wide-area Internet routing, each network passes a routing path advertisement to its neighboring networks, which then append to it information about themselves and forward the updated advertisement to their neighbors. For security, the route advertisements must be authenticated; intermediate networks must be incapable of modifying routes except according to the protocol (that is, by appending their names to already-received advertisements). Likewise, in the context of secure resource delegation for distributed systems, users need to delegate their share of resources to other users, who may then re-delegate to other users by including their own resources in

---

* Any opinions, findings, and conclusions or recommendations expressed in this material are those of the authors and do not necessarily reflect the views of the NSF.
** Supported in by a DAAD postdoc fellowship.
*** Supported in part by NSF grants ANR-0129617 and CCR-0208842.
[†] Supported in part by NSF grant 0313241.
[‡] Supported by a NSF Graduate Research Fellowship.

the pool. In many of these applications, it is desirable that delegation is possible without parties having to share any cryptographic keys and that the authenticity of any information received through a series of delegations is verifiable based only on the identity of the first party in the chain.

To directly address these needs, we present a new cryptographic primitive called Append-Only Signatures (AOS). An AOS scheme enables the extension of signed messages and update of the corresponding signatures, without requiring possession of the signing key. That is, any party given an AOS signature $\text{Sig}[M_1]$ on message M_1 can compute $\text{Sig}[M_1\|M_2]$ for any message M_2, where $M_1\|M_2$ is the concatenation of M_1 and M_2. The verifier of the final signature needs the initial signer's public key but does not need to know the public keys or any other information from intermediate signers except the message data appended. Clearly, such a scheme cannot be secure according to the standard notion of security for signatures. Instead, we define an AOS scheme to be secure if it is infeasible to forge signatures of messages that are not obtained by extending already-signed messages. A formal definition appears in Section 2.

In Section 3 we present several provably secure AOS schemes, offering different tradeoffs of flexibility and efficiency. Our first construction shows a generic approach to building AOS schemes from any standard digital signature scheme using certificate chains. The construction works as follows: The secret and public keys for the AOS scheme are obtained by running the key generator for $\mathcal{SIG}$. For any message $M = M_1\|M_2\|\cdots\|M_n$, each M_i being a symbol in some predetermined message space, the AOS signature of M is defined as a sequence of n public keys $pk_1, pk_2, \cdots, pk_n$ (generated using the key generator for $\mathcal{SIG}$) and a sequence of n certificates binding the message symbols to these public keys. The ith certificate in the chain binds the message symbol M_i to the corresponding public key pk_i and is signed using the secret key, sk_{i-1}, corresponding to pk_{i-1}. The secret key, sk_0, of the AOS scheme signs the first certificate in the chain while the secret key sk_n (corresponding to the last public key), is revealed as part of the AOS signature and is used for appending new symbols to M. We observe that if the message space is small enough, we can make use of "weaker", and more efficient, signature schemes without compromising the security of the resulting AOS scheme. Using aggregation techniques of [2, 10] one can reduce the length of the signature by a factor of two.

We also present a more efficient construction of AOS for applications in which the message space is constant size and the total number of append operations performed is also constant. This construction is based on a seemingly stronger assumption (than that of the existence of signature schemes) and makes use of pseudorandom generators and collision-resistant hash functions (CRHFs).

RELATION TO HIERARCHICAL IDENTITY-BASED SIGNATURES. Identity-Based Signature (IBS) schemes, due to Shamir [14], are signature schemes in which the identity of the signer (for example, her email address) plays the role of his public key. Such schemes assume the existence of a trusted authority that holds a master public-private key pair that is used to assign secret keys to users based on their identities. Anyone can verify signatures on messages signed by a user knowing only the master public key and the identity of that user. Hierarchical IBS (HIBS) schemes, proposed by Gentry and Silverberg [4], are identity-based signature schemes in which users are arranged in a hierarchy and a user at any level in this hierarchy can delegate secret keys to her descendants based on their identities and her own secret key. To verify the signature created by any user, one needs to know only the identity of the user (and her position in the hierarchy) and the public key of the root user.

HIBS can be implemented using certificate chains (as suggested in [4]) and the resulting construction bears a strong resemblance to the certificate-based construction of AOS we give in this paper. Upon closer examination, we find that the similarity between the two constructions is not accidental: it is an artifact of the close relationship between the two primitives themselves—AOS and HIBS are, in fact, tightly equivalent. This means that (a) there exist generic transformations from any HIBS scheme into a corresponding AOS scheme and, likewise, from any AOS scheme into a corresponding HIBS scheme; and (b) these transformations are extremely efficient (the derived scheme is as efficient as the scheme being derived from) and highly security-preserving (an adversary attacking the derived scheme can be transformed into an adversary attacking the original one, losing only a constant factor in efficiency and query complexity). Section 4 gives details.

A benefit of this equivalence is that it considerably simplifies the notion of HIBS and makes security analysis for HIBS schemes less onerous: AOS is simpler than HIBS, and, for any HIBS scheme, it is typically easy to find an equivalent AOS scheme whose security properties carry over to the corresponding HIBS scheme. For example, our security proof for certificate-based AOS translates to a security proof for certificate-based HIBS (originally proposed in [4]). Although this construction of HIBS was known prior to our work, it was never analyzed in the literature, and, to the best of our knowledge, we give the first proof of security for it. Furthermore, our construction of AOS based on pseudorandom generators and CRHFs yields a novel approach to designing HIBS and can be useful for some restricted scenarios (for example, in a constant-depth hierarchy wherein each user signs messages from a constant-size message space). We remark that both these constructions yield HIBS schemes in the standard model and neither involves the use of computationally intensive bilinear maps (this is in contrast with some recent results on HIBS [3]).

APPLICATION TO SECURE ROUTING. In Section 5 we discuss an important real-life application of AOS in internet routing security.

RELATED WORK. Append-only signatures belong to a general class of primitives called *algebraic signatures*. Informally, an algebraic signature scheme allows the creation of signatures on a message M using the signatures on some known messages, $M_1, M_2, \ldots, M_n$, and the public key, provided M can be obtained from the known messages using some prespecified set of (n-ary) operations, say $\mathcal{O} = \{f_1, f_2, \cdots, f_m\}$. That is, given the signatures, $\text{sig}[M_1], \ldots, \text{sig}[M_n]$ and the public key, it is easy to compute $\text{sig}[f_i(M_1,\ldots,M_n)]$ for any $f_i \in \mathcal{O}$. In our setting, each f_i has arity 1 and appends some fixed message symbol M_i to an input message M. Security for algebraic signatures is defined in a manner similar to our approach to security of AOS (that is, it should be hard to forge signatures of messages that cannot be obtained by applying the operations in $\mathcal{O}$ to already-signed messages). Examples of algebraic signatures studied in the literature include transitive signatures by Micali and Rivest [12], homomorphic signatures by Johnson, Molnar, Song and Wagner [7], and graph-based algebraic signatures by Hevia and Micciancio [5].

Although no obvious relation exists between our primitive and any of the previously studied algebraic signature primitives, we do note that some of the techniques we use in our constructions parallel prior techniques. For example, our construction of AOS schemes using CRHFs can be viewed as a special instance of graph-based algebraic signature schemes studied in [5] (although the set of update operations considered there are different from the append operation that we consider).

2 Append-Only Signatures

Informally, append-only signatures (AOS) are signatures that enable the public extension of existing signatures. That is, any party given an AOS signature Sig on a message $(M_1,\ldots,M_n)$ can compute an AOS signature on any message $(M_1,\ldots,M_n,M_{n+1})$. (As in the introduction, one could represent the message $(M_1,\ldots,M_n)$ as the string $M_1||\ldots||M_n$ which better captures the idea of appending. However, since we want to differentiate between the message "A"||"B" and the message symbol "AB", we prefer to think of messages as n-tuples. That is, in our example, we have the two different tuples (A,B) and (AB)). Besides the append operation, AOS is the same as ordinary signatures. That is, given only an AOS signature on the message $(M_1,\ldots,M_n)$ it should be infeasible to forge an AOS signature on any message not having $(M_1,\ldots,M_n)$ as a prefix.

Let AOS.MSpace be any set of symbols (for example, $\{0,1\}$ or $\{0,1\}^*$). A *message of length n* is an n tuple of symbols written as $M[1..n] = (M_1, M_2, \ldots, M_n)$. The special case of $n = 0$ is the empty message, denoted ε. We use the symbol $\sqsubseteq$ to denote the prefix relation over the messages. An append-only signature (AOS) scheme with respect to the message space AOS.MSpace is a collection of three polynomial-time algorithms: a setup algorithm (AOS.Setup), an append algorithm (AOS.Append), and a verify algorithm (AOS.Vfy), defined as follows:

- AOS.Setup takes the security parameter as input and outputs a pair of keys: the public key AOS.pk and the secret key $\text{Sig}[\varepsilon]$, which is the signature on the empty message ε.
- AOS.Append takes the public key AOS.pk, a signature on a message $M[1..n-1] = (M_1,\ldots M_{n-1})$, of length $n-1$, and a symbol $M_n \in$ AOS.MSpace and produces a signature on the message $M[1..n] = (M_1,\ldots,M_n)$.
- AOS.Vfy takes the public key AOS.pk, a message $M[1..n]$, and a signature sig, and returns either true or false.

All algorithms can be randomized. Additionally, the scheme should have the property that for any pair (AOS.pk, Sig[ε]) and any message $M[1..n]$, the signature sig obtained by iteratively appending $M_1, \ldots, M_n$ to Sig[ε] should be accepted by AOS.Vfy. Appendig the symbols one-by-one should be the only way of generating a signatures on the message $M[1..n]$. This fact ensures history independence of AOS: that is, no party, given an AOS signature, can tell whether the signature was created by the owner of the secret key or whether it passed through multiple parties that appended symbols at every step. History independence is a useful property to have in most applications, as already highlighted in previous work on algebraic signatures [7].

Definition 1. Let $\mathcal{AOS}$ = (AOS.Setup, AOS.Append, AOS.Vfy) be an AOS scheme, let k be the security parameter, and let $\mathcal{A}$ be an adversary. We consider the experiment:

Experiment $\mathbf{Exp}^{\text{aos-uf-cma}}_{\mathcal{AOS},\mathcal{A}}(k)$

$MSGSet \leftarrow \emptyset$; (AOS.pk, Sig[ε]) $\xleftarrow{\$}$ AOS.Setup(1^k)
$(M[1..n], \text{sig}) \xleftarrow{\$} \mathcal{A}^{\text{AOSSIGN}(\cdot)}(\text{AOS.pk})$
if AOS.Vfy(AOS.pk, $M[1..n]$, sig) = true
and $\forall J[1..j] \sqsubseteq M[1..n] : J[1..j] \notin MSGSet$
then return 1 else return 0

Oracle AOSSIGN($M[1..n]$)
  Add $M[1..n]$ to $MSGSet$
  return EXTRACT($M[1..n]$)

Oracle EXTRACT($M[1..i]$) // defined recursively
  if $i = 0$ then return Sig[ε]
  else if Sig[$M[1..i]$] = defined
    then return Sig[$M[1..i]$]
    else Sig[$M[1..i]$] $\xleftarrow{\$}$ AOS.Append(AOS.pk, $M[1..i-1]$, EXTRACT($M[1..i-1]$), M_i)
  return Sig[$M[1..i]$]

The aos-uf-cma-advantage of an adversary $\mathcal{A}$ in breaking the security of the scheme $\mathcal{AOS}$ is defined as $\mathbf{Adv}^{\text{aos-uf-cma}}_{\mathcal{AOS},\mathcal{A}}(k) = \Pr[\mathbf{Exp}^{\text{aos-uf-cma}}_{\mathcal{AOS},\mathcal{A}}(k) = 1]$, and $\mathcal{AOS}$ is said to be *unforgeable under chosen message attacks* (aos-uf-cma secure) if the above advantage is a negligible function in k for all polynomial-time adversaries $\mathcal{A}$.

Note that $\mathcal{A}$ is given access to the oracle AOSSIGN($\cdot$), not to the oracle EXTRACT($\cdot$) (the latter is used internally by AOSSIGN($\cdot$) to create intermediate signatures).

3 Efficient AOS Constructions

We briefly sketch our constructions for AOS in this section. More details (including proofs of all theorems) can be found in [9].

CERTIFICATE-BASED APPEND-ONLY SIGNATURES. We present an efficient construction of a provably-secure AOS scheme based on a public-key signature scheme. Let $\mathcal{SGN}$ = (SGN.G, SGN.S, SGN.V) be a signature scheme with a space of public keys SGN.PKSpace and message space SGN.MSpace = AOS.MSpace $\times$

SGN.PKSpace. That is, messages to be signed by $\mathcal{SGN}$ are tuples of the form (M, pk), where $M \in$ AOS.MSpace and $pk \in$ SGN.PKSpace. An AOS signature Sig of $M[1..n]$ is a tuple $(pk_1, sig_1, \ldots, pk_n, sig_n, sk_n)$, where for $1 \leq i \leq n$, (pk_i, sk_i) are random public/secret key pairs of the public-key signature scheme $\mathcal{SGN}$ and sig_i is a signature on the tuple (M_i, pk_i) under the secret key sk_{i-1}. The signature sig_0 is signed with the secret key sk_0, which is the signature of ε (the master secret key). Our AOS scheme $\mathcal{AOS}1$ with message space AOS.MSpace is specified as follows:

- AOS.Setup(1^k): Run SGN.G(1^k) to generate a pair (pk_0, sk_0) and returns it as AOS public/secret key pair.
- AOS.Append(AOS.pk, Sig$[M[1..n]], M_{n+1}$): Parse Sig as $(pk_1, sig_1, \ldots, pk_n, sig_n, sk_n)$. Run SGN.G($1^k$) to generate a pair (sk_{n+1}, pk_{n+1}). Compute $sig_{n+1} \leftarrow$ SGN.S$_{sk_n}(M_{n+1}, pk_{n+1})$. Return $(pk_1, sig_1, \ldots, pk_{n+1}, sig_{n+1}, sk_{n+1})$.
- AOS.Vfy(AOS.pk, $M[1..n]$, Sig): Parse Sig as $(pk_1, sig_1, \ldots, pk_n, sig_n, sk_n)$. Set pk_0 to be the master public key AOS.pk. For $i = 1..n-1$ verify that SGN.V($pk_{i-1}, sig_i, (M_i, pk_i)$) = true. If any of the verifications fail, return false. If all the verifications succeed, verify that (sk_n, pk_n) is a valid secret key/public key pair (by signing and veryfing a signature on a random message under sk_n).

Theorem 2. *The AOS scheme $\mathcal{AOS}1$ is aos-uf-cma secure assuming that the public-key signature scheme $\mathcal{SGN}$ is unforgeable under choosen message attacks.*

SHORTER SIGNATURES VIA AGGREGATION. An aggregate signature scheme, $\mathcal{ASGN} = $ (ASGN.G, ASGN.S, ASGN.AGG, ASGN.V), allows the aggregation of n signatures on n distinct messages from n distinct users into a single signature. Its verification algorithm, ASGN.V($n, \cdot$), takes an aggregated signature, n messages, and n public keys and verifies that the n users signed the n messages. When using the certificate-based construction of AOS from Section 3, we can use sequential signature aggregation to shrink the size of the signature (without significantly decreasing security or efficiency). To be more precise, the length of an AOS signature of a message of length n can be condensed to one signature of $\mathcal{ASGN}$, n public keys of $\mathcal{ASGN}$, and one secret key of $\mathcal{ASGN}$. We note that there are two known signature aggregation techniques. The first scheme, given in [2], is based on bilinear maps. The second scheme (only supporting sequential aggregation) is from [10] and can be based on homomorphic trapdoor permutations (such as RSA). Both aggregation schemes are in the random oracle model.

AOS VIA HASH TREES. If the number of symbols in the alphabet AOS.MSpace is small, AOS can be efficiently implemented using hash trees [11]. This approach suffers from dramatic complexity blowup as the size of the message space increases, but uses only secret-key primitives and provides good security guarantees. We believe that this construction is useful in computationally constrained applications. Here we show how to construct an AOS scheme $\mathcal{AOS}2$ with message space AOS.MSpace = $\{0,1\}$ and message length restricted to d. The construction uses a pseudorandom generator $G : \{0,1\}^k \to \{0,1\}^{2k}$ and a collision-resistant hash function $H : \{0,1\}^k \times \{0,1\}^k \to \{0,1\}^k$. We denote by $G_i : \{0,1\}^k \to \{0,1\}^k$ the i-th k-bit component of G for $i \in \{0,1\}$.

Consider the graph T depicted in the left part of Figure 1, whose nodes are denoted as shown on the figure (the upper part (UT) in round brackets and the lower part (LT) in square brackets). In general, the graph T has d levels in both upper and lower parts. For any node $u = \langle v_1, \ldots, v_j \rangle$ from the upper part of the graph, we define the complement of u, denoted Comp(u), to be the minimal set of nodes in LT $- \{\tilde{\varepsilon}\}$ such that every path from ε to $\tilde{\varepsilon}$ passes through exactly one node from $\{u\} \cup$ Comp(u). An example of a complement set is given on the right half of Figure 1.

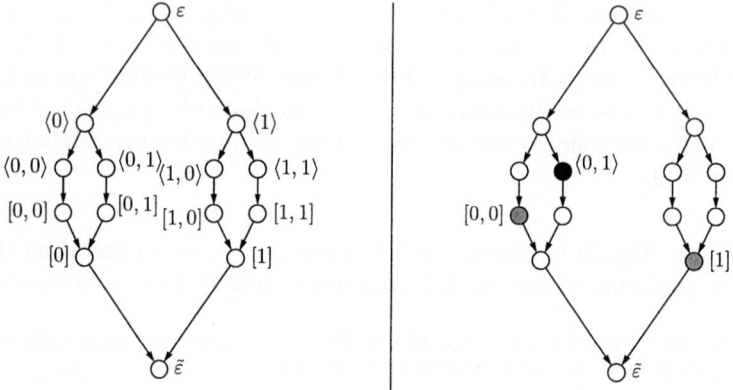

Fig. 1. Structure of the hash-tree construction for $d = 2$. The diagram on the left depicts the hash tree. The diagram on the right highlights the node $u = \langle 0, 1 \rangle$ (shown in black) and the set of its complements, Comp(u) (shown in gray)

In $\mathcal{AOS}2$, a message $M = (M_1, \ldots, M_n)$ is assosiated with a node $u = \langle M_1, \ldots, M_n \rangle$ from the upper part of the graph. Each node u is assosiated with a k-bit value key(u), which is called the "key" of the message. The setup algorithm assigns these values in a top-down manner, starting from the root node ε. Initially, the root key key(ε) is chosen at random from $\{0,1\}^k$. Keys of all the other nodes in the upper part of T are obtain by repeated application of G: keys of left and right children of a node u are G_0(key(u)) and G_1(key(u)), respectively. Keys of the nodes on the highest lower level are obtained by applying G_0 to their parents' keys; keys of the other nodes in the lower part are obtained by applying H to their parents' keys. The setup algorithm outputs key($\tilde{\varepsilon}$) as the public key and key(ε) as the secret key of AOS.

The signature of a node u consists of the keys in the set $\{u\} \cup$ Comp(u). Note that given a signature of u, one can compute the keys of all the descendants of u including the last node $\tilde{\varepsilon}$. Verification of a signature is done by computing the keys of all the descendants of u and comparing the obtained key($\tilde{\varepsilon}$) with the public key. The append algorithm, given a signature of $u = \langle M_1, \ldots, M_n \rangle$ and a bit M_{n+1}, computes the key of $u' = \langle M_1, \ldots, M_{n+1} \rangle$ (which is a child of u) and the keys of all the nodes in Comp(u') (which are descendants of u and Comp(u)). It returns these keys as a signature on $(M_1, \ldots, M_{n+1})$.

Theorem 3. If $G(\cdot)$ is a secure pseudorandom generator, $G_0(\cdot)$, $G_1(\cdot)$, are secure one-way functions and $H(\cdot, \cdot)$, $G_0(\cdot)$, and $G_1(\cdot)$ are all collision-resistant hash functions, then $\mathcal{AOS}2$ is aos-uf-cma secure.

AOS VIA ONE-TIME SIGNATURES. We observe that we can combine the ideas of the certificate-based AOS and the hash-tree AOS to gain a more efficient append-only signature scheme when the message space is small. Assume the message space AOS.MSpace consists of m elements, where m is a constant. Then we can use our certificate-based construction $\mathcal{AOS}1$ instantiated with an m-time signature scheme. m-time signatures can be efficiently constructed using hash-trees (see [5, 13] for the definition and efficient constructions of one-time and m-time signatures). In addition, the security proof of $\mathcal{AOS}1$ guarantees unforgeability if $\mathcal{SGN}$ is at least an |AOS.MSpace|-time signature scheme.

COMPACT AOS. In the full version of the paper [9] we show how to use a recent technique of Boneh, Boyen and Goh [1] to get an AOS scheme in which the signature size is proportional to the square root of the length of the message.

4 Relations Between HIBS and AOS

In this section, we show that the concepts of AOS and Hierarchical Identity-based Signatures (HIBS) are in fact equivalent.

We start with a formal defintion of HIBS. Let HIBS.IDSpace be any set of identities (typically $\{0,1\}^*$). A *hierarchical identity of length* n is an n-tuple of identities from HIBS.IDSpace, written as $I[1..n] = (I_1, I_2, \ldots, I_n)$. The *root identity* is denoted as $I[1..0]$ or ε. Again we use the symbol $\sqsubseteq$ to denote the prefix relation over the set of hierarchical identities.

A HIBS scheme over identity space HIBS.IDSpace is made up of four (possibly randomized) algorithms: a setup algorithm HIBS.Setup, a key delegation algorithm HIBS.KeyDel, a signing algorithm HIBS.Sign and a verification algorithm HIBS.Vfy.

Definition 4. Given a HIBS scheme $\mathcal{HIBS}$ = (HIBS.Setup, HIBS.KeyDel, HIBS.Sign, HIBS.Vfy), security parameter k and adversary $\mathcal{A}$ consider the following experiment:

Experiment $\mathbf{Exp}_{\mathcal{HIBS},\mathcal{A}}^{\text{hibs-uf-cma}}(k)$
  $IDSet \leftarrow \emptyset$
  $(\text{HIBS.pk}, \text{HIBS.SK}[\varepsilon]) \leftarrow \text{HIBS.Setup}(1^k)$
  $(I[1..n], M, \text{sig})$
    $\leftarrow \mathcal{A}^{\text{Corrupt}(\cdot), \text{Sign}(\cdot, \cdot)}(\text{HIBS.pk})$
  if $\text{HIBS.Vfy}(I[1..n], M, \text{sig}) = \text{true}$
   and $\forall\, j \leq n\ I[1..j] \notin IDSet$
   and $(I[1..n], M) \notin MSGSet$
   then return 1 else return 0

Oracle $\text{Corrupt}(I[1..n])$
  $IDSet \leftarrow IDSet \cup \{I[1..n]\}$
  return $\text{Extract}(I[1..n])$

Oracle $\text{Sign}(I[1..n], M)$
  $MSGSet \leftarrow MSGSet \cup \{(I[1..n], M)\}$
  $sk \leftarrow \text{Extract}(I[1..n])$
  return $\text{HIBS.Sign}(sk, I[1..n], M)$

Oracle $\text{Extract}(I[1..i])$ // defined recursively
  if $i = 0$ return $\text{HIBS.SK}[\varepsilon]$
  else if $\text{HIBS.SK}[I[1..i]] = $ defined
    then return $\text{HIBS.SK}[I[1..i]]$
    else $sk \leftarrow \text{HIBS.KeyDel}(\text{HIBS.pk}, I[1..i-1], \text{Extract}(I[1..i-1]), I_i)$
  return sk

The hibs-uf-cma-advantage of an adversary $\mathcal{A}$ in breaking the security of the scheme $\mathcal{HIBS}$ is defined as $\mathbf{Adv}_{\mathcal{HIBS},\mathcal{A}}^{\text{hibs-uf-cma}}(k) = \Pr[\mathbf{Exp}_{\mathcal{HIBS},\mathcal{A}}^{\text{hibs-uf-cma}}(k) = 1]$, and $\mathcal{HIBS}$ is said to be *existentially unforgeable under chosen message attacks* (hibs-uf-cma secure) if the above advantage is a negligible function in k for all polynomial time adversaries $\mathcal{A}$.

CONSTRUCTING AOS FROM HIBS. We set AOS.MSpace = HIBS.IDSpace and associate an AOS message $(M_1, \ldots, M_n)$ of length n with the hierarchical identity $I[1..n] = (M_1, \ldots, M_n)$ of depth n. We then define the signature of this message as the secret key $\text{HIBS.SK}[I[1..n]]$ of $I[1..n]$. Given the above analogy between signatures of messages and secret keys of hierarchical identities, we construct an AOS scheme given a HIBS scheme as follows. Appending to a given signature in $\mathcal{AOS}$ is done using key delegation in $\mathcal{HIBS}$. The verification of an AOS signature $\text{HIBS.SK}[I[1..n]]$ is done by signing a random message $M \in \text{HIBS.MSpace}$ under the secret key $\text{HIBS.SK}[I[1..n]]$ and verifying that the resulting signature is valid. A formal construction is given in the full version [9].

Theorem 5. *If the HIBS scheme $\mathcal{HIBS} = (\text{HIBS.Setup}, \text{HIBS.KeyDel}, \text{HIBS.Sign}, \text{HIBS.Vfy})$ is hibs-uf-cma secure, then the above AOS scheme is aos-uf-cma secure.*

CONSTRUCTING HIBS FROM AOS. A naive approach to building a HIBS scheme from an AOS scheme would be as follows: for any hierarchical identity $I[1..n]$, define $\text{HIBS.SK}[[]I[1..n]]$ as the AOS signature on $I[1..n]$ and the HIBS signature created with $\text{HIBS.SK}[[]I[1..n]]$ on message M as the AOS signature formed by appending M to $\text{HIBS.SK}[[]I[1..n]]$. However, it can be shown that such a scheme is insecure. Our tweak is to insert a unique identifier to separate identities and messages. Let $\mathcal{AOS} = (\text{AOS.Setup}, \text{AOS.Append}, \text{AOS.Vfy})$ be an AOS scheme with message space AOS.MSpace. Let HIBS.IDSpace and HIBS.MSpace be subsets of AOS.MSpace such that there is some symbol Δ from the AOS message space which is not a valid identity for the HIBS scheme (Δ can still be in the

HIBS message space). Then we can construct a HIBS scheme with identity space HIBS.IDSpace and message space HIBS.MSpace as follows:

Construction 6. $\mathcal{HIBS}$ = (HIBS.Setup, HIBS.KeyDel, HIBS.Sign, HIBS.Vfy):

- HIBS.Setup(1^k): Run the AOS.Setup(1^k) to generate a pair (AOS.pk, Sig[ε]); output it as the master public/private key pair for $\mathcal{HIBS}$.
- HIBS.KeyDel(HIBS.pk, HIBS.SK[$I[1..n]$], I_{n+1}): The delegation algorithm interprets HIBS.SK[$I[1..n]$] as an $\mathcal{AOS}$ signature of $I[1..n]$. It appends to the signature a symbol I_{n+1} and outputs the resulting signature as the secret key of $I[1..n+1]$.
- HIBS.Sign(HIBS.pk, HIBS.SK[I_n], M): The signing algorithm for $\mathcal{HIBS}$ interprets HIBS.SK[$I[1..n]$] as an $\mathcal{AOS}$ signature of $I[1..n]$. It appends a symbol Δ to HIBS.SK[$I[1..n]$] and then appends the message M to the resulting AOS signature to get the final signature sig.
- HIBS.Vfy(HIBS.pk, $I[1..n]$, M, sig): The verification algorithm for $\mathcal{HIBS}$ verifies if sig is a valid AOS signature of $(I_1, \ldots, I_n, \Delta, M)$.

The following theorem is proven in the version [9]:

Theorem 7. If the AOS scheme $\mathcal{AOS}$ = (AOS.Setup, AOS.Append, AOS.Vfy) is aos-uf-cma secure, then the HIBS scheme $\mathcal{HIBS}$ from Construction 6 is hibs-uf-cma secure.

5 Applications

An important application of AOS is in the construction of secure routing protocols for the Internet. The Border Gateway Protocol (BGP), which is the primary routing protocol used today in the Internet, has some well-known security weaknesses which require cryptographic solutions. While there have been many proposals for securing BGP in the past [8, 6], each must develop its own cryptographic constructions due to the lack of any primitive designed specifically for this application. In the discussion below, we briefly describe Internet routing and explain how our primitive is useful for ensuring important security requirements in BGP.

The Internet is composed of various autonomous systems (ASes), each having control over some portion of the IP address space. BGP is the protocol used to spread information about the routes to all IP addresses in this network of ASes. Initially, all ASes advertise the IP addresses they own to their neighboring ASes. Upon receipt of such advertisements, each neighbor records this information, appends itself to the advertised route and sends the new information further down. The process repeats with the next AS in the chain and eventually, all ASes learn a route to the originating AS (In case an AS receives two or more routes to the same IP address, it selects one of them based on some local policy). Authenticity of route announcements is essential for ensuring the correct behaviour of BGP for otherwise, malicious ASes can play havoc with the Internet traffic. For example, if an AS truncates the route in some advertisement or modifies it selectively, it could convince its neighbors to forward all their traffic to it, which it could then modify or drop at will (incidents of this nature have indeed occured in the recent past [6]).

Append-only Signatures are a useful tool in addressing this problem. Suppose that an AS R_0 wishes to announce routes for some IP prefix it owns. It first generates an AOS public-private key pair, distributes the public key AOS.pk throughout the network (this can be done through a PKI as in [8, 6]) and to every neighboring AS R_{i_1}, sends the usual BGP information along with the AOS signature AOS.Append(AOS.pk, Sig[ε], R_{i_1}). In order to continue the advertisement process, R_{i_1} sends to each of its own neighbors R_{i_2} a BGP announcement containing the route (R_0, R_{i_1}) and the signature AOS.Append(AOS.pk, Sig[R_{i_1}], R_{i_2}). In other words, R_0 appends the label of its neighbor R_{i_1} into the AOS signature chain and R_{i_1} further appends the label of R_{i_2} into it. The advertisement process continues in this manner until all ASes in the network receive information about a route to R_0. Each recipient can verify the validity of the announced route using the public key AOS.pk. If the AOS scheme is secure, then all that a malicious AS can do now is to append one of its neighbors into the AOS signature chain (since each R_i can check that the AS it receives a route from was the last to be appended before R_i). In practice, the number of path advertisements received by an AS from any given source AS is extremely small: as observed in real routing data [6], the odds that an AS receives more than 15 path advertisements coming from the same source are about 1 in a 1000. This enables us to use m-time signature schemes (with $m = 15$) for an efficient AOS with reasonable security guarantee. For more details and other applications of AOS, see [9].

6 Final Remarks and Open Problems

FINALIZATION OF AOS SIGNATURE. A property of append-only signature schemes which might be needed by some applications is the ability to "finalize" the signature, that is, to modify the signature of a message in a way that prohibits any further appending. The general solution to this problem is to use a special symbol Θ (from the message space) to denote the end of the message. When one wants to finalize the signature of some message, he should append Θ to the signature. Messages that contain symbol Θ in the middle of the message (not as the last symbol) are therefore considered to be invalid.

RESTRICTED AOS. In AOS, anyone can append and verify signatures. In certain scenarios, however, one may want to restrict the ability to append messages to a limited group of users. Still, anyone should be able to verify the signatures. We call this extension of AOS *Restricted Append-Only Signatures* (RAOS). Using a symmetric encryption scheme we show in the full version [9] how to modify a given AOS scheme to get an RAOS scheme.

SHORTER AOS SIGNATURES. Given that wide-area routing protocols propagate a large number of messages, compact signatures are desirable. Thus we raise an open problem of whether it is possible to build an AOS scheme with constant signature length (in both message length and maximal message length). This problem is equivalent to building a HIBS scheme where secret keys of the users have constant length (in the depth of the given user in the hierarchy and in the maximal depth of the hierarchy).

Acknowledgments

We thank Mihir Bellare (for suggesting an improvement to the proof of Theorem 2) and Daniele Micciancio (for useful insight about the definition of AOS). Thanks also to the anonymous reviewers for helpful comments.

References

1. D. Boneh, X. Boyen and E.-J. Goh. Hierarchical identity based encryption with constant size ciphertext. In *Proceedings of EUROCRYPT 2005*, LNCS, 2005.
2. Dan Boneh, Craig Gentry, Ben Lynn, and Hovav Shacham. Aggregate and verifiably encrypted signatures from bilinear maps. In *Proceedings of EUROCRYPT 2003*, volume 2656 of *LNCS*, pages 416–432, 2003.
3. S. S. M. Chow, L. C. K. Hui, S. M. Yiu, and K. P. Chow. Secure hierarchical identity based signature and its application. In *Proceedings of ICICS 2004*, pages 480–494, 2004.
4. Craig Gentry and Alice Silverberg. Hierarchical id-based cryptography. In *Proceedings of ASIACRYPT 2002*, volume 2501 of *LNCS*, pages 548–566, 2002.
5. Alejandro Hevia and Daniele Micciancio. The provable security of graph-based one-time signatures and extensions to algebraic signature schemes. In *Proceedings of ASIACRYPT 2002*, volume 2501 of *LNCS*, pages 379 – 396, 2002.
6. Yih-Chun Hu, Adrian Perrig, and Marvin Sirbu. SPV: secure path vector routing for securing BGP. In *Proceedings of the ACM SIGCOMM*, pages 179–192, 2004.
7. Robert Johnson, David Molnar, Dawn Xiaodong Song, and David Wagner. Homomorphic signature schemes. In *Proceedings of CT-RSA 2002*, volume 2271 of *LNCS*, pages 244–262, 2002.
8. Stephen Kent, Charles Lynn, and Karen Seo. Secure border gateway protocol (S-BGP). In *IEEE Journal on Selected Areas in Communications*, 18(4):582–592, 2000.
9. Eike Kiltz, Anton Mityagin, Saurabh Panjwani and Barath Raghavan. Append-Only Signatures. Full version. http://eprint.iacr.org/2005/124
10. Anna Lysyanskaya, Silvio Micali, Leonid Reyzin, and Hovav Shacham. Sequential aggregate signatures from trapdoor permutations. In *Proceedings of EUROCRYPT 2004*, volume 3027 of *LNCS*, pages 74–90, 2004.
11. Ralph C. Merkle. A digital signature based on a conventional encryption function. In *Proceedings of CRYPTO'87*, volume 293 of *LNCS*, pages 369–378, 1988.
12. Silvio Micali and Ronald L. Rivest. Transitive signature schemes. In *Proceedings of CT-RSA 2002*, volume 2271 of *LNCS*, pages 236–243, 2002.
13. Leonid Reyzin and Natan Reyzin. Better than biba: Short one-time signatures with fast signing and verifying. In *Proceedings of 7th Australasian Conference ACSIP*, 2002.
14. Adi Shamir. Identity-based cryptosystems and signature schemes. In *Proceedings of CRYPTO'84*, volume 196 of *LNCS*, pages 47–53, 1985.

Hierarchical Group Signatures

Mårten Trolin and Douglas Wikström

Royal Institute of Technology (KTH),
Nada, SE-100 44 Stockholm, Sweden
{marten, dog}@nada.kth.se

Abstract. We introduce the notion of *hierarchical group signatures*. This is a proper generalization of group signatures, which allows multiple group managers organized in a tree with the signers as leaves. When opening a signature a group manager only learns to which of its subtrees, if any, the signer belongs.

We provide definitions for the new notion and construct a scheme that is provably secure given the existence of a family of trapdoor permutations. We also present a construction which is relatively practical, and prove its security in the random oracle model under the strong RSA assumption and the DDH assumption.

1 Introduction

Consider the notion of group signatures introduced by Chaum and van Heyst [13]. A group member can compute a signature that reveals nothing about the signer's identity except that he is a member of the group. On the other hand the group manager can always reveal the identity of the signer.

An application for group signatures is anonymous credit cards. The cardholder wishes to preserve his privacy when he pays a merchant for goods, i.e., he is interested in unlinkability of payments. The bank must obviously be able to extract the identity of a cardholder from a payment or at least an identifier for an account, to be able to debit the account. To avoid fraud, the bank, the merchant, and the cardholder all require that a cardholder cannot pay for goods without holding a valid card. To solve the problem using group signatures we let the bank be the group manager and the cardholders be signers. A cardholder signs a transaction and hands it to the merchant. The merchant then hands the signed transaction to the bank, which debits the cardholder and credits the merchant. Since signatures are unlinkable, the merchant learns nothing about the cardholder's identity. The bank on the other hand can always extract the cardholder's identity from a valid signature and debit the correct account.

The above scenario is somewhat simplified since normally there are many banks that issue cards of the same brand which are processed through the same payment network. The payment network normally works as an administrator and routes transactions to several independent banks. Thus, the merchant hands a payment to the payment network which hands the payment to the issuing bank. We could apply group signatures here as well by making the payment network act as the group manager. The network would then send the extracted identity to the issuing bank. Another option is to set up several independent group signatures schemes, one for each issuer. In the first approach, the payment network learns the identity of the customer, and in the second approach the merchant learns which bank issued the customer's card. A better solution would reveal nothing except what is absolutely necessary to each party. The merchant needs to be convinced that the credit card is valid, the payment network must be able to route the payment to the correct card issuer, and the issuer must be able to determine the identity of the cardholder.

In this extended abstract we introduce and investigate the notion of *hierarchical group signatures*. These can be employed to solve the above problem. When using a hierarchical group signature scheme there is not one single group manager. Instead there are several group managers organized in a tree, i.e., each group manager either manages a group of signers or a group of group managers. In the original notion the group manager can always identify the signer of a message, but nobody else can distinguish between signatures by different signers. The corresponding property for hierarchical group signatures is more complicated. When opening a signature from a signer in its subtree, a group manager learns to which of the subtrees directly below it the signer belongs. Signatures from other signers are indistinguishable. Hence a group manager on the level directly above the signers can identify its signers, whereas group managers higher in the hierarchy only learns to which subtree the signer belongs.

When we use hierarchical group signatures to construct anonymous credit cards for the more realistic setting we let the payment network be the root manager that manages a set of group managers, i.e., the issuing banks, and we let the cardholders be signers. The credit card application also demonstrates what kind of responsibility model is likely to be used with a hierarchical group signature scheme. With a valid signature on a transaction, the merchant has a valid demand on the payment network. If the payment network has a signature that can be shown to belong to a certain bank, the network has a valid demand on that bank. Thus, it is in the network's interest to open the signatures it receives from merchants, and it is in the issuing banks' interest to open the signatures they receive from the network.

1.1 Previous Work

The concept of group signatures was first introduced by Chaum and van Heyst [13] in 1991. This and the group signature schemes that followed [14, 7] all have the property that the complexity of the scheme grows with the number of participants. In [11] Camenisch and Stadler presented a system where the key does not grow with the number of participants. This system, however, relies on a non-standard number-theoretic assumption. The assumption was actually found to be incorrect and modified in [2]. An efficient system whose security rests on the strong RSA assumption and the Diffie-Hellman decision assumption was presented by Camenisch and Michels in 1998 [10]. This system was improved in [1]. The currently most efficient scheme that is secure under standard assumptions is [8]. More efficient schemes do exist [6,9], but they are based on bilinear maps and thus relies on less well-studied assumptions for security.

A related notion is *traceable signatures* introduced by Kiayias et al. [19], where signatures belonging to a member can be opened, or traced, in a distributed way without revealing the group secret.

Bellare et al. [4] give a definitional framework for group signatures for static groups, i.e., when the set of members cannot be changed after the initial setup. The paper also contains a scheme based on general methods in this setting. Kiayias and Yung [20] define security for dynamic groups and prove that a modification of [1] is secure under these definitions. Independently, Bellare et al. [5] extend the definitions of [4] in a similar way to handle dynamic groups, and present a scheme that is secure under general assumptions.

In [2] the concepts of *multi-group signatures* and *subgroup signatures* are described, and in [21] a system for hierarchical multi-groups is given. It may be worthwhile to consider the differences between these concepts and hierarchical signatures introduced here. Subgroup signatures make it possible for an arbitrary number i of signers to produce a joint signature which can be verified to stem from i distinct group members. Multi-group signature schemes allow a signer who is a member of two groups to produce a signature that shows membership of either both groups or just one of them. In hierarchical multi-groups a signer who is a member of a supergroup with subgroups can produce a signature that reveals membership either of the supergroup or of a subgroup of his choice. However, the opening procedure is not hierarchical, i.e., there are no group managers for the subgroups.

1.2 Notation

Throughout the text, κ denotes a security parameter. A function $f : \mathbb{N} \to [0,1]$ is said to be negligible if for each $c > 0$ there exists a $\kappa_0 \in \mathbb{N}$ such that $f(\kappa) < \kappa^{-c}$ for $\kappa_0 < \kappa \in \mathbb{N}$. We write $\emptyset$ to denote both the empty set and the empty string. If T is a tree we denote by $\mathcal{L}(T)$ its set of leaves and by $\mathcal{V}(T)$ the set of all vertices. We write G_q for the unique subgroup of order q of $\mathbb{Z}_p^*$ for a prime $p = 2q+1$. In the ElGamal cryptosystem a secret key is a randomly generated $x \in \mathbb{Z}_q$ and the public key is $y = g^x$. To encrypt message $m \in G_q$, $r \in \mathbb{Z}_q$ is chosen randomly and the cryptotext is given by $(u,v) = E_y(m,r) = (g^r, y^r m)$. To decrypt a cryptotext $D_x(u,v) = u^{-x}v = m$ is computed. We denote by $N = PQ$ an RSA module for two strong primes P and Q, and let QR_N be the subgroup of squares in $\mathbb{Z}_N^*$ with generators $\mathbf{g}$ and $\mathbf{h}$. The adversaries in this paper are modeled as polynomial time Turing machines with *non-uniform* auxiliary advice string. We denote the set of such adversaries by PPT*.

2 Hierarchical Group Signatures

In this section we discuss the notion of hierarchical group signatures. We begin by describing the parties of a hierarchical group signature system. Then we proceed by giving formal definitions.

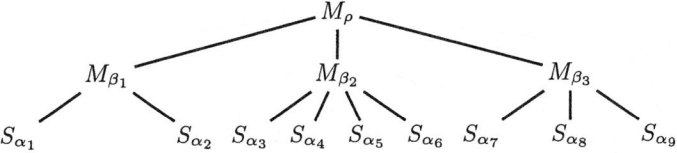

Fig. 1. A tree of group managers and signers, where $\rho = \{\beta_1, \beta_2, \beta_3\}$, $\beta_1 = \{\alpha_1, \alpha_2\}$, $\beta_2 = \{\alpha_3, \alpha_4, \alpha_5, \alpha_6\}$, and $\beta_3 = \{\alpha_7, \alpha_8, \alpha_9\}$

There are two types of parties: signers denoted S_α for α in some index set $\mathcal{I}$, and group managers denoted M_α for indices α described below. The parties form a tree T, where the signers are leaves and the group managers are inner nodes. The indices of the group managers are formed as follows. If a group manager manages a set of signers $\beta \subset \mathcal{I}$ we denote it by M_β. This corresponds to M_β having S_α for $\alpha \in \beta$ as children. If a group manager M_γ manages a set of group managers $\{M_{\beta_1}, \ldots, M_{\beta_l}\}$ we denote it by M_γ where $\gamma = \{\beta_1, \ldots, \beta_l\}$. This corresponds to M_γ having M_{β_i} for $i = 1, \ldots, l$ as children. Let M_ρ denote the root group manager. We define the root group manager to be at depth 0 and assume that all leaves in the tree are at the same depth δ. Figure 1 illustrates a tree of parties.

Note that standard group signatures correspond to having a single group manager $M_{\{1,\ldots,l\}}$ that manages all signers $S_1, \ldots, S_l$.

2.1 Definition of Security

Bellare et al. [4] give a definition of a group signature scheme, but more importantly they argue that two properties of group signatures, full anonymity and full traceability, imply any reasonable security requirements one can expect from a group signature scheme. We follow their definitional approach closely.

Definition 1 (Hierarchical Group Signature). *A hierarchical group signature scheme* $\mathcal{HGS} = $ (HKg, HSig, HVf, HOpen) *consists of four polynomial-time algorithms*

1. *The randomized* key generation algorithm HKg *takes as input* $(1^\kappa, T)$, *where T is a tree of size polynomially bounded in the security parameter κ with all leaves at the same depth, and outputs a pair of maps* hpk, hsk $: \mathcal{V}(T) \to \{0,1\}^*$. *For each node (or leaf) α,* hpk(α) *is the public key and* hsk(α) *is the secret key.*
2. *The randomized* signature algorithm HSig *takes as input a message m, a tree T, a public key* hpk, *and a secret signing key* hsk(α), *and returns a signature of m.*
3. *The deterministic* signature verification algorithm HVf *takes as input a tree T, a public key* hpk, *a message m and a candidate signature σ of m and returns either 1 or 0.*

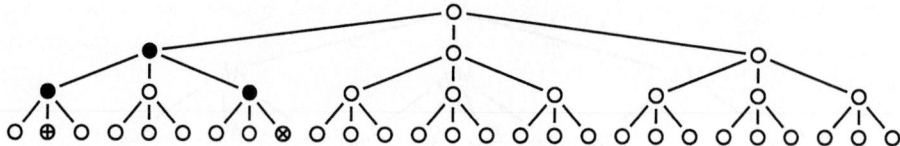

Fig. 2. Nodes in black represent group managers able to distinguish between signatures by $S_{\alpha^{(0)}}$ and $S_{\alpha^{(1)}}$, the two marked leaves

4. *The deterministic opening algorithm* HOpen *takes as input a tree* T, *a public key* hpk, *a secret opening key* hsk(β), *a message* m, *and a candidate signature* σ. *It outputs an index* $\alpha \in \beta$ *or* $\bot$.

We need to define what we mean by security for a hierarchical group signature scheme. We begin with anonymity. Consider Figure 2, where two signers $S_{\alpha^{(0)}}$ and $S_{\alpha^{(1)}}$ are marked. Assume that it is known that a message has been signed by one of them. Then any group manager on the path leading from $S_{\alpha^{(0)}}$ or $S_{\alpha^{(1)}}$ to their first common ancestor can determine which of them signed the message. In the figure those group managers are marked with black. In the definition of anonymity we capture the property that unless the adversary corrupts one of these group managers, it cannot determine whether $S_{\alpha^{(0)}}$ or $S_{\alpha^{(1)}}$ signed the message, even if the adversary is given the private keys of all signers and is allowed to select $\alpha^{(0)}$, $\alpha^{(1)}$ and the message itself.

We define Experiment 1 to formalize these ideas. Throughout the experiment the adversary has access to an HOpen$(T, \text{hpk}, \text{hsk}(\cdot), \cdot, \cdot)$ oracle. At the start of the experiment the adversary is given the public keys of all parties and the private keys of all signers. Then it can adaptively ask for the private keys of the group managers. At some point it outputs the indices $\alpha^{(0)}$ and $\alpha^{(1)}$ of two leaves and a message m. The HSig$(\cdot, T, \text{hpk}, \text{hsk}(\cdot))$ oracle computes the signature of m using the private key hsk$(\alpha^{(b)})$ and hands it to the adversary. The adversary finally outputs a guess d of the value of b. If the scheme is anonymous the probability that $b = d$ should be negligibly close to $1/2$ when b is a randomly chosen bit. The labels corrupt, choose and guess below distinguish between the phases of the experiment.

Experiment 1 (Hierarchical Anonymity, $\text{Exp}_{\mathcal{HGS},A}^{\text{anon}-b}(\kappa, T)$).

$(\text{hpk}, \text{hsk}) \leftarrow \text{HKg}(1^\kappa, T)$; $s_{\text{state}} \leftarrow (\text{hpk}, \text{hsk}(\mathcal{L}(T)))$; $\mathcal{C} \leftarrow \emptyset$; $\alpha \leftarrow \emptyset$;
Do
$\quad \mathcal{C} \leftarrow \mathcal{C} \cup \{\alpha\}$
$\quad (s_{\text{state}}, \alpha) \leftarrow A^{\text{HOpen}(T, \text{hpk}, \text{hsk}(\cdot), \cdot, \cdot)}(\text{corrupt}, s_{\text{state}}, \text{hsk}(\alpha))$
While $(\alpha \in \mathcal{V}(T) \setminus \mathcal{C})$
$(s_{\text{state}}, \alpha^{(0)}, \alpha^{(1)}, m) \leftarrow A^{\text{HOpen}(T, \text{hpk}, \text{hsk}(\cdot), \cdot, \cdot)}(\text{choose}, s_{\text{state}})$
$\sigma \leftarrow \text{HSig}(m, T, \text{hpk}, \text{hsk}(\alpha^{(b)}))$
$d \leftarrow A^{\text{HOpen}(T, \text{hpk}, \text{hsk}(\cdot), \cdot, \cdot)}(\text{guess}, s_{\text{state}}, \sigma)$

Let B be the set of nodes on the paths from $\alpha^{(0)}$ and $\alpha^{(1)}$ up to their first common ancestor α_t excluding $\alpha^{(0)}$ and $\alpha^{(1)}$ but including α_t, i.e., the set of nodes $\alpha_l^{(0)}, \alpha_l^{(1)}$, $l = t, \ldots, \delta - 1$, such that

$$\alpha^{(0)} \in \alpha_{\delta-1}^{(0)} \in \alpha_{\delta-2}^{(0)} \in \ldots \in \alpha_{t+1}^{(0)} \in \alpha_t \ni \alpha_{t+1}^{(1)} \ni \ldots \ni \alpha_{\delta-2}^{(1)} \ni \alpha_{\delta-1}^{(1)} \ni \alpha^{(1)} \;.$$

If $B \cap C \neq \emptyset$ or if A asked its $\mathsf{HOpen}(T, \mathrm{hpk}, \mathrm{hsk}(\cdot), \cdot, \cdot)$ oracle a question $(\alpha_l^{(0)}, m, \sigma)$ or $(\alpha_l^{(1)}, m, \sigma)$ return 0. Otherwise return d.

Consider the above experiment with a depth one tree T and root ρ. In that case we may assume that $\mathrm{hsk}(\rho)$ is never handed to the adversary, since the adversary fails in that case anyway. Similarly the $\mathsf{HOpen}(T, \mathrm{hpk}, \mathrm{hsk}(\cdot), \cdot, \cdot)$ oracle reduces to the Open oracle in [4]. Thus, our experiment reduces to the experiment for full anonymity given in [4] where the adversary gets the secret keys of all signers, but only the public key of the group manager.

Next we consider how the notion of full traceability can be defined in our setting. Full traceability as defined in [4] is similar to security against chosen message attacks (CMA-security) as defined by Goldwasser, Micali and Rivest [18] for signatures. The only essential difference is that the group manager must always be able to open a signature and identify the signer. In our setting this amounts to the following. Given a signature deemed valid by the HVf algorithm, the root should always be able to identify the child directly below it of which the signer is a descendent. The child should have the same ability for the subtree of which it is a root and so on until the child itself is a signer.

Again we define an experiment consisting of two phases. The adversary is given the secret keys of all group managers. Then the adversary adaptively chooses a set of signers to corrupt. Then in a second phase the adversary outputs a message and a signature. If the output amounts to a signature deemed valid by HVf and the signer cannot be traced, or if the signature is traced to a non-corrupted signer, the adversary has succeeded and the experiment outputs 1. Otherwise it outputs 0. Thus, the distribution of the experiment should be negligibly close to 0 for all adversaries if the scheme is secure.

Experiment 2 (Hierarchical Traceability, $\mathrm{Exp}_{\mathcal{HGS},A}^{\mathrm{trace}}(\kappa, T)$).

$(\mathrm{hpk}, \mathrm{hsk}) \leftarrow \mathsf{HKg}(1^\kappa, T); \; s_{\mathrm{state}} \leftarrow (\mathrm{hpk}, \mathrm{hsk}(\mathcal{V}(T) \backslash \mathcal{L}(T)); \; \mathcal{C} \leftarrow \emptyset, \; \alpha \leftarrow \emptyset;$
Do
$\quad \mathcal{C} \leftarrow \mathcal{C} \cup \{\alpha\}$
$\quad (s_{\mathrm{state}}, \alpha) \leftarrow A^{\mathsf{HSig}(\cdot, T, \mathrm{hpk}, \mathrm{hsk}(\cdot))}(\mathrm{corrupt}, s_{\mathrm{state}}, \mathrm{hsk}(\alpha))$
While $(\alpha \in \mathcal{V}(T) \setminus \mathcal{C})$
$(m, \sigma) \leftarrow A^{\mathsf{HSig}(\cdot, T, \mathrm{hpk}, \mathrm{hsk}(\cdot))}(\mathrm{guess}, s_{\mathrm{state}})$

If $\mathsf{HVf}(T, \mathrm{hpk}, m, \sigma) = 0$ return 0. Define $\alpha_0 = \rho$ and $\alpha_l = \mathsf{HOpen}(T, \mathrm{hpk}, \mathrm{hsk}(\alpha_{l-1}), m, \sigma)$ for $l = 1, \ldots, \delta$. If $\alpha_l = \bot$ for some $0 < l \leq \delta$ return 1. If $\alpha_\delta \notin \mathcal{C}$ and the $\mathsf{HSig}(\cdot, T, \mathrm{hpk}, \mathrm{hsk}(\cdot))$ oracle did not get a question (m, α_δ) return 1. Otherwise return 0.

Consider the experiment above with a depth one tree. This corresponds to giving the adversary the secret key of the group manager, and letting it adaptively choose additional signing keys. This is precisely the setting of [4].

The advantage of the adversary is defined in the natural way by setting $\mathbf{Adv}^{anon}_{\mathcal{HGS},A}(\kappa,T) = |\Pr[\mathbf{Exp}^{anon-0}_{\mathcal{HGS},A}(\kappa,T) = 1] - \Pr[\mathbf{Exp}^{anon-1}_{\mathcal{HGS},A}(\kappa,T) = 1]|$ and $\mathbf{Adv}^{trace}_{\mathcal{HGS},A}(\kappa,T) = \mathbf{Exp}^{trace}_{\mathcal{HGS},A}(\kappa,T)$.

Definition 2 (Security of Hierarchical Group Signatures). *A hierarchical group signature scheme* $\mathcal{HGS} = $ (HKg, HSig, HVf, HOpen) *is secure if for all trees* T *of polynomial size in* κ *with all leaves at the same depth, and all* $A \in \text{PPT}^*$, $\mathbf{Adv}^{trace}_{\mathcal{HGS},A}(\kappa,T) + \mathbf{Adv}^{anon}_{\mathcal{HGS},A}(\kappa,T)$ *is negligible.*

3 Our Constructions

We construct two hierarchical group signature schemes, one under general assumptions, and one under standard assumptions in the random oracle model. Both require a trusted key generator at the start of the protocol. The two constructions are different, but based on similar ideas. In this extended abstract we only give the main ideas behind our constructions. Detailed descriptions and proofs of our claims are given in the full paper [26].

3.1 Our Approach

All known group signatures are based on the idea that the signer encrypts a secret of some sort using the group manager's public key, and then proves that the resulting cryptotext is on this special form. The security of the cryptosystem used implies anonymity, since no adversary can distinguish cryptotexts of two distinct messages if they are encrypted using the *same* public key. We generalize this approach.

First we consider the problem of forwarding partial information on the identity of the signer to group managers without leaking information. Each group manager M_β is given a secret key sk_β and a public key pk_β of a cryptosystem. We also give each signer S_α a public key pk_α that is used to identify the signer. Each signer is associated in the natural way with the path $\alpha_0, \alpha_1, \ldots, \alpha_\delta$ from the root $\rho = \alpha_0$ to the leaf $\alpha = \alpha_\delta$ in the tree T of group managers and signers. To compute a signature, the signer computes as part of the signature a chain

$$(C_0, C_1, \ldots, C_{\delta-1}) = \left(E_{\text{pk}_{\alpha_0}}(\text{pk}_{\alpha_1}), E_{\text{pk}_{\alpha_1}}(\text{pk}_{\alpha_2}), \ldots, E_{\text{pk}_{\alpha_{\delta-1}}}(\text{pk}_{\alpha_\delta})\right) .$$

Note that each cryptotext C_l in the list encrypts the public key $\text{pk}_{\alpha_{l+1}}$ used to form the following cryptotext. The particular structure of the chain and the fact that all leaves are on the same depth in the tree ensures that a group manager M_β on depth l can try to open a signature by decrypting C_l, i.e., it computes $\text{pk} = D_{\text{sk}_\beta}(C_l)$. If $\alpha_l = \beta$, then $\text{pk} = \text{pk}_{\alpha_{l+1}}$. Thus, if M_β manages signers, it learns the identity of the signer S_α, and if it manages other group managers it learns the identity of the group manager below it in the tree which (perhaps indirectly) manages the signer S_α.

Now suppose that $\alpha_l \neq \beta$, so $\text{pk} \neq \text{pk}_{\alpha_{l+1}}$. What does M_β, or indeed any outsider, learn about the identity of the signer S_α? It clearly does not learn anything from a cryptotext C_l about the encrypted cleartext, as long as the cryptosystem is semantically secure. However, if the cryptotext C_{l+1} somehow indicates which public key was used to form it, M_β, or any outsider, can simply look at C_{l+1} and recover the cleartext of C_l. This means that it can look at the chain of cryptotexts and extract information on the identity of the signer. We conclude that using the approach above, we need a cryptosystem which not only hides the cleartext, but also hides the public key used to form the cryptotext. Such a cryptosystem is called an *anonymous* cryptosystem [3].

Next we consider the problem of ensuring hierarchical traceability. This problem consists of two parts. We must ensure chosen message security to avoid that an illegitimate signer is able compute a valid signature at all. However, the more difficult problem is to ensure that the signer S_α not only formed $(C_0, \ldots, C_{\delta-1})$ as described above for some public keys $\text{pk}_{\alpha_0}, \ldots, \text{pk}_{\alpha_\delta}$, but also that the public keys used correspond to the unique path $\alpha_0, \alpha_1, \ldots, \alpha_\delta$ from the root $\rho = \alpha_0$ to the leaf $\alpha = \alpha_\delta$ corresponding to the signer S_α. This is the main obstacle to construct an efficient hierarchical group signature scheme.

3.2 A Construction Under General Assumptions

We sketch the construction under general assumptions. To achieve hierarchical anonymity we employ the cryptosystem of Goldwasser and Micali [17] and prove that this cryptosystem is anonymous. To achieve traceability we use the group signature scheme of Bellare et al. [4] and a non-interactive adaptive zero-knowledge unbounded simulation sound proof (NIZK) as defined and constructed for any language in NP in Feige, Lapidot and Shamir [16], Sahai [23], and De Santis [24]. Both constructions are provably secure under the existence of a trapdoor permutation family.

The signer proves using the NIZK that the chain of cryptotexts is formed correctly, and the group signature scheme ensures that only legitimate signers can form a signature, without losing anonymity. The group signature also allows us to use a semantically secure cryptosystem for the chain (cf. [4,8]), since any query to the HOpen oracle obviously can be answered correctly by the simulator if we know the full identity of the signer, i.e., we use a variant of the double-cryptotext trick of Naor and Yung [22]. The following theorem is proved in the full version [26].

Theorem 1. *If there exists a family of trapdoor permutations, then there exists a secure hierarchical group signature scheme.*

3.3 A Construction Under the DDH Assumption and the Strong RSA Assumption

To achieve hierarchical anonymity in the practical construction we employ the ElGamal cryptosystem, which is semantically secure under the DDH assumption. It is easy to see that ElGamal is also anonymous, as long as a fixed group is used for each security parameter. Thus, each group manager M_β holds a secret key x_β and a public key $y_\beta = g^{x_\beta}$, and the chain of cryptotexts is on the form

$$((u_0, v_0), \ldots, (u_{\delta-1}, v_{\delta-1})) = (E_{y_{\alpha_0}}(y_{\alpha_1}), \ldots, E_{y_{\alpha_{\delta-1}}}(y_{\alpha_\delta})) \ .$$

To achieve chosen message security we employ the Fiat-Shamir heuristic to turn an identification scheme into a signature scheme. The secret key of a signer S_α is a Cramer-Shoup [15] signature $\sigma_\alpha = \text{Sig}^{\text{cs}}(y_{\alpha_1}, \ldots, y_{\alpha_{\delta-1}})$ of the public keys corresponding to the path $\alpha_0, \alpha_1, \ldots, \alpha_\delta$ from the root $\rho = \alpha_0$ to the leaf $\alpha = \alpha_\delta$. The Cramer-Shoup scheme is provably secure under the strong RSA assumption.

To form a signature of a message m the signer first computes a commitment $C(\sigma_\alpha)$ of the signature σ_α. Then it computes an honest verifier zero-knowledge public coin proof $\pi(m)$ that the cryptotexts $((u_0, v_0), \ldots, (u_{\delta-1}, v_{\delta-1}))$ form a chain and that $C(\sigma_\alpha)$ hides a signature of the list $(y_{\alpha_1}, \ldots, y_{\alpha_{\delta-1}})$ of public keys used to form the chain of cryptotexts. The proof is given in the random oracle model and the message m to be signed is given as a prefix to every query to the random oracle. Thus, the complete signature is given by

$$(E_{y_{\alpha_0}}(y_{\alpha_1}), \ldots, E_{y_{\alpha_{\delta-1}}}(y_{\alpha_\delta}), C(\sigma_\alpha), \pi(m)) \ .$$

Intuitively, this means that if a signer S_α can produce a valid signature, we can by rewinding extract a signature of the list of public keys corresponding to the path from the root to the signer. Thus, a signature can only be formed if the signer is legitimate and if it has formed the chain correctly. Denote the hierarchical group signature scheme sketched above by $\mathcal{HGS}$. We prove the following theorem in the full version [26].

Theorem 2. *The hierarchical signature scheme $\mathcal{HGS}$ is secure under the DDH assumption and the strong RSA assumption in the random oracle model.*

Efficiency Analysis. The complexity of the protocol is largely determined by the proof sketched in the next subsection. This protocol has soundness $1 - O(\delta 2^{-\kappa'})$, where κ' is a secondary security parameter. Using standard computational tricks we estimate the complexity of the protocol to correspond to roughly $\kappa'(\delta + 3)$ general exponentiations modulo a κ-bit integer. If we set $\delta = 3$ and $\kappa' = 160$ this corresponds to less than 1000 general exponentiations. The size of a signature is about 1 Mb. The full version [26] contains a more detailed analysis.

Construction of the Proof of Knowledge. The main obstacle to find an efficient hierarchical group signature scheme following our approach is how to prove efficiently that $C(\sigma_\alpha)$ is a commitment of a signature σ_α of the list of public keys $(y_{\alpha_1}, \ldots, y_{\alpha_{\delta-1}})$ used to form the chain $((u_0, v_0), \ldots, (u_{\delta-1}, v_{\delta-1}))$. We construct a reasonably practical honest verifier zero-knowledge public coin proof for this relation by carefully selecting and combining a variety of cryptographic primitives and techniques. Due to the complexity of this protocol we can only sketch the main ideas of this protocol in this extended abstract. Details are given in the full version [26].

Let $q_0, \ldots, q_3$ be primes such that $q_i = 2q_{i+1} + 1$ for $i = 0, 1, 2$. A list of such primes is called a Cunningham chain and exists under mild assumptions on the distribution of primes. There is a subgroup $G_{q_{i+1}} \subset \mathbb{Z}_{q_i}^*$ of order q_{i+1} for $i = 0, 1, 2$. Denote by g_i and y_i fixed and independently chosen generators of G_{q_i} for $i = 1, 2, 3$, i.e., $\log_{g_i} y_i$ is not known to any party in the protocol. Thus, we can form a commitment of a value $y_\alpha \in G_{q_3}$ in three ways, as

$$(y_3^{t'''} g_3^{s'''}, y_3^{s'''} y_\alpha) \;,\quad (y_2^{t''} g_2^{s''}, y_2^{s''} g_2^{y_\alpha}) \;,\quad \text{and} \quad (y_1^{t'} g_1^{s'}, y_1^{s'} g_1^{g_2^{y_\alpha}}) \;,$$

where $t''', s''' \in \mathbb{Z}_{q_3}$, $t'', s'' \in \mathbb{Z}_{q_2}$, and $t', s' \in \mathbb{Z}_{q_1}$ are randomly chosen. By extending the ideas of Stadler [25] we can give a reasonably practical cut-and-choose proof that the elements hidden in two such commitments are identical.

Recall that the collision-free Chaum-Heijst-Pfitzmann [12] hash function is defined by $H^{\mathsf{CHP}} : \mathbb{Z}_{q_2}^\delta \to G_{q_2}$, $H^{\mathsf{CHP}} : (z_1, \ldots, z_\delta) \mapsto \prod_{l=1}^\delta h_l^{z_l}$, where the bases $h_1, \ldots, h_\delta \in G_{q_2}$ are randomly chosen, i.e., no party knows a non-trivial representation of $1 \in G_{q_2}$ in these elements.

We employ ElGamal over G_{q_3}. This means that the public keys $y_{\alpha_1}, \ldots, y_{\alpha_\delta}$ belong to G_{q_3}. Although it is not trivial, the reader should not find it too hard to imagine that Stadler-techniques can be used to prove that the public keys used for encryption are identical to values hidden in a list of commitments formed as

$$((\mu_0, \nu_0), \ldots, (\mu_{\delta-1}, \nu_{\delta-1})) = ((y_2^{t_0''} g_2^{s_0''}, y_2^{s_0''} h_1^{y_{\alpha_1}}), \ldots, (y_2^{t_{\delta-1}''} g_2^{s_{\delta-1}''}, y_2^{s_{\delta-1}''} h_\delta^{y_{\alpha_\delta}})) \;.$$

The importance of this is that if we take the product of the commitments we get a commitment of $H^{\mathsf{CHP}}(y_{\alpha_1}, \ldots, y_{\alpha_\delta})$, i.e.,

$$\left(\prod_{i=0}^{\delta-1} \mu_i, \prod_{i=0}^{\delta-1} \nu_i \right) = \left(y_2^{t''} g_2^{s''}, y_2^{s''} \prod_{i=1}^{\delta} h_i^{y_{\alpha_i}} \right) \;, \qquad (1)$$

for some $t'', s'' \in \mathbb{Z}_{q_2}$. Thus, at this point we have devised a way for the signer to verifiably commit to the hash value of the keys it used to form the chain of cryptotexts. This is a key step in the construction.

Recall that the signer commits to a Cramer-Shoup signature σ_α of the list of public keys it uses to form the chain of cryptotexts. This signature scheme uses an RSA-modulus N and elements from the subgroup QR_N of squares in $\mathbb{Z}_N^*$, and it is parameterized by two collision-free hash functions. The first hash function is used to compute a message digest of the message to be signed, i.e., the list $(y_{\alpha_1}, \ldots, y_{\alpha_\delta})$ of public keys. Above we have sketched how the signer can verifiably form a commitment of the H^{CHP} hash value of this message, so it is only natural that we let this be the first of the two hash functions in the signature scheme. However, in the signature scheme the message digest lives in the exponent of an element in QR_N in the signature scheme. To move the hash value up in the exponent and to change group from G_{q_1} to QR_N, the signer forms two commitments

$$\left(y_1^{t'} g_1^{s'}, y_1^{s'} g_1^{H^{\mathsf{CHP}}(y_{\alpha_1},\ldots,y_{\alpha_\delta})}\right) \text{ and } \mathbf{h}^t \mathbf{g}^{H^{\mathsf{CHP}}(y_{\alpha_1},\ldots,y_{\alpha_\delta})} .$$

Then it gives a cut-and-choose proof that the exponent in the left commitment equals the value committed to in the product (1). It also proves that the exponents in the two commitments are equal. Thus, at this point the signer has proved that it holds a commitment over QR_N of the hash value of the public keys it used to form the chain of cryptotexts.

The second hash function used in the Cramer-Shoup signature scheme is applied to a single element in QR_N. Since H^{CHP} is not collision-free on such inputs, we use the Shamir hash function $H^{\mathsf{Sh}}_{(\mathbf{g},N)} : \mathbb{Z} \to \mathrm{QR}_N$, $x \mapsto \mathbf{g}^x \bmod N$ instead. Using similar techniques as explained above the signer evaluates the hash function and moves the result into the exponent, by two Stadler-like cut-and-choose proofs.

Given the two hash values in the exponents of two commitments, standard techniques can be used to prove that the commitment $C(\sigma_\alpha)$ is a commitment of the Cramer-Shoup signature σ_α of the list of public keys used to form the chain of cryptotexts.

4 Conclusion

We have introduced and formalized the notion of hierarchical group signatures and given two constructions. The first is provably secure under general assumptions, whereas the second is provably secure under the DDH assumption and the strong RSA assumption in the random oracle model. The latter is practical, i.e., it can be implemented and run on modern workstations, bit it is still slow. Both require a trusted key generator. Thus, an interesting open problem is to eliminate these deficiencies.

Acknowledgments

We thank Johan Håstad for excellent advice. We also thank Torsten Ekedahl for explaining the heuristic view on the existence of Cunningham chains. We thank Ronald Cramer for providing us with references. We had valuable discussions on anonymity of cryptosystems with Gustav Hast.

References

1. G. Ateniese, J. Camenisch, M. Joye, and G. Tsudik. A practical and provably secure coalition-resistant group signature scheme. In *CRYPTO 2000*, volume 1880 of *LNCS*, 2000.
2. G. Ateniese and G. Tsudik. Some open issues and directions in group signatures. In *Financial Cryptography '99*, volume 1648 of *LNCS*, 1999.
3. M. Bellare, A. Boldyreva, A. Desai, and D. Pointcheval. Key-privacy in public-key encryption. In *ASIACRYPT 2001*, volume 2248 of *LNCS*, 2001.
4. M. Bellare, D. Micciancio, and B. Warinschi. Foundations of group signatures: Formal definitions, simplified requirements, and a construction based on general assumptions. In *EUROCRYPT 2003*, volume 2656 of *LNCS*, 2003.
5. M. Bellare, H. Shi, and C. Zhang. Foundations of group signatures: The case of dynamic groups. In *RSA-CT 2005*, volume 3376 of *LNCS*, 2005.
6. D. Boneh, X. Boyen, and H. Shacham. Short group signatures. In *CRYPTO 2004*, volume 3152 of *LNCS*, 2004.
7. J. Camenisch. Efficient and generalized group signatures. In *EUROCRYPT'97*, volume 1233 of *LNCS*, 1997.
8. J. Camenisch and J. Groth. Group signatures: Better efficiency and new theoretical aspects. In *SCN 2004*, volume 3352 of *LNCS*, 2005.
9. J. Camenisch and A. Lysyanskaya. Signature schemes and anonymous credentials from bilinear maps. In *CRYPTO 2004*, volume 3152 of *LNCS*, 2004.
10. J. Camenisch and M. Michels. A group signature scheme with improved effiency. In *ASIACRYPT'98*, volume 1514 of *LNCS*, 1999.
11. J. Camenisch and M. Stadler. Efficient group signature schemes for large groups. In *CRYPTO'97*, volume 1294 of *LNCS*, 1997.
12. D. Chaum, E. van Heijst, and B. Pfitzmann. Cryptographically strong undeniable signatures, unconditionally secure for the signer. In *CRYPTO'91*, volume 576 of *LNCS*, 1991.
13. D. Chaum and E. van Heyst. Group signatures. In *EUROCRYPT'91*, volume 547 of *LNCS*, 1991.
14. L. Chen and T.P. Pedersen. New group signature schemes. In *EUROCRYPT'94*, volume 950 of *LNCS*, 1994.
15. R. Cramer and V. Shoup. Signature schemes based on the strong RSA assumption. In *6th CCS*, 1999.
16. U. Feige, D. Lapidot, and A. Shamir. Multiple non-interactive zero-knowledge proofs under general assumptions. *SIAM Journal of Computing*, 29(1), 1999.
17. S. Goldwasser and S. Micali. Probabilistic encryption. *Journal of Computer and System Sciences*, 28(2), 1984.
18. S. Goldwasser, S. Micali, and R. Rivest. A digital signature scheme secure against adaptive chosen-message attacks. *SIAM Journal of Computing*, 17(2), 1988.
19. A. Kiayias, Y. Tsiounis, and M. Yung. Traceable signatures. In *EUROCRYPT 2004*, volume 3027 of *LNCS*, 2004.
20. A. Kiayias and M. Yung. Group signatures: Provable security, efficient constructions and anonymity from trapdoor-holders. Cryptology ePrint Archive, Report 2004/076, 2004. http://eprint.iacr.org/2004/076.
21. S. Kim, S. Park, and D. Won. Group signatures for hierarchical multigroups. In *ISW'97*, volume 1396 of *LNCS*, 1998.
22. M. Naor and M. Yung. Public-key cryptosystems provably secure against chosen ciphertext attacks. In *22nd STOC*, 1990.

23. A. Sahai. Non-malleable non-interactive zero-knowledge and adaptive chosen-ciphertext security. In *40th FOCS*, 1999.
24. A. De Santis, G. Di Crescenzo, R. Ostrovsky, G. Persiano, and A. Sahai. Robust non-interactive zero knowledge. In *CRYPTO 2001*, volume 2139 of *LNCS*, 2001.
25. M. Stadler. Publicly verifiable secret sharing. In *EUROCRYPT'96*, volume 1070 of *LNCS*, 1996.
26. M. Trolin and D. Wikström. Hierarchical group signatures. Cryptology ePrint Archive, Report 2004/076, 2004. http://eprint.iacr.org/2004/311.

Designated Verifier Signature Schemes: Attacks, New Security Notions and a New Construction

Helger Lipmaa[1], Guilin Wang[2], and Feng Bao[2]

[1] Cybernetica AS and University of Tartu, Estonia
[2] Institute for Infocomm Research (I^2R), Singapore

Abstract. We show that the signer can abuse the disavowal protocol in the Jakobsson-Sako-Impagliazzo designated-verifier signature scheme. In addition, we identify a new security property—non-delegatability—that is essential for designated-verifier signatures, and show that several previously proposed designated-verifier schemes are delegatable. We give a rigorous formalisation of the security for designated-verifier signature schemes, and propose a new and efficient designated-verifier signature scheme that is provably unforgeable under a tight reduction to the Decisional Diffie-Hellman problem in the non-programmable random oracle model, and non-delegatable under a loose reduction in the programmable random oracle model. As a direct corollary, we also get a new efficient conventional signature scheme that is provably unforgeable under a tight reduction to the Decisional Diffie-Hellman problem in the non-programmable random oracle plus common reference string model.

Keywords: Designated verifier signature scheme, non-delegatability, non-programmable random oracle model, signature scheme.

1 Introduction

In 1996, Jakobsson, Sako and Impagliazzo introduced the concept of designated-verifier signature (DVS) schemes [JSI96]. A DVS scheme makes it possible for a prover Signy to convince a designated verifier Desmond that she has signed a statement so that Desmond cannot transfer the signature to a third party Trevor. This is achieved since Desmond himself can efficiently simulate signatures that are indistinguishable from Signy's signatures. Moreover, in a disavowable DVS scheme, Signy can prove to Trevor that a simulated signature was not created by Desmond, while she can not disavow her own signatures. This is possible only when Signy's and Desmond's signatures are computationally but not perfectly indistinguishable.

We point out weaknesses in the designated-verifier signature schemes of [JSI96, SKM03, SBWP03, SWP04, LV04]. Of these schemes, the JSI scheme from [JSI96] is the only disavowable DVS scheme. However, we show that in the JSI scheme, a malicious Signy can generate signatures exactly from the same distribution as Desmond and thus the JSI scheme is perfectly non-transferable and thus also not disavowable. Our attack against the DVS schemes from [SKM03, SBWP03, SWP04, LV04] is not an attack according to the definitions of the designated verifier signatures in these papers, although it is an attack according to the original informal definition of [JSI96]: namely, we show that Signy can delegate her signing ability—with respect to a fixed designated verifier Desmond—to a third party Trevor, without revealing her secret key or making it possible for Trevor to sign with respect to other designated verifiers. This delegation property, while desirable in some settings (e.g., proxy DVS schemes), is extremely undesirable in many other settings and must therefore be considered as a serious weakness of a DVS scheme.

By pointing out the described flaws in these designated verifier signature scheme, we arrive to a stronger security notion for DVS that includes two novel requirements: (a) most importantly, *non-delegatability*: there exists an efficient knowledge extractor that can extract either Signy's secret key or Desmond's secret key, when given oracle access to an adversary who can create valid signatures with a high probability (this property is not shared by the DVS schemes from [SKM03, SBWP03, SWP04, LV04]), and (b) secure disavowability: if the DVS scheme has a disavowal protocol, it must be the case that Signy cannot disavow signatures, given by herself (this property is not shared by the DVS scheme from [JSI96]).

Non-delegatability of a DVS means that a valid designated-verifier signature constitutes a proof of knowledge of either Signy's or Desmond's secret key. Now, for conventional signatures, ability to sign is conceptually equal to the knowledge of the secret key. Therefore, a valid signature does not be a proof of knowledge. Now, as it is known from [KW03], one can construct conventional signature schemes whose unforgeability is proven by giving a tight reduction to an underlying cryptographic problem; this is achieved by specially avoiding the use of proofs of knowledge. However, in the case of a DVS scheme, we can also avoid proofs of knowledge in the proof of unforgeability, but not in the proof of non-delegatability. Therefore, even if we have a proof that a DVS scheme is unforgeable (w.r.t. any verifier), we cannot directly derive from that that this scheme is also non-delegatable. Therefore, in some sense, a (non-delegatable) DVS is a more "complex" notion than a conventional signature scheme.

It is not difficult to show that the DVS scheme from [JSI96] is secure—more precisely, unforgeable, non-delegatable, computationally non-transferable and securely disavowable—after a trivial fix of just adding some additional variables under the used hash value, by following the usual proof of knowledge methodology. We do not present these proofs in this paper: while it is straightforward to prove these results, the corresponding proofs do not really give an insight to the just pointed out difference between unforgeability and non-delegatability.

Instead, we propose a new DVS scheme, DVS-KW, based on the provably secure signature scheme of Katz and Wang [KW03], where the signer presents a designated-verifier proof that his public key is a Decisional Diffie-Hellman (DDH) tuple. We prove that DVS-KW is unforgeable by providing a tight reduction to the underlying cryptographic problem (DDH) in the non-programmable random oracle (NPRO) model. The NPRO model is is known to be strictly weaker than the random oracle (RO) model [Nie02] and thus the unforgeability of DVS-KW in the NPRO model is interesting by itself, especially since the unforgeability proof of the original Katz-Wang signature scheme relies heavily on the programmability of the random oracle. We also prove non-delegatability of DVS-KW, though this proof is in the programmable random oracle model and has a larger security degradation due to the involved proof-of-knowledge property. More precisely, we show that if some forger can create valid signatures with probability $\varepsilon > \kappa$ where κ is the knowledge error, then there exists a knowledge extractor that extracts one of the two secret keys in time, dominated by $56/\kappa$ oracle queries to the forger.

DVS-KW can be seen as a proof of concept, showing how to design DVS schemes that have a tight reduction in the unforgeability proof and are still non-delegatable. Moreover, DVS-KW is more efficient than the JSI scheme from [JSI96], and DVS-KW does not allow the signer to disavow simulated signatures; the latter property makes DVS-KW attractive in many applications. (Recall also that we broke the disavowability of the JSI scheme.) At this moment, the most efficient secure disavowable designated-verifier signature scheme seems to be the corrected JSI DVS scheme, while the most efficient secure designated-verifier signature scheme seems to be the DVS-KW scheme.

We also show the existence of an efficient conventional signature scheme that is unforgeable under a tight reduction to the Decisional Diffie-Hellman problem in the NPRO+CRS (common reference string) model. In this model, all parties will additionally have access to the common reference string that corresponds to Desmond's public key in DVS-KW. The importance of this result is that signature schemes, secure in the plain model, are considerably slower than this scheme and/or are secure under incompatible and often less studied assumptions. Therefore, if one wants to avoid the programmable random oracle model—where one can construct very efficient signature schemes—one might want to use the new scheme.

2 Preliminaries

Let $\mathcal{G}_q$ be a finite, cyclic group of prime order q in which the group operation is represented multiplicatively; furthermore, let g be a generator of $\mathcal{G}$. The most common setting is as follows: let p, q be two large primes such that $q|(p-1)$, then $\mathcal{G}_q$ a multiplicative subgroup of $\mathbb{Z}_p^*$ of order q, and g a generator of $\mathcal{G}_q$. Other settings (e.g., using elliptic curves) are possible. A distinguishing algorithm $\mathcal{A}$ is said to (τ, ε)-break DDH (Decisional Diffie-Hellman) in group $\mathcal{G}_q$ if $\mathcal{A}$ runs in time at most τ and furthermore $\mathrm{Adv}_\mathcal{G}^{\mathrm{ddh}}(\mathcal{A}) := |\Pr[x, y, z \leftarrow_r \mathbb{Z}_q : \mathcal{A}(g, g^x, g^y, g^z) = 1] - \Pr[x, y \leftarrow_r \mathbb{Z}_q : \mathcal{A}(g, g^x, g^y, g^{xy}) = 1]| \geq \varepsilon$, where the probability is taken over the choice of random variables and the coin tosses of $\mathcal{A}$. We say that $\mathcal{G}$ is a (τ, ε)-DDH group if no algorithm (τ, ε)-breaks DDH in $\mathcal{G}$.

A *designated-verifier signature scheme* [JSI96] is a tuple of probabilistic algorithms (Gen, Sign, Simul, Vrfy) over a message space $\mathcal{M}$, such that: (a) The key-generation algorithm Gen outputs a public key pk and a secret key sk; (b) The signing algorithm Sign takes as input signer's secret key sk_S, designated verifier's public key pk_D and a message $m \in \mathcal{M}$ and returns signature σ; (c) The simulation algorithm Simul takes as input signer's public key pk_S, designated verifier's secret key sk_D and a message $m \in \mathcal{M}$ and returns signature σ; (d) Verification algorithm Vrfy takes as input signer's public key pk_S, designated verifier's public key pk_D, a message $m \in \mathcal{M}$, and a signature σ and returns accept or reject. In some of the existing designated verifier schemes, the verification algorithm must have access to the designated verifier's secret key sk_D. We call such a designated-verifier signature scheme *privately verifiable*. We make the standard correctness requirement: for all $(\mathsf{sk}_S, \mathsf{pk}_S)$ and $(\mathsf{sk}_D, \mathsf{pk}_D)$ output by Gen and for all $m \in \mathcal{M}$ we have $\mathsf{Vrfy}_{\mathsf{pk}_S, \mathsf{pk}_D}(\mathsf{Sign}_{\mathsf{sk}_S, \mathsf{pk}_D}(m)) = \mathsf{Vrfy}_{\mathsf{pk}_S, \mathsf{pk}_D}(\mathsf{Sign}_{\mathsf{sk}_D, \mathsf{pk}_S}(m)) =$ accept. We say that a signature σ is *valid* if $\mathsf{Vrfy}_{\mathsf{pk}_S, \mathsf{pk}_D}(\sigma) = $ accept.

3 Previous DVS Schemes and Their Security

Jakobsson-Sako-Impagliazzo Disavowable DVS Scheme [JSI96]. Let p, q and $\mathcal{G}_q$ be as described in Sect. 2. Assume that Signy and Desmond have the Diffie-Hellman key pairs $(x_S, y_S = g^{x_S} \mod p)$ and $(x_D, y_D = g^{x_D} \mod p)$, respectively. Assume that H_q is a random oracle mapping to $\mathbb{Z}_q$. (Note: If $m \notin \mathbb{Z}_p$ then m must be hashed by using a full-domain hash, modelled by a random oracle. We will ignore this issue throughout this paper.) In $\mathsf{Sign}_{\mathsf{sk}_S, \mathsf{pk}_D}(m)$, Signy sets $s \leftarrow m^{x_S} \mod p$, selects three random numbers $w, t, r \leftarrow_r \mathbb{Z}_q$, and computes $G \leftarrow g^r \mod p$, $M \leftarrow m^r \mod p$, $h \leftarrow H_q(g^w y_D^t \mod p, G, M)$ and $z \leftarrow r + (h + w)x_S \mod q$. Then, Signy sends the signature $\sigma := (s, P)$, where $P = (w, t, G, M, z)$, to the designated verifier, Desmond. In $\mathsf{Simul}_{\mathsf{sk}_D, \mathsf{pk}_S}(m, s)$, by selecting three random numbers $z, \alpha, \beta \leftarrow_r \mathbb{Z}_q$, Desmond creates $P = (w, t, G, M, z)$, for any message m and any $s \leftarrow \mathcal{G}_q$, as follows: $(G, M) \leftarrow (g^z y_S^{-\beta} \mod p, m^z s^{-\beta} \mod p)$, $h \leftarrow H_q(g^\alpha \mod p, G, M)$, $w \leftarrow \beta - h \mod q$, $t \leftarrow (\alpha - w)x_D^{-1} \mod q$. He sets $\sigma \leftarrow (s, P)$. In $\mathsf{Vrfy}_{\mathsf{pk}_S, \mathsf{pk}_D}(m; s, w, t, G, M, z)$, the verifier computes $h \leftarrow H_q(g^w y_D^t \mod p, G, M)$ and checks whether $G = g^z y_S^{-(h+w)} \mod p$ and $M = m^z s^{-(h+w)} \mod p$. The JSI scheme can be made more communication-efficient by transferring h (instead of G and M) to Desmond. Then the verifier must check that $h = H_q(g^w y_D^t \mod p, g^z \cdot y_S^{-(h+w)} \mod p, m^z \cdot s^{-(h+w)} \mod p)$. This version is security-wise equivalent to the original scheme but somewhat more efficient.

Our Attack: First, a honest Signy generates valid signatures only for $s = m^{x_S}$ while Desmond can generate valid signatures for any $s \in \mathbb{Z}_p^*$. That is, knowing x_S, a honest Signy generates valid designated-verifier proof P only for $s = m^{x_S}$, while knowing x_D, Desmond generates valid designated-verifier proofs P for any $s \in \mathbb{Z}_p^*$. Thus it suffices to have a disavowal where Signy proves in non-interactive zero-knowledge that $s \neq m^{x_S}$. Next, we show that Signy can also compute valid signatures for any $\bar{s} \in \mathbb{Z}_p^*$, therefore, Signy can create signatures from the same distribution as Desmond and thus, the JSI scheme is perfectly non-transferable. This means that there exists no disavowal protocol for the JSI scheme at all.

Here is how Signy does it. Signy computes a signature $(\bar{s}; w, t, G, \bar{M}, z)$ for a message m, with $\bar{s} \neq m^{x_S}$, as follows. She selects four random numbers $w, t, r, \bar{r} \leftarrow_r \mathbb{Z}_q$ and then sets $c \leftarrow g^w y_D^t \mod p$, $G \leftarrow g^r \mod p$, $\bar{M} \leftarrow m^{\bar{r}} \mod p$, $h \leftarrow H_q(c, G, \bar{M})$, $z \leftarrow r + (h + w)x_S \mod q$ and $\bar{s} \leftarrow m^{x_S} \cdot m^{(r-\bar{r})/(h+w)} \mod q$ mod p. After that, Signy sends the message-signature pair $(m, \bar{s})$ with $\bar{\sigma} = (\bar{s}, P = (w, t, G, \bar{M}, z))$, to Desmond. Clearly, $\mathsf{Vrfy}_{\mathsf{pk}_S, \mathsf{pk}_D}(m, \bar{\sigma}) = $ accept so Desmond will believe that $\bar{s}$ is Signy's signature for message m. In later disputes, however, Signy can convince a third party (e.g., a judge) that $\bar{s}$ was simulated by Desmond, by using a standard disavowal protocol to show that $\log_g y_S \neq \log_m \bar{s}$. This attack does not result in a signature forgery, it just shows that the JSI scheme is not disavowable.

There are two intuitive countermeasures to avoid this attack. First, Signy provides an additional proof of knowledge that $\log_m M = \log_g G$. However, this increases the signature length. Second, include s (together with pk_S and pk_D) to the input of the hash function. This turns out to be is sufficient, related discussion can be found in Sect. 6.

Saeednia-Kremer-Markowitch Privately Verifiable DVS Scheme [SKM03]. Let p, q and $\mathcal{G}_q$ be as defined in Sect. 2. In the SKM scheme, Signy and Desmond have the Diffie-Hellman key pairs $(x_S, y_S = g^{x_S} \bmod p)$ and $(x_D, y_D = g^{x_D} \bmod p)$, respectively. Assume that $H_q(\cdot)$ is a random oracle mapping to $\mathbb{Z}_q$. In $\mathsf{Sign}_{\mathsf{sk}_S, VK_D}(m)$, Signy selects two random numbers $k \leftarrow_r \mathbb{Z}_q$, $t \leftarrow_r \mathbb{Z}_q^*$, and then computes the signature $\sigma = (h, d, t)$ by setting $c \leftarrow y_D^k \bmod p$, $h \leftarrow H_q(m, c)$ and $d \leftarrow kt^{-1} - h \cdot x_S \bmod q$. In $\mathsf{Simul}_{\mathsf{sk}_D, \mathsf{pk}_S}(m)$, Desmond picks two random numbers $\bar{d} \leftarrow_r \mathbb{Z}_q$, $\bar{r} \leftarrow_r \mathbb{Z}_q^*$, and then computes $\sigma = (h, d, t)$ as follows: for $c \leftarrow g^{\bar{d}} y_S^{\bar{r}} \bmod p$, set $h \leftarrow H_q(m, c)$, $d \leftarrow h \cdot \bar{d} \cdot \bar{r}^{-1} \bmod q$ and $t \leftarrow \bar{r} \cdot (x_D \cdot h)^{-1} \bmod q$. In $\mathsf{Vrfy}_{\mathsf{pk}_S, \mathsf{sk}_D}(m; h, d, t)$, the verifier accepts iff $h = H_q(m, (g^d y_S^h)^{t x_D} \bmod p)$.

Our attack: We show that the knowledge of $y_{SD} := g^{x_S \cdot x_D} \bmod p$ is sufficient to generate both a valid signature and to verify it, and therefore, this scheme is delegatable (both in the sense of signing and verifying). On the one hand, given y_{SD}, one can verify whether a message-signature pair (m, h, d, t) is valid for the designated verifier Desmond by checking whether $h = H_q(m, (y_D^d y_{SD}^h)^t \bmod p)$. On the other hand, given y_{SD}, anybody can produce a valid designated-verifier signature $\sigma = (h, d, t)$ for any message m by selecting two random numbers $\bar{d} \leftarrow_r \mathbb{Z}_q$, $\bar{r} \leftarrow_r \mathbb{Z}_q^*$, and then setting $c \leftarrow y_D^{\bar{d}} y_{SD}^{\bar{r}} \bmod p$, $h \leftarrow H_q(m, c)$, $d \leftarrow h \cdot \bar{d} \cdot \bar{r}^{-1} \bmod q$ and $t \leftarrow \bar{r} h^{-1} \bmod q$. The resulting signature $\sigma = (h, d, t)$ is accepted by the verifier, since $(g^d y_S^h)^{t x_D} \bmod p = (g^{h \cdot \bar{d} \cdot \bar{r}^{-1}} y_S^h)^{\bar{r} \cdot h^{-1} x_D} = y_D^{\bar{d}} y_{SD}^{\bar{r}} \bmod p = c$. In particular, this means that this scheme is perfectly non-transferable. (The authors of [SKM03] get the same result by a complicated analysis of probability theory.) It might again seem that one can remedy this situation by including t under the hash. However, in the simulation, Desmond needs to choose t after fixing h, and therefore this fix does not work.

Steinfeld-Wang-Pieprzyk DVS. Assume again that p, q and $\mathcal{G}_q$ are as defined in Sect. 2. At least the first privately-verifiable DVS scheme, $\mathsf{SchUDVS}_1$, from [SWP04] is also delegatable. Recall that there, a designated-verifier signature of message m is equal to (r, u, K), for $r \leftarrow H_q(m, u)$, and the verifier accepts only when $K = (u \cdot y_S^r)^{x_D} \bmod p$. But if Signy publishes $y_{SD} \leftarrow y_D^{x_S}$, anybody can produce valid signatures by setting $u \leftarrow g^k \bmod p$ and $K \leftarrow y_D^k \cdot y_{SD}^r \bmod p$ for random $k \leftarrow_r \mathbb{Z}_q$. Thus, also this scheme is delegatable.

Steinfeld-Bull-Wang-Pieprzyk and Laguillaumie-Vergnaud DVS. The SBWP designated-verifier signature scheme from [SBWP03] is based on a group pair $(\mathcal{G}_1, \mathcal{G}_2)$ with bilinear pairing $\langle \cdot, \cdot \rangle$ from $\mathcal{G}_1^2$, with $\mathrm{ord}\,\mathcal{G}1 = q$, to $\mathcal{G}_2$. (See [SBWP03] for more details.) Let m be the message to be signed, $H_{\mathcal{G}_1}$—a random oracle with outputs from $\mathcal{G}_1$, and let g be a generator of $\mathcal{G}_1$. Assume that Signy's key-pair is $(x_S, y_S = g^{x_S})$

and that Desmond's key-pair is $(x_D, y_D = g^{x_D})$ for randomly chosen $x_S, x_D \leftarrow \mathbb{Z}_q$. Signy's designated-verifier signature on a message m is $\sigma \leftarrow \langle H_{\mathcal{G}_1}(m)^{x_S}, y_D \rangle$. Desmond simulates the signature by setting $\sigma \leftarrow \langle H_{\mathcal{G}_1}(m)^{x_D}, y_S \rangle$. The signature is verified by checking that $\sigma = \langle H_{\mathcal{G}_1}(m)^{x_D}, y_S \rangle$. Again, if the value of $y_{SD} = g^{x_S x_D} \bmod q$ is compromised, any entity who gets to know y_{SD} can also produce a valid σ by computing $\sigma \leftarrow \langle H_{\mathcal{G}_1}(m), y_{SD} \rangle$. Therefore, this scheme is also delegatable. The Laguillaumie-Vernaud [LV04] DVS scheme is delegatable for the same reason.

4 DVS Security Definitions

In the original paper [JSI96], a designated-verifier signature was required to convince the designated verifier Desmond that the signer Signy has signed the message, in a way that Desmond is able to simulate signature that is indistinguishable from the real signatures, even by a verifier who has access to the designated verifier's private key. A more formal definition was given in say [SBWP03]. We will first repeat formal definitions of unforgeability and non-transferability and then give a definition of the new notion, non-delegatability.

A designated-verifier signature scheme must satisfy at least the next two conditions: (a) Unforgeability: signatures are verifiable by the designated verifier Desmond who rejects it when the signature was not signed by himself or Signy, and (b) Non-transferability: given a message-signature pair (m, σ), that is accepted by the designated-verifier, and without access to the secret key of the signer, it is computationally infeasible to determine whether the message was signed by the signer, or the signature was simulated by the designated verifier.

In the following, Ω denotes the space from which the random oracle H is selected; definition without a random oracle is analogous. Depending on the situation, we will have either $\Omega = \Omega_{\text{npro}}$ to be the set of all non-programmable random oracles [Nie02] or $\Omega = \Omega_{\text{ro}}$ to be the set of all random oracles with proper input and output domains.

Let $\Delta = (\text{Gen}, \text{Sign}, \text{Simul}, \text{Vrfy})$ be a designated-verifier signature scheme with the message space $\mathcal{M}$. We say that Δ is *perfectly non-transferable* if $\text{Sign}_{\text{sk}_S, \text{pk}_D}(m) = \text{Simul}_{\text{sk}_D, \text{pk}_S}(m)$ as distributions for every $(\text{pk}_S, \text{sk}_S) \leftarrow \text{Gen}$, $(\text{pk}_D, \text{sk}_D) \leftarrow \text{Gen}$, $H_q \leftarrow \Omega$ and $m \leftarrow \mathcal{M}$. Analogously, one can define statistically non-transferable and computationally non-transferable schemes. An adversial forging algorithm $\mathcal{F}$ is said to $(\tau, q_h, q_s, \varepsilon)$-*forge* Δ if $\mathcal{F}$ runs in time at most τ, makes at most q_h hash queries and in total at most q_s signing and simulation queries, and furthermore

$$\text{Adv}_\Delta^{\text{forge}}(\mathcal{F}) := \Pr \begin{bmatrix} (\text{pk}_S, \text{sk}_S) \leftarrow \text{Gen}; (\text{pk}_D, \text{sk}_D) \leftarrow \text{Gen}; H \leftarrow \Omega; \\ (m, \sigma) \leftarrow \mathcal{F}^{\text{Sign}_{\text{sk}_S, \text{pk}_D}(\cdot), \text{Simul}_{\text{sk}_D, \text{pk}_S}(\cdot), H(\cdot)}(\text{pk}_S, \text{pk}_D) : \\ \sigma \notin \Sigma(m) \wedge \text{Vrfy}_{\text{pk}_S, \text{pk}_D}(m, \sigma) \neq \text{reject} \end{bmatrix} \geq \varepsilon ,$$

where $\Sigma(m)$ is the set of signatures received either from $\text{Sign}_{\text{sk}_S, \text{pk}_D}(m)$ or from $\text{Simul}_{\text{sk}_D, \text{pk}_S}(m)$. A designated-verifier signature scheme is $(\tau, q_h, q_s, \varepsilon)$-*unforgeable* if no forger can $(\tau, q_h, q_s, \varepsilon)$-forge it. In the case a DVS scheme is only computationally non-transferable, it is important that $\Sigma(m)$ also includes signatures received from

$\mathsf{Simul}_{\mathsf{sk}_D,\mathsf{pk}_S}(m)$. If a scheme is perfectly non-transferable then an access to the Simul oracle will not help the forger.

Delegatability. The definition of unforgeability does not cover the case when the signer, without disclosing her secret key sk_S, delegates her signing rights (w.r.t. to a concrete designated verifier Desmond) to $\mathcal{F}$ by disclosing some side information $y_{SD} := f_S(\mathsf{sk}_S, \mathsf{pk}_D)$, that helps the latter to produce valid signatures. Analogously, the designated verifier might delegate his signature simulating capability by disclosing some—potentially different—side information $y'_{SD} := f_D(\mathsf{sk}_D, \mathsf{pk}_S)$. This implies that when Desmond sees a valid signature σ that is not generated by himself, he can only conclude that σ is generated by somebody who knows either y_{SD} or y'_{SD}. In some scenarios, Signy may reveal or be forced to reveal the value of y_{SD} to a third party Trevor so that Trevor can generate valid Signy's signatures for Desmond. Such a delegation is essentially different from the situation where Signy reveals her secret key x_S to Trevor, since knowledge of y_{SD} allows Trevor to sign messages designated only to Desmond, and not to anybody else. Therefore, Signy might be more willing to give out the value y_{SD} than her secret key sk_S. This is not an issue in the case of conventional signature schemes where non-delegatability follows from unforgeability.

We will next give a longer explanation why delegatability is bad. First of all, in the original definition of designated-verifier proofs [JSI96], it was said that a proof of the truth of some statement Φ is a designated-verifier proof if it is a proof that either Φ is true or the signer knows Desmond's secret key. This intuitive requirement is clearly not satisfied by delegatable DVS schemes, where a signer proves that either Φ is true or she knows y_{SD} or she knows y'_{SD}.

Moreover, delegatability is undesirable in many applications of the designated-verifier signature scheme. We will give two examples. First, in an hypothetical e-voting protocol where voters sign messages by using a designated-verifier signature scheme (with the tallier being the designated verifier Desmond), knowing that this information can only be used to vote in this concrete election, a voter Signy could be motivated to give a copy of y_{SD} to the coercer for a moderate sum of money. On the other hand, since Signy might use sk_S in many other applications, she might not be willing to send sk_S to the coercer for any imaginable "moderate" amount of money. Second, assume that Signy is a subscriber to an e-library, maintained by Desmond, and that she identifies herself by using a designated-verifier signature scheme so that Desmond could not sell Signy's access history to third parties. If the DVS scheme is delegatable, Signy could however send y_{SD} to a non-subscriber who could then also start enjoying Desmond's service. Since y_{SD} is used only in this application, Signy could be happily willing to do that. On the other hand, if the DVS were not delegatable, Signy would have to reveal her secret key. Finally, it may happen that Signy and Desmond also participate in some other protocols where y_{SD} or y'_{SD} is revealed legitimately and thus the attacker can gain access to either of these values. Note also that a DVS scheme with delegatability is somewhat similar to the proxy signatures, except that in the case of proxy signatures, (a) the verifier can distinguish between messages, signed by Signy and a proxy, and (b) the signatures are universally verifiable.

The preceding discussion motivates the next definition. It basically says that a non-delegatable DVS scheme is a non-interactive system of proofs of knowledge of either

sk_S or sk_D. Here, $\mathcal{F}_m$ denotes $\mathcal{F}$ with m as its input, and oracle calls are counted as one step. More precisely, let $\kappa \in [0,1]$ be the knowledge error. We say that Δ is (τ, κ)-*non-delegatable* if there exists a black-box knowledge extractor $\mathcal{K}$ that, for every algorithm $\mathcal{F}$ and for every valid signature σ, satisfies the following condition: For every $(\mathsf{pk}_S, \mathsf{sk}_S) \leftarrow \mathsf{Gen}$, $(\mathsf{pk}_D, \mathsf{sk}_D) \leftarrow \mathsf{Gen}$ and message m, if $\mathcal{F}$ produces a valid signature on m with probability $\varepsilon > \kappa$ then, on input m and on access to the oracle $\mathcal{F}_m$, $\mathcal{K}$ produces either sk_S or sk_D in expected time $\frac{\tau}{\varepsilon - \kappa}$ (without counting the time to make the oracle queries). (Here, $\mathcal{F}$'s probability is taken over the choice of her random coins and over the choice of $H_q \leftarrow \Omega$.)

5 New DVS Scheme with Tight Reduction to DDH in NPRO

Let p, q and $\mathcal{G}_q$ be as defined in Sect. 2. Let $g_1, g_2 \in \mathcal{G}_q$ be two elements such that nobody knows the mutual discrete logarithms of g_1 and g_2. Following the ideas from [KW03], in the next DVS-KW DVS scheme, we let Signy to prove Desmond that $(g_1, g_2, y_{1S}, y_{2S})$ is a Decisional Diffie-Hellman tuple, where $x_i \leftarrow_r \mathbb{Z}_q$ is i's private key and $\mathsf{pk}_i := (g_1, g_2, y_{1i}, y_{2i})$ is i's public key with $y_{1i} = g_1^{x_i}$ and $y_{2i} = g_2^{x_i}$. This proof is made designated-verifier by using the same trick as in the JSI scheme [JSI96], and non-interactive by using a non-programmable random oracle [Nie02] H_q with outputs from $\mathbb{Z}_q$. In particular, the random oracle H_q can be chosen at the same stage as other system parameters, g_1 and g_2. This is an interesting result by itself since in [KW03], H_q had to be a *programmable* random oracle for their security proof to go through. The description of the full DVS-KW scheme follows:

$\mathsf{Sign}_{\mathsf{sk}_S, \mathsf{pk}_D}(m)$: Signy generates random $r, w, t \leftarrow \mathbb{Z}_q$, and sets $a_1 \leftarrow g_1^r \bmod p$, $a_2 \leftarrow g_2^r \bmod p$, $c \leftarrow g_1^w y_{1D}^t \bmod p$, $h \leftarrow H_q(\mathsf{pk}_S, \mathsf{pk}_D, a_1, a_2, c, m)$ and $z \leftarrow r + (h+w)x_S \bmod q$. She outputs the signature $\sigma \leftarrow (w, t, h, z)$.

$\mathsf{Simul}_{\mathsf{sk}_D, \mathsf{pk}_S}(m)$: By selecting three random numbers $z, \alpha, \beta \leftarrow_r \mathbb{Z}_q$, Desmond creates $\sigma = (w, t, h, z)$ for any message m as follows: $(a_1, a_2) \leftarrow (g_1^z y_{1S}^{-\beta} \bmod p, g_2^z y_{2S}^{-\beta} \bmod p)$, $h \leftarrow H_q(\mathsf{pk}_S, \mathsf{pk}_D, a_1, a_2, g_1^\alpha \bmod p, m)$, $w \leftarrow \beta - h \bmod q$, $t \leftarrow (\alpha - w)x_D^{-1} \bmod q$.

$\mathsf{Vrfy}_{\mathsf{pk}_S, \mathsf{pk}_D}(m; w, t, h, z)$: The verifier checks whether $h = H_q(\mathsf{pk}_S, \mathsf{pk}_D, g_1^z y_{1S}^{-(h+w)} \bmod p, g_2^z y_{2S}^{-(h+w)} \bmod p, g_1^w y_{1D}^t \bmod p, m)$.

Security: First, clearly this scheme is correct and perfectly non-transferable. Second, we can prove the next result that is a generalisation of a proof from [KW03]:

Theorem 1. *Let $\mathcal{G}$ be a (τ', ε')-DDH group with $|\mathcal{G}| = q$ such that exponentiation in $\mathcal{G}$ takes time $t_{\exp}$. Then the above designated-verifier signature scheme is $(\tau, q_h, q_s, \varepsilon)$-unforgeable in the non-programmable random oracle model, where $\tau \leq \tau' - (3.2q_s + 5.6)t_{\exp}$ and $\varepsilon \geq \varepsilon' + q_s q_h q^{-2} + q^{-1} + q_h q^{-2}$.*

Proof. Assume that we have an algorithm $\mathcal{F}$, running in time at most τ and making at most q_h hash queries and in total at most q_s signing and simulation queries, which forges a new message/signature or a new message/simulated signature pair with probability at least ε. We use $\mathcal{F}$ to construct an algorithm $\mathcal{A}$ running in time τ' which solves the DDH problem with probability ε'. The stated result follows.

Algorithm $\mathcal{A}$ is given as input a tuple $(g_1, g_2, y_{1D}, y_{2D})$; its goal is to determine whether this represents a "random tuple" or a "DDH tuple". To this end, it sets $\mathsf{pk}_D = (g_1, g_2, y_{1D}, y_{2D})$, sets Signy's public key pk_S to be equal to a random DDH tuple $\mathsf{pk}_S = (g_1, g_2, y_{1S} = g_1^{x_S}, y_{2S} = g_2^{x_S})$ for which she chooses the corresponding secret key $x_S \leftarrow \mathbb{Z}_q$, and runs $\mathcal{F}$ on input $(\mathsf{pk}_S, \mathsf{pk}_D)$. Algorithm $\mathcal{A}$ simulates the signing oracle and the random oracle for $\mathcal{F}$ as follows:

Hash queries. In response to a query $H_q(\mathsf{pk}_S, \mathsf{pk}_D, a_1, a_2, c, m)$, algorithm $\mathcal{A}$ responds with $h \leftarrow H_q(\mathsf{pk}_S, \mathsf{pk}_D, a_1, a_2, c, m)$. Additionally, if this query was not made before, $\mathcal{A}$ stores the tuple $(\mathsf{pk}_S, \mathsf{pk}_D, a_1, a_2, c, m)$.

Signing and simulation queries. If $\mathcal{F}$ asks either for a signature or a simulation on message m, then $\mathcal{A}$ attempts to sign pk_S. That is, $\mathcal{A}$ chooses random $r, w, t \leftarrow \mathbb{Z}_q$, sets $a_1 \leftarrow g_1^r \bmod p$, $a_2 \leftarrow g_2^r \bmod p$ and $c \leftarrow g_1^w y_{1D}^t \bmod p$. If H_q had previously been queried on input $H_q(\mathsf{pk}_S, \mathsf{pk}_D, a_1, a_2, c, m)$ then $\mathcal{A}$ aborts (with output 0); otherwise, $\mathcal{A}$ sets $h \leftarrow H_q(\mathsf{pk}_S, \mathsf{pk}_D, a_1, a_2, c, m)$ and outputs the signature (w, t, h, z), where $z \leftarrow r + (h + w) x_S \bmod q$.

At some point, $\mathcal{F}$ outputs its forgery $(\bar{m}, \bar{\sigma} = (\bar{w}, \bar{t}, \bar{h}, \bar{z}))$, where we assume that $\bar{\sigma}$ was not previously the response to a query $\mathsf{Sign}_{\mathsf{sk}_S}(\bar{m})$ or $\mathsf{Simul}_{\mathsf{sk}_D}(\bar{m})$. Set $\bar{a}_1 \leftarrow g_1^{\bar{z}} y_{1S}^{-(\bar{h}+\bar{w})} \bmod p$, $\bar{a}_2 \leftarrow g_2^{\bar{z}} y_{2S}^{-(\bar{h}+\bar{w})} \bmod p$, $\bar{c} \leftarrow g_1^{\bar{w}} y_{1D}^{\bar{t}} \bmod p$. Now, if $H_q(\mathsf{pk}_S, \mathsf{pk}_D, \bar{a}_1, \bar{a}_2, \bar{c}, \bar{m}) = \bar{h}$ (i.e., $\mathsf{Vrfy}_{\mathsf{pk}}(\bar{m}, \bar{\sigma}) = 1$), then $\mathcal{A}$ outputs 1; otherwise, $\mathcal{A}$ outputs 0.

We now analyse the probability that $\mathcal{A}$ outputs 1. If $(g_1, g_2, y_{1D}, y_{2D})$ is a Diffie-Hellman tuple, then $\mathcal{A}$ provides a perfect simulation for $\mathcal{F}$ except for the possibility of an abort. An abort can occur during $\mathcal{A}$'s simulation of any of the signing queries; in the simulation of any particular signing query, it is not hard to see that the probability of abort is at most q_h/q^2. Thus the overall probability that $\mathcal{A}$ aborts is at most $q_s q_h/q^2$. This means that $\mathcal{A}$ outputs a forgery (and hence $\mathcal{A}$ outputs 1) with probability at least $\varepsilon - q_s q_h/q^2$. On the other hand, if $(g_1, g_2, y_{1D}, y_{2D})$ is a random tuple then with probability $1 - q^{-1}$ it is not a Diffie-Hellman tuple. In this case, for any query $H_q(\mathsf{pk}_S, \mathsf{pk}_D, a_1, a_2, c, m)$ made by $\mathcal{F}$ there is at most q possible values of (w, t, z) such that the verification succeeds. Thus, $\mathcal{F}$ outputs a forgery (and hence $\mathcal{A}$ outputs 1) with probability at most $q^{-1} + q_h q^{-2}$. Putting everything together, we see that $\mathsf{Adv}_{\mathcal{G}}^{\mathsf{ddh}}(\mathcal{A}) \geq \varepsilon - q_s q_h q^{-2} - q^{-1} - q_h q^{-2} = \varepsilon'$. The running time of $\mathcal{A}$ includes the running time of $\mathcal{F}$ and is otherwise dominated by two exponentiations and one multi-exponentiation that are performed for each query to the signing oracle plus those done in generating Signy's key and verifying the output of $\mathcal{F}$. Assuming as in [KW03] that a two-exponent multi-exponentiation takes time $1.2 t_{\mathsf{exp}}$ gives the result of the theorem. □

On NPRO Versus RO Model. The non-programmable random oracle model is known to be strictly weaker than the random oracle model [Nie02]. This is not surprising looking at the corresponding security proofs, e.g., at the security proof in [KW03] and at the proof of Thm. 1. In the former, the adversary does not know the secret key, and therefore is forced to program the random oracle to be able to answer successfully to the signature queries. In the latter, the adversary knows Signy's secret key, and knowing this, can answer successfully to the signature and simulation queries without a need to program the random oracle. A conceptual difference between the two models is that proofs in the RO model work for the "best case" (showing that for every forger, there exists a function H_q such that the signature scheme is unforgeable), while proofs in the NPRO model work for the "average case" (showing that the signature scheme is unforgeable for a randomly chosen function $H_q \leftarrow \Omega_{\mathsf{npro}}$, independent of the forger). Given such arguments, we think that unforgeability in the NPRO model is an important property of DVS-KW.

Conventional Signature Scheme with Tight Reduction to DDH in the NPRO+CRS Model. From DVS-KW, one can build a conventional signature scheme with the same properties (tight reduction to the DDH problem in the non-programmable random oracle model) if one additionally assumes the common reference string (CRS) model. More precisely, in this case, the CRS is equal to a pair (y_{1D}, y_{2D}), drawn randomly from the set $\{(g_1^{x_D}, g_2^{x_D}) : x_D \in \mathbb{Z}_q\}$. It is assumed that nobody will use the corresponding secret key to simulate a DVS; in our case, this is easily achieved since CRS can be a random DDH tuple. On the other hand, for the proof to go through, $\mathcal{A}$ is allowed to generate the CRS herself. This scheme is about twice less efficient than the Katz-Wang scheme, but its security does not rely on programmable random oracles. On the other hand, this scheme is more efficient than the currently known signature schemes that are unforgeable in the plain model, and/or has a more standard underlying assumption.

Non-delegatability of DVS-KW. Next, we will give a proof of the non-delegatability of DVS-KW. Because of the structure of our proof, we expect that an arbitrary DVS scheme would have a similar time bound in its security proof.

Theorem 2. *Assume that for some $m \in \mathcal{M}$, $\mathcal{F}$ can produce valid signatures in time τ and with probability ε. Then DVS-KW is $(56\tau/\varepsilon, \frac{1}{q})$-non-delegatable in the programmable random oracle model.*

Proof. Assume that $\varepsilon > \kappa = 1/q$. We must show there exists a knowledge extractor $\mathcal{K}$ that on input σ and on black-box oracle access to $\mathcal{F}$ can produce either sk_S or sk_D in expected time $\tau' \leq 56\tau/\varepsilon$ and with probability 1.

Let $\mathcal{F}_m$ be a forger with input m. Consider two executions of $\mathcal{F}_m$ by $\mathcal{K}$ on the same random input of $\mathcal{F}_m$. In both cases, $\mathcal{K}$ executes $\mathcal{F}_m$ step-by-step, except that $\mathcal{K}$ returns different random values (h versus h') as the answer to the hash query $H_q(\mathsf{pk}_S, \mathsf{pk}_D, a_1, a_2, c, m)$. Since (a_1, a_2, c) are under the hash, their values must be equal in both runs. If both signatures are valid, one must have $z - (h+w)x_S \equiv z' - (h'+w')x_S \mod q$ and $w + tx_D \equiv w' + t'x_D \mod q$, where (w, t, h, z) is the first signature and (w', t', h', z'), $h \neq h'$, is the second signature. Now we have two cases. If $w \not\equiv w' \mod q$ then one can find $x_D \leftarrow (w-w')/(t'-t) \mod q$. If $w \equiv w' \mod q$ then $h + w \not\equiv h' + w' \mod q$ and thus one can find $x_S \leftarrow (z-z')/(h-h'+w-w') \mod q$. Now, assume Rewind is an algorithm that, given an oracle access to $\mathcal{F}_m$, in time τ_R, for some τ_R, produces two different valid signatures (w, t, h, z) and (w', t', h', z') on m, such that $h \neq h'$ but $(a_1, a_2, c) = (a'_1, a'_2, c')$. Then one can compute either x_S

or x_D with probability 1. Thus, given that algorithm Rewind runs in expected time $56/\varepsilon$ (counting every access to $\mathcal{F}_m$ as one step), we have proven the theorem.

Next, let us describe the algorithm Rewind. We are given a forger $\mathcal{F}_m$ who returns a signature (w, t, h, z) that may or may not be correct. We are given that the probability of a correct answer taken over $\mathcal{F}_m$ coins and the choice of h is at least ε. We want to find correct answers to two different h-values for a given (a_1, a_2, c, m) as efficiently as possible. The idea is to run the prover, and use rewinding to try to make him answer two different challenges correctly. But to run him, we need to supply random coins. Although we know that the average success probability is ε, we do not know that $\mathcal{F}_m$ is equally successful with any random input. To get a better view of this, let H be a matrix with a row for each possible set of random coins for $\mathcal{F}_m$, and one column for each possible challenge value. Write 1 in an entry if $\mathcal{F}_m$ answers correctly with the corresponding random choices and challenge, and 0 otherwise. Using $\mathcal{F}_m$ as black-box, we can probe any entry we want in H, and our goal can be rephrased to: find two 1's in the same row. What we know is that ε equals the fraction of 1-entries in H. Now, Rewind uses an algorithm from [DF02] to find such 1 entries in time $56/\varepsilon$. □

6 On Disavowability

In a few proposed designated-verifier signature schemes [JSI96], given a pair (m, σ), Signy and Desmond cannot prove to a third party, even when they join their forces, which one of them generated σ. In some other schemes—that we call *disavowable*—Signy can prove that (a) she signed messages that she really signed, and (b) she has not signed signatures, simulated by Desmond. This is achieved by decomposing the signature σ of a message m in two non-empty parts, s ("undeniable signature") and P ("designated-verifier proof"), where for every message m and for every possibly incorrect $\bar{s}$, the designated verifier can produce a simulated proof $\bar{P}$ such that the distribution of $(\bar{s}, \bar{P})$ is computationally indistinguishable from the distribution of the real signature (s, P). Therefore, in this case, Simul takes an additional argument $\bar{s}$, and one requires also the existence of four additional protocols, Confirm, ConfirmVrfy, Disavowal and DisavowalVrfy. To model our attack on the JSI scheme we must also define "secure disavowability". The idea behind such definition is to prove that if $\mathsf{Vrfy}_{\mathsf{pk}_S,\mathsf{pk}_D}(m, s) = \mathsf{accept}$ and on input (m, s), an oracle machine $\mathcal{F}$ can, with high probability, produce a δ such that $\mathsf{DisavowalVrfy}_{\mathsf{pk}_S,\mathsf{pk}_D}(m; \sigma, \delta) = \mathsf{success}$ then there exists a knowledge extractor that can use $\mathcal{F}$ to recover sk_D.

The JSI scheme [JSI96] was claimed to be disavowable—although, as shown earlier in this paper, it is not—while most of the subsequent DVS schemes provide perfect non-transferability and thus are not disavowable. (If a DVS scheme is perfectly non-transferable then there is no trapdoor information that Signy might use to prove that a message was or was not signed by her. Thus, to make it able for Signy to disavow a signature, a DVS scheme must be computationally but not perfectly non-transferable.) However, even in [JSI96], disavowability was not seen as a feature of the DVS schemes and disavowability was never formally defined; their scheme, being based on Chaum's undeniable signature, just happens to have this property.

Corrected JSI Scheme: JSI+. As mentioned before, the original JSI scheme, is secure including pk_S, pk_D and $s = m^{xs}$ under the hash. Since we want this scheme to be disavowable, $\mathsf{Simul}_{\mathsf{sk}_D,\mathsf{pk}_S}$ also gets an additional input $s \in \mathbb{Z}_p$. Clearly, JSI+ is computationally but not perfectly non-transferable. Unforgeability, non-delegatability and secure disavowability of this scheme can be proven by using standard cryptographic tools, although most probably either in the RO model or with looser reductions. One of the reasons is that since DVS-KW is *perfectly* non-transferable, in its unforgeability proof $\mathcal{A}$ could answer signing and simulation queries in the same manner. Therefore, it was sufficient for $\mathcal{A}$ to know only one of sk_S and sk_D to simulate both kind of queries without using a programmable random oracle. It is not a priori clear how to achieve the same in the case of JSI+.

Further Work and Acknowledgments. We feel that this paper raises interesting questions about the relationship between different non-standard security models; it is an interesting open question to study the necessity of these models in any concrete case. We would like to thank anonymous referees for useful comments. The first author was partially supported by the Estonian Science Foundation.

References

[DF02] Ivan Damgård and Eiichiro Fujisaki. An Integer Commitment Scheme Based on Groups with Hidden Order. In Yuliang Zheng, editor, *Advances on Cryptology — ASIACRYPT 2002*, volume 2501 of *Lecture Notes in Computer Science*, pages 125–142, Queenstown, New Zealand, December 1–5, 2002. Springer-Verlag.

[JSI96] Markus Jakobsson, Kazue Sako, and Russell Impagliazzo. Designated Verifier Proofs and Their Applications. In Ueli Maurer, editor, *Advances in Cryptology — EUROCRYPT '96*, volume 1070 of *Lecture Notes in Computer Science*, pages 143–154, Saragossa, Spain, May 12–16, 1996. Springer-Verlag.

[KW03] Jonathan Katz and Nan Wang. Efficiency Improvements for Signature Schemes with Tight Security Reductions. In *10th ACM Conference on Computer and Communications Security*, pages 155–164, Washington, D.C., USA, October 27–31, 2003. ACM Press.

[LV04] Fabien Laguillaumie and Damien Vergnaud. Designated Verifier Signatures: Anonymity and Efficient Construction from Any Bilinear Map. In Carlo Blundo and Stelvio Cimato, editors, *Security in Communication Networks, 4th International Conference, SCN 2004*, volume 3352 of *Lecture Notes in Computer Science*, pages 105–119, Amalfi, Italy, September 8–10, 2004. Springer Verlag.

[Nie02] Jesper Buus Nielsen. Separating Random Oracle Proofs from Complexity Theoretic Proofs: The Non-committing Encryption Case. In Moti Yung, editor, *Advances in Cryptology — CRYPTO 2002, 22nd Annual International Cryptology Conference*, volume 2442 of *Lecture Notes in Computer Science*, pages 111–126, Santa Barbara, USA, August 18–22, 2002. Springer-Verlag.

[SBWP03] Ron Steinfeld, Laurence Bull, Huaxiong Wang, and Josef Pieprzyk. Universal Designated-Verifier Signatures. In Chi Sung Laih, editor, *Advances on Cryptology — ASIACRYPT 2003*, volume 2894 of *Lecture Notes in Computer Science*, pages 523–542, Taipei, Taiwan, November 30–December 4, 2003. Springer-Verlag.

[SKM03] Shahrokh Saeednia, Steve Kremer, and Olivier Markowitch. An Efficient Strong Designated Verifier Signature Scheme. In Jong In Lim and Dong Hoon Lee, editors, *Information Security and Cryptology - ICISC 2003*, volume 2971 of *Lecture Notes in Computer Science*, pages 40–54, Seoul, Korea, November 27–28, 2003. Springer-Verlag.

[SWP04] Ron Steinfeld, Huaxiong Wang, and Josef Pieprzyk. Efficient Extension of Standard Schnorr/RSA Signatures into Universal Designated-Verifier Signatures. In Feng Bao, Robert H. Deng, and Jianying Zhou, editors, *Public Key Cryptography 2004*, volume 2947 of *Lecture Notes in Computer Science*, pages 86–100, Singapore, March 1–4, 2004. Springer-Verlag.

Single-Key AIL-MACs from Any FIL-MAC

Ueli Maurer and Johan Sjödin

Department of Computer Science,
Swiss Federal Institute of Technology (ETH), Zurich,
CH-8092 Zurich, Switzerland
{maurer, sjoedin}@inf.ethz.ch

Abstract. We investigate a general paradigm for constructing arbitrary-input-length (AIL) MACs from fixed-input-length (FIL) MACs, define the waste as the relevant efficiency parameter of such constructions, and give a simple and general security proof technique applicable to very general constructions. We propose concrete, essentially optimal constructions for practical use, Chain-Shift (CS) and Chain-Rotate (CR), and prove their security. They are superior to the best previously known construction, the NI-construction proposed by An and Bellare: Only one rather than two secret keys are required, the efficiency is improved, and the message space is truly AIL, i.e., there is no upper bound on the message length. The generality of our proof technique is also illustrated by giving a simple security proof of the NI-construction and several improvements thereof.

Keywords: Message authentication code (MAC), arbitrary-input-length (AIL), variable-input-length (VIL), fixed-input-length (FIL).

1 Introduction

1.1 Message Authentication Codes (MACs)

Authenticity is a fundamental security requirement for data transmissions. A well-known technique for authenticating messages is to use a so-called *message authentication code* (MAC), an important symmetric-key cryptographic primitive with widespread use in many practical applications. A MAC is a function family $H := \{h_k : \mathcal{M} \to \mathcal{T}\}_{k \in \mathcal{K}}$, where $\mathcal{M}$ is the message space, $\mathcal{T}$ the tag space, and $\mathcal{K}$ the key space. Two communicating parties who share a secret key k can authenticate a message m, sent over an insecure channel, by computing a tag $\tau = h_k(m)$, which is sent together with the message. The message need not be encrypted. The receiver accepts the message if and only if the received pair (m', τ') satisfies $\tau' = h_k(m')$. We refer to h_k as an *instantiation* of H.

The message space is an important parameter of a MAC scheme. In most applications one needs to authenticate messages of potentially *arbitrary-input-length* (AIL), i.e., $\mathcal{M} = \{0,1\}^*$. Many proposed MACs are not AIL-MACs since there is an upper bound on the message length, i.e., the message space is the set $\mathcal{M} = \{0,1\}^{\leq N}$ of all bit strings of length at most N. We refer to such MACs as having *variable-input-length* (VIL). The constant N is typically large enough to be of little practical concern. A MAC which has $\mathcal{M} = \{0,1\}^L$ for a constant L is referred to as having *fixed-input-length* (FIL). In this paper we address the problem of constructing (VIL- and) AIL-MACs from any FIL-MAC.

1.2 Previous Work

Constructing VIL- or AIL-primitives from FIL-primitives have been addressed in many papers. A well-known example is the Merkle-Damgård [5,7] iteration method for constructing AIL collision-resistant functions from FIL collision-resistant functions. The question of constructing VIL- or AIL- pseudo random functions (PRFs) based on any FIL-PRF has received substantial attention, see for example the CBC-MAC [4,6,10] and the XOR-MAC [3] (which are PRFs and thus trivially also MACs). Other examples of AIL-MAC constructions are the hash function-based MACs like NMAC and HMAC [2].

A central goal in cryptography is to prove the security of cryptographic schemes under as weak assumptions as possible. In the context of constructing VIL- or AIL-MACs, a natural assumption (much weaker than the PRF assumption) is that the underlying FIL-primitive is a secure FIL-MAC. In 1997 Naor and Reingold [8] constructed a FIL-PRF from any FIL-MAC, but with high cost. While this FIL-PRF could in principle be used in some well-known construction of an AIL-MAC from any FIL-PRF (e.g. the CBC-MAC), it would be impractical. Their question whether a FIL-PRF can be obtained from any FIL-MAC at low cost is still open to date. In 1999 the problem of constructing VIL-MACs from FIL-MACs was proposed and investigated by An and Bellare [1]. They showed that the CBC-MAC is insecure under this weaker assumption for the FIL-primitive. They also presented the *nested iterated* (NI) construction, the first practical construction of a VIL-MAC based on any FIL-MAC (see Fig. 3). As we will see, the NI-construction leaves room for improvements.

1.3 MAC Constructions and Important Design Criteria

Throughout this paper, let $G := \{g_k : \{0,1\}^L \to \{0,1\}^\ell\}_{k \in \{0,1\}^\kappa}$ denote a FIL-MAC, with *compression* $b := L - \ell > 0$. We consider a general type of construction C, which uses G to construct a MAC $C^G := \{C^{g_k} : \mathcal{M} \to \{0,1\}^\ell\}_{k \in \{0,1\}^\kappa}$, where $\mathcal{M}$ is either AIL (i.e., $\{0,1\}^*$) or VIL (i.e., $\{0,1\}^{\leq N}$). The instantiation C^{g_k} is constructed by invoking g_k several times in a black-box manner. To be more precise, let us describe the computation of the tag $\tau = C^{g_k}(m)$ for an n-bit message m. In a pre-processing step m is encoded into a bit string m' of length (denoted by) $\lambda(n)$, for instance by padding m and appending information about its length. The processing step is best described with a buffer initialized with m', where each call to g_k fetches (and deletes) some L bits and writes back the ℓ-bit result to the buffer. This reduces the number of bits in the buffer (by b bits) with each call to g_k. As soon as the number of bits is less than L, the content of the buffer is returned as the tag τ. To obtain an ℓ-bit output, an appropriate encoding is used such that $\lambda(n) = t(n) \cdot b + \ell$ for some $t(n)$. Note that $t(n)$ is exactly the number of calls to g_k required to compute τ, and that τ is the last output of g_k. The function $t(\cdot)$ is referred to as the *application* function of $\mathrm{C}^{\cdot}$. A particular construction can thus be described by the encoding function mapping m to m' and by the scheme by which the L-bit blocks are fetched.

In a more general variant of such a construction, several (say 2) instantiations g_{k_1} and g_{k_2} of G can be used to build an instantiation $\mathrm{C}^{g_{k_1}, g_{k_2}}$ of the MAC $\mathrm{C}^{G,G} := \{\mathrm{C}^{g_{k_1}, g_{k_2}} : \mathcal{M} \to \{0,1\}^{\ell}\}_{k_1, k_2 \in \{0,1\}^{\kappa}}$ (with key space $(\{0,1\}^{\kappa})^2$). The only difference in the computation of the tag, described above, is that for each L-bit block that is fetched, the instantiation to be invoked needs to be specified. For such schemes $t^i(n)$ (with $i \in \{1, 2\}$) denotes the number of calls needed to g_{k_i} in order to compute the tag of an n-bit message, and $t(n) := t^1(n) + t^2(n)$.

Note that the key space of $\mathrm{C}^{G,G}$ is twice the size of the key space of C^G. We refer to $\mathrm{C}^{\cdot}$ as a single-key construction and to $\mathrm{C}^{\cdot,\cdot}$ as a 2-key construction. We now discuss the main design criteria for the constructions:

Number of Keys: We will propose single-key constructions (like $\mathrm{C}^{\cdot}$) for practical use and see that there is essentially no reason for considering multiple-key constructions (like $\mathrm{C}^{\cdot,\cdot}$).

Efficiency: The efficiency can be measured in the number of *applications* $t(n)$ of the FIL-MAC, or equivalently in terms of the *waste* $w(n) := \lambda(n) - n = t(n) \cdot b + \ell - n$, i.e., the amount by which pre-processing expands the message.

Type of Processing: It is desirable that a message can be processed *on-line*, i.e., as the message bits arrive, without knowing the message length in advance. Moreover, it is desirable that the computation of the tag τ can be *parallelized*, i.e., sped up by a factor of roughly c (over the construction using one processor) when c processors are available.

Message Space: As we will see, it turns out that no bound on the message length is necessary, and therefore our focus is on AIL-MAC constructions.

1.4 Contributions of This Paper

The purpose of this paper is to investigate systematically a natural and general paradigm for constructing (VIL- or) AIL-MACs from FIL-MACs, a problem introduced by An and Bellare [1]. Our proof technique, applicable to a very general type of construction, turns out to be insightful for constructing (VIL- and) AIL-MACs from FIL-MACs. We propose concrete, essentially optimal AIL-MAC constructions for practical use, Chain-Shift (CS) and Chain-Rotate (CR) (see Fig. 1 and Fig. 2), and prove their security. They use a single key, have constant waste, allow for on-line and parallel processing of the messages, and their security reduction is essentially tight.

The only previously known (practical) VIL-MAC construction, the NI-construction (see Fig. 3), uses two keys, has an upper bound of 2^b on the message length, and is not optimal in terms of the number of applications to the FIL-MAC (especially not for short messages). In Sect. 4.2 we give a simple security proof (using our proof technique) and several improvements of the NI-construction.

2 Preliminaries

2.1 Notation and Definitions

Let $\{0,1\}^L$ denote the set of all bit strings of length L, $\{0,1\}^{\leq N}$ the set of all bit strings of length at most N, $\{0,1\}^*$ the set of all bit strings, and $[n] := \{1,\ldots,n\}$ (with $[0] := \emptyset$). If M is a set, $\#M$ denotes its cardinality. For $x, y \in \{0,1\}^*$, let $|x|$ denote the length of x (in bits), $x\|y$ the concatenation of x and y, $\langle n \rangle_b$ a b-bit encoding of a positive integer $n \leq 2^b$, $x[i]$ the ith bit of x, and $x[i,j] := x[i]\|x[i+1]\|\cdots\|x[j]$ for $1 \leq i < j \leq |x|$. Furthermore, let $\text{RR}(\cdot)$ denote the operator on bit strings that rotates the input by one position to the right, i.e., $\text{RR}(x) := x[L]\|x[1, L-1]$. For a sequence S of elements, $|S|$ denotes its length, and S_i the sequence of the first $i \leq |S|$ elements of S. An encoding $\sigma : \{0,1\}^* \to \{0,1\}^*$ is called *prefix-free* if there are no three strings $x, x', y \in \{0,1\}^*$ such that $x \neq x'$, $|y| \geq 1$, and $\sigma(x)\|y = \sigma(x')$. A *suffix-free* encoding is an encoding which becomes prefix-free if the bit-order of the output is reversed. A *non-trivial collision* for a function f is a pair $x \neq x'$ of inputs for which $f(x) = f(x')$. If $\mathcal{E}$ denotes an event, $\bar{\mathcal{E}}$ denotes the complementary event.

2.2 Security Definition for MACs

A forger F for a MAC $H := \{h_k : \mathcal{M} \to \mathcal{T}\}_{k \in \mathcal{K}}$ has oracle access to $h_k(\cdot)$ (for which k is chosen uniformly at random from $\mathcal{K}$) and can thus learn the tag values for some adaptively chosen messages $m_1, m_2, \ldots$. It then returns a *forgery* (m, τ), i.e., a message m together with a tag τ. The forger F is considered successful if $h_k(m) = \tau$. The only constraint on m is that it must be *new*, i.e., different from all previous messages $m_1, m_2, \ldots$. We refer to a forger F of this kind as a (t, q, μ, ε)-forger, where t, q, and μ are upper bounds on the running time, the number of messages (or oracle queries), respectively the total length (in bits) of the oracle queries including the forgery message m, and ε is a lower bound on the success probability. Informally, a MAC is considered secure against *existential forgery* under an *adaptive chosen-message attack*, if there is no (t, q, μ, ε)-forger, even for very high values of t, q, and μ, and a very small value of ε. A forger for a FIL-MAC will be denoted simply as a (t, q, ε)-forger, since the parameter μ is determined by q and the message-input-length L, i.e., $\mu = (q+1)L$.

To prove the security of a MAC based on a FIL-MAC one shows that the existence of a (t, q, μ, ε)-forger F for the MAC implies the existence of a (t', q', ε')-forger F' for the FIL-MAC, where t', q', and ε' are functions of t, q, μ, and ε. In all our security proofs F is called only once by F'. Therefore, the running time of F' is essentially that of F, i.e., $t' \approx t$, with some small overhead that is obvious from the construction of F'. We will therefore not bother to explicitly compute the running time of forgers, as this complicates the analysis unnecessarily without providing more insight. Therefore we drop the time parameter t in the sequel.

3 Single-Key AIL-MACs Based on Any FIL-MAC

3.1 FIL-MAC Forgers Based on an AIL-MAC Forger (Single Key)

Let F be a (q, μ, ε)-forger for a MAC C^G, i.e., if k is chosen uniformly at random from $\{0,1\}^\kappa$, and F is allowed at most q oracle queries to C^{g_k} of total length at most μ (including the length of the forgery message), then F returns a valid forgery (m, τ) with probability at least ε. We refer to $F \circ C^{g_k}$ as the process in which F's queries to C^{g_k} are computed and returned to F, and where F's forgery (m, τ) is verified by computing $C^{g_k}(m)$. Let us consider the random variables occurring at the interface to g_k (in the process $F \circ C^{g_k}$). Let z_i denote the ith input to g_k and let $y_i := g_k(z_i)$. The sequences $\mathbf{Z} := (z_1, z_2, \ldots)$ and $\mathbf{Y} := (y_1, y_2, \ldots)$ are thus naturally defined. Note that as soon as the key k and the random coins of F are fixed, all values in $\mathbf{Z}$ and $\mathbf{Y}$ are determined, and also whether F is successful or not. Let $\mathcal{E}$ denote the event that F is successful. Without loss of generality we assume that F's forgery message m is distinct from F's oracle queries. Thus $\mathcal{E}$ occurs if and only if $C^{g_k}(m) = \tau$.

A FIL-MAC forger F' for G simulates $F \circ C^{g_k}$ with the help of F and its oracle access to g_k. At some query z_i to g_k it stops the simulation and returns a forgery (z', τ') for g_k (without making any other oracle queries to g_k). Such a forger is characterized by the moment it stops (i.e., i) and the way it produces its forgery. We refer to this as the *strategy* s of F' and let F'_s denote the corresponding forger.

The most simple strategy is the *naïve* strategy s_{na}. $F'_{s_{\text{na}}}$ stops the simulation of $F \circ C^{g_k}$ at the very last query $\mathbf{z}$ to g_k (i.e., $\mathbf{z}$ is the last entry in $\mathbf{Z}$). Then it returns $(\mathbf{z}, \tau)$ as a forgery, where τ is the forgery tag of F's forgery (m, τ) for C^{g_k}. $F'_{s_{\text{na}}}$ is successful if the following two conditions hold. First, $\mathcal{E}$ occurs, i.e., $C^{g_k}(m) = \tau$ (and thus $g_k(\mathbf{z}) = \tau$ by definition of C), and second $\mathbf{z}$ is new, i.e., $\mathbf{z}$ is only the last entry in $\mathbf{Z}$. Let $\mathcal{E}_{\text{new}}$ denote the event that $\mathbf{z}$ is new. Thus $F'_{s_{\text{na}}}$ is successful whenever $\mathcal{E} \wedge \mathcal{E}_{\text{new}}$ occurs.

Imagine that there is a set $\mathcal{S}$ of strategies, such that whenever $\bar{\mathcal{E}}_{\text{new}}$ occurs there exists at least one strategy $s \in \mathcal{S}$ for which F'_s is successful. We refer to such a set $\mathcal{S}$ as *complete* for the construction. Obviously, the set $\mathcal{S} \cup \{s_{\text{na}}\}$ has the property that whenever $\mathcal{E}$ occurs, there is at least one strategy $s \in \mathcal{S} \cup \{s_{\text{na}}\}$ for which F'_s is successful. Thus an overall strategy of F' is to pick its strategy uniformly at random from $\mathcal{S} \cup \{s_{\text{na}}\}$. Its success probability is at least the probability that $\mathcal{E}$ occurs divided by $\#\mathcal{S} + 1$, since the choice of strategy is independent of $\mathcal{E}$. As F''s number of oracle queries is $|\mathbf{Z}|$, which is a random variable, it is convenient to introduce the following function.

Definition 1. *The* expansion *function e of a construction C is defined as*

$$e(\tilde{q}, \tilde{\mu}) := \max\left\{\sum_{i=1}^{\tilde{q}} t(n_i) : n_1, \ldots, n_{\tilde{q}} \in \mathbb{N}_0,\ n_1 + \cdots + n_{\tilde{q}} \leq \tilde{\mu}\right\},$$

where $t(\cdot)$ is the application function of C.

It follows that $|\mathbf{Z}| \leq e(q+1, \mu)$, since there are at most $q+1$ queries of total length at most μ to C^{g_k} in $F \circ C^{g_k}$. In general $\#\mathcal{S}$ is a function of $e(q+1, \mu)$.

Proposition 1. *The existence of a complete set $\mathcal{S}$ for a construction C and a (q, μ, ε)-forger F for C^G implies the existence of a (q', ε')-forger F' for G, where $q' = e(q+1, \mu)$ and $\varepsilon' = \frac{\varepsilon}{\#\mathcal{S}+1}$.*

Proof. F' picks its strategy s uniformly at random from $\mathcal{S} \cup \{s_{\text{na}}\}$. Let $\mathcal{E}'$ denote the event that F' is successful, and let $\mathcal{E}$ and $\mathcal{E}_{\text{new}}$ be defined as above.

$$\Pr[\mathcal{E}'] \geq \underbrace{\Pr[\mathcal{E}' \mid \mathcal{E} \wedge \mathcal{E}_{\text{new}}]}_{\geq 1/(\#\mathcal{S}+1)} \cdot \Pr[\mathcal{E} \wedge \mathcal{E}_{\text{new}}] + \underbrace{\Pr[\mathcal{E}' \mid \bar{\mathcal{E}}_{\text{new}}]}_{\geq 1/(\#\mathcal{S}+1)} \cdot \underbrace{\Pr[\bar{\mathcal{E}}_{\text{new}}]}_{\geq \Pr[\mathcal{E} \wedge \bar{\mathcal{E}}_{\text{new}}]} \geq \underbrace{\frac{\Pr[\mathcal{E}]}{\#\mathcal{S}+1}}_{=\varepsilon/(\#\mathcal{S}+1)}$$

$=: \varepsilon'$ □

3.2 Deterministic Strategies

An important class of strategies for F' are the deterministic strategies. A deterministic strategy s is characterized by a pair (i, f), where $i \in [e(q+1, \mu)]$ is an index and f a function mapping $(\mathbf{Z}_i, \mathbf{Y}_{i-1})$ to some value $\hat{y}_i \in \{0, 1\}^\ell$ (which can be seen as a prediction of y_i). To be more precise, the corresponding forger F'_s stops (the simulation of $F \circ C^{g_k}$) at query z_i and returns $(z_i, \hat{y}_i)$ as a forgery.[1] The forger is successful if $\hat{y}_i = y_i$ and if z_i is new, i.e., not contained in the sequence $\mathbf{Z}_{i-1}$. Next follow three particular sets of strategies, which will be used in the sequel:

- Let $s_{i,y}$ (with $y \in \{0, 1\}^\ell$) denote the strategy of stopping at query z_i and returning (z_i, y) as a forgery. Note that whenever the event occurs that g_k outputs y, i.e., when y is an entry in $\mathbf{Y}$, then there is at least one strategy $s \in \mathcal{S}_y := \{s_{i,y} \mid i \in [e(q+1, \mu)]\}$ for which F'_s is successful. We have

$$\#\mathcal{S}_y = e(q+1, \mu). \tag{1}$$

- Let $s_{\text{coll},i,j}$ (with $i > j$) denote the strategy of stopping at query z_i and returning (z_i, y_j) as a forgery. Note that whenever a non-trivial collision for g_k occurs, i.e., $\alpha, \beta \in [|\mathbf{Z}|]$ satisfying $z_\alpha \neq z_\beta$ and $y_\alpha = y_\beta$, then there is at least one strategy $s \in \mathcal{S}_{\text{coll}} := \{s_{\text{coll},i,j} \mid i, j \in [e(q+1, \mu)], i > j\}$ for which F'_s is successful. The cardinality of $\mathcal{S}_{\text{coll}}$ is

$$\#\mathcal{S}_{\text{coll}} = e(q+1, \mu)^2/2 - e(q+1, \mu)/2. \tag{2}$$

- Let $s_{\text{coll2},i,j,a,\text{left}}$ (with $a \in \{0, 1\}$ and $i > j$) denote the strategy of stopping at input z_i and returning $(z_i, a\|y_j[1, \ell-1])$ as a forgery, and let $s_{\text{coll2},i,j,a,\text{right}}$ denote the strategy of stopping at input z_i and returning $(z_i, y_j[2, \ell]\|a)$ as a forgery. Note that whenever the event occurs that there are $\alpha, \beta \in [|\mathbf{Z}|]$ satisfying $z_\alpha \neq z_\beta$ and $g_k(z_\alpha)[2, \ell] = g_k(z_\beta)[1, \ell-1]$, then there is a strategy $s \in \mathcal{S}_{\text{coll2}} := \{s_{\text{coll2},i,j,a,d} \mid i, j \in [e(q+1, \mu)], i > j, a \in \{0, 1\}, d \in \{\text{left}, \text{right}\}\}$ for which F'_s is successful. The cardinality of $\mathcal{S}_{\text{coll2}}$ is

$$\#\mathcal{S}_{\text{coll2}} = 2 \cdot e(q+1, \mu)^2 - 2 \cdot e(q+1, \mu). \tag{3}$$

---

[1] If $i > |\mathbf{Z}|$ the forger aborts.

3.3 The Chain-Shift (CS) Construction

The CS-construction uses any FIL-MAC $G := \{g_k : \{0,1\}^{b+\ell} \to \{0,1\}^\ell\}_{k \in \{0,1\}^\kappa}$ with compression $b \geq \ell$, to construct an AIL-MAC $\text{CS}^G := \{\text{CS}^{g_k} : \{0,1\}^* \to \{0,1\}^\ell\}_{k \in \{0,1\}^\kappa}$. For a message $m \in \{0,1\}^*$ of length $n := |m|$, the tag $\tau = \text{CS}^{g_k}(m)$ is computed according to the following recursion (as depicted in Fig. 1). Parse m into a sequence of b-bit blocks $m_1, \ldots, m_{t-1}$ and a $(b-\ell)$-bit block m_t, such that $m_1 \| \cdots \| m_t = m \| 10^\nu$ for some $\nu \in \{0, \ldots, b-1\}$:

$$y_0 := 0^\ell, \; y_i := g_k(y_{i-1} \| m_i) \text{ for } i \in [t-1], \text{ and } \tau := g_k(1^\ell \| y_{t-1} \| m_t).$$

The waste $w(n) = t(n) \cdot b + \ell - n = \lceil (n+1+\ell)/b \rceil \cdot b + \ell - n \leq L + \ell$ is upper bounded by a constant, and on-line processing is possible.

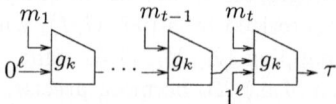

Fig. 1. The Chain-Shift (CS) construction

Theorem 1.[2] *A (q, μ, ε)-forger F for CS^G implies a (q', ε')-forger F' for G, where $q' = \lfloor \frac{\mu}{b} \rfloor + 2(q+1)$ and $\varepsilon' = \frac{\varepsilon}{q'^2/2 + 3q'/2 + 1}$.*

Proof. We apply Proposition 1 and show that $\mathcal{S} := \mathcal{S}_{\text{coll}} \cup \mathcal{S}_{0^\ell} \cup \mathcal{S}_{1^\ell}$ is complete for CS by proving that whenever the last entry $\mathbf{z}$ in $\mathbf{Z}$ is not new, then there is a non-trivial collision in g_k, or an output from g_k that equals 0^ℓ or 1^ℓ.

Assume that $\mathbf{z}$ is not new. Let $\tilde{z}_1, \ldots, \tilde{z}_t$ denote the sequence of queries to g_k resulting from the last query m_β to CS^{g_k}. Note that m_β is the forgery message of F and thus new. Since $\tilde{z}_t = \mathbf{z}$ is not new, $\tilde{z}_t$ must have been an earlier query to g_k, resulting from some query m_α (with $\alpha \leq \beta$) to CS^{g_k}. Let $\tilde{z}'_1, \ldots, \tilde{z}'_{t'}$ denote the sequence of queries to g_k in the computation of $\text{CS}^{g_k}(m_\alpha)$. There are three cases to distinguish, depending on the index $i \in [t']$ for which $\tilde{z}_t = \tilde{z}'_i$.

At the end of the chain $(\tilde{z}_t = \tilde{z}'_{t'})$: First, we note that this can not be the case if $\alpha = \beta$, since in that case $\tilde{z}'_{t'}$ is not an earlier occurring query. Thus we have (the non-trivial collision) $m_\alpha \neq m_\beta$ satisfying $\text{CS}^{g_k}(m_\alpha) = \text{CS}^{g_k}(m_\beta)$. Without loss of generality assume that $t' \geq t$. Now, either there exist an index $j \in [t-1]$ such that $\tilde{z}_{t-j} \neq \tilde{z}'_{t'-j}$ and $\tilde{z}_{t-j+1} = \tilde{z}'_{t'-j+1}$, i.e., a non-trivial collision in g_k occurs (since $\tilde{z}_{t-j+1} = \tilde{z}'_{t'-j+1}$ implies $g_k(\tilde{z}_{t-j}) = g_k(\tilde{z}'_{t'-j})$) or $\tilde{z}'_{t'-t+1} = \tilde{z}_1 = 0^\ell \| v$ for some $v \in \{0,1\}^b$, which implies $g_k(\tilde{z}'_{t'-t}) = 0^\ell$ (with $t' - t \geq 1$ since $m_\alpha \neq m_\beta$).

---

[2] An and Bellare point out in [1] that the security loss of roughly $(\mu/b)^2$ is unavoidable for iterative constructions of this nature. It is shown using birthday attacks illustrated by Preneel and Van Oorschot [11].

In the middle of the chain ($\tilde{z}_t = \tilde{z}'_i$ with $1 < i < t'$): We have $1^\ell \| v = \tilde{z}_t = \tilde{z}'_i = g_k(\tilde{z}'_{i-1}) \| v$, for some $v \in \{0,1\}^b$. Thus g_k outputs 1^ℓ.

At the beginning of the chain ($\tilde{z}_t = \tilde{z}'_1$ and $t' > 1$): This case is obviously impossible, since $\tilde{z}_t = 1^\ell \| v \neq 0^\ell \| v' = \tilde{z}'_1$ for any $v, v' \in \{0,1\}^b$.

By definition of $e(q+1, \mu)$, there is a sequence $n_1, \ldots, n_{q+1} \in \mathbb{N}_0$ such that:

$$e(q+1, \mu) = \sum_{i=1}^{q+1} t(n_i) \leq \left\lfloor \frac{\mu + (q+1)L}{b} \right\rfloor \leq \left\lfloor \frac{\mu}{b} + 2(q+1) \right\rfloor =: q'.$$

Thus $\#\mathcal{S} + 1 \leq (q'^2/2 - q'/2) + q' + q' + 1 \leq q'^2/2 + 3q'/2 + 1$ by (1) and (2). □

Improving the Waste for Short Messages. We improve the efficiency of the CS-construction for $n := |m| < rb$, where $r \in \mathbb{N}_0$ is a design parameter. This is relevant (see for example [9]). The computation of the tag τ is redefined for messages m of length shorter than rb as follows. Parse m into a sequence of b-bit blocks $m_1, \ldots, m_t$ such that $m_1 \| \ldots \| m_t = m \| 10^\nu$ where $\nu \in \{0, \ldots, b-1\}$:

$$y_0 := \langle t \rangle_\ell, \; y_i := g_k(y_{i-1} \| m_i) \text{ for } i \in [t], \text{ and } \tau := y_t.$$

Now, $t(n) = \lceil (n+1)/b \rceil$ if $n < rb$ (and $t(n) = \lceil (n+1+\ell)/b \rceil$ if $n \geq rb$). The proof that $\mathcal{S}_{\text{coll}} \cup \mathcal{S}_{0^\ell} \cup \mathcal{S}_{1^\ell} \cup (\cup_{i=1}^r \mathcal{S}_{\langle i \rangle_\ell})$ is complete for the construction is omitted. The only modification of Theorem 2 is thus that $\epsilon' = \frac{\epsilon}{q'^2/2 + (3/2+r)q'+1}$, i.e., the reduction is essentially as tight (as for $r = 0$) for reasonable r's.

Parallelizing the CS-Construction. We modify the CS-construction to allow $c \geq 1$ processors to compute the tag in parallel, achieving a speed up by a factor of roughly c for long messages. The tag τ of an n-bit message m is computed according to the following recursion:

1. If $c \leq \lceil (n+1)/b \rceil$ then set $c' := c$, and else set $c' := \lceil (n+1)/b \rceil$.
2. Parse m into $m_1 \| \cdots \| m_{c't} = m \| 10^\nu$, where $m_1, \ldots, m_{c't}$ are b-bit blocks and $\nu \in \{0, \ldots, c'b - 1\}$. Set $m_{i,j} := m_{i+(j-1)c'}$ for $i \in [c']$ and $j \in [t]$.
3. Set $y_{i,0} := 0^\ell$, and compute $y_{i,j} := g_k(y_{i,j-1} \| m_{i,j})$ for $i \in [c']$ and $j \in [t]$.
4. Return $\tau := \text{CS}^{g_k}(y_{1,t} \| \cdots \| y_{c',t})$.[3]

---

[3] The construction can be further parallelized by replacing step 4 as follows. For simplicity assume $b = \ell$ (the generalization to $b \geq \ell$ is straight forward). Apply g_k to every pair of adjacent blocks in $(y_{1,t}, \cdots, y_{c',t})$, resulting in a new sequence of $\lceil c'/2 \rceil$ blocks, and repeat this until a single block y is obtained. Then set $\tau := g_k(1^\ell \| y)$.

By setting $c := \infty$ this construction is *fully parallelized* (FP) (here meaning that the computation time is in $\Theta(\log(n))$ when arbitrary many processors are available) with $w(n) \in \Theta(n)$. From a theoretical viewpoint it would be interesting to see whether FP single-key AIL-MAC constructions with $w(n) \in \Theta(1)$ exists. There are FP single-key AIL-MAC constructions with $w(n) \in \Theta(\log(n))$ and FP 2-key AIL-MAC constructions with $w(n) \in \Theta(1)$.

The waste remains constant and the on-line property is preserved. We omit the proof that $\mathcal{S} = \mathcal{S}_{\text{coll}} \cup \mathcal{S}_{0^\ell} \cup \mathcal{S}_{1^\ell}$ is complete for the construction, as it is similar to the proof that $\mathcal{S}$ is complete for the CS-construction.

3.4 The Chain-Rotate (CR) Construction

The purpose of presenting this single-key AIL-MAC construction is twofold. First, it shows that constant waste and on-line processing is possible (even) with compression $b < \ell$. Second, it demonstrates the generality of our proof technique.

The CR-construction transforms any FIL-MAC $G := \{g_k : \{0,1\}^{b+\ell} \to \{0,1\}^\ell\}_{k \in \{0,1\}^\kappa}$ into an AIL-MAC $\text{CS}^G := \{\text{CS}^{g_k} : \{0,1\}^* \to \{0,1\}^\ell\}_{k \in \{0,1\}^\kappa}$. The tag $\tau = \text{CR}^{g_k}(m)$ of an n-bit message m is computed as follows (see Fig. 2). Parse m into a sequence of b-bit blocks $m_1, \ldots, m_t$ such that $m_1 \| \cdots \| m_t = m \| 10^\nu$ with $\nu \in \{0, \ldots, b-1\}$. If $t > 1$, set $y_0 := 0^\ell$ and else set $y_0 := 1^\ell$:

$$y_i := g_k(y_{i-1} \| m_i) \text{ for } i \in [t-1] \text{ and } \tau := g_k(\text{RR}(y_{t-1} \| m_t)).$$

The waste is $w(n) = \lceil (n+1)/b \rceil \cdot b + \ell - n \leq L$, and on-line processing is possible.

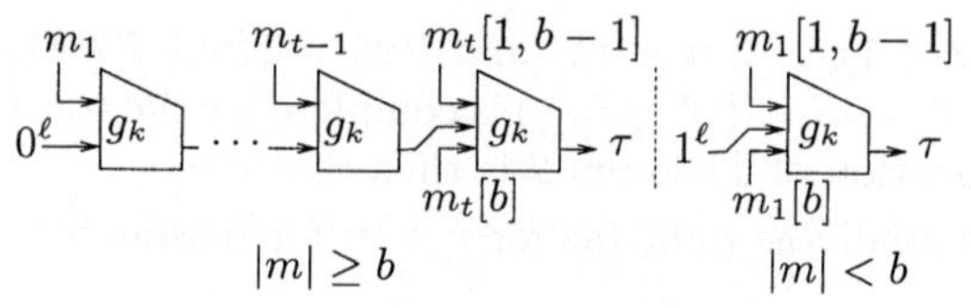

Fig. 2. The Chain-Rotate (CR) construction

Theorem 2. *A (q, μ, ε)-forger F for CR^G implies a (q', ε')-forger F' for G, where $q' = \lfloor \frac{\mu}{b} \rfloor + q + 1$ and $\varepsilon' = \frac{\varepsilon}{5q'^2/2 + 3q'/2 + 1}$.*

Proof (Sketch). We apply Proposition 1. With a case study similar to that for the CS (here omitted), one shows that $\mathcal{S} := \mathcal{S}_{\text{coll}} \cup \mathcal{S}_{\text{coll2}} \cup \mathcal{S}_{0^\ell} \cup \mathcal{S}_{1^\ell} \cup \mathcal{S}_{0^{\ell-1}1} \cup \mathcal{S}_{01^{\ell-1}}$ is complete for CR. There exist $n_1, \ldots, n_{q+1} \in \mathbb{N}_0$ such that:

$$e(q+1, \mu) = \sum_{i=1}^{q+1} t(n_i) \leq \sum_{i}^{q+1} \left\lceil \frac{n_i + 1}{b} \right\rceil \leq \left\lfloor \sum_{i=1}^{q+1} \frac{n_i + b}{b} \right\rfloor \leq \left\lfloor \frac{\mu}{b} \right\rfloor + q + 1 =: q'.$$

As a consequence, $\#\mathcal{S} + 1 \leq 5q'^2/2 + 3q'/2 + 1$ by (1), (2), and (3). □

Parallelizing the CR-Construction. The CR-construction can be parallelized in a similar way as the CS-construction. Just replace CS by CR in step 4 of the corresponding paragraph of Sect. 3.3. As for the CR-construction the set $\mathcal{S} := \mathcal{S}_{\text{coll}} \cup \mathcal{S}_{\text{coll2}} \cup \mathcal{S}_{0^\ell} \cup \mathcal{S}_{1^\ell} \cup \mathcal{S}_{0^{\ell-1}1} \cup \mathcal{S}_{01^{\ell-1}}$ is complete for the construction.

4 Comparison with the NI-Construction and Variations

The security analysis described in Sect. 3.1 can be generalized to multiple-key constructions. Motivated by the NI-construction (see Fig. 3), we consider constructions $C^{\cdot,\cdot}$ (using two instantiations g_{k_1} and g_{k_2} of G to construct an instantiation $C^{g_{k_1},g_{k_2}}$ of the MAC $C^{G,G}$), where one of the instantiations (say g_{k_2} without loss of generality) is invoked at the end of the computation. We prove the security of the NI-construction and give several improvements thereof.

4.1 FIL-MAC Forgers Based on an AIL-MAC Forger (2 Keys)

Let F denote a (q, μ, ε)-forger for the MAC $C^{G,G}$. As before let $F \circ C^{g_{k_1},g_{k_2}}$ denote the process in which for each query $\tilde{m}$ issued by F, the corresponding tag $C^{g_{k_1},g_{k_2}}(\tilde{m})$ is computed and returned to F, and once F returns a forgery (m, τ), the forgery is verified by computing $C^{g_{k_1},g_{k_2}}(m)$. Let $\mathbf{Z}^i := (z_1^i, z_2^i, \ldots)$ and $\mathbf{Y}^i := (y_1^i, y_2^i, \ldots)$ be the sequence of inputs respectively outputs occurring at the interface to instantiation g_{k_i} (for $i \in \{1, 2\}$).

The FIL-MAC forger F' simulates $F \circ C^{g_{k_1},g_{k_2}}$ by letting its own oracle simulate one of the instantiations g_{k_i} (say the instantiation *under attack*) and by choosing a random key for the other, but stops the simulation at some query z_j^i to its oracle and returns a forgery (without making any further query to any FIL-MAC instantiation). This is equivalent to first instantiating g_{k_1} and g_{k_2} (by choosing the keys k_1, k_2 uniformly at random) and then letting F' specify which instantiation to attack, i.e., consider as its own oracle, after which the key to the other instantiation is revealed to F'. We adopt this view. Any such forger is characterized by its *strategy*, i.e., which instantiation it attacks (i.e., i), the moment it stops (i.e., j), and the way it produces its forgery.

Let s_{na} denote the naïve strategy described in Sect. 3.1, with the only modification that the second instantiation, g_{k_2} is put under attack (recall that the tag τ is an output of g_{k_2}). $F'_{s_{\text{na}}}$ stops at the very last query $\mathbf{z}$ to g_{k_2} and returns $(\mathbf{z}, \tau)$ as a forgery. Of course F' is successful if the following two conditions hold. First, $\mathcal{E}$ occurs, i.e., $C^{g_{k_1},g_{k_2}}(m) = \tau$ (and thus $g_{k_2}(\mathbf{z}) = \tau$),[4] and second $\mathcal{E}_{\text{new}}$ holds, i.e., $\mathbf{z}$ is new for g_{k_2} or equivalently $\mathbf{z}$ is only the last entry in $\mathbf{Z}^2$.

Imagine as before, that a *complete* set of strategies $\mathcal{S}$ exists, i.e., a set for which whenever $\bar{\mathcal{E}}_{\text{new}}$ occurs, there exists a strategy $s \in \mathcal{S}$ for which F'_s is successful. Then an overall strategy of F' is to pick its strategy uniformly at random from $\mathcal{S} \cup \{s_{\text{na}}\}$. Its success probability is at least the probability that $\mathcal{E}$ occurs (i.e., ε) divided by the number $\#\mathcal{S} + 1$ of strategies, since the choice of strategy is independent of the event $\mathcal{E}$. Since F''s number of queries to its oracle is upper bounded by $\max\{|\mathbf{Z}^1|, |\mathbf{Z}^2|\}$, which is a random variable, it is convenient to introduce the expansion function for each instantiation, i.e., for $i \in \{1, 2\}$, let $e^i(\tilde{q}, \tilde{\mu}) := \max\{\sum_{j=1}^{\tilde{q}} t^i(n_j) : n_1, \ldots, n_{\tilde{q}} \in \mathbb{N}_0, n_1 + \cdots + n_{\tilde{q}} \leq \tilde{\mu}\}$. Thus $|\mathbf{Z}^i| \leq e^i(q+1, \mu)$. Proposition 1 generalizes as follows.

---

[4] We assume w.l.o.g. that F's forgery message m is distinct from its oracle queries.

Proposition 2. *The existence of a complete set S for a construction $C^{\cdot,\cdot}$ and a (q, μ, ε)-forger F for $C^{G,G}$ implies a (q', ε')-forger F' for G, where $q' = \max(e^1(q+1, \mu), e^2(q+1, \mu))$ and $\varepsilon' = \frac{\varepsilon}{\#S+1}$.*

A *deterministic* strategy s is now characterized by a triple of values (i, j, f), where i denotes the instantiation to attack, z_j^i the moment to stop, and f a function mapping $(\mathbf{Z}_j^i, \mathbf{Y}_{j-1}^i)$ to some value $\hat{y}_j^i \in \{0,1\}^\ell$. The pair $(z_j^i, \hat{y}_j^i)$ is the forgery of F'_s. The sets of deterministic strategies introduced in Sect. 3.2 is naturally defined for each instantiation. Let S_y^i, S_{coll}^i, and S_{coll2}^i denote the corresponding sets for the ith instantiation.

4.2 The NI-Construction

The NI-construction [1] transforms any FIL-MAC G into a VIL-MAC $\text{NI}^{G,G} := \{\text{NI}^{g_{k_1}, g_{k_2}} : \{0,1\}^{\leq 2^b} \to \{0,1\}^\ell\}_{k_1, k_2 \in \{0,1\}^\kappa}$. For a message $m \in \{0,1\}^{\leq 2^b}$ of length $n := |m|$, the tag $\tau = \text{NI}^{g_{k_1}, g_{k_2}}(m)$ is computed according to the following recursion (as illustrated in Fig. 3). Break m into $t - 1 = \lceil n/b \rceil$ blocks $\{m_i\}_{i=1}^{t-1}$ of length b, where m_{t-1} is padded with zeroes if necessary, and set $m_t := \langle n \rangle_b$:

$$y_0 := 0^\ell, \; y_i := g_{k_1}(y_{i-1} \| m_i) \text{ for } i \in [t-1], \text{ and } \tau := g_{k_2}(y_{t-1} \| m_t). \quad (4)$$

The waste is $w(n) = t(n) \cdot b + \ell - n = \lceil \frac{n}{b} + 1 \rceil \cdot b + \ell - n \leq L + b$, and on-line processing is possible. Note that the message space is VIL, due to $m_t := \langle n \rangle_b$.

Fig. 3. The nested iterated (NI) construction

Theorem 3. *A (q, μ, ε)-forger F for $\text{NI}^{G,G}$ implies a (q', ε')-forger F' for G, where $q' = \lfloor \frac{\mu}{b} \rfloor + q + 1$ and $\varepsilon' = \frac{\varepsilon}{q'^2/2 - q'/2 + 1}$.*

Proof. We show that S_{coll}^1 is complete for $\text{NI}^{\cdot,\cdot}$ by proving that whenever the last entry $\mathbf{z}$ in $\mathbf{Z}^2$ is not new, then a non-trivial collision in g_{k_1} occurs. Let m_β denote the forgery message of F. Since $\mathbf{z}$ is not new, there is a query m_α (issued by F and different from m_β) with same input to g_{k_2}. Thus we have $|m_\alpha| = |m_\beta|$, and (the non-trivial collision) $m_\alpha \neq m_\beta$ satisfying $\text{NI}^{g_{k_1}, g_{k_2}}(m_\alpha) = \text{NI}^{g_{k_1}, g_{k_2}}(m_\beta)$. Since $m_\alpha \neq m_\beta$, all corresponding intermediate values in the computation chains can not be the same. As a consequence a non-trivial collision in g_{k_1} occurs.

By definition of $e^1(q+1, \mu)$, there is a sequence $n_1, \ldots, n_{q+1} \in \mathbb{N}_0$ such that:

$$e^2(q+1, \mu) \leq e^1(q+1, \mu) = \sum_{i=1}^{q+1} t^1(n_i) \leq \left\lfloor \sum_{i=1}^{q+1} \frac{n_i + b - 1}{b} \right\rfloor \leq \left\lfloor \frac{\mu}{b} + q + 1 \right\rfloor =: q'.$$

Thus $\#S_{\text{coll}}^1 + 1 \leq q'^2/2 - q'/2 + 1$ by (2). Proposition 2 concludes the proof. □

Improvements on the NI-Construction

1. By replacing $y_0 := 0^\ell$ with a message block, the waste decreases by ℓ bits, the security reduction is slightly tighter, and the on-line property is of course preserved. The security proof is identical to that of the NI-construction.
2. The block $m_t := \langle n \rangle_b$, encoding the message length, is superfluous. It can be replaced by a message block with appropriate padding. This decreases the waste of the construction, improves the tightness of the reduction, lifts the message space to AIL, and preserves the on-line property. To be more precise the message m is parsed into a sequence of b-bit blocks $m_1, \ldots, m_t$ such that $m_1 \| \ldots \| m_t = m \| 10^\nu$ with $\nu \in \{0, \ldots, b-1\}$ and processed according to (4). It is straight forward to see that $\mathcal{S}^1_{\text{coll}} \cup \mathcal{S}^1_{0^\ell}$ is complete for the construction.
3. If the block encoding the message length is used as the first block instead of the last or if any other prefix-free encoding of the message into blocks is used, the two keys can actually be replaced by a single key. By choosing an appropriate prefix-free encoding (for example the one on page 126 in [6]) the message space can be lifted to AIL, at the cost of having $w(n) \in \Theta(\log(n))$. We conjecture that linear waste, i.e., $w(n) \in \Theta(n)$ is needed for the on-line property. It is easy to verify that $\mathcal{S}_{\text{coll}} \cup \mathcal{S}_{0^\ell}$ is complete for the construction.

5 Conclusions

A general paradigm for constructing VIL- and AIL-MACs from any FIL-MAC was presented. The design goals were minimal key-length, optimal waste, as well as suitability for on-line and parallel processing of the messages. Our single-key AIL-MAC constructions, CS and CR, have constant waste, allow for on-line and parallel processing of the message, and have essentially tight security reductions.

References

1. J. H. An and M. Bellare. Constructing VIL-MACs from FIL-MACs: Message authentication under weakened assumptions. In *Advances of Cryptology — CRYPTO '99*, volume 1666 of *LNCS*, pages 252–269. Springer-Verlag, 1999.
2. M. Bellare, R. Canetti, and H. Krawczyk. Keying hash functions for message authentication. In *Advances of Cryptology — CRYPTO '96*, volume 1109 of *LNCS*, pages 1–15. Springer-Verlag, 1996.
3. M. Bellare, J. Guérin, and P. Rogaway. XOR MACs: New methods for message authentication using finite pseudorandom functions. In *Advances of Cryptology — CRYPTO '95*, volume 963 of *LNCS*, pages 15–28. Springer-Verlag, 1995.
4. M. Bellare, J. Kilian, and P. Rogaway. The security of the cipher block chaining message authentication code. In *Journal of Computer and System Sciences*, 61(3):362–399, 2000.
5. I. Damgård. A design principle for hash functions. In *Advances in Cryptology — CRYPTO '89*, volume 435 of *Lecture Notes in Computer Science*, pages 416–427. Springer-Verlag, 1990.

6. U. Maurer. Indistinguishability of random systems. In *Advances of Cryptology — EUROCRYPT '02*, volume 2332 of *LNCS*, pages 110–132. Springer-Verlag, 2002.
7. R. Merkle. A certified digital signature. In *Advances in Cryptology — CRYPTO '89*, volume 435 of *LNCS*, pages 218–232. Springer-Verlag, 1990.
8. M. Naor and O. Reingold. From unpredictability to indistinguishability: A simple construction of pseudo-random functions from MACs (extended abstract). In *Advances in Cryptology — CRYPTO '98*, volume 1462 of *LNCS*, pages 267–282. Springer-Verlag, 1998.
9. S. Patel. An efficient MAC for short messages. In *Selected Areas in Cryptography*, volume 2595 of *LNCS*, pages 352–368. Springer-Verlag, 2003.
10. E. Petrank and C. Rackoff. CBC MAC for real-time data sources. In *Journal of Cryptology*, 13(3):315–338, 2000.
11. B. Preneel and P. C. van Oorschot, MDx-MAC and building fast MACs from hash functions. In *Advances in Cryptology — CRYPTO '95*, volume 953 of *LNCS*, pages 1–14. Springer-Verlag, 1995.

The Efficiency and Fairness of a Fixed Budget Resource Allocation Game

Li Zhang

Hewlett-Packard Labs, 1501 Page Mill Road,
Palo Alto, CA 94304, USA
l.zhang@hp.com

Abstract. We study the resource allocation game in which price anticipating players compete for multiple divisible resources. In the scheme, each player submits a bid to a resource and receives a share of the resource according to the proportion of his bid to the total bids. Unlike the previous study (e.g. [5]), we consider the case when the players have budget constraints, i.e. each player's total bids is fixed. We show that there always exists a Nash equilibrium when the players' utility functions are strongly competitive. We study the efficiency and fairness at the Nash equilibrium. We show the tight efficiency bound of $\Theta(1/\sqrt{m})$ for the m player balanced game. For the special cases when there is only one resource or when there are two players with linear utility functions, the efficiency is 3/4. We extend the classical notion of envy-freeness to measure fairness. We show that despite a possibly large utility gap, any Nash equilibrium is 0.828-approximately envy-free in this game.

1 Introduction

We study the performance of a mechanism for allocating multiple *divisible* resources to strategic and selfish players. We consider the price anticipating allocation scheme in which each player submits a bid to a resource and receives a fraction of his bid to the total bids submitted to that resource. Compared to, for example, auction based schemes [4,19,14], such commodity based scheme is simple to implement and is highly responsive to the player's need. It has been implemented for allocating computing resources in several distributed systems [2,9] and analyzed in [10,5,3]. One important constraint we consider in this paper is that each player has a budget constraint, i.e. a player's total bids is fixed. As we explain later, the natural condition on the budget adds difficulty for applying the existing technique. We also consider general utility functions instead of restricting to network game as considered in [17,5].

Similar to the previous work, we analyze the scheme as a noncooperative game and focus on the game's Nash equilibrium, the stable state at which no player has incentive to change his behavior. We first answer the question about the existence of Nash equilibrium in such a game. It is easy to construct an example to show that even when the utility functions satisfy the standard assumptions, there may still not exist an equilibrium. We show that an additional requirement of *strong competitiveness* can guarantee the existence of Nash equilibria. Informally, the players are strongly competitive if for each resource, there are at least two players who always desire it. However, unlike some similar game without budget constraint, the Nash equilibria are no longer unique.

For the performance at the Nash equilibrium, we first consider the efficiency, or price of anarchy according to [13], of the Nash equilibrium, defined as the ratio of the social welfare at the worst equilibrium to the social optimum. We show that with the budget constraint, the efficiency can be unfortunately low, at $\Theta(1/\sqrt{m})$ for m players. For the two special cases when there is only one resource and when there are only two players with linear utility functions, we show the tight efficiency bound of $3/4$. Interestingly, in both cases, the efficiency is $3/4$ although the proofs are completely different. In particular, the bound of $3/4$ in one resource case is a property of concave functions, which is probably interesting in its own right.

Another important criterion for evaluating a mechanism is its fairness. However, unlike the common agreement on the importance of efficiency metric, there are multiple definitions on fairness, such as max-min fairness [8] and proportional fairness [6,7]. In this paper, we evaluate the fairness by extending the notion of envy-freeness commonly used in the study of fair division [18,1]. An allocation is called *envy-free* if no one feels someone else gets a share more than he does. We extend this notion to approximately envy-free. Roughly speaking, an allocation is called c-approximately envy-free if everyone feels he gets at least c times of others share. We are able to show that if the players have the same budget, then any allocation at the Nash equilibrium is $2\sqrt{2} - 2 \approx 0.828$ approximately envy-free. This shows that while the utility gap between two players at an equilibrium might be big, the gap is bounded by a constant if we evaluate the player's utility from any single player's view.

In the previous work, there is usually no constraint on the budget so one can bid as much as one can as long as the marginal profit gain is positive. Such assumption about infinite supply of money is not realistic in practice. It is often the case that the player receives a limited endowment. The fixed budget can also be a result of multi-stage decision process in which the fund is reserved to cope the future need. The budget constraint has been considered in [10] with only empirical results.

The budget constraint, however, adds difficulty to analyzing those games. One typical technique used in analyzing Nash equilibria is to relate the equilibrium condition to the optimality condition of some potential function [16,12,11] or some convex program [17,5]. Intuitively, such connection is established by that at a Nash equilibrium, the marginal profit for each player on each resource is equally 0 and is therefore equivalent to the first order condition at the solution of some optimization problem. However, this is no longer true with budget constraint as the Lagrangian multiplier condition only stipulates that each player has equal margin on the resources he bids on but those margins may vary from player to player. Actually, there may exist multiple Nash equilibria when adding budget constraints, which is different from its counterpart without budget constraint.

2 Definitions

Price Anticipating Allocation Scheme. Suppose that there are m players (or users) and n resources (or machines) where each resource is continuously divisible. An *allocation scheme* $\omega = (\mathbf{r}_1, \ldots, \mathbf{r}_m)$, where $\mathbf{r}_i = (r_{i1}, \cdots, r_{in})$ with r_{ij} representing player i's share of resource j, satisfies that for any $1 \leq i \leq m$ and $1 \leq j \leq n$, $r_{ij} \geq 0$ and $\sum_{i=1}^{m} r_{ij} \leq 1$. Let Ω denote the set of all the allocation schemes.

Each player i has a finite budget $X_i > 0$. Let $X = \sum_i X_i$ be the total budget. In the *price anticipating allocation mechanism*, the resources are allocated as follows: a player submits a bid to each resource under the constraint that the total bids sum to the player's budget. Let x_{ij} denote the bid of player i on resource j, where $x_{ij} \geq 0$ and $\sum_{j=1}^{n} x_{ij} = X_i$. The bidding vector of player i is $\mathbf{x}_i = (x_{i1}, \ldots, x_{in})$. Let $Y_j = \sum_{i=1}^{m} x_{ij}$ be the total bids on resource j. Then, the fraction of player i gets from resource j is $r_{ij} = \frac{x_{ij}}{Y_j}$ if $Y_j > 0$ and 0 otherwise.

Utility function. The player's utility $U_i(r_{i1}, \cdots, r_{in}) : [0,1]^n \to R$ is a function of the fraction of each resource assigned to the player. We assume common properties about the utility functions: for each i, U_i is non-negative, differentiable, non-decreasing, and concave. One special family is *linear utility function*: $U_i(r_{i1}, \cdots, r_{in}) = w_{i1} r_{i1} + \cdots + w_{in} r_{in}$ for $w_{ij} \geq 0$. For linear utility functions, we also call w_{ij} the *weight* of resource j to the player i. We sometimes abuse the notation and let $U_i(\mathbf{x}_1, \ldots, \mathbf{x}_m)$ denote the utility of player i under the bidding vectors $\mathbf{x}_i$ for player i. Note that while $U_i(\mathbf{r})$ is differentiable and concave, $U_i(\mathbf{x})$ may not — in fact, it is not even continuous at the points where $x_{ij} = 0$ for some j and for all $1 \leq i \leq m$.

Let U_{ij} denote the partial derivative of U_i with respect to the j-th variable[1]. We say a resource j is *non-satiated* to player i if for any $\mathbf{r}$, $U_{ij}(\mathbf{r}) > 0$. Denote by Γ_j the set of players to whom the resource j is non-satiated. For example, for linear utility function, $\Gamma_j = \{i \,|\, w_{ij} > 0\}$. A game is called *strongly competitive* if for any j, $|\Gamma_j| > 1$. For linear utility functions, a game is strongly competitive if for any j, there exist $i \neq k$ such that $w_{ij}, w_{kj} > 0$.

Nash equilibrium. The bidding vectors $\mathbf{x}_1, \ldots, \mathbf{x}_m$ is a *Nash equilibrium* if for any $1 \leq i \leq m$ and for any legitimate bidding vector $\mathbf{x}'_i$, $U_i(\mathbf{x}_1, \ldots, \mathbf{x}_i, \ldots, \mathbf{x}_m) > U_i(\mathbf{x}_1, \ldots, \mathbf{x}'_i, \ldots, \mathbf{x}_m)$. At a Nash equilibrium, each player's bidding vector is the best response to the other's biddings. It represents a stable state at which no player has incentive to change his strategy. Denote by Ω_0 the set of all the allocation schemes at the Nash equilibria.

For an allocation scheme $\omega \in \Omega$, denote by $U(\omega) = \sum_i U_i(\mathbf{r}_i)$ the social welfare under ω. Let $U^* = \max_{\omega \in \Omega} U(\omega)$ denote the maximal social welfare. The efficiency at ω is defined as $\pi(\omega) = \frac{U(\omega)}{U^*}$. When $\Omega_0 \neq \emptyset$, the *efficiency* of a game Q is defined to be $\pi(Q) = \min_{\omega \in \Omega_0} \pi(\omega)$. While the definition of

---

[1] At the boundary, the derivative is understood as from one side.

efficiency is standard, there are multiple ways to define fairness. We extend the concept of envy-freeness [18] and measure the fairness of an allocation scheme ω by $\rho(\omega) = \min_{i \neq j} \frac{U_i(\mathbf{r}_i)}{U_i(\mathbf{r}_j)}$. When $\rho(\omega) \geq 1$, the scheme is known as an envy-free allocation scheme. We call a game c-*approximately envy-free* if for any $\omega \in \Omega_0$, $\rho(\omega) \geq c$. This shall be contrasted to the definition of utility gap $\min_{i \neq j} \frac{U_i(\mathbf{r}_i)}{U_j(\mathbf{r}_j)}$.

3 The Existence of Nash Equilibrium

The differentiability and concavity alone are not sufficient to guarantee the existence of the Nash equilibrium. This can be shown by a simple example of two players and two resources. For example, let $U_1(r_1, r_2) = r_1$ and $U_2(r_1, r_2) = r_1 + r_2$. Then player 1 should never bid on resource 2 because it has no value to him. Now, player 2 has to put a positive bid on resource 2 in order to claim the resource, but there is no lower limit, resulting the non-existence of the Nash equilibrium.[2] Clearly, this happens whenever there is a resource that is "wanted" by exactly one player, or when the game is not strongly competitive. As we shall show, this is the only case when the Nash equilibrium may not exist.

We use Rosen's theorem, a generalization of Nash's theorem, for proving the existence of Nash equilibrium. According to [15], a *concave game* is a game in which each player's strategy domain is a compact convex set, and each user's utility function is a concave function for any fixed choice of the other player's strategies. In [15], it is shown that

Theorem 1 (Rosen [15]). *An equilibrium exists for any concave game.*

Our game is almost a concave game except at the domain boundary where there is some j, $x_{ij} = 0$ for all $1 \leq i \leq m$. As noted before, the utility function is discontinuous at those points. This discontinuity causes the problem in the above example and prevents us from applying Rosen's theorem. To overcome this difficulty, by a similar technique used in [5], we perturb the game to fix the discontinuity and then argue that the limiting point is a Nash equilibrium of the original game, with the help of the additional assumption that the game is strongly competitive.

Now consider a perturbed game Q^ε in which each player's payoff function is $U_i^\varepsilon(\mathbf{x}) = U_i\left(\frac{x_{i1}}{\varepsilon + Y_1}, \cdots, \frac{x_{in}}{\varepsilon + Y_n}\right)$. It is easily verified that U_i^ε is differentiable, concave, and non-decreasing in x_{ij}'s, for $1 \leq j \leq n$. The domain of player i's strategy is the set $\Omega_i = \{(x_{i1}, \ldots, x_{in}) \mid \sum_{j=1}^n x_{ij} = X_i, \ x_{ij} \geq 0\}$, which is clearly a compact convex set. Therefore, by Rosen's theorem, there exists a Nash equilibrium of game Q^ε. Let $\omega^\varepsilon = (x_{ij}^\varepsilon)$ be any equilibrium of Q^ε. Now let $\varepsilon \to 0$. Since the strategy space is compact, there exist an infinite sequence that converge to a limiting point. Suppose the limiting point is ω, i.e. there is a sequence $\varepsilon_k \to 0$ and $\omega^{\varepsilon_k} \to \omega$. Clearly, ω is a legitimate strategy. We shall show that ω is a Nash equilibrium of the original game $Q = Q^0$.

---

[2] In fact, there does not even exist mixed strategy Nash equilibrium in this case.

Let us consider only those ε's in the converging sequence. In what follows, a constant means a number that is solely determined by the system parameters, m, n, U_i's, and is independent of ε. Let $Y_j^\varepsilon = \sum_{i=1}^m x_{ij}^\varepsilon$, and $z_{ij}^\varepsilon = Y_j^\varepsilon - x_{ij}^\varepsilon$. Then, $\frac{\partial U_i^\varepsilon}{\partial x_{ij}^\varepsilon}(\omega^\varepsilon) = U_{ij}\left(\frac{x_{ij}^\varepsilon}{\varepsilon+Y_j^\varepsilon}\right)\frac{\varepsilon+z_{ij}^\varepsilon}{(\varepsilon+Y_j^\varepsilon)^2}$.

By definition, the bidding vectors $\mathbf{x}_1,\ldots,\mathbf{x}_m$ is a Nash equilibrium if and only if each player's strategy is the best response to the group's bids. Since U_i is differentiable and concave, and Ω_i is convex, the optimality condition is that there exists $\lambda_i^\varepsilon > 0$ such that

$$\frac{\partial U_i^\varepsilon}{\partial x_{ij}^\varepsilon} \begin{cases} = \lambda_i^\varepsilon \text{ if } x_{ij}^\varepsilon > 0, \text{ and} \\ < \lambda_i^\varepsilon \text{ if } x_{ij}^\varepsilon = 0. \end{cases} \tag{1}$$

It is easy to see, from our condition on U_i and by the compactness of the domain, that

Lemma 1. *Let $r = (r_1,\cdots,r_n)$. For any $\varepsilon > 0$, there exists a constant $C > 0$, such that $U_{ij}(r) < C$ for any $r_j \geq \varepsilon$; and for any $1 \leq i \leq m$, there exists a constant $D_i > 0$ such that for $i \in \Gamma_j$, $U_{ij}(r) > D_i$.*

For an n-tuple $(r_1, r_2, \ldots, r_n)$, let $(\mathbf{r}_{-j}; s)$ denote the n-tuple where r_j is replaced by s. In addition, let $\mathbf{0}$ and $\mathbf{1}$ denote the n-tuple with all 0's and all 1's, respectively. For $\varepsilon > 0$ and $\delta < 1$, set $C_\varepsilon = \max(\max_{i,j} U_{ij}(\mathbf{0}_{-j};\varepsilon), 1)$ and $D = \min_i D_i$. Clearly, $C_{\varepsilon_1} \geq C_{\varepsilon_2}$ for $\varepsilon_1 < \varepsilon_2$.

We first upper bound the marginal profit of ω^ε.

Lemma 2. *There exists a constant $M_1 > 0$ such that for sufficiently small ε, $\lambda_i^\varepsilon < M_1$ for any $1 \leq i \leq m$.*

Proof. At the equilibrium, the player i has to bid at least $\frac{X_i}{n}$ on some resource, say k. By optimality condition (1), $\lambda_i^\varepsilon = U_{ik}^\varepsilon(r_{i1},\cdots,r_{in})\frac{\varepsilon+z_{ik}^\varepsilon}{(\varepsilon+Y_k^\varepsilon)^2}$.

For any given x_{ik}^ε, $\frac{\varepsilon+z_{ik}^\varepsilon}{(\varepsilon+Y_k^\varepsilon)^2} = \frac{\varepsilon+z_{ik}^\varepsilon}{(\varepsilon+z_{ik}^\varepsilon+x_{ik}^\varepsilon)^2}$ achieves maximum when $\varepsilon + z_{ik}^\varepsilon = x_{ik}^\varepsilon$. Thus, when $\varepsilon < X$,

$$\lambda_i^\varepsilon \leq U_{ik}^\varepsilon(r_{i1},\cdots,r_{in})\frac{1}{4x_{ik}^\varepsilon} \leq U_{ik}^\varepsilon(\mathbf{r}_{-k}; \frac{X_i}{n(\varepsilon+X)})\frac{n}{4X_i}$$
$$\leq U_{ik}^\varepsilon(\mathbf{r}_{-k}; \frac{X_i}{2nX})\frac{n}{4X_i} \leq C_{d_i/2n}\frac{n}{4X_i}.$$

Let $d_{min} = \min_i \frac{X_i}{X} > 0$. Set $M_1 = C_{d_{min}/2n}\frac{n}{4d_{min}X}$. By Lemma 1, we have that $\lambda_i^\varepsilon \leq M_1$. □

We then lower bound the total bids on each resource.

Lemma 3. *For a strongly competitive game, there exists a constant $c_0 > 0$ such that for sufficiently small ε and for any j, $Y_j^\varepsilon \geq c_0$.*

Proof. We first consider the case when there is some player $k \in \Gamma_j$ not bidding on resource j, i.e. $x_{kj}^\varepsilon = 0$. Then, $\lambda_k = U_{kj}(\mathbf{r}_k)\frac{\varepsilon+Y_j^\varepsilon-x_{kj}^\varepsilon}{(\varepsilon+Y_j^\varepsilon)^2} = \frac{U_{kj}(\mathbf{r}_k)}{\varepsilon+Y_j^\varepsilon}$.

Since $\lambda_k \leq M_1$, we have that $Y_j^\varepsilon > \frac{U_{kj}(\mathbf{r}_k)}{M_1} - \varepsilon > \frac{D}{M_1} - \varepsilon$. Thus, if we set $c_0 = \frac{D}{4M_1}$, when $\varepsilon < c_0$, we have that $Y_j^\varepsilon > c_0$.

For the other case, we assume that for all $k \in \Gamma_j$, $x_{kj}^\varepsilon > 0$. Then,

$$\lambda_k = U_{kj}(\mathbf{r}_k)\frac{\varepsilon + Y_j^\varepsilon - x_{kj}^\varepsilon}{(\varepsilon + Y_j^\varepsilon)^2} \geq D\frac{\varepsilon + Y_j^\varepsilon - x_{kj}^\varepsilon}{(\varepsilon + Y_j^\varepsilon)^2}.$$

Let $\ell = |\Gamma_j|$. Then $\ell \geq 2$ by the assumption that the game is strongly competitive. Thus

$$\sum_{k \in \Gamma_j} \lambda_k \geq D \sum_{k \in \Gamma_j} \frac{\varepsilon + Y_j^\varepsilon - x_{kj}^\varepsilon}{(\varepsilon + Y_j^\varepsilon)^2} = D\frac{\ell\varepsilon + \ell Y_j^\varepsilon - \sum_{k \in \Gamma_j} x_{kj}^\varepsilon}{(\varepsilon + Y_j^\varepsilon)^2}$$

$$\geq D\frac{\ell\varepsilon + (\ell-1)Y_j^\varepsilon}{(\varepsilon + Y_j^\varepsilon)^2} \geq D(\ell-1)\frac{1}{\varepsilon + Y_j^\varepsilon}.$$

By that $\lambda_k < M_1$, we have that $D(\ell-1)\frac{1}{\varepsilon+Y_j^\varepsilon} < \ell M_1$. Thus $Y_j^\varepsilon > D(\ell-1)/(\ell M_1) - \varepsilon \geq D/(2M_1) - \varepsilon$. Again, $Y_j^\varepsilon > c_0$ if $\varepsilon < c_0$. □

We are now ready for the main lemma.

Lemma 4. *If the game is strongly competitive, then for any $\delta > 0$, when ε is sufficiently small,*

$$\left|\frac{\partial U_i(x)}{\partial x_{ij}}(\omega) - \frac{\partial U_i^\varepsilon(x)}{\partial x_{ij}}(\omega^\varepsilon)\right| \leq \delta.$$

Proof. Clearly, $\frac{\partial U_i(x)}{\partial x_{ij}}(\omega) = U_i(r_{ij})\frac{z_{ij}}{Y_j^2}$ and $\frac{\partial U_i^\varepsilon(x)}{\partial x_{ij}}(\omega^\varepsilon) = U_i(r_{ij}^\varepsilon)\frac{z_{ij}^\varepsilon}{(\varepsilon+Y_j^\varepsilon)^2}$.

The lemma follows immediately by $z_{ij}^\varepsilon \to z_{ij}$ and $Y_j^\varepsilon \to Y_j$, and that $Y_j^\varepsilon \geq c_0$, for some constant $c_0 > 0$. □

Thus, we have the following theorem.

Theorem 2. *ω is a Nash equilibrium if Q is strongly competitive.*

Proof. (By contradiction) Suppose it were not true, then the optimality condition (1) is violated for some player i. There are two possibilities.

1. There are j, k, where $j \neq k$, such that $x_{ij}, x_{ik} > 0$ and $\frac{\partial U_i}{\partial x_{ij}} \neq \frac{\partial U_i}{\partial x_{ik}}$. By Lemma 4, we know that for sufficiently small ε, the following holds: $x_{ij}^\varepsilon > 0, x_{ik}^\varepsilon > 0$, and $\frac{\partial U_i^\varepsilon}{\partial x_{ij}}(\omega^\varepsilon) \neq \frac{\partial U_i^\varepsilon}{\partial x_{ik}}(\omega^\varepsilon)$. This contradicts with that ω^ε is a Nash equilibrium of Q^ε. Now, we assume that $\lambda_i = \frac{\partial U_i}{\partial x_{ij}}(\omega)$ for any $x_{ij} > 0$.

2. There is j where $x_{ij} = 0$ and $\frac{\partial U_i}{\partial x_{ij}}(\omega) > \lambda_i$. By the same reasoning, we again derive contradiction.

Hence, ω is a Nash equilibrium of Q. □

We remark that there may exist multiple Nash equilibria. For example, consider the case with two players and two resources where $X_1 = X_2 = 1$, and $U_1 = \alpha r_1 + (1-\alpha)r_2$, $U_2 = (1-\alpha)r_1 + \alpha r_2$ for $0 < \alpha < 1$. As shown in [3], there exist three Nash equilibria whenever $\alpha > (2\sqrt{2}+2)/4 \approx 0.854$,.

4 Efficiency

In this section, we study the efficiency of Nash equilibria. If we allow arbitrary utility function and budget, the efficiency at Nash equilibria can be unboundedly low, for example, when there are players with high budget but very low utility. Therefore, we assume that the players' utilities are related with their budgets by requiring that $U_i(\mathbf{0}) = 0$ and $X_i/U_i(\mathbf{1}) = \delta$ for some constant δ and for $1 \leq i \leq n$. In other words, we assume that each player's utility is proportional to their budget if they can own the whole system. We refer to such game a *balanced* game. We first show asymptotic bounds for the general case and then present tight bounds for two special cases.

For large number of players, the efficiency can be unfortunately low, in the order of $\frac{1}{\sqrt{m}}$ for m players, as shown in the following theorem. Let $v_i = U_i(\mathbf{1})$ and $v = \sum_i v_i$.

Theorem 3. *For an m-player balanced game Q, $\pi(Q) \geq \max(\frac{v}{mU^*}, \frac{U^*}{2v}) \geq \frac{1}{\sqrt{2m}}$, and the bound is asymptotically tight as there exists an equilibrium w such that $\pi(w) = O(\frac{1}{\sqrt{m}})$.*

Proof. We first construct an upper bound. Consider a system with $m = n^2 + n$ players and n resources. In addition, each player has a linear utility function. Of m players, there are n^2 who have the same weights on all the resources, i.e. $1/n$ on each resource (recall the definition of weight in Section 2). The other n players have weight 1 on each different resource and 0 on all the other resources. Clearly $U^* = n$. It is easy to verify the following allocation is an equilibrium: the first n^2 players evenly distribute their bids among all the resources, the other n player put all the bid on their respective favorite resource. At this equilibrium, the total bids on each resource is $n+1$. Each of the first n^2 players receives $\frac{1}{n}\frac{1/n}{n+1} = \frac{1}{n^2(n+1)}$ on each resource, summing up to a total utility of $n^3 \cdot \frac{1}{n^2(n+1)} < 1$. The other n players each receives $\frac{1}{n+1}$ on their favorite resource, summing up to a total utility of $n \cdot \frac{1}{n+1} < 1$. Therefore, the total utility of the equilibrium is < 2, while the social optimum is $n = \Theta(\sqrt{m})$.

Now we show the lower bound. Let $X = \sum_i X_i = \delta v$. First, we show that each player i gets at least $\frac{X_i}{X} v_i$. For any existing allocations, a player can distribute his bid proportional to the amount of total bids already allocated to the resources. Let y_j denote the total bids of the players other than i on resource j. We can then assign $x_{ij} = \frac{y_j X_i}{Y}$, where $Y = \sum_j y_j = X - X_i$. Under such scheme, player i's share of resource j is $r_{ij} = \frac{x_{ij}}{x_{ij}+y_j} = X_i/(X_i + Y) = X_i/X$. Thus the utility of the player i is at least $U_i(X_i/X, X_i/X, \ldots, X_i/X) > U_i(\mathbf{1})X_i/X = X_i v_i/X$ by concavity of U_i. Therefore, $U \geq \sum_i X_i v_i/X = \sum_i v_i^2/v \geq v/m$ by Cauchy-Schwarz inequality. Thus, $U/U^* \geq v/(mU^*)$.

Now we show that $U \geq (U^*)^2/(2v)$. Let r_{ij}^* denote player i's share of resource j at the social optimum, and $U^* = \sum_i U_i(r_{i1}^*, \ldots, r_{in}^*)$. Let $Y_j = \sum_i x_{ij}$, the total bids on resource j at the Nash equilibrium. Let $B_i = \sum_j r_{ij}^* Y_j$ and $z_{ij} =$

$r_{ij}^* Y_j X_i / B_i$. Clearly, $\sum_j z_{ij} = X_i$. Consider the bidding where player i submits bid z_{ij} to resource j. Then $r_{ij} = \frac{z_{ij}}{Y_j - x_{ij} + z_{ij}} \geq \frac{z_{ij}}{Y_j + z_{ij}} = \frac{r_{ij}^* X_i}{B_i + X_i r_{ij}^*} \geq \frac{r_{ij}^* X_i}{X_i + B_i}$.

Since at the equilibrium, player i's strategy is the best response to the group. We have that $U_i(\mathbf{x}) \geq U_i(\mathbf{z}) \geq U_i(r_{ij}^* X_i/(X_i + B_i)) \geq \frac{X_i}{X_i + B_i} U_i(r_{ij}^*)$. Therefore, $U = \sum_{i=1}^m U_i \geq \sum_{i=1}^m U_i(r_{ij}^*) \frac{X_i}{X_i + B_i}$. Further, $\sum_i B_i = \sum_i \sum_j r_{ij}^* Y_j = \sum_j Y_j \sum_i r_{ij}^* = \sum_j Y_j = X$.

Let $W_i = U_i(r_{ij}^*)$. We now minimize $\sum_{i=1}^m \frac{W_i X_i}{X_i + B_i}$ under the constraints that $\sum_{i=1}^m B_i = X$, and $B_i \geq 0$. Let $\zeta = \min_i \frac{\sqrt{X_i W_i}}{\sum_j \sqrt{X_j W_j}} X$. We relax the above constraints to that $\sum_{i=1}^m B_i = X$, and $B_i \geq -X_i + \zeta$.

The domain remains a convex set and the minimization function is a convex function on the domain. By Lagrangian multiplier method, the minimum is achieved when setting $B_i = \frac{2X\sqrt{X_i W_i}}{\sum_{j=1}^m \sqrt{X_j W_j}} - X_i > -X_i + \zeta$. Thus

$$U \geq \sum_{i=1}^m \frac{\sqrt{X_i W_i} \sum_{j=1}^m \sqrt{X_j W_j}}{2X} = \frac{(\sum_{i=1}^m \sqrt{X_i W_i})^2}{2X}.$$

Since $X_i = \delta v_i \geq \delta W_i$, we have that $U \geq \frac{\delta(\sum_{i=1}^m W_i)^2}{2X} = \frac{(U^*)^2}{2v}$. Hence, $U/U^* \geq \frac{U^*}{2v}$. Combining with $U/U^* \geq v/(mU^*)$, we have that $U/U^* \geq 1/\sqrt{2m}$ as $\max(v/(mU^*), U^*/2v)$ is minimized when $U^* = v\sqrt{2/m}$. □

In the following, we show tight bounds for two special cases.

One resource. In this case, the Nash equilibrium is the trivial one: everyone bids all of his budget on the resource, and player i receives the fraction of $X_i/X = v_i/v$. Recall that $v_i = U_i(1)$, $v = \sum_i v_i$, and $X_i/v_i = \delta$. At the equilibrium, $U = \sum_i U_i(v_i/v)$. On the other hand, $U^* = \max_{r_i} \sum_i U_i(r_i)$, where $r_i \geq 0$ and $\sum_i r_i = 1$. Surprisingly, the efficiency is at least $3/4$ under such a "dumb" allocation scheme.

Theorem 4. *Let $U_1, \cdots, U_m : [0,1] \to \mathbb{R}$ be non-negative, non-decreasing concave functions with $U_i(1) = v_i$. Let $v = \sum_i v_i$. Then for any $r_i \geq 0$ that satisfy $\sum_i r_i = 1$, $\sum_{i=1}^m U_i(v_i/v) \geq 3/4 \cdot \sum_{i=1}^m U_i(r_i)$.*

Proof. Write $a_i = U_i(r_i)$. Let f_i be the piecewise linear function with $f_i(0) = 0$, $f_i(r_i) = a_i$, and $f_i(1) = 1$. Clearly, $f(r_i) = U_i(r_i)$. Further $U_i(x) \geq f(x)$ for $x \in [0,1]$ by concavity. Write $s_i = v_i/v$. Thus, it suffices to show that $\sum_{i=1}^m f_i(s_i) \geq 3/4 \cdot \sum_{i=1}^m a_i$.

If $r_i \leq s_i$, then $f_i(s_i) = a_i + \frac{v_i - a_i}{1 - r_i}(s_i - r_i)$, and when $r_i > s_i$, $f_i(s_i) = \frac{s_i a_i}{r_i}$.

Let $r_i \leq s_i$ for $1 \leq i \leq k$ and $r_i > s_i$ for $k < i \leq m$. Then,

$$\sum_i f_i(s_i) = \sum_{i=1}^k \left(a_i + \frac{v_i - a_i}{1 - r_i}(s_i - r_i) \right) + \sum_{i=k+1}^m \frac{a_i s_i}{r_i}$$

$$= \sum_{i=1}^m a_i + \sum_{i=1}^k \frac{v_i - a_i}{1 - r_i}(s_i - r_i) + \sum_{i=k+1}^m \left(\frac{a_i s_i}{r_i} - a_i \right).$$

Let $\Delta = (\sum_{i=1}^{k} \frac{v_i - a_i}{1 - r_i}(s_i - r_i) + \sum_{i=k+1}^{m} (\frac{a_i s_i}{r_i} - a_i)) / \sum_i a_i$. We observe that Δ is minimized when $a_i = v_i$ for all $1 \leq i \leq m$: it is when the denominator is maximized, and every individual term in the numerator is minimized. Thus, Δ is at least

$$\sum_{i=k+1}^{m} (v_i s_i / r_i - v_i)/v = \sum_{i=k+1}^{m} (v_i^2/(vr_i) - v_i)/v. \tag{2}$$

Write $w = \sum_{i=k+1}^{m} v_i$. (2) is minimized when $r_i = \frac{v_i}{w}$, and the minimum value is $\frac{w^2/v - w}{v} \geq -1/4$. Thus, $\sum_i f_i(s_i) / \sum_i a_i = 1 + \Delta \geq 3/4$. □

The bound of $3/4$ is tight even when $m = 2$ by setting $U_1(x) = 1$ and $U_2(x) = x$. As a special case, when $v_i = 1$ for all i, we have that for any m non-negative, non-decreasing concave functions $f_i(x)$ with $f_i(1) = 1$, $\sum_i f_i(1/m) \geq 3/4 \sum_i f_i(r_i)$, for any $r_i \geq 0$ and $\sum_i r_i = 1$,

Two player, linear utility function. In this case, we can obtain again the same efficiency of $3/4$, but curiously with completely different proof.[3] The proof is by reducing the problem further to when there are only two resources. The details will appear in the full version of the paper.

Theorem 5. *For any two player, two resource game with linear utility functions, $\pi(Q) \geq 3/4$, and the bound is tight in the worst case.*

5 Fairness

It is important for a resource allocation scheme to be "fair". There are many standards to evaluate fairness [18, 1]. In one definition, one may consider whether any player gets a more than average share. According to Theorem 3, each player i's utility is at least $U_i(1)/m$, or in other words, each player gets at least $1/m$-th of the maximum utility he can possibly get from the system. Thus, in this sense the scheme is fair.

Another possibility is to compare the utility between two players. We can easily construct an example so that one player has utility $1/m$ while some player has 1. Or $U_i(\mathbf{x}_i) \leq U_j(\mathbf{x}_j)/m$. In this standard, the scheme is highly unfair as it creates large utility gap.

Here, we consider fairness by extending the notion of envy-freeness [18], a concept central in the economical study of fair division. Different from the utility gap measure, envy-freeness, as defined in Section 2, measures how much each player likes (or envies) other's share compared to his own. For a meaningful discussion, we assume that all the players have the same budget, or $X_i = 1$ for

---

[3] The constant $3/4$ also appears in [17] in analyzing the efficiency of selfish routing and in [5] when there is no budget constraint. The bound is tight in all the four cases, but all those proofs do not have apparent connection. We do not know if it is just a coincidence or there is a deeper connection.

all the i's. Unlike the efficiency analysis, we do not require $U_i(\mathbf{1})$ proportional to the player's budget. In general, an equilibrium allocation is not envy-free. But it is not very far from being envy free according to the following theorem.

Theorem 6. *When $X_i = 1$ for all the i's, any Nash equilibrium is at least c-approximately envy-free for $c = 2\sqrt{2} - 2 \approx 0.828$. The bound is tight in the worst case.*

Proof. Suppose that at the Nash equilibrium ω, $x_1, \ldots, x_n$ are the allocation of bids of a player A, and $y_1, \ldots, y_n$ are the total bids by the other players. Let $r_i = x_i/(x_i + y_i)$. We need to show that for all $z_1, \ldots, z_n$ with $0 \leq z_i \leq y_i$ and $\sum_i z_i = 1$, we have that $U_i(\mathbf{r}) \geq c \cdot U_i\left(\frac{z_1}{x_1 + y_1}, \ldots, \frac{z_n}{x_n + y_n}\right)$.

First, we show that we can assume A has a linear utility function. Otherwise, we consider the linear utility function $\overline{U}(s_1, \ldots, s_n) = \sum_i U_i'(r_1, \ldots, r_n) s_i$. Because $\frac{\partial \overline{U}}{\partial x_i}(\mathbf{r}) = U_i'(\mathbf{r}) \frac{y_i}{(x_i + y_i)^2} = \frac{\partial U}{\partial x_i}(\mathbf{r})$, the bids $x_1, \ldots, x_n$ is still A's best response to the group bids $y_1, \ldots, y_n$ under the utility function $\overline{U}$. In addition, the change of A's utility function does not affect the other players. Hence, ω remains a Nash equilibrium of the changed utility function. In addition, by concavity of U, we have that

$$U(\mathbf{s}) \leq U(\mathbf{r}) + \sum_i U_i'(r_1, \ldots, r_n)(s_i - r_i) = U(\mathbf{r}) + \overline{U}(\mathbf{s}) - \overline{U}(\mathbf{r}). \quad (3)$$

Thus, if $\overline{U}(\mathbf{r}) \geq c \cdot \overline{U}(\mathbf{s})$, then

$$U(\mathbf{r}) = \overline{U}(\mathbf{r}) + (U(\mathbf{r}) - \overline{U}(\mathbf{r})) \geq c \cdot \overline{U}(\mathbf{s}) + (U(\mathbf{r}) - \overline{U}(\mathbf{r}))$$
$$\geq c(U(\mathbf{s}) - U(\mathbf{r}) + \overline{U}(\mathbf{r})) + (U(\mathbf{r}) - \overline{U}(\mathbf{r}))$$
$$= c \cdot U(\mathbf{s}) + (1 - c)(U(\mathbf{r}) - \overline{U}(\mathbf{r})) \geq c \cdot U(\mathbf{s}).$$

Therefore, without loss of generality, we assume that A has linear utility function $U(\mathbf{s}) = w_1 s_1 + \cdots + w_n s_n$, and $U_A = U(\mathbf{r})$. Let λ_i be the margin on the resources i for A. By optimality condition, for any i, j, $\lambda_i \geq \lambda_j$ if $x_i > 0$ and the equality holds when $x_j > 0$. The price p_i of the resources $\frac{w_i}{x_i + y_i}$ on each resource. We first note that

$$\frac{\lambda_i}{p_i} = \frac{w_i y_i/(x_i + y_i)^2}{w_i/(x_i + y_i)} = \frac{y_i}{x_i + y_i} \leq 1. \quad (4)$$

Now suppose that $\mathbf{z} = (z_1, \cdots, z_n)$ maximizes $V = \sum_i p_i z_i$ subject to that $0 \leq z_i \leq y_i$, and $\sum_i z_i = 1$. We now show that $U_A \geq (2\sqrt{2} - 2)V$. Intuitively, the proof is done by comparing the utility of A on each resource under two allocations. When $x_i > z_i$, we collect the "residual" utility to compensate those i's where $x_i < z_i$. Formally, consider b_{ij} for $1 \leq i, j \leq n$ that satisfy

$$b_{ii} = \min(x_i, z_i), \quad \sum_j b_{ij} = x_i, \quad \sum_i b_{ij} = z_j, \quad b_{ij} \geq 0.$$

Since $\sum_i x_i = \sum_i z_i$, there always exist b_{ij}'s satisfying the above condition.

$$U_A = \sum_i p_i x_i = \sum_i (p_i \sum_j b_{ij}) = \sum_i (p_i b_{ii} + \sum_{i \neq j} p_i b_{ij})$$

$$\geq \sum_i (p_i b_{ii} + \sum_{i \neq j} \lambda_i b_{ij}), \text{ by (4)}.$$

$$\geq \sum_i (p_i b_{ii} + \sum_{i \neq j} \lambda_j b_{ij}), \text{ since } b_{ij} > 0 \text{ implies } x_i > 0.$$

$$= \sum_j (\sum_{i \neq j} \lambda_j b_{ij} + p_j b_{jj}) = \sum_j (\lambda_j \sum_{i \neq j} b_{ij} + p_j b_{jj}).$$

Let $u_j = \lambda_j \sum_{i \neq j} b_{ij} + p_j b_{jj}$, and $v_j = p_j z_j$. We distinguish two cases.

- When $z_j \leq x_j$, then $b_{jj} = z_j$ and for all $i \neq j$, $b_{ij} = 0$. Thus, $u_j = p_j z_j = v_j$.
- When $z_j > x_j$, then $b_{jj} = z_j$ and $\sum_{i \neq j} b_{ij} = z_j - x_j$. Thus $u_j = p_j x_j + \lambda_j (z_j - x_j)$. Therefore, $c_j = u_j/(p_j z_j) \geq (x_j + (z_j - x_j)\frac{y_j}{x_j + y_j})/z_j = \frac{x_j^2}{z_j(x_j + y_j)} + \frac{y_j}{x_j + y_j}$. Clearly, c_j is minimized when z_j is maximized, i.e $z_j = y_j$, and the minimum value is: $\frac{x_j^2}{y_j(x_j + y_j)} + \frac{y_j}{x_j + y_j} = \frac{x_j^2 + y_j^2}{y_j(x_j + y_j)}$. Now, fix y_j, c_j is minimized when $x_j = (\sqrt{2} - 1)y_j$ and the minimum value is $2\sqrt{2} - 2$.

In both cases, $u_j \geq (2\sqrt{2} - 2)v_j$. Thus, $\sum_i u_i \geq (2\sqrt{2} - 2) \sum_i p_i z_i$. Hence, the scheme is c-approximately envy-free for $c = 2\sqrt{2} - 2 \approx 0.828$.

For the upper bound, consider the case with $s + 2$ players and 2 resources. Let $\alpha = 2/(s+1)$. Suppose that the utility functions are $U_1 = \alpha r_1 + (1-\alpha)r_2$, $U_2 = r_1$, and for $3 \leq i \leq s+2$, $U_i = r_2$. Clearly, at the equilibrium, player 2 bids $(1, 0)$, and the player 3 to $s+2$ bid $(0, 1)$. By equalizing the margin on the two resources, we derive that the bids of player 1 are $\frac{2\beta - \alpha}{\alpha(1+\beta)}$ and $\frac{2\alpha + \alpha\beta - 2\beta}{\alpha(1+\beta)}$, where $\beta = \frac{\alpha}{\sqrt{(1-\alpha)(2-\alpha)}}$. When $s \to +\infty$, i.e. $\alpha \to 0$, the bids of player 1 approaches $(\sqrt{2} - 1, 2 - \sqrt{2})$, and $U_1(\mathbf{x}_1)/U_1(\mathbf{x}_2) \to 2\sqrt{2} - 2$. □

6 Conclusion

In this paper, we study the existence and performance of Nash equilibrium of a price anticipating resource allocation game. We consider general utility functions with the budget constraint. We show the existence of Nash equilibrium for strongly competitive game and the bounds for efficiency and fairness of such a game. It would be interesting to design efficient algorithms to achieve the equilibrium and to study the specific utility functions that arise from practice, e.g. those discussed in [3].

Acknowledgments. The author would like to thank Kevin Lai, Lars Rasmusson, Michal Feldman, Fang Wu, and Bernardo Huberman for their many helpful discussions. The author would also like to thank the anonymous reviewers for their useful comments.

References

1. S. J. Brams and A. D. Taylor. *Fair Division: From Cake-cutting to Dispute Resolution*. Cambridge University Press, 1996.
2. B. N. Chun and D. E. Culler. Market-based proportional resource sharing for clusters. Technical Report CSD-1092, University of California at Berkeley, Computer Science Division, January 2000.
3. M. Feldman, K. Lai, and L. Zhang. A price-anticipating resource allocation mechanism for distributed shared clusters. In *Proceedings of ACM Conference on Electronic Commerce*, 2005.
4. D. Ferguson, Y. Yemimi, and C. Nikolaou. Microeconomic algorithms for load balancing in distributed computer systems. In *International Conference on Distributed Computer Systems*, pages 491–499, 1988.
5. R. Johari and J. N. Tsitsiklis. Efficiency loss in a network resource allocation game. *Mathematics of Operations Research*, 2004.
6. F. P. Kelly. Charging and rate control for elastic traffic. *European Transactions on Telecommunications*, 8:33–37, 1997.
7. F. P. Kelly and A. Maulloo. Rate control in communication networks: Shadow prices, proportional fairness and stability. *Operational Res. Soc.*, 49:237–252, 1998.
8. Y. Korilis and A. Lazar. Why is flow control hard: optimality, fairness, partial and delayed information. In *Proceedings of 2nd ORSA Telecommunications Conference*, 1992.
9. K. Lai, L. Rasmusson, S. Sorkin, L. Zhang, and B. A. Huberman. Tycoon: a distributed market-based resource allocation system. Manuscript, http://www.hpl.hp.com/research/idl/papers/tycoon, 2004.
10. R. T. Maheswaran and T. Basar. Nash equilibrium and decentralized negotiation in acutioning divisible resources. *Group Decision and Negotiation*, 12:361–395, 2003.
11. I. Milchtaich. Congestion games with player-specific payoff functions. *Games and Economic Behavior*, 13:111–124, 1996.
12. D. Monderer and L. S. Sharpley. Potential games. *Games and Economic Behavior*, 14:124–143, 1996.
13. C. Papadimitriou. Algorithms, games, and the Internet. In *Proceedings of 33rd Annual ACM Symposium on Theory of Computing*, pages 749–753, 2001.
14. O. Regev and N. Nisan. The POPCORN market – an online market for computational resources. In *Proceedings of 1st International Conference on Information and Computation Economies*, pages 148–157, 1998.
15. J. B. Rosen. Existence and uniqueness of equilibrium points for concave N-person games. *Econometrica*, 33(3):520–534, 1965.
16. R. W. Rosenthal. A class of games possessing pure-strategy Nash equilibria. *International Journal of Game Theory*, 2:65–67, 1973.
17. T. Roughgarden and E. Tardos. How bad is selfish routing? *Jounral of the ACM*, 49(2):236–259, 2002.
18. H. R. Varian. Equity, envy, and efficiency. *Journal of Economic Theory*, 9:63–91, 1974.
19. C. A. Waldspurger, T. Hogg, B. A. Huberman, J. O. Kephart, and W. S. Stornetta. Spawn: A distributed computational economy. *Software Engineering*, 18(2):103–117, 1992.

Braess's Paradox, Fibonacci Numbers, and Exponential Inapproximability

Henry Lin[1,*], Tim Roughgarden[2,**], Éva Tardos[3,***], and Asher Walkover[4]

[1] UC Berkeley, Computer Science Division,
Soda Hall, Berkeley, CA 94720
henrylin@cs.berkeley.edu

[2] Department of Computer Science, Stanford University,
462 Gates Building, Stanford, CA 94305
tim@cs.stanford.edu

[3] Cornell University, Department of Computer Science,
Upson Hall, Ithaca, NY 14853
eva@cs.cornell.edu

[4] Google Inc., Mountain View, CA
walkover@gmail.com

Abstract. We give the first analyses in multicommodity networks of both the worst-case severity of Braess's Paradox and the price of anarchy of selfish routing with respect to the maximum latency. Our first main result is a construction of an infinite family of two-commodity networks, related to the Fibonacci numbers, in which both of these quantities grow exponentially with the size of the network. This construction has wide implications, and demonstrates that numerous existing analyses of selfish routing in single-commodity networks have no analogues in multicommodity networks, even in those with only two commodities. This dichotomy between single- and two-commodity networks is arguably quite unexpected, given the negligible dependence on the number of commodities of previous work on selfish routing.

Our second main result is an exponential upper bound on the worst-possible severity of Braess's Paradox and on the price of anarchy for the maximum latency, which essentially matches the lower bound when the number of commodities is constant.

Finally, we use our family of two-commodity networks to exhibit a natural network design problem with intrinsically exponential (in)approximability: while there is a polynomial-time algorithm with an exponential approximation ratio, subexponential approximation is unachievable in polynomial time (assuming $P \neq NP$).

---

* Supported by a UC Berkeley Research Fellowship.
** Supported in part by ONR grant N00014-04-1-0725 and DARPA grant W911NF-04-9-0001.
*** Supported in part by NSF grant CCR-032553, NSF grant 0311333, and ONR grant N00014-98-1-0589.

1 Introduction

Selfish Routing and the Price of Anarchy. A recent trend in theoretical computer science is to analyze the extent to which equilibria in a noncooperative game approximate a social optimum. The most popular measure is the *price of anarchy* [15] (also called the *coordination ratio* [13]), defined as the worst-case ratio between the objective function value of a Nash equilibrium of a game and that of an optimal solution. The price of anarchy is thus defined relative to a game and to an objective function.

In this paper, we will study one of the most popular models for price of anarchy analyses, the so-called *selfish routing* model. Selfish routing is a mathematical model of how noncooperative agents route traffic in a network with congestion. Formally, the game takes place in a directed multicommodity flow network, where each edge possesses a continuous, nondecreasing latency function that models how the performance of an edge degrades as it becomes increasingly congested. The traffic in the network is assumed to comprise a large number of independent network users, so that each individual has negligible impact on the experience of others. Under this assumption, equilibria—*flows at Nash equilibrium*—are naturally defined as the multicommodity flows in which all traffic travels only on minimum-latency paths.

As in most noncooperative games, flows at Nash equilibrium are inefficient, in the sense that they need not optimize natural objective functions. One such objective function, the average latency incurred by traffic, has been extensively studied. Beginning with Roughgarden and Tardos [21] and continuing with studies of ever-increasing generality [4, 6, 16, 18, 22], exact worst-case bounds on the price of anarchy with respect to the average latency have been established under a wide variety of different assumptions.

The Maximum Latency Objective. As in many combinatorial optimization problems, permitting an objective function to average the cost of different users can be problematic from a fairness perspective. Specifically, to attain or approximate a flow that minimizes the average latency, some users may need to be sacrificed to very costly paths, in order to reduce the congestion encountered by others. This unfairness inherent in the average latency measure motivates modifying the objective function to be more attuned to those users on the most costly paths. Arguably, the most obvious way to accomplish this is to aspire toward minimizing the *maximum* latency incurred by any user.

Compared to the average latency objective, considerably less is known about the price of anarchy relative to the maximum latency. The first paper on the topic is by Weitz [25], whose results we will review below. Most relevant for us is a paper by Roughgarden [19], where only the special case of single-commodity networks, networks in which all traffic shares the same source and destination, were considered. The main result of [19] states that, if latency functions are allowed to be arbitrary continuous, nondecreasing functions, then the (worst-case) price of anarchy with respect to the maximum latency objective in single-commodity networks with at most n vertices is precisely $n - 1$.

Roughgarden [19] also made two conjectures about this price of anarchy in multicommodity networks. The *weak conjecture* of [19] asserts that in multicommodity networks, this price of anarchy can be bounded by a function of the number of vertices, edges, and commodities in the network. As a point of contrast, simple examples show that no such bound is possible for the price of anarchy relative to the average latency, unless additional structure is imposed on the network latency functions [21]. The *strong conjecture* of [19] states that the price of anarchy with respect to the maximum latency remains $n - 1$ in multicommodity networks. This conjecture was motivated in part by the provable equivalence of single-commodity and multicommodity networks for the price of anarchy relative to the average latency [6, 18].

Braess's Paradox. Braess's Paradox is the following counterintuitive fact: adding additional edges to a network with selfish routing can make all of the traffic worse off. First discovered in 1968 [3], Braess's Paradox has since motivated a vast number of follow-up papers; see [17] for a survey. Almost all existing work on the paradox confines attention to close variations on or analogues of Braess's original example in a four-node network. Only recently have larger, more severe versions of Braess's Paradox been discovered. Specifically, Roughgarden [17] defined an infinite family of networks, beginning with Braess's original example, that shows that adding edges to a single-commodity network with n vertices can increase the latency experienced by all of the traffic (and hence the maximum latency) by a factor of $\lfloor n/2 \rfloor$. It was also shown in [19] that no greater increase is possible in single-commodity networks, but the proof of this crucially used the combinatorial structure of flows at Nash equilibrium in such networks. Once again the story was left incomplete: are the networks of [17] the worst examples possible, or are more severe versions of Braess's Paradox lurking in the richer landscape of multicommodity networks?

Our Results. In this paper, we establish nearly matching upper and lower bounds on both the price of anarchy with respect to the maximum latency and on the worst-possible severity of Braess's Paradox in multicommodity networks. Our results resolve both of the conjectures in [19]—one in the affirmative, one in the negative—and also give the first demonstration that Braess's Paradox is provably more severe in multicommodity networks than in single-commodity ones. Specifically, our two main results are the following.

- We give a parameterized construction, based on the Fibonacci numbers, that shows that adding one edge to a two-commodity network with n vertices can increase the latency of all traffic by a $2^{\Omega(n)}$ factor.
- We prove that the price of anarchy with respect to the maximum latency in networks with k commodities, n vertices, and m edges is $2^{O(\min\{kn, m \log n\})}$.

The construction used to prove the first result has wide implications. In particular, for all existing approximation-type analyses of selfish routing that were only known to hold for single-commodity networks [14, 17, 19], this construction rules out any reasonable extension to multicommodity networks, even those with only two commodities. For example, removing one edge from a single-commodity network can only decrease the maximum (or average) latency of a Nash flow by a factor of 2 [14], while our construction shows that a single edge removal can cause an exponential improvement in the average and the maximum latency (even with only two commodities). This dichotomy between single- and two-commodity networks is somewhat unexpected, given the negligible role that the number of commodities has played in previous work in this area [6, 18, 21].

The first result easily implies a lower bound of $2^{\Omega(n)}$ on the price of anarchy for the maximum latency in multicommodity networks, as an optimal flow has the option of ignoring edges that are causing Braess's Paradox. By the same reasoning, the second result implies that adding any number of edges to a network with k commodities, n vertices, and m edges can only increase the maximum latency by a $2^{O(\min\{kn, m \log n\})}$ factor. Our upper and lower bounds on both the price of anarchy and on the worst-possible severity of Braess's Paradox are thus essentially tight for networks with a constant number of commodities.

Finally, we consider the following network design problem, motivated by the goal of detecting and avoiding Braess's Paradox: given a network, find the subnetwork with the smallest maximum latency. Using our family of two-commodity networks and ideas from the gap reductions of [17] that apply to the single-commodity version of the problem, we prove that there is no polynomial-time algorithm for this network design problem with subexponential approximation ratio (assuming $P \neq NP$). Since our upper bound on the price of anarchy trivially implies that an exponential performance guarantee is achievable, this network design problem is a rare example of a natural optimization problem with intrinsically exponential approximability.

Further Related Work. There have been numerous price of anarchy analyses in the past few years. Study of the original load-balancing model of Koutsoupias and Papadimitriou [13] continues unabated; see [7, 10] for surveys. A survey of the selfish routing model studied here, including results on the price of anarchy, can be found in [20]. Other noncooperative games have also been studied recently from a price of anarchy perspective, including facility location games [8, 24], network design games [1, 9], and resource allocation games [12].

As noted above, Weitz [25] was the first to study the price of anarchy of selfish routing under the maximum latency objective. Weitz [25] noted that, for single-commodity networks and classes of restricted latency functions, the price of anarchy for the maximum latency is no more than that for the average latency objective. For example, a theorem of Roughgarden and Tardos [21] bounding the price of anarchy with respect to the average latency objective then implies that the price of anarchy for the maximum latency in single-commodity networks with linear latency functions is at most $4/3$, and a matching lower bound is furnished by the original form of Braess's Paradox [3, 25]. However, upper bounds on the price of anarchy with respect to the maximum latency objective do not imply upper bounds on the price of anarchy with respect to the average latency objective: for example, the price of anarchy for the maximum latency objective is at most $n - 1$ in single-commodity networks with arbitrary latency functions [19], while the price of anarchy for the average latency can be arbitrarily large even in two-node, two-link networks [21].

Weitz [25] also gave a family of networks that shows that this price of anarchy is $\Omega(n)$ for multicommodity networks with n vertices and linear latency functions. Concurrently with Roughgarden [19], Correa, Schulz, and Stier Moses [5] studied the maximum latency objective from several different perspectives. The results of [5] mostly concern the computational complexity of computing an optimal solution and the extent to which multiple objective functions can be simultaneously optimized, and are disjoint from those in [19] and in the present work.

2 Preliminaries

The Model. We now describe our model of selfish routing, following Roughgarden and Tardos [21]. We will study a multicommodity flow network, described by a directed graph $G = (V, E)$ and k source-destination vertex pairs $(s_1, t_1), \ldots, (s_k, t_k)$. We denote by r_i the amount of traffic that wishes to travel from the source s_i to the destination t_i—the *traffic rate*. The graph G can contain parallel edges, but we can exclude self-loops. We will denote the s_i-t_i paths of G by $\mathcal{P}_i$. We assume that $\mathcal{P}_i$ is non-empty for all i, and define $\mathcal{P} = \cup_{i=1}^{k} \mathcal{P}_i$.

A *flow* is a nonnegative vector indexed by $\mathcal{P}$. By f_e we mean the amount $\sum_{P \in \mathcal{P} : e \in P} f_P$ of flow that traverses edge e. With respect to a network G and a vector r of traffic rates, a flow is *feasible* if $\sum_{P \in \mathcal{P}_i} f_P = r_i$ for all commodities i.

We assume that the network G suffers from congestion effects, and to model this we give edge e a nonnegative, continuous, nondecreasing *latency function* ℓ_e that describes the time needed to traverse the edge as a function of the edge congestion f_e. Given a flow f, the latency ℓ_P of a path P is the sum of the latencies of the edges in the path: $\ell_P(f) = \sum_{e \in P} \ell_e(f_e)$. We will call a triple of the form (G, r, ℓ) an *instance*.

Our objective function is the maximum latency incurred by a flow, defined formally by $M(f) = \max_{P \in \mathcal{P} : f_P > 0} \ell_P(f)$. With respect to an instance (G, r, ℓ), a flow that minimizes $M(\cdot)$ over all feasible flows will be called *optimal*. Since the feasible flows of an instance form a compact subset of Euclidean space and $M(\cdot)$ is a continuous function, every instance admits an optimal flow.

Flows at Nash Equilibrium. We next define the flows that we expect to arise from selfish routing. Assuming that all network users have negligible size and want to minimize the latency experienced, we expect all users to travel on paths with minimum-possible latency. We formalize this in the next definition.

Definition 1. *A flow f feasible for (G, r, ℓ) is at Nash equilibrium, or is a Nash flow, if for every $i \in \{1, 2, \ldots, k\}$ and two paths $P_1, P_2 \in \mathcal{P}_i$ with $f_{P_1} > 0$,*

$$\ell_{P_1}(f) \le \ell_{P_2}(f).$$

Happily, Nash flows always exist, and all Nash flows of an instance have equal maximum latency.

Proposition 1. *Let (G, r, ℓ) be an instance.*

(a) There is at least one Nash flow for (G, r, ℓ).
(b) If $f, \tilde{f}$ are Nash flows for (G, r, ℓ), then $M(f) = M(\tilde{f})$.

Proposition 1 is classical; for example, it follows from arguments of Beckmann, McGuire, and Winsten [2].

Definition 1 implies that in a Nash flow, all of the traffic of a given commodity experiences a common latency. We will sometimes use the notation $L_i(G, r, \ell)$ to denote the common latency of the ith commodity's traffic in a Nash flow for (G, r, ℓ); analogously to Proposition 1(b), this is well defined (i.e., independent of the particular Nash flow).

We will also benefit from the following alternative definition of a Nash flow, which was first noted by Smith [23]. It is an easy consequence of Definition 1.

Proposition 2. *A flow f feasible for (G, r, ℓ) is at Nash equilibrium if and only if*

$$\sum_{e \in E} \ell_e(f_e) f_e \leq \sum_{e \in E} \ell_e(f_e) \tilde{f}_e \qquad (1)$$

for every flow $\tilde{f}$ that is feasible for (G, r, ℓ).

The Price of Anarchy. We now formalize what we mean by the price of anarchy. As noted in the introduction, it is the ratio of the objective function values of a flow at Nash equilibrium and an optimal flow. If (G, r, ℓ) is an instance, then the *price of anarchy of* (G, r, ℓ), denoted $\rho(G, r, \ell)$, is the ratio $M(f)/M(f^*)$, where f is a Nash flow and f^* is an optimal flow. Proposition 1 ensures that the price of anarchy of an instance is well defined provided $M(f^*) > 0$. If $M(f^*) = 0$, then f^* is also a flow at Nash equilibrium and we define the price of anarchy of the instance to be 1.

Finally, the price of anarchy $\rho(\mathcal{I})$ of a collection $\mathcal{I}$ of instances is defined in the obvious way:

$$\rho(\mathcal{I}) = \sup_{(G,r,\ell) \in \mathcal{I}} \rho(G, r, \ell).$$

3 Braess's Paradox in Multicommodity Networks

In this section, we prove that Braess's Paradox can be much more severe in multicommodity networks than in single-commodity networks. In fact, there will be a "phase transition" of sorts: the worst-case severity of Braess's Paradox is polynomial in single-commodity instances, but exponential in two-commodity instances. The family of instances that we construct in this section will also serve as a starting point for our inapproximability results in Section 5.

We will begin this section by formally stating the properties of our construction in Theorem 3 below. Prior to detailing this construction and proving Theorem 3, we will discuss its many consequences for multicommodity networks.

Our family of two-commodity instances is closely related to the *Fibonacci numbers*. Recall that for a nonnegative integer p, the pth Fibonacci number F_p is defined as follows: $F_0 = 0$, $F_1 = 1$, and $F_p = F_{p-2} + F_{p-1}$ for $p \geq 2$. It is well known that $F_p \approx c \cdot \phi^p$ as $p \to \infty$, where $c \approx 0.4472$ and $\phi \approx 1.618$ is the golden ratio.

We can now state the main result of this section.

Theorem 3. *There is an infinite family $\{(G^p, r^p, \ell^p)\}_{p=1}^{\infty}$ of instances with the following properties:*

(a) (G^p, r^p, ℓ^p) has two commodities and $O(p)$ vertices and edges as $p \to \infty$;
(b) for p odd, $L_1(G^p, r^p, \ell^p) = F_{p-1} + 1$ and $L_2(G^p, r^p, \ell^p) = F_p$;

(c) for p even, $L_1(G^p, r^p, \ell^p) = F_p + 1$ and $L_2(G^p, r^p, \ell^p) = F_{p-1}$;
(d) for all p, there is a subgraph H^p of G^p with one less edge than G^p that satisfies $L_1(H^p, r^p, \ell^p) = 1$ and $L_2(H^p, r^p, \ell^p) = 0$.

Theorem 3 has a number of implications. We begin by noting two immediate corollaries of the theorem.

Corollary 1. *Adding a single edge to an n-vertex two-commodity instance can increase the latency of all traffic by a $2^{\Omega(n)}$ factor as $n \to \infty$.*

Corollary 2. *If $\mathcal{I}_n$ is the set of instances with at most n vertices, then $\rho(\mathcal{I}_n) = 2^{\Omega(n)}$ as $n \to \infty$.*

Furthermore, Corollary 1 trivially implies that for every $k \geq 2$, adding a single edge to an n-vertex k-commodity instance can increase the latency of all traffic by a $2^{\Omega(n)}$ factor as $n \to \infty$.

Theorem 3 and Corollaries 1 and 2 show that a number of previously established properties of single-commodity instances do not carry over to multicommodity networks. In particular, the following statements are known to hold in single-commodity instances.

(1) Adding one edge to a single-commodity instance can only increase the maximum or average latency of a Nash flow by a factor of 2 [14].
(2) Adding any number of edges to an n-vertex single-commodity instance can only increase the maximum or average latency of a Nash flow by a factor of $\lfloor n/2 \rfloor$ [17].
(3) The price of anarchy with respect to maximum latency in an n-vertex single-commodity instance is at most $n - 1$ [19].

Theorem 3 and Corollaries 1 and 2 demonstrate that all of these statements utterly fail to extend to multicommodity networks, even to those with only two commodities. This dichotomy stands in contrast to other work on selfish routing, such as bounds on the price of anarchy with respect to the average latency objective function, where there is provably no separation between single-commodity and multicommodity instances [18, 6].

We now give the construction of the family of instances claimed in Theorem 3. We begin by defining the graph G^p for $p \geq 1$, see Figure 1. We will describe the construction only for p odd; the construction for even p is similar. We begin with two paths, which we will call P_1 and P_2. The $(p+3)$-vertex path P_2, drawn vertically in Figure 1, is $s_2 \to w_0 \to w_1 \to \cdots \to w_p \to t_2$. The $(p+4)$-vertex path P_1, drawn horizontally in Figure 1, is $s_1 \to a \to w_1 \to v_1 \to \cdots \to v_p \to t_1$. We also add the following edges between the two paths:

- (a, w_i) for all positive even i;
- (v_i, w_i) for all positive even i;
- (s_2, v_i) for all odd i at most $p - 2$;
- (w_i, v_i) for all odd i.

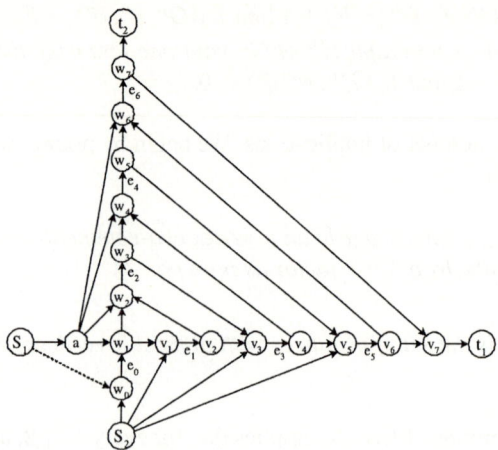

Fig. 1. Construction of the instance (G^p, r^p, ℓ^p) when $p = 7$. Dotted edge (s_1, w_0) is the "extra edge". Edges with non-constant latency functions are labelled

Finally, we complete G^p by adding what we will call an *extra edge*, defined as the edge (s_1, w_0).

For all p, the traffic rate vector r^p will be $r_1^p = r_2^p = 1$. To complete the construction, we therefore need only describe the edge latency functions. All edges will either possess a constant latency function, or a latency function that approximates a step function. We next introduce notation for the latter function type. For a positive integer i and a positive real number δ, f_δ^i will denote a continuous, nondecreasing function satisfying $f_\delta^i(x) = 0$ for $x \leq 1$ and $f_\delta^i(x) = F_i$ for $x \geq 1 + \delta$. (The function f_δ^i can be defined arbitrarily on $(1, 1 + \delta)$, provided it is continuous and nondecreasing.)

For $i \in \{0, 1, \ldots, p-1\}$, we define the edge e_i to be (w_i, w_{i+1}) if i is even and (v_i, v_{i+1}) if i is odd. (See Figure 1.) We now define the latency functions ℓ^p for G^p as follows, where δ is sufficiently small (to be chosen later): for each $i > 0$, edge e_i receives the latency function $\ell^p(x) = f_\delta^i(x)$, edge e_0 receives the latency function $\ell^p(x) = f_\delta^1(x)$, edge (s_1, a) receives the latency function $\ell^p(x) = 1$, and all other edges receive the latency function $\ell^p(x) = 0$.

With the construction in hand, we now turn toward proving Theorem 3 for odd p (the arguments for even p are similar). Part (a) is obvious. Part (d) is easy to see: if H^p is obtained from G^p by removing the extra edge (s_1, w_0) and f is the flow that routes one unit of traffic on both P_1 and P_2, then f is at Nash equilibrium for (G^p, r^p, ℓ^p), showing that $L_1(H^p, r^p, \ell^p) = 1$ and $L_2(H^p, r^p, \ell^p) = 0$. (See Figure 2.)

To finish the proof of Theorem 3 (for p odd), we need only prove part (b). We will accomplish this via a sequence of lemmas, the first of which requires some further definitions. First, we will say that a flow f, feasible for (G^p, r^p, ℓ^p), *floods* the instance if $f_{e_i} \geq 1 + \delta$ for all $i \in \{0, 1, \ldots, p-1\}$. Thus if f floods (G^p, r^p, ℓ^p), all edge latencies are at their maximum, as in Figure 3. Second, we introduce notation for some of the paths of G^p. For i even, Q_i will denote the unique s_1-t_1 path which traverses edge e_i before any odd labelled edges, and includes no other edge of P_2. For i odd,

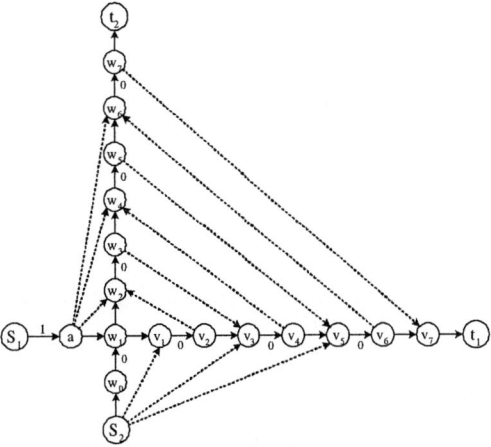

Fig. 2. Nash flow in (H^p, r^p, ℓ^p), with $p = 7$, where H^p is G^p with the extra edge (s_1, w_0) removed. Solid edges carry flow, dotted edges do not. Edge latencies are with respect to the Nash flow. Unlabelled edges have zero latency

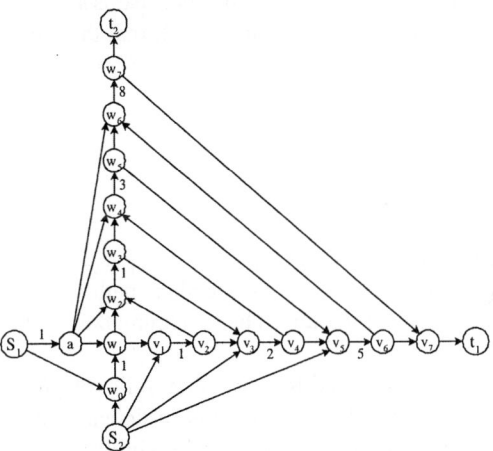

Fig. 3. Nash flow in (G^p, r^p, ℓ^p), with $p = 7$. Solid edges carry flow, dotted edges do not. Edge latencies are with respect to the Nash flow. Unlabelled edges have zero latency

Q_i will denote the unique s_2-t_2 path which traverses edge e_i before any even labelled edges, and includes no other edge of P_1. We will call the paths $Q_0, \ldots, Q_{p-1}$, together with the "axis-aligned" paths P_1 and P_2, the *short paths*. The next lemma justifies this terminology, at least for flows that flood the instance (G^p, r^p, ℓ^p).

Lemma 1. *If f floods (G^p, r^p, ℓ^p) with p odd, then:*

(a) $\ell_P(f) \geq F_{p-1} + 1$ for every s_1-t_1 path P, and equality holds for short paths;
(b) $\ell_P(f) \geq F_p$ for every s_2-t_2 path P, and equality holds for short paths.

We will only prove part (b) of Lemma 1, as the proof of part (a) is similar. In the proof, we will use the following lemma about Fibonacci numbers, which is easy to verify.

Lemma 2. *Let j and p be odd positive integers with $j < p$, and I the even numbers between j and p. Then, $F_j + \sum_{i \in I} F_i = F_p$.*

We now prove Lemma 1.

Proof of Lemma 1: Let P be an s_2-t_2 path. Let j be the largest odd number such that $e_j \in P$, or 0 if there is no such number. We only need to prove the case where $j > 0$, since the $j = 0$ and $j = 1$ cases are the same. If $j > 0$, then P contains e_j and also e_i for all even i between j and p. Since f floods (G^p, r^p, ℓ^p), Lemma 2 implies that $\ell_P(f) \geq F_p$. Moreover, this inequality holds with equality for short paths. ∎

Our final lemma states that routing flow on short paths suffices to flood the instance (G^p, r^p, ℓ^p). For the statement of the lemma, recall that the parameter δ controls how rapidly the non-constant latency functions of (G^p, r^p, ℓ^p) increase as the amount of flow on the edge exceeds one.

Lemma 3. *For all p odd and δ sufficiently small, there is flow f, with $f_P > 0$ only for short paths P, that floods (G^p, r^p, ℓ^p).*

Proof. Define the flow f as follows. First, for $i = 0, 1, \ldots, p-1$, route $2^{-(i+1)}$ units of flow (of the appropriate commodity) on the short path Q_i. This routes strictly less than one unit of flow of each commodity. The remaining flow is then routed on the short paths P_1 and P_2.

To complete the proof, we need to show that f floods (G^p, r^p, ℓ^p)—that $f_{e_i} \geq 1 + \delta$ for all $p \in \{0, 1, \ldots, p-1\}$ provided δ is sufficiently small. We will prove this inequality only for i odd; the argument for even i is similar.

The second commodity uses edge e_i only in the short path Q_i, on which it routes $2^{-(i+1)}$ units of flow. The first commodity uses edge e_i in all of its flow paths except for the short paths Q_j for j even and greater than i. The total amount of flow on e_i is thus at least

$$2^{-(i+1)} + 1 - \sum_{j \geq 0} 2^{-(i+2+2j)} = 1 + 2^{-(i+1)} - \frac{4}{3} \cdot 2^{-(i+2)} > 1 + 2^{-(i+3)}.$$

Thus, provided $\delta \leq 2^{-(p+3)}$, f floods (G^p, r^p, ℓ^p), and the proof is complete. ∎

Theorem 3(b) now follows immediately from Definition 1, Lemma 1, and Lemma 3.

4 Upper Bounds on the Price of Anarchy

We now turn toward proving upper bounds on the price of anarchy and, as a consequence, on the worst-possible severity of Braess's Paradox. We will aim for an upper bound that matches the lower bound of Theorem 3, and will largely succeed in this goal.

We begin by proving a very weak bound on the price of anarchy, a bound that depends on parameters other than the size of the network. While not interesting in its own right, this bound will play a crucial role in later proofs in this section.

Lemma 4. *Let f be a Nash flow and f^* a feasible flow for an instance (G, r, ℓ), where G has m edges. For every edge e of G with $f_e > f_e^*$,*

$$\ell_e(f_e) \leq \frac{m \sum_i r_i}{f_e - f_e^*} \cdot \max_e \ell_e(f_e^*). \tag{2}$$

Proof of Lemma 4: Let $F \subseteq E$ denote the edges e of G for which $f_e > f_e^*$. Using inequality (1) in Proposition 2 and the fact that $\ell_e(f_e^*) \leq M(f^*)$ whenever $f_e^* > 0$, we can derive the following crude bound:

$$\sum_{e \in F} \ell_e(f_e)(f_e - f_e^*) \leq \sum_{e \in E \setminus F} \ell_e(f_e)(f_e^* - f_e)$$

$$\leq \left(\max_e \ell_e(f_e^*)\right) \sum_{e \in E \setminus F} (f_e^* - f_e)$$

$$\leq \max_e \ell_e(f_e^*) \cdot m \cdot \sum_i r_i.$$

The lemma now follows easily. ∎

We next use Lemma 4 as a bootstrap for deriving upper bounds on the price of anarchy that depend only on the size of the network. We will accomplish this as follows. For an arbitrary instance, we will set up a linear program, with edge latencies as variables, that maximizes the price of anarchy among instances that are "basically equivalent" to the given instance. We will define our notion of equivalence so that Lemma 4 ensures that the linear program has a bounded maximum, and will then analyze the vertices of the feasible region of the linear program to derive the following bound.

Theorem 4. *If (G, r, ℓ) is an instance with n vertices and m edges, then*

$$\rho(G, r, \ell) = 2^{O(m \log n)}.$$

Before implementing the proof approach outlined above, we state a proposition that bounds the maximum size of the optimal value of a linear program with a constraint matrix with entries in $\{-1, 0, 1\}$.

Proposition 5. *Let A be an $m \times n$ matrix with entries in $\{0, \pm 1\}$ and at most α non-zero entries in each row. Let b be a real-valued m-vector, and let the linear program $\max x_i$ subject to $Ax \leq b$ have a finite maximum. Then, this maximum is at most $n\alpha^n \|b\|_\infty$, where $\|b\|_\infty$ denotes $\max_j |b_j|$.*

Proof of Theorem 4: Let (G, r, ℓ) be an instance with n vertices and m edges. Let f and f^* be Nash and optimal flows for (G, r, ℓ), respectively. We aim to show that $\rho(G, r, \ell) = 2^{O(m \log n)}$.

We begin by performing some preprocessing on the instance (G, r, ℓ). First, if $f_e = f_e^* = 0$ for some edge e, then that edge can be removed from the instance without affecting its ρ-value. We can therefore assume that $f_e^* > 0$ or $f_e > 0$ for every edge e. Second, we can assume that $\ell_e(0) = 0$ whenever $f_e^* = 0$. To see why, note that replacing the latency function $\ell_e(x)$ of such an edge by the function equal to (e.g.) $\min\{x/f_e, 1\} \cdot \ell_e(x)$ leaves the Nash flow unaffected while only decreasing the maximum latency of f^* and hence increasing the ρ-value of the instance. Combining these two assumptions, we can assume without loss of generality $\ell_e(f_e^*) \leq M(f^*)$ for every edge e of G.

We now set up a linear program that attempts to further transform the latency functions to make the ρ-value of the given instance as large as possible. In the linear program, the flow amounts $\{f_e\}$ and $\{f_e^*\}$, as well as the latencies $\{\ell_e(f_e^*)\}$ with respect to f^*, will be held fixed. There will be a nonnegative variable $\hat{\ell}_e(f_e)$ representing the latency of edge e with respect to the flow f. So that the new latency functions are nondecreasing, we impose the following linear constraints, which we call *monotonicity constraints*:

- For all edges e with $f_e = f_e^*$, $\hat{\ell}_e(f_e) = \ell_e(f_e^*)$.
- For all edges e with $f_e < f_e^*$, $\hat{\ell}_e(f_e) \leq \ell_e(f_e^*)$.
- For all edges e with $f_e > f_e^*$, $\hat{\ell}_e(f_e) \geq \ell_e(f_e^*)$.

Additionally, we will insist that the (fixed) flow f be at Nash equilibrium with respect to the (variable) latencies $\{\hat{\ell}_e(f_e)\}$. There are several ways that this requirement can be encoded with linear constraints. For this proof, we will be content with the following naive approach: for every commodity i, and every pair of paths $P, \tilde{P} \in \mathcal{P}_i$ for which $f_e^{(i)} > 0$ for all $e \in P$, we insist that $\sum_{e \in P} \hat{\ell}_e(f_e) \leq \sum_{e \in \tilde{P}} \hat{\ell}_e(f_e)$ in our linear program. Since this linear program has a small number of variables, we will not be hampered by its potentially massive number of constraints.

By construction, our constraints ensure the following: for every feasible solution $\{\hat{\ell}(f_e)\}$, there is an instance $(G, r, \hat{\ell})$ with continuous, nondecreasing latency functions $\hat{\ell}$, so that these latency functions interpolate their two prescribed values and f is a Nash flow for $(G, r, \hat{\ell})$. Consider the objective function $\max \hat{\ell}_e(f_e)$ for an edge e. Our key claim is that the resulting linear program is not unbounded. For edges e with $f_e \leq f_e^*$, the claim is obvious from the constraints. For edges e with $f_e > f_e^*$, the claim follows from Lemma 4 and the fact that all parameters on the right-hand side of the bound (2) are fixed in the linear program.

Since the maximum of the above linear program is bounded, we can apply Proposition 5. In our linear program, there are a total of m variables, of which each constraint contains at most $2n$. The right-hand side of each constraint is either a 0 or a term of the form $\ell_e(f_e^*)$. By our preprocessing step, $\ell_e(f_e^*) \leq M(f^*)$ for all edges e. Hence, Proposition 5 implies that the maximum of the linear program is at most

$mn^{O(m)} \cdot M(f^*)$. Hence, returning to the original instance (G, r, ℓ), we must have $\ell_e(f_e) \leq mn^{O(m)} \cdot M(f^*)$ for all edges e. Since a flow path of f can contain only n edges, we can conclude that $\rho(G, r, \ell) \leq nmn^{O(m)} = 2^{O(m \log n)}$. ∎

When the number of commodities is small (e.g., a constant), we can improve the bound to $2^{O(kn)}$ using a different way to encode the constraint that f must be at Nash equilibrium with respect to $\{\hat{\ell}_e(f_e)\}$.

Theorem 6. *If (G, r, ℓ) is an instance with n vertices and k commodities, then*

$$\rho(G, r, \ell) = 2^{O(kn)}.$$

Proof. To prove our bound, we start with the linear program described in Theorem 4. We leave the objective and monotonicity constraints the same, but replace the constraints that ensure f is a Nash equilibrium for $\{\hat{\ell}_e(f_e)\}$. To ensure the latencies $\{\hat{\ell}_e(f_e)\}$ define a Nash equilibrium for f, we introduce an auxiliary variable $\hat{d}^i(v)$ for each commodity i and for every vertex $v \in V$ reachable from that commodity's source s_i, which will represent the length of the shortest path from s_i to v, with respect to the latencies $\{\hat{\ell}_e(f_e)\}$. Now, we define the following constraints:

- $\hat{d}_i(s_i) = 0$, for all commodities i.
- $\hat{d}_i(u) + \hat{\ell}_e(f_e) = \hat{d}_i(v)$, for all edges $e = (u, v)$ and commodities i with $f_e^{(i)} > 0$.
- $\hat{d}_i(u) + \hat{\ell}_e(f_e) \leq \hat{d}_i(v)$, for all edges $e = (u, v)$ and commodities i.

To complete the proof, we need to show:

(a) A set of latencies $\{\hat{\ell}_e(f_e)\}$ is feasible for our linear program if and only if it defines a Nash equilibrium for f.
(b) The latency variables can be removed, yielding a linear program with kn variables, that can be bounded by $2^{O(kn)}$.

To prove the forward direction of (a), it is easy to see that with our constraints, for every commodity i, each path $P \in \mathcal{P}_i$, with $f_e^{(i)} > 0$ for all $e \in P$, must have length precisely equal to $\hat{d}_i(t_i)$. Furthermore, no path in $P \in \mathcal{P}_i$ may have length strictly less than $\hat{d}_i(t_i)$. Therefore, for every commodity i and every $P, \tilde{P} \in \mathcal{P}_i$, with $f_e^{(i)} > 0$ for all $e \in P$, $\sum_{e \in P} \hat{\ell}_e(f_e) \leq \sum_{e \in \tilde{P}} \hat{\ell}_e(f_e)$, and our latencies define a Nash equilibrium for f. To prove the other direction of (a), note that for any set of latencies $\hat{\ell}_e(f_e)$ defining a Nash equilibrium for f, we can define $\hat{d}_i(v)$ to be the length of the shortest path from s_i to v, and we have a feasible solution for our linear program.

With (a) proven, we know that the maximum value of our linear program is precisely the maximum edge length that occurs in any Nash equilibrium for flow f. As before, note that Lemma 4 implies our linear program is bounded, and we can apply Proposition 5 to bound the maximum value of the linear program.

Before applying Proposition 5 however, we first eliminate the latency variables, which allows us to prove a better bound. Note that for any edge $e = (u, v)$ with $f_e^i > 0$ for some commodity i, $\hat{d}_i(u) + \hat{\ell}_e(f_e) = \hat{d}_i(v)$, so we can replace any occurrence of $\hat{\ell}_e(f_e)$ in our linear program with $\hat{d}_i(v) - \hat{d}_i(u)$. Furthermore, for any other edges, $f_e = 0$, and it must be the case that $\hat{\ell}_e(f_e) \le M(f^*)$. For these edges, an optimal solution must assign some value $x_e \le M(f^*)$ to the latencies $\hat{\ell}_e(f_e)$, and thus for these edges, we can substitute the $\hat{\ell}_e(f_e)$ variable with the constant x_e value used in the optimal solution. With these substitutions, we have not changed the optimal value of our linear program, and we are only left with $O(kn)$ variables. Moreover, there are still a constant number of variables per constraint, and each entry of b is still bounded by $M(f^*)$. Therefore, applying Proposition 5 bounds the price of anarchy by $2^{O(kn)}$.

Corollary 2 shows that Theorem 6 is essentially tight for a constant number of commodities.

5 Exponential Inapproximability for Network Design

In this section, we will show that a network design problem that is naturally motivated by Braess's Paradox has intrinsically exponential approximability. The problem, which we call MULTICOMMODITY NETWORK DESIGN (MCND), is as follows.

> Given a (multicommodity) instance (G, r, ℓ), find a subgraph H of G that minimizes $M(H, r, \ell)$.

By $M(H, r, \ell)$, we mean the maximum latency of a Nash flow for (H, r, ℓ) (well defined by Proposition 1). MCND is tantamount to detecting and avoiding Braess's Paradox. For single-commodity instances, this problem was studied in [17].

The *trivial algorithm* is defined as the algorithm that always returns the entire graph G—the algorithm that always punts on trying to detect Braess's Paradox. The following was proved in [17]: the trivial algorithm is an $\lfloor n/2 \rfloor$-approximation algorithm for the special case of single-commodity instances; and for every $\epsilon > 0$, no $(\lfloor n/2 \rfloor - \epsilon)$-approximation algorithm exists (assuming $P \ne NP$). Here, we will succeed in proving analogues of these results for multicommodity networks, where the best-possible approximation ratio is inherently exponential.

First, we note that since Theorems 4 and 6 imply limits on the largest possible increase in the maximum latency due to Braess's Paradox, they also translate to an upper bound on the trivial algorithm.

Proposition 7. *The trivial algorithm is a $2^{O(\min\{kn, m \log n\})}$-approximation algorithm for* MCND.

Much more interesting is the next result, which states that there is *no* polynomial-time algorithm with subexponential approximation ratio (assuming $P \ne NP$).

Theorem 8. *Assuming $P \ne NP$, there is no $2^{o(n)}$-approximation algorithm for* MCND.

The proof of Theorem 8 combines ideas from the gap reductions of [17] for the single-commodity version of MCND with the family of two-commodity instances described in Section 3. Because of space constraints, we will content ourselves here with a high-level overview of the proof.

Recall that in an instance of the NP-complete problem PARTITION, we are given q positive integers $\{a_1, a_2, \ldots, a_q\}$ and seek a subset $S \subseteq \{1, 2, \ldots, q\}$ such that $\sum_{j \in S} a_j = \frac{1}{2} \sum_{j=1}^{q} a_j$ [11–SP12]. The idea of the reduction is to start with an instance (G^p, r^p, ℓ^p) of the form described in Section 3, and to replace the extra edge (s_1, w_0) with a collection of parallel edges representing an instance $\mathcal{I} = \{a_1, \ldots, a_p\}$ of PARTITION. We will give these edges latency functions that simulate "capacities", with an edge representing an integer a_j of $\mathcal{I}$ receiving capacity proportional to a_j. The proof then break down into three parts. First, if too many of these parallel edges are removed from the network, there will be insufficient remaining capacity to send flow cheaply. To implement this, we must also augment the latency functions of the edges $e_0, \ldots, e_{p-1}$ of Section 3 to have effective capacities. Second, if too few of the parallel edges are removed, the excess of capacity results in a bad flow at Nash equilibrium similar to that of Figure 3. Finally, these two cases can be avoided if and only if $\mathcal{I}$ is a "yes" instance of PARTITION, in which case removing the appropriate collection of parallel edges results in a network that admits a good Nash equilibrium similar to that of Figure 2.

References

1. E. Anshelevich, A. Dasgupta, É. Tardos, and T. Wexler. Near-optimal network design with selfish agents. In *Proceedings of the 35th Annual ACM Symposium on the Theory of Computing (STOC)*, pages 511–520, 2003.
2. M. Beckmann, C. B. McGuire, and C. B. Winsten. *Studies in the Economics of Transportation*. Yale University Press, 1956.
3. D. Braess. Über ein Paradoxon aus der Verkehrsplanung. *Unternehmensforschung*, 12:258–268, 1968.
4. C. K. Chau and K. M. Sim. The price of anarchy for non-atomic congestion games with symmetric cost maps and elastic demands. *Operations Research Letters*, 31(5):327–335, 2003.
5. J. R. Correa, A. S. Schulz, and N. E. Stier Moses. Computational complexity, fairness, and the price of anarchy of the maximum latency problem. In *Proceedings of the 10th Conference on Integer Programming and Combinatorial Optimization (IPCO)*, volume 3064 of *Lecture Notes in Computer Science*, pages 59–73, 2004.
6. J. R. Correa, A. S. Schulz, and N. E. Stier Moses. Selfish routing in capacitated networks. *Mathematics of Operations Research*, 29(4):961–976, 2004.
7. A. Czumaj. Selfish routing on the Internet. In J. Leung, editor, *Handbook of Scheduling: Algorithms, Models, and Performance Analysis*, chapter 42. CRC Press, 2004.
8. N. Devanur, N. Garg, R. Khandekar, V. Pandit, and A. Saberi. Price of anarchy, locality gap, and a network service provider game. Unpublished manuscript, 2003.
9. A. Fabrikant, A. Luthra, E. Maneva, C. H. Papadimitriou, and S. J. Shenker. On a network creation game. In *Proceedings of the 22nd ACM Symposium on Principles of Distributed Computing (PODC)*, pages 347–351, 2003.
10. R. Feldmann, M. Gairing, T. Lücking, B. Monien, and M. Rode. Selfish routing in non-cooperative networks: A survey. In *Proceedings of the Conference on Mathematical Foundations of Computer Science (MFCS)*, volume 2747 of *Lecture Notes in Computer Science*, pages 21–45, 2003.
11. M. R. Garey and D. S. Johnson. *Computers and Intractability: A Guide to the Theory of NP-Completeness*. Freeman, 1979.

12. R. Johari and J. N. Tsitsiklis. Efficiency loss in a network resource allocation game. *Mathematics of Operations Research*, 29(3):407–435, 2004.
13. E. Koutsoupias and C. H. Papadimitriou. Worst-case equilibria. In *Proceedings of the 16th Annual Symposium on Theoretical Aspects of Computer Science (STACS)*, pages 404–413, 1999.
14. H. Lin, T. Roughgarden, and É. Tardos. A stronger bound on braess's paradox. In *Proceedings of the 15th Annual Symposium on Discrete Algorithms (SODA)*, pages 333–334, 2004.
15. C. H. Papadimitriou. Algorithms, games, and the Internet. In *Proceedings of the 33rd Annual ACM Symposium on the Theory of Computing (STOC)*, pages 749–753, 2001.
16. G. Perakis. The price of anarchy when costs are non-separable and asymmetric. In *Proceedings of the 10th Conference on Integer Programming and Combinatorial Optimization (IPCO)*, volume 3064 of *Lecture Notes in Computer Science*, pages 46–58, 2004.
17. T. Roughgarden. Designing networks for selfish users is hard. In *Proceedings of the 42nd Annual Symposium on Foundations of Computer Science*, pages 472–481, 2001.
18. T. Roughgarden. The price of anarchy is independent of the network topology. *Journal of Computer and System Sciences*, 67(2):341–364, 2003.
19. T. Roughgarden. The maximum latency of selfish routing. In *Proceedings of the 15th Annual Symposium on Discrete Algorithms (SODA)*, pages 973–974, 2004.
20. T. Roughgarden. *Selfish Routing and the Price of Anarchy*. MIT Press, 2005.
21. T. Roughgarden and É. Tardos. How bad is selfish routing? *Journal of the ACM*, 49(2):236–259, 2002.
22. T. Roughgarden and É. Tardos. Bounding the inefficiency of equilibria in nonatomic congestion games. *Games and Economic Behavior*, 47(2):389–403, 2004.
23. M. J. Smith. The existence, uniqueness and stability of traffic equilibria. *Transportation Research*, 13B:295–304, 1979.
24. A. Vetta. Nash equilibria in competitive societies, with applications to facility location, traffic routing and auctions. In *Proceedings of the 43rd Annual Symposium on Foundations of Computer Science (FOCS)*, pages 416–425, 2002.
25. D. Weitz. The price of anarchy. Unpublished manuscript, 2001.

Weighted Automata and Weighted Logics*

Manfred Droste[1] and Paul Gastin[2]

[1] Institut für Informatik, Universität Leipzig,
Augustusplatz 10-11, D-04109 Leipzig, Germany
droste@informatik.uni-leipzig.de
[2] LSV, CNRS UMR 8643 & ENS de Cachan 61,
Av. du Président Wilson, F-94235 Cachan Cedex, France
Paul.Gastin@lsv.ens-cachan.fr

Abstract. Weighted automata are used to describe quantitative properties in various areas such as probabilistic systems, image compression, speech-to-text processing. The behaviour of such an automaton is a mapping, called a formal power series, assigning to each word a weight in some semiring. We generalize Büchi's and Elgot's fundamental theorems to this quantitative setting. We introduce a weighted version of MSO logic and prove that, for commutative semirings, the behaviours of weighted automata are precisely the formal power series definable with our weighted logic. We also consider weighted first-order logic and show that aperiodic series coincide with the first-order definable ones, if the semiring is locally finite, commutative and has some aperiodicity property.

1 Introduction

Büchi's and Elgot's fundamental theorems [3, 9] established the coincidence of regular languages with languages definable in monadic second-order logic. At the same time, Schützenberger [22] investigated finite automata with weights and characterized their behaviours as rational formal power series. Both of these results have inspired a wealth of extensions and further research, cf. [24, 21, 15, 2] for surveys and monographs, and also led to recent practical applications, e.g. in verification of finite-state programs (model checking, [17, 1, 16]), in digital image compression [5, 11, 13, 12] and in speech-to-text processing [19, 20, 4].

It is the goal of this paper to introduce a logic with weights and to show that the behaviours of weighted finite automata are precisely the series definable in our weighted monadic second-order logic. This can be viewed as a quantitative version of Büchi's and Elgot's classical (qualitative) results.

The syntax of our weighted logics incorporates weights taken from a semiring K, just as done for weighted automata in order to model a variety of applications and situations. The semantics of a weighted logic formula will be a formal power

---

* Work partly supported by the DAAD-PROCOPE project Temporal and Quantitative Analysis of Distributed Systems.

series with values in K. We restrict negation to atomic formulas since in general there is no natural complement operation in the semiring that would allow us to define the semantics of negation elementwise. In comparison to classical MSO-logic, this is not a restriction, since we include disjunction, conjunction, existential and universal quantifications in our syntax. Thus, we obtain the classical MSO-logics by letting $K = \mathbb{B}$, the 2-element Boolean algebra.

Even for the semiring of natural numbers or the tropical semiring it turns out that neither universal first-order nor universal second-order quantification of formulas preserve recognizability. Therefore, for the syntax of restricted MSO-logic we exclude universal second-order quantification, and we permit universal first-order quantification only for formulas whose semantics takes finitely many values in K. Moreover, if we allow existential set quantifications only to occur at the beginning of a fomula, we arrive at restricted existential MSO-logic.

Now we give a summary of our results. First we show for any commutative semiring K that the behaviours of weighted automata with values in K are precisely the series definable by sentences of our restricted MSO-logic, or, equivalently, of our restricted existential MSO-logic. Second, if the semiring K is locally finite, we obtain that the semantics of all sentences of our full weighted MSO-logic are representable by weighted automata. Locally finite semirings were investigated in [6]; they form a large class of semirings including e.g. all finite semirings, the max-min-semiring employed for capacity problems of networks, and all Boolean algebras. Thus we obtain Büchi's and Elgot's theorems as a particular consequence. Moreover, if the semiring K is a field or locally finite and is given in some effective way, then the constructions in our proofs yield effective conversions of sentences of our weighted logics to weighted automata, and viceversa, and we obtain also decision procedures.

Finally, we investigate weighted first-order logic. As is well-known, the first-order definable languages are precisely the starfree languages which in turn coincide with the the aperiodic ones [23, 18]. Aperiodic and starfree formal power series were introduced and investigated in [6]. Easy examples show that even if the semiring K is finite, series definable in our weighted first-order logic need not be aperiodic. However, we obtain that the aperiodic series coincide with the first-order definable ones, if the semiring is locally finite, commutative and both addition and multiplication satisfy a certain aperiodicity property. Such semirings include again all Boolean algebras, but also quite different ones like the truncated max-plus semiring.

We believe that the present paper opens a new research road. Obviously, one could try to extend our results to other structures such as trees or traces. One could also try to define weighted temporal logics and study not only expressiveness but also decidability and complexity of natural problems such as quantitative model checking.

Full proofs are available in the technical report [7].

2 Weighted Automata

We give basic definitions and properties of semirings, formal power series and weighted automata. For background, we refer the reader to [2, 15, 21].

A *semiring* is a structure $(K, +, \cdot, 0, 1)$ where $(K, +, 0)$ is a commutative monoid, $(K, \cdot, 1)$ is a monoid, multiplication distributes over addition, and $0 \cdot x = x \cdot 0 = 0$ for each $x \in K$. If the multiplication is commutative, we say that K is *commutative*. If the addition is idempotent, then the semiring is called *idempotent*. Important examples include

- the natural numbers $(\mathbb{N}, +, \cdot, 0, 1)$ with the usual addition and multiplication,
- the Boolean semiring $\mathbb{B} = (\{0,1\}, \vee, \wedge, 0, 1)$,
- the tropical semiring Trop = $(\mathbb{N} \cup \{\infty\}, \min, +, \infty, 0)$ (also known as min-plus semiring), with min and $+$ extended to $\mathbb{N} \cup \{\infty\}$ in the natural way,
- the arctical semiring Arc = $(\mathbb{N} \cup \{-\infty\}, \max, +, -\infty, 0)$,
- the semiring $([0,1], \max, \cdot, 0, 1)$ which can be used to compute probabilities,
- the semiring of languages $(\mathcal{P}(A^*), \cup, \cap, \emptyset, A^*)$.

If K is a semiring and $n \in \mathbb{N}$, then $K^{n \times n}$ comprises all $(n \times n)$-matrices over K. With usual matrix multiplication $(K^{n \times n}, \cdot)$ is a monoid.

A formal power series is a mapping $S : A^* \to K$. It is usual to write (S, w) for $S(w)$. The set $\mathrm{Supp}(S) := \{w \in A^* \mid (S, w) \neq 0\}$ is called the *support* of S, and $\mathrm{Im}(S) = \{(S, w) \mid w \in A^*\}$ is the *image* of S. The set of all formal power series over K and A is denoted by $K\langle\!\langle A^* \rangle\!\rangle$. Now let $S, T \in K\langle\!\langle A^* \rangle\!\rangle$. The *sum* $S + T$ and the *Hadamard product* $S \odot T$ are both defined pointwise:

$$(S+T, w) := (S, w) + (T, w) \text{ and } (S \odot T, w) := (S, w) \cdot (T, w) \quad (w \in A^*).$$

For $L \subseteq A^*$, we define the *characteristic series* $\mathbb{1}_L : A^* \to K$ by $(\mathbb{1}_L, w) = 1$ if $w \in L$, and $(\mathbb{1}_L, w) = 0$ otherwise.

Now we turn to weighted automata. We fix a semiring K and an alphabet A. A *weighted finite automaton* over K and A is a quadruple $\mathcal{A} = (Q, \lambda, \mu, \gamma)$ where Q is a finite set of states, $\mu : A \to K^{Q \times Q}$ is the transition weight function and $\lambda, \gamma : Q \to K$ are weight functions for entering and leaving a state, respectively. Here $\mu(a)$ is a $(Q \times Q)$-matrix whose (p, q)-entry $\mu(a)_{p,q} \in K$ indicates the weight of the transition $p \xrightarrow{a} q$. Then μ extends uniquely to a monoid homomorphism (also denoted by μ) from A^* into $(K^{Q \times Q}, \cdot)$.

The *weight* of a path $P : q_0 \xrightarrow{a_1} q_1 \longrightarrow \ldots \longrightarrow q_{n-1} \xrightarrow{a_n} q_n$ in $\mathcal{A}$ is the product $\mathrm{weight}(P) := \lambda(q_0) \cdot \mu(a_1)_{q_0, q_1} \cdots \mu(a_n)_{q_{n-1}, q_n} \cdot \gamma(q_n)$. This path has label $a_1 \ldots a_n$. The *weight* of a word $w = a_1 \ldots a_n \in A^*$ in $\mathcal{A}$, denoted $(||\mathcal{A}||, w)$, is the sum of $\mathrm{weight}(P)$ over all paths P with label w. One can check that $(||\mathcal{A}||, w) = \lambda \cdot \mu(w) \cdot \gamma$ with usual matrix multiplication, considering λ as a row vector and γ as a column vector. If $w = \varepsilon$, we have $(||\mathcal{A}||, \varepsilon) = \lambda \cdot \gamma$. This defines a formal power series $||\mathcal{A}|| : A^* \to K$ called the *behavior* of $\mathcal{A}$. A formal power series $S \in K\langle\!\langle A^* \rangle\!\rangle$ is called *recognizable*, if there exists a weighted finite automaton $\mathcal{A}$ such that $S = ||\mathcal{A}||$. Then we also call $\mathcal{A}$ or (λ, μ, γ) a *representation* of S. We let $K^{\mathrm{rec}}\langle\!\langle A^* \rangle\!\rangle$ be the collection of all recognizable formal power series over K and A.

Now let $h : A^* \to B^*$ be a homomorphism. If $T \in K\langle\!\langle B^*\rangle\!\rangle$, then $h^{-1}(T) := T \circ h \in K\langle\!\langle A^*\rangle\!\rangle$. That is, $(h^{-1}(T), w) = (T, h(w))$ for each $w \in A^*$. We say that h is *non-erasing*, if $h(a) \neq \varepsilon$ for any $a \in A$. In this case, for $S \in K\langle\!\langle A^*\rangle\!\rangle$, define $h(S) : B^* \to K$ by $(h(S), v) := \sum_{w \in h^{-1}(v)}(S, w)$ ($v \in B^*$), noting that the sum is finite since h is non-erasing. Recognizability is preserved by inverse morphisms and non-erasing morphisms [8, 2].

We say $S : A^* \to K$ is a *recognizable step function*, if $S = \sum_{i=1}^n k_i \cdot \mathbb{1}_{L_i}$ for some $n \in \mathbb{N}, k_i \in K$ and recognizable languages $L_i \subseteq A^*$ ($i = 1, \ldots, n$). As is well-known, any recognizable step function is a recognizable power series.

3 Weighted Logics

In this section, we introduce our weighted logics and study its first properties. We fix a semiring K and an alphabet A. For each $a \in A$, P_a denotes a unary predicate symbol. The syntax of formulas of the *weighted MSO-logic* is given by

$$\varphi ::= k \mid P_a(x) \mid \neg P_a(x) \mid x \leq y \mid \neg(x \leq y) \mid x \in X \mid \neg(x \in X)$$
$$\mid \varphi \vee \psi \mid \varphi \wedge \psi \mid \exists x.\varphi \mid \exists X.\varphi \mid \forall x.\varphi \mid \forall X.\varphi$$

where $k \in K$ and $a \in A$. We denote by $\mathrm{MSO}(K, A)$ the collection of all such weighted MSO-formulas φ.

Now we turn to the definition of the semantics of formulas $\varphi \in \mathrm{MSO}(K, A)$. We let $\mathrm{Free}(\varphi)$ be the set of all free variables of φ. Let $w = a_1 \ldots a_n \in A^*$ with $a_i \in A$. We also write $w(i) = a_i$ ($1 \leq i \leq n$). The length of w is $|w| = n$.

Let $\mathcal{V}$ be a finite set of first-order and second-order variables. A $(\mathcal{V}, w)$-assignment σ is a function mapping first-order variables in $\mathcal{V}$ to elements of $\{1, \ldots, |w|\}$ and second-order variables in $\mathcal{V}$ to subsets of $\{1, \ldots, |w|\}$. If x is a first-order variable and $i \in \{1, \ldots, |w|\}$ then $\sigma[x \to i]$ is the $(\mathcal{V} \cup \{x\}, w)$-assignment which assigns x to i and acts like σ on all other variables. Similarly, $\sigma[X \to I]$ is defined for $I \subseteq \{1, \ldots, |w|\}$.

As usual, a pair (w, σ) where σ is a $(\mathcal{V}, w)$-assignment will be encoded using an extended alphabet $A_\mathcal{V} = A \times \{0, 1\}^\mathcal{V}$. More precisely, we will write a word over $A_\mathcal{V}$ as a pair (w, σ) where w is the projection over A and σ is the projection over $\{0, 1\}^\mathcal{V}$. Now, σ represents a *valid* assignment over $\mathcal{V}$ if for each first-order variable $x \in \mathcal{V}$, the x-row of σ contains exactly one 1. In this case, we identify σ with the $(\mathcal{V}, w)$-assignment such that for each first-order variable $x \in \mathcal{V}$, $\sigma(x)$ is the position of the 1 on the x-row, and for each second-order variable $X \in \mathcal{V}$, $\sigma(X)$ is the set of positions carrying a 1 on the X-row. Clearly, the language $N_\mathcal{V} = \{(w, \sigma) \in A_\mathcal{V}^* \mid \sigma \text{ is a valid } \mathcal{V}\text{-assignment}\}$ is recognizable. We simply write $A_\varphi = A_{\mathrm{Free}(\varphi)}$ and $N_\varphi = N_{\mathrm{Free}(\varphi)}$.

Definition 3.1. *Let $\varphi \in \mathrm{MSO}(K, A)$ and $\mathcal{V}$ be a finite set of variables containing $\mathrm{Free}(\varphi)$. The semantics of φ is a formal power series $[\![\varphi]\!]_\mathcal{V} \in K\langle\!\langle A_\mathcal{V}^* \rangle\!\rangle$. Let $(w, \sigma) \in A_\mathcal{V}^*$. If σ is not a valid $\mathcal{V}$-assignment, then we put $[\![\varphi]\!]_\mathcal{V}(w, \sigma) = 0$. Otherwise, we define $[\![\varphi]\!]_\mathcal{V}(w, \sigma) \in K$ inductively as follows:*

$$[\![k]\!]_\mathcal{V}(w,\sigma) = k \qquad\qquad [\![P_a(x)]\!]_\mathcal{V}(w,\sigma) = \begin{cases} 1 & \text{if } w(\sigma(x)) = a \\ 0 & \text{otherwise} \end{cases}$$

$$[\![x \leq y]\!]_\mathcal{V}(w,\sigma) = \begin{cases} 1 & \text{if } \sigma(x) \leq \sigma(y) \\ 0 & \text{otherwise} \end{cases} \quad [\![x \in X]\!]_\mathcal{V}(w,\sigma) = \begin{cases} 1 & \text{if } \sigma(x) \in \sigma(X) \\ 0 & \text{otherwise} \end{cases}$$

$$[\![\neg\varphi]\!]_\mathcal{V}(w,\sigma) = \begin{cases} 1 & \text{if } [\![\varphi]\!]_\mathcal{V}(w,\sigma) = 0 \\ 0 & \text{if } [\![\varphi]\!]_\mathcal{V}(w,\sigma) = 1 \end{cases} \quad \text{if } \varphi \text{ is of the form } P_a(x),\, (x \leq y) \text{ or } (x \in X).$$

$$[\![\varphi \vee \psi]\!]_\mathcal{V}(w,\sigma) = [\![\varphi]\!]_\mathcal{V}(w,\sigma) + [\![\psi]\!]_\mathcal{V}(w,\sigma)$$

$$[\![\varphi \wedge \psi]\!]_\mathcal{V}(w,\sigma) = [\![\varphi]\!]_\mathcal{V}(w,\sigma) \cdot [\![\psi]\!]_\mathcal{V}(w,\sigma)$$

$$[\![\exists x.\varphi]\!]_\mathcal{V}(w,\sigma) = \sum_{1 \leq i \leq |w|} [\![\varphi]\!]_{\mathcal{V} \cup \{x\}}(w, \sigma[x \to i])$$

$$[\![\exists X.\varphi]\!]_\mathcal{V}(w,\sigma) = \sum_{I \subseteq \{1,\ldots,|w|\}} [\![\varphi]\!]_{\mathcal{V} \cup \{X\}}(w, \sigma[X \to I])$$

$$[\![\forall x.\varphi]\!]_\mathcal{V}(w,\sigma) = \prod_{1 \leq i \leq |w|} [\![\varphi]\!]_{\mathcal{V} \cup \{x\}}(w, \sigma[x \to i])$$

$$[\![\forall X.\varphi]\!]_\mathcal{V}(w,\sigma) = \prod_{I \subseteq \{1,\ldots,|w|\}} [\![\varphi]\!]_{\mathcal{V} \cup \{X\}}(w, \sigma[X \to I])$$

where we fix some order on the power set of $\{1, \ldots, |w|\}$ so that the last product is defined even if K is not commutative. We simply write $[\![\varphi]\!]$ for $[\![\varphi]\!]_{\text{Free}(\varphi)}$.

Note that if φ is a sentence, i.e. has no free variables, then $[\![\varphi]\!] \in K\langle\!\langle A^* \rangle\!\rangle$. We give several examples of possible interpretations for weighted formulas:

I. Let $K = (\mathbb{N}, +, \cdot, 0, 1)$ and assume φ does not contain constants $k \in \mathbb{N}$. We may interpret $[\![\varphi]\!](w, \sigma)$ as the number of proofs we have that (w, σ) satisfies formula φ. Indeed, for atomic formulas the number of proofs is clearly 0 or 1, depending on whether φ holds for (w, σ) or not. Now if e.g. $[\![\varphi]\!](w, \sigma) = m$ and $[\![\psi]\!](w, \sigma) = n$, the number of proofs that (w, σ) satisfies $\varphi \vee \psi$ should be $m + n$ (since any proof suffices), and for $\varphi \wedge \psi$ it should be $m \cdot n$ (since we may pair the proofs of φ and ψ arbitrarily). Similarly, the semantics of the existential and universal quantifiers can be interpreted.

II. The formula $\exists x.P_a(x)$ counts how often a occurs in the word. Here how often depends on the semiring: e.g. Boolean semiring, natural numbers, integers modulo 3, ...

III. Consider the probability semiring $K = ([0,1], \max, \cdot, 0, 1)$ and the alphabet $A = \{a_1, \ldots, a_n\}$. Assume that each letter a_i has a reliability k_i. Then, the series assigning to a word its reliability can be given by the first-order formula $\forall x. \bigvee_{1 \leq i \leq n}(P_{a_i}(x) \wedge k_i)$.

IV. Let K be an arbitrary Boolean algebra $(B, \vee, \wedge, \bar{}, 0, 1)$. In this case, sums correspond to suprema, and products to infima. Here we can define the semantics of $\neg\varphi$ for an arbitrary formula φ by $[\![\neg\varphi]\!](w, \sigma) := \overline{[\![\varphi]\!](w, \sigma)}$. Then clearly $[\![\varphi \wedge \psi]\!] = [\![\neg(\neg\varphi \vee \neg\psi)]\!]$, $[\![\forall x.\varphi]\!] = [\![\neg(\exists x.\neg\varphi)]\!]$ and $[\![\forall X.\varphi]\!] =$

$\llbracket \neg(\exists X.\neg\varphi)\rrbracket$. This may be interpreted as a multi-valued logics. In particular, if $K = \mathbb{B}$, the 2-valued Boolean algebra, our semantics coincides with the usual semantics of unweighted MSO-formulas, identifying characteristic series with their supports.

Let K be a semiring and A an alphabet. Observe that if $\varphi \in \mathrm{MSO}(K,A)$, we have defined a semantics $\llbracket\varphi\rrbracket_\mathcal{V}$ for each finite set of variables $\mathcal{V}$ containing Free(φ). Now we show that these semantics' are consistent with each other.

Proposition 3.2. *Let $\varphi \in \mathrm{MSO}(K,A)$ and $\mathcal{V}$ a finite set of variables containing Free(φ). Then $\llbracket\varphi\rrbracket_\mathcal{V}(w,\sigma) = \llbracket\varphi\rrbracket(w,\sigma_{|\,\mathrm{Free}(\varphi)})$ for each $(w,\sigma) \in A^*_\mathcal{V}$ such that σ is a valid $\mathcal{V}$-assignment. In particular, $\llbracket\varphi\rrbracket$ is recognizable iff $\llbracket\varphi\rrbracket_\mathcal{V}$ is recognizable.*

Proof (sketch). The first claim can be shown by induction on φ. For the final claim, let $\llbracket\varphi\rrbracket$ be recognizable, and consider the projection $\pi : A_\mathcal{V} \to A_\varphi$. For $(w,\sigma) \in A^*_\mathcal{V}$, we have $\pi(w,\sigma) = (w,\sigma_{|\,\mathrm{Free}(\varphi)})$. Hence, $\llbracket\varphi\rrbracket_\mathcal{V} = \pi^{-1}(\llbracket\varphi\rrbracket) \odot \mathbb{1}_{N_\mathcal{V}}$ is recognizable by standard preservation results, cf. [2]. The converse can be shown similarly. □

Now let $Z \subseteq \mathrm{MSO}(K,A)$. A series $S : A^* \to K$ is called *Z-definable*, if there is a sentence $\varphi \in Z$ such that $S = \llbracket\varphi\rrbracket$. The main goal of this paper is the comparison of Z-definable with recognizable series, for suitable fragments Z of $\mathrm{MSO}(K,A)$. Crucial for this will be closure properties of recognizable series under the constructs of our weighted logic. However, first we will show that $K^{\mathrm{rec}}\langle\!\langle A^*\rangle\!\rangle$ is in general not closed under universal quantification.

Example 3.3. Let $K = (\mathbb{N},+,\cdot,0,1)$. Then $\llbracket\forall x.2\rrbracket(w) = 2^{|w|}$ and $\llbracket\forall y\forall x.2\rrbracket(w) = (2^{|w|})^{|w|} = 2^{|w|^2}$. Clearly, the series $\llbracket\forall x.2\rrbracket$ is recognizable by the weighted automaton (Q,λ,μ,γ) with $Q = \{1\}$, $\lambda_1 = \gamma_1 = 1$ and $\mu_{1,1}(a) = 2$ for all $a \in A$. However, $\llbracket\forall y\forall x.2\rrbracket$ is not recognizable. Suppose there was an automaton $\mathcal{A}' = (Q',\lambda',\mu',\gamma')$ with behavior $\llbracket\forall y\forall x.2\rrbracket$. Let $M = \max\{|\lambda'_p|, |\gamma'_p|, |\mu'(a)_{p,q}| \mid p,q \in Q', a \in A\}$. Then $(\|\,\mathcal{A}'\,\|,w) \leq |Q'|^{|w|+1} \cdot M^{|w|+2}$ for any $w \in A^*$, a contradiction with $(\|\,\mathcal{A}'\,\|,w) = 2^{|w|^2}$.

A similar argument applies also for the tropical and the arctical semirings. Observe that in all these cases, $\llbracket\forall x.2\rrbracket$ has infinite image.

Example 3.4. Let $K = (\mathbb{N},+,\cdot,0,1)$. Then $\llbracket\forall X.2\rrbracket(w) = 2^{2^{|w|}}$ for any $w \in A^*$, and as above $\llbracket\forall X.2\rrbracket$ is not recognizable due to its growth. Again, this counterexample also works for the tropical and the arctical semirings.

The examples show that unrestricted universal quantification is too strong to preserve recognizability. This motivates the following definition. We will call a formula $\varphi \in \mathrm{MSO}(K,A)$ *restricted*, if it contains no universal set quantification of the form $\forall X.\psi$, and whenever φ contains a universal first-order quantification $\forall x.\psi$, then $\llbracket\psi\rrbracket$ is a recognizable step function. We let $\mathrm{RMSO}(K,A)$ comprise all restricted formulas of $\mathrm{MSO}(K,A)$. Furthermore, let $\mathrm{REMSO}(K,A)$ contain all restricted *existential* MSO-formulas φ, i.e. φ is of the form $\varphi = \exists X_1,\ldots,X_n.\psi$ with $\psi \in \mathrm{RMSO}(K,A)$ containing no set quantification.

We let $K^{\mathrm{rmso}}\langle\!\langle A^*\rangle\!\rangle$ (resp. $K^{\mathrm{remso}}\langle\!\langle A^*\rangle\!\rangle$) contain all series $S \in K\langle\!\langle A^*\rangle\!\rangle$ which are definable by some sentence in RMSO(K, A) (resp. in REMSO(K, A)). The main result of this paper is the following theorem. It will be proved in sections 4 and 5.

Theorem 3.5. *Let K be a commutative semiring and A an alphabet. Then*

$$K^{\mathrm{rec}}\langle\!\langle A^*\rangle\!\rangle = K^{\mathrm{rmso}}\langle\!\langle A^*\rangle\!\rangle = K^{\mathrm{remso}}\langle\!\langle A^*\rangle\!\rangle.$$

4 Definable Series Are Recognizable

In all of this section, let K be a semiring and A an alphabet. We wish to show that if K is commutative, then all RMSO-definable series $[\![\varphi]\!]$ over K and A are recognizable. We proceed by induction over the structure of RMSO-formulas.

Lemma 4.1. *Let $\varphi, \psi \in \mathrm{MSO}(K, A)$.*

(a) If φ is atomic or the negation of an atomic formula, then $[\![\varphi]\!]$ is recognizable.
(b) If $[\![\varphi]\!]$ and $[\![\psi]\!]$ are recognizable, then $[\![\varphi \vee \psi]\!]$ is recognizable. If in addition K is commutative, then $[\![\varphi \wedge \psi]\!]$ is also recognizable.
(c) If $[\![\varphi]\!]$ is recognizable, then $[\![\exists x.\varphi]\!]$ and $[\![\exists X.\varphi]\!]$ are recognizable series.

Proof (sketch). (a) Automata with behavior $[\![\varphi]\!]$ can easily be given explicitly.

(b) Let $\mathcal{V} = \mathrm{Free}(\varphi) \cup \mathrm{Free}(\psi)$. By definition, we have $[\![\varphi \vee \psi]\!] = [\![\varphi]\!]_\mathcal{V} + [\![\psi]\!]_\mathcal{V}$ and $[\![\varphi \wedge \psi]\!] = [\![\varphi]\!]_\mathcal{V} \odot [\![\psi]\!]_\mathcal{V}$. Hence the result follows from Proposition 3.2 and since sum and Hadamard product preserve recognizability of series [8, 2].

(c) Let $\mathcal{V} = \mathrm{Free}(\exists X.\varphi)$ and consider the projection $\pi : A^*_{\mathcal{V} \cup \{X\}} \to A^*_\mathcal{V}$. We can show that $[\![\exists X.\varphi]\!](w, \sigma) = \pi([\![\varphi]\!]_{\mathcal{V} \cup \{X\}})(w, \sigma)$. Now, $[\![\varphi]\!]_{\mathcal{V} \cup \{X\}}$ is recognizable by Proposition 3.2 and since non-erasing projections preserve recognizability we get $[\![\exists X.\varphi]\!]$ recognizable. The case $\exists x.\varphi$ can be dealt with similarly. □

The most interesting and new case here arises from universal quantification.

Lemma 4.2. *Let K be commutative and $\varphi \in \mathrm{MSO}(K, A)$ such that $[\![\varphi]\!]$ is a recognizable step function. Then $[\![\forall x.\varphi]\!]$ is recognizable.*

Proof. Let $\mathcal{W} = \mathrm{Free}(\varphi)$ and $\mathcal{V} = \mathrm{Free}(\forall x.\varphi) = \mathcal{W} \setminus \{x\}$. We may write $[\![\varphi]\!] = \sum_{j=1,\ldots,n} k_j \cdot \mathbb{1}_{L_j}$ with $n \in \mathbb{N}$, $k_j \in K$ and recognizable languages $L_j \subseteq A^*_\mathcal{W}$ ($j = 1, \ldots, n$) forming a partition of $A^*_\mathcal{W}$.

First, we assume that $x \in \mathcal{W}$. Let $\widetilde{A} = A \times \{1, \ldots, n\}$. A word in $(\widetilde{A}_\mathcal{V})^*$ will be written (w, ν, σ) where $(w, \sigma) \in A^*_\mathcal{V}$ and $\nu \in \{1, \ldots, n\}^*$ is interpreted as a mapping from $\{1, \ldots, |w|\}$ to $\{1, \ldots, n\}$. Let $\widetilde{L}$ be the set of $(w, \nu, \sigma) \in (\widetilde{A}_\mathcal{V})^*$ such that for all $i \in \{1, \ldots, |w|\}$ and $j \in \{1, \ldots, n\}$ we have $\nu(i) = j$ implies $(w, \sigma[x \to i]) \in L_j$. Observe that for each $(w, \sigma) \in A^*_\mathcal{V}$ there is a unique ν such that $(w, \nu, \sigma) \in \widetilde{L}$ since the L_j form a partition of $A^*_\mathcal{W}$.

It can be shown that $\widetilde{L}$ is a recognizable word language. Hence there is a deterministic automaton $\widetilde{\mathcal{A}}$, with state set Q, say, recognizing $\widetilde{L}$. Now we obtain a weighted automaton $\mathcal{A}$ with the same state set by adding weights to the transitions of $\widetilde{\mathcal{A}}$ as follows: If $(p,(a,j,s),q)$ is a transition in $\widetilde{\mathcal{A}}$ with $p,q \in Q$, $(a,s) \in A_\mathcal{V}$ and $1 \leq j \leq n$, we let this transition in $\mathcal{A}$ have weight k_j, i.e. $\mu_\mathcal{A}(a,j,s)_{p,q} = k_j$. All triples which are not transitions in $\widetilde{\mathcal{A}}$ get weight 0. Also, the initial state of $\widetilde{\mathcal{A}}$ gets initial weight 1 in $\mathcal{A}$, all non-initial states of $\widetilde{\mathcal{A}}$ get initial weight 0, and similarly for the final states and final weights.

Clearly, since $\widetilde{\mathcal{A}}$ is deterministic and accepts $\widetilde{L}$, the weight of $(w,\nu,\sigma) \in \widetilde{L}$ in $\mathcal{A}$ is $\prod_{1 \leq j \leq n} k_j^{|\nu^{-1}(j)|}$, and the weight of $(w,\nu,\sigma) \in \widetilde{A}^* \setminus \widetilde{L}$ in $\mathcal{A}$ is 0. Now let $h : (\widetilde{A}_\mathcal{V})^* \to A_\mathcal{V}^*$ be the projection mapping (w,ν,σ) to (w,σ). Then for any $(w,\sigma) \in A_\mathcal{V}^*$ and the unique ν such that $(w,\nu,\sigma) \in \widetilde{L}$ we obtain

$$h(\|\mathcal{A}\|)(w,\sigma) = \sum_\rho \|\mathcal{A}\|(w,\rho,\sigma) = \|\mathcal{A}\|(w,\nu,\sigma) = \prod_{1 \leq j \leq n} k_j^{|\nu^{-1}(j)|}.$$

Now we have

$$[\![\forall x.\varphi]\!](w,\sigma) = \prod_{1 \leq i \leq |w|} [\![\varphi]\!](w,\sigma[x \to i]) = \prod_{1 \leq j \leq n} k_j^{|\nu^{-1}(j)|}$$

where the last equality holds due to the form of φ. Hence $[\![\forall x.\varphi]\!] = h(\|\mathcal{A}\|)$ which is recognizable [8, 2].

Now assume that $x \notin \mathcal{W}$, so that $\mathcal{V} = \mathcal{W}$. Let $\varphi' = \varphi \wedge (x \leq x)$. So $[\![\varphi']\!]$ is recognizable by Lemma 4.1, and clearly $[\![\varphi]\!]_{\mathcal{V} \cup \{x\}} = [\![\varphi']\!]_{\mathcal{V} \cup \{x\}}$. Thus $[\![\forall x.\varphi]\!]_\mathcal{V} = [\![\forall x.\varphi']\!]_\mathcal{V}$ which is recognizable by what we showed above. □

Now the following result is immediate by Lemmata 4.1 and 4.2.

Theorem 4.3. *Let K be a commutative semiring, A an alphabet and $\varphi \in \mathrm{RMSO}(K,A)$. Then $[\![\varphi]\!] \in K^{\mathrm{rec}}\langle\!\langle A^* \rangle\!\rangle$ is recognizable.*

Next we turn to decidability questions. Employing decidability results from the theory of formal power series and our previous constructions, we can show:

Proposition 4.4. *Let K be a computable field, and let $\varphi \in \mathrm{MSO}(K,A)$. It is decidable whether φ is restricted, and in this case one can effectively compute a weighted automaton $\mathcal{A}_\varphi$ for $[\![\varphi]\!]$.*

Corollary 4.5. *Let K be a computable field, and let $\varphi, \psi \in \mathrm{RMSO}(K,A)$. Then it is decidable whether $[\![\varphi]\!] = [\![\psi]\!]$. It is also decidable whether $[\![\varphi]\!]$ and $[\![\psi]\!]$ differ only for finitely many words.*

Proof. By Proposition 4.4, the series $[\![\varphi]\!], [\![\psi]\!]$ and hence also $[\![\varphi]\!] - [\![\psi]\!] = [\![\varphi]\!] + (-1) \cdot [\![\psi]\!]$ are effectively recognizable. By [2–Propositions VI.1.1, VI.1.2], it is decidable whether such a series equals 0, or whether its support is finite. □

5 Recognizable Series Are Definable

In all of this section let K be a semiring and A an alphabet. We wish to show that if K is commutative, then all recognizable series are REMSO-definable. For this, the concept of an unambiguous MSO-formula will be useful. The class of *unambiguous* formulas in $MSO(K, A)$ is defined inductively as follows: All atomic formulas of the form $P_a(x)$, $x \leq y$ or $(x \in X)$, and their negations are unambiguous. If φ, ψ are unambiguous, then $\varphi \wedge \psi$, $\forall x.\varphi$ and $\forall X.\varphi$ are also unambiguous. If φ, ψ are unambiguous and $\mathrm{Supp}(\llbracket \varphi \rrbracket) \cap \mathrm{Supp}(\llbracket \psi \rrbracket) = \emptyset$, then $\varphi \vee \psi$ is unambiguous. Let φ be unambiguous and $\mathcal{V} = \mathrm{Free}(\varphi)$. If for any $(w, \sigma) \in A_\mathcal{V}^*$ there is at most one element $i \in \{1, \ldots, |w|\}$ such that $\llbracket \varphi \rrbracket_{\mathcal{V} \cup \{x\}}(w, \sigma[x \to i]) \neq 0$, then $\exists x.\varphi$ is unambiguous. If for any $(w, \sigma) \in A_\mathcal{V}^*$ there is at most one subset $I \subseteq \{1, \ldots, |w|\}$ such that $\llbracket \varphi \rrbracket_{\mathcal{V} \cup \{X\}}(w, \sigma[X \to I]) \neq 0$, then $\exists X.\varphi$ is unambiguous.

Proposition 5.1. *Let $\varphi \in \mathrm{MSO}(K, A)$ be unambiguous. We may also regard φ as a classical MSO-formula defining the language $\mathcal{L}(\varphi) \subseteq A_\varphi^*$. Then, $\llbracket \varphi \rrbracket = \mathbb{1}_{\mathcal{L}(\varphi)}$ is a recognizable step function.*

Next we show that, conversely, classical MSO-formulas can be transformed into unambiguous formulas.

Lemma 5.2. *For each classical MSO-formula φ not containing set quantifications (but possibly including atomic formulas of the form $(x \in X)$) we can effectively construct two unambiguous $\mathrm{MSO}(K, A)$-formula φ^+ and φ^- such that $\llbracket \varphi^+ \rrbracket = \mathbb{1}_{\mathcal{L}(\varphi)}$ and $\llbracket \varphi^- \rrbracket = \mathbb{1}_{\mathcal{L}(\neg \varphi)}$.*

Proof (sketch). We may assume (using also conjunction and universal quantification in our syntax or as abbreviations) that in φ negations are applied only to atomic formulas. Now we proceed by induction, giving only the formulas for some cases: $(\varphi \vee \psi)^- = \varphi^- \wedge \psi^-$, $(\varphi \vee \psi)^+ = \varphi^+ \vee (\varphi^- \wedge \psi^+)$, $(\exists x.\varphi)^- = \forall x.\varphi^-$ and $(\exists x.\varphi)^+ = \exists x.(\varphi^+(x) \wedge \forall y.((x \leq y) \vee (\neg(x \leq y) \wedge \varphi^-(y))))$. □

Theorem 5.3. *Let K be commutative. Then $K^{\mathrm{rec}}\langle\!\langle A^* \rangle\!\rangle \subseteq K^{\mathrm{remso}}\langle\!\langle A^* \rangle\!\rangle$.*

Proof (sketch). Let $\mathcal{A} = (Q, \lambda, \mu, \gamma)$ be a weighted automaton over A. For each triple $(p, a, q) \in Q \times A \times Q$ choose a set variable $X_{p,a,q}$, and let $\mathcal{V} = \{X_{p,a,q} \mid p, q \in Q, a \in A\}$. We choose an enumeration $X_1, \ldots, X_m$ of $\mathcal{V}$ with $m = |Q|^2 \cdot |A|$. In the formulas below we use classical macros such as $\mathrm{partition}(X_1, \ldots, X_m)$, $(y = x + 1)$, $\min(y)$ and $\max(y)$. Define the unambiguous formula

$$\psi(X_1, \ldots, X_m) := \mathrm{partition}(X_1, \ldots, X_m)^+ \wedge \bigwedge_{p,a,q} \forall x.((x \in X_{p,a,q}) \to P_a(x))^+$$

$$\wedge \; \forall x \forall y. \Big((y = x+1) \to \bigvee_{p,q,r \in Q, a,b \in A} (x \in X_{p,a,q}) \wedge (y \in X_{q,b,r}) \Big)^+.$$

Let $w = a_1 \ldots a_n \in A^+$. One can show that there is a bijection between the set of paths in $\mathcal{A}$ over w and the set of $(w, \mathcal{V})$-assignments σ satisfying ψ. Consider now the formula $\varphi(X_1, \ldots, X_m)$ defined by

$$\psi(X_1,\ldots,X_m) \wedge \Big(\bigwedge_{p,a,q} \forall x.(\neg(x \in X_{p,a,q}) \vee ((x \in X_{p,a,q}) \wedge \mu(a)_{p,q})) \Big) \wedge$$
$$\exists y.\Big(\min(y) \wedge \bigvee_{p,a,q}(y \in X_{p,a,q}) \wedge \lambda_p\Big) \wedge \exists z.\Big(\max(z) \wedge \bigvee_{p,a,q}(z \in X_{p,a,q}) \wedge \gamma_q\Big).$$

Let ρ be a path in $\mathcal{A}$ over w and let σ_ρ be the associated $(w,\mathcal{V})$-assignment. We can show that $[\![\varphi]\!]_\mathcal{V}(w,\sigma_\rho) = \text{weight}(\rho)$. Let $\xi = \exists X_1 \cdots \exists X_m.\varphi(X_1,\ldots,X_m)$. For $w \in A^+$ we can show that $([\![\xi]\!],w) = (\|\mathcal{A}\|,w)$. Let $\zeta = (\lambda \cdot \gamma) \wedge \forall x.\neg(x \leq x)$. Then, $\|\mathcal{A}\| = [\![\zeta \vee \xi]\!] \in K^{\text{remso}}\langle\!\langle A^* \rangle\!\rangle$. □

Now Theorem 3.5 is immediate by Theorems 4.3 and 5.3.

Observe that the proof of Theorem 5.3 is constructive, i.e. given a weighted automaton $\mathcal{A}$, we effectively obtain an REMSO(K,A)-sentence φ with $[\![\varphi]\!] = \|\mathcal{A}\|$. Using this, from the theory of formal power series (cf. [21, 15, 2]) we immediately obtain undecidablility results for the semantics of weighted MSO-sentences. For instance, it is undecidable whether a given REMSO-sentence φ over $\mathbb{Q}$, the field of rational numbers, and an alphabet A, satisfies $\text{Supp}([\![\varphi]\!]) = A^*$. Also, by a result of Krob [14], the equality of given recognizable series over the tropical semiring is undecidable. Hence, the equality of two given REMSO(Trop, A)-sentences is also undecidable.

6 Locally Finite Semirings

In section 3 we gave examples of semirings K showing that the results of Theorem 3.5 and 4.3 in general do not hold for arbitrary MSO(K,A)-sentences. In contrast, here we wish to show that for a large class of semirings K, all MSO(K,A)-formulas have a recognizable semantics.

A semiring K is called *locally finite*, if each finitely generated subsemiring of K is finite. For example, any Boolean algebra $(B,\vee,\wedge,0,1)$ is locally finite. The max-min semiring $\mathbb{R}_{\max,\min} = (\mathbb{R}_+ \cup \{\infty\}, \max, \min, 0, \infty)$ of positive reals, used in operations research for maximum capacity problems of networks, is locally finite. In fact, more generally, any distributive lattice $(L, \vee, \wedge, 0, 1)$ with smallest element 0 and largest element 1 is a locally finite semiring, cf. [6] for further basic properties. Examples of infinite but locally finite fields are provided by the algebraic closures of the finite fields $\mathbf{Z}/p\mathbf{Z}$ for any prime p. We can show (see [7]):

Theorem 6.1. *Let K be a locally finite commutative semiring and A an alphabet. Then $K^{rec}\langle\!\langle A^* \rangle\!\rangle = K^{mso}\langle\!\langle A^* \rangle\!\rangle$.*

Again, given an MSO(K,A)-formula φ, we can effectively construct a weighted automaton $\mathcal{A}$ over K and A_φ such that $\|\mathcal{A}\| = [\![\varphi]\!]$. As a consequence of this and of corresponding decidability results given in the full version of [6] for recognizable series over locally finite semirings, we immediately obtain:

Corollary 6.2. *Let K be a locally finite commutative semiring and A an alphabet. It is decidable whether two given MSO(K,A)-formulas φ and ψ satisfy $[\![\varphi]\!] = [\![\psi]\!]$; whether a given MSO$(K,A)$-formula φ satisfies $\text{Supp}([\![\varphi]\!]) = A_\varphi^*$.*

7 Weighted First-Order Logic

In this section, we investigate weighted first-order logic and the relationship to aperiodic series. Most of our results will require additional assumptions on the semiring K.

Let K be a semiring and A an alphabet. A formula $\varphi \in \mathrm{MSO}(K, A)$ is called a (weighted) first-order formula, if φ does not contain any set variable. We let $\mathrm{FO}(K, A)$ contain all first-order formulas and $\mathrm{RFO}(K, A)$ all restricted first-order formulas over K and A. The collections of series definable by these formulas are denoted $K^{fo}\langle\langle A^*\rangle\rangle$ and $K^{rfo}\langle\langle A^*\rangle\rangle$, respectively.

As is well-known, the first-order definable languages are precisely the starfree languages which in turn coincide with the the aperiodic ones [23, 18]. Aperiodic and starfree formal power series were introduced and investigated in [6]. Recall that a monoid M is said to be *aperiodic*, if there exists some $m \geq 0$ such that $x^m = x^{m+1}$ for all $x \in M$. We call a monoid M *weakly aperiodic*, if for each $x \in M$ there exists $m \geq 0$ such that $x^m = x^{m+1}$. Clearly, a finite monoid is aperiodic iff it is weakly aperiodic.

A series $S: A^* \to K$ is called *aperiodic*, if there exists a representation $S = (Q, \lambda, \mu, \gamma)$ with $\mu(A^*)$ aperiodic. Observe that then there exists some $m \geq 0$ such that for all $w \in A^*$ we have $\mu(w^m) = \mu(w^{m+1})$ and hence $(S, w^m) = (S, w^{m+1})$. The collection of all aperiodic series over K and A will be denoted $K^{aper}\langle\langle A^*\rangle\rangle$.

Now we turn to the relationship between aperiodic and FO-definable series. First we show that even if K is finite and commutative, in general we do not have $K^{aper}\langle\langle A^*\rangle\rangle = K^{fo}\langle\langle A^*\rangle\rangle$.

Example 7.1. Let $K = \mathbb{Z}/2\mathbb{Z}$, the field with two elements, and $S = [\![\exists x.1]\!]$. Then $S(w) = |w|$ mod 2 for any $w \in A^*$. Hence S is not aperiodic since otherwise we would obtain some $m \geq 1$ such that $S(a^m) = S(a^{m+1})$ ($a \in A$), a contradiction. Note that here the monoid $(K, \cdot)$ is idempotent, and $(K, +)$ is not aperiodic.

Example 7.2. Let K be the tropical semiring and $T = [\![\forall x.1]\!]$. Then $T(w) = |w|$ for all $w \in A^*$, so T is not aperiodic. Note that $(\mathbb{N} \cup \{-\infty\}, \max)$ is idempotent, but $(\mathbb{N} \cup \{-\infty\}, +)$ is not weakly aperiodic.

These examples indicate that in order to achieve the inclusion $K^{fo}\langle\langle A^*\rangle\rangle \subseteq K^{aper}\langle\langle A^*\rangle\rangle$, we need some aperiodicity assumption both for $(K, +)$ and $(K, \cdot)$.

We call a semiring K *weakly bi-aperiodic*, if both $(K, +)$ and $(K, \cdot)$ are weakly aperiodic. If K is also commutative, then in particular K is locally finite. Clearly, any idempotent monoid is weakly aperiodic. Thus the weakly bi-aperiodic semirings include all semirings in which both addition and multiplication are idempotent, and this class of semirings properly contains (cf. [10]) the class of all distributive lattices $(L, \vee, \wedge, 0, 1)$ with smallest element 0 and greatest element 1. There are further examples:

Example 7.3. Let $0 < d \in \mathbb{R}$. We let $\mathbb{R}_{\max}^d$ be the real max$-$plus semiring truncated at d, i.e. $\mathbb{R}_{\max}^d = ([0, d] \cup \{-\infty\}, \max, +_d, -\infty, 0)$ with $x +_d y := x + y$

if $x + y \leq d$, and $x +_d y := d$ if $x + y \geq d$. This semiring is weakly bi-aperiodic, and $(\mathbb{R}^d_{max}, +_d)$ is weakly aperiodic but not aperiodic.

We can show (see [7]):

Theorem 7.4. *Let K be a commutative weakly bi-aperiodic semiring, and A an alphabet. Then $K^{aper}\langle\langle A^*\rangle\rangle = K^{fo}\langle\langle A^*\rangle\rangle = K^{rfo}\langle\langle A^*\rangle\rangle$.*

References

1. A. Arnold. *Finite Transition Systems*. International Series in Computer Science. Prentice Hall, 1994.
2. J. Berstel and Ch. Reutenauer. *Rational Series and Their Languages*, volume 12 of *EATCS Monographs in Theoretical Computer Science*. Springer Verlag, 1988.
3. J.R. Büchi. Weak second-order arithmetic and finite automata. *Z. Math. Logik Grundlagen Math.*, 6:66–92, 1960.
4. A.L. Buchsbaum, R. Giancarlo, and J.R. Westbrook. On the determinization of weighted finite automata. *SIAM Journal on Computing*, 30(5):1502–1531, 2000.
5. K. Culik and J. Kari. Image compression using weighted finite automata. *Computer and Graphics*, 17:305–313, 1993.
6. M. Droste and P. Gastin. On aperiodic and star-free formal power series in partially commuting variables. In *Proccedings of FPSAC'00*, pages 158–169. Springer, 2000. Full version available as a Research Report, LSV, ENS de Cachan, France, 2005.
7. M. Droste and P. Gastin. Weighted automata and weighted logics. Research Report LSV-05-02, ENS de Cachan, France, 2005. 23 pages.
8. S. Eilenberg. *Automata, Languages and Machines*, volume A. Academic Press, New York, 1974.
9. C.C. Elgot. Decision problems of finite automata design and related arithmetics. *Trans. Amer. Math. Soc.*, 98:21–52, 1961.
10. S. Golan. *Semirings and their Applications*. Kluwer Academic Publisher, 1999.
11. U. Hafner. *Low Bit-Rate Image and Video Coding with Weighted Finite Automata*. PhD thesis, Universität Würzburg, Germany, 1999.
12. Z. Jiang, B. Litow, and O. de Vel. Similarity enrichment in image compression through weighted finite automata. In *COCOON'00*, number 1858 in Lecture Notes in Computer Science, pages 447–456. Springer Verlag, 2000.
13. F. Katritzke. *Refinements of data compression using weighted finite automata*. PhD thesis, Universität Siegen, Germany, 2001.
14. D. Krob. The equality problem for rational series with multiplicities in the tropical semiring is undecidable. *International Journal of Algebra and Computation*, 4(3):405–425, 1994.
15. W. Kuich and A. Salomaa. *Semirings, Automata, Languages*, volume 6 of *EATCS Monographs in Theoretical Computer Science*. Springer Verlag, 1986.
16. R.P. Kurshan. *Computer-Aided Verification of Coordinating Processes*. Princeton Series in Computer Science. Princeton University Press, 1994.
17. K. McMillan. *Symbolic Model Checking*. Kluwer Academic Publishers, 1993.
18. R. McNaughton and S. Papert. *Counter-free automata*. MIT Press, Cambridge, 1971.
19. M. Mohri. Finite-state transducers in language and speech processing. *Computational Linguistics*, 23:269–311, 1997.

20. M. Mohri, F. Pereira, and M. Riley. The design principles of a weighted finite-state transducer library. *Theoretical Computer Science*, 231:17–32, 2000.
21. A. Salomaa and M. Soittola. *Automata-Theoretic Aspects of Formal Power Series*. Texts and Monographs in Computer Science. Springer Verlag, 1978.
22. M.P. Schützenberger. On the definition of a family of automata. *Information and Control*, 4:245–270, 1961.
23. M.P. Schützenberger. On finite monoids having only trivial subgroups. *Information and Control*, 8:190–194, 1965.
24. W. Thomas. Languages, automata and logic. In G. Rozenberg and A. Salomaa, editors, *Handbook of Formal Languages, Vol. 3*, pages 389–485. Springer Verlag, 1997.

Restricted Two-Variable FO+MOD Sentences, Circuits and Communication Complexity

Pascal Tesson[1,*] and Denis Thérien[2,**]

[1] Département d'Informatique et de Génie Logiciel, Université Laval
pascal.tesson@ift.ulaval.ca
[2] School of Computer Science, McGill University
denis@cs.mcgill.ca

Abstract. We obtain a logical characterization of an important class of regular languages, denoted $\mathcal{DO}$, and of its most important subclasses in terms of two-variable sentences with ordinary and modular quantifiers but in which all modular quantifiers lie outside the scope of ordinary quantifiers. The result stems from a new decomposition of the variety of monoids **DO** in terms of iterated block products.

This decomposition and the ensuing logical characterization allows us to shed new light on recent results on regular languages which are recognized by bounded-depth circuits with a linear number of wires and regular languages with small communication complexity.

1 Introduction

Descriptive complexity uses logical sentences to define languages. For instance, one can view the sentence

$$\exists x \exists y \exists z (x < y < z \wedge Q_a x \wedge Q_b y \wedge Q_c z)$$

as defining the set of words in which there are positions x, y, z with $x < y < z$ holding the letters a, b and c respectively, i.e. the set $\Sigma^* a \Sigma^* b \Sigma^* c \Sigma^*$. A large amount of research has sought to understand the expressive power of such sentences. In particular, the celebrated result of McNaughton and Papert [6] shows that languages definable by a first-order sentence which, as the one above, use only the order predicate $<$ are exactly the star-free regular languages. By a result of Schützenberger, these are also the regular languages recognizable by an aperiodic (or group-free) monoid and we can thus decide whether a language is **FO** definable. Kamp [4] further showed that these **FO** sentences could be restricted to use only three variables (provided that they can be reused). Much later, Thérien and Wilke characterized languages definable by **FO** formulas using only *two* variables (**FO₂**) [16] as languages recognized by monoids in **DA**.

---

[*] Part of this research took place while the author was at the University of Tübingen, supported by the von Humboldt Foundation.
[**] Research supported in part by NSERC, FQRNT and the von Humboldt Foundation.

When **FO** is augmented with modular quantifiers (checking whether a property holds for $i \pmod p$ of the positions in the string), all languages recognizable by solvable monoids can further be described. In this case too, three variables suffice to obtain full expressivity. The power of **FO + MOD** two-variable sentences was studied in [11, 12] and the results depended crucially on showing that every such sentence was equivalent to one where no ordinary quantifier occurs within the scope of a modular quantifier. In this paper, we handle the other extreme case, that is sentences in which no modular quantifier lies in the scope of an ordinary quantifier. This restriction is meaningful since we show that these are provably less expressive than general two-variable sentences and can provide a logical characterization for important subclasses of regular languages.

The key to our new results is a decomposition of the variety of monoids **DO** (precise definitions will be given in Section 2) in terms of weakly-iterated block products, an idea put forward in [12]. That result is interesting in its own right but our motivation comes from two recent results in complexity theory. We show that a regular language is definable by a two-variable sentence in which no modular quantifier lies in the scope of another quantifier iff it is recognized by a monoid in the class $\mathbf{DO} \cap \overline{\mathbf{Ab}}$. This happens to be *precisely* the class of regular languages that can be recognized by ACC^0-circuits with a linear number of wires [5] and by $O(\log n)$-cost communication protocols [15]. In contrast, languages defined by a two-variable sentence without this restriction can be computed by ACC^0-circuits using a linear number of *gates* and we believe that they are the only regular languages with this property. The logical characterization of languages recognized by monoids in $\mathbf{DO} \cap \overline{\mathbf{Ab}}$ sheds new light on the circuit complexity and communication complexity upper bounds, highlighting the interplay between logic, algebra, complexity and formal languages in this context.

In Section 2, we introduce the algebraic tools needed for our discussion. In Section 3, we decompose subvarieties of **DO** using weakly iterated block-products and use this to obtain a logical characterization of the corresponding regular languages in Section 4. Finally, we discuss the applications of these results to circuit complexity and communication complexity in Section 5.

2 Finite Monoids and Regular Languages

We sketch here the algebraic background needed for our discussion. For a more thorough overview of algebraic automata theory, we refer the reader to e.g. [7]. A *monoid* M is a set with a binary associative operation (which we denote multiplicatively) and a distinguished identity element 1_M. For an alphabet Σ, the set of finite words Σ^* forms a monoid under concatenation and with identity element ϵ. For a finite monoid M and language $K \subseteq \Sigma^*$, we say that K is recognized by M if there exists a homomorphism $h : \Sigma^* \to M$ and a subset $T \subseteq M$ such that $K = h^{-1}(T)$. A simple variant of Kleene's Theorem states that a language is regular iff it can be recognized by some finite monoid. In fact, all monoids considered in this paper, with the exception of the free monoid Σ^*, will be finite. It is useful to point out that if K and K' are recognized by M and

M' respectively then K's complement is recognized by M and both $K \cup K'$ and $K \cap K'$ are recognized by $M \times M'$.

A class of finite monoids is a *variety* if it is closed under direct product, homomorphic images and submonoids. In particular, we will deal with the variety of solvable groups $\mathbf{G_{sol}}$, p-groups $\mathbf{G_p}$, Abelian groups $\mathbf{Ab}$, Abelian p-groups $\mathbf{Ab_p}$, aperiodic or group-free monoids $\mathbf{A}$, semilattices $\mathbf{SL}$, i.e. monoids satisfying $x^2 = x$ and $xy = yx$, and the variety $\mathbf{DO}$ of monoids M which satisfy the identity $(xy)^\omega (yx)^\omega (xy)^\omega = (xy)^\omega$ for all $x, y \in M$. In fact, ω can be chosen as the smallest integer such that $x^{2\omega} = x^\omega$ for all $x \in M$.

Furthermore, for any variety of groups $\mathbf{H}$, we denote as $\overline{\mathbf{H}}$ the variety of monoids whose subgroups all lie in $\mathbf{H}$. For any variety $\mathbf{V}$ of monoids, we will denote as $L(\mathbf{V})$ the class of languages that can be recognized by a monoid of $\mathbf{V}$. These *varieties of languages* are fairly robust and in particular are closed under Boolean operations. We focus on varieties of the form $\mathbf{DO} \cap \overline{\mathbf{H}}$ and we give here a description of the corresponding regular languages.

We say that words $x, y \in \Sigma^*$ are M-equivalent if $h(x) = h(y)$ for any homomorphism $h : \Sigma^* \to M$. Let $\alpha(x) \subseteq \Sigma$ be the set of letters occurring in x. For $a \in \alpha(x)$, the *a-left decomposition* of x is the factorization $x = x_0 a x_1$ with $a \notin \alpha(x_0)$. The *a-right decomposition* is defined symmetrically. For a finite group G we define the congruence $\sim_{n,k}^G$ on Σ^* with $|\Sigma| = n$ by induction on $n + k$. First, $x \sim_{n,0}^G y$ iff x, y are G-equivalent. Next, we let $x \sim_{n,k}^G y$ iff:

1. $x \sim_{n,k-1}^G y$;
2. $\alpha(x) = \alpha(y)$;
3. For any $a \in \alpha(x) = \alpha(y)$, if $x = x_0 a x_1$ and $y = y_0 a y_1$ are the a-left decompositions of x and y then $x_0 \sim_{n-1,k}^G y_0$ and $x_1 \sim_{n,k-1}^G y_1$;
4. For any $a \in \alpha(x) = \alpha(y)$, if $x = x_0 a x_1$ and $y = y_0 a y_1$ are the a-right decompositions of x and y then $x_0 \sim_{n,k-1}^G y_0$ and $x_1 \sim_{n-1,k}^G y_1$.

This equivalence relation is well-defined since $|\alpha(x_0)| < |\alpha(x)|$ in (3) and $|\alpha(x_1)| < |\alpha(x)|$ in (4). These congruences allow us to describe languages in $L(\mathbf{DO} \cap \overline{\mathbf{H}})$ and this will be crucial in the proof of Theorem 3.

Theorem 1 ([15]). *Let K be a language in Σ^* with $|\Sigma| = n$. Then K is in $L(\mathbf{DO} \cap \overline{\mathbf{H}})$ iff K is the union of $\sim_{n,k}^G$-classes for some $k \in \mathbb{N}$ and $G \in \mathbf{H}$.*

The following simple fact will also prove useful for Theorem 3 and Lemma 4:

Lemma 1 (e.g. [7]). *A language $K \in \Sigma^*$ can be recognized by a semilatice iff it is in the Boolean algebra generated by the languages $\Sigma^* a \Sigma^*$. Furthermore, u, v are M-equivalent for any semilattice M iff $\alpha(u) = \alpha(v)$.*

Let M and N be finite monoids. As in [11] we denote the operation of M as $+$ and its identity element as 0 to distinguish it from the operation of N. A left-action of N on M is a function mapping pairs $(n, m) \in N \times M$ to $nm \in M$ and satisfying $n(m_1 + m_2) = nm_1 + nm_2$, $n_1(n_2 m) = (n_1 n_2)m$, $n0 = 0$ and $1m = m$. Right actions are defined symmetrically. If we have both a right and a left-action of N on M that further satisfy $n_1(mn_2) = (n_1 m)n_2$, we can construct the bilateral semidirect product $M**N$ which we define as the monoid with elements in $M \times N$ and multiplication defined as $(m_1, n_1)(m_2 n_2) = (m_1 n_2 + n_1 m_2, n_1 n_2)$. This operation is associative and $(0, 1)$ acts as an identity for it.

For varieties $\mathbf{V}, \mathbf{W}$, we denote $\mathbf{V}\square\mathbf{W}$ the variety generated by all semidirect products $M**N$ with $M \in \mathbf{V}$, $N \in \mathbf{W}$. For varieties $\mathbf{U}, \mathbf{V}, \mathbf{W}$ we always have $(\mathbf{U}\square\mathbf{V})\square\mathbf{W} \subseteq \mathbf{U}\square(\mathbf{V}\square\mathbf{W})$ but the containment is strict in general. Block-product decompositions of varieties traditionally use the stronger bracketing but [11] showed the relevance of the weak bracketing, particularly in relation to two-variable logical sentences.

The languages recognized by $\mathbf{V}\square\mathbf{W}$ can be conveniently described in terms of languages recognized by $\mathbf{V}$ and $\mathbf{W}$. For a monoid $N \in \mathbf{W}$, an N-transduction τ is a function determined by two homomorphisms $h_l, h_r : \Sigma^* \to N$ and mapping words in Σ^* to words in $(N \times \Sigma \times N)^*$. For a word $w = w_1 \ldots w_n \in \Sigma^*$ we set $\tau(w) = \tau(w_1)\tau(w_2)\ldots\tau(w_n)$ with $\tau(w_i) = (h_l(w_1 \ldots w_{i-1}), w_i, h_r(w_{i+1} \ldots w_n))$. For a language $K \subseteq (N \times \Sigma \times N)^*$, let $\tau^{-1}(K) = \{w \in \Sigma^* : \tau(w) \in K\}$.

Theorem 2. *[9, 7] A regular language lies in $L(\mathbf{V}\square\mathbf{W})$ iff it is the Boolean combination of languages in $L(\mathbf{W})$ and languages $\tau^{-1}(K)$ for some $K \in \mathbf{V}$ and N-transduction τ with $N \in \mathbf{W}$.*

A *pointed word* is a pair (w, p) consisting of a word $w \in \Sigma^*$ and a pointer p with $1 \leq p \leq |w|$. A *pointed language* $\dot{K}$ is a set of such structures and, as in [17], we extend the notion of monoid recognizability to these languages: $\dot{K}$ is recognized by M if there are homomorphisms $h_l, h_r : \Sigma^* \to M$ and a set of triples $T \subseteq (M \times \Sigma \times M)$ such that $\dot{K} = \{(w, p) : (h_l(w_1 \ldots w_{p-1}), w_p, h_r(w_{p+1} \ldots w_{|w|})) \in T\}$. For a variety $\mathbf{V}$ we will denote as $P(\mathbf{V})$ the set of pointed languages recognized by a monoid in $\mathbf{V}$. Abusing our terminology, it will be convenient to think of ordinary words in Σ^* as pointed words with $p = 0$ and thus view $L(\mathbf{V})$ as a subset of $P(\mathbf{V})$.

3 A Weak Block Product Decomposition of DO

In this section, we characterize each variety $\mathbf{DO} \cap \overline{\mathbf{H}}$ for a variety of groups $\mathbf{H}$ using weakly iterated block products. The idea is similar to that of [11] but we also need the combinatorial description of the corresponding regular languages.

Theorem 3. *Let $\mathbf{H}$ be a variety of finite groups, and $\mathbf{V_H}$ be the smallest variety such that $\mathbf{H} \subseteq \mathbf{V_H}$ and $\mathbf{V_H}\square\mathbf{SL} \subseteq \mathbf{V_H}$, then $\mathbf{V_H} = \mathbf{DO} \cap \overline{\mathbf{H}}$.*

Proof. Clearly, $\mathbf{DO} \cap \overline{\mathbf{H}}$ contains $\mathbf{H}$ and we also need to show that it is closed under block product with $\mathbf{SL}$. First, we claim that $\mathbf{DO}\square\mathbf{SL} = \mathbf{DO}$. Let $M \in \mathbf{DO}$ and $N \in \mathbf{SL}$. As we did earlier, we denote the multiplication of M additively (even though M is not necessarily commutative). Since $M \in \mathbf{DO}$ there exists an integer k such that $k(v + w) + k(w + v) + k(v + w) = k(v + w)$ and $(2k)v = kv$ for all $v, w \in M$ and we prove that any bilateral semidirect product $M**N$ satisfies the identity $(xy)^{2k}(yx)^{2k}(xy)^{2k} = (xy)^{2k}$.

Let $x = (m_1, n_1)$ and $y = (m_2, n_2)$ be arbitrary elements of $M * *N$. By definition of the bilateral semidirect product, we have $xy = (m_1 n_2 + n_1 m_2, n_1 n_2)$ and so $(xy)^{2k} = (z, (n_1 n_2)^{2k})$ where

$$z = m_1 n_2 (n_1 n_2)^{2k-1} + n_1 m_2 (n_1 n_2)^{2k-1} + \ldots + (n_1 n_2)^{2k-1} n_1 m_2$$

Since N is commutative and idempotent ($n^2 = n$), this is simply:

$$z = m_1(n_1 n_2) + n_1 m_2(n_1 n_2) + \ldots + (n_1 n_2) m_2$$
$$= m_1(n_1 n_2) + n_1 m_2(n_1 n_2) + (2k-1)[(n_1 n_2) m_1(n_1 n_2) + (n_1 n_2) m_2(n_1 n_2)]$$
$$+ (n_1 n_2) m_2 n_1 + (n_1 n_2) m_2$$

By the same argument $(xy)^{2k}(yx)^{2k}(xy)^{2k}$ is the pair $(z', n_1 n_2)$ with

$$z' = m_1(n_1 n_2) + n_1 m_2(n_1 n_2) + (2k-1)\left[(n_1 n_2) m_1(n_1 n_2) + (n_1 n_2) m_2(n_1 n_2)\right]$$
$$+ (2k)[(n_1 n_2) m_2(n_1 n_2) + (n_1 n_2) m_1(n_1 n_2)]$$
$$+ (2k-1)\left[(n_1 n_2) m_1(n_1 n_2) + (n_1 n_2) m_2(n_1 n_2)\right] + (n_1 n_2) m_2 n_1 + (n_1 n_2) m_2.$$

If we let $v = (n_1 n_2) m_1(n_1 n_2)$ and $w = (n_1 n_2) m_2(n_1 n_2)$, then the middle part of the product is simply $(2k-1)(v+w) + 2k(w+v) + (2k-1)(v+w)$ and since M satisfies $k(v+w) + k(w+v) + k(v+w) = k(v+w)$ and $(2k)v = kv$, this sum is equal to $(k-1)(v+w) + k(v+w) + (k-1)(v+w) = (3k-2)(v+w) = (2k-2)(v+w)$. This gives

$$z' = m_1(n_1 n_2) + n_1 m_2(n_1 n_2) + (2k-2)[(n_1 n_2) m_1(n_1 n_2)$$
$$+ (n_1 n_2) m_2(n_1 n_2)] + (n_1 n_2) m_2 n_1 + (n_1 n_2) m_2$$
$$= z$$

and so $(xy)^{2k}(yx)^{2k}(xy)^{2k} = (xy)^{2k}$ as claimed.

Furthermore, folklore results show that if the variety **W** is aperiodic (i.e. contains no non-trivial groups), then for any variety **V** the groups lying in **V**□**W** are exactly those in **V**. In particular, **SL** is aperiodic so $(\mathbf{DO} \cap \overline{\mathbf{H}})\square\mathbf{SL} \subseteq \overline{\mathbf{H}}$ and by the first part of our argument $(\mathbf{DO} \cap \overline{\mathbf{H}})\square\mathbf{SL} \subseteq \mathbf{DO} \cap \overline{\mathbf{H}}$.

Let $\mathbf{V_0} = \mathbf{H}$ and $\mathbf{V_{i+1}} = \mathbf{V_i}\square\mathbf{SL}$: to complete our proof we now show that $\mathbf{DO} \cap \overline{\mathbf{H}}$ is contained in any variety containing **H** and closed under block product with **SL**. We do so by proving that any monoid of $\mathbf{DO} \cap \overline{\mathbf{H}}$ lies in some $\mathbf{V_i}$. We use the family $\sim^\mathbf{H}$ of congruences for $\mathbf{DO} \cap \overline{\mathbf{H}}$ and show that for each alphabet Σ with $|\Sigma| = n$, any $k \geq 0$, any group G and any word $w \in \Sigma^*$, the language $L_{n,k}^{(w)} = \{v \in \Sigma^* : v \sim_{n,k}^G w\}$ is recognized by some M in $\mathbf{V_{n+k}}$. We argue by induction on $n+k$: if $n = 0$ the claim is trivial. If $k = 0$ then L_w is the set of words that are G-equivent to w and this can be recognized by a direct product of t copies of G where t is the number of homomorphisms from Σ^* to G.

Suppose $n, k > 0$: to check if $v \sim_{n,k}^G w$, we need to first verify that $\alpha(v) = \alpha(w)$ and, as stated in Lemma 1, this can be done using some monoid in **SL**. Now, let $w = w_l a w_r$ and $v = v_l a v_r$ be the a-left-decompositions of w and v.

We claim that there is an **SL**-transduction τ and a language $K' \in L(\mathbf{V_{n+k-1}})$ such that $\tau^{-1}(K')$ is the set of words v such that $v_l \sim^G_{n-1,k} w_l$. Consider the two-element monoid $U_1 = \{0,1\}$ in **SL** with multiplication given by $0 \cdot x = x \cdot 0 = 0$ and $1 \cdot x = x \cdot 1 = x$. Let $h_l = h_r : \Sigma^* \to U_1$ be the homomorphism mapping a to 0 and every other letter to the identity 1. Thus, $h_l(v) = 0$ iff $a \in \alpha(v)$ and the transduction τ defined by h_l and h_r maps v to the sequence of triples

$$(1, v_1, 0)\ldots(1, v_i, 0)(0, v_{i+1}, 0)\ldots(0, v_j, 0)(0, v_{j+1}, 1)\ldots(0, v_n, 1)$$

where v_i and v_j are the first and last occurrence of a in v (if any such occurrence exists). We are trying to check if $v_l \sim^G_{n-1,k} w_l$ and we know by induction that the language $L^{(w_l)}_{n-1,k}$ is in $L(\mathbf{V_{n+k-1}})$, i.e. there exists $M \in \mathbf{V_{n+k-1}}$, a homomorphism $h : \Sigma^* \to M$ and a subset $T \subseteq M$ with $L^{(w_l)}_{n-1,k} = h^{-1}(T)$. Let $h' : (U_1 \times \Sigma \times U_1)^* \to M$ be the homomorphism which maps triples $(t,b,t') \in (U_1 \times \Sigma \times U_1)$ to $h'(t,b,t') = \begin{cases} h(b) & \text{if } t=1 \text{ and } b \neq a; \\ 1_M & \text{otherwise.} \end{cases}$

One can verify that for any $v \in \Sigma^*$ with the a-left decomposition $v_l a v_r$ we now get $h'(\tau(v)) = h'(\tau(v_l)) = h(v_l)$ because all the triples of $\tau(v)$ beyond that prefix are mapped to 1_M. Let $K' = h'^{-1}(T)$: we have $K' \in L(\mathbf{V_{n+k-1}})$ and $\tau^{-1}(K') = \{v \in \Sigma^* : v_l \in L^{(w_l)}_{n-1,k}\}$ as we had required.

Similarly, we can recognize words such that $w_r \sim^G_{n,k-1} v_r$ with the help of a U_1-transduction. We argue symmetrically about the right-decompositions and conclude that $L^{(w)}_{n,k} \in L(\mathbf{V_{n+k}})$.

4 An Application to Two-Variable Sentences

We refer the reader to e.g. [9] for a thorough overview of logical descriptions of regular languages and their relation to algebraic automata theory and circuit complexity. As we suggested in our introduction, we view finite words over the finite alphabet Σ as logical structures. We construct logical formulas using the order predicate $<$, the "content" predicates $\{Q_a | a \in \Sigma\}$, constants **t** (true) and **f** (false), Boolean connectives, a set of variables $\{x_1, \ldots, x_n\}$, existential and universal quantifiers as well as modular quantifiers $\exists^{i \bmod m} x$. The atomic formulas are either **t**, **f**, $x_i < x_j$ or $Q_a x_i$. A *word structure* over alphabet Σ and variable set $\mathcal{V} \subseteq \{x_1, \ldots, x_n\}$ is a pair $(w, \boldsymbol{p})$ consisting of a word $w \in \Sigma^*$ and a set of pointers $\boldsymbol{p} = (p_{i_1}, \ldots, p_{i_k})$ with $1 \leq p_i \leq |w|$ which associate each variable $x_{i_j} \in \mathcal{V}$ with a position p_{i_j} in the string. A *simple extension* of a word structure $(w, \boldsymbol{p})$ over $\Sigma, \mathcal{V}$ is a word structure $(w, \boldsymbol{p}')$ over $\Sigma, (\mathcal{V} \cup \{x_{i_k+1}\})$ such that $x_{i_{k+1}} \notin \mathcal{V}$ and $p_{i_j} = p'_{i_j}$ for $1 \leq j \leq k$. We can now formally define the semantics of our formulas in a natural way. If $w = w_1 \ldots w_t$ is a word, we have

$(w, \boldsymbol{p}) \models Q_a x_i$ \qquad if $w_{p_i} = a$;
$(w, \boldsymbol{p}) \models x_i < x_j$ \qquad if $p_i < p_j$;
$(w, \boldsymbol{p}) \models \exists x(\phi(x_k))$ \qquad if there exists an extension $(w, \boldsymbol{p}')$ of $(w, \boldsymbol{p})$ such that $(w, \boldsymbol{p}') \models \phi(x_k)$;
$(w, \boldsymbol{p}) \models \exists^{i \bmod m} x_k(\phi(x_k))$ if there exists i modulo m extensions $(w, \boldsymbol{p}')$ of $(w, \boldsymbol{p})$ such that $(w, \boldsymbol{p}') \models \phi(x_k)$;

We omit for space the obvious definition of the semantics of the Boolean connectives and of the universal quantifiers.

If ϕ is a sentence, i.e. a formula with no free variable, we denote as $L_\phi \subseteq \Sigma^*$ the language $L_\phi = \{w : (w, \emptyset) \models \phi\}$. Similarly, it will be useful for our purposes to consider the special case of formulas with a single free variable. Such a formula naturally defines a set of word structures (w, p) with $1 \leq p \leq |w|$, i.e. a *pointed language*. For any formula ϕ having a single free variable and Φ a class of such formulas, we will denote as P_ϕ the pointed language $P_\phi = \{(w, p) : (w, p) \models \phi\}$ and $P(\Phi)$ the class of all P_ϕ with $\phi \in \Phi$.

We will denote as **FO**, **MOD**, **MOD^p**, **FO + MOD** and **FO + MOD^p** the class of respectively first-order sentences (no modular quantifier), modular sentences (no existential or universal quantifier), modular sentences with only mod p quantifiers and so on. The expressive power of such sentences has been investigated thoroughly and all existing results are algebraic in nature: languages definable by sentences of the fragment Γ are exactly those in $L(\mathbf{V_\Gamma})$ for some appropriate variety $\mathbf{V_\Gamma}$. (see [10] for elements of a meta-explanation). In particular, we have $L(\mathbf{FO}) = L(\mathbf{A})$ [6, 8], $L(\mathbf{MOD}) = L(\mathbf{G_{sol}})$, $L(\mathbf{MOD^p}) = L(\mathbf{G_p})$, $L(\mathbf{FO + MOD}) = L(\overline{\mathbf{G_{sol}}})$ [13].

While it is natural to construct logical sentences using a new variable for each quantifier, one can just as well write sentences that reuse variables. For instance, we gave earlier a three-variable sentence defining the language $\Sigma^* a \Sigma^* b \Sigma^* c \Sigma^*$. It is also definable by the two-variable sentence $\exists x (Q_a x \wedge \exists y (Q_b y \wedge x < y \wedge \exists x (Q_c x \wedge y < x)))$. The semantics of the sentence are perfectly unambiguous (see [12] for a formal discussion). We denote $\mathbf{FO_k}$, $\mathbf{FO + MOD_k}$ and so on the class of sentences using only k variables. Surprisingly, three variables suffice to describe any language in $L(\mathbf{FO + MOD})$ [4, 3, 12]. If only one variable is allowed, it is easy to show that our expressive power is dramatically reduced and, for instance, that $L(\mathbf{FO_1}) = L(\mathbf{SL})$ and $L(\mathbf{MOD_1}) = L(\mathbf{Ab})$. The expressiveness of $\mathbf{FO_2}$ or $\mathbf{FO + MOD_2}$ is trickier to understand [16, 12] but also admits algebraic characterizations. In order to study $\mathbf{FO + MOD_2}$, [12] show that every such sentence can be rewritten so that no existential or universal quantifier appears in the scope of a modular quantifier. We will show that this is not just an artefact of the proof: at the other end of the spectrum, two-variable sentences in which no modular quantifier appears in the scope of an ordinary quantifier are provably less expressive than general $\mathbf{FO + MOD_2}$ sentences.

Definition 1. *Let Σ be an alphabet and $\Phi = \{\phi_1, \ldots \phi_k\}$ be a set formulas over Σ with one free variable, say x, and a single bound variable y (possibly bound* by more than one quantifier). A recycling Φ-substitution σ over Σ is a function mapping two-variable sentences over the alphabet 2^Φ (the power set of Φ) to two-variable sentences over Σ as follows: each occurrence of the predicate $Q_S x$ with $S \subseteq \Phi$ is replaced by the conjunction $\bigwedge_{\phi_i \in S} \phi_i$ and each occurrence of $Q_S y$ is replaced by the similar formula in which the roles of x and y are interchanged.

Note that recycling substitutions preserve the two-variable property. If Γ is a class of two-variable sentences and Λ is a class of formulas with one bound and at most one free variable we will denote by $\Gamma \circ \Lambda$ the class of sentences which are Boolean combinations of *sentences* in Λ and of sentences obtained by applying to a sentence ψ of Γ a recycling substitution in which all formulas in Φ lie in Λ.

Lemma 2. *Let σ be a substitution and for any $w = w_1 \ldots w_n$ in Σ^* let $\sigma^{-1}(w)$ be the word $u_1 \ldots u_n$ over the alphabet 2^Φ with $u_i = \{\phi_j : (w,i) \models \phi_j\}$. Then $w \models \sigma(\psi)$ iff $\sigma^{-1}(w) \models \psi$.*

The proof is omitted but is straightforward. The function mapping w to $\sigma^{-1}(w)$ is of course quite similar to the notion of transduction which we introduced in Section 2 and we now prove an equivalent of the "block product/substitution principle" of [17] which formalizes the idea:

Lemma 3. *Let Γ be a class of $\mathbf{FO + MOD_2}$ sentences and Λ a class of $\mathbf{FO + MOD_2}$ formulas with one free variable. If $\mathbf{V}, \mathbf{W}$ are monoid varieties such that $L(\Gamma) = L(\mathbf{V})$ and $P(\Lambda) = P(\mathbf{W})$, then $L(\Gamma \circ \Lambda) = L(\mathbf{V} \square \mathbf{W})$.*

Proof. Since $L(\mathbf{V}\square\mathbf{W})$ is closed under Boolean combinations, the left-to-right containment follows if we show that for any $\psi \in \Gamma$ and any Λ-substitution σ we have $L(\sigma(\psi)) \in L(\mathbf{V}\square\mathbf{W})$. Let w be some word in Σ^* and $\Phi = \{\phi_1, \ldots \phi_k\}$ be the formulas used by σ. Since $P(\mathbf{W}) = P(\Lambda)$, the pointed languages $P_{\phi_j} : \{(w,i) : (w,i) \models \phi_j\}$ can be recognized by monoids $M_1, \ldots, M_k$ in $\mathbf{W}$ and $M = M_1 \times \ldots \times M_k$ recognizes any Boolean combination of them. We can therefore construct an M-transduction τ such that $\tau(w_i)$ completely determines the set $\{\phi_j : (w,i) \models \phi_j\}$. Since we assume that $L(\psi)$ is recognized by a monoid N in $\mathbf{V}$, we get that $L(\sigma(\psi)) = \tau^{-1}(K)$ for some $K \subseteq (M \times \Sigma \times M)^*$ also recognized by N. Hence, $L(\sigma(\psi)) \in L(\mathbf{V}\square\mathbf{W})$.

For the right-to-left containment, we need to show that any language of $L(\mathbf{V}\square\mathbf{W})$ can be described by a sentence of $\Gamma \circ \Lambda$ and we proceed similarly. If τ is an M-transduction for some $M \in \mathbf{W}$ then for any triple $(m_1, a, m_2) \in M \times \Sigma \times M$, the pointed language $T_{(m_1,a,m_2)} = \{(w,i) : \tau(w_i) = (m_1, a, m_2)\}$ is in $P(\mathbf{W})$ and is thus definable by some formula $\phi_{(m_1,a,m_2)}$ in $P(\Lambda)$. Any language $K \subseteq (M \times \Sigma \times M)^*$ in $L(\mathbf{V})$ is definable by some sentence $\psi_K \in \Gamma$. Now the set of words such that $\tau(w) \in K$ is defined by the sentence obtained from ψ_K by a recycling substitution using the formulas[1] $\phi_{(m_1,a,m_2)}$.

Let $\mathbf{FF_1}$ be the class of $\mathbf{FO}$ formulas with one free and one quantified variable.

---

[1] Note that at any position i, exactly one of the $\phi_{(m_1,a,m_2)}$ is true and we can thus identify the domain of $\sigma^{-1}(w_i)$ with the set $M \times \Sigma \times M$.

Lemma 4. $P(\mathbf{FF_1}) = P(\mathbf{SL})$.

Proof (sketch). Each $\mathbf{FF_1}$ formula can be rewritten as a Boolean combination of formulas of the form $\exists y((x * y) \wedge Q_a y)$ with $* \in \{<, >, =\}$. By Lemma 1, the pointed language defined by such a formula is recognized by some $M \in \mathbf{SL}$ since we are simply asking whether the letter a occurs somewhere before x (if $*$ is $>$), at x (if $*$ is $=$) or after x (if $*$ is $<$). Conversely, if a pointed language K is recognized by some $M \in \mathbf{SL}$ then by Lemma 1 membership of (w, p) in K depends on the letter w_p and on the set of letters occurring before and after the pointer. Thus, K can be defined by an $\mathbf{FF_1}$ formula.

Theorem 4. *Let Σ be a finite alphabet. A language $L \subseteq \Sigma^*$ is*

(1) definable by a $\mathbf{FO}+\mathbf{MOD_2}$ sentence in which no modular quantifier appears in the scope of an ordinary quantifier iff $M(L) \in \mathbf{DO} \cap \overline{\mathbf{G_{sol}}}$;
(2) definable by a $\mathbf{FO}+\mathbf{MOD_2}$ sentence in which no modular quantifier appears in the scope of another quantifier iff $M(L) \in \mathbf{DO} \cap \overline{\mathbf{Ab}}$;
(3) definable by a $\mathbf{FO}+\mathbf{MOD_2}$ sentence in which all modular quantifiers have prime moduli and no Mod_p-quantifier appears in the scope of an ordinary quantifier or of a Mod_q quantifier for p, q distinct primes iff $M(L) \in \mathbf{DO} \cap \mathbf{G_{nil}}$;

Proof. The proofs are applications of Lemma 3. By [11], we have $L(\mathbf{MOD_2}) = L(\mathbf{G_{sol}})$. Let $\mathbf{D_k}$ be the class $\mathbf{FO}+\mathbf{MOD_2}$ sentences in which no modular quantifier appears in the scope of an ordinary quantifier and in which the nesting depth of ordinary quantifiers is k. To complete our argument, it suffices to note that sentences in $\mathbf{D_k}$ are exactly those which can be obtained from $\mathbf{MOD_2}$ sentences by applying successively k $\mathbf{FF_1}$ substitutions. By applying the block-product substitution principle and Lemma 4 we get that $L(\mathbf{D_k}) = L((\ldots(\mathbf{G_{sol}}\square\mathbf{SL})\ldots\square\mathbf{SL}))$ (where the product has length k) and by Theorem 3 the union of the varieties on the right is exactly $\mathbf{DO} \cap \overline{\mathbf{G_{sol}}}$.

To obtain, the second and third parts of our theorem we simply need to recall that $L(\mathbf{MOD_1}) = L(\mathbf{Ab})$ and that $L(\mathbf{G_{nil}})$ is the class of languages definable by $\mathbf{MOD_2}$ sentences satisfying the restriction given in (3) [11, 13].

The regular language $K = (b*ab*a)^*b\Sigma^*$ is defined by the sentence $\exists x(Q_b x \wedge \exists^{0 \mod 2} y(x < y \wedge Q_a x))$. We claim that K cannot be recognized by some M in $\mathbf{DO}$ and, thus defined by a $\mathbf{FO}+\mathbf{MOD_2}$ sentence in which the modular quantifiers lie outside the scope of the ordinary quantifiers. Indeed, let h be the recognizing morphism and consider $x = h(ab)$ and $y = h(a)$: we have for some ω that $h[(aba)^\omega (baa)^\omega (aba)^\omega] = h[(aba)^\omega]$ in M but this is impossible since $(aba)^\omega (baa)^\omega (aba)^\omega$ is in K while $(aba)^\omega$ is not.

The variety $\mathbf{DO}$ is decidable since it is defined by a simple identity and it is also easy to decide whether the subgroups of such a monoid lie in $\mathbf{Ab}, \mathbf{G_p}, \mathbf{G_{nil}}$ or $\mathbf{G_{sol}}$. Moreover the proof of Theorem 4 actually shows a tight correspondence between the depth of ordinary quantifiers in the $\mathbf{FO}+\mathbf{MOD_2}$ formulas with the depth of the product $(\ldots(\mathbf{H}\square\mathbf{SL})\square\mathbf{SL})\ldots\square\mathbf{SL}$ and each of these varieties is also decidable, as long as $\mathbf{H}$ is. The argument requires machinery beyond the scope of this paper but the key ideas can be found in [17].

5 A New Perspective on Communication Complexity and Circuit Complexity Upper Bounds

Our decomposition of $\mathbf{DO} \cap \overline{\mathbf{Ab}}$ in terms of block products and the ensuing logical characterization of the corresponding languages allow us to shed new light on two complexity problems recently resolved in [5, 15].

In [5], one considers Boolean circuits of bounded depth. A circuit C_n is an acyclic digraph: the nodes of in-degree 0 are *input* nodes labeled with a Boolean variable x_i with $1 \leq i \leq n$ or a Boolean constant, the other nodes are labeled by one of the Boolean functions $\{\wedge, \vee, Mod_m\}$. We also assume that there is a single *output* node, i.e. a node of out-degree 0. A circuit C_n with n inputs computes a function $C_n : \{0,1\}^n \to \{0,1\}$ in the obvious way (note that the output of a Mod_m gate is 1 if the sum of its inputs is 0 modulo m). The *depth* of C_n is the longest path from an input to the output node. The wire-size and the gate-size of C_n are, respectively, the number of edges and nodes in the graph. By extension a family of circuits $C = (C_n)_{n \geq 0}$ computes a function $C : \{0,1\}^* \to \{0,1\}$ (i.e. defines a language of $\{0,1\}^*$) and we then think of the depth and size of C as a function of n. The class of languages which can be recognized by such circuits in polynomial-size and bounded-depth is denoted ACC^0.

To process a non-Boolean input $w \in \Sigma^*$ with such circuits we can for instance require that the circuit be given a simple Boolean encoding of each letter w_i. In [5], it was shown that the regular languages definable by an ACC^0 circuit using only a linear number of wires are exactly those recognized by monoids in $\mathbf{DO} \cap \overline{\mathbf{Ab}}$. The result relies on a communication complexity result: suppose Alice and Bob are given a word $w = w_1 w_2 \ldots w_{2n}$ where Alice knows only the odd-indexed w_i and Bob only the even-indexed ones. What is then the minimal number of bits that Alice and Bob need to exchange in the worst case to determine whether w lies in some regular language L? If L is recognized by some M in $\mathbf{DO} \cap \overline{\mathbf{Ab}}$ this can be done with $O(\log n)$ bits of communication but requires $\Theta(n)$ bits otherwise [15]. It is interesting to note that the circuit complexity and communication complexity upper bounds follow simply from the block-product decomposition of this variety:

Theorem 5. *If every $K \in L(\mathbf{V})$ has a circuit with a linear number of wires then so does any $K' \in L(\mathbf{V} \square \mathbf{SL})$. Similarly, if each $K \in L(\mathbf{V})$ has communication complexity $O(\log n)$ then so does any $K' \in L(\mathbf{V} \square \mathbf{SL})$.*

Proof (sketch). Bilardi and Preparata show in [2] that there exists a linear-size circuit computing Prefix-Suffix-OR i.e. a circuit with n Boolean inputs $x_1, \ldots, x_n$ and $2n$ outputs $p_1, \ldots, p_n, s_1, \ldots, s_n$ with $p_i = \bigvee_{j=1}^{i-1} x_j$ and $s_i = \bigvee_{j=i+1}^{n} x_j$.

It suffices to show that for any language K recognized by a monoid of $\mathbf{V}$ and any M-transduction τ with $M \in \mathbf{SL}$, the language $\tau^{-1}(K)$ has a linear size circuit. By Lemma 1, the value of $\tau(w_i)$ is determined entirely by the letter w_i and the sets $\alpha(w_1 \ldots w_{i-1})$ and $\alpha(w_{i+1} \ldots w_n)$ of letters occurring in the prefix and suffix around w_i. By using $|\Sigma|$ copies of the Prefix-Suffix-OR circuit, we can therefore build a circuit with a linear number of wires which on input $w_1 \ldots w_n$

produces n blocks of $k = \log(|M|^2 \cdot |\Sigma|)$ outputs such that the ith block encodes the value of $\tau(w_i)$. These values can now simply be fed into a circuit recognizing K which, by assumption, uses only a linear number of wires.

For the communication complexity problem, Alice and Bob begin by exchanging the location of the first and last occurrence of a letter that they have access to. This requires $O(\log n)$ bits of communication and this information is enough for each player to privately compute $\tau(w_i)$ for any w_i he has access to.

In fact, the result in [2] is more general: for a regular language L, there exists a circuit which computes Prefix-Suffix-L iff L is piecewise-testable, i.e. recognized by a monoid in the variety **J**. This well known variety contains **SL** and is in fact the largest variety of aperiodics satisfying the equality **DO**□**V** = **DO** [1].

Corollary 1. *Every language in $L(\mathbf{DO} \cap \overline{\mathbf{Ab}})$ can be defined by a linear-wire-size family of ACC^0 circuits and has two-party communication complexity $O(\log n)$.*

Proof. For any $K \in L(\mathbf{Ab})$, membership of $w \in K$ depends solely on the number $|w|_a$ of occurrences of each letter a in w modulo some integer m [7]. Computing $|w|_a$ modulo m, can clearly be done with a single Mod_m gates with n input wires and by a communication protocol of cost $2 \log |m|$. Thus, any K in $L(\mathbf{Ab})$ has linear-wire-size circuits and communication complexity $O(1)$. Our statement then follows from Theorem 5 and the decomposition result of Theorem 3.

We should note that the same idea can be used to show that for every language L recognized by $M \in \mathbf{DO} \cap \overline{\mathbf{G}_{nil}}$ there exists k such that the k-party communication complexity of L is $O(1)$ (see [14] for a discussion of the multi-party communication model).

Alternatively, one can view the linear-wire-size circuits for $\mathbf{DO} \cap \overline{\mathbf{Ab}}$ as evaluating a two-variable formula with no modular quantifier nested in another quantifier. First the circuit evaluates the most deeply nested subformulas with one free variable, say x. These are $\mathbf{FF}_1$ formulas and thus Boolean combinations of formulas such as $\phi(x) = \exists y(x < y \wedge Q_a y)$ and the Prefix-Suffix-OR construction allows us to simultaneously compute the value of $\phi(x)$ for *each* value of x using only $O(n)$ wires. These results are used in the next step to compute, for *all* positions y the value of formulas such as $\psi(y) = \exists x(x < y \wedge \phi(x))$. At the output level, we evaluate the value of a modular quantifier $\exists^{i \bmod m} x \chi(x)$. It suffices to feed the n values of $\chi(x)$ obtained in the previous step into a Mod_m gate.

In general, recognizing a language definable by an $\mathbf{FO} + \mathbf{MOD}_2$ sentence in which modular quantifiers *are* allowed in the scope of other quantifiers will require an ACC^0 circuit with a super-linear number of wires. However, these languages can still be recognized using only $O(n)$ *gates*. The strategy to build such a circuit is similar: we are given, at some stage in our circuit the values for each x of a formula $\phi(x)$ and want to compute for each y the value of a formula of the form $\psi(y) = \exists^{i \bmod m} x(x < y \wedge \phi(x))$. For each y, this can of course be computed by a *single* Mod_m gate in which we feed the results of $\phi(x)$ for each $x < y$. The sub-circuit computing all values $\psi(y)$ from the values $\phi(x)$ will use only n gates but $\Omega(n^2)$ wires. In fact, we conjecture that the regular languages recognized by an ACC^0 circuit with a linear number of gates are exactly those definable in **FO + MOD$_2$** and that, among these, the fine line between linear-gate-size and linear-wire-size is thus captured exactly by the ability to pull modular quantifiers out of the scope of any other quantifier.

It is also very natural to try and characterize regular languages which can be computed by linear-gate-size restricted ACC^0 circuits where we allow only Mod_m gates (i.e. CC^0 circuits), or only $\wedge, \vee$ gates (i.e. AC^0 circuits). We believe that these regular languages are exactly those in **MOD$_2$** and **FO$_2$** respectively. In particular, we conjecture that every regular language recognized by an AC^0 circuit with a linear number of gates can also be recognized by an AC^0 circuit with a linear number of wires. Resolving this question for CC^0 circuits amounts to proving a lower bound of $\omega(n)$ for the number of gates in CC^0 circuits computing the AND of n bits or the word problem of a non-solvable group. To the best of our knowledge, the state-of-the-art lower bound state that CC^0 circuits for AND requires $\Omega(n)$ non-input gates, a world away from the suspected $\Omega(c^n)$. As a first step, it would an interesting start to establish an $\omega(n)$ lower bound for the number of wires in such circuits.

References

1. J. Almeida and P. Weil. Profinite categories and semidirect products. *J. Pure and Applied Algebra*, 123:1–50, 1998.
2. G. Bilardi and F. Preparata. Characterization of associative operations with prefix circuits of constant depth and linear size. *SIAM J. Comput.*, 19(2):246–255, 1990.
3. N. Immerman and D. Kozen. Definability with a bounded number of bound variables. *Information and Computation*, 83(2):121–13, 1989.
4. J. A. W. Kamp. *Tense Logic and the Theory of Linear Order*. PhD thesis, University of California, Berkeley, 1968.
5. M. Koucký, P. Pudlák, and D. Thérien. Bounded-depth circuits: separating wires from gates. In *Symposium on Theory of Computing (STOC'05)*, 2005.
6. R. McNaughton and S. Papert. *Counter-Free Automata*. MIT Press, Cambridge, Mass., 1971.
7. J.-E. Pin. Syntactic semigroups. In G. R. et A. Salomaa, editor, *Handbook of language theory*, volume 1, chapter 10, pages 679–746. Springer Verlag, 1997.
8. M. P. Schützenberger. On finite monoids having only trivial subgroups. *Information and Control*, 8(2):190–194, Apr. 1965.
9. H. Straubing. *Finite Automata, Formal Logic and Circuit Complexity*. Boston: Birkhauser, 1994.
10. H. Straubing. On the logical description of regular languages. In *Proc. of the 5th Latin American Theoretical Informatics Conf. (LATIN '02)*, 2002.
11. H. Straubing and D. Thérien. Weakly iterated block products of finite monoids. In *Proc. of the 5th Latin American Theoretical Informatics Conf. (LATIN '02)*, 2002.
12. H. Straubing and D. Thérien. Regular languages defined by generalized first-order formulas with a bounded number of bound variables. *Theory of Computing Systems*, 36(1):29–69, 2003.
13. H. Straubing, D. Thérien, and W. Thomas. Regular languages defined by generalized quantifiers. *Information and Computation*, 118:289–301, 1995.

14. P. Tesson. *Computational Complexity Questions Related to Finite Monoids and Semigroups*. PhD thesis, McGill University, 2003.
15. P. Tesson and D. Thérien. Complete classifications for the communication complexity of regular languages. *Theory of Comput. Syst.*, 2004. To appear.
16. D. Thérien and T. Wilke. Over words, two variables are as powerful as one quantifier alternation. In *Proc. 30th ACM Symposium on the Theory of Computing*, pages 256–263, 1998.
17. D. Thérien and T. Wilke. Nesting until and since in linear temporal logic. *Theory Comput. Syst.*, 37(1):111–131, 2004.

Suitable Curves for Genus-4 HCC over Prime Fields: Point Counting Formulae for Hyperelliptic Curves of Type $y^2 = x^{2k+1} + ax$

Mitsuhiro Haneda[1], Mitsuru Kawazoe[2], and Tetsuya Takahashi[2]

[1] Sharp Corporation
[2] Faculty of Liberal Arts and Sciences,
Osaka Prefecture University,
1-1 Gakuen-cho Sakai Osaka 599-8531 Japan
{kawazoe, takahasi}@las.osakafu-u.ac.jp

Abstract. Computing the order of the Jacobian group of a hyperelliptic curve over a finite field is very important to construct a hyperelliptic curve cryptosystem (HCC), because to construct secure HCC, we need Jacobian groups of order in the form $l \cdot c$ where l is a prime greater than about 2^{160} and c is a very small integer. But even in the case of genus two, known algorithms to compute the order of a Jacobian group for a general curve need a very long running time over a large prime field. In this article, we give explicit formulae of the order of Jacobian groups for hyperelliptic curves over a finite prime field of type $y^2 = x^{2k+1} + ax$, which allows us to search suitable curves for HCC. By using these formulae, we can find many suitable curves for genus-4 HCC and show some examples.

1 Introduction

Let C be a hyperelliptic curve of genus g over $\mathbb{F}_q$, J_C the Jacobian variety of C and $J_C(\mathbb{F}_q)$ the Jacobian group of C which is the set of $\mathbb{F}_q$-rational points of J_C. Then $J_C(\mathbb{F}_q)$ is a finite abelian group and we can construct a public-key-cryptosystem by using DLP on it. This cryptosystem is called "hyperelliptic curve cryptosystem (HCC)". In particular, HCC obtained by using a hyperelliptic curve of genus g is called "genus-g HCC". To construct secure HCC, the order of $J_C(\mathbb{F}_q)$ is required to be in the form $l \cdot c$ where l is a prime greater than about 2^{160} and c is a very small integer. We call a hyperelliptic curve "suitable for HCC" if its Jacobian group has such a suitable order. The advantage of HCC to an elliptic curve cryptosystem (ECC) is that we can construct a cryptosystem at the same security level as an elliptic one by using a smaller defining field. More precisely, we need a 160-bit field to construct a secure ECC, but for a genus-g HCC with $g \geq 2$, we only need about $(160/g)$-bit field. This comes from the fact that the order of the Jacobian group of a hyperelliptic curve defined over an N-bit field is about (Ng)-bit. The merit of using higher genus HCC is its short operand size; less than 64-bit for genus ≥ 3; its implementation on 64-bit CPUs does not need multi-precision arithmetic and it can encrypt very fast (cf. [12]).

Due to Gaudry [8], there is a variant of the index calculus attack which can solve the discrete logarithm problem in the Jacobian of hyperelliptic curves. As a result of it, it is recommended that the genus should be taken less than five to construct a secure HCC. In AsiaCrypt2003, Thériault [23] improved Gaudry's algorithm and recently Gaudry-Thomé [11] and Nagao [21] independently improved Thériault's result. The running time of their algorithms is asymptotically better than the original index calculus and the Rho method for curves of genus ≥ 3. In particular Gaudry-Thomé showed by computer experiments on the genus 3 case that even for rather small sizes, their method is better than Pollard Rho algorithm and a genus-3 HCC should have a key size about 12 % larger than an elliptic curve cryptosystem for an equivalent security. This implies that a genus-3 HCC on a field of less than 64-bit may be insecure in comparison with 160-bit ECC in the near future. For genus-4 HCC, we also need a larger key size and hence a larger field, but it still has its advantage that it can be implemented on 64bit CPU because it has enough margins to enlarge a field size under 64bit. So the search for suitable curves for genus-4 HCC is still important.

As in the case of ECC, to get a fast algorithm for adding points on the Jacobian group and to give an efficient way to produce a suitable curve for HCC are very important to construct HCC. For the first problem, we have already got many good results. See [10][16][17] for genus two, [15] for genus three and [22] for genus four. For the second problem, we need to calculate the order of the Jacobian group. However, for HCC over prime fields, there are few results in this direction even for the genus two case (cf. [9]) and there is no efficient algorithm for the case of genus greater than two.

There is a known algorithm to construct a curve with complex multiplication (CM) whose Jacobian group has a 160-bit prime factor. But this algorithm is efficient only for genus two at this moment. For genus three, only a few examples of suitable curves are constructed by this method [24]. For a special curve of type $y^2 + y = x^n$ over a prime field $\mathbb{F}_p$, Buhler-Koblitz [2] obtained a point counting algorithm.

If the Jacobian of a curve is obtained as a reduction mod p of an abelian variety of CM type over $\mathbb{Q}$, the calculation of the order of the Jacobian group is much easier. For such curves, the problem is reduced to finding an element of norm p in a CM field. Note that such an element is defined up to a root of unity.

In this article, we study characteristic polynomials of curves defined by $y^2 = x^{2k+1} + ax$, $a \in \mathbb{F}_p$ and give explicit formulae giving the order of Jacobian groups of those curves. The Jacobian of such curve is obtained as a reduction mod p of an abelian variety of CM type over $\mathbb{Q}$. Our tactics for calculating the order of the Jacobian group of the curve $C : y^2 = x^{2k+1} + ax$ is to calculate $|C(\mathbb{F}_{p^r})|$ for $r = 1, 2, \ldots, g$ directly. Since the curve C is obtained as a quotient of a Fermat curve, $|C(\mathbb{F}_{p^r})|$ is expressed by Jacobi sums over $\mathbb{F}_{p^r}$. But the explicit calculation of Jacobi sums over $\mathbb{F}_{p^r}$ for $r \geq 2$ and large p is not easy. In this article, we prove Theorem 1; it reduces the calculation of $|C(\mathbb{F}_{p^r})|$ to Jacobi sums over $\mathbb{F}_p$. And as a result of it, we can determine the characteristic polynomial of the p-th power Frobenius endomorphism without ambiguity over unit elements. Moreover, we can give a necessary condition for such curves to be suitable for HCC. By using our formulae, we show that the case $k = 4$ produces suitable curves for genus-4 HCC when $p \equiv 1 \pmod{16}$ and give some examples of such curves. We can obtain explicit formulae for the case $k = 2$ and 3. But for the case $k = 3$, the genus three case, it is shown that the Jacobian splits over the base field and hence cannot produce suitable curves. So we omit the formulae for that case. For the case $k = 2$, we should note that it was already appeared in [7] and explicit formulae are given but under some condition on a. Combining our formula and the result in [7], we have a complete result for $k = 2$.

Our result gives a family of hyperelliptic curves with closed formula for the order of the Jacobian and a necessary condition on the parameter to be suitable for genus 4-HCC.

2   The Characteristic Polynomial and the Order of the Jacobian Group

Let p be an odd prime, $\mathbb{F}_q$ a finite field of order $q = p^r$ and C a hyperelliptic curve of genus g defined over $\mathbb{F}_q$. Then the defining equation of C is given as $y^2 = f(x)$ where $f(x)$ is a polynomial in $\mathbb{F}_q[x]$ of degree $2g + 1$.

Let J_C be the Jacobian variety of a hyperelliptic curve C. We denote the group of $\mathbb{F}_q$-rational points on J_C by $J_C(\mathbb{F}_q)$ and call it the Jacobian group of C. Let $\chi_q(t)$ be the characteristic polynomial of the q-th power Frobenius endomorphism of C. We call $\chi_q(t)$ for C the characteristic polynomial of C and denote it by $\chi(t)$ for the convenience. Then, it is well-known that the order $|J_C(\mathbb{F}_q)|$ is given by

$$|J_C(\mathbb{F}_q)| = \chi(1). \tag{1}$$

Due to Mumford [19], every point on $J_C(\mathbb{F}_q)$ can be represented by a pair $\langle u(x), v(x) \rangle$ where $u(x)$ and $v(x)$ are polynomials in $\mathbb{F}_q[x]$ with $\deg v(x) < \deg u(x) \leq g$ such that $u(x)$ divides $f(x) - v(x)^2$. The identity element of the addition law is represented by $\langle 1, 0 \rangle$. By using this representation of points on $J_C(\mathbb{F}_q)$, we obtain an algorithm for adding two points on $J_C(\mathbb{F}_q)$. This algorithm was firstly given by Cantor [3] in general and has been improved for genus 2, 3 and 4 by many people [10][16][17][22].

In the following, for a generator g of $\mathbb{F}_p^\times$, we denote $\text{Ind}_g\, a = k$ when $a = g^k$, $k = 0, 1, \ldots, p - 1$.

3   Jacobstahl Sum and the Key Theorem

For two characters ψ, ψ' of $\mathbb{F}_{p^r}^\times$, the Jacobi sum $J_r(\psi, \psi')$ is defined by

$$J_r(\psi, \psi') = \sum_{t \in \mathbb{F}_{p^r}} \psi(t)\psi'(1 - t).$$

For the convenience we use the notation $K_r(\psi) = \psi(4)J_r(\psi,\psi)$. When $r = 1$, we drop the subscript J_r and K_r. For properties of Jacobi sums, see [1].

Let ℓ be a positive integer and p a prime such that $p \equiv 1 \pmod{2\ell}$. Let ψ_2 be a character of order 2 on a finite field $\mathbb{F}_{p^r}$. For an element a in $\mathbb{F}_p$,

$$\phi_{\ell,r}(a) := \sum_{x \in \mathbb{F}_{p^r}} \psi_2(x^{\ell+1} + ax)$$

is called a "Jacobstahl sum". It is easy to see that for a hyperelliptic curve defined by an equation $y^2 = x^{\ell+1} + ax$ over $\mathbb{F}_p$,

$$|C(\mathbb{F}_{p^r})| = p^r + 1 + \phi_{\ell,r}(a) \tag{2}$$

where $|C(\mathbb{F}_{p^r})|$ denotes the number of rational points of C over $\mathbb{F}_{p^r}$.

The following properties of $\phi_{\ell,r}$ enables us to reduce the calculation of $\phi_{\ell,r}(a)$ to the case $p^r \equiv 1 \pmod{2\ell}$.

Lemma 1. *(1) For $d = (\ell, p^r - 1)$, $\phi_{\ell,r}(a) = \phi_{d,r}(a)$.*
(2) If $p^r - 1 \equiv \ell \pmod{2\ell}$, then $\phi_{\ell,r}(a) = 0$.

Then we have the following theorem. This is the key theorem in our results.

Theorem 1. *Let p be a prime such that $p \equiv 1 \pmod{2\ell}$ for some positive integer ℓ. For $a \in \mathbb{F}_p$,*

$$\phi_{\ell,r}(a) = (-1)^{r-1}\hat{\psi}(-1)\hat{\psi}^{\ell+1}(a)\sum_{j=0}^{\ell-1}\hat{\psi}^{2j}(a)K(\psi^{2j+1})^r$$

where $\hat{\psi}$ is a character of $\mathbb{F}_{p^r}^{\times}$ of order 2ℓ and $\psi = \hat{\psi} \circ N_{\mathbb{F}_{p^r}/\mathbb{F}_p}$.

Proof. Since $\hat{\psi}^{\ell} = \psi_2$,

$$\phi_{\ell,r}(a) = \sum_{x \in \mathbb{F}_{p^r}} \hat{\psi}^{\ell}(x)\hat{\psi}^{\ell}(x^{\ell} + a) = \sum_{x \in \mathbb{F}_{p^r}} \hat{\psi}(x^{\ell})\hat{\psi}^{\ell}(x^{\ell} + a).$$

By the equality

$$\sum_{j=0}^{\ell-1}\hat{\psi}^{2j}(x) = \begin{cases} 0 & \hat{\psi}^2(x) \neq 1 \\ \ell & \hat{\psi}^2(x) = 1 \end{cases}$$

and the fact each fiber of the map $x \mapsto x^{\ell}$ has ℓ elements, we have

$$\phi_{\ell,r}(a) = \sum_{x \in \mathbb{F}_{p^r}} \hat{\psi}(x)\hat{\psi}^{\ell}(x + a)\sum_{j=0}^{\ell-1}\hat{\psi}^{2j}(x).$$

By the change of variable $x \to -x$ and $x \to -ax$,

$$\phi_{\ell,r}(a) = \hat{\psi}(-1)\hat{\psi}^{1+\ell}(a) \sum_{x \in \mathbb{F}_{p^r}} \hat{\psi}(x)\hat{\psi}^{\ell}(1-x) \sum_{j=0}^{\ell-1} \hat{\psi}^{2j}(ax)$$

$$= \hat{\psi}(-1)\hat{\psi}^{1+\ell}(a) \sum_{j=0}^{\ell-1} \hat{\psi}^{2j}(a) \sum_{x \in \mathbb{F}_{p^r}} \hat{\psi}^{2j+1}(x)\hat{\psi}^{\ell}(1-x)$$

$$= \hat{\psi}(-1)\hat{\psi}^{1+\ell}(a) \sum_{j=0}^{\ell-1} \hat{\psi}^{2j}(a) J_r(\hat{\psi}^{2j+1}, \hat{\psi}^{\ell}).$$

Since $J_r(\hat{\psi}^{2j+1}, \hat{\psi}^{\ell}) = \hat{\psi}^{2j+1}(4) J_r(\hat{\psi}^{2j+1}, \hat{\psi}^{2j+1}) = K_r(\hat{\psi}^{2j+1})$, we get the formula

$$\phi_{\ell,r}(a) = \hat{\psi}(-1)\hat{\psi}^{1+\ell}(a) \sum_{j=0}^{\ell-1} \hat{\psi}^{2j}(a) K_r(\hat{\psi}^{2j+1}).$$

It follows from the Hasse-Davenport relation (cf. [1–Theorem 11.5.2]) that for any integer m, $K_r(\hat{\psi}^m) = (-1)^{r-1} K_1(\psi^m)^r$. Hence our Theorem. □

Remark 1. Theorem 1 is a new result. Combining this theorem with the following well-known fact, we get the formula of $\chi(t)$ for the curve $C : y^2 = x^{2k+1} + ax$.

Theorem 2. *Let C be a hyperelliptic curve of genus g over $\mathbb{F}_p$. Assume $\chi(t)$ for C is decomposed as $\chi(t) = \prod_{i=1}^{2g}(t - \alpha_i)$. Then $|C(\mathbb{F}_{p^r})| = p^r + 1 - \sum_{i=1}^{2g} \alpha_i^r$.*

4 Explicit Formula for $y^2 = x^5 + ax$

Let p be an odd prime and C a hyperelliptic curve defined by an equation $y^2 = x^5 + ax$ over $\mathbb{F}_p$. In [7], the explicit formulae of the order of $J_C(\mathbb{F}_p)$ are given for all cases except for the only one case $p \equiv 1 \pmod 8$ with $\left(\frac{a}{p}\right) = -1$. Here we show the explicit formula for the remaining case.

Theorem 3. *Let p be a prime such that $p \equiv 1 \pmod 8$ and C a hyperelliptic curve defined by an equation $y^2 = x^5 + ax$ over $\mathbb{F}_p$. Put $f = (p-1)/8$. Write p as $p = c^2 + 2d^2$ where $c \equiv 1 \pmod 4$ and $2d \equiv -(a^f + a^{3f})c \pmod p$. Then the characteristic polynomial of p-th power Frobenius map for C is given by the following formula: $\chi(t) = t^4 + (-1)^f 4dt^3 + 8d^2t^2 + (-1)^f 4dpt + p^2$. In particular, $|J_C(\mathbb{F}_p)| = 1 + (-1)^f 4d + 8d^2 + (-1)^f 4dp + p^2$.*

Proof. This follows from Theorem 1, Theorem 2 and the formula for $K(\psi)$. (See [1]). □

This formula provides us a faster algorithm to compute the order of the Jacobian than in [7].

Remark 2. All formulae for $\chi(t)$ in this paper are obtained in the same way. Since we have not enough space, we describe the proof only for Theorem 4 which is the most important result of this article and omit the proofs for other formulae.

5 Remark on $y^2 = x^7 + ax$

Unfortunately the Jacobian of a hyperelliptic curve of type $y^2 = x^7 + ax$ splits over $\mathbb{F}_p$ splits over $\mathbb{F}_p$, because for $k \equiv 0 \pmod{3}$, one has a degree 3 covering from $y^2 = x^{2k+1} + ax$ to $Y^2 = X^{2k/3+1} + aX$, given by $(x, y) \mapsto (x^3, xy)$. Therefore the characteristic polynomial is reducible over $\mathbb{Z}$ and this curve is not suitable for HCC.

6 Explicit Formula for $y^2 = x^9 + ax$

Let p be an odd prime and C a hyperelliptic curve defined by an equation $y^2 = x^9 + ax$ over $\mathbb{F}_p$.

6.1 The Case of $p \equiv 1 \pmod{16}$

Let p be a prime such that $p \equiv 1 \pmod{16}$. We fix a generator g of $\mathbb{F}_p^\times$. Put $f = (p-1)/16$ and $\alpha = g^{(p-1)/16}$. Then there exist integers x, u, v, w such that

$$p = x^2 + 2(u^2 + v^2 + w^2),\ 2xv = u^2 - 2uw - w^2,\ x \equiv 1 \pmod{8},$$
$$x + u(\alpha + \alpha^7) + v(\alpha^2 - \alpha^6) + w(\alpha^3 + \alpha^5) \equiv 0 \pmod{p}, \quad (3)$$
$$2v^2 - x^2 \equiv -(u^2 + 2uw - w^2)(\alpha^2 - \alpha^6) \pmod{p}.$$

It is known that the above x, u, v, w are uniquely determined.

Let $\chi(t) = t^8 - s_1 t^7 + s_2 t^6 - s_3 t^5 + s_4 t^4 - s_3 p t^3 + s_2 p^2 t^2 - s_1 p^3 t + p^4$ be the characteristic polynomial of C. Then by using the above notation, we have the following theorems.

Theorem 4. s_1, s_2, s_3 and s_4 are given by the following tables.

$\mathrm{Ind}_g a \pmod{16}$	s_1
1, 7	$(-1)^f 8w$
9, 15	$(-1)^{f+1} 8w$
3, 5	$(-1)^{f+1} 8u$
11, 13	$(-1)^f 8u$
2, 14	$(-1)^{f+1} 8v$
6, 10	$(-1)^f 8v$
8	$(-1)^{f+1} 8x$
0	$(-1)^f 8x$
4, 12	0

$\mathrm{Ind}_g a \pmod{16}$	s_2
1, 7, 9, 15	$32w^2 + 16xv$
3, 5, 11, 13	$32u^2 - 16xv$
2, 6, 10, 14	$32v^2$
0, 8	$4p + 24x^2 - 16v^2$
4, 12	$-4p + 8x^2 + 16v^2$

$\mathrm{Ind}_g\, a \pmod{16}$	s_3
1, 7	$(-1)^{J+1}8(pu - 4(u^3 + w^3 + u^2w - 3uw^2))$
9, 15	$(-1)^{J}8(pu - 4(u^3 + w^3 + u^2w - 3uw^2))$
3, 5	$(-1)^{J+1}8(pw + 4(u^3 - w^3 + 3u^2w + uw^2))$
11, 13	$(-1)^{J}8(pw + 4(u^3 - w^3 + 3u^2w + uw^2))$
2, 14	$(-1)^{J+1}(8pv + 64v^3 - 32x^2v)$
6, 10	$(-1)^{J}(8pv + 64v^3 - 32x^2v)$
8	$(-1)^{J+1}(24px + 32x^3 - 64xv^2)$
0	$(-1)^{J}(24px + 32x^3 - 64xv^2)$
4, 12	0

$\mathrm{Ind}_g\, a \pmod{16}$	s_4
1, 7, 9, 15	$32u^4 + 32w^4 + 64u^2w^2 - 64puw + 128u^3w - 128uw^3$
3, 5, 11, 13	$32u^4 + 32w^4 + 64u^2w^2 + 64puw + 128u^3w - 128uw^3$
2, 6, 10, 14	$2p^2 + 16x^4 + 64v^4 - 16px^2 - 64x^2v^2 + 32pv^2$
0, 8	$6p^2 + 16x^4 + 64v^4 + 48px^2 - 64x^2v^2 - 32pv^2$
4, 12	$6p^2 + 16x^4 + 64v^4 - 16px^2 - 64x^2v^2 - 32pv^2$

Proof. Here we show the outline of the proof. Let x, u, v, w be integers which satisfy the condition (3). Then for the character ψ which maps α to $e^{2\pi i/16}$,

$$K(\psi) = -x + v\sqrt{2} + iu\sqrt{2 - \sqrt{2}} - iw\sqrt{2 + \sqrt{2}}.$$

From this result, similar descriptions for $K(\psi^{2j+1})$ can be easily obtained. Hence by using equations (1), (2) and Theorem 2, we have the theorem. □

Corollary 1. *If $a^{1/8} \in \mathbb{F}_p$, the characteristic polynomial of C is given by*

$$\chi(t) = \left(t^4 - s_1 t^3/2 + (s_2/2 - s_1^2/8)t^2 - s_1 pt/2 + p^2\right)^2.$$

In particular, if $a^{1/8} \in \mathbb{F}_p$, it is not suitable for HCC.

We look at the case when $a^{1/8} \notin \mathbb{F}_p$. Since $\left(\frac{-1}{p}\right) = 1$, if a is square, then there is an element $b \in \mathbb{F}_p$ such that $b^2 = -a$. Then $x^9 + ax$ factors into $x(x^4 + b)(x^4 - b)$ and we have that $|J_C(\mathbb{F}_p)|$ is divided by at least 4. Moreover if $a^{1/4} \in \mathbb{F}_p$, $|J_C(\mathbb{F}_p)|$ is divided by at least 16.

If a is not a square, it is possible to obtain a Jacobian group whose order is in the form $2l$ where l is prime.

6.2 The Case of $p \equiv 7 \pmod{16}$

Let p be a prime such that $p \equiv 7 \pmod{16}$. Then there exist integers x, u, v, w such that

$$p = x^2 + 2(u^2 + v^2 + w^2), \quad 2xv = u^2 - 2uw - w^2,$$
$$x \equiv 1 \pmod{8}, \quad u \equiv v \equiv w \equiv 1 \pmod{2}. \tag{4}$$

Let $\chi(t) = t^8 - s_1 t^7 + s_2 t^6 - s_3 t^5 + s_4 t^4 - s_3 p t^3 + s_2 p^2 t^2 - s_1 p^3 t + p^4$ be the characteristic polynomial of C. Then, for a fixed generator g of $\mathbb{F}_p^\times$, we have the following theorem.

Theorem 5. *The characteristic polynomial of C is determined by the following formula. (1) $s_1 = s_3 = 0$, (2) $s_2 = (-1)^{\mathrm{Ind}_g a}(4p - 8x^2 - 16v^2)$, (3) $s_4 = 6p^2 + 16x^4 + 64v^4 - 16px^2 - 64x^2v^2 - 32pv^2$.*

Remark 3. There is some ambiguity with respect to u, w and the sign of v. But it does not affect to determine the characteristic polynomial of C.

Corollary 2. *If a is square, the characteristic polynomial of C is given by*

$$\chi(t) = 5(t^4 + 4xt^3 + \delta t^2 + 4xpt + p^2)(t^4 - 4xt^3 + \delta t^2 - 4xpt + p^2), \quad \delta = 2p + 4x^2 - 8v^2.$$

In particular, if a is square, it is not suitable for HCC.

We look at the case when a is not a square. From Theorem 5, we have that $|J_C(\mathbb{F}_p)|$ is divided by at least 2^7.

6.3 The Case of $p \not\equiv 1, 7 \pmod{16}$

Theorem 6. *If $p \equiv 3, 11 \pmod{16}$, then the characteristic polynomial of C is given by $\chi(t) = (t^4 + (-1)^{\mathrm{Ind}_g a} p^2)^2$.*

Theorem 7. *Assume that $p \equiv 5, 13 \pmod{16}$. Then the characteristic polynomial of C is given by the following formula. (1) If $\mathrm{Ind}_g a \not\equiv 0 \pmod 2$, then $\chi(t) = t^8 + p^4$, (2) if $\mathrm{Ind}_g a \equiv 0 \pmod 4$, then $\chi(t) = (t^4 + p^2)^2$, (3) if $\mathrm{Ind}_g a \equiv 2 \pmod 4$, then $\chi(t) = (t^2 - p)^2(t^2 + p)^2$.*

Theorem 8. *Assume that $p \equiv 9 \pmod{16}$. Then the characteristic polynomial of C is given by the following formula. (1) If $\mathrm{Ind}_g a \not\equiv 0 \pmod 2$, then $\chi(t) = t^8 + p^4$, (2) if $\mathrm{Ind}_g a \equiv 2 \pmod 4$, then $\chi(t) = (t^4 + p^2)^2$, (3) if $\mathrm{Ind}_g a \equiv 4 \pmod 8$, then $\chi(t) = (t^2 - p)^4$, (4) if $\mathrm{Ind}_g a \equiv 0 \pmod 8$, then $\chi(t) = (t^2 + p)^4$.*

Theorem 9. *If $p \equiv 15 \pmod{16}$, then the characteristic polynomial of C is given by $\chi(t) = (t^2 + p)^4$.*

In particular, for $p \not\equiv 1, 7 \pmod{16}$, C is a supersingular curve which is not recommended to use for HCC.

6.4 Which Parameter Is Suitable for HCC?

From the above results, all the cases which can produce suitable curves for HCC are the followings: (1) $p \equiv 1 \pmod{16}$ with a not a square, (2) $p \equiv 1 \pmod{16}$ with a square but $a^{1/4} \notin \mathbb{F}_p$, (3) $p \equiv 1 \pmod{16}$ with $a^{1/4} \in \mathbb{F}_p$ but $a^{1/8} \notin \mathbb{F}_p$, (4) $p \equiv 7 \pmod{16}$ with a not a square. In each case, the best possible order is in the form (1) $2l$, (2) $4l$, (3) $2^4 l$ and (4) $2^7 l$ where l is a prime.

Now we consider splitting of the Jacobian over extension fields. Let $\chi(t)$ be the characteristic polynomial of a hyperelliptic curve C over $\mathbb{F}_p$. In the following, we denote by $\chi_{p^r}(t)$ the characteristic polynomial of the p^r-th power Frobenius endomorphism of C as a curve over $\mathbb{F}_{p^r}$.

Proposition 1. *Let $\chi(t) = t^8 - s_1 t^7 + s_2 t^6 - s_3 t^5 + s_4 t^4 - s_3 p t^3 + s_2 p^2 t^2 - s_1 p^3 t + p^4$ be the characteristic polynomial of a hyperelliptic curve C of genus four over $\mathbb{F}_p$. If $\chi(t)$ is irreducible over $\mathbb{Q}$, then the followings hold: (1) $\chi_{p^2}(t)$ is a product of two polynomials of degree four if and only if $s_1 = s_3 = 0$, (2) $\chi_{p^4}(t)$ is a product of four polynomials of degree two if and only if $\chi_{p^2}(t)$ is a product of two polynomials of degree four.*

Proof. Let γ be a root of $\chi(t)$. We have only to show that if $[\mathbb{Q}(\gamma^2) : \mathbb{Q}] = 4$, $s_1 = s_3 = 0$. Since $\gamma \bar\gamma = p$, we can show $(\gamma^2 + \bar\gamma^2)^2 \in \mathbb{Q}$ under the assumption $[\mathbb{Q}(\gamma^2) : \mathbb{Q}] = 4$. This implies $s_1 = s_3 = 0$. □

By Proposition 1 and Theorem 4, we have that in the case $p \equiv 1 \pmod{16}$ with a not a square, $\chi_{p^2}(t)$ is irreducible and therefore its DLP cannot be reduced to DLP of HCC of genus two over $\mathbb{F}_{p^2}$ nor of ECC over $\mathbb{F}_{p^4}$.

7 Examples of Suitable Curves for Genus-4 HCC

In this section, we describe how to search suitable curves for genus-4 HCC of type $y^2 = x^9 + ax$ and show the result of search. Based on the argument in 6.4, we only treat the case of $p \equiv 1 \pmod{16}$.

7.1 LLL Algorithm

Let p be a prime such that $p \equiv 1 \pmod{16}$. We describe the algorithm to determine $|J_C(\mathbb{F}_p)|$. For a given p, if we obtain x, u, v and w in (3), we can determine the order of $J_C(\mathbb{F}_p)$ and check its suitability. So the main part of the algorithm is determining x, u, v, w in (3). To determine x, u, v, w, we use the LLL algorithm.

Let α_i, $i = 1, 2, \ldots, 7$ be positive integers such that $0 \le \alpha_i < p$ and $\alpha_i \equiv g^{(p-1)i/16} \pmod{p}$. Let $\zeta \in \mathbb{C}$ be a primitive 16th root of unity and P a prime ideal over (p) in the integer ring $\mathcal{O}_K$ of $K = \mathbb{Q}(\zeta + \zeta^7)$. A $\mathbb{Z}$-basis $\{b_0, b_1, b_2, b_3\}$ of P is given by

$$b_0 = p, \quad b_1 = \zeta + \zeta^7 - \alpha_1 - \alpha_7,$$
$$b_2 = \zeta^2 - \zeta^6 - \alpha_2 + \alpha_6, \quad b_3 = \zeta^3 + \zeta^5 - \alpha_3 - \alpha_5. \tag{5}$$

For this basis, any entry of the Gram matrix with respect to an inner product $\langle u, v \rangle = \operatorname{Tr}_{K/\mathbb{Q}}(u\bar v)$ is an integer. Put $c_1 = -\alpha_1 - \alpha_7$, $c_2 = -\alpha_2 + \alpha_6$, $c_3 = -\alpha_3 - \alpha_5$. Then each entry of the Gram matrix is given as follows: $\langle b_0, b_0 \rangle = 4p^2$, $\langle b_0, b_i \rangle = 4pc_i$ $(1 \le i \le 3)$, $\langle b_i, b_j \rangle = 4c_i c_j$ $(1 \le i \ne j \le 3)$, $\langle b_i, b_i \rangle = 8 + 4c_i^2$ $(1 \le i \le 3)$. Then the LLL algorithm for the Gram matrix works and we can determine x, u, v and w in (3) by using the following algorithm. (For the details on the LLL algorithm, see [4] for example.)

Algorithm[1]

Input	p: a prime ($p \equiv 1 \pmod{16}$)
Output	x, u, v, w satisfying (3)

(Step 1-5: Finding $\beta \in \mathcal{O}_K$, $N_{K/\mathbb{Q}}(\beta) = p$.)

Step 1 $g \leftarrow$ a generator of $\mathbb{F}_p^\times$.
Step 2 $\mathbf{b} = (b_0, b_1, b_2, b_3) \leftarrow$ a $\mathbb{Z}$-basis (5) of $\mathcal{O}_K$.
Step 3 $G \leftarrow$ the Gram matrix for $\mathbf{b}$.
Step 4 $H = (h_{ij}) \leftarrow$ a transformation matrix obtained by the LLL algorithm for G.
Step 5 $\beta \leftarrow \sum_{i=0}^{3} b_i h_{0i}$
Step 6 Determine x, u, v, w by $\beta\tau(\beta) = x + u(\zeta + \zeta^7) + v(\zeta^2 - \zeta^6) + w(\zeta^3 + \zeta^5)$ and (3) where τ is an automorphism of $\mathbb{Q}(\zeta + \zeta^7)$ given by $\zeta \mapsto \zeta^3$.
Step 7 Return x, u, v, w.

This algorithm can be easily implemented and we can compute $|J_C(\mathbb{F}_p)|$ of C defined by $y^2 = x^9 + ax$ in a very short time, e.g. only 0.01 seconds over 41bit fields.

7.2 Examples of Suitable Curves

We can obtain many suitable curves for HCC by varying p and a.

Table 1. Search results

search range (r, s) for $r < p < s$	the number of primes[2] s.t. $\|J_C(\mathbb{F}_p)\| = 2 \cdot (\text{prime})$	time[sec]
$(2^{41}, 2^{41} + 10^6)$	714	267.054
$(2^{60}, 2^{60} + 10^6)$	349	293.421

Here we show some examples of suitable curves for HCC obtained by our algorithm.

Table 2. An Example of a suitable curve for genus-4 HCC over 41-bit fields

$y^2 = x^9 + 29x$, $p = 1759218504481$ (41-bit)	
s_1	4722688
s_2	14617568463136
s_3	29894897984637227312
s_4	46358542553945186095112704
$\|J_C(\mathbb{F}_p)\|$	$2 \cdot 4789034620376653463540859489797855263219497047089$ (162-bit)

---

[1] This algorithm does not always give β with $N_{K/\mathbb{Q}}(\beta) = p$ theoretically. But the vector produced by this algorithm had norm p in all experiments we tried.
[2] Here we count the number of primes p such that $|J_C(\mathbb{F}_p)| = 2 \cdot (\text{prime})$ for at least one a.

Table 3. An Example of a suitable curve for genus-4 HCC over 60-bit fields

$y^2 = x^9 + 1953125x$, $p = 1152921504606851137$(61-bit)	
s_1	564124384
s_2	13459706070512888832
s_3	$-74918598083954270881163363$2
s_4	80166212866222609539893111807958016
$\|J_C(\mathbb{F}_p)\|$	2·88342353195694586762722787339232382489941057864943049973024832482813964$9$(239-bit)

All computation were done on a system with Pentium 4 1.6GHz.

7.3 Notes on Security

All examples in Table 2 are not weak against Frey-Rück attack[6]. To see this, one can easily check that a large prime factor of $|J_C(\mathbb{F}_p)|$ does not divide $p^r - 1$, $r = 1, 2, \ldots, 4^3 \lfloor \log^2 p \rfloor$.

From the result of Duursma, Gaudry and Morain [5], an automorphism of large order can be exploited to accelerate the Pollard's rho algorithm. If there is an automorphism of order m, we can get a speed up of $\sqrt{m}$. The order of any automorphism of $y^2 = x^9 + ax$ is at most 16. So the Pollard's rho algorithm for these curves can be improved only by a factor 4.

As we saw in 6.4, J_C in Table 2 and 3 cannot split over $\mathbb{F}_{p^2}$ and $\mathbb{F}_{p^4}$. So examples in Table 2 cannot be reduced to genus-2 HCC of about 80-bit and ECC of about 160-bit. And an example in Table 3 cannot be reduced to genus-2 HCC of about 120-bit and ECC of about 240-bit.

8 Conclusions

We gave explicit formulae of the order of Jacobian groups for hyperelliptic curves of type $y^2 = x^{2k+1} + ax$ over a finite prime field and gave a necessary condition that the curves are suitable for genus-4 HCC. By using these formulae, we searched suitable curves for HCC among the above curves and found many suitable curves for genus-4 HCC.

References

1. B. C. Berndt, R. J. Evans and K. S. Williams, Gauss and Jacobi Sums, Canadian Mathematical Society Series of Monographs and Advanced Texts **21**, A Wiley-Interscience Publication, 1998,
2. J. Buhler and N. Koblitz, *Lattice Basis Reduction, Jacobi Sums and Hyperelliptic Cryptosystems*, Bull. Austral. Math. Soc. **58** (1998), pp. 147–154,
3. D. G. Cantor, *Computing in the Jacobian of hyperelliptic curve*, Math. Comp. **48** (1987), pp. 95–101,
4. H. Cohen, A Course in Computational Algebraic Number Theory, Graduate Texts in Mathematics **138**, Springer, 1996,

5. I. Duursma, P. Gaudry and F. Morain, *Speeding up the Discrete Log Computation on Curves with Automorphisms*, Advances in Cryptology – ASIA CRYPT '99, Springer-Verlag LNCS 1716, 1999, pp. 103–121,
6. G. Frey and H.-G. Rück, *A Remark Concerning m-divisibility and the Discrete Logarithm in the Divisor Class Group of Curves*, Math. Comp. **62**, No.206 (1994) pp. 865–874,
7. E. Furukawa, M. Kawazoe and T. Takahashi, *Counting Points for Hyperelliptic Curves of type $y^2 = x^5 + ax$ over Finite Prime Fields*, Selected Areas in Cryptography(SAC2003), Springer LNCS 3006, pp. 26–41,
8. P. Gaudry, *An algorithm for solving the discrete logarithm problem on hyperelliptic curves*, EUROCRYPT 2000, Springer LNCS 1807, 2000, pp. 19–34,
9. P. Gaudry and E. Schost, *Construction of Secure Random Curves of Genus 2 over Prime Fields*, EUROCRYPT 2004, Springer LNCS 3027, 2004, pp. 239–256,
10. P. Gaudry and R. Harley, *Counting Points on Hyperelliptic Curves over Finite Fields*, ANTS-IV, Springer LNCS 1838, 2000, pp. 297–312.
11. P. Gaudry, N. Thériault and E. Thomé, *A Double Large Prime Variation for Small Genus Hyperelliptic Index Calculus*, Cryptology ePrint Archive, Report 2004/153, 2004, http://eprint.iacr.org/
12. M. Gonda,K. Matsuo, K. Aoki, J. Chao and S. Tsujii, *Improvements of addition algorithm on genus 3 hyperelliptic curves and their implementations*, Proc. of SCIS2004, IEICE Japan, January 2004, pp. 995-1000,
13. R. H. Hudson and K. S. Williams, *Binomial Coefficients and Jacobi Sums*, Trans. Amer. Math. Soc. **281** (1984), pp. 431–505,
14. N. Koblitz, Algebraic Aspects of Cryptography, Algorithms and Computation in Mathematics Vol. 3, Springer-Verlag, 1998,
15. J. Kuroki, M. Gonda, K. Matsuo, J. Chao and S. Tsujii, *Fast Genus Three Hyperelliptic Curve Cryptosystems*, Proc. of SCIS2002, IEICE Japan, January 2002, pp. 503–507,
16. T. Lange, *Efficient Arithmetic on Genus 2 Hyperelliptic Curves over Finite Fields via Explicit Formulae*, Cryptology ePrint Archive, Report 2002/121, 2002, http://eprint.iacr.org/,
17. K. Matsuo, J. Chao and S. Tsujii, *Fast Genus Two Hyperelliptic Curve Cryptosystem*, ISEC2001-31, IEICE, 2001,
18. K. Matsuo, J. Chao and S. Tsujii, *An improved baby step giant step algorithm for point counting of hyperelliptic curves over finite fields*, ANTS-V, Springer-Verlag LNCS 2369, 2002, pp. 461–474,
19. D. Mumford, Tata Lectures on Theta II, Progress in Mathematics **43**, Birkhäuser, 1984,
20. K. Nagao, *Improving Group Law Algorithms for Jacobians of Hyperelliptic Curves*, In W. Bosma ed., ANTS IV, Springer-Verlag LNCS 1838, pp. 439–448,
21. K. Nagao, *Improvement of Thériault Algorithm of Index Calculus for Jacobian of Hyperelliptic Curves of Small Genus*, Cryptology ePrint Archive, Report 2004/161, 2004, http://eprint.iacr.org/,
22. J. Pelzl, T. Wollinger and C. Paar, *Low Cost Security: Explicit Formulae for Genus 4 Hyperelliptic Curves*, Selected Areas in Cryptography (SAC2003), Springer-Verlag LNCS 3006, pp. 1–16,
23. N. Thériault, *Index Calculus Attack for Hyperelliptic Curves of Small Genus*, in Advances in Cryptology – AsiaCrypt 2003, Lecture Notes in Computer Science 2894 (2003), Springer, pp. 75–92,
24. A. Weng, *Hyperelliptic CM-curves of genus 3*, Journal of the Ramanujan Mathematical Society **16**, No. 4, 2001, pp.339-372.

Solvability of a System of Bivariate Polynomial Equations over a Finite Field

(Extended Abstract)

Neeraj Kayal

Indian Institute of Technology, Kanpur
and
National University of Singapore

Abstract. We investigate the complexity of the following polynomial solvability problem: Given a finite field $\mathbb{F}_q$ and a set of polynomials

$$f_1(x,y), f_2(x,y), \cdots, f_n(x,y), g(x,y) \in \mathbb{F}_q[x,y]$$

determine the $\mathbb{F}_q$-solvability of the system

$$f_1(x,y) = f_2(x,y) = \cdots = f_n(x,y) = 0 \text{ and } g(x,y) \neq 0$$

We give a *deterministic* polynomial-time algorithm for this problem.

1 Motivation

The need to find solutions of polynomial congruences and to count their number over various fields has appeared in many important works in the history of number theory, algebra and geometry. For example, Fermat's Last Theorem asserts that for any integer $n \geq 3$ the following bivariate system has no solution over the field $\mathbb{Q}$ of rational numbers.

$$x^n + y^n = 1 \quad \text{and} \quad xy \neq 0$$

The general algorithmic problem of the decidability of solvability of such systems over $\mathbb{Q}$ remains a major open problem in algorithmic number theory. Here we examine the computational complexity of determining the existence of a solution to a system of bivariate polynomial equations over a given finite field $\mathbb{F}_q$. For a general polynomial system in n variables, the problem is known to be efficiently solvable when the system is linear and is **NP**-complete otherwise, even when the field is as small as 2 and the degree of each polynomial is bounded by 2.

Our focus is on *deterministic* algorithms for the problem of solvability of a system of bivariate polynomials over a given finite field $\mathbb{F}_q$. Huang and Wong, [HW99] give a *randomized* polynomial time algorithm for this problem when the field is a prime field. We give a *deterministic* polynomial-time algorithm for this problem over *any* given finite field $\mathbb{F}_q$. The techniques used herein also give an efficient *deterministic* approximation algorithm for the closely related problem of counting the number of solutions to a given bivariate system.

Problem - Existence of solution to a general bivariate system
Input. A finite field $\mathbb{F}_q$, polynomials $f_1(x,y), f_2(x,y), \cdots, f_m(x,y) \in \mathbb{F}_q[x,y]$ and a polynomial $g(x,y) \in \mathbb{F}_q[x,y]$.
Question. Does there exist a point $(a,b) \in \mathbb{F}_q \times \mathbb{F}_q$ such that

$$f_i(a,b) = 0 \text{ for all } 1 \leq i \leq m$$

$$\text{and } g(a,b) \neq 0$$

We give an efficient deterministic algorithm for this problem. As an application of this result, we immediately get an efficient deterministic algorithm for recognizing permutation functions, wherein a rational function $f(x) = \frac{g(x)}{h(x)} \in \mathbb{F}_q(x)$ is said to be a permutation function if the corresponding map $f : \mathbb{F}_q \mapsto \mathbb{F}_q$ defined by $a \mapsto f(a)$ is total and bijective. Permutation functions have found numerous applications, including cryptography [LM83]. Indeed, the RSA cryptosystem [RSA78] is one such application. A survey of permutation functions and their applications can be found in the articles by Lidl and Mullen [LM88], [LM93]. Ma and Gathen showed that this problem is in **ZPP**, and had posed as an open problem the deterministic complexity of recognizing permutation functions :

> "Besides the problem *PRIMES* of recognizing prime numbers, it seems to be the only natural decision problem in **ZPP** unknown to be in **P**." - [MG94]

For polynomials, Lenstra [Len05] showed that this problem is in **P** by using a classification of permutation polynomials, this classification itself being based on the classification of finite simple groups. The problem of recognizing permutation functions can be easily seen to reduce to the bivariate system solvability problem above by noting that a rational function $f(x) = \frac{g(x)}{h(x)} \in \mathbb{F}_q(x)$ is a permutation function if and only if the system

$$h(x)(g(x)h(y) - h(x)g(y)) = 0, \ x - y \neq 0$$

has no solution over $\mathbb{F}_q$. Thus our deterministic algorithm for bivariate systems immediately gives a deterministic test for permutation functions. Moreover, the proof of correctness of our algorithm is completely elementary.

Determining the existence of solutions to polynomial equations and finding these solutions is closely related to the problem of polynomial factorization. In spite of a lot of effort by several researchers, an efficient deterministic algorithm for the general polynomial factorization problem remains elusive. In the first part of this paper we examine the problem of deterministically factoring bivariate polynomials over a given finite field $\mathbb{F}_q$. We devise an algorithm that partially factors a given bivariate polynomial. We will later on describe precisely what the output of this partial factorization algorithm looks like. This algorithm is itself of independent interest. We then apply this factoring algorithm to solve the problem of existence of solution to systems of equations as mentioned above.

2 Some Definitions and Notation

2.1 Resultant and Discriminant

For two bivariate polynomials $f(x,y)$ and $g(x,y)$ in $\mathbb{F}_q[x,y]$ we will denote by Resultant$_x(f(x,y), g(x,y)) \in \mathbb{F}_q[y]$ the resultant of the two polynomials with respect to the variable x. We will denote by $\Delta_x(f(x,y)) \in \mathbb{F}_q[y]$ the discriminant of the polynomial $f(x,y)$ with respect to the variable x.

2.2 Nice Bivariate Polynomials

Definition 1. *A bivariate polynomial $f(x,y) \in \mathbb{F}_q[x,y]$ of total degree n is nice if $f(x,0)$ is squarefree and of degree n.*

Note that the coefficient of x^i of a nice polynomial $f(x,y)$ as a polynomial in y has degree no more than $n-i$, in particular the leading coefficient of $f(x,y)$ with respect to x is in $\mathbb{F}_q$.

Also observe that a nice polynomial $f(x,y) \in \mathbb{F}_q[x,y]$ remains nice over any extension field $\mathbb{K}$ of $\mathbb{F}_q$ and that any factor of a nice polynomial is also a nice polynomial. We will see that the problem of general bivariate factoring can be reduced to factoring a nice bivariate polynomial.

2.3 Absolutely Irreducible Polynomial

Definition 2. *A bivariate polynomial $h(x,y) \in \mathbb{F}[x,y]$ is said to be absolutely irreducible if it is irreducible over $\mathbb{F}$ and remains irreducible over the algebraic closure $\bar{\mathbb{F}}$ of $\mathbb{F}$.*

For example $(y^2 - x^3) \in \mathbb{F}_7[x,y]$ is absolutely irreducible whereas $(y^2 + x^2) \in \mathbb{F}_7[x,y]$ is irreducible over $\mathbb{F}_7$ but factors into $(y + \sqrt{-1}x)(y - \sqrt{-1}x)$ over the extension $\mathbb{F}_{7^2} = \mathbb{F}_7(\sqrt{-1})$ and hence is not absolutely irreducible over $\mathbb{F}_7$. The importance of absolutely irreducible polynomials stems from a classic theorem of Weil which states that every non-singular absolutely irreducible curve of degree d over a finite field $\mathbb{F}_q$ has a lot of (at least $q - d^2\sqrt{q}$) $\mathbb{F}_q$-rationals points. This has been generalized by Bach [Bac93] to *all* absolutely irreducible curves.

By the degree of a bivariate polynomial $h(x,y) \in \mathbb{F}[x,y]$ we will mean the total degree of $h(x,y)$. Moreover $h(x,y)$ has a unique factorization over the algebraic closure $\bar{\mathbb{F}}$ of $\mathbb{F}$. Now collect all the elements of $\bar{\mathbb{F}}$ that occur as the coefficient of some term $x^i y^j$ in some irreducible factor of $h(x,y)$ over $\bar{\mathbb{F}}$. Since this is a finite set, all these coefficients lie in some finite extension $\mathbb{K}$ of $\mathbb{F}$. We will call the smallest such extension field $\mathbb{K}$ the splitting field of $h(x,y)$. We will denote by $dim_{\mathbb{F}}(h(x,y))$ the dimension of the splitting field of $h(x,y)$ over $\mathbb{F}$.

Definition 3. *We will call a polynomial $h(x,y) \in \mathbb{F}_q[x,y]$ a (n,d)-uniform polynomial over $\mathbb{F}_q$ if all its $\mathbb{F}_q$-irreducible factors have the same degree n and the same dimension d over $\mathbb{F}_q$.*

Our first goal is to extend the distinct degree factorization algorithm of Gao, Kaltofen and Lauder [GKL04] to split a given polynomial $h(x, y) \in \mathbb{F}_q[x, y]$ into a product of uniform polynomials.

Theorem 1. *[Uniform factoring] There exists a deterministic algorithm that on input a polynomial $h(x, y) \in \mathbb{F}_q[x, y]$ outputs*

$$< (h_1(x,y), n_1, d_1), (h_2(x,y), n_2, d_2), \cdots (h_k(x,y), n_k, d_k) >$$

such that
$$h(x,y) = h_1(x,y)h_2(x,y)\cdots h_k(x,y)$$
where each $h_i(x, y)$ is a (n_i, d_i)-uniform polynomial over $\mathbb{F}_q$
Moreover the algorithm runs in time $poly(deg(h)) \log q$.

Note that the output of the algorithm of Theorem 1 is a refinement of the distinct degree factorization of $h(x, y)$ over $\mathbb{F}_q$. It also tells us about the irreducibility and the absolute irreducibility of $h(x, y)$ over $\mathbb{F}_q$. In the next section we prove some preliminary lemmas needed for the algorithm in Theorem 1.

3 Splitting of Nice $\mathbb{F}_q$-Irreducible Polynomials in Extensions of $\mathbb{F}_q$

Let $\mathbb{K}$ be a field extension of the finite field $\mathbb{F}_q$. Let $\phi \in Gal_{\mathbb{K}/\mathbb{F}_q}$ be an automorphism of $\mathbb{K}$. We extend ϕ to $\mathbb{K}[x, y]$ as follows

Definition 4. *Let $\phi \in Gal_{\mathbb{K}/\mathbb{F}_q}$ be an automorphism of $\mathbb{K}$. Define the map $\phi : \mathbb{K}[x, y] \mapsto \mathbb{K}[x, y]$ as*

$$\phi(f(x,y)) = \sum_{1 \leq k,l \leq n} \phi(a_{kl}) x^k y^l$$

$$\text{where} \quad f(x,y) = \sum_{1 \leq k,l \leq n} a_{kl} x^k y^l$$

Observe that the map $\phi : \mathbb{K}[x, y] \mapsto \mathbb{K}[x, y]$ is an automorphism of the ring $\mathbb{K}[x, y]$ that fixes the subring $\mathbb{F}_q[x, y]$. In particular,

- $\phi(f(x,y) + g(x,y)) = \phi(f(x,y)) + \phi(g(x,y))$
- $\phi(f(x,y)g(x,y)) = \phi(f(x,y))\phi(g(x,y))$

We now define an equivalence relation on $\mathbb{K}[x, y]$ induced by such automorphisms of $\mathbb{K}[x, y]$.

Definition 5. *Let $f(x, y) \in \mathbb{K}[x, y]$ be any bivariate polynomial. Then $\phi(f(x, y))$ for any $\phi \in Gal_{\mathbb{K}/\mathbb{F}_q}$ is said to be a conjugate of $f(x, y)$ over $\mathbb{F}_q$. When the underlying field $\mathbb{F}_q$ is clear from context, we will simply say that $\phi(f(x, y))$ is a conjugate of $f(x, y)$.*

Observe that conjugacy is an equivalence relation on $\mathbb{K}[x,y]$. Now consider a nice $\mathbb{F}_q$-irreducible polynomial $h(x,y) \in \mathbb{F}_q[x,y]$. Let $\mathbb{K} \supseteq \mathbb{F}_q$ be a finite field extension of $\mathbb{F}_q[x,y]$. How does $h(x,y)$ factor over $\mathbb{K}$? We claim that all the $\mathbb{K}$-irreducible factors of $h(x,y)$ in $\mathbb{K}$ are in fact conjugates of each other. In particular, all the $\mathbb{K}$-irreducible factors of $h(x,y)$ are of equal degree.

Claim 2. *Let $h(x,y) \in \mathbb{F}_q[x,y]$ be a nice irreducible polynomial of total degree n. Let $\mathbb{K}$ be any finite field extension of $\mathbb{F}_q$. If $f_1(x,y) \in \mathbb{K}[x,y]$ and $f_2(x,y) \in \mathbb{K}[x,y]$ are any two factors of $h(x,y)$ that are irreducible over the extension field $\mathbb{K}$, then $f_1(x,y)$ and $f_2(x,y)$ are conjugates over the base field $\mathbb{F}_q$.*

Proof. For a polynomial $f(x,y) \in \mathbb{K}[x,y]$, define $H_f \leq Gal_{\mathbb{K}/\mathbb{F}_q}$ to be the subgroup of $Gal_{\mathbb{K}/\mathbb{F}_q}$ consisting of automorphisms in $Gal_{\mathbb{K}/\mathbb{F}_q}$ that fix $f(x,y)$. Since the galois groups of finite extensions of finite fields are cyclic groups, H_f must be a normal subgroup of $Gal_{\mathbb{K}/\mathbb{F}_q}$.

Let $f(x,y) \in \mathbb{K}[x,y]$ be a factor of $h(x,y)$ which is irreducible over $\mathbb{K}$. Let the set of distinct cosets of H_f in $Gal_{\mathbb{K}/\mathbb{F}_q}$ be

$$Gal_{\mathbb{K}/\mathbb{F}_q}/H_f = \{H_f\phi_1, H_f\phi_2, \cdots H_f\phi_t\}$$

Then $\phi_1(f(x,y)), \phi_2(f(x,y)), \cdots \phi_t(f(x,y))$ are all the distinct conjugates of $f(x,y)$. We claim that the unique factorization of $h(x,y)$ into irreducible polynomials over $\mathbb{K}$ is simply the product of all these distinct conjugates of $f(x,y)$. That is,

$$h(x,y) = \prod_{H_f\phi \in Gal_{\mathbb{K}/\mathbb{F}_q}/H_f} \phi(f(x,y)) \quad (1)$$

Let ϕ be any automorphism in $Gal_{\mathbb{K}/\mathbb{F}_q}$. Now

$$f(x,y)|h(x,y) \Rightarrow \exists g(x,y) \in \mathbb{K}[x,y] \text{ such that } h(x,y) = f(x,y)g(x,y)$$

Applying ϕ to both sides, $\phi(h(x,y)) = \phi(f(x,y))\phi(g(x,y))$

or, $h(x,y) = \phi(f(x,y))\phi(g(x,y))$

$$\Rightarrow \phi(f(x,y))|h(x,y)$$

By the same reasoning $\phi(f(x,y)) \in \mathbb{K}[x,y]$ is irreducible over $\mathbb{K}$ for if any $g(x,y) \in \mathbb{K}[x,y]$, $deg(g(x,y)) < deg(\phi(f(x,y))) = deg(f(x,y))$ divides $\phi(f(x,y))$ then $\phi^{-1}(g(x,y))$ divides $f(x,y)$, contradicting the irreducibility of $f(x,y)$ over $\mathbb{K}$. Thus any conjugate of $f(x,y)$ is also an irreducible factor of $h(x,y)$. Moreover, $f(x,y)$ being irreducible over $\mathbb{K}$, is coprime to all conjugates distinct from itself. Thus the rhs of equation (1) divides $h(x,y)$. Moreover the rhs of equation (1) is fixed by all the automorphisms in $Gal_{\mathbb{K}/\mathbb{F}_q}$. Since finite extensions of finite fields are normal extensions, so any polynomial in $\mathbb{K}[x,y]$ that is fixed by all the automorphisms in $Gal_{\mathbb{K}/\mathbb{F}_q}$ is in fact a polynomial in $\mathbb{F}_q[x,y]$. Hence the rhs of equation (1) is in fact a polynomial in $\mathbb{F}_q[x,y]$ that divides $h(x,y)$. By the irreducibility of $h(x,y)$ over $\mathbb{F}_q$, we deduce that equation (1) is indeed the unique factorization of $h(x,y)$. Thus all the irreducible factors of $h(x,y)$ over $\mathbb{K}$ are precisely all the distinct conjugates of $f(x,y)$.

Now consider an irreducible polynomial $h(x,y) \in \mathbb{F}_q[x,y]$ that factors in the algebraic closure of $\mathbb{F}_q$. What is the splitting field of $h(x,y)$? Can we put a bound on the dimension of the splitting field over $\mathbb{F}_q$? Assuming that $h(x,y)$ is a nice polynomial, the following proposition shows that if $t(x)$ is an irreducible factor of $h(x,0)$, then the splitting field of $h(x,y)$ is a subfield of the finite field $\mathbb{F}_q[z]/\langle t(z) \rangle$. In particular, if $h(x,0)$ has a root $\alpha \in \mathbb{F}_q$, then $h(x,y)$ must be absolutely irreducible.

Proposition 3. *Let $h(x,y) \in \mathbb{F}_q[x,y]$ be a nice irreducible polynomial of total degree n. Let $dim_{\mathbb{F}_q}(h(x,y)) = d$. Also let $t(z) \in \mathbb{F}_q[z]$ be an irreducible factor of $h(z,0)$. Then $d | deg(t(z))$ and $h(x,y)$ breaks into absolutely irreducible factors over $\mathbb{K} := \mathbb{F}_q[z]/\langle t(z) \rangle$, each irreducible factor over $\mathbb{K}$ being of degree $m = \frac{n}{d}$.*

Proof. Let $f(x,y) \in \mathbb{K}[x,y]$ be an irreducible factor of $h(x,y)$ in $\mathbb{K}[x,y]$. Suppose if possible that $f(x,y)$ is not absolutely irreducible but breaks further over some finite extension $\mathbb{L} \supset \mathbb{K}$.

Let H_f be as in claim 2. By claim 2

$$h(x,y) = \prod_{H_f \phi \in G/H_f} \phi(f(x,y)) \quad (2)$$

Let $\alpha \in \mathbb{K}$ be a root of the polynomial $t(z)$. Now

$$(x-\alpha) \mid (h(x,0) = \prod_{H_f \phi \in G/H_f} \phi(f(x,0)))$$

$$\Rightarrow \exists \phi \in Gal_{\mathbb{K}/\mathbb{F}_q} \text{ such that } (x-\alpha)|\phi(f(x,0))$$

$$\Rightarrow (x-\beta)|f(x,0) \text{ where } \beta = \phi^{-1}(\alpha)$$

Note that $\beta = \phi^{-1}(\alpha) \in \mathbb{K}$ is also a root of the polynomial $t(z)$.

By claim 2 the irreducible factors of $f(x,y)$ over $\mathbb{L}$ are all conjugates. Let $f_1(x,y)$ be an irreducible factor of $f(x,y)$ over $\mathbb{L}$ such that $(x-\beta)$ divides $f_1(x,0)$. Let $\psi \in Gal_{\mathbb{L}/\mathbb{K}}$ be such that $\psi(f_1(x,y))$ is another irreducible (over $\mathbb{L}$) factor of $f(x,y)$ distinct from $f_1(x,y)$. Now since $(x-\beta)|f_1(x,0)$, we must also have $(x-\psi(\beta)) = (x-\beta)|\psi(f_1(x,0))$. This implies that $(x-\beta)^2$ divides $f(x,0)$ which is a contradiction since $h(x,0)$ and hence $f(x,0)$ are squarefree.

Thus the irreducible factors of $h(x,y)$ over $\mathbb{K}$ are absolutely irreducible. Hence there exists a subfield $\mathbb{F} \subseteq \mathbb{K}$ which is the splitting field of $h(x,y)$. Therefore $d = [\mathbb{F} : \mathbb{F}_q]$ divides $deg(t(z)) = [\mathbb{K} : \mathbb{F}_q] = [\mathbb{K} : \mathbb{F}][\mathbb{F} : \mathbb{F}_q]$.

By the definition of the splitting field of $h(x,y)$, the coefficients occuring in $f(x,y)$ lie in the field $\mathbb{F}$ and do not all lie in some proper subfield of $\mathbb{F}$. Hence $\mathbb{F}$ is precisely the subfield of $\mathbb{K}$ which is fixed by every automorphism in H_f. So

$$d = [\mathbb{F} : \mathbb{F}_q] = ord(Gal_{\mathbb{K}/\mathbb{F}_q}/H_f)$$

Further $ord(Gal_{\mathbb{K}/\mathbb{F}_q}/H_f)$ is the number of distinct absolutely irreducible factors of $h(x,y)$. Since all the irreducible factors of $h(x,y)$ are of the same degree, say m, we have

$$m.ord(Gal_{\mathbb{K}/\mathbb{F}_q}/H_f) = deg(h(x,y))$$

$$\Rightarrow m.d = n$$
$$\Rightarrow m = \frac{n}{d}$$

In summary, if $h(x,y) \in \mathbb{F}_q[x,y]$ is a nice polynomial that is $\mathbb{F}_q$-irreducible and $t_1(z), t_2(z) \in \mathbb{F}_q[z]$ are any two $\mathbb{F}_q$-irreducile factors of $h(z,0)$, then the degree of an irreducible factor of $h(x,y)$ over $\mathbb{K}_1 := \mathbb{F}_q[z]/\langle t_1(z)\rangle$ is the same as the degree of an irreducible factor of $h(x,y)$ over $\mathbb{K}_2 := \mathbb{F}_q[z]/\langle t_2(z)\rangle$. Moreover, this degree is $m = \frac{n}{d}$, where the dimension d of $h(x,y)$ was defined indepently of the choice of any factor of $h(z,0)$. This observation will be the key to our uniform-factoring algorithm.

4 Algorithm for Uniform Factoring

Given a polynomial $f(x,y) \in \mathbb{F}_q[x,y]$, the first step of our algorithm is to invoke the algorithm in [GKL04] on $f(x,y)$ thereby obtaining a distinct-degree factorization of $f(x,y)$. So now we can assume that the input to our algorithm is a nice bivariate polynomial $h(x,y) \in \mathbb{F}_q[x,y]$ and an integer n, such that each $\mathbb{F}_q$ irreducible factor of $h(x,y)$ has the same degree n.

If we could construct an irreducible polynomial $t(z) \in \mathbb{F}_q[z]$ of degree n in deterministic polynomial time, we could simply invoke the distinct-degree factorization algorithm on $h(x,y)$ over the field $\mathbb{F}_q[z]/\langle t(z)\rangle$ and obtain all the distinct dimension factors as well. No deterministic algorithm is currently known for constructing irreducible polynomials. Fortunately we can get around this problem by working over the ring $R \stackrel{\text{def}}{:=} \mathbb{F}_q[z]/\langle h(z,0)\rangle$.

Our algorithm from this point onwards is the similar to that in [GKL04] but with one key difference - we work over the ring R whereas [GKL04] work over the Berlekamp subalgebra of R. The proof of correctness is also similar and we omit the details due to lack of space.

5 Solvability of a Bivariate Polynomial System

Recall the problem we started with - given a set of polynomials $f_1, f_2, \cdots, f_m, g \in \mathbb{F}_q[x,y]$ does there exist a point $(a,b) \in \mathbb{F}_q \times \mathbb{F}_q$ such that

$$f_1(a,b) = f_2(a,b) = \cdots = f_m(a,b) = 0 \ , \ g(a,b) \neq 0 \ ?$$

5.1 Overview of the Algorithm

In this section we apply the uniform factoring algorithm devised above to the problem of determining the solvability of a bivariate polynomial system. We now give an overview of the algorithm. If the size of the field q is very small ($< d^5m$), we can determine the solvability in $O(d^{10}m)$ time by simplying trying all possible values of x and y. Henceforth we shall assume that $q > d^5$.

There is an initial preprocessing stage, the effect of which is to make *all* the given polynomials nice and separable with respect to x. For now assume that all the input polynomials are *nice*. Our algorithm is recursive and we now consider two special base cases of this recursion.

Special case: $m \geq 2$ and $\text{Resultant}_x(f_1, f_2) \neq 0$. Suppose that $m \geq 2$ and consider $f_1(x,y), f_2(x,y) \in \mathbb{F}_q[x,y]$. Further suppose that

$$\rho(y) \stackrel{\text{def}}{=} \text{Resultant}_x(f_1, f_2) \neq 0$$

In this case for any point $(a,b) \in \mathbb{F}_q \times \mathbb{F}_q$ satisfying $f_1(a,b) = f_2(a,b) = 0$, it must be the case that b is a root of $\rho(y)$. We will say that y is algebraic over $\mathbb{F}_q$ and in this case we can devise a deterministic polynomial-time algorithm that not only determines the existence of a solution to the given system but also determines the exact number of distinct solutions to the system.

Special case: $m = 1$. Here we are looking for $\mathbb{F}_q$ solutions to the system

$$f_1(x,y) = 0 \text{ and } g(x,y) \neq 0$$

By removing those factors from $f_1(x,y)$ that are common to both $f_1(x,y)$ and $g(x,y)$, we can assume that $\text{Resultant}_x(f_1, g) \neq 0$.

We use the uniform bivariate factoring algorithm devised above to determine whether $f_1(x,y)$ has an absolutely irreducible factor or not. Suppose that $f_1(x,y)$ does indeed have an absolutely factor $l(x,y) \in \mathbb{F}_q[x,y]$. Then Weil's theorem [Bac93] assures us of the existence of a large (about q) number of $\mathbb{F}_q$-solutions to the equation $l(x,y) = 0$ and hence to the equation $f_1(x,y) = 0$. Moreover, Bezout's theorem assures us that only a few ($\leq d^2$) of these solutions to $f_1(x,y) = 0$ are common to $g(x,y) = 0$. So if $f_1(x,y)$ does have an absolutely irreducible factor and if the size of the field q is large enough ($\geq d^5$) then the existence of a solution to the above system is guaranteed.

Otherwise none of the $\mathbb{F}_q$-irreducible factors of $f_1(x,y)$ are absolutely irreducible and hence by lemma 3 it follows that every $\mathbb{F}_q$-point (a,b) on the curve $f_1(x,y) = 0$ is a repeated point. In particular (a,b) is also a zero of $f_2(x,y) \stackrel{\text{def}}{=} \frac{\partial f_1}{\partial x}$. The preprocessing stage assures us that $\Delta_x(f_1) = \text{Resultant}_x(f_1, f_2) \neq 0$. This boils our problem down to the first special case.

General case. We now return to the general problem where $m \geq 2$ and $\text{Resultant}_x(f_1, f_2) = 0$. Compute $l(x,y) = \gcd_x(f_1, f_2)$ which must be nontrivial since the resultant is zero. Then we can express the given polynomials f_1 and f_2 as

$$f_1(x,y) = l(x,y) f_{11}(x,y) \ , \ f_2(x,y) = l(x,y) f_{21}(x,y)$$

where $\text{Resultant}_x(f_{11}, f_{21}) \neq 0$. Thus the given system has a $\mathbb{F}_q$-solution if and only if either the system

$$f_{11} = f_{21} = f_3 = f_4 = \cdots = f_m = 0 \ , g \neq 0$$

or the system
$$l = f_3 = f_4 = \cdots = f_m = 0 , g \neq 0$$
has a $\mathbb{F}_q$-solution. The first system is the first special case considered above. The $\mathbb{F}_q$-solvability of the second system is determined recursively. Thus if the time taken by the algorithm is denoted by $T(m,d,q)$ then

$$T(m,d,q) = T(m-1,d,q) + \text{poly}(md\log q)$$
$$\text{Or,}\quad T(m,d,q) = T(1,d,q) + \text{poly}(md\log q)$$
$$\text{Or,}\quad T(m,d,q) = \text{poly}(md\log q)$$

5.2 Handling the Case $m \geq 2$ and $\text{Resultant}_x(f_1, f_2) \neq 0$

In this section, we present the details of how to deterministically determine the solvability of a bivariate system of this form. Compute

$$\rho(y) \stackrel{\text{def}}{=} \gcd(y^q - y, \text{Resultant}_x(f_1, f_2)) \neq 0$$

which has degree at most d^2. Suppose that $(a,b) \in \mathbb{F}_q \times \mathbb{F}_q$ is a solution to the system $f_1(x,y) = f_2(x,y) = 0$. This implies that

$$\text{Resultant}_x(f_1(x,b), f_2(x,b)) = 0$$

(Note that the converse is not true since $f_1(x,b)$ and $f_2(x,b)$ may share a common root which does not lie in $\mathbb{F}_q$ but in some finite extension of $\mathbb{F}_q$). Thus b is an $\mathbb{F}_q$-root of $\text{Resultant}_x(f_1, f_2)$ and hence must be a root of $\rho(y)$. Let the roots of $\rho(y)$ be $b_1, b_2, \cdots, b_l \in \mathbb{F}_q$. Then the given system has a solution if and only if there exists a $b \in \{b_1, b_2, \cdots, b_l\}$ such that

$$g(x,b) \neq 0 \pmod{h(x)}$$

where
$$h(x) \stackrel{\text{def}}{=} \gcd(f_1(x,b), f_2(x,b), \cdots, f_m(x,y), x^q - x)$$

We can check for the existence of such a b without needing to actually factor $\rho(y)$ by working over the ring $R \stackrel{\text{def}}{=} \mathbb{F}_q[y]/\langle \rho(y)\rangle$. We simply view $f_1, f_2, \cdots, f_m, g$ as univariate polynomials in the variable x over the ring R and compute $h(x) \in R[x]$, $h(x) \stackrel{\text{def}}{=} \gcd_x(f_1(x), f_2(x), \cdots, f_m(x), x^q - x)$. (This gcd can be computed by the standard Euclidean algorithm applied over the ring R with just one minor modification - if the leading coefficient $\alpha \in R$ of a divisor polynomial $b(x) \in R[x]$ is not a unit of R, then the zero-divisorness of α allows us to factor the ring R into a direct sum of rings R_1 and R_2. Having obtained this factorization, we can carry out the division in each of these rings R_1 and R_2 and combine the result using chinese remaindering to get the result in the original ring R). Then the given system has an $\mathbb{F}_q$-solution if and only if

$$g(x) \neq 0 \pmod{h(x)} \quad \text{(over } R\text{)}$$

5.3 Preprocessing

The job of the preprocessing stage is to make the given polynomials nice with respect to x.

Preprocessing: Input and Output specifications
Input. A finite field $\mathbb{F}_q$ of characteristic p, a set of bivariate polynomials

$$f_1(x,y), f_2(x,y), \cdots, f_m(x,y), f_{m+1}(x,y) \in \mathbb{F}_q[x,y]$$

of total degree at most d. Further $q \geq 2d^2m$.
Output. Polynomials $\hat{f}_1(x,y), \hat{f}_2(x,y), \cdots, \hat{f}_m(x,y), \hat{h}(x,y) \in \mathbb{F}_q[x,y]$ of total degree at most d^2 such that each $\hat{f}_i(x,y)$ is a nice and separable polynomial with respect to x. Moreover the number of solutions to the system

$$f_1(x,y) = f_2(x,y) = \cdots = f_m(x,y) = 0 \ , \ f_{m+1}(x,y) \neq 0$$

is the same as the number of solutions to the system

$$\hat{f}_1(x,y) = \hat{f}_2(x,y) = \cdots = \hat{f}_m(x,y) = 0 \ , \ \hat{f}_{m+1}(x,y) \neq 0$$

In the first part of the preprocessing stage we make each of the given polynomials square-free by replacing each polynomial by its radical and then doing a substitution of the following form:

$$\hat{f}_i(x,y) = f_i(x, y + ax)$$

Suppose that $f_i(x,y)$ has total degree d_i then note that the coefficient of x^{d_i} in $\hat{f}_i$ is a polynomial t_i in a of degree at most d_i. Thus we need to chose an $a \in \mathbb{F}_q$ such that

$$t(a) \stackrel{\text{def}}{:=} \prod_{1 \leq i \leq m+1} t_i(a) \neq 0$$

$t(a)$ is a polynomial of degree at most dm and thus we can certainly find such an a in deterministic polynomial time. Thus in what follows we will assume that each $f_i(x,y)$ is squarefree and monic in x.

Making the polynomials separable with respect to x requires a little more work. Let $l \stackrel{\text{def}}{:=} \lfloor \log_p d \rfloor$. The main idea is that if $f_i(x,y)$ is not separable with respect to x (i.e. $\Delta_x(f(x,y)) = 0$) then we can deterministically compute a factorization of $f_i(x,y)$ of the form

$$f_i(x,y) = \prod_{0 \leq k \leq l} f_{ik}(x^{p^k}, y)$$

where each $f_{ik}(x,y)$ is a polynomial in x^{p^k} but not a polynomial $x^{p^{k+1}}$.

For a polynomial $f(x,y) = \sum_{i,j} a_{ij} x^i y^j \in \mathbb{F}_q[x,y]$ let us define $f^{(p)}(x,y)$ to be the polynomial $f^{(p)}(x,y) \stackrel{\text{def}}{:=} \sum_{i,j} a_{ij}^p x^i y^j$. Note that for any point $(a,b) \in \mathbb{F}_q \times \mathbb{F}_q$,

$$f(a,b) = 0 \Leftrightarrow f^{(p)}(a^p, b^p) = 0$$

That is, the set of zeroes of $f(x,y)$ is exactly the same as the set of zeroes of the polynomial $f^{(p)}(x^p, y^p)$.

Corresponding to every input polynomial

$$f_i(x,y) = \prod_{0 \leq k \leq l} f_{ik}(x^{p^k}, y)$$

we output the polynomial

$$\hat{f}_i(x,y) \stackrel{\text{def}}{=} \prod_{0 \leq k \leq l} f_{ik}^{(p^{l-k})}(x, y^{p^{l-k}})$$

It can be shown that the number of $\mathbb{F}_q$-solutions to this new system is the same as the number of $\mathbb{F}_q$-solutions to the original system. We omit the proof due to lack of space.

6   Conclusion

A very general problem in algorithmic number theory is the following: let

$$f_1, f_2, \cdots, f_m \in \mathbb{F}_q[x_1, x_2, \cdots, x_n]$$

be a collection of m dense polynomials in n variables over the finite field $\mathbb{F}_q$. Let $Q_1, Q_2, \cdots, Q_n$ be some sequence of quantifiers, i.e. each quantifier is either " $\exists$ " or " $\forall$ ". What is the computational complexity of determining the truth value of the following statement:

$$Q_1 x_1 \in \mathbb{F}_q \cdots Q_n x_n \in \mathbb{F}_q \quad f_1 = f_2 = \cdots = f_m = 0$$

when the number n of variables is bounded? (For unbounded number of variables, the problem is **PSPACE**-complete and bounding the number of alternations between " $\exists$ " and " $\forall$ ", we get the different levels of the polynomial hierarchy).

Generalizing the techniques of this paper quite a bit further, it has recently been shown by the author that the existential version of this problem in which all the quantifiers are existential can be solved deterministically in time polynomial in the size of the input. Beyond this, we do not know an efficient solution for even the Π_2-version of this problem with 2 variables -

Open Question. Is there a random polynomial time algorithm for determining the truth of the statement

$$\forall x \in \mathbb{F}_q \; \exists y \in \mathbb{F}_q \; f(x,y) = 0 \;?$$

Acknowledgements

The author is grateful to Alan Lauder, Manindra Agrawal and Nitin Saxena for giving many helpful comments on an earlier version of this paper that improved the presentation. The author also wishes to thank Igor Shparlinski and

Hendrik Lenstra for pointing out the existence of a classification of permutation polynomials and the derived algorithm for recognizing permutation functions in deterministic polynomial time.

References

[GKL04] Shuhong Gao, Erich Kaltofen and Alan Lauder. *Deterministic distinc-degree factorization of polnomials over finite fields.* Journal of Symbolic Computing, Volume 38, Number 6, 2004. pages 1461-70.

[HW99] Ming-Deh Huang and Yiu-Chung Wong. *Solvability of systems of polynomial congruences modulo a large prime.* Computational Complexity Volume 8, Number 3, 1999. pages 227-257.

[LM88] R. Lidl and G. L. Mullen. *When does a polynomial over a finite field permute the elements of the field?* American Mathematical Monthly 95, 1988. 243-246.

[LM93] R. Lidl and G. L. Mullen. *When does a polynomial over a finite field permute the elements of the field?, II* American Mathematical Monthly 100, 1993. 71-74.

[LM83] Rudolf Lidl and Winfried B. Muller. *Permutation Polynomials in RSA cryptosystems.* In D. Chaum, editor, Proceedings CRYPTO 83. pages 293-301.

[MG94] Keju Ma and Joachim Von Zur Gathen. *The computational complexity of recognizing permutation functions.* Computational Complexity, Volume 5, Number 1, 1995. pages 76-97.

[Len05] Hendrik Lenstra. *Private Communication, 2005.*

[RSA78] Ronald L. Rivest, Adi Shamir and Leonard M. Adleman. *A method for obtaining digital signatures and public-key cryptosystems.* Communications of the ACM, 21(2), 1978. pages 120-126.

[Bac93] E. Bach. *Weil bounds for singular curves.* Applicable Algebra in Engineering, Communication and Computing 7, 1996. pages 289-298.

Cache-Oblivious Planar Shortest Paths

Hema Jampala* and Norbert Zeh*

Faculty of Computer Science, Dalhousie University,
6050 University Ave, Halifax, NS B3H 1W5, Canada
{jampala, nzeh}@cs.dal.ca

Abstract. We present an efficient cache-oblivious implementation of the shortest-path algorithm for planar graphs by Klein et al., and prove that it incurs no more than $\mathcal{O}(\frac{N}{B^{1/2-\epsilon}} + \frac{N}{B} \log N)$ block transfers on a graph with N vertices. This is the first cache-oblivious algorithm for this problem that incurs $o(N)$ block transfers.

1 Introduction

The *single-source shortest-path* (SSSP) problem is a fundamental combinatorial optimization problem with numerous applications. Let $G = (V, E)$ be a directed graph with vertex set V and edge set E; let $s \in V$ be a distinguished vertex, called the *source vertex*; and let $\omega : E \to \mathbb{R}^+$ be an assignment of non-negative real weights to the edges of G. The SSSP-problem is to find, for every vertex $v \in V$, the distance $d(v)$ from s to v, that is, the weight of a minimum-weight (shortest) path from s to v. This problem is well-studied in the RAM-model. The classical algorithm for this problem is Dijkstra's algorithm [15], which has seen many improvements (e.g.,[19, 21, 22, 23]). In particular, on planar graphs, much progress has been made: Frederickson [17] proposes an algorithm that takes $\mathcal{O}(N\sqrt{\log N})$ time, pioneering the idea to use separator decompositions to speed up shortest-path computations. Klein et al. [19] present a non-trivial refinement of Frederickson's approach that uses a hierarchy of nested separator decompositions to solve SSSP in planar directed graphs in linear time.

More recently, the SSSP-problem has been studied in memory hierarchy models, which take the varying access times of different levels of cache, main memory, and disk into account. Such algorithms can be *cache aware* or *cache-oblivious*. The former require knowledge of the parameters of the different levels of memory and often explicitly transfer data between the different levels; the latter are oblivious of these parameters, but help the default paging algorithm by laying out the data appropriately and accessing it in a local fashion.

The most widely used model for the design of *cache-aware* algorithms is the I/O-model of Aggarwal and Vitter [1]. This model assumes a memory hierarchy consisting of two levels: the lower level has size M; data is transferred

---

* Research supported by the Natural Sciences and Engineering Research Council of Canada and by the Canadian Foundation for Innovation.

between the two levels in blocks of B consecutive data items. The complexity of an algorithm is the number of blocks transferred (I/Os). Algorithms that perform a small number of I/Os are usually referred to as *I/O-efficient* algorithms. The strength of the I/O-model is its simplicity, while it still adequately models the situation when the I/Os between two levels of the memory hierarchy dominate the running time of the algorithm, which is often the case in large-scale applications. Complexities that arise often in I/O-efficient algorithms are the sorting bound, $\text{sort}(N) = \Theta(\frac{N}{B} \log_{M/B} \frac{N}{B})$ I/Os [1, 18], the permutation bound, $\text{perm}(N) = \Theta(\min(N, \text{sort}(N)))$ I/Os [1], and the scanning bound, $\text{scan}(N) = \Theta(N/B)$ I/Os.

Solving interesting problems using cache-aware algorithms for *multi-level* hierarchies is cumbersome, because it is necessary to tune the algorithms to the sizes and block sizes of all levels of memory. *Cache-oblivious* algorithms provide an elegant solution to this problem. They are designed to be I/O-efficient without knowing M or B; that is, they are formulated in the RAM-model and analyzed in the I/O-model, assuming that the memory transfers are performed by an optimal offline paging algorithm. Since the analysis holds for any block and memory sizes, it holds for *all* levels of a multi-level memory hierarchy (see [18] for details). Thus, the cache-oblivious model elegantly combines the simplicity of the I/O-model with a coverage of the entire memory hierarchy. The bounds for sorting and scanning are the same as in the I/O-model [9, 18], whereas $\text{perm}(N) = \text{sort}(N)$ in the cache-oblivious model [11]. Since any internal-memory algorithm is by definition cache-oblivious, but usually incurs a substantial number of block transfers, we refer to an algorithm as *cache-oblivious* in this paper if it is cache-oblivious in the sense of the definition and incurs $o(T(N))$ block transfers, where $T(N)$ is the computation time of the best internal-memory algorithm.

Previous work on graph algorithms for memory hierarchies has focused mainly on *I/O-efficient* algorithms, motivated by a number of large-scale applications that have to deal with massive graphs, such as geographic information systems, web modelling, and data mining of phone call databases. The obtained results include a large number of algorithms for planar graphs, such as $\mathcal{O}(\text{sort}(N))$-I/O algorithms for computing connected components [13], minimum spanning trees [13], and strongly connected components [8]; breadth-first search (BFS) [3] and undirected depth-first search (DFS) [5]; single-source shortest paths [3]; and topological sorting [6, 7]. Directed planar DFS is studied in [8], and an optimal $\mathcal{O}(N^2/B)$-I/O all-pairs shortest path algorithm is presented in [4].

Recently, a number of cache-oblivious graph algorithms have been obtained for general graphs, including algorithms for computing connected components and minimum spanning trees [2], directed breadth-first search and depth-first search [2], undirected breadth-first search [12], and undirected shortest paths [12, 14]. All these algorithms are obtained from I/O-efficient algorithms for these problems by designing cache-oblivious data structures that can replace the cache-aware ones in these algorithms. This strategy does not seem to work for most of the specialized algorithms for planar graphs, mentioned above: Their dependence on B is not hidden in data structures; and many of them exploit that computation in internal memory is free in the I/O-model, by performing $\Omega(BN)$ computation in main memory. In a multi-level hierarchy, this extra computation may incur block transfers at lower cache levels, thereby leading to an excessive number of block transfers. The internal-memory computation can often be reduced. For example, for shortest paths, it can be reduced to $\mathcal{O}(n \cdot \text{polylog}(B))$, using results from [16]; but these algorithms are not easily made cache-oblivious. Nevertheless, a number of cache-oblivious algorithms for planar graphs exist. Using cache-oblivious data structures from [2, 10], the I/O-efficient algorithms for connectivity, biconnectivity, and minimum spanning trees [13], and for topologically sorting planar directed acyclic graphs [7] can be made cache-oblivious, without asymptotically increasing their complexities.

2 New Result

We make the first progress towards solving the SSSP-problem cache-obliviously in planar graphs, by analyzing the number of block transfers incurred by a cache-efficient implementation of the algorithm of [19]. We assume that a suitable multi-level separator decomposition of the graph is given. We consider finding such a decomposition the central open problem in the design of cache-oblivious algorithms for planar graphs, because separators also have played a central role in the design of I/O-efficient algorithms for planar graphs. In Sec. 7, we suggest one approach that may lead to a cache-oblivious algorithm for this problem. Our result is summarized in the following theorem.

Theorem 1. *The SSSP-problem in a planar directed graph G with N vertices and non-negative edge weights can be solved using a cache-oblivious algorithm that incurs $\mathcal{O}\left(\frac{N}{B^{1/2-\epsilon}} + \frac{N}{B}\log N\right)$ block transfers, for any constant $\epsilon > 0$, provided that a suitable multi-level separator decomposition of G is given.*

In this paper, we assume that every vertex in G has in-degree at most 2 and out-degree at most 2. A simple transformation described, for instance, in [20] achieves this. The discussion of the algorithm is divided into several sections. In Sec. 3, we discuss the necessary terminology regarding graph separators. We outline the linear-time SSSP-algorithm of [19] in Sec. 4. We show in Sec. 5 how to implement this algorithm in a cache-efficient manner. The analysis of the algorithm is provided in Sec. 6. Open problems are discussed in Sec. 7.

3 Separator Decompositions

The algorithm of [19], as many other SSSP-algorithms for planar graphs, uses a separator decomposition to organize its computation. Next we define the required partition and discuss its representation.

Definition. For a planar graph $G = (V, E)$ and an integer parameter $r > 0$, an *r-partition* of G is a pair $(S, \{G_1, G_2, \ldots, G_h\})$ with the following properties:

(i) S is a subset of V of size $\mathcal{O}(N/\sqrt{r})$. (ii) The graphs $G_1, G_2, \ldots, G_h$ are edge-disjoint subgraphs of G whose union is G and such that any two such graphs G_i and G_j share only vertices in S. (iii) Every graph G_i, $1 \le i \le h$, contains at most r edges and at most $\sqrt{r}$ vertices from S. We call the vertices in S *separator vertices*. The set of vertices in S that are contained in a graph G_i are the *boundary* of G_i.

Given a vector $\mathbf{r} = (r_0, r_2, \ldots, r_k)$, a *recursive $\mathbf{r}$-partition* $\mathcal{P}$ of G consists of a sequence of partitions of G, $(\mathcal{P}_0, \mathcal{P}_1, \ldots, \mathcal{P}_k)$, with the following properties: (i) Each $\mathcal{P}_i$ is an r_i-partition of G. (ii) For two consecutive partitions $\mathcal{P}_i = (S_i, \{G_1, G_2, \ldots, G_s\})$ and $\mathcal{P}_{i+1} = (S_{i+1}, \{G'_1, G'_2, \ldots, G'_t\})$, $S_i \supset S_{i+1}$ and, for every graph G_j in $\mathcal{P}_i$, there exists a graph G'_ℓ in $\mathcal{P}_{i+1}$ such that $G_j \subseteq G'_\ell$.

In this paper, a recursive $\mathbf{r}$-partition will always satisfy $r_0 = 1$ and $r_k = |E|$; that is, the lowest level of the partition splits G into its edges, and the highest level consists of the entire graph.

Representation. A recursive $\mathbf{r}$-partition $\mathcal{P}$ can be represented quite naturally by a rooted tree. The root of the tree is the single subgraph in $\mathcal{P}_k$, which is G. For a node corresponding to a graph R in $\mathcal{P}_i$, $i > 0$, its children represent the graphs in $\mathcal{P}_{i-1}$ that are subgraphs of R. The leaves represent the edges of G. We call this tree the *partition tree* of $\mathcal{P}$. The subgraphs of G corresponding to the nodes in this tree are referred to as *regions*. The edges of G, which correspond to the leaves of the tree, are *atomic regions*, as they cannot be split any further. The *level* of a region R is the index i such that R is a graph in $\mathcal{P}_i$. We define ancestry of regions to be the ancestry of the corresponding nodes in the partition tree.

4 Outline of the Algorithm

The algorithm of [19] is a variant of Dijkstra's algorithm, implemented using a hierarchy of priority queues associated with the regions of a recursive $\mathbf{r}$-partition of G (see Alg. 1). The entries in the priority queue $Q(R)$ of region R are the children of R; the priority of such a child R' is the minimum tentative distance of the source vertices of all unrelaxed edges in R'. Initially, the priorities of the regions containing out-edges of s are 0 and those of the rest are $+\infty$. The algorithm chooses the child with minimal priority from $Q(G)$ and recurses on this child. If the child is non-atomic, it repeats this process; otherwise it relaxes the edge (u, v) in this region and updates the priorities of all regions containing the out-edges of v. Procedure Shortest-Paths is called on G until all edges are relaxed, that is, the priority of G itself is $+\infty$.

If every invocation of this procedure on a non-atomic region were to return after the first recursive call it makes on one of its children, this algorithm would be Dijkstra's algorithm. However, the procedure returns from a recursive call on a level-i region R only after all edges in the region have been relaxed or α_i recursive calls have been made on children of R, for a suitable parameter α_i.

By keeping the algorithm focused on a region for some time, once a recursive call has been made on this region, Klein et al. limit the number of priority queue

```
Shortest-Paths(R)
 1  ▷ R is the region on which the current invocation operates.
 2  j ← level(R)
 3  if j = 0
 4     then Relax the edge (u, v) in R, that is, set d(v) = min(d(v), d(u) + w(u, v)).
 5          if this changes the distance of v from s
 6             then Change the priority of every out-edge (v, w) of v to d(v).
 7                  Change the priority of every ancestor of (v, w) whose priority is
                    greater than d(v) to d(v).
 8          Change the priority of (u, v) to ∞.
 9     else i ← 0
10          while the minimum entry in Q(R) is not ∞ and $i < \alpha_j$
11             do R' ← Delete-Min(Q(R))
12                Shortest-Paths(R')
13                Let the priority of R' in Q(R) be the minimum priority in Q(R').
14                i ← i + 1
```

Algorithm 1. Outline of the shortest-path algorithm. This procedure is called repeatedly with the whole graph G as the argument until all edges are relaxed

operations on large priority queues (that is, priority queues attached to regions at higher levels) and obtain a linear time bound for their algorithm. The other advantage this approach has is that it avoids "jumping around randomly"; that is, once the algorithm focuses on a subgraph that fits into cache, it stays focused on this subgraph for a while. The computation inside the subgraph does not incur any block transfers after the whole subgraph has been loaded into the cache. To take full advantage of this, however, the algorithm has to be implemented using cache-oblivious data structures, and the parameters in the algorithm have to be chosen differently, which increases the amount of computation to $\mathcal{O}(N \log N)$. In Sec. 7, we argue that, while a reduction of the log-factor may be possible, it cannot be eliminated entirely.

5 Cache-Oblivious Implementation

In [19], Algorithm 1 is used more or less verbatim to obtain a linear running time. The updates of the priorities of all relevant regions in Lines 6 and 7 are performed using a so-called GlobalUpdate operation. This operation traverses the path from region (v, w) to the lowest common ancestor of regions (u, v) and (v, w) in the partition tree and performs the updates described in Lines 6 and 7.

A cache-efficient algorithm cannot use this strategy because the updates of atomic regions alone would require $\Theta(N)$ block transfers in the worst case. But we can exploit that the recursive calls in the algorithm correspond to a traversal of the partition tree: The correct priority of a region R has to be known only by the time the traversal visits its parent R' because only then this information is required to extract the correct child to be visited from $Q(R')$. In order to reach R', we have to traverse the path from the current region (u, v) to R'. Thus, if the relaxation of edge (u, v) affects the priority of R, we can carry this update along the path from (u, v) to R' and apply it to $Q(R')$ immediately before the next Delete-Min operation is to be performed on $Q(R')$. The details are as follows:

Every invocation I on a region R collects all updates on atomic regions outside of R that are triggered by descendants of I and passes them to its parent invocation I' when it returns. Let R' be the region of I'. Then I' inspects the updates received from I. Every update on an atomic region outside of R' is scheduled to be passed to the parent of I'. Updates on atomic regions inside R' are scheduled to be sent to the appropriate children of R' when the next invocations on these children are made. When an invocation on R' makes a recursive call on one of these children, R'', the updates scheduled to be sent to R'' are applied to $Q(R'')$ and then scheduled to be sent to the appropriate children of R'', based on which child contains the edge affected by each update.

Line 5 presents a similar problem, with a similar solution: We cannot afford to access vertex v to test whether the relaxation of edge (u, v) decreases $d(v)$. To avoid this, we store $d(v)$ with both in-edges (u_1, v) and (u_2, v) of v. When one in-edge, say (u_1, v), decreases the distance, it informs the other in-edge, (u_2, v), by sending an Update-Target message to (u_2, v). Again, the correct distance of v has to be known to region (u_2, v) only when the next invocation on this region is made; that is, Update-Target messages can be delivered in the same fashion as Update messages for the out-edges of v.

We implement this strategy using the following data structures: We use a stack S to collect and pass updates to ancestors of the current region. We associate a cache-oblivious buffered repository tree (BRT) $B(R)$ [2] with every region, which we use to collect all updates to be sent from R to its children. The tree $B(R)$ serves yet another purpose: Since the updates to be performed on $Q(R)$ are in fact updates on atomic regions, and it is too costly to identify the child of R affected by every update, we store atomic regions in $Q(R)$. When retrieving the minimum entry (u, v) from $Q(R)$, we have to (1) determine the child R' of R that contains edge (u, v), in order to make a recursive call on R', and (2) remove all edges in R' from $Q(R)$, in order to effectively set the priority of R' to $+\infty$. Using the BRT, we can achieve both: Assume that the leaves of the partition tree are numbered left to right and that every internal node has been labelled with the interval of numbers of its descendant leaves. (This can be achieved in a preprocessing step.) We associate with every leaf of a BRT $B(R)$, which corresponds to a region R', the interval associated with R' and with every internal node the union of the intervals of its children. This is sufficient to decide, for every update on an atomic region, to which leaf in the BRT it should be sent and to use any atomic region contained in a child R' as the key for an Extract operation that identifies R' and retrieves all up-

CO-Shortest-Paths(R, U)
1. ▷ R is the region on which the current invocation operates; U is an array of Update operations provided by the parent invocation.
2. $j \leftarrow \text{level}(R)$
3. **if** $j = 0$
4.   **then** Update $d(u)$ and $d(v)$ using the minimum priorities of the Update and Update-Target operations in U, respectively.
5.   Relax the edge (u, v) in R, that is, set $d(v) = \min(d(v), d(u) + \omega(u, v))$.
6.   **if** Line 5 changes the distance of v from s
7.   **then** Insert an Update$((v, w), d(v))$ operation into S, for every out-edge (v, w) of v; mark this operation as inserted by region R.
8.   Insert an Update-Target$((u', v), d(v))$ operation into S, where (u', v) is the other in-edge of v; mark this operation as inserted by region R.
9.   Change the priority of (u, v) to ∞ and return region (u, v) with priority ∞, for insertion into the parent's priority queue.
10.   **else** Perform the updates in U on $Q(R)$ and insert them into $B(R)$.
11. $i \leftarrow 0$
12.   **while** the minimum entry in $Q(R)$ is not ∞ and $i < \alpha_j$
13.   **do** $(u, v) \leftarrow$ Delete-Min$(Q(R))$
14.   Identify the child R' of R that contains edge (u, v) and extract all updates on descendants of R' from $B(R)$.
15.   Store these updates in an array U' and delete all regions affected by these updates from $Q(R)$.
16.   CO-Shortest-Paths(R', U'); let (x, y) be the edge returned by this recursive call, and let p be its priority.
17.   Insert edge (x, y) into $Q(R)$, with priority p, and into $B(R)$.
18.   Extract all updates from S that have been inserted by R'. Re-insert the updates on edges outside of R and mark them as inserted by R. Apply all other updates to $Q(R)$ and insert them into $B(R)$.
19. $i \leftarrow i + 1$
20.   Return the minimum-priority entry in $Q(R)$ for insertion into the parent's priority queue.

Algorithm 2. Cache-oblivious implementation of the shortest-path algorithm

dates on descendants of R'. Once the updates on R' are retrieved, they can be deleted from $Q(R)$. The details of the cache-oblivious implementation are shown in Alg. 2.

From our discussion, it follows that the modifications do not change the sequence of edge relaxations performed by the algorithm. Hence, the algorithm remains correct. We summarize this in the following lemma:

Lemma 1. *Procedure CO-Shortest-Paths terminates with the label $d(v)$ of every vertex v set to its distance from s in G.*

6 Analysis

In this section, we analyze the number of block transfers incurred by the algorithm. One of the keys to minimizing the number of block transfers is an appropriate layout of the data structures in memory. We describe this layout in Sec. 6.1. We also describe the paging strategy we use in our analysis; the optimal paging strategy cannot do worse than ours. In Sec. 6.2, we show that our paging strategy incurs at most the number of block transfers stated in Thm. 1.

6.1 Memory Layout and Space Bound

The partition, including the associated data structures, is laid out in memory as follows: We number the nodes of the partition tree in a depth-first manner. For every node representing a region R of size r, we allocate space for $Q(R)$, $B(R)$, and a buffer space $H(R)$ of size $3a^{1/2a}r^{1-1/2a}$, for some constant $1 < a \le 2$ defined in Sec. 6.2, which holds the buffers associated with the nodes in $B(R)$. If we refer to every region by its number, we arrange these structures in the order $Q(1), B(1), H(1), Q(2), B(2), H(2), \ldots$. The space for stack S succeeds these data structures and buffer spaces for the regions in the partition.

The space allocated for every data structure is fixed, that is, the memory layout is static. Preallocating space for the BRT's and priority queues is easy because we can bound their sizes by $\mathcal{O}(r^{1/a})$ and $\mathcal{O}(r^{1-1/2a})$, respectively. For the BRT, excluding buffers, this follows because we will show in Sec. 6.2 that the number of children of R is $\mathcal{O}(r^{1-1/a})$, and $B(R)$ contains one leaf per child. For the priority queue, we observe that an edge (atomic region) stored in $Q(R)$ is either incident to a boundary vertex of a child of R or it is the value returned by the last recursive call to a child of R. The number of entries of the first type is bounded by the number of edges incident to boundary vertices of children of R, which is $\mathcal{O}(r^{1-1/2a})$. For entries of the second type, we observe that each such entry is removed before the next recursive call to the child that inserted the entry. Hence, there can be at most one such entry per child of R in $Q(R)$, that is, at most $\mathcal{O}(r^{1-1/a}) = \mathcal{O}(r^{1-1/2a})$ entries. In order to bound the buffer space required by the BRT, we exploit that, similar to $Q(R)$, $B(R)$ cannot store updates on more than $\mathcal{O}(r^{1-1/2a})$ atomic regions, which allows us to use a variant of the BRT with limited buffer space. This variant is discussed in the next paragraph.

Another way to look at the above space bound is that every level uses space proportional to the number of boundary vertices at the next lower level. Since the number of boundary vertices at level 0 is $\mathcal{O}(N)$, and the number of boundary vertices per level decreases by at least a constant factor as we proceed towards the root (see Sec. 6.2), we obtain the following lemma:

Lemma 2. *The layout of the partition tree, including priority queues, BRT's, and buffer spaces associated with its nodes, uses linear space.*

A BRT with Limited Buffer Space. In order to limit the buffer space for every BRT, we exploit that, if there is more than one update pending on an atomic region, it suffices to perform the update with minimum priority. We partition the buffer space for the BRT of R into three regions of size $t = a^{1/2a} r^{1-1/2a}$, called the *sorted, unsorted*, and *root* buffer space. The root buffer space holds the root buffer. The unsorted buffer space holds buffers for all non-root nodes as in [2]. The sorted buffer space stores additional buffers, one per leaf, whose entries are sorted by the atomic regions they affect. As will become clear from the following discussion, the sorted buffer space can never overflow. As long as neither the root buffer nor the unsorted buffer space overflows, we operate on the BRT as described in [2], except that an Extract operation needs to read out an additional leaf buffer in the sorted buffer space.

When either the root buffer or the unsorted buffer space overflows, we sort the contents of these buffer spaces and then merge them with the contents of the sorted buffer space. If there are duplicate entries in the resulting list, we keep only the one with minimum priority. Hence, the resulting list has size at most t and completely fits into the sorted buffer space.

The cost of all BRT operations, excluding the cost of compacting the buffers when the root buffer or unsorted buffer space overflows, remains $\mathcal{O}\bigl(\frac{\log N}{B}\bigr)$ for Insert operations and $\mathcal{O}(\log N + K/B)$ for Extract operations. Next we argue that the cost of buffer compaction is $\mathcal{O}\bigl(\frac{\log t}{B}\bigr)$ amortized per element, if $t \geq B$: The cost of sorting the contents of the unsorted and root buffer space is $\mathcal{O}(\text{sort}(t'))$, where t' is the number of elements in these two buffer spaces. The rest of the compaction takes two scans of $\mathcal{O}(t)$ data, which requires $\mathcal{O}(t/B)$ block transfers. This is $\mathcal{O}\bigl(\frac{\log t}{B}\bigr)$ amortized per element if we can prove that $t' \geq t/\log t$. To do so, observe that every element, as it is propagated down the tree, requires us to allocate one memory cell at each level in the BRT, $\log t$ in total. Hence, it takes $t' \geq t/\log t$ elements to make the unsorted or root buffer space overflow.

A Paging Strategy. We permanently keep the top two blocks of S in cache. For every invocation on a region of size at most B, we load the part of the data structure into memory that corresponds to the region and its descendants. We call such an invocation *small*. The same argument that establishes the linear space bound in Lem. 2 implies that the size of this portion of the data structure is $\mathcal{O}(B)$; and it is stored consecutively in memory. Hence, such an invocation costs $\mathcal{O}(1)$ block transfers, and the descendant invocations do not incur any further block transfers, except for stack operations. For any invocation on a region of size greater than B, we load the first block of $Q(R)$ and the last block of the root buffer of $B(R)$ into cache. Thus, excluding the cost of priority queue, BRT, and stack operations, each such invocation costs $\mathcal{O}(1)$ I/Os. We call such an invocation *large*.

6.2 Counting Block Transfers

It remains to analyze the number of block transfers performed by our paging algorithm. We start by choosing suitable region sizes $r_1, r_2, \ldots, r_k$ and parameters $\alpha_1, \alpha_2, \ldots, \alpha_k$. In order to obtain the complexity stated in Thm. 1, we choose

$$r_j = a^j \cdot 2^{a^j/(a-1)} \quad \text{and} \quad \alpha_j = 2^{a^{j-1}/2},$$

where $a = 1/(1-\epsilon) > 1$. The height of the recursive partition is the minimum k such that $r_k \geq 3N$; thus, the height of the partition is no more than $\log_a((a-1)\log 3N) = \mathcal{O}(\log\log N)$. We can divide the block transfers incurred by the algorithm into 4 groups, depending on which part of the algorithm triggers them:

$T_1(N)$: The cost of loading buffer blocks for large invocations and complete data structures for small invocations. We call these block transfers *invocation swaps*.
$T_2(N)$: The cost of priority queue operations and Insert operations on BRT's.
$T_3(N)$: The cost of Extract operations on BRT's.
$T_4(N)$: The cost of stack operations.

We prove that $T_1(N) = \mathcal{O}(\frac{N}{B^{1/2-\epsilon}})$, $T_2(N) = \mathcal{O}(\frac{N}{B}\log N)$, $T_3(N) = \mathcal{O}(\frac{N}{B^{1/2-\epsilon}} + \frac{N}{B}\log\log N)$, and $T_4(N) = \mathcal{O}(\frac{N}{B}\log\log N)$. Summing these four terms, we obtain the following lemma:

Lemma 3. *Algorithm 2 incurs* $\mathcal{O}(\frac{N}{B^{1/2-\epsilon}} + \frac{N}{B}\log N)$ *block transfers.*

Next we sketch how to derive the above bounds for $T_1(N)$ through $T_4(N)$. Details will be provided in the full paper.

Invocation Swaps. It suffices to count the number of invocations whose parent regions have size at least B because each such invocation incurs $\mathcal{O}(1)$ block transfers, and these are the only invocations that incur block transfers. Klein et al. [19] classify invocations as *truncated* or *non-truncated*. The former is an invocation that returns because all edges in the region have been relaxed. The latter returns because α_j recursive calls on children of the current region have been made. We denote the number of truncated level-j invocations by S'_j and the total number of level-j invocations by S_j. Using a non-trivial charging scheme, Klein et al. prove that the number of truncated level-j invocations is

$$S'_j \leq \sum_{i \geq j} \mathcal{O}(\beta_{ij} \cdot N/\sqrt{r_i}),\text{[1]}$$

where $\beta_{ij} = \prod_{k=j+1}^{i} \alpha_k$. Each non-truncated invocation at level j gives rise to α_j invocations at level $j-1$. Hence, the total number of invocations at level j is

$$S_j \leq \frac{S_{j-1}}{\alpha_j} + S'_j.$$

Note that all level-0 invocations are truncated, that is, $S_0 = S'_0$. Using our choice of parameters α_i, we have $\beta_{ij} = \frac{2^{a^i/(2(a-1))}}{2^{a^j/(2(a-1))}}$. Substituting this in the expressions for S'_j and S_j, we obtain that $S'_j = \mathcal{O}(N/\sqrt{r_j})$ and $S_j = \mathcal{O}((j+1)a^{j/2}N/\sqrt{r_j})$. Since the r_j increase doubly exponentially, we have $\sum_{i \geq j} S_i = \mathcal{O}(S_j)$. Hence,

---

[1] They prove $S'_j \leq \sum_{i \geq j} \mathcal{O}(\beta_{ij} \cdot Nf(r_i)/r_i)$, where $f(r_i)$ is a bound on the boundary size of every level-i region. We use optimal planar separators; hence, $f(r_i) = \mathcal{O}(\sqrt{r_i})$.

it suffices to argue that, for the maximum j_0 such that $r_{j_0} \leq B$, we obtain $S_{j_0} = \mathcal{O}(\frac{N}{B^{1/2-\epsilon}})$. For this j_0, we have $r_{j_0+1} = a^{j_0+1} \cdot 2^{a^{j_0+1}/(a-1)} > B$, that is, $r_{j_0} \geq (r_{j_0+1}/a)^{1/a} = \Omega(B^{1/a})$. On the other hand, we have $j_0 \leq \log_a((a-1)\log B) = \mathcal{O}(\log \log B)$ because $r_{j_0} \leq B$. Hence, $T_1(N) = \mathcal{O}(\sum_{i \geq j_0} S_i) = \mathcal{O}\left(\frac{N\sqrt{\log B} \log \log B}{B^{1/2a}}\right) = \mathcal{O}\left(\frac{N}{B^{1/2-\epsilon}}\right)$.

Priority Queue Operations and Insertions into BRT's. The cost of a priority queue operation or an Insert operation on a BRT of a level-j region is $\mathcal{O}(\frac{\log r_j}{B})$. The number of priority queue operations is bounded by the number of insertions into the BRT. Hence, it suffices to bound the cost of the latter.

Every insertion into a BRT is the result of the relaxation of an edge (u, v). Such a relaxation triggers updates of the priorities of both out-edges of v and their ancestors and an Update-Target operation on the other in-edge of v. We argue about the cost of updates on the out-edges; the cost of the Update-Target operations can be bounded in a similar manner. In the worst case, the lowest common ancestor of edge (u, v) and an out-edge (v, w) of v is the root. Then the update of (v, w) traverses all levels in the partition and triggers one insertion at every level. Hence, the cost per level-0 invocation is at most $\sum_{j \geq 0} \mathcal{O}(\frac{\log r_j}{B})$, which is $\mathcal{O}(\frac{\log r_k}{B}) = \mathcal{O}(\frac{\log N}{B})$ because the graph sizes r_j increase doubly exponentially. Hence, the total cost of priority queue operations and BRT insertions is $T_2(N) = \mathcal{O}(\frac{S_0}{B} \log N) = \mathcal{O}(\frac{N}{B} \log N)$.

Extractions from BRT's. Every Extract operation on the BRT of a level-j region costs $\mathcal{O}(\log r_i + K/B)$ block transfers, where K is the number of extracted elements. Every extracted element must have been inserted before, and we have argued that every level-0 invocation triggers at most three insertions per level. The number of extract operations at level j equals the number of invocations at level $j-1$. Hence, $T_3(N) = \mathcal{O}(\sum_{i > j_0} S_{i-1} \log r_i) + \mathcal{O}(\frac{S_0}{B} \log \log N)$. Using similar arguments as the ones we used to bound the cost of invocation swaps, this simplifies to $T_3(N) = \mathcal{O}(S_{j_0} \log B^a) + \mathcal{O}(\frac{N}{B} \log \log N) = \mathcal{O}(\frac{N}{B^{1/2-\epsilon}} + \frac{N}{B} \log \log N)$.

Stack Operations. Every update incurs at most a constant number of stack operations at every level. Hence, the cost of all stack operations is $T_4(N) = \mathcal{O}(\frac{S_0}{B} \log \log N) = \mathcal{O}(\frac{N}{B} \log \log N)$.

7 Open Problems

The most interesting open questions are the following: (1) Can the required separator decomposition be computed cache-obliviously? (2) Can the $\mathcal{O}(\frac{N}{B^{1/2-\epsilon}})$-term in the complexity of the algorithm be reduced to $o(N/B^{1/2})$? (3) Can the $\mathcal{O}(\frac{N}{B} \log N)$-term be reduced to $\mathcal{O}(N/B)$ (the equivalent of the linear time bound obtained in [19])?

To answer question (1), we believe that the contraction-based separator algorithm of [20] can be extended to compute the desired partition: At every contraction level, use the (by a constant factor suboptimal) separator produced at the previous level to compute a BFS-tree of the current level. Then use this BFS-tree to obtain an optimal separator for the current level (which is constant-factor suboptimal for the next level), and iterate.

The answer to question (2) may be yes, but not using purely separator-based ideas because these algorithms trade-off the number of times a region is loaded into cache against a certain amount of wasteful computation in each region. While the I/O-model can ignore the latter, a cache-oblivious algorithm cannot and must therefore balance this trade-off, which is what our choice of parameters in the algorithm of [19] achieves. The answer to question (3) is most definitely no because this would violate the $\Omega(\mathrm{perm}(N))$ lower bound for shortest paths [13], which is $\Omega(\mathrm{sort}(N))$ in the cache-oblivious model.

References

1. A. Aggarwal and J. S. Vitter. The input/output complexity of sorting and related problems. *Comm. ACM*, pp. 1116–1127, 1988.
2. L. Arge, M. A. Bender, E. Demaine, B. Holland-Minkley, and J. I. Munro. Cache-oblivious priority queue and graph algorithm applications. *Proc. 34th STOC*, pp. 268–276, 2002.
3. L. Arge, G. S. Brodal, and L. Toma. On external-memory MST, SSSP and multi-way planar graph separation. *J. Alg.*, 53:186–206, 2004.
4. L. Arge, U. Meyer, and L. Toma. External memory algorithms for diameter and all-pairs shortest-paths on sparse graphs. *Proc. 31st ICALP*, LNCS 3142, pp. 146–157. Springer-Verlag, 2004.
5. L. Arge, U. Meyer, L. Toma, and N. Zeh. On external-memory planar depth first search. *J. Graph Alg. and Appl.*, 7(2):105–129, 2003.
6. L. Arge and L. Toma. Simplified external memory algorithms for planar DAGs. *Proc. 9th SWAT*, LNCS 3111, pp. 493–503. Springer-Verlag, 2004.
7. L. Arge, L. Toma, and N. Zeh. I/O-efficient algorithms for planar digraphs. *Proc. 15th SPAA*, pp. 85–93. 2003.
8. L. Arge and N. Zeh. I/O-efficient strong connectivity and depth-first search for directed planar graphs. *Proc. 44th FOCS*, pp. 261–270, 2003.
9. G. S. Brodal and R. Fagerberg. Cache oblivious distribution sweeping. *Proc. 29th ICALP*, LNCS 2380, pp. 426–438. Springer-Verlag, 2002.
10. G. S. Brodal and R. Fagerberg. Funnel heap—a cache oblivious priority queue. *Proc. 13th ISAAC*, LNCS 2518, pp. 219–228. Springer-Verlag, 2002.
11. G. S. Brodal and R. Fagerberg. On the limits of cache-obliviousness. *Proc. 35th STOC*, pp. 307–315, 2003.
12. G. S. Brodal, R. Fagerberg, U. Meyer, and N. Zeh. Cache-oblivious data structures and algorithms for undirected breadth-first search and shortest paths. *Proc. 9th SWAT*, LNCS 3111, pp. 480–492. Springer-Verlag, 2004.
13. Y.-J. Chiang, M. T. Goodrich, E. F. Grove, R. Tamassia, D. E. Vengroff, and J. S. Vitter. External-memory graph algorithms. *Proc. 6th SODA*, pp. 139–149, 1995.
14. R. A. Chowdhury and V. Ramachandran. Cache-oblivious shortest paths in graphs using buffer heap. *Proc. 16th SPAA*, pp. 245–254, 2004.
15. E. W. Dijkstra. A note on two problems in connection with graphs. *Num. Math.*, 1:269–271, 1959.

16. J. Fakcharoenphol and S. Rao. Planar graphs, negative weight edges, shortest paths, near linear time. *Proc. 42nd FOCS*, pp. 232–241, 2001.
17. G. N. Frederickson. Fast algorithms for shortest paths in planar graphs, with applications. *SIAM J. Comp.*, 16:1004–1022, 1987.
18. M. Frigo, C. E. Leiserson, H. Prokop, and S. Ramachandran. Cache-oblivious algorithms. *Proc. 40th FOCS*, pp. 285–297, 1999.
19. P. Klein, S. Rao, M. Rauch, and S. Subramanian. Faster shortest path algorithms for planar graphs. *J. Comp. Sys. Sci.*, 55:3–23, 1997.
20. A. Maheshwari and N. Zeh. I/O-optimal algorithms for planar graphs using separators. *Proc. 13th SODA*, pp. 372–381, 2002.
21. S. Pettie and V. Ramachandran. Computing shortest paths with comparisons and additions. *Proc. 13th SODA*, pp. 267–276, 2002.
22. M. Thorup. Undirected single source shortest paths with positive integer weights in linear time. *J. ACM*, 46:362–394, 1999.
23. M. Thorup. Floats, integers, and single source shortest paths. *J. Alg.*, 35:189–201, 2000.

Cache-Aware and Cache-Oblivious Adaptive Sorting

Gerth Stølting Brodal[1,*], Rolf Fagerberg[2,**], and Gabriel Moruz[1]

[1] BRICS[***], Department of Computer Science,
University of Aarhus, IT Parken,
Åbogade 34, DK-8200 Århus N, Denmark
{gerth, gabi}@daimi.au.dk
[2] Department of Mathematics and Computer Science,
University of Southern Denmark,
Campusvej 55, DK-5230 Odense M, Denmark
rolf@imada.sdu.dk

Abstract. Two new adaptive sorting algorithms are introduced which perform an optimal number of comparisons with respect to the number of inversions in the input. The first algorithm is based on a new linear time reduction to (non-adaptive) sorting. The second algorithm is based on a new division protocol for the GenericSort algorithm by Estivill-Castro and Wood. From both algorithms we derive I/O-optimal cache-aware and cache-oblivious adaptive sorting algorithms. These are the first I/O-optimal adaptive sorting algorithms.

1 Introduction

1.1 Adaptive Sorting

A well known fact concerning sorting is that optimal sorting algorithms perform $\Theta(n \log n)$ comparisons [9–Section 9.1]. However, in practice there are many cases where the input sequences are already nearly sorted, i.e. have low disorder according to some measure [16, 19]. In such cases one can hope for a sorting algorithm to be faster.

In order to quantify the disorder of input sequences, several *measures of presortedness* have been proposed, e.g. see [11, 16, 18]. One of the most commonly considered measures is Inv, the number of inversions in the input, defined by $Inv(X) = |\{(i,j) \mid i < j \wedge x_i > x_j\}|$ for a sequence $X = (x_1, \ldots, x_N)$. Other examples of measures include: $Runs$, the number of boundaries between ascending subsequences; Max, the largest difference between the ranks of an element

---

[*] Supported by the Carlsberg Foundation (contract number ANS-0257/20) and the Danish Natural Science Foundation (SNF).
[**] Partially supported by the Danish Natural Science Foundation (SNF).
[***] Basic Research in Computer Science, www.brics.dk, funded by the Danish National Research Foundation.

in the input and the sorted sequence; *Dis*, the largest distance determined by an inversion. A sorting algorithm is denoted *adaptive* if the time complexity is a function dependent on the size as well as the presortedness of the input sequence [19]. For an overview concerning adaptive sorting, see e.g. the survey by Estivill-Castro and Wood [13].

Manilla [18] introduced the concept of optimality of an adaptive sorting algorithm in the comparison model. An adaptive sorting algorithm S is optimal with respect to some measure of presortedness $\mathcal{D}$, if for some constant $c > 0$ and for all inputs X, the time complexity $T_A(X)$ satisfies

$$T_A(X) \leq c \cdot \max(N, \log|\text{below}(X, \mathcal{D})|),$$

where below$(X, \mathcal{D})$ is the number of permutations of the input sequence Y for which $\mathcal{D}(Y) \leq \mathcal{D}(X)$ and $\log x$ denotes $\log_2 x$. By the usual information theoretic lower bound, this is asymptotically the best possible. In particular, an adaptive sorting algorithm that is optimal with respect to the measure *Inv* performs $\Theta(N(1 + \log(1 + Inv/N)))$ comparisons [15].

1.2 The I/O Model and the Cache-Oblivious Model

Traditionally, the RAM model has been used in the design and analysis of algorithms. It consists of a CPU and an infinite memory, where all memory accesses are assumed to take equal time. However, this model is not always adequate in practice, due to the memory hierarchy found on modern computers. Modern computers have several memory levels, each level having smaller size and access time than the next one. Typically, a desktop computer contains CPU registers, L1, L2, and L3 caches, main memory and hard-disk. The access time increases from one cycle for registers and level 1 cache to around 10, 100 and 10,000,000 cycles for level 2 cache, main memory and disk, respectively. Therefore, the I/Os of the disk often become a bottleneck with respect to the running time of a given algorithm, and the number of I/Os, not CPU cycles, should be minimized.

Several models have been proposed to capture the effect of memory hierarchies. The most successful of these is the *I/O model*, introduced by Aggarwal and Vitter [1]. It models a simple two-level memory hierarchy consisting of a fast memory of size M and a slow infinite memory. The data transfers between the slow and fast memory are performed in *blocks* of size B of consecutive data. The I/O complexity of an algorithm is the number of transfers it performs between the slow and the fast memories. A comprehensive list of I/O efficient algorithms for different problems have been proposed, e.g. see the surveys by Vitter [21] and Arge [2]. Among the fundamental results concerning the I/O model is that sorting a sequence of size N requires $\Theta(\frac{N}{B} \log_{\frac{M}{B}} \frac{N}{B})$ I/Os [1].

The I/O model assumes that the size M of the fast memory and the block size B are known, which does not always hold in practice. Moreover, as the modern computers have multiple memory levels with different sizes and block sizes, different parameters are required at the different memory levels. Frigo et al. [14] proposed the *cache-oblivious model*, which is similar to the I/O model, but assumes no knowledge about M and B. In short, a cache-oblivious algorithm is an algorithm described in the RAM model, but analyzed in the I/O model with an analysis valid for any values of M and B. The power of this model is that if a cache-oblivious algorithm performs well on a two-level memory hierarchy with arbitrary parameters, it performs well between all the consecutive levels of a multi-level memory hierarchy.

Many problems have been addressed in the cache-oblivious model (see the surveys by Arge et al. [3], Brodal [6], and Demaine [10]). Among these there are several optimal cache-oblivious sorting algorithms. Frigo et al. [14] gave two optimal cache-oblivious algorithms for sorting: *Funnelsort* and a variant of *Distributionsort*. Brodal and Fagerberg [7] introduced a simplified version of Funnelsort, *Lazy Funnelsort*. The I/O complexity of all these sorting algorithms is $O(\frac{N}{B} \log_{\frac{M}{B}} \frac{N}{B})$. All these algorithms require a *tall cache* assumption, i.e. $M = \Omega(B^{1+\varepsilon})$ for a constant $\varepsilon > 0$. In [8] it is shown that a tall cache-assumption is required for all optimal cache-oblivious sorting algorithms.

1.3 Results and Outline of Paper

In Section 2 we apply the lower bound technique from [4] to obtain lower bounds on the number of I/Os for comparison based sorting algorithms that are adaptive with respect to different measures of presortedness.

In Section 3 we present a linear time reduction from adaptive sorting to general (non-adaptive) sorting, directly implying comparison optimal and I/O-optimal cache-aware and cache-oblivious algorithms with respect to measure *Inv*.

In Section 4 we describe a cache-aware generic sorting algorithm, *cache-aware GenericSort* based on *GenericSort*, introduced in [12], and characterize its I/O adaptiveness. Section 5 introduces a cache-oblivious version of GenericSort.

In Section 6 we introduce a new greedy division protocol for GenericSort, interesting in its own right due to its simplicity. We prove that the resulting algorithm, *GreedySort*, is comparison optimal with respect to measure *Inv*. We show that using our division protocol we obtain both cache-aware and cache-oblivious algorithms that are optimal with respect to *Inv*.

In the remainder of this paper, sorted means sorted in increasing order.

2 I/O Lower Bounds

In this section we show lower bounds on the number of I/Os performed by comparison based sorting algorithms that are adaptive with respect to several measures of presortedness.

Theorem 1. *A comparison based sorting algorithm performs* $\Omega(\frac{N}{B}(1+\log_{\frac{M}{B}}(1+\frac{Inv}{N})))$ *I/Os for sorting input sequences of size N and Inv inversions, assuming $M = \Omega(B^2)$.*

Measure of presortedness	I/Os	Comparisons [13]
Dis	$\Omega\left(\frac{N}{B}\left(1+\log_{\frac{M}{B}}(1+Dis)\right)\right)$	$\Omega(N(1+\log(1+Dis)))$
Exc	$\Omega\left(\frac{N}{B}\left(1+Exc\log_{\frac{M}{B}}(1+Exc)\right)\right)$	$\Omega(N+Exc\log(1+Exc))$
Enc	$\Omega\left(\frac{N}{B}\left(1+\log_{\frac{M}{B}}(1+Enc)\right)\right)$	$\Omega(N(1+\log(1+Enc)))$
Inv	$\Omega\left(\frac{N}{B}\left(1+\log_{\frac{M}{B}}\left(1+\frac{Inv}{N}\right)\right)\right)$	$\Omega\left(N\left(1+\log\left(1+\frac{Inv}{N}\right)\right)\right)$
Max	$\Omega\left(\frac{N}{B}\left(1+\log_{\frac{M}{B}}(1+Max)\right)\right)$	$\Omega(N(1+\log(1+Max)))$
Osc	$\Omega\left(\frac{N}{B}\left(1+\log_{\frac{M}{B}}\left(1+\frac{Osc}{N}\right)\right)\right)$	$\Omega\left(N\left(1+\log\left(1+\frac{Osc}{N}\right)\right)\right)$
Reg	$\Omega\left(\frac{N}{B}\left(1+\log_{\frac{M}{B}}(1+Reg)\right)\right)$	$\Omega(N(1+\log(1+Reg)))$
Rem	$\Omega\left(\frac{N}{B}\left(1+Rem\log_{\frac{M}{B}}(1+Rem)\right)\right)$	$\Omega(N+Rem\log(1+Rem))$
Runs	$\Omega\left(\frac{N}{B}\left(1+\log_{\frac{M}{B}}(1+Runs)\right)\right)$	$\Omega(N(1+\log(1+Runs)))$
SMS	$\Omega\left(\frac{N}{B}\left(1+\log_{\frac{M}{B}}(1+SMS)\right)\right)$	$\Omega(N(1+\log(1+SMS)))$
SUS	$\Omega\left(\frac{N}{B}\left(1+\log_{\frac{M}{B}}(1+SUS)\right)\right)$	$\Omega(N(1+\log(1+SUS)))$

Fig. 1. Lower bounds on the number of I/Os and the number of comparisons

Proof. Consider an adaptive sorting algorithm A and some input sequence X of size N. Let $T_A(X)$ and $I/O_A(X)$ denote the number of comparisons and the number of I/Os performed by a comparison based sorting algorithm A for sorting an input sequence X respectively.

Recall that $below(X, Inv)$ denotes the set of all permutations Y for the input sequence with $Inv(Y) \le Inv(X)$. Consider the decision tree of A (see e.g. [9–Section 9.1]) restricted to the inputs in $below(X, Inv)$. The tree has at least $|below(X, Inv)|$ leaves and therefore A performs at least $\log|below(X, \mathcal{D})|$ comparisons in the worst case. Therefore, for any sequence X, there is a sequence $Y \in below(X, Inv)$, such that $\log|below(X, Inv)| \le T_A(Y)$.

Using the decision tree translation by Arge et al. [4–Theorem 1] we get:

$$\log(|below(X, Inv)|) \le N\log B + \max_{Y\in below(X, Inv)} I/O_A(Y)\left(B\log\left(\frac{M}{B}\right)+3B\right).$$

Since $\log(|below(X, Inv)|) = \Omega(N(1+\log(1+\frac{Inv}{N})))$ [15], we obtain that $\max_{Y\in below(X, Inv)} I/O_A(Y) = \Omega(\frac{N}{B}(1+\log_{\frac{M}{B}}(1+\frac{Inv}{N})))$, given $M = \Omega(B^2)$. □

Using a similar technique we obtain lower bounds on the number of I/Os for other measures of presortedness, assuming that $M = \Omega(B^2)$. Figure 1 lists these lower bounds. For definitions of the different measures, refer to [13].

3 GroupSort

In this section we describe a reduction to derive *Inv* adaptive sorting algorithms from non-adaptive sorting algorithms. The reduction is cache-oblivious and requires $O(N)$ comparisons and $O(N/B)$ I/Os.

The basic idea is to distribute the input sequence into a sequence of buckets $S_1, \ldots, S_k$ each of size at most $32(Inv/N)^2$, where the elements in bucket S_i are all smaller than or equal to the elements in S_{i+1}. Each S_i is then sorted independently by a non-adaptive cache-oblivious sorting algorithm [7,14]. During the construction of the buckets $S_1, \ldots, S_k$ some elements might fail to get inserted into an S_i and are instead inserted into a *fail set* F. It will be guaranteed that at most half of the elements are inserted into F. The fail set F is sorted recursively and merged with the sequence of sorted buckets.

The S_i buckets are constructed by scanning the input left-to-right by inserting an element x into the rightmost bucket S_k if $k = 1$ or $x \geq \min(S_k)$ and otherwise inserting x in F. During the construction we generate increasing bucket capacities $\beta_j = 2 \cdot 4^j$, which will be used for $\alpha_j = N/(2 \cdot 2^j)$ insertions into F. If during construction $|S_k| > \beta_j$, the bucket S_k is split into two buckets S_k and S_{k+1} by computing its median using the cache-oblivious selection algorithm from [5] and distributing its elements relatively to the median. This ensures $|S_i| \leq \beta_j$ for $1 \leq i \leq k$. We maintain the invariant $|S_k| \geq \beta_j/2$ if there are at least two buckets by repeatedly concatenating the two last buckets after an increment of i. Since $\beta_{j-1} = \beta_j/4$, this ensures $\beta_j/2 \leq |S_k| \leq \frac{3}{4}\beta_j$ after this concatenation process. If only one bucket remains, then $|S_k| \leq \frac{3}{4}\beta_j$.

The pseudo-code of the reduction is given in Figure 2. We assume that $S_1, \ldots, S_k$ are stored consecutively in an array by storing the start index and the minimum element from each bucket on a separate stack, i.e. the concatenation of S_{k-1} and S_k can be done implicitly in $O(1)$ time. The fail set F is stored as a list of subsets $F_1, \ldots, F_j$, where F_i stores the elements inserted into F while the bucket size is β_i. Similarly $F_1, \ldots, F_j$ are stored consecutively in an array.

Theorem 2. *GroupSort is cache-oblivious and is comparison optimal and I/O-optimal with respect to Inv, assuming $M = \Omega(B^2)$.*

Proof. Consider the last bucket capacity β_j and fail set size α_j. Each element x inserted into the fail set F_j induces in the input sequence at least $\beta_j/2$ inversions, since $|S_k| \geq \beta_j/2$ when x is inserted into F_j and all elements in S_k appeared before x in the input and are larger than x.

For $i = \lceil \log \frac{Inv}{N} \rceil + 1$, we have $\alpha_i \cdot \frac{\beta_i}{2} = \frac{N}{2 \cdot 2^i} \cdot \frac{2 \cdot 4^i}{2} \geq Inv$, i.e. F_i is guaranteed to be able to store all failed elements. This immediately leads to $j \leq \lceil \log \frac{Inv}{N} \rceil + 1$, and $\beta_j = 2 \cdot 4^j \leq 32 \left(\frac{Inv}{N}\right)^2$. The fail set F has size at most $\sum_{i=1}^{j} \alpha_i = \sum_{i=1}^{j} N/(2 \cdot 2^i) \leq N/2$.

Taking into account that the total size of the fail sets is at most $N/2$, the number of comparisons performed by GroupSort is given by the following recurrence:

procedure GroupSort(X)
Input: Sequence $X = (x_1, \ldots, x_N)$
Output: Sequence X sorted
begin
    $S_1 = (x_1); F_1 = (); \beta_1 = 8; \alpha_1 = N/4; j = 1; k = 1;$
    for $i = 2$ **to** N
        if $k = 1$ **or** $x_i \geq \min(S_k)$
            append(S_k, x_i);
            if $|S_k| > \beta_j$
                $(S_k, S_{k+1}) = \text{split}(S_k); k = k+1;$
        else
            append(F_j, x_i);
            if $|F_j| > \alpha_j$
                $\beta_{j+1} = \beta_j \cdot 4; \alpha_{j+1} = \alpha_j/2; j = j+1;$
                while $k > 1$ **and** $|S_k| < \beta_j/2$
                      $S_{k-1} = \text{concat}(S_{k-1}, S_k); k = k-1;$
    $S = \text{concat}(\text{sort}(S_1), \text{sort}(S_2), \ldots, \text{sort}(S_k));$
    $F = \text{concat}(F_1, F_2, \ldots, F_j);$
    GroupSort(F);
    $X = \text{merge}(S, F);$
end

Fig. 2. Linear time reduction to non-adaptive sorting

$$T(N) = T\left(\frac{N}{2}\right) + \sum_{i=1}^{k} T_{\text{Sort}}(|S_i|) + O(N),$$

where the $O(N)$ term accounts for the bucket splittings and the final merge of S and F. The $O(N)$ term for splitting buckets follows from that when a bucket with β_j elements is split then at least $\beta_j/4$ elements in a bucket have been inserted since the most recent bucket splitting or increase in bucket capacity, and we can charge the splitting of the bucket to these recent $\beta_j/4$ elements.

Since $T_{\text{Sort}}(N) = O(N \log N)$ and each $|S_i| \leq \beta_j = O((\frac{Inv}{N})^2)$ the number of comparisons performed by GroupSort is:

$$T(N) = T\left(\frac{N}{2}\right) + O\left(N\left(1 + \log\left(1 + \left(\frac{Inv}{N}\right)^2\right)\right)\right).$$

Since F is a subsequence of the input, Inv for the recursive call is at most Inv for the input. As $\sum_{i=0}^{\infty} \frac{N}{2^i} \log \frac{Inv}{N/2^i} = N \log \frac{Inv}{N} \sum_{i=0}^{\infty} \frac{1}{2^i} + N \sum_{i=0}^{\infty} \frac{i}{2^i}$, it follows that GroupSort performs $T(N) = O\left(N\left(1 + \log\left(1 + \frac{Inv}{N}\right)\right)\right)$ comparisons, which is optimal.

The cache-oblivious selection algorithm from [5] performs $O(N/B)$ I/Os and the cache-oblivious sorting algorithms [7, 14] perform $O(\frac{N}{B} \log_{\frac{M}{B}} \frac{N}{B})$ I/Os for $M = \Omega(B^2)$. Since GroupSort otherwise does sequential access to the input and data structures, we get that GroupSort is cache-oblivious and the number of I/Os performed is given by the recurrence:

$$I/O(N) = I/O\left(\frac{N}{2}\right) + O\left(\frac{N}{B}\left(1 + \log_{\frac{M}{B}}\left(1 + \left(\frac{Inv}{N}\right)^2 \cdot \frac{1}{B}\right)\right)\right).$$

It follows that GroupSort performs $O(\frac{N}{B}(1 + \log_{\frac{M}{B}}(1 + \frac{Inv}{N})))$ I/Os provided $M = \Omega(B^2)$, which by Theorem 1 is I/O-optimal. □

Pagh et al. [20] gave a related reduction for adaptive sorting on the RAM model. Their reduction assumes that a parameter q is provided such that the number of inversions is at most qN. A valid q is found by selecting increasing values for q such that the running time doubles for each iteration. In the cache oblivious setting the doubling approach fails, since the first q value should depend on the unknown parameter M. We circumvent this limitation of the doubling technique by selecting the increasing β_j values internally in the reduction.

4 Cache-Aware GenericSort

Estivill-Castro and Wood [12] introduced a generic sorting algorithm, *GenericSort*, as a framework for adaptive sorting algorithms. It is a generalization of Mergesort, and is described using a generic division protocol, i.e. an algorithm for splitting an input sequence into two or more subsequences. The algorithm works as follows: consider an input sequence X; if X is sorted then the algorithm returns; if X is "small", then X is sorted using some alternate non-adaptive sorting algorithm; otherwise, X is divided according to the division protocol and the resulting subsequences are recursively sorted and merged.

In this section we modify GenericSort to achieve a generic I/O-adaptive sorting algorithm. Consider an input sequence $X = (x_1, \ldots, x_N)$ and some division protocol DP such that DP splits the input in $s \geq 2$ subsequences of roughly equal sizes in a single scan, visiting each element of the input exactly once. To avoid testing whether X is sorted before applying the division protocol, we derive a new division protocol DP' by modifying DP to identify the longest sorted prefix of X: we scan the input sequence until we find some i such that $x_i < x_{i-1}$. Denote $S = (x_1, \ldots, x_{i-1})$ and $X' = (x_i, \ldots, x_N)$. We apply DP to X', recursively sort the resulting s subsequences, and finally merge them with S. The adaptive bounds for GenericSort proved in [12-Theorem 3.1] are not affected by these modifications, and we have the following theorem.

Theorem 3. *Let $\mathcal{D}$ be a measure of presortedness, d and s constants, $0 < d < 2$, and DP a division protocol that splits some input sequence of size N into s subsequences of size at most $\lceil \frac{N}{s} \rceil$ each using $O(N)$ comparisons.*

- *the modified GenericSort performs $O(N \log N)$ comparisons in the worst case;*

- if for all sequences X, the division of a suffix of X into $X_1, \ldots, X_s$ by DP satisfies that $\sum_{j=1}^{s} \mathcal{D}(X_j) \leq d \lfloor \frac{s}{2} \rfloor \cdot \mathcal{D}(X)$, then the modified GenericSort performs $O\left(N\left(1 + \log(1 + \mathcal{D}(X))\right)\right)$ comparisons.

We now describe a cache-aware version of the modified GenericSort provided that the division protocol DP works in a single scan of the input. Let T be the recursion tree of GenericSort using the new division protocol DP'. We obtain a new tree T' by contracting T top-down such that every node in T' corresponds to a subtree of height $O(\log_s(M/B))$ in T and each node in T' has a fanout of at most m, where $m = \Theta(M/B)$. There are $O(m)$ sorted prefixes for every node in T'. In cache-aware GenericSort, for each node of T' we scan its input sequence and distribute the elements accordingly to one of the $O(m)$ output sequences. Each output sequence is a linked list of blocks of size $\Theta(B)$. If the size of the input sequence is at most M, then we sort it in internal memory, hence performing $O(N/B)$ I/Os. Theorem 4 gives a characterization of the adaptiveness of cache-aware GenericSort in the I/O model. It is an I/O version of Theorem 3.

Theorem 4. *Let $\mathcal{D}$ be a measure of presortedness, d and s constants, $0 < d < 2$ and $s \leq \frac{M}{2B}$, and DP a division protocol that splits some input sequence of size N into s subsequences of size at most $\lceil \frac{N}{s} \rceil$ each using $O(\frac{N}{B})$ I/Os. If DP performs the splitting in one scan visiting each element of the input exactly once, then:*

- *cache-aware GenericSort performs $O(\frac{N}{B} \log_{\frac{M}{B}} \frac{N}{B})$ I/Os in the worst case;*
- *if for all sequences X, the division of a suffix of X into $X_1, \ldots, X_s$ by DP satisfies that $\sum_{j=1}^{s} \mathcal{D}(X_j) \leq d \lfloor \frac{s}{2} \rfloor \cdot \mathcal{D}(X)$, then cache-aware GenericSort performs $O\left(\frac{N}{B}\left(1 + \log_{\frac{M}{B}}(1 + \mathcal{D}(X))\right)\right)$ I/Os.*

Proof. We analyze the I/Os performed at the nodes of T' separately for the nodes having input sizes less than or equal to M and greater than M.

At a node with input X and $|X| > M$, $O(m + |X|/B) = O(|X|/B)$ I/Os are performed to read the input and to write to the at most $m - 1$ sorted output prefixes and m sequences to be recursively sorted. If we charge $O(1/B)$ I/Os per element in the input this will pay for the I/Os required at the node.

At a node with input X and $|X| \leq M$, $O(1 + |X|/B)$ I/Os are performed. These I/Os can be charged to the parent node, since at the parent we will already charge $O(1 + |X|/B)$ I/Os to write the output X.

By Theorem 3 we have that the sum of the depths in T reached by the elements in the input X is bounded by $O(N(1 + \log(1 + \mathcal{D}(X))))$. Since each node in T' spans $\Theta(\log \frac{M}{B})$ levels from T, we get that cache-aware GenericSort performs $O(\frac{N}{B} + N(1 + \log(1 + \mathcal{D}(X)))/(B \log \frac{M}{B})) = O(\frac{N}{B}(1 + \log_{\frac{M}{B}}(1 + \mathcal{D}(X))))$ I/Os, where the N/B term counts for the I/Os at the root of T'. □

The power of cache-aware GenericSort lies in its generality, meaning that using different division protocols we obtain sorting algorithms that are I/O adaptive with respect to different measures of presortedness. For example, using the straight division protocol, we achieve I/O optimality with respect to *Runs*. Using the odd-even division protocol, we obtain an algorithm that is I/O optimal with respect to *Dis* and *Max*. Furthermore, the different division protocols can be combined as shown in [13] in order to achieve I/O optimality with respect to more measures of presortedness.

5 Cache-Oblivious GenericSort

We give a cache-oblivious algorithm that achieves the same adaptive bounds as the cache-aware GenericSort introduced in Section 4. It works only for division protocols that split the input into two unsorted subsequences. It is based on a modification of the k-merger used in FunnelSort [7, 14].

A k-merger is a binary tree stored using the recursive van Emde Boas layout. The edges contain buffers of variable sizes and the nodes are binary mergers. The tree and the buffer sizes are recursively defined: consider an output sequence of size k^3 and h the height of the tree. We split the tree at level $\frac{h}{2}$ yielding $k^{\frac{1}{2}}+1$ subtrees, each of size $O(k^{\frac{1}{2}})$. The buffers at this level have sizes $k^{\frac{3}{2}}$. See [7] for further details.

Consider DP division protocol that scans the input a single time and DP' the modified DP as introduced in Section 4. Each node of the k-merger corresponds to a node in the recursion tree of GenericSort using DP' as the division protocol. Therefore, each node has a fanout of three and becomes a ternary merger. The resulting unsorted sequences are pushed in the buffers to the children, while the sorted prefix is stored as a list of memory chunks of size $O(N^{\frac{2}{3}})$ for an input buffer of size N.

Our algorithm uses a single $N^{\frac{1}{3}}$-merger. It fills the buffers in a top-down fashion and then merges the resulted sorted subsequences in a bottom-up manner. The $N^{\frac{1}{3}}$ output buffers at the leaves of the k-merger are sorted using a non-adaptive I/O-optimal cache oblivious sorting algorithm [7, 14].

Lemma 1. *The $N^{\frac{1}{3}}$-merger and the sorted subsequences use $O(N)$ space.*

Proof. Consider the $N^{\frac{1}{3}}$-merger and an input sequence of size N. The total size of the inner buffers is $O(N^{\frac{2}{3}})$ [14]. The memory chunks storing the sorted subsequences use $O(N)$ space because there are $N^{\frac{1}{3}}$ nodes in the merger and the size of a single memory chunk is $O(N^{\frac{2}{3}})$. Adding the input sequence, we conclude that the $N^{\frac{1}{3}}$-merger and the sorted subsequences take $O(N)$ space together. □

Lemma 2. *Cache-oblivious GenericSort and cache-aware GenericSort have the same comparison and I/O complexity, for division protocols that split the input into two subsequences.*

Proof. Consider $\ell = \frac{1}{3}\log N$ the height of the $N^{\frac{1}{3}}$-merger of the cache-oblivious GenericSort.

We first prove that cache-aware and cache-oblivious GenericSort have the same comparison complexity. For some element x_i let d_i be its depth in the recursion tree of the GenericSort using DP' as a division protocol. If $d_i \leq \ell$ then x_i reaches the same level in the recursion tree of cache-oblivious GenericSort, because the two algorithms have the same recursion trees at the top ℓ levels. If $d_i > \ell$ then the number of comparisons performed by cache-oblivious GenericSort for x_i is $O(\log N) = O(d_i)$ because $d_i > l = \Omega(\log N)$.

We analyze the number of I/Os used by cache-aware and cache-oblivious GenericSort. Consider an element x_i that reaches level d_i in the recursion tree of cache-aware GenericSort.

If $d_i < \ell$ then x_i is placed in a sorted prefix at a node in the $N^{\frac{1}{3}}$-merger. In this case, cache-oblivious GenericSort spends linear I/Os when the size of the input reaches $O(M)$ because the $N^{\frac{1}{3}}$-merger together with the sorted subsequences take linear space by Lemma 1. Taking into account that the height of the $N^{\frac{1}{3}}$-merger is $O(\log(M/B))$ due to the tall cache assumption, it follows that $O(1 + d_i/(\log(M/B)))$ I/Os are performed by cache-oblivious GenericSort for getting x_i to its sorted subsequence.

If $d_i > \ell$ then x_i reaches an output buffer of the $N^{\frac{1}{3}}$-merger, where it is sorted using an optimal cache-oblivious sorting algorithm. In this case the number of I/Os performed for the sorting involving x_i is still $O(1/B + d_i/(B \log(M/B)))$, because both the $N^{\frac{1}{3}}$-merger and the optimal sorting algorithms require $O(1/B + d_i/(B \log(M/B)))$ I/Os for the sorting involving x_i, since $d_i = \Theta(\log N)$.

We obtain that the number of I/Os performed by cache-oblivious GenericSort is $O\left(\frac{N}{B} + \frac{\sum_{i=1}^{n} d_i}{B \log(M/B)}\right)$. Cache-aware GenericSort performs $O\left(\frac{N}{B} + \frac{\sum_{i=1}^{n} d_i}{B \log \frac{M}{B}}\right)$ I/Os too because the fanout of the nodes in the recursion tree is $O(\log \frac{M}{B})$. We conclude that cache-aware GenericSort and cache-oblivious GenericSort have the same I/O complexity. □

6 GreedySort

We introduce *GreedySort*, a sorting algorithm based on GenericSort using a new division protocol, *GreedySplit*. The protocol is inspired by a variant of the Kim-Cook division protocol, which was introduced and analyzed in [17]. Our division protocol achieves the same adaptive performance with respect to *Inv*, but is simpler and moreover facilitates cache-aware and cache-oblivious versions. It may be viewed as being of a greedy type, hence the name. We first describe GreedySort and its division protocol and then prove that it is optimal with respect to *Inv*. GreedySplit partitions the input sequence X into three subsequences S, Y, and Z, where S is sorted and Y and Z have balanced sizes, i.e. $|Z| \leq |Y| \leq |Z| + 1$. In one scan it builds an ascending subsequence S of the input in a greedy fashion and at the same time distributes the remaining elements in two subsequences, Y and Z, using an odd-even approach.

Lemma 3. *GreedySplit splits an input sequence X in the three subsequences S, Y and Z, where S is sorted and $Inv(X) \geq \frac{5}{4} \cdot (Inv(Y) + Inv(Z))$.*

Proof. Let $X = (x_1, \ldots x_N)$. By construction S is sorted. Consider an inversion in Y, $y_i > y_j$, $i < j$ and i_1 and j_1 the indices in X of y_i and y_j respectively. Due to the odd-even construction of Y and Z, there exists an $x_k \in Z$ such that in the original sequence X we have $i_1 < k < j_1$.

We prove that there is one inversion between x_k and at least one of x_{i_1} and x_{j_1}, for any $i_1 < k < j_1$. Indeed, if $x_{i_1} > x_k$, we get an inversion between x_{i_1} and x_k. If $x_{i_1} \leq x_k$, we get an inversion between x_{j_1} and x_k, because we assume that $y_i > y_j$ which yields $x_{i_1} > x_{j_1}$. Let $z_{i_1}, \ldots, z_{j-1}$ be all the elements from Z which appear between y_i and y_j in the original sequence. We know that there exists at least an inversion between $z_{\lfloor i+j \rfloor/2}$ and y_i or y_j. The inversion $(y_i, z_{\lfloor (i+j)/2 \rfloor})$ can be counted for two different pairs in Y, $(y_i, y_{i+2\lfloor (j-i)/2 \rfloor})$ and $(y_i, y_{i+1+2\lfloor (j-i)/2 \rfloor})$. Similarly, the inversion $(z_{\lfloor (i+j)/2 \rfloor, j})$ can be counted for two different pairs in Y. Taking into account that the inversions involving elements of Y and elements of Z appear in X, but neither in Y nor Z, we have that $Inv(X) \geq Inv(Y) + Inv(Z) + Inv(Y)/2$. In a similar manner we obtain $Inv(X) \geq Inv(Y) + Inv(Z) + Inv(Z)/2$. Summing the two equations we obtain $Inv(X) \geq \frac{5}{4}(Inv(Y) + Inv(Z))$. □

Theorem 5. *GreedySort performs $O(N(1 + \log(1 + Inv(X)/N)))$ comparisons to sort a sequence X of size N, i.e. it is comparison optimal with respect to Inv.*

Proof. Similar to [17], we first prove the claimed bound for the upper levels of recursion where the total number of inversions is greater than $N/4$ and then prove that the total number of comparisons for the remaining levels is linear. Let $Inv_i(X)$ denote the total number of inversions in the subsequences at the ith level of recursion. By Lemma 3, $Inv_i(X) \leq \left(\frac{4}{5}\right)^i Inv(X)$.

We want to find the first level ℓ of the recursion for which $\left(\frac{4}{5}\right)^\ell Inv(X) \leq \frac{N}{4}$, which yields $\ell = \left\lceil \frac{\log(4 Inv(X)/N)}{\log(5/4)} \right\rceil$.

At each level of recursion GreedySort performs $O(N)$ comparisons. Therefore at the first ℓ levels of recursion the total number of comparisons performed is $O(\ell \cdot N) = O(N(1 + \log(1 + Inv(X)/N))))$. We now prove that the remaining levels perform a linear number of comparisons.

Let $|(X, i)|$ denote the total size of Ys and Zs at the ith level of recursion. As each element in Y and Z is obtained as a result of an inversion in the sequence X, we have $|(X, i)| \leq Inv_{i-1}(X)$. Using Lemma 3 we obtain: $|(X, \ell+i)| \leq Inv_{\ell+i-1}(X) \leq \left(\frac{4}{5}\right)^{i-1} \cdot \left(\frac{4}{5}\right)^\ell \cdot Inv(X) \leq \left(\frac{4}{5}\right)^{i-1} \frac{N}{4}$. Taking into account that the sum of the $|(X, \ell+i)|$s is $O(N)$ and that at each level $\ell+i$ we perform a linear number of comparisons with respect to $|(X, \ell+i)|$, it follows that the total number of comparisons performed at the lower levels of the recursion tree is $O(N)$. We conclude that GreedySort performs $O(N(1 + \log(1 + \frac{Inv}{N})))$ comparisons. □

We derive both cache-aware and cache-oblivious algorithms by using our greedy division protocol in both the cache-aware and the cache-oblivious

GenericSort frameworks described in Sections 4 and 5. In both cases the division protocol considered does not identify the longest prefix of the input, but simply apply the greedy division protocol. We prove that these new algorithms, *cache-aware GreedySort* and *cache-oblivious GreedySort* achieve the I/O-optimality with respect to *Inv* under the tall cache assumption $M = \Omega(B^2)$.

Theorem 6. *Cache-aware GreedySort and cache-oblivious GreedySort are I/O-optimal with respect to Inv, provided that $M = \Omega(B^2)$.*

Proof. From Theorem 5 the average number of levels of recursion for an element is $O(1 + \log(1 + Inv/N))$. In Theorem 4 each element is charged $O(\frac{1}{B})$ I/Os for every $\Theta(\log \frac{M}{B})$ levels. This implies that cache-aware GreedySort performs $\Theta(\frac{N}{B}(1 + \log_{\frac{M}{B}}(1 + \frac{Inv}{N})))$ I/Os, which is optimal by Theorem 1. Similar observations apply to cache-oblivious GreedySort based on the proof of Lemma 2. □

References

1. A. Aggarwal and J. S. Vitter. The input/output complexity of sorting and related problems. *Communications of the ACM*, 31(9):1116–1127, 1988.
2. L. Arge. External memory data structures. In J. Abello, P. M. Pardalos, and M. G. C. Resende, editors, *Handbook of Massive Data Sets*.
3. L. Arge, G. S. Brodal, and R. Fagerberg. Cache-oblivious data structures. In D. Mehta and S. Sahni, editors, *Handbook of Data Structures and Applications*, page 27. CRC Press, 2004.
4. L. Arge, M. Knudsen, and K. Larsen. A general lower bound on the I/O-complexity of comparison-based algorithms. In *Proc. of Workshop on Algorithms and Data Structures*, 1993.
5. M. Blum, R. W. Floyd, V. Pratt, R. L. Rivest, and R. E. Tarjan. Time bounds for selection. *J. Comput. Syst. Sci.*, 7:448–461, 1973.
6. G. S. Brodal. Cache-oblivious algorithms and data structures. In *Proc. 9th Scandinavian Workshop on Algorithm Theory*, Lecture Notes in Computer Science. Springer Verlag, Berlin, 2004.
7. G. S. Brodal and R. Fagerberg. Cache oblivious distribution sweeping. In *Proc. 29th International Colloquium on Automata, Languages, and Programming*, pages 426–438. Springer Verlag, Berlin, 2002.
8. G. S. Brodal and R. Fagerberg. On the limits of cache-obliviousness. In *Proc. 35th Annual ACM Symposium on Theory of Computing*, pages 307 315, 2003.
9. T. H. Cormen, C. E. Leiserson, R. L. Rivest, and C. Stein. *Introduction to Algorithms, 2nd Edition*. MIT Press, 2001.
10. E. Demaine. Cache-oblivious algorithms and data structures. *Lecture Notes from the EEF Summer School on Massive Data Sets*, 2002.
11. V. Estivill-Castro and D. Wood. A new measure of presortedness. *Information and Computation*, 83(1):111–119, 1989.
12. V. Estivill-Castro and D. Wood. Practical adaptive sorting. In *Advances in Computing and Information - Proc. of the International Conference on Computing and Information*, pages 47–54. Springer-Verlag, 1991.
13. V. Estivill-Castro and D. Wood. A survey of adaptive sorting algorithms. *ACM Computing Surverys*, 24(4):441–475, 1992.

14. M. Frigo, C. E. Leiserson, H. Prokop, and S. Ramachandran. Cache oblivious algorithms. In *40th Ann. IEEE Symp. on Foundations of Computer Science*, pages 285–298, 1999.
15. L. J. Guibas, E. M. McCreight, M. F. Plass, and J. R. Roberts. A new representation of linear lists. In *Proc. 9th Ann. ACM Symp. on Theory of Computing*, pages 49–60, 1977.
16. D. E. Knuth. *The Art of Computer Programming. Vol 3, Sorting and searching*. Addison-Wesley, 1973.
17. C. Levcopoulos and O. Petersson. Splitsort – an adaptive sorting algorithm. *Information Processing Letters*, 39(1):205–211, 1991.
18. H. Manilla. Measures of presortedness and optimal sorting algorithms. *IEEE Trans. Comput.*, 34:318–325, 1985.
19. K. Mehlhorn. *Data structures and algorithms. Vol. 1, Sorting and searching*. Springer, 1984.
20. A. Pagh, R. Pagh, and M. Thorup. On adaptive integer sorting. In *Proc. 12th Annual European Symposium on Algorithms*, volume 3221, pages 556–567. 2004.
21. J. S. Vitter. External memory algorithms and data structures: Dealing with massive data. *ACM Computing Surveys*, 33(2):209–271, 2001.

Simulated Annealing Beats Metropolis in Combinatorial Optimization

Ingo Wegener*

FB Informatik, LS2, Univ. Dortmund, Germany
`ingo.wegener@uni-dortmund.de`

Abstract. The Metropolis algorithm is simulated annealing with a fixed temperature. Surprisingly enough, many problems cannot be solved more efficiently by simulated annealing than by the Metropolis algorithm with the best temperature. The problem of finding a natural example (artificial examples are known) where simulated annealing outperforms the Metropolis algorithm for all temperatures has been discussed by Jerrum and Sinclair (1996) as "an outstanding open problem." This problem is solved here. The examples are instances of the well-known minimum spanning tree problem. Moreover, it is investigated which instances of the minimum spanning tree problem can be solved efficiently by simulated annealing. This is motivated by the aim to develop further methods to analyze the simulated annealing process.

1 Introduction

Simple randomized search heuristics like randomized local search (RLS), the Metropolis algorithm (MA), simulated annealing (SA), evolutionary algorithms (EA), and genetic algorithms (GA) find many applications. One cannot hope that they outperform sophisticated problem-specific algorithms for well-studied problems. They are easy to implement and good alternatives if one does not know efficient problem-specific algorithms and if one shies away from developing a clever algorithm. They are the tool of choice in black-box optimization where the problem instance is hidden from the algorithm. And they are useful as parts of hybrid algorithms combining general search principles with problem-specific modules.

Hence, it is interesting to understand the working principles behind these heuristics. The aim is to analyze the expected optimization time and the success probability within a given time bound of heuristics applied to specific problems. Up to now there are not many of such results. One reason is that the heuristics are not designed to support their analysis (in contrast to many problem-specific

---

* Supported in part by the Deutsche Forschungsgemeinschaft (DFG) as part of the Collaborative Research Center "Computational Intelligence" (SFB 531) and by the German-Israeli Foundation (GIF) in the project "Robustness Aspects of Algorithms".

algorithms). To simplify the problem many authors have first investigated quite artificial problems hoping to develop methods which can be used in many other situations.

Here, we are interested in simulated annealing and the Metropolis algorithm (which can be defined as SA with a fixed temperature). Both algorithms are defined in Section 2. It is an obvious question how to use the freedom to choose a cooling schedule for SA and whether this option is essential. Little is known about this leading Jerrum and Sinclair (1996, page 516) to the following statement: "It remains an outstanding open problem to exhibit a natural example in which simulated annealing with any non-trivial cooling schedule provably outperforms the Metropolis algorithm at a carefully chosen fixed value of α." In their paper, α is the temperature. The notion of a "natural example" is vague, but the known examples are obviously artificial. Sorkin (1991) has proven the considered effect for a so-called fractal energy landscape. The chaotic behavior of this function asks for different temperatures in different phases of the search. The artificial example due to Droste, Jansen, and Wegener (2001) allows a simpler analysis.

Jerrum and Sorkin (1998) have analyzed the Metropolis algorithm for the graph bisection problem. They focus the interest on problems from combinatorial optimization: "Unfortunately no combinatorial optimization problem that has been subjected to rigorous theoretical analysis has been exhibited this phenomenon: those problems that can be solved efficiently by simulated annealing can be solved just as effectively by 'annealing' at a single carefully selected temperature. A rigorous demonstration that annealing is provably beneficial for some natural optimization problems would rate as a significant theoretical advance."

Our problem of choice is the minimum spanning tree problem (MSTP) which is contained in all textbooks on combinatorial optimization and should be accepted as "natural optimization problem." It should be obvious that SA cannot beat MA for each problem instance. E. g., for graphs where all edge weights equal 1 the frozen MA (at temperature 0) cannot be beaten by SA. In Section 3, we describe the notion of efficiency for randomized search heuristics and, in Section 4, we describe simple instances of the MSTP where SA outperforms MA. The underlying graphs will be so-called connected triangles (CT), see Figure 1.

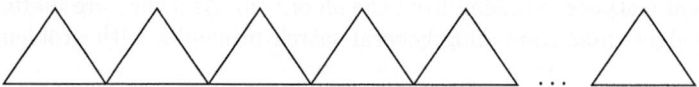

Fig. 1. Graphs called connected triangles.

The idea is to produce examples as simple as possible. This allows proofs which can be taught in introductory courses on randomized search heuristics. Afterwards, we try to understand which instances of the MSTP can be solved efficiently by SA and MA, only by SA, or by none of them. Weights $w_1, \ldots, w_m$ are called $(1 + \varepsilon)$-separated if $w_i > w_j$ implies $w_i \geq (1 + \varepsilon) \cdot w_j$. For each $\varepsilon(m) = o(1)$ there are graphs with $(1 + \varepsilon(m))$-separated weights such that SA cannot attack them efficiently (Section 5). For each constant $\varepsilon > 0$, SA can attack all graphs with $(1 + \varepsilon)$-separated weights efficiently (Section 6). These results imply that SA outperforms MA on a much larger class of graphs than the connected triangles discussed in Section 4. We finish with some conclusions.

It should be obvious that we do not hope that SA or MA beats the well-known algorithms due to Kruskal and to Prim. Again we like to transfer a statement of Jerrum and Sorkin (1998) from minimum bisections to minimum spanning trees (MSTs): "Our main contribution is not, then, to provide a particularly effective algorithm for the minimum bisection problem ..., but to analyze the performance of a popular heuristic applied to a reasonably realistic problem in combinatorial optimization."

2 Metropolis Algorithm, Simulated Annealing, and Minimum Spanning Trees

An instance of the MSTP consists of an undirected graph $G = (V, E)$ with n vertices and m edges and a weight $w(e)$ for each edge e. Weights are positive integers. The problem is to find an edge set E' connecting all vertices with minimal total weight. The edges are numbered and edge sets are described as characteristic vectors, i.e., $x \in \{0,1\}^m$ describes the set of edges e_i where $x_i = 1$. This formalization is well-suited for MA and SA.

We describe the Metropolis algorithm with temperature T for minimization problems on $\{0,1\}^m$. The first search point x is chosen in some way discussed later. Each round of an infinite loop consists of local change and selection.

Local change: Let x be the current search point. Choose $i \in \{1, \ldots, m\}$ uniformly at random and flip x_i, i.e., let $x' = (x'_1, \ldots, x'_m)$ where $x'_j = x_j$, if $j \neq i$, and $x'_i = 1 - x_i$.

Selection of the new current search point with respect to a fitness function f:
if $f(x') \leq f(x)$: select x',
if $f(x') > f(x)$: select x' with probability $\exp\{-(f(x')-f(x))/T\}$, otherwise select x.

We have to discuss some details in order to ensure that our results are not based on too special choices. Randomized search heuristics do not produce a certificate that a search point is optimal. Therefore, the algorithm contains an infinite loop, but the run time is defined as the number of rounds until an optimal search point is produced. A round cannot be performed in time $O(1)$ but quite efficiently and people have agreed to count the number of rounds.

We choose 1^m as starting point. This is similar to the choice 0^m for the maximum matching problem (Sasaki and Hajek (1988)) and the maximum clique problem (Jerrum (1992)). The starting points are the worst legal solutions. This choice of the starting point implies that we cannot apply the lower bound technique for MA due to Sasaki (1991) which ensures only the existence of some bad starting point. It would be an alternative to start with a search point chosen uniformly at random. For many graphs, we then choose a non-legal solution (an unconnected graph) and the fitness function has to contain hints directing the search to legal search points. It is not difficult to obtain similar results in this situation. However, most papers on MA and SA only work on legal search points.

We have chosen the fitness function f where $f(x) = \infty$ for search points x describing unconnected graphs and where $f(x)$ is the total weight of all chosen edges if x describes a connected graph. Unconnected graphs are never accepted as current search points. This again is in accordance with Sasaki and Hajek (1988) and Jerrum (1992). All search points are legal solutions in the graph bisection problem and therefore Jerrum and Sorkin (1993, 1998) start with randomly chosen search points.

We follow Sasaki and Hajek (1988) and Jerrum (1992) in allowing only 1-bit neighborhoods. Neumann and Wegener (2004) have analyzed RLS with 1-bit and 2-bit flips (RLS equals the frozen MA at temperature $T = 0$) and a simple EA for the MSTP. These algorithms do not select new search points which are worse than the old one. Hence, their search strategy is completely different from the strategy applied by MA and SA that have to accept sometimes worsenings to find the optimum. Flips of two bits allow to include an edge into a tree and to exclude simultaneously an edge of the newly created cycle. RLS and the simple EA find an MST in an expected number of $O(m^2(\log m + \log w_{\max}))$ steps, where $w_{\max}$ denotes the maximal weight. Note that we are not looking for a "best" algorithm for the MSTP. The main idea of an elitist EA is to reject worsenings and to escape from local optima by non-local steps. The main idea of MA and SA is to work with very local steps and to escape from local optima by accepting worsenings. The situation here is similar to the case of maximum matchings where also flips of 2 bits are helpful to shorten augmenting paths, compare Sasaki and Hajek (1988) who analyze SA with 1-bit flips only and Giel and Wegener (2003) who analyze RLS with 1-bit and 2-bit flips and a simple EA.

Finally, we introduce SA based on a cooling schedule $T(t)$. The initial temperature $T(1)$ may depend on m and the largest possible weight w_{max}. The temperature $T(t)$ applied by the selection operator in step t equals $\alpha^{t-1} \cdot T(1)$, where $\alpha < 1$ is a constant which may depend on m and an upper bound on w_{max}. This cooling schedule does not include any knowledge about the problem instance. We use a kind of "continuous cooling", other possibilities are longer phases with a constant temperature or dynamic cooling schedules that depend on the success rate (where a step is called successful if x' is selected) or the rate of f-improving steps.

3 Efficiency Measures

There are many well-known convergence results on MA and SA. We want to distinguish "efficient behavior" from non-efficient one. The first idea is to define efficiency as expected polynomial time. We think that this is not a good choice. There may be a small probability of missing a good event for temperatures in some interval $[T_1, T_2]$. For temperatures smaller than T_1 it may be very unlikely that the good event happens. This may cause a superpolynomial or even exponential expected run time although the run time is polynomially bounded with overwhelming probability.

Definition 1. *Let A be a randomized search heuristic (RSH) running for a polynomial number of $p(m)$ rounds and let $s(m)$ be the success probability, i.e., the probability that A finds an optimal search point within this phase. A is called*

- *successful, if $s(m) \geq 1/q(m)$ for some polynomial $q(m)$,*
- *highly successful, if $s(m) \geq 1 - 1/q(m)$ for some polynomial $q(m)$, and*
- *successful with overwhelming probability, if $s(m) = 1 - e^{-\Omega(m^\varepsilon)}$ for some $\varepsilon > 0$.*

One can be satisfied with successful RSHs, since then multistart variants not depending on p and q are successful with overwhelming probability and have an expected polynomial run time. An RSH is called unsuccessful if, for each polynomial p, the success probability within $p(m)$ steps is $o(m^{-k})$ for each constant k. This implies a superpolynomial expected optimization time. Moreover, multistart variants do not help.

4 Simulated Annealing Beats Metropolis on Some Simple Graphs

Our plan is to present simple graphs where SA beats MA for each temperature. The graphs should allow proofs as simple as possible. The idea behind the chosen graphs is the following. The problem to compute an MST on graphs with many two-connected components is separable, i.e., an MST consists of MSTs on the two-connected components. We investigate graphs where each two-connected component can be handled easily by MA with a well-chosen temperature, but different components need different temperatures. To keep the analysis easy the components have constant size. This implies that, for high temperatures, each component can be optimized, but the solutions are not stable. They are destroyed from time to time and then reconstructed. Therefore, it is unlikely that all the components are optimized simultaneously. SA can handle these graphs efficiently.

As announced, we investigate connected triangles (CT), see Figure 1, with $m = 6n$ edges. The number of triangles equals $2n$ and the number of vertices equals $4n + 1$. The weight profile (w_1, w_2, w_3) of a triangle is simply the ordered vector of the three edge weights. We investigate CTs with n triangles with weight profile $(1, 1, m)$ and n triangles with weight profile (m^2, m^2, m^3). The unique MST consists of all edges of weight 1 or m^2.

Theorem 1. *The probability that the Metropolis algorithm applied to CTs with n triangles with weight profile $(1, 1, m)$ and n triangles with weight profile (m^2, m^2, m^3) computes the MST within e^{cm} steps (c a positive constant which is small enough) is bounded above by $e^{-\Omega(m)}$, i.e., MA is unsuccessful on these instances.*

Proof. We distinguish the cases of high temperature $(T \geq m)$ and low temperature $(T < m)$.

The low temperature case is easy. We do not care about the triangles with weight profile $(1, 1, m)$. For each other triangle, MA accepts the exclusion of the first flipping edge. By Chernoff bounds, with probability $1 - 2^{-\Omega(m)}$, we obtain $\Omega(m)$ triangles where the first spanning tree contains the heavy edge. In order to obtain the MST it is necessary to include the missing edge of weight m^2. If this edge is chosen to flip, the probability of selecting the new search point equals $e^{-m^2/T} \leq e^{-m}$. Hence, the success probability within $e^{m/2}$ steps is $e^{-\Omega(m)}$.

In the high temperature case, we do not care about the heavy triangles. For the light triangles, we distinguish between complete triangles (the search point chooses all three edges), optimal triangles (the two weight-1 edges are chosen), and bad triangles. The status of each triangle starts with "complete" and follows a Markov chain with the following transition probabilities:

	complete	optimal	bad
complete	$1 - 3/m$	$1/m$	$2/m$
optimal	$\frac{1}{m} \cdot e^{-m/T}$	$1 - \frac{1}{m} \cdot e^{-m/T}$	0
bad	$\frac{1}{m} \cdot e^{-1/T}$	0	$1 - \frac{1}{m} \cdot e^{-1/T}$

Let X_t be the number of optimal triangles after time step t, i.e., $X_0 = 0$. We are waiting for the first point of time t when $X_t = n$. Obviously, $|X_{t+1} - X_t| \leq 1$. Moreover,

$$\text{Prob}(X_{t+1} = a + 1 \mid X_t = a) \leq \frac{n - a}{m}$$

since it is necessary to flip the heaviest edge in one of the at most $n - a$ complete triangles, and

$$\text{Prob}(X_{t+1} = a - 1 \mid X_t = a) = \frac{a}{m} \cdot e^{-m/T} \geq \frac{a}{3m}$$

since $T \geq m$ and since it is necessary to flip the heaviest edge in one of the optimal triangles and to accept the new search point. Since we are interested in lower bounds, we use the upper bound for the probability of increasing a and the lower bound for the probability of decreasing a. Ignoring steps not changing a, we obtain the following transition probabilities for the new Markov chain Y_t:

$$\text{Prob}(Y_{t+1} = a - 1 \mid Y_t = a) = \frac{a/(3m)}{a/(3m) + (n - a)/m} = \frac{a}{3n - 2a}.$$

There has to be a phase where the Y-value increases from $(10/11)n$ to n without reaching $(9/11)n$. In such a phase the probability of decreasing steps is bounded below by $\frac{(9/11)n}{3n - (18/11)n} = \frac{3}{5}$. Applying results on the gambler's ruin problem, the probability that one phase starting at $a = (10/11)n$ and finishing at $a = (9/11)n$ or $a = n$ stops at $a = n$ is bounded above by

$$((3/2)^{n/11} - 1)/((3/2)^{2n/11} - 1) = e^{-\Omega(m)}$$

since the probability of decreasing steps is at least by a factor of $3/2$ larger than the probability of increasing steps. Hence, the probability of finding the MST within e^{cm} steps, $c > 0$ small enough, is bounded by $e^{-\Omega(m)}$. □

Theorem 2. *Let p be a polynomial and let the cooling schedule be described by $T(1) = m^3$ and $\alpha = 1 - 1/(cm)$ for some constant $c > 0$. If c is large enough, the probability that simulated annealing applied to CTs with n $(1, 1, m)$-triangles and n (m^2, m^2, m^3)-triangles computes the MST within $3cm \ln m$ steps is bounded below by $1 - 1/p(m)$.*

Proof. We only investigate the search until the temperature drops below 1. This phase has a length of at most $3cm \ln m$ steps and contains two subphases where the temperature is in the interval $[m^2, m^{5/2}]$ or in the interval $[1, m^{1/2}]$. The length of each subphase is at least $(c/4)m \ln m$.

If $T \leq m^{5/2}$, the probability of including an edge of weight m^3 is bounded above by $e^{-m^{1/2}}$. Each run where such an event happens is considered as unsuccessful. If $T \in [m^2, m^{5/2}]$ and an (m^2, m^2, m^3)-triangle is optimal, this triangle remains optimal unless the event considered above happens. Applying Chernoff bounds to each edge and choosing c large enough, the probability of not flipping edges of each triangle at least $c'' \log m$ times is bounded by m^{-k}, $c'' > 0$ and k arbitrary constants. This is a second source of bad behavior. Now, we investigate one (m^2, m^2, m^3)-triangle and the steps flipping one of its edges. For each complete or bad triangle, there is a chance that it turns into optimal within the next two steps concerning this triangle. This happens if the right two edges flip in the right order (probability $1/9$) and the inclusion of the edge with weight m^2 is accepted (probability $e^{-m^2/T} \geq e^{-1}$). The probability of not having a good pair among the at least $(c''/2) \log m$ step pairs, can be made much smaller than m^{-k} by choosing c'' large enough. Altogether, the probability that the first subphase does not finish with MSTs on all (m^2, m^2, m^3)-triangles can be made smaller than $1/(3p(m))$.

The same calculations for $T \in [1, m^{1/2}]$ and the $(1, 1, m)$-triangles show that the probability that the second subphase does not finish with MSTs on all $(1, 1, m)$-triangles can be made smaller than $1/(3p(m))$. Finally, the probability that an (m^2, m^2, m^3)-triangle has turned from optimal into non-optimal after the first subphase is smaller than $1/(3p(m))$. This proves the theorem. □

We have proved that SA is highly successful for the considered graph instances. It is easy to choose a cooling schedule such that SA is even successful with overwhelming probability, e.g., $T(1) = m^3$ and $\alpha = 1 - 1/m^2$. See Neumann and Wegener (2004) to compare SA with simple evolutionary algorithms.

This section contains the result announced in the title of the paper. In the remaining sections, we investigate which graphs can be handled efficiently by MA and SA, only by SA, or by none of them.

5 Connected Triangles with the Same Weight Profile

It is interesting to understand how much different weights have to differ such that MA or SA are able to construct efficiently an MST. For this reason, we investigate graphs consisting of connected triangles in more detail. In this section, we consider the case of n CTs with the same weight profile $(w, w, (1+\varepsilon(m)) \cdot w)$ where $\varepsilon(m) > 0$. We distinguish the cases where $\varepsilon(m)$ is bounded below by a positive constant ε and the case where $\varepsilon(m) = o(1)$.

Theorem 3. *If $\varepsilon(m) \geq \varepsilon > 0$, MA with an appropriate temperature finds the MST on CTs with n $(w, w, (1+\varepsilon(m)) \cdot w)$-triangles in expected polynomial time and is successful with overwhelming probability.*

Proof. A good temperature has to fulfil two properties:

- It has to be low enough to distinguish w-edges effectively from $(1+\varepsilon) \cdot w$-edges.
- It has to be high enough to allow the inclusion of a w-edge in expected polynomial time.

We choose $\gamma := 3/\varepsilon$ and $T := w/(\gamma \cdot \ln m)$. The probability to accept the inclusion of a w-edge equals $e^{-w/T} = m^{-\gamma}$ while the corresponding probability for a $((1+\varepsilon(m)) \cdot w)$-edge equals $m^{-\gamma \cdot (1+\varepsilon(m))} \leq m^{-\gamma - 3}$. We analyze the success probability of a phase of length $m^{\gamma+2}$ starting with an arbitrary connected graph. The event to accept the inclusion of a heavy edge is considered as an unsuccessful phase. The probability of this event is bounded above by $1/m$. Following the lines of the proof of Theorem 2 we have for each triangle with overwhelming probability $\Omega(m^{\gamma+1})$ steps flipping an edge of this triangle which we partition into $\Omega(m^{\gamma+1})$ pairs of consecutive steps. The probability that a complete or bad triangle is turned within such two steps into an optimal one is $\Omega(m^{-\gamma})$. Hence, with overwhelming probability, all triangles turn into optimal during this phase and with probability at least $1 - 1/m$ none of them is turned into non-optimal. Hence, the expected number of phases is $O(1)$ and the probability that a sequence of m phases is unsuccessful is exponentially small. □

It is obvious how to tune the parameters in order to get improved run times. We omit such calculations which do not need new ideas. SA finds the MST in polynomial time with a probability exponentially close to 1 if it starts with $T(1) := w/(\gamma \cdot \ln m)$ and has a cooling schedule that cools down the temperature sufficiently slow. This follows in the same way as Theorem 3.

Theorem 4. *If $\varepsilon(m) = o(1)$, MA and SA are unsuccessful on CTs with n $(w, w, (1+\varepsilon(m)) \cdot w)$-triangles.*

Proof. First, we investigate MA. The search starts with n complete triangles and each one has a probability of $2/3$ to be turned into a bad one before it is turned into an optimal one. With overwhelming probability, at least $n/2$ bad triangles are created where the missing w-edge has to be included in order to be able to turn it into an optimal triangle. The probability of including a w-edge within a polynomial number of $p(m)$ steps is bounded above by $p(m) \cdot e^{-w/T}$. This is bounded below by $\Omega(m^{-k})$ only if $e^{-w/T} = \Omega(m^{-\gamma})$ for some constant $\gamma > 0$. Hence, we can assume that $T \geq w/(\gamma \cdot \ln m)$ for some constant $\gamma > 0$.

Let $p^*(T)$ be the probability of accepting the inclusion of a w-edge and $p^{**}(T)$ the corresponding probability for a $((1+\varepsilon(m)) \cdot w)$-edge. Since $T \geq w/(\gamma \cdot \ln m)$ and $\varepsilon(m) = o(1)$,

$$\begin{aligned} p^*(T)/p^{**}(T) &= e^{-w/T} \cdot e^{(1+\varepsilon(m)) \cdot w/T} \\ &= e^{\varepsilon(m) \cdot w/T} \\ &\leq e^{\varepsilon(m) \cdot \gamma \cdot \ln m} \\ &= m^{\varepsilon(m) \cdot \gamma}. \end{aligned}$$

Choosing m large enough, this gets smaller than any m^δ, $\delta > 0$. It will turn out that this advantage of w-edges against $((1 + \varepsilon(m)) \cdot w)$-edges is too small. The stochastic process behind MA can be described by the parameters b (number of bad triangles) and c (number of complete triangles). We use the potential function $2b + c$ which starts with the value n and has the value 0 for the MST. The value of the potential function changes in the following way:

- It increases by 1 if a complete triangle turns into a bad one or an optimal one turns into a complete one. The probability of the first event equals $2c/m$, since we have to flip one of the two light edges of one of the complete triangles. The probability of the second event equals $p^{**}(T) \cdot (n-b-c)/m$ since we have to flip the heavy edge in one of the $n-b-c$ optimal triangles and to accept this flip.
- It decreases by 1 if a complete triangle turns into an optimal one (probability c/m) or a bad triangle turns into a complete one (probability $p^*(T) \cdot b/m$).
- It remains unchanged, otherwise.

Since we are interested in lower bounds on the optimization time, we can ignore all non-accepted steps, i.e., all steps not changing the potential. If $b \leq n^{1/2}$ and m is large enough, the probability that an accepted step increases the potential is at least $3/5$. This claim is equivalent to

$$\frac{2c/m + p^{**}(T) \cdot (n-b-c)/m}{2c/m + p^{**}(T) \cdot (n-b-c)/m + c/m + p^*(T) \cdot b/m} \geq \frac{3}{5}$$

which is equivalent to

$$2c + p^{**}(T) \cdot (n-b-c) \geq \frac{9}{5}c + \frac{3}{5} \cdot p^{**}(T) \cdot (n-b-c) + \frac{3}{5}p^*(T) \cdot b$$

and

$$\frac{1}{5}c + \frac{2}{5}p^{**}(T) \cdot (n-b-c) \geq \frac{3}{5} \cdot p^*(T) \cdot b.$$

This is obviously true if $c \geq 3 \cdot b$. Otherwise, $n - b - c \geq n - 4b \geq n - 4n^{1/2}$ and it is sufficient to show that

$$2 \cdot p^{**}(T) \cdot (n - 4n^{1/2}) \geq 3 \cdot p^*(T) \cdot n^{1/2}$$

or

$$p^*(T)/p^{**}(T) \leq \frac{2}{3}(n^{1/2} - 4).$$

We have shown that this holds for large enough m, since $n = \Omega(m)$. The claim for MA follows now from results on the gambler's ruin problem. The probability to start with a potential of $n^{1/2}/2$ and to reach the value 0 before the value $n^{1/2}$ is exponentially small. Finally, we investigate a polynomial number of $p(m)$ steps of SA. Let d be chosen such that $p(m) \leq m^d$. We claim that it is unlikely that the potential drops below $n^{1/2}/4$ within m^d steps. With overwhelming probability, we produce a bad triangle. Therefore, it is necessary to accept the inclusion of a w-edge. Hence, as seen above, only steps where the temperature is at least $w/(\gamma \cdot \ln m)$ for some appropriate constant $\gamma > 0$ have to be considered. However, the analysis of MA treats all these temperatures in the same way. The probability to start with a potential of $n^{1/2}/2$ and to reach the value $n^{1/2}/4$ before $(3/4)n^{1/2}$ is still exponentially small. □

The proof also shows that SA with an arbitrary cooling schedule is unsuccessful in the considered situation.

6 Simulated Annealing Is Successful for $(1+\varepsilon)$-Separated Weights

We have seen in Theorem 4 that MA and even SA are unsuccessful on certain graphs if we allow that different weights may differ by a factor of $1+\varepsilon(m)$ where $\varepsilon(m)$ is an arbitrary function such that $\varepsilon(m) = o(1)$. Here, we prove that SA is highly successful on all graphs if the different weights differ at least by a factor of $1 + \varepsilon$ for some constant $\varepsilon > 0$.

Before proving this result, we repeat some well-known facts about MSTs. Let $E_1, \ldots, E_r$ be the partition of the edge set E such that all edges in E_i have the same weight W_i and $W_1 > \cdots > W_r$. Let $c_i, 1 \leq i \leq r+1$, be the number of connected components of $G_i := (V, E_i \cup \cdots \cup E_r)$. Each MST contains exactly $a_i := c_{i+1} - c_i$ E_i-edges such that the chosen edges from $E_i \cup \cdots \cup E_r$ span the connected components of G_i. A set E_i^* of a_i E_i-edges is called optimal if $G_i^* := (V, E_i^* \cup E_{i+1} \cup \cdots \cup E_r)$ has the same connected components as G_i. An MST contains exactly the edges of optimal sets $E_1^*, \ldots, E_r^*$. The set E_i^* is not necessarily uniquely defined. The idea of the proof is the following. There is some point of time $t_i, 1 \leq i \leq r+1$, such that, with large probability, the following holds. After step t_i, no inclusion of an edge from $E_1 \cup \cdots \cup E_i$ is accepted and at step t_i the current search point has chosen among all E_j-edges, $1 \leq j \leq i-1$, an optimal subset E_j^*. This implies that after step t_i no edges from $E_1 \cup \cdots \cup E_{i-1}$ are included (the first property) or excluded (this would destroy the connectedness of the graph described by the search point). Moreover, no edges from E_i are included and we hope to exclude enough E_i-edges until step t_{i+1} such that then the search point chooses an optimal set E_i^* of E_i-edges. Note that after time step t_i the set of chosen E_i-edges is always a superset of an optimal set E_i^* since, otherwise, the considered graph would be unconnected. Finally, the properties imply that at step t_{r+1} the search point describes an MST.

Theorem 5. *Let the weights of the edges be bounded by 2^m and $(1+\varepsilon)$-separated for some constant $\varepsilon > 0$, i.e., $w_i > w_j$ implies $w_i \geq (1+\varepsilon) \cdot w_j$. SA with an appropriate cooling schedule is highly successful when searching for an MST on such graphs.*

Proof. Let $T(1) := 2^m$, $\gamma := 8/\varepsilon$, α be the cooling factor such that it takes $m^{\gamma+7}$ steps to decrease the temperature from T to $T/(1+\varepsilon/2)$, and β be defined by $(1+\varepsilon/2)^\beta = 2$. Then we set $t_{r+1} := 2\beta m^{\gamma+8}$. Until step t_{r+1}, the temperature has dropped (far) below $1/m$. Our claim is that, with a probability of $1-O(1/m)$, the search point at step t_{r+1} describes an MST.

To follow the proof strategy discussed above let $t_i, 1 \leq i \leq r$, be the earliest point of time when $T(t_i) \leq W_i/((1+\varepsilon) \cdot \gamma \cdot \ln m)$. The probability of accepting the inclusion of an edge of weight W_i after step t_i is bounded above by $m^{-\gamma-8}$. During the next $m^{\gamma+7}$ steps, with overwhelming probability, there are $O(m^{\gamma+6})$ steps flipping a specified edge and the probability to accept this edge at least once is $O(1/m^2)$. Afterwards, the temperature has dropped by a factor of $1/(1+\varepsilon/2)$. The probability to accept this edge is then bounded by $m^{-\gamma-12}$ and the probability to accept the edge during the next $m^{\gamma+7}$ steps is $O(1/m^5)$. This argumentation can be continued implying that the probability to accept the inclusion of the considered edge after step t_i is $O(1/m^2)$. Hence, with probability $1 - O(1/m)$, it holds that, for all i, edges of weight W_i are not included after step t_i. In the following, we assume that this event holds.

We assume that at step t_i the search point chooses optimal sets $E_1^*, \ldots, E_{i-1}^*$ and a superset E_i' of an optimal set E_i^*. This is obviously true for $i = 1$. We analyze the steps $t_i, \ldots, t_i + m^{\gamma+7} - 1$. The probability to accept an edge with weight $w \leq W_{i+1}$ in one step is bounded below by $m^{-\gamma-4}$ during this phase. By our assumption, we do not include edges of weight $w \geq W_i$. Let $b_i := |E_i'| - |E_i^*|$ at step t_i. As long as $|E_i'| > |E_i^*|$ there are at least $|E_i'| - |E_i^*|$ candidate E_i-edges whose exclusion is possible. The exclusion of such an edge is only accepted if this edge lies on a cycle. Either the edge lies on a cycle or there is a missing edge of weight $w \leq W_{i+1}$ whose inclusion creates a cycle containing the considered E_i-edge. If no cycle with an E_i-edge exists, the probability of creating such a cycle in the next step is at least $m^{-\gamma-5}$. If a cycle with an E_i-edge exists, the probability to destroy the cycle by excluding an E_i-edge is at least $1/m$ (there may be more than one E_i-edge on the cycle). Let us assume that we do not exclude b_i E_i-edges within the considered $m^{\gamma+7}$ steps. Let s be the number of steps in this phase where a cycle with an E_i-edge exists.

If $s \geq m^{3/2}$, then the probability of less than $b_i \leq m$ steps excluding an E_i-edge on the cycle is exponentially small. If $s < m^{3/2}$, then the probability that among the at least $m^{\gamma+7} - m^{3/2}$ steps without a cycle with E_i-edges there are less than $m^{3/2}$ steps creating such a cycle is exponentially small. Hence, with overwhelming probability, enough E_i-edges are excluded and the claim holds for step t_{i+1}.

Altogether, with a probability of $1 - O(1/m)$, SA has found an MST after $O(m^{\gamma+8})$ steps. □

It is easy to see that we can generalize the result to weights up to $2^{p(m)}$ for a polynomial p. The run time increases by a factor of $O(p(m)/m)$. It is possible to tune the parameters to obtain better run times. However, the purpose of Theorem 4 and Theorem 5 was to identify the border (with respect to quotients of different weights) between cases where SA is highly successful and cases where SA can be unsuccessful. With respect to these aims we have obtained optimal results. It is easy to generalize our results to prove that SA is always highly successful if one is interested in $(1 + \varepsilon)$-optimal spanning trees. It remains an open problem to find other sufficient conditions implying that MA or SA is successful or unsuccessful on the MSTP.

7 Conclusions

The paper contributes to the theory of randomized search heuristics, in particular, the Metropolis algorithm and simulated annealing. The problem to present a natural example from combinatorial optimization where simulated annealing beats the Metropolis algorithm is solved by investigating the problem of computing minimum spanning trees. Moreover, the minimal factor between different weights to guarantee that simulated annealing finds minimum spanning trees efficiently is determined.

References

1. Droste, S., Jansen, T., and Wegener, I. (2001). Dynamic parameter control in simple evolutionary algorithms. FOGA'2000. Foundations of Genetic Algorithms 6 (Eds. Martin, W. N. and Spears, W. M.), 275–294. Morgan Kaufmann.
2. Giel, O. and Wegener, I. (2003). Evolutionary algorithms and the maximum matching problem. Proc. of 20th Symp. on Theoretical Aspects of Computer Science (STACS), LNCS 2607, 415–426.
3. Jerrum, M. (1992). Large cliques elude the Metropolis process. Random Structures and Algorithms 3, 347–359.
4. Jerrum, M. and Sinclair, A. (1996). The Markov chain Monte Carlo method. An approach to approximate counting and integration. Ch. 12 of Hochbaum, D. (Ed.). *Approximation Algorithms for NP-hard Problems*, 482–522. PWS Publishing Company.
5. Jerrum, M. and Sorkin, G. B. (1993). Simulated annealing for graph bisection. Proc. of 37th Symp. Foundations of Computer Science (FOCS), 94–103.

6. Jerrum, M. and Sorkin, G. B. (1998). The Metropolis algorithm for graph bisection. Discrete Applied Mathematics 82, 155–175.
7. Neumann, F. and Wegener, I. (2004). Randomized local search, evolutionary algorithms, and the minimum spanning tree problem. Proc. of Genetic and Evolutionary Computation. GECCO 2004. LNCS 3102, 713–724.
8. Sasaki, G. (1991). The effect of the density of states on the Metropolis algorithm. Information Processing Letters 37, 159–163.
9. Sasaki, G. and Hajek, B. (1988). The time complexity of maximum matching by simulated annealing. Journal of the ACM 35, 387–403.
10. Sorkin, G. B. (1991). Efficient simulated annealing on fractal energy landscapes. Algorithmica 6, 367–418.

Online Interval Coloring and Variants

Leah Epstein[1,*] and Meital Levy[2]

[1] Department of Mathematics, University of Haifa, 31905 Haifa, Israel
lea@math.haifa.ac.il
[2] School of Computer Science, Tel-Aviv University, Israel
levymeit@post.tau.ac.il

Abstract. We study interval coloring problems and present new upper and lower bounds for several variants. We are interested in four problems, online coloring of intervals with and without bandwidth and a new problem called lazy online coloring again with and without bandwidth. We consider both general interval graphs and unit interval graphs. Specifically, we establish the difference between the two main problems which are interval coloring with and without bandwidth. We present the first non-trivial lower bound of **3.2609** for the problem with bandwidth. This improves the lower bound of 3 that follows from the tight results for interval coloring without bandwidth presented in [9].

1 Introduction

We study online interval coloring problems. In the basic problem intervals are presented one by one and the online algorithm must assign each interval a color before the next interval arrives where two intersecting intervals can not be colored by the same color. We are also interested in the case where every interval has an associated bandwidth in [0,1], this problem was first introduced by Adamy and Erlebach [1]. A set of intervals can be assigned the same color c, if for any point p, on the real line, the sum of the bandwidths of intervals colored c and containing p, does not exceed 1. We refer to a coloring satisfying the above condition as a *proper coloring*.

As mentioned in [1], the interval coloring problem with bandwidth arises in many applications. Most of the applications come from networks field.

Consider a network with a line topology that consists of links, where each link has channels of constant capacity. This can be either an all-optical WDM (wavelength-division multiplexing) network or an optical network supporting SDM (space-division multiplexing). A connection request is from one network node a to another node b has a bandwidth associated with it. The set of requests assigned to a channel must not exceed the capacity of the channel on any of the links on the path $[a, b]$. The goal is to minimize the number of channels (colors) used. A connection request from a to b corresponds to an interval $[a, b]$ with the respective bandwidth requirement and the goal is to minimize the number of required channels to serve all requests.

Another network related application is that if the requests have constant duration c, and we have to serve all requests as fast as possible. With respect to our online interval

---
* Research supported by Israel Science Foundation (grant no. 250/01).

coloring problem with bandwidth, the colors correspond to time slots, and the total number of colors corresponds to the schedule length.

The last example comes from scheduling, a requested job has a duration and resource requirement during its execution. Jobs (intervals) arrive online and must be assigned to a machine (color) immediately. All the machines have the same capabilities and the objective is to minimize the number of machines used.

The unweighted problem is equivalent to coloring an interval graph, where each interval corresponds to a node and an edge between two nodes appears if the corresponding intervals intersect. Interval graphs are perfect, therefore the chromatic number of the graph is the maximum clique size [6], which represents a point where the most intervals intersect. It can be elaborated for the bandwidth case, if we refer to the maximum clique size as the maximum weighted clique. Each node has the weight of the related interval, i.e., its bandwidth, and the clique size is the sum of weights of the clique.

We study online coloring problems in terms of competitive analysis. Thus we compare an online algorithm to an optimal offline algorithm OPT that knows the complete sequence of intervals in advance. In this paper, we make use of two types of competitive ratios. The *absolute competitive ratio* and the *asymptotic competitive ratio*.

Let $B(\sigma)$ (or B, if the sequence σ is clear from the context), be the cost of algorithm B on the request sequence σ. An algorithm A is $\mathcal{R}$-competitive (with respect to the absolute competitive ratio) if for every sequence σ, $A(\sigma) \leq \mathcal{R} \cdot OPT(\sigma)$. The absolute competitive ratio of an algorithm is the infimum value of $\mathcal{R}$ such that the algorithm is $\mathcal{R}$-competitive.

The asymptotic competitive ratio for an online algorithm A is defined to be

$$\mathcal{R}_A^\infty = \limsup_{n \to \infty} \sup_\sigma \left\{ \frac{A(\sigma)}{OPT(\sigma)} \middle| OPT(\sigma) = n \right\}.$$

Most results in this paper hold for both definitions of competitive ratio. We mention the cases where a result holds for only one definition.

Coloring interval graphs has been intensively studied, Kierstead and Trotter [9] constructed an online algorithm which uses at most $3\omega - 2$ colors where ω is the maximum clique size of the interval graph. They also presented a matching lower bound of $3\omega - 2$ on the number of colors in a coloring of an arbitrary online algorithm. As mentioned above, the chromatic number of interval graphs equals to the size of a maximum clique. This means that the optimal offline algorithm can color every interval graph with ω colors.

Much research has been done analyzing the performance of the simple First Fit algorithm. An upper bound of 40 on the competitive ratio was proven in [7], and later on an upper bound of 25.72 was presented by Kierstead and Qin [8]. In a recent study [11], a competitive ratio of 10 was proved. Chrobak and Slusarek [3] show a lower bound close to 4.5 on the competitive ratio of First Fit.

Coloring intervals with bandwidth was first posed in 2003 in [1], they presented an online algorithm with a competitive ratio of 195. In [10] Narayanaswamy presented a new algorithm with a competitive ratio of 10.

Motivated by The Maximum Resource Bin Packing Problem [2], we introduce a new problem called Lazy Interval Coloring. As opposed to the regular interval coloring problem, we seek to use as many colors as possible. If all bandwidths are 1, a new color may be used for an interval, only if it intersects intervals with all the previously used colors. We also consider the bandwidth case, where intervals have bandwidth in [0,1]. For both problems, a newly presented interval must be colored by a used color if a proper coloring can be achieved. An application for this problem can be any of the applications mentioned before, where using additional colors (channels/time slots/machines) can improve the quality of service. The scheduler (who assigns colors) has the purpose of using as many colors as possible. However, to avoid the usage of too many colors, the instructions of a scheduler (given by the boss, who pays for the equipment) are not to use a new color unless it is absolutely necessary.

Our Results: We show that introducing bandwidth to interval coloring makes the problem harder. We present the first non-trivial lower bound of **3.2609**, which improves the lower bound of 3 for interval coloring without bandwidth. Recall that for the problem without bandwidth, the bound 3 is tight. We also show that bandwidth makes lazy interval coloring a much harder problem.

In this work we consider resource augmentation for the problem of interval coloring with and without bandwidth. We show that there exists an online algorithm that when uses twice as much resource for each color, can perform as good as the optimal offline algorithm. We also present a matching lower bound.

For the bandwidth case we study two interesting cases. In the first case, the online algorithm may use twice as much capacity for each color as the offline algorithm, for which we present an online algorithm that uses at most 3 times the number of colors that the optimal offline algorithm uses. In the seconds case, we present an online algorithm that uses at most as many colors as OPT in the worst case, where each color has a capacity of 4.

Another interesting variant is to restrict the class of intervals, so all intervals are of the same length. This type of graph is called "Unit Interval Graph". For the interval coloring problem we show that First Fit uses at most $2\omega - 1$ colors and that the analysis is tight. We also show a lower bound of $\frac{3}{2}$ on the competitive ratio of any online algorithm.

For interval coloring with bandwidth for unit interval graphs, we present several algorithms, the best algorithm has $3\frac{1}{2}$ competitive ratio. We also present a lower bound of 2. For the asymptotic competitive ratio, the bounds become 3.17778 and 1.831.

For lazy interval coloring, we prove that for general instances any online algorithm performs arbitrarily bad. However if all intervals have the same length we present upper and lower bounds of 2. When introducing bandwidth to the lazy interval coloring problem, we show that any online algorithm is arbitrarily bad compared to the maximum weighted clique (even for unit interval graphs). We summarize our results and previous results in the following table.

Remark: All omitted proofs can be found in the full version of the paper.

2 Preliminaries

A weighted interval graph G, is a graph where each node corresponds to an interval. The weight of the node is the bandwidth of the interval related to it. If two intervals

Table 1. Results obtained in this paper and previous work. For each case, a single entry means that the results hold for both asymptotic and absolute competitive ratios. If two entries exist, the first one is the asymptotic competitive ratio and the second one stands for the absolute competitive ratio

	Interval Graph		Unit Interval Graph	
	LB	UB	LB	UB
Interval Coloring	3 [9]	3 [9]	$\frac{3}{2}$	2
Interval Coloring with Bandwidth	**3.2609**	10 [10]	1.831	2 3.17778 3.5
Lazy Interval Coloring	∞		2	2
Lazy Interval Coloring with Bandwidth	∞		∞	

intersect, there is an edge between their related nodes in G. Recall that we denote the optimal coloring of the offline algorithm by OPT.

Let $\omega(G)$ denote the size of the maximum cardinality clique in G (ω for short), i.e., ignoring the weights. Let $\omega^*(G)$ (ω^* for short) denote the largest weighted clique in G. A weighted clique is the sum of the weights of the vertices in a clique. Note that for the interval coloring problem with bandwidth we have $OPT \geq \lceil \omega^* \rceil$

Below we give a generalized presentation of the algorithm of Kierstead and Trotter [9]. We use specific cases of the generalized algorithm for the variant of resource augmentation in the next sections. We present the algorithm using notations similar to these of [10].

Let $\sigma = v_1, \ldots, v_n$ be the list of vertices of G, in the order of arrival. Algorithm $KT_{l,b}$ is defined for inputs σ such that, $b(v_i) \in (0, b]$. The algorithm partitions the intervals (i.e. the vertices of G) into sets A_m (for integer values of m, such that $m \geq 1$). We use C_m to denote the set of colors dedicated to A_m. Every set A_m is colored using First Fit, independently of other sets. Therefore the colors have the property $C_x \cap C_y = \emptyset$ for $x \neq y$. A critical point, q, in interval $v_i \in A_{m(v_i)}$, is a point where $\omega^*(G_{m(v_i)-1}(v_i) \cup \{v_i\}) > (m(v_i) - 1) \cdot l$. Since $v_i \in A_{m(v_i)}$, there is at least one such point for every interval in $A_{m(v_i)}$.

Algorithm 1. $KT_{l,b}$

On a new interval v_i:

1: For every integer $m \geq 1$, let $V_m(v_i)$ and $E_m(v_i)$ be the following subsets of $V(G)$ and $E(G)$ respectively.
  $V_m(v_i) = \{v_j \in V(G) : j < i, m(v_j) \leq m\}$;
  $E_m(v_i) = \{(u,v) \in E(G) : u, v \in V_m(v_i)\}$;
  $G_m(v_i) \cup \{v_i\} = G(V_m(v_i) \cup v_i, E_m(v_i) \cup \{(u, v_i) \in E(G) : u \in V_m(v_i)\})$
  $\omega_i^*(H) = $ The maximum weighted clique in graph H containing interval v_i
2: Let $G_m(v_i) = G(V_m(v_i), E_m(v_i))$
3: $m(v_i) = $ the smallest m such that $\omega_i^*(G_m(v_i) \cup \{v_i\}) \leq m \cdot l$.
4: $A_{m(v_i)} \Leftarrow A_{m(v_i)} \cup \{v_i\}$
5: Color v_i considering only the intervals of $A_{m(v_i)}$ using First Fit on colors of $C_{m(v_i)}$.

Lemma 1. *Given an interval v_i, let $m = m(v_i)$. For the set A_m and every critical point $q \in v_i$, the total bandwidth at q of intervals in A_m does not exceed $b + l$.*

Lemma 2. *For every m, $\omega^*(A_m) \leq 2(b+l)$.*

Proof. Proof by contradiction, assume that there is a weighted clique of more than $2(b+l)$ in A_m obtained at point p_j. By the previous lemma, this point is not a critical point of any interval in A_m. For every interval $v_i \in A_m$, where $p_j \in v_i$, there is a critical point either to the right of p_j or the left of p_j or both. Denote the closest critical point (of any interval of A_m which contains p_j) to the left of p_j, q_l and the closest critical point to the right of p_j, q_r. Since there is at least one critical point for every v_i, either $q_l \in v_i$ or $q_r \in v_i$ or both for every $v_i \in A_m$ (since the critical point of v_i cannot be in the interval (q_l, q_r). But this means that either q_l or q_r have a total bandwidth of more than $b + l$. By Lemma 1, this is not possible. Note that either q_l or q_r or both must exist. If one of q_r or q_l does not exist, we get the contradiction at the point that exists.

Lemma 3. *If all intervals have the same bandwidth, b, and l is divisible by b, for every m, $\omega^*(A_m) \leq 2l$.*

Lemma 4. *(i) The largest value of m ever used in $KT_{l,b}$ is $\lceil \frac{\omega^*}{l} \rceil$*
(ii) The coloring of $KT_{l,b}$ is at most $\lceil \frac{\omega^*}{l} \rceil (\max_m FF(C_m))$, where $FF(C_m)$ denotes the coloring of the First Fit algorithm on the set C_m of intervals that were presented online.

Proof. **(i):** For a maximum weighted clique of ω^* and for every interval $v_i \in \sigma$, $\omega^*(G_{\lceil \frac{\omega^*}{l} \rceil}(v_i) \cup \{v_i\}) \leq \omega^* \leq \lceil \frac{\omega^*}{l} \rceil \cdot l$ **(ii):** By (i) the largest value of m is at most $\lceil \frac{\omega^*}{l} \rceil$. For each m, A_m is colored by First Fit using the related colors of C_m (last step of the algorithm).

2.1 Variants of Interval Coloring

Resource Augmentation in Interval Coloring: In the resource augmentation approach, the online algorithm is given more resources than the offline algorithm. Interval coloring is a natural problem to consider with resource augmentation. In this case, in the online coloring, the total bandwidth of intersecting intervals with the same color can exceed 1. The allowed maximum bandwidth of the intersecting intervals in the online coloring will be denoted by B. For the analysis we use the concept of competitive ratio.

Equal length intervals in Interval Coloring: If the intervals must be of equal length, the associated graph is called a unit interval graph. Recognition of unit interval graphs has been studied in [13, 4, 12]. It was also studied in the context of interval selection in [5]. For simplicity, we use intervals of length 1 in some of the proofs of this paper.

3 Interval Coloring

3.1 Resource Augmentation

In [9] Kierstead and Trotter give a lower and upper bound of 3. The main goal when resource augmentation is allowed, is to find a value of B for which there exists an algorithm with a competitive ratio 1. We argue that for $B = 2$, the online algorithm presented in [9] uses $\max\{1, OPT - 1\}$ colors. We also present a matching lower bound of $\max\{1, OPT - 1\}$ on the number of colors used by any online algorithm.

Theorem 1. *An adaptation of the algorithm of Kierstead and Trotter [9] can be used for interval coloring with $B = 2$, and uses at most $\max\{1, OPT - 1\}$ colors.*

Proof. **Algorithm:** $KT_{1,1}$ without bandwidth is exactly the the algorithm of Kierstead and Trotter.

According to Lemma 3 every set A_m has a maximum bandwidth of at most 2. Since $B = 2$, First Fit can color each A_m by a unique color. Moreover, by the definition of sets A_1 and A_2, the total bandwidth of intervals in $A_1 \cup A_2$ does not exceed 2. We use the same color for $m = 1, 2$, and one color for each other value of m. By Lemma 4, we get a coloring which uses $\lceil \omega^* \rceil - 1 \leq OPT - 1$ colors, if $OPT \geq 1$ and otherwise it uses a single color.

We prove that the above bound cannot be improved. Clearly if $OPT = 1$ the algorithm also uses at least one color, therefore we need to show a lower bound of $OPT - 1$. The lower bound construction holds only for the absolute competitive ratio.

Theorem 2. *There exists an infinite sequence of values of α, such that there exists an input for which $OPT = \alpha$ and any online algorithm for interval coloring with $B = 2$, uses at least $OPT - 1$ colors.*

3.2 Unit Interval Graphs

In the following 2 theorems, we show that First Fit uses at most $2\omega - 1$ colors for unit interval graphs and that the analysis is tight.

Theorem 3. *First Fit uses at most $2\omega - 1$ colors for coloring unit interval graphs.*

Proof. In this version, since all the intervals are of equal length, if the maximum clique equals to ω, then there is no interval that intersects with more than $2\omega - 2$ different intervals. Assume by contradiction that there exists an interval $I = [x, x + 1]$ that does intersect with more than $2\omega - 2$ different intervals. Since all intervals are of unit length, every interval that intersects I, must contain either the point x or $x + 1$ (or both). Therefore either the point x or the point $x + 1$ is contained in more than $\omega - 1$ intervals not including interval I. This contradicts the fact that the maximum clique size does not exceeds ω. Therefore First Fit uses at most $2\omega - 1$ colors.

Theorem 4. *There exist unit interval graphs on which First Fit uses exactly $2\omega - 1$ colors.*

Proof. We show a family of instances, where k is the size of largest clique on which First Fit uses exactly $2k - 1$ colors to color it. We define the sequence of intervals in phases.

Phase i ($1 \leq i \leq 2k - 1$) contains $2k - i$ intervals that receive the same color by First Fit. The largest clique size after phase i is $\lfloor \frac{i}{2} \rfloor + 1$. The intervals of phase i are

$[i-1+2^{-i+1}+2j, i-1+2^{-i+1}+2j+1]$, for $0 \le j < 2k - i$. All intervals have length 1, and all consecutive intervals of one phase have fixed distances of 1 between them.

We prove the following claims.

1. The largest clique after phase $2\ell + 1$ is of size $\ell + 1$, and in ranges of the form $(2j, 2j + 2^{-2\ell})$, the total requested bandwidth is at most ℓ.
2. All intervals of phase i receive the color i.

Proof of 1: We prove the claim by induction. After phase 1 ($\ell = 0$), the largest clique is clearly of size 1 (the intervals are non-intersecting). The ranges $(2j, 2j + 1)$ are empty since the intervals of phase 1 start at odd points. Assume now that the claim holds for $\ell = s - 1$ and prove for $\ell = s$. Phases $2s$ and $2s + 1$ introduce two sets of intervals. To show the first part of the claim we need to show that the overlap between the intervals does not overlap with areas where the largest clique is s. Non overlapping parts of new intervals may increase the size of the largest clique by 1. By definition, the overlap interval between intervals of these two phases are intervals of the form $[2j + 2^{-2s}, 2j + 2^{1-2s}]$. Using the inductive hypothesis, intervals of the form $(2j, 2j + 2^{2-2s})$ have bandwidth request of size $s - 1$ only, and therefore the largest clique in these intervals after phase $2s + 1$ does not exceed $s + 1$. To prove the second part of the claim, the interval $(2j, 2j + 2^{-2s})$ is not a part of the overlap between phases $2s$ and $2s + 1$, therefore its bandwidth request increases by at most 1, and becomes at most s.

Proof of 2: We prove that an interval of phase i intersects with intervals of all smaller colors. If i is odd, then its left endpoint intersects with all intervals of even colors, and its right endpoint with all intervals of odd colors. If i is even, its left endpoint intersects with intervals of all odd colors, and its right endpoint with even colors. This can be easily verified by the intervals definitions.

Theorem 5. *Any online algorithm for unit interval graphs has a competitive ratio of at least $\frac{3}{2}$*

4 Online Coloring of Intervals with Bandwidth

Interval coloring with bandwidth was recently studied by [1] and [10]. Adamy and Erlebach [1] gave a 195-competitive ratio algorithm and Narayanaswamy [10] gave a 10-competitive ratio. However, is this case really harder? In the theorem below we answer that question affirmatively. We give a lower bound which is strictly higher than the upper bound for the problem of interval coloring without bandwidth presented in [9]. To prove the lower bound we adapt the lower bound on classical interval coloring given in [9]. In that paper a lower bound of of $3\omega - 2$ colors is shown, for inputs where $OPT = \omega$. Since interval graphs are perfect, these are exactly inputs where the largest clique has size ω. Note that two intervals whose width is strictly larger than $\frac{1}{2}$ cannot have the same color. Therefore the same lower bound can be applied not only for intervals of width 1, but for intervals of arbitrary widths in $(\frac{1}{2}, 1]$. In this case, let q be the largest number of intersecting intervals of width in $(\frac{1}{2}, 1]$, then we immediately get a lower bound of $3q - 2$. Finally, we make another adaptation to the lower bound, namely, we make use of the following lemma.

Lemma 5. *The lower bound $3q - 2$ on the number of colors holds even if q is given in advance.*

With this, we are ready to prove our main theorem of this section.

Theorem 6. *Any deterministic online algorithm for interval coloring with bandwidth has competitive ratio of at least 3.2609*

Proof. Let α be a constant rational number fixed later and let t be an integer such that αt is integer (there are infinitely many such values of t). Let $\varepsilon = \frac{1}{2\alpha t+1}$. The first phase of requests is a large number $T = N!(2\alpha t + 1)$ of identical requests (for a large enough integer N), all of bandwidth ε. The requests are for a long enough interval. All future requests will be given within this interval, and therefore they all intersect the initial requests. The sequence either stops here (and has $OPT = N!$) or continues with a second phase which contains requests which are all of bandwidth $1 - \frac{k}{2\alpha t+1} = 1 - k\varepsilon$, for some $t \le k \le \alpha t$.

Furthermore, these requests all have bandwidth larger than $\frac{1}{2}$, therefore if they are packed independently from the first phase, they are treated as requests of width 1. Adding the first phase means that colors that were used for at least $k + 1$ intervals, cannot be used again in the second phase, and other colors can be used again. Intervals of the second phase are introduced as in Lemma 5 so that the optimal number to color them is w, and the number of colors used in this phase is $3w$. If the bandwidth of these intervals is $1 - k\varepsilon$, then we use $w = \frac{T}{k}$, so that an optimal coloring uses w colors, and each color is used for k intervals of the first phase. We denote by X_i ($1 \le i \le \alpha t$) the number of colors used in the first phase for exactly i intervals. The algorithm has no reason to color less than t intervals with one color, since a color used for t intervals can always be used again, therefore $X_i = 0$ for $i < t$. If a color is used for more than αt intervals, it will not be used again, so we can assume that in such a case, the color is used for $\frac{1}{\varepsilon}$ intervals. We denote by Y the number of colors that are used for this maximum number of intervals. We have $T = Y(2\alpha t + 1) + \sum_{i=t}^{\alpha t} i \cdot X_i$. If there is no second phase, $OPT = T\varepsilon$ and $ALG = Y + \sum_{i=1}^{2\alpha t} X_i$. Otherwise, we compute the number of colors used by the algorithm for a specific choice of $t \le k \le \alpha t$. We get

$$ALG = Y + \sum_{i=k+1}^{\alpha t} X_i + 3\frac{T}{k} = Y + \sum_{i=k+1}^{\alpha t} X_i + 3 \cdot OPT.$$

Let C be the competitive ratio of ALG. We have $Y + \sum_{i=1}^{\alpha t} X_i \le C \cdot T\varepsilon$ and for every $t \le k \le \alpha t$, $Y + \sum_{i=k+1}^{j} X_i \le (C - 3)\frac{T}{k}$. We multiply the first inequality by t, the last inequality (i.e., the second inequality for $k = \alpha t$) by $\alpha t + 1$, and all other inequalities by 1 (i.e., the second inequality for all other values of k). We sum them and get $T = \sum_{i=t}^{\alpha t} i X_i + (2\alpha t + 1)Y \le C \cdot T \frac{t}{2\alpha t+1} + (C - 3) \sum_{i=t}^{\alpha t} \frac{T}{2\alpha t+1} + (C - 3)T$. Letting t tend to infinity, we have $\sum_{i=t}^{\alpha t} \frac{T}{2\alpha t+1} \to \ln \alpha$. We get $C \ge \frac{3\ln\alpha+4}{\frac{1}{2\alpha}+\ln\alpha+1}$.

Solving in Maple, we see that for an appropriate choice of α this gives a lower bound of 3.2609.

4.1 Resource Augmentation

In this section we consider two interesting possible values of B. For $B = 4$ we provide an online algorithm that can color with $OPT - 3$ colors. For $B = 2$ we provide an online algorithm that uses $3OPT - 2$ colors. The lower bound on the competitive ratio for $B = 2$ is at least 1 as was shown for the case of interval coloring without bandwidth.

Proposition 1. *An adaptation of the algorithm of Kierstead and Trotter [9] can be used for interval coloring with $B = 4$, and uses at most $\max\{1, OPT - 3\}$ colors.*

Proof. We use **Algorithm:** $KT_{1,1}$ with bandwidth. According to Lemma 2 every A_m has a maximum bandwidth of at most 4. By Lemma 4 the coloring of $KT_{1,1}$ with bandwidth, is $\omega^* \cdot 1 = \omega^*$. Since $B = 4$, First Fit can color each A_m by a unique color. Moreover, by the definition of sets $A_1, \ldots, A_4$ bandwidth of intervals in $A_1 \cup A_2 \cup A_3 \cup A_4$ does not exceed 4. We use the same color for $m = 1, 2, 3, 4$, and one color for each other value of m. By Lemma 4, we get a coloring which uses $\lceil \omega^* \rceil - 3 \leq OPT - 3$ colors. If $m \leq 4$ we get a coloring using a single color.

Proposition 2. *There exists an online algorithm for interval coloring with bandwidth with $B = 2$, that uses at most 3OPT-1 colors.*

4.2 Unit Interval Graphs

For this version we present three algorithms.

1. First Fit.
2. 2-First Fit. Perform an online partition of the intervals into two subsequences according to the bandwidth of the intervals. One subsequence for intervals with bandwidth b such that $b \leq \frac{1}{2}$ and the other for intervals with bandwidth b such that $b > \frac{1}{2}$. Apply First Fit on each subsequence separately with disjoint sets of colors.
3. Odd-Even bin packing. Scale the real line into integers and assume all intervals are of unit length .

Perform an online partition of the intervals into two subsequences called *evens* and *odds*. Each interval intersects an integer point. If an interval is exactly between two integers, assign it to the left integer point, and otherwise there is a unique integer point. If the integer point is an even number assign the interval into the evens subsequence, otherwise assign it to the odds subsequence. Apply the best online bin packing algorithm known separately for the odds subsequence and for the evens subsequence using two disjoint sets of colors. Each class of intervals that was assigned to an integer point is handled as an instance of a bin packing problem.

In the following we show that algorithms First Fit, 2-First Fit and Odd-Even bin packing have an absolute competitive ratios of at most the values 8, 6, 3.5 respectively.

Intensive research has been done analyzing the performance of the simple First Fit algorithm for the problem of interval coloring. Adamy and Erlebach [1] argue that First Fit is arbitrarily bad when introducing bandwidth. In the following theorem we show that on unit interval graph the competitive ratio of first fit is constant.

Theorem 7. *(i) Algorithm First Fit has a competitive ratio of 8 for unit interval graphs with bandwidth. (ii) Algorithm First Fit has a competitive ratio of 4 for unit interval graphs with bandwidth if each interval has a bandwidth of at most $\frac{1}{2}$.*

Proposition 3. *Algorithm 2-First Fit uses at most 6ω colors.*

Proof. 2-First Fit uses different sets of colors for intervals of bandwidth in $(0, \frac{1}{2}]$ and in $(\frac{1}{2}, 1]$. By Theorem 7 part *(ii)* First Fit on intervals with bandwidth of at most $\frac{1}{2}$ has a competitive ratio of 4. By Theorem 3 First Fit for intervals with bandwidth that exceeds $\frac{1}{2}$, the competitive ratio is 2. Combining these competitive ratios we get a competitive ratio of at most 6.

Proposition 4. *(i) Algorithm Odd Even bin packing has an absolute competitive ratio of 3.5 for coloring unit interval graphs with bandwidth, using First Fit as the online bin packing algorithm. (ii) The asymptotic competitive ratio of the algorithm Odd Even bin packing is at most 3.17778 using the algorithm Harmonic++ of Seiden [14].*

Proof. We claim that the odds subsequence can be split into different classes, such that an interval intersects all intervals within its class, but no other intervals. Each class is represented by an odd number and it contains all the intervals that were assigned to that odd integer by the algorithm. Same argument holds for the evens subsequence. Note that, since all intervals of the same class intersect, each class can be viewed as an instance to the online bin packing problem. For the first part we use the fact that First Fit for bin packing has competitive ratio of at most 1.75 with respect to the absolute measure [15]. Since we use it with two sets of colors, we get a competitive ratio of at most 3.5. For the second part we use the Harmonic++ algorithm of Seiden [14] and therefore get 3.17778.

Theorem 8. *Any online algorithm for unit interval graph with bandwidth has an absolute competitive ratio of at least 2. Any online algorithm for unit interval graph with bandwidth has an asymptotic competitive ratio of at least 1.831.*

Proof. We prove the lower bound of 2, the proof of the lower bound 1.831 is omitted. We introduce two identical intervals $I_1 = I_2 = [1, 2]$ and have bandwidth $\frac{1}{3}$. If they are assigned distinct colors then already $ALG = 2$ and $OPT = 1$, and we are done. Otherwise all future intervals have bandwidth $\frac{2}{3}$ and intersect with the previous intervals. This means that no future interval has the same color as the first two. Two further intervals are first given, $I_3 = [\frac{1}{5}, \frac{6}{5}]$ and $I_4 = [\frac{9}{5}, \frac{14}{5}]$. If they receive distinct colors, we introduce the interval $I_5 = [1, 2]$ which gets a fourth color. It is possible to color using two colors only, coloring I_1, I_3, I_4 with one color and I_2, I_5 with a second color. If they receive the same color, we introduce two intervals $I_6 = [\frac{3}{5}, \frac{8}{5}], I_7 = [\frac{7}{5}, \frac{12}{5}]$, which must receive two new colors. The total number of colors used is again 4, while it is possible to color using only two colors, one color for I_1, I_3, I_7, and a second color for I_2, I_4, I_6.

5 Lazy Online Interval Coloring

Motivated by The Maximum Resource Bin Packing Problem, we introduce a new problem called Lazy Online Interval Coloring. In this problem the objective is to use as many colors as possible. A newly presented interval can be colored by a new color only if it intersects intervals with all the previously used colors.

Theorem 9. *Any online algorithm for the problem Lazy Online Interval Coloring is arbitrarily bad.*

5.1 Unit Interval Graphs

Proposition 5. *Any online algorithm for the Lazy Online Interval Coloring with equal length intervals uses at least $\frac{OPT+1}{2}$ colors.*

Proof. In this version, since all the intervals are of equal length, if the maximum size clique equals to ω, then there is no interval that intersects with more than $2\omega - 2$ different intervals (see proof of Theorem 3). Therefore OPT can only use as much as $2\omega - 1$ colors. Since any online coloring uses at least ω we get a coloring of at least $\frac{OPT+1}{2}$ colors.

Theorem 10. *There is an upper bound of $\frac{OPT+1}{2}$ on the coloring of any online algorithm for the Lazy Online Interval Coloring problem on unit interval graphs.*

5.2 Lazy Online Coloring of Intervals with Bandwidth

In this case we show that no algorithm is competitive even on unit interval graphs.

Theorem 11. *Any online algorithm for lazy online coloring of intervals with bandwidth and equal length intervals is arbitrarily bad.*

Proof. For an integer $D > 0$, we show that the competitive ratio is at least D. We break up the construction of the sequence of intervals into $2D - 1$ phases. In the initial phase (phase 0) we provide two identical intervals of bandwidth 1 which are requests for $[0, 1]$. These are colored by two colors by any online algorithm. We denote these colors by 1 and 2. For the next phases, it suffices to show that for every phase i, we increase the number of colors used by OPT from $i + 1$ to $i + 2$ while the number of colors used by the online algorithm remains two.

Phase i: In the first step we present intersecting intervals of bandwidth $\delta = \frac{1}{2D}$. The amount of intervals never exceeds $2D$ so the total requested bandwidth is at most 2. The online algorithm can only color these intervals with colors 1 and 2. These intervals are presented such that all intervals colored by 1 by the online algorithm are slightly shifted to the left with respect to all intervals that are colored 2. Moreover all the intervals presented here intersect. We present exactly $2i + 1$ intervals in the following way.

Let $I_1 = [a, a + 1]$ be the rightmost interval colored by 1 and let $I_2 = [d, d + 1]$ be the leftmost interval colored by 2. If there is no interval colored 1 we say that I_1 is empty and If there is no interval colored 2 we say that I_2 is empty. For $0 < \varepsilon \ll \frac{1}{12D}$ a new interval, I, is presented as follows.

1. If both I_1 and I_2 are empty (presentation of the first interval) then $I = [5i - 1, 5i]$.
2. If only I_1 is empty, $I = [d - \varepsilon, d + 1 - \varepsilon]$
3. If only I_2 is empty, $I = [a - 1 + \varepsilon, a + \varepsilon]$
4. If I_1 and I_2 are not empty then, $I = [\frac{d+a}{2}, \frac{d+a}{2} + 1]$, this is an interval of length 1, located halfway between I_1 and I_2 and intersecting all previously presented intervals of this step.

After $2i + 1$ intervals there are at least $i + 1$ intervals with the same color. In the next step, an interval of bandwidth 1 that intersects these $i + 1$ intervals of the same color and does not intersect any other interval is presented. The online algorithm has to color this interval by the second color and therefore does not increase the number of colors. An offline algorithm can color the $i + 1$ intervals with $i + 1$ colors it used before. The next interval can be colored by a new color $i + 2$ since it intersects intervals that are colored with $i + 1$ distinct colors.

Acknowledgment. We would like to thank an anonymous referee for pointing out an error in an earlier version of this paper.

References

1. U. Adamy and T. Erlebach. Online coloring of intervals with bandwidth. In *Proc. of te First International Workshop on Approximation and Online Algorithms (WAOA2003)*, pages 1–12, 2003.
2. J. Boyar, L. Epstein, L. M. Favrholdt, J. S. Kohrt, K. S. Larsen, M. M. Pedersen, and S. Wøhlk. The maximum resource bin packing problem. manuscript.
3. M. Chrobak and M. Ślusarek. On some packing problems relating to dynamical storage allocation. *RAIRO Journal on Information Theory and Applications*, 22:487–499, 1988.
4. D. G. Corneil, H .Kim, S. Natarajan, S. Olariu, and A P. Sprague. Simple linear time recognition of unit interval graphs. *Information Processing Letters*, 55(2):99–104, 1995.
5. T. Erlebach and F. C. R. Spieksma. Interval selection: applications, algorithms, and lower bounds. *J. Algorithms*, 46(1):27–53, 2003.
6. T. R. Jensen and B. Toft. *Graph coloring problems*. Wiley, 1995.
7. H. A. Kierstead. The linearity of first-fit coloring of interval graphs. *SIAM Journal on Discrete Mathematics*, 1(4):526–530, 1988.
8. H. A. Kierstead and J. Qin. Coloring interval graphs with First-Fit. *SIAM Journal on Discrete Mathematics*, 8:47–57, 1995.
9. H. A. Kierstead and W. T. Trotter. An extremal problem in recursive combinatorics. *Congressus Numerantium*, 33:143–153, 1981.
10. N. S. Narayanaswamy Dynamic storage allocation and online colouring interval graphs. In *Proc of the 10th Annual International Conference on Computing and Combinatorics (COCOON2004)*, pages 329–338, 2004.
11. S. Pemmaraju, R. Raman, and K. Varadarajan. Buffer minimization using max-coloring. In *Proc. of the Fifteenth Annual ACM-SIAM Symposium on Discrete Algorithms (SODA 2004)*, pages 562–571, 2004.
12. I. Rabinovitch. The scott-suppes theorem on semiorders. *J. Math. Psych.*, 15:209–212, 1977.
13. D. Scott and P. Suppes. Foundational aspects of theories of measurement. *J. Symbolic Logic*, 23:113–128, 1958.
14. S. S. Seiden. On the online bin packing problem. *Journal of the ACM*, 49(5):640–671, 2002.
15. D. Simchi-Levi. New worst-case results for the bin-packing problem. *Naval Res. Logist.*, 41(4):579–585, 1994.

Dynamic Bin Packing of Unit Fractions Items

Wun-Tat Chan[1,*], Tak-Wah Lam[1], and Prudence W.H. Wong[2,**]

[1] Department of Computer Science,
University of Hong Kong, Hong Kong
{wtchan, twlam}@cs.hku.hk
[2] Department of Computer Science,
University of Liverpool, UK
pwong@csc.liv.ac.uk

Abstract. This paper studies the dynamic bin packing problem, in which items arrive and depart at arbitrary time. We want to pack a sequence of unit fractions items (i.e., items with sizes $1/w$ for some integer $w \geq 1$) into unit-size bins such that the maximum number of bins used over all time is minimized. Tight and almost-tight performance bounds are found for the family of any-fit algorithms, including first-fit, best-fit, and worst-fit. We show that the competitive ratio of best-fit and worst-fit is 3, which is tight, and the competitive ratio of first-fit lies between 2.45 and 2.4985. We also show that no on-line algorithm is better than 2.428-competitive. This result improves the lower bound of dynamic bin packing problem even for general items.

1 Introduction

Bin packing problem has been studied since the early 70's and different variants of the problem continue to attract researchers attentions (see the survey [6, 9, 10]). In the classical bin packing problem, we want to pack a sequence of items each with size in the range $(0, 1]$ into unit-size bins using the minimum number of bins. One of the generalizations of the problem is known as the *dynamic bin packing problem* [8], in which items arrive and depart at arbitrary time. The objective is to minimize the maximum number of bins used over all time. In this paper, we study dynamic bin packing of *unit fractions items*. A unit fraction item has size of the form $1/w$ for some integer $w \geq 1$. We analyze the performance of the family of any-fit algorithms, which includes first-fit, best-fit and worst-fit, and provide tight and almost-tight performance bounds. Our lower bound on dynamic bin packing of unit fractions items even improves the lower bound of Coffman et al. [8] on dynamic bin packing of general items.

There is a long history of results for the classical bin packing problem and its variants [6, 9, 10]. Most of the previous works considered the "static" bin packing

---

* This research was supported in part by Hong Kong RGC Grant HKU-5172/03E.
** This research was supported in part by Nuffield Foundation Grant NAL/01004/G.

where items will not depart. In this model, the off-line bin packing problem is NP-hard [11]. In the on-line setting, each item must be assigned to a bin, without knowledge of the subsequent items. Moreover, no migration of items are allowed, i.e., items are not allowed to move from one bin to another. The performance is measured in terms of competitive ratio (see [3] for a survey). The current best upper bound is due to Seiden [14], who proved that the algorithm HARMONIC++ has a competitive ratio at most 1.58889[1]. The current best lower bound is due to van Vliet [15] who showed that no on-line algorithm can achieve a competitive ratio less than 1.54014.

In many real applications, item sizes are often not arbitrary real numbers in $(0,1]$. Bar-Noy et al. [2] initiated the study of the *unit fractions bin packing problem* (UFBP), a restricted version of the classical bin packing problem in which all sizes are of the form $1/w$ for some integer $w \geq 2$. In the on-line setting, they gave an on-line algorithm with a competitive ratio $1+O(1/\sqrt{H})$, where H denotes the sum of sizes of all items. Note that this algorithm is asymptotically optimal. Bin packing with other restricted form of item sizes includes divisible item sizes [7] (where each possible item size can be divided by the next smaller item size) and discrete item sizes [5] (where possible item sizes are $\{1/k, 2/k, \cdots, j/k\}$ for some $1 \leq j \leq k$).

Dynamic bin packing is a generalization of the classical bin packing problem introduced by Coffman et al. [8]. The problem assumes that items may depart at arbitrary time, and the objective is to minimize the maximum number of bins used over all time. It was shown in their paper that the on-line algorithm first-fit has a competitive ratio lies between 2.75 and 2.897, and no on-line algorithm can achieve a competitive ratio better than 2.5. Note that these results assume a very general optimal off-line algorithm, which can re-pack the items. Coffman et al. also gave an improved lower bound of 2.388 when the off-line algorithm is not allowed to re-pack the items. Ivkovic and Lloyd [12] studied an even more general problem called the *fully dynamic bin packing problem*, where migration of items are allowed, and gave a 1.25-competitive on-line algorithm for this problem.

This paper studies the problem of dynamic bin packing of unit fractions items, the main contribution are several very close upper and lower bounds (see Table 1). We show that any-fit algorithms, which include first-fit, best-fit and worst-fit are 3-competitive. We further show that the performance of best-fit and worst-fit are indeed tight, i.e., they cannot be better than 3-competitive. On the other hand, we show that first fit has a better performance, its competitive ratio lies between 2.45 and 2.4985. In addition, we prove that no on-line algorithm can be better than 2.428-competitive. This result improves the lower bound of 2.388 of Coffman et al. [8] on dynamic bin packing for general items.

There is a problem related to UFBP, called the *windows scheduling problem* (WS) [1, 2, 4], as pointed out by Bar-Noy et al. [2]. Similar to UFBP, the input of WS is a sequence of items, each with a *window* represented by an inte-

---

[1] Seiden [14] pointed out that the previous best algorithm HARMONIC+1 by Richey [13] has competitive ratio at least 1.59217 rather than the claimed 1.58872.

Table 1. Summary of results

Algorithms	Upper bounds	Lower bounds
First-fit	2.4985	2.45
Best-fit	3	3
Worst-fit	3	3
Any-fit	3	2.428
Any on-line algorithms	–	2.428

ger. Each item represents a piece of information to be broadcast to all clients. It is assumed that all items are of the same length, which takes the same amount of time to broadcast. The objective of WS is to use the minimum number of broadcast channels to broadcast each item periodically such that the duration between two consecutive broadcasts of the same item must not exceed the window of that item. By letting the bins as broadcast channels and the reciprocal of item sizes as windows, UFBP can be considered as a special case of WS, and hence the lower bound result on UFBP applies to WS. (Note that the upper bound on UFBP does not carry over to WS.) Chan and Wong [4] considered the dynamic version of WS, in which items may also depart. They gave a 5-competitive algorithm and showed that no on-line algorithm can be better than 2-competitive. The lower bound of dynamic bin packing of unit fractions item in this paper improves the lower bound for the dynamic version of WS to 2.428.

The rest of the paper is organized as follows. Section 2 gives some definitions. Section 3 analyzes the performance of the family of any-fit algorithms. This includes upper and lower bounds for first-fit (Sections 3.1 and 3.2, respectively), and upper and lower bounds for best-fit and worst-fit (Section 3.3). Finally, Section 4 gives a lower bound for any on-line algorithm.

2 Preliminaries

In this section, we give a precise definition of the dynamic unit fractions bin packing problem and the necessary notations for further discussion. We are to pack a sequence of items into bins of unit-capacity. Items arrive and depart at arbitrary time. We denote the i-th item by m_i and its arrival time by a_i. Each item m_i comes with a size s_i which is a reciprocal of an integer, i.e., $s_i = 1/w_i$ for some integer $w_i \geq 1$. When item m_i arrives at a_i, it must be assigned to a bin immediately. At any time, the *load* of a bin is the total size of items that are currently assigned to that bin and have not yet departed, and this load must be at most 1 because of unit-capacity. Migration is not allowed in the sense that once an item is assigned to a bin, it cannot be moved to another bin. The objective is to minimize the maximum number of bins used over all time.

As with previous work, we measure the performance of an on-line algorithm in terms of a competitive ratio. Given a sequence σ of items and an on-line bin packing algorithm $\mathcal{A}$, let $\mathcal{A}(\sigma, t)$ denote the number of bins used by $\mathcal{A}$ at time t.

We say that $\mathcal{A}$ is c-competitive if there exists a constant k such that for any input sequence σ, we have $\max_t \mathcal{A}(\sigma, t) \leq c \cdot \max_t \mathcal{O}(\sigma, t) + k$, where $\mathcal{O}$ is the optimal off-line algorithm.

In this paper, we consider several on-line algorithms: any-fit, first-fit, best-fit, and worst-fit. When an item arrives, all these algorithms pack the item into an occupied bin as long as there exists such a bin that can accommodate the item; a new bin is only used when no occupied bins can accommodate the item. The algorithms differ in the rule used to choose the occupied bin for the newly arrived item. To describe the rules of these algorithms, we first define a way to label the occupied bins at a specific time.

At any time t, suppose that there are n occupied bins. For any bin X among these n bins, let $f(x) \leq t$ be the latest time X turns from empty to non-empty. At time t, we label these n non-empty bins using integers $1, 2, \ldots, n$ such that the label of bin X is less than that of bin Y if $f(X) \leq f(Y)$. Notice that the labels of bins change over time.

When a new item m_i arrives, if there is any occupied bin with load no more than $1 - 1/w_i$, the algorithms assign m_i to one of these bins as follows:

Any-fit (AF) assigns m_i to any of these bins arbitrarily.
First-fit (FF) assigns m_i to the one with the smallest label at a_i.
Best-fit (BF) assigns m_i to the heaviest load one; ties are broken arbitrarily.
Worst-fit (WF) assigns m_i to the lightest load one; ties are broken arbitrarily.

3 Performance of the Family of Any-Fit Algorithms

In this section we analyze the performance of any-fit algorithms. In Sections 3.1 and 3.2, we give an upper bound of 2.4985 and a lower bound of 2.45 for the competitive ratio of FF. Then in Section 3.3, we show that both BF and WF cannot be better than 3-competitive and then give the matching upper bounds.

3.1 Upper Bound for First-Fit

Before we analyze the upper bound of FF, let us have some definitions. Consider any positive integers x and y. Suppose that we pack a bin using items of sizes $1, 1/2, \ldots, 1/x$ only. We want to define the notion of the minimum load that such a bin must have in order that an additional item of size $1/y$ cannot be packed into the bin. We define a function $\alpha(x, y)$ to capture this notion. Formally,

$$\alpha(x, y) = \min_{1 \leq j \leq x \,\&\, n_j \geq 0} \{n_1 + n_2/2 + \ldots n_x/x \mid n_1 + n_2/2 + \ldots n_r/x > 1 - 1/y\}.$$

For example, when $x = 4$ and $y = 3$, we have $\alpha(4, 3) = 3/4$ and correspondingly $n_1 = 0$, $n_2 = 1$, $n_3 = 0$ and $n_4 = 1$.

With respect to a particular input σ, we define a sequence of integer pairs (b_i, r_i) as follows. Let b_1 denote the maximum number of bins used by FF over all time.

Suppose the smallest item that FF ever packs into a bin with label b_1 is of size $1/r_1$. We define b_i and r_i for $i \geq 2$ as follows. Let $b_i < b_{i-1}$ be the largest integer such that FF ever packs an item of size smaller than $1/r_{i-1}$ into a bin with label b_i. The size of the smallest item that FF ever packs into a bin with label b_i is denoted $1/r_i$. Let k be the largest value of i that b_i and r_i can be defined. Notice that $b_1 > b_2 > \ldots > b_k$ and $r_1 < r_2 < \ldots < r_k$.

Now we are ready for the analysis. Consider the time instance t_k when FF packs an item X of size $1/r_k$ into a bin B with label b_k. Since k is the largest index of b_i that can be defined, no item with size smaller than $1/r_k$ has ever been packed into any bin, in particular the bins with label from 1 to $b_k - 1$. Together with the fact that FF packs X into B but not bins with labels 1 to $b_k - 1$, we can conclude that at time t_k, each of the bins with labels from 1 to $b_k - 1$ must have a load at least $\alpha(r_k, r_k)$. Including item X, the total load of all bins at t_k is at least $1/r_k + (b_k - 1) \cdot \alpha(r_k, r_k)$. Let

$$\ell_k = 1/r_k + (b_k - 1) \cdot \alpha(r_k, r_k).$$

Next, for any integer $1 \leq i \leq k-1$, consider the time instance t_i when FF packs an item X_i of size $1/r_i$ into a bin with label b_i. By the definition of b_i and r_i, we can use a similar argument as before to show that: (1) each of the bins with labels from 1 to b_k must have load at least $\alpha(r_k, r_i)$; (2) for any integer p with $k > p \geq i+1$, each of the bins with labels from $b_{p+1}+1$ to b_p must have load at least $\alpha(r_p, r_i)$; and (3) each of the bins with labels from $b_{i+1}+1$ to $b_i - 1$ must have load at least $\alpha(r_i, r_i)$. Including item X_i, the total load of all bins at time t_i is at least $1/r_i + (b_i - b_{i+1} - 1) \cdot \alpha(r_i, r_i) + \sum_{p=i+1}^{k-1}(b_p - b_{p+1}) \cdot \alpha(r_p, r_i) + b_k \cdot \alpha(r_k, r_i)$. For $1 \leq i \leq k-1$, let

$$\ell_i = 1/r_i + (b_i - b_{i+1} - 1) \cdot \alpha(r_i, r_i) + \sum_{p=i+1}^{k-1}(b_p - b_{p+1}) \cdot \alpha(r_p, r_i) + b_k \cdot \alpha(r_k, r_i).$$

Let $\ell = \max_{1 \leq i \leq k} \ell_i$. The number of bins used by the optimal off-line algorithm is at least ℓ. On the other hand, the maximum number of bins used by FF is b_1. Below we show that $b_1 < 2.4985\ell + 1.337$, which implies the following theorem.

Theorem 1. *First-fit is* 2.4985-*competitive.*

To prove Theorem 1, we assume $k \geq 5$. The case for $k < 5$ can be proved similarly and will be given in the full paper. Depending on the values of r_i's, we consider the following six sub-cases: **Case 1:** $r_1 \geq 2$; **Case** i, for $2 \leq i \leq 5$: $r_1 = 1, r_2 = 2, \cdots, r_{i-1} = i-1$, and $r_i \geq i+1$; and **Case 6:** $r_1 = 1, r_2 = 2, \cdots, r_5 = 5$. We analyze the relationship between b_1 and ℓ case by case in each of the following lemmas.

Lemma 1. *If* $r_1 \geq 2$, *then* $b_1 < 2\ell + 1$.

Proof. Since $\alpha(x, y) > 1 - 1/y$ for any integers x and y, we have $\ell_1 > 1/r_1 + (b_1 - b_2 - 1)(1 - 1/r_1) + \sum_{p=2}^{k-1}(b_j - b_{j+1})(1 - 1/r_1) + b_k(1 - 1/r_1) = 1/r_1 + (b_1 - 1)(1 - 1/r_1)$. By simple arithmetic, we have $b_1 < \ell_1 r_1/(r_1 - 1) + (r_1 - 2)/(r_1 - 1) < 2\ell + 1$; the latter inequality holds because $r_1 \geq 2$ and $\ell_1 \leq \ell$.

Lemma 2. *If $r_1 = 1$ and $r_2 \geq 3$, then $b_1 < 2.4445\ell + 1$.*

Proof. Notice that for any integer x, $\alpha(x, 1)$ is the minimum value in the form of $n_1 + n_2/2 + n_3/3 + \ldots + n_x/x$ that is greater than 0; therefore, $\alpha(x, 1)$ must be equal to $1/x$. Then, we have $\ell_1 = 1/1 + (b_1 - b_2 - 1) \cdot \alpha(1,1) + \sum_{p=2}^{k-1}(b_p - b_{p+1}) \cdot \alpha(r_p, 1) + b_k \cdot \alpha(r_k, 1) > (b_1 - b_2) + (b_2 - b_3)/r_2$.

Next, by the definition that $\alpha(x, y) \geq 1 - 1/y$ for any integers x and y, we have $\ell_2 = 1/r_2 + (b_2 - b_3 - 1) \cdot \alpha(r_2, r_2) + \sum_{p=3}^{k-1}(b_p - b_{p+1}) \cdot \alpha(r_p, r_2) + b_k \cdot \alpha(r_k, r_2) > (b_2 - 1)(1 - 1/r_2)$. Similarly, we have $\ell_3 > (b_3 - 1)(1 - 1/r_3)$. By solving the three inequalities, we have $b_1 < 22\ell/9 + 1 < 2.4445\ell + 1$.

Lemma 3. *If $r_1 = 1$, $r_2 = 2$, and $r_3 \geq 4$, then $b_1 < 2.4792\ell + 1.25$.*

Proof. Recall that $\alpha(x, 1) > 1/x$ for any integer x. We have $\ell_1 > (b_1 - b_2) + (b_2 - b_3)/2 + (b_3 - b_4)/r_3$. By the fact that $\alpha(2,2) = 1$ and $\alpha(x, 2) > 1/2$ for any integer x, we have $\ell_2 = 1/2 + (b_2 - b_3 - 1) \cdot \alpha(2,2) + \sum_{p=3}^{k-1}(b_p - b_{p+1}) \cdot \alpha(r_p, 2) + b_k \cdot \alpha(r_k, 2) > (b_2 - b_3 - 1) + b_3/2$. We can also prove that $\ell_3 > (b_3 - 1)(1 - 1/r_3)$ and $\ell_4 > (b_4 - 1)(1 - 1/r_4)$. By solving the four inequalities, we have $b_1 < 119\ell/48 + 5/4 < 2.4792\ell + 1.25$, and the lemma follows.

Lemma 4. *If $r_1 = 1$, $r_2 = 2$, $r_3 = 3$, and $r_4 \geq 5$, then $b_1 < 2.4942\ell + 1.3167$.*

Proof. Using the same approach, we have

$\ell_1 > (b_1 - b_2) + (b_2 - b_3)/2 + (b_3 - b_4)/3 + (b_4 - b_5)/r_4$,
$\ell_2 > (b_2 - b_3 - 1) + (b_3 - b_4)(2/3) + b_4/2$, $\quad \{\because \alpha(3,2) = 2/3\}$
$\ell_3 > ((b_3 - b_4 - 1)(5/6) + 2b_4/3$, $\quad \{\because \alpha(3,3) = 5/6\}$
$\ell_4 > (b_4 - 1)(1 - 1/r_4)$, and
$\ell_5 > (b_5 - 1)(1 - 1/r_5)$.

By solving the five inequalities, we have $b_1 < 2993\ell/1200 + 79/60 < 2.4942\ell + 1.3167$, and the lemma follows.

Lemma 5. *If $r_i = i$ for all $1 \leq i \leq 4$, and $r_5 \geq 6$, then $b_1 < 2.49345\ell + 1.3325$.*

Proof. Using the same approach, we have

$\ell_1 > (b_1 - b_2) + (b_2 - b_3)/2 + (b_3 - b_4)/3 + (b_4 - b_5)/4 + (b_5 - b_6)/r_5$,
$\ell_2 > (b_2 - b_3 - 1) + (b_3 - b_4)(2/3) + (b_4 - b_5)(7/12) + b_5/2$, $\{\because \alpha(4,2) = 7/12\}$
$\ell_3 > (b_3 - b_4 - 1)(5/6) + (b_4 - b_5)(3/4) + 2b_5/3$, $\quad \{\because \alpha(4,3) = 3/4\}$
$\ell_4 > ((b_4 - b_5 - 1)(5/6) + b_5(3/4)$, $\quad \{\because \alpha(4,4) = 5/6\}$
$\ell_5 > (b_5 - 1)(1 - 1/r_5)$, and
$\ell_6 > (b_6 - 1)(1 - 1/r_6)$.

Solving these inequalities, we have $b_1 < 2.4935\ell + 1.3325$, and the lemma follows.

Lemma 6. *If $r_i = i$ for all $1 \leq i \leq 5$, then $b_1 < 2.4985\ell + 1.337$.*

The proof of Lemma 6 uses a similar approach as in Lemmas 1 to 5, and the details will be given in the full paper. By Lemmas 1 to 6, Theorem 1 follows.

In fact, we conjecture that the worst case happens when $r_i = i$ for all $1 \leq i \leq k$. In that case, the computed competitive ratio is approaching 2.48.

3.2 Lower Bound for First-Fit

In this section we give a lower bound for FF by constructing an adversary sequence of items such that the maximum number of bins used by FF is at least 2.45 times that used by the optimal off-line algorithm. For any positive integers x and y, define $\beta(x, y)$ to be the minimum number of items of size $1/x$ in a bin such that an additional item of size $1/y$ cannot be packed into the bin. Formally,

$$\beta(x, y) = \min_{z \in \mathcal{Z}^+} \{z \mid z/x > 1 - 1/y\},$$

i.e., $\beta(x, y) = 1 + x - \lceil x/y \rceil$. For example, if $x = 4$ and $y = 3$, then $\beta(x, y) = 3$.

Let n be an integer and let $D = n!$. The adversary sequence consists of n stages. In each stage, some items released in the previous stage depart and a number of new items of the same size are released. The choices of which items to depart depend on how FF packs the items in previous stages. In Stage 1, Dn items of size $1/n$ are released. FF packs all Dn items into D bins, and each bin is fully packed.

For subsequent stages, i.e., Stage i, for $2 \leq i \leq n$, the adversary targets to maintain an invariant on how FF packs the items: At the beginning of Stage i,

- each occupied bin contains only items of the same size; and
- a bin that contains items of size $1/x$ contains $\beta(x, n - i + 2)$ items.

The invariant holds at the beginning of Stage 2 because each occupied bin contains $\beta(n, n) = n$ items of size $1/n$. Stage i consists of two steps.

1. For each occupied bin, if it contains items of size $1/x$, we arbitrarily choose $\beta(x, n - i + 2) - \beta(x, n - i + 1)$ items and let them depart, in other words, there are $\beta(x, n-i+1)$ items remained. Let D_i be the sum of item size for all the departed items. We will prove later that D_i is an integer (see Lemma 7).
2. Next, $D_i(n-i+1)$ items of size $1/(n-i+1)$ are released. Since each bin with item of size $1/x$ contains $\beta(x, n - i + 1)$ items, none of the newly released items can be packed into any occupied bin. Therefore, FF will use D_i empty bins to pack all these items, each bin contains $n-i+1 = \beta(n-i+1, n-i+1)$ items. Thus, the invariant also holds at the beginning of Stage $i + 1$.

Define $D_1 = D$, which is the number of empty bins used in Stage 1. From the above discussion, we can see that the number of empty bins required for items of size $1/(n-i+1)$ in Stage i is D_i. Then, at the beginning of Stage i, there will be D_j bins each with items of size $1/(n - j + 1)$ for all $1 \leq j \leq i - 1$. Therefore, the sum of the size of all departed items in Stage i satisfies:

$$D_i = \sum_{j=1}^{i-1} \left\{ \frac{D_j(\beta(n - j + 1, n - i + 2) - \beta(n - j + 1, n - i + 1))}{n - j + 1} \right\}.$$

Lemma 7. *D_i is an integer multiple of $(n - i + 1)$ for $1 \leq i \leq n$.*

Proof. We prove by induction a stronger claim that D_i is an integer multiple of $(n-i+1)!$ for $1 \leq i \leq n$. It is clear that $D_1 = D$ is an integer multiple of $n!$. Suppose D_i is an integer multiple of $(n-i+1)!$ for $1 \leq i \leq k$. We have $D_{k+1} = \sum_{j=1}^{k}(D_j/(n-j+1))(\beta(n-j+1, n-k+1) - \beta(n-j+1, n-k))$. Since the function β gives an integer output and $D_j/(n-j+1)$ is an integer multiple of $(n-j)!$, the summation gives an integer multiple of $(n-k)!$, which completes the induction.

The following lemmas give the performance of FF on the adversary sequence.

Lemma 8. *There exists some integer n such that the maximum number of bins used by FF is at least $2.45D$.*

Proof. After Stage n, FF uses $\sum_{i=1}^{n} D_i$ bins. We carry out the analysis on the value of $\sum_{i=1}^{n} D_i$ by actually computing the value of $\sum_{i=1}^{n} D_i$ with different values of n. We find that the increase in n generally leads to an increase in $\sum_{i=1}^{n} D_i$, though not monotonically. By letting $n = 21421$, and computing the values of D_i, we have $\sum_{i=1}^{n} D_i > 2.45D$.

Lemma 9. *The optimal off-line algorithm uses at most D bins at any time.*

Proof. We give an algorithm $\mathcal{O}$ to pack the items in the adversary such that $\mathcal{O}$ uses at most D bins over all time. In this proof permanent items refer to the items remain after Stage n and temporary items refer to the items depart in Stage n or before.

The algorithm $\mathcal{O}$ runs as follows. In each stage, when there are new items released, $\mathcal{O}$ packs the new items using the minimum number of empty bins such that a bin contains only permanent items, or temporary items that will depart in the same stage. We claim that $\mathcal{O}$ uses exactly D bins after each stage. In the initial stage, $\mathcal{O}$ packs the $D = n!$ permanent items into $(n-1)!$ bins and the $(n-1)n!$ temporary items to another $(n-1)(n-1)!$ bins. Totally, there are $n! = D$ occupied bins.

We prove that in each subsequent Stage i, for $2 \leq i \leq n$, the departed items produce D_i empty bin and $\mathcal{O}$ uses the D_i empty bins to pack the $D_i(n-i+1)$ items of size $1/(n-i+1)$ released. First, the number of empty bins produced in Stage i equals $\sum_{j=1}^{i-1} \lfloor D_j(\beta(n-j+1, n-i+2) - \beta(n-j+1, n-i+1))/(n-j+1) \rfloor$, which is equal to D_i because by Lemma 7, the term $D_j/(n-j+1)$ is an integer. Second, among the $D_i(n-i+1)$ items of size $1/(n-i+1)$ released in Stage i, the total size of those items that will depart in Stage p, for $i+1 \leq p \leq n$, is $D_i(\beta(n-j+1, n-p+2) - \beta(n-j+1, n-p+1))/(n-i+1)$, which is an integer because by Lemma 7, $D_i/(n-i+1)$ is an integer. Thus, we show that $\mathcal{O}$ can use D_i empty bins to pack all $D_i(n-i+1)$ items of size $1/(n-i+1)$. In other words, $\mathcal{O}$, and thus the optimal off-line algorithm, uses at most D bins at any time, and the lemma follows.

By Lemmas 8 and 9, the following theorem holds.

Theorem 2. *FF is at least 2.45-competitive.*

3.3 Performance of Other Any-Fit Algorithms

We show that BF and WF have a worse performance than FF, precisely, we show that BF and WF cannot be better than 3-competitive. On the other hand, we give the matching upper bounds. We prove in the Appendix that AF, including BF and WF, is 3-competitive.

Theorem 3. *Any-fit is 3-competitive.*

We give an adversary for WF as follows. Let k be an arbitrarily large integer constant and let $w = 2k$. The sequence contains $5k$ items, $m_1, m_2, \ldots, m_{5k}$, with m_i arriving at time i. There are three different sizes of the items: (1) $s_i = 1/2$ for $i = 1, 3, \ldots, 4k-1$; (2) $s_i = 1/w$ for $i = 2, 4, \ldots, 4k$; and (3) $s_i = 1$ for $i = 4k+1, 4k+2, \ldots, 5k$. All items of size $1/2$ depart at time $4k$ while items of size $1/w$ and 1 never depart. In the Appendix we show that the maximum number of bins used by WF is at least $3k$ and by the optimal off-line algorithm is at most $k+1$. For any $0 < \epsilon \le 3/2$, setting $k = 3/\epsilon - 1$ results in the competitive ratio $3k/(k+1) > (3-\epsilon)$. Hence, we have the following theorem.

Theorem 4. *Worst-fit is no better than 3-competitive.*

Next, we give an adversary for BF. Let k be an arbitrarily large integer constant and let $w = 2k$. The adversary sequence consists of $2k$ stages, each lasts for 4 time units. Precisely, Stage i spans from time $4i+1$ to $4i+4$. There are three different sizes of items: $1/w$, $1/2$ and 1, all items of size $1/2$ will depart at some time while items of size $1/w$ and 1 never depart. Before Stage 0, two items are released, one with size $1/2$, and the other with size $1/w$. The stages proceed as follows. **Stage i**, for $0 \le i \le 2k-2$: At time $4i+1$, i items of size $1/2$ are released. At time $4i+2$, one more item of size $1/2$ is released. At time $4i+3$, all items of size $1/2$ released before time $4i+2$ depart, including those released at time $4i+1$ and the one released in Stage $i-1$. At time $4i+4$, a single item with size $1/w$ is released. **Stage $2k-1$**: At time $4(2k-1)+3$, the item with size $1/2$ released in Stage $(2k-2)$ departs. At time $4(2k-1)+4$, k items of size 1 are released. In the Appendix, we show that the maximum number of bins used by BF is at least $3k$ and that by the optimal off-line algorithm is at most $k+1$. For any $0 < \epsilon \le 3/2$, setting $k = 3/\epsilon - 1$ results in the competitive ratio $3k/(k+1) > (3-\epsilon)$. Hence, we have the following theorem.

Theorem 5. *Best-fit is no better than 3-competitive.*

4 General Lower Bound

We give an adversary sequence of items such that the maximum number of bins used by any on-line algorithm is at least 2.428 times that used by the optimal off-line algorithm. First, we need the following notion. For any positive integers x and y, define $\lambda(x, y)$ to be the maximum number of items of size $1/x$ that can be packed into a bin containing an item of size $1/y$. Formally,

$$\lambda(x,y) = \max_{z \in \mathcal{Z}^+}\{z \mid z/x \le 1 - 1/y\},$$

i.e., $\lambda(x,y) = x - \lceil x/y \rceil$. For example, if $x=4$ and $y=3$, then $\lambda(x,y)=2$.

Consider any on-line algorithm $\mathcal{A}$. Let n be an integer and let $F = n!(n-1)!$. The adversary sequence consists of n stages and has the following properties: In each stage, some items released in the previous stage depart and a number of items of the same size are released. The arrival of items in the adversary sequence ensures that at the end of each stage, $\mathcal{A}$ has to use some additional bins to pack the items released in that stage. We are going to define a sequence of numbers F_i for $1 \le i \le n$, which is related the number of additional bins required in Stage i.

In Stage 1, Fn items of size $1/n$ are released. The algorithm $\mathcal{A}$ uses at least F bins to pack the Fn items. If $\mathcal{A}$ uses more than F bins, all items in bins other than the first F bins depart. We define F_1 to be F.

In each of the subsequent stages, i.e., Stage i, for $2 \le i \le n$, there are three steps. (1) For each occupied bin, all its items except the smallest one depart. (2) Let R_i be the total size of the items remained. (We will prove later, in the proof of Lemma 11, that R_i is indeed an integer.) The adversary then releases $(F - R_i)(n - i + 1)$ items of size $1/(n - i + 1)$. At this point, the total size of all items not yet departed, including those released in previous stages, is F. (3) Define

$$F_i = F - \sum_{j=1}^{i-1} F_j \left(\frac{1}{n-j+1} + \frac{\lambda(n-j+1, n-i+1)}{n-i+1} \right).$$

If $\mathcal{A}$ uses more than F_i additional bins in Stage i, all items packed into the additional bins other than the first F_i additional bins depart. Roughly speaking, F_i is the minimum number of additional bins required in Stage i; the term $\lambda(n-j+1, n-i+1)$ reflects the maximum number of items released in Stage i that can be packed into an occupied bin which was an additional bin in Stage j. We will prove this formally in Lemma 11. We first prove a property of F_i.

Lemma 10. F_i is an integer multiple of $(n-i+1)!(n-i)!$ for $1 \le i \le h$.

Proof. We prove the lemma by induction. It is clear that $F_1 = F$ is an integer multiple of $n!(n-1)!$. Suppose F_i is an integer multiple of $(n-i+1)!(n-i)!$ for $1 \le i \le k$. We have $F_{k+1} = F - \sum_{j=1}^{k} F_j(1/(n-j+1) + \lambda(n-j+1, n-i+1)/(n-k))$. Since the function λ gives an integer output and $F_j/((n-j+1)(n-k))$ is an integer multiple of $(n-k)!(n-k-1)!$ as $k \ge j$, the summation gives an integer multiple of $(n-k)!(n-k-1)!$, which completes the induction.

Lemma 11. *For $1 \le i \le n$, $\mathcal{A}$ uses no less than F_i additional bins in Stage i.*

Proof. In Stage 1, it is clear that $\mathcal{A}$ uses at least $F_1 = F$ additional bins. We show by induction that in Step (2) of Stage i for $2 \le i \le n$, $\mathcal{A}$ also uses at least F_i additional bins. Assume that it is true for $i = k$. Before Step (2) of Stage $k+1$, $\mathcal{A}$ already has F_j bins containing a single item of size $1/(n-j+1)$ for $1 \le j \le k$. Therefore, $R_{k+1} = \sum_{j=1}^{k} F_j/(n-j+1)$. We can see that R_{k+1} is an integer as

F_j is an integer multiple of $(n-j+1)!(n-j)!$. The number of items of size $1/(n-k)$ released is $(F - R_{k+1})(n-k) = (F - \sum_{j=1}^{k} F_j/(n-j+1))(n-k)$. The number of items that can be packed into the occupied bins is equal to $\sum_{j=1}^{k} F_j \cdot \lambda(n-j+1, n-k)$. Therefore, the number of additional bins needed in Stage $k+1$ is at least

$$F - \sum_{j=1}^{k} \frac{F_j}{n-j+1} - \sum_{j=1}^{k} \frac{F_j \cdot \lambda(n-j+1, n-k)}{n-k}$$

$$= F - \sum_{j=1}^{k} F_j \left(\frac{1}{n-j+1} + \frac{\lambda(n-j+1, n-k)}{n-k} \right) = F_{k+1}.$$

This induction is completed.

Lemma 12. *There exists some integer n such that the maximum number of bins used by $\mathcal{A}$ is at least $2.428F$.*

Proof. After Stage n, $\mathcal{A}$ uses $\sum_{i=1}^{n} F_i$ bins. We carry out the analysis on the value of $\sum_{i=1}^{n} F_i$ by actually computing the value of $\sum_{i=1}^{n} F_i$ with different values of n. We find that the increase in n results an increase in $\sum_{i=1}^{n} F_i$ monotonically. In particular, by letting $n = 12794$, and computing the values of F_i, we have $\sum_{i=1}^{n} F_i > 2.428F$.

Theorem 6. *Any on-line algorithm is at least 2.428-competitive.*

Proof. Similar to Lemma 9, we can show that the optimal off-line algorithm uses at most F bins. Together with Lemma 12, the theorem follows.

References

1. A. Bar-Noy and R. E. Ladner. Windows scheduling problems for broadcast systems. *SIAM J. Comput.*, 32(4):1091–1113, 2003.
2. A. Bar-Noy, R. E. Ladner, and T. Tamir. Windows scheduling as a restricted version of bin packing. In J. I. Munro, editor, *SODA*, pages 224–233. SIAM, 2004.
3. A. Borodin and R. El-Yaniv. *Online Computation and Competitive Analysis*. Cambridge University Press, 1998.
4. W.-T. Chan and P. W. H. Wong. On-line windows scheduling of temporary items. In R. Fleischer and G. Trippen, editors, *ISAAC*, volume 3341 of *Lecture Notes in Computer Science*, pages 259–270. Springer, 2004.
5. E. G. Coffman, Jr., C. Courcoubetis, M. R. Garey, D. S. Johnson, P. W. Shor, R. R. Weber, and M. Yannakakis. Bin packing with discrete item sizes, Part I: Perfect packing theorems and the average case behavior of optimal packings. *SIAM J. Discrete Math.*, 13:38–402, 2000.
6. E. G. Coffman, Jr., G. Galambos, S. Martello, and D. Vigo. Bin pakcing approximation algorithms: Combinatorial analysis. In *Handbook of Combinatorial Optimization*. Kluwer Academic Publishers, 1998.
7. E. G. Coffman, Jr., M. Garey, and D. Johnson. Bin packing with divisible item sizes. *Journal of Complexity*, 3:405–428, 1987.

8. E. G. Coffman, Jr., M. R. Garey, and D. S. Johnson. Dynamic bin packing. *SIAM J. Comput.*, 12(2):227–258, 1983.
9. E. G. Coffman, Jr., M. R. Garey, and D. S. Johnson. Bin packing approximation algorithms: A survey. In D. S. Hochbaum, editor, *Approximation Algorithms for NP-Hard Problems*, pages 46–93. PWS, 1996.
10. J. Csirik and G. J. Woeginger. On-line packing and covering problems. In A. Fiat and G. J. Woeginger, editors, *On-line Algorithms—The State of the Art*, volume 1442 of *Lecture Notes in Computer Science*, pages 147–177. Springer, 1996.
11. M. R. Garey and D. S. Johnson. *Computers and Intractability: A Guide to the Theory of NP-Completeness*. W. H. Freeman, San Francisco, 1979.
12. Z. Ivkovic and E. L. Lloyd. Fully dynamic algorithms for bin packing: Being (mostly) myopic helps. *SIAM J. Comput.*, 28(2):574–611, 1998.
13. M. B. Richey. Improved bounds for harmonic-based bin packing algorithms. *Discrete Applied Mathematics*, 34:203–227, 1991.
14. S. S. Seiden. On the online bin packing problem. *J. ACM*, 49(5):640–671, 2002.
15. A. van Vliet. An improved lower bound for on-line bin packing algorithms. *Inf. Process. Lett.*, 43(5):277–284, 1992.

Appendix

Proof of Theorem 3: Consider any input seqeunce σ. Suppose AF uses at most n bins. The proof is based on two notions. (1) Let t_1 be a time instance such that AF uses n occupied bins, and n_1 be the number of occupied bins with item of size 1 at time t_1. Notice that the optimal off-line algorithm uses at least n_1 occupied bins at time t_1. (2) Let n_2 be the largest integer such that AF packs an item m of size $1/2$ or less into an empty bin with label n_2 and suppose this happens at t_2. At time t_2, all bins with labels smaller than n_2 have load greater than $1/2$; otherwise, AF can pack m to one of these bins. In that case, the optimal off-line algorithm uses at least $\lceil n_2/2 \rceil$ bins at time t_2. By the definition of n_2, at time t_1, every bin with label greater than n_2 contains an item of size greater than $1/2$, i.e., 1. Hence, we have $n \leq n_1 + n_2$. On the other hand, the optimal off-line algorithm uses at least $\max\{n_1, \lceil n_2/2 \rceil\}$ bins. By simple arithmetic, we have $n \leq 3\max\{n_1, \lceil n_2/2 \rceil\}$ (the worst case happens when $n_1 = \lceil n_2/2 \rceil$).

Proof of Theorem 4: We first describe how WF packs the $5k$ items in the adversary. WF packs the item m_{2j-1} of size $1/2$ and m_{2j} of size $1/w$ to the same bin with label j, for $1 \leq j \leq 2k$. After all items of size $1/2$ depart, there are $2k$ occupied bins; k more bins are needed for the items of size 1. Therefore, WF uses $3k$ bins.

On the other hand, the optimal off-line algorithm can use a single bin to pack all the items of size $1/w$ and k bins to pack the items of size $1/2$. This packing uses $k+1$ bins. After all the items of size $1/2$ depart, the k bins can be used to pack the items of size 1. Thus, the optimal off-line algorithm uses at most $k+1$ bins, and the competitive ratio of WF is at least $3k/(k+1)$. For any $0 < \epsilon \leq 3/2$, picking k to be $3/\epsilon - 1$ implies that WF is no better than $(3-\epsilon)$-competitive.

Proof of Theorem 5: We first describe how BF packs the items in the adversary. We claim that for $1 \leq i \leq 2k-1$, at the beginning of Stage i, BF uses $i+1$

bins, one of them has load $1/2 + 1/w$, and the other i bins each has load $1/w$. The base case for Stage 1 can be verified easily. Suppose the claim is true for some $i \geq 1$. Consider what happens in Stage i. At time $4i+1$, BF packs each of the i new items into the i bins with load $1/w$. All the $i+1$ occupied bins now have load $1/2 + 1/w$. The item of size $1/2$ released at time $4i+2$ must then be packed into a new bin. After the departure of items of size $1/2$ at time $4i+3$, we are left with a bin with load $1/2$ and $i+1$ bins each with load $1/w$. When the item of size $1/w$ is released at time $4i+4$, BF packs it into the bin with load $1/2$. Then, at the beginning of Stage $i+1$, BF uses $i+2$ bins where $i+1$ of them have load $1/w$ and one has load $1/2 + 1/w$, and the claim follows. Finally, in Stage $2k-1$, BF needs k more bins, and thus uses at least $2k + k = 3k$ bins.

On the other hand, the optimal off-line algorithm can reserve a single bin for the $2k$ items of size $1/w$. At any time, there are at most $2k$ items of size $1/2$ which can be packed into k bins. In the final stage, all these items depart and the k bins can be used for the items of size 1. Hence, the optimal off-line algorithm uses at most $k+1$ bins, and the competitive ratio of BF is at least $3k/(k+1)$. For any $0 < \epsilon \leq 3/2$, picking k to be $3/\epsilon - 1$ implies that BF is no better than $(3-\epsilon)$-competitive.

Reordering Buffer Management for Non-uniform Cost Models*

Matthias Englert and Matthias Westermann

Department of Computer Science,
RWTH Aachen, D-52056 Aachen, Germany
{englert, marsu}@cs.rwth-aachen.de

Abstract. A sequence of objects which are characterized by their color has to be processed. Their processing order influences how efficiently they can be processed: Each color change between two consecutive objects produces non-uniform cost. A reordering buffer which is a random access buffer with storage capacity for k objects can be used to rearrange this sequence in such a way that the total cost are minimized. This concept is useful for many applications in computer science and economics.

We show that a reordering buffer reduces the cost of each sequence by a factor of at most $2k - 1$. This result even holds for cost functions modeled by arbitrary metric spaces. In addition, a matching lower bound is presented. From this bound follows that each strategy that does not increase the cost of a sequence is at least $(2k - 1)$-competitive.

As main result, we present the deterministic Maximum Adjusted Penalty (MAP) strategy which is $O(\log k)$-competitive. Previous strategies only achieve a competitive ratio of k in the non-uniform model. For the upper bound on MAP, we introduce a basic proof technique. We believe that this technique can be interesting for other problems.

1 Introduction

Frequently, a number of tasks has to be processed and their processing order influences how efficiently they can be processed. Hence, a reordering buffer can be expedient to influence the processing order. This concept is useful for many applications in computer science and economics. In the following and in Sect. 1.2 we give some examples.

In computer graphics, a rendering system displays a 3D scene which is composed of primitives. In current rendering systems a significant factor for the performance are the state changes performed by the graphics hardware. A state change occurs when two consecutively rendered primitives differ in their attribute values, e.g., in their texture or shader program. These state changes slow down a rendering system. Note that the duration of a state change is non-uniform and heavily depends on the attribute values of the primitive causing this state change, e.g., textures and shader programs vary significantly in size which has

---

* Supported by the DFG grant WE 2842/1.

a great impact on the state change. To reduce the cost of the state changes, a reordering buffer can be included between application and graphics hardware. This reordering buffer, which is a random access buffer with limited memory capacity, can be used to rearrange the incoming sequence of primitives online in such a way that the cost of the state changes are minimized.

Hard disks consist of one or more rotating platters. A read/write head is positioned above the rotating surface of each platter. The position of a head determines which cylinder can be accessed. The latency of an access is mainly induced by the movement of the head to the respective cylinder. The latencies are the dominating factor for the performance of a hard disk. This can be modeled by a non-uniform metric, e.g., the line metric: Accesses are categorized according to their destination cylinder, and the cost are defined as the distance between start and destination cylinder. A reordering buffer can be used to rearrange the incoming sequence of accesses online in such a way that latencies are minimized. This problem is known as disk scheduling (see, e.g., [1]).

File servers are high-capacity storage devices which each computer in a network can access to retrieve files. A file on a server is denoted as open, if it is ready to be accessed. Otherwise, it is denoted as closed. By technical reasons, the number of open files on a server is limited. The overhead induced by the opening and closing processes is a significant factor for the performance of a file server. Note that the cost of an opening or closing process is non-uniform and depends on the characteristics of the involved file. This overhead can be minimized by preceding a file server with a reordering buffer. If several open files are allowed, this scenario is a generalization of the rendering scenario. In addition to the possibility to choose a file that has to be opened next, there is the possibility to choose a file that has to be closed in the case that the maximum number of files is open. This scenario is equivalent to the classical paging problem (see, e.g., [2]) if the reordering buffer has storage capacity for only one file.

1.1 The Model

An *input sequence* $\sigma = \sigma_1 \sigma_2 \cdots$ of objects which are only characterized by a specific attribute has to be processed. To simplify matters, we suppose that the objects are characterized by their color, and, for each object σ_i, let $c(\sigma_i)$ denote the color of σ_i. A *reordering buffer* which is a random access buffer with storage capacity for k objects can be used to rearrange the input sequence in the following way.

The *current input object* σ_i, i.e., the first object of σ that is not handled yet, can be stored in the reordering buffer, or objects currently stored in the reordering buffer can be removed. These removed objects result in an *output sequence* $\sigma_{\pi^{-1}} = \sigma_{\pi^{-1}(1)} \sigma_{\pi^{-1}(2)} \cdots$ which is a partial permutation of σ. Let the *current output object* denote the object that was last assigned to the output sequence. We suppose that the reordering buffer is initially empty and, after processing the whole input sequence, the buffer is empty again.

For each color c, we are given *weight* b_c. Cost b_c are produced for each color change to color c, i.e., for two consecutive objects $\sigma_{\pi^{-1}(i)}$ and $\sigma_{\pi^{-1}(i+1)}$ of the output sequence, we define the cost function $d(\sigma_{\pi^{-1}(i)}, \sigma_{\pi^{-1}(i+1)}) = b_{c(\sigma_{\pi^{-1}(i+1)})}$, if $c(\sigma_{\pi^{-1}(i)}) \neq c(\sigma_{\pi^{-1}(i+1)})$, and $d(\sigma_{\pi^{-1}(i)}, \sigma_{\pi^{-1}(i+1)}) = 0$, otherwise. Then, the goal is to minimize the *cost* $C^A(\sigma) = \sum_i d(\sigma_{\pi^{-1}(i)}, \sigma_{\pi^{-1}(i+1)})$ *of a management strategy A*. Note that this models the presented application examples well except disk scheduling (see also Sect. 1.3).

The notion of an online strategy is intended to formalize the realistic scenario, where the strategy does not have knowledge about the whole input sequence in advance. The online strategy has to serve the input sequence σ one after the other, i.e., a new object is not issued before there is a free slot in the reordering buffer. Online strategies are typically evaluated in a competitive analysis. In this kind of analysis the cost of the online strategy are compared with the cost of an optimal offline strategy. For a given sequence σ, let $C^{\mathrm{OP}}(\sigma)$ denote the minimum cost produced by an optimal offline strategy. An online strategy is denoted as α-*competitive* if it produces cost at most $\alpha \cdot C^{\mathrm{OP}}(\sigma) + \kappa$, for each sequence σ, where κ is a term that does not depend on σ. The value α is also called the *competitive ratio* of the online strategy.

W.l.o.g., we only consider *lazy* strategies, i.e., strategies that fulfill the following two properties:

- If an object with the same color as the current output object is stored in the reordering buffer, a lazy strategy does not make a color change.
- If the current input object can be stored in the reordering buffer, a lazy strategy does not remove an object from the reordering buffer.

Note that every (in particular every optimal offline) strategy can be transformed into a lazy strategy without increasing the cost.

1.2 Previous Work

Web caching with request reordering extends the classic paging model by allowing reordering of requests under the constraint that a request is delayed by no longer than a predetermined number of time steps (see, e.g., [3,4]). Albers [3] presents a deterministic strategy that achieves an optimal competitive ratio of $k+1$, where k denotes the storage capacity of the cache. Feder et al. [4] introduce a randomized strategy that achieves an asymptotically optimal competitive ratio of $\Theta(\log k)$.

The uniform case of our problem is studied in, e.g., [5,6]. In the uniform model, for each color c, weight $b_c = 1$, i.e., just the number of color changes is considered. Räcke, Sohler and Westermann [5] show that several standard strategies are unsuitable for a reordering buffer, i.e., the competitive ratio of the First In First Out and Least Recently Used strategy is $\Omega(\sqrt{k})$ and the competitive ratio of the Most Common First strategy is $\Omega(k)$, where k denotes the buffer size. Further, the deterministic Bounded Waste strategy is presented and it is proven that this strategy achieves a competitive ratio of $O(\log^2 k)$ in the uniform model. Kohrt and Pruhs [6] present a polynomial-time offline algorithm that achieves a constant approximation ratio. However, their goal is to maximize the number of saved color changes. Note that a constant approximation of the minimal number of color changes in the output sequence is preferable, if it is possible to save a large number of color changes.

Krokowski et al. [7] examine the previously mentioned rendering application in an uniform version, i.e., just the number of state changes is considered. They use a small reordering buffer to rearrange the incoming sequence of primitives online in such a way that the number of state changes is minimized. Due to its simple structure and its low memory requirements this method can easily be implemented in software or even hardware. In their experimental evaluation this method typically reduces the number of state changes by an order of magnitude and the rendering time by roughly 30%. Note that the studied strategies do not consider the individual cost of a state change. A conclusion is that there is a lack of efficient strategies that consider these individual cost.

In the painting shop of a car plant, a sequence of cars bodies traverses the final layer painting where each car body is painted with its own top coat. If two consecutive cars have to be painted in different colors then a color change is required which causes set-up cost. In addition to the color change cost, further important non-uniform cost arise, e.g., the individual accessing times of the parking slots for the car bodies. These costs can be minimized by preceding the final layer painting with a reordering buffer. In several practical work, heuristic strategies for reordering buffers are evaluated by simulation (see, e.g., [8]). Efficient strategies for reordering buffer are considered to be a major problem of operating a paint shop.

1.3 Results and Further Work

In Sect. 2, the possible gain of a reordering buffer is investigated. We show that a reordering buffer of size k reduces the cost of each sequence by a factor of at most $2k - 1$. This result holds for online and offline strategies and even for cost functions modeled by arbitrary metric spaces, i.e., the cost function $d(\sigma_{\pi^{-1}(i)}, \sigma_{\pi^{-1}(i+1)})$ can be any positive and symmetric function obeying the triangle inequality. In addition, a matching lower bound is presented. From this basic upper bound follows immediately that each strategy that does not increase the cost of an input sequence is at least $(2k - 1)$-competitive. In particular, the simple online strategy that does no reordering at all is already $(2k - 1)$-competitive. This shows the poor performance of some strategies. For example, Yeh et al. [9] give a lower bound of $2k - 1$ on the competitive ratio of the disk scheduling strategies Shortest Seek Time First and Look in the line metric model.

In Sect. 3, we show a lower bound of k on the competitive ratio of the Bounded Waste (BW) strategy in our model, where k denotes the size of the reordering buffer. The BW strategy is introduced by Räcke, Sohler and Westermann [5] for the uniform case of our problem: For each color c, weight $b_c = 1$, i.e., just the number of color changes is considered. Note that this lower bound even holds for the case that BW takes the non-uniform cost into account, i.e., BW is aware of the individual cost b_c, for each color c. In Sect. 4, the deterministic Maximum Adjusted Penalty (MAP) strategy is presented. We show that the MAP strategy is $O(\log k)$-competitive, where k denotes the size of the reordering buffer. Note that, although MAP is equivalent to BW in the uniform case of our model, our analysis provides a better result. Currently, we only know a trivial lower bound of 5/3 on the competitive ratio of any deterministic strategy.

For the upper bound on MAP we introduce the following basic proof technique: First, it is shown that MAP with buffer size k is 4-competitive against an optimal offline strategy with buffer size $k/4$. Finally, it is proven that an optimal offline strategy with buffer size $k/4$ is $O(\log k)$-competitive against an optimal offline strategy with buffer size k. We believe that this technique can be interesting for other problems.

Our non-uniform scenario can be modeled by the following star-like metric space: $d(x,y) = (b_x + b_y)/2$, if $x \neq y$, and $d(x,y) = 0$, otherwise. Above, we conclude that there is a lack of efficient strategies for the disk scheduling problem, i.e., the line metric space. We consider to transfer this technique from star-like to line metric spaces.

2 Basic Upper Bound

In this section, we show that a reordering buffer of size k reduces the cost of each sequence by a factor of at most $2k-1$. This result holds for online and offline strategies and even for cost functions modeled by arbitrary metric spaces. Note that this result is tight. Fix the two colors c_1 and c_2 with weights $b_{c_1} = b_{c_2} = 1$. The input sequence $\sigma = (c_1 c_2)^k$ of length $2k$ can obviously be reordered to $\sigma_{\pi^{-1}} = c_1^k c_2^k$ with a reordering buffer of size k. The cost of σ is $C(\sigma) = 2k-1$, and the cost of $\sigma_{\pi^{-1}}$ is $C(\sigma_{\pi^{-1}}) = 1$. Hence, $C(\sigma) = (2k-1) \cdot C(\sigma_{\pi^{-1}}) \geq (2k-1) \cdot C^{\text{OP}}(\sigma)$, where $C^{\text{OP}}(\sigma)$ denotes the cost of an optimal offline strategy using a reordering buffer of size k.

Theorem 1. *For every metric space (M,d), and every input sequence $\sigma = \sigma_1 \cdots \sigma_l$, with $\sigma_i \in M$,*

$$C(\sigma) \leq (2k-1) \cdot C^{\text{OP}}(\sigma) \ ,$$

where $C(\sigma)$ denotes the cost of σ, and $C^{\text{OP}}(\sigma)$ denotes the cost of an optimal offline strategy using a reordering buffer of size k.

Proof. Fix an input sequence $\sigma = \sigma_1 \cdots \sigma_l$. Let $\sigma_{\pi^{-1}} = \sigma_{\pi^{-1}(1)} \cdots \sigma_{\pi^{-1}(l)}$ denote the output sequence of an optimal offline strategy using a reordering buffer of size k. We define a *subsequence* $I_r^s = \sigma_{\pi^{-1}(r)} \cdots \sigma_{\pi^{-1}(s)}$, if $r \leq s$, and $I_r^s = \sigma_{\pi^{-1}(s)} \cdots \sigma_{\pi^{-1}(r)}$, otherwise. Let $C(I_r^s)$ denote the cost of this subsequence, i.e., $C(I_r^s) = \sum_{j=\min\{r,s\}}^{\max\{r,s\}-1} d(\sigma_{\pi^{-1}(j)}, \sigma_{\pi^{-1}(j+1)})$.

Due to the triangle inequality, $d(\sigma_i, \sigma_{i+1}) \leq C(I_{\pi(i)}^{\pi(i+1)})$. Thus,

$$C(\sigma) = \sum_{i=1}^{l-1} d(\sigma_i, \sigma_{i+1}) \leq \sum_{i=1}^{l-1} C(I_{\pi(i)}^{\pi(i+1)}) \ .$$

Fix two consecutive objects $\sigma_{\pi^{-1}(j)}$ and $\sigma_{\pi^{-1}(j+1)}$ of $\sigma_{\pi^{-1}}$. We show that these objects do only occur in at most $2k-1$ of the subsequences above. If

$\sigma_{\pi^{-1}(j)}$ and $\sigma_{\pi^{-1}(j+1)}$ are part of a subsequence $I_{\pi(i)}^{\pi(i+1)}$, one of the following two cases is true:

1. $\pi(i) \leq j$ and $\pi(i+1) \geq j+1$
2. $\pi(i+1) \leq j$ and $\pi(i) \geq j+1$

In case (1), σ_i is one of the first j objects and σ_{i+1} is not under the first j objects of $\sigma_{\pi^{-1}}$. In case (2), it is the other way around.

Obviously, the following observation can be made.

Observation 2. *For each input sequence* $\sigma = \sigma_1 \cdots \sigma_l$, *the output sequence of a reordering buffer of size* k *is a permutation* $\sigma_{\pi^{-1}} = \sigma_{\pi^{-1}(1)} \cdots \sigma_{\pi^{-1}(l)}$ *of* σ, *with* $\pi^{-1}(i) < i + k$, *for each* i. *In addition, each such permutation can be generated using a reordering buffer of size* k.

The observation shows that only one of the first $i + k - 1$ objects of σ can be placed at the i-th position of $\sigma_{\pi^{-1}}$. Thus, we conclude for case (1) that $i \leq j + k - 1$. In the same way, we conclude for case (2) the even stronger inequality $i + 1 \leq j + k - 1$.

In the following, we consider case (1). The above conclusions provide that σ_{i+1} must be one of the first $j + k$ objects of σ. But, σ_{i+1} is not one of the first j objects of $\sigma_{\pi^{-1}}$. Recall that the observation shows that the first j objects of $\sigma_{\pi^{-1}}$ have to be under the first $j + k - 1$ objects of σ. Hence, at most k objects of the first $j + k$ objects of σ cannot be under the first j objects of $\sigma_{\pi^{-1}}$. It follows that case (1) is true for at most k different subsequences. Obviously, case (2) can be addressed analogously. It follows that case (2) is true for at most $k - 1$ different subsequences.

Hence,

$$C(\sigma) \leq \sum_{i=1}^{l-1} C(I_{\pi(i)}^{\pi(i+1)}) \leq (2k-1) \sum_{i=1}^{l-1} d(\sigma_{\pi^{-1}(i)}, \sigma_{\pi^{-1}(i+1)}) = (2k-1) \cdot C^{\mathrm{OP}}(\sigma) \ ,$$

since at most $2k - 1$ subsequences are containing the two objects $\sigma_{\pi^{-1}(i)}$ and $\sigma_{\pi^{-1}(i+1)}$. □

3 Lower Bound for the BW Strategy

The Bounded Waste (BW) strategy is introduced in [5] for the uniform case of our model: For each color c, weight $b_c = 1$, i.e., just the number of color changes is considered. BW chooses one color as the active color, and continues to remove objects of this active color from the reordering buffer until all objects in the buffer have a color different from the active color. Then a new active color has to be chosen. For this purpose, a counter P_c, which is initially set to zero, is assigned to each color c. At each color change, the counter of each color c is increased by the number of objects of color c currently stored in the buffer. Then a color c' with maximal counter $P_{c'}$ is chosen as the new active color and $P_{c'}$ is reset to zero.

The following theorem shows a lower bound of k, where k denotes the size of the reordering buffer, on the competitive ratio of BW in our non-uniform model. Note that this lower bound even holds for the case that BW takes the non-uniform weights of the colors into account.

Theorem 3. *The competitive ratio of BW is at least k, where k denotes the size of the reordering buffer.*

Proof. Fix one expensive color x with weight $b_x = 1$, and several inexpensive colors $c_1, \ldots c_l$ with weights $b_{c_1} = \cdots = b_{c_l} = \varepsilon$. The input sequence $\sigma = \sigma_1 \cdots \sigma_l$ with $l = k \cdot (k+1)$ is defined as follows: σ_i is of color x, if i is divisible by $(k+1)$, and of color c_i, otherwise. Obviously, it is possible to produce an output sequence $\sigma^{op}_{\pi-1}$ with cost $C(\sigma^{op}_{\pi-1}) = 1 + (k^2 - 1) \cdot \varepsilon$ by aggregating objects of the expensive color.

Each pair of objects in σ has different colors, except the pairs where both objects have the color x. However, BW will never hold two objects of color x in the reordering buffer at the same time. Assume this statement holds until some step i in which an object of color x arrives. If each pair of objects in the buffer of BW has different colors, the values of the counters are exclusively based on the number of steps the corresponding objects are stored in the buffer. The object of color x is after $k-1$ steps the oldest object in the buffer. Hence, this object is removed from the buffer in the next step, i.e., one step before the next object of color x arrives.

The cost of the produced output sequence $\sigma^{on}_{\pi-1}$ is $C(\sigma^{on}_{\pi-1}) = C(\sigma) = k + k^2 \cdot \varepsilon$. Since the input sequence σ can be iterated and ε can be chosen arbitrarily small, this yields a lower bound of k on the competitive ratio. Of course, BW could take the non-uniform weights of the colors into account, i.e., the counter P_c is increased by the number of objects of color c currently stored in the buffer times the weight of color c. However, this proof does not dependent on this decision. □

4 The MAP Strategy

In this section, we present the Maximal Adjusted Penalty (MAP) strategy. We show that the MAP strategy is $O(\log k)$-competitive, where k denotes the size of the reordering buffer.

MAP chooses one color as the active color, and removes at each time step one object of this active color from the reordering buffer until all objects in the buffer have a color different from the active color. Then a new active color has to be chosen. For this purpose, a penalty counter P_c, which is initially set to zero, is assigned to each color c. MAP chooses a color c as the new active color with $P_c - k \cdot b_c \geq P_{c'} - k \cdot b_{c'}$, for each color c'. The counters are updated after a new active color is chosen. Suppose a step in which a color change from color x to color y occurs. Let n_c denote the number of objects of color c stored in the buffer at the beginning of this step. Then each counter P_c is increased by $n_c \cdot b_y$ and counter P_x is reset to zero.

The MAP strategy does not need to know the weights of all colors in advance. It is sufficient to provide the weight b_c when the first object of color c arrives. In addition, a counter P_c can be deleted if no objects of color c are stored in the buffer. Hence, each step can be performed in time $O(k)$, since at most k counters are active at the same time.

Theorem 4. *The MAP strategy is $O(\log k)$-competitive, where k denotes the size of the reordering buffer.*

Proof. The proof consists of two parts.

1. First, we prove that MAP with buffer size k is 4-competitive against an optimal offline strategy with buffer size $h = k/4$.
2. Finally, we show that an optimal offline strategy with buffer size h is $O(\log k)$-competitive against an optimal offline strategy with buffer size k.

Together, this yields the theorem.

Part 1. Fix an input sequence σ and a lazy optimal offline strategy OPT. MAP has a reordering buffer of size k and OPT has a reordering buffer of size h. We exclude the last k color changes of MAP. Hence, it can be assumed that there are k objects in the buffer of MAP at any time. Under this assumption we show that MAP is 4-competitive against OPT. This yields part 1, since the last k color changes of MAP produce at most cost $k \cdot \max_{\text{color } c}\{b_c\}$.

Color changes of MAP and OPT are denoted as online and offline color changes, respectively. An *online (offline) c-interval* starts with an online (offline) color change from color c to a different color and ends right before the next online (offline) color change from c to a different color (the first online (offline) c-interval starts with the first step). Each object of color c falls into exactly one online and one offline c-interval, and it enters and leaves the buffer of the respective strategy in the same c-interval. Also each step i falls into exactly one online and one offline c-interval, and these intervals are denoted as *active* at step i.

Now, we introduce counters to which k-times the cost of each online color change is assigned. For each color c, and each online c-interval I, the two counters $w_c^{\text{on},I}$ and $\hat{w}_c^{\text{on},I}$ are introduced. $w_c^{\text{on},I}(i)$ and $\hat{w}_c^{\text{on},I}(i)$ denote the value of $w_c^{\text{on},I}$ and $\hat{w}_c^{\text{on},I}$ at the beginning of step i, respectively. The counters are initially set to zero, and they are monotonously increasing. A counter is denoted as active at step i, if the according online c-interval I is active at step i. Otherwise, the counter is denoted as inactive. For simplicity, we just write $w_c^{\text{on}}(i)$ and $\hat{w}_c^{\text{on}}(i)$ to denote the active counters for color c at the beginning of step i. Inactive counters do not change their value.

Fix a step i. In the following, we describe how k-times the cost of an online color change is distributed among the counters. Let $n_c^{\text{on}}(i)$ denote the number of objects with color c in the buffer of MAP at the beginning of step i. For simplicity, we just write n_c^{on}, if the step is fixed. Note that $\sum_{\text{color } c} n_c^{\text{on}} = k$. Suppose there is an online color change to color c' in step i. Then $k \cdot b_{c'}$ has to be assigned to the counters. For each color c, w_c^{on} is increased by $n_c^{\text{on}} \cdot b_{c'}$. In total, we assign $\sum_{\text{color } c} n_c^{\text{on}} \cdot b_{c'} = k \cdot b_{c'}$ to active counters.

But we prevent a counter w_c^{on} from becoming larger than $k \cdot b_c$. This restriction might cause that nothing or only a part of the value $n_c^{\text{on}} \cdot b_{c'}$ is really assigned to w_c^{on}. The remaining part, i.e, the part that would lead to a counter w_c^{on} larger than $k \cdot b_c$, is assigned to $\hat{w}_{c'}^{\text{on}}$ instead. Note that w_c^{on} equals the P_c counter in MAP, as long as $P_c \leq k \cdot b_c$. Otherwise, w_c^{on} remains on the value $k \cdot b_c$, but P_c is further increased. Note in addition that, for each color c' and each online c'-interval I, the counter $\hat{w}_{c'}^{\text{on},I}$ is increased in at most one step.

Since we have assigned k-times the produced cost to counters, we can express the cost of MAP $C^{\text{on}} = (1/k) \cdot \sum_{\text{color } c} \sum_{\text{on. c-int. } I} (W_c^{\text{on},I} + \hat{W}_c^{\text{on},I})$, where $W_c^{\text{on},I}$ and $\hat{W}_c^{\text{on},I}$ denote the final, i.e., maximum, value of the counters $w_c^{\text{on},I}$ and $\hat{w}_c^{\text{on},I}$, respectively.

In addition, for each color c and each online c-interval I, the counter $w_c^{\text{op},I}$ is introduced. $w_c^{\text{op},I}(i)$ denotes the value of $w_c^{\text{op},I}$ at the beginning of step i. The counters are initially set to zero, and they are monotonously increasing. A counter is denoted as active at step i, if the according online c-interval I is active at step i. Otherwise, the counter is denoted as inactive. For simplicity, we just write $w_c^{\text{op}}(i)$ to denote the active counter for color c at the beginning of step i. Inactive counters do not change their value.

Fix a step i. Let $n_c^{\text{op}}(i)$ denote the number of objects with color c in the buffer of OPT at the beginning of step i. For simplicity, we just write n_c^{op}, if the step is fixed. Note that $\sum_{\text{color } c} n_c^{\text{op}} = h$. Suppose there is an online color change to color c' in step i. For each color c, w_c^{op} is increased by $n_c^{\text{op}} \cdot b_{c'}$. In total, we assign $\sum_{\text{color } c} n_c^{\text{op}} \cdot b_{c'} = h \cdot b_{c'}$ to active counters.

Hence, we yield a new possibility to express the cost of MAP $C^{\text{on}} = (1/h) \cdot \sum_{\text{color } c} \sum_{\text{on. c-int. } I} W_c^{\text{op},I}$, where $W_c^{\text{op},I}$ denotes the final, i.e., maximum, value of the counter $w_c^{\text{op},I}$.

For each color c, we show the following *main inequality*

$$4k \cdot C_c^{\text{op}} + \sum_{\text{on. c-int. } I} (4W_c^{\text{op},I} - (W_c^{\text{on},I} + \hat{W}_c^{\text{on},I})) \geq \sum_{\text{on. c-int. } I} (W_c^{\text{on},I} + \hat{W}_c^{\text{on},I}) ,$$

where C_c^{op} denotes the total cost produced by offline color changes to color c. Summing up over all colors, we yield $4k \cdot C^{\text{op}} + 4h \cdot C^{\text{on}} - k \cdot C^{\text{on}} \geq k \cdot C^{\text{on}}$. Hence, $4C^{\text{op}} \geq C^{\text{on}}$. This yields part 1.

We distinguish between two kinds of online c-intervals. An online c-interval I is denoted as *problematic*, if $4W_c^{\text{op},I} < 2(W_c^{\text{on},I} + \hat{W}_c^{\text{on},I})$. Otherwise, I is denoted as *non-problematic*. Now, we show the following inequality

$$4k \cdot C_c^{\text{op}} \geq 2 \sum_{\text{prob. c-int. } I} (W_c^{\text{on},I} + \hat{W}_c^{\text{on},I}) .$$

Obviously, the main inequality can be concluded with the help of the above inequality.

The following lemma provides an upper bound on $W_c^{\text{on},I}$ and $\hat{W}_c^{\text{on},I}$. Then, an upper bound on the number of problematic c-intervals is shown. These two results together complete part 1 of the proof.

Lemma 5. *For each color c and each online c-interval I,*

$$\hat{W}_c^{on,I} \leq W_c^{on,I} \leq k \cdot b_c .$$

Proof. Due to the cost assignment for $w_c^{on,I}$, $W_c^{on,I} \leq k \cdot b_c$. Suppose that $\hat{W}_c^{on,I} > 0$. Recall that $\hat{w}_c^{on,I}$ is increased in at most one step. We consider the step i in which $\hat{w}_c^{on,I}$ is increased due to the online color change to color c in the online c-interval I. In this step, for each color c', $P_c - k \cdot b_c \geq P_{c'} - k \cdot b_{c'}$.

Now, we distinguish the following two cases.

– Suppose that $W_c^{on,I} = k \cdot b_c$.
  The color change in step i produces cost b_c. Hence, $k \cdot b_c$ is assigned to the counters in this step. Even if the whole value $k \cdot b_c$ is assigned to $\hat{w}_c^{on,I}$, $\hat{W}_c^{on,I} \leq k \cdot b_c = W_c^{on,I}$, since $\hat{w}_c^{on,I}$ is increased in at most one step.
– Suppose that $W_c^{on,I} < k \cdot b_c$.
  In this case, $P_c = w_c^{on}(i)$ in step i. In fact, for each color c', $P_{c'} = w_{c'}^{on}(i)$ at the beginning of step i, since $w_{c'}^{on}(i) < k \cdot b_{c'}$. If, for some color $c' \neq c$, $w_{c'}^{on}(i) = k \cdot b_{c'}$, MAP would have chosen color c' as new active color in step i. Of course, for a color $c' \neq c$, the active counter $w_{c'}^{on}$ can reach its limit $k \cdot b_{c'}$ in this step. $w_{c'}^{on}$ exceeds its limit by $x_{c'} = \max\{0, w_{c'}^{on}(i) + n_{c'}^{on}(i) \cdot b_c - k \cdot b_{c'}\}$. Due to MAP, $w_{c'}^{on}(i) - k \cdot b_{c'} \leq w_c^{on}(i) - k \cdot b_c$. Hence, $x_{c'} \leq \max\{0, w_c^{on}(i) + n_{c'}^{on}(i) \cdot b_c - k \cdot b_c\}$.
  Let V denote the set of all colors c' with $x_{c'} > 0$. If $V = \emptyset$, $\hat{W}_c^{on,I} = 0$. Otherwise,

$$\hat{W}_c^{on,I} = \sum_{c' \in V} x_{c'} \leq |V| \cdot (w_c^{on}(i) - k \cdot b_c) + \sum_{c' \in V} n_{c'}^{on}(i) \cdot b_c$$

$$\leq w_c^{on}(i) - k \cdot b_c + \sum_{c' \in V} n_{c'}^{on}(i) \cdot b_c \leq w_c^{on}(i) \leq W_c^{on,I} ,$$

since $w_c^{on}(i) - k \cdot b_c < 0$.

This finishes the proof of the lemma. □

The beginning of an offline c-interval I, before the offline color change to color c occurs, is denoted as increasing phase, since the number of objects of color c in the offline buffer is monotonously increasing. The remaining part of I, after the offline color change to color c, is denoted as decreasing phase, since the number of objects of color c in the offline buffer is monotonously decreasing.

Lemma 6. *At most one problematic online c-interval starts in an offline c-interval.*

Proof. Fix a problematic online c-interval I. From Lem. 5 follows $4W_c^{op,I} < 4W_c^{on,I}$. Hence, there exists at least one step i in I with $n_c^{op}(i) < n_c^{on}(i)$. Let I' denote the offline c-interval in which I starts.

Suppose I starts in the increasing phase of I'. Let start(I) denote the first step of interval I. Then $n_c^{op}(\text{start}(I)) \geq 0 = n_c^{on}(\text{start}(I))$. Hence, no step i with

$n_c^{\mathrm{op}}(i) < n_c^{\mathrm{on}}(i)$ can exist in this increasing phase, since every arriving object of color c is stored in the offline buffer and no objects of color c are removed from the offline buffer.

Consider the decreasing phase of I'. If n_c^{on} is decreased, then n_c^{op} is decreased by the same amount. Hence, if there exists a step i with $n_c^{\mathrm{op}}(i) < n_c^{\mathrm{on}}(i)$ in the decreasing phase of I', the offline c-interval I' ends before the end of the problematic online c-interval I. □

The total number of offline c-intervals is C_c^{op}/b_c. We can exclude the first offline c-interval, if c is the color of the first object in the output sequence of OPT, since the total produced cost by MAP in the only problematic interval starting in this offline c-interval can be bounded by a term independent of σ. Note that we exclude, for only one color c, an offline c-interval. From the lemma above it follows that the total number of problematic online c-intervals is at most C_c^{op}/b_c. Then

$$2 \sum_{\text{prob. } c\text{-int. } I} (W_c^{\mathrm{on},I} + \hat{W}_c^{\mathrm{on},I}) \leq 2 \sum_{\text{prob. } c\text{-int. } I} k \cdot b_c + k \cdot b_c$$
$$\leq 4k \cdot b_c \cdot C_c^{\mathrm{op}}/b_c = 4k \cdot C_c^{\mathrm{op}} .$$

This completes part 1 of the proof.

Part 2. It remains to show, that an optimal offline strategy with buffer size $h = k/4$ is $O(\log k)$-competitive against an optimal offline strategy with buffer size k. Fix an input sequence σ. For each step i, let $n_c^h(i)$ denote the number of objects of color c in the buffer of size h and let $n_c^k(i)$ denote the number of objects of color c in the buffer of size k.

Fix a lazy optimal offline strategy LARGE for the reordering buffer of size k. The offline strategy SMALL for the reordering buffer of size h chooses a new active color c, with $n_c^h(i) \geq n_c^k(i)/4$. Note that there exists always such a color, since $\sum_{\text{color } c} n_c^h(i) = \sum_{\text{color } c} n_c^k(i)/4$.

Large (small) c-intervals are defined for LARGE (SMALL) according to the definition of online and offline c-intervals. The definitions of increasing and decreasing phases apply to large c-intervals, too. Note that the total number of large (small) c-intervals is order of the total number of color changes to color c of LARGE (SMALL).

For every color c, we show that there are at most $O(\log k)$ small c-intervals in one large c-interval. Then, $C_c^h \leq O(\log k) \cdot C_c^k$, where C_c^h (C_c^k) denotes the cost of LARGE (SMALL) for color c. This yields part 2 of the proof.

Fix a color c. In the following, we only consider small c-intervals that are completely contained in a large c-interval. In addition, we exclude a small c-interval, if LARGE performs a color change to color c during this interval. Hence, in total at most two small c-intervals are excluded for every large c-interval. The remaining small c-intervals are completely contained either in an increasing or in a decreasing phase of a large c-interval.

The following lemma shows that there are at most $O(\log k)$ small c-intervals in a large c-interval, since the buffer size of LARGE is k.

Lemma 7. Let start(I) and end(I) denote the first and last step of a small c-interval I, respectively.

- In an increasing phase: $n_c^k(\text{end}(I)) \geq (5/4)n_c^k(\text{start}(I))$.
- In a decreasing phase: $n_c^k(\text{start}(I)) \geq (5/4)n_c^k(\text{end}(I))$.

Proof. We only prove the inequality for the increasing phase. The inequality for the decreasing phase can be addressed analogously. Fix a small c-interval I. Let i be the step in I at which SMALL performs a color change to color c. Due to SMALL, $n_c^h(i) \geq n_c^k(i)/4$.

At least $n_c^h(i)$ objects of color c arrive in I. Since I lies in an increasing phase, these objects are not removed from the buffer of LARGE during I. At the end of I, the $n_c^k(\text{start}(I))$ objects of color c stored in the buffer of LARGE before I, are still there, and at least $n_c^h(i) \geq n_c^k(i)/4 \geq n_c^k(\text{start}(I))/4$ new objects of color c are added. Hence, at least $(5/4)n_c^k(\text{start}(I))$ objects of color c are stored in the buffer of LARGE at the end of I. □

This completes part 2 of the proof. □

References

1. Teorey, T., Pinkerton, T.: A comparative analysis of disk scheduling policies. Communications of the ACM **15** (1972) 177–184
2. Fiat, A., Karp., R.M., Luby, M., McGeoch, L.A., Sleator, D.D., Young, N.E.: Competitive paging algorithms. Journal of Algorithms **12** (1991) 685–699
3. Albers, S.: New results on web caching with request reordering. In: Proceedings of the 16th ACM Symposium on Parallel Algorithms and Architectures (SPAA). (2004) 84–92
4. Feder, T., Motwani, R., Panigrahy, R., Seiden, S., van Stee, R., Zhu, A.: Combining request scheduling with web caching. Theoretical Compuer Science **324** (2004) 201–218
5. Räcke, H., Sohler, C., Westermann, M.: Online scheduling for sorting buffers. In: Proceedings of the 10th European Symposium on Algorithms (ESA). (2002) 820–832
6. Kohrt, J., Pruhs, K.: A constant approximation algorithm for sorting buffers. In: Proceedings of the 6th Latin American Symposium on Theoretical Informatics (LATIN). (2004) 193–202
7. Krokowski, J., Räcke, H., Sohler, C., Westermann, M.: Reducing state changes with a pipeline buffer. In: Proceedings of the 9th International Fall Workshop Vision, Modeling, and Visualization (VMV). (2004) 217–224
8. Gutenschwager, K., Spieckermann, S., Voss, S.: A sequential ordering problem in automotive paint shops. International Journal of Production Research **42** (2004) 1865–1878
9. Yeh, T., Kuo, C., Lei, C., Yen, H.: Competitive analysis of on-line disk scheduling. Theory of Computing Systems **31** (1998) 491–506

Combining Intruder Theories*

Yannick Chevalier and Michaël Rusinowitch

[1] IRIT Université Paul Sabatier, France
ychevali@irit.fr
[2] LORIA-INRIA-Lorraine, France
rusi@loria.fr

Abstract. Most of the decision procedures for symbolic analysis of protocols are limited to a fixed set of algebraic operators associated with a fixed intruder theory. Examples of such sets of operators comprise XOR, multiplication/exponentiation, abstract encryption/decryption. In this paper we give an algorithm for combining decision procedures for arbitrary intruder theories with disjoint sets of operators, provided that solvability of ordered intruder constraints, a slight generalization of intruder constraints, can be decided in each theory. This is the case for most of the intruder theories for which a decision procedure has been given. In particular our result allows us to decide trace-based security properties of protocols that employ any combination of the above mentioned operators with a bounded number of sessions.

1 Introduction

Recently many procedures have been proposed to decide insecurity of cryptographic protocols in the Dolev-Yao model w.r.t. a finite number of protocol sessions [2, 5, 18]. Among the different approaches the symbolic ones [16, 10, 4] are based on reducing the problem to constraint solving in a term algebra. This reduction has proved to be quite effective on standard benchmarks and also was able to discover new flaws on several protocols [4].

However while most formal analysis of security protocols abstracts from low-level properties, i.e., certain algebraic properties of encryption, such as the multiplicative properties of RSA or the properties induced by chaining methods for block ciphers, many real attacks and protocol weaknesses rely on these properties. For attacks exploiting the XOR properties in the context of mobile communications see [7]. Also the specification of *Just Fast Keying* protocol (an alternative to IKE) in [1] employs a set constructor that is idempotent and commutative and a Diffie-Hellman exponentiation operator with the property $(g^y)^z = (g^z)^y$.

In this paper we present a general procedure for deciding security of protocols in presence of algebraic properties. This procedure relies on the combination of constraint solving algorithm for disjoint intruder theories, provided that solvability of ordered intruder constraints, a slight generalization of intruder constraints,

---

* Supported by IST-2001-39252 AVISPA, ACI SATIN, ACI-Jeune Chercheur JC9005.

can be decided in each theory. Such combination algorithm already exists for solving E-unification problems [19, 3]. We have extended it in order to solve intruder constraints on disjoint signatures. This extension is non trivial since intruder deduction rules allow one to build *contexts* above terms and therefore add some second-order features to the *standard* first-order E-unification problem.

Our approach is more modular than the previous ones and it allows us to decide interesting intruder theories that could not be considered before by reducing them to simpler and independent theories. For instance it allows one to combine the exponentiation with abelian group theory of [17] with the Xor theory of [8].

Related works. Recently several protocol decision procedures have been designed for handling algebraic properties in the Dolev-Yao model [15, 6, 11, 8]. These works have been concerned by fixed equational theories corresponding to a fixed intruder power. A couple of works only have tried to derive generic decidability results for *class* of intruder theories. For instance, in [12] Delaune and Jacquemard consider the class of *public collapsing* theories. These theories have to be presented by rewrite systems where the right-hand side of every rule is a ground term or a variable, which is a strong restriction.

2 Motivation

Combination of algebraic operators. We consider in this section the Needham–Schroeder Public-Key protocol. This well-known protocol is described in the Alice and Bob notation by the following sequence of messages, where the comma denotes a pairing of messages and $\{M\}K_a$ denotes the encryption by the public key K_a of A.

$$A \to B : \{N_a, A\}K_b$$
$$B \to A : \{N_a, N_b\}K_a$$
$$A \to B : \{N_b\}K_b$$

Assume now that the encryption algorithm follows El-Gamal encryption scheme. The public key of A is defined by three publicly-available parameters: a modulus p_a, a base g_a and the proper public key $(g_a)^a \mod p_a$. The private key of A is a. Denoting $\exp_p$ the exponentiation modulo p, and with new nonces k_1, k_2 and k_3 we can rewrite the protocol as:

$$A \to B : \exp_{p_b}(g_b, k_1), (N_a, A) \oplus \exp_{p_b}(\exp_{p_b}(g_b, b), k_1)$$
$$B \to A : \exp_{p_a}(g_a, k_2), (N_a, N_b) \oplus \exp_{p_a}(\exp_{p_a}(g_a, a), k_2)$$
$$A \to B : \exp_{p_b}(g_b, k_3), (N_b) \oplus \exp_{p_b}(\exp_{p_b}(g_b, b), k_3)$$

In this simple example we would like to model the group properties of the Exclusive-or ($\oplus$), the associativity of exponential ($(x^y)^z = x^{y \times z}$), the group properties of the exponents. Several works have already been achieved toward taking into account these algebraic properties for detecting attacks on a bounded number of sessions. However none of these works handles the analysis of protocols combining several algebraic operators like the example above. The algorithm given in this paper will permit to decide the trace-based security properties of such protocols.

Examples of intruder theories. A convenient way to specify intruder theories in the context of cryptographic protocols is by giving a set L of *deduction rules* describing how the intruder can construct new messages from the ones she already knows and a set of *equational laws* $\mathcal{E}$ verified by the functions employed in messages. We give here two examples of intruder theories. Some other theories are given in [9].

Abelian group theory. This intruder may treat messages as elements of an abelian group. We assume here there is only one such group and that the composition law is $\cdot \times \cdot$, the inverse law is $i(\cdot)$ and the neutral element is denoted 1.

$$L_\times \begin{cases} \to 1 \\ x \to i(x) \\ x,y \to x \times y \end{cases} \qquad \mathcal{E}_\times \begin{cases} (x \times y) \times z = x \times (y \times z) \\ x \times y = y \times x \\ 1 \times x = x \\ x \times i(x) = 1 \end{cases}$$

Dolev Yao with explicit destructors. The intruder is given with a pairing operator and projections to retrieve the components of a pair. There are symmetric encryption $(\mathrm{se}(\_,\_))$ and decryption $(\mathrm{sd}(\_,\_))$ operators. For conciseness we omit the public-key encryption specification.

$$L_{DY} \begin{cases} x,y \to \langle x,y \rangle \\ x \to \pi_1(x) \\ x \to \pi_2(x) \\ x,y \to \mathrm{se}(x,y) \\ x,y \to \mathrm{sd}(x,y) \end{cases} \qquad \mathcal{E}_{DY} \begin{cases} \pi_1(\langle x,y \rangle) = x \\ \pi_2(\langle x,y \rangle) = y \\ \mathrm{sd}(\mathrm{se}(x,y),y) = x \end{cases}$$

3 Terms and Subterms

We consider an infinite set of free constants C and an infinite set of variables $\mathcal{X}$. For all signatures $\mathcal{G}$ (i.e. a set of function symbols with arities), we denote by $T(\mathcal{G})$ (resp. $T(\mathcal{G}, \mathcal{X})$) the set of terms over $\mathcal{G} \cup C$ (resp. $\mathcal{G} \cup C \cup \mathcal{X}$). The former is called the set of ground terms over $\mathcal{G}$, while the latter is simply called the set of terms over $\mathcal{G}$. Variables are denoted by x, y, v, terms are denoted by s, t, u, and finite sets of terms are written $E, F, ...$, and decorations thereof, respectively.

A *constant* is either a free constant or a function symbol of arity 0. Given a term t we denote by $\mathrm{Var}(t)$ the set of variables occurring in t and by $\mathrm{Cons}(t)$ the set of constants occurring in t. We denote by $\mathrm{Atoms}(t)$ the set $\mathrm{Var}(t) \cup \mathrm{Cons}(t)$. We denote by $\mathcal{A}$ the set of all constants and variables. A substitution σ is an involutive mapping from $\mathcal{X}$ to $T(\mathcal{G}, \mathcal{X})$ such that $\mathrm{Supp}(\sigma) = \{x \mid \sigma(x) \neq x\}$, the *support* of σ, is a finite set. The application of a substitution σ to a term t (resp. a set of terms E) is denoted $t\sigma$ (resp. $E\sigma$) and is equal to the term t (resp. E) where all variables x have been replaced by the term $x\sigma$. A substitution σ is *ground* w.r.t. $\mathcal{G}$ if the image of $\mathrm{Supp}(\sigma)$ is included in $T(\mathcal{G})$.

In this paper, we consider 2 disjoint signatures $\mathcal{F}_1$ and $\mathcal{F}_2$, and 2 consistent equational theories $\mathcal{E}_1$ and $\mathcal{E}_2$ on $\mathcal{F}_1$ and $\mathcal{F}_2$, resp. We denote by $\mathcal{F}$ the union of the signatures $\mathcal{F}_1$ and $\mathcal{F}_2$, $\mathcal{E}$ the union of the theories $\mathcal{E}_1$ and $\mathcal{E}_2$. A term t in $\mathrm{T}(\mathcal{F}_1, \mathcal{X})$ (resp. in $\mathrm{T}(\mathcal{F}_2, \mathcal{X})$) is called a *pure 1-term* (resp. a pure 2-term).

The *syntactic subterms* of a term t are defined recursively as follows and denoted $\mathrm{Sub}_{\mathrm{syn}}(t)$. If t is a variable or a constant then $\mathrm{Sub}_{\mathrm{syn}}(t) = \{t\}$. If $t = f(t_1, \ldots, t_n)$ then $\mathrm{Sub}_{\mathrm{syn}}(t) = \{t\} \cup \bigcup_{i=1}^n \mathrm{Sub}_{\mathrm{syn}}(t_i)$. The *positions* in a term t are defined recursively as usual (*i.e.* as sequences of integers), ϵ being the empty sequence. We denote by $t_{|p}$ the syntactic subterm of t at position p. We denote by $t[p \leftarrow s]$ the term obtained by replacing in t the syntactic subterm $t_{|p}$ by s. We denote by $\mathrm{Sign}(\cdot)$ the function that associates to each term $t \notin \mathcal{C} \cup \mathcal{X}$ the signature ($\mathcal{F}_1$, or $\mathcal{F}_2$) of its symbol at position ϵ. For $t \in \mathcal{C} \cup \mathcal{X}$ we define $\mathrm{Sign}(t) = \bot$, with $\bot$ a new symbol. The term s is *alien* to u if $\mathrm{Sign}(s) \neq \mathrm{Sign}(u)$.

Factors. We define the set of *factors* of a term t, and denote $\mathrm{Factors}(t)$, the set of maximal syntactic subterms of t that are either alien to t or atoms and different from t. In particular $\mathrm{Factors}(t) = \emptyset$ for $t \in \mathcal{A}$.

Subterms. We now define the notion of *subterm values*. Given a term t, the set of its subterm values is denoted by $\mathrm{Sub}(t)$ and is defined recursively by: $\mathrm{Sub}(t) = \{t\} \cup \bigcup_{u \in \mathrm{Factors}(t)} \mathrm{Sub}(u)$. For a set of terms E, $\mathrm{Sub}(E)$ is defined as the union of the subterms values of the elements of E.

As an example consider $\mathcal{F}_1 = \{\oplus, 0\}$ and $\mathcal{F}_2 = \{f\}$ where f has arity 1. Then $\mathrm{Sub}(a \oplus (b \oplus 0)) = \{a \oplus (b \oplus 0), a, b, 0\}$. On the other hand $\mathrm{Sub}(f(b \oplus c)) = \{f(b \oplus c), b \oplus c, b, c\}$. This shows the difference with the notion of *syntactic subterms*. In the rest of this paper and unless otherwise indicated, *the notion of subterm will refer to subterm values*.

Congruences and ordered rewriting. We shall introduce the notion of *ordered rewriting* [13], which is a useful technique that has been utilized (e.g. [3]) for proving the correctness of combination of unification algorithms.

Let $<$ be a simplification ordering on $\mathrm{T}(\mathcal{G})$ [1] assumed to be total on $\mathrm{T}(\mathcal{G})$ and such that the minimum for $<$ is a constant $c_{\min} \in \mathcal{C}$. Given a possibly infinite set of equations $\mathcal{O}$ on the signature $\mathrm{T}(\mathcal{G})$ we define the ordered rewriting relation $\to_\mathcal{O}$ by $s \to_\mathcal{O} s'$ iff there exists a position p in s, an equation $l = r$ in $\mathcal{O}$ and a substitution τ such that $s = s[p \leftarrow l\tau]$, $s' = s[p \leftarrow r\tau]$, and $l\tau > r\tau$.

It has been shown (see [13]) that by applying the *unfailing completion procedure* [14] to a set of equations $\mathcal{H}$ we can derive a (possibly infinite) set of equations $\mathcal{O}$ such that:

1. the congruence relations $=_\mathcal{O}$ and $=_\mathcal{H}$ are equal on $\mathrm{T}(\mathcal{F})$.
2. $\to_\mathcal{O}$ is convergent (i.e. terminating and confluent) on $\mathrm{T}(\mathcal{F})$.

We shall say that $\mathcal{O}$ is an *o-completion* of $\mathcal{H}$. The relation $\to_\mathcal{O}$ being convergent on ground terms we can define $(t){\downarrow}_\mathcal{O}$ as the unique normal form of the ground term t for $\to_\mathcal{O}$. Given a ground substitution σ we denote by $(\sigma){\downarrow}_\mathcal{O}$ the substitution with the same support such that for all variables $x \in \mathrm{Supp}(\sigma)$ we have

---

[1] By definition $<$ satisfies for all $s, t, u \in \mathrm{T}(\mathcal{G})$ $s < t[s]$ and $s < u$ implies $t[s] < t[u]$.

$x(\sigma)\downarrow_\mathcal{O} = (x\sigma)\downarrow_\mathcal{O}$. A substitution σ is *normal* if $\sigma = (\sigma)\downarrow_\mathcal{O}$. We will denote by R an o-completion of $\mathcal{E} = \mathcal{E}_1 \cup \mathcal{E}_2$. We denote by C_{spe} the set containing the constants in $\mathcal{F}$ and c_{min}.

4 Protocols, Intruders and Constraint Systems

Security of a given protocol is assessed with respect to a class of environments in which the protocol is executed. Dolev and Yao have described the environment not in terms of possible attacks on the protocol but by the deduction an intruder attacking a protocol execution is able to perform.

In Subsection 4.1 we define an extension of Dolev-Yao model to arbitrary operators for modeling the possible deductions of the intruder. In Subsection 4.2 we define the protocol semantics for an execution within an hostile environment controlled by the intruder and in Subsection 4.3 we describe how we represent this execution by a constraint system.

4.1 Intruder Systems

We shall model messages as ground terms and intruders deduction rules as rewrite rules on sets of messages representing the knowledge of an intruder. An intruder derives new messages from a given (finite) set of messages by applying intruder rules. Since we assume some equational axioms $\mathcal{H}$ are verified by functions symbols in the signature, all these derivations have to be considered *modulo* the equational congruence $=_\mathcal{H}$ generated by these axioms.

An intruder deduction rule in our setting is specified by a term t in some signature $\mathcal{G}$. Given values for the variables of t the intruder is able to generate the corresponding instance of t.

Definition 1. *An* intruder system $\mathcal{I}$ *is given by a triple* $\langle \mathcal{G}, T, \mathcal{H} \rangle$ *where $\mathcal{G}$ is a signature, $T \subseteq \mathrm{T}(\mathcal{G}, \mathcal{X})$ and $\mathcal{H}$ is a set of axioms between terms in $\mathrm{T}(\mathcal{G}, \mathcal{X})$. To each $t \in T$ we associate a deduction rule $\mathrm{L}^t : \mathrm{Var}(t) \to t$ and $\mathrm{L}^{t,g}$ denotes the set of ground instances of the rule L^t:*

$$\mathrm{L}^{t,g} = \{l \to r \mid \exists \sigma, \text{ground substitution on } \mathcal{G} \text{ s.t. } l = \mathrm{Var}(t)\sigma \text{ and } r =_\mathcal{H} t\sigma\}$$

The set of rules $\mathrm{L}_\mathcal{I}$ is defined as the union of the sets $\mathrm{L}^{t,g}$ for all $t \in T$.

Each rule $l \to r$ in $\mathrm{L}_\mathcal{I}$ defines an intruder deduction relation $\to_{l \to r}$ between finite sets of terms. Given two finite sets of terms E and F we define $E \to_{l \to r} F$ if and only if $l \subseteq E$ and $F = E \cup \{r\}$. We denote $\to_\mathcal{I}$ the union of the relations $\to_{l \to r}$ for all $l \to r$ in $\mathrm{L}_\mathcal{I}$ and by $\to_\mathcal{I}^*$ the transitive closure of $\to_\mathcal{I}$. We simply denote by $\to$ the relation $\to_\mathcal{I}$ when there is no ambiguity about $\mathcal{I}$.

For instance we can define $\mathcal{I}_\times = \langle \{\times, \mathrm{i}, 1\}, \{x \times y, \mathrm{i}(x), 1\}, \mathcal{E}_\times \rangle$ and we have $a, b, c \to_{\mathcal{I}_\times} a, b, c, c \times a$ by applying the rule $c, a \to c \times a \in \mathrm{L}^{x \times y, g}$.

A *derivation* D of length n, $n \geq 0$, is a sequence of steps of the form $E_0 \to_\mathcal{I} E_0 \cup \{t_1\} \to_\mathcal{I} \cdots \to_\mathcal{I} E_n$ with finite sets of ground terms $E_0, \ldots E_n$, and ground terms $t_1, \ldots, t_n$, such that $E_i = E_{i-1} \cup \{t_i\}$ for every $i \in \{1, \ldots, n\}$. The term t_n is called the *goal* of the derivation. We define $\overline{E}^{\mathcal{I}}$ to be equal to the set $\{t \mid \exists F \text{ s.t. } E \to_{\mathcal{I}}^{*} F \text{ and } t \in F\}$ i.e. the set of terms that can be derived from E. If there is no ambiguity on the deduction system $\mathcal{I}$ we write $\overline{E}$ instead of $\overline{E}^{\mathcal{I}}$.

Let $\mathcal{O}$ be an o-completion of $\mathcal{H}$. We will assume from now that all the deduction rules generate terms that are normalized by $\to_{\mathcal{O}}$ and the goal and the initial set are in normal form for $\to_{\mathcal{O}}$. It can be shown [9] that this is not restrictive for our main decidability result.

Given a set of terms $T \subseteq T(\mathcal{G}, \mathcal{X})$ we define the set of terms $\langle T \rangle$ to be the minimal set such that $T \subseteq \langle T \rangle$ and for all $t \in \langle T \rangle$ and for all substitutions σ with image included in $\langle T \rangle$, we have $t\sigma \in \langle T \rangle$. Hence terms in $\langle T \rangle$ are built by composing terms in T iteratively. We can prove easily that the intruder systems $\mathcal{I} = \langle \mathcal{G}, T, \mathcal{H} \rangle$ and $\mathcal{J} = \langle \mathcal{G}, \langle T \rangle, \mathcal{H} \rangle$ define the same sets of derivable terms, i.e. for all E we have $\overline{E}^{\mathcal{I}} = \overline{E}^{\mathcal{J}}$.

We want to consider the union of the 2 intruder systems $\mathcal{I}_1 = \langle \mathcal{F}_1, T_1, \mathcal{E}_1 \rangle$ and $\mathcal{I}_2 = \langle \mathcal{F}_2, T_2, \mathcal{E}_2 \rangle$. In particular we are interested in the derivations obtained by using $\to_{\mathcal{I}_1} \cup \to_{\mathcal{I}_2}$. It can be noticed that $\langle T_1 \cup T_2 \rangle = \langle \langle T_1 \rangle \cup \langle T_2 \rangle \rangle$. Hence by the remarks above the derivable terms using $\langle T_1 \cup T_2 \rangle$ or $\langle T_1 \rangle \cup \langle T_2 \rangle$ are the same. For technical reason it will be more convenient to use $\langle T_1 \rangle \cup \langle T_2 \rangle$ for defining the union of 2 intruder systems:

Definition 2. *The* union *of the two intruder systems* $\langle \mathcal{F}_1, T_1, \mathcal{E}_1 \rangle, \langle \mathcal{F}_2, T_2, \mathcal{E}_2 \rangle$ *is the intruder system* $\mathcal{U} = \langle \mathcal{F}, \langle T_1 \rangle \cup \langle T_2 \rangle, \mathcal{E} \rangle$.

A derivation $E_0 \to_{\mathcal{U}} E_0 \cup \{t_1\} \to_{\mathcal{U}} \cdots \to_{\mathcal{U}} E_n$ of intruder system $\mathcal{U}$ is *well-formed* if for all $i \in \{1, \ldots, n\}$ we have $t_i \in \text{Sub}(E_0 \cup \{t_n\})$; in other words every message generated by an intermediate step either occurs in the goal or in the initial set of messages. In the following lemma the derivations refer to the intruder system $\mathcal{U} = \langle \mathcal{F}, \langle T_1 \rangle \cup \langle T_2 \rangle, \mathcal{E} \rangle$. For the proof see [9]:

Lemma 1. *A derivation of minimal length starting from E of goal t is well-formed.*

4.2 Protocol Analysis

In this subsection we describe how protocols are modelled. In the following we only model a single session of the protocol since it is well-known how to reduce several sessions to this case. Our semantics follows the one by [12].

In Dolev-Yao model the intruder has complete control over the communication medium. We model this by considering the intruder *is* the network. Messages sent by honest agents are sent directly to the intruder and messages received by the honest agents are always sent by the intruder. From the intruder side a finite execution of a protocol is the interleaving of a finite sequence of messages she has to send and a finite sequence of messages she receives (and add to her knowledge).

We also assume that the interaction of the intruder with one agent is an atomic step. The intruder sends a message m to an honest agent, this agent tests the validity of this message and responds to it. Alternatively an agent may initiate an execution and in this case we assume it reacts to a dummy message c_{min} sent by the intruder.

A *step* is a triplet $(\text{RECV}(x); \text{SEND}(s); \text{COND}(e))$ where $x \in \mathcal{X}$, $s \in \text{T}(\mathcal{G}, \mathcal{X})$ and e is a set of equations between terms of $\text{T}(\mathcal{G}, \mathcal{X})$. The meaning of a step is that upon receiving message x, the honest agent checks the equations in e and sends the message s. An execution of a protocol is a finite sequence of steps.

Example 1. Consider the following toy protocol where K is a symmetric key initially known by A only:

$$A \to B : \{M \oplus B\}_K$$
$$B \to A : B$$
$$A \to B : K$$
$$B \to A : M$$

Assuming the algebraic properties of $\oplus$, symmetric encryption $\text{se}(,)$ and symmetric decryption $\text{sd}(,)$ we model this protocol as:

$\text{RECV}(v_1); \text{SEND}(\text{se}(M \oplus B, K)); \text{COND}(v_1 = c_{min})$
$\text{RECV}(v_2); \text{SEND}(B); \text{COND}(\emptyset)$
$\text{RECV}(v_3); \text{SEND}(K); \text{COND}(v_3 = B)$
$\text{RECV}(v_4); \text{SEND}(\text{sd}(v_2, v_4) \oplus B); \text{COND}(v_2 = \text{se}(x, v_4,))$
$\text{RECV}(v_5); \text{SEND}(c_{min}); \text{COND}(v_5 = M)$

Note that in our setting we can model that at some step i the message must match the pattern t_i by adding an equation $v_i \stackrel{?}{=} t_i$ as a condition for this step.

In order to define whether an execution of a protocol is feasible we must first define when a substitution σ satisfies a set of equations $\mathcal{S}$.

Definition 3. *(Unification systems) Let $\mathcal{H}$ be a set of axioms on $\text{T}(\mathcal{G}, \mathcal{X})$. An $\mathcal{H}$-Unification system $\mathcal{S}$ is a finite set of equations in $\text{T}(\mathcal{G}, \mathcal{X})$ denoted by $(t_i \stackrel{?}{=} u_i)_{i \in \{1,\ldots,n\}}$. It is satisfied by a ground substitution σ, and we note $\sigma \models \mathcal{S}$, if for all $i \in \{1, \ldots, n\}$ $t_i\sigma =_\mathcal{H} u_i\sigma$.*

Let $\mathcal{I} = \langle \mathcal{G}, T, \mathcal{H} \rangle$ be an intruder system. A *configuration* is a couple $\langle P, N \rangle$ where P is a finite sequence of steps and N is a set of ground terms (the knowledge of the intruder). From the configuration $\langle (\text{RECV}(x); \text{SEND}(s); \text{COND}(e)) \cdot P, N \rangle$ a transition to (P', N') is possible iff there exists a ground substitution σ such that $x\sigma \in \overline{N}$, $\sigma \models e$, $N' = N \cup \{s\sigma\}$ and $P' = P\sigma$. Trace based-security properties like secrecy can be reduced to the *Execution feasibility* problem:

Execution feasibility

    Input: an initial configuration $\langle P, N_0 \rangle$
    Output: SAT iff there exists a reachable configuration $\langle \emptyset, M \rangle$

4.3 Constraints Systems

We express the execution feasibility of a protocol by a constraint problem $\mathcal{C}$.

Definition 4. *(Constraint systems)* Let $\mathcal{I} = \langle \mathcal{G}, T, \mathcal{H} \rangle$ be an intruder system. An $\mathcal{I}$-Constraint system $\mathcal{C}$ is denoted: $((E_i \triangleright v_i)_{i \in \{1,\ldots,n\}}, \mathcal{S})$ and it is defined by a sequence of couples $(E_i, v_i)_{i \in \{1,\ldots,n\}}$ with $v_i \in \mathcal{X}$ and $E_i \subseteq \mathrm{T}(\mathcal{G}, \mathcal{X})$ for $i \in \{1,\ldots,n\}$ and $E_{i-1} \subseteq E_i$ for $i \in \{2,\ldots,n\}$ and by an $\mathcal{H}$-unification system $\mathcal{S}$. It is deterministic iff for all $i \in \{1,\ldots,n\}$, $\mathrm{Var}(E_i) \subseteq \{v_1,\ldots,v_{i-1}\}$

An $\mathcal{I}$-Constraint system $\mathcal{C}$ is satisfied by a ground substitution σ if for all $i \in \{1,\ldots,n\}$ we have $v_i \sigma \in \overline{E_i \sigma}$ and if $\sigma \models \mathcal{S}$. We denote that a ground substitution σ satisfies a constraint system $\mathcal{C}$ by $\sigma \models_{\mathcal{I}} \mathcal{C}$.

Constraint systems are denoted by $\mathcal{C}$ and decorations thereof. Note that if a substitution σ is a solution of a constraint system $\mathcal{C}$, by definition of constraints and of unification systems the substitution $(\sigma){\downarrow}_\mathcal{O}$ is also a solution of $\mathcal{C}$ (where $\mathcal{O}$ is an o-completion of H). In the context of cryptographic protocols the inclusion $E_{i-1} \subseteq E_i$ means that the knowledge of an intruder does not decrease as the protocol progresses: after receiving a message an honest agent will respond to it. This response can be added to the knowledge of an intruder who listens all communications.

The condition defining the *deterministic* constraint systems expresses that a message to be sent at some step i should be built from previously received messages recorded in the variables v_j for $j < i$ and from the initial knowledge.

Example 2. We model the protocol of Example 1 by the following constraint system. First we gather all conditions in a unification system $\mathcal{S}$

$$\mathcal{S} = \left\{ v_1 \stackrel{?}{=} c_{\min}\ ,\ v_3 \stackrel{?}{=} B\ ,\ v_2 \stackrel{?}{=} \mathrm{se}(x, v_4)\ ,\ v_5 \stackrel{?}{=} M \right\}$$

The protocol execution for intruder $\mathcal{I}$ with initial knowledge $\{c_{\min}\}$ is then expressed by the constraint system:

$$\begin{aligned}
\mathcal{C} = ((\ & c_{\min} \triangleright v_1, \\
& c_{\min}, \mathrm{se}(M \oplus B, K) \triangleright v_2, \\
& c_{\min}, \mathrm{se}(M \oplus B, K), B \triangleright v_3, \\
& c_{\min}, \mathrm{se}(M \oplus B, K), B, K \triangleright v_4), \\
& c_{\min}, \mathrm{se}(M \oplus B, K), B, K, \mathrm{sd}(v_2, v_4) \oplus B \triangleright v_5, \mathcal{S})
\end{aligned}$$

The *deterministic* condition imposes to write the last message $\mathrm{sd}(v_2, v_4)$ instead of x though both are equivalent with respect to satisfiability.

The decision problems we are interested in are the *satisfiability* and the *ordered satisfiability* of intruder constraint systems.

Satisfiability

Input: an $\mathcal{I}$-constraint system $\mathcal{C}$
Output: SAT iff there exists a substitution σ such that: $\sigma \models_{\mathcal{I}} \mathcal{C}$.

In order to be able to combine solutions of constraints in component theories to get a solution for the full theory these solutions have to satisfy some ordering constraints too. Intuitively, this is to avoid introducing cycle when building a global solution. With respect to this use we can always assume $c_{\min}$ is the minimum of $\prec$ in the following definition:

Ordered Satisfiability

Input: an $\mathcal{I}$-constraint system $\mathcal{C}$, X the set of all variables and C the set of all free constants occurring in $\mathcal{C}$ and a linear ordering $\prec$ on $X \cup C$.

Output: SAT iff there exists a substitution σ such that:
$$\begin{cases} \sigma \models_{\mathcal{I}} \mathcal{C} \\ \forall x \in X \text{ and } \forall c \in C, x \prec c \text{ implies } c \notin \text{Sub}_{\text{syn}}(x\sigma) \end{cases}$$

The main result of this paper is the following modularity result:

Theorem 1. *If the ordered satisfiability problem is decidable for two intruders $\langle \mathcal{F}_1, T_1, \mathcal{E}_1 \rangle$ and $\langle \mathcal{F}_2, T_2, \mathcal{E}_2 \rangle$ for disjoint signatures $\mathcal{F}_1$ and $\mathcal{F}_2$ then the satisfiability problem is decidable for deterministic constraint systems for the intruder $\mathcal{U} = \langle \mathcal{F}, \langle T_1 \rangle \cup \langle T_2 \rangle, \mathcal{E} \rangle$.*

This result is obtained as a direct consequence of the next section where we give an algorithm for solving $\mathcal{U}$-constraints using algorithms for solving *ordered satisfiability* for intruders $\langle \mathcal{F}_1, T_1, \mathcal{E}_1 \rangle$ and $\langle \mathcal{F}_2, T_2, \mathcal{E}_2 \rangle$.

5 Combination of Decision Procedures

We introduce Algorithm 1 for solving satisfiability of constraint systems for the union $\mathcal{U}$ of two intruders systems $\mathcal{I}_1 = \langle \mathcal{F}_1, T_1, \mathcal{E}_1 \rangle$ and $\mathcal{I}_2 = \langle \mathcal{F}_2, T_2, \mathcal{E}_2 \rangle$ with disjoint signatures $\mathcal{F}_1$ and $\mathcal{F}_2$. The completeness of Algorithm 1 is sketched below, and the proofs (for completeness and soundness) are fully detailed in [9]. Let us explain this algorithm:

Step 2 The algorithm input is a $\mathcal{U}$-Constraint system $(\mathcal{D}, \mathcal{S})$. An equational system $\mathcal{S}$ is *homogeneous* if for all $u \stackrel{?}{=} v \in \mathcal{S}$, u and v are both pure 1-terms or both pure 2-terms. It is well-known that equational systems can be transformed into equivalent (w.r.t. satisfiability) homogeneous systems. Thus we can assume that $\mathcal{S}$ is homogeneous without loss of generality.

Step 3 abstracts every subterm t of $\mathcal{C}$ by a new variable $\psi(t)$. A choice of ψ such that $\psi(t) = \psi(t')$ will lead to solutions that identify t and t'.

Step 4 assign non-deterministically a signature to the root symbol of the subterms of $\mathcal{C}$ instantiated by a solution. The choice $th(\psi(t)) = 0$ corresponds to the situation where t gets equal to a free constant.

Steps 5–8 choose and order non-deterministically the intermediate subterms in derivations that witness that the solution satisfies the constraints in $\mathcal{D}$.

Step 9 defines a constraint problem $\mathcal{C}'$ collecting the previous choices on subterms identification, subterms signatures and derivation structures.

Step 10 splits the problem $\mathcal{S}'$ in two pure subproblems.

Step 11 splits non-deterministically the problem $\mathcal{D}'$, that is we select for each $E \triangleright v$ in $\mathcal{D}'$ an intruder system to solve it.

Step 12 guesses an ordering on variables: this ordering will preclude the value of a variable from being a subterm of the value of a smaller variable. This is used to avoid cycles in the construction of the solution.

Step 13 solves independently the 2 pure subproblems obtained at steps 10–11. In $\mathcal{C}_i$ the variables q with $th(q) \neq i$ will be considered as constants.

Algorithm 1. Combination Algorithm

1: **Solve**$_\mathcal{U}(\mathcal{C})$
2: **Let** $\mathcal{C} = ((E_i \triangleright v_i)_{i \in \{1,\ldots,n\}}, \mathcal{S})$ with $\mathcal{S}$ homogeneous.
3: **Choose** ψ an application from $\mathrm{Sub}(\mathcal{C})$ to $\mathcal{X} \setminus \mathrm{Var}(\mathcal{C})$ and let $Q = \psi(\mathrm{Sub}(\mathcal{C}))$
4: **Choose** a theory $th(q) \in \{0, 1, 2\}$ for all $q \in Q$
5: **for** $i = 1$ to n **do**
6:     **Choose** $Q_i \subseteq Q$
7:     **Choose** a linear ordering over the elements of Q_i say $(q_{i,1},\ldots,q_{i,k_i})$
8: **end for**
9: **Let** $\mathcal{C}' = (\mathcal{D}', \mathcal{S}')$ where
$$\begin{cases} \mathcal{S}' = \mathcal{S} \cup \left\{ z \stackrel{?}{=} \psi(z) \mid z \in \mathrm{Sub}(\mathcal{C}) \right\} \\ \mathcal{D}' = \Delta_1, \ldots, \Delta_i, \ldots \Delta_n \end{cases}$$
and $\Delta_i = (K_i, Q_i^{<j} \triangleright q_{i,j})_{j \in \{1,\ldots,k_i\}}, (K_i, Q_i \triangleright \psi(v_i))$ with
$$\begin{cases} K_i = \psi(E_i) \cup \bigcup_{j=1}^{i-1} Q_j \\ Q_i^{<j} = q_{i,1}, q_{i,2}, \ldots, q_{i,j-1} \end{cases}$$
10: **Split** $\mathcal{S}'$ into $\mathcal{S}_1, \mathcal{S}_2$ such that $\mathcal{S}' = \mathcal{S}_1 \cup \mathcal{S}_2$ and:
$$\begin{cases} \mathcal{S}_1 = \left\{ z \stackrel{?}{=} z' \in \mathcal{S}' \mid z, z' \text{ are pure 1-terms} \right\} \\ \mathcal{S}_2 = \left\{ z \stackrel{?}{=} z' \in \mathcal{S}' \mid z, z' \text{ are pure 2-terms} \right\} \end{cases}$$
11: **Split** non-deterministically $\mathcal{D}'$ into $\mathcal{D}_1, \mathcal{D}_2$
12: **Choose** a linear ordering $\prec$ over Q.
13: **Solve** $\mathcal{C}_i = (\mathcal{D}_i, \mathcal{S}_i)$ for intruder $\mathcal{I}_i$ with linear ordering $\prec$ for $i \in \{1, 2\}$
14: **if** both are satisfied **then**
15:     **Output:** SATISFIED
16: **end if**

We assume $\mathrm{C}_{\mathrm{spe}} \subseteq \mathrm{Sub}(\mathcal{C})$. Recall that R is the rewrite system associated to $\mathcal{E} = \mathcal{E}_1 \cup \mathcal{E}_2$. We say a normal substitution σ is *bound* if for all variables x with $x\sigma \neq x$ and for all $t \in \mathrm{Sub}(x\sigma)$ there exists $u \in \mathrm{Sub}(\mathcal{C})$ such that $(u\sigma){\downarrow}_R = t$. A key proposition is:

Proposition 1. *If $\mathcal{C}$ is a satisfiable constraint system there exists a bound substitution σ such that $\sigma \models \mathcal{C}$. Moreover $\mathrm{Sub}((\mathrm{Sub}(\mathcal{C})\sigma){\downarrow}_R) = (\mathrm{Sub}(\mathcal{C})\sigma){\downarrow}_R$.*

5.1 Completeness of the Algorithm

Proposition 2. *If $\mathcal{C}$ is satisfiable then there exists $\mathcal{C}_1$ and $\mathcal{C}_2$ satisfiable at Step 13 of the algorithm.*

Proof. First let us prove that the 11 first steps of the algorithm preserve satisfiability. Assume $\mathcal{C}$ is satisfiable. By Proposition 1 there exists a normal bound substitution σ which satisfies $\mathcal{C}$. Define ψ to be a function from $\text{Sub}(\mathcal{C})$ to a set of variables Q such that $\psi(t) = \psi(t')$ if and only if $(t\sigma)\downarrow_R = (t'\sigma)\downarrow_R$. Thus by Proposition 1 there exists a bijection ϕ from Q to $\text{Sub}((\text{Sub}(\mathcal{C})\sigma)\downarrow_R)$. We let $th(q) = i$ if $\text{Sign}(\phi(q)) = \mathcal{F}_i$ and $th(q) = 0$ if $\text{Sign}(\phi(q)) = \bot$. By the construction of $\mathcal{S}'$ and the choice of ψ we can extend σ on Q by $q\sigma = (\psi^{-1}(q)\sigma)\downarrow_R$ for all $q \in Q$.

For each $i \in \{1, \ldots, n\}$ by Lemma 1 we can consider a well-formed derivation $\mathcal{D}_i$ starting from $F_i = (E_i\sigma)\downarrow_R$ and of goal $g_i = v_i\sigma$:

$$\mathcal{D}_i : F_i \to_{\mathcal{U}} F_i \cup \{r_{i,1}\} \to_{\mathcal{U}} \cdots \to_{\mathcal{U}} F_i \cup \{r_{i,1}, \ldots, r_{i,k_i}\} \to_{\mathcal{U}} F_i \cup \{r_{i,1}, \ldots, r_{i,k_i}, g_i\}$$

We have $\text{Sub}(F_i, g_i) \subseteq \text{Sub}((\text{Sub}(\mathcal{C}\sigma))\downarrow_R)$. Since the derivation is well-formed we have $\{r_{i,1}, \ldots, r_{i,k_i}\} \subseteq \text{Sub}(F_i, g_i)$. By Proposition 1, $\text{Sub}((\text{Sub}(\mathcal{C}\sigma))\downarrow_R) = (\text{Sub}(\mathcal{C})\sigma)\downarrow_R$. Thus the function ϕ^{-1} is defined for each $r_{i,j}$. Let $q_{i,j} = \phi^{-1}(r_{i,j})$ and Q_i be the sequence of the $q_{i,j}$.

The algorithm will non-deterministically produce a $\mathcal{C}'$ corresponding to these choices and satisfied by σ (extended over Q by $\psi(t)\sigma = (t\sigma)\downarrow_R$) by construction. Since $\mathcal{S}$ is satisfiable, following the lines of F. Baader and K. Schulz [3] permits to prove that $\mathcal{S}_1$ and $\mathcal{S}_2$ are satisfiable with a linear constant restriction $\prec$ chosen such that $q \prec q'$ implies $q'\sigma$ is not a subterm of $q\sigma$.

We choose the sequence of constraints in $\mathcal{D}_1$ (resp. $\mathcal{D}_2$) to be the subsequence of constraints $F \triangleright q$ from $\mathcal{D}'$ such that the corresponding transition in the solution was performed by a rule in $\text{L}^{u,g}$ with $\text{Sign}(u) = \mathcal{F}_1$ (resp. $\mathcal{F}_2$). By construction these two systems are satisfiable.

From the soundness and completeness of Algorithm 1 we can derive our main result on the combination of two intruders. It can be easily generalized to n intruders over disjoint signatures $\mathcal{F}_1, \ldots, \mathcal{F}_n$.

The main drawback of the combination algorithm that we have presented here is that it requires the solvability of general constraints from sub-theories. However the decision procedures which already exist for fixed intruder theories are limited to *deterministic* constraint systems. Fortunately we have been able to show ([9]) that our combination algorithm can be adapted so that it suffices to decide the solvability of *deterministic* constraints systems in sub-theories.

6 Conclusion

We have proposed an algorithm for combining decision procedures for intruder constraints on disjoint signatures. This algorithm allows for a modular treatment of algebraic operators in protocol analysis and we believe that it will contribute to a better understanding of complexity issues in the domain. Since constraint satisfiability is required only from the intruder sub-theories the approach should permit one to handle more complex protocols than with alternative techniques.

References

1. M. Abadi, B. Blanchet, and C. Fournet. Just Fast Keying in the Pi Calculus. In David Schmidt, editor, *Proceedings of ESOP'04*, volume 2986 of *Lecture Notes on Computer Science*, pages 340–354, Barcelona, Spain, 2004. Springer Verlag.
2. R. Amadio, D. Lugiez, and V. Vanackère. On the symbolic reduction of processes with cryptographic functions. *Theor. Comput. Sci.*, 290(1):695–740, 2003.
3. F. Baader and K. U. Schulz. Unification in the union of disjoint equational theories. combining decision procedures. *J. Symb. Comput.*, 21(2):211–243, 1996.
4. D. Basin, S. Mödersheim, and L. Viganò. An On-The-Fly Model-Checker for Security Protocol Analysis. In Einar Snekkenes and Dieter Gollmann, editors, *Proceedings of ESORICS'03*, LNCS 2808, pages 253–270. Springer-Verlag, 2003.
5. M. Boreale. Symbolic trace analysis of cryptographic protocols. In *Proceedings of the 28th ICALP'01*, LNCS 2076, pages 667–681. Springer-Verlag, Berlin, 2001.
6. M. Boreale and M. Buscemi. Symbolic analysis of crypto-protocols based on modular exponentiation. In *Proceedings of MFCS 2003*, volume 2747 of *Lecture Notes in Computer Science*. Springer, 2003.
7. N. Borisov, I. Goldberg, and D. Wagner. Intercepting mobile communications: the insecurity of 802.11. In *Proceedings of MOBICOM 2001*, pages 180–189, 2001.
8. Y. Chevalier, R. Kuesters, M. Rusinowitch, and M. Turuani. An NP Decision Procedure for Protocol Insecurity with XOR. In *Proceedings of the Logic In Computer Science Conference, LICS'03*, June 2003.
9. Y. Chevalier and M. Rusinowitch. Combining intruder theories. Technical report, INRIA, 2005. http://www.inria.fr/rrrt/rr-5495.html.
10. Y. Chevalier and L. Vigneron. A Tool for Lazy Verification of Security Protocols. In *Proceedings of the Automated Software Engineering Conference (ASE'01)*. IEEE Computer Society Press, 2001.
11. H. Comon-Lundh and V. Shmatikov. Intruder Deductions, Constraint Solving and Insecurity Decision in Presence of Exclusive or. In *Proceedings of the Logic In Computer Science Conference, LICS'03*, pages 271–280, 2003.
12. S. Delaune and F. Jacquemard. A decision procedure for the verification of security protocols with explicit destructors. In *Proceedings of the 11th ACM Conference on Computer and Communications Security (CCS'04)*, pages 278–287, Washington, D.C., USA, October 2004. ACM Press.
13. N. Dershowitz and J-P. Jouannaud. Rewrite systems. In *Handbook of Theoretical Computer Science, Volume B*, pages 243–320. Elsevier, 1990.
14. J. Hsiang and M. Rusinowitch. On word problems in equational theories. In *ICALP*, volume 267 of *Lecture Notes in Computer Science*, pages 54–71. Springer, 1987.
15. C. Meadows and P. Narendran. A unification algorithm for the group Diffie-Hellman protocol. In *Workshop on Issues in the Theory of Security (in conjunction with POPL'02)*, Portland, Oregon, USA, January 14-15, 2002.

16. J. Millen and V. Shmatikov. Constraint solving for bounded-process cryptographic protocol analysis. In *ACM Conference on Computer and Communications Security*, pages 166–175, 2001.
17. J. Millen and V. Shmatikov. Symbolic protocol analysis with an abelian group operator or Diffie-Hellman exponentiation. *Journal of Computer Security*, 2005.
18. M. Rusinowitch and M. Turuani. Protocol insecurity with finite number of sessions is NP-complete. In *Proceedings of CSFW 2001*. IEEE, 2001.
19. M. Schmidt-Schauß. Unification in a combination of arbitrary disjoint equational theories. *J. Symb. Comput.*, 8(1/2):51–99, 1989.

Computationally Sound Implementations of Equational Theories Against Passive Adversaries

Mathieu Baudet[1], Véronique Cortier[2], and Steve Kremer[1]

[1] LSV/ CNRS UMR 8643 & INRIA Futurs projet SECSI & ENS Cachan, France
{baudet, kremer}@lsv.ens-cachan.fr
[2] Loria/CNRS UMR 7503 & INRIA Lorraine projet Cassis, France
cortier@loria.fr

Abstract. In this paper we study the link between formal and cryptographic models for security protocols in the presence of a passive adversary. In contrast to other works, we do not consider a fixed set of primitives but aim at results for an arbitrary equational theory. We define a framework for comparing a cryptographic implementation and its idealization w.r.t. various security notions. In particular, we concentrate on the computational soundness of static equivalence, a standard tool in cryptographic pi calculi. We present a soundness criterion, which for many theories is not only sufficient but also necessary. Finally, we establish new soundness results for the exclusive OR and a theory of ciphers and lists.

1 Introduction

Today's ubiquity of computer networks increases the need for theoretic foundations for cryptographic protocols. For more than twenty years now, two communities separately developed two families of models. Both views have been very useful in increasing the understanding and quality of security protocol design. On the one hand *formal* or *logical* models have been developed, based on the seminal work of Dolev and Yao [9]. These models view cryptographic operations in a rather abstract and idealized way. On the other hand *cryptographic* or *computational* models [10] are closer to implementations: cryptographic operations are modeled as algorithms manipulating bit-strings. Those models cover a large class of attacks, namely all those implementable by a probabilistic polynomial-time Turing machine.

The advantage of formal models is that security proofs are generally simpler and suitable for automatic procedures, even for complex protocols. Unfortunately, the high degree of abstraction and the limited adversary power raise serious questions regarding the security offered by such proofs. Potentially, justifying symbolic proofs with respect to standard computational models has tremendous benefits: protocols can be analyzed using automated tools and still benefit from the security guarantees of the computational model.

Recently, a significant research effort has been directed at linking these two approaches. In their seminal work [3], Abadi and Rogaway prove the computational soundness of formal (symmetric) encryption in the case a passive attacker. Since then, many results [5, 11, 12] have been obtained. Notably, Backes *et al.* [5] prove the soundness of a rich language including digital signatures, public-key and symmetric key encryption in the presence of an active attacker. Laud [11] presents an automated procedure for computationally sound proofs of confidentiality in the case of an active attacker and symmetric encryption when the number of sessions is bounded.

Each of these results considers a fixed set of primitives, *e.g.* symmetric or public-key encryption. In this paper, we aim at presenting general results for arbitrary equational theories, such as encryption, but also less studied ones, *e.g.* groups or exclusive OR. We concentrate on *static equivalence*, a now standard notion originating from the applied pi calculus [2]. Intuitively, static equivalence asks whether an attacker can distinguish between two tuples of terms, by exhibiting an equation which holds on one tuple but not on the other. This provides an elegant means to express security properties against passive attackers. Moreover there exist exact [1] and approximate [8] algorithms to decide static equivalence for a large family of equational theories.

Our first contribution is a general framework for comparing formal and computational models in the presence of a passive attacker. We define the notions of *soundness* and *faithfulness* of a cryptographic implementation w.r.t. equality, static equivalence and deducibility. Soundness holds when each formal proof has a computational interpretation. Faithfulness is the converse, *i.e.* the formal model does not provide false attacks.

Our second contribution is a sufficient criterion for soundness w.r.t static equivalence: intuitively the usual computational semantics of terms has to be indistinguishable to an idealized one. We also provide a general definition of patterns for arbitrary equational theories that encompasses the notion usually defined for symmetric and public encryption. Those patterns allow us to characterize a large class of theories for which our soundness criterion is necessary.

Our third contribution consists in applying our framework to obtain two novel soundness results. The first theory deals with the exclusive OR. Interestingly, our proof reflects the unconditional security (in the information-theoretic sense) of the One-Time Pad encryption scheme. Second we consider a theory of symmetric encryption and lists. In some sense, the result is similar to the one of Abadi and Rogaway [3]. However, we consider deterministic, length-preserving, symmetric encryption schemes a.k.a. ciphers. To the best of our knowledge, this is the first result on such schemes, whose specificity is that decryption always succeeds.

Outline of the paper. In the next section, we introduce our abstract and concrete models together with the notions of indistinguishability. We then define the notions of soundness and faithfulness and illustrate some consequences of soundness w.r.t. static equivalence on groups. In Section 4, we define the ideal semantics of abstract terms, present our soundness criterion and also show that for a large family of interesting equational theories, the soundness criterion is a necessary condition. As an illustration (Section 5), we prove the soundness for the theories modeling exclusive OR, as well as ciphers and lists. We then conclude and give directions for future work. Note that, due to lack of space most proofs have been omitted; those can be found in the extended version [7].

2 Modeling Cryptographic Primitives with Abstract Algebras

In this section we introduce some notations and set our abstract and concrete models.

2.1 Abstract Algebras

Our abstract models—which we call *abstract algebras*—consist of term algebras defined on a first-order signature with sorts and equipped with equational theories.

Specifically a *signature* $(\mathcal{S}, \mathcal{F})$ is made of a set of *sorts* $\mathcal{S} = \{s, s_1 \ldots\}$ and a set of *symbols* $\mathcal{F} = \{f, f_1 \ldots\}$ together with arities of the form $\mathrm{ar}(f) = s_1 \times \ldots \times s_k \to s$, $k \geq 0$. Symbols that take $k = 0$ arguments are called *constants*; their arity is simply written s. We fix an infinite set of *names* $\mathcal{N} = \{a, b \ldots\}$ and an infinite set of *variables* $\mathcal{X} = \{x, y \ldots\}$. We assume that names and variables are given with sorts. The set of *terms of sort s* is defined inductively by

$$T ::= \quad \text{term of sort } s$$
$$\mid x \quad \text{variable } x \text{ of sort s}$$
$$\mid a \quad \text{name } a \text{ of sort s}$$
$$\mid f(T_1, \ldots, T_k) \quad \text{application of symbol } f \in \mathcal{F}$$

where for the last case, we further require that T_i is a term of some sort s_i and $\mathrm{ar}(f) = s_1 \times \ldots \times s_k \to s$. As usual, we write $\mathrm{var}(T)$ and $\mathrm{names}(T)$ for the set of variables and names occurring in T respectively. A term is *ground* or *closed* iff it has no variables.

Substitutions are written $\sigma = \{x_1 = T_1, \ldots, x_n = T_n\}$ with domain $\mathrm{dom}(\sigma) = \{x_1, \ldots, x_n\}$. We only consider *well-sorted* substitutions, that is, substitutions $\sigma = \{x_1 = T_1, \ldots, x_n = T_n\}$ for which x_i and T_i have the same sort. σ is *closed* iff all of the T_i are closed. We extend the notation $\mathrm{names}(.)$ from terms to substitutions in the obvious way. The application of a substitution σ to a term T is written $\sigma(T) = T\sigma$.

Symbols in $\mathcal{F}$ are intended to model cryptographic primitives, whereas names in $\mathcal{N}$ are used to model nonces *i.e.* concretely random numbers. The abstract semantics of symbols is described by an equational theory E, that is an equivalence relation (also written $=_E$) which is stable by application of contexts and well-sorted substitutions of variables. We further require that E is stable under substitution of names. All the equational theories that we consider in this paper satisfy these properties. For instance, symmetric and deterministic encryption is modeled by the theory E_{enc} generated by the classical equation $E_{\mathrm{enc}} = \{\mathrm{dec}(\mathrm{enc}(x, y), y) = x\}$.

2.2 Frames, Deducibility and Static Equivalence

Following [2,1], a *frame* is an expression $\varphi = \nu \tilde{a}.\sigma$ where $\tilde{a}$ is a set of *bound (or restricted) names* and σ is a well-sorted substitution. Intuitively, frames represent sequences of messages learned by an attacker during the execution of a protocol.

For simplicity we only consider frames $\nu \tilde{a}.\sigma$ which restrict *every* name occurring in σ, that is $\tilde{a} = \mathrm{names}(\sigma)$. In other words, names a must be disclosed *explicitly* by adding a mapping $x_a = a$ to the substitution. Thus we tend to assimilate frames and their underlying substitutions.

A term T is *deducible* from a closed frame φ, written $\varphi \vdash_E T$ iff there exists a term M with $\text{var}(M) \subseteq \text{dom}(\varphi)$ and $\text{names}(M) \cap \text{names}(\varphi) = \emptyset$ such that $M\varphi =_E T$. Consider for instance the theory E_{enc} and the frame $\varphi_1 = \nu k_1, k_2, k_3, k_4. \{x_1 = \text{enc}(k_1, k_2), x_2 = \text{enc}(k_4, k_3), x_3 = k_3\}$: the name k_4 is deducible from φ_1 since $\text{dec}(x_2, x_3)\varphi_1 =_{E_{\text{enc}}} k_4$ but neither k_1 nor k_2 are deducible.

Deducibility is not always sufficient to account for the knowledge of an attacker. E.g. it lacks partial information on secrets. This is why the notion of static equivalence is used. Two closed frames φ_1 and φ_2 are *statically equivalent*, written $\varphi_1 \approx_E \varphi_2$, iff (i) $\text{dom}(\varphi_1) = \text{dom}(\varphi_2)$, (ii) for all terms M, N with variables included in $\text{dom}(\varphi_i)$ and using no names occurring in φ_1 or φ_2, $M\varphi_1 =_E N\varphi_1$ is equivalent to $M\varphi_2 =_E N\varphi_2$.

For instance, the two frames $\nu k. \{x = \text{enc}(0, k)\}$ and $\nu k. \{x = \text{enc}(1, k)\}$ are statically equivalent with respect to E_{enc}, whereas the two frames $\nu k. \{x = \text{enc}(0, k), y = k\}$ and $\nu k, k'. \{x = \text{enc}(0, k'), y = k\}$ are not.

2.3 Concrete Semantics

We now give terms and frames a concrete semantics, parameterized by an implementation of the primitives. Provided a set of sorts S and a set of symbols $\mathcal{F}$ as above, a $(S, \mathcal{F})$-*computational algebra* A consists of

- a non-empty set of bit-strings $[\![s]\!]_A \subseteq \{0,1\}^*$ for each sort $s \in S$;
- a computable function $f_A : [\![s_1]\!]_A \times \ldots \times [\![s_k]\!]_A \to [\![s]\!]_A$ for each $f \in \mathcal{F}$ with $\text{ar}(f) = s_1 \times \ldots \times s_k \to s$;
- a computable congruence $=_{A,s}$ for each sort s, in order to check the equality of elements in $[\![s]\!]_A$ (the same element may be represented by different bit-strings); by congruence, we mean a reflexive, symmetric, transitive relation such that $e_1 =_{A,s_1} e'_1, \ldots, e_k =_{A,s_k} e'_k \Rightarrow f_A(e_1, \ldots, e_k) =_{A,s} f_A(e'_1, \ldots, e'_k)$ (in the remaining we often omit s and write $=_A$ for $=_{A,s}$);
- an effective procedure to draw random elements from $[\![s]\!]_A$; we denote such a drawing by $x \xleftarrow{R} [\![s]\!]_A$; the drawing may not follow a uniform distribution, but no $=_{A,s}$-equivalence class should have probability 0.

Assume a fixed $(S, \mathcal{F})$-computational algebra A. We associate to each closed frame $\varphi = \{x_1 = T_1, \ldots, x_n = T_n\}$ a distribution $\psi = [\![\varphi]\!]_A$, of which the drawings $\hat{\psi} \xleftarrow{R} \psi$ are computed as follows:

1. for each name u of sort s appearing in $T_1, \ldots, T_n$, draw a value $\hat{a} \xleftarrow{R} [\![s]\!]_A$;
2. for each x_i ($1 \leq i \leq n$) of sort s_i, compute $\widehat{T_i} \in [\![s_i]\!]_A$ recursively on the structure of terms: $\widehat{f(T'_1, \ldots, T'_m)} = f_A(\widehat{T'_1}, \ldots, \widehat{T'_m})$;
3. return the value $\hat{\psi} = \{x_1 = \widehat{T_1}, \ldots, x_n = \widehat{T_n}\}$.

Such values $\phi = \{x_1 = e_1, \ldots, x_n = e_n\}$ with $e_i \in [\![s_i]\!]_A$ are called *concrete frames*. We extend the notation $[\![\cdot]\!]_A$ to (sets of) closed terms in the obvious way. We also generalize the notation to terms or frames with variables, by specifying the concrete values for all of them: $[\![\cdot]\!]_{A,\{x_1=e_1,\ldots,x_n=e_n\}}$. Notice that when a term or a frame contains no names, the translation is deterministic; in this case, we use the same notation to denote the distribution and its unique value.

(Families of) distributions over concrete frames benefit from the usual notion of cryptographic indistinguishability. Let us note $\eta \geq 0$ the complexity parameter. Intuitively, two families (ψ_η) and (ψ'_η) of distributions over concrete frames are *indistinguishable*, written $(\psi_\eta) \approx (\psi'_\eta)$, iff no probabilistic polynomial-time adversary $\mathcal{A}$ can guess whether he is given a sample from ψ_η or ψ'_η with a probability significantly greater than $\frac{1}{2}$. Rigorously, we ask the *advantage* of $\mathcal{A}$,

$$\mathrm{Adv}^{\mathrm{IND}}(\mathcal{A}, \eta, \psi_\eta, \psi'_\eta) = \mathbb{P}[\widehat{\psi} \xleftarrow{R} \psi_\eta; \mathcal{A}(\eta, \widehat{\psi}) = 1] - \mathbb{P}[\widehat{\psi} \xleftarrow{R} \psi'_\eta; \mathcal{A}(\eta, \widehat{\psi}) = 1]$$

to be a *negligible* function of η, that is, to remain eventually smaller than any η^{-n} ($n > 0$) for sufficiently large η.

3 Relating Abstract and Computational Algebras

In the previous section we have defined abstract and computational algebras. We now relate formal notions such as equality, (non-)deducibility and static equivalence to their computational counterparts, *i.e.* equality, one-wayness and indistinguishability.

3.1 Soundness and Faithfulness

We introduce the notions of sound, *resp.* faithful, computational algebras with respect to the formal relations studied here: equality, static equivalence and deducibility. In the remaining of the paper we only consider families of computational algebras (A_η) such that each required operation on algebras is feasible by a (uniform) polynomial-time algorithm in the complexity parameter η. We also require that for every sort s, either there exists no name of sort s, or the probability of collision of two random elements in $[\![s]\!]_{A_\eta}$, $\mathbb{P}[e_1, e_2 \xleftarrow{R} [\![s]\!]_{A_\eta}; e_1 =_{A_\eta} e_2]$, is negligible.

Specifically a family of computational algebras (A_η) is

- $=_E$-*sound* iff for every closed terms T_1, T_2 of the same sort, $T_1 =_E T_2$ implies that $\mathbb{P}[e_1, e_2 \xleftarrow{R} [\![T_1, T_2]\!]_{A_\eta}; e_1 \neq_{A_\eta} e_2]$ is negligible;
- $=_E$-*faithful* iff for every closed terms T_1, T_2 of the same sort, $T_1 \neq_E T_2$ implies that $\mathbb{P}[e_1, e_2 \xleftarrow{R} [\![T_1, T_2]\!]_{A_\eta}; e_1 =_{A_\eta} e_2]$ is negligible;
- $\approx_E$-*sound* iff for every closed frames φ_1, φ_2 with the same domain, $\varphi_1 \approx_E \varphi_2$ implies that $([\![\varphi_1]\!]_{A_\eta}) \approx ([\![\varphi_2]\!]_{A_\eta})$;
- $\approx_E$-*faithful* iff for every closed frames φ_1, φ_2 of the same domain, $\varphi_1 \not\approx_E \varphi_2$ implies that there exists a polynomial-time adversary $\mathcal{A}$ for distinguishing concrete frames, such that $1 - \mathrm{Adv}^{\mathrm{IND}}(\mathcal{A}, \eta, [\![\varphi_1]\!]_{A_\eta}, [\![\varphi_2]\!]_{A_\eta})$ is negligible;
- $\not\vdash_E$-*sound* iff for every closed φ and T, $\varphi \not\vdash_E T$ implies that for each polynomial-time adversary $\mathcal{A}$, $\mathbb{P}[\phi, e \xleftarrow{R} [\![\varphi, T]\!]_{A_\eta}; \mathcal{A}(\phi) =_{A_\eta} e]$ is negligible;
- $\not\vdash_E$-*faithful* iff for every closed φ and T, $\varphi \vdash_E T$ implies that there exists a polynomial-time adversary $\mathcal{A}$ such that $1 - \mathbb{P}[\phi, e \xleftarrow{R} [\![\varphi, T]\!]_{A_\eta}; \mathcal{A}(\phi) =_{A_\eta} e]$ is negligible.

Sometimes, it is possible to prove stronger notions of soundness that hold without restriction on the computational power of adversaries. In particular, (A_η) is *unconditionally* $=_E$-*sound* iff for every closed terms T_1, T_2 of the same sort, $T_1 =_E T_2$ implies that $\mathbb{P}[e_1, e_2 \xleftarrow{R} [\![T_1, T_2]\!]_{A_\eta}; e_1 =_{A_\eta} e_2] = 1$; *unconditionally* $\approx_E$-*sound* iff for every closed frames φ_1, φ_2 with the same domain, $\varphi_1 \approx_E \varphi_2$ implies $(\![\varphi_1]\!]_{A_\eta}) = (\![\varphi_2]\!]_{A_\eta})$; *unconditionally* $\not\vdash_E$-*sound* iff for every closed φ and T s.t. $\varphi \not\vdash_E T$, the distributions for φ and T are independent: for all ϕ_0, e_0, $\mathbb{P}[\phi, e \xleftarrow{R} [\![\varphi, T]\!]_{A_\eta}; \phi = \phi_0 \text{ and } e = e_0] = \mathbb{P}[\phi \xleftarrow{R} [\![\varphi]\!]_{A_\eta}; \phi = \phi_0] \times \mathbb{P}[e \xleftarrow{R} [\![T]\!]_{A_\eta}; e = e_0]$.

Generally, (unconditional) $=_E$-soundness is given by construction. Indeed true formal equations correspond to the expected behavior of primitives and should hold in the concrete world with overwhelming probability. The other criteria are however more difficult to fulfill. Therefore it is often interesting to restrict frames to *well-formed* ones in order to achieve soundness or faithfulness: for instance Abadi and Rogaway [3] do forbid encryption cycles (*c.f.* Section 5.2).

It is worth noting that the notions introduced above are not independent.

Proposition 1. *Let (A_η) be a $=_E$-sound family of computational algebras. Then (A_η) is $\not\vdash_E$-faithful. If moreover (A_η) is $=_E$-faithful, then it is also $\approx_E$-faithful.*

For many interesting theories, we have that $\approx_E$-soundness implies all the other notions of soundness and faithfulness. As an illustration, let us consider an arbitrary theory which includes keyed hash functions.

Proposition 2. *Let (A_η) be a family of $\approx_E$-sound computational algebras. Assume that free binary symbols $\mathsf{h}_s : s \times \mathsf{Key} \to \mathsf{Hash}$ are available for every sort s, and the sorts Hash and Key have infinitely many names. Then (A_η) is $=_E$-faithful and $\not\vdash_E$-sound. Besides, if the implementations for the h_s are collision-resistant, then (A_η) is $=_E$-sound, $\approx_E$-faithful and $\not\vdash_E$-faithful.*

3.2 $\approx_E$-Soundness Implies Classical Assumptions on Groups

Inspired by the work of Rivest on pseudo-freeness [14], we now study some consequences of $\approx_E$-soundness on *groups*. Let E_G be the equational theory modeling a free group G with exponents taken over a free commutative ring A. Assume a $\approx_{E_G}$-sound family of computational algebras (A_η). Then the static equivalence $\nu g, a, b.\{x_1 = g, x_2 = g^a, x_3 = g^b, x_4 = g^{a \cdot b}\} \approx_{E_G} \nu g, a, b, c.\{x_1 = g, x_2 = g^a, x_3 = g^b, x_4 = g^c\}$ implies the hardness of the decisional Diffie-Hellman problem for this implementation.

In a similar way we prove that $\approx_{E_G}$-soundness implies the hardness of RSA. More details can be found in [7].

4 A Sufficient (and Often Necessary) Criterion for $\approx_E$-Soundness

We now present useful results for proving $\approx_E$-soundness properties in general. Notably, we provide a sufficient criterion for $\approx_E$-soundness in Section 4.1 and prove it necessary under additional assumptions in Section 4.2.

4.1 Ideal Semantics and $\approx_E$-Soundness Criterion

Given an implementation of the primitives, what we called the concrete semantics maps every closed frame φ to a distribution $[\![\varphi]\!]_{A_\eta}$ in the expected way. We now define the *ideal semantics* of a φ, intuitively as the uniform distribution over sequences of bit-strings (in the appropriate space) that pass all the formal tests verified by φ.

Given a closed frame φ, let us write $\text{eq}_E(\varphi)$ for the set of tests that are true in φ: $\text{eq}_E(\varphi) = \{(M, N) \mid \text{var}(M) \cup \text{var}(N) \subseteq \text{dom}(\phi), (\text{names}(M) \cup \text{names}(N)) \cap \text{names}(\varphi) = \emptyset \text{ and } M\varphi =_E N\varphi\}$. Notice that $\varphi \approx_E \varphi'$ iff $\text{eq}_E(\varphi) = \text{eq}_E(\varphi')$.

We say that (A_η) *has uniform distributions* iff for every η and every sort s, $[\![s]\!]_{A_\eta}$ is a finite set, $=_{A_\eta,s}$ is the usual equality and, the distribution associated to s by A_η is the uniform one over $[\![s]\!]_{A_\eta}$.

Definition 1 (Ideal semantics). *Let (A_η) be an unconditionally $=_E$-sound family of computational algebras, having uniform distributions. Let $\varphi = \{x_1 = t_1, \ldots, x_n = t_n\}$ be a closed frame and s_i the sort of x_i. The ideal semantics $[\![\varphi]\!]^{ideal}_{A_\eta}$ of φ is the uniform distribution over the finite (non-empty) set of concrete frames:*

$$\{\{x_1 = e_1, \ldots, x_n = e_n\} \mid (e_1, \ldots, e_n) \in [\![s_1]\!]_{A_\eta} \times \cdots \times [\![s_n]\!]_{A_\eta} \text{ and}$$
$$\forall (M, N) \in \text{eq}_E(\varphi) \cdot [\![M]\!]_{A_\eta, \{x_1=e_1, \ldots, x_n=e_n\}} = [\![N]\!]_{A_\eta, \{x_1=e_1, \ldots, x_n=e_n\}}\}$$

For instance, let $\varphi = \nu n_1, n_2.\{x_1 = n_1, x_2 = n_2\}$ with n_1 and n_2 of sort s. Then $\text{eq}_E(\varphi) \subseteq \{(M, N) \mid M =_E N\}$ implies that $[\![\varphi]\!]^{ideal}_{A_\eta}$ is simply the uniform distribution over $[\![s]\!]_{A_\eta} \times [\![s]\!]_{A_\eta}$. A more general definition of the ideal semantics, which does not restrict (A_η) to uniform distributions is given in [7].

We can now state our $\approx_E$-soundness criterion: intuitively, the two semantics, concrete and ideal, should be indistinguishable.

Theorem 1 ($\approx_E$-soundness criterion). *Let (A_η) be an unconditionally $=_E$-sound family of computational algebras. Assume that for every closed frame φ it holds that $([\![\varphi]\!]_{A_\eta}) \approx ([\![\varphi]\!]^{ideal}_{A_\eta})$. Then (A_η) is $\approx_E$-sound.*

4.2 Patterns Revisited

Patterns have been introduced by Abadi and Rogaway [3] and used in subsequent work [12, 6] as a way to define computationally sound formal equivalences. Typically frames are mapped to patterns by replacing non-decipherable terms by boxes $\square$. Two frames are then equivalent iff they yield the same pattern (up to renaming of names). For example, the pattern associated to the frame $\varphi_1 = \{x_1 = \text{enc}(\text{enc}(k_4, k_3), k_1), x_2 = \text{enc}(k_1, k_2), x_3 = k_2\}$ is $\{x_1 = \text{enc}(\square, k_1), x_2 = \text{enc}(k_1, k_2), x_3 = k_2\}$.

In this section we propose a general, novel definition of patterns and study some of their properties. We then use these properties to prove that our soundness criterion is necessary in many cases.

Definition 2. *A closed frame φ is a pattern if each of its subterms is deducible from φ.*

Equivalently a pattern is a closed frame of the form $\varphi = \{x_1 = C_1[a_1, \ldots, a_m], \ldots, x_n = C_n[a_1, \ldots, a_m]\}$ where the $C_1 \ldots C_n$ are closed (not necessarily linear) contexts and the $a_1 \ldots a_m$ are distinct deducible names: $\varphi \vdash_E a_i$. For example, φ_1 as defined above is not a pattern, while $\varphi_2 = \{x_1 = \mathsf{enc}(n_1, k_1), x_2 = \mathsf{enc}(k_1, k_2), x_3 = k_2\}$ is.

The following proposition finitely characterizes the equations verified by a pattern.

Proposition 3. *Let $\varphi = \{x_1 = C_1[a_1, \ldots, a_m], \ldots, x_n = C_n[a_1, \ldots, a_m]\}$ be a pattern, using the notations above. For each a_i, let ζ_{a_i} be a term such that $\mathrm{var}(\zeta_{a_i}) \subseteq \{x_1, \ldots, x_n\}$, $\mathrm{names}(\zeta_{a_i}) \cap \mathrm{names}(\varphi) = \emptyset$ and $\zeta_{a_i}\varphi =_E a_i$. Then every equation which holds in φ is a logical consequence (in the first-order theory of equality) of E and the equations $x_j = C_j[\zeta_{a_1}, \ldots, \zeta_{a_m}]$.*

Interestingly the concrete and the ideal semantics of patterns often coincide.

Proposition 4. *Let (A_η) be an unconditionally $=_E$-sound family of computational algebras, having uniform distributions. Let φ be a pattern. The concrete and the ideal semantics of φ yield the same family of distributions: for all η, $[\![\varphi]\!]_{A_\eta} = [\![\varphi]\!]_{A_\eta}^{ideal}$.*

The idea of the proof is that, using the finite characterization of $\mathrm{eq}_E(\varphi)$ (Proposition 3), one can draw a bijection between the drawing of nonces and the eligible values for the ideal semantics.

A theory E *admits patterns* iff for every closed frame φ, there exists a (not necessarily unique) pattern $\overline{\varphi}$ such that $\varphi \approx_E \overline{\varphi}$. In practice many theories useful in cryptography satisfy this property, e.g. the theories considered in Section 5. Note that we have proved *en passant* that $\approx_E$ is decidable for equational theories that admit patterns and for which $=_E$ is decidable, provided the construction of patterns is effective. Indeed, given two frames φ_1 and φ_2, we associate to each of them one of its statically equivalent pattern $\overline{\varphi_1}$ and $\overline{\varphi_2}$, respectively. It is then straightforward to check whether $\overline{\varphi_1}$ and $\overline{\varphi_2}$ are equivalent using the finite characterization of $\mathrm{eq}_E(\overline{\varphi_i})$ by Proposition 3.

The following theorem states that our soundness criterion is actually very tight: whenever a theory admits patterns, our criterion is a necessary condition.

Theorem 2. *Assume that the theory E admits patterns. Let (A_η) be a family of computational algebras, such that (A_η) has uniform distributions, is $\approx_E$- and unconditionally $=_E$-sound. Then the soundness criterion of Theorem 1 is satisfied: for every closed frame φ, $([\![\varphi]\!]_{A_\eta}) \approx ([\![\varphi]\!]_{A_\eta}^{ideal})$.*

5 Examples

We now apply the framework of Sections 3 and 4 to establish two novel $\approx_E$-soundness results, concerning the theory of exclusive OR and that of ciphers and lists.

5.1 Exclusive OR

We study the soundness and faithfulness problems for the usual theory and implementation of the exclusive OR (XOR).

The formal model consists of a single sort $Data$, an infinite number of names, the infix symbol $\oplus : Data \times Data \to Data$ and two constants $0, 1 : Data$. Terms are equipped with the equational theory $E_\oplus$ generated by:

$$x \oplus y = y \oplus x \qquad x \oplus x = 0$$
$$(x \oplus y) \oplus z = x \oplus (y \oplus z) \qquad x \oplus 0 = x$$

As an implementation, we define the computational algebras A_η, $\eta \geq 0$: the concrete domain $[\![Data]\!]_{A_\eta}$ is $\{0,1\}^\eta$ equipped with the uniform distribution; $\oplus$ is interpreted by the usual XOR function over $\{0,1\}^\eta$, $[\![0]\!]_{A_\eta} = 0^\eta$, $[\![1]\!]_{A_\eta} = 1^\eta$.

In this setting, statically equivalent frames enjoy an algebraic characterization. Indeed, let φ and φ' be two frames with names$(\varphi) \cup$ names$(\varphi') \subseteq \{a_1, \ldots, a_n\}$ and dom$(\varphi) =$ dom$(\varphi') = \{x_1, \ldots, x_m\}$. We associate to φ a $(m+1) \times (n+1)$-matrix $\alpha = (\alpha_{i,j})$ over the two element field $\mathbb{F}_2$: the 0-th row of α is $(1, 0 \ldots 0)$ and for $1 \leq i \leq m, 1 \leq j \leq n$ (resp. $j=0$) $\alpha_{i,j}$ is the number of occurrences of a_j (resp. of 1) in $\varphi(x_i)$, taken modulo 2. In the same way, a matrix α' is associated to φ'. Using classical manipulations on matrices, it is easy to show that $\varphi \approx_{E_\oplus} \varphi'$ iff the two associated matrices α and α' have the same image, that is $\alpha(\mathbb{F}_2^{n+1}) = \alpha'(\mathbb{F}_2^{n+1})$.

This characterization is the key point of our main result for the theory of XOR.

Theorem 3. *The usual implementation of the XOR theory is unconditionally $=_{E_\oplus}$-, $\approx_{E_\oplus}$- and $\not\vdash_{E_\oplus}$-sound. It is also $=_{E_\oplus}$-, $\approx_{E_\oplus}$- and $\not\vdash_{E_\oplus}$-faithful.*

This result is comparable to the work of Bana [6], who shows the unconditional soundness of the One-Time Pad encryption in a setting similar to that of Abadi and Rogaway [3]. In some sense our result is more precise as we model the XOR symbol itself and not a particular use of it.

5.2 Symmetric, Deterministic, Length-Preserving Encryption and Lists

We now detail the example of symmetric, deterministic and length-preserving encryption schemes. Such schemes, also known as *ciphers* [13], are widely used in practice, the most famous examples being DES and AES.

Our formal model consists of a set of sorts $\mathcal{S} = \{Data, List_0, List_1 \ldots List_n \ldots\}$, an infinite number of names for every sort $Data$ and $List_n$, $n \neq 0$, and the symbols:

$$\begin{aligned}
enc_n, dec_n &: List_n \times Data \rightarrow List_n &&\text{encryption, decryption} \\
cons_n &: Data \times List_n \rightarrow List_{n+1} &&\text{list constructor} \\
head_n &: List_{n+1} \rightarrow Data &&\text{head of a list} \\
tail_n &: List_{n+1} \rightarrow List_n &&\text{tail of a list} \\
nil &: List_0 \quad 0,1 : Data &&\text{empty list, constants}
\end{aligned}$$

We consider the equational theory E_{sym} generated by (for every $n \geq 0$)

$$\begin{aligned}
dec_n(enc_n(x,y),y) &= x & cons_n(head_n(x), tail_n(x)) &= x \\
enc_n(dec_n(x,y),y) &= x & enc_0(nil, x) &= nil \\
head_n(cons_n(x,y)) &= x & dec_0(nil, x) &= nil \\
tail_n(cons_n(x,y)) &= y
\end{aligned}$$

When oriented from left to right, the equations E_{sym} form an (infinite) convergent rewriting system, written $\mathcal{R}$. The equations $enc_n(dec_n(x,y),y) = x$ are characteristic of length-preserving encryption schemes. Indeed, encryption and decryption functions under each key then form a pair of mutually inverse bijections. The concrete meaning of sorts and symbols is given by the computational algebras A_η, $\eta > 0$, defined as follows:

- the carrier sets are $[\![Data]\!]_{A_\eta} = \{0,1\}^\eta$ and $[\![List_n]\!]_{A_\eta} = \{0,1\}^{n\eta}$ equipped with the uniform distribution and the usual equality relation;
- enc_n, dec_n are implemented by a cipher for data of size $n\eta$ and keys of size η (we discuss the required cryptographic assumptions later);
- $[\![nil]\!]_{A_\eta}$ is the empty bit-string, $[\![cons_n]\!]_{A_\eta}$ is the usual concatenation, $[\![0]\!]_{A_\eta} = 0^\eta$, $[\![1]\!]_{A_\eta} = 1^\eta$, $[\![head_n]\!]_{A_\eta}$ returns the η first digits of bit-strings (of size $(n+1)\eta$) whereas $[\![tail_n]\!]_{A_\eta}$ returns the last $n\eta$ digits.

Obviously, the above implementation is unconditionally $=_{E_{sym}}$-sound. Before studying the $\approx_{E_{sym}}$-soundness, we need to characterize statically equivalent frames. Specifically we show that this theory admits patterns, in the sense of Section 3.

Proposition 5. *Let φ be a closed frame. There exists a pattern $\overline{\varphi}$ such that $\varphi \approx_{E_{sym}} \overline{\varphi}$.*

Proof (outline). We associate a pattern to any frame φ by the following procedure:

1. normalize φ using the rules $\mathcal{R}$ (the result is still denoted φ);
2. while φ is not a pattern, repeat: find any subterm T of the form $T = enc_n(U, V)$, $T = dec_n(U, V), T = head_n(V)$ or, $T = tail_n(V)$, with $\varphi \not\vdash_{E_{sym}} V$ and replace T everywhere in φ by a fresh name a of the appropriate sort.

We prove in [7] that this procedure always terminates on a pattern statically equivalent to the initial frame.

We now study the $\approx_{E_{sym}}$-soundness problem under realistic cryptographic assumptions. Classical assumptions on ciphers include the notions of super pseudo-random permutation (SPRP) and several notions of indistinguishability (IND-Pi-Cj, $i, j = 0, 1, 2$). In particular, IND-P1-C1 denotes the indistinguishability against lunchtime chosen-plaintext and chosen-ciphertext attacks. These notions and the relations between them have been studied notably in [13].

Initially, the SPRP and IND-P1-C1 assumptions apply to (block) ciphers specialized to plaintexts of a given size. Interestingly, this is not sufficient to imply $\approx_{E_{sym}}$-soundness for frames which contain plaintexts of heterogeneous sizes, encrypted under the same key. Thus we introduce a strengthened version of IND-P1-C1, applying to a *collection* of ciphers $(\mathcal{E}_{\eta,n}, \mathcal{D}_{\eta,n})$, where η is the complexity parameter and $n \geq 0$ is the number of blocks of size η contained in plaintexts and ciphertexts.

We define the ω-IND-P1-C1 assumption by considering the following experiment $\mathcal{G}_\eta$ involving a 2-stage adversary $\mathcal{A} = (\mathcal{A}_1, \mathcal{A}_2)$:

- first a key k is randomly chosen from $\{0,1\}^\eta$;
- (Stage 1) $\mathcal{A}_1$ is given access to the encryption oracles $\mathcal{E}_{\eta,n}(\cdot, k)$ and the decryption oracles $\mathcal{D}_{\eta,n}(\cdot, k)$; it outputs two plaintexts $m_0, m_1 \in \{0,1\}^{n_0\eta}$ for some n_0, and possibly some data d;
- (Stage 2) a random bit $b \in \{0,1\}$ is drawn; $\mathcal{A}_2$ receives the data d, the *challenge ciphertext* $c = \mathcal{E}_{\eta,n_0}(m_b, k)$ and outputs a bit b';
- $\mathcal{A}$ is successful in $\mathcal{G}_\eta$ iff $b = b'$ and it has never submitted m_0 or m_1 to an encryption oracle, nor c to a decryption oracle.

Define the *advantage* of $\mathcal{A}$ as: $\text{Adv}_{\mathcal{A}}^{\omega\text{-IND-P1-C1}}(\eta) = 2 \times \mathbb{P}[\mathcal{A} \text{ is successful in } \mathcal{G}_\eta] - 1$. The ω-*IND-P1-C1* assumption holds for $(\mathcal{E}_{\eta,n}, \mathcal{D}_{\eta,n})$ iff the advantage of any probabilistic polynomial-time adversary is negligible. It holds for the *inverse* of the encryption scheme, iff it holds for the collection of ciphers $(\mathcal{D}_{\eta,n}, \mathcal{E}_{\eta,n})$.

As in previous work [3, 12, 4, 11], we restrict frames to those with only atomic keys and no encryption cycles. Specifically a closed frame φ *has only atomic keys* if for all subterms $\text{enc}_n(u,v)$ and $\text{dec}_n(u,v)$ of φ, v is a name. Given two (atomic) keys k_1 and k_2, we say that k_1 *encrypts* k_2 *in* φ, written $k_1 >_\varphi k_2$, iff there exists a subterm U of φ of the form $U = \text{enc}_n(T, k_1)$ or $U = \text{dec}_n(T, k_1)$ such that k_2 appears in T *not used as a key*, i.e. k_2 appears in T at a position which is not the right-hand argument of a $\text{enc}_{n'}$ or a $\text{dec}_{n'}$. An *encryption cycle* is a tuple $k_1 \ldots k_m$ such that $k_1 >_\varphi \ldots >_\varphi k_m >_\varphi k_1$.

The effect of the condition "not used as a key" is to allow considering more terms as free of encryption cycles, for instance $\text{enc}_n(\text{enc}_n(a,k), k)$. This improvement is already suggested in [3].

We now state our $\approx_{E_{\text{sym}}}$-soundness theorem. A closed frame is *well-formed* iff its $\mathcal{R}$-normal form has only atomic keys, contains no encryption cycles and uses no head and tail symbols.

Theorem 4 ($\approx_{E_{\text{sym}}}$-**soundness**). *Let φ_1 and φ_2 be two well-formed frames of the same domain. Assume that the concrete implementations for the encryption and its inverse satisfy both the ω-IND-P1-C1 assumption. If $\varphi_1 \approx_{E_{\text{sym}}} \varphi_2$ then $([\![\varphi_1]\!]_{A_n}) \approx ([\![\varphi_2]\!]_{A_n})$.*

Note on the cryptographic assumptions. Cryptographic assumptions of Theorem 4 may appear strong compared to existing work on passive adversaries [3, 12]. Nevertheless if φ_1 and φ_2 contain no decryption symbols, our proofs are easily adapted to work when the encryption scheme is ω-IND-P1-C0 only, where ω-IND-P1-C0 is defined similarly to ω-IND-P1-C1 except that the adversary has no access to the decryption oracle.

Also, it is possible to recover the classical assumptions IND-P1-C1 by modeling the ECB mode (Electronic Code Book). Let us add two symbols $\text{enc} : Data \times Data \to Data$ and $\text{dec} : Data \times Data \to Data$, and define the symbols enc_n and dec_n (formally and concretely) recursively by

$$\text{enc}_{n+1}(x,y) = \text{cons}_n(\text{enc}(\text{head}_n(x), y), \text{enc}_n(\text{tail}_n(x), y)) \quad \text{and}$$
$$\text{dec}_{n+1}(x,y) = \text{cons}_n(\text{dec}(\text{head}_n(x), y), \text{dec}_n(\text{tail}_n(x), y)).$$

Define well-formed frames as those of which the normal forms contain no encryption cycles. The $\approx_{E_{\text{sym}}}$-soundness property holds for well-formed frames as soon as the implementations for enc and dec are both IND-P1-C1, or equivalently [13] enc is SPRP.

6 Conclusion and Future Work

In this paper we developed a general framework for relating formal and computational models of security protocols in the presence of a passive attacker. These are the first results on abstract models allowing arbitrary equational theories. We define the soundness and faithfulness of cryptographic implementations w.r.t. abstract models. We also provide a soundness criterion which for a large number of theories—those that admit a general notion of patterns—is not only sufficient but also necessary. Finally, we provide new soundness results for the exclusive OR and a theory of ciphers and lists.

As future work, we foresee to study the soundness of other theories. An interesting case would be the combination of the two theories considered in this paper: in a theory combining XOR, ciphers and lists, one can precisely model the *Cipher Block Chaining* (CBC) mode, which is commonly used with block ciphers such as DES or AES. Another ambitious extension is to consider the case of an active attacker.

Acknowledgments. This work has been partially supported by the ACI-SI Rossignol, the ACI JC 9005 and the RNTL project PROUVÉ 03V358 and 03V360.

References

1. M. Abadi and V. Cortier. Deciding knowledge in security protocols under equational theories. In *Proc. 31st International Colloquium on Automata, Languages and Programming (ICALP'04)*, volume 3142 of *LNCS*, pages 46–58, 2004.
2. M. Abadi and C. Fournet. Mobile values, new names, and secure communications. In *Proc. 28th Annual ACM Symposium on Principles of Programming Languages (POPL'01)*, pages 104–115, 2001.
3. M. Abadi and P. Rogaway. Reconciling two views of cryptography (the computational soundness of formal encryption). In *Proc. 1st IFIP International Conference on Theoretical Computer Science (IFIP–TCS'00)*, volume 1872 of *LNCS*, pages 3–22, 2000.
4. M. Backes and B. Pfitzmann. Symmetric encryption in a simulatable Dolev-Yao style cryptographic library. In *Proc. 17th IEEE Computer Science Foundations Workshop (CSFW'04)*, pages 204–218, 2004.
5. M. Backes, B. Pfitzmann, and M. Waidner. A composable cryptographic library with nested operations. In *Proc. 10th ACM Conference on Computer and Communications Security (CCS'03)*, 2003.
6. G. Bana. *Soundness and Completeness of Formal Logics of Symmetric Encryption*. PhD thesis, University of Pennsylvania, 2004.
7. M. Baudet, V. Cortier, and S. Kremer. Computationally sound implementations of equational theories against passive adversaries. Research Report 2005/074, Cryptology ePrint Archive, Mar. 2005. 28 pages.
8. B. Blanchet. Automatic proof of strong secrecy for security protocols. In *Proc. 25th IEEE Symposium on Security and Privacy (SSP'04)*, pages 86–100, 2004.
9. D. Dolev and A. C. Yao. On the security of public key protocols. *IEEE Transactions on Information Theory*, IT-29(12):198–208, 1983.
10. S. Goldwasser and S. Micali. Probabilistic encryption. *Journal of Computer and System Sciences*, 28:270–299, 1984.
11. P. Laud. Symmetric encryption in automatic analyses for confidentiality against active adversaries. In *Proc. IEEE Symposium on Security and Privacy (SSP'04)*, pages 71–85, 2004.
12. D. Micciancio and B. Warinschi. Completeness theorems for the Abadi-Rogaway logic of encrypted expressions. *Journal of Computer Security*, 12(1):99–129, 2004.
13. D.H. Phan and D. Pointcheval. About the security of ciphers (semantic security and pseudo-random permutations). In *Proc. Selected Areas in Cryptography (SAC'04)*, volume 3357 of *LNCS*, pages 185–200, 2004.
14. R.L. Rivest. On the notion of pseudo-free groups. In *Proc. 1st Theory of Cryptography Conference (TCC'04)*, volume 2951 of *LNCS*, pages 505–521, 2004.

Password-Based Encryption Analyzed

Martín Abadi[1] and Bogdan Warinschi[2]

[1] Computer Science Department, University of California, Santa Cruz
[2] Computer Science Department, Stanford University

Abstract. The use of passwords in security protocols is particularly delicate because of the possibility of off-line guessing attacks. We study password-based protocols in the context of a recent line of research that aims to justify symbolic models in terms of more concrete, computational ones. We offer two models for reasoning about the concurrent use of symmetric, asymmetric, and password-based encryption in protocol messages. In each of the models we define a notion of equivalence between messages and also characterize when passwords are used securely in a message or in a set of messages. Our new definition for the computational security of password-based encryption may be of independent interest. The main results of this paper are two soundness theorems. We show that under certain (standard) assumptions about the computational implementation of the cryptographic primitives, symbolic equivalence implies computational equivalence. More importantly, we prove that symbolically secure uses of passwords are also computationally secure.

1 Introduction

Passwords and other weak secrets sometimes serve as cryptographic keys in security protocols and elsewhere (e.g., [5, 14, 16, 21]). The use of weak secrets is particularly delicate because of the possibility of off-line guessing attacks. In such an attack, data that depends on a weak secret is used in checking guesses of the values of the weak secret. Consider, for example, a protocol where two parties exchange the encryption c of some fixed message, say Ok, under a shared password pwd. If pwd is picked from a relatively small dictionary, then an attacker that obtains a transcript of the protocol execution can mount the following off-line attack. It decrypts the ciphertext c with the passwords in the dictionary, one by one, until the result of the decryption is the text Ok. The password used for this last decryption is likely to be pwd. Guessing attacks such as this one are passive, in the sense that they do not require interaction with the protocol participants, so they are hard to detect. A guessing attack may however be carried out after an active attack, relying on the messages exchanged in the course of the active attack.

Early research on the design and analysis of protocols based on weak secrets focused on techniques for defending against guessing attacks (e.g. [13]). These techniques basically aim to ensure that plaintexts encrypted under passwords do not contain redundancy that can later be used to verify a password guess. While this is a helpful guideline, its informal application need not guarantee security. As experience demonstrates (e.g., [23]), conjecturing the security of a protocol, or arguing it only heuristically, is not sufficient. Instead, and this is the goal pursued by recent research on the subject, the security of protocols should be rigorously analyzed. Models for carrying out such analyzes have been designed using two different, yet related approaches.

The first approach, known as the symbolic or formal methods approach, adopts an abstract view of executions. Messages are modeled as elements of a term algebra constructed with symbolic operations that represent various cryptographic primitives. Parties operate on terms using a limited number of inference rules, sometimes generically known as the Dolev-Yao rules. The rules reflect a common understanding of the security of cryptographic primitives. For example, they say that the message encrypted in a ciphertext can be recovered only if the appropriate decryption key is known. Quite often, proofs that rely on these rules can be mechanized. Work done on symbolic models for password-based protocol has concentrated on extending the Dolev-Yao rules to guessing attacks [7–9, 18]. Typical formalisms enrich standard symbolic models with an operation that represents encryption under passwords, and they offer a careful account of when a password guess can be verified from a given set of terms (presumably a transcript of a protocol execution). The resulting decision procedure has been automated [7, 8, 18]; a corresponding decision problem has been shown NP-complete [9]. Unfortunately, as remarked by authors of prior work [18], it is quite difficult to determine if a set of formal criteria for the existence of guessing attacks is exhaustive. Hence, the possibility remains that a formal analysis would miss some attacks, and unsoundly conclude that a protocol is secure when in fact it is not.

The second approach, known as the computational approach, uses a concrete (bit level) representation, for protocol executions. The attacker is modeled as a powerful, arbitrary probabilistic polynomial-time Turing machine. Although proofs with this approach tend to be lengthy, difficult, and tedious, it is generally accepted that it provides strong guarantees. For the case of password-based protocols, work with the computational approach seems to have focused almost exclusively on the important use of passwords for authenticated key exchange. This work includes designing models and giving provably secure constructions [4, 6, 10, 11, 15]. Surprisingly, the security of password-based encryption as a stand-alone primitive has not been addressed.

A recent line of research aims to justify the abstractions made by symbolic methods with respect to computational models (e.g., [1, 2, 17, 20, 22]) via soundness theorems. These theorems typically state that, under certain assumptions on the implementation of cryptographic primitives, symbolic security proofs imply security in the computational model. The applications of soundness theorems are quite appealing: simple reasoning techniques and automatic tools, specific to the symbolic setting, can be used to carry out proofs that guarantee strong, computational security.

This paper is a first exploration on the subject of computationally sound symbolic analysis for protocols based on passwords. We concentrate on off-line guessing attacks, because they are the main original concern in the analysis of those protocols and because they appear mostly orthogonal to the standard active attacks. Our framework is an extension of the framework introduced by Abadi and Rogaway [1] to asymmetric and password-based encryption. That framework focuses, as an initial step, on passive attacks; in that respect, it is a good match for our purposes, since off-line guessing attacks are passive by definition (even if they may occur in conjunction with active attacks).

We introduce a language of expressions whose elements are abstract representations of the messages sent during protocol executions. The cryptographic primitives that we consider are symmetric encryption, asymmetric encryption, and encryption that uses passwords as keys. To these expressions we attach two different semantics. The first is symbolic; it is based on an extension of the classical Dolev-Yao inference rules to include password-based encryption. The second is computational; it is based on concrete implementations of the encryption operations. In current protocols, password-based encryption typically serves for achieving authenticity rather than secrecy properties, despite the use of the term "encryption" (which we preserve for historical reasons). Accordingly, our semantics do not require that the encryption of a plaintext under a password ensure the secrecy of the plaintext. Further, we give symbolic and computational definitions for expression equivalence (when two expressions convey the same information to an adversary) and for secure use of passwords (which expressions do not leak passwords despite guessing attacks). The main results of our paper are soundness theorems that link the two models. We prove that if two expressions are equivalent symbolically then they are equivalent computationally. We also prove that if an expression hides a password symbolically then it hides the password computationally.

In Section 2 we give the syntax of the language of expressions. As a counterpart, we introduce a computational setting in Section 4. We define expression equivalence and password hiding, symbolically and computationally, in Sections 3 and 5, respectively. In Section 6 we give our main results; as an example, we show an application to the EKE protocol [5]. We conclude in Section 7. Because of space constraints, we leave many details and proofs to a longer version of this paper.

2 Syntax

In this section we define the language of expressions Exp. We consider messages constructed from bits and cryptographic keys by using pairing, symmetric and asymmetric encryption, as well as encryption that employs passwords as keys. In what follows, Bool is the set of bits $\{0, 1\}$. Keys is the set of cryptographic keys; it is the union of the disjoint sets SKeys, EKeys, DKeys, and Passwd which contain symbols for symmetric keys, asymmetric encryption keys, asymmetric decryption keys, and passwords, respectively. We write EncKeys for SKeys ∪ EKeys ∪ Passwd, the set of keys that can be used for encryption; and write $(\cdot)^{-1}$: SKeys ∪ EKeys → SKeys ∪ DKeys for a bijection that maps an encryption key to the associated decryption key. We usually follow the convention that $K_1^s, K_2^s, \ldots$ represent symmetric keys, $K_1^e, K_2^e, \ldots$ asymmetric encryption keys, and $K_1^d, K_2^d, \ldots$ the corresponding asymmetric decryption keys. In this paper we concentrate on the simple setting where expressions use a single password symbol for encryption, so the set Passwd contains a single element W. This setting is sufficient for analyzing multiple concurrent runs of the execution of a protocol between principals that share a password; with some complications, our approach extends to the general case where multiple passwords are used simultaneously.

The set Exp of formal expressions is defined by the grammar:

Exp ::= Bool | Passwd | EKeys | DKeys | SKeys | (Exp, Exp) | $\{$Exp$\}_{\text{EncKeys}}$

For example, expression $\{K^s\}_W, \{(0,0)\}_{K^s}$ represents the encryption of symmetric key K^s under the password W paired with the encryption of $(0,0)$ under key K^s. The expression $\{K^e\}_{K^e}, \{K^s\}_{K^e}, \{0\}_{K^s}$ represents the encryption of public key K^e under itself, paired with the encryption of symmetric key K^s under K^e and the encryption of the bit 0 under K^s. As we do here, we omit parenthesis when there is no risk of ambiguity or when ambiguity is harmless.

An important subset of Exp is that of *acyclic expressions*. Acyclicity was introduced in previous work [1] for expressions that use only symmetric encryption. Here we generalize this notion to deal also with asymmetric encryption. Given an expression $E \in$ Exp we build the following directed graph. The nodes of the graph are pairs of encryption and decryption keys $(K, K^{-1}) \in$ SKeys $\times$ SKeys $\cup$ EKeys $\times$ DKeys for which at least one of the components appears in E. We add an edge between nodes (K_1, K_1^{-1}) and (K_2, K_2^{-1}), and say that K_1 encrypts K_2, if there exists $E' \in$ Exp such that $\{E'\}_{K_1}$ occurs in E and K_2^{-1} occurs in E'. We say that an expression $E \in$ Exp is acyclic if its associated graph is acyclic. For example, the two expressions $\{K_1^e\}_{K_1^e}$ and $(\{K_1^e\}_{K_1^s}, \{K_1^s\}_{K_1^e})$ are acyclic, and the three expressions $\{K_1^s\}_{K_1^s}$, $(\{K_1^s\}_{K_2^s}, \{K_2^s\}_{K_1^s})$, and $(\{K_1^d\}_{K_1^s}, \{K_1^s\}_{K_1^e})$ are not.

3 A Symbolic Model for Expressions

In this section we introduce a symbolic semantics for expressions in Exp. Intuitively, the semantics of an expression is a *pattern* that represents the information that an adversary learns by observing the expression. With this interpretation, we give an equivalence relation on the set of expressions that identifies expressions that convey the same information to the adversary (extending [1]). Furthermore, we use the symbolic semantics to give a characterization of expressions that do not leak a password.

Symbolic semantics and expression equivalence. The inference rules that an adversary can use for deriving new information are formalized by the *entailment* relation $M \vdash N$, which is the least relation that satisfies:

1. $M \vdash 0$ and $M \vdash 1$,
2. $M \vdash M$,
3. if $M \vdash N_1$ and $M \vdash N_2$ then $M \vdash (N_1, N_2)$,
4. if $M \vdash (N_1, N_2)$ then $M \vdash N_1$ and $M \vdash N_2$,
5. if $M \vdash N$ and $M \vdash K$ then $M \vdash \{N\}_K$, for $K \in$ EncKeys,
6. if $M \vdash \{N\}_K$ and $M \vdash K^{-1}$ then $M \vdash N$, for $K \in$ SKeys $\cup$ EKeys,
7. if $M \vdash \{N\}_W$ then $M \vdash N$, for $W \in$ Passwd.

Most of the rules are self-explanatory: for example, they allow an adversary to pair messages that it knows (rule (3)), recover the components of a pair (rule (4)), and compute the encryption of a message M under a certain key K, provided the adversary can compute both M and K (rule (5)). Rule (6) is the standard rule of Dolev-Yao deduction systems: an adversary can decrypt a ciphertext if it has the right decryption key. For instance, we have: $K_1^d, \{W\}_{K_1^e} \vdash W$ and $\{\{K_1^s\}_{K_2^s}, \{K_1^d\}_{K_2^s}, \{W\}_{K_1^e}, K_2^s \vdash W$.

Rule (7) shows that our definitions make a pessimistic (but perhaps realistic!) assumption about the secrecy of plaintexts encrypted under a password. For instance, we have $K_1^d, \{\{0, K_2^s\}_{K_1^e}\}_W \vdash K_2^s$. As indicated in the introduction, this assumption is compatible with current uses of passwords for authentication. However, none of the rules allows the recovery of W by simply observing encryptions of messages under W.

Patterns are elements of the language Pat obtained by extending the language Exp with symbols that represent undecryptable (symmetric and asymmetric) ciphertexts. We let $\text{Undec} = \{\square^s, \square^a\}$. The set of patterns is defined by the grammar:

$$\text{Pat} ::= \text{Exp} \mid \text{Undec} \mid (\text{Pat}, \text{Pat}) \mid \{\text{Pat}\}_{\text{EncKeys}}$$

The pattern $p(M, T)$ represents what an adversary can see in an expression $M \in \text{Exp}$ using for decryption the keys in $T \subseteq \text{Keys}$. It is defined inductively by:

$$p(M, T) = M \quad \text{for } M \in \text{Bool} \cup \text{Keys}$$
$$p((M, N), T) = (p(M, T), p(N, T))$$
$$p(\{M\}_W, T) = \{p(M, T)\}_W$$
$$p(\{M\}_{K^s}, T) = \begin{cases} \{p(M, T)\}_{K^s} & \text{if } K^s \in T \\ \square^s & \text{otherwise} \end{cases}$$
$$p(\{M\}_{K^e}, T) = \begin{cases} \{p(M, T)\}_{K^e} & \text{if } K^d \in T \\ \square^a & \text{otherwise} \end{cases}$$

We let $\textit{recoverable}(M) = \{K \in \text{Keys} \mid M \vdash K\}$ be the set of keys that can be recovered from an expression M. The pattern associated to M is the pattern computed from M given the set of keys recoverable from M, that is: $\textit{pattern}(M) = p(M, \textit{recoverable}(M))$. For instance, in the case of the expression $\{K_1^s\}_W, \{K_1^e\}_W, \{\{K_2^s\}_{K_1^s}\}_{K_3^s}, \{0\}_{K_1^e}, K_2^s, \{\{K_1^d\}_{K_3^s}, 0\}_{K_1^s}$, the recoverable keys are K_1^s, K_1^e, and K_2^s, and the pattern is $\{K_1^s\}_W, \{K_1^e\}_W, \square^s, \square^a, K_2^s, \{\square^s, 0\}_{K_1^s}$.

We use patterns for defining equivalence of expressions: two expressions are equivalent if they have the same pattern. Much as in previous work, this equivalence relation can be a little too restrictive, so we relax it by using key renaming functions. A key renaming function is a bijection on the set Keys that preserves the types of keys: it maps passwords to passwords, asymmetric encryption (decryption) keys to asymmetric encryption (respectively decryption) keys, and symmetric keys to symmetric keys.

Definition 1. $M \equiv N$ if and only if $\textit{pattern}(M) = \textit{pattern}(N)$, and $M \cong N$ if and only if there exists a key renaming σ such that $M \equiv N\sigma$.

For example, we have $\{0\}_{K_1^s} \cong \{1\}_{K_2^s}$ and $\{0\}_{K_1^e} \cong \{1\}_{K_2^e}$. These equivalences reflect the standard assumption that symmetric and asymmetric encryption hide plaintexts. We also have $\{0\}_{K^s} \not\cong \{0\}_{K^e}$: symmetric and asymmetric ciphertexts can in principle be distinguished. Coming to passwords, we have $\{0\}_W \cong \{0\}_W$ and $\{0\}_W \not\cong \{1\}_W$: password-based encryptions of different known plaintexts are inequivalent. On the other hand, we have $\{K_1^s\}_W \cong \{K_2^s\}_W$: encryptions of random keys with a password cannot be distinguished. Finally, in contrast, we have $(\{K_1^s\}_W, \{0\}_{K_1^s}) \not\cong (\{K_2^s\}_W, \{1\}_{K_2^s})$: if keys encrypted with a password are used elsewhere, then the two resulting expressions may not be equivalent anymore.

Secure use of passwords, symbolically. Next we identify a set of expressions in which a password is used securely, that is, the password is not subject to a guessing attack. Our definition is in two steps. First we introduce patterns with variables. Then we say that an expression uses passwords securely if its pattern can be obtained from a pattern with variables by instantiating the variables in a certain *appropriate* way.

Let $\mathsf{Var} = \{x_1, x_2, \ldots\}$ be a set of variables. The set $\mathsf{Pat}[\mathsf{Var}]$ of patterns with variables from Var is defined by the grammar:

$$\mathsf{Pat}[\mathsf{Var}] ::= \mathsf{Bool} \mid \mathsf{EKeys} \mid \mathsf{DKeys} \mid \mathsf{SKeys} \mid \mathsf{Undec} \mid (\mathsf{Pat}[\mathsf{Var}], \mathsf{Pat}[\mathsf{Var}]) \mid$$
$$\{\mathsf{Pat}\}_{\mathsf{EKeys}} \mid \{\mathsf{Pat}\}_{\mathsf{SKeys}} \mid \{\mathsf{Var}\}_{\mathsf{Passwd}}$$

Informally, in a pattern with variables, a password may appear only as an encryption key, and only be used for encrypting variables. For example, $(\{x_1\}_W, \{(\{x_2\}_W, 0)\}_{K^s})$ is in $\mathsf{Pat}[\mathsf{Var}]$, but $(W, \{x_1\}_W)$ and $\{W\}_{K^s}$ are not. Intuitively, the variables mark places in an expression where we can place concrete subexpressions.

For security, we should ensure that these subexpressions do not offer redundancy that could permit a guessing attack. The subexpressions that we consider benign (in this sense) are ciphertexts and keys that do not themselves appear elsewhere in the pattern. More precisely, an instantiation of a pattern with variables into a pattern is appropriate if variables are mapped to one of the symbols $\square^s$ or $\square^a$ or to (symmetric or asymmetric) encryption keys that do not appear elsewhere in the pattern. For example, the pattern with variables $\{x_1\}_W, \{(\{x_2\}_W, 0)\}_{K^s}, K^s$ has occurrences of K^s, so it cannot be instantiated to $\{K^s\}_W, \{(\{\square^s\}_W, 0)\}_{K^s}, K^s$. On the other hand, it can be instantiated to $\{\square^a\}_W, \{(\{K^e\}_W, 0)\}_{K^s}, K^s$ and to $\{K^e\}_W, \{(\{\square^s\}_W, 0)\}_{K^s}, K^s$ via the appropriate instantiations $[x_1 \mapsto \square^a, x_2 \mapsto K^e]$ and $[x_1 \mapsto K^e, x_2 \mapsto \square^s]$. Hence, we define:

Definition 2. *Let $p \in \mathsf{Pat}[\mathsf{Var}]$. A mapping $\sigma : \mathsf{Var} \to \mathsf{Pat}$ is appropriate for p if for all $x \in \mathsf{Var}$ it holds that $\sigma(x) \in \mathsf{SKeys} \cup \mathsf{EKeys} \cup \{\square^s, \square^a\}$ and, if $\sigma(x)$ is a key $K \in \mathsf{SKeys} \cup \mathsf{EKeys}$, then neither K nor K^{-1} occur in p.*

Definition 3. *An expression $E \in \mathsf{Exp}$ hides passwords symbolically if there exist $p \in \mathsf{Pat}[\mathsf{Var}]$ and a mapping $\sigma : \mathsf{Var} \to \mathsf{Pat}$ appropriate for p such that $\mathrm{pattern}(E) = p\sigma$.*

For example, the expression $\{\{(0,1)\}_{K^e}\}_W, \{(\{K^e\}_W, 0)\}_{K^s}, K^s$ hides passwords symbolically: its pattern is $\{\square^a\}_W, \{(\{K^e\}_W, 0)\}_{K^s}, K^s$ which, as noted above, can be obtained from a pattern with variables via an appropriate instantiation. On the other hand, neither $\{0\}_W$ nor $\{(K_1^s, K_2^s)\}_W$ hide passwords symbolically. The former is subject to the attack we sketched in the introduction. The same attack may apply to the latter if any kind of fixed delimiters are used to implement pairing. This possibility cannot be ruled out *a priori*, and is in fact quite reasonable, so we chose to consider this expression insecure. Further, $(\{K^s\}_W, \{0\}_{K^s})$ does not hide passwords symbolically either. Although W encrypts the symmetric key K^s (potentially a random string), and therefore the same attack does not seem to apply, the key K^s is also used for encrypting a fixed plaintext, which allows a simple guessing attack: an adversary decrypts the first part with a possible password, then uses the result for decrypting the second part in order to check the password guess. It might appear that the same attack does not apply to

$\{K_1^s\}_W, \{K_2^s\}_{K_1^s}$, since here the key K_1^s is used for encrypting another symmetric key. We consider this expression insecure because the symmetric encryption scheme may well provide a mechanism for ensuring that decryptions succeed only if the appropriate key is used, as in the case of authenticated encryption (e.g. [3]), thus offering an indirect way to check a password guess.

Our definitions are at the same level of abstraction as those found in the literature on formal analysis of guessing attacks. However, those tend to be, at least superficially, in a somewhat different style. They also model (symmetric and asymmetric) deterministic encryption, while we focus on probabilistic (symmetric and asymmetric) encryption, because this is the standard kind of encryption used in modern cryptography. We expect that a secure expression in the sense defined in this paper is also secure against the symbolic guessing attacks captured by previous work.

4 Computational Security of Encryption Schemes

A password-based encryption scheme Π^p is given by a pair of polynomial-time algorithms $(\mathcal{E}^p, \mathcal{D}^p)$ for encryption and decryption, respectively. The scheme is used for encrypting messages in a set $\mathsf{Plaintext}(\Pi^p) \subseteq \{0,1\}^*$ under passwords from a dictionary $\mathsf{D} \subseteq \{0,1\}^*$. The messages may be chosen according to a probability distribution, part of a distribution ensemble (a parameterized family of distributions). Thus, for generality, we partition the set of plaintexts and the set of passwords according to a security parameter: $\mathsf{Plaintext}(\Pi^p) = \cup_\eta \mathsf{Plaintext}(\Pi^p)_\eta$ and $\mathsf{D} = \cup_\eta \mathsf{D}_\eta$. Furthermore, we require that dictionaries can be sampled efficiently: each dictionary D comes with a probabilistic polynomial-time algorithm that, for security parameter η, returns a sample w from D_η; we write this $w \xleftarrow{R} \mathsf{D}_\eta$. For each η, the encryption function takes as input a password $pwd \in \mathsf{D}_\eta$ and a plaintext $m \in \mathsf{Plaintext}(\Pi^p)_\eta$ and returns an encryption $\mathcal{E}^p(pwd, m)$ of m under pwd. The decryption function $\mathcal{D}^p$ takes as input a password pwd and a ciphertext c and returns the decryption $\mathcal{D}^p(pwd, c)$ of c using pwd. For any security parameter η, any $m \in \mathsf{Plaintext}_\eta$, and $pwd \in \mathsf{D}_\eta$, the equality $m = \mathcal{D}^p(pwd, \mathcal{E}^p(pwd, m))$ must hold.

Before this work, it appears that the security of password-based encryption had not been defined from a computational perspective. We aim to fill this gap. Our definition captures the idea that, given the encryptions of one or more plaintexts under a password, it should be hard to recover the password—and, as suggested in the introduction, our definition does not capture any possible, additional authenticity or secrecy properties. A common assumption is that passwords are selected from a relatively small dictionary that is likely to be known to an adversary; the attack sketched in the introduction indicates that, unless the plaintexts are selected from a distribution with sufficient entropy, there is no hope for the password to be secure. Therefore, in our definition, the plaintexts are chosen according to distributions. Moreover, the distributions are parameterized by a security parameter; we require that it be hard to recover the password asymptotically.

For instance, let us consider a protocol where two parties have exchanged a session key k (for a security parameter η), without authentication, and wish to use a shared password pwd for authenticating k. For this purpose, one party might encrypt a predefined message, say Ok, under k, with a symmetric encryption algorithm $\mathcal{E}^s$, then encrypt it further under pwd, and transmit the result $\mathcal{E}^p(pwd, \mathcal{E}^s(k, \mathsf{Ok}))$. The other party would first decrypt the message that it receives using pwd and then k. It would accept k as valid only if the result of this last decryption is Ok. Ideally, $\mathcal{E}^s(k, \mathsf{Ok})$ should not expose redundancy, so that it can be safely encrypted under pwd. The security of pwd can be guaranteed only for large values of η: for small values, an adversary that sees $\mathcal{E}^p(pwd, \mathcal{E}^s(k, \mathsf{Ok}))$ can check a password guess pwd' by decrypting with pwd' and then breaking the inner encryption—a feasible task for small values of η.

The following definition of security uses an adversary A that has access to an encryption oracle $\mathcal{E}^p(pwd, Dist)$. At each query to the oracle, the oracle samples a string d according to distribution $Dist$ and returns $\mathcal{E}^p(pwd, d)$, the encryption of d under pwd. Intuitively, the definition says that A cannot tell which of w_0 and w_1 (two possible values of pwd) is used for creating encryptions of plaintexts selected according to $Dist$.

Definition 4. *A dictionary D^0 is a subdictionary of D if $\mathsf{D}^0_\eta \subseteq \mathsf{D}_\eta$ for all η. A dictionary D^0 is a singleton dictionary if $|\mathsf{D}^0_\eta| = 1$ for all η. Let $\Pi^p = (\mathcal{E}^p, \mathcal{D}^p)$ be a password-based encryption scheme. We say that Π^p securely encrypts distribution ensemble $Dist = (Dist_\eta)_\eta$ using passwords from dictionary D, if for any probabilistic polynomial-time adversary A, and any singleton subdictionaries D^0 and D^1 of D,*

$$\mathsf{Adv}_{\Pi^p, Dist, A}(\eta) \stackrel{\text{def}}{=} \Pr\left[w_0 \stackrel{R}{\leftarrow} \mathsf{D}^0_\eta, w_1 \stackrel{R}{\leftarrow} \mathsf{D}^1_\eta : A^{\mathcal{E}^p(w_1, Dist_\eta)}(\eta, w_0, w_1) = 1\right] - \Pr\left[w_0 \stackrel{R}{\leftarrow} \mathsf{D}^0_\eta, w_1 \stackrel{R}{\leftarrow} \mathsf{D}^1_\eta : A^{\mathcal{E}^p(w_0, Dist_\eta)}(\eta, w_0, w_1) = 1\right]$$

is negligible (as a function of the security parameter η).

(Recall that a function is negligible if it is smaller than the inverse of any polynomial for all sufficiently large inputs.)

In defining the syntax of password-based encryption, we do not require that the encryption function be randomized. Interestingly, randomization and security appear to be somewhat in conflict for password-based encryption. In order to explain this observation, let us write $\mathcal{E}^p(pwd, m, r)$ for the encryption of m under password pwd with random coins r. Consider an adversary A with access to an encryption oracle as in the definition above, but now with the (reasonable) capability of obtaining several encryptions of the same plaintext using different random coins. When A queries the encryption oracle twice, it obtains ciphertexts $c_0 = \mathcal{E}^p(w_b, m, r_0)$ and $c_1 = \mathcal{E}^p(w_b, m, r_1)$ for some $b \in \{0, 1\}$, some plaintext m, and some fresh random coins r_0 and r_1. Suppose that $b = 0$, without loss of generality. When A decrypts c_0 and c_1 with w_0, it obtains m twice. For b to remain secret, it also must be the case that $\mathcal{D}^p(w_1, \mathcal{E}^p(w_0, m, r_0)) = \mathcal{D}^p(w_1, \mathcal{E}^p(w_0, m, r_1))$. In this sense, the use of the random coins is trivial.

In addition to password-based encryption schemes, we rely on symmetric and asymmetric encryption schemes. As usual, a symmetric or asymmetric encryption scheme consists of algorithms $(\mathcal{K}, \mathcal{E}, \mathcal{D})$ for key generation, encryption, and decryption. We require that these satisfy a variant of the standard notion of semantic security [12], called *type-0 security*. This notion was previously introduced for the case of symmetric encryption [1] and extends to the case of asymmetric encryption. We leave precise definitions and constructions for a longer version of this paper.

An *encryption suite* is a triple $\Pi = (\Pi^a, \Pi^s, \Pi^p)$ with an asymmetric encryption scheme Π^a, a symmetric encryption scheme Π^s, and a password-based encryption scheme Π^p. We say that an encryption suite is secure if it provides type-0 secure asymmetric and symmetric encryption schemes, and its password-based encryption scheme securely encrypts keys and ciphertexts:

Definition 5. *An encryption suite* $\Pi = (\Pi^a, \Pi^s, \Pi^p)$ *is secure if* $\Pi^a = (\mathcal{K}^a, \mathcal{E}^a, \mathcal{D}^a)$ *and* $\Pi^s = (\mathcal{K}^s, \mathcal{E}^s, \mathcal{D}^s)$ *are type-0 secure encryption schemes and* Π^p *securely encrypts distribution ensembles* sym key, sym ciphertext, asym key, *and* asym ciphertext *defined by the algorithms below:*

sym key(η)	sym ciphertext(η)	asym key(η)	asym ciphertext(η)
$(k,k) \xleftarrow{R} \mathcal{K}^s(\eta)$	$(k,k) \xleftarrow{R} \mathcal{K}^s(\eta)$	$(pk, sk) \xleftarrow{R} \mathcal{K}^a(\eta)$	$(pk, sk) \xleftarrow{R} \mathcal{K}^a(\eta)$
As a sample	As a sample	As a sample	As a sample
return k	return $\mathcal{E}^s(k, 0)$	return pk	return $\mathcal{E}^a(pk, 0)$

5 A Computational Model for Expressions

In this section we give a computational interpretation to expressions in the form of ensembles of probability distributions, and give computational definitions for expression equivalence and password hiding.

For an encryption suite Π and a dictionary $\mathsf{D} = \cup_\eta \mathsf{D}_\eta$, we associate with each expression $M \in \mathsf{Exp}$ a distribution $[\![M]\!]_{\Pi[\eta], \mathsf{D}}$ on strings of bits, and thereby an ensemble $[\![M]\!]_{\Pi, \mathsf{D}}$. The definition is inductive:

- Each key symbol K that occurs in M is mapped to a string $\tau(K)$, via the key generation algorithms of Π^s and Π^a for symmetric and asymmetric keys, respectively, and by selecting at random from D_η for passwords.
- The formal bits 0 and 1 are mapped to standard string representations for them.
- The image of a pair (M, N) is obtained by concatenating the images of M and N.
- The image of a formal encryption $\{M\}_K$ is obtained by calculating $\mathcal{E}^{t(K)}_{\tau(K)}(x)$, where x is the image of M and $t(K) \in \{a, s, p\}$ selects the type of encryption.

Definition 6. *Two ensembles D^0 and D^1 are indistinguishable ($D^0 \approx D^1$) if for any probabilistic polynomial-time algorithm A,*

$$\mathrm{Adv}^{dist}_{D^0, D^1, A}(\eta) = \Pr\left[x \xleftarrow{R} D^0_\eta : A(x, \eta) = 1\right] - \Pr\left[x \xleftarrow{R} D^1_\eta : A(x, \eta) = 1\right]$$

is negligible (as a function of the security parameter η).

Definition 7. *The expressions $E_0, E_1 \in \mathsf{Exp}$ are computationally equivalent if their associated distribution ensembles are indistinguishable, that is,* $[\![E_0]\!]_{\Pi, \mathsf{D}} \approx [\![E_1]\!]_{\Pi, \mathsf{D}}$.

Definition 8. *Let Π be an arbitrary encryption suite and let D be a dictionary. An expression $E \in \mathsf{Exp}$ hides passwords in D computationally if for all singleton dictionaries D^0 and D^1, subdictionaries of D, it holds that* $[\![E]\!]_{\Pi, \mathsf{D}^0} \approx [\![E]\!]_{\Pi, \mathsf{D}^1}$.

In this definition, intuitively, an adversary is given two singleton dictionaries and a sample from the distribution associated with the expression E. This sample is created by using one of the two singleton dictionaries, and the goal of the adversary is to determine which. The expression hides passwords computationally if the adversary has only a negligible chance of success.

6 Soundness Theorems

Our soundness theorems link the symbolic definitions for expression equivalence and secure use of passwords to their computational counterparts. The theorem on expression equivalence can be regarded as an extension of the main theorem of Abadi and Rogaway [1] to the richer language of expressions of this paper.

Theorem 1 (Soundness for expression equivalence). *Let Π be a secure encryption suite and let D be a dictionary. For any two acyclic expressions $E_0, E_1 \in$ Exp we have that $E_0 \cong E_1$ implies $[\![E_0]\!]_{\Pi,\mathsf{D}} \approx [\![E_1]\!]_{\Pi,\mathsf{D}}$.*

Our main theorem says that, under certain hypotheses, if the use of passwords is secure symbolically, then it is also secure computationally.

Theorem 2 (Soundness of password hiding). *Let Π be a secure encryption suite and let D be a dictionary. For any acyclic expression $E \in$ Exp if E hides passwords symbolically then E hides passwords in D computationally.*

A question that we do not investigate in this paper is under what conditions the converses of Theorems 1 and 2 hold. However, it seems quite likely that the techniques and the assumptions for proving completeness of symbolic equivalence for the case of symmetric encryption (e.g. [19]) extend to the setting of this paper.

As an example, we show how to apply our results in the case of the influential Encrypted Key Exchange (**EKE**) protocol [5]. In the language Exp, the flows of the protocol between parties A and B that share a password W are as follows.

1. A generates an asymmetric key pair (K_1^e, K_1^d) and sends $\{K_1^e\}_W$ to B.
2. B decrypts this message using W. Then B generates a symmetric key K_1^s and sends $\{\{K_1^s\}_{K_1^e}\}_W$ to A.
3. At this point the parties share the key K_1^s, and check if the protocol was executed as expected: A generates a symmetric key K_A^s and sends $\{K_A^s\}_{K_1^s}$ to B.
4. Upon receiving this message, B obtains K_A^s, generates a new symmetric key K_B^s, and sends $\{(K_A^s, K_B^s)\}_{K_1^s}$ to A. (In the original protocol, K_A^s and K_B^s are random nonces; for simplicity we model these nonces as random symmetric keys.)
5. A decrypts this message and checks that the first component of the resulting pair is K_A^s. If so, it obtains K_B^s, sends $\{K_B^s\}_{K_1^s}$ to B, and terminates successfully.
6. Finally, B decrypts this last message, verifies that it contains the key K_B^s it previously sent to A, and if so, it terminates successfully.

A transcript of the execution of the protocol is given by the expression:

$$E = \{K_1^e\}_W, \{\{K_1^s\}_{K_1^e}\}_W, \{K_A^s\}_{K_1^s}, \{(K_A^s, K_B^s)\}_{K_1^s}, \{K_B^s\}_{K_1^s}$$

Since $pattern(E) = \{K_1^e\}_W, \{\square^a\}_W, \square^s, \square^s, \square^s$ is the instantiation of a pattern in Pat[Var] with an appropriate mapping, by definition, E hides the password symbolically. It follows from Theorem 2 that E also hides the password computationally. Informally, this means that for any probabilistic polynomial-time adversary, the probability that the adversary can determine correctly which of two passwords w_0 and w_1 was used in a given protocol execution is negligible. Once we have Theorem 2, the proof of this fact via the formal definitions is much simpler than a computational proof from scratch.

7 Conclusions

In this paper we investigate the use of password-based encryption schemes in protocols from the perspective of a recent line of research aimed at bridging the gap between the symbolic and computational views of cryptography. We give symbolic and computational interpretations to the elements of a language of formal expressions built using symmetric, asymmetric, and password-based encryption. We then prove that symbolic accounts of expression equivalence and password hiding imply strong, computational formulations of the same properties. We base our results on a new computational security definition for password-based encryption, which may be of independent interest.

Off-line guessing attacks, as typically considered in the literature, are inherently passive: an adversary, with some data about a protocol execution, analyzes the data in an attempt to obtain information about the password in use. Our definitions and theorems focus strictly on the data analysis, and do not consider how the data is obtained. Thus, we neither address nor exclude the possibility that the adversary may play a role in protocol executions, perhaps mounting standard active attacks, and obtaining data from interactions with other participants. For protocols that do not rely on passwords, research on the relations between symbolic and computational models has recently dealt with active attacks (e.g., [2, 20]). In further work, it may be worthwhile to integrate the results of that research with the present analysis of password-based encryption.

Acknowledgements. We thank Bruno Blanchet, Cédric Fournet, and Phil Rogaway for helpful discussions on password-based protocols, and thank Mathieu Baudet and Nathan Whitehead for suggesting improvements to the presentation of this paper. This research was partly carried out while the second author was at the University of California at Santa Cruz, and it was partly supported by the National Science Foundation under Grants CCR-0204162, CCR-0208800, and ITR-0430594.

References

1. M. Abadi and P. Rogaway. Reconciling two views of cryptography (The computational soundness of formal encryption). *Journal of Cryptology*, 15(2):103–127, 2002.
2. M. Backes, B. Pfitzmann, and M. Waidner. A composable cryptographic library with nested operations. In *Proceedings of the 10th ACM Conference on Computer and Communications Security*, pages 220–330, 2003.

3. M. Bellare and C. Namprempre. Authenticated encryption: Relations among notions and analysis of the generic composition paradigm. In *Advances in Cryptology — ASIACRYPT 2000*, volume 1976 of *LNCS*, pages 531–545. Springer-Verlag, 2000.
4. M. Bellare, D. Pointcheval, and P. Rogaway. Authenticated key exchange secure against dictionary attacks. In *Advances in Cryptology – EUROCRYPT 2000*, volume 1807 of *LNCS*, pages 139–155. Springer-Verlag, 2000.
5. S. M. Bellovin and M. Merritt. Encrypted key exchange: Password-based protocols secure against dictionary attacks. In *Proceedings of the 1992 IEEE Symposium on Security and Privacy*, pages 72–84, 1992.
6. V. Boyko, P. MacKenzie, and S. Patel. Provably secure password-authenticated key exchange using Diffie-Hellman. In *Advances in Cryptology – EUROCRYPT 2000*, volume 1807 of *LNCS*, pages 156–171. Springer-Verlag, 2000.
7. R. Corin, J. M. Doumen, and S. Etalle. Analysing password protocol security against off-line dictionary attacks. Technical report TR-CTIT-03-52, Centre for Telematics and Information Technology, Univ. of Twente, The Netherlands, 2003.
8. R. Corin, S. Malladi, J. Alves-Foss, and S. Etalle. Guess what? Here is a new tool that finds some new guessing attacks (extended abstract). In *IFIP WG 1.7 and ACM SIGPLAN Workshop on Issues in the Theory of Security (WITS)*, pages 62–71, 2003.
9. S. Delaune and F. Jacquemard. A theory of dictionary attacks and its complexity. In *Proc. of the 17th IEEE Computer Security Foundations Workshop (CSFW 2004)*, pages 2–15, 2004.
10. R. Gennaro and Y. Lindell. A framework for password-based authenticated key exchange. In *Advances in Cryptology – EUROCRYPT 2003*, volume 2656 of *LNCS*, pages 524–543. Springer-Verlag, 2003.
11. O. Goldreich and Y. Lindell. Session key generation using human passwords only. In *Advances in Cryptology – CRYPTO 2001*, volume 2139 of *LNCS*, pages 403–432. Springer-Verlag, 2001.
12. S. Goldwasser and S. Micali. Probabilistic encryption. *Journal of Computer and System Sciences*, 28:270–299, 1984.
13. L. Gong. Verifiable-text attacks in cryptographic protocols. In *INFOCOM '90*, pages 686–693, 1990.
14. L. Gong, T. M. A. Lomas, R. M. Needham, and J. H. Saltzer. Protecting poorly chosen secrets from guessing attacks. *IEEE Journal on Selected Areas in Communications*, 11(5):648–656, 1993.
15. J. Katz, R. Ostrovsky, and M. Yung. Practical password-authenticated key exchange provably secure under standard assumptions. In *Advances in Cryptology – EUROCRYPT 2001*, volume 2045 of *LNCS*, pages 475–494. Springer-Verlag, 2001.
16. J. Kohl and C. Neuman. RFC 1510: The Kerberos network authentication service (V5). Web page at ftp://ftp.isi.edu/in-notes/rfc1510.txt, 1993.
17. P. Laud. Symmetric encryption in automatic analyses for confidentiality against active adversaries. In *Proc. of 2004 IEEE Symposium on Security and Privacy*, pages 71–85, 2004.
18. G. Lowe. Analysing protocols subject to guessing attacks. *Journal of Computer Security*, 12(1):83–98, 2004.
19. D. Micciancio and B. Warinschi. Completeness theorems for the Abadi-Rogaway logic of encrypted expressions. *Journal of Computer Security*, 12(1):99–129, 2004.
20. D. Micciancio and B. Warinschi. Soundness of formal encryption in the presence of active adversaries. In *Theory of Cryptography Conference (TCC 2004)*, volume 2951 of *LNCS*, pages 133–151. Springer-Verlag, 2004.

21. G. Miklau and D. Suciu. Controlling access to published data using cryptography. In *Proceedings of 29th International Conference on Very Large Data Bases – VLDB 2003*, pages 898–909. Morgan Kaufmann Publishers, 2003.
22. J. Mitchell, A. Ramanathan, A. Scedrov, and V. Teague. A probabilistic polynomial-time calculus for analysis of cryptographic protocols. *Electronic Notes in Theoretical Computer Science*, 45, 2001.
23. S. Patel. Number theoretic attacks on secure password schemes. In *Proc. of the IEEE Symposium on Research in Security and Privacy*, pages 236–247, 1997.

On the Cover Time of Random Geometric Graphs

Chen Avin and Gunes Ercal

Computer Science Department,
University of California, Los Angeles,
Los Angeles, CA 90095-1596, USA
{avin, ercal}@cs.ucla.edu

Abstract. The cover time of graphs has much relevance to algorithmic applications and has been extensively investigated. Recently, with the advent of ad-hoc and sensor networks, an interesting class of random graphs, namely *random geometric graphs*, has gained new relevance and its properties have been the subject of much study. A random geometric graph $\mathcal{G}(n,r)$ is obtained by placing n points uniformly at random on the unit square and connecting two points iff their Euclidean distance is at most r. The phase transition behavior with respect to the radius r of such graphs has been of special interest. We show that there exists a critical radius r_{opt} such that for any $r \geq r_{\text{opt}}$ $\mathcal{G}(n,r)$ has optimal cover time of $\Theta(n \log n)$ with high probability, and, importantly, $r_{\text{opt}} = \Theta(r_{\text{con}})$ where r_{con} denotes the critical radius guaranteeing asymptotic connectivity. Moreover, since a disconnected graph has infinite cover time, there is a phase transition and the corresponding threshold width is $O(r_{\text{con}})$. We are able to draw our results by giving a tight bound on the electrical resistance of $\mathcal{G}(n,r)$ via the power of certain constructed flows.

1 Introduction

The *cover time* C_G of a graph G is the expected time taken by a simple random walk on G to visit all nodes in G. This property has much relevance to algorithmic applications [1, 2, 3, 4, 5], and methods of bounding the cover time of graphs have been thoroughly investigated [6, 7, 8, 9, 10, 11]. Several bounds on the cover times of particular classes of graphs have been obtained with many positive results [8, 9, 12, 13, 14].

A random geometric graph (RGG) is a graph $\mathcal{G}(n,r)$ resulting from placing n points uniformly at random on the unit square[1] and connecting two points iff their Euclidean distance is at most r. While these graphs have traditionally been studied in relation to subjects such as statistical physics and hypothesis testing [15], random geometric graphs have gained new relevance with the advent of ad-hoc and sensor networks [16, 17] as they are a model of such networks.

---
[1] We focus on the 2-dimensional, see section 6 for discussion.

Sensor networks have strict energy and memory constraints and in many cases are subject to high dynamics, created by failures, mobility and other factors. Thus, purely deterministic algorithms have disadvantages for such networks as they need to maintain data structures and have expensive recovery mechanism. Recently, questions regarding the random walk properties of such networks have been of interest especially due to the locality, simplicity, low-overhead and robustness to failures of the process [18, 19, 20]. In particular random walk techniques have been proposed for gossiping in random geometric graphs [1], for information collection and query answering [21, 5] and even for routing [22, 23].

In ad-hoc and sensor networks, interference grows with increased communication radius. So, for a desirable property P of random geometric graphs, one wants to find a tight upper bound on the smallest radius r_P, that will guarantee that P holds with high probability. The radius r_P is called *critical radius* if P exhibits a sharp threshold, the difference between the smallest radius for which the property holds with high probability and the radius for which the property holds with very low probability goes to zero as $n \to \infty$. The critical radius for connectivity, r_{con}, has been of special interest, and it has been shown that if $\pi r^2 \geq \pi r_{con}^2 = \frac{\log n + \gamma_n}{n}$ then $\mathcal{G}(n,r)$ is connected with probability going to one as $n \to +\infty$ iff $\gamma_n \to +\infty$ [24, 25].

In this paper we study the existence of a critical radius r_{opt} that will guarantee with high probability that $\mathcal{G}(n,r)$ with $r \geq r_{opt}$ has *optimal cover-time*. That is cover time of $\Theta(n \log n)$ [26], the same order as the complete graph. We show that such a threshold does exist, and, surprisingly, occurs at a radius $r_{opt} = \Theta(r_{con})$.

1.1 Discussion of Our Results and Techniques

Our main result can be formalized as follows:

Theorem 1 (Cover Time of RGG). *For $c > 1$, if $r^2 \geq \frac{c8 \log n}{n}$, then w.h.p.[2] $\mathcal{G}(n,r)$ has cover time $\Theta(n \log n)$. If $r^2 \leq \frac{\log n}{\pi n}$, then $\mathcal{G}(n,r)$ has infinite cover time with positive probability (bounded away from zero).*

The main contribution of this paper is in giving new tight theoretical bounds on the cover time and sharp threshold width associated with cover time for random geometric graphs. Our results improve upon bounds on the cover time obtained through bounding the mixing-time and spectral gap of random geometric graphs [27, 20, 19], as cover time can be bounded by the spectral gap [9]. In particular, the spectral gap method only guarantees optimal cover time of $\mathcal{G}(n,r)$ for $r = \Theta(1)$.

Aside from that, our results also have important implications for applications. Corollaries to our results are that both the *partial cover time* [5], which is the expected time taken by a random walk to visit a constant fraction of the nodes, and the *blanket time* [28], which is the expected time taken by a random walk

---

[2] Event $\mathcal{E}_n$ occurs with high probability if probability $P(\mathcal{E}_n)$ is such that $\lim_{n \to \infty} P(\mathcal{E}_n) = 1$.

to visit all nodes with frequencies according to the stationary distribution, are optimal for random geometric graphs. This demonstrates both the efficiency and quality of random walk approaches and certain token-management schemes for some ad-hoc and sensor networks [29, 1, 5].

In a recent related work Goel et al. [30] have proved that any monotonic property of random geometric graphs has a sharp threshold and have bounded the threshold width. While for general graphs optimality of cover time is not a monotonic property (see full version [31]), it follows from our result that optimality of cover time is monotonic for $\mathcal{G}(n,r)$ and has a threshold width of $O(r_{\text{con}})$.

The method that we used to derive our result is by bounding the electrical resistance of $\mathcal{G}(n,r)$, which bounds the cover time by the following result of Chandra et al. [8]: for any graph with n nodes and m edges, where R is the the electrical resistance of the graph:

$$mR \leq \text{cover time} \leq O(mR \log n) \tag{1}$$

In turn, we bound the resistance R of $\mathcal{G}(n,r)$ by bounding the power of a unit flow as permitted by Thomson's Principle which we formalize later. For any pair of points u and v, we construct a flow c in such a manner that the power of the flow satisfies $P(c) = O(\frac{n}{m}) = O(\frac{1}{\delta_{\text{avg}}})$ where δ_{avg} denotes the average degree of a node in $\mathcal{G}(n,r)$. Since $R \leq P(c)$ the above flow together with (1) establish to be sufficient for $\mathcal{G}(n,r)$ to have optimal cover time.

To construct a flow from u to v, we partition the nodes into contour layers based on distance from u and expanding outward until the midpoint between u and v, then from the midpoint line onward contracting towards v in a mirror fashion. The idea of using contour layers that expand with distance from a point is similar to the layering ideas used by Chandra et al. [8] for meshes and originally by Doyle and Snell [32] for infinite grids. Layers in our case can be visualized as slices of an isosceles right triangle along the hypotenuse that connects u and v. The flow can thus be thought of as *moving through* consecutive layers, with the total flow on the edges connecting consecutive layers being 1. Just as the variance of a probability function is minimized for the uniform distribution, we minimize the power by allocating flow almost uniformly along the set of edges used between layer l and layer $l+1$.

The construction of the above flow is based on "nice" properties of random geometric graphs, such as the uniformity of nodes distribution and the regularity of node degree. We formalize this "niceness" using the notion of a *geo-dense* graph: a geometric graph (random or deterministic) with close to uniform node density across the unit square. In *geo-dense* graphs there are no large areas that fail to contain a sufficient number of nodes. To construct the flow we define *bins* as equal size areas that partition the unit square. These bins are used as the building blocks of our layered flow: nodes in neighboring bins are in the same clique, and only edges between neighboring bins contribute to the flow. Finally, We show that *random* geometric graphs are in fact *geo-dense* for radius on the order of $\Theta(r_{\text{con}})$. Note however that *geo-dense* graphs are not necessary dense graphs in the graph theoretic meaning, i.e have $\Theta(n^2)$ edges. For example RGG are *geo-dense* even with $\Theta(n \log n)$ edges.

1.2 Related Work

There is a vast body of literature on cover times and on geometric graphs, and to attempt to summarize all of the relevant work would not do it justice. We have already mentioned some of the related results previously, however, here we would like to highlight the related literature that has been most influential to our result, namely that of Chandra et al. [8] and Doyle and Snell [32].

The work of Doyle and Snell [32] is a seminal work regarding the connection between random walks and electrical resistance. In particular, they proved that while the infinite 2-dimensional grid has infinite resistance, for any $d \geq 3$ the resistance of the d-dimensional grid is bounded from above, and these results were established to be sufficient in re-proving Pólya's beautiful result that a random walk on the infinite 2-dimensional grid is recurrent whereas a random walk on the infinite d-dimensional grid for any $d \geq 3$ is transient. In obtaining this result, essentially thors bounded the power of a unit current flow from the origin out to infinity and found that the power diverges for the 2-dimensional case and converges for every dimension greater than two. The authors used a layering argument, namely partitioning nodes into disjoint contour layers based on their distance from the origin, and the rate of growth of consecutive layers can be seen as the crucial factor yielding the difference between the properties of the different dimensions. Later, Chandra et al. [8] proved the tight relation between commute time and resistance, and used that relationship to extend Doyle and Snell's result by bounding the cover time of the *finite* d-dimensional mesh by computing the power and resistance via an expanding contour layers argument. Together with the tight lower bound of Zuckerman [10], they showed that the 2-dimensional torus has cover time of $\Theta(n \log^2 n)$, and for $d \geq 3$ the d-dimensional torus has an optimal cover time of $\Theta(n \log n)$.

While this paper deals with random geometric graphs there are striking similarities between $\mathcal{G}(n,r)$ and a more familiar family of random graphs, the *Bernoulli* graphs $\mathcal{B}(n,p)$ in which each edge is chosen independently with probability p [33]. For example, for critical probability $p_{\text{con}} = \pi r_{\text{con}}^2 = \frac{\log n + \gamma_n}{n}$, $\mathcal{B}(n,p)$ is connected with probability going to one as $n \to +\infty$ iff $\gamma_n \to +\infty$, and both classes of graphs have sharp thresholds for monotone properties [33]. Regarding cover time, Jonasson [12] and Cooper and Frieze [14] gave tight bounds on the cover time and an interesting aspect of our result is that we add another similarity and both classes of graphs have optimal cover time around the same threshold for connectivity. Yet, despite the similarities between $\mathcal{G}(n,r)$ and $\mathcal{B}(n,p)$, *Bernoulli* graphs are not appropriate models for connectivity in wireless networks since edges are introduced independent of the distance between nodes. In wireless networks the event of edges existing between i and j and between j and k is *not* independent of the event of an edge existing between k and i. There are other notable differences between $\mathcal{G}(n,r)$ and $\mathcal{B}(n,p)$ as well. For example, the proof techniques for the above results for $\mathcal{G}(n,r)$ are very different than the proof techniques for the respective results for $\mathcal{B}(n,p)$. Interestingly, whereas the proof of [14] for optimality of cover time in Bernoulli graphs of $\Theta(\log n)$ average degree depends on the property that Bernoulli graphs do *not* have small cliques (and, in particular that small cycles are sufficiently far apart), in the case of random geometric graphs the existence of many small cliques uniformly distributed over the unit square like bins is essential in our analysis.

Another recent result with a bin-based analysis technique for random geometric graphs is that of Muthukrishnan and Pandurangan [34]. However, as their technique uses large overlapping bins where the overlap is explicitly stated to be essential and there is no direct utilization of cliques.

2 Bounding the Cover Time via Resistance

For a graph $G = (V,E)$ with $|V| = n, |E| = m$, the electrical network $\mathcal{E}(G)$ is obtained by replacing each edge $e \in E$ with a 1 Ohm resistor, and this is the network we analyze when we speak of the resistance properties of G. For $u, v \in V$ let R_{uv} be the *effective resistance* between u and v: the voltage induced between u and v by passing a current flow of one ampere between them. Let R be the electrical resistance of G: the maximum effective resistance between any pair of nodes [32].

Let H_{uv} be the *hitting time*, the expected time for a random walk starting at u to arrive to v for the first time, and let C_{uv} be the *commute time*, the expected time for a random walk starting at u to first arrive at v and then return to u. Chandra et al. [8] proved the following equality for the commute time C_{uv} in terms of the effective resistance R_{uv}:

Theorem 2. *For any two vertices u and v in G the commute time $C_{uv} = 2mR_{uv}$*

Using this direct relation between resistance and random walks and Matthews' theorem [6] they introduced the bound of (1) on the cover time for G.

Let $H_{\max}$ be the maximum hitting time over all pairs of nodes in G. Since $H_{uv} \leq C_{uv}$ it follows that $H_{\max} \leq \max_{u,v \in V} C_{uv} = 2mR$. In [5] it has been shown that the partial cover time can be bounded by $H_{\max}$, so combining:

$$\text{partial cover time} \leq O(mR) \qquad (2)$$

Thus, by bounding the resistance R we may obtain tight bounds on the cover time C_G through (1) and on the partial cover time through (2).

A powerful method used to bound resistance is by bounding the power of a current flow in the network. The following definitions and propositions from the literature [8, 32, 35] help to formalize this method.

Definition 1 (Power of a flow). *Given an electrical network (V, E, ρ), with resistance $\rho(e)$ for each edge e, a flow c from a source u to a sink v is a function from $V \times V$ to $\mathbb{R}$, having the property that $c(x,y) = 0$ unless $\{x,y\} \in E$, and c is anti-symmetric, i.e., $c(x,y) = -c(y,x)$. The net flow out of a node will be denoted $c(x) = \sum_{y \in V} c(x,y)$ and $c(x) = 0$ if $x \neq u,v$. The flow along an edge e*

is $c(e) = |c(u,v)|$. The power $P(c)$ in a flow is $P(c) = \sum_{e \in E} \rho(e)c^2(e)$. A flow is a current flow if it satisfies Kirchoff's voltage law, i.e., for any directed cycle $x_0, x_1, \ldots, x_{k-1}, x_0$, $\sum_{i=0}^{k-1} c(x_i, x_{i+1 \bmod k}) \cdot \rho(x_i, x_{i+1 \bmod k}) = 0$.

Proposition 1. *[Thomson Principle [32, 35]] For any electrical network (V, E, ρ) and flow c with only one source u, one sink v, and $c(u) = -c(v) = 1$ (i.e a **unit flow**), we have $R_{uv} \leq P(c)$, with equality when the flow is a current flow.*

Finally,

Proposition 2. *[Rayleigh's Short/Cut Principle [32]] Resistance is never raised by lowering the resistance on an edge, e.g. by "shorting" two nodes together, and is never lowered by raising the resistance on an edge, e.g. by "cutting" it.*

3 The Cover Time and Resistance of Geometric Graphs

Before proving Theorem 1 about *random* geometric graphs we are going to prove a more general Theorem about geometric graphs. A **geometric graph** is a graph $G(n,r) = (V,E)$ with $n = |V|$ such that the nodes of V are embedded into the unit square with the property that $e = (u,v) \in E$ if and only if $d(u,v) \leq r$ (where $d(u,v)$ is the Euclidean distance between points u and v). We say that a geometric graph (either random or deterministic) is **geo-dense** if every square bin of area at least $A = r^2/8$ (in the unit square) has $\Theta(nA) = \Theta(nr^2)$ nodes.

Theorem 3. *A geometric graph $G(n,r)$ that is geo-dense and has $r = \Theta(\sqrt{\frac{\log n}{n}})$ has optimal cover time of $\Theta(n \log n)$, optimal partial cover time of $\Theta(n)$, and optimal blanket time of $\Theta(n \log n)$.*

Let $G(n,r)$ be a geometric graph that is *geo-dense*. We will prove Theorem 3 using the bound on the cover time from Eq. (1) and by bounding the resistance between any two points u, v in $G(n,r)$. Let V be the set of nodes of $G(n,r)$ and $\delta(v)$ denote the degree (i.e number of neighbors) of $v \in V$

Claim 1. $\forall v \in V \ \delta(v) = \Theta(nr^2)$

Proof. First note that the *geo-dense* property guarantees that if we divide the unit square into square bins of size $\frac{r}{\sqrt{2}} \times \frac{r}{\sqrt{2}}$ each, then the number of nodes in every bin will be $\Theta(nr^2)$. Since, for every bin, the set of nodes in the bin forms a clique, and every node $v \in V$ is in some bin, we have that $\delta(v) = \Omega(nr^2), \forall v \in V$. Similarly, when we divide the area into bins of size $r \times r$ every node may be connected to the nodes of at most nine bins (that is its own bin and the bordering bins), and we have that $\delta(v) = \Theta(nr^2), \forall v \in V$. □

Thus, since we showed that $m = |E| = \Theta(n^2r^2)$, if the resistance R of $G(n,r)$ is $O(\frac{n}{m}) = O(\frac{1}{nr^2})$ then we are done.

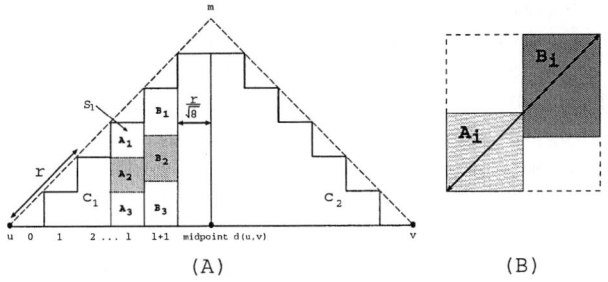

Fig. 1. $T(u,v)$ and the flow c between u and v in $G(n,r)$

Theorem 4. *The resistance R_{uv} between $u, v \in V$ is $\Theta(\frac{1}{nr^2} + \frac{\log(d(u,v)/r)}{n^2 r^4})$.*

Proof. The proof of the upper bound will be by bounding the power of a unit flow c that we construct between u and v.

Let $T(u,v)$ be an isosceles right triangle such that the line (u,v) is the hypotenuse. It is clear that such a triangle which lies inside the unit square must exist. We divide our flow c into two disjoint flows c_1 and c_2 where c_1 carries a unit flow from u up to the line perpendicular to the *midpoint* of $d(u,v)$ in increasing layer size, and c_2 forwards the flow in decreasing layer size up to v which is the only sink. By symmetry we can talk only about c_1 since the construction of c_2 mirrors that of c_1 and $P(c) = P(c_1 + c_2) = 2P(c_1)$ since the flows are disjoint. To construct the flow in c_1 we divide the line $(u, midpoint(u,v))$ into $d(u,v)\sqrt{2}/r$ segments of size $r/\sqrt{8}$, and number them from 0 to $d(u,v)\sqrt{2}/r - 1$ (see Fig 1 (A))[3]. Let S_l be the largest rectangle of width $r/\sqrt{8}$ included in the intersection of the area perpendicular to the lth segment and $T(u,v)$. S_l will define the lth layer in our flow. Note that the area of S_l is $lr^2/8$ and contains l squares of area $r^2/8$, each of them containing $\Theta(nr^2)$ nodes by the *geo-dense* property.

Let $V_l \subseteq V$ be the set of nodes in layer l. $V_0 = u$, and for $l > 0$ a node v is in layer l if and only if it is located inside S_l. It follows that $|V_l| = \Theta(nr^2 l)$. Edges in our flow are only among edges $e = (x,y)$ s.t. $x \in V_l$ and $y \in V_{l+1}$, and all other edges have zero flow. In particular, the set of edges E_l that carries flow from layer l to layer $l+1$ in c_1 is defined as follows: for the case $l = 0$, E_0 contains all the edges from u to nodes in V_1, noting that $|F_0| = |V_1| = \Theta(nr^2)$ since $u \cup V_1$ is a clique (i.e the maximum $d(u,x), x \in V_1$ is r). This allows us to make the flow uniform such that each node in V_1 has incoming flow of $1/|V_1|$ and for each edge $e \in E_0$, $c_1(e) = 1/|E_0|$. For $l > 0$ (see again Fig. 1 (A)) we divide S_l into l equal squares $A_1, A_2, \ldots A_l$ each of size $r^2/8$. Let V_{A_i} be the set of nodes contained in the area A_i. We then divide S_{l+1} into l equal sized rectangles $B_1, B_2 \ldots B_l$ and define V_{B_i} similarly, with B_i touching A_i for each i.

---

[3] Assume for simplicity the expression divides nicely, if not, the proof holds by adding one more segment that will end at the midpoint and overlap with the previous segment.

Now let $E_l = \{(x,y) | x \in V_{A_i}$ and $y \in V_{B_i}\}$. Note again that since, for each i, the maximum $d(x,y)$ between nodes in A_i and nodes in B_i is r (see Fig. 1 (B)), $V_{A_i} \cup V_{B_i}$ is a clique (as the worst case distance occurs between the first two layers). So, the number of edges crossing from A_i to B_i is $|V_{A_i}||V_{B_i}| = \Theta(n^2 r^4)$ by *geo-dense* property. The clique construction allows us to easily maintain the uniformity of the flow such that into each node in V_{B_i} the total flow is $1/l|V_{B_i}|$, and each edge carries a flow of $\Theta(1/n^2 r^4 l) = \Theta(1/E_l)$. All other edges have no flow. Now we compute the power of c:

$$R_{uv} \leq \sum_{e \in c} c(e)^2 = \sum_{e \in c_1} c_1(e)^2 + \sum_{e \in c_2} c_2(e)^2 =$$

$$= 2 \sum_{l=0}^{\sqrt{2}d(u,v)/r} \sum_{e \in E_l} c_1(e)^2 = 2\frac{1}{|E_0|} + 2 \sum_{l=1}^{\sqrt{2}d(u,v)/r} \frac{1}{|E_l|}$$

$$= 2O(\frac{1}{nr^2}) + 2O(\frac{1}{n^2 r^4}) \sum_{l=1}^{\sqrt{2}d(u,v)/r} \frac{1}{l}$$

$$= O(\frac{1}{nr^2} + \frac{\log(d(u,v)/r)}{n^2 r^4})$$

To prove the lower bound we again follow in the spirit of [32] and use the "Short/Cut" Principle. We partition the graph into $\lfloor d(u,v)/r \rfloor + 1$ partitions by drawing $\lfloor d(u,v)/r \rfloor$ squares perpendicular to the line (u,v), where the first partition P_0 is only u itself and the lth partition P_l is the area of the lth square excluding the $(l-1)^{th}$ square area. The last partition contains all the nodes outside the last square including v (see Fig 2 (A)). We are shorting all vertices in the same partition (see Fig. 2 (B)), and following the reasoning of the upper bound, let m_l be the number of edges between partition l and $l+1$. m_0 is $\Theta(nr^2)$ and for $l > 0$, $m_l = \Theta(n^2 r^4 l)$, so

$$R_{uv} \geq \sum_{l=0}^{\lfloor d(u,v)/r \rfloor} \frac{1}{m_l}$$

$$= \Omega(\frac{1}{nr^2}) + \sum_{l=1}^{\lfloor d(u,v)/r \rfloor} \Omega(\frac{1}{n^2 r^4 l})$$

$$= \Omega(\frac{1}{nr^2} + \frac{\log(d(u,v)/r)}{n^2 r^4}) \qquad \square$$

Corollary 1. *The resistance R of $G(n,r)$ is $\Theta(\frac{1}{nr^2} + \frac{\log(\sqrt{2}/r)}{n^2 r^4})$.*

This follows directly from the fact that $\max d(u,v) \leq \sqrt{2}$. Now we can prove Theorem 3.

Proof (of Theorem 3). Remember that $m = \Theta(n^2 r^2)$, so all we need is $R = O(n/m) = O(1/nr^2)$ and then the cover time bound will follow by (1), the

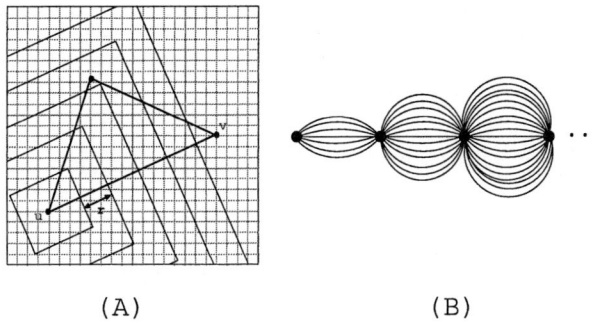

Fig. 2. Lower bound for R_{uv} on the $\mathcal{G}(n,r)$

partial cover time bound will follow from (2), and the blanket time will follow from [28] and the $\log n$ order difference between the cover time and maximum hitting time. In order to have $R = \Theta(\frac{1}{nr^2})$ we want that $\frac{\log(\sqrt{2}/r)}{n^2 r^4} = O(\frac{1}{nr^2})$, which means $\frac{\log(1/r)}{nr^2} \leq \alpha$ for some constant α. Taking $r^2 = \frac{c\log n}{n}$, for a constant β, we get $\frac{\log(n/\beta \log n)}{\beta 2 \log n} = \frac{1}{2\beta} - \frac{\log(\beta \log n)}{2\beta \log n} \leq \frac{1}{2\beta}$. □

4 Cover Time and Resistance of $\mathcal{G}(n,r)$

After Proving Theorem 3, in order to prove Theorem 1 all we need to show is that for $c > 1$, $r^2 = \frac{c8 \log n}{n}$ is sufficient to guarantee with high probability that $\mathcal{G}(n,r)$ is *geo-dense*. Note however that the second part of the theorem follows directly from [25] since if $\mathcal{G}(n,r)$ is disconnected with positive probability bounded away from zero when $r^2 \leq \frac{\log n}{\pi n}$, then it has infinite cover time with at least the same probability.

To prove the *geo-dense* property for $\mathcal{G}(n,r)$ we utilize the following lemma which seems to be folklore [28] although we include a proof in the full version since we have not found a reference including a proof of the minimum condition.

Lemma 2 (Balls in Bins). *For a constant $c > 1$, if one throws $n \geq cB \log B$ balls uniformly at random into B bins, then w.h.p both the minimum and the maximum number of balls in any bin is $\Theta(\frac{n}{B})$.*

And, the following lemma easily follows from the Balls in Bins Lemma:

Lemma 3 (Node Density). *For constants $c > 1$ and $a \geq 1$, if $r^2 = \frac{ca \log n}{n}$ then w.h.p any area of size r^2/a in $\mathcal{G}(n,r)$ has $\Theta(c \log n)$ nodes.*

Proof. Let an area of r^2/a be a bin. If we divide the unit square into such equal size bins we have $B = \frac{n}{c \log n}$ bins. For the result to follow we check that Lemma 2 holds by showing that $n \geq c'B \log B$ for some constant $c' > 1$:

$$B \log B = \frac{n}{c \log n} \log(\frac{n}{c \log n})$$
$$= \frac{n}{c \log n}(\log(n) - \log(c \log n))$$
$$= \frac{n}{c} - (\frac{n}{c \log n})(\log(c \log n))$$
$$\leq n/c$$

□

Now combining the results of Lemmas (2) and (3) we can prove Theorem 1

Proof (of Theorem 1). Clearly from Lemma 3 for $c > 1$, $r^2 = \frac{c8 \log n}{n}$ satisfies the geo-dense property w.h.p, and since r^2 is also $\Theta(\frac{\log n}{n})$ the result follows from Theorem 3. □

Corollary 2. *For $c > 1$, if $r^2 \geq \frac{c8 \log n}{n}$, then w.h.p $\mathcal{G}(n,r)$ has optimal partial cover time $\Theta(n)$ and optimal blanket time $\Theta(n \log n)$.*

5 Deterministic Geometric Graphs

As an example of other applications of our results consider the following: for an integer k, let the **k-fuzz** [32] of a graph G be the graph G_k obtained from G by adding an edge xy if x is at most k hops away from y in G. In particular, let $G_1(n)$ denote the 2-dimensional grid of n nodes, and let $G_k(n)$ be the k-fuzz of $G_1(n)$. It is known that the cover time of $G_1(n)$ is $\Theta(n \log^2 n)$. and so we ask what is the minimum k s.t. G_k has an optimal cover time of $\Theta(n \log n)$. Using the results of Theorem 4 and the lower bound method of Zuckerman [10] we can prove the following (see [31]):

Theorem 5. *For any constant k, the cover time of $G_k(n)$ is $\Theta(k^{-2} n \log^2 n)$.*

Corollary 3. *$G_k(n)$ has Cover Time of $\Theta(n \log n)$ if $k = \gamma_n$ and $\lim_{n \to \infty} \frac{\log n}{\gamma_n^2} \leq c$ for some constant c.*

This means that if each node is neighbor with his $\Omega(\log n)$ closest neighbors, the cover time of the 2-dimensional grid becomes optimal.

6 Conclusions

We have shown that for a two dimensional random geometric graph $\mathcal{G}(n,r)$, if the radius r_{opt} is chosen just on the order of guaranteeing asymptotic connectivity then $\mathcal{G}(n,r)$ has optimal cover time of $\Theta(n \log n)$ for any $r \geq r_{opt}$. We present a similar proof for 1-dimensional random geometric graphs in [31]. We find that the critical radius guaranteeing optimal cover time is $r_{opt} = \Omega(\frac{1}{\sqrt{n}})$ for such graphs, whereas the critical radius guaranteeing asymptotic connectivity is $r_{con} = \frac{\log n}{n}$. So, unlike the 2-dimensional case, we have $r_{opt} = \omega(r_{con})$.

Our proof techniques can be generalized to the d-dimensional random geometric graph $\mathcal{G}^d(n,r)$, yielding that for any given dimension d, $r_{\text{opt}} = \Theta(r_{\text{con}})$ with correspondingly optimal cover time. However, both grow exponentially with d which seems to be a consequence of a separation between average degree and minimum degree for higher dimensions rather than just an artifact of our method. Nevertheless, the case of dimension $d = 2$ is considered to be the hardest one [36]. This can intuitively be seen from the mesh results. The case for $d = 1$ (i.e the cycle) is easy to analyze. For $d > 2$ the cover time of the d-dimensional mesh is optimal [8], and we can show that for any k the cover time of the k-fuzz is also optimal. On the other hand, as we showed earlier, the cover time of the k–fuzz in 2 dimensions (i.e. $G_k(n)$) for constant k is not optimal making this the most interesting case.

Acknowledgment. The authors would like to thank Shailesh Vaya and Eli Gafni for helpful discussions and David Dayan-Rosenman and the anonymous reviewers for their comments and corrections. The first author acknowledges partial support from ONR (MURI) grant #N00014-00-1-0617 and from the Department of Communication System Engineering at Ben-Gurion University, Israel.

References

1. Kempe, D., Dobra, A., Gehrke, J.: Gossip-based computation of aggregate information. In: Proc. of the 44th Annual IEEE Symposium on Foundations of Computer Science. (2003) 482–491
2. Gkantsidis, C., Mihail, M., Saberi, A.: Random walks in peer-to-peer networks. In: in Proc. 23 Annual Joint Conference of the IEEE Computer and Communications Societies (INFOCOM). (2004)
3. Wagner, I.A., Lindenbaum, M., Bruckstein, A.M.: Robotic exploration, brownian motion and electrical resistance. Lecture Notes in Computer Science **1518** (1998) 116–130
4. Jerrum, M., Sinclair, A.: The markov chain monte carlo method: an approach to approximate counting and integration. In: Approximations for NP-hard Problems, Dorit Hochbaum ed. PWS Publishing, Boston, MA (1997) 482–520
5. Avin, C., Brito, C.: Efficient and robust query processing in dynamic environments using random walk techniques. In: Proc. of the third international symposium on Information processing in sensor networks. (2004) 277–286
6. Matthews, P.: Covering problems for Brownian motion on spheres. Ann. Probab. **16** (1988) 189–199
7. Aldous, D.J.: Lower bounds for covering times for reversible Markov chains and random walks on graphs. J. Theoret. Probab. **2** (1989) 91–100
8. Chandra, A.K., Raghavan, P., Ruzzo, W.L., Smolensky, R.: The electrical resistance of a graph captures its commute and cover times. In: Proc. of the 21st annual ACM symposium on Theory of computing. (1989) 574–586
9. Broder, A., Karlin, A.: Bounds on the cover time. J. Theoret. Probab. **2** (1989) 101–120
10. Zuckerman, D.: A technique for lower bounding the cover time. In: Proc. of the twenty-second annual ACM symposium on Theory of computing. (1990) 254–259

11. Aleliunas, R., Karp, R.M., Lipton, R.J., Lovász, L., Rackoff, C.: Random walks, universal traversal sequences, and the complexity of maze problems. In: 20th Annual Symposium on Foundations of Computer Science (San Juan, Puerto Rico, 1979). IEEE, New York (1979) 218–223
12. Jonasson, J.: On the cover time for random walks on random graphs. Comb. Probab. Comput. **7** (1998) 265–279
13. Jonasson, J., Schramm, O.: On the cover time of planar graphs. Electronic Communications in Probability **5** (2000) 85–90
14. Cooper, C., Frieze, A.: The cover time of sparse random graphs. In: Proc. of the fourteenth Annual ACM-SIAM Symposium on Discrete Algorithms (SODA-03). (2003) 140–147
15. Penrose, M.D.: Random Geometric Graphs. Volume 5 of Oxford Studies in Probability. Oxford University Press (2003)
16. Estrin, D., Govindan, R., Heidemann, J., Kumar, S.: Next century challenges: Scalable coordination in sensor networks. In: Proc. of the ACM/IEEE International Conference on Mobile Computing and Networking. (1999) 263–270
17. Pottie, G.J., Kaiser, W.J.: Wireless integrated network sensors. Communications of the ACM **43** (2000) 51–58
18. Goel, A., Rai, S., Krishnamachari, B.: Monotone properties have sharp thresholds in random geometric graphs. STOC slides (2004) http://www.stanford.edu/sanat/slides/thresholdsstoc.pdf.
19. Avin, C., Ercal, G.: Bounds on the mixing time and partial cover of ad-hoc and sensor networks. In: Proceedings of the 2nd European Workshop on Wireless Sensor Networks (EWSN 2005). (2005) 1–12
20. Boyd, S., Ghosh, A., Prabhakar, B., Shah, D.: Gossip and mixing times of random walks on random graphs. Unpublished. http://www.stanford.edu/boyd/reports/gossip_gnr.pdf (2004)
21. Sadagopan, N., Krishnamachari, B., Helmy, A.: Active query forwarding in sensor networks (acquire). To appear Elsevier journal on Ad Hoc Networks (2003)
22. Braginsky, D., Estrin, D.: Rumor routing algorthim for sensor networks. In: Proc. of the 1st ACM Int. workshop on Wireless sensor networks and applications. (2002) 22–31
23. Servetto, S.D., Barrenechea, G.: Constrained random walks on random graphs: routing algorithms for large scale wireless sensor networks. In: Proc. of the 1st Int. workshop on Wireless sensor networks and applications. (2002) 12–21
24. Penrose, M.D.: The longest edge of the random minimal spanning tree. The Annals of Applied Probability **7** (1997) 340–361
25. Gupta, P., Kumar, P.R.: Critical power for asymptotic connectivity in wireless networks. In Stochastic Analysis, Control, Optimization and Applications: A Volume in Honor of W. H. Fleming (1998) 547–566
26. Feige, U.: A tight lower bound on the cover time for random walks on graphs. Random Structures and Algorithms **6** (1995) 433–438
27. Rai, S.: The spectrum of a random geometric graph is concentrated. http://arxiv.org/PS_cache/math/pdf/0408/0408103.pdf (2004)
28. Winkler, P., Zuckerman, D.: Multiple cover time. Random Structures and Algorithms **9** (1996) 403–411
29. Dolev, S., Schiller, E., Welch, J.: Random walk for self-stabilizing group communication in ad-hoc networks. In: Proc. of the 21st IEEE Symposium on Reliable Distributed Systems (SRDS'02), IEEE Computer Society (2002) 70

30. Goel, A., Rai, S., Krishnamachari, B.: Sharp thresholds for monotone properties in random geometric graphs. In: Proc. of the thirty-sixth annual ACM symposium on Theory of computing. (2004) 580–586
31. Avin, C., Ercal, G.: On the cover time of random geometric graphs. Technical Report 040050, UCLA (2004) ftp://ftp.cs.ucla.edu/tech-report/2004-reports/040050.pdf.
32. Doyle, P.G., Snell, J.L.: Random Walks and Electric Networks. Volume 22. The Mathematical Association of America (1984)
33. Bollobás, B.: Random Graphs. Academic Press, Orlando, FL (1985)
34. Muthukrishnan, S., Pandurangan, G.: The bin-covering technique for thresholding random geometric graph properties. In: Proc. of the ACM-SIAM Symposium on Discrete Algorithms, to appear. (2005)
35. Synge, J.L.: The fundamental theorem of electrical networks. Quarterly of Applied Math. (1951) 113–127
36. Aldous, D., Fill, J.: Reversible markov chains and random walks on graphs. Unpublished. http://stat-www.berkeley.edu/users/aldous/RWG/ book.html (1997)

On the Existence of Hamiltonian Cycles in Random Intersection Graphs

Charilaos Efthymiou[1] and Paul G. Spirakis[1,2,*]

[1] Computer Engineering and Informatics Department, Patras University, Greece
euthimio@ceid.upatras.gr
[2] Research Academic Computer Technology Institute (R.A.C.T.I.), Greece
spirakis@cti.gr

Abstract. Random Intersection Graphs is a new class of random graphs introduced in [5], in which each of n vertices randomly and independently chooses some elements from a universal set, of cardinality m. Each element is chosen with probability p. Two vertices are joined by an edge iff their chosen element sets intersect. Given n, m so that $m = n^\alpha$, for any real α different than one, we establish here, for the first time, tight lower bounds $p_0(n, m)$, on the value of p, as a function of n and m, above which the graph $G_{n,m,p}$ is almost certainly Hamiltonian, i.e. it contains a Hamilton Cycle almost certainly. Our bounds are tight in the sense that when p is asymptotically smaller than $p_0(n, m)$ then $G_{n,m,p}$ almost surely has a vertex of degree less than 2. Our proof involves new, nontrivial, coupling techniques that allow us to circumvent the edge dependencies in the random intersection model. Interestingly, Hamiltonicity appears well below the general thresholds, of [4], at which $G_{n,m,p}$ looks like a usual random graph. Thus our bounds are much stronger than the trivial bounds implied by those thresholds.

Our results strongly support the existence of a threshold for Hamiltonicity in $G_{n,m,p}$.

1 Introduction

E. Marczewski in [8] proved that every graph can be represented by a list of sets where each vertex corresponds to a set and the edges to non-empty intersections of sets. Consider that each vertex chooses randomly the members of its corresponding set from a universal set, each independently of the others. The probability space that is created is the space of random intersection graphs, $G_{n,m,p}$, where n is the number of vertices, m the cardinality of a universal set of elements and p the probability that each vertex chooses each of the elements from the universal set. The random intersection graph model was first introduced by M. Karoński, E. Scheinerman and K. Singer-Cohen in [5]. A rigorous definition of the model of random intersection graphs follows:

---

[*] This work has been partially supported by the IST/FET Programme of the European Union under contract number 6FP 001907 (DELIS).

Definition 1. *Let n, m be positive integers, $0 \leq p \leq 1$. The random intersection graph $G_{n,m,p}$ is a probability space over the set of graphs on the vertex set $\{1,\ldots,n\}$ where each vertex is assigned a random subset from a fixed set of m elements. An edge arises between two vertices when their sets have at least a common element. Each random subset assigned to a vertex is determined by*

$$\mathbf{Pr}\left[\text{vertex } i \text{ chooses element } j\right] = p$$

with these events mutually independent.

In this work, we find an asymptotically tight lower bound for the probability p of the random intersection graph model $G_{n,m,p}$ so that the graph will be *Hamiltonian* almost surely. By "almost surely" we mean with probability tending to 1 as $n \to \infty$.

Definition 2. *Consider an undirected graph $G(V, E)$ where V is the set of vertices and E the set of edges. This graph contains a hamiltonian cycle if there is a simple cycle that contains each vertex in V.*

The derivation of the lower bound for p is made in two steps. First, we establish a stochastic order relations between the random intersection graph model $G_{n,m,p}$ and the random graph model $G_{n,\hat{p}}$. In particular, we find a function $\hat{p}$, of the probability p, such that if $G_{n,\hat{p}}$ has an *increasing property* [1], say $\mathcal{A}$, with probability P, then w.h.p. $G_{n,m,p}$, has the same property, $\mathcal{A}$, with probability at least P. Then, we calculate the value p, so that no vertex in $G_{n,m,p}$ has degree less than 2, w.h.p.. Finally, because of the stochastic order relation we have established, we conclude that when all degrees in $G_{n,m,p}$ are greater than 2, w.h.p. the graph is Hamitlonian a.s. Note that p will be a function of n, m.

As mentioned in [12], when we study the properties of $G_{n,m,p}$ for large n there are two "parameters" to adjust m, p. When m is very small compared to n, the model is not particularly interesting and when m is exceedingly large (compared to n) the behavior of $G_{n,m,p}$ is essentially the same as the Erdös-Rényi model of random graphs (see [4]). If we take $m = \lfloor n^\alpha \rfloor$, for fixed real $\alpha > 0$, then there is some deviation from the standard models, while allowing for a natural progression from sparse to dense graphs.

The probability for an edge e to arise in $G_{n,m,p}$ is $\mathbf{Pr}\left[e\right] = 1 - (1-p^2)^m$. We will assume $mp^2 \to 0$ i.e. $\mathbf{Pr}\left[e\right] \to 0$ as $n \to \infty$ (we have a sparse graph).

1.1 Previous Work

The model of random intersection graphs $G_{n,m,p}$ was first introduced by M. Karoński, E. Scheinerman and K. Singer-Cohen in [5] where they explored the evolution of random intersection graphs by studying the thresholds for the appearance and disappearance of small induced subgraphs. Also, J. A. Fill, E. R.

---

[1] A graph property $\mathcal{A}$ is increasing iff given that $\mathcal{A}$ holds for a graph $G(V, E)$ then $\mathcal{A}$ holds for any $G(V, E')$: $E' \supseteq E$.

Scheinerman and K. Singer Cohen in [4] proved an equivalence theorem relating the evolution of $G_{n,m,p}$ and $G_{n,p}$, in particular they proved that when $m = n^\alpha$ where $\alpha > 6$, the total variation distance between the graph random variables has limit 0. S. Nikoletseas, C. Raptopoulos and P. Spirakis in [10] studied the existence and the efficient algorithmic construction of close to optimal independent sets in random intersection graphs. No previous work is known to us for the existence of hamiltonian cycles in $G_{n,m,p}$.

There is a vast literature regarding the existence and the discovery of Hamiltonian cycles in random graphs $G_{n,p}$ and random regular graphs $G_{n,r}$ (n nodes with degree r each). The best result, known to us, for the existence of Hamiltonian cycles in random graphs, $G_{n,p}$, first proved by Komlós, Szeméredy in [6] and Korshunov [7], is stated in the following theorem (from [2]) .

Theorem 1. *(Komlós, Szeméredy, Korshunov) Let* $\omega(n) \to \infty$, $p = (1/n)(\log n + \log \log n + \omega(n))$. *Then almost every* $G_{n,p}$ *is Hamiltonian.*

1.2 Our Results and Techniques

The main practical difference between random intersection graphs $G_{n,m,p}$ and the random graphs $G_{n,p}$ is that in the first model, randomness is located on the vertices while in $G_{n,p}$ on the edges. Note, also, that in $G_{n,m,p}$ the existence of an edge affects the probability of existence of its adjacent edges. In this paper, we approximate the set of edges of an instance of $G_{n,m,p}$ with the set of edges of a $G_{n,\hat{p}}$ such that the second set of edges is a subset of the first one. This approximation is achieved by using weak coupling, i.e. we consider a discreet step process that creates, uniformly, an instance of $G_{n,m,p}$ and $G_{n,\hat{p}}$ simultaneously. In each step of this process it is decided which vertices choose a specific element from the universal set of elements. The choice or not of an element by some vertex depends on the degree of this vertex in the instance of $G_{n,p'}$, for a suitable p', that is created during that step. Each non-isolated vertex is considered to have chosen the corresponding element (coupling). The dependence between the degrees of the vertices of $G_{n,p'}$ is dealt either by using Poisson approximation for the number of non-isolated vertices in $G_{n,p'}$ or by setting the degrees of each vertex independently from the others, which under some conditions (see [9]), satisfied in our case, leads to the creation of an instance of $G_{n,p'}$ in a uniform manner. We finally get a $G_{n,m,p}$ by forming cliques between the vertices that chose the same elements or we get a $G_{n,\hat{p}}$ by taking the union of all the instances of $G_{n,p'}$ that are created in each step.

Having established the stochastic order relation between the random graph models $G_{n,m,p}$ and $G_{n,\hat{p}}$ we show that the moment the vertices of degree less than 2 disappear in $G_{n,m,p}$ is, asymptotically, the moment that the $G_{n,\hat{p}}$ becomes hamiltonian, thus the $G_{n,m,p}$ also becomes hamiltonian. We use Theorem 1 to argue when a random graph $G_{n,p}$ is hamiltonian almost surely. The following theorem is the main result of our paper.

Theorem 9. *Let $m = \lfloor n^\alpha \rfloor$, and C_1, C_2 be sufficiently large constants. If*

$$p \geq C_1 \frac{\log n}{m} \quad \text{for } 0 < \alpha < 1 \quad \text{or}$$
$$p \geq C_2 \sqrt{\frac{\log n}{nm}} \quad \text{for } \alpha > 1$$

then almost all $G_{n,m,p}$ are Hamiltonian. Our bounds are asymptotically tight.

In fact, all our results, conditioned on the Hamiltonicity threshold of $G_{n,p}$, hold with probability at least $1 - n^{-c}$, for some fixed $c > 0$.

2 On the Existence of Degree Less Than 2

If graph G is hamiltonian, then all its vertices must have degree greater than one. In this section we calculate bounds on the probability p for the existence (or not) of vertices with degree less than 2. To prove this, we use the first and second moment methods (see [1]).

Theorem 2. *Let $G_{n,m,p}$ be a random intersection graph. When $\alpha > 1$ and $p = C\sqrt{\frac{\log n}{nm}}$, if $C < 1$ then w.h.p. there is at least a vertex with degree less than 2, if $C > 1$ then w.h.p. the vertices with degree less than 2 disappear. When $\alpha \leq 1$ and $p = C\frac{\log n}{m}$, if $1 - \alpha < C < 2 - \alpha$ then w.h.p there is at least a vertex with degree less than 2 and if $C > 2 - \alpha$ vertices with degree less than 2 disappear.*

Proof. Let p_0 and p_1 be the probability for a vertex to have degree 0 or 1 respectively and $p_{l,k}^{i,j} = \mathbf{Pr}\left[deg(i) = l | deg(j) = k\right]$ for $i, j \in \{1, \ldots, n\}$, in an instance of $G_{n,m,p}$. Let $X_i, i = 1, \ldots, n$ be indicator random variables such that

$$X_i = \begin{cases} 1 & \text{if vertex } i \text{ has degree less than two} \\ 0 & \text{otherwise} \end{cases}$$

It holds that $\mathbf{Pr}\left[X_i = 1\right] = (p_0 + p_1)$ where $p_0 = \left(1 - p + p(1-p)^{(n-1)}\right)^m$ and $p_1 = nmp^2(1-\epsilon)\left(1 - p + p(1-p)^{(n-2)}\right)^m$ (see full paper [3] for the derivations). Let also $X = \sum_{i=1}^n X_i$. From the Chebyshev inequality it holds

$$\mathbf{Pr}\left[X = 0\right] \leq r = \frac{\Delta}{E^2[X]}$$

where the quantity Δ is defined as follows

$$\Delta = \sum_{i \sim j} \mathbf{Pr}\left[X_i X_j\right]$$

The symbol $i \sim j$ stands for "for all i, j such that $i \neq j$". Towards calculating Δ first note that

$$\mathbf{Pr}\left[X_i X_j\right] = p_0 \left(p_{0,0}^{i,j} + p_{0,1}^{i,j}\right) + p_1 \left(p_{0,1}^{i,j} + p_{1,1}^{i,j}\right)$$

The calculations and asymptotic bounds regarding $p_{k,l}^{i,j}$ can be found in [3]. Finally we get that $\Delta \leq n^2(p_0+p_1)^2(1+nmp^2)e^{-mp(1-e^{-np})+mnp^3}$ and consequently

$$r \leq (1+nmp^2)e^{-mp(1-e^{-np})+mnp^3}$$

By linearity of expectation we get

$$E[X] = nE[X_i] = n(p_1+p_0)$$
$$\simeq n(1+nmp^2)e^{-nmp^2}(1-\epsilon)$$

If $nmp^2 = C\log n$ then $E[X] = O(\frac{\log n}{n^{C-1}})$ and $r \leq O(\frac{\log n}{n^C})$. Taking $C < 1$ we have $E[X] \to \infty$ and $r \to 0$ which, by the Second momend method, leads to conclusion that w.h.p. there are vertices of degree less than 2. On the other hand, if $C > 1$ then $E[X] \to 0$, and by the first moment method, w.h.p. there are no vertices with degree less than 2.

In the case where $\alpha \leq 1$ matters are similar, but now we must have $np \to \infty$. The derivations of the result can be found in the full version of the paper, [3]. ◇

3 Weak Coupling

Let $\mathcal{I}(n,m,p)$ be the following process, that creates an instance of $G_{n,m,p}$: There are n vertices and m elements. In step i each vertex chooses element l_i with probability p and independently from the others. Form a clique with the vertices that chose l_i. Repeat the above step for $i = 1,\ldots,m$. The union of all cliques is the $G_{n,m,p}$.

We use coupling to establish a stochastic order relation between $G_{n,m,p}$ and $G_{n,\hat{p}}$. Coupling is made by using different processes, but similar to $\mathcal{I}(n,m,p)$. After each step of the processes the result can be interpreted either as a choice of an element from a set of vertices, as in $\mathcal{I}$, or as the choice of some edges, each independently of the others. In our case we use two processes, the $\mathcal{C}_1(n,m,p)$ and $\mathcal{C}_2(n,m,p)$.

Definition 3. *Process $\mathcal{C}_1(n,m,p)$. The process $\mathcal{C}_1(n,m,p)$ consists of m steps. Choose a value p' from the Normal distribution with mean $\bar{p} = p/(n-1)$ and variance $\bar{p}\bar{q}/(2N)$ truncated to the unit interval $(0,1)$, where $\bar{q} = 1 - \bar{p}$ and $N = \binom{n}{2}$. There are n vertices and in step i, vertex j chooses a random integer d_j which is binomially distributed, i.e. as in $B(n-1,p')$. If in step i, $d_j > 0$ then vertex j is considered to have choosen element l_i.*

Under conditions, which we will analyze in detail, the vector $d = (d_1,\ldots,d_n)$, formed in step i, is w.h.p. a degree sequence of a random graph $G_{n,p/(n-1)}$. We choose an instance of $G_{n,p/(n-1)}$ uniformly among those that have degree sequence d. The union of the graphs $G_{n,p/(n-1)}$ that are created in each step is an instance of $G_{n,\hat{p}}$, where $\hat{p} = 1 - (1-p/(n-1))^m$ (see subsection 3.1). The union of cliques between vertices that chose the same elements is an instance of $G_{n,m,p(1-\epsilon)}$, where ϵ is a non-negative constant.

Definition 4. *Process* $C_2(n,m,p)$ *The process* $C_2(n,m,p)$ *consists of m steps. There are n vertices. In step i: (a) with probability $1-(1-p)^{n-1}$, form an instance, H, of $G_{n,p/(n-1)}$ among the n vertices. If $deg_H(j) > 0$ for vertex j then vertex j is considered to have chosen element l_i. (b) Else, choose uniformly one vertex from the set of vertices. With probability $np(1-\epsilon)$, where $\epsilon < np$ is a positive real, this vertex chooses element l_i, otherwise it does not.*

Under conditions, which we will see in detail, the choice of the elements from the vertices can be considered independent for each vertex. The union of cliques between vertices that chose the same elements is an instance of $G_{n,m,p(1-\epsilon)}$, where ϵ non-negative constant. The union of the graphs $G_{n,p/(n-1)}$ that are created during $C_2(n,m,p)$, is an instance of $G_{n,\hat{p}}$, where $\hat{p} = E[1-(1-p/(n-1))^{m^*}]$ and m^* is a random variable following the Binomial distribution $B(n, 1-(1-p)^{n-1})$ that indicates the number of $G_{n,p/(n-1)}$ instances that were created during $C_2(n,m,p)$ (see subsection 3.2).

Corollary 1. *After the execution of either $C_1(n,m,p)$ or $C_2(n,m,p)$ the instance of $G_{n,\hat{p}}$ is a subgraph of $G_{n,m,p(1-\epsilon)}$.*

3.1 Analysis of $C_1(n,m,p)$

We present, here some results from the work of McKay and Wormald in [9] that will be useful to understand $C_1(n,m,p)$.

Consider a random vector $d \in I_n$, where $I_n = \{0,1,\ldots,n-1\}^n$ and d has components $d_1,\ldots,d_n$. Define $m(d) = \sum_{i=1}^{n} d_i$ and $E_n = \{d \in I_n \mid m(d) \text{ is even}\}$. We also have $\bar{d} = m(d)/n$, $\lambda = \bar{d}/(n-1)$ and $\gamma_2 = (n-1)^{-2}\sum_{i=1}^{n}(d_i - \bar{d})^2$. We create d according to the following models:

Weighted Even Sum binomial models, $\mathcal{E}_p$ and $\mathcal{E}'_p$ Model $\mathcal{E}_p$ has domain E_n, each component d_i is independently distributed according to binomial distribution, B(n-1,p), with the restriction that $m(d)$ is even. The model $\mathcal{E}'_p$ is constructed from $\mathcal{E}_p$ by weighting each $d \in E_n$ with weight depending on $m(d)$, such that $\frac{1}{2}m$ has distribution $B(N,p)$, where $N = \binom{n}{2}$.
Integrated model $\mathcal{I}_p$ Model $\mathcal{I}_p$ has domain E_n. To generate a vector d in this model we first choose a value p' from the normal distribution with mean p and variance $pq/(2N)$, where $q = 1-p$, truncated to the unit interval $(0,1)$. Then we generate a variant of model $\mathcal{E}_{p'}$
Graph model $\mathcal{D}_p$ Model $\mathcal{D}_p$ has domain E_n. A variant d in this model is the degree sequence of a random graph with n vertices and each edge is selected with probability p.

Theorem 3. *(McKay-Wormald) If $\omega(n)\log n/n^2 \le pq \le o(n^{-1/2})$ then there is a set-valued function $R_P(n) \subseteq E_n$ and a real function $\delta(n) \to 0$ such that the following condition is satisfied*

1. For each $d \in R_p(n)$ there is a number δ_d such that $|\delta_d| \le \delta(n)$ and

$$P_{D_p}(d) = P_{\mathcal{E}'_p}(d) exp\left\{\frac{1}{4} - \frac{\gamma_2^2}{4\lambda^2(1-\lambda)^2} + \delta_d\right\}$$

2. In each of the models $\mathcal{E}_p$ and D_p, we have $\mathbf{Pr}\left[R_p(n)\right] = 1 - n^{-\omega(n)}$

Theorem 4. *(McKay-Wormald)* If $\omega(n) \cdot \log n/n^2 \le pq \le o(n^{-1/2})$ then $\gamma_2 = \lambda(1-\lambda)(1+o(1))$ with probability $1 - n^{\omega(n)}$ for both $\mathcal{E}'_p$ and D_p models.

Theorem 5. *(McKay-Wormald)* If $pqN \to \infty$ and $y = o((pqN)^{(1/6)})$, then

$$P_{I_p}(d) = P_{\mathcal{E}'_p}(d)\left(1 + O\left(\frac{1+|y|^3}{\sqrt{pqN}}\right)\right)$$

uniformly over all $d \in E_n$ such that $|\frac{1}{2}m - pN| \le y\sqrt{pqN}$

Corollary 2. *If $\omega(n) \log n/n^2 \le pq \le o(n^{-1/2})$, where $q = 1-p$, we can approximate the model D_p by the model I_p. More particularly, for fixed reals $\epsilon > 0$ and ϵ' and sufficiently large n, for the random vector d we have that w.h.p.*

$$Pr_{D_p}(d) = Pr_{I_p}(d)(1+\epsilon') \qquad \text{where } |\epsilon'| \le \epsilon$$

Proof. See full version of this paper [3]. ◇

The next theorem shows that $\mathcal{C}_1(n,m,p)$ generates either a $G_{n,m,p(1-\epsilon)}$, with ϵ a non-negative constant, or a $G_{n,\hat{p}}$, under some conditions regarding p.

Theorem 6. *If $\omega(n) \log n/n^2 \le \bar{p}\bar{q} \le o(n^{-1/2})$, where $\bar{p} = p/(n-1)$, $\bar{q} = 1-\bar{p}$ for sufficiently large n, the execution of process $\mathcal{C}_1(n,m,p)$ results in either an instance of a random intersection graph $G_{n,m,p(1-\epsilon)}$, where ϵ is small non-negative number, or an instance of random graph $G_{n,\hat{p}}$ where $\hat{p} = 1 - (1-p/(n-1))^m$.*

Proof. For $\bar{p} = p/(n-1)$, $\bar{q} = 1-\bar{p}$ and $N = \binom{n}{2}$, if $\bar{p}\bar{q}N = \omega(n)\log n$, then, by Corollary 2, in step i of the process $\mathcal{C}_1(n,m,p)$ the vector d is w.h.p. the degree sequence of a random graph $G_{n,p/(n-1)}$. We choose one graph uniformly among those that have degree sequence d. Taking the union of all $G_{n,p/(n-1)}$ created in each step, we have a random graph $G_{n,\hat{p}}$ where $\hat{p} = 1 - (1-p/(n-1))^m$.

After each step of $\mathcal{C}_1(n,m,p)$, an instance of $G_{n,p/(n-1)}$ is created w.h.p. and each non-isolated vertex, in step i is considered to have chosen element l_i. This happens with probability $1-(1-p/(n-1))^{n-1}$. By the fact that $(n-1)p/(n-1) = p \to 0$ and using the binomial expansion of $(1-p/(n-1))^{(n-1)}$ it holds:

$$\mathbf{Pr}\left[\text{vertex } i \text{ is non isolated in } G_{n,\bar{p}}\right] = p(1-\epsilon)$$

where $0 \le \epsilon \le p$. Note that $\epsilon \to 0$ as $n \to \infty$, thus given a fixed value for ϵ we can find a sufficiently large n_0 such that $\mathbf{Pr}\left[\text{vertex } i \text{ is non isolated in } G_{n,\bar{p}}\right] \ge p(1-\epsilon)$ when $n \ge n_0$. But this probability is the probability for a vertex to choose element l_i in step i of $\mathcal{C}_1(n,m,p)$. The choice of each element l_i is independent for each vertex because each vertex's degree is set independently from the others' (we used model I_p to create the instance of the degree sequence). ◇

3.2 Analysis of C_2

The process $C_2(n, m, p)$ consists of m steps. There are n vertices. In step i: (a) with probability $1 - (1 - p)^{n-1}$, form a instance, H, of $G_{n,p/(n-1)}$ among the n vertices. If $deg_H(j) > 0$ for vertex j then vertex j is considered to have chosen element l_i. (b) Else, choose uniformly one vertex from the set of vertices. With probability $np(1-\epsilon)$, where $\epsilon < np$ is a positive real, this vertex chooses element l_i, otherwise it does not.

Someone could object saying that in random intersection graphs each vertex should choose each element independently of the others. In $C_2(n, m, p)$, this condition is violated, because the event that a vertex has degree greater than zero affects also other vertices' degrees. However, we will show that this dependence is weak for some range of probability p, and its effects weakens as n gets large.

We approximate the distribution of the number of non-isolated vertices in each step of $C_2(n, m, p)$ by the Poisson distribution. To show that the approximation is precise we use the *Stein-Chen Method* to bound the error in the approximation. The following theorem states the *Stein-Chen Method* (see [11]).

Theorem 7. *(Stein-Chen Method) Let X_i, $i = 1, \ldots, n$ be random indicator variables such that*

$$X_i = \begin{cases} 1 & \text{with probability } \lambda_i \\ 0 & \text{with probability } 1 - \lambda_i \end{cases}$$

and $W = \sum_{i=1}^{n} X_i$. Also, let V_i be any random variable that is distributed as the conditional distribution of $\sum_{j \neq i} X_j$ given that $X_i = 1$, $i = 1, \ldots, n$. That is, for each i, V_i is such that

$$\mathbf{Pr}\left[V_i = k\right] = \mathbf{Pr}\left[\sum_{j \neq i} X_j = k | X_i = 1\right], \quad \text{for all } k$$

Setting $\lambda = \sum_{i=1}^{n} \lambda_i$, for any set A of nonnegative integers we have

$$\left|\mathbf{Pr}\left[W \in A\right] - \sum_{i \in A} e^{-\lambda} \lambda^i / i!\right| \leq min(1, 1/\lambda) \sum_{i-1}^{n} \lambda_i \cdot E[|W - V_i|]$$

Lemma 1. *Let $G_{n,p}$ be a random graph and W be random variable that indicates the number of non-isolated vertices in that graph. If $n^2 p \to 0$ then the distribution of the random variable W can be approximated by a Poisson distribution with mean $\lambda = n\left(1 - (1-p)^{n-1}\right)$, more specifically*

$$\left|\mathbf{Pr}\left[W = i\right] - e^{-\lambda} \lambda^i / i!\right| \leq n^2 p$$

Proof. Let $G_{n,p}$ be a random graph and $X_1, \ldots, X_n$ be indicator random variables such that:

$$X_i = \begin{cases} 1 & \text{if vertex } i \text{ is non-isolated in } G_{n,p} \\ 0 & \text{otherwise} \end{cases}$$

Let $W = \sum_{i=1}^{n} X_i$, W is a random variable that counts the non-isolated vertices in $G_{n,p}$. We show that W is Poisson distributed with mean $\lambda = n\left(1 - (1-p)^{n-1}\right)$.

For the proof we use the *Stein-Chen Method* (Theorem 7). First calculate $E[|V_i - W|]$, V_i is a random variable that counts the non-isolated vertices in $G_{n,p}$ given that vertex i is non-isolated, vertex i is not counted by V_i. To simplify the calculations, we will use coupling.

Coupling. Let $G_{n,p}^1$ and $G_{n,p}^2$ be two random graphs. In G^1, all edges but these that are adjacent to vertex i, are selected with probability p, independently of each other. Each edge that is selected in G^1 is also selected in G^2. Then, for the graph G^2 set the degree of vertex i according to binomial distribution, $B(n-1,p)$ but with the restriction that the degree should be greater than zero. In G^1, the vertex i chooses the same edges as in G^2, with probability $1 - (1-p)^{(n-1)}$ otherwise it remains isolated.

W is the number of non-isolated vertices in G^1 while the number of non-isolated vertices in G^2 is $1 + V_i$. If vertex i is non-isolated in G^1 then $W = V_i + 1$, otherwise $V_i \geq W$. We have

$$E[|W - V_i|] = \mathbf{Pr}\left[deg_{G^1}(i) > 0\right] + E[V_i - W | deg_{G^1}(i) = 0]\mathbf{Pr}\left[deg_{G^1}(i) = 0\right] \quad (1)$$

In the full version of this paper [3] we calculate $E[V_i - W | deg_{G^1}(i) = 0]$ and we get

$$E[V_i - W | deg_{G^1}(i) = 0] \leq \frac{1}{1 - (n-1)p}$$

We also have that

$$\mathbf{Pr}\left[deg_{G^1}(i) > 0\right] = 1 - (1-p)^{(n-1)} \leq np \quad (2)$$

Finally, by equation 1, $E[|V_i - W|] \leq np + \frac{1}{1-np}$.
From Theorem 7 we get

$$\left|\mathbf{Pr}\left[W = i\right] - e^{-\lambda}\lambda^i/i!\right| = min(1, 1/\lambda) \sum_{i=1}^{n} \lambda_i E[|W - V_i|]$$

Because of equation 2 and the assumption that $n^2 p \to 0$ as $n \to \infty$, for sufficiently large n, we have that $min(1, 1/\lambda) = 1$ Thus:

$$\left|\mathbf{Pr}\left[W = i\right] - e^{-\lambda}\lambda^i/i!\right| \leq n^2 p(np + \frac{1}{1-np})$$
$$= n^2 p(1 + o(1)) \to 0 \quad \text{as } n \to \infty \quad \diamond$$

It is well known that the binomial distribution, $B(n,p)$, under some range of p, can be approximated by the *Poisson* distribution with mean $\lambda = np$. This is stated in detail by the following lemma.

Lemma 2. *Let Q be a random variable which is binomially distributed, $B(n,p)$. If $np \to 0$ as $n \to \infty$ the distribution of the random variable Q can be approximated by a Poisson distribution with mean $\lambda = np$*

$$\left|\mathbf{Pr}\left[Q = i\right] - e^{-\lambda}\lambda^i/i!\right| \leq np^2$$

Proof. See full version of paper [3]. ◇

From lemmas 1 and 2 we can deduce that the non isolated vertices in a $G_{n,p}$ (under conditions that we study in the following corollary), can be regarded, (with a negligible error) as independent. The following Corollary states it in detail.

Corollary 3. *Let W be the number of non-isolated vertices in an instance of $G_{n,p}$ and Q be a random variable that is binomially distributed, $B(n, 1-(1-p)^{(n-1)})$. If $n^2 p \to 0$ as $n \to \infty$ then for sufficiently large n*

$$|\mathbf{Pr}\,[W = i] - \mathbf{Pr}\,[Q = i]| \leq n^2 p + n^3 p^2 (1 - \epsilon) \to 0$$

where ϵ is a fixed real positive and $i \in \{0, \ldots, n\}$.

Proof. See full version of paper [3] ◇

Theorem 8. *If $n^2 p \to 0$ as $n \to \infty$, for sufficiently large n, the execution of the process $\mathcal{C}_2(n, m, p)$ results in either an instance of a random intersection graph $G_{n,m,p(1-\epsilon)}$, where ϵ is a small number, or an instance of random graph $G_{n,\hat{p}}$ where $\hat{p} = E[1 - (1 - p/(n-1))^{m^*}]$. m^* is a random variable binomially distributed as in $B(m, 1 - (1-p)^{n-1})$.*

Proof. The case of $G_{n,\hat{p}}$ is easily shown. In each step of $\mathcal{C}_2(n, m, p)$ an instance of $G_{n,p/(n-1)}$ is created with probability $1 - (1-p)^{n-1}$. So, after the m-th step of the process the probability of an edge to be created, i.e. $\hat{p}$, is the probability of the edge to be created during, at least, one step of the process. Let m^* be the number of instances of $G_{n,p/n-1}$ that are created during $\mathcal{C}_2(n, m, p)$ then, conditioning on m^*, we get

$$\hat{p} = \sum_{i=0}^{m} \mathbf{Pr}\,[m^* = i]\left(1 - (1 - \frac{p}{n-1})^i\right)$$
$$= E[1 - (1 - \frac{p}{n-1})^{m^*}]$$

On the other hand, if an instance of $G_{n,p/(n-1)}$ is created, from Corollary 3, we see that the number of non-isolated vertices in this step tend to be distributed as if each was chosen independently from the others. The probability for a vertex i to be non-isolated, in $G_{n,p/(n-1)}$, is $\mathbf{Pr}\,[degree(i) > 0] = 1 - (1 - \frac{p}{n-1})^{(n-1)}$. Using the binomial expansion of $(1 - p/n)^n$ and the fact that $p \to 0$, we get

$$\mathbf{Pr}\,[degree(i) > 0] = p(1-\epsilon) \quad \text{where } 0 \leq \epsilon \leq p$$

For fixed ϵ we can find n_0 such that $\mathbf{Pr}\,[degree(i) > 0] \geq p(1-\epsilon)$ if $n \geq n_0$. The probability for a vertex j to choose element l_i is thus

$$\mathbf{Pr}\,[j \text{ chooses } l_i] = \frac{1}{n} np(1-\epsilon)(1-p)^{(n-1)} + \left(1 - (1-p)^{n-1}\right) p(1-\epsilon)$$
$$= p(1-\epsilon)$$

Considering the above we indeed form an instance of $G_{n,m,p(1-\epsilon)}$. ◇

4 The Emergence of a Hamiltonian Cycle

In this section, we show that almost every random intersection graph $G_{n,m,p}$ becomes hamiltonian, asymptotically, at the moment that every vertex in the graph acquires degree greater than 1. To prove this we use the stochastic order relations we established in Section 3.

Case $0 < \alpha < 1$. By Theorem 2, if $p = C\frac{\log n}{m}$, for $C > 2 - \alpha$, an instance of $G_{n,m,p}$ has every vertex with degree greater than one w.h.p.. Also, note that for this p, $\frac{p}{n-1}$ satisfies the conditions of $C_1(n,m,p)$. Thus, using process $C_1(n,m,p)$, where $p = C'\frac{\log n}{m}$ and $C'(1-\epsilon) > 2 - \alpha$ we create an instance of $G_{n,m,p(1-\epsilon)}$ and an instance of $G_{n,\hat{p}}$, where $\hat{p} = 1 - [1 - p/(n-1)]^m$ or

$$\hat{p} = 1 - (1 - p/(n-1))^m = \frac{mp}{n}(1 - \epsilon')$$
$$= (1 - \epsilon')C'\frac{\log n}{n}$$

where $0 \le \epsilon' \le \frac{mp}{n}$. The second equality is derived from the fact that $\frac{mp}{n} \to 0$ and the binomial expansion of $(1 - p/(n-1))^m$. Note that $\epsilon' \to 0$ as $n \to \infty$. We can fix ϵ' and take sufficiently large n so as $\hat{p} \ge \frac{mp}{n}(1 - \epsilon')$. Substituting p we get $\hat{p} = K\frac{\log n}{n}$, for sufficiently large C', we get $K > 1$. By Theorem 1 and Corollary 1 the instances of $G_{n,\hat{p}}$ and $G_{n,m,p(1-\epsilon)}$ are almost surely Hamiltonian.

Case $\alpha > 1$. By Theorem 2, if $p = C\sqrt{\frac{\log n}{nm}}$, for $C > 1$ an instance of $G_{n,m,p}$ has every vertex with degree greater than one w.h.p.. We also note that in this case $p/(n-1)$ satisfies the conditions of $C_2(n,m,p)$. Thus using the process $C_2(n,m,p)$ where $p = C'\sqrt{\frac{\log n}{nm}}$ and $C'(1-\epsilon) > 1$ we can create either $G_{n,m,p(1-\epsilon)}$ or $G_{n,\hat{p}}$, where $\hat{p} = E[1 - (1 - p/(n-1))^{m^*}]$.

The random variable m^*, indicates the number of instances of $G_{n,p/(n-1)}$ that are created during $C_2(n,m,p)$. In the full version of paper [3] we find that $\hat{p} = mp^2(1-\epsilon')$. Substituting p we get $\hat{p} = K\frac{\log n}{n}$, for sufficiently large C', we get $K > 1$. By Theorem 1 and Corollary 1 the instances of $G_{n,\hat{p}}$ and $G_{n,m,p(1-\epsilon)}$ are almost surely Hamiltonian. In view of Theorem 2 our bounds are asymptotically tight. Thus, we have proved our main result:

Theorem 9. *Let $m = \lfloor n^\alpha \rfloor$, C_1 and C_2 be sufficiently large positive constants. If*

$$p \ge C_1 \frac{\log n}{m} \quad \text{for } 0 < \alpha < 1$$
$$p \ge C_2 \sqrt{\frac{\log n}{nm}} \quad \text{for } \alpha > 1$$

then almost all $G_{n,m,p}$ is Hamiltonian. Our bounds are asymptotically tight.

Note: The probability of the events that Theorem 9 implies, is at least, that of Theorem 1 multiplied by the probability of the similarity between the models D_p and I_p. This, in fact, gives stronger probability bounds than simple convergence to certainty as n grows.

References

1. N. Alon and J. H. Spencer *The Probabilistic Method*, Second Edition, John Wiley & Sons, Inc, 2000.
2. B. Bollobas, *Random Graphs*, Academic Press. Inc., New York, 1985.
3. C. Efthymiou, P.G. Spirakis *On the Existence of Hamiltonian Cycles in Random Intersection Graphs (full version)*, R.A.C.T.I. Technical Report TR2005/04/02, http://www.cti.gr. Also, DELIS Technical Report, http://delis.upb.de/docs/?subproject=2.
4. J. A. Fill, E. R. Scheinerman and K. B. Singer-Cohen *Random intersection graphs when $m = \omega(n)$: an equivalence theorem relating the evolution of the $G(n,m,p)$ and $G(n,p)$ models*, Random Structures and Algorithms 16, 156-176, (2000).
5. M. Karoński, E. R. Scheinerman and K. Singer-Cohen *On Random Intersection Graphs: The Subgraph Problem* Combinatorics, Probability and Computing, 8 131-159, 1999.
6. J. Komlós, and Szemerédi, *Limit Distributions for the existence of Hamilton cycles in a random graph*, Discrete Math. 43, 55-63, 1983.
7. A. D. Korshunov *Solution of a problem of P. Erdös and A. Rényi on Hamilton Cycles in non-oriented graphs*, Metody Diskr. Anal. Teoriy Upr. Syst. Sb. Trubov Novosibrirsk 31, 17-56, 1977.
8. E. Marczewski, *Sur deux propriétés des classes d'ensemble*, Fun. Math. 33, 303-307 (1945).
9. B. D. McKay and N. C. Wormald, *The degree sequence of a random graph. I. The models*, Random Structures and Algorithms 11 (1997), 97-117
10. S. Nikoletseas, C. Raptopoulos and P. Spirakis *The existence and Efficient construction of Large Independent Sets in General Random Intersection Graphs*, in the Proc. of the 31st International Colloquium on Automata, Languages and Programming, Lecture Notes in Computer Science Vol. 3142 (Springer Verlag) pp 1029-1040, 2004.
11. S. M. Ross, Stochastic Processes, 2nd Edition. *John Wiley and Sons, Inc.*, 1995.
12. K. Singer, *Random Intersection Graphs*, Ph.D. thesis, The Johns Hopkins University, 1995.

Optimal Cover Time for a Graph-Based Coupon Collector Process

Nedialko B. Dimitrov* and C. Greg Plaxton**

University of Texas at Austin,
1 University Station C0500,
Austin, Texas 78712–0233
{ned, plaxton}@cs.utexas.edu

Abstract. In this paper we study the following covering process defined over an arbitrary directed graph. Each node is initially uncovered and is assigned a random integer rank drawn from a suitable range. The process then proceeds in rounds. In each round, a uniformly random node is selected and its lowest-ranked uncovered outgoing neighbor, if any, is covered. We prove that if each node has in-degree $\Theta(d)$ and out-degree $O(d)$, then with high probability, every node is covered within $O(n + \frac{n \log n}{d})$ rounds, matching a lower bound due to Alon. Alon has also shown that, for a certain class of d-regular expander graphs, the upper bound holds no matter what method is used to choose the uncovered neighbor. In contrast, we show that for arbitrary d-regular graphs, the method used to choose the uncovered neighbor can affect the cover time by more than a constant factor.

1 Introduction

One of the most commonly discussed stochastic processes in computer science is the so-called coupon collector process [7]. In that process, there are n distinct coupons and we proceed in rounds, collecting one uniformly random coupon (with replacement) in each round. Are $O(n)$ rounds sufficient to collect all of the coupons? Put differently, is picking coupons with replacement as efficient, to within a constant factor, as picking them without replacement? No, it is a well-known fact that with high probability the number of rounds required to collect all of the coupons is $\Theta(n \log n)$.

This shortcoming has motivated Adler et al. [1] and Alon [2] to study a similar graph-based covering process. The nodes of the graph nodes represent the coupons and covering a node represents collecting a coupon. In each round, a uniformly random node w is selected. If an uncovered neighbor of w exists, choose one such uncovered neighbor and cover it. We refer to this process as process CC.

Process CC can use a variety of different *covering methods* to decide which uncovered neighbor to cover. If our ultimate goal is to minimize cover time, certainly the most powerful covering method available is an offline method with knowledge of the entire

---

* Supported by an MCD Fellowship from the University of Texas at Austin.
** Supported by NSF Grants CCR–0310970 and ANI–0326001. Also affiliated with Akamai Technologies, Inc., Cambridge, MA 02142.

sequence of node selections and with infinite computing power. We refer to this powerful cover time minimizing version of process CC as process MIN. To achieve our $O(n)$ goal, it is natural to consider $\log n$-regular graphs since the work of Alon implies process MIN has an expected cover time of $\Omega(n + \frac{n \log n}{d})$ rounds on d-regular graphs [2].

1.1 Logarithmic-Degree Graphs

Another natural version of process CC — in which the covering method chooses a uniformly random uncovered neighbor, if any — was studied by Adler et al. [1] and by Alon [2]. We refer to this version of process CC as process UNI. Alon shows that for logarithmic-degree Ramanujan expander graphs, process UNI completes in $O(n)$ time, matching the lower bound for process MIN.

Adler et al. show that for the hypercube, which has a weak expansion property but is not an expander, process UNI takes $O(n)$ time, also matching the lower bound for process MIN [1]. They also show that for arbitrary logarithmic-degree graphs, process UNI completes in $O(n \log \log n)$ time. Furthermore, Adler et al. present an application of process UNI to load balancing in a hypercubic distributed hash table (DHT).

A process that is intuitively similar to process UNI is one where we initially assign a rank to each node using a uniformly random permutation of the nodes, and the covering method covers the minimum-rank uncovered neighbor, if any. We refer to this permutation-based version of process CC as process P-RANK. In this paper, we show that process P-RANK completes in $O(n)$ time on arbitrary logarithmic-degree graphs.

In fact, we analyze a more general and local version of process CC in which each node initially chooses a uniformly random rank in a suitable range, and the covering method covers the minimum-rank uncovered neighbor of the selected node. (We assume that the nodes are numbered from 1 to n, and that ties in rank are broken in favor of the lower-numbered node.) We refer to this random rank version of process CC as process R-RANK. We show that the more general and local process R-RANK completes in $O(n)$ time on arbitrary logarithmic-degree graphs.

1.2 Results for General Graphs

Alon shows that process MIN on any d-regular graph has expected cover time at least $n - \frac{n}{d} + \frac{n}{d} \ln(\frac{n}{d})$ [2]. Alon also shows that process UNI completes in time $n + (1 + o(1)) \frac{n \ln n}{d}$ for random nearly d-regular graphs. Alon further shows that on any (n, d, λ)-expander graph the expected cover time of process UNI is at most $n + n(\frac{\lambda}{d})^2 (\ln n + 1)$. In particular, this implies that on Ramanujan graphs process UNI completes in $(1 + o(1))n$ time, matching the lower bound for process MIN.

If our goal is to maximize cover time, certainly the most powerful covering method available is an offline adversary with knowledge of the entire sequence of node selections and with infinite computing power. We refer to this powerful cover time maximizing version of process CC as process MAX. Alon notes that the upper bounds for expanders hold even if after every round an adversary "is allowed to shift the uncovered nodes to any place he wishes, keeping their number." In particular, this shows that on Ramanujan graphs, the cover time for process MAX matches the cover time for process MIN, up to constant factors. In effect, the covering method does not matter for this class of graphs.

Another previously studied variant of process CC favors covering the selected node. In this variant, we check — immediately after selecting a uniformly random node — if the selected node is uncovered. If it is, we cover it and move to the next selection. Only otherwise do we consider the neighbors of the selected node. We refer to the selection-biased variants of processes UNI, P-RANK, and R-RANK as UNI', P-RANK', and R-RANK', respecively.

Adler et al. show that for all d-regular graphs, processes UNI and UNI' finish in $O(n + n(\log n)(\log d)/d)$ time[1]. They also show that for random d-regular graphs only $O(n + \frac{n \log n}{d})$ steps are needed. Furthermore, they exhibit an application of process UNI' to load balancing in DHTs.

All of the results matching Alon's lower bound for process MIN presented prior to this work have used some expansion properties of the underlying graph. In contrast, our proof techniques do not require the underlying graph to have any particular structure. Thus, we show the following general result: for directed graphs, with self-loops but no parallel edges, where each node has in-degree at least δ_{in} and at most Δ_{in}, and out-degree at most Δ_{out}, both process R-RANK and process R-RANK' cover all nodes in $O(n \max(\Delta_{in}\Delta_{out}/\delta_{in}^2, (\log n)/\delta_{in}))$ rounds with high probability. This result matches Alon's lower bound for $\delta_{in} = \Delta_{in} = \Delta_{out} = \Theta(d)$, and is thus optimal under these conditions.

Furthermore, Alon's results for Ramanujan graphs raise the question whether there is any separation between the cover times for process MAX and process MIN. In other words, are there any graphs for which the choice of covering method matters? We define a weakly adversarial process, process A-RANK, that is similar to process P-RANK. In process A-RANK, instead of picking a uniformly random permutation, an adversary is initally allowed to fix the permutation used to assign ranks to the nodes. We then proceed as in process P-RANK. In addition, we define the selection-biased variant of process A-RANK as process A-RANK'. We establish that there exists a logarithmic-degree graph on which process A-RANK and process A-RANK' each take $\omega(n)$ rounds to complete. This implies that in general there is separation between the cover times of process MIN and process MAX. In other words, the covering method does matter. Due to space limitations, the proofs of our $\omega(n)$ lower bounds for processes A-RANK and A-RANK' are omitted from this paper; these proofs may be found in [4].

1.3 Proof Outline

The proof of our theorem is inspired by the delay sequence argument used by Ranade for the analysis of a certain packet routing problem on the butterfly [8] (see also [6]). In a delay sequence argument, we identify certain combinatorial structures that exist whenever the random process lasts for a long time. Then, we show that the probability any of these structures exist is small. This in turn implies an upper bound on the running time of the random process.

There are significant differences between our proof and that of Ranade. For example, in our problem, the connection between the running time and the length of a delay sequence is not clear-cut, while in the butterfly routing problem analyzed by Ranade, the length of the delay sequence is equal to the running time. But let us begin by giving the notion of a delay sequence in our problem.

Consider the node that was covered last, say w_1. Why wasn't w_1 covered earlier? It was not covered earlier because at the last opportunity to cover w_1 — that is, the last selection in w_1's neighborhood — we covered some other node, w_2, instead. In such a case we consider w_1 to be delayed by w_2. Similarly, w_2 may be delayed by some node w_3, et cetera, until finally we reach a node w_k that is not delayed, i.e., w_k is covered at the first opportunity. The sequence of nodes $w_1, \ldots, w_k$ corresponds to our notion of a delay sequence.

In analyzing process R-RANK, we find it useful to first analyze a much simpler process, process SELECT, in which we repeatedly select a uniformly random node, never covering anything. After establishing several lemmas for the simpler process, we proceed to analyzing process R-RANK. This is the bulk of the proof, and includes a technical lemma to work around the difficulties in linking cover time to delay sequence length. Finally, we reduce process R-RANK' to process R-RANK to show that the same bounds hold.

Due to space limitations, we omit our analysis of process R-RANK' from the present paper. In [4], we analyze process R-RANK' via a reduction from process R-RANK; in addition, we establish the existence of a logarithmic-degree graph on which processes A-RANK and A-RANK' each take $\omega(n)$ rounds to complete, establishing that the covering method does matter.

The rest of this paper is structured as follows. In Section 2, we present a number of useful definitions and lemmas related to standard probability distributions. In Section 3, we analyze process SELECT. In Section 4, we analyze process R-RANK using the results in Section 3.

2 Preliminaries

We use the term ℓ-sequence to refer to a sequence of length ℓ. For any ℓ-sequence σ of elements of a given type, and any element x of the same type, we let $\sigma : x$ denote the $(\ell + 1)$-sequence obtained by appending element x to σ.

For any nonnegative integer n and probability p, we use the notation $X \sim \text{Bin}(n, p)$ to denote that the random variable X has a binomial distribution with n trials and success probability p. Similarly, we write $X \sim \text{Geo}(p)$ to indicate that the random variable X has a geometric distribution with success probability p, and we write $X \sim \text{NegBin}(r, p)$ to indicate that the random variable X has a negative binomial distribution with r successes and success probability p. Due to space limitations, we include proofs for only two of the lemmas stated in this section. The other lemmas follow from simple arguments involving independence of random variables or tail bounds for the binomial distribution [4].

Lemma 1. *Let p denote an arbitrary probability, let ℓ denote an arbitrary nonnegative integer, and let $X \sim \text{NegBin}(\ell, p)$. For any integer j such that $1 \leq j \leq \ell$, let p_j denote an arbitrary probability such that $p_j \geq p$, let $Y_j \sim \text{Geo}(p_j)$, and let $Y = \sum_{1 \leq j \leq \ell} Y_j$. Then for any nonnegative integer i, $\Pr(X \geq i) \geq \Pr(Y \geq i)$.*

Proof. Note that if $p_j = p$ for all j, then the random variables X and Y have the same distribution. Furthermore, increasing any of the p_j's can only decrease Y. □

Lemma 2. *For any nonnegative integers r and n, and any probability p, we have $\Pr(X < r) = \Pr(Y > n)$, where $X \sim \text{Bin}(n,p)$ and $Y \sim \text{NegBin}(r,p)$.*

Proof. The random variables X and Y can be seen as different views of the same experiment where we successively flip coins with probability of success p. With Y, we ask "How many flips are required for r successes?" With X, we ask "How many successes are in the first n flips?" In this experiment, the event of seeing less than r successes in the first n flips ($X < r$) corresponds to the event that we have to wait more than n flips for the first r successes ($Y > n$). This gives the result. □

Lemma 3. *For any integer $r \geq 2$, $\Pr(X \geq 2E[X]) = \Pr(X \geq 2r/p) \leq \exp(-r/8)$, where $X \sim \text{NegBin}(r,p)$.*

Proof. Let $j = \lfloor \frac{2r}{p} \rfloor - 1$ and let $Y \sim \text{Bin}(j,p)$. By Lemma 2, we know that $\Pr(X \geq \frac{2r}{p}) \leq \Pr(X \geq \lfloor \frac{2r}{p} \rfloor) = \Pr(X > \lfloor \frac{2r}{p} \rfloor - 1) = \Pr(Y < r) = \Pr(Y \leq r - 1)$.

$$\Pr\left(Y \leq \frac{jp}{2}\right) = \Pr\left(Y \leq r - (\eta + 1)\frac{p}{2}\right)$$
$$= \Pr(Y \leq r - 1)$$

where $\frac{2r}{p} = \lfloor \frac{2r}{p} \rfloor + \eta$ and the last equality holds because $0 < (\eta+1)\frac{p}{2} < 1$.

Recall the Chernoff bounds in the form $\Pr(Y \leq (1-\lambda)jp) \leq \exp(-\lambda^2 jp/2)$ for $0 < \lambda < 1$ (see [3, 5]).

We apply this bound with $\lambda = \frac{1}{2}$ to get

$$\Pr(Y \leq r - 1) = \Pr(Y \leq jp/2)$$
$$\leq \exp(-jp/8)$$
$$\leq \exp\left(\frac{-2r + (\eta+1)p}{8}\right)$$
$$\leq \exp(-r/8)$$

where η is as previously defined and the last inequality holds because $r \geq 2$. □

Lemma 4. *Let p be an arbitrary probability and let X be the sum of n independent Bernoulli variables $X_1, \ldots, X_n$, where X_j has success probability $p_j \geq p$. Then $\Pr(X \leq np/2) \leq \exp(-np/12)$.*

Proof. The result follows from Chernoff bounds (see, e.g., [3, 5]). □

Lemma 5. *Suppose we repeatedly throw balls independently and uniformly at random into n bins, and let the random variable X denote the number of throws required for every bin to receive at least n balls. Then X is $O(n^2)$ with high probability, that is, with failure probability that is an arbitrary inverse polynomial in n.*

Proof. The result follows from Lemma 4. □

3 Process SELECT

Throughout the remainder of the paper, we fix an arbitrary directed graph $G = (V, E)$ where $|V| = n > 0$. We say that an event holds "with high probability" if the probability that it fails to occur is upper bounded by an arbitrary inverse polynomial in n. We let δ_{in}, Δ_{in}, and Δ_{out} denote the minimum in-degree, maximum in-degree, and maximum out-degree of any node, respectively. For ease of exposition, we assume throughout the paper that $\delta_{\text{in}} > 0$. The edge set E is allowed to contain loops but not parallel edges. For any node v, we define $\Gamma_{\text{in}}(v)$ as $\{w \mid (w, v) \in E\}$. For any sequence of edges $\sigma = (u_1, v_1), \ldots, (u_\ell, v_\ell)$, we define the two sequences of nodes $\text{src}(\sigma) = u_1, \ldots, u_\ell$ and $\text{dst}(\sigma) = v_1, \ldots, v_\ell$.

In this section, we analyze a simple stochastic process, process SELECT, defined as follows. Initially, we fix a positive integer r and independently assign each node in V a uniformly random integer rank between 1 and r. Process SELECT then proceeds in an infinite number of rounds, indexed from 1. In each round, one node is selected uniformly at random, with replacement. The following definitions are central to our analysis of this process.

A node sequence is said to be *rank-sorted* if the associated sequence of node ranks is nondecreasing.

For any node sequence σ, we inductively define a nonnegative integer $\text{duration}(\sigma)$ and a node sequence $\text{select}(\sigma)$ as follows. If σ is empty, then $\text{duration}(\sigma)$ is 0 and $\text{select}(\sigma)$ is empty. Otherwise, σ is of the form $\tau : v$ for some shorter node sequence τ and node v. Let i denote the the least i such that $i > \text{duration}(\tau)$ and the node selected in round i belongs to $\Gamma_{\text{in}}(v)$. Let u denote the node selected in round i. Then we define $\text{duration}(\sigma)$ as i, and $\text{select}(\sigma)$ as $\text{select}(\tau) : u$.

Lemma 6. *For any ℓ-sequence of distinct nodes σ, $\Pr(\sigma \text{ is rank-sorted}) = \binom{\ell+r-1}{\ell} r^{-\ell}$.*

Proof. There are $\binom{\ell+r-1}{\ell}$ ways that ranks can be assigned to the ℓ distinct nodes so that the resulting ℓ-sequence is rank-sorted. The result follows since each such assignment occurs with probability $r^{-\ell}$. □

Lemma 7. *For any ℓ-sequence of nodes $\sigma = v_1, \ldots, v_\ell$ and any nonnegative integer i, we have*
$\Pr(\text{duration}(\sigma) = i) \le \Pr(X \ge i)$*, where $X \sim \text{NegBin}\left(\ell, \frac{\delta_{\text{in}}}{n}\right)$.*

Proof. We proceed by proving that

$$\Pr(\text{duration}(\sigma) = i) = \Pr\left(\sum_{k=1}^{\ell} Y_k = i\right)$$

where $Y_k \sim \text{Geo}\left(\frac{d_k}{n}\right)$ and d_k denotes the in-degree of v_k. The desired bound then follows by Lemma 1.

We prove the foregoing claim by induction on ℓ. If $\ell = 0$, the claim holds since $\text{duration}(\sigma) = \sum_{k=1}^{\ell} Y_k = 0$.

For $\ell > 0$, we let τ denote the node sequence $v_1, \ldots, v_{\ell-1}$ and assume inductively that

$$\Pr(\text{duration}(\tau) = i) = \Pr\left(\sum_{k=1}^{\ell-1} Y_k = i\right).$$

Thus,

$$\Pr(\text{duration}(\sigma) = i) = \sum_{j=0}^{i-1} \Pr(\text{duration}(\tau) = j) \cdot \Pr(\text{duration}(\sigma) - \text{duration}(\tau)$$
$$= i - j \mid \text{duration}(\tau) = j)$$
$$= \sum_{j=0}^{i-1} \Pr(\text{duration}(\tau) = j) \cdot \Pr(\text{duration}(\sigma) - \text{duration}(\tau)$$
$$= i - j)$$
$$= \sum_{j=0}^{i-1} \Pr(\text{duration}(\tau) = j) \cdot \Pr(Y_\ell = i - j)$$
$$= \sum_{j=0}^{i-1} \Pr\left(\sum_{k=1}^{\ell-1} Y_k = j\right) \cdot \Pr(Y_\ell = i - j)$$
$$= \Pr\left(\sum_{k=1}^{\ell} Y_k = i\right).$$

The second equality holds because each selection is independent of previous selections. The third equality holds because the waiting time to obtain a selection in $\Gamma_{\text{in}}(v_\ell)$ is distributed as Y_ℓ. □

Lemma 8. *For any ℓ-sequence of edges σ, $\Pr(\text{select}(\text{dst}(\sigma)) = \text{src}(\sigma)) \leq \delta_{\text{in}}^{-\ell}$.*

Proof. We proceed by induction on ℓ. For $\ell = 0$, $\Pr(\text{select}(\text{dst}(\sigma)) = \text{src}(\sigma)) = 1 = \delta_{\text{in}}^0$ since we have assumed that $\delta_{\text{in}} > 0$.

For $\ell > 0$, σ can be written in the form $\tau : (u, v)$, where we inductively assume that the claim of the lemma holds for τ. Let A denote the event that the first node selected in $\Gamma_{\text{in}}(v)$ after round $\text{duration}(\text{dst}(\tau))$ is u. We have

$$\Pr(\text{select}(\text{dst}(\sigma)) = \text{src}(\sigma)) = \Pr(\text{select}(\text{dst}(\tau)) = \text{src}(\tau)) \cdot \Pr(A \mid \text{select}(\text{dst}(\tau)) = \text{src}(\tau))$$
$$= \Pr(\text{select}(\text{dst}(\tau)) = \text{src}(\tau)) \cdot \Pr(A)$$
$$\leq \delta_{\text{in}}^{-\ell}.$$

The second step follows from the independence of the events A and $\text{select}(\text{dst}(\sigma')) = \text{src}(\sigma')$. (These two events are independent since each selection is independent of previous selections.) The third step follows from the induction hypothesis and the observation that $\Pr(A)$ is equal $1/\Gamma_{\text{in}}(v)$, which is at most $1/\delta_{\text{in}}$. □

Lemma 9. *For any ℓ-sequence of edges σ and nonnegative integer i, the events $A =$ "$\mathrm{dst}(\sigma)$ is rank-sorted", $B =$ "$\mathrm{duration}(\mathrm{dst}(\sigma)) = i$", and $C =$ "$\mathrm{select}(\mathrm{dst}(\sigma)) = \mathrm{src}(\sigma)$" are mutually independent.*

Proof. Note that event A depends only on the rank assignments, while events B and C depend only on the selections. Thus event A is independent of events B and C. Below we argue that events B and C are independent.

Let $\sigma = (u_1, v_1), \ldots, (u_\ell, v_\ell)$ and let σ_j denote the length-j prefix of σ, $0 \le j \le \ell$. Define a selection to be j-*special*, $1 \le j \le \ell$, if it is the first selection after round $\mathrm{duration}(\sigma_{j-1})$ in $\Gamma_{\mathrm{in}}(v_j)$. A selection is *special* if it is j-special for some j. Note that event B depends only on the timing of the special events; in particular, B occurs if and only if the ℓ-special selection occurs in round i. Suppose we run process SELECT, but at each step, instead of revealing the selected node, we reveal only whether the selection is special. This information is sufficient to determine the unique i for which B occurs, but does not bias the distribution of $\mathrm{select}(\mathrm{dst}(\sigma))$. Since event C only depends on $\mathrm{select}(\mathrm{dst}(\sigma))$, it is independent of B. □

Lemma 10. *Let σ be an ℓ-sequence of edges so that the nodes of $\mathrm{dst}(\sigma)$ are distinct, let $X \sim \mathrm{NegBin}\left(\ell, \frac{\delta_{\mathrm{in}}}{n}\right)$, let i be a nonnegative integer, and let events A, B, and C be defined as in the statement of Lemma 9. Then $\Pr(A \cap B \cap C) \le \binom{\ell+r-1}{\ell} \cdot \Pr(X \ge i) \cdot (r\delta_{\mathrm{in}})^{-\ell}$.*

Proof. By Lemma 6, $\Pr(A) \le \binom{\ell+r-1}{\ell} r^{-\ell}$. By Lemma 7, $\Pr(B) \le \Pr(X \ge i)$. By Lemma 8, $\Pr(C) \le \delta_{\mathrm{in}}^{-\ell}$. The claim then follows by Lemma 9. □

4 Process R-RANK

In the section we analyze an augmented version of process SELECT, referred to as Process R-RANK, in which we maintain a notion of a "covered subset" of the nodes. Initially, all of the nodes are uncovered. Process R-RANK then proceeds in rounds in exactly the same manner as process SELECT, except that in any given round, if one or more outgoing neighbors of the selected node are uncovered, we cover the uncovered outgoing neighbor with minimum rank. (As indicated in Section 1, ties are broken according to some arbitrary numbering of the nodes.)

Note that process R-RANK simply augments process SELECT by also covering nodes; rank assignment and selections are performed in exactly the same manner in the two processes. Thus all of the definitions and lemmas presented in Section 3 are applicable to process R-RANK. The following additional definitions are useful for our analysis of process R-RANK.

The *cover time* of process R-RANK is defined as the number of rounds required to cover all of the nodes.

We inductively define the notion of a *linked* sequence of edges. For ℓ equal to 0 or 1, any ℓ-sequence of edges is linked. For $\ell > 1$, an ℓ-sequence of edges of the form $\sigma : (u,v) : (u',v')$ is linked if the $(\ell-1)$-sequence $\sigma : (u,v)$ is linked and (u,v') belongs to E.

For any node v, we define parent(v) as follows. Let i denote the round in which node v is covered. If i is the first round in which some node in $\Gamma_{\text{in}}(v)$ is selected, then parent(v) is defined to be nil. Otherwise, parent(v) is the node covered in the first round prior to round i in which the selected node belongs to $\Gamma_{\text{in}}(v)$.

We inductively define the notion of a *chronological* sequence of nodes as follows. Any ℓ-sequence of nodes with $\ell \leq 1$ is chronological. An ℓ-sequence of nodes of the form $\sigma : v : v'$ is chronological if $\sigma : v$ is chronological and node v is covered before node v'.

We inductively define the notion of an *active* node sequence as follows. The empty node sequence is active. A singleton node sequence consisting of the node v is active if parent(v) = nil. An ℓ-sequence of nodes of the form $\sigma : v : v'$ is active if $\sigma : v$ is active and parent(v') = v.

We call an ℓ-sequence of edges σ *active* if dst(σ) is active and select(dst(σ)) = src(σ).

We call an ℓ-sequence of edges σ *i-active* if it is active and either $\ell = i = 0$ or $\ell > 0$, σ is of the form $\sigma : (u, v)$, and v is the node covered in round i.

Lemma 11. *For any nonnegative integer ℓ, there are at most $n\Delta_{\text{out}}^{\ell} \Delta_{\text{in}}^{\ell-1}$ linked ℓ-sequences of edges.*

Proof. We proceed by induction on ℓ, treating $\ell = 0$ and $\ell = 1$ as the base cases. For $\ell = 0$, the empty sequence is the only linked 0-sequence, and the claim holds since $n/\Delta_{\text{in}} \geq 1$. (Note that Δ_{in} is at most n since we do not allow parallel edges.) For $\ell = 1$, the number of linked 1-sequences is at most $|E| \leq n\Delta_{\text{out}}$.

Now let ℓ be greater than 1 and inductively assume that the number of linked $(\ell-1)$-sequences of edges is at most $n\Delta_{\text{out}}^{\ell-1} \Delta_{\text{in}}^{\ell-2}$. Recall that any linked ℓ-sequence of edges is of the form $\sigma : (u, v) : (u', v')$ where the $(\ell - 1)$-sequence of edges $\sigma : (u, v)$ is linked and (u, v') belongs to E. Observe that for any linked $(\ell - 1)$-sequence of edges $\sigma : (u, v)$, there are at most Δ_{out} nodes v' such that (u, v') belongs to E, and for each such choice of v', there are at most Δ_{in} nodes u' such that (u', v') belongs to E. Thus the number of linked ℓ-sequences is at most $\Delta_{\text{out}} \Delta_{\text{in}}$ times the number of linked $(\ell - 1)$-sequences, and the desired bound follows from the induction hypothesis. □

Lemma 12. *Suppose we run two instances of process R-RANK in parallel using the same random ranks and the same sequence of random selections, but in the second instance, we allow an arbitrary subset of the covered nodes to be uncovered after each round. Then the cover time of the first instance is at most the cover time of the second instance.*

Proof. By a straightforward induction on the number of rounds, at all times, the set of covered nodes in the first instance contains the set of covered nodes in the second instance. The claim of the lemma follows. □

Lemma 13. *For any rank assignment, the expected cover time of process R-RANK is $O(n^2)$.*

Proof. It follows from Lemma 5 that the cover time is $O(n^2)$ with high probability since in that time each vertex is selected at least n times, implying that all of its neighbors are covered.

We can then consider a modified version of process R-RANK in which the infinite sequence of rounds is partitioned into epochs of $O(n^2)$ rounds, and where at the end of each epoch, if the nodes are not all covered, all nodes are uncovered before proceeding to the next epoch. Since each epoch covers all the nodes with high probability, the expected cover time of this modified version of process R-RANK is $O(n^2)$. By Lemma 12, for any rank assignment, the expected cover time of process R-RANK is $O(n^2)$. □

Lemma 14. *Assume that v is the node covered in round i and let u be the node selected in round i. Then there is an i-active edge sequence σ terminating in edge (u, v) and such that* $\mathrm{duration}(\mathrm{dst}(\sigma)) = i$.

Proof. Observe that u belongs to $\Gamma_{\mathrm{in}}(v)$. Furthermore, if $\mathrm{parent}(v) = \mathrm{nil}$, then the singleton node sequence v is active with $\mathrm{duration}(v) = i$. Thus the singleton edge sequence $\sigma = (u, v)$ is i-active with $\mathrm{duration}(\mathrm{dst}(\sigma)) = i$.

We prove the claim by induction on i. For $i = 1$, we have $\mathrm{parent}(v) = \mathrm{nil}$ and so the claim follows by the observations of the previous paragraph.

For $i > 1$, if $\mathrm{parent}(v) = \mathrm{nil}$, the claim once again follows from the foregoing observations. Otherwise, $\mathrm{parent}(v) = v'$ where v' is the node covered in round j with $j < i$. Let u' denote the node selected in round j. Since $j < i$, we can inductively assume that there is a j-active edge sequence, call it τ, terminating in edge (u', v') and such that $\mathrm{duration}(\mathrm{dst}(\tau)) = j$. Since τ is active, the node sequence $\mathrm{dst}(\tau)$ is active and $\mathrm{select}(\mathrm{dst}(\tau)) = \mathrm{src}(\tau)$. Let $\sigma = \tau : (u, v)$. Thus $\mathrm{src}(\sigma) = \mathrm{src}(\tau) : u$ and $\mathrm{dst}(\sigma) = \mathrm{dst}(\tau) : v$. Since $\mathrm{parent}(v) = v'$ and $\mathrm{dst}(\tau)$ is an active node sequence terminating in node v', $\mathrm{dst}(\sigma)$ is active. Since $\mathrm{duration}(\mathrm{dst}(\tau)) = j$, $\mathrm{select}(\mathrm{dst}(\tau)) = \mathrm{src}(\tau)$, u was selected in round i, and i is the least integer greater that j such that the node selected in round i belongs to $\Gamma_{\mathrm{in}}(v)$, we have $\mathrm{duration}(\mathrm{dst}(\sigma)) = i$ and $\mathrm{select}(\mathrm{dst}(\sigma)) = \mathrm{src}(\sigma)$. Since $\mathrm{dst}(\sigma)$ is active and $\mathrm{select}(\mathrm{dst}(\sigma)) = \mathrm{src}(\sigma)$, σ is active. Since σ is active and v is the node covered in round i, σ is i-active. Thus the edge sequence σ satisfies all of the requirements of the lemma. □

Lemma 15. *Any active node sequence is rank-sorted, chronological, and consists of distinct nodes.*

Proof. Note that any chronological node sequence consists of distinct nodes. Thus, in what follows, it is sufficient to prove that any active node sequence is rank-sorted and chronological.

We proceed by induction on the length of the sequence. For the base case, note that any node sequence of length 0 or 1 is rank-sorted and chronological. For the induction step, consider an active node sequence σ of the form $\tau : v : v'$. Since σ is active, $\tau : v$ is active and $\mathrm{parent}(v') = v$. Since $\tau : v$ is active, the induction hypothesis implies that it is also rank-sorted and chronological. Since $\mathrm{parent}(v') = v$, $\mathrm{rank}(v) \leq \mathrm{rank}(v')$ and v is covered before v'. Hence σ is rank-sorted and chronological. □

Lemma 16. *For any nonempty active edge sequence σ, if the last edge in σ is (u, v), then v is the node covered in round $\mathrm{duration}(\mathrm{dst}(\sigma))$ and node u is selected in the same round.*

Proof. We prove the claim by induction on the length of the active edge sequence σ.

If σ consists of a single edge (u, v), then by the definition of an active edge sequence, the singleton node sequence $\mathrm{dst}(\sigma)$ is active and $\mathrm{select}(\mathrm{dst}(\sigma)) = \mathrm{src}(\sigma)$. Since $\mathrm{dst}(\sigma)$ is active, $\mathrm{parent}(v) = \mathrm{nil}$, that is, v is the node covered in the first round in which a node in $\Gamma_{\mathrm{in}}(v)$ is selected, which is round $\mathrm{duration}(\mathrm{dst}(\sigma))$. Since $\mathrm{select}(\mathrm{dst}(\sigma)) = \mathrm{src}(\sigma)$, node u is selected in the same round.

Now assume that σ is an active edge sequence of the form $\tau : (u, v)$, where τ is of the form $\tau' : (u', v')$. Since σ is active, the node sequence $\mathrm{dst}(\sigma)$ is active and $\mathrm{select}(\mathrm{dst}(\sigma)) = \mathrm{src}(\sigma)$. It follows that $\mathrm{dst}(\tau)$ is active and $\mathrm{select}(\mathrm{dst}(\tau)) = \mathrm{src}(\tau)$, that is, τ is also active. Since τ is active and shorter than σ, we can inductively assume that v' is the node covered in round $\mathrm{duration}(\mathrm{dst}(\tau))$ and node u' is selected in the same round. Since $\mathrm{dst}(\sigma)$ is active, $\mathrm{parent}(v) = v'$, that is, v is the node covered in the first round after round $\mathrm{duration}(\mathrm{dst}(\tau))$ in which a node in $\Gamma_{\mathrm{in}}(v)$ is selected. Applying the definition of $\mathrm{duration}(\mathrm{dst}(\sigma))$, we conclude that v is the node covered in round $\mathrm{duration}(\mathrm{dst}(\sigma))$. Since $\mathrm{select}(\mathrm{dst}(\sigma)) = \mathrm{src}(\sigma)$, node u is selected in the same round. □

Lemma 17. *If σ is an active sequence of edges, then σ is linked.*

Proof. We proceed by induction on the length of σ. If the length of σ is 0 or 1, then σ is linked by definition.

Now assume that σ is an edge sequence of the form $\tau : (u, v)$, where τ is of the form $\tau' : (u', v')$ and σ is active. Since σ is active, $\mathrm{dst}(\sigma)$ is active. Since $\mathrm{dst}(\sigma)$ is active, $\mathrm{dst}(\tau)$ is also active. Since $\mathrm{dst}(\tau)$ is active and τ is shorter than σ, we can inductively assume that τ is linked. Therefore, in order to establish that σ is linked, it is sufficient to prove that (u', v) is an edge. Since $\mathrm{dst}(\sigma)$ is active, $\mathrm{parent}(v) = v'$. Hence, letting i denote the round in which node v is covered, we find that v' is the node covered in the first round prior to round i in which the selected node belongs to $\Gamma_{\mathrm{in}}(v)$. By Lemma 16, v' is covered in a round in which node u' is selected. Thus u' belongs to $\Gamma_{\mathrm{in}}(v)$, that is, (u', v) is an edge, as required. □

Lemma 18. *If an edge sequence σ is i-active, then $\mathrm{duration}(\mathrm{dst}(\sigma)) = i$.*

Proof. If σ is empty, then the claim holds since $i = 0$ and $\mathrm{duration}(\mathrm{dst}(\sigma)) = 0$. Otherwise, σ is of the form $\tau : (u, v)$, and by the definition of an i-active edge sequence, v is the node covered in round i. By Lemma 16, v is the node covered in round $\mathrm{duration}(\mathrm{dst}(\sigma))$, so $\mathrm{duration}(\mathrm{dst}(\sigma)) = i$. □

Lemma 19. *For any ℓ-sequence of edges σ, and any nonnegative integer i, the probability that σ is i-active is at most $\binom{\ell+r-1}{\ell} \cdot \Pr(X \geq i) \cdot (r\delta_{\mathrm{in}})^{-\ell}$, where $X \sim \mathrm{NegBin}\left(\ell, \frac{\delta_{\mathrm{in}}}{n}\right)$.*

Proof. If the nodes in $\mathrm{dst}(\sigma)$ are not all distinct, then $\Pr(\sigma \text{ is } i\text{-active}) = 0$ by Lemma 15 and the claimed inequality holds since the right-hand side is nonnegative.

Now assume that $\mathrm{dst}(\sigma)$ consists of distinct nodes, and let events A, B, and C be as defined in the statement of Lemma 9. Below we prove that if σ is i-active, then events A, B, and C all occur. The claimed inequality then follows by Lemma 10.

Assume that σ is i-active. Thus event B occurs by Lemma 18. Furthermore, σ is active, so $\mathrm{dst}(\sigma)$ is active and event C occurs by the definition of an active edge sequence. Since $\mathrm{dst}(\sigma)$ is active, event A occurs by Lemma 15. $\square$

Lemma 20. *For any nonnegative integers i and ℓ, the probability that some ℓ-sequence of edges is i-active is at most*

$$n\Delta_{\mathrm{out}}^{\ell}\Delta_{\mathrm{in}}^{\ell-1}\binom{\ell+r-1}{\ell}\frac{\Pr(X\geq i)}{(r\delta_{\mathrm{in}})^{\ell}}$$

where $X \sim \mathrm{NegBin}\left(\ell, \frac{\delta_{\mathrm{in}}}{n}\right)$.

Proof. By Lemma 17, if an edge sequence σ is not linked, then $\Pr(\sigma \text{ is } i\text{-active}) = 0$. A union bound then implies that the probability some ℓ-sequence of edges is i-active is at most the number of linked ℓ-sequences of edges multiplied by the maximum probability that any particular ℓ-sequence is i-active. The desired inequality then follows by Lemmas 11 and 19. $\square$

Lemma 21. *For nonnegative integers i, ℓ, and r such that $i \geq 64n\max(\Delta_{\mathrm{out}}\Delta_{\mathrm{in}}/\delta_{\mathrm{in}}^2, (\ln n)/\delta_{\mathrm{in}})$ and $r \geq \min(\lceil 2e^2\Delta_{\mathrm{out}}\Delta_{\mathrm{in}}/\delta_{\mathrm{in}}\rceil, \ell)$, we have*

$$\Delta_{\mathrm{out}}^{\ell}\Delta_{\mathrm{in}}^{\ell-1}\binom{\ell+r-1}{\ell}\frac{\Pr(X\geq i)}{(r\delta_{\mathrm{in}})^{\ell}} \leq \exp(-i\delta_{\mathrm{in}}/(32n))$$

where $X \sim \mathrm{NegBin}\left(\ell, \frac{\delta_{\mathrm{in}}}{n}\right)$.

Proof. First, we show that the LHS of the claimed inequality is a nonincreasing function of r.

It is sufficient to prove that the expression $\binom{\ell+r-1}{\ell}r^{-\ell}$ is a nonincreasing function of r. Fix ℓ and let $f(r)$ denote the preceding expression. Note that

$$\frac{f(r+1)}{f(r)} = \frac{r+\ell}{r}\left(\frac{r}{r+1}\right)^{\ell}$$

$$= \left(1+\frac{\ell}{r}\right)\left(1+\frac{1}{r}\right)^{-\ell}$$

$$\leq 1,$$

where the last inequality holds since the binomial theorem implies $(1+\frac{1}{r})^{\ell} \geq 1+\frac{\ell}{r}$.

Since we have established that the LHS of the claimed inequality is a nonincreasing function of r, we can assume in what follows that $r = \min(\lceil 2e^2\Delta_{\mathrm{out}}\Delta_{\mathrm{in}}/\delta_{\mathrm{in}}\rceil, \ell)$.

Let us rewrite the LHS of the claimed inequality as $\lambda \cdot \Pr(X \geq i)$, where

$$\lambda = \Delta_{\mathrm{out}}^{\ell}\Delta_{\mathrm{in}}^{\ell-1}\binom{\ell+r-1}{\ell}(r\delta_{\mathrm{in}})^{-\ell}$$

$$\leq \Delta_{\mathrm{out}}^{\ell}\Delta_{\mathrm{in}}^{\ell}\left(\frac{e(\ell+r-1)}{\ell r\delta_{\mathrm{in}}}\right)^{\ell}$$

$$\leq \left(\frac{e\Delta_{\mathrm{out}}\Delta_{\mathrm{in}}(\ell+r)}{\ell r\delta_{\mathrm{in}}}\right)^{\ell}. \qquad (1)$$

We begin by establishing two useful upper bounds on λ, namely, Equations (2) and (4) below.

If $r = \lceil 2e^2 \Delta_{out} \Delta_{in}/\delta_{in} \rceil$, then since since $r = \min(\lceil 2e^2 \Delta_{out} \Delta_{in}/\delta_{in} \rceil, \ell)$, we have $r \leq \ell$. Substituting the value of r into Equation (1), we find that

$$\lambda \leq \left(\frac{e(\ell + r)}{2e^2 \ell} \right)^\ell$$

$$\leq \left(\frac{2e\ell}{2e^2 \ell} \right)^\ell$$

$$\leq e^{-\ell}. \tag{2}$$

If $r = \ell$, then Equation (1) implies

$$\lambda \leq \left(\frac{2e \Delta_{out} \Delta_{in}}{\ell \delta_{in}} \right)^\ell. \tag{3}$$

Let $h(\ell)$ denote the natural logarithm of the RHS of Equation (3), that is, $h(\ell) = \ell \ln(2e \Delta_{out} \Delta_{in}/(\ell \delta_{in}))$. Using elementary calculus, it is straightforward to prove that the derivative of $h(\ell)$ with respect to ℓ is positive for $\ell < 2 \Delta_{out} \Delta_{in}/\delta_{in}$, is 0 when $\ell = 2\Delta_{out} \Delta_{in}/\delta_{in}$, and is negative for $\ell > 2\Delta_{out} \Delta_{in}/\delta_{in}$. It follows that $h(\ell) \leq h(2\Delta_{out}\Delta_{in}/\delta_{in}) = 2\Delta_{out}\Delta_{in}/\delta_{in}$. Since ln is monotonic, the RHS of Equation (3) is also maximized when $\ell = 2\Delta_{out}\Delta_{in}/\delta_{in}$. Combining this result with Equation (2), we find that for any r

$$\lambda \leq \exp(2\Delta_{out}\Delta_{in}/\delta_{in}). \tag{4}$$

(Note that $\exp(2\Delta_{out}\Delta_{in}/\delta_{in}) \geq 1$ and Equation (2) implies $\lambda \leq 1$ when $r = \lceil 2e^2 \Delta_{out}\Delta_{in}/\delta_{in} \rceil$.)

We are now ready to proceed with the proof of the lemma. We consider the two cases $\ell > \lceil i\delta_{in}/(2n) \rceil$ and $\ell \leq \lceil i\delta_{in}/(2n) \rceil$ separately.

If $\ell > \lceil i\delta_{in}/(2n) \rceil$, then $\ell > 2ec \max(\Delta_{out}\Delta_{in}/\delta_{in}, \ln n)$ where $c = 16/e > e$. Thus $\ell > \lceil 2e^2 \Delta_{out}\Delta_{in}/\delta_{in} \rceil$ and so $r = \lceil 2e^2 \Delta_{out}\Delta_{in}/\delta_{in} \rceil$. It follows from Equation (2) that $\lambda \leq e^{-\ell} \leq \exp(-i\delta_{in}/(2n)) \leq \exp(-i\delta_{in}/(64n))$, and hence the claim holds since $\Pr(X \geq i) \leq 1$.

Now assume that $\ell \leq \lceil i\delta_{in}/(2n) \rceil$. Let $Y \sim \text{NegBin}(\lfloor \frac{i\delta_{in}}{2n} \rfloor, \frac{\delta_{in}}{n})$ and $Z \sim \text{NegBin}(\lfloor \frac{i\delta_{in}}{2n} \rfloor - \ell, \frac{\delta_{in}}{n})$. By the definition of the negative binomial distribution, $\Pr(Y \geq i) = \Pr(X + Z \geq i)$. And, since Z is nonnegative, $\Pr(X + Z \geq i) \geq \Pr(X \geq i)$. Thus

$$\Pr(X \geq i) \leq \Pr(Y \geq i). \tag{5}$$

Since $E[Y] \leq \frac{i}{2}$ and $\lfloor i\delta_{in}/(2n) \rfloor \geq \lfloor 32 \max(\Delta_{out}\Delta_{in}/\delta_{in}, \ln n) \rfloor > 2$, Lemma 3 implies $\Pr(Y \geq i) \leq \Pr(Y \geq 2E[Y]) \leq \exp\left(\frac{-i\delta_{in}}{16n} + \frac{1}{8}\right)$. The claim follows since

$$\lambda \cdot \Pr(X \geq i) \leq \exp\left(\frac{2\Delta_{\text{out}}\Delta_{\text{in}}}{\delta_{\text{in}}}\right) \cdot \Pr(Y \geq i)$$

$$\leq \exp\left(\frac{-i\delta_{\text{in}}}{16n} + \frac{1}{8} + \frac{2\Delta_{\text{out}}\Delta_{\text{in}}}{\delta_{\text{in}}}\right)$$

$$\leq \exp\left(\frac{-i\delta_{\text{in}}}{32n} + \frac{1}{8}\right)$$

$$\leq \exp\left(\frac{-i\delta_{\text{in}}}{64n}\right).$$

(The first step follows from Equations (4) and (5). For the third step and fourth steps, note that the assumption $i \geq 64n\max(\Delta_{\text{out}}\Delta_{\text{in}}/\delta_{\text{in}}^2, (\ln n)/\delta_{\text{in}})$ implies $i\delta_{\text{in}}/(32n) \geq 2\Delta_{\text{out}}\Delta_{\text{in}}/\delta_{\text{in}}$ and $i\delta_{\text{in}}/(64n) \geq 1/8$, respectively.) □

Lemma 22. *If $r \geq \min(\lceil 2e^2\Delta_{\text{out}}\Delta_{\text{in}}/\delta_{\text{in}}\rceil, n)$, then every active edge sequence is, with high probability, $O(n\max(\Delta_{\text{out}}\Delta_{\text{in}}/\delta_{\text{in}}^2, (\log n)/\delta_{\text{in}}))$-active.*

Proof. Let c denote an arbitrary positive real greater than or equal to 1, and let i denote the positive integer $\lceil 64cn\max(\Delta_{\text{out}}\Delta_{\text{in}}/\delta_{\text{in}}^2, (\ln n)/\delta_{\text{in}})\rceil$.

For any nonnegative integer j, let p_j denotes the probability that there is a j-active edge sequence. Any j-active edge sequence σ is active, so the associated node sequence $\text{dst}(\sigma)$ is active. It follows from Lemma 15 that any j-active sequence has length at most n. In other words, $\ell \leq n$ for any j-active ℓ-sequence of edges. Furthermore, if $j > 0$ then the length of a j-active sequence is nonzero. Since any j-active ℓ-sequence of edges satisfies $\ell \leq n$, the condition $r = \min(\lceil 2e^2\Delta_{\text{out}}\Delta_{\text{in}}/\delta_{\text{in}}\rceil, n)$ allows us to apply Lemmas 20 and 21. Applying these two lemmas, together with a union bound, we obtain $p_j \leq n^2 \exp(-j\delta_{\text{in}}/(64n))$ for $j > i$.

Let p denote the probability that there is a j-active edge sequence for some $j \geq i$. By a union bound, $p \leq \sum_{j \geq i} p_j$. Using the upper bound on p_j derived in the preceding paragraph, we find that p is upper bounded by an infinite geometric sum with initial term $n^2 \exp(-i\delta_{\text{in}}/(64n))$ and ratio $\exp(-\delta_{\text{in}}/(64n))$. Thus

$$p = O((n^3/\delta_{\text{in}})\exp(-i\delta_{\text{in}}/(64n)))$$
$$= O(n^3 \exp(-c\max(\Delta_{\text{out}}\Delta_{\text{in}}/\delta_{\text{in}}, \log n)))$$
$$= O(n^{3-c}).$$

By setting c to a sufficiently large positive constant, we can drive p below any desired inverse polynomial threshold. The claim of the lemma follows. □

Lemma 23. *If $r \geq \min(\lceil 2e^2\Delta_{\text{out}}\Delta_{\text{in}}/\delta_{\text{in}}\rceil, n)$, then the cover time of process R-RANK is, with high probability, $O(n\max(\Delta_{\text{out}}\Delta_{\text{in}}/\delta_{\text{in}}^2, (\log n)/\delta_{\text{in}}))$. The same asymptotic bound holds for the expected cover time.*

Proof. The high probability claim is immediate from Lemmas 14 and 22. The bound on the expected cover time then follows by Lemma 13. □

Theorem 1. *If both Δ_{in} and Δ_{out} are $O(\delta_{\text{in}})$, then there is an r in $O(\delta_{\text{in}})$ such that the cover time of process R-RANK is $O(n + \frac{n\log n}{\delta_{\text{in}}})$ with high probability. The same asymptotic bound holds for the expected cover time.*

Proof. Immediate from Lemma 23. □

The result of Theorem 1 matches the lower bound proved by Alon for process MIN and is thus optimal [2].

Note that as r tends to infinity, the behavior of process R-RANK converges to that of process P-RANK. Thus, the bounds of Theorem 1 also hold for process P-RANK.

References

1. M. Adler, E. Halperin, R. Karp, and V. Vazirani. A stochastic process on the hypercube with applications to peer-to-peer networks. In *Proceedings of the 35th Annual ACM Symposium on Theory of Computing*, pages 575–584, 2003.
2. N. Alon. Problems and results in extremal combinatorics, II. Manuscript, 2004.
3. N. Alon and J. H. Spencer. *The Probabilistic Method*. Wiley, New York, NY, 1991.
4. Nedialko B. Dimitrov and C. Greg Plaxton. Optimal cover time for a graph-based coupon collector process. Technical Report TR–05–01, Department of Computer Science, University of Texas at Austin, January 2005.
5. Stasys Jukna. *Extremal Combinatorics*, pages 224–225. Springer, 2001.
6. F. T. Leighton. *Introduction to Parallel Algorithms and Architectures: Arrays, Trees, and Hypercubes*, pages 547–556. Morgan-Kaufmann, San Mateo, CA, 1991.
7. R. Motwani and P. Raghavan. *Randomized Algorithms*. Cambridge University Press, Cambridge, UK, 1995.
8. A. G. Ranade. How to emulate shared memory. *Journal of Computer and System Sciences*, 42:307–326, 1991.

Stability and Similarity of Link Analysis Ranking Algorithms *

Debora Donato[1], Stefano Leonardi[2], and Panayiotis Tsaparas[3]

[1] Universita di Roma,"La Sapienza",
donato@dis.uniroma1.it
[2] Universita di Roma,"La Sapienza",
leon@dis.uniroma1.it
[3] University of Helsinki,
tsaparas@cs.helsinki.fi

Abstract. Recently, there has been a surge of research activity in the area of *Link Analysis Ranking*, where hyperlink structures are used to determine the relative *authority* of Web pages. One of the seminal works in this area is that of Kleinberg [15], who proposed the HITS algorithm. In this paper, we undertake a theoretical analysis of the properties of the HITS algorithm on a broad class of random graphs. Working within the framework of Borodin et al. [7], we prove that on this class (a) the HITS algorithm is stable with high probability, and (b) the HITS algorithm is similar to the INDEGREE heuristic that assigns to each node weight proportional to the number of incoming links. We demonstrate that our results go through for the case that the expected in-degrees of the graph follow a power-law distribution, a situation observed in the actual Web graph [9]. We also study experimentally the similarity between HITS and INDEGREE, and we investigate the general conditions under which the two algorithms are similar.

1 Introduction

In the past years there has been increasing research interest in the analysis of the Web graph for the purpose of improving the performance of search engines. The seminal works of Kleinberg [15] and Brin and Page [8] introduced the area of *Link Analysis Ranking*, where hyperlink structures are used to rank the results of search queries. Their work was followed by a plethora of modifications, generalizations and improvements (see [7] and references within). As a result, today there exists a wide range of Link Analysis Ranking (LAR) algorithms, many of which are variations of each other.

The multitude of LAR algorithms creates the need for a formal framework for assessing and comparing their properties. Borodin et al., introduced such a theoretical framework in [7]. In this framework an LAR algorithm is defined as a function from a class of graphs of size n to an n-dimensional real vector that assigns an *authority weight* to each node in the graph. The nodes are ranked in decreasing order of their

---

* Partially supported by the EU under contract 001907 (DELIS) and 33555 (COSIN), and by the Italian MIUR under contract ALGO-NEXT.

weights. Borodin et al. [7] define various properties of LAR algorithms. In this work we focus on *stability* and *similarity*. Stability considers the effect of small changes in the graph to the output of an LAR algorithm. Similarity studies how close the outputs of two algorithms are on the same graph.

Borodin et al. [7] considered the question of stability and similarity over an unrestricted class of graphs. They studied a variety of algorithms, and they proved that no pair of these algorithms is similar, and almost all algorithms are unstable. It appears that the class of all possible graphs is too broad to allow for positive results. This raises naturally the question whether it is possible to prove positive results if we restrict ourselves to a smaller class of graphs. Since the explosion of the Web, various stochastic models have been proposed for the Web graph [4, 5, 16, 3]. The model we consider, which was proposed by Azar et al. [4], is the following: assume that every node i in the graph comes with two parameters a_i and h_i which take values in $[0, 1]$. For some node i, the value h_i can be thought of as the probability of node i to be a good *hub*, while the value a_i is the probability of the node i to be a good *authority*. We then generate an edge from i to j with probability proportional to $h_i a_j$. We will refer to this model as the *product model*, and the corresponding class of graphs as the class of *product graphs*. The product graph model generalizes the traditional random graph model of Erdös and Rèny [13] to include graphs where the *expected* degrees follow specific distributions. This is of particular interest since it is well known [16, 9] that the in-degrees of the nodes in the Web graph follow a power law distribution.

Our contribution. In this paper we study the behavior of the HITS algorithm, proposed by Kleinberg [15], on the class of product graphs. The study of HITS on product graphs was initiated by Azar et al. [4] who showed that under some assumptions the HITS algorithm returns weights that are very close to the authority parameters. We formalize the findings of Azar et al. [4] in the framework of Borodin et al. [7]. We extend the definitions of stability and similarity for classes of random graphs, and we demonstrate the link between stability and similarity. We then prove that, with high probability, under some restrictive assumptions, the HITS algorithm is stable on the class of product graphs, and similar to the INDEGREE heuristic that ranks pages according to their in-degree. This similarity result is the main contribution of the paper. The implication of the result is that on product graphs, with high probability, the HITS algorithm reduces to simple in-degree count. We show that our assumptions are general enough to capture graphs where the expected degrees follow a power law distribution as the one observed on the real Web. We also analyze the correlation between INDEGREE and HITS on a large sample of the Web graph. The experimental analysis reveals that similarity between HITS and INDEGREE can also be observed on the real Web. We conclude with a discussion on the conditions that guarantee similarity of HITS and INDEGREE for the class of all possible graphs.

2 Related Work

Link Analysis Ranking Algorithms: Let P be a collection of n Web pages that need to be ranked. This collection may be the whole Web, or a query dependent subset of the Web. We construct the underlying *hyperlink graph* $G = (P, E)$ by creating a node for each Web page in the collection, and a directed edge for each hyperlink between two pages. The input to a LAR algorithm is the $n \times n$ adjacency matrix W of the graph G. The output of the algorithm is an n-dimensional *authority weight vector* w, where w_i, the i-th coordinate of w, is the authority weight of node i.

We now describe the two LAR algorithms we consider in this paper: the INDEGREE algorithm, and the HITS algorithm. The INDEGREE algorithm is the simple heuristic that assigns to each node weight equal to the number of incoming links in the graph G. The HITS algorithm was proposed by Kleinberg [15] in the seminal paper that introduced the hubs and authorities paradigm. In this framework, every page can be thought of as having a *hub* and an *authority* weight. Let h and a denote the n-dimensional hub and authority weight vectors. Kleinberg proposed an iterative algorithm, termed HITS, for computing the vectors h and a; the algorithm is essentially a power method computation of the principle eigenvectors of the matrices WW^T and W^TW respectively. These are the principal *singular vectors* of the matrix W. The HITS algorithm returns the vector a, the right singular vector of matrix W.

Independently from Kleinberg, Brin and Page developed the celebrated PAGERANK algorithm [8], which outputs the stationary distribution of a random walk on the Web graph. The works of Kleinberg [15] and Brin and Page [8] were followed by numerous modifications and extensions (see [7] and references within). Of particular interest is the SALSA algorithm by Lempel and Moran [18], which performs a random walk that alternates between hubs and authorities.

Theoretical study of LAR algorithms: Borodin et al. [7], in the paper that introduced the theoretical framework for the analysis of LAR algorithms, considered various algorithms, including HITS, SALSA, INDEGREE, and variants of HITS defined in their paper. They proved that, on the class of all possible graphs, no pair of algorithms is similar, and only the INDEGREE algorithm is stable. They also defined the notion of *rank stability* and *rank similarity*, where they considered the ordinal rankings induced by the weight vectors. The same results carry over in this case. Their work was extended by Lempel and Moran [19], and Lee and Borodin [17]. The stability of HITS and PAGERANK has also been studied elsewhere [22, 6].

The product graph model: Product graphs (also known as random graphs with given expected degrees) were first considered as a model for the Web graph by Azar et al. [4]. The undirected case, where the $h_i = a_i$ and the edges are undirected, has been studied extensively [20, 10, 11, 12]. The focus of these works is on the case where the parameters follow a power law distribution, as it is the case with most real-life networks.

3 The Theoretical Framework

In this section we review the definitions of Borodin et al. [7], and we extend them for classes of random graphs. Let $\mathcal{G}_n$ denote the set of all possible graphs of size n. The size of a graph is the number of nodes in the graph. Let $\overline{\mathcal{G}}_n \subseteq \mathcal{G}_n$ denote a collection of graphs in $\mathcal{G}_n$. Following the work of Borodin et al. [7], we define a link analysis algorithm $\mathcal{A}$ as a function $\mathcal{A} : \overline{\mathcal{G}}_n \to \mathbb{R}^n$ that maps a graph $G \in \overline{\mathcal{G}}_n$ to an n-dimensional real vector. The vector $\mathcal{A}(G)$ is the authority weight vector produced by the algorithm

$\mathcal{A}$ on graph G. The weight vector $\mathcal{A}(G)$ is normalized under some chosen norm L, that is, the algorithm maps the graphs in $\overline{\mathcal{G}}_n$ onto the unit L-sphere. Typically, the weights are normalized under some L_p norm. The L_p norm of a vector w is defined as $\|w\|_p = \left(\sum_{i=1}^n |w_i|^p\right)^{1/p}$.

Distance measures: In order to compare the behavior of different algorithms, or the behavior of the same algorithm on different graphs, Borodin et al. [7] defined various distance measures between authority weight vectors. The distance functions we consider are defined using the L_q norm. The d_q distance between two weight vectors w_1, w_2 is defined as follows.

$$d_q(w_1, w_2) = \min_{\gamma_1, \gamma_2 \geq 1} \|\gamma_1 w_1 - \gamma_2 w_2\|_q$$

The constants γ_1 and γ_2 serve the purpose of alleviating differences due to different normalization factors. When using distance d_q we will assume that the vectors are normalized in the L_q norm. In this paper we consider mainly the d_2 distance measure. We can prove that the $d_2(a, b) = \|a - b\|$, and thus the d_2 distance is a metric. The proof appears in the full version of the paper.

Similarity: Borodin et al. [7] give the following general definition of similarity for any distance function d and any normalization norm L. In the following we define $M_n(d, L) = \sup_{\|w_1\| = \|w_2\| = 1} d(w_1, w_2)$ to be the maximum distance between any two n-dimensional vectors with unit norm $L = \|\cdot\|$.

Definition 1. *Algorithms $\mathcal{A}_1$ and $\mathcal{A}_2$ are (L, d)-similar on the class $\overline{\mathcal{G}}_n$ if as $n \to \infty$*

$$\max_{G \in \overline{\mathcal{G}}_n} d(\mathcal{A}_1(G), \mathcal{A}_2(G)) = o(M_n(d, L))$$

Consider now the case that the class $\overline{\mathcal{G}}_n$ is a class of random graphs, generated according to some random process. That is, we define a probability space $\langle \overline{\mathcal{G}}_n, \mathcal{P} \rangle$, where $\mathcal{P}$ is a probability distribution over the class $\overline{\mathcal{G}}_n$. We extend the definition of similarity on the class $\overline{\mathcal{G}}_n$ as follows.

Definition 2. *Algorithms $\mathcal{A}_1$ and $\mathcal{A}_2$ are (L, d)-similar with high probability on the class of random graphs $\overline{\mathcal{G}}_n$ if for a graph G drawn from $\overline{\mathcal{G}}_n$, as $n \to \infty$*

$$d(\mathcal{A}_1(G), \mathcal{A}_2(G)) = o(M_n(d, L))$$

with probability $1 - o(1)$.

We note that when we consider (L_q, d_q)-similarity we have that $M_n(d_q, L_q) = \Theta(1)$. Furthermore, if the distance function d is a metric, or a near metric[1], then the transitivity property holds. It is easy to show that if algorithms $\mathcal{A}_1$ and $\mathcal{A}_2$ are similar

---

[1] A near metric [14] is a distance function that is reflexive, and symmetric, and there exists a constant c independent of n, such that for all $k > 0$, and all vectors $u, w_1, w_2, \ldots, w_k, v$, $d(u, v) \leq c(d(u, w_1) + d(w_1, w_2) + \cdots + d(w_k, v))$.

(with high probability), and algorithms $\mathcal{A}_2$ and $\mathcal{A}_3$ are similar (with high probability), then algorithms $\mathcal{A}_1$ and $\mathcal{A}_3$ are also similar (with high probability).

Stability: Let $\overline{\mathcal{G}}_n$ be a class of graphs, and let $G = (P, E)$ and $G' = (P, E')$ be two graphs in $\overline{\mathcal{G}}_n$. The *link distance* d_ℓ between graphs G and G' is defined as $d_\ell(G, G') = |(E \cup E') \setminus (E \cap E')|$ That is, $d_\ell(G, G')$ is the minimum number of links that we need to add and/or remove so as to change one graph into the other.

Given a class of graphs $\overline{\mathcal{G}}_n$, let $\mathcal{C}_k(G) = \{G' \in \overline{\mathcal{G}}_n : d_\ell(G, G') \leq k\}$ denote the set of all graphs that have link distance at most k from graph G. Borodin et al. [7] give the following generic definition of stability.

Definition 3. *An algorithm $\mathcal{A}$ is (L, d)-stable on the class of graphs $\overline{\mathcal{G}}_n$ if for every fixed positive integer k, we have as $n \to \infty$*

$$\max_{G \in \overline{\mathcal{G}}_n} \max_{G' \in \mathcal{C}_k(G)} d(\mathcal{A}(G), \mathcal{A}(G')) = o(M_n(d, L))$$

Given a class of random graphs $\overline{\mathcal{G}}_n$ we define stability with high probability as follows.

Definition 4. *An algorithm $\mathcal{A}$ is (L, d)-stable with high probability on the class of random graphs $\overline{\mathcal{G}}_n$ if for every fixed positive integer k, for a graph G drawn from $\overline{\mathcal{G}}_n$ we have as $n \to \infty$*

$$\max_{G' \in \mathcal{C}_k(G)} d(\mathcal{A}(G), \mathcal{A}(G')) = o(M_n(d, L))$$

with probability $1 - o(1)$.

Stability and Similarity: The following lemma shows the connection between stability and similarity. The lemma is a generalization of a lemma by Borodin et al. [7]. The proof appears in the full version of the paper.

Lemma 1. *Let d be a metric or near metric distance function, L a norm, and $\overline{\mathcal{G}}_n$ a class of random graphs. If algorithm $\mathcal{A}_1$ is (L, d)-stable with high probability on the class $\overline{\mathcal{G}}_n$, and algorithm $\mathcal{A}_2$ is (L, d)-similar to $\mathcal{A}_1$ with high probability on the class $\overline{\mathcal{G}}_n$, then $\mathcal{A}_2$ is (L, d)-stable with high probability on the class $\overline{\mathcal{G}}_n$.*

4 Stability and Similarity on the Class of Product Graphs

The class of product graphs $\mathcal{G}_n^p(\boldsymbol{h}, \boldsymbol{a})$ (or, for brevity, $\mathcal{G}_n^p$) is defined with two parameters $\boldsymbol{h}$ and $\boldsymbol{a}$, which are two n-dimensional real vectors, with h_i and a_i taking values in $[0, 1]$. These can be thought of as the *latent* hub and authority vectors. A link is generated from node i to node j with probability $h_i a_j$.

Let $G \in \mathcal{G}_n^p$, and let W be the adjacency matrix of the graph G. The matrix W can be written as $W = \boldsymbol{h}\boldsymbol{a}^T + R$, where R is a random matrix, such that

$$R[i, j] = \begin{cases} -h_i a_j & \text{with probability } 1 - h_i a_j \\ 1 - h_i a_j & \text{with probability } h_i a_j \end{cases}$$

We refer to matrix R as the *rounding* matrix, that rounds the entries of M to 0 or 1. We can think of the matrix W as a perturbation of the matrix $M = \mathbf{h}\mathbf{a}^T$ by the rounding matrix R. The matrix M is a rank-one matrix. If we run HITS on the matrix M (assuming a small modification of the algorithm so that it runs on weighted graphs), the algorithm will reconstruct the latent vectors $\mathbf{a}$ and $\mathbf{h}$, which are the singular vectors of matrix M. Note also that if we run the INDEGREE algorithm on the matrix M (assuming again that we take the weighted in-degrees), the algorithm will also output the latent vector $\mathbf{a}$. So, on rank-one matrices the two algorithms are identical. The question is how the addition of the rounding matrix R affects the output of the two algorithms. We will show that it has only a small effect, and the two algorithms remain similar.

More formally, let LATENT denote the (imaginary) LAR algorithm which, for any graph G in the class $\mathcal{G}_n^p(\mathbf{h}, \mathbf{a})$, outputs the vector $\mathbf{a}$. We will show that both HITS and INDEGREE are similar to LATENT with high probability. This implies that the two algorithms are similar with high probability. Furthermore, we will show that it also implies the stability of the HITS algorithm.

4.1 Mathematical Tools

We now introduce some mathematical tools that we will use for the remaining of this section.

Perturbation Theory: Perturbation theory studies how adding a perturbation matrix E to a matrix M affects the eigenvalues and eigenvectors of M. Let G and G' be two graphs, and let W and W' denote the respective adjacency matrices. The matrix W' can be written as $W' = W + E$, where E is a matrix with entries in $\{-1, 0, 1\}$. The entry $E[i, j]$ is 1 if we add a link from i to j, and -1 if we remove a link from i to j. Therefore, we can think of the matrix W' as a perturbation of the matrix W by a matrix E. Note that if we assume that only a constant number of links is added and removed, then both the Frobenius and the L_2 norms of E are bounded by a constant.

We now introduce an important lemma that we will use in the following. The proof of the lemma appears in the full version of the paper.

Lemma 2. *Let W be a matrix, and let $W + E$ be a perturbation of the matrix. Let $\mathbf{u}$ and $\mathbf{v}$ denote the left and right principal singular vectors of the matrix W, and $\mathbf{u}'$ and $\mathbf{v}'$ the principal singular vectors of the perturbed matrix. Let σ_1, σ_2 denote the first and second singular values of the matrix W. If $\sigma_1 - \sigma_2 = \omega(\|E\|_2)$, then $\|\mathbf{u}' - \mathbf{u}\|_2 = o(1)$ and $\|\mathbf{v}' - \mathbf{v}\|_2 = o(1)$.*

Norms of random matrices: We also make use of the following theorem for concentration bounds on the L_2 norm of random symmetric matrices. We state the theorem as it appears in [1].

Theorem 1. *Given an $m \times n$ matrix A and any $\epsilon > 0$, let $\widehat{A}$ be any random matrix such that for all i, j: $E[\widehat{A}_{ij}] = A_{ij}$, $Var(\widehat{A}_{ij}) \leq \sigma^2$, and $|\widehat{A}_{ij} - A_{ij}| \leq K$, where*

$$K = \left(\frac{4\epsilon}{4+3\epsilon}\right)^3 \frac{\sigma\sqrt{m+n}}{\log^3(m+n)}$$

For any $\alpha > 0$, and $m + n \geq 20$, with probability at least $1 - (m+n)^{-\alpha^2}$,

$$\|\widehat{A} - A\|_2 < (2 + \alpha + \epsilon)\sigma\sqrt{m+n}$$

Chernoff bounds: We will make use of standard Chernoff bounds. The following theorem can be found in the textbook of Motwani and Raghavan [21].

Theorem 2. *Let $X_1, X_2, \ldots, X_n$ be independent Poisson trials such that, for $1 \leq i \leq n$, $\Pr[X_i = 1] = p_i$, where $0 \leq p_i \leq 1$. Let $X = \sum_{i=1}^n X_i, \mu = E[X] = \sum_{i=1}^n p_i$. Then, for $0 < \delta \leq 1$, we have that*

$$\Pr[X < (1-\delta)\mu] < \exp(-\mu\delta^2/2) \tag{1}$$
$$\Pr[X > (1+\delta)\mu] < \exp(-\mu\delta^2/4) \tag{2}$$

4.2 Conditions for the Stability of HITS

We first provide general conditions for the stability of the HITS algorithm. Let $\mathcal{G}_n^\sigma$ denote the class of graphs with adjacency matrix W that satisfies $\sigma_1(W) - \sigma_2(W) = \omega(1)$. The proof of the following theorem follows directly from Lemma 2, and the fact that the perturbation matrix E has L_2 norm bounded by a constant.

Theorem 3. *The HITS algorithm is (L_2, d_2)-stable on the class of graphs $\mathcal{G}_n^\sigma$.*

Theorem 3 provides a sufficient condition for the stability of HITS on general graphs and it will be useful when considering stability on the class of product graphs. The class $\mathcal{G}_n^\sigma$ is actually a subset of the class defined by the result of Ng et al. [22]. Translating their result in the framework of Borodin et al. [7], they prove that the HITS algorithm is stable on the class of graphs with $\sigma_1(W)^2 - \sigma_2(W)^2 = \omega(\sqrt{d})$, where d is the maximum out-degree.

4.3 Similarity of HITS and LATENT

We now turn our attention to product graphs, and we prove that HITS and LATENT are similar on this class. A result of similar spirit is shown in the work of Azar et al. [4]. We make the following assumption for the vectors a and h.

Assumption 1. *For the class $\mathcal{G}_n^p(h, a)$, the latent vectors a and h satisfy $\|a\|_2 \|h\|_2 = \omega(\sqrt{n})$.*

As we show below, Assumption 1 places a direct lower bound on the principal singular value of the matrix $M = ha^T$. Also, let $A = \sum_{i=1}^n a_i$, denote the sum of the authority values, and let $H = \sum_{j=1}^n h_j$ the sum of the hub values. Since the values are positive, we have $A = \|a\|_1$ and $H = \|h\|_1$. The product HA is equal to expected number of edges in the graph. We have that $HA \geq \|a\|_2 \|h\|_2$, thus, from Assumption 1, $HA = \omega(\sqrt{n})$. This implies that the graph is not too sparse.

Lemma 3. *The algorithms HITS and LATENT are (L_2, d_2)-similar with high probability on the class $\mathcal{G}_n^p$, subject to Assumption 1.*

Proof. The singular vectors of the matrix M are the L_2 unit vectors $a_2 = a/\|a\|_2$ and $h_2 = h/\|h\|_2$. The matrix M can be expressed as $M = h_2^T \|h\|_2 \|a\|_2 a_2$. Therefore, the principal singular value of M is $\sigma_1 = \|h\|_2 \|a\|_2 = \omega(\sqrt{n})$. Since M is rank-one, $\sigma_i = 0$, for all $i = 2, 3, \ldots, n$. Therefore, for matrix M we have that $\sigma_1 - \sigma_2 = \omega(\sqrt{n})$.

Matrix R is a random matrix, where each entry is a independent random variable with mean 0, and maximum value and variance bounded by 1. Using Theorem 1, we observe that $K = 1$, and $\sigma = 1$. Setting $\epsilon = 1$ and $\alpha = 1$, we get that $Pr[\|R\|_2 \leq 8\sqrt{n}] \geq 1 - o(1/n)$, thus $\|R\|_2 = O(\sqrt{n})$ with high probability.

Therefore, we have that $\sigma_1 - \sigma_2 = \omega(\|R\|_2)$ with probability $1 - o(1)$. If w_2 is the right singular vector of matrix W normalized in the L_2 norm, then, using Lemma 2, we have that $\|w_2 - a_2\|_2 = o(1)$ with probability $1 - o(1)$. □

Assumption 1 guarantees also the stability of HITS on $\mathcal{G}_n^p$. The proof follows from the fact that if $G \in \mathcal{G}_n^p$, then $G \in \mathcal{G}_n^\sigma$, with high probability.

Theorem 4. *The HITS algorithm is (L_2, d_2)-stable with high probability on the class of graphs $\mathcal{G}_n^p$, subject to Assumption 1.*

4.4 Similarity of INDEGREE and LATENT

We now consider the (L_q, d_q)-similarity of INDEGREE and LATENT, for all $1 \leq q < \infty$. Again, let $A = \sum_{i=1}^n a_i$, and let $H = \sum_{j=1}^n h_j$. Also, let d denote the vector of the INDEGREE algorithm before any normalization is applied. That is, d_i is the in-degree of node i. For some node i, we have that

$$d_i = \sum_{j=1}^n W[j,i] = \sum_{j=1}^n M[j,i] + \sum_{j=1}^n R[j,i]$$

We have that $\sum_{j=1}^n M[j,i] = Ha_i$. Furthermore, let $r_i = \sum_{j=1}^n R[j,i]$, and let $r = [r_1, \ldots, r_n]^T$. Vector d can be expressed as $d = Ha + r$.

We first prove the following auxiliary lemma.

Lemma 4. *For every $q \in [1, \infty)$, if $H\|a\|_q = \omega(n^{1/q} \ln n)$, then $\|r\|_q = o(H\|a\|_q)$ with high probability.*

Proof. For the following we will use $\|\cdot\|$ to denote the L_q norm, for some $q \in [1, \infty)$. We will prove that $\|r\| = o(H\|a\|)$ with probability at least $1 - 1/n$. We have assumed that $H\|a\| = \omega(n^{1/q} \ln n)$, so it is sufficient to show that $\|r\| = O(n^{1/q} \ln n)$, or equivalently that for all $1 \leq i \leq n$, $|r_i| = O(\ln n)$ with probability at least $1 - 1/n^2$. Note that $r_i = d_i - Ha_i$, so essentially we need to bound the deviation of d_i from its expectation.

We partition the nodes into two sets S and B. Set S contains all nodes such that $Ha_i = O(\ln n)$, that is, nodes with "small" expected in-degree, and set B contains all nodes such that $Ha_i = \omega(\ln n)$, that is, node with "big" expected in-degree.

Consider a node $i \in S$. We have that $Ha_i \leq c \ln n$, for some constant c. Using Theorem 2, Equation 2, we set $\delta = k \ln n/(Ha_i)$, where k is a constant such that $k \geq \sqrt{8c}$, and we get that $Pr[d_i - Ha_i \geq k \ln n] \leq exp(-2 \ln n)$. Therefore, for all nodes in S we have that $|r_i| = O(\ln n)$ with probability at least $1 - 1/n^2$. This implies that $\sum_{i \in S} |r_i|^q = O(n \ln^q n) = o(H^q \|a\|^q)$, with probability $1 - 1/n$.

Consider now a node $i \in B$. We have that $Ha_i = \omega(\ln n)$, thus, $Ha_i = (\ln n)/s(n)$, where $s(n)$ is a function such that $s(n) = o(1)$. Using Theorem 2, we set $\delta = k\sqrt{s(n)}$, where k is a constant such that $k \geq \sqrt{8}$, and we get that $Pr[|d_i - Ha_i| \geq \delta Ha_i] \leq exp(-2 \ln n)$. Therefore, for the nodes in B, we have that $|r_i| = o(Ha_i)$ with probability at least $1 - 1/n^2$. Thus, $\sum_{i \in B} |r_i|^q = o(H^q \|a\|^q)$, with probability $1 - 1/n$.

Putting everything together we have that $\|r\|^q = \sum_{i \in S} |r_i|^q + \sum_{i \in B} |r_i|^q = o(H^q \|a\|^q)$, with probability $1 - 2/n$. Therefore, $\|r\| = o(H\|a\|)$ with probability $1 - 2/n$. This concludes our proof. □

We are now ready to prove the similarity of INDEGREE and LATENT. The following lemma follows from Lemma 4. The details of the proof appears in the full version of the paper.

Lemma 5. *For every $q \in [1, \infty)$, the INDEGREE and LATENT algorithms are (L_q, d_q)-similar with high probability on the class $\mathcal{G}_n^p$, when the latent vectors a and h satisfy $H\|a\|_q = \omega(n^{1/q} \ln n)$.*

We now make the following assumption for vectors a and h.

Assumption 2. *For the class $\mathcal{G}_n^p(h, a)$, the latent vectors a and h satisfy $H\|a\|_2 = \omega(\sqrt{n} \ln n)$.*

Assumption 2 implies that the expected number of edges in the graph satisfies $HA = \omega(\sqrt{n} \ln n)$. Note that we can satisfy Assumption 2 by requiring $HA = \omega(n \ln n)$, that is, the graph is dense enough. We can satisfy both Assumption 1 and 2 by requiring that $\sigma_1(M) = \|h\|_2 \|a\|_2 = \omega(\sqrt{n} \ln n)$.

The INDEGREE and LATENT algorithms are (L_2, d_2)-similar subject to Assumption 2. The following theorem follows from the transitivity property of similarity.

Theorem 5. *The HITS and INDEGREE algorithms are (L_2, d_2)-similar with high probability on the class $\mathcal{G}_n^p$, subject to Assumptions 1 and 2.*

4.5 Power Law Graphs

A discrete random variable X follows a power law distribution with parameter α, if $Pr[X = x] \propto x^{-\alpha}$. Closely related to the power-law distribution is the Zipfian distribution, also known as Zipf's law [24]. Zipf's law states that the r-th largest value of the random variable X is proportional to $r^{-\beta}$. It can be proved [2] that if X follows a Zipfian distribution with exponent β, then it also follows a power law distribution with parameter $\alpha = 1 + 1/\beta$. We will now prove that Assumptions 1 and 2 are general enough to include graphs with *expected* in-degrees that follow Zipf's law with parameter $\beta < 1$.

Without loss of generality we assume that $a_1 \geq a_2 \geq \cdots \geq a_n$. For some constant $c \leq 1$ the i-th authority value is defined as $a_i = ci^{-\beta}$, for $\beta < 1$. This implies a power law distribution on the *expected* in-degrees with exponent $\alpha > 2$. This is typical for most real-life graphs. The exponent of the in-degree distribution for the Web graph is

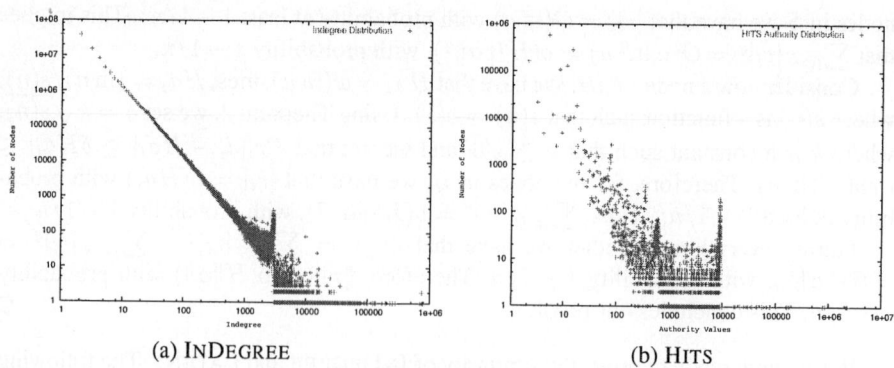

Fig. 1. INDEGREE and HITS distributions on the Web graph

2.1 [9]. For the hub values we assume that $h_i = \Theta(1)$, for all $1 \leq i \leq n$. Therefore, we have that $H = \Theta(n)$, and $\|h\|_2 = \Theta(\sqrt{n})$. Furthermore, it is easy to show that for $\beta < 1$, $\|a\|_2^2 = \sum_{i=1}^{n} \frac{c}{i^{2\beta}} = \omega(1)$.

Therefore, $\|a\|_2 \|h\|_2 = \omega(\sqrt{n})$, and $H\|a\|_2 = \omega(n)$, thus satisfying Assumptions 1 and 2. Therefore, we can conclude that HITS and INDEGREE are similar with high probability when the expected degrees follow a power law distribution. Note that on this graph we have that the expected number of edges is $HA = \omega(n \ln n)$.

5 Experimental Analysis

In this section we study experimentally the similarity of HITS and INDEGREE on a large sample of the Web. We analyze a sample of 136M vertices and about 1,2 billion edges of the Web graph collected in 2001 by the WebBase project[2] at Stanford. Figures 1(a) and 1(b) show the distributions of the INDEGREE and HITS authority values. The in-degree distribution, as it is well known, follows a power law distribution. The HITS authority weights also follow a "fat" power law distribution in the central part of the plot. Table 1 summarizes our findings on the relationship between INDEGREE and HITS. Since we only have a single graph and not a sequence of graphs, the distance measures are not very informative, so we also compute the correlation coefficient between the two weight vectors. We observe a strong correlation between the authority weights of HITS and the in-degrees, while almost no correlation between the hub weights and the out-degrees.

Table 1. Similarity between HITS and INDEGREE

	authority/in-degree	hub/out-degree
d_2 distance	0.36	1.23
correlation coefficient	0.93	0.005

---

[2] http://www-diglib.stanford.edu/~testbed/doc2/WebBase/

Similar trends are observed for the d_2 distance, where the distance between hub weights and out-degrees is much larger than that between authority weights and in-degrees. These results suggest that although the Web, as expected, is not a product graph, the HITS authority weights can be well approximated by the in-degrees.

6 Similarity of HITS and INDEGREE

In this section we study the general conditions under which the HITS and INDEGREE algorithms are similar. Consider a graph $G \in \mathcal{G}_n$ and the corresponding adjacency matrix W. Let $\sigma_1 \geq \sigma_2 \geq \ldots \geq \sigma_n$ be the singular values of W, and let $\boldsymbol{a}_1, \ldots, \boldsymbol{a}_n$ and $\boldsymbol{h}_1, \ldots, \boldsymbol{h}_n$ denote the right (authority) and left (hub) singular vectors respectively. All vectors are unit vectors in the L_2 norm. The HITS algorithm outputs the vector $\boldsymbol{a} = \boldsymbol{a}_1$. Let $\boldsymbol{w}$ denote the output of the INDEGREE algorithm (normalized in L_2). Also, let $H_i = \sum_{j=1}^{n} h_i(j)$ be the sum of the entries of the i-th hub vector. We can prove the following proposition. The proof appears in the full version of the paper.

Proposition 1. *For a graph $G \in \overline{\mathcal{G}}_n$, the d_2 distance between HITS and INDEGREE is*

$$d_2(\boldsymbol{a}, \boldsymbol{w}) = \sqrt{\left(\frac{\sigma_2 H_2}{\sigma_1 H_1}\right)^2 + \cdots + \left(\frac{\sigma_n H_n}{\sigma_1 H_1}\right)^2} \qquad (3)$$

We now study the conditions under which $d_2(\boldsymbol{a}, \boldsymbol{w}) = o(1)$. Since the values of $\boldsymbol{h}_1$ are positive, we have that $H_1 = \|\boldsymbol{h}_1\|_1$, and $1 \leq H_1 \leq \sqrt{n}$. For every $i > 1$, we have that $|H_i| \leq \|\boldsymbol{h}_i\|_1$ and $|H_i| \leq \sqrt{n}$. The following conditions guarantee the similarity of HITS and INDEGREE: (a) $\sigma_2/\sigma_1 = o(1/\sqrt{n})$, and there exists a constant k such that $\sigma_{k+1}/\sigma_1 = o(1/n)$; (b) $H_1 = \Theta(\sqrt{n})$, and $\sigma_2/\sigma_1 = o(1)$, and there exists a constant k such that $\sigma_{k+1}/\sigma_1 = o(1/n)$; (c) $H_1 = \Theta(\sqrt{n})$, and $\sigma_2/\sigma_1 = o(1/\sqrt{n})$.

Assume now that $|H_i|/(\sigma_1 H_1) = o(1)$, for all $i \geq 2$. One possible way to obtain this bound is to assume that $\sigma_1 = \omega(\sqrt{n})$, or that $H_1 = \Theta(\sqrt{n})$ and $\sigma_1 = \omega(1)$. Then, we can obtain the following characterization of the distance between HITS and INDEGREE. From Equation (3) we have that $d_2(\boldsymbol{a}, \boldsymbol{w}) = o\left(\sqrt{\sigma_2^2 + \cdots + \sigma_n^2}\right)$. Let $W_1 = \sigma_1 \boldsymbol{h}_1 \boldsymbol{a}_1^T$ denote the rank-one approximation of W. The matrix $R = W - W_1$ is called the residual matrix, and it has singular values $\sigma_2, \ldots, \sigma_n$. We have that

$$d_2(\boldsymbol{a}, \boldsymbol{w}) = o\left(\|W - W_1\|_F\right) \quad \text{and} \quad d_2(\boldsymbol{a}, \boldsymbol{w}) = o\left(\sqrt{\|W\|_F^2 - \|W\|_2^2}\right) \qquad (4)$$

Equation (4) says that the similarity of HITS and INDEGREE algorithms depends on the Frobenius norm of the residual matrix. Furthermore, the similarity of the HITS and INDEGREE algorithms depends on the difference between the Frobenius and the spectral (L_2) norm of matrix W. The L_2 norm measures the strength of the strongest linear trend in the matrix, while the Frobenius norm captures the sum of the strengths of all linear trends in the matrix [1]. The similarity of the HITS and INDEGREE algorithms depends upon the contribution of the strongest linear trend to the sum of linear trends.

7 Conclusions

In this paper we studied the behavior of the HITS algorithm on the class of product graphs. We proved that under some assumptions the HITS algorithm is stable, and it is similar to the INDEGREE algorithm. Our assumptions include graphs with expected degrees that follow a power law distribution.

Our work opens a number of interesting directions for future work. First, it would be interesting to determine a necessary condition for the stability of the HITS algorithm. Also, it would be interesting to study the stability and similarity of other LAR algorithms on product graphs, such as the PAGERANK and the SALSA algorithms. Finally, it would be interesting to study other classes of random graphs [5, 16].

References

1. D. Achlioptas and F. McSherry. Fast computation of low rank matrix approximations. In *ACM Symposium on Theory of Computing (STOC)*, 2001.
2. L. A. Adamic and B. A. Huberman. Zipf's law and the internet. *Glottometrics*, 3:143–150, 2002.
3. W. Aiello, F. R. K. Chung, and L. Lu. Random evolution in massive graphs. In *IEEE Symposium on Foundations of Computer Science*, pages 510–519, 2001.
4. Y. Azar, A. Fiat, A. Karlin, F. McSherry, and J. Saia. Spectral analysis of data. In *Proceedings of the 33rd Symposium on Theory of Computing (STOC 2001)*, Greece, 2001.
5. A.-L. Barabasi and R. Albert. Emergence of scaling in random networks. *Science*, 286:509–512, 1999.
6. M. Bianchini, M. Gori, and F. Scarselli. Pagerank: A circuital analysis. In *Proceedings of the Eleventh International World Wide Web (WWW) Conference*, 2002.
7. A. Borodin, G. O. Roberts, J. S. Rosenthal, and P. Tsaparas. Link Analysis Ranking: Algorithms, Theory, and Experiments. *ACM Transactions on Internet Technology*, 05(1), 2005.
8. S. Brin and L. Page. The anatomy of a large-scale hypertextual Web search engine. In *Proceedings of the 7th International World Wide Web Conference*, Brisbane, Australia, 1998.
9. A. Broder, R. Kumar, F. Maghoul, P. Raghavan, S. Rajagopalan, R. Stata, A. Tomikns, and W. Wiener. Graph structure in the Web. In *Proceedings of WWW9*, 2000.
10. F. Chung and L. Lu. Connected components in random graphs with given degree sequences. *Annals of Combinatorics*, 6:125–145, 2002.
11. F. Chung and L. Lu. The average distances in random graphs with given expected degrees. *Internet Mathematics*, 1:91–114, 2003.
12. F. Chung, L. Lu, and V. Vu. Eigenvalues of random power law graphs. *Annals of Combinatorics*, 7:21–33, 2003.
13. P. Erdös and A. Rènyi. On the evolution of random graphs. *Publ. Math. Inst. Hung. Acad. Sci.*, 5:17–61, 1960.
14. R. Fagin, R. Kumar, and D. Sivakumar. Comparing top k lists. In *Proceedings of the ACM-SIAM Symposium on Discrete Algorithms (SODA)*, 2003.
15. J. Kleinberg. Authoritative sources in a hyperlinked environment. In *Proceedings of the Ninth Annual ACM-SIAM Symposium on Discrete Algorithms*, pages 668–677, 1998.
16. R. Kumar, P. Raghavan, S. Rajagopalan, D. Sivakumar, A. Tomkins, and E. Upfal. Stochastic models for the web graph. In *Proceedings of the 41st Annual Symposium on Foundations of Computer Science*, 2000.

17. H. C. Lee and A. Borodin. Perturbation of the hyperlinked environment. In *Proceedings of the Ninth International Computing and Combinatorics Conference*, 2003.
18. R. Lempel and S. Moran. The stochastic approach for link-structure analysis (SALSA) and the TKC effect. In *Proceedings of the 9th International World Wide Web Conference*, 2000.
19. R. Lempel and S. Moran. Rank stability and rank similarity of link-based web ranking algorithms in authority connected graphs. In *Second Workshop on Algorithms and Models for the Web-Graph (WAW 2003)*, 2003.
20. M. Mihail and C. H. Papadimitriou. On the eigenvalue power law. In *Proceedings of the 6th International Workshop on Randomization and Approximation Techniques*, 2002.
21. R. Motwani and P. Raghavan. *Randomized Algorithms*. Cambridge University Press, Cambridge, England, June 1995.
22. A. Y. Ng, A. X. Zheng, and M. I. Jordan. Link analysis, eigenvectors, and stability. In *Proceedings of the International Joint Conference on Artificial Intelligence (IJCAI)*, 2001.
23. G. W. Stewart and J. Sun. *Matrix Perturbation Theory*. Academic Press, 1990.
24. G. K. Zipf. *Human Behavior and the principle of least effort*. Addison-Wesley, 1949.

Up-to Techniques for Weak Bisimulation

Damien Pous

ENS Lyon

Abstract. Up-to techniques have been introduced to enhance the bisimulation proof method for establishing bisimilarity results. While up-to techniques for strong bisimilarity are well understood, in the weak case they come as a collection of unrelated results, and lack a unified presentation. We propose a uniform and modular theory of up-to techniques for weak bisimulation that captures existing proof technology and introduces new techniques. Some proofs rely on non trivial – and new – commutation results based on termination guarantees.

Introduction

Bisimilarity is a widely used behavioural equivalence in concurrency theory. It can be seen as the finest extensional equivalence that enjoys a natural formulation and nice mathematical properties. Bisimilarity can be defined as the greatest *bisimulation*. Given a *labelled transition system* (LTS), allowing one to write transitions between states of the form $P \xrightarrow{\alpha} P'$ (meaning that a state P can perform an action α and evolve to P'), we say that a relation $\mathcal{R}$ between states is a bisimulation whenever the leftmost diagram below holds: if P and Q are related by $\mathcal{R}$ and $P \xrightarrow{\alpha} P'$, there is Q' such that $Q \xrightarrow{\alpha} Q'$ and $\mathcal{R}$ relates P' and Q', and symmetrically for the transitions of Q.

$$
\begin{array}{ccc}
P & \mathcal{R} & Q \\
\alpha\downarrow & & \downarrow\alpha \\
P' & \mathcal{R} & Q'
\end{array}
\qquad
\begin{array}{ccc}
P & \mathcal{R} & Q \\
\alpha\downarrow & & \downarrow\alpha \\
P' & \mathcal{F}(\mathcal{R}) & Q'
\end{array}
\qquad
\begin{array}{ccc}
P & \mathcal{R} & Q \\
\alpha\downarrow & & \downarrow\alpha \\
P' & \mathcal{S} & Q'
\end{array}
\qquad
\begin{array}{ccc}
P & \mathcal{R} & Q \\
\alpha\downarrow & & \downarrow\alpha \\
P' & \mathcal{R} & Q'
\end{array}
$$

Bisimulation is the most popular technique to establish bisimilarity results: to prove that P and Q are bisimilar (written $P \sim Q$), exhibit a bisimulation $\mathcal{R}$ such that $P \mathcal{R} Q$. *Up-to techniques for bisimulation* have been introduced to alleviate the task of bisimulation proofs, by working with smaller relations. The proof scheme is shown on the second diagram above: a correct up-to technique is given by a function $\mathcal{F}$ from relations to relations such that if we prove that $\mathcal{R}$ 'evolves to' $\mathcal{F}(\mathcal{R})$, then we know that $\mathcal{R} \subseteq \sim$. The advantage is that $\mathcal{R}$ need not be a bisimulation (and can be 'much smaller' than a bisimulation). The notion of evolution of relations (depicted on the third diagram, where $\mathcal{R}$ evolves to $\mathcal{S}$ — its informal meaning is made precise below) serves as the basis of [8], where a general theory of up-to techniques for bisimulation is presented. The corresponding framework gives a unified and modular view of known up-to techniques, that can be combined together to yield powerful proof techniques for bisimilarity.

Up to now, we have implicitly been referring to *strong* bisimulation. When analysing nontrivial systems, however, one is often interested in the *weak* version, where a special action, called τ, is isolated, and the game of bisimulation is redefined by abstracting over τ transitions (τ is treated as a *silent* action, while other actions are *visible*). In the weak version of the bisimulation game, as shown on the rightmost diagram above, Q responds to $P \xrightarrow{\alpha} P'$ by performing an $\xrightarrow{\alpha}$ transition: this means that Q can do zero or several silent steps before and after the transition along α, or even not move at all in the case where $\alpha = \tau$ (and symmetrically when Q offers a challenge). One might then want to follow the same path as above: redefine the evolution of relations, and look for some functions $\mathcal{F}$ that yield correct up-to proof techniques for weak bisimilarity (written $\approx$). An important motivation for doing so is that in general, weak bisimulation proofs tend to be much larger than strong bisimulation proofs, so that having up-to techniques for the weak case is at least as important as in the strong case.

Unfortunately, in the weak case, irregularities appear, the paradigmatic example being given by the unsoundness of the 'weak bisimulation up to weak bisimilarity' proof technique. We recall the counterexample, from [9]. We suppose that the reader is familiar with CCS, and define $\mathcal{R} \triangleq \{(\tau.a, 0)\}$. Let us show that $\mathcal{R}$ is a weak bisimulation up to $\approx$, i.e., that $\mathcal{R}$ evolves to $\approx \mathcal{R} \approx$ (we use juxtaposition to denote relation composition). The right process, 0, cannot move. The only move the left process can do is a τ transition to a, to which the right process answers by no move, and we get the pair $(a, 0)$. Now since we are reasoning up to $\approx$, and since $a \approx \tau.a$, we are allowed to replace this pair with $(\tau.a, 0)$, and we are back in $\mathcal{R}$. Nevertheless, we obviously cannot conclude that $\tau.a$ and 0 are bisimilar processes.

Novel and useful proof techniques have been introduced to circumvent this difficulty [9, 3], notably based on the expansion preorder [1], that allows one to avoid situations where one can 'undo a τ transition' as in the example above. However, as we have experienced in a recent study [4], in some cases reasoning up to expansion is not possible. The intuitive reason can be formulated as follows: when a process P expands a process Q, P has to be more efficient (in terms of internal computations, represented by silent transitions) than Q *at every step*. Typically, expansion is a well suited relation to get rid of intermediate computation steps that do not affect the behaviour of the system. However, it is common (in particular, it is the case in [4]) that along such transitions, an increased efficiency is achieved at the cost of some initial computation. Because of its 'very controlled' nature, expansion fails in handling this kind of pre-calculation techniques.

In the present work, we develop a theory of up-to techniques for weak bisimulation that enjoys nice properties in terms of generality and modularity, and we introduce new useful proof techniques for weak bisimilarity that can be used in that framework.

We start by adapting the work of [7] to the weak case, yielding the notion of *monotonic function* over relations. We explore the class of monotonic functions, and argue that it is too restrictive. We are thus led to relax the notion of monotonicity, and introduce *weakly monotonic* functions, for which up-to techniques can be applied only to reason about visible actions (those that cannot be undone by $\approx$). We then show under which conditions monotonic and weakly monotonic functions can be combined together to obtain sound proof techniques. The resulting framework gives a unified and modular account of existing technology for weak bisimulation proofs. Beyond that, we validate some proof principles, such as 'up to bisimilarity and transitivity on visible actions', that to our knowledge had not been proposed before.

We then attack the question of finding alternatives to the expansion relation to handle τ transitions in weak bisimulation proofs. We propose an *up to controlled bisimulation* technique. The notion of controlled bisimulation intuitively captures the idea of avoiding 'going back in time' in bisimulation proofs. We introduce *relaxed expansion*, a co-inductively defined relation that is a controlled bisimulation and is coarser than expansion. We also propose two new proof principles for which the control on τ steps exploits a different kind of argument, based on termination guarantees. The corresponding correctness proofs are best formulated as rewriting results, that are technically difficult and may be of interest *per se*; we therefore describe them in that setting in a dedicated section. All our results have been formally checked using the Coq proof assistant [7]. For the lack of space, most are omitted in this extended abstract. They can be found in [5].

Outline of the paper. In Sect. 1, we introduce some necessary background and show where the approach of [8] breaks when adapted to the weak case. We develop our theory of up-to techniques for weak bisimulation in Sect. 2, introducing monotonic and weakly monotonic functions. In Sect. 3 we introduce controlled simulations and present new up-to techniques based on this notion. The correctness of some of these techniques is supported by the proofs given in Sect. 4, which are formulated in the setting of commutation results. We give final remarks in Sect. 5.

1 The Problem of "Weak Bisimulation Up to"

1.1 Labelled Transition Systems, Relations, Evolution

We consider a labelled transition system (LTS) $(\mathcal{P}, \mathcal{L}, \rightarrow)$, with domain $\mathcal{P}$, *labels* or *actions* in $\mathcal{L}$ and transition relation $\rightarrow \subseteq \mathcal{P} \times \mathcal{L} \times \mathcal{P}$. The elements of $\mathcal{P}$ are called *processes* and are denoted by P, Q. We distinguish a *silent action*, $\tau \in \mathcal{L}$. We let α, β (resp. a, b) range over actions, in $\mathcal{L}$ (resp. *visible actions*, in $\mathcal{L}\backslash\{\tau\}$). We write $P \xrightarrow{\alpha} Q$ when $(P, \alpha, Q) \in \rightarrow$ (so that $P \xrightarrow{a} Q$ stands for a transition of P along a visible action a).

We let $\mathcal{R}, \mathcal{S}, \mathcal{B}, \mathcal{E}$ range over binary relations (simply called *relations* in the sequel) on processes, and denote respectively by $\mathcal{R}^+, \mathcal{R}^=, \mathcal{R}^*$ the transitive, reflexive, transitive and reflexive closure of the relation $\mathcal{R}$. $P \mathcal{R} Q$ stands for $(P, Q) \in \mathcal{R}$. The composition of two relations $\mathcal{R}$ and $\mathcal{S}$, written $\mathcal{RS}$, is defined by $\mathcal{RS} \triangleq \{(P, Q) \text{ s.t. } P \mathcal{R} T \text{ and } T \mathcal{S} Q \text{ for some process } T\}$. We will also need the inverse of a relation: $\mathcal{R}^{-1} \triangleq \{(P, Q) \text{ s.t. } Q \mathcal{R} P\}$. $\mathcal{I}$ will denote the identity relation. We say that $\mathcal{R}$ *contains* $\mathcal{S}$ (alternatively, that $\mathcal{S}$ is contained in $\mathcal{R}$), written $\mathcal{S} \subseteq \mathcal{R}$, if $P \mathcal{S} Q$ implies $P \mathcal{R} Q$. A relation $\mathcal{R}$ *terminates* if there is no infinite sequence $P_1, P_2 \ldots$ such that $\forall i, P_i \mathcal{R} P_{i+1}$.

Definition 1.1 (weak transitions). *The* weak transition relation, *written* $\overset{\alpha}{\Longrightarrow}$, *is defined as the reflexive transitive closure of* $\overset{\tau}{\rightarrow}$ *when* $\alpha = \tau$, *and the composition* $\overset{\tau}{\Longrightarrow} \overset{a}{\rightarrow} \overset{\tau}{\Longrightarrow}$ *for* $\alpha = a \in \mathcal{L} \setminus \{\tau\}$.

Definition 1.2 (evolution). *Let* α *be an action and* $\mathcal{R}, \mathcal{S}$ *two relations. We say that* $\mathcal{R}$ α-*evolves to* $\mathcal{S}$, *if whenever* $P \mathcal{R} Q$, $P \overset{\alpha}{\rightarrow} P'$ *implies* $Q \overset{\alpha}{\Longrightarrow} Q'$ *and* $P' \mathcal{S} Q'$ *for some* Q'. *Given two relations* $\mathcal{R}$ *and* $\mathcal{S}$, *we say that:*

- $\mathcal{R}$ *evolves to* $\mathcal{S}$, *denoted by* $\mathcal{R} \rightarrowtail \mathcal{S}$, *if* $\mathcal{R}$ α-*evolves to* $\mathcal{S}$ *for all* $\alpha \in \mathcal{L}$,
- $\mathcal{R}$ *evolves silently to* $\mathcal{S}$, *denoted by* $\mathcal{R} \overset{\tau}{\rightarrowtail} \mathcal{S}$, *if* $\mathcal{R}$ τ-*evolves to* $\mathcal{S}$,
- $\mathcal{R}$ *evolves visibly to* $\mathcal{S}$, *denoted by* $\mathcal{R} \overset{v}{\rightarrowtail} \mathcal{S}$, *if* $\mathcal{R}$ a-*evolves to* $\mathcal{S}$ *for all* $a \in \mathcal{L} \setminus \{\tau\}$.

Our notion of evolution is the 'asymmetric' version of *progression* in [8]: $\mathcal{R}$ progresses to $\mathcal{S}$ in the sense of [8] iff $\mathcal{R}$ evolves to $\mathcal{S}$ and $\mathcal{R}^{-1}$ evolves to $\mathcal{S}^{-1}$.

In the following, we build a theory of up-to techniques to reason about simulations. This leads to simpler developments, and we show at the end of each section how to use the results to obtain proof techniques for bisimulation.

In the definition below, and in the remainder of the paper, we implicitly refer to *weak* relations. There are several equivalent definitions of bisimilarity. The following directly gives the standard way to prove a bisimilarity result between two processes P and Q: exhibit a bisimulation $\mathcal{R}$ containing the pair (P, Q).

Definition 1.3 (simulation, bisimulation, expansion). *Let* $\mathcal{R}$ *be a relation,* $\mathcal{R}$ *is a* simulation *(resp.* silent simulation*) if* $\mathcal{R} \rightarrowtail \mathcal{R}$ *(resp.* $\mathcal{R} \overset{\tau}{\rightarrowtail} \mathcal{R}$*).* $\mathcal{R}$ *is a* bisimulation *if* $\mathcal{R}$ *and* $\mathcal{R}^{-1}$ *are simulations. Two processes* P *and* Q *are* bisimilar, *written* $P \approx Q$, *if* $P \mathcal{R} Q$ *for some bisimulation* $\mathcal{R}$.

Expansion, *denoted by* $\succsim$, *is the largest relation such that* $\succsim^{-1}$ *is a simulation, and, whenever* $P \succsim Q$,

1. $P \overset{\tau}{\rightarrow} P'$ *implies* $Q \overset{\tau}{\rightarrow} Q'$ *and* $P' \succsim Q'$ *for some* Q', *or* $P' \succsim Q$;
2. $P \overset{a}{\rightarrow} P'$ *implies* $Q \overset{a}{\Longrightarrow} Q'$ *and* $P' \succsim Q'$ *for some* Q'.

1.2 The Difficulty in the Weak Case

We now adapt the theory of up-to techniques of [8] to the weak case, and show where the difficulties arise. We let $\mathcal{F}, \mathcal{G}$ range over *functions* from relations to relations. We say that $\mathcal{F}$ *contains* $\mathcal{G}$, written $\mathcal{G} \subseteq \mathcal{F}$, if $\mathcal{G}(\mathcal{R}) \subseteq \mathcal{F}(\mathcal{R})$ for any relation $\mathcal{R}$. Given a relation $\mathcal{S}$, we define *identity* ($\mathcal{U}$), *constant-to-$\mathcal{S}$* ($\tilde{\mathcal{S}}$), *$\mathcal{S}$-left-chaining* ($\mathcal{S}\bullet$) and *$\mathcal{S}$-right-chaining* ($\bullet\mathcal{S}$) as follows:

$$\mathcal{U}(\mathcal{R}) \triangleq \mathcal{R} \qquad \tilde{\mathcal{S}}(\mathcal{R}) \triangleq \mathcal{S} \qquad \mathcal{S}\bullet(\mathcal{R}) \triangleq \mathcal{S}\mathcal{R} \qquad \bullet\mathcal{S}(\mathcal{R}) \triangleq \mathcal{R}\mathcal{S}$$

We define four *constructors*, i.e., functions from functions to functions: *composition* ($\circ$), *union* ($\cup$), *iteration* ($*$) and *chaining* ($\frown$), as follows:

$$(\mathcal{F} \circ \mathcal{G})(\mathcal{R}) \triangleq \mathcal{F}(\mathcal{G}(\mathcal{R})) \qquad\qquad (\mathcal{F}^0)(\mathcal{R}) \triangleq \mathcal{R}$$
$$(\mathcal{F} \cup \mathcal{G})(\mathcal{R}) \triangleq \mathcal{F}(\mathcal{R}) \cup \mathcal{G}(\mathcal{R}) \qquad\qquad (\mathcal{F}^{n+1})(\mathcal{R}) \triangleq \mathcal{F}^n(\mathcal{R}) \cup \mathcal{F}(\mathcal{F}^n(\mathcal{R}))$$
$$(\mathcal{F}^\frown\mathcal{G})(\mathcal{R}) \triangleq \mathcal{F}(\mathcal{R})\mathcal{G}(\mathcal{R}) \qquad\qquad (\mathcal{F}^*)(\mathcal{R}) \triangleq \bigcup_{n \geq 0} \mathcal{F}^n(\mathcal{R})$$

Definition 1.4 (monotonicity). *A function $\mathcal{F}$ is monotonic if $\mathcal{R} \subseteq \mathcal{S}$ entails $\mathcal{F}(\mathcal{R}) \subseteq \mathcal{F}(\mathcal{S})$ and the following conditions hold:*

(1) $\begin{cases} \mathcal{R} \overset{\tau}{\rightarrowtail} \mathcal{S} \\ \mathcal{R} \subseteq \mathcal{S} \end{cases} \Rightarrow \mathcal{F}(\mathcal{R}) \overset{\tau}{\rightarrowtail} \mathcal{F}(\mathcal{S})$
(2) $\begin{cases} \mathcal{R} \rightarrowtail \mathcal{S} \\ \mathcal{R} \subseteq \mathcal{S} \end{cases} \Rightarrow \mathcal{F}(\mathcal{R}) \overset{v}{\rightarrowtail} \mathcal{F}(\mathcal{S})$

This slightly strengthens the notion of *respectfulness* found in [8], in which the two kinds of transitions are treated uniformly. While the results of this section would hold using respectful functions, we will need this separation between silent and visible actions in Sect. 2.2.

Proposition 1.5 (correctness of monotonic functions). *Let $\mathcal{F}$ be a monotonic function. If $\mathcal{R} \rightarrowtail \mathcal{F}(\mathcal{R})$, then $\mathcal{F}^*(\mathcal{R})$ is a simulation.*

This proposition ensures that a monotonic function provides a sound up-to technique: whenever we can prove that $\mathcal{R}$ evolves to $\mathcal{F}(\mathcal{R})$, then $\mathcal{R}$ is contained in $\mathcal{F}^*(\mathcal{R})$, which is a simulation. We now exhibit some monotonic functions, and show how to combine them to obtain more powerful techniques.

Lemma 1.6. *Let $\mathcal{S}$ be a simulation, $\mathcal{U}$, $\tilde{\mathcal{S}}$, $\bullet\mathcal{S}$ and $\succsim\bullet$ are monotonic functions.*

In the sequel, we will say that a constructor *respects* a predicate P over functions, if, given arguments that satisfy P, it returns a function satisfying P.

Lemma 1.7. *Constructors $\circ$, $\cup$ and $*$ respect monotonicity.*

We can now apply our framework to reason about bisimulation relations, and revisit a result from [9]. We show that the proof becomes elementary.

Theorem 1.8. *If $\mathcal{R} \rightarrowtail \succsim \mathcal{R}^= \approx$ and $\mathcal{R}^{-1} \rightarrowtail \succsim (\mathcal{R}^{-1})^= \approx$, then $\mathcal{R} \subseteq \approx$.*

Proof. Using the previous results, $\mathcal{F}(\mathcal{R}) \triangleq \succsim \mathcal{R}^= \approx$ is monotonic, and $\mathcal{F}^*(\mathcal{R})$ and $\mathcal{F}^*(\mathcal{R}^{-1})$ are simulations. Then $\approx \mathcal{F}^*(\mathcal{R})$ and $\mathcal{F}^*(\mathcal{R}^{-1}) \approx$ are simulations. We check that $(\approx \mathcal{F}^*(\mathcal{R}))^{-1} = \mathcal{F}^*(\mathcal{R}^{-1})\approx$, so that $\mathcal{R} \subseteq \approx \mathcal{F}^*(\mathcal{R}) \subseteq \approx$. □

The transitivity problem. The $\approx$-left-chaining function is not monotonic. As a consequence, the chaining constructor does not respect monotonicity in general. Indeed, when trying to prove the monotonicity of $\approx\bullet$, we lack some hypotheses about silent transitions to close the corresponding diagram.

2 A Smooth Theory for the Weak Case

2.1 A Weaker Notion of Monotonicity

When looking at the counterexample given in the Introduction, we can observe that the problem is related to silent transitions: unlike visible transitions, they can be cancelled by $\approx$. We now exploit this observation to relax the definition of monotonicity, which leads to a smoother theory, where reasoning *up to weak bisimilarity* is allowed, but on visible actions only.

Definition 2.1 (weak monotonicity). *A function $\mathcal{F}$ is weakly monotonic if $\mathcal{R} \subseteq \mathcal{S}$ entails $\mathcal{F}(\mathcal{R}) \subseteq \mathcal{F}(\mathcal{S})$ and the following conditions hold:*

(1) $\mathcal{R} \stackrel{\tau}{\rightarrowtail} \mathcal{R} \Rightarrow \mathcal{F}(\mathcal{R}) \stackrel{\tau}{\rightarrowtail} \mathcal{F}(\mathcal{R})$ (2) $\begin{cases} \mathcal{R} \stackrel{\tau}{\rightarrowtail} \mathcal{R},\, \mathcal{R} \stackrel{v}{\rightarrowtail} \mathcal{S} \\ \mathcal{S} \stackrel{\tau}{\rightarrowtail} \mathcal{S},\, \mathcal{R} \subseteq \mathcal{S} \end{cases} \Rightarrow \mathcal{F}(\mathcal{R}) \stackrel{v}{\rightarrowtail} \mathcal{F}(\mathcal{S})$

The main difference w.r.t. Definition 1.4 is in clause (1): instead of respecting silent evolutions, a weakly monotonic function has to respect silent simulations. On the visible side (2), we suppose that $\mathcal{R}$ and $\mathcal{S}$ are silent simulations. The immediate consequence of these modifications appears in the following result: the up-to function may only be used on visible evolutions, and the candidate relation $\mathcal{R}$ has to be a silent simulation.

Proposition 2.2 (correctness of weakly monotonic functions). *Let $\mathcal{F}$ be weakly monotonic. If $\mathcal{R} \stackrel{\tau}{\rightarrowtail} \mathcal{R}$, and $\mathcal{R} \stackrel{v}{\rightarrowtail} \mathcal{F}(\mathcal{R})$, then $\mathcal{F}^*(\mathcal{R})$ is a simulation.*

Now we study the class of weakly monotonic functions: the following lemma ensures that the functions given by Lemma 1.6 can be used in the setting of weakly monotonic functions. Furthermore, weakly monotonic functions can be composed using the most important constructors:

Lemma 2.3. *Any monotonic function is weakly monotonic. Composition ($\circ$), union ($\cup$), iteration ($*$) and chaining ($\frown$) respect weak monotonicity.*

The closure under the chaining constructor naturally suggests the use of interesting up-to techniques, and in particular *up to transitivity*, given by $\mathcal{F}(\mathcal{R}) = \mathcal{R}^*$, and *up to weak bisimilarity*, using $\mathcal{F}(\mathcal{R}) = \approx \mathcal{R} \approx$.

2.2 Combining Monotonicity and Weak Monotonicity

In introducing weakly monotonic functions, we have restricted the use of up-to techniques to visible steps. We show how to develop further this approach by combining a monotonic function and a weakly monotonic function so as to employ constrained up-to techniques on silent steps, and full-fledged up-to techniques on visible steps.

Proposition 2.4 (unified up-to technique). *Let $\mathcal{F}$ be monotonic and $\mathcal{G}$ be weakly monotonic, and suppose further that $\mathcal{F} \subseteq \mathcal{G}$.*
If $\mathcal{R} \stackrel{\tau}{\rightarrowtail} \mathcal{F}(\mathcal{R})$ and $\mathcal{R} \stackrel{v}{\rightarrowtail} \mathcal{G}(\mathcal{R})$, then $(\mathcal{G}^*)^*(\mathcal{R})$ is a simulation.

In this result, we have to iterate $\mathcal{G}$ twice for technical reasons (see [5] for details). The following theorem is the counterpart of Theorem 1.8 in the richer setting we have introduced:

Theorem 2.5. *If* $\begin{cases} \mathcal{R} \stackrel{\tau}{\rightarrowtail} \widetilde{\succsim} \mathcal{R}^= \approx \\ \mathcal{R} \stackrel{v}{\rightarrowtail} (\mathcal{R} \cup \approx)^* \end{cases}$ *and* $\begin{cases} \mathcal{R}^{-1} \stackrel{\tau}{\rightarrowtail} \widetilde{\succsim} \mathcal{R}^{-1=} \approx \\ \mathcal{R}^{-1} \stackrel{v}{\rightarrowtail} (\mathcal{R}^{-1} \cup \approx)^* \end{cases}$ *then $\mathcal{R} \subseteq \approx$.*

We thus have a modular theory of up-to techniques for weak bisimulation that follows the approach for the strong case in [8]. Technically, the main improvement over previous works is the ability to exploit weaker hypotheses when reasoning about visible steps: for instance, *up to transitivity* ($\mathcal{R} \stackrel{v}{\rightarrowtail} \mathcal{R}^*$) and *up to weak bisimilarity* ($\mathcal{R} \stackrel{v}{\rightarrowtail} \approx \mathcal{R} \approx$) techniques entail valid proof methods.

3 Beyond Expansion

3.1 Controlled Relations

In this section, we enrich our framework with the possibility to use alternatives to $\succsim$ (which is the best we can do using Theorem 2.5) to handle τ transitions in bisimulation proofs. We define a class of relations that are controlled w.r.t. silent transitions, meaning that they prevent silent steps from being cancelled in an up-to bisimulation game.

The left-chaining functions associated to such relations are not weakly monotonic, and we thus have to depart from the theory we have developed so far. Roughly, a controlled relation is defined as a relation that induces a correct proof technique when used as a left-chaining up-to technique. The following technical definition introduces a uniform way to plug a non weakly monotonic left-chaining function into our setting.

Definition 3.1 (controlled relation). *We says that $\mathcal{B}$ is a* controlled relation *if the following holds for all relations $\mathcal{R}$, $\mathcal{S}$:*

(1) $\mathcal{R} \stackrel{\tau}{\rightarrowtail} \mathcal{B}^\star \mathcal{R} \Rightarrow \mathcal{B}^\star \mathcal{R} \stackrel{\tau}{\rightarrowtail} \mathcal{B}^\star \mathcal{R}$
(2) $\begin{cases} \mathcal{R} \stackrel{\tau}{\rightarrowtail} \mathcal{B}^\star \mathcal{R} \\ \mathcal{R} \stackrel{v}{\rightarrowtail} \mathcal{S}, \quad \mathcal{S} \stackrel{\tau}{\rightarrowtail} \mathcal{S} \end{cases} \Rightarrow \mathcal{B}^\star \mathcal{R} \stackrel{v}{\rightarrowtail} \mathcal{B}^\star \mathcal{S}.$

Remark 3.2. Note that a controlled relation need not be a simulation. However, by taking $\mathcal{R} = \mathcal{S} = \mathcal{I}$, we see that if $\mathcal{B}$ is controlled, then $\mathcal{B}^\star$ is a simulation. Also, the union of two controlled relations is not necessarily a controlled relation. Thus, this does not a priori induce a notion of *controlled bisimilarity*.

We say that $\mathcal{B}$ is a *controlled bisimulation* if it is a controlled relation contained in bisimilarity.

We now show how controlled relations can be used in simulation proofs.

Definition 3.3 (transparency). *Given a relation $\mathcal{B}$ and a function $\mathcal{F}$, $\mathcal{F}$ is $\mathcal{B}$-transparent if $\mathcal{F}(\mathcal{B}^\star \mathcal{R}) \subseteq \mathcal{B}^\star \mathcal{F}(\mathcal{R})$ for any relation $\mathcal{R}$.*

$\mathcal{F}$ is transparent *if it is $\mathcal{B}$-transparent for any relation $\mathcal{B}$.*

Proposition 3.4 (up to controlled relation). *Let $\mathcal{F}$ and $\mathcal{G}$ be two functions, and $\mathcal{B}$ a relation such that: $\mathcal{B}$ is a controlled relation, $\mathcal{F}$ is monotonic and $\mathcal{B}$-transparent, $\mathcal{G}$ is weakly monotonic. Suppose moreover that $\mathcal{G}$ contains $\mathcal{F}$ and $\mathcal{B}^\star\bullet$. If $\mathcal{R} \stackrel{\tau}{\rightarrowtail} \mathcal{B}^\star \mathcal{F}(\mathcal{R})$ and $\mathcal{R} \stackrel{v}{\rightarrowtail} \mathcal{G}(\mathcal{R})$, then $(\mathcal{G}^\star)^*(\mathcal{R})$ is a simulation.*

Lemma 3.5. *The identity and all $\mathcal{S}$-right-chaining functions are transparent. If $\mathcal{B} \subseteq \mathcal{S}$ then the constant-to-$\mathcal{S}$ function is $\mathcal{B}$-transparent.*

The composition, union and iteration constructors respect $\mathcal{B}$-transparency.

In practise, we will work with $\mathcal{S} = \approx$ and require that $\mathcal{B} \subseteq \approx$, so that condition $\mathcal{B} \subseteq \mathcal{S}$ will be satisfied.

Also notice that $\succsim \bullet$, the expansion-left-chaining function, is not transparent in general. This hence prevents us from encompassing the up to expansion proof technique in the statement of the following theorem.

Theorem 3.6. *Let $\mathcal{B}$ be a controlled bisimulation.*

If $\begin{cases} \mathcal{R} \stackrel{\mathcal{I}}{\rightarrowtail} \mathcal{B}^{\star}\mathcal{R}^{=} \approx \\ \mathcal{R} \stackrel{v}{\rightarrowtail} (\mathcal{R} \cup \approx)^{\star} \end{cases}$ *and* $\begin{cases} \mathcal{R}^{-1} \stackrel{\mathcal{I}}{\rightarrowtail} \mathcal{B}^{\star}\mathcal{R}^{-1=} \approx \\ \mathcal{R}^{-1} \stackrel{v}{\rightarrowtail} (\mathcal{R}^{-1} \cup \approx)^{\star} \end{cases}$ *then* $\mathcal{R} \subseteq \approx$.

This theorem is the counterpart of Theorem 2.5 using a controlled bisimulation instead of $\succsim$. A refined version of this result, in which two distinct controlled bisimulations are used for the silent evolutions of $\mathcal{R}$ and $\mathcal{R}^{-1}$, also holds. This can be useful in particular because the class of controlled bisimulations is not closed under union, as explained in Remark 3.2.

The remainder of the section is devoted to the construction of controlled relations.

3.2 Relaxed Expansion

Definition 3.7 (relaxed expansion). *A relation $\mathcal{E}$ is a* relaxed expansion *if whenever $P \mathrel{\mathcal{E}} Q$,*

1. *$P \stackrel{\tau}{\rightarrow} P'$ implies $Q \stackrel{\tau}{\rightarrow} Q'$ and $P' \mathrel{\mathcal{E}} Q'$ for some Q' or $P' \mathrel{\mathcal{E}} Q$,*
2. *$P \stackrel{a}{\rightarrow} P'$ implies $Q \stackrel{a}{\rightarrow}\stackrel{\tau}{\twoheadrightarrow} Q'$ and $P' \mathrel{\mathcal{E}} Q'$ for some Q'.*

Relaxed expansion, denoted by $\succsim$, is the union of all relaxed expansions $\mathcal{E}$ such that $\mathcal{E}^{-1}$ is a simulation.

When $P \succsim Q$ and $P \stackrel{a}{\rightarrow} P'$, Q has to do immediately a transition along a, but then can do as many silent transitions as necessary. The intuition behind the definition of relaxed expansion is that, using this possibility, Q can do some 'preliminary internal computation' in order to be able to remain faster than P until the next visible action.

Lemma 3.8. $\succsim$ *is a relaxed expansion, and we have:* $\succsim \subsetneq \succapprox \subsetneq \approx$.

Proof. The first point and the inclusions are straightforward. We illustrate the strictness of the inclusions using CCS processes: $a.b \succapprox a.\tau.b$ holds but not $a.b \succsim a.\tau.b$, and $a \approx \tau.a$ holds but not $a \succapprox \tau.a$. □

Theorem 3.9. *A relaxed expansion is a controlled relation. $\succsim$ is a controlled bisimulation.*

In general, $\succsim$ is not a congruence: for instance, in CCS, $a.b \succsim a.\tau.b$ holds but not $\bar{a} \mid a.b \succsim \bar{a} \mid a.\tau.b$. We remark that $\succsim$ is very close to *almost weak bisimilarity*, defined in [9]; the definition of $\succsim$ only fits better to our setting.

3.3 Introducing Termination Guarantees

We now show how to obtain controlled relations using termination guarantees. The theorems below follow from general results about commuting diagrams, presented in Sect. 4. Their proofs are thus deferred to that section.

Theorem 3.10. *Let $\mathcal{B}$ be a relation such that $\mathcal{B} \rightarrowtail \mathcal{B}^{+}$ and $\mathcal{B}$ terminates. Then $\mathcal{B}$ is a controlled relation.*

Theorem 3.11. *Let $\mathcal{B}$ be a relation such that $\mathcal{B} \rightarrowtail \mathcal{B}^*$ and $\mathcal{B}^+ \xrightarrow{\tau} {}^+$ terminates. Then $\mathcal{B}$ is a controlled relation.*

Unlike $\gtrapprox$, where the control on silent moves is fixed by the co-inductive definition of the relation, in these two results we start with a relation that roughly respects the – too permissive – weak bisimulation game, and constrain it a posteriori, in such a way that it cannot cancel silent steps indefinitely. For example, the erroneous up-to relation $\mathcal{B} = \{(a, \tau.a)\}$ is rejected because $\mathcal{B}$ evolves to $\mathcal{I} = \mathcal{B}^0$, and $\mathcal{B}^+ \xrightarrow{\tau}{}^+ = \{(a, a)\}$ obviously does not terminate.

There are processes that are not related by $\gtrapprox$, but by a relation satisfying the conditions of the previous theorems: consider $(a \mid (\nu b)b, \tau.a)$ or $(a + a, \tau.a)$.

Like for controlled relations, there is no direct way to define the greatest relation satisfying the requirements in Theorems 3.10 and 3.11, the main reason being that the union of terminating relations does not terminate in general. Also remark that the termination of $\mathcal{B}^+ \xrightarrow{\tau}{}^+$ does not entail the termination of $\mathcal{B}$ or $\xrightarrow{\tau}$. Theorem 3.11 can thus be applied to systems exhibiting infinite chains of τ transitions (e.g., π or CCS with replication).

We can use the up-to techniques we have defined previously to show the evolution condition in the above theorems ($\mathcal{B} \rightarrowtail \mathcal{B}^+$ or $\mathcal{B} \rightarrowtail \mathcal{B}^*$). However one has to be careful, because the simulation relation obtained with these techniques is $\mathcal{F}^*(\mathcal{B})$. Depending on $\mathcal{F}$, this relation may be reflexive, which discards Theorem 3.10, or just quite complex, so that proving the termination of $\mathcal{F}^*(\mathcal{B})$ or $\mathcal{F}^*(\mathcal{B})^+ \xrightarrow{\tau}{}^+$ may be delicate.

4 Results About Commuting Diagrams

In this section, we work in the more general setting of *diagrams*, commonly found in rewriting theory. In addition to $\mathcal{R}, \mathcal{S}$ we let $\rightarrow, \hookrightarrow$ and $\rightsquigarrow$ range over relations. As before, $\rightarrow^+$ (resp. $\twoheadrightarrow$) is the transitive (resp. reflexive transitive) closure of $\rightarrow$. We shall say that four relations $(\mathcal{R}, \rightarrow, \mathcal{S}, \hookrightarrow)$ form a *diagram*, denoted $(\mathcal{R}, \rightarrow) \gg (\mathcal{S}, \hookrightarrow)$, if whenever $P \mathrel{\mathcal{R}} Q$ and $P \rightarrow P'$, there is Q' such that $P' \mathrel{\mathcal{S}} Q'$ and $Q \hookrightarrow Q'$ (in our proofs, we shall sometimes adopt the usual graphical notation for diagrams). We say that two relations $\mathcal{R}$ and $\rightarrow$ *commute* if $(\mathcal{R}, \rightarrow) \gg (\mathcal{R}, \rightarrow)$. Notice that a relation $\mathcal{R}$ is a simulation iff $\mathcal{R}$ commutes with $\xrightarrow{\alpha}$ for all $\alpha \in \mathcal{L}$.

4.1 A First Termination Argument

Lemma 4.1. *Let $\mathcal{B}, \rightarrow$ be two relations such that $\mathcal{B}$ terminates. If $(\mathcal{B}, \rightarrow) \gg (\mathcal{B}^+, \twoheadrightarrow)$, then $\mathcal{B}^+$ and $\twoheadrightarrow$ commute.*

Remark 4.2. The commutation hypothesis $(\mathcal{B}, \rightarrow) \gg (\mathcal{B}^+, \twoheadrightarrow)$ cannot be weakened to $(\mathcal{B}, \rightarrow) \gg (\mathcal{B}^*, \twoheadrightarrow)$, or to "whenever $P \mathrel{\mathcal{B}} Q$ and $P \rightarrow P'$, $P' = Q$ or there is Q' such that $P' \mathrel{\mathcal{B}^+} Q'$ and $Q \twoheadrightarrow Q'$". Indeed, if we define

$$\mathcal{B} \triangleq \{(2,3),(3,4),(1,0)\}$$
$$\rightarrow \triangleq \{(3,2),(2,1),(1,0)\}$$

$$0 \underset{\mathcal{B}}{\leftarrow\!\!\!\leftarrow} 1 \leftarrow 2 \overset{\mathcal{B}}{\underset{}{\rightleftarrows}} 3 \xrightarrow{\mathcal{B}} 4$$

$\mathcal{B}$ terminates and satisfies the two alternative hypotheses; 2 $\mathcal{B}^*$ 4 and 2 $\to$ 1, but there is no i s.t. 4 $\twoheadrightarrow$ i and 1 $\mathcal{B}^*$ i.

A similar result: "if $\mathcal{B}$ terminates and $(\mathcal{B}, \to) \gg (\mathcal{B}^+, \twoheadrightarrow)$, then $\mathcal{B}^*$ and $\twoheadrightarrow$ commute" is given in [10–Exercise 1.3.2]. However we are interested in showing the stronger result below, in which diagrams can be composed with other relations (this is necessary to obtain controlled simulations).

Lemma 4.3. *Let $\mathcal{B}, \to, \hookrightarrow$ be three relations such that $\mathcal{B}$ terminates. If $(\mathcal{B}, \to) \gg (\mathcal{B}^+, \twoheadrightarrow)$ and $(\mathcal{B}, \hookrightarrow) \gg (\mathcal{B}^+, \twoheadrightarrow\hookrightarrow)$, then $\mathcal{B}^+$ and $\twoheadrightarrow\hookrightarrow$ commute.*

Proof. By induction over the well-founded relation $\mathcal{B}^{-1}$ with the predicate $\phi(P')$: "For all P, Q such that $P \twoheadrightarrow\hookrightarrow P'$ and P $\mathcal{B}^+$ Q, there is Q' such that $Q \twoheadrightarrow\hookrightarrow Q'$ and P' $\mathcal{B}^+$ Q'". □

Proposition 4.4. *Let $\mathcal{B}, \to, \hookrightarrow, \mathcal{R}, \mathcal{S}, \rightsquigarrow$ be six relations such that $\mathcal{B}$ terminates. If*
$$\begin{cases} (\mathcal{B}, \to) \gg (\mathcal{B}^+, \twoheadrightarrow) \\ (\mathcal{B}, \hookrightarrow) \gg (\mathcal{B}^+, \twoheadrightarrow\hookrightarrow) \end{cases} \text{and} \begin{cases} (\mathcal{R}, \to) \gg (\mathcal{B}^*\mathcal{R}, \twoheadrightarrow) \\ (\mathcal{R}, \hookrightarrow) \gg (\mathcal{B}^*\mathcal{S}, \twoheadrightarrow\rightsquigarrow) \end{cases}$$
then $(\mathcal{B}^*\mathcal{R}, \twoheadrightarrow\hookrightarrow) \gg (\mathcal{B}^*\mathcal{S}, \twoheadrightarrow\rightsquigarrow)$.

We can now give the first deferred proof from the previous section:

Proof (of Theorem 3.10).

1. Suppose $\mathcal{R} \stackrel{\tau}{\rightarrowtail} \mathcal{B}^*\mathcal{R}$, we apply Proposition 4.4, taking $\stackrel{\tau}{\to}$ for $\to$, and the identity relation for $\hookrightarrow$, $\rightsquigarrow$, and $\mathcal{S}$.
2. Suppose furthermore $\mathcal{R} \stackrel{v}{\rightarrowtail} \mathcal{S}$ and $\mathcal{S} \stackrel{\tau}{\rightarrowtail} \mathcal{S}$. Lemma 4.1 ensures that $\mathcal{B}^+$ is a silent simulation. We close the diagram marked with a (*) below with a simple induction.

$$\begin{array}{ccc} & \mathcal{B} & \\ a \downarrow & (H) & \downarrow a \\ & \mathcal{B}^+ & \\ \tau \downarrow & (\text{Lem. } 4.1) & \downarrow \tau \\ & \mathcal{B}^+ & \end{array} \qquad \begin{array}{ccc} & \mathcal{R} & \\ a \downarrow & (H) & \downarrow a \\ & \mathcal{S} & \\ \tau \downarrow & (*) & \downarrow \tau \\ & \mathcal{S} & \end{array}$$

We then apply Proposition 4.4, using $\stackrel{\tau}{\to}$ for $\to$, $\stackrel{a}{\twoheadrightarrow}\stackrel{\tau}{\twoheadrightarrow}$ for $\hookrightarrow$ and $\rightsquigarrow$. □

4.2 A Generalisation of Newman's Lemma

Lemma 4.5. *Let $\mathcal{B}, \to, \mathcal{R}$ be three relations such that $\mathcal{B}^+\to^+$ terminates. If $(\mathcal{B}, \to) \gg (\mathcal{B}^*, \twoheadrightarrow)$ and $(\mathcal{R}, \to) \gg (\mathcal{B}^*\mathcal{R}, \twoheadrightarrow)$, then $\mathcal{B}^*\mathcal{R}$ and $\twoheadrightarrow$ commute.*

Proof. It suffices to prove $(\mathcal{B}^*\mathcal{R}, \to) \gg (\mathcal{B}^*\mathcal{R}, \twoheadrightarrow)$: the commutation result then follows by a simple induction. We use an induction over the well-founded order induced by the termination of $\mathcal{B}^+\to^+$, with the predicate $\phi(P)$: "For all P', Q such that $P \to P'$ and P $\mathcal{B}^*\mathcal{R}$ Q, there is Q' such that $Q \twoheadrightarrow Q'$ and P' $\mathcal{B}^*\mathcal{R}$ Q'" (IH_1). Then we do a second induction on the derivation of P $\mathcal{B}^*\mathcal{R}$ Q (IH_2). From the first hypothesis, we get P_n such that the leftmost diagram below holds (we show the interesting case where $P_0 \to^+ P_n$). We use the internal induction to obtain Q_1 in the central diagram; this is possible since any process P'' such that $P_0 \ \mathcal{B}^+ \to^+ P''$ satisfies $P \ \mathcal{B}^+ \to^+ P''$: the external induction hypothesis is preserved. Finally, using a third induction on the derivation $P_1 \twoheadrightarrow P_n$, we close the diagram by applying $n-1$ times the external induction hypothesis (all processes between P_1 and P_n satisfy $P \ \mathcal{B}^+ \to^+ P_i$).

$$
\begin{array}{ccccc}
P & \mathcal{B} & P_0 & \mathcal{B}^\star \mathcal{R} & Q \\
\downarrow & (H) & \downarrow & & \downarrow \\
& & P_1 & & \\
\downarrow & & \downarrow & & \\
P' & \mathcal{B}^\star & P_n & &
\end{array}
\qquad
\begin{array}{ccccc}
P & \mathcal{B} & P_0 & \mathcal{B}^\star \mathcal{R} & Q \\
\downarrow & & \downarrow & (IH_2) & \downarrow \\
& & P_1 & \mathcal{B}^\star \mathcal{R} & Q_1 \\
\downarrow & & \downarrow & & \\
P' & \mathcal{B}^\star & P_n & &
\end{array}
\qquad
\begin{array}{ccccc}
P & \mathcal{B} & P_0 & \mathcal{B}^\star \mathcal{R} & Q \\
\downarrow & & \downarrow & & \downarrow \\
& & P_1 & \mathcal{B}^\star \mathcal{R} & Q_1 \\
\downarrow & & \downarrow & (IH_1)^{n-1} & \downarrow \\
P' & \mathcal{B}^\star & P_n & \mathcal{B}^\star \mathcal{R} & Q'
\end{array}
$$

$\square$

By taking $\mathcal{R} = \mathcal{I}$ in this lemma, we obtain the following corollary:

Corollary 4.6. *Let $\mathcal{B}, \to$ be two relations such that $\mathcal{B}^+ \to^+$ terminates. If $(\mathcal{B}, \to) \gg (\mathcal{B}^\star, \twoheadrightarrow)$, then $\mathcal{B}^\star$ and $\twoheadrightarrow$ commute.*

By taking $\mathcal{B} = \to$, we get Newman's lemma: "Local confluence and termination entail confluence". A different generalisation of this confluence lemma to commutation can be found in [2–Lemma 4.26]. However, the latter result is weaker than ours since it requires the termination of $\mathcal{B} \cup \to$, and thus the termination of both $\mathcal{B}$ and $\to$.

Remark 4.7 (up-to techniques and commuting diagrams). The previous corollary admits a direct and elegant proof using the decreasing diagram techniques of van Oostrom et al. [2–Theorem 4.25]. The details of this proof are given in [5]. However, results like Lemma 4.5 and Proposition 4.8 cannot be proved within the setting of [2], because they express properties beyond 'pure commutation'.

Fournet [3] and others have been using results from [2] to validate up-to techniques for *barbed equivalences*. This is not directly comparable to the present work, since in that setting, commutation results apply directly (visible actions are not taken into account). Moreover, these works do not exploit results based on termination guarantees on the relations between processes.

Proposition 4.8. *Let $\mathcal{B}, \to, \mathcal{R}, \hookrightarrow, \mathcal{S}, \rightsquigarrow$ be six relations s.t. $\mathcal{B}^+ \to^+$ terminates. If*
$$\begin{cases} (\mathcal{B}, \to) \gg (\mathcal{B}^\star, \twoheadrightarrow) \\ (\mathcal{B}, \hookrightarrow) \gg (\mathcal{B}^\star, \twoheadrightarrow\hookrightarrow) \end{cases} \text{and} \begin{cases} (\mathcal{R}, \to) \gg (\mathcal{B}^\star \mathcal{R}, \twoheadrightarrow) \\ (\mathcal{R}, \hookrightarrow) \gg (\mathcal{B}^\star \mathcal{S}, \twoheadrightarrow\rightsquigarrow) \end{cases}$$
then $(\mathcal{B}^\star \mathcal{R}, \twoheadrightarrow\hookrightarrow) \gg (\mathcal{B}^\star \mathcal{S}, \twoheadrightarrow\rightsquigarrow)$.

Like in the proof of Theorem 3.10, we use Proposition 4.8 and Corollary 4.6 to establish Theorem 3.11.

5 Concluding Remarks

Applications of our proof techniques. We have analysed two example systems where existing methods do not really help in establishing bisimilarity results, while the techniques we have presented are applicable, and indeed simplify the proofs. One of these examples comes from the work reported in [4] (which presents a direct bisimilarity proof). For lack of space, we do not present these here; the interested reader can refer to the long version of this paper [5]. More experience on case studies has to be developed in order to have a better understanding of how our techniques can be best combined, and how to tune the distinction between visible and internal computation steps.

A theorem prover formalisation of our results. All results in this paper have been formally checked in the Coq proof assistant [7], and the descriptions of the proofs we give actually closely follow the proof scripts (available from [6]). This is of particular interest for the proofs in Sect. 4, which require non trivial and error-prone reasoning, especially when reasoning about nested inductions.

Results about decreasing diagrams. Due to the presence of labelled transitions, results about decreasing diagrams from [2] are not applicable directly in our setting. We plan to study how the theory of [2] can be adapted to keep track of visible actions. This could be a way to provide an abstract approach for the definition of 'up to transitivity' techniques based on termination guarantees.

Acknowledgements. We would like to thank Davide Sangiorgi for his comments and suggestions, and Daniel Hirschkoff for helpful discussions and a great help during the redaction process.

References

1. S. Arun-Kumar and M. Hennessy. An efficiency preorder for processes. *Acta Informatica*, 29(9):737–760, 1992.
2. M. Bezem, J. W. Klop, and V. van Oostrom. Diagram techniques for confluence. *Information and Computation*, 141(2):172–204, 1998.
3. C. Fournet. *The Join-Calculus: a Calculus for Distributed Mobile Programming*. PhD thesis, Ecole Polytechnique, 1998.
4. D. Hirschkoff, D. Pous, and D. Sangiorgi. An Efficient Abstract Machine for Safe Ambients. Technical Report 2004-63, LIP – ENS Lyon, 2004. An extended abstract appeared in the proceedings of COORDINATION'05.
5. D. Pous. Up-to Techniques for Weak Bisimulation. Technical Report 2005-16, LIP – ENS Lyon, 2005.
6. D. Pous. Web appendix of this paper, 2005. Available at http://perso.ens-lyon.fr/damien.pous/upto.
7. INRIA projet Logical. The Coq proof assistant. http://coq.inria.fr/.
8. D. Sangiorgi. On the Bisimulation Proof Method. *Mathematical Structures in Computer Science*, 8:447–479, 1998.
9. D. Sangiorgi and R. Milner. The problem of "Weak Bisimulation up to". In *Proc. CONCUR '92*, volume 630 of *Lecture Notes in Computer Science*, pages 32–46. Springer Verlag, 1992.
10. TeReSe. *Term Rewriting Systems*. Cambridge University Press, 2003.

Petri Algebras

Eric Badouel[1], Jules Chenou[2], and Goulven Guillou[3]

[1] INRIA, IRISA, Campus Universitaire de Beaulieu,
F35042 Rennes Cedex, France
ebadouel@irisa.fr
[2] Faculté des Sciences, Université de Douala,
B.P. 24157 Douala, Cameroon
chenouj@yahoo.fr
[3] Université de Bretagne Occidentale, E3883,
B.P.817, 29285 Brest Cedex, France
goulven.guillou@univ-brest.fr

Abstract. The firing rule of Petri nets relies on a residuation operation for the commutative monoid of natural numbers. We identify a class of residuated commutative monoids, called Petri algebras, for which one can mimic the token game of Petri nets to define the behaviour of generalized Petri net whose flow relation and place contents are valued in such algebraic structures. We show that Petri algebras coincide with the positive cones of lattice-ordered commutative groups and constitute the subvariety of the (duals of) residuated lattices generated by the commutative monoid of natural numbers. We introduce a class of nets, termed lexicographic Petri nets, that are associated with the positive cones of the lexicographic powers of the additive group of real numbers. This class of nets is universal in the sense that any net associated with some Petri algebras can be simulated by a lexicographic Petri net. All the classical decidable properties of Petri nets however are undecidable on the class of lexicographic Petri nets. Finally we turn our attention to bounded nets associated with Petri algebras and show that their dynamics can be reformulated in term of MV-algebras.

1 Introduction

The Petri net model is a graphical and mathematical modeling tool that, since its introduction in the early sixties, have come to play a pre-eminent role in the formal study of concurrent discrete-event dynamic systems. A Petri net $(P, T, Pre, Post)$ consists of a finite set P of places, a finite set T of transitions (disjoint from P), and flow relations $Pre, Post : P \times T \to \mathbb{N}$. Places can contain some tokens representing the resources available in this place for the current configuration. A configuration of a Petri net is given as a vector $M : P \to \mathbb{N}$, called marking, indicating the number of tokens available in each place. Tokens are consumed and produced by the firing of transitions according to the so-called token game

$M\,[t\rangle\,M' \Leftrightarrow (\forall p \in P)\,M(p) \sqsupseteq Pre(p,t) \wedge M'(p) = (M(p)-Pre(p,t))+Post(p,t)$

The token game of Petri net says that in order for a transition t to fire in marking M it should be the case that each place contains enough resources as it is expressed by the condition $M(p) \sqsupseteq Pre(p,t)$ where $\sqsubseteq$ is the usual order relation on $\mathbb{N}$. Then the firing of transition t proceeds in two stages : a consumption of resources ($Pre(p,t)$ tokens are removed from place p) followed by a production of resources ($Post(p,t)$ tokens are added to place p). The notation $M\,[t\rangle\,M'$ expresses the fact that transition t is allowed to fire in marking M and that firing t in marking M produces the new marking M'. Numerous techniques, supported and automated by software tools, can be used to verify that some required properties are met for systems specified using Petri nets. For instance reachability, coverability, place-boundedness, deadlock and liveness can be decided on the class of Petri nets [13].

Numerous extensions of this basic model of Petri nets have been introduced over the years. Some of them are high level nets that allow for more compact representations but do not increase the expressive power of Petri nets: these high level nets can be unfolded into equivalent, even though in general much larger, Petri nets. Some extensions however change more dramatically the semantics of the original model. For instance timing constraints may be added, as in timed Petri nets or stochastic Petri nets for the purpose of enabling performance analysis. With continuous Petri nets the discrete state transition rule is replaced by a notion of trajectory using a continuum of intermediate states. In Fuzzy Petri nets one has a possibilistic measure of the firing of a transition in the given marking thus enabling to deal with uncertainty. Our purpose in this paper is to put forward an axiomatisation of the token game of Petri nets. More precisely we identify a class of commutative residuated monoids, called Petri algebras, for which one can mimic the token game of Petri nets to define the behaviour of generalized Petri nets whose flow relations and place contents are valued in such algebraic structures. The sum and its associated residuation capture respectively how resources within places are produced and consumed through the firing of a transition. The class of usual Petri nets is associated with the commutative monoid of natural numbers. We show that Petri algebras coincide with the positive cones of lattice-ordered commutative groups and constitute the subvariety of the (duals of) residuated lattices generated by the commutative monoid of natural numbers. The basic Petri net model is thus associated with the generator of the variety of Petri algebras which shows that these extended nets share all algebraic properties of Petri nets, in particular they have the same equational and inequational theory. We however exhibit a Petri algebra whose corresponding class of nets is strictly more expressive than the class of Petri nets, i.e. their class of marking graphs is strictly larger. More precisely, we introduce a class of nets, termed lexicographic Petri nets, that are associated with the positive cones of the lexicographic powers of the additive group of real numbers. This class of nets is proved to be universal in the sense that any net associated with some Petri algebra can be simulated by a lexicographic Petri net. All the classical decidable properties of Petri nets however (termination, covering, boundedness, structural boundedness, accessibility, deadlock, liveness ...) are proved to be undecidable on the class of lexicographic Petri nets. Finally we turn our attention to bounded nets associated with Petri algebras and show that their dynamics can be reformulated in term of MV-algebras.

2 An Axiomatisation of the Token Game

In order to obtain an axiomatisation of the token game of Petri nets we represent the marking of a net as a map $M : P \to \bigsqcup_{p \in P} A_p$ that associates with each place $p \in P$ the local value of the current configuration $M(p) \in A_p$ in this place. Content of places are resources that are consumed and produced according to the token game. Thus we assume that each place $p \in P$ is associated with a commutative divisibility monoid $A_p = (A_p, \oplus, 0)$, i.e. a monoid such that

$$\text{the relation } a \sqsupseteq b \Leftrightarrow \exists c \cdot a = b \oplus c \text{ is an order relation} \quad (1)$$

The constant 0 represents the absence of resource and the binary operator $\oplus$ the accumulation of resources in places. Immediate consequences of condition (1) are the following:

$$a \oplus b \sqsupseteq a, b$$
$$0 \sqsubseteq a$$
$$a \oplus b = 0 \Rightarrow a = b = 0$$

Moreover we need to have a residuation operation $\ominus$ such that $a \ominus b$ represents the residual resource obtained by substracting b from a when $b \sqsubseteq a$. Thus the following should hold true:

$$b \sqsubseteq a \Rightarrow a = (a \ominus b) \oplus b \quad (2)$$

Usual Petri nets corresponds to the situation where, for every place p, $A_p = (\mathbb{N}, +, 0)$ is the commutative monoid of natural numbers with the truncated difference $n \ominus m = \max(0; n - m)$ as residuation. This operation is characterized by the universal property that for every natural numbers n, m and p

$$n + m \sqsupseteq p \Leftrightarrow n \sqsupseteq p \ominus m$$

Up to the reversal of the order relation, it is a *commutative residuated monoid* i.e. a commutative monoid $(A, \oplus, 0)$ with an order relation $\leq$ and a residuation operation $\ominus$ which is a right adjoint to the addition, in the sense that

$$a \oplus b \leq c \Leftrightarrow a \leq c \ominus b \quad (3)$$

It follows immediately from this definition that a commutative monoid is residuated if and only if its addition is order preserving in each argument and the inequation $a \oplus b \leq c$ has a largest solution for a (namely $c \ominus b$). In particular the residual is uniquely determined by the addition and the order relation.

When the monoid is a divisibility monoid the order relation itself is defined in terms of the addition and thus the whole structure is characterized by its monoid reduct.

Proposition 1. *Let $(A, \oplus, 0, \sqsubseteq)$ be a commutative monoid where the neutral element is also the least element for the order relation, we assume that this monoid is co-residuated in the sense that there exists a residuation operation $\ominus$ such that*

$$a \oplus b \sqsupseteq c \Leftrightarrow a \sqsupseteq c \ominus b \qquad (4)$$

then the following conditions are equivalent

(i) It is a divisibility monoid: $a \sqsupseteq b \Leftrightarrow \exists c \cdot a = b \oplus c$
(ii) It is an upper semi-lattice with: $a \sqcup b = (a \ominus b) \oplus b$
(iii) $b \sqsubseteq a \Rightarrow a = (a \ominus b) \oplus b$

Definition 2. *A* Petri pre-structure *is a commutative monoid equipped with a residuation operation $(M, \oplus, 0, \ominus)$ satisfying the conditions (1) and (4).*

The firing of a transition proceeds in two stages: a consumption of resources in the input places followed by a production of resources in the output places. More precisely, the transition relation $M\,[t\rangle\,M'$ stating that transition t can fire in marking M and leads, when it is fired, to the new marking M' is given by:

$$M\,[t\rangle\,M' \Leftrightarrow \forall p \in P \;\; M(p) \sqsupseteq Pre(p,t) \wedge M'(p) = (M(p) \ominus Pre(p,t)) \oplus Post(p,t)$$

A net is called *homogeneous* if all the algebras A_p are identical. We will stick to homogeneous nets until Section 3 where it will be noticed that the "multi-sorted" case adds in fact no extra generality. By the way we also restrict our attention in this paper to commutative algebras. With non commutative monoids it would be possible [1] for example to take fifo nets [11] into account.

For any non empty sequence of transitions $u = a_0 \ldots a_{n-1} \in T^+$ we let $M\,[u\rangle\,M'$ state the existence of markings $M = M_0, M_1, \ldots, M_n = M'$ such that $M_i\,[a_i\rangle\,M_{i+1}$ for every $0 \leq i < n$. Moreover we set $M\,[\varepsilon\rangle\,M$ where $\varepsilon \in E^*$ is the empty sequence and M an arbitrary marking. We use $M\,[u\rangle$ (respectively $[u\rangle\,M'$) as a shorthand for $\exists M'\;M\,[u\rangle\,M'$ (resp. $\exists M\;M\,[u\rangle\,M'$). If $a, b \in T$ are transitions in a (n usual) Petri net we have the following equivalences (using the vectorial notations $\mathcal{P}(t) = (\mathcal{P}(p,t); p \in P) \in \mathbb{N}^P$ for $\mathcal{P} \in \{Pre, Post\}$)

$$\begin{aligned} M\,[ab\rangle &\Leftrightarrow M \sqsupseteq Pre(a) \text{ and } (M - Pre(a)) + Post(a) \sqsupseteq Pre(b) \\ &\Leftrightarrow M \sqsupseteq \max(Pre(a); Pre(a) + (Pre(b) - Post(a))) \\ &\Leftrightarrow M \sqsupseteq Pre(a) + \max(0; Pre(b) - Post(a)) \\ &\Leftrightarrow M \sqsupseteq Pre(a) \oplus (Pre(b) \ominus Post(a)) \end{aligned}$$

This suggests to let $Pre(uv) = Pre(u) \oplus (Pre(v) \ominus Post(u))$ for any sequences $u, v \in T^*$ and symmetrically $Post(uv) = (Post(u) \ominus Pre(v)) \oplus Post(v)$. For these definitions to make sense however, it remains to show that they do not depend upon the specific chosen decomposition $w = uv$; otherwise stated, the product defined on $A \times A$ by $(x, y) \otimes (x', y') = (x \oplus (x' \ominus y), (y \ominus x') \oplus y')$ should be associative.

Theorem 3. *For any Petri pre-structure, the following conditions are equivalent:*
  (i) Operation $\otimes$ is associative,
  (ii) the identity $(b \oplus c) \ominus a = (b \ominus (a \ominus c)) \oplus (c \ominus a)$ holds,
  (iii) the monoid is cancellable: $a \oplus b = a \oplus c \Rightarrow b = c$, and
  (iv) the identity $(a \oplus b) \ominus b = a$ holds.

Definition 4. *A Petri algebra is a Petri pre-structure with a cancellable monoid reduct.*

Corollary 5. *Petri algebras satisfy the following equivalence*

$$a \sqsupseteq b \oplus c \quad \Leftrightarrow \quad a \sqsupseteq b \text{ and } a \ominus b \sqsupseteq c \qquad (5)$$

Identity (ii) of Theorem 3 is an internalization of (5) using the axiomatization of the order relation: $a \sqsubseteq b \Leftrightarrow a \ominus b = 0$.

Let us consider a net over a Petri algebra A, then we can inductively define the applications $Pre, Post : P \times T^* \to A$ by letting $\varphi(p, u) = (Pre(p, u), Post(p, u))$ where $\varphi(p, -) : T^* \to A \times A$ is the unique monoid morphism such that the images $\varphi(p, t) = (Pre(p, t), Post(p, t))$ of the generators $t \in T$ be given by the flow relations of the net. Then the following holds:

$$Pre(p, \varepsilon) = Post(p, \varepsilon) = 0$$
$$Pre(p, uv) = Pre(p, u) \oplus (Pre(p, v) \ominus Post(p, u))$$
$$Post(p, uv) = (Post(p, u) \ominus Pre(p, v)) \oplus Post(p, v)$$

Theorem 6. *The generalized transition relation $M [u\rangle M'$ stating the existence of a sequence u of transitions leading from M to M' is given by any of the three following equivalent conditions*
  1. $\forall p \in P$ $M(p) \sqsupseteq Pre(p, u)$ and $M'(p) = (M(p) \ominus Pre(p, u)) \oplus Post(p, u)$
  2. $\forall p \in P$ $M'(p) \sqsupseteq Post(p, u)$ and $M(p) = (M'(p) \ominus Post(p, u)) \oplus Pre(p, u)$
  3. $\forall p \in P$ $M(p) \sqsupseteq Pre(p, u)$; $M'(p) \sqsupseteq Post(p, u)$ and $M(p) \ominus Pre(p, u) = M'(p) \ominus Post(p, u)$

We have so far identified the set of conditions that should be fulfilled by Petri algebras so that we can play the token game and the resulting firing rule is associative. To sum up, these structures are duals of commutative residuated lattices whose joins and meets are given by the formulas $a \sqcup b = a \ominus (a \ominus b)$ and $a \sqcap b = b \oplus (a \ominus b)$. Moreover this lattice is integral in the sense that the neutral element for the sum is also the least element of the lattice. Finally the underlying monoid is cancellable and this condition is equivalent to the identity $(a \oplus b) \ominus b = a$.

Using [3,9] we can conclude that Petri algebras coincide with the (duals of) *integral, cancellative and commutative GMV-algebras*. These algebras form a sub-variety of the variety of residuated lattices and the following result is a direct consequence of [10–Theorem 5.6 and corollaries].

Theorem 7. *Petri algebras coincide with the positive cones of lattice-ordered abelian groups. Moreover lattice-ordered abelian groups constitute the subvariety of lattice-ordered groups generated by the group $\mathbb{Z}$ of integer, and their positive cones (i.e. Petri algebras) is the subvariety of residuated lattices generated by $\mathbb{N}$.*

3 Lexicographic Petri Nets

We define a (generalized) Petri net as a structure $\mathcal{N} = (P, T, Pre, Post, M_0)$ where P is a finite set of places with a Petri algebra A_p associated with each place $p \in P$, T is a finite set of transitions disjoint from P and $Pre, Post : P \times T \to \bigsqcup_{p \in P} A_p$, the flow relations, are such that $\forall p \in P \; \forall t \in T \quad Pre(p,t), Post(p,t) \in A_p$. A marking is a map $M : P \to \bigsqcup_{p \in P} A_p$ that associates with each place $p \in P$ the local value of the current configuration $M(p) \in A_p$ in this place. M_0 is some fixed marking, called the *initial marking*. The transition relation $M\,[t\rangle\,M'$ stating that transition t can fire in marking M and leads, when it is fired, to the new marking M' is given by:

$$M\,[t\rangle\,M' \Leftrightarrow \forall p \in P \quad M(p) \sqsupseteq Pre(p,t) \wedge M'(p) = (M(p) \ominus Pre(p,t)) \oplus Post(p,t)$$

This relation can be extended inductively to sequences $u \in T^*$ of transitions by letting $M\,[\varepsilon\rangle\,M$ for every marking M and $M\,[t \cdot u\rangle\,M'$ if and only if there exists some marking M'' such that $M\,[t\rangle\,M''$ and $M''\,[u\rangle\,M'$ for every $t \in T$ and $u \in T^*$. The set of *reachable markings* is $Reach(\mathcal{N}) = \{M \mid \exists u \in T^* \; M_0\,[u\rangle\,M\}$, and the *marking graph* of a generalized net $\mathcal{N} = (P, T, Pre, Post, M_0)$ is the labelled graph $\Gamma_\mathcal{N} = (V, \Lambda, v_0)$ whose set of vertices is given by the set $V = Reach(\mathcal{N})$ of reachable markings with $v_0 = M_0$ and whose set of arcs $\Lambda \subseteq V \times T \times V$ is the restriction of the transition relation to the set of reachable markings: $\Lambda = \{(M, t, M') \mid M, M' \in V \wedge M\,[t\rangle\,M'\}$. Two generalized Petri nets are termed *equivalent* when they have isomorphic marking graphs.

We immediately see that a place p whose type A_p is a sub-algebra of a product of Petri algebras ($A_p \subseteq A_1 \times \cdots \times A_n$) can be replaced by n places $p_1, \ldots, p_n$ with respective types $A_1, \ldots, A_n$ without changing the marking graph (at least up to isomorphism). A classical result of universal algebra says that any algebra of a variety is a sub-direct product of sub-directly irreducible algebras. Thus we can assume without loss of generality that all algebras A_p are sub-directly irreducible algebras in the variety of Petri algebras. Now any $M(p)$ belongs to the sub-algebra of A_p generated by the set $\{M_0(p)\} \cup \bigcup_{t \in T} \{Pre(p,t), Post(p,t)\}$. Thus:

Theorem 8. *Every generalized Petri net is equivalent to a generalized Petri net all of whose types are sub-directly irreducible and finitely generated Petri algebras.*

Let $Irr(V)$ denote the set of sub-directly irreducible algebras of a variety V, then if V is a subvariety of W one has $Irr(W) \cap V = Irr(V)$; using the fact that the sub-directly irreducible commutative GMV-algebras are chains (totally ordered sets) we deduce that

Proposition 9. *sub-directly irreducible Petri algebras are chains.*

An algebra is sub-directly irreducible if and only if it admits a least non trivial congruence [4]. Now we know [5,6] that the congruences of Petri algebras are in bijective correspondance with their convex sub-monoids. On the one hand we can associate each congruence θ of a Petri algebra A with the class of the neutral element which is a convex sub-monoid $M_\theta = [0]_\theta$ of A. Conversely we associate each such monoid M to the congruence $\theta_M = \{(a,b) \in A^2 \mid b \ominus a, a \ominus b \in M\}$. The correspondances $\theta \mapsto M_\theta$ and $M \mapsto \theta_M$ are inverses to each other and they establish an isomorphism between the lattice of congruences of A and the lattice of the convex sub-monoids of A. Moreover for every $a \in A$, the principal congruence generated by the equation $a = 0$ corresponds to the convex sub-monoid generated by a. A Petri algebra is then sub-directly irreducible if and only if it admits a least non trivial convex sub-monoid. Let us assume that A is a totally ordered Petri algebra. Let

$$M(x) = \{y \in A \mid \exists k \in \mathbb{N} \cdot y \sqsubseteq k \cdot x = \underbrace{x \oplus \cdots \oplus x}_{k \text{ times}}\}$$

denote the principal convex sub-monoid generated by $x \in A$. $M(x)$ is non-trivial if and only if $x \neq 0$. Now if x is some element of a convex sub-monoid M of A one necessarily has $M(x) \subseteq M$; thus a minimal convex sub-monoid is principal and is generated by any of its non null elements. Since A is totally ordered and $x \leq y \Rightarrow M(x) \subseteq M(y)$ we deduce that A admits at most one minimal non trivial sub-monoid. $M(x)$ is minimal if and only if $y \ll x \Rightarrow y = 0$ where relation $\ll$ is given by $y \ll x \Leftrightarrow \forall k \in \mathbb{N} \cdot k \cdot y \sqsubset x$. Otherwise stated $y \ll x$ if and only if $y \sqsubset x$ and $M(y)$ is strictly included in $M(x)$. Therefore A has no non trivial minimal sub-monoid if and only if for every $x \in A \setminus \{0\}$ one can find some $y \in A \setminus \{0\}$ such that $y \ll x$. Under that condition one can form an infinite strictly decreasing chain thus proving that the order relation $\ll$ is not well-founded. Conversely if this order is well-founded then any non empty subset of A, and thus in particular $A \setminus \{0\}$ if A is not trivial, admits a least element for this order which shows the existence of a minimal non trivial sub-monoid. We thus have established the following:

Theorem 10. *A Petri algebra is sub-directly irreducible if and only if it is a chain and the order relation* $y \ll x \Leftrightarrow \forall k \in \mathbb{N} \cdot k \cdot y \sqsubset x$ *is well-founded.*

The lexicographic product $\mathbb{G} \circ \mathbb{H}$ of two ordered groups $\mathbb{G}$ and $\mathbb{H}$ is the product group $\mathbb{G} \times \mathbb{H}$ equipped with the lexicographic order relation:

$$(x,y) \leq_{\mathbb{G} \circ \mathbb{H}} (x',y') \Leftrightarrow x <_{\mathbb{G}} x' \text{ or } (x = x' \text{ and } y \leq_{\mathbb{H}} y')$$

If $\mathbb{G}$ and $\mathbb{H}$ are simply ordered abelian groups then the same holds for their lexicographic product. This product is associative and we can define inductively

$L_n(\mathbb{G}) = (\mathbb{G}^n)^+$ for every simply ordered abelian group $\mathbb{G}$ and integer $n \in \mathbb{N}$ by letting $\mathbb{G}^0 = \{0\}$ be the trivial group and $\mathbb{G}^{n+1} = \mathbb{G}^n \circ \mathbb{G}$, and where $\mathbb{G}^+$ denote the positive cone of group $\mathbb{G}$. The group $\mathbb{G}^n$ naturally embedds into $\mathbb{G}^m$ when $n \leq m$; the projective limit of this sequence of embeddings is the group $\mathbb{G}^\omega$ whose elements are the infinite sequences of elements in $\mathbb{G}$, with componentwise composition and the lexicographic order relation defined as follows: $u \leq_{lex} v \Leftrightarrow u <_{lex} v$ or $u = v$ where $u <_{lex} v \Leftrightarrow \exists n \in \mathbb{N} \, \forall m \leq n \ u_m = v_m$ and $u_n <_\mathbb{G} v_n$. The inductive limit, or "union" $\bigcup_{n<\omega} \mathbb{G}^n$, is the subgroup of $\mathbb{G}^\omega$ consisting of the sequences u of finite support $(supp(u) = \sup\{k \in \mathbb{N} \mid u_k \neq 0\} < \omega)$ with $\mathbb{G}^n$ identified with the subgroup of $u \in \mathbb{G}^\omega$ such that $supp(u) \leq n$.

Definition 11. *The set $Lex(\mathbb{G})$ of lexicographic Petri nets based on a totally ordered abelian group $\mathbb{G}$ is the set of (homogeneous) generalized Petri net of type $(\mathbb{G}^\omega)^+$. $Lex(\mathbb{G},n) \subseteq Lex(\mathbb{G})$ is the set of n-dimensional lexicographic Petri nets with type $L_n(\mathbb{G}) = (\mathbb{G}^n)^+ \subseteq (\mathbb{G}^\omega)^+$, i.e. all flow arc inscriptions and initial place contents, and hence all place contents in every accessible marking, are elements in $(\mathbb{G}^n)^+$.*

If $\mathcal{K}$ and $\mathcal{L}$ are subclasses of generalized Petri nets we let $\mathcal{K} \lesssim \mathcal{L}$ when every net in $\mathcal{K}$ is equivalent to some net in $\mathcal{L}$. This is a pre-order relation, we let $\approx$ denote its associated equivalence relation and $\underset{\not\approx}{\lesssim}$ the corresponding strict relation: $\mathcal{K} \underset{\not\approx}{\lesssim} \mathcal{L}$ when every net in $\mathcal{K}$ is equivalent to some net in $\mathcal{L}$ but there exists some net in $\mathcal{L}$ not equivalent to any net in $\mathcal{K}$. Notice that $Lex(\mathbb{G},n) \lesssim Lex(\mathbb{H},m)$ when $\mathbb{G} \subseteq \mathbb{H}$ and $n \leq m$; and that $Lex(\mathbb{Z},1)$ is the class of Petri nets.

Lemma 12. *Any finitely generated sub-directly irreducible Petri algebra A is isomorphic to a sub-algebra of the positive cone of some finite power of the additive group of real numbers: $A \subseteq (\mathbb{R}^n)^+$.*

By Theorem 8 we deduce the following result.

Theorem 13. *Every generalized Petri net is equivalent to some lexicographic Petri net, more precisely : $GenPetri \approx Lex(\mathbb{R})$*

We provide an example showing that $Lex(\mathbb{Z},1) \underset{\not\approx}{\lesssim} Lex(\mathbb{Z},2)$, i.e. that lexicographic Petri nets based on the group of integers of dimension 2 are already strictly more expressive than the class of Petri nets. Let us consider the net of type $L_2 = (\mathbb{Z} \circ \mathbb{Z})^+ = \{(n,m) \mid (n=0 \text{ and } m \geq 0) \text{ or } (n>0 \text{ and } m \in \mathbb{Z})\}$.

$$\boxed{a} \xleftarrow{(1,0)} (1,0) \xrightarrow{(0,1)} \boxed{b}$$

From the initial marking $(1,0)$ transition a can fire once $(1,0) [a\rangle (0,0)$ and transition b can fire an infinite number of time leading to the infinite firing sequence $(1,0) [b\rangle (1,-1) [b\rangle (1,-2) \ldots [b\rangle (1,-n) \ldots$ and there are no other transitions in the marking graph of the net. Suppose there exists some Petri net with an isomorphic marking graph. Since transition b can fire an infinite number of time, and

$$M_0 [b^n\rangle M_n \Rightarrow \forall p \in P\ M_n(p) = M_0(p) - n \times (Pre(p,b) - Post(p,b))$$

we deduce that for every place p it is the case that $Post(p,b) \sqsupseteq Pre(p,b)$ and thus $M_n(p) \sqsupseteq M_0(p)$. By monotony of the firing rule, any transition that can fire in the initial marking M_0 can also fire in any of the markings M_n obtained by firing b. Transition a is in contradiction with this property. Thus

Proposition 14. $Petri = Lex(\mathbb{Z}, 1) \precsim GenPetri$

It can also be shown that $Lex(\mathbb{Z}, n) \precsim Lex(\mathbb{Z}, m) \precsim Lex(\mathbb{Z})$ for $n < m < \omega$, and $Lex(\mathbb{Z}, n) \approx Lex(\mathbb{Q}, n) \precsim Lex(\mathbb{R}, n)$.

It appears to be difficult to obtain strict extensions of the class of Petri nets that preserve all of its decidable properties. Many of these extensions, like the class of Petri nets with inhibitor arcs, are indeed Turing-powerful. We recall that an inhibitor arc from a place p to a transition t (one such arc is depicted in Fig. 1) is intended to inhibit the firing of transition t as long as place p is not empty.

Theorem 15. *Lexicographic Petri nets are a strict extension of the class of Petri nets with inhibitor arcs. Thus Reachability, Coverability, Place-boundedness, Boundedness, Deadlock and Liveness are undecidable for the class of lexicographic Petri nets.*

The translation of a Petri net with inhibitor arcs $\mathcal{N}$ into an equivalent lexicographic Petri net $\overline{\mathcal{N}}$, illustrated in Fig. 1, consists in splitting every place p with

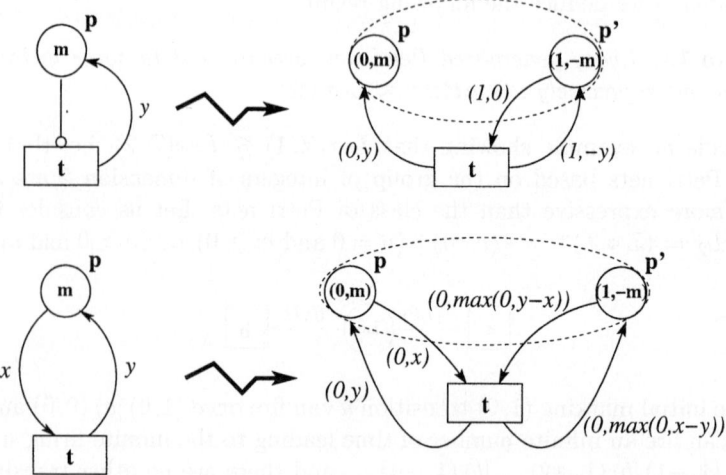

Fig. 1. A translation from Petri nets with inhibitor arcs into lexicographic Petri nets

initial marking $m \in \mathbb{N}$ of the original net into two places denoted p and p' with initial markings $(0, m) \in (\mathbb{Z} \circ \mathbb{Z})^+$ and $(1, -m) \in (\mathbb{Z} \circ \mathbb{Z})^+$ respectively.

4 Bounded Nets

A net is bounded if we can find an upper bound on the possible values of places in any accessible marking. Let us start our study on the algebraization of the dynamic of bounded nets by the following observation.

Proposition 16. *Any non trivial commutative Petri algebra is an unbounded lattice.*

However we can enforce boundedness by modifying the rule of the token game. Let us consider first the case of the usual Petri nets: assume that each place $p \in P$ is associated with a capacity $k_p \in \mathbb{N}$ and that we want to ensure that the value of a place p of a Petri net be bounded from above by its capacity k_p. For that purpose we modify the firing rule as follows (where all computations are performed in $\mathbb{Z}$)

$$M[t\rangle M' \Leftrightarrow \forall p \in P \begin{cases} M(p) \sqsupseteq Pre(p, t) \wedge (M(p) - Pre(p, t)) + Post(p, t) \sqsubseteq k_p \\ M'(p) = (M(p) - Pre(p, t)) + Post(p, t) \end{cases}$$

this rule can be reformulated as:

$$M[t\rangle M' \Leftrightarrow \forall p \in P \begin{cases} Pre(p, t) \sqsubseteq M(p) \sqsubseteq (k_p + Pre(p, t)) - Post(p, t) \\ M'(p) = (M(p) - Pre(p, t)) + Post(p, t) \end{cases}$$

Petri algebras are the positive cones G^+ of lattice-ordered abelian groups $G = (G, +, 0, \sqcup, \sqcap)$. It is an algebra with the following operations: the sum (restriction of the group operation $x \oplus y = x + y$), and the truncated difference ($x \ominus y = (x - y) \sqcup 0$).

Let $k \sqsupseteq 0$ be some element of this positive cone; we suppose that it is a *strong unit* in the sense that $\forall g \in G \, \exists n \in \mathbb{N} \cdot n \cdot k \sqsupseteq g$. We can by modifying the firing rule ensure that the values of places stay within the interval $I = [0, k] = \{g \in G \mid 0 \sqsubseteq g \sqsubseteq k\}$. This interval with induced order is a bounded lattice that can be equipped with the following operations: the truncated sum ($x \boxplus y = (x + y) \sqcap k$), the truncated difference ($x \ominus y = (x - y) \sqcup 0$), the product ($x \bullet y = ((x + y) - k) \sqcup 0$), the implication ($x \to y = ((y + k) - x) \sqcap k$), and the negation ($\neg x = (x \to 0) = k \ominus x$).

Such a structure is called an MV-algebra. MV-algebras are generalizations of boolean algebras used in the algebraic analysis of Lukasiewicz infinite-valued propositional logic and this class of algebras admits several equivalent definitions [7, 12]. We then state that the firing relation of a so-called bounded net associated with some MV-algebras is given by:

$$M[t\rangle M' \Leftrightarrow \forall p \in P \begin{cases} Pre(p, t) \sqsubseteq M(p) \sqsubseteq Post(p, t) \to Pre(p, t) \\ M'(p) = (M(p) \ominus Pre(p, t)) \boxplus Post(p, t) \end{cases}$$

The boolean algebra $2 = \{0,1\}$ is an MV-algebra where $x \boxplus y = x \sqcup y$ and $x \bullet y = x \sqcap y$. Let P be the set of places of a net and $B = 2^P = \wp(P)$ the corresponding product structure, the preceding firing rule can be reformulated as:

$$M [a\rangle M' \Leftrightarrow M \supseteq Pre(a) \wedge M \cap Post(a) \subseteq Pre(a) \wedge M' = (M \setminus Pre(a)) \cup Post(a)$$

which is the usual firing rule of 1-safe nets. If we replace the boolean algebra $2 = \{0,1\}$ by the interval $[0,1]$ of the additive group of real numbers, i.e. with $x \boxplus y = min(1, x+y)$ and $x \bullet y = max(0, x+y-1)$ then we obtain the firing rule of some kind of 1-safe "fuzzy" nets :

$$M [a\rangle M' \Leftrightarrow \forall p \in P \begin{cases} Pre(a,p) \sqsubseteq M(p) \sqsubseteq Pre(a,p) + 1 - Post(a,p) \\ M'(p) = min(1, M(p) + Post(a,p) - Pre(a,p)) \end{cases}$$

Mundici [12] proved that MV-algebras coincide with $[0, k]$ intervals of abelian lattice-ordered groups where k is a strong unit. More precisely if $G = (G, +, 0)$ is an abelian lattice-ordered group with strong unit k then $\Gamma(G, k) = (A, \boxplus, 0, \bullet, k, \neg)$ where $A = [0, k] = \{x \in G \mid 0 \sqsubseteq x \sqsubseteq k\}$, $x \boxplus y = (x+y) \sqcap k$, $x \bullet y = ((x+y) - k) \sqcup 0$, and $\neg x = k - x$ is an MV-algebra such that the restriction of the order relation of the group on the unit interval $[0, k]$ coincides with the order relation of the MV-algebra: $x \sqsubseteq y \Leftrightarrow x \ominus y = 0$ (where $x \ominus y = x \bullet \neg y = (x - y) \sqcup 0$). Moreover Γ extends into an equivalence between the respective categories, i.e. it induces a bijective correspondence between isomorphism classes of abelian lattice-ordered groups with strong unit and isomorphism classes of MV-algebras. Now we have seen that Petri algebras corresponds bijectively, up to isomorphism, to the positive cones of abelian lattice-ordered groups, we thus have a Petri algebra canonically associated with each MV-algebra. The following result shows that, by using complementary places, we can simulate a bounded Petri net by a generalized Petri net defined on the associated Petri algebra.

Theorem 17. *Any bounded net can be simulated by a generalized Petri net.*

The translation of a bounded net $\mathcal{N}$ into an equivalent generalized Petri net $\overline{\mathcal{N}}$, illustrated in Fig. (2), consists in splitting every place p with initial marking $m \in A_p$ of the bounded net into two places denoted p and p' with initial markings $m \in (\mathbb{G}_p)^+$ and $k_p - m \in (\mathbb{G}_p)^+$ respectively where $(\mathbb{G}_p, k_p)$ is the lattice-ordered abelian group with strong unit associated with the MV-algebra $A_p \simeq \Gamma(\mathbb{G}_p, k_p)$.

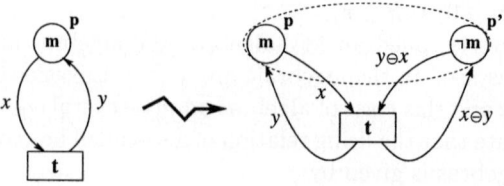

Fig. 2. Translation from bounded nets to generalized Petri nets

5 Conclusion

In this paper we have put forward an axiomatization of the token game of Petri nets by identifying a class of commutative residuated monoids, called Petri algebras, for which one can generalize the rule of token game of Petri nets to define the behaviour of generalized Petri net whose flow relation and place contents are valued in such algebras. In this way we have put the basis for a uniform presentation of various families of Petri nets by recasting them as particular instances of a generic class of Petri nets parametric in algebraic structures representing some concrete notion of resources. We thus have followed the line of research best illustrated in [8], a special issue of *Advances in Petri nets*, dedicated to the development of uniform approaches to Petri nets. However the present approach, centered on the notion of resources, is probably too concrete to be of practical interest in many situations. For instance even though one can describe continuous Petri nets in this framework the obtained semantics is too much extensional. We see two directions that can be used in order to derived more abstract representations for the behaviour of these generic nets. First, one can abstract of the flow arc inscriptions. Such an inscription takes its value in some algebra of abstract properties. A precondition then appears as a guard stating that some property has to be satisfied by the resources contained in the corresponding place and a postcondition is interpreted as adding resources to enforce some property. Second, one can abstract on the firing relation itself by giving the measure in some adequate semiring of the "firability" of a transition in some marking.

Acknowledgement. *The first author thanks Nikolaos Galatos for fruitful discussions on totally ordered commutative GMV algebras. The second author was supported by a grant from the University of Douala, Cameroon.*

References

1. E. Badouel, J. Chenou. Nets Enriched over Closed Monoidal Structures. In Proc. ICATPN'03, Eindhoven, Lecture Notes in Computer Science vol. 2679 (2003), 64-81.
2. E. Badouel, J. Chenou, G. Guillou. Petri Algebras. Inria Research Report 5355, November 2004. http://www.inria.fr/rrrt/rr-5355.html
3. P. Bahls, J. Cole, N. Galatos, P. Jipsen, C. Tsinakis. Cancellative residuated lattices. *Algebra Universalis* 12:42 (2003), 1-24.
4. G. Birkhoff. *Lattice Theory.* Third edition, AMS Colloquium Publications, vol. XXV (American Mathematical Society, Providence, 1967).
5. K. Blount. On the structure of residuated lattices. Ph. D. Thesis, Dept. of Mathematics, (Vanderbilt University, Nashville, Tennessee, 1999).
6. K. Blount, C. Tsinakis. The structure of Residuated Lattices, International Journal of Algebra and Computation (to appear).
7. R. Cignoli, I. D'Ottaviano, D. Mundici. Algebraic foundations of many-valued reasoning. Trends in Logic-studia Logica Library 7. (Kluwer Academic Publishers, Dordrecht, 2000).

8. H. Ehrig, G. Juhas, J. Padberg, G. Rozenberg (Eds.). *Unifying Petri Nets.* Advances in Petri Nets. Volume 2128 of Lecture Notes in Computer Science (2001).
9. N. Galatos. *Varieties of Residuated Lattices.* Ph. D. Thesis, Dept. of Mathematics, (Vanderbilt University, Nashville, Tennessee, 2003).
10. P. Jipsen, C. Tsinakis. A survey of Residuated Lattices. In *Ordered Algebraic Structures*, J. Martinez, editor (Kluwer Academic Publishers, Dordrecht, 2002), 19-56.
11. G. Memmi, A. Finkel. An introduction to fifo nets - monogeneous nets: a subclass of fifo nets. *Theoretical Computer Science* 35 (1985), 191-214.
12. D. Mundici. Interpretation of AF C*-algebras in Lukasiewicz sentential calculus. Journal of Functional Analysis 65:1 (1986), 15-63.
13. T. Murata. Petri Nets: Properties, Analysis and Applications. Proceedings of the IEEE, Vol. 77, No. 4 (1989), 541-580.

A Finite Basis for Failure Semantics

Wan Fokkink[1,2] and Sumit Nain[3]

[1] Vrije Universiteit Amsterdam,
Department of Theoretical Computer Science,
De Boelelaan 1081a, 1081 HV Amsterdam, The Netherlands
[2] CWI, Department of Software Engineering,
PO Box 94079, 1090 GB Amsterdam, The Netherlands
wanf@cs.vu.nl
[3] BRICS, Department of Computer Science, Aalborg University,
Fr. Bajersvej 7E, 9220 Aalborg Ø, Denmark
nain@cs.aau.dk

Abstract. We present a finite ω-complete axiomatization for the process algebra BCCSP modulo failure semantics, in case of a finite alphabet. This solves an open question by Groote [12].

1 Introduction

Labeled transition systems model processes by explicitly describing their states and their transitions from state to state, together with the actions that produce these transitions. Several notions of behavioral equivalence have been proposed, with the aim to identify those states of labeled transition systems that afford the same observations.

Van Glabbeek [10, 11] presented the linear time - branching time spectrum of behavioral equivalences for finitely branching, concrete, sequential processes. In this paper we focus on *failure semantics* [6, 7], which distinguishes a process by its "failure pairs", which consist of a finite (partial) trace together with a set of actions that cannot be executed at the ultimate state of this trace. Other semantics in the spectrum are based on (bi)simulation and on (decorated) traces. Figure 1 depicts the linear time - branching time spectrum, where a directed edge from one equivalence to another means that the source of the edge is finer than the target.

Van Glabbeek [10, 11] studied the semantics in his spectrum in the setting of the process algebra BCCSP, which contains only basic process algebraic operators from CCS and CSP, but is sufficiently powerful to express all finite synchronization trees. Van Glabbeek gave (sound and complete) axiomatizations for semantics in the spectrum, meaning that two *closed* BCCSP terms (i.e., terms that do not contain variables) can be equated if and only if they are equivalent.

An axiomatization E is ω-*complete* when an equation can be derived from E if (and only if) all its closed instantiations can be derived from E. In theorem proving applications, it is convenient if an axiomatization has this property, because it means that proofs by (structural) induction can be avoided in favor

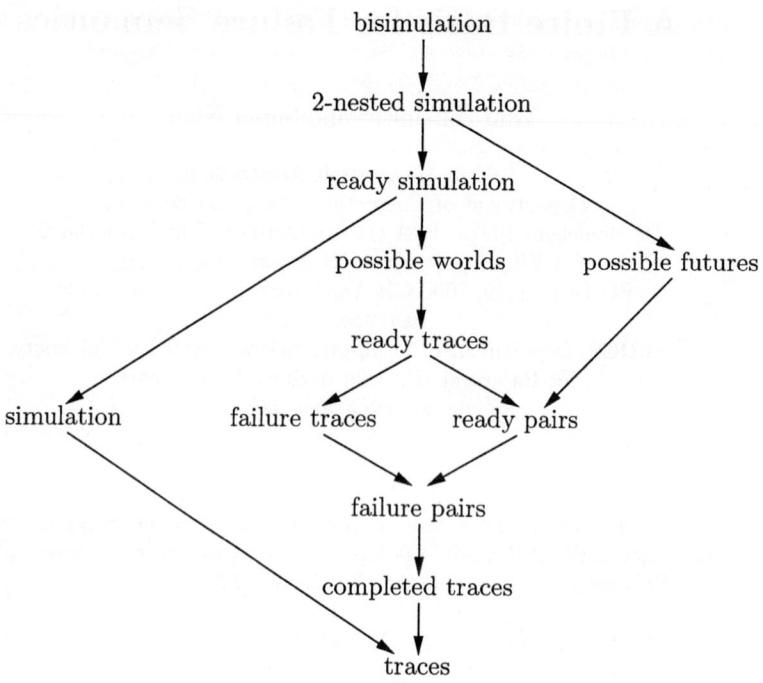

Fig. 1. The linear time - branching time spectrum

of purely equational reasoning; see [16]. In [14] it was argued that ω-completeness is desirable for the partial evaluation of programs.

Notable examples of ω-incomplete axiomatizations in the literature are the $\lambda K\beta\eta$-calculus (see [27]) and the equational theory of CCS [20]. Therefore laws such as commutativity of parallelism, which are valid in the initial model but which cannot be derived, are often added to the latter equational theory. For such extended equational theories, ω-completeness results were presented in the setting of CCS [22] and ACP [8]. Another negative result, for basic process algebra with the binary Kleene star, was reported in [2]: semantics no coarser than completed trace equivalence and no finer than ready simulation equivalence have no finite (sound and complete) axiomatization, so by default no finite ω-complete axiomatization.

A number of positive and negative results regarding finite ω-complete axiomatizations for BCCSP occur in the literature. Moller [22] proved that the finite axiomatization for BCCSP modulo bisimulation equivalence is ω-complete. Groote [12] presented a similar result for completed trace equivalence, for trace equivalence (in case of an alphabet with more than one element), and for readiness, failure and failure trace equivalence (in case of an infinite alphabet). Blom, Fokkink and Nain [4] proved that in case of an infinite alphabet, BCCSP modulo ready trace equivalence does not have a finite (sound and complete) axiomatization. Aceto, Fokkink and Ingólfsdóttir [3] proved a similar negative result for 2-nested simulation equivalence, independent of the cardinality of the alphabet.[1] Fokkink and Nain [9] showed that in case of a finite alphabet with more than one element, BCCSP modulo any semantics no coarser than readiness equivalence and no finer than possible worlds equivalence does not have a finite ω-complete axiomatization.

A *basis* of an equational theory is a set of axioms from which all equations in the theory can be derived. The existence of a finite basis for an equational theory is a classic topic of study in universal algebra (see, e.g., [19]), dating back to Lyndon [17]. Murskiĭ [24] proved that "almost all" finite algebras (namely all quasi-primal ones) are finitely based, while in [23] he presented an example of a three-element algebra that has no finite basis. Henkin [15] showed that the algebra of naturals with addition and multiplication is finitely based, while Gurevič [13] showed that after adding exponentiation the algebra is no longer finitely based. McKenzie [18] settled Tarski's Finite Basis Problem in the negative, by showing that the general question whether a finite algebra is finitely based is undecidable.

In this paper we present a finite ω-complete axiomatization for BCCSP modulo failure semantics. This provides a positive answer to an open question from [12]. The axiomatization consists of the standard axioms A1-4 for bisimulation, two standard axioms F1-2 for failure semantics, and a new axiom F3 that requires a finite alphabet. The latter axiom was obtained by considering *cover equations*, which aim to obtain a full coverage of the equational theory for (in this case) failure semantics. The central idea is that it is sufficient to only consider equations of the form $at + u \approx u$ and $x + u \approx u$ (where a denotes an action, x a variable, and t, u BCCSP terms). We classified the sound equations of this form, which we call *cover equations*. Now one can proceed in two ways. Either one can determine an infinite family of cover equations that obstructs a finite basis; this approach we took in [9] to prove the absence of a finite basis for a range of process semantics. Or one can determine a finite basis among the cover equations; this approach is followed in the current paper. We only present A1-4 and F1-3 together with a proof that this axiomatization is ω-complete, as the full classification of cover equations is quite involved and not needed for the proof.

We also present a proof that A1-4 together with F1-2 are ω-complete in case of an infinite alphabet. This proof is considerably simpler than the proof in [12], and moreover it is a sub-proof of the proof that A1-4 together with F1-3 are ω-complete in case of a finite alphabet. Last but not least, our axioms F1-2 are simplifications with respect to the axioms for failure equivalence as presented in [11, 12].

Groote [12] also asked whether in case of a finite alphabet, BCCSP modulo failure trace semantics has a finite ω-complete axiomatization. This question remains open.

---

[1] In case of an infinite alphabet, occurrences of action names in axioms should be interpreted as variables, as else most of the axiomatizations mentioned in this paragraph would be infinite.

2 Preliminaries

Syntax of BCCSP. BCCSP(A) is a basic process algebra for expressing finite process behavior. Its syntax consists of closed (process) terms p,q that are constructed from a constant $\mathbf{0}$, a binary operator $\_+\_$ called *alternative composition*, and unary *prefix* operators $a\_$, where a ranges over a nonempty set A of actions, called the *alphabet* (with typical elements a, b, c). Open terms t, u can moreover contain variables from a countably infinite set V (with typical elements x, y, z). A (closed) substitution maps variables in V to (closed) terms. For every term t and substitution σ, the term $\sigma(t)$ is obtained by replacing every occurrence of a variable x in t by $\sigma(x)$.

Transition rules. Intuitively, closed terms represent finite process behaviors, where $\mathbf{0}$ does not exhibit any behavior, $p+q$ is the nondeterministic choice between the behaviors of p and q, and ap executes action a to transform into p. This intuition is captured, in the style of Plotkin [26], by the transition rules below, which give rise to A-labeled transitions between closed terms.

$$\frac{}{ax \xrightarrow{a} x} \qquad \frac{x \xrightarrow{a} x'}{x+y \xrightarrow{a} x'} \qquad \frac{y \xrightarrow{a} y'}{x+y \xrightarrow{a} y'}$$

The *depth* of a term t, denoted by $depth(t)$, is the maximal number of transitions in sequence that t can exhibit. It is defined by: $depth(\mathbf{0}) = 0$, $depth(x) = 0$, $depth(t+u) = \max\{depth(t), depth(u)\}$, and $depth(at) = depth(t) + 1$.

For a closed term p, $\mathcal{I}(p)$ denotes the set of actions a for which there exists a transition $p \xrightarrow{a} p'$.

Definition 1. *A pair $(a_1 \cdots a_k, B)$ with $B \subseteq A$ and $k \geq 0$ is a* failure pair *of p_0 if $p_0 \xrightarrow{a_1} p_1 \cdots \xrightarrow{a_k} p_k$ with $\mathcal{I}(p_k) \cap B = \emptyset$. Two closed terms p and q are* failure equivalent, *denoted by $p \sim_F q$, if they have exactly the same failure pairs.*

Failure equivalence is a *congruence* for BCCSP(A), meaning that $p_1 \sim_F q_1$ and $p_2 \sim_F q_2$ implies $ap_1 \sim_F aq_1$ for $a \in A$ and $p_1 + p_2 \sim_F q_1 + q_2$. This follows from the fact that the transition rules above are in the failure format from [5].

Axiomatization. An *(equational) axiomatization* E for BCCSP(A) is a collection of equations $t \approx u$. We write $E \vdash t \approx u$ if this equation can be derived from the equations in E using the standard rules of equational logic, and $E \vdash F$ if $E \vdash t \approx u$ for all $t \approx u \in F$. An axiomatization E is *sound* modulo an equivalence $\sim$ on closed terms if $(E \vdash p \approx q) \Rightarrow p \sim q$, and it is *complete* modulo $\sim$ if $p \sim q \Rightarrow (E \vdash p \approx q)$, for all closed terms p and q. An axiomatization E is ω-*complete* if for each equation $t \approx u$ with $E \vdash \sigma(t) \approx \sigma(u)$ for all closed substitutions σ, we have $E \vdash t \approx u$.

The core axioms A1-4 below from [20] are sound and complete for BCCSP(A) modulo bisimulation equivalence [25], which is the finest semantics in the linear time - branching time spectrum (see Figure 1).

A1 $x + y \approx y + x$
A2 $(x + y) + z \approx x + (y + z)$
A3 $x + x \approx x$
A4 $x + 0 \approx x$

In the remainder of this paper, process terms are considered modulo A1-4.

We use *summation* $\sum_{i=1}^{k} t_i$ or $\sum_{i \in \{1,\ldots,k\}} t_i$, with $k \geq 0$, to denote $t_1 + \cdots + t_k$, where the empty sum denotes $\mathbf{0}$. Each process term is of the form $\sum_{i=1}^{k} a_i t_i + \sum_{j=1}^{\ell} x_j$. The $a_i t_i$ and x_j are called the *summands* of this process term.

As binding convention, alternative composition and summation bind weaker than prefixing.

3 A Finite Basis for Failure Semantics

3.1 Cover Equations

The central idea is that for bisimulation semantics, and thus for all process semantics in the linear time - branching time spectrum, axiom A3 is sound. So if an equation $t \approx u$ is sound, then $u + t \approx t$ and $t + u \approx u$ are sound too; and from the last two equations one can derive $t \approx u$, using axiom A1. Hence, by A2 and A4, it is sufficient to only consider sound equations of the form $at + u \approx u$ and $x + u \approx u$. We call these the *cover equations*. We present three lemmas that limit the form that cover equations can have.

Lemma 1. *If $t \approx u$ is sound modulo $\sim_F$, then t and u have the same variable summands.*

Proof. Let $x \in V$ be a summand of t. We define $\sigma(x) = a^{depth(u)+1}\mathbf{0}$ for some $a \in A$, and $\sigma(y) = \mathbf{0}$ for $y \neq x$. Then $(a^{depth(u)+1}, A)$ is a failure trace of $\sigma(t)$, so it must be a failure trace of $\sigma(u)$. This implies that x is a summand of u. □

Remark 1. Lemma 1 fails for trace equivalence. Namely, let $A = \{a\}$. Then $x + ax \approx ax$ is sound modulo trace equivalence. Lemma 1 does hold for completed trace equivalence.

Lemma 2. *If $at + u + bv \approx u + bv$ with $a \neq b$ is sound modulo $\sim_F$, then $at + u \approx u$ is sound modulo $\sim_F$.*

Proof. Since $at + u + bv \approx u + bv$ is sound modulo $\sim_F$ and $a \neq b$, for each closed substitution σ:

- $\mathcal{I}(\sigma(at + u)) = \mathcal{I}(\sigma(u))$, and
- each failure pair $(ab_1 \cdots b_k, B)$ of $a\sigma(t)$ is a failure pair of $\sigma(u)$.

So $at + u \approx u$ is sound modulo $\sim_F$. □

Lemma 3. *If $t + x \approx u + x$ is sound modulo $\sim_F$, and x is not a summand of $t + u$, then $t \approx u$ is sound modulo $\sim_F$.*

Proof. Suppose that $t \approx u$ is not sound modulo $\sim_F$; we prove that then $t + x \approx u + x$ is not sound modulo $\sim_F$.

$\sigma(t) \not\sim_F \sigma(u)$ for some closed substitution σ. Without loss of generality we can assume that some failure pair $(b_1 \cdots b_k, B_0)$ of $\sigma(t)$ is not a failure pair of $\sigma(u)$. The (fixed) set B_0 will play a crucial role in the construction below. We distinguish two cases.

1. $k = 0$.
   Then $\mathcal{I}(\sigma(t)) \neq \mathcal{I}(\sigma(u))$. Let $\sigma'(x) = \mathbf{0}$ and $\sigma'(y) = \sigma(y)$ for $y \neq x$. Since x is not a summand of $t + u$, $\mathcal{I}(\sigma'(t+x)) = \mathcal{I}(\sigma(t))$ and $\mathcal{I}(\sigma'(u+x)) = \mathcal{I}(\sigma(u))$. Then $\mathcal{I}(\sigma'(t+x)) \neq \mathcal{I}(\sigma'(u+x))$, and so $\sigma'(t+x) \not\sim_F \sigma'(u+x)$.

2. $k > 0$.
   We define a substitution σ' such that $\sigma'(y) = \sigma(y)$ for $y \neq x$, and $\sigma'(x)$ has the same failure pairs $(c_1 \cdots c_\ell, B_0)$ as $\sigma(x)$ for $\ell < k$, while it does not have failure pairs $(c_1 \cdots c_k, B_0)$. We obtain $\sigma'(x)$ from $\sigma(x)$ by replacing subterms ap at depth $k - 1$ by $\mathbf{0}$ if $a \notin B_0$, or by $aa\mathbf{0}$ if $a \in B_0$. That is,

$$\sigma'(x) = chop_{k-1}(\sigma(x))$$

where

$$chop_m(\mathbf{0}) = \mathbf{0}$$
$$chop_m(p + q) = chop_m(p) + chop_m(q)$$
$$chop_0(ap) = \begin{cases} \mathbf{0} & \text{if } a \notin B_0 \\ aa\mathbf{0} & \text{if } a \in B_0 \end{cases}$$
$$chop_{m+1}(ap) = a\, chop_m(p)$$

We prove the two desired properties concerning the failure pairs of $chop_m(p)$, for $m \geq 0$ and closed terms p. Let $c_1, \ldots, c_{m+1}$ range over A.

I For $\ell \leq m$, p and $chop_m(p)$ have the same failure pairs $(c_1 \cdots c_\ell, B_0)$.
   We use induction on ℓ. *Base case:* Since the summands of $chop_0(p)$ are $aa\mathbf{0}$ with $a \in \mathcal{I}(p) \cap B_0$, $\mathcal{I}(p) \cap B_0 = \emptyset$ if and only if $\mathcal{I}(chop_0(p)) \cap B_0 = \emptyset$. *Inductive case:* Let $\ell + 1 \leq m$. By induction, for closed terms q, q and $chop_{m-1}(q)$ have the same failure pairs $(c_2 \cdots c_{\ell+1}, B_0)$. Since $m > 0$, the transitions of $chop_m(p)$ are $chop_m(p) \xrightarrow{c_1} chop_{m-1}(p')$ for $p \xrightarrow{c_1} p'$. Hence p and $chop_m(p)$ have the same failure pairs $(c_1 \cdots c_{\ell+1}, B_0)$.

II $chop_m(p)$ does not have failure pairs $(c_1 \cdots c_{m+1}, B_0)$.
   We use induction on m. *Base case:* Since the summands of $chop_0(p)$ are $aa\mathbf{0}$ with $a \in \mathcal{I}(p) \cap B_0$, $chop_0(p)$ does not have a failure pair (c_1, B_0). *Inductive case:* By induction, for closed terms q, $chop_m(q)$ does not have failure pairs $(c_2 \cdots c_{m+2}, B_0)$. Since the transitions of $chop_{m+1}(p)$ are $chop_{m+1}(p) \xrightarrow{c_1} chop_m(p')$ for $p \xrightarrow{c_1} p'$, it follows that $chop_{m+1}(p)$ does not have failure pairs $(c_1 \cdots c_{m+2}, B_0)$.

We proceed to relate the failure pairs of $\sigma(t_0)$ and $\sigma'(t_0)$, for terms t_0.

III For $\ell < k$, $\sigma(t_0)$ and $\sigma'(t_0)$ have the same failure pairs $(c_1 \cdots c_\ell, B_0)$.
    We apply induction on ℓ. *Base case:* Clearly, $\sigma(x) \cap B_0 = \emptyset$ if and only if $\sigma'(x) \cap B_0 = \emptyset$, and moreover $\sigma(y) = \sigma'(y)$ for $y \neq x$. This implies that $\mathcal{I}(\sigma(t_0)) \cap B_0 = \emptyset$ if and only if $\mathcal{I}(\sigma'(t_0)) \cap B_0 = \emptyset$. *Inductive case:*

Let $\ell + 1 < k$. We prove for each summand of t_0 that applying σ or σ' gives rise to the same failure pairs $(c_1 \cdots c_{\ell+1}, B_0)$. By property I, $\sigma(x)$ and $\sigma'(x)$ have the same failure pairs $(c_1 \cdots c_{\ell+1}, B_0)$. Moreover, $\sigma(y) = \sigma'(y)$ for $y \neq x$. Furthermore, by induction, for each summand $c_1 t_1$ of t_0, $\sigma(t_1)$ and $\sigma'(t_1)$ have the same failure pairs $(c_2 \cdots c_{\ell+1}, B_0)$; so $\sigma(c_1 t_1)$ and $\sigma'(c_1 t_1)$ have the same failure pairs $(c_1 \cdots c_{\ell+1}, B_0)$.

IV If t_0 does not have the summand x, then $\sigma(t_0)$ and $\sigma'(t_0)$ have the same failure pairs $(c_1 \cdots c_k, B_0)$.

We prove for each summand of t_0 that applying σ or σ' gives rise to the same failure pairs $(c_1 \cdots c_k, B_0)$. $\sigma(y) = \sigma'(y)$ for $y \neq x$. Furthermore, by property III, for each summand $c_1 t_1$ of t_0, $\sigma(t_1)$ and $\sigma'(t_1)$ have the same failure pairs $(c_2 \cdots c_k, B_0)$; so $\sigma(c_1 t_1)$ and $\sigma'(c_1 t_1)$ have the same failure pairs $(c_1 \cdots c_k, B_0)$.

Recall that $(b_1 \cdots b_k, B_0)$ is a failure pair of $\sigma(t)$ and not of $\sigma(u)$. Since x is not a summand of $t + u$, by property IV, $(b_1 \cdots b_k, B_0)$ is a failure pair of $\sigma'(t)$ and not of $\sigma'(u)$. Since $k > 0$, $(b_1 \cdots b_k, B_0)$ is a failure pair of $\sigma'(t + x)$. By property II, $\sigma'(x) = chop_{k-1}(\sigma(x))$ does not have the failure pair $(b_1 \cdots b_k, B_0)$, so $(b_1 \cdots b_k, B_0)$ is not a failure pair of $\sigma'(u + x)$. Hence $\sigma'(t + x) \not\sim_F \sigma'(u + x)$.

We conclude that $t + x \approx u + x$ is not sound modulo $\sim_F$. $\square$

Remark 2. The condition in Lemma 3 that x is not a summand of $t + u$ is essential. For instance, $x + x \approx \mathbf{0} + x$ is sound modulo $\sim_F$, but $x \approx \mathbf{0}$ is not.

Remark 3. There exist equivalences in between bisimulation and partial traces that are a congruence for BCCSP(A), but for which Lemma 2 and/or Lemma 3 fail. So these lemmas have to be proved for each equivalence in the linear time - branching time spectrum individually.

3.2 ω-Complete Axiomatizations for Failure Semantics

We present two axioms for failure semantics.[2]

$$\begin{aligned} \text{F1} \quad & a(x + y) + ax + a(y + z) \approx ax + a(y + z) \\ \text{F2} \quad & a(x + by) + a(x + by + bz) \approx a(x + by + bz) \end{aligned}$$

It is not hard to see that F1 and F2 are sound for BCCSP(A) modulo $\sim_F$.

Theorem 1. *[10] A1-4+F1-2 is complete for BCCSP(A) modulo $\sim_F$.*

Theorem 2. *[12] If $|A| = \infty$, then A1-4+F1-2 is ω-complete.*

In case of a finite alphabet, A1-4+F1-2 is not ω-complete. Let $A = \{a_1, \ldots, a_n\}$ for some $n > 0$. Then the following axiom is sound for BCCSP(A) modulo $\sim_F$.

---

[2] Van Glabbeek [11] and Groote [12] presented a somewhat more complicated version of F2. In [11] it takes the form $a(bx + u) + a(by + v) \approx a(bx + by + u) + a(by + v)$.

F3$_n$ $a(x + \sum_{i=1}^n a_i z_i) + a(x + y + \sum_{i=1}^n a_i z_i) \approx a(x + y + \sum_{i=1}^n a_i z_i)$

F3$_n$ cannot be derived from A1-4+F1-2. This follows from the fact that A1-4+F1-2 are sound for BCCSP(A) modulo $\sim_F$ in case of an infinite alphabet, while F3$_n$ is not.

Theorem 3. *If $A = \{a_1, \ldots, a_n\}$, then A1-4+F1-2+F3$_n$ is ω-complete.*

The main aim of this paper is to present a proof of Theorem 3. Furthermore, although Theorems 1 and 2 have already been proved in earlier papers, we provide new proofs here. First of all, since axiom F2 is presented in a simpler form here than in [11, 12], strictly speaking Theorems 1 and 2 have not been proved in [10, 12], but are immediate corollaries from results in those papers and the fact that the earlier axiom $a(bx + u) + a(by + v) \approx a(bx + by + u) + a(by + v)$ can be derived from A1-4+F1-2. More important, Theorems 1 and 2 follow immediately from our proof of Theorem 3. So we obtain the new proofs for free, and moreover they are considerably simpler than the old proofs. In particular, the new proofs are fully equational, and more direct than the old proofs, which involve term rewriting and normal forms.

By abuse of notation, we let a finite set $X \subset V$ denote the term $\sum_{x \in X} x$. From now on, X, Y (possibly subscripted) denote finite subsets of V.

Proof. We derive all equations $t \approx u$ that are sound modulo $\sim_F$, by induction on $\max\{depth(t), depth(u)\}$. Clearly $u + t \approx t$ and $t + u \approx u$ are sound too; and from the last two equations one can derive $t \approx u$. So it suffices to derive all sound equations of the form $at + u \approx u$ and $x + u \approx u$. In view of Lemma 1, soundness of $x + u \approx u$ implies that x is a summand of u, so that $x + u \approx u$ follows from A3. So it suffices to derive all sound equations of the form $at + u \approx u$. In view of Lemmas 2 and 3, we can take u to be of the form $\sum_{j \in J} au_j$. (Note that the equations to-be-proved, which are thus obtained from the original equation $t \approx u$, involve terms with a depth $\leq \max\{depth(t), depth(u)\}$; so the induction order is respected.)

Hence, to prove Theorems 1, 2 and 3, it suffices to prove that each equation $at + \sum_{j \in J} au_j \approx \sum_{j \in J} au_j$ that is sound modulo $\sim_F$ can be derived from A1-4+F1-2 in case of an infinite alphabet, and from A1-4+F1-2+F3$_n$ in case of a finite alphabet with n elements. Consider such a sound equation modulo $\sim_F$:

$$a(X + \sum_{i \in I} b_i t_i) + \sum_{j \in J} a(Y_j + \sum_{k \in K_j} c_k u_k) \approx \sum_{j \in J} a(Y_j + \sum_{k \in K_j} c_k u_k) \quad (1)$$

As said before, we apply induction on the depth of the terms in the equation. In the base case of the induction, $I = \emptyset$ and $K_j = \emptyset$ for $j \in J$.

The main idea of the proof will be to restrict the syntactic form that the summands in equation (1) can have, by exploiting the fact that it is sound modulo $\sim_F$. Thus the application of a suitable closed substitution can provide information on syntactic relations between the summands.

Suppose, towards a contradiction, that $X \not\subseteq \cup_{j \in J} Y_j$. Let $x \in X \backslash (\cup_{j \in J} Y_j)$. Let σ be a closed substitution with $\sigma(x) = a^\ell \mathbf{0}$ where ℓ is greater than the depth of the terms in (1), and $\sigma(y) = \mathbf{0}$ for all $y \neq x$. Then clearly $(a^{\ell+1}, A)$ is a failure pair of $\sigma(a(X + \sum_{i \in I} b_i t_i))$, but not of $\sigma(\sum_{j \in J} a(Y_j + \sum_{k \in K_j} c_k u_k))$ for $j \in J$, contradicting the soundness of (1). Hence $X \subseteq \cup_{j \in J} Y_j$.

Suppose, towards a contradiction, that $\{b_i \mid i \in I\} \not\subseteq \cup_{j \in J} \{c_k \mid k \in K_j\}$. Let $b \in \{b_i \mid i \in I\} \setminus \cup_{j \in J} \{c_k \mid k \in K_j\}$. Let σ be the closed substitution with $\sigma(y) = \mathbf{0}$ for all $y \in V$. Then clearly $(ab, \emptyset)$ is a failure pair of $\sigma(a(X + \sum_{i \in I} b_i t_i))$, but not of $\sigma(a(Y_j + \sum_{k \in K_j} c_k u_k))$ for $j \in J$, contradicting the soundness of (1). Hence $\{b_i \mid i \in I\} \subseteq \cup_{j \in J} \{c_k \mid k \in K_j\}$.

Since (1) is sound, clearly

$$X + \sum_{i \in I} b_i t_i + \sum_{j \in J}(Y_j + \sum_{k \in K_j} c_k u_k) \approx \sum_{j \in J}(Y_j + \sum_{k \in K_j} c_k u_k) \tag{2}$$

is also sound modulo $\sim_F$. Let $L_j = \{k \in K_j \mid c_k \in \{b_i \mid i \in I\}\}$. By Lemmas 2 and 3, (2) implies that

$$\sum_{i \in I} b_i t_i + \sum_{j \in J} \sum_{k \in L_j} c_k u_k \approx \sum_{j \in J} \sum_{k \in L_j} c_k u_k \tag{3}$$

is sound modulo $\sim_F$. Hence, by induction on depth (or, in the base case, because $I = \emptyset$ and $L_j = \emptyset$ for $j \in J$), it can be derived from A1-4, F1-2 and, in case of a finite alphabet with n elements, F3$_n$.

We consider two cases.

1. $\{b_i \mid i \in I\} \neq A$.
   Suppose, towards a contradiction, that for all $j \in J$, either $Y_j \not\subseteq X$ or $\{c_k \mid k \in K_j\} \not\subseteq \{b_i \mid i \in I\}$. Let $c \in A \setminus \{b_i \mid i \in I\}$. Let σ be the closed substitution with $\sigma(x) = \mathbf{0}$ for $x \in X$ and $\sigma(y) = c$ for $y \notin X$. Then clearly $(a, A \setminus \{b_i \mid i \in I\})$ is a failure pair of $\sigma(a(X + \sum_{i \in I} b_i t_i))$, but not of $\sigma(a(Y_j + \sum_{k \in K_j} c_k u_k))$ for $j \in J$, contradicting the soundness of (1). Hence there is a $j_0 \in J$ such that $Y_{j_0} \subseteq X$ and $\{c_k \mid k \in K_{j_0}\} \subseteq \{b_i \mid i \in I\}$. We start with the term $\sum_{j \in J} a(Y_j + \sum_{k \in K_j} c_k u_k)$. Since $Y_{j_0} \subseteq X \subseteq \cup_{j \in J} Y_j$ and $K_{j_0} = L_{j_0}$, we can use F1 to convert the summand $a(Y_{j_0} + \sum_{k \in K_{j_0}} c_k u_k)$ into $a(X + \sum_{j \in J} \sum_{k \in L_j} c_k u_k)$. By (3) we convert this into $a(X + \sum_{i \in I} b_i t_i + \sum_{j \in J} \sum_{k \in L_j} c_k u_k)$. Finally, since $c_k \in \{b_i \mid i \in I\}$ for $k \in L_j$ and $j \in J$, we can use F2 to convert this into $a(X + \sum_{i \in I} b_i t_i)$. Thus we have derived (1).

2. $\{b_i \mid i \in I\} = A$. Note that in this case A is finite.
   We start with the term $\sum_{j \in J} a(Y_j + \sum_{k \in K_j} c_k u_k)$. Using F1 we can create the summand $a(\sum_{j \in J} Y_j + \sum_{j \in J} \sum_{k \in K_j} c_k u_k)$. Note that $K_j = L_j$ for $j \in J$, because $\{b_i \mid i \in I\} = A$. So we can use (3) to convert this summand into $a(\sum_{j \in J} Y_j + \sum_{i \in I} b_i t_i + \sum_{j \in J} \sum_{k \in K_j} c_k u_k)$. Finally, since $X \subseteq \cup_{j \in J} Y_j$ and $\{b_i \mid i \in I\} = A$, we can use F3$_{|A|}$ to convert this into $a(X + \sum_{i \in I} b_i t_i)$. Thus we have derived (1).

We note that in case of an infinite alphabet, the axioms F3$_n$ were not used in the derivation of (1). □

References

1. L. Aceto, W.J. Fokkink, R.J. van Glabbeek, and A. Ingólfsdóttir. Axiomatizing prefix iteration with silent steps. *Information Computation*, 127(1):26–40, 1996.
2. L. Aceto, W.J. Fokkink, and A. Ingólfsdóttir. A menagerie of non-finitely based process semantics over BPA*: From ready simulation to completed traces. *Mathematical Structures in Computer Science*, 8(3):193–230, 1998.
3. L. Aceto, W.J. Fokkink, and A. Ingólfsdóttir. 2-nested simulation is not finitely equationally axiomatizable. In *Proceedings 18th Symposium on Theoretical Aspects of Computer Science (STACS'01)*, Dresden, LNCS 2010, pp. 39–50. Springer, 2001.
4. S.C.C. Blom, W.J. Fokkink, and S. Nain. On the axiomatizability of ready traces, ready simulation and failure traces. In *Proceedings 30th Colloquium on Automata, Languages and Programming (ICALP'03)*, Eindhoven, LNCS. Springer, 2003. To appear.
5. B. Bloom, W.J. Fokkink, and R.J. van Glabbeek. Precongruence formats for decorated trace semantics. *ACM Transactions on Computational Logic*, 5(1):26–78, 2004.
6. S.D. Brookes, C.A.R. Hoare, and A.W. Roscoe. A theory of communicating sequential processes. *Journal of the ACM*, 31(3):560–599, 1984.
7. R. De Nicola and M.C.B. Hennessy. Testing equivalences for processes. *Theoretical Computer Science*, 34(1/2):83–133, 1984.
8. W.J. Fokkink and S.P. Luttik. An ω-complete equational specification of interleaving. In *Proceedings 27th Colloquium on Automata, Languages and Programming (ICALP'00)*, Geneva, LNCS 1853, pp. 729–743. Springer, 2000.
9. W.J. Fokkink and S. Nain. On finite alphabets and infinite bases: from ready pairs to possible worlds. In *Proceedings 7th Conference on Foundations of Software Science and Computation Structures (FOSSACS'04)*, Barcelona, LNCS 2987, pp. 182-194. Springer, 2004.
10. R.J. van Glabbeek. The linear time – branching time spectrum. In *Proceedings 1st Conference on Concurrency Theory (CONCUR'90)*, Amsterdam, LNCS 458, pp. 278–297. Springer 1990
11. R.J. van Glabbeek. The linear time – branching time spectrum I. The semantics of concrete, sequential processes. In J.A. Bergstra, A. Ponse, and S.A. Smolka, eds, *Handbook of Process Algebra*, pp. 3–99. Elsevier, 2001.
12. J.F. Groote. A new strategy for proving ω-completeness with applications in process algebra. In *Proceedings 1st Conference on Concurrency Theory (CONCUR'90)*, Amsterdam, LNCS 458, pp. 314–331. Springer, 1990.
13. R. Gurevič. Equational theory of positive natural numbers with exponentiation is not finitely axiomatizable. *Annals of Pure and Applied Logic*, 49:1–30, 1990.
14. J. Heering. Partial evaluation and ω-completeness of algebraic specifications. *Theoretical Computer Science*, 43:149–167, 1986.
15. L. Henkin. The logic of equality. *American Mathematical Monthly*, 84(8):597–612, 1977.
16. A. Lazrek, P. Lescanne, and J.-J. Thiel. Tools for proving inductive equalities, relative completeness, and ω-completeness. *Information and Computation*, 84(1):47–70, 1990.
17. R.C. Lyndon. Identities in two-valued calculi. *Transactions of the American Mathematical Society*, 71:457–465, 1951.
18. R.N. McKenzie. Tarski's finite basis problem is undecidable. *International Journal of Algebra and Computation*, 6(1):49–104, 1996.

19. R.N. McKenzie, G. McNulty, and W. Taylor. *Algebras, Varieties, Lattices.* Wadsworth & Brooks/Cole, 1987.
20. R. Milner. *Communication and Concurrency.* Prentice Hall, 1989.
21. R. Milner. A complete axiomatisation for observational congruence of finite-state behaviours. *Information and Computation,* 81(2):227–247, 1989.
22. F. Moller. *Axioms for Concurrency.* PhD thesis, University of Edinburgh, 1989.
23. V.L. Murskiĭ. The existence in the three-valued logic of a closed class with a finite basis having no finite complete system of identities. *Doklady Akademii Nauk SSSR,* 163:815–818, 1965. In Russian.
24. V.L. Murskiĭ. The existence of a finite basis of identities, and other properties of "almost all" finite algebras. *Problemy Kibernetiki,* 30:43–56, 1975. In Russian.
25. D.M.R. Park. Concurrency and automata on infinite sequences. In *Proceedings 5th GI (Gesellschaft für Informatik) Conference,* Karlsruhe, LNCS 104, pp. 167–183. Springer, 1981.
26. G.D. Plotkin. A structural approach to operational semantics. Report DAIMI FN-19, Aarhus University, 1981.
27. G.D. Plotkin. The λ-calculus is ω-incomplete. *Journal of Symbolic Logic,* 39:313–317, 1974.

Spatial Logics for Bigraphs[*]

Giovanni Conforti[1,3], Damiano Macedonio[2,3], and Vladimiro Sassone[3]

[1] Università di Pisa
[2] Università Ca' Foscari di Venezia
[3] University of Sussex

Abstract. Bigraphs are emerging as an interesting model for concurrent calculi, like CCS, pi-calculus, and Petri nets. Bigraphs are built orthogonally on two structures: a hierarchical place graph for locations and a link (hyper-)graph for connections. With the aim of describing bigraphical structures, we introduce a general framework for logics whose terms represent arrows in monoidal categories. We then instantiate the framework to bigraphical structures and obtain a logic that is a natural composition of a place graph logic and a link graph logic. We explore the concepts of separation and sharing in these logics and we prove that they generalise some known spatial logics for trees, graphs and tree contexts.

1 Introduction

To describe and reason about structured, distributed, dynamic resources is one of the main goals of global computing research. Recently, many *spatial logics*, in different contexts, have been studied to fulfill this goal. The term 'spatial,' as opposed to 'temporal,' refers to the use of modal operators inspecting the structure of the terms in the considered model. Spatial logics are usually equipped with a separation/composition binary operator that *splits* a term into two parts, in order to 'talk' about them separately. Looking closely, we observe that the notion of *separation* is interpreted differently in different logics. In 'separation' logics [18], it is used to reason about dynamic update of heap-like structures, and it is *strong* in that it forces names of resources in separated components to be disjoint. As a consequence, term composition is usually partially defined. In static spatial logics (e.g., for, trees [2], graphs [4] or trees with hidden names [5]), the separation/composition does not require any constraint on terms, and names are usually shared between separated parts. Similarly in dynamic spatial logics (for, e.g., ambients [6] or π-calculus [1]), where the separation is intended only for location in space. Context tree logic, introduced in [3], integrates the first approach above with a spatial logic for trees. The result is a logic able to express properties of tree-shaped structures (and contexts) with pointers, and it is used as an assertion language for Hoare-style program specifications in a tree memory model.

---

[*] Research partially supported by the EU projects: IHP 'Marie Curie **DisCo**' HPMT-CT-2001-00290, FET-GC '**MIKADO**' IST-2001-32222, and FET-GC '**MyThS**' IST-2001-32617.

Bigraphs [12, 14] are an emerging model for structures in global computing, which can be instantiated to model several well-known examples, including CCS [17], π-calculus [12], and Petri nets [16]. Bigraphs consist essentially of two graphs sharing the same nodes. The first graph, the *place graph*, is tree structured and expresses a hierarchical relationship on nodes (viz. locality in space and nesting of locations). The second graph, the *link graph*, is an hyper-graph and expresses a generic *"many-to-many"* relationship among nodes (e.g. data link, sharing of a channel). The two structures are orthogonal, so links between nodes can cross locality boundaries. Thus, bigraphs make clear the difference between structural separation (i.e., separation in the place graph) and name separation (i.e., separation on the link graph).

In this paper we introduce a spatial logic for bigraphs as a natural composition of a place graph logic (for tree contexts) and a link graph logic (for name linkings). The main point is that a resource has a spatial structure as well as a link structure associated to it. Suppose for instance to be describing a tree-shaped distribution of resources in locations. We may use formulae like $PC(A)$ and $PC_x(A)$ to describe a resource in an unnamed location, respectively location x, of 'type' PC (e.g. a computer) whose contents satisfy A. We can then write $PC(T) \otimes PC(T)$ to characterise terms with two unnamed PC resources whose contents satisfy the tautological formula (i.e., with anything inside). Using named locations, as e.g. in $PC_a(T) \otimes PC_b(T)$, we are able to express name separation, i.e., that names a and b are different. Furthermore, using link expressions we can force name-sharing between resources with formulae like:

$$PC_a(\mathsf{in}_c \otimes T) \overset{c}{\otimes} PC_b(\mathsf{out}_c \otimes T)$$

This describes two PC with different names, a and b, sharing a link on a distinct name c, which models, e.g., a communication channel. Name c is used as input (in) for the first PC and as an output (out) for the second PC. No other names are shared and c cannot be used elsewhere inside the PCs.

A bigraphical structure is, in general, a context with several holes and open links that can be filled by composition. This means that the logic can describe contexts for resources at no additional cost. We can then express formulae like $PC_a(T \otimes HD(id_1))$ that describes a modular computer PC, where id_1 represents a 'pluggable' hole in the hard disc HD. Contextual resources have many important applications. In particular, the contextual nature of bigraphs is useful to specify reaction rules, but it can also be used as a general mechanism to describe contexts of bigraphical data structures (cf. [8, 10]).

As bigraphs are establishing themselves as a truly general (meta)model of global systems, our bigraph logic, *BiLog*, aims at achieving the same generality as a description language: as bigraphs specialise to particular models, we expect BiLog to specialise to powerful logics on these. In this sense, the contribution of this paper is to propose BiLog as a unifying language for the description of global resources. We will explore this path in future work, fortified by the positive preliminary results obtained for semistructured data [8] and CCS [9].

2 An Informal Introduction to Bigraphs

Bigraphs formalise distributed systems by focusing on two of their main characteristics: locality and interconnections. A bigraph consists of a set of *nodes*, which may be nested in a hierarchical tree structure (the so-called *place graph*), and have ports that may be connected to each other (and to names) by *links* (the so-called *link graph*). Place graphs express locality, i.e., the physical arrangement of the nodes. Link graphs are hypergraphs and formalise connections among nodes. The orthogonality of the two structures dictates that nestings impose no constrain upon interconnections.

The bigraph G of Fig. 1 represents a system where people and things interact. We imagine two offices with employees logged on PCs. Every entity is represented by a node, shown with bold outlines, and every node is associated with a *control* (either PC, U, R1, R2). Controls represent kinds of nodes, and have fixed *arities* that determine their number of ports. Control PC marks nodes representing computers, and its arity is 3: in clockwise order, these ports represent a keyboard interacting with an employee U, a LAN to an other PC and open to the outside network, and a plug connecting the computer to the electrical mains of office R. Employees U may communicate with each other via the upper port in the picture. The nesting of nodes (place graph) is shown by the inclusion of nodes into each other; the connections (link graph) are drawn like lines.

At the top level of the nesting structure sit the *regions*. In Fig. 1 there is one sole region (the dotted box). Inside nodes there may be 'context' *holes*, drawn as shaded boxes, which are uniquely identified by ordinals. In figure the hole marked by 1 represents the possibility for another user U to get into office R1 and sit in front of a PC. The hole marked by 2 represents the possibility to plug a subsystem inside office R2.

Place graphs can be seen as *arrows* over a symmetric monoidal category whose objects are finite ordinals. We write $P : m \to n$ to indicate a place graph P with m holes and n regions. In Fig. 1, the place graph of G is of type $2 \to 1$. Given place graphs P_1, P_2, their composition $P_1 \circ P_2$ is defined only if the holes of P_1 are as many as the regions of P_2, and amounts to *filling* holes with regions, according to the number each carries. The tensor product $P_1 \otimes P_2$ is not commutative, as it 'renumbers' regions and holes 'from left to right'.

Link graphs are arrows of a partial monoidal category whose objects are (finite) sets Λ of names, that we assume to be *denumerable*. A link graph is an arrow $X \to Y$,

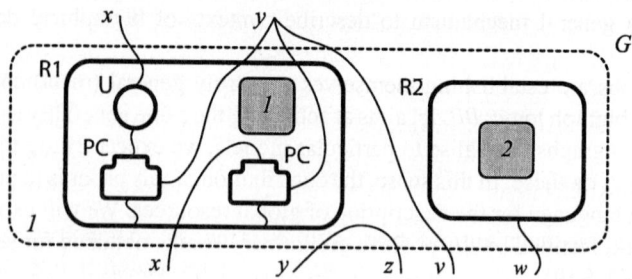

Fig. 1. A bigraph $G : \langle 2, \{x, y, z, v, w\}\rangle \to \langle 1, \{x, y\}\rangle$

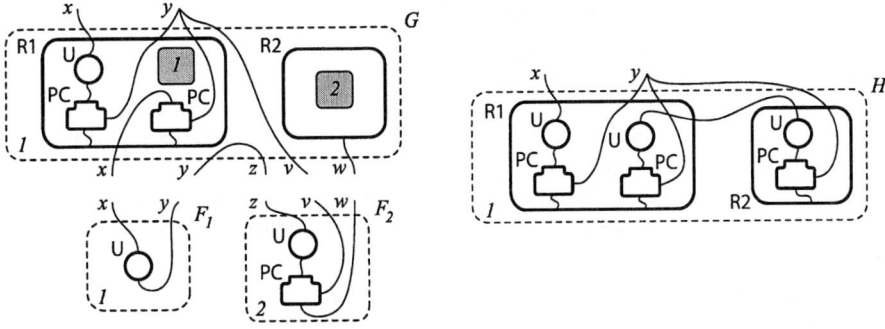

Fig. 2. Bigraphical composition, $H \equiv G \circ (F_1 \otimes F_2)$

with $X, Y \subseteq \Lambda$. The set X represents the *inner* names (drawn at the bottom of the bigraph) and Y represents the set of *outer* names (drawn on the top). The link graph connects ports to names or to *edges*, in any finite number. A link to a name is *open*, i.e., it may be connected to other nodes as an effect of composition. A link to an edge (represented in Fig. 1 by a line between nodes) is *closed*, as it cannot be further connected to ports. Thus, edges are *private*, or hidden, connections. The composition of link graphs $W \circ W'$ corresponds to *linking* the inner names of W with the corresponding outer names of W' and forgetting about their identities. As a consequence, the outer names of W' (resp. inner names of W) are not necessarily inner (resp. outer) names of $W \circ W'$, and the link graphs can perform substitution and renaming. The tensor product of link graphs is defined in the obvious way only if their inner (resp. outer) names are disjoint.

Combining ordinals with names we obtain *interfaces*, i.e., pairs $\langle m, X \rangle$ where m is an ordinal and X is a set of names. Combining the notion of place graph and link graphs on the same nodes we obtain the notion of bigraphs, i.e., arrows $G : \langle m, X \rangle \to \langle n, Y \rangle$.

Fig. 2 represents a more complex situation. At the top left-hand side is the system of Fig. 1, At the bottom left-hand side F_1 represents a user U ready to interact with a PC or with some other users, F_2 represents a user logged on its laptop, ready to communicate with other users. The system with F_1 and F_2 represents the tensor product $F = F_1 \otimes F_2$. The right-hand side of Fig. 2 represents the composition $G \circ F$. The idea is to insert F into the context G. The operation is partially defined, since it requires the inner names and the number of holes of G to match the outer names and the number of regions of F, respectively. Shared names create the new links between the two structures.

3 BiLog: Syntax and Semantics

As place and link graphs are arrows of a (partial) monoidal category, we first introduce a logic having monoidal categories as models and then we adapt it to model the orthogonal structures of place and link graphs. Each of these is expressive enough to model and generalise (e.g. by means of contexts) well-known spatial logics. Finally we apply the logic to model the whole structure of abstract bigraphs.

Table 1. BiLog($M, \otimes, \epsilon, \Theta, \equiv, \tau$)

$\Omega ::=$	$\text{id}_I \mid \ldots$	a constant formula for every Ω s.t. $\tau(\Omega)$			
$A, B ::=$	**F**	false	$A \Rightarrow B$	implication	
	id	identity	Ω	Constant for a simple term	
	$A \otimes B$	tensor product	$A \circ B$	composition	
	$A \multimap B$	left comp. adjunct	$A \multimapdot B$	right comp. adjunct	
	$A \otimes\!\!- B$	left prod. adjunct	$A -\!\!\otimes B$	right prod. adjunct	

$G \models \mathbf{F}$	$\stackrel{def}{=}$	never	
$G \models A \Rightarrow B$	$\stackrel{def}{=}$	$G \models A$ implies $G \models B$	
$G \models \Omega$	$\stackrel{def}{=}$	$G \equiv \Omega$	
$G \models \mathbf{id}$	$\stackrel{def}{=}$	$\exists I. G \equiv \text{id}_I$	
$G \models A \otimes B$	$\stackrel{def}{=}$	$\exists G_1, G_2. G \equiv G_1 \otimes G_2$ and $G_1 \models A$ and $G_2 \models B$	
$G \models A \circ B$	$\stackrel{def}{=}$	$\exists G_1, G_2. G \equiv G_1 \circ G_2$ and $\tau(G_1)$ and $G_1 \models A$ and $G_2 \models B$	
$G \models A \multimap B$	$\stackrel{def}{=}$	$\forall G'. G' \models A$ and $\tau(G')$ and $(G' \circ G) \downarrow$ implies $G' \circ G \models B$	
$G \models A \multimapdot B$	$\stackrel{def}{=}$	$\tau(G)$ implies $\forall G'. G' \models A$ and $(G \circ G') \downarrow$ implies $G \circ G' \models B$	
$G \models A \otimes\!\!- B$	$\stackrel{def}{=}$	$\forall G'. G' \models A$ and $(G' \otimes G) \downarrow$ implies $G' \otimes G \models B$	
$G \models A -\!\!\otimes B$	$\stackrel{def}{=}$	$\forall G'. G' \models A$ and $(G \otimes G') \downarrow$ implies $G \otimes G' \models B$	

The models are categories built on a (possibly partial) monoid $(M, \otimes, \epsilon)$, whose elements are dubbed *interfaces* and denoted by I, J. The elements of a BiLog model are arrows on the corresponding (partial) monoid. Given a set of term constructors Θ, ranged over by Ω, the arrows are defined by the term language $G ::= G \otimes G' \mid G \circ G' \mid \Omega$. Each Ω in Θ has a type $\Omega : I \to J$. For each interface I, we assume a distinguished construct $\text{id}_I : I \to I$. The types of constructors, together with the obvious rules [9] determine the type of each term. Terms of type $\epsilon \to J$ are called *ground*.

We consider terms up to a structural congruence $\equiv$, which subsumes the axioms of monoidal categories [9]. Later on, the congruence will be refined to model specialised structures, such as place graphs or bigraphs. All axioms are required to hold only when both sides are well typed. Throughout the paper, when using = or $\equiv$ we imply that both sides are defined and we write $(G) \downarrow$ to say that G is defined.

BiLog internalises the term constructors in the style of the ambient logic [6]. Constructors are represented in the logic as constant formulae, while tensor product and composition are expressed by connectives. We thus have two binary spatial operators. This contrasts with other spatial logics, which have only one: ambient-like logics, with parallel composition $A \mid B$, Separation Logic [18], with separating conjunction $A * B$, and Context Tree Logic [3], with application $K(P)$. Our logic is parameterised on a transparency predicate τ, reflecting that not every term can be directly observed in the logic: some are opaque and do not allow inspection of their contents. We will see that when all terms are observable (i.e. $\tau(G)$ for all G), logical equivalence corresponds to $\equiv$. Otherwise, it can be less discriminating. We assume that id_I and ground terms are always transparent, and τ preserves $\equiv$, hence $\otimes$ and $\circ$, in particular.

The logic BiLog($M, \otimes, \epsilon, \Theta, \equiv, \tau$) is formally defined in Table 1 and the meaning of formulae is given in terms of a satisfaction relation. It features a logical constant

Ω for each *transparent* construct Ω. The satisfaction of logical constants is simply the congruence to the corresponding constructor. The *horizontal decomposition* formula $A \otimes B$ is satisfied by a term that can be decomposed as the tensor product of terms satisfying A and B respectively. The degree of separation enforced by $\otimes$ between terms plays a fundamental role in the various fragments of the logic (notably link graph and place graph). The *vertical decomposition* formula $A \circ B$ is satisfied by terms that can be seen as the composition of terms satisfying A and B. We shall see that both connectives correspond in some cases to well known spatial connectives. We define the *left* and *right adjuncts* for composition and tensor to express extensional properties. The left adjunct $A \multimapdotinv B$ expresses the property of a term to satisfy B whenever inserted in a context satisfying A. Similarly, the right adjunct $A \multimap B$ expresses the property of a context to satisfy B whenever filled with a term satisfying A. A similar description for $\otimes\!\!-$ and $-\!\!\otimes$, the adjoints of $\otimes$. Observe that these collapse if the tensor is commutative in the model.

Derived Operators and Logical Properties. In Table 2 we outline some interesting operators that can be derived in BiLog. The operators constraining the interfaces are self-explanatory. The 'dual' operators have the following semantics: $A \ominus B$ is satisfied by terms G such that for every possible decomposition $G_1 \otimes G_2$ either $G_1 \models A$ or $G_2 \models B$. For instance, $A \ominus A$ describes terms where A is true in (at least) one part of each $\otimes$-decomposition. Similarly, the composition $A \bullet B$ expresses structural properties universally quantified on every $\circ$-decomposition. Both these connectives are useful to specify security properties or types. The adjuncts work as usual. The formulae $A^{\exists \otimes}$, $A^{\forall \otimes}$, $A^{\exists \circ}$, and $A^{\forall \circ}$ correspond to quantifications on the horizontal/vertical structure of terms. The equality between interfaces $I = J$ is easily derivable using $\otimes$ and $\otimes\!\!-$.

We can extend the idea of *sublocation* ($\sqsubseteq$) defined in [7] to our terms. The inductive definition of $\sqsubseteq$ specifies that $G \sqsubseteq G$, and $G' \sqsubseteq G$ if either $G \equiv G_1 \otimes G_2$, with $G' \sqsubseteq G_1$ (and symmetrically $G' \sqsubseteq G_2$) or $G \equiv G_1 \circ G_2$, with $\tau(G_1)$ and $G' \sqsubseteq G_2$. Exploiting this relation between ground terms, we define a *somewhere* modality. Intuitively, we say that a term satisfies $\diamondsuit A$ whenever one of its sublocations satisfies A. Quite surprisingly, $\diamondsuit A$ is expressible in the logic, as described in [9].

The lemma below states that the relation $\models$ is well defined w.r.t. the congruence and that the interfaces for transparent terms can be observed.

Lemma 3.1 (Type and Congruence preservation).
For every couple of term G, G', it holds: $G \models A$ and $G \equiv G'$ implies $G' \models A$.
For every term G, it holds: $G \models A_{I \to J}$ if and only if $G : I \to J$, $G \models A$, and $\tau(G)$.

BiLog induces a logical equivalence $=_L$ on terms in the usual sense, that is $G_1 =_L G_2$ if $G_1 \models A$ implies $G_2 \models A$ and vice versa, for every formula A.

Theorem 3.2 (Logical equivalence and congruence). *If the transparency predicate is always true, then for every term G, G', it holds: $G =_L G'$ if and only if $G \equiv G'$.*

Place Graph Logic (PGL). Place graphs are essentially ordered lists of regions hosting unordered labelled trees with holes. The labels of the trees correspond to controls K belonging to the fixed signature $\mathcal{K}$. We consider the monoid $(\omega, +, 0)$ of finite ordinals

Table 2. Derived Operators

$\mathbf{T}, \wedge, \vee, \Leftrightarrow, \Leftarrow, \neg$		Classical operators
A_I	$\stackrel{def}{=} A \circ id_I$	Constraining the source to be I
$A_{\to J}$	$\stackrel{def}{=} id_J \circ A$	Constraining the target to be J
$A_{I\to J}$	$\stackrel{def}{=} (A_I)_{\to J}$	Constraining the type to be $I \to J$
$A \circ_I B$	$\stackrel{def}{=} A \circ id_I \circ B$	Composition with interface I
$A \multimap_J B$	$\stackrel{def}{=} A_{\to J} \multimap B$	Contexts with J as target guarantee
$A \multimapinv_I B$	$\stackrel{def}{=} A_I \multimapinv B$	Composing with terms with I as source guarantee
$A \ominus B$	$\stackrel{def}{=} \neg(\neg A \otimes \neg B)$	Dual of tensor product
$A \bullet B$	$\stackrel{def}{=} \neg(\neg A \circ \neg B)$	Dual of composition
$A \leftarrowtail B$	$\stackrel{def}{=} \neg(\neg A \multimapinv \neg B)$	Dual of composition left adjunct
$A \rightarrowtail B$	$\stackrel{def}{=} \neg(\neg A \multimap \neg B)$	Dual of composition right adjunct
$A^{\exists\otimes}$	$\stackrel{def}{=} \mathbf{T} \otimes A \otimes \mathbf{T}$	Some horizontal term satisfies A
$A^{\forall\otimes}$	$\stackrel{def}{=} \mathbf{F} \ominus A \ominus \mathbf{F}$	Every horizontal term satisfies A
$A^{\exists\circ}$	$\stackrel{def}{=} \mathbf{T} \circ A \circ \mathbf{T}$	Some vertical term satisfies A
$A^{\forall\circ}$	$\stackrel{def}{=} \mathbf{F} \bullet A \bullet \mathbf{F}$	Every vertical term satisfies A
$I = J$	$\stackrel{def}{=} \mathbf{T} \otimes (id_\epsilon \wedge id_I \otimes\!\!-\, id_J)$	Equality between interfaces
$\Diamond A$	$\stackrel{def}{=} (\mathbf{T} \circ A)_\epsilon$	Somewhere modality (on ground terms)
$\boxempty A$	$\stackrel{def}{=} \neg\Diamond\neg A$	Anywhere modality (on ground terms)

m, n. Interfaces here represent the number of holes and regions of place graphs. Place graph terms are generated from the set $\Theta = \{1 : 0 \to 1, id_n : n \to n, join : 2 \to 1, \gamma_{m,n} : m + n \to n + m, K : 1 \to 1$ for $K \in \mathcal{K}\}$. The main structural term is K, that represents a region containing a single node with a hole inside. Other simple terms are *placings*, representing trees $m \to n$ with no nodes; the constructor 1 represents a barren region; *join* is a mapping of two regions into one; $\gamma_{m,n}$ is a permutation that interchanges the first m regions with the following n. The structural congruence $\equiv$ for place graph terms is refined by the usual axioms for symmetry of $\gamma_{m,n}$ and by the place axioms that essentially turn the operation *join* $\circ$ ($\_ \otimes \_$) in a commutative monoid with neutral element 1. Hence, the places generated by composition and tensor product from $\gamma_{m,n}$ are *permutations*. A place graph is *prime* if it has type $I \to 1$ (i.e., with a single region).

Defined the transparency predicate τ on each control in $\mathcal{K}$, the Place Graph Logic PGL($\mathcal{K}, \tau$) is $BiLog(\omega, +, 0, \equiv, \mathcal{K} \cup \{1, join, \gamma_{m,n}\}, \tau)$. We assume the τ to be true for *join* and $\gamma_{m,n}$. It follows from Theorem 3.2 that PGL can describe place graphs precisely. The logic resembles a propositional spatial tree logic, like [2]. The main differences are that PGL models contexts of trees and that the tensor product is not commutative, allowing us to model the order of regions. We can define a commutative separation using **join** and the tensor product, the *parallel composition* $A \mid B \stackrel{def}{=} \textbf{join} \circ (A_{\to 1} \otimes B_{\to 1})$. This separation is purely structural, and corresponds at term level to *join* $\circ (P \otimes P')$ that is a total operation on all prime place graphs.

We show that BiLog restricted to prime ground place graphs (with the always-true transparency predicate) is equivalent to the propositional spatial tree logic of [2] (STL in the following). The logic STL expresses properties of unordered labelled trees T

Table 3. Encoding STL in PGL over prime ground place graphs

Trees into Prime Ground Place Graphs	
$[\![\, 0 \,]\!] \stackrel{def}{=} 1 \qquad [\![\, a[T] \,]\!] \stackrel{def}{=} K(a) \circ [\![\, T \,]\!]$	$[\![\, T_1 \mid T_2 \,]\!] \stackrel{def}{=} join \circ ([\![\, T_1 \,]\!] \otimes [\![\, T_2 \,]\!])$
STL formulae into PGL formulae	
$[\![\, 0 \,]\!] \stackrel{def}{=} 1$	$[\![\, a[A] \,]\!] \stackrel{def}{=} \mathbf{K}(a) \circ_1 [\![\, A \,]\!]$
$[\![\, \mathbf{F} \,]\!] \stackrel{def}{=} \mathbf{F}$	$[\![\, A@a \,]\!] \stackrel{def}{=} \mathbf{K}(a) \circ_{-1} [\![\, A \,]\!]$
$[\![\, A \Rightarrow B \,]\!] \stackrel{def}{=} [\![\, A \,]\!] \Rightarrow [\![\, B \,]\!]$	$[\![\, A \mid B \,]\!] \stackrel{def}{=} [\![\, A \,]\!] \mid [\![\, B \,]\!]$
$[\![\, A \triangleright B \,]\!] \stackrel{def}{=} ([\![\, A \,]\!] \mid \mathbf{id}_1) \circ_{-1} [\![\, B \,]\!]$	

constructed from the empty tree 0, the labelled node containing a tree $a[T]$, and the parallel composition of trees $T_1 \mid T_2$. It is a static fragment of the ambient logic [6] characterised by propositional connectives, spatial connectives (i.e., $0, a[A], A \mid B$), and their adjuncts (i.e., $A@a, A \triangleright B$).

In Table 3 we encode the tree model of STL into prime ground place graphs, and STL operators into PGL operators. We assume a bijective encoding between labels and controls, and associate every label a with a distinct control $K(a)$. The monoidal properties of parallel composition are guaranteed by the symmetry and unit axioms of *join*. The equations are self-explanatory once we remark that: *(i)* the parallel composition of STL is the structural commutative separation of PGL; *(ii)* tree labels can be represented by the corresponding controls of the place graph; and *(iii)* location and composition adjuncts of STL are encoded in terms of the left composition adjunct, as they add logically expressible contexts to the tree. This encoding allows us to prove the following.

Theorem 3.3 (Encoding STL). *For each tree T and formula A of STL we have that $T \models_{SL} A$ if and only if $[\![\, T \,]\!] \models ([\![\, A \,]\!])_{0 \to 1}$.*

Differently from STL, PGL can also describe structures with several holes and regions. In [8] we show how PGL describes contexts of tree-shaped semistructured data. In particular multi-contexts can be useful to specify properties of web-services. Consider for instance a function taking two trees and returning the tree obtained by merging their roots. Such function is represented by the term *join*, which solely satisfies the formula **join**. Similarly, the function that takes a tree and encapsulates it inside a node *labelled* by K, is represented by the term K and captured by the formula **K**. Moreover, the formula $\mathbf{join} \circ (\mathbf{K} \otimes (\mathbf{T} \circ \mathbf{id}_1))$ expresses all contexts of form $2 \to 1$ that place their first argument inside a K node and their second one as a sibling of such node.

Link Graph Logic (LGL). For Λ a denumerable set of names, we consider the monoid of interfaces $(\mathcal{P}_{fin}(\Lambda), \uplus, \emptyset)$, where $\mathcal{P}_{fin}(\_)$ is the finite powerset operator and $\uplus$ is the union on disjoint pairs of sets and undefined otherwise. The structures that arise from such a monoid are the link graphs discussed in §2. They can describe nominal resources common in many areas, such as object identifiers, location names in memory structures, channel names, and ID attributes in XML documents.

Wiring terms are a structured way to map a set of inner names X into a set of outer names Y. They are generated by the constructors: $/a : \{a\} \to \emptyset$ and $^a/_X : X \to a$. The closure $/a$ hides the inner name a in the outer face. The substitution $^a/_X$ associates all the names in the set X to the name a. We denote wirings by ω, substitutions by σ, τ, and *renamings* (i.e., bijective substitutions) by α, β. Substitutions can be specialised in:

$$a \stackrel{def}{=} {}^a/_\emptyset; \qquad a \leftarrow b \stackrel{def}{=} {}^a/_{\{b\}}; \qquad a \Leftarrow b \stackrel{def}{=} {}^a/_{\{a,b\}}.$$

Constructor a represents the introduction of a name a, term $a \leftarrow b$ the renaming of b to a, and finally $a \Leftarrow b$ links (or fuse) a and b in the name a.

Given a signature $\mathcal{K}$ of controls K with corresponding ports $ar(K)$ we generate link graphs from wirings and the constructor $K_{\vec{a}} : \emptyset \to \vec{a}$ with $\vec{a} = a_1, \ldots, a_k$, $K \in \mathcal{K}$, and $k = ar(K)$; $K_{\vec{a}}$ represents a resource of kind K with named ports $\vec{a}$. Any ports may be connected to other node ports via wiring compositions.

The structural congruence $\equiv$ for link graphs is refined with obvious axioms for links, modelling α-conversion and extrusion of closed names, cf. [9]. We assume the transparency predicate τ to be true for wiring constructors.

Given the transparency τ for each control in $\mathcal{K}$, the Link Graph Logic LGL($\mathcal{K}, \tau$) is $BiLog(\mathcal{P}_{fin}(\Lambda), \uplus, \emptyset, \equiv, \mathcal{K} \cup \{/a, {}^a/_X\}, \tau)$. By Theorem 3.2, LGL describes the link graphs precisely. The logic expresses structural spatiality for resources and strong spatiality (separation) for names, and it can therefore be viewed as a generalisation of Separation Logic for contexts and multi-ports locations. On the other side the logic can describe resources with local (hidden/private) names between resources, and in this sense the logic is a generalisation of Spatial Graph Logic [4], considering the edges as resources.

In LGL the formula $A \otimes B$ describes a decomposition into two *separate* link graphs (i.e., sharing no resources, names, nor connections) satisfying respectively A and B. Observe that in this case, horizontal decomposition inherits the commutativity property from the monoidal tensor product. If we want a name a to be shared between separated resources, we need the sharing to be made explicit, and the sole way to do that is through the link operation. We therefore need a way to first separate the names occurring in two wirings in order to apply the tensor, and then link them back together.

As a shorthand if $W : X \to Y$ and $W' : X' \to Y'$ with $Y \subset X'$, we write $[W']W$ for $(W' \otimes id_{X' \setminus Y}) \circ W$ and if $\vec{a} = a_1, \ldots, a_n$ and $\vec{b} = b_1, \ldots, b_n$, we write $\vec{a} \leftarrow \vec{b}$ for $a_1 \leftarrow b_1 \otimes \ldots \otimes a_n \leftarrow b_n$ (and similarly for $\vec{a} \Leftarrow \vec{b}$). It is possible to derive from the tensor product a product with sharing on $\vec{a}$. Given $G : X \to Y$ and $G' : X' \to Y'$ with $X \cap X' = \emptyset$, we choose a list $\vec{b}$ (with the same length as $\vec{a}$) of fresh names. The composition with sharing $\vec{a}$ is:

$$G \overset{\vec{a}}{\otimes} G' \stackrel{def}{=} [\vec{a} \Leftarrow \vec{b}](([\vec{b} \leftarrow \vec{a}] \circ G) \otimes G')$$

By extending this sharing to all names we can define the parallel composition $G \mid G'$ as a total operation. However, such an operator does not behave "well" with respect to the composition, as shown in [15]. In addition a direct inclusion of a corresponding connective in the logic would impact the satisfaction relation by expanding the finite horizontal decompositions to the boundless possible name-sharing decompositions.

As a matter of fact, without name quantification it is not possible to build formulae that explore a link, since the latter has the effect of hiding names. For this task, we employ the name variables $x_1, \ldots, x_n$ and a fresh name quantification in the style of Nominal Logic [19].

$$G \models \mathsf{N} x_1, \ldots, x_n. A \stackrel{def}{=} \exists a_1 \ldots a_n \notin fn(G) \cup fn(A). G \models A\{x_1, \ldots x_n \leftarrow a_1 \ldots a_n\}$$

Using fresh name quantification we can define a notion of $\vec{a}$-linked name quantification for fresh names, whose purpose is to identify names that are linked to $\vec{a}$:

$$\vec{a} \mathbf{L} \vec{x}. A \stackrel{def}{=} \mathsf{N} \vec{x}. ((\vec{a} \Leftarrow \vec{x}) \otimes \mathbf{id}) \circ A.$$

The formula above expresses that variables in $\vec{x}$ denote in A names that are linked in the term to $\vec{a}$, and the role of $(\vec{a} \Leftarrow \vec{x})$ is to link the fresh names $\vec{x}$ with $\vec{a}$, while **id** deals with names not in $\vec{a}$. We also define a *separation-upto*, namely the decomposition in two terms that are separated apart from the link on the specific names in $\vec{a}$, which crosses the separation line.

$$A \stackrel{\vec{a}}{\otimes} B \stackrel{def}{=} \vec{a} \mathbf{L} \vec{x}. (((\vec{x} \leftarrow \vec{a}) \otimes \mathbf{id}) \circ A) \otimes B.$$

The idea of the formula above is that the shared names $\vec{a}$ are renamed in fresh names $\vec{x}$, so that the product can be performed and finally $\vec{x}$ is linked to $\vec{a}$ in order to actually have the sharing. The corresponding parallel composition operator is not directly definable using separation-upto, since we do not know a priori the name shared in arbitrary decompositions. However, we will show that a careful encoding is possible for the parallel composition of spatial logics with nominal resources.

We show that LGL can be seen as a contextual (and multi-edge) version of Spatial Graph Logic (SGL) [4]. The logic SGL expresses properties of directed edge labelled graphs G built from the empty graph *nil*, the edge labelled a from x to y nodes $a(x, y)$, the parallel composition of graphs $G_1 \mid G_2$, and the binding for local names of nodes $(vx)G$. We consider a $\mathcal{K}$ such that: there is a bijective function associating every edge label a to a distinct control $K(a)$ and the arity of every control is 2 (the ports represent the starting and arrival node respectively). The resulting link graphs can be interpreted as contextual edge labelled graphs and the resulting class of ground link graphs is isomorphic to the graph model of SGL. In Table 4 we encode the graphs modelling SGL into ground link graphs and SGL formulae into LGL formulae. The encoding is parametric on a finite set X of names containing the free names of the graph under consideration. Observe that when we force the outer face of the graphs to be a fixed finite set X, the encoding of parallel composition is simply the separation-upto $\vec{a}$, where $\vec{a}$ is a list of all the elements in X. Notice also how local names are encoded into name closures (and identity).

Theorem 3.4 (Encoding SGL). *For each graph G, finite set X containing $fn(G)$, and formula ϕ of the propositional fragment of SGL, we have that $G \models_{GL} \phi$ if and only if $[\![G]\!]_X \models ([\![\phi]\!]_X)_{\emptyset \to X}$.*

In LGL is also possible to encode the Separation Logics on heaps: names used as identifiers of location will be forcibly separated by tensor product, while names used for pointers will be shared/linked.

Bigraphs as a Model for BiLog. We combine the structures of link graphs and place graphs to generate all *(abstract pure) bigraphs* of [12]. We take as monoid the product of link and place interfaces, i.e. $(\omega \times \mathcal{P}_{fin}(\Lambda), \otimes, \epsilon)$ where $\langle m, X \rangle \otimes \langle n, X \rangle \stackrel{def}{=} \langle m + n, X \uplus Y \rangle$

Table 4. Encoding Propositional SGL in LGL over two ported ground link graphs

Spatial Graphs into Two-ported Ground Link Graphs	
$[\![\, nil\,]\!]_X \stackrel{def}{=} X$	$[\![\, a(x,y)\,]\!]_X \stackrel{def}{=} K(a)_{x,y} \otimes X \setminus \{x,y\}$
$[\![\, G \mid G'\,]\!]_X \stackrel{def}{=} [\![\, G\,]\!]_X \stackrel{\vec{a}}{\otimes} [\![\, G'\,]\!]_X$	$[\![\, (\nu x)G\,]\!]_X \stackrel{def}{=} ((/x \otimes id_{X\setminus\{x\}}) \circ [\![\, G\,]\!]_{\{x\}\cup X})) \otimes (\{x\} \cap X)$
SGL formulae into LGL formulae	
$[\![\, \mathbf{nil}\,]\!]_X \stackrel{def}{=} X$	$[\![\, a(x,y)\,]\!]_X \stackrel{def}{=} \mathbf{K}(a)_{x,y} \otimes (X \setminus \{x,y\})$
$[\![\, \mathbf{F}\,]\!]_X \stackrel{def}{=} \mathbf{F}$	$[\![\, \phi \Rightarrow \psi\,]\!]_X \stackrel{def}{=} [\![\, \phi\,]\!]_X \Rightarrow [\![\, \psi\,]\!]_X$
$[\![\, \phi \mid \psi\,]\!]_X \stackrel{def}{=} [\![\, \phi\,]\!]_X \stackrel{\vec{a}}{\otimes} [\![\, \psi\,]\!]_X$	

and $\epsilon \stackrel{def}{=} \langle 0, \emptyset \rangle$. We will use X for $\langle 0, X \rangle$ and n for $\langle n, \emptyset \rangle$. As constructors for bigraphical terms we have the union of place and link graph constructors apart from the controls $K : 1 \to 1$ and $K_{\vec{a}} : \emptyset \to \vec{a}$, which are replaced by the new *discrete ion* constructor, which we note $K_{\vec{a}} : 1 \to \langle 1, \vec{a} \rangle$; this is a prime bigraph containing a single node with ports named $\vec{a}$ and an hole inside. Bigraphical terms thus are defined w.r.t. a control signature $\mathcal{K}$ and a set of names Λ, cf. [15] for details.

PGL excels at expressing properties of *unnamed* resources, i.e., resources accessible only by following the structure of the term. On the other hand, LGL characterises names and their links to resources, but it has no notion of locality. A combination of them ought to be useful to model spatial structures, either private or public. BiLog promises to be a good (contextual) spatial logic for (semi-structured) resources with nominal links, thanks to bigraphs' orthogonal treatment of locality and connectivity. To witness this we have proved in [9] that also the recently proposed Context Logic for Trees [3] can be encoded into bigraphs. The idea of the encoding is to extend the one of STL with contexts and identified nodes. Essentially, in [9] we show that the model of [3] is a particular class of prime bigraphs with one port for each node and a number of holes and regions limited to one. Since [3] is more general than separation logic, and is used to reason about programs that manipulate tree structured memories, it is possible to generalise separation logic as well.

4 Conclusion and Future Work

In this paper we moved a first step towards describing global resources by focusing on bigraphs. Our final objective is to design a general dynamic logic able to cope uniformly with all the models bigraphs have been proved useful for, as of today these include CCS [17], pi-calculus [12] and Petri-nets [13, 16]. We introduced BiLog, a logic for bigraphs (and more generally for monoidal categories), with two main spatial connectives: composition and tensor product. Our main technical results are the embedding and comparison with other spatial logics previously studied. Moreover, we have shown that BiLog is expressive enough to internalise the somewhere modality. In §3 we observed that the induced logical equivalence can be forced to coincide with the structural congruence of terms. This property is fundamental in order to describe, query and reason about bigraphical data structures. For a more detailed discussion we refer to [8].

In [9] we study how BiLog can deal with dynamics. A natural solution is to add a temporal *next step* modality basically describing bigraphs that can compute (*react*) according to a Bigraphical Reactive System [12]. When the transparency predicate τ enables the inspection of 'dynamic' controls, BiLog is '*intensional*' in the sense of [11], namely it can observe internal structures. In several cases, notably the bigraphical system describing CCS [17], this can be to the extent that the next step modality can be expressed directly by using the static fragment of BiLog. Notice that τ specifies what structure the logic can directly observe, while the next step modality, along with the spatial connectives, allows to deduce the structure by observing the behaviour. It would be interesting to investigate how the transparency predicate influences the expressiveness and intentionality of significant fragments of the dynamic logic.

The 'separation' plays differently in various fragments of the logic. For instance, in the case of *Place Graph Logic*, where the model is the class of bigraphs without names, the separation is purely structural and coincides with the notion of parallel composition in Spatial Tree Logic. The separation in the *Link Graph Logic* is disjointness of nominal resources. Finally, for *Bigraph Logic* it is a combination that can be seen as separation in a structured term with nominal resources (e.g. the trees with pointers of [3] and trees with hidden names [5]). In the paper we have not addressed logical operators for hidden names (e.g. ®, **H**, © of ambient logic). We can encode them easily using, in particular, Ʋ and /a. The decidability of BiLog is an open question, we are working on extending the results of [2], and we are isolating decidable fragments of BiLog.

We are currently developing a proof theory for Bilog in order to complete the robust logical setting provided by the model theory presented here. Besides aiming at a generalise existing proof systems, this will allow direct comparisons between BiLog and other spatial logics also from the proof-theoretic point of view.

Acknowledgment. We would like to thank Philippe Bidinger, Annalisa Bossi, Rohit Chadha, Murdoch Gabbay, Giorgio Ghelli, Robin Milner, Peter O'Hearn and all the anonymous referees for useful comments and discussions.

References

1. L. Caires and L. Cardelli. A spatial logic for concurrency (Part I). In *Proc. of TACS*, volume 2215 of *LNCS*, pages 1–37. Springer-Verlag, 2001.
2. C. Calcagno, L. Cardelli, and A. D. Gordon. Deciding validity in a spatial logic for trees. In *Proc. of TLDI*, 2003.
3. C. Calcagno, P. Gardner, and U. Zarfaty. A context logic for tree update. In *Proc. of LRPP*, 2004; revised version to appear in *POPL*, 2005.
4. L. Cardelli, P. Gardner, and G. Ghelli. A spatial logic for querying graphs. In *Proc. of ICALP*, volume 2380 of *LNCS*, page 597. Springer-Verlag, 2002.
5. L. Cardelli, P. Gardner, and G. Ghelli. Manipulating trees with hidden labels. In *Proc. of FOSSACS*, volume 2620 of *LNCS*, pages 216–232. Springer-Verlag, 2003.
6. L. Cardelli and A. D. Gordon. Ambient logic. To appear in Mathematical Structures in Computer Science.
7. L. Cardelli and A. D. Gordon. Anytime, anywhere: Modal logics for mobile ambients. In *Proc. of POPL*. ACM Press, 2000.

8. G. Conforti, D. Macedonio, and V. Sassone. Bigraphical logics for XML. In *Proc. of SEBD*, 2005. To appear.
9. G. Conforti, D. Macedonio, and V. Sassone. BiLog: spatial logics for bigraphs. Computer Science Report 2005:02, University of Sussex, 2005.
10. T. Hildebrandt and J.W. Winther. Bigraphs and (Reactive) XML, an XML-centric model of computation. IT University of Copenhagen Technical Report TR-2005-26, 2005.
11. D. Hirschkoff. An extensional spatial logic for mobile processes. In *Proc. of CONCUR*, volume 3170 of *LNCS*, pages 325–339. Springer-Verlag, 2004.
12. O. H. Jensen and R. Milner. Bigraphs and mobile processes (revised). Technical Report UCAM-CL-TR-580. University of Cambridge, 2004.
13. J. J. Leifer and R. Milner. Transition systems, link graphs and petri nets. Technical Report UCAM-CL-TR-598. University of Cambridge, 2004.
14. R. Milner. Bigraphical reactive systems. In *Proc. of CONCUR*, volume 2154 of *LNCS*, pages 16–35. Springer-Verlag, 2001.
15. R. Milner. Axioms for bigraphical structure. Technical Report UCAM-CL-TR-581. University of Cambridge, 2004.
16. R. Milner. Bigraphs for petri-nets. In *Lectures on Concurrency and Petri Nets: Advances in Petri Nets*, pages 686–701. Springer-Verlag, 2004.
17. R. Milner. Pure bigraphs. Technical Report UCAM-CL-TR-614. University of Cambridge, 2005.
18. P. O'Hearn, J. C. Reynolds, and H. Yang. Local reasoning about programs that alter data structures. In *Proc. of CSL*, volume 2142 of *LNCS*, pages 1–19. Springer-Verlag, 2001.
19. A. M. Pitts. Nominal logic: A first order theory of names and binding. In *Proc. of TACS*, volume 2215 of *LNCS*, pages 219–242. Springer-Verlag, 2001.

Completely Non-malleable Schemes

(Extended Abstract)

Marc Fischlin*

Institute for Theoretical Computer Science, ETH Zürich, Switzerland
marc.fischlin@inf.ethz.ch
http://www.fischlin.de/

Abstract. An encryption scheme is non-malleable if the adversary cannot transform a ciphertext into one of a related message under the given public key. Although providing a very strong security property, some application scenarios like the recently proposed key-substitution attacks yet show the limitations of this notion. In such settings the adversary may have the power to transform the ciphertext *and* the given public key, possibly without knowing the corresponding secret key of her own public key. In this paper we therefore introduce the notion of completely non-malleable cryptographic schemes withstanding such attacks. We show that classical schemes like the well-known Cramer-Shoup DDH encryption scheme become indeed insecure against this stronger kind of attack, implying that the notion is a strict extension of chosen-ciphertext security. We also prove that, unless one puts further restrictions on the adversary's success goals, completely non-malleable schemes are hard to construct (as in the case of encryption) or even impossible (as in the case of signatures). Identifying the appropriate restrictions we then show how to modify well-known constructions like RSA-OAEP and Fiat-Shamir signatures yielding practical solutions for the problem in the random oracle model.

1 Introduction

According to the seminal paper by Dolev et al. [7] an encryption scheme is called non-malleable if giving a ciphertext to an adversary does not significantly help this adversary to produce a ciphertext of a related message under the same public key. Analogous requirements can be formulated for other cryptographic primitives like signatures or commitments. While this definition of non-malleability is already quite strong and suffices in most settings it yet leaves open if there are cases where refined notions are needed and, if so, whether they can be achieved at all.

---

* This work was supported by the Emmy Noether Programme Fi 940/1-1 of the German Research Foundation (DFG). Part of this work done while visiting University of California, San Diego, USA.

Motivation. A possible stronger definition of non-malleability, introduced here as complete non-malleability, basically allows the adversary to transform the public key as well. That is, in case of encryption the adversary may output a ciphertext of a related message *under an adversarial chosen public key*. For this, the adversary does not even need to know the matching secret key to the chosen public key.

Our initial interest in completely non-malleable schemes stems from the area of (regular) non-malleable commitments. Previous constructions of such non-malleable commitments usually require a common reference string [4, 8, 5, 9], or are rather theoretical in terms of efficiency [7, 1]. Coming up with an efficient non-malleable commitment protocol in the plain model is still an open problem.

Early in cryptography it has been observed that efficient commitment schemes can be derived from encryption schemes. To commit, the sender creates a key pair and sends a ciphertext of the message together with the public key. To decommit, the sender transmits the message with the random bits used to create the ciphertext, or simply sends the secret key (if appropriate). Now, if the encryption scheme was *completely* non-malleable then the resulting commitment scheme in this basic construction would be non-malleable in the ordinary sense. And the derived commitment scheme would be non-interactive and would not rely on public parameters either.

In addition to the application to commitment schemes, it turns out that, recently, the problem of complete non-malleability also appeared in similar flavors in related areas like signatures or hash functions [3, 13, 12]. For example, Blake-Wilson and Menezes [3] show how to deploy unknown key-share attacks to show weaknesses in the station-to-station key agreement protocol. In their case, the adversary is given a signature s for message m under some public verification key vk and her task is to find a different key pair (sk^*, vk^*) such that s is also a valid signature for m under vk^*.

Our Results. In this work we discuss the issue of complete non-malleability for public-key encryption and signatures. We first show that most of the well-known encryption and signature schemes fall prey to complete non-malleability attacks. Specifically, we propose attacks against the Cramer-Shoup DDH encryption scheme, RSA-OAEP and signatures of the Fiat-Shamir type like Schnorr signatures (of which only the first one appears in this version). This shows that the security notion of complete non-malleability is not covered by chosen-ciphertext security and by unforgeability against chosen-message attacks, respectively.

Then we give a formal framework for complete non-malleability of public-key encryption and signatures. There are two major differences to the basic definition of non-malleability. First, the adversary's goal in the definition of [7] for encryption is to relate the original secret message m to a chosen message m^* via a relation $R(m, m^*)$. Here we extend the relation to include the given public key pk. For message-only relations it remains for example unclear if it is easy to modify a ciphertext of some message m under some RSA-based non-malleable encryption with random RSA-exponent e into a ciphertext of the related message $m^* = m+e$ under the same public key. We answer this in the affirmative, showing that this is indeed easy for general schemes. Namely, we present a scheme which is non-malleable for relations over messages, but for which the adversary can easily produce a ciphertext c^* of a message m^* under pk such that a specific relation $R(pk, m, m^*)$ is satisfied. We stress that the adversary does not even take advantage of the possibility to select her own public key for this attack.

Our separating example for relations $R(m, m^*)$ over messages shows that (regular) non-malleable commitments constructed by means of encryption schemes in the common reference string model (as in [5]) may not provide adequate security for the classical Internet auction example. In the auction case the users' bids are encrypted with a public key published in the reference string. Now, an adversarial bidder may be able to transform such a sealed bid of an honest user into one which is related via this public key, and may thus overbid this user easily with a reasonably small amount (e.g., by $m^* = m + e$).

The second, and more significant extension of the [7] framework for encryption is that the adversary now has the power to tamper the public key. Consequently, the relations now also range over the given public key pk, the adversarial chosen public key pk^* and, for sake of generality, also over adversary's ciphertext. Similarly, for signatures we let the relation include the given verification key vk, the adversarial key vk^*, message m^* and signature s^*.

Concerning constructions of completely non-malleable schemes, the bad news is that schemes *for general relations* are hard to derive or even impossible. We show that there are relations where complete non-malleability cannot be proven via black-box proofs for both encryption and signatures. Even worse, for more complex relations we prove that completely non-malleable signature schemes do not exist at all.

On the positive side, we can show that practical schemes like RSA-OAEP and Fiat-Shamir signatures can be made completely non-malleable *in the random oracle model* (while the basic versions do not achieve this goal, not even in the random oracle model). Security holds for a broad class of relations which, roughly, excludes only such relations for which we are able to show our unconditional impossibility results. Also, our solutions are essentially as efficient as the original schemes, thus giving us complete non-malleability almost for free.

However, we remark that the completely non-malleable versions of the schemes above are proven secure in the random oracle model only. A closer look reveals why this model provides a useful countermeasure: Random oracles are by nature highly non malleable constructs, because outputs of related inputs are completely uncorrelated and because all users in the system use the *same* hash function oracle as a common anchor. The advantage of giving security of these schemes in terms of complete non-malleability, even in the random oracle model, is that security now follows for a vast number of attacks, including for example so-called key-substitution and strong-unforgeability attacks. That is, any attack where the adversary's goal can be cast through such relations provably fails; extra security proofs become obsolete. An interesting open question is whether there are secure schemes in the plain model for interesting relations or not.

Organization. To provide some intuition about the power of complete non-malleability attack we start with the attack on the Cramer-Shoup encryption scheme in Section 2. Then we define completely non-malleable schemes formally in Section 3. Because of the complexity of the topic we mainly focus on the definitions. Our impossibility and positive results are outlined in Section 4, further details appear in the full version.

2 Attack on Cramer-Shoup Encryption Scheme

The Cramer-Shoup encryption scheme [6] is semantically secure against adaptive chosen-ciphertext attacks under the decisional Diffie-Hellman assumption. It is thus also non-malleable (in the classical sense) with respect to such attacks.

Key Generation: The public key is given by the description of a group $\mathcal{G}_q$ of prime order q for which the decisional Diffie-Hellman problem is believed to be intractable, two random generators g_1, g_2 of this group as well as c, d and h where

$$c = g_1^{x_1} g_2^{x_2}, \quad d = g_1^{y_1} g_2^{y_2}, \quad h = g_1^{z_1} g_2^{z_2}$$

for random values $x_1, x_2, y_1, y_2, z_1, z_2 \leftarrow \mathbb{Z}_q$. The public key also contains a collision-intractable hash function H. The secret key is $(x_1, x_2, y_1, y_2, z_1, z_2)$.

Encryption: To encrypt a message $m \in \mathcal{G}_q$ pick a random $r \leftarrow \mathbb{Z}_q$ and compute

$$u_1 = g_1^r, \quad u_2 = g_2^r, \quad e = h^r m, \quad \alpha = H(u_1, u_2, e), \quad v = c^r d^{r\alpha}$$

The ciphertext is given by (u_1, u_2, e, v).

Decryption: To decrypt a ciphertext (u_1, u_2, e, v) compute $\alpha = H(u_1, u_2, e)$ and verify that $v = u_1^{x_1 + \alpha y_1} u_2^{x_2 + \alpha y_2}$. If so, then output $m = e/u_1^{z_1} u_2^{z_2}$.

The attack showing that the scheme fails to provide complete non-malleability now proceeds as follows. Given a public key $(\mathcal{G}, g_1, g_2, H, c, d, h)$ and a ciphertext (u_1, u_2, e, v) first recompute $\alpha = H(u_1, u_2, e)$. With overwhelming probability $\alpha \neq 0 \mod q$ and we can invert α in $\mathbb{Z}_q^*$; else, if a random ciphertext maps to 0 with noticeable probability, collisions for H could be found easily. Next compute

$$u_1^* = u_1^2, \quad u_2^* = u_2^2, \quad e^* = e^2, \quad \alpha^* = H(u_1^*, u_2^*, e^*), \quad v^* = v^{2\alpha^*/\alpha}$$

and finally prepare the public key as

$$c^* = c^{\alpha^*/\alpha}, \quad d^* = d, \quad h^* = h.$$

A simple calculation shows that

$$v^* = v^{2\alpha^*/\alpha} = c^{2r\alpha^*/\alpha} d^{2r\alpha\alpha^*/\alpha} = (c^*)^{2r} (d^*)^{2r\alpha^*}$$

Hence, the tuple (u_1^*, u_2^*, e^*, v^*) is a valid ciphertext of $m^* = m^2 \in \mathcal{G}$ under randomness $r^* = 2r \bmod q$ and public key $(\mathcal{G}, g_1, g_2, H, c^*, d^*, h^*)$.[1]

The attack shows that the encryption scheme cannot be used as a non-malleable commitment scheme, as explained in the introduction. With this attack the adversary would be able to open her commitment correctly with $(2r, m^2)$ after seeing the decommitment (m, r) of the original sender. Analogously, if the adversary is given the original secret key $(x_1, x_2, y_1, y_2, z_1, z_2)$ she can modify it to $(x_1\alpha^*/\alpha, x_2\alpha^*/\alpha, y_1, y_2, z_1, z_2)$. We stress that the scheme still satisfies its designated security goal of chosen-ciphertext security.

3 Definitions

In this section we define completely non-malleable public-key encryption and signature schemes. Our approach follows the line of Dolev at el. [7] and also investigates the non-malleability question of an encryption or signature scheme merely with respect to itself. Achieving non-malleability between different schemes is in general impossible, even in the basic case.

3.1 Encryption

An obvious problem with defining completely non-malleable encryption schemes lies in the adversary's possibility to choose her own public key and the uniqueness of ciphertexts. With a fake, yet valid-looking public key the adversary might be able to produce ciphertexts which can be decrypted ambiguously. We consider this to be a characteristic of the encryption scheme, and not an issue of complete non-malleability. Specifically, we allow the adversary to produce such phony keys if the scheme supports it, i.e., if one cannot distinguish good keys from fake ones. We note that, for the application to non-malleable commitments as explained in the introduction, verifying the validity of keys is for example necessary.

Relations. As mentioned in the introduction, regular non-malleability says that it is hard to transform a given ciphertext of message m into one of a related message m^* under the same key. There, related messages are designated according to an efficiently computable (probabilistic) algorithm R which basically takes the messages m and m^* as input.[2] But here we are interested in more general attacks where, as in the examples of non-malleable commitments or key-substitution attacks on signatures, finding a related public key pk^* or ciphertext c^* to the given

---

[1] At first glance it seems that replacing h by $h^* = h^a$ (or similar substitutions), and leaving the other ciphertext components untouched, would work as well. But then the adversary would encrypt a message $m^* = e/(h^*)^r = mh^{r(1-a)}$. This, however, would be a random message (over the choice of r) and would be thus unlikely to be related to m in a reasonable way.

[2] The definition in [7] lets the relations include another string chosen by the adversary, mainly to deal with the case of symmetric encryption schemes. All our positive and negative results for public-key encryption and signatures remain valid for this extension.

key pk may be considered a success. Hence, we let the relations in general also depend on pk and pk^*, c^*.

Our approach of allowing the relation to depend on other parameters than the messages introduces an interesting issue for non-malleable encryption schemes in the "ordinary" sense. In the original definition of [7] the relation $R(m, m^*)$ does not range over the user's public key pk. Hence, it remains unclear if it is infeasible to find a ciphertext of a message m^* to a given ciphertext of some unknown m such that m^* is related to m via the public key pk for such schemes. We discuss this in more detail in the final version, presenting an example which is malleable if the relation includes the public key, but which is provably non-malleable if the relations are defined over messages only.

To capture both the original definition of relations over messages only and the more general approach including public keys, we look at classes $\mathcal{R}$ of relations and define complete non-malleability with respect to such classes. The class for the basic definition then spans over relations $R(pk, m, pk^*, m^*, c^*) = R_0(m, m^*) \wedge pk = pk^*$, for example.

Message Distributions. We assume that the distribution of the user's message is determined according to some efficiently computable probabilistic algorithm M from some class $\mathcal{M}$. The message distribution M may depend on the given public key. Dolev at al. [7] let the adversary and the simulator determine the message distribution after seeing the public key and having queried the decryption oracle in a preprocessing phase. This can be subsumed in our model by letting these two algorithms output some parameter μ before the ciphertext is created. Unless stated differently all our results, positive and negative ones, remain valid in the setting where the adversary and simulator select such values; yet, we usually do not include them here for sake of simplicity.

Attack Model. In the first stage of the actual attack the adversary $\mathcal{A}$ is given a public key pk and access to a decryption oracle $\text{DEC}(sk, \cdot)$, where $(sk, pk) \leftarrow \text{KGEN}(1^k)$ have been produced by the key generator. The adversary also gets a description of the relation R and the message distribution M. A message m is sampled according to the distribution $M(pk) \in \mathcal{M}$ and encrypted under pk to ciphertext $c \leftarrow \text{ENC}(pk, m; r)$. The adversary starts the attack on the ciphertext c, the decryption oracle and some information about the message m in form of the value $h \leftarrow \text{hist}(m)$ of an efficiently computable probabilistic function hist. This function can be formally regarded of part of the distribution M. The adversary finally outputs a public key pk^*, possibly for a different yet polynomially related security parameter, and a ciphertext c^*.

Let $\pi_{\text{enc}}(\mathcal{A}, M, R)$ be the probability that $(pk, c) \neq (pk^*, c^*)$ and that there exists some m^*, r^* such that $c^* = \text{ENC}(pk^*, m^*; r^*)$ and $R(pk, m, pk^*, m^*, c^*)$ for the relation R. We call this a related-ciphertext attack. Here, as usual for non-malleability definitions, R may implicitly depend on the encryption scheme itself and some security parameter. However, we do not demand that $m \neq m^*$; it suffices to produce a different key/ciphertext pair.

As explained in the introduction, the usage of the encryption scheme as a commitment may result in different attacks and success goals, e.g., the adver-

	$\mathcal{A}$ gets pk, c, oracle DEC$(sk,\cdot)$ and ...	$\mathcal{S}$ gets pk [and possibly oracle DEC$(sk,\cdot)$] and...
$\pi_{\text{enc}}^{(\prime)}$	$\mathcal{A}$ outputs pk^*, c^*	$\mathcal{S}$ outputs pk', c', m', r'
$\pi_{\text{open}}^{(\prime)}$	$\mathcal{A}$ outputs pk^*, c^*, then m^*, r^* after m, r	$\mathcal{S}$ outputs pk', c', m', r'
$\pi_{\text{sk-open}}^{(\prime)}$	$\mathcal{A}$ outputs pk^*, c^*, then sk^* after sk	$\mathcal{S}$ outputs pk', c', m', r', sk'

Fig. 1. Overview of Attack and Simulation Modes for Encryption

sary may be obliged to actually open her ciphertext after seeing the opening of the original ciphertext. Therefore, let $\pi_{\text{open}}(\mathcal{A}, \mathsf{M}, \mathsf{R})$ denote the probability that $\mathcal{A}$ after the first stage, on input α^* and values m, r, also returns m^*, r^* such that $c^* = \text{ENC}(pk^*, m^*; r^*)$ and $\mathsf{R}(pk, m, pk^*, m^*, c^*) = 1$. This is called a related-opening attack. Write $\pi_{\text{sk-open}}(\mathcal{A}, \mathsf{M}, \mathsf{R})$ for the probability that $\mathcal{A}$ for input α^* and the secret key sk returns sk^* such that DEC$(sk^*, c^*) = m^*$ and $\mathsf{R}(pk, m, pk^*, m^*, c^*) = 1$ in a so-called related-key-opening attack. The three cases are described informally in the middle column in Figure 1.

Simulation Model. To capture the idea of the user's ciphertext not helping to produce a ciphertext of a related message we define a simulator $\mathcal{S}$ which is supposed to be as successful as the adversary but without seeing the ciphertext. $\mathcal{S}$ gets as input a public key pk and descriptions of the relation and the message distribution, but does not get access to a decryption oracle. Then, a message m is sampled according to $\mathsf{M}(pk)$ and algorithm $\mathcal{S}$ receives $h \leftarrow \text{hist}(m)$ as input.

Depending on the adversary's attack mode, the simulator's task becomes increasingly challenging such that a successful simulator for a security level automatically constitutes a simulator for a lower level. Precisely, the simulator is supposed to output a key pk', a ciphertext c', a message m' and randomness r' (if the adversary runs a related-ciphertext or a related-opening attack),[3] and a key pk', a ciphertext c', a message m', a random string r' and a secret key sk' (if the adversary runs a related-key-opening attack). Again, see Figure 1 for an overview.

Concerning the auxiliary power of the simulator there are two possibilities. One version is to give the simulator, like the adversary, additional access to the decryption oracle. We call this an *assisted* simulator. This reflects the approach that the simulator should have comparable power as the adversary. The other possibility is to deny the simulator access to DEC. We call such simulators *stand-alone* simulators. This approach follows the definition of [7].

Although the definition with assisted simulators appears to be more intuitive at first, it is not clear that giving the simulator access to DEC captures the "right flavor" of complete non-malleability. The additional power may for

---

[3] For some of our negative results we use a milder requirement and let the simulator only output pk', c'. This even strengthens these results.

example allow to prove schemes to be secure which are completely malleable in a natural sense. While this question has somewhat been settled for chosen-ciphertext security, where this additional power is acceptable, our separation of complete non-malleability from chosen-ciphertext security means that these arguments cannot be transfered without precautions. Instead, a conservative approach for designing schemes is therefore to rely on stand-alone simulators, as it suffices for our solutions in the random oracle model for example. We note that our impossibility results hold for both cases, although in a slightly weaker sense for assisted simulators.

Let both $\pi'_{enc}(\mathcal{S}, \mathsf{M}, \mathsf{R})$ and $\pi'_{open}(\mathcal{S}, \mathsf{M}, \mathsf{R})$ denote the probability that $c' = \text{ENC}(pk', m'; r')$ and that $\mathsf{R}(pk, m, pk', m', c') = 1$ in the first and second simulation experiment, respectively. Similarly, $\pi'_{sk\text{-}open}(\mathcal{S}, \mathsf{M}, \mathsf{R})$ stands for the probability that $c' = \text{ENC}(pk', m'; r')$, $m' = \text{DEC}(sk', c')$ and $\mathsf{R}(pk, m, pk', m', c') = 1$ in the third simulation experiment.

Definition 1. *A public-key encryption scheme is completely non-malleable (for stand-alone or assisted simulator) with respect to kind $\in \{enc, open, sk\text{-}open\}$, distribution class $\mathcal{M}$ and relation class $\mathcal{R}$, if for any adversary $\mathcal{A}$ there exists a (stand-alone or assisted) simulator $\mathcal{S}$ such that for any distribution $\mathsf{M} \in \mathcal{M}$ and any relation $\mathsf{R} \in \mathcal{R}$ the absolute difference $|\pi_{kind}(\mathcal{A}, \mathsf{M}, \mathsf{R}) - \pi'_{kind}(\mathcal{S}, \mathsf{M}, \mathsf{R})|$ is negligible.*

In the sequel, when speaking of completely non-malleable encryption schemes we refer to related-ciphertext attacks and $\pi_{enc}(\mathcal{A}, \mathsf{M}, \mathsf{R})$, $\pi'_{enc}(\mathcal{S}, \mathsf{M}, \mathsf{R})$. The definitions for completely non-malleable encryption (and signatures in the next section) can be extended in a straightforward way to the random oracle model.

3.2 Signatures

The attack scenario for completely non-malleable signature schemes resembles the setting of adaptive chosen-message attacks known from regular signature schemes.

Discussion. Defining the attack model for completely non-malleable signature schemed as outlined above, it seems that the adversary can always generate a new signature under a new public key, i.e., the adversary can naturally generate a new key pair and sign some message with the self-generated secret key. As explained, this attack can be confined as in the example of unknown-key attacks [3] where the adversary is supposed to find a matching key pair for a given message and a given signature. Here we do not restrict the adversary's goal in such a way. First, we do not want to give up generality and exclude certain application scenarios, e.g., signatures encrypted together with the message under a malleable encryption scheme, where the message is not known but the signature may still be transformable by permeating the malleable ciphertext. Second, if the adversary can trivially output a signature, i.e., without relying on the original signature, then this does not violate the idea of (complete) non-malleability and we should therefore be able to prove this formally as well.

	$\mathcal{A}$ gets vk, oracle SIG$(sk, \cdot)$ and ...	$\mathcal{S}$ gets vk and ...
$\pi_{\text{sig}}^{(\prime)}$	$\mathcal{A}$ outputs vk^*, m^*, s^*	$\mathcal{S}$ outputs vk', m', s'

Fig. 2. Overview of Attack and Simulation Mode for Signatures

Attack and Simulation Model. At the outset of the complete non-malleability attack the adversary $\mathcal{A}$ gets as input the description of the relation R and a verification key vk, generated together with the secret signing key sk by KGEN(1^k). The adversary is then allowed to query a signature oracle SIG($sk, \cdot$) about messages of her choice. For definitional reasons we let the signature oracle prepend the verification key vk and the message m to each signature reply s for such a query. The adversary finally outputs some verification key vk^*, a message m^* and some signature s^*. Define $\pi_{\text{sig}}(\mathcal{A}, \mathsf{R})$ as the probability that s^* is a valid signature for m^* under vk^*, i.e., VF(vk^*, m^*, s^*) = 1, that (vk^*, m^*, s^*) is different from any previously given answer (vk, m, s) of the signature oracle, and that R(vk, vk^*, m^*, s^*) holds for relation R from the class $\mathcal{R}$.

The simulator only gets vk and the relation as input and is supposed to output a triple (vk', m', s') without having oracle access to SIG($sk, \cdot$). Let $\pi'_{\text{sig}}(\mathcal{S}, \mathsf{R})$ be the probability that s' is a valid signature for m' under vk' and that R(vk, vk', m', s') is satisfied. The attack and simulation model is outlined in Figure 2.

Similar to the encryption case one could also distinguish between standalone simulators (as defined here) and assisted simulators (which additionally get access to the signature oracle). In the latter case one would have to unorthodoxly extend the model to allow the adversary to ask for a "challenge signature" which the simulator is denied. We do not follow this approach here as our negative results would hold for this case as well, and our constructions in the random oracle already work for stand-alone simulators.

Security Definition. The idea is now to say that for any adversary there is a simulator such that the success probabilities differ only insignificantly. But with this definition a signature scheme could be completely non-malleable and yet be insecure in the sense of unforgeability, e.g., if it is easy to derive the secret key from the verification key. Therefore, we also throw in the mild assumption that the signature scheme must be unforgeable under key-only attacks, i.e., it must be infeasible on input vk (but no signature oracle) to find some message together with a valid signature under vk.

Definition 2. *A signature scheme is completely non-malleable for relation class $\mathcal{R}$ if it is existentially unforgeable under key-only attacks and if for any adversary $\mathcal{A}$ there exists a simulator $\mathcal{S}$ such that for any relation $\mathsf{R} \in \mathcal{R}$ the absolute difference $|\pi_{\text{sig}}(\mathcal{A}, \mathsf{R}) - \pi'_{\text{sig}}(\mathcal{S}, \mathsf{R})|$ is negligible.*

We briefly discuss some consequences of the definition, showing that the definition is powerful to reflect the notions of strong unforgeability (i.e., where the adversary is also considered victorious if she finds a new signature under the original verification key to a message previously signed by the signature oracle) or key-substitution attacks (where the adversary tries to find another key vk^* to a valid triple vk, m, s), both under adaptive chosen-message attacks. For this, let $R_{\text{str-unf}}(vk, vk^*, m^*, s^*)$ be the relation such that $R_{\text{str-unf}}(vk, vk^*, m^*, s^*) = 1$ iff $vk = vk^*$; let $R_{\text{key-sub}}$ be the relation such that $R_{\text{key-sub}}(vk, vk^*, m^*, s^*) = 1$ iff $\text{VF}(vk, m^*, s^*) = 1$. The proof is omitted.

Proposition 1. *Let* $(\text{KGEN}, \text{SIG}, \text{VF})$ *be a signature scheme which is completely non-malleable with respect to* $\mathcal{R} \ni R_{\text{str-unf}}$ *resp.* $\mathcal{R} \ni R_{\text{key-sub}}$. *Then the scheme is strongly unforgeable under adaptive chosen-message attacks resp. secure against key-substitution attacks.*

4 Summary of Results

In this section we summarize our (positive and negative) results. For better comprehensibility the results are stated in an informal way. The formal results and technical details can be found in the the full version.

Regular Non-Malleability and Relations over Messages Only. We show that extending the relations in the definition of [7] for regular non-malleability, i.e., where the adversary does not tamper the public key, to include the given public key pk (in addition to the messages m, m^*) can be fatal to security:

Theorem 1 (informal). *There is an encryption scheme which is non-malleable with respect to* $\mathcal{R}_{msg} = \{R(m, m^*)\}$ *but which is malleable with respect to some relation* $R_{pk}(pk, m, m^*)$.

Hardness of Constructions for General Relations. Here we discuss our negative results for constructions of completely non-malleable schemes where, in contrast to the previous case, the adversary is allowed to output another key pk^*. We show that there are relations for which completely non-malleable schemes are hard to construct. Although we prove this result for a specific set of "bad" relations, we note that the implication carries over to any class where such relations can be "somehow embedded" in relations of the class.

Theorem 2 (informal). *Public-key encryption schemes which are completely non-malleable according to black-box stand-alone simulators and general relations, do not exist.*

Note that the previous theorem assumes that the simulator is stand-alone. For assisted simulators, which are granted access to DEC, we can show the same result for relations which are efficiently computable relative to an oracle. We note that the black-box simulator does not have access to this oracle directly, but only through the relation. This corresponds to the case that the simulator can efficiently compute the relation (via black-box access) but is denied the description of the relation.

Theorem 3 (informal). *Public-key encryption schemes which are completely non-malleable according to black-box assisted simulators and general relations (relative to an oracle), do not exist.*

The results about encryption easily transfers to signatures:

Proposition 2 (informal). *Signature schemes which are completely non-malleable according to black-box simulations for general relations, do not exist.*

Yet, for signatures we can show that completely non-malleable systems for general relations are impossible at all, even when allowing non-black-box constructions or if the simulator depends on the relation.

Theorem 4 (informal). *There do not exist completely non-malleable signature schemes with respect to general relations.*

Constructions in the Random Oracle Model. On the positive side, solutions in the random oracle for completely non-malleable schemes exist. And while OAEP encryption [2] and Fiat-Shamir signatures [11] provably do not have this property, slight variations of these schemes work. The basic idea to simply include the public encryption or signature key, respectively, to each hash function evaluation. We append the term "with public-key hashing" to such modified schemes:

Proposition 3 (informal). *RSA-OAEP with public-key hashing is completely non-malleable with respect to stand-alone simulators and any relations, in the random oracle model.*

A similar result holds for Fiat-Shamir signatures:

Proposition 4 (informal). *Fiat-Shamir signatures with public-key hashing are completely non-malleable with respect to general relations (except for essentially those relations, for which the unconditional impossibility results of Theorem 4 holds), in the random oracle model.*

In both cases the proofs rely on the original results [2, 10, 14] about the security against regular chosen-ciphertext attacks and chosen-message attacks.

Acknowledgments

We would like to thank Yevgeniy Dodis, Alejandro Hevia, Bogdan Warinschi and the reviewers for helpful input.

References

1. Boaz Barak. *Constant-Round Coin-Tossing With a Man in the Middle or Realizing the Shared Random String Model.* Proceedings of the Annual Symposium on Foundations of Computer Science (FOCS) 2002. IEEE Computer Society Press, 2002.
2. Mihir Bellare and Phillip Rogaway. *Optimal Asymmetric Encryption — How to Encrypt with RSA.* Advances in Cryptology — Eurocrypt'94, Volume 950 of Lecture Notes in Computer Science, pages 92–111. Springer-Verlag, 1995.
3. Simon Blake-Wilson and Alfred Menezes. *Unknown Key-Share Attacks on the Station-to-Station (STS) Protocol.* Public-Key Cryptography (PKC)'99, Volume 1560 of Lecture Notes in Computer Science, pages 154–170. Springer-Verlag, 1999.

4. G. Di Crescenzo, Y. Ishai, and R. Ostrovsky. *Non-interactive and Non-Malleable Commitment*. Proceedings of the Annual Symposium on the Theory of Computing (STOC) 1998, pages 141–150. ACM Press, 1998.
5. G. Di Crescenzo, J. Katz, R. Ostrovsky, and A. Smith. *Efficient And Non-Interactive Non-Malleable Commitment*. Advances in Cryptology — Eurocrypt 2001, Volume 2045 of Lecture Notes in Computer Science, pages 40–59. Springer-Verlag, 2001.
6. Ronald Cramer and Victor Shoup. *A Practical Public Key Cryptosystem Provably Secure Against Adaptive Chosen Ciphertext Attacks*. Advances in Cryptology — Crypto'98, Volume 1462 of Lecture Notes in Computer Science, pages 13–25. Springer-Verlag, 1998.
7. D. Dolev, C. Dwork, and M. Naor. *Non-Malleable Cryptography*. SIAM Journal on Computing, 30(2):391–437, 2000.
8. Marc Fischlin and Roger Fischlin. *Efficient Non-Malleable Commitment Schemes*. Advances in Cryptology — Crypto 2000, Volume 1880 of Lecture Notes in Computer Science, pages 414–432. Springer-Verlag, 2000.
9. Marc Fischlin and Roger Fischlin. *The Representation Problem Based on Factoring*. Topics in Cryptology — Cryptographer's Track, RSA Conference (CT-RSA) 2002, Volume 2271 of Lecture Notes in Computer Science, pages 96–113. Springer-Verlag, 2002.
10. E. Fujisaki, T. Okamoto, David Pointcheval, and Jacques Stern. *RSA-OAEP is Secure Under the RSA Assumption*. Advances in Cryptology — Crypto 2001, Volume 2139 of Lecture Notes in Computer Science. Springer-Verlag, 2001.
11. A. Fiat and A. Shamir. *How to Prove Yourself: Practical Solutions to Identification and Signature Schemes*. Advances in Cryptology — Crypto'86, Volume 263 of Lecture Notes in Computer Science, pages 186–194. Springer-Verlag, 1986.
12. Burton Kaliski. *On Hash Function Firewalls in Signature Schemes*. Topics in Cryptology — Cryptographer's Track, RSA Conference (CT-RSA) 2002, Volume 2271 of Lecture Notes in Computer Science, pages 1–16. Springer-Verlag, 2002.
13. Alfred Menezes and Nigel Smart. *Security of Signature Schemes in a Multi-User Setting*. Designs, Codes and Cryptography, Volume 33, pages 261–274. Springer-Verlag, 2004.
14. David Pointcheval and Jacques Stern. *Security Arguments for Digital Signatures and Blind Signatures*. Journal of Cryptology, 13(3):361–396, 2000.

Boneh-Franklin Identity Based Encryption Revisited

David Galindo

Institute for Computing and Information Sciences,
Radboud University Nijmegen,
P.O.Box 9010 6500 GL, Nijmegen, The Netherlands
d.galindo@cs.ru.nl

Abstract. The first practical identity based encryption (IBE) scheme was proposed by Boneh and Franklin in [BF03]. In this work we point out that there is a flawed step in the security reduction exhibited by the authors. Fortunately, it is possible to fix it without changing the scheme or the underlying assumption.

In the second place, we introduce a variant of the seminal IBE scheme which allows a more efficient security reduction. This variant is simpler, and has more compact ciphertexts than Boneh-Franklin's proposal, while keeping the computational cost.

Finally, we observe that the flawed step pointed out here is present in several works, and that our techniques can be applied to obtain tighter reductions for previous relevant schemes.

Keywords: provable security, identity-based encryption, exact security, bilinear maps.

1 Introduction

The concept of Identity Based Encryption (IBE) was proposed by Shamir in [Sha84], aimed at simplifying certificate management in e-mail related systems. The idea is that an arbitrary string such as an e-mail address or a telephone number could serve as a public key for an encryption scheme. Once a user U receives a communication encrypted using its identity ID_U, the user authenticates itself to a Private Key Generation Center (KGC) from which it obtains the corresponding private key d_{ID_U}.

The problem was not satisfactorily solved until the work by Boneh and Franklin [BF03]. They proposed formal security notions for IBE systems and designed a fully functional secure IBE scheme using bilinear maps. The security is based on a variant of the Computational Diffie-Hellman assumption, called Bilinear Diffie-Hellman assumption. This scheme and the tools developed in its design have been successfully applied in numerous cryptographic settings, transcending by far the identity based cryptography framework.

On the other hand, an important concern when exhibiting a security reductionist proof is that of the efficiency of the reduction. One of the goals pursued is to preserve as much as possible the strength of the underlying hard problem which is used in the protocol's design. An inefficient security reduction would imply the use of larger key sizes to attain a given security level.

Our contributions. In the first place, we show there is a flawed step in the security reduction exhibited in [BF03] for the scheme proposed for chosen ciphertext security. Fortunately, the reduction can be changed without modifying the original scheme or the underlying hard problem used to state the security. The efficiency of the new security reduction is a bit worse than the previous one. This is just another example in which a well-known and widely used construction turns out to have an unnoticed flawed security reduction.

In the second place, we modify the scheme by Boneh and Franklin towards obtaining a more efficient security reduction. Indeed, it is possible to show a tighter security reduction for a modified scheme which uses one less random oracle. The new proposal also presents more compact ciphertexts than the original scheme.

Finally, since Boneh-Franklin IBE scheme has been used as a building block for numerous protocols, the corrections and/or improvements we present here are likely to be applied to further schemes. For instance, this is the case for the schemes in [GS02, HL02, Gen03, AP03, YFDL04, CC05].

2 Preliminaries

We begin by fixing some notation. If A is a non-empty set, then $x \leftarrow A$ denotes that x has been uniformly chosen in A. If A is a finite set, then $|A|$ denotes its cardinality.

2.1 Definitions for IBE Schemes

Identity based encryption (IBE). An IBE is specified by four probabilistic polynomial time (PPT) algorithms:

Setup takes a security parameter 1^ℓ and returns the system parameters params and master-key. The system parameters include the description of sets $\mathcal{M}, \mathcal{C}$, which denote the set of messages and ciphertexts respectively. params is publicly available, while the master-key is kept secret by the KGC.

Extract takes as inputs params, master-key and an arbitrary string $\mathsf{ID} \in \{0,1\}^*$ and returns a private key d_ID to the user with identity ID. This must be done over a secure channel, since d_ID enables to decrypt ciphertexts under the identity ID.

Encrypt takes as inputs params, $\mathsf{ID} \in \{0,1\}^*$ and $M \in \mathcal{M}$. It returns a ciphertext $C \in \mathcal{C}$.

Decrypt takes as inputs params, $C \in \mathcal{C}$ and a private key d_ID, and it returns $M \in \mathcal{M}$ or rejects.

Chosen ciphertext security. An IBE scheme is said to have indistinguishability against an adaptive chosen ciphertext attack (IND-ID-CCA) if any PPT algorithm $\mathcal{A}$ has a negligible advantage in the following game:

Setup The challenger takes a security parameter 1^ℓ and runs the Setup algorithm. It gives params to the adversary. It keeps the master-key to itself.
Phase 1 The adversary issues queries of the form
- Extraction query $\langle \mathsf{ID}_i \rangle$. The challenger runs algorithm Extract to generate the private key d_i corresponding to ID_i. It sends d_i to the adversary.
- Decryption query $\langle \mathsf{ID}_i, C_i \rangle$. The challenger generates the private key d_i. It then runs Decrypt to decrypt C_i under ID_i.

These queries may be asked adaptively, that is, each query may depend on the answers obtained to the previous queries.
Challenge The adversary outputs equal length plaintexts $M_0, M_1 \in \mathcal{M}$ and an identity ID_{ch}. The only constraint is that the private key for ID_{ch} was not requested in Phase 1. The challenger picks $\beta \leftarrow \{0,1\}$ and sets $C = \mathsf{Encrypt}(\mathsf{params}, \mathsf{ID}_{ch}, M_\beta)$. It sends C to the adversary.
Phase 2 The adversary issues extraction and decryption queries as in Phase 1, with the restriction $\langle \mathsf{ID}_i \rangle \neq \langle \mathsf{ID}_{ch} \rangle$ and $\langle \mathsf{ID}_i, C_i \rangle \neq \langle \mathsf{ID}_{ch}, C \rangle$.
Guess The adversary outputs a guess $\beta' \in \{0,1\}$.
Such an adversary is called an IND-ID-CCA adversary $\mathcal{A}$, and its advantage is defined as $\mathsf{Adv}^{\mathsf{CCA}}_{\mathcal{E},\mathcal{A}}(1^\ell) = |\Pr[\beta = \beta'] - 1/2|$.

Similarly, indistinguishability against passive adversaries (IND-ID-CPA) can also be defined. In this case, the game between the challenger and the adversary is similar to the IND-ID-CCA case, but disallowing decryption queries. The advantage of an adversary in this game is defined as $\mathsf{Adv}^{\mathsf{CPA}}_{\mathcal{E},\mathcal{A}}(1^\ell) = |\Pr[\beta = \beta'] - 1/2|$.

Definition 1. *An IBE system $\mathcal{E}$ is secure under chosen ciphertext attacks (resp. chosen plaintext attacks) if for any probabilistic polynomial time IND-ID-CCA (resp. IND-ID-CPA) adversary $\mathcal{A}$ the function $\mathsf{Adv}^{\mathsf{CCA}}_{\mathcal{E},\mathcal{A}}(1^\ell)$ (resp. $\mathsf{Adv}^{\mathsf{CPA}}_{\mathcal{E},\mathcal{A}}(1^\ell)$) is negligible.*

2.2 Bilinear Maps and Bilinear Groups

Let $\mathbb{G}_1, \mathbb{G}_2$ and $\mathbb{G}_T$ be finite abelian groups in which the discrete logarithm is believed to be hard. We use additive notation for $\mathbb{G}_1, \mathbb{G}_2$ whereas multiplicative notation is used for $\mathbb{G}_T$. Thus, $\mathbb{G}_1^* = \mathbb{G}_1 \setminus \{O_1\}$ and $\mathbb{G}_T^* = \mathbb{G}_T \setminus \{1_T\}$, where O_1 and 1_T are the identity elements in $\mathbb{G}_1$ and $\mathbb{G}_T$ respectively. By a *pairing* or *bilinear map* we will refer to a non-degenerate bilinear function $\widehat{e} : \mathbb{G}_1 \times \mathbb{G}_2 \to \mathbb{G}_T$. In some protocols the existence of a computable isomorphism $\psi : \mathbb{G}_2 \to \mathbb{G}_1$ is assumed. In particular, this implies that $\psi(aP_2) = a\psi(P_2)$. By a *bilinear group* we refer to a tuple $(\mathbb{G}_1, \mathbb{G}_2, \mathbb{G}_T, \widehat{e}, \psi)$ with the properties described above.

Bilinear maps are usually implemented using the Weil or modified Tate pairings on an elliptic curve. In general, the elements in $\mathbb{G}_1$ allow more compact representation than those in $\mathbb{G}_2$. In the following it is assumed that $|\mathbb{G}_1| = |\mathbb{G}_2| = |\mathbb{G}_T| = p$, where p is prime; $\mathbb{G}_1, \mathbb{G}_2$ are cyclic groups generated by P_1, P_2

respectively and $\psi(P_2) = P_1$. In this context, the map $\widehat{e}$ is non-degenerate if and only if $\widehat{e}(P_1, P_2) \neq 1_{G_T}$. We refer to [BF03] for further details.

Bilinear Diffie-Hellman (BDH) Problem on $(\mathbb{G}_1, \mathbb{G}_2)$. Given $aP_2, bP_2 \in \mathbb{G}_2^*$ and $cP_1 \in \mathbb{G}_1^*$, where $P_2 \leftarrow \mathbb{G}_2^*$, $P_1 = \psi(P_2)$, $a, b, c \leftarrow \mathbb{Z}_p^*$; compute $W = \widehat{e}(P_1, P_2)^{abc} \in \mathbb{G}_T$.

We say that an algorithm $\mathcal{B}$ (t, ε) breaks BDH on $(\mathbb{G}_1, \mathbb{G}_2)$ if it runs in time at most t and has advantage at least ε, that is,

$$\Pr[\mathcal{B}(P_2, aP_2, bP_2, cP_1) = \widehat{e}(P_1, P_2)^{abc})] \geq \varepsilon,$$

where the probability is taken over the random choices of the parameters, and the random bits of $\mathcal{B}$.

Bilinear Decision Diffie-Hellman (BDDH) Problem on $(\mathbb{G}_1, \mathbb{G}_2)$. Let $aP_2, bP_2 \in \mathbb{G}_2^*$, $cP_1 \in \mathbb{G}_1^*$, and $T \leftarrow \mathbb{G}_T$, where $P_2 \leftarrow \mathbb{G}_2^*$, $P_1 = \psi(P_2)$, $a, b, c \leftarrow \mathbb{Z}_p^*$. We say that an algorithm $\mathcal{B}$ (t, ε) breaks BDDH on $(\mathbb{G}_1, \mathbb{G}_2)$ if it runs in time at most t and

$$\left|\Pr[\mathcal{B}(P_2, aP_2, bP_2, cP_1, \widehat{e}(P_1, P_2)^{abc}) = 1] - \Pr[\mathcal{B}(P_2, aP_2, bP_2, cP_1, T) = 1]\right| \geq \varepsilon,$$

where the probability is computed over the random choices of the parameters, and the random bits of $\mathcal{B}$. Hereafter, the distribution on the left side is called *BDH distribution* and is denoted by $\mathcal{P}_{BDH}$, while the distribution on the right is called *random BDH distribution* and is denoted by $\mathcal{R}_{BDH}$.

3 Security Proof of Boneh-Franklin Identity Based Encryption Scheme Revisited

In this section we consider the identity based encryption (IBE) scheme by Boneh and Franklin [BF03]. In the first place, we point out and fix a flaw in the security reduction given by the authors. In repairing the proof, we do not need to change the security assumption neither the specification of the scheme. However, the security reduction is a bit worse than the original one.

3.1 Boneh-Franklin IBE Scheme

We will not directly use the original description of the BF scheme, because it is phrased with bilinear group pairs where $\mathbb{G}_1 = \mathbb{G}_2$, so we must adapt their scheme to the more general case $\mathbb{G}_1 \neq \mathbb{G}_2$. In choosing how to use $\mathbb{G}_1$ and $\mathbb{G}_2$, we preferred to minimize the length of the ciphertexts. This means we use $\mathbb{G}_2$ as the set of private keys and then ciphertexts are elements in $\mathbb{G}_1^* \times \{0,1\}^n$. Here follows the description of the BF scheme, which is called Full-Ident in [BF03].

> **Full-Ident**
>
> **Setup.** Let $(\mathbb{G}_1, \mathbb{G}_2, \mathbb{G}_T, \hat{e}, \psi)$ a bilinear group. Choose a generator $P_2 \leftarrow \mathbb{G}_2$ and set $P_1 = \psi(P_2)$. Next pick $s \leftarrow \mathbb{Z}_p^*$ and set $Q_{pub} = sP_2 \in \mathbb{G}_2^*$, $P_{pub} = sP_1 \in \mathbb{G}_1^*$. Choose cryptographic hash functions $H_1 : \{0,1\}^* \to \mathbb{G}_2^*$, $H_2 : \mathbb{G}_T \to \{0,1\}^n$, $H_3 : \{0,1\}^n \times \{0,1\}^n \to \mathbb{Z}_p^*$, $H_4 : \{0,1\}^n \to \{0,1\}^n$. The message space is $\mathcal{M} = \{0,1\}^n$ and the ciphertext space is $\mathcal{C} = \mathbb{G}_1^* \times \{0,1\}^n \times \{0,1\}^n$.
>
> **Extract.** For a given string $\mathsf{ID} \in \{0,1\}^*$, compute $Q_{\mathsf{ID}} = H_1(\mathsf{ID}) \in \mathbb{G}_2^*$ and set the private key d_{ID} to be $d_{\mathsf{ID}} = sQ_{\mathsf{ID}} \in \mathbb{G}_2^*$.
>
> **Encrypt.** To encrypt $M \in \{0,1\}^n$ under identity ID, compute $Q_{\mathsf{ID}} = H_1(\mathsf{ID}) \in \mathbb{G}_2^*$, choose $\sigma \leftarrow \{0,1\}^n$, set $r = H_3(\sigma, M) \in \mathbb{Z}_p^*$ and finally
>
> $$C = \langle rP_1, \sigma \oplus H_2(g_{\mathsf{ID}}^r), M \oplus H_4(\sigma) \rangle \quad \text{where} \quad g_{\mathsf{ID}} = \hat{e}(P_{pub}, Q_{\mathsf{ID}}) \in \mathbb{G}_T.$$
>
> **Decrypt.** Let $C = \langle U, V, W \rangle \in \mathcal{C}$ be a ciphertext under the identity ID. To decrypt C using the private key $d_{\mathsf{ID}} \in \mathbb{G}_2^*$ do:
> 1. Compute $V \oplus H_2(\hat{e}(U, d_{\mathsf{ID}})) = \sigma$.
> 2. Compute $W \oplus H_4(\sigma) = M$.
> 3. Set $r = H_3(\sigma, M)$. Check that $U = rP$. If not, reject the ciphertext.
> 4. Output M.

This completes the description of Full-Ident. This IBE scheme is sound since

$$\hat{e}(U, d_{\mathsf{ID}}) = \hat{e}(rP_1, sQ_{\mathsf{ID}}) = \hat{e}(P_1, Q_{\mathsf{ID}})^{sr} = \hat{e}(P_{pub}, Q_{\mathsf{ID}})^r = g_{\mathsf{ID}}^r.$$

In [BF03] it is proven that the above scheme is IND-ID-CCA secure under the BDH assumption in the Random Oracle model. That scheme uses Fujisaki and Okamoto transformation [FO99] from a one-way encryption scheme into an IND-CCA encryption scheme in the ROM (we refer to [BDPR98] for public key encryption security notions). If we denote by $E_{\mathsf{pk}}(M, r)$ the encryption of M using the random bits r under the public key pk, the transformation by Fujisaki and Okamoto is the hybrid scheme[1]

$$E_{\mathsf{pk}}^{hy}(M) = \langle E_{\mathsf{pk}}(\sigma, H_3(\sigma, M)), H_4(\sigma) \oplus M \rangle \tag{1}$$

where σ is random and H_3, H_4 are random oracles. To decrypt (C_1, C_2), one first obtains σ' decrypting C_1 using the original scheme, next computes M' and finally checks if $E_{\mathsf{pk}}(\sigma', H_3(\sigma', M)) = C_1$. If this is so, outputs M; otherwise outputs reject.

Two additional schemes are needed in order to exhibit the security proof in [BF03]. These schemes are not IBE schemes but merely public key encryption schemes. They are called BasicPub and BasicPub^{hy}. Here follows the description of

---

[1] In the case where the symmetric encryption scheme is the one-time pad.

BasicPub

KeyGen. Let $(\mathbb{G}_1, \mathbb{G}_2, \mathbb{G}_T, \widehat{e}, \psi)$ a bilinear group. Choose a generator $P_2 \leftarrow \mathbb{G}_2$ and set $P_1 = \psi(P_2)$. Next pick $s \leftarrow \mathbb{Z}_p^*, Q \leftarrow \mathbb{G}_2^*$ and set $Q_{pub} = sP_2 \in \mathbb{G}_2^*$, $P_{pub} = sP_1 \in \mathbb{G}_1^*$. Choose $H_2 : \mathbb{G}_T \to \{0,1\}^n$. Then $\mathcal{M} = \{0,1\}^n$ and $\mathcal{C} = \mathbb{G}_1^* \times \{0,1\}^n$. The public key is

$$\mathsf{pk} = \langle p, \mathbb{G}_1, \mathbb{G}_2, \mathbb{G}_T, \widehat{e}, P_1, P_2, P_{pub}, Q_{pub}, Q, H_2 \rangle.$$

The private key is $\mathsf{sk} = d = sQ \in \mathbb{G}_2^*$.
Encrypt. To encrypt $M \in \{0,1\}^n$ choose $r \leftarrow \mathbb{Z}_p^*$ and set the ciphertext to be

$$C = \langle rP_1, M \oplus H_2(g^r) \rangle \quad \text{where} \quad g = \widehat{e}(P_{pub}, Q) \in \mathbb{G}_T$$

Decrypt. Let $C = \langle U, V \rangle \in \mathcal{C}$ be a ciphertext under the public key pk. To decrypt C using the private key $d \in \mathbb{G}_2^*$, compute $V \oplus H_2(\widehat{e}(U,d)) = M$.

Finally, the scheme $\mathsf{BasicPub}^{hy}$ is the result of applying Fujisaki-Okamoto transformation (1) to the above scheme. The security reduction for Full-Ident scheme under the BDH assumption follows the diagram below

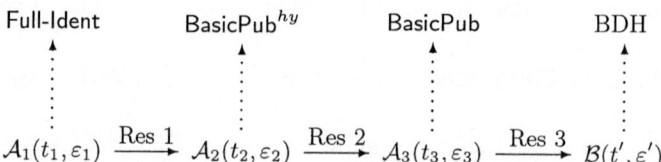

The following results are shown in [BF03]. Hereafter, q_E, q_D, q_{H_i} denote the number of extraction, decryption and random oracle H_i queries respectively.

Result 1. Let $\mathcal{A}_1$ an IND-ID-CCA adversary that has advantage ε_2 against Full-Ident making at most q_E, q_D and q_{H_1} queries. Then there is an IND-CCA adversary $\mathcal{A}_2$ that has advantage at least $\frac{\varepsilon_1}{e(1+q_E+q_D)}$ against $\mathsf{BasicPub}^{hy}$. Its running time is $t_2 \leq t_1 + c_{\mathbb{G}_2}(q_D + q_{H_1} + q_E)$, where $c_{\mathbb{G}_2}$ denotes the time of computing a random multiple in $\mathbb{G}_2$.

Result 2. Let $\mathcal{A}_2$ an IND-CCA adversary that has advantage ε_2 against $\mathsf{BasicPub}^{hy}$ making at most q_D, q_{H_3} and q_{H_4} queries. Then there is an IND-CPA adversary $\mathcal{A}_3$ that has advantage at least $\frac{1}{2(q_{H_3}+q_{H_4})}[(\varepsilon_2+1)(1-2/p)^{q_D}-1]$ against BasicPub. Its running time is $t_3 \leq t_2 + \mathcal{O}((q_{H_3}+q_{H_4}) \cdot (n + \log p))$.

Result 3. Let $\mathcal{A}_3$ an IND-CPA adversary that has advantage ε_3 against BasicPub making at most q_{H_2} queries. Then there is an algorithm $\mathcal{B}$ breaking the BDH problem on $(\mathbb{G}_1, \mathbb{G}_2)$ with advantage at least $\frac{2\varepsilon_3}{q_{H_2}}$ and running time $t' \approx t_3$.

In order to come up with the total concrete security, we can bound any q_{H_i} with a single q_H, and assume that $q_E = q_D$, since extraction and decryption operations have roughly the same computational complexity. Then, taking the above reductions, we obtain that the BF scheme is $(t_1, q_H, q_D, \varepsilon_1)$ IND-ID-CCA secure if the BDH problem on $(\mathbb{G}_1, \mathbb{G}_2)$ is

$$\left(t_1 + c_{\mathbb{G}_1}(2q_D + q_H)) + 2q_H(n + \log p), \frac{\varepsilon_1}{8eq_H^2 q_D}\right)\text{-secure.} \qquad (2)$$

Therefore, the security reduction is far from tight, mainly because of the $q_H^2 q_D$ factor relating the advantages against the scheme and the underlying problem.

3.2 A Flaw in the Security Reduction

In this section we point out a flaw in the reduction used to state Result 1, which is Lemma 4.6 in [BF03].

The goal of that reduction is to construct an IND-CCA adversary $\mathcal{B}$ with advantage $\varepsilon/e(1 + q_E + q_D)$ against BasicPub^{hy} by using an IND-ID-CCA adversary $\mathcal{A}$ with advantage ε against Full-Ident. $\mathcal{B}$ receives a public key
$$K_{pub} = \langle p, \mathbb{G}_1, \mathbb{G}_2, \mathbb{G}_T, \hat{e}, P_1, P_2, P_{pub}, Q, Q_{pub}, H_2, H_3, H_4\rangle$$
from its challenger. Then $\mathcal{B}$ simulates the challenger for $\mathcal{A}$ as follows:

Setup $\mathcal{B}$ gives $\mathcal{A}$ the parameters $\langle p, \mathbb{G}_1, \mathbb{G}_2, \mathbb{G}_T, \hat{e}, P_1, P_2, P_{pub}, Q_{pub}, H_1, H_2, H_3, H_4\rangle$, where H_1 is an oracle controlled by $\mathcal{B}$ as indicated in the following:

H_1-queries To respond to $\mathcal{A}$ queries, algorithm $\mathcal{B}$ maintains a list H_1^{list} of tuples $\langle \text{ID}_i, Q_i, b_i, c_i\rangle$ as explained below. When $\mathcal{A}$ queries H_1 at an unrepeated ID_i, $\mathcal{B}$ generates a random coin $c_i \in \{0, 1\}$ such that $\Pr[c_i = 0] = \delta$, and a random $b_i \leftarrow \mathbb{Z}_p^*$. If $c_i = 0$ it computes $Q_i = b_i P_2 \in \mathbb{G}_2^*$, and if $c_i = 1$ it computes $Q_i = b_i Q \in \mathbb{G}_2^*$. Finally, $\mathcal{B}$ adds the tuple $\langle \text{ID}_i, Q_i, b_i, c_i\rangle$ to the H_1^{list} and sends $H_1(\text{ID}_i) = Q_i$ to $\mathcal{A}$.

The idea is that tuples with $c_i = 0$ enable $\mathcal{B}$ to answer private key queries for identity ID_i, while $\mathcal{B}$ can only take profit of $\mathcal{A}$'s advantage when $\mathcal{A}$ chooses a challenge identity ID_{ch} such that $c_{ch} = 1$.

Phase 1 - Extraction queries When $\mathcal{A}$ asks for the private key associated to ID_i, $\mathcal{B}$ runs the algorithm for responding H_1-queries and gets $H_1(\text{ID}_i) = Q_i$, where $\langle \text{ID}_i, Q_i, b_i, c_i\rangle$ is the corresponding entry in H_1^{list}. If $c_i = 1$ then $\mathcal{B}$ aborts the game and the attack against BasicPub^{hy} failed. Otherwise, $c_i = 0$ and therefore $Q_i = b_i P_2$. It turns out that d_i can be computed as $d_i := b_i Q_{pub}$, since by definition $d_i = sQ_i$. Finally, $\mathcal{B}$ gives d_i to algorithm $\mathcal{A}$.

Phase 1 - Decryption queries $\mathcal{B}$ answers to a decryption query $\langle \text{ID}_i, C_i\rangle$ as follows. It runs H_1-queries algorithm and let $\langle \text{ID}_i, Q_i, b_i, c_i\rangle \in H_1^{\text{list}}$. If $c_i = 0$, then $\mathcal{B}$ retrieves the private key d_i and decrypts C_i using the decryption algorithm. If $c_i = 1$, then $Q_i = b_i Q$. Recall that the unknown private key is $d_i = sQ_i = sb_i Q$. Set $C_i' = \langle b_i U_i, V_i, W_i\rangle$, where $C_i = \langle U_i, V_i, W_i\rangle$. Then, the authors claim that the Full-Ident decryption of C_i is *equal* to the BasicPub^{hy} decryption of C_i'. The reason given is that

$$\hat{e}(b_i U_i, d) = \hat{e}(b_i U_i, sQ) = \hat{e}(U_i, sb_i Q) = \hat{e}(U_i, sQ_i) = \hat{e}(U_i, d_i),$$

which implies that the values σ and M obtained by decrypting C_i with Full-Ident and by decrypting C'_i with BasicPub^{hy} are equal. However, BasicPub^{hy} will output the reject symbol when decrypting C'_i with overwhelming probability. To see this, remember that $b_i U_i = b_i r_i P_1$, and at least $b_i \leftarrow \mathbb{Z}_p^*$, which implies that $b_i r$ is uniformly random in $\mathbb{Z}_p^*$. On the other hand, we have that H_3 is a random oracle not controlled by $\mathcal{B}$. These facts imply that $H_3(\sigma, M) \neq b_i r$ with probability $1 - 1/p$, and therefore the decryption algorithm of BasicPub^{hy} will reject the ciphertext. Thereby, we can not use the decryption oracle for BasicPub^{hy} to decrypt ciphertexts under any ID$_i$ such that $H_1(\text{ID}_i) \neq Q$. Therefore, the reduction in [BF03] is not valid.

3.3 Fixing the Security Reduction

Due to the ciphertext integrity checking in FO transformation [FO99], we can only answer decryption queries $\langle \text{ID}_i, C_i \rangle$ such that:

- $H_1(\text{ID}_i) = b_i P_2$, since we can use the private key d_i, or
- $H_1(\text{ID}_i) = Q$, since in this case, the decryption of C_i under such ID$_i$ is equal to the decryption of C_i by BasicPub^{hy}.

This remark enables us to fix the flawed reduction shown above. In the following we describe the new answers delivered by $\mathcal{B}$.

Setup As in Section 3.2.

H_1-queries Before initializing H_1^{list}, $\mathcal{B}$ selects at random $j \leftarrow \{1, \ldots, q_{H_1}\}$. When $\mathcal{A}$ queries H_1 at ID$_i$, algorithm $\mathcal{B}$ proceeds as follows: if $i \neq j$, it picks $b_i \leftarrow \mathbb{Z}_p^*$, sets $Q_i = b_i P_2$, adds $\langle \text{ID}_i, Q_i, b_i \rangle$ to the list and gives back Q_i to $\mathcal{A}$. If $i = j$, it sets $Q_j = Q$, adds $\langle \text{ID}_i, Q_i, * \rangle$ to the list and sends Q_j to $\mathcal{A}$. Here $*$ denotes a special symbol. Note that the outputs of H_1 are uniformly distributed in $\mathbb{G}_2^*$ and independent of $\mathcal{A}$'s current view, since Q is unknown to $\mathcal{A}$ and is uniformly distributed in $\mathbb{G}_2^*$.

Phase 1 - Extraction queries When $\mathcal{A}$ asks for the private key for ID$_i$, $\mathcal{B}$ runs the algorithm for responding H_1-queries and gets $H_1(\text{ID}_i) = Q_i$, where $\langle \text{ID}_i, Q_i, b_i \rangle$ is the corresponding entry in H_1^{list}. If $i = j$, then $\mathcal{B}$ aborts the game and the attack against BasicPub^{hy} failed. Otherwise, it sets $d_i := b_i Q_{pub}$. Finally, $\mathcal{B}$ gives d_i to algorithm $\mathcal{A}$.

Phase 1 - Decryption queries $\mathcal{B}$ answers to a decryption query $\langle \text{ID}_i, C_i \rangle$ as follows. It runs H_1-queries algorithm and let $\langle \text{ID}_i, Q_i, b_i \rangle \in H_1^{\text{list}}$. If $i \neq j$, then $\mathcal{B}$ retrieves the private key d_i and decrypts C_i using the decryption algorithm. If $i = j$, then $Q_i = Q$, and the decryption of $\langle \text{ID}_j, C_j \rangle$ is the same as the decryption of C_j under BasicPub^{hy}. Then, $\mathcal{B}$ asks its challenger to decrypt C_j and relays the answer to $\mathcal{A}$.

Challenge $\mathcal{A}$ outputs a public key ID$_{\text{ch}}$ and two equal length plaintexts M_0, M_1. Algorithm $\mathcal{B}$ proceeds as follows. If ID$_{\text{ch}} \neq $ ID$_j$, it aborts the game and the attack against BasicPub^{hy} failed. Otherwise, it sends M_0, M_1 to its own challenger and gets back C, the encryption of M_β for a random bit β under BasicPub^{hy}. Finally, $\mathcal{B}$ relays C to $\mathcal{A}$, which is an also encryption of M_β under ID$_{\text{ch}}$ for Full-Ident.

Phase 2 - Extraction queries Algorithm $\mathcal{B}$ proceeds as in Phase 1, except for the extraction query for $\mathsf{ID}_{\mathsf{ch}}$, which is rejected.

Phase 2 - Decryption queries Algorithm $\mathcal{B}$ proceeds as in Phase 1, except for the decryption query $\langle \mathsf{ID}_{\mathsf{ch}}, C_\beta \rangle$, which is rejected.

Guess Algorithm $\mathcal{A}$ outputs a guess β' for β. $\mathcal{B}$ outputs β' as its guess.

Using this algorithm $\mathcal{B}$, we are able to state the following:

Result 4. *Let $\mathcal{A}$ an* IND-ID-CCA *adversary that has advantage ε against* Full-Ident *making at most q_E, q_D and q_{H_1} queries. Then there is an* IND-CCA *adversary $\mathcal{B}$ that has advantage at least $\frac{\varepsilon}{q_{H_1}}\left(1 - \frac{q_E}{q_{H_1}}\right) \approx \frac{\varepsilon}{q_{H_1}}$ against* BasicPub^{hy}. *Its running time is $t_2 \leq t_1 + c_{\mathbb{G}_2}(q_D + q_{H_1} + q_E)$, where $c_{\mathbb{G}_2}$ denotes the time of computing a random multiple in $\mathbb{G}_2$.*

Proof: The proof appears in the full version of the paper. □

Therefore, joining Results 2, 3 and 4, an IND-ID-CCA advantage ϵ_1 against Full-Ident is turned into an algorithm with advantage roughly $\varepsilon_1/(q_H^3)$ in solving the BDH problem. Compared to the original flawed reduction, where the advantage obtained against BDH was roughly $\varepsilon_1/(q_H^2 q_D)$, the new reduction is a bit worse, since in general $q_D \ll q_H$. In the next section we show a modification of Full-Ident which allows a tighter security reduction.

4   A New Identity Based Encryption Scheme with Improved Tightness

In this section we design a new IBE scheme using the scheme Basic-Ident from the previous section and a second general transformation also due to Fujisaki and Okamoto [FO00]. This conversion starts from an IND-CPA encryption scheme and builds an IND-CCA scheme in the ROM. If we denote by $E_{\mathsf{pk}}(M, r)$ the encryption of M using the random bits r under the public key pk, with set of messages $\mathcal{M} = \{0,1\}^n$, set of coins $\mathcal{R}$ and set of ciphertexts $\mathcal{C}$, the new transformation is the scheme

$$E_{\mathsf{pk}}^{hyNew}(M) = E_{\mathsf{pk}}(M||r, H(M||r)) \qquad (3)$$

where $M||r \in \{0,1\}^{n-k_0} \times \{0,1\}^{k_0}$ and $H : \{0,1\}^* \to \mathcal{R}$ is a hash function. Then, $\mathcal{M}^{hyNew} = \{0,1\}^{n-k_0}$, $\mathcal{R}^{hyNew} = \{0,1\}^{k_0}$ and $\mathcal{C}^{hyNew} = \mathcal{C}$. To decrypt C, one first obtains $M'||r'$ using the original decryption algorithm, and next checks if $E_{\mathsf{pk}}(M'||r', H(M'||r')) = C$. If this is so, outputs M; otherwise outputs reject.

Let us describe the new IBE scheme thereby obtained.

NewFull-Ident

Setup. Let $(\mathbb{G}_1, \mathbb{G}_2, \mathbb{G}_T, \hat{e}, \psi)$ a bilinear group. Choose a generator $P_2 \leftarrow \mathbb{G}_2$ and set $P_1 = \psi(P_2)$. Next pick $s \leftarrow \mathbb{Z}_p^*$ and set $Q_{pub} = sP_2 \in \mathbb{G}_2^*$, $P_{pub} = sP_1 \in \mathbb{G}_1^*$. Choose hash functions $H_1 : \{0,1\}^* \to \mathbb{G}_2^*$, $H_2 : \mathbb{G}_T \to \{0,1\}^n$ and $H_3 : \{0,1\}^* \to \mathbb{Z}_p^*$. Now $\mathcal{M} = \{0,1\}^{n-k_0}$, $\mathcal{C} = \mathbb{G}_1^* \times \{0,1\}^n$ and params $= \langle p, \mathbb{G}_1, \mathbb{G}_2, \mathbb{G}_T, \hat{e}, P_1, P_2, P_{pub}, P_{pub}, Q_{pub}, H_1, H_2, H_3 \rangle$.
The master-key is $s \in \mathbb{Z}_p^*$.

Extract. For a given ID $\in \{0,1\}^*$, compute $Q_{\mathsf{ID}} = H_1(\mathsf{ID}) \in \mathbb{G}_2^*$ and set $d_{\mathsf{ID}} = sQ_{\mathsf{ID}} \in \mathbb{G}_2^*$ where s is the master key.

Encrypt. To encrypt $M \in \{0,1\}^{n-k_0}$ under ID, compute $Q_{\mathsf{ID}} = H_1(\mathsf{ID}) \in \mathbb{G}_2^*$, choose $\sigma \leftarrow \{0,1\}^{k_0}$, set $r = H_3(M, \sigma) \in \mathbb{Z}_p^*$ and finally

$$C = \langle rP_1, (M||\sigma) \oplus H_2(g_{\mathsf{ID}}^r) \rangle \quad \text{where} \quad g_{\mathsf{ID}} = \hat{e}(P_{pub}, Q_{\mathsf{ID}}) \in \mathbb{G}_T$$

Decrypt. Let $C = \langle U, V \rangle \in \mathcal{C}$ be a ciphertext under the public key ID. To decrypt C using the private key $d_{\mathsf{ID}} \in \mathbb{G}_2^*$ do:
1. Compute $V \oplus H_2(\hat{e}(U, d_{\mathsf{ID}})) = M||\sigma$.
2. Parse $M||\sigma$ and compute $r = H_3(M, \sigma)$. Check that $U = rP$. If not, reject the ciphertext.
4. Output M.

On the basis of the proof sketched in the previous section, we define in a similar fashion a public key encryption scheme NewBasicPub^{hy}, which is obtained applying the conversion from expression (3) to Basic-Pub. Then the following results hold:

Result 5. *Let $\mathcal{A}_1$ an IND-ID-CCA adversary with advantage ε_1 against NewFull-Ident making at most q_E private key extraction queries, q_D decryption queries and q_{H_1} hash queries. Then there is an IND-CCA adversary $\mathcal{A}_2$ that has advantage at least $\frac{\varepsilon}{q_{H_1}}\left(1 - \frac{q_E}{q_{H_1}}\right) \approx \frac{\varepsilon}{q_{H_1}}$ against NewBasicPub^{Hy}. Its running time is $t_2 \leq t_1 + c_{\mathbb{G}_1}(q_D + q_{H_1} + q_E)$.*

Proof: Use the same reduction as for Result 4 in Section 3.3.

Result 6. *Let $\mathcal{A}_2$ an IND-CCA adversary with advantage ε_2 against NewBasicPub^{hy} making at most q_D decryption queries and at most q_{H_2} hash queries. Then there is an IND-CPA adversary $\mathcal{A}_3$ that has advantage at least*

$$\left(\varepsilon_2 - q_{H_2} \cdot 2^{-(k_0-1)}\right)\left(1 - \frac{1}{p}\right)^{q_D} \approx \varepsilon_2$$

against BasicPub. Its running time is $t_3 \leq t_2 + q_{H_2}(T_{\mathsf{BasicPub}} + \log p)$, where T_{BasicPub} is the running time of Encrypt algorithm in BasicPub.

Proof: This result is obtained as a special case of Theorem 5.4 in [FO00].

Finally, taking into account these new reductions, we obtain that NewFull-Ident scheme is $(t_1, q_H, q_D, \varepsilon_1)$ IND-ID-CCA secure if the BDH problem on $(\mathbb{G}_1, \mathbb{G}_2)$ is

$$\left(t_1 + c_{\mathbb{G}_1}(2q_D + q_H) + q_H \mathcal{O}(\log^3 p + \log p), \frac{\varepsilon_1}{q_H^2}\right)\text{-secure}$$

The last expression has been simplified replacing any of the hash queries q_{H_i} by q_H and setting $q_D = q_E$. Then, we get rid of a q_H factor in the BDH advantage with respect to the reduction in expression (2).

Compared to Full-Ident scheme, which is the result of using FO transformation in expression (1), the NewFull-Ident scheme presents several advantages:

- It provides more compact ciphertexts. In fact, Full-Ident scheme adds a n-bits component to a Basic-Ident ciphertext to get chosen ciphertext security, while NewFull-Ident achieves this preserving Basic-Ident ciphertext's structure.
- It presents a tighter security reduction to the BDH problem.
- It uses one less hash function than Full-Ident.

We can obtain a second tightness improvement using a stronger assumption, namely, the BDDH assumption. In this case, we have the following result:

Result 7. *Let $\mathcal{A}_3$ an IND-CPA adversary that has advantage ε_3 against BasicPub making at most q_{H_2} hash queries. Then there is an algorithm $\mathcal{B}$ breaking the BDDH problem on $(\mathbb{G}_1, \mathbb{G}_2)$ with advantage roughly ε_3 and running time $t' \approx t_3$.*

Proof: The proof appears in the full version of the paper. □

With this second tightness improvement, we obtain that NewFull-Ident scheme is $(t_1, q_H, q_D, \varepsilon_1)$ IND-ID-CCA secure if the BDDH problem on $(\mathbb{G}_1, \mathbb{G}_2)$ is

$$\left(t_1 + c_{\mathbb{G}_1}(2q_D + q_H) + q_H \mathcal{O}(\log^3 q + \log q), \frac{\varepsilon_1}{q_H}\right)\text{-secure}$$

Then, we get rid of a q_H factor in the security reduction at the cost of relying on a stronger assumption.

5 Conclusions

In this work, we have shown there is a flawed step in the security reduction exhibited in [BF03] for the so called Boneh-Franklin IBE scheme. We have provided a new reduction without modifying the original scheme neither the underlying hard problem used to state the security.

In the second place, we have proposed a new IBE scheme slightly changing the original scheme. The proposal presents a tighter reduction than BF scheme, uses one less random oracle and has more compact ciphertexts.

Finally, we point out that it is still an open problem to design an IND-ID-CCA IBE scheme with a tight security reduction under a reasonable assumption either in the standard or the random oracle models.

Acknowledgements. The author is grateful to Javier Herranz and Paz Morillo for useful comments on an early draft of this paper. The author also acknowledges the anonymous referees' comments.

References

[AP03] S. AlRiyami and K.G. Paterson. Certificateless public key cryptography. In *ASIACRYPT 2003*, volume 2894 of *LNCS*, pages 452–473, 2003. Full version available at http://eprint.iacr.org/.

[BDPR98] M. Bellare, A. Desai, D. Pointcheval, and P. Rogaway. Relations among notions of security for public-key encryption schemes. In - *CRYPTO 1998*, volume 1462 of *LNCS*, pages 26–45, 1998.

[BF03] D. Boneh and M. Franklin. Identity-Based encryption from the Weil pairing. *SIAM Journal of Computing*, 32(3):586–615, 2003. This is the full version of an extended abstract of the same title presented at *Crypto'01*.

[CC05] Z. Cheng and R. Comley. Efficient certificateless public key encryption. Cryptology ePrint Archive, Report 2005/012, 2005.

[FO99] E. Fujisaki and T. Okamoto. Secure integration of asymmetric and symmetric encryption schemes. In - *CRYPTO 1999*, volume 1666 of *LNCS*, pages 537–554, 1999.

[FO00] E. Fujisaki and T. Okamoto. How to enhance the security of public-key encryption at minimum cost. *IEICE Trans. Fund.*, E83-9(1):24–32, 2000.

[Gen03] C. Gentry. Certificate-based encryption and the certificate revocation problem. In *EUROCRYPT 2003*, volume 2656 of *LNCS*, pages 272–293, 2003.

[GS02] C. Gentry and A. Silverberg. Hierarchical ID-based cryptography. In *ASIACRYPT 2002*, volume 2501 of *LNCS*, pages 548–566, 2002.

[HL02] J. Horwitz and B. Lynn. Toward hierarchical identity-based encryption. In - *EUROCRYPT 2002*, volume 2332 of *LNCS*, pages 466–481, 2002.

[Sha84] A. Shamir. Identity-based cryptosystems and signature schemes. In *CRYPTO 1984*, volume 196 of *LNCS*, pages 47–53, 1985.

[YFDL04] D. Yao, N. Fazio, Y. Dodis, and A. Lysyanskaya. Id-based encryption for complex hierarchies with applications to forward security and broadcast encryption. In *Proceedings of the 11th ACM CCS*, pages 354–363. ACM Press, 2004.

Single-Database Private Information Retrieval with Constant Communication Rate

Craig Gentry and Zulfikar Ramzan

DoCoMo Communications Laboratories USA, Inc.
{cgentry, ramzan}@docomolabs-usa.com

Abstract. We present a single-database private information retrieval (PIR) scheme with communication complexity $\mathcal{O}(k+d)$, where $k \geq \log n$ is a security parameter that depends on the database size n and d is the bit-length of the retrieved database block. This communication complexity is better asymptotically than previous single-database PIR schemes. The scheme also gives improved performance for practical parameter settings whether the user is retrieving a single bit or very large blocks. For large blocks, our scheme achieves a constant "rate" (e.g., 0.2), even when the user-side communication is very low (e.g., two 1024-bit numbers). Our scheme and security analysis is presented using general groups with hidden smooth subgroups; the scheme can be instantiated using composite moduli, in which case the security of our scheme is based on a simple variant of the "Φ-hiding" assumption by Cachin, Micali and Stadler [2].

1 Introduction

PROBLEM STATEMENT AND BACKGROUND. Private Information Retrieval (PIR) schemes allow a user to retrieve the ith bit of an n-bit database, without revealing to the database the value of i. The "trivial" solution is for the user to retrieve the entire database, but this approach may incur enormous communication costs. A good PIR scheme, on the other hand, should have considerably lower (certainly sub-linear) communication complexity. Private Block Retrieval (PBR) is a natural and more practical extension of PIR in which, instead of retrieving only a single bit, the user retrieves a d-bit block that begins at index i.

PIR and PBR have been studied extensively; here, we only mention the work most relevant to us. The notion of PIR was introduced by Chor et al. [5], who focused on the *information-theoretic* case, where one requires that the user's query give absolutely no information about i. They proved that if only a single database is used, then n bits must be communicated. On the other hand, if the database is replicated in k servers, and if the user is allowed to give a separate query to each server, one can construct a PIR scheme with k user queries each being $\mathcal{O}(n^{1/k})$-bits and k single-bit server responses.[1] However, to ensure user privacy in the multi-server setting, the servers must be trusted not to collude.

---

[1] Currently, the lowest asymptotic total communication complexity for information-theoretic PIR is $\mathcal{O}(n^{\log \log k/k \log k})$ [1].

Chor et al. also introduced PBR. They showed that any PIR scheme with $\alpha_k(n)$-bit queries and $\beta_k(n)$-bit responses can be converted into a PBR scheme for d-bit blocks with $\alpha_k(n)$-bit queries and $d\beta_k(n)$-bit responses. This means that, for a constant $k \geq 2$ of servers, the above information-theoretic PIR scheme can be converted into a PBR scheme with an asymptotically constant "rate" of $1/k$ – i.e., the ratio of bits retrieved (i.e., d) versus total communication complexity (i.e., $kd + \mathcal{O}(n^{1/k})$) tends towards $1/k$ as n and d increase appropriately. Increasing the rate to 1 in the information-theoretic setting seems difficult.

Chor and Gilboa studied the problem of whether one could achieve better communication complexity for multi-server PIR by using computational assumptions [4]. Subsequently, Kushilevitz and Ostrovsky showed that one can achieve *single database* PIR under the Quadratic Residuosity assumption with communication $2^{\mathcal{O}\sqrt{\log n \log l_m}}$, where l_m is the bit length of a composite modulus m. Like all current single-database PIR schemes, the server needs $\Omega(n)$ computation to generate a query response. Since the number field sieve [10] can factor an l_m-bit composite number in time $2^{\mathcal{O}(1) l_m^{1/3} (\log l_m)^{2/3}}$ (and hence solve quadratic residuosity), and since it seems reasonable that the server should need at least as much computation to break user privacy as to generate a response, one should set $l_m = \Omega(\log^{3-o(1)} n)$ to ensure adequate security.

Cachin, Micali, and Stadler [2] constructed the first single-database PIR scheme with *poly-logarithmic* communication complexity (about $\mathcal{O}(\log^8 n)$ for their suggested parameters), addressing an open problem left by Kushilevitz and Ostrovsky. The security of their scheme (CMS) is based on the "Φ-hiding" assumption – roughly, that is hard to distinguish which of two primes divides $\phi(m)$ for composite modulus m. Essentially, the scheme works as follows. Each index $j \in [1, n]$ is mapped to a distinct prime p_j. To recover bit b_i from database $B = b_1 \cdots b_n$, the user sends a composite (hard-to-factor) modulus m such that p_i divides $\phi(m)$ and a generator $x \in \mathbb{Z}_m^*$ with order divisible by p_i. The server sends back $r = x^P \pmod{m}$ for $P = \prod_j p_j^{b_j}$. The user concludes that $b_i = 1$ if r is a p-residue modulo m; otherwise, $b_i = 0$. The communication complexity of (this simplified version of) CMS is $3l_m$ to recover 1 database bit. Again, $l_m = \Omega(\log^{3-o(1)} n)$ for adequate security, though [2] recommends an even larger value of l_m ($\mathcal{O}(\log^8 n)$).

Recently, Lipmaa [11] gave a PBR scheme with stated $\Theta(l_m \cdot \log^2 n + d \cdot \log n)$ communication complexity for d-bit blocks, where again $l_m = \Omega(\log^{3-o(1)} n)$. Thus, Lipmaa's scheme has a better "rate" – namely $1/(\log n)$ – than CMS for large blocks. In fact, as we describe in the full version of this paper, one can apply Chor et al.'s [5] abovementioned conversion from PIR to PBR to Lipmaa's scheme to get a PBR scheme with rate arbitrarily close to 1. However, for Lipmaa's scheme to achieve a good rate in practice, n and d must be quite large (on the order of gigabits and megabits, respectively) before they begin to offset the large one-time cost represented by the $l_m \cdot \log^2 n$ term.

OUR RESULTS. We present a single-database PBR scheme that has, to the best of our knowledge, the lowest asymptotic communication complexity of

$\Theta(k+d)$. The scheme is somewhat similar to CMS [2], but the scheme is described (and its security proven) with respect to general groups that have "hidden subgroups" of smooth order. Our scheme also transforms the CMS technique to *maximize* the number of database bits the user can recover from a short server response. The essential technique is to associate each *block* of bits with a distinct *small* prime (or power of a small prime), rather than allocating a (largish) prime to each bit. The database's response protocol uses the Chinese Remainder Theorem to encode each database chunk modulo its associated prime power. To decode, the user computes a discrete logarithm, but in a subgroup whose order is *smooth* – i.e., a product of small primes. We can carry out this step efficiently in a (somewhat surprising) *constructive* application of the Pohlig-Hellman method [14]. In the full version of the paper, we show that our scheme is secure against generic attacks even when $k = \mathcal{O}(\log n)$ and when the rate of the scheme approaches 1. We provide an oblivious transfer scheme with similar performance characteristics by using the Naor-Pinkas transformation[13].

We describe an instantiation of our scheme that, like CMS, uses a (e.g., 1024-bit) composite modulus m. In CMS as described above, a user sends a $2l_m$-bit query and gets back a l_m-bit response that allows the user to retrieve a single bit; in our scheme, with the same communication complexity, the user can recover $c \cdot l_m$ bits for $c < 1/4$; this is a fairly high constant "rate" – i.e., the communication of the PBR scheme is only a small constant times more than the communication needed to transmit the block with *no privacy at all*. This instantiation has the best known asymptotic communication complexity $\Theta(\log^{3-o(1)} n, d)$ in terms of n and d among single-database PIR schemes and has the lowest complexity for most practical parameters (until it is overtaken by the modified version of Lipmaa's scheme with rate approaching 1). However, this instantiation does not perform as well as our scheme *could* perform according to the generic group model, since it is vulnerable to the number field sieve unless $k = \Omega(\log^{3-o(1)} n)$ and to Coppersmith's attack [6,7] when $c \geq 1/4$. We speculate on approaches to instantiating the scheme that may achieve better performance.

2 Preliminaries

In the sequel, n denotes the database size in bits. If S is a set of elements, and D is a sampleable probability distribution on S, we let $s \xleftarrow{D} S$ denote the process of picking an element s from S according to distribution D. Throughout, π will denote a prime power. We say that an integer m $\varPhi$-hides π if π divides $\phi(m)$.

If A is an algorithm, we let $A(\cdot, \ldots, \cdot)$ denote that A may take one or more inputs. By $Pr[y \leftarrow A(x) : b(y)]$, we denote the probability that $b(y)$ is true after y was generated by A on input x. By $A^{(B)}(\cdot)$, we denote an algorithm that can make oracle queries to B. For $a, b \in \mathbb{Z}$ with $a \leq b$, let $[a, b]$ denote the set of integers between a and b inclusive. Let $[b]$ denote $[1, b]$.

Now, we define polylogarithmic private information retrieval as in [2].

Definition 1 (Polylogarithmic CPIR). *Let $Q(\cdot,\cdot,\cdot)$, $D(\cdot,\cdot,\cdot)$ and $R(\cdot,\cdot,\cdot,\cdot,\cdot)$ be polynomial-time algorithms. We say that (Q, D, R) is a fully polylogarithmic CPIR scheme if there exists constants $a, b, c, d > 0$ such that:*

- *(Correctness)* $\forall n \in \mathbb{N}$, $\forall B \in \{0,1\}^n$, $\forall i \in [1,n]$, and $\forall k' \in \mathbb{N}$,

$$\Pr[(q,s) \xleftarrow{R} Q(n,i,1^{k'}); r \xleftarrow{R} D(B,q,1^{k'}) : R(n,i,(q,s),r,1^{k'}) = B_i] > 1 - 2^{-ak'}.$$

- *(User Privacy)* $\forall n \in \mathbb{N}$, $\forall i,j \in [1,n]$, $\forall k' \in \mathbb{N}$ such that $2^{k'} > n^b$, and $\forall 2^{ck'}$-gate circuits A,

$$\left|\Pr[(q,s) \xleftarrow{R} Q(n,i,1^{k'}) : A(n,q,1^{k'}) = 1] - \Pr[(q,s) \xleftarrow{R} Q(n,j,1^{k'}) : A(n,q,1^k)\right.$$

$$\left. = 1]\right| < 2^{-dk'}.$$

Here a, b, c, d are the fundamental constants of the CPIR scheme; B is the contents of the database; D is the database's response algorithm; Q is the user's query-generating algorithm; R is the user's response reconstruction algorithm; q is the user's actual query; s is the user's secret (associated with q); r is the database's response; and k' is a security parameter.

Notice that we have mentioned two security parameters – namely, $k' > b \log n$ above, and k in the Introduction (which may be, e.g., the bit-length of a composite modulus). The two parameters are related by $k = \mathcal{O}(f(k'))$ for some polynomial f. For example, for the modulus-based instantiation, we may have $k = \max\{1024, Ck'^3\}$ for some constant C to ensure that no $(2^{ck'} = poly(n))$-gate circuits A (e.g., a circuit running NFS) can break user privacy with probability $1/poly(n)$. Against generic attacks, $k = k'$ suffices to ensure user privacy. In short, the security parameter k' is useful because it ensures (above) that no algorithms A that are polynomial in n can break user privacy, while allowing us to separately define the security parameter k in the "common parlance" of a particular instantiation. (For example, for cryptosystems related to factoring, the security parameter k is typically defined as the modulus bit-length, even though such schemes have only $\exp(\mathcal{O}(1)k^{1/3}(\log k)^{2/3})$ security against NFS.)

3 Our General Private Block Retrieval Scheme

We now describe our PIR scheme using general groups with hidden smooth-order subgroups; afterwards, once the essential strategy of our scheme has been laid out, we will describe the computational assumption on which user privacy is based (which, by then, will seem relatively natural).

First, we give a high-level description of the scheme. The scheme has some public parameters known to all users, including the database size n, an integer parameter ℓ, a set of $t = \lceil n/\ell \rceil$ (small) distinct prime numbers $\{p_1, \ldots, p_t\}$, and a set $\mathcal{S} = \{\pi_1, \ldots, \pi_t\}$ of prime powers $\pi_i = p_i^{c_i}$, where $c_i = \lceil \ell / \log_2 p_i \rceil$ (i.e., so that $p_i^{c_i} \geq 2^\ell$). The server partitions the database B into t blocks $B = C_1 \| C_2 \| \cdots \| C_t$ of size at most ℓ. In our scheme, the user will retrieve the entire ℓ-bit block that contains its desired bit. Each block C_i is associated to a prime power π_i. Using the Chinese Remainder Theorem, the server can express the entire database B as an integer e that satisfies $e \equiv C_i (\bmod \pi_i)$, where the ℓ-bit block C_i is treated as an integer satisfying $0 \le C_i < 2^\ell \le \pi_i$. Notice that to retrieve C_i, it suffices to retrieve $e(\bmod \pi_i)$.

Roughly speaking, to query the value of $e(\bmod \pi_i)$, the user generates an appropriate cyclic group $G = \langle g \rangle$ with order $|G| = q\pi_i$ for some suitable integer q. It sends (G, g) to the server and keeps q private. Notice that G contains a subgroup H of order π_i, and that $h = g^q$ is a generator of H. (For technical reasons, in the actual scheme below, $\langle g \rangle$ may be a proper subgroup of G.)

The server responds with $g_e = g^e \in G$. The user then obtains $e(\bmod \pi_i)$ by setting $h_e = g_e^q \in H$ and performing a (tractable) discrete logarithm computation: $\log_h h_e \equiv e(\bmod \pi_i)$. This discrete logarithm computation, which occurs entirely in the subgroup H of order $p_i^{c_i}$, can actually be quite efficient if p_i is small. Correctness is demonstrated below. Now, we give a more precise description of the general scheme.

For some parameter choices, the user can select G such that $|G|$ is divisible by multiple π_i's. In this case, the user can recover multiple ℓ-bit blocks (note that this does not contradict the security requirements for PIR schemes). However, for simplicity, we focus on the single-block case.

SPECIFICATION OF THE SCHEME. Let B be an n-bit database. Let $f_1(x, y)$ and $f_2(x, y)$ be functions. Let $k' = \Theta(\log n)$ and $k = f_2(k', \log n)$ be security parameters. Set $\ell = \lfloor f_1(k, \log n) \rfloor$ and $t = \lceil n/\ell \rceil$. For primes $\mathcal{P} = \{p_1, \ldots, p_t\}$, set $\pi_i = p_i^{c_i}$ for $c_i = \lceil \ell/(\log_2 p_i) \rceil$, and $\mathcal{S} = \{\pi_i\}$. Let $\mathcal{G}_i$ be the set of cyclic groups whose order is a number in $[2^k, 2^{k+1}]$ that is divisible by π_i. Let D_i be a distribution under which elements of $\mathcal{G}_i$ can be efficiently sampled. We assume that for $G \xleftarrow{D_i} \mathcal{G}_i$, each $g \in G$ has a unique "normal" representation. (We will discuss the security considerations involved in choosing k', f_1, f_2 and $\{D_i\}$ later.)

Query Generation: Given input $(n, f_1, f_2, \mathcal{S}, \{D_i\}, 1^{k'})$, the user determines the index i of its desired block, and generates a query for block C_i as follows:

1. Generate $G \xleftarrow{D_i} \mathcal{G}_i$ and a uniformly random "quasi-generator" g of G – i.e., g is a random element of G such that $GCD(|G : \langle g \rangle|, \prod_{j=1}^t p_j) = 1$;
2. Output query (G, g); keep $q = |\langle g \rangle|/\pi_i$ private; store $h = g^q$ for future use.

Database Response Generation: Given the input $(B, f_1, f_2, \mathcal{S}, G, g, 1^{k'})$, the server responds to the user's query as follows:

1. Express each ℓ-bit database block C_j (after appending zeros to C_t if needed) as a number in $[0, 2^\ell - 1]$ in the obvious fashion;
2. Set e to be the smallest positive integer such that $e \equiv C_j \pmod{\pi_j}$ for all j;
3. Output the response $g_e = g^e \in G$.

Note that steps 1 and 2 are independent of the query, and can be precomputed.

Response Retrieval: Given the input $(\pi_i, g_e, G, q, h, 1^{k'})$, the user retrieves block C_i as follows:

1. Compute $h_e = g_e^q$;
2. Compute C_i as the discrete logarithm $\log_h h_e$ within the subgroup $H \subset G$ of order $\pi_i = p_i^{c_i}$ using Pohlig-Hellman.

Notice that we need p_i to be small (unlike CMS) for the discrete logarithm computation using Pohlig-Hellman to be efficient. Fortunately, as we show below, the Prime Number Theorem will help us ensure that $\max\{p_i\}$ is small, and that response retrieval is efficient.

CORRECTNESS OF RESPONSE RETRIEVAL. Let $e_{\pi_i} \in [0, \pi_i - 1]$ satisfy $e_{\pi_i} \equiv e \pmod{\pi_i}$; observe that e_{π_i} is equal to C_i. So, it suffices to show that e_{π_i} is the discrete logarithm of h_e with respect to base h. Write $e = e_{\pi_i} + \pi_i \cdot E$, for some $E \in \mathbb{Z}$. Now:

$$h_e = g_e^{|\langle g \rangle|/\pi_i} = g^{e|\langle g \rangle|/\pi_i} = g^{e_{\pi_i}|\langle g \rangle|/\pi_i} = h^{e_{\pi_i}}.$$

Remark 1. The above scheme has some similarities to CMS, particularly if one instantiates the group G using a composite modulus m. However, for recovering *blocks* of data (a more realistic scenario anyway), our scheme is much more communication efficient; the server's $(\log m)$-bit response uses the Chinese Remainder Theorem to give the user ℓ bits instead of 1 bit. Later, we will see that ℓ can equal $(\log m)/C$ for reasonably small constant C.

CHOOSING THE SET $\mathcal{P}$ WISELY. Recall that $\mathcal{P} = \{p_1, \ldots, p_t\}$ is the set of primes that the scheme uses; let p_t be the largest. As mentioned above, p_t must be reasonably small to ensure efficient response retrieval. Also, since we must have $\log|G| \geq \max\{\pi_i\} \geq p_t$, the size of p_t also affects communication complexity. The following result of Rosser and Schoenfeld related to the Prime Number Theorem [15] gives an upper bound on p_t.

Theorem 1 (Rosser and Schoenfeld). *For $t > 20$, let $\mathcal{P} = \{p_1, \ldots, p_t\}$ be the first t primes, with p_t the largest. Then, $p_t < t(\ln t + \ln \ln t - 1/2)$.*

For technical reasons in the security proof, we need $p_1 \geq 2t$. Nonetheless, in terms of n and ℓ, we easily get that $p_t < 16(n/\ell) \log_2(n/\ell)$ suffices. For the performance analysis below, we assume for convenience that ℓ is chosen so that $2^\ell \geq p_t$.

COMPUTATIONAL COMPLEXITY. The dominant component of the querier's computation is in computing the discrete logarithm of h_e for base h. This step involves solving c_i discrete logarithm sub-problems in groups of order p_i for $p_i^{c_i} \in [2^\ell, 2^\ell p_t]$. Assuming that each sub-problem involves $\sqrt{p_i}$ group operations – e.g., using baby-step giant-step – the entire discrete logarithm problem requires about $c_i \sqrt{p_i}$ group operations. Considering the curve $y^x = 2^\ell p_t$ for $y \leq p_t$, we see that $x\sqrt{y} = (\sqrt{y}/\log y)(x \log y) = (\sqrt{y}/\log y)(\log(2^\ell p_t))$ takes its maximum at $y = p_t$. As a very rough upper bound, $\sqrt{p_t}/\log p_t < 2\sqrt{n/\ell}$ and $\log(2^\ell p_t) < 2\ell$, so the querier's computation is no more than $4\sqrt{n\ell}$ group operations, where ℓ must be less than $\log|G|$ (which will be polylogarithmic in n). This does not seem unreasonable given that the database's computation in single-database PIR is *unavoidably linear* in n (since otherwise the database has not included every database bit in the computation, which would imply that it knows at least one bit that the user did not request).

The dominant component of the database's computation is in computing $g^e \bmod m$. This requires (roughly) $\log e$ group operations. Since e is a number modulo $\prod_{i=1}^{t} \pi_i$, we have $\log e \leq \sum_{i=1}^{t} \log \pi_i$. Since, $p_i \leq 2^\ell$ for all i, $\pi_i = p_i^{c_i} < 2^{2\ell}$ for all $c_i = \lceil \ell/(\log p_i) \rceil$. Thus, we have $\sum_{i=1}^{t} \log \pi_i < 2\ell t = 2\ell \lceil n/\ell \rceil$ – i.e., the database needs $\Theta(n)$ group operations, which is about the best we can hope for in single-database PIR.

COMMUNICATION COMPLEXITY. Suppose that the group G and any element of G can be described in $l_G = \Omega(\log |G|)$ bits. (For example, the group generated by g modulo m for composite modulus m can be described in $\mathcal{O}(\log \phi(m))$ bits.) Then, the total communication complexity is $3l_G$. The size of l_G depends, in part, on security considerations pertaining to the particular instantiation of our general scheme; so, we obviously cannot give a general upper bound for l_G. Here, we merely note that, in terms of the scheme's *correctness*, the only constraint on $|G|$ is that it be divisible by (and, hence, at least as large as) π_i. Above, we saw that when $2^\ell > p_t$, $\pi_i < 2^{2\ell}$ for all i. Thus, if we set $\ell = \lceil \log p_t \rceil$, then $\max\{\log \pi_i\} < 2\ell < 4 \log p_t < 8 \log n$. Thus, the mechanics of the scheme do not prevent $\log |G| = \Theta(\log n)$ or $l_G = \Theta(\log n)$.

We stress that l_G may need to be larger to ensure user privacy. However, in our analysis of the scheme's security in the generic group model in Section 6, we find that generic attacks do not prevent our scheme with $l_G = \Theta(\log n)$ from having the security required by CMS's definition of polylogarithmic PIR; any attack that forces l_G to be larger must exploit the encoding of the group or its elements.

PRIVATE BLOCK RETRIEVAL. In our scheme, the user already recovers ℓ-bit blocks. This scheme can be converted, using the general transformation described in [5], into a scheme that recovers d ℓ-bit blocks with total communication complexity $(2+d)l_G$, as follows. The user generates a query (G, g) for the ℓ-bit block beginning with index i. To allow the user to retrieve the ℓ-bit block with index $i + x\ell$ for $x \in [0, d-1]$, the server temporarily relabels the database, giving the database bit with index j (for $j \in [n]$) the "temporary index" $j - x\ell (\bmod n)$; it then responds to the user's query (G, g) using the temporary indices, rather than the actual ones. The "rate" of our scheme – i.e., the ratio of the number of bits that the user retrieves over the total communication complexity – is $d\ell/(d+2)l_G$, which approaches ℓ/l_G as d increases. We will see that our general scheme is secure against generic group attacks for ℓ/l_G arbitrarily close to 1. When we instantiate the scheme using Φ-hiding and a composite modulus m in the natural way, however, an attack by Coppersmith [7] forces $\ell/l_G < 1/4$.

OBLIVIOUS TRANSFER. Naor and Pinkas [13] describe how to construct 1-out-of-n OT scheme from a PIR scheme for n-bit databases and $\log n$ invocations of a 1-out-of-2 OT scheme. Since the transformation is generic, we omit the details, except to mention that 1-out-of-2 OT can be accomplished fairly efficiently through the ElGamal encryption scheme. If k'' is the bit-length of group elements in the ElGamal group (e.g., $k'' = 160$), the transformation only adds $6k''(\log n)$ bits to our PIR scheme, regardless of the block size d.

4 Our General Computational Assumption

In our PIR scheme, the server is given not only a description of G (and generator g), but also a promise that one of the prime powers in S – i.e., the one associated to the user's target block index i – actually divides $|G|$. For our PIR scheme to be user-private, the server should be unable to distinguish which of π_0 or π_1 divides $|G|$ – or, equivalently, to distinguish whether the "smooth" subgroup H hidden inside G has order π_0 or π_1. So, our computational assumption is roughly that, given (π_0, π_1, G) and the promise that π_b divides $|G|$ for one $b \in \{0,1\}$, it is computationally hard (if G is generated appropriately) to distinguish the value of b, even if π_0 and π_1 are not "much smaller" than $|G|$, and even if π_0 and π_1 are "special" integers such as powers of small primes. We formalize this assumption in terms of the following problem.

Definition 2 (The Decision Subgroup Problem). *Let ℓ be an integer and k a parameter. Let $\pi_0, \pi_1 \in [2^\ell, 2^{2\ell} - 1]$ be distinct integers. Let $\mathcal{G}_i$ be the set of cyclic groups whose order is a number in $[2^k, 2^{k+1}]$ that is divisible by π_i. Let D_i be a distribution on $\mathcal{G}_i$. We say that algorithm A has advantage ϵ against the $(\ell, k, \pi_0, \pi_1, D_0, D_1)$-Decision Subgroup Problem if*

$$\left| \Pr\left[b \xleftarrow{R} \{0,1\}, G_b \xleftarrow{D_b} \mathcal{G}_b : A(G_b, \ell, k, \pi_0, \pi_1, \{D_i\}, \{\mathcal{G}_i\}) = b \right] - \frac{1}{2} \right| \geq \epsilon.$$

A solves the problem if it guesses b correctly.

In our PIR scheme, we want the above problem to be hard for each pair $\pi_{i_0}, \pi_{i_1} \in S$, a set of prime powers. Thus, we state our computational assmption as follows.

Definition 3 (The (Extended) Decision Subgroup Assumption). *Let $f_1(x,y)$ and $f_2(x,y)$ be functions. Let S be a set of $t \geq 2$ powers of distinct primes. The (f_1, f_2, S)-Extended Decision Subgroup Assumption is that there exist constants $b, c, d > 0$ such that, for all $n \in \mathbb{N}$ and all $k' > b \log n$ with $\ell = \lfloor f_1(k', \log n) \rfloor$ and $k = f_2(k', \log n)$, there exist efficiently sampleable distributions $\{D_i : i \in [t]\}$ such that, for all $i_0, i_1 \in [t]$, all circuits $\mathcal{A}$ with $(2^{ck'} + t \cdot f_2(k', \log n) \cdot C_{\{\mathcal{D}_i\}})$ gates have advantage at most $2^{-dk'}$ against the $(\ell, k, \pi_{i_0}, \pi_{i_1}, D_{i_0}, D_{i_1})$-Decision Subgroup Problem, where $C_{\{\mathcal{D}_i\}}$ is an upper bound on the circuit complexity of a group multiplication in groups drawn according to D_i for $i \in [t]$.*

5 Security Proof for Our PIR Scheme

We base the security of our scheme on the extended decision subgroup assumption. The proof is done in the standard model.

Theorem 2. *Suppose that a circuit $\mathcal{A}$ with $2^{ck'}$ gates can break user privacy with advantage $2^{-dk'}$. Then, there is an $\mathcal{A}'$ with $\mathcal{O}(2^{ck'} + t \cdot f_2(k', \log n) \cdot C_{\{\mathcal{D}_i\}})$ gates that solves the extended decision subgroup problem with advantage $\frac{1}{5} 2^{-dk'}$.*

Proof. Suppose that the privacy condition fails for (Q, D, R). Then for all $b, c, d > 0$, there exist $n, k' > b \log n$, $B \in \{0,1\}^n$, block indices $i \neq j$, and a circuit $\mathcal{A}$ with $2^{ck'}$ gates, such that $|\alpha_{i,0^t} - \alpha_{j,0^t}| \geq 2^{-dk'}$, where:

$$\alpha_{i,\overline{v}} \triangleq \Pr[((G,g),q) \leftarrow Q(i,T) : \mathcal{A}((G, g^{\prod_{x=1}^{t} p_x^{v_x}}), T) = 1],$$

$$\alpha_{j,\overline{v}} \triangleq \Pr[((G,g),q) \leftarrow Q(j,T) : \mathcal{A}((G, g^{\prod_{x=1}^{t} p_x^{v_x}}), T) = 1],$$

where $\overline{v}$ is a t-element integer vector and 0^t is the zero vector, and where we define $T = (n, f_1, f_2, \mathcal{S}, \{D_i\}, 1^{k'})$ for convenience. We now define two probabilities representing $\mathcal{A}$'s output when g is chosen uniformly at random from G (as opposed to being a random quasi-generator of G):

$$\beta_{i,\overline{v}} \triangleq \Pr[G \xleftarrow{D_i} \mathcal{G}_i; g \xleftarrow{R} G : \mathcal{A}((G, g^{\prod_{x=1}^{t} p_x^{v_x}}), T) = 1],$$

$$\beta_{j,\overline{v}} \triangleq \Pr[G \xleftarrow{D_j} \mathcal{G}_j; g \xleftarrow{R} G : \mathcal{A}((G, g^{\prod_{x=1}^{t} p_x^{v_x}}), T) = 1].$$

Let $\overline{e_x}$ be the unit vector in dimension x. If $p_x^{v_x}$ is greater than $2^{f_2(k, \log n)+1}$ (the maximum possible group order), then $\alpha_{i,\overline{v}} = \alpha_{i,\overline{v}-\overline{e_x}}$ since the distributions of the element given to $\mathcal{A}$ are identical. Let $\overline{v_0}$ be s.t. $\epsilon_{\alpha\alpha} = |\alpha_{i,\overline{v_0}} - \alpha_{j,\overline{v_0}}|$ is maximal and s.t. $p_x^{v_{0,x}} \leq 2^{f_2(k,\log n)+1}$ for all $x \in [t]$. Set $\epsilon_{\beta\beta} \triangleq |\beta_{i,\overline{v_0}} - \beta_{j,\overline{v_0}}|$. Then, $\mathcal{A}'$ can solve the decision subgroup problem instance for prime powers π_i, π_j with advantage $\epsilon_{\beta\beta}$ simply by generating random $g \in G$ and passing $(G, g^{\prod_{x=1}^{t} p_x^{v_{0,x}}})$ to $\mathcal{A}$, and then outputting "i" if $\mathcal{A}$ outputs 1 and outputting "j" otherwise.

Let $\overline{w} \geq \overline{v}$ denote $\forall x \in [t], w_x \geq v_x$. We express $\beta_{i,\overline{v}}$ in terms of $\{\alpha_{i,\overline{w}} : \overline{w} \geq \overline{v}\}$ by noting that choosing an element of G uniformly at random is equivalent to choosing a uniformly random quasi-generator and then exponentiating it by $\prod_{x=1}^{t} p_x^{w_x - v_x}$ with probability $(\prod_{x=1}^{t} \frac{p_x - 1}{p_x})/(\prod_{x=1}^{t} p_x^{w_x - v_x})$. We obtain:

$$\beta_{i,\overline{v}} = \left(\prod_{x=1}^{t} \frac{p_x - 1}{p_x}\right) \left(\sum_{\overline{w} \geq \overline{v}} \alpha_{i,\overline{w}}/(\prod_{x=1}^{t} p_x^{w_x - v_x})\right).$$

Since $p_1 > 2t$, $\sum_{\overline{w} \geq \overline{v_0}} \alpha_{i,\overline{w}}/(\prod_{x=1}^{t} p_x^{w_x - v_{0,x}}) < \epsilon_{\alpha\alpha} \prod_{x=1}^{t} \frac{p_x}{p_x - 1} < \epsilon_{\alpha\alpha} e^{t/(p_1 - 1)} < \epsilon_{\alpha\alpha}\sqrt{e}$. By the triangle inequality, $|\beta_{i,\overline{v_0}} - \beta_{j,\overline{v_0}}| \geq (\epsilon_{\alpha\alpha} - (\sqrt{e} - 1)\epsilon_{\alpha\alpha})(\prod_{x=1}^{t} \frac{p_x - 1}{p_x}) \geq \epsilon_{\alpha\alpha}(2 - \sqrt{e})/(\sqrt{e}) > \epsilon_{\alpha\alpha}/5$. So, $\mathcal{A}$ has $\epsilon_{\beta\beta} \geq \epsilon_{\alpha\alpha}/5 \geq (1/5)2^{-dk'}$ advantage against the Decision Subgroup Problem for (π_i, π_j). The circuit complexity of $\mathcal{A}'$ is basically that of $\mathcal{A}$, plus that needed to compute $g^{\prod_{x=1}^{t} p_x^{v_{0,x}}}$. □

6 Lessons from the Generic Group Model

To gain confidence in our computational assumption, we can consider the Decision Subgroup Problem's vulnerability to generic attacks. The following theorem, which is quite similar to a result by Damgard and Koprowski [8] on root extraction in generic groups, roughly states that, as long as the distributions D_0 and D_1 each tend to output a group whose order is divisible by a large evenly-distributed prime, the Decision Subgroup Problem is hard against generic attacks. In other words, the security of the Decisional Subgroup Problem against generic attacks depends less on the value of $|H|$ (the order of the subgroup hidden in G) than it does on the distribution of $|G:H|$.

Theorem 3. *Let A be a generic algorithm for solving the Decision Subgroup Problem on $(\ell, k, \pi_0, \pi_1, D_0, D_1)$ that makes at most m oracle queries. Let S be a set of bit strings of cardinality at least $2^{2\ell}$. For group G_i, let $\theta(G_i)$ be the largest prime divisor of $|G_i|$ that does not divide $\pi_0 \pi_1$. Let $\alpha(D_i) = \max_q \{\Pr[\theta(G_i) = q \mid G_i \xleftarrow{D_i} \mathcal{G}_i]\}$. Let $\beta(D_i, M)$ be the probability that $\theta(G_i) \leq M$ for distribution D_i; let $\beta(D, M) = \max\{\beta(D_0, M), \beta(D_1, M)\}$. Now, randomly choose $b \xleftarrow{R} \{0,1\}$, $G_b \xleftarrow{D_b} \mathcal{G}_b$, and a random mapping $\sigma_b : G_b \to S$. Then,*

$$\left| \Pr\left[A^{(O)}(S, \ell, k, \pi_0, \pi_1, D_0, D_1) = b \right] - \frac{1}{2} \right| \leq m^2 \left(m\alpha(D) + \beta(D, M) + \frac{1}{M} \right)/2,$$

where the probability is over the random bits of the oracle and A.

Proof. See full version of this paper.

Let's choose parameters to give Theorem 3 meaning. Suppose $2^k/\max\{\pi_i\} \geq 2^{k'}$, and define D_i to choose $|G_i|$ as follows: choose a uniformly random prime q from $[2^{k'}, 2^{k'+1}] \setminus P$ (where P is the set of primes dividing $\pi_0 \pi_1$) and an integer d from the set of integers in the interval $[2^k/q\pi_i, 2^{k+1}/q\pi_i]$ whose prime divisors are all less than q; set $|G_i| = \pi_i q d$. Then, by the Prime Number Theorem, $\alpha(D) \approx 2^{-k'+\log k'} \ln 2$. If we set $M = 2^{k'}$, then $\beta(D, M) = 0$. Once we insert these values into Theorem 3, we find that a generic algorithm for solving the Decision Subgroup Problem for such D_i takes $\Omega(2^{(k'-\log k')/3})$ oracle queries. Thus, when $2^k/\max\{\pi_i\} \geq 2^{k'}$, the Extended Decision Subgroup Assumption is absolutely true in the generic group model.

Now, let's consider how well our scheme could perform, if generic attacks were the only security concern. First, consider the rate of our scheme. If we set $k = \lceil k' + \ell + \log p_t \rceil$, then $2^k/\max\{\pi_i\} \geq 2^{k'}$ as required above, while the rate $f_1(k', \log n)/f_2(k', \log n) = k/\ell$ can be arbitrarily close to 1. Also, since $k' = b \log n$, $\log p_t = \mathcal{O}(\log n)$, and ℓ can be chosen to be $\mathcal{O}(\log n)$, k can also be purely logarithmic in n. Thus, generic attacks do not prevent our scheme from achieving an optimal rate (approaching 1) for blocks, and minimal communication $\mathcal{O}(\log n)$ for private bit retrieval.

7 Instantiating Groups with Hidden Smooth Subgroups

Up to this point, we have discussed our PIR scheme and its performance and security properties in a general way, without discussing in detail how to instantiate the group G securely. One way to instantiate G is using a composite modulus, as in [2]. For example, to construct a modulus m that Φ-hides π, one may choose a random "semi-safe" prime $Q_0 = 2q_0\pi + 1$ for prime q_0 and a random semi-safe prime $Q_1 = 2dq_1 + 1$ for prime q_1 and d chosen uniformly from a large interval, and set $m = Q_0 Q_1$. Then, m should have good uniformity properties, even modulo the primes dividing π.

Cachin, Micali and Stadler [2] note that when a divisor $\pi \geq m^{1/4}$ of $(Q_0 - 1)$ is known, however, it is easy to decide whether π divides $\phi(m)$; in particular, given $m = Q_0 Q_1$ and divisor $\pi \geq m^{1/4}$ of $(Q_0 - 1)$, one can factor m using Coppersmith's method [7], [6] – a lattice-based attack. An abundance of work relating to Coppersmith's method has appeared in the literature, the most recent being May's Eurocrypt 2005 paper [12], which provides a unifying framework for most of the results. His Corollary 14 applies to the Φ-hiding situation; it states:

Corollary 14 (A. May). *Let $f(x) \in \mathbb{Z}[X]$ be a polynomial of degree δ. Let m be a composite number of unknown factorization with divisor $Q_0 \geq m^\beta$. Then, we can find all points $x_0 \in \mathbb{Z}$ satisfying $f(x_0) = Q_0$ in time polynomial in $\log m$ and δ if $|x_0| \leq m^{\beta^2}$.*

Setting $\beta = 1/2$ and $f(x) = \pi x + 1$, the algorithm will give us the divisor $Q_0 = \pi c + 1$ in polynomial time, since $c \approx Q_0/\pi < m^{1/4}$ for $\pi > m^{1/4}$. As May notes, this $|x_0| \leq m^{1/4}$ bound occurs frequently in the literature on Coppersmith's method. Since the algorithm works well (in polynomial time) when $\pi > m^{1/4}$, one might expect that the algorithm's performance declines only gradually – e.g., so that for $\pi > m^{1/5}$, the algorithm (while not polynomial-time) would be only slightly super-polynomial, perhaps because of the inefficiency of lattice reduction. However, this is not true; when $(\log m)/(\log \pi)$ is larger than 4, the target vector (i.e., the one that would help us factor m) is not even the shortest vector in the lattice; thus, even perfect lattice reduction algorithms would not, by themselves, make the attack work.

These considerations give us confidence that, as long as $(\log m)/(\log \pi) > 4$ (perhaps by a "comfortable" margin), then the Φ-hiding assumption, as outlined above, is hard. Thus, it seems plausible that the bit-length of m only needs to be a constant factor greater than $\log \pi$. This allows our PIR scheme to achieve constant rate when instantiated with groups modulo composite numbers. A drawback of using composite moduli is that, as mentioned before, we need $\log m = \Omega(\log^{3-o(1)} n)$, due to the number field sieve [10]. This makes our PIR scheme somewhat communication-inefficient for short block sizes d, even though it is most efficient among the single-database PIR schemes that currently exist.

Of course, it would be preferable to instantiate our scheme using groups for which the number field sieve is inapplicable if this could be done securely. For example, one might try elliptic curve groups. However, algorithms exist to find the orders of elliptic curves over finite fields; when we try using the compositum of finite fields, we seem to be reverting back to a factorization problem. Class groups are another interesting alternative, since currently the best known algorithms for attacking class groups (e.g., determining their order) have quadratic-sieve-type complexity. Unfortunately, in our scheme, the user generating the group must know its order for response retrieval; currently, there are no efficient algorithms that would allow the user to generate a class group with known partially-smooth order, as required by our scheme.

8 Conclusion and Open Problems

We described single-database computational block retrieval schemes based on the decision subgroup problem with communication complexity $\mathcal{O}(k+d)$, where d is the size of the block to be retrieved and k is the security parameter. Asymptotically, this is about as good as one might expect since there is only an additive communication overhead of the security parameter k. Indeed, our scheme has better asymptotic performance compared to previous schemes.

We leave it as an open problem to construct an instantiation of our scheme that achieves rate arbitrarily close to 1, while circumventing Coppersmith's attack. Clearly, based on our analysis of the Decision Subgroup Problem in the generic group model, any attack that prevents the scheme from achieving rate close to 1 must exploit the encoding of the elements.

Acknowledgments. We thank Phil Mackenzie, David Woodruff, Helger Lipmaa, Yuval Ishai, and the anonymous referees for fruitful comments.

References

[1] A. Beimel, Y. Ishai, E. Kushilevitz, and J. F. Raymond. *Breaking the $O(n^{1/(2k-1)})$ Barrier for Information-Theoretic Private Information Retrieval*, FOCS 2002.
[2] C. Cachin, S. Micali, M. Stadler, *Computational Private Information Retrieval with Polylogarithmic Communication*, Eurocrypt 1999.
[3] Y. Chang. *Single-Database Private Information Retreival with Logarithmic Communication*, ACISP 2004.
[4] B. Chor and N. Gilboa, *Comput. Private Information Retrieval*, STOC 1997.
[5] B. Chor, E. Kushilevitz, O. Goldreich, and M. Sudan, *Private Information Retrieval*, Journal of the ACM, 45, 1998. Earlier version in FOCS 95.
[6] D. Coppersmith, *Finding a Small Root of a Bivariate Integer Equation; Factoring with High Bits Known*, Eurocrypt 1996.
[7] D. Coppersmith, *Finding a Small Root of a Univ. Mod. Equation*, Eurocrypt 1996.
[8] I. Damgard and M. Koprowski, *Generic Lower Bounds for Root Extraction and Signature Schems in General Groups*, Eurocrypt 2002.
[9] E. Kushilevits and R. Ostrovsky, *Replication is not needed: single database, computationally private information Retrieval.* FOCS 1997.
[10] A.K. Lenstra and H.W. Lenstra, Jr., (eds.), *The Development of the Number Field Sieve*, Lecture Notes in Mathematics 1554, Springer-Verlag, 1995.

[11] H. Lipmaa, *An Oblivious Transfer Protocol with Log-Squared Communication.* Cryptology ePrint Archive, 2004/063.
[12] A. May, *A Tool Kit for Finding Small Roots of Bivariate Polynomials over the Integers*, Eurocrypt 2005.
[13] M. Naor and B. Pinkas, *Obl. Transfer and Polynomial Evaluation*, STOC 1999.
[14] S.C. Pohlig and M. Hellman. *An Improved Algorithm for Computing Logarithms Over GF(p) and its Crypt. Significance,* IEEE Trans. Inf. Th. IT-24 (1978).
[15] J.B. Rosser and L. Schoenfeld, *Sharper Bounds for Chebyshev Functions $\theta(x)$ and $\psi(x)$*, Math. Comput. 29, 243-269, 1975.
[16] J.P. Stern, *A New and Efficient All or Nothing Disclosure of Secrets Protocol,* Asiacrypt 1998.

Concurrent Zero Knowledge in the Public-Key Model*

Giovanni Di Crescenzo[1] and Ivan Visconti[2]

[1] Telcordia, Piscataway, NJ, USA
giovanni@research.telcordia.com
[2] Dip. di Inf. ed Appl., Univ. di Salerno, Baronissi, Italy
visconti@dia.unisa.it

Abstract. The concurrent setting for Zero-Knowledge protocols is very challenging as it requires protocols to remain secure even when several parties execute the same protocol concurrently. Indeed, it has been proved that achieving concurrent security for (black-box-simulation) zero-knowledge protocols in standard models requires a non-constant number of rounds, thus severely limiting efficiency. As a result, a few models with additional setup or network assumptions have been introduced to present constant-round concurrently-secure zero-knowledge protocols for all languages in $\mathcal{NP}$.

In this paper we consider the bare public-key model, which is known to have very minimal setup assumptions, and we present the first constant round and concurrently secure zero-knowledge argument for any languages in $\mathcal{NP}$, under standard intractability assumptions. In fact, our protocol requires 4 rounds and is therefore round-optimal, is a proof of knowledge, and is time-efficient, in the sense that it is based on a tranformation that does not require any expensive $\mathcal{NP}$ reduction from prover or verifier. One 5-round variant of our protocol can be based on the minimal assumption of the existence of a one-way function.

1 Introduction

The classical notion of a zero knowledge proof (a proof that reveals no additional information other than the theorem being true, even to malicious verifiers) was introduced in [20] and, since its introduction, it has been widely used to prove the security of distributed protocols for several applications. Motivated by the use of such protocols in networks like the Internet, several researchers recently realized the need of extending the security properties of zero-knowledge protocols to multi-party and asynchronous settings. In particular, the notion of concurrent zero-knowledge considers the case where several concurrent executions of the same protocol can take place and a malicious adversary may control the scheduling of the messages and corrupt multiple provers or verifiers in order to vi-

---

* Copyright Telcordia. The second author releases his portion of the copyright to Springer-Verlag. Part of the second author's work done while being a post-doctoral fellow at the Dép. d'Inf. of the Ecole Normale Supérieure in Paris, France; and part supported by NoE ECRYPT under contract IST-2002-507932.

olate the soundness or zero-knowledge properties. This notion is being studied in a (standard) model without additional setup infrastructures or network assumptions, where several protocols have been proposed [26, 21, 25], but super-constant lower bounds on the round complexity have been given (see, e.g. [6]). As these bounds severely limit the applicability of these protocols and this notion, other models are being studied to achieve efficient, and, in particular, constant-round concurrent zero-knowledge protocols. Specifically, a model with synchronous network assumptions is studied in [8]; the timing model [15, 16, 18] makes other assumptions on the network asynchronicity; the preprocessing model [11] requires an interactive preprocessing stage involving all parties; the common random string model [3], and its various extensions, require a trusted party or a trusted public auxiliary string. The model that seems to have the minimal set-up or network assumptions is the bare public-key (BPK) model [5], where verifiers register their public key in a public file during a set-up stage, and there is no interactive preprocessing stage, trusted third party, trusted string, or assumption on the asynchronicity of the network. In this model, both the concurrent soundness and zero-knowledge notions are harder to achieve than their non-concurrent variants, as noted in [22], who discussed four distinct and increasingly stronger soundness notions: one-time, sequential, concurrent and resettable soundness. Indeed, the constant-round concurrent zero-knowledge (in fact, resettable zero-knowledge, a stronger notion from [5]) protocols in the BPK model presented in [5, 22] only enjoy sequential soundness while it is conjectured that they do not satisfy concurrent soundness. Recently, [12] shows constant-round concurrently sound resettable zero-knowledge arguments in the BPK model, under non-standard assumptions on the hardness of computational problems against subexponential adversaries; and [13] shows 3-round resettable zero-knowledge arguments in a public-key model requiring stronger set-up assumptions.

Our results. We obtain the first constant-round argument in the BPK model with concurrent security (both soundness and zero-knowledge) under standard intractability assumptions. Our protocol is based on number-theoretic hardness assumptions, and is time-efficient, in the sense that it is based on a tranformation not requiring inefficient $\mathcal{NP}$ reductions. Moreover, it has desirable round-complexity and security in the BPK model, with respect to the black-box simulation model, as:

1. it is a 4-round protocol and therefore optimal among concurrently sound or concurrent zero-knowledge arguments (see [22, 19]), unless $\mathcal{NP}$ is in $\mathcal{BPP}$;
2. it is a concurrent zero-knowledge argument of knowledge and the stronger notion of resettable zero-knowledge arguments of knowledge only exists for trivial relations (see [5]);
3. it is concurrently sound and the stronger notion of resettably sound zero-knowledge protocols can only be obtained for languages in $\mathcal{BPP}$ (see [22, 1]).

One 5-round variant of our protocol can be based on the minimal assumption of the existence of a one-way function. Both our constructions are arguments of knowledge and crucially use equivocal commitment schemes in the BPK model that withstand certain types of malleability attacks.

2 Definitions

Model description. The BPK model can be seen as a relaxed version of two well-known models in Cryptography: the Public-Key Infrastructure model, (one main difference being in that the BPK model does not deal with certification issues and therefore we do not discuss certification authorities) and the Preprocessing model (one main difference being in that in the BPK model the preprocessing is reduced to users non-interactively posting public keys on a public file). Formally, this model assumes that: 1) there exists a public file F that is a collection of records, each containing a public key; 2) an (honest) prover is an interactive deterministic polynomial-time Turing machine that takes as input a security parameter 1^n, F, an n-bit string x, such that $x \in L$, for some language L, an auxiliary input y, a reference to an entry of F and a random tape; 3) an (honest) verifier V is an interactive deterministic polynomial-time Turing machine that works in the following two stages: (a) on input a security parameter 1^n and a random tape, V generates a key pair (pk, sk) and stores the public key pk in one entry of the file F; (b) later, V takes as input the secret key sk, a statement $x \in L$ and a random string, and outputs "accept" or "reject" after performing an interactive protocol with a prover; 4) the first interaction between a prover and a verifier starts after all verifiers have completed their first stage.

Malicious provers in the BPK model. Let s be a positive polynomial. We say that $P^\star$ is an *s-concurrent malicious* prover if it is a probabilistic polynomial-time Turing Machine that, on input 1^n and PK, can perform the $s(n)$ interactive protocols with V as follows: 1) if $P^\star$ is already running i protocols $0 \leq i < s(n)$ he can choose a new statement x_i to be proved and start a new protocol with V with $x_i \in L$ as statement; 2) he can output a message for any running protocol, receive immediately the response from V and continue. (We assume that each message is unambiguously associated with only one of the protocols.) Given an s-concurrent malicious prover $P^\star$ and an honest verifier V, an *s-concurrent attack* is performed as follows: 1) the first stage of V is run on input 1^n and a random string to obtain pair (pk, sk); 2) $P^\star$ is run on input 1^n and pk so to obtain an n-bit string x_1; 3) whenever $P^\star$ starts a new protocol choosing an n-bit string x_i, V uses inputs x_i, a new random string r_i and sk, and interacts with $P^\star$.

Malicious verifiers in the BPK model. We say that $V^\star$ is an *s-concurrent malicious* verifier if it is probabilistic polynomial-time Turing Machine that, on input 1^n and PK, can perform the following $s(n)$ interactive protocols with P: 1) if $V^\star$ is already running i protocols $0 \leq i < s(n)$ he can decide the i-th protocol to be started with P; 2) he can output a message for any running protocol, receive immediately the next message from P and continue. Given an s-concurrent malicious verifier $V^\star$ and an honest prover P, an *s-concurrent attack* is performed as follows: 1) in its first stage, $V^\star$, on input 1^n and a random string, generates a public file F; 2) $V^\star$ is run on input 1^n and F so to start the first protocol with P; 3) whenever $V^\star$ starts a new protocol, P uses a new statement, a new random string, and interacts with $V^\star$.

We now define concurrently sound and zero-knowledge protocols in the BPK model.

Definition 1. *Given a language $L \in \mathcal{NP}$, and its corresponding relation R_L, we say that a pair $\langle P, V \rangle$ is* complete *for L, if for all n-bit strings $x \in L$ and any witness y such that $(x, y) \in R_L$, the probability that V, at the end of the interaction with P on input y, outputs "reject", is negligible in n. We say that $\langle P, V \rangle$ is* concurrently-sound *over L if, for any false statement "$x \in L$", for all positive polynomials s, for all s-concurrent malicious provers $P^\star$, the probability that in an execution of an s-concurrent attack V outputs "accept" for such a statement is negligible in n. Finally, we say that $\langle P, V \rangle$ is* concurrently-zero-knowledge *over L if for all positive polynomials s, for any s-concurrent malicious verifiers $V^\star$, there exists a probabilistic polynomial-time algorithm $S_{V^\star}$, called the* simulator, *such that for all $x_1, \ldots, x_{s(n)} \in L$, the probability distributions $\{\mathtt{view}_{V^\star}^P(\bar{x})\}$ and $\{S_{V^\star}(\bar{x})\}$ are computationally indistinguishable, where $\{\mathtt{view}_{V^\star}^P(\bar{x})\}$ is the distribution of the transcript seen by $V^\star$ on its input tape (i.e., $\bar{x} = x_1, \ldots, x_{s(n)}$), random tape and communication tape during its interaction with P.*

3 Equivocal Commitments in the BPK Model

Crucial tools towards our main results are equivocal commitment schemes in the BPK model that maintain security properties against certain concurrent adversaries. We now describe these schemes after reviewing the notions of (conventional) commitment schemes and equivocal commitment schemes.

Commitment schemes. Informally speaking, a *commitment scheme* (sen, rec) is a two-phase interactive protocol between two probabilistic polynomial time parties sen and rec, called the sender and the receiver, respectively, such that the following is true. sen commits to a bit b in the first phase (called the commitment phase); in the second phase (called the decommitment phase) sen convinces rec of the value of the bit b committed in the first phase. A commitment scheme has three requirements. First, if sen and rec behave honestly, then at the end of the decommitment phase rec is convinced that sen committed to bit b with high probability (this is the *correctness* requirement). Then, no matter which polynomial-time computable strategy rec* uses in the commitment phase, rec* is not able to guess such a bit with probability significantly better than its a priori probability before such phase (this is the *hiding* property). Finally, for any strategy played by a polynomial-time sen*, the probability that sen* can later decommit both as 0 and as 1 is negligible (this is the *binding* property).

Equivocal commitment schemes. Informally speaking, a commitment scheme (sen,rec) is *equivocal* if there exists an efficient equivocator algorithm returning a transcript leading to a faked commitment key such that: (a) this commitment can be decommitted both as 0 and as 1, (the *key equivocation* property) and (b) the transcript is indistinguishable from a real execution (the *transcript indistinguishability* property). The advantage of the equivocator algorithm over a real committer is usually realized by rewinding of the receiver, or auxiliary information about a common reference string. Equivocal commitment schemes are a strict generalization of trapdoor commitment schemes [4] and and have been already applied in several works towards the construction of various types of commitments and zero-knowledge proofs. In particular, [11] uses equivocal commitment schemes to construct concurrent zero-knowledge protocols in the preprocessing model.

Constructing equivocal commitments in the BPK model. Following [11], a first natural step towards our goal of constructing concurrently zero-knowledge and concurrently sound protocols in the BPK model could be that of constructing an equivocal commitment scheme in this model. As we will see, this both requires some work and turns out not be enough due to a subtle malleability attack by a concurrent adversary. A first candidate for such a construction could be the perfectly-hiding commitment scheme in [24], which we now describe.

Let p, q be primes such that $p = 2q+1$ and let G_q denote the only subgroup of Z_p^* of order q. We note that it can be efficiently decided whether an integer a is in G_q, by checking that $a^q \equiv 1 \bmod p$. Moreover, any element of G_q different from 1 generates such a subgroup. For any $a, b \in G_q$, if $b \neq 1$ the *discrete logarithm of a in base b* is the integer x such that $b^x = a \bmod p$. In order to commit to an integer $z \in Z_q$, the committer uniformly chooses $r \in Z_q$, computes com $= g^z h^r \bmod p$, and sends com to the receiver. In order to decommit com as z, the committer sends the pair (z, r) to the receiver, who checks that com $= g^z h^r \bmod p$. The hiding property of this scheme follows from the fact that com is uniformly distributed in G_q for any z; the binding property follows from the fact that if a committer is able to successfully decommit a string com both as z and z' with $z' \neq z$, then he can be used to compute the discrete logarithm of h in base g. A simulator recovering, either through rewinding or as an auxiliary input, the value α such that $g^\alpha = h$, can open a commitment com $= g^z h^r \bmod p$ of z as z' by sending pair $(z', r' = r + (z - z')\alpha^{-1} \bmod q)$ as a decommitment.

One way to use this scheme in the BPK model goes as follows. The receiver sets her public key equal to values (p, q, g, h) chosen as above, and proves, using well-known 3-round witness-indistinguishable proofs of knowledge (e.g., [27]), that he knows the discrete logarithm of h modulo p. The committer computes the commitment key com and the decommitment key (z, r) as above. While the commitment is equivocal as an equivocator can extract α from the proof of knowledge, it is unclear how to prove the binding property because of the conflicting requirements of computing the discrete logarithm of h from a dishonest committer, and simulating the proof of knowledge to the dishonest committer.

Achieving the binding property. A first idea to fix the above problem, based on an approach of [23], is to extend the previously discussed commitment scheme so that the receiver's public key has two integers h_0, h_1 instead of one. The receiver's public key is then a tuple (p, q, g, h_0, h_1) where p, q, g are as before, $h_0, h_1 \in G_q$, and $h_i = g^{\alpha_i} \bmod p$, for $i = 0, 1$ and some α_0, α_1. In this case, the receiver proves, using well-known 3-round witness-indistinguishable proofs of knowledge (e.g., [7]), that he knows the discrete logarithm modulo p of one among h_0, h_1. A commitment to a string z can be obtained by randomly choosing $u, v \in Z_q$, and computing com $= g^u h_0^v h_1^{z-v} \bmod p$ and the decommitment is implemented by sending (u, v, z) and verifying that com $= g^u h_0^v h_1^{z-v} \bmod p$. We will refer to this scheme as EQ_{dlog}.

The correctness and hiding properties of EQ_{dlog} directly follow from the analogue properties of the scheme in [24]. To see that the scheme enjoys the equivocation property, consider an equivocator that extracts one of the two values α_0, α_1 from the receiver's proof of knowledge, and then computes a commitment com as com $= g^u h_0^v h_1^{z-v} \bmod p$. Later, on input one of the two trapdoors α_0, α_1, it can compute a decommitment (u', v', z') as $z' \in Z_q$ as follows:

1. on input α_0 such that $g^{\alpha_0} = h_0$, set $v' = z' + v - z \bmod q$ and $u' = u + \alpha_0(z - z') \bmod q$;
2. on input α_1 such that $g^{\alpha_1} = h_1$, set $v' = v$ and $u' = u + \alpha_1(z - z') \bmod q$.

It can be seen that in both cases (u', v', z') is a valid decommitment of com as z' and that the distribution of the tuples (com, u', v', z') is perfectly indistinguishable from that produced by a honest committer on input z'.

To see that EQ_{dlog} is computationally binding, we show an efficient algorithm DLbreak that can randomly choose $b \in \{0, 1\}$ and use α_b to compute the prover's messages in the witness-indistinguishable proofs of knowledge. Later, for any two different decommitments $(u, v, z), (u', v', z')$ of a commitment com (i.e., $z \neq z'$), it can succeed in computing α_{1-b} on input α_b, with probability at least $1/2$. Specifically, we consider the following two cases:

1. $v = v' \bmod q$: notice that $z \neq z' \bmod q$ and thus $z' - v' \neq z - v \bmod q$; since $g^{u+\alpha_0 v+\alpha_1(z-v)} = g^{u'+\alpha_0 v'+\alpha_1(z'-v')} \bmod p$, if $b = 0$ then DLbreak computes $\alpha_1 = (u - u')(z' - v' - z + v)^{-1} \bmod q$;
2. $v \neq v' \bmod q$: since $g^{u+\alpha_0 v+\alpha_1(z-v)} = g^{u'+\alpha_0 v'+\alpha_1(z'-v')} \bmod p$, if $b = 1$ then DLbreak computes $\alpha_0 = (u' - u + \alpha_1(z' - v' - z + v))(v - v')^{-1} \bmod q$.

A variant based on any one-way function. A variant of EQ_{dlog}, denoted as EQ_{oway}, can be based on the (minimal) assumption of the existence of a one-way function f. We now only sketch the main ideas behind scheme EQ_{oway}. The receiver computes a public key $pk = (y_0, y_1)$, where $y_j = f(x_j)$, for x_j randomly chosen in $\{0, 1\}^n$ and $j = 0, 1$. The receiver proves, using a 3-round witness-indistinguishable system (such as a parallel repetition of the protocol with soundness error $1/2$ from [2]), that he knows at least one of (x_0, x_1). The committer uses the same language and the same atomic proof system with soundness error $1/2$, as follows. For this protocol, there exists a simulator that, on input $b \in \{0, 1\}$, can compute a first message from the prover and an answer to the verifier's challenge, if equal to b. We use the first message returned by the simulator as a commitment key and the answer as a decommitment key. We can prove the following

Theorem 1. EQ_{dlog} *(resp. EQ_{oway}) is an equivocal commitment scheme in the BPK model for which the correctness, hiding, and equivocation property hold unconditionally, and the binding property holds assuming the hardness of computing discrete logarithms modulo primes p of the form $p = 2q + 1$, for q prime (resp., assuming the existence of a one-way function).*

Malleability attacks during concurrent executions. Both schemes from Theorem 1 are not enough to construct concurrently sound and zero-knowledge arguments in the BPK model, as their properties may not hold under concurrent attacks when the scheme is used as a subprotocol in a larger protocol. Assume, for instance, that we use a constant-round honest-verifier zero-knowledge proof system for an $\mathcal{NP}$ language L, denoted as BL, such as the one in [2], and implement the commitment scheme necessary in this protocol by using, say, scheme EQ_{dlog}. The resulting protocol has 4 messages: $m1$ (containing the first message from the proof of knowledge in EQ_{dlog}), $m2$ (containing the second message from the proof of knowledge and the commitment keys for BL computed as com in EQ_{dlog}), $m3$ (containing the third message from the proof of knowledge in EQ_{dlog} and the challenge message from BL), and $m4$ (containing the decommitments as in EQ_{dlog} and the answer message from BL). While one would expect the hiding and equivocation property of the commitment to still hold, this may not be true for the binding property. In fact, we show an example of two concurrent executions between a malicious prover $P^\star$ and two incarnations of the same verifier V, V', where it is unclear how to prevent a malleability attack from $P^\star$ in the execution with V using the public key and the proof of knowledge from the execution with V'. Specifically, the schedule decided by a malicious prover $P^\star$ is $(m1, m1', m2, m3, m2', m3', m4, m4')$, and $P^\star$ might decide to compute the commitments in $m2$ with some malleability attack over $m1'$ and the public key of V'. Now, assume $P^\star$ tries to prove a false statement to V; we then see that the known approach in proving that $P^\star$ only succeeds with negligible probability, fails. Indeed, in trying to extract two different decommitments from the proof given by $P^\star$ on a false statement to V, the rewind performed on $P^\star$ while interacting with V and possibly changing message $m3$, also rewinds the proof given by V' to $P^\star$ and may change message $m2'$ (as $m2'$ is carefully scheduled by $P^\star$ to happen after $m3$). As a consequence, the secret key α_b extracted from $P^\star$ could always be the same used by V' in its witness indistinguishable proof of knowledge, just due to a successful malleability attack, and thus not violating the witness indistinguishability of the proof of knowledge used.[1]

Our atomic commitment scheme. Even if we will not use EQ_{dlog} or EQ_{oway} as a black box in our main construction, we will crucially use their components and properties. To avoid the above malleability attack, we cannot use known round-efficient non-malleable commitment schemes, as some of these solutions do not preserve their security when executed in a concurrent setting (e.g., [17]), or are designed in the stronger public random string model (e.g., [10]). Instead,

---

[1] We note that a similar problem applies to (unrefereed) protocols in [28], that consequently do not seem to satisfy concurrent soundness.

we observe that we do not need the full power of non-malleability, and avoid the above attacks by requiring that the committer, after computing an equivocal commitment key eqcom and its decommitment key eqdec as in scheme EQ_{dlog} (resp., EQ_{oway}), additionally commits to eqdec at commitment stage and using a statistically-binding commitment scheme, that can be implemented as a simple application of the El-Gamal encryption scheme (resp., starting from any one-way function). To prevent the loss of the equivocation property, we extend an idea of [9] so that it can be applied to the BPK model, again using statistically-binding commitments. Formally, we define an atomic commitment scheme (sen,rec) using discrete logarithms (the analogue description using one-way functions is similarly obtained), as follows. The common input to sen and rec is public key (p, q, g, h_0, h_1), the private input of sen is the bit or string z to be committed.

Commitment Phase:
1. sen checks that $p = 2q + 1$, that p, q are primes, that g is a generator of G_q and that h_0, h_1 are in G_q.
2. sen commits to z by randomly choosing $u, v \in Z_q$ and by computing the commitment com = (eqcom, sbcom_0, sbcom_1), where eqcom = $g^u h_0^v h_1^{z-v} \bmod p$, sbcom_{1-b} is a statistically-binding commitment of a random chosen string while sbcom_b is a statistically-binding commitment of (u, v) for $b \in \{0, 1\}$.

Decommitment Phase:
1. sen decommits z by sending (u, v, z) and the decommitment that corresponds to the statistically-binding commitment sbcom_b (instead sbcom_{1-b} is not opened).
2. rec returns z if eqcom = $g^u h_0^v h_1^{z-v} \bmod p$ and one between sbcom_0 and sbcom_1 has been correctly opened and corresponds to a commitment of (u, v).

We note that a polynomial-time malicious committer cannot change the committed bit unless he both knows two valid decommitments for opening eqcom and commits to them in sbcom_b and sbcom_{1-b} at *commitment phase*. In particular, the previously described concurrent scheduling does not help a malicious committer in computing the decommitted value at decommitment phase.

4 Concurrent Zero Knowledge and Soundness

We now present our concurrently sound and concurrently zero-knowledge argument system for any $\mathcal{NP}$-language in the BPK model. Most importantly, it only uses hardness assumptions with respect to polynomial-time adversaries, while [12] obtains the same result (in fact, a stronger one) with hardness assumptions against subexponential-time adversaries. We obtain the following

Theorem 2. *Let L be a language in $\mathcal{NP}$. Assuming the intractability of the DDH problem for integers of the form $p = 2q + 1$, for p, q primes, in the BPK model, there exists (constructively) a 4-round concurrently sound and concurrently zero-knowledge argument system for L. Moreover, this protocol is also an argument of knowledge for the relation R_L associated with L.*

The above result is obtained using tools from commitment scheme EQ_{dlog}. If, instead, we start from EQ_{oway}, we can reduce the assumption to the existence of a one-way function, but we obtain a 5-round and less efficient protocol.

Construction of our argument system. Our protocol (P, V), depicted in Figure 1, uses the following tools:

1. The atomic commitment scheme (sen,rec) constructed in Section 3 to compute commitments $\text{atcom} = (\text{eqcom}, \text{sbcom}_0, \text{sbcom}_1)$.
2. A 3-round witness indistinguishable proof of knowledge π_v given by the verifier in order to prove knowledge of one of the two secret keys corresponding to his public key. We further need a *partial witness-independence* property from π_v: the message sent at its first round should have distribution independent from the value of any witness for the theorem proved. We can obtain such a protocol using [27, 7]. (For the variant based on one-way functions we can use [2], requiring 4 rounds.)
3. A 3-round public-coin honest-verifier zero-knowledge argument of knowledge π_p for any $\mathcal{NP}$-complete language L with the following special structure and property. We denote by (a, c, z) the three messages exchanged by prover P and verifier V and assume that (a, c, z) is an l-times parallel iteration of a basic protocol with soundness $1/2$, and a single-bit challenge. Then we assume that π_p satisfies a *special witness-extraction* property (specifically, an extractor, given two accepting transcripts (a, c, z) and (a, c', z') with $c' \neq c$, can compute a witness y for x). We can obtain such a protocol from [2].

Properties of our argument system. The completeness follows by the completeness of protocols π_p, π_v and the correctness of scheme (sen,rec). Witness extraction also follows from the witness extraction of protocol π_p, the completeness of π_v and the binding property of (sen,rec). For the concurrent zero-knowledge property (when a concurrent malicious verifier is considered), we show a simulator S that extracts the secret keys from the arguments of knowledge π_v given by the adversarial verifier V^* and later uses them in order to compute equivocal commitments. Note that since S performs a rewind, he could run in exponential-time when concurrent executions are performed. We show that after an expected polynomial number of rewinds S can complete the simulation without any further rewinds and thus run in expected polynomial-time.

The most interesting part of our proof is verifying the concurrent soundness property. Assume by contradiction that the protocol is not concurrently sound, thus there exists a malicious s-concurrent prover P^* that in a concurrent attack succeeds with non-negligible probability in completing a proof for a false statement "$x \in L$". We show a probabilistic polynomial-time algorithm $\mathcal{A}$ that, using black-box access to P^*, breaks the discrete logarithm assumption.

We now describe algorithm $\mathcal{A}$. On input as challenge a discrete logarithm instance (p, q, g, h) $\mathcal{A}$ randomly chooses $\alpha \in Z_q$, $b \in \{0, 1\}$, and the session number $s_j \in \{1, \ldots, s\}$ (trying to guess the index of the proof in which P^* will succeed in cheating), and stores public key $\text{pk}_i = (p, q, g, h_0, h_1)$, where $h_b = g^\alpha$, $h_{1-b} = h$. Note that $\mathcal{A}$ knows the secret key corresponding to pk_i.

Common input: the public file F, n-bit string $x \in L$, index i specifying the entry $\text{pk}_i = (p, q, g, h_0, h_1)$ of F.
P's private input: a witness y for $x \in L$.
V's private input: a secret key α.
V-P-PoK (rounds 1,2,3): V and P engage in the 3-round witness-indistinguishable protocol π_v in which V proves knowledge of α such that $g^\alpha = h_0 \bmod p$ or $g^\alpha = h_1 \bmod p$.
P-round-2:
  1. compute l atomic commitments $\text{atcom}_c = (\text{atcom}^1, \ldots, \text{atcom}^l)$ along with the decommitment $\text{atdec}_c = (\text{atdec}^1, \ldots, \text{atdec}^l)$ of l random bits $c_1, \ldots, c_l$ by using public key pk_i;
  2. compute the first round message a of protocol π_p for proving "$x \in L$";
  3. send atcom_c, a to V.
V-round-3:
  1. compute the second l-bit message c of protocol π_p;
  2. send c to P.
P-round-4:
  1. compute the third message z of protocol π_p for "$x \in L$" by using as challenge $\hat{c} = c \oplus (c_1 | \cdots | c_l)$;
  2. send $z, \text{atdec}_c, \hat{c}$ to V.
V-decision: accept if atdec_c is a valid decommitment of atcom_c as $\hat{c} \oplus c$ on public key pk_i, and $(x, a, \hat{c}, z)$ is an accepting transcript for π_p.

Fig. 1. The four-round concurrently sound and concurrently zero-knowledge argument of knowledge for any $\mathcal{NP}$ language in the BPK model

For all the proof sessions, $\mathcal{A}$ interacts with an s-concurrent malicious prover P^* during a concurrent attack. In particular in all but the s_j-th session, $\mathcal{A}$ works as an honest verifier, by using α to run the prover's algorithm of π_v. In correspondence with the s_j-th session, $\mathcal{A}$ uses the following rewind strategy.

1. The first time that $\mathcal{A}$ runs the session, he works as an honest verifier. After the 4-th round $\mathcal{A}$ logs the accepting transcript $(a, \hat{c}, z)$ of the honest-verifier zero-knowledge proof π_p given by the prover. Then $\mathcal{A}$ rewinds the prover so that round V-round-3 can be played again.
2. $\mathcal{A}$ plays round V-round-3, by choosing random challenges $c' \neq c$ until he obtains an accepting transcript $(a, \hat{c}', z')$ for protocol π_p by the prover.

If $\hat{c} \neq \hat{c}'$, by the special witness extraction property of the honest-verifier zero-knowledge proof of knowledge π_p, $\mathcal{A}$ obtains a witness for $x \in L$ and this contradicts the assumption that "$x \notin L$". If instead $\hat{c} = \hat{c}'$ then for each $t \in \{1, \ldots, l\}$ such that the t-th bit of c' is different from the t-th bit of c, the corresponding t-th commitment atcom^t in atcom_c has been opened as both 0 and 1. We note that the probability that this happens because P^* reveals two decommitments of any among commitments $\text{sbcom}_0, \text{sbcom}_1$ contained in atcom_c is exponentially small because the commitment scheme used is statistically binding. Therefore $P^\star$ opens in two different ways commitment eqcom. Then $\mathcal{A}$ selects the pairs that correspond to different openings and for each of them, say the t-th, $\mathcal{A}$ runs DLbreak on input α, eqcom^t and the t-th pair of decommitments. For $j = 0, 1$, let $sk_{i,j}$ be such that $g^{sk_{i,j}} = h_j \bmod p$ and $sk_{i,b} = \alpha$ in pk$_i$. In case the output of DLbreak is at least once secret key $sk_{i,1-b}$, then $\mathcal{A}$ obtains the discrete logarithm of h in base g and thus we reach a contradiction.

Therefore we have only to deal with the case that DLbreak always outputs $sk_{i,b}$, that is the same secret key that $\mathcal{A}$ knows. Notice that $\mathcal{A}$ uses the secret key as witness to compute the third round of the witness indistinguishable proofs of knowledge. Let m be the number of sessions played until the end of session s_j. We now use hybrid arguments to contradict the assumed witness indistinguishability of the proof of knowledge. We know that using m times $sk_{i,0}$, DLbreak outputs $sk_{i,0}$ and using m times $sk_{i,1}$, DLbreak outputs $sk_{i,1}$. Then there must be $\gamma \in \{1, \ldots, m\}$ such that by using $sk_{i,0}$ for the first $\gamma - 1$ times and $sk_{i,1}$ for the last $m - \gamma$ times, the witness used in the γ-th proof is the same that will be extracted by DLbreak. Now we distinguish two different cases, according to whether the γ-th proof has been completed before or after P-round-2 of session s_j:

1. "before": here the γ-th proof is not affected by rewinds performed so far, and we can use this session to break the witness indistinguishability of π_v;
2. "after": here, rewinds to $P^\star$ may rewind the γ-th proof. Then we can flip a bit j to use one of the two $sk_{i,j}$ as witness in the γ-th proof and the same discrete logarithm is also given as input to DLbreak; as assumed so far, the same discrete logarithm will be also given as output by DLbreak. However, because of the statistical binding property of the commitment scheme sbcom, except with exponentially small probability, all values decommitted by $P^\star$ at step P-round-4 of the γ-th proof are already fixed at step P-round-2 of the same proof. Furthermore, at this step $P^\star$ has seen at most step V-round-1 of the γ-th proof, and, by the partial witness-independence property of π_v, the message sent by $\mathcal{A}$ at this step has distribution independent from bit j. Therefore, except with exponentially small probability, $P^\star$ can guess j with probability at most $1/2$, which contradicts our assumptions so far.

Acknowledgments. We thank Pino Persiano and Yunlei Zhao for many interesting discussions on zero knowledge protocols in various models.

References

1. B. Barak, O. Goldreich, S. Goldwasser, and Y. Lindell. Resettably-Sound Zero-Znowledge and its Applications. In Proc. of IEEE FOCS 01.
2. M. Blum. How to Prove a Theorem So No One Else Can Claim It. In *Proceedings of the International Congress of Mathematicians*, 1986.
3. M. Blum, A. De Santis, S. Micali, and G. Persiano. Non-Interactive Zero-Knowledge. *SIAM J. on Computing*, 20(6):1084–1118, 1991.
4. J. Brassard, D. Chaum, and C. Crepéau. Minimum Disclosure Proofs of Knowledge. *Journal of Computer and System Science*, 37(2):156–189, 1988.

5. R. Canetti, O. Goldreich, S. Goldwasser, and S. Micali. Resettable Zero-Knowledge. In Proc. of ACM STOC 2000.
6. R. Canetti, J. Kilian, E. Petrank, and A. Rosen. Black-Box Concurrent Zero-Knowledge Requires $\omega(\log n)$ Rounds. In Proc. of ACM STOC 2001.
7. A. De Santis, G. Di Crescenzo, G. Persiano, and M. Yung. On Monotone Formula Closure of SZK. In Proc. of IEEE FOCS 1994.
8. G. Di Crescenzo. Concurrent Zero-Knowledge without Complexity Assumptions. In Proc. of Cocoon 2000, LNCS, Springer-Verlag.
9. G. Di Crescenzo. Equivocable and Extractable Commitment Schemes. In Proc. of Security in Communication Networks 2002, LNCS, Springer-Verlag.
10. G. Di Crescenzo, J. Katz, R. Ostrovsky and A. Smith. Efficient and Non-Interactive Non-Malleable Commitment. In Proc. of Eurocrypt 2001, LNCS, Springer-Verlag.
11. G. Di Crescenzo and R. Ostrovsky. On Concurrent Zero-Knowledge with Preprocessing. In Proc. of CRYPTO 99, LNCS, Springer-Verlag.
12. G. Di Crescenzo, G. Persiano, and I. Visconti. Constant-Round Resettable Zero Knowledge with Concurrent Soundness in the Bare Public-Key Model. In Proc. of CRYPTO 04, LNCS, Springer-Verlag.
13. G. Di Crescenzo, G. Persiano, and I. Visconti. Improved Setup Assumptions for 3-Round Resettable Zero Knowledge. In Proc. of Asiacrypt '04, LNCS, Springer-Verlag.
14. D. Dolev, C. Dwork, and M. Naor. Non-Malleable Cryptography. *SIAM J. on Computing*, 30(2):391–437, 2000.
15. C. Dwork, M. Naor, and A. Sahai. Concurrent Zero-Knowledge. In Proc. of ACM STOC 2001.
16. C. Dwork and A. Sahai. Concurrent Zero-Knowledge: Reducing the Need for Timing Constraints. In Proc. of CRYPTO 98, LNCS, Springer-Verlag.
17. M. Fischlin and R. Fischlin, Efficient Non-Malleable Commitment Schemes. In Proc. of CRYPTO 2000, LNCS, Springer-Verlag.
18. O. Goldreich. Concurrent Zero-Knowledge with Timing, Revisited. In Proc. of ACM STOC 2002.
19. O. Goldreich and H. Krawczyk. On the Composition of Zero-Knowledge Proof Systems. *SIAM J. on Computing*, 25(1): 169–192 (1996).
20. S. Goldwasser, S. Micali, and C. Rackoff. The Knowledge Complexity of Interactive Proof-Systems. *SIAM J. on Computing*, 18(6):186–208, 1989.
21. J. Kilian and E. Petrank. Concurrent and Resettable Zero-Knowledge in Poly-Logarithmic Rounds. In Proc. of ACM STOC 2001.
22. S. Micali and L. Reyzin. Soundness in the Public-Key Model. In Proc. of CRYPTO 01, LNCS, Springer-Verlag.
23. M. Naor and M. Yung. Public-key Cryptosystems Provably Secure against Chosen Ciphertext Attacks. In Proc. of STOC 1990, 427–437.
24. T. Pedersen. Non-interactive and information-theoretic secure verifiable secret sharing. In Proc. of CRYPTO 91, LNCS, Springer-Verlag.
25. M. Prabhakaran, A. Rosen, and A. Sahai. Concurrent Zero-Knowledge with Logarithmic Round Complexity. In Proc. of IEEE FOCS 2002.
26. R. Richardson and J. Kilian. On the Concurrent Composition of Zero-Knowledge Proofs. In Proc. of EUROCRYPT 99, LNCS, Springer-Verlag.
27. C. P. Schnorr. Efficient Signature Generation for Smart Cards. *Journal of Cryptology*, 4(3):239–252, 1991.
28. Y. Zhao. Concurrent/Resettable Zero-Knowledge With Concurrent Soundness in the Bare Public-Key Model and Its Applications. Cryptology ePrint Archive, Report 2003/265, last update June 5th, 2004.

A Faster Combinatorial Approximation Algorithm for Scheduling Unrelated Parallel Machines*

Martin Gairing, Burkhard Monien**, and Andreas Woclaw

Faculty of Computer Science,
Electrical Engineering and Mathematics,
University of Paderborn, Fürstenallee 11,
33102 Paderborn, Germany
{gairing, bm, wocland}@uni-paderborn.de

Abstract. We consider the problem of scheduling n independent jobs on m unrelated parallel machines without preemption. Job i takes processing time p_{ij} on machine j, and the total time used by a machine is the sum of the processing times for the jobs assigned to it. The objective is to minimize makespan. The best known approximation algorithms for this problem compute an optimum fractional solution and then use rounding techniques to get an integral 2-approximation.

In this paper we present a combinatorial approximation algorithm that matches this approximation quality. It is much simpler than the previously known algorithms and its running time is better. This is the first time that a combinatorial algorithm always beats the interior point approach for this problem. Our algorithm is a generic minimum cost flow algorithm, without any complex enhancements, tailored to handle unsplittable flow. It pushes unsplittable jobs through a two-layered bipartite generalized network defined by the scheduling problem. In our analysis, we take advantage from addressing the approximation problem directly. In particular, we replace the classical technique of solving the LP-relaxation and rounding afterwards by a completely integral approach. We feel that this approach will be helpful also for other applications.

1 Introduction

We consider the scheduling problem where n independent jobs have to be assigned to a set of m unrelated parallel machines without preemption. Processing

---

* This work has been partially supported by the DFG-Sonderforschungsbereich 376 Massive Parallelität: Algorithmen, Entwurfsmethoden, Anwendungen, by the European Union within the 6th Framework Programme under contract 001907 (DELIS) and by the DFG Research Training Group GK-693 of the Paderborn Institute for Scientific Computation (PaSCo).
** Parts of this work were done while the author was visiting Università di Roma La Sapienza at Rome and the University of Texas at Dallas.

job i on machine j takes time p_{ij}. For each machine j, the total time used by machine j is the sum of processing times p_{ij} for the jobs that are assigned to machine j. The makespan of a schedule is the maximum total time used by any machine. The objective is to find a schedule (assignment) that minimizes makespan. This problem has many applications. Typically, they arise in the area of scheduling multiprocessor computers and industrial manufacturing systems (see [18, 28]).

Related Work. There is a large amount of literature on scheduling independent jobs on parallel machines (a collection of several approximation algorithms can be found in [10]). A good deal of these publications concentrate on scheduling jobs on unrelated machines. Horowitz and Sahni [11] presented a (non-polynomial) dynamic programming algorithm to compute a schedule with minimum makespan. Lenstra et al. [17] proved that unless $\mathcal{P} = \mathcal{NP}$, there is no polynomial-time approximation algorithm for the optimum schedule with approximation factor less than $\frac{3}{2}$. They also presented a polynomial-time 2-approximation algorithm. This algorithm computes an optimal fractional solution and then uses rounding to obtain a schedule for the discrete problem with approximation factor 2. Shmoys and Tardos [23] generalized this technique to obtain the same approximation factor for the generalized assignment problem. They also generalized the rounding technique to hold for any fractional solution.

The fractional unrelated scheduling problem can also be formulated as a *generalized maximum flow* problem, where the network is defined by the scheduling problem and the capacity of some edges, that corresponds to the makespan, is minimized. This generalized maximum flow problem is a special case of *linear programming*. Using techniques of Kapoor and Vaidya [14] and by exploiting the special structure of the problem, an optimum fractional solution can be found with the interior point algorithm of Vaidya [27] in time $O(|E|^{1.5}|V|^2 \log(U))$, where U denotes the maximal p_{ij}.

In contrast to the linear programming methods, the aforementioned generalized maximum flow problem can also be solved with a purely *combinatorial* approach. Here, the makespan minimization is done by binary search. Computing generalized flows has a rich history, going back to Dantzig [2]. The first combinatorial algorithms for the generalized maximum flow problem were exponential time augmenting path algorithms by Jewell [13] and Onaga [19]. Truemper [26] showed that the generalized maximum flow problem and the *minimum cost flow* problem are closely related. More specifically, he transformed a generalized maximum flow problem into some minimum cost flow problem by setting the cost of an edge to be the logarithm of the gain from the generalized maximum flow problem. Goldberg et al. [6] designed the first polynomial-time combinatorial algorithms for the generalized maximum flow problem. Their algorithms were further refined and improved by Goldfarb, Jin and Orlin [7] and later by Radzik [22]. Radzik's algorithm is so far the fastest combinatorial algorithm with a running time of $O(|E||V|(|E| + |V|\log|V|)\log U)$. In order to minimize makespan, this algorithm has to be called at most $O(\log(nU))$ times.

There exist fast fully polynomial-time approximation schemes for computing a fractional solution [4, 12, 20, 21, 25]. Using the rounding technique from [23], this leads to a $(2+\varepsilon)$-approximation for the discrete problem. The approximation schemes can be divided into those that approximate generalized maximum flows [4, 21, 25] and those that directly address the scheduling problem [12, 20].

Unrelated machine scheduling is a very important problem and many heuristics and exact methods have been proposed. Techniques used here range from combinatorial approaches with partial enumeration to integer programming with branch-and-bound and cutting planes. For a selection we refer to [18, 24, 28] and references therein.

Finding a discrete solution for the unrelated scheduling problem can be formulated as an *unsplittable* generalized maximum flow problem. Several authors [3, 15, 16] have studied the unsplittable flow problem for usual flow networks. Kleinberg [15] formulated the problem of finding a solution with minimum makespan for the restricted scheduling problem as an unsplittable flow problem. Here the restricted scheduling problem is a special case of our problem, in which each job i has some weight w_i, each machine j has some speed s_j and $p_{ij} = \frac{w_i}{s_j}$ or $p_{ij} = \infty$ holds for all i, j. Gairing et al. [5] exploited the special structure of the network, gave a 2-approximation algorithm for the restricted scheduling problem based on preflow-push techniques and also an algorithm for computing a Nash equilibrium for the restricted scheduling problem on identical machines.

Contribution. The algorithm presented in this paper computes an assignment for the unrelated scheduling problem with makespan at most twice the optimum. We prove that a 2-approximative schedule can be computed in $O(m^2 A \log(m) \log(nU))$ time, where A is the number of pairs (i,j) with $p_{ij} \neq \infty$. This is better than the previously known best time bounds of Vaidya's [27] and Radzik's [22] algorithms. In particular, this is the first time that a combinatorial algorithm always beats the interior point approach for this problem.

An essential element of our approximation algorithm is the procedure Unsplittable-Blocking-Flow from [5]. This procedure was designed to solve the unsplittable maximum flow problem in a bipartite network, which is defined by the *restricted* scheduling problem. In this paper the connection to flow is more tenuous. We solve an unsplittable flow problem in a *generalized* bipartite network, which is defined by the *unrelated* scheduling problem. The generalized flow problem can be transformed to a minimum cost flow problem. Our algorithm uses the primal-dual approach combined with a gain scaling technique to obtain a polynomial running time. To compute a flow among the edges with zero reduced cost it uses the procedure Unsplittable-Blocking-Flow from [5] in the inner loop.

Given some candidate value for the makespan, our algorithm finds an approximate solution for the generalized flow problem in the two-layered bipartite network. Throughout execution the algorithm always maintains an integral assignment of jobs to machines. Each assignment defines a partition of the machines into underloaded, medium loaded and overloaded machines. Our overloaded machines are heavily overloaded, that is, their load is at least twice as large as the candidate makespan.

The main idea of our algorithm is to utilize the existence of overloaded machines in conjunction with the fact that we are looking for an approximate integral solution. We use this idea twice. On the one hand this allows us to show an improved lower bound on the makespan of an optimum schedule and thus to overcome the $(1 + \varepsilon)$ error usually induced by the gain scaling technique. On the other hand this is also used to reduce the number of outer loops to $O(m \log m)$, which is the main reason for the substantial running time improvement. Our algorithm is a generic minimum cost flow algorithm without any complex enhancements for generalized flow computation. Overloaded and underloaded machines are treated as sources and sinks, respectively. The height of a node is its minimum distance to a sink. In our algorithm the admissible network, used for the unsplittable maximum flow computation, consists only of edges and nodes which are on shortest paths from overloaded machines with minimum height to underloaded machines. This modification to the primal-dual approach is important to show the improved lower bound on the makespan of an optimum schedule.

Our algorithm is simpler and faster than the previously known algorithms. For the unrelated scheduling problem we have replaced the classical technique, i.e., computing first a fractional solution and rounding afterwards, by a completely integral approach. Our algorithm takes advantage from addressing the approximation problem directly. In particular, this allows us to benefit from an unfavorable preliminary assignment. We feel that this might be helpful also in other applications.

Identifying the connection to flow might be the key for obtaining combinatorial (approximation) algorithms for problems for which solving the LP-relaxation and rounding is currently the (only) alternative. Our techniques and results do not improve upon the approximation factor for the unrelated scheduling problem, however, we expect more exciting improvements for other hard problems.

Comparison of Running Times. We compare the running time of our algorithm with the so far fastest algorithms of Vaidya [27] and Radzik [22]. Both of the former approaches have been designed to solve the fractional generalized maximum flow problem on a graph with node set V and edge set E. Rounding the fractional solution yields the 2-approximation.

Technique and running time for computing a 2-approximative schedule:

- $O(|E|^{1.5}|V|^2 \log(U))$: Interior Point approach for generalized flow problem and rounding [27]
- $O(|E||V|(|E|+|V|\log|V|)\log U \log(nU))$: Combinatorial algorithm for generalized flow problem and rounding [22]
- $O(m^2 A \log(m) \log(nU))$: The integral approach presented in this paper

To compare these bounds, note that in our bipartite network $A = |E| = O(nm)$ and $|V| = n + m$. Our algorithm is linear in A. It clearly outperforms the previous algorithms if $n + m = o(A)$. In the case $A = \Theta(n + m)$ our algorithm is better by a factor of $\Omega(\frac{(n+m)^{0.5}}{\log(n)\log(m)})$ than Vaidya's algorithm and by a factor of $\Omega(\log U)$ faster than Radzik's algorithm. This is the first time that a combinatorial algorithm always beats the interior point approach for this problem. The heuristics [18, 24, 28] consider instances where $A = \Theta(nm)$. In this case our algorithm outperforms both former approaches by a factor almost linear in n.

The $(1+\varepsilon)$-approximation algorithms for the generalized maximum flow problem in [4, 21, 25] have all running time $\widetilde{O}(\log \varepsilon^{-1}|E|(|E| + |V| \log \log U))$, where the $\widetilde{O}()$ notation hides a factor polylogarithmic in $|V|$. Again, an extra factor of $O(\log(nU))$ is needed for the makespan minimization. This running time is not always better than ours. The fastest approximation scheme that directly addresses the scheduling problem is due to Jansen and Porkolab [12] and has a running time of $O(\varepsilon^{-2}(\log \varepsilon^{-1})mn \min\{m, n \log m\} \log m)$. Clearly, for constant ϵ this algorithm is faster than our algorithm. However, for ϵ in the order of $\frac{1}{m}$ and $\log(U) = O(n)$ their running times become comparable.

Roadmap. The rest of the paper is organized as follows. In Section 2, we introduce notation and model. Section 3.1 presents our approximation algorithm and Section 3.2 shows the analysis.

2 Notation

2.1 The Scheduling Problem

We consider the problem of scheduling a set J of n independent *jobs* on a set M of m *machines*. The *processing time* of job i on machine j is denoted by p_{ij}. Define the $n \times m$ matrix of processing times $\mathbf{P}$ in the natural way. Throughout the paper we assume that p_{ij} is either an integer or ∞ for all $i \in J$ and $j \in M$. Define $U = \max_{i \in J, j \in M}\{p_{ij} \neq \infty\}$. Furthermore, define A as the number of pairs (i, j) with $p_{ij} \neq \infty$. An *assignment* of jobs to machines is denoted by a function $\alpha : J \mapsto M$. We denote $\alpha(i) = j$ if job i is assigned to machine j. For any assignment α, the *load* δ_j on machine j for a matrix of processing times $\mathbf{P}$ is the sum of processing times for the jobs that are assigned to machine j, thus $\delta_j(\mathbf{P}, \alpha) = \sum_{i \in J, \alpha(i)=j} p_{ij}$. We omit $\mathbf{P}$ in the notation of δ_j if $\mathbf{P}$ is clear from the context.

Define the *makespan* of an assignment α for a processing time matrix $\mathbf{P}$, denoted $\mathsf{Cost}(\mathbf{P}, \alpha)$, as the maximum load on a machine, hence $\mathsf{Cost}(\mathbf{P}, \alpha) = \max_{j \in M} \delta_j(\alpha)$. Associated with a matrix of processing times $\mathbf{P}$ is the *optimum makespan*, which is the least possible makespan of an assignment α, that is $\mathsf{OPT}(\mathbf{P}) = \min_\alpha \mathsf{Cost}(\mathbf{P}, \alpha)$. Following Graham's notation [9], our problem is equivalent to $R||C_{\max}$.

2.2 Generalized Maximum Flows and Minimum Cost Flows

The generalized maximum flow problem is a generalization of the maximum flow problem, where each edge (i, j) has some gain factor μ_{ij}. If f_{ij} units of flow are sent from node i to node j along edge (i, j), then $\mu_{ij} f_{ij}$ units arrive at j. More specifically, let $G = (V, E)$ be a directed graph of the generalized flow problem, $\mu : E \mapsto \mathbb{R}^+$ a gain function, and s and t source and sink node, respectively. Furthermore, there is a capacity function on the edges. A generalized flow $f : E \mapsto \mathbb{R}$ is a function on the edges that satisfies the capacity and antisymmetry constraints on all edges, and the conservation constraints $\sum_{(j,i)\in E} \mu_{ji} f_{ji} - \sum_{(i,j)\in E} f_{ij} = 0$ on all nodes $i \in V \setminus \{s,t\}$. The value of the flow f is defined as the amount of flow into the sink. Among all generalized flows of maximum value, the goal is to find one that minimizes the flow out of the source.

The fractional version of the scheduling problem can be converted into a generalized maximum flow problem [20]. In order to check whether a fractional schedule of length w exists, one can construct a bipartite graph with nodes representing jobs and machines and introduce an edge from machine node i to job node j with gain $1/p_{ij}$. There is a source which is connected to all the machine nodes with edges of unit gain and capacity w, and the job nodes are connected to a sink with edges of unit gain and unit capacity. A generalized flow in this network that results in an excess of n at the sink corresponds to a solution of the fractional scheduling problem. If the maximum excess that can be generated at the sink is below n, then the fractional scheduling problem is infeasible, i.e., the current value of w is too small.

Truemper [26] established a relationship between the generalized maximum flow problem and the minimum cost flow problem. In his construction, he defined the cost for each arc in the minimum cost flow problem as the logarithm of the gain in the generalized maximum flow problem. In order to transform the generalized maximum flow problem to a minimum cost flow problem with integral arc costs, a gain rounding technique can be used (see e.g. [25]). Gains are rounded down to integer powers of some base $b > 1$. The rounded gain of each residual arc (i,j) is defined as $\gamma_{ij} = b^{c_{ij}}$ where $c_{ij} = \lfloor \log_b \mu_{ij} \rfloor$. Antisymmetry is maintained by setting $\gamma_{ij} = 1/\gamma_{ji}$ and $c_{ij} = -c_{ji}$. The cost of arc (i,j) in the resulting minimum cost flow problem equals c_{ij}. Using a potential function $\pi : V \mapsto \mathbb{R}^+$, the reduced costs c_{ij}^π of an arc (i,j) are defined as $c_{ij}^\pi = c_{ij} - \pi(i) + \pi(j)$ (see [1]). The PRIMAL-DUAL approach [1] for minimum cost flows can be used to compute a generalized maximum flow (see e.g. [25]). The PRIMAL-DUAL approach preserves the *reduced cost optimality condition*, i.e., $c_{ij}^\pi \geq 0$ for each edge (i,j) in the residual network. Because of the rounding, an optimum solution of the minimum cost flow problem gives only a $(1+\epsilon)$-approximation of the generalized (fractional) maximum flow problem. Using techniques from [23], the fractional solution can be transformed to an integral solution. This approach leads to a $(2+\epsilon)$-approximation algorithm for the scheduling problem.

2.3 Our Model

We also formulate the scheduling problem as a generalized maximum flow problem. However, we use a different construction as in [20]. We construct a bipartite graph with nodes representing jobs and machines. There is an arc from job node i to machine node j with unit capacity and gain $\mu_{ij} = p_{ij}$ if $p_{ij} \leq w$. The parameter w will be determined by binary search. Each job node i has supply 1. A generalized flow f is a solution to the fractional version of the scheduling problem, if in f all supplies are sent to the machines. In this case, we call f a *feasible flow*. A generalized flow in such a network creates excess on the machine nodes. An excess on machine j corresponds to the load on machine j. Define $\delta_j(\mathbf{P}, f)$ as the load on machine j under the generalized flow f with gains defined by $\mathbf{P}$. If we require that the supply of each job is sent to exactly one machine, then we get an integral solution to the scheduling problem. In this case, we call f a *generalized unsplittable flow* and f corresponds to an assignment α, i.e., assigning job i to machine j corresponds to sending one unit of flow along edge (i,j). We are interested in finding a generalized unsplittable flow f such that the maximum excess over all machines is at most $2w$. This is not always possible, however, if we can't find such a flow, we can still derive the lower bound $\mathsf{OPT}(\mathbf{P}) \geq w + 1$.

Following the construction from Section 2.2, we formulate this generalized maximum unsplittable flow problem as a minimum cost flow problem. For the gain rounding, we choose $b = (1 + z)$ where $z = \frac{1}{m}$. If (i, j) is an edge from job node i to machine node j then the cost c_{ij} and the rounded gain γ_{ij} is defined by $c_{ij} = \lfloor \log_b(p_{ij}) \rfloor$, and $\gamma_{ij} = b^{c_{ij}}$. For any path W, we define $\gamma(W) = \prod_{(i,j) \in W} \gamma_{ij}$. In the same way we define $\gamma(K)$ for some cycle K. In the following, denote $\mathbf{C} = (c_{ij})$ and $\boldsymbol{\Gamma} = (\gamma_{ij})$. In order to solve the minimum cost flow problem we use the well known PRIMAL-DUAL approach [1].

For a given assignment α, a positive integer w and a matrix of processing times $\mathbf{P}$, we now define the residual network $G_\alpha(w)$ (Definition 1) and we partition the machines, with respect to their loads, into three subsets (Definition 2).

Definition 1. *Let α be an assignment and $w \in \mathbb{N}$. We define a directed bipartite graph $G_\alpha(w) = (V, E_\alpha(w))$ where $V = M \cup J$ and each machine is represented by a node in M, whereas each job defines a node in J. Furthermore, $E_\alpha = E_\alpha^1 \cup E_\alpha^2$ with $E_\alpha^1 = \{(j, i) : j \in M, i \in J, \alpha(i) = j, p_{ij} \leq w\}$ and $E_\alpha^2 = \{(i, j) : j \in M, i \in J, \alpha(i) \neq j, p_{ij} \leq w\}$.*

Definition 2. *Let $w \in \mathbb{N}$ and α be an assignment. We partition the set of machines M into three subsets:*
$$M^-(\alpha) = \{j : \delta_j(\mathbf{P}, \alpha) \leq w\}$$
$$M^0(\alpha) = \{j : w + 1 \leq \delta_j(\mathbf{P}, \alpha) \leq 2w\}$$
$$M^+(\alpha) = \{j : \delta_j(\mathbf{P}, \alpha) \geq 2w + 1\}$$

In our setting, at each time, nodes from M^- can be interpreted as sink nodes, whereas nodes from M^+ as source nodes.

We now give a lemma that generalizes the path decomposition theorem to generalized flows. The proof of a similar decomposition theorem can be found in [8]. Note, that a fractional generalized flow on a path is defined as a flow that fulfills the flow conservation constraints on the inner nodes. Similarly, a generalized flow on a cycle fulfills the flow conservation constraints on all nodes in the cycle except one.

Lemma 1 (Decomposition theorem). *Let f and g be two generalized feasible flows in $G = (J \cup M, E)$. Then g equals f plus fractional flow: on some directed* cycles in G_f, and on some directed paths in G_f with end points in M and with the additional property that no end point of some path is also the starting point of some other path.

2.4 Unsplittable Blocking Flows

Our approximation algorithm will make use of the algorithm UNSPLITTABLE-BLOCKING-FLOW introduced in [5]. UNSPLITTABLE-BLOCKING-FLOW was designed for a restricted scheduling problem on identical machines. Here, each job i has some weight w_i and is only allowed to use a subset A_i of the machines. This is a special case of the unrelated scheduling problem considered in this paper, where $p_{ij} = w_i$ if $j \in A_i$ and $p_{ij} = \infty$ otherwise. Given an assignment α and an integer w, UNSPLITTABLE-BLOCKING-FLOW(α, w) computes an assignment β, where there is no path from $M^+(\beta)$ to $M^-(\beta)$ in $G_\beta(w)$.

We use UNSPLITTABLE-BLOCKING-FLOW for arbitrary processing times p_{ij}. In order to make clear that UNSPLITTABLE-BLOCKING-FLOW runs on the original processing times (p_{ij}) we include $\mathbf{P}$ in the parameter list. Furthermore, we allow UNSPLITTABLE-BLOCKING-FLOW only to reassign jobs according to some graph $G_\alpha^0(w)$, which can be any subgraph of $G_\alpha(w)$. These adaptations do not influence the correctness and the running time of algorithm UNSPLITTABLE-BLOCKING-FLOW.

Lemma 2 and Theorem 1 are derived from [5] and state properties of algorithm UNSPLITTABLE-BLOCKING-FLOW that are used in the discussion of our approximation algorithm.

Let $G_\alpha^0(w)$ be any subgraph of $G_\alpha(w)$. Let β be the assignment computed by UNSPLITTABLE-BLOCKING-FLOW$(\alpha, G_\alpha^0(w), \mathbf{P}, w)$. In this call jobs are reassigned by pushing them through edges of $G_\alpha^0(w)$. We define $G_\beta^0(w)$ as the graph that results from $G_\alpha^0(w)$ after this reassignments.

Lemma 2 ([5, Lemma 4.2]). *Let β be the assignment computed by* UNSPLITTABLE-BLOCKING-FLOW$(\alpha, G_\alpha^0(w), \mathbf{P}, w)$. *Then*
(a) $j \in M^-(\alpha) \Rightarrow \delta_j(\mathbf{P}, \beta) \geq \delta_j(\mathbf{P}, \alpha)$
(b) $j \in M^0(\alpha) \Rightarrow w + 1 \leq \delta_j(\mathbf{P}, \beta) \leq 2w$
(c) $j \in M^+(\alpha) \Rightarrow \delta_j(\mathbf{P}, \beta) \leq \delta_j(\mathbf{P}, \alpha)$.

Theorem 1 ([5, Lemma 4.4/Theorem 4.5]). UNSPLITTABLE-BLOCKING-FLOW$(\alpha, G_\alpha^0(w), \mathbf{P}, w)$ *takes time $O(mA)$ and computes an assignment β, having the property, that there is no path from $M^+(\beta)$ to $M^-(\beta)$ in $G_\beta^0(w)$.*

3 Approximation Algorithm

We now present our approximation algorithm, UNSPLITTABLE-TRUEMPER, which will be used to compute an assignment α where $\mathsf{Cost}(\mathbf{P}, \alpha) \leq 2 \cdot \mathsf{OPT}(\mathbf{P})$. We always maintain an *unsplittable* flow, i.e., an integral solution. We loose a factor of 2 by allowing some gap for the machine loads. The special structure of our algorithm allows us to compensate the error, introduced by the gain scaling technique, by a better lower bound on $\mathsf{OPT}(\mathbf{P})$. We stop the computation as soon as we get this better lower bound. This improves also the running time.

3.1 Algorithm Unsplittable-Truemper

We formulate the scheduling problem as a generalized maximum unsplittable flow problem with rounded gain factors as described in Section 2.2. In order to solve this generalized unsplittable flow problem we use the PRIMAL-DUAL approach for computing a minimum cost flow [1]. Our algorithm maintains the reduced cost optimality condition. In our setting this means that it does not create negative cost cycles in the residual network. In order to achieve this, UNSPLITTABLE-TRUEMPER iteratively computes a shortest path graph $G_\alpha^0(w)$, which we define below, and uses UNSPLITTABLE-BLOCKING-FLOW to compute a blocking flow on this shortest path graph. While the costs in UNSPLITTABLE-TRUEMPER refer to the rounded processing times, it operates on the original processing times. It is important to note, that both the costs as well as the original processing times are integer. Because of Theorem 1, there is no path from a machine from M^+ to a machine from M^- in $G_\alpha^0(w)$ after termination of UNSPLITTABLE-BLOCKING-FLOW. We stop this procedure, when we can either derive a *good* lower bound on OPT(**P**) (see Theorem 2) or we found an assignment α with $M^+ = \emptyset$.

UNSPLITTABLE-TRUEMPER($\alpha, \mathbf{P}, \mathbf{C}, w$)

Input: assignment α with each job i assigned to a machine from $B(i)$,
matrix of processing times **P**, matrix of edge costs **C**,
positive integer w
Output: assignment β

// $G_\alpha(w)$ is the graph corresponding to α and w.
$\pi := 0$;
while $\exists$ machine in M^+ with a path to some machine in M^- in $G_\alpha(w)$
 and $\forall u \in M^+ : \pi(u) < \log_b(m)$
{
 determine shortest path distances $d(\cdot)$ from all nodes to the set
 of sinks M^- in $G_\alpha(w)$ with respect to the reduced costs c_{ij}^π;
 update $\pi := \pi + d$;
 define $M_{\min}^+$ as the set of machines from M^+ with minimum distance
 to a node in M^- with respect to the costs c_{ij};
 define $G_\alpha^0(w)$ as the admissible graph, consisting only of edges on
 shortest paths from $M_{\min}^+$ to M^- in $G_\alpha(w)$;
 $\beta :=$ UNSPLITTABLE-BLOCKING-FLOW($\alpha, G_\alpha^0(w), \mathbf{P}, w$);
 update $\alpha := \beta$;
}
return α;

We now describe our algorithm in more detail. UNSPLITTABLE-TRUEMPER starts with an assignment α. In α, each job $i \in J$ is assigned to some machine $j \in B(i)$, where its processing time is minimum, i.e., $B(i) = \{j \in M : p_{ij} \leq p_{ik}, \forall k \in M\}$. Arc capacities are given by **P** whereas arc costs are given by **C** (as defined in

Section 2). Furthermore, UNSPLITTABLE-TRUEMPER gets as input an integer w. Assignment α and integer w define a graph $G_\alpha(w)$ as in Definition 1, and a partition of the machines as in Definition 2. At all times, UNSPLITTABLE-TRUEMPER maintains a total assignment, that is all jobs are always assigned to some machine. If a job gets unassigned from a machine, it is immediately assigned to some other machine.

Our algorithm iteratively computes shortest path distances $d(u)$ from each node u to the set of sinks M^-, with respect to the reduced costs c_{ij}^π. Then π is updated, such that all arcs on shortest paths have zero reduced costs. For each node $u \in M$, $\pi(u)$ never decreases. After the update of π, $\pi(u)$ holds the minimum distance from u to M^- for each node u with respect to the costs c_{ij}. We define $M^+_{\min}$ as the set of machines from M^+ with minimum distance to a node in M^- with respect to the costs c_{ij}. Note, that $M^+_{\min}$ consists of all machines $u \in M^+$ where $\pi(u)$ is minimum. $G^0_\alpha(w)$ is then defined as the admissible graph, consisting only of edges on shortest paths from $M^+_{\min}$ to M^- in $G_\alpha(w)$. We will see in Section 3.2 that this is essential for our algorithm. Note, that $G^0_\alpha(w)$ consists only of arcs with zero reduced costs. Afterwards, UNSPLITTABLE-BLOCKING-FLOW is applied to the admissible graph $G^0_\alpha(w)$. It reassigns jobs from the admissible graph, such that after UNSPLITTABLE-BLOCKING-FLOW returns, there is no longer a path from a machine in $M^+_{\min}$ to a machine in M^- in the admissible graph $G^0_\alpha(w)$. Therefore, $\min\{\pi(u); u \in M^+\}$ increases in the next iteration of the while loop. The residual network $G_\alpha(w)$ is then updated accordingly. The while-loop terminates when there exists no machine from M^+ with a path to a machine from M^- in $G_\alpha(w)$ or there exists a machine $u \in M^+$ with $\pi(u) \geq \log_b(m)$.

3.2 Analysis

We now analyze the behavior of our algorithm. The main result in this section is Theorem 2. A call of UNSPLITTABLE-TRUEMPER$(\alpha, \mathbf{P}, \mathbf{C}, w)$ terminates if $M^+(\alpha) = \emptyset$. In this case, we know that $\mathsf{Cost}(\mathbf{P}, \alpha) \leq 2w$. We will see, that we can take also some advantage from an assignment α which is still unfavorable, i.e., for which $M^+(\alpha) \neq \emptyset$ holds.

The reduced cost optimality condition $c_{ij}^\pi \geq 0$ holds for all $(i,j) \in E_\alpha(w)$ during the whole computation. It implies $\gamma(K) \geq 1$ for each cycle K in $G_\alpha(w)$. This property does not necessarily hold for every path. Lemma 3 is of crucial importance in our analysis. It shows that $\gamma(W) \geq 1$ holds for every path W connecting some node from $M^+(\alpha)$ to any other node from M in $G_\alpha(w)$. For proving this result, we need that $G^0_\alpha(w)$ was defined only by shortest paths from nodes in $M^+_{\min}$ to nodes in M^-.

Lemma 3. UNSPLITTABLE-TRUEMPER *maintains the property, that for each path W in $G_\alpha(w)$ from any machine in M^+ to any other machine in M, we have $\gamma(W) \geq 1$.*

The following lemma will be used to derive a lower bound on $\mathsf{OPT}(\mathbf{P})$.

Lemma 4. *Let (G, Γ) denote a generalized maximum unsplittable flow problem defined by network G and matrix of processing times Γ. Let f be a generalized feasible unsplittable flow in (G, Γ), and let $s, t \in \mathbb{R}^+$. Suppose $\forall u \in M : \delta_u(\Gamma, f) \geq s$, and $\exists \hat{u} \in M : \delta_{\hat{u}}(\Gamma, f) \geq s + t$, and for each cycle K in G_f, $\gamma(K) \geq 1$. If on every path W in G_f from $\hat{u}$ to any other machine $u \in M$, $\gamma(W) \geq 1$, then $\mathsf{OPT}(\Gamma) \geq s + \frac{t}{m}$.*

Theorem 2. UNSPLITTABLE-TRUEMPER *takes time $O(m^2 A \log(m))$. Furthermore, if* UNSPLITTABLE-TRUEMPER$(\alpha, \mathbf{P}, \mathbf{C}, w)$ *terminates with $M^+ \neq \emptyset$ then $\mathsf{OPT}(\mathbf{P}) \geq w + 1$.*

We will now show how to use UNSPLITTABLE-TRUEMPER to approximate a schedule with minimum makespan. We do series of calls to UNSPLITTABLE-TRUEMPER$(\alpha, \mathbf{P}, \mathbf{C}, w)$ where, by a binary search on $w \in [1, nU]$, we identify the smallest w such that a call to UNSPLITTABLE-TRUEMPER$(\alpha, \mathbf{P}, \mathbf{C}, w)$ returns an assignment with $M^+ = \emptyset$. Afterwards we have identified a parameter w, such that UNSPLITTABLE-TRUEMPER$(\alpha, \mathbf{P}, \mathbf{C}, w)$ returns an assignment where $M^+ \neq \emptyset$ and UNSPLITTABLE-TRUEMPER$(\alpha, \mathbf{P}, \mathbf{C}, w+1)$ returns with $M^+ = \emptyset$.

Theorem 3. UNSPLITTABLE-TRUEMPER *can be used to compute a schedule α with $\mathsf{Cost}(\mathbf{P}, \alpha) \leq 2 \cdot \mathsf{OPT}(\mathbf{P})$ in time $O(m^2 A \log(m) \log(nU))$.*

Proof. We use UNSPLITTABLE-TRUEMPER as described above. Let β_1 be the assignment returned by UNSPLITTABLE-TRUEMPER$(\alpha, \mathbf{P}, \mathbf{C}, w)$ where $M^+ \neq \emptyset$. Let β_2 be the assignment returned by UNSPLITTABLE-TRUEMPER$(\alpha, \mathbf{P}, \mathbf{C}, w+1)$ where $M^+ = \emptyset$. From β_1 we follow by Theorem 2 that $\mathsf{OPT}(\mathbf{P}) \geq w+1$ and in β_2 we have $\mathsf{Cost}(\mathbf{P}, \beta_2) \leq 2(w+1)$. Thus, $\mathsf{Cost}(\mathbf{P}, \beta_2) \leq 2 \cdot \mathsf{OPT}(\mathbf{P})$. It remains to show the running time of $O(m^2 A \log(m) \log(nU))$. Due to Theorem 2, one call to UNSPLITTABLE-TRUEMPER takes time $O(m^2 A \log(m))$. The binary search contributes a factor $\log(nU)$. This completes the proof of the theorem. □

Acknowledgments. We would like to thank Thomas Lücking for many fruitful discussions and helpful comments.

References

1. R.K. Ahuja, T.L. Magnanti, and J.B. Orlin. *Network Flows: Theory, Algorithms, and Applications*. Prentice Hall, 1993.
2. G. Dantzig. *Linear Programming and Extensions*. Princeton University Press, Princeton, New York, 1963.
3. Y. Dinitz, N. Garg, and M.X. Goemans. On the single-source unsplittable flow problem. *Combinatorica*, 19(1):17–41, 1999.
4. L. Fleischer and K. D. Wayne. Fast and simple approximation schemes for generalized flow. *Mathematical Programming*, 91(2):215–238, 2002.
5. M. Gairing, T. Lücking, M. Mavronicolas, and B. Monien. Computing Nash equilibria for scheduling on restricted parallel links. In *Proceedings of the 36th Annual ACM Symposium on the Theory of Computing (STOC'04)*, pages 613–622, 2004.

6. A.V. Goldberg, S.A. Plotkin, and E. Tardos. Combinatorial algorithms for the generalized circulation problem. *Math. of Operations Research*, 16:351–379, 1991.
7. D. Goldfarb, Z. Jin, and J.B. Orlin. Polynomial-time highest-gain augmenting path algorithms for the generalized circulation problem. *Math. of Operations Research*, 22:793–802, 1997.
8. M. Gondran and M. Minoux. *Graphs and Algorithms*. Wiley, 1984.
9. R.L. Graham. Bounds for certain multiprocessor anomalies. *Bell System Technical Journal*, 45:1563–1581, 1966.
10. D.S. Hochbaum. *Approximation Algorithms for NP-hard Problems*. PWS Publishing Co., 1996.
11. E. Horowitz and S. Sahni. Exact and approximate algorithms for scheduling nonidentical processors. *Journal of the ACM*, 23(2):317–327, 1976.
12. K. Jansen and L. Porkolab. Improved approximation schemes for scheduling unrelated parallel machines. *Math. of Operations Research*, 26(2):324–338, 2001.
13. W.S. Jewell. Optimal flow through networks with gains. *Operations Research*, 10:476–499, 1962.
14. S. Kapoor and P.M. Vaidya. Fast algorithms for convex quadratic programming and multicommodity flows. In *Proceedings of the 18th Annual ACM Symposium on Theory of Computing (STOC'86)*, pages 147–159, 1986.
15. J. Kleinberg. Single-source unsplittable flow. In *Proceedings of the 37th Annual Symposium on Foundations of Computer Science (FOCS'96)*, pages 68–77, 1996.
16. S.G. Kolliopoulos and C. Stein. Approximation algorithms for single-source unsplittable flow. *SIAM Journal on Computing*, 31:919–946, 2002.
17. J.K. Lenstra, D.B. Shmoys, and E. Tardos. Approximation algorithms for scheduling unrelated parallel machines. *Mathematical Programming*, 46:259–271, 1990.
18. E. Mokotoff and P. Chrétienne. A cutting plane algorithm for the unrelated parallel machine scheduling problem. *European Journal of Operational Research*, 141:515–525, 2002.
19. K. Onaga. Dynamic programming of optimum flows in lossy communication nets. *IEEE Transactions on Circuit Theory*, 13:308–327, 1966.
20. S.A. Plotkin, D.B. Shmoys, and E. Tardos. Fast approximation algorithms for fractional packing and covering problems. *Math. of Operations Research*, 20(2):257–301, 1995.
21. T. Radzik. Faster algorithms for the generalized network flow problem. *Math. of Operations Research*, 23:69–100, 1998.
22. T. Radzik. Improving time bounds on maximum generalised flow computations by contracting the network. *Theoretical Computer Science*, 312(1):75–97, 2004.
23. D.B. Shmoys and E. Tardos. An approximation algorithm for the generalized assignment problem. *Mathematical Programming*, 62:461–474, 1993.
24. F. Sourd. Scheduling tasks on unrelated machines: Large neighborhood improvement procedures. *Journal of Heuristics*, 7:519–531, 2001.
25. E. Tardos and K. D. Wayne. Simple generalized maximum flow algorithms. In *Proceedings of the 6th Integer Programming and Combinatorial Optimization Conference (IPCO'98)*, pages 310–324, 1998.
26. K. Truemper. On max flows with gains and pure min-cost flows. *SIAM Journal on Applied Mathematics*, 32(2):450–456, 1977.
27. P.M. Vaidya. Speeding up linear programming using fast matrix multiplication. In *Proceedings of the 30th Annual Symposium on Foundations of Computer Science (FOCS'89)*, pages 332–337, 1989.
28. S. L. van de Velde. Duality-based algorithms for scheduling unrelated parallel machines. *ORSA Journal on Computing*, 5(2):182–205, 1993.

Polynomial Time Preemptive Sum-Multicoloring on Paths

Annamária Kovács

Max-Planck Institut für Informatik,
Stuhlsatzenhausweg 85,
66123 Saarbrücken, Germany
panni@mpi-sb.mpg.de

Abstract. The *preemptive Sum-Multicoloring (pSMC)* problem is a scheduling problem where pairwise conflicting jobs are represented by a conflict graph. The time demands of jobs are given by integer weights on the nodes. The goal is to schedule the jobs in such a way that the *sum* of their finish times is minimized. We give an $\mathcal{O}(n \cdot \min(n, \log p))$ time algorithm for pSMC on paths and cycles, where n is the number of nodes and p is the largest time demand. This is the first polynomial algorithm for this problem. It answers the question raised in [8] about the hardness of this problem. In this respect our result identifies a gap between binary-tree conflict graphs – where the question is NP-hard – and paths. Furthermore, our time bound gets very close to that of $\mathcal{O}(n \cdot \log p / \log \log p)$ for the *non-preemptive* SMC on paths [8]. A detailed version of this paper is available at [3].

1 Introduction

In the paper we consider the *preemptive Sum-Multicoloring (pSMC)* problem. In spite of the name "multicoloring", we prefer to discuss it from a scheduling point of view. The input is given by a simple undirected *conflict graph* $G(V, E)$, where the nodes model different jobs, and by integer weights on the nodes $x : V \to \mathbb{N}^+$, i.e., *time demands* for each job. The edges of the graph represent pairwise conflicts between certain jobs, meaning that they cannot be processed at the same time, e.g., due to some non-shareable resource they use. The goal is to schedule the jobs, i.e., to determine a function $\Phi : V \to 2^\mathbb{N}$ that assigns a set of time units to each job, so that the sets assigned to conflicting jobs do not intersect, and $\forall v \in V : |\Phi(v)| = x(v)$. Let $f(v) = \max \Phi(v)$ be the finish time of node v. In the Sum-Multicoloring problems, the *average* finish time, or equivalently, the *sum* of finish times $\sum_{v \in V} f(v)$ has to be minimized. In the *non-preemptive* version of SMC *(npSMC)* the assigned sets $\Phi(v)$ must be contiguous, whereas in *preemptive* SMC they are arbitrary.

Scheduling jobs with pairwise conflicts is a classical problem, with applications in diverse areas like traffic intersection control, VLSI routing, session management in local area networks, operating systems, etc. (see [7]). From the jobs' point of view, it is a natural goal to minimize the average completion time.

The concern of this paper is the pSMC problem on path conflict graphs. The path has n nodes, and the nodes are numbered by $1,...,n$, from left to right. Let $p := \max_{1 \leq i \leq n} x(i)$ be the largest demand in the input. We provide the first polynomial algorithm for pSMC on paths. It runs in time $\mathcal{O}(n \cdot \min(n, \log p))$. Moreover, our result easily carries over to cycle conflict graphs.

Related work. For a comprehensive history of the SMC problems see, e.g., [2, 8]. The Sum-Multicoloring problem was introduced by Bar-Noy et al. in [7], as a generalization of the *Sum-Coloring* problem (the latter was studied by Kubicka in [5]). Even the Sum-Coloring problem was shown to be NP-hard on several conflict graph classes, like e.g., on planar graphs [2]; it is hard to approximate within a constant factor $c > 1$ on bipartite graphs [1].

The SMC problem on trees was studied by Halldórsson et al. in [8] and some of the results were generalized to partial k-trees in [2]. In [8] an $\mathcal{O}(n^2)$ and an $\mathcal{O}(np)$ algorithm was provided for the *non-preemptive* SMC on trees. The first one runs in time $\mathcal{O}(n \cdot \log p / \log \log p)$ on paths. For the *preemptive* SMC on trees a PTAS was given. The hardness of pSMC on trees or paths was posed as an open question. Marx [6] showed, that pSMC is NP-hard on binary trees, even for polynomially bounded p. Thus, the preemptive problem on trees turned out to be essentially harder than the non-preemptive version. In [4] we gave a pseudo-polynomial algorithm, running in time $\mathcal{O}(n^3 p)$ for pSMC on paths.

Our result. We provide an $\mathcal{O}(n \cdot \min(n, \log p))$ time algorithm for the pSMC problem on paths which immediately implies an $\mathcal{O}(n \cdot \min(n, \log p))$ time solution for cycles. This is the first polynomial algorithm for this problem. It improves substantially on our previous $\mathcal{O}(n^3 p)$-time result, and answers the question raised in [8], whether pSMC is efficiently solvable on paths. Thus, we identify a major difference between the solvability of the problem on binary trees – where it is NP-hard [6] – and graphs of maximum degree 2. More interestingly, compared to the results on the *non-preemptive* SMC problem in [8], our result gets very close to the time-bound of $\mathcal{O}(n \cdot \log p / \log \log p)$ for paths, whereas for very large p, our $\mathcal{O}(n^2)$ bound matches the $\mathcal{O}(n^2)$ bound for npSMC on paths (or trees in general). Our method is an improved version of the algorithm in [4]. The core of this improvement is a result about the structure of some "nice" optimal schedules, which we will briefly sketch here.

Without loss of generality, we add nodes 0 and $n + 1$ to the path, with demands $x(0) = x(n+1) = 0$. We denote by $\langle i, j \rangle$ the subpath $i, ..., j$. In a fixed schedule Φ, we distinguish so called local minimum *(loc-min)*, and local maximum *(loc-max)* nodes on the path: a loc-min node has smaller finish time, resp. a loc-max has larger finish time than any of its two neighbors. The rest is called *stair* node. Restricted to optimal schedules with maximum value of $\sum_{v \in V} f^2(v)$, the following hold: The schedule of a loc-min node determines the schedule of the stairs on its two sides in a greedy fashion, and a loc-max node receives the first time units idle on both sides. The hardness of this problem lies in the fact that loc-mins are not always *compact*, i.e., in general $f(i) \neq x(i)$ for a loc-min i. Let's denote by $r(i)$ and $\ell(i)$ the first nodes to the right and to the left of loc-min i with finish time less than $f(i)$ (see Fig. 1). The main idea of [4]

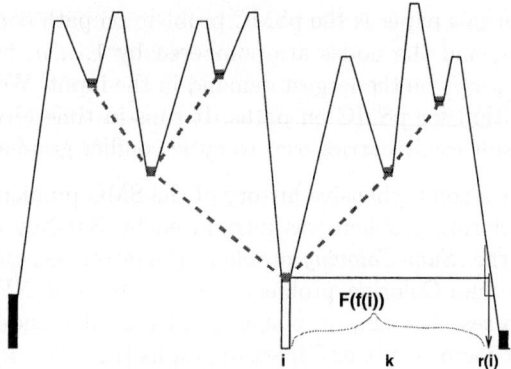

Fig. 1. Minimum-tree of a block

was that any "simultaneous permutation" of the time units in $\Phi(i)$ and $\Phi(r(i))$ results in the same optimum sum on the subpath strictly between i and $r(i)$. As a further consequence, $f(i)$ alone determines the optimum $F(f(i)) := \sum_{j=i+1}^{r(i)-1} f(j)$ on this subpath. This observation facilitates a dynamic programming algorithm: Two scheduled subpaths are glued at a loc-min, proceeding from short subpaths to longer ones, from large loc-mins to smaller ones. However, on the subpath $\langle i+1, r(i)-1 \rangle$ the optimum $F(f(i))$ for every possible $f(i)$ has to be determined (since $f(i)$ depends also on what will be glued to the left of i).

Our new result is that the possible changes in $f(i)$ cannot modify the structure on $\langle i, r(i) \rangle$, i.e., the order of the finish times remains the same. Moreover, by increasing $f(i)$ to $f(i)+\Delta$, we increase the finish time of all the loc-mins and decrease the finish time of all the loc-max's on $\langle i+1, r(i)-1 \rangle$ by Δ. In total we decrease $F(f(i))$ by Δ. The optimum on $\langle i+1, r(i)-1 \rangle$ is a simple linear function of the form $F(f(i)) = C - f(i)$. The only question for the dynamic algorithm is the constant C in this function and the range of possible $f(i)$ values.

What is the explanation for this determined structure on $\langle i, r(i) \rangle$? A subpath between nearest compact loc-mins will be called a *block*. (Note that blocks are easy to connect.) Inside a block, the loc-mins are non-compact, and there must be "large" differences between their demands. They form a binary tree (see Fig. 1), and the finish times on growing levels of this tree grow exponentially. In other words, if there are many loc-mins of similar demands, not separated by nodes of smaller demands, then some (e.g. the odd ones) will be compact. The exponential growth within a block implies that for fixed i there are just a constant number of candidates for $r(i)$, and just a few candidates to be the loc-min finished next, i.e., the child of i in the minimum-tree.

The fast growth within a block is quite intuitive. However, the order of finish times must remain the same already on the lowest levels above $f(i)$, meaning roughly, that even the smallest differences between finish times should exceed potential changes in $f(i)$. The proof of these results is lengthy and technical; it makes use of a handful of simple operations, and a tricky induction argument.

Overview. In Section 2 we characterize the type of optimal schedules we will look for, based on [4]. There we also try to provide more insight into the spirit of our improvement. Section 3 contains the key result that facilitates a polynomial-time algorithm. The proofs can be found at [3]. In Section 4 we present our fast algorithm. We conclude with a few words about cycle conflict graphs, and potential future work.

2 Definitions and Basic Facts

We will refer to time units as to **levels**. Given a schedule Φ, we say that i is **black** on level a, if $a \in \Phi(i)$, and i is **white** on level a if $a \notin \Phi(i)$. For $i < j$ we will also say, that the ordered pair (i, j) is **black-white, black-black,...** etc. on level a. Note that $(i, i + 1)$ cannot be black-black on any level.

Definition 1. *An (i, j) pair has a conflict or is conflicting on level a, if i and j are of the same parity and (i, j) is black-white, or white-black, or i and j are of opposite parity and (i, j) is black-black, or white-white.*

Obviously, if (i, j) is conflicting on level a, then $\exists k \in \langle i, j-1 \rangle$ s.t. $(k, k+1)$ is white-white on level a.

Definition 2. *For any ordered pair of nodes (i, j) and any fixed schedule Φ, we may consider the number of levels under some fixed level Γ, where (i, j) is black-black, white-black, white-white and black-white, respectively. We call the 4-tuple of these numbers the scheme of (i, j) under Γ.*

Definition 3. *Given a schedule, we say that node i is*
  *a **loc-min**, if $f(i-1) > f(i)$ and $f(i) < f(i+1)$, or $i = 0$, or $i = n+1$;*
  *a **loc-max**, if $f(i-1) < f(i)$ and $f(i) > f(i+1)$;*
  *a **stair** otherwise, in particular a **stair-up**, if $f(i-1) < f(i) < f(i+1)$, and a **stair-down**, if $f(i-1) > f(i) > f(i+1)$;*
  compact, *if $f(i) = x(i)$. Note that 0 and $n+1$ are compact.*
  *If both i and j are compact with no compact nodes between them, then we call $\langle i, j \rangle$ or (i, j) a **block**.*

We provide with notation the following 4 nodes concerning a loc-min i :

Definition 4. *If i is a loc-min in a schedule, then we denote by $r(i)$ the first node to the right of i s.t. $f(r(i)) < f(i)$ and we denote by $\ell(i)$ the first node to the left of i s.t. $f(\ell(i)) < f(i)$. Furthermore, $R(i) := r(i) - 1$ and $L(i) := \ell(i) + 1$.*

Note that $R(i)$ ($L(i)$) is either a stair-down (stair-up) or a loc-max.
We would like to restrict our attention to schedules of the simplest possible structure. In [4] we have shown that for any instance of the pSMC problem, a schedule exists with properties (P1)–(P5) below. For example, among schedules of minimum $\sum_{i=1}^{n} f(i)$ those of maximum $\sum_{i=1}^{n} f^2(i)$ fulfil (P1)–(P5).

(P1) the sum of finish times $\sum_{i=1}^{n} f(i)$ is minimum over all proper schedules;

(P2) stairs and loc-max's are scheduled "greedily": e.g., a stair-up i is black on the first possible levels where $i-1$ is white; a loc-max m is black on the first levels where both $m-1$ and $m+1$ are white;

(P3) if i is a non-compact loc-min and $\ell(i) \leq k \leq r(i)$ then i has no conflict with k on the level $f(i)$ and on any level where i is white; therefore there are no compact nodes inside $\langle \ell(i), r(i) \rangle$, since any node is white either on level $f(i)$ or on the levels where i is white.

(P4) if i is a non-compact loc-min and k is a loc-min $i < k < r(i)$, then there is no level where k (is black and) has conflict on both sides (with $\ell(k)$ and with $r(k)$).

(P5) if $i < j$ are loc-mins, $j < r(i)$ and $f(i) = f(j)$, then i and j are compact.

Definition 5. *A schedule Φ is called **optimal schedule** or a **solution** if properties (P1)-(P5) hold.*

Corollary 1. *The schedule of a loc-min determines the schedules of stairs leading up to the next loc-max's on both sides. If, e.g., i is a stair-up, then $f(i) = x(i-1) + x(i)$.*

Corollary 2. *Let i be a non-compact, even loc-min. (P2), (P3) and (P4) imply the following on $\langle i, r(i) \rangle$: If $(i, r(i))$ have no conflict on level b, then exactly the odd or exactly the even nodes are black on level b. If $(i, r(i))$ have a conflict on level b, then there is a loc-max m, s.t. on $\langle i, m-1 \rangle$ exactly the even nodes, on $\langle m+1, r(i) \rangle$ exactly the odd nodes are black on level b.*

Corollary 3. *Let i be a non-compact loc-min. $(i, r(i))$ is black-white on level $f(i)$, so i and $r(i)$ are of different parity, and $i + 3 \leq r(i)$ (resp. $\ell(i) \leq i - 3$).*

Definition 6. *Let i be a non-compact loc-min. We denote by $pit(i, r(i))$ the loc-min of smallest finish time between i and $r(i)$. If there is no loc-min between them, then $pit(i, r(i)) = \emptyset$ and $top(i, r(i))$ denotes the unique loc-max in $\langle i, r(i) \rangle$. The nodes $pit(\ell(i), i)$ and $top(\ell(i), i)$ are defined analogously.*

Due to (P5), the $pit()$ is well-defined. Note that if $k = pit(i, r(i))$ then $\langle i+1, \ell(k) \rangle$ is a series of stair-ups and $\langle r(k), r(i) - 1 \rangle$ is a series of stair-downs, and (P3) implies that $k = pit(i, r(i))$ is non-compact.

If g is a compact loc-min, then – instead of $\langle g, r(g) \rangle$ –, the block $\langle g, h \rangle$ is of relevance, where h is the nearest compact node. Blocks can be regarded as maximal $(i, r(i))$ or $(\ell(i), i)$ pairs, and we can define $pit(g, h)$ similarly to Definition 6.

If $\langle g, h \rangle$ is a block then the (non-compact) loc-mins inside form a binary tree in the following way: $i = pit(g, h)$ is the root of the tree, and its two children are $pit(\ell(i), i)$ and $pit(i, r(i))$, etc. The loc-max's can be regarded as the leaves of this tree. We will call this tree the **minimum-tree** of (g, h) (see Fig. 1). We say that the root i is on the **1st floor**, the children of i are on the **2nd floor**, etc. of the minimum-tree. Having the minimum-tree fixed, the root i is the only loc-min that can have (black) levels conflicting with both $\ell(i)$ and $r(i)$, and so $f(i)$ is not automatically determined by (P3), (P4). But once we have $f(i)$ ($\Phi(i)$), it determines the schedule of all the other loc-mins (by (P3), (P4)), consequently of all other nodes in the block (by (P2)).

Suppose we know that i is a non-compact loc-min and $j = r(i)$. In [4] we have shown that the optimum sum of finish times $F := \sum_{k=i+1}^{r(i)-1} f(k)$ depends only on $f(i)$. This fact is based on the following simple observations: The optimum on $\langle i+1, r(i) - 1 \rangle$ depends only on the scheme of $(i, r(i))$ and not on the exact schedule of i and $r(i)$. And, since there are no conflicting white-white levels of $(i, r(i))$ under $f(i)$, fixing the scheme of $(i, r(i))$ boils down to fixing the number of conflicting black-black levels under $f(i)$ or equivalently, to fixing $f(i)$ itself.

Let α denote the number of levels where $(i, r(i))$ is black-black. In this paper we consider the optimum F as a function of α. Our goal is to show that for the range of feasible α values, the order of finish times on $\langle i, r(i) \rangle$ is fixed. As a consequence, $F(\alpha)$ is a simple linear function, namely $F(\alpha - \Delta) = F(\alpha) - \Delta$. This result in turn, allows us to modify our previous algorithm: in the previous version, for each potential $(i, r(i))$ we had to test the $F(\alpha)$ value for each α one by one. Now it is sufficient to compute dynamically the domain and one value of $F(\alpha)$ for each possible $(i, r(i))$ (resp. $(\ell(i), i)$).

From a more global perspective, observe that inside a block $\langle i, j \rangle$, or inside $\langle i, r(i) \rangle$, there is one more loc-max than loc-mins. On the whole, we will derive that increasing the finish time of the root in a minimum-tree by Δ increases every other loc-min and decreases every loc-max by the same value, resulting in an improvement of Δ on the whole block. However, each increase must mean exchanging a black level below the finish time for a white level above the finish time of a loc-min, both having conflicts on both sides. This condition bounds from above the potential values of Δ in a recursive manner. The *rung* of a stair is the topmost set of contiguous black levels of the stair. The rungs of consecutive stairs partition the set of levels above the loc-min.

Definition 7. *Let s be a stair-up (stair-down) in a fixed solution Φ. The **rung of s** is the set of levels $S := [f(s-1)+1, f(s)]$, (resp. $S := [f(s+1)+1, f(s)]$). For simplicity, we use the respective capital letters to denote rungs.*

Definition 8. *We say that s is a 2nd **stair** resp. S is a 2nd **rung** if $s - 1$ or $s + 1$ is a loc-min. We call a stair or rung **high stair**, resp. **high rung** otherwise. Let i be a non-compact loc-min, $c = L(i)$ and $d = R(i)$. Now C and D are the rungs of c and d, respectively, and $f(i) \in C \cap D$. We say that i (or $f(i)$) **ends in C on the left and in D on the right**.*

3 The Main Theorem

Let $\langle i, j \rangle$ be a subpath, and $x(i), x(i+1), ..., x(j)$ a fixed (partial) instance of the pSMC problem. Regard all cases when this is part of an instance on a longer path, and all solutions (if existent) where i is a *non-compact* loc-min and $j = r(i)$.

Let's take all $f(i)$ that ever occur in such a situation, and fix one solution on $\langle i, r(i) \rangle$ for each of them. We can only talk about fixed solutions modulo permutation of the levels below $f(i)$, but with fixed finish times on $\langle i, r(i) - 1 \rangle$.

From here on we discuss the case of potential $(i, r(i))$ pairs. Case $(\ell(i), i)$ can be handled symmetrically. Instead of considering the optimum as a function of $f(i)$, we will look at it as follows: Let α be the number of black-black conflicts of $(i, r(i))$, and $F(\alpha)$ be the sum of finish times on $\langle i+1, r(i) - 1 \rangle$ in the corresponding fixed solution Φ_α. Since $\alpha = x(i) + x(r(i)) - f(i)$ (see (P3)), this makes just a tiny technical difference. Those α values that correspond to an occurence of $f(i)$ constitute the **domain** $\mathcal{D}_F$ of $F(\alpha)$.

We fix a level $\Gamma := \min(x(i) + x(i+1), x(r(i)) + x(r(i) - 1))$. We claim that Γ separates the set of levels into two parts: For any of the fixed solutions on $\langle i, r(i) \rangle$, $f(i) \leq \Gamma$, and $f(pit(i, r(i))) > \Gamma$. (The latter holds, because i and $r(i)$ are of different parity, and $pit(i, r(i))$ ends in rungs of the same parity.) In the rest of the paper Γ denotes this level.

For fixed α, the optimum is indifferent to the permutation (exchange) of levels below Γ (or even to change of roles of i and r(i)). Since $f(pit(i, r(i))) > \Gamma$, such a permutation does not interfere with finish times inside the subpath.

Decreasing $f(i)$ by Δ corresponds to the following change in the scheme of $(i, r(i))$ below Γ: α increases by Δ, and equivalently, the number of white-white conflicts increases by Δ, the number of black-white levels and white-black levels both decrease by Δ. Theorem 1 serves as an illustration of what we would like to show in general, and as the base step of an induction proof of the general case:

Theorem 1. *Let i, $r(i)$, α, etc. be as defined above. If for all $\alpha \in \mathcal{D}_F$, $pit(i, r(i)) = \emptyset$ in Φ_α, then*
(I.) *we may assume wlog. that $top(i, r(i))$ is the same node m in every Φ_α and*
(II.) $F(\alpha - \Delta) = F(\alpha) - \Delta$.

We will state a similar theorem about the general case, i.e. when for some α in Φ_α there is a $pit(i, r(i)) \neq \emptyset$. In the full version [3] we show that independent of $f(i)$, the solutions on $\langle i, r(i) \rangle$ have the same minimum-tree. The reason for this is roughly as follows: Within $\langle i, r(i) \rangle$ the loc-mins end in different (pairs of) rungs, and the sizes of these rungs are large compared to possible differences in $f(i)$. Consequently, the demands of nodes in $\langle i, r(i) \rangle$ differ too much to be alternative candidates for, e.g., $pit(i, r(i))$. We could not exclude a few potential exceptions to this rule (although we did not find any example for such an exception), but fortunately these might appear only on the first two floors of a minimum-tree. For example, we have more than one candidate for $pit(i, r(i))$ when $r(i)$ is compact. The notions of *irregular* vs. *regular* partial solutions on $\langle i, r(i) \rangle$ were introduced in order to handle the possible exceptional solutions separately [3].

Let $\mathcal{D}'_F \subseteq \mathcal{D}_F$ be the set of α for which a regular solution exists, and let Φ_α be regular for $\alpha \in \mathcal{D}'_F$. Now we state the generalized theorem:

Theorem 2. *Let i, $r(i)$, α, Φ_α, etc. and the partial instance on $\langle i, r(i) \rangle$ be as defined above.*

(I.) *Either* $m := top(i, r(i))$ *can be fixed to be the same node; or* $k := pit(i, r(i))$ *is the same, $L(k)$ is the same, resp. $R(k)$ is the same node in every* Φ_α ($\alpha \in \mathcal{D}'_F$).
(II.) $F(\alpha - \Delta) = F(\alpha) - \Delta$ *on* $\mathcal{D}'_F$.

Theorem 3 below characterizes the candidate nodes for $k, L(k)$, and $R(k)$. Recall that the potential finish time $f_\alpha(k)$ of any possible $k = pit(i, r(i))$ is determined by α (see (P3) and (P4)).

Theorem 3. *Suppose that the unique $pit(i, r(i))$ of Theorem 2 is known to be even. Let $i < c$ and $d < r(i)$ be the lowest even stairs s.t. $\exists \alpha$, $\exists k \in \langle c+2, d-2 \rangle$ so that k is even, and $f_\alpha(k) \in C \cap D$. Then $L(k) = c$, $R(k) = d$, and $k = pit(i, r(i))$ in any regular partial solution. Moreover, for all $\alpha \in \mathcal{D}'_F$ there is no even $k' \neq k$ s.t. $f_\alpha(k') \in C \cap D$. Similar statement holds for odd k.*

Corollary 4. *Let $x(i), ..., x(r(i))$ be a given (partial) instance. When restricted to regular solutions, there is at most one even and at most one odd node in $\langle i, r(i) \rangle$ to be a candidate for $pit(i, r(i))$, independent of α.*

4 The Algorithm

Let $x(1), x(2), ..., x(n)$ be an instance of the pSMC problem on the path (or cycle) $1, 2, ..., n$. Recall that $p = \max_i x(i)$. Theorem 2 allows us to get rid of the factor p that was there in the $\mathcal{O}(n^3 p)$ algorithm of [4]. By the following lemmas we can improve on the factor n^3 and obtain an algorithm that runs in $\mathcal{O}(n \cdot \min(n, \log p))$ time. The lemmas are quite intuitive: they reflect the fact that within one block the finish times exhibit exponential growth.

In a fixed schedule, let the **rungs of a loc-min** mean the maximal sets of contiguous black or contiguous white levels of the loc-min. Let $I_1, I_2, ..., I_\eta$ be the rungs of a loc-min i and $S_1, S_2, ..., S_\zeta$ ($\zeta \geq 0$) be the rungs of any number of consecutive stair-ups $i = s_0, s_1, s_2, ..., s_\zeta$ following i. Now all of these sets I_τ and S_ρ are disjoint and $[1, f(s_\zeta)] = \bigcup_{\tau=1}^{\eta} I_\tau \cup \bigcup_{\rho=1}^{\zeta} S_\rho$.

Lemma 1. *If η and ζ are numbers of rungs as described above, then $\eta + \zeta < 5 \log f(s_\zeta)$.*

Lemmas 2 and 3 imply that for given non-compact i, there is a constant number of candidate nodes for $r(i)$ and $pit(i, r(i))$:

Definition 9. *Let*
$\chi(i) := \min(x(i) + x(i+1), 2x(i));$
$\chi(r(i)) := \min(x(r(i)) + x(r(i) - 1), 2x(r(i))).$

Lemma 2. *If i is a loc-min, then $x(r(i)) \leq \chi(i)$.*

Lemma 3. *Let i be a non-compact loc-min. A constant $\mathcal{C} \leq 30$ exists, such that if i is even,*

(a.) there are $\leq \mathcal{C}$ odd nodes l, and $\leq \mathcal{C}$ even nodes k in $\langle i, r(i) \rangle$ s.t. $x(l) \leq \chi(i)$, resp. $x(k) \leq \chi(r(i))$.
(b.) there are $\leq \mathcal{C}$ even nodes $k \in \langle i, r(i) \rangle$, s.t. $x(k) \leq 2x(i)$.

We did not provide the definition of *irregular* partial solutions on $\langle i, r(i) \rangle$ in this short version. We will also omit the details concerning irregular solutions from the algorithm description. We include here two hints, why these potential exceptions to Theorem 2 do not blow up the running time. On the one hand, the following claim, together with Lemma 3, imply that possible $pit(i, r(i))$ nodes of irregular solutions can be found in constant time:

Claim 1. Suppose i is even, and $k = pit(i, r(i))$ in an irregular solution. If k is odd, then $x(k) \leq \chi(i)$; if k is even, then $x(k) \leq \chi(r(i))$.

On the other hand, recall that irregular partial solutions on $\langle i, r(i) \rangle$ may appear only if i is on $\leq$2nd floor of a minimum-tree. Although we have to deal with irregular solutions in the dynamic programming algorithm, fortunately they do not propagate: We have to take the minimum function of $F(\alpha)$ functions resulted by irregular solutions, which may be a non-linear function due to breaks or holes in the domain. However, we can forget these solutions as soon as we stepped down at least two floors in the minimum-tree. This implies, that we will never have to deal with more than $\approx \mathcal{C}^2$ breaks in the optimum function.

Proposition 1. *For any node i in a solution, unless i has compact neighbours on both sides, $f(i) \leq 4x(i)$.*

The algorithm has two phases. In Phase 1. the blocks, and the minimum-tree of each block are determined. In Phase 2. the schedule is determined.

Phase 1. We group the nodes into $\log p$ groups: nodes of demand in $[2^{t-1}, 2^t - 1]$ belong to group $\mathcal{G}_t$. First we traverse the path from right to left, and from a node k in $\mathcal{G}_t$ we set pointers to the nearest nodes of each of $\mathcal{G}_{t+2}, \mathcal{G}_{t+1}, \mathcal{G}_t, ..., \mathcal{G}_1$ to the right of k. After that we do the same from left to right. This preprocessing takes $\mathcal{O}(n \cdot \min(n, \log p))$ time.

We test for every node $1 \leq i \leq n$, if it can be a *non-compact* loc-min, and every potential $r(i)$ (and $\ell(i)$). Let i be a fixed even node. By Corollary 3 and Lemma 2, $r(i)$ must be an odd node s.t. $x(r(i)) \leq \chi(i)$. According to Lemma 3, we have to test at most $\mathcal{C}$ odd nodes of demand $\leq \chi(i)$, both on the left and on the right of i. However, we may stop earlier if we find a node of demand $< x(i)/4$. Because of the preprocessing, finding these $2\mathcal{C}$ candidate nodes for all i, takes overall $\mathcal{O}(n)$ time. We partition all these candidate $(i, r(i))$ and $(\ell(i), i)$ pairs into sets depending on their distances. In the testing process, we have to determine the optimum dynamically, and therefore we proceed from pairs of short distances to pairs of long ones. We start with the possible $r(i) = i + 3$ and $\ell(i) = i - 3$ values for every $i \in \langle 1, n \rangle$.

Suppose we have i and $r(i)$ fixed. Let α and $F(\alpha)$ be as defined in Section 3. Now we have to compute $F(\alpha)$ dynamically, for every potential $(i, r(i))$ (and $(\ell(i), i)$) pairs. When we set the domain $\mathcal{D}'_F$ of the possible α values, we must take care that each α and $F(\alpha)$ can be realized on $\langle i, r(i)\rangle$, and that *all* realizable α are considered. It is straightforward to show that the nonnegative part of $[x(i) + x(r(i)) - \Gamma, \min(x(r(i)), x(i) - 1)]$ is an enclosing interval of all α. Later we cut this domain as required by the connecting step of the dynamic algorithm.

Next, we search for $k = pit(i, r(i))$ or $m = top(i, r(i))$. Corollary 4 says that there is not more than one even and one odd potential $pit(i, r(i))$. As for the number of possible $top(i, r(i))$ nodes, there are obviously always ≤ 2 even and ≤ 2 odd candidates, since $x(m-1) > x(m-3) > \ldots$, resp. $x(m+1) > x(m+3) > \ldots$, must hold. The potential k or m nodes we find as described in Claim 2:

Claim 2. For given i and $r(i)$ it takes $\mathcal{O}(\min(n, \log p))$ steps to find the ≤ 2 candidates for $pit(i, r(i))$ or ≤ 4 candidates for $top(i, r(i))$ in a regular solution.

Sketch of proof. We proceed upwards along the stairs on the two sides of $\langle i, r(i)\rangle$. The potential finish time $f(k)$ of a loc-min k, is calculated by summing up the number of (potential) black levels of k: say k is even, these are the levels, where on at least one side there is an even rung (see (P3)). Finally, $f(k)$ is the level where the sum totals to $x(k)$. By Theorem 3, the even candidate is the node k of the least possible $f(k)$. Consequently, $x(k)$ is minimum over all demands *inside* $\langle L(k), R(k)\rangle$. The starting value of the summation is the number of black levels below Γ, this depends on α. Large α values yield small $f(k)$ values, and we search for minimum $f(k)$, so we fix $\alpha := \min(x(r(i)), x(i) - 1)$.

In order to find potential nodes k of minimum $x(k)$, we can check the linked lists of each group $\mathcal{G}_t$, starting nodes might be the stairs on the left. Applying Lemma 3 (b.) to k implies that in the lowest group that is nonempty between stairs on the two sides, there are at most $2\mathcal{C}$ nodes. So, by Lemma 1 we find the candidate k in $\mathcal{O}(\min(n, \log p))$ steps. If stairs on the two sides meet before we find a possible $pit(i, r(i))$, then the topmost stairs are candidates for $top(i, r(i))$.

If we have found an appropriate node k of minimum $x(k)$, we have to restrict the domain of α values to those for which $f(k)$ ends in an even rung on both sides. Should there be two different *pairs* of even rungs, where $f(k)$ might end, the higher pair does not provide optimal solutions by Theorem 3. Thus, the restricted set of α values is an interval. □

First suppose that the process of Claim 2 yields possible $top(i, r(i))$ nodes. Such $(i, r(i))$ pairs constitute the starting steps in the dynamic process. It is trivial to check, which subinterval of the starting domain of α values allows that m is a local maximum (it is easy to show that for $m = top(i, r(i))$ all of the starting α are realizable). Moreover, each potential m provides a function $F_m(\alpha) = \alpha + C_m$, where C_m is a constant, computable in $\mathcal{O}(\min(n, \log p))$ time along with the procedure of Claim 2. Finally, $F(\alpha) = \alpha + C$ where $C = \min_m C_m$.

Second, suppose we have a candidate $k = pit(i, r(i))$. How do we compute the optimum function $F_k(\alpha)$ for this k? Claim 2 settled a restricted interval domain for α s.t. $f(k)$ ends in a fixed pair of even rungs. We only search for "hereditarily" regular solutions. In this case the optimum function on $\langle \ell(k) + 1, k - 1\rangle$ has the form $F'(\beta) = \beta + C'$ and the optimum function on $\langle k + 1, r(k) - 1\rangle$ has the form $F''(\gamma) = \gamma + C''$, where β and γ are the numbers of black-black conflicts of

$(\ell(k), k)$ resp. $(k, r(k))$ below $f(k)$. These functions we computed earlier. Now we further restrict the domain of α so that the β and γ values stay within the domains of F', resp. F''. Since we take intersection of intervals, we have the following:

(1) the restricted domain of $F(\alpha)$ is an interval $[\alpha_0, \alpha_0 + \Delta]$.
(2) the corresponding finish times range from $f_0(k)$ to $f_0(k) - \Delta$.
(3) β ranges from β_0, to $\beta_0 + \Delta$ and γ ranges from γ_0, to $\gamma_0 + \Delta$.

The stairs $\langle i+1, \ell(k) \rangle$ and $\langle r(k), r(i) - 1 \rangle$ have constant finish times, that total up to some S (see Corollary 1).

So, we have $F_k(\alpha_0 + \delta) = F'(\beta_0 + \delta) + f_0(k) - \delta + F''(\gamma_0 + \delta) + S = \beta_0 + \delta + C' + f_0(k) - \delta + \gamma_0 + \delta + C'' + S = \alpha_0 + \delta + C_k$. That is, $F_k(\alpha) = \alpha + C_k$, where $C_k := \beta_0 + C' + f_0(k) + \gamma_0 + C'' + S - \alpha_0$.

The even k, and the odd l candidate for $pit(i, r(i))$ yield the optimum functions $F_k(\alpha)$ and $F_l(\alpha)$. Theorem 2 makes sure, that the function of smaller constant (C_k or C_l) is the better one overall.

The last steps of this dynamic process are calculating the optimum on $\langle i, r(i) \rangle$, ($\langle \ell(i), i \rangle$) when i is compact. We also include here the case when $\langle i, j \rangle$ is a block and $x(i) = x(j)$. Suppose that node i is fixed, then node $r(i)$ (in this broader sense) may be a node of arbitrary parity, s.t. $x(r(i)) \leq x(i)$. By Lemma 3 (a.), it is enough to test $\leq 2\mathcal{C}$ such nodes of both parity in both directions to be $r(i)$.

A natural variant of Theorem 3 holds for compact i. The 2 potential k (or 4 potential $m = top(i, r(i))$) nodes can be found as described in Claim 2. Since (P4) does not apply to such a k, $f(i)$ does not determine $f(k)$. Let k be fixed and even, and let $f_0(k)$ be the potential finish time of k as determined by $f(i) = x(i)$ like in Claim 2. The starting domain of possible $f(k)$ values is $f(k) \in [x(k), f_0(k)]$. This domain is trimmed further so that $f(k)$ ends in rungs of some even $L(k)$ and $R(k)$. (There are ≤ 3 different pairs of $L(k)$ and $R(k)$ to be considered.) We fix $L(k)$ and $R(k)$ as well. Now the possible $f(k)$ values constitute an interval $\mathcal{I}$.

We have the optimum functions $F'_k(\beta)$ on $\langle \ell(k) + 1, k - 1 \rangle$ and $F''_k(\gamma)$ on $\langle k+1, r(k)-1 \rangle$. The domains of these two functions correspond to two intervals of possible $f(k)$ values. We take the intersection of these two intervals and $\mathcal{I}$. The maximum possible $f(k)$ will provide the optimum sum of finish times on $\langle i+1, j \rangle$ for this $k, L(k)$ and $R(k)$.

We conclude Phase 1. with searching for (potential) blocks (i, j) (resp. (j, i)) assuming that $f(i) \geq f(j)$. Let i be fixed. Since $x(r(i)) = x(i)$ was allowed, we may assume that $r(i) \leq j$. Claim 3 shows that for a fixed $(i, r(i))$ pair, any possible j is within the first $< 2 \log p$ nodes to the right of $r(i)$.

Claim 3. Let (i, j) be a block in a solution Φ, s.t. $x(i) > x(j)$. Let $r(i) = s_\eta < \ldots < s_2 < s_1 < j$ be the series of consecutive stair-downs leading from $r(i)$ to j. Then $\forall \tau \; x(s_{\tau+2}) \geq 2x(s_\tau)$.

Let $F(i, j)$ denote the optimum sum of finish times on $\langle i+1, j \rangle$. Calculating $F(i, j)$ is straightforward from the optimum on $\langle i, r(i) \rangle$. We can define a graph $G(V, E)$ on the nodes $V := \{0, 1, \ldots, n, n+1\}$ having $\leq 8 \cdot \mathcal{C} \cdot n \min(2 \log p, n)$ edges. An edge leads between i and j if (i, j) is a potential block. The edge can be weighted by $F(i,j)$. Finally, we determine the optimum and the blocks on $\langle 0, n+1 \rangle$ dynamically: if i is compact, then $F(i)$, the optimum on $\langle 0, i \rangle$ is determined by $F(i) = \min\{F(i-t) + F(i-t, i) \mid (i-t, i) \in E\}$. The running time is proportional to the number of edges in G.

Phase 2. Now we have the blocks, and the minimum-tree for each block in an optimal solution. Let (i, j) be a block of t nodes and $x(i) > x(j)$. First we schedule the stairs $\langle r(i), j \rangle$. Next we schedule $k = pit(i, r(i))$ and the stairs $\langle i, \ell(k) \rangle$ and $\langle r(k), r(i) \rangle$, and so on. (The root k of the minimum-tree has black levels conflicting with both sides of the block (determined by $f(k)$). These levels can be arbitrary, e.g. the lowest possible such levels.)

The scheduling procedure is basically the same as described in Claim 2. The preemption levels in $\Phi(k)$ are obtained by way of merging the finish-times of stairs on the two sides. Lemma 1 implies that a node has $\mathcal{O}(\min(t, \log p))$ preemption levels. Scheduling the whole block takes $\mathcal{O}(\min(t^2, t \log p))$ time.

5 Concluding Remarks

Since on cycles there is always at least one compact node i, the overlapping path $i, i+1, ..., i+n, i+n+1 = i(\bmod n)$ constitutes one or more blocks. Let's take the first two nonempty groups of $\mathcal{G}_1, ..., \mathcal{G}_t$. By Lemma 3, there must be a compact node among the first $\leq 2\mathcal{C}$ nodes of these groups. Testing these possible starting nodes one by one, yields the same $\mathcal{O}(n \cdot \min(n, \log p))$ time bound.

Due to the observation, that the *total* number of different preemption levels is $\leq 2n$ (see [4]), the lower bound for pSMC on paths is $\mathcal{O}(n)$. If we may assign a certain preemption level to a whole set of nodes at once (e.g., in $\langle i, m \rangle$ the odd jobs should be started, and even jobs should be stopped at time a), then it is not excluded that by way of a better analysis or an improved algorithm the time-bound could be reduced. It could also be investigated, what our results imply for conflict graphs with just one, or very few nodes of higher degree.

Acknowledgements. I would like to thank Katalin Friedl and Ulrich Meyer for their remarks and suggestions during the writing of this paper.

References

1. A. Bar-Noy and G. Kortsarz. The minimum color-sum of bipartite graphs. *Journal of Algorithms*, 28:339–365, 1998.
2. M. M. Halldórsson and G. Kortsarz. Tools for multicoloring with applications to planar graphs and partial k-trees. *Journal of Algorithms*, 42(2):334–366, 2002.
3. A. Kovács. Fast preemtive sum-multicoloring on paths. Extended version: http://www.mpi-sb.mpg.de/~panni/pSMC_long.ps.
4. A. Kovács. Sum-multicoloring on paths. In *Proc. 21st Ann. Symp. on Theo. Aspects of Comp. Sci. (STACS)*, volume 2996 of *LNCS*, pages 68–80. Springer, 2004.

5. E. Kubicka. *The Chromatic Sum of a Graph*. PhD thesis, Western Michigan University, 1989.
6. D. Marx. The complexity of tree multicolorings. In *Proc. 27th Intl. Symp. Math. Found. Comput. Sci. (MFCS)*, LNCS. Springer, 2002.
7. A. Bar-Noy M. M. Halldórsson G. Kortsarz H. Shachnai and R. Salman. Sum multicoloring of graphs. *Journal of Algorithms*, 37:422–450, 2000.
8. M. M. Halldórsson G. Kortsarz A. Proskurowski R. Salman H. Shachnai and J. A. Telle. Multi-coloring trees. *Information and Computation*, 180(2):113–129, 2002.

The Generalized Deadlock Resolution Problem

Kamal Jain[1], MohammadTaghi Hajiaghayi[2], and Kunal Talwar[1]

[1] Microsoft Research
{kamalj, kunal}@microsoft.com
[2] MIT Computer Science and Artificial Intelligence Laboratory,
hajiagha@csail.mit.edu

Abstract. In this paper we initiate the study of the AND-OR directed feedback vertex set problem from the viewpoint of approximation algorithms. This AND-OR feedback vertex set problem is motivated by a practical deadlock resolution problem that appears in the development of distributed database systems[1]. This problem also turns out be a natural generalization of the directed feedback vertex set problem. Awerbuch and Micali [1] gave a polynomial time algorithm to find a minimal solution for this problem. Unfortunately, a minimal solution can be arbitrarily more expensive than the minimum cost solution. We show that finding the minimum cost solution is as hard as the directed Steiner tree problem (and thus $\Omega(log^2 n)$ hard to approximate). On the positive side, we give algorithms which work well when the number of writers (AND nodes) or the number of readers (OR nodes) are small.

We also consider a variant that we call *permanent deadlock resolution* where we cannot specify an execution order for the surviving processes; they should get completed even if they were scheduled adversarially. When all processes are writers (AND nodes), we give an $O(\log n \log \log n)$ approximation for this problem.

Finally we give an LP-rounding approach and discuss some other natural variants.

1 Introduction

One of the best ways to understand deadlocks in databases is the dining philosophers problem. Say there are five philosophers sitting on a circular table to eat spaghetti, with a fork between every two of them. Each philosopher needs two forks to eat. But everybody grabs the fork on the right, hence everybody has one fork and waiting for another to be freed. This wait will be never ending unless one of the philosophers gave up and freed up his fork. This never ending is an example of a *deadlock*. Picking up a philosopher who can give up on eating the spaghetti is an example of *deadlock resolution*. Now suppose that these philosophers have different likings for the spaghetti and hence different inherent cost of giving up eating it. In this case we want to pick the philosopher who likes spaghetti the least. This is called the *minimum cost deadlock resolution* problem.

In databases, philosophers correspond to independent agents e.g., transactions and processes. Forks correspond to shared resources, e.g., shared memory. Eating spaghetti

---
[1] Thanks to Ondrej Such from Microsoft for asking an algorithm for this problem.

corresponds to actions which these independent agents want to perform on the shared resources e.g., reading or writing a memory location. So in general besides asking for two forks these philosophers may ask for two spoons too, while they have grabbed only one each. These spoons and forks can be of different kinds (e.g., plastic or metallic). In general demands for resources can be very complicated and it can be represented by a monotonic binary function, called *demand function*. A demand function takes a vector of resources as an input and outputs whether it can satisfy the demand or not.

When a process does not get all the resources to satisfy its demand then it has to wait. Like any other protocol involving waiting, there is a risk of deadlock. There are ways to avoid deadlock, like putting a total order on all the resources and telling to the users to ask them in the same order. In big or distributed databases, such solutions are difficult to implement. Moreover such a solution works when the demand functions consist of only ANDs. In essence deadlocks do happen and they need to be resolved at a small cost. In practice one of the convenient solution is to time out on wait, i.e., if it takes too long for a transaction to acquire further resources then it aborts and frees up the resources held so far. This solution does not have any guarantee on the cost incurred. For notational convenience, aborting a transaction will also be referred as *killing* it. We assume that there is an associated cost of killing a process (this cost can also be the cost of restarting it). The cost of a solution is the total cost of all the processes killed. For the minimum cost deadlock resolution problem we want to kill the least expensive set of processes to resolve the deadlock.

An instance of a generalized deadlock detection problem is captured by a *waits-for-graph* (WFG) on transactions. An old survey by Knapp [17] mentions many relevant models of WFG graphs. In the AND model, formally defined by Chandy and Misra [5], transactions are permitted to request a set of resources. A transaction is blocked until it gets all the resources it has requested. In the OR model, formally defined by Chandy et al. [6], a request for numerous resources are satisfied by granting any requested resource, such as satisfying a read request for a replicated data item by reading any copy of it. In a more generalized AND-OR model, defined by Gray et al. [12] and Herman et al. [15], requests of both kinds are permitted. A node making an AND request is called an AND node and a node making an OR request is called an OR node. An advantage of using both these kinds of nodes is that one can express[2] arbitrary demand functions e.g., if a philosopher wants any one fork and any one spoon then we can create two sub-agents for this philosopher, one responsible for getting a fork and the other for getting a spoon. This philosopher then becomes an AND node and the two sub-agents become two OR nodes. From the perspective of algorithm design, detecting deadlocks in all these models is not a difficult task (see e.g. [11,21,23]). The difficult task is to resolve it once detected and that too at a minimum cost (for some heuristics and surveys on the generalized AND-OR model see e.g. [1,3,4,14]). In the next section we formally model the problem as an AND-OR directed feedback vertex set problem.

Often it may not be possible for the deadlock resolving algorithm to specify a schedule for the remaining processes, and when the cost of calling the deadlock resolution algorithm is large (as one would expect in a distributed setting), we would like that

---

[2] This expression may be of exponential size. See [17] for more models of waits-for-graphs.

no matter in what order the surviving transactions are scheduled, they do not deadlock again. This motivates the *permanent deadlock resolution* problem. For the case when the transactions are all writers (the AND only case), we show a polynomial-time approximation algorithm for the problem.

1.1 Our Results

When all the nodes are OR nodes then the problem can be solved in polynomial time via strongly connected components decomposition. But the problem quickly becomes at least as hard as the set-cover problem even in the presence of a single AND node. Our reduction has deadlock cycles of length 3 capturing the special case mentioned by Jim Gray, who says in practice deadlocks happen because of cycles of length 2 or 3. We give an $O(n_a \log(n_o))$ factor approximation algorithm, where n_o is the number of OR nodes and n_a is the number of AND nodes. On the other hand if all the nodes are AND nodes, the problem is the well-studied directed feedback vertex set problem. There are approximation algorithms with polylog approximation factor for this problem due to Leighton-Rao [18] and Seymour [22]. We generalize those algorithms to destroy all the *handles* at a pivot vertex. We define handles later in the paper and there we also show that destroying handles is a more general problem than destroying cycles. We use this generalization as a subroutine to develop an $O(n_o \log(n_a) \log \log(n_a))$ factor approximation algorithm.

From the hardness point of view, we show that the problem is as hard as the directed Steiner tree problem, which was shown to be hard to approximate better than a factor of $O(log^{2-\epsilon} n)$ by Halperin and Krauthgamer [13], and has no known polynomial time polylogarithmic approximation algorithm. One difficulty in designing an approximation algorithm for our problem is that we do not know any good LP relaxation. The natural LP relaxation itself is at least as hard as the directed Steiner tree problem, even for the case of one OR node. It will be interesting to interpret our algorithms in terms of LP rounding. We do that in case there is one (or a constant number of) OR nodes. The size of this LP is exponential in the number of OR nodes.

For the *permanent deadlock resolution* problem, we show that the case with only AND nodes is reducible to the feedback vertex set problem in mixed graphs. Acyclicity implies schedulability for both undirected and directed graphs - acyclic undirected graphs have leaves and acyclic directed graphs have sinks. Corresponding theorem for mixed graphs is not clear. We develop a corresponding theorem for bipartite mixed graphs. This leads to an $O(\log n \log \log n)$ approximation algorithm for this problem. We leave open the approximability of this problem in the general case.

This problem was also studied in theoretical computer science by Awerbuch and Micali [1]. In their paper, they mentioned that the ideal goal is to kill a set of processes with minimum cost, but the problem is a generalization of feedback vertex set and seems very hard. Thus they gave a distributed algorithm for finding a minimal solution. Unfortunately, a minimal solution can be arbitrarily more expensive than the minimum cost solution. We study this problem from approximation algorithm point of view. We are excited with the fact that the problem has such a rich mathematical structure. It allows use of many results, which were discovered after the paper due to Awerbuch and Micali. In this paper we try to find a proper place for the problem in the vast area of approximation algorithms. We show that this problem blends naturally with feedback vertex and arc set problems. From hardness point of view it blends naturally with the directed Steiner tree and set cover problems. In the discussion section we mention an alternative approach to design approximation algorithms for the directed Steiner tree problem. This approach is suggested by interpreting our algorithm for the case of one OR node in terms of linear programming. This approach does not seem to be based upon the standard LP for the directed Steiner tree problem, which some researchers suspect to have integrality ratio worse than polylogarithmic.

Due to interest of space, we omit several proofs in this extended abstract and defer them to the journal version of this paper.

2 Problem Definition and Preliminary Results

All the graphs in this paper are directed without loops or multiple edges, unless stated otherwise. Our graph terminology is as follows. A graph G is represented by $G = (V, E)$, where V (or $V(G)$) is the set of vertices (or nodes) and E (or $E(G)$) is the set of edges. We denote an edge e from u to v by (u, v), and we call it an *outgoing* edge for u and an *incoming* edge for v. We say node u can *reach* node v (or equivalently v is *reachable* from u) if there is a path from u to v in the graph. We shall use the notation $u \leadsto v$ to denote that v is reachable from u. We define n to be the number of vertices of a graph when this is clear from context. We denote the maximum out-degree by Δ_{out} and the maximum in-degree by Δ_{in}. We assume that the node set V is partitioned into two sets V_a and V_o. Nodes in V_a and V_o are referred to as AND nodes and OR nodes respectively. We let $n_a = |V_a|$ and $n_o = |V_o|$. With this terminology we now define the wait-for-graphs (WFG).

Each node of a wait-for-graph, $G = (V, E)$, represents a transaction. An edge (u, v) denotes that transaction u has made a request for a resource currently held by transaction v. There are two kinds of nodes. An AND node represents a transaction which has made an AND request on a set of resources, which are held by other transactions. An OR node represents a transaction which has made an OR request on a set of resources. Without loss of generality we assume that a transaction is allowed to make only one request. If a transaction makes multiple requests then we can create a sub-transaction for each request and put the necessary dependency edges. Each transaction has an associated weight. We denote the weight of a transaction u by w_u.

An AND transaction can be scheduled if it gets all the resources it has requested. An OR transaction can be scheduled if it gets at least one of the resources it has requested. Once a transaction is scheduled, it gives up all its locks, potentially allowing other processes to get scheduled. A wait-for-graph is called *deadlock free* if there exist an ordering of the transactions in which they can be executed successfully. If no such ordering exist then we say that the graph has a deadlock. The *minimum cost generalized deadlock resolution problem (GDR)* is to kill the minimum weight set of transactions to free up the resources held by them so that the remaining transactions are deadlock free. In other words, there exists an order on the remaining transactions so that for each AND transaction, each of its children is either killed or can be completed before it and for each OR transaction at least one of its children is either killed or can be completed before it.

2.1 Some Special Cases

We show some simple propositions which give us some intuition about the problem.

Proposition 1. *The GDR problem when there is no OR node has an approximation algorithm with ratio $O(\log n \log \log n)$.*

Proposition 2. *The GDR problem with all OR nodes can be solved in polynomial time.*

In fact, we can strengthen Proposition 2 as follows:

Proposition 3. *The GDR problem, when the reachability graph on the AND nodes is a directed acyclic graph, can be solved in polynomial time.*

Proposition 4. *The GDR problem with uniform weights and $O(\log n)$ AND nodes can be solved in polynomial time.*

Using ideas of Propositions 3 and 4, we can show the following theorem.

Theorem 1. *The GDR problem with uniform weights and n_a AND nodes has an $O(n_a)$-approximation algorithm.*

3 Hardness Results and Natural LP

In this section, we consider the hardness of the GDR problem. First, we show a simple approximation preserving reduction from the set cover problem to this problem. Recall that the set cover problem is to find a minimum collection $\mathcal{C}$ of sets from a family $\mathcal{F} \subseteq 2^U$, such that $\mathcal{C}$ covers U, i.e. $\cup_{S \in \mathcal{C}} S = U$. From the results of Lund and Yannakakis [20] and Feige [10], it follows that no polynomial time algorithm approximates the set cover problem better than a factor of $\ln n$ unless $NP \subseteq \text{DTIME}(n^{\log \log n})$. Our reduction then implies a similar hardness for the GDR problem. There is no similar inapproximability result known for the directed feedback vertex set problem.

Theorem 2. *There exists an approximation preserving reduction from (unweighted) set cover to GDR with only one AND node.*

It is worth mentioning that in the reduction of Theorem 2, there is only one AND node whose weight is $m + 1$ and the rest of the vertices are OR nodes with weight one. Moreover, the one AND node of high weight can be replaced by $m + 1$ AND nodes of unit weight placed "in parallel". Thus the uniform weight case is also hard to approximate better than a factor of $\Omega(\log n)$.

Now the question is that whether it is possible to get a better inapproximability result. To answer this question, we use a recent result of Halperin and Krauthgamer [13] on the inapproximability of the *directed Steiner tree* problem. In the directed Steiner tree problem, given a directed graph $G = (V, E)$, a *root* $r \in V$ and a set of *terminals* $T \in V$, our goal is to find a minimum subset $E' \subseteq E$ such that in graph $G' = (V, E')$ there is a path from r to every $t \in T$. Halperin and Krauthgamer [13] show that the *directed Steiner tree* problem is hard to approximate better than a factor of $\Omega(\log^2 n)$, unless

$NP \subseteq \text{ZTIME}(n^{\text{polylog } n})$. So far, no polynomial-time polylogarithmic approximation algorithm is known for this problem. We show a similar non-approximability result in Theorem 3 for GDR by giving an approximation preserving reduction from directed Steiner tree.

Theorem 3. *There exists an approximation preserving reduction from directed Steiner tree to GDR.*

Proof. We consider an instance of directed Steiner tree given by a directed graph $G = (V, E)$, a set of terminals $T \subseteq V$ and a root node $r \in V$. The goal is to find a minimum cost subset E' of edges containing a path from r to every terminal $t \in T$. The reduction is as follows. For each vertex $v \in V - \{r\}$, we create an OR node v of weight ∞^3 in our GDR instance. For r, we create an OR node r of weight zero. In addition, we have an AND node a of weight ∞ which has an edge (a, t) for each $t \in T$ and an edge (v, a) for each $v \in V$. For each edge $e \in E$, we put an AND-OR gadget shown in Figure 1, with the weight of each node as shown in the figure. Recall that a is the global AND node introduced before and o_e and a_e are new OR and AND nodes corresponding to e respectively. Intuitively, using an edge e in the Steiner tree corresponds to killing the OR node o_e in this gadget.

Next we show that the cost of an optimum Steiner tree is equal to the minimum cost of nodes to be killed such that the remaining graph is deadlock-free. First consider a Steiner tree S in G. We kill all OR nodes corresponding to edges in S. For each edge $e = (u, v) \in S$, killing o_e allows v to be complete after u. Thus, first complete node r, then complete nodes according to the directed Steiner tree. Since the Steiner tree solution contains a path to each terminal, we can complete all terminals. Now, after completing all terminals, we can complete the global AND node a and then complete every other node in the graph.

On the other hand, since the only nodes with finite weight are the OR nodes corresponding to edges and the node corresponding to root r, any feasible solution of finite weight for GDR kills only such nodes. It is easy to check that the set of edges for which the OR nodes are killed contain a directed Steiner tree. □

Again, we might replace each node of weight ∞ with several nodes of unit weight, say $|E(G)|$, in order to reduce the directed Steiner tree problem to the uniform weighted case.

3.1 Natural LP and Hardness

We end this section by considering a natural LP for the GDR problem, which is a generalization of the LP for feedback vertex set (see e.g. [9]). We say a set of nodes H forms a *Minimal Deadlocked Structure (MDS)* if

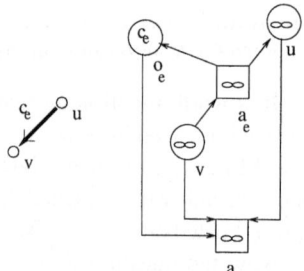

Fig. 1. Edge $e = (u, v)$ in graph G and its AND-OR gadget in the new instance of GDR

---

[3] As usual, the ∞ weights can be replaced by a (polynomially) large weight.

1. For any OR node $u \in H$, all its outneighbors are in H.
2. For any AND node $u \in H$, at least one of its outneighbors is in H.
3. H is minimal[4] amongst sets satisfying (1) and (2).

We now write a linear program (called *LP 1*) is as follows: minimize $\sum_{v \in V} w_v x_v$ such that $\sum_{v \in H} x_v \geq 1$ for any *MDS* H and $x_v \geq 0$ for all $v \in V$

Clearly an integral solution to this linear program is a feasible solution to the underlying GDR instance and hence this is a relaxation. However, this linear program can potentially have exponentially many constraints. Note that if the graph G does not have any OR node, MDSs are exactly the minimal directed cycles and our LP is the same as the LP considered in [18,22,9] for applying region growing techniques for the feedback vertex set problem. In this special case of feedback vertex set, this LP has a simple separation oracle which enables us to solve it using Ellipsoid method. However, we now show that even the separation oracle for LP 1 is as hard as the directed Steiner tree problem.

Theorem 4. *The separation oracle for LP 1 is as hard as solving directed Steiner tree.*

Proof. A separation oracle for LP 1 solves the following problem: given a vector $\vec{x}$, is there an MDS H for which $\sum_{v \in H} x_v < 1$. We shall reduce the directed Steiner tree problem to this problem

We consider an instance of directed Steiner tree: given a root r and a set of terminals T in a directed graph $G = (V, E)$, is there is Steiner tree of weight at most 1 (by scaling). Without loss of generality we assume G is a directed acyclic graph (DAG), since the directed Steiner tree problem on DAGs is as hard as the one on general directed graphs (see e.g. [7]). Also without loss of generality assume we have weights on vertices instead of edges (again the two problems are equivalent). Now we are ready to demonstrate the reduction. For each vertex $v \in V$, we place an AND node v with x_v equal to its weight in the Steiner instance. For each edge (u, v) in G, we place an edge (v, u) in our new graph. In addition, we add an OR node with $x_o = 0$ which has an outgoing edge (o, t) for each terminal $t \in T$ and an incoming edge (r, o) (r is the root node). Call the new graph G'. It is easy to check that $H \cup \{o\}$ is an MDS in G' if and only if H is a directed Steiner tree in G. Hence the claim follows. □

As shown by Jain et. al. [16], for these kinds of problems optimizing LP 1 is equivalent to solving the separation oracle problem. Furthermore, these reductions are approximation preserving. Thus if we can optimize LP 1 within some factor then we can solve its separation oracle for the same factor. Hence by Theorem 4, we can solve the directed Steiner tree problem within the same factor.

Corollary 1. *Optimizing LP 1 is at least as hard as the directed Steiner tree problem.*

Finally, notice that finding the integrality gap of LP 1 is an interesting open problem.

---

[4] With respect to set inclusion.

4 Approximation Algorithms

In this section, we give an $O(\min\{n_a \log n, n_o \log n \log \log n\})$ algorithm for this problem, where n_a is the number of AND nodes and n_o is the number of OR nodes in the instance. Thus, when either of n_a or n_o is small, the problem is well approximable.

In subsection 4.1 we show how to use region growing to solve a slight generalization of feedback vertex set. We use this to get an $O(n_o \log n \log \log n)$ algorithm in subsection 4.2. In subsection 4.3, we give an $O(n_a \log n)$ approximation algorithm for the problem. Thus the better of these two algorithms gives the claimed performance guarantee.

4.1 Handle Removal Algorithm

In this section, we consider the following handle removal problem which plays an important role in the algorithm for the case of few OR nodes (see Section 4.2): Given a directed graph G, and a designated vertex r, delete the smallest number (weight) of vertices such that the remaining graph has no cycles reachable from r. For ease of exposition, we shall replace each vertex by a pair of vertices joined by an edge, and transfer the weight to this edge. The edges in the original graph are given an infinite weight. The problem then reduces to finding the smallest cost set of edges whose removal eliminates all cycles reachable from r.

We shall write this problem as an integer program, and consider its linear programming relaxation. We first formally define a *handle*.

Definition 1. *Let $H = (r = u_0, u_1, \ldots, u_k)$ be a simple path in G. We call H a* handle *if for some $p : 0 \leq p < k$, there is an edge (u_k, u_p). We refer to u_p as the* pivot *of the handle H. The edges on the path along with the edge $\{u_k, u_p\}$ constitute the edges of the handle.*

Let $\mathcal{H}$ be the set of all handles in G. We can write the following linear programming relaxation for this problem: minimize $\sum_{e \in E} w_e x_e$ such that $\sum_{e \in H} x_e \geq 1$ for all $H \in \mathcal{H}$ and $x_e \geq 0$ for all $e \in E$.

Note that the above linear program has an exponential number of constraints. The separation oracle for this LP requires us to find a violated handle in a given fractional solution. Note that we can find in polynomial time, for each $u \in V$, the smallest cycle passing through u, and the shortest path from r to u. The shortest handle in the graph is then just the minimum, over all u, of the sum of the above two quantities. Thus the LP has a polynomial time separation oracle, and hence can be solved by Ellipsoid method.

Given a solution to this linear program, we shall now argue that the techniques used by Seymour [22] and Even et.al. [9] for the feedback arc set problem apply here to give an $O(\log n \log \log n)$-approximation algorithm to the problem. Given a graph G, and a non negative length function x_e on the edges, we can define the shortest path function d_x on the vertices of G. A modification of the algorithm of Seymour implies the following theorem:

Theorem 5. *Given a weighted graph G, a special vertex r and non-negative length function x_e, let $W = \sum_e w_e x_e$. There exists a set of edges C such that:*

- $\sum_{e \in C} w_e \leq O(\log n \log \log n) \cdot W$
- For any vertex v with $d_x(r, v) \geq \frac{1}{4}$, C is an r-v cut.
- For any pair of vertices u and v such that $d_x(u, v) \geq \frac{1}{4}$, C contains either a u-v cut or a v-u cut.

We now argue that the rounding described in the theorem applied to a feasible LP solution, gives a feasible solution to the handle removal problem.

Claim. Let x be a feasible solution to the linear program above and d be the shortest path function defined accordingly. Then for any handle H with pivot u, either $d(r, u) \geq \frac{1}{4}$, Or there exists $v \in H$ such that $d(u, v) \geq \frac{1}{4}$.

From the above claim, and theorem 5, it follows that the handle removal problem is approximable within a factor of $O(\log n \log \log n)$.

4.2 Few OR Nodes Algorithm

Using the algorithm for the handle removal problem in section 4.1, we are now ready to prove the following theorem.

Theorem 6. *There is an $O(n_o \log(n_a) \log \log(n_a))$-approximation algorithm for the Generalized Deadlock Resolution Problem.*

Proof. We first assume that the OR nodes in the graph have infinite cost, and thus are all scheduled. We shall give an $O(\alpha n_o)$ solution where α is the approximability of the handle removal problem.

Let u be the first OR node to be scheduled. Since this node is scheduled, one of its outneighbors, say v, is killed/scheduled before any other OR node. Since no cycle of AND nodes can be scheduled, no such cycle reachable from v survives in the optimum solution. OPT thus includes a solution to the handle removal problem with root v. Let O_v be the optimum of the handle removal problem with root v, when all OR nodes are removed. Hence $OPT \geq \min_{u \in V_O} \min_{v:(u,v) \in E} O_v$.

Our algorithm is as follows. Using an α-approximation algorithm for the handle removal problem, we compute solutions to handle removal problems rooted at $\{v : (u, v) \in E, u \in V_O\}$. We pick the cheapest of these and kill the nodes in this solution. The cost of killing these nodes is at most αOPT. Now the OR node u can be scheduled, and consequently some more nodes can be scheduled. We remove all such nodes along with their incoming edges, and recur. In the base case, when there are no OR nodes, we have the feedback vertex set problem, which is also approximable within α (by a simple reduction to the handle removal problem). Thus we get obtain $(n_o + 1)\alpha$ approximation.

We now show how to remove the assumption about OR node removal. To each OR node u in the graph, we add a new outneighbor u_a which is an AND node with cost equal to the original OR node. We add another AND node v of infinite cost to the graph, with edges to all the original vertices of the graph. Finally, we add an edge from u_a to the vertex v, and increase the costs of all OR nodes to infinity.

For any solution to the original instance that kills an OR node u, we can get a solution to the new instance by killing u_a. This lets us schedule u instead of killing u. Moreover, after scheduling/killing all original nodes, the node v can be scheduled, after which any unkilled u_a can be scheduled. Finally, the cost of the new solution is the same as the original one.

A solution to the new instance immediately gives a solution to the original instance: kill OR node u whenever the new solution killed u_a. It is easy to see that this transformation preserves feasibility and cost. □

4.3 Few AND Nodes Algorithm

In this section, we present an $O(n_a \log n)$-approximation algorithm for this problem. We note that in the reduction of set cover to generalized deadlock resolution (mentioned in Theorem 2), we have only one AND node and thus our result is tight in this case. However, in the reduction of directed Steiner tree to this problem, the number of AND nodes is linear and the best non-approximability result is in $\Omega(\log^2 n)$.

The algorithm is as follows. We start with the original graph G and in each iteration we update it. More precisely, if in an iteration graph G does not have any AND node, we can obtain the optimal solution for G by the procedure mentioned in Proposition 2 (and thus we stop). Otherwise, for each AND node a whose outgoing edges are $(a, c_1), (a, c_2), \cdots, (a, c_{\Delta_{out}})$ in graph G and all c_i's, $1 \leq i \leq \Delta_{out}$, are OR nodes, we construct the following hitting set instance (note that the hitting set problem is the dual of the set cover problem). For each c_i, $1 \leq i \leq \Delta_{out}$, we form a set S_i which contains all OR nodes reachable via OR nodes from c_i (i.e. paths from c_i to S_i do not use any AND nodes). Now, the collection $\mathcal{C}$ contains all sets $S_i \subseteq S$, where S is the set of all OR nodes. Using the $(1+\ln \Delta_{out}) = O(\log n)$ approximation for hitting set, we obtain a set S_a^* of weight w_a^* of OR nodes which hit every set. Let $W_a = \min\{w_a, w_a^*\}$ (w_a is the weight of node a). Choose the AND node a with minimum W_a over all AND nodes. Kill AND node a or all the OR nodes in the corresponding hitting set solution (the one with minimum weight). Clear graph G, i.e., remove every AND/OR node which can be completed after killing the aforementioned nodes, and repeat the above iteration for G. The final solution contain all AND/OR nodes killed during the iterations.

We finish by showing that

Theorem 7. *The above algorithm kills a set of AND/OR nodes such that the remaining graph is deadlock free and the weight of the solution is at most $(1 + \ln \Delta_{out})n_a + 1 = O(n_a \log n)$ times optimum.*

Proof. The correctness of the solution can be easily seen from the description of the algorithm. Thus, we only show the approximation factor here. To this end, we prove that in each iteration, except the case in which there is no AND node, we kill nodes of total weight at most $(1 + \ln \Delta_{out})$ times optimum weight for the updated graph G in that iteration. In the last iteration, we kill nodes of total weight at most OPT according to the description of the algorithm. Using these facts and that OPT in each iteration is at most the original optimum, we obtain the desired approximation factor.

Consider an optimum solution and let a be the first AND node which is completed or killed in the optimum resolution. Thus either we have killed a or we have completed a by killing at least one OR node from the OR nodes reachable from each of its children. Hence for at least one AND node, the weight of the solution to the corresponding hitting set instance is at most the weight of optimum. Since the approximation factor of hitting set is $1+\ln \Delta_{out}$ and we try all AND nodes and then take the minimum, the total weight of the killed nodes is at most $(1 + \ln \Delta_{out})$ times optimum, as desired. □

5 Permanent Deadlock Resolution

Here we consider another version of the deadlock resolution problem where it is not possible for the algorithm to specify a feasible schedule on the remaining processes. In particular, we want to kill enough processes, such that if the remaining processes try to acquire locks in any order, they cannot deadlock. We then say that the remaining processes are *adversarially schedulable*.

We consider the special case of this problem when all processes are writers (AND nodes). In this case, we show that this problem can be reduced to the feedback vertex set problem on mixed graphs (i.e. graphs with both directed and undirected edges). Since this problem yields to the same techniques as those used for feedback vertex set of directed graphs, we get an $O(\log n \log \log n)$-approximation.

We are given a set of resources R and a set of processes P, each holding a lock on some subset of resources, and waiting to get locks on another subset of resources. We construct a bipartite mixed graph as follows: create a vertex v_r for every resource r with infinite cost, and a vertex v_p for every process p. Whenever process p holds the lock on resource r, we add a directed edge from v_p to v_r. Moreover, we add an undirected edge between v_p and $v_{r'}$ whenever process p is waiting to get a lock on resource r'.

Theorem 8. *An instance is adversarially schedulable if and only if the corresponding graph is acyclic.*

Proof. We first argue that greedily schedulability implies acyclicity. Assume the contrary, and let the graph have a cycle $p_1, r_1, p_2, r_2, \ldots, p_k, r_k, p_1$. Now consider the schedule in which p_i grabs a lock on r_i (or already holds it, in case the edge is directed). Note that p_i waits for a lock on r_{i-1} and p_1 waits on r_k. this entails a cyclic dependency amongst processes $p_1, \ldots, p_k$: p_i cannot finish unless p_{i-1} finishes and releases r_{i-1}. This configuration is therefore deadlocked. Since we have shown how to reach a deadlocked state from the initial state, the initial state was not adversarially schedulable, which contradicts the assumption. Hence the claim follows.

Now suppose that the graph is acyclic. We claim that the initial configuration is adversarially schedulable. Suppose not. Then there is a sequence of lock acquisition that lead to a deadlocked configuration. Clearly, a deadlocked configuration corresponds to processes $p_1, p_2, \ldots, p_k$ such that p_{i+1} is waiting for p_i to release some resource r_i. Since p_i holds r_i in this configuration, (p_i, r_i) must be directed/undirected edge in the graph. Moreover, since p_{i+1} is waiting for r_i, (r_i, p_{i+1}) is an undirected edge in the graph. However, we have just shown that $p_1, r_1, p_2, r_2, \ldots, p_k, r_k, p_1$ is a cycle in G, which contradicts the acyclicity of G. Thus the claim follows. □

Theorem 9. *The permanent deadlock resolution problem for AND nodes has an $O(\log n \log \log n)$ approximation algorithm.*

Acknowledgement

We would like to thank, Ondrej Such, a developer at Microsoft for asking his question on the Algorithm's email list. The first author would also like to thanks Joseph Cheriyan and Laci Lovasz for many initial and fruitful discussions. He would also like to thank David Shmoys for a very short but helpful discussion. The last author would like to thank Christos Papadimitriou and Joe Hellerstein for very useful discussions.

References

1. B. AWERBUCH AND S. MICALI, *Dynamic deadlock resolution protocols*, in The 27th Annual Symposium on Foundations of Computer Science, 1986, 196–207.
2. R. BAR-YEHUDA, D. GEIGER, J. NAOR, AND R. M. ROTH, *Approximation algorithms for the feedback vertex set problem with applications to constraint satisfaction and Bayesian inference*, SIAM J. Comput., 27 (1998), 942–959.
3. G. BRACHA AND S. TOUEG, *A distributed algorithm for generalized deadlock detection*, in Proceedings of the 3rd annual ACM symposium on Principles of distributed computing, ACM Press, 1984, 285–301.
4. K. M. CHANDY AND L. LAMPORT, *Distributed snapshots: determining global states of distributed systems*, ACM Transactions on Computer Systems (TOCS), 3 (1985), 63–75.
5. K. M. CHANDY AND J. MISRA, *A distributed algorithm for detecting resource deadlocks in distributed systems*, in Proceedings of the 1st ACM SIGACT-SIGOPS symposium on Principles of distributed computing, ACM Press, 1982, 157–164.
6. K. M. CHANDY, J. MISRA, AND L. M. HAAS, *Distributed deadlock detection*, ACM Transactions on Computer Systems (TOCS), 1 (1983), 144–156.
7. M. CHARIKAR, C. CHEKURI, T.-Y. CHEUNG, Z. DAI, A. GOEL, S. GUHA, AND M. LI, *Approximation algorithms for directed Steiner problems*, J. Algorithms, 33 (1999), 73–91.
8. J. CHERIYAN, H. J. KARLOFF, AND Y. RABANI, *Approximating directed multicuts*, in The 42th Annual Symposium on Foundations of Computer Science, 2001, 348–356.
9. G. EVEN, J. NAOR, B. SCHIEBER, AND M. SUDAN, *Approximating minimum feedback sets and multicuts in directed graphs*, Algorithmica, 20 (1998), 151–174.
10. U. FEIGE, *A threshold of $\ln n$ for approximating set cover*, J. ACM, 45 (1998), 634–652.
11. M. FLATEBO AND A. K. DATTA, *Self-stabilizing deadlock detection algorithms*, in Proceedings of the '92 ACM annual conference on Communications, ACM Press, 1992, 117–122.
12. J. GRAY, P. HOMAN, R. OBERMARCK, AND H. KORTH, *A straw man analysis of probability of waiting and deadlock*, in Proceedings of the 5th Internafional Conference on Distributed Data Management and Computer Networks, 1981.
13. E. HALPERIN AND R. KRAUTHGAMER, *Polylogarithmic inapproximability*, in The 35th Annual ACM Symposium on Theory of Computing (STOC'03), 2003, 585–594.
14. J.-M. HELARY, C. JARD, N. PLOUZEAU, AND M. RAYNAL, *Detection of stable properties in distributed applications*, in Proceedings of the 6th PODC, ACM Press, 1987, 125–136.
15. T. HERMAN AND K. M. CHANDY, *A distributed procedure to detect and/or deadlock*, Tech. Rep. TR LCS-8301, Dept. of Computer Sciences, Univ. of Texas, 1983.
16. K. JAIN, M. MAHDIAN, AND M. R. SALAVATIPOUR, *Packing Steiner trees*, in The 14th Annual ACM-SIAM Symposium on Discrete Algorithms (SODA'03), 2003, 266–274.
17. E. KNAPP, *Deadlock detection in distributed databases*, ACM Computing Surveys (CSUR), 19 (1987), 303–328.

18. T. LEIGHTON AND S. RAO, *Multicommodity max-flow min-cut theorems and their use in designing approximation algorithms*, J. ACM, 46 (1999), 787–832.
19. R. J. LIPTON AND R. E. TARJAN, *Applications of a planar separator theorem*, SIAM J. Comput., 9 (1980), 615–627.
20. C. LUND AND M. YANNAKAKIS, *On the hardness of approximating minimization problems*, J. Assoc. Comput. Mach., 41 (1994), 960–981.
21. K. MAKKI AND N. PISSINOU, *Detection and resolution of deadlocks in distributed database systems*, in Proceedings of the 4th international conference on Information and knowledge management, ACM Press, 1995, 411–416.
22. P. D. SEYMOUR, *Packing directed circuits fractionally*, Combinatorica, 15 (1995), 281–288.
23. H. WU, W.-N. CHIN, AND J. JAFFAR, *An efficient distributed deadlock avoidance algorithm for the and model*, IEEE Transactions on Software Engineering, 28 (2002), 18–29.

Facility Location in Sublinear Time

Mihai Bădoiu[1], Artur Czumaj[2,*], Piotr Indyk[1], and Christian Sohler[3,**]

[1] MIT Computer Science and Artificial Intelligence Laboratory,
Stata Center, Cambridge, Massachusetts 02139, USA
{mihai, indyk}@theory.lcs.mit.edu
[2] Department of Computer Science,
New Jersey Institute of Technology,
Newark, NJ 07102, USA
czumaj@cis.njit.edu
[3] Heinz Nixdorf Institute and Computer Science Department,
University of Paderborn, D-33102 Paderborn, Germany
csohler@uni-paderborn.de

Abstract. In this paper we present a randomized constant factor approximation algorithm for the problem of computing the optimal *cost* of the metric Minimum Facility Location problem, in the case of uniform costs and uniform demands, and in which every point can open a facility. By exploiting the fact that we are approximating the optimal cost without computing an actual solution, we give the first algorithm for this problem with running time $O(n \log^2 n)$, where n is the number of metric space points. Since the size of the representation of an n-point metric space is $\Theta(n^2)$, the complexity of our algorithm is *sublinear* with respect to the input size.

We consider also the general version of the metric Minimum Facility Location problem and we show that there is no $o(n^2)$-time algorithm, even a randomized one, that approximates the optimal solution to within any factor. This result can be generalized to some related problems, and in particular, the cost of minimum-cost matching, the cost of bichromatic matching, or the cost of $n/2$-median cannot be approximated in $o(n^2)$-time.

1 Introduction

The design of algorithms operating on massive data sets has received a lot of attention in recent years. The practical motivation of this study is that polynomial-time algorithms that are efficient in relatively small inputs, may become impractical for input sizes of several gigabytes. For example, when we consider approximation algorithms for clustering problems in metric spaces then they typically

---

* Research supported in part by NSF grant ITR-CCR-0313219.
** Research supported in part by DFG grant Me 872/8-2 and by the IST program of EC under contract no. IST-2002-001-907 (DELIS).

have $\Omega(n^2)$ running time where n is the number of input points. Clearly, such a running time is not feasible for massive data sets. But for many problems — like the facility location problem considered in this paper — such a running time is provably unavoidable. Surprisingly, these lower bounds do not necessarily hold when one wants to estimate the *cost* of an optimal solution. In this paper we will indeed show that one can find a constant factor approximation algorithm for the metric uncapacitated facility location problem with uniform costs and in which every point can open a facility, that runs in $O(n \log^2 n)$ time, that is, in time *sublinear* in the input size.

Our approach is motivated by the fact that in many applications it suffices to know an *approximate cost* of the facility location problem rather than to find an approximate solution to the facility location problem. Let us consider the example that a company wants to invest money and it can relate the cost of the facility location problem to the possible return on investment. Then it would first solve an instance of the problem for every market to find out the most profitable one. In such a situation it is sufficient to know the return on investment before one decides which market to enter. It is not (yet) necessary to know how to achieve it. Finally, when one knows which market to enter one only has to compute a solution to a single instance of the problem. Therefore, if one could approximate the *cost* of an optimal solution significantly faster than finding such a particular approximate solution this would significantly speed up the market analysis.

Similar arguments hold for another popular application of facility location algorithms, that of clustering data sets. In particular, it is good to know if the data can be "well-clustered" before actually attempting to find the clustering.

1.1 Our Results

In this paper we consider the metric Minimum Facility Location problem with uniform opening costs and demands, and in which every point can open a facility. We give a randomized $O(1)$-approximation algorithm for this problem that runs in time $O(n \log^2 n)$, where n is the number of metric space points. Since the size of the representation of an n-point metric space is $\Theta(n^2)$, the complexity of our algorithm is *sublinear* with respect to the input size. No $o(n^2)$-time approximation algorithm for this problem was known before. It has been known that any constant factor approximation algorithm that returns not only the cost, but also a solution itself, requires the running time of $\Omega(n^2)$ [14].

Next, we prove that if the set of facilities and the cities (points that are to be connected to the facilities) are allowed to be disjoint, then any, *even randomized*, approximation algorithm for the cost of the Minimum Facility Location that guarantees any bounded approximation ratio for the cost, requires time $\Omega(n^2)$. This bound holds even when the opening costs and demands are uniform. Furthermore, our proof can be extended to the problems of estimating the *cost of minimum-cost matching*, the *cost of bi-chromatic matching*, and the *cost of k-median* for $k = n/2$; all these problems require $\Omega(n^2)$ to estimate the cost of their optimal solution to within any factor. We feel that these results demonstrate that most optimization problems for metric instances do not have sublinear-time algorithms even to estimate well the cost of the optimal solution; results like our sublinear-time algorithm for a $O(1)$-factor approximation of the cost of the optimum solution for the metric uniform Minimum Facility Location problem are rare (see however, [4, 6, 7]).

1.2 Our Techniques

Our analysis of a sublinear-time algorithm consists of two principal steps: we first prove the existence of an appropriated estimator for the cost of the Minimum Facility Location problem and then we show how such an estimator can be approximated in time $O(n \log^2 n)$. Our estimator is obtained by extending the primal-dual approach from [12]: for each point we define an approximation of the contribution of that point to the total cost, and then we prove that the sum of the contributions for all the points approximates the cost of the Minimum Facility Location problem. An important property of our estimator is that it can be efficiently approximated by *adaptive sampling*. We first prove that the individual value of an estimator for any single point can be efficiently approximated by sampling with the running time depending on the value of the estimator, and then we apply another adaptive sampling scheme to efficiently approximate the sum of the estimators. A similar approach has been used in recent sublinear-time algorithms for estimating the cost of the minimum spanning tree problem in [2] and [4].

1.3 Definition of the Problem

The formal definition of the general form of the (Metric) *Minimum Facility Location* problem is as follows: We are given a metric (P, D), and a subset $\mathcal{F} \subseteq P$ of *facilities*. For each facility $v \in \mathcal{F}$, we are given a nonnegative *cost* $\mathfrak{f}(v)$, and for each point $u \in P$, a nonnegative *demand* $d(u)$. The problem consists of finding a set $F \subseteq \mathcal{F}$, so as to minimize

$$\sum_{v \in F} \mathfrak{f}(v) + \sum_{u \in P} d(u) \cdot D(u, F) ,$$

where $D(u, F) = \min_{v \in F} D(u, v)$.

In this paper we focus on the variant of the facility location problem with $\mathcal{F} = P$ and in which the costs as well as the demands are uniform. That is, for each $v \in \mathcal{F}$, $\mathfrak{f}(v) = c$ for some $c > 0$, and for each $u \in P$, $d(u) = 1$. Observe that we can assume that $c = 1$, if we re-scale the given metric, by dividing all the distances by c. In what follows, we will refer to this variant of the facility location problem as *uniform*.

The key property of our formulation, is that we are interested in computing the cost of the optimal solution, without computing a solution itself. Thus, in what follows, our task is to approximate the value:

$$\min_{F \subseteq P} |F| + \sum_{u \in P} D(u, F) .$$

In the final part of the paper we also consider a more general variant of the problem when P and $\mathcal{F}$ do not have to be the same. We prove in Theorem 2 that in that case there is no hope to obtain a sublinear-time algorithm.

1.4 Previous Work

The *Minimum Facility Location* problem is one of the most extensively studied problems in combinatorial optimization. The problem is known to be $\mathcal{NP}$-hard and the first constant factor approximation algorithm was given by Shmoys et al. [13]. Several other approximation algorithms are given in [1, 3, 8]. The best approximation ratio of 1.52, is due to Madhian, Ye, and Zhang [10], while the best lower bound of 1.463 for the approximation ratio is due to Guha and Khuller [5].

The first constant factor approximation algorithm with almost linear running time (that is, the running time of $O(n^2 \log n)$) was given by Jain and Vazirani [9]; Mettu and Plaxton [12] gave a simple $O(n^2)$-time constant approximation ratio algorithm. Thorup [14] considered the facility location problem in metric spaces defined by a graph. If the underlying graph has m edges, then even though the metric space is of size $\Theta(n^2)$, Thorup gives a constant-factor approximation algorithm running in time $\tilde{O}(m)$; this is a sublinear time for sparse graphs. On the other hand, it has been shown [14] that for general metric spaces, any constant factor approximation algorithm, even a randomized one, requires running time of $\Omega(n^2)$. Notice that this does not exclude the possibility of approximating the *cost* of the Minimum Facility Location problem in sublinear time, in particular, in time $O(n \operatorname{polylog}(n))$.

2 Estimating the Cost of Uniform Minimum Facility Location

In this section we present an $O(n \log^2 n)$ time algorithm that approximates the cost of the Minimum Facility Location in the uniform case, that is, when the costs as well as the demands are uniform.

2.1 Preliminaries

Let (P, D) be a metric with a point set $P = \{p_1, \ldots, p_n\}$. For any point $p_i \in P$, and for any $r \geq 0$, we denote by $B(p_i, r)$ the set of points in P which are at distance at most r from p_i. For each i, $1 \leq i \leq n$, let $r_i > 0$ be the number satisfying

$$\sum_{p \in B(p_i, r_i)} (r_i - D(p_i, p)) = 1 .$$

Observe that the value $\sum_{p \in B(p_i, r)} (r - D(p_i, p))$ is continuous and strictly monotonically increasing with r. Thus, there exists a unique value r_i satisfying the above equality. Moreover, for any i, $1 \leq i \leq n$, we have $1/n \leq r_i \leq 1$.

We begin with a lemma that establishes the relation between the value of r_i and the size of $B(p_i, r_i)$.

Lemma 1. *For every i, with $1 \leq i \leq n$, we have $\frac{1}{|B(p_i, r_i)|} \leq r_i \leq \frac{2}{|B(p_i, r_i/2)|}$.*

Proof. By the definition of r_i, we have $\sum_{p \in B(p_i, r_i)} (r_i - D(p_i, p)) = 1$, which implies $\sum_{p \in B(p_i, r_i)} r_i \geq 1$, and thus $r_i \geq 1/|B(p_i, r_i)|$. The other inequality follows directly from the following,

$$1 = \sum_{p \in B(p_i, r_i)} (r_i - D(p_i, p)) \geq \sum_{p \in B(p_i, r_i/2)} (r_i - D(p_i, p)) \geq |B(p_i, r_i/2)| \cdot r_i/2. \quad \square$$

MP algorithm. In our analysis we will use a simple approximation algorithm for the Minimum Facility Location problem due to Mettu and Plaxton [12]; we will refer to that algorithm as the *MP algorithm*.

1. Compute the value of r_i for every $p_i \in P$.
2. Sort the input such that $r_1 \leq r_2 \leq \cdots \leq r_n$.
3. For $i = 1$ to n: if there is no open facility in $B(p_i, 2r_i)$ then open the facility at p_i.

Mettu and Plaxton [12] proved that this simple algorithm will return a set of open facilities for which the total cost is at most 3 times the minimum.

2.2 Cost Estimation

In this section, we show that the sum of the radii approximates the optimal cost of the facility location to within a constant factor. Our analysis uses the relation between the sum $\sum_{p_i \in P} r_i$ and the cost of optimal solution and that of the solution obtained by the MP algorithm discussed above.

Let C_{OPT} be the cost of an optimal solution. Let also F_{MP} be the set of facilities computed by the MP algorithm. For this solution given by the MP algorithm, we define C_{MP}, C_{MP}^c, and C_{MP}^f to be the total cost, the connection cost, and the facility cost respectively.

The following lemma shows that the sum of the radii estimates well C_{OPT}.

Lemma 2. $\frac{1}{4} \cdot C_{OPT} \leq \sum_{p_i \in P} r_i \leq 6 \cdot C_{OPT}$.

Proof. We first prove the lower bound that $C_{OPT} \leq 4 \cdot \sum_{p_i \in P} r_i$ and then the upper bound that $\sum_{p_i \in P} r_i \leq 6 \cdot C_{OPT}$.

Lower bound: Since in the MP algorithm for every $p_i \in P$ there is an open facility within distance at most $2r_i$ (for if not, then the algorithm would open the facility at p_i), we get that $2 \sum_{p_i \in P} r_i \geq C_{MP}^c$.

It remains to show that $\sum_{p_i \in P} r_i$ is an upper bound for C_{MP}^f. We first observe that every $p_i \in P$ is contained in at most one ball $B(p_j, r_j)$, for some $p_j \in F_{MP}$. Indeed, if $p_i \in B(p_j, r_j) \cap B(p_k, r_k)$ for some $p_j, p_k \in F_{MP}$, $j < k$, then since

$r_j \leq r_k$, we would have $p_j \in B(p_k, 2r_k)$. But this implies that the MP algorithm would not open the facility at p_k, a contradiction.

This observation yields:

$$\sum_{p_i \in P} r_i \geq \sum_{p_j \in F_{MP}} \sum_{p_k \in B(p_j, r_j)} r_k . \tag{1}$$

Next, we observe that if $p_j \in F_{MP}$ and $p_k \in B(p_j, r_j)$, then we must have $r_j \leq 2r_k$. Indeed, for if not, then we would have $B(p_k, 2r_k) \subseteq B(p_k, r_j) \subseteq B(p_k, r_j + D(p_j, p_k)) \subseteq B(p_j, 2r_j)$, and thus the MP algorithm would not open the facility at p_j, a contradiction. This observation can be now combined with (1) to conclude:

$$\sum_{p_i \in P} r_i \geq \sum_{p_j \in F_{MP}} \sum_{p_k \in B(p_j, r_j)} r_k \geq \sum_{p_j \in F_{MP}} \sum_{p_k \in B(p_j, r_j)} r_j/2$$

$$= \tfrac{1}{2} \cdot \sum_{p_j \in F_{MP}} r_j \cdot |B(p_j, r_j)| \geq \tfrac{1}{2} \cdot \sum_{p_j \in F_{MP}} 1 = \tfrac{1}{2} \cdot C_{MP}^f ,$$

where the second inequality follows from the fact that $r_j \geq 1/|B(p_j, r_j)|$ (Lemma 1). Thus, we have $2 \cdot \sum_{p_i \in P} r_i \geq C_{MP}^c/2 + C_{MP}^f/2 \geq C_{MP}/2 \geq C_{OPT}/2$.

Upper bound: Next, we show that the sum of the radii is not much bigger than the cost of optimal solution. Before we proceed, we introduce one definition from [12]. For a set $X \subseteq P$ and a point $p_i \in P$, we define

$$charge(p_i, X) = D(p_i, X) + \sum_{p_j \in X} \max\{0, r_j - D(p_i, p_j)\} .$$

Mettu and Plaxton proved [12] that $C_{MP} = \sum_{p_i \in P} charge(p_i, F_{MP})$.

Now we are ready to prove that $\sum_{p_i \in P} r_i \leq 2 \cdot C_{MP}$ what will imply that $\sum_{p_i \in P} r_i \leq 6 \cdot C_{OPT}$. We have,

$$2 \cdot C_{MP} = 2 \cdot \sum_{p_i \in P} charge(p_i, F_{MP})$$

$$\geq 2 \cdot \left(\sum_{p_i \in F_{MP}} r_i + \sum_{p_j \in P \setminus F_{MP}} \max\{r_{\delta(j)}, D(p_j, p_{\delta(j)})\} \right) ,$$

where $\delta(j)$ denotes the index of the facility in F_{MP} that is closest to p_j. We want to show

$$2 \cdot \left(\sum_{p_i \in F_{MP}} r_i + \sum_{p_j \in P \setminus F_{MP}} \max\{r_{\delta(j)}, D(p_j, p_{\delta(j)})\} \right) \geq \sum_{p_i \in P} r_i .$$

We will show that $r_j \leq D(p_j, p_{\delta(j)}) + r_{\delta(j)}$, which immediately implies the above inequality because then $\max\{r_{\delta(j)}, D(p_j, p_{\delta(j)})\} \geq r_j/2$. Assume $r_j > D(p_j, p_{\delta(j)}) + r_{\delta(j)}$. In this case we have $B(p_{\delta(j)}, r_{\delta(j)}) \subseteq B(p_j, r_j)$. We get

$$\sum_{p \in B(p_j, r_j)} (r_j - D(p_j, p)) \geq \sum_{p \in B(p_{\delta(j)}, r_{\delta(j)})} (r_j - D(p_j, p))$$

$$> \sum_{p \in B(p_{\delta(j)}, r_{\delta(j)})} (r_{\delta(j)} - D(p_{\delta(j)}, p)) = 1 .$$

This is a contradiction because the definition of r_j requires

$$\sum_{p \in B(p_j, r_j)} (r_j - D(p_j, p)) = 1 .$$

To summarize, we have proven that $2 \cdot C_{MP} \geq \sum_{p_i \in P} r_i$, and now the lower bound follows from the fact that $C_{MP} \leq 3 \cdot C_{OPT}$ [12]. □

2.3 Estimating the Cost of the Facility Location Problem

From the previous section we know that to approximate the cost of the facility location problem it suffices to estimate the sum $\sum_i r_i$ of the radii $r_1, \ldots, r_n$ of the points $p_1, \ldots, p_n$. A standard approach to this problem would be to sample a set of s points (for a suitable s), determine (possibly approximately) their radii, and then output n times their average radius as an approximation for $\sum_i r_i$. However, this approach cannot lead to a sublinear-time algorithm for the following reason. In general, the time to determine the radius of a point in $\Omega(n)$. For example, this might be the case when the radius is constant, because there is only a constant number of points within the radius. Therefore, to certify that a point has constant radius the algorithm must be able to certify that no more than a constant number of points are within the radius. This task cannot be done in $o(n)$ time (even if one aims at an approximation and uses randomization). We also note that, in general, $s = \Omega(n)$, if we need a constant factor approximation of $\sum_i r_i$. This follows from standard Chernoff-Hoeffding bounds (which are essentially tight in this setting) and the fact that the average radius can be as small as $1/n$. Therefore, this standard sampling approach would not give us a sublinear time algorithm.

In the following we will show that an *adaptive sampling* algorithm can estimate the size of r_i in $O(r_i n \log n)$ time (recall that $r_i < 1$). We start with a constant size sample of points and determine their average radius. If our sample is too small we double it and continue until we have found a sample of sufficient size. For the analysis we will parameterize the sample size s by the average value of the r_i. Combining this with the running time of the adaptive algorithm leads to a sublinear algorithm. Details follow in the next two subsections.

2.4 Estimating r_i

In this section we present an algorithm that for a given i, in time $O(r_i n \log n)$ approximate the value of r_i to within a constant factor, with high probability.

Let us fix i. Our approach of estimating the value of r_i is by approximating the value of r for which $B(p_i, r)$ contains approximately $1/r$ points. This is formalized in the following lemma.

Lemma 3. *Let j_0 be the maximum integer j, with $1 \leq j \leq \log n$, such that $|B(p_i, 2^{-j})| \geq 2^j$. Then, we have $2^{-(j_0+1)} \leq r_i \leq 2^{-j_0+1}$.*

Proof. We will use Lemma 1. By our assumption about j_0 we have $|B(p_i, 2^{-(j_0+1)})| < 2^{j_0+1}$ and $|B(p_i, 2^{-j_0})| \geq 2^{j_0}$. The first inequality implies that for any $r < 2^{-(j_0+1)}$, $|B(p_i, r)| \leq |B(p_i, 2^{-(j_0+1)})| < 2^{j_0+1} < 1/r$. This bound together with the lower bound in Lemma 1 yield that $r_i \geq 2^{-(j_0+1)}$. On the other hand, the inequality $|B(p_i, 2^{-j_0})| \geq 2^{j_0}$ implies that for any $r > 2^{-j_0+1}$, $|B(p_i, r/2)| \geq |B(p_i, 2^{-j_0})| \geq 2^{j_0} > 2/r$. Therefore, by the upper bound in Lemma 1 we must have $r_i \leq 2^{-j_0+1}$. □

Lemma 3 implies that in order to estimate r_i, it suffices to estimate the value of j_0. Our algorithm to estimate j_0 runs as follows: We begin with setting $j = \log n$, and then we are decreasing j by one until for the first time $|B(p_i, 2^{-j})| \geq 2^j$. Since computing $|B(p_i, 2^{-j})|$ exactly requires $\Omega(n)$ time, we only approximate $|B(p_i, 2^{-j})|$ by *random sampling*. This reduces the running time. At each step, we pick uniformly at random, and with replacement, $K_j = c 2^{-j} n \log n$ sample points to estimate the value of $|B(p_i, 2^{-j})|$, where c is a sufficiently large constant. Let N_j be the number of sample points that are inside the ball $B(p_i, 2^{-j})$. We return $\beta_j = n N_j / K_j$ as the estimator of $|B(p_i, 2^{-j})|$.

In the following three lemmas we first analyze the quality of the estimator β_j and then discuss the running time of this sampling scheme.

Lemma 4. *If $j \geq j_0 + 2$, then $\mathbf{Pr}[\beta_j \geq 2^j] < 1/poly(n)$.*

Proof. Since $j \geq j_0 + 2$, it follows that $B(p_i, 2^{-j}) \subseteq B(p_i, 2^{-(j_0+1)})$. Let q be the probability that a randomly chosen sample point is in $B(p_i, 2^{-j})$. We have $q \leq |B(p_i, 2^{-(j_0+1)})|/n$. By the choice of j_0, we have $|B(p_i, 2^{-(j_0+1)})| < 2^{j_0+1}$, and thus $q < 2^{j_0+1}/n \leq 2^{j-1}/n$.

The expected number of sample points that fall inside $B(p_i, 2^{-j})$ is $\mathbf{E}[N_j] = qK_j < \frac{c \log n}{2}$. Applying the Chernoff bound, we obtain

$$\mathbf{Pr}[\beta_j \geq 2^j] = \mathbf{Pr}[N_j \geq c \log n] < 1/poly(n) \ . \qquad \square$$

Lemma 5. *If $j \leq j_0 - 1$, then $\mathbf{Pr}[\beta_j \geq 2^j] > 1 - 1/poly(n)$.*

Proof. Since $j \leq j_0 - 1$, it follows that $|B(p_i, 2^{-j})| \geq |B(p_i, 2^{-j_0})| \geq 2^{j_0} \geq 2^{j+1}$. Let q be the probability that a randomly chosen sample point is in $B(p_i, 2^{-j})$. We have that $q \geq 2^{j+1}/n$.

The expected number of sample points that fall inside $B(p_i, 2^{-j})$ is $\mathbf{E}[N_j] = q K_j \geq 2 c \log n$. Applying the Chernoff bound, we obtain

$$\mathbf{Pr}[\beta_j \geq 2^j] = \mathbf{Pr}[N_j \geq c \log n] > 1 - 1/poly(n) \ . \qquad \square$$

Lemma 6. *The described procedure estimates the value of r_i to within a constant factor in time $O(r_i n \log n)$, with high probability.*

Proof. Let j'_0 be the estimated value of j_0. By Lemmas 4 and 5, it follows that with high probability, $j_0 \leq j'_0 \leq j_0 + 1$. If we use the value $r'_i = 2^{-j'_0}$ as an estimation of r_i, then by Lemma 3 we obtain that $r_i/2 \leq r'_i \leq 4r_i$.

Moreover, with high probability, the running time of the procedure is at most $\sum_{j=j_0}^{\log n} O(K_j) = O(r_i \, n \log n)$. □

2.5 Estimating the Sum of the Radii

In this section we show how to estimate $\sum_i r_i$ in time almost linear in n. Let us first assume that we know the cost of the solution c, and we sample a set of s points independently and uniformly at random, where $s = \Theta(\frac{n}{c} \log n)$. Since by Lemma 6, the running time to estimate a radius r_i is $O(r_i \, n \log n)$, the total expected running time of the algorithm is

$$\mathbf{E}[\text{time}] = s \cdot \mathbf{E}[\text{one step}] = s \cdot O(\tfrac{1}{n} \cdot \sum_i r_i \, n \log n) = O(n \log^2 n) \ .$$

Let x_i, for $i \in \{1, 2, \ldots, s\}$, be the radii of the sample points taken by the algorithm. We have

$$\mathbf{E}[x_i] = \frac{\sum_j r_j}{n} \ .$$

Let $S = \sum_{i=1}^{s} x_i$ and hence, $\mathbf{E}[S] = \frac{s \cdot \sum_i r_i}{n} = \frac{\Theta(\frac{n}{c} \log n) \cdot \sum_i r_i}{n} = \Theta\left(\frac{\sum_i r_i}{c} \cdot \log n\right) = \Theta(\log n)$. Let $\epsilon > 0$ be arbitrary. Our goal is to use the value of S as the estimator of $\frac{n}{s} \sum_i r_i$. To show the quality of this estimator we will bound $\mathbf{Pr}[|S - \mathbf{E}[S]| \geq \epsilon \cdot \mathbf{E}[S]]$. By using the fact that $0 \leq x_i \leq 1$ for every i, we apply a variant of the Hoeffding inequality, see [11, Theorem 2.3], to obtain

$$\mathbf{Pr}[S \geq (1+\epsilon) \cdot \mathbf{E}[S]] \leq e^{-\frac{\epsilon^2 \cdot \mathbf{E}[S]}{2(1+\epsilon/3)}} \ ,$$

$$\mathbf{Pr}[S \leq (1-\epsilon) \cdot \mathbf{E}[S]] \leq e^{-\frac{1}{2} \cdot \epsilon^2 \cdot \mathbf{E}[S]} \ .$$

This immediately implies the following bound for any $0 < \epsilon \leq 1$,

$$\mathbf{Pr}[|S - \mathbf{E}[S]| \geq \epsilon \cdot \mathbf{E}[S]] \leq 2 \, e^{-\Theta(\epsilon^2 \cdot \mathbf{E}[S])} = 2 \, e^{-\Theta(\epsilon^2 \cdot \log n)} \ .$$

We now show how to remove the assumption that we know the cost of the solution. We run the algorithm in phases: we start in the first phase by "guessing" $c = n$, because we know that the cost of the optimal solution is not bigger than n. If $S < \frac{s}{n} \cdot c$, then we start a new phase with estimated cost $c/2$, and so on. If $S \geq \frac{s}{n} \cdot c$, we return $S \cdot n/s$ as the approximation of the cost. The probability that the algorithm ends in a bad phase (when S far away from $\frac{s}{n} \cdot c$) is low, because $\mathbf{Pr}[S \geq (1+\epsilon) \cdot \mathbf{E}[S]] < 1/\text{poly}(n)$, as shown above. Since we need to have at least one facility in a solution, we have $c \geq 1$, therefore we have at most a logarithmic number of phases.

Note that we only get a constant slowdown by running these phases to guess c, because the last phase, for the smallest c, dominates the running time of all the other phases. Thus we obtain the following theorem.

Theorem 1. *There exists a constant factor approximation algorithm for the uniform case of the Minimum Facility Location problem which runs in time $O(n \log^2 n)$ with high probability.*

3 Lower Bounds: Estimating the Cost in the General Case of the Uniform Minimum Facility Location Problem Requires $\Omega(n^2)$ Time (Even for Randomized Algorithms)

In this section, we consider a general case of the Minimum Facility Location problem in which we do not impose the restriction that $\mathcal{F} = P$ (that is, we allow only for a subset of points to be able to open a facility). We focus again on the uniform case, and the goal is to minimize the following cost:

$$\min_{F \subseteq \mathcal{F}} \left(|F| + \sum_{p \in P} d(p, F) \right) .$$

Our main result is the following theorem.

Theorem 2. *For any $\varrho \geq 1$, every approximation algorithm (even a randomized one) with approximation ratio ϱ for the cost of the Minimum Facility Location problem as defined above requires time $\Omega(n^2)$.*

Proof. We show the existence of two instances of the metric spaces which are undistinguishable by any $o(n^2)$-time algorithms and such that the cost of the

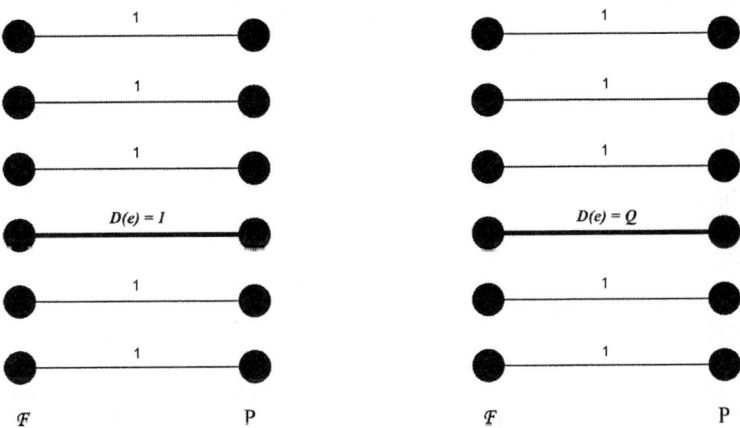

Fig. 1. Two metric spaces undistinguishable by any $o(n^2)$-time algorithms whose costs of the Minimum Facility Location differ by factor ϱ. The perfect matching connecting $\mathcal{F}$ with P is selected at random and the edge e is selected as a random edge from the matching. We set $Q = 2n(\varrho - 1) + 2$. The distances not shown are all equal to $n^3 \varrho$

Minimum Facility Location in one instance is greater than ϱ times than the one in the other instance (see Fig. 1).

Let us consider the metric space with $2n$ points: n points in P and n points in $\mathcal{F}$. Take a random perfect matching $\mathbb{M}$ between the points in P and $\mathcal{F}$, and choose an edge $e \in \mathbb{M}$ at random. Now, we define the distances in $(P \cup \mathcal{F}, D)$ according to the following:

- for any $e^* \in \mathbb{M} \setminus \{e\}$, $D(e^*) = 1$,
- $D(e)$ is either 1 or $Q = 2n(\varrho - 1) + 2$, and
- for any pair of points x, y not connected by an edge from $\mathbb{M}$, $D(x, y) = n^3 \varrho$.

It is easy to see that both instances define properly a metric space $(P \cup \mathcal{F}, D)$. Furthermore, that for such problem instances, the solution to the Minimum Facility Location will open all facilities and the cost of the Minimum Facility Location problem will depend on the choice of $D(e)$: if $D(e) = Q$ then the cost will be $2n - 1 + Q > 2n\varrho$, and if $D(e) = 1$, then the cost will be $2n$. Hence, any ϱ-factor approximation algorithm for the matching problem must distinguish between these two problem instances. However, this requires to find if there is an edge of length Q, and this is known to require time $\Omega(n^2)$, even if a randomized algorithm is used. □

3.1 Extensions

It is not difficult to see that almost an identical proof will also work for estimating the cost of *minimum-cost matching*, the cost of *minimum-cost bi-chromatic matching*, and also the cost of *k-median* for $k = n/2$; all these problems require $\Omega(n^2)$ to estimate the cost of their optimal solution to within any factor. No such lower bounds have been previously known.

Theorem 3. *For any $\varrho \geq 1$, every approximation algorithm (even a randomized one) with approximation ratio ϱ for each of the following problems requires time $\Omega(n^2)$:*

- *estimating the cost of minimum-cost matching for a set of n points in a metric space,*
- *estimating the cost of minimum-cost bi-chromatic matching for a set of n points in a metric space,*
- *estimating the cost of metric k-median for $k = n/2$.*

References

1. M. Charikar and S. Guha. Improved combinatorial algorithms for the facility location and k-median problems. *Proceedings of the 40th IEEE Symposium on Foundations of Computer Science (FOCS)*, pp. 378–388, 1999.
2. B. Chazelle, R. Rubinfeld, and L. Trevisan. Approximating the minimum spanning tree weight in sublinear time. *Proceedings of the 28th Annual International Colloquium on Automata, Languages and Programming (ICALP)*, pp. 190–200, 2001.

3. F. A. Chudak. Improved approximation algorithms for uncapacitated facility location. *Proceedings of the 6th International Integer Programming and Combinatorial Optimization Conference (IPCO)*, pp. 180–194, 1998.
4. A. Czumaj and C. Sohler. Estimating the weight of metric minimum spanning trees in sublinear time. *Proceedings of the 36th Annual ACM Symposium on Theory of Computing (STOC)*, pp. 175–183, 2004.
5. S. Guha and S. Khuller. Greedy strikes back: Improved facility location algorithms. *Journal of Algorithms*, 31(1): 228–248, 1999.
6. P. Indyk. Sublinear time algorithms for metric space problems. *Proceedings of the 31st Annual ACM Symposium on Theory of Computing (STOC)*, pp. 428–434, 1999.
7. P. Indyk. A sublinear time approximation scheme for clustering in metric spaces. *Proceedings of the 40th IEEE Symposium on Foundations of Computer Science (FOCS)*, pp. 154–159, 1999.
8. K. Jain, M. Mahdian, and A. Saberi. A new greedy approach for facility location problems. *Proceedings of the 34th Annual ACM Symposium on Theory of Computing (STOC)*, pp. 731–740, 2002.
9. K. Jain and V. Vazirani. Approximaton algorithms for metric facility location and k-median problems using the primal-dual schema and lagrangian relaxation. *Journal of the ACM*, 48(2): 274–296, 2001.
10. M. Mahdian, Y. Ye, and J. Zhang. Improved approximation algorithms for metric facility location problems. *Proceedings of the 5th International Workshop on Approximation Algorithms for Combinatorial Optimization(APPROX)*, pp. 229–242, 2002.
11. C. McDiarmid. Concentration. In M. Habib, C. McDiarmid, J. Ramirez-Alfonsin, and B. Reed, editors, *Probabilistic Methods for Algorithmic Discrete Mathematics*, Algorithms and Combinatorics, pp. 195–247. Springer-Verlag, Berlin, 1998.
12. R. R. Mettu and C. G. Plaxton. The online median problem. *SIAM Journal on Computing*, 32(3): 816–832, 2003.
13. D. B. Shmoys, E. Tardos, and K. Aardal. Approximation algorithms for facility location problems. *Proceedings of the 29th Annual ACM Symposium on Theory of Computing (STOC)*, pp. 265–274, 1997.
14. M. Thorup. Quick k-median, k-center, and facility location for sparse graphs. *SIAM Journal on Computing*, 34(2):405–432, 2005.

The Complexity of Stochastic Rabin and Streett Games[*,**]

Krishnendu Chatterjee[1], Luca de Alfaro[2], and Thomas A. Henzinger[1,3]

[1] EECS, University of California, Berkeley, USA
[2] CE, University of California, Santa Cruz, USA
[3] EPFL, Switzerland
{c_krish, tah}@eecs.berkeley.edu, luca@soe.ucsc.edu

Abstract. The theory of graph games with ω-regular winning conditions is the foundation for modeling and synthesizing reactive processes. In the case of stochastic reactive processes, the corresponding stochastic graph games have three players, two of them (System and Environment) behaving adversarially, and the third (Uncertainty) behaving probabilistically. We consider two problems for stochastic graph games: the *qualitative* problem asks for the set of states from which a player can win with probability 1 (*almost-sure winning*); the *quantitative* problem asks for the maximal probability of winning (*optimal winning*) from each state. We show that for Rabin winning conditions, both problems are in NP. As these problems were known to be NP-hard, it follows that they are NP-complete for Rabin conditions, and dually, coNP-complete for Streett conditions. The proof proceeds by showing that pure memoryless strategies suffice for qualitatively and quantitatively winning stochastic graph games with Rabin conditions. This insight is of interest in its own right, as it implies that controllers for Rabin objectives have simple implementations. We also prove that for every ω-regular condition, optimal winning strategies are no more complex than almost-sure winning strategies.

1 Introduction

A stochastic graph game [5] is played on a directed graph with three kinds of states: player-1, player-2, and probabilistic states. At player-1 states, player 1 chooses a successor state; at player-2 states, player 2 chooses a successor state; and at probabilistic states, a successor state is chosen according to a given probability distribution. The result of playing the game forever is an infinite path through the graph. If there are no probabilistic states, we refer to the game as a *2-player graph game*; otherwise, as a *2½-player graph game*. There has been a long history of using 2-player graph games for modeling and synthesizing reactive processes [1, 14, 16]: a reactive system and its environment represent the

---

[*] This research was supported in part by the ONR grant N00014-02-1-0671, the AFOSR MURI grant F49620-00-1-0327, and the NSF grant CCR-0225610.
[**] Full proofs are available in [2].

two players, whose states and transitions are specified by the states and edges of a game graph. Consequently, 2 1/2-player graph games provide the theoretical foundation for modeling and synthesizing processes that are both reactive and stochastic [9, 15].

For the modeling and synthesis (or "control") of reactive processes, one traditionally considers ω-regular winning conditions, which naturally express the temporal specifications and fairness assumptions of transition systems [11]. This paper focuses on the complexity of solving 2 1/2-player graph games with respect to two important normal forms of ω-regular winning conditions: *Rabin conditions* and *Streett conditions* [17]. Rabin and Streett conditions are dual (i.e., complementary), and their practical relevance stems from the fact that their form corresponds to the form of fairness conditions for transition systems.

In the case of 2-player graph games, where no randomization is involved, a fundamental determinacy result ensures that, given an ω-regular winning condition, at each state, either player 1 has a strategy to ensure that the condition holds, or player 2 has a strategy to ensure that the condition does not hold [10]. Thus, the problem of solving 2-player graph games consists in finding the set of *winning states*, from which player 1 can ensure that the condition holds. This problem is known to be in NP ∩ coNP for parity conditions, to be NP-complete for Rabin conditions [8], and consequently, to be coNP-complete for Streett conditions. The proofs of inclusion in NP rely on the existence of pure (i.e., deterministic) memoryless winning strategies, which act as polynomial witnesses. The existence of pure memoryless winning strategies is also of independent interest, as such strategies can be simply and effectively implemented by a controller. Note that for Streett conditions, winning strategies in general require memory.

In the case of 2 1/2-player graph games, where randomization is present in the transition structure, the notion of winning needs to be clarified. Player 1 is said to *win surely* if she has a strategy that guarantees to achieve the winning condition against all player-2 strategies. While this is the classical notion of winning in the 2-player case, it is less meaningful in the presence of probabilistic states, because it makes all probabilistic choices adversarial (it treats them analogously to player-2 choices). To adequately treat probabilistic choice, we consider the *probability* with which player 1 can ensure that the winning condition is met. We thus define two solution problems for 2 1/2-player graph games: the *qualitative* problem asks for the set of states from which player 1 can ensure winning with probability 1; the *quantitative* problem asks for the maximal probability with which player 1 can ensure winning from each state (this probability is called the *value* of the game at a state) [7]. Correspondingly, we define *almost-sure winning strategies*, which enable player 1 to win with probability 1 whenever possible, and *optimal strategies*, which enable player 1 to win with maximal probability. The main result of this paper is that, in 2 1/2-player graph games, both the qualitative and the quantitative solution problems are NP-complete in the case of Rabin conditions, and coNP-complete in the case of Streett conditions. The NP-hardness for Rabin conditions follows from the NP-hardness of 2-player games with Rabin conditions [8]; we establish the membership in NP. Both questions are known to be in NP ∩ coNP for the more restrictive, self-dual case of parity conditions [4, 13, 18], whose exact complexity is an important open problem.

Our proof of membership in NP for stochastic Rabin games relies on establishing the existence of pure memoryless almost-sure winning and optimal strategies. The corresponding result for stochastic parity games has been proved only recently [4, 13, 18], and these proofs rely on the self-duality of parity conditions. For Rabin conditions, a new proof approach is required. First, we show the existence of pure memoryless almost-sure winning strategies in stochastic Rabin games by a reduction from $2\frac{1}{2}$-player games to 2-player. The reduction preserves the ability of player 1 to win with probability 1, but it does not preserve the maximal probability of winning. The proof technique is combinatorial and uses graph-theoretic arguments to account for the fact that Rabin conditions are not closed under complementation. Second, to show the existence of pure memoryless optimal strategies in stochastic Rabin games, we partition the game graph into value classes, each consisting of states where the value of the game is identical. We prove that if the players play according to optimal strategies, then the game leaves every intermediate value class (in which the value is neither 0 nor 1) with probability 1. We then use the qualitative result on almost-sure winning to establish the existence of pure memoryless optimal strategies.

We emphasize that, as mentioned earlier, the existence of pure memoryless strategies is relevant in its own right, as such strategies consist in mappings that associate with each player-1 state a unique successor, without need for randomization or memory; such mappings are easily implemented in controllers. Furthermore, our techniques lead us to a more general result, which states a strong connection between certain qualitative and quantitative games: we show that for every ω-regular winning condition in a $2\frac{1}{2}$-player game graph, if a restricted family of strategies suffices for almost-sure winning, then it suffices also for optimality. Hence future research on $2\frac{1}{2}$-player games with ω-regular conditions can focus on qualitatively (i.e., almost-sure) winning strategies, and our result generalizes these strategies to quantitatively winning (i.e., optimal) strategies.

2 Definitions

We consider several classes of turn-based games, namely, two-player turn-based probabilistic games ($2\frac{1}{2}$-player games), two-player turn-based deterministic games (2-player games), and Markov decision processes ($1\frac{1}{2}$-player games).

Game graphs. A *turn-based probabilistic game graph* ($2\frac{1}{2}$-*player game graph*) $G = ((S, E), (S_1, S_2, S_\bigcirc), \delta)$ consists of a directed graph (S, E), a partition $(S_1, S_2, S_\bigcirc)$ of the finite set S of states, and a probabilistic transition function δ: $S_\bigcirc \to \mathcal{D}(S)$, where $\mathcal{D}(S)$ denotes the set of probability distributions over the state space S. The states in S_1 are the *player-1* states, where player 1 decides the successor state; the states in S_2 are the *player-2* states, where player 2 decides the successor state; and the states in $S_\bigcirc$ are the *probabilistic* states, where the successor state is chosen according to the probabilistic transition function δ. We assume that for $s \in S_\bigcirc$ and $t \in S$, we have $(s,t) \in E$ iff $\delta(s)(t) > 0$, and we often write $\delta(s,t)$ for $\delta(s)(t)$. For technical convenience we assume that every state in the graph (S, E) has at least one outgoing edge. For a state $s \in S$, we write $E(s)$ to denote the set $\{t \in S \mid (s,t) \in E\}$ of possible successors.

A set $U \subseteq S$ of states is called δ-*closed* if for every probabilistic state $u \in U \cap S_\bigcirc$, if $(u,t) \in E$, then $t \in U$. The set U is called δ-*live* if for every nonprobabilistic state $s \in U \cap (S_1 \cup S_2)$, there is a state $t \in U$ such that $(s,t) \in E$. A δ-closed and δ-live subset U of S induces a *subgame graph* of G, indicated by $G \upharpoonright U$.

The *turn-based deterministic game graphs* (*2-player game graphs*) are the special case of the $2\frac{1}{2}$-player game graphs with $S_\bigcirc = \emptyset$. The *Markov decision processes* ($1\frac{1}{2}$-*player game graphs*) are the special case of the $2\frac{1}{2}$-player game graphs with $S_1 = \emptyset$ or $S_2 = \emptyset$. We refer to the MDPs with $S_2 = \emptyset$ as *player-1 MDPs*, and to the MDPs with $S_1 = \emptyset$ as *player-2 MDPs*.

Plays and strategies. An infinite path, or *play*, of the game graph G is an infinite sequence $\omega = \langle s_0, s_1, s_2, \ldots \rangle$ of states such that $(s_k, s_{k+1}) \in E$ for all $k \in \mathbb{N}$. We write Ω for the set of all plays, and for a state $s \in S$, we write $\Omega_s \subseteq \Omega$ for the set of plays that start from the state s.

A *strategy* for player 1 is a function $\sigma: S^* \cdot S_1 \to \mathcal{D}(S)$ that assigns a probability distribution to all finite sequences $\boldsymbol{w} \in S^* \cdot S_1$ of states ending in a player-1 state (the sequence represents a prefix of a play). Player 1 follows the strategy σ if in each player-1 move, given that the current history of the game is $\boldsymbol{w} \in S^* \cdot S_1$, she chooses the next state according to the probability distribution $\sigma(\boldsymbol{w})$. A strategy must prescribe only available moves, i.e., for all $\boldsymbol{w} \in S^*$, $s \in S_1$, and $t \in S$, if $\sigma(\boldsymbol{w} \cdot s)(t) > 0$, then $(s,t) \in E$. The strategies for player 2 are defined analogously. We denote by Σ and Π the set of all strategies for player 1 and player 2, respectively.

Once a starting state $s \in S$ and strategies $\sigma \in \Sigma$ and $\pi \in \Pi$ for the two players are fixed, the outcome of the game is a random walk $\omega_s^{\sigma,\pi}$ for which the probabilities of events are uniquely defined, where an *event* $\mathcal{A} \subseteq \Omega$ is a measurable set of paths. Given strategies σ for player 1 and π for player 2, a play $\omega = \langle s_0, s_1, s_2, \ldots \rangle$ is *feasible* if for every $k \in \mathbb{N}$ the following three conditions hold: (1) if $s_k \in S_\bigcirc$, then $(s_k, s_{k+1}) \in E$; (2) if $s_k \in S_1$, then $\sigma(s_0, s_1, \ldots, s_k)(s_{k+1}) > 0$; and (3) if $s_k \in S_2$ then $\pi(s_0, s_1, \ldots, s_k)(s_{k+1}) > 0$. Given two strategies $\sigma \in \Sigma$ and $\pi \in \Pi$, and a state $s \in S$, we denote by $\mathrm{Outcome}(s, \sigma, \pi) \subseteq \Omega_s$ the set of feasible plays that start from s given strategies σ and π. For a state $s \in S$ and an event $\mathcal{A} \subseteq \Omega$, we write $\mathrm{Pr}_s^{\sigma,\pi}(\mathcal{A})$ for the probability that a path belongs to $\mathcal{A}$ if the game starts from the state s and the players follow the strategies σ and π, respectively. In the context of player-1 MDPs we often omit the argument π, because Π is a singleton set.

We classify strategies according to their use of randomization and memory. The strategies that do not use randomization are called pure. A player-1 strategy σ is *pure* if for all $\boldsymbol{w} \in S^*$ and $s \in S_1$, there is a state $t \in S$ such that $\sigma(\boldsymbol{w} \cdot s)(t) = 1$. We denote by $\Sigma^P \subseteq \Sigma$ the set of pure strategies for player 1. A strategy that is not necessarily pure is called *randomized*. Let M be a set called *memory*. A player-1 strategy can be described as a pair of functions: a *memory-update* function $\sigma_u \colon S \times \mathtt{M} \to \mathtt{M}$ and a *next-move* function $\sigma_m \colon S_1 \times \mathtt{M} \to \mathcal{D}(S)$. The strategy (σ_u, σ_m) is *finite-memory* if the memory M is finite. We denote by Σ^F the set of finite-memory strategies for player 1, and by Σ^{PF} the set of *pure finite-memory* strategies; that is, $\Sigma^{PF} = \Sigma^P \cap \Sigma^F$. The strategy (σ_u, σ_m) is *memoryless* if $|\mathtt{M}| = 1$; that is, the next move does not depend on the history of the play but only on the current state. A memoryless player-1 strategy can be represented as a function $\sigma \colon S_1 \to \mathcal{D}(S)$. A *pure memoryless strategy* is a pure strategy that is memoryless. A pure memoryless strategy for player 1 can be represented as a function $\sigma \colon S_1 \to S$. We denote by Σ^M the set of memoryless strategies for player 1, and by Σ^{PM} the set of pure memoryless strategies; that is, $\Sigma^{PM} = \Sigma^P \cap \Sigma^M$. Analogously we define the corresponding strategy families $\Pi^P, \Pi^F, \Pi^{PF}, \Pi^M$, and Π^{PM} for player 2.

Given a finite-memory strategy $\sigma \in \Sigma^F$, let G_σ be the game graph obtained from G under the constraint that player 1 follows the strategy σ. The corresponding definition G_π for a player-2 strategy $\pi \in \Pi^F$ is analogous, and we write $G_{\sigma,\pi}$ for the game graph obtained from G if both players follow the finite-memory strategies σ and π, respectively. Observe that given a $2\frac{1}{2}$-player game graph G and a memoryless player-1 strategy σ, the result G_σ is a player-2 MDP. Similarly, for a player-1 MDP G and a memoryless player-1 strategy σ, the result G_σ is a Markov chain. Hence, if G is a $2\frac{1}{2}$-player game graph and the two players follow memoryless strategies σ and π, the result $G_{\sigma,\pi}$ is a Markov chain. These observations will be useful in the analysis of $2\frac{1}{2}$-player games.

Objectives. An *objective* for a player consists of an ω-regular set of *winning plays* $\Phi \subseteq \Omega$ [17]. In this paper we study zero-sum games [9,15], where the objectives of the two players are complementary; that is, if the objective of one player is Φ, then the objective of the other player is $\Omega \setminus \Phi$. We consider ω-regular objectives specified in Rabin or Streett normal forms. For a play $\omega = \langle s_0, s_1, s_2, \ldots \rangle$, let $\mathrm{Inf}(\omega)$ be the set $\{\, s \in S \mid s = s_k \text{ for infinitely many } k \geq 0 \,\}$ of states that occur infinitely often in ω. We use colors to define objectives independent of game graphs. For a set C of colors, we write $[\![\cdot]\!] \colon C \to 2^S$ for a function that maps each color to a set of states. Inversely, given a set $U \subseteq S$ of states, we write $[U] = \{\, c \in C \mid [\![c]\!] \cap U \neq \emptyset \,\}$ for the set of colors that occur in U. Note that a state can have multiple colors.

A *Rabin objective* is specified as a set $P = \{(e_1, f_1), \ldots, (e_d, f_d)\}$ of pairs of colors $e_i, f_i \in C$. Intuitively, the Rabin condition P requires that for some $1 \leq i \leq d$, all states of color e_i be visited finitely often and some state of color f_i be visited infinitely often. Let $[\![P]\!] = \{(E_1, F_1), \ldots, (E_d, F_d)\}$ be the corresponding set of so-called *Rabin pairs*, where $E_i = [\![e_i]\!]$ and $F_i = [\![f_i]\!]$ for all $1 \leq i \leq d$. Formally, the set of winning plays is $\mathrm{Rabin}(P) = \{\, \omega \in \Omega \mid \exists\, 1 \leq i \leq d.\ (\mathrm{Inf}(\omega) \cap E_i = \emptyset \wedge \mathrm{Inf}(\omega) \cap F_i \neq \emptyset)\,\}$. Without loss of generality, we require that $\left(\bigcup_{i \in \{1,2,\ldots,d\}} (E_i \cup F_i)\right) = S$. The *parity* (or *Rabin-chain*) objectives are the special case of Rabin objectives such that $E_1 \subset F_1 \subset E_2 \subset F_2 \ldots \subset E_d \subset F_d$. A *Streett objective* is again specified as a set $P = \{(e_1, f_1), \ldots, (e_d, f_d)\}$

of pairs of colors. The Streett condition P requires that for each $1 \leq i \leq d$, if some state of color f_i is visited infinitely often, then some state of color e_i be visited infinitely often. Formally, the set of winning plays is $\text{Streett}(P) = \{ \omega \in \Omega \mid \forall\, 1 \leq i \leq d.\ (\text{Inf}(\omega) \cap E_i \neq \emptyset \ \vee\ \text{Inf}(\omega) \cap F_i = \emptyset) \}$, for the set $[\![P]\!] = \{(E_1, F_1), \ldots, (E_d, F_d)\}$ of so-called *Streett pairs*. Note that the Rabin and Streett objectives are dual; i.e., the complement of a Rabin objective is a Streett objective, and vice versa. Moreover, every parity objective is both a Rabin objective and a Streett objective.

Sure winning, almost-sure winning, and optimality. Given a player-1 objective Φ, a strategy $\sigma \in \Sigma$ is *sure winning* for player 1 from a state $s \in S$ if for every strategy $\pi \in \Pi$ for player 2, we have $\text{Outcome}(s, \sigma, \pi) \subseteq \Phi$. The strategy σ is *almost-sure winning* for player 1 from the state s for the objective Φ if for every player-2 strategy π, we have $\Pr_s^{\sigma,\pi}(\Phi) = 1$. The sure and almost-sure winning strategies for player 2 are defined analogously. Given an objective Φ, the *sure winning set* $\langle\!\langle 1 \rangle\!\rangle_{sure}(\Phi)$ for player 1 is the set of states from which player 1 has a sure winning strategy. The *almost-sure winning set* $\langle\!\langle 1 \rangle\!\rangle_{almost}(\Phi)$ for player 1 is the set of states from which player 1 has an almost-sure winning strategy. The sure winning set $\langle\!\langle 2 \rangle\!\rangle_{sure}(\Omega \setminus \Phi)$ and the almost-sure winning set $\langle\!\langle 2 \rangle\!\rangle_{almost}(\Omega \setminus \Phi)$ for player 2 are defined analogously. It follows from the definitions that for all $2\frac{1}{2}$-player game graphs and all objectives Φ, we have $\langle\!\langle 1 \rangle\!\rangle_{sure}(\Phi) \subseteq \langle\!\langle 1 \rangle\!\rangle_{almost}(\Phi)$. Computing sure and almost-sure winning sets and strategies is referred to as the *qualitative* analysis of $2\frac{1}{2}$-player games [7].

Given ω-regular objectives $\Phi \subseteq \Omega$ for player 1 and $\Omega \setminus \Phi$ for player 2, we define the *value* functions $\langle\!\langle 1 \rangle\!\rangle_{val}$ and $\langle\!\langle 2 \rangle\!\rangle_{val}$ for the players 1 and 2, respectively, as the following functions from the state space S to the interval $[0, 1]$ of reals: for all states $s \in S$, let $\langle\!\langle 1 \rangle\!\rangle_{val}(\Phi)(s) = \sup_{\sigma \in \Sigma} \inf_{\pi \in \Pi} \Pr_s^{\sigma,\pi}(\Phi)$ and $\langle\!\langle 2 \rangle\!\rangle_{val}(\Omega \setminus \Phi)(s) = \sup_{\pi \in \Pi} \inf_{\sigma \in \Sigma} \Pr_s^{\sigma,\pi}(\Omega \setminus \Phi)$. In other words, the value $\langle\!\langle 1 \rangle\!\rangle_{val}(\Phi)(s)$ gives the maximal probability with which player 1 can achieve her objective Φ from state s, and analogously for player 2. The strategies that achieve the value are called *optimal*: a strategy σ for player 1 is *optimal* from the state s for the objective Φ if $\langle\!\langle 1 \rangle\!\rangle_{val}(\Phi)(s) = \inf_{\pi \in \Pi} \Pr_s^{\sigma,\pi}(\Phi)$. The optimal strategies for player 2 are defined analogously. Computing values is referred to as the *quantitative* analysis of $2\frac{1}{2}$-player games. The set of states with value 1 is called the *limit-sure winning set* [7]. For $2\frac{1}{2}$-player game graphs with ω-regular objectives the almost-sure and limit-sure winning sets coincide [3].

Let $\mathcal{C} \in \{P, M, F, PM, PF\}$ and consider the family $\Sigma^\mathcal{C} \subseteq \Sigma$ of special strategies for player 1. We say that the family $\Sigma^\mathcal{C}$ *suffices* with respect to a player-1 objective Φ on a class $\mathcal{G}$ of game graphs for *sure winning* if for every game graph $G \in \mathcal{G}$ and state $s \in \langle\!\langle 1 \rangle\!\rangle_{sure}(\Phi)$, there is a player-1 strategy $\sigma \in \Sigma^\mathcal{C}$ such that for every player-2 strategy $\pi \in \Pi$, we have $\text{Outcome}(s, \sigma, \pi) \subseteq \Phi$. Similarly, the family $\Sigma^\mathcal{C}$ *suffices* with respect to the objective Φ on the class $\mathcal{G}$ of game graphs for *almost-sure winning* if for every game graph $G \in \mathcal{G}$ and state $s \in \langle\!\langle 1 \rangle\!\rangle_{almost}(\Phi)$, there is a player-1 strategy $\sigma \in \Sigma^\mathcal{C}$ such that for every player-2 strategy $\pi \in \Pi$, we have $\Pr_s^{\sigma,\pi}(\Phi) = 1$; and for *optimality*, if for every game graph $G \in \mathcal{G}$ and state $s \in S$, there is a player-1 strategy $\sigma \in \Sigma^{\mathcal{C}}$ such that $\langle\!\langle 1 \rangle\!\rangle_{val}(\Phi)(s) = \inf_{\pi \in \Pi} \Pr_s^{\sigma,\pi}(\Phi)$.

For sure winning, the $1\frac{1}{2}$-player and $2\frac{1}{2}$-player games coincide with 2-player (deterministic) games where the random player (who chooses the successor at the probabilistic states) is interpreted as an adversary, i.e., as player 2. Theorem 1 and Theorem 2 state the classical determinacy results for 2-player and $2\frac{1}{2}$-player game graphs with ω-regular objectives.

Theorem 1 (Qualitative determinacy [8, 10]**).** *For all 2-player game graphs and Rabin or Streett objectives Φ, we have $\langle\!\langle 1 \rangle\!\rangle_{sure}(\Phi) \cap \langle\!\langle 2 \rangle\!\rangle_{sure}(\Omega \setminus \Phi) = \emptyset$ and $\langle\!\langle 1 \rangle\!\rangle_{sure}(\Phi) \cup \langle\!\langle 2 \rangle\!\rangle_{sure}(\Omega \setminus \Phi) = S$. Moreover, on 2-player game graphs, the family of pure memoryless strategies suffices for sure winning with respect to Rabin objectives, and the family of pure finite-memory strategies suffices for sure winning with respect to Streett objectives.*

Theorem 2 (Quantitative determinacy [12]**).** *For all $2\frac{1}{2}$-player game graphs, all Rabin or Streett objectives Φ, and all states s, we have $\langle\!\langle 1 \rangle\!\rangle_{val}(\Phi)(s) + \langle\!\langle 2 \rangle\!\rangle_{val}(\Omega \setminus \Phi)(s) = 1$.*

3 Qualitative Analysis

We show that the pure memoryless strategies suffice for almost-sure winning with respect to Rabin objectives on $2\frac{1}{2}$-player game graphs. The result is achieved by a reduction to 2-player Rabin games. The reduction also allows us to apply algorithms for solving 2-player Rabin games to the qualitative analysis of $2\frac{1}{2}$-player Rabin games. Furthermore, in the next section, we will use the existence of pure memoryless almost-sure winning strategies to prove the existence of pure memoryless optimal strategies.

End components of MDPs. We review some facts about *end components* [6] which are needed for the further development of the paper. We consider player-1 MDPs and hence strategies for player 1. Let $G = ((S, E), (S_1, S_2, S_\bigcirc), \delta)$ with $S_2 = \emptyset$ be a $1\frac{1}{2}$-player game graph.

Definition 1 (End components). *A set $U \subseteq S$ of states is an* end component *if U is δ-closed and the subgame graph $G \upharpoonright U$ is strongly connected.*

We denote by $\mathcal{E} \subseteq 2^S$ the set of all end-components of G. The next lemma states that, under every strategy (memoryless or not), with probability 1 the set of states visited infinitely often along a play is an end component. This lemma allows us to derive conclusions on the (infinite) set of plays in an MDP by analyzing the (finite) set of end components in the MDP. In particular, the lemma implies that to show that a set $\{(E_1, F_1), \ldots, (E_d, F_d)\}$ of Rabin pairs is satisfied with probability 1, it suffices to show that for each reachable end component U, there exists an $1 \leq i \leq d$ such that $U \cap E_i = \emptyset$ and $U \cap F_i \neq \emptyset$. To state the lemma, for $s \in S$ and $U \subseteq S$, we define $\Omega_s^U = \{\,\omega \in \Omega_s \mid \mathrm{Inf}(\omega) = U\,\}$.

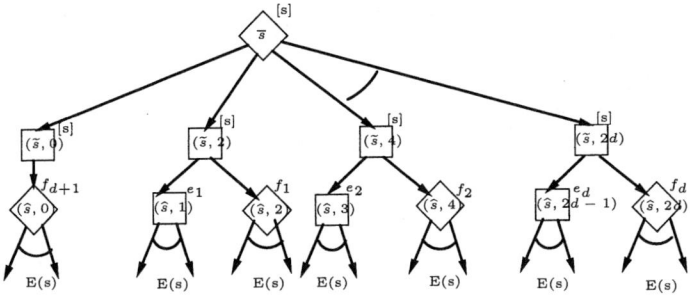

Fig. 1. Gadget for the reduction of $2\frac{1}{2}$-player Rabin games to 2-player Rabin games

Lemma 1. [6] *For all states $s \in S$ and strategies $\sigma \in \Sigma$, $\Pr_s^\sigma(\bigcup_{U \in \mathcal{E}} \Omega_s^U) = 1$.*

Reduction. Given a $2\frac{1}{2}$-player game graph $G = ((S,E),(S_1,S_2,S_\bigcirc),\delta)$, a set $C = \{e_1, f_1, \ldots, e_d, f_d\}$ of colors, and a color map $[\cdot] \colon S \to 2^C \setminus \emptyset$, we construct a 2-player game graph $\overline{G} = ((\overline{S}, \overline{E}), (\overline{S}_1, \overline{S}_2), \overline{\delta})$ together with a color map $[\cdot] \colon \overline{S} \to 2^{\overline{C}} \setminus \emptyset$ for the extended color set $\overline{C} = C \cup \{e_{d+1}, f_{d+1}\}$. The construction is specified as follows. For every nonprobabilistic state $s \in S_1 \cup S_2$, there is a corresponding state $\overline{s} \in \overline{S}$ such that (1) $\overline{s} \in \overline{S}_1$ iff $s \in S_1$, and (2) $[\overline{s}] = [s]$, and (3) $(\overline{s}, \overline{t}) \in \overline{E}$ iff $(s,t) \in E$. Every probabilistic state $s \in S_\bigcirc$ is replaced by the gadget shown in Figure 1. In the figure, diamond-shaped states are player-2 states (in $\overline{S}_2$), and square-shaped states are player-1 states (in $\overline{S}_1$). From the state $\overline{s}$ with $[\overline{s}] = [s]$, the players play the following 3-step game in $\overline{G}$. First, in state $\overline{s}$ player 2 chooses a successor $(\widetilde{s}, 2k)$, for $k \in \{0, 1, \ldots, d\}$. For every state $(\widetilde{s}, 2k)$, we have $[(\widetilde{s}, 2k)] = [s]$. For $k > 1$, in state $(\widetilde{s}, 2k)$ player 1 chooses from two successors: state $(\widehat{s}, 2k-1)$ with $[(\widehat{s}, 2k-1)] = e_k$, or state $(\widehat{s}, 2k)$ with $[(\widehat{s}, 2k)] = f_k$. The state $(\widetilde{s}, 0)$ has only one successor $(\widehat{s}, 0)$, with $[(\widehat{s}, 0)] = f_{d+1}$. Note that no state in $\overline{S}$ is labeled by the new color e_{d+1}, that is, $[e_{d+1}] = \emptyset$. Finally, in each state $(\widehat{s}, j)$ the choice is between all states $\overline{t}$ such that $(s,t) \in E$, and it belongs to player 1 if k is odd, and to player 2 if k is even.

We consider the $2\frac{1}{2}$-player game played on the graph G with the Rabin condition $P = \{(e_1, f_1), \ldots, (e_d, f_d)\}$ for player 1. Let $\overline{U}_1$ and $\overline{U}_2$ be the sure winning sets for players 1 and 2, respectively, in the constructed 2-player game graph $\overline{G}$ with the modified Rabin condition $\overline{P} = \{(e_1, f_1), \ldots, (e_{d+1}, f_{d+1})\}$ for player 1. Define the sets U_1 and U_2 in the original $2\frac{1}{2}$-player game graph G by $U_1 = \{s \in S \mid \overline{s} \in \overline{U}_1\}$ and $U_2 = \{s \in S \mid \overline{s} \in \overline{U}_2\}$. From the determinacy of 2-player Rabin games (Theorem 1), it follows that $\overline{U}_1 = \overline{S} \setminus \overline{U}_2$, and hence $U_1 = S \setminus U_2$.

Lemma 2. *In the $2\frac{1}{2}$-player game graph G with the Rabin condition P for player 1, there exists a pure memoryless strategy σ for player 1 such that for all player-2 strategies π and all states $s \in U_1$, we have $\Pr_s^{\sigma,\pi}(\mathrm{Rabin}(P)) = 1$.*

Proof. We define a pure memoryless strategy σ for player 1 in the game G from a strategy $\overline{\sigma}$ in the game $\overline{G}$ as follows: for all states $s \in S_1$, if $\overline{\sigma}(\overline{s}) = \overline{t}$, then set $\sigma(s) = t$. Consider a pure memoryless sure winning strategy $\overline{\sigma}$ in the game $\overline{G}$ from every state $\overline{s} \in \overline{U}_1$. Our goal is to establish that σ is an almost-sure winning strategy from every state in U_1.

For the Rabin objective Rabin(P), let the set Rabin pairs be $[\![P]\!] = \{(E_1, F_1), (E_2, F_2), \ldots, (E_d, F_d)\}$. A strongly connected component (s.c.c.) W in a graph G_1 is winning for player 1, if there exists $i \in \{1, 2, \ldots, d\}$ such that $W \cap F_i \neq \emptyset$ and $W \cap E_i = \emptyset$; otherwise W is winning for player 2. If G_1 is a MDP, then an end component W in G_1 is winning for player 1, if there exists $i \in \{1, 2, \ldots, d\}$ such that $W \cap F_i \neq \emptyset$ and $W \cap E_i = \emptyset$; otherwise W is winning for player 2.

We prove that every end component in the player-2 MDP $(G \upharpoonright U_1)_\sigma$ is winning for player 1. It would follow from Lemma 1 that σ is an almost-sure winning strategy. We argue that if there is an end component W in $(G \upharpoonright U_1)_\sigma$ that is winning for player 2, then we can construct an s.c.c. in the subgraph $(\overline{G} \upharpoonright \overline{U}_1)_{\overline{\sigma}}$ that is winning for player 2, which is impossible because $\overline{\sigma}$ is a sure winning strategy for player 1 from the set $\overline{U}_1$ in the 2-player Rabin game $\overline{G}$. Let W be an end component in $(G \upharpoonright U_1)_\sigma$ that is winning for player 2. We denote by $\overline{W}$ the set of states in the gadget of states in W. Hence for all $i \in \{1, 2, \ldots, d\}$ we have if $F_i \cap W \neq \emptyset$, then $W \cap E_i \neq \emptyset$. Let us define the set $I = \{i_1, i_2, \ldots, i_j\}$ such that $E_{i_k} \cap W \neq \emptyset$. Thus for all $i \in (\{1, 2, \ldots, d\} \setminus I)$ we have $F_i \cap W = \emptyset$. Note that $I \neq \emptyset$, as every state has at least one color. We now construct a sub-game in $\overline{G}_{\overline{\sigma}}$ as follows:

1. For a state $\overline{s} \in \overline{W} \cap \overline{S}_2$ keep all the edges $(\overline{s}, \overline{t})$ such that $\overline{t} \in \overline{W}$.
2. For a state $\overline{s} \in \overline{W} \cap \overline{S}_\bigcirc$ the sub-game is defined as follows:
   – At state $\overline{s}$ choose the edges to state $(\widetilde{s}, 2i)$ such that $i \in I$.
   – For a state $s \in W$, let $dis(s, W \cap E_i)$ denote the shortest distance (BFS distance) from s to $W \cap E_i$ in the graph of $(G \upharpoonright W)_\sigma$. At state $(\widehat{s}, 2i)$, which is a player 2 state, player 2 chooses a successor $\widehat{s}_1$ such that $dis(s_1, W \cap E_i) < dis(s, W \cap E_i)$ (i.e., shorten distance to the set $\overline{W} \cap E_i$ in $\overline{G}$).

We now prove that every terminal s.c.c. is winning for player 2 in the subgame thus constructed in $(\overline{G} \upharpoonright \overline{W})_{\overline{\sigma}}$, where $\overline{W}$ is the set of states in the gadget of states in W. Consider any arbitrary terminal s.c.c. $\overline{Y}$ in the subgame constructed in $(\overline{G} \upharpoonright \overline{W})_{\overline{\sigma}}$. It follows from the construction that for every $i \in (\{1, 2, \ldots, d\} \setminus I)$, we have $F_i \cap \overline{Y} = \emptyset$. Suppose for a $i \in I$ we have $F_i \cap \overline{Y} \neq \emptyset$, we show that $E_i \cap \overline{Y} \neq \emptyset$. There are two cases:

1. If there is at least one state $(\widetilde{s}, 2i)$ such that the strategy $\overline{\sigma}$ chooses the successor $(\widehat{s}, 2i - 1)$, then $E_i \cap \overline{Y} \neq \emptyset$, since $[(\widetilde{s}, 2i - 1)] = e_i$.
2. Else for every state $(\widetilde{s}, 2i)$ the strategy for player 1 chooses the successor $(\widehat{s}, 2i)$. At state $(\widehat{s}, 2i)$, which is a player 2 state, player 2 chooses a successor $\widehat{s}_1$ that shortens distance to the set $\overline{Y} \cap E_i$. Hence the terminal s.c.c. $\overline{Y}$ must contain a state $\overline{s}$ such that $[\overline{s}] = e_i$. Hence $E_i \cap \overline{Y} \neq \emptyset$.

We argue that for every probabilistic state $s \in S_\bigcirc \cap U_1$, all of its successors are in U_1. Otherwise, player 2 in the state $\overline{s}$ of the game $\overline{G}$ can choose the successor $(\tilde{s}, 0)$ and then a successor to its winning set $\overline{U}_2$, which contradicts the assumption that the strategy $\overline{\sigma}$ is a sure winning strategy for player 1 in the game $\overline{G}$ from $\overline{U}_1$. It follows from Lemma 1 that for all strategies π, for all states $s \in U_1$, with probability 1 the set of states visited infinitely often along the play $\omega_s^{\sigma,\pi}$ is an end component in U_1. Since every end component in $(G \upharpoonright U_1)_\sigma$ is winning for player 1 the strategy σ is an almost-sure winning strategy for player 1 from U_1. ∎

Lemma 3. *In the $2\frac{1}{2}$-player game graph G with the Rabin condition P for player 1, there exists a finite-memory strategy π for player 2 such that for all player-1 strategies σ and all states $s \in U_2$, we have $\Pr_s^{\sigma,\pi}(\Omega \setminus \mathrm{Rabin}(P)) > 0$.*

From Lemma 2, it follows that $U_1 \subseteq \langle\!\langle 1 \rangle\!\rangle_{almost} \mathrm{Rabin}(P)$. From Lemma 3, it follows that $\langle\!\langle 1 \rangle\!\rangle_{almost} \mathrm{Rabin}(P) \subseteq U_1$. Therefore $U_1 = \langle\!\langle 1 \rangle\!\rangle_{almost} \mathrm{Rabin}(P)$. The proof of Lemma 2 also establishes the existence of pure memoryless almost-sure winning strategies for Rabin objectives.

Theorem 3. *The family of pure memoryless strategies suffices for almost-sure winning with respect to Rabin objectives on $2\frac{1}{2}$-player game graphs.*

4 Quantitative Analysis

We extend sufficiency results for families of strategies from almost-sure winning to optimality with respect to all ω-regular objectives. In the following, we fix a $2\frac{1}{2}$-player game graph G. Given an ω-regular objective Φ, for every real $r \in \mathbb{R}$ the *value class* with value r is $\mathrm{VC}(r) = \{s \in S \mid \langle\!\langle 1 \rangle\!\rangle_{val}(\Phi)(s) = r\}$. Proposition 1 states that there exist optimal strategies for player 1 such that they never choose an edge to a lower value class.

Proposition 1. *For all ω-regular objectives Φ, there exists an optimal strategy σ for player 1 such that for all $w \in S^*$, $s \in S_1$, and $t \in S$, if $\langle\!\langle 1 \rangle\!\rangle_{val}(\Phi)(t) < \langle\!\langle 1 \rangle\!\rangle_{val}(\Phi)(s)$, then $\sigma(w \cdot s)(t) = 0$.*

Definition 2 (Boundary probabilistic states). *Given an ω-regular objective Φ, a probabilistic state $s \in S_\bigcirc$ is a boundary probabilistic state if there exists a successor $t \in E(s)$ such that $\langle\!\langle 1 \rangle\!\rangle_{val}(\Phi)(t) \neq \langle\!\langle 1 \rangle\!\rangle_{val}(\Phi)(s)$. Observe that for every boundary probabilistic state s, there exist $t_1, t_2 \in E(s)$ such that $\langle\!\langle 1 \rangle\!\rangle_{val}(\Phi)(t_1) < \langle\!\langle 1 \rangle\!\rangle_{val}(\Phi)(s)$ and $\langle\!\langle 1 \rangle\!\rangle_{val}(\Phi)(t_2) > \langle\!\langle 1 \rangle\!\rangle_{val}(\Phi)(s)$.*

Lemma 4. *Consider a $2\frac{1}{2}$-player game G with an ω-regular objective Φ. Given a value class $\mathrm{VC}(r)$ with $0 < r < 1$, let $B(r)$ be the set of boundary probabilistic states in the value class $\mathrm{VC}(r)$. Convert each state in $B(r)$ into a sink state that is winning for player 1. Let the new game be G'. Then player 1 wins almost-surely from all states in the subgame with game graph $G' \upharpoonright \mathrm{VC}(r)$ and objective Φ.*

Proof. Assume that player 1 does not win almost-surely from every state in $G' \upharpoonright \mathrm{VC}(r)$. Then there exists a state where player 2 wins with positive bounded probability. It follows from Corollary 1 of [7] that there exist a non-empty set $U \subseteq \mathrm{VC}(r)$ such that that player 2 wins almost-surely from U in $G' \upharpoonright \mathrm{VC}(r)$. Consider an optimal strategy σ that never chooses an edge with positive probability to a lower value class (such a strategy exists from Proposition 1). Since player 2 wins almost-surely from U it follows that for every state $s \in U \cap S_1$, for every successor t of s in $\mathrm{VC}(r)$ we have $t \in U$. It follows that every move of the strategy σ exists in U. Hence player 2 wins almost-surely from U against σ. This is a contradiction to the assumption that $r > 0$ and that σ is an optimal strategy. ∎

Definition 3 (Qualitatively optimal strategies). *A strategy σ is qualitatively optimal for player 1, for an ω-regular objective Φ, if the following conditions hold: (a) for every state $s \in \langle\!\langle 1 \rangle\!\rangle_{almost}(\Phi)$, the strategy σ is almost-sure winning, and (b) for every state $s \in \mathrm{VC}(r)$ such that $0 < r < 1$, there is a constant $c > 0$ such that $\inf_{\pi \in \Pi} \mathrm{Pr}_s^{\sigma,\pi}(\Phi) \geq c$.*

Lemma 4 shows that in every value class, if the boundary probabilistic states are assumed to be winning for player 1, then player 1 wins almost-surely. We call such an almost-sure winning strategy a *conditional* almost-sure winning strategy. We compose conditional almost-sure winning strategies in value classes to obtain an optimal strategy. If a strategy σ is conditional almost-sure winning, it follows that for all player-2 strategies π that are optimal against σ, the play $\omega_s^{\sigma,\pi}$ reaches the boundary probabilistic states with positive probability, for $s \in \mathrm{VC}(r)$ and $r > 0$. From every boundary probabilistic state the game proceeds to a higher value class with positive probability. An induction on the number of value classes yields Lemma 5.

Lemma 5. *For every ω-regular objective Φ, if a player-1 strategy σ is almost-sure winning from every state $s \in \langle\!\langle 1 \rangle\!\rangle_{almost}(\Phi)$, and is conditionally almost-sure winning from every state $s \notin \langle\!\langle 2 \rangle\!\rangle_{almost}(\Omega \setminus \Phi)$, then σ is qualitatively optimal for Φ.*

Definition 4 (Locally optimal strategies). *A strategy σ is locally optimal for player 1, for an ω-regular objective Φ, if for all $w \in S^*$, $s \in S_1$, and $t \in S$, if $\langle\!\langle 1 \rangle\!\rangle_{val}(\Phi)(t) < \langle\!\langle 1 \rangle\!\rangle_{val}(\Phi)(s)$, then $\sigma(w \cdot s)(t) = 0$.*

Note that by definition, a conditional almost-sure winning strategy is locally optimal. The following Lemma generalizes Lemma 5.3 of [4]. Theorem 4 follows from Lemma 6. Since pure memoryless strategies suffice for almost-sure winning with respect to Rabin objectives on $2\frac{1}{2}$-player game graphs (Theorem 3), Theorem 5 is immediate from Theorem 4.

Lemma 6. *Consider a $2\frac{1}{2}$-player game G with an ω-regular objective Φ for player 1. Let σ be a finite-memory strategy such that σ is both qualitatively optimal and locally optimal for Φ. Then σ is an optimal strategy for Φ from all states of G.*

Theorem 4. *If a family Σ^C of strategies suffices for almost-sure winning with respect to an ω-regular objective Φ on $2\frac{1}{2}$-player game graphs, then Σ^C suffices for optimality with respect to Φ on $2\frac{1}{2}$-player game graphs.*

Theorem 5. *The family of pure memoryless strategy suffices for optimality with respect to Rabin objectives on $2\frac{1}{2}$-player game graphs.*

The existence of pure memoryless optimal strategies for $2\frac{1}{2}$-player game graphs with Rabin objectives, and of polynomial-time algorithms for computing the values of MDPs with Streett objectives [2], establishes that the $2\frac{1}{2}$-player games with Rabin objectives can be decided (qualitatively and quantitatively) in NP. The NP-hardness follows from the hardness of 2-player Rabin games.

Theorem 6. *Given a $2\frac{1}{2}$-player game graph G, an objective Φ for player 1, a state s of G, and a rational r, the complexity of determining whether $\langle\!\langle 1 \rangle\!\rangle_{val}(\Phi)(s) \geq r$ is as follows: NP-complete if Φ is a Rabin objective; coNP-complete if Φ is a Streett objective; and in NP $\cap$ coNP if Φ is a parity objective.*

References

1. J.R. Büchi and L.H. Landweber. Solving sequential conditions by finite-state strategies. *Transactions of the AMS*, 138:295–311, 1969.
2. K. Chatterjee, L. de Alfaro, and T.A. Henzinger. The complexity of stochastic Rabin and Streett games. Technical Report UCB/CSD-3-1355, UC Berkeley, 2004.
3. K. Chatterjee, M. Jurdziński, and T.A. Henzinger. Simple stochastic parity games. In *CSL'03*, volume 2803 of *LNCS*, pages 100–113. Springer, 2003.
4. K. Chatterjee, M. Jurdziński, and T.A. Henzinger. Quantitative stochastic parity games. In *SODA'04*, pages 114–123. SIAM, 2004.
5. A. Condon. The complexity of stochastic games. *Information and Computation*, 96:203–224, 1992.
6. L. de Alfaro. *Formal Verification of Probabilistic Systems*. PhD Thesis, Stanford University, 1997.
7. L. de Alfaro and T.A. Henzinger. Concurrent ω-regular games. In *LICS'00*, pages 141–154. IEEE Computer Society, 2000.
8. E.A. Emerson and C. Jutla. The complexity of tree automata and logics of programs. In *FOCS'88*, pages 328–337. IEEE Computer Society, 1988.
9. J. Filar and K. Vrieze. *Competitive Markov Decision Processes*. Springer, 1997.
10. Y. Gurevich and L. Harrington. Trees, automata, and games. In *STOC'82*, pages 60–65. ACM, 1982.
11. Z. Manna and A. Pnueli. *The Temporal Logic of Reactive and Concurrent Systems: Specification*. Springer, 1992.
12. D.A. Martin. The determinacy of Blackwell games. *Journal of Symbolic Logic*, 63:1565–1581, 1998.
13. A.K. McIver and C.C. Morgan. Games, probability, and the quantitative μ-calculus $qm\mu$. In *LPAR'02*, volume 2514 of *LNAI*, pages 292–310. Springer, 2002.
14. A. Pnueli and R. Rosner. On the synthesis of a reactive module. In *POPL'89*, pages 179–190. ACM, 1989.

15. T.E.S. Raghavan and J.A. Filar. Algorithms for stochastic games—a survey. *ZOR—Methods and Models of Operations Research*, 35:437–472, 1991.
16. P.J. Ramadge and W.M. Wonham. Supervisory control of a class of discrete-event processes. *SIAM Journal of Control and Optimization*, 25:206–230, 1987.
17. W. Thomas. Languages, automata, and logic. In *Handbook of Formal Languages*, volume 3 (*Beyond Words*), pages 389–455. Springer, 1997.
18. W. Zielonka. Perfect-information stochastic parity games. In *FoSSaCS'04*, volume 2987 of *LNCS*, pages 499–513. Springer, 2004.

Recursive Markov Decision Processes and Recursive Stochastic Games

Kousha Etessami[1] and Mihalis Yannakakis[2]

[1] LFCS, School of Informatics, University of Edinburgh
[2] Department of Computer Science, Columbia University

Abstract. We introduce Recursive Markov Decision Processes (RMDPs) and Recursive Simple Stochastic Games (RSSGs), and study the decidability and complexity of algorithms for their analysis and verification. These models extend Recursive Markov Chains (RMCs), introduced in [EY05a, EY05b] as a natural model for verification of probabilistic procedural programs and related systems involving both recursion and probabilistic behavior. RMCs define a class of denumerable Markov chains with a rich theory generalizing that of stochastic context-free grammars and multi-type branching processes, and they are also intimately related to probabilistic pushdown systems. RMDPs & RSSGs extend RMCs with one controller or two adversarial players, respectively. Such extensions are useful for modeling nondeterministic and concurrent behavior, as well as modeling a system's interactions with an environment.

We provide upper and lower bounds for deciding, given an RMDP (or RSSG) A and probability p, whether player 1 has a strategy to force termination at a desired exit with probability at least p. We also address "qualitative" termination, where $p = 1$, and model checking questions.

1 Introduction

Markov Decision Processes (MDPs) are a fundamental formalism for modeling control optimization problems in sequential stochastic environments. They have found widespread applications in many fields (see, e.g., [Put94, FS02]). They have also been studied extensively in recent years for verification of probabilistic systems. Stochastic games generalize MDPs with multiple players, and in their 2-player zero-sum version are also known as Competitive MDPs (see [FV97]). Simple Stochastic Games (SSGs) [Con92] are a special class of 2-player zero-sum stochastic games, where the goal of one player is to reach a given terminal state, while the other aims to avoid it. SSGs generalize parity games and other important games for model checking, and the termination problem for finite SSGs already presents a well-known algorithmic challenge: it is in NP ∩ coNP, but no P-time algorithm is known ([Con92]).

Recursive Markov Chains (RMCs) were introduced and studied in our earlier work ([EY05a, EY05b]) as a natural model of probabilistic procedural programs and systems exhibiting both recursion and probabilistic behavior. There we provided strong upper and lower bounds for both reachability and ω-regular model checking questions for RMCs. Informally, a RMC consists of a (finite) collection of finite state Markov chains that can call each other in a potentially recursive manner. RMCs define a class of denumerable Markov chains with a rich theory generalizing that of Stochastic Context-Free Grammars (SCFGs) (see, e.g., [MS99]) and Multi-Type Branching Processes ([Har63]), both of which correspond to 1-exit RMCs: RMCs in which each component Markov chain has 1 terminating exit state where it can return control back to a component that called it. RMCs are also intimately related to probabilistic Pushdown Systems (pPDSs), which have also been studied recently in connection to verification of probabilistic programs ([EKM04, BKS05]).

For verification, it is natural and useful to extend RMCs with nondeterministic choice, where some states are controlled by the system while others exhibit probabilistic behavior. Indeed, finite MDPs have been studied extensively for verification of probabilistic systems, and optimized verification tools already exist for them (see, e.g.,[CY98, Var85, dAKN$^+$00, Kwi03]; [Kwi03] is a recent survey). SSGs extend MDPs with a second (adversarial) player. Like non-probabilistic game graphs, they can also be used to model and analyze the interactions between a controlled (but probabilistic) system and an (adversarial) environment.

In this paper we focus on precisely such extensions of RMCs: we introduce *Recursive Markov Decision Processes* (RMDPs) and *Recursive Simple Stochastic Games* (RSSGs), which define natural classes of countable MDPs and SSGs, respectively. In the stochastic dynamic programming literature, MDPs are studied under many different reward criteria, such as average reward, discounted reward, etc. Our focus here is on verification of probabilistic systems, and for this purpose we study RMDPs and RSSGs under reachability criteria which are central to any analysis like model checking. In particular, we ask the *quantitative termination* question: given an RMDP (or RSSG) A and a probability p, is there a strategy for the controller where (regardless of the strategy used by the adversary, in the case of RSSGs) the process terminates at a desired exit with probability at least p (or at most p)? We also ask the *qualitative* question of whether the controller has a strategy to force termination with probability 1. Lastly, we address model checking questions.

Our positive results apply primarily to 1-exit RMDPs and 1-exit RSSGs, which correspond to controlled and game extensions, respectively, of both SCFGs and Multi-Type Branching Processes (MT-BPs). Branching processes are an important class of stochastic processes, dating back to the early work of Galton and Watson in the 19th century (they studied the single-type case, a subcase of 1-exit 1-entry 1-component RMCs), and continuing in the 20th century in the work of Kolmogorov, Sevastianov, Harris and others for MT-BPs and beyond (see, e.g., [Har63]). These have been used to model a wide variety of applications, including in population genetics ([Jag75]), nuclear chain reactions, and RNA modeling in computational biology (based on SCFGs) ([SBH$^+$94]). SCFGs are also fundamental models in statistical natural language processing (see, e.g., [MS99]). 1-exit RMDPs correspond to a controlled version of MT-BPs (and SCFGs): the reproduction of some types can be controlled, while the dynamics of other types is probabilistic as in ordinary MT-BPs. This model would also be suitable for analysis of population dynamics under worst-case (or best-case) assumptions for some types and probabilistic assumptions for others. Such controlled MT-BPs can be readily translated to 1-entry, 1-exit RMDPs, where the number of components is bounded by the number of types (a reverse translation is possible, but will not in general preserve the number of components, i.e., 1-entry, 1-exit RMDPS with a bounded number of components are more general than MT-BPs with a bounded number of types). Thus, our results on 1-exit RMDPs apply, among other things, to such controlled MT-BPs; these do not appear to have been studied in the rich Branching Process literature. Indeed, even some basic algorithmic problems about SCFGs and MT-BPs had received limited attention prior to our work in [EY05a, EY05b].

We now outline our main results in this paper:

- We show that the Least Fixed Point solution of certain systems of nonlinear min/max equations captures optimal termination probabilities for 1-exit RMDPs & 1-exit RSSGs. These equations generalize linear Bellman's equations for finite MDPs (see, e.g., [Put94, FV97]) and also generalize the monotone systems of nonlinear equations for RMCs that we studied in ([EY05a]).
- We show a quite nontrivial *Stackless & Memoryless (S&M) Determinacy* result for 1-exit RSSG termination, whereas we observe this fails badly even for 2-exit RMDPs (namely, optimal strategies of any kind do not always exist for 2-exit RMDP termination; one must make do with ϵ-optimal strategies).
- Using the equations, we show that quantitative termination for 1-exit RMDPs and 1-exit RSSGs is decidable in PSPACE. This matches our PSPACE upper bound for the special case of 1-exit RMCs in [EY05a] and, as shown there, it can not be improved without resolving a long standing open problem in the complexity of numerical computation, namely the square-root sum problem.
- Using S&M-determinacy, we show qualitative termination (where $p = 1$) can be decided in NP for 1-exit RMDPs, and in $\Sigma_2^P \cap \Pi_2^P$ for 1-exit RSSGs.
- For the special case of *linearly recursive* 1-exit RMDPs (RSSGs), we show that the exact optimal, and rational, termination probabilities can be computed in polynomial time (in NP∩co-NP, respectively).
- Lastly, and unfortunately, we show that for multi-exit RMDPs & RSSGs the situation is far worse: even qualitative termination for general RMDPs is undecidable, even when the number of exits in bounded by a fixed constant and the RMDP is restricted to be linearly-recursive. It is even undecidable, for any fixed $\epsilon > 0$, to distinguish whether the optimal value is 1 or $< \epsilon$. So optimal probabilities can not be approximated in a strong sense, with any resources. Furthermore, we show undecidability applies already to qualitative model checking of 1-exit RMDPs, against regular or LTL properties. Our undecidability results are derived from classic and recent undecidability results for Probabilistic Finite Automata (PFA) [Paz71, CL89, BC03]. We show PFAs can be viewed as essentially a special case of multi-exit RMDPs.

Related work. Both MDPs and Stochastic Games have a vast literature, dating back to Bellman and Shapley (see, e.g., [Put94, FS02, FV97]). MDPs are studied in both finite state and infinite state variants. Verification of finite state MDPs, also called concurrent Markov chains, has been studied for a long time (see, e.g., [CY98, CY95, Var85, HSP83]). [CY98] provides efficient algorithms for ω-regular model checking of finite MDPs. Model checking tools like PRISM contain optimized implementations of branching-time model checkers for finite MDPs (see, e.g., [dAKN+00, Kwi03]).

Our earlier work [EY05a, EY05b] developed the basic theory of RMCs and studied efficient algorithms for both their reachability analysis and model checking. We showed, among many results, that qualitative model checking of ω-regular properties for 1-exit RMCs can be decided in polynomial time in the size of the RMC, and that quantitative model checking of RMCs can be done in PSPACE in the size of the RMC. As mentioned, 1-exit RMCs correspond to both MT-BPs and SCFGs (see, e.g., [Har63] and [MS99]), while general RMCs are intimately related to probabilistic Pushdown Systems (pPDSs). Model checking questions for pPDSs, for both linear and branching time properties, have also been recently studied in [EKM04, BKS05]. RMDPs and RSSGs are natural extensions of RMCs, introducing nondeterministic and game behavior. Countable state MDPs are studied extensively in the MDP literature (see, e.g., [Put94, FS02]), but the concise representations afforded by RMDPs and its algorithmic properties, appear not to have been studied prior to our work.

2 Basics

A *Recursive Simple Stochastic Game (RSSG)*, A, is a tuple $A = (A_1, \ldots, A_k)$, where each *component graph* $A_i = (N_i, B_i, Y_i, En_i, Ex_i, \mathtt{pl}_i, \delta_i)$ consists of:

- A set N_i of *nodes*. Let $N = \cup_{i=1}^k N_i$ be the (disjoint) union of all nodes of A.
- A distinguished subset of *entry* nodes $En_i \subseteq N_i$, and a disjoint subset of *exit* nodes $Ex_i \subseteq N_i$. Let $En = \cup_{i=1}^k En_i$ and $Ex = \cup_{i=1}^k Ex_i$.
- A set B_i of *boxes*. Let $B = \cup_{i=1}^k B_i$ be the (disjoint) union of all boxes of A.
- A mapping $Y_i : B_i \mapsto \{1, \ldots, k\}$ that assigns to every box (the index of) of a component. Let $Y = \cup_{i=1}^k Y_i$ be the map $Y : B \mapsto \{1, \ldots, k\}$ where $Y|_{B_i} = Y_i$, for $1 \leq i \leq k$.
- To each box $b \in B_i$, we associate a set of *call ports*, $Call_b = \{(b, en) \mid en \in En_{Y(b)}\}$, and a set of *return ports*, $Return_b = \{(b, ex) \mid ex \in Ex_{Y(b)}\}$. Let $Call^i = \cup_{b \in B_i} Call_b$ and let $Call = \cup_{i=1}^k Call^i$ denote all calls in A. Similarly, define $Return^i$ and $Return$.
- We let $Q_i = N_i \cup Call^i \cup Return^i$, denote collectively the nodes, call ports, and return ports, We will use the term *vertex* of A_i to refer to elements of Q_i. We let $Q = \bigcup_{i=1}^k Q_i$ be the set of all vertices of the RSSG A.
- A mapping $\mathtt{pl}_i : Q_i \mapsto \{0, 1, 2\}$ that assigns to every vertex a player (Player 0 represents "chance" or "nature"). We assume $\mathtt{pl}_i(ex) = 0$ for all $ex \in Ex_i$. Let $\mathtt{pl} = \cup_{i=1}^k \mathtt{pl}_i$ denote $\mathtt{pl} : Q \mapsto \{0, 1, 2\}$ where $\mathtt{pl}|_{Q_i} = \mathtt{pl}_i$, for $1 \leq i \leq k$.
- A transition relation $\delta_i \subseteq (Q_i \times (\mathbb{R} \cup \{\bot\}) \times Q_i)$, where for each tuple $(u, x, v) \in \delta_i$, the source $u \in (N_i \setminus Ex_i) \cup Return^i$, the destination $v \in$

$(N_i \setminus En_i) \cup Call^i$, and x is either (i) a real number $p_{u,v} \in [0,1]$ (the transition probability) if $\text{pl}(u) = 0$, or (ii) $x = \bot$ if $\text{pl}(u) = 1$ or 2. For computational purposes we assume that the given probabilities $p_{u,v}$ are rational. Furthermore they must satisfy the consistency property: for every $u \in \text{pl}^{-1}(0)$, $\sum_{\{v' | (u, p_{u,v'}, v') \in \delta_i\}} p_{u,v'} = 1$, unless u is a call port or exit node, neither of which have outgoing transitions, in which case by default $\sum_{v'} p_{u,v'} = 0$. Let $\delta = \cup_i \delta_i$ be the set of all transitions of A.

An RSSG A defines a global denumerable Simple Stochastic Game (SSG) $M_A = (V = V_0 \cup V_1 \cup V_2, \Delta, \text{pl})$ as follows. The global *states* $V \subseteq B^* \times Q$ of M_A are pairs of the form $\langle \beta, u \rangle$, where $\beta \in B^*$ is a (possibly empty) sequence of boxes and $u \in Q$ is a *vertex* of A. More precisely, the states $V \subseteq B^* \times Q$ and transitions Δ are defined inductively as follows:

1. $\langle \epsilon, u \rangle \in V$, for $u \in Q$. (ϵ denotes the empty string.)
2. if $\langle \beta, u \rangle \in V$ & $(u, x, v) \in \delta$, then $\langle \beta, v \rangle \in V$ and $(\langle \beta, u \rangle, x, \langle \beta, v \rangle) \in \Delta$.
3. if $\langle \beta, (b, en) \rangle \in V$ & $(b, en) \in Call_b$, then $\langle \beta b, en \rangle \in V$ & $(\langle \beta, (b, en) \rangle, 1, \langle \beta b, en \rangle) \in \Delta$.
4. if $\langle \beta b, ex \rangle \in V$ & $(b, ex) \in Return_b$, then $\langle \beta, (b, ex) \rangle \in V$ & $(\langle \beta b, ex \rangle, 1, \langle \beta, (b, ex) \rangle) \in \Delta$.

Item 1 corresponds to the possible initial states, item 2 corresponds to control staying within a component, item 3 is when a new component is entered via a box, item 4 is when control exits a box and returns to the calling component. The mapping $\text{pl} : V \mapsto \{0, 1, 2\}$ is given as follows: $\text{pl}(\langle \beta, u \rangle) = \text{pl}(u)$ if u is in $Q \setminus (Call \cup Ex)$, and $\text{pl}(\langle \beta, u \rangle) = 0$ if $u \in Call \cup Ex$. The set of vertices V is partitioned into V_0, V_1, and V_2, where $V_i = \text{pl}^{-1}(i)$.

We consider M_A with various *initial states* of the form $\langle \epsilon, u \rangle$, denoting this by M_A^u. Some states of M_A are *terminating states* and have no outgoing transitions. These are states $\langle \epsilon, ex \rangle$, where ex is an exit node.

An RSSG where $V_2 = \emptyset$ ($V_1 = \emptyset$) is called a maximizing (minimizing, respectively) *Recursive Markov Decision Process* (RMDP); an RSSG where $V_1 \cup V_2 = \emptyset$ is called a *Recursive Markov Chain* (RMC) ([EY05a, EY05b]); an RSSG where $V_0 \cup V_2 = \emptyset$ is called a *Recursive Graph* ([AEY01]); an RSSG where $V_0 = \emptyset$ is called a *Recursive Game Graph* (see [ATM03, Ete04]). We use *1-exit RSSG* to refer to RSSGs where every component has 1 exit. W.l.o.g., we can assume every component has 1 entry, because multi-entry RSSGs can be transformed to equivalent 1-entry RSSGs with polynomial blowup (similar to RSM transformations [AEY01]). This is decidedly not so for exits: 1-exit RSSGs form an important sub-class of RSSGs and are the main focus of our upper bounds. We shall call a RSSG (RMDP, RMC, etc.) *linearly-recursive* (denoted lr-RSSG, etc.) if there in no path of transitions in any component from any return port to a call port. lr-RMCs are much easier to analyse than general RMCs: reachability probabilities are rational and both reachability analysis and model checking can be performed with the same complexity as for finite Markov chains, using the decomposed Newton's method [EY05a] and techniques we developed in [EY05a, EY05b] (although lr-RMCs were not mentioned explicitly in [EY05a, EY05b]).

A basic goal is to answer termination questions for RSSGs: *"Does player 1 have a strategy to force the game to terminate at ex (i.e., reach state $\langle \epsilon, ex \rangle$),* starting at $\langle \epsilon, u \rangle$, with probability $\geq p$, regardless of how player 2 plays?". A *strategy* σ for player i, $i \in \{1, 2\}$, is a function $\sigma : V^* V_i \mapsto V$, where, given the history $ws \in V^* V_i$ of play so far, with $s \in V_i$ (i.e., it is player i's turn to play a move), $\sigma(ws) = s'$ determines the next move of player i, where $(s, \bot, s') \in \Delta$. (We could also allow randomized strategies.)

Let Ψ_i denote the set of all strategies for player i. A pair of strategies $\sigma \in \Psi_1$ and $\tau \in \Psi_2$ induce in a straightforward way a Markov chain $M_A^{\sigma,\tau} = (V^*, \Delta')$, whose set of states is the set V^* of histories. Given initial vertex u, a final exit ex in the same component, and a $k \geq 0$, let $q_{(u,ex)}^{k,\sigma,\tau}$ be the probability that, in $M_A^{\sigma,\tau}$, starting at initial state $\langle \epsilon, u \rangle$, we will reach a state $w\langle \epsilon, ex \rangle$ in at most k "steps" (i.e., where $|w| \leq k$). Let $q_{(u,ex)}^{*,\sigma,\tau} = \lim_{k \to \infty} q_{(u,ex)}^{k,\sigma,\tau}$ be the probability of ever terminating at ex, i.e., reaching $\langle \epsilon, ex \rangle$ (the limit exists: the sequence is monotonically non-decreasing & bounded by 1). Let $\mathbf{q}_{(u,ex)}^k = \max_{\sigma \in \Psi_1} \min_{\tau \in \Psi_2} q_{(u,ex)}^{k,\sigma,\tau}$ and let $\mathbf{q}_{(u,ex)}^* = \sup_{\sigma \in \Psi_1} \inf_{\tau \in \Psi_2} q_{(u,ex)}^{*,\sigma,\tau}$. Next, for a strategy $\sigma \in \Psi_1$, let $q_{(u,ex)}^{k,\sigma} = \min_{\tau \in \Psi_2} q_{(u,ex)}^{k,\sigma,\tau}$, and let $q_{(u,ex)}^{*,\sigma} = \inf_{\tau \in \Psi_2} q_{(u,ex)}^{*,\sigma,\tau}$. Lastly, given instead a strategy $\tau \in \Psi_2$, let $q_{(u,ex)}^{k,\cdot,\tau} = \max_{\sigma \in \Psi_1} q_{(u,ex)}^{k,\sigma,\tau}$, and let $q_{(u,ex)}^{*,\cdot,\tau} = \sup_{\sigma \in \Psi_1} q_{(u,ex)}^{*,\sigma,\tau}$.

From very general determinacy results (eg. Martin's "Blackwell determinacy" [Mar98]) it follows that the games M_A are *determined*, meaning that $\sup_{\sigma \in \Psi_1} \inf_{\tau \in \Psi_2} q_{(u,ex)}^{*,\sigma,\tau} = \inf_{\tau \in \Psi_2} \sup_{\sigma \in \Psi_1} q_{(u,ex)}^{*,\sigma,\tau}$. Of course, finite SSGs are even *memorylessly determined* ([Con92]), meaning that the strategies of either player can be restricted to *memoryless strategies* which ignore the history prior to the current position, without harming the optimal outcome. As we shall see, 1-exit RSSGs exhibit memoryless determinacy in an even stronger sense, namely, the strategy is also independent of the call stack. This fails badly for multi-exit RMDPs, as we will see. We are interested in the following questions:

(1) The *qualitative* termination problem: Is $\mathbf{q}_{(u,ex)}^* = 1$?
(2) The *quantitative* termination problems: Given $r \in [0, 1]$, is $q_{(u,ex)}^* \geq r$? Is $q_{(u,ex)}^* = r$? Or we may wish to compute or approximate probabilities $q_{(u,ex)}^*$.

More generally, we can ask model checking questions, where, given a Σ-labeling of vertices, and e.g., an LTL formula φ over Σ, we ask what is the supremum probability with which player 1 can force the satisfaction of property φ? We refrain from formal definitions due to space (see,e.g., [CY98, EY05b]). Our results for model checking will be negative: undecidability, stemming from the undecidability of termination problems for general RMDPs.

3 Systems of Nonlinear Min-max Equations for 1-Exit RSSGs

We generalize the monotone nonlinear system of equations for RMCs ([EY05a]) to monotone nonlinear min-max systems for 1-exit RSSGs, whose Least Fixed Point yields the desired probabilities $q_{(u,ex)}^*$. Let us use a variable $x_{(u,ex)}$ for each unknown $q^*_{(u,ex)}$. We will often find it convenient to index the variables $x_{(u,ex)}$ according to a fixed order (say lexicographical), so we can refer to them also as $x_1, \ldots, x_n$, with each $x_{(u,ex)}$ identified with x_j for some j. In this way we obtain a vector of variables: $\mathbf{x} = (x_1 \; x_2 \ldots x_n)^T$.

Definition 1. *Given 1-exit RSSG $A = (A_1, \ldots, A_k)$, we define a system of polynomial/min-max equations, S_A, over the variables $x_{(u,ex)}$, where $u \in Q_i$ and $ex \in Ex_i$, for $1 \leq i \leq k$. The system contains one equation of the form $x_{(u,ex)} = P_{(u,ex)}(\mathbf{x})$, for each variable $x_{(u,ex)}$. There are 5 cases to distinguish, based on what "type" of vertex u is:*

1. *Type I: $u = ex$. In this case: $x_{(ex,ex)} = 1$.*
2. *Type II: $\mathtt{pl}(u) = 0 \; \& \; u \in (N_i \setminus \{ex\}) \cup \text{Return}^i$: $x_{(u,ex)} = \sum_{\{v|(u,p_{u,v},v)\in\delta\}} p_{u,v} x_{(v,ex)}$. (If u has no outgoing transitions, this equation is by definition $x_{(u,ex)} = 0$.)*
3. *Type III: $u = (b, en)$ is a call port: $x_{((b,en),ex)} = x_{(en,ex')} \cdot x_{((b,ex'),ex)}$, where $ex' \in Ex_{Y(b)}$ is the unique exit of $A_{Y(b)}$.*
4. *Type IV: $\mathtt{pl}(u) = 1 \; \& \; u \in (N_i \setminus \{ex\}) \cup \text{Return}^i$: $x_{(u,ex)} = \max_{\{v|(u,\bot,v)\in\delta\}} x_{(v,ex)}$. (If u has no outgoing transitions, we define $\max(\emptyset) = 0$.)*
5. *Type V: $\mathtt{pl}(u) = 2$ and $u \in (N_i \setminus \{ex\}) \cup \text{Return}^i$: $x_{(u,ex)} = \min_{\{v|(u,\bot,v)\in\delta\}} x_{(v,ex)}$. (If u has no outgoing transitions, we define $\min(\emptyset) = 0$.)*

In vector notation, we denote $S_A = (x_j = P_j(\mathbf{x}) \mid j = 1, \ldots, n)$ by: $\mathbf{x} = P(\mathbf{x})$.

Given 1-exit RSSG A, we can easily construct $\mathbf{x} = P(\mathbf{x})$ in linear time. We now identify a particular solution to $\mathbf{x} = P(\mathbf{x})$, called the *Least Fixed Point* (LFP) solution, which gives precisely the termination game values. For vectors $\mathbf{x}, \mathbf{y} \in \mathbb{R}^n$, define the partial-order $\mathbf{x} \preceq \mathbf{y}$ to mean $x_j \leq y_j$ for every coordinate j. For $D \subseteq \mathbb{R}^n$, we call a mapping $H : \mathbb{R}^n \mapsto \mathbb{R}^n$ *monotone* on D, if: for all $\mathbf{x}, \mathbf{y} \in D$, if $\mathbf{x} \preceq \mathbf{y}$ then $H(\mathbf{x}) \preceq H(\mathbf{y})$. Define $P^1(\mathbf{x}) = P(\mathbf{x})$, and define $P^k(\mathbf{x}) = P(P^{k-1}(\mathbf{x}))$, for $k > 1$. Let $\mathbf{q}^* \in \mathbb{R}^n$ denote the n-vector $q^*_{(u,ex)}$ (using the same indexing as used for $\mathbf{x}$). For $k \geq 0$, let $\mathbf{q}^k$ denote, similarly, the n-vector $q^k_{(u,ex)}$. Let $\mathbf{0}$ ($\mathbf{1}$) denote the n-vector consisting of 0 (respectively, 1) in every coordinate. Define $\mathbf{x}^0 = \mathbf{0}$, and for $k \geq 1$, define $\mathbf{x}^k = P(\mathbf{x}^{k-1}) = P^k(\mathbf{0})$.

Theorem 1. *Let $\mathbf{x} = P(\mathbf{x})$ be the system S_A associated with 1-exit RSSG A.*

1. *$P : \mathbb{R}^n \mapsto \mathbb{R}^n$ is monotone on $\mathbb{R}^n_{\geq 0}$. Hence, for $k \geq 0$, $\mathbf{0} \preceq \mathbf{x}^k \preceq \mathbf{x}^{k+1}$.*
2. *For all $k \geq 0$, $\mathbf{q}^k \preceq \mathbf{x}^{k+1} \preceq \mathbf{q}^{2^k}$.*
3. *$\mathbf{q}^* = P(\mathbf{q}^*)$. In other words, $\mathbf{q}^*$ is a fixed point of the map P.*
4. *For all $k \geq 0$, $\mathbf{x}^k \preceq \mathbf{q}^*$.*
5. *For all $\mathbf{q}' \in \mathbb{R}^n_{\geq 0}$, if $\mathbf{q}' = P(\mathbf{q}')$, then $\mathbf{q}^* \preceq \mathbf{q}'$.*
   In other words, $\mathbf{q}^*$ is the Least Fixed Point, LFP(P), of $P : \mathbb{R}^n_{\geq 0} \mapsto \mathbb{R}^n_{\geq 0}$.
6. *$\mathbf{q}^* = \lim_{k \to \infty} \mathbf{x}^k = \lim_{k \to \infty} \mathbf{q}^k$.*

The proofs are omitted due to space. They are similar to those of an analogous theorem in [EY05a] for nonlinear systems associated with RMCs, but some parts are substantially more tricky because of the players. We sketch here the idea for part (5). Consider any fixpoint $\mathbf{q}'$ of the equations, i.e., $\mathbf{q}' = P(\mathbf{q}')$. Let τ' be the (S&M) strategy for player 2 that always picks, at any state $\langle \beta, u \rangle$, for vertex $u \in \mathtt{pl}^{-1}(2)$, the particular successor v of u such that $v = \arg\min_{\{v | (u, \bot, v) \in \delta\}} \mathbf{q}'_{(v, ex)}$ (breaking ties, say, lexicographically). Then we prove a lemma stating that, for all strategies $\sigma \in \Psi_1$ of player 1, and for all $k \geq 0$, $\mathbf{q}^{k, \sigma, \tau'} \preceq \mathbf{q}'$. The lemma implies that $\mathbf{q}^{*, \sigma, \tau'} = \lim_{k \to \infty} \mathbf{q}^{k, \sigma, \tau'} \preceq \mathbf{q}'$. This holds for any strategy $\sigma \in \Psi_1$. Therefore, $\sup_{\sigma \in \Psi_1} q^{*, \sigma, \tau'}_{(u, ex)} \leq \mathbf{q}'_{(u, ex)}$, for every vertex u. Thus, by the determinacy of RSSG games, it follows that $\mathbf{q}^*_{(u, ex)} = \inf_{\tau \in \Psi_2} \sup_{\sigma \in \Psi_1} q^{*, \sigma, \tau}_{(u, ex)} \leq \sup_{\sigma \in \Psi_1} q^{*, \sigma, \tau'}_{(u, ex)} \leq \mathbf{q}'_{(u, ex)}$, for all vertices u. In other words, $\mathbf{q}^* \preceq \mathbf{q}'$. □

4 S&M Determinacy

We now identify a very restricted kind of strategy that suffices as an optimal strategy in 1-exit RSSGs. Call a strategy *Stackless & Memoryless (S&M)* if it is not only independent of the history of the game, but also independent of the current call stack, i.e., only depends on the current vertex. (See also [ATM03], where such strategies are called *modular* strategies.)

Corollary 1. *In every 1-exit RSSG termination game, player 2 (the minimizer) has an optimal S&M strategy.*

Proof. Consider the strategy τ' in the proof of part (5) of Theorem 1, chosen not for just any fixed point $\mathbf{q}'$, but for $\mathbf{q}^*$ itself. □

Far less trivially, we establish next that player 1 (the maximizer) also has an optimal S&M strategy and thus the game is *S&M-determined*, meaning both players have optimal S&M strategies. (Note that the game is not symmetric with respect to the two players.)

Theorem 2. *Every 1-exit RSSG termination game is S&M-determined.*

Although the statement is intuitive, the proof is quite nontrivial and delicate; the full proof is given in the full paper. We sketch the approach here. By Corollary 1, we only need to show that player 1 has an optimal S&M strategy. Let σ be any S&M strategy for player 1, and let $\mathbf{q}^{*, \sigma} = \inf_{\tau \in \Psi_2} \mathbf{q}^{*, \sigma, \tau}$. If $\mathbf{q}^{*, \sigma}$ is a fixpoint of the equations then it follows that it is the least fixpoint and σ is optimal. On the other hand, it can be shown that $\mathbf{q}^{*, \sigma}$ satisfies all the equations except possibly for some type IV equations. We argue that if u is such a vertex (belonging to player 1) whose equation is violated, then switching to another strategy σ' where u picks another successor leads to a strictly better strategy than σ (for any strategy of player 2). This is the heart of the proof. We parameterize the game with respect to the value t at vertex u, and we express the optimal values of the other vertices z (for all strategies τ of player 2) as functions $f_z(t)$. We then carefully analyze the properties of these functions, which are power series in t

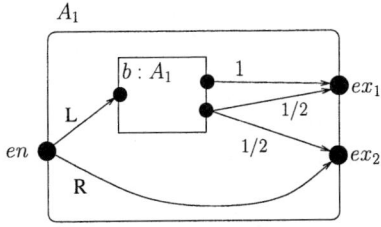

Fig. 1. *2-exit RMDPs:* no optimal strategy exists for terminating at ex_1

with non-negative coefficients, and we analyze their fixpoints, and conclude that switching the choice at vertex u leads to a strategy σ' that has at least as great value as σ at every vertex, and strictly better at u. We repeat the process until we arrive at a S&M strategy σ^* whose probabilities satisfy all the equations, and hence it is optimal. We refer to the full paper for the details. □

Already for 2-exit RMDP termination, not only are there no optimal S&M strategies for player 1, there are in general no optimal strategies at all! Figure 1 illustrates this. In this 2-exit maximizing RMDP the *supremum* probability of terminating at exit ex_1 starting from en is 1. However, no strategy player 1 achieves this. Specifically, for $n \geq 0$, the strategy $L^n R$ terminates at ex_1 with probability $(1 - \frac{1}{2^n})$. Note that any S&M strategy for player 1 would yield probability 0 of terminating at ex_1, so such strategies are the worst possible.

5 Termination Problems for 1-Exit RMDPs & RSSGs

Using Corollary 1 and Theorem 2, and results from [EY05a], we can show the following results for qualitative termination of 1-exit RMDPs and 1-exit RSSGs:

Theorem 3.
1. *We can decide in P-time if the value of a 1-exit RSSG termination game (and optimal termination probability in a maximizing or minimizing 1-exit RMDP) is exactly 0.*
2. *We can decide in NP whether the maximum probability of termination in a maximizing 1-exit RMDP is exactly 1, and in coNP whether the minimum probability of termination in a minimizing 1-exit RMDP is exactly 1.*
3. *Deciding whether a 1-exit RSSG termination game has value 1 is in $\Sigma_2^P \cap \Pi_2^P$.*
4. *For 1-exit* $\mathtt{lr}$*-RMDPs we can compute the exact optimal (rational) termination probability in P-time, and for 1-exit* $\mathtt{lr}$*-RSSGs we can compute the exact optimal (rational) value of the termination game in NP∩co-NP.*

Part (1) is done via a fixpoint algorithm; parts (2) and (3) involve guessing the optimal S&M strategies and verifying the optimality with the appropriate complexity using techniques from [EY05a]; part (4) exploits the fact that for 1-exit $\mathtt{lr}$-RMDPs & $\mathtt{lr}$-RSSGs the non-linear system $\mathbf{x} = P(\mathbf{x})$ can be decomposed into linear parts that can be solved sequentially by linear programming (and by guessing strategies for RSSGs); see the full paper. □

We next show that quantitative termination questions for 1-exit RMDPs and 1-exit RSSGs can be answered in PSPACE by appealing to the deep algorithms for deciding the *Existential Theory of Reals*, ***ExTh***(ℝ). A first-order sentence in the theory of reals is formed from quantifiers and boolean connectives over a vocabulary with "atomic predicates" of the form: $f_i(\mathbf{x})\Delta 0$, where the f_i, are multi-variate polynomials with rational coefficients over the variables $\mathbf{x} = x_1, \ldots, x_n$, and Δ is any comparison operator: $=, <$, or $\leq$. The *existential theory of reals*, ***ExTh***(ℝ), consists of prenex sentences: $\exists x_1, \ldots, x_n R(x_1, \ldots, x_n)$, where R is a boolean combination of "atomic predicates". Beginning with Tarski, algorithms for deciding the theory of reals and fragments such as ***ExTh***(ℝ) have been deeply investigated. Current, it is known that ***ExTh***(ℝ) can be decided in PSPACE [Can88, Ren92, BPR96]. Furthermore, it can be decided in exponential time where the exponent depends (linearly) only on the number of variables; hence for a fixed number of variables the time is polynomial.

Suppose we want to decide whether a vector $\mathbf{c} = [c_1, \ldots, c_n]^T$ of rational numbers is $LFP(P)$, where $\mathbf{x} = P(\mathbf{x})$ is the system of equations for a given 1-exit RSSG. Consider the sentence:

$$\varphi \equiv \exists x_1, \ldots, x_n \bigwedge_{i=1}^{n} (P_i(x_1, \ldots, x_n) = x_i) \wedge \bigwedge_{i=1}^{n} (x_i = c_i)$$

φ is true iff $\mathbf{c} = P(\mathbf{c})$. For type I, II, and III nodes, P_i is a polynomial. It remains to show how to encode, in arithmetic, the predicate "$P_i(x_1, \ldots, x_n) = x_i$" in the case (IV) where $P_i(x_1, \ldots, x_n) = \max_{j \in J} x_j$, and in the case (V) where $P_i(x_1, \ldots, x_n) = \min_{j \in J} x_j$, for some subset $J \subseteq \{1, \ldots, n\}$. For type IV nodes, note that $x_i = \max_{j \in J} x_j$ iff $\bigwedge_{j \in J} x_i \geq x_j \wedge (\bigvee_{j \in J} x_i \leq x_j)$. Likewise, for type V nodes, $x_i = \min_{j \in J} x_j$ iff $\bigwedge_{j \in J} x_i \leq x_j \wedge (\bigvee_{j \in J} x_i \geq x_j)$. Thus, we can encode the predicates $x_i = P_i(x_1, \ldots, x_n)$ as a boolean combination of quantifier-free predicates, and we can encode the sentence φ in ***ExTh***(ℝ). To guarantee that $\mathbf{c} = \mathrm{LFP}(P)$, we need in addition to check the following sentence:

$$\psi \equiv \exists x_1, \ldots, x_n \bigwedge_{i=1}^{n} (P_i(x_1, \ldots, x_n) = x_i) \wedge \bigwedge_{i=1}^{n} (0 \leq x_i) \wedge \bigvee_{i=1}^{n} (x_i < c_i)$$

ψ is false iff there is no solution $\mathbf{z} \in \mathbb{R}_{\geq 0}^n$ to $\mathbf{x} = P(\mathbf{x})$ such that $\mathbf{c} \not\leq \mathbf{z}$. Hence, to decide whether $\mathbf{c} = \mathrm{LFP}(P)$, we only need two queries to ***ExTh***(ℝ). Namely, we check that φ is true, and hence $\mathbf{c} = P(\mathbf{c})$, and that ψ is false, and hence $\mathbf{c} = \mathrm{LFP}(P)$. If we only want to check an inequality $q_j^* \leq c_j$, then let φ' be φ with the last conjunction of equations replaced by $\bigwedge_{i=1}^{n}(0 \leq x_i) \wedge (x_j \leq c_j)$. Applying the results on ***ExTh***(ℝ), we obtain the following:

Theorem 4. *Given a 1-exit RSSG A and a vector of rational probabilities* $\mathbf{c}$, *there is a PSPACE algorithm to decide whether* $\mathbf{q}^* = \mathbf{c}$, *as well as to decide whether* $\mathbf{q}_j^* \Delta c_j$, *for any comparison operator* Δ. *Moreover, the running time is* $O(|A|^{O(1)} \cdot 2^{O(n)})$ *where n is the number of variables in* $\mathbf{x} = P(\mathbf{x})$. *Hence the running time is polynomial if n is bounded.*

Since $0 \leq \text{LFP}(P) \leq 1$, we can use such queries to $\textbf{\textit{ExTh}}(\mathbb{R})$ in a "binary search" to "narrow in" on the value of each coordinate of $\text{LFP}(P)$. Via obvious modifications of sentences like ψ, we can gain one extra bit of precision on the exact value of each c_i with one extra query to $\textbf{\textit{ExTh}}(\mathbb{R})$. This yields:

Theorem 5. *Given 1-exit RSSG A and a number j in unary, there is an algorithm that approximates every coordinate of $\mathbf{q}^*$ to within j bits of precision in PSPACE. Moreover, the running time is $O(j \cdot |A|^{O(1)} \cdot 2^{O(n)})$, where n is the number of variables in $\mathbf{x} = P(\mathbf{x})$.*

6 Multi-exit RMDP Termination and 1-Exit RMDP Model Checking: Undecidability

We next show strong undecidability results for RMDPs, and thus for RSSGs.

Theorem 6. *Given a multi-exit linearly-recursive RMDP, A, entry en and exit ex, it is undecidable whether $q^*_{(en,ex)} = 1$. This is so even when the number of exits in each component of A is bounded by a fixed constant. Furthermore, there is no algorithm that approximates the probability $q^*_{(en,ex)}$ within any constant (multiplicative) factor. In particular:*

1. For any fixed rational ϵ with $0 < \epsilon < 1$, given 1r-RMDP A with only one component such that either $q^*_{(en,ex)} > 1 - \epsilon$ or $q^*_{(en,ex)} < \epsilon$, it is undecidable to distinguish which is the case.
2. For any fixed rational ϵ with $0 < \epsilon < 1$, given a 1r-RMDP A with only two components such that either $q^*_{(en,ex)} = 1$ or $q^*_{(en,ex)} < \epsilon$, it is undecidable to distinguish which is the case.

We have two proofs: one is a reduction from the halting problem for 2-counter machines. The second is a reduction from the emptiness problem for Probabilistic Finite Automata (PFAs). The latter reduction is simpler and connects RMDPs to the well-studied area of PFAs, allowing us to leverage extensive research in that area [Paz71, CL89, BC03]. We show that PFAs are, in effect, a special case of RMDPs. Recall, a PFA is a FA whose transitions from a state on a given input are probabilistic. The PFA emptiness problem is to decide for a given PFA A and threshold λ, whether there is a word accepted by A with probability $> \lambda$. It is undecidable in strong ways ([Paz71, CL89, BC03]). Our reduction constructs from a PFA A a 1r-RMDP whose termination probability under the optimal strategy is precisely the supremum probability of acceptance of any word by A. Please see the full paper for details. With a modified construction we show:

Theorem 7. *Qualitative & quantitative LTL model checking for 1-exit 1r-RMDPs is undecidable; this holds even for a fixed property; moreover, the optimum probability of the property can't be approximated within any constant factor.*

Acknowledgement. Thanks to Amir Pnueli for asking us about 1r-RMCs.

References

[AEY01] R. Alur, K. Etessami, and M. Yannakakis. Analysis of recursive state machines. In *Proc. of 13th CAV*, pages 304–313, 2001.

[ATM03] R. Alur, S. La Torre, and P. Madhusudan. Modular strategies for recursive game graphs. In *TACAS*, volume 2619 of *LNCS*, pages 363–378, 2003.

[BC03] V. Blondel and V. Canterini. Undecidable problems for probabilistic automata of fixed dimension. *Theory of Comput. Sys.*, 36:231–245, 2003.

[BKS05] T. Brázdil, A. Kučera, and O. Stražovský. Decidability of temporal properties of probabilistic pushdown automata. In *Proc. of STACS'05*, 2005.

[BPR96] S. Basu, R. Pollack, and M. F. Roy. On the combinatorial and algebraic complexity of quantifier elimination. *J. ACM*, 43(6):1002–1045, 1996.

[Can88] J. Canny. Some algebraic and geometric computations in PSPACE. In *Prof. of 20th ACM STOC*, pages 460–467, 1988.

[CL89] A. Condon and R. Lipton. The complexity of space bounded interactive proofs. In *Proc. of 30th IEEE FOCS*, 1989.

[Con92] A. Condon. The complexity of stochastic games. *Inf. & Comp.*, 96(2):203–224, 1992.

[CY95] C. Courcoubetis and M. Yannakakis. The complexity of probabilistic verification. *Journal of the ACM*, 42(4):857–907, 1995.

[CY98] C. Courcoubetis and M. Yannakakis. Markov decision processes and regular events. *IEEE Trans. on Automatic Control*, 43(10):1399–1418, 1998.

[dAKN+00] L. de Alfaro, M. Kwiatkowska, G. Norman, D. Parker, and R. Segala. Symbolic model checking of probabilistic processes using MTBDDs and the kronecker representation. In *Proc. of 6th TACAS*, pages 395–410, 2000.

[EKM04] J. Esparza, A. Kučera, and R. Mayr. Model checking probabilistic pushdown automata. In *Proc. of 19th IEEE LICS'04*, 2004.

[Ete04] K. Etessami. Analysis of recursive game graphs using data flow equations. In *Proc. 5th VMCAI*, vol. 2937 *LNCS*, pages 282–296. Springer, 2004.

[EY05a] K. Etessami and M. Yannakakis. Recursive markov chains, stochastic grammars, and monotone systems of non-linear equations. In *Proc. of 22nd STACS'05*. Springer, 2005. (Tech. Report, U. Edinburgh, June 2004).

[EY05b] K. Etessami and M. Yannakakis. Algorithmic verification of recursive probabilistic state machines. In *Proc. 11th TACAS*, vol. 3440 of LNCS, 2005.

[FS02] E. Feinberg and A. Shwartz, editors. *Handbook of Markov Decision Processes*. Kluwer, 2002.

[FV97] J. Filar and K. Vrieze. *Competitive Markov Decision Processes*. Springer, 1997.

[Har63] T. E. Harris. *The Theory of Branching Processes*. Springer-Verlag, 1963.

[HSP83] S. Hart, M. Sharir, and A. Pnueli. Termination of probabilistic concurrent programs. *ACM Trans. Prog. Lang. & Sys.*, 5(3):356–380, 1983.

[Jag75] P. Jagers. *Branching Processes with Biological Applications*. Wiley, 1975.

[Kwi03]   M. Kwiatkowska. Model checking for probability and time: from theory to practice. In *18th IEEE LICS*, pages 351–360, 2003.
[Mar98]   D. A. Martin. Determinacy of Blackwell games. *J. Symb. Logic*, 63(4):1565–1581, 1998.
[MS99]   C. Manning and H. Schütze. *Foundations of Statistical Natural Language Processing*. MIT Press, 1999.
[Paz71]   A. Paz. *Introduction to Probabilistic Automata*. Academic Press, 1971.
[Put94]   M. L. Puterman. *Markov Decision Processes*. Wiley, 1994.
[Ren92]   J. Renegar. On the computational complexity and geometry of the first-order theory of the reals, parts I-III. *J. Symb. Comp.*, 13(3):255–352, 1992.
[SBH+94]   Y. Sakakibara, M. Brown, R Hughey, I.S. Mian, K. Sjolander, R. Underwood, and D. Haussler. Stochastic context-free grammars for tRNA modeling. *Nucleic Acids Research*, 22(23):5112–5120, 1994.
[Var85]   M. Vardi. Automatic verification of probabilistic concurrent finite-state programs. In *Proc. of 26th IEEE FOCS*, pages 327–338, 1985.

Decidability in Syntactic Control of Interference

J. Laird

Dept. of Informatics, University of Sussex, UK
jiml@sussex.ac.uk

Abstract. We investigate the decidability of observational equivalence and approximation in "Syntactic Control of Interference" (SCI). By associating denotations of terms in an inequationally fully abstract model of finitary basic SCI with multitape finite state automata, we show that observational approximation is not decidable (even at first order), but that observational equivalence is decidable for all terms. We then consider the same problems for basic SCI extended with non-local control in the form of backwards jumps. We show that both observational approximation and observational equivalence are decidable in this language by describing a fully abstract games model in which strategies are regular languages.

1 Introduction

Reynolds' *Syntactic Control of Interference* [18] is a prototypical functional-imperative language in which covert *interference* between functions and their arguments is prevented by the use of an affine typing discipline. By eliminating phenomena such as aliasing, interference control should make it easier to predict program behaviour. Here, we shall investigate decidability of properties of observational approximation and equivalence for terms over finite datatypes.

The full SCI system contains a notion of passive type (as do related languages with interference control such as SCIR [15]); the programs (and contexts) typable at purely passive types are essentially those of PCF, and equivalence at such types is therefore not decidable [9]. Consequently, we restrict attention to the active types or *basic* SCI, which may be thought of as a common core for interference controlled languages. Basic SCI is essentially a subsystem of Idealized Algol, for which a hierarchy of decidability results has recently been developed, based on "algorithmic game semantics" [3]: associating the denotation of each term as a strategy in a fully abstract game semantics with a formal language. These techniques have been used to show that observational equivalence and approximation in (finitary) Idealized Algol are decidable at third-order types — by showing that the denoting strategies are recognized by deterministic pushdown automata [14] — but not at fourth order [13]. The latter undecidability result depends crucially on the nesting of function calls, which is not possible in SCI.

These results provide motivation and a methodological basis for investigating observational equivalence in basic SCI, but one obstacle is that game semantics appears to be "too sequential" to capture this directly. (McCusker and Wall have described a games model [12], but the associated full abstraction result depends on a quotient operation.) However, basic SCI has an appealingly simple semantics based on sets and relations, described by Reddy [17], and investigated further by McCusker, who proved that it is fully abstract [10]. We show that the denotations of first-order terms in this semantics correspond to *multitape deterministic finite state automata*, as introduced by Rabin and Scott in 1959 [16]. It is straightforward to show that containment of 2-tape deterministic FSA is undecidable [7]. By contrast, the problem of decidability of equivalence for all multitape deterministic FSA remained open for thirty years, before being resolved (affirmatively) by Harju and Karhumäki [4]. These results have direct implications for basic SCI: observational equivalence (equivalence of denotations) is decidable at first order but observational approximation (inclusion of denotations) is not (even though all first-order terms of Idealized Algol are typable in SCI (up to β-equivalence) — the point being that the possible *observations* of first-order terms are more restricted in SCI). Although denotations of terms at higher-order types do not correspond directly to multitape automata, we show that there is a "definable monomorphism" from every type to a first-order type, and that equivalence is therefore decidable at *all* types.

The undecidability of observational approximation (and associated results, such as the undecidability of equivalence in a non-deterministic variant of the language) raises the question of whether deciding such properties in the presence of interference control is always difficult. The problem in basic SCI is in some respects analogous to the situation in PCF: whilst it is a sequential language, it is not *observably sequential*; different sequentializations of a program may approximate each other (for example, $\lambda x.\lambda y.x; y$ is equivalent in SCI to $\lambda x.\lambda y.y; x$). In the case of PCF, Cartwright and Felleisen [2] have shown that adding simple non-local control operators results in a language ("observably sequential PCF") with a finitely presentable fully abstract model, for which equivalence and approximation are therefore decidable. Pursuing the analogy between PCF and SCI, we obtain an observably sequential version of the latter by adding backwards jumps to labelled points. We give a regular game semantics for this language and show that it is inequationally fully abstract (although it contains compact elements which are not definable) and that observational approximation is therefore decidable.

2 Basic SCI

The syntax of basic SCI [18] is, in essence, the same as that of Idealized Algol [19] but the typing rules are more restrictive. In our finitary version of the language we assume a base type $\underline{n}$ (of expressions) for each natural number n, containing the mumerals $0, \ldots, n-1$, and generate types using the constructors $\to$ (affine function-space) and & (additive product). The depth or order of a type (and the corresponding closed terms) is defined: $o(\underline{n}) = 0$, $o(S \& T) = \max\{o(S), o(T)\}$, $o(S \to T) = \max\{o(S) + 1, o(T)\}$.

Typing judgements are based on the affine λ-calculus with pairing:

$$\frac{}{x{:}T, \Gamma \vdash x{:}T} \qquad \frac{\Gamma, x{:}S, \Delta \vdash M{:}T}{\Gamma, \Delta, x{:}S \vdash M{:}T}$$

$$\frac{\Gamma \vdash M{:}S \quad \Gamma \vdash N{:}T}{\Gamma \vdash \langle M, N\rangle{:}S\&T} \quad \frac{\Gamma, x{:}S \vdash M{:}T}{\Gamma \vdash \lambda x.M{:}S \to T} \quad \frac{\Gamma \vdash M{:}S \to T \quad \Delta \vdash N{:}S}{\Gamma, \Delta \vdash M\,N{:}T}$$

For the sake of concision, we work with a lean syntax which can be easily sugared up to more closely resemble that of Idealized Algol. We write T^n for the n-fold product of copies of T (setting $T^0 = \underline{1}$). We represent the type $\mathtt{var}[n]$ of imperative variables with values in $\underline{n}$ as the type $\underline{n}\&\underline{1}^n$ (the product of the types of its "methods", assignment and dereferencing).[1] We extend the term formation rules with the following constants (we omit parallel composition as it is observationally equivalent to sequential composition):

Projections $\pi_l : A\&B \to A$, $\pi_r : A\&B \to B$ yielding $\pi_i : A^n \to A$ for $i < n$.
Numerals $\mathtt{i} : \underline{n}$ for $i < n$.
Case statements $\mathtt{case}_{n,m} : \underline{n}\&\underline{m}^n \to \underline{m}$. from which we derive $\mathtt{case}_{n,T} : \underline{n}\&T^n \to T$ for arbitrary T.
Loops $\mathtt{while0} : \underline{m+1} \to \underline{m}$. We write Ω for $\mathtt{while0}\ 0$.
New variable declaration $\mathtt{new} : (\mathtt{var}[n] \to \underline{m}) \to \underline{m}$.

If $M : \underline{1}$, we write $\mathtt{case}\,\langle M, N\rangle$ as $M; N$, and if $M : \underline{0}$ we write $\mathtt{abort}\,M$ for $\mathtt{case}\,\langle M, 0\rangle$. Given $M : \mathtt{var}[n]$, we may write $!M$ for $\pi_l M$ and $M := N$ for $\mathtt{case}\,\langle N, \pi_r M\rangle$.

The "small-step" operational semantics is based on the evaluation contexts:

$$E[\cdot] ::= [\_] \mid E[\_]\,M \mid \pi_i\,E[\_] \mid \mathtt{case}\,E[\_] \mid \mathtt{case}\,\langle E[\_], M\rangle \mid \mathtt{while0}\,E[\_]$$

and the following reductions for pairs $M, \mathcal{E}$ of a term (of ground type) and a store (a set of pairs $\langle a, v\rangle$ of location names and natural numbers).

$$\begin{aligned}
E[(\lambda x.M)\,N], \mathcal{E} &\longrightarrow E[M[N/x]], \mathcal{E} \\
E[\pi_i\,\langle M_1, M_2\rangle] &\longrightarrow M_i, \mathcal{E} \\
E[\mathtt{case}\,\langle i, M\rangle], \mathcal{E} &\longrightarrow E[\pi_i\,M] \\
E[\mathtt{while0}\,M] &\longrightarrow E[\mathtt{case}\,\langle M, \langle \mathtt{while0}\,M, 0, \ldots, 0\rangle\rangle] \\
E[\mathtt{new}\,M], \mathcal{E} &\longrightarrow E[M\,a], \mathcal{E} \cup \{\langle a, 0\rangle\} \quad a \notin \pi_1(\mathcal{E}) \\
E[\pi_l\,a], \mathcal{E} &\longrightarrow E[v], \mathcal{E} \quad \langle a, v\rangle \in \mathcal{E} \\
E[\pi_i\,(\pi_l\,a)], \mathcal{E} &\longrightarrow E[0], \mathcal{E}[a \mapsto i]
\end{aligned}$$

We write $M \Downarrow$ if evaluation of $M, \varnothing$ terminates, and so define standard notions of observational approximation and equivalence:
$M \precsim N$ if $C[M] \Downarrow$ implies $C[N] \Downarrow$, and $M \simeq N$ if $M \precsim N$ and $N \precsim M$.

---

[1] So this version of the language implicitly contains "bad variables". This does not affect the results given here for basic SCI: approximation is shown to be undecidable for the var-free types, whilst McCusker has shown that observational equivalence in basic SCI with bad variables is conservative over equivalence in the language without bad variables [11].

2.1 Fully Abstract Semantics of Basic SCI

We briefly describe the fully abstract model of basic SCI, based on sets and relations, which we shall use to prove our results. This is essentially a version of Reddy's semantics [17], simplified and proved to be fully abstract by McCusker [10]. We may present this model as a typed **BCK**-algebra [5] with products:

- A collection of objects and a set $\mathrm{el}(\alpha)$ of elements for each object. In the relational model, objects are sets and $\mathrm{el}(\alpha) = \mathcal{P}(\alpha)$.
- Objects $\alpha \to \beta$ and $\alpha\&\beta$ for each pair of objects α, β. In the relational model, we define $\alpha \to \beta$ to be the set $\alpha^* \times \beta$ (where α^* is the set of finite words over α) and $\alpha\&\beta$ to be the disjoint union $\alpha + \beta = \{e^l \mid e \in \alpha\} \cup \{e^r \mid e \in \beta\}$. So in the relational model, elements of $\alpha \to \beta$ are relations between α^* and β, and elements of $\alpha\&\beta$ correspond to pairs of elements from α and β.
- An application operation — a function $\cdot : \mathrm{el}(\alpha \to \beta) \times \mathrm{el}(\alpha) \to \mathrm{el}(\beta)$. In the relational model $X \cdot Y = \{z \in \beta \mid \exists \boldsymbol{y} \in Y^*. \langle \boldsymbol{y}, z \rangle \in X\}$.
- For each α, β, γ, elements $\mathbf{B} \in \mathrm{el}((\beta \to \gamma) \to (\alpha \to \beta) \to \alpha \to \gamma)$, $\mathbf{C} \in \mathrm{el}((\alpha \to \beta \to \gamma) \to \beta \to \alpha \to \gamma)$, $\mathbf{T} \in \mathrm{el}(\alpha \to \beta \to \alpha\&\beta)$, $\mathbf{P} \in \mathrm{el}((\alpha \to \beta)\&(\alpha \to \gamma) \to \alpha \to \beta\&\gamma)$, $\pi_l \in \mathrm{el}(\alpha\&\beta \to \alpha)$ and $\pi_r \in \mathrm{el}(\alpha\&\beta \to \beta)$ satisfying the axioms:
$\mathbf{B} \cdot x \cdot y \cdot z = x \cdot (y \cdot z)$,
$\mathbf{C} \cdot x \cdot y \cdot z = x \cdot z \cdot y$
$\pi_l \cdot (\mathbf{T} \cdot x \cdot y) = x$ and $\pi_r \cdot (\mathbf{T} \cdot x \cdot y) = y$
$\mathbf{P} \cdot (\mathbf{T} \cdot x \cdot y) \cdot z = \mathbf{T} \cdot (x \cdot z) \cdot (y \cdot z)$
(note that we may define $\mathbf{K} \in \mathrm{el}(\alpha \to \beta \to \alpha)$ to be $\mathbf{B} \cdot (\mathbf{B} \cdot \pi_r) \cdot \mathbf{T}$). In the relational model we define e.g. $\mathbf{B} \in \mathrm{el}((\beta \to \gamma) \to (\alpha \to \beta) \to \alpha \to \gamma) =$

$$\{\langle\langle \boldsymbol{b}, c\rangle, \langle e, \langle \boldsymbol{a}, c\rangle\rangle\rangle \mid \boldsymbol{a} \in \alpha^*, \boldsymbol{b} \in \beta^*, c \in \gamma, e \in (\alpha^* \times \beta)^*. e{\upharpoonright}\alpha = \boldsymbol{a} \wedge e{\upharpoonright}\beta = \boldsymbol{b}\}$$

BCK-algebras yield an interpretation of *derivations* in the affine λ-calculus (as standard combinatory algebras do for the λ-calculus), and the extension with products is straightforward. Thus to interpret basic SCI, we define $[\![\underline{n}]\!] = \lceil n \rceil = \{i \mid i < n\}$, and assign meanings to the remaining constants as follows:

- $[\![\mathtt{case}_{m,n}]\!] = \{\langle i^l j^{ir}, j\rangle \mid i < m \wedge j < n\}$,
- $[\![\mathtt{while0}]\!] = \{\langle 0^*(n+1), 0\rangle \mid s \in \{i < n \mid i \neq 0\}^*\}$,
- $[\![\mathtt{new}]\!] = \{\langle\langle(0^l)^*s, j\rangle, j\rangle \mid s \in \{(0^i)^l(i^r)^* \mid i < n\}^*\}$

Concretely, this interpretation is equivalent to Reddy's and McCusker's (for closed terms), and therefore fully abstract.

Theorem 1 (McCusker [10]). $M \lesssim N$ *if and only if* $[\![M]\!] \subseteq [\![N]\!]$.

Note that the semantics does not have the "finite definability" property: there are compact elements which are not the denotation of any term. In particular, we may interpret a non-deterministic choice between any two terms of the same type as the union of their denotations.

3 Multitape Automata and SCI Terms

Our definition of deterministic multitape finite state automata is tailored for establishing a correspondence with the relational model. However, it is straightforward to show that it yields the same class of languages as the original [16].

A deterministic n-tape automaton $\alpha = (C_0, C_1, \ldots, C_n, F, s_0, \delta)$ over an alphabet $\Sigma = \{0, \ldots, k\}$ consists of disjoint sets of states $C_0, C_1, \ldots, C_n, F$, an initial state s_0 and a (partial) transition function $\delta : (\Sigma \cup \{\varepsilon\}) \times S \rightharpoonup S$, where $S = C_0 \cup \ldots \cup C_n \cup F$. C_0 consists of states in which (only) a ε-transition may be performed, C_i of states in which a symbol is read from tape i (for $0 < i \leq n$), and F of final states. We say that α accepts a tuple of tapes if it consumes all of them to reach a final state. Formally, for each final state f, we may say that the set of pairs $A_f \subseteq (\Sigma^*)^n \times S$ of tuples and states accepted in state f is the least set such that $\langle\langle\varepsilon, \ldots, \varepsilon\rangle, f\rangle \in A_f$, and if $\langle\langle t_1, \ldots, t_n\rangle, s\rangle \in A_f$, $s' \in C_i$ and $\delta(m, s') = s$ then if $i = 0$ then $\langle\langle t_1, \ldots, t_n\rangle, s'\rangle \in A_f$ and if $i > 0$ then $\langle\langle t_1, \ldots, mt_i, \ldots, t_n\rangle, s'\rangle \in A_f$.

We write $\mathcal{L}(\alpha)$ for the set $\{x \mid \exists f \in F.\langle x, s_0\rangle \in A_f\}$ of tuples accepted by α.

There is a natural connection between multitape FSA and SCI terms at first-order types, based on the fact that the denotations of the latter are sets of tuples of words over a finite alphabet.

Proposition 1. *For any 2-tape deterministic FSA α over the alphabet $\{0, \ldots, m-1\}$ there is a (closed) term $M_\alpha : \underline{m} \to \underline{m} \to \underline{1}$ such that $\langle s, t\rangle \in L(\alpha) \Leftrightarrow \langle s, \langle t, 0\rangle\rangle \in [\![M_\alpha]\!]$.*

Proof. Every deterministic n-tape FSA is equivalent to one with no ε transitions, so we assume C_0 is empty. We number the states of α as $0, \ldots, k-1$, with 0 being initial. We assume terms $\mathtt{state} : \underline{k} \to \underline{3}$ such that $[\![\mathtt{state}\, s]\!] = 0$ if $s \in F$ and $[\![\mathtt{state}\, s]\!] = i$ if $s \in C_i$, and $\mathtt{tr} : \underline{k} \to \underline{k}^{\underline{m}}$ such that $[\![\pi_i\,(\mathtt{tr}\, j)]\!] = \delta(j, i)$. M_α may then be defined as follows:

$$\lambda x.\lambda y.\mathtt{new}\,\lambda s.\mathtt{while0}\,(\mathtt{case}\,\langle\mathtt{state}\,!s, \langle 1, s := \mathtt{case}\langle x, \mathtt{tr}\,!s\rangle, s := \mathtt{case}\langle y, \mathtt{tr}\,!s\rangle\rangle\rangle)$$

So we may show undecidability of inclusion of denotations in the fully abstract model, and hence of observational approximation in SCI via the following result, which may be proved via an encoding of Post's correspondence problem.

Proposition 2. *[7] Inclusion of deterministic 2-tape FSA is undecidable.*

Corollary 1. *Observational approximation in SCI is undecidable at first-order.*

Since $M \lesssim N$ if and only if M or $N \simeq N$, this entails that observational equivalence (w.r.t. may-testing) in SCI with erratic choice is undecidable at first order. The denotations of first-order terms *without loops* are finite sets, and therefore approximation for such terms is decidable at first-order. However, we can simulate any 2-tape FSA as a term of type $(\underline{1}^n \to \underline{1}) \to (\underline{1}^n \to \underline{1}) \to \underline{1}$ (as in Proposition 1) without using the $\mathtt{while0}$ constant, and so approximation is undecidable at second order for the loop-free language.

The situation with respect to *equivalence* of multitape automata (and hence, as we shall show, observational equivalence of SCI) is different.

Theorem 2 (Harju and Karhumäki [4]). *Equivalence of n-tape deterministic finite state automata is decidable for all n.*

To use this result to show decidability of SCI equivalence, we prove a converse to Proposition 1: the denotation of every first-order term corresponds to the language accepted by a multitape automaton.

Proposition 3. *For each term $x_1 : T_1, \ldots, x_n : T_n \vdash M : \underline{m}$, where each T_i is a product of ground types, there is a deterministic n-tape FSA with final states $F = f_0, \ldots, f_{m-1}$ which accepts $\langle s_1, \ldots, s_n \rangle$ in final state f_i if and only if $\langle \langle s_1, \ldots, s_n \rangle, i \rangle \in [\![M]\!]$.*

Proof. We may show using a reducibility argument that every first-order term is reducible to a $\beta\pi$-normal form (i.e. no subterms of the form $(\lambda x.P)Q$ or $\pi_i \langle M, N \rangle$). So we assume that M is $\beta\pi$-normal, and define an n-tape deterministic automaton $(C_0^M, \ldots, C_n^M, F, s_0^M, \delta^M)$ with the additional property that if $T_i = \underline{m_1} \& \ldots \& \underline{m_k}$ then for any state $s \in C_i^M$, if $\delta^M(v^i, s)$ and $\delta^M(u^j, s)$ are both defined then $i = j$.

- If $M = i$, then $C_j^M = \varnothing$ for all j, $s_0^M = f_i$ and $\delta^M = \varnothing$.
- If $M = \pi_j(x_i)$, then $C_i^M = \{s\}$, $C_j^M = \varnothing$ for $j \neq i$, $s_0^M = s$,
  $\delta(l^k, s_0) = f_l$ if $k = j$, and $\delta(m, s)$ is undefined otherwise.
- If $M = \text{case}\,\langle L, N_0, \ldots, N_{k-1}\rangle$ then $C_i^M = C_i^L + (C_i^{N_0} + \ldots + C_i^{N_{k-1}})$,
  $s_0^M = (s_0^L)^l$ and
  $\delta^M(m, s^l) = (s_0^{N_i})^{ir}$ if $\delta^L(m, s) = f_i$, $\delta^M(m, s^l) = (\delta^L(m, s))^l$ otherwise,
  $\delta^M(m, s^{ir}) = f_j$ if $\delta^{N_i}(m, s) = f_j$, $\delta^M(m, s^{ir}) = (\delta^{N_i}(m, s))^{ir}$ otherwise.
- If $M = \text{while0}\,N$ then $C_i^M = C_i^N$ for each i, $s_0^M = s_0^N$,
  $\delta^M(m, s) = s_0^N$ if $\delta^N(m, s) = f_0$ and $\delta^M(m, s) = \delta^N(m, s)$, otherwise.
- If $M = \text{new}\,\lambda x_j : \text{var}[n].N$, then $C_0^M = (C_0^N \cup C_i^N) \times \lceil n \rceil$ and $C_k^M = C_k^M \times \lceil n \rceil$
  for $k \notin \{0, j\}$, $s_0^M = \langle s_0^N, 0 \rangle$ and
  $\delta^M(\varepsilon, \langle s, i\rangle) = \langle s', i \rangle$ if $s \in C_j^N$ and $\delta^N(i^l, s) = s'$,
  $\delta^M(\varepsilon, \langle s, i\rangle) = \langle s', j \rangle$ if $s \in C_j^N$ and $\delta(0^{ir}, s)) = s'$,
  $\delta^M(m, \langle s, i\rangle) = \langle \delta^N(m, s), i \rangle$, otherwise.

Corollary 2. *Observational equivalence in basic SCI is decidable at first-order.*

At higher types, denotations no longer consist of tuples of words and so we cannot decide observational equivalence between terms at these types simply by constructing multitape automata which recognize their denotations. However, we will show that we can associate a first-order term (and hence a multitape automaton) to each higher-order term in a way which reflects observational equivalence, which is therefore decidable at all types.

Definition 1. *A definable monomorphism between types S and T is a term $\text{mono} : S \to T$ such that $[\![\text{mono}]\!] \cdot e = [\![\text{mono}]\!] \cdot e'$ implies $e = e'$.*

By full abstraction, for any $M, N : S$ we have $M \simeq N$ iff $[\![M]\!] = [\![N]\!]$ iff $[\![\mathtt{mono}]\!] \cdot [\![N]\!] = [\![\mathtt{mono}]\!] \cdot [\![N]\!]$ iff $\mathtt{mono}\, N \simeq \mathtt{mono}\, N$. Thus we can prove that equivalence is decidable for terms of type S by showing that there is a definable monomorphism from S to some first-order type.

First, we observe that we may restrict attention to *functional* types — i.e. types which do not contain any instances of the product construction. A *definable retraction* between types S and T is a pair of terms $\mathtt{in} : S \to T$ and $\mathtt{out} : T \to S$ such that $[\![\lambda x.\mathtt{in}\,(\mathtt{out}\, x)]\!] = [\![\lambda x.x]\!]$. Clearly, $\mathtt{in}$ is a monomorphism. We note also that if there is a definable retraction from R to R' and from S to S' then there are definable retractions from $S \to R$ to $S' \to R'$ and from $S\&R$ to $S'\&R'$.

Proposition 4. *Given functional types S, T there is a functional type $\mathrm{prod}(S, T)$ (of order $\max\{o(S), o(T), 1\}$) such that $S\&T$ is a definable retract of $\mathrm{prod}(S, T)$.*

Proof. is by induction on $o(S) + o(T)$. For the base case, we have $S = \underline{m}$, $T = \underline{n}$ for some m, n. We define $\mathrm{prod}(S, T) = \underline{2} \to \overline{\max\{m, n\}}$, $\mathtt{in} = \lambda x.\lambda y.\mathtt{case}\, \langle y, x \rangle$ and $\mathtt{out} = \lambda x.\langle x\, 0, x\, 1\rangle$. For the inductive case, suppose $T = U \to V$. Then $S\&T \trianglelefteq (U \to S)\&(U \to V) \cong U \to (S\&V) \trianglelefteq U \to \mathrm{prod}(S, V) =_{df} \mathrm{prod}(S, T)$.

Corollary 3. *For every type T there exists a functional type $\overline{T}$ of order at most $o(T) + 1$ and a definable retraction from T to $\overline{T}$.*

We now define monomorphisms from higher-order to lower-order functional types. They are based on the observation that we may uniquely represent a sequence of tuples of sequences $\langle s_{11}, \ldots, s_{1n}\rangle \ldots \langle s_{k1}, \ldots, s_{kn}\rangle$ as a tuple of sequences $\langle s_{11}@\ldots @s_{k1}, \ldots, s_{1n}@\ldots @s_{kn}\rangle$, where @ is a symbol not occurring in any of the s_{ij}.

Lemma 1. *For any types S, T, there is a definable monomorphism from $(\underline{n} \to S) \to T$ to $\underline{n+1} \to S \to T$.*

Proof. Assuming a term $\mathtt{eq} : \underline{n}\&\underline{n} \to \underline{2}$ such that $[\![\mathtt{eq}\,\langle \mathtt{n}, \mathtt{n}\rangle]\!] = \{0\}$ and $[\![\mathtt{eq}\,\langle \mathtt{n}, \mathtt{m}\rangle]\!] = \bot$ if $n \neq m$, we define $\mathtt{mono1} : ((\underline{n} \to S) \to T) \to (\underline{n+1}) \to S \to T$:

$$\lambda f.\lambda x.\lambda y. f\,(\lambda z.(\mathtt{while0}\,(\mathtt{case}\,\langle x, \langle\langle \mathtt{eq}\,\langle \mathtt{i}, z\rangle \mid i < n\rangle, 1\rangle\rangle); y)$$

The denotation of $\mathtt{mono1}$ relates each element $\langle\langle j_1, s_1\rangle \ldots \langle j_m, s_m\rangle, t\rangle$ to the unique element $\langle j_1 n \ldots n j_m n, \langle s_1 \ldots s_n, t\rangle\rangle$. Clearly this is a 1-1 relation, and therefore a monomorphism.

Lemma 2. *For any type $((S \to T) \to U) \to V$ there is a definable monomorphism into $(S\&\underline{1}) \to (T \to U) \to V$.*

Proof. We define $\mathtt{mono2} : (((S \to T) \to U) \to V) \to (S\&\underline{1}) \to (T \to U) \to V =$

$$\lambda f.\lambda x.\lambda g.f\,\lambda y.(g\,(\pi_2(x); (y\,\pi_1(x))))$$

The denotation of mono2 relates each element

$$\langle\langle s_{11}, t_{11}\rangle \ldots \langle s_{1m_1}, t_{1m_1}\rangle, u_1\rangle \ldots \langle\langle s_{n1}, t_{n1}\rangle \ldots \langle s_{nm_n}, t_{nm_n}\rangle, u_n\rangle, v\rangle$$

to the unique element

$$\langle 0^r s^l{}_{11} \ldots 0^r s^l{}_{1m_1} \ldots 0^r s^l{}_{n1} \ldots 0^r s^l{}_{nm_n}, \langle\langle t_{11} \ldots t_{1m_1}, u_1\rangle \ldots \langle t_{n1} \ldots t_{nm_n}, u_n\rangle, v\rangle\rangle$$

Proposition 5. *For any functional type R of order $n+2$ there is a functional type $\widehat{R}$ of order $n+1$ and a definable monomorphism from R to $\widehat{R}$.*

Proof. By induction on the size of R. For the induction step, if R is of order 2 then it is isomorphic to a type of the form $(\underline{n} \to S) \to T$. By Lemma 1 there is a definable isomorphism from R to $\underline{n+1} \to S \to T$ and hence — using the induction hypothesis — to $\underline{n+1} \to \widehat{S \to T}$.

If $o(R) > 2$ then R is isomorphic to a type of the form $((S \to T) \to U) \to V$. By Lemma 2 there is a definable monomorphism from R to $S\&\underline{1} \to (T \to U) \to V$ and hence — using the induction hypothesis — to $\overline{S\&\underline{1}} \to \widehat{(T \to U)} \to V$.

Theorem 3. *Observational equivalence in finitary basic SCI is decidable.*

Proof. We use Proposition 5 to show there is a definable monomorphism from each type of SCI into a first-order type.

4 Observably Sequential SCI

We will now show that observational approximation is decidable in an "observably sequential" version of SCI containing a simple form of non-local control in the form of backwards jumps to labelled program points.[2] We extend the syntax of SCI with the constant label : $(\underline{0} \to \underline{0}) \to \underline{1}$, and the operational semantics with the evaluation contexts label $E[\_]$ and label $\lambda k.E[\_]$ and the reduction:

$$E[\text{label } \lambda k.E'_k[k]], \mathcal{E} \longrightarrow E[0], \mathcal{E}$$

(where we write $E_k[\_]$ for an evaluation context which does not bind k, and assume that all substitutions are capture avoiding).

We may use label to distinguish SCI-equivalent terms such as $x; y$ and $y; x$. More generally, we can express Cartwright and Felleisen's catch operators [2]. For each n, $\text{catch}_n : (T_0 \to \ldots \to T_{n-1} \to \underline{m}) \to \underline{m+n}$ returns i if its argument is strict in its ith argument, or $n + j$ if it is non-strict with value j. We define $\text{catch}_0 = \lambda x.x$, and
$\text{catch}_{n+1} = \lambda f.\text{new } \lambda x.(\text{label } \lambda k.(x := ((\text{catch}_n \ (f \ (\text{abort } k))) + 1); k); !x)$.

---

[2] This breaks the equivalence between parallel and sequential composition.

We will now give a fully abstract games model of observably sequential SCI, and show that denotations may be represented as regular languages. It is similar to Abramsky and McCusker's semantics of Idealized Algol [1], based on Hyland-Ong dialogue games [6]. Since only one thread of computation relating to each argument may be "open" at a time in SCI programs, we may omit explicit justification pointers from our model. As in McCusker and Wall's semantics of SCI [12] (and the author's semantics of linearly used continuations [8]) we add an equivalence relation $\sim$ "interference" to the notion of HO-arena in order to indicate which moves may not occur in the same thread. We allow the interpretation of backwards jumps by abandoning the notion of questions and answers, and hence the "bracketing condition".

A SCI *arena* is a directed acyclic graph $(M_A, \vdash_A)$ — in the form of a set of nodes or *moves* M_A, and a set of directed edges (or *enabling relation*) $\vdash_A \subseteq M_A \times M_A$ — with a labelling function $\lambda_A : M_A \to \{O, P\}$, partitioning the moves between Player and Opponent, and an equivalence relation $\sim \subseteq M_A \times M_A$ such that:

- Root nodes (the *initial moves* M_A^I) are Opponent moves, and there is no edge between two Opponent moves or two Player moves.
- If $m \vdash n$ and $m' \vdash n$ then $m \sim m'$.

For any sequence s of moves, we define a subsequence $\mathsf{open}(s)$ — the "stack of open moves" — containing at most one move from each $\sim$ equivalence class.
$\mathsf{open}(\varepsilon) = \varepsilon$,
$\mathsf{open}(sm) = tm$, if $tm' \sqsubseteq \mathsf{open}(s)$ and $m' \sim m$,
$\mathsf{open}(sm) = \mathsf{open}(s)m$, otherwise.

By definition, the stack of open moves contains at most one enabler for each move — which we may designate its *justifier* — and so we may unambiguously define the *view* $\ulcorner \mathsf{open}(s) \urcorner$ as follows:[3]
$\ulcorner \varepsilon \urcorner = \varepsilon$ and $\ulcorner m \urcorner = m$
$\ulcorner smtn \urcorner = \ulcorner s \urcorner mn$ if $m \vdash n$.

The set L_A of *legal* sequences of the arena A consists of finite alternating sequences of moves of A containing at most one initial move (well-openedness), in which every non-initial move is preceded by an enabling move, which occurs in the view of the stack of open moves (visibility), and which satisfy the *non-interference condition*. Essentially this requires that two potentially interfering moves cannot be "called" from non-interfering parts of the game.

Definition 2. *A sequence t satisfies the non-interference condition if for any $sb, s'b' \sqsubseteq t$ and moves a, a' ocurring (respectively) in $\ulcorner \mathsf{open}(s) \urcorner$ and $\ulcorner \mathsf{open}(s') \urcorner$ and having the same justifier: if $b \sim b'$ and $\lambda(a) = \overline{\lambda(b)}$ then $a \sim a'$.*

We define a **BCK**-algebra with products in which the objects are arenas, with product and function-space given by the following constructions:

---

[3] We do not distinguish Player and Opponent views — the view of any sequence is that of the participant about to move.

- $A\&B = (M_A + M_B, [\lambda_A, \lambda_B], \vdash_A + \vdash_B, (\sim_A + \sim_B) \cup ((M_A^I)^l \times (M_B^I)^r) \cup ((M_B^I)^l \times (M_A^I)^r))$
- $A \to B = (M_A + M_B, [\lambda_A, \lambda_B], (\vdash_A + \vdash_B) \cup ((M_A^I)^l \times (M_B^I)^r), \sim_A + \sim_B)$.

The elements of A are the deterministic strategies on A (non-empty and even-branching subsets of L_A). Application of $\sigma : A \to B$ to $\tau : A$ is defined $\sigma \cdot \tau = \{s \restriction B \mid s \in \sigma \land s \restriction A \in \tau^*\}$. The combinators $\mathbf{B}, \mathbf{C}, \mathbf{T}, \mathbf{P}$ and π_i are interpreted as *copycat strategies*.

Proposition 6. *The $\mathbf{BCK}$-algebra of SCI-arenas and strategies is well-defined.*

Proof. The key point is to show that application is well-defined — that the set of legal sequences is *compositional*: i.e. if $s \in L_{A \to B}$ and $s \restriction A \in (L_A)^*$ then $s \restriction B \in L_B$. The proof that $s \restriction B \in L_B$ satisfies the non-interference condition is based on techniques used to prove compositionality of innocence in HO-games.

We obtain a model of SCI by setting the denotation of the ground type $\underline{n}$ to be the arena with a single initial move enabling n different Player moves. Thus $\underline{0} \to \underline{0}$ is isomorphic to $\underline{1}$, and we define the denotation of label to be this isomorphism. The remaining constants of SCI are interpreted as in the game semantics of Idealized Algol [1]. Proof of soundness and adequacy uses standard techniques and follows those of similar results for games models.

Proposition 7. *For any closed term $M : \underline{1}$, $M \Downarrow$ if and only if $\llbracket M \rrbracket \neq \bot$.*

We now observe that denotational equivalence and approximation are decidable because the strategy denoting each term is a *regular language*. The key point is that the set of legal sequences over each SCI type-object is itself regular.

Lemma 3. *For any finitary SCI type T, the set $L_{\llbracket T \rrbracket}$ is regular.*

Proof. We define a FSA in which each state is a pair consisting of a non-repetitive (and hence bounded) legal sequence t and a function from M_A to sets of non-repetitive sequences over M_A. The initial state is the pair of the empty sequence and the constantly $\emptyset$ function, and the transition function is defined so that having read the legal sequence of moves r, the FSA is in the state $\langle \text{open}(r), f \rangle$, where $f(m)$ is the set of views of prefixes of r which contain an occurrence of m which is in open(r): this information is sufficient to determine the legality of the next move to be read.

Hence the copycat strategies interpreting the $\mathbf{B}, \mathbf{C}, \mathbf{K}, \mathbf{P}$ and π combinators are regular. Following [3], we show that the application of one regular strategy to another is regular, and that the interpretations of case, while0, label and new are regular.

Proposition 8. *The denotation of each term of observably sequential SCI is regular.*

4.1 Full Abstraction

We will now prove that our interpretation of observably sequential SCI is (inequationally) fully abstract. This is the case despite the fact that there are finitary strategies in the model which are not the denotations of terms (and may in fact exhibit interfering behaviour). For example, the strategy on the arena $(\underline{1} \to \underline{1} \to \underline{1}) \to \underline{2}$ — corresponding to the Idealized Algol term $\lambda f.\mathtt{new}\, \lambda x.((f\, (x := 1))\, x := 0); !x$ — which runs its argument once, and returns one if the last argument tested by f was the leftmost one and 0 otherwise.

It is sufficient to establish full abstraction for functional types, since every type is a definable retract of such a type (as in Proposition 4). We further simplify the set of types for which we need to prove full abstraction to the *zero* types: functional types generated from the atomic type $\underline{0}$.

Proposition 9. *Every functional type is definably isomorphic to a zero type.*

Proof. Define the type $\underline{0}_k$ for each $k \geq 0$ by $\underline{0}_0 = \underline{0}$ and $\underline{0}_{k+1} = \underline{0} \to \underline{0}_k$. The arenas denoting $\underline{0}_k$ and $\underline{k}$ are clearly isomorphic; these isomorphisms are definable as $\mathtt{catch}_k : \underline{0}_k \to \underline{k}$ and $\lambda x.\lambda \mathbf{y}.\mathtt{case}\, \langle x, \langle y_0, \ldots, y_{k-1} \rangle \rangle : \underline{k} \to \underline{0}_k$.

We now show that sufficient "observations" are definable to prove full abstraction. For any sequence s, let $|s|$ be the *multiset* of moves occurring in s.

Lemma 4. *Let T be a functional type. For any sequence $s \in L_{[\![T]\!]}$, there exists a term $M(s) : T$ such that $s \in [\![M(s)]\!]$ and if $t \in [\![M(s)]\!]$ then $|t| \sqsubseteq |s|$.*

Proof. By Corollary 9 we may assume T is a zero type. We define $M(s)$ by induction on the length of s. For the induction case, suppose $s = m_1 m_2 r$ for some r. First we suppose that the move m_2 does not occur in r, and define $M(s)$ by induction on the *arity* of T.

- For the base case, suppose $T = (\boldsymbol{S} \to \underline{0}) \to \underline{0}$. For each i, $s \upharpoonright [\![S_i]\!] = t_{i1} t_{i2} \ldots t_{in_i}$, where each t_{ij} is a well-opened legal sequence on $[\![S_i]\!]$ which is shorter than s. Thus we may define $x_i : \mathtt{var}[n+2] \vdash N_i : S_i = x_i := x_i + 1; \mathtt{case}\, \langle !x_i, \langle \Omega, M(t_1), \ldots, M(t_n), \Omega \rangle \rangle$ for each i, and

$$M(s) = \lambda f.\mathtt{new}\, \lambda x_1 \ldots \mathtt{new}\, \lambda x_n.(f\, N_1 \ldots N_n)$$

- If T has arity greater than one, it is (definably) isomorphic to a type of the form $\boldsymbol{R} \to \boldsymbol{S} \to (\boldsymbol{U} \to \underline{0}) \to \underline{0}$, where m_2 is the initial move in $[\![\boldsymbol{U} \to \underline{0}]\!]$, so we assume that it has this form. We may show that if $tb \sqsubseteq s$, where b is the initial move in S, then there exists a unique i such that the initial move in U_i occurs in $\ulcorner \mathtt{open}(tb) \urcorner$. By the non-interference condition, for any two occurrences of b in s, this index i is the same. Hence we may define a legal sequence $\hat{s}$ on the arena $[\![\boldsymbol{R} \to (\boldsymbol{V} \to \underline{0}) \to \underline{0}]\!]$ — where $V_i = S \to U_i$, and $V_j = U_i$ if $j \neq i$ — by relabelling each move from S in s as the corresponding move in V_i. We may then define

$$M(s) = \lambda \boldsymbol{x}.\lambda y.\lambda \boldsymbol{z}.(M(\hat{s})\, \boldsymbol{x})\, \lambda \boldsymbol{a}.\boldsymbol{z}\, a_1 \ldots a_{i-1}(a_i\, y) a_{i+1} \ldots a_n$$

If m_2 does occur in r — i.e. $s = m_1 m_2 r_1 m_2 r_2$, where m_2 does not occur in r_1, then the sequence $m_1 m_2 r_1 n$ in $\underline{0} \to T$ (where n is the move from $\underline{0}$) is no longer than s, whilst $m_1 m_2 r_2$ is shorter than s. Thus we may define

$$M(s) = \lambda \boldsymbol{x}.(\texttt{label } \lambda k.(M(m_1 m_2 r_1 n)\, k\boldsymbol{x})); (M(m_1 m_2 r_2)\, \boldsymbol{x})$$

Proposition 10. *For any closed $M, N : T$, $[\![M]\!] \subseteq [\![N]\!]$ iff $M \not\gtrsim N$.*

Proof. From left to right (soundness) this is standard. For the converse, suppose $[\![M]\!] \not\subseteq [\![N]\!]$ and let smn be a minimal length sequence in $[\![M]\!] - [\![N]\!]$. By prefixing an initial move $q \in M_1$ and postfixing its "answer" a, we obtain a legal sequence $qsmna$ in $[\![T \to \underline{1}]\!]$. By Lemma 4, there exists $L : T \to \underline{1}$ such that $qsmna \in [\![L]\!]$ (hence $[\![L\, M]\!] \neq \bot$) and if $t \in [\![L]\!]$ then $|t| \subseteq |qsmna|$.

Suppose $[\![L\, N]\!] \neq \bot$. Then there exists $qta \in [\![L]\!]$ such that $t \in [\![N]\!]$. By minimality of smn, $s \in [\![N]\!]$, and hence by determinacy of $[\![L]\!]$ and $[\![N]\!]$, $qsm \sqsubseteq qt$, and so $|qsm| \subseteq |qta| \subseteq |qsmna|$ by definition of $[\![L]\!]$. But then $t = smn$, contradicting the assumption that $smn \notin [\![N]\!]$.

Hence $[\![L\, M]\!] \neq \bot$ and $[\![L\, N]\!] = \bot$, and $L\, M \Downarrow$, $L\, N \Uparrow$ as required.

Theorem 4. *Observational equivalence and approximation are decidable in finitary observably sequential SCI.*

References

1. S. Abramsky and G. McCusker. Linearity, Sharing and State: a fully abstract game semantics for Idealized Algol with active expressions. In P.W. O'Hearn and R. Tennent, editors, *Algol-like languages*. Birkhauser, 1997.
2. R. Cartwright and M. Felleisen. Observable sequentiality and full abstraction. In *Proceedings of POPL '92*, 1992.
3. D. Ghica and G. McCusker. The regular language semantics of second-order Idealised Algol. *Theoretical Computer Science (To appear)*, 2003.
4. T. Harju and J. Karhumäki. The equivalence problem of multitape finite automata. *Theoretical Computer Science*, 78:347–355, 1991.
5. J. R. Hindley. *Basic Simple Type Theory*. Cambridge University Press, 1997.
6. J. M. E. Hyland and C.-H. L. Ong. On full abstraction for PCF: I, II and III. *Information and Computation*, 163:285–408, 2000.
7. E. Kinber. The inclusion problem for some classes of deterministic multitape automata. *Theoretical Computer Science*, 26:62–73, 1983.
8. J. Laird. Game semantics and Linear CPS interpretation. *Theoretical Computer Science*, 333:199–224, 2005.
9. R. Loader. Finitary PCF is undecidable. *Annals of Pure and Applied Logic*, 2000.
10. G. McCusker. A fully abstract relational model of Syntactic Control of Interference. In *Proceedings of Computer Science Logic '02*, number 2471 in LNCS. Springer, 2002.
11. G. McCusker. On the semantics of the bad-variable constructor in Algol-like languages. In *Proceedings of MFPS XIX*, ENTCS, 2003. To appear.
12. G. McCusker and M. Wall. Categorical and game semantics for SCIR. In the proceedings of Games for Logics and Programming Languages, 2004.

13. A. Murawski. On program equivalence in languages with ground-type references. In *Proceedings of LICS '03*. IEEE Press, 2003.
14. A. Murawski and I. Walukiewicz. Third-order Idealized Algol with iteration is decidable. In *Proceedings of FoSSACS '05*, number 3411 in LNCS, pages 202–218. Springer, 2005.
15. P. W. O'Hearn, A.J. Power, M. Takeyama and R.D. Tennent. Syntactic control of interference revisited. *Theoretical Computer Science*, 228(1-2):211–252, 1999.
16. M. Rabin and D. Scott. Finite automata and their decision problems. *IBM Journal of Research and Development*, 3:114–125, 1959.
17. U. S. Reddy. Global state considered unnecesary: Object-based semantics for interference-free imperative programs. *Lisp and Symbolic Computation*, 9(1), 1996.
18. J. Reynolds. Syntactic Control of Interference. In *Conf. Record 5th ACM Symposium on Principles of Programming Languages*, pages 39–46, 1978.
19. J. Reynolds. The essence of Algol. In *Algorithmic Languages*, pages 345–372. North Holland, 1981.

Idealized Algol with Ground Recursion, and DPDA Equivalence*

A.S. Murawski[1], C.-H.L. Ong[1], and I. Walukiewicz[2]

[1] Oxford University Computing Laboratory,
Parks Road, Oxford OX1 3QD, UK
[2] CNRS, Université Bordeaux-1, 351, Cours de la Libération,
33 405, Talence, France

Abstract. We prove that observational equivalence of $\mathsf{IA}_3 + \mathbf{Y}_0$ (3rd-order Idealized Algol with 0th-order recursion) is equivalent to the DPDA Equivalence Problem, and hence decidable. This completes the classification of decidable fragments of Idealized Algol. We also prove that observational approximation of $\mathsf{IA}_1 + \mathbf{Y}_0$ is undecidable by reducing the DPDA Containment Problem to it.

1 Introduction

Observational equivalence is an extensional notion of program equivalence. Two program phrases are *observationally equivalent* if one can be replaced by the other in any program without causing any observable difference to the computational outcome. Reynolds's Idealized Algol (IA) is an elegant and compact programming language that combines imperative programming with high-order features, mediated by a simple type theory. Observational equivalence in IA is in general undecidable even when ground types are finite sets. This paper is concerned with the question of decidability of observational equivalence for appropriate fragments of IA.

We begin with a quick review of IA. Ground types of IA, which are ranged over by β, are *exp* (expressions), *com* (commands) and *var* (assignable variables). Types of IA, ranged over by θ, θ' etc., are generated from ground types by the function space constructor $\theta \to \theta'$. The *order* of a type is defined by $\mathsf{ord}(\beta) = 0$ and $\mathsf{ord}(\theta \to \theta') = \max(\mathsf{ord}(\theta) + 1, \mathsf{ord}(\theta'))$. *Finitary* Idealized Algol, IA_f, is just recursion-free Idealized Algol over finite ground types. We can extend IA_f with iteration by adding the rule

$$\frac{\Gamma \vdash M : exp \quad \Gamma \vdash N : com}{\Gamma \vdash \mathbf{while}\ M\ \mathbf{do}\ N : com}$$

and with general recursion by adding the rule

$$\frac{\Gamma, x : \theta \vdash M : \theta}{\Gamma \vdash \mu x^\theta . M : \theta}.$$

---

* Work supported by the UK EPSRC (GR/R88861/01), European Research Training Network GAMES and St John's College, Oxford.

We call a term an *ith-order term* provided its typing derivation uses exclusively judgments of the shape $\Gamma \vdash M : \theta$ where the types of the free identifiers in Γ have order less than i and $\mathsf{ord}(\theta) \leq i$. The collection of ith-order $\mathsf{IA_f}$ (resp. $\mathsf{IA_f} + \mathbf{while}$) terms will be denoted by IA_i (resp. $\mathsf{IA}_i + \mathbf{while}$). We write $\mathsf{IA}_i + \mathbf{Y}_j$ to refer to ith-order $\mathsf{IA_f}$ terms with recursion that can be typed using the recursion rule in which $\mathsf{ord}(\theta) \leq j$. We tabulate all known results on the complexity of observational equivalence of β-normal terms in IA [1, 2, 3, 4]:

	pure	+**while**	+$\mathbf{Y}_0$	+$\mathbf{Y}_1$
IA_1	CONP	PSPACE	?	= $\mathsf{IA}_1 + \mathbf{Y}_0$
IA_2	PSPACE	PSPACE	?	undecidable
IA_3	EXPTIME	EXPTIME	?	undecidable
IA_4	undecidable	undecidable	undecidable	undecidable

The same results as above also hold for observational approximation.

This paper addresses the cases marked with question marks. In $\mathsf{IA}_i + \mathbf{Y}_0$ only programs of ground type can call themselves recursively. For example, while-loops **while** b **do** c can then be defined by

$$b : exp,\ c : com,\ z \vdash \mu z^{com}.\ \mathbf{ifzero}\ b\ skip\ (c; z).$$

We show that observational equivalence in $\mathsf{IA}_i + \mathbf{Y}_0$ is decidable for $i = 1, 2, 3$ by giving a reduction to the DPDA Equivalence Problem (recently proved decidable by Sénizergues [8]). This does not tell us much about the complexity, though. At the moment it is only known that the complexity of DPDA Equivalence is bounded by a primitive recursive function [5]. We also show that already for $i = 1$ observational equivalence is at least as hard as DPDA Equivalence. In consequence, no advance on the complexity of the former problem can be made without an advance on the latter. Another result is that observational approximation in $\mathsf{IA}_i + \mathbf{Y}_0$ is undecidable for $i = 1, 2, 3$, because the undecidable DPDA Containment Problem [6] can be reduced to it.

Let us comment on the relationship of our results to a recent paper [4] showing decidability of $\mathsf{IA}_3 + \mathbf{while}$. In that paper a simpler language was considered but it was translated to a weaker form of pushdown automata. This was essential to get a precise complexity bound. In this paper we model a richer language but, for reasons explained above, we do not concern ourselves with complexity, hence our constructions are not designed to optimize the size of the resulting automata. In [4] most constructions relied on parallel composition of automata. This is not possible here as DPDAs are not closed under that operation.

2 Game Semantics: Complete Plays

We assume familiarity with the treatment of game semantics of IA as presented, for example, in [7]. Recall that the *multiplicative composition* $\sigma;_m \tau : A \multimap C$

of two strategies $\sigma : A \multimap B$ and $\tau : B \multimap C$ is defined by parallel composition with hiding by letting the strategies interact in the shared subgame B and subsequently hiding the B-moves.

Let $\sigma : A \Rightarrow B$ and $\tau : B \Rightarrow C$. Recall that $A \Rightarrow B = !A \multimap B$ and $B \Rightarrow C = !B \multimap C$. In order to compose the strategies, one first defines $\sigma^\dagger : !A \multimap !B$ by

$$\sigma^\dagger = \{ s \in L_{!A \multimap !B} \mid \text{for all initial } m,\ s \upharpoonright m \in \sigma \},$$

where $s \upharpoonright m$ stands for the subsequence of s (pointers included) whose moves are hereditarily justified by m. Then $\sigma; \tau : A \Rightarrow C$ is taken to be $\sigma^\dagger ;_m \tau$.

Given a set σ of positions on G we write $\mathcal{L}(\sigma)$ for the set of the underlying sequences of moves from M_G. For a given strategy σ, we write $\text{comp}\,\sigma$ for the set of its non-empty plays in which the number of questions matches the number of answers; such plays are called *complete*.

Remark 1. We will be interested in the $\dagger$ construction when $B = [\![\theta]\!]$ and $\text{ord}(\theta) \leq 1$. Then $\sigma^\dagger$ can be characterized explicitly as follows.

- $B = [\![\beta]\!]$: Then $\sigma^\dagger = (\text{comp}\,\sigma)^* \sigma$, i.e. $\sigma^\dagger$ simply iterates σ.
- $B = [\![\beta_1 \to \beta_0]\!]$: Then the switching conditions in the game $!A \multimap !B$ imply that a new copy of σ can be started each time σ is finished (as above) and, additionally, after each question q_1 from β_1. We can thus capture $K = \{\epsilon\} \cup \text{comp}\,(\sigma^\dagger)$ by the equation below.

$$K = \{\varepsilon\} \cup \bigcup \{ q_0 U q_1 K a_1 U \cdots q_1 K a_1 U a_0 K \mid q_0 U q_1 a_1 U \cdots q_1 a_1 U a_0 \in \text{comp}\,\sigma \},$$

where U's stand for (possibly empty and possibly different) segments of moves from A.

The operational semantics of IA can be found in [7]. We write $M \Downarrow$ if the closed term M reduces to *skip*. Recall that two terms $\Gamma \vdash M_1, M_2 : \theta$ are *observationally equivalent*, written $\Gamma \vdash M_1 \cong M_2$, if for any context $C[-]$ such that $C[M_1]$ and $C[M_2]$ are closed terms of type *com*, we have $C[M_1]\Downarrow$ iff $C[M_2]\Downarrow$. Similarly, M_1 *observationally approximates* M_2, written $\Gamma \vdash M_1 \lesssim M_2$, just if for all contexts satisfying the properties above, $C[M_1]\Downarrow$ implies $C[M_2]\Downarrow$.

Theorem 1 ([7]). $\Gamma \vdash M_1 \lesssim M_2$ *iff* $\text{comp}\,[\![\Gamma \vdash M_1]\!] \subseteq \text{comp}\,[\![\Gamma \vdash M_2]\!]$. *Consequently,* $\Gamma \vdash M_1 \cong M_2$ *iff* $\text{comp}\,[\![\Gamma \vdash M_1]\!] = \text{comp}\,[\![\Gamma \vdash M_2]\!]$.

3 Simple Terms and Pointer-Free Representation

Suppose $\theta = \theta_1 \to \cdots \to \theta_n \to \beta$. Then we write $\text{tail}(\theta) = \beta$. Given $\Gamma \vdash M : \theta$, depending on whether $\text{tail}(\beta)$ is *com*, *exp* or *var* respectively, we define the *sets of sequences of moves* $(\![\Gamma \vdash M : \theta]\!)$ by the following decompositions:

$$\mathcal{L}(\text{comp}\,[\![\Gamma \vdash M : \theta]\!]) = run \cdot (\![\Gamma \vdash M]\!) \cdot done$$

$$\mathcal{L}(\text{comp}\,[\![\Gamma \vdash M : \theta]\!]) = q \cdot \sum_{j=0}^{max} ((\![\Gamma \vdash M]\!)_j \cdot j)$$

$$\mathcal{L}(\text{comp}\,[\![\Gamma \vdash M : \theta]\!]) = \sum_{j=0}^{max} write(j) \cdot (\![\Gamma \vdash M]\!)_j^w \cdot ok + read \cdot \sum_{j=0}^{max} ((\![\Gamma \vdash M]\!)_j^r \cdot j).$$

It will turn out convenient to define automata accepting $(\![\cdots]\!)$ instead of $\mathcal{L}(\text{comp}\,[\![\cdots]\!])$, because then it will not be necessary to interpret hiding in many cases (an operation under which DCFLs are not closed in general). Since $(\![\cdots]\!)$ are sets of sequences of moves, they do not always represent $\text{comp}\,[\![\cdots]\!]$ faithfully because they ignore pointers. Nevertheless, we are going to identify a sufficiently rich class of terms for which $\text{comp}\,[\![\cdots]\!]$ can be recovered from $(\![\cdots]\!)$.

First, we note that to establish decidability it suffices to consider β-normal terms only. This does not solve the pointer problem though. To address it we replace the application rule with its multiplicative version (left) and contraction (right):

$$\frac{\Gamma \vdash M : \theta \to \theta' \quad \Delta \vdash N : \theta}{\Gamma, \Delta \vdash MN : \theta'} \qquad \frac{\Gamma, x_1 : \theta, x_2 : \theta \vdash M : \theta'}{\Gamma, x : \theta \vdash M[x/x_1, x/x_2] : \theta'}.$$

All β-normal terms in $\mathsf{IA}_3 + \mathbf{Y}_0$ are typable if we allow the above rules for $\text{ord}(\theta) \le 2$. We will call a β-normal $\mathsf{IA}_3 + \mathbf{Y}_0$ term *simple* if it can be typed by using the contraction rule only for θ such that $\text{ord}(\theta) < 2$. For instance, $\lambda f.f(\lambda x.f(\lambda y.x))$ is *not* simple. Note that pointer reconstruction is uniquely defined for all moves *except* third-order questions (pointers for answers can be reconstructed uniquely thanks to the bracketing condition; first-order moves must point to the unique initial move; finally, because of Visibility, second-order questions must point to the necessarily unique first-order question in the appropriate O-view). As made precise in the lemma below, for simple terms, we can still recover $\text{comp}\,[\![\cdots]\!]$ from $(\![\cdots]\!)$.

Lemma 1. *Suppose $\Gamma \vdash M : \theta$ is simple. If $sq_3 \in [\![\Gamma \vdash M : \theta]\!]$, and q_3 is a third-order question then q_3's justifier in sq_3 is the last unanswered enabler of q_3 in s.*

Proof. By induction on the structure of simple terms. The crucial point is that if $\sigma : A \multimap B$ satisfies the Lemma for $B = [\![\theta]\!]$ such that $\text{ord}(\theta) \le 1$, so does $\sigma^\dagger$. This follows from the description in Remark 1.

On the other hand, simple terms are good representatives of β-normal terms. Any β-normal term of $\mathsf{IA}_3 + \mathbf{Y}_0$ can be typed by extending a typing derivation of a simple term with a number of applications of the contraction rule for θ of order 2 followed by a number of applications of the λ-abstraction rule.

4 G-Automata

We are going to use a variant of deterministic pushdown automata. Their states will be divided into O-states and P-states and the stack will be modified and inspected only during ϵ-moves.

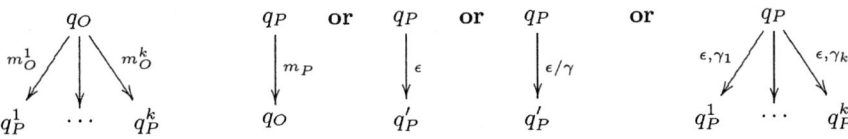

Fig. 1. Transitions in G-automata

- In O-states the automaton will only be able to read an O-move from the input without inspecting or changing the stack; after the transition the state will always change to a P-state.
- In P-states the automaton will either: read a *single* P-move without modifying the stack and move to an O-state, or perform an ϵ-move and go to a P-state possibly modifying the stack.

The constraints are summarized in Figure 1 and captured formally below.

Definition 1. *Let G be a game. A G-automaton $\mathcal{A}$ is a tuple $\langle Q, \Omega, i, F, \delta \rangle$ such that*

- $Q = Q^O + Q^P$ *is the set of states partitioned into Q^O and Q^P, called the sets of O-states and P-states respectively;*
- Ω *is the stack alphabet;*
- $i \in Q^P$ *is the initial state, i does not have any incoming transitions;*
- $F \subseteq Q^P$ *is the set of final states, final states do not have any outgoing transitions;*
- $\delta : [Q^O \rightharpoonup (M_G^O \rightharpoonup Q^P)] + [Q^P \rightharpoonup ((M_G^P \times Q^O) + Q^P + (\Omega \times Q^P) + (\Omega \rightharpoonup Q^P))]$ *is the transition function.*

The interpretation of a transition $\delta(q_O) \in M_G^O \rightharpoonup Q^P$ is that the automaton reads a letter $a \in M_G^O$ from the input and changes the state to $\delta(q_O)(a)$. Similarly for $\delta(q_P) = (b, q_O) \in M_G^P \times Q^O$, the automaton expects to see b as the input and then changes its state to q_O. On a transition $\delta(q_P) = q'_P$ the automaton changes its state but does not consume the input or change the stack. On a transition $\delta(q_P) = (\gamma, q'_P)$ the automaton changes the state and pushes γ on the stack. On a transition $\delta(q_P) \subseteq \Omega \rightharpoonup Q^P$ the automaton makes a pop move and changes its state depending on the letter that was popped. We will use arrow notation for transitions as in Figure 1. Note that slash and comma denote push and pop operations, respectively.

We write $L(\mathcal{A})$ for the language that $\mathcal{A}$ accepts by final state and the empty stack. It is easy to see that G-automata are DPDAs. They have a particularity that they look at the stack only when doing a pop move. We make this restriction to simplify the definitions that follow. It is not difficult to see that for every deterministic pushdown automaton there is an equivalent one with this property.

By a G-automaton *configuration*, we mean a pair qw, where q is a state and $w \in \Omega^*$ is the content of the stack.

Definition 2. *We say that a G-automaton is* productive *if every non-initial reachable configuration is productive, i.e. the configuration occurs in some accepting run of the automaton.*

Lemma 2. *For every G-automaton there is an equivalent productive automaton.*

Proof. Let $\mathcal{A} = \langle Q, \Omega, i, F, \delta \rangle$ be a G-automaton. Consider the function $\beta : \Omega^* \to \mathcal{P}(Q)$ which for all contents of the stack $w \in \Omega^*$ returns the set of states q such that the configuration qw is productive. We will first modify $\mathcal{A}$ so that it keeps the current value of β in its state. Note that it is easy to update β after push-transitions, because $\beta(wa)$ depends only on $\beta(w)$ and a: $q \in \beta(wa)$ iff there exists q' such that $q' \in \beta(w)$ and when $\mathcal{A}$ starts with a on the stack in state q it can end up in q' with empty stack (states q' with this property can be precomputed at the very beginning of the construction). In order to handle pop-transitions, we simply force the automaton to push the old value of β in addition to the pushed stack symbol. Finally, it now suffices to suppress all transitions that are not consistent with β to obtain a productive automaton.

Remark 2. (i) The productiveness of the automaton will play an important role in our constructions for several reasons. One is that when $\mathcal{A}$ is productive we can be sure that when running as a subautomaton of some automaton construction (as in the proof of Theorem 2), it will only use the symbols it pushed (it cannot try to make a pop operation on the empty stack as that would be an unproductive configuration). Another consequence of productiveness is that the automaton cannot enter an accepting state while the stack is not empty. Indeed, since there are no outgoing transitions from an accepting state, the resulting configuration would not be productive. Further, as the automaton stack is necessarily empty whenever it reaches a final state, we also know that when it finishes the stack will be exactly as before it has started.

(ii) From an O-state there are in principle several input letters from M_G^O that can make the automaton advance. However, because of productiveness, in general, not every playable O-move can label an outgoing transition from the O-state. From a P-state there is only one input letter from M_G^P that the automaton is prepared to read. We have this asymmetry because G-automata are designed to accept strategies. Thus, an automaton works in cycles: it reads a letter from M_G^P, then a letter from M_G^O and then does some internal manipulations on the stack.

Definition 3. *We say that an automaton is* careful *if whenever it reaches a configuration qw after reading the sequence of moves s then the sequence of open second-order questions in s appears in the stack w. More formally, there is a function $\pi_\mathcal{A} : \Omega \to M_G \cup \{\epsilon\}$ such that $\pi_\mathcal{A}^*(w)$ is the sequence of open second-order questions in s.*

The final definition of this section makes it precise what kind of automata we want to construct. Below, by $\mathcal{A}(F')$ we will denote the automaton $\mathcal{A}$ with F' as the set of accepting states.

Definition 4. *We say that a tuple of G-automata $\langle \mathcal{A}^1, \cdots, \mathcal{A}^n \rangle$, where each $\mathcal{A}^j = \langle Q^j, \Omega^j, i^j, F^j, \delta^j \rangle$, is fully productive for $\Gamma \vdash M : \theta$ just if each $\mathcal{A}^j$ is productive and careful; further*

- *suppose $tail(\theta) = com$: we have $n = 1$ and $L(\mathcal{A}^1) = (\Gamma \vdash M : \theta)$*
- *suppose $tail(\theta) = exp$: we have $n = 1$, $F^1 = \oplus_{j=0}^{max} F_j$ and $L(\mathcal{A}^1(F_j)) = (\Gamma \vdash M : \theta)_j$ for any $0 \le j \le max$;*
- *in case $tail(\theta) = var$: we have $n = max + 2$, $F^1 = \oplus_{j=0}^{max} F_j$ and $L(\mathcal{A}^1(F_j)) = (\Gamma \vdash M : \theta)_j^r$ for any $0 \le j \le max$; for each $0 \le k \le max$, we have $L(\mathcal{A}^{k+2}) = (\Gamma \vdash M : \theta)_k^w$.*

5 Modelling Simple Terms

Here we focus on simple terms. The extension to other β-normal terms is discussed in Section 6.

Theorem 2. *For any simple $\mathsf{IA}_3 + \mathbf{Y}_0$ term $\Gamma \vdash M : \theta$ there exists a fully productive tuple of $[\![\Gamma \vdash \theta]\!]$-automata for $\Gamma \vdash M : \theta$.*

Proof. We use structural induction. Whenever our constructions fail to preserve productiveness, we simply appeal to Lemma 2 to obtain an equivalent productive automaton.

Thanks to the equalities below and the fact that productive G-automata compose well (see Remark 2 (i))

$$(\Gamma \vdash \lambda x.M) = (\Gamma, x \vdash M) \qquad (\Gamma \vdash M; N) = (\Gamma \vdash M) \cdot (\Gamma \vdash N)$$
$$(\Gamma \vdash !M)_j = (\Gamma \vdash M)_j^r \qquad (\Gamma \vdash M := N) = \sum_{j=0}^{max}((\Gamma \vdash N)_j \cdot (\Gamma \vdash M)_j^w)$$
$$(\Gamma \vdash \text{if } M \text{ then } N_1 \text{ else } N_2) = (\Gamma \vdash M)_0 \cdot (\Gamma \vdash N_2) + (\sum_{j=1}^{max}(\Gamma \vdash M)_j) \cdot (\Gamma \vdash N_1)$$

the corresponding cases are easy. We can simply appeal to the inductive hypothesis and construct the new automata by connecting suitable final states with suitable initial states with ϵ-transitions. Note that this does not violate determinism as final states do not have outgoing transitions. Equivalently, one could "glue" final states with initial ones suitably. If one performs the constructions for reachable final states only, productiveness will be preserved. The remaining cases are treated as follows:

The case of $\Gamma \vdash \text{new } X \text{ in } M : \beta$. Suppose $\beta = com$ and $\langle \mathcal{A} \rangle$ is fully productive for $\Gamma, X : var \vdash M : \beta$. Let us construct $max + 1$ copies of $\mathcal{A}$ denoted by $\mathcal{A}^0, \cdots, \mathcal{A}^{max}$. We will use superscripts to refer to their states. The unique automaton $\mathcal{B}$ in the fully productive tuple for $\Gamma \vdash \text{new } X \text{ in } M$ will consist of all these copies, the idea being that the index of the copy corresponds to the interim value stored in the variable X. The new initial state will be i^0, i.e. the initial state of $\mathcal{A}^0$. The set of final states will be $H = \bigcup_{h=0}^{max} F^h$, i.e. the set of final states from all the copies $\mathcal{A}^h$ of $\mathcal{A}$. We make the following changes to the copies of $\mathcal{A}$. Because $\mathcal{A}$ is productive, we can assume that each $\mathcal{A}$-transition using the P-moves $read_X$ or $write(k)_X$ ($0 \leq k \leq max$) originating from the distinguished copy of var is followed by transitions on $i_X^1, \cdots, i_X^d$ and ok_X respectively, where $\{i^1, \cdots, i^d\} \subseteq \{0, \cdots, max\}$. To construct $\mathcal{B}$ we redirect transitions in the various copies of $\mathcal{A}$ as shown below

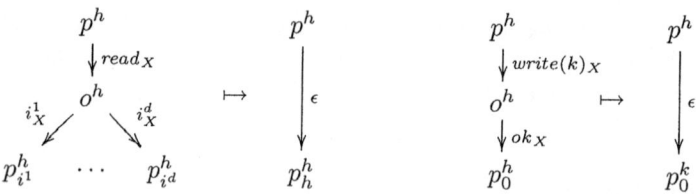

where $0 \leq h \leq max$. These transformations redirect only the transition from p^h, while the transitions from o^h remain as they were. In the first case the transformation is performed only if $h = i^c$ for some $1 \leq c \leq d$. Otherwise, the transition from p^h is just erased. Note that in the second case the new transition connects $\mathcal{A}^h$ to $\mathcal{A}^k$. After all the transitions on P-moves are dealt with, we delete the transitions from o^h. It is clear that the $\mathcal{B}$ is careful if $\mathcal{A}$ is. Using Lemma 2 we can make it productive. If $\beta = exp$ the same construction can be performed to give us a fully productive tuple for $\Gamma \vdash$ **new** X **in** M. To define the required partition of states $H = \oplus_{j=0}^{max} H_j$ we simply take $H_j = \bigcup_{h=0}^{max} F_j^h$, where $F = \oplus_{j=0}^{max} F_j$ is the partition given by the inductive hypothesis. The case of $\beta = var$ combines the previous two cases.

The case of $\Gamma \vdash \mu x^\beta.M : \beta$. Suppose $\beta = com$ and $\langle \mathcal{A} \rangle$ is a fully productive tuple for $\Gamma, x : \beta \vdash M : \beta$, where $\mathcal{A} = \langle Q, \Omega, i, F, \delta \rangle$. The unique automaton $\mathcal{B}$ in the fully productive tuple for $\Gamma \vdash \mu x.M$ will be constructed from two copies $\mathcal{A}^0, \mathcal{A}^1$ of $\mathcal{A}$. Intuitively, $\mathcal{A}^1$ will be used to process recursive calls from M, whereas $\mathcal{A}^0$ will correspond to the base copy of M. Accordingly, the initial state will be i^0 and the final states will be those from F^0. The stack alphabet of $\mathcal{B}$ will, in addition to the stack alphabet of $\mathcal{A}$, contain two copies of the set of O-states of $\mathcal{A}$. By productiveness of $\mathcal{A}$, we can assume that each transition on run_x is followed by a transition on $done_x$. In order to define $\mathcal{B}$, for each $h = 0, 1$ and for each block of the shape $p^h \xrightarrow{run_x} o^h \xrightarrow{done_x} p^h_{done}$ we erase the transition from p^h and add the transitions: $p^h \xrightarrow{\epsilon/o^h} i^1$ and $f^1 \xrightarrow{\epsilon,o^h} p^h_{done}$ for any $f \in F$. The case of $\beta = exp$ is similar but, additionally, we have to pass on the result of the recursive call. The notion of a productive tuple will make that easy. Suppose $F = \oplus_{j=0}^{max} F_j$. Then we first create two copies $\mathcal{A}^0, \mathcal{A}^1$ of $\mathcal{A}$ and for $h = 0, 1$ and for each block

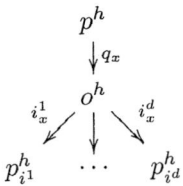

we erase the transition from p^h and add the transitions: $p^h \xrightarrow{\epsilon/o^h} i^1$ and $f_{i^c}^1 \xrightarrow{\epsilon,o^h} p_{i^c}^h$ for any $1 \leq c \leq d$ and $f_{i^c} \in F_{i^c}$. The initial state of the new automaton is i^0, the final states are those in $F^0 = \oplus_{j=0}^{max} F_j^0$. The remaining case of $\beta = var$ is more tedious but completely analogous to the previous two. Given a fully productive tuple $\langle \mathcal{A}^1, \cdots, \mathcal{A}^{max+2} \rangle$ for $\Gamma, x \vdash M$, blocks $write(j)_x ok_x$ are replaced with calls to a copy of $\mathcal{A}^{j+2}$ as for com, while blocks $read_x j_x$ are replaced by calls to a copy of $\mathcal{A}^1$. It is easy to see that the obtained automaton is careful. After the above constructions we can apply Lemma 2 to make the resulting automaton productive.

The case of $fM_1 \cdots M_k$. For demonstration we assume that $k = 1, M_1 = M$ and $fM : com$ and f does not occur in M, i.e. $f : \theta \to com, \Gamma \vdash fM : com$. Let us write r_f, d_f and r, d for the $[\![com]\!]$-moves on the left and right respectively. Observe that because fM is interpreted by interaction of the identity strategy (corresponding to f) with $([\![\Gamma \vdash M]\!])^\dagger$ we have $\text{comp}\,[\![f : \theta \to com, \Gamma \vdash fM : com]\!] = r\,r_f\,(\{\epsilon\} \cup L)\,d_f\,d$, where $L = \phi(\text{comp}\,[\![\Gamma \vdash M : \theta]\!]^\dagger)$ and ϕ is the (injective) renaming map which acts like identity on moves from $[\![\Gamma]\!]$ and maps $[\![\theta]\!]$-moves to their copies in $[\![\theta \to com]\!]$. Below we examine two representative examples in detail and give the precise shape of ϕ. We use subscripts to refer to (moves from) various copies of com.

- Suppose $\Gamma \vdash M : com_0$ and $f : com_{f,1} \to com_f, \Gamma \vdash fM : com$. Then $L = \text{comp}\,[\![\Gamma \vdash M]\!]^\dagger[r_0 \mapsto r_{f,1}, d_0 \mapsto d_{f,1}]$.
- Suppose $\Gamma \vdash M : com_1 \to com_0$ and $f : (com_{f,2} \to com_{f,1}) \to com_f, \Gamma \vdash fM : com$. Then $L = \text{comp}\,[\![\Gamma \vdash M]\!]^\dagger[r_0 \mapsto r_{f,1}, d_0 \mapsto d_{f,1}, r_1 \mapsto r_{f,2}, d_1 \mapsto d_{f,2}]$.

Consequently, the main difficulty is the construction of an automaton accepting $\text{comp}\,(\sigma^\dagger) \cup \{\epsilon\}$, where $\sigma = [\![\Gamma \vdash M : \theta]\!]$ and $\text{ord}(\theta) \leq 1$. What is constructed in the following is not a $[\![\Gamma \vdash \theta]\!]$ automaton (the initial state, which is an O-state, will coincide with the final one and it will have both incoming and outgoing transitions), but the crucial fact is that it becomes one when the transitions on r_f and d_f (which will become the new initial and final moves respectively) are added. We consider the case of $\theta = com_1 \to com_0$ for illustration.

Suppose $\langle \mathcal{A} \rangle$ is fully productive for $\Gamma \vdash M : \theta$, $\mathcal{A} = \langle Q, \Omega, i, F, \delta \rangle$ and $\text{ord}(\theta) \leq 1$. Recall from Remark 1 that in this case $\dagger$ iterates as well as stacks copies of the original strategy and a new copy can start either if the previous one finished or after r_1. To model that, we create two copies $\mathcal{A}^0, \mathcal{A}^1$ of $\mathcal{A}$ with the aim of delegating the copies started after r_1 to $\mathcal{A}^1$. The outermost iterated copies will be processed by $\mathcal{A}^0$.

First we define a few new states: i^0_{new} (which will be the initial and final state of the automaton) and $(\circ, q), (\bullet, q)$, where $q \in Q^0 + Q^1 + \{i^0_{new}\}$. We make the following additions to $\mathcal{A}^0, \mathcal{A}^1$:

- $i^0_{new} \xrightarrow{r_0} (\bullet, i^0_{new}) \xrightarrow{\epsilon/(i^0_{new}, r_0)} i^0$ and $f^0 \xrightarrow{\epsilon, (i^0_{new}, r_0)} (\circ, i^0_{new}) \xrightarrow{d_0} i^0_{new}$ for any $f \in F$.
- Observe that r_1 is always followed by d_1 in σ. By productiveness of $\mathcal{A}$, we can assume that each transition $p^h \xrightarrow{r_1} o^h$ ($h = 0, 1$) is followed by $o^h \xrightarrow{d_1} p_0^h$. Then we add $o^h \xrightarrow{r_0} (\bullet, o^h) \xrightarrow{\epsilon/(o^h, r_0)} i^1$. This makes it possible for the automaton to start processing a new copy of σ, the return state is saved on the stack. To process call returns, we add $f^1 \xrightarrow{\epsilon, (o^h, r_0)} (\circ, o^h) \xrightarrow{d_0} o^h$ for any $f \in F$.

Note that in addition to return addresses in the form of O-states we have also arranged for the questions r_0 to be pushed on the stack. They will remain there as long as they are not answered in the corresponding position. These questions are redundant for accepting $\text{comp}(\sigma^\dagger) \cup \{\epsilon\}$, but are necessary for the resulting automaton to be careful. Since r_0 will be substituted by $r_{f,1}$, the questions r_0 are exactly the second-order questions contributed by f. The case of $fM_1 \cdots M_k$ is analogous, except that we need to apply $\dagger$ to $\sum_{i=1}^k [\![\Gamma \vdash M_i]\!]$.

The contraction rule. The contraction rule for free identifiers of type θ, where $\text{ord}(\theta) = 0, 1$, is interpreted simply by identifying moves from the two copies of $[\![\theta]\!]$. In general this might lead to nondeterminism, but thanks to the structure of our automata this can never happen in our case. There is no problem with P-moves as from each P-state the automaton can read at most one letter. Suppose that a nondeterminism arises from some O-state reachable from the initial configuration. This means that there exists a position s such that the automaton can read both $s\, o_1$ and $s\, o_2$, where o_1 and o_2 are the O-moves from two different copies of θ. However, this contradicts that fact that at most one of $s\, o_1$ and $s\, o_2$ can satisfy Visibility (because only one first-order question is O-visible in s). Consequently, contraction can be interpreted without loss of determinacy.

6 Representing Pointers

We are going to introduce a representation of pointers for simple terms by following Lemma 1. Consider a position sq_3, where q_3 is a third-order question. We define $\pi(s, q_3) = k - j + 2$ where q_2^j is the last enabler of q_3 in the sequence $q_2^1, \cdots, q_2^k$ of all second-order unanswered questions in s (written in the order they appear in s).

Definition 5. *Suppose $\sigma = [\![\Gamma \vdash M : \theta]\!]$, where $\Gamma \vdash M : \theta$ is an $\mathsf{IA}_3 + \mathbf{Y}_0$ term. The language $\mathcal{P}(\sigma)$ over the alphabet $M_{[\![\Gamma \vdash \theta]\!]} + \{\text{check}\}$ is defined by $\mathcal{P}(\sigma) = \{\, s\, \text{check}^{\pi(s, q_3)} \mid sq_3 s' \in \text{comp}\,\sigma \text{ for some } s' \,\}$.*

Lemma 3. *For any simple term $\Gamma \vdash M : \theta$ there exists a DPDA accepting $\mathcal{L}(\operatorname{comp}\sigma) \cup \mathcal{P}(\sigma)$, where $\sigma = [\![\Gamma \vdash M : \theta]\!]$.*

Proof. By Theorem 2 there is a fully productive tuple for $\Gamma \vdash M : \theta$. By adding transitions on the missing initial and final moves, we can construct an automaton $\mathcal{A}$ accepting $\mathcal{L}(\operatorname{comp}\sigma)$. Though $\mathcal{A}$ is not, strictly speaking, a G-automaton, it is nonetheless a DPDA that is productive and careful. (The definitions of productive and careful do carry over in this case.) Next, we will construct a DPDA accepting $\mathcal{L}(\operatorname{comp}\sigma) \cup \mathcal{P}(\sigma)$. Suppose that the automaton $\mathcal{A}$ reads the sequence sq_3 where q_3 is a third-order question. Note that q_3 is always a P-move, so it is uniquely determined by s and σ. As $\mathcal{A}$ is productive, we know that sq_3 can be extended to a complete position. It remains to take care of *check* letters. As $\mathcal{A}$ is careful, all second-order questions that are open in sq_3 will be stored on the stack. Hence, in order to accept the right number of *check*'s after reading s, the automaton can move to a fresh state (reading the letter *check*) and enter a new mode in which it will repeatedly pop the stack reading *check* whenever the topmost stack symbol contains a second-order question and ϵ otherwise. It accepts exactly after it encounters the first enabler of q_3.

Now that we can represent the semantics of simple terms with pointers, the semantics of other β-normal terms can be obtained by renaming, because in order to type such terms we only need to perform contraction at order 2 and λ-abstraction. The former is done by relabelling (in the same manner as at order 0 or 1; the previous argument that determinism is preserved remains valid), the latter amounts to identity in the games setting. Consequently, we get:

Theorem 3. *For any $\mathsf{IA}_3 + \mathbf{Y}_0$ term $\Gamma \vdash M : \theta$ in β-normal form there exists a DPDA accepting $\mathcal{L}(\operatorname{comp}\sigma) \cup \mathcal{P}(\sigma)$, where $\sigma = [\![\Gamma \vdash M : \theta]\!]$. Hence, the theorem also holds for any $\mathsf{IA}_3 + \mathbf{Y}_0$ term.*

By Lemma 1, $\mathcal{L}(\operatorname{comp}\sigma) \cup \mathcal{P}(\sigma)$ is a faithful representation of $\operatorname{comp}\sigma$. Thus, by Theorem 3 and the decidability of DPDA equivalence [8], we get

Theorem 4. *Observational equivalence is decidable for terms in $\mathsf{IA}_3 + \mathbf{Y}_0$.*

7 Hardness

A DPDA [6] is a tuple $\mathcal{B} = \langle Q, \Sigma, \Gamma, \delta, q_0, Z_0 \rangle$, where $\delta : Q \times (\Sigma \cup \{\epsilon\}) \times \Gamma \rightharpoonup Q \times \Gamma^*$. Additionally, whenever $\delta(q, a, X)$ is defined for some $a \in \Sigma$, $\delta(q, \epsilon, X)$ must be undefined. We consider acceptance by empty stack (initially the stack contains Z_0) and write $N(\mathcal{B})$ for the language accepted by $\mathcal{B}$. For simplicity, we further assume that $\mathcal{B}$ can either pop the stack or push one symbol onto it, i.e. if $\delta(q, a, X) = (q', \alpha)$ then $\alpha = \epsilon$ or $\alpha = \alpha_0 \alpha_1$ ($\alpha_0, \alpha_1 \in \Gamma$) and $\alpha_0 = X$. Any DPDA can be easily converted into this form.

We identify values of type exp with Σ. Consider the game $G = [\![exp]\!] \Rightarrow [\![com]\!]$ so that we have $M_G = \{q\} \cup \Sigma \cup \{r, d\}$. Given $L \subseteq \Sigma^*$ define $\widehat{L} \subseteq M_G^*$ by $\widehat{L} = \{\, rqx_1 \cdots qx_n d \mid x_1 \cdots x_n \in L \,\}$. Note that $\widehat{L_1} = \widehat{L_2}$ iff $L_1 = L_2$.

Lemma 4. *For any DPDA $\mathcal{B}$ there exists a term $x : exp \vdash M_\mathcal{B} : com$ such that $\mathcal{L}(\text{comp}\,[\![x : exp \vdash M_\mathcal{B} : com]\!]) = \widehat{N(\mathcal{B})}$.*

Proof. Push transitions are simulated by recursive calls, pop moves by call returns. Before each call, the symbol to be pushed is stored in the variable TOP which is used to initialize the local copy of X after the call. Take $M_\mathcal{B}$ to be

$x : exp \vdash \textbf{new } Q := q_0,\ TOP := Z_0,\ CH \textbf{ in}$
$\quad \mu z^{com}.\ \textbf{new } POP := 0,\ X := !TOP \textbf{ in}$
$\quad\quad \textbf{while (not }!POP\textbf{) do}$
$\quad\quad\quad (\textbf{if } \delta(!Q, \epsilon, !X) = (q', \alpha) \textbf{ then}$
$\quad\quad\quad\quad (Q := q';$
$\quad\quad\quad\quad \textbf{if } \alpha = \epsilon \textbf{ then } POP := 1 \textbf{ else } ((TOP := \alpha_1); z))$
$\quad\quad\quad \textbf{else}$
$\quad\quad\quad\quad (CH := x;$
$\quad\quad\quad\quad \textbf{if } \delta(!Q, !CH, !X) = (q', \alpha) \textbf{ then}$
$\quad\quad\quad\quad\quad (Q := q';$
$\quad\quad\quad\quad\quad \textbf{if } \alpha = \epsilon \textbf{ then } POP := 1 \textbf{ else } ((TOP := \alpha_1); z))$
$\quad\quad\quad\quad \textbf{else } \Omega_{com})) : com$

Proposition 1. *For any DPDAs $\mathcal{B}_1, \mathcal{B}_2$ we have $N(\mathcal{B}_1) \subseteq N(\mathcal{B}_2)$ iff $M_{\mathcal{B}_1} \sqsubseteq_{\approx} M_{\mathcal{B}_2}$.*

Observe that the term used in Lemma 4 is from $\mathsf{IA}_1 + \mathbf{Y}_0$. Since the Containment Problem for DPDAs is undecidable [6] we have:

Corollary 1. *Observational approximation is undecidable for $\mathsf{IA}_1 + \mathbf{Y}_0$ terms; observational equivalence is at least as hard as DPDA Equivalence.*

Our results complete the classification of decidable and undecidable fragments of IA. The exact complexity of observational equivalence for β-normal terms is known in most cases, as shown in the table in Section 1. The complexity of $\mathsf{IA}_i + \mathbf{Y}_0$ depends on the complexity of DPDA Equivalence, which is not known at present. Our construction, when suitably optimized, yields DPDAs that are doubly exponentially larger than the given term in β-normal form. It is also not yet understood how the presence of β-redexes affects the complexity.

References

1. Ong, C.-H. L.: Observational equivalence of 3rd-order Idealized Algol is decidable. In: Proceedings of LICS (2002) 245–256
2. Murawski, A. S.: On program equivalence in languages with ground-type references. In: Proceedings of LICS (2003) 108–117
3. Murawski, A. S.: Games for complexity of second-order call-by-name programs. Theoretical Computer Science, to appear.
4. Murawski, A. S., Walukiewicz, I.: Third-order Idealized Algol with iteration is decidable. In: Proceedings of FOSSACS. LNCS **3441** (2005) 202–218

5. Stirling, C.: Deciding DPDA equivalence is primitive recursive. In: Proceedings of ICALP. LNCS **2380** (2002) 821–832
6. Hopcroft, J. E., Ullman, J. D.: Introduction to Automata Theory, Languages and Computation. Addison-Wesley (1979)
7. Abramsky, S., McCusker, G.: Linearity, sharing and state: a fully abstract game semantics for Idealized Algol with active expressions. In O'Hearn, P.W., Tennent, R.D., eds.: Algol-like languages, Birkhaüser (1997) 297–329
8. Sénizergues, G.: L(A)=L(B)? decidability results from complete formal systems. Theoretical Computer Science **251(1-2)** (2001) 1–166

From Primal-Dual to Cost Shares and Back: A Stronger LP Relaxation for the Steiner Forest Problem

Jochen Könemann[1], Stefano Leonardi[2], Guido Schäfer[2], and Stefan van Zwam[3]

[1] Department of Combinatorics and Optimization,
University of Waterloo, Canada
jochen@uwaterloo.ca
[2] Dipartimento di Informatica e Sistemistica,
Università di Roma "La Sapienza", Italy
{leon, schaefer}@dis.uniroma1.it
[3] Department of Mathematics and Computer Science,
Eindhoven University of Technology, The Netherlands
s.h.m.v.zwam@student.tue.nl

Abstract. We consider a game-theoretical variant of the Steiner forest problem, in which each of k users i strives to connect his terminal pair (s_i, t_i) of vertices in an undirected, edge-weighted graph G. In [1] a natural primal-dual algorithm was shown which achieved a 2-approximate budget balanced cross-monotonic cost sharing method for this game.

We derive a new linear programming relaxation from the techniques of [1] which allows for a simpler proof of the budget balancedness of the algorithm from [1]. Furthermore we show that this new relaxation is strictly stronger than the well-known *undirected cut relaxation* for the Steiner forest problem.

We conclude the paper with a negative result, arguing that no cross-monotonic cost sharing method can achieve a budget balance factor of less than 2 for the Steiner tree and Steiner forest games. This shows that the results of [1, 2] are essentially tight.

1 Introduction

In the *Steiner forest problem* we are given an undirected graph $G = (V, E)$, with vertex set V and edge set E, a non-negative cost function $c : E \to \mathbb{R}^+$ on the edges of G, and a set of $k > 0$ terminal pairs $R = \{(s_1, t_1), \ldots, (s_k, t_k)\} \subseteq V \times V$. A feasible solution for a Steiner forest instance is a forest $F \subseteq E$ such that vertices s_j and t_j are in the same tree of F for all $1 \leq j \leq k$. The objective is to find a feasible solution F of smallest total cost $c(F) := \sum_{e \in F} c(e)$. The *Steiner tree problem* is a special case of the Steiner forest problem that consists of connecting a set of terminals $R \subseteq V$ to a root vertex $r \in V$ in the cheapest possible way.

Computing minimum-cost Steiner trees and forests is NP-hard [3] and APX-complete [4,5] and therefore, neither of the two problems admits a polynomial-time approximation scheme unless P=NP. The best known algorithm for the Steiner forest problem, due to Agrawal, Klein and Ravi [6] and generalized by Goemans and Williamson [7], uses the primal-dual schema. The algorithms in [6,7] achieve an approximation ratio of $(2 - 1/k)$.

In this paper we will consider the following natural game-theoretic version of the Steiner forest problem: each terminal pair $(s_j, t_j) \in R$ is associated with a player j that wants to establish a connection between s_j and t_j. Player j derives a (privately known) utility value u_j from an existing connection between its terminals.

A *cost sharing method* ξ is an algorithm that, given a subset $Q \subseteq R$ of the players, computes a Steiner forest of cost $c(Q)$ satisfying the connectivity requirements of all players in Q. Moreover, for each player $j \in Q$ it determines a non-negative cost share $\xi_Q(j)$. We say that a cost sharing method is α-*budget balanced* if

$$\frac{1}{\alpha} \cdot c(Q) \leq \sum_{j \in Q} \xi_Q(j) \leq \mathsf{opt}_Q.$$

The first inequality says that at least a $1/\alpha$ fraction of the total cost of servicing the users in Q is recovered by the sum of the cost shares of the users in Q. The second inequality establishes fairness in that the sum of all cost shares is not allowed to exceed the optimum cost of servicing the users in Q, denoted opt_Q. This second inequality is often referred to as *competitiveness*. In this paper we will be interested in cost sharing methods ξ that are computable in polynomial time. Therefore, the cost $c(Q)$ of servicing a set of players Q will necessarily exceed the cost opt_Q of an optimum solution for some instances. This also means that there is no hope to achieve budget balance, i.e., $\alpha = 1$.

An important class of cost sharing methods are the *cross-monotonic* methods. ξ is cross-monotonic if, for any two sets Q and S such that $Q \subseteq S$, and any player $j \in Q$ we have $\xi_S(j) \leq \xi_Q(j)$. In other words, the cost share of any player under the given cost sharing method does not increase if the player set increases. The importance of cross-monotonic cost sharing methods stems from a result by Moulin and Shenker [8]: any budget balanced cross-monotonic cost sharing method can be turned into a budget balanced group-strategyproof mechanism.

Despite the recent interest in computational game theory, examples for combinatorial optimization problems that possess a cross-monotonic cost sharing method are still few: Moulin and Shenker [8] gave a cross-monotonic cost sharing method for problems whose optimal cost function is a sub-modular function of the set Q. However, this condition does not hold for many important network design problems such as Steiner trees and facility location. Jain and Vazirani [2] showed a 2-budget balanced and cross-monotonic cost sharing method for Steiner trees. Pál and Tardos [9] later obtained a 3-budget balanced and group-strategyproof cost sharing method for facility location and also provided a 15-budget balanced solution to the single sink rent-or-buy problem.

In a recent paper, Immorlica, Mahdian and Mirrokni [10] show that combinatorial problems that are well-behaved with respect to their approximability may prove hard when looking for approximately budget balanced cross-monotonic cost sharing methods. Their lower bounds are achieved by using cross-monotonicity only. The authors left open the issue of finding a lower bound on the budget balance factor for the Steiner tree problem.

Our contribution. Both Jain and Vazirani [2] and Pál and Tardos [9] show that the computed cost shares form a feasible solution for the dual of a linear programming relaxation for the problem. Proving competitiveness of the methods can therefore be reduced to an application of weak duality. The budget balance factor corresponds to the performance guarantee of the underlying primal-dual algorithm.

In a recent paper [1], Könemann, Leonardi and Schäfer depart from this line. The authors present a cost sharing method KLS which is an adaptation of the primal-dual algorithm AKR for Steiner forests due to Agrawal, Klein and Ravi [6]. The cost shares computed by KLS are proven to be 2-budget balanced. They do not, however, correspond to a feasible dual solution for any of the known Steiner forest duals.

The obvious question left open by [1], that we answer affirmatively in this paper, is this: Is there an alternate Steiner forest LP formulation such that the cost shares computed by KLS correspond to a feasible dual solution? If so, how does this new LP relaxation relate to the standard undirected-cut LP relaxation?

Theorem 1. *There is a linear programming relaxation (LC-D) for the Steiner forest problem whose optimum solution value is at most the cost of any feasible Steiner forest for the given instance. (LC-D) is strictly stronger than the well-known undirected-cut relaxation for Steiner forests. The dual solution computed by KLS is feasible for (LC-D).*

The algorithms in [6,7] are based on the classical undirected cut formulation for Steiner forests [11]. The integrality gap of this relaxation is known to be $(2 - 1/k)$ and the results in [6,7] are therefore tight. Our lifted-cut dual relaxation is strictly stronger than the classical undirected cut formulation. There are instances in which the dual solution achieved by our relaxation provides a much better approximation of the optimum than the undirected-cut dual relaxation. On the other hand, there are instances in which both relaxations achieve an integrality gap of $(2 - 1/k)$.

Secondly, one naturally wonders whether there is a $(2 - \epsilon)$- budget balanced and cross-monotonic cost sharing method for Steiner trees and forests. The answer to this question is negative, and holds for *any* cross-monotonic cost sharing method for these games, including those taking exponential time:

Theorem 2. *There is no $(2 - \epsilon)$-budget balanced, cross-monotonic cost sharing method for Steiner trees for any $\epsilon > 0$.*

This lower bound shows that the results in [1,2] are essentially best possible. Hence there is no hope to obtain a $(2 - \epsilon)$-budget-balanced cross-monotonic cost sharing scheme even if a linear programming relaxation with such integrality gap existed. Prior to our work, the only upper bound known for the budget balance factor of a cross-monotonic cost sharing method for the Steiner tree and forest games was the IP/LP gap of the bidirected cut relaxation [10] which is at most 8/9.

2 A Cost Sharing Method for Steiner Forests

We review the cross-monotonic cost sharing method for Steiner forests as given in [1]. The algorithm is similar to the primal-dual algorithm AKR due to Agrawal, Klein and Ravi [6]. In this section, we first state the standard LP formulation on which AKR is based, then review AKR, and finally describe the cost sharing algorithm KLS from [1].

Let $S \subseteq V$. We define $\delta(S)$ to be the set of all edges that have exactly one endpoint in S and we also let $R(S)$ be the set of terminal pairs in R that are separated by S, i.e., $R(S) := \{(s,t) \in R : |\{s,t\} \cap S| = 1\}$. We use $r(S)$ for the cardinality of $R(S)$. A subset $S \subseteq V$ is a *Steiner cut* if $r(S) \geq 1$. Let $\mathcal{S}$ be the set of all Steiner cuts.

Consider a Steiner cut $S \in \mathcal{S}$. Any feasible solution F for a given Steiner forest instance must *cross* this cut at least once, i.e., $|\delta(S) \cap F| \geq 1$. This gives rise to the following integer programming formulation for the Steiner forest problem: We have a variable x_e for each edge $e \in E$ which has value 1 if e is part of the resulting forest and 0 otherwise.

$$\text{opt}_{\text{IP}} := \min \sum_{e \in E} c(e) \cdot x_e \qquad \text{(IP)}$$

$$\text{s.t.} \sum_{e \in \delta(S)} x_e \geq 1 \qquad \forall S \in \mathcal{S} \qquad (1)$$

$$x_e \in \{0,1\} \qquad \forall e \in E \ .$$

The dual of the linear programming relaxation (LP) of (IP) has a variable y_S for all Steiner cuts $S \in \mathcal{S}$. There is a constraint for each edge $e \in E$ that limits the total dual assigned to sets $S \in \mathcal{S}$ that contain exactly one endpoint of e to be at most the cost $c(e)$ of the edge.

$$\text{opt}_{\text{D}} := \max \sum_{S \in \mathcal{S}} y_S \qquad \text{(D)}$$

$$\text{s.t.} \sum_{S \in \mathcal{S}: e \in \delta(S)} y_S \leq c(e) \qquad \forall e \in E \qquad (2)$$

$$y_S \geq 0 \qquad \forall S \in \mathcal{S} \ .$$

AKR is a *primal-dual* algorithm. That is, it constructs both a feasible and integral primal solution for (LP) and a feasible dual solution for (D). The algorithm starts with an infeasible primal solution and reduces the degree of infeasibility as it progresses. At the same time, it creates a dual feasible packing of sets of largest possible total value. The algorithm raises dual variables of certain subsets of vertices. The final dual solution is maximal in the sense that no single set can be raised without violating a constraint of type (2).

We can think of an execution of AKR as a process over time. Let x^τ and y^τ be the primal incidence vector and feasible dual solution at time τ. We use F^τ to denote the forest corresponding to x^τ. Initially, $x_e^0 = 0$ for all $e \in E$ and $y_S^0 = 0$ for $S \in \mathcal{S}$. In the following we say that an edge $e \in E$ is *tight* if the corresponding constraint (2) holds with equality.

Assume that the forest F^τ at time τ is infeasible. We use $\bar{F}^\tau$ to denote the subgraph of G that is induced by the tight edges for dual y^τ. A connected component S of $\bar{F}^\tau$ is *active* iff S separates at least one terminal pair, i.e., iff $S \in \mathcal{S}$. Let $\mathcal{C}^\tau$ be the set of all active connected components of $\bar{F}^\tau$ at time τ. AKR raises the dual variables for all sets in $\mathcal{C}^\tau$ uniformly at all times $\tau \geq 0$.

Suppose now that two active connected components S_1 and S_2 *collide* at time τ in the execution of AKR. In other words, there are terminals $u \in S_1$ and $v \in S_2$ such that a path between u and v becomes tight as a consequence of increasing y_{S_1} and y_{S_2}. If this happens, we add the path to F^τ and continue. S_1 and S_2 are part of the same connected component of $\bar{F}^{\tau'}$ for $\tau' > \tau$.

The following is the main result of [6]:

Theorem 3. *Suppose that algorithm* AKR *outputs a forest F and a feasible dual solution $\{y_S\}_{S \in \mathcal{S}}$. Then $c(F) \leq (2 - 1/k) \cdot \sum_{S \in \mathcal{S}} y_S \leq (2 - 1/k) \cdot \mathrm{opt}_R$, where opt_R is the minimum-cost of a Steiner forest for the given input instance with terminal set R.*

We next describe the modifications that are necessary to turn AKR into a cross-monotonic cost sharing algorithm. We use KLS to refer to this algorithm.

Define the *time of death* $\mathsf{d}(s,t)$ for each terminal pair $(s,t) \in R$ as

$$\mathsf{d}(s,t) := \frac{1}{2} \cdot c(s,t) \,, \tag{3}$$

where $c(s,t)$ denotes the cost of the minimum-cost s,t-path in G. We assume for ease of presentation that each vertex $v \in V$ has at most one terminal on it. This assumption is without loss of generality since we can replace each vertex in V by a sufficient number of copies and link these copies by zero-cost edges. We extend the death time notion to individual terminals and define $\mathsf{d}(s) = \mathsf{d}(t) = \mathsf{d}(s,t)$ for terminals $s, t \in R$.

Using the notation introduced above we obtain KLS by modifying the definition of $\mathcal{C}^\tau$. We say that a connected component S of $\bar{F}^\tau$ is *active* at time τ if it contains at least one terminal $v \in S$ with death time at least τ, i.e., S is active iff there exists a $v \in S$ with $\mathsf{d}(v) \geq \tau$. KLS grows all active connected components in $\mathcal{C}^\tau$ uniformly at all times $\tau \geq 0$. Observe that this way KLS also raises dual variables of connected components in $\mathcal{C}^\tau$ that do not correspond to Steiner cuts. In what follows we denote by $\mathcal{N}$ the set of *non-Steiner cuts*: $\mathcal{N} := \{S \subseteq V : S \notin \mathcal{S}, S \cap R \neq \emptyset\}$. Furthermore, we let $\mathcal{U} := \mathcal{S} \cup \mathcal{N}$ be the set of all Steiner and non-Steiner cuts.

The intuition behind KLS is that a terminal pair (s,t) is active for the time it would take s and t to connect in the absence of all other terminals. Therefore its activity time is independent of other terminal pairs and this is crucial to achieve cross-monotonicity.

For a terminal $v \in R$ and for $\tau \le \mathsf{d}(v)$ we let $S^\tau(v)$ be the connected component in $\bar{F}^\tau$ that contains v. Also let $a^\tau(v)$ be the number of terminals in $S^\tau(v)$ whose death time is at least τ. The cost share of terminal vertex $v \in R$ is defined as

$$\xi_R(v) := \int_{\tau=0}^{\mathsf{d}(v)} \frac{1}{a^\tau(v)} \, d\tau \ . \tag{4}$$

Furthermore, we define $\xi_R(s,t) := \xi_R(s) + \xi_R(t)$ for all $(s,t) \in R$.

Theorem 4. ξ is a cross-monotonic cost sharing method that is 2-budget balanced.

A proof of this theorem was presented in [1]. There, one of the major difficulties was to show that ξ is *competitive*, i.e., that the sum of the cost shares is at most the optimal cost, opt_R. Since we share the entire dual produced during the execution of KLS among the terminal pairs in R, proving competitiveness is equivalent to showing that the dual solution $\{y_S\}_{S \in \mathcal{U}}$ satisfies $\sum_{S \in \mathcal{U}} y_S \le \mathrm{opt}_R$. If y were a feasible solution to (D) this would follow immediately from weak duality. Here, however, we cannot apply this argument, since KLS also raises dual variables of non-Steiner cuts. Subsequently, we present an alternative, LP-based proof for the competitiveness of ξ.

3 Lifted-Cut LP Relaxation for Steiner Forests

Recall that we let $R = \{(s_1, t_1), \ldots, (s_k, t_k)\}$ be the set of terminal pairs in our instance. Without loss of generality, we assume in the following that $\mathsf{d}(s_1, t_1) \le \ldots \le \mathsf{d}(s_k, t_k)$. We define a precedence order $\prec$ on R by letting $(s_i, t_i) \prec (s_j, t_j)$ iff $i \le j$ and we extend this order to terminal vertices by letting $s_1 \prec t_1 \prec s_2 \prec t_2 \prec \ldots \prec s_k \prec t_k$. For ease of notation we assume that $v \prec v$ for all $v \in R$.

Consider a terminal w and let $\bar{w}$ be w's mate in the Steiner forest instance (i.e., $(w, \bar{w}) \in R$). We let $\mathcal{S}_w \subseteq \mathcal{S}$ be the set of Steiner cuts that separate w and $\bar{w}$ and for which $(w, \bar{w})$ is the highest ranked such terminal pair:

$$\mathcal{S}_w := \{ S \in \mathcal{S} : w \in R(S), \ v \prec w \text{ for all } v \in R(S) \} \ .$$

We also let $\mathcal{N}_w \subseteq \mathcal{N}$ be the set of all non-Steiner cuts containing w and $\bar{w}$ where $(w, \bar{w})$ is the terminal pair of highest rank:

$$\mathcal{N}_w := \{ S \in \mathcal{N} : \{w, \bar{w}\} \subseteq S \cap R, \ (v, \bar{v}) \prec (w, \bar{w}) \text{ for all } (v, \bar{v}) \in S \cap R \} \ .$$

Recall that we define $\mathcal{U} := \mathcal{S} \cup \mathcal{N}$ as the set of all Steiner and non-Steiner cuts. We then say that a terminal $w \in R$ is *responsible* for a cut $S \in \mathcal{U}$ if $S \in \mathcal{S}_w \cup \mathcal{N}_w$.

The dual of the lifted-cut relaxation for the Steiner forest problem is as follows:

$$\text{opt}_{\text{LC-D}} := \max \sum_{S \in \mathcal{U}} y_S \qquad \text{(LC-D)}$$

$$\text{s.t.} \sum_{S \in \mathcal{U}:\ e \in \delta(S)} y_S \leq c(e) \qquad \forall e \in E \quad (5)$$

$$\sum_{S \in \mathcal{S}_w} y_S + \sum_{S \in \mathcal{N}_w} y_S \leq \text{d}(w) \qquad \forall w \in R \quad (6)$$

$$y_S \geq 0 \qquad \forall S \in \mathcal{U}.$$

Notice that a feasible solution to (LC-D) may assign positive values to non-Steiner cuts $S \in \mathcal{N}$. The constraints of type (6) are necessary as the objective function value of (LC-D) would be unbounded in their absence.

The linear programming dual of (LC-D) has variables x_e for every edge $e \in E$ and variables x_w for every terminal $w \in R$:

$$\text{opt}_{\text{LC-P}} := \min \sum_{e \in E} c(e) \cdot x_e + \sum_{w \in R} \text{d}(w) \cdot x_w \qquad \text{(LC-P)}$$

$$\text{s.t.} \sum_{e \in \delta(S)} x_e + x_w \geq 1 \qquad \forall S \in \mathcal{S}_w, \forall w \in R \quad (7)$$

$$\sum_{e \in \delta(S)} x_e + x_w + x_{\bar{w}} \geq 1 \qquad \forall S \in \mathcal{N}_w, \forall w \in R \quad (8)$$

$$x_e, x_w \geq 0 \qquad \forall e \in E, \forall w \in R.$$

Let $\{x_e, x_w\}_{e \in E, w \in R}$ be an integral solution that is feasible for (LC-P). We argue that this solution gives rise to a feasible Steiner forest with cost not exceeding the objective function value. Define $F := \{e \in E : x_e = 1\}$. The total cost $c(F)$ of F is $\sum_{e \in E} c(e) \cdot x_e$. F is not necessarily a feasible Steiner forest since there might exist a Steiner cut $S \in \mathcal{S}$ with no crossing edge, i.e., $\delta(S) \cap F = \emptyset$. Let $S \in \mathcal{S}_w$ be such a set and let $\bar{w}$ be the mate of w. Constraint (7) for S and w implies that $x_w = 1$ in this case. Next consider the complement $\bar{S} = V \setminus S$. It can be seen that $\bar{w}$ is responsible for $\bar{S}$ and hence, $\bar{S} \in \mathcal{S}_{\bar{w}}$. As no edge crosses $\bar{S}$, we must have $x_{\bar{w}} = 1$. Therefore, we can add all edges along the shortest $w, \bar{w}$-path to F at a cost of $2\text{d}(w, \bar{w})$.

The following lemma relates the cost of any feasible solution for the given Steiner forest instance to the objective function value of an optimal solution for (LC-P).

Lemma 1. *Let F be a feasible solution for the underlying Steiner forest instance. We can then construct a solution x that is feasible for (LC-P) and satisfies:*

$$\sum_{e \in E} c(e) \cdot x_e + \sum_{w \in R} \mathsf{d}(w) \cdot x_w \le c(F) \ .$$

In particular, this implies that $\mathsf{opt}_{LC\text{-}D} = \mathsf{opt}_{LC\text{-}P} \le \mathsf{opt}_R$.

Proof. Let T be a tree in F. We use $E(T)$ and $V(T)$ to refer to the edges and vertices of T, respectively. We construct a solution x that is feasible for (LC-P) and show that for each tree $T \in F$

$$\sum_{e \in E(T)} c(e) \cdot x_e + \sum_{w \in R \cap V(T)} \mathsf{d}(w) \cdot x_w \le c(T) \ .$$

The lemma then follows by summing over all trees in F.

Consider a tree $T \in F$. Let $(w, \bar{w})$ be the terminal pair that is responsible for the non-Steiner cut $V(T)$. Moreover, let P denote the unique $w, \bar{w}$-path in T. We set $x_e := \frac{1}{2}$ for each edge $e \in E(P)$ and $x_e := 1$ for each edge $e \in E(T) \setminus E(P)$. Moreover, we assign $x_w = x_{\bar{w}} := \frac{1}{2}$ and $x_r := 0$ for all terminals $r \in (R \cap V(T)) \setminus \{w, \bar{w}\}$. By definition (3) of death time, $\mathsf{d}(w, \bar{w}) \le \frac{1}{2} c(P)$. Thus, the objective value for x on T is $c(T) - c(P)/2 + \mathsf{d}(w, \bar{w}) \le c(T)$.

It remains to be shown that x is feasible for (LC-P). We show for each tree T in F and for all $v \in R \cap V(T)$ that x satisfies the cut-requirements of constraints (7) and (8) for sets $S \in \mathcal{S}_v \cup \mathcal{N}_v$.

Consider a cut $S \in \mathcal{S}_v$ for some $v \in R \cap V(T)$. If $v \in \{w, \bar{w}\}$, constraint (7) holds since S intersects P and $x_v = \frac{1}{2}$. Now let $v \notin \{w, \bar{w}\}$. As $S \in \mathcal{S}_v$ and $v \prec w$, by assumption, it follows that either $\{w, \bar{w}\} \subseteq S$ or $\{w, \bar{w}\} \cap S = \emptyset$. We also have $\bar{v} \notin S$. As T connects v and $\bar{v}$, it can be seen that S either intersects at least one edge e of T that is not on P (and hence $x_e = 1$) or it intersects at least two edges e_1 and e_2 on P (and therefore $x_{e_1} = x_{e_2} = \frac{1}{2}$). Thus, constraint (7) holds in this case as well.

Next consider a non-Steiner cut $S \in \mathcal{N}_v$ for terminal $v \in R \cap V(T)$. If $v \notin \{w, \bar{w}\}$ then $\{w, \bar{w}\} \cap S = \emptyset$ and S crosses at least one edge of T that is not on P or at least two edges of P. Hence constraint (8) holds. Otherwise, S may cross no edge of T but $x_w + x_{\bar{w}} = 1$ and thus (8) is satisfied. □

Running algorithm KLS on terminal set R yields a cost-share $\xi_R(s, t)$ for all $(s, t) \in R$. It also returns a dual solution y such that $\sum_{(s,t) \in R} \xi_R(s, t) = \sum_{S \in \mathcal{U}} y_S$. It is easy to verify that y is feasible for (LC-D). Lemma 1 therefore yields an alternate proof of the competitiveness of KLS:

Corollary 1. $\sum_{(s,t) \in R} \xi_R(s, t) = \sum_{S \in \mathcal{U}} y_S \le \mathsf{opt}_{LC\text{-}D} \le \mathsf{opt}_R$.

The next lemma shows that (LC-D) is at least as strong as the standard LP dual (D).

Lemma 2. *Let* $\{y_S\}_{S \in \mathcal{S}}$ *be a feasible dual solution for (D). Then there is a feasible dual solution* $\{y'_S\}_{S \in \mathcal{U}}$ *for (LC-D) with* $\sum_{S \in \mathcal{S}} y_S \le \sum_{S \in \mathcal{U}} y'_S$. *This implies that* $\mathsf{opt}_D \le \mathsf{opt}_{LC\text{-}D}$.

Proof. Let y be a feasible solution for (D). The sets $\mathcal{S}_w$ for terminals $w \in R$ form a partition of $\mathcal{S}$: $\mathcal{S} = \bigcup_{w \in R} \mathcal{S}_w$. First, we argue that we can assume that y is *symmetric* in the following sense. y is symmetric if for all $(s,t) \in R$:

$$l_s := \sum_{S \in \mathcal{S}_s} y_S = \sum_{S \in \mathcal{S}_t} y_S =: l_t \ .$$

Suppose that this equality does not hold for some $(s,t) \in R$ and, without loss of generality, assume that $l_s > l_t$. Then, let $S \in \mathcal{S}_s$ be a set with $y_S > 0$.

Consider the set $\bar{S} = V \setminus S$ and observe that this set is a Steiner cut as well. Moreover, S and $\bar{S}$ trivially separate the same terminal pairs in R, i.e., $R(S) = R(\bar{S})$. It therefore follows that t must be the responsible terminal for $\bar{S}$ and hence $\bar{S} \in \mathcal{S}_t$. Finally, notice that $\delta(S) = \delta(\bar{S})$ and hence we can increase $y_{\bar{S}}$ and decrease y_S at the same rate without violating any of the constraints of type (2). Continuing this procedure will lead to a symmetric dual y that is feasible for (D).

Now define $y'_S = y_S$ if S is a Steiner cut and let $y'_S = 0$ otherwise. y' clearly satisfies all constraints of type (5). We will now show that y' also satisfies all constraints of type (6) and this will finish the proof of the lemma.

Assume for the sake of contradiction that y' violates constraint (6) for some terminal $w \in R$: $l_w + \sum_{S \in \mathcal{N}_w} y'_S > \mathsf{d}(w)$. Since $y'_S = 0$ for all non-Steiner cuts $S \in \mathcal{N}_w$ we therefore must have $l_w > \mathsf{d}(w) = c(P)/2$, where $c(P)$ is the cost of a minimum-cost $w, \bar{w}$-path in G.

Using the symmetry of y we know that $l_w = l_{\bar{w}}$ and hence we must have $l_w + l_{\bar{w}} > c(P)$. On the other hand, adding the constraints of type (2) for all edges $e \in P$ yields

$$l_w + l_{\bar{w}} \leq \sum_{S \in \mathcal{S}} |\delta(S) \cap P| \cdot y_S = \sum_{e \in P} \sum_{S \in \mathcal{S}: e \in \delta(S)} y_S \leq c(P) \ ,$$

and this is a contradiction. □

The dual of the lifted-cut relaxation is strictly stronger than the standard LP dual (D). The proof of the following lemma is omitted.

Lemma 3. *There exist instances for which* $\mathsf{opt}_D < \mathsf{opt}_{LC\text{-}D}$.

Lemmas 1, 2, and 3 together with Corollary 1 finish the proof of Theorem 1. Unfortunately, as with the undirected cut formulation for Steiner forests, the IP/LP gap of the lifted-cut relaxation is about 2 for certain instances. The proof is omitted.

Lemma 4. *There exist instances for which* $\mathsf{opt}_R/\mathsf{opt}_{LC\text{-}D} = 2 - 2/(k+1)$.

4 A Lower Bound for the Steiner Tree Game

The tools used in this section are adaptations of those used in [10]. In particular we consider any given cross-monotonic cost sharing method ξ for the Steiner tree game and show that there is an instance of the game where the sum of the cost shares of all players is considerably smaller than the cost of an optimum solution. Instead of using a probabilistic argument similar to the one described in [10], we use a more direct (but ultimately equivalent) proof based on convex combinations.

The family of instances used in our proof resembles the one used for the facility location lower bound in [10]. We construct an undirected graph $G = (V, E)$. In this graph, there are k pairwise disjoint classes A_i, $i = 1, \ldots, k$, each of which contains m vertices. Every one of these vertices corresponds to a player who wants to connect this vertex with the root. The set of all players that have a vertex associated with them in A_i is denoted by $\mathcal{A}_i$. The set of all players is $\mathcal{R} := \bigcup_{i=1}^{k} \mathcal{A}_i$.

Let $\mathcal{B}$ be a set containing all sets with exactly one element from each of the A_i, i.e., $\mathcal{B} := \{\{a_1, \ldots, a_k\} : a_i \in A_i, i = 1, \ldots, k\}$. For each set $B \in \mathcal{B}$, we introduce a unique vertex f_B with distance 1 to all vertices in B. The distance to the vertices not in B is, by triangle inequality, equal to 3. Finally, every vertex f_B is connected to the root r, with edges of length 3.

The following lemma argues that we may assume that ξ is *symmetric* in the following sense: Consider a subset $\mathcal{T} \subseteq \mathcal{R}$. We then may assume that $\xi_\mathcal{T}(c) = \xi_\mathcal{T}(d)$ for any two players $c, d \in \mathcal{A}_i \cap \mathcal{T}$ and for any $1 \leq i \leq k$.

Lemma 5. *Suppose that there is an α-budget balanced cost sharing method for the Steiner tree game. Then, given an arbitrary player subset $\mathcal{T} \subseteq \mathcal{R}$, there also is an α-budget balanced cost sharing method with $\xi_\mathcal{T}(c) = \xi_\mathcal{T}(d)$ for all $c, d \in \mathcal{T} \cap \mathcal{A}_i$ and for all $1 \leq i \leq k$. Moreover, for all $c \in \mathcal{T} \cap \mathcal{A}_i$ and for all $d \in \mathcal{A}_i \setminus \mathcal{T}$, we may assume that $\xi_\mathcal{T}(c) = \xi_{(\mathcal{T}\setminus\{c\})\cup\{d\}}(d)$.*

Proof. Let $\tilde{\xi}$ be an α-budget balanced cost sharing method for the Steiner tree game. Pick an index i and a set of players $\mathcal{T} \subseteq \mathcal{R}$. For a player $c \in \mathcal{A}_i \cap \mathcal{T}$ we let $a_c \in A_i$ be the (original) terminal associated with c. Now consider a permutation π_i of the players in class $\mathcal{A}_i$. The terminal of c under permutation π_i is $a_{\pi_i(c)}$.

Define the map π by letting $\pi(c) = \pi_i(c)$ iff $c \in \mathcal{A}_i$. For a set of players $\mathcal{T}$, we define $\pi(\mathcal{T}) := \{d \in \mathcal{R} : \exists c \in \mathcal{T} : d = \pi(c)\}$. Let Π be the set of all $(m!)^k$ possible maps that arise in this way. The cost sharing method ξ is then given by

$$\xi_\mathcal{T}(c) = \sum_{\pi \in \Pi} \frac{1}{(m!)^k} \tilde{\xi}_{\pi(\mathcal{T})}(\pi(c))$$

for all $c \in \mathcal{R}$.

The important observation here is that the cost sharing method $\tilde{\xi}$ works on vertices. So if we swap the terminals associated with two players in $\mathcal{T}$, the cost shares for these players will be swapped as well. In other words, for any two permutations the algorithm will be presented with the same set of terminals, but the players associated with these terminals may have changed.

Notice that, for a player $c \notin \mathcal{T}$, the value $\xi_\mathcal{T}(c)$ is 0 as $\pi(c) \notin \pi(\mathcal{T})$ for all $\pi \in \Pi$. It is now not difficult to see that, for all $1 \leq i \leq k$ and for any two players

$c, d \in A_i \cap T$, we have $\xi_T(c) = \xi_T(d)$ as we average over all player permutations. It remains to show that $\tilde{\xi}$ is cross-monotonic and α-budget balanced.

Consider adding a player d to set T. We have to argue that the cost share of an individual player cannot increase. For a player $c \in T$ we see that

$$\xi_{T \cup \{d\}}(c) = \sum_{\pi \in \Pi} \frac{1}{(m!)^k} \tilde{\xi}_{\pi(T \cup \{d\})}(\pi(c)) \leq \sum_{\pi \in \Pi} \frac{1}{(m!)^k} \tilde{\xi}_{\pi(T)}(\pi(c)) = \xi_T(c) .$$

This follows since $\pi(T \cup \{d\}) = \pi(T) \cup \{\pi(d)\}$ and hence the cross-monotonicity of $\tilde{\xi}$ can be applied to each term.

Now we show α-budget balance. To this end we must specify which solution is returned by the algorithm. If we denote with S^π the solution returned by cost sharing method $\tilde{\xi}$ when run on set $\pi(T)$, we return the solution $S \in \{S^\pi : \pi \in \Pi\}$ with cost $c(S) = \min_{\pi \in \Pi} c(S^\pi)$.

Of course this solution is not necessarily feasible for the original player set, but because of the symmetry of the instance there is a graph isomorphism that maps the solution back to a feasible one without changing the cost.

Now we can write

$$\sum_{c \in T} \xi_T(c) = \sum_{c \in T} \sum_{\pi \in \Pi} \frac{1}{(m!)^k} \tilde{\xi}_\pi(T)(\pi(c)) = \sum_{\pi \in \Pi} \frac{1}{(m!)^k} \sum_{c \in T} \tilde{\xi}_\pi(T)(\pi(c))$$
$$\geq \sum_{\pi \in \Pi} \frac{1}{(m!)^k} \frac{1}{\alpha} \cdot c(S^\pi) \geq \sum_{\pi \in \Pi} \frac{1}{(m!)^k} \frac{1}{\alpha} \cdot c(S) = \frac{1}{\alpha} \cdot c(S) .$$

Competitiveness can be proven using a similar line of reasoning: the cost of the optimal solution must be the same in any permutation. With that, the proof is complete. □

Now suppose we are given a symmetric cost sharing method ξ. From this point on we will identify players and vertices to avoid complication of notation. Ask the algorithm for cost shares for a subset of players $\{a_1, \ldots, a_k\}$ where $a_i \in A_i$. By construction of the graph, all these terminals can connect to vertex $f_{\{a_1,\ldots,a_k\}}$ at cost 1, at which point they are only 3 units away from the root. Hence there is a solution of cost $k+3$ for this subset. Competitiveness states that

$$\sum_{j=1}^k \xi_{\{a_1,\ldots,a_k\}}(a_j) \leq \mathrm{opt}_{\{a_1,\ldots,a_k\}} \leq k+3 .$$

Therefore there must be at least one index i such that $\xi_{\{a_1,\ldots,a_k\}}(a_i) \leq (k+3)/k$. By Lemma 5, this holds for every set $\{a_1, \ldots, a_{i-1}, c, a_{i+1}, \ldots, a_k\}$ where $c \in A_i$ and the other a_j are the same as before.

For this index i we consider the instance with subset $Q := \{a_1, \ldots, a_k\} \cup A_i$. We bound the sum of the cost shares for this set as follows:

$$\sum_{c \in Q} \xi_Q(c) = \sum_{c \in A_i} \xi_Q(c) + \sum_{j \neq i} \xi_Q(a_j)$$

$$\leq \sum_{c \in A_i} \xi_{\{a_1,\ldots,a_{i-1},c,a_{i+1},\ldots,a_k\}}(c) + \sum_{j \neq i} \xi_{\{a_1,\ldots,a_{i-1},a_{i+1},\ldots,a_k\}}(a_j) \quad (9)$$

$$\leq m \cdot \frac{k+3}{k} + k + 2 \ . \quad (10)$$

The inequality in (9) is due to cross-monotonicity: the cost share of a player over a subset of Q cannot be smaller than the cost share for that player over Q. We know that a set of players never pays more than the cost of the optimal tree connecting these players. For set $\{a_1, \ldots, a_{i-1}, a_{i+1}, \ldots, a_k\}$, which has one player from $k-1$ of the sets A_j, there is a solution of $3 + k - 1 = k + 2$. This provides an upper bound on the rightmost term of (9). For the leftmost term, we argued above that $\xi_{\{a_1,\ldots,a_{i-1},c,a_{i+1},\ldots,a_k\}}(c) \leq (k+3)/k$.

Due to the large amount of symmetry in the instance, we can in fact describe the optimal solution. The proof of the following lemma is omitted.

Lemma 6. *The optimal solution for connecting the players in a set Q, as defined above, to the root has cost $2m + k + 1$.*

Combining Lemma 6 with Inequality (10), we can now prove Theorem 2.

Proof (Theorem 2). The ratio between the cost shares of players in the subset Q as defined above and the cost of the network they use can be bounded as follows:

$$\frac{\sum_{c \in Q} \xi_Q(c)}{c(Q)} \leq \frac{\sum_{c \in Q} \xi_Q(c)}{\text{opt}_Q} \leq \frac{m\frac{k+3}{k} + k + 2}{2m + k + 1} = \frac{k^2 + 4k + 2}{2k^2 + k + 1} \ ,$$

where the last equality holds if we choose $m = k^2$. This ratio tends to $\frac{1}{2}$ as $k \to \infty$, which completes the proof. □

References

1. Könemann, J., Leonardi, S., Schäfer, G.: A group-strategyproof mechanism for Steiner forests. In: Proceedings of the Sixteenth Annual ACM-SIAM Symposium on Discrete Algorithms. (2005) 612–619
2. Jain, K., Vazirani, V.V.: Applications of approximation algorithms to cooperative games. In: Proceedings of the Thirty-Third Annual ACM Symposium on Theory of Computing. (2001) 364–372
3. Garey, M.R., Johnson, D.S.: Computers and Intractability: A Guide to the Theory of NP-completeness. Freeman, San Francisco (1979)
4. Arora, S., Lund, C., Motwani, R., Sudan, M., Szegedy, M.: Proof verification and intractability of approximation problems. In: Proceedings of the Thirty-Third IEEE Symposium on Foundations of Computer Science. (1992) 210–214

5. Bern, M., Plassmann, P.: The Steiner problems with edge lengths 1 and 2. Information Processing Letters **32** (1989) 171–176
6. Agrawal, A., Klein, P., Ravi, R.: When trees collide: An approximation algorithm for the generalized Steiner problem in networks. SIAM Journal on Computing **24** (1995) 445–456
7. Goemans, M.X., Williamson, D.P.: A general approximation technique for constrained forest problems. SIAM Journal on Computing **24** (1995) 296–317
8. Moulin, H., Shenker, S.: Strategyproof sharing of submodular costs: budget balance versus efficiency. Economic Theory **18** (2001) 511–533
9. Pál, M., Tardos, É.: Group strategyproof mechanisms via primal-dual algorithms. In: Proceedings of the Forty-Fourth Annual IEEE Symposium on Foundations of Computer Science. (2003) 584–593
10. Immorlica, N., Mahdian, M., Mirrokni, V.S.: Limitations of cross-monotonic cost sharing schemes. In: Proceedings of the Sixteenth Annual ACM-SIAM Symposium on Discrete Algorithms. (2005) 602–611
11. Aneja, Y.P.: An integer linear programming approach to the Steiner problem in graphs. Networks **10** (1980) 167–178

How Well Can Primal-Dual and Local-Ratio Algorithms Perform?

Allan Borodin, David Cashman, and Avner Magen

Department of Computer Science,
University of Toronto
{bor, cashman, avner}@cs.toronto.edu

Abstract. We define an algorithmic paradigm, the stack model, that captures most primal-dual and local-ratio algorithms for approximating covering and packing problems. The stack model is defined syntactically and without any complexity limitations. Hence our approximation bounds are independent of the P vs NP question. We provide tools to bound the performance of primal dual and local ratio algorithms and supply a $(\log n + 1)/2$ inapproximability result for set-cover, a $4/3$ inapproximability for min steiner tree, and a 0.913 inapproximability for interval scheduling on two machines.

1 Introduction

The primal dual and local ratio schemas for approximation algorithms are two fundamental algorithm design techniques. The use of the primal dual schema is pervasive, having been applied to give good approximation algorithms for several basic NP-hard combinatorial optimization problems including set cover, vertex cover, numerous network design problems, facility location and k-median, steiner forest, and many others. The origins of the primal dual schema can be found in the Bar Yehuda and Even [7] algorithm for the weighted vertex cover problem. The re-introduction of the enhanced primal dual schema leading to its current importance is due to the works of Agarwal, Klein and Ravi [1] and Goemans and Williamson [15].

The list of problems tackled successfully by primal-dual and local-ratio algorithm is long and impressive. For many of these problems the methods achieve the best known approximation (see [24, 6] and references therein). But just how far can these methods go? The goal of this paper is to shed some light on this question. Our first step in achieving this is to put the two algorithmic schemas under a larger umbrella – a syntactic computational model capturing both, devoid of complexity considerations and that is related to the LP formulation via the input model environment it works over. This model is what we call the *stack model* and will be described in detail in Section 2. We then show hardness of approximation results within this model for a few prominent problems in which the primal-dual schema was applied before, sometimes achieving the best bound.

For the clarity of our exposition we will use the set-cover problem as a specific running example. We shall focus our attention on what is considered to be the common method of the primal-dual schema for covering problems. Let us be more specific: our abstraction of this method corresponds to what Williamson [24] calls the "primal dual algorithm with reverse delete step". The more restricted version of the method which was the first to appear is called "basic primal-dual algorithm", can be simulated by the *priority model* of [11]. In the priority model, an algorithm considers each input item once (according to some ordering) and immediately makes an irrevocable decision (i.e. accept or reject) about each input item. For the simulation to be possible, the input items are in correspondence with the 0/1 variables constituting the problem, and the input items contain information about their cost and about the constraints they appear in. In the set-cover example, the items are the sets and the natural representation of a set is its weight and the list of elements it contains, namely the cost and the constraints of the natural LP relaxation. We elaborate on this somewhat subtle issue when we discuss the specific applications we deal with. Returning to priority algorithms, there is some natural restriction on what orderings are permissible and we shall argue that the primal dual schema always leads to a permissible ordering. Roughly speaking, the permissible orderings are those that are induced by any (not necessarily computable) mapping of all possible input items into $\mathbb{R}$. Also, the ordering in priority algorithms can be established afresh at every iteration of the algorithm.

A substantial leap forward in terms of the applicability of the primal-dual approach was made at the beginning of the nineties, namely the introduction of the "reverse delete step" [15, 17, 21]. Roughly speaking, this is a phase that guarantees that the output of the algorithm is a *minimal* feasible solution (w.r.t set inclusion). This property is shown to be crucial in the analyses of problems such as minimum Steiner tree, the feedback vertex set problem, generalized steiner forests, and others. As is shown in [24] the analysis may sometime depend not only on the minimality of the produced solution, but also on the exact process that ensures this minimality. The analogous leap in the world of abstract models leads us to the definition of the *stack model*. Here, instead of irrevocably rejecting/accepting an item, the algorithm decides whether to reject or to *push* an item onto a stack. Then the elements in the stack are popped and removed when this is possible.

The *stack model* also captures the local-ratio schema as described in the survey of Bar-Yehuda et al [6]. This should not be taken as a surprise as Bar-Yehuda and Rawitz [9] have demonstrated that there is an equivalence between the primal dual and local-ratio schemas. In fact, in this extended abstract we use their observation to argue that the *stack model* simulates regular local-ratio algorithms, without a concrete description of such a simulation.

In contrast to the extensive use of primal-dual and local-ratio for covering problems, the only (albeit important) use of the primal dual schema for packing problems was the approximation of the *bandwidth allocation problem* in Bar-Noy et al [5]. As was shown in [5], the local ratio technique can *optimally* solve the one machine weighted interval selection problem whereas the negative result in [11] precludes the possibility of any constant approximation to the problem within the priority model. This provably shows that it is sometimes necessary to have a second "clean-up" stage to obtain good approximation bounds.

It is important to note that, while most of the known applications of primal-dual fall under "primal-dual with reverse delete", there are several applications that do not fit within the stack model framework. Of these, one of the most prominent examples is Jain and Vazirani's use of primal dual method for the approximation of metric facility location and k-median problems [16]. Notice that these problems are not covering or packing problems, and therefore inherently in need of a more specialized algorithmic paradigm. A more basic example is the steiner tree algorithm for the special class of quasi-bipartite graphs [18]. Their method provides a $3/2 + \epsilon$ approximation by combining local search with primal dual and using the so called *bidirected* LP relaxation as a starting point. (This method utilizes a non-standard input representation by considering each edge as two directed edges and hence automatically falls outside of our model.) Another example of a primal-dual/local-ratio algorithm that does not fit our model are the so called *fractional* local-ratio/primal-dual algorithms which first solve the LP, and use the solution as a guide to ordering items [8] and [10].

2 The Stack Model and Its Ancestor the Primal-Dual

Consider an integer program for a covering problem, and assume that the variables to the problem are the natural 0/1 variables of the problem, and that the constraint matrix A and the cost vector c are nonnegative. In a nutshell, what a primal dual schema does is to consider an LP relaxation of this IP and the dual maximization problem and (i) starts with the feasible dual solution $y = 0$ (ii) increases y continuously until a dual constraint(s) becomes tight (iii) in this case the corresponding primal variable(s) enter the solution (iv) continues as long as the primal solution generated so far is not feasible (v) goes over the primal variables in the solution in a reverse order and removes them whenever feasibility is maintained. Notice that we haven't specified exactly how the dual vector y is increased, and we choose to stay at this level of generality. Returning to our set-cover example, the dual variables in the natural LP relaxation are elements of the ground set. We increase all the dual variables that do not occur in a tight equation (uncovered elements) uniformly, until a new set (dual constraint) becomes tight, in which case this set joins the solution. Step (v), the reverse-delete, removes a set if it is not needed, while following the reverse order.

We now get to the critical observation that later leads to the abstraction of the model and to our lower bounds. Notice that the above process induces, at each iteration, an ordering on the primal variables that has not entered the solution yet. If at any point an adversary announces that a primal variable that has not yet become tight *has never existed*, the algorithm has no way to recognize such a change as it has no effect on the algorithm's history. An adversarial approach can be developed from this limitation of the algorithm and may lead to inapproximability results.

This sets the stage for the stack model. We consider a two-pass algorithm, in which the first pass, the push phase, resembles an adaptive priority algorithm, as defined in [11], and the second pass, the pop phase, is a simple greedy phase that is not under the control of the algorithm. Here is a precise description. The algorithm first orders the items by some valid ordering. This is just a total ordering of all possible input items. In our example, one can imagine ordering by cardinality of sets, or more generally by any function f_1 on the characteristic 0/1 vectors and the weight of the sets. Next, it looks at the first item, and either pushes it onto the stack or rejects it. It then may decide to change the ordering using the information of the first item (defining f_2 after looking at the first set). Again, the first (unseen) item in this ordering is either pushed or rejected. This process continues for at most n iterations, [1] at which point the algorithm enters the "pop phase" in which items are popped from the stack (last in, first out). The solution to be produced is a subset of the items on the stack and is defined as follows. Each popped item will be rejected if the items that were popped and accepted, together with the items still on the stack constitutes a feasible solution; otherwise it is accepted. It is easy to see that if, at the beginning of the pop phase, we have a feasible solution, we will have one at the end, and further, that solution is minimal w.r.t set inclusion.

Having defined the stack model, we are ready to make the central claim of this section : Every primal dual algorithm under the scheme described above can be simulated by a stack algorithm. It is here that we insist that the input representation of an item contains its cost and the objects that correspond to the constraints that include it. Indeed, regardless of the way y is increased, we get a primal variable that matches a dual constraint that was just made tight. The variable(s) made tight is the one (are those) minimizing the expression $f(i) = c_i - \sum_{j:a_{ij}>0} a_{ij} y_j$. But, since the information about item i includes c_i and the positive a_{ij}, and since the algorithm may keep track of y we conclude that $f : [n] \mapsto \mathbb{R}$ is a valid ordering function as it is a function on the items. The first item is taken [2] and the stack algorithm pushes it and continues. The pop phase is clearly a reflection of step (v) in the primal dual schema.

If the ordering used during the push phase is determined at the beginning and is not changed during the process, we call this a *fixed order* stack algorithm. Without this restriction we say the ordering is *adaptive*.

We now consider the modifications needed in order to define a stack model for packing problems. Here too, we have a push phase and a pop phase where the push phase proceeds like a fixed or adaptive priority algorithm in determining which items to place on the stack. The pop phase starts with the empty set as its existing solution, and then adds popped items to the existing solution unless the addition of the item renders the current solution infeasible.

---

[1] As we shall see later, the stack algorithm can end the push phase once a feasible solution exists on the stack, as the items past this point are bound to get deleted.

[2] If some items become tight simultaneously this is simulated by taking those items one after the other.

We briefly indicate how we can derive negative results for approximation ratios in the (adaptive ordering) stack model. An adversary initially presents a large set of potential input items. As the computation proceeds, the adversary is allowed to delete input items. We note that the adversary is allowed this license since the ordering functions depend only on the item and any previously considered items. Eventually the algorithm will consider all input items that have not been deleted and that will end the push phase. Then since the pop phase is completely determined, we can calculate the cost/profit of the stack algorithm and compare it to an optimal solution for the actual set of items (i.e. the input items not deleted by the adversary and hence considered by the algorithm). For the case of fixed-order stack algorithms, the adversary can often construct the initial set of potential items in such a way that some "difficult structure" exists within a subset of the items no matter how the algorithm chose its order.

The reader may notice that in moving from the primal-dual methodology to the abstraction we describe, the linear structure of the (relaxed) problem disappears. One may also ask what role does the integrality gap play in this discussion? As was observed [24], the ratio guarantee of primal dual algorithms is never better than the integrality gap for *the specific linear program relaxation used by the algorithm*. However, *any* formulation of the problem as an LP with positive coefficients can lead to a primal dual algorithm, which in turn provides an ordering of the elements that gives rise to a stack algorithm. In other words the abstraction helps us to understand the power of primal dual algorithms with respect to every LP relaxation with positive coefficients under the (nontrivial) condition that the variables are the "natural" ones, and that the input representation "knows" the cost of an item and the constrains involving it. Specifically, to show an inapproximability result for a suggested LP relaxation we are required to adjust the input-representation when we give the stack lower bound. In vertex cover, for example, we may look at the LP relaxation that uses the inequalities that require variables of the vertices in an odd cycle of size l to sum up to $(l+1)/2$ (on top of the regular inequalities). Notice that this type of tightening is a result of a lift and project method and was discussed in the context of hardness result in [4]. If we obtain a lower bound for an input representation in which a vertex knows all the edges *and* all the odd cycles containing it, this would imply a lower bound for PD algorithm applied to this relaxation!

3 The Minimum Set Cover Problem

We consider the classical minimum set cover problem. We are given a family of subsets $S_1, S_2, \ldots, S_m$ of a ground set U of size n. In addition, each set S_i is associated with a cost c_i. The goal is to minimize the cost of a collection of the S_i that cover every element in U. The unweighted version of the problem seeks to minimize the number of subsets chosen.

The well known natural greedy algorithm for set cover selects sets with minimum cost per size ratio, and continues recursively on the remaining set of uncovered items with the remaining sets. The natural greedy algorithm yields an $H(n)$

approximation. Notice that this algorithm is a priority algorithm, and hence a special case of a stack algorithm.

Raz and Safra [19] showed that if $P \neq NP$, then for some constant $c > 0$, set cover cannot be approximated within a ratio of $c \log n$ in polynomial time. Using a stronger complexity assumption, namely that $NP \not\subset DTIME(n^{\log \log n})$, Feige was able to show an almost tight bound $(1-\epsilon)H(n)$. But these hardness results do not apply to our model as the stack algorithm is not restricted to run in polynomial time. Furthermore,, stack algorithms may be nonuniform, in the sense of allowing a different algorithm for each n, the the number of input items.

In [2], a $(\log n + 1)/2$ inapproximability bound for the weaker priority algorithm model is shown. Without restating the proof given there, we note that a stack model could provide an optimal solution to the instance that is used in that paper to demonstrate the priority lower bound. Hence, we will need a somewhat more complex set-up to prove the same lower bound for the stack algorithm. We note that a somewhat similar construction is used [23] to show that the integrality gap of the natural LP relaxation for set cover is $\Omega(\log n)$.

We define the ground set U to be the cube $\{0,1\}^{\log n}$. The initial input to the algorithm consists of $2(n-1)$ sets, S_v^b where $v \in \{0,1\}^{\log n} \setminus \{0\}$, and $b \in \{0,1\}$. Set S_v^b contains all points x for which $\langle v, x \rangle = b$ where $\langle \cdot, \cdot \rangle$ is the inner product over GF_2. The sets can be viewed as the inverse images of 0 (1) of all the nontrivial Fourier characters of the boolean cube. We note that for any v, S_v^0 and S_v^1 are a pair of complementary sets.

We require a simple lemma relating the combinatorial properties of sets in our system to the algebraic properties of the vectors that define them:

Lemma 1. *For any set of linearly independent vectors $\{v_1, \ldots, v_k\}$ and any choice of $b_1, \ldots, b_k$, the number of elements of U left uncovered by $\{S_{v_1}^{b_1}, \ldots, S_{v_k}^{b_k}\}$ are selected is exactly $n/2^k$. In particular, any family of sets whose corresponding vectors are linearly independent does not cover U completely.*

Proof. The elements covered by the above sets are $S_{v_1}^{b_1} \cup \ldots \cup S_{v_k}^{b_k}$, hence the uncovered ones are $S_{v_1}^{1-b_1} \cap \ldots \cap S_{v_k}^{1-b_k}$. Therefore, x is uncovered iff it is a solution of the system $\langle v_1, x \rangle = 1 - b_1, \ldots \langle v_k, x \rangle = 1 - b_k$. Since the v_i are linearly independent, the co-dimension of the solution space of uncovered elements is k and so it contains $|F_2|^{l-k}$ elements and the first part of the lemma follows. The second part is obvious.

Using the set-system above, we will show that no stack algorithm can achieve a set cover smaller than $\log(n+1)$, while the optimal set cover is of size 2. At stage i, the algorithm chooses a set $S_{v_i}^{b_i}$ to look at next. Recall that the adversary may remove sets from the input as the algorithm proceeds.

At each step i, $i < \log n$, suppose the algorithm accepts $S_{v_i}^{b_i}$. Then, the adversary removes all the sets S_v^b with $v \in \mathrm{span}\{v_1, \ldots, v_i\}$. Notice that this strategy ensures that as long as $i \leq \log n$, the set of vectors defining the sets in the stack are linearly independent. In particular, for any vector v, S_v^0 and S_v^1 cannot both be on the stack.

In the other case in which the algorithm rejects the set $S_{v_i}^{b_i}$, we argue that the algorithm may fail to generate *any* set cover. Indeed, as a response the adversary simply deletes all unseen inputs except for $S_{v_i}^{1-b_i}$. Note that $v_i \notin$ span$\{v_1, \ldots, v_{i-1}\}$ so this is a valid strategy. But even if the algorithm takes that last set, the stack still contains sets with corresponding linearly independent vectors $\{v_1, \ldots, v_i\}$, which by Lemma 1 does not form a cover. But of course one exists as the input contains the two complementary sets $\{S_{v_i}^{b_0}, S_{v_i}^{b_1}\}$. This argument holds also for $i = \log n$.

Now, assuming that the algorithm continues to accept, after the $(\log n)$-th step, the algorithm has accepted $\log n$ sets whose defining vectors are linearly independent, leaving exactly $n/2^{\log n} = 1$ uncovered element. The adversary will delete all sets *except* for $S_{v^*}^0$ and $S_{v^*}^1$, where $v^* = \sum_{i=1}^{\log n} v_i$. (these sets were not removed before as $v^* \notin$ span$\{v_1, \ldots, v_{\log n-1}\}$). At this point our the "game" is over, and the input to the algorithm is determined. The stack contains $\log n$ linearly independent sets and there are the two remaining sets that the algorithm must consider. Clearly, the sets $S_{v^*}^0$ and $S_{v^*}^1$ constitute a set cover of size 2. It remains to argue that the algorithm must take $\log n + 1$ sets. Since only one element is left uncovered before the two complementary sets are considered, one of the sets is contained in the union of the sets on the stack. This means that this set will be rejected in the pop phase and we may as well assume the algorithm rejects it. The algorithm now has $\log n + 1$ sets on the stack.

We claim that all of these sets must survive the pop phase. Indeed, by the special choice of v^* the vectors $v_1, \ldots v_l, v^*$ are in general position, that is no $\log n$ of them are linearly dependent. Since, by the independence of $v_1, \ldots v_{\log n}$ there may be only one dependency between the $\log n + 1$ vectors, and since summing *all* of them is a dependency, we know that there are no other dependencies, and so the vectors are indeed in general position. We use the corollary again to deduce that no strict sub-family of sets may cover the ground set, and in particular no set can be removed in the pop phase. We have established:

Theorem 1. *No stack algorithm can achieve an approximation ratio better than* $\log(n + 1)/2$ *for the unweighted minimum set cover problem, where n is the number of elements in the ground set to be covered.*

4 The Steiner Tree Problem

In the Steiner tree problem we are given a weighted graph and a set of distinguished *terminal* vertices T. The goal is to choose a subset of edges that connect all terminal vertices, so as to minimize the sum of the weights. The primal-dual algorithm that achieves a 2-approximation uses the natural LP relaxation in which the IP constraints dictate that every cut separating T is crossed by an edge. The input representation we consider is with edges as items. The information in an item is its weight and the identity of the vertices of the edge. Also, the names of the vertices is known and whether they are terminals or not. Notice that this allows us to know all the separating sets an edge crosses. We later mention that a lower bound under a much richer input representation is possible, and will discuss the resulting implication. The reverse delete step ensures that if an edge is not needed it will not be taken. In fact, here this step is easily seen to be essential, and the analysis depends on the fact that the resulting set is minimal w.r.t set inclusion. In this section we show that stack algorithms cannot approximate the problem with better than a 4/3 factor.

In [22], it is shown that unless Co-$RP = NP$, no approximation ratio better than about 1.007 is possible for the Steiner Tree problem. The best known algorithm for Steiner tree, due to Robins and Zelikovsky [20] achieves approximation ratio 1.55 and is not a local-ratio/primal-dual algorithm. For the class of quasi-bipartite graphs (i.e. graphs not containing edges between Steiner nodes), their algorithm achieves a ratio of about 1.28, slightly better than our lower bound of 4/3 for any stack algorithm. We note that our proof applies to quasi-bipartite graphs. In some sense this shows that the algorithm in [20] is superior to any primal-dual approach. We now turn to our result.

Theorem 2. *No stack algorithm can achieve a worst-case approximation ratio better than 4/3 for the (unweighted) Steiner tree problem.*

Proof. Our proof is motivated by the priority lower bound of Davis and Impagliazzo [13]. Initially we consider the complete bipartite graph $G = <T \cup S, R \times S>$, where $T = \{t_1, t_2, t_3\}$ are the terminals and $S = \{s_1, s_2\}$ are the steiner nodes. The algorithm is given the set of terminals and steiner nodes in advance, but is not provided with a list of edges. After each stage of the algorithm, the adversary has the option of deleting any edges that have not yet been seen.

The main ingredient of the proof is the following lemma.

Lemma 2. *In the push phase, consider the first time t that the algorithm has seen two edges connected to some s_j assuming it has not rejected any edges before time t. Call this event D. Then the solution produced by the algorithm must contain these two edges.*

Proof. The lemma is not vacuous : for the graph to have a steiner tree, at least one of the steiner node must have degree at least 2. These edges must be pushed since at time t there is no way to preclude the possibility that the graph is a path of length 4 (this is still consistent with the input so far) and therefore all edges must be taken, and a reject is impossible.

To show that these two edges must be used in the solution produced by the algorithm, we should show that there is no edge e in this pair that will be not rejected in the pop phase. Assume otherwise, then by the definition of the pop phase, there must be a cycle C containing e, so that e is pushed last among all edges in C. But since the graph is bipartite, C must contain two edges out of the other steiner node. This leads to a contradiction, since the later of these edges to be considered must come after e in the order.

Observe that the stars rooted at s_1 and s_2 both have cost 3, while the other steiner trees must use both steiner nodes, and will have cost 4. Note that if an edge adjacent to either steiner node is missing (i.e. deleted by the adversary) then the star rooted at the other steiner node will be the only Steiner tree with cost 3. Here then is the lower bound strategy. The adversary lets the algorithm run, and waits until either the algorithm rejects an edge, or event D happens.

Suppose that the algorithm rejects an edge before event D happens, say the one connected to s_1. Then the adversary simply deletes an unseen edge adjacent to s_2. Such an edge must exist, since event D has not occurred yet. This makes the star rooted at s_1 the unique optimal solution, but the algorithm has just rejected an edge in that star. Hence, in this case the algorithm can achieve a cost of at best 4, for a ratio of 4/3.

Now, suppose that event D happens first. The algorithm pushes the two edges connected to, say, s_1. At this point the adversary deletes the third edge adjacent to s_1, leaving the star rooted at s_2 as the only optimal solution. But by Lemma 2, the algorithm cannot return this unique optimal solution, since it is bound to take both edges connected to s_1 in the solution. This again leads to an approximation lower bound of 4/3.

By a somewhat more involved example and analysis we can prove a 7/6-approximation lower bound when we allow the algorithm to not only know in advance the nodes of the graph, but also the edges of the graph. The only thing not known in advance is the edge costs. This relates quite nicely to the discussion about what LP relaxations we may consider: when all the information is known but the cost vector, then *any* LP relaxation with the natural variables and with positive constraint coefficients is captured by the model.

5 Packing Problems: Scheduling

As was mentioned in the introduction, Bar Noy et al [5] provide the only use to date of local-ratio/primal dual for packing problems. Several problems are discussed there, the most general one being the NP-hard bandwidth allocation problem. Here we concentrate on one (polynomial time solvable) special case, namely weighted interval scheduling on 2 identical machines ($WISP_2$), and show that 0.913 is an upper bound to the approximation of this problem by a fixed order (packing) stack algorithm. This should be contrasted with the optimal local ratio algorithm for one machine interval scheduling ($WISP_1$) and the 2/3-approximation supplied in [5] for $WISP_2$. We also note that an optimal algorithm for $WISP_m$ can be obtained by a time $O(n^m)$ dynamic programming algorithm or a time $O(n^2(n-m))$ min cost max flow based algorithm [3].

For the interval scheduling problem (and the more general bandwidth allocation problem) the natural input representation is that the input items are intervals represented by their start times, finish times and weights (profits). This representation is thus good "against" the natural LP formulation in which the constraints are the bounds on the number of intervals that can be scheduled at any time instance t.

For our packing results, we are thus far only able to provide bounds for fixed order stack algorithms. This, however, does capture the one "meta-algorithm" that Bar Noy et al use to solve various cases of the bandwidth allocation problem. The fixed order there is determined by non-decreasing finishing times which is also the order used for the optimal greedy algorithm for unweighted interval scheduling, and for the one pass algorithms of Erlebach and Spieksma [14]. Obviously our bound does not require the algorithm to use this ordering.

To provide a bound for a fixed-order model (be it a priority algorithm, backtracking algorithm [12], or stack algorithm), it is often useful to provide an initial input set and claim that regardless of the ordering of elements in that set, some combinatorial property must apply to a subset of the initial set. This is analogous to the Ramsey phenomena in graphs; i.e. in every colouring we can say something about one of the colour classes. Restricted to such a (now ordered) subset, we are able to bound the quality of the algorithm (see [12] Theorem 4.1).

We start by describing *forbidden configurations* of the input with respect to an ordering. The first forbidden configuration consists of intervals I, J, K, $I \cap J \cap K \neq \emptyset$, that appear in that order and with profits x, y, z respectively, such that $y < x, z$. We claim that if there is such a configuration, the best approximation ratio achievable is

$$\max\{x/(x+y), (x+y)/(x+z), (y+z)/(x+z)\}. \qquad (1)$$

To see the claim, consider the action of the algorithm on input that contains (at most) only I, J and K. If the algorithm decides to reject an interval, that interval becomes the last interval of the input. This leads to approximation ratios $0, x/(x+y)$ or $(x+y)/(x+z)$ when rejecting the first, second or third items respectively. In the more interesting case, all intervals are accepted to the stack, and in the pop phase I will be rejected as it will be popped when J and K are already in the solution. Therefore the algorithm achieves $y + z$ while $x + z$ is possible, and the bound 1 follows.

Another forbidden configuration we consider is the following. There are four intervals I_1, I_2, J, K of profits x, x, y, z respectively. I_1 and I_2 are disjoint, and $I_i \cap J \cap K \neq \emptyset$ for $i = 1, 2$. Also assume that $y < 2x, z$ and that the order is I_1, I_2, J, K or K, J, I_1, I_2. By a slightly more careful yet similar analysis, we show that this configuration implies a lower bound of

$$\max\{1/2, 2x/(2x+y), (2x+y)/(2x+z), z/(y+z), (y+z)/(2x+z)\}. \qquad (2)$$

We now describe the initial input set for our proof: There are two overlapping intervals, one with profit a and the other with profit b, and three intervals, each with profit c. Each c-interval overlaps the a and b intervals, but does not overlap any other c-interval. Finally, for each c-interval, there are three "child" intervals (for a total of 9), each with profit d. Each d-interval overlaps its parent c-interval, and the a and b intervals, but no other intervals. We will later fix the profit values, but for now we only require that $a > b > c > d$ and $2d > b$.

Now, a d or a c interval appearing between two other intervals of greater profit corresponds to configuration I above. A b-interval appearing between an a-interval and two c-intervals, or a c-interval appearing between a b-interval and two d-intervals corresponds to configuration II above. By appropriately substituting the a, b, c, d values for x, y, z in equations 1 and 2, we obtain the following bound on the approximation factor:

$$\max \{a/(a+d), (a+c)/(a+b), (a+d)/(a+c), \\ (c+d)/(b+c), 2c/(2c+b), (2c+b)/(2c+a), \\ (a+b)/(2c+a), 2d/(2d+c), (2d+c)/(2d+b)\}\} \quad (3)$$

Lemma 3. *Any ordering selected by a fixed ordering stack algorithm will result in at least one of the forbidden configurations described above.*

The proof of the lemma is omitted from this extended abstract. Setting $(a, b, c, d) = (10, 8, 3\sqrt{30} - 10, 5)$, and substituting into (3) gives:

Theorem 3. *For interval scheduling on two machines, no fixed ordering stack algorithm can achieve a constant approximation factor better than .913.*

We remark that the above can be extended to k machines, leading to a $1-O(1/k)$ inapproximability result. This is not satisfying, as the upper bound achieved by local ratio [5] is $\frac{1}{2-1/k}$ for $k \geq 2$ which limits to $1/2$ for large k.

6 Discussion and Conclusion

We have presented a syntactic model that captures the standard use of primal dual/local ratio algorithms in the context of covering and packing problems. Our framework exposes limits of these paradigms and hence hopefully suggests new ways that modifications of these algorithmic techniques can be applied so as to obtain better approximation guarantees while still maintaining the syntactic and computational simplicity of the basic methods.

For example, our analysis of the interval scheduling problem does not preclude the possibility of close to optimal and efficient (e.g. $O(n \log n)$ time) algorithms for a large but fixed number of processors. For the more general bandwidth allocation problem, we would like to be able to derive a stack algorithm that can yield a ratio better than $1/2$ for small bandwidths. We have also seen the dependency of these methods on the input representation which corresponds to the constraints used in an LP relaxation of the problem. For the natural representations of set cover, steiner tree and bandwidth allocation/interval allocation we can derive limitations on the approximation ratio of such algorithms. But our bounds suggest that further improvements can be made even assuming we stay within the natural input representation. Our stack framework also suggests some natural extensions to the known primal-dual/local-ratio paradigm; for example, allowing the stack algorithm to make irrevocable acceptances during the push phase. The framework also encourages us to think of other reasonable ways to order input items.

References

1. A. Agrawal, P. Klein, and R. Ravi. When trees collide: An approximation algorithm for the generalized steiner problem on networks. *SICOMP*, 24:440–465, 1995.
2. S. Angelopoulos and A. Borodin. The power of priority algorithms for facility location and set cover, 2002.

3. E. M. Arkin and E. L. Silverberg. Scheduling jobs with fixed start and end times. *Disc. Appl. Math*, 18:1–8, 1987.
4. S. Arora, B. Bollobás, and L. Lovász. Proving integrality gaps without knowing the linear program. In *Proceedings of the 43rd Annual IEEE Conference on Foundations of Computer Science*, pages 313–322, 2002.
5. A. Bar-Noy, R. Bar-Yehuda, A. Freund, J. Naor, and B. Schieber. A unified approach to approximating resource allocation and scheduling. *JACM*, 48(5):1069–1090, 2001.
6. R. Bar-Yehuda, A. Bendel, A. Freund, and D. Rawitz. Local ratio: A unified framework for approxmation algorithms in memoriam: Shimon even 1935-2004. *Computing Surveys*, 36:422–463, 2004.
7. R. Bar-Yehuda and S. Even. A linear time approximation algorithm for the weighted vertex cover problem. *Journal of Algorithms*, 2:198–203, 1981.
8. R. Bar-Yehuda, M. M. Halldorsson, J. Naor, H. Shachnai, and I. Shapira. Scheduling split intervals. In *Proceedings of the 13th ACM-SIAM Symposium on Discrete Algorithms (SODA)*, pages 732–741, 2002.
9. R. Bar-Yehuda and D. Rawitz. On the equivalence between the primal-dual schema and the local ratio technique. In *4th International Workshop on Approximation Algorithms for Combinatorial Optimization Problems, APPROX*, pages 24–35, 2001.
10. R. Bar-Yehuda and D. Rawitz. Using fractional primal-dual to schedule split intervals with demands, 2004.
11. A. Borodin, M. N. Nielsen, and C. Rackoff. (Incremental) priority algorithms. In *Proceedings of the 13th Annual ACM-SIAM Symposium on Discrete Algorithms*, 2002.
12. M. Alekhnovich A. Borodin, J. Buresh-Oppenheim, R. Impagliazzo, A. Magen, and T. Pitassi. Toward a model for backtracking and dynamic programming. *Unpublished manuscript*, 2004.
13. S. Davis and R. Impagliazzo. Models of greedy algorithms for graph problems. In *Proceedings of the 15th ACM-SIAM Symposium on Discrete Algorithms*, 2004.
14. T. Erlebach and F.C.R. Spieksma. Interval selection: Applications, algorithms, and lower bounds. *Technical Report 152, Computer Engineering and Networks Laboratory, ETH*, October 2002.
15. M. X. Goemans and D.P. Williamson. A general approximation technique for constrained forest problems. *SICOMP*, 24:296–317, 1995.
16. K. Jain and V. Vazirani. Approximation algorithms for the metric facility location problem and k-median problem using the primal-dual schema and lagrangian relaxation. *JACM*, 48:274–299, 2001.
17. P. Klein and R. Ravi. When cycles collapse: A general approximation technique for constrained two-connectivity problems. In *Proceedings of the Third MPS Conference on Integer Programming and Combinatorial Optimization*, pages 39–55, 1993.
18. Rajagopalan and Vazirani. On the bidirected cut relaxation for the metric steiner tree problem. In *SODA*, pages 742–751, 1999.
19. R. Raz and S. Safra. A sub-constant error-probability low-degree test, and sub-constant error-probability pcp characterization of np. In *Proceedings of the 29th Annual ACM Symposium on the Theory of Computing*, pages 475–484, 1997.
20. G. Robins and A. Zelikovsky. Improved steiner tree approximation in graphs. In *SODA*, pages 770–779, 2001.
21. H. Saran, V. Vazirani, and N. Young. A primal-dual approach to approximation algorithms for network steiner problems. In *Proceedings of the Indo-US workshop on Cooperative Research in Computer Science*, pages 166–168, 1992.

22. Martin Thimm. On the approximability of the steiner tree problem. In *Lecture Notes in Computer Science*, page 678. Springer-Verlag Heidelberg, 2001.
23. V. V. Vazirani. *Approximation algorithms*. Springer-Verlag New York, Inc., 2001.
24. D. P. Williamson. The primal-dual method for approximation algorithms. *Mathematical Programming, Series B*, 91(3):447–478, 2002.

Approximating MAX kCSP - Outperforming a Random Assignment with Almost a Linear Factor

Gustav Hast

Department of Numerical Analysis and Computer Science,
Royal Institute of Technology, 100 44 Stockholm, Sweden
ghast@nada.kth.se

Abstract. An instance of MAX kCSP consists of weighted k-ary constraints acting over a set of Boolean variables. The objective is to find an assignment to the Boolean variables such that the total weight of satisfied constraints is maximized. In this paper we provide a probabilistic polynomial time approximation algorithm that $c_0 k(\log k)^{-1} 2^{-k}$-approximates MAX kCSP, for a constant $c_0 > 0$.

1 Introduction

An instance of MAX kCSP consists of a set of Boolean variables and a set of weighted constraints, each acting over a k-tuple of the variables. A solution is an assignment of the Boolean variables, and its value is the sum of the weights of the satisfied constraints. The objective is to find a solution with as large value as possible. The MAX kCSP problem is a natural generalization of many well-known combinatorial maximization problems, such as MAX CUT and MAX 3SAT.

It is well-known that it is NP-hard to solve MAX kCSP optimally. Therefore, it is interesting to investigate if an efficient algorithm can find a good approximate solution. In order to measure the quality of an approximation algorithm, we say that an algorithm r-approximates a maximization problem if the ratio of the value of the solution found by the algorithm, and the value of an optimal solution, is at least r for all instances. If the algorithm is probabilistic the value of the solution may be an expected value over the random choices that are made by the algorithm.

Trevisan [10] used a linear relaxation algorithm in order to 2^{1-k}-approximate MAX kCSP. He observed that the hardest instances consist of only conjunctions of literals, where a literal is a variable or a negated variable. The problem MAX kCONJSAT consists of such instances. Trevisan then showed how to 2^{1-k}-approximate MAX kCONJSAT. Note that a random assignment satisfies a single conjunction of length k with probability 2^{-k}. This implies that a random assignment 2^{-k}-approximate MAX kCONJSAT, and thus the Trevisan algorithm outperforms a random assignment with a factor of two. Recently, Hast

[5] produced a $2^{1.54-k}$-approximation of MAX kCSP by utilizing a semi-definite relaxation approach. Essentially, it combined already known algorithms for MAX 2CONJSAT, MAX 3CONJSAT and MAX 4CONJSAT with a technique to reduce large conjunctions into smaller ones.

If a MAX kCSP instance is known to be satisfiable, then it can be $(k+1)2^{-k}$-approximated by using an algorithm by Trevisan [11]. The technique by Trevisan reduces constraints, with at most k accepting inputs, into linear constraints. A random assignment is then picked from the set of assignments that adhere to the produced linear constraints. This ensures that each constraint is accepted with probability at least $(k+1)2^{-k}$, and thus the algorithm is a $(k+1)2^{-k}$-approximation.

In this paper we show that it is possible to $c_0 k(\log k)^{-1}2^{-k}$-approximate general MAX kCSP, for a constant $c_0 > 0$. Our algorithm is the first that outperforms a random assignment with an increasing factor for larger values of k. It is interesting to see that we can match, up to a logarithmic factor, the approximation ratio of Trevisan for satisfiable instances, even though we use very different methods.

The power of a probabilistical checkable proof where the verifier asks k bits is closely connected with the approximability of MAX kCSP. The PCP of Samorodnitsky and Trevisan [9] and the enhancement of Engebretsen and Holmerin [3] shows that it is NP-hard to approximate MAX kCSP within $2^{\sqrt{2k-2}+1/2-k}$. Our algorithm implies the following inclusion:

$$\text{PCP}_{c,s}[\log, k] \subseteq \text{P}, \text{ for any } c/s > \frac{\log k}{c_0 k}2^k ,$$

where c is the completeness and s the soundness of a verifier that uses a logarithmic number of random bits and asks k questions.

1.1 Our Method

An instance of MAX kALLEQUAL consists of a collection of weighted constraints. Each constraint is a k-tuple of literals and a constraint is satisfied if all its literals have the same value, i.e., all are true or all are false. A simple reduction shows that an r-approximation of MAX kALLEQUAL can be turned into an $r/2$-approximation of MAX kCONJSAT, and thus also into an $r/2$-approximation of MAX kCSP. In this paper we approximate MAX kALLEQUAL in two steps. The first step is to produce an unbalanced solution. For such a solution, constraints tend to either have many literals that are true or many that are false. We find such an unbalanced solution by using semi-definite relaxation techniques [2, 13]. We produce an assignment that is biased towards this solution. This is done by, for some $\alpha \in [0,1]$, assigning a variable according to the unbalanced solution with probability $(1+\alpha)/2$, and negating the value with probability $(1-\alpha)/2$.

It can be shown that as long as α is chosen appropriately such a biased random assignment makes the value of all literals in a constraint equal with much higher probability than if an unbiased random assignment is used. Too see why this happen we look at a constraint consisting of eight literals of which the first seven have the same value but the last one has the opposite value according to the unbalanced solution. A random assignment makes all literals equal with probability $2^{-7} \approx 0.0078$. For a biased random assignment the constraint is satisfied if we assign the first seven literals according to the unbalanced solution but the last literal is negated. This happens with probability $\left(\frac{1+\alpha}{2}\right)^7 \left(\frac{1-\alpha}{2}\right)$. By choosing $\alpha = 3/4$ this probability is larger than 0.049 making it more than six times more probable to satisfy the constraint using a biased random assignment compared with using an unbiased random assignment.

Let us briefly explain from where the approximation ratio of our algorithm for MAX kALLEQUAL comes from. Assume that we have a MAX kALLEQUAL instance and let W be the total weight of its constraints and let OPT be the value of an optimal solution to the instance. The expected value of a random assignment is $2^{-k}W$, thus if OPT $\leq W/k$ then a random assignment achieves an approximation ratio of $k2^{-k}$. If OPT $> W/k$ then the optimal solution turns out to be unbalanced in our measure. If we pick an assignment that is biased towards this solution we achieve an approximation ratio of $k2^{-k}$. However, we do not know the optimal solution and instead we find an unbalanced solution using the approximation algorithm by Charikar and Wirth [2]. In this process we loose a factor of $\log k$ and thus the approximation ratio gets to be $\Omega\left(k(\log k)^{-1}2^{-k}\right)$.

1.2 Overview of the Paper

In Section 2 we define problems relevant to this paper. The next section contains the formal description of our MAX kALLEQUAL algorithm along with an analysis of its approximation ratio. In Section 4 we evaluate our algorithm numerically and give approximation ratios for MAX kCSP, $k = 5\ldots 100$. In the last section we identify a PCP class as a subset of P.

2 Definitions

An instance of the MAX kCSP problem consists of a set $\{C_1,\ldots,C_m\}$ of constraints with associated non-negative weights $\{w_1,\ldots,w_m\}$ and a set of Boolean variables $X = \{x_1,\ldots,x_n\}$. Each clause C_i consists of a Boolean function f_i of arity $h \leq k$ and a size h tuple of Boolean variables $(x_{i_1},\ldots,x_{i_h})$ where $x_{i_j} \in X$. A solution is an assignment to X and the objective value of the solution is the sum of the weights of the satisfied clauses. A clause $C_i = (f_i, (x_{i_1},\ldots,x_{i_h}))$ is satisfied if and only if $f_i(x_{i_1},\ldots,x_{i_h})$ is true.

Let $\overline{X}$ define the set $\{\overline{x} : x \in X\}$. MAX kCONJSAT is a special type of MAX kCSP where each constraint is a conjunction of at most k literals from $X \cup \overline{X}$.

MAX kALLEQUAL is also a special type of MAX kCSP. Each constraint is a tuple of at most k literals from $X \cup \overline{X}$, and it is satisfied if and only if all literals have the same value.

The sum of all weights in an instance, $\sum_{i=1}^{m} w_i$, is denoted by W. We use OPT to refer to the value of an optimal solution to an instance and APP is used to refer to the value of an approximative solution. We also use the normalized values app = APP/W and opt = OPT/W.

3 An Approximation Algorithm for MAX kALLEQUAL

In this section we describe and analyze our approximation algorithm for MAX kALLEQUAL. After the analysis we show some consequences for the approximability of MAX kCONJSAT and MAX kCSP.

3.1 Algorithm Description

If some of the constraints of a MAX kALLEQUAL are not of size k, then these constraints are padded into size k using auxiliary variables, where each new variable only appears once in the instance. This will not effect the satisfiability of the instance. Thus, we can assume that all constraints are of length k. The algorithm is shown in Figure 1.

Input: A set of Boolean variables $\{x_1, \ldots x_n\}$ and a set of all equal constraints $\{C_1, \ldots C_m\}$ with corresponding weights $\{w_1, \ldots w_m\}$.

1. (EQ_2-**gadget**) Each constraint $C_i = z_{i_1} \equiv z_{i_2} \ldots \equiv z_{i_k}$ is transformed into $k(k-1)/2$ equality constraints, $z_{i_1} \equiv z_{i_2}$, $z_{i_1} \equiv z_{i_3}$, $\ldots z_{i_{k-1}} \equiv z_{i_k}$. Each constraint is given weight w_i.
2. (**Solve** MAX 2ALLEQUAL) Use the Charikar and Wirth [2] algorithm in order to satisfy as much weight of the equality constraints as possible. Let b_i be the value of variable x_i in the produced solution.
3. (**Biased random assignment**) For $i := 1, \ldots, n$: assign x_i according to

$$x_i := \begin{cases} b_i \text{ with probability } (1+\alpha)/2 \\ \overline{b_i} \text{ with probability } (1-\alpha)/2 \end{cases},$$

with $\alpha = 1/\sqrt{k}$.

Fig. 1. Algorithm **ALG**: a MAX kALLEQUAL algorithm

3.2 Algorithm Analysis

Theorem 1. *For a constant $c > 0$, MAX kALLEQUAL, for $k \geq 2$, can in probabilistical polynomial time be approximated within a factor of $ck(\log k)^{-1}2^{-k}$.*

The proof of Theorem 1 is based on Lemmas 2 and 4.

Given an assignment to a MAX kALLEQUAL instance we call a constraint unbalanced if either many literals are true or many literals are false. Given an assignment and a constraint C_i, γ_i is defined such that $k/2 + \gamma_i$ literals are true and $k/2 - \gamma_i$ are false. Thus, γ_i is a function depending on an assignment but due to notational convenience we do not make this dependency explicit. For a fixed assignment, we let γ_i^2 be a measure of how unbalanced C_i is. The following lemma shows that if there exists a good solution to a MAX kALLEQUAL instance, then we are able to find an assignment which makes constraints of large weight unbalanced. We do this by first transforming the MAX kALLEQUAL instance into a MAX 2ALLEQUAL instance, and then solving it using the Charikar-Wirth algorithm.

Lemma 2. *Let I be an instance of* MAX kALLEQUAL *with total weight W and assume that the value of an optimal solution is* opt $\cdot W$ *where $k \cdot $ opt ≥ 3. Let $d = c_{cw}(k \cdot \text{opt} - 1)/(4 \log k)$, where c_{cw} is a positive constant defined by the Charikar-Wirth algorithm. Then a solution can be produced in polynomial time such that*

$$dk \leq \mathrm{E}\left[\frac{1}{W} \sum_{i:|\gamma_i|>\sqrt{k}/2} w_i \gamma_i^2\right], \qquad (1)$$

where the expectation is taken over the random choices of the Charikar-Wirth algorithm.

Proof. Each constraint of k literals is transformed into $k(k-1)/2$ equality constraints of arity two. If the original constraint is satisfied, then all new equality constraints are satisfied as well. If $k/2+\gamma$ literals are true and $k/2-\gamma$ false, then the number of equality constraints that are not satisfied is $(k/2+\gamma)(k/2-\gamma) = k^2/4 - \gamma^2$. Thus, at least $k(k-1)/2 - k^2/4$ constraints are satisfied for any assignment.

A MAX kALLEQUAL instance is transformed into a MAX 2ALLEQUAL instance according to the first step of Algorithm **ALG**. An optimal solution that satisfies weight opt $\cdot W$ in the original MAX kALLEQUAL instance then satisfies clauses of weight

$$\text{opt}\binom{k}{2}W + (1-\text{opt})\left(\binom{k}{2}-\frac{k^2}{4}\right)W = \left(\binom{k}{2}-\frac{k^2}{4}+\text{opt}\frac{k^2}{4}\right)W,$$

in the MAX 2ALLEQUAL instance. We define that a solution of weight $W_{eq}(1/2+\delta)$ has *gain* δ, where $W_{eq} = Wk(k-1)/2$ is the total weight of the MAX 2ALLEQUAL instance. The gain is a measure of how much better a solution is compared to a random assignment. We let δ^* denote the optimal gain. From

$$W_{eq}\left(\frac{1}{2}+\delta^*\right) \geq \left(\binom{k}{2}-\frac{k^2}{4}+\text{opt}\frac{k^2}{4}\right)W$$

it is not hard to derive the following lower bound

$$\delta^* \geq \frac{k \cdot \text{opt} - 1}{2(k-1)}.$$

We have the following Lemma due to Charikar and Wirth.

Lemma 3 (Charikar and Wirth [2]). *If δ^* is the optimum gain of a MAX CUT instance, ApproxMaxQP returns a solution whose expected gain is at least*

$$c_{cw}\left(\frac{\delta^*}{\log(1/\delta^*)}\right),$$

for a constant $c_{cw} > 0$.

The proof of Lemma 3 shows that $c_{cw} = 1/64$ is a valid choice. *ApproxMaxQP* is a probabilistic polynomial time algorithm and the above lemma is still valid if we instead of a MAX CUT instance consider the more general case of a MAX 2ALLEQUAL instance.

We analyze the logarithmic factor that is lost when using the Charikar-Wirth algorithm. The assumption from Lemma 2 implies that $k \cdot \text{opt} - 1 \geq 2$, thus

$$\log(1/\delta^*) \leq \log \frac{2(k-1)}{k \cdot \text{opt} - 1} < \log k .$$

We run the algorithm of Charikar and Wirth. Lemma 3 implies that the expected gain δ of the produced solution can be lower bounded

$$\delta \geq c_{cw} \frac{\delta^*}{\log(1/\delta^*)} > c_{cw} \frac{k \cdot \text{opt} - 1}{2(k-1)\log k} . \qquad (2)$$

Remember that γ_i is the value such that $k/2+\gamma_i$ of the literals in constraint C_i are true and $k/2-\gamma_i$ are false according to the solution produced by the Charikar-Wirth algorithm. The weight of satisfied equality constraints corresponding to constraint C_i is then

$$\left(\binom{k}{2} - \left(\frac{k}{2}+\gamma_i\right)\left(\frac{k}{2}-\gamma_i\right)\right)w_i = \left(\frac{k(k-1)}{4} + \left(\gamma_i^2 - \frac{k}{4}\right)\right)w_i .$$

We sum the contribution of each constraint in order to get the total weight of satisfied equality constraints. Thus,

$$W_{eq}\left(\frac{1}{2}+\delta\right) = E\left[\sum_{i=1}^m w_i\left(\frac{k(k-1)}{4} + \left(\gamma_i^2 - \frac{k}{4}\right)\right)\right]$$

which implies that

$$\binom{k}{2}\delta W = E\left[\sum_{i=1}^m w_i\left(\gamma_i^2 - \frac{k}{4}\right)\right] . \qquad (3)$$

We derive from (2) a lower bound for the weighted sum of γ_i^2 which concludes the proof of the lemma:

$$\mathrm{E}\left[\frac{1}{W}\sum_{i:|\gamma_i|>\sqrt{k}/2} w_i\gamma_i^2\right] \geq \mathrm{E}\left[\frac{1}{W}\sum_{i:|\gamma_i|>\sqrt{k}/2} w_i\left(\gamma_i^2 - \frac{k}{4}\right)\right]$$

$$\geq \mathrm{E}\left[\frac{1}{W}\sum_{i=1}^{m} w_i\left(\gamma_i^2 - \frac{k}{4}\right)\right] = \binom{k}{2}\delta$$

$$\geq \frac{k(k-1)}{2}c_{cw}\frac{k\cdot\mathrm{opt}-1}{2(k-1)\log k} \geq \frac{kc_{cw}(k\cdot\mathrm{opt}-1)}{4\log k},$$

where the equality follows from (3). □

Lemma 4. *Let I be an instance of* MAX kALLEQUAL *with total weight W and $\{b_1,\ldots b_n\}$ is a solution such that*

$$dk \leq \frac{1}{W}\sum_{i:|\gamma_i|>\sqrt{k}/2} w_i\gamma_i^2,$$

for $d \geq 1$. Then the solution defined by

$$x_i := \begin{cases} b_i & \text{with probability } (1+\alpha)/2 \\ \overline{b}_i & \text{with probability } (1-\alpha)/2 \end{cases},$$

with $\alpha = 1/\sqrt{k}$ has an expected value of at least $e^{2\sqrt{d}-1/2}2^{-k}W$.

Proof. Consider a constraint C_i that has $|\gamma_i| > \sqrt{k}/2$. Let AE_i be the event that all literals in C_i get the same value after the biased random assignment.

$$\Pr[AE_i] = \Pr[\text{all literals true}] + \Pr[\text{all literals false}]$$

$$= \left(\frac{1+\alpha}{2}\right)^{\frac{k}{2}+\gamma_i}\left(\frac{1-\alpha}{2}\right)^{\frac{k}{2}-\gamma_i} + \left(\frac{1-\alpha}{2}\right)^{\frac{k}{2}+\gamma_i}\left(\frac{1+\alpha}{2}\right)^{\frac{k}{2}-\gamma_i}$$

$$> 2^{-k}\left((1+\alpha)^{\frac{k}{2}+|\gamma_i|}(1-\alpha)^{\frac{k}{2}-|\gamma_i|}\right)$$

$$= 2^{-k}\left((1+\alpha)^{2|\gamma_i|}(1-\alpha^2)^{\frac{k}{2}-|\gamma_i|}\right)$$

$$> 2^{-k}e^{2\alpha|\gamma_i|-\frac{k}{2}\alpha^2}.$$

The following claim validates the last inequality. The proof of the claim can be found in [6].

Claim. For $1 \geq \alpha > 0$ and $2|\gamma_i| \geq k\alpha$,

$$(1+\alpha)^{2|\gamma_i|}(1-\alpha^2)^{\frac{k}{2}-|\gamma_i|} > e^{2\alpha|\gamma_i|-\frac{k}{2}\alpha^2}.$$

We calculate a lower bound on the expected weight of constraints that either have all literals true or all literals false.

$$\sum_{i:|\gamma_i|>\sqrt{k}/2} w_i \Pr[AE_i] > 2^{-k} \sum_{i:|\gamma_i|>\sqrt{k}/2} w_i e^{2\alpha|\gamma_i|-\frac{k}{2}\alpha^2} \quad (4)$$

We let $s_i = \gamma_i^2$ and rewrite the above lower bound:

$$2^{-k} e^{-\frac{k}{2}\alpha^2} \sum_{i:\sqrt{s_i}>\sqrt{k}/2} w_i e^{2\alpha\sqrt{s_i}} \ .$$

We consider the terms $e^{2\alpha\sqrt{s_i}}$ as functions of s_i and calculate their second derivative:

$$\frac{\partial}{\partial^2 s_i} e^{2\alpha\sqrt{s_i}} = \left(\alpha - \frac{1}{2\sqrt{s_i}}\right) \alpha s_i^{-1} e^{2\alpha\sqrt{s_i}} \ .$$

We see that the second derivative is positive as long as $\alpha > \frac{1}{2\sqrt{s_i}}$, thus $e^{2\alpha\sqrt{s_i}}$ is convex for all terms of the sum because $\sqrt{s_i} > \sqrt{k}/2$ and $\alpha = 1/\sqrt{k}$. The condition in the lemma gives a lower bound to $\sum_{i:\sqrt{s_i}>\sqrt{k}/2} w_i s_i$. Thus, using Jensen's inequality we conclude that (4) is minimized if all values of $|\gamma_i|$ over the threshold are equal.

We let W_+ be the weight of all conjunctions meeting the threshold condition, $W_+ = \sum_{i:|\gamma_i|>\sqrt{k}/2} w_i$. We will see that the worst case happens if $W_+ = W$, but for now we are general and let $W_+ = xW$ where x is a value between 0 and 1. In order to minimize (4), the value of all $|\gamma_i|$ above the threshold should be equal and by the assumption of the lemma at least $\sqrt{dk/x}$. We apply this to (4) and get the expected weight of constraints that either have all literals true or all literals false.

$$2^{-k} \sum_{i:|\gamma_i|>\sqrt{k}/2} w_i e^{\frac{2|\gamma_i|}{\sqrt{k}}-\frac{1}{2}} \geq 2^{-k} xW e^{2\sqrt{d/x}-1/2} \ . \quad (5)$$

We calculate the derivative of the above expression

$$\frac{\partial}{\partial x} 2^{-k} xW e^{2\sqrt{d/x}-1/2} = (1-\sqrt{d/x}) 2^{-k} e^{2\sqrt{d/x}-1/2} W \ .$$

As $d \geq 1$, we see that the derivative is non-positive for $x \in (0,1]$. Thus, (5) is minimized by setting $x = 1$ and thus the expected weight of satisfied constraints is at least $2^{2\sqrt{2d}-1/2} 2^{-k} W$ which concludes the proof of Lemma 4. □

We are now ready to prove Theorem 1 by using Lemma 2 and Lemma 4.

Proof (Theorem 1). Let I be an instance of MAX kALLEQUAL and let opt be the normalized value of an optimal solution. Set $c = (c_1/c_{cw} + 1)^{-1}$, where c_1

is a positive constant yet to be defined. If opt $\leq \log k/(ck)$, then a random assignment achieves the following approximation ratio:

$$\frac{\text{APP}}{\text{OPT}} \geq \frac{2^{-k}W}{\log k/(ck)W} = \frac{ck}{\log k}2^{-k} .$$

Thus, we only need to consider if opt $> \log k/(ck)$. Assume that for some $r > 1$ we have that opt $= r\log k/(ck)$. We note that $k \cdot \text{opt} \geq \log k/c > 3$ and apply Lemma 2 with

$$d = c_{cw}\frac{k \cdot \text{opt} - 1}{4\log k} = c_{cw}\frac{r/c - 1}{4} = c_{cw}\frac{r(c_1/c_{cw} + 1) - 1}{4} > rc_1/4 .$$

We can now apply Lemma 4 with an expected value of $d = rc_1/4$. However, if $d < 1$ then Lemma 4 does not give anything and thus the expected value of the solution could be zero. The following function expresses a lower bound for the expected value of the solution

$$\begin{cases} 0 & \text{if } 0 \leq d < 1 \\ e^{2\sqrt{d}-1/2}2^{-k}W & \text{if } d \geq 1 \end{cases} .$$

It is not convex and thus we cannot apply Jensen's inequality. However, the following linear function is a lower bound for that function:

$$c_2(d-1)2^{-k}W \text{ , where } c_2 = \min_{d>1}\frac{e^{2\sqrt{d}-1/2}}{d-1} \approx 9.5 . \tag{6}$$

By applying the lower bound (6) we get that APP, the expected value of the solution, is at least $c_2(rc_1/4 - 1)2^{-k}W$. By setting $c_1 = 4/c_2 + 4$ we ensure that APP $\geq r2^{-k}W$ and thus the approximation ratio is

$$\frac{\text{APP}}{\text{OPT}} > \frac{r2^{-k}W}{r\log k/(ck)W} = \frac{ck}{\log k}2^{-k} ,$$

which concludes the proof of Theorem 1. □

3.3 Consequences for MAX kCONJSAT and MAX kCSP

By using a simple observation we derive from Theorem 1 the following result on the approximability of MAX kCONJSAT. The proof can be found in [6].

Theorem 5. *There exists a constant $c_0 > 0$ such that MAX kCONJSAT, for $k \geq 2$, can in probabilistical polynomial time be approximated within a factor of $c_0 k(\log k)^{-1}2^{-k}$.*

By using the observation of Trevisan [10], we conclude that an algorithm for MAX kCONJSAT implies an approximation algorithm for MAX kCSP with the same approximation ratio. Thus, as a consequence of Theorem 5 we get our main theorem:

Theorem 6. *There exists a constant $c_0 > 0$ such that MAX kCSP, for $k \geq 2$, can in probabilistical polynomial time be approximated within a factor of $c_0 k(\log k)^{-1}2^{-k}$.*

4 Numerical Approximation Ratios for Fixed Values of k

In this section we give approximation ratios for MAX kCSP for values of $5 \leq k \leq 100$. The ratios are obtained using numerical methods and are presented in Figure 2. For $k \leq 4$ there are known good approximation algorithms, which outperform our algorithm with a broad margin [4, 7, 12].

In this section we use an algorithm slightly different from Algorithm **ALG** which is presented in Figure 1. First we use Zwick's outward rotation algorithm [13] in order to approximate the MAX 2ALLEQUAL instance instead of the Charikar-Wirth algorithm. This choice is made because an exact approximation ratio can numerically be calculated from [13], if the optimal gain is known. In the last step of Algorithm **ALG** we use a somewhat different value of the bias α. Given the solution of the MAX 2ALLEQUAL instance, it is possible to calculate the optimal value of the parameter α. However, our calculation show that by setting

$$\alpha = \sqrt{\frac{1}{k} + \left(2 - \frac{2}{k}\right)\delta} \;,$$

where δ is the expected gain of the approximate solution, we obtain an approximation ratio very close to the ratio that would be obtained if we used the optimal

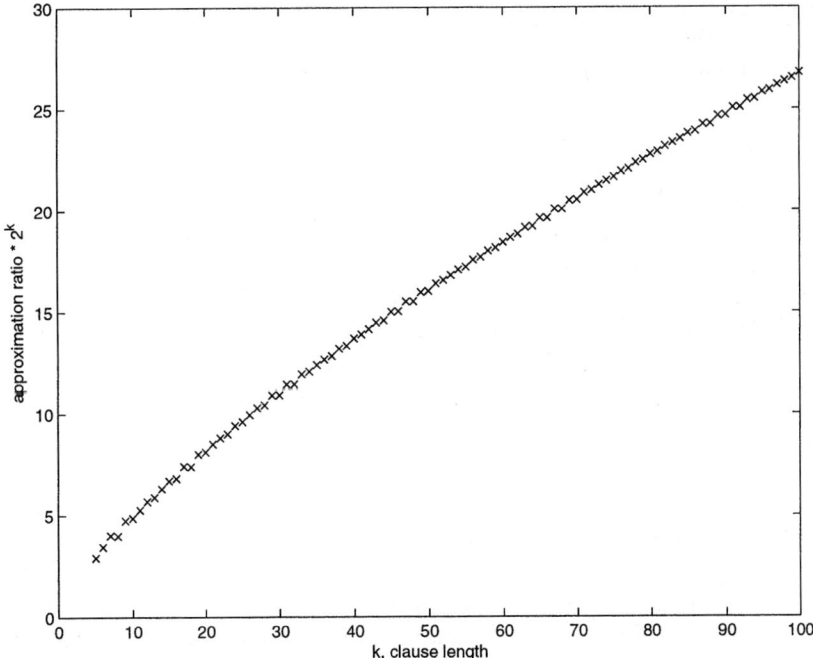

Fig. 2. Approximation ratios for MAX kCSP

value of α. We choose this suboptimal value to ensure easier reproducibility of the graph in Figure 2.

We can heuristically show why this value of α is suitable. It seems that the worst case is if the solution makes all constraints equally unbalanced, thus the value of $|\gamma_i|$ is equal for all constraints. In the proof of Lemma 2 we saw that

$$\binom{k}{2}\delta W = \mathrm{E}\left[\sum_{i=1}^{m} w_i \left(\gamma_i^2 - \frac{k}{4}\right)\right].$$

This implies that the worst case is if $|\gamma_i| = \sqrt{\delta k(k-1)/2 + k/4}$.

In the proof of Lemma 4 we showed that the probability that all literals in a constraint get the same value after a biased random assignment is at least $2^{-k}e^{2\alpha|\gamma_i|-\frac{k}{2}\alpha^2}$. It is easy to show that this expression is maximized by setting $\alpha = 2|\gamma_i|/k$. Inserting the worst case value for $|\gamma_i|$ in this expression we see that α is equal to the value we use.

For each value of k we calculate the approximation ratio in the following way. For each possible optimal gain, using a suitable discretization, we calculate the expected gain δ of the solution produced by the algorithm of Zwick. We then produce a linear program where the objective value is equal to the expected normalized objective value of the solution produced by the biased random assignment. The linear program has $k+1$ variables $y_0, y_1, \ldots y_k$, where y_i indicates the total weight of constraints with exactly i literals true. There are also constants $d_0, \ldots d_k$, where $d_i = \frac{k(k-1)}{2} - i(k-i)$ is the number of pair of literals that are equal in a constraint with exactly i literals true. We let b_i be the probability that a constraint with exactly i literals true will be all equal after the biased random assignment. This value depends on our choice of α.

$$\min \sum_{i=0}^{k} b_i y_i, \text{ given } \sum_{i=0}^{k} y_i = 1 \text{ and } \sum_{i=0}^{k} d_i y_i = \frac{k(k-1)}{2}\left(\frac{1}{2} + \delta\right)$$

By solving the linear program we get a numerical value of the approximation ratio of the MAX kALLEQUAL algorithm. By dividing this value with two we get the approximation ratio of the MAX kCSP algorithm.

Our algorithm approximates MAX 5CSP within $2.91 \cdot 2^{-5}$. The general $2^{1.54-k} = 2.90 \cdot 2^{-k}$ approximation algorithm in [5] do better for small values of k and actually $3.68 \cdot 2^{-5}$ approximates MAX 5CSP. However, for values of $k \geq 6$ our algorithm achieves the best known approximation ratio for MAX kCSP.

5 Relation with PCP Classes

We use the standard notation and definitions in connection with probabilistical checkable proofs, PCPs. For those unfamiliar with these, we refer to the work of Bellare et al. [1] which also contains the history of PCPs. The complexity class $\mathrm{PCP}_{c,s}[\log, q]$ contains all languages that have a verifier with completeness c, soundness s, which uses only a logarithmic number of random bits and asks at most q (adaptive) questions.

The power of PCPs asking k questions is closely related to the approximability of MAX kCONJSAT as can be seen by the following theorem:

Theorem 7 (Trevisan [10]). *If, for some $r \leq 1$, MAX kCONJSAT is deterministically r-approximable in polynomial time then $\text{PCP}_{c,s}[\log, k] \subseteq \text{P}$ for any $c/s > 1/r$.*

Our MAX kCONJSAT algorithm is probabilistic, thus it is not immediate that the above theorem can be applied. There are two steps in our algorithm that are probabilistic, the rounding in the Charikar-Wirth algorithm and the biased random assignment. We believe that the Charikar-Wirth algorithm can be derandomized using the method of Mahajan and Ramesh [8], but we have not looked into the details. The biased random assignment can be derandomized using conditional probabilities. However, in Theorem 7 we only need an algorithm that give a lower bound of the optimal value, not an approximate solution. We get such a lower bound directly from the semidefinite program of the Charikar-Wirth algorithm. Thus, we have the following theorem.

Theorem 8. $\text{PCP}_{c,s}[\log, k] \subseteq \text{P}$ *for any* $c/s > \frac{\log k}{c_0 k} 2^k$, *where $c_0 > 0$ is the constant in Theorem 5.*

Acknowledgment. I am very grateful for the help that Johan Håstad gave me when analyzing Algorithm **ALG** and in preparing this paper.

References

1. Mihir Bellare, Oded Goldreich, and Madhu Sudan. Free bits, PCPs, and nonapproximability - towards tight results. *SIAM Journal on Computing*, 27(3):804–915, 1998.
2. Moses Charikar and Anthony Wirth. Maximizing quadratic programs: Extending Grothendieck's inequality. In *Proceedings of the 45th Annual IEEE Symposium on Foundations of Computer Science*, pages 54–60, 2004.
3. Lars Engebretsen and Jonas Holmerin. More efficient queries in PCPs for NP and improved approximation hardness of maximum CSP. In *Proceedings of STACS 2005, Lecture Notes in Computer Science 3404*, pages 194–205, 2005.
4. Venkatesan Guruswami, Daniel Lewin, Madhu Sudan, and Luca Trevisan. A tight characterization of NP with 3 query PCPs. In *Proceedings of the 39th Annual IEEE Symposium on Foundations of Computer Science*, pages 8–17, 1998.
5. Gustav Hast. Approximating Max kCSP using random restrictions. In *Proceedings of APPROX 2004, Lecture Notes in Computer Science 3122*, pages 151–162, 2004.
6. Gustav Hast. *Beating a Random Assignment*. PhD thesis, Royal Institute of Technology, 2005.
7. Michael Lewin, Dror Livnat, and Uri Zwick. Improved rounding techniques for the MAX 2-SAT and MAX DI-CUT problems. In *Proceedings of 9th IPCO, Lecture Notes in Computer Science 2337*, pages 67–82, 2002.

8. Sanjeev Mahajan and H. Ramesh. Derandomizing approximation algorithms based on semidefinite programming. *SIAM Journal on Computing*, 28(5):1641–1663, 1999.
9. Alex Samorodnitsky and Luca Trevisan. A PCP characterization of NP with optimal amortized query complexity. In *Proceedings of the 32nd ACM Symposium on Theory of Computing*, pages 191–199, 2000.
10. Luca Trevisan. Parallel approximation algorithms by positive linear programming. *Algorithmica*, 21(1):72–88, 1998.
11. Luca Trevisan. Approximating satisfiable satisfiability problems. *Algorithmica*, 28(1):145–172, 2000.
12. Uri Zwick. Approximation algorithms for constraint satisfaction problems involving at most three variables per constraint. In *Proceedings of the 9th Annual ACM-SIAM Symposium on Discrete Algorithms*, pages 201–210, 1998.
13. Uri Zwick. Outward rotations: a tool for rounding solutions of semidefinite programming relaxations, with applications to max cut and other problems. In *Proceedings of the 31st annual ACM symposium on Theory of computing*, pages 679–687, 1999.

On Dynamic Bit-Probe Complexity

Corina E. Pătrașcu[1] and Mihai Pătrașcu[2]

[1] Harvard University
patrascu@fas.harvard.edu
[2] MIT
mip@mit.edu

Abstract. This paper presents several advances in the understanding of dynamic data structures in the bit-probe model:

- We improve the lower bound record for dynamic language membership problems to $\Omega((\frac{\lg n}{\lg \lg n})^2)$. Surpassing $\Omega(\lg n)$ was listed as the first open problem in a survey by Miltersen.
- We prove a bound of $\Omega(\frac{\lg n}{\lg \lg \lg n})$ for maintaining partial sums in $\mathbb{Z}/2\mathbb{Z}$. Previously, the known bounds were $\Omega(\frac{\lg n}{\lg \lg n})$ and $O(\lg n)$.
- We prove a surprising and tight upper bound of $O(\frac{\lg n}{\lg \lg n})$ for predecessor problems. We use this to obtain the same upper bound for dynamic word and prefix problems in group-free monoids.

1 Introduction

Bit-probe complexity can be considered a fundamental measure of computation. When analyzing space-bounded algorithms (branching programs), it is usually preferred to cell-probe[1] complexity. In data structures, cell-probe complexity is used more frequently, but the machine independence and overall cleanness of the bit-probe measure have made it a persistent object of study since the dawn of theoretical computer science. Nonetheless, many of the most fundamental questions are not yet understood. In this paper, we address this on several fronts.

Record Lower Bound. We prove a lower bound of $\Omega((\frac{\lg n}{\lg \lg n})^2)$ for dynamic connectivity. This problem asks to maintain an undirected graph, under insertion and deletion of edges, and queries asking whether two nodes are in the same connected component. The best upper bound is $O(\lg^2 n \cdot (\lg \lg n)^3)$ [1], so our lower bound is optimal up to doubly logarithmic factors.

Our lower bound is the highest known bound for an explicit dynamic language membership problem. The previous record was $\Omega(\lg n)$, shown in [2]. A survey on cell probe complexity by Miltersen [3] lists improving this bound as the first open

---

[1] Of course, the bit-probe model is an instantiation of the cell-probe model with one-bit cells. For conciseness, however, we shall use "cell-probe" in a more restricted sense, to refer to the cell-probe model with cells of $\Theta(\lg n)$ bits.

problem among three major challenges for future research. Our lower bound is based on the recent technique of Pătraşcu and Demaine [4], which proved the first $\Omega(\lg n)$ bounds in the cell-probe model. While our contribution is mostly a series of technical tricks necessary for the bit-probe model, it should be noted that in no way is our $\widetilde{\Omega}(\lg^2 n)$ bound a mere echo of an $\Omega(\lg n)$ bound in the cell-probe model. Indeed, $\Omega(\frac{\lg n}{\lg \lg n})$ bounds in the cell-probe model have been known for one and a half decades (including for dynamic connectivity), but the bit-probe record has remained just the trivially higher $\Omega(\lg n)$. To our knowledge, our bound is the first to show a quasi-optimal $\widetilde{\Omega}(\lg n)$ separation between the cell-probe and bit-probe complexity, when the cell-probe complexity is superconstant.

Lower Bound for Maintaining Partial Sums. One of the most fundamental data-structure problems is maintaining partial sums. This problems asks to maintain an array $A[1..n]$ under an UPDATE(k, x) operation, which changes $A[k] \leftarrow x$, and a SUM(k) operation, which asks for a partial sum $\sum_{i=1}^{k} A[i]$. Because we are concerned with the bit-probe model, it is most natural to consider the case when all operations are done in $\mathbb{Z}/2\mathbb{Z}$, i.e. we are interested in prefix parity. This can be generalized to any fixed cyclic group $\mathbb{Z}/k\mathbb{Z}$, without any change in the bounds discussed below. The only known upper bound is the classic $O(\lg n)$, based on a balanced binary tree. It is widely believed that this is optimal.

Fredman [5] proved a lower bound of $\Omega(\frac{\lg n}{\lg \lg n})$ by considering the following greater-than problem. First, the algorithm is given a number $a \in [n]$, and can write t_u bits. Then, the algorithm is given another number $b \in [n]$, and it must determine whether $b > a$ through t_q bit probes. Observe that there is a trivial reduction from this problem to the partial sums problem, with one update and one query. It is quite tempting to believe that one cannot improve past the trivial upper bound $t_u = t_q = O(\lg n)$, since, in some sense, this is the complexity of "writing down" a. However, we show below that Fredman's bound is optimal for the greater-than problem. Therefore, one needs a different strategy to improve the lower bound for maintaining partial sums.

It is natural to approach the problem in the framework of [4], which could prove an $\Omega(\lg n)$ bound in the cell-probe model (of course, the group was $\mathbb{Z}/n\mathbb{Z}$ in that case, matching the cell size). Applying the tricks we developed for dynamic connectivity will only reprove the old $\Omega(\frac{\lg n}{\lg \lg n})$ bound. For reasons discussed below, obtaining $\Omega(\lg n)$ seems quite difficult.

We can, nonetheless, prove an $\Omega(\frac{\lg n}{\lg \lg \lg n})$ bound, which is only a triply-logarithmic factor away from the upper bound! We find the technique of this bound at least as interesting as the result itself. The proof uses a variation of the chronogram technique of Fredman and Saks [6]. However, it is well known that the classic chronogram technique can only prove $\Omega(\frac{\lg n}{\lg \lg n})$. We present a small, yet very important variation, which brings a considerable strengthening of this technique: with this improvement, the chronogram technique can prove lower bounds of $\Omega(\lg n)$ in the cell-probe model. To fully appreciate this development, one must remember that the chronogram technique was virtually the only available technique for proving dynamic lower bounds before the work of [4].

At the same time, obtaining a logarithmic bound in the cell-probe model was viewed as one of the most important problems in data-structure lower bounds. It is now quite surprising to find that the answer has always been this close.

Upper Bound for Predecessor Problems. As mentioned already, we can achieve an $O(\frac{\lg n}{\lg \lg n})$ upper bound for Fredman's greater-than problem. In fact, we can achieve the same bound for several predecessor-type problems. Consider the classic predecessor problem: maintain a dynamic set S, under queries to determine (some information about) the predecessor in S of a given number. Unfortunately, we cannot determine the actual predecessor in $o(\lg n)$, because the output itself has this many bits of entropy. But we can recover some constant amount of information about the predecessor (a stored "color"), which proves to be enough for many purposes. For our classification of dynamic prefix problems, it is important to generalize this slightly to the colored k-predecessor problem: the query asks for the colors of the k predecessors of a given number. Here k is part of the definition of the problem, so it is a constant.

In the interest of dynamic prefix problems, we also study a more unusual stabbing problem. The problem itself is an interesting trick to circumvent finding an exact predecessor in many cases. We have to maintain a dynamic set $S = \{b_1, b_2, \dots\}$ under the following query operation: given $j \in (b_i, b_{i+1})$, the query determines a value in (b_i, b_{i+1}), which is only a function of b_i and b_{i+1}, but not of j or i. Imagine that the elements of S break $[n]$ into segments. The query must then produce a representative for the segment stabbed by j, which is inside the segment, but is independent of the actual choice of j. Adding or removing an element merges or splits segments. The representatives of these affected segments may change arbitrarily, but those of any other segments must remain the same (because they are only functions of the end-points). The segment representative has $\lg n$ bits, so it is maybe surprising that one can be found in $O(\frac{\lg n}{\lg \lg n})$ time. Of course, the query cannot actually write down the representative, but the representative is determined by the query's input and its bit probes.

A tight tradeoff between t_u and t_q was recently given for the greater-than problem [7]. We can match this tradeoff for the more general predecessor problems that we consider. More details will be given in the full version of this paper.

Dynamic Word and Prefix Problems. Dynamic prefix problems are defined like the partial sums problem, except that all additions take place in an arbitrary finite monoid. The word problem is identical to the prefix problem, except that queries only ask for the sum of the entire array, not an arbitrary prefix. The problem is defined by the monoid, so the monoid is considered fixed (and constants may depend on it). The aim is to understand the complexity of the problem in terms of the structure of the monoid. This line of research was inspired by the intense study of parallel word problems, which eventually led to a complete classification. Both in the parallel and in the dynamic case, it can be seen that many fundamental problems are equivalent to word and prefix problems for certain classes of monoids. Examples include partial sums modulo some value, the predecessor problem, and the priority queue problem. In the full version, we also show that existential range queries in one dimension are captured by a class of monoids. In general, we would expect any fundamental problem of a certain one-dimensional flavor to be represented, making word problems an interesting avenue for complexity-theoretic research.

The seminal paper of Frandsen, Miltersen and Skyum [8] achieved tight bounds for many classes of monoids, both in the bit-probe and in the cell-probe models, but the classification is incomplete in both cases. In this paper, we further the classification for the bit-probe model in two ways. First, our lower bound for partial sums in $\mathbb{Z}/k\mathbb{Z}$ applies to the prefix problem in any monoid containing groups, and for the word problem in monoids containing a certain kind of externally noncommutative cycles [8–Theorem 2.5.1]. Second, we derive an upper bound of $O(\frac{\lg n}{\lg \lg n})$ for the word and prefix problems in group-free monoids. This uses the same algebraic toolkit as used by [8] in the cell-probe model, but our application needs several interesting algorithmic ideas to handle the idiosyncrasies of the bit-probe model.

2 Lower Bound for Dynamic Connectivity

Theorem 1. *Maintaining dynamic connectivity requires $\Omega((\frac{\lg n}{\lg \lg n})^2)$ bit probes, amortized per operation, in the average case of a certain input distribution.*

We first describe the dynamic graph used in our lower-bound construction; refer to Figure 1. The vertex set is roughly given by an integer grid of size $\sqrt{n} \times \sqrt{n}$. The edge set is given by a series of permutation boxes. A permutation box connects the nodes in a column to the nodes in the next column arbitrarily, according to a given permutation in $S_{\sqrt{n}}$. Notice that the permutations decompose the graph into a collection of $\sqrt{n}$ paths. As the paths evolve horizontally, the y coordinates change arbitrarily at each point due to the permutations. In addition to this, there is a special test vertex to the left. This is connected to a subset of the vertices in the first column.

We now describe the hard sequence of operations. The shape of the graph allows us to implement a partial sums problem over $S_{\sqrt{n}}$. The partial sums macro-operations are implemented as follows:

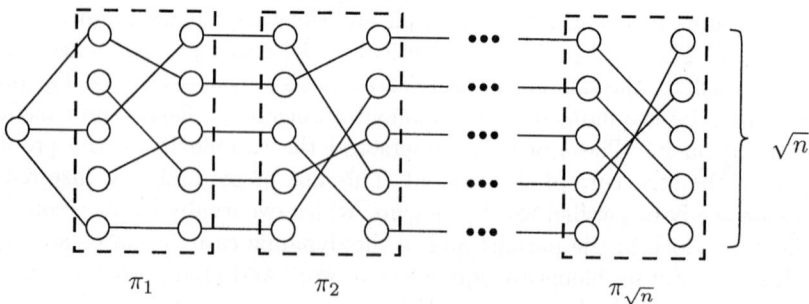

Fig. 1. Our graphs can be viewed as a sequence of $\sqrt{n}$ permutation boxes

UPDATE(i, π): sets $\pi_i = \pi$. This is done by removing all edges in permutation box i and inserting new edges corresponding to the new permutation π. This uses $O(\sqrt{n})$ elementary operations.

SUM(i): returns $\sigma = \pi_1 \circ \cdots \circ \pi_i$. We use $O(\lg n)$ phases, each one guessing a bit of $\sigma(j)$ for all j. Phase k begins by removing all edges incident to the test node. Then, we add edges from the test vertex to all vertices in the first column, whose row number has a one in the k-th bit. Then, we test connectivity of all vertices from the i-th column and the test node, respectively. This determines the k-th bit of $\sigma(j)$ for all j. In total, SUM uses $O(\sqrt{n} \lg n)$ elementary operations.

This construction differs from the one used in the cell-probe model. There, it was possible to prove a lower bound for a VERIFY operation, which verifies a given partial sum, rather than compute one from scratch. This does not seem possible in the bit-probe model. The reason is quite technical: the "separators" used for the cell-probe proof are too large. Unfortunately, known lower bounds show that those separators are actually optimal, so the approach fails altogether. In the bit-probe model, we are forced to revert to the SUM operation. Interestingly, we can still prove a good lower bound even if we blow up the queries by a $\lg n$ factor: the cost is just one of the $\lg \lg n$ factors in the denominator.

To fully specify the hard sequence, we must still say how UPDATEs and SUMs are generated. We choose $m = n^\varepsilon$ macro-operations randomly and independently ($\varepsilon > 0$ can be an arbitrary small constant). Because queries are $O(\lg n)$ time more expensive, each operation is chosen to be a query with probability $\frac{1}{\lg n}$. The arguments of each operation (i, possibly π) are picked uniformly at random. Let t be the running time of a connectivity operation. The total expected running for m operations is $O(t\sqrt{n}(1 - \frac{1}{\lg n})m + t\sqrt{n}\lg n \cdot \frac{1}{\lg n}m) = O(mt\sqrt{n})$. We shall prove an $\Omega(m\sqrt{n}(\frac{\lg n}{\lg \lg n})^2)$ lower bound, implying the desired bound on t.

Proof of the Lower Bound. Following [4], we consider a tree whose leaves represent the entire sequence of operations in time order. Any bit probe is characterized by when it occurs and when the bit was last written. These times are given by two leaves in the tree. For every node in the tree, we will lower bound the number of bit probes with the write time in the subtree of the node, and the read time in a subtree of a right sibling of the node. We then add up these lower bounds for all nodes, to obtain the overall bound. The correctness of this strategy follows by two observations. First, we never double count a bit probe: it is only counted for the node immediately under the lowest common ancestor of the read and write times. Second, we can sum the lower bounds because they all hold in the average case, under the same distribution.

We choose the branching factor of the time-tree to be $\Theta(\lg n)$, as suggested by the following intuition. Queries are fewer than updates by a logarithmic factor (because they are expensive). For maximal hardness, we must show that essentially all information in a series of updates is extracted. To do this, we must reuse most queries $\Theta(\lg n)$ times. Most nodes have $\Theta(\lg n)$ right siblings. If the node has L leaves under it, we expect $\Theta(L)$ updates in its subtree, and

$\Theta(\lg n \cdot L \frac{1}{\lg n}) = \Theta(L)$ queries in the subtrees of its right siblings, which allows all information to be extracted. Thus, a query is effectively reused $\Theta(\lg n)$ times on each level – in the lower bound for all left siblings of its ancestor.

Lemma 2. *Consider two adjacent intervals of operations, the left one of size L, and the right one of size $\Theta(L \lg n)$. Let r be the number of bit probes executed in the right interval, and c the number of probes from the right interval, which access a bit whose value after the left interval differs from its value before the left interval. Then $E[\lg \binom{r}{c}] = \Omega(L\sqrt{n} \lg n)$.*

Proof. Assume that we are told the entire sequence of macro-operations, except the ones in the left interval. Given this information, the sequence of answers to the SUM queries from the right interval has entropy $\Omega(L\sqrt{n} \lg n)$; see [4– Lemmas 5.3 and 5.4]. In essence, we expect $\Theta(L)$ updates in the left interval, and $\Theta(L)$ queries in the right interval. Further, we expect these to interleave almost perfectly, causing the answers to $\Omega(L)$ queries to be independent.

We now propose the following encoding for the answers to the queries: encode r, c, and the subset of the r bit probes which touch a bit changed during the left interval, and not yet changed in the right interval. This takes $E[\lg \binom{r}{c}] + O(\lg n)$ bits on average. Note that we only encode a c-subset of $[r]$; this identifies the relevant bit probes, but doesn't, for example, identify the addresses they probe. To decode, we first simulate the data structure's behavior before the left interval (we know everything from the past). We then simulate it for the right interval; implicitly we will recover the answers to the queries. This simulation needs to recover the values of all probed bits. If the bit location was written during the simulation of the right interval, we already know the value. Otherwise, we check whether the bit probe is in the set of c special bit probes. In this is true, the value is the negation of what it was before the left interval began; otherwise, it is the old value from before the left interval. □

Now consider an arbitrary level of the tree. We apply the lemma by setting the left interval to the subtree of a node, and the right interval to the union of the subtrees of its right siblings. We do this only for nodes which are in the first half of their siblings, so that the right interval is $\Theta(\lg n)$ times bigger. For some node i, we obtain $E[\lg \binom{r_i}{c_i}] = \Omega(L_i \sqrt{n} \lg n)$. Let us sum this up for all nodes on a level. Clearly, $\sum L_i = \frac{m}{2} = \Theta(m)$. By linearity of expectation, the left hand side becomes $E[\lg \prod \binom{r_i}{c_i}]$. It is well known that $\prod \binom{r_i}{c_i} \leq \binom{\sum r_i}{\sum c_i}$. Now, $\sum r_i = \Theta(T \lg n)$, where T is the total number of bit probes. This is because each bit probe is counted at most once for every sibling of its ancestor on the current level. Let $\sum c_i = C_\ell$, where ℓ is the current level. We can now obtain $E[\lg \binom{\gamma T \lg n}{C_\ell}] = \Omega(m\sqrt{n} \lg n)$, for an appropriate constant γ. Now let us sum over all levels ℓ. As explained in the beginning, $\sum C_\ell \leq T$. Then, we obtain $E[\lg \binom{\gamma T \lg n \cdot \lg n / \lg \lg n}{T}] = \Omega(m\sqrt{n} \lg n \frac{\lg n}{\lg \lg n})$, which implies $E[T] O(\lg \lg n) = \Omega(m\sqrt{n} \frac{\lg^2 n}{\lg \lg n})$.

This completes the proof of our $\Omega((\frac{\lg n}{\lg \lg n})^2)$ lower bound. As explained already, one of the $\lg \lg n$ factors is lost because we need to increase the complexity of queries by a $\lg n$ factor, which translates into a higher branching factor for the tree, and a depth of just $O(\frac{\lg n}{\lg \lg n})$. The other $\lg \lg n$ factor is lost from our encoding. Intuitively, Lemma 2 should state that $c = \Omega(L\sqrt{n} \lg n)$. This is because the right interval needs $\Omega(L\sqrt{n} \lg n)$ bits of information, and, intuitively, this can only come from so many bit probes. Unfortunately, the fact that a bit was not modified during the left interval is also information, even if it is not clear how such information could ever be used in an upper bound. The factor of $\lg \binom{r}{c} \approx c \lg \lg n$ seems very hard to avoid.

3 Lower Bound for Partial Sums in Cyclic Groups

Let us review the ideas behind the chronogram method, at an intuitive level. First, we generate a sequence of random updates, ended by one random query. Looking back in time from the query, we partition the updates into exponentially growing epochs: for a certain t, epoch i contains the t^i operations immediately *before* epoch $i - 1$. We then argue that for all i, the query needs to read at least one cell from epoch i with constant probability. This is done as follows. Clearly, information about epoch i cannot be reflected in higher epochs (those occurred back in time). The first $i - 1$ epochs contain $O(t^{i-1})$ updates. Assume the cell-probe complexity of each update is bounded by t_u. Then, during the first $i - 1$ epochs, only $O(t^{i-1} t_u)$ cells are written. If $t = C \cdot t_u$ for a sufficiently large constant C, these cells can only include complete information about a fraction of the t^i updates in epoch i. If a random query forces us to learn information about a random update from epoch i, we are forced to read a cell from epoch i, because this information is not available in any other epoch. This gives $\Omega(1)$ probes in expectation to every epoch, so the lower bound on the query time is given by the number of epochs that we can construct, i.e. $t_q = \Omega(\lg n / \lg t_u)$. A tradeoff of this form was indeed obtained by [9], and is the highest tradeoff obtained by the chronogram method. Of course, these calculations assumed that update inputs and query outputs match the cell size.

We now describe the ideas needed to improve the $\lg t_u$ factor in the denominator. Intuitively, the analysis done by the chronogram technique is overly pessimistic, in that it assumes all cells written in the first $i - 1$ epochs concentrate on epoch i, encoding a maximum amount of information about it. In the setup from above, this may actually be tight, up to constant factors, because the data structure knows the division into epochs, and can build a strategy based on it. However, we can randomize the construction of epochs to foil such strategies. We generate a random number of updates, followed by one query; since the data structure cannot anticipate the number of updates, it cannot base its decisions on a known epoch pattern. Due to this randomization, we intuitively expect each update to write $O(t_u/t_q)$ cells about a random epoch (as before, t_q is roughly the number of epochs). Then, we should take $t = C \frac{t_u}{t_q}$, and obtain bounds of the form $t_q \lg \frac{t_u}{t_q} = \Omega(\lg n)$, implying $\max\{t_u, t_q\} = \Omega(\lg n)$.

Unfortunately, formalizing the intuition that the information written by updates "splits" between epochs seems to lead to elusive information theoretic arguments. To circumvent this, we need a second idea: we can look at cell reads, as opposed to cell writes. Reads have more graspable information theoretic properties, dictated by the time when the cell was last written, and the time when the read occurs. We can actually show that in expectation, $O(t_u/t_q)$ of the reads made by each update obtain information about a random epoch. Then, regardless of how many cells are written, subsequent epochs can only encode little information about epoch i, because very little information was read by the updates in the first place.

One case in which the above reasoning works is the partial sums problem in the cell-probe model (with the group $\mathbb{Z}/n\mathbb{Z}$), for which the tradeoff is tight. Our bit-probe lower bound from below contains all ideas needed by this proof. This is the largest cell-probe tradeoff known, and a different proof was described recently in [4]. In the bit-probe model, we achieve a slightly weaker bound, because $O(t_u/t_q)$ bit probes into an epoch give more than this number of bits of information. As for the connectivity bound, we must also account for the choice of the relevant $O(t_u/t_q)$ probes from the total of t_u.

Theorem 3. *For the partial-sums problem in $\mathbb{Z}/2\mathbb{Z}$, let t_u be the amortized bit-probe complexity of an update, and t_q the bit-probe complexity of a query. Then in the average case of a certain input distribution, $t_q(\lg \frac{t_u}{t_q} + \lg \lg t_q) = \Omega(\lg n)$. In particular, $\max\{t_u, t_q\} = \Omega(\frac{\lg n}{\lg \lg \lg n})$.*

Let $M = n^\varepsilon$, for any small $\varepsilon > 0$. We generate $2M$ operations UPDATE(i, x), each with random $i \in [n], x \in \mathbb{Z}/2\mathbb{Z}$. Now pick m uniformly at random from $\{M, \ldots, 2M\}$. After m operations, we insert a query to a random index. Looking back in time from the query, divide that last M operations into epochs. Epoch i has the t^i operations preceding epoch $i-1$, where t will be specified later ($t \geq 2$). The number of epochs is $\Theta(\lg_t M) = \Theta(\lg_t n)$. Let P_i be the number of bit probes from epochs 1 through $i-1$, touching bits written during epoch i.

Lemma 4. *If i is chosen uniformly at random, $E[\frac{P_i}{t^{i-1}}] = O(\frac{t_u}{\lg_t n})$.*

Proof. We say that a bit probe has span j if $t^{j-1} \leq w - r < t^j$, where r and w are the read and write times (i.e. the index of the operation during which they occur). Let C_j be the total number of bit probes of span j that the $2M$ updates would execute in the absence of an interleaving query. Fix i to some arbitrary value, and let f be the last operation in epoch i. Observe that f is a random variable depending on m. We are interested in bit probes with $w \in (f - t^i, f], r \in (f, m]$. All such bit probes must have span at most $i + 1$, because $m - f < 2t^{i-1}$. For all j, if $r \geq f + t^{j-1}$, the bit probe must have span at least j. We count all probes which satisfy these necessary conditions, thus upper bounding P_i.

Fix the $2M$ updates arbitrarily. Because m is not known to the data structure, the first m updates behave as if no query ever happened. As m is chosen randomly, we are interested in the probes of span j from a random segment of size t^j.

There are $\Omega(M)$ choices for the position of this segment, and we choose uniformly at random between them. Then, the expected number of span-j operations that are relevant is $O(t^j \frac{C_j}{M})$. For $j = i$ or $j = i+1$, we can upper bound the segment size by $m - f = O(t^{i-1})$. Then, $E_m[P_i] = O(t^{i-1} \frac{C_i + C_{i+1}}{M}) + \sum_{j<i} O(t^j \frac{C_j}{M})$, so $E_m[\frac{P_i}{t^{i-1}}] = O(\frac{C_i}{M} + \frac{C_{i+1}}{M} + \sum_{j<i} \frac{C_j}{M \cdot t^{i-j}})$.

Observe that $\sum_i E_m[\frac{P_i}{t^{i-1}}] \leq \sum_i (2 + \sum_{k=1}^{\infty} t^{-k}) \cdot O(\frac{C_i}{M}) = \sum_i O(\frac{C_i}{M})$. By definition, $\sum_i \frac{C_i}{M}$ is the amortized running time of an update. When updates are chosen uniformly at random, $E[\sum_i \frac{C_i}{M}] \leq t_u$. Thus, if i is chosen uniformly at random, $E[\frac{P_i}{t^{i-1}}] = O(\frac{t_u}{\lg_t n})$. □

Now pick an epoch i at random. We show that with constant probability, the query needs to probe a bit from epoch i. Then, by linearity of expectation, $t_q = \Omega(\lg_t n)$. By the Markov bound, there is a $1 - \delta$ probability that for the chosen i, $E[\frac{P_i}{t^{i-1}}] = O(\frac{t_u}{\lg_t n})$. Here the expectation is over the choice of m, and the random updates, but i is already fixed; $\delta > 0$ is an arbitrarily small constant. If this relation does not hold, we make no claim. Otherwise, let p be the probability that a random query accesses a cell from epoch i. We want to prove that p is bounded away from 0. Then, be the union bound, there is a constant probability that a query needs a bit probe into epoch i.

Pick t^i queries independently at random, and imagine that each is run starting with the state of the data structure after m updates. Given the choice of the queries, and the random updates from all epoch except i, the entropy of the answers to all queries is $\Omega(t^i)$; see [4–Lemmas 5.3 and 5.4]. This is intuitive, since the t^i query indices are expected to interleave with the indices updated in epoch i, so each query recovers one more bit of information about the unknown updates. We now propose an encoding for the answers to these queries.

First, we encode which bit probes executed in epochs $[1, i-1]$ read bits that were changed in epoch i. Let $R_{<i}$ be the total number of bit probes made by epochs $[1, i-1]$. Since m is chosen randomly, $E[R_{<i}] = O(t^{i-1} t_u)$. This part of the encoding takes $O(\lg P_i + \lg R_{<i}) + \lg \binom{R_{<i}}{P_i} = O(P_i \lg \frac{R_{<i}}{P_i})$ bits. Since $(x, y) \mapsto x \lg \frac{y}{x}$ is convex, the expected size is $O(E[P_i] \lg \frac{E[R_{<i}]}{E[P_i]}) = O(t^{i-1} \frac{t_u}{\lg_t n} \lg \frac{t_u}{t_u / \lg_t n}) = O(t^{i-1} \frac{t_u}{\lg_t n} \lg \lg_t n)$.

The second part of the encoding identifies which of the t^i queries reads at least one bit-probe from epoch i. If this number is Q, this part takes $\lg \binom{t^i}{Q} = O(Q \lg \frac{t^i}{Q})$ bits. By convexity, the expected size is $O(pt^i \lg \frac{t^i}{pt^i}) = O(t^i p \lg \frac{1}{p})$. For the queries which read a bit from epoch i, we explicitly encode their answers; this takes pt^i bits in expectation.

It is easy to see that the encoding actually works. We first simulate the data structure up until the beginning of epoch i. Then, we simulate it after epoch i. This can be done because we know which bit probes read later get a changed bit because of epoch i. At the end, we simulate all queries which do not read bits from epoch i. For the rest of the queries, we know the answers.

The total size of the encoding is $O(t^i p \lg \frac{1}{p}) + O(t^{i-1} \frac{t_u}{\lg_t n} \lg \lg_t n)$. This must be $\Omega(t^i)$, from which we obtain $p \lg \frac{1}{p} = \Omega(1) - O(\frac{1}{t} \cdot \frac{t_u}{\lg_t n} \lg \lg_t n)$. Now choose t satisfying the equation $t = C \cdot \frac{t_u}{\lg_t n} \lg \lg_t n$, for a large enough constant C. For this t, we obtain $p \lg \frac{1}{p} = \Omega(1)$, hence $p = \Omega(1)$. The lower bound is $t_q = \Omega(\lg_t n)$, from which we obtain $t_q = \Omega(\lg n / (\lg \frac{t_u}{t_q} + \lg \lg t_q))$.

4 Upper Bounds for Predecessor Problems

We only sketch the solution for the segment representatives problem. The other problems are discussed in the full version of this paper. We construct a trie with branching factor $B = \Theta(\frac{\lg n}{\lg \lg n})$, in which each element is represented by a root-to-leaf path. A node is marked active if any leaf under it is in the current set. Among the children of the node, the minimum and maximum active nodes are specially marked. In the full version of this paper, we describe how to maintain these markings in $O(\frac{\lg n}{\lg \lg n})$ per insertion and deletion. It remains to implement queries. The main idea is to find the lowest common ancestor of the predecessor and the successor of the query point, and the children corresponding to the predecessor and the successor. Once this is known, it is easy to determine a canonical representative. First, take the bits corresponding to the LCA, which are common to the query, the predecessor, and the successor. Then, take the $\lg B$ bits giving the child corresponding to the successor. Finally, pad with zeroes. This computation of the representative takes zero time, since all elements are already known.

To find the lowest common ancestor of the predecessor and successor, we proceed as follows. Traverse the leaf-to-root path from the query until an active node is found. Scan the children of the node. If there is both an active node to the left and one to the right of the query, we have found the answer. Otherwise, assume by symmetry there is an active node to the left, so we have only found the predecessor. We continue searching up the tree, until we find a node which is *not* marked as maximum among its siblings. At that point, we scan the right siblings, finding the next active one, which corresponds to the successor.

5 Group-Free Monoids

Theorem 5. *The dynamic prefix problem in any group-free monoid has bit-probe complexity* $\Theta(\frac{\lg n}{\lg \lg n})$.

We use a corollary of the Krohn-Rhodes decomposition [10], which was also used in the cell-probe case [8]. Unfortunately, our application of it is considerably more involved due to the idiosyncrasies of the bit-probe model: we go through considerable lengths to avoid an exact predecessor query.

Theorem 6 ([10]). *Let M be a finite nontrivial group-free monoid. One of the following holds:*

1. $M \setminus \{1\} = \langle a \rangle = \{a, a^2, \ldots, a^k = a^{k+1}\}$;
2. M is left simple, i.e. $\forall a, b \in M \setminus \{1\}, ab = a$;
3. $M = V \cup T$, with V, T proper submonoids of M, and $T \setminus \{1\}$ a left ideal of M (i.e. $(\forall) a \in M, b \in T : ab \in T$).

We prove our upper bound by induction on the size of M. However, we need a stronger induction hypothesis, which assumes a solution for a slightly harder problem. We call this the prefix problem with breakpoints. Consider an array $A[1..n]$, in which an element can either be an element of M, or a breakpoint, denoted ⓑ. The update operation can change any position of A to any element from $M \cup \{ⓑ\}$. A query on an arbitrary position i must return the composition of $A[j+1], \ldots, A[i]$, where j is the predecessor of i in the set of breakpoints. For uniformity, we say $A[0] = A[n+1] = ⓑ$. Also, we assume breakpoints do not appear in consecutive positions. This can easily be arranged by doubling the array, and inserting an identity at every other position.

Assume by induction that the predecessor problem with breakpoints in all group-free monoids of size less than $|M|$ has complexity $O(\frac{\lg n}{\lg \lg n})$. Now apply the decomposition theorem. Due to space constraints, cases 1 and 2 are discussed in the full version of this paper.

Case 3 - $M = T \cup V$. Since $|T|, |V| \leq |M| - 1$, we have, by induction, solutions for the prefix problem with breakpoints, both running in $O(\frac{\lg n}{\lg \lg n})$. To obtain a data structure for M, we use the following components:

- a prefix structure for V, built over the array $A^V[1..n]$. The values of A^V are defined as follows: if $A[i] \in V$, $A^V[i] = A[i]$; otherwise ($A[i] \in T \setminus V$ or $A[i] = ⓑ$), let $A^V[i] = ⓑ$.
- a structure for finding segment representatives with the dynamic set being $B = \{i \mid A^V[i] = ⓑ\}$. Denote the elements $B = \{b_1, b_2, \ldots\}$.
- a prefix structure for T, built over the array A^T. If $A[i] = ⓑ$, let $A^T[i] = ⓑ$. The other elements are represented in a less straightforward way. For any segment (b_i, b_{i+1}), let $A^T[\text{repr}(b_i + 1)] = \bigoplus_{j=b_i+1}^{b_{i+1}} A[j]$. Note that all but $A[b_{i+1}]$ are elements from V, but the composition is in T because of the ideal property. All elements which are not segment representatives are 1_M.
- a simple array C. For any segment (b_i, b_{i+1}), $C[\text{repr}(b_i+1)] = A[b_{i+1}]$. Other values of C are ignored.

We claim that we can support the following operation: given any element $j \in (b_i, b_{i+1})$, compute the composition of the entire segment in A^V. Clearly, we can recover the composition of the elements between $b_i + 1$ and $j - 1$, by running a prefix query in A^V. In addition, we can define a monoid W, which is "the reverse" of V: $(\forall) a, b : a \oplus_W b = b \oplus_V a$. Since $|W| = |V| \leq m - 1$, we can construct a prefix structure on W by the induction hypothesis. We maintain $A^W[n - i + 1] = A^V[i], (\forall) i$. Now, by running a prefix query in A^W at position $n - j + 1$, we are effectively running a suffix query in A^V. By composing the prefix and the suffix, we obtain the composition of the entire segment.

We can now easily support a prefix query to position i. First, we run a prefix query in A^T up to $\text{repr}(i) - 1$. This will give the desired partial sum up to the predecessor of i in B. To get the last part of the prefix, which consists only of elements from V', we simply ask for the prefix up to i in A^V. This works because the predecessor from B of i is the most recent breakpoint in A^V.

Updates are done in two steps: first we change the old value to 1_M, and then we change this to the new value. We distinguish the following cases:

- both the old and new values are in V (possibly 1_M). We update $A^V[i]$, and compute the composition of the entire segment containing i. Then, we compose this with $C[\text{repr}(i)]$, and we update $A^T[\text{repr}(i)]$ to this new value.
- an element from $T \setminus V$ is replaced by 1_M. We update A^V and remove i from B, which merges two segments. We remove the old segment representatives from A^T (set $A^T[\text{repr}(i-1)] = A^T[\text{repr}(i+1)] = 1_M$). Then, we add the representative for the new segment. The value of $B[\text{repr}(i)]$ is given by the old value $B[\text{repr}(i+1)]$, corresponding to the right segment; $A^T[\text{repr}(i)]$ is obtained by recomputing the composition of i's segment, as above.
- 1_M is replaced by an element in $T\setminus V$. We update $A^V[i]$ to ⓑ, and insert i into B, thus splitting a segment. We first remove the old segment's representative, setting $A^T[\text{repr}(i)] = 1_M$, and then we add the two new representatives. The values in B are obvious: $B[\text{repr}(i-1)]$ gets the new value of $A[i]$, and $B[\text{repr}(i+1)]$ gets the old $B[\text{repr}(i)]$ from the unsplit segment. The values in A^T are obtained by recomputing the compositions of the two segments.
- ⓑ is replaced by 1_M. We update $A^T[i]$ to 1_M. Changing the other structures is identical to the case when an element from $T \setminus V$ is removed.
- 1_M is replaced by ⓑ. We propagate the change to $A^T[i]$. Changing the other structures is identical to the case when an element from $T \setminus V$ is added.

Acknowledgements. We are grateful to Thore Husfeldt for many helpful suggestions. Some of the results in this paper originate in a project of the first author for a research course at Harvard. She is grateful to Nick Rogers for useful comments made on that occasion.

References

1. Thorup, M.: Near-optimal fully-dynamic graph connectivity. In: Proc. 32nd ACM Symposium on Theory of Computing (STOC). (2000) 343–350
2. Miltersen, P.B., Subramanian, S., Vitter, J.S., Tamassia, R.: Complexity models for incremental computation. Theor. Comp. Sci. **130** (1994) 203–236 Also STACS'93.
3. Miltersen, P.B.: Cell probe complexity - a survey. In: Advances in Data Structures Workshop, FSTTCS (1999)
4. Pătrașcu, M., Demaine, E.D.: Logarithmic lower bounds in the cell-probe model. *arXiv:cs.DS/0502041*. Based on publications from SODA'04 and STOC'04 (2004)
5. Fredman, M.L.: The complexity of maintaining an array and computing its partial sums. Journal of the ACM **29** (1982) 250–260
6. Fredman, M.L., Saks, M.E.: The cell probe complexity of dynamic data structures. In: Proc. 21st ACM Symposium on Theory of Computing (STOC). (1989) 345–354

7. Mortensen, C.W., Pagh, R., Pătraşcu, M.: On dynamic range reporting in one dimension. In: Proc. 37th ACM Symposium on Theory of Computing (STOC). (2004) to appear
8. Frandsen, G.S., Miltersen, P.B., Skyum, S.: Dynamic word problems. Journal of the ACM **44** (1997) 257–271 See also FOCS'93.
9. Alstrup, S., Husfeldt, T., Rauhe, T.: Marked ancestor problems. In: Proc. 39th IEEE Symposium on Foundations of Computer Science (FOCS). (1998) 534–543
10. Krohn, K., Rhodes, J.: Algebraic theory of machines I. Prime decomposition theorem for finite semigroups and machines. Trans. AMS **116** (1965) 450–464

Time-Space Lower Bounds for the Polynomial-Time Hierarchy on Randomized Machines

Scott Diehl* and Dieter van Melkebeek**

University of Wisconsin-Madison, Madison WI 53706, USA
{sfdiehl, dieter}@cs.wisc.edu

Abstract. We establish the first polynomial-strength time-space lower bounds for problems in the linear-time hierarchy on randomized machines with bounded two-sided error. We show that for any integer $\ell > 1$ and constant $c < \ell$, there exists a positive constant d such that QSAT_ℓ cannot be computed by such machines in time n^c and space n^d, where QSAT_ℓ denotes the problem of deciding the validity of a Boolean first-order formula with at most $\ell - 1$ quantifier alternations. Corresponding to $\ell = 1$, we prove that for any constant $c < \phi \approx 1.618$, there exists a positive constant d such that the set of Boolean tautologies cannot be decided by a randomized machine with one-sided error in time n^c and space n^d.

1 Introduction

Proving lower bounds remains one of the most challenging tasks in computational complexity. Satisfiability, the seminal NP-complete problem, is particularly unyielding in this respect. While we believe that any algorithm for satisfiability takes time linear exponential in the number of variables in the formula, we have been unable to prove super-linear time lower bounds on random access machines despite several decades of effort. Additionally, problems complete for higher levels of the polynomial-time hierarchy, while not receiving as much attention, have also resisted nontrivial time lower bounds.

A few years ago, Fortnow [5] realized that if we restrict the work space that a machine can use to solve satisfiability, then we can establish nontrivial lower bounds. Fortnow's technique has its roots in earlier work by Kannan [6] and has been further developed in recent years [8,9]. For example, Fortnow and Van Melkebeek derived the following time-space lower bound for nondeterministic linear time, which gives the same lower bound for satisfiability due to the tight connection between satisfiability and nondeterministic linear time.

Theorem 1 (Fortnow-Van Melkebeek [4]). *Let $\phi \doteq (\sqrt{5} + 1)/2 \approx 1.618$ denote the golden ratio. For any constant $c < \phi$ there exists a positive constant*

---

* Supported by NSF Career award CCR-0133693.
** Partially supported by NSF Career award CCR-0133693.

d such that nondeterministic linear time cannot be simulated by deterministic random-access machines running in time n^c and space n^d.

Fortnow and Van Melkebeek also considered higher levels of the linear-time hierarchy and managed to prove the following time-space lower bound.

Theorem 2 (Fortnow-Van Melkebeek [4]). *For any integer $\ell \geq 2$ and constant $c < \ell$, there exists a positive constant d such that $\Sigma_\ell \text{TIME}[n]$ cannot be simulated by deterministic random-access machines running in time n^c and space n^d.*

The same tight relationship between nondeterministic linear time and satisfiability also exists between $\Sigma_\ell \text{TIME}[n]$ and QSAT_ℓ, the problem of deciding the validity of a given Boolean first-order formula with at most $\ell - 1$ quantifier alternations. Thus, the time-space lower bound for $\Sigma_\ell \text{TIME}[n]$ of Theorem 2 holds for QSAT_ℓ as well.

In this paper, we establish time-space lower bounds for the same problems on *randomized* machines with two-sided error bounded away from $1/2$.

Theorem 3 (Main Theorem). *For any integer $\ell \geq 2$ and constant $c < \ell$, there exists a positive constant d such that $\Sigma_\ell \text{TIME}[n]$ cannot be simulated by randomized random-access machines with two-sided error bounded away from $1/2$ running in time n^c and space n^d.*

Observe that our bounds essentially match those for the deterministic simulations given in Theorem 2, the only difference being the exact dependence of d on c. As in the deterministic case, Theorem 3 implies the same time-space lower bounds for QSAT_ℓ.

One can also view the instantiation of our Main Result for $\ell = 2$ as an analog of Theorem 1. Note that Theorem 1 relates to the question of P versus NP. We know the trivial inclusion that $\text{P} \subseteq \text{NP}$ and do not believe the converse but fail to prove that conjecture. Thus, research evolved towards time-space lower bounds to achieve partial negative results for the converse, as in [5, 8, 4, 9]. Similarly for BPP, we know that $\text{BPP} \subseteq \Sigma_2^p$ but we do not believe the converse. In this work, we turn to time-space lower bounds to achieve nontrivial negative results about the converse.

We also further strengthen Theorem 3 by showing similar lower bounds for problems that can be decided by Σ_ℓ machines in linear time and space n^a for some constant $a < 1$.

We note that Theorem 3 establishes the first polynomial-strength time-space lower bounds for two-sided error randomized simulations of the polynomial-time hierarchy. By time-space lower bounds of "polynomial strength" we mean time lower bounds of the form $\Omega(n^c)$ for some constant $c > 1$ under nontrivial space upper bounds. Previous works establish randomized time-space lower bounds but they either consider problems believed not to be in the polynomial-time hierarchy, or the time lower bounds involved are only slightly super-linear. Allender et al.'s [1] time-space lower bounds for problems in the counting hierarchy on probabilistic machines with unbounded error fall within the first category. Beame et al. [2] give a (nonuniform) time-space lower bound for a problem in P based on a binary quadratic form, which falls in the second category.

At first glance, it might seem that the known results about space-bounded derandomization let us derive time-space lower bounds on randomized machines as immediate corollaries to the known time-space lower bounds on deterministic machines. In particular, assuming that we can solve satisfiability on a randomized machine in logarithmic space and time n^c, Nisan's deterministic simulation [11] yields a deterministic algorithm for satisfiability that runs in polylogarithmic space and polynomial time. However, even for $c = 1$, the degree of the latter polynomial is far too large for this simulation to yield a contradiction with Theorem 1 as we would like. Thus, we need a more delicate approach for the randomized setting.

Our proofs follow the paradigm of indirect diagonalization. The technique establishes a desired separation by contradiction – assuming the separation does not hold, we derive a sequence of progressively unlikely inclusions of complexity classes until we reach one that contradicts a known diagonalization result. Kannan [6] used the paradigm avant la lettre to investigate the relationship between deterministic linear time and nondeterministic linear time. All of the recent work on time-space lower bounds for satisfiability and problems higher up in the polynomial-time hierarchy [5, 8, 4, 9] follow it as well. Allender et al. [1] employed the technique to establish time-space lower bounds for problems in the counting hierarchy.

The critical ingredient that allows us to apply the paradigm to two-sided error randomized algorithms for problems in the polynomial-time hierarchy is a time and space efficient simulation of randomized computations in the second level of the polynomial-time hierarchy with very few guess bits. The latter follows from a careful combination of Nisan's partial space-bounded derandomization [10] and a version of Lautemann's proof that BPP $\subseteq \Pi_2^p$ [7].

We point out that earlier work implies lower bounds for (the complements of) the above problems on randomized machines with *one-sided error*. This follows because the lower bound arguments for conondeterministic linear time on *deterministic* machines typically also work on *nondeterministic* machines, of which randomized machines with one-sided error are special cases. However, the bounds become weaker. For example, the bound on c in Theorem 1 reduces from the golden ratio to $\sqrt{2}$.

Theorem 4 (Fortnow-Van Melkebeek [4]). *For any constant $c < \sqrt{2}$ there exists a positive constant d such that conondeterministic linear time cannot be simulated by randomized random-access machines with one-sided error running in time n^c and space n^d.*

Using ideas from the proof of our main result, we manage to strengthen Theorem 4 so that the bound matches the one in Theorem 1.

Theorem 5. *For any constant $c < \phi$, there exists a positive constant d such that conondeterministic linear time cannot be simulated by randomized random-access machines with one-sided error running in time n^c and space n^d.*

A similar strengthening holds for the analog of Theorem 2.

2 Preliminaries

Most of the notation we use is standard [3, 12]. For a detailed description of the machine model we use, we refer the reader to [9]. We adopt the convention that time and space functions refer to constructible functions from natural numbers to natural numbers. Our results ultimately apply to computations with polynomial time and space bounds, which certainly meet these conditions.

We introduce some additional terminology to reason about randomized computation. In particular, we use the notation BPTISP$[t, s]$ to refer to the class of languages recognized by randomized machines using time t and space s with error bounded by $\frac{1}{3}$ on both sides.

Our arguments involve Σ_k^p and Π_k^p computations in which the numbers of bits guessed at each stage are bounded by explicitly given small functions. To this end, we use the following notation to describe such computations:

Definition 1. *Given a complexity class $\mathcal{C}$ and a function f, we define the class $\exists^f \mathcal{C}$ to be the set of languages that can be described as*

$$\{x | \exists y \in \{0,1\}^{O(f(|x|))} P(x,y)\},$$

where P is a predicate accepting a language in the class $\mathcal{C}$ when its complexity is measured in terms of $|x|$ (not $|x| + |y|$). We analogously define $\forall^f \mathcal{C}$.

For example, $\exists^f \text{DTIME}[n]$ and $\forall^f \text{DTIME}[n]$ are subsets of NP and coNP for $f(n) = n^{O(1)}$. The requirement that the complexity of P be measured in terms of $|x|$ allows us to express the running times simply in terms of the original input length, which is a more natural notion for our arguments.

We also make use of the standard divide-and-conquer approach for speeding up space bounded computation by introducing alternations. Namely, by splitting up the computation tableau of a DTISP$[T, S]$ computation into $b > 0$ equal size blocks, we obtain

$$\text{DTISP}[T, S] \subseteq \exists^{bS} \forall^{\log b} \text{DTISP}[T/b, S] \subseteq \Sigma_2 \text{TIME}[bS + T/b]. \tag{1}$$

If we choose b to optimize the running time of the resulting Σ_2 computation, the result is

$$\text{DTISP}[T, S] \subseteq \exists^{\sqrt{TS}} \exists^{\log T} \subseteq \Sigma_2 \text{TIME}[\sqrt{TS}]. \tag{2}$$

When used within the framework of time-space lower bounds, it is not always desirable to choose the block size to optimize the running time in this way, as exhibited by [4]. Therefore, most of our arguments make use of (1) for some unspecified b, and then set b later to yield the strongest results.

Finally, we need a standard diagonalization result from which we can derive contradictions. The following lemma states that for a fixed number of alternations, if we switch from universal to existential initial states and allow for a little more time, we can compute something we couldn't compute before.

Lemma 1 (Folklore). *Let ℓ be a positive integer and t a time function. Then*

$$\Sigma_\ell \text{TIME}[t] \not\subseteq \Pi_\ell \text{TIME}[o(t)].$$

3 Framework of Earlier Deterministic Results

In this section, we provide an overview of the arguments used to establish Theorem 1 since our approach uses the same general framework. Specifically, both arguments follow the paradigm of indirect diagonalization, which can be divided into the following general steps:

1. Assume the inclusion that we wish to show does not hold. For example, $\Sigma_2\text{TIME}[n] \subseteq \text{BPTISP}[t, s]$.
2. Using the hypothesis, derive inclusions of complexity classes which are increasingly unlikely.
3. Eventually one of these inclusions contradicts a known diagonalization result, proving the desired result.

Let us step through a weaker instantiation of Theorem 1 as an example. Namely, we prove the result of [8] that $\text{NTIME}[n] \not\subseteq \text{DTISP}[n^c, n^{o(1)}]$ for $c < \sqrt{2}$ following the outline described above. Therefore, the first step is to assume that

$$\text{NTIME}[n] \subseteq \text{DTISP}[n^c, n^{o(1)}]. \quad (3)$$

Consider the class $\text{DTISP}[T, T^{o(1)}]$ for some polynomial T, say $T(n) = n^2$. Using (2), we can speed up this computation by introducing alternations, resulting in a $\Sigma_2\text{TIME}[T^{1/2+o(1)}]$ simulation. Now observe that we can use the hypothesis (3) to collapse Σ_2 to NP, eliminating one alternation at the small cost of raising the running time of the simulation to the power of c. Since DTISP is closed under complement, this process gives the inclusion

$$\text{DTISP}[T, T^{o(1)}] \subseteq \text{coNTIME}[T^{c/2+o(1)}, T^{o(1)}]. \quad (4)$$

If the cost of removing an alternation by this technique is less than the speedup we gained by its introduction, then (4) is a more unlikely inclusion than the hypothesis, which could lead to a contradiction with Lemma 1.

To find the values of c which yield a contradiction, consider the hypothesis (3) padded to time T. Combining this with (4), we can conclude

$$\text{NTIME}[T] \subseteq \text{DTISP}[T^c, T^{o(1)}] \subseteq \text{coNTIME}[T^{c^2/2+o(1)}],$$

which contradicts Lemma 1 so long as $c^2/2 < 1$. This proves the desired result.

One can obtain stronger results by applying these arguments recursively. Specifically, once the speedup of (1) is applied, the final stage involved in the resulting Σ_2 simulation is itself a space bounded computation taking less time than what we started with. Therefore, we can obtain a contradiction for larger c if we recursively apply the same arguments to further speed up this computation. Doing so in an economical way (with respect to alternations) and choosing the block numbers optimally at each recursive step yields Theorem 1.

4 Lautemann's Proof and Derandomization

For this paper, we wish to derive a contradiction from the assumption

$$\Sigma_2 \text{TIME}[n] \subseteq \text{BPTISP}[t,s]. \tag{5}$$

Taking a cue from the deterministic results, we would like to figure out a way to transfer the hypothesis (5) into a statement giving a strong collapse of the polynomial-time hierarchy, such as $\Sigma_2 \text{TIME}[n] \subseteq \Pi_2 \text{TIME}[f(n)]$ for some small function $f(n)$. We can then use such a collapse to eliminate alternations introduced by applying the speedup of (1). The focus of this section is the derivation of such a statement.

Lautemann's proof that $\text{BPP} \subseteq \Sigma_2^p \cap \Pi_2^p$ is the first tool which helps us accomplish this task. The proof relies on the large gap in sizes of witness sets for a BPP algorithm which uses R bits to accept a language L with sufficiently small error. When $x \in L$, the witness set is large enough so that for most sets of R shifts, the union of the R shifted witness sets covers the entire universe of possible witnesses; when $x \notin L$, the witness set is so small that no set of R shifts covers the universe. These complementary conditions can be expressed by a Σ_2^p predicate. Since BPP is closed under complement, this shows that $\text{BPP} \subseteq \Sigma_2^p \cap \Pi_2^p$. Specifically, we are interested in the Π_2^p side of the inclusion for our results.

Theorem 6 (Lautemann [7]). *Let L be a language recognized by a randomized machine M with error bounded on both sides by $1/R$ that runs in time T, space S (when provided two-way access to the random bits), and uses R random bits. Then*

$$L \in \forall^{R^2} \exists^R \text{DTISP}[RT, S + \log R]. \tag{6}$$

Theorem 6 is a natural candidate to derive the desired strong collapse of the polynomial-time hierarchy from (5). However, the key question is if the inclusion given by Theorem 6 is efficient enough to allow for a sufficiently strong collapse. To answer this, first note the requirement that the error be $1/R$ does not pose a problem, since this can be achieved by taking the majority vote of $O(\log R) = O(\log T)$ repetitions of a standard BPP algorithm. Thus, we can transform an arbitrary BPP computation into an equivalent one satisfying the conditions of Theorem 6 with only a logarithmic time blowup. This is good news since we would be in trouble if the theorem required error exponentially small in R – in general, the running time of the amplified algorithm would be at least quadratic.

On the other hand, the Π_2 simulation resulting in an application of Theorem 6 must run the randomized machine on R different shifts of a witness so its running time is a factor of R greater than that of the randomized machine. Since, in general, R can be as large as T, this slow-down is too much for our arguments to compensate. Therefore, an additional ingredient is needed.

That ingredient exploits the fact that the hypothesis (5) gives simulations by *space-bounded* BPP computations. In that setting, we know of techniques to reduce the number of random bits used, which in turn reduces the overhead in the Π_2 simulation given by Theorem 6. The means by which we achieve the needed reduction in randomness is the space-bounded derandomization of Nisan [10].

Theorem 7 (Nisan [10]). *Any randomized algorithm running in time T and space S with error ϵ can be simulated by one running in time $O(T)$ and space $O(S \log T)$, which uses only $O(S \log T)$ random bits and has error $\epsilon + 2^{-S}$. If two-way access to the random bits is allowed, the space requirement is reduced to $O(S)$.*

Note that we do not apply Theorem 7 to deterministically simulate the randomized algorithm. Instead, we use it to reduce the randomness required by a BPTISP$[T, S]$ algorithm to $O(S \log T)$ with no blowup in time. If we subsequently apply Theorem 6, we only incur a blowup of $O(S \log T)$, which is acceptable for polynomial T and small S. Therefore, we can conclude:

Theorem 8.

$$\text{BPTISP}[T, S] \subseteq \forall^{(S \log T)^2} \exists^{S \log T} \text{DTISP}[TS \log^2 T, S]. \tag{7}$$

Proof. Let A be the randomized time T, space S procedure for recognizing $L \in$ BPTISP$[T, S]$. We first take the majority of $O(\log T)$ independent repetitions of A to reduce the error to $1/T^2$, increasing the running time to $O(T \log T)$. We then apply the derandomization provided by Theorem 7 to obtain a procedure A' taking time $O(T \log T)$, using $O(S \log T)$ random bits, and using space $O(S)$ when allowed two-way access to the random bits. Additionally, A' has error at most $1/T^2 + 2^{-S}$. Since we can assume without loss of generality that $S \leq T \leq 2^S$ and $S = \omega(1)$, we have that $1/T^2 + 2^{-S} = o(1/(S \log T))$. We now apply Theorem 6 to derive (7). □

This result gives exactly the efficient inclusion of space bounded randomized classes in the polynomial time hierarchy that we need. Combining it with the hypothesis (5), we derive the inclusion

$$\Sigma_2 \text{TIME}[n] \subseteq \forall^{(s \log t)^2} \exists^{s \log t} \text{DTISP}[ts \log^2 t, s], \tag{8}$$

which gives the desired strong collapse for small enough t and s.

5 Main Result

We now use the techniques discussed in the previous sections to complete our indirect diagonalization argument for the proof of Theorem 3. Using (8) as a starting point, we derive a series of inclusions giving stronger and stronger collapses of the polynomial-time hierarchy, towards the end of contradicting Lemma 1. Specifically, we derive inclusions of the form $\Sigma_2 \text{TIME}[n] \subseteq \Pi_2 \text{TIME}[f(n)]$ for smaller and smaller $f(n)$, eventually hoping to find such an inclusion for $f(n) = o(n)$. Given that the hypothesis allows a simulation of $\Sigma_2 \text{TIME}[n]$ by

BPTISP$[t, s]$, we rely on finding a sequence of inclusions for the latter to give these collapses. The following process derives the first such inclusion giving an improvement over (8):

1. First, we use Theorem 8 to derive a Π_2 simulation of a BPTISP$[t, s]$ computation.
2. Since the quantifiers of the Π_2 simulation derived in the previous step are over a small number of bits, the dominant term in the running time comes from the final space bounded deterministic computation. Therefore, we can apply the speedup of (1) to the final deterministic computation to achieve a simulation taking less time, whose quantifiers look like $\forall\exists\exists\forall$, i.e., a Π_3 computation.
3. We now eliminate an alternation from this Π_3 computation in two steps. We first apply the hypothesis to simulate the computation represented by the last two quantifiers (a Σ_2 computation) by a BPTISP computation.
4. Finally, we complete the collapse by applying Theorem 8 to simulate this BPTISP computation by a Π_2 computation. Merging the two initial universal quantifiers gives us a Π_2TIME$[f(n)]$ simulation of BPTISP$[t, s]$ for small $f(n)$, depending on the choice of t and s.

For small enough t and s, we have derived a stronger collapse of Σ_2 to Π_2 than that given by (8). We can achieve even stronger collapses by viewing the above procedure as a process that takes as input a BPTISP computation and returns a Π_2 simulation which possibly takes less time. If we recursively apply this procedure to the BPTISP computation in step 4 instead of Theorem 8, we can complete the collapse to Π_2 while possibly further speeding up the computation, providing a stronger collapse. Running this recursive argument more and more times yields the sequence of collapses we require.

The following lemma formalizes the above idea. Specifically, we consider the hypothesis (5) for polynomial t and s, namely $t = n^c, s = n^d$, and derive the running time of the resulting Π_2 simulation in terms c, d, and k, the number of times the argument is recursively applied.

Lemma 2. *Suppose that*

$$\Sigma_2\text{TIME}[n] \subseteq \text{BPTISP}[n^c, n^d] \tag{9}$$

for some constants $c \geq 1$ and $d > 0$ where $c + 2d \leq 2$. Then for any functions T and S and positive integer k such that $2d \leq f_k$,

$$\text{BPTISP}[T, S] \subseteq \Pi_2\text{TIME}\left[\left((TS^2)^{f_k} + (n + S^2)^{c+2d}\right)\text{polylog}(T + n)\right],$$

where

$$f_k = \left(\tfrac{c+2d}{2}\right)^k. \tag{10}$$

Lemma 2 gives the sequence of collapses which, for certain values of c and d, leads to a contradiction with Lemma 1. We now derive the resulting time-space lower bound for Σ_2TIME$[n]$.

Theorem 9. *For any constant $c < 2$, there exists a positive constant d such that $\Sigma_2\text{TIME}[n]$ cannot be simulated by randomized random-access machines with error bounded away from $1/2$ running in time n^c and space n^d.*

Proof. For $c < 1$, the Theorem holds for any d by standard techniques.

We prove the case for $c \geq 1$ via indirect diagonalization. Suppose, by way of contradiction, that
$$\Sigma_2\text{TIME}[n] \subseteq \text{BPTISP}[n^c, n^d], \tag{11}$$
for some constant $d > 0$ to be determined later. Then for any time function $\tau(n) \geq n$, the hypothesis and Lemma 2 give us the inclusions
$$\Sigma_2\text{TIME}[\tau] \subseteq \text{BPTISP}[\tau^c, \tau^d]$$
$$\subseteq \Pi_2\text{TIME}\left[\left((\tau^{c+2d})^{f_k} + (\tau^{2d} + n)^{c+2d}\right)\text{polylog}(\tau)\right]$$
when $c + 2d \leq 2$ and $2d \leq f_k$. In the case that $2d \leq f_k$, the dominant power of τ will be $2f_{k+1}$, allowing us to simplify the running time of the Π_2 machine to big-O of
$$(\tau^{2f_{k+1}} + n^{c+2d})\text{polylog}(\tau).$$
Choosing a sufficiently large polynomial τ, we can further simplify this to
$$\tau^{2f_{k+1}}\text{polylog}(\tau).$$
Therefore, we have shown
$$\Sigma_2\text{TIME}[\tau] \subseteq \Pi_2\text{TIME}[\tau^{2f_{k+1}}\text{polylog}(\tau)]. \tag{12}$$
The inclusion (12) gives a contradiction with Lemma 1 for any k with $f_{k+1} < 1/2$. Note that $f_k \to 0$ as $k \to \infty$ if $c + 2d < 2$. Therefore, all that remains is to show that the latter condition is compatible with the other ones, i.e., that we can pick a constant $d > 0$ and an integer $k > 0$ such that
$$c + 2d < 2, \tag{13}$$
$$2d \leq f_k, \text{ and} \tag{14}$$
$$f_{k+1} < 1/2. \tag{15}$$

It remains to show that d and k can be chosen to satisfy these constraints. For any c and d satisfying (13), consider choosing k to be the smallest integer such that (15) is satisfied. For such a choice, we can see that $f_k \geq 1/2$, so $d \leq 1/4$ satisfies (14). Therefore, choosing d such that $d \leq \min(1/4, \frac{2-c}{2})$ and then calculating k as described above yields a d and k satisfying all of the constraints. □

We point out that the dependence of d on c in Theorem 9 differs from the deterministic setting. The proofs of Theorem 1 and Theorem 2 show that as c approaches 1 from above, d approaches 1 from below. In the proof of Theorem 9, however, the strategy described for choosing d yields a value approaching $1/4$

from below as c approaches 1 from above. Although we have not optimized our arguments to obtain the largest value of d possible, a smaller value than in the deterministic case seems inherent to our approach.

The proof of Theorem 9 generalizes to higher levels of the linear-time hierarchy, as stated in our Main Theorem. In this setting, we can use the hypothesis that $\Sigma_\ell \text{TIME}[n] \subseteq \text{BPTISP}[n^c, n^d]$ along with Theorem 8 to eliminate more than one alternation at the same cost of removing one alternation in the setting of Theorem 9. This lets us eliminate the alternations introduced by many recursive applications of (1), achieving a greater speedup and allowing contradictions for values of c less than ℓ.

Through a tight connection to nondeterministic linear time, the results of Theorem 1 extend to satisfiability and many other NP-complete problems. This connection also exists between $\Sigma_\ell \text{TIME}[n]$ and Σ_ℓ-complete problems such as QSAT_ℓ, allowing us to extend our time-space lower bounds to randomized computations of such problems.

Corollary 1. *For any integer $\ell \geq 2$ and constant $c < \ell$, there exists a positive constant d such that QSAT_ℓ cannot be solved by randomized random-access machines with error bounded away from 1/2 running in time n^c and space n^d.*

Paying close attention to the space used by the simulations in our proofs, we actually obtain time-space lower bounds for $\Sigma_2 \text{TIME}[n, n^a]$ for certain values of $a < 1$.

Theorem 10. *For any integer $\ell \geq 2$ and any positive constants c and a with $c < 1 + (\ell - 1)a$, there exists a constant $d > 0$ such that $\Sigma_\ell \text{TISP}[n, n^a] \nsubseteq \text{BPTISP}[n^c, n^d]$.*

6 Other Results

In this section, we show how to extend the golden ratio result of Fortnow and Van Melkebeek for deterministic machines (Theorem 1) to randomized machines with one-sided error, yielding Theorem 5. As we will argue, Theorem 5 follows from the next extension of Theorem 1 to a a slightly stronger class of machines.

Theorem 11. *For any constant $c < \phi$ there exist positive constants d and b such that cononodeterministic linear time cannot be simulated by nondeterministic random-access machines which run in time n^c, space n^d, and nondeterministically guess only n^b bits.*

Nisan's space-bounded derandomization [10] given in Theorem 7 allows us to reduce the number of random bits used by a randomized machine with one-sided error running in time T and space S to $O(S \log T)$ without significantly increasing the time or space used. Since a randomized machine with one-sided error is also a special type of nondeterministic machine, we can view such a derandomized machine as a space bounded nondeterministic machine which guesses only

$O(S \log T)$ bits. Thus, the time-space lower bounds of Theorem 5 for conondeterministic linear time on randomized machines with one-sided error follow as a corollary to Theorem 11.

In order to generalize Theorem 1 to nondeterministic machines that guess a bounded number of bits, we derive an analog to Lemma 2 which gives an unlikely inclusion of $\exists^B \mathrm{DTISP}[T,S]$ into NTIME. Recall that in Section 5, we observed that if the number of bits guessed in a $\Pi_2 \mathrm{TISP}[T,S]$ computation is much less than the running time T, then the entire computation can be sped up by applying (1) to the final deterministic stage of the computation. The same observation allows us to speed up a $\exists^B \mathrm{DTISP}[T,S]$ computation when T dominates B. We accomplish the latter through an extension of Lemma 3.1 from [4], which gives a speedup of $\mathrm{DTISP}[T,S]$ on nondeterministic machines under the hypothesis that $\mathrm{coNTIME}[n] \subseteq \mathrm{NTIME}[n^c]$.

Lemma 3. *Suppose that*

$$\mathrm{coNTIME}[n] \subseteq \mathrm{NTIME}[n^c]$$

for some constant $c \geq 1$. Then for any functions T, S, and B and any integer $k \geq 0$

$$\exists^B \mathrm{DTISP}[T,S] \subseteq \mathrm{NTIME}[(T \cdot S^k)^{f_k} + (n+B+S)^{c^k}],$$

where f_k is given by,

$$f_0 = 1$$
$$f_{k+1} = c \cdot f_k/(1+f_k). \qquad (16)$$

We now describe how to use Lemma 3 to prove Theorem 11. First assume that $\mathrm{coNTIME}[n] \subseteq \exists^{n^b} \mathrm{DTISP}[n^c, n^d]$ for b and d to be determined later. Padding this assumption and applying Lemma 3 yields the inclusion

$$\mathrm{coNTIME}[\tau] \subseteq \mathrm{NTIME}[\tau^{(c+kd)f_k} + (n+\tau^b+\tau^d)^{c^k}] \qquad (17)$$

for any time function $\tau(n) \geq n$. Letting τ be a large enough polynomial and choosing small enough values for b and d, (17) forms a contradiction to Lemma 1 as long as there is a k such that $cf_k < 1$. Since the sequence $(f_k)_k$ defined by (16) decreases monotonically to $c-1$ for $c < 2$, this is the case if and only if $c(c-1) < 1$. Since $c(c-1) = 1$ defines the golden ratio ϕ, this establishes Theorem 11.

We point out that a similar strengthening as Theorem 5 holds for the analog of Theorem 2.

7 Further Research

The techniques discussed in this work allow us to establish time-space lower bounds for two-sided error randomized simulations of the polynomial-time hierarchy at the second level and higher. They do not seem to extend to the first level in a straightforward way. This is mainly due to the fact that the assumption NTIME[n] $\subseteq$ BPTISP[t, s] doesn't seem to allow the collapsing of alternations in an efficient manner. Thus, establishing time-space lower bounds for satisfiability on randomized machines with two-sided error remains open.

Improving the quantitative results of this paper is another direction needing further work. Very recently, Williams [13] used a bootstrapping argument to improve the lower bounds for NTIME[n] on deterministic machines. He was able to boost the bound on the exponent c in Theorem 1 from the golden ratio $\phi \approx 1.618$ to 1.732, and we have been able to boost it further to 1.759. Using the same technique, we can improve our lower bound for coNTIME[n] on randomized machines with one-sided error, boosting the bound on the exponent c in Theorem 5 from the golden ratio to 1.759. However, we have been unable to apply the technique to improve our main result, the lower bounds for Σ_ℓTIME[n] on randomized machines with two-sided error, as given in Theorem 3.

Acknowledgements

We would like to thank Bess Berg and the anonymous referees for helpful comments.

References

1. E. Allender, M. Koucky, D. Ronneburger, S. Roy, and V. Vinay. Time-space tradeoffs in the counting hierarchy. In *CCC*, pages 295-302. IEEE, 2001.
2. P. Beame, M. Saks, X. Sun, and E. Vee. Time-space tradeoff lower bounds for randomized computation of decision problems. *Journal of the ACM*, 50(2):154-195, 2003.
3. J. Balcázar, J. Díaz, J. Gabarró. *Structural Complexity I*, volume 11 of *EATCS Monographs on Theoretical Computer Science*. Springer-Verlag, 1995.
4. L. Fortnow and D. van Melkebeek. Time-space tradeoffs for nondeterministic computation. In *CCC*, pages 2-13. IEEE, 2000.
5. L. Fortnow. Time-space tradeoffs for satisfiability. *Journal of Computer and System Sciences*, 60:337-353, 2000.
6. R. Kannan. Towards separating nondeterminism from determinism. *Mathematical Systems Theory*, 17:29-45, 1984.
7. C. Lautemann. BPP and the polynomial hierarchy. *Information Processing Letters*, 17(4):215-217, 1983.
8. R. Lipton and A. Viglas. On the complexity of SAT. In *FOCS*, pages 459-464. IEEE, 1999.
9. D. van Melkebeek. Time-Space Lower Bounds for NP-Complete Problems. In G. Paun, G. Rozenberg, and A. Salomaa, editors, *Current Trends in Theoretical Computer Science*, pages 265-291. World Scientific, 2004.
10. N. Nisan. Pseudorandom generators for space-bounded computation. *Combinatorica*, 12(4):449-461, 1992.
11. N. Nisan. RL $\subseteq$ SC. *Computational Complexity*, 4:1-11, 1994.
12. C. Papadimitriou. *Computational Complexity*. Addison-Wesley, 1994.
13. R. Williams. Better Time-Space Lower Bounds for SAT and Related Problems. To appear in *CCC*. IEEE, 2005.

Lower Bounds for Circuits with Few Modular and Symmetric Gates

Arkadev Chattopadhyay[1] and Kristoffer Arnsfelt Hansen[2]

[1] School of Computer Science, McGill University
achatt3@cs.mcgill.ca
[2] Department of Computer Science, University of Aarhus, Denmark
arnsfelt@daimi.au.dk

Abstract. We consider constant depth circuits augmented with few modular, or more generally, arbitrary symmetric gates. We prove that circuits augmented with $o(\log^2 n)$ symmetric gates must have size $n^{\Omega(\log n)}$ to compute a certain (complicated) function in **ACC⁰**.

This function is also hard on the average for circuits of size $n^{\epsilon \log n}$ augmented with $o(\log n)$ symmetric gates, and as a consequence we can get a pseudorandom generator for circuits of size m containing $o(\sqrt{\log m})$ symmetric gates.

For a composite integer m having r distinct prime factors, we prove that circuits augmented with s MOD_m gates must have size $n^{\Omega(\frac{1}{s} \log^{\frac{1}{r-1}} n)}$ to compute MAJORITY or MOD_l, if l has a prime factor not dividing m. For proving the latter result we introduce a new notion of representation of boolean function by polynomials, for which we obtain degree lower bounds that are of independent interest.

1 Introduction

Strong size lower bounds have been obtained for several classes of circuits. In particular *constant depth* circuits (**AC⁰** circuits) require exponential size to compute even simple functions such as PARITY [1,2,3,4]. More generally, if we also allow gates computing MOD_q for a prime power q we have exponential lower bounds for computing MAJORITY, and if l has a prime divisor not dividing q also for computing MOD_l [5,6]. If we however allow gates computing MOD_m for an arbitrary integer m, we have no nontrivial lower bounds. The corresponding circuit class **ACC⁰** is the smallest natural circuit class, for which we have no nontrivial lower bounds. Another very interesting class of constant depth circuits (containing **ACC⁰**) is the class of constant depth threshold circuits (**TC⁰** circuits), i.e constant depth circuits built entirely from MAJORITY gates.

Viewing **TC⁰** circuits as **AC⁰** circuits augmented with MAJORITY gates instead, raises a natural question: Can lower bounds be proved if we limit the amount of MAJORITY gates used. This question was answered affirmatively in a series of papers. Aspnes et al [7] proved that **AC⁰** circuits with a MAJORITY gate on top, i.e **MAJ** ∘ **AC⁰** circuits, and in fact even **AC⁰** circuits with a single MAJORITY gate anywhere in the circuit, require exponential size to compute the MOD_2 function. Beigel et al [8] improved this lower bound to allow for $o(\log n)$ MAJORITY gates, and Beigel [9] showed that the lower bound also holds for **AC⁰** circuits augmented with $n^{o(1)}$ MAJORITY gates. Finally Barrington and Straubing [10] generalized this lower bound, showing that it also holds for circuits computing the MOD_m function for any constant m. In fact, the proofs of these lower bounds show that **AC⁰** circuits of size $2^{n^{o(1)}}$ augmented with $n^{o(1)}$ MAJORITY gates must differ from the MOD_m function on a constant fraction all inputs of length n.

The analogous question for **ACC⁰** is: Can lower bounds be proved if we limit the amount of MOD_m gates used. Note that although these functions *can* be computed by small **TC⁰** circuits, the above lower bounds for **AC⁰** circuits augmented with few MAJORITY gates are not strong enough to be applied. Indeed, these lower bounds are proved for circuit *computing* the MOD_m functions.

Another way of viewing **TC⁰**, is as **AC⁰** augmented with gates computing arbitrary symmetric functions, since every such function is computable by a depth 2 threshold circuit and MAJORITY is obviously symmetric. Thus, regarding **TC⁰**, we could also ask, if lower bounds can be proved if we limit the amount of symmetric functions used. Also note that since the MOD_m functions are symmetric, an affirmative answer would answer both questions.

Our most powerful lower bound is a superpolynomial lower bound for circuits containing $o(\log^2 n)$ arbitrary symmetric gates. Combining a technique of Beigel [9] for reducing the number of symmetric gates with a lower bound for **MAJ** ∘ **SYM** ∘ **AC⁰** circuits by Hansen and Miltersen [11], one can immediately obtain superpolynomial lower bounds for **AC⁰** circuits augmented with up to $o(\log \log n)$ MOD_m gates. Our significantly stronger result, is obtained by instead generalizing the results of Håstad and Goldmann [12], Razborov and Wigderson [13] and Hansen and Miltersen [11]. We now give the definition of the functions we prove the lower bounds for. The so-called Generalized Inner Product [14] is defined as $GIP_{n,k} = (MOD_2)_n \circ AND_k$, i.e. the MOD_2 function of n conjunctions each consisting of k variables. Let further $f_{n,k} = (MOD_2)_n \circ AND_k \circ (MOD_2)_n$, i.e. $GIP_{n,k}$ with each input replaced by the MOD_2 function of n variables.

Theorem 1. *Any* **AC⁰** *circuit augmented with* $o(\log^2 n)$ *arbitrary symmetric gates computing* $f_{n,\log n}$ *must have size* $n^{\Omega(\log n)}$.

It follows as a corollary of the following theorem.

Theorem 2. *Any* **MAJ** ∘ **ANY**$_s$ ∘ **SYM** ∘ **ANY**$_t$ *circuit computing* $GIP_{n,t+1}$ *must have size* $2^{\Omega(\frac{n}{st2^t})}$. *For constants* $\delta > 0$ *and* $\gamma > 0$ *such that* $\delta + \gamma < 1$, *if* $t = \delta \log n$ *and* $s = n^\gamma$, *any* **MAJ** ∘ **ANY**$_s$ ∘ **SYM** ∘ **AC⁰** *circuit computing* $f_{n,t+1}$ *must have size* $n^{\Omega(\log n)}$.

For circuits augmented with MOD_m gates we are able to give lower bounds for computing much simpler functions than that used for proving Theorem 1. In particular are all these functions symmetric, and the results are thus incomparable.

Theorem 3. *Let m be a positive integer with $r \geq 2$ distinct prime factors. Any $\mathbf{AC^0}$ circuit augmented with s MOD_m gates require size $n^{\Omega(\frac{1}{s}\log^{\frac{1}{r-1}} n)}$ to compute MAJORITY or MOD_l, if l has a prime factor not dividing m.*

This complements the mentioned results about circuits with few MAJORITY gates: We cannot compute MOD_m with few MAJORITY gates and we cannot compute MAJORITY with few MOD_m gates. Note also that the lower bounds are for the same functions and under the same conditions as the mentioned lower bounds for circuits with MOD_q gates for prime powers q.

Previously, Hansen and Miltersen [11] used results on *weak representation* of boolean function by polynomials over $\mathbf{Z}_m$ as introduced and proved by Green [15] to obtain exponential lower bounds for circuits augmented with just one MOD_m gate. Theorem 3 is proved, by greatly generalizing this technique. We introduce a new notion of representation of boolean functions be polynomials over $\mathbf{Z}_m$ for which we prove lower bounds and finally apply these to obtain circuit lower bounds.

Beigel and Maciel [16] showed how to convert a $\mathbf{MAJ} \circ \mathbf{OR} \circ \mathbf{MOD}_m$ circuit into a $\mathbf{MAJ} \circ \mathbf{OR}_{O(1)} \circ \mathbf{MOD}_m$ circuit with only a polynomial increase in size, which they could then apply to results by Krause and Pudlák [17] to obtain lower bounds for $\mathbf{MAJ} \circ \mathbf{OR} \circ \mathbf{MOD}_m$ circuits. Applying their technique to our results we immediately obtain the following as a corollary to Theorem 2.

Theorem 4. *Any $\mathbf{MAJ} \circ \mathbf{OR} \circ \mathbf{MOD}_m \circ \mathbf{ANY}_t$ circuit computing $GIP_{n,t+1}$ must have size $2^{\Omega(\frac{n}{t2^t})}$. Any $\mathbf{MAJ} \circ \mathbf{OR} \circ \mathbf{MOD}_m \circ \mathbf{AC^0}$ circuit computing $f_{n,\log n}$ must have size $n^{\Omega(\log n)}$.*

Viola [18] showed that by combining and modifying the proofs of [13] and [11] one can obtain a function that is hard on the average for $\mathbf{SYM} \circ \mathbf{AC^0}$ circuits of size $n^{\epsilon \log n}$. Viola used his result for constructing a pseudorandom generator for these circuits, using the construction of Nisan and Wigderson [19], and generalized it to circuits with a *constant* number of symmetric gates using the technique of Beigel [9]. Adding our constructions, one can prove that the same function is hard on the average for $\mathbf{ANY}_{n^\gamma} \circ \mathbf{SYM} \circ \mathbf{AC^0}$ circuits of size $n^{\epsilon \log n}$ for $\gamma < 1$, and thus also for $\mathbf{AC^0}$ circuits of size $n^{\epsilon \log n}$ with $\gamma \log n$ symmetric gates. This gives a pseudorandom generator for circuits of size m with $o(\sqrt{\log m})$ symmetric gates. Subsequently to this result, Viola showed how to improve this to $o(\log m)$ symmetric gates [18].

1.1 Organisation of Paper

In Sect. 2 we introduce the notation and previous results we will use in our proofs. In Sect. 3 we prove Theorem 1 and 2. In Sect. 4 we prove Theorem 3 as well as our degree lower bounds used for this.

2 Preliminaries

2.1 Constant Depth Circuits

We consider circuits built from families of unbounded fanin gates. Inputs are allowed to be boolean variables and their negations as well as the constants 0 and 1. Let $x_1, \ldots, x_n$ be boolean inputs. For a positive integer m, the MOD_m function is 1 if and only if $\sum_{i=1}^n x_i \not\equiv 0 \pmod{m}$. Generally, for $A \subset \mathbf{Z}_m$ the MOD_m^A function is 1 is and only $\sum_{i=1}^n x_i \in A \pmod{m}$. The MAJORITY function is 1 if and only if $\sum_{i=1}^n x_i \geq \frac{n}{2}$. A function is symmetric if and only if it only depends on the sum of its inputs.

Let **AND** and **OR** denote the families of unbounded fanin AND and OR gates. Let $\mathbf{MOD}_m$, **MAJ**, **SYM** and **ANY** denote the families of MOD_m, MAJORITY, all symmetric gates and *any* kind of gate, respectively. If G is a family of boolean gates and $\mathcal{C}$ is a family of circuits we let $G \circ \mathcal{C}$ denote the class of circuits consisting of a gate from G taking circuits from $\mathcal{C}$ as inputs. If we need to specify a specific bound on the fanin of some of the gates, this will be specified by a subscript.

$\mathbf{AC^0}$ is the class of functions computed by constant depth circuits built from AND and OR gates. $\mathbf{ACC^0}$ is the the analogous class of functions computed when we also allow unbounded fanin MOD_m gates for constants m, and similarly is $\mathbf{TC^0}$ the class of functions computed when we instead allow unbounded fanin MAJORITY gates.

2.2 Multiparty Communication Complexity

We consider the "Number on the Forehead" model of multiparty communication as introduced by Chandra, Furst and Lipton [20]. Here k players wish to evaluate a boolean function $f(x_1, \ldots, x_k)$ where $x_i \in \{0,1\}^n$ and player i knows all input except x_i. They exchange messages according to a fixed protocol by broadcasting, and we are interested in the number of bits that must be exchanged in order to evaluate f. We let $C_\epsilon(f)$ denote the minimal number of bits that must be exchanged in the worst case, for a deterministic protocol that computes f correctly with probability at least $1 - \epsilon$ when x is chosen uniformly at random.

Babai, Nisan and Szegedy [14] studied the k-party communication complexity of the Generalized Inner Product, and proved $C_{\frac{1}{2}-\epsilon}(\text{GIP}_{n,k}) = \Omega(\frac{n}{4^k} + \log \epsilon)$. We state an improvement due to Chung and Tetali [21].

Theorem 5. $C_{\frac{1}{2}-\epsilon}(GIP_{n,k}) = \Omega(\frac{n}{2^k} + \log \epsilon)$.

Håstad and Goldmann [12] observed that depth 2 threshold circuit can be evaluated efficiently by a multiparty protocol if the fanin of the bottom gates is restricted. We observe that essentially the same protocol works for depth 3 circuits if we restrict the fanin of both the top and bottom gates.

Proposition 6. *Let f be a boolean function computed by a $\mathbf{ANY}_s \circ \mathbf{SYM} \circ \mathbf{ANY}_t$ circuit of size S. Then there is a $t+1$ player protocol computing f with $1 + st \log S$ bits exchanged.*

Proof. Since the bottom fanin of the circuit is less than the number of players, every $\mathbf{ANY}_t$ gate can be evaluated by at least one player. In the protocol we fix a partition of these gates to the players, such that every player can evaluate all assigned gates. Now for each of the symmetric gates, the first t players compute the sum of the assigned inputs and send the results separately to player $t+1$. Player $t+1$ can then compute the output of the circuit and communicate the result with 1 extra bit of communication. □

2.3 The Discriminator Lemma and the Switching Lemma

Let C be a circuit taking n inputs and f a boolean function on n variables. We say that C is an ϵ-discriminator for f if $\Pr[C(x) = 1 | f(x) = 1] - \Pr[C(x) = 1 | f(x) = 0] \geq \epsilon$. The so-called Discriminator Lemma by Hajnal et al [22], states that if a circuit with a MAJORITY gate at the output computes a boolean function f, then one of the inputs to the output gate is an ϵ-discriminator for f.

Lemma 7. *Let f be a boolean function computed by a circuit C with a MAJORITY gate as the output gate, and let $C_1, \ldots, C_s$ be the subcircuits of C whose output gates are the inputs to the output of C. Then for some i, C_i is an $\frac{1}{s}$-discriminator for f.*

A *restriction* on a set V of boolean variables is a map $\rho : V \to \{0, 1, \star\}$. It acts on a boolean function $f : V \to \{0, 1\}$, creating a new boolean function f_ρ on the set of variables for which $\rho(x) = \star$, obtained by substituting $\rho(x)$ for $x \in V$ whenever $\rho(x) \neq \star$. The variables x for which $\rho(x) = \star$ are called *free*; the other variables *set*. Let R_n^l denote the set of all restriction ρ leaving l of n variables free.

A decision tree is a binary tree, where the internal nodes are labeled by variables and leafs are labeled by either 0 or 1. On a given input x, its value is the value of the leaf reached by starting at the root, and at any internal node labeled by x_i proceeding to the left child if $x_i = 0$ and to the right child otherwise. We will use the following version of Håstads Switching Lemma due to Beame [23].

Lemma 8. *Let f be a DNF formula in n variables with terms of length at most r. Let $l = pn$ and pick ρ uniformly at random from R_n^l. Then the probability that f_ρ does not have a decision tree of depth at most d is less than $(7pr)^d$.*

The advantage of using Beame's switching lemma is that it directly gives us a decision tree. If we convert a decision tree into a DNF, we have that all terms are *mutually contradictory*, i.e. we can view it as a sum of terms, instead as an OR of AND's. This will allow us to absorb the sum into a symmetric gate.

2.4 Representation by Polynomials

For composite m there are several different ways of defining representation by polynomials. To obtain our lower bounds we will define new a notion which we will call *weak generalized* representation. Let f be a boolean function in n variables, and let P be a polynomial in n variables over $\mathbf{Z}_m$.

- P is a *strong* representation of f if $f(x) = P(x)$ for all $x \in \{0,1\}^n$.
- P is a *one-sided* representation of f if $f(x) = 0 \Leftrightarrow P(x) \equiv 0 \pmod{m}$ for all $x \in \{0,1\}^n$.
- P is a *weak* representation of f if $P \not\equiv 0$ and $P(x) \not\equiv 0 \pmod{m} \Rightarrow f(x) = 1$ for all $x \in \{0,1\}^n$.
- P is a *generalized*[1] representation of f if there is a set $S \subset \mathbf{Z}_m$ such that $f(x) = 1 \Leftrightarrow P(x) \in S$.
- P is a *weak generalized* representation of f if there is a set $S \subset \mathbf{Z}_m$ and an $\bar{x} \in \{0,1\}^n$ such that $P(\bar{x}) \in S$ and that for all $x \in \{0,1\}^n$ we have $P(x) \in S \Rightarrow f(x) = 1$.

The minimal degree of a polynomial satisfying the above properties will be called the strong, one-sided, weak, generalized and weak generalized MOD$_m$ degree, respectively. Note that a strong representation is also a one-sided representation. A one-sided representation is also a weak representation as well as a generalized representation, and these are both weak generalized representations. For a weak generalized representation we can assume that $|S| = 1$, in fact, if P is a weak generalized representation there exist $a \in \mathbf{Z}_m$ such that $P - a$ is a weak generalized representation with respect to $\{0\}$ of the same boolean function.

For convenience we will also consider a representation of a boolean function by more than one polynomial. Let f be as before and let $P_1, \ldots, P_s$ be polynomials in n variables over $\mathbf{Z}_m$. We say $P_1, \ldots, P_s$ is a *simultaneous weak representation* of f if there is an $\bar{x} \in \{0,1\}^n$ such that $P_i(\bar{x}) \not\equiv 0 \pmod{m}$ for all i, and if it holds that whenever $P_i(x) \not\equiv 0 \pmod{m}$ for all i, we have that $f(x) = 1$. The s-simultaneous weak MOD$_m$ degree of f is the minimum over all choices of polynomials $P_1, \ldots, P_s$ over $\mathbf{Z}_m$ representing f of the maximal degree of $P_1, \ldots, P_s$.

We will several times use the following well known lemma. (cf. [24]).

Lemma 9. *Let $q = p^k$ for a prime p, let P be a polynomial of degree d in n variables over $\mathbf{Z}_q$ and let $S \subseteq \mathbf{Z}_q$. Then there exists another polynomial P' of degree at most $(q-1)d$ in n variables over $\mathbf{Z}_p$ such that $P(x) \in S \Rightarrow P'(x) = 1$ and $P(x) \notin S \Rightarrow P'(x) = 0$ for all $x \in \{0,1\}$.*

Thus if $q = p^k$ for a prime p, the strong MOD$_p$ degree of f is at most $(q-1)$ times the generalized MOD$_q$ degree of f, and similarly the weak MOD$_p$ degree of f is at most $(q-1)$ times the weak generalized MOD$_q$ degree of f.

The following lemma shows, that s-simultaneous weak degree and weak generalized degree are essentially the same, when s is a constant.

Lemma 10. *Let m be a positive integer and let $m = q_1 \cdots q_t$ be the factorization in prime powers $q_i = p_i^{k_i}$, let $m' = p_1 \cdots p_t$ and let f be a boolean function. The weak generalized MOD$_{m'}$ degree of f is at most $s(q-1)$ times the s-simultaneous weak MOD$_m$ degree of f, where q is the largest prime power factor of m.*

---

[1] This notion was actually called *weak* representation in [24], but we prefer to reserve this name for the representation introduced by Green [15], which is analogous to the weak degree of a voting polynomial defined by Aspnes et al [7].

On the other hand, the $(m-1)$-simultaneous weak MOD_m degree of f is at most as large as the weak generalized MOD_m degree of f.

3 Circuits with Few Symmetric Gates

Proof (of Theorem 2). Let C be a $\mathbf{MAJ} \circ \mathbf{ANY}_s \circ \mathbf{SYM} \circ \mathbf{ANY}_t$ circuit of size S which computes $f = \mathrm{GIP}_{n,t+1}$. From Lemma 7 we have a $\mathbf{ANY}_s \circ \mathbf{SYM} \circ \mathbf{ANY}_t$ subcircuit C_i that is an $\frac{1}{S}$-discriminator for f, i.e. $\Pr[C_i(x) = 1 \mid f(x) = 1] - \Pr[C_i(x) = 1 \mid f(x) = 0] \geq \frac{1}{S}$. From the Binomial Theorem we have $\Pr[f(x) = 0] - \Pr[f(x) = 1] = (1-2^{-t})^n$ and thus $\Pr[f(x) = 0] = \frac{1}{2} + \frac{(1-2^{-t})^n}{2}$. For $a,b \in \{0,1\}$ let $P_{ab} = \Pr[C_i(x) = a \mid f(x) = b]$. We thus get $\Pr[C_i(x) = f(x)] = P_{11}\Pr[f(x)=1] + P_{00}\Pr[f(x)=0] = \frac{1}{2} + \frac{1}{2}(P_{11}-P_{10}) + \frac{(1-2^{-t})^n}{2}(1-P_{11}-P_{10}) \geq \frac{1}{2} + \frac{1}{2S} - \frac{(1-2^{-t})^n}{2} \geq \frac{1}{2} + \frac{1}{4S}$, where we assume without loss of generality that $S \leq \frac{1}{2}e^{\frac{n}{2^t}}$. Combining this with Theorem 5 and Proposition 6 we have $1 + st \log S = \Omega(\frac{n}{2^t} + \log \frac{1}{4S})$ and we can conclude that $S = 2^{\Omega(\frac{n}{st2^t})}$.

For the second part, let C be a depth $d+3$ $\mathbf{MAJ} \circ \mathbf{ANY}_s \circ \mathbf{SYM} \circ \mathbf{AC}^0$ circuit of size $n^{\epsilon \log n}$ computing $f_{n,t+1}$. Let $m = n^2(t+1)$. We choose a random restriction $\rho \in R_m^{m^{\frac{2}{3}}}$, and argue that for sufficiently small ϵ, with positive probability, $f_{n,t+1}$ contains $\mathrm{GIP}_{n,t+1}$ as a subfunction and the function computed by C_ρ is also computed by a $\mathbf{MAJ} \circ \mathbf{ANY}_s \circ \mathbf{SYM} \circ \mathbf{AND}_t$ circuit of size $n^{\epsilon \log n} 2^t = n^{\delta + \epsilon \log n}$. The statement then follows from the first part.

The probability that ρ fails the first requirement is at most $n(t+1)$ times the probability that a random subset of size $m^{\frac{2}{3}}$ does not intersect a fixed subset of size n (corresponding to the inputs of the MOD_2 functions substituted in $\mathrm{GIP}_{n,t+1}$). This happens with probability at most $2^{-n^{\Omega(1)}}$.

For the other part, view ρ as a composition of several restrictions $\rho_1, \ldots, \rho_d$ where $\rho_i \in R_{m_{i-1}}^{m_i}$ and $m_i = m\left(m^{\frac{1}{3d}}\right)^{-i}$. Assume that after having applied the first $i-1$ restrictions, that all functions computed at level $i-1$ of C are computed by decision trees of depth at most t, and hence DNFs with terms of size at most t. Assume without loss of generality that the gates at level i are OR gates, and thus also computable by decision trees of depth at most t. By Lemma 8, the probability that the function computed by such an OR gate can not be computed by a decision tree of depth t after applying ρ_i is at most $\left(7 \frac{m_i}{m_{i-1}} t\right)^t = \left(7m^{-\frac{1}{3d}} \delta \log n\right)^{\delta \log n} = n^{-\Omega(\log n)}$. Thus for ϵ sufficiently small, ρ will convert all gates at level d into decision trees of depth at most t. By rewriting these into DNFs with *mutually contradictory* terms, allows us to simply directly feed these terms into the symmetric gate above, resulting in a $\mathbf{MAJ} \circ \mathbf{ANY}_s \circ \mathbf{SYM} \circ \mathbf{AND}_t$ circuit computing the same function. □

Proof (of Theorem 1). Let C be a $\mathbf{AC}^0$ circuit of size S augmented with s symmetric gates. Let $g_1, \ldots, g_s$ denote the symmetric gates, such that there is no path from the output of g_j to an input of g_i if $i < j$. For $\alpha \in \{0,1\}^s$, let C_i^α

be the $\mathbf{SYM} \circ \mathbf{AC^0}$ subcircuit of C with g_i as output, where each g_j for $j < i$ is replaced by the constant α_j. Similarly, let C^α be the $\mathbf{AC^0}$ circuit obtained from C where every g_i is replaced by α_i. Note that since the subcircuit of C with output g_i contain at most $i-1$ other symmetric gates than g_i, we have at most $\sum_{i=1}^s 2^{i-1} = 2^s - 1$ different $\mathbf{SYM} \circ \mathbf{AC^0}$ subcircuits of C. We can now compute the same function as C by a $\mathbf{OR}_{2^s} \circ \mathbf{AND}_{s+1} \circ \mathbf{SYM} \circ \mathbf{AC^0}$ circuit of size $O(2^s S)$ constructed as follows: The output OR gate is fed by ANDs corresponding to all $\alpha \in \{0,1\}^s$. The AND gates takes C^α as input, as well as C_i^α if $\alpha_i = 1$ and $\neg C_i^\alpha$ (which is also a $\mathbf{SYM} \circ \mathbf{AC^0}$ circuit) otherwise. If $s = o(\log^2 n)$ Theorem 2 gives that $S = n^{\Omega(\log n)}$. □

4 Circuits with Few Modular Gates

4.1 Degree Lower Bounds

Barrington and Tardos obtained the following lower bound on the generalized degree of the OR function.

Theorem 11 ([24]). *Let m be a positive integer with $r \geq 2$ distinct prime factors, and let q be the smallest maximal prime power of m. The generalized degree of the OR function on n variables is at least $\left(\left(\frac{1}{q-1} - o(1)\right) \log n\right)^{\frac{1}{r-1}}$.*

As a corollary to this we obtain the same lower bounds for the weak generalized degree of the MAJORITY and ¬MAJORITY functions.

Theorem 12. *Let m be a positive integer with $r \geq 2$ distinct prime factors, and let q be the smallest maximal prime power of m. The weak generalized MOD_m degree of the MAJORITY and ¬MAJORITY functions on n variables is at least $\left(\left(\frac{1}{q-1} - o(1)\right) \log n\right)^{\frac{1}{r-1}}$.*

We will use the following lower bound on the degree of weak representation due to Green [15] in a crucial way for proving Lemma 15.

Theorem 13. *Let m and l be positive relative prime integers. The weak MOD_m degree of the MOD_l and $\neg \mathrm{MOD}_l$ functions on n variables is at least $\left|\frac{n}{2(l-1)}\right|$.*

Corollary 14. *Let l be a positive integer and let $q = p^a$ for a prime p not dividing l. The weak generalized MOD_q degree of the MOD_l and $\neg \mathrm{MOD}_l$ functions on n variables is at least $\frac{1}{(q-1)} \lfloor \frac{n}{2(l-1)} \rfloor$.*

For a subset $S \subseteq \{1, \ldots, n\}$, let $\chi(S) \in \{0,1\}^n$ denote its characteristic vector. Conversely for $x \in \{0,1\}^n$, let $\sigma(x) \subseteq \{1, \ldots, n\}$ be the set of indices where $x_i = 1$.

Lemma 15. *Let P be a polynomial of degree d in n variables over $\mathbf{Z}_q$ where $q = p^a$ for a prime p, and let l be a positive integer not divisible by p. If $k \geq 1$*

satisfies $n \geq 2(l-1)\left(k + (q-1)\sum_{i=1}^{d}(d+1-i)\binom{k}{i}\right)$, then there exists pairwise disjoint nonempty sets $S^1, \ldots, S^k \subseteq \{1, \ldots, n\}$ such that for every $y \in \{0,1\}^k$ we have $P(\sum_{i=1}^{k} y_i \chi(S^i)) \equiv P(0) \pmod{q}$ and furthermore we have $|S^i| \not\equiv 0 \pmod{l}$ for all i.

Proof. Assume without loss of generality that $P(0) = 0$. We will find the sets S^i inductively with $|S^i| \leq s_i$, where $s_j = 2(l-1)\left(1 + (q-1)\sum_{i=1}^{d}\binom{j-1}{i-1}(d+1-i)\right)$. First pick a set S of $s_1 = 2(l-1)(d(q-1)+1)$ variables. If we consider the polynomial obtained from P by substituting 0 for all variables not in S, we obtain a polynomial which can not be a weak generalized representation of $\neg \text{MOD}_l$ with respect to the set $\{0\}$, by Corollary 14. Thus there is a subset $S^1 \subseteq S$ such that $P(\chi(S^1)) = 0 = P(0)$ and $\neg \text{MOD}_l(\chi(S^1)) = 0$, that is $|S^1| \not\equiv 0 \pmod{l}$.

In the general case, assume for $j < k$, that we have already found sets $S^1, \ldots, S^j$, where $|S^i| \leq s_i$ and $|S^i| \not\equiv 0 \pmod{l}$ for all $i \leq j$ and we have $P(\sum_{i=1}^{t} y_i \chi(S^i)) = 0 = P(0)$ for all $y \in \{0,1\}^j$. Pick a set S of size s_{j+1} of the remaining variables.

For any $y \in \{0,1\}^j$, let P_y be the polynomial obtained from P be substituting y_i for all variables in S^i for all i, and further substituting 0 for all remaining variables not in S. Let P'_y be the polynomial over $\mathbf{Z}_p$, obtained using Lemma 9, that is a strong representation of the boolean function of which P_y is a generalized representation with respect to $\{0\}$. That is $P'_y(x) \equiv 0 \pmod{p} \Leftrightarrow P_y(x) \not\equiv 0 \pmod{q}$ and $P'_y(x) \equiv 1 \pmod{p} \Leftrightarrow P_y(x) \equiv 0 \pmod{q}$.

Let $R = \prod_{y \in \{0,1\}^j} P'_y$. Observe that R only take the values 0 and 1 modulo p, and that $R(x) \equiv 1 \pmod{p}$ if and only if $P'_y(x) \equiv 1 \pmod{p}$ for all y, that is, if and only if $P_y(x) \equiv 0 \pmod{q}$ for all y. If R is not a weak representation of $\neg \text{MOD}_l$, we could thus find a new set S^{j+1} as before. But since the degree of R is $2^j(q-1)d$, this does not give our desired bound on the number of variables s_{j+1} required for this contradiction. However, exactly as in [24] we can use inclusion-exclusion to construct an equivalent polynomial R', i.e $R'(x) \not\equiv 0$ if and only if $P_y(x) \equiv 0 \pmod{q}$ for all y, of degree $(q-1)\sum_{i=0}^{d-1}\binom{j}{i}(d-i)$. From Theorem 13 and the choice of s_{j+1} we have that R' is not a weak representation of $\neg \text{MOD}_l$. We can thus find $S^{j+1} \subseteq S$ such that $R'(\chi(S^{j+1})) \not\equiv 0 \pmod{p}$ and $\neg \text{MOD}_l(\chi(S^{j+1})) = 0$. It follows that $P_y(\chi(S^{j+1})) = 0$ for all y and $|S^{j+1}| \not\equiv 0 \pmod{l}$. To allow the induction to go through, we need that $n \geq \sum_{j=1}^{k} s_j$, which is exactly the requirement stated. □

Theorem 16. *Let m be a positive integer with $r \geq 2$ distinct prime factors, let q be the smallest maximal prime power factor of m and let p be a prime not dividing m. For all $a \in \mathbf{Z}_p$ the weak generalized MOD_m degree of the $\text{MOD}_p^{\{a\}}$ and $\neg \text{MOD}_p^{\{a\}}$ functions on n variables is at least $\left(\left(\frac{1}{2(p-1)^2(q-1)} - o(1)\right) \log n\right)^{\frac{1}{r-1}}$.*

Proof. The idea of the proof is to succesively use Lemma 15 to convert a given representation into another representation (on fewer variables) that depend on less prime factors, and finally use Corollary 14.

Let $n = n(m,d)$ denote the maximal number of variables, for which there is a weak generalized representation over $\mathbf{Z}_m$ of degree d, for any of the $\mathrm{MOD}_p^{\{a\}}$ and $\neg\mathrm{MOD}_p^{\{a\}}$ functions. We need to prove that $\log n(m,d) \leq (2(p-1)^2(q-1)+o(1))d^{r-1}$. Let $m = q_1 m_1$ where q_1 is a maximal prime power divisor of m different from q.

Assume that P is a polynomial in n variables of degree d over $\mathbf{Z}_m$ which is a weak generalized representation of f with respect to $\{0\}$, where f is either $\mathrm{MOD}_p^{\{a\}}$ or $\neg\mathrm{MOD}_p^{\{a\}}$ for some $a \in \mathbf{Z}_p$. By assumption there exists $\bar{x} \in \{0,1\}^n$ such that $P(\bar{x}) \equiv 0 \pmod{m}$ and $f(\bar{x}) = 1$. If $|\sigma(\bar{x})| < \frac{n}{2}$ let P' be the polynomial obtained from P by setting the variables indexed by $\sigma(\bar{x})$ to 1. Otherwise, if $|\sigma(\bar{x})| \geq \frac{n}{2}$ we can let P' be the polynomial where variables x_i are substituted with $1 - x_i$ if $i \in \sigma(\bar{x})$ and otherwise set to 0, and modify the following proof accordingly. In either case the number n' of variables in P' is at least $\frac{n}{2}$.

For a given integer k, let $k' = (p-1)k$ and assume that n' is at least $2(p-1)\left(k' + (q_1-1)\sum_{i=1}^d (d+1-i)\binom{k'}{i}\right)$. Then using Lemma 15 we can find pairwise disjoint nonempty sets $S'^1, \ldots, S'^{k'} \subseteq \{1, \ldots, n'\}$ such that for every $y \in \{0,1\}^{k'}$ we have $P'(\sum_{i=1}^{k'} y_i \chi(S'^i)) \equiv P'(0) \equiv 0 \pmod{q_1}$ and furthermore we have $|S'^i| \not\equiv 0 \pmod{p}$ for all i. Choosing the most occurring residue $b \in \mathbf{Z}_p \setminus \{0\}$ among $|S'^i|$ modulo p and extending the sets to $\{1, \ldots, n\}$, we have pairwise disjoint nonempty sets $S^1, \ldots, S^k \subseteq \{1, \ldots, n\}$ such that $P(\bar{x} + \sum_{i=1}^k y_i \chi(S^i)) \equiv P(\bar{x}) \equiv 0 \pmod{q_1}$ for every $y \in \{0,1\}^k$, and $|S^i| \equiv b$ for all i.

In case f is $\mathrm{MOD}_p^{\{a\}}$, we have $P(\bar{x} + \sum_{i=1}^k y_i \chi(S^i)) \equiv 0 \pmod{m} \Rightarrow |\sigma(\bar{x})| + \sum_{i=1}^k y_i |S^i| \not\equiv a \pmod{p} \Rightarrow \sum_{i=1}^k y_i \not\equiv b^{-1}(a - |\sigma(\bar{x})|) \pmod{p}$. In case f is $\neg\mathrm{MOD}_p^{\{a\}}$ we have $P(\bar{x} + \sum_{i=1}^k y_i \chi(S^i)) \equiv 0 \pmod{m} \Rightarrow |\sigma(\bar{x})| + \sum_{i=1}^k y_i |S^i| \equiv a \pmod{p} \Rightarrow \sum_{i=1}^k y_i \equiv 0 \pmod{p}$. Since $P(\bar{x} + \sum_{i=1}^k y_i \chi(S^i)) \equiv 0 \pmod{m_1} \Rightarrow P(\bar{x} + \sum_{i=1}^k y_i \chi(S^i)) \equiv 0 \pmod{m}$, we thus have that $P(\bar{x} + \sum_{i=1}^k y_i \chi(S^i))$ is a weak generalized MOD_{m_1} representation of either $\mathrm{MOD}_p^{\{b^{-1}(a-|\sigma(\bar{x})|)\}}$ or $\neg\mathrm{MOD}_p^{\{0\}}$. Thus for $k = n(m_1, d) + 1$ we must have $n' < 2(p-1)\left(k' + (q_1-1)\sum_{i=1}^d (d+1-i)\binom{k'}{i}\right)$.

If $r = 2$ then $m_1 = q$ and from Corollary 14 we have that $k \leq 2(p-1)((q-1)d+1)$. Using $n(m,d) \leq O\left(d2^{k'}\right)$ gives $\log n(m,d) \leq O(\log d) + (p-1)k \leq \left(2(p-1)^2(q-1) + o(1)\right)d$.

If $r > 2$, we have by induction that $\log k \leq (2(p-1)^2(q-1) + o(1))d^{r-2}$. Using $n(m,d) \leq O(k'^d)$ gives $\log n(m,d) \leq O(1) + d(\log(p-1) + \log k) \leq (2(p-1)^2(q-1) + o(1))d^{r-1}$. □

Corollary 17. *Let m be a positive integer with $r \geq 2$ distinct prime factors, let q be the smallest maximal prime power factor of m and let l be a positive integer having a prime factor p not dividing m.*

Then the weak generalized MOD_m degree of the $\mathrm{MOD}_l^{\{a\}}$ and $\neg\mathrm{MOD}_l^{\{a\}}$ functions on n variables is at least $\left(\left(\frac{1}{2(p-1)^2(q-1)} - o(1)\right)\log n\right)^{\frac{1}{r-1}}$.

4.2 Circuit Lower Bounds

Proof (of Theorem 3). Let C be a depth d **AC⁰** circuit of size $n^{\frac{\epsilon}{s}\log^{\frac{1}{r-1}}n}$ containing s MOD_m gates $g_1,\ldots,g_s$ computing a function f. Assume there is no path from the output of g_j to an input of g_i if $i < j$. For $\alpha \in \{0,1\}^s$ let C_i^α be the $\text{MOD}_m \circ \mathbf{AC^0}$ subcircuit of C with g_i as output, where every g_j for $j < i$ is replaced by the constant α_j. Similarly, let C^α be the **AC⁰** circuit obtained from C where every g_i is replaced by α_i. We will choose a random restriction $\rho \in R_n^{\sqrt{n}}$ and show that for every $\delta > 0$ we can choose ϵ sufficiently small such that with high probability we can obtain polynomials p_i^α and q^α, of degree $\frac{\delta}{s}\log^{\frac{1}{r-1}}n$, such that $C_{i,\rho}^\alpha(x) = 1 \Leftrightarrow p_i^\alpha(x) \not\equiv 0 \pmod{m}$ and $C_\rho^\alpha(x) = q^\alpha(x)$ for all x.

If we obtain this we can construct a simultaneous weak representation, using $s+1$ of the polynomials, of either f_ρ or $\neg f_\rho$ as follows: Pick a *maximal* set G of the MOD_m gates that are 1 at the same time for some input x in C_ρ and define α such that $\alpha_i = 1 \Leftrightarrow g_i \in G$. If there exist x such that all the gates in G evaluate to 1 on x and at the same time $C_\rho(x) = 1$, then $\{p_i^\alpha \mid g_i \in G\} \cup \{q^\alpha\}$ is a simultaneous weak representation of f_ρ. Otherwise $\{p_i^\alpha \mid g_i \in G\}$ is a simultaneous weak representation of $\neg f_\rho$.

Note that if f is MOD_l then f_ρ is $\text{MOD}_l^{\{a\}}$ for some a. If f is MAJORITY and the number of 0 and 1 assigned by ρ differ by at most 1 (which happens with probability $\Omega(n^{-\frac{1}{2}})$), we can fix at most 1 extra variable such that f_ρ will compute MAJORITY. In any case we can pick δ sufficiently small and obtain a contradiction to the degree lower bounds in Theorem 12 and Corollary 17, using Lemma 10.

The polynomials are obtained similarly as in the proof of Theorem 2 by applying a series of restrictions $\rho_1,\ldots,\rho_d$, where ρ_i leaves $n\left(n^{\frac{1}{2d}}\right)^{-i}$ variables free, simultaneously on all of the at most $2^{s+1}-1$ different circuits defined above. Here again the crucial step is to feed the terms of a DNF directly to the MOD_m gates, thus obataining a polynomial over $\mathbf{Z}_m$. □

Acknowledgement. We thank Emanuele Viola for sending his paper [18].

The first author is supported by a postgraduate scholarship from the Natural Sciences and Engineering Research Council (NSERC) of Canada and research grants of Prof. Denis Thérien. The second author is supported by BRICS, Basic Research in Computer Science, a Centre of the Danish National Research Foundation.

References

1. Furst, M., Saxe, J.B., Sipser, M.: Parity, circuits, and the polynomial-time hierarchy. Mathematical Systems Theory **17** (1984) 13–27
2. Ajtai, M.: Σ_1^1-formulae on finite structures. Annals of Pure and Applied Logic **24** (1983) 1–48
3. Håstad, J.: Computational limitations of small-depth circuits. MIT Press (1987)

4. Yao, A.C.C.: Separating the polynomial–time hierarchy by oracles. In: 26th Annual Symposium on Foundations of Computer Science, IEEE Computer Society Press (1985) 1–10
5. Razborov, A.A.: Lower bounds for the size of circuits of bounded depth with basis $(\wedge, \oplus)$. Mathematical Notes of the Academy of Science of the USSR **41** (1987) 333–338
6. Smolensky, R.: Algebraic methods in the theory of lower bounds for Boolean circuit complexity. In: 19th Annual ACM Symposium on Theory of Computing. (1987) 77–82
7. Aspnes, J., Beigel, R., Furst, M.L., Rudich, S.: The expressive power of voting polynomials. Combinatorica **14** (1994) 135–148
8. Beigel, R., Reingold, N., Spielman, D.A.: PP is closed under intersection. Journal of Computer and System Sciences **50** (1995) 191–202
9. Beigel, R.: When do extra majority gates help? Polylog(n) majority gates are equivalent to one. Computational Complexity **4** (1994) 314–324
10. Barrington, D.A.M., Straubing, H.: Complex polynomials and circuit lower bounds for modular counting. Computational Complexity **4** (1994) 325–338
11. Hansen, K.A., Miltersen, P.B.: Some meet-in-the-middle circuit lower bounds. In: 29th International Symposium on Mathematical Foundations of Computer Science. Volume 3153 of Lecture Notes in Computer Science., Springer (2004) 334–345
12. Håstad, J., Goldmann, M.: On the power of small-depth threshold circuits. Computational Complexity **1** (1991) 113–129
13. Razborov, A., Wigderson, A.: $n^{\Omega(\log n)}$ lower bounds on the size of depth-3 threshold circuits with AND gates at the bottom. Information Processing Letters **45** (1993) 303–307
14. Babai, L., Nisan, N., Szegedy, S.: Multiparty protocols, pseudorandom generators for logspace, and time-space trade-offs. Journal of Computer and System Sciences **45** (1992) 204–232
15. Green, F.: A complex-number fourier technique for lower bounds on the mod-m degree. Computational Complexity **9** (2000) 16–38
16. Beigel, R., Maciel, A.: Upper and lower bounds for some depth-3 circuit classes. Computational Complexity **6** (1997) 235–255
17. Krause, M., Pudlák, P.: On the computational power of depth-2 circuits with threshold and modulo gates. Theoretical Computer Science **174** (1997) 137–156
18. Viola, E.: Pseudorandom bits for constant-depth circuits with few arbitrary symmetric gates. In: 20th Annual IEEE Conference on Computational Complexity, IEEE Computer Society Press (2005) (to appear).
19. Nisan, N., Wigderson, A.: Hardness vs randomness. Journal of Computer and System Sciences **49** (1994) 149–167
20. Chandra, A.K., Furst, M.L., Lipton, L.: Multi-party protocols. In: 15th Annual ACM Symposium on Theory of Computing, ACM Press (1983) 94–99
21. Chung, F.R.K., Tetali, P.: Communication complexity and quasi randomness. SIAM Journal on Discrete Mathematics **6** (1993) 110–123
22. Hajnal, A., Maass, W., Pudlák, P., Szegedy, M., Turán, G.: Threshold circuits of bounded depth. Journal of Computer and System Sciences **46** (1993) 129–154
23. Beame, P.: A switching lemma primer. Technical Report UW CSE 95 07-01, Department of Computer Science and Engineering, University of Washington (1994) Availible online at www.cs.washington.edu/homes/beame.
24. Tardos, G., Barrington, D.A.M.: A lower bound on the mod 6 degree of the or function. Computational Complexity **7** (1998) 99–108

Discrete Random Variables over Domains*

M.W. Mislove

Tulane University, New Orleans, LA 70118

Abstract. In this paper we explore discrete random variables over domains. We show that these lead to a continuous endofunctor on the categories RB (domains that are retracts of bifinite domains), and FS (domains where the identity map is the directed supremum of deflations finitely separated from the identity). The significance of this result lies in the fact that there is no known category of continuous domains that is closed under the probabilistic power domain, which forms the standard approach to modeling probabilistic choice over domains. The fact that RB and FS are cartesian closed and also are closed under the discrete random variables power domain means we can now model, e.g., the untyped lambda calculus extended with a probabilistic choice operator, implemented via random variables.

1 Introduction

Domain theory, perhaps the most widely used method for assigning denotational meanings to programming languages, has recently seen its influence broaden to other areas of computation and mathematics. It provides a wide range of constructors for modeling data types, nondeterminism, functional programming, and several other constructs needed in semantics. It also admits a number of cartesian closed categories, the fundamental objects needed to model the lambda calculus. Even probabilistic computation admits a model in the theory, although truth to tell, this particular constructor has proven to be very difficult to unravel. Of particular interest is the question

Is there a cartesian closed category of domains that is closed under the probabilistic power domain?

There have been many attempts to resolve this, but the most we know to date is contained in [9], where it is shown that the probabilistic power domain of a finite tree is in RB, that the probabilistic power domain of a finite reversed tree is in FS, and that RB is closed under the probabilistic power domain if the probabilistic power domain of every finite poset is in RB. But, other than finite trees, the only finite posets whose probabilistic power domain is known to be in RB is the class of flat posets, whose probabilistic power domains are bounded complete (the continuous analogs of Scott domains).

---

* This work supported by the US National Science Foundation and the US Office of Naval Research.

We do not contribute to settling this question here, but we do provide an alternative construction—what we call the *power domain of discrete random variables*, which we show defines a continuous endofunctor on the category RB, as well as on FS and on CDOM, the category of coherent domains.

Objects in RB are retracts of *bifinite domains*, those that can be expressed as bilimits of finite posets under embedding–projection pairs. This category is cartesian closed, and it also is closed under the various power domains for nondeterminism [4]. With the addition of a mechanism to model probabilistic choice, RB provides virtually all the tools required to support semantic modeling. Furthermore, playing off results of Varacca [20, 21], we show that the formation of the power domain of discrete random variables over RB yields a monad that enjoys a distributive law with respect to each of the power domain monads, and this in turn implies that each of these power domain monads lifts to a monad on the category RB that are also algebras for the discrete random variable power domain monad. In short, we can now form domain-theoretic models of computation that respect the laws of discrete random variables as well as any of the laws we choose for nondeterminism: angelic, demonic or convex choice.

The outline of the rest of the paper is as follows. In the next section we provide some background about domains and about the constructions we need. We then review briefly a construction by Varacca [21] which inspired our work, and that Varacca calls the *Plotkin indexed valuations*. In the following section, we investigate bag domains—domain-theoretic models for multisets, which are at the heart of our reformulation of Varacca's construction, and we then show how our approach recaptures Varacca's construction. Next we present the main result of the paper, which is to define that what we call the power domain of discrete random variables, and we show it defines a continuous endofunctor on RB, FS and on CDOM. In the final section, we discuss further work along this line, including how to construct Varacca's other examples of indexed valuations. We also discuss the relationship between a random variable approach to modeling probabilistic computation and one based directly on probability distributions.

1.1 Background

We begin with some basic results about partial orders, and about domains in particular. A general reference for this material is [1] or [3].

A subset A of a partially ordered set P is *directed* if A has an upper bound for each of its finite subsets. A mapping between partially ordered sets is *Scott continuous* if it preserves the order and the suprema of those directed sets that have a supremum. A *directed complete partial order (dcpo)* is a partially ordered set in which each directed set has a least upper bound. A *cpo* is a dcpo with a least element $\bot$.

If P is a partial order and $x, y \in P$, then we say x is *way-below* y ($x \ll y$) if whenever $A \subseteq P$ is directed and has a supremum, if $y \sqsubseteq \sqcup A$, then $x \sqsubseteq a$ for some $a \in A$. A poset P is *continuous* if $\Downarrow y = \{x \in P \mid x \ll y\}$ is directed and $y = \sqcup \Downarrow y$ for each $y \in P$. A *domain* is a continuous dcpo.

An *abstract basis* is a pair $(B, \ll)$ where $\ll$ is a transitive relation on B satisfying the *interpolation property*:

$$F \ll x \ \& \ F \subseteq B \text{ finite} \Rightarrow (\exists y \in B) \ F \ll y \ll x.$$

By $F \ll x$ we mean $z \ll x \ \forall z \in F$. If $(B, \ll)$ is an abstract basis, then $I \subseteq B$ is a *round ideal* if I is a $\ll$-lower set, and $x \in I \Rightarrow (\exists y \in I) \ x \ll y$. The *round-ideal completion* of an abstract basis $(B, \ll)$ is the family of round ideals, ordered by inclusion. This forms a domain, where $I \ll J$ iff $(\exists x \ll y \in B) \ I \subseteq \Downarrow x \subseteq \Downarrow y \subseteq J$. In fact, every domain P is isomorphic to the round-ideal completion of an abstract basis, namely P is isomorphic to the round ideal completion of $(P, \ll)$ under the mapping sending a point x to $\Downarrow x$, whose inverse is the mapping that sends a round ideal to its supremum.

One of the fundamental results about dcpos is that the family of Scott continuous maps between two dcpos is another dcpo in the pointwise order. Since it's easy to show that the finite product of a family of continuous posets is another such, and the one-point poset is a terminal object, a central question is which categories of dcpos and Scott continuous maps are cartesian closed. This is true of DCPO, but not of DOM, the category of domains. Still, there are several full subcategories of DOM that are cartesian closed. Among the notable such categories are:

BCD *Bounded complete domains*, in which every non-empty subset has a greatest lower bound.
RB Domains which are retracts of *bifinite domains*, themselves bilimits of families of finite posets under embedding-projection maps; these are pairs of Scott continuous mappings $e: P \to Q$ and $p: Q \to P$ satisfying $p \circ e = 1_P$ and $e \circ p \leq 1_Q$.
FS Domains D satisfying the property that the identity map is the directed supremum of selfmaps $f: D \to D$ each *finitely separated* from the identity: i.e., for each f there is a finite subset $M_f \subseteq D$ with the property that, for each $x \in D$, there is some $m \in M_f$ with $f(x) \leq m \leq x$.

Actually, BCD is a full subcategory of RB, which in turn is a full subcategory of FS, and FS is a maximal ccc of domains. An interesting (some might say frustrating) open question is whether RB and FS are equal. The objects in all of these categories are *coherent*,[1] but the category CDOM of coherent domains and Scott continuous maps is not a ccc.

We also recall some facts about categories. A *monad* on a category A is a *triple* (T, μ, η) where $T: \mathsf{A} \to \mathsf{A}$ is an endofunctor, and $\mu: T^2 \longrightarrow T$ and $\eta: 1_\mathsf{A} \longrightarrow T$ are natural transformations satisfying the laws:

$$\mu \circ T\mu = \mu \circ \mu_T \quad \text{and} \quad \mu \circ \eta_T = T = \mu \circ T\eta$$

If (T, μ, η) is a monad, then a *T-algebra* is a pair (a, h), where $a \in \mathsf{A}$ and $h: Ta \to a$ is an A-morphism satisfying $h \circ \eta_a = 1_a$ and $h \circ Th = h \circ \mu_a$. For

---
[1] A domain is *coherent* if its Lawson topology is compact; cf. [1].

example, each of a power domains, $\mathcal{P}_L(P), \mathcal{P}_U(P)$ and $\mathcal{P}_C(P)$ define monads on DCPO (cf. [4]), whose algebras are ordered semilattices; another example is the *probabilistic power domain* $\mathbb{V}(P)$ (cf. [7]), whose algebras satisfy equations that characterize the probability measures over P.

One goal of domain theory is to provide a setting in which all of the constructors needed to model a given programming language can be combined. If the aim is to model both nondeterminism and probabilistic choice, then one needs to combine the appropriate nondeterminism monad with the probabilistic power domain monad, in order that the laws of each constructor be preserved in the resulting model. This is the function of a *distributive law*, which is a natural transformation $d\colon ST \longrightarrow TS$ between monads S and T on A satisfying several identities—cf. [2]. The significance of distributive laws is a theorem of Beck [2], which says there is a distributive law of S over T if and only if T lifts to a monad on the category of S-algebras. Unfortunately, it was shown by Plotkin and Varacca [20] that there is no distributive law of $\mathbb{V}$ over $\mathcal{P}_X$, or of $\mathcal{P}_X$ over $\mathbb{V}$ for any of the nondeterminism monads $\mathcal{P}_X$. This led to the work we report on next.

2 Indexed Valuations

We now recall some of the work of Varacca [21] that was motivated by problems associated with trying to model *both* nondeterminism and probabilistic choice. Once it was shown that there is no distributive law between $\mathbb{V}$ and any of the nondeterminism monads, Varacca realized that weakening one of the laws of probabilistic choice could result in a monad that enjoys a distributive law with respect to a monad for nondeterminism. For $0 < p < 1$ and A a domain element, the law in question is

$$pA + (1-p)A = A \qquad (1)$$

which he weakened in three ways:

$$pA + (1-p)A \sqsupseteq A \qquad (2)$$
$$pA + (1-p)A \sqsubseteq A \qquad (3)$$
$$pA + (1-p)A \text{ and } A \text{ unrelated by order.} \qquad (4)$$

He called the monad he constructed satisfying (2) the *Hoare indexed valuations*, the one satisfying (3) the *Smyth indexed valuations* and the one satisfying (4), a non-relation, the *Plotkin indexed valuations*. We exploit this last construction—the so-called *Plotkin indexed valuations* over a domain—in defining our power domain of discrete random variables.

2.1 Plotkin Indexed Valuations

An *indexed valuation* over the poset P is a tuple $(r_i, p_i)_{i \in I}$ where I is an index set,[2] each $r_i \geq 0$ is a non-negative real number and $p_i \in P$ for each $i \in I$.

---

[2] For our discussion, we can assume I is always finite.

Two indexed valuations satisfy $(r_i,p_i)_I \simeq_1 (s_j,q_j)_J$ if $|I| = |J|$ and there is a permutation $\phi \in S(|I|)$[3] with $r_{\phi(i)} = s_i$ and $p_{\phi(i)} = q_i$ for each i. If we let $I' = \{i \in I \mid r_i \neq 0\}$ and similarly for J, then $(r_i,p_i)_I \simeq_2 (s_j,q_j)_J$ if $(r_i,p_i)_{I'} \simeq_1 (s_j,q_j)_{J'}$, and we let $\simeq$ denote the equivalence relation on indexed valuations generated by $\simeq_2$. For an indexed valuation $(r_i,p_i)_I$, we let $\langle r_i,p_i\rangle_I$ denote the equivalence class modulo $\simeq$.

Next, let $\overline{\mathbb{R}_{\geq 0}}$ denote the extended non-negative real numbers, with the usual order. Then for a domain P, Varacca defines a relation $\ll_P$ on the family $\{\langle r_i,p_i\rangle_I \mid r_i \in \overline{\mathbb{R}_{\geq 0}}\ \&\ p_i \in P\}$ of indexed valuations over P by

$$\langle r_i,p_i\rangle_I \ll_P \langle s_j,q_j\rangle_J \quad \text{iff} \quad (|I'|=|J'|)\,(\exists \phi \in S(|I'|)) \tag{5}$$
$$r_i < s_{\phi(i)}\ \&\ p_i \ll_P q_{\phi(i)}\ (\forall i \in I').[4]$$

Note that we can "add" indexed valuations $\langle r_i,p_i\rangle_I$ and $\langle s_j,q_j\rangle_J$ by simply concatenating the tuples and taking the equivalence class of the resulting $I \cup J$-tuple. This forms a continuous operation on indexed valuations that is commutative, by construction. We also can let $\overline{\mathbb{R}_+}$ act on $\langle r_i,p_i\rangle_I$ by $r \cdot \langle r_i,p_i\rangle_I = \langle r \cdot r_i,p_i\rangle_I$. Varacca's main result for the family of Plotkin indexed valuations is:

Theorem 1 (Varacca [20]).

- *If P is a continuous poset, then the family of Plotkin indexed valuations ordered by $\ll_P$ as defined in (5) is an abstract basis. The family $IV_P(P)$, the round ideal completion of the Plotkin indexed valuations, satisfies the following family of equations:*

  (1) $A \oplus B = B \oplus A$ \qquad *(2)* $A \oplus (B \oplus C) = (A \oplus B) \oplus C$
  (3) $A \oplus \underline{0} = A$ \qquad *(4)* $0A = \underline{0}$
  (5) $1A = A$ \qquad *(6)* $p(A \oplus B) = pA \oplus pB$
  (7) $p(qA) = (pq)A$ \qquad *where $p,q \in \mathbb{R}_+$ and $A,B,C \in IV_P(P)$.*

- *The Plotkin indexed valuation defines the object level of a functor which is monadic over* **DOM**, *and each of the power domain monads satisfies a distributive law with respect to the Plotkin indexed valuations monad.*

A corollary of this result is that the composition $\mathcal{P}_P \circ IV_P$ defines a monad on **CDOM**, the category of coherent domains, whose algebras satisfy the laws listed in Theorem 1 and the laws of the Plotkin power domain:

(i) $X + Y = Y + X$ \quad (ii) $X + X = X$ \quad (iii) $X + (Y + Z) = (X + Y) + Z$

In other words, $\mathcal{P}_P(IV_P(P))$ is the initial domain semilattice algebra over P that also satisfies the laws listed in Theorem 1.

---
[3] $S(n)$ denotes the permutation group on n.
[4] Note that $r < s$ iff $r \ll s$ for $r,s \in \mathbb{R}$.

3 Bag Domains

In this section we develop some results that are fundamental for our main construction. The details of these results are contained in [15]. The construction of *bag domains*—domains whose elements are *bags* or *multisets* from an underlying domain, originated in the work of Vickers [22], and also have been considered by Johnstone [5, 6]. Those works were inspired by work in database theory, and the goals of their work was to capture the abstract categorical nature of the construction. Here we present results along the same line, but we provide a more direct construction that allows us to analyze the internal structure of the objects more closely. It also allows us to capture the constructions of Varacca [20] more concretely. We begin with a simple result about posets:

Definition 1. *Let P be a poset and let $n \in \mathbb{N}$. For $\phi \in S(n)$, define a mapping $\phi \colon P^n \to P^n$ by $\phi(d)_i = d_{\phi^{-1}(i)}$. Then ϕ permutes the components of d according to ϕ's permutation of the indices $i = 1, \ldots, n$. Next, define a preorder $\sqsubseteq_n$ on P^n by*

$$d \sqsubseteq_n e \quad \text{iff} \quad (\exists \phi \in S(n))\ \phi(d) \leq e \quad \text{iff} \quad d_{\phi^{-1}(i)} \leq e_i\ (\forall i = 1\ldots, n). \tag{6}$$

Finally, we define the equivalence relation $\equiv$ on P^n by

$$\equiv\ =\ \sqsubseteq_n \cap \sqsupseteq_n, \tag{7}$$

and we note that $(P^n/\!\equiv, \sqsubseteq_n)$ [5] is a partial order. We denote by $[d]$ the image of $d \in P^n$ in $P^n/\!\equiv$.

Lemma 1. *Let P be a poset and let $n \in \mathbb{N}$. Then the following are equivalent:*

1. *$d \sqsubseteq_n e$,*
2. *$(\exists \phi \in S(n))(\forall i = 1, \ldots, n)\ d_i \leq e_{\phi(i)}$, for $i = 1, \ldots, n$.*
3. *$\uparrow\!\{\phi(d) \mid \phi \in S(n)\} \supseteq\ \uparrow\!\{\phi(e) \mid \phi \in S(n)\}$.*

Proof. For (i) implies (ii), we note that, if $\phi \in S(n)$ satisfies $d_{\phi^{-1}(i)} \leq e_i$, then $d_i \leq e_{\phi(i)}$, for each $i = 1, \ldots, n$, so (ii) holds. Next, (ii) implies $\phi^{-1}(e) \in\ \uparrow d$, and then $\psi(e) \in\ \uparrow\!\{(\phi(d) \mid \phi \in S(n)\}$ for each $\psi \in S(n)$ by composing permutations, from which (iii) follows. Finally, (iii) implies (i) is clear.

We also need a classic result due to M.-E. Rudin [10]

Lemma 2 (Rudin). *Let P be a poset and let $\{\uparrow F_i \mid i \in I\}$ be a filter basis of non-empty, finitely-generated upper sets. Then there is a directed subset $A \subseteq \bigcup_i F_i$ with $A \cap F_i \neq \emptyset$ for all $i \in I$.*

Next, let P be a dcpo and let $n > 0$. We can apply the lemma above to derive the following:

---

[5] We use $\sqsubseteq_n$ to denote the $\sqsubseteq/\!\equiv$, the partial order on $P^n/\!\equiv$ induced by the pre-order defined above.

Proposition 1. *Let P be a dcpo, and let $n > 0$.*

- *If $A \subseteq P^n/\equiv$ is directed, then there is a directed subset $B \subseteq \bigcup_{[a] \in A} \{\phi(a) \mid \phi \in S(n)\}$ satisfying*

$$\uparrow\{\phi(\sqcup B) \mid \phi \in S(n)\} = \bigcap_{a \in A} \uparrow\{\phi(a) \mid \phi \in S(n)\} \quad \text{and} \quad [\sqcup B] = \sqcup A. \qquad (8)$$

- *In particular, $(P^n/\equiv, \sqsubseteq)$ is a dcpo, and the mapping $x \mapsto [x]\colon P^n \to P^n/\equiv$ is Scott continuous.*

Proof. If $A \subseteq P^n/\equiv$ is directed, then Lemma 1 implies that $\{\bigcup_{\phi \in S(n)} \uparrow \phi(a) \mid [a] \in A\}$ is a filter basis of finitely generated upper sets, and so by Lemma 2 there is a directed set $B \subseteq \bigcup_{[a] \in A}\{\phi(a) \mid \phi \in S(n)\}$ with $B \cap \{\phi(a) \mid \phi \in S(n)\} \neq \emptyset$ for each $[a] \in A$. Since $B \subseteq P^n$ is directed, we have $x = \sqcup B$ exists. If $[a] \in A$, then $B \cap \{\phi(a) \mid \phi \in S(n)\} \neq \emptyset$ means there is some $\phi \in S(n)$ with $\phi(a) \in B$, so $\phi(a) \leq x$ by Lemma 1. Hence $a \sqsubseteq x$ for each $[a] \in A$, so $[x]$ is an upper bound for A.

We also note that, since $\sqcup B = x$,

$$\bigcap_{b \in B} \uparrow\{\phi(b) \mid \phi \in S(n)\} = \uparrow\{\phi(x) \mid \phi \in S(n)\}.$$

Indeed, the right hand side is clearly contained in the left hand side since $b \leq x$ for all $b \in B$. On the other hand, if y is in the left hand side, then $b \sqsubseteq y$ for each $b \in B$. Now, since $S(n)$ is finite, there is some $\phi \in S(n)$ and some cofinal subset $B' \subseteq B$ with $\phi(b) \leq y$ for each $b \in B'$. But then $\sqcup B' = \sqcup B$, and so $\sqcup\{\phi(b) \mid b \in B'\} = \phi(x)$, from which we conclude that $\phi(x) \leq y$. Thus y is in the right hand side, so the sets are equal.

Now, if $y \in P^n$ satisfies $a \sqsubseteq y$ for each $[a] \in A$, then since $B \subseteq \bigcup_{[a] \in A}\{\phi(a) \mid \phi \in S(n)\}$, it follows that $b \sqsubseteq y$ for each $b \in B$. Then $y \in \bigcap_{b \in B} \uparrow\{\phi(b) \mid \phi \in S(n)\} = \uparrow\{\phi(x) \mid \phi \in S(n)\}$, and so $x \sqsubseteq y$. Thus $[x] = \sqcup A$ in the order $\sqsubseteq_n$. This also shows the $\bigcap_{[a] \in A} \uparrow\{\phi(a) \mid \phi \in S(n)\} = \uparrow\{\phi(x) \mid \phi \in S(n)\}$

It is clear now that $P^n/\equiv$ is a dcpo, and the argument we just gave shows that directed sets $B \subseteq P^n$ satisfy $[\sqcup B] = \sqcup_{b \in B}[b]$. ∎

Proposition 2. *Let P be a domain and let $n \in \mathbb{N}$. Then*

1. *$(P^n/\equiv, \sqsubseteq_n)$ is a domain. Moreover, $[d_i]_n \ll [e_i]_n$ iff $(\exists \phi \in S(n))\, \phi((d_i)_n) \ll (e_i)_n$.*
2. *If P is RB or FS, then so is $P^n/\equiv$.*
3. *If P is coherent, then so is $P^n/\equiv$.*

Proof. **$P^n/\equiv$ is a domain:** Proposition 1 shows that $(P^n/\equiv, \sqsubseteq_n)$ is directed complete and that the quotient map is Scott continuous. To characterize the way-below relation on $P^n/\equiv$, let $x, y \in P^n$ with $x \ll y$. Then $x_i \ll y_i$ for each $i = 1, \ldots, n$, and it follows that $\phi(x) \ll \phi(y)$ for each $\phi \in S(n)$. If $A \subseteq P^n/\equiv$ is directed and $[y] \sqsubseteq_n \sqcup A$, then there is some $\phi \in S(n)$ with $\phi(y) \leq z$, where

$[z] = \sqcup A$. Then Proposition 1 shows there is a directed set $B \subseteq \cup_{[a] \in A} \uparrow \{\phi(a) \mid \phi \in S(n)\}$ with $\sqcup B \equiv z$. Hence, there is some $\psi \in S(n)$ with $\psi(y) \leq \sqcup B$. Since $\psi(x) \ll \psi(y)$, it follows that there is some $b \in B$ with $\psi(x) \leq b$, so $[x] \sqsubseteq_n [b]$. Hence $[x] \ll [y]$ in $D^n/\equiv$.

We have just shown that $x \ll y$ in P^n implies that $[x] \ll [y]$ in $P^n/\equiv$. Since D^n is a domain, $\downarrow y$ is directed with $y = \sqcup \downarrow y$, and so the same is true for $\downarrow [y] \in D^n/\equiv$. Thus $P^n/\equiv$ is a domain.

$P^n/\equiv$ **is RB if P is:** Now suppose the P is in RB. Then, by Theorem 4.1 of [8] there is a directed family $f_k: P \to P$ of Scott continuous maps with $1_P = \sqcup_k f_k$ and $f_k(P)$ finite for each $k \in K$. Then the mappings $(f_k)^n: P^n \to P^n$ also form such a family, showing P^n is in RB.

Next, given $k \in K, x \in P^n$ and $\phi \in S(n)$, we have $\phi(f_k^n(x)) = f_k^n(\phi(x))$ since f_k^n is f_k acting on each component of x. It follows that there is an induced map $[f_k^n]: P^n/\equiv \to P^n/\equiv$ satisfying $[f_k^n]([x]) = [f_k^n(x)]$, and this map is continuous since $[\ ]$ is a quotient map. Finally, $[f_k^n](P^n/\equiv)$ is finite since $f_k^n(P^n)$ is finite, and that $\sqcup_k [f_k^n] = 1_{P^n/\equiv}$ follows from $\sqcup_k f_k^n = 1_{P^n}$. Thus, $P^n/\equiv$ is RB is P is.

$P^n/\equiv$ **is FS if P is:** Continuing, the domain P is FS if there is a directed family of selfmaps $f_k: P \to P$ satisfying $\sqcup_k f_k = 1_P$ and for each $k \in K$, there is some finite $M_k \subseteq P$ with $f_k(x) \leq m_x \leq x$ for some $m_x \in M_k$, for each $x \in P$. The remainder is similar to the case of RB.

$P^n/\equiv$ **is coherent if P is:** Last, we consider coherent domains. Recall a domain is coherent if the Lawson topology is compact, where the Lawson topology has for a basis the family of sets $\{U \setminus \uparrow F \mid F \subseteq P \text{ finite } \& \ U \text{ Scott open}\}$. Now, if $x \in P^n$, then $\{\phi(x) \mid \phi \in S(n)\}$ is finite, and so if $F \subseteq P^n/\equiv$ is finite, then $[\uparrow F]^{-1} = \cup_{[x] \in F} \uparrow \{\phi(x) \mid \phi \in S(n)\}$ is finitely generated. It follows that $[\ ]: P^n \to P^n/\equiv$ is Lawson continuous, so if P is coherent, then so are P^n and $P^n/\equiv$.

Definition 2. *For a domain P, we let $\mathcal{B}_n(P) = P^n/\equiv$ denote the domain of n-bags over P. We also let $\mathcal{B}(P) = \oplus_n \mathcal{B}_n(P)$ denote the coalesced sum[6] of the $\mathcal{B}_n(P)$, where we identify $P^0 \equiv \underline{0}$.*

Theorem 2.

1. *$\mathcal{B}_n$ defines a continuous endofunctor on the categories DCPO, DOM, CDOM, RB and FS for each $n \in \mathbb{N}$.*
2. *$\mathcal{B}$ defines a continuous endofunctor on DCPO and DOM, as well as on $\text{CDOM}_\perp$, $\text{RB}_\perp$ and $\text{FS}_\perp$.[7]*
3. *In fact, $\mathcal{B}$ defines the monad of commutative monoids over each of the categories listed in 2.*

---

[6] The coalesced sum of cpos takes their disjoint union and identifies their least elements.

[7] If A is a category of dcpos, the $A_\perp$ denotes the subcategory of cpos in A and maps that preserve least elements.

Proof. $\mathcal{B}_n$ is defined as a composition of constructors that define continuous endofunctors on each of the categories, so it is continuous.

For $\mathcal{B}$, we must add the countable coalesced sum operator, which it is easy to show leaves each of the indicated categories invariant.

The fact that $\mathcal{B}(P)$ is a monoid follows by defining $\langle p_i \rangle_m \oplus \langle q_j \rangle_n = \langle r_k \rangle_{m+n}$, where $r_k = p_k$ for $k \leq m$ and $r_k = q_{m+k}$ otherwise, with $P^0 \equiv \underline{0}$ as the identity.

Example 1. If $P = \{\bot, a, b, \top\}$ is the four-element lattice with a and b incomparable, then $P^2/\equiv$ is not in BCD: the pair $[a, \bot], [b \bot]$ has $[a, b]$ and $[\top, \bot]$ as minimal upper bounds.[8]

3.1 Reconstructing $IV_P(P)$

We now use our results on bag domains to reconstruct Varacca's Plotkin indexed valuations. The facts we need are contained in the following.

Definition 3. *We define the family of* real n-bags *over a domain* P *by* $\mathcal{B}_n^{\mathbb{R}}(P) = \mathcal{B}_n(\overline{\mathbb{R}_{\geq 0}} \times P)/\{[r_i, d_i]_n \mid (\exists i)\ r_i = 0\}$. *We also define* $\mathcal{B}^{\mathbb{R}}(P) = \oplus_n \mathcal{B}_n^{\mathbb{R}}(P)$.

Theorem 3. $\mathcal{B}_n^{\mathbb{R}}$ *and* $\mathcal{B}^{\mathbb{R}}$ *define continuous endofunctors on* CDOM, RB *and on* FS. *Moreover,* $\mathcal{B}^{\mathbb{R}}$ *defines a monad on each of these categories whose algebras satisfy the laws (1) – (7) of Theorem 1.*

Proof. $\mathcal{B}_n^{\mathbb{R}}(P)$ is the quotient of $\mathcal{B}_n(\overline{\mathbb{R}_{\geq 0}} \times P)$ modulo a Scott-closed subset, which shows $\mathcal{B}_n^{\mathbb{R}}(P)$ is in CDOM, RB or FS if P is. The definition of $\mathcal{B}^{\mathbb{R}}(P)$ implies $\underline{0}$ is the least element, which implies $\mathcal{B}^{\mathbb{R}}$ leaves these categories invariant.

Given $f \colon P \to Q$, we define $\mathcal{B}_n^{\mathbb{R}}(f)([r_i, d_i]_n) = [r_i, f(d_i)]_n$, and then $\mathcal{B}^{\mathbb{R}}(f) = \oplus_n \mathcal{B}_n^{\mathbb{R}}(f)$. Clearly $\mathbb{R}_+$ acts on $\mathcal{B}_n^{\mathbb{R}}(P)$, and the last part follows from the fact that these Scott-closed subsets are invariant with respect to this action.

Theorem 4. *For a domain* P, $IV_P(P) \simeq \mathcal{B}^{\mathbb{R}}(P)$.

Proof. This follows by noting that the mapping $\langle r_i, p_i \rangle_n \mapsto [r_i, p_i]_n$ and $\underline{0} \mapsto \underline{0}$ defines a bijection that takes $\ll_P$ on $IV_P(P)$ to $\ll$ on $\mathcal{B}^{\mathbb{R}}(P)$.

Corollary 1. *Each of the power domain monads* $\mathcal{P}_X$ *lifts to a monad on* $\mathcal{B}^{\mathbb{R}}$-*algebras.*

Proof. Varacca showed that there is a distributive law of IV_P over $\mathcal{P}_X$ in [20], and this implies that $\mathcal{P}_X$ lifts to a monad on the class of $\mathcal{B}^{\mathbb{R}}$-algebras by Beck's Theorem [2] and by Theorem 4. In fact, we can easily recover the distributive law that Varacca obtains in [21]: the distributive law $d \colon \mathcal{B}^{\mathbb{R}} \mathcal{P}_X \longrightarrow \mathcal{P}_X \mathcal{B}^{\mathbb{R}}$ can be defined as $d_P([r_i, X_i]_n) = \langle [r_i, x_i]_n \mid x_i \in X_i \in \mathcal{P}_X(P) \rangle$. The result follows.

---

[8] Thanks to one of the anonymous referees for this example.

4 Random Variables over Domains

We now show how to construct the power domain of discrete random variables over a domain. Recall that a *random variable* is a function $f: (X, \mu) \to (Y, \Sigma)$ where (X, μ) is a probability space, (Y, Σ) is a measure space, and f is a measurable function, which means $f^{-1}(A)$ is measurable in X for every $A \in \Sigma$, the specified σ-algebra of subsets of Y. Most often random variables take their values in $\mathbb{R}$, equipped with the usual Borel σ-algebra. For us, X will be a countable, discrete space, and Y will be a domain, where Σ will be the Borel σ-algebra generated by the Scott-open subsets.

Given a random variable $f: X \to Y$, the usual approach is to "push the probability measure μ forward" onto Y by defining $f \cdot \mu(A) = \mu(f^{-1}(A))$ for each measurable subset A of Y. But this defeats one of the attractions of random variables: namely, that there may be several points $x \in X$ which f takes to the same value $y \in Y$. This is 'attractive' because it means that the random variable f makes distinctions that the probability measure $f \cdot \mu$ does not, and we would like to exploit this fact. Varacca makes exactly this point in his work [20, 21], a point he justifies by showing how to distinguish the random variable f from the probability measure $f \cdot \mu$ operationally. We return to this point later. For the moment, we define our power domain of random variables.

Definition 4. *For a domain P, we define the* power domain of discrete random variables over P *to be the subdomain*

$$\mathbb{RV}(P) = \bigcup_n \{[r_i d_i]_n \mid n \geq 1 \ \& \ \sum_i r_i \leq 1\} \cup \{\underline{0}\} \subseteq \mathcal{B}^{\mathbb{R}}(P).$$

Remark 1. We can think of a discrete random variable over P as a formal sum $\sum_{i \leq n} r_i \delta_{x_i}$ where some of the x_i can be repeated. But, the order from $\mathcal{B}^{\mathbb{R}}(P)$ distinguishes, for example, $\frac{1}{2} \delta_x \oplus \frac{1}{2} \delta_x$ from δ_x, while these two would be identified as probability measures.

We now come to our main result.

Theorem 5. $\mathbb{RV}$ *defines a continuous endofunctor on* **CDOM**, **RB** *and* **FS**.

Proof. $\mathbb{RV}$ is obtained by restricting $\mathcal{B}^{\mathbb{R}}$ in the "real components" to ones whose sum is at most 1, which defines a Scott-closed subset of $\mathcal{B}^{\mathbb{R}}(P)$. Hence $\mathbb{RV}(P)$ is in each of the indicated categories if P is. Since continuous maps $f: P \to Q$ are extended to $\mathcal{B}^{\mathbb{R}}(P)$ by $\mathcal{B}^{\mathbb{R}}(f)[r_i, p_i] = [r_i, f(p_i)]$ and since the elements in $\mathbb{RV}(P)$ are those in $\mathcal{B}^{\mathbb{R}}(P)$ whose real components sum to at most 1; it follows that $\mathcal{B}^{\mathbb{R}}(f)$ preserves this property in $\mathcal{B}^{\mathbb{R}}(Q)$. The result follows.

Corollary 2. *Each of the power domain monads $\mathcal{P}_X$ lifts to a monad on $\mathbb{RV}$-algebras.*

Proof. The distributive law given in the proof of Corollary 1 clearly restricts to one for $\mathbb{RV}$.

This corollary means we can solve domain equations such as $P \simeq \mathcal{P}_X \circ \mathbb{RV}(P)$ for each of the power domain monads $\mathcal{P}_X$. The resulting domain P will be a $\mathcal{P}_X$-algebra and simultaneously a $\mathbb{RV}$-algebra.

One might also ask about the relationship between our construction and the traditional probabilistic power domain over a domain. The following provides the answer.

Theorem 6. *If P is a domain, then there is an epimorphism $Flat \colon \mathbb{RV}(P) \to \mathbb{V}(P)$, the domain of valuations over P.*

Proof. The mapping $Flat$ is defined by $Flat([r_i, d_i]_n) = \sum_{i \leq n} r_i \delta_{d_i}$, where in $\mathbb{V}(P)$, summands with the same support are identified. This is easily seen to define a continuous map. It is an epimorphism of domains because the simple valuations are dense [7], and clearly they are the range of $Flat$.

5 Summary and Future Work

We have presented a power domain for discrete random variables, and shown that it enjoys a distributive law with respect to each of the power domain monads. Moreover, our construction defines a continuous endofunctor on the cartesian closed categories **RB** and **FS**, as well as on the category **CDOM**. This is where our results on bag domains have their payoff, since trying to devise these last results using abstract bases would be much more difficult. Varacca actually presents three separate constructions, as indicated in Section 2. Our methods can be adopted to recapture each of them; a discussion of the Hoare indexed valuations from our approach is presented in [15]. Since no similar result is known to hold for the probabilistic power domain, our construction provides an alternative for modeling probabilistic choice on domains.

One issue we haven't discussed is what sort of operational intuition there is for random variables. Again, we rely on Varacca, who showed in [21] that, for a simple state-based language supporting nondeterminism and probabilistic choice, probabilistic schedulers could distinguish distinct programs in his model. This is similar to refusal testing in CSP: one tests a process at each place where a probabilistic choice is made. In the more traditional approach using probabilistic bisimulation, such as in [16], one tests processes at the end of their computation, not at each stage. This provides a viable, albeit more complicated method of assigning behaviors to programs.

Another issue not discussed here is whether one can bring Shannon's information theory into the picture [18]. This is based on bringing entropy into play; there are some very interesting results about domains and entropy in Martin's recent work [12], a line we plan to explore. A particularly appealing issue here is defining an order on random variables over a domain relative to which entropy forms a measurement. If Martin's work is any indication, this will probably be a fairly difficult issue to resolve.

Acknowledgment. We wish to thank the anonymous referees for many valuable suggestions.

References

1. Abramsky, S. and A. Jung, "Doman Theory," in: Handbook of Logic in Computer Science, S. Abramsky and D. M. Gabbay and T. S. E. Maibaum, editors, Clarendon Press, 1994, pp. 1—168.
2. Beck, J., *Distributive laws*, in: *Semian on Triples and Categorical Homology Theory*, 1969, pp. 119–140.
3. Gierz, G., K. H. Hofmann, K. Keimel, J. Lawson, M. Mislove and D. Scott, "Continuous Lattices and Domains," Cambridge University Press, 2003.
4. Hennessy, M. and G. D. Plotkin, *Full abstraction for a simple parallel programming language*, Lecture Notes in Computer Science **74** (1979), pp. 108–120.
5. Johnstone, P. T., *Partial products, bag domains and hyperlocal toposes*, LMS Lecture Notes Series **77** (1992), pp. 315–339.
6. Johnstone, P. T., *Variations on a bagdomain theme*, Theoretical Computer Science **136** (1994), pp. 3–20.
7. Jones, C., "Probabilistic Nondeterminism," PhD Dissertation, University of Edinburgh, Scotland, 1989.
8. Jung, A., "Cartesian Closed Categories of Domains," CWI Tracts **66** (1989), Centrum voor Wiskunde en Informatica, Amsterdam.
9. Jung, A. and R. Tix, *The problematic probabilistic power domain*, Electronic Notes in Theoretical Computer Science **13** (1999).
 http://www.elsevier.com/locate/entcs/volume13.html
10. Lawson, J. D., *The upper interval topology, property $\mathcal{M}$ and compactness*, Electronic Notes in Theoretical Computer Science **13** (1998).
 http://www.elsevier.com/locate/entcs/volume13.html
11. Lowe, G., "Probabilities and Priorities in Timed CSP," DPhil Thesis, Oxford University, 1993.
12. Martin, K. *Entropy as a fixed point*, Proceedings of ICALP 2004, LNCS **3142**, 2004.
13. Mislove, M. *Algebraic posets, algebraic cpos and models of concurrency*, Proceedings of the Oxford Symposium on Topology, G. M. Reed, A. W. Ros-coe and R. Wachter, editors, Oxford University Press, 75 – 111.
14. Mislove, M. *Nondeterminism and probabilistic choice: Obeying the laws*, Lecture Notes in Computer Science **1877** (2000), pp. 350–364.
15. Mislove, M. *Monoids over domains*, submitted to MSCS, 2005.
16. Mislove, M., J. Ouaknine and J. B. Worrell, *Axioms for probability and nondeterminism*, Proceedings of *EXPRESS 2003*, Electronic Notes in Theoretical Computer Science **91(3)**, Elsevier.
17. Morgan, C., et al, *Refinement-oriented probability for CSP*, Technical Report PRG-TR-12-94, Oxford University Computing Laboratory, 1994.
18. Shannon, C., *A mathematical theory of information*, Bell Systems Technical Journal **27** (1948), pp. 379–423 & 623–656.
19. Tix, R., "Continuous D-Cones: Convexity and Powerdomain Constructions," PhD Thesis, Technische Universität Darmstadt, 1999.
20. Varacca, D., *The powerdomain of indexed valuations*, Proceedings 17th IEEE Symposium on Logic in Computer Science (LICS 2002), IEEE Press, 2002.
21. Varacca, D., "Probability, Nondeterminism and Concurrency: Two Denotational Models for Probabilistic Computation," PhD Dissertation, Aarhus University, Aarhus, Denmark, 2003.
22. Vickers, S. *Geometric theories and databases*, LMS Lecture Notes Series **77** (1992), pp. 288–314.

An Accessible Approach to Behavioural Pseudometrics

With an Application to Probabilistic Systems

Franck van Breugel[1,*], Claudio Hermida[2], Michael Makkai[3,*], and James Worrell[4,**]

[1] York University, Department of Computer Science,
4700 Keele Street, Toronto, Ontario M3J 1P3, Canada
franck@cs.yorku.ca
[2] Instituto Superior Técnico, Department of Mathematics,
Av. Rovisco Pais, 1049-001, Lisbon, Portugal
chermida@math.ist.utl.pt
[3] McGill University, Department of Mathematics and Statistics,
805 Sherbrooke Street West, Montreal, Quebec H3A 2K6, Canada
makkai@math.mcgill.ca
[4] Tulane University, Department of Mathematics,
6823 St Charles Avenue, New Orleans LA 70118, USA
jbw@math.tulane.edu

Abstract. Behavioural pseudometrics are a quantitative analogue of behavioural equivalences. They provide robust models for those concurrent systems in which quantitative data plays a crucial role. In this paper, we show how behavioural pseudometrics can be defined coalgebraically. Our results rely on the theory of accessible categories. We apply our results to obtain a robust model for probabilistic systems.

1 Introduction

The study of concurrent systems with *quantitative data*, such as probabilities, time, costs or rewards, has led to various extensions of the classical notion of *transition system* and its associated notions of behavioural equivalences, with emphasis on *bisimilarity*. However, such discrete notions (states are either behaviourally equivalent or they are not) sit uneasily with the presence of quantitative data. For instance, if some of the probabilities associated with the transitions change slightly—the probabilities are often obtained experimentally and, hence, are usually approximations—states that used to be behaviourally equivalent may not be anymore or vice versa. In short, behavioural equivalences like probabilistic bisimilarity are not *robust*.

---

[*] Supported by the Natural Sciences and Engineering Research Council of Canada.
[**] Supported by the US Office of Naval Research.

To address this problem, Giacalone, Jou and Smolka [18] suggested to exploit pseudometrics that assign a distance, a real number between 0 and 1, to each pair of states of a system. Such a pseudometric yields a quantitative notion of behavioural equivalence in that the distances between states express the similarity of the behaviour of those states. The smaller the distance, the more the states behave alike. In particular, the distance between states is 0 if they are behaviourally indistinguishable. Recently, there has been a renewed interest in such *behavioural pseudometrics* (see, for example, [6, 9, 10, 11, 12, 13, 14, 15, 19]).

Three different approaches have been put forward to define behavioural pseudometrics. Desharnais, Gupta, Jagadeesan and Panangaden [13] defined a behavioural pseudometric in terms of a real-valued interpretation of a modal *logic*. Van Breugel and Worrell [6] gave a *coalgebraic* definition of a behavioural pseudometric and showed that it coincides with the logical definition. Desharnais et al. [14] also defined a behavioural pseudometric as the greatest *fixed point* of a monotone function on a complete lattice. Each approach has its advantages and disadvantages. The coalgebraic approach is completely driven by a single functor. This allows for a uniform approach and a general theory, which are two of the advantages of the coalgebraic approach. Having a logical, a coalgebraic and a fixed point characterisation of a behavioural pseudometric allows us to exploit the advantages of all three approaches.

There are two types of behavioural pseudometrics: those that *discount the future* and those that do not. In the discounted setting, differences in behaviour in the farther future have less impact on the distance, whereas in the undiscounted setting, all differences are weighted equally (see, for example, [11] for more details). Pseudometrics that do not discount the future usually provide quantitative information different from those that discount the future (even the induced topologies often differ). Both types of behavioural pseudometrics can be defined by means of the logical approach and the fixed point approach. So far, the coalgebraic approach has only been shown to be applicable to the discounted setting. In this paper, we show that behavioural pseudometrics that do not discount the future can be defined coalgebraically as well. Furthermore, we show that the three approaches give rise to the same behavioural pseudometric for (a generalisation of) probabilistic transition systems.

Let us provide some more details of the coalgebraic approach to define a behavioural pseudometric for a class of systems. This approach is completely driven by a functor $\mathcal{F} : \mathbb{CM}et_1 \to \mathbb{CM}et_1$ on the category $\mathbb{CM}et_1$ of 1-bounded complete metric spaces and nonexpansive functions. The functor $\mathcal{F}$ should be such that the $\mathcal{F}$-coalgebras represent the systems of interest and such that a *terminal $\mathcal{F}$-coalgebra* exists. An $\mathcal{F}$-coalgebra consists of a space S, which represents the state space of the system, and a function $t : S \to \mathcal{F}(S)$, which captures the transitions of the system. The terminal $\mathcal{F}$-coalgebra can be viewed as a universal system, as it contains all possible behaviours. Let us denote the state space of this universal system by T. Then for every $\mathcal{F}$-coalgebra with state space S, there exists a unique function $h : S \to T$ that preserves and reflects behaviour. That is, h only maps states s_1 and s_2 to the same state in T if s_1 and s_2 are behaviourally equivalent. Since T is a metric space, its distance function can be transferred to the set of states S along the function h. That is, the behavioural distance of states s_1 and s_2 is defined as the distance of $h(s_1)$ and $h(s_2)$ in T. Since h preserves and reflects behaviour, states only have distance 0 if they are behaviourally indistinguishable.

De Vink and Rutten [27] have shown that *probabilistic transition systems* can be represented as coalgebras (see also [6]). To prove the existence of a terminal coalgebra, they exploit the metric terminal coalgebra theorem of Turi and Rutten [26]. This theorem is only applicable to functors that give rise to behavioural pseudometrics that discount the future. More generally, all known techniques to solve *recursive equations* of the form $X = \mathcal{F}(X)$ require the functor $\mathcal{F}$ to discount the future (note that this is indeed more general since a terminal $\mathcal{F}$-coalgebra solves the equation). Such recursive equations play a key role in semantics (see, for example, [3]). In this paper, we show that for many functors $\mathcal{F}$ that are used in semantics, in particular those that do not discount the future, a terminal $\mathcal{F}$-coalgebra exists and, hence, the equation $X = \mathcal{F}(X)$ can be solved.

The rest of this paper is organised as follows. In Section 2 we present a terminal coalgebra theorem, which is shown to be applicable to the category $\mathbb{C}\mathrm{Met}_1$ in Section 3. In Section 4 and 5 we study some functors used in semantics that do not discount the future. In Section 6 we apply our results to give a coalgebraic definition of a behavioural pseudometric for probabilistic transition systems that does not discount the future. Furthermore, we show that the coalgebraic definition coincides with a logical and a fixed point definition. We assume that the reader is familiar with category theory, metric topology and measure theory (see, for example, [23, 17, 5]).

2 A Terminal Coalgebra Theorem for Accessible Categories

Numerous terminal coalgebra theorems can be found in the literature. As far as we know, only one of them, due to Turi and Rutten [26], is applicable to metric spaces. In this section, we present a terminal coalgebra theorem for accessible categories. This theorem is implicit in the work of Makkai and Paré [24] (see also [4, Theorem 1.2]). As we will see in Section 3, the theorem is applicable to metric spaces. Before we present the theorem, we introduce the reader to those concepts and results of the theory of coalgebra and the theory of accessible categories that we will use in the rest of this paper.

Definition 1. *Let $\mathbb{C}$ be a category. Let $\mathcal{F} : \mathbb{C} \to \mathbb{C}$ be a functor. An $\mathcal{F}$-coalgebra consists of an object C in $\mathbb{C}$ together with a morphism $f : C \to \mathcal{F}(C)$ in $\mathbb{C}$. An $\mathcal{F}$-homomorphism from $\mathcal{F}$-coalgebra $\langle C, f \rangle$ to $\mathcal{F}$-coalgebra $\langle D, g \rangle$ is a morphism $h : C \to D$ in $\mathbb{C}$ such that $\mathcal{F}(h) \circ f = g \circ h$. The $\mathcal{F}$-coalgebras and $\mathcal{F}$-homomorphisms form a category. If this category has a terminal object, then this object is called a terminal $\mathcal{F}$-coalgebra.*

For more details about the theory of coalgebra, we refer the reader to, for example, the tutorial [21] of Jacobs and Rutten.

Next, we turn to accessible categories and accessible functors.

Definition 2. *A category $\mathbb{C}$ is accessible if it is κ-accessible for some infinite regular cardinal κ.*

We will not provide the definition of κ-accessible category since we do not need this definition in our paper. It can be found in, for example, [24, Definition 2.1.3]. To prove that categories are accessible, either we will exploit the "logical" characterisation of accessible categories as proposed by Makkai and Paré in [24, § 3.2] or we will make use of the following result.

Proposition 1. (Theorem 5.1.6 of [24]) *If the categories $\mathbb{C}$ and $\mathbb{D}$ are accessible, then the category $\mathbb{C} \times \mathbb{D}$ is accessible.*

Definition 3. *Let κ be an infinite regular cardinal. A functor $\mathcal{F} : \mathbb{C} \to \mathbb{D}$ is κ-accessible if $\mathbb{C}$ and $\mathbb{D}$ are κ-accessible and $\mathcal{F}$ preserves κ-filtered colimits. A functor $\mathcal{F} : \mathbb{C} \to \mathbb{D}$ is accessible if it is κ-accessible for some infinite regular cardinal κ.*

We will also not introduce the definition of κ-filtered colimit because we do not need this definition in our paper either. It can be found in, for example, [24, page 13]. To prove that a functor is accessible, we will either prove that the functor preserves all colimits or we will exploit one of the following results.

Proposition 2. (page 32 of [24]) *If the functors $\mathcal{F} : \mathbb{C} \to \mathbb{D}$ and $\mathcal{G} : \mathbb{D} \to \mathbb{E}$ are accessible, then the functor $\mathcal{G} \circ \mathcal{F} : \mathbb{C} \to \mathbb{E}$ is accessible.*

Proposition 3. (Theorem 5.1.6 of [24]) *If the functors $\mathcal{F} : \mathbb{C}_1 \to \mathbb{D}_1$ and $\mathcal{G} : \mathbb{C}_2 \to \mathbb{D}_2$ are accessible, then the functor $\mathcal{F} \times \mathcal{G} : \mathbb{C}_1 \times \mathbb{C}_2 \to \mathbb{D}_1 \times \mathbb{D}_2$ is accessible.*

Proposition 4. (Proposition 2.4.8 of [24]) *Let $\mathbb{C}$ and $\mathbb{D}$ be accessible categories. Any functor $\mathcal{F} : \mathbb{C} \to \mathbb{D}$ which has a left or right adjoint is accessible.*

The main point of working with accessible categories in this paper is the existence of a terminal coalgebra for accessible functors.

Theorem 1. *If the category $\mathbb{C}$ is accessible and complete and the functor $\mathcal{F} : \mathbb{C} \to \mathbb{C}$ is accessible, then a terminal $\mathcal{F}$-coalgebra exists.*

For more details about theory of accessible categories we refer the reader to the textbook [24] of Makkai and Paré and the textbook [1] of Adámek and Rosický.

3 The Category of Complete Metric Spaces Is Accessible

Since the category $\mathbb{T}op$ of topological spaces and continuous functions is not accessible as shown, for example, in [1, Example 2.3(1)], it may come as a surprise that the category $\mathbb{C}\mathbb{M}et_1$ of 1-bounded complete metric spaces and nonexpansive functions is accessible, as we will show below. However, since the category $\mathbb{B}an$ of Banach spaces and contractions is accessible as demonstrated in, for example, [24, § 3.4], the accessibility of the category $\mathbb{C}\mathbb{M}et_1$ may not be that surprising after all.

Definition 4. *A metric space is a pair (X, d_X) consisting of a set X and a distance function $d_X : X \times X \to [0, \infty)$ satisfying*

1. *for all $x, y \in X$, $d_X(x, y) = 0$ if and only if $x = y$,*
2. *for all $x, y \in X$, $d_X(x, y) = d_X(y, x)$,*
3. *for all $x, y, z \in X$, $d_X(x, z) \leq d_X(x, y) + d_X(y, z)$.*

Instead of (X, d_X) we often write X and we denote the distance function of a metric space X by d_X. A metric space X is 1-bounded if

4. *for all $x, y \in X$, $d_X(x, y) \leq 1$.*

A metric space X is complete if

5. *each Cauchy sequence in X is convergent in X.*

A sequence $(x_n)_{n \in \mathbb{N}}$ in X is Cauchy if

$$\forall \epsilon > 0 \, \exists N \in \mathbb{N} \, \forall m > N \, \forall n > N \, d_X(x_m, x_n) < \epsilon.$$

A sequence $(x_n)_{n \in \mathbb{N}}$ in X is convergent in X if

$$\exists x \in X \, \forall \epsilon > 0 \, \exists N \in \mathbb{N} \, \forall n > N \, d_X(x_n, x) < \epsilon.$$

A function $f : X \to Y$ is nonexpansive if for all $x, y \in X$,

$$d_Y(f(x), f(y)) \leq d_X(x, y).$$

The category $\mathbb{C}\mathbb{M}et_1$ has 1-bounded complete metric spaces as objects and nonexpansive functions as morphisms. One can prove that $\mathbb{C}\mathbb{M}et_1$ is accessible exploiting the "logical" characterisation of accessible categories of Makkai and Paré [24, § 3.2] (see also [1, Section 5.B]).

Theorem 2. *The category $\mathbb{C}\mathbb{M}et_1$ is accessible.*

4 The Kantorovich Functor Is Accessible

As we will see in Section 6, probabilistic nondeterminism can be modelled by means of the set of tight Borel probability measures on a metric space endowed with the Kantorovich metric. Below, we will show that the corresponding Kantorovich functor is accessible.

First, let us review the notion of a tight Borel probability measure on a metric space. Let X be a metric space and let the set of open sets $O(X)$ be the smallest set of subsets of X which contains $\{y \in X \mid d_X(x,y) < r\}$ for each $x \in X$ and $r \in [0, \infty)$, and which is closed under unions. The set $B(X)$ of Borel sets is the smallest set of subsets of X which contains $O(X)$ and which is closed under countable intersections and countable unions. Then $\emptyset \in B(X)$, if $B \in B(X)$ then $X \setminus B \in B(X)$, and for each sequence $(B_n)_n$ in $B(X)$, $\bigcup_n B_n \in B(X)$ (see, for example, [25, Theorem I.1.3]), that is, $B(X)$ is a σ-field. A Borel probability measure on X is a function $\mu : B(X) \to [0,1]$ satisfying $\mu(X) = 1$ and for each sequence $(B_n)_n$ of pairwise disjoint sets in $B(X)$, $\mu(\bigcup_n B_n) = \sum_n \mu(B_n)$. A Borel probability measure on X is tight if for each $\epsilon > 0$ there exists a compact subset K_ϵ of X such that $\mu(X \setminus K_\epsilon) < \epsilon$. We write $\mathcal{K}(X)$ for the set of tight Borel probability measures on a metric space X. This set is endowed with the Kantorovich metric.

Definition 5. *The Kantorovich metric $d_{\mathcal{K}(X)} : \mathcal{K}(X) \times \mathcal{K}(X) \to [0, \infty]$ is defined by*

$$d_{\mathcal{K}(X)}(\mu, \nu) = \sup \left\{ \int f \, d\mu - \int f \, d\nu \mid f \in X \to [0,1] \text{ is nonexpansive} \right\}.$$

If X is a 1-bounded metric space, then so is $\mathcal{K}(X)$. Furthermore, if X is complete, then also $\mathcal{K}(X)$ is complete (see, for example, [16, Theorem 2.5.25]). Given a 1-bounded complete metric space X, the 1-bounded complete metric space $\mathcal{K}(X)$ together with the binary operation mapping the tight Borel probability measures μ and ν to the tight Borel probability measure $\frac{1}{2} \cdot \mu + \frac{1}{2} \cdot \nu$ form a metric mean-value algebra.

Definition 6. *A metric mean-value algebra is a pair $\langle X, \oplus \rangle$ consisting of a 1-bounded complete metric space X and a choice operation $\oplus : X \times X \to X$ satisfying*

1. *for all $x \in X$, $x \oplus x = x$,*
2. *for all $x, y \in X$, $x \oplus y = y \oplus x$,*
3. *for all $v, w, x, y \in X$, $(v \oplus w) \oplus (x \oplus y) = (x \oplus w) \oplus (v \oplus y)$, and*
4. *for all $v, w, x, y \in X$, $d_X(v \oplus w, x \oplus y) \leq \frac{d_X(v,x) + d_X(w,y)}{2}$.*

A function $f : X \to Y$ is choice preserving if $f(x \oplus_X y) = f(x) \oplus_Y f(y)$ for all $x, y \in X$.

The above definition is a metric analogue of the notion of mean-value algebra [20] given by Heckmann.

The category $MV(\mathbb{CM}\text{et}_1)$ has metric mean-value algebras as objects and nonexpansive and choice preserving functions as morphisms. The metric mean-value algebras provide a finitary description of the algebras of the Kantorovich monad which will enable us to prove that the category of these algebras is accessible (again exploiting the "logical" characterisation of accessible categories) and, hence, that the Kantorovich functor itself is accessible.

Theorem 3. *The category $MV(\mathbb{C}Met_1)$ is accessible.*

The operation mapping a 1-bounded complete metric space X to the metric mean-value algebra $\langle \mathcal{K}(X), \frac{1}{2} \cdot - + \frac{1}{2} \cdot - \rangle$ can be extended to a functor $\mathcal{K}: \mathbb{C}Met_1 \to MV(\mathbb{C}Met_1)$ as follows.

Definition 7. *Let $f : X \to Y$ be a nonexpansive function. The function $\mathcal{K}(f) : \mathcal{K}(X) \to \mathcal{K}(Y)$ is defined by $\mathcal{K}(f)(\mu)(B) = \mu(f^{-1}(B))$.*

The forgetful functor $\mathcal{U} : MV(\mathbb{C}Met_1) \to \mathbb{C}Met_1$ maps each metric mean-value algebra to the underlying 1-bounded complete metric space. The functors $\mathcal{K}$ and $\mathcal{U}$ are related as follows.

Theorem 4. *$\mathcal{K}$ is a left adjoint for $\mathcal{U}$.*

From Theorem 2, 3 and 4 and Proposition 2 and 4 we can derive

Corollary 1. *The functor $\mathcal{U} \circ \mathcal{K} : \mathbb{C}Met_1 \to \mathbb{C}Met_1$ is accessible.*

Instead of $\mathcal{U} \circ \mathcal{K}$ we will just write $\mathcal{K}$ in the sequel.

Similarly, we can show that the Hausdorff functor $\mathcal{H} : \mathbb{C}Met_1 \to \mathbb{C}Met_1$ that maps a 1-bounded complete metric space to the set of nonempty and compact subsets of the space endowed with the Hausdorff metric is an accessible functor. This functor is often used to model nondeterminism.

5 Other Accessible Functors

After having shown that the Kantorovich functor is accessible, we introduce some other functors that are often used in semantics and we show that they are accessible as well.

The identity functor $\mathrm{Id}_{\mathbb{C}Met_1} : \mathbb{C}Met_1 \to \mathbb{C}Met_1$ preserves all colimits and, hence, is accessible. Let X be a 1-bounded complete metric space. We denote the constant functor mapping each 1-bounded complete metric space to the space X by $X : \mathbb{C}Met_1 \to \mathbb{C}Met_1$. We write 1 for a singleton metric space (which is unique up to isomorphism). The functor $1 : \mathbb{C}Met_1 \to \mathbb{C}Met_1$ is an example of a constant functor. Obviously, these constant functors preserve all colimits and, hence, are accessible.

The coproduct $X + Y$ of X and Y in $\mathbb{C}Met_1$ consists of the disjoint union of the sets underlying the 1-bounded complete metric spaces X and Y endowed with the following distance function.

Definition 8. *Let X and Y be 1-bounded complete metric spaces. The distance function $d_{X+Y} : (X+Y) \times (X+Y) \to [0,1]$ is defined by*

$$d_{X+Y}(u,v) = \begin{cases} d_X(u,v) & \text{if } u, v \in X \\ d_Y(u,v) & \text{if } u, v \in Y \\ 1 & \text{otherwise.} \end{cases}$$

This operation can be extended to a bifunctor $+ : \mathbb{C}\mathrm{Met}_1 \times \mathbb{C}\mathrm{Met}_1 \to \mathbb{C}\mathrm{Met}_1$. Since the bifunctor $+$ is left adjoint to the diagonal (see, for example, [23, page 85]), we can conclude from Theorem 2 and Proposition 1 and 4 that the functor $+$ is accessible.

The copower $A \cdot X$ of a set A and a 1-bounded complete metric space X consists of the A-fold disjoint union of the set underlying X endowed with the following distance function.

Definition 9. *Let A be a set and X a 1-bounded complete metric space. The distance function $d_{A \cdot X} : A \cdot X \times A \cdot X \to [0,1]$ is defined by*

$$d_{A \cdot X}(x_a, y_b) = \begin{cases} d_X(x,y) & \text{if } a = b \\ 1 & \text{otherwise.} \end{cases}$$

The power X^A of a set A and a 1-bounded complete metric space X consists of the A-indexed Cartesian product of the set underlying X endowed with the following distance function.

Definition 10. *Let A be a set and X a 1-bounded complete metric space. The distance function $d_{X^A} : X^A \times X^A \to [0,1]$ is defined by*

$$d_{X^A}(\langle x_a \rangle_a, \langle y_a \rangle_a) = \sup_{a \in A} d_X(x_a, y_a).$$

Given a set A, the operations $A \cdot -$ and $-^A$ can be extended to functors $A \cdot - : \mathbb{C}\mathrm{Met}_1 \to \mathbb{C}\mathrm{Met}_1$ and $-^A : \mathbb{C}\mathrm{Met}_1 \to \mathbb{C}\mathrm{Met}_1$. Since the functors $A \cdot -$ and $-^A$ form an adjunction (see, for example, [23, page 88]), we can conclude from Theorem 2 and Proposition 4 that both functors are accessible.

Several other functors, including the product and tensor product functor, are accessible as well. The hom functor is not accessible.

6 A Behavioural Pseudometric for Probabilistic Transition Systems

The results of the foregoing sections are applied in this section to give a coalgebraic definition of a behavioural pseudometric for (a generalisation of) probabilistic transition systems. The behavioural pseudometric does not discount the future. Furthermore, we give a fixed point characterisation and a logical characterisation of the behavioural pseudometric.

Definition 11. *A probabilistic transition system is a triple $\langle S, A, (\pi_a)_{a \in A} \rangle$ consisting of*

- *a set S of states,*
- *a set A of actions, and*
- *for each $a \in A$, a function $\pi_a : S \times S \to [0,1]$ satisfying $\sum_{s' \in S} \pi_a(s, s') \leq 1$ for each $s \in S$.*

Next, we introduce a family of functors $\mathcal{P}^A$, such that each probabilistic transition system $\langle S, A, (\pi_a)_{a \in A}\rangle$ can be represented as a $\mathcal{P}^A$-coalgebra.

Definition 12. *The functor $\mathcal{P}^A : \mathbb{C}\mathrm{Met}_1 \to \mathbb{C}\mathrm{Met}_1$ is defined by*

$$\mathcal{P}^A = -^A \circ \mathcal{K} \circ + \circ \langle 1, \mathrm{Id}_{\mathbb{C}\mathrm{Met}_1}\rangle,$$

where A is a set.

Proposition 5. *Each probabilistic transition system $\langle S, A, (\pi_a)_{a \in A}\rangle$ can be represented as a $\mathcal{P}^A$-coalgebra.*

From the fact that the functors $-^A$, $\mathcal{K}$, $+$, 1 and $\mathrm{Id}_{\mathbb{C}\mathrm{Met}_1}$ are accessible and Proposition 2 and 3, we can conclude that the functor $\mathcal{P}^A$ is accessible. Since the category $\mathbb{C}\mathrm{Met}_1$ is accessible (Theorem 2) and complete (see, for example, [2, Chapter 4]), we can deduce from Theorem 1 the following result.

Proposition 6. *For each set A, a terminal $\mathcal{P}^A$-coalgebra exists.*

For the rest of this section, we fix a $\mathcal{P}^A$-coalgebra $\langle S, \pi\rangle$. To simplify the presentation, we assume that the space S of states is compact.

Definition 13. *Let $\langle T^A, \iota\rangle$ be a terminal $\mathcal{P}^A$-coalgebra. The distance function $d_c : S \times S \to [0, 1]$ is defined by*

$$d_c(s_1, s_2) = d_{T^A}(h(s_1), h(s_2)),$$

where h is the unique $\mathcal{P}^A$-homomorphism from the $\mathcal{P}^A$-coalgebra $\langle S, \pi\rangle$ to the terminal $\mathcal{P}^A$-coalgebra $\langle T^A, \iota\rangle$.

Since terminal objects are unique up to isomorphism, the definition of the distance function d_c does not depend on which terminal $\mathcal{P}^A$-coalgebra is chosen. The distance function d_c satisfies the condition 2–4 of Definition 4 but not condition 1. It does satisfy the following weaker condition: 1'. for all $s \in S$, $d(s, s) = 0$. Therefore, d_c is called a 1-bounded pseudometric. Note that in a pseudometric space, different elements may have distance 0. In the behavioural pseudometric d_c, distance 0 captures the behavioural equivalence probabilistic bisimilarity [22].

Theorem 5. *For all s_1, $s_2 \in S$, $d_c(s_1, s_2) = 0$ iff s_1 and s_2 are probabilistic bisimilar.*

Next, we characterise the behavioural pseudometric as the greatest fixed point of a monotone function on a complete lattice. This approach was first proposed by Desharnais et al. [14]. We consider those distance functions on the set underlying the 1-bounded compact metric space S that satisfy condition 1' and condition 2–4 of Definition 4 and order them as follows.

Definition 14. *The relation $\sqsubseteq$ on 1-bounded pseudometrics on S is defined by*

$$d_1 \sqsubseteq d_2 \text{ if } d_1(s_1, s_2) \geq d_2(s_1, s_2) \text{ for all } s_1, s_2 \in S.$$

The set of 1-bounded pseudometrics on S endowed with the order $\sqsubseteq$ forms a complete lattice (see, for example, [14, Lemma 3.2]).

Definition 15. *Let d be a 1-bounded pseudometric on S. The distance function $\Delta(d) : S \times S \to [0,1]$ is defined by*

$$\Delta(d)(s_1, s_2) = \sup_{a \in A} \left(\inf \left\{ \int d\, d\mu \mid \mu \in \pi(s_1)_a \otimes \pi(s_2)_a \right\} \right),$$

where $\pi(s_1)_a \otimes \pi(s_2)_a$ denotes the set of Borel probability measures on the product space $(1+S)^2$ with marginals $\pi(s_1)_a$ and $\pi(s_2)_a$, that is, those Borel probability measures μ such that for all $B \in \mathcal{B}(1+S)$, $\mu(B \times (1+S)) = \pi(s_1)_a(B)$ and $\mu((1+S) \times B) = \pi(s_2)_a(B)$.

Note that $\Delta(d)$ is also a 1-bounded pseudometric on S. Furthermore, Δ is monotone, that is, if $d_1 \sqsubseteq d_2$ then $\Delta(d_1) \sqsubseteq \Delta(d_2)$. According to Tarski's fixed point theorem, a monotone function $f : X \to X$ on a complete lattice X has a greatest fixed point. Hence, we can conclude that Δ has a greatest fixed point d_f.

Theorem 6. $d_c = d_f$.

We conclude this section with a logical characterisation of the behavioural pseudometric, as first given by Desharnais et al. in [13]. We introduce a logic that shows similarities with the modal logic that characterises probabilistic bisimilarity [22] (see [15] for a detailed discussion).

Definition 16. *The logic $\mathcal{L}$ is defined by*

$$\varphi ::= \text{true} \mid \neg \varphi \mid \langle a \rangle \varphi \mid \varphi \ominus q \mid \varphi \wedge \varphi$$

where $a \in A$ and $q \in \mathbb{Q} \cap [0,1]$.

Next, we provide a real-valued interpretation of the logic. For each formula φ and state s, the real number $\varphi(s)$ provides a quantitative measure of the validity of φ in s. The larger $\varphi(s)$, the more likely it is that φ holds in s.

Definition 17. *For each $\varphi \in \mathcal{L}$, the function $\varphi : S \to [0,1]$ is defined by*

$$\text{true}(s) = 1$$
$$(\neg \varphi)(s) = 1 - \varphi(s)$$
$$(\langle a \rangle \varphi)(s) = \int \varphi \, d\pi(s)_a$$
$$(\varphi \ominus q)(s) = \max\{\varphi(s) - q, 0\}$$
$$(\varphi \wedge \psi)(s) = \varphi(s) \min \psi(s)$$

Given the logic and its real-valued interpretation, we can define a behavioural pseudometric as follows.

Definition 18. *The distance function $d_\ell : S \times S \to [0,1]$ is defined by*

$$d_\ell(s_1, s_2) = \sup_{\varphi \in \mathcal{L}} \varphi(s_1) - \varphi(s_2).$$

The above definition of d_ℓ can be seen as a logical characterisation of d_c since the pseudometrics d_ℓ and d_c coincide.

Theorem 7. $d_c = d_\ell$.

The proof of the above theorem shows some similarities with the proof of [8, Theorem 42] in which Van Breugel and Worrell show that a coalgebraic and a logical characterisation of a behavioural pseudometric that discounts the future coincide. However, the proof of the above theorem also includes some new ingredients. These ingredients allow us to remove from [8, Theorem 42] the assumption that the set A of actions is finite.

7 Conclusion

Let us briefly highlight our main contributions, before discussing related and future work. We have shown that an accessible endofunctor on an accessible and complete category has a terminal coalgebra. This terminal coalgebra theorem was implicit in [24] (see also [4]). Furthermore, we have shown that the category $\mathbb{CM}et_1$ is accessible (it is well-known to be complete) and that many functors used in semantics are accessible. As a consequence, for most functors $\mathcal{F}$ used in semantics, a terminal $\mathcal{F}$-coalgebra exists and, hence, the equation $X = \mathcal{F}(X)$ can be solved, even if $\mathcal{F}$ does not discount the future. As an application of our developed theory, we have presented a coalgebraic definition of a behavioural pseudometric that does not discount the future for probabilistic transition systems. Furthermore, we have provided a fixed point characterisation and a logical characterisation of the behavioural pseudometric. As far as we know, we are the first to relate a fixed point definition to the corresponding coalgebraic and logical definitions. We have also shown that there is no need to restrict to finite action sets when relating the different approaches.

In this paper we have made fruitful use of the theory of accessible categories [1, 24]. Our coalgebraic definition of the behavioural pseudometric is inspired by the work of De Vink and Rutten [27] and Van Breugel and Worrell [6]. The logical characterisation and the fixed point characterisation are based on work of Desharnais et al. [13, 14].

The theory developed in this paper has already been applied to a large class of timed transition systems. The details will be presented in [7].

We are confident that the results developed in this paper can also be exploited to provide coalgebraic characterisations of the fixed point and logical definitions of behavioural pseudometrics given in [9, 10, 11, 12].

Acknowledgements

The authors thank Jiří Adámek, Michael Barr and the referees for their helpful comments.

References

1. J. Adámek and J. Rosický. *Locally Presentable and Accessible Categories*, Cambridge University Press, 1994.
2. M.A. Arbib and E.G. Manes. *Arrows, Structures, and Functors: the categorical imperative*, Academic Press, 1975.
3. J.W. de Bakker and E.P. de Vink. *Control Flow Semantics*, The MIT Press, 1996.
4. M. Barr. Terminal coalgebras in well-founded set theory. *Theoretical Computer Science*, 114(2):299–315, 1993.
5. P. Billingsley. *Probability and Measure*, John Wiley & Sons, 1995.
6. F. van Breugel and J. Worrell. Towards quantitative verification of probabilistic transition systems. In *Proceedings of ICALP*, volume 2076 of *LNCS*, pages 421–432, 2001. Springer-Verlag.
7. F. van Breugel and J. Worrell. A behavioural pseudometric for metric labelled transition systems. 2005.
8. F. van Breugel and J. Worrell. A behavioural pseudometric for probabilistic transition systems. *Theoretical Computer Science*, 331(1):115–142, 2005.
9. L. de Alfaro. Quantitative verification and control via the mu-calculus. In *Proceedings of CONCUR*, volume 2761 of *LNCS*, pages 102–126, 2003. Springer-Verlag.
10. L. de Alfaro, M. Faella, and M. Stoelinga. Linear and branching metrics for quantitative transition systems. In *Proceedings of ICALP*, volume 3142 of *LNCS*, pages 97–109, 2004. Springer-Verlag.
11. L. de Alfaro, T.A. Henzinger, and R. Majumdar. Discounting the future in systems theory. In *Proceedings of ICALP*, volume 2719 of *LNCS*, pages 1022–1037, 2003. Springer-Verlag.
12. Y. Deng, T. Chothia, C. Palamidessi, and J. Pang. Metrics for action-labelled quantitative transition systems. In *Proceedings of QAPL*, ENTCS, 2005. Elsevier.
13. J. Desharnais, V. Gupta, R. Jagadeesan, and P. Panangaden. Metrics for labeled Markov systems. In *Proceedings of CONCUR*, volume 1664 of *LNCS*, pages 258–273, 1999. Springer-Verlag.
14. J. Desharnais, V. Gupta, R. Jagadeesan, and P. Panangaden. The metric analogue of weak bisimulation for probabilistic processes. In *Proceedings of LICS*, pages 413–422, 2002. IEEE.
15. J. Desharnais, V. Gupta, R. Jagadeesan, and P. Panangaden. Metrics for labelled Markov processes. *Theoretical Computer Science*, 318(3):323–354, 2004.
16. G.A. Edgar. *Integral, Probability, and Fractal Measures*, Springer-Verlag, 1998.
17. R. Engelking. *General Topology*, Heldermann Verlag, 1989.
18. A. Giacalone, C.-C. Jou, and S.A. Smolka. Algebraic reasoning for probabilistic concurrent systems. In *Proceedings of PROCOMET*, pages 443–458, 1990. North-Holland.
19. V. Gupta, R. Jagadeesan, and P. Panangaden. Approximate reasoning for real-time probabilistic processes. In *Proceedings of QEST*, pages 304–313, 2004. IEEE.
20. R. Heckmann. Probabilistic domains. In *Proceedings of CAAP*, volume 787 of *LNCS*, pages 142–156, 1994. Springer-Verlag.
21. B. Jacobs and J.J.M.M. Rutten. A tutorial on (co)algebras and (co)induction. *Bulletin of the EATCS*, 62:222–259, 1997.
22. K.G. Larsen and A. Skou. Bisimulation through probabilistic testing. *Information and Computation*, 94(1):1–28, 1991.

23. S. Mac Lane. *Categories for the Working Mathematician*, Springer-Verlag, 1971.
24. M. Makkai and R. Paré. *Accessible Categories: The Foundation of Categorical Model Theory*, American Mathematical Society, 1989.
25. K.R. Parthasarathy. *Probability Measures on Metric Spaces*, Academic Press, 1967.
26. D. Turi and J.J.M.M. Rutten. On the foundations of final coalgebra semantics: non-well-founded sets, partial orders, metric spaces. *Mathematical Structures in Computer Science*, 8(5):481–540, 1998.
27. E.P. de Vink and J.J.M.M. Rutten. Bisimulation for probabilistic transition systems: a coalgebraic approach. *Theoretical Computer Science*, 221(1/2):271–293, 1999.

Noisy Turing Machines

Eugene Asarin[1] and Pieter Collins[2]

[1] LIAFA, Université Paris 7 / CNRS, case 7014,
2 place Jussieu, 75251 Paris Cedex 05, France
Eugene.Asarin@liafa.jussieu.fr
[2] Centrum voor Wiskunde in Informatica,
P.O. Box 94079, 1090 GB Amsterdam, The Netherlands
Pieter.Collins@cwi.nl

Abstract. Turing machines exposed to a small stochastic noise are considered. An exact characterisation of their ($\approx \Pi_2^0$) computational power (as noise level tends to 0) is obtained. From a probabilistic standpoint this is a theory of large deviations for Turing machines.

1 Introduction

Computers are always subjected to faults and component failures, and even random changes of memory bits caused by to cosmic rays or neutrons flipping memory cells [1]. From the practical viewpoint these phenomena are particularly important for computers operating in hostile environments, such as aboard a spacecraft [2]. In the present paper we adopt a more theoretical and abstract approach to this issue and study how small random perturbations can affect the computational power of popular computational models, in our case Turing machines (TMs).

As far as we know, the pioneering paper considering influence of infinitesimal noise on computational models was Puri's [3], where the author introduces the infinitesimally perturbed semantics and solves the reachability problem for timed automata. Fränzle [4] applies a version of Puri's noise to hybrid systems, and argues that such a noise leads to a sort of "practical decidability". The immediate predecessor of the present paper is [5] where computational power is analysed for infinitesimally perturbed hybrid systems and TMs. The main result of [5] is the Π_1^0 completeness of reachability or acceptance problems for such machines. It is important to notice that all the papers cited above considered a non-deterministic noise of a bounded (and tending to zero) magnitude, with no probabilistic assumptions.

The influence of a small stochastic noise on computational models has been considered in [6] for finite-state models, and in [7] for neural networks. Other related work concerns the behaviour of dynamical systems under a small stochastic noise, known as the theory of large deviations. A good reference is [8].

In the present paper we consider TMs exposed to a small stochastic noise, or in other words large deviations for TMs. We give an exact characterisation of their computational power in terms of classes of arithmetic hierarchy (see [9]).

The rest of the paper is organised as follows. In Sec. 2 we introduce Noisy Turing Machines (NTMs) and several versions of "computability" by such machines. In Sec. 3 we explore NTMs with a noise level $\varepsilon > 0$ and establish some basic properties, such as continuity and computability of acceptance probability, decidability of the halting problem etc. In Sec. 4 we describe several interesting NTMs used in the subsequent sections. This section also gives a flavour of "noisy programming" and reasoning about noisy programs. The main technical results of the paper are established in Sec. 5 and 6 where we explore the computational power of NTMs for a noise level *tending to zero*. Such a "limit computational power" turns out to be stronger than that of a TM; we give its precise characterisation.

2 The Model

We consider a standard multi-tape TM augmented by an additional parameter giving the noise level. Formally, a *Noisy Turing Machine* (NTM) is a tuple

$$\mathcal{M}_\varepsilon = (Q, \Sigma, \Gamma, N, \rho, \varepsilon, q_0, q_\top, q_\bot), \tag{1}$$

where Q is the set of states, Σ is the input alphabet (not containing the special blank symbol $\sqcup$), $\Gamma \supset \Sigma \cup \{\sqcup\}$ is the tape alphabet, N is the number of tapes, $\rho : Q \times \Gamma^N \to Q \times \Gamma^N \times \{L, R, S\}^N$ is the transition function, ε is the noise level, q_0 is the initial state, $q_\top$ the accepting state and $q_\bot$ the rejecting state.

Every tape is infinite in both directions. Initially the ith tape contains a word $w_i \in \Sigma^*$ completed by two infinite tails of blank symbols $\sqcup$, and the ith tape head is pointing to the first symbol of w_i. Every computation step performed by an NTM consists of two stages:

At the *noisy stage* the tapes are exposed to a noisy environment, which changes each symbol on the tape independently with probability $\varepsilon \ll 1$. A changed symbol takes any other value in the tape alphabet with equal probability.

At the *progress stage* the computation proceeds as follows. Starting in state q^-, the machine reads the symbol on each tape, giving an N-tuple $s^- \in \Gamma^N$. If $\rho(q^-, s^-) = (q^+, s^+, m^+)$, the machine changes to state q^+, writes $(s^+)_i$ on the ith tape, and shifts the ith tape head left if $(m^+)_i = L$, right if $(m^+)_i = R$, and does not move it if $(m^+)_i = S$. Whenever the machine arrives at $q_\top$ or at $q_\bot$ it halts.

We are interested in the probabilities $\mathbb{P}(\mathcal{M}_\varepsilon(w) \downarrow)$, $\mathbb{P}(\mathcal{M}_\varepsilon(w) = \top)$ and $\mathbb{P}(\mathcal{M}_\varepsilon(w) = \bot)$ that, for a given noise level ε and a given input word w, the NTM $\mathcal{M}$ halts, accepts or rejects, respectively. We are even more interested in the behaviour of those probabilities as $\varepsilon \to 0$.

Definition 1. *An NTM $\mathcal{M}$ is*

- *lim-halting if $\forall x \in \Sigma^*$, $\lim_{\varepsilon \to 0} \mathbb{P}(\mathcal{M}_\varepsilon(x) \downarrow) = 1$, i.e the limit probability to halt is 1;*

- almost sure (a.s.)-halting *if for any $x \in \Sigma^*$ and any $\varepsilon > 0$ the probability to halt is 1*;
- converging *if $\forall x \in \Sigma^*$, $\lim_{\varepsilon \to 0} \mathbb{P}\left(\mathcal{M}_\varepsilon(x) = \top\right)$ exists.*

Clearly if an NTM is a.s.-halting, then it is lim-halting. The former two properties seem restrictive, but in Sec. 3 it will be shown that any NTM is equivalent to an a.s.-halting one, in the sense that the acceptance probabilities are equal.

The "limit computational power" of an NTM captures the behaviour of a machine operating in an environment which is almost, but not entirely, noise-free.

Definition 2. *An NTM $\mathcal{M}$ lim-generates a function $p : \Sigma^* \to [0,1]$ if it is lim-halting, converging, and for any $x \in \Sigma^*$, the limit probability to accept it is $p(x)$:*

$$\forall x \in \Sigma^*, \lim_{\varepsilon \to 0} \mathbb{P}\left(\mathcal{M}_\varepsilon(x) = \top\right) = p(x).$$

An NTM $\mathcal{M}$ lim-decides a set $S \subset \Sigma^*$ if it lim-generates its characteristic function.

Notice that in order to lim-decide a set S, an NTM should satisfy a 0-1 law:

$$\lim_{\varepsilon \to 0} \mathbb{P}\left(\mathcal{M}_\varepsilon(x) = \top\right) = \begin{cases} 1, & \text{if } x \in S; \\ 0, & \text{if } x \notin S. \end{cases}$$

A weaker notion of computability considering lim sup (or lim inf) rather than lim is suitable for non-converging machines.

Definition 3. *An NTM $\mathcal{M}$ lim sup-generates a function $p : \Sigma^* \to [0,1]$ if it is lim-halting and for any $x \in \Sigma^*$, the upper limit probability to accept it is $p(x)$:*

$$\forall x \in \Sigma^*, \limsup_{\varepsilon \to 0} \mathbb{P}\left(\mathcal{M}_\varepsilon(x) = \top\right) = p(x).$$

An NTM $\mathcal{M}$ lim sup-decides a set $S \subset \Sigma^*$ if for any $x \in \Sigma^*$ it lim sup-generates its characteristic function.

The question whether this can be really considered as a computation is left to a philosophically-minded reader.

2.1 If It Halts Without Noise

We start the study of noisy machines with the easy case when a machine without noise halts on an input x.

Theorem 1. *If $\mathcal{M}(x) = \top$ then $\lim_{\varepsilon \to 0} \mathbb{P}\left(\mathcal{M}_\varepsilon(x) = \top\right) = 1$. Symmetrically, if $\mathcal{M}(x) = \bot$ then $\lim_{\varepsilon \to 0} \mathbb{P}\left(\mathcal{M}_\varepsilon(x) = \bot\right) = 1$.*

Proof. Consider the first case, $\mathcal{M}(x) = \top$; the other case is similar. Let τ be the computation time of $\mathcal{M}$ on the input x. For any $\delta > 0$ take an $\varepsilon < \delta \tau^{-2}$.

A normal computation (without noise) of $\mathcal{M}$ on x uses at most τ tape cells during τ time units. When noise is added the probability for at least one of those cells being perturbed during τ time units cannot exceed $\tau \cdot T \cdot \varepsilon < \delta$. The computation of $\mathcal{M}$ on x leading to acceptance is then unaffected by the noise, hence the NTM $\mathcal{M}_\varepsilon$ accepts x with probability at least $1-\delta$. Since δ is arbitrary, $\lim_{\varepsilon \to 0} \mathbb{P}\left(\mathcal{M}_\varepsilon(x) = \top\right) = 1$. □

Corollary 1. *If a TM $\mathcal{M}$ decides a set $S \subset \Sigma^*$ then its NTM version lim-decides the same set S.*

3 General Properties: $\varepsilon > 0$

In this section we explore NTMs and acceptance by NTMs for a positive noise level $\varepsilon > 0$. This paves the way for the characterisation of the limit behaviour of NTMs as $\varepsilon \to 0$ in subsequent sections.

3.1 Automaton

For an NTM $\mathcal{M}$ described by a tuple (1), we can abstract away the memory tapes to obtain an automaton

$$\mathcal{A} = (Q, \rho', q_0, q_\top, q_\bot),$$

where the transition relation $\rho' \subset Q \times Q$ is obtained from $\rho : Q \times \Gamma^N \to Q \times \Gamma^N \times \{L, R, S\}^N$ by projection. Any allowable sequence of transitions of $\mathcal{A}$ is a possible sequence of transitions for $\mathcal{M}$, since it is always possible that the element at the tape head changes just before the step to enable the desired transition.

We say a state q of $\mathcal{M}$ is a *looping state* if for every possible sequence of transitions starting at q, no halting state is reached. In this case, the probability that $\mathcal{M}$ halts given that it reaches q is zero.

Theorem 2. *For any NTM $\mathcal{M}$, there exists an effectively constructible a.s.-halting NTM $\mathcal{M}'$ with the same acceptance probability for any input $x \in \Sigma^*$ and any $\varepsilon > 0$.*

Proof. We construct $\mathcal{M}'$ by deleting all looping states from $\mathcal{M}$, and replacing all transitions leading to looping states with a transition leading to $q_\bot$. Then clearly, $\mathbb{P}\left(\mathcal{M}'_\varepsilon(x) = \top\right) = \mathbb{P}\left(\mathcal{M}_\varepsilon(x) = \top\right)$.

It remains to show that $\mathbb{P}\left(\mathcal{M}'_\varepsilon(x) \downarrow\right) = 1$. Since $\mathcal{M}'$ has no looping states, for any state q of $\mathcal{M}'$, there is a sequence of transitions leading to a halting state in at most $k = |Q|$ steps. This sequence of transitions occurs for the noisy machine with probability at least $(\varepsilon/|\Sigma|)^k$. Therefore the probability that the machine halts after nk steps is at least $1 - \left(1 - (\varepsilon/|\Sigma|^k)\right)^n$. Hence $\mathcal{M}'$ halts with probability 1. □

We can therefore replace any NTM with one that halts with probability 1 for every $\varepsilon > 0$ without changing the acceptance probability. This means that, unlike ordinary TMs, we need only consider NTMs which almost surely halt on any input. In the rest of this paper, we assume that all NTMs have no looping states.

3.2 Continuity

Theorem 3. $\mathbb{P}\left(\mathcal{M}_\varepsilon(w) = \top\right)$ *is continuous with respect to ε for $\varepsilon > 0$.*

Proof. Let $p(w, \varepsilon, t)$ be the probability that $\mathcal{M}$ halts in time t in state $q_\top$, and $q(w, \varepsilon, t)$ be the probability that $\mathcal{M}$ halts in time t in state $q_\bot$. Then clearly $p(w, \varepsilon, t)$ and $q(w, \varepsilon, t)$ are continuous as functions of $\varepsilon > 0$, since they depend on finite computations. Let $r(w, \varepsilon, t) = 1 - p(w, \varepsilon, t) - q(w, \varepsilon, t)$. Then by our standing assumption of almost-sure halting, $r(w, \varepsilon, t) \to 0$ as $t \to \infty$.

It is easy to see that

$$p(w, \varepsilon, t) < \mathbb{P}\left(\mathcal{M}_\varepsilon(w) = \top\right) < p(w, \varepsilon, t) + r(w, \varepsilon, t) = 1 - q(w, \varepsilon, t). \quad (2)$$

To prove continuity of $\mathbb{P}\left(\mathcal{M}_\varepsilon(w) = \top\right)$ at ε, take t such that $r(w, \varepsilon, t) < \delta/3$. For ε' sufficiently close to ε, both $|p(w, \varepsilon', t) - p(w, \varepsilon, t)| < \delta/3$ and $r(w, \varepsilon', t) < 2\delta/3$. Then, using (2), we obtain $|\mathbb{P}\left(\mathcal{M}_\varepsilon(w) = \top\right) - \mathbb{P}\left(\mathcal{M}_{\varepsilon'}(w) = \top\right)| < \delta$. Hence $\mathbb{P}\left(\mathcal{M}_\varepsilon(w) = \top\right)$ is continuous. □

3.3 Computability

We now consider ε-perturbed machines for a fixed rational ε. By computability of a real number x (see [10]), we mean that given an error bound $\delta > 0$, there is a TM which computes an approximation to x with an error of at most δ.

Theorem 4. $\mathbb{P}\left(\mathcal{M}_\varepsilon(w) = \top\right)$ *is computable as a function of $\mathcal{M}$, rational $\varepsilon > 0$ and w.*

Proof. Let $p(w, \varepsilon, t)$, $q(w, \varepsilon, t)$ and $r(w, \varepsilon, t)$ be as in the proof of Theorem 3. By simulating all possible runs of the NTM of length at most t and computing the probability of each, we can compute p, q and r exactly. Since $r(w, \varepsilon, t) \to 0$ as $t \to \infty$, we can take t sufficiently large so that $r(t, \varepsilon, t) < \delta$, and so $|\mathbb{P}\left(\mathcal{M}_\varepsilon(w) = \top\right) - p(w, \varepsilon, t)| < \delta$. □

4 Some Gadgets

We now describe some generally useful NTMs and their properties.

4.1 Measuring Time

The first gadget is a TIMER. Its construction is very simple: it is just a TM with one tape (initially blank), whose head goes right at every step. If it sees a non-blank cell it stops.

The following lemma establishes that, when subjected to an ε-noise, TIMER is capable to measure approximately a lapse of $\varepsilon^{-1/2}$ time units.

Lemma 1. *Let τ be the time before the TIMER_ε stops. Then for any a, b with $a < 1/2 < b$, the following estimates hold:*

1. $\mathbb{P}\left(\tau < \varepsilon^{-a}\right) = O(\varepsilon^{1-2a})$;
2. $\mathbb{P}\left(\tau > \varepsilon^{-b}\right) = O(\varepsilon^{d})$ for any $0 < d < 1/2$.

Proof. First we estimate the probability of the event E_1 that $\tau < \varepsilon^{-a}$. This probability can be majorated by the probability of the event E_2 that during $\lfloor \varepsilon^{-a} \rfloor$ time units at least one of the first $\lfloor \varepsilon^{-a} \rfloor$ cells on the tape has been altered by the noise. For each cell and each step the probability to be altered is ε, which gives an upper bound

$$\mathbb{P}(E_1) \leq \mathbb{P}(E_2) \leq \lfloor \varepsilon^{-a} \rfloor \cdot \lfloor \varepsilon^{-a} \rfloor \cdot \varepsilon = O(\varepsilon^{1-2a}).$$

In the sequel we will omit $\lfloor \cdot \rfloor$ and $\lceil \cdot \rceil$ symbols and make all the computations as if all the powers of ε considered were integer numbers.

The event E_3 that $\tau > \varepsilon^{-b}$, implies either the event E_4 that none of the first ε^{-b} cells have been modified before the timer scans them, or the event E_5 that at least one of the first ε^{-b} cells has been modified at least twice in time ε^{-b}. Hence $\mathbb{P}(E_3) \leq \mathbb{P}(E_4) + \mathbb{P}(E_5)$.

E_4 is a conjunction of $\varepsilon^{-b} \cdot \varepsilon^{-b}/2$ independent events with probabilities $1-\varepsilon$; each event means that a cell has not been perturbed at a time instant. Hence

$$\mathbb{P}(E_4) = (1-\varepsilon)^{\varepsilon^{-2b}/2} = O(\exp(-\varepsilon^{1-2b})).$$

In particular, if $b > 1/2$, $\mathbb{P}(E_4) = o(\varepsilon^n)$ for any $n > 0$.

The event E_5 is a disjunction of independent $n = \varepsilon^{-b}$ events. Each of those events is that a particular cell has been perturbed at least twice during ε^{-b} time units. Hence

$$\mathbb{P}(E_5) \leq n \left(1 - (1-\varepsilon)^n - n\varepsilon(1-\varepsilon)^{n-1}\right) = O(n^3 \varepsilon^2) = O(\varepsilon^{2-3b}).$$

Therefore, $\mathbb{P}(E_3) = O(\varepsilon^{2-3b})$ for $1/2 < b < 2/3$. Since $\mathbb{P}(E_3)$ is a decreasing function of b, we must have $\mathbb{P}(E_3) = O(\varepsilon^d)$ for any $0 < d < 1/2$. □

We say an event E occurs with *high probability* if there exists $d > 0$ such that $\mathbb{P}(E) = 1 - O(\varepsilon^d)$. Similarly, it occurs with *low probability* if $\mathbb{P}(E) = O(\varepsilon^d)$.

We remark that it would be easier to build a timer measuring ε^{-1} lapse of time. Such a timer is an NTM staying in place and observing one cell, until its contents is modified. Unfortunately, such a timer would be rather useless, because during such a long time the contents of all the cells on the tape becomes completely random.

We prefer the $\varepsilon^{-1/2}$ timer described at the beginning of this section because during such time the probability of perturbation on a small zone of tape or of a "double error" in the same position of two tapes is low. In the next subsection we formalise these properties, and explain how they allow for reliable computations of duration $\varepsilon^{-1/2}$ and even more. We will then be able to use these constructions to build $O(\varepsilon^{-c})$ TIMERS for $1/2 < c < 1$.

4.2 Tossing Coins

By letting two TIMERS race each other to find a non-blank symbol, we can generate random bits.

A RANDOMBIT machine has two tapes. Two TIMERS are launched concurrently on both tapes. If the first one stops before the second one, the result is $\bot$, if the second one stops before the result is $\top$. In the highly improbable case of a tie, the TIMERS are restarted. The following result is straightforward, but important; it shows that NTMs can produce random bits.

Lemma 2. *The* RANDOMBIT$_\varepsilon$ *terminates almost surely and returns* $\bot$ *and* $\top$ *(or 0 and 1) with probabilities 1/2 each. Its computation time is bounded above by* ε^{-b} *(with* $b > 1/2$*) with probability* $1 - O(\varepsilon^d)$ *for any* $0 < d < 1/2$.

Notice that the RANDOMBIT$_\varepsilon$ gadget can also be started some time T after the beginning, and can be run continuously to generate a succession of random bits.

4.3 Memory

Even on time scales of order $\varepsilon^{-1/2}$ generated by a TIMER, the behaviour of a noisy version of a regular TM $\mathcal{M}$ has unacceptably high errors. To obtain correct execution with high probability of a time interval of order ε^{-a}, we run all computations of $\mathcal{M}$ on a multi-tape MEMORY with error correction.

For computations taking $O(\varepsilon^{-a})$ time with $a < 2/3$ we can guarantee error-freedom with high probability by taking three identical copies of the tape and making the same computation on all of them. If at some moment a disagreement between the three tapes is observed, it is immediately corrected by a majority vote. This procedure allows to correct single errors, while Lemma 3 ensures that double errors are highly improbable.

By using more tapes, we can, in fact, construct MEMORY which is error-free with high probability on time intervals $O(\varepsilon^{-c})$ for any $c < 1$.

Lemma 3. *Let* $\mathcal{M}_\varepsilon$ *be an NTM running on a three-tape* MEMORY *for a time period* $O(\varepsilon^{-a})$ *using space* $O(\varepsilon^{-b})$. *Then the probability of incorrect execution of* $\mathcal{M}_\varepsilon$ *is of order* $O(\varepsilon^{2-2a-b})$.

Proof. Incorrect execution can only occur if two cells with the same coordinate are perturbed in time period $\tau = O(\varepsilon^{-a})$. The probability of such a "double error" in a given cell is $O(\varepsilon^{2(1-a)})$, hence the probability of a double error in any of b cells is $O(\varepsilon^{2-2a-b})$.

4.4 Programming NTMs

Armed with a TIMER, a RANDOMBIT and (fairly) reliable MEMORY, we can start to program NTMs by running ordinary TMs on a MEMORY, using a TIMER to halt the computation before the noise-induced errors become too high.

A simple, but very useful gadget is a COUNTER machine. This machine stores an integer n in binary form in a three-tape MEMORY. The COUNTER spends all its time incrementing its value, which asymptotically grows as $n \sim t/\log t$.

Using a COUNTER, we can construct a DELAY gadget. When this gadget is activated, it copies the time n contained in the counter, and computes some (easy to compute) function $f(n) \sim n^s$. It then waits until the COUNTER reaches $f(n)$, emits a signal, and goes back to sleep.

Using the COUNTER and DELAY gadgets, we can construct an improved version of a TIMER. We run an $\sim \varepsilon^{-1/2}$ timer as usual, but when this stops, we activate a DELAY gadget with $f(n) \sim n^{2c}$. The program continues running until the DELAY gadget deactivates. This new TIMER(c) gadget stops in time $\sim \varepsilon^{-c}$ with high probability.

We can use TIMER(c) and RANDOMBIT to construct a RANDOMNUMBER. By storing successively generated bits in a MEMORY, we generate an increasing sequence of rationals r_i converging to a real number r which is uniformly distributed in $[0, 1]$. By using a TIMER(c), we can generate $\sim \varepsilon^{1/2-c}$ digits of r.

4.5 Oscillators

An OSCILLATOR is a gadget which stores a binary digit in a "register" variable where it is unaffected by the noise. (Formally, we construct a register by taking a set of states $Q \times \{0, 1\}$.) When the OSCILLATOR is halted by a TIMER, it stops in $q_\top$ if the register holds 1, and in $q_\bot$ if the register holds 0.

A simple oscillator which changes register state at every step is not very interesting; the limiting acceptance probability is $1/2$. By using a DELAY, we can hold the value of the register for a period $[m, f(m) \sim m^c]$; long enough for its value to be seen when the TIMER halts.

Lemma 4. *Let $\mathcal{M}$ be an OSCILLATOR which uses a delay to switch state at times $n_i = f(n_{i-1})$ with $f(n) \geq n^c$, and which halts when a TIMER stops. Then M halts almost surely, but $\mathbb{P}(\mathcal{M}_\varepsilon = \top)$ does not converge as $\varepsilon \to 0$.*

Proof. Choose a, b such that $a < 1/2 < b < 2/3$, $b/a < c$ and $a + b < 1$, and let $d = 1 - a - b$. For any given n, we can find $\varepsilon < 1$ such that $[\varepsilon^{-a}, \varepsilon^{-b}] \subset [n, n^c]$. If oscillator switches at times n_i, then $n_{i+1} \geq n_i^c$. Hence, there is a sequence ε_i with $\varepsilon_i \to 0$ as $i \to \infty$ such that $[\varepsilon_i^{-a}, \varepsilon_i^{-b}] \subset [n_i, n_{i+1}]$.

The TIMER halts at $\tau \in [\varepsilon_n^{-a}, \varepsilon_n^{-b}]$ with high probability. Further, in this time the MEMORY ensures correct execution with high probability. Hence for noise levels ε_{2i}, the probability that the timer halts with the register in state 0 tends to 1 as $n \to \infty$, and for noise levels ε_{2i+1} timer halts with the register in state 1 with high probability. Thus $\mathbb{P}(\mathcal{M}_\varepsilon = \top)$ oscillates between 0 and 1 as $\varepsilon \to 0$, and does not converge as $\varepsilon \to 0$. □

5 Decisional Power of NTMs

In this section we address the capabilities of NTMs for deciding sets $S \subset \Sigma^*$. The main result of this section is

Main Result 1. *A set $S \subset \Sigma^*$ is lim-decidable if and only if it is Δ_2^0. A set $S \subset \Sigma^*$ is lim sup-decidable if and only if it is Π_2^0.*

The upper complexity bounds follow from Theorems 9 and 11 in the next section.

5.1 Deciding Recursively Enumerable Sets

The following result illustrates how converging NTMs can be stronger than ordinary TMs, and solve, for example, the halting problem.

Theorem 5. *For any recursively enumerable (Σ_1^0) set S there exists an NTM $\mathcal{N}$ which* lim*-decides S.*

Proof. Recall that S is Σ_1^0 if there is a TM $\mathcal{M}$ such that $\mathcal{M}(w)$ halts if, and only if, $w \in S$. Given such a TM, we construct an NTM $\mathcal{N}_\varepsilon$ with 4 tapes to lim-decide S. On tapes 1, 2 and 3 we have a MEMORY store on which we run $\mathcal{M}$, correcting errors by majority vote. On tape 4, we run a TIMER gadget.

The computation terminates in the accepting state if $\mathcal{M}$ runs successfully and reaches its halting state. The computation terminates in the rejecting state if the TIMER stops.

Since the TIMER stops almost surely, and does so with high probability in time $\tau \in [\varepsilon^{-a}, \varepsilon^{-b}]$, the computation performed by $\mathcal{M}$ runs successfully with high probability, terminating in state $q_\top$ if $w \in S$ (as long as ε is small enough, namely such that ε^{-a} exceeds the computation time of $\mathcal{M}(w)$), and halts in state $q_\bot$ if $w \notin S$. □

5.2 Deciding Δ_2^0 Sets

We now strengthen the result of Theorem 5 to show that NTMs can lim-decide Δ_2^0 sets. Recall that a set S is Δ_2^0 if both S and its complement are Π_2^0.

A particularly useful characterisation of a Π_2^0 set, similar to Büchi acceptance by ω-automata, can be given in terms of *signalling* TMs. A signalling TM has no halting states, but instead a distinguished set of *signalling states* $Q_s \subset Q$. A set S is Π_2^0 if there is a signalling TM $\mathcal{M}$ which enters states from Q_s infinitely often if, and only if, $w \in S$.

Theorem 6. *For any Δ_2^0 set S there exists an NTM which* lim*-decides S.*

Proof. Let $\mathcal{M}^1$ be a TM such that $\mathcal{M}^1(w)$ emits a signal infinitely often iff $w \in S$, and $\mathcal{M}^0$ a TM such that $\mathcal{M}^0(w)$ emits a signal infinitely often iff $w \notin S$.

We can run $\mathcal{M}^1$ and $\mathcal{M}^0$ in parallel with a register variable. Whenever $\mathcal{M}^i$ emits a signal, we store i in the register. If $w \in S$, then eventually $\mathcal{M}^0$ emits no more signals, but $\mathcal{M}^1$ continues to do so, and the register sticks to 1. Conversely, if $w \notin S$, then eventually the register contains 0. In both cases, after some time $T(w)$ the register's content never changes and is equal to $\chi_S(w)$.

We lim-decide S by an NTM $\mathcal{N}$ running $\mathcal{M}^1$ and $\mathcal{M}^0$ as described above. Computation is terminated when a TIMER stops, the outcome is the register's content. Whenever ε is small enough to ensure that the TIMER stops after time $T(w)$, we can easily see that $\mathcal{N}_\varepsilon$ operates correctly with high probability and outputs $\chi_S(w)$. Hence $\mathcal{N}$ lim decides S. □

5.3 Deciding Π_2^0 Sets

We now investigate the computational power of NTMs without restriction of convergence.

Theorem 7. *For any Π_2^0 set S there exists an NTM which* lim sup-*decides S.*

Proof. Let $\mathcal{M}$ be a TM such that $\mathcal{M}(w)$ emits a signal infinitely often iff $w \in S$. To ensure that signals are noticed in the presence of noise, we combine $\mathcal{M}$ with a DELAY which activates when $\mathcal{M}$ emits a signal, setting a register to 1. Whenever DELAY deactivates, it resets the register to 0. As usual, the computation is halted whenever the TIMER stops.

Assuming correct operation of $\mathcal{M}$ and DELAY, which occurs with high probability, the register variable is equal to 1 on time intervals $[\tau_i, \tau_i^c]$ infinitely often if $w \in S$, and is eventually equal to 0 if $w \notin S$. Using the argument from the proof of Lemma 4 we can see that $\mathcal{N}$ lim sup-decides S. □

6 Generating Probability Functions

In this section, we investigate the functions which can be generated as the acceptance probability of an NTM as $\varepsilon \to 0$. We shall prove the following result:

Main Result 2. *A function $p : \Sigma^* \to [0;1]$ can be* lim-*generated by a converging NTM if and only if it is $0'$-computable. A function $p : \Sigma^* \to [0;1]$ is* lim sup-*generated by an NTM if and only if it is upper $0'$-semicomputable.*

6.1 Generating $0'$-Computable Probabilities

Recall that a function is called $0'$-computable if it can be computed by a TM using an oracle for the halting problem. Equivalently, p is $0'$-computable iff $\{(r, w) \in \mathbb{Q} \times \Sigma^* \mid r < p(w)\}$ and $\{(r, w) \in \mathbb{Q} \times \Sigma^* \mid r > p(w)\}$ are Π_2^0-sets.

Theorem 8. *Let $p : \Sigma^* \to [0, 1]$ be a $0'$-computable function. There exists a converging NTM $\mathcal{S}$ that* lim-*generates the function p.*

Proof. Since $\{(r, w) \in \mathbb{Q} \times \Sigma^* \mid r < p(w)\}$ is Π_2^0, there is a TM $\mathcal{M}_<$ such that $\mathcal{M}_<(r, w)$ emits infinitely many signals if, and only if, $r < p(w)$.

We now aim to extend this computation to real numbers. Suppose we have access to an increasing sequence of rationals r_i converging to a real number r. We run $\mathcal{M}_<$ sequentially on inputs (r_i, w), starting $\mathcal{M}_<(r_n, w)$ after each computation $\mathcal{M}_<(r_i, w)$ (with $i < n$) has emitted $n - i$ signals.

Suppose $r < p(w)$. Then $r_i < p(w)$ for all i, so $\mathcal{M}_<(r_i, w)$ emits infinitely many signals for all i. Conversely, if $r > p(w)$, then $r_i \geq p(w)$ for some i, and $\mathcal{M}_<(r_i, w)$ emits only finitely many signals before looping indefinitely.

We can therefore construct a TM $\mathcal{M}'_<$ which, given $w \in \Sigma^*$ and n digits of r, emits infinitely many signals if $r < p(w)$, and finitely many signals if $r > p(w)$ and n is sufficiently large. Similarly, we can construct a TM $\mathcal{M}'_>$ which, given n elements of a decreasing sequence converging to r, emits infinitely many signals if $r > p(w)$, and finitely many signals if $r < p(w)$ and n is sufficiently large.

We now construct an NTM $\mathcal{N}$ to lim-generate the function p. We use a RANDOMNUMBER gadget to generate the binary approximants r_i to a random variable r uniformly distributed in $[0,1]$. Notice that the distribution of the r_i is independent of the noise ε. We use $\mathcal{M}'_<$ and $\mathcal{M}'_>$ to compute $r < p(w)$ and $r > p(w)$. The computation is halted by a TIMER(c) with $1/2 < c < 2/3$ to ensure that RANDOMNUMBER generates sufficiently many bits, but that the MEMORY is still error-free with high probability.

Fix $w \in \Sigma^*$ and n, and suppose $p(w) \notin [r_n, r_n + 1/2^n]$, which occurs with probability $1 - 1/2^n$. We claim that after a fixed time T, independent of r, the value of the register does not change. In the case $r_n + 1/2^n < p(w)$, then $r_i < p(w)$ for all i, so $\mathcal{M}'_<$ emits infinitely many signals, whereas $\mathcal{M}_>(r_n, w)$ emits only finitely many signals, so after some time $T(r_n)$, machine $\mathcal{M}_>(r_n, w)$ does not emit further signals. Since there are only finitely many possible values of r_n, we can choose T independently of r_n. The case $r_n > p(w)$ is similar.

Using the same argument as in Theorem 6, we see that with high probability, $\mathcal{N}_\varepsilon$ accepts if $r_n < p(w)$ and rejects if $p(w) < r_n + 1/2^n$. Hence $\mathbb{P}(\mathcal{N}_\varepsilon(w) = \top) \in [r_n, r_n + 1/2^n]$, and since n is arbitrary, $\mathbb{P}(\mathcal{N}_\varepsilon(w) = \top) = p(w)$. □

To prove that Theorem 8 gives a precise characterisation of the computational power of a converging NTM, we analyse the limit as $\varepsilon \to 0$.

Theorem 9. $\lim_{\varepsilon \to 0} \mathbb{P}(\mathcal{M}_\varepsilon(w) = \top)$ is $0'$-computable for any converging NTM.

Proof. The function $f(\varepsilon, w)$ given by $\mathbb{P}(\mathcal{M}_\varepsilon(w) = \top)$ is computable when ε is rational, and converges as $\varepsilon \to 0$ for all $w \in \Sigma^*$. By definition,

$$r < \lim_{\varepsilon \to 0} f(\varepsilon, w) \iff \exists s > r, \exists q > 0, \forall \varepsilon < q, (\neg(s > f(\varepsilon, w))).$$

The inequality $(s > f(\varepsilon, w))$ belongs to the class Σ^0_1. We deduce that the relation $(r < \lim_{\varepsilon \to 0} f(\varepsilon, w))$ belongs to the class Σ^0_2, and hence is $0'$-recursively enumerable. Symmetrically, the inequality $(r > \lim_{\varepsilon \to 0} f(\varepsilon, w))$ is also $0'$-recursively enumerable. Hence $\lim_{\varepsilon \to 0} f(\varepsilon, w)$ is $0'$-computable. □

6.2 Generating Upper $0'$-Semicomputable Probabilities

Recall that a function p is upper $0'$-semicomputable if the set $\{(r, w) \in \mathbb{Q} \times \Sigma^* \mid r > p(w)\}$ is of class Σ^0_2, so that $\{(r, w) \in \mathbb{Q} \times \Sigma^* \mid r \leq p(w)\}$ is of class Π^0_2.

Theorem 10. *Let $p : \Sigma^* \to [0, 1]$ be an upper $0'$-semicomputable function. There exists an NTM $\mathcal{S}$ that lim sup-generates p.*

Proof. We use the machine $\mathcal{M}'_<(r, w)$ and the RANDOMNUMBER from the proof of Theorem 8, and combine this with DELAY as in the proof of Theorem 7, and a TIMER(c) to halt the computation.

Computation proceeds by running $\mathcal{M}'_<$, and starting a DELAY whenever $\mathcal{M}_<$ emits a signal. If $\mathcal{M}_<$ emits signals infinitely often, then every $r_n < p(w)$, so $r \leq p(w)$, and if $\mathcal{M}_<(r_n, w)$ loops for some r_n, then $r_n \geq p(w)$, so $r \geq p(w)$.

The rest of the proof follows that of Theorems 7 and 8, and is omitted. □

Theorem 11. $\limsup_{\varepsilon \to 0} \mathbb{P}(\mathcal{M}_\varepsilon(w) = \top)$ *is upper $0'$-semicomputable.*

Proof. As in the proof of Theorem 9, it is easy to show that $\limsup_{\varepsilon \to 0} f(\varepsilon, w)$ is upper $0'$-semicomputable whenever $f : (0,1) \times \Sigma^* \to [0;1]$ is continuous and computable. □

7 Concluding Remarks

We have described a class of randomly perturbed Turing machines and studied their computational properties. We have shown that in the limit of infinitesimal noise, these machines can be programmed to lim-decide Δ_2^0, and lim sup-decide Π_2^0 sets. It is interesting to compare this result with [5], where a small nondeterministic noise led to a Π_1^0 computational power only. We have also given a characterisation of the acceptance probability distributions which can be generated. As a future work we are planning to explore how sensitive are these results to the choice of a computational model (discrete, hybrid or analog) and of a stochastic noise model.

References

1. McKee, W., Bergman, D., Nguyen, N., Aton, T., Block, L., Huynh, V., McAdams, H., Smith, E., McPherson, J., Janzen, J., Ondrusek, J., Hyslop, A., Russell, D., Coy, R.: Cosmic ray neutron induced upsets as a major contributor to the soft error rate of current and future generation DRAMs. In: Proceedings of the International Reliability Physics Symposium. (1996)
2. Howard, J.J., Hardage, D.: Spacecraft environments interactions: Space radiation and its effects on electronic systems. Technical Report TP-1999-209373, NASA (1999) http://trs.nis.nasa.gov/archive/00000502/01/tp209373.pdf.
3. Puri, A.: Dynamical properties of timed automata. Discrete Event Dynamic Systems **10** (2000) 87–113
4. Fränzle, M.: Analysis of hybrid systems: An ounce of realism can save an infinity of states. In Flum, J., Rodríguez-Artalejo, M., eds.: Computer Science Logic (CSL'99). Volume 1683 of LNCS., Springer-Verlag (1999) 126–140
5. Asarin, E., Bouajjani, A.: Perturbed Turing machines and hybrid systems. In: Proceedings of the 16th Annual IEEE Symposium on Logic in Computer Science, IEEE Computer Society (2001) 269
6. Delyon, B., Maler, O.: On the effects of noise and speed on computations. Theoretical Computer Science **129** (1994) 279–291
7. Maass, W., Orponen, P.: On the effect of analog noise in discrete-time analog computations. Neural Computation **10** (1998) 1071–1095
8. Freidlin, M., Wentzell, A.: Random perturbations of dynamical systems. Springer-Verlag, New York (1984)
9. Rogers, H.: Theory of Recursive Functions and Effective Computability. McGraw-Hill (1967)
10. Weihrauch, K.: Computable analysis. Springer-Verlag, Berlin (2000)

A Better Approximation Ratio for the Vertex Cover Problem[*]

George Karakostas[**]

Department of Computing and Software, McMaster University
karakos@mcmaster.ca

Abstract. We reduce the approximation factor for Vertex Cover to $2 - \Theta(\frac{1}{\sqrt{\log n}})$ (instead of the previous $2 - \Theta(\frac{\log \log n}{\log n})$, obtained by Bar-Yehuda and Even [3], and by Monien and Speckenmeyer [11]). The improvement of the vanishing factor comes as an application of the recent results of Arora, Rao, and Vazirani [2] that improved the approximation factor of the sparsest cut and balanced cut problems. In particular, we use the existence of two big and well-separated sets of nodes in the solution of the semidefinite relaxation for balanced cut, proven in [2]. We observe that a solution of the semidefinite relaxation for vertex cover, when strengthened with the triangle inequalities, can be transformed into a solution of a balanced cut problem, and therefore the existence of big well-separated sets in the sense of [2] translates into the existence of a big independent set.

1 Introduction

One of the most well-studied problems in combinatorial optimization is the vertex cover (VC) problem: given a graph $G = (V, E)$, we look for a minimum size subset of vertices such that for every $(u, v) \in E$, at least one of u, v belongs to this subset. In the *weighted* version of VC, each vertex has an integral weight, and we are looking for the minimum total weight subset of vertices with the property above.

Since the complexity of VC has been heavily studied since Karp's original proof of its NP-completeness [9], the related bibliography is vast and cannot be covered, of course, in this introductory note. We mention here that VC is known to be APX-complete [12], and moreover it cannot be approximated within a factor of 1.36 [6], unless P=NP. A 2-approximation on the other hand can be trivially obtained by taking all the vertices of a maximal matching in the graph.

Improving this simple 2-approximation algorithm has been a quite non-trivial task. The best approximation algorithms known before this work were published 20 years ago by Bar-Yehuda and Even [3], and by Monien and Speckenmeyer [11].

---

[*] A preliminary version of this work appeared as a McMaster University Technical Report CAS-04-05-GK and ECCC Report TR04-084, September/October 2004.
[**] Research supported by an NSERC Discovery grant.

They achieved an approximation factor of $2 - \frac{\ln \ln n}{2 \ln n}$, where n is the number of vertices. If Δ is the maximum degree of the graph, Halperin [8] showed that a factor of $2 - (1 - o(1))\frac{2 \ln \ln \Delta}{\ln \Delta}$ can be achieved by using the semidefinite programming (SPD) relaxation of VC.

In this work we use a stronger SDP relaxation to improve the approximation factor achieved in polynomial time to $2 - \Theta(\frac{1}{\sqrt{\log n}})$. We observe that the introduction of all the so-called *triangle inequalities* to the standard SDP relaxation of VC is, in fact, very similar to the balanced cut SDP relaxation used by Arora, Rao, and Vazirani [2]. Then we use one of the main results of [2], which asserts that in the solution of this SDP, there are two big and well-separated vertex subsets. At the same time, we show that edges that were not covered by a trivial initial rounding are too big to have both of their endpoints in either of these two sets. Hence, one of these two big subsets has to be a big independent set, which can be excluded. We show this process first for the unweighted VC, and then we show how it can be extended to the weighted case in a straight-forward manner.

Hence the main idea in this improvement of the approximation factor is the transformation of the 'classic' SDP formulation of VC to a formulation that corresponds to a balanced cut problem with the addition of the 'antipodal' points of the original points. This appears to be a very general technique that can find application to other problems, and bring the Arora, Rao, Vazirani [2] improved approximation factor for balanced cuts to other contexts as well. Indeed, subsequently to our work, Agarwal, Charikar, Makarychev, and Makarychev [1] used this stronger SDP formulation and the negative symmetric metric that it implies, together with a clever iterative application of the separation algorithm of [2], and several other ideas like volume arguments to improve the approximation factor of several problems (MIN UNCUT, MIN 2CNF DELETION, DIRECTED BALANCED SEPARATOR, DIRECTED SPARSEST CUT) to $O(\sqrt{\log n})$.

2 The Unweighted Case

The following is a semidefinite-programming relaxation of unweighted Vertex Cover (VC) for a graph $G = (V, E)$ with n nodes:

$$\min \sum_{i=1}^{n} \frac{1 + v_0 v_i}{2} \text{ s.t.} \tag{SDP}$$

$$(v_0 - v_i)(v_0 - v_j) = 0, \quad \forall (i,j) \in E \tag{1}$$

$$(v_i - v_j)(v_i - v_k) \geq 0, \quad \forall i, j, k \in V \cup \{0\} \tag{2}$$

$$v_i^2 = 1, \quad \forall i \in V \cup \{0\} \tag{3}$$

where $v_i \in \mathbb{R}^{n+1}$. Constraints (2) are *triangular inequalities*, which must be satisfied by the vertex cover. In an 'integral' solution of (SDP) (which would correspond to a vertex cover of G), vertices that are picked coincide with v_0, while vertices that are not picked coincide with $-v_0$. In general though, an optimal solution of (SDP) will not be 'integral'.

In fact one can strengthen this SDP relaxation for VC by adding all the so called triangle inequalities:

$$\min \sum_{i=1}^{n} \frac{1+v_0 v_i}{2} \text{ s.t.}$$

$$(v_0 - v_i)(v_0 - v_j) = 0, \qquad \forall (i,j) \in E$$
$$(v_i - v_j)(v_i - v_k) \geq 0, \qquad \forall i,j,k \in V \cup \{0\}$$
$$(v_i + v_j)(v_i - v_k) \geq 0, \qquad \forall i,j,k \in V \cup \{0\}$$
$$(v_i + v_j)(v_i + v_k) \geq 0, \qquad \forall i,j,k \in V \cup \{0\}$$
$$v_i^2 = 1, \qquad \forall i \in V \cup \{0\}$$

This relaxation is in fact equivalent to the following relaxation: We add n more 'shadow' points to (SDP) so that for every unit vector v_i, $i = 1, \ldots, n$ we add unit vector v_i' which is the antipodal of v_i, i.e., $v_i v_i' = -1$, $\forall i$. Let V' be the set of shadow points. Note that in an integral solution of (SDP), exactly half (n) of the points in $V \cup V'$ coincide with v_0 and the other half coincide with $-v_0$. Therefore the following must hold

$$\sum_{i,j \in V \cup V'} |v_i - v_j|^2 = 4n^2$$

where every pair (i, j) appears only once in the sum. (Hence the set $V \cup V'$ is 1/2-spread in the terminology of [2]). In addition, the triangular inequalities (2) must also hold when we extend V with V'. Hence we have the following strengthened SDP:

$$\min \sum_{i=1}^{n} \frac{1+v_0 v_i}{2} \text{ s.t.} \qquad \text{(SDP')}$$

$$(v_0 - v_i)(v_0 - v_j) = 0, \qquad \forall (i,j) \in E \qquad (4)$$
$$(v_i - v_j)(v_i - v_k) \geq 0, \qquad \forall i,j,k \in V \cup V' \cup \{0\} \qquad (5)$$
$$v_i^2 = 1, \qquad \forall i \in V \cup V' \cup \{0\} \qquad (6)$$
$$v_i v_i' = -1, \qquad \forall i \in V \qquad (7)$$
$$\sum_{i,j \in V \cup V'} |v_i - v_j|^2 = 4n^2 \qquad (8)$$

where $v_i, v_i' \in \mathbb{R}^d$ for some $d \gg \log n$. Constraint (8) is in fact unnecessary since it is always satisfied by a set of points and their antipodals, but we include it in order to point out that this relaxation defines a *spread metric* as defined in [2]. Now we can use results of [2] to find an approximate VC.

For any $\varepsilon > 0$, we define the following two sets of graph vertices:

$$S_1 := \{v \in V : v_0 v > \varepsilon\}$$
$$S_2 := \{v \in V \cup V' : -\varepsilon \leq v_0 v \leq \varepsilon\}$$

For now, we concentrate our attention on S_2. Note that in S_2 we have included also shadow points. In fact, note that if $v_i \in V$ belongs to S_2 then its shadow $v_i' \in V'$ belongs to S_2 as well, and vice-versa. In other words, S_2 contains both original points and their shadows.

Lemma 1.
$$\sum_{i,j \in S_2} |v_i - v_j|^2 = 4|S_2|^2$$

Proof. Note that for a particular pair $i, j \in S_2 \cap V$ we have $v_i v_j' = v_i' v_j = -v_i v_j$. So if we group the summation terms according to pairs of vertices $i, j \in S_2 \cap V$, we get the lemma, due to cancellation of terms.

Let $\Delta, \sigma > 0$ be two parameters to be determined later. Let u be a random unit vector, and let

$$S_u := \{v \in S_2 : uv \geq \frac{\sigma}{\sqrt{d}}\}$$
$$T_u := \{v \in S_2 : uv \leq -\frac{\sigma}{\sqrt{d}}\}$$

Since $v_i = -v_i'$, it is easy to prove the following

Lemma 2. *If $v_i \in S_u$ for some $v_i \in V$, then $v_i' \in T_u$, and vice-versa, if $v_i' \in T_u$, then $v_i \in S_u$. The same holds with the roles of S_u, T_u interchanged.*

As a result of Lemma 2, $S_u \cup T_u$ contains only pairs of points in V with their shadow points, and each such pair is separated between S_u, T_u, and $|S_u| = |T_u|$. Moreover, the following easy fact also holds:

Lemma 3.
$$\left.\begin{array}{l} v_i \in S_u, v_j \in T_u, |v_i - v_j|^2 \leq \Delta \\ v_i' \in S_u, v_j \in T_u, |v_i' - v_j|^2 \leq \Delta \\ v_i \in S_u, v_j' \in T_u, |v_i - v_j'|^2 \leq \Delta \\ v_i' \in S_u, v_j' \in T_u, |v_i' - v_j'|^2 \leq \Delta \end{array}\right\} \Rightarrow \left\{\begin{array}{l} v_j' \in S_u, v_i' \in T_u, |v_i' - v_j'|^2 \leq \Delta \\ v_j' \in S_u, v_i \in T_u, |v_i - v_j'|^2 \leq \Delta \\ v_j \in S_u, v_i' \in T_u, |v_i' - v_j|^2 \leq \Delta \\ v_j \in S_u, v_i \in T_u, |v_i - v_j|^2 \leq \Delta \end{array}\right.$$

Let $c' > 0$ be another parameter which will be defined later. We modify the procedure SET-FIND of [2] as follows:

- If $|S_u| < 2c'|S_2|$ or $|T_u| < 2c'|S_2|$ then we HALT (just like in [2]).
- Otherwise, pick any $x \in S_u, y \in T_u$ such that $|x - y|^2 \leq \Delta$. Then, because of Lemma 3, the corresponding pair of antipodal points $y' \in S_u, x' \in T_u$ also satisfy $|x' - y'|^2 \leq \Delta$. Delete x, x', y, y'. Repeat until no such x, y can be found.

Note that initially T_u contains the antipodal points of S_u (Lemma 2), and every deletion eliminates two points from each of S_u, T_u, and these four actually form two (a point in V, its shadow point in V') pairs. Therefore, in the end, the remaining points in S_u are *exactly* the antipodal points of T_u (or both S_u, T_u are empty). As in [2], $|x - y|^2 > \Delta$, $\forall x \in S_u, y \in T_u$. One can define the parameters c', σ so that, initially, S_u, T_u are big with high probability:

Lemma 4. [Lemma 4 in [2]] *For every positive $c < 1/3$, there are $c', \sigma > 0$ such that the probability (over the choice of u) is at least $c/8$ that the initial sets S_u, T_u defined above have size at least $2c'|S_2|$.*

Proof. From Lemma 1 and application of Lemma 4 of [2].

In fact, since S_u initially (and throughout the running of the algorithm) contains the antipodal points of T_u, $|S_u| = |T_u| = |S_2|/2$ before the algorithm starts running no matter which u we choose, therefore $c' = 1/4$.

One of the main results of [2] is to show that, with high probability over u, not many points are deleted before SET-FIND terminates. Note that the points removed form a matching (at every step, x is matched to y, and x' is matched to y'). Theorem 5 in [2] shows that, with $\Delta = O(\log^{-2/3} n)$, the probability that SET-FIND removes a matching of size $c'|S_2|$ is $o(1)$. Hence the final S_u, T_u of SET-FIND have size $\geq c'|S_2|$ with probability $\Omega(1)$, and $|S_u| = |T_u|$. In what follows, we assume that S_u, T_u are the big final sets we get with high probability from SET-FIND.

Lemma 5. *If $\varepsilon \leq \Delta/4$, then there is no edge $(i,j) \in E$ such that $v_i, v_j \in V$ belong both to S_u or both to T_u.*

Proof. W.l.o.g. suppose that there is $(i,j) \in E$ such that $v_i, v_j \in S_u$. Then their shadow (antipodal) points belong to T_u, i.e., $v'_i, v'_j \in T_u$. Since $v_i, v_j \in S_2$ and constraint (4) holds, we have that

$$v_i v_j = v_0 v_i + v_0 v_j - 1 \leq -(1 - 2\varepsilon). \tag{9}$$

Since $v'_i \in T_u$ and $v_j \in S_u$ are not deleted in SET-FIND, $|v'_i - v_j|^2 > \Delta$, or, equivalently, $|v_i + v_j|^2 > \Delta$. This implies that

$$v_i v_j > -1 + \frac{\Delta}{2}. \tag{10}$$

But (9) and (10) together imply that $\varepsilon > \Delta/4$ which contradicts the hypothesis.

From now on we set $\varepsilon := \Delta/4 > 0$. Since $|S_u| = |T_u| \geq c'|S_2|$, and the two sets contain antipodal points, one of them (w.l.o.g. let's assume that this is S_u), contains at least $\frac{c'|S_2|}{2}$ points from V. Let I be this set of points from V. Lemma 5 implies that I is an *independent set* of G of size at least $c_0|S_2|$, where $c_0 := c'/2 > 0$. We return the set $S := S_1 \cup (S_2 \setminus (I \cup V'))$ as our vertex cover.

Lemma 6. *S is a vertex cover of G.*

Proof. If there is $(i,j) \in E$ with $v_i, v_j \in V \setminus (S_1 \cup S_2)$, we have (by the definition of S_1, S_2) that $v_0 v_i < -\varepsilon$ and $v_0 v_j < -\varepsilon$, which implies that $v_0 v_i + v_0 v_j - 1 < -1 - 2\varepsilon$. Then constraint (4) implies that $v_i v_j < -1 - 2\varepsilon$, a contradiction. Also, since I is an independent set, not both of v_i, v_j can belong to it. If $v_i \in I$ and $v_j \in V \setminus (S_1 \cup S_2)$, then $v_0 v_i \leq \varepsilon$ and $v_0 v_j < -\varepsilon$, therefore constraint (4) implies that $v_i v_j < -1$, a contradiction. We conclude that every edge must have at least one of its endpoints in S.

Our main result is the following

Theorem 1. $|S| \leq (2 - \Theta(\frac{1}{\log^{2/3} n})) VC(G).$

Proof. We follow the analysis of Halperin [8]. From (SDP') and the definition of S_1, S_2 we have that

$$VC(G) \geq |S_1| \frac{1+\varepsilon}{2} + |S_2 \setminus V'| \frac{1-\varepsilon}{2}$$

or, equivalently,

$$|S_1| \leq \frac{2 \cdot VC(G)}{1+\varepsilon} - |S_2 \setminus V'| \frac{1-\varepsilon}{1+\varepsilon}. \quad (11)$$

Hence

$$|S| = |S_1| + |S_2 \setminus V'| - |I| \stackrel{(11)}{\leq} \frac{2}{1+\varepsilon} VC(G) + |S_2 \setminus V'| (\frac{2\varepsilon}{1+\varepsilon} - c_0).$$

Note that for $\Delta = \Theta(\log^{-2/3} n)$, $\frac{2\varepsilon}{1+\varepsilon} = \Theta(\log^{-2/3} n) < c_0$, for big enough n. Therefore,

$$|S| \leq \frac{2}{1+\varepsilon} VC(G) = (2 - \Theta(\log^{-2/3} n)) \cdot VC(G).$$

Very recently, J. Lee proved that the SET-FIND algorithm of [2] can also be used to obtain their stronger result [10], i.e., Δ can be as big as $\Theta(1/\sqrt{\log n})$. Therefore we can get the following strengthening of Theorem 1:

Theorem 2. $|S| \leq (2 - \Theta(\frac{1}{\sqrt{\log n}})) VC(G).$

Theorem 2 can be somewhat strengthened by noticing that in the proof of Theorem 1 we just need to pick Δ so that $\frac{2\varepsilon}{1+\varepsilon} < c_0$, and therefore [2] and [10] imply that if $x := 1/\Delta^2$, it is enough for x to be the solution of equation

$$\frac{x}{\log x} = c \log n$$

where $c > 0$ is a constant (cf. [5] for more details on solving this equation through Lambert's W function).

3 The Weighted Case

The following is a semidefinite-programming relaxation of weighted Vertex Cover (VC) for a graph $G = (V, E)$ with n nodes:

$$\min \sum_{i=1}^{n} w_i \cdot \frac{1 + v_0 v_i}{2} \text{ s.t.} \tag{WSDP}$$

$$(v_0 - v_i)(v_0 - v_j) = 0, \quad \forall (i,j) \in E$$
$$(v_i - v_j)(v_i - v_k) \geq 0, \quad \forall i, j, k \in V \cup \{0\}$$
$$v_i^2 = 1, \quad \forall i \in V \cup \{0\}$$

where w_i is the *integral* weight of node i. Let $W := \sum_{i=1}^{n} w_i$.

In order to apply the methods of Section 2, we solve (SDP') with the weights incorporated in the objective function, and replace every v_i by w_i copies of v_i (v_i' is also replaced by w_i copies of v_i'). In fact we don't need to do this replacement in practice, but this mental experiment is helpful in order to see how the unweighted case applies here, too. Note that this new set of vectors still satisfies the triangular inequalities, and Lemmata 4 through 6 in Section 2 apply here as well with $n := W$. Note that SET-FIND can be made to run in polynomial time in this case (recall that we don't really do the replacement of v_i with w_i, all we need to do is to keep track of how much weight remains for each node after each matching). Now Theorem 1 (and hence Theorem 2) can be proven in the same way as before, if we replace the cardinality of sets $|\cdot|$ with their weights $w(\cdot)$.

4 Open Problems

Obviously one of the biggest open problems in theoretical computer science is the exact determination of the approximability of VC. There is a big gap between the hardness and the approximability results. Unfortunately, the SET-FIND procedure of [2] is limited to a gap of at most $\Theta(1/\sqrt{\log n})$ by the embedding of the $\log n$-dimensional hypercube: in this case any two subsets of linear size are closer than $\Theta(1/\sqrt{\log n})$ (this simple fact was pointed out to us by James R. Lee).

We couldn't extend our techniques to other problems related to VC, for example the maximum independent set problem (IS), and we don't know whether this is possible (Halperin's [8] techniques, on the contrary, can be applied to IS). Another extension of VC is the vertex cover problem in hypergraphs. We don't know how to extend our techniques to this problem as well. Therefore we leave the application of the results above to these and other problems as an open question.

Finally, we point out that we don't know what the integrality gap of the strengthened SDP relaxation (SDP') used above is. A weaker formulation, that doesn't contain all the triangle inequalities but is equivalent to Schrijver's θ' function [7], was proven to have an integrality gap of $2 - \varepsilon$ for any constant $\varepsilon > 0$ by Charikar [4]. It would be interesting to show the same result for the stronger SDP.

Acknowledgements. I am grateful to Sanjeev Arora for reading an earlier draft of this work and for bringing [10] to my attention.

References

1. A. Agarwal, M. Charikar, K. Makarychev, and Y. Makarychev. $O(\sqrt{\log n})$ approximation algorithms for Min UnCut, Min 2CNF Deletion, and directed cut problems. In 37th STOC, 2005 (to appear).
2. S. Arora, S. Rao, and U. Vazirani. Expander flows, geometric embeddings and graph partitioning. In Proc. of 36th STOC, pp. 222–231, 2004.
3. R. Bar-Yehuda and S. Even. A local-ratio theorem for approximating the weighted vertex cover problem. *Annals of Discrete Mathematics*, **25**, pp. 27–45, 1985.
4. M. Charikar. On semidefinite programming relaxations for graph coloring and vertex cover. In Proc. of 13th SODA, pp. 616–620, 2002.
5. R. M. Corless, G. H. Gonnet, D. E. G. Hare, D. J. Jeffrey, and D. E. Knuth. On the Lambert W Function. *Advances in Computational Mathematics*, vol. 5, pp. 329–359, 1996.
6. I. Dinur and S. Safra. On the importance of being biased (1.36 hardness of approximating Vertex-Cover). *Annals of Mathematics*, to appear. Also in Proc. of 34th STOC, 2002.
7. M. Goemans and J. Kleinberg. The Lovász Theta Function and a Semidefinite Programming Relaxation of Vertex Cover. *SIAM Journal on Discrete Mathematics* **11**(2), pp. 196–204, 1998.
8. E. Halperin. Improved approximation algorithms for the vertex cover problem in graphs and hypergraphs. *SIAM J. on Computing*, **31**(5), pp. 1608–1623, 2002. Also in Proc. of 11th SODA, pp. 329–337, 2000.
9. R. Karp. Reducibility among combinatorial problems. in R. E. Miller and J. W. Thatcher (eds.) *Complexity of Computer Computations*, Plenum Press, NY, pp. 85–103.
10. J. R. Lee. On distance scales, embeddings, and efficient relaxations of the cut cone. SODA 2005 (to appear), see http://www.eecs.berkeley.edu/~jrl.
11. B. Monien and E. Speckenmeyer. Ramsey numbers and an approximation algorithm for the vertex cover problem. *Acta Informatica*, **22**, pp. 115–123, 1985.
12. C. E. Papadimitriou and M. Yannakakis. Optimization, approximation, and complexity classes. *JCSS*, **43**(3), pp. 425–440, 1991.

Stochastic Steiner Trees Without a Root

Anupam Gupta[1] and Martin Pál[2,*]

[1] Dept. of Computer Science,
Carnegie Mellon University, Pittsburgh PA 15213
anupamg@cs.cmu.edu
[2] DIMACS, Rutgers University, Piscataway, NJ
mpal@acm.org

Abstract. This paper considers the Steiner tree problem in the model of *two-stage stochastic optimization with recourse*. This model, the focus of much recent research [11, 16, 8, 18], tries to capture the fact that many infrastructure planning problems have to be solved in the presence of uncertainty, and that we have make decisions knowing merely market forecasts (and not the precise set of demands); by the time the actual demands arrive, the costs may be higher due to inflation.

In the context of the Stochastic Steiner Tree problem on a graph $G = (V, E)$, the model can be paraphrased thus: on Monday, we are given a probability distribution π on subsets of vertices, and can build some subset E_M of edges. On Tuesday, a set of terminals D materializes (drawn from the same distribution π). We now have to buy edges E_T so that the set $E_M \cup E_T$ forms a Steiner tree on D. The goal is to minimize the expected cost of the solution.

We give the first constant-factor approximation algorithm for this problem. To the best of our knowledge, this is the first $O(1)$-approximation for the stochastic version of a *non sub-additive problem*. In fact, algorithms for the *unrooted* stochastic Steiner tree problem we consider are powerful enough to solve the Multicommodity Rent-or-Buy problem, itself a topic of recent interest [3, 7, 15].

1 Introduction

Real world planning problems often have a significant component of uncertainty. For instance, when designing networks, the precise demand patterns and future costs of building capacity are often unknown to begin with, and only become clear as time progresses. However, with our increasing ability to collect statistical data, and the development of sophisticated and realistic forecast models, the paradigm of stochastic optimization has gained much traction. Indeed, we can now aim to solve a wider class of problems: given not a single input, but a distribution over inputs, we want to find a solution that is good in expectation (taken with respect to the randomness in the model).

In this paper, we study the problem of connecting a group of terminals by a Steiner tree in a stochastic setting. In the classical Steiner tree problem, we are given an undirected graph $G = (V, E)$ with edge costs c_e, and a group of terminals $g = \{t_1, t_2, \ldots,$

---

* Supported by ONR grant N00014-98-1-0589 (at Cornell University) and NSF grant EIA 02-05116 (at DIMACS).

$t_k\}$; the goal is to find a subset E' of edges of minimum cost that connects all these terminals. We consider this problem when the group g is not deterministically given in advance; instead, it is given by a random variable Γ, with $\Pr[\Gamma = g]$ being the probability that we will be required to build a network that connects a particular group $g \subseteq V$ of terminals. As sketched in the abstract, we work the model of two-stage stochastic optimization with recourse.

- In the first stage, we assume to have (some) knowledge of the distribution of the random variable Γ. Armed with this information, we construct a network $F^0 \subseteq E$ of edges bought as the first *anticipatory* part of the solution.
- In the second stage, we learn a group $g \subseteq V$ of terminals that is a realization of the random variable Γ. We have to purchase an additional *augmenting* set $F^1(g)$ of edges to ensure that $F^0 \cup F^1(g)$ connects the terminals of g. The problem is interesting when the edges bought in the second stage have a higher cost (due to inflation, or because the second phase has to be built on short notice). We use $\sigma > 1$ to denote the *inflation factor* by which the edges are more expensive.

Our goal is to minimize the expected cost of the two-stage solution. If we define $c(F) = \sum_{e \in F} c_e$, and denote the first and second stage solutions $F^0 \in E$ and $F^1 : 2^V \mapsto 2^E$ to minimize

$$c(F^0) + \mathbf{E}_\Gamma[\sigma \cdot c(F^1(\Gamma))]. \tag{1.1}$$

Our results. The main quantitative result of this paper is the following:

Theorem 1. *There is a 12.6-approximation algorithm for the two-stage stochastic (unrooted) Steiner tree problem.*

Note that while the stochastic Steiner tree problem has been considered in previous papers [11, 8, 9], their model is subtly but significantly different. All these works make the crucial assumption that the there is a *fixed root* r, and the goal is to connect the group g to the root r. This assumption, while a trifling detail in the deterministic case, turns out to make a big difference in the stochastic setting, requiring us to develop new techniques. For example, a fact used in one way or another by all previous results was that the first stage solution F^0 in the rooted case can be assumed to be a connected tree containing the root; this is just not true in the unrooted case: in fact, insisting on a connected first stage network may cost arbitrarily more than the optimum solution. Indeed, our result is the first approximation algorithm given for a problem that is not sub-additive, and requires us to interpret and use cost-sharing ideas in a novel way.

Specifying the distribution. The distribution π of the random variable Γ is an object whose size may be exponential in $|V|$. Our algorithm does not need to know π explicitly; it only requires access to a *sampling oracle* that can upon request supply independent samples drawn from the distribution π.

Related work. As already mentioned, several papers studied the *rooted version* of the stochastic Steiner tree problem. Immorlica et al. [11] give a $O(\log n)$ approximation in the independent decisions model, while [8] and [9] give constant approximation algorithms for the oracle and scenario models respectively. Karger and Minkoff [13] and

Hayrapetyan et al. [10] study the *maybecast* problem, where one is to output a single tree T, to minimize the expected size of the smallest subtree of T spanning a random set of terminals. While technically this is also a stochastic problem, the recourse action is fixed, and the randomness is present only in the objective function.

Gupta et al. [8] give a simple boosted sampling framework to convert an algorithm for a deterministic minimization problem to an algorithm for its stochastic counterpart. Their framework relies crucially on two ingredients: the deterministic version of the problem at hand has to be *subadditive*, and have an approximation algorithm that admits a *strict* cost sharing function. Since the unrooted Steiner tree problem is not sub-additive (i.e., if T_1 is a solution for terminal set g_1, and T_2 for g_2, then $T_1 \cup T_2$ may not be a solution for $g_1 \cup g_2$), we cannot apply their techniques directly here.

The general area of stochastic optimization is studied heavily in the operations research community, dating back to the seminal works of Dantzig [5] and Beale [2] in the 1950s; the books [4, 12] and monograph [17] could serve as introduction for the interested reader. Much of the work related to combinatorial optimization problems in this area has been concerned with finding and characterizing optimal solutions either for restricted classes of inputs or with algorithms without polynomial running times guarantees. Recently, there has been some work on taking solutions to stochastic linear programs and rounding those to obtain approximation algorithms for the stochastic problems [18]; however, it is not clear how to apply those techniques to the Steiner tree problem.

The Boosted Sampling Framework. Gupta et al. [8] propose the *Boosted Sampling* framework of Figure 1 to solve any two-stage stochastic problem Π where the set Γ of demand points is stochastic.

One would naturally expect that in the case of stochastic Steiner tree, the deterministic algorithm of Step 2 would build a Steiner tree on the set of terminals $g_1 \cup g_2 \cup \cdots \cup g_\sigma$. In fact, if the support of Γ was on sets that all contained the fixed root r, [8] shows that this is enough to obtain an 3.55-approximation algorithm for stochastic Steiner tree.

Unfortunately, building a Steiner tree fails in the unrooted case. For an example, consider two groups g_1 and g_2 that are very far apart relative to their diameter; assume that $\Pr[\Gamma = g_i] \cdot \sigma$ is large. In this case, the optimum solution must connect up each group g_i in the first stage to avoid high second stage cost, but it should not build a link between g_1 and g_2 (to make F_0 span $g_1 \cup g_2$) if it wants to avoid a high first stage cost. On the other hand, if the two groups are interspersed in the same region

---

1: *Boosted Sampling:* Take $\lfloor \sigma \rfloor$ independent samples $g_1, g_2, \ldots, g_{\lfloor \sigma \rfloor}$ from the sampling oracle for Γ.
2: *Building First Stage Solution:* Use an algorithm $\mathcal{A}$ to find a solution to the *deterministic equivalent* of the problem Π on the groups $g_1, g_2, \ldots, g_{\lfloor \sigma \rfloor}$. Use this solution as the first stage solution to the stochastic problem.
3: *Building Recourse:* Once the group g of required terminals materializes, use an *augmenting algorithm* $\text{Aug}_\mathcal{A}$ to augment the first stage solution to a valid solution that satisfies g.

Fig. 1. Algorithm Boost-and-Sample(Π)

of the graph, the optimum solution may benefit from link sharing and hence build a single Steiner tree spanning both groups. Hence it seems natural to suggest that the algorithm $\mathcal{A}$ should build a forest ensuring that each group lies within a single connected component; different groups may or may not be in the same component. As it turns out, building a *Steiner Forest* on the groups g_i is a suitable deterministic equivalent of stochastic unrooted Steiner tree; however, proving this requires a lot more work.

To this end, we have to show that the main theorem of [8] which relates the performance of the boosted sampling framework to the notion of *strictness* (which we define shortly) of certain cost-sharing functions can be proved in our case, even though our problem is not sub-additive. The proof of this is simple, and we will sketch it in Section 2. We then define the cost-shares in Section 3, and prove them to be strict in 4.

2 Notation and Preliminaries

Let $G = (V, E)$ be an undirected weighted graph with weigths c_e on the edges. A *network* is simply a subset of the edges. We say that a network F is *feasible* for (or *connects*) a group of terminals $g = \{t_1, t_2, \ldots, t_k\}$, if all the terminals of g lie in the same connected component of F. The cost of a network F is simply the sum of costs of its edges; that is $c(F) = \sum_{e \in F} c_e$.

In the *Steiner Forest* problem, given a weighted undirected graph G and a list of groups of terminals $\mathcal{D} = \{g_1, g_2, \ldots, g_n\}$ with each $g_i = \{t_{i1}, \ldots, t_{ik_i}\}$, we want to construct a network F of minimum cost that is feasible for each group g_i. For a set $\mathcal{D}$ of terminal groups, let $\mathsf{Sols}(\mathcal{D})$ denote the set of networks that are feasible for each of the groups in $\mathcal{D}$, and let $\mathsf{OPT}(\mathcal{D})$ be the network in $\mathsf{Sols}(\mathcal{D})$ of minimum cost. An algorithm $\mathcal{A}$ is an α-approximation algorithm for the Steiner Forest problem, if for any set D of terminal groups, it finds a network $F_\mathcal{D} \in \mathsf{Sols}(\mathcal{D})$ of cost at most $\alpha \, \mathsf{cost}(\mathsf{OPT}(\mathcal{D}))$.

Given a group g of terminals and an existing network $F \subseteq E$, the goal of an *augmenting algorithm* is buy a set of extra edges F' so that $F \cup F'$ is a network that connects the group g. For instance, given a network $F_\mathcal{D} \in \mathsf{Sols}(\mathcal{D})$ that connects each of the groups in $\mathcal{D}$, and a new group $g \notin \mathcal{D}$, the augmenting algorithm $\mathsf{Aug}_\mathcal{A}$ seeks to find a set of edges F' of minimum cost so that $F_\mathcal{D} \cup F' \in \mathsf{Sols}(\mathcal{D} \cup \{g\})$.

Definition 1. *A cost-sharing function ξ is a function that, for any instance $(G, \mathcal{D})$ of the Steiner forest problem, assigns a non-negative real number $\xi(G, \mathcal{D}, g_i)$ to every participating group $g_i \in \mathcal{D}$.*

We shall drop a reference to the graph G, if clear from the context. *Note that the cost sharing function assigns shares to groups, and not to the individual terminals.*

Since the above definition is so general, let us specify some properties of these functions that we would like to get. A cost-sharing function ξ is *competitive* if $\sum_{g \in \mathcal{D}} \xi(\mathcal{D}, g) \leq \mathsf{cost}(\mathsf{OPT}(\mathcal{D}))$ holds for any Steiner forest instance $(G, \mathcal{D})$. Thus, competitive cost-shares serve as a lower bound on the cost of the optimal solution. The following notion is crucial to the development of the paper, and implicitly places lower bounds on the cost-shares themselves.

Definition 2. *A cost sharing function ξ is β-strict with respect to an algorithm $\mathcal{A}$, if there exists an augmenting algorithm $\text{Aug}_\mathcal{A}$, such that for any set of demand groups $\mathcal{D}$ and any group $g \notin \mathcal{D}$,*

$$\text{cost}(\text{Aug}_\mathcal{A}(\mathcal{A}(\mathcal{D}), g)) \leq \beta \xi(\mathcal{D} + g, g). \tag{2.2}$$

Remark 1. There is a fine distinction between the notion of strictness we use here and strictness as defined in [7, 8]. In [7], strictness was defined only for augmentations with groups of size 2; in this paper, we allow for groups of larger sizes. However, the strictness in [8] is stronger than our notion, and allows for multiple group augmentations; the question of proving strictness by this definition remains open despite much effort.

Given all these definitions, we can now state the the following theorem, which can be derived from the proof of [8–Theorem 3.1]. We defer its proof to a full version of the paper.

Theorem 2. *Suppose that $\mathcal{A}$ is an α-approximation algorithm for deterministic Steiner forest. Then, the boosted sampling algorithm of Figure 1 is an $(\alpha + \beta)$-approximation algorithm for unrooted stochastic Steiner tree whenever there is a cost-sharing function ξ that is β-strict with respect to $\mathcal{A}$ and single group augmentations.*

3 The Algorithm $\mathcal{A}$ and the Cost Shares ξ

In this section we review the Steiner forest algorithm of [7], although the algorithm of Becchetti et al. [3] would serve our purpose equally well. Both algorithms are extensions of the algorithm of Agarwal, Klein, and Ravi (AKR) [1], and Goemans and Williamson (GW) [6], and are designed to "build a few extra edges" over and above the AKR-GW algorithms, while keeping the overall cost of the solution within a constant factor of the cost of the optimum. We also describe our cost-sharing method.

Recall that we are given a graph $G = (V, E)$ and a set $\mathcal{D}$ of groups $g_1, \ldots, g_n$ of terminals, where each group $g_i = \{t_{i1}, t_{i2}, \ldots, t_{ik_i}\} \subseteq V$. Before defining our algorithm, we review the LP relaxation and the corresponding LP dual of the Steiner forest problem that was used in [6]:

$$\min \sum_e c_e x_e \quad \text{(SF-LP)} \qquad \max \sum_S f(S) y_S \quad \text{(SF-DP)}$$
$$x(\delta(S)) \geq f(S) \quad \forall S \subseteq V \qquad \sum_{S \subseteq V: e \in \delta(S)} y_S \leq c_e \tag{3.3}$$
$$x_e \geq 0 \qquad\qquad y_S \geq 0,$$

where $f(S)$ is equal to 1 if S separates g_i for some i (that is, if both $S \cap g_i$ and $(V - S) \cap g_i$ is nonempty), and is 0 otherwise. Note that variables y_S for sets S that do not separate any group are not contributing to the dual objective function, they still play an important role in our algorithm.

We now describe a general way to define primal-dual algorithms for the Steiner forest problem. As is standard for the primal-dual approach, the algorithm with maintain a feasible (fractional) dual, initially the all-zero dual, and a primal integral solution (a set of edges), initially the empty set. The algorithm will terminate with a feasible Steiner forest, which will be proved approximately optimal with the dual solution (which is a lower bound on the optimal cost by weak LP duality). The algorithms of [1, 6] arise as a particular instantiation of the following algorithm. Our presentation is closer to [1], where the "reverse delete step" of Goemans and Williamson [6] is implicit; this version of the algorithm is more suitable for our analysis.

Our algorithm has a notion of *time*, initially 0 and increasing at a uniform rate. At any point in time, some terminals will be *active* and others *inactive*. All terminals are initially active and eventually become inactive. At any point of time, the vertex set is also partitioned into *clusters*, which can again be either active or inactive. In our algorithm, a cluster will be one or more connected components (w.r.t. the currently built edges). Initially, each vertex is a cluster by itself, and the active clusters are just the terminals. We will consider different rules by which demands and clusters become active or inactive, which we describe shortly. To maintain dual feasibility, whenever the constraint (3.3) for some edge e between two clusters S and S' becomes tight (i.e., first holds with equality), the clusters are *merged* and replaced by the cluster $S \cup S'$. We raise dual variables of active clusters until there are no more such clusters.

We have not yet specified how an edge can get built. Towards this end, let us define a (time-varying) equivalence relation $\mathcal{R}$ on the set of terminals. Initially, all terminals lie in their own equivalence class; these classes will only merge with time. When two active clusters are merged, we merge the equivalence classes of all active terminals in the two clusters. Since inactive terminals cannot become active, this rule ensures that all active terminals in a cluster are in the same equivalence class. (Note that if an active cluster merges with an inactive one, this merging of equivalence classes does not happen.)

We build enough edges to maintain the following invariant: the terminals in the same equivalence class are connected by built edges. This clearly holds at the beginning, since the equivalence classes are all singletons. When two active clusters meet, the invariant ensures that, in each cluster, all active terminals lie in a common connected component. To maintain the invariant, we join these two components by adding a path between them. Building such paths without incurring a large cost is simple but somewhat subtle; Agrawal et al. [1] (and implicitly, Goemans and Williamson [6]) show how to do this. We refer the reader to [1] for details of this procedure, instead of repeating it here. Specifying the rule by which clusters are deemed active or inactive now gives us two different algorithms:

1. **Algorithm** GW$(G, \mathcal{D})$: A terminal $t_{ij} \in g_i$ is active if the current cluster containing it does not contain the entire group g_i. A cluster is active as long as it contains at least one active demand. This implementation of the algorithm is equivalent to the algorithms of Agrawal et al. [1] and Goemans and Williamson [6].
2. **Algorithm** Timed(G, D, T): This algorithm takes as an additional input a function $T : V \to \mathbb{R}_{\geq 0}$ which assigns a *stopping time* to each vertex. (We can also view T as a vector with coordinates indexed by V.) A vertex j is active at time τ if $j \in D$ and $\tau \leq T(j)$. (T is defined for vertices not in D for future convenience, but such values are irrelevant, and can be imagined to be set to 0 for the rest of the paper.) As before, a cluster is said to be active if at least one demand in it is active.

To get a feeling for $\mathsf{Timed}(G, D, T)$, consider the following procedure: run the algorithm $\mathsf{GW}(G, \mathcal{D})$ and set $T_\mathcal{D}(j)$ to be the time at which vertex j becomes inactive during this execution. (If $j \notin D$, then $T_\mathcal{D}(j)$ is set to zero.) Since a vertex stays active for exactly the same duration of time in the two algorithms $\mathsf{GW}(G, \mathcal{D})$ and $\mathsf{Timed}(G, D, T_\mathcal{D})$, the two algorithms clearly have identical outputs. Similarly, if for each $t_{ij} \in g_i$ we set $T(t_{ij}) = \max_{t, t' \in g_i} d_G(t, t')$, we obtain the algorithm of Könemann et al. [14].

It turns out that the Timed algorithm gives us a nice principled way to force the GW algorithm to build additional edges: run the Timed algorithm with a vector of demand activity times that is larger than what is naturally induced by the GW algorithm.

The Algorithm $\mathcal{A}$: The algorithm **Algorithm $\mathcal{A}(G, \mathcal{D})$** that we use to build the first stage solution is

1: Run $\mathsf{GW}(G, \mathcal{D})$, and let $T_\mathcal{D}(v)$ be the time at which v becomes inactive.
2: Run $\mathsf{Timed}(G, D, \gamma T_\mathcal{D})$—the timed algorithm with the above time vector $T_\mathcal{D}$ scaled up by a parameter $\gamma \geq 1$—and output the resulting forest $F_\mathcal{D}$.

A technical point: when $\gamma > 1$, algorithm $\mathcal{A}$ may raise the dual variables of vertex sets that do not separate any group, and hence do not contribute to the value of the dual objective function. However, this will not hinder our analysis. The fact that $F_\mathcal{D}$ is a feasible Steiner network for $\mathcal{D}$ is easily verified, using the fact that the terminals of each group became inactive at the same time $T_\mathcal{D}(g_i)$ (equal to $T_\mathcal{D}(t_{ij})$ for any $t_{ij} \in g_i$) when g_i became connected, and that $\gamma \geq 1$. We now define the cost shares ξ.

The Cost Shares ξ: We want the cost share of a group g_i of users to account for the growth of components that contain only terminals from g_i. Let $a(g_i, \tau)$ be the number of active clusters in the execution of $\mathsf{GW}(G, \mathcal{D})$ that contain a terminal from g_i but *do not* contain any active terminals outside g_i. We define the cost share of g_i to be

$$\xi(\mathcal{D}, g_i) = \int a(g_i, \tau) \, d\tau, \tag{3.4}$$

where the integral is over the entire execution of the algorithm. Note that the cost shares defined by Equation (3.4) do not account for the full cost of the dual solution y, as the cost of growth of clusters with active demands from more than one group more than one active demand is not reflected at all. We could fix this by dividing the cost of growing mixed clusters among participating groups in some way; however, we do not see how to use this to improve our approximation ratio.

Augmentation Algorithm $\mathsf{Aug}_\mathcal{A}$: A practical augmenting algorithm $\mathsf{Aug}_\mathcal{A}$ would simply contract all edges of $F_\mathcal{D}$, and then find an approximate Steiner tree on the terminals of g in this contracted graph $G/F_\mathcal{D}$. However, in order to bound the second stage cost, we build a specific Steiner tree on g in $G/F_\mathcal{D}$, and argue that the cost of *this* tree can be bounded by $\beta \xi(\mathcal{D} + g, g)$ for some $\beta \in \mathbb{R}$. The construction of this tree is implicit in the proof of Theorem 4, and can be found efficiently in polynomial time if required. In the following, we let $\mathsf{Aug}_\mathcal{A}$ be the algorithm that constructs this implicit tree. Our main technical result is thus the following.

Theorem 3. *For any $\gamma > 2$, $\mathcal{A}$ is a $\alpha = (\gamma+1)$-approximation for the Steiner network problem, and ξ is a $\beta = (4\gamma/(\gamma-2))$-strict cost sharing method with respect to the algorithms $\mathcal{A}$ and $\mathrm{Aug}_{\mathcal{A}}$.*

Proof. The fact that $\mathcal{A}$ is a $(\gamma+1)$-approximation can be proved along the lines of [3–Lemma 3.1] (We postpone the details to the full paper). The proof of strictness (Theorem 4) is the analytical heart of this paper, and is given in the following section.

4 Proving Strictness

Our analysis follows a fairly natural line of analysis that was also used in [7]. We start by fixing a set $\mathcal{D}$ of demand groups, and a group $g \notin \mathcal{D}$. To prove strictness of our cost shares, we compare two executions of the GW algorithm: the inflated algorithm $\mathcal{A}(G, \mathcal{D})$ on the set of groups $\mathcal{D}$ that results in the forest $F_{\mathcal{D}}$, and the uninflated algorithm $\mathrm{GW}(G, \mathcal{D} + g)$ which is responsible for computing the cost share $\xi(\mathcal{D}+g, g)$.

Recall that we have to show that g can be connected in $F_{\mathcal{D}}$ with cost at most $O(\xi(\mathcal{D}+g,g))$. We prove this in the following theorem, which also implicitly describes the augmenting algorithm $\mathrm{Aug}_{\mathcal{A}}$. In the rest of the discussion, we will assume that $\gamma > 2$.

Theorem 4. *There is a tree F' in the graph $G/F_{\mathcal{D}}$ that spans all terminals of g and has cost at most $4\gamma/(\gamma-2)\,\xi(\mathcal{D}+g,g)$. The tree F' can be constructed in polynomial time.*

The main difficulty in proving Theorem 4 arises from the fact that the two executions $\mathcal{A}(G,\mathcal{D})$ and $\mathrm{GW}(G, \mathcal{D}+g)$ may be very different. Hence it is not immediately clear how to relate the cost of augmenting the forest $F_{\mathcal{D}}$ produced by the former by the cost share $\xi(\mathcal{D}+g,g)$ computed by the latter. To make a direct comparison possible, we work through some transformations that allow us to find a mapping between dual variables in these two executions. In the grand finale, we produce a tree $\mathbb{T}$ that spans terminals of g, and show that a $1/\beta$ fraction of its edges is covered by dual variables corresponding to the cost share of g, which will complete the proof. Let us introduce some time vectors to facilitate this comparison.

- Let $T_{\mathcal{D}}$ be the time vector obtained by running $\mathrm{GW}(G,\mathcal{D})$. Recall that $F_{\mathcal{D}}$ is the forest constructed by $\mathrm{Timed}(G,\mathcal{D},\gamma T_{\mathcal{D}})$; we also let $\mathcal{R}_{\mathcal{D}}$ be the equivalence relation constructed by the latter algorithm.
- Let $T_{\mathcal{D}+g}$ be the time vector generated by the execution $\mathrm{GW}(G, \mathcal{D}+g)$ and let $\tau = T_{\mathcal{D}+g}(g)$ be the time when the terminals of g got connected in this execution.
- Let T be the vector obtained by truncating $T_{\mathcal{D}+g}$ at time τ. That is, $T(v) = \min(\tau, T_{\mathcal{D}+g}(v))$ for $v \in V$. (The intuition for T is loosely this: we do not care about time after g has been connected, and this truncation captures this fact.)
- Finally, let T_{-g} be the vector T with g "taken out", that is, $T_{-g}(v) = T(v)$ if $v \notin g$, and $T_{-g}(v) = 0$ if $v \in g$. Let $\mathcal{R}_{-g}$ be the equivalence relation constructed by the execution $\mathrm{Timed}(G, \mathcal{D}, \gamma T_{-g})$.

A side-by-side comparison of the executions $\mathsf{GW}(G, \mathcal{D})$ and $\mathsf{GW}(G, \mathcal{D} + g)$ shows that for all $v \in V$,
$$T_{\mathcal{D}}(v) \geq T_{-g}(v); \tag{4.5}$$
the simple inductive proof is omitted. The forest constructed by $\mathsf{Timed}(G, \mathcal{D}, \gamma T_{-g})$ will serve us as a proxy for the forest $F_{\mathcal{D}}$ created by $\mathsf{Timed}(G, \mathcal{D}, \gamma T_{\mathcal{D}})$; intuitively, since T_{-g} is smaller than $T_{\mathcal{D}}$, it should also produce a forest with fewer edges. We will make this intuition precise in Lemma 1 below.

To state the lemma in a general form that will be useful later, we need some more notation. For two weighted graphs G and G' on the same vertex set V, we write $G' \leq G$ if the shortest path distance between any pair of vertices (u,v) in G' is no more than their distance in G. For a graph $G = (V, E)$ and a set $F \subseteq (V \times V)$, the graph $G' = G/F$ is a *contraction* of G, and is obtained by adding a zero-cost edge in G between every pair $(u,v) \in F$. Since $\mathcal{R} \subseteq V \times V$, we can define $G/\mathcal{R}$ in the same way. It immediately follows that if G' is a contraction of G, then $G' \leq G$. For time vectors, let $T \leq T'$ denote coordinate-wise inequality (and hence we can rewrite (4.5) as $T_{-g} \leq T_{\mathcal{D}}$).

Lemma 1 ([7]). *Let $G \leq G'$ be two weighted graphs and $T \leq T'$ be two time vectors. Then, for the equivalence relations $\mathcal{R}$ and $\mathcal{R}'$ produced by the executions $\mathsf{Timed}(G, \mathcal{D}, T)$ and $\mathsf{Timed}(G', \mathcal{D}, T')$, it holds that $\mathcal{R} \subseteq \mathcal{R}'$.*

A Simpler graph H: We now define a simpler graph $H = G/\mathcal{R}_{-g}$; this graph H will act as a proxy for $G/F_{\mathcal{D}}$ in the following sense. For two vertices u, v connected by a zero-cost path in H, we know that u and v are connected by a path in $F_{\mathcal{D}}$. This is because the inequality $T_{-g} \leq T_{\mathcal{D}}$ used with Lemma 1 implies that $\mathcal{R}_{-g} \subseteq \mathcal{R}_{\mathcal{D}}$; now the invariant maintained by the algorithm Timed implies that there is a path connecting u and v in $F_{\mathcal{D}}$ whenever $(u,v) \in \mathcal{R}_{\mathcal{D}}$.

Thus, to prove Theorem 4, it suffices to exhibit a tree $\mathbb{T}$ in H that spans all terminals of g, and has cost at most $4\gamma/(\gamma - 2)\xi(\mathcal{D} + g, g)$. By the properties of the graph H, it then follows that the network $\mathbb{T} \cup F_{\mathcal{D}}$ is feasible for the group g.

Note that each equivalence class of $\mathcal{R}_{-g}$ can also be thought of as a single (super)-vertex of the graph H; this view may be more useful in some contexts. To complete the correspondence between the two views, let us extend the definition of a time vector to supernodes in the natural way: if w_C is an equivalence class of the relation $\mathcal{R}_{-g}$, we let $T(w_c) = \max_{v_i \in C} T(v_i)$; this allows us to talk about running the Timed algorithm on H with the vector T.

4.1 The Tree $\mathbb{T}$ Spanning Terminals of g

We will obtain the desired Steiner tree on the group g in H by considering the execution of the algorithm $\mathsf{Timed}(H, \mathcal{D} + g, T)$; we denote this execution by $\mathcal{E}$. Recall that the time vector T was defined to ensure that in the execution $\mathsf{Timed}(G, \mathcal{D} + g, T)$ on the original graph G, the terminals of g eventually merge into a single equivalence class of the respective relation $\mathcal{R}$. Since the graph H is a contraction of G, it follows from Lemma 1 that the terminals of g must end up in the same equivalence class in $\mathcal{E}$, and hence in the same connected component of the forest constructed by $\mathcal{E}$. There is a unique minimal tree that spans the terminals of g in this forest; let $\mathbb{T}$ denote this tree.

Since $\mathbb{T}$ was constructed by the execution $\mathcal{E}$, all of its edges must be fully tight with the dual grown in $\mathcal{E}$. Our plan of attack is to show that the dual variables corresponding to the terminals of g account for a significant fraction of this dual, and hence the cost share of g must be large enough to pay for a $1/\beta$ fraction of the tree. To pursue this plan, we introduce the following notion of layers as in [7]; this terminology is just a convenient way of talking about "dual moats".

In an execution of an algorithm, a *layer* (C, I) corresponds to an active cluster C whose dual variable y_C has been growing during the time interval $I = [\tau_1, \tau_2)$; the *thickness* of this layer is $|I| = \tau_2 - \tau_1$. A layering $\mathcal{L}$ of an execution is a set of layers such that, for every time τ and every active cluster C, there is exactly one layer $(C, I) \in \mathcal{L}$ such that $\tau \in I$.

Lonely layers: A layer (C, I) is *lonely*, if it does not contain any active terminals except terminals belonging to g. Thus, the cost share of g can be expressed as the total thickness of lonely layers in any layering of $\mathsf{Timed}(G, \mathcal{D} + g, T)$. Using Lemma 1, we can argue that the total thickness of lonely layers in the execution $\mathcal{E}$ is no more than in $\mathsf{Timed}(G, \mathcal{D} + g, T)$ (see [7] for details). Hence the total thickness of lonely layers in the execution $\mathcal{E}$ is a lower bound on the cost share of g.

We lower bound the thickness of lonely layers by arguing that the thickness of non-lonely layers intersecting $\mathbb{T}$ is significantly smaller than the length of $\mathbb{T}$: since all of $\mathbb{T}$ has to be covered, this leaves a considerable fraction of the tree to be covered by lonely layers. Hence our overall goal can be reduced to giving an upper bound on the thickness of non-lonely layers that intersect the tree $\mathbb{T}$.

To get a hold on this quantity, we proceed to compare a layering $\mathcal{L}$ of the execution $\mathcal{E}$—recall that $\mathcal{E} = \mathsf{Timed}(H, \mathcal{D} + g, T)$—with a layering $\mathcal{L}'$ of its inflated counterpart $\mathcal{E}' = \mathsf{Timed}(H, \mathcal{D}, \gamma T_{-g})$. We construct a mapping that maps every non-lonely layer $\ell = (C, I) \in \mathcal{L}$ to a distinct layer $\ell' = (C', \gamma I) \in L'$ that is γ times thicker. (Note that lonely layers do not have a natural counterpart, as the terminals of g do not appear at all in the execution $\mathcal{E}'$.) To ensure the existence of such a mapping, we align the two layerings to satisfy the following property: if $(C, I) \in \mathcal{L}$ and $(C', I') \in \mathcal{L}'$ with $\gamma I \cap I' \neq \emptyset$, then $I' = \gamma I$. (I.e., $I' = [\gamma \tau_1, \gamma \tau_2)$ and $I = [\tau_1, \tau_2)$.) This condition can easily be imposed by repeatedly *splitting* layers of $\mathcal{L}$ and $\mathcal{L}'$, that is, replacing an offending layer $(C, [\tau_1, \tau_2))$ by two layers $(C, [\tau_1, \hat{\tau}))$ and $(C, [\hat{\tau}, \tau_2))$ for a suitably chosen $\hat{\tau} \in [\tau_1, \tau_2)$.

Mapping non-lonely layers of $\mathcal{L}$ to layers of $\mathcal{L}'$: Every non-lonely layer $\ell = (C, [\tau_1, \tau_2))$ must contain a terminal $t \in C$ such that $t \notin g$, that was active in the interval $[\tau_1, \tau_2)$. Since $T_{-g} \leq T_{\mathcal{D}}$, the terminal t must have been active in the interval $[\gamma \tau_1, \gamma \tau_2)$ in the execution $\mathcal{E}'$, and hence there is a unique layer $\ell' = (C', [\gamma \tau_1, \gamma \tau_2))$ such that $t \in C'$. We thus map ℓ to ℓ'. A layer ℓ may contain multiple active terminals outside g; in that case, pick one of them arbitrarily.

The following two lemmas supply us with all the ammunition we will need to finish our argument. In the next lemma, let $V(\mathbb{T})$ denote the vertex set of the tree $\mathbb{T}$. The proofs are omitted due to lack of space.

Lemma 2. *The mapping from non-lonely layers of $\mathcal{L}$ to layers of $\mathcal{L}'$ is one to one; that is, distinct layers of $\mathcal{L}$ map to distinct layers of $\mathcal{L}'$.*

Lemma 3. *Let $\ell = (C, I) \in \mathcal{L}$ be a non-shared layer, such that $V(\mathbb{T}) \cap C \neq \emptyset$. Then, for its corrsponding layer $\ell' = (C', \gamma I)$ we have that $V(\mathbb{T}) \cap C \neq \emptyset$.*

4.2   The Book Keeping

Let L and N denote the total thickness of lonely and non-lonely layers that intersect the tree $\mathbb{T}$. Note that we count every layer only once, irrespective of how many edges of $\mathbb{T}$ it cuts. We can express the total length of the tree as

$$|\mathbb{T}| = L + N + X, \qquad (4.6)$$

where X represents the "extra" contributions of layers that intersect $\mathbb{T}$ more than once. (If a lonely layer intersects $\mathbb{T}$ in three edges, it is counted once in L and twice in X).

At any time instant τ, consider all the active clusters in the execution $\mathcal{E}$ that have a non-empty intersection with the tree $\mathbb{T}$. We claim that any such cluster C "carves out" a connected portion of the tree $\mathbb{T}$, that is, $C \cap \mathbb{T}$ is a connected graph. Hence if we construct a graph with a node for every cluster intersecting $\mathbb{T}$ and an edge between every pair of clusters connected by a direct path along $\mathbb{T}$, this graph will also be a tree. The number of layers intersecting $\mathbb{T}$ is equal to the number of nodes in this graph; the number of times each layer intersects $\mathbb{T}$ is equal to the degree of the corresponding vertex in this graph. Since the average vertex degree in a tree is at most 2, the number of intersections is at any time bounded by twice the number of layers intersecting $\mathbb{T}$. Integrating over the course of the execution $\mathcal{E}$, we obtain that

$$L + N + X \leq 2(L + N). \qquad (4.7)$$

A non-lonely layer ℓ is considered *wasted* if ℓ intersects $\mathbb{T}$, but its image ℓ' does not. According to Lemma 3, this happens only if $\mathbb{T}$ is fully contained inside ℓ'. Let W denote the total thickness of wasted layers. The total thickness of layers of $\mathcal{L}'$ intersecting $\mathbb{T}$ is a lower bound on the length of $\mathbb{T}$. Since the image of every non-lonely layer ℓ that intersects $\mathbb{T}$ and is not wasted also intersects $\mathbb{T}$, and because images of distinct layers do not overlap, we get the following lower bound on the length of $\mathbb{T}$.

$$\gamma(N - W) \leq |\mathbb{T}|. \qquad (4.8)$$

The final piece of our argument is the following claim: for every layer that is wasted, there must be a lonely layer growing at the same time, and hence $W \leq L$. To see this claim, suppose that a non-lonely layer $\ell = (C, I)$ intersects $\mathbb{T}$ but is wasted—hence for its inflated image $\ell' = (C', \gamma I)$, we have $V(\mathbb{T}) \subseteq C'$. Since ℓ intersects $\mathbb{T}$, there must be a terminal $t \in g$ such that $t \notin C$. We now claim that during the interval I, the terminal t must have been a part of a lonely cluster. Indeed, suppose not; let t be inside a non-lonely layer $\ell_1 = (C_1, I)$ with some other active terminal $t_1 \notin g$. But then, by Lemma 3, the inflated image $\ell'_1 = (C'_1, \gamma I)$ of this layer ℓ_1 must contain some vertex of $\mathbb{T}$, and since $V(\mathbb{T}) \subseteq C'$, the layers ℓ' and ℓ'_1 have a nonempty intersection. This is possible only if ℓ' and ℓ'_1 are the same inflated layer, which contradicts Lemma 2, as the clearly distinct layers ℓ and ℓ_1 would then map to the same layer $\ell' = \ell'_1$. Thus,

$$W \leq L. \tag{4.9}$$

Combining the inequalities (4.6–4.9), we obtain $(\gamma - 2)|\mathbb{T}| \leq 4\gamma L$, thus proving Theorem 4.

References

1. Ajit Agrawal, Philip Klein, and R. Ravi. When trees collide: an approximation algorithm for the generalized steiner problem on networks. *SIAM J. Comput.*, 24(3):440–456, 1995. (Preliminary version in *23rd STOC*, 1991).
2. E. M. L. Beale. On minimizing a convex function subject to linear inequalities. *J. Roy. Statist. Soc. Ser. B.*, 17:173–184; discussion, 194–203, 1955. (Symposium on linear programming.).
3. Luca Becchetti, Jochen Könemann, Stefano Leonardi, and Martin Pál. Sharing the cost more efficiently: Improved approximation for multicommodity rent-or-buy. In *Proceedings of the 16th Annual ACM-SIAM Symposium on Discrete Algorithms*, 2005.
4. John R. Birge and François Louveaux. *Introduction to stochastic programming*. Springer Series in Operations Research. Springer-Verlag, New York, 1997.
5. George B. Dantzig. Linear programming under uncertainty. *Management Sci.*, 1:197–206, 1955.
6. Michel X. Goemans and David P. Williamson. A general approximation technique for constrained forest problems. *SIAM J. Comput.*, 24(2):296–317, 1995. (Preliminary version in *5th SODA*, 1994).
7. Anupam Gupta, Amit Kumar, Martin Pál, and Tim Roughgarden. Approximation via cost sharing: A simple approximation algorithm for the multicommodity rent or buy problem. In *Proceedings of the 44th Annual IEEE Symposium on Foundations of Computer Science*, pages 606–615, 2003.
8. Anupam Gupta, Martin Pál, R. Ravi, and Amitabh Sinha. Boosted sampling: Approximation algorithms for stochastic optimization. In *Proceedings of the 36th Annual ACM Symposium on Theory of Computing*, 2004.
9. Anupam Gupta, R. Ravi, and Amitabh Sinha. An edge in time saves nine: Lp rounding approximation algorithms. In *Proceedings of the 45th Annual IEEE Symposium on Foundations of Computer Science*, 2004.
10. Ara Hayrapetyan, Chaitanya Swamy, and Éva Tardos. Network design for information networks. In *ACM-SIAM Symposium on Discrete Algorithms*, 2005.
11. Nicole Immorlica, David Karger, Maria Minkoff, and Vahab Mirrokni. On the costs and benefits of procrastination: Approximation algorithms for stochastic combinatorial optimization problems. In *Proceedings of the 15th Annual ACM-SIAM Symposium on Discrete Algorithms*, 2004.
12. Peter Kall and Stein W. Wallace. *Stochastic programming*. Wiley-Interscience Series in Systems and Optimization. John Wiley & Sons Ltd., Chichester, 1994.
13. David R. Karger and Maria Minkoff. Building steiner trees with incomplete global knowledge. In *Proceedings of the 41st Annual Symposium on Foundations of Computer Science*, pages 613–623, 2000.

14. Jochen Könemann, Stefano Leonardi, and Guido Schäffer. A group-strategyproof mechanism for steiner forests. In *Proceedings of the 16th Annual ACM-SIAM Symposium on Discrete Algorithms*, 2005.
15. Amit Kumar, Anupam Gupta, and Tim Roughgarden. A constant factor approximation algorithm for the multicommodity rent-or-buy problem. In *Proceedings of the 43rd Annual Symposium on Foundations of Computer Science*, 2002.
16. R. Ravi and Amitabh Sinha. Hedging uncertainty: Approximation algorithms for stochastic optimization problems. In *Proceedings of the 10th International Conference on Integer Programming and Combinatorial Optimization (IPCO)*, 2004. GSIA Working Paper 2003-E68.
17. R. Schultz, L. Stougie, and M. H. van der Vlerk. Two-stage stochastic integer programming: a survey. *Statist. Nederlandica*, 50(3):404–416, 1996.
18. David Shmoys and Chaitanya Swamy. Stochastic optimization is (almost) as easy as deterministic optimization. In *Proceedings of the 45th Annual IEEE Symposium on Foundations of Computer Science*, 2004.

Approximation Algorithms for the Max-coloring Problem*

Sriram V. Pemmaraju and Rajiv Raman

The University of Iowa, Iowa City, IA 52242, USA
{sriram, rraman}@cs.uiowa.edu

Abstract. Given a graph $G = (V, E)$ and positive integral vertex weights $w : V \rightarrow \mathbf{N}$, the *max-coloring problem* seeks to find a proper vertex coloring of G whose color classes $C_1, C_2, \ldots, C_k$, minimize $\sum_{i=1}^{k} max_{v \in C_i} w(v)$. The problem arises in scheduling conflicting jobs in batches and in minimizing buffer size in dedicated memory managers.

In this paper we present three approximation algorithms and one inapproximability result for the max-coloring problem. We show that if for a class of graphs $\mathcal{G}$, the classical problem of finding a proper vertex coloring with fewest colors has a c-approximation, then for that class $\mathcal{G}$ of graphs, max-coloring has a $4c$-approximation algorithm. As a consequence, we obtain a 4-approximation algorithm to solve max-coloring on perfect graphs, and well-known subclasses such as chordal graphs, and permutation graphs. We also obtain constant-factor algorithms for max-coloring on classes of graphs such as circle graphs, circular arc graphs, and unit disk graphs, which are not perfect, but do have a constant-factor approximation for the usual coloring problem. As far as we know, these are the first constant-factor algorithms for all of these classes of graphs. For bipartite graphs we present an approximation algorithm and a matching inapproximability result. Our approximation algorithm returns a coloring whose weight is within $\frac{8}{7}$ times the optimal. We then show that for any $\epsilon > 0$, it is impossible to approximate max-coloring on bipartite graphs to within a factor of $(\frac{8}{7} - \epsilon)$ unless $P = NP$. Thus our approximation algorithm yields an optimum approximation factor. Finally, we also present an exact sub-exponential algorithm and a PTAS for max-coloring on trees.

1 Introduction

The *max-coloring problem* takes as input a vertex-weighted graph $G = (V, E)$ with weight function $w : V \rightarrow \mathbf{N}$. The problem requires that we find a proper vertex coloring of G whose color classes $C_1, C_2, \ldots, C_k$, minimize the sum of the weights of the heaviest vertices in the color classes, that is, $\sum_{i=1}^{k} max_{v \in C_i} w(v)$. When all the weights are one, this problem reduces to the classical problem of

---

* This research is partially supported by the National Science Foundation Grant DMS-0213305.

finding a proper vertex coloring of a graph using fewest possible colors. For any color class C of G, we will use $weight(C)$ to denote $\max\{w(v) \mid v \in C\}$. The weight of a coloring $C_1, C_2, \ldots, C_k$ is then $\sum_{i=1}^{k} weight(C_i)$.

The max-coloring problem arises in two distinct applications. In one application the max-coloring problem models the problem of minimizing the total buffer size needed for memory management in wireless protocol stacks like GPRS or 3G [7] and in digital signal processing applications [2]. In general, programs that run with stringent memory or timing constraints use a dedicated memory manager that provides better performance than the general purpose memory management of the operating system. The most commonly used memory manager design for this purpose is the *segregated buffer pool*. The problem of minimizing the total size of the buffer pool corresponds to the max-coloring problem.

A second application of max-coloring arises in the scheduling of jobs with conflicts in a multiprocessor environment. In systems in which jobs require exclusive access to certain resources, a fundamental problem is of scheduling jobs onto processors such that jobs requiring access to the same resource are not scheduled together. The problem of scheduling jobs in conflict to processors can be modeled as a graph coloring problem. When jobs have different processing times, this is modeled as a generalized coloring problem on vertex weighted graphs. One such generalization that models the problem of scheduling conflicting jobs in batches to minimize the *makespan* or the time to complete all the jobs in the system corresponds to the max-coloring problem.

Our Results. Although graph coloring is hopelessly hard to approximate on general graphs, the underlying conflict graphs that arise in applications have more structure, and this structure can be exploited to obtain efficient exact or approximation algorithms for max-coloring. However, the max-coloring problem is hard even on instances where the coloring problem can be solved in polynomial time. In [7], the authors prove that max-coloring is NP-hard on interval graphs, even though there is a simple greedy algorithm for the usual coloring problem [1]. [7] also presents a 2-approximation for the max-coloring problem on interval graphs. For other classes of graphs, very little seems to be known about how to solve the max-coloring problem efficiently, either exactly or approximately. In this paper we present three approximation algorithms and one inapproximability result. We show that for any hereditary[1] class of graphs $\mathcal{G}$, if the usual vertex coloring problem has a c-approximation, then max-coloring has a $4c$-approximation on $\mathcal{G}$. One implication is that there is a 4-approximation algorithm to solve max-coloring on perfect graphs. For bipartite graphs we present an approximation algorithm and a matching inapproximability result. Our approximation algorithm always returns a coloring whose weight is within $\frac{8}{7}$ times the optimal and following this we show that for any $\epsilon > 0$, it is impossible to approximate max-coloring on bipartite graphs to with a factor of $(\frac{8}{7} - \epsilon)$ unless

---

[1] A class $\mathcal{G}$ of graphs is hereditary, if for any $G \in \mathcal{G}$, every induced subgraph of G is also in $\mathcal{G}$.

$P = NP$. Thus our approximation algorithm yields an optimum approximation factor. Finally, we also present an exact sub-exponential algorithm and a PTAS for trees. The max-coloring problem on trees was also studied by Guan and Zhu [3] where the authors present a polynomial time algorithm that finds an optimal max-coloring of a given tree, that uses exactly r colors for a fixed constant r.

2   Max-coloring Trees

The max-coloring problem has turned out to be surprisingly difficult even for trees. Though we believe that the problem can be solved in polynomial time, the two best algorithms we have are (i) a sub-exponential exact algorithm and (ii) a PTAS. We present these in this section. Our first observation is on the distribution of weights of color classes in an optimal max-coloring of bipartite graphs.

Lemma 1. *Let G be a bipartite graph. Let $\{C_1, C_2, \ldots, C_k\}$ be an optimal max-coloring of G with $w_i = weight(C_i)$ and $w_1 \geq w_2 \geq \cdots \geq w_k$. Then, we have that $w_i \geq \sum_{j=i+1}^{k} w_j$, $i = 1, \cdots, k-1$.*

Proof. If $w_i < \sum_{j=i+1}^{k} w_j$, then the subgraph induced by vertices in $\cup_{j=i}^{k} C_j$ can be colored with two colors with weight at most $2w_i$. This coloring has weight less than the weight of $\{C_1, C_2, \ldots, C_k\}$, a contradiction. □

Corollary 1. *Let G be a bipartite graph. Let $\{C_1, \cdots, C_k\}$ be an optimal max-coloring of G with $w_i = weight(C_i)$ and $w_1 \geq w_2 \geq \cdots \geq w_k$, we have that $\frac{w_i}{2} \geq w_{i+2}$, for $i = 1, \cdots, k-2$, and hence, $w_1 \geq 2^{\lfloor (i-1)/2 \rfloor} \cdot w_i$.*

Since the weights of the color classes decrease rapidly, we can expect that the max-color number of a tree may not be too high. Let $\chi_{mc}(G)$ denote the *max-color number* of a graph G, the minimum number of colors required in a minimum cost max-coloring. We now state three upper bounds on χ_{mc}.

Lemma 2. *Let T be an n-vertex tree with maximum degree Δ. Let W denote the ratio of the weight of the heaviest vertex to the weight of the least heavy vertex. Then, (i) $\chi_{mc}(T) \leq \Delta + 1$, (ii) $\chi_{mc}(T) \leq \lceil \log_2 n \rceil + 1$, and (iii) $\chi_{mc}(T) \leq \lceil \log_2 W \rceil + 1$.*

Proof. Let $k = \chi_{mc}(T)$ and let $\{C_1, \cdots, C_k\}$ be the color classes in an optimal max-coloring of T. Let $w_i = weight(C_i)$ and without loss of generality assume that $w_1 \geq w_2 \geq \cdots \geq w_k$.

(i) Suppose $\chi_{mc}(T) > \Delta + 1$. For each vertex v in C_k, we can find a color class C_i, $i < k$ such that v is not adjacent to any vertex in C_i. We can thus move each vertex in C_k to a lower color class thus decreasing the coloring weight, a contradiction. Note that this upper bound holds in general for any graph G.

(ii) For each $i > 1$, we can assume without loss of generality that every vertex $v \in C_i$ has a neighbor in C_j, for every $j < i$.

For each vertex $v \in C_1$, let $T(v)$ denote the rooted tree with one vertex, namely v. For each $v \in C_i$, $i > 1$, define $T(v)$ as the tree rooted at v, such that (i) the children of v in $T(v)$ are exactly the neighbors of v in T belonging to color classes $C_1, C_2, \ldots, C_{i-1}$, and (ii) for each child u of v, the subtree of $T(v)$ rooted at u is simply $T(u)$. For each i, $1 \leq i \leq k$, let $S_i = \min\{|T(v)| \mid v \in C_i\}$. In other words, S_i is the size of a smallest tree $T(v)$ rooted at a vertex v in C_i. Then,

$$S_1 = 1$$

$$S_i \geq \sum_{j=1}^{i-1} S_j + 1, \text{ for each } i > 1$$

This implies that $S_i \geq 2^{i-1}$, $1 \leq i \leq k$. Using the fact that $S_k \leq n$, we get $\chi_{mc} = k \leq \lfloor \log_2 n \rfloor + 1$.

(iii) Let $\ell = \min\{t \in \mathbf{N} \mid \text{ for all } v \in V(T), w(v) \geq w_1/2^t\}$. Therefore, $\ell = \lceil \log_2 W \rceil$. Recall that W is the ratio of the weights of the heaviest vertex to the lightest vertex. Consider the collection of disjoint intervals $\mathcal{I} = \{I_0, I_1, \ldots, I_{\ell-1}\}$, where $I_i = [\frac{w_1}{2^{i+1}}, \frac{w_1}{2^i})$, for $i = 1, \ldots, \ell - 1$ and let $I_0 = [\frac{w_1}{2}, w_1]$. Because of the choice of ℓ, for each vertex $v \in V(T)$, $w(v)$ belongs to exactly one interval I_j. Let $V_j = \{v \in V(T) \mid w(v) \in I_j\}$, $j = 0, 1, \ldots, \ell - 1$. We say that a vertex v *contributes* to a color class C_i if $v \in C_i$, and $w(v) = \max\{w(u) \mid u \in C_i\}$. The *contribution* of an interval I_j is the maximum number of vertices in V_j that contribute to distinct color classes.

Corollary 1 tells us that $w_i \geq 2 \cdot w_{i+2}$ for $i = 1, \cdots k - 2$. This immediately implies that no interval I_j, $j = 1, 2, \ldots, \ell - 1$ has a contribution of more than two. Now suppose that intervals $I_{i_1}, I_{i_2}, \ldots, I_{i_t}$, $0 \leq i_1 < i_2 < \cdots < i_t \leq \ell - 1$, is the sequence of all intervals in $\mathcal{I}$, each of whose contribution is two. We now claim that for any pair of consecutive intervals I_p, $p = i_j$ and I_q, $q = i_{j+1}$, where $j < t$, there is an interval in $\{I_{p+1}, I_{p+2}, \ldots, I_{q-1}\}$ with contribution zero. If we can show this claim, then we can charge the "extra" contribution of each I_{i_j} to an interval between I_{i_j} and $I_{i_{j+1}}$, whose contribution is zero. Since there are ℓ intervals and since the contribution of I_{i_t} is at most two, there is a total contribution of at most $\ell + 1$, implying that there are at most $\ell + 1$ color classes.

We prove the above claim by contradiction, assuming that the contribution of every interval in $\{I_{p+1}, I_{p+2}, \ldots, I_{q-1}\}$ is one. Let $\{x_p, x_{p+1}, \ldots, x_q\} \cup \{y_p, y_q\}$ be vertices such that (i) for each $j = p, p+1, \ldots, q$, $x_j \in V_j$ and x_j contributes to some color class and (ii) for each $j \in \{p, q\}$, $y_j \in V_j$ and x_j and y_j contribute to distinct color classes. Since $x_j \in V_j$, $w(x_j) \geq \frac{w_1}{2^{j+1}}$, $j = p, p+1, \ldots, q$. Also, since $y_q \in V_q$, $w(y_q) \geq \frac{w_1}{2^{q+1}}$. Therefore,

$$\sum_{j=p}^{q} w(x_j) + w(y_q) \geq \sum_{j=p}^{q} \frac{w_1}{2^{j+1}} + \frac{w_1}{2^{q+1}}$$

$$= w_1 \frac{2^{q-p+1} - 1}{2^{q+1}} + \frac{w_1}{2^{q+1}}$$

$$= \frac{w_1}{2^p} > w(y_p)$$

This contradicts Lemma 1 and proves the claim. □

There are simple examples that show that the bounds in Lemma 2 are all tight [6].

Since the number of colors are at most $\lceil \log n \rceil + 1$, this immediately gives a simple sub-exponential time algorithm. Try all $\lceil \log n \rceil + 1$ possible colors for each vertex, and return a feasible coloring of minimum weight. This algorithm runs in $O(n^{\log n + 1})$ time.

Now we show that if the given tree has a constant number of distinct vertex weights, we can find an optimal max-coloring in polynomial time. We deal with the case of constant number of distinct weights via the solution to a problem called FEASIBLE k-COLORING.

FEASIBLE k-COLORING

INPUT: A tree T with weight function $w : V \rightarrow \mathbf{N}$, and a positive integer sequence $(W_1, W_2, \cdots, W_k)$ of positive integers, satisfying $W_1 \geq W_2 \geq \cdots \geq W_k$.
OUTPUT: Either a coloring of the tree into color classes $A_1, \cdots, A_k$, such that for all $v \in A_i$, $w(v) \leq W_i$ or if such a coloring does not exist, a report that no such feasible coloring exists.

There is a simple dynamic programming algorithm for solving FEASIBLE k-COLORING on trees in $O(nk)$ time [6].

The main idea underlying our PTAS is the reduction of the number of distinct weights of the vertices down to a constant. We then pick candidates for the weights of the color classes and for each such choice, using the algorithm for FEASIBLE k-COLORING, we test if there is a legal coloring of the tree with the chosen weights for the color classes.

We are given a tree T, with weight function $w : V \rightarrow \mathbf{N}$ and an $\epsilon > 0$. Let $c > 0$ be an integer such that $(2 \log c + 3)/c \leq \epsilon$, and let $\alpha = (W-1)/c$ where W is the maximum weight of any vertex. Let $I_1, I_2, \cdots, I_c$ be a partition of the range $[1, W)$, where $I_i = [1 + (i-1)\alpha, 1 + i \cdot \alpha)$, $1 \leq i \leq c$. Let T' be a tree that is identical to T, except in its vertex weights. The tree T' has vertex weights $w' : V \rightarrow \mathbf{N}$ defined by the rule: for any $v \in V$, if $w(v) \in I_j$ then $w'(v) = 1 + (j-1) \cdot \alpha$ and if $w(v) = W$, then $w'(v) = W$. In other words, except for vertices with maximum weight W, all other vertices have their weights "rounded" down. As a result T' has $c+1$ distinct vertex weights. Now let OPT' denote the weight of an optimal max-coloring of T' and let $\mathcal{C}' = C'_1, C'_2, \ldots, C'_k$ be the color classes corresponding to OPT'. Since the weights of vertices have fallen in going from T to T', clearly $OPT' \leq OPT$. If we use the coloring $\mathcal{C}'$ for T, we get a coloring whose weight is at most $OPT' + k\alpha$. Substituting $(W-1)/c$ for α and noting that $W \leq OPT'$, we obtain that weight of $\mathcal{C}'$ used as a coloring for T' is at most $(1 + \frac{k}{c})OPT'$ We now show that given the distribution of vertex weights of T', $k = O(\log c)$. If $k = 2$ we are done, so assume that $k \geq 3$. To see this first observe that the weights of last three color classes C'_k, C'_{k-1}, and C'_{k-2} cannot all be identical, by Lemma 1. Also, observe that the possible vertex weights of T' are $1, 1 + \alpha, 1 + 2\alpha, \ldots$. Therefore, $weight(C'_{k-2}) \geq 1 + \alpha$. From Corollary 1, we obtain

$$1 + \alpha \leq weight(C'_{k-2}) \leq \frac{W}{2^{\lfloor (k-3)/2 \rfloor}}.$$

Solving this for k yields $k \leq 2\log_2(c) + 3$. Therefore, by our choice of c, we have
$$\frac{k}{c} \leq \frac{2\log_2(c) + 3}{c} \leq \epsilon.$$

Thus $(1+\epsilon)OPT'$ is an upper bound on the weight of $\mathcal{C}'$ used as a coloring for T. Since $OPT' \leq OPT$, we see that the weight of $\mathcal{C}'$ used as a coloring for T is at most $(1+\epsilon)OPT$.

To construct OPT' in polynomial time, for each $k = 1, \ldots, 2\lceil \log_2 c \rceil + 3$, we generate all $O(c^k)$ possible sequences of weights and call algorithm FEASIBLE k-COLORING for each subsequence and pick the coloring with the minimum weight. This gives OPT'. Each solution to FEASIBLE k-COLORING takes $O(nk)$ time, and we have $O(c^k)$ sequences, for $k = 1, \ldots, 2\lceil \log_2 c \rceil + 3$. Using the fact that $(2\log_2 c + 3)/c \leq \epsilon$, a little bit of algebra yields a running time that is linear in n and exponential in $1/\epsilon$.

3 Max-coloring Bipartite Graphs

This section presents an $\frac{8}{7}$-approximation algorithm for the max-coloring problem on bipartite graphs, followed by a hardness of approximation result that shows that for any $\epsilon > 0$, there is no $(\frac{8}{7} - \epsilon)$-approximation algorithm unless $P = NP$. Thus our approximation algorithm produces an optimal approximation ratio.

One feature of our approximation algorithm is that it uses at most 4 colors, even though though an optimal max-coloring of an n-vertex bipartite graph may need a $\Omega(n)$ colors [6]. Our PTAS for the max-coloring problem on trees relied on the fact that the FEASIBLE k-COLORING problem on trees can be solved in polynomial time for any k. However, FEASIBLE k-COLORING is NP-complete for bipartite graphs for $k \geq 3$ [5]. This has forced us to use a different approach for bipartite graphs. Another difference between max-coloring on trees and max-coloring on bipartite graphs is that in contrast to the $O(\log n)$ upper bound on the number of colors used by an optimal max-coloring for an n-vertex tree, there are simple examples of n-vertex bipartite graphs G with $\chi_{mc}(G) \geq n/2$ [6].

Our $(\frac{8}{7} - \epsilon)$-hardness result for max-coloring bipartite graphs is via a gap introducing reduction from the PRE-COLORING EXTENSION problem [5].

3.1 An $\frac{8}{7}$-Approximation Algorithm

First note that since bipartite graphs are 2-colorable, Lemma 1 holds and hence if an optimal max-coloring of a bipartite graph uses a large number of colors, the contribution of all but the first few color classes must be quite small. We can use this to our advantage and develop an algorithm that tries to find a *good* approximation to the weights of the first few color classes. We run three algorithms, A_2, A_3, and A_4, that use 2, 3 and 4 colors respectively. The color classes produced by algorithm A_i, $2 \leq i \leq 4$, are denoted $\{A_1^i, A_2^i, \cdots\}$, and the weights of the corresponding color classes are denoted $\{a_1^i, a_2^i, \cdots\}$. We start with a description of algorithm A_2.

Algorithm $A_2(G, w)$
1. For each connected component G_i of G do
2.   Color G_i with colors 1 and 2, such that a vertex with maximum weight is colored 1.

The fact that A_2 is a 2-approximation immediately follows from the fact that $weight(A_2) \leq 2w_1$, and $w_1 \leq OPT$. We encode this result in the following lemma.

Lemma 3. $weight(A_2) \leq 2w_1$

In an optimum coloring, the weight of the first color class, w_1 is fixed. By using more colors, OPT may gain an advantage because it can then push *heavy* vertices into *lower* color classes. We now introduce algorithm A_3 which constructs a 3-coloring of G such that the weight of the second color class is minimized.

Algorithm $A_3(G, w)$
1. Let S be a *maximal* independent set of G picked by examining vertices in non-increasing weight order.
2. Use Algorithm A_2 to color $G \setminus S$.
3. Rename colors 1 and 2, as colors 2 and 3 respectively.
4. Color S with color 1.

Lemma 4. $weight(A_3) \leq w_1 + 2w_2$.

Proof. In algorithm A_3, $a_1^3 = w_1$. Since S is a maximal independent set selected in non-increasing weight order, the weight of the second color class of OPT, w_2 cannot be smaller than the weight of any vertex in $G \setminus S$. Hence, $w_2 \geq a_2^3$. Since $a_3^3 \leq a_2^3$, it follows that $weight(A_3) = a_1^3 + a_2^3 + a_3^3 \leq w_1 + w_2 + w_2 = w_1 + 2w_2$. □

The greedy strategy employed by algorithm A_3 in selecting the first color class causes a_2^3 to be no larger than w_2. However, it might cause a_3^3 to be significantly larger than w_3. We rectify this situation by introducing algorithm A_4 that uses four colors to color G.

Algorithm $A_4(G, w)$
1. For all w^* such that there is a $u \in V$, with $w(u) = w^*$ do
2.   Partition the vertices of G into two parts
$$P_1 = \{v \in V \mid w(v) > w^*\|, \text{ and}$$
$$P_2 = \{v \in V \mid w(v) \leq w^*\}.$$
3. Use algorithm A_2 to color P_2.
4. Rename colors 1 and 2 as 3 and 4 respectively.
5. Use algorithm A_2 to color P_1.
6. Return the coloring with minimum weight, over all choices of w^*.

Lemma 5. $weight(A_4) < w_1 + w_2 + 2w_3$

Proof. Since the weight of every vertex in G is used for the threshold w^*, in some iteration of A_4, $w^* = w_3$. At this point, A_4 partitions the vertex set such that $P_1 = \{v \mid w(v) > w_3\}$ and $P_2 = \{v \mid w(v) \leq w_3\}$. In this iteration, A_4 colors P_1 with weight at most $w_1 + w_2$, and colors P_2 with weight at most $2w_3$. Since A_4 returns the coloring with minimum weight, over all choices of w^*, it follows that $weight(A_4) \leq w_1 + w_2 + 2w_3$. □

The final algorithm, which we call Bipartite Max-Color runs A_2, A_3, A_4, and returns the minimum weight coloring.

Theorem 1. *Algorithm* Bipartite Max-Color *is a $\frac{8}{7}$-approximation for the max-coloring problem on bipartite graphs.*

Proof. Let $w(B)$ denote the weight of the coloring produced by algorithm Bipartite Max-Color. From Lemmas 3, 4, and 5, we know that $w(B) \leq 2w_1$, $w(B) \leq w_1 + 2w_2$, $w(B) \leq w_1 + w_2 + 2w_3$. Now, multiplying the first inequality by 1, the second inequality by 2, the third inequality by 4 and adding, we get

$$7 \cdot w(B) \leq 8 \cdot (w_1 + w_2 + w_3) \leq 8 \cdot OPT$$ □

3.2 An ($\frac{8}{7} - \epsilon$)-Hardness Reduction

We now show that the 8/7-approximation produced by the above algorithm is optimal. We do this by showing a matching hardness result via a reduction from the PRE-COLORING EXTENSION problem on bipartite graphs. The PRE-COLORING EXTENSION problem for general graphs is defined below.

PRE-COLORING EXTENSION

INPUT: A graph $G = (V, E)$, with $r \geq \chi(G)$, a subset $P \subseteq V$, and a proper assignment $c : P \to \{1, \cdots, r\}$ of colors to vertices in P.

QUESTION: Is there an extension of the proper vertex coloring of P to a proper vertex coloring of G, using colors from $\{1, \cdots, r\}$?

In [5], Kratochvil proved that PRE-COLORING EXTENSION is NP-complete for planar bipartite graphs even when the color bound $r = 3$. We now show a simple gap introducing reduction from PRE-COLORING EXTENSION on bipartite graphs with $r = 3$ to max-coloring on bipartite graphs.

Theorem 2. *For any $\epsilon > 0$, there is no $(8/7 - \epsilon)$-approximation algorithm for max-coloring on bipartite graphs, unless P=NP.*

Proof. The reduction is from PRE-COLORING EXTENSION. Let the given instance of PRE-COLORING EXTENSION consist of a bipartite graph $G = (V_1, V_2, E)$, a subset $P \subseteq V_1 \cup V_2$, and a proper assignment $c : P \to \{1, 2, 3\}$ of colors to vertices in P. We transform G into a vertex-weighted bipartite graph $G' = (V_1', V_2', E')$ as follows. Add four new vertices, x_1, x_2, y_1, and y_2 to G. Let $X = \{x_1, x_2\}$,

$Y = \{y_1, y_2\}$, $V_1' = V_1 \cup X$, and $V_2' = V_2 \cup Y$. To each vertex $v \in P$, assign a weight $w(v)$ using the rule: $w(v) = 2^{3-i}$ if $c(v) = i$, for each $i \in \{1, 2, 3\}$. If $v \in (V_1 \cup V_2) - P$, set $w(v) = 1$. The new vertices are assigned weights as follows: $w(x_1) = w(y_1) = 4$ and $w(x_2) = w(y_2) = 2$. The edge set E' of G' contains some additional edges between the new vertices and the old.

$$E' = E \cup \{\{x_i, y\} | y \in P \cap V_2', \text{ and } w(y) < w(x_i)\} \cup$$
$$\{\{y_i, x\} | x \in P \cap V_1' \text{ and } w(x) < w(y_i)\} \cup \{\{x_1, y_2\}\} \cup \{\{x_2, y_1\}\}.$$

This completes the description of G'.

Now suppose that the coloring of P can be extended to a proper 3-coloring $c : V_1 \cup V_2 \to \{1, 2, 3\}$ of G. Start with the coloring c and extend this to a proper vertex coloring of G' by assigning colors to the new vertices as follows: $c(x_1) = c(y_1) = 1$ and $c(x_2) = x(y_2) = 2$. Observe that this indeed produces a proper coloring of G'. To see that the weight of this coloring on G' has weight at most 7, note that the weight of the coloring in G' restricted to the vertices of G is at most 7, and since the coloring above of the vertices $\{x_1, x_2, y_1, y_2\}$ does not increase this cost, we are done.

Now suppose that G does not have a pre-coloring extension. We show by contradiction that in this case G' does not have a proper vertex coloring of weight less than 8. So suppose that there is a proper vertex coloring $c' : V_1' \cup V_2' \to \{1, 2, \ldots\}$ of weight less than 8. Without loss of generality, assume that in this coloring, the color classes are labeled in non-increasing order of their weight. Therefore, all vertices of weight 4 are in color class 1. This includes vertices x_1 and y_1 and this forces all vertices of weight 2 to be excluded from color class 1. Since color class 1 has weight 4, to prevent the total weight of the coloring from reaching 8, all vertices of weight 2 have to be included in color class 2. This includes vertices x_2 and y_2, and so this color class is also non-empty. Therefore the total weight of color classes 1 and 2 is 6. Since c' is a coloring of G' of weight less than 8, it must be the case that color class k, for each $k \geq 4$, is empty. This means that c' is a 3-coloring of G'. Furthermore, it is a 3-coloring of G that respects the pre-coloring of P. This contradicts the assumption that G has no pre-coloring extension and therefore we have that any proper vertex coloring of G' has weight at least 8.

If for some $\epsilon > 0$, there were an $(\frac{8}{7} - \epsilon)$-approximation algorithm for max-coloring bipartite graphs, then using the above polynomial time transformation from G to G', we could distinguish between positive and negative instances of PRE-COLORING EXTENSION on bipartite graphs with $r = 3$. This is not possible unless $P = NP$. □

4 Max-coloring on Arbitrary Graphs

Let $\mathcal{G}$ be a hereditary class of graphs for which the minimum vertex coloring problem has a c-approximation. In other words, there is a polynomial time algorithm A that takes a graph $G \in \mathcal{G}$ as input and returns a proper vertex coloring of G using at most $c \cdot \chi(G)$ colors. In this section, we present an $4c$-approximation algorithm, that we call GeomFit, for the max-coloring problem on the class of graphs $\mathcal{G}$. The algorithm is inspired by the algorithm of Halldórsson, et. al. [4] for sum-coloring interval and comparability graphs. GeomFit will repeatedly use A as a black box to obtain "good" vertex colorings of portions of the input graph. For ease of exposition, below we describe GeomFit assuming that $c = 1$.

GeomFit(G, w)
   1. Let $i = 0$, $l_i = 0$
   2. While $G \neq \emptyset$ do
      3. Set $c_i = 2^i$
      4. Let $G_i = mkc(G, c_i)$
      5. Color G_i optimally using colors $l_i + 1, \cdots, l_i + c_i$
      6. Set $l_{i+1} = l_i + c_i$, $i = i + 1$.
      7. Set $G = G \setminus G_i$.
   8. End While

A round of the algorithm corresponds to an iteration of the while loop. Suppose that each round is labeled with the value of i at the beginning of that round. For some integer $t > 0$, suppose that the algorithm executes rounds $0, 1, \cdots, t-1$, after which the graph is entirely colored. In each round i, $0 \leq i < t$, the algorithm calls the subroutine $mkc(G, c_i)$, that returns a maximal c_i-colorable subgraph of G, obtained by examining vertices in non-increasing order of weight. Here G is the subgraph of the input graph induced by the not yet colored vertices and $c_i = 2^i$. When called, the subroutine $mkc(G, c_i)$ starts with an empty set S and processes each vertex v of G, in non-increasing order of weight. The subroutine tests if $G[S \cup \{v\}]$ is c_i-colorable or not and if it is, it adds v to S, and proceeds to the next vertex in G. To perform this test, $mkc(G, c_i)$ calls the algorithm A that returns a minimum vertex coloring of G.

Lemma 6. *If GeomFit uses t rounds to color G, then $\chi(G) > c_{t-2}$.*

Proof. In round $t - 2$, the algorithm picks a maximal c_{t-2} colorable subgraph of G. If G were c_{t-2}-colorable, then all of it would have been picked up in round $t - 2$ or earlier. Since we used one more round to color G, it must mean that $\chi(G) > c_{t-2}$. □

Without loss of generality, suppose that OPT uses numbers $1, 2, \ldots$ for colors such that color classes are numbered in non-increasing order of weight. Now observe that color classes created in round i by GeomFit are all heavier than color classes created in round $i + 1$. Without loss of generality, assume that the color classes created in each round of GeomFit are numbered in non-increasing order of weight. Let $color_{OPT}(v)$ denote the color assigned to vertex v in OPT. Now using the color classes of OPT we define a pairwise disjoint collection of vertex subsets of G, $\{V_0, \cdots, V_{t-1}\}$, where $V_i = \{v \in G | c_{i-1} < color_{OPT}(v) \leq c_i\}$, $i = 0, \cdots, t - 1$. For the definition to make sense, we assume that $c_{-1} = 0$. Since V_{t-1} contains vertices colored $c_{t-2} + 1, c_{t-2} + 2, \ldots, c_{t-1}$ by OPT, from Lemma

6, it follows that $V_{t-1} \neq \emptyset$. Now we state and prove a critical inequality that follows from the greedy choice of a subgraph in each round of GeomFit. Let W_i denote the weight of color class $c_{i-1}+1$ in OPT. Note that color class $c_{i-1}+1$ is a subset of V_i and by our labeling convention, it is a heaviest color class in V_i. Similarly, let R_i denote the weight of color class l_i+1 created by GeomFit. Note that this is a heaviest color class created in round i be GeomFit. Also note that $l_i = \sum_{j=0}^{i-1} c_j = c_i - 1$ and therefore color class l_i+1 is simply color class c_i.

Lemma 7. $R_i \leq W_i$, for $i = 0, 1, \cdots, t-1$.

Proof. Since R_0 and W_0 are equal to the maximum weight vertex in G, the lemma holds for $i = 0$. By the greedy choice employed in selecting G_0, we ensure that for any other independent set S of G, the maximum weight of a vertex in $G \setminus S$ is at least as large as the maximum weight vertex in $G \setminus G_0$. This ensures that $R_1 \leq W_1$. By the same reasoning, since in round $i-1$, we greedily select a maximal c_{i-1} colorable subgraph of OPT, and $V_1 \cup V_2 \cup \cdots V_{i-1}$ is c_{i-1} colorable, it follows that $R_i \leq W_i$. □

Theorem 3. *Let $\mathcal{G}$ be a hereditary class of graphs on which the minimum vertex coloring problem can be solved in polynomial time. Algorithm GeomFit is a 4-approximation algorithm for the max-coloring problem on $\mathcal{G}$.*

Proof. The weight of the max-coloring produced by GeomFit is bounded above by

$$weight(\text{GeomFit}) \leq \sum_{i=0}^{t-1} c_i \cdot R_i \leq \sum_{i=0}^{t-1} c_i \cdot W_i$$

The first inequality follows from the fact that in each round i, GeomFit uses at most c_i colors and a heaviest color class in round i has weight R_i. The second inequality follows from Lemma 7.

We obtain a lower bound on OPT as follows. The set V_0 contains one color class and this has weight W_0. Now consider a set V_i, $1 \leq i \leq t-2$. It contains one color class of weight W_i and the remaining color classes have weight at least W_{i+1}. Recall that V_i has color classes labeled $c_{i-1}+1, c_{i-1}+2, \ldots, c_i$ and therefore $weight(V_i) \geq W_i + (c_{i-1}-1)W_{i+1}$.

$$OPT \geq \sum_{i=0}^{t-1} weight(V_i) \geq W_0 + \sum_{i=1}^{t-2}(W_i + (c_{i-1}-1)W_{i+1}) + W_{t-1}$$

$$= W_0 + W_1 + \sum_{i=0}^{t-3} c_i W_{i+2}.$$

Therefore, $4 \cdot OPT \geq 4W_0 + 4W_1 + \sum_{i=0}^{t-3} 4c_i W_{i+2} = 4W_0 + 4W_1 + \sum_{i=2}^{t-1} c_i W_i$. This lower bound on $4 \cdot OPT$ is larger than the upper bound on $weight(\text{GeomFit})$ above. Therefore, $weight(\text{GeomFit}) \leq 4 \cdot OPT$. □

Now suppose that $\mathcal{G}$ is a hereditary class of graphs that has a c-approximation algorithm A for the minimum vertex coloring problem. A $4c$-approximation algorithm for max-coloring on graphs in $\mathcal{G}$ is obtained by modifying GeomFit slightly. In Step (4), the algorithm computes a maximal $\lfloor c \cdot c_i \rfloor$-colorable subgraph. Correspondingly, in Step (5), G_i is colored using colors $l_i + 1, \cdots, l_i + \lfloor c \cdot c_i \rfloor$. The analysis of this modified GeomFit proceeds in a manner similar to the $c = 1$ case. For details, see [6].

Theorem 4. *Let $\mathcal{G}$ be a hereditary class of graphs on which the minimum vertex coloring problem has a c-approximation algorithm. Algorithm* GeomFit *is a $4c$-approximation algorithm for the max-coloring problem on $\mathcal{G}$.*

The choice of $c_i = 2^i$ in GeomFit gave us an approximation factor of $4c$. This approximation factor can be improved to $3.5c$ by running GeomFit twice, once by setting $c_i = 2^i$, and once by setting $c_i = \lfloor 1.5 \times 2^i \rfloor$ and returning the coloring with smaller weight. More generally, setting $c_i = \lfloor \alpha q^i \rfloor$, where α is chosen uniformly at random from a certain range and q is an appropriately chosen constant, may yield a furthur improvement in the approximation ratio.

Acknowledgments. We would like to thank the anonymous referees for suggestions that led to a simplified proof of Theorem 1, and also for pointing out the work done in [3].

References

1. M.C. Golumbic. *Algorithmic graph theory and perfect graphs*. Academic Press, NY, 1980.
2. R. Govindarajan and S. Rengarajan. Buffer allocation in regular dataflow networks: An approach based on coloring circular-arc graphs. In *Proceedings of the 2nd International Conference on High Performance Computing*, 1996.
3. D.J. Guan and Xuding Zhu. A coloring problem for weighted graphs. *Information Processing Letters*, 61:77–81, 1997.
4. Magnús M. Halldórsson, Guy Kortsarz, and Hadas Shachnai. Sum coloring interval and k-claw free graphs with application to scheduling dependent jobs. *Algorithmica*, 37(3):187–209, 2003.
5. J. Kratochvil. Precoloring extensions with a fixed color bound. *Acta Mathematica Universitatsis Comenianae*, 62:139–153, 1993.
6. S. V. Pemmaraju and R. Raman. Approximation algorithms for the max-coloring problem. http://www.cs.uiowa.edu/~sriram/papers/tbPerfectFull.ps.
7. S.V. Pemmaraju, R. Raman, and K. Varadarajan. Buffer minimization using max-coloring. In *Proceedings of The ACM-SIAM Symposium on Discrete Algorithms (SODA)*, pages 562–571, 2004.

Tight Lower Bounds for Query Processing on Streaming and External Memory Data

Martin Grohe[1], Christoph Koch[2], and Nicole Schweikardt[1]

[1] Institut für Informatik, Humboldt-Universität Berlin, Germany
{grohe, schweika}@informatik.hu-berlin.de
[2] Database Group, Universität des Saarlandes, Saarbrücken, Germany
koch@cs.uni-sb.de

Abstract. We study a clean machine model for external memory and stream processing. We show that the number of scans of the external data induces a strict hierarchy (as long as work space is sufficiently small, e.g., polylogarithmic in the size of the input). We also show that neither joins nor sorting are feasible if the product of the number $r(n)$ of scans of the external memory and the size $s(n)$ of the internal memory buffers is sufficiently small, e.g., of size $o(\sqrt[5]{n})$. We also establish tight bounds for the complexity of XPath evaluation and filtering.

1 Introduction

It is generally assumed that databases have to reside in external, inexpensive storage because of their sheer size. Current technology for external storage systems (disks and tapes) presents us with a reality that performance-wise, a small number of *sequential scans* of the data is strictly preferable over random data accesses. Indeed, the combined latencies and access times of moving to a certain position in external storage are by orders of magnitude greater than actually reading a small amount of data once the read head has been placed on its starting position.

Database engines rely on main memory buffers for assuring acceptable performance. These are usually small compared to the size of the externally stored data. Database technology – in particular query processing technology – has developed around this notion of memory hierarchies with layers of greatly varying sizes and access times. There has been a wealth of research on query processing and optimization along these lines (cf. e.g. [27, 14, 32, 22]). It seems that the current technologies scale up to current user expectations, but on closer investigation it may appear that our theoretical understanding of the problems involved – and of optimal algorithms for these problems – is not quite as developed.

Recently, data stream processing has become an object of study by the data management community (e.g. [15]) but from the viewpoint of database theory, this is, in fact, a special case of the query processing problem on data in external storage where we are limited to a single scan of the input data.

In summary, it appears that there are a variety of data management and query processing problems in which a comparably small but efficiently accessible main memory buffer is available and where accessing external data is costly and is best performed by sequential read/write scans. This calls for an appropriate formal model that captures the essence of external memory and stream processing. In this paper, we study such a model, which employs a Turing machine with one *external memory tape* (external tape for short) and a number *internal memory tapes* (internal tapes for short). The external tape initially holds the input; the internal tapes correspond to the main memory buffers of a database management system and are thus usually small compared to the input.

As computational resources for inputs of size n, we study the space $s(n)$ available on the internal tapes and the number $r(n)$ of scans of (or, random accesses to) the external tape, and we write $ST(r, s)$ to denote the class of all problems solvable by (r, s)-bounded Turing machines, i.e., Turing machines which comply to the resource bounds $r(n)$ and $s(n)$ on inputs of size n.

Formally, we model the number of scans, respectively the number of random accesses, by the number of reversals of the Turing machine's read/write head on the external tape. The number of reversals of the read/write head on the internal tapes remains unbounded. The reversals done by a read/write head are a clean and fundamental notion [8], but of course real external storage technology based on disks does not allow to reverse their direction of rotation. On the other hand, we can of course simulate k forward scans by $2k$ reversals in our machine model — and allowing for forward as well as backward scans makes our *lower* bound results even stronger.

As we allow the external tape to be both read and written to, the external tape can be viewed, for example, as modeling a hard disk. By closely watching reversals of the external tape head, anything close to random I/O will result in a very considerable number of reversals, while a full sequential scan of the external data can be effected cheaply. We will obtain strong lower bounds in this paper that show that even if the external tape (whose size we do not put a bound on) may be written to and re-read, certain bounds cannot be improved upon. For our matching upper bounds, we will usually not write to the external tape. Whenever one of our results requires writing to the external tape, we will explicitly indicate this.

The model is similar in spirit to the frameworks used in [18, 19], but differs from the previously considered *reversal complexity* framework [8]. Reversal complexity is based on Turing machines with a single read/write tape and the overall number of reversals of the read/write head the main computational resource. In our notion, only the number of reversals on the external tape is bounded, while reversals on the internal tapes are free; however, the space on the internal tapes is considered to be a limited resource.[1]

---

[1] The justification for this assumption is simply that accessing data on disks is currently about five to six orders of magnitude slower than accessing main memory. For that reason, processor cycles and main memory access times are often neglected when estimating query cost in relational query optimizers, where cost measures are often exclusively based on the amount of expected page I/O as well as disk latency and access times. Moreover, by taking buffer space rather than running time as a parameter, we obtain more robust complexity classes that rely less on details of the machine model (see also [31]).

Apart from formalizing the $ST(r,s)$ model, we study its properties and locate a number of data management problems in the hierarchy of $ST(\cdot,\cdot)$ classes. Our technical contributions are as follows:

- We prove a reduction lemma (Lemma 4.1) which allows easy lower bound proofs for certain problems.
- We prove a hierarchy (Corollary 4.11 and Theorem 4.10), stating for each fixed number k that $k+1$ scans of the external memory tape are strictly more powerful than k scans of the external memory tape.
- We consider machines where the product of the number of scans of the external memory tape, $r(n)$, and internal memory tape size, $s(n)$, is of size $o(\frac{n}{\log n})$, where n is the input size, and show that *joins* cannot be computed by (r,s)-bounded Turing machines (cf., Lemma 4.4).
- We show that the *sorting* problem cannot be solved with $(o(\sqrt[5]{n}), O(\sqrt[5]{n}))$-bounded Turing machines that are not allowed to write intermediate results to the external memory tape (cf., Corollary 4.9).
- We show (cf., Theorem 4.5) that for some *XQuery* queries, filtering is impossible for machines with $r(T) \cdot s(T) \in o(\frac{n}{\log n})$, where n is the size of the input XML document T.
- We show (cf., Corollary 5.5) that for some *Core XPath* [12] queries, filtering is impossible for machines with $r(T) \cdot s(T) \in o(d)$, where d denotes the *depth* of the input XML document T. Furthermore, we show that the lower bound on Core XPath is *tight* in that we give an algorithm that solves the Core XPath filtering problem with a single scan of the external data (zero reversals) and $O(d)$ buffer space.

The primary technical machinery that we use for obtaining lower bounds is that of *communication complexity* (cf. [21]). Techniques from communication complexity have been used previously to study queries on streams [4, 6, 2, 3, 5, 23, 24, 18]. The work reported on in [4] addresses the problem of determining whether a given relational query can be evaluated scalably on a data stream or not at all. In comparison, we ask for tight bounds on query evaluation problems, i.e. we give algorithms for query evaluation that are in a sense worst-case optimal. As we do, the authors of [6] study XPath evaluation; however, they focus on *instance data complexity* while we study worst-case bounds. This allows us to find strong and tight bounds for a greater variety of query evaluation problems. Many of our results apply beyond stream processing in a narrow sense to a more general framework of queries on data in external storage. Also, our worst-case bounds apply for *any* evaluation algorithm possible, that is, our bounds are not in terms of complexity classes closed under reductions that allow for nonlinear expansions of the input (such as LOGSPACE) as is the case for the work on the complexity of XPath in [12, 13, 28].

Lower bound results for a machine model with *multiple* external memory tapes (or harddisks) are presented in [17]. In the present paper, we only consider a single external memory tape, and are consequently able to show (sometimes exponentially) stronger lower bounds.

Due to space limitations we had to defer detailed proofs of our results to the full version of this paper [16] which extends the present paper by an appendix that contains proofs of all results presented here.

2 Preliminaries

In this section we fix some basic notation concerning trees, streams, and query languages. We write $\mathbb{N}$ for the set of non-negative integers. If M is a set, then 2^M denotes the set of all subsets of M. Throughout this paper we make the following convention: Whenever the letters r and s denote functions from $\mathbb{N}$ to $\mathbb{N}$, then these functions are *monotone*, i.e., we have $r(x) \leqslant r(y)$ and $s(x) \leqslant s(y)$ for all $x, y \in \mathbb{N}$ with $x \leqslant y$.

Trees and Streams. We use standard notation for trees and streamed trees (i.e. *documents*). In particular, we write $Doc(T)$ to denote the XML document associated with an XML document tree T. An example is given in Figure 1.

Query Languages. By $Eval(\cdot, \cdot)$ we denote the evaluation function that maps each tuple (Q, T), consisting of a query Q and a tree T to the corresponding query result. Let $\mathcal{Q}$ be a query language and let $\mathcal{T}_1 \subseteq \mathit{Trees}_\tau$ and $\mathcal{T}_2 \subseteq \mathcal{T}_1$. We say that $\mathcal{T}_2$ *can be filtered from* $\mathcal{T}_1$ *by a $\mathcal{Q}$-query* if, and only if, there is a query $Q \in \mathcal{Q}$ such that the following is true for all $T \in \mathcal{T}_1$: $T \in \mathcal{T}_2 \iff Eval(Q, T) \neq \emptyset$.

We assume that the reader is familiar with first-order logic (*FO*) and monadic second-order logic (*MSO*). An *FO-* or *MSO-sentence* (i.e., a formula without any free variable) specifies a Boolean query, whereas a formula with exactly one free first-order variable specifies a unary query, i.e., a query which selects a set of nodes from the underlying input tree.

It is well-known [9, 30] that the MSO-definable Boolean queries on binary trees are exactly the (Boolean) queries that can be defined by finite (deterministic or nondeterministic) bottom-up tree automata. An analogous statement is true about MSO on unranked trees and unranked tree automata [7].

Theorem 4.5 in section 4 gives a lower bound on the worst case complexity of the language *XQuery*. As we prove a lower bound for *one* particular XQuery query, we do not give a formal definition of the language but refer to [33].

Apart from *FO*, *MSO*, and XQuery, we also consider a fragment of the XPath language, Core XPath [12, 13]. As we will prove not only lower, but also upper bounds for Core XPath, we give a precise definition of this query language in [16]. An example of a Core XPath query is

/descendant::*[child::A and child::B]/child::*,

which selects all children of descendants of the root node that (i.e., the descendants) have a child node labeled A and a child node labeled B.

Core XPath is a strict fragment of XPath [12], both syntactically and semantically. It is known that Core XPath is in LOGSPACE w.r.t. data complexity and P-complete w.r.t. combined complexity [13]. In [12], it is shown that Core XPath can be evaluated in time $O(|Q| \cdot |D|)$, where $|Q|$ is the size of the query and $|D|$ is the size of the XML data. Furthermore, every Core XPath query is equivalent to a unary MSO query on trees (cf., e.g., [11]).

Communication complexity. To prove basic properties and lower bounds for our machine model, we use some notions and results from *communication complexity*, cf., e.g., [21].

Let A, B, C be sets and let $F : A \times B \to C$ be a function. In Yao's [34] basic model of communication two players, Alice and Bob, jointly want to evaluate $F(x, y)$, for input values $x \in A$ and $y \in B$, where Alice only knows x and Bob only knows y. The two players can exchange messages according to some fixed protocol $\mathcal{P}$ that depends on F, but not on the particular input values x, y. The exchange of messages starts with Alice sending a message to Bob and ends as soon as one of the players has enough information on x and y to compute $F(x, y)$.

$\mathcal{P}$ is called a *k-round protocol*, for some $k \in \mathbb{N}$, if the exchange of messages consists, for each input $(x, y) \in A \times B$, of at most k rounds. The *cost* of $\mathcal{P}$ on input (x, y) is the number of bits communicated by $\mathcal{P}$ on input (x, y). The *cost of* $\mathcal{P}$ is the *maximal* cost of $\mathcal{P}$ over all inputs $(x, y) \in A \times B$. The *communication complexity of F*, comm-compl(F), is defined as the minimum cost of $\mathcal{P}$, over all protocols $\mathcal{P}$ that compute F. For $k \geq 1$, the *k-round communication complexity of F*, comm-compl$_k(F)$, is defined as the minimum cost of $\mathcal{P}$, over all k-round protocols $\mathcal{P}$ that compute F.

Many powerful tools are known for proving lower bounds on communication complexity, cf., e.g., [21]. In the present paper we will use the following basic lower bound for the problem of deciding whether two sets are disjoint.

Definition 2.1. For $n \in \mathbb{N}$ let the function $\text{Disj}_n : 2^{\{1,\ldots,n\}} \times 2^{\{1,\ldots,n\}} \to \{0, 1\}$ be given via

$$\text{Disj}_n(X, Y) := \begin{cases} 1, & \text{if } X \cap Y = \emptyset \\ 0, & \text{otherwise.} \end{cases}$$

□

Theorem 2.2 (cf., e.g., [21]). *For every $n \in \mathbb{N}$, comm-compl(Disj_n) $\geq n$.*

3 Machine Model

We consider Turing machines with (1) an input tape, which is a read/write tape and will henceforth be called "external memory tape" or "external tape", for short, (2) an arbitrary number u of work tapes, which will henceforth be called "internal memory tapes" or "internal tapes", for short, and, if needed, (3) an additional write-only output tape.

Let M be such a Turing machine and let ρ be a run of M. By rev(ρ) we denote the number of times the external memory tape's head changes its direction in the run ρ. For $i \in \{1, \ldots, u\}$ we let space(ρ, i) be the number of cells of internal memory tape i that are used by ρ.

The class $ST(r,s)$ for strings

Definition 3.1 ($ST(r,s)$ for strings). Let $r: \mathbb{N} \to \mathbb{N}$ and $s: \mathbb{N} \to \mathbb{N}$.

(a) A Turing machine M is (r,s)-*bounded*, if every run ρ of M on an input of length n satisfies the following conditions:
(1) ρ is finite, (2) $1 + \text{rev}(\rho) \leq r(n)$,[2] and (3) $\sum_{i=1}^{u} \text{space}(\rho, i) \leq s(n)$, where u is the number of internal tapes of M.

(b) A string-language $L \subseteq \Sigma^*$ belongs to the class $ST(r,s)$ (resp., $NST(r,s)$), if there is a deterministic (respectively, nondeterministic) (r,s)-bounded Turing machine which accepts exactly those $w \in \Sigma^*$ that belong to L.

(c) A function $f: \Sigma^* \to \Sigma^*$ belongs to the class $ST(r,s)$, if there is a deterministic (r,s)-bounded Turing machine which produces, for each input string $w \in \Sigma^*$, the string $f(w)$ on its write-only output tape. □

For classes R and S of functions, we let $ST(R,S) := \bigcup_{r \in R, s \in S} ST(r,s)$.

If $k \in \mathbb{N}$ is a constant, then we write $ST(k,s)$ instead of $ST(r,s)$, where r is the function with $r(x) = k$ for all $x \in \mathbb{N}$. We freely combine these notations and use them for $NST(\cdot, \cdot)$ instead of $ST(\cdot, \cdot)$, too.

If we think of the external memory tape of an (r,s)-bounded Turing machine as representing the incoming stream, stored on a hard disk, then admitting the external memory tape's head to reverse its direction might not be very realistic. But as we mainly use our model to prove lower bounds, it does not do any harm either, since the reversals can be used to simulate *random access*. Random access can be introduced explicitly into our model as follows: A *random access Turing machine* is a Turing machine M which has a special internal memory tape that is used as *random access address tape*, i.e., on which only binary strings can be written. Such a binary string is interpreted as a positive integer specifying an external memory address, that is, the position index number of a cell on the external tape (we think of the external tape cells being numbered by positive integers). The machine has a special state q_{ra}. If q_{ra} is entered, then in one step the external memory tape head is moved to the cell that is specified by the number on the random access address tape, and the content of the random access address tape is deleted.

Definition 3.2. Let $q, r, s: \mathbb{N} \to \mathbb{N}$. A random access Turing machine M is (q, r, s)-*bounded*, if it is (r,s)-bounded (in the sense of an ordinary Turing machine) and, in addition, every run ρ of M on an input of length n involves at most $q(n)$ random accesses. □

Noting that a random access can be simulated with at most 2 changes of the direction of the external memory tape head, one immediately obtains:

---

[2] It is convenient for technical reasons to add 1 to the number $\text{rev}(\rho)$ of changes of the head direction. As defined here, $r(n)$ bounds the number of sequential scans of the external memory tape rather than the number of changes of head directions.

Lemma 3.3. *Let $q, r, s : \mathbb{N} \to \mathbb{N}$. If a problem can be solved by a (q, r, s)-bounded random access Turing machine, then it can also be solved by an $(r + 2q, O(s))$-bounded Turing machine.*

In the subsequent parts of this paper, we will concentrate on ordinary Turing machines (without random access). Via Lemma 3.3, all results can be transferred from ordinary Turing machines to random access Turing machines.

The class $ST(r,s)$ for trees. We make an analogous definition to $ST(r,s)$ on strings for trees. This definition is given in detail in [16].

4 Lower Bounds for the ST Model

A reduction lemma. The following lemma provides a convenient tool for showing that a problem L does not belong to $ST(r,s)$. The lemma's assumption can be viewed as a *reduction* from the problem $Disj_n(\cdot, \cdot)$ to the problem L.

Lemma 4.1. *Let Σ be an alphabet and let $\lambda : \mathbb{N} \to \mathbb{N}$ such that the following is true: For every $n_0 \in \mathbb{N}$ there is an $n \geqslant n_0$ and functions $f_n, g_n : 2^{\{1,\ldots,n\}} \to \Sigma^*$ such that for all $X, Y \subseteq \{1, \ldots, n\}$ the string $f_n(X)g_n(Y)$ has length $\leqslant \lambda(n)$.*

Then we have for all $r, s : \mathbb{N} \to \mathbb{N}$ with $r(\lambda(n)) \cdot s(\lambda(n)) \in o(n)$, that there is no (r, s)-bounded deterministic Turing machine which accepts a string of the form $f_n(X)g_n(Y)$ if, and only if, $X \cap Y = \emptyset$.

Disjointness. Every n-bit string $x = x_1 \cdots x_n \in \{0,1\}^n$ specifies a set $S(x) := \{i : x_i = 1\} \subseteq \{1, \ldots, n\}$. Let L_{Disj} consist of those strings $x\#y$ where x and y specify disjoint subsets of $\{1, \ldots, n\}$, for some $n \geqslant 1$. That is,

$$L_{Disj} := \{\ x\#y\ :\ \text{ex. } n \geqslant 1 \text{ with } x, y \in \{0,1\}^n \text{ and } S(x) \cap S(y) = \emptyset\ \}.$$

From Lemma 4.1 one easily obtains

Proposition 4.2. *Let $r : \mathbb{N} \to \mathbb{N}$ and $s : \mathbb{N} \to \mathbb{N}$. If $r(n) \cdot s(n) \in o(n)$, then $L_{Disj} \notin ST(r,s)$.*

The bound given by Proposition 4.2 is tight, as it can be easily seen that $L_{Disj} \in ST(r,s)$ for all $r, s : \mathbb{N} \to \mathbb{N}$ with $r(n) \cdot s(n) \in \Omega(n)$.

Joins. Let τ be the set of tag names { rels, rel1, rel2, tuple, no1, no2, 0, 1 }. We represent a pair (A, B) of finite relations $A, B \subseteq \mathbb{N}^2$ as a τ-tree $T(A, B)$ whose associated XML document $Doc(T(A,B))$ is a Σ_τ-string of the following form: For each number $i \in \mathbb{N}$ let $Bin(i) = b^{(i)}_{\ell_i} \cdots b^{(i)}_0$ be the binary representation of i. For each tuple $(i, j) \in \{1, \ldots, n\}^2$ let $Doc(i, j) :=$

$\langle\text{tuple}\rangle\ \langle\text{no1}\rangle\ \langle b^{(i)}_{\ell_i}/\rangle\ \cdots\ \langle b^{(i)}_0/\rangle\ \langle/\text{no1}\rangle\ \langle\text{no2}\rangle\ \langle b^{(j)}_{\ell_j}/\rangle\ \cdots\ \langle b^{(j)}_0/\rangle\ \langle/\text{no2}\rangle\ \langle/\text{tuple}\rangle\ .$

For each finite relation $A \subseteq \mathbb{N}^2$ let $t_1, \ldots, t_{|A|}$ be the lexicographically ordered list of all tuples in A. We let $Doc(A) := Doc(t_1) \cdots Doc(t_{|A|})$. Finally, we let

$$Doc(T(A,B)) := \langle\text{rels}\rangle \langle\text{rel1}\rangle Doc(A) \langle/\text{rel1}\rangle \langle\text{rel2}\rangle Doc(B) \langle/\text{rel2}\rangle \langle/\text{rels}\rangle.$$

It is straightforward to see that the string $Doc(T(A,B))$ has length $O((|A| + |B|) \cdot \log n)$, if $A, B \subseteq \{1, \ldots, n\}^2$.

We write $A \bowtie_1 B$ to denote the join of A and B on their first component, i.e., $A \bowtie_1 B := \{(x,y) : \exists z\, A(z,x) \wedge B(z,y)\}$. We let

$$\mathcal{T}_{Rels} := \{T(A,B) : A,B \subseteq \mathbb{N}^2,\ A,B \text{ finite}\}$$
$$\mathcal{T}_{EmptyJoin} := \{T(A,B) \in \mathcal{T}_{Rels} : A \bowtie_1 B = \emptyset\}$$
$$\mathcal{T}_{NonEmptyJoin} := \{T(A,B) \in \mathcal{T}_{Rels} : A \bowtie_1 B \neq \emptyset\}.$$

Lemma 4.3. $\mathcal{T}_{NonEmptyJoin}$ *can be filtered from* $\mathcal{T}_{Rels}$ *by an XQuery query.*

Lemma 4.4. *Let* $r, s : \text{Trees}_\tau \to \mathbb{N}$.
If $r(T) \cdot s(T) \in o\bigl(\frac{size(T)}{\log(size(T))}\bigr)$, *then* $\mathcal{T}_{EmptyJoin} \notin ST(r,s)$.

From Lemma 4.4 and Lemma 4.3 we immediately obtain a lower bound on the worst-case data complexity for filtering relative to an XQuery query:

Theorem 4.5. *The tree-language* $\mathcal{T}_{EmptyJoin}$
(a) can be filtered from $\mathcal{T}_{Rels}$ *by an XQuery query,*
(b) does not belong to the class $ST(r,s)$, *whenever* $r, s : \text{Trees}_\tau \to \mathbb{N}$ *with*

$$r(T) \cdot s(T) \in o\Bigl(\frac{size(T)}{\log(size(T))}\Bigr).$$

Remark 4.6. Let us note that the above bound is "almost tight" in the following sense: The problem of deciding whether $A \bowtie_1 B = \emptyset$ and, in general, all *FO*-definable problems belong to $ST(1,n)$ – in its single scan of the external memory tape, the Turing machine simply copies the entire input on one of its internal memory tapes and then evaluates the *FO*-sentence by the straightforward LOGSPACE algorithm for *FO*-model-checking (cf. e.g. [1]). □

Sorting. By KEYSORT, we denote the problem of sorting a set S of tuples $t = (K,V)$ consisting of a key K and a value V by their keys. Let $ST^-(r,s)$ denote the class of all problems in $ST(r,s)$ that can be solved without writing to the external memory tape. Then,

Theorem 4.7. *Let* $r, s : \mathbb{N} \to \mathbb{N}$. *If* KEYSORT *is in* $ST^-(r,s)$, *then computing the natural join* $A \bowtie B$ *of two finite relations* A, B *is in*

$$ST^-\bigl(r(n^2) + 2,\, s(n^2) + O(\log n) + O(\max_{t \in A \cup B} |t|)\bigr).$$

Remark 4.8. Given that the size of relations A and B is known (which is usually the case in practical database management systems DBMS), the algorithm given in the previous proof can do a merge-join without additional scans after the sort run and without a need to buffer more than one tuple. This is guaranteed even if both relations may contain many tuples with the same join key – in current implementations of the merge join in DBMS, this may lead to grass-roots swapping. The (substantial) *practical* drawback of the join algorithm of the proof of Theorem 4.7, however, is that much larger relations A', B' need to be sorted: indeed $|A'| = |A| * |B|$. □

Corollary 4.9.

(a) *Let* $r, s : \mathbb{N} \to \mathbb{N}$ *such that* $r(n^2) \cdot (s(n^2) + \log n) \in o\left(\frac{n}{\log n}\right)$.
  Then, KEYSORT $\notin ST^-(r, s)$.

(b) KEYSORT $\notin ST^-\left(o(\sqrt[5]{n}), O(\sqrt[5]{n})\right)$.

It is straightforward to see that by using MergeSort, the sorting problem can be solved using $O(\log n)$ scans of external memory provided that *three* external memory tapes are available. (In [17], this logarithmic bound is shown to be tight, for arbitrarily many external tapes.) Corollary 4.9 gives an exponentially stronger lower bound for the case of a single external tape.

A hierarchy based on the number of scans

Theorem 4.10. *For every fixed $k \geq 1$,*

$$ST(k, O((\log k) + \log n)) \cap NST(1, O(k \cdot \log n)) \not\subseteq ST\left(k-1, o\left(\frac{\sqrt{n}}{k^5 (\log n)^3}\right)\right).$$

The proof of this theorem is based on a result due to Duris, Galil and Schnitger [10]. Theorem 4.10 directly implies

Corollary 4.11. *For every fixed $k \in \mathbb{N}$ and all classes S of functions from $\mathbb{N}$ to $\mathbb{N}$ such that $O(\log n) \subseteq S \subseteq o\left(\frac{\sqrt{N}}{(\lg n)^3}\right)$ we have $ST(k, S) \subsetneq ST(k+1, S)$.*

Remark 4.12. On the other hand, of course, the hierarchy collapses if internal memory space is at least linear in the size of the input: For every $r : \mathbb{N} \to \mathbb{N}$ and for every $s : \mathbb{N} \to \mathbb{N}$ with $s(n) \in \Omega(n)$, we have

$$ST(r, s) \subseteq ST(1, n + s(n)) \quad \text{and} \quad ST(r, O(s(n))) = \text{DSPACE}(O(s(n))).$$

5 Tight Bounds for Filtering and Query Evaluation on Trees

Lower Bound. We need the following notation: We fix a set τ of tag names via $\tau := \{\text{ root, left, right, blank }\}$. Let T_1 be the τ-tree from Figure 1. Note that

T_1 has a unique leaf v_1 labeled with the tag name "left". For any arbitrary τ-tree T we let $T_1(T)$ be the τ-tree rooted at T_1's root and obtained by identifying node v_1 with the root of T and giving the label "left" to this node. Now, for every $n \geq 2$ let T_n be the τ-tree inductively defined via $T_n := T_1(T_{n-1})$. It is straightforward to see that T_n has exactly $2n$ leaves labeled "blank". Let $x_1, \ldots, x_n, y_n, \ldots, y_1$ denote these leaves, listed in *document order* (i.e., in the order obtained by a pre-order depth-first left-to-right traversal of T_n). For an illustration see Figure 2.

We let $\tau_{01} := \tau \cup \{0, 1\}$. For all sets $X, Y \subseteq \{1, \ldots, n\}$ let $T_n(X, Y)$ be the τ_{01}-tree obtained from T_n by replacing, for each $i \in \{1, \ldots, n\}$, (*) the label "blank" of leaf x_i by the label 1 if $i \in X$, and by the label 0 otherwise and (*) the label "blank" of leaf y_i by the label 1 if $i \in Y$, and by the label 0 otherwise.

We let

$$\mathcal{T}_{Sets} := \{ T_n(X, Y) \; : \; n \geq 1, \; X, Y \subseteq \{1, \ldots, n\} \},$$
$$\mathcal{T}_{Disj} := \{ T_n(X, Y) \in \mathcal{T}_{Sets} \; : \; X \cap Y = \emptyset \},$$
$$\mathcal{T}_{NonDisj} := \{ T_n(X, Y) \in \mathcal{T}_{Sets} \; : \; X \cap Y \neq \emptyset \}.$$

Lemma 5.1. *(a) There is a Core XPath query Q such that the following is true for all τ-trees $T \in \mathcal{T}_{Sets}$: $\mathit{Eval}(Q, T) \neq \emptyset \iff T \in \mathcal{T}_{NonDisj}$.*
(b) There is a FO-sentence φ such that the following is true for all τ-trees T: $T \models \varphi \iff T \in \mathcal{T}_{NonDisj}$.

Lemma 5.2. *Let $r, s : \mathit{Trees}_\tau \to \mathbb{N}$.*
If $r(T) \cdot s(T) \in o(\mathit{depth}(T))$, then $\mathcal{T}_{NonDisj} \notin ST(r, s)$.

From Lemma 5.1 and Lemma 5.2 we directly obtain a lower bound on the worst-case data complexity of Core XPath filtering:

Theorem 5.3. *The tree-language $\mathcal{T}_{NonDisj}$*
(a) can be filtered from $\mathcal{T}_{Sets}$ by a Core XPath query,

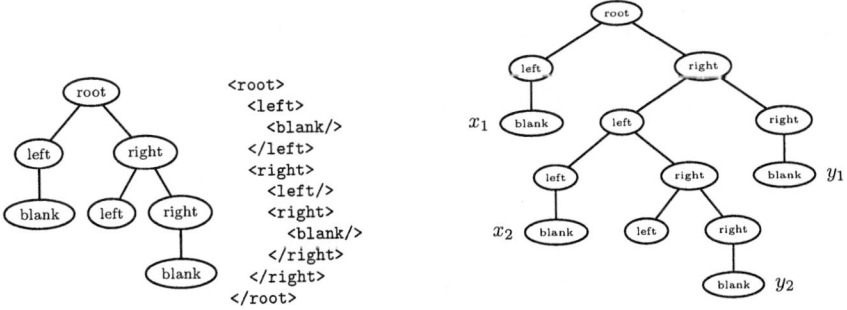

Fig. 1. A τ-tree T_1 and its XML document $\mathit{Doc}(T_1) \in \Sigma_\tau^*$ with tag names $\tau := \{\text{root}, \text{left}, \text{right}, \text{blank}\}$

Fig. 2. Tree T_2 and nodes x_1, x_2, y_1, y_2

(b) is definable by an FO-sentence (and therefore, also definable by a Boolean MSO query and recognizable by a tree automaton), and
(c) does not belong to the class $ST(r, s)$, whenever $r, s : \mathit{Trees}_\tau \to \mathbb{N}$ with $r(T) \cdot s(T) \in o(\mathit{depth}(T))$.

In the following subsection we match this lower bound with the corresponding upper bound.

Upper bounds. Recall that a tree-language $\mathcal{T} \subseteq \mathit{Trees}_\tau$ is definable by an MSO-sentence if, and only if, it is recognizable by an unranked tree automaton, respectively, if, and only if, the language $\{\mathit{BinTree}(T) : T \in \mathcal{T}\}$ of associated binary trees is recognizable by an ordinary (ranked) tree automaton (cf., e.g., [7, 9, 30]).

Theorem 5.4 (implicit in [25, 29]). Let $\mathcal{T} \subseteq \mathit{Trees}_\tau$ be a tree-language. If $\mathcal{T}$ is definable by an MSO-sentence (or, equivalently, recognizable by a ranked or an unranked finite tree automaton), then $\mathcal{T} \in ST\bigl(1, \mathit{depth}(\cdot) + 1\bigr)$.

Recall that every Core XPath query is equivalent to a unary MSO query. Thus a Core XPath filter can be phrased as an MSO sentence on trees. From the Theorems 5.4 and 5.3 we therefore immediately obtain a tight bound for Core XPath filtering:

Corollary 5.5. (a) Filtering from the set of unranked trees with respect to every fixed Core XPath query Q belongs to $ST\bigl(1, O(\mathit{depth}(\cdot))\bigr)$.
(b) There is a Core XPath query Q such that, for all $r, s : \mathit{Trees}_\tau \to \mathbb{N}$ with $r(T) \cdot s(T) \in o\bigl(\mathit{depth}(T)\bigr)$, filtering w.r.t. Q does not belong to $ST(r, s)$.

Next, we provide an upper bound for the problem of computing the set $\mathit{Eval}(Q, T)$ of nodes in an input tree T matching a unary MSO (or Core XPath) query Q. We first need to clarify what this means, because writing the subtree of each matching node onto the output tape requires a very large amount of internal memory (or a large number of head reversals on the external memory tape), and this gives us no appropriate characterization of the difficulty of the problem. We study the problem of computing, for each node matched by Q, its index in the tree, in the order in which they appear in the document $\mathit{Doc}(T)$. We distinguish between the case where these indexes are to be written to the output tape in ascending order and the case where they are to be output in descending (i.e., reverse) order.

Theorem 5.6 (implicit in [26, 20]). For every unary MSO or Core XPath query Q, the problem of computing, for input trees T, the nodes in $\mathit{Eval}(Q, T)$
(a) in ascending order belongs to $ST\bigl(3, O(\mathit{depth}(\cdot))\bigr)$.
(b) in reverse order belongs to $ST\bigl(2, O(\mathit{depth}(\cdot))\bigr)$.

Note that this bound is tight: From Corollary 5.5(c) we know that, for some Core XPath query Q, not even filtering (i.e., checking whether $\mathit{Eval}(Q, T)$ is empty) is possible in $ST(r, s)$ if $r(T) \cdot s(T) \in o\bigl(\mathit{depth}(T)\bigr)$.

References

1. S. Abiteboul, R. Hull, and V. Vianu. *Foundations of Databases*. Addison-Wesley, 1995.
2. G. Aggarwal, M. Datar, S. Rajagopalan, and M. Ruhl. On the streaming model augmented with a sorting primitive. In *Proc. FOCS'04*, pages 540–549.
3. N. Alon, Y. Matias, and M. Szegedy. The space complexity of approximating the frequency moments. *Journal of Computer and System Sciences*, 58:137–147, 1999.
4. A. Arasu, B. Babcock, T. Green, A. Gupta, and J. Widom. "Characterizing Memory Requirements for Queries over Continuous Data Streams". In *Proc. PODS'02*, pages 221–232, 2002.
5. B. Babcock, S. Babu, M. Datar, R. Motwani, and J. Widom. Models and issues in data stream systems. In *Proc. PODS'02*, pages 1–16.
6. Z. Bar-Yossef, M. Fontoura, and V. Josifovski. "On the Memory Requirements of XPath Evaluation over XML Streams". In *Proc. PODS'04*, pages 177–188, 2004.
7. A. Brüggemann-Klein, M. Murata, and D. Wood. "Regular Tree and Regular Hedge Languages over Non-ranked Alphabets: Version 1, April 3, 2001". Technical Report HKUST-TCSC-2001-05, Hong Kong Univ. of Science and Technology, 2001.
8. J.-E. Chen and C.-K. Yap. "Reversal Complexity". *SIAM J. Comput.*, **20**(4):622–638, Aug. 1991.
9. J. Doner. "Tree Acceptors and some of their Applications". *Journal of Computer and System Sciences*, 4:406–451, 1970.
10. P. Duris, Z. Galil, and G. Schnitger. Lower bounds on communication complexity. *Information and Computation*, 73:1–22, 1987. Journal version of STOC'84 paper.
11. G. Gottlob and C. Koch. "Monadic Datalog and the Expressive Power of Web Information Extraction Languages". *Journal of the ACM*, **51**(1):74–113, 2004.
12. G. Gottlob, C. Koch, and R. Pichler. "Efficient Algorithms for Processing XPath Queries". In *Proc. VLDB 2002*, pages 95–106, Hong Kong, China, 2002.
13. G. Gottlob, C. Koch, and R. Pichler. "The Complexity of XPath Query Evaluation". In *Proc. PODS'03*, pages 179–190, San Diego, California, 2003.
14. G. Graefe. "Query Evaluation Techniques for Large Databases". *ACM Computing Surveys*, **25**(2):73–170, June 1993.
15. T. J. Green, G. Miklau, M. Onizuka, and D. Suciu. "Processing XML Streams with Deterministic Automata". In *Proc. ICDT'03*, 2003.
16. M. Grohe, C. Koch, and N. Schweikardt. Tight lower bounds for query processing on streaming and external memory data. Technical report CoRR cs.DB/0505002, 2005. Full version of ICALP'05 paper.
17. M. Grohe and N. Schweikardt. Lower bounds for sorting with few random accesses to external memory. In *Proc. PODS*, 2005. To appear.
18. M. Henzinger, P. Raghavan, and S. Rajagopalan. Computing on data streams. In *External memory algorithms*, volume 50, pages 107–118. DIMACS Series In Discrete Mathematics And Theoretical Computer Science, 1999.
19. J. E. Hopcroft and J. D. Ullman. Some results on tape-bounded Turing machines. *Journal of the ACM*, **16**(1):168–177, 1969.
20. C. Koch. "Efficient Processing of Expressive Node-Selecting Queries on XML Data in Secondary Storage: A Tree Automata-based Approach". In *Proc. VLDB 2003*, pages 249–260, 2003.
21. E. Kushilevitz and N. Nisan. *Communication Complexity*. Cambridge Univ. Press, 1997.
22. U. Meyer, P. Sanders, and J. Sibeyn, editors. *Algorithms for Memory Hierarchies*, volume 2832 of *Lecture Notes in Computer Science*. Springer-Verlag, 2003.

23. J. Munro and M. Paterson. Selection and sorting with limited storage. *Theoretical Computer Science*, 12:315–323, 1980.
24. S. Muthukrishnan. Data streams: algorithms and applications. In *Proc. 14th SODA*, pages 413–413, 2003.
25. A. Neumann and H. Seidl. "Locating Matches of Tree Patterns in Forests". In *Proc. 18th FSTTCS, LNCS 1530*, pages 134–145, 1998.
26. F. Neven and J. van den Bussche. "Expressiveness of Structured Document Query Languages Based on Attribute Grammars". *J. ACM*, **49**(1):56–100, Jan. 2002.
27. R. Ramakrishnan and J. Gehrke. *Database Management Systems*. McGraw-Hill, 2002.
28. L. Segoufin. "Typing and Querying XML Documents: Some Complexity Bounds". In *Proc. PODS'03*, pages 167–178, 2003.
29. L. Segoufin and V. Vianu. "Validating Streaming XML Documents". In *Proc. PODS'02*, 2002.
30. J. Thatcher and J. Wright. "Generalized Finite Automata Theory with an Application to a Decision Problem of Second-order Logic". *Math. Syst. Theory*, **2**(1):57–81, 1968.
31. P. van Emde Boas. "Machine Models and Simulations". In J. van Leeuwen, editor, *Handbook of Theoretical Computer Science*, volume 1, chapter 1, pages 1–66. Elsevier Science Publishers B.V., 1990.
32. J. Vitter. External memory algorithms and data structures: Dealing with massive data. *ACM Computing Surveys*, 33(2):209–271, June 2001.
33. World Wide Web Consortium. "XQuery 1.0 and XPath 2.0 Formal Semantics. W3C Working Draft (Aug. 16th 2002), 2002. `http://www.w3.org/XML/Query`.
34. A. Yao. Some complexity questions related to distributive computing. In *Proc. 11th STOC*, pages 209–213, 1979.

Decidability and Complexity Results for Timed Automata via Channel Machines

Parosh Aziz Abdulla[1], Johann Deneux[1], Joël Ouaknine[2], and James Worrell[3]

[1] Department of Computer Systems,
Uppsala University, Sweden
{parosh, johannd}@it.uu.se
[2] Oxford University Computing Laboratory, UK
joel@comlab.ox.ac.uk
[3] Department of Mathematics, Tulane University, USA
jbw@math.tulane.edu

Abstract. This paper is concerned with the language inclusion problem for timed automata: given timed automata $\mathcal{A}$ and $\mathcal{B}$, is every word accepted by $\mathcal{B}$ also accepted by $\mathcal{A}$? Alur and Dill [3] showed that the language inclusion problem is decidable if $\mathcal{A}$ has no clocks and undecidable if $\mathcal{A}$ has two clocks (with no restriction on $\mathcal{B}$). However, the status of the problem when $\mathcal{A}$ has one clock is not determined by [3]. In this paper we close this gap for timed automata over infinite words by showing that the one-clock language inclusion problem is undecidable. For timed automata over finite words, building on our earlier paper [20], we show that the one-clock language inclusion problem is decidable with non-primitive recursive complexity. This reveals a surprising divergence between the theory of timed automata over finite words and over infinite words. Finally, we show that if ε-transitions or non-singular postconditions are allowed, then the one-clock language inclusion problem is undecidable over both finite and infinite words.

1 Introduction

An execution of a real-time system can be modelled as a *timed word* consisting of a sequence of events and their associated timestamps, and properties of such systems can be expressed as languages of timed words. Timed automata were introduced by Alur and Dill [3] to specify languages of timed words, and have since been extensively studied by many researchers. In particular, timed automata have been used as the foundation for several verification algorithms and tools (see [8] for a survey).

One of the most fundamental results about timed automata is the undecidability of the language inclusion problem: 'Given timed automata $\mathcal{A}$ and $\mathcal{B}$, is every word accepted by $\mathcal{B}$ also accepted by $\mathcal{A}$?' This problem is undecidable irrespective of whether one considers automata over finite words or automata over infinite words with a Büchi acceptance condition. In this context it is natural to seek subclasses of timed automata, with reduced expressive power, for which the language inclusion problem is decidable [5, 7, 8, 4, 13, 19, 20]. In this paper we consider subclasses parameterized by the number of clocks an automaton has. In particular, we consider the *n-clock language inclusion problem* in which $\mathcal{A}$ is allowed n clocks but where no restriction is placed on $\mathcal{B}$.

A close analysis of the proof of the undecidability of language inclusion in [3] reveals that the tightest possible formulation of their result is that the two-clock language inclusion problem is undecidable. On the other hand, from the decidability of language emptiness for timed automata, also proved in [3], it follows that the zero-clock language inclusion problem is (PSPACE-complete) decidable. This leaves an interesting open question about the status of the one-clock language inclusion problem. In fact, many interesting specifications can be expressed by automata with a single clock, or parallel combinations thereof. This is particularly so for *alternating timed automata* [16, 21]. For instance, every formula of *Metric Temporal Logic* [6, 4] can be translated into an alternating timed automaton with a single clock [21]. The model checking problem then corresponds to language inclusion.

Recently, using techniques from the theory of well-quasi-ordered transition systems [11], we showed that over finite words the one-clock language inclusion problem is decidable [20]. However, while finite words are sufficient to capture safety properties, to capture liveness or fairness properties it is most natural to consider automata over infinite words. The main result of this paper is that, for timed automata over infinite words (with Büchi acceptance conditions) the one-clock language inclusion problem is undecidable. This reveals a surprising divergence between the theory of timed automata over finite words and over infinite words. We also show that over finite words the one-clock language inclusion problem has non-primitive recursive complexity. Finally, language inclusion becomes undecidable over both finite and infinite words if ε-transitions are allowed or if reset clocks have nonsingular postconditions (as in [9, 12]).

We use *channel machines* [10] as a convenient middleware between Turing machines and timed automata. This allows us to develop a schematic approach to proving undecidability and complexity results for various classes of timed automata. In each case we show how to encode a certain class of channel computations as a timed language, whose complement can be recognized by a timed automaton of a certain type.

Related Work. The non-primitive recursive complexity of language inclusion over finite words and the undecidability of language inclusion over finite words with ε-transitions have recently and independently been proved by Lasota and Walukiewicz [16]. They have also concurrently discovered the undecidability of universality for one-clock Büchi timed automata [17]. Like us, they make use of channel machines in their work, although via a different encoding of channel histories as timed words.

Alur, La Torre and Madhusudan [7] consider automata with *perturbed clocks* whose rates may vary; they show that for every automaton with a single perturbed clock there is an equivalent deterministic timed automaton. It follows that the language inclusion problem is decidable for this class of automata.

Laroussinie, Markey and Schnoebelen [15] classify the complexity of deciding language emptiness for timed automata with one, two and three clocks respectively.

2 Timed Automata and Timed Words

Let Σ be a finite alphabet and write Σ^ε for $\Sigma \cup \{\varepsilon\}$, where $\varepsilon \notin \Sigma$. Let $\mathbb{R}_+$ be the set of non-negative reals. A *timed event* is a pair (t, a), where $t \in \mathbb{R}_+$ is called the *timestamp* of the *event* $a \in \Sigma$. A *timed word* is a finite or infinite sequence $u = (t_0, a_0)(t_1, a_1)(t_2, a_2)\ldots$ of timed events whose sequence of timestamps $t_0 t_1 t_2 \ldots$ is non-decreasing and is either finite or diverges to infinity. (This last assumption rules out so-called *Zeno words*.) We say that a timed word is *strictly monotonic* if its sequence of timestamps is strictly monotonic increasing. We write $T\Sigma^*$ for the set of finite timed words over alphabet Σ and $T\Sigma^\omega$ for the set of infinite timed words over alphabet Σ.

Let X be a finite set of clocks, denoted x, y, z, etc. We define the set $\Phi(X)$ of clock constraints over X via the following grammar (here $k \in \mathbb{N}$ is a non-negative integer).

$$\phi ::= x < k \mid x \leqslant k \mid \phi \wedge \phi \mid \neg\phi \ .$$

Definition 1. *A* timed automaton *is a tuple* $\mathcal{A} = (\Sigma, S, S_0, F, X, E)$, *where*

- Σ *is a finite alphabet of events,*
- S *is a finite set of control states,*
- $S_0 \subseteq S$ *is a set of initial control states,*
- $F \subseteq S$ *is a set of accepting control states,*
- X *is a finite set of clocks, and*
- $E \subseteq S \times S \times \Phi(X) \times \Sigma^\varepsilon \times 2^X \times \Phi(X)$ *is a finite set of edges. An edge $(s, s', \phi, \alpha, R, \phi')$ allows α-labelled transition from s to s', provided the precondition ϕ on clocks is met. Afterwards, the clocks in R are nondeterministically reset to values satisfying the postcondition ϕ', and all other clocks remain unchanged.*

A *clock valuation* of $\mathcal{A}$ is a function $\nu : X \to \mathbb{R}_+$. If $\delta \in \mathbb{R}_+$, we let $\nu + \delta$ be the clock valuation such that $(\nu + \delta)(x) = \nu(x) + \delta$ for all $x \in X$. A *global state* of $\mathcal{A}$ is a pair (s, ν), where $s \in S$ is a control state and ν is a clock valuation. Write $Q = S \times (\mathbb{R}_+)^X$ for the set of global states.

Automaton $\mathcal{A}$ induces an $(\mathbb{R}_+ \times \Sigma^\varepsilon)$-labelled transition relation on the set of global states as follows: write $(s, \nu) \xrightarrow{\delta, \alpha} (t, \nu')$ iff there is an edge $(s, t, \phi, \alpha, R, \phi')$ in E such that $\nu + \delta$ satisfies ϕ, ν' satisfies ϕ' and $(\nu + \delta)(x) = \nu'(x)$ for all $x \notin R$.

A *run* of $\mathcal{A}$ is a finite or infinite sequence of transitions

$$(s_0, \nu_0) \xrightarrow{\delta_0, \alpha_0} (s_1, \nu_1) \xrightarrow{\delta_1, \alpha_1} (s_2, \nu_2) \xrightarrow{\delta_2, \alpha_2} \ldots \tag{1}$$

where $s_0 \in S_0$ is an initial control state and $\nu_0(x) = 0$ for all $x \in X$. We require that an infinite run contain infinitely many transitions labelled from Σ and that $\sum_{i=0}^\infty \delta_i$ be infinite.

A finite run is *accepting* if the last control state in the run is accepting. An infinite run is accepting if infinitely many control states in the run are accepting. Let $\alpha_{i_0}\alpha_{i_1}\alpha_{i_2}\ldots$ be the sequence of non-ε-labels occurring in an accepting run and let $t_j = \sum_{i=0}^{j} \delta_i$. Then the timed word $(t_{i_0}, \alpha_{i_0})(t_{i_1}, \alpha_{i_1})(t_{i_2}, \alpha_{i_2})\ldots$ is said to be *accepted by* $\mathcal{A}$. We write $L_f(\mathcal{A})$ for the set of finite timed words accepted by $\mathcal{A}$ and $L_\omega(\mathcal{A})$ for the set of infinite timed words accepted by $\mathcal{A}$.

Remark 1. Definition 1 represents quite a general model of timed automata. We will adopt the convention that, unless otherwise specified, a given timed automaton has no ε-transitions, and has *singular postconditions*, i.e., for each edge $(s, t, \phi, \alpha, R, \phi')$, if clock valuations ν and ν' both satisfy ϕ', then $\nu(x) = \nu'(x)$ for all $x \in R$.

3 Hardness Results over Finite Words

In [20] we showed that it is decidable whether $L_f(\mathcal{B}) \subseteq L_f(\mathcal{A})$ for an arbitrary timed automaton $\mathcal{B}$ and a one-clock automaton $\mathcal{A}$. In this section we show that this problem has non-primitive recursive complexity and is undecidable if $\mathcal{A}$ is allowed ε-transitions or non-singular postconditions. We prove these results by reduction from the reachability problem for channel machines.

A *channel machine* [1, 10, 22] consists of a finite-state automaton acting on an unbounded fifo channel (or buffer). More precisely, a channel machine is a tuple $\mathcal{C} = (S, s_0, M, \Delta)$, where S is a finite set of *control states*, $s_0 \in S$ is the *initial control state*, M is a finite set of *messages*, and $\Delta \subseteq S \times L \times S$ is the transition relation over label set $L = \{m!, m? : m \in M\}$.

A *global state* of $\mathcal{C}$ is a pair (s, x), where $s \in S$ is the control state and $x \in M^*$ is the contents of the channel. The rules in Δ induce an L-labelled transition relation on the set of global states as follows: $(s, m!, t) \in \Delta$ yields a transition $(s, x) \xrightarrow{m!} (t, x \cdot m)$ that writes $m \in M$ to the tail of the channel, and $(s, m?, t) \in \Delta$ yields a transition $(s, m \cdot x) \xrightarrow{m?} (t, x)$ that reads $m \in M$ from the head of the channel. If we only allow the transitions indicated above, then we call $\mathcal{C}$ an *error-free* channel machine.

We also consider channel machines that operate with *insertion errors*. Given $x, y \in M^*$, write $x \sqsubseteq y$ if x can be obtained from y by deleting any number of letters, e.g. sub $\sqsubseteq$ s<u>t</u>u<u>bb</u>orn, as indicated by the underlining. Following [22] we introduce *insertion errors* by extending the transition relation on global states with the following clause: if $(s, x) \xrightarrow{\alpha} (t, y)$, $x' \sqsubseteq x$ and $y \sqsubseteq y'$, then $(s, x') \xrightarrow{\alpha} (t, y')$. Dually, we define *lossy channel machines* by adding by a clause: if $(s, x) \xrightarrow{\alpha} (t, y)$, $x \sqsubseteq x'$ and $y' \sqsubseteq y$, then $(s, x') \xrightarrow{\alpha} (t, y')$.

A *computation* of $\mathcal{C}$ is a finite or infinite sequence of transitions between global states $(s_0, x_0) \xrightarrow{\alpha_0} (s_1, x_1) \xrightarrow{\alpha_1} (s_2, x_2) \xrightarrow{\alpha_2} \cdots$.

The *control-state reachability problem* asks, given a channel machine $\mathcal{C} = (S, s_0, M, \Delta)$ and a control state $t \in S$, whether there is a computation of $\mathcal{C}$ starting in global state (s_0, ε) and ending in global state (t, ε). It is well-known that the control-state reachability problem for error-free channel machines is undecidable[1]. Next we show how to reduce the control-state reachability problem for error-free channel machines to the universality problem (which is a special case of language inclusion) for various classes of timed automata.

Let $\mathcal{C} = (S, s_0, M, \Delta)$ and $t \in S$ be an instance of the control-state reachability problem. Given this data, let $\Sigma = \{m!, m? : m \in M\} \cup \{\checkmark\}$ be a finite alphabet. We encode the finite control of $\mathcal{C}$ as an untimed automaton (i.e., a timed automaton with no clocks) $\mathcal{A}_{cont}$ over alphabet Σ. $\mathcal{A}_{cont}$ is just the underlying control automaton of $\mathcal{C}$ with a $\checkmark$-labelled self-transition added to every control state, with s_0 as the initial control state and $t \in S$ as the only accepting control state. Let L_{cont} denote the timed language $L_f(\mathcal{A}_{cont})$.

Definition 2. Let $L_{chan} \subseteq T\Sigma^*$ consist of those timed words u such that:

1. u is strictly monotonic.
2. u contains a $\checkmark$-event at time zero, and thereafter consecutive $\checkmark$-events are separated by one time unit.
3. Every $m!$-event in u is followed one time unit later by an $m?$-event.
4. Every $m?$-event is preceded one time unit earlier by an $m!$-event.

Clauses 3 and 4 capture the channel discipline: every message written to the channel is read from the channel one time unit later, and every message that is read from the channel was written to the channel one time unit earlier. The one-to-one unit-time-delayed correspondence between read and write events ensures that messages are read from the channel in the order that they were written to the channel. The requirement that every message written to the channel is eventually read corresponds to the fact that we consider computations that end with an empty channel. The $\checkmark$-events in L_{chan} have no particular significance other than to facilitate the encoding below.

Proposition 1. $\mathcal{C}$ has an error-free computation from (s_0, ε) to (t, ε) iff $L_{cont} \cap L_{chan} \neq \emptyset$.

Let $\overline{\mathcal{A}_{cont}}$ denote the complement of $\mathcal{A}_{cont}$ as an untimed automaton. It is clear that $\overline{\mathcal{A}_{cont}}$ is also the complement of $\mathcal{A}_{cont}$ with respect to timed languages, i.e., $L_f(\overline{\mathcal{A}_{cont}}) = T\Sigma^* - L_{cont}$. Now suppose that $\mathcal{A}_{chan}$ is a timed automaton such that $L_f(\mathcal{A}_{chan}) = T\Sigma^* - L_{chan}$. From Proposition 1 it holds that $\overline{\mathcal{A}_{cont}} \cup \mathcal{A}_{chan}$ is universal (i.e. accepts every timed word) iff $\mathcal{C}$ has no error-free computation from (s_0, ε) to (t, ε). Since the control-state reachability problem is undecidable for error-free channel machines, it follows that the universality problem is undecidable for any class of timed automata that is closed under unions and can capture the complement of L_{chan}.

---

[1] The usual formulation of the problem asks whether there is a computation from (s_0, ε) to (t, x) for some $x \in M^*$. It is straightforward to reduce this problem to the formulation above.

3.1 Two Clocks

We show how to define a timed automaton $\mathcal{A}_{chan}$ with two clocks such that $L_f(\mathcal{A}_{chan}) = T\Sigma^* - L_{chan}$. We define $\mathcal{A}_{chan}$ as the disjunction of several automata, each of which accepts the set of words that fail to satisfy a particular clause in the definition of L_{chan}. The interesting clauses here are 3 and 4.

Automaton $\mathcal{A}_1$, below, accepts those timed words in which some $m!$-event is not followed one time unit later by an $m?$-event, i.e., those words that fail to satisfy Clause 3 in Definition 2. Automaton $\mathcal{A}_2$ (both left-hand and right-hand components) accepts those timed words in which some $m?$-event is not preceded one time unit earlier by *any* event. Note that a strictly monotonic timed word satisfying Clause 2 in Definition 2 fails to satisfy Clause 4 if and only if it is either accepted by $\mathcal{A}_2$ or contains an α-event followed one time unit later by an $m?$-event, with $\alpha \neq m!$. It is straightforward to capture this last condition with a one-clock timed automaton. In fact $\mathcal{A}_2$ is the only component of $\mathcal{A}_{chan}$ that uses two clocks.

Thus we obtain a new proof of Alur and Dill's classical result [3].

Theorem 1. *The universality problem for timed automata with two clocks is undecidable.*

3.2 ε-Transitions

By allowing ε-transitions, we can replace the left-hand component of automaton $\mathcal{A}_2$, above, with the following automaton which uses only one clock. The ε-transition and the $m?$-transition in $\mathcal{A}_3$ are separated by exactly one time unit; this prevents any visible event from preceding this occurrence of $m?$ by one time unit.

Theorem 2. *The universality problem for the class of timed automata with one clock and ε-transitions is undecidable.*

3.3 Non-singular Postconditions

Instead of ε-transitions we can consider non-singular postconditions for clock resets. In this case we can replace automaton $\mathcal{A}_3$ with the following one-clock timed automaton, where the $\checkmark$-labelled edge non-deterministically resets x to a value strictly less than 1.

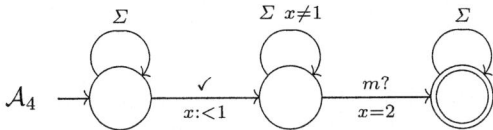

Theorem 3. *The universality problem for the class of timed automata with one clock and with non-singular postconditions is undecidable.*

3.4 Complexity

The control-state reachability problem for lossy channel machines was shown to be decidable, in contrast to the error-free case, by Abdulla and Jonsson [2]. Later Schnoebelen [22] proved that it has non-primitive recursive complexity.

Proposition 2. *The control-state reachability problem for channel machines with insertion errors has non-primitive recursive complexity.*

Proof. Given a channel machine $\mathcal{C} = (S, s_0, M, \Delta)$, define a new transition relation $\Delta^{\mathrm{op}} \subseteq S \times \{m!, m? : m \in M\} \times S$ by

$$\Delta^{\mathrm{op}} = \{(s, m!, t) : (t, m?, s) \in \Delta\} \cup \{(s, m?, t) : (t, m!, s) \in \Delta\} \ .$$

Notice that there is a transition from global state (s, x) to global state (t, y) under Δ iff there is a transition from $(t, rev(y))$ to $(s, rev(x))$ under Δ^{op}, where $rev : M^* \to M^*$ reverses the order of a finite string. Thus there is computation with lossiness errors from (s, ε) to (t, ε) under Δ iff there is a computation with insertion errors from (t, ε) to (s, ε) under Δ^{op}. This observation allows us to reduce the control-state reachability problem for channel machines with lossiness errors to the control-state reachability problem for channel machines with insertion errors. □

Define the timed language $L_{ins} \subseteq T\Sigma^*$ to consist of those timed words satisfying Clauses 1-3 in Definition 2. Thus for a word in L_{ins}, every $m!$-event is followed one time unit later by an $m?$-event, but every $m?$-event need not be preceded one time unit earlier by an $m!$-event. This corresponds to a channel with insertion errors.

Proposition 3. $\mathcal{C}$ *has a computation with insertion errors starting in state (s_0, ε) and ending in state (t, ε) iff $L_{cont} \cap L_{ins} \neq \emptyset$.*

Note that we can express $T\Sigma^* - L_{ins}$ as the language of a one-clock timed automaton. This automaton incorporates $\mathcal{A}_1$ in Section 3.1, but not $\mathcal{A}_2$. Thus we obtain

Theorem 4. *The universality problem for the class of timed automata with a single clock has non-primitive recursive complexity.*

4   Universality for One-Clock Büchi Automata

In this section we prove the undecidability of the following universality problem: 'Given a one-clock timed automaton $\mathcal{A}$ (without ε-transitions and with singular postconditions) does $L_\omega(\mathcal{A}) = T\Sigma^\omega$?'.

As in Section 3 the idea behind the proof is to encode the computations of a certain type of channel machine as a timed language. To this end, say that a non-terminating computation of a channel machine is *space-bounded* if there exists $N \in \mathbb{N}$ such that the number of messages stored on the channel during the computation never exceeds N. We define a timed language $L_{bound} \subseteq T\Sigma^\omega$ encoding space-bounded computations of a channel with insertion errors. We capture the space bound by requiring an upper bound on the number of events per time unit for each word $u \in L_{bound}$. Besides space-boundedness the other key ingredient in our proof is the notion of *alternation*, defined below.

Define a channel machine $\mathcal{C}$ to be *alternating* if we can partition the set of control states into two classes, called *read states* and *write states* respectively, such that each edge is of the form $(s, m?, t)$ with s a read state and t a write state, or of the form $(s, m!, t)$ with s a write state and t a read state. Then any computation of $\mathcal{C}$ consists of an alternating sequence of read transitions and write transitions.

The *recurrent-state problem* for alternating channel machines is as follows. Given an alternating channel machine $\mathcal{C} = (S, s_0, M, \Delta)$, does there exist $x \in M^*$ such that $\mathcal{C}$ has a non-terminating computation starting in global state (s_0, x) and visiting s_0 infinitely often? The following proposition is a relatively straightforward reduction from the halting problem for Turing machines.

Proposition 4. *The recurrent-state problem for error-free alternating channel machines is undecidable.*

The *space-bounded recurrent-state problem* for alternating channel machines with insertion errors asks if a given alternating channel machine has a space-bounded computation, possibly with insertion errors, passing infinitely often through the initial control state. The following proposition, which is reminiscent of a result of Mayr [18] on lossy counter machines, asserts that this problem is undecidable.

Proposition 5. *The space-bounded recurrent-state problem for alternating channel machines with insertion errors is undecidable.*

Proof. Given an alternating channel machine $\mathcal{C}$, we claim that $\mathcal{C}$ has a space-bounded recurrent computation with insertion errors iff it has an error-free recurrent computation. Then Proposition 5 follows from Proposition 4. Indeed, since $\mathcal{C}$ is alternating, at any point in a computation the total number of insertion errors up to that point is within one of the current size of the channel minus the size of the channel at the start of the computation. Thus any error-free computation of C is space-bounded, and any space-bounded computation of C with insertion errors is eventually error-free (the space bound gives an upper bound on the total number of insertion errors). □

Definition 3. *Given a strictly monotonic timed word* $u = (t_0, a_0)(t_1, a_1)\ldots$, *define* density$(u) = \sup\{j - i : t_j - t_i \leq 1\}$. *The density of a timed word measures the maximum number of events in any time unit along the word.*

Theorem 5. *The universality problem for one-clock Büchi timed automata is undecidable.*

The proof is by reduction from the space-bounded recurrent-state problem for alternating machines with insertion errors. Given an alternating channel machine $C = (S, s_0, M, \Delta)$, we define a one-clock Büchi timed automaton $\mathcal{A}$ such that C has a space-bounded recurrent computation with insertion errors iff $\mathcal{A}$ is non-universal.

Let $\Sigma = \{m!, m? : m \in M\} \cup \{\checkmark\}$ be a finite alphabet. We encode the finite control of C as a Büchi timed automaton $\mathcal{A}_{cont}$ with no clocks over alphabet Σ. $\mathcal{A}_{cont}$ is just the underlying control automaton of C with a $\checkmark$-labelled self-transition added to every control state and with s_0 as the initial control state and only accepting control state. Let L_{cont} denote the timed language $L_\omega(\mathcal{A}_{cont})$.

Next we capture the behaviour of a space-bounded channel with insertion errors using a timed language L_{bound} over alphabet Σ.

Definition 4. L_{bound} *consists of those timed words* u *satisfying:*

1. u *is strictly monotonic and contains infinitely many non-$\checkmark$-events.*
2. *There is a $\checkmark$-event at time zero, and thereafter consecutive $\checkmark$-events are separated by one time unit.*
3. *For every $m!$-event in u there is an $m?$-event one time unit later.*
4. *For every $m?$-event in u there is a $n!$-event one time unit later, for some $n \in M$.*
5. density$(u) < \infty$.

As with the corresponding clause in Definition 2, Clause 3 captures the channel discipline: every message sent is received. The channel has insertion errors because not every $m?$-event is necessarily preceded one time unit earlier by an $m!$-event. On the other hand, Clause 4 has nothing to do with the channel discipline. However its presence, together with Clauses 2 and 3, ensures that for every event of $u \in L_{bound}$ there is an event exactly one time unit later. (This fact will play a significant role later.) Since we are dealing with alternating channel machines, the imposition of Clause 4 will prove to be no restriction when we seek to match words in L_{bound} with channel computations. Finally, Clause 5 corresponds to the space-boundedness of the channel.

Proposition 6. C *has a space-bounded recurrent computation with insertion errors iff* $L_{cont} \cap L_{bound} \neq \emptyset$.

Proof. ($\Leftarrow$) Let $u \in L_{cont} \cap L_{bound}$. We show how to recover a space-bounded recurrent computation of $\mathcal{C}$ from u. Since $u \in L_{cont}$, the automaton $\mathcal{A}_{cont}$, which represents the finite control of $\mathcal{C}$, has a run

$$(s_0, \nu_0) \xrightarrow{\delta_0, \alpha_0} (s_1, \nu_1) \xrightarrow{\delta_1, \alpha_1} (s_2, \nu_2) \xrightarrow{\delta_2, \alpha_2} \ldots \quad (2)$$

on u. Let $\alpha_{i_0} \alpha_{i_1} \alpha_{i_2} \ldots$ be the sequence of non-$\checkmark$-events in u. Then we obtain a recurrent computation of $\mathcal{C}$

$$(s_{i_0}, x_0) \xrightarrow{\alpha_{i_0}} (s_{i_1}, x_1) \xrightarrow{\alpha_{i_1}} (s_{i_2}, x_2) \xrightarrow{\alpha_{i_2}} \ldots$$

where $x_j \in M^*$ is the sequence of messages that occur as read events in the unit time interval $(t_{i_{j-1}}, t_{i_{j-1}} + 1]$, where, by convention, $i_{-1} = 0$. Since $u \in L_{bound}$, Clause 3 in Definition 4 ensures that this is a legitimate computation of $\mathcal{C}$, albeit with insertion errors. Since u has finite density this computation is space-bounded.

($\Rightarrow$) We have already observed that if $\mathcal{C}$ has a space-bounded recurrent computation with insertion errors, then it has a space-bounded recurrent error-free computation. The trace of channel events along such an error-free computation can easily be encoded as a word in L_{bound} as we now explain. Since $\mathcal{C}$ is alternating, there is a number $N \in \mathbb{N}$ such that the size of the channel is either N or $N-1$ at any point in the computation. When any message is written to the channel, the machine performs exactly $2N-1$ (read and write) operations before that message is read off the channel. We transform the sequence of read and write events along a computation into a timed word u by putting exactly $1/(2N-1)$ time units between consecutive events. This automatically guarantees that Clauses 3–5 in Definition 4 hold. Finally, adding $\checkmark$-events at integer times yields a timed word in L_{bound}. □

Similarly to the development in Section 3, the undecidability of the universality problem will follow from Proposition 6 provided that we can define a one-clock timed automaton $\mathcal{A}_{bound}$ such that $L_\omega(\mathcal{A}_{bound}) = T\Sigma^\omega - L_{bound}$. We define $\mathcal{A}_{bound}$ to be the disjunction of several automata, corresponding to the different clauses in the definition of L_{bound}. It is straightforward, for each clause 1–4, to define an automaton that accepts precisely the timed words that fail to satisfy that clause. Below we define two automata $\mathcal{A}_{inc}$ and $\mathcal{A}_{dec}$ such that, if a timed word u already satisfies 1–4, then it is accepted by $\mathcal{A}_{inc}$ or $\mathcal{A}_{dec}$ precisely if it fails Clause 5, i.e., it has infinite density.

First we recall the following simple proposition about real numbers.

Proposition 7. *If* $\mathbf{x} = \langle x_n : n \in \mathbb{N} \rangle$ *is a sequence of real numbers in the open interval* $(0,1)$ *that takes on infinitely many values, then* $\mathbf{x}$ *has either a strictly increasing subsequence or a strictly decreasing subsequence.*

Let $u = (t_0, a_0)(t_1, a_1)(t_2, a_2) \ldots$ be a timed word satisfying Clauses 1–4 in Definition 4. Then for every event of u there is an event exactly one time unit later. Thus u has infinite density iff $\{frac(t_i) : i \in \mathbb{N}\}$ is infinite. By Proposition 7, this holds iff the sequence $\langle frac(t_i) : i \in \mathbb{N} \rangle$ has either a strictly increasing subsequence or a strictly decreasing subsequence. We define an automaton $\mathcal{A}_{inc}$ that accepts u iff $\langle frac(t_i) : i \in \mathbb{N} \rangle$ has a strictly increasing subsequence, and an automaton $\mathcal{A}_{dec}$ that accepts u iff $\langle frac(t_i) : i \in \mathbb{N} \rangle$ has a strictly decreasing subsequence.

Consider a run of $\mathcal{A}_{inc}$ (depicted below) on $u = (t_0, a_0)(t_1, a_1)(t_2, a_2)\ldots$. Let t_{i_j} be the timestamp of the transition that resets clock x for the j-th time. Notice that either $t_{i_{j+1}} = t_{i_j} + 1$ or $frac(t_{i_{j+1}}) > frac(t_{i_j})$. The Büchi condition ensures that the second eventuality holds infinitely often in the run, and so the sequence $frac(t_{i_j})$ has a strictly increasing subsequence. Thus, among those timed words u satisfying Clauses 1–4 in Definition 4, $\mathcal{A}_{inc}$ accepts precisely those for which $\langle frac(t_i) : i \in \mathbb{N} \rangle$ has a strictly increasing subsequence. (Notice the importance of the fact that for each event in u there is an event one time unit later.)

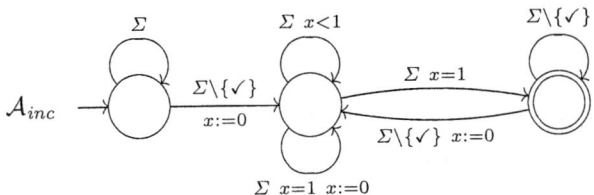

$\mathcal{A}_{dec}$ (depicted below) operates in a similar manner to $\mathcal{A}_{inc}$ except that it accepts those words $u = (t_0, a_0)(t_1, a_1)(t_2, a_2)\ldots$ for which $\langle frac(t_i) : i \in \mathbb{N} \rangle$ has a strictly decreasing subsequence.

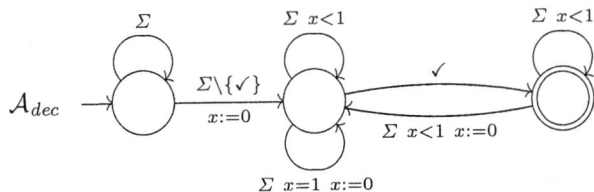

5 Remarks and Future Work

The main result of this paper is that the universality problem for one-clock Büchi timed automata is undecidable. A closely related problem concerns the decidability of the satisfiability and model checking problems for Metric Temporal Logic (MTL) [6, 4]. This logic is known to be undecidable under an *interval semantics*. However, under a *point-based semantics*—i.e., interpreting the logic over timed words—the satisfiability problem is open, cf. [21].

In [21] we show how to translate an MTL formula into an *alternating timed automaton* with a single clock. In terms of alternating automata, the present paper shows that the emptiness problem for one-clock alternating automata with co-Büchi acceptance conditions is undecidable. However one can express MTL formulas using a particularly simple acceptance condition—the so-called *weak parity acceptance condition*[2]. It does not seem possible to capture the timed language L_{bound} by an MTL formula, or, more generally, a one-clock alternating automata with weak parity acceptance conditions. So it remains possible that the satisfiability problem for MTL is decidable, and we are currently investigating this question.

References

1. P. Abdulla and B. Jonsson. Undecidable Verification Problems with Unreliable Channels. *Information and Computation* 130:71–90, 1996.
2. P. Abdulla and B. Jonsson. Model checking of systems with many identical timed processes. *Theoretical Computer Science* 290(1):241-264, 2003.
3. R. Alur and D. Dill. A Theory of Timed Automata. *Theoretical Computer Science* 126:183–235, 1994.
4. R. Alur, T. Feder and T. A. Henzinger. The Benefits of Relaxing Punctually. *Journal of the ACM* 43:116–146, 1996.
5. R. Alur, L. Fix and T. A. Henzinger. Event-clock automata: a determinizable class of timed automata. *Theoretical Computer Science* 211(1-2):253–273, 1999.
6. R. Alur and T. A. Henzinger. Real-time Logics: Complexity and Expressiveness. *Information and Computation* 104:35–77, 1993.
7. R. Alur, S. La Torre and P. Madhusudan. Perturbed Timed Automata. *Proc. HSCC 05*, LNCS 3414, Springer-Verlag, 2005.
8. R. Alur and P. Madhusudan. Decision problems for timed automata: A survey, *4th Intl. School on Formal Methods for Computer, Communication, and Software Systems: Real Time*, LNCS 3185, Springer-Verlag, 2004.
9. P. Bouyer, C. Dufourd, E. Fleury and A. Petit. Updatable timed automata. *Theoretical Computer Science* 321(2-3):291-345, 2004.
10. G. Cécé, A. Finkel and S. Purushothaman Iyer. Unreliable Channels are Easier to Verify Than Perfect Channels. *Information and Computation* 124:20–31, 1996.
11. A. Finkel and P. Schnoebelen. Well-Structured Transition Systems Everywhere! *Theoretical Computer Science* 256(1-2):63–92, 2001.
12. T. A. Henzinger, P.W. Kopke, A. Puri and P. Varaiya. What's Decidable About Hybrid Automata? *Journal of Computer and System Sciences* 57:94–124, 1998.
13. T. A. Henzinger, Z. Manna, and A. Pnueli. What Good Are Digital Clocks? *Proc. ICALP 92*, LNCS 623, Springer-Verlag, 1992.
14. J. Hopcroft and J. Ullman. *Introduction to Automata Theory, Languages, and Computation.* Addison-Wesley, 1979.
15. F. Laroussinie, N. Markey and P. Schnoebelen. Model Checking Timed Automata with One or Two Clocks. *Proc. CONCUR 04*, LNCS 3170, Springer-Verlag, 2004.
16. S. Lasota and I. Walukiewicz. Alternating Timed Automata. *Proc. of FOSSACS 05*, LNCS 3441, Springer-Verlag, 2005.
17. S. Lasota and I. Walukiewicz. Personal communication, 2005.

---

[2] In the untimed case an alternating automaton with a Büchi or co-Büchi acceptance condition can be translated into an alternating automaton with a weak parity acceptance condition, however this translation does not apply in the timed case.

18. R. Mayr. Undecidable problems in unreliable computations. *Theoretical Computer Science* 1-3(297):337-354, 2003.
19. J. Ouaknine and J. Worrell. Universality and Language Inclusion for Open and Closed Timed Automata. *Proc. HSCC 03*, LNCS 2623, Springer-Verlag, 2003.
20. J. Ouaknine and J. Worrell. On the Language Inclusion Problem for Timed Automata: Closing a Decidability Gap. *Proc. LICS 04*, IEEE, 2004.
21. J. Ouaknine and J. Worrell. On the Decidability of Metric Temporal Logic. *Proc. LICS 05*, IEEE, 2005.
22. P. Schnoebelen. Verifying Lossy Channel Systems has Nonprimitive Recursive Complexity. *Information Processing Letters* 83(5):251-261, 2002.

Congruences for Visibly Pushdown Languages*

Rajeev Alur[1], Viraj Kumar[2], P. Madhusudan[2], and Mahesh Viswanathan[2]

[1] University of Pennsylvania, Philadelphia, PA, USA
alur@cis.upenn.edu
[2] University of Illinois at Urbana-Champaign, Urbana, IL, USA
{kumar, madhu, vmahesh}@cs.uiuc.edu

Abstract. We study congruences on words in order to characterize the class of visibly pushdown languages (VPL), a subclass of context-free languages. For any language L, we define a natural congruence on words that resembles the syntactic congruence for regular languages, such that this congruence is of finite index if, and only if, L is a VPL. We then study the problem of finding canonical minimal deterministic automata for VPLs. Though VPLs in general do not have unique minimal automata, we consider a subclass of VPAs called k-module single-entry VPAs that correspond to programs with recursive procedures without input parameters, and show that the class of well-matched VPLs do indeed have unique minimal k-module single-entry automata. We also give a polynomial time algorithm that minimizes such k-module single-entry VPAs.

1 Introduction

The class of *visibly pushdown languages* (VPL), introduced in [1], is a subclass of context-free languages accepted by pushdown automata in which the input letter determines the type of operation permitted on the stack. Visibly pushdown languages are closed under all boolean operations, and problems such as inclusion, that are undecidable for context-free languages, are decidable for VPL. VPLs are relevant to several applications that use context-free languages such as the model-checking of software programs using their pushdown models [1, 2, 3]. Recent work has shown applications in other contexts: in modeling semantics of effects in processing XML streams [4], in game semantics for programming languages [5], and in identifying larger classes of pushdown specifications that admit decidable problems for infinite games on pushdown graphs [6].

Our main result in this paper is a characterization of the class of VPLs in terms of congruences on strings. It is well known that the *syntactic congruence*, which is defined as $w_1 \approx w_2$ when for every u, v, $uw_1v \in L$ if and only if $uw_2v \in L$, has finite index precisely for languages L that are regular. Our central

---

* This research was partially supported by ARO URI award DAAD19-01-1-0473, NSF awards CCR-0306382 and CCF 04-29639, and DARPA/AFOSR MURI award F49620-02-1-0325.

thesis is that for VPLs L, when we restrict our attention to *well-matched* words w_1 and w_2 (i.e., words where every push transition has a corresponding pop transition and vice versa), the syntactic congruence has finite index. Moreover, for languages consisting only of well-matched words, if the syntactic congruence on well-matched words has finite index then the language is a VPL. For languages containing strings that are not well-matched, we need some additional conditions only because no congruence on well-matched words can *saturate* such a language. Our characterization of VPLs is a natural generalization of the Myhill-Nerode theorem for regular languages— when restricted to languages that do not require any push or pop operations, our congruence coincides with the right congruence defined by Myhill and Nerode [7, 8].

One important consequence of the congruence based characterization of regular (word) languages and regular tree languages is that for any regular language there is a unique minimum state deterministic automaton recognizing the language, which can also be constructed efficiently [8,9]. For VPLs, however, we show that in general there is no unique minimum state recognizer. Thus, while our characterization yields the construction of a *canonical* deterministic acceptor for VPLs, it may not in general be minimal. An implicit consequence of the results in [1] is that VPLs have canonical deterministic pushdown automata. It is shown in [1] that with any language L, a language of trees called *stack trees*, can be associated such that L is a VPL exactly when the corresponding set of stack trees form a regular tree language. The unique minimal bottom-up tree automaton accepting the language of stack trees can then be translated to a canonical deterministic visibly pushdown automaton. However, since bottom-up tree automata can only be translated into deterministic pushdown automata with exponentially more states, the implicit construction in [1] does not result in necessarily small deterministic VPAs.

Visibly pushdown automata are a natural model for programs with recursive procedure calls and finite data types. Such programs are called Boolean programs in the literature on software model checking [3]. When modeling a program as a visibly pushdown automaton, the natural structure the model assumes is one where the machine's states are partitioned into k modules, one for each procedure in the program. As one expects, these modules are such that from a state in a module, a sequence of calls and returns to other modules results in a state of the same module. Moreover, if the programs modeled are such that the calls to modules have no input parameters (or if a function is modeled separately for each possible value of its input parameters), then the visibly pushdown automaton assumes additional structure, namely that every call results in going to a unique state in the module corresponding to the call. We call such structured VPAs k-module single-entry VPAs (k-SEVPAs). They correspond roughly to the model of *recursive state machines* with a single entry per module [10].

Though visibly pushdown languages in general do not have unique minimum-state recognizers, partitioning the calls into the modules they correspond to fixes enough additional structure that there is a minimum-state k-SEVPA that respects the partition and accepts the language. More precisely, we show that for any partition of the call-alphabet into k-sets, there is a unique minimum-state k-SEVPA accepting any well-matched VPL L. If $k = 0$ (that is, there are no calls), the result is equivalent to the Myhill-Nerode theorem for regular languages. The characterization of this unique minimal k-SEVPA is done via a set of $k+1$ congruences on words. We also present an algorithm which, given any deterministic k-SEVPA accepting a well-matched language, minimizes it in $O(n^3)$ time, where n is the size of the original machine.

The rest of the paper is organized as follows. We first recall the definitions of visibly pushdown languages and visibly pushdown automata in Section 2. Our main result characterizing visibly pushdown languages in terms of language theoretic congruences is presented in Section 3. We also show that VPLs, in general, do not have unique minimum state deterministic recognizers. In Section 4, we define the notion of how partitions on calls define k-module single-entry VPAs and prove that every (well-matched) VPL has a unique minimal k-SEVPA accepting it. We also present an example of a family of languages for which the minimal 1-module machine is super polynomial in the size of the smallest visibly pushdown automaton recognizing it. Conclusions and open problems are presented in Section 5.

2 Preliminaries

In this section, we recall definitions of *visibly pushdown automata* and *visibly pushdown languages*, and introduce some notation that we will use in the rest of the paper.

Pushdown Alphabet. A *pushdown alphabet* is a tuple $\widehat{\Sigma} = (\Sigma_{\text{call}}, \Sigma_{\text{ret}}, \Sigma_{\text{int}})$ that comprises three *disjoint* finite alphabets— Σ_{call} is a finite set of *calls*, Σ_{ret} is a finite set of *returns*, and Σ_{int} is a finite set of *internal actions*. For any such $\widehat{\Sigma}$, let $\Sigma = \Sigma_{\text{call}} \cup \Sigma_{\text{ret}} \cup \Sigma_{\text{int}}$. In the paper we will use $u, v, u_1, \ldots$ for strings in Σ^*, $c, c_1, c_i, \ldots$ for elements of Σ_{call}, $r, r_1, r_i, \ldots$ for elements of Σ_{ret}, and $i, i_1, i_j, \ldots$ for elements of Σ_{int}.

Visibly Pushdown Automata. For visibly pushdown automata, unlike the case of pushdown automata, it turns out that deterministic VPAs are as powerful as a non-deterministic VPAs [1]. In light of this, we will only consider deterministic VPAs. A *visibly pushdown automaton* (VPA) on finite strings over $\widehat{\Sigma} = (\Sigma_{\text{call}}, \Sigma_{\text{ret}}, \Sigma_{\text{int}})$ is a tuple $M = (Q, q_0, \Gamma, \delta, Q_F)$ where Q is a finite set of states, $q_0 \in Q$ is the initial state, Γ is a finite stack alphabet that contains a special bottom-of-stack symbol $\bot$, $\delta = \delta_{\text{call}} \cup \delta_{\text{ret}} \cup \delta_{\text{int}}$ is the transition function, where $\delta_{\text{call}} : Q \times \Sigma_{\text{call}} \to Q \times (\Gamma \setminus \{\bot\})$, $\delta_{\text{ret}} : Q \times \Sigma_{\text{ret}} \times \Gamma \to Q$, and $\delta_{\text{int}} : Q \times \Sigma_{\text{int}} \to Q$, and $Q_F \subseteq Q$ is a set of final states.

If $\delta_{\text{call}}(q, c) = (q', \gamma)$, where $c \in \Sigma_{\text{call}}$ and $\gamma \neq \bot$, there is a *push-transition* from q on input c where on reading c, γ is pushed onto the stack and the control changes from state q to q'; we denote such a transition by $q \xrightarrow{c/\gamma} q'$. Similarly, if $\delta_{\text{ret}}(q, r, \gamma) = q'$, there is a *pop-transition* from q on input r where γ is read from the top of the stack and popped (if the top of the stack is $\bot$, then it is read but not popped), and the control changes from q to q'; we denote such a transition by $q \xrightarrow{r/\gamma} q'$. If $\delta_{\text{int}}(q, i) = q'$, there is an *internal-transition* from q on input i where on reading i, the state changes from q to q'; we denote such a transition by $q \xrightarrow{i} q'$. Note that there are no stack operations on internal transitions.

Acceptance. A *stack* is a non-empty finite sequence over Γ ending in the bottom-of-stack symbol $\bot$. The set of all stacks is denoted as $St = (\Gamma \setminus \{\bot\})^* \cdot \{\bot\}$. A *configuration* is a pair (q, σ) such that q is a state and $\sigma \in St$. The transition function of a VPA can be used to define how the configuration of the machine changes in a single step: we say $\delta((q, \sigma), a) = (q', a')$ [1] if one of the following holds:

1. If $a \in \Sigma_{\text{call}}$ then there exists $\gamma \in \Gamma$ such that $\delta_{\text{call}}(q, a) = (q', \gamma)$ and $\sigma' = \gamma \cdot \sigma$
2. If $a \in \Sigma_{\text{ret}}$, then there exists $\gamma \in \Gamma$ such that $\delta_{\text{ret}}(q, a, \gamma) = q'$ and either $\gamma \neq \bot$ and $\sigma = \gamma \cdot \sigma'$, or $\gamma = \bot$ and $\sigma = \sigma' = \bot$
3. If $a \in \Sigma_{\text{int}}$ is an internal action, then $\delta_{\text{int}}(q, a) = q'$ and $\sigma' = \sigma$

The transitive closure of the single-step transition function, which we also denote by δ, can be easily defined in the standard inductive manner. For a stack $\sigma \in St$, we define the function $\delta_\sigma : Q \times \Sigma^* \to Q$ as $\delta_\sigma(q, u) = q'$ whenever $\delta((q, \sigma), u) = (q', \sigma')$ for some $\sigma' \in St$.

A string $u \in \Sigma^*$ is *accepted* by VPA M if $\delta_\bot(q_0, u) \in Q_F$. The *language* of M, $L(M)$, is the set of strings accepted by M.

Visibly Pushdown Languages. A language over finite strings $L \subseteq \Sigma^*$ is a *visibly pushdown language* (VPL) with respect to $\widehat{\Sigma}$ (a $\widehat{\Sigma}$-VPL) if there is a VPA M over $\widehat{\Sigma}$ such that $L(M) = L$.

Matched calls and returns. Let $MR(\widehat{\Sigma})$ denote the set of all strings where every return has a matched call before it, i.e. $u \in MR(\widehat{\Sigma})$ if for every prefix u' of u, the number of return symbols in u' is at most the number of call symbols in u'. Similarly, let $MC(\widehat{\Sigma})$ denote the set of all strings where every call has a matching return after it, i.e. $u \in MC(\widehat{\Sigma})$ if for every suffix u' of u, the number of call symbols in u' is at most the number of return symbols in u'. The set of *well-matched* strings over $\widehat{\Sigma}$ is $WM(\widehat{\Sigma}) = MR(\widehat{\Sigma}) \cap MC(\widehat{\Sigma})$.

A $\widehat{\Sigma}$-VPL L is said to be *well-matched* if $L \subseteq WM(\widehat{\Sigma})$.

Remark 1. For every $w \in WM(\widehat{\Sigma})$, there is a unique matching between call and return symbols such that every call-symbol always precedes its matching return-symbol and the substring w' between a matching pair of call and return symbols is a well-matched string.

---

[1] We abuse notation and use δ for both the transition function of the automaton and the single step transition function on configurations.

3 Congruence Based Characterization of VPLs

In this section we present a congruence based characterization of when a language over $\widehat{\Sigma}$ is a visibly pushdown language. Before presenting the characterization for general VPLs, we first consider the case of VPLs that have only well-matched words.

3.1 Well-Matched Visibly Pushdown Languages

For a language L over the pushdown alphabet $\widehat{\Sigma} = (\Sigma_{\text{call}}, \Sigma_{\text{ret}}, \Sigma_{\text{int}})$, consider the following congruence on well-matched words:

$$w_1 \approx w_2 \text{ iff } \forall u, v \in \Sigma^*, \, uw_1v \in L \text{ iff } uw_2v \in L$$

Recall that this is the standard syntactic congruence restricted to well-matched words over $\widehat{\Sigma}$. For example, if $\widehat{\Sigma} = (\{c\}, \{r\}, \emptyset)$ and $L = \{c^n.r^n \mid n \geq 0\}$, then there are only two equivalence classes that $\approx$ defines: $\{c^n r^n \mid n \geq 0\}$ and the complement of this set with respect to $WM(\widehat{\Sigma})$.

Analogous to the case of regular languages, the finiteness of the number of equivalence classes of the syntactic congruence (on well-matched words) provides a precise characterization of *well-matched* VPLs.

Theorem 1. *L is a well-matched $\widehat{\Sigma}$-VPL iff $\approx$ (as defined above) has finitely many equivalence classes.*

Proof. Suppose L is a $\widehat{\Sigma}$-VPL and $M = (Q, q_0, \Gamma, \delta, Q_F)$ is a VPA over $\widehat{\Sigma}$ with (unique) initial state q_0 such that $L(M) = L$. Every well-matched string w defines a function $f_w : Q \to Q$ as follows: $f_w(q) = \delta_\bot(q, w)$. Define the following equivalence on well-matched strings:

$$w_1 \approx_M w_2 \text{ iff } f_{w_1} = f_{w_2}$$

Observe that $\approx_M$ has finitely many equivalence classes (bounded by $|Q|^{|Q|}$). We will show that $\approx_M$ is a refinement of $\approx$, thus establishing that $\approx$ is also of finite index. Consider $w_1 \approx_M w_2$. Then for any $u, v \in \Sigma^*$, we know

$$\begin{aligned}\delta((q_0, \bot), uw_1v) &= \delta(\delta(\delta((q_0, \bot), u), w_1), v) \\ &= \delta(\delta(\delta((q_0, \bot), u), w_2), v) \text{ since } f_{w_1} = f_{w_2} \\ &= \delta((q_0, \bot), uw_2v)\end{aligned}$$

Hence $uw_1v \in L$ iff $uw_2v \in L$, and so $w_1 \approx w_2$. Thus $\approx_M$ is a refinement of $\approx$. Observe that this proof does not rely on L being a well-matched language.

To prove the converse, consider a language L such that $\approx$ is of finite index. We construct a deterministic (but incomplete[2]) VPA that recognizes L and whose

---

[2] A VPA is incomplete if the transition function δ is not total. An incomplete VPA can be easily modified to yield a VPA with at most one extra "dead" state to which all undefined transitions go.

states are the equivalence classes of $\approx$. Consider a string with no unmatched returns $u = w_1 c_1 w_2 c_2 \cdots c_k w_{k+1} \in MR(\widehat{\Sigma})$, where $c_1, \ldots c_k$ are the unmatched call symbols in u, and $w_1, \ldots w_{k+1}$ are well-matched strings between the unmatched call symbols. The automaton we construct will maintain the following invariant: after reading the string $u \in MR(\widehat{\Sigma})$, the state of the machine will be $[w_{k+1}]_\approx$ and the stack will be $([w_k]_\approx, c_k)([w_{k-1}]_\approx, c_{k-1}) \cdots ([w_1]_\approx, c_1) \bot$.

The formal construction of VPA $M = (Q, q_0, \Gamma, \delta, Q_F)$ is as follows: $Q = \{[w]_\approx \mid w \in WM(\widehat{\Sigma})\}$, $q_0 = [\epsilon]_\approx$, $\Gamma = \{\bot\} \cup (Q \times \Sigma_{\text{call}})$, and $Q_F = \{[w]_\approx \mid w \in L\}$. The transition function δ is defined as follows.

- $[w]_\approx \xrightarrow{i} [wi]_\approx$ for every $i \in \Sigma_{\text{int}}$
- $[w]_\approx \xrightarrow{c/([w]_\approx, c)} [\epsilon]_\approx$ for every $c \in \Sigma_{\text{call}}$
- $[w]_\approx \xrightarrow{r/([w']_\approx, c)} [w'cwr]_\approx$ for every $r \in \Sigma_{\text{ret}}$

The above machine has no pop transitions when $\bot$ is the only symbol on the stack. Observe that the definitions of Q_F and δ are sound because $\approx$ saturates L [3] and $\approx$ is a congruence with respect to well-matched words. Further, it is easy to verify that the above invariant is maintained. Thus, after reading a *well-matched* word w, the automaton will be in the state $[w]_\approx$ and hence $L = L(M) \cap WM(\widehat{\Sigma})$. Since $WM(\widehat{\Sigma})$ is a VPL, and VPLs are closed under intersection, the result follows. □

3.2 General Visibly Pushdown Languages

For visibly pushdown languages that are not necessarily well-matched, $\approx$ being of finite index is not sufficient. This is because $\approx$ is no longer a congruence that saturates the VPL. We need to define two additional congruences on strings— one that will capture the behavior of a state when the stack only has $\bot$, and one that will capture the behavior when the stack has more than one element. The reason we need to distinguish the cases of the stack having only $\bot$ and that of the stack having additional elements, is because symbols in Σ_{ret} behave differently. In the first case, elements of Σ_{ret} are like internal actions which leave the stack unchanged, and in the second case they result in the stack being popped.

For a language L over $\widehat{\Sigma}$, define the following congruences.

For $u_1, u_2 \subset \Sigma^*$, $u_1 \equiv u_2$ iff $\forall v \in MR(\widehat{\Sigma}). u_1 v \in L$ iff $u_2 v \in L$
For $u_1, u_2 \in MC(\widehat{\Sigma})$, $u_1 \sim_0 u_2$ iff $\forall v \in \Sigma^*. u_1 v \in L$ iff $u_2 v \in L$

Intuitively, the congruence $\equiv$ says that the two strings u_1 and u_2 cannot be distinguished by experiments ($v \in MR(\widehat{\Sigma})$) that do not examine the stacks reached on u_1 and u_2. The congruence $\sim_0$ is only defined on strings where every call is matched. Thus, after reading such a word, any VPA will only have $\bot$ on the stack. Starting from such configurations, as was observed earlier, return symbols

---

[3] An equivalence $\equiv$ saturates L iff either $[w]_\equiv \cap L = \emptyset$ or $[w]_\equiv \subseteq L$, for any equivalence class $[w]_\equiv$ of $\equiv$.

behave like internal actions, and the congruence is the usual Myhill-Nerode right congruence. We now present the main theorem of this paper.

Theorem 2. *L is a $\widehat{\Sigma}$-VPL iff $\approx$, $\equiv$ and $\sim_0$ all have finite index.*

Proof. For a VPL L, let $M = (Q, q_0, \Gamma, \delta, Q_F)$ be a VPA recognizing L. In the proof of Theorem 1, we already showed that $\approx$ will have finite index. Define the following two equivalences over words in Σ^*:

$$u_1 \equiv^M u_2 \text{ iff } \delta_\perp(q_0, u_1) = \delta_\perp(q_0, u_2)$$
$$u_1 \sim_0^M u_2 \text{ iff } \delta_\perp(q_0, u_1) = \delta_\perp(q_0, u_2)$$

It can be shown that $\equiv^M$ refines $\equiv$, and $\sim_0^M$ refines $\sim_0$ when restricted to $MC(\widehat{\Sigma})$ (proof skipped in the interests of space). Hence, both $\equiv$ and $\sim_0$ have finitely many equivalence classes.

For the converse, we show that L is a VPL by once again constructing a VPA M whose states are equivalence classes of the congruences we have defined, but the construction is a bit more involved. The main intuition behind the construction is to ensure that the following invariant is maintained after M has read a string $u \in \Sigma^*$

- If $u \in MC(\widehat{\Sigma})$ then the state of M is $[u]_{\sim_0}$ and the stack is $\perp$.
- If $u = vc_1w_1\cdots c_kw_k$, where $v \in MC(\widehat{\Sigma})$, each $w_j \in WM(\widehat{\Sigma})$, and each $c_j \in \Sigma_{\text{call}}$, then M is in state $([u]_\equiv, [w_k]_\approx)$ and the stack is $([w_{k-1}]_\approx, c_k) \cdots ([w_1]_\approx, c_2)([v]_{\sim_0}, c_1)\perp$.

The formal construction of M is as follows. $M = (Q, q_0, \Gamma, \delta, Q_F)$ where $Q = \{[u]_{\sim_0} \mid u \in MC(\widehat{\Sigma})\} \cup \{([u]_\equiv, [w]_\approx) \mid u \in \Sigma^*, w \in WM(\widehat{\Sigma})\}$; $q_0 = [\epsilon]_{\sim_0}$; $\Gamma = Q \times \Sigma_{\text{call}} \cup \{\perp\}$; $Q_F = \{[u]_{\sim_0} \mid u \in L\} \cup \{([u]_\equiv, [w]_\approx) \mid u \in L\}$; and δ is defined as follows:

- $[u]_{\sim_0} \xrightarrow{i} [ui]_{\sim_0}$ for every $i \in \Sigma_{\text{int}}$
- $[u]_{\sim_0} \xrightarrow{c/([u]_{\sim_0},c)} ([uc]_\equiv, [\epsilon]_\approx)$ for every $c \in \Sigma_{\text{call}}$
- $[u]_{\sim_0} \xrightarrow{r/\perp} [ur]_{\sim_0}$ for every $r \in \Sigma_{\text{ret}}$
- $([u]_\equiv, [w]_\approx) \xrightarrow{i} ([ui]_\equiv, [wi]_\approx)$ for every $i \in \Sigma_{\text{int}}$
- $([u]_\equiv, [w]_\approx) \xrightarrow{c/(([u]_\equiv,[w]_\approx),c)} ([uc]_\equiv, [\epsilon]_\approx)$ for every $c \in \Sigma_{\text{call}}$
- $([u]_\equiv, [w]_\approx) \xrightarrow{r/([u']_{\sim_0},c)} [u'cwr]_{\sim_0}$ for every $r \in \Sigma_{\text{ret}}$
- $([u]_\equiv, [w]_\approx) \xrightarrow{r/(([u']_\equiv,[w']_\approx),c)} ([u'cwr]_\equiv, [w'cwr]_\approx)$ for every $r \in \Sigma_{\text{ret}}$

The correctness of the construction relies on the intuition outlined earlier and is skipped in the interests of space. □

Remark 2. Note that in the case where $\Sigma_{\text{call}} = \Sigma_{\text{ret}} = \emptyset$ (i.e. for regular languages), the machine M constructed in Theorem 2 is the unique minimum-state

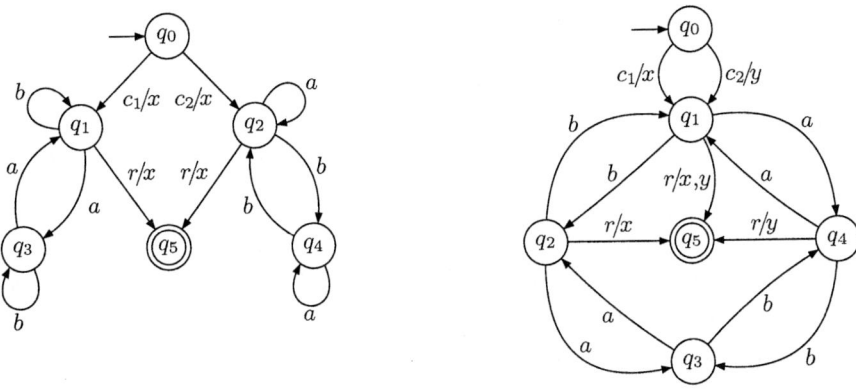

Fig. 1. Two non-isomorphic minimum-state VPAs

automaton for L because the only reachable states will be of the form $[w]_{\sim_0}$, where $w \in \Sigma^*$.

Despite the above remark, the VPA M constructed in Theorem 2 need not be a *minimum-state* $\widehat{\Sigma}$-VPA accepting L. Furthermore,

Proposition 1. *There are VPLs that have no unique minimum-state VPA accepting them.*

To illustrate the above proposition, consider the VPAs in Figure 1. Let $\widehat{\Sigma} = (\{c_1, c_2\}, \{r\}, \{a, b\})$. Let $L = c_1 L_1 r + c_2 L_2 r$, where L_1 is the regular language over $\{a, b\}$ such that the number of a's is even, and L_2 is the regular language over $\{a, b\}$ such that the number of b's is even. The figure shows two *non-isomorphic* minimum-state $\widehat{\Sigma}$-VPAs accepting L. In both machines, the initial state is q_0 and the following transitions have been omitted in the figure for readability: in both machines, every call-transition not shown is of the form $q \xrightarrow{c_j/z} q_2$, and every other transition not shown goes to state q_2.

Notice that the first machine consists of two distinct "modules", one recognizing L_1 and one recognizing L_2, and the call symbol c_1 or c_2 determines which module is "invoked". In contrast, the second machine consists of a single recognizer for both L_1 and L_2, and this module is invoked regardless of the call symbol. As this example illustrates, it is not clear when splitting the task of recognition into distinct modules reduces the total number of states in the VPA. In the following section, we consider a restricted class of VPAs for which the partition of the VPA into modules has already been provided and the call-symbol determines which module is to be invoked. The task then is to minimize the number of states of the automaton, while preserving the given partition of states into modules.

4 k-Module Single-Entry Visibly Pushdown Automata

In this section we show that the class of well-matched VPLs have *unique* minimum-state k-*module* single-entry automata (k-SEVPA). As mentioned in the introduction, these automata are motivated by models of programs with finite data-types, and are similar to single-entry recursive state machines [10] (see [11] for a precise comparison).

k-*SEVPAs.* Let $\{\Sigma_{\text{call}}^j\}_{j=1}^k$ be a partition of Σ_{call}. A VPA $M = (Q, q_0, \Gamma, \delta, Q_F)$ is a k-module single-entry VPA with respect to $\{\Sigma_{\text{call}}^j\}_{j=1}^k$ if there is a partition $\{Q_j\}_{j=0}^k$ of Q and distinguished states $q_j \in Q_j$ for every $j = 1, \ldots, k$ such that:

1. $Q_F \subseteq Q_0$, $q_0 \in Q_0$;
2. $\Gamma = \{\bot\} \cup (Q \times \Sigma_{\text{call}})$;
3. if $q \xrightarrow{i} q'$ for some $i \in \Sigma_{\text{int}}$, then $\exists j.\ q, q' \in Q_j$;
4. if $q \xrightarrow{c/(q,c)} q'$ for some $c \in \Sigma_{\text{call}}^j$, then $q' = q_j$;
5. if $q' \xrightarrow{r/(q,c)} q''$ for some $c \in \Sigma_{\text{call}}$, then $\exists j.\ q, q'' \in Q_j$.

Intuitively, Q_0 is the base module (corresponding to the 'main' module of a program), and the transition relation is such that a call leads to a unique state in the module corresponding to the call (in models of programs, this state will be the initial control state of the function called), and upon return will return to the calling module. Such automata are no less expressive than VPAs: as Theorem 3 below shows, for any partition of call symbols, any well-matched VPL is accepted by some k-SEVPA.

We use the abbreviation k-SEVPA for such machines and explicitly denote them as $M = ((Q, q_0, \Gamma, \delta, Q_F), \{\Sigma_{\text{call}}^1, \ldots, \Sigma_{\text{call}}^k\}, \{Q_0, \ldots, Q_k\}, \{q_1, \ldots, q_k\})$. For example, for the first VPA in Figure 1, given the partition $\Sigma_{\text{call}} = \{\{c_1\}, \{c_2\}\}$, there is a partition of Q as $\{Q_0, Q_1, Q_2\}$, where $Q_0 = \{q_0, q_5\}$, $Q_1 = \{q_1, q_3\}$, and $Q_2 = \{q_2, q_4\}$ witnessing the fact that this is a 2-SEVPA. Similarly, for the second VPA, given the partition $\Sigma_{\text{call}} = \{\{c_1, c_2\}\}$, there is a partition of Q as $\{Q_0, Q_1\}$, where $Q_0 = \{q_0, q_5\}$ and $Q_1 = \{q_1, q_2, q_3, q_4\}$ establishing that it is a 1-SEVPA.

Remark 3. The automaton constructed in Theorem 1 is a 1-SEVPA. However, in general, this VPA will be much bigger than the smallest 1-SEVPA. The reason for this is similar to the reason why the finite automaton constructed for regular languages from the syntactic congruence is much larger than that obtained from the syntactic right congruence. Thus, in order to characterize the minimal k-SEVPAs, we need a new congruence that partitions words like the Myhill-Nerode right congruence does for regular languages.

In our construction of the minimal k-SEVPA for a language L, we will use $k + 1$ congruences. To model the case when the stack only has $\bot$, we will use $\sim_0$ (defined in Section 3.2 on strings with matched calls). For the case when the stack has additional symbols, we need k new congruences that make use of the fact that the states of the machine are partitioned into k modules identified by the call symbol. Given a k-SEVPA $M = (M', \{\Sigma_{\text{call}}^j\}_{j=1}^k, \{Q_j\}_{j=0}^k, \{q_j\}_{j=1}^k)$ accepting a language L over $\widehat{\Sigma}$, define the following congruences on well-matched strings: for every $j = 1, \ldots, k$,

$$w_1 \sim_j w_2 \text{ iff } \forall u, v \in \Sigma^* \forall c \in \Sigma_{\text{call}}^j.\ ucw_1v \in L \text{ iff } ucw_2v \in L$$

Since $\sim_j$'s will be used to define states when the stack has more than just $\bot$, when defining the equivalence we only need to consider contexts where there is an unmatched call. We are ready to present the main theorem of this section.

Theorem 3. *For any well-matched $\widehat{\Sigma}$-VPL L and any partition $\{\Sigma_{\text{call}}^j\}_{j=1}^k$ of Σ_{call}, there is a unique (upto isomorphism) minimum-state k-SEVPA for L with respect to this partition.*

Proof. We first show that given any partition $\{\Sigma_{\text{call}}^j\}_{j=1}^k$ of Σ_{call}, there is a k-SEVPA M that recognizes L. We construct M using the equivalences $\{\sim_j\}_{j=1}^k$ and $\sim_0$ (defined in Section 3.2 on strings with matched calls). We then show that this machine M is the unique minimum-state k-SEVPA that recognizes L. The construction of M relies on the observation that $\sim_0$ and $\sim_j$'s all have finite index if L is a VPL. From Theorems 1 and 2, we know that when L is a VPL, $\sim_0$ and $\approx$ are of finite index. Since $\approx$ is a refinement of $\sim_j$ for every j, it follows that all $\sim_j$'s are also of finite index. For $0 \leq j \leq k$, we use the notation $[u]_j$ to denote the equivalence class of $\sim_j$ containing u.

The formal construction of $M = ((Q, q_0, \Gamma, \delta, Q_F), \{\Sigma_{\text{call}}^j\}_{j=1}^k, \{Q_j\}_{j=0}^k, \{q_j\}_{j=1}^k)$ is: $Q_0 = \{[u]_0 \mid u \in MC(\widehat{\Sigma})\}$, and for every $j = 1, \ldots, k$, $Q_j = \{[w]_j \mid w \in WM(\widehat{\Sigma})\}$. For every $j \geq 0$, $q_j = [\epsilon]_j$, and $Q_F = \{[u]_0 \mid u \in L\}$. The transition function δ is given as follows.

- For every $i \in \Sigma_{\text{int}}$ and $j \geq 0$, $[u]_j \xrightarrow{i} [ui]_j$
- For every $c \in \Sigma_{\text{call}}^{j'}$ and $j \geq 0$, $[u]_j \xrightarrow{c/([u]_j, c)} [\epsilon]_{j'}$
- For every $r \in \Sigma_{\text{ret}}$ and $j, j' \geq 0$, $[w]_j \xrightarrow{r/([u]_{j'}, c)} [ucwr]_{j'}$
- For every $r \in \Sigma_{\text{ret}}$, $[u]_0 \xrightarrow{r/\bot} [ur]_0$

Q_F is well-defined because $\sim_0$ is an equivalence that saturates L. The transition function is consistent because $u_1 i \sim_j u_2 i$ whenever $u_1 \sim_j u_2$ and $i \in \Sigma_{\text{int}}$, and because when $u_1 \sim_{j'} u_2$ and $w_1 \sim_j w_2$, $u_1 cw_1 r \sim_{j'} u_2 cw_2 r$ for every $j > 0, j' \geq 0$ and $c \in \Sigma_{\text{call}}^j$, $r \in \Sigma_{\text{ret}}$. Thus, the above machine is well defined. Further observe that the following invariant is maintained during the execution: after reading a string u

- If $u \in MC(\widehat{\Sigma})$ then the state of M is $[u]_0$ and the stack is $\bot$.
- If $u = vc_1 w_1 \ldots c_l w_l$, where $v \in MC(\widehat{\Sigma})$, each $c_j \in \Sigma_{\text{call}}^{m_j}$ and each $w_j \in WM(\widehat{\Sigma})$, then the state of M is $[w_l]_{m_l}$ and the stack is $([w_{l-1}]_{m_{l-1}}, c_l) \ldots ([w_1]_{m_1}, c_2)([v]_0, c_1)\bot$.

Hence, if a string reaches a final state, we are guaranteed that the stack only has $\perp$, and recognizes L. The formal proof of correctness is skipped.

Consider any k-SEVPA $M' = ((Q', q'_0, \Gamma', \delta', Q'_F), \{\Sigma^j_{\text{call}}\}^k_{j=1}, \{Q'_j\}^k_{j=0}, \{q'_j\}^k_{j=1})$ recognizing L. We show that M is the unique minimum-state k-SEVPA by demonstrating a homomorphism from M' to M. In other words, we construct an onto function $f : \bigcup_{j \geq 0} Q'_j \to \bigcup_{j \geq 0} Q_j$ having the following properties.

1. $f(q'_j) = q_j$ for every $j \geq 0$
2. For any $i \in \Sigma_{\text{int}}$, if $p' \xrightarrow{i}_{M'} q'$ then $f(p') \xrightarrow{i}_M f(q')$
3. For any $c \in \Sigma_{\text{call}}$, if $p' \xrightarrow{c/(p',c)}_{M'} q'$ then $f(p') \xrightarrow{c/(f(p'),c)}_M f(q')$
4. For any $r \in \Sigma_{\text{ret}}$, if $p' \xrightarrow{r/(s',c)}_{M'} q'$ then $f(p') \xrightarrow{r/(f(s'),c)}_M f(q')$

Thus, we will be able to conclude that $|\cup Q'_j| \geq |\cup Q_j|$, and if $|\cup Q'_j| = |\cup Q_j|$ then f witnesses an isomorphism between M and M'.

The homomorphism f from M' to M is defined as follows:

$$f(q') = \begin{cases} [u]_0 & \text{if } \exists u \in MC(\widehat{\Sigma}). \, \delta'_\perp(q'_0, u) = q' \\ [w]_j & \text{if } \exists u \in \Sigma^*, c \in \Sigma^j_{\text{call}}, w \in WM(\widehat{\Sigma}). \, \delta'_\perp(q'_0, ucw) = q' \end{cases}$$

Note that $f(q'_j) = [\epsilon]_j$ for every $0 \leq j \leq k$. Observe that f maps states of Q'_j to the equivalence classes of $\sim_j$ for every $0 \leq j \leq k$. We need to show that f is well defined, i.e., f is indeed a function and does not map a state of M' to two different states of M. This follows from the following lemma.

Lemma 1. *If $u_1, u_2 \in MC(\widehat{\Sigma})$ are such that $\delta'_\perp(q'_0, u_1) = \delta'_\perp(q'_0, u_2)$, then $u_1 \sim_0 u_2$. In addition, for well-matched strings w_1 and w_2 and every $j = 1, \ldots, k$, if $\delta'_\perp(q'_j, w_1) = \delta'_\perp(q'_j, w_2)$ then $w_1 \sim_j w_2$.*

The proof of the above lemma is similar to the proofs in Theorems 1 and 2 where we show our congruences to have finite index. Thus, f is indeed a function. Further, f is clearly onto. Also, f preserves initial state and distinguished states q_j, by definition. It preserves the transitions of M' because $\sim_0$ and $\sim_j$'s are congruences. This completes the proof that there is a unique minimum-state k-SEVPA. □

While the above theorem shows that each (well-matched) VPL has a unique k-SEVPA with respect to a given partition of Σ_{call}, the constructed machine may be much bigger than the smallest VPA recognizing the language because in a k-SEVPA, each module is constrained to have a unique "entry" (an entry is the destination of a push-transition). The presence of multiple entries can greatly reduce the size of the VPA as the following proposition shows.

Proposition 2. *For positive integers m, n, there is a family of well-matched VPLs $L_{m,n}$ such that the smallest VPA recognizing $L_{m,n}$ has at most $O(nm)$ states, while the smallest 1-SEVPA recognizing $L_{m,n}$ has at least n^m states.*

As the following theorem states, there is an efficient algorithm to minimize k-SEVPAs. The algorithm is omitted due to lack of space, but can be found along with the proof of correctness and complexity analysis in [11].

Theorem 4. *Given a k-SEVPA M with respect to a partition $\{\Sigma_{\text{call}}^j\}_{j=1}^k$ of Σ_{call} accepting a well-matched language L, the unique minimum-state k-SEVPA with respect to $\{\Sigma_{\text{call}}^j\}_j$ that accepts L can be computed in time $O(n^3)$, where n is the size M.*

5 Conclusions

We presented a characterization of VPLs in terms of congruences on strings of finite index and gave constructions of canonical automata recognizing visibly pushdown languages. We showed that while VPLs in general do not have unique minimum-state deterministic recognizers, the class of well-matched VPLs do have unique minimal k-module single-entry deterministic visibly pushdown automata (k-SEVPAs) for any fixed partition of the call symbols.

Our constructions of visibly pushdown automata based on congruences can, in general, result in automata with exponentially more states than a smallest deterministic visibly pushdown automaton recognizing the language. A characterization and construction of visibly pushdown automata that are at most polynomial in the size of the smallest automaton recognizing a language is an interesting open problem.

We presented a minimization algorithm for k-SEVPAs that runs in time $O(n^3)$. The computational complexity of the problem of constructing the smallest k-SEVPA given any visibly pushdown automaton (not necessarily k-module) is open, and would be interesting to investigate.

Acknowledgements. We would like to thank a referee for strengthening the lower bound for Proposition 2.

References

1. Alur, R., Madhusudan, P.: Visibly pushdown languages. In: Proceedings of STOC '04, ACM Press (2004) 202–211
2. Alur, R., Etessami, K., Madhusudan, P.: A temporal logic of nested calls and returns. In: Proceedings of TACAS '04. LNCS 2988, Springer (2004) 467–481
3. Ball, T., Rajamani, S.: Bebop: A symbolic model checker for boolean programs. In: SPIN 2000 Workshop on Model Checking of Software. LNCS 1885. Springer (2000) 113–130
4. Pitcher, C.: Visibly pushdown expression effects for XML stream processing. In: Programming Language Technologies for XML. (2005) 1–14
5. Murawski, A., Walukiewicz, I.: Third-order idealized algol with iteration is decidable. In: FOSSACS. 3441 (2005) 202–218
6. Löding, C., Madhusudan, P., Serre, O.: Visibly pushdown games. In: Proceedings of FSTTCS'04. LNCS (2004)

7. Nerode, A.: Linear automaton transformations. In: Proc. AMS. Volume 9. (1958) 541–544
8. Hopcroft, J.E., Ullman, J.D.: Introduction to Automata Theory, Languages and Computation. Addison Wesley (1979)
9. Hopcroft, J.E.: An $n \log n$ algorithm for minimizing the states in a finite automaton. In: The Theory of Machines and Computations. Acad. Press (1971) 189–196
10. Alur, R., Benedikt, M., Etessami, K., Godefroid, P., Reps, T., Yannkakis, M.: Analysis of recursive state machines. ACM Transactions on Programming Languages and Systems (to appear) (2005)
11. Alur, R., Kumar, V., Madhusudan, P., Viswanathan, M.: Congruences for visibly pushdown languages. Technical Report UIUCDCS-R-2005-2565, UIUC (2005)

Approximation Algorithms for Euclidean Group TSP

Khaled Elbassioni, Aleksei V. Fishkin, Nabil H. Mustafa, and René Sitters

Max-Planck-Institut für Informatik, Saarbrücken, Germany
{elbassio, avf, nmustafa, sitters}@mpi-sb.mpg.de

Abstract. In the Euclidean group *Traveling Salesman Problem* (TSP), we are given a set of points P in the plane and a set of m connected regions, each containing at least one point of P. We want to find a tour of minimum length that visits at least one point in each region. This unifies the TSP with Neighborhoods and the Group Steiner Tree problem. We give a $(9.1\alpha + 1)$-approximation algorithm for the case when the regions are disjoint α-fat objects with possibly varying size. This considerably improves the best results known, in this case, for both the group Steiner tree problem and the TSP with Neighborhoods problem. We also give the first $O(1)$-approximation algorithm for the problem with intersecting regions.

1 Introduction

A salesman wants to meet a set of potential buyers. Each buyer indicates a set of potential locations where he or she can meet the buyer. The salesman would like to minimize the total length of the tour required to meet all the potential buyers. How to construct such a tour? This problem is a generalization of the classical Traveling Salesman Problem (TSP), and hence is NP-hard. More formally, the problem we study can be stated as follows.

Euclidean Group TSP. Given a set P of points in the Euclidean plane, and m subsets $\{S_1, \ldots, S_m\}$ of P, one has to construct a tour on a set $P' \subseteq P$ such that P' contains at least one point from each subset (group) S_i. The objective is to minimize the length of the tour.

This model unifies two important separate bodies of research – the Group Steiner Tree problem, and the Euclidean TSP with Neighborhoods problem, which arise in VLSI design [RW90], and routing-related applications [Mit00]. We describe these problems in more detail.

Group Steiner Tree. Given a graph G on n vertices with non-negative weights on the edges, and m subsets of vertices, the group Steiner tree problem calls for computing a sub-tree of G which contains at least one vertex from each subset (group), and whose total length is minimized.

Slavik [Sla97] presented an $O(k)$-approximation algorithm for the Group Steiner Problem in the metric case, where k is the maximum group size. Using probabilistic tree embeddings together with randomized rounding, Garg

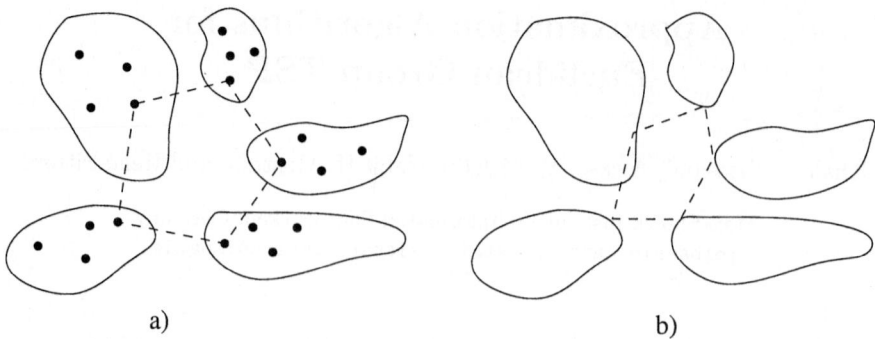

Fig. 1. a) a tour in Euclidean Group TSP b) a tour in Euclidean TSP with Neighborhoods

et al. [GKR00] obtained a randomized $O(\log^3 n \log m)$-approximation algorithm for the general case. On the negative side, it was observed (see e.g. [Sla96]) that the problem is a direct generalization of the set covering problem, and hence is NP-hard to approximate within a factor of $o(\log m)$.

A natural question is whether the above approximation ratio can be improved, if we consider restricted versions of the problem. For example, when the metric is Euclidean and each group is induced by a geometric object, such as a disk in the plane.

Euclidean TSP with Neighborhoods (TSPN). Given m connected geometric regions $S_1, \ldots, S_m$ in the Euclidean plane, find a minimum length tour that hits each region at least once.

The TSPN is a generalization of the classical Euclidean TSP which is known to be NP-hard [GGJ76, Pap77]. A 3/2-approximation algorithm, that works for any metric, was given by Christofides [Chr76]. Arora and Mitchell [Aro98, Mit99] independently obtained a $(1+\varepsilon)$- approximation algorithm.

The TSPN was first studied by Arkin and Hassin [AH94]. They presented constant-factor approximations for the case where the geometric regions are translates of disjoint convex polygons, and for disjoint unit disks. For the general problem, Mata and Mitchell [MM95], and later Gudmundsson and Levcopoulos [GL00], gave an $O(\log m)$-approximation algorithm. Dumitrescu and Mitchell [DM03] gave an $O(1)$-approximation algorithm for intersecting unit disks. For disjoint varying-sized convex fat regions de Berg et al. [dBGK+02] present a sophisticated algorithm with approximation ratio $12000\alpha^3$, where α is a measure of fatness of the regions. Their algorithm uses Slavik's algorithm [Sla97] for the Group Steiner tree problem, mentioned above. On the hardness side, Safra and Schwartz [SS03] showed that approximating Euclidean TSPN within $(2-\varepsilon)$ is NP-hard.

By viewing each geometric region S_i as an infinite set of points, the TSPN becomes a special case of the Euclidean Group TSP. In this context, it is natural to study the Euclidean Group TSP in which the point sets are separated by geometric objects, such as disks or more generally fat objects (see Figure 1).

Our Results. As mentioned above, our problem relates to both the Group Steiner Tree and the TSP with neighborhoods problems. Our algorithms improve and give new results in both areas.

In Section 3 we consider the problem with disjoint regions and give a $(9.1\alpha + 1)$-approximation algorithm for the problem where groups are enclosed by non-intersecting α-fat objects of arbitrary size. This improves the previous best result on the TSP with neighborhoods problem in several ways. First, we dramatically reduce the previous best approximation factor of $12,000\alpha^3$ [dBGK+02]. Second, our groups are arbitrary point sets $S_1, \ldots, S_m$, separated by fat objects in the Euclidean plane, whereas previous results only deal with the (continuous) case where each set S_i is the (infinite) set of points belonging to an object. We also do not require the objects to be convex. Furthermore, our algorithm yields an $O(\alpha/\sqrt{d})$-approximate solution for any dimension d. In contrast, it was shown in [SS03] that TSP with neighborhoods in $\mathbb{R}^3$ is unlikely to be approximable within $O(\log^{1/2} n)$, and thus, for $d \geq 3$, it is unlikely that there is a approximation factor independent of both n and α.

In Section 4 we consider the variant of the problem in which the instances are defined by m sets of points in the Euclidean plane and a set of m disks. Disks may intersect and each disk defines a group consisting of the enclosed points. Clearly, the intersection between disks admits much more complicated systems of subsets than what we can get from the non-intersecting sets. We present an $O(1)$-approximation algorithm for this problem.

2 Preliminaries

We consider instances of the Euclidean Group TSP in which the sets S_i are contained in geometric regions in the plane. Formally, we define a *region* as a connected and closed subset of the Euclidean plane $\mathbb{R}^2$. An instance of our problem is given by a finite set of points P and m subsets $S_1, \ldots, S_m$ of P. The subsets have the property that there exist regions $O_1, O_2, \ldots, O_m$ such that $S_i = P \cap O_i$. A solution is given by a tour (spanning tree) on a subset P' of P such that $P' \cap S_i \neq \emptyset$. The objective is to minimize the length of the tour. We denote an optimal tour (spanning tree) by OPT, and its length by |OPT|.

In this paper we restrict to so called *fat* regions. The definition of fatness we use here was introduced by Van der Stappen [Sta94] and used by De Berg et al. [dBGK+02] in their paper on the TSP with neighborhoods problem.

Definition 1. *An object $O \subseteq \mathbb{R}^2$ is said to be α-fat if for any disk Θ which does not fully contain O and whose center lies in O, the area of the intersection of O and Θ is at least $1/\alpha$ times the area of Θ.*

Notice for example that the plane $\mathbb{R}^2$ has fatness 1, a halfspace has fatness 2 and a disk has fatness 4. We define the *size* of an object as the diameter of it's smallest enclosing disk.

Lemma 1 (Packing Lemma). *The length of the shortest path connecting c disjoint α-fat objects in $\mathbb{R}^2$ is at least $(c/\alpha - 1)\pi S/4$, where S is the size of the smallest object.*

Proof. Consider a path T that connects the c objects and let the center of a disk with diameter S follow this path. At the point where the path touches a certain object, the disk intersects its boundary and hence at least an $1/\alpha$ fraction of the disk at that point intersects the object. The total area covered by the moving disk must be at least c/α times the area $\pi S^2/4$ of the disk. On the other hand, it is easy to see that the total area covered by a disk that follows a continuous path T in $\mathbb{R}^2$ is at most $\pi S^2/4 + S|T|$. Combining the upper and lower bound on the area we get,

$$c\pi S^2/(4\alpha) \leq \pi S^2/4 + S|T| \Rightarrow |T| \geq (c/\alpha - 1)\pi S/4.$$
□

3 Varying-Sized Objects

In this section, we consider the case where the object $O_1, \ldots, O_m$ are pairwise disjoint. Further, we assume that all objects have fatness at most α. However, we do not put any restriction on the size of the objects, i.e., we do not assume that objects have comparable sizes. The problem reduces to geometric TSP when each set S_i consists of a single point.

There exists a simple $(m-1)$-approximation algorithm that we denote by GREEDY. We define the *distance* between a point p and a set X as $d(p, X) = \min_{x \in X} d(p, x)$.

Algorithm GREEDY:
(1) Pick the points $p_i \in S_i$ ($i = 1 \ldots m$) that minimize $\sum_{j=2}^m d(p_1, p_j)$.
(2) For all $j \geq 2$, select twice the edge (p_1, p_j) and construct a tour by short cutting the edges.

Lemma 2. *Algorithm GREEDY gives an $(m-1)$-approximate solution for Group TSP.*

Proof. Any TSP-tour contains two edge disjoint paths from S_1 to S_i for all $i \in \{2, \ldots, m\}$. Therefore, $(m-1)|\text{OPT}| \geq 2\sum_{i=2}^m d(p_1, p_i)$, which is at most the length of the tour constructed by the algorithm. □

By δ_i we denote the *diameter* of the point set S_i, i.e. the largest distance between any two points in S_i. Notice that δ_i is at most the size of its enclosing object. The following is an immediate corollary of Lemma 1.

Corollary 1. *The length of the shortest path connecting c of the given sets is at least $(c/\alpha - 1)\pi\delta/4$, where δ is the minimum of the diameters of the sets.*

Algorithm $\mathcal{A}$:
(1) Order the point sets by their diameter $\delta_1 \leq \delta_2 \leq \ldots \leq \delta_m$. Pick any p_1 in S_1. For $i = 2$ up to m pick the point p_i in S_i that minimizes $d(p_i, \{p_1, \ldots, p_{i-1}\})$, i.e. pick the point that is closest to the already chosen points.
(2) Construct a $(1+\epsilon)$-approximate TSP tour T on this set of m points.
(3) Output the minimum of T and the tour constructed by algorithm GREEDY.

The second step can be done efficiently for any $\epsilon > 0$ using techniques from [Aro98] and [Mit99].

Theorem 1. *Algorithm $\mathcal{A}$ gives a $(1+\epsilon)(9.1\alpha+1)$-approximate solution for the group TSP with non-intersecting α-fat neighborhoods.*

Proof. We assume $m - 1 > 9.1\alpha + 1$ since we can use GREEDY for smaller values of m. Denote the set of points chosen by $\mathcal{A}$ as $P' = \{p_1, \ldots, p_m\}$. Let $p_i^* \in \{p_1, \ldots, p_{i-1}\}$ be the point at minimum distance from p_i and denote the distance by x_i.

Consider some optimal solution OPT and fix an orientation of this tour. We choose some number $c \in \{1, \ldots, m\}$ and define T_i as the part of this directed tour that connects exactly c sets and starts from the point in S_i. Let t_i be the length of path T_i.

We choose $c = \lceil \alpha(4/\pi + 1) \rceil$. Notice that $c \in \{1, \ldots, m\}$ is satisfied by the assumption in the first line of the proof. (By choosing c a bit smaller we can get a marginal improvement of the approximation ratio. Since this would make the proof more complicated we omit this here.) Consider some $i \in \{1, \ldots, m\}$ and let $S_{h(i)}$ be a set with smallest diameter among those from the c sets on the path T_i. Then, by Corollary 1 and the choice of c we have

$$t_i \geq (c/\alpha - 1)\pi\delta_{h(i)}/4 \geq \delta_{h(i)}. \qquad (1)$$

Since S_i is on this path T_i and we ordered the sets by their diameter we have $1 \leq h(i) \leq i$. We distinguish two cases.

If $h_i = i$, meaning that S_i has smallest diameter, then by (1) we have

$$t_i \geq \delta_i. \qquad (2)$$

Otherwise, if $h(i) < i$, then we argue as follows. Since the algorithm picked point p_i we know that the distance from any point in S_i to the point $p_{h(i)}$ (which is chosen before p_i) is at least x_i. Hence, the distance from any point in S_i to any point in $S_{h(i)}$ is at least $x_i - \delta_{h(i)}$, implying $t_i \geq x_i - \delta_{h(i)}$. Together with (1) this yields

$$t_i \geq \max\{\delta_{h(i)}, x_i - \delta_{h(i)}\} \geq x_i/2. \qquad (3)$$

We will construct a tour on the set of points $P' = \{p_1, \ldots, p_m\}$ chosen by the algorithm, using the bounds (2) and (3). Let H be the set of indices i for which $t_i \geq \delta_i$ and let OPT$_H$ be the smallest tour through the points $\{p_i | i \in H\}$. Clearly,

$$|\text{OPT}_H| \leq |\text{OPT}| + 2\sum_{i \in H} \delta_i \leq |\text{OPT}| + 2\sum_{i \in H} t_i.$$

Let $\bar{H} = \{1, \ldots, m\} \setminus H$. Then, by (3) we know that for any $i \in \bar{H}$, the length of the edge (p_i, p_i^*) equals $x_i \leq 2t_i$. We add this edge twice for any $i \in \bar{H}$ to the tour OPT$_H$. Clearly, the resulting graph is Eulerian. Moreover, it is connected since $p_1 \in H$, and for any i we have $p_i^* = p_j$ for some $j < i$. The total length of the Eulerian graph is

$$|\text{OPT}_H| + 2\sum_{i \in \bar{H}} d(p_i, p_i^*) \leq |\text{OPT}| + 2\sum_{i \in H} t_i + 2\sum_{i \in \bar{H}} 2t_i \leq |\text{OPT}| + 4\sum_{i=1}^{m} t_i.$$

When we take the sum over all t_i then every edge is counted $c-1$ times, implying $(c-1)|\text{OPT}| = \sum_{i=1}^{m} t_i$. Substituting the value of c we conclude that the tour given by our algorithm has length at most

$$(1+\epsilon)(1 + 4(c-1))|\text{OPT}| < (1+\epsilon)(9.093\alpha + 1)|\text{OPT}|.$$

□

The algorithm and proof apply directly to the Euclidean TSP with Neighborhoods problem under the weak assumption that, given the points $\{p_1, \ldots, p_{i-1}\}$, we can efficiently find the point p_i in the infinite set of points S_i that minimizes $d(p_i, \{p_1, \ldots, p_{i-1}\})$.

Corollary 2. *Algorithm $\mathcal{A}$ gives a $(1+\epsilon)(9.1\alpha + 1)$-approximate solution for the TSP with neighborhoods problem.*

3.1 Higher Dimensions

The generalization of the definitions and lemma's of the previous section to higher dimensions is straightforward.

Definition 2. *An object $O \subseteq \mathbb{R}^d$ is said to be α-fat if for any d-dimensional sphere D which does not fully contain O and whose center lies in O, the volume of the intersection of O and D is at least $1/\alpha$ times the volume of D.*

We denote the volume of a d-dimensional sphere with radius r by $V_d(r)$.

Lemma 3. *If the center of a d-dimensional sphere with radius r follows a path T in $\mathbb{R}^d$, then the volume covered by the sphere is at most $|T|V_{d-1}(r) + V_d(r)$.*

Lemma 4. *The length of the shortest path connecting c disjoint α-fat objects is at least $(c/\alpha - 1)V_d(r)/V_{d-1}(r)$, where r is half the size of the smallest object.*

The volume of a d-dimensional sphere with radius r is

$$V_d(r) = \frac{\pi^{d/2} r^d}{\Gamma(\frac{d+2}{2})},$$

where Γ is the well-known gamma function. For $d \geq 3$ we get

$$\frac{V_d(r)}{V_{d-1}(r)} = r\sqrt{\pi}\frac{\Gamma((d+2)/2)}{\Gamma((d+1)/2)} > r\sqrt{\pi d/2}.$$

For small values of K we can simply get a K-approximation as described in the previous section. If we choose c such that $(c/\alpha - 1)\sqrt{\pi d/2} = 1$, then the proof of Theorem 1 applies here without any adjustment.

Theorem 2. *Algorithm $\mathcal{A}$ is an $O(\alpha/\sqrt{d})$-approximation algorithm for the TSP in $\mathbb{R}^d$ with non-intersecting α-fat neighborhoods.*

Notice that the approximation factor decreases in the dimension for constant α. However, for bounded objects, α grows exponentially in d. For example, $\alpha = 2^d$ for a d-dimensional sphere.

Safra and Schwartz [SS03] showed that TSP with neighborhoods in $\mathbb{R}^3$ is unlikely to be approximable within $O(\log^{1/2} n)$. Hence, there is little hope to improve our result for $d \geq 3$ to a ratio independent of the fatness α.

4 Intersecting Objects

In this section, we consider the case when the objects defining the sets $S_1, \ldots, S_m$ are intersecting disks of the same radius r. We denote these disks by $\mathcal{D} = \{D_1, \ldots, D_m\}$, and their centers by c_i. Then $S_i = P \cap D_i$, and assume that $P = S_1 \cup \ldots \cup S_m$.

A subset $P' \subseteq P$ is called a *hitting pointset* for $\mathcal{D}$ if $P' \cap S_i \neq \emptyset$ for $i = 1, \ldots, m$ and a *minimal hitting pointset* if for every $x \in P'$ there exists an $i \in [m]$ such that $(P' \setminus \{x\}) \cap S_i = \emptyset$. A minimal hitting set can be found by the natural greedy algorithm: Set $P' = P$, and keep deleting points from P' as long as it is still a hitting set. A (square) box B is called *a covering box* for the set of disks $\mathcal{D}$ if B contains a hitting pointset for $\mathcal{D}$, and a *minimum covering box* if it has the smallest size amongst all such covering boxes. Since a minimum covering box is determined by two points of P on its boundary, there are only $O(n^2)$ such candidates. By enumerating over all such boxes, and verifying if they contain a hitting set, one can compute a minimum covering box.
Consider the following algorithm for the Group TSP problem on the sets $S_1, \ldots, S_m$. Note that the last step can be done efficiently for any $\epsilon > 0$ using techniques from [Aro98] and [Mit99].

Algorithm $\mathcal{B}$:
(1) Compute a minimum covering box B of $\mathcal{D}$.
(2) Find a minimal hitting pointset $P' \subseteq P$ for $\mathcal{D}$ inside B.
(3) Compute a $(1 + \epsilon)$-approximate TSP tour on P'.

To analyze the performance of the algorithm, we need the following lemma.

Lemma 5. *Let B be a box of diameter L that contains $P = \{p_1, \ldots, p_n\}$. Let $\mathcal{D} = \{D_1, \ldots, D_n\}$ be a collection of disks of radius r, such that (i) each point $p \in P$ is contained in exactly one disk $D(p) \in \mathcal{D}$ (ii) each disk D contains exactly one point of P. Then there exists a tour T on P with length at most $f(L/r)L$, where $f(L/r)$ is defined in (5). In particular, $f(6) \leq 113$.*

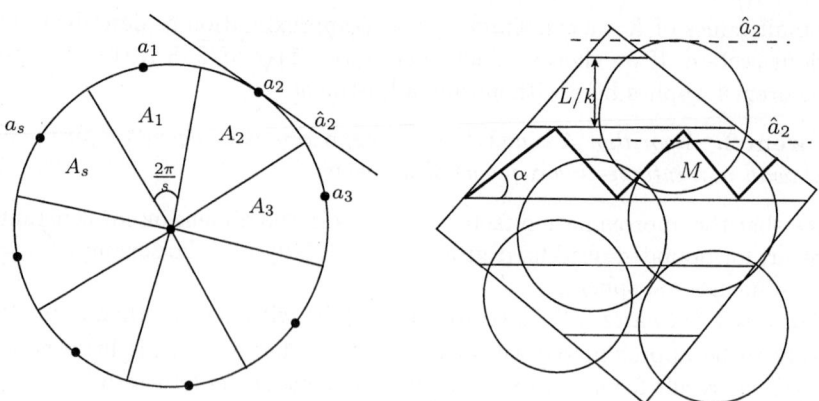

Fig. 2. (a) Partitioning disks into cones. (b) Partitioning B into k strips in direction $\hat{a}_2$. The longest path is M (bold) if each angle is at most α

Proof. Fix constants $s, k > 0$. Denote the center of disk D_i by c_i. Set s equally-spaced identical 'direction' points on each circle, say $a_1, \ldots, a_s$. Partition each disk D_i into s identical cones $A_1, \ldots, A_s$ where $A_j(D_i) = \{q \in D_i : \pi/s \le \angle c_i q \, c_i a_j \le \pi/s\}$. Denote by $\hat{a}_j$ the direction of the tangent to the circles at point a_j. See Figure 2(a).

Now partition P into s subsets $P_1, \ldots, P_s$, where P_j contains all the points $p \in P$ that lie in the cone $A_j(D(p))$ of the disk containing p. We will first construct a path on all the points in each P_j separately, and then connect these paths together.

Fix any set P_j. Partition B into k strips of equal width (each of width at most L/k), in the direction $\hat{a}_j$. W.l.o.g., assume $\hat{a}_j$ is horizontal (one can always rotate everything to get this). The situation is shown in Figure 2 (b). Let P_j^i be the set of points belonging to the i-th strip. Assume P_j^i is sorted along the x-coordinates. Now construct a path M by connecting all the points in P_j^i in this linear sequence.

Claim. The path M constructed above has length at most $L/\cos\alpha$, where

$$\alpha \le \max\{ \sin^{-1}\frac{L}{rk}, \frac{\pi}{2s} + \frac{\pi}{4} - \frac{1}{2}\sin^{-1}(\cos\frac{\pi}{s} - \frac{L}{rk}) \} \qquad (4)$$

Proof. Let $P_j^i = \langle r_1, \ldots, r_m \rangle$. We first bound the angle that each $r_l r_{l+1}$ and $r_l r_{l-1}$ makes with the horizontal line passing through r_l. The case of $r_l r_{l-1}$ is symmetric, so we only consider the first case. One can assume that r_{l+1} is below r_l (otherwise, consider the angle made by $r_{l+1} r_l$ with line passing through r_{l+1}; both angles are the same and now r_l lies below r_{l+1}).

We would like to place points r_l and r_{l+1} such that the angle of the edge $r_l r_{l+1}$ with the horizontal is maximized. By assumption, r_l and r_{l+1} are contained in two disks of radius r containing only their respective points. It is not hard to see that there are two possible worse-case choices, as shown in Figure 3. The first

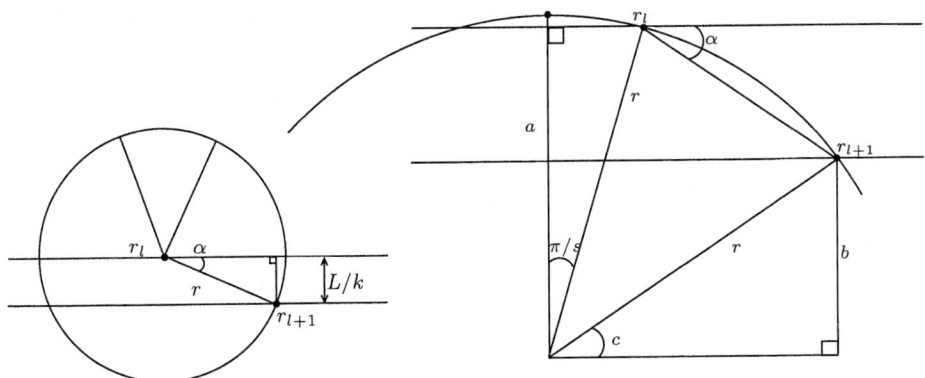

Fig. 3. The two cases for bounding α

case is when r_l is the center of its enclosing disk $D(r_l)$, and r_{l+1} lies (almost) on the boundary of $D(r_l)$, see Figure 3(a). Clearly, the required angle $\alpha \leq \sin^{-1}\frac{L}{rk}$. The second case is shown in Figure 3(b). By elementary geometry, we get

$$a = r\cos\frac{\pi}{s}, \quad b = a - \frac{L}{k}, \quad \sin c = \frac{b}{r}$$

from which one can derive that

$$\alpha \leq \frac{\pi}{2s} + \frac{\pi}{4} - \frac{1}{2}\sin^{-1}\frac{b}{r}.$$

Therefore, the maximum angle that any edge $r_l r_{l+1}$ makes with the horizontal is given by Equation 4. Finally, the weight of M is maximized if each edge makes the maximum angle (at most α) with the horizontal, as shown in Figure 2 (b). Then $M \leq L/\cos\alpha$, as required. □

The above claim bounds the weight of the spanning path (constructed by simply sorting the points by x-coordinates and connecting them in this order) of points lying in cone A_j and the i-th strip. There are at most sk such paths, and they can be connected together into a tour T at the cost of additional total length of $sk \cdot L$. Therefore the constructed tour has a total length of at most $L \cdot f(L/r)$, where

$$f(\frac{L}{r}) = \min_{s,k}\{sk(1 + \sec\alpha(s, k, L/r))\}, \tag{5}$$

and where $\alpha(s, k, L/r) = \max\{\sin^{-1}\frac{L}{rk}, \frac{\pi}{2s} + \frac{\pi}{4} - \frac{1}{2}\sin^{-1}(\cos\frac{\pi}{s} - \frac{L}{rk})\}$. In particular, with $L/r = 6$, and setting $k = 6.5$ and $s = 4.5$, the tour has length at most $113L$. □

Theorem 3. *Algorithm $\mathcal{B}$ is a $O(1)$-approximation algorithm for the Group TSP problem.*

Proof. Since P' is a hitting set for $\mathcal{D}$, it is enough to show that there exists a tour T on the set P' whose total cost is within $O(1)$ of the optimum for $\mathcal{D}$ [1]. To P' we can associate a subset of the disks $\mathcal{D}' \subseteq \mathcal{D}$ with the property that $|P' \cap D| = 1$ for all $D \in \mathcal{D}'$ and $|\{D \in \mathcal{D}' : x \in D\}| = 1$ for all $x \in P'$. The set $\mathcal{D}'$ can be found as follows. By the minimality of P', for every point $x \in P'$ there exists a disk $D(x) \in \mathcal{D}$ such that $D \cap P' = \{x\}$. Let $\mathcal{D}' = \{D(x) : x \in P'\}$.

For a disk $D \in \mathcal{D}'$, let us denote by $x(D)$ the (unique) point of P' contained inside D. Let $\mathcal{I} \subseteq \mathcal{D}'$ be a *maximal independent set* of disks in $\mathcal{D}'$, i.e. a maximal collection of pairwise disjoint disks. If every maximal independent set in $\mathcal{D}'$ has size at most 2, we assume that $\mathcal{I}$ consists of two disks in $\mathcal{D}'$, the distance δ between which is the largest. Let $\text{OPT}_\mathcal{I}$ be an optimal Group TSP tour on $\mathcal{I}$, when we are allowed to use any of the points lying inside the disks of $\mathcal{I}$, and let $\text{OPT}'_\mathcal{I}$ an optimal TSP tour on the set of points $\{x(I) : I \in \mathcal{I}\}$. Clearly, $|\text{OPT}_\mathcal{I}| \leq |\text{OPT}|$. We consider three cases:

Case 1: $|\mathcal{I}| \geq 3$: By the maximality of $\mathcal{I}$, every disk in $\mathcal{D}' \setminus \mathcal{I}$ must intersect some disk in $\mathcal{I}$. Let $\{\mathcal{D}_I \subseteq \mathcal{D}' \setminus \mathcal{I} : I \in \mathcal{I}\}$ be a partition of $\mathcal{D}' \setminus \mathcal{I}$, such that, for $I \in \mathcal{I}$, $\mathcal{D}_I$ contains only disks intersecting I. For $I \in \mathcal{I}$, let OPT_I be an optimal TSP tour on the pointset $S_I = \{x(D) : D \in \mathcal{D}_I\} \cup \{x(I)\}$ contained in the partition of disks intersecting I. To define the tour T, take the union T' of $\text{OPT}_\mathcal{I}$ and $\bigcup_{I \in \mathcal{I}} \text{OPT}_I$ to obtain an Eulerian connected graph T' on the pointset P'. Finally, we use short-cutting on an Eulerian tour in T' to get a TSP tour T on P' (see Figure 4).

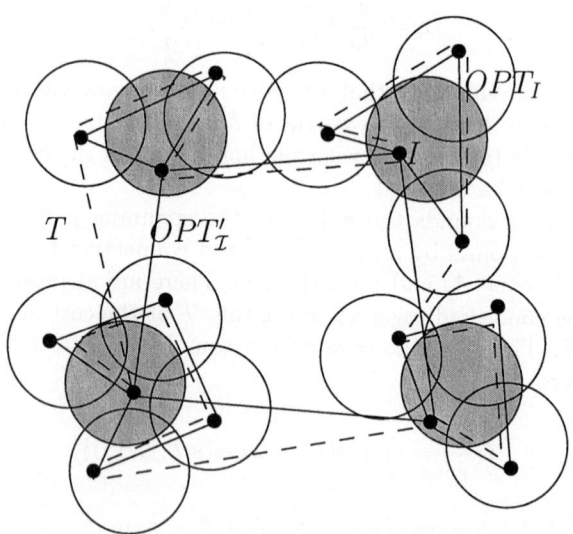

Fig. 4. Constructing an approximate tour as in the proof of Theorem 3

---

[1] The constants in the approximation factor can be improved by a more complicated analysis, which we omit from this extended abstract.

Claim. $|T| \leq (1 + (4 + 6f(6))/0.14)|\text{OPT}|$, where $f(\cdot)$ is defined by (5).

Proof. Note that, for $I \in \mathcal{I}$, all the points of S_I lie inside a box of diameter at most $6r$. Thus applying Lemma 5, we conclude that there is a tour on S_I of size at most $f(6) \cdot 6r$. In particular, $|\text{OPT}_I| \leq 6f(6) \cdot r$. On the other hand, by Lemma 1, and using $\alpha = 4$ for disks, one can derive that $|\text{OPT}_\mathcal{I}| \geq 0.14r|\mathcal{I}|$. Furthermore, by connecting by a double edge, for each disk $I \in \mathcal{I}$, the point picked by the optimal solution $\text{OPT}_\mathcal{I}$ to the point $x(I)$, we can construct a TSP tour on the set $\{x(I) \ ; I \in \mathcal{I}\}$ of total length at most $\text{OPT}_\mathcal{I} + 4r|\mathcal{I}|$. Thus it follows that $|\text{OPT}_{\mathcal{I}'}| \leq |\text{OPT}_\mathcal{I}| + 4r|\mathcal{I}|$. Combining these together, we get

$$|T| \leq |\text{OPT}'_\mathcal{I}| + \sum_{I \in \mathcal{I}} |\text{OPT}_I| \leq |\text{OPT}_\mathcal{I}| + (4 + 6f(6))r|\mathcal{I}|$$

$$\leq (1 + \frac{4 + 6f(6)}{0.14})|\text{OPT}_\mathcal{I}| \leq (1 + \frac{4 + 6f(6)}{0.14})|\text{OPT}|,$$

and our claim follows. □

Case 2: $|\mathcal{I}| = 2$ and $\delta > 2r$: Note that δ is a lower bound on $|\text{OPT}_\mathcal{I}|$. Then $r|\mathcal{I}| \leq |\text{OPT}_\mathcal{I}|$ and, similar to Case 1, we can construct a tour T of length at most $(1 + (4 + 6f(6))/0.14)|\text{OPT}|$.

Case 3: $|\mathcal{I}| = 2$ and $\delta \leq 2r$, or $|I| = 1$: Let L be the size of the minimum covering box B. Note that $\text{OPT} \geq L$ because (i) any box containing OPT is a covering box, and (ii) the smallest such box has width or height at most $|\text{OPT}|$. It is easy to see, in this case, that all the points of P' lie inside a box of size at most $6r$. In particular, this implies that $L \leq 6r$. By Lemma 5, we can construct a tour T on the pointset P' whose total length does not exceed $f(6) \cdot L \leq f(6)|\text{OPT}|$. □

References

[AH94]   E. M. Arkin and R. Hassin. Approximation algorithms for the geometric covering salesman problem. *Discrete Applied Mathematics*, 55(3):197–218, 1994.

[Aro98]   S. Arora. Nearly linear time approximation schemes for euclidean TSP and other geometric problems. *J. ACM*, 45(5):1–30, 1998.

[Chr76]   N. Christofides. Worst-case analysis of a new heuristic for the traveling salesman problem. Technical report, GSIA, Carnegie-Mellon University, 1976.

[dBGK+02]   M. de Berg, J. Gudmundsson, M.J. Katz, C. Levcopoulos, M.H. Overmars, and A. F. van der Stappen. TSP with Neighborhoods of varying size. In *Proceedings 10th Annual European Symposium on algorithms (ESA)*, pages 187–199, 2002.

[DM03]   A. Dumitrescu and J.S.B. Mitchell. Approximation algorithms for TSP with neighborhoods in the plane. *J. Algorithms*, 48(1):135–159, 2003.

[GGJ76]   M. R. Garey, R. L. Graham, and D. S. Johnson. Some NP-complete geometric problems. In *Proceedings 8th Annual ACM Symposium on the Theory of Computing (STOC)*, 1976.

[GKR00]   N. Garg, G. Konjevod, and R. Ravi. A polylogarithmic approximation algorithm for the Group Steiner Tree Problem. *J. Algorithms*, 37(1):66–84, 2000.

[GL00]   J. Gudmundsson and C. Levcopoulos. Hardness result for TSP with neighborhoods, 2000. Technical Report LU-CS-TR:2000-216, Department of Computer Science, Lund University, Sweden.

[Mit99]   J.S.B. Mitchell. Guillotine subdivions approximate polygonal subdivisons: A simple polynomial-time approximation scheme for geometric TSP, k-MST and related problems. *SICOMP*, 28(4):1298–1309, 1999.

[Mit00]   J.S.B. Mitchel. *Handbook of Computational Geometry*, chapter Geometric shortest paths and network optimization, pages 633–701. Elsevier, North-Holland, Amsterdam, 2000.

[MM95]   C.S. Mata and J.S.B. Mitchell. Approximation algorithms for geometric tour and network design problems (extended abstract). In *Proceedings 11th Ann. ACM Symposium on Computational Geometry*, pages 360–369, 1995.

[Pap77]   C. H. Papadimitriou. The Euclidean traveling salesman problem is NP-complete. *Theoretical Computer Science*, 4(3):237–244, 1977.

[RW90]   G. Reich and P. Widmayer. Beyond steiner's problem: a VLSI oriented generalization. In *WG '89: Proceedings of the fifteenth international workshop on Graph-theoretic concepts in computer science*, pages 196–210. Springer-Verlag New York, Inc., 1990.

[Sla96]   Petr Slavik. A tight analysis of the greedy algorithm for set cover. In *STOC '96: Proceedings of the twenty-eighth annual ACM symposium on Theory of computing*, pages 435–441, 1996.

[Sla97]   P. Slavik. The errand scheduling problem. Technical report, March 14 1997. Technical Report, SUNY, Buffalo, USA.

[SS03]   S. Safra and O. Schwartz. On the complexity of approximating TSP with Neighborhoods and related problems. In *Proceedings 11th Annual European Symposium on algorithms (ESA)*, volume 2832 of *Lecture Notes in Computer Science*, pages 446–458. Springer, 2003.

[Sta94]   A. F. van der Stappen. *Motion Planning amidst Fat Obstacles*. Ph.d. dissertation, Utrecht University, Utrecht, Netherlands, 1994.

Influential Nodes in a Diffusion Model for Social Networks

David Kempe[1,*], Jon Kleinberg[2,**], and Éva Tardos[2,***]

[1] Department of Computer Science,
University of Southern California
dkempe@usc.edu
[2] Department of Computer Science, Cornell University
kleinber@cs.cornell.edu, eva@cs.cornell.edu

Abstract. We study the problem of maximizing the expected spread of an innovation or behavior within a social network, in the presence of "word-of-mouth" referral. Our work builds on the observation that individuals' decisions to purchase a product or adopt an innovation are strongly influenced by recommendations from their friends and acquaintances. Understanding and leveraging this influence may thus lead to a much larger spread of the innovation than the traditional view of marketing to individuals in isolation.

In this paper, we define a natural and general model of influence propagation that we term the *decreasing cascade model*, generalizing models used in the sociology and economics communities. In this model, as in related ones, a behavior spreads in a cascading fashion according to a probabilistic rule, beginning with a set of initially "active" nodes. We study the *target set selection* problem: we wish to choose a set of individuals to target for initial activation, such that the cascade beginning with this active set is as large as possible in expectation. We show that in the decreasing cascade model, a natural greedy algorithm is a $1 - 1/e - \varepsilon$ approximation for selecting a target set of size k.

1 Introduction

Suppose that we are trying to market a product, or promote an idea, innovation or behavior, within a population of individuals. In order to do so, we can "target" individuals; for instance, this "targeting" could take the form of offering free samples of the product, demonstrating an innovation, or explaining an idea (such as the consequences of drug use to teenagers). An important question is

---

[*] This research was supported by an Intel Graduate Fellowship and an NSF Graduate Research Fellowship.
[**] Supported in part by a David and Lucile Packard Foundation Fellowship and NSF grants 0311333 and 0329064.
[***] Supported in part by NSF ITR grant CCR-0325453, NSF grant CCR-0311333, and ONR grant N00014-98-1-0589.

then whom we should target. Clearly, if there were no interaction between the individuals, this would be straightforward: the effect on each targeted individual could be determined in isolation, and we could choose the set of individuals with largest (expected) revenue or reach. However, individuals do not exist in a vacuum; rather, they form complex social networks based on a multitude of different relations and interactions. By virtue of these interactions, they influence each other's decisions in adopting a product or behavior.

Research in the area of *viral marketing* [1, 2, 3, 4, 5] takes advantage of these social network effects, based on the premise that targeting a few key individuals may lead to strong "word-of-mouth" effects, wherein friends recommend a product to their friends, who in turn recommend it to others, and so forth, creating a cascade of recommendations. In this way, decisions can spread through the network from a small set of initial adopters to a potentially much larger group. Given a probabilistic model for the way in which individuals influence one another, the *influence maximization problem* consists in determining a set A of k individuals yielding the largest expected cascade.

The influence maximization problem has been proposed and studied by Domingos and Richardson [2, 5], who gave heuristics for the problem in a very general descriptive model of influence propagation. In recent work [6], we obtained provable performance guarantees for approximation algorithms in several simple, concrete, but extensively studied models from mathematical sociology (see, e.g., [7, 8, 9] for comprehensive introductions to this area).

In this paper, we show that the influence maximization problem can be approximated in a very general model that we term the *decreasing cascade model*. The analysis techniques from our earlier work [6] rely on the concrete forms of influence used in that paper, and we show that they cannot be applied to the general model considered here. We therefore develop a more general framework, which we believe will be of interest in its own right, for reasoning about dynamic processes in network models such as these.

1.1 The Decreasing Cascade Model

Throughout this paper, we call individuals (nodes) *active* if they have adopted the product, and *inactive* otherwise. We assume that once a node becomes active, it will remain so forever (see [6] for a discussion on how this assumption can be lifted). We focus on *cascade models* that capture the dynamics of recommendations in a step-by-step fashion: when a node u first becomes active, say at time t, it is considered *contagious*. It has one chance of influencing each previously inactive neighbor v. A successful attempt will cause v to become active in the next time step $t + 1$. If multiple neighbors of v become active at time t, then their activation attempts are sequenced in an arbitrary order, but we assume that they all happen within time step t. After a node u has made all its attempts at influencing other nodes, it remains active, but is now *non-contagious*. The process terminates when there are no more contagious nodes.

In order to fully describe the model, we need to specify the probability of success for node u's attempt at activating v. In the simplest *independent cascade* model [3], this probability is a constant $p_v(u)$, independent of the history of the process. In general, however, v's propensity for being activated may change as a function of which of its neighbors have already attempted (and failed) to influence it; if S denotes the set of v's neighbors that have already attempted to influence v, then u's *success probability* is denoted by $p_v(u, S)$. For this model to be well-defined, we also need to assume *order-independence*: if all nodes from a set T try to influence v, then the order in which their attempts are made does not affect the probability of v being active in the end. Formally, if $u_1, \ldots, u_r$, and $u'_1, \ldots, u'_r$ are two permutations of T, and $T_i = \{u_1, \ldots, u_{i-1}\}$ as well as $T'_i = \{u'_1, \ldots, u'_{i-1}\}$, then order-independence means that

$$\prod_{i=1}^{r}(1 - p_v(u_i, S \cup T_i)) = \prod_{i=1}^{r}(1 - p_v(u'_i, S \cup T'_i))$$

for all sets S disjoint from T.

From the point of view of influence maximization, we start by *targeting* a set A of individuals for activation at time 1, making them contagious. Afterwards, the process unfolds as described above, until there are no more contagious nodes; we say that the process *quiesces*. Note that this happens after at most $n + 1$ rounds. At that point, we have some set $\varphi(A)$ of active nodes, which is is a random variable. The goal is to choose A so as to maximize the expected size $\sigma(A) := \mathrm{E}[|\varphi(A)|]$ of this final set of active nodes. Due to the computational difficulty of this goal (see the discussion below), we will consider approximation algorithms: for a constant c, we wish to choose a set A for which $\sigma(A)$ is at least $\frac{1}{c}$ times as large as $\sigma(A^*)$ for *any* set A^* of k nodes. The quantity c is thus the approximation guarantee of the algorithm.

The *order-independent cascade model* is very general — it specifies how each node influences each other node, and how the influence is "attenuated" by previous interactions a node has had. It is also equivalent in a precise sense to a generalization of Granovetter's threshold model [10] for social networks (see Section 3).

In general, it is NP-hard to approximately maximize the size $\sigma(A)$ of the final active set to within $n^{1-\varepsilon}$, for any $\varepsilon > 0$. The inapproximability follows from a straightforward reduction, e.g., from VERTEXCOVER, and can already be shown in the case of a *hard threshold* model [11, 12, 13], where a node v is activated if at least a fixed fraction (say, $1/2$) of its neighbors are active; this corresponds to $p_v(u, S)$ being 0 if S contains fewer than half of v's neighbors, and 1 otherwise.

Thus, we study here a natural restriction that we term the *decreasing cascade model*. In the decreasing cascade model, the functions $p_v(u, S)$ are non-increasing in S, i.e., $p_v(u, S) \geq p_v(u, T)$ whenever $S \subseteq T$. Intuitively, this restriction states that a contagious node's probability of activating some $v \in V$ decreases if more nodes have already attempted to activate v, and v is hence more "marketing-saturated". The decreasing cascade model contains the *independent cascade model* [3] as a special case, and even for the independent cascade model, maximizing $\sigma(A)$ is NP-hard [6]; in fact, the proof in [6] shows that it is NP-hard to approximate within $1 - 1/e + \varepsilon$ for any $\varepsilon > 0$.

2   An Approximation Algorithm

In this paper, we analyze the following simple greedy algorithm (Algorithm 1.) for influence maximization. The approximation guarantee for this algorithm is the main theorem of this paper:

Algorithm 1. Greedy Approximation Algorithm

1: Start with $A = \emptyset$
2: **for** $i = 1$ to k **do**
3:   Let v_i be a node (approximately) maximizing the *marginal gain* $\sigma(A \cup \{v\}) - \sigma(A)$.
4:   Set $A \leftarrow A \cup \{v_i\}$.
5: **end for**

Theorem 1. *Let A^* be the the set maximizing $\sigma(\cdot)$ among all sets of k nodes.*

1. *If the optimal v_i is chosen in each iteration, then the greedy algorithm is a $(1 - 1/e)$-approximation, i.e., the set A found by the algorithm satisfies $\sigma(A) \geq (1 - 1/e) \cdot \sigma(A^*)$.*
2. *If the node v_i is a $1 - \varepsilon$ approximate best node in each iteration, then the greedy algorithm is a $(1 - 1/e - \varepsilon')$-approximation, where ε' depends on ε polynomially.*

Before proceeding with the proof of Theorem 1, a few words are in order about determining the node v_i in the **for** loop of the algorithm. Even in the simple independent cascade model, it is not clear how to evaluate $\sigma(A)$ exactly, or whether this can be done in polynomial time; in fact, we consider the question of evaluating $\sigma(A)$ an interesting direction for further research. However, the cascade process has the property that it can be efficiently simulated, simply by running the probabilistic rule for influence propagation until quiescence (which, as noted above, will occur within at most $n + 1$ rounds). By repeatedly simulating the cascade process and sampling $\varphi(A)$, we can compute arbitrarily close approximations to $\sigma(A)$. A straightforward calculation shows that with a number of simulations polynomial in ε, δ, and n, one can obtain a $1 \pm \varepsilon$ approximation to $\sigma(A)$, with probability at least $1 - \delta$. This approximate evaluation of $\sigma(A)$ in turn is enough to find an element v whose marginal gain $\sigma(A \cup \{v\}) - \sigma(A)$ is within a factor of $1 - \varepsilon'$ of maximal.

The idea for the proof of Theorem 1 is to show that $\sigma(A)$ is a monotone and submodular function of A. The property of submodularity formally means that $\sigma(S \cup \{w\}) - \sigma(S) \geq \sigma(T \cup \{w\}) - \sigma(T)$ whenever $S \subseteq T$. Informally, this is known as the "diminishing returns condition": the return derived from investing in node w diminishes as the size of the total investment (set) increases.

These properties of $\sigma(A)$ are sufficient to prove the desired approximation guarantee, for we can apply a well-known theorem of Nemhauser, Wolsey and Fischer. The first part of the theorem below is due to Nemhauser, Wolsey and Fischer [14,15]; the generalization can be obtained by straightforward modifications to the proof.

Theorem 2. *Let f be a non-negative, monotone, submodular function on sets.*

1. *The greedy algorithm, which always picks the element v with largest marginal gain $f(S\cup\{v\})-f(S)$, is a $(1-1/e)$-approximation algorithm for maximizing f on k-element sets S.*
2. *A greedy algorithm which always picks an element v within $1-\varepsilon$ of the largest marginal gain results in a $1-1/e-\varepsilon'$ approximation, for some ε' depending polynomially on ε.*

Given Theorem 2, in order to prove Theorem 1 (or its approximate version), it is sufficient to establish the following result:

Theorem 3. *For the decreasing cascade model, $\sigma(A)$ is a monotone and submodular function of A.*

Remark. The proof of the $(1 - 1/e)$ approximation guarantee in [6] was based on the same outline. In order to establish submodularity for the independent cascade and linear threshold models of [6], it was shown that for both models, it is possible to define distributions over directed graphs with the following property: for any set S of nodes, the probability that $\varphi(A) = S$ under the influence model is equal to the probability that the nodes of S are exactly the ones reachable from A in a graph chosen according to the corresponding distribution. Submodularity then follows readily from the fact that the number of reachable nodes in a fixed graph is a submodular function of the set of source nodes.

The decreasing cascade model is more general than the models considered in [6]. In Section 5, we give an instance which provably has no corresponding distribution on graphs. Therefore, the proof for submodularity becomes more intricate, and we have to consider the dynamics of the process in a more detailed way.

Most of the rest of this paper will be concerned with the proof of Theorem 3. We first introduce a generalized version of Granovetter's threshold model [10] in Section 3, as a useful reparametrization of the probability space. Using this threshold model, we then give the proof of Theorem 3 in Section 4.

3 The General Threshold Model

Recall that the notion of order-independence, as defined in Section 1.1, postulates that for a given set S of nodes trying to influence node v, the order in which these attempts are made does not affect the probability that v will be active once all the nodes in S have made their attempts. For the proof of Theorem 3, we require a stronger version of this statement: namely that even if the activation of nodes, or some activation attempts, are deferred for many time steps, the ultimate distribution over active sets remains the same.

It is not clear how to argue this fact directly from the definition of the cascade model, and we therefore introduce the general threshold model, a natural generalization of Granovetter's linear threshold model [10]. The linear threshold model has been the foundation for a large body of work in sociology; see, e.g., [8, 16, 17, 18, 19, 20, 21]; its generalization was introduced in [6]. While the General threshold model is a natural model in its own right, in this work, we are most interested in it as a reparametrization of the cascade model. Indeed, Lemma 1 proves that the two models are equivalent.

In the *general threshold model* [6], each node v has a monotone *activation function* $f_v : 2^V \to [0, 1]$, and a threshold θ_v, chosen independently and uniformly at random from the interval $(0, 1]$. A node v becomes active at time $t + 1$ if $f_v(S) \geq \theta_v$, where S is the set of nodes active at time t. Again, the process starts with the activation of a select set A at time 1.

The threshold model focuses more on the "cumulative effect" of a node set S's influence on v, instead of the individual attempts of nodes $u \in S$. The perhaps somewhat surprising fact is that for any activation functions $f_v(\cdot)$, we can define corresponding success probabilities $p_v(\cdot, \cdot)$ such that the distribution over final active sets $\varphi(A)$ is identical under both models, for all sets A.

Specifically, given success probabilities $p_v(u, S)$, we define the activation functions

$$f_v(S) = 1 - \prod_{i=1}^{r}(1 - p_v(u_i, S_i)), \tag{1}$$

where $S = \{u_1, u_2, \ldots, u_r\}$, and $S_i = \{u_1, \ldots, u_{i-1}\}$. That f_v is well defined follows from the order-independence assumption on the $p_v(u, S)$. Conversely, given activation functions f_v, we define success probabilities

$$p_v(u, S) = \frac{f_v(S \cup \{u\}) - f_v(S)}{1 - f_v(S)}. \tag{2}$$

It is straightforward to verify that the activation functions defined via Equation (1) satisfy Equation (2), and the success probabilities defined via Equation (2) satisfy Equation (1).

Lemma 1. *Assume that the success probabilities $p_v(u, S)$ and activation functions $f_v(S)$ satisfy Equation (2). Then, for each node set T and each time t, the probability that exactly the nodes of set T are active at time t is the same under the order-independent cascade process with success probabilities $p_v(u, S)$ and the general threshold process with activation functions $f_v(S)$.*

Proof. We show, by induction, a slightly stronger statement: namely that for each time t and any pair (T, T'), the probability that exactly the nodes of T are active at time t, and exactly those of T' are active at time $t + 1$, is the same under both views. By summing over all sets T', this clearly implies the lemma.

At time $t = 0$, the inductive claim holds trivially, as the probability is 1 for the pair $(\emptyset, A)$ and 0 for all other pairs, for both processes. For the inductive step to time t, we first condition on the event that the nodes of T are active at time $t - 1$, and those of T' at time t.

Consider a node $v \notin T'$. Under the cascade process, v will become active at time $t+1$ with probability $1 - \prod_{i=1}^{r}(1 - p_v(u_i, T \cup T'_i))$, where we write $T' \setminus T = \{u_1, \ldots, u_r\}$ and $T'_i = \{u_1, \ldots, u_{i-1}\}$. Under the threshold process, node v becomes active at time $t+1$ iff $f_v(T) < \theta_v \leq f_v(T')$. Because node v is not active at time t, and by the Principle of Deferred Decisions, θ_v is uniformly distributed in $(f_v(T), 1]$ at time t, so the probability that v becomes active is $\frac{f_v(T') - f_v(T)}{1 - f_v(T)}$. Substituting Equation (1) for $f_v(T)$ and $f_v(T')$, a simple calculation shows that

$$\frac{f_v(T') - f_v(T)}{1 - f_v(T)} = 1 - \prod_{i=1}^{r}(1 - p_v(u_i, T \cup T'_i)).$$

Thus, each individual node becomes active with the same probability under both processes. As both the thresholds θ_v and activation attempts are independent for distinct nodes, the probability for any set T'' to be the set of active nodes at time $t+1$ is the same under both processes. Finally, as the probability distribution over active sets T'' is the same conditioned on any pair (T, T') of previously active sets, the overall distribution over pairs (T', T'') is the same in both the cascade and threshold processes.

Lemma 1, which was stated without proof in [6], shows that the threshold model is a non-trivial reparametrization of the cascade model. In a natural way, it allows us to make all random choices at time 0, before the process starts. An alternate way of attempting to pre-flip all coins, for instance by providing a sequence of random numbers from $[0, 1]$ for use in deciding the success of activation attempts, would not preserve order-independence.

The nice thing about this view is that it makes a strong generalization of the notion of order-independence an almost trivial feature of the model. To formulate this generalization, we allow each node v a finite *waiting time* τ_v, meaning that when v's criterion for activation has been met at time t (i.e., an influence attempt was successful in the cascade model, or $f_v(S) \geq \theta_v$ in the threshold model), v only becomes active at time $t+\tau_v$. Notice that when $\tau_v = 0$ for all nodes, this is the original threshold/cascade model.

Lemma 2. *Under the general threshold model, the distribution $\varphi(A)$ over active sets at the time of quiescence is the same regardless of the waiting times τ_v. This even holds conditioned upon any random event $\mathcal{E}$.*

Proof. We prove the stronger statement that for every choice of thresholds θ_v, and every vector τ of waiting times τ_v, the set S_τ of nodes active at the time of quiescence is the same as the set S_0 of nodes active at quiescence when all waiting times are 0. This will clearly imply the claim, by integrating over all thresholds that form the event $\mathcal{E}$. So from now on, fix the thresholds θ_v.

Let $A_{0,t}$ denote the set of nodes active at time t when all waiting times are 0, and $A_{\tau,t}$ the set of nodes active at time t with waiting times τ. A simple inductive proof using the monotonicity of the activation functions f_v shows that $A_{\tau,t} \subseteq A_{0,t}$ for all times t, which, by setting t to be the time of quiescence of the process with waiting times τ, implies that $S_\tau \subseteq S_0$.

Assume now that $S_\tau \neq S_0$, and let $T = S_0 \setminus S_\tau \neq \emptyset$. Among the nodes in T, let v be one that was activated earliest in the process without waiting times, i.e., $T \cap A_{0,t} = \emptyset$, and $v \in A_{0,t+1}$ for some time t. Because v was activated, we know that $\theta_v \leq f_v(A_{0,t})$, and by definition of v, no previously active nodes are in T, i.e., $A_{0,t} \subseteq S_\tau$. But then, the monotonicity of f_v implies that $\theta_v \leq f_v(S_\tau)$, so v should be active in the process with waiting times τ, a contradiction.

4 Proof of Theorem 3

The monotonicity is an immediate consequence of Lemma 3 below, applied with $V = V'$ and $p'_v(u, S) = p_v(u, S)$ for all S, v, u. So we focus on submodularity for the remainder of the proof. We have to show that, whenever $A \subseteq A'$, we have $\sigma(A \cup \{w\}) - \sigma(A) \geq \sigma(A' \cup \{w\}) - \sigma(A')$, for any node $w \notin A'$.

The basic idea of the proof is to characterize $\sigma(A \cup \{w\}) - \sigma(A)$ in terms of a *residual process* which targets only the node w, and has appropriately modified success probabilities (similarly for $\sigma(A' \cup \{w\}) - \sigma(A')$). To show that these residual processes indeed have the same distributions over final active sets $\varphi(\{w\})$ as the original processes, we use Lemma 2.

Given a node set B, we define the *residual process* on the set $V \setminus B$: the success probabilities are $p_v^{(B)}(u, S) := p_v(u, S \cup B)$, and the only node targeted is w, targeted at time 1. Let $\varphi_B(w)$ denote the set of nodes active at the time of quiescence of the residual process; notice that this is a random variable. We claim that, conditioned on the event that $[\varphi(A) = B]$, the variable $\varphi_B(w)$ has the same distribution as the variable $\varphi(A \cup \{w\}) \setminus \varphi(A)$.

In order to prove this fact, we focus on the threshold interpretation of the process, and assign node w a waiting time of $\tau_w = n + 1$. By Lemma 2, this view does not change the distribution of $\varphi(A \cup \{w\}) \setminus \varphi(A)$. Then, w is the only contagious node at time $n + 1$, and by the conditioning, the other active (but non-contagious) nodes are those from B. This implies that only nodes from $V \setminus B$ will make activation attempts after time $n + 1$. By using the same order of activation attempts, and the same coin flips for each pair $u, v \in V \setminus B$, a simple inductive proof on the time t shows that the set S of nodes is active in the residual process at time t if and only if the set $S \cup B$ is active in the original process at time $n + t$. In particular, this shows that the two random variables have the same distributions.

Having shown this equivalence, we want to compare the expected sizes of $\varphi_B(w)$ and $\varphi_{B'}(w)$, when $B \subseteq B'$. We write $\sigma_B(w) = \mathrm{E}\left[|\varphi_B(w)|\right]$, as well as $\sigma_{B'}(w) = \mathrm{E}\left[|\varphi_{B'}(w)|\right]$. First off, notice that the node set $V \setminus B$ of the former process is a superset of $V \setminus B'$. Furthermore, for all nodes u, v and node sets S, the decreasing cascade condition implies that

$$p_v^{(B)}(u, S) = p_v(u, S \cup B) \geq p_v(u, S \cup B') = p_v^{(B')}(u, S).$$

Lemma 3 below proves the intuitively obvious fact that the combination of a larger ground set of nodes and larger success probabilities results in a larger set of activated nodes, i.e.,

$$\sigma_w(B) \geq \sigma_w(B') \tag{3}$$

Finally, we can rewrite the expected number of active nodes as

$$\begin{aligned}
\sigma(A \cup \{w\}) - \sigma(A) &= \sum_B \sigma_w(B) \cdot \text{Prob}[\varphi(A) = B] \\
&= \sum_B \sum_{B' \supseteq B} \sigma_w(B) \cdot \text{Prob}[\varphi(A) = B, \varphi(A') = B'] \\
&\geq \sum_B \sum_{B' \supseteq B} \sigma_w(B') \cdot \text{Prob}[\varphi(A) = B, \varphi(A') = B'] \\
&= \sum_{B'} \sigma_w(B') \cdot \text{Prob}[\varphi(A') = B'] \\
&= \sigma(A' \cup \{w\}) - \sigma(A').
\end{aligned}$$

The inequality followed by applying Inequality (3) under the sum. In both of the steps surrounding the inequality, we used that $\text{Prob}[\varphi(A) = B, \varphi(A') = B'] = 0$ whenever $B \not\subseteq B'$, by the monotonicity of the cascade process. This completes the proof of submodularity. ■

Lemma 3. *Let $V' \subseteq V$, and assume that $p'_v(u, S) \leq p_v(u, S)$ for all nodes $u, v \in V$ and all sets S. If $A' \subseteq A$ are the targeted sets for cascade processes on V' and V, then the expected size of the active set at the end of the process on V is no smaller than the corresponding expected size for the process on V'.*

Proof. This claim is most easily seen in the threshold view of the process. Equation (1) shows that the activation functions f'_v, f_v corresponding to the success probabilities $p'_v(u, S)$ and $p_v(u, S)$ satisfy $f'_v(S) \leq f_v(S)$, for all nodes v and sets S. Then, for any fixed thresholds θ_v, a simple inductive proof on time steps t shows that the set of active nodes in the former process (with functions f'_v) is always a subset of the set of active notes in the latter one (with functions f_v). Since the inequality thus holds for every point of the probability space, it holds in expectation.

5 Distributions over Graphs

As mentioned briefly before, the outline of the proof of the $(1 - 1/e)$ approximation guarantee in [6] was the same as here. However, a simpler technique was used to show the submodularity of $\sigma(A)$.

This technique can be most easily understood in the case of the independent cascade model, where each activation attempt of a node u on a node v succeeds independently with probability $p_v(u)$. By the definition of the process, a node v is active in the end if it is reachable from one of the initially targeted nodes by a chain of successful activation attempts. If we consider a graph G that contains a directed arc (u, v) iff u's activation attempt on v succeeded, then it follows that a node v is active iff it is reachable in G from the targeted set

A. Due to the independence of activation attempts, and by the Principle of Deferred Decisions, the graph G can be generated by including each arc (u, v) independently with probability $p_v(u)$. As the set of nodes reachable from a given set A is a submodular function of A, and the expected size of the activated set is a non-negative linear combination (over all possible graphs G) of these functions, the function $\sigma(A)$ is shown to be submodular.

This technique can be applied whenever the influence model allows for a corresponding distribution on directed graphs G — the fact that we included each arc independently did not matter. In fact, [6] uses this technique to show submodularity in two other, less obvious, cases. In this section, we give an instance of the decreasing cascade model for which there is no distribution over graphs resulting in the same activation probabilities. This example shows that the techniques used to show submodularity of $\sigma(A)$ in [6] cannot be applied for the more general decreasing cascade model.

Our example has five nodes. Node v could potentially be influenced by four nodes $u_1, \ldots, u_4$. The first two nodes to try activating v have a probability of $\frac{1}{2}$ each to succeed, whereas all subsequent attempts fail. The influences are thus $p_v(u_i, S) = \frac{1}{2}$ whenever $|S| < 2$, and $p_v(u_i, S) = 0$ otherwise. Notice that this is indeed an instance of the decreasing cascade model, and order independent.

Assume, for contradiction, that there is a distribution on graphs such that node v is reachable from a set S with the same probability that S will activate v in the cascade model. For any set $S \subseteq \{1, 2, 3, 4\}$, let q_S denote the probability that in this distribution over graphs, exactly the edges from u_i to v for $i \in S$ are present. Because with probability $\frac{1}{4}$, v does not become active even if all u_i are, we know that $q_\emptyset = \frac{1}{4}$. If u_1, u_2, u_3 are active, then v is also active with probability $\frac{3}{4}$, so the edge (u_4, v) can never be present all by itself (if it were, then the set $\{u_1, u_2, u_3, u_4\}$ together would have higher probability of reaching v than the set $\{u_1, u_2, u_3\}$). Thus, we have that $q_{\{i\}} = 0$ for all i. The same argument shows that $q_{\{i,j\}} = 0$ for all i, j.

Thus, the only non-empty edge sets with non-zero probabilities can be those of size three or four. If node u_1 is the only active node, then v will become active with probability $\frac{1}{2}$, so the edge (u_1, v) is present with probability $\frac{1}{2}$. Hence, $q_{\{1,2,3\}} + q_{\{1,2,4\}} + q_{\{1,3,4\}} + q_{\{1,2,3,4\}} = \frac{1}{2}$, while $q_{\{1,2,3\}} + q_{\{1,2,4\}} + q_{\{1,3,4\}} + q_{\{2,3,4\}} + q_{\{1,2,3,4\}} = 1 - q_\emptyset = \frac{3}{4}$. Therefore, $q_{\{2,3,4\}} = \frac{1}{4}$, and a similar argument for nodes u_2, u_3, u_4 gives that $q_S = \frac{1}{4}$ for each set S of cardinality 3. But then, the total probability mass on edge sets is at least $\frac{5}{4}$, as there are four such sets S, and the empty set also has probability $\frac{1}{4}$. This is a contradiction, so there is no such distribution over graphs.

6 Conclusions

In this paper, we have presented and analyzed a simple greedy algorithm for maximizing the spread of influence in a general model of social influence termed the decreasing cascade model. The proof centered on showing that the expected number of influenced nodes is a monotone and submodular function of the targeted set, which required new techniques beyond those used in previous work, including a non-trivial reparametrization of the probability space.

An interesting direction for future work is to investigate which are the most general influence models for which provable approximation guarantees can be achieved. A conjecture in [6], which is as of yet unresolved, states that whenever the activation functions f_v of the general threshold process of Section 3 are monotone and submodular at each node v, so is $\sigma(A)$.

Another direction for future work concerns the evaluation of the function $\sigma(A)$. At this point, we do not know if the function can be evaluated exactly in polynomial time, even for the simplest influence models.

References

1. Brown, J., Reinegen, P.: Social ties and word-of-mouth referral behavior. Journal of Consumer Research **14** (1987) 350–362
2. Domingos, P., Richardson, M.: Mining the network value of customers. In: Proc. 7th Intl. Conf. on Knowledge Discovery and Data Mining. (2001) 57–66
3. Goldenberg, J., Libai, B., Muller, E.: Using complex systems analysis to advance marketing theory development: Modeling heterogeneity effects on new product growth through stochastic cellular automata. Academy of Marketing Science Review (2001)
4. Goldenberg, J., Libai, B., Muller, E.: Talk of the network: A complex systems look at the underlying process of word-of-mouth. Marketing Letters **12** (2001) 211–223
5. Richardson, M., Domingos, P.: Mining knowledge-sharing sites for viral marketing. In: Proc. 8th Intl. Conf. on Knowledge Discovery and Data Mining. (2002) 61–70
6. Kempe, D., Kleinberg, J., Tardos, E.: Maximizing the spread of influence in a social network. In: Proc. 9th Intl. Conf. on Knowledge Discovery and Data Mining. (2003) 137–146
7. Rogers, E.: Diffusion of innovations. 4th edn. Free Press (1995)
8. Valente, T.: Network Models of the Diffusion of Innovations. Hampton Press (1995)
9. Wasserman, S., Faust, K.: Social Network Analysis. Cambridge University Press (1994)
10. Granovetter, M.: Threshold models of collective behavior. American Journal of Sociology **83** (1978) 1420–1443
11. Berger, E.: Dynamic monopolies of constant size. Journal of Combinatorial Theory Series B **83** (2001) 191–200
12. Morris, S.: Contagion. Review of Economic Studies **67** (2000) 57–78
13. Peleg, D.: Local majority voting, small coalitions, and controlling monopolies in graphs: A review. In: 3rd Colloquium on Structural Information and Communication. (1996) 170–179
14. Cornuejols, G., Fisher, M., Nemhauser, G.: Location of bank accounts to optimize float. Management Science **23** (1977) 789–810
15. Nemhauser, G., Wolsey, L., Fisher, M.: An analysis of the approximations for maximizing submodular set functions. Mathematical Programming **14** (1978) 265–294
16. Macy, M.: Chains of cooperation: Threshold effects in collective action. American Sociological Review **56** (1991) 730–747

17. Macy, M., Willer, R.: From factors to actors: Computational sociology and agent-based modeling. Annual Review of Sociology **28** (2002) 143–166
18. Schelling, T.: Micromotives and Macrobehavior. Norton (1978)
19. Watts, D.: A simple model of fads and cascading failures. Technical Report 00-12-062, Santa Fe Institute Working Paper (2000)
20. Young, H.P.: Individual Strategy and Social Structure: An Evolutionary Theory of Institutions. Princeton University Press (1998)
21. Young, H.P.: The diffusion of innovations in social networks. Technical Report 02-14-018, Santa Fe Institute Working Paper (2002)

An Optimal Bound for the MST Algorithm to Compute Energy Efficient Broadcast Trees in Wireless Networks

Christoph Ambühl[*]

Department of Computer Science,
The University of Liverpool
christoph@csc.liv.ac.uk

Abstract. Computing energy efficient broadcast trees is one of the most prominent operations in wireless networks. For stations embedded in the Euclidean plane, the best analytic result known to date is a 6.33-approximation algorithm based on computing an Euclidean minimum spanning tree. We improve the analysis of this algorithm and show that its approximation ratio is 6, which matches a previously known lower bound for this algorithm.

1 Introduction

Multi-hop wireless networks [14] require neither fixed, wired infrastructure nor predetermined interconnectivity. In particular, ad hoc networks [11, 17] are the most popular type of multi-hop wireless networks. An ad hoc wireless network is built of a bunch of radio stations. The links between them are established in a wireless fashion using the radio transmitters and receivers of the stations.

In order to send a message from a station s to a station t, station s needs to emit the message with enough power such that t can receive it. In the model, the power P_s required by a station s to transmit data to station t must satisfy the inequality

$$\frac{P_s}{\text{dist}(s,t)^\alpha} > \gamma. \tag{1}$$

The term $\text{dist}(s,t)$ denotes the distance between s and t, and $\alpha \geq 1$ is the **distance-power gradient**, and $\gamma \geq 1$ is the **transmission-quality** parameter. In an ideal environment (i.e. in the empty space) it holds that $\alpha = 2$ but it may vary from 1 to more than 6 depending on the environment conditions of the location of the network (see [19]).

In ad hoc networks, a power value is assigned to each station. These values, according to Equation (1), determine the so-called **range** of each station. The range of a station s is the area in which stations can receive all messages sent by s.

---

[*] This paper was written while the author worked at the Istituto Dalle Molle di Studi sull' Intelligenza Artificiale in Manno, Switzerland and was supported by the Swiss National Science Foundation project 200021-100539/1, "Approximation Algorithms for Machine Scheduling Through Theory and Experiments". It also contains ideas obtained while the author stayed at Università degli Studi di Roma "Tor Vergata" in Italy and was supported by the European Union under the RTN Project ARACNE.

Using the ranges, one can determine the so-called **transmission graph** $G = (S, A)$. The vertex set S is the set of stations, and the directed edge from s to t is in A if and only if t is within the range of s.

All stations in the range of a station i can receive messages sent by i. The minimal range needed for station i to establish all its out-going connections in G is therefore

$$r_G(i) := \max_{j \in \Gamma_G(i)} \text{dist}(i, j). \tag{2}$$

where $\Gamma_G(i)$ denotes the set of out-neighbors of station i in G. The total power needed to establish all connections in G is therefore

$$\text{power}(G) := \sum_{i \in V} \gamma \cdot r_G(i)^\alpha, \tag{3}$$

Since the value of γ does not influence the relative quality of the solutions, we assume $\gamma = 1$ for the rest of the paper. In this paper we address the following problem:

Problem 1 (Energy Efficient Broadcast Tree Problem (EEBT)). Let S be a set of stations represented by points from the Euclidean plane. That is, the distance function becomes $\text{dist}(s, t) := |st|$, where $|st|$ is the Eulidean distance between s and t. One of the stations is called the source station s. The goal is to find the transmission graph G which minimizes $\text{power}(G)$ and contains a directed spanning tree rooted at s (a branching from s).

The relevance of this problem is due to the fact that any transmission graph satisfying the above property allows the source station to perform a **broadcast** operation. Broadcast is a task initiated by the source station to transmit a message to all stations in the wireless network: This task constitutes a major part of real life multi-hop radio networks [1, 2, 7].

The **EEBT** Problem is known to be $\mathcal{NP}$-hard [4, 3]. Furthermore, if the dist function is arbitrary, the problem cannot be approximated with a logarithmic factor unless $\mathcal{P} = \mathcal{NP}$ [10]. The currently best approximation algorithm for the **EEBT** Problem is as follows.

Algorithm 1 (MSTALG). *The input of the algorithm is a set of stations S represented by points in the Euclidean plane. One of the stations is designated as the source. The algorithm first computes the Euclidean minimum spanning tree (EMST) of the point set S. Then the EMST is turned into a directed EMST by directing all the edges such that there exists a directed path from the source station to all other stations.*

In [21], Wan, Călinescu, Li, and Frieder claimed that **MSTALG** is a 12-approximation. Unfortunately, there is a small error in their paper. The correct analysis yields an approximation ratio of 12.15, as stated by Klasing, Navarra, Papadopoulos, and Perennes in [13]. Independently, Clementi, Crescenzi, Penna, Rossi, and Vocca showed an approximation ratio of 20 for **MSTALG** [4]. Recently, Flammini, Klasing, Navarra, and Perennes [8] showed that **MSTALG** is a 7.6-approximation algorithm. And even more recently, Navarra proved an approximation ratio of 6.33 [18]. In this paper, we show that **MSTALG** is a 6-approximation for all $\alpha \geq 2$. This matches the lower bound given in [4] and [21].

Experimental studies reported in [6, 15] show that for most instances, the approximation ratio of **MSTALG** is much better than 6. In [16], exact algorithms for **EEBT** have been studied. The **EEBT** problem falls into the class of so-called **range assignment problems**: Find a transmission range assignment such that the corresponding transmission graph G satisfies a given connectivity property Π, and power(G) is minimized (see for example [12, 7]). In [5], the reader may find an exhaustive survey on previous results related to range assignment problems.

Theorem 1. *Let S be a set of points from the unit disk around the origin, with the additional property that the origin is in S. Let $e_1, e_2, \ldots, e_{|S|-1}$ be the edges of the Euclidean minimum spanning tree of S. Then*

$$\mu(S) := \sum_{i=1}^{|S|-1} |e_i|^2 \leq 6.$$

Theorem 1 is the main theorem of this paper. Together with the next lemma, it proves that **MSTALG** is a 6-approximation algorithm for the **EEBT** problem.

The problem of giving upper bounds for $\mu(S)$ has already been looked at independently of the **EEBT** problem. Already in 1968, Gilbert and Pollack [9] gave a upper bound of $8\pi/\sqrt{3}$, based on a technique very similar to the one used by Wan et al in [21]. In 1989, Steele gave a bound of 16 based on space filling curves [20].

Lemma 1. *A bound on $\mu(S)$ automatically implies the same bound on the approximation ratio of* **MSTALG** *for $\alpha \geq 2$.*

Up to a few differences concerning the station at the origin of the unit disk, this lemma has already been proven in [21] to obtain the 12.15-approximation. We therefore skip its proof. For the case $\alpha < 2$, Clementi et al have shown that the MST algorithm does not provide a constant approximation ratio [4].

We now sketch the proof of the $\mu(S) \leq 12.15$ bound given in [21]. It works as follows. The cost of each edge e of the MST is represented by a geometric shape called diamonds, shown in Figure 1 on the bottom left. Diamonds consist of two isosceles

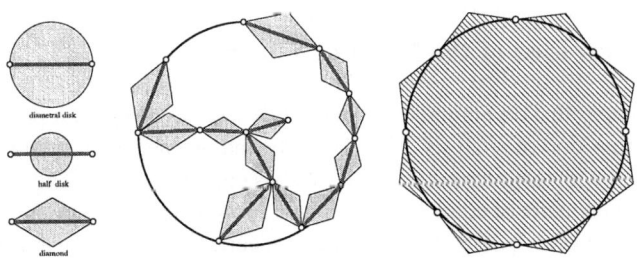

Fig. 1. Proof idea of previous bounds

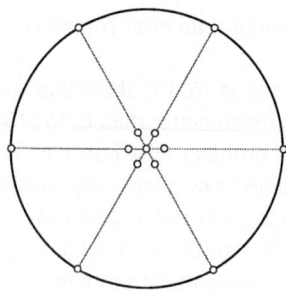

Fig. 2. A worst case example for **MSTALG**

triangles with an angle of $\frac{2}{3}\pi$. The area of a diamond for an edge e with length $|e|$ is $\lambda \cdot |e|^2$, with $\lambda = \sqrt{3}/6$. Diamonds are considered being open sets. It can be shown that if one puts these diamonds along the edges of an MST as shown in the middle of Figure 1, they do not intersect. It can further be shown that the area of the polygon shown on the right of Figure 1 is an upper bound on the area that can be covered by the diamonds along the MST edges. The area of this polygon is 12.15λ. Therefore one can conclude $\mu(S) \leq 12.15\lambda/\lambda = 12.15$.

Similar bounds can be obtained using diametral disks ($\mu(S) \leq 40$) or half disks ($\mu(S) \leq 20$) [3]. In both cases, one has to give an upper bound on the area generated by these shapes. In the case of diametral disks, this is done using the fact that in any point, at most five diametral disks can intersect. This gives only a very crude bound, which leads to a very crude bound on $\mu(S)$. On the other hand, open half disks do not intersect. But since they are smaller than diamonds, the bound provided by them is worse.

Concerning lower bounds on $\mu(S)$, there is a point set S that attains $\mu(S) = 6$. It is a regular 6-gone with one point in the middle. A lower bound on the approximation ratio of **MSTALG** is shown in Figure 2 [21,4]. The length of the edges of the MST shown in Figure 2 are ε and $1 - \varepsilon$, respectively. We have $opt(S) = 1$ and $power(G) = \varepsilon^2 + 6 \cdot (1 - \varepsilon)^2$. Hence for $\varepsilon \to 0$, the ratio between the two becomes 6. This lower bound holds for all values of α. Our analysis will give a matching upper bound for this lower bound for $\alpha \geq 2$. As already stated earlier, **MSTALG** does not have a constant approximation ratio for the case $\alpha < 2$ [4].

2 The Main Idea of the Proof of Theorem 1

Among the shapes that do not intersect, diamonds seem to be the best possible geometric shape for this kind of analysis. For a better bound, we need to use larger shapes and we need to deal with the intersection of the shapes more accurately. The shapes used for our new bound are pairs of equilateral triangles. As depicted in Figure 3, the equilateral triangles intersect heavily. We will give a quite accurate bound on the area generated by them.

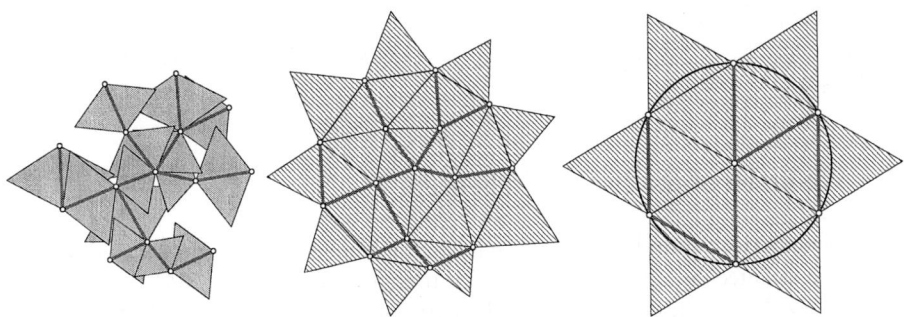

Fig. 3. The total area of the equilateral triangles on the left is bounded by the hatched area in the middle. The point set that maximizes the hatched area is the star shown on the right

A high level description of the proof of our bound is the following. Consider a point set S with n points. Hence, the MST will have $n - 1$ edges and therefore, there will be $2(n - 1)$ equilateral triangles representing the cost of the MST. Let A_{MST} be the total area generated by these triangles.

In order to obtain an upper bound on A_{MST}, let c be the number of edges of the convex hull of S. By triangulating S, we end up with a planar graph G with n vertices, e edges, and f facets. Let t be the number of triangles of the triangulation. Then the following three equations hold.

$$\begin{vmatrix} f &= t + 1 \\ 3t + c &= 2e \\ n + f - e &= 2 \end{vmatrix}$$

The first one simply states that the number of facets is equal to the number of triangles plus the infinite facet. For the second one, we add up the number of edges of all facets. For triangles, this is 3, whereas for the infinite facet, it is c. Since every edge is part of exactly two facets, this sums up to $2e$. The last equation is the Descartes-Euler polyhedral formula [22]. If we solve the system for t, we obtain $t = 2(n - 1) - c$. Hence, if we add c equilateral triangles along the convex hull of S as depicted in the center of Figure 3, the number of triangles becomes equal to the number of triangles generated by the MST, as shown on the left side of the figure. Let A_{TRI} be the total area of the triangles within the convex hull of S plus the c additional triangles along the convex hull.

The main idea of the proof is to show that $A_{\text{MST}} \leq A_{\text{TRI}}$. To get an intuitive understanding of it, consider a point set S obtained from the hexagonal grid for which all edges of the triangulation of its convex hull have the same length. In this case, all triangles that are involved in A_{MST} and A_{TRI} are congruent. Furthermore, since their number is equal, it holds $A_{\text{MST}} = A_{\text{TRI}}$. Intuitively, if the edges of the triangulation have different lengths, A_{MST} will be smaller compared to A_{TRI} since the MST will be composed mainly of small edges.

We then conclude the proof by showing that A_{TRI} is maximized by the star configuration depicted on the right of Figure 3. The area of the star is 6λ. Therefore we get $\mu(S) \leq 6\lambda/\lambda = 6$.

3 A Sketch of the Proof of Theorem 1

First, we introduce some notations. The area of an equilateral triangle with side length s will be denoted by $\triangle(s)$. The area of a triangle with edge lengths a, b, and c is denoted by $\triangle(a, b, c)$. Every edge can be partitioned lengthwise into two **half edges**. Both half edges are incident to the same vertices, but each of them is incident to only one facet. Slightly abusing notations, we call the largest side of an obtuse triangle its **hypothenuse**.

Consider the two triangles incident to an edge e. Let α and β be the two angles opposite e in the two triangles. We will call β the **opposite angle of** α.

Consider the MST of S and the Delaunay triangulation of the convex hull of S. Remember that the MST edges are also edges of the Delaunay triangulation. Now choose any edge e of the triangulation. Consider the unique cycle that is formed by adding e to the MST. This cycle and its (finite) interior is called a **pocket**. The triangles of the Delaunay triangulation within a pocket are called **pocket triangles**. The **area of a pocket** is the total area of all pocket triangles. The edge e is called the **door** of the pocket. All MST edges of the cycle are called **border edges**. Those in the interior are called **interior edges**. If e is an MST edge, the pocket will be called an **empty pocket**. Here, e is a border edge and the door at the same time. Empty pockets have area 0.

Note that the door of a pocket is incident to exactly one pocket triangle. If this triangle is obtuse and the door is its hypothenuse, the pocket is called an **obtuse pocket**, otherwise we call it an **acute pocket**.

The **MST-triangles of a pocket** P is the following set of triangles. Every half edge which is part of the MST and incident to a pocket triangle of P generates an MST-triangle. An MST-triangle of a half edge of length l is an equilateral triangle with side length l.

Obviously, both half edges of an interior edge belong to the pocket. On the other hand, only one half edge of a border edge belongs to the pocket. The **MST-area of a pocket** is the sum of the areas of all the MST-triangles. The MST-area of an empty pocket is $\triangle(e)$.

Figure 4 shows a pocket. The door of the pocket is the dashed line. Its border consists of all the edges of the MST connecting the two end points of the pocket. The largest of these edges is denoted by b_1. There are four inner edges. Note that the inner edges have two MST-triangles attached, one for each half edge, whereas the border edges have only one. The area of the pocket consists of the interior of the pocket. Because b_1 is part of the MST whereas the door is not, b_1 is never longer then the door.

Lemma 2. *In a acute pocket with largest border edge b, the difference between MST-area and pocket area is bounded by $\triangle(|b|)$.*

The proof of Lemma 2 is quite complicated. We therefore only give a sketch of it towards the end of this section. Using this lemma, one can prove Lemma 3. Due to lack

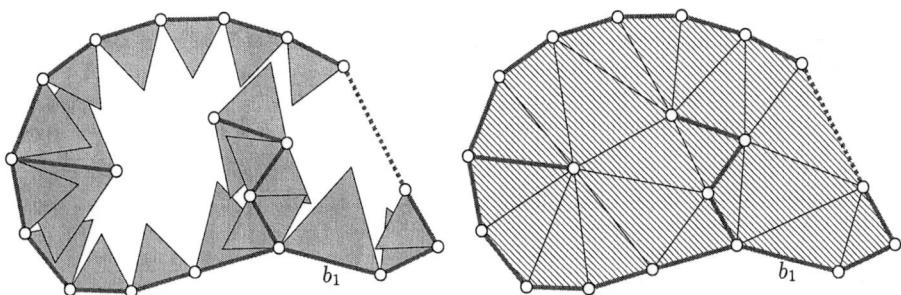

Fig. 4. A pocket with its MST-triangles on the left and the pocket triangles on the right. Note that there are 22 MST-triangles and 21 pocket triangles

to space, also this proof is omitted here. Lemma 3 in turn leads directly to the proof of Theorem 1.

Lemma 3. *Consider a pocket formed by an edge e. Then its MST-area can be bounded by the area of the pocket plus the area of a set of equilateral triangles whose side lengths are bounded by 1 and add up to $|e|$.*

Proof of Theorem 1. Consider the pockets whose doors are the edges of the convex hull of S. The sum of the MST-areas of all these pockets is equal to the total MST-area generated by S. Using Lemma 3, we can conclude that the total MST-area is bounded by the area of a so-called sun. A sun is defined by a convex set T from the unit disk, with the additional property that all edges of the convex hull are bounded by 1. The **area of a sun** is the convex hull of T plus, for each edge e of the convex hull of T, the area of an equilateral triangle with side length $|e|$.

Observe that the MST-area of S is $\mu(S) \cdot \sqrt{3}/2$. Hence we just need to prove that the area of a sun is bounded by $6\sqrt{3}/2$, which is exactly the area of a sun produced by a regular hexagon.

A point set T maximizing the area of its sun has all points on the unit circle. This holds since by moving a point towards the unit circle, the area of the sun increases. The area of a sun with all points on the unit circle can be partitioned as indicated in Figure 5(a). Each part consists of the triangle formed by the origin and an edge of the convex hull, plus the corresponding equilateral triangle. The area of each part can be expressed in terms of the angle ρ the first triangle forms at the origin. It is

$$f(\rho) := \sin\left(\frac{\rho}{2}\right)\left(\cos\left(\frac{\rho}{2}\right) + 2 \cdot \sin\left(\frac{\rho}{2}\right)\frac{\sqrt{3}}{2}\right) = \frac{1}{2}sin(\rho) + \sqrt{3} \cdot sin\left(\frac{\rho}{2}\right)^2.$$

Because we assumed that the edges of the convex hull are bounded by 1, the angle ρ must be between 0 and $\frac{\pi}{3}$. Note that $f(\frac{\pi}{3}) = \sqrt{3}/2$. In order to prove that the sun area is maximized by a regular hexagon, observe from Figure 5(b) that $f(\rho)/\rho$, restricted to the range $0 \le \rho \le \frac{\pi}{3}$, is maximized for $\rho = \frac{\pi}{3}$. □

In the remainder of this section, we describe the main ideas of the proof of Lemma 2.

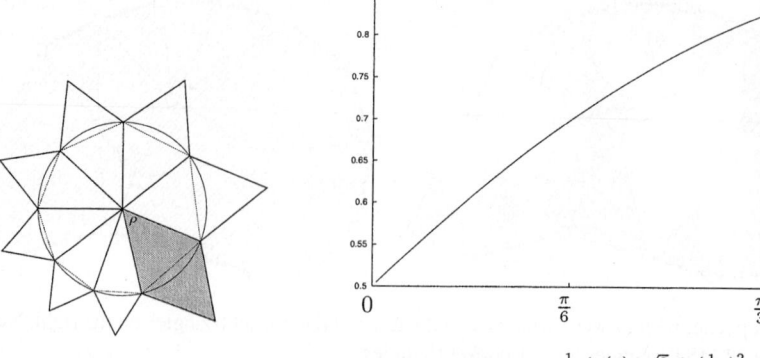

Fig. 5(a). A sun **Fig. 5(b).** $f(\rho)/\rho = \frac{\frac{1}{2}sin(\rho)+\sqrt{3}sin(\frac{1}{2}\rho)^2}{\rho}$

Lemma 4. *In a pocket, the number of MST-triangles exceeds the number of pocket triangles by one.*

Proof. Let n, e, b, t, be the number of nodes, edges, border edges (not including the door), and pocket triangles of a pocket. Let us first count the MST-triangles. Since the edges of a pocket form a tree, there are $n-1$ edges of the MST involved in the pocket. Each border edge produces one MST-triangle, whereas each inner edges produces two. Therefore their number is $2n - 2 - b$. Let us now count the pocket triangles. The Descartes-Euler polyhedral formula gives $n + (t + 1) - e = 2$, where the additional 1 is the face outside the pocket. By double counting the half edges, we obtain $2e = 3t + (b+1)$. Here, the additional 1 stands for the door of the pocket. Solving for t, we get that the number of pocket triangles is $t = 2n - 2 - b - 1$. □

The **extended pocket area (EP-area)** of a pocket is defined as the area of the pocket triangles plus an additional equilateral triangle with side length b, where b is the longest border edge of the pocket. This additional triangle is called **door triangle**. We call the union of the pocket triangles and the door triangle **EP-triangles**. The **net-area** of the pocket is defined as its EP-area minus its MST-area. Using the above definition, Lemma 2 can be rewritten as

Lemma 2 (reformulated) *The net-area of an acute pocket is non-negative.*

Let $V \subseteq S$ be the set of vertices that are part of the pocket, i.e., all the vertices that are inside or at the border of the pocket. Let $G = (V, E)$ be the weighted graph obtained by adding all the edges of the triangulation of the pocket, including the border edges and the door. The weight of the edge $e = uv$, $u, v \in S$, is denoted by $w(e)$ and its value is $|uv|$. Observe that the EP-, MST-, and net-area of a pocket are just a sum of triangle areas. Therefore, using Heron's formula [22]

$$\triangle(a,b,c) = \frac{1}{4}\sqrt{-a^4 - b^4 - c^4 + 2a^2b^2 + 2a^2c^2 + 2b^2c^2}$$

for the area of a triangle with side lengths a, b, and c, their areas can all be expressed in terms of the weighted planar graph G. What is more, defining EP-, MST-, and net-area in terms of weighted planar graphs allows to define them even if the planar graph does not have an embedding in the plane. The next lemma takes full advantage of this fact.

Lemma 5. *Let G' be a graph obtained from G by setting all weights to the same value. Then the net-area of G' is 0.*

Proof. If all weights of G' are equal, all the triangles involved in the EP-area and the MST-area of G' are equilateral and have the same side length. By Lemma 4, the number of EP-triangles is equal to the number of MST-triangles. Since both the EP-area and the MST-area consist of the same number of congruent triangles, we can conclude that their area is equal. □

We will now define a continuous process that turns G into a graph in which all edges have the same weight. During the process, only the weights of the edges are altered, whereas the combinatorial structure of G remains unchanged. The process is designed in such a way that the net-area of G decreases monotonically. This property, together with Lemma 5, proves that the net-area of G is non-negative.

Let $w_{\min}$ and $w_{\max}$ be the length of the smallest and the largest edge in G. The process will be described by a set of graphs $G(m)$, $m \in \mathbb{R}$, $w_{\min} \leq m \leq w_{\max}$. We start with $G = G(w_{\max})$ and end with $G(w_{\min})$, in which all weights will be $w_{\min}$.

The complete proof is quite involved and therefore deferred to the full version of this paper. In the remainder of the paper, we sketch the proof for a special case. Namely, we assume that all pocket triangles in G are acute. In this case, the process can be described very easily: Let $w(e)$ and $w_m(e)$ be the weight of edge e in G and $G(m)$ respectively. Then $w_m(e) = \min(m, w(e))$. That is, in every stage of the process, all maximal edges are decreased simultaneously until all edges have the same weight.

During this process only maximal edges are decreased. Hence, the ordering of the edges in terms of length remains unchanged during the process. Therefore the MST of G remains valid in all $G(m)$.

It is easy to see that during the process, the area of the pocket triangles of G decrease monotonically. This holds only because we assumed that the pocket triangles are acute.

We need to show that the net-area decreases. Hence, we have to show that in every $G(m)$, the decrease of the pocket area is at least as large as the decrease of the MST-area. To do this, we will partition $G(m)$ into so-called **chains** for which we will prove that their total net-area decreases monotonically.

Consider a graph $G(m)$ for fixed m. Chains are defined in terms of a graph Q. The vertex set of Q is the set of triangles in $G(m)$ plus the door triangle. Each maximal MST edge e of $G(m)$ creates the following set of edges in Q. If we remove e, the MST is divided into two subtrees. Let $R(e)$ be the ring of triangles that separates the two subtrees. For any pair of adjacent triangles in $R(e)$, we add an edge in Q. Note that if e is a border edge, the door triangle is also part of the ring and it is connected with the triangle incident to e and the pocket triangle incident to the door, as shown on the right of Figure 5. This completes the definition of the graph Q.

The **chains** are defined as the connected components of Q. Let Q' be a chain. We can define the area, the MST-area, and the net-area of Q' as follows. The **area of Q'** is equal to the sum of the areas of all triangles of Q'. Concerning the MST-area, note that every MST half edge h of $G(m)$ and its corresponding MST-triangle can be assigned

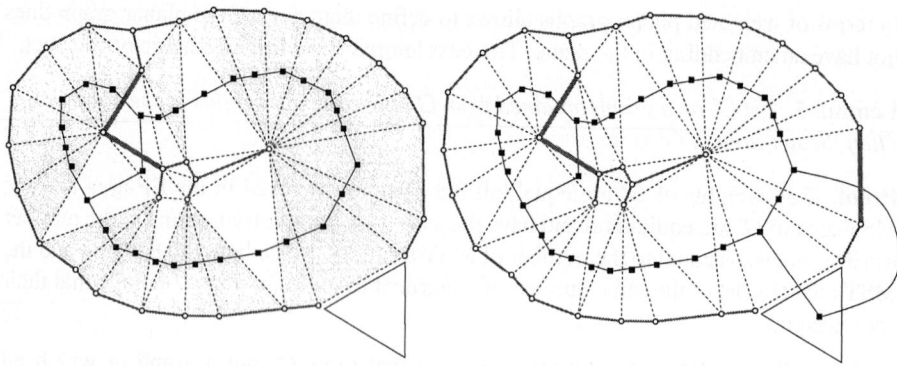

Fig. 6. Two chains in a graph G plus the door triangle. During the process, the chain on the left appears when there are two maximal MST edges, whereas the one on the right appears when there are three maximal MST edges. The maximal edges are the thick MST edges

to a unique chain, namely the one that contains the unique triangle h is incident to. The **MST-area of** Q' is equal to the sum of the areas of all MST-triangles belonging to the chain. The **net-area of** Q' is its area minus its MST-area. Since the net-area of a pocket is equal to the sum of the net-areas of all its chains, all we have to do to complete the proof is to show that the net-area of a chain decreases monotonically.

Remember that in $G(m)$, only the maximal MST edge decrease. Therefore only the MST-triangles of maximal MST edges decrease. In what follows, we will show that the decrease of the area of the chain makes up for the decrease of these **maximal MST-triangles**.

Some chains contain only a single triangle and no maximal MST-triangles. For these chains, it is obvious that the net-area decreases. Let us now look at chains that contain maximal MST-triangles. We need to consider two cases.

In case (i), we assume that the door triangle is not part of the chain. The **border of** Q' is the cycle in G with smallest area that contains all the triangles of Q'. The triangles from Q' that are incident to a border edge are called **border triangles**. On the left of Figure 6, the border triangles are shaded.

Let us now change Q' as follows. Let e be a maximal MST edge belonging to Q'. Assume its two incident triangles are q_1 and q_2. Add two new vertices h_1 and h_2 to Q'. They represent the half edges of e. Then remove the edge q_1q_2 from Q' and add q_1h_1 and q_2h_2. If we do this for all maximal MST edges e, Q' becomes a tree. Let d_i be the number of vertices of Q' with degree i, let n and e be the number of vertices and edges, respectively. From $e = n - 1$ (since Q' is a tree), $n = d_1 + d_2 + d_3$ and $2e = 3d_3 + 2d_2 + d_1$ (by double counting), we get $d_3 = d_1 - 2$.

The vertices of degree three represent equilateral triangles with side length m in $G(m)$. Let D_3 be the set of these triangles. The vertices of degree one represent the maximal MST half edges in $G(m)$. Let D_1 be the set of their corresponding MST-triangles. The area of the triangles in D_1 and D_3 decreases in the same way. Hence, the decrease of all but two triangles from D_1 is made up by the triangles in D_3. To complete the proof, one can show that the decrease of the border triangles of the chain makes up for the decrease of the remaining two maximal MST-triangles. Due to lack of space, we skip this proof here.

For case (ii), assume that the door triangle belongs to Q'. Apply the edge splitting of the previous case to Q'. This time, remove all the leafs adjacent to vertex representing the door triangle. We can do the same analysis as in the previous case to find that if the number of leafs in Q' is d_1, then the number of degree three vertices is $d_3 = d_1 - 2$. But this time, one of the leafs is the door triangle, which is an equilateral triangle with side length m. Hence there are as many half edges as equilateral triangles. Hence, if m decreases, the MST-area of the chain decreases at least as much as the area of the chain.

Acknowledgements

I am grateful for interesting discussions with Andrea Clementi, Miriam di Ianni, Gianluca Rossi, Andreas Paffenholz, Ingo Schurr, Emo Welzl, and Birgitta Weber.

References

1. R. Bar-Yehuda, O. Goldreich, and A. Itai. On the Time Complexity of Broadcast Operations in Multi-Hop Radio Networks: An Exponential Gap Between Determinism and Randomization. *Journal of Computer and Systems Science*, 45:104–126, 1992.
2. R. Bar-Yehuda, A. Israeli, and A. Itai. Multiple Communication in Multi-Hop Radio Networks. *SIAM Journal on Computing*, 22:875–887, 1993.
3. A.E.F. Clementi, P. Crescenzi, P. Penna, G. Rossi, and P. Vocca. A Worst-case Analysis of an MST-based Heuristic to Construct Energy-Efficient Broadcast Trees in Wireless Networks. Technical Report 010, University of Rome "Tor Vergata", Math Department, 2001.
4. A.E.F. Clementi, P. Crescenzi, P. Penna, G. Rossi, and P. Vocca. On the Complexity of Computing Minimum Energy Consumption Broadcast Subgraphs. In *Proceedings of the 18th Annual Symposium on Theoretical Aspects of Computer Science (STACS)*, pages 121–131, 2001.
5. A.E.F. Clementi, G. Huiban, P. Penna, G. Rossi, and Y.C. Verhoeven. Some Recent Theoretical Advances and Open Questions on Energy Consumption in Ad-Hoc Wireless Networks. In *Proceedings of the 3rd Workshop on Approximation and Randomization Algorithms in Communication Networks (ARACNE)*, pages 23–38, 2002.
6. A.E.F. Clementi, G. Huiban, P. Penna, G. Rossi, and Y.C. Verhoeven. On the Approximation Ratio of the MST-based Heuristic for the Energy-Efficient Broadcast Problem in Static Ad-Hoc Wireless Networks. In *3rd Workshop on Wireles, Mobile and Ad-Hoc Networks (WMAN) in the Proceedigs of the 17th International Parallel and Distributed Precessing Symposium (IPDPS)*, 2003.
7. A. Ephremides, G.D. Nguyen, and J.E. Wieselthier. On the Construction of Energy-Efficient Broadcast and Multicast Trees in Wireless Networks. In *Proceedings of the 19th Annual Joint Conference of the IEEE Computer and Communications Societies (INFOCOM)*, pages 585 594, 2000.
8. M. Flammini, R. Klasing, A. Navarra, and S. Perennes. Improved approximation results for the minimum energy broadcasting problem. *Proceedings of the 2004 joint workshop on Foundations of mobile computing*, 2004.
9. E. N. Gilbert and H. O. Pollak. Steiner minimal trees. *SIAM Journal on Applied Mathematics*, 16(1):1–29, 1968.

10. S. Guha and S. Khuller. Improved Methods for Approximating Node Weighted Steiner Trees and Connected Dominating Sets. *Information and Computation*, 150:57–74, 1999.
11. Z. Haas and S. Tabrizi. On Some Challenges and Design Choices in Ad-Hoc Communications. In *Proceedings of the IEEE Military Communication Conference (MILCOM)*, 1998.
12. L. M. Kirousis, E. Kranakis, D. Krizanc, and A. Pelc. Power Consumption in Packet Radio Networks. *Theoretical Computer Science*, 243:289–305, 2000.
13. R. Klasing, A. Navarra, A. Papadopoulos, and S. Perennes. Adaptive broadcast consumption (abc), a new heuristic and new bounds for the minimum energy broadcast routing problem. pages 866–877, 2004.
14. G.S. Lauer. *Packet radio routing*, chapter 11 of *Routing in communication networks*, M. Streenstrup (ed.), pages 351–396. Prentice-Hall, 1995.
15. A. Navarra M. Flammini and S. Perennes. The "real" approximation factor of the mst heuristic for the minimum energy broadcasting. pages 22–31, 2005.
16. R. Montemanni and L.M. Gambardella. Exact algorithms for the minimum power symmetric connectivity problem in wireless networks. *Computers and Operations Research, to appear*.
17. R. Montemanni, L.M. Gambardella, and A.K. Das. Mathematical models and exact algorithms for the min-power symmetric connectivity problem: an overview. *In Handbook on Theoretical and Algorithmic Aspects of Sensor, Ad Hoc Wireless, and Peer-to-Peer Networks. Jie Wu ed., CRC Press, to appear*.
18. A. Navarra. Tighter bounds for the minimum energy broadcasting problem. pages 313–322, 2005.
19. K. Pahlavan and A. Levesque. *Wireless Information Networks*. Wiley-Interscience, 1995.
20. J. M. Steele. Cost of sequential connection for points in space. *Operations Research Letters*, 8(3):137–142, 1989.
21. P.J. Wan, G. Călinescu, X.Y. Li, and O. Frieder. Minimum-Energy Broadcast Routing in Static Ad Hoc Wireless Networks. In *Proceedings of the 20th Annual Joint Conference of the IEEE Computer and Communications Societies (INFOCOM)*, pages 1162–1171, 2001.
22. E. W. Weisstein. *MathWorld–A Wolfram Web Resource*. http://mathworld.wolfram.com.

New Approaches for Virtual Private Network Design

Friedrich Eisenbrand[1,*], Fabrizio Grandoni[2],
Gianpaolo Oriolo[3], and Martin Skutella[4,**]

[1] Max-Planck-Institut für Informatik,
Stuhlsatzenhausweg 85, D-66123 Saarbrücken, Germany
eisen@mpi-sb.mpg.de
[2] Università di Roma "La Sapienza",
Dipartimento di Informatica,
Via Salaria 113, 00198 Roma, Italy
grandoni@di.uniroma1.it
[3] Università di Roma "Tor Vergata",
Dipartimento di Ingegneria dell'Impresa,
Via del Politecnico 1, 00165, Roma, Italy
oriolo@disp.uniroma2.it
[4] Universität Dortmund, Fachbereich Mathematik,
44221 Dortmund, Germany
martin.skutella@uni-dortmund.de

Abstract. *Virtual private network design* is the following NP-hard problem. We are given a communication network, represented as a weighted graph with thresholds on the nodes which represent the amount of flow that a node can send to and receive from the network. The task is to reserve capacities at minimum cost and to specify paths between every ordered pair of nodes such that all valid traffic-matrices can be routed along the corresponding paths.

Recently, this network design problem has received considerable attention in the literature. It is motivated by the fact that the exact amount of flow which is exchanged between terminals is not known in advance and prediction is often illusive. The main contributions of this paper are as follows:

- Using Hu's 2-commodity flow theorem, we provide a new lower bound on the cost of an optimum solution.
- Using this lower bound we reanalyze a simple routing scheme which has been described in the literature many times and provide a considerably stronger upper bound on its approximation ratio.
- We present a new randomized approximation algorithm for which, in contrast to earlier approaches from the literature, the resulting solution does not have tree structure.

---

* Partially supported by the DFG as part of the Transregional Collaborative Research Center "Automatic Verification and Analysis of Complex Systems" (SFB/TR 14 AVACS).
** Partially supported by the DFG Focus Program 1126, "Algorithmic Aspects of Large and Complex Networks", grant no. SK 58/4-1.

– Finally we show that a combination of our new algorithm with the simple routing scheme yields a randomized algorithm with expected performance ratio 3.55 for virtual private network design. This is a considerable improvement of the previously best known approximation results (5.55 STOC'03, 4.74 SODA'05).

1 Introduction

Consider a communication network which is represented by an undirected graph $G = (V, E)$ with edge costs $c : E \longrightarrow \mathbb{R}_+$. Within this network there is a set of terminals $T \subseteq V$ which want to communicate with each other. However, the exact amount of traffic between pairs of terminals is not known in advance. Instead, each terminal $v \in T$ has an associated input and output threshold $b_{in}(v) \in \mathbb{Z}_{\geq 0}$ and $b_{out}(v) \in \mathbb{Z}_{\geq 0}$. A *traffic matrix* $D \in \mathbb{Q}_{\geq 0}^{T \times T}$ is *valid*, if it respects the lower and upper bounds on the incoming and outgoing traffic of the terminals, i.e., if the following holds for each terminal $i \in T$

$$\sum_{j \in T, j \neq i} D(i,j) \leq b_{out}(i) \quad \text{and} \quad \sum_{j \in T, j \neq i} D(j,i) \leq b_{in}(i).$$

The *(asymmetric) Virtual Private Network Design Problem* (VPND) defined by G, c, T and b consists of finding capacities $u(e)$, $e \in E$, and paths P_{ij} for each ordered pair $(i, j) \in T \times T$ such that the following conditions hold:

i) All valid traffic matrices can be routed without exceeding the installed capacities where all traffic from terminal i to terminal j is routed along path P_{ij}.
ii) The total cost of the capacity reservation $\sum_{e \in E} u(e)\, c(e)$ is minimal.

A reservation of capacities $u : E \longrightarrow \mathbb{R}_+$ is a *tree reservation*, if the subgraph of G induced by the edges $e \in E$ with $u(e) > 0$ is a tree. A general reservation is sometimes referred to as a *graph reservation*.

The virtual private network design problem is NP-hard by the following reduction from the Steiner tree problem [9]. Given an instance of the Steiner tree problem, pick a terminal v which has to be connected with the other terminals in a Steiner tree. This terminal is assigned thresholds $b_{in}(v) := 0$ and $b_{out}(v) := 1$. All other terminals of the Steiner tree instance have input threshold one and output threshold zero. It is easy to see that a minimum cost Steiner tree yields an optimum reservation for this VPND-instance.

The virtual private network design problem has independently been defined by Fingerhut et al. [8] in the context of broadband ATM networks, and by Duffield et al. [5] (VPN "hose" model). Since then, it has been studied by various authors in several variations which we discuss next. In the following list, the last variant (*AsymG*) is the one which we refer to as VPND.

(*SymT*) Symmetric thresholds, tree reservation: In this variant, each terminal $i \in T$ has only one threshold $b(i)$, which is an upper bound on the cumulative amount of traffic that terminal i can send or receive. The task is to find an optimal tree reservation, which supports all valid traffic matrices. Gupta et al. [9] show that $(SymT)$ is polynomially solvable.

$(SymG)$ Symmetric thresholds, graph reservation: This variant is defined in the same way as $(SymT)$, except that the capacity reservation can be arbitrary and not necessarily a tree. Gupta et al.[9] present a 2-approximation for $(SymG)$. It is not known whether $SymG$ is NP-hard.

$(BalT)$ Balanced thresholds, tree reservation: The thresholds are *balanced*, which means that $\sum_{v \in T} b_{in}(v) = \sum_{v \in T} b_{out}(v)$. The reservation has to be a tree. Italiano et al. [13] show that this variant can be solved in polynomial time.

$(AsymT)$ Asymmetric thresholds, tree reservation: This problem is NP-hard [9]. Constant approximation algorithms are presented in [9, 10, 6]. Interestingly, while the algorithm in [9] is deterministic, the other algorithms are randomized and seem difficult to de-randomize.

$(AsymG)$ Asymmetric thresholds, graph reservation: This is the VPND problem defined above. We have seen that this problem is NP-hard. The randomized approximation results presented in [10, 6] in fact compare the computed tree solution to an optimal graph reservation. The current best approximation algorithm is the one in [6] which achieves an expected performance ratio 4.74.

Simplifying Assumptions and a Lower Bound

Following [10], we make some simplifying assumptions without loss of generality. By duplicating nodes, we can assume that each terminal is either a *sender* s, with $b_{out}(s) = 1$ and $b_{in}(s) = 0$, or a *receiver* r, with $b_{out}(r) = 0$ and $b_{in}(r) = 1$. This simplifying assumption is feasible as long as we make sure that the selected paths in our solution between copies of a terminal v and copies of a terminal u are all equal. Let $\mathcal{S}$ and $\mathcal{R}$ be the set of senders and the set of receivers, respectively. Let $S = |\mathcal{S}|$ and $R = |\mathcal{R}|$ denote the corresponding cardinalities. The algorithms presented in this paper can easily be adapted such as to run in polynomial time even when the thresholds are not polynomially bounded and to satisfy the consistence property described above. Moreover, by symmetry reasons, we always assume $R \geq S$.

We can now interpret VPND as follows. Let $B = (\mathcal{S} \cup \mathcal{R}, E^B)$ be the complete bipartite graph with nodes partitioned into senders and receivers. We have to reserve capacities on the edges of G and we have to specify a set of paths $\mathcal{P}$ in graph G containing one path P_{sr} for each edge $sr \in E^B$ such that each bipartite matching of B can be routed along these paths. In other words, for each edge $e \in E$, the reservation $u(e)$ has to satisfy the following condition:

$$|\{P_{rs} \in \mathcal{P} \mid e \in P_{rs} \text{ and } rs \in M\}| \leq u(e) \quad \text{for each matching } M \text{ in } B. \quad (1)$$

Notice that for a fixed set of paths $\mathcal{P}$, an optimal reservation of capacity is the component-wise minimal u satisfying (1). (In particular, given $\mathcal{P}$, the integral capacity $u(e)$ of edge e can be obtained by a maximum bipartite matching computation.) Thus, a solution to VPND can be encoded by only specifying a set of paths $\mathcal{P}$ in G.

The cost of a bipartite matching between senders and receivers in the metric closure of G is obviously a lower bound on OPT, the value of an optimum solution to the VPND-instance. We denote the shortest path distance between nodes u and v of G by $\ell(u,v)$. Thus, if edges (r,s) in B are assigned weights $\ell(r,s)$, then the cost of any matching in B is a lower bound on OPT. This lower bound is used in the analyses of all previous constant factor approximation algorithms for VPND.

Lemma 1 ([10]). *Let $B = (\mathcal{S} + \mathcal{R}, E^B)$ be the complete bipartite graph on the senders and receivers with edge weights $\ell : E^B \to \mathbb{R}_+$ given by the shortest path distances in the graph G. Then, the weight of any matching in B is a lower bound on OPT.*

Contribution of This Paper

The design of good approximation results usually requires two main ingredients: Cleverly constructed algorithms and thoroughly chosen lower bounds on the optimum such that the quality of the computed solutions can be assessed in terms of the lower bounds. We considerably advance the state of the art of approximating VPND by making contributions to both ingredients.

In Section 2 we present a new lower bound which generalizes and thus strengthens the one stated in Lemma 1. We prove that the weight of *any matching* (not necessarily bipartite) on the union of the senders and at most S receivers is at most OPT. The edge-weights in the matching are again the shortest path distances in G. This new lower bound relies on an interesting interrelation between a special case of VPND and 2-commodity flows. Its proof is based on an application of Hu's 2-commodity flow theorem [11].

In Section 3 we employ the new lower bound in order to show that the following simple algorithm achieves performance ratio $1 + R/S$: Find a vertex $v \in V$ whose shortest path tree to the union of senders and receivers is of minimal cost; cumulatively install a capacity of one on each shortest path. One interesting consequence of this result is that $(BalG)$, VPND with balanced thresholds and graph reservation, has a 2-approximation. Our result improves upon the 3-approximation of Italiano et al. [13] for this problem and generalizes the 2-approximation for $(SymG)$ by Gupta et al. [9].

In Section 4 we present a new randomized algorithm for VPND. The algorithm chooses a random subset of receivers and connects each sender via its own Steiner tree to this subset. The remaining receivers are then connected to the randomly chosen subset of receivers by shortest paths. Due to the Steiner trees for each individual receiver, the resulting solution has in general no tree structure. In contrast to our new approach, the previous algorithm by Gupta et al. [10] and its refinement in [6] construct only one 'high-bandwidth' core which is a Steiner tree with high capacity. In particular, all previous approximation algorithms for VPND produce tree solutions.

Finally, we can show that our new algorithm in combination with the simple algorithm from above yields a 3.55 randomized approximation algorithm. The previously best known algorithm [6] achieves performance ratio 4.74.

Related Work

As discussed above, VPND can be seen as a generalization of the Steiner tree problem. The currently best known approximation ratio for the Steiner tree problem is $\rho < 1.55$ [15]. A related problem is *buy at bulk* network design (see, e.g. [1,2]). In this problem, there is a fixed demand d_{ij} between each pair of nodes in the graph, specifying the amount of flow which has to be sent from i to j. The costs of the capacities however is a concave function on the amount purchased, which reflects "economies of scale". Gupta et al. [10] consider the single source buy-at-bulk network design problem and present a constant factor approximation algorithm.

Another important issue in this context is to cope with edge failures [3]. Italiano et al. [14] consider the problem of restoring the network, when at most one edge in a tree-solution to VPND might fail and provide a constant factor approximation algorithm.

Recently, Hurkens, Keijsper and Stougie [12] considered the problem $(SymT)$ in the special case when the given network is a ring. It is conjectured that $(SymG)$ can be solved in polynomial time. It is in fact conjectured that an optimal solution to $(SymT)$ is also an optimal solution to $(SymG)$. Hurkens et al. [12] show that this conjecture holds for ring networks. The authors also describe an integer programming formulation for VPND, which proves that a fractional version can be solved in polynomial time. This fractional version allows to specify several paths for each sender-receiver pair and requires the fraction for each of these paths, which describes how the commodity has to be split.

2 A New Lower Bound via Hu's 2-Commodity Flow Theorem

This section is devoted to proving a new lower bound on the cost of an optimal solution to VPND. Generalizing Lemma 1, we prove that the cost of an arbitrary (not necessarily bipartite) matching between terminals in $S \cup \mathcal{R}'$ is at most OPT, for any subset of receivers $\mathcal{R}' \subseteq \mathcal{R}$ of cardinality $|\mathcal{R}'| = S$. The proof of this result is based on Hu's classical 2-commodity flow theorem [11].

Theorem 1 (Hu's 2-commodity flow theorem). *Let $G = (V, E)$ be a graph and let $\{s_1, r_1\}, \{s_2, r_2\}$ be pairs of vertices of G; let $u: E \longrightarrow \mathbb{R}_+$ be a capacity function on the edges and let $d_1, d_2 \in \mathbb{R}_+$. Then, there exists a (fractional) 2-commodity flow of value d_1, d_2 if and only if the cut condition is satisfied. Moreover, if all edge capacities are integral, then a half-integral flow exists.*

The cut condition requires that $u(\delta(X)) \geq d_1 \chi_1(X) + d_2 \chi_2(X)$ for each $X \subseteq V$. Here $\delta(X)$ denotes the cut induced by X. Moreover, for $i = 1, 2$, we set $\chi_i(X) = 1$ if the cut $\delta(X)$ separates s_i from r_i and $\chi_i(X) = 0$, otherwise.

Corollary 1. *Let $G = (V, E)$ be an undirected graph with edge capacity function $u: E \to \mathbb{R}_+$ and $s_1, s_2, r_1, r_2 \in V$. In the following, all demand values* are equal to 1. If there exists a feasible 2-commodity flow for terminal pairs $\{s_1, r_1\}, \{s_2, r_2\}$ and for terminal pairs $\{s_1, r_2\}, \{s_2, r_1\}$, then there also exists a feasible 2-commodity flow for terminal pairs $\{s_1, s_2\}, \{r_1, r_2\}$.

Proof. In the case of unit demands, the cut condition requires that, for all $X \subseteq V$, the capacity $u(\delta(X))$ of the cut induced by X must be at least the number of terminal pairs which are separated by the cut. It thus remains to prove that the cut condition holds for the terminal pairs $\{s_1, s_2\}, \{r_1, r_2\}$ if it holds for $\{s_1, r_1\}, \{s_2, r_2\}$ and for $\{s_1, r_2\}, \{s_2, r_1\}$.

Consider an arbitrary $X \subseteq V$. If the corresponding cut separates neither $\{s_1, s_2\}$ nor $\{r_1, r_2\}$, nothing needs to be shown. If $\delta(X)$ separates one terminal pair, say $\{s_1, s_2\}$, then it separates either $\{s_1, r_1\}$ or $\{s_2, r_1\}$ since s_1 and s_2 lie on different sides of the cut. In particular, the capacity of the cut is at least 1. Finally, if $\delta(X)$ separates both terminal pairs $\{s_1, s_2\}, \{r_1, r_2\}$, then it either separates $\{s_1, r_1\}$ and $\{s_2, r_2\}$ or it separates $\{s_1, r_2\}$ and $\{s_2, r_1\}$. In both cases it follows that the capacity of the cut is at least 2.

We remark that Corollary 1 is no longer true if we replace "2-commodity flow" by "integral 2-commodity flow". Even, Itai, and Shamir show that finding an integer 2-commodity flow is NP-hard [7]. On the other hand, Hu's result states that there always exists a half-integral flow in this case. For a more detailed account of results we refer to Schrijver's book [16–Chapter 71].

Before we formulate and prove our new lower bound for VPND, we first state a general technique for deriving such lower bounds.

Lemma 2. *Let $\mathcal{S}_1, \ldots, \mathcal{S}_k$ be a partition of $\mathcal{S}$ and let $\mathcal{R}_1, \ldots, \mathcal{R}_k$ be a partition of $\mathcal{R}$. We denote the VPND-instance on graph G with senders $\mathcal{S}_i$ and receivers $\mathcal{R}_i$ by I_i. Then,*

$$\sum_{i=1}^{k} OPT_i \leq OPT ,$$

where OPT_i is the cost of an optimal solution to instance I_i.

We are now ready to prove the main result of this section.

Theorem 2. *Let $\mathcal{R}' \subseteq \mathcal{R}$ be an arbitrary subset of cardinality $|\mathcal{R}'| = S$ and let M be a matching in the complete graph on $\mathcal{S} \cup \mathcal{R}'$. Then,*

$$\sum_{vw \in M} \ell(v, w) \leq OPT.$$

Proof. Let $\mathcal{S} = \{s_1, s_2, \ldots, s_S\}$ and $\mathcal{R}' = \{r_1, r_2, \ldots, r_S\}$. It suffices to prove the claim for perfect matchings M. Suppose that M consists of edges

$$s_1 s_2, \ s_3 s_4, \ldots, s_{2k-1} s_{2k}, \ \text{and} \ r_1 r_2, \ r_3 r_4, \ldots, r_{2k-1} r_{2k},$$
$$\text{and} \ s_{2k+1} r_{2k+1}, \ s_{2k+2} r_{2k+2}, \ldots, s_S r_S \ .$$

Consider the following partition of $\mathcal{S}$ and $\mathcal{R}'$ into $S - k$ subsets $\mathcal{S}_i$ and $\mathcal{R}'_i$ each:

$$\mathcal{S}_i = \{s_{2i-1}, s_{2i}\}, \ \mathcal{R}'_i = \{r_{2i-1}, r_{2i}\}, \qquad \text{for } 1 \leq i \leq k,$$
$$\mathcal{S}_i = \{s_i\}, \ \mathcal{R}'_i = \{r_i\}, \qquad \text{for } 2k+1 \leq i \leq S.$$

By Lemma 2, the sum of OPT_i for the VPND-instances I_i with senders S_i and receivers $\mathcal{R}'_i$ is a lower bound on OPT. Thus we only need to prove that

$$\ell(s_{2i-1}, s_{2i}) + \ell(r_{2i-1}, r_{2i}) \leq OPT_i \text{ for each } 1 \leq i \leq k.$$

For $1 \leq i \leq k$, any solution to instance I_i yields a reservation of capacities that supports 2-commodity flows with unit demands for terminal pairs $\{s_{2i-1}, r_{2i-1}\}$, $\{s_{2i}, r_{2i}\}$ and for $\{s_{2i-1}, r_{2i}\}$, $\{s_{2i}, r_{2i-1}\}$. By Corollary 1, it must also support a 2-commodity flow for terminal pairs $\{s_1, s_2\}$, $\{r_1, r_2\}$. Therefore, the cost of this solution is at least $\ell(s_1, s_2) + \ell(r_1, r_2)$. This concludes the proof.

3 The Quality of a Simple Routing Scheme

Consider the following simple VPND algorithm: Select the node with the cheapest shortest path tree to the union of senders and receivers and reserve one unit of capacity along each shortest path. The effect of installing capacities along shortest paths is cumulative. In other words, if k shortest paths share the same edge, the algorithm assigns k units of capacity to that edge. Moreover, the shortest paths can be computed with a consistent tie-breaking rule such that the edges with nonzero capacity form a tree.

This algorithm produces an optimal tree reservation in the symmetric case ($SymT$) [9] and in the balanced case ($BalT$) [13]. In the symmetric case, Gupta et al. [9] show that the tree produced by the algorithm is a 2-approximate solution to the optimum graph reservation. Italiano et al. [13] show that, in the balanced case, the produced tree is a 3-approximate solution to the optimum graph reservation.

In this section, we apply our new lower-bound result to show that this algorithm produces a tree-solution whose cost is within a factor of $1 + R/S$ of the optimum graph reservation cost. As a consequence, also ($BalG$) can be approximated within a factor of two.

In [6] the following inequality is shown which follows from Lemma 1 and which we will use in the proof of the theorem below:

$$\sum_{s \in S} \sum_{r \in \mathcal{R}} \ell(s, r) \leq R \, OPT. \tag{2}$$

We are now ready to bound the approximation ratio provided by this simple routing scheme.

Theorem 3. *The above described algorithm achieves a performance ratio* $1 + R/S$.

Proof. Let $G^m = (\mathcal{R} \cup S, E^m)$ be the metric closure of $\mathcal{R} \cup S$, i.e., the complete graph on $\mathcal{R} \cup S$ with edge weight $\ell(u, v)$ given by the shortest path distance between u and v in G. We show that there exists a node $u \in \mathcal{R} \cup S$ such that the cost of its star satisfies

$$\sum_{v \in \mathcal{R} \cup \mathcal{S}} \ell(u, v) \leq (1 + R/S)\, OPT.$$

If $R = S$, then the edges of E^m can be partitioned into $2S - 1$ perfect matchings. By Theorem 2, the weight of each matching is a lower bound on OPT. Since each edge is contained in exactly two stars of G^m, there must exist one star, whose weight is at most $(2\,(2S-1)/(2S))\, OPT < 2\, OPT$.

Suppose for the remainder of the proof that $R > S$. In the following we denote by $\mathcal{M}_\mathcal{S}$ the set of (possibly not perfect) matchings of the senders in G^m and by $\mathcal{M}_\mathcal{R}$ the matchings of at most S receivers. Theorem 2 implies the inequality

$$\ell(M_\mathcal{S}) + \ell(M_\mathcal{R}) \leq OPT \qquad \text{for each } M_\mathcal{S} \in \mathcal{M}_\mathcal{S},\, M_\mathcal{R} \in \mathcal{M}_\mathcal{R}, \qquad (3)$$

where $\ell(E') := \sum_{uv \in E'} \ell(u,v)$ for any $E' \subseteq E^m$. In consideration of (3), we distinguish two cases.

First case: $\ell(M_\mathcal{S}) \leq OPT/2$ for each $M_\mathcal{S} \in \mathcal{M}_\mathcal{S}$.
Consider the subgraph $G_\mathcal{S}^m$ of G^m which is induced by the senders. The edges of $G_\mathcal{S}^m$ can be partitioned into S matchings. On the other hand one has

$$\sum_{s' \in \mathcal{S}} \sum_{s \in \mathcal{S}} \ell(s', s) = 2\, \ell(E_\mathcal{S}^m) \qquad (4)$$

and thus $\sum_{s' \in \mathcal{S}} \sum_{s \in \mathcal{S}} \ell(s', s) \leq S \cdot OPT$. This means that the average weight of a complete star in $G_\mathcal{S}^m$ is at most OPT. Let s' be a random sender. Inequality (2) implies that $E[\sum_{r \in \mathcal{R}} \ell(s', r)] \leq (R/S)\, OPT$. Together with the above discussion this implies for a random sender s'

$$E\left[\sum_{s \in \mathcal{S}} \ell(s', s) + \sum_{r \in \mathcal{R}} \ell(s', r) \right] \leq (1 + R/S)\, OPT. \qquad (5)$$

Second case: $\ell(M_\mathcal{S}) > OPT/2$ for some maximum weight matching $M_\mathcal{S} \in \mathcal{M}_\mathcal{S}$.
Let $G_\mathcal{R}^m$ denote the subgraph of G^m which is induced by the receivers. We will show below that the cost of any matching $\widetilde{M}$ in $G_\mathcal{R}^m$ is bounded by $(R/S)\, OPT/2$. Since the edges of $G_\mathcal{R}^m$ can be partitioned into R matchings, we can then argue in a similar manner as above that $E[\sum_{r \in \mathcal{R}} \ell(r', r)] \leq (R/S)\, OPT$ for a random receiver r'. Together with (2) this implies

$$E\left[\sum_{s \in \mathcal{S}} \ell(s, r') + \sum_{r \in \mathcal{R}} \ell(r, r') \right] \leq (1 + R/S)\, OPT. \qquad (6)$$

It remains to bound the cost of any matching $\widetilde{M}$ in $G_\mathcal{R}^m$. First assume that S is even. Theorem 2 implies $\ell(M_\mathcal{R}) \leq OPT/2$ for each matching $M_\mathcal{R} \in \mathcal{M}_\mathcal{R}$ of at most S receivers. As a consequence, the average cost of an edge in a matching $\widetilde{M}$ is at most $(OPT/2)/(S/2) = OPT/S$. Since $\widetilde{M}$ has at most $R/2$ edges, we get $\ell(\widetilde{M}) \leq (R/S)\, OPT/2$ for any matching $\widetilde{M}$ in $G_\mathcal{R}^m$.

It remains to consider the case that S is odd. Then there is a sender s^* which is missed by the maximum cost matching M_S of G_S^m. Theorem 2 yields

$$\ell(M_{\mathcal{R}}) + \ell(s^*, r_1^*) \leq OPT/2 \tag{7}$$

for each matching $M_{\mathcal{R}} \in \mathcal{M}_{\mathcal{R}}$ and $r_1^* \in \mathcal{R}$ which is not matched by $M_{\mathcal{R}}$. Since $R > S$, there exists another receiver r_2^* which is missed by $M_{\mathcal{R}}$. By the triangle inequality one has $\ell(r_1^*, r_2^*) \leq \ell(s^*, r_1^*) + \ell(s^*, r_2^*)$. As a result we get

$$\ell(M_{\mathcal{R}}) + 1/2\,\ell(r_1^*, r_2^*) \leq OPT/2 \tag{8}$$

for each matching $M_{\mathcal{R}} \in \mathcal{M}_{\mathcal{R}}$ and receivers r_1^*, r_2^* which are missed by $M_{\mathcal{R}}$. This implies that the average weight of an edge of $\widetilde{M}$ is bounded by OPT/S and thus $\ell(\widetilde{M}) \leq (R/S)\,OPT/2$.

4 A New Algorithm for VPND

In Section 3 we described an algorithm which guarantees a good approximation ratio for R close to S. In this section we present a better approximation algorithm for the case that R is sufficiently larger than S. As the algorithm by Gupta et al. [10], this algorithm is based on Steiner-tree computations. However, in contrast to the algorithm in [10], it does not construct a "high bandwidth core", which is a small Steiner tree with high capacity, which collects and distributes the demands from outside, and routes them along its high capacity paths. Instead, we proceed by constructing Steiner trees for each sender to a previously sampled subset of receivers, and by connecting the other receivers along their shortest paths to the sampled subset.

Algorithm 1.

(1) Partition $\mathcal{R}$ into S subsets uniformly at random. Among the non-empty subsets in the partition, select one subset $\mathcal{R}'$ uniformly at random.
(2) For each sender $s \in \mathcal{S}$, compute a ρ-approximate Steiner tree $T(s)$ on $\{s\} \cup \mathcal{R}'$, and add one unit of capacity to each edge of $T(s)$.
(3) Add one unit of capacity along a shortest path between each receiver $r \in \mathcal{R}$ and $\mathcal{R}'$.

It is interesting to note here that the thereby produced solution is not a tree solution. Though an optimal tree-solution is a constant-factor approximation to an optimal graph-solution, it is known [9] that an optimal solution to $(AsymT)$ is in general not an optimal solution to VPND. All previous constant factor approximation algorithms for VPND produce tree reservations [10, 6].

It remains to specify the path between each sender-receiver pair (s, r). Assume that the shortest paths are computed in a consistent way. Let r^* be the receiver in $\mathcal{R}'$ which is closest to r. The path P_{sr} between s and r is obtained by concatenating the (simple) path between s and r^* in $T(s)$ with the shortest path between r^* and r.

Before we proceed with the analysis of Algorithm 1, we state a corollary of Lemma 2. Here, given a subset V' of nodes, we denote the cost of an optimum Steiner tree on terminal set V' by $st(V')$.

Corollary 2 ([10]). *Let $\mathcal{R}_1, \ldots, \mathcal{R}_S$ be a partition of $\mathcal{R}$ into S (disjoint) subsets. Consider an arbitrary perfect matching between S and this family of subsets. Let $\mathcal{R}(s)$ be the subset matched with sender s. Then,*

$$\sum_{s \in S} st(\{s\} \cup \mathcal{R}(s)) \leq OPT.$$

Theorem 4. *Algorithm 1 is a $(2+\rho)/(1-e^{-R/S})$ approximation algorithm for VPND.*

Theorem 4 is a straightforward consequence of the following lemmas.

Lemma 3. *For a uniformly chosen random sender s',*

$$E[st(\{s'\} \cup \mathcal{R}')] \leq \frac{OPT}{S(1-e^{-R/S})}.$$

Proof. Consider the following random process. For each receiver r, we *assign* r to a sender s chosen uniformly at random. Let $\mathcal{R}(s)$ be the subset of receivers assigned to s. Note that the subsets $\mathcal{R}(s)$ partition $\mathcal{R}$ into S (possibly empty) subsets. Thus, by Corollary 2, $\sum_{s \in S} st(\{s\} \cup \mathcal{R}(s)) \leq OPT$. This means that, for the random sender s', $E[st(\{s'\} \cup \mathcal{R}(s'))] \leq OPT/S$. Let A denote the event that $\mathcal{R}(s')$ is empty. By elementary probability theory,

$$E[st(\{s'\} \cup \mathcal{R}(s'))] = P(A)\, E[st(\{s'\} \cup \mathcal{R}(s')) \mid A]$$
$$+ P(\overline{A})\, E[st(\{s'\} \cup \mathcal{R}(s')) \mid \overline{A}].$$

Now observe that $P(\overline{A}) = 1 - (1-1/S)^R \geq 1 - e^{-R/S}$. Moreover $E[st(\{s'\} \cup \mathcal{R}(s')) \mid A] = E[st(\{s'\})] = 0$. Thus

$$E[st(\{s'\} \cup \mathcal{R}(s')) \mid \overline{A}] = \frac{E[st(\{s'\} \cup \mathcal{R}(s'))]}{P(\overline{A})} \leq \frac{OPT}{S(1-e^{-R/S})}.$$

The claim follows by observing that, given $\overline{A}$, $\mathcal{R}(s')$ and $\mathcal{R}'$ are identically distributed. Thus

$$E[st(\{s'\} \cup \mathcal{R}')] = E[st(\{s'\} \cup \mathcal{R}(s')) \mid \overline{A}] \leq \frac{OPT}{S(1-e^{-R/S})}.$$

This implies the following upper bound on the expected cost of capacity which is installed in the second step of the algorithm.

Lemma 4. *The expected cost of the capacity installed in the second step of Algorithm 1 is at most*

$$\rho\, OPT/(1-e^{-R/S}).$$

The cost of the third step can be bounded as in the following lemma. A proof can be found in the full version of this paper.

Lemma 5. *The expected cost of the capacity installed in the third step of Algorithm 1 is at most*
$$2 OPT/(1 - e^{-R/S}).$$

In Section 3 we described a $(1 + R/S)$ approximation algorithm. The factors $1 + R/S$ and $(2 + \rho)/(1 - e^{-R/S})$ are equal for $R/S = 2.78\ldots < 2.79$. Note that $1 + R/S$ is increasing in R/S and $(2 + \rho)/(1 - e^{-R/S})$ is decreasing in R/S. It follows that a combination (taking the minimum cost solution) of the simple routing scheme of section 3 and Algorithm 1 has an expected approximation guarantee of 3.79.

Theorem 5. *The combination (taking the cheaper solution) of the simple routing scheme and Algorithm 1 is an expected 3.79 approximation algorithm for* VPND.

4.1 Computing the Optimum Steiner Tree

The performance ratios of the simple scheme and Algorithm 1 meet roughly at $R/S = 2.78$. In this case, the sampled set $\mathcal{R}'$ of receivers has expected constant size. An optimum Steiner tree on a graph with n nodes and t terminals can be computed in $O(3^t n + 2^t n^2 + n^3)$ time with the Dreyfus-Wagner algorithm [4]. This suggests the following variant of Algorithm 1.

In the second step of Algorithm 1 compute an *optimal* Steiner tree whenever $|\mathcal{R}'| \leq \log n$, where n is the number of nodes in (the original graph) G. A similar technique was already applied in [6].

Clearly, this modification is a polynomial time algorithm whose expected approximation guarantee is not worse than the one of Algorithm 1. In particular, if $R/S \geq \log \log n$, the approximation achieved is

$$(2 + \rho)/(1 - e^{-R/S}) \leq (2 + \rho)/(1 - 1/\log n) = 2 + \rho + o(1) < 3.55.$$

What can be said about the approximation guarantee if $R/S \leq \log \log n$? In that case, the expected size of $\mathcal{R}'$ is $1 + (R - 1)/S < 1 + \log \log n$. The probability that the size of $\mathcal{R}'$ exceeds $\log n$ is, by Markov's inequality, at most $(1 + \log \log n)/\log n$. In this unlikely event however, we can estimate the outcome of the combination of the thereby computed solution with the solution computed by the simple routing scheme described in section 3.

The next theorem is proved in a similar way as the main theorem in [6]. The proof can be found in the full version of this paper.

Theorem 6. *The combination (taking the cheaper solution) of the above described modification of Algorithm 1 and the simple routing scheme described in section 3 is an expected 3.55-approximation algorithm for* VPND.

References

1. M. Andrews and L. Zhang. The access network design problem. In *Proceedings of the 39th Annual Symposium on Foundations of Computer Science(FOCS-98)*, pages 40–49, Los Alamitos, CA, Nov.8–11 1998. IEEE Computer Society.
2. B. Awerbuch and Y. Azar. Buy-at-bulk network design. In *38th Annual Symposium on Foundations of Computer Science*, pages 542–547, Miami Beach, Florida, 20–22 Oct. 1997. IEEE.
3. G. Brightwell, G. Oriolo, and F.B. Shepherd. Reserving resilient capacity in a network. *SIAM J. Discrete Math.*, 14(4):524–539, 2001.
4. S.E. Dreyfus and R.A. Wagner. The Steiner problem in graphs. *Networks*, 1:195–207, 1971/72.
5. N.G. Duffield, P. Goyal, A. Greenberg, P. Mishra, K.K. Ramakrishnan, and J.E. van der Merive. A flexible model for resource management in virtual private networks. In *Proceedings of the conference on Applications, technologies, architectures, and protocols for computer communication*, pages 95–108. ACM Press, 1999.
6. F. Eisenbrand and F. Grandoni. An improved approximation algorithm for virtual private network design. In *Proceedings of the sixteenth annual ACM-SIAM symposium on Discrete algorithms, SODA 05*, pages 928–932, 2005.
7. S. Even, A. Itai, and A. Shamir. On the complexity of timetable and multicommodity flow problems. *SIAM Journal on Computing*, 5(4):691–703, 1976.
8. J.A. Fingerhut, S. Suri, and J. S. Turner. Designing least-cost nonblocking broadband networks. *Journal of Algorithms*, 24(2):287–309, 1997.
9. A. Gupta, J. Kleinberg, A. Kumar, R. Rastogi, and B. Yener. Provisioning a virtual private network: a network design problem for multicommodity flow. In *ACM Symposium on the Theory of Computing*, pages 389–398, 2001.
10. A. Gupta, A. Kumar, and T. Roughgarden. Simpler and better approximation algorithms for network design. In ACM, editor, *Proceedings of the Thirty-Fifth ACM Symposium on Theory of Computing, San Diego, CA, USA, June 9–11, 2003*, pages 365–372, New York, NY, USA, 2003. ACM Press.
11. T. Hu. Multi-commodity network flows. *Operations Research*, 11:344–360, 1963.
12. C. Hurkens, J. Keijsper, and L. Stougie. Virtual private network design: A proof of the tree routing conjecture on ring networks. In *Proceedings of the eleventh Conference on Integer Programming and Combinatorial Optimization, IPCO XI*, 2005. to appear.
13. G. Italiano, S. Leonardi, and G. Oriolo. Design of networks in the hose model. In *Proceedings of ARACNE 2002*, pages pp 65–76, 2002.
14. G.F. Italiano, R. Rastogi, and B. Yener. Restoration algorithms for virtual private networks in the hose model. In *Proceedings of the 21st Annual Joint Conference of the IEEE Computer and Communications Society (INFOCOM-02)*, volume 1 of *Proceedings IEEE INFOCOM 2002*, pages 131–139. IEEE Computer Society, 2002.
15. G. Robins and A. Zelikovsky. Improved steiner tree approximation in graphs. In *Proceedings of the eleventh annual ACM-SIAM Symposium on Discrete Algorithms, SODA'00*, pages 770–779, 2000.
16. A. Schrijver. *Combinatorial optimization. Polyhedra and efficiency (3 volumes)*. Algorithms and Combinatorics 24. Berlin: Springer., 2003.

Hadamard Tensors and Lower Bounds on Multiparty Communication Complexity

Jeff Ford and Anna Gál*

Dept. of Computer Science,
University of Texas at Austin, Austin, TX 78712-1188, USA
{jeffford, panni}@cs.utexas.edu

Abstract. We develop a new method for estimating the discrepancy of tensors associated with multiparty communication problems in the "Number on the Forehead" model of Chandra, Furst and Lipton. We define an analogue of the Hadamard property of matrices for tensors in multiple dimensions and show that any k-party communication problem represented by a Hadamard tensor must have $\Omega(n/2^k)$ multiparty communication complexity. We also exhibit constructions of Hadamard tensors, giving $\Omega(n/2^k)$ lower bounds on multiparty communication complexity for a new class of explicitly defined Boolean functions.

1 Introduction

Communication complexity was introduced by Yao [23] in 1979. Two players wish to compute $f(x, y)$. One player knows x, and the other knows y. Both have unlimited computational power. The communication complexity of f is the number of bits they must exchange on an arbitrary input in order to determine the value of f. This model and many of its variants have been widely studied [14]. Communication complexity arguments have been used to derive results in circuit complexity and in other computational models.

We consider the multiplayer model of Chandra, Furst, and Lipton [7] usually called the "Number on the Forehead" model. With k players, the input is partitioned into k parts: $x_1, \ldots, x_k$. The i-th player has access to every x_j except x_i. The Number on the Forehead model is stronger than the 2-party model, and sometimes the overlap between the players' inputs can be used to obtain surprising upper bounds (e.g. [18, 17]). This model is harder to analyze than the 2-party model, and very few lower bounds are known. On the other hand, lower bounds in this model have many applications in complexity theory, including constructions of pseudorandom generators for space bounded computation, universal traversal sequences, and time-space tradeoffs [2], as well as circuit complexity lower bounds [13, 16, 18].

---

* Supported in part by NSF CAREER Award CCR-9874862, NSF Grant CCF-0430695 and an Alfred P. Sloan Research Fellowship.

The largest known lower bounds for explicit functions are of the form $\Omega(n/2^k)$ where k is the number of players, and n is the number of bits each player misses. The first bounds of this form were given by Babai, Nisan and Szegedy [2] for the "quadratic character of the sum of coordinates" (QCS) function. They also gave an $\Omega(n/4^k)$ lower bound for the "generalized inner product" (GIP) function that was later improved to $\Omega(n/2^k)$ by Chung and Tetali [10]. Chung [9] and Raz [19] generalized the method of [2] to give a sufficient condition for a function to have $\Omega(n/2^k)$ multiparty communication complexity. Raz [19] also obtained $\Omega(\sqrt{n}/2^k)$ lower bounds for a new function based upon matrix multiplication over GF(2). Babai, Hayes and Kimmel [3] obtained further examples of functions with $\Omega(n/2^k)$ multiparty communication complexity. All of these lower bounds were obtained by estimating discrepancy, and so they also hold in the distributional and randomized communication complexity models.

The known bounds all decrease exponentially as the number of players grows, becoming trivial for $k > \log n$. It is a major open problem, with important implications in circuit complexity, to prove nontrivial lower bounds on multiparty communication problems for a large number of players. The class ACC^0, defined by Barrington [4], consists of languages recognized by constant depth, unbounded fan-in polynomial size circuit families with AND, OR, NOT and MOD_m gates for a fixed m. By the results of [24, 5, 13], families of functions that belong to ACC^0 can be computed by multiparty protocols with polylogarithmic (in n) communication by a polylogarithmic (in n) number of players (where n is the number of bits each player misses). Separating ACC^0 from other complexity classes (e.g. NP) is a major open problem, and a sufficiently large multiparty communication complexity lower bound would resolve it.

As proved by Chor and Goldreich [8], any Boolean function defined by a Hadamard matrix has $\Omega(n)$ 2-party communication complexity. Their proof uses a lemma by Lindsey (see [11] p. 88) that estimates the largest possible sum of entries in a submatrix of a Hadamard matrix. Lindsey's lemma implies upper bounds on the discrepancy of functions defined by Hadamard matrices and "nearly" Hadamard matrices. Babai, Nisan and Szegedy [2] generalized the proof of Lindsey's lemma to obtain upper bounds on the discrepancy of tensors associated with certain multiparty communication problems. The lower bounds that followed (e.g. [9, 10, 19, 3]) all used this approach. These papers did not consider generalizing the Hadamard property to tensors. In fact, [10] mentions that it is not clear how to generalize Hadamard matrices to tensors.

In this paper we propose a generalization of the Hadamard property of matrices to tensors of arbitrary dimension. We show that any k-party communication problem represented by a Hadamard tensor must have $\Omega(n/2^k)$ multiparty communication complexity. We construct families of Hadamard tensors, giving $\Omega(n/2^k)$ lower bounds for a new class of explicitly defined Boolean functions. Our Hadamard property is stronger than the sufficient condition of Chung [9] and Raz [19] for $\Omega(n/2^k)$ bounds, and could yield larger than $\Omega(n/2^k)$ lower bounds. There are no matching upper bounds known for functions represented by Hadamard tensors. We show how the Chung-Raz condition and some previous lower bounds fit into a "nearly" Hadamard framework. We believe that Hadamard tensors may also be of independent interest.

Our approach is based upon a new general upper bound on the discrepancy of tensors in terms of the largest possible value achievable by multiplying a collection of lines of the tensor by -1 and taking the sum of the entries of the resulting tensor. We refer to this value as the *weight*. This measure has been analyzed for matrices (see e.g. [1, 20]), and the corresponding matrix problem is sometimes called the "switching lights game". Generalizing the switching lights game to tensors was previously suggested in [10]. As far as we know, the general upper bound we give for the discrepancy of a tensor in terms of its weight is new. We also show that this upper bound is not too much larger than the actual discrepancy. Thus, the weight will give good bounds and may be easier to use than directly computing discrepancy. Since our lower bounds are based on discrepancy, they also hold in the distributional and randomized models.

2 Preliminaries

In the k-party model of Chandra, Furst and Lipton [7], k players with unlimited computational power wish to compute the value of a function $f : X_1 \times \cdots \times X_k \to \{-1, 1\}$ on input $\boldsymbol{x} = (x_1, \ldots, x_k)$. Usually we assume that $X_1 = \ldots = X_k = \{0, 1\}^n$. The function f is known to each player, and player P_i gets all of the input except $x_i \in X_i$. Players communicate by broadcasting messages, so all players receive all messages. If each player misses n bits of input, then $n + 1$ bits of communication is sufficient: Player P_2 broadcasts x_1, and then player P_1 who now has the entire input broadcasts the answer.

Definition 2.1. *The deterministic k-party communication complexity of f (denoted $C(f)$) is the number of bits communicated by the players on the worst input $\boldsymbol{x}$ using the best protocol for computing f.*

Definition 2.2. *Let μ be a probability distribution over the input of f. The bias achieved by a protocol P is defined as $|Pr[P(\boldsymbol{x}) = f(\boldsymbol{x})] - Pr[P(\boldsymbol{x}) \neq f(\boldsymbol{x})]|$, where $\boldsymbol{x}$ is chosen according to the distribution μ.*

The ϵ-distributional communication complexity of f (denoted $C_{\epsilon,\mu}(f)$) is the number of bits communicated by the players on the worst input $\boldsymbol{x}$ using the best protocol for computing f that achieves bias at least ϵ under the distribution μ. When μ is the uniform distribution we abbreviate to $C_\epsilon(f)$.

Definition 2.3. *[2] A subset $Z_i \subseteq X_1 \times \cdots \times X_k$ is called a* cylinder *in the i-th dimension, if membership in Z_i does not depend on the i-th coordinate, that is for every $(x_1, \ldots, x_i, \ldots, x_k) \in Z_i$ and every $x'_i \in X_i$ we have $(x_1, \ldots, x'_i, \ldots, x_k) \in Z_i$ as well. A subset $Z \subseteq X_1 \times \cdots \times X_k$ is called a* cylinder intersection *if it can be represented as $Z = \cap_{i=1}^{k} Z_i$, where each Z_i is a cylinder in the i-th dimension.*

A protocol can be thought of as reducing the space of possible inputs at each step until all the remaining possibilities give the same output. A message from player P_i winnows the input space, but not along the i-th dimension. Thus it causes the space of possible inputs to be intersected with a cylinder in the i-th dimension. After each message the consistent inputs form a cylinder intersection.

Definition 2.4. *The* discrepancy *of f on the cylinder intersection Z (denoted $\mathrm{Disc}_Z(f)$) is defined by*

$$\mathrm{Disc}_Z(f) = |\Pr[(\boldsymbol{x} \in Z) \wedge (f(\boldsymbol{x}) = 1)] - \Pr[(\boldsymbol{x} \in Z) \wedge (f(\boldsymbol{x}) \neq 1)]| \,,$$

where $\boldsymbol{x}$ is chosen according to the uniform distribution. The discrepancy *of f (denoted $\mathrm{Disc}(f)$) is the maximum of $\mathrm{Disc}_Z(f)$ over all cylinder intersections Z.*

Since $\mathrm{Disc}(f)$ is defined with respect to the uniform distribution, and the output of f is from $\{-1, 1\}$, we have the following:

$$\mathrm{Disc}_Z(f) = |\sum_{\boldsymbol{x} \in Z} f(\boldsymbol{x})| \,/\, |X_1 \times \ldots \times X_k| \,.$$

Lemma 2.1. *[2] For any function $f : X_1 \times X_2 \times \cdots \times X_k \to \{-1, 1\}$, $C(f) \geq \log_2(1/\mathrm{Disc}(f))$ and $C_\epsilon(f) \geq \log_2(\epsilon/\mathrm{Disc}(f))$.*

3 A General Upper Bound on Discrepancy

Problems in 2-party communication complexity can be represented as matrices with rows labeled by the possible inputs for player P_1 and columns labeled by the possible inputs for player P_2. An entry in the matrix at location (x, y) is given by $f(x, y)$.

A multiparty communication complexity problem can be represented by a tensor, the multidimensional analogue of a matrix. Each dimension of the tensor is labeled by the piece of input missed by a player. That is, the i-th dimension of the tensor is indexed by the elements of X_i. We denote by $A(x_1, \ldots, x_k)$ the entry of the k-dimensional tensor A at location $(x_1, \ldots, x_k)$. For tensor A_f representing function f we have $A_f(x_1, \ldots, x_k) = f(x_1, \ldots, x_k)$. If $|X_1| = \ldots = |X_k| = N$, we say that the tensor has *order N*.

Definition 3.1. *Given a tensor A in k dimensions, a* line *of A is any vector formed by fixing all but one coordinate of A. A* face *of A is any $(k-1)$-dimensional tensor formed by fixing one coordinate of A.*

A tensor of order N has N entries in each line and N^{k-1} entries in each face. It has N^{k-1} lines and N faces along each of the k dimensions.

Definition 3.2. *Let A be a tensor with ± 1 entries. We say that a line of the tensor A is* flipped *if each entry in that line is multiplied by -1.*

Definition 3.3. *We say that a tensor is* cylindrical *in the i-th dimension, if it does not depend on the i-th coordinate x_i.*

If a tensor is cylindrical in the i-th dimension, the entries of any given line along the i-th dimension are identical, and the corresponding N faces are identical. Thus, a k-dimensional cylindrical tensor can be specified by a $(k-1)$-dimensional tensor (specifying the face that is repeated N times).

Definition 3.4. *We define the* excess *of a tensor A (denoted $S(A)$) to be the sum of its entries; that is, $S(A) = \sum_{x \in X_1 \times \cdots \times X_k} A(x)$.*

Lemma 3.1. *(implicit in [10])* $\mathrm{Disc}(f) = \max S(A_f \circ C_1 \circ \ldots \circ C_k)/N^k$, *where A_f is the ± 1 tensor representing f, and each C_i is a 0/1 tensor which is cylindrical in the i-th dimension. ($A_f \circ C_1 \circ \ldots \circ C_k$ denotes the entrywise product of the tensors $A, C_1, \ldots, C_k$.)*

Proof. Let $Z_i \subseteq X_1 \times \cdots \times X_k$ be a cylinder in the i-th dimension, and let C_i be the 0/1 tensor representing the characteristic function of the cylinder Z_i. Then C_i is cylindrical in the i-th dimension. Conversely, every 0/1 tensor which is cylindrical in the i-th dimension represents the characteristic function of some cylinder in the i-th dimension. The lemma immediately follows from the definitions and our notation. □

Definition 3.5. *We define the* weight *of a tensor A (denoted $W(A)$) to be the largest possible excess of a tensor A' where A' can be obtained from A by flipping an arbitrary collection of lines (in any direction). Note that the order in which the flips are performed does not matter.*

Alternatively, $W(A)$ can be described as $W(A) = \max S(A \circ T_1 \circ \ldots \circ T_k)$, where each T_i is a ± 1 tensor which is cylindrical in the i-th dimension. ($A \circ T_1 \circ \ldots \circ T_k$ denotes the entrywise product of the tensors $A, T_1, \ldots, T_k$.)

Theorem 3.1. $\mathrm{Disc}(f) \leq W(A_f)/N^k$, *where N is the order of the tensor A_f representing f.*

Proof. For $i = 1, \ldots, k$, let C_i be an arbitrary 0/1 tensor which is cylindrical in the i-th dimension. We inductively define related ± 1 tensors $\hat{C}_i$ and T_i. For each $i = 1, \ldots, k$, we define a $(k-1)$-dimensional ± 1 tensor $\hat{C}_i$, where the i-th coordinate is left out. For example, $\hat{C}_1$ is a $(k-1)$-dimensional tensor that depends on the $k-1$ coordinates $x_2, \ldots, x_k$. To simplify notation, we will denote the entries of these tensors by $\hat{C}_i(x)$, with the understanding that for $\hat{C}_i$, x_i is not used for indexing. For example, $\hat{C}_1(x)$ stands for $\hat{C}_1(x_2, \ldots, x_k)$.

We define $\hat{C}_1$ as follows: $\hat{C}_1(x) = \mathrm{sign}(\sum_{x_1} A_f(x) \cdot C_2(x) \cdots C_k(x))$. In other words, to obtain $\hat{C}_1$, we collapse the k dimensional tensor $A_f \circ C_2 \circ \ldots \circ C_k$ to a $k-1$ dimensional tensor by summing the entries of each line along the first dimension and taking the sign of each line sum as an entry of $\hat{C}_1$. (If a given line sums to a negative number, the corresponding entry in $\hat{C}_1$ is -1, otherwise it is 1.) We use $\hat{C}_1$ to define the ± 1 tensor T_1, which is k-dimensional, and cylindrical in the first dimension. T_1 is obtained by taking N copies of $\hat{C}_1$ and using them as the faces of T_1 (along the first dimension).

Assume that $T_1, \ldots, T_{i-1}$ are already defined. We define $\hat{C}_i$ as follows:

$$\hat{C}_i(\boldsymbol{x}) = \text{sign}(\sum_{x_i} A_f(\boldsymbol{x}) \cdot T_1(\boldsymbol{x}) \cdots T_{i-1}(\boldsymbol{x}) \cdot C_{i+1}(\boldsymbol{x}) \cdots C_k(\boldsymbol{x}))$$

Once $\hat{C}_i$ is defined we use it to obtain T_i which is k-dimensional, and cylindrical in the i-th dimension. T_i is obtained by taking N copies of $\hat{C}_i$ and using them as the faces of T_i (along the i-th dimension).

First we show $S(A_f \circ C_1 \circ C_2 \circ \ldots \circ C_k) \leq S(A_f \circ T_1 \circ C_2 \circ \ldots \circ C_k)$. When we replace C_1 by T_1, the contribution of each line of the tensor $A_f \circ C_1 \circ C_2 \circ \ldots \circ C_k$ (along the first dimension) is replaced by a nonnegative value at least as large as the absolute value of the sum of the entries of the original line. To see this, notice that by definition, $\hat{C}_1$ and T_1 contain the signs of the sum of the entries of the corresponding lines of $A_f \circ C_2 \circ \ldots \circ C_k$. (If the sum is 0, we use 1 for the sign.) Obtaining $A_f \circ T_1 \circ C_2 \circ \ldots \circ C_k$ corresponds to multiplying each entry of a given line of $A_f \circ C_2 \circ \ldots \circ C_k$ by the sign of the sum of the entries of that line. Recall that each C_i is cylindrical, thus the lines of C_1 along the x_1 coordinate are constants (all 0 or all 1). If all entries of a given line of C_1 are 0 then the corresponding line of $A_f \circ C_1 \circ C_2 \circ \ldots \circ C_k$ did not contribute anything to the sum, while after the replacement it contributes a nonnegative value. For the lines of C_1 that are constant 1, the contribution of the corresponding line of $A_f \circ C_1 \circ \ldots \circ C_k$ is replaced by its absolute value. Thus, we never decrease the total sum. Similarly, at each inductive step above, we maintain that $S(A_f \circ T_1 \circ \ldots \circ T_{i-1} \circ C_i \circ \ldots \circ C_k) \leq S(A_f \circ T_1 \circ \ldots \circ T_i \circ C_{i+1} \circ \ldots \circ C_k)$. It follows that $S(A_f \circ C_1 \circ \ldots \circ C_k) \leq S(A_f \circ T_1 \circ \ldots \circ T_k)$. By Lemma 3.1 and the definition of $W(A_f)$ the theorem follows. □

The following simple example shows that the discrepancy $\text{Disc}(f)$ can be strictly smaller than $W(A_f)/N^k$. Let $k = 2$, and f be the parity function, i.e., f is 1 if the number of 1's among the input bits is even, and -1 otherwise. Then the discrepancy $\text{Disc}(f) = 1/4$, while $W(A_f)/N^2 = 1$. To see this, note that in the matrix corresponding to the parity function the sum of entries in any rectangle is at most $N^2/4$. On the other hand, it is possible to flip the lines of the matrix so that we obtain the all 1 matrix. (Theorem 8 in [10] appears to claim that $\text{Disc}(f) = W(A_f)/N^k$. However, this seems to be a mistake in notation, and they in fact prove Lemma 3.1.)

The following theorem shows that the discrepancy can not be too much smaller than the bound given by the weight. Thus, using the weight for bounding discrepancy will give good bounds.

Theorem 3.2.
$$\text{Disc}(f) \geq W(A_f)/(2^k N^k).$$

Proof. Consider the lines used to generate $W(A_f)$. Partition the entries of A_f into 2^k groups according to whether they were flipped by the lines along each of the k dimensions. Along each dimension the entries flipped by the lines form a cylinder, as do the unflipped entries. Thus the partition splits the entries of A_f

into 2^k cylinder intersections. At least one of these cylinder intersections has entry sum with absolute value at least $W(A_f)/2^k$. Using that cylinder intersection in the discrepancy definition gives discrepancy at least $W(A_f)/(2^k N^k)$. □

It is known that $W(A) \geq N^{3/2}/\sqrt{2}$ for any N by N matrix A with ± 1 entries (see Theorem 5.1 in [1]; see also [6] (c.f. [15])). We show the following extension of that result:

Theorem 3.3. $W(A) \geq N^{k-\frac{1}{2}}/\sqrt{2}$, for any k dimensional ± 1 tensor A of order N.

Proof. Consider the set of matrices formed by fixing all but the first two dimensions of A. Each of the matrices has weight at least $N^{3/2}/\sqrt{2}$. They do not intersect, so their lines can be flipped independently giving a tensor weight at least $N^{k-2}(N^{3/2}/\sqrt{2})$. □

A standard probabilistic argument shows that there are tensor with weight $O(\sqrt{k}N^{k-\frac{1}{2}})$. Proving similar upper bounds on the weight of explicitly defined tensors would yield lower bounds of the form $\Omega(n)$ on multiparty communication complexity, for any number of players. Thus, estimating the weight of tensors can potentially give close to optimal bounds on the discrepancy, and on the multiparty communication complexity of the corresponding functions.

4 Hadamard Tensors

An N by N matrix with ± 1 entries is called a *Hadamard matrix* if the inner product of any two of its distinct rows is 0. It is equivalent to state the condition for columns: The product of any two distinct rows is 0 if and only if the product of any two distinct columns is 0.

The Hadamard property is invariant under the arbitrary flipping of lines. Thus, Lindsey's lemma (see [11] p. 88) gives the following well known statement:

Lemma 4.1. *For any Hadamard matrix A of order N, $W(A) \leq N^{3/2}$.*

We define the *product of t lines* (along the same dimension) of a tensor as the sum of entries in their entrywise product. For example, if $l_1, \ldots, l_t$ are lines along the first dimension, then their product is $\sum_{x_1} l_1(x_1) \cdots l_t(x_1)$.

Let A be a k-dimensional tensor of order N with ± 1 entries. For each of the first $k-1$ dimensions $i = 1, \ldots, k-1$, choose two distinct indices $y_i, z_i \in X_i$. Picking exactly one of y_i or z_i for each $i = 1, \ldots, k-1$ gives a point in $X_1 \times \ldots \times X_{k-1}$, and each such point specifies a line of A along the last coordinate x_k. There are 2^{k-1} possible choices for the selection described above, and since for each $i = 1, \ldots, k-1$, $y_i \neq z_i$, we get 2^{k-1} distinct lines this way. We say that the tensor A is Hadamard, if the product of any 2^{k-1} lines chosen in this way is 0. More formally, we define Hadamard tensors as follows:

Definition 4.1. *Let A be a k-dimensional tensor of order N with ± 1 entries. We say that A is a Hadamard tensor if for any $y_1, z_1 \in X_1, \ldots, y_{k-1}, z_{k-1} \in X_{k-1}$ such that $y_i \neq z_i$ for $i = 1, \ldots, k-1$, the following holds:*

$$\sum_{x_k \in X_k} \prod_{x_1 \in \{y_1, z_1\}, \ldots, x_{k-1} \in \{y_{k-1}, z_{k-1}\}} A(x_1, x_2, \ldots, x_k) = 0 .$$

When $k = 2$ this definition is identical to the definition of Hadamard matrices.

Lemma 4.2. *Let A^{x_i} denote the face of A obtained by fixing the i-th coordinate to the value x_i. Let $k \geq 3$. A is a k-dimensional Hadamard tensor if and only if for any $i \neq k$ and $y_i \neq z_i$ the entrywise product of two faces $A^{y_i} \circ A^{z_i}$ is a $(k-1)$-dimensional Hadamard tensor.*

Proof. Without loss of generality, let $i = k - 1$. We need to show that for any $y_1, z_1 \in X_1, \ldots, y_{k-2}, z_{k-2} \in X_{k-2}$ such that $y_i \neq z_i$ for $i = 1, \ldots, k-2$, the following holds:

$$\sum_{x_k \in X_k} \prod_{x_1 \in \{y_1, z_1\}, \ldots, x_{k-2} \in \{y_{k-2}, z_{k-2}\}} A^{y_{k-1}} \circ A^{z_{k-1}}(x_1, \ldots, x_{k-2}, x_k) = 0 .$$

But $A^{y_{k-1}} \circ A^{z_{k-1}}(x_1, \ldots, x_{k-2}, x_k) = \prod_{x_{k-1} \in \{y_{k-1}, z_{k-1}\}} A(x_1, x_2, \ldots, x_k)$, and the statement directly follows from Definition 4.1. The proof in the reverse direction is similar. □

Since the k-th coordinate plays a special role in the definition of a Hadamard tensor, we can say that the definition is given with respect to the k-th dimension. It is not hard to see (using Lemma 4.2) that, just as for matrices, if a tensor is Hadamard with respect to one dimension, then it is Hadamard with respect to any other dimension. We leave the proof of this statement for the full version of the paper.

Lemma 4.3. *Let A' be a tensor obtained from a Hadamard tensor A by flipping a collection of lines. Then A' is a Hadamard tensor.*

Proof. This follows by induction from the characterization of Hadamard tensors given by Lemma 4.2. The result holds for matrices since after flipping a row or column any row or column product that was 0 remains 0. Suppose the result holds for tensors of dimension $k - 1$. Consider any face product $A^{y_i} \circ A^{z_i}$ of a k-dimensional Hadamard tensor A. Flipping a line of A may miss A^{y_i} and A^{z_i} entirely, intersect both in one entry, or flip an entire line of A^{y_i} or A^{z_i}. In the first case the face product is unaffected. In the second case the face product is unchanged since the corresponding entry is negated twice. In the third case the face product has a line flipped. By the induction hypothesis this is still a Hadamard tensor. □

4.1 The Discrepancy of Hadamard Tensors

In light of Theorem 3.1, we can prove upper bounds on the discrepancy of any tensor A by proving upper bounds on $W(A)$. Let $W_k(N)$ denote the largest possible value of $W(A)$ if A is a k-dimensional Hadamard tensor of order N.

Lemma 4.4. *Let A be a k-dimensional Hadamard tensor of order N. Then*

$$(W(A))^2 \leq N^{2k-1} + N^{k+1}(W_{k-1}(N)) \,.$$

Proof. Let A' be the k-dimensional tensor obtained from A by flipping a collection of lines that achieves maximal excess, that is $W(A) = S(A')$. By Lemma 4.3, A' is a Hadamard tensor, and by Lemma 4.2 the entrywise product of any two distinct faces of A' is a Hadamard tensor in $k-1$ dimensions. Thus, we have the following estimates (using the Cauchy-Schwartz inequality).

$$(S(A'))^2 = \left(\sum_{x \in X_1 \times \cdots \times X_k} A'(x) \right)^2 \leq N^{k-1} \sum_{x_1, \ldots, x_{k-1}} \left(\sum_{x_k} A'(x) \right)^2$$

$$= N^{k-1} \left(N^k + \sum_{i \neq j} \sum_{x_1, \ldots, x_{k-1}} A'(x_1, \ldots, x_{k-1}, i) A'(x_1, \ldots, x_{k-1}, j) \right)$$

$$\leq N^{k-1}(N^k + (N^2 - N)(W_{k-1}(N))) \leq N^{2k-1} + N^{k+1}(W_{k-1}(N))) \,. \square$$

Theorem 4.1. *Let A be a k-dimensional Hadamard tensor of order N. Then $W(A) \leq \phi N^{k-(1/2^{k-1})}$ where $\phi = (1+\sqrt{5})/2$.*

Proof. Follows by induction using Lemma 4.1 and Lemma 4.4. $\square$

Theorem 4.2. *Let $f : (\{0,1\}^n)^k \to \{1,-1\}$ be a function represented by a Hadamard tensor. Then $\mathrm{Disc}(f) \leq \phi N^{-1/2^{k-1}}$ where $\phi = (1+\sqrt{5})/2$.*

Proof. Follows from Theorem 4.1 and Theorem 3.1. $\square$

By the results of [2] (see Lemma 2.1) this yields the following:

Theorem 4.3. *Let $f : (\{0,1\}^n)^k \to \{1,-1\}$ be a function represented by a Hadamard tensor. Then $C(f) = \Omega(n/2^k)$, and $C_\epsilon(f) = \Omega((n/2^k) + \log_2 \epsilon)$.*

4.2 Constructions of Hadamard Tensors

Let $x_1, \ldots, x_k$ be n-bit strings. Consider each of these strings as an element of the finite field $\mathrm{GF}(2^n)$, representing the field elements as univariate polynomials over $\mathrm{GF}(2)$ modulo a fixed irreducible polynomial of degree n. (In this representation the i-th bit ($0 \leq i \leq n-1$) of a given n-bit string indicates whether the corresponding polynomial $p(a)$ contains the term a^i.)

Let χ_S stand for the function obtained by raising -1 to the parity of the bits with coordinates in S, such that χ_S is 1 when the parity is even, and -1 when the parity is odd. It is not hard to see that for any $x, y \in \{0,1\}^n$,

$$\chi_S(x)\chi_S(y) = \chi_S(x+y), \tag{1}$$

where $+$ represents addition in $\mathrm{GF}(2^n)$. (In fact the χ_S are the additive characters of $\mathrm{GF}(2^n)$.) By the definition of χ_S, $\chi_S(x)\chi_S(y) = \chi_S(x \oplus y)$, viewing x and y as strings and taking bitwise XOR, which is the same as $\chi_S(x+y)$ using addition in the field.

Definition 4.2. *Given a function $f : \{0,1\}^n \to \{1,-1\}$, we define the function $\mathrm{FFM}_f^{n,k} : (\{0,1\}^n)^k \to \{1,-1\}$ by*

$$\mathrm{FFM}_f^{n,k}(x_1,\ldots,x_k) = f(x_1 \cdot x_2 \cdot \ldots \cdot x_k),$$

where $x_1 \cdot x_2 \cdot \ldots \cdot x_k$ denotes the product of the field elements $x_1,\ldots,x_k$, and f is applied to the n-bit string representing the resulting field element.

For $S \subseteq \{0,1,\ldots,n-1\}$, we denote by $\mathrm{FFM}_S^{n,k}$ the function $\mathrm{FFM}_{\chi_S}^{n,k}$.

"FFM" is an abbreviation for "Finite Field Multiplication".

Theorem 4.4. *For every $\emptyset \neq S \subseteq \{0,1,\ldots,n-1\}$, the k-dimensional tensor associated with $\mathrm{FFM}_S^{n,k}$ is Hadamard.*

We need the following technical lemma:

Lemma 4.5. *For any k and for any $y_1, z_1, \ldots, y_k, z_k \in \mathrm{GF}(2^n)$ with $y_1 \neq z_1, \ldots, y_k \neq z_k$,*

$$\sum_{x_1 \in \{y_1,z_1\},\ldots,x_k \in \{y_k,z_k\}} x_1 x_2 \cdots x_k \neq 0$$

Proof. The proof is by induction. For distinct y_1 and z_1, $y_1 + z_1$ is nonzero since in $\mathrm{GF}(2^n)$ each element is its own additive inverse. Suppose the statement holds for $k-1$. Let $y_k, z_k \in \mathrm{GF}(2^n)$ with $y_k \neq z_k$.

$$\sum_{x_1 \in \{y_1,z_1\},\ldots,x_k \in \{y_k,z_k\}} x_1 x_2 \cdots x_k$$

$$= y_k \sum_{x_1 \in \{y_1,z_1\},\ldots,x_{k-1} \in \{y_{k-1},z_{k-1}\}} x_1 x_2 \cdots x_{k-1} +$$

$$z_k \sum_{x_1 \in \{y_1,z_1\},\ldots,x_{k-1} \in \{y_{k-1},z_{k-1}\}} x_1 x_2 \cdots x_{k-1}$$

$$= (y_k + z_k) \sum_{x_1 \in \{y_1,z_1\},\ldots,x_{k-1} \in \{y_{k-1},z_{k-1}\}} x_1 x_2 \cdots x_{k-1}$$

Since $y_k + z_k$ is nonzero (because in $\mathrm{GF}(2^n)$ each element is its own additive inverse), and the sum is nonzero by the induction hypothesis, this is nonzero. □

Proof. (of Theorem 4.4) Consider the following sum from Definition 4.1:

$$\sum_{x_k} \prod_{x_1 \in \{y_1, z_1\}, \ldots, x_{k-1} \in \{y_{k-1}, z_{k-1}\}} \chi_S(x_1 x_2 \cdots x_k)$$

By (1) this is the same as

$$\sum_{x_k} \chi_S \left(\sum_{x_1 \in \{y_1, z_1\}, \ldots, x_{k-1} \in \{y_{k-1}, z_{k-1}\}} x_1 x_2 \cdots x_k \right)$$

$$= \sum_{x_k} \chi_S \left(x_k \sum_{x_1 \in \{y_1, z_1\}, \ldots, x_{k-1} \in \{y_{k-1}, z_{k-1}\}} x_1 x_2 \cdots x_{k-1} \right)$$

As shown in Lemma 4.5, the inner sum evaluates to a non-zero field element, so for some fixed non-zero w, we obtain $\sum_{x_k} \chi_S(x_k w) = \sum_{x_k} \chi_S(x_k) = 0$. □

By Theorem 4.3 we immediately obtain the following:

Theorem 4.5. *For every $\emptyset \neq S \subseteq \{0, 1, \ldots, n-1\}$, $C(\mathrm{FFM}_S^{n,k}) = \Omega(n/2^k)$, and $C_\epsilon(\mathrm{FFM}_S^{n,k}) = \Omega(n/2^k + \log_2 \epsilon)$.*

Although all finite fields of order 2^n are isomorphic, it is necessary to specify exactly which one is being used to obtain explicit constructions of Boolean functions this way. The deterministic algorithm developed by Shoup [21] can be used to construct an irreducible polynomial of degree n for any given n. Thus the family of Boolean functions associated with the tensors FFM_S belongs to the complexity class P. Note also that the polynomial $x^n + x^{n/2} + 1$ is irreducible over $\mathrm{GF}(2)$ when n is of the form $n = 2 \cdot 3^m$ (Theorem 1.1.28 in [22]). Assuming that an irreducible polynomial of degree n is given, we can show that the corresponding Boolean function can be computed by depth $O(\log k)$ ACC circuits. We leave the proof of this for the full version of the paper.

4.3 Relaxations of the Hadamard Property

Raz [19] considered the function defined as follows: each part of the input $x_i \in \{0,1\}^n$ is interpreted as a $\sqrt{n}$ by $\sqrt{n}$ matrix with $0, 1$ entries. The function is defined by the bit in the upper left corner of the matrix obtained by taking the product (over $\mathrm{GF}(2)$) of the k matrices. Raz [19] proved that this function has (probabilistic) k-party communication complexity $\Omega(\sqrt{n}/2^k)$. The tensor associated with this function is not Hadamard, but we can show that it contains a subtensor of order $2^{\sqrt{n}}$ which is Hadamard. Thus, our methods give $\Omega(\sqrt{n}/2^k)$ lower bounds on the k-party communication complexity of the function.

Chung [9] and Raz [19] state a sufficient condition for a function to have $\Omega(n/2^k)$ multiparty communication complexity (generalizing the method of [2]). We can show that satisfying the condition of [9] and [19] is equivalent to being nearly Hadamard in the following, relaxed sense: Instead of requiring that all the products of the 2^{k-1}-tuples of lines selected according to the Hadamard definition are 0, it is enough to require that the products are small on average; e.g. that the sum of the squares of the line products is small. The tensor corresponding to the "generalized inner product" (GIP) function of [2] is nearly Hadamard in this relaxed sense, but it is not Hadamard. For the tensor corresponding to the "quadratic character of the sum of coordinates" (QCS) function of [2] we can show that each (nontrivial) product of the selected 2^{k-1} tuples of lines is small (at most $2^k\sqrt{N}$). We leave the proof for the full version of the paper. Note that the property we prove for QCS is stronger than the condition required in [9, 19], but weaker than the Hadamard property.

Grolmusz [12] proved an $O(kn/2^k)$ upper bound on the multiparty communication complexity of GIP, showing that the $\Omega(n/2^k)$ lower bounds for GIP cannot be significantly improved. There are no similar upper bounds known for any of the functions that we presented as examples of Hadamard tensors. The examples of Hadamard tensors we give and the QCS function are candidates for having $\Omega(n/poly(k))$ multiparty communication complexity.

References

[1] N. Alon, J. H. Spencer, "The Probabilistic Method", Wiley-Interscience, 2000.
[2] L. Babai, N. Nisan, M. Szegedy, "Multiparty Protocols, Pseudorandom Generators for Logspace, and Time-Space Trade-Offs", *JCSS*, 45(2):204-232, 1992.
[3] L. Babai, T. P. Hayes, P. G. Kimmel, "The Cost of the Missing Bit: Communication Complexity with Help", *Proc. 30th ACM STOC*, 673-682, 1998.
[4] D. Barrington, "Bounded-width polynomial size branching programs recognize exactly those languages in NC_1", *JCSS*, 38(1):150-164, 1989.
[5] R. Beigel, J. Tarui, "On ACC", *Proc. 32nd IEEE FOCS*, 783-792, 1991.
[6] M. R. Best, "The Excess of a Hadamard Matrix", *Indag. Math.*, 39(5):357-361, 1977.
[7] A. Chandra, M. Furst, R. Lipton: "Multiparty protocols", *Proc. 15th ACM STOC*, 94-99, 1983.
[8] B. Chor, O. Goldreich, "Unbiased Bits from Sources of Weak Randomness and Probabilistic Communication Complexity", *SIAM J. Comp.* 17:230-261, 1988.
[9] F. Chung, "Quasi-Random Classes of Hypergraphs", *Random Structures and Algorithms*, 1(4):363-382, 1990.
[10] F. Chung, P. Tetali, "Communication complexity and quasi randomness", *SIAM J. Discrete Math.*, 6(1):110-123, 1993.
[11] P. Erdős, J. H. Spencer, "Probabilistic methods in combinatorics", Academic Press, 1974.
[12] V. Grolmusz, "The BNS Lower Bound for Multi-Party Protocols is Nearly Optimal", *Information and Computation*, 112:51-54, 1994.
[13] J. Håstad, M. Goldmann, "On the power of small depth threshold circuits", *Computational Complexity*, 113-129, 1991.
[14] E. Kushilevitz, N. Nisan, "Communication complexity", Cambridge, 1997.
[15] J. H. van Lint, R. M. Wilson, "A Course in Combinatorics", Cambridge, 1992.
[16] N. Nisan, A. Wigderson, "Rounds in communication complexity revisited" *SIAM J. Comp.*, 22:211-219, 1993.

[17] P. Pudlák, "Unexpected upper bounds on the complexity of some communication games", *Proc. ICALP'94*, 1-11, 1994.
[18] P. Pudlák, V. Rödl, J. Sgall, "Boolean circuits, tensor ranks and communication complexity", *SIAM J. Comp.*, 26:605-633, 1997.
[19] R. Raz, "The BNS-Chung criterion for multiparty communication complexity", *Computational Complexity*, 9(2):113-122, 2000.
[20] J. Spencer, "Ten lectures on the probabilistic method", Soc. for Industrial and Applied Math., 1987.
[21] V. Shoup, "New Algorithms for Finding Irreducible Polynomials over Finite Fields", *Mathematics of Computation*, 54:435-447, 1990.
[22] J.H. van Lint, "Introduction to Coding Theory", Springer-Verlag, 1998.
[23] A. Yao, "Some complexity questions related to distributed computing", *Proc. 11th ACM STOC*, 209-213, 1979.
[24] A. Yao, "On ACC and threshold circuits", *Proc. 31st IEEE FOCS*, 619-627, 1990.

Lower Bounds for Lovász-Schrijver Systems and Beyond Follow from Multiparty Communication Complexity*

Paul Beame[1], Toniann Pitassi[2], and Nathan Segerlind[1]

[1] Computer Science and Engineering,
University of Washington, Seattle, WA 98195-2350
[2] Computer Science Department, University of Toronto,
Toronto, ON M5S 1A4

Abstract. We prove that an $\omega(\log^3 n)$ lower bound for the three-party number-on-the-forehead (NOF) communication complexity of the set-disjointness function implies an $n^{\omega(1)}$ size lower bound for tree-like Lovász-Schrijver systems that refute unsatisfiable CNFs. More generally, we prove that an $n^{\Omega(1)}$ lower bound for the $(k+1)$-party NOF communication complexity of set-disjointness implies a $2^{n^{\Omega(1)}}$ size lower bound for all tree-like proof systems whose formulas are degree k polynomial inequalities.

1 Introduction

Linear programming, the problem of optimizing a linear objective function over the points of a given polyhedron, was shown to be polynomial-time solvable over the rationals by Khachian [15]. When integrality constraints are added, however, the resulting integer linear programming problem becomes NP-hard. Many algorithms for such problems attempt to apply efficiencies from rational linear programming to the integral case.

One of the most powerful of such approaches is to begin with the polytope defined by the original linear program without integrality constraints and systematically pare down the polytope by repeatedly refining the linear program with "cutting planes" that remove only nonintegral solutions until we are left with the convex hull of the integral solutions. These are local methods in which the initial polytope Q (expressed by the natural cutting planes constraints) is transformed through a sequence of local operations to smaller and smaller polytopes (each contained in the original one), until the integral hull of Q is reached. (At this point, rational linear programming will find the correct solution.) For decision problems, this sequence terminates with the empty polytope if and only if the initial polytope contains no integral points.

---

* Paul Beame's research was supported by NSF grants CCR-0098066 and ITR-0219468. Toniann Pitassi's research was supported by an Ontario Premiere's Research Excellence Award, an NSERC grant, and the Institute for Advanced Study where this research was done. Nathan Segerlind's research was supported by NSF Postdoctoral Fellowship DMS-0303258 and done while at the Institute for Advanced Study.

One such method is that of Gomory-Chvátal cuts [6] which derives each new cutting plane as a linear combination and shift of existing facet constraints. There are even more subtle methods available, particularly in the case of 01-programming, which is also NP-complete. In a seminal paper, Lovász and Schrijver [16] introduced a variety of cutting planes methods that derive new cutting planes by first "lifting" the inequalities to higher degree polynomial inequalities (in particular quadratic inequalities) and then "projecting" them down to linear inequalities using polynomial identities and the fact that $x^2 = x$ for $x \in \{0, 1\}$. These systems are now known as *Lovász-Schrijver systems (LS)*.

It may be too costly to apply these techniques to pare all the way down to the integral hull. However, even applying a smaller number of rounds of the procedure can often lead to a smaller polytope that has good approximability ratio, one for which the best nonintegral solution is not too far away from the best integral solution, so that by rounding we can achieve a good approximation to the optimal value.

Two complexity measures are commonly studied for Lovász-Schrijver and related cutting planes proof systems: *size* and *rank*. Intuitively, rank is the number of intermediate polytopes that must be passed through before arriving at the integral hull. In [16] it was shown that for any (relaxed) polytope P, if the rank of P is d, then the optimization and decision problems for P can be solved exactly deterministically in time $n^{O(d)}$. This very nice algorithmic property of Lovász-Schrijver systems makes them especially appealing for solving or approximating NP-hard optimization problems via linear programming. A variety of rank lower bounds for exact solution are known, even for the case of unsatisfiable systems [4, 8, 11, 7, 12]. Moreover, interesting bounds on the ranks required for good approximations to vertex cover [1] and MaxSAT [5] have been obtained. This, in turn, implies inapproximability results for these problems for *any* polynomial-time algorithm based on rank.

While there is a rich and growing body of results concerning rank, very little is known about the size of LS proofs. Informally, the size of an LS procedure with respect to some polytope P is the smallest number of hyperplanes defining all of the polytopes that we need to pass through before arriving at the integral hull. Clearly size lower bounds imply rank lower bounds, and even tree-size lower bounds imply rank lower bounds, but the converse is not known to be true. The one unconditional (tree-like) size lower bound known for LS [12] is for a family of polytopes for which decision and optimization are trivial and for which the integral hull has a trivial derivation in Chvátal's cutting planes proof system.

Problems in which the facets represent clauses of a CNF formula and a decision algorithm for 01-programming yields a propositional proof system are particularly important to analyze. Proving (tree-like) size lower bounds for such polytopes was given as one of the main open problems in [12]. The only LS size lower bounds known at present for such polytopes formulas are conditional results. First, it is an easy observation that NP $\neq$ coNP implies superpolynomial LS size lower bounds for some family of unsatisfiable CNF formulas. It has also been shown by [19, 9, 10] that these lower bounds also hold under other natural complexity assumptions.

In this paper we develop a new method for attacking size lower bounds for LS and for systems that generalize LS. Our main result is a proof that lower bounds on the 3-

party communication complexity of set disjointness (in the number-on-forehead model) imply lower bounds on the size of tree-like LS proofs for a particular family of unsatisfiable CNF formulas. We also generalize this result to a much more powerful family of proof systems known as semantic LS^k, where lines are now degree k polynomial inequalities. All versions of LS are special cases of LS^2, and Chvátal's Cutting Planes proof system is a special case of LS^1.

More generally, we show that proving lower bounds on the $(k+1)$-party communication complexity of set disjointness implies lower bounds on the size of tree-like semantic LS^k proofs. By a natural extension of the ideas in [2] one can show that the $(k+1)$-party set disjointness problem is "complete" for the $(k+1)$-party communication complexity class $(k+1)$-NP^{cc} and a lower bound showing that it is not in $(k+1)$-RP^{cc} would already given excellent lower bounds for LS^k proofs. Such a result is already known in the case $k = 1$ [2] (and was used in [13] to derive tree-like size lower bounds for Chvátal's Cutting Planes system) and set disjointness is one of the most well-studied problems in communication complexity.

Our proof can be seen as a generalization of [13] to arbitrary k but the extension requires a number of new ideas and a substantially more complicated argument that includes a detailed analysis of large sets of vertex-disjoint paths in expander graphs.

2 Definitions

2.1 Multiparty Communication Complexity and Set Disjointness

The k-party number-on-the-forehead (NOF) model of communication complexity computes functions (or relations) of input vectors $(x_1, \ldots, x_k) \in X_1 \times \ldots \times X_k$ distributed among k parties, such that party $i \in [k]$ sees all x_j for all $j \in [k], j \neq i$.

The k-party set disjointness problem $\mathrm{DISJ}_{k,n} : (\{0,1\}^m)^k \to \{0,1\}$ is defined by $\mathrm{DISJ}_{k,n}(\vec{x}) = 1$ iff there is some $j \in [n]$ such that $x_{i,j} = 1$ for all $i \in [k]$. (We follow standard terminology although it might be more appropriate to call this set intersection rather than disjointness.)

A $(0, \epsilon)$-error k-party NOF communication protocol for set disjointness is a protocol that for every disjoint input produces output 0 and for intersecting inputs outputs 1 with probability at least $1 - \epsilon$.

It is conjectured that for any $k \geq 2$ the k-party set disjointness problem requires nearly linear randomized NOF communication complexity. This conjecture is equivalent showing that nondeterministic k-party communication complexity can be almost optimally separated from randomized k-party communication complexity. The conjecture is proven for $k = 2$ [14], but the best known lower bound for $k \geq 3$ is $\Omega(\log n)$ for general models and $\Omega(n^{1/k})$ for more restricted models [3].

2.2 Threshold Logics and the Complexity of a Search Problem

The two most prevalent classes of threshold logics are Gomory-Chvátal cutting planes [6], and matrix cuts, defined by Lovász and Schrijver [16]. These proof systems, CP LS, LS_0, and LS_+, are special cases of more general *semantic* threshold logic proof systems.

A *k-threshold formula* over Boolean variables $x_1, \ldots, x_n$ is a formula of the form $\sum_j \gamma_j m_j \geq t$, where γ_j, t are integers, and for all j, m_j is a multilinear monomial of degree at most k. The *size* of a k-threshold formula is the sum of the sizes of γ_j and t, written in binary notation.

Let f_1, f_2, g be k-threshold formulas in the variables $\vec{x}$. We say that g *is semantically entailed* by f_1 and f_2 if for every 0/1 assignment to $\vec{x}$ that satisfies both f_1 and f_2, g is also satisfied.

Let f be an unsatisfiable CNF formula over $x_1, \ldots, x_n$, and let $t_1, \ldots, t_m$ be the underlying set of clauses of f, written as 1-threshold inequalities. A **Th(k)** *refutation of* f, $\mathcal{P}$, is a sequence of k-threshold formulas, $L_1, \ldots, L_q$, where each L_j is one of the inequalities t_i, $i \in [m]$, or is semantically entailed by two formulas L_i and $L_{i'}$ with $i, i' < j$, and the final formula L_q is $0 \geq 1$. The *size* of $\mathcal{P}$ is the sum of the sizes of all k-threshold formulas occurring in $\mathcal{P}$. The proof is *tree-like* if the underlying directed acyclic graph, representing the implication structure of the proof, is a tree. (That is, every formula in the proof, except for the formulas from f, is used at most once as an antecedent of an implication.)

CP refutations are a special case of **Th(1)** semantic refutations, and thus lower bounds for tree-like **Th(1)** semantic refutations imply similar lower bounds for tree-like CP. (This was already shown in [13].)

As mentioned earlier, since we can assume that any of the Lovász-Schrijver systems can be assumed to have fan-in two, it follows that any of the systems LS_0, LS and LS^+ can easily be converted into **Th(2)** semantic refutations with at most a polynomial increase in size, and if the original proof is tree-like, so is the semantic refutation. Thus, lower bounds for tree-like **Th(2)** semantic refutations imply similar lower bounds for all tree-like Lovász-Schrijver systems.

Let f be an unsatisfiable CNF formula. We will be interested in the following search problem, $Search_f$ associated with f: given a truth assignment α, find a clause from f which is falsified by α. The model for this computation is a decision tree whose nodes evaluate polynomial threshold functions:

A *k-threshold decision tree* is a rooted, directed tree whose vertices are labeled with k-threshold functions and edges are labeled with either 0 or 1. The leaves of the tree are labeled with clauses of f. A k-threshold decision tree solves $Search_f$ in the obvious way: start at the root and evaluate the threshold function; follow the edge that is consistent with the value of the threshold function; continue until the computation reaches a leaf and output the associated clause. The size S of a k-threshold decision tree is the sum of the sizes of all threshold formulas in the tree, where the coefficients are written in binary. The depth of a k-threshold decision tree is the depth of the underlying tree.

Theorem 1. *Suppose that f has a tree-like* **Th(k)***-semantic refutation of size S. Then there exists a $k+1$-party 0-error randomized NOF communication complexity protocol for $Search_f$ (over any partition of the variables into k groups) that communicates $O(\log^3 S)$ bits and produces an answer with probability at least $1 - 1/n$.*

Further, if all k-threshold formulas in the **Th(k)**-*semantic refutation have coefficients bounded by a polynomial in n, then the 0-error randomized communication complexity can reduced to $O(\log S (\log \log n)^2)$ or the protocol can be made deterministic using $O(\log S \log n)$ bits.*

Proof (Sketch). First, following ideas similar to the degree 1 case in [13], we recursively search the proof tree using the $\frac{1}{3}$-$\frac{2}{3}$ trick to derive a k-threshold decision tree for $Search_f$ of depth $O(\log S)$ and size $O(S)$. Then, adapting arguments from [18], we show that any relation computed by a shallow k-threshold decision tree can also be efficiently computed by a $k + 1$ player communication complexity protocol (number-on-forehead model), over any partition of the variables.

2.3 k-Fold Tseitin Formulas

Our hard examples are based on the well-known Tseitin graph formulas. Let $G = (V, E)$ be any connected, undirected graph and let $\vec{c} \in \{0, 1\}^V$. The *Tseitin formula for G with respect to charge vector $\vec{c}$*, $TS(G, \vec{c})$, has variables $Vars(G) = \{y_e \mid e \in E\}$. The formula states that for every vertex $v \in V$, the parity of the edges incident with v is equal to the charge, c_v, at node v. It is expressed propositionally as the conjunction of the clauses obtained by expanding $\oplus_{e \ni v} y_e = c_v$ for each $v \in V$. For a graph with maximum degree d, each clause is of width $\leq d$ and the number of clauses is $\leq |V| 2^d$.

$TS(G, \vec{c})$ is satisfiable if and only if $\sum_{v \in V} c_v$ is even. For odd $\vec{c}$, $Search_{TS(G,\vec{c})}$ takes a 0/1 assignment α to $Vars(G)$ and outputs a clause of $TS(G, \vec{c})$ that is violated. In particular, a solution to $Search_{TS(G,\vec{c})}$ will produce a vertex v such that the parity equation associated with vertex v is violated by α.

To make the search problem hard for k-party NOF communication protocols (and thus, by Theorem 1, hard for $k - 1$-threshold decision trees) we modify $TS(G, \vec{c})$ by replacing each variable y_e by the conjunction of k variables, $\bigwedge_{i=1}^{k} y_e^i$, and expanding the result into clauses. We call the resulting k-fold Tseitin formula, $TS^k(G, \vec{c})$, and its variable set, $Vars^k(G) = \{y_e^i \mid e \in E, i \in [k]\}$.

For a fixed graph G and different odd-charge vectors $\vec{c} \in \{0, 1\}^{V(G)}$, the various problems $Search_{TS^k(G,\vec{c})}$ are very closely related. Define $\text{ODDCHARGE}^k(G)$ to be the k-party NOF communication search problem which takes as input an odd charge vector $\vec{c} \in \{0, 1\}^{V(G)}$, seen by all players, and an assignment α to $Vars^k(G)$, in which player i sees all values but the assignment α_e^i to y_e^i for $e \in E(G)$, and requires that the players output a vertex v that is a solution to $Search_{TS^k(G,\vec{c})}$.

3   Reduction from Set Disjointness to ODDCHARGE

We give a sequence of reductions to show that for a suitably chosen graph G, an efficient k-party NOF communication complexity protocol for $\text{ODDCHARGE}^k(G)$ will imply an efficient 1-sided error randomized k-party NOF protocol for the set disjointness relation.

We apply the Valiant-Vazirani argument to show that, without loss of generality, it suffices to derive a 1-sided error protocol for a version of set disjointness in which the input has intersection size 0 or size 1, and the job of the players is to distinguish between these two cases. We call this promise problem *zero/one set disjointness*.

Our reduction from zero/one set disjointness to $\text{ODDCHARGE}^k(G)$ goes via an intermediate problem, $\text{EVENCHARGE}^k(G)$, which is the exact analog of $\text{ODDCHARGE}^k(G)$ except that the input charge vector $\vec{c}$ is even rather than odd and the requirement is *either* to find a charge violation or to determine that no charge violation exists.

The reduction from EVENCHARGE$^k(G)$ to ODDCHARGE$^k(G)$, which is similar in spirit to a reduction of Raz and Wigderson [20], works by planting a single randomly chosen additional charge violation. This yields a protocol for EVENCHARGE$^k(G)$ that works well on average for each class of inputs with a given number of charge violations.

The most difficult part of our argument is the reduction from zero/one set disjointness to EVENCHARGE$^k(G)$ for suitable graphs G. The key idea is that for even $\vec{c}$, charge violations of $TS^k(G,\vec{c})$ come in pairs: Given an instance $\vec{x} \in (\{0,1\}^m)^k$ of zero/one set disjointness, using the public coins, the players randomly choose an even charge vector $\vec{c}$ and m vertex-disjoint paths in G, $p_1, \ldots, p_m$, for each $j \in [m]$, the players plant the $x_{1,j}, \ldots, x_{k,j}$ as the assignment along each edge of path p_j, in a random solution that otherwise meets the chosen charge constraint. By construction, a charge violation can occur only at the endpoints of a path and only if there is an intersection in the set disjointness problem.

It is tricky to ensure that the resulting problem looks sufficiently like a random instance of EVENCHARGE$^k(G)$ with either 0 or 2 charge violations so that we can apply the average case properties of the protocol for EVENCHARGE$^k(G)$. This places major constraints on the graph G and in particular requires that $m \leq n^{1/3}/\log n$ where $|V(G)| = n$. The bulk of the work is in showing that a small number of specific properties: rapid mixing, modest degree, and high girth – properties all met by a family of expanders constructed in [17] – are sufficient.

Distributions on labeled graphs. For the rest of the paper in the Tseitin tautologies we will use a family of graphs H_n that is the union of two edge-disjoint graphs on the same set of n vertices $[n]$, G_n and T_n. G_n will be a Δ-regular expander graph of the form defined by Lubotzky, Phillips, and Sarnak [17] for $\Delta = \Theta(\log n)$. Since $\overline{G_n}$ has degree $> n/2$, there is a spanning tree T_n of maximum degree 2 (a Hamiltonian path) in $\overline{G_n}$. Clearly H_n also has maximum degree $\Theta(\log n)$ and thus $TS^k(H_n, \vec{c})$ has size $n^{O(k)}$.

Let H_n be such a graph and let $\vec{c}$ be an even charge vector. We define $Sol(H_n, \vec{c})$ to be the set of all 0/1 assignments to the edges of H_n so that for each vertex $v \in [n]$, the parity of edges incident with v is equal to c_v. A uniform random distribution over $Sol(H_n, \vec{c})$ can be obtained by first selecting 0/1 values uniformly at random for all edges in G_n and then choosing the unique assignment to the edges of T_n that fulfill the charge constraints given by $\vec{c}$.

Given a bit value b associated with an edge $e \in G_n$, we can define a uniform distribution $\mathcal{L}^k = \mathcal{L}_k(b)$ over the corresponding variables y_e^i, $i \in [k]$. Such an assignment is chosen randomly from $\mathcal{L}_k$ on input b by the following experiment. If $b = 1$ then set all variables associated with edge e, y_e^i, $i \in [k]$ to 1. Otherwise if $b = 0$, set the vector $(\vec{y}_e)_{i \in [k]}$ by choosing uniformly at random from the set of $2^k - 1$ not-all-1 vectors.

Definition 1. *For any $t \geq 0$ let $\mathcal{D}_t$ be a distribution given by the following experiment on input $H_n = G_n \cup T_n$.*
1. *Choose an even charge vector $\vec{c} \in \{0,1\}^n$ uniformly at random.*
2. *Choose some $\beta \in Sol(H_n, \vec{c})$ uniformly at random.*

3. For each $e \in G_n$, select the values for the vector $(y_e)_{i \in [k]}$ from $\mathcal{L}_k(\beta_e)$ and for each $e \in T_n$, set $y_e^i = \beta_e$ for all $i \in [k]$.
4. Select a random subset $U \subseteq [n]$ of $2t$ vertices and produce charge vector $\vec{c}^U$ from $\vec{c}$ by toggling all bits c_v for $v \in U$.
5. Return the pair $(\alpha, \vec{c}^U)$ where α is the boolean assignment to the variables y_e^i, $i \in [k]$, $e \in H_n$.

Reduction from EVENCHARGE to ODDCHARGE

Lemma 1. *Let G be any connected graph on n vertices and let $\Delta(G)$ be the maximum degree in G. Suppose that Π_{odd} is a randomized k-party NOF protocol for $\text{ODDCHARGE}^k(G)$ that produces an answer with probability at least $1 - \epsilon$, is correct whenever it produces an answer, and uses at most s bits of communication. Then there is a randomized k-party NOF protocol Π_{even} for $\text{EVENCHARGE}^k(G)$ that uses $s + \Delta(G)$ bits of communication and has the following performance:*

$$\Pr_{(\alpha, \vec{c}) \in \mathcal{D}_0} [\Pi_{even}(\alpha, \vec{c}) = \text{true}] = 1$$

$$\Pr_{(\alpha, \vec{c}) \in \mathcal{D}_t} [\Pi_{even}(\alpha, \vec{c}) \in Err(\alpha, \vec{c})] \geq 2/3 - \epsilon \text{ for } t \geq 1.$$

Proof. Let Π_{odd} be a protocol for $\text{ODDCHARGE}^k(G)$ and assume that $V(G) = [n]$. We give a protocol Π_{even} for $\text{EVENCHARGE}^k(G)$. On input $(\alpha, \vec{c})$ and random public string r: Using r, choose a random vertex $v \in [n]$. Check whether the parity equation associated with vertex v is satisfied by α using at most $\Delta(G)$ bits of communication. If it is not, return v. Otherwise, create an odd charge vector, $\vec{c}^{\{v\}}$, which is just like $\vec{c}$ except that the value of c_v is toggled. Now run Π_{odd} on input $(\vec{c}^{\{v\}}, \alpha)$. If Π_{odd} returns the planted error v or if Π_{odd} does not return a value then return "true"; if Π_{odd} returns $u \neq v$, output u.

Suppose that $(\alpha, \vec{c}) \in \mathcal{D}_0$. Then α satisfies all charges specified by $\vec{c}$, so when Π_{odd} returns a vertex the above protocol must output "true" because Π_{odd} has one-sided error–that is, Π_{odd} will only return a vertex u when there is an error on the parity equation associated with u. Now suppose that $(\alpha, \vec{c}) \in \mathcal{D}_t$ so exactly $2t$ parity equations are violated. If the random vertex v does not satisfy its parity constraints, then the algorithm is correct. The remaining case is when v satisfies the parity equation and in this case we call Π_{odd} on a pair $(\alpha, \vec{c}^{\{v\}})$ where exactly $2t + 1$ parity equations are violated.

We show the probability bound separately for each $T \in [n]^{(2t+1)}$. Because the events $Err(\alpha, \vec{c}) = T$ partition the probability space, this proves the claim. By symmetry, for $T \in [n]^{(2t+1)}$ and any function g with codomain T, we have that $\Pr_{\alpha, \vec{c}, v}[g(\alpha, \vec{c}^{\{v\}}) = v \mid Err(\alpha, \vec{c}^{\{v\}}) = T] = 1/(2t + 1)$ since it is equally likely for $\vec{c} = \vec{c}^{\{v\}}$ to be generated as $\vec{c}^{\{u\}}$ for any $u \in T$. Thus we obtain:

$$\Pr_{\alpha, \vec{c}, v}[\Pi_{even}(\alpha, \vec{c}^{\{v\}}) \text{ errs} \mid Err(\alpha, \vec{c}^{\{v\}}) = T]$$

$$= \Pr_{\alpha, \vec{c}, v}[\Pi_{odd}(\alpha, \vec{c}^{\{v\}}) = v \text{ or } \Pi_{odd}(\alpha, \vec{c}^{\{v\}}) \text{ is not defined} \mid Err(\alpha, \vec{c}^{\{v\}}) = T]$$

$$\leq 1/(2t + 1) + \epsilon \leq 1/3 + \epsilon \quad \text{for } t \geq 1.$$

Reduction from Zero/One Set Disjointness to EVENCHARGE: We now show how to use a k-party NOF communication complexity protocol Π_{even} for EVENCHARGE$^k(H_n)$ as guaranteed by Lemma 1 to produce a k-party NOF protocol for the zero/one set disjointness problem which uses the following definition.

Definition 2. *Let $P_l^{(m)}$ be the set of all sequences of m vertex-disjoint length l paths in G_n.*

Lemma 2. *Let $m = n^{1/3}/\log n$. For sufficiently large n and for any even charge vector $\vec{c}$, if there is a probabilistic k-party NOF communication complexity protocol, Π_{even} for* EVENCHARGE$^k(H_n)$ *using s bits, satisfying the conditions in Lemma 1 for $\mathcal{D}_0$ and $\mathcal{D}_1$, then there is a randomized $(0, 1/3 + \epsilon + o(1))$ error k-party NOF communication complexity protocol Π_{01disj} for zero/one set disjointness on input $\vec{x} \in (\{0,1\}^m)^k$ that uses s bits of communication.*

Proof. Let $\vec{x}$ be an instance of zero/one set disjointness. Protocol Π_{01disj} will call Π_{even} on the graph H_n, on a pair $(\alpha, \vec{c})$ chosen according to the following distribution/experiment:
1. On input $\vec{x}$ with public coins r:
   (a) Using public coins r, choose a random even charge vector $\vec{c} \in \{0,1\}^n$.
   (b) Using public coins r, choose a sequence of m vertex-disjoint length l paths, $p_1, \ldots p_m$ uniformly at random from $P_l^{(m)}$.
   (c) Using the public coins r, choose $\beta \in Sol(H_n - \bigcup_{j=1}^m p_j, \vec{c})$
2. For all edges $e \in H_n$, all players other than player i compute α_e^i as follows:
   (a) If $e \in p_j$ for $j \in [m]$, set $\alpha_e^i = x_{i,j}$
   (b) If $e \in G_n$ and $e \notin \bigcup_{j=1}^m p_j$, choose the vector $\alpha_e^1 \ldots \alpha_e^k$ according to the distribution $\mathcal{L}_k(\beta_e)$.
   (c) For the remaining edges $e \in T_n$, set all variables α_e^i for $i \in [k]$ equal to β_e.
3. Return $(\alpha, \vec{c})$

We write $\mathcal{R}(\vec{x})$ to denote the distribution on assignment/charge pairs produced by reduction Π_{01disj} when given an input $\vec{x}$. The following lemma, proven in section 4, has the main technical argument and shows that for $t = |\cap \vec{x}| \in \{0, 1\}$, although $\mathcal{R}(\vec{x})$ is not the same as $\mathcal{D}_t$, $\mathcal{R}(\vec{x})$ is close to the distribution $\mathcal{D}_t$ in the ℓ_1 norm.

Lemma 3. *Let $\vec{x} \in (\{0,1\}^m)^k$ and $|\cap \vec{x}| = 1$. Then $||\mathcal{R}(\vec{x}) - \mathcal{D}_1||_1$ is $o(1)$.*

Protocol Π_{01disj} will output 0 if Π_{even} returns "true" and 1 otherwise. If $\cap \vec{x} = \emptyset$, by the above construction, the support of $\mathcal{R}(\vec{x})$ is contained in that of $\mathcal{D}_0$ and thus on $\mathcal{R}(\vec{x})$, Π_{even} must answer "true" and the vector $\vec{x}$ is correctly identified as being disjoint. In the case that $\cap \vec{x}$ contains exactly one element, $\Pr[\Pi_{01disj}(\vec{x})) = 0] \geq 2/3 - \epsilon - o(1)$. This completes the proof of the Lemma 2.

Reduction from Set disjointness to Zero/One Set disjointness

Lemma 4. *If there is an $(0, \epsilon)$ randomized NOF protocol for the k-party zero-one-promise set-disjointness problem that uses s bits of communication where ϵ is a constant < 1, then there is a $(0, \frac{1}{3})$ randomized NOF protocol for the k-party set-disjointness problem that uses $O(s \log n)$ bits of communication.*

Naturally, our starting point is the well-known result of Valiant and Vazirani [21].

Lemma 5 (Valiant-Vazirani). *Let a be a positive integer. Fix a nonempty $S \subseteq \{0,1\}^a$, and choose $w_1, \ldots w_a \in \{0,1\}^a$ independently and uniformly. With probability at least $1/4$, there exists $j \in \{0, \ldots, a\}$ so that $|\{x \in S \mid \forall i \leq j,\ x \cdot w_i = 0\}| = 1$.*

Proof (of Lemma 4). Let Π be the protocol for the promise problem. Set $a = \lceil \log n \rceil$. Using public coins, independently and uniformly choose $w_1, \ldots w_l \in \{0,1\}^a$. For $j \in \{0, \ldots a\}$, the players run the protocol Π, using the following rule for evaluating the input $x_{i,r}$ for $i \in [k], r \in [m]$: interpret r as a vector in $\{0,1\}^a$, and replace the value of $x_{i,r}$ by zero if for some $j' \leq j$, $w_{j'} \cdot r \neq 0$, and use the value $x_{i,r}$ if for all $j' \leq j$, $w_{j'} \cdot r = 0$. If the protocol Π returns 1, the players halt and output 1, otherwise, the players proceed to round $j + 1$. If no intersection is found after all $a + 1$ rounds, the players announce that the inputs are disjoint.

Clearly, this protocol uses $O(s \log n)$ bits of communication, and by the 0-error property of Π on disjoint inputs, it never outputs 1 when the inputs are disjoint. When the inputs are non-disjoint, the Valiant-Vazirani construction ensures that with probability at least $1/4$, at some round j the protocol Π is used on an input with a unique intersection, and therefore, conditioned on this event, the correct answer is returned with probability at least $1 - \epsilon$. Therefore, the correct answer is returned with probability at least $\frac{1}{4} - \frac{\epsilon}{4}$. Because ϵ is bounded away from 1 and the error is one-sided, a constant number of repetitions decreases the probability of error to $1/3$.

Combining the reductions

Theorem 2. *Let $k \geq 2$ and $m = n^{1/3}/\log n$. For each n there is an odd charge vector $\vec{c} \in \{0,1\}^n$ such that for any $\epsilon < 1/2$ the size of any tree-like* **Th(k-1)** *refutation of $TS^k(H_n, \vec{c})$ is at least $2^{\Omega((R^k_\epsilon(\mathrm{DISJ}_{k,m})/\log n)^{1/3})}$. Further if the coefficients in the* **Th(k-1)** *refutations are bounded by a polynomial in n then the refutation size must be at least $2^{\Omega(R^k_\epsilon(\mathrm{DISJ}_{k,m})/(\log n (\log \log n)^2))}$ or at least $2^{\Omega(D^k_\epsilon(\mathrm{DISJ}_{k,m})/\log^2 n)}$.*

Proof (Sketch). By Theorem 1 and the definition of $\mathrm{ODDCHARGE}^k(H_n)$, if for every $\vec{c} \in \{0,1\}^n$ there is tree-like **Th(k-1)** refutation of $TS^k(H_n, \vec{c})$ of size at most S, then there is a $1/n$-error randomized k-party NOF communication complexity protocol for $\mathrm{ODDCHARGE}^k(H_n)$ in which at most $O(\log^3 S)$ bits are communicated. By sending one more bit the players can check that the answer is correct and only output it in this case. Then applying Lemmas 1, 2, and 4 in turn yields an error $1/3$ randomized k-party NOF protocol for $\mathrm{DISJ}_{k,m}$ of complexity $O(\log^3 S \log n + \log^2 n)$ bits in total. Applying a similar reduction using the other parts of Theorem 1 yields the claimed result.

In the full paper we prove that the same lower bounds as Theorem 2 hold for *every* odd charge vector $\vec{c} \in \{0,1\}^n$.

4 Proximity of Distributions $\mathcal{D}_1$ and $\mathcal{R}(\vec{x})$ When $|\cap \vec{x}| = 1$

In this section we prove Lemma 3 that for $|\cap \vec{x}| = 1$ the distributions $\mathcal{R}(\vec{x})$ and $\mathcal{D}_1$ are close in the ℓ_1 norm. Let $\mu_{\mathcal{D}_1}$ and $\mu_{\mathcal{R}(\vec{x})}$ be their associated probability measures.

We will show that for all but a set of $(\alpha, \vec{c})$ with $\mu_{\mathcal{D}_1}$ measure $o(1)$, $\mu_{\mathcal{D}_1}(\alpha, \vec{c}) = (1 \pm o(1))\mu_{\mathcal{R}(\vec{x})}(\alpha, \vec{c})$.

Given an instance of the set disjointness variables, $\vec{x} = (\{0,1\}^m)^k$, for $j \in [m]$ we say that the *color* of j is the tuple $(x_{1,j}, \ldots, x_{k,j}) \in \{0,1\}^k$. By construction, the assignment $\mathcal{R}(\vec{x})$ produced by R on this instance has color $(x_{1,j}, \ldots, x_{k,j})$ on each edge of the path p_j.

Definition 3. *Given an ordered sequence of paths $\vec{p} \in P_l^{(m)}$, an $\vec{x} \in (\{0,1\}^m)^k$, and an assignment α, write $\chi(\alpha_{\vec{p}}) = \vec{x}$ if and only if every edge on path p_j has color $(x_{1,j}, \ldots, x_{k,j})$ for every $j \in [m]$.*

We first observe that for any $(\alpha, \vec{c})$ with $|\text{Err}(\alpha, \vec{c})| = 2$ the probability $\mu_{\mathcal{D}_t}(\alpha, \vec{c})$ depends only on the number of edges $e \in G_n$ having color 1^k in α.

Definition 4. *Let $\phi(a, b) = 2^{-a}(2^k - 1)^{-(a-b)}$.*

Lemma 6. *For any $(\alpha, \vec{c})$ with $|\text{Err}(\alpha, \vec{c})| = 2t$ and $m_1 = |\{e \in E(G_n) \mid \alpha_e = 1^k\}|$, $\mu_{\mathcal{D}_1}(\alpha, \vec{c}) = \phi(|E(G_n)|, m_1)/(2^{n-1}\binom{n}{2t})$.*

Proof. Let $U = \text{Err}(\alpha, \vec{c})$. The probability under $\mathcal{D}_1$ that U is chosen to be flipped is $1/\binom{n}{2t}$ and, given U, all of the 2^{n-1} even charge vectors $\vec{c}^U$ are equally likely. Conditioned on these events, the chance that α labels the edges for the randomly selected element of $Sol(H_n, \vec{c})$ is $2^{-|E(G_n)|}(2^k - 1)^{-(|E(G_n)|-m_1)}$.

Definition 5. *For $U \subset V$ with $|U| = 2$ let $P_l^{(m)}(U)$ be the set of all elements of $P_l^{(m)}$ that have a path whose endpoints are U.*

Now consider the measure $\mu_{\mathcal{R}(\vec{x})}(\alpha, \vec{c})$. Let $\{i\} = \cap \vec{x} \subseteq [n]$, $U = \text{Err}(\alpha, \vec{c})$ with $|U| = 2$, and $m_1 = |\{e \in E(G_n) \mid \alpha_e = 1^k\}|$. By the definition of R,

$$\mu_{\mathcal{R}(\vec{x})}(\alpha, \vec{c}) = \Pr_{\vec{p} \in P_l^{(m)}}[\text{Ends}(p_i) = \text{Err}(\alpha, \vec{c}) \wedge \chi(\alpha_{\vec{p}}) = \vec{x}]$$

$$\times \Pr_{\vec{c}' \in \{0,1\}^n,\ \alpha' \in \mathcal{L}_k(Sol(H_n - \vec{p}, \vec{c}'))}[\alpha' = \alpha_{G_n - \vec{p}} \text{ and } \vec{c}' = \vec{c}]$$

$$= \Pr_{\vec{p} \in P_l^{(m)}}[\text{Ends}(p_i) = \text{Err}(U)] \times \Pr_{\vec{p} \in P_l^{(m)}(U)}[\chi(\alpha_{\vec{p}}) = \vec{x}]$$

$$\times \phi(|E(G_n)| - ml, m_1 - l)/2^{n-1}.$$

Observe that p_i is a uniformly chosen element of P_l and we can analyze the first term using the following property of random paths on LPS expanders proven in the full paper.

Lemma 7. *For $u \neq v \in V(G_n)$ and $l \geq c_1 \log n / \log \log n$,*
$$\Pr_{p \in P_l}[\text{Ends}(p) = \{u, v\}] = (1 \pm o(1))/\binom{n}{2}.$$

Thus $\mu_{\mathcal{R}(\vec{x})}(\alpha, \vec{c}) = (1 \pm o(1)) \dfrac{\phi(|E(G_n)| - ml, m_1 - l)}{\binom{n}{2} 2^{n-1}} \cdot \Pr_{\vec{p} \in P_l^{(m)}(U)} [\chi(\alpha_{\vec{p}}) = \vec{x}]$

$= (1 \pm o(1)) \dfrac{\mu_{\mathcal{D}_1}(\alpha, \vec{c})}{\phi(ml, l)} \cdot \Pr_{\vec{p} \in P_l^{(m)}(U)} [\chi(\alpha_{\vec{p}}) = \vec{x}].$

It follows that we will obtain the desired result if we can show that for all but a $o(1)$ measure of $(\alpha, \vec{c})$ under $\mu_{\mathcal{D}_1}$,

$$\Pr_{\vec{p} \in P_l^{(m)}(U)} [\chi(\alpha_{\vec{p}}) = \vec{x}] = (1 \pm o(1)) \phi(ml, l) = (1 \pm o(1)) 2^{-ml} (2^k - 1)^{-(m-1)l}$$

where $U = \mathrm{Err}(\alpha, \vec{c})$. In the case that this happens, we say that $(\alpha, \vec{c})$ is *well-distributed for $\vec{x}$*.

Using the second moment method we prove the following lemma which shows that for all but a $o(1)$ measure of $(\alpha, \vec{c})$ under $\mu_{\mathcal{D}_1}$, $(\alpha, \vec{c})$ is indeed well-distributed for $\vec{x}$. The detailed proof is given in the full paper; the proof uses the fact that $\Theta(\log n)$-degree LPS expanders have $O(\log n / \log \log n)$ mixing time and $\Omega(\log n / \log \log n)$ girth.

Lemma 8. *Let $m \leq n^{1/3}/\log n$ and $l = 2\lceil c_1 \log n / \log \log n \rceil$ and $\vec{x} \in (\{0,1\}^m)^k$ with $|\cap \vec{x}| = 1$. For almost all $U \subset [n]$ with $|U| = 2$,*
$$\Pr_{(\alpha, \vec{c}) \in \mathcal{D}_1}[(\alpha, \vec{c}) \text{ is well-distributed for } \vec{x} \mid \mathrm{Err}(\alpha, \vec{c}) = U] = 1 - o(1).$$

Lemma 3 follows from this almost immediately.

Proof (of Lemma 3). Let $\vec{x} \in (\{0,1\}^m)^k$ and $|\cap \vec{x}| = 1$. By Lemma 8 and the preceding argument, for all but a set B of U that forms $o(1)$ fraction of all subsets $[n]$ of size 2, $\Pr_{(a, \vec{c}) \in \mathcal{D}_1}[\mu_{\mathcal{R}(\vec{x})}(\alpha, \vec{c}) = (1 \pm o(1)) \mu_{\mathcal{D}_1}(\alpha, \vec{c}) \mid \mathrm{Err}(\alpha, \vec{c}) = U] = 1 - o(1)$. By Lemma 7, $\Pr_{(\alpha, \vec{c}) \in \mathcal{D}_1}[\mathrm{Err}(\alpha, \vec{c}) \in B] = o(1)$. Therefore by summing over distinct choices of U, we obtain that with probability $1 - o(1)$ over $(\alpha, \vec{c}) \in \mathcal{D}_1$, $\mu_{\mathcal{R}(\vec{x})}(\alpha, \vec{c}) = (1 \pm o(1)) \mu_{\mathcal{D}_1}(\alpha, \vec{c})$. This is equivalent to the desired conclusion that $||\mathcal{D}_1 - \mathcal{R}(\vec{x})||_1$ is $o(1)$.

5 Discussion

There are a couple of interesting open problems related to our work beyond the natural problem of the communication complexity of DISJ_k. First, does semantic LS^k have a separation oracle, as LS does? This is closely related to whether or not LS^k is automatizable and we conjecture that the answer to both questions is negative. Secondly, is it possible to extend our lower bounds to other tautologies that would imply inapproximability results for polynomial-time LS^k-based algorithms? (For example, if we could prove superpolynomial lower bounds for tree-like LS^k proofs of random 3CNF formulas, this would imply inapproximability results for LS^k-based linear programming algorithms for MaxSAT [5].)

Finally we would like to point out a connection between our main result and the complexity of disjoint NP pairs. An open question in complexity theory is whether or not all pairs of disjoint NP sets can be separated by a set in P. This is known to be false under the assumption P $\neq$ UP and also by the assumption P $\neq$ NP $\cap$ coNP. It is an open question whether or not it is implied by P $\neq$ NP. Let us consider the same question with respect to communication complexity rather than polynomial time: can every pair of relations with small nondeterministic k-party communication complexity be separated by a small probabilistic/deterministic protocol? In [20] the answer is shown to be unconditionally false for $k = 2$. In particular, they give a pair of disjoint properties on $3m$-vertex graphs G, a matching on $2m$ vertices of G and an independent set of $2m + 1$ vertices of G, and show that this pair cannot be separated by any small probabilistic/deterministic protocol. In this paper, we have shown that for any k, the question is still false, under k-$\mathsf{RP}^{cc} \neq k$-NP^{cc}.

Acknowledgements

We are indebted to Avi Wigderson for helpful discussions and insights.

References

1. S. Arora, B. Bollobás, and L. Lovász. Proving integrality gaps without knowing the linear program. In *Proceedings 43nd Annual Symposium on Foundations of Computer Science*, pages 313–322, Vancouver, BC, November 2002. IEEE.
2. L. Babai, P. Frankl, and J. Simon. Complexity classes in communication complexity theory. In *27th Annual Symposium on Foundations of Computer Science*, pages 337–347, Toronto, Ontario, October 1986. IEEE.
3. P. Beame, T. Pitassi, N. Segerlind, and A. Wigderson. A direct sum theorem for corruption and the multiparty NOF communication complexity of set disjointness. In *Proceedings Twentieth Annual IEEE Conference on Computational Complexity*, San Jose, CA, June 2005.
4. A. Bockmayr, F. Eisenbrand, M.E. Hartmann, and A.S. Schulz. On the Chvatal rank of polytopes in the 0/1 cube. *Discrete Applied Mathematics*, 98(1-2):21–27, 1999.
5. J. Buresh-Oppenheim, N. Galesi, S. Hoory, A. Magen, and T. Pitassi. Rank bounds and integrality gaps for cutting planes procedures. In *Proceedings 44th Annual Symposium on Foundations of Computer Science*, pages 318–327, Boston, MA, October 2003. IEEE.
6. V. Chvátal. Edmonds polytopes and a hierarchy of combinatorial problems. *Discrete Mathematics*, 4:305–337, 1973.
7. V. Chvátal, W. Cook, and M. Hartmann. On cutting-plane proofs in combinatorial optimization. *Linear Algebra and its Applications*, 114/115:455–499, 1989.
8. W. Cook, C. R. Coullard, and G. Turan. On the complexity of cutting plane proofs. *Discrete Applied Mathematics*, 18:25–38, 1987.
9. S. Dash. *On the matrix cuts of Lovász and Schrijver and their use in Integer Programming*. PhD thesis, Department of Computer Science, Rice University, March 2001.
10. S. Dash. An exponential lower bound on the length of some classes of branch-and-cut proofs. In W. Cook and A. S. Schulz, editors, *IPCO*, volume 2337 of *Lecture Notes in Computer Science*, pages 145–160. Springer-Verlag, 2002.
11. F. Eisenbrand and A. S. Schulz. Bounds on the Chvatal rank of polytopes in the 0/1-cube. *Combinatorica*, 23(2):245–261, 2003.

12. D. Grigoriev, E.A. Hirsch, and D.V. Pasechnik. Complexity of semi-algebraic proofs. In *(STACS) 2002: 19th Annual Symposium on Theoretical Aspects of Computer Science*, volume 2285 of *Lecture Notes in Computer Science*, pages 419–430, Antibes, France, February 2002. Springer-Verlag.
13. R. Impagliazzo, T. Pitassi, and A. Urquhart. Upper and lower bounds on tree-like cutting planes proofs. In *9th Annual IEEE Symposium on Logic in Computer Science*, pages 220–228, Paris, France, 1994.
14. B. Kalyanasundaram and Georg Schnitger. The probabilistic communication complexity of set intersection. In *Proceedings, Structure in Complexity Theory, Second Annual Conference*, pages 41–49, Cornell University, Ithaca, NY, June 1987. IEEE.
15. L. G. Khachian. A polynomial time algorithm for linear programming. *Doklady Akademii Nauk SSSR, n.s.*, 244(5):1093–1096, 1979. English translation in *Soviet Math. Dokl.* 20, 191–194.
16. L. Lovasz and A. Schrijver. Cones of matrices and set-functions and 0-1 optimization. *SIAM J. Optimization*, 1(2):166–190, 1991.
17. A. Lubotzky, R. Phillips, and P. Sarnak. Ramanujan graphs. *Combinatorica*, 8(3):261–277, 1988.
18. N. Nisan. The communication complexity of threshold gates. In V.S.D. Mikl'os and T. Szonyi, editors, *Combinatorics: Paul Erdös is Eighty, Volume I*, pages 301–315. Bolyai Society, 1993.
19. P. Pudlák. Lower bounds for resolution and cutting plane proofs and monotone computations. *Journal of Symbolic Logic*, 62(3):981–998, September 1997.
20. R. Raz and A. Wigderson. Monotone circuits for matching require linear depth. *Journal of the ACM*, 39(3):736–744, July 1992.
21. L. Valiant and V. Vazirani. NP is as easy as detecting unique solutions. *Theoretical Computer Science*, pages 85–93, 1986.

On the l-Ary GCD-Algorithm in Rings of Integers

Douglas Wikström

Royal Institute of Technology (KTH)
KTH, Nada, S-100 44 Stockholm, Sweden

Abstract. We give an l-ary greatest common divisor algorithm in the ring of integers of any number field with class number 1, i.e., factorial rings of integers. The algorithm has a quadratic running time in the bit-size of the input using naive integer arithmetic.

1 Introduction

The greatest common divisor (GCD) of two integers a and b is the largest integer d such that d divides both a and b. The problem of finding the GCD of two integers efficiently is one of the oldest problems studied in number theory. The corresponding problem can be considered for two elements α and β in any factorial ring R. Then $\lambda \in R$ is a GCD of α and β if it divides both elements, and whenever $\lambda' \in R$ divides both α and β it also holds that λ' divides λ. A precise understanding of the complexity of different GCD algorithms gives a better understanding of the arithmetic in the domain under consideration.

1.1 Previous Work

The Euclidean GCD algorithm is well known. The basic idea of Euclid is that if $|a| \geq |b|$, then $|a \bmod b| < |b|$. Since we always have $\gcd(a, b) = \gcd(a \bmod b, b)$, this means that we can replace a with $a \bmod b$ without changing the GCD. Swapping the order of a and b does not change the GCD, so we can repeatedly reduce $|a|$ or $|b|$ until one becomes zero, at which point the other equals the GCD of the original inputs. In a more general setting with α and β in a factorial ring R, Euclid's idea works essentially unchanged if there exists a valuation $v : R \to \mathbb{R}^+$ with the following properties for $\alpha, \beta \in R$. There exists $\gamma, \delta \in R$ with $\alpha = \gamma\beta + \delta$ and $\delta = 0$ or $v(\delta) < v(\beta)$, if $\alpha\beta \neq 0$ then $v(\alpha) < v(\alpha\beta)$. Rings for which there exists such a valuation are called Euclidean. If in an algebraic ring $v(\alpha) = |N\alpha|$, where $N\alpha$ is the algebraic norm of α, the ring is called norm-Euclidean. Most algebraic rings are not even Euclidean. If we also want the Euclidean algorithm to terminate there must be a constant k such that $\{\alpha \mid v(\alpha) < k\}$ is finite.

All is however not lost. Kaltofen and Rolletschek [5] devise a GCD algorithm with quadratic running time for the ring of integers in any quadratic number field. Their approach is based on the idea to find an integer j such that $N(j\alpha \bmod$

$\beta) < N\beta$. This is always possible with $|j|$ bounded essentially by the square root of the discriminant. We are not aware of any generalization of this approach to more general rings of integers.

Interestingly, there are alternative approaches to compute the GCD. These are generalizations of Stein's binary GCD algorithm [9], which is particularly well suited for implementation on computers. It is based on the following facts.

$$\begin{aligned}
\gcd(a,b) &= 2\gcd(a/2, b/2) && \text{if } a \text{ and } b \text{ are even,} \\
\gcd(a,b) &= \gcd(a/2, b) && \text{if } a \text{ is even and } b \text{ is odd, and} \\
\gcd(a,b) &= \gcd((a-b)/2, b) && \text{if } a \text{ and } b \text{ are odd.}
\end{aligned}$$

One may always apply one of the rules to reduce the size of elements, while preserving the GCD. Thus, by simply shifting and subtracting integers, the GCD of two integers can be computed. Weilert [11] generalizes this algorithm to the Gaussian integers. Damgård and Skovbjerg Frandsen [3,4] independently also generalize the binary algorithm to the Eisenstein and Gaussian integers.

Sorenson [8] give the l-ary algorithm for computing the GCD of integers, which generalizes the binary algorithm. The l-ary algorithm is based on the result by Minkowski that given a and b one can find c and d such that $ca+db = 0 \bmod l$ for an integer l, where $|c|$ and $|d|$ essentially are bounded by $\sqrt{l}$. Thus, in each iteration the larger of a and b is replaced by $(ca+db)/l$. This is an analog to the binary algorithm, in that in each iteration the size of the largest integer is reduced roughly by a factor $2\sqrt{l}/l$. The details of this algorithm is slightly more involved than the binary algorithm, since the linear expression does not preserve the GCD. Sorenson also constructs a parallel version of his algorithm. Weilert [10] generalizes also this algorithm to the Gaussian integers.

Agarwal and Skovbjerg Frandsen [2] introduce an algorithm related to both the binary and the l-ary algorithms for computing GCD in several complex quadratic rings. It is interesting to note that one of the rings they consider is not Euclidean.

Wikström [13] generalizes the l-ary approach to compute the GCD in the ring of integers in the octic cyclotomic field. This is the first l-ary GCD algorithm in a non-quadratic ring, and the main inspiration to the current work.

The binary or l-ary GCD algorithm in the ring of integers $\mathbb{Z}[\zeta_m]$ of the cyclotomic number fields $\mathbb{Q}(\zeta_m)$ for $m = 2, 3, 4, 8$ can be "translated" to compute the corresponding power residue symbol. Shallit and Sorenson [7] give a binary algorithm for computing the Jacobi symbol. Weilert [11] generalizes the idea to compute the quartic residue symbol. Independently, both Damgård and Skovbjerg Frandsen [3,4] and Wikström [12] generalize the idea to compute the cubic and quartic residue symbols. Wikström [13] also uses the idea to compute octic residue symbols.

1.2 Contribution

We give a GCD algorithm in the ring of integers $\mathcal{O}_K$ of any number field K with class number 1, i.e., rings of integers with unique factorization. Our result is non-uniform in the sense that we, for each ring, assume that we already know an integral basis, fundamental units, and several constants derived from these. The running time of the algorithm is quadratic in the bit-size of the input.

As far as we know, the only previous generic GCD algorithm with quadratic running time is given by Kaltofen and Rolletschek [5], and this is only applicable to quadratic rings. The algorithm in [5] is in some sense "almost Euclidean", whereas our algorithm generalizes the l-ary algorithm [8]. As explained in the introduction, l-ary algorithms have appeared in the literature for specific rings, but the present work is the first to give a generic description of this approach.

We are confident that our algorithm can be "translated" to compute the m-th power residue symbol in the ring of integers of the m-th cyclotomic number field $\mathbb{Q}(\zeta_m)$ if it has class number 1.

Proofs of all claims are given in [14], but we try to convey the main ideas used in each proof here.

2 Notation

We denote the number of elements in a finite set A by $\#(A)$. We write $\mathbb{Z}$, $\mathbb{Q}$, $\mathbb{R}$, and $\mathbb{C}$ for the rational integers, the rational numbers, the real numbers, and the complex numbers. The imaginary unit is denoted by $i = \sqrt{-1}$. We denote the complex absolute value by $|\cdot| : \mathbb{C} \to \mathbb{R}$, where $|a+bi| = \sqrt{a^2 + b^2}$. We write $\overline{\alpha}$ to denote the complex conjugate of an $\alpha \in \mathbb{C}$.

Throughout the paper we use K to denote a number field with class number 1, and we use $\mathcal{O}_K$ to denote its ring of integers. Since $\mathbb{Q}$ is a perfect field $K/\mathbb{Q}$ is a separable extension. The ring $\mathcal{O}_K$ has an integral basis which we denote by $\omega_1, \ldots, \omega_g$, since $\mathcal{O}_K$ is the integral closure of $\mathbb{Z}$ which is a principal ideal domain. This means that $\mathcal{O}_K = \mathbb{Z}\omega_1 + \ldots + \mathbb{Z}\omega_g$ and $K = \mathbb{Q}\omega_1 + \ldots + \mathbb{Q}\omega_g$. We write $\mathcal{O}_K^*$ to denote the units of $\mathcal{O}_K$, i.e., the invertible elements. The corresponding notation is used also for other domains. We use $\varepsilon_1, \ldots, \varepsilon_h$ to denote a maximal set of independent fundamental units in $\mathcal{O}_K$. We denote the group of roots of unity by $\mu(K)$. We denote by $G = \mathrm{Hom}_\mathbb{Q}(K, \mathbb{C})$ the set of $\mathbb{Q}$-embeddings of K into $\mathbb{C}$, i.e., isomorphisms of K, which keep $\mathbb{Q}$ fixed. This implies that $g = \#(G)$. We assume throughout that $g \geq 2$. We use multiplicative notation for the action of an element $\sigma \in \mathrm{Hom}_\mathbb{Q}(K, \mathbb{C})$, i.e., $\sigma : \alpha \mapsto \alpha^\sigma$. We denote by $N\alpha = \prod_{\sigma \in G} \alpha^\sigma$ the algebraic norm of α. For $\alpha \in \mathcal{O}_K^*$, we have $N\alpha \in \mathbb{Z}$. We use the term irreducible only for non-units. One source for the above facts is Neukirch [6].

The naive complexity model we use in this paper stipulates that addition or subtraction of positive integers x and y takes time $O(\log x + \log y)$, and multiplication, integer division or computing remainders takes time $O(\log x \log y)$.

3 Preliminary Results

Before we describe the algorithm and analyze it we need to generalize the results given in [13].

3.1 Balanced Elements

Consider the absolute value of the algebraic norm, $|N\alpha| = \prod_{\sigma \in G} |\alpha^\sigma|$, of an element $\alpha \in \mathcal{O}_K$. It is given by the product of the absolute values of the conjugates of α. The quotient $|\alpha^\sigma|/|\alpha^{\sigma'}|$ of two such absolute values can be arbitrarily large for elements with a fixed absolute norm $|N\alpha|$. However, it follows from Dirichlet's Unit Theorem that there exists an associate β of α for which the absolute values $|\beta^\sigma|$ are roughly the same size. This is an important observation, since it allows us to establish a weak triangle inequality. Informally, we could say that we can balance the complex absolute values of the algebraic conjugates of α. We use the following definition.

Definition 1 (Δ-Balanced Element) *We say that a non-zero $\alpha \in K$ is Δ-balanced if $|\alpha^\sigma| \leq \Delta|\alpha^{\sigma'}|$ for all $\sigma, \sigma' \in \mathrm{Hom}_\mathbb{Q}(K, \mathbb{C})$.*

Note that α is Δ-balanced precisely when all of its conjugates are Δ-balanced, and that the requirement is equivalent to $\frac{1}{\Delta}|\alpha^\sigma| \leq |\alpha^{\sigma'}|$ for all $\sigma, \sigma' \in G$.

3.2 A Weak Triangle Inequality

It would be nice if given $\alpha, \beta \in K$, we had $|N(\alpha + \beta)| \leq c\max\{|N\alpha|, |N\beta|\}$ for a constant $c \in \mathbb{R}$, i.e., some type of "triangle inequality". Unfortunately, for almost all K there is no such law. Instead we show that there exists a triangle inequality for balanced elements.

Theorem 1 (Triangle Inequality for Δ-Balanced Elements) *Let α and β be Δ-balanced elements in K, and set $g = \#(\mathrm{Hom}_\mathbb{Q}(K, \mathbb{C}))$. Then*

$$|N(\alpha + \beta)| \leq 2^g \Delta^{g-1} \max\{|N\alpha|, |N\beta|\} \ .$$

The idea of the proof is to expand the product $|N(\alpha+\beta)| = |\prod_{\sigma \in G}(\alpha^\sigma + \beta^\sigma)|$ as a sum, apply the triangle inequality for the complex absolute value, and bound each term using the fact that α and β are balanced.

Remark 1 If the conjugates of α can be organized in pairs of complex conjugates one can give a slightly tighter inequality as is done in [13].

3.3 Linear Combinations

In this section we construct the cofactors of the l-ary approach, but first we exhibit a large set of elements with relatively small norm. Let $l \in \mathbb{Z}$, $l > 0$, denote a constant to be determined later and define the set

$$S_l = \left\{ \sum_{j=1}^{g} a_j \omega_j \ \bigg| \ 0 \leq a_j \leq \sqrt{l} + 1 \right\} \ .$$

Each $\sigma \in G$ may be described as a $\mathbb{Z}$-linear map in the basis $\omega_1, \ldots, \omega_g$. We denote the matrix corresponding to this map by $f_\sigma = (f^\sigma_{k,j})_{1 \leq k,j \leq g}$, and define the constant $c_\omega = \max_{1 \leq k, j \leq g, \sigma \in G}\{|f^\sigma_{k,j}\omega_k|\}$. We have the following result.

Lemma 1 *Let $\gamma, \gamma' \in S_l$. Then for all $\sigma \in G$*

$$|(\gamma - \gamma')^\sigma| \leq g^2 c_\omega(\sqrt{l}+1) \ , \quad \text{and} \quad \#(S_l) > l^{g/2} \ .$$

The first part of the lemma follows by the linearity of σ, application of the triangle inequality, and the fact that elements in S_l have small positive coefficients. The second part follows by counting.

Denote by T_l the set of pairwise differences $T_l = \{\gamma - \gamma' \mid \gamma, \gamma' \in S_l\}$. We show that for any Δ-balanced elements $\alpha, \beta \in \mathcal{O}_K$, we can find elements $\gamma, \delta \in T_l$ such that $l \mid (\gamma\alpha + \delta\beta)$ and still keep $|N(\gamma\alpha + \delta\beta)|$ relatively small. More precisely we define $C_{\text{lin}}(l) = (g^2 c_\omega(\sqrt{l}+1))^g \Delta^{g-1}$ and have the following theorem.

Theorem 2 *Let α and β be Δ-balanced elements in $\mathcal{O}_K$. Then there exists $\gamma, \delta \in T_l$, with $(\gamma, \delta) \neq (0,0)$, such that $l \mid (\gamma\alpha + \delta\beta)$, and*

$$|N(\gamma\alpha + \delta\beta)| \leq C_{\text{lin}}(l) \max\{|N\alpha|, |N\beta|\} \ .$$

The idea of the proof is as follows. The existence of the γ and δ follows by the pigeon-hole principle. The bound follows by an argument similar to that in the proof of Theorem 1, except that we apply Lemma 1 to bound the norm of the cofactors γ and δ.

In the following we need a notation to identify the cofactors guaranteed to exist by the theorem. We write $\gamma_{\alpha,\beta}$ and $\delta_{\alpha,\beta}$ for a pair of cofactors in T_l such that $l \mid (\gamma_{\alpha,\beta}\alpha + \delta_{\alpha,\beta}\beta)$.

3.4 Spurious Factors

Sorenson [8] notes that $\gcd(a,b) = \gcd(ca + db, b)$ may not hold for rational integers $a, b, c, d \in \mathbb{Z}$. A similar problem arises for algebraic integers. Fortunately, the following straightforward lemma explains this completely.

Lemma 2 *Let α, β, γ, and δ lie in $\mathcal{O}_K$. Then $\gcd(\alpha, \beta) \mid \gcd(\gamma\alpha + \delta\beta, \beta)$ and $(\gcd(\gamma\alpha + \delta\beta, \beta)/\gcd(\alpha,\beta)) \mid \gamma$.*

3.5 Approximating the Norm of a Δ-Balanced Element

The norm of an element gives in some sense the "size" of the element. Unfortunately, the way the norm is defined requires multiplication of integers, which takes time $O(n^2)$ in the naive arithmetic model. This is far too expensive to be done in each iteration of our algorithm, since we are looking for an algorithm that has a total running time of $O(n^2)$. It is natural to try to approximate the norm, but since elements can have small norm but large representation, i.e., be unbalanced, there may be much cancellation during the computation of the norm.

We consider a weaker estimate of the size of an element, which we call $N_+ : K \to \mathbb{R}$, and prove some useful results about this function. We do not know how to compute this function quickly, but in contrast to the norm it can be approximated within a constant factor in linear time for elements in $\mathcal{O}_K$.

Definition 2 *Define $N_+ : K \to \mathbb{R}$ by $N_+\alpha = \max_{\sigma \in G}\{|\alpha^\sigma|\}$.*

It is not hard to see that N_+ approximates the norm N arbitrarily badly, but it turns out to be useful anyway. The next lemma says that if an element is Δ-balanced, then N_+ is essentially a good approximation of the norm N.

Lemma 3 *Let $\alpha \in K$ be Δ-balanced. Then $\sqrt[g]{|N\alpha|} \leq N_+\alpha \leq \Delta \sqrt[g]{|N\alpha|}$.*

For the lemma to be useful there must be a way to balance an element α without computing its norm $N\alpha$, but we ignore this issue for now. Instead we introduce a function N'_+ which approximates N_+ within a constant factor.

Definition 3 *Define $N'_+ : K \to \mathbb{R}$ by $N'_+\alpha = \max_{1 \leq j \leq g}\{|a_j|\}$ for an element $\alpha \in K$ given by $\alpha = \sum_{j=1}^{g} a_j \omega_j$ with $a_j \in \mathbb{Q}$.*

The function N'_+ can obviously be evaluated in linear time in the bit-size of the input when $\alpha \in \mathcal{O}_K$, since then $a_j \in \mathbb{Z}$. Next we show that it approximates N_+ within a constant factor.

Denote by $K_\mathbb{C}^*$ the direct product $\prod_{\sigma \in G} \mathbb{C}^*$, and denote by $\psi : K \to K_\mathbb{C}^*$ the map given by $\psi : \alpha \mapsto (\alpha^\sigma)_{\sigma \in G}$. We consider K as a g-dimensional $\mathbb{Q}$-vector space, where elements are represented in the basis $\omega_1, \ldots, \omega_g$. Then the image $\psi(K)$ is an isomorphic $\mathbb{Q}$-vector space from the $\mathbb{Q}$-linearity of the homomorphisms. Denote by $(\psi_{\sigma,j})_{\sigma \in G, 1 \leq j \leq g}$ the complex valued matrix which represents the map $\psi : K \to \psi(K)$ expressed in the basis $\omega_1, \ldots, \omega_g$. Denote by $(\psi'_{j,\sigma})_{\sigma \in G, 1 \leq j \leq g}$ the complex valued matrices corresponding to the map $\psi^{-1} : \psi(K) \to K$ expressed in the canonical orthonormal basis $\{e_\sigma\}_{\sigma \in G}$ for $K_\mathbb{C} = \prod_{\sigma \in G} \mathbb{C}$. Define $\Gamma = g \max_{1 \leq j \leq g, \sigma \in G}\{|\psi_{\sigma,j}|, |\psi'_{j,\sigma}|\}$. The lemma below follows straightforwardly from the linearity of ψ and its inverse.

Lemma 4 *Let $\alpha \in K$. Then $\frac{1}{\Gamma} N_+ \alpha \leq N'_+ \alpha \leq \Gamma N_+ \alpha$.*

Corollary 1 *Let $\alpha \in K$ be Δ-balanced. Then*

$$\frac{1}{\Gamma} \sqrt[g]{|N\alpha|} \leq N'_+ \alpha \leq \Gamma \Delta \sqrt[g]{|N\alpha|} \ .$$

3.6 Balancing Elements

In this section we prove a result that allows us to balance elements in $\mathcal{O}_K$ efficiently. Recall the statement of Dirichlet's Unit Theorem. It considers a number field K which has r real embeddings and s pairs of conjugates of complex embeddings of K in $\mathbb{C}$, and says that the group of units $\mathcal{O}_K^*$ is the direct product of the group of roots of unity, $\mu(K)$, and a free abelian group of rank $r + s - 1$. The theorem itself is not strong enough for our purposes, but we can extract a useful result from the construction used in its proof. We follow the exposition given in Neukirch [6], but use slightly different notation.

We have already defined the map $\psi : K \to K_\mathbb{C}^*$. Denote by $\text{vlog} : K_\mathbb{C}^* \to \prod_{\sigma \in G} \mathbb{R}$ the map given by $\text{vlog} : (z_\sigma)_{\sigma \in G} \mapsto (\log |z_\sigma|)_{\sigma \in G}$. Conjugation $F : z \mapsto \bar{z}$ in $\mathbb{C}$ induces involutions. In $K_\mathbb{C}^*$ it acts by $F(z_\sigma)_{\sigma \in G} = (\overline{z_{\bar{\sigma}}})_{\sigma \in G}$ and in $\prod_{\sigma \in G} \mathbb{R}$

it acts by $F(x_\sigma)_{\sigma \in G} = (x_{\bar\sigma})_{\sigma \in G}$. We define $K_\mathbb{R}^*$ and $\left[\prod_{\sigma \in G} \mathbb{R}\right]^+$ to be the vector spaces consisting of fixed points of F in $K_\mathbb{C}^*$ and $\prod_{\sigma \in G} \mathbb{R}$ respectively. If σ is a real embedding, it is clearly fixed by F, and the complex embeddings comes in pairs. Since there are s pairs of complex embeddings we see that $\left[\prod_{\sigma \in G} \mathbb{R}\right]^+$ is isomorphic to $\mathbb{R}^{r+s}$.

Define $N_\mathbb{C} : K_\mathbb{C} \to \mathbb{C}$, $N_\mathbb{C} : (z_\sigma)_{\sigma \in G} \mapsto \prod_{\sigma \in G} z_\sigma$ and $\text{Tr}_\mathbb{R} : \prod_{\sigma \in G} \mathbb{R} \to \mathbb{R}$, $\text{Tr}_\mathbb{R} : (x_\sigma)_{\sigma \in G} \mapsto \sum_{\sigma \in G} x_\sigma$. It is shown in [6] that the following diagram commutes.

$$\begin{array}{ccccc} K^* & \xrightarrow{\psi} & K_\mathbb{R}^* & \xrightarrow{\text{vlog}} & \left[\prod_{\sigma \in G} \mathbb{R}\right]^+ \\ \downarrow N & & \downarrow N_\mathbb{C} & & \downarrow \text{Tr}_\mathbb{R} \\ \mathbb{Q}^* & \longrightarrow & \mathbb{R}^* & \xrightarrow{\log} & \mathbb{R} \end{array}$$

Consider the following subgroups.

$$\begin{array}{ll} \mathcal{O}_K^* = \{\varepsilon \in \mathcal{O}_K \mid N\varepsilon = \pm 1\} & \text{the units,} \\ S = \{y \in K_\mathbb{R}^* \mid N_\mathbb{C} y = \pm 1\} & \text{the norm one surface, and} \\ H = \{x \in \left[\prod_{\sigma \in G} \mathbb{R}\right]^+ \mid \text{Tr}_\mathbb{R}(x) = 0\} & \text{the trace zero hyperplane.} \end{array}$$

The commutative diagram above induces the homomorphisms

$$\mathcal{O}_K^* \xrightarrow{\psi} S \xrightarrow{\text{vlog}} H \ .$$

Denote by λ the composed map $\lambda = \text{vlog} \circ \psi$, and let $L = \lambda(\mathcal{O}_K^*)$ be the image of the units in $\left[\prod_{\sigma \in G} \mathbb{R}\right]^+$. Recall the definition of a lattice.

Definition 4 *A lattice in an $\mathbb{R}$-vector space V is a subgroup $L = E_1 \mathbb{Z} + \ldots + E_h \mathbb{Z}$, where $E_1, \ldots, E_h$ are linearly independent vectors in V. It is called complete if $E_1, \ldots, E_h$ is a basis for V.*

It is proved in [6] that the group of roots of unity, $\mu(K)$, is isomorphic to the kernel of λ and that L is a complete lattice in the $(r+s-1)$-dimensional vector space H. Dirichlet's theorem follows from this. Let $h = r + s - 1$. To define the fundamental units $\varepsilon_1, \ldots, \varepsilon_h$ we pick a basis $E_1, \ldots, E_h$ for L, and define $\varepsilon_j = \lambda^{-1}(E_j)$. We also define $E_j' = gE_j$.

Suppose we map an element α into $\left[\prod_{\sigma \in G} \mathbb{R}\right]^+$ using λ. Let $x = (x_\sigma) = \lambda(\alpha)$. Then it is not hard to see that α is balanced when all x_σ are of roughly the same size. Another way to phrase this is that the orthogonal projection of x onto H is close to the origin. If we multiply α by ε_j, the image $\lambda(\alpha \varepsilon_j)$ is translated by the vector E_j, i.e., $\lambda(\alpha \varepsilon_j) = \lambda(\alpha) + \lambda(\varepsilon_j) = x + E_j$. To balance an element we want to find some integer combination of the vectors $E_1, \ldots, E_h$ that translates x close to the origin, since this corresponds to multiplying α by the fundamental units $\varepsilon_1, \ldots, \varepsilon_h$. We can always write $x = \sum_{j=1}^h r_j E_j$, with $r_j \in \mathbb{R}$, since L is a complete lattice, i.e., $E_1, \ldots, E_h$ is a $\mathbb{R}$-basis for H. Then we pick integers close to these real coefficients. Below we prove two lemmas that allow us to give a simple algorithm for balancing elements that is easy to analyze.

Recall that $\{e_\sigma\}_{\sigma \in G}$ denotes the canonical orthonormal basis for the space $\prod_{\sigma \in G} \mathbb{R}$. We define the max-norm $\|\cdot\| : H \to \mathbb{R}$ in terms of this basis by $\|\sum_{\sigma \in G} x_\sigma e_\sigma\| = \max_{\sigma \in G}\{|x_\sigma|\}$. It is intuitively clear that if an element $x \in H$ is sufficiently far from the origin, we may reduce its max-norm $\|x\|$ by an additive term t by translating it by a bounded element in the lattice L. We define a constant $c_E = \max_{r_1,\ldots,r_h \in [-1/2,1/2]} \|\sum_{j=1}^h r_j E'_j\|$, and turn this into a precise statement as follows. Denote by $A(t)$ the set $\{x \in H \mid x = \sum_{j=1}^h r_j E'_j, r_j \in \mathbb{R}, \|x\| \leq t + c_E\}$, and define

$$w(t) = \frac{1}{2} + \max_{\sum_{j=1}^h r_j E'_j \in A(t)} \{|r_j|\},$$

where $r_j \in \mathbb{R}$. We prove the following result.

Lemma 5 *Let $t > 0$ and let $x \in H$ be an element such that $\|x\| > t + c_E$. Then there exists $k_1, \ldots, k_h \in \mathbb{Z}$ with $|k_j| \leq w(t)$ such that*

$$\left\| x + \sum_{j=1}^h k_j E'_j \right\| \leq \|x\| - t \ .$$

Choose t such that $\frac{\Gamma^2}{2^t} < \frac{1}{2}$ and define the constants $\Phi = w(gt)$ and $\Delta = 2^{\frac{2}{g}(gt+c_E)}$. We translate the above lemma from the space H back to $\mathcal{O}_K$ and take care of the lack of precision in our approximation of N_+. This gives the following lemma.

Lemma 6 *If α in $\mathcal{O}_K$ is not Δ-balanced, then there exists $k_1, \ldots, k_h \in \mathbb{Z}$ with $|k_j| \leq \Phi$ such that*

$$N'_+\left(\alpha \prod_{j=1}^h \varepsilon_j^{k_j}\right) < \frac{1}{2} N'_+ \alpha \ .$$

The idea of the proof is the following. Suppose α is not balanced and consider the element $\beta = \alpha^g/N\alpha$. Note that β is "normalized" in the sense that $x = \lambda(\beta) \in H$ (we may have $\beta \notin \mathcal{O}_K$ though). Up to a constant factor, the element α is balanced if and only if β is balanced. This implies that $x \in H$ is far from the origin. We then apply Lemma 5 to translate x closer to the origin. Since we do this using the basis $E'_1, \ldots, E'_h$ this translates to multiplying β by a product of the fundamental units ϵ_j. When this is no longer possible, x is close to the origin, which implies that β, and thus α, are balanced.

4 The Algorithm

In this section we describe the algorithm. We divide it into subroutine calls to improve readability.

4.1 Subroutines

Consider the set of non-unit elements that divide some δ in the set of cofactors T_l. This set is clearly infinite, since each element in $\mathcal{O}_K$ has an infinite number of associates. This makes it natural to consider the following set instead

$$F_l = \{\pi \in \mathcal{O}_K \;:\; \pi \mid l \text{ or } \pi \mid \delta \in T_l, \text{ and } \pi \text{ is } \Delta\text{-balanced and irreducible}\}\;.$$

The set F_l is bounded and we denote its elements by $F_l = \{\pi_1, \ldots, \pi_{s_F}\}$. We write $F_l \nmid \alpha$ to denote the fact that $\pi \nmid \alpha$ for all $\pi \in F_l$. For clarity we state trial division as an algorithm below.

Algorithm 1 (Extract Small Factors)

```
SMALL(α)
Input : α ∈ O_K.
Output : (α', (k_1, ..., k_{s_F})), where α = α' ∏_{j=1}^{s_F} π_j^{k_j} and F_l ∤ α'.
The algorithm is the trivial one. Find α', and k_j by trial division.
```

Lemma 7 *Let $\alpha \in \mathcal{O}_K$ and suppose $(\alpha', (k_1, \ldots, k_{s_F})) = \text{SMALL}(\alpha)$. Then the running time of the SMALL-algorithm on input α is $O(n(1 + \log \frac{|N\alpha|}{|N\alpha'|}))$.*

Note that $\pi \mid \alpha$ if and only if $N\pi \mid \frac{N\pi}{\pi}\alpha$. Since $\frac{N\pi}{\pi} \in \mathcal{O}_K$ this reduces, in linear time, trial division in $\mathcal{O}_K$ to trial division in $\mathbb{Z}$.

Next we consider the problem of Δ-balancing elements. The algorithm below repeatedly applies Lemma 6 to find an increasingly balanced associate of the input. When this is no longer possible, we know that the current associate is Δ-balanced.

Algorithm 2 (Balance Element)

```
BALANCE(α)
Input : α ∈ O_K.
Output : a Δ-balanced associate β of α.
β ← α
Do
    α ← β
    For (k_1, ..., k_h) ∈ [-Φ, Φ]^h Do
        If N'_+ (β ∏_{j=1}^h ε_j^{k_j}) < ½ N'_+ β Then
            β ← β ∏_{j=1}^h ε_j^{k_j}
        End If
    End For
While N'_+ β < N'_+ α
Return α
```

Lemma 8 *The output of the BALANCE-algorithm is Δ-balanced, and the algorithm runs in time $O(n(1 + \log(\max_{\sigma, \sigma' \in G} \frac{|\alpha^\sigma|}{|\alpha^{\sigma'}|})))$.*

Identify $\mathcal{O}_K/(l)$ with the set of representatives $\sum_{j=1}^{g} a_j \omega_j$, with $0 \leq a_j < l$. Then let $(\gamma_{\alpha',\beta'}, \delta_{\alpha',\beta'})_{\alpha',\beta' \in \mathcal{O}_K/(l)}$ be the table of elements from T_l guaranteed to exist by Theorem 2, i.e., elements such that $l \mid (\gamma_{\alpha',\beta'}\alpha' + \delta_{\alpha',\beta'}\beta')$. For clarity we state finding the cofactors as an algorithm.

Algorithm 3 (Find γ and δ)

GAMMADELTA(α, β)
Input : $\alpha, \beta \in \mathcal{O}_K$.
Output : $(\gamma, \delta) \in T_l^2$, such that $l \mid (\gamma\alpha + \delta\beta)$.
Compute $\alpha' = \alpha \bmod (l)$ and $\beta' = \beta \bmod (l)$. Then output $(\gamma_{\alpha',\beta'}, \delta_{\alpha',\beta'})$.

Lemma 9 *The algorithm is correct and runs in time $O(n)$.*

4.2 Greatest Common Divisor

Finally, we are ready to give the algorithm for computing a greatest common divisor of two elements α and β in $\mathcal{O}_K$. The special case where one of the inputs is zero is treated in the first two lines. Then we extract all small factors of both inputs and store these. This allows us to determine all small factors in a GCD. Then we make sure that β is balanced. Consider now the while-loop of the algorithm. In each iteration α is balanced. This ensures that when we compute $N'_+\alpha$ and $N'_+\beta$ the results are in fact approximations of $\sqrt[n]{|N\alpha|}$ and $\sqrt[n]{|N\beta|}$. This gives us a good idea of which of the elements is the larger. The if-statement swaps α and β such that the norm of α is larger or at least within a constant factor of the norm of β. Then a linear expression is formed using the special

Algorithm 4 (Greatest Common Divisor)

GCD(α, β)
Input : $\alpha, \beta \in \mathcal{O}_K$, with either α or β non-zero.
Output : The greatest common divisor of α and β.
If $\alpha = 0$ Return β
If $\beta = 0$ Return α
$(\alpha, (k_1, \ldots, k_{s_F})) \leftarrow$ SMALL(α)
$(\beta, (k'_1, \ldots, k'_{s_F})) \leftarrow$ SMALL(β)
$\beta \leftarrow$ BALANCE(β)
While $\alpha \neq 0$ Do
    $\alpha \leftarrow$ BALANCE(α)
    If $N'_+\alpha < N'_+\beta$ Then
        $(\alpha, \beta) \leftarrow (\beta, \alpha)$
    End If
    $(\gamma, \delta) \leftarrow$ GAMMADELTA(α, β)
    $(\alpha, \cdot) \leftarrow$ SMALL$((\gamma\alpha + \delta\beta)/l)$
Done
Return $\beta \prod_{j=1}^{s_F} \pi_j^{\min\{k_j, k'_j\}}$

---

cofactors of bounded norm, and the result is divided by l. This reduces the norm of α. During the iterations of the while-loop spurious factors from the set F_l may be introduced into the current β. These are removed in the subroutine call, and the output is formed in the obvious way.

In each iteration, α, perhaps after swapping with β, is replaced by the expression $(\gamma\alpha + \delta\beta)/l$. We must obviously choose l large enough such that $|N((\gamma\alpha+\delta\beta)/l)| < |N\alpha|$. But, we must also take into account the lack of exactness in the approximation N'_+ of the norm used when deciding which of α and β is the larger. For simplicity we choose l such that the norm of α is guaranteed to be reduced by a factor of two in each iteration. More precisely we choose l as the smallest integer that satisfies the inequality $C_{\text{lin}}(l)/l^g < 1/(2\Gamma^{2g}\Delta^g)$. We can choose l to satisfy this inequality since $C_{\text{lin}}(l) = O(l^{g/2})$.

5 Analysis

In this section we prove the correctness of the algorithm and bound its running time. To simplify the exposition we denote by α_j and β_j, and α'_j and β'_j the values of α and β before and after the if-statement in the jth iteration of the while-loop.

Lemma 10 *The jth iteration, $j > 1$, runs in time $O\bigl(n(1+\log \frac{|N\alpha_j|\cdot|N\beta_j|}{|N\alpha_{j+1}|\cdot|N\beta_{j+1}|})\bigr)$.*

To see why the lemma is true, note that from the triangle inequality of the complex absolute value follows that in each iteration, $\max\{|\alpha_{j+1}^\sigma|\}$ can only be a constant factor larger than $\max\{|(\alpha'_j)^\sigma|\}$. This means that if α_{j+1} is very unbalanced, then $|N\alpha_{j+1}|$ must also be much smaller than $|N\alpha'_j|$. The lemma then follows from Lemma 7 and Lemma 8.

Theorem 3 *Algorithm 4 computes the greatest common divisor of its inputs in time $O(n^2)$ in the bit-size n of its input using naive arithmetic in $\mathbb{Z}$.*

The proof of correctness is straightforward except from the handling of spurious factors. Since all small factors are removed from both inputs and stored before the while-loop, any small factors found in the while-loop can safely be discarded. By Lemma 2, replacing α by $(\gamma\alpha+\delta\beta)/l$ preserves the GCD up to small factors. Since all small factors are removed by the call to the SMALL-algorithm the GCD of α and β is preserved and the output of the algorithm is correct. The bound of the running time is explained as follows. We have chosen l such that the absolute norm of one of the elements is reduced at least by a factor $1/2$ in each iteration. Since the norm is an integer and the algorithm halts when $\alpha = 0$, the algorithm executes at most $d = O(n)$ iterations. The subroutine calls made outside of the while-loop can be done in time $O(n^2)$. The running time of each iteration is bounded in Lemma 10. Thus, it remains to argue that the combined execution time $\sum_{j=2}^{d} O(n\log \frac{|N\alpha_j|\cdot|N\beta_j|}{|N\alpha_{j+1}|\cdot|N\beta_{j+1}|})$ of all subroutine calls sum to $O(n^2)$, but this follows by calculation.

6 On the Existence of Practical Algorithms

In this paper we focus on conceptual simplicity, and not on minimizing the constants in the running time. In particular the three subroutines SMALL, BALANCE, and GAMMADELTA are trivial brute force algorithms. Thus, an interesting line of future research is to device more efficient specialized versions of these subroutines, e.g., it should be possible to use lattice reduction techniques to balance elements. Another line of research is to exploit specific properties of $\mathcal{O}_K$ when $\#(G)$ is relatively small. In addition to the complex rings mentioned in the introduction, this seems possible for some real quadratic rings [1].

Acknowledgments

I wish to thank my advisor Johan Håstad for excellent advise. Without his help I would still be struggling. I also thank Torsten Ekedahl who essentially played the role of an extra advisor during this work.

References

1. S. Agarwal, *Personal communication*, November, 2004.
2. S. Agarwal, G. Skovbjerg Frandsen, *Binary GCD Like Algorithms for Some Complex Quadratic Rings*, ANTS 2004, LNCS 3076, pp. 57-71, 2004.
3. I. Damgård, G. Skovbjerg Frandsen, *Efficient Algorithms for gcd and Cubic Residuosity in the Ring of Eisenstein Integers*, BRICS Technical Report, ISSN 0909-0878, BRICS RS 03-8, 2003.
4. I. Damgård, G. Skovbjerg Frandsen, *Efficient algorithms for GCD and cubic residuosity in the ring of Eisenstein integers*, Fundamentals of computation theory, LNCS 2751, pp. 109-117, 2003 (revised version to appear in Journal of Symbolic Computation).
5. E. Kaltofen, H. Rolletschek, *Computing greatest common divisors and factorizations in quadratic number fields*, Mathematics of Computation, 53(188):697-720, 1989.
6. J. Neukirch, *Algebraic Number Theory*, ISBN 3-540-65399-6, Springer-Verlag Berlin, 1999.
7. J. Shallit, J. Sorenson, *A binary algorithm for the Jacobi symbol*, ACM SIGSAM Bulletin, 27 (1), pp. 4-11, 1993.
8. J. Sorenson, *Two Fast GCD Algorithms*, Journal of Algorithms, 16(1):110-144, 1994.
9. J. Stein, *Computational problems associated with Racah algebra*, Journal of Computational Physics No. 1, pp. 397-405, 1969.
10. A. Weilert, *Asymptotically fast GCD computation in $\mathbb{Z}[i]$*, In Algorithmic number theory (Leiden, 2000), LNCS 1838, pp. 595-613, 2000.
11. A. Weilert, *$(1+i)$-ary GCD computation in $\mathbb{Z}[i]$ as an analogue to the binary GCD algorithm*, Journal of Symbolic Computation, 30(5):605-617, 2000.

12. D. Wikström, *On the Security of Mix-Nets and Related Problems*, Licentiate thesis, Nada, KTH, TRITA-NA-04-06, ISSN: 0348-2952, ISRN KTH/NA/R--04/06--SE, ISBN 91-7283-717-9, May, 2004.
13. D. Wikström, *On the l-Ary GCD-Algorithm and Computing Residue Symbols*, Technical Report, Nada, KTH, Royal Institute of Technology, TRITA-NA-04-39, ISSN: 0348-2952, ISRN KTH/NA/R--04/39--SE, November, 2004.
14. D. Wikström, *On the l-Ary GCD-Algorithm in Rings of Integers*, Technical Report, Nada, KTH, Royal Institute of Technology, TRITA-NA-05-15, ISSN: 0348-2952, ISRN KTH/NA/R--05/15--SE, April, 2005.

A Fully Abstract Encoding of the π-Calculus with Data Terms*

(Extended Abstract)

Michael Baldamus[1,**], Joachim Parrow[2], and Björn Victor[2]

[1] Linnaeus Centre for Bioinformatics
[2] Department of Information Technology, Uppsala University, Sweden

Abstract. The π-calculus with data terms (πT) extends the pure π-calculus by data constructors and destructors and allows data to be transmitted between agents. It has long been known how to encode such data types in π, but until now it has been open how to make the encoding *fully abstract*, meaning that two encodings (in π) are semantically equivalent precisely when the original πT agents are semantically equivalent. We present a new type of encoding and prove it to be fully abstract with respect to may-testing equivalence. To our knowledge this is the first result of its kind, for any calculus enriched with data terms. It has particular importance when representing security properties since attackers can be regarded as may-test observers. Full abstraction proves that it does not matter whether such observers are formulated in π or πT, both are equally expressive in this respect. The technical new idea consists of achieving full abstraction by encoding data as table entries rather than active processes, and using a firewalled central integrity manager to ensure data security.

1 Introduction

The increasingly complicated mechanisms to guarantee secure communications have spurred the development of appropriate formal description techniques. In this paper we study a prototypical such formalism, a π-calculus enriched with data terms (πT), and show how it can be encoded in the more fundamental π-calculus [13] while preserving full abstraction. This means that two processes in πT are equivalent precisely when their encodings are equivalent. Although encodings between such calculi has proved a rich field of study this is the first result of its kind. We achieve it by designing the encoding in a different way from what is usually done. The full proof is very technical and we here outline the main ideas, which are quite general and can in principle be applied to many similar calculi.

πT extends the basic π-calculus by including constructors and deconstructors for data terms. In this the calculus resembles a high level parallel programming language, and specifications of security protocols can be made in a familiar operational style. It

---

* Work supported by European Union project PROFUNDIS, Contract No. IST-2001-33100.
** To whom correspondence should be addressed (*http://www.lcb.uu.se/~michaelb*).

can be seen as a generalisation of the spi-calculus [3] (which extends π with primitives for encryption and decryption) in that arbitrary constructors and destructors are allowed. It can also be seen as a simplification of the applied π-calculus by Abadi and Fournet [1], since it does not admit equations over the data terms.

A key idea when encoding data was expressed already in the first papers on the π-calculus: a data structure is represented in π as a process with the ability to interact along designated ports in order to carry out operations. In such a manner all known kinds of data structures, certainly including those of πT, can be encoded.

The problem with this kind of encoding appears when we consider a useful notion of semantic equality between πT processes. We shall here as a prime example look at *testing* equivalence. The main idea is that two processes are behaviourally equivalent precisely when they can satisfy the same *tests*. Tests are arbitrary processes of the calculus, which of course is expressive enough to describe not only security protocols but also the potential attackers on them. So if we know of two processes that they are testing equivalent we know that they will behave similarly in all conceivable environments. In an earlier paper [4] we have proved for the spi calculus that a process satisfies a test if and only if the encoding of the process satisfies the encoding of the test. The same result, as we will show in this paper, holds for an encoding from πT to π.

Unfortunately, with the traditional kind of encoding from πT to π the testing equivalences in πT and π turn out to be different. The reason is, briefly put, that the encodings of two πT processes satisfy the same tests *only if the tests are encodings of πT tests*. But in the π-calculus there are also tests which are not encodings of any πT test, and some of these may be able to discriminate between the encodings of otherwise equivalent processes. Formally *full abstraction*, the result that two πT processes are equivalent precisely when their encodings are equivalent, has proved elusive, not only for πT but for all similar calculi, and for similar operationally defined equivalences.

In this paper we exhibit an encoding from πT to π and prove it to be fully abstract. A key ingredient is that we use a central so called *integrity manager* (**M**) which stores all data values generated throughout a computation. All access to data must go through a level of indirection at **M**, which only allows accesses that adheres to the protocols for interacting with data. Thus, the previously dangerous π-calculus processes that are not encodings of πT processes are rendered impotent.

The remaining sections of this paper are organised as follows: Section 2 introduces πT. Section 3 presents the encoding of πT into the polyadic π-calculus. The reader may refer to [5] for some parts that have to be left out here due to a lack of space. Section 4 states the full abstraction result. It also gives an idea of its long and complex proof. The full proof can be found in [5]. Section 5 discusses related work. Section 6 concludes the paper with some final remarks, in particular about the lack of compositionality of the encoding due to the global integrity manager.

2 Background: π-Calculus, πT-Calculus

2.1 The πT-Calculus

As always in π-like calculi, an infinite set of *names* is assumed to be given. This set is here typically ranged over by lower-case letters from the middle of the alphabet, such as n. Sets of names are typically ranged over by upper-case letters from the middle of the alphabet, such as N. As usual in spi-calculus-like extensions of the π-calculus we distinguish names and *variables*. The set of variables is also assumed to be infinite. It is here typically ranged over by lower-case letters from the end of the alphabet, such as x. We designate finite vectors by means of the $\tilde{}$-symbol. The length of any finite vector $\widetilde{X}$ is denoted by $|\widetilde{X}|$.

The πT-calculus is further characterised by assuming a set of function symbols from which data terms can be formed, distinguishing dedicated *constructors* and *deconstructors* (with minor restrictions, see below). Predictably, constructors are used to build data terms and deconstructors to take them apart. Both the set of constructors and the set of deconstructors are assumed to be finite. Constructors are typically ranged over by f, deconstructors by d. Each constructor f is assumed to have a fixed, finite *arity* $\mathrm{ar}[f]$.

Data terms are then given as follows:

$$T, \ldots ::= n \mid x \mid f(T_1, \ldots, T_{\mathrm{ar}[f]})$$

For the sake of simplicity, no type discipline is assumed for data terms, only the constructor arities must be respected. The set of all names that occur in any given vector $\widetilde{T}$ of data terms is denoted by $\mathrm{nm}[\widetilde{T}]$. Deconstructors must not occur within data terms. They may only be applied via the let construct introduced below. A *value* is a data term without any variables. Values are typically ranged over by V and W.

Agent expressions are then given as follows:

$$P, \ldots ::= \mathbf{0} \mid T(x).P \mid \overline{T}\langle U\rangle.P \mid P \mid Q \mid (\nu n)\, P \mid\, !P$$
$$\mid\ \mathrm{if}\ T = U\ \mathrm{then}\ P\ \mathrm{else}\ Q \mid \mathrm{let}\ x = d(\widetilde{T})\ \mathrm{in}\ P$$

The $\mathbf{0}$ constant denotes the inert process. An input prefix $T(x).P$ can be performed in a context where T evaluates to some name n. Then a value, e.g. V, is received over the channel denoted by n and the process continues as P where x is substituted by V. An output prefix $\overline{T}\langle U\rangle.P$ can also be performed in a context where T evaluates to some name n and U to some value V. Then V is sent via the channel denoted by n and the process continues as P. Here n and T are called the *subject*, and x and V the *object* of the input or output. The next three operators are standard in pi-calculus-like calculi: $P \mid Q$ behaves like P and Q acting concurrently; $(\nu n)\, P$ behaves like P where n is local; $!P$ behaves like infinitely many copies of P put in parallel. The conditional and the let have their usual meaning. In πT, let is the only place where deconstructors may be applied.

Input prefixing, let, and restriction are *binders*: $T(x).P$ and let $x = d(\widetilde{T})$ in P bind the variable x in P; $(\nu n)\, P$ binds the name n in P. The set of names occurring free (non-bound) in a process expression P is denoted by $\mathrm{fn}[P]$ and similarly for a vector $\widetilde{P} = P_1 \ldots P_n$ of process expressions $\mathrm{fn}[\widetilde{P}]$ means the union of all $\mathrm{fn}[P_i]$.

We do not distinguish between expressions that differ only up to alpha conversion of bound names and variables. Given any data term or process expression H, $H\{\widetilde{x} := \widetilde{T}\}$ denotes the term after simultaneously substituting each x_i by T_i. Substitution always entails implicit alpha-conversion of bound names in H such that there is no capture of any free names in $\widetilde{T}$. A *process* is an agent expression without free variables.

We often omit trailing $\mathbf{0}$ suffixes, writing $x\langle y\rangle$ for $x\langle y\rangle.\mathbf{0}$. We often also use an "inline" notation for restriction, and e.g. write $\bar{x}\langle \nu\, y\rangle.P$ as shorthand for $(\nu\, y)\,\bar{x}\langle y\rangle.P$. Further, we use the standard notation $\prod_{i\in\{j_1,\dots,j_k\}} P_i = P_{j_1} \mid \dots \mid P_{j_k}$.

Deconstructor Equations. *Deconstructor equations* describe how deconstructors act upon values. To this end, we need to introduce *value patterns*:

$$G ::= x \mid f(G_1, \dots, G_{\mathrm{ar}[f]})$$

A deconstructor equation is then of the form $d(\widetilde{G}) \triangleq x$ where x must occur in $\widetilde{G}$. There may be several, but only finitely many equations for each deconstructor. Since there are finitely many deconstructors, there are only finitely many deconstructor equations.

Operational Semantics. Abadi's and Fournet's semantics for the applied π-calculus [1] use active substitutions and rely heavily on structural congruence rules. Our semantics, using the **let** construct for deconstruction, is more direct and does not utilise structural congruence. The rules for scope opening are similar to the variant in [1] giving the finest bisimulation equivalence relation; we are dealing with may testing where this finesse does not matter.

Actions are of one of three forms:

$n(V)$: Input of value V on channel n where V is bound to x as shown below.
$(\nu\, M)\,\bar{n}\langle V\rangle$: Output of value V on channel n where the names in M are extruded as private names. The SOS clauses ensure that we always have $M \subseteq \mathrm{fn}(V)$.
τ: Silent action.

The individual clauses are as shown below. The treatment of replication follows [17]. The clauses for **if** _ **then** _ **else** are slightly non-standard since they entail a τ-action, just like in [1]. This is advantageous for our purposes of considering may-testing. The missing symmetric clauses for interleaving and closure are left implicit.

$$n(x).P \xrightarrow{n(V)} P\{x := V\} \qquad \frac{P \xrightarrow{n(V)} P' \quad Q \xrightarrow{(\nu\, M)\,\bar{n}\langle V\rangle} Q' \quad \mathrm{fn}[P] \cap M = \emptyset}{P \mid Q \xrightarrow{\tau} (\nu\, M)(P' \mid Q')}$$

$$\bar{n}\langle V\rangle.P \xrightarrow{(\nu\, \emptyset)\,\bar{n}\langle V\rangle} P$$

$$\frac{P \xrightarrow{\alpha} P' \quad \mathrm{bn}[\alpha] \cap \mathrm{fn}[Q] = \emptyset}{P \mid Q \xrightarrow{\alpha} P' \mid Q} \qquad \frac{P \xrightarrow{(\nu\, M)\,\bar{n}\langle V\rangle} P' \quad n \neq m' \quad m' \in \mathrm{nm}[V] \setminus M}{(\nu\, m')\,P \xrightarrow{(\nu\, M + m')\,\bar{n}\langle V\rangle} P'}$$

$$\frac{P \xrightarrow{\alpha} P' \quad n \notin \mathrm{nm}[\alpha]}{(\nu\, n)\,P \xrightarrow{\alpha} (\nu\, n)\,P'} \qquad \frac{\mathbf{if}\ T = T\ \mathbf{then}\ P\ \mathbf{else}\ Q \xrightarrow{\tau} P}{\mathbf{if}\ T = U\ \mathbf{then}\ P\ \mathbf{else}\ Q \xrightarrow{\tau} Q} \qquad \frac{P \xrightarrow{\alpha} P' \quad \mathrm{bn}[\alpha] \cap \mathrm{fn}[P] = \emptyset}{!P \xrightarrow{\alpha} P' \mid !P}$$

$$\frac{P \xrightarrow{n(V)} P_1' \quad P \xrightarrow{(\nu\, M)\,\bar{n}\langle V\rangle} P_2' \quad \mathrm{fn}[P] \cap M = \emptyset}{!P \xrightarrow{\tau} ((\nu\, M)(P_1' \mid P_2')) \mid !P} \qquad \frac{d(\widetilde{G}) \triangleq x \quad \widetilde{G}\sigma = \widetilde{V}}{\mathbf{let}\ y = d(\widetilde{V})\ \mathbf{in}\ P \xrightarrow{\tau} P\{y := x\sigma\}}$$

We denote by $\mathrm{nm}[\alpha]$ the set of names that syntactically occur in any action α; we denote by $\mathrm{bn}[\alpha]$ the set of bound names of α: $\mathrm{bn}[n(x)] = \emptyset$, $\mathrm{bn}[(\nu\, M)\,\bar{n}\langle T\rangle] = M$, $\mathrm{bn}[\tau] = \emptyset$; $(\nu\, \{n_1, \dots, n_k\})$ stands for $(\nu\, n_1) \dots (\nu\, n_k)$, $k \geq 0$. Also, we denote mappings from variables to values by σ, and by $\widetilde{T}\sigma$ the vector that results from applying σ to each element of $\widetilde{T}$.

2.2 Polyadic π-Calculus

In the next section we will encode πT into the polyadic π-calculus. The specific dialect that we use is derived from πT via three simple modifications: First, the sets of constructors and deconstructors are assumed to be empty; second, nondeterministic CCS-like choice of the form $P + Q$ and polyadic input and output prefixes of the form $a(\widetilde{x}).P$ or $\overline{a}\langle \widetilde{b} \rangle$, respectively, are admitted, where a and b range over both names and variables; third, *process constants* are admitted. They are typically ranged over by upper-case letters from the beginning of the alphabet, such as A. *Process constant definitions* are of the form $A(\widetilde{x}) \stackrel{\Delta}{=} P$ where the free variables of P must be from $\widetilde{x}$. There must be a unique definition for each process constant. The number of actual parameters in each instantiation of any process constant must be the same as the number of formal parameters in the constant's definition. The necessary modifications of the above SOS clauses and the accompanying notion of action are standard. Here, we just give the semantics of the nondeterministic choice operator:

$$\frac{P \stackrel{\alpha}{\to} P'}{P + Q \stackrel{\alpha}{\to} P'} \qquad \frac{Q \stackrel{\alpha}{\to} Q'}{P + Q \stackrel{\alpha}{\to} Q'}$$

As α ranges over all actions including τ, $+$ is the ordinary CCS-like nonderministic choice operator.

3 The Encoding of the πT-Calculus

In this section we present the encoding of πT-processes into the polyadic π-calculus. As in previous work [4], the main issue is encoding values.

The idea is to let encoded processes operate on encoded values only via *value identifiers* (IDs). To this end, an *integrity manager* **M** is set up, which maintains tables of the existing IDs and the values these correspond to. Whenever an encoded process wants to operate on one or more encoded values, it must send a specific request carrying the value IDs to **M**. **M** will check whether it already knows the IDs, i.e., whether the IDs correspond to any actual values: if that is the case, then **M** will perform the operation; if not, then the request will deadlock. If the operation requires generating a new value, **M** will do so, and it will also generate a new ID that it will associate with the value. Completion will be signalled on a dedicated channel that has been supplied when the request was issued. The channel carries an ID which refers to the result of the operation, if there was any.

M thus shields encoded processes from any malformed complex values that might be sent to them by π-observers which are not encodings of πT processes. Moreover, as πT- and π-processes have the same pure synchronisation capabilities, π-observers are unable to do anything to encoded πT-processes that encoded πT-observers cannot already do. The only requirement is that no encoded πT-process may ever get connected to a "fake" integrity manager. This is assured by (a) restricting the channels that are used for interacting with **M** and (b) setting up a uni-directional firewall **F** via which **M** can receive requests from an external observer via duplicate, non-restricted channels.

The encoding of πT processes is constructed using one encoding function $(\!|\cdot|\!)$ for processes as such (section 3.1), and one encoding function $(\!|\cdot|\!)_r$ of values occurring in the processes (section 3.2), parameterised by a location r used to represent the value. $\mathbf{M}$ is presented in section 3.3.

The full encoding of a $\pi\mathrm{T}$ process P is thus $[\![P]\!] = (\nu\,\mathrm{K}_\mathbf{M})(\langle\!| P |\!\rangle \mid \mathbf{M} \mid \mathbf{F})$, and our main theorem
$$P \approx_{\mathrm{may}} Q \text{ iff } [\![P]\!] \approx_{\mathrm{may}} [\![Q]\!].$$

The encoded processes communicate with $\mathbf{M}$ using a simple protocol over the free names of $\mathbf{M}$, which we denote by $\mathrm{K}_\mathbf{M}$. Below we describe their use and the parameters passed over them.

input : (i, r) i is the ID of the channel the process wants to input a value over; r is a fresh name over which the process will receive the value (itself an ID).

output : (i, j, r) i is the ID of the channel the process wants to output a value over; j is the ID of the value to be sent; r is a fresh name which will be used for synchronisation when the output has been performed by $\mathbf{M}$.

apply$_E$: $(\widetilde{i}, r)$ (for each deconstructor equation $E : d(G_1, \ldots, G_k) = x$, where $|\widetilde{i}| = k$). $\widetilde{i}$ are the IDs of the actual components of the value pattern; r is a fresh name where the ID of the value corresponding to x (if any) can be received by the process.

r : (i) (for each value occurring in the process). Each value used in the process is represented by a fresh name r, where the process can receive the ID of the value, e.g. to pass in the above operations.

The encoded values register with $\mathbf{M}$ by communicating with it over two additional channels:

new : (n, r) n is a $\pi\mathrm{T}$ name occurring in the source process; r is a name where the ID of the name can be received.

new$_f$: $(\widetilde{x}, r)$ (for each constructor f, where $|\widetilde{x}| = \mathrm{ar}[f]$). $\widetilde{x}$ are the actual components of the constructor f; r is a name where the ID of the constructed value $f(\widetilde{x})$ can be received.

3.1 Encoding Processes

The encoding of processes is homomorphic for $\mathbf{0}, |, (\nu)$ and $!$ operators:

$$\langle\!| 0 |\!\rangle = 0 \quad \langle\!| P \mid Q |\!\rangle = \langle\!| P |\!\rangle \mid \langle\!| Q |\!\rangle \quad \langle\!| (\nu\,n)\,P |\!\rangle = (\nu\,n)\langle\!| P |\!\rangle \quad \langle\!|\, !\,P |\!\rangle = !\langle\!| P |\!\rangle$$

Input and output prefixes (below) encode the subject, request its ID, and contacts $\mathbf{M}$ for doing the actual input or output. In the case of input, the response channel s is used for performing the input, binding the object x, while in the case of output, $\mathbf{M}$ performs the output and responds with a pure synchronisation. r, s, t, i and j are fresh.

$$\langle\!| T(x).P |\!\rangle = (\nu\,r)(\langle\!| T |\!\rangle_r \mid r(i).\overline{input}\langle i, \nu\,s\rangle.s(x).\langle\!| P |\!\rangle)$$
$$\langle\!| \overline{T}\langle U\rangle.P |\!\rangle = (\nu\,r,s)(\langle\!| T |\!\rangle_r \mid \langle\!| U |\!\rangle_s \mid r(i).s(j).\overline{output}\langle i, j, \nu\,t\rangle.t().\langle\!| P |\!\rangle)$$

The conditional retrieves the IDs of the encodings of the compared values and tests them for identity. ($\mathbf{M}$ ensures that identical values have the same ID.) r, s, i and j are fresh.

$$\langle\!| \text{if } T = U \text{ then } P \text{ else } Q |\!\rangle = (\nu\,r,s)(\langle\!| T |\!\rangle_r \mid \langle\!| U |\!\rangle_s \mid r(i).s(j).\text{if } i = j \text{ then } \langle\!| P |\!\rangle \text{ else } \langle\!| Q |\!\rangle)$$

$\mathbf{M} = (\nu\, lock, unlock, put, getId, getAlias, [\forall f \in \mathbb{F}.\, put_f, getId_f, getArgIds_f], lookUp)($
  $\mathbf{EmptyValTbl}(put, getId, getAlias, [\forall f \in \mathbb{F}.getId_f, getArgIds_f], lookUp)$
  $|\quad \mathbf{NameRegistrar}(lock, unlock, put, getId, lookUp, new)$
  $|\quad \prod_{f \in \mathbb{F}} \mathbf{ConsTermRegistrar}_f(lock, unlock, put_f, getId_f, new_f)$
  $|\quad \mathbf{InputInterpreter}(lock, unlock, getAlias, input)$
  $|\quad \mathbf{OutputInterpreter}(lock, unlock, getAlias, lookUp, output)$
  $|\quad \prod_{E \in \mathbb{E}} \mathbf{EquationInterpreter}_E(lock, unlock, [\forall f \in \mathbb{F}.getArgIds_f], apply_E)$
  $|\quad \mathbf{Mutex}(lock, unlock))$

Fig. 1. Defining equation of integrity manager and empty value table

The let processes use a deconstructor. Each deconstructor d of arity k has a set of associated deconstructor equations $\mathbb{E}_{d,k}$. Each such equation E has a handler in $\mathbf{M}$, contacted over $apply_E$. In the encoding of let below (left), all such handlers are applied in parallel (line 2), and the first one to complete (line 3) locks the others out (using the one-time lock u on line 1). $\tilde{r}, \tilde{i}$ and u are fresh.

$(\!|\text{let } x = d(T_1,\ldots,T_k) \text{ in } P|\!) =$
$\quad (\nu\, r_1,\ldots,r_k)($
$\quad (\!|T_1|\!)_{r_1} | \ldots | (\!|T_k|\!)_{r_k}$
$\quad |\; r_1(i_1).\ldots r_k(i_k).$
$\quad (\nu\, u)($
$\quad\quad \overline{u}\langle\rangle \qquad\qquad\qquad\qquad (1)$
$\quad |\; \prod_{E \in \mathbb{E}_{d,k}}($
$\quad\quad \overline{apply_E}\langle i_1,\ldots,i_k,\nu\, r\rangle.\quad (2)$
$\quad\quad r(x).u().(\!|P|\!))) \qquad\qquad (3)$

$(\!|f(T_1,\ldots,T_{\text{ar}[f]})|\!)_r$
$\quad = (\nu\, r_1,\ldots,r_{\text{ar}[f]})($
$\quad\quad (\!|T_1|\!)_{r_1} | \ldots | (\!|T_{\text{ar}[f]}|\!)_{r_{\text{ar}[f]}} \quad (4)$
$\quad |\; r_1(i_1).\ldots r_{\text{ar}[f]}(i_{\text{ar}[f]}). \quad (5)$
$\quad\quad \overline{new_f}\langle i_1,\ldots,i_{\text{ar}[f]},r\rangle) \quad (6)$
$(\!|n|\!)_r = \overline{new}\langle n,r\rangle$
$(\!|x|\!)_r = \overline{r}\langle x\rangle$

3.2 Encoding Value Terms

Value terms come in three kinds: constructor terms, names, and variables. Their encoding is above (right). Terms using a constructor f encode their component values (line 4), and supply their IDs (line 5) to the new_f handler (line 6). Names are encoded by calling the new handler of $\mathbf{M}$, while occurrences of variables will always be substituted at runtime by an ID, which will then be returned.

3.3 Integrity Manager M

$\mathbf{M}$ is shown in Figure 1. The names $input$, $output$, new, new_f for all $f \in \mathbb{F}$, and $apply_E$ for all $E \in \mathbb{E}$ are free in $\mathbf{M}$. They may be used by π-calculus observers to interact with any encoded process, but thanks to the firewall, they can not interfere in the communication between the encoded processes and $\mathbf{M}$.

$\mathbf{M}$ uses an initially empty value table (**EmptyValTbl**) to maintain the correspondence between values and their value IDs. Each name in the table has an *alias*, used for the actual input and output, such that $\mathbf{M}$ can monitor all values passed, making sure they are value IDs. For each constructor in the table, the IDs of its subcomponents are maintained. The **Mutex** is used as a mutual exclusion lock for the table. (The reader may refer to [5] for parts that have to be left out here due to a lack of space.)

Additional components of **M** insert new values into the tables, if necessary:

NameRegistrar. The first time a name is used in the encoded process, the name registrar adds its corresponding ID and alias to the name table. Later uses of the name only result in a lookup; the ID and alias is maintained.

ConsTermRegistrar$_f$ (for each $f \in \mathbb{F}$). The constructor term registrars perform the corresponding function for values built by constructors f; the ID of the value and its components is maintained, and together the name and constructor term registrars ensure that identical values have the same ID.

The remaining components of **M** are (1) the handlers for input and output, **InputInterpreter** and **OutputInterpreter**, which ensure that all names used as channels, and values passed over them, have appropriate table entries and (2) the deconstructor equation handlers **EquationInterpreter$_E$**, used to match a deconstructor application against deconstructor equations.

Value Table. IDs of names and constructors are added to the value table by communication over put and put_f, respectively; they can be retrieved over $getId$ and $getId_f$. The $lookUp$ is used to verify that an ID is in the table. Subcomponents of constructors can be retrieved over $getArgIds$.

The table is built by appending table cells to the initial empty table. Such a cell is either a **NameEntry** or a **ConsTermEntry$_f$** ($f \in \mathbb{F}$). All requests to retrieve information have among their parameters the names r^+, r^-, which are used for the response. Each cell checks if it should handle the request, and if not, passes it on to the next cell. If no cell handles the request, the **EmptyValTbl** eventually signals on r^-.

Name Registrar. Name records maintain the IDs and *aliases* of names. The latter are the channels used internally for communication, to make sure all objects passed are value IDs. The *name registrar* first checks if the supplied name is a value ID, supplied by a "malicious" observer: this is an error, and the requester gets no reply. If it is not a value ID, it checks if the name is in the name table; in that case it returns its ID, otherwise it adds the name together with its new ID and alias to the name table, adds the ID to the ID table (since all IDs used must be there), and responds with the ID.

Constructor Term Registrars. The constructor registrars for a constructor f is a variation of the name registrar: it checks that each component value is in the ID table (and thus is a valid value in the encoding), and if they are, either returns the ID of the constructed term (if it is already there) or puts in an association between the constructed term and its (new) ID, puts the ID in the ID table, and returns it.

Equation Interpreters. Given an equation $E : d(G_1, \ldots, G_{\mathrm{ar}[E]}) \stackrel{\Delta}{=} x$, an interpreter for E is given a vector of $\mathrm{ar}[E]$ actual value IDs, and tries to match each value ID against the corresponding value pattern G_i of the equation. If it succeeds, the value ID corresponding to x of the equation is returned over the result channel r. If it fails, no result is given.

Given $E : d(G_1, \ldots, G_{\mathrm{ar}[E]}) \stackrel{\Delta}{=} x$,

EquationInterpreter$_E(lock, unlock, [\forall f \in \mathbb{F}.getArgIds_f], lookUp, apply_E) \stackrel{\Delta}{=}$
$\,!\ \overline{apply_E}(i_1, \ldots, i_{\mathrm{ar}[E]}, r).lock().\overline{lookUp}\langle i_1, \nu\, s_1^+, \nu\, s_1^-\rangle$
$\quad(\ s_1^-().\overline{unlock}\langle\rangle$
$\quad + s_1^+().$
$\quad\vdots$

$\overline{lookUp}\langle i_{\mathrm{ar}[E]}, \nu\, s_{\mathrm{ar}[E]}^+, \nu\, s_{\mathrm{ar}[E]}^-\rangle.$
$(\ s_{\mathrm{ar}[f]}^-().\overline{unlock}\langle\rangle$
$+ s_{\mathrm{ar}[f]}^+().(\!| G_1, i_1, \ldots, G_{\mathrm{ar}[E]}, i_{\mathrm{ar}[E]}, \emptyset |\!) \cdots)$

where the $(\!| H_1, j_1, \ldots, H_k, j_k, S |\!)$-construct, $k \geq 0$, handles the matching of actual component values against components of the equation. Its definition is recursive.

1. If $k = 0$, then $(\!| S |\!) = \overline{unlock}\langle\rangle.\overline{r}\langle x\rangle$.
2. If $k \geq 1$ and $H_1 = z$ for some variable $z \notin S$,
   $(\!| H_1, j_1, \ldots, H_k, j_k, S |\!) = (\!| H_2, j_2, \ldots, H_k, j_k, S \cup \{z\} |\!)\{z := j_1\}$
3. If $k \geq 1$ and $H_1 = z$ for some variable $z \in S$,
   $(\!| H_1, j_1, \ldots, H_k, j_k, S |\!) = \mathbf{if}\ z = j_1\ \mathbf{then}\ (\!| H_2, j_2, \ldots, H_k, j_k, S |\!)$
   $\qquad\qquad\qquad\qquad\qquad \mathbf{else}\ \overline{unlock}\langle\rangle$
4. If $k \geq 1$ and $H_1 = f(H_1', \ldots, H_{\mathrm{ar}[f]}')$ for some constructor f and arguments H_1', $\ldots, H_{\mathrm{ar}[f]}'$, then:
   $(\!| H_1, j_1, \ldots, H_k, j_k, S |\!) = \overline{getArgIds_f}\langle j_1, \nu\, t^+, \nu\, t^-\rangle.$
   $\qquad(\ t^-().\overline{unlock}\langle\rangle$
   $\qquad + t^+(j_1', \ldots, j_{\mathrm{ar}[f]}').$
   $\qquad (\!| H_1', j_1', \ldots, H_{\mathrm{ar}[f]}', j_{\mathrm{ar}[f]}', H_2, j_2, \ldots, H_k, j_k, S |\!))$

where $j_1', \ldots, j_{\mathrm{ar}[f]}'$ are fresh.

Communication Interpreters. The communication interpreters handle the *input* and *output* requests to M. Both look up the *alias* of the subject channel in the name table; the input handler returns this so the encoded input prefix can perform the input and bind the object variable in the correct context; the output handler looks up the object ID and performs the output on behalf of the encoded output prefix, and synchronises on the result channel when done.

InputInterpreter(
$\quad lock, unlock, getAlias, input$
$) \stackrel{\Delta}{=}\ !\ input(i, r).lock().$
$\quad \overline{getAlias}\langle i, \nu\, s^+, \nu\, s^-\rangle.$
$\quad (\ s^-().\overline{unlock}\langle\rangle$
$\quad + s^+(a).\overline{unlock}\langle\rangle.a(x).\overline{r}\langle x\rangle)$

OutputInterpreter(
$\quad lock, unlock, getAlias, lookUp, output$
$) \stackrel{\Delta}{=}\ !\ \overline{output}\langle i, j, r\rangle.lock().$
$\quad \overline{getAlias}\langle i, \nu\, s^+, \nu\, s^-\rangle.$
$\quad (\ s^-().\overline{unlock}\langle\rangle$
$\quad + s^+(a).\overline{lookUp}\langle j, \nu\, t^+, \nu\, t^-\rangle$
$\quad\quad (\ t^-().\overline{unlock}\langle\rangle$
$\quad\quad + t^+().\overline{unlock}\langle\rangle.\overline{a}\langle j\rangle.\overline{r}\langle\rangle)))$

3.4 Firewall

For every free name n of **M**, we introduce a distinct fresh name n' via which external observers may interact with **M** and thus indirectly also with any encoded πT-process. The firewall **F** receives requests on the channels named with those fresh names and encodes them to requests on the corresponding internal channels.

4 Full Abstraction

The relevance of may-testing equivalence for analysing security protocols has been stated elsewhere (see e.g. [3]). For this reason, we keep this section relatively technical apart from stating an outline of the long and complex proof of the full abstraction result. Recall the full encoding of a πT process P is $[\![P]\!] = (\nu \, \mathrm{K_M})((\!|P|\!) \mid \mathbf{M} \mid \mathbf{F})$.

Definition 1.
1. *An* observer *is a process that may use a distinguished name* $. *An action on channel* $ *is a* success signal.
2. *A* test *is a parallel composition* $P \mid O$ *of a process P and an observer O.*
3. *A process P may* pass *a test $P \mid O$ if some sequence of τ-steps of $P \mid O$ has a state in which O signals success. Formally, we denote this property by P may O.*
4. *Any two processes P and Q are* may-testing equivalent *if P may O if and only if Q may O for every observer O – in other words, the tests P and Q may pass are the same. We denote this property by $P \approx_{\mathrm{may}} Q$.*

Theorem 2. *Given any πT-processes P and Q, $P \approx_{\mathrm{may}} Q$ iff $[\![P]\!] \approx_{\mathrm{may}} [\![Q]\!]$.*

Proof. (Outline) The proof has two main steps:

1. This step consists of proving that may-tests are preserved, that is to say, proving that the encodings of two πT processes satisfy the same tests if the tests are encodings of πT tests. This result is analogous to earlier work [4] but, at the same time, it is by far more difficult. The reason is that, unlike [4], the global context in the form of the integrity manager has to be taken into account. The solution consists of working with a set-theoretical abstraction of the integrity manager that allows us to consider only factions of it as we exploit the syntax-directed way in which the pure encoding $(\!|\_|\!)$ is defined. In this way we are able to re-instantiate the concept of an ancestor relation introduced in [4] to prove both a forward and a backward operational correspondence between unencoded and encoded process-observer couplings. The preservation of may tests is then an easy corollary once we also have an operational correspondence between abstract and concrete integrity management.

2. This step mainly consists of using the result from Step 1 in obtaining a π-calculus trace characterisation of πT-may-testing equivalence. This property is reminiscent of trace characterisations of may-testing equivalence in other settings (see e.g. [10]). All what is left is then to take the firewall into account to obtain the final full-abstraction result.

The full proof can be found in [5].

5 Related Work

Arbitrary equations between data terms are the most crucial feature of the applied π-calculus that is lacking from πT. Also, we require that deconstructors are only used in let-expressions. πT is therefore in between the applied π-calculus and the spi calculus in that it resembles Borgström, Briais and Nestmann's parameterised spi calculus [7]. These simplifications entail a loss of expressiveness when describing security protocols. The advantage is that we can still give a straightforward Structural Operational Semantics (SOS) for agents without having to first define a complex evaluation semantics for value terms. This simplifies our proofs to a great extent.

Full abstraction in the kind of encoding considered here is difficult to achieve. Milner's encoding of the λ-calculus [12] is not fully abstract. Sangiorgi [16] demonstrates full abstraction for an encoding from the higher order (HO) π-calculus. The main difference is that in HOπ the language does not contain an equality test for HO values. In πT there is such a test for data terms. This means that in the encoding the interior of a term must, so to speak, be open for inspection, and this makes our encoding and proof very different.

The applied π-calculus, the spi calculus [3], Burrows, Abadi and Needham's logic of authentication [9], which is nowadays known as BAN logic, and a number of other modelling and analysis techniques for security protocols rest on the Dolev-Yao assumption [11]. This means that the underlying crypto-system is considered unbreakable, so that the protocol logic on top of that is the only concern. A central aspect of all of these lines of work is that an analysis makes it necessary to explicitly represent the knowledge that an observer of a supposedly secure system can accumulate over time. This situation is similar in mobile process approaches: Observer knowledge has there been incorporated in notions of bisimulation and testing within the framework of the spi calculus; Borgström and Nestmann give a good overview of bisimulation for the spi calculus in [8]; Boreale, De Nicola and Pugliese have, besides treating bisimulation, shown how to take account of may-testing for the spi calculus in terms of knowledge-enriched traces [6]. In contrast to all of them our characterisation seems to be unique in that all observer knowledge is internalised. That is to say, it appears on the calculus level, not on the meta-level. In this sense it could be considered much closer to traditional process algebra methodology than the above-mentioned approaches.

6 Conclusion

We have presented a new encoding from the value-enriched mobile process calculus πT into the polyadic π-calculus. The cornerstone of the translation is an integrity manager that acts as a clearing house for all operations that involve translated values. It is protected by a firewall so that it cannot be impersonated by a hostile environment. Similar techniques were used in [2]. This encoding solves the long open full abstraction problem for πT-like calculi.

Our encoding is not compositional since it has the integrity manager and the firewall as a global context. While compositionality has been put forward as a criterion for whether an encoding from one calculus to the other is good [15], it can be argued (see [14]) that these criteria are too strong for practical purposes, and that by allowing a top-level context (but keeping the inner encoding compositional), many practically or theoretically motivated encodings turn out to be "good".

One way to interpret our result is that the π-calculus can provide as much "security" in the sense of protected data values as the πT-calculus and its instantiations such as the spi calculus. The means to achieve this are an integrity manager and a firewall. Thus we could claim to have proved that integrity management plus firewalling is a viable security philosophy. Modulo possible termonological differences this belief may actually be common; our contribution is to prove the correctness of a formal statement of it.

References

[1] M. Abadi and C. Fournet. Mobile Values, New Names, and Secure Communication. In *Principles of Programming Languages*, pages 104–115. ACM, 2001.

[2] M. Abadi, C. Fournet, and G. Georges. Secure Implementation of Channel Abstractions. In *Logic in Computer Science*, pages 105–116. IEEE, 1998.

[3] M. Abadi and A. Gordon. A Calculus for Cryptographic Protocols: The Spi Calculus. *Information and Computation*, 148(1):1–70, 1999.

[4] M. Baldamus, J. Parrow, and B. Victor. Translating Spi Calculus to π-Calculus Preserving May-Tests. In *Logic in Computer Science*, pages 22–31. IEEE, 2004.

[5] M. Baldamus, J. Parrow, and B. Victor. A Fully Abstract Encoding of the π-Calculus with Data Terms. Technical Report 2005-004, Department of Information Technology, Uppsala University, February 2005.

[6] M. Boreale, R. De Nicola, and R. Pugliese. Proof Techniques for Cryptographic Processes. *SIAM Journal on Computing*, 31(3):947–986, 2002.

[7] J. Borgström, S. Briais, and U. Nestmann. Symbolic Bisimulation in the Spi Calculus. In *Concurrency Theory*, LNCS 3170, pages 161–176, 2004. Concur conference proceedings.

[8] J. Borgström and U. Nestmann. On Bisimulations for the Spi Calculus. Technical Report IC/2003/34, EPFL I&C, 2003.

[9] A. Burrows, M. Abadi, and R. Needham. A logic of authentication. *Proceedings of the Royal Society of London A*, 426:233–271, 1989.

[10] R. De Nicola and M. Hennessy. Testing Equivalences for Processes. *Theoretical Computer Science*, 34:83–133, 1984.

[11] D. Dolev and A. Yao. On the Security of Public-Key Protocols. *IEEE Transactions on Information Technology*, 29(2):198–208, 1983.

[12] R. Milner. Functions as Processes. *Mathematical Structures in Computer Science*, 2(2):119–141, 1992.

[13] R. Milner, J. Parrow, and D. Walker. A Calculus of Mobile Processes, Parts I/II. *Information and Computation*, 100:1–77, 1992.

[14] U. Nestmann. What Is a 'Good' Encoding of Guarded Choice? *Information and Computation*, 156:287–319, 2000.

[15] C. Palamidessi. Comparing the Expressive Power of the Synchronous and the Asynchronous π-Calculus. In *Principles of Programming Languages*. ACM, 1997.

[16] D. Sangiorgi. From π-Calculus to Higher-Order π-Calculus – and Back. In *Theory and Practice of Software Development*, LNCS 668, pages 151–161. Springer, 1993.

[17] D. Sangiorgi and D. Walker. *The π-Calculus: A Theory of Mobile Processes*. Cambridge University Press, 2003.

Orthogonal Extensions in Structural Operational Semantics

(Extended Abstract)

MohammadReza Mousavi and Michel A. Reniers

Department of Computer Science,
Eindhoven University of Technology,
NL-5600MB Eindhoven, The Netherlands

Abstract. In this paper, we give novel and more liberal notions of operational and equational conservativity for language extensions. We motivate these notions by showing their practical application in existing formalisms. Based on our notions, we formulate and prove meta-theorems that establish conservative extensions for languages defined using Structural Operational Semantics (SOS).

Keywords: Formal Semantics, Structural Operational Semantics (SOS), Conservative Extension, Operational Conservativity, Equational Conservativity, Orthogonality.

1 Introduction

Programming languages and process calculi have been subject to constant extensions. It is often crucial to make sure that such extensions do not change the intuition behind the old subset, or said otherwise, the extensions are *conservative*. In the context of languages with Structural Operational Semantics (SOS) [13], this topic has been studied in depth in [1, 3, 5, 10, 15, 17]. This research has resulted in meta-theorems proving sufficient conditions for an extension to be *operationally* and/or *equationally conservative*. In the remainder, we mostly refer to [5] which gives the most detailed account of the problem and subsumes almost all previous results. We do not treat multi-sorted and variable binding signatures, addressed in [5, 10], in this paper.

So far, *operational conservativity* has only allowed for extensions that consistently deny the addition of any new behavior to the old syntax. One can imagine that an extension which grants a new behavior consistently to the old syntax can also be considered safe or "conservative". This phenomenon occurs quite often in practice. For example, designers of many timed extensions of existing formalisms (e.g., the timed process algebras of [2, 9, 14]) have decided to add timed behavior homogenously to the terms from the old syntax. Unfortunately, it turns out that the existing definitions and their corresponding meta-theorems come short of any formal result about such extensions.

In this paper, we present a more liberal notion of operational conservativity, called *orthogonality*, which caters for both possibilities (i.e., denying some types of behavior from the old syntax while granting some other types). We show that our notion is useful in the aforementioned cases where the old notions cannot be used. We formulate orthogonality meta-theorems for languages with Structural Operational Semantics and prove them correct.

In [15], *equational conservativity* is considered in the setting where a new set of axioms is added to an existing set. Then, the extension is called *equationally conservative* if it induces exactly the same derivable closed equalities on the old syntax as the original equational theory. In this paper, we remove the requirement for including the old set of axioms in the extended equational theory. This relaxation is motivated by the fact that in many extensions, such as those of [2, 14], for some axioms, only all closed derivable equalities on the old syntax are kept and the axioms themselves are removed. Operational conservativity is usually considered as a means for equational conservativity and we show that our notion of orthogonality leads to equational conservativity in the same way as operational conservativity does (no matter which notion of equational conservativity is chosen, the traditional notion or the relaxed one).

The rest of this paper is structured as follows. Section 2 gives the basic definitions about Structural Operational Semantics, Transition System Specification (TSS) and equational theory. Section 3 presents the notions of operational and equational conservativity and gives sufficient conditions for proving operational conservativity. Orthogonality and related notions are defined in Section 4. Subsequently, Section 5 defines sufficient conditions for orthogonality. In the same section, we also present theorems establishing the link between orthogonality and equational conservativity. Finally, Section 6 summarizes the results and presents future directions. In each section, we provide abstract and concrete examples from the area of process algebra to motivate the definitions and illustrate the results. Due to lack of space, we could not present all of the results and the proofs of the theorems in this extended abstract. Interested readers can find these in the full version of this paper [11].

2 Preliminaries

2.1 Structural Operational Semantics

Structural Operational Semantics [13] is a logical way of defining operational semantics which has found lots of applications in different areas of computer science. A semantic specification in the style of SOS, called a *Transition System Specification (TSS)*, consists of a number of deduction rules which specify the possibility of a transition (in the conclusion of the rules) in terms of the (im)possibility of other transition (in the premises). Predicates on states are other possible ingredients of TSS's which can both be defined in the conclusion of the rules and used in the premises. Predicates can always be coded as transitions with dummy right-hand sides (cf. [16]) and thus, we do not complicate

Definition 3 (Stratification [8]). A TSS tss is stratified by a function $\mathcal{S}$ from closed positive formulae to an ordinal if and only if for all deduction rules in tss of the following form:

$$\frac{\{t_i \xrightarrow{l_i} t'_i | i \in I\} \quad \{t_j \xrightarrow{l_j} \not{|} j \in J\}}{t \xrightarrow{l} t'}$$

and for all closed substitutions σ, $\forall_{i \in I} \mathcal{S}(\sigma(t_i \xrightarrow{l_i} t'_i)) \leq \mathcal{S}(\sigma(t \xrightarrow{l} t'))$ and $\forall_{j \in J}$ $\mathcal{S}(\sigma(t_j \xrightarrow{l_j} t'')) < \mathcal{S}(\sigma(t \xrightarrow{l} t))$, for all terms t''. If the measure decreases also from the conclusion to the positive premises, then tss is *strictly stratified* by $\mathcal{S}$.

The following example illustrates the concepts defined in this section.

Example 1 (Minimal Process Algebra (*MPA*)). Consider the following deduction rules defined on a signature with a constant δ, a family of unary operators $a.\_$ (for all $a \in A$, where A is a given set of atomic actions) and a binary operator $\_ + \_$. The labels of transitions are $a \in A$.

$$\text{(a)} \frac{}{a.x \xrightarrow{a} x} (a \in A) \quad \text{(c0)} \frac{x \xrightarrow{a} x'}{x + y \xrightarrow{a} x'} \quad \text{(c1)} \frac{y \xrightarrow{a} y'}{x + y \xrightarrow{a} y'}$$

This TSS (called tss_m in the remainder) is supposed to define a transition relation for the Minimal Process Algebra (*MPA*) of [2], simplified here by removing the concept of termination, which we use as our running example in the remainder. Deduction rules of *MPA* are (strictly) stratified using a measure of size on the terms in the source of formulae and it defines a unique transition relation by all possible interpretations. The following transitions are among those: $tss_m \models (a.\delta) + \delta \xrightarrow{a} \delta$ and $tss_m \models a.(\delta + a.\delta) \xrightarrow{a} \delta + a.\delta$.

2.2 Equational Theory

Equational theories play a central role in process algebras. They capture the basic intuition behind the algebra, and the models of the algebra are expected to respect this intuition (e.g., the models induced by operational semantics). To establish a reasonable link between the operational model and the equational theory of the algebra, a notion of behavioral equality is needed. This notion captures when two syntactic terms show the "same behavior" and thus they should belong to the same equivalence class. There is a spectrum of notions of behavioral equality in the literature [7]. We take the notion of strong bisimilarity [12], denoted by $\underline{\leftrightarrow}$, as the notion of behavioral equivalence, but as we show in the extended version of this paper [11], our results are valid for a wide range of notions in this spectrum.

Getting back to the equational side of the algebra, the notion of behavioral equivalence should ideally coincide with the closed derivations of the equational theory. One side of this coincidence is captured by the soundness theorem which states that all closed derivations of the equational theory are indeed valid with

the presentation with their formal treatment. Next, we formalize the rest of the concepts mentioned above.

Definition 1 (Term and Substitution). We assume that the set of *process terms*, denoted by $T(\Sigma)$ with typical members $t, t', t_i, \ldots$, is inductively defined on a given set of *variables* $V = \{x, y, \ldots\}$ and a signature Σ. A *signature* contains a number of *function symbols* (composition operators: $f, g, \ldots$) with fixed *arities*. Function symbols with arity 0 are called *constants*. *Closed terms*, denoted by $C(\Sigma)$ with typical members $p, q, p_i, \ldots$, are terms that do not contain variables. The set of variables appearing in term t is denoted by $vars(t)$.

A *(closed) substitution* σ replaces variables in a term with other (closed) terms. *The set of terms generated by a set of terms S, denoted by $G(S)$, is the set of all terms $t' = \sigma(t)$, for some $t \in S$ and some σ such that $\forall_{x \in V} \sigma(x) \in S$. A set of terms S covers Σ-terms, if $C(\Sigma) \subseteq G(S)$.*

A transition system specification, defined below, is a logical way of defining a transition relation on (closed) terms.

Definition 2 (Transition System Specification (TSS)). A *transition system specification* is a tuple (Σ, L, D) where Σ is a signature, L is a set of *labels* (with typical members $l, l', l_0, \ldots$) and D is a set of deduction rules. For all $l \in L$, and $t, t' \in T(\Sigma)$ we define that $(t, l, t') \in \rightarrow$ and $(t, l) \not\rightarrow$ are *formulae* (positive and negative, respectively). To avoid any confusion, note that $\ldots \in \rightarrow$ and $\ldots \not\rightarrow$ are used as a syntactic notation and are not intended to denote the set-theoretic membership at this point. The notion of closed is lifted from terms to formulae in the natural way. A *deduction rule* $dr \in D$, is defined as a tuple (H, c) where H is a set of formulae and c is a positive formula. The formula c is called the *conclusion* and the formulae from H are called *premises*. A deduction rule with label l in its conclusion is called an *l-rule*.

Formulae $(t, l, t') \in \rightarrow$ and $(t, l) \not\rightarrow$ are denoted by the more intuitive notations $t \xrightarrow{l} t'$ and $t \not\xrightarrow{l}$, respectively. We refer to t as the source of both formulae and to t' as the target of the first one. A deduction rule (H, c) is denoted by $\frac{H}{c}$ in the remainder.

Different interpretations of the transition relation (the set of closed positive formulae) induced by a TSS are given in the literature. In [6], an extensive overview of alternative interpretations is provided. We formulate and prove our main results in such a general way that they remain independent from the chosen interpretation and can be adopted for several existing ones. In cases where we need an explicit transition relation, we assume that this transition relation is uniquely defined by the corresponding TSS using one of the interpretations given in [6]. In such cases, we use the notation $tss \models \phi$ to denote that a closed positive formula ϕ is in the transition relation induced by tss.

One criterium that guarantees the existence and uniqueness of a transition relation associated with a TSS is the following concept of (strict) stratification, which we use for other purposes in this paper, as well.

respect to the particular notion of behavioral equality. The other side of the coincidence, called completeness, phrases that all induced behavioral equalities are derivable from the equational theory, as well. These concepts are formalized in the remainder.

Definition 4 (Equational Theory). *An* equational theory *or* axiomatization *(Σ, E) is a set of equalities E on a signature Σ of the form $t = t'$, where $t, t' \in T(\Sigma)$. A closed instance $p = p'$, for some $p, p' \in C(\Sigma)$, is derivable from E, denoted by $E \vdash p = p'$ if and only if it is in the smallest congruence relation induced by the equalities of E.*

An equational theory (Σ, E) is sound *with respect to a TSS tss (also on signature Σ) if and only if for all $p, p' \in C(\Sigma)$, if $E \vdash p = p'$, then it holds that $tss \vdash p \leftrightarrow p'$. It is* complete *if the implication holds in the other direction.*

An equational theory E on Σ eliminates *function symbols from $\Sigma' \subseteq \Sigma$ if and only if for all $p \in C(\Sigma)$ there exists a term $p' \in C(\Sigma \setminus \Sigma')$ such that $E \vdash p = p'$.*

The following example illustrates the idea of equational theory.

Example 2 (*MPA: Equational Theory*). Consider the Minimal Process Algebra of Example 1. The following is an axiomatization of *MPA* [2].

$$x + y = y + x \qquad x + (y + z) = (x + y) + z \qquad x + x = x \qquad x + \delta = x$$

It is well-known that this axiomatization is sound and complete with respect to tss_m given in Example 1. The following are examples of derivable equalities from the above axiomatization: $(a.\delta) + \delta = a.\delta$ and $(a.\delta) + a.\delta = a.\delta$.

3 Operational and Equational Conservativity

In this section, we define different concepts regarding language extensions. To extend a language defined by a TSS, one may have to combine an existing signature with a new one. However, not all signatures can be combined into one as the arities of the function symbols may clash. To prevent this, we define two signatures to be *consistent* when they agree on the arity of the shared function symbols. Henceforth, we always assume that extended and extending TSS's are consistent. The following definition formalizes the concept of operational extension.

Definition 5 (Extension of a TSS). *Consider TSS's $tss_0 = (\Sigma_0, L_0, D_0)$ and $tss_1 = (\Sigma_1, L_1, D_1)$. The extension of tss_0 with tss_1, denoted by $tss_0 \cup tss_1$, is defined as $(\Sigma_0 \cup \Sigma_1, L_0 \cup L_1, D_0 \cup D_1)$.*

Next, we define when an extension of a TSS is called operationally conservative.

Definition 6 (Operational Conservativity [15]). *Consider TSS's $tss_0 = (\Sigma_0, L_0, D_0)$ and $tss_1 = (\Sigma_1, L_1, D_1)$. If $\forall_{p \in C(\Sigma_0)} \forall_{p' \in C(\Sigma_0 \cup \Sigma_1)} \forall_{l \in L_0 \cup L_1} tss_0 \cup tss_1 \models$*

$p \xrightarrow{l} p' \Leftrightarrow tss_0 \models p \xrightarrow{l} p'$, then $tss_0 \cup tss_1$ is an *operationally conservative extension* of tss_0.

Note that in the above definition, the labels and the targets of the transitions are taken from the extended TSS and thus, any new transition of the old syntax, even with a new label or a new target is prohibited. The following example illustrates the idea of extending TSS's and the concept of operational conservativity.

Example 3 (Timed *MPA*: Operational Semantics). Consider the following deduction rules (divided into three parts) which are defined on a signature with two constants δ and $\underline{\delta}$, a unary function symbol $\underline{\sigma}.\_$, two families of unary function symbols $a.\_$ and $\underline{a}.\_$ (for all $a \in A$) and a binary function symbol $\_+\_$. The set of labels of the TSS is $A \cup \{1\}$ (for $1 \notin A$).

$$\text{(ua)} \frac{}{a.x \xrightarrow{a} x} \quad \text{(td)} \frac{}{\underline{\sigma}.x \xrightarrow{1} x} \tag{1}$$

$$\text{(tc0)} \frac{x \xrightarrow{1} x' \quad y \xrightarrow{1} y'}{x+y \xrightarrow{1} x'+y'} \quad \text{(tc1)} \frac{x \xrightarrow{1} x' \quad y \xnrightarrow{1}}{x+y \xrightarrow{1} x'} \quad \text{(tc2)} \frac{y \xrightarrow{1} y' \quad x \xnrightarrow{1}}{x+y \xrightarrow{1} y'} \tag{2}$$

$$\text{(ta)} \frac{}{a.x \xrightarrow{1} a.x} \quad \text{(d)} \frac{}{\delta \xrightarrow{1} \delta} \tag{3}$$

The above TSS, which we call tss_t defines the aspect of timing in terms of new time transitions $\xrightarrow{1}$ and it is added in [2] to tss_m in Example 1 to define a relative-discrete-time extension of *MPA*. The intuition behind the new underlined function symbols ($\underline{a}.\_$ and $\underline{\sigma}.\_$) is that they are not delayable in time and should take their (respectively action and time) transitions immediately. Addition of the first and/or the second parts of the above TSS (each or both) to tss_m results in an operationally conservative extension of the latter as the newly added transitions will be restricted to the new syntax. (Note that in the first and second parts, there is no rule about timed transition of constants in the old syntax.) This claim can be checked formally as an instance of a meta-theorem in the rest of this section. However, the addition of part (3) violates the conservativity of the extension as it adds time transitions ($\xrightarrow{1}$) to the behavior of terms from the old syntax. For example, in combination with the first two parts, it allows for transitions such as $tss_m \cup tss_t \models a.\delta \xrightarrow{1} a.\delta$ and $tss_m \cup tss_t \models (a.\delta)+\delta \xrightarrow{1} (a.\delta)+\delta$, all of which are prohibited by the original TSS and thus are considered harmful from the operational conservativity point of view.

Next, we formulate sufficient conditions to prove operational conservativity. But before that, we need a few auxiliary definitions.

Definition 7 (Source Dependency). All variables appearing in the source of the conclusion of a deduction rule are called *source dependent*. A variable of a deduction rule is *source dependent* if it appears in a target of a premise of which all the variables of the source are source dependent. A premise is *source dependent* when all the variables appearing in it are source dependent. A deduction rule is *source dependent* when all its variables are. A TSS is *source dependent* when all its rules are.

Definition 8 (Reduced Rules). For a deduction rule $d = (H, c)$, the reduced rule with respect to a signature Σ is defined by $\rho(d, \Sigma) \doteq (H', c)$ where H' is the set of all premises from H which have a Σ-term as a source.

Theorem 1. *(Operational Conservativity Meta-Theorem [5]) Given two TSS's $tss_0 = (\Sigma_0, L_0, D_0)$ and $tss_1 = (\Sigma_1, L_1, D_1)$, $tss_0 \cup tss_1$ is an operationally conservative extension of tss_0 if:*

1. *tss_0 is source dependent;*
2. *for all $d \in D_1$ at least one of the following holds:*
   (a) the source of the conclusion has a function symbol in $\Sigma_0 \setminus \Sigma_1$, or
   (b) $\rho(d, \Sigma_0)$ has a source-dependent positive premise $t \xrightarrow{l} t'$ such that $l \notin \Sigma_0$ or $t' \notin T(\Sigma_0)$.

The following definition formalizes the concept of equational conservativity.

Definition 9 (Equational Conservativity). An equational theory E_1 on signature Σ_1 is an equationally conservative extension of E_0 on Σ_0 if and only if $\Sigma_0 \subseteq \Sigma_1$ and for all $p, p' \in C(\Sigma_0)$, $E_0 \vdash p = p' \Leftrightarrow E_1 \vdash p = p'$.

It is worth mentioning that the above definition is more liberal than the similar notion in [15] in that there, it is required that the same axioms are included in the extended equational theory (i.e., $E_0 \subseteq E_1$). In practice, some process algebras do not keep the same axioms when extending the formalism while they make sure that the closed instantiations of the old axioms with old terms indeed remain derivable (see for example, [2, 14] and Example 4 in the remainder). Hence, we believe that the restriction imposed by [15] unnecessarily limits the applicability of the theory. If, for any reason, one chooses the more restricted notion of [15], the theorems concerning equational conservativity in this paper remain valid.

Example 4 (Timed *MPA*: Equational Theory). Consider the TSS resulting from extending tss_m of Example 1 with (all three parts) of the timed extension defined in Example 3. The following are a set of sound and complete axioms (w.r.t. strong bisimilarity) for this TSS:

$$x + y = y + x \quad x + (y + z) = (x + y) + z \quad x + x = x \quad \delta = \underline{\sigma}.\delta$$

$$x + \underline{\delta} = x \quad (\underline{\sigma}.x) + \underline{\sigma}.y = \underline{\sigma}.(x + y) \quad a.x = (\underline{a}.x) + \underline{\sigma}.a.x \quad (a.x) + \delta = a.x$$

The above axiomatization underscores the fact we mentioned before. Namely, the axioms of the old system do not hold in the new system (e.g., $(\underline{a}.x) + \delta \neq \underline{a}.x$ as an instance of $x + \delta = x$) but all closed instantiations of the old axioms by the old syntax are derivable from the new set of axioms.

It can be checked that the above axiomatization of timed MPA is indeed an equationally conservative extension of the axiomatization of MPA in the sense of Definition 9. Thus, if one considers operational conservativity as a means to equational conservativity, this example already suggests the need for an extension of Definition 6. In other words, we believe that the transitions added by the extension are quite innocent and harmless to the intuition behind the original semantics, for they are added uniformly to the old syntax without changing the old behavior or violating previously valid equalities. In the next section, we formalize our idea of orthogonal extensions which caters for extensions of the above type.

4 Orthogonality

In this section, we define the notion of orthogonality and an instance of this notion, called *granting extensions*, which can be checked syntactically.

Definition 10 (Orthogonal Extension). Consider TSS's $tss_0 = (\Sigma_0, L_0, D_0)$ and $tss_1 = (\Sigma_1, L_1, D_1)$. The TSS $tss_0 \cup tss_1$ is an *orthogonal* extension of tss_0 when first, $\forall_{p,p' \in C(\Sigma_0)} \forall_{l \in L_0} \ tss_0 \cup tss_1 \models p \xrightarrow{l} p' \Leftrightarrow tss_0 \models p \xrightarrow{l} p'$ and second, and $\forall_{p,p' \in C(\Sigma_0)} \ tss_0 \cup tss_1 \models p \leftrightarrow p' \Leftrightarrow tss_0 \models p \leftrightarrow p'$.

Note that it immediately follows from the above definition that orthogonality is a preorder, i.e., a reflexive and transitive relation, on TSS's. Besides strong bisimilarity, our results in the this paper are valid for orthogonality with respect to most other notions of behavioral equivalence in the literature (cf. [11]). The notion of *operational conservativity up to ϕ-equivalence* of [15, 3] can be seen as a variant of orthogonality which only has the second condition. This and other variants of the notion of orthogonality can also be useful and our results can be used to establish meta-theorems for these notions. To our knowledge, beyond operational conservativity results (e.g., in [15]), no systematic study of these notions (including meta-theorems guaranteeing them) has been carried out.

Corollary 1. An operationally conservative extension is an orthogonal extension.

Corollary 1 addresses operational conservativity as an extreme case of orthogonality which denies all new transitions from the old syntax; the other extreme is an extension which grants all new behavior to the old syntax. However, for such an extension to be orthogonal, these transitions should be made to equivalent terms from the old syntax. In particular, if we allow for self transitions, we are able to prove orthogonality with respect to many notions of behavioral equivalence. The following definitions and the subsequent theorem substantiate these concepts.

Definition 11 (Granting Extension). Consider TSS's $tss_0 = (\Sigma_0, L_0, D_0)$ and $tss_1 = (\Sigma_1, L_1, D_1)$ with disjoint labels. We call $tss_0 \cup tss_1$ a *granting extension* of tss_0 when first, $\forall_{p,p'\in C(\Sigma_0)} \forall_{l\in L_0}\ tss_0 \vDash p \xrightarrow{l} p' \Leftrightarrow tss_0 \cup tss_1 \vDash p \xrightarrow{l} p'$ and second, $\forall_{p\in C(\Sigma_0)} \forall_{p'\in C(\Sigma_0\cup\Sigma_1)} \forall_{l\in L_1\setminus L_0}\ tss_0 \cup tss_1 \vDash p \xrightarrow{l} p' \Leftrightarrow p = p'$.

The above definition states that granting extensions keep the old transitions on the old terms intact and only add self transitions with all of the new labels to old terms. This definition does not make any statement about the transitions on the new terms, i.e., terms from $T(\Sigma_0 \cup \Sigma_1) \setminus T(\Sigma_0)$.

We are doubtful whether any meaningful relaxation of Definition 11 would be at all possible that allows for anything coarser than syntactic equality on the old terms involved in (the left- or the right-hand side of) the new transitions and still can be captured by simple syntactic checks. This suggests that to formulate syntactic criteria for proving orthogonality, we have to resort to one of the two extremes (operational conservativity or granting extensions). We admit that combining these two extremes is interesting. This is partly possible by exploiting the the transitivity of orthogonal extension relation. This way, one can interleave the application of granting and operational conservativity theorems (cf. Example 5). We propose an alternative method of combining operationally conservative and granting extensions in [11]. Next, we show that granting extensions are indeed orthogonal.

Theorem 2. *For TSS's tss_0 and tss_1, if tss_1 is a granting extension of tss_0 then tss_1 is an orthogonal extension of tss_0.*

5 Meta-theorems

In this section, we seek sufficient conditions for establishing orthogonality and equational conservativity.

We start with defining sufficient conditions to prove an extension to be granting. Hence, we need to define when a deduction rule proves (only) self transitions. We use unification as a means to this end.

Definition 12 (Unification). *A term t is unifiable with t' using σ, denoted by $t \approx_\sigma t'$ if and only if $\sigma(t) = \sigma(t')$. The set of unifiers of t and t' is defined by $U(t,t') = \{\sigma \mid t \approx_\sigma t'\}$. The set of unifiers of a set of pairs is defined as the intersection of the sets of unifiers of each pair. The set of unifiers of an empty set is defined to include all substitutions.*

The set of unifiers of a positive formula $t \xrightarrow{l} t'$ is defined as the set of unifiers of t and t'. Unification also naturally extends to a set of positive formulae, again, using intersection.

Next, we characterize the set of rules that induce self transitions. This is done by only allowing for unifiable (positive) formulae in the premises and the conclusion of a rule and further, by forcing the unification of the conclusion to follow from that of the premises.

Definition 13 (Source Preserving Rules). *A deduction rule $\frac{H}{c}$ without negative premises is source preserving if $U(H) \neq \emptyset$ and $U(H) \subseteq U(c)$. A TSS is source* *preserving* if all its deduction rules are. For a source preserving TSS, the set of *unified conclusions* contains conclusions of the deduction rules with their unifiers applied to them.

Source-preserving rules are safe for the purpose of proving self transitions. However, there might be other rules in the extending TSS that can be harmful in that they may prove other types of transition for old terms. This may be prevented by forcing the other (non-source-preserving) rules to have negative or non-unifiable positive premises addressing the old syntax. The following definition gives sufficient conditions for an extension to be granting.

Definition 14 (Granting Criteria). Consider a TSS $tss = (\Sigma, L, D)$ stratified by $\mathcal{S}$. It *grants* L_0 *transitions on* Σ_0-*terms*, if $tss = tss_0 \cup tss_1$ (with $tss_x = (\Sigma_x, L_x, D_x)$ for $x \in \{0, 1\}$) such that:

1. tss_0 is strictly stratified by $\mathcal{S}$, it is source preserving and for all $l \in L_0$, the set containing sources of unified conclusions of l-rules covers Σ_0-terms, and
2. for all deduction rules $d \in D_1$ at least one of the following holds:
   (a) d has a function symbol from $\Sigma_1 \setminus \Sigma_0$ in the source of its conclusion, or
   (b) $\rho(d, \Sigma_0)$ has a negative source-dependent premise with a label in L_1, or
   (c) $\rho(d, \Sigma_0)$ has a positive source-dependent premise $t \xrightarrow{l} t'$ with $l \in L_1$ and $U(t, t') = \emptyset$.

The first condition in the above definition is dedicated to proving self transitions from the syntax of Σ_0, and the the second one takes care of preventing Σ_0-terms from performing other types of transitions while allowing other terms to do so.

Theorem 3. *(Granting Meta-theorem) Consider source-dependent TSS's $tss_0 = (\Sigma_0, L_0, D_0)$ and $tss_1 = (\Sigma_1, L_1, D_1)$. If tss_1 grants L_1 transitions on Σ_0-terms and $L_0 \cap L_1 = \emptyset$ then $tss_0 \cup tss_1$ is a granting extension of tss_0.*

The following example applies our meta-theorem to obtain orthogonality of relative-discrete-time extension of *MPA*.

Example 5 (Timed *MPA*: Orthogonality). Consider the tss_m of *MPA* in Example 1 and tss_t of Example 3. TSS tss_t can be decomposed into the following three parts: $tss_0 \doteq (\{\underline{a}._-, \underline{\delta}\}, A, \{(\mathbf{ua}), (\mathbf{td})\})$, $tss_1 \doteq (\{\delta, a._-, \_ + \_\}, \{1\}, \{(\mathbf{tc0}), (\mathbf{ta}), (\mathbf{d})\})$ and $tss_2 \doteq (\{\_ + \_\}, \{1\}, \{(\mathbf{tc1}), (\mathbf{tc2})\})$.

It follows from Definition 13 that tss_1 is source preserving since:

1. the conclusions of (**ta**) and (**d**) are unifiable using any substitution, hence using the unifiers of the empty set of premises,
2. and the conclusion of (**tc0**) is unifiable using the unifiers of the premises, i.e., those that evaluate x and x' to the same term and y and y' to the same term.

It then follows from Definition 14 that $tss_1 \cup tss_2$ grants time transitions over *MPA* terms since

1. tss_1 is strictly stratified using a simple measure of size on terms, it is source preserving as shown before, and by applying unifiers to the source of conclusion of **(tc0)**, **(ta)** and **(d)**, i.e., the set $\{x + y, a.x, \delta\}$, we can cover the syntax of *MPA*,
2. in tss_2, deduction rules **(tc1)** and **(tc2)** have source-dependent negative premises with label 1 (note that **(tc1)** and **(tc2)** are the same as their reduced versions).

From Theorem 3, it follows that the extension of tss_m with $tss_1 \cup tss_2$ is a granting extension, hence an orthogonal extension. Furthermore, the extension of $tss_m \cup tss_1 \cup tss_2$ with tss_0 is conservative, hence orthogonal, following Theorem 1. Since orthogonality is a preorder, we conclude that $tss_m \cup tss_t$ is an orthogonal extension of tss_m.

The following theorem establishes the link between orthogonality and equational conservativity. It is very similar to the theorem stated in [16, 17] about the relation between operational and equational conservativity. The theorem states that a sound axiomatization of an operationally conservative extension cannot induce new equalities on the old syntax.

Theorem 4. *(Equational Conservativity Theorem) Consider TSS's $tss_0 = (\Sigma_0, L_0, D_0)$ and $tss_1 = (\Sigma_1, L_1, D_1)$ where tss_1 is an orthogonal extension of tss_0. Also let E_0 be a sound and complete axiomatization of tss_0 and E_1 be a sound axiomatization of tss_1. If $\forall_{p,p' \in C(\Sigma_0)} E_0 \vdash p = p' \Rightarrow E_1 \vdash p = p'$ then E_1 is an equational conservative extension of E_0.*

Finally, the last theorem establishes sufficient conditions for a sound equationally conservative extension to be a complete equational theory for the extended language.

Theorem 5. *(Elimination Theorem) Consider TSS's $tss_0 = (\Sigma_0, L_0, D_0)$ and $tss_1 = (\Sigma_1, L_1, D_1)$ where tss_1 is an orthogonal extension of tss_0. Also let E_0 and E_1 be sound axiomatizations of tss_0 and tss_1, respectively. If E_0 is also complete for tss_0, E_1 is an equational conservative extension of E_0 and E_1 eliminates terms from $\Sigma_1 \setminus \Sigma_0$, then E_1 is complete for tss_1.*

A typical line of reasoning starts with taking an orthogonal extension and a sound axiomatization thereof, and proving equational conservativity using Theorem 4. Then, by proving an elimination result for the newly introduced operators, one can get completeness of the axiomatization following Theorem 5.

6 Conclusions

In this paper, we defined a more relaxed notion of operational conservativity, called orthogonality which allows for non-destructive extension of the behavior of the old language. We gave a meta-theorem providing sufficient conditions for this notion. Also, we presented a slightly more general notion of equational conservativity and established the link between these two notions.

Extending the theory presented in this paper with the concept of variable binding is an straightforward extension along the lines of [5]. The second enhancement of our work concerns operational extensions that require a translation of labels (using a kind of abstraction function). Finally, investigating the possibility of other realizations of orthogonality is an interesting subject for future research.

References

1. L. Aceto, W. J. Fokkink, and C. Verhoef. Structural operational semantics. In *Handbook of Process Algebra*, Chapter 3, pages 197–292. Elsevier, 2001.
2. J. C. M. Baeten. Embedding untimed into timed timed process algebra: the case for explicit termination. *MSCS*, 13(4):589–618, 2003.
3. J. C. M. Baeten and C. Verhoef. Concrete Process Algebra. In *Handbook of Logic in Computer Science*, volume 4, pages 149–268. Oxford University Press, 1995.
4. R. Bol and J. F. Groote. The meaning of negative premises in transition system specifications. *JACM*, 43(5):863–914, 1996.
5. W. J. Fokkink and C. Verhoef. A conservative look at operational semantics with variable binding. *I&C*, 146(1):24–54, 1998.
6. R. J. van Glabbeek. The meaning of negative premises in transition system specifications II. *JLAP*, 60-61:229–258, 2004.
7. R.J. van Glabbeek. The linear time - branching time spectrum I. In *Handbook of Process Algebra*, Chapter 1, pages 3–100. Elsevier, 2001.
8. J. F. Groote. Transition system specifications with negative premises. *TCS*, 118(2):263–299, 1993.
9. G. Leduc and L. Leonard. A timed LOTOS supporting a dense time domain and including new timed operators. In *Proceedings of FORTE'92*, pages 87–102. North-Holland, 1993.
10. C.A. Middelburg. An alternative formulation of operational conservativity with binding terms. *JLAP*, 55(1/2):1–19, 2003.
11. M.R. Mousavi and M. A. Reniers. Orthogonal Extensions in Structural Operational Semantics. Technical Report, Dept. of Computer Science, Eindhoven Univ. of Tech., 2005.
12. D. M. Park. Concurrency and automata on infinite sequences. In *Proceedings of the 5th GI Conference*, volume 104 of *LNCS*, pages 167–183. Springer, 1981.
13. G. D. Plotkin. A structural approach to operational semantics. *JLAP*, 60:17–139, 2004.
14. J. J. Vereijken. *Discrete Time Process Algebra*. PhD thesis, Department of Computer Science, Eindhoven University of Technology, 1997.
15. C. Verhoef. A general conservative extension theorem in process algebra. In *Proceedings of PROCOMET'94*, pages 274–302. Elsevier, 1994.
16. C. Verhoef. A congruence theorem for structured operational semantics with predicates and negative premises. *Nordic Journal of Computing*, 2(2):274–302, 1995.
17. C. Verhoef, L. Aceto, and W. Fokkink. Conservative extension in structural operational semantics. *BEATCS*, 69:110–132, 1999.

Basic Observables for a Calculus for Global Computing

Rocco De Nicola[1], Daniele Gorla[2], and Rosario Pugliese[1]

[1] Dipartimento di Sistemi e Informatica, Università di Firenze
[2] Dipartimento di Informatica, Università di Roma "La Sapienza"

Abstract. We introduce a foundational language for modelling applications over global computers whose interconnection structure can be explicitly manipulated. Together with process distribution, mobility, remote operations and asynchronous communication through distributed data spaces, the language provides constructs for explicitly modelling inter-node connections and for dynamically establishing and removing them. For the proposed language, we define natural notions of extensional observations and study their closure under operational reductions and/or language contexts to obtain *barbed congruence* and *may testing* equivalence. For such equivalences, we provide alternative characterizations in terms of a labelled *bisimulation* and a *trace* equivalence that can be used for actual proofs.

1 Introduction

In the last decade, we have witnessed the birth of many calculi and kernel languages intended to support programming of global systems and to provide formal tools for reasoning over them. These formalisms in general provide constructs and mechanisms, at different abstraction levels, for modelling the execution contexts of the network where applications roam and run, for coordinating and monitoring the use of resources, for expressing process communication and mobility, and for specifying and enforcing security policies. However, much research effort has been devoted to studying the impact of different communication and mobility paradigms, but little attention has been devoted to the modelling of the actual network underlying global computers as such. Usually, the model of the network implicitly originates from other linguistic choices concerning, e.g., the mobility paradigm. All foundational languages proposed in the literature either model the network as an evolving graph of fully connected nodes [17,9,26,1] or model it as an evolving forest of trees [7,14,24,8]. In our view, both approaches do not convincingly model global computers (the Internet is neither a clique nor a forest of trees) and lack of flexibility ('sharing of resources' is difficult to control and requires complex modelling).

Here, we want to develop the semantic theory of a new model that takes its origin from two formalisms with opposite objectives, namely the programming language X-KLAIM [2] and the π-calculus [23]. The former one is a full fledged programming language based on KLAIM [9], while the latter one is the generally recognized minimal common denominator of calculi for mobility. The resulting model has been called TKLAIM (*topological* KLAIM); it retains the main features of KLAIM (distribution, remote operations, process mobility and asynchronous communication through distributed data spaces), but extends it with new constructs to flexibly model the interconnection structure underlying a net. tKLAIM permits explicit creation of inter-node connections and their destruction. Connections are essential to perform remote operations: these are possible only if the node where they are initiated and the target one are directly connected.

For the proposed formalism, we introduce two abstract semantics, *barbed congruence* and *may testing*, that are obtained as the closure under operational reductions and/or language contexts of the extensional equivalences induced by what we consider basic *observables* for global computers. For deciding the observables to use, we have been struggling with the following ones:

  i. a specific site is up and running (i.e., it provides a datum of any kind)
  ii. a specific information is available in (at least) a site,
  iii. a specific information is present at a specific site.

Other calculi for global computers make use of (barbed) congruences induced by similar observables: for example, Ambient uses barbs that are somehow related to *i.*; the barbs in Dπ-calculus instead, are strongly related to *iii.*. Within our framework, it can be proved that, by closing observations under any tKLAIM context, the three observables all yield the same congruence. This is already an indication of the robustness of the resulting semantic theories. Moreover, the observables are powerful enough to yield interesting theories also when considering lower-level features, such as failures [11].

Of course, the step that comes next after defining equivalence as context closure is determining some alternative characterizations that would permit to better appreciate their discriminating power and to devise proof techniques that avoid universal quantification over contexts (that would render equivalence checking very hard).

In this paper, we concentrate on the barbed and may equivalences induced by the first basic observable (*a site is up and running*) and establish their correspondence with a bisimulation-based and a trace-based equivalence. To this aim, we introduce a labelled transition system for tKLAIM (with labels indicating the performed action) and, on top of it, we define alternative characterizations of barbed congruence and may testing in terms of (non-standard) labelled *bisimilarity* and *trace* equivalence, resp.. The actual development of the alternative characterizations, although performed along the lines of similar results for CCS [20, 4] and π-calculus [23] had to face problems raised by process distribution and mobility, by the explicit use of connections and by asynchrony.

2   The Process Language tKLAIM

The syntax of tKLAIM is reported in Table 1. We assume the existence of two countable and disjoint sets: *names*, ranged over by $l, l', \ldots, u, \ldots, x, y, \ldots$, and *process variables*, ranged over by $X, Y, \ldots$. Names provide the abstract counterpart of the set of *communicable* objects and can be used as localities or variables; notationally, we prefer letters $l, l', \ldots$ when we want to stress the use of a name as a locality, and $x, y, \ldots$ when we want to stress the use of a name as a variable. We will use u for variables and localities.

Nets, ranged over by $N, M, \ldots$, are finite collections of nodes and inter-node connections. A *node* is a pair $l :: C$, where locality l is the address of the node and C is the (parallel) component located at l. *Components*, ranged over by $C, D, \ldots$, can be either

Table 1. tKlaim Syntax

Nets:	$N ::= \mathbf{0} \mid l :: C \mid \{l_1 \leftrightarrow l_2\} \mid (\nu l)N \mid N_1 \| N_2$
Components:	Processes:
$C ::= \langle l \rangle \mid P \mid C_1 \mid C_2$	$P ::= \mathbf{nil} \mid a.P \mid P_1 \mid P_2 \mid X \mid \mathbf{rec}\, X.P$
Actions:	
$a ::= \mathbf{in}(!x)@u \mid \mathbf{in}(u_2)@u_1 \mid \mathbf{out}(u_2)@u_1 \mid \mathbf{eval}(P)@u \mid \mathbf{new}(l) \mid \mathbf{conn}(u) \mid \mathbf{disc}(u)$	

processes or data, denoted by $\langle l \rangle$. *Connections* are pairs of node addresses $\{l_1 \leftrightarrow l_2\}$ stating that the nodes at address l_1 and l_2 are directly and bidirectionally connected. In $(\nu l)N$, name l is private to N; the intended effect is that, if one considers the term $M \| (\nu l)N$, then locality l of N cannot be referred from within M.

Processes, ranged over by $P, Q, R, \ldots$, are the tKlaim active computational units and may be executed concurrently either at the same locality or at different localities. They are built from the inert process **nil** and from the basic actions by using prefixing, parallel composition and recursion. *Actions* permit removing/adding data from/to node repositories (actions **in** and **out**), activating new threads of execution (action **eval**), creating new nodes (action **new**), and establishing and removing connections (actions **conn** and **disc**). Notice that $\mathbf{in}(l)@l'$ differs from $\mathbf{in}(!x)@l'$ in that the former evolves only if datum $\langle l \rangle$ is present at l', while the latter accepts any datum. Indeed, $\mathbf{in}(l)@l'$ is a form of *name matching operator* reminiscent of Linda's [16] pattern-matching.

Names occurring in tKlaim processes and nets can be *bound*. More precisely, prefix $\mathbf{in}(!x)@u.P$ binds x in P; prefix $\mathbf{new}(l).P$ binds l in P, and, similarly, net restriction $(\nu l)N$ binds l in N; finally, $\mathbf{rec}\, X.P$ binds X in P. A name that is not bound is called *free*. The sets $fn(\cdot)$ and $bn(\cdot)$ of free and bound names of a term, respectively, are defined accordingly. The set $n(\cdot)$ of names of a term is the union of its free and bound names. As usual, we say that two terms are *alpha-equivalent* if one can be obtained from the other by renaming bound names. We shall say that a name u is fresh for $\_$ if $u \notin n(\_)$. In the sequel, we shall work with terms whose bound names are all distinct and different from the free ones.

tKlaim operational semantics relies on a structural congruence and a reduction relation. The *structural congruence*, $\equiv$, is formally defined in [10] and identifies nets which intuitively represent the same net. It is inspired to π-calculus's structural congruence (see, e.g., [23]): it states that '$\|$' and '$|$' are monoidal operators with $\mathbf{0}$ and **nil** as identity elements, it equates alpha-equivalent nets, it regulates commutativity of restrictions, and it allows to freely fold/unfold recursive processes. Moreover, the following laws are crucial in our setting:

(Clone)   (Self)   (BiDir)
$l :: C_1 | C_2 \equiv l :: C_1 \| l :: C_2$ $l :: \mathbf{nil} \equiv \{l \leftrightarrow l\}$ $\{l_1 \leftrightarrow l_2\} \equiv \{l_2 \leftrightarrow l_1\}$

(RNode)   (Ext)
$(\nu l)N \equiv (\nu l)(N \| l :: \mathbf{nil})$ $N \| (\nu l)M \equiv (\nu l)(N \| M) \quad \text{if } l \notin fn(N)$

(Clone) turns the parallel composition of co-located components into a parallel between nodes; (Self) states that nodes are self-connected; (BiDir) states that connections are

Table 2. tKLAIM Operational Semantics

(R-Out)
$l_1 :: \mathbf{out}(l)@l_2.P \parallel \{l_1 \leftrightarrow l_2\} \longmapsto l_1 :: P \parallel \{l_1 \leftrightarrow l_2\} \parallel l_2 :: \langle l \rangle$

(R-Eval)
$l_1 :: \mathbf{eval}(P_2)@l_2.P_1 \parallel \{l_1 \leftrightarrow l_2\} \longmapsto l_1 :: P_1 \parallel \{l_1 \leftrightarrow l_2\} \parallel l_2 :: P_2$

(R-In)
$l_1 :: \mathbf{in}(!x)@l_2.P \parallel \{l_1 \leftrightarrow l_2\} \parallel l_2 :: \langle l \rangle \longmapsto l_1 :: P[l/x] \parallel \{l_1 \leftrightarrow l_2\}$

(R-Match)
$l_1 :: \mathbf{in}(l)@l_2.P \parallel \{l_1 \leftrightarrow l_2\} \parallel l_2 :: \langle l \rangle \longmapsto l_1 :: P \parallel \{l_1 \leftrightarrow l_2\}$

(R-New)
$l :: \mathbf{new}(l').P \longmapsto (\nu l')(l :: P \parallel l' :: \mathbf{nil})$

(R-Conn)
$l_1 :: \mathbf{conn}(l_2).P \parallel l_2 :: \mathbf{nil} \longmapsto l_1 :: P \parallel \{l_1 \leftrightarrow l_2\}$

(R-Disc)
$l_1 :: \mathbf{disc}(l_2).P \parallel \{l_1 \leftrightarrow l_2\} \longmapsto l_1 :: P \parallel l_2 :: \mathbf{nil}$

(R-Par)
$$\frac{N_1 \longmapsto N_1'}{N_1 \parallel N_2 \longmapsto N_1' \parallel N_2}$$

(R-Res)
$$\frac{N \longmapsto N'}{(\nu l)N \longmapsto (\nu l)N'}$$

(R-Struct)
$$\frac{N \equiv M \longmapsto M' \equiv N'}{N \longmapsto N'}$$

bidirectional; (Ext) is the standard π-calculus rule for scope extension. Finally, (RNode) states that any restricted name can be used as the address of a node; indeed, we consider restricted names as private network addresses, whose corresponding nodes can be activated and deactivated on demand. In the sequel, we shall assume that each restricted name does correspond to an actual node. This assumption is justified by law (RNode).

The reduction relation is given in Table 2. In (R-Out) and (R-Eval), the existence of a connection between the nodes source and target of the action is necessary to place the spawned component. Notice that existence of the connection can only be checked at run-time: an approach like [17] does not fit well in a global computing setting because it relies on a typing mechanism that would require to statically know the whole net. (R-In) and (R-Match) additionally require the existence of a matching datum in the target node. (R-Match) states that action $\mathbf{in}(l)@l_2$ consumes exactly the datum $\langle l \rangle$ at l_2, while (R-In) states that action $\mathbf{in}(!\,x)@l_2$ can consume any $\langle l \rangle$ at l_2; l will then replace the free occurrences of x in the continuation of the process performing the action. (R-New) states that execution of action $\mathbf{new}(l')$ adds a restriction over l' to the net, while creating a node with address l'. Finally, (R-Conn) and (R-Disc) deal with activation/deactivation of connections. In the first case, we need to ensure that the connected nodes do exist; in the second case, we need to check existence of the connection to be removed.

tKLAIM adopts a LINDA-like [16] communication mechanism: communication is asynchronous and data are anonymous. Indeed, no synchronization takes place between (sending and receiving) processes, because their interactions are mediated by nodes, that act as data repositories. For the sake of simplicity, we only consider monadic data, but the semantic theories we develop could be smoothly extended to deal with tuples of data and with a full-blown LINDA-like *pattern matching* mechanism.

3 Observables, Closures and Equivalences

In this section we present both a linear time and a branching time equivalence that yield sensible semantic theories for TKLAIM. The approach we follow relies on the definition of an *observation* (also called *barb*) that intuitively formalises the possible interactions of a process. We use observables to define equivalence relations that identify those nets that cannot be taken apart by any basic observation along reductions in any execution context. As usual, $\Longrightarrow$ denotes the reflexive and transitive closure of $\longmapsto$ and $\widetilde{l}$ denotes a possibly empty set of names.

Definition 1 (Barbs and Contexts). *Predicate* $N \downarrow l$ *holds true if and only if* $N \equiv (\nu\widetilde{l})(N' \parallel l :: \langle l' \rangle)$ *for some* $\widetilde{l}$, N' *and* l' *such that* $l \notin \widetilde{l}$. *Predicate* $N \Downarrow l$ *holds true if and only if* $N \Longrightarrow N'$ *for some* N' *such that* $N' \downarrow l$. *A context* $C[\cdot]$ *is a* TKLAIM *net with an occurrence of a hole* $[\cdot]$ *to be filled in with any net. Formally,*

$$C[\cdot] \quad ::= \quad [\cdot] \quad | \quad N \parallel C[\cdot] \quad | \quad (\nu l)C[\cdot]$$

We have chosen the basic observables by taking inspiration from those used for the asynchronous π-calculus [23]. One may wonder if our choice is "correct" and argue that there are other alternative notions of basic observables that seem quite natural, as we have discussed in the Introduction. In the full paper [10], we prove that the congruences induced by these alternative observables do coincide. This means that our results are quite independent from the observable chosen and vindicates our choice. Now, we say that a binary relation $\mathcal{R}$ between nets is

- *barb preserving*, if $N \mathcal{R} M$ and $N \Downarrow l$ imply $M \Downarrow l$;
- *reduction closed*, if $N \mathcal{R} M$ and $N \longmapsto N'$ imply $M \Longrightarrow M'$ and $N' \mathcal{R} M'$, for some M';
- *context closed*, if $N \mathcal{R} M$ implies $C[N] \mathcal{R} C[M]$, for every context $C[\cdot]$.

Our touchstone equivalences should at the very least relate nets with the same observable behaviour; thus, they must be barb preserving. However, an equivalence defined only in terms of this property would be too weak: indeed, the set of barbs of a net may change during computations or when interacting with an external environment. Moreover, for the sake of compositionality, our touchstone equivalences should also be congruences. These requirements lead us to the following definitions.

Definition 2 (May testing). $\simeq$ *is the largest symmetric, barb preserving and context closed relation between nets.*

Definition 3 (Barbed congruence). $\cong$ *is the largest symmetric, barb preserving, reduction and context closed relation between nets.*

We want to remark that the above definition of barbed congruence is the standard one, see [18, 23]. May testing is, instead, usually defined in terms of *observers*, *experiments* and *success of an experiment* [13]. In [10], we prove that, if we let $\simeq'$ denote the equivalence on TKLAIM nets defined a lá [13], the two definitions do coincide. Moreover, the inclusions between our touchstone equivalences reflect the inclusions that hold in the π-calculus, since also in our setting may testing, differently from barbed congruence, ignores the branching structure of a process. A pair of nets proving that $\cong \subset \simeq$ can be obtained from the CCS terms $a_1.(a_2 + a_3)$ and $a_1.a_2 + a_1.a_3$, that are may testing equivalent but not barbed congruent, by implementing the non-deterministic choice ('+') through parallel composition.

The problem with the definitions of barbed congruence and may testing is that context closure makes it hard to prove equivalences due to the universal quantification over contexts. In the following section, we shall provide two alternative characterisations of $\cong$ and $\simeq$, as a *bisimulation-based* and as a *trace-based* equivalence, respectively.

4 Alternative Characterisations

4.1 A Labelled Transition System

In order to provide more tractable characterisations of our touchstone equivalences, we introduce a labelled transition system (LTS) to make apparent the action a net is willing to perform in order to evolve. The *labelled transition relation*, $\xrightarrow{\alpha}$, is defined as the least relation over nets induced by the inference rules in Table 3. Labels take the form

$$\alpha ::= \tau \mid l_1 \frown l_2 \mid (\widetilde{\nu l})\langle l\rangle @ l_1 : l_2 \mid l_1 : \triangleright l_2 \mid l_1 : (\widetilde{\nu l})l \triangleleft l_2 \mid l_1 : l_2 \mid l_1 : \neg l_2$$

We let $bn(\alpha)$ be $\widetilde{l}$ if $\alpha = (\widetilde{\nu l})\langle l\rangle @ l_1 : l_2$ or $\alpha = l_1 : (\widetilde{\nu l})l \triangleleft l_2$, and be $\emptyset$ otherwise; $fn(\alpha)$ and $n(\alpha)$ are defined accordingly.

Let us now explain the intuition behind the labels of the LTS and some key rules. Label α in $N \xrightarrow{\alpha} N'$ can be

τ : this means that N may perform a reduction step to become N' (see Proposition 1).
$l_1 \frown l_2$: this means that in N there is a direct connection between nodes l_1 and l_2 (see (LTS-LINK)).
$(\widetilde{\nu l})\langle l\rangle @ l_1 : l_2$: this means that in N there is a datum $\langle l\rangle$ located at l_1 and a connection $\{l_1 \leftrightarrow l_2\}$; the datum is available for processes located at l_2 (see (LTS-DATUM), (LTS-OFFER) and (LTS-LINK)). Moreover, according to whether $\widetilde{l} = \{l\}$ or $\widetilde{l} = \emptyset$, we also know if N restricts l or not (see (LTS-OPEN)).
$l_1 : \triangleright l_2$: this means that in N there is a process located at l_1 willing to send a component at l_2 (see (LTS-OUT) and (LTS-EVAL)[1]). For the sending to take place, a direct connection between such nodes is needed (see (LTS-SEND)).
$l_1 : (\widetilde{\nu l})l \triangleleft l_2$: this means that in N there is a process located at l_1 willing to retrieve a (possibly fresh) datum $\langle l\rangle$ at l_2 (see (LTS-IN), (LTS-MATCH) and (LTS-BIN)). For the actual retrieval, a direct connection between such nodes and a proper datum at l_2 are needed (see (LTS-COMM)).
$l_1 : l_2$: this means that in N there is a process located at l_1 willing to activate a connection with l_2 (see (LTS-CONN)). For the actual activation, the net must contain a node with address l_2, as pointed out by label $l_2 \frown l_2$ (see (LTS-EST) and (SELF)).

---

[1] It should not be surprising that actions **out** and **eval** yield the same label. Of course, the two actions should be taken apart for security reasons because accepting processes for execution is more dangerous than accepting data. However, in our setting, an external observer has not enough power to notice any difference: in both cases, it can just observe that a packet is sent.

$l_1 : \neg l_2$: this means that in N there is a process located at l_1 willing to deactivate a connection with l_2 (see (LTS-DISC)). For the actual deactivation, the net must contain the connection $\{l_1 \leftrightarrow l_2\}$ (see (LTS-REM)).

The last four kinds of labels describe 'intentions' of a process running in the net. Thus, (LTS-OUT) should be read as: "process $\mathbf{out}(l)@l_2.P$ running at l_1 is willing to send a component at l_2; whenever the execution context provides the connection needed, l_1 will host process P for execution and will run in a net where the connection $\{l_1 \leftrightarrow$

Table 3. A Labelled Transition System

(LTS-OUT)
$l_1 :: \mathbf{out}(l)@l_2.P \xrightarrow{l_1 : \triangleright l_2} l_1 :: P \parallel \{l_1 \leftrightarrow l_2\} \parallel l_2 :: \langle l \rangle$

(LTS-EVAL)
$l_1 :: \mathbf{eval}(P_2)@l_2.P_1 \xrightarrow{l_1 : \triangleright l_2} l_1 :: P_1 \parallel \{l_1 \leftrightarrow l_2\} \parallel l_2 :: P_2$

(LTS-IN)
$l_1 :: \mathbf{in}(!x)@l_2.P \xrightarrow{l_1 : l \triangleleft l_2} l_1 :: P[l/x] \parallel \{l_1 \leftrightarrow l_2\}$

(LTS-MATCH)
$l_1 :: \mathbf{in}(l)@l_2.P \xrightarrow{l_1 : l \triangleleft l_2} l_1 :: P \parallel \{l_1 \leftrightarrow l_2\}$

(LTS-NEW)
$l :: \mathbf{new}(l').P \xrightarrow{\tau} (\nu l')(l :: P \parallel l' :: \mathbf{nil})$

(LTS-CONN)
$l_1 :: \mathbf{conn}(l_2).P \xrightarrow{l_1 : l_2} l_1 :: P \parallel \{l_1 \leftrightarrow l_2\}$

(LTS-DISC)
$l_1 :: \mathbf{disc}(l_2).P \xrightarrow{l_1 : \neg l_2} l_1 :: P \parallel l_2 :: \mathbf{nil}$

(LTS-LINK) $\{l_1 \leftrightarrow l_2\} \xrightarrow{l_1 \frown l_2} \mathbf{0}$

(LTS-DATUM) $l_1 :: \langle l \rangle \xrightarrow{\langle l \rangle @ l_1 : l_1} \mathbf{0}$

(LTS-OFFER)
$$\frac{N_1 \xrightarrow{\langle l \rangle @ l_2 : l_2} N_1' \qquad N_2 \xrightarrow{l_1 \frown l_2} N_2'}{N_1 \parallel N_2 \xrightarrow{\langle l \rangle @ l_2 : l_1} N_1' \parallel N_2'}$$

(LTS-BIN)
$$\frac{N \xrightarrow{l_1 : l \triangleleft l_2} N' \qquad l \notin fn(N)}{N \xrightarrow{l_1 : (\nu l) l \triangleleft l_2} N' \parallel l :: \mathbf{nil}}$$

(LTS-PAR)
$$\frac{N_1 \xrightarrow{\alpha} N_2 \qquad bn(\alpha) \cap fn(N) = \emptyset}{N_1 \parallel N \xrightarrow{\alpha} N_2 \parallel N}$$

(LTS-SEND)
$$\frac{N_1 \xrightarrow{l_1 : l_2} N_1' \qquad N_2 \xrightarrow{l_1 \frown l_2} N_2'}{N_1 \parallel N_2 \xrightarrow{\tau} N_1' \parallel N_2'}$$

(LTS-COMM)
$$\frac{N_1 \xrightarrow{l_1 : l \triangleleft l_2} N_1' \qquad N_2 \xrightarrow{\langle l \rangle @ l_2 : l_1} N_2'}{N_1 \parallel N_2 \xrightarrow{\tau} N_1' \parallel N_2'}$$

(LTS-EST)
$$\frac{N_1 \xrightarrow{l_1 : l_2} N_1' \qquad N_2 \xrightarrow{l_2 \frown l_2} N_2'}{N_1 \parallel N_2 \xrightarrow{\tau} N_1' \parallel N_2'}$$

(LTS-REM)
$$\frac{N_1 \xrightarrow{l_1 : \neg l_2} N_1' \qquad N_2 \xrightarrow{l_1 \frown l_2} N_2'}{N_1 \parallel N_2 \xrightarrow{\tau} N_1' \parallel N_2'}$$

(LTS-RES)
$$\frac{N \xrightarrow{\alpha} N' \qquad l \notin n(\alpha)}{(\nu l)N \xrightarrow{\alpha} (\nu l)N'}$$

(LTS-OPEN)
$$\frac{N \xrightarrow{\langle l \rangle @ l_2 : l_1} N' \qquad l \notin \{l_1, l_2\}}{(\nu l)N \xrightarrow{(\nu l)\langle l \rangle @ l_2 : l_1} N'}$$

(LTS-STRUCT)
$$\frac{N \equiv N_1 \qquad N_1 \xrightarrow{\alpha} N_2 \qquad N_2 \equiv N'}{N \xrightarrow{\alpha} N'}$$

$l_2\}$ does exist and the datum $\langle l \rangle$ is placed at l_2". (LTS-EVAL), (LTS-IN), (LTS-MATCH), (LTS-CONN) and (LTS-DISC) should be interpreted similarly.

(LTS-OPEN) signals extrusion of bound names; as in some presentation of the π-calculus, this rule is used to investigate the capability of processes to export bound names, rather than to actually extend the scope of bound names.

Notice that the LTS of Table 3 may appear unnecessarily complicated as a tool to define the operational semantics of TKLAIM: consider, e.g., the right hand side of the rules for **out/in/eval**, or rule (LTS-BIN) (used to signal that a received name is fresh for the receiving net). Nevertheless, it is adequate as a tool to establish alternative, more tractable, characterisations of the touchstone equivalences we are interested in. Indeed, the complications in the operational rules of Table 3 resemble those arisen in [25] when defining an 'equivalent' LTS depending on the reduction semantics of a calculus. However, in [25] only simple calculi are considered and it would be interesting to investigate if the approach can be satisfactory extended to TKLAIM. Finally, the LTS is 'correct' w.r.t. the actual operational semantics of TKLAIM, $\longmapsto$, as stated by the following Proposition.

Proposition 1. $N \longmapsto M$ if and only if $N \xrightarrow{\tau} M$.

4.2 Bisimulation Equivalence

We now characterize barbed congruence by using the labels of the LTS instead of the universal quantification over contexts; in this way, we obtain an alternative characterization of $\cong$ in terms of a labelled *bisimilarity*. As a matter of notation, we let

$$\chi ::= \tau \mid l_1 \frown l_2 \mid (\widetilde{vl})\langle l \rangle @ l_1 : l_2$$

Moreover, $\Rightarrow$ stands for $\xrightarrow{\tau}^*$, $\xRightarrow{\alpha}$ stands for $\Rightarrow \xrightarrow{\alpha} \Rightarrow$, and $\xRightarrow{\hat{\alpha}}$ stands for $\Rightarrow$, if $\alpha = \tau$, and for $\xRightarrow{\alpha}$, otherwise.

Definition 4 (Bisimilarity). *A symmetric relation $\mathcal{R}$ between TKLAIM nets is a (weak) bisimulation if, for each $N \mathcal{R} M$, it holds that:*

1. $N \xrightarrow{\chi} N'$ *implies that* $M \xRightarrow{\hat{\chi}} M'$ *and* $N' \mathcal{R} M'$, *for some M'*;
2. $N \xrightarrow{l_1 : \triangleright l_2} N'$ *implies that* $M \parallel \{l_1 \leftrightarrow l_2\} \Rightarrow M'$ *and* $N' \mathcal{R} M'$, *for some M'*;
3. $N \xrightarrow{l_1 : l \triangleleft l_2} N'$ *implies that* $M \parallel \{l_1 \leftrightarrow l_2\} \parallel l_2 :: \langle l \rangle \Rightarrow M'$ *and* $N' \mathcal{R} M'$, *for some M'*;
4. $N \xrightarrow{l_1 : l_2} N'$ *implies that* $M \parallel l_2 :: \textbf{nil} \Rightarrow M'$ *and* $N' \mathcal{R} M'$, *for some M'*;
5. $N \xrightarrow{l_1 : \neg l_2} N'$ *implies that* $M \parallel \{l_1 \leftrightarrow l_2\} \Rightarrow M'$ *and* $N' \mathcal{R} M'$, *for some M'*.

Bisimilarity, $\approx$, is the largest bisimulation.

Bisimilarity requires that labels of the form $(\widetilde{vl})\langle l \rangle @ l_1 : l_2$ or $l_1 \frown l_2$ must be replied to with the same label (possibly with some additional τ-step). This is necessary since such labels describe the structure of the net (its data and connections) and, to be equivalent, two nets must have at least the same structure. Labels different from χ only express intentions and are handled differently. For example, the intention of sending a component, say $N \xrightarrow{l_1 : \triangleright l_2} N'$, can be simulated by a net M (in a context where l_1 and l_2 are connected) through the execution of some τ-steps that lead to some M' equivalent to N'. Indeed, since we want our bisimulation to be a congruence, a context that provides a connection between the source and the target nodes of the sending action must not tell apart N and M. Similar considerations also hold for the last three items of Definition 4.

Notice that labels of the form $l_1 : (\nu l)l \triangleleft l_2$ are not necessary for the definition of bisimulation. Indeed, they exactly work like labels $l_1 : l \triangleleft l_2$ with the extra information that l is fresh for the receiving net; for the bisimulation game this information is useless, while it will be of fundamental importance when considering trace-based equivalence.

Remarkably, though in TKLAIM processes can occur as arguments in process actions (**eval**), the LTS and the bisimulation we developed do not use labels containing processes. Thus, the bisimulation relies only on a standard quantification over names (in the input case) and we strongly conjecture that it is decidable, under proper assumptions: techniques similar to those in [21] could be used here. Moreover, the presence of rule (LTS-STRUCT) in the LTS does not compromise the tractability of $\approx$; obviously, (LTS-STRUCT) can be dropped, if one is prepared to have more rules in the LTS.

We can now present our first main result, whose proof is in the full paper [10].

Theorem 1 (Alternative Characterization of Barbed Congruence). $\approx \ = \ \cong$.

4.3 Trace Equivalence

In this section, we develop an alternative characterization of may testing. For some well-known process calculi, may testing coincides with trace equivalence [13, 3, 5]; in this section, we show how a similar result is obtained also in the setting of TKLAIM.

The idea behind trace equivalence is that N and M are related if and only if the sets of their traces coincide. Put in another form, if N exhibits a sequence of visible actions σ, then M must exhibit σ as well, and vice versa. In an asynchronous setting [5], this requirement must be properly weakened, since the discriminating power of asynchronous contexts is weaker: in the asynchronous π-calculus, for example, contexts cannot observe input actions.

To carry out proofs, we found it convenient to introduce a *complementation function* $\overline{\cdot}$ over visible labels (i.e. labels different from τ), ranged over by ϕ, such that

$$\overline{l_1 \frown l_2} \triangleq l_1 : \triangleright l_2 \qquad \overline{l_1 : \triangleright l_2} \triangleq l_1 \frown l_2 \qquad \overline{l_1 : (\widetilde{\nu l})\,l \triangleleft l_2} \triangleq (\widetilde{\nu l})\langle l \rangle @ l_2 : l_1$$

$$\overline{l_1 : l_2} \triangleq l_2 \frown l_1 \qquad \overline{l_1 : \neg l_2} \triangleq l_1 \frown l_2 \qquad \overline{(\widetilde{\nu l})\langle l \rangle @ l_2 : l_1} \triangleq l_1 : (\widetilde{\nu l})\,l \triangleleft l_2$$

Because of the interplay between free and bound names (bound names are always associated to nodes, see rule (RNODE)), we need to distinguish reception of a free name from reception of a bound name (that must be fresh for the receiving net). Similarly to the π-calculus [3, 5], this can be done by exploiting *bound input* labels, $l_1 : (\nu l)l \triangleleft l_2$, generated by rule (LTS-BIN) (that also adds a node with address l because of law (RNODE)). Finally, we let σ to range over (possibly empty) sequences of visible actions, i.e.

$$\sigma \ ::= \ \epsilon \ \mid \ \phi \cdot \sigma$$

where ϵ denotes the empty sequence of actions and '$\cdot$' represents concatenation. As usual, $N \stackrel{\epsilon}{\Rightarrow}$ denotes $N \Rightarrow$ and $N \stackrel{\phi \cdot \sigma}{\Longrightarrow}$ denotes $N \stackrel{\phi}{\Rightarrow} \stackrel{\sigma}{\Rightarrow}$.

The naive formulation of trace equivalence such as "$N \stackrel{\sigma}{\Rightarrow}$ if and only if $M \stackrel{\sigma}{\Rightarrow}$" is too strong in an asynchronous setting: for example, it would distinguish $l \ ::$

Basic Observables for a Calculus for Global Computing   1235

Table 4. The Ordering Relation on Traces

(L1)	$\sigma \cdot (\widetilde{vl})\sigma' \leq \sigma \cdot (\widetilde{vl})(\beta \cdot \sigma')$	if $(\widetilde{vl})\sigma' \neq UNDEF$
(L2)	$\sigma \cdot (\widetilde{vl})(\phi \cdot \gamma \cdot \sigma') \leq \sigma \cdot (\widetilde{vl})(\gamma \cdot \phi \cdot \sigma')$	if $(\widetilde{vl})(\phi \cdot \gamma \cdot \sigma') \neq UNDEF$
(L3)	$\sigma \cdot (\widetilde{vl})\sigma' \leq \sigma \cdot (\widetilde{vl})(\gamma \cdot \overline{\gamma} \cdot \sigma')$	if $(\widetilde{vl})\sigma' \neq UNDEF$
(L4)	$\sigma \cdot l : \triangleright l \cdot \phi \cdot \sigma' \leq \sigma \cdot \phi \cdot \sigma'$	if $l \in \Upsilon(\phi)$
(L5)	$\sigma \cdot l : \triangleright l \cdot \phi \cdot \sigma' \leq \sigma \cdot \phi \cdot l : \triangleright l \cdot \sigma'$	if $l \notin bn(\phi)$
(L6)	$\sigma \cdot \phi' \cdot \sigma' \leq \sigma \cdot \phi \cdot \sigma'$	if $(\phi, \phi') \in \Psi$
(L7)	$\sigma \cdot l_1 : (\widetilde{vl}) l \triangleleft l_1 \cdot \sigma' \leq \sigma \cdot l_2 : (\widetilde{vl}) l \triangleleft l_1 \cdot \sigma'$	
(L8)	$\sigma \cdot l_2 : (\widetilde{vl}) l \triangleleft l_1 \cdot \sigma' \leq \sigma \cdot l_1 : (\widetilde{vl}) l \triangleleft l_1 \cdot l_1 : \triangleright l_2 \cdot \sigma'$	

In law (L1), β stands for either $l_1 : \triangleright l_2$ or $l_1 : l \triangleleft l_2$ or $l_1 : l_2$ or $l_1 : \neg l_2$.
In laws (L2) and (L3), γ stands for either $l_1 : \triangleright l_2$ or $l_1 : l \triangleleft l_2$.
In law (L4), function $\Upsilon(\cdot)$ is defined as follows: $\Upsilon(l_1 : (\widetilde{vl})l \triangleleft l_2) = \Upsilon(l_1 : l_2) = \{l_1, l_2\}$ and $\Upsilon(l_1 \curvearrowright l_2) = \{l_1\}$ and $\Upsilon(l_1 : \triangleright l_2) = \Upsilon((\widetilde{vl}) \langle l \rangle @ l_1 : l_2) = \{l_2\}$.
In law (L6), relation Ψ is $\{(l_1 : \neg l_2, l_1 : \triangleright l_2), (l_1 : \triangleright l_2, l_1 : \neg l_2), (l_2 : \triangleright l_2, l_1 : l_2)\}$.

$\mathbf{in}(!x)@l_1.\mathbf{in}(!y)@l_2$ and $l :: \mathbf{in}(!y)@l_2.\mathbf{in}(!x)@l_1$, which are indeed may testing equivalent. Like in [5], a weaker trace-based equivalence can be defined as follows.

Definition 5 (Trace Equivalence). $\asymp$ *is the largest symmetric relation between* TKLAIM *nets such that, whenever* $N \asymp M$, *it holds that* $N \stackrel{\sigma}{\Longrightarrow}$ *implies* $M \stackrel{\sigma'}{\Longrightarrow}$, *for some* $\sigma' \leq \sigma$.

The crux is to identify a proper ordering on the traces such that may testing is exactly captured by $\asymp$. The ordering $\leq$ is the least reflexive and transitive relation induced by the laws in Table 4. The first three laws have been inspired by [5], while the last five ones are strictly related to inter-node connections. The intuition behind $\sigma' \leq \sigma$ is that, if a context can interact with a net that exhibits σ, then the context can interact with any net that exhibits σ' as well. The ordering $\leq$ relies on the function $(\widetilde{vl})\sigma$, that is used in laws (L1), (L2) and (L3) when moving/removing a label of the form $l_1 : (vl) l \triangleleft l_2$. In this case, the information that l is a fresh received value must be kept in the remaining trace. The formal definition is

$$(\widetilde{vl})\sigma \triangleq \sigma \qquad \text{if } \widetilde{l} \cap fn(\sigma) = \emptyset$$

$$(vl)(\phi \cdot \sigma) \triangleq \begin{cases} l_1 : (vl) l \triangleleft l_2 \cdot \sigma & \text{if } \phi = l_1 : l \triangleleft l_2 \text{ and } l \notin \{l_1, l_2\} \\ \phi \cdot (vl)\sigma & \text{if } l \notin n(\phi) \text{ and } (vl)\sigma \neq UNDEF \\ UNDEF & \text{otherwise} \end{cases}$$

Further explanations can be found in the full paper [10].

The rules in Table 4 can be explained as follows. (L1) states that labels representing intentions cannot be directly observed; at most, their effect can be somehow observed. (L2) states that the execution of an input/output/migration can be delayed along computations without being noticed by any observer. (L3) states that two adjacent 'complementary' actions can be deleted. (L4) states that an action involving l as source or target node always enables sending actions from l to l; because, in all these cases, a node at address l exists. Function $\Upsilon(\cdot)$ is needed to restrict applicability of (L4) only to the cases needed to prove Theorem 2. (L5) states that, if a sending action from l to l is enabled after an action ϕ, then the action can take place before ϕ, since the node at l was already present; clearly, this is possible only if l is not bound by ϕ. (L6) states that some intentions are interchangeable; indeed, since the complementation function is not injective, the same observer may enable different kinds of process actions. (L7) states that, if a process located at l_2 can retrieve a datum from l_1, then processes located at l_1 can retrieve such datum as well. Finally, (L8) states that, if a process located at l_1 can retrieve a datum $\langle l \rangle$ locally and then migrate at l_2, then processes located at l_2 can retrieve $\langle l \rangle$ remotely.

Remarkably, may testing in the (synchronous/asynchronous) π-calculus [3, 5] cannot distinguish bound names from free ones; thus, a bound name can be replaced with any name in a trace. This is *not* the case here: indeed, bound names can always be considered as addresses of nodes, while free names cannot. This makes a difference for an external observer; thus, a law like

$$\sigma \cdot \langle l' \rangle @ l_1 : l_2 \cdot (\sigma'[^{l'}/l]) \leq \sigma \cdot (\nu l) \langle l \rangle @ l_1 : l_2 \cdot \sigma'$$

(that, mutatis mutandis, holds for the π-calculus [3, 5]) does not hold for tKLAIM.

We can now state our second major result; detailed proofs are in [10].

Theorem 2 (Alternative Characterization of May Testing). $\asymp \: = \: \simeq$.

5 Conclusions and Related Work

We have introduced tKLAIM, a foundational language that provides constructs to explicitly model and dynamically establish/remove inter-node connections, and some associated semantic theories. In a companion paper [11], we have applied the theory to a few examples that illustrate usability.

We believe that, although tKLAIM can be somehow encoded in the π-calculus, the introduction of the former is justified by at least two reasons. First, tKLAIM clearly enlightens the key features we want to model such as distribution and mobility of processes, and inter-node connections; an encoding of such features in the π-calculus would hide them within complex process structures. Second, a convincing encoding should enjoy 'reasonable' properties, like those pointed out in [22]. We believe this is *not* the case. For example, in [12] we developed an intuitive encoding of a tKLAIM's sub-calculus into the asynchronous π-calculus that does not preserve convergence. We are now working on proving that this is not incidental and is due to the check of existence of the target of a communication that is performed in tKLAIM and not in the π-calculus. We conjecture that a divergence free encoding does not exist.

Related Work. To our knowledge, no alternative characterization of may testing in terms of a trace-based equivalence has ever been given for a distributed language with process mobility. Bisimulation-based equivalences for calculi relying on a flat net topology are developed in [1, 17]; such equivalences are mainly derived from bisimulation equivalences for the π-calculus and its variants. Bisimulation-based equivalences for calculi relying on a hierarchical net topology are developed in [19, 6, 8]. Although these bisimulations are inspired by Sangiorgi's *context bisimulation* [23] and, thus, exploit universal quantification over processes, they yield proof techniques that are usable in practice.

Finally, the most closely related work is [15]; there, a distributed version of the π-calculus is presented where nodes are connected through links that can fail during the computation. A bisimulation-based proof technique is used to establish properties of systems. However, differently from our approach, the authors only consider links that can fail and do not model dynamic connections establishment.

Acknowledgements. We thank the anonymous referees for their useful comments.

References

1. R. M. Amadio. On modelling mobility. *Theor. Comp. Sci.*, 240(1):147–176, 2000.
2. L. Bettini, R. De Nicola, G. Ferrari, and R. Pugliese. Interactive Mobile Agents in X-KLAIM. In *Proc. of the 7th WETICE*, pages 110–115. IEEE, 1998.
3. M. Boreale and R. De Nicola. Testing equivalences for mobile processes. *Information and Computation*, 120:279–303, 1995.
4. M. Boreale, R. De Nicola, and R. Pugliese. Basic observables for processes. *Information and Computation*, 149(1):77–98, 1999.
5. M. Boreale, R. De Nicola, and R. Pugliese. Trace and testing equivalence on asynchronous processes. *Information and Computation*, 172:139–164, 2002.
6. M. Bugliesi, S. Crafa, M. Merro, and V. Sassone. Communication and Mobility Control in Boxed Ambients. To appear in *Information and Computation*.
7. L. Cardelli and A. D. Gordon. Mobile ambients. *TCS*, 240(1):177–213, 2000.
8. G. Castagna and F. Zappa Nardelli. The Seal Calculus Revisited: contextual equivalence and bisimilarity. In *Proc. of FSTTCS'02*, volume 2556 of *LNCS*, pages 85–96.
9. R. De Nicola, G. Ferrari, and R. Pugliese. KLAIM: a Kernel Language for Agents Interaction and Mobility. *IEEE Transactions on Software Engineering*, 24(5):315–330, 1998.
10. R. De Nicola, D. Gorla, and R. Pugliese. Basic observables for a calculus for global computing. Tech. Rep. 07/2004, Dip. Informatica, Università di Roma "La Sapienza", 2004.
11. R. De Nicola, D. Gorla, and R. Pugliese. Global Computing in a Dynamic Network of Tuple Spaces. In *Proc. of COORDINATION'05*, volume 3454 of *LNCS*, pages 157–172.
12. R. De Nicola, D. Gorla, and R. Pugliese. On the Expressive Power of KLAIM-based Calculi. *Proc. of EXPRESS'04*, ENTCS 128(2):117–130.
13. R. De Nicola and M. Hennessy. Testing equivalence for processes. *TCS*, 34:83–133, 1984.
14. C. Fournet, G. Gonthier, J.-J. Lévy, L. Maranget, and D. Rémy. A calculus of mobile agents. In *Proc. of CONCUR '96*, volume 1119 of *LNCS*, pages 406–421. Springer, 1996.
15. A. Francalanza and M. Hennessy. A Theory of System Behaviour in the Presence of Node and Link Failures. Tech. Rep. cs01:2005, Univ. of Sussex.
16. D. Gelernter. Generative communication in linda. *TOPLAS*, 7(1):80–112. ACM, 1985.
17. M. Hennessy, M. Merro, and J. Rathke. Towards a behavioural theory of access and mobility control in distributed systems. In *FoSSaCS '03*, volume 2620 of *LNCS*, pages 282–299.
18. K. Honda and N. Yoshida. On reduction-based process semantics. *TCS*, 152(2), 1995.
19. M. Merro and F. Zappa Nardelli. Bisimulation proof methods for mobile ambients. In *Proc. of ICALP'03*, volume 2719 of *LNCS*, pages 584–598. Springer, 2003.
20. R. Milner and D. Sangiorgi. Barbed bisimulation. In *Proc. of ICALP '92*, volume 623 of *LNCS*, pages 685–695. Springer, 1992.

21. U. Montanari and M. Pistore. Finite state verification for the asynchronous pi-calculus. In *Proc. of TACAS'99*, volume 1579 of LNCS, pages 255–269. Springer, 1999.
22. C. Palamidessi. Comparing the Expressive Power of the Synchronous and the Asynchronous π-calculi. *Mathematical Structures in Computer Science*, 13(5):685–719, 2003.
23. D. Sangiorgi and D. Walker. The π-calculus: a Theory of Mobile Processes. *Cambridge University Press*, 2001.
24. A. Schmitt and J.-B. Stefani. The m-calculus: a higher-order distributed process calculus. *SIGPLAN Not.*, 38(1):50–61, 2003.
25. P. Sewell. From Rewrite Rules to Bisimulation Congruences. In *Proc. of CONCUR'98*, volume 1466 of *LNCS*, pages 269–284. Springer, 1998.
26. P. Sewell, P. Wojciechowski, and B. Pierce. Location independence for mobile agents. In *Proc. of ICCL*, volume 1686 of *LNCS*. Springer, 1999.

Compositional Verification of Asynchronous Processes via Constraint Solving

Giorgio Delzanno[1] and Maurizio Gabbrielli[2]

[1] Dip. di Informatica e Scienze dell'Informazione, Università di Genova,
via Dodecaneso 35, 16146 Genova, Italy
giorgio@disi.unige.it

[2] Dip. di Scienze dell'Informazione, Università di Bologna,
Mura Anteo Zamboni, 7 40127 Bologna, Italy
gabbri@cs.unibo.it

Abstract. We investigate the foundations of a constraint-based compositional verification method for infinite-state systems. We first consider an asynchronous process calculus which is an abstract formalization of several existing languages based on the blackboard model. For this calculus we define a constraint-based symbolic representation of finite computations of a compositional model based on traces. The constraint system we use combines formulas of integer arithmetics with equalities over uninterpreted functions in which satisfiability is decidable. The translation is inductively defined via a CLP program. Execution traces of a process can be compositionally obtained from the solutions of the answer constraints of the CLP encoding. This way, the task of compositional verification can be reduced to constraint computing and solving.

1 Introduction

Compositional verification of infinite state systems introduces several problems along two main axis. On one hand, in order to apply verification methods developed for finite state systems we need suitable abstractions to finitely represent infinite sets of states. On the other hand, compositionality is usually quite difficult to achieve already in the case of finite state systems, as it requires semantics structures which are often rather complicated. Nevertheless, when considering some important classes of infinite state systems, for example those arising in distributed computing, compositional verification is almost mandatory as the environment in which software agents operate cannot be fixed in advance.

In this paper we deal with this problem by proposing a new technique which combines classical compositional semantics based on sequences (or traces) with constraint programming. More specifically, we consider an asynchronous process calculus which is an abstract formalization of several existing languages based on the blackboard model like [4, 11, 16], where processes communicates and synchronize by posting and retrieving messages from a global, common store. For this process calculus we define a symbolic representation, in terms of constraints, of a compositional model based on traces, thus reducing the task of (compositional) verification to constraint solving.

Technically, our approach is based on the following steps. First of all, we define a compositional model based on traces of basic actions on the store which describes correctly the sequences of stores obtained in a *finite computation*. This follows a standard approach to trace based models of concurrent languages (e.g. see [14,8]), even though our technical treatment is different. We represent the traces arising in the compositional model by using the constraint system C_{E+Z} which combines linear integer arithmetics with uninterpreted function symbols. Arithmetic constraints allow us to describe the addition and deletion of a message to the store at a given time instant. Uninterpreted symbols are used to relate store configurations in different time instants. Indeed, they can be used to represent variables which depend on other variables as in the formula $s_a(i+1) = s_a(i) + 1$ which can be used to represent the addition of a message a to the store at time i. Following from the results on the combination of decision procedures of [15], we know that the quantifier-free satisfiability problem for constraints in C_{E+Z} is decidable. In order to deal with finite computations of recursive process definitions, the above mentioned constraint system is embedded into a Constraint Logic Programming (CLP) language [13]. Then we compositionally translate a process of our algebra into a CLP program P and a goal G so as to extract the traces describing the semantics of the process from the constraints computed by the evaluation of G in P. In this setting a recursive process definition naturally translates into a CLP recursive clause.

The resulting encoding has several benefits. Firstly, we introduce an implicit, compact representation of finite computations, which are described by C_{E+Z} formulas and which can be obtained explicitly by taking the solutions of these constraints. Moreover, we can use logical operators to represents the operators on sequences which are the semantic counterpart of the syntactic operators of the language, thus obtaining a compositional construction. Notably, the interleaving of sequences which models the syntactic parallel composition, can be simply defined in terms of conjunction of constraints, provided that we select the solutions of the resulting constraint with some care. Secondly, we obtain a natural compositional model by translating processes into CLP programs, where we exploit the and-compositionality of the computed answer constraint of CLP programs. Thirdly, we can combine CLP systems with solvers like CVCL [5] and MathSAT [7] to define compositional verification procedures based on our symbolic semantics. Indeed, since C_{E+Z} constraints can also be used to express initial and final conditions on the store, the verification of a property $\mathcal{P}$ for a process P can be reduced to the existence of a satisfiable answer constraint for the CLP translation of P in conjunction with the constraint defining $\mathcal{P}$.

Related Work The use of sequences (or traces) in the semantics of concurrent languages is not new: compositional models based on traces have been defined for a variety of concurrent languages, ranging from dataflow languages [14] and imperative ones [8] to (timed) concurrent constraint programming [3] and Linda-like languages (expressed in terms of process algebras) [2]. However, all the existing approaches consider sequences explicitly and do not take into account the issue of defining a suitable language for expressing and manipulating them implicitly. As a consequence, the resulting models cannot be used directly to define automatic verification tools. On the other hand, our approach introduces a constraint system for expressing and manipulating sequences, where the semantic parallel composition operator can be simply seen in terms of conjunction of constraints. This allows us to express symbolically the semantic of a process and to reason on it without the need to generate explicitly all the sequences, thus reducing the task of property verification to constraint solving. Finally, while in other CLP-based verification methods like [9] arithmetic constraints are used to represent infinite sets of states, in our approach we combine them with uninterpreted function symbols to represent set of traces.

Plan of the paper In Section 2 we present a process algebra for asynchronous processes. In Section 3 we present the logic language $\text{CLP}(C_{E+Z})$. In Section 3 we present the encoding from processes to $\text{CLP}(C_{E+Z})$ programs. Finally, in Section 5 we discuss related and future work.

2 An Asynchronous Process Algebra

The calculus that we consider in this paper is an abstract formalization of several existing concurrent languages based on the blackboard model, e.g., [4, 11, 16]. The calculus is equipped with two basic operations $out(a)$ and $in(a)$ for adding and removing a message from a common global store. Then we have the usual parallel composition and (internal) choice operators. We also have a construct which allows to test for presence of a message, allowing different continuations depending on the result of the test. This construct, which cannot be simulated by the basic operations, is needed in order to obtain Turing completeness. Infinite behaviors are expressed by allowing the recursive definition of process constants.

Definition 1 (The language). Given a finite set of messages Msg, with typical elements $a, b \ldots$, and a set of process constants K with typical elements $p, q, \ldots$, the syntax of process terms is given by the following grammar:

$$P, Q ::= \alpha.P \mid \beta?P : Q \mid P||Q \mid P + Q \mid p \mid halt$$
$$\alpha ::= out(a) \mid in(a)$$
$$\beta ::= inp(a) \mid rdp(a)$$

where we assume that for each process constant p there exists a single definition $p =_{def} P$ where P is a process term and we assume guarded recursion. A *closed process* is a pair $\langle P, D \rangle$ where P is a process term and D is a set of definitions for all the constants in P.

Since the *out* operation is non blocking the communication is asynchronous: a process that wants to communicate with another one simply adds a message to the global store and then proceeds in its computation. The process that wants to receive a message can retrieve it from the global store by performing an *in*

Table 1. The transition system (symmetric rules omitted)

R1 $\langle out(a).P, \mathcal{M} \rangle \rightarrow \langle P, \mathcal{M} \oplus \{a\} \rangle$

R2 $\langle in(a).P, \mathcal{M} \rangle \rightarrow \langle P, \mathcal{M} \ominus \{a\} \rangle$ provided $a \in \mathcal{M}$

R3 $\langle rdp(a)?P : Q, \mathcal{M} \rangle \rightarrow \langle P, \mathcal{M} \rangle$ provided $a \in \mathcal{M}$
$\langle rdp(a)?P : Q, \mathcal{M} \rangle \rightarrow \langle Q, \mathcal{M} \rangle$ provided $a \notin \mathcal{M}$

R4 $\langle inp(a)?P : Q, \mathcal{M} \rangle \rightarrow \langle P, \mathcal{M} \ominus \{a\} \rangle$ provided $a \in \mathcal{M}$
$\langle inp(a)?P : Q, \mathcal{M} \rangle \rightarrow \langle Q, \mathcal{M} \rangle$ provided $a \notin \mathcal{M}$

R5 $\langle P + Q, \mathcal{M} \rangle \rightarrow \langle P, \mathcal{M} \rangle$

R6 $\dfrac{\langle P, \mathcal{M} \rangle \rightarrow \langle P', \mathcal{M}' \rangle}{\langle P \parallel Q, \mathcal{M} \rangle \rightarrow \langle P' \parallel Q, \mathcal{M}' \rangle}$

R7 $\langle p, \mathcal{M} \rangle \longrightarrow \langle P, \mathcal{M} \rangle$ provided $(p =_{def} P) \in D$

operation, which is blocking: if (an occurrence of) the required message is not present in the store then the computation suspends, possibly being resumed later when the message is available. The process $inp(a)?P : Q$ tests for the presence of (an occurrence of) message a in the current store: if present, (an occurrence of) a is removed and the process continues as P, otherwise the process continues as Q. The process $rdp(a)?P : Q$ behaves similarly with the exception that rdp only tests for the presence of a without removing it. The parallel operator is modeled in terms of interleaving, as usual. As for the choice, we consider here the internal (or local) choice, where the environment cannot influence the choice. In the following we will work modulo the congruence relation $halt \equiv halt \parallel halt$.

Operational Semantics The *operational semantics* is formally described by a transition system $T = (Conf, \longrightarrow)$. Configurations (in $Conf$) are pairs consisting of a process and a *multiset* of messages $\mathcal{M}$ representing the common store. We will use use $\ominus$ and $\oplus$ to denote multiset union and difference, respectively. The transition relation $\longrightarrow$ is the least relation satisfying the rules R1-R7 in Table 1, where we omit definitions since these do not change during the computation. So $\langle P, \mathcal{M} \rangle \longrightarrow \langle Q, \mathcal{M}' \rangle$ means that the process P with the store $\mathcal{M}$ and a given set of declarations D can perform a transition step yielding the process Q and the store $\mathcal{M}'$. A computation is a sequence of configurations $C_1 C_2 \ldots C_k$ such that $C_i \longrightarrow C_{i+1}$ for any $i : 1, \ldots, k-1$. In the following $Store$ will denote the set of possible stores, i.e. multiset over Msg, while $Store^*$ will indicate the set of finite sequences over $Store$. The observables are obtained by projecting runs over sequences in $Store^*$.

Definition 2. For any program $\langle P, D \rangle$, we define

$$\mathcal{O}(\langle P, D \rangle) = \{\mathcal{M}_1 \cdot \ldots \cdot \mathcal{M}_n \mid \langle P_1, \mathcal{M}_1 \rangle \rightarrow \ldots \rightarrow \langle P_n, \mathcal{M}_n \rangle \not\rightarrow, \ P = P_1\}$$

Furthermore, we define the observable of *terminating computations* as follows.

Definition 3. For any program $\langle P, D \rangle$, we define

$$\mathcal{O}_h(\langle P, D \rangle) = \{\mathcal{M}_1 \cdot \ldots \cdot \mathcal{M}_n \mid \langle P_1, \mathcal{M}_1 \rangle \to \ldots \to \langle halt, \mathcal{M}_n \rangle, \ P = P_1\}$$

For instance, let $P = out(b).in(a).halt$. Then, $\emptyset \cdot \{b\}$ is in $\mathcal{O}(Q)$ but not in $\mathcal{O}_h(Q)$, whereas $\{a\} \cdot \{a,b\} \cdot \{b\} \in \mathcal{O}(Q) \cap \mathcal{O}_h(Q)$.

2.1 Denotational Semantics

We can compositionally characterize the observables using the set $\mathcal{A}^*$ of sequences built on $\mathcal{A} = \{in(a), out(a), inp(a), rdp(a), \overline{inp}(a), \overline{rdp}(a) \mid a \in Msg\}$. To define the semantics of parallel, we consider the *shuffle operator* of [12]. The shuffling of s and t in $\mathcal{A}^*$, denoted $s \odot t$, is the set of all sequences of the form

$$s_1 \cdot t_1 \cdot \ldots \cdot s_k \cdot t_k$$

where $k > 0$, $s = s_1 \cdot \ldots \cdot s_k$, $t = t_1 \cdot \ldots \cdot t_k$, and the sequences s_i and t_i, $1 \leq i \leq k$, can be of arbitrary length (including the empty sequence). The operator $\odot$ associates less than concatenation; it can be extended to sets of sequences as follows: $S \odot T = \{s \odot t \mid s \in S, t \in T\}$.
As an example, $in(a) \cdot in(b) \odot in(c) = \{in(c) \cdot in(a) \cdot in(b), in(a) \cdot in(c) \cdot in(b)\}$.

Definition 4. The denotational semantics is the least function $\mathcal{D}$ from programs to $\wp(\mathcal{A}^*)$ which satisfies the equations in Table 2.

As an example, consider the processes $p = inp(a)?q : r$, $q = out(b).out(c).halt$ and $r = out(c).out(b).halt$. Then, the denotational semantics $\mathcal{D}[\![p]\!]$ of p contains the sequences $inp(a) \cdot out(b) \cdot out(c)$ and $\overline{inp}(a) \cdot out(c) \cdot out(b)$.

From the denotational semantics we can reconstruct the observables by resorting to the partial map $eval \in [\mathcal{A}^* \times Store \to Store^*]$ defined in Table 3. Going back to the previous example, $eval(inp(a) \cdot out(b) \cdot out(c), \emptyset)$ is undefined, since the evaluation of $inp(a)$ in the empty store does not succeed, whereas

Table 2. Denotational semantics for process terms with definitions in D

E1 $\mathcal{D}[\![halt]\!] = \{\epsilon\}$

E2 $\mathcal{D}[\![out(a).P]\!] = \{out(a) \cdot s \mid s \in \mathcal{D}[\![P]\!]\}$

E3 $\mathcal{D}[\![in(a).P]\!] = \{in(a) \cdot s \mid s \in \mathcal{D}[\![P]\!]\}$

E4 $\mathcal{D}[\![rdp(a)?P : Q]\!] = \{rdp(a) \cdot s \mid s \in \mathcal{D}[\![P]\!]\} \cup \{\overline{rdp}(a) \cdot s \mid s \in \mathcal{D}[\![Q]\!]\}$

E5 $\mathcal{D}[\![inp(a)?P : Q]\!] = \{inp(a) \cdot s \mid s \in \mathcal{D}[\![P]\!]\} \cup \{\overline{inp}(a) \cdot s \mid s \in \mathcal{D}[\![Q]\!]\}$

E6 $\mathcal{D}[\![P \parallel Q]\!] = \mathcal{D}[\![P]\!] \odot \mathcal{D}[\![Q]\!]$

E7 $\mathcal{D}[\![P + Q]\!] = \mathcal{D}[\![P]\!] \cup \mathcal{D}[\![Q]\!]$.

E6 $\mathcal{D}[\![p]\!] = \mathcal{D}[\![B]\!]$ provided $(p =_{def} B) \in D$

Table 3. The *eval* mapping

$$eval(\epsilon, \mathcal{M}) = \mathcal{M}$$

$$eval(out(a) \cdot s, \mathcal{M}) = \mathcal{M} \cdot eval(s, \mathcal{M} \oplus \{a\})$$

$$eval(in(a) \cdot s, \mathcal{M}) = \begin{cases} \mathcal{M} \cdot eval(s, \mathcal{M} \ominus \{a\}) & \text{if } a \in \mathcal{M} \\ \mathcal{M} & \text{otherwise} \end{cases}$$

$$eval(inp(a) \cdot s, \mathcal{M}) = \mathcal{M} \cdot eval(s, \mathcal{M} \ominus \{a\}) \quad \text{if } a \in \mathcal{M}$$

$$eval(rdp(a) \cdot s, \mathcal{M}) = \mathcal{M} \cdot eval(s, \mathcal{M}) \quad \text{if } a \in \mathcal{M}$$

$$eval(\overline{rdp}(a) \cdot s, \mathcal{M}) = eval(\overline{inp}(a) \cdot s, \mathcal{M}) = \mathcal{M} \cdot eval(s, \mathcal{M}) \quad \text{if } a \notin \mathcal{M}$$

$eval(\overline{inp}(a) \cdot out(c) \cdot out(b), \emptyset) = \emptyset \cdot \{c\} \cdot \{b, c\}$. Notice that in the definition of the *eval* map the treatment of $in(a)$ and $inp(a)$ is different: in fact, in case the message a is not present in the store the evaluation of $in(a)$ suspends, hence the result of the *eval* is simply the store $\mathcal{M}$, without any further continuation. On the other hand, if the message a is not present in the store the evaluation of $inp(a)$ does not suspend and follows the alternative branch which, in our sequences, is indicated by the $\overline{inp}(a)$ construct (see equation E5 in Table 2). The following result states the correctness of the denotational semantics with respect to the notion of observables of Def. 2.

Theorem 1. *For any* P, $\mathcal{O}(P) = \{eval(s, \mathcal{M}) \mid s \in \mathcal{D}[\![P]\!], \mathcal{M} \in Store\}$.

3 The Logic Language CLP(C_{E+Z})

In this section we introduce the CLP(C_{E+Z}) language, obtained as a specific instance of the CLP scheme [13] by considering constraints defined over the combination of the first order theories of *equality over uninterpreted functions* and of *integer arithmetics*.

The combined constraint system C_{E+Z} is defined as follows. C_{E+Z} constraints are quantifier free formulas built over the signature $\Sigma_E \cup \Sigma_Z$, where Σ_E is a set of functions symbols, and $\Sigma_Z = \{0, 1, \ldots, +, -, <, \leq\}$ is the usual signature of integer arithmetics. An example of C_{E+Z} constraint is the formula $x \leq y \wedge f(x) = g(y) + 1 \wedge f(g(x) + 1) \leq 2$. Thus in C_{E+Z} we can represent arithmetics over variables that depend on other variables. The theory of equality is the Σ_E-theory with no axioms, whereas the theory of arithmetics is defined as the set of Σ_Z-sentences that are true in the standard interpretation of constants, function and predicate symbols in Σ_Z. From the general results of Nelson and Oppen [15], quantifier-free satisfiability is decidable in the combined theory $\mathcal{T}_{E+Z}$. Solvers like CVCL [5] and MathSAT [7] implement decision procedures to check satisfiability of these constraints.

Let Π be a finite set of predicate symbols (program predicates), disjoint from Σ_{E+Z} and let $\mathcal{V}$ be a denumerable set of variables. The language CLP(C_{E+Z}) is defined on top of C_{E+Z} as follows.

Definition 5 (Clauses, Goals, and Programs). A *clause* is an implicitly universally quantified formula of the form

$$A_0 \Leftarrow \varphi \wedge A_1 \wedge \ldots \wedge A_n,$$

and a *goal* is an implicitly existentially quantified formula of the form

$$\varphi \wedge A_1 \wedge \ldots \wedge A_n,$$

where φ is a C_{E+Z} constraint over $\mathcal{V}$, $A_i = p_i(x_1^i, \ldots, x_{n_i}^i)$ is an atom with $p_i \in \Pi$, $0 \le i \le n$, and $x_j^k \in \mathcal{V}$ for $j, k \ge 0$. A *program* is a set of clauses.

We use $\Rightarrow_P$ to denote the usual notion of CLP derivation relation (see [13]) for goals and (renamings of) clauses taken from a CLP(C_{E+Z}) program P. A derivation step replaces an atomic formula A in a goal G with the body B of a clause $A' \Leftarrow B$ whose head A' can be unified with A. The constraint resulting from unification in conjunction with those in B and in G must be satisfiable. $\Rightarrow_P^*$ denotes its reflexive and transitive closure. CLP(C_{E+Z}) programs allow to compute *answer constraints* [10]. Formally, we have the following definition.

Definition 6. The *answer constraint semantics* of P and G is defined as

$$\mathcal{A}(P, G) = \{\varphi \mid G \Rightarrow_P^* \varphi, \varphi \text{ is a } C_{E+Z} \text{ constraint}\}.$$

Thus an answer constraint is the conjunction of constraints which are left when all atomic goals have been resolved using program clauses.

4   Encoding Processes in CLP(C_{E+Z})

First of all, given the set of messages $Msg = \{a_1, \ldots, a_k\}$, we build the signature $\Sigma_E = \{s_1, \ldots, s_k\}$ (contained in Σ_{E+Z} by definition). The term $s_i(t)$ is used to represent the number of occurrences of message a_i in the store at time t. The effect of an action like $out(a_i)$ executed at time x can then be described via the clause like that defining the *store atom* $out(a_i, x)$ of Fig. 4. Thus in the rest of the paper we will focus our attention on the subclass of C_{E+Z} constraints defined as follows.

Definition 7 (Store Constraints). A *store atom* is a formula $\alpha(a_i, x)$ with $\alpha \in \{out, in, inp, rdp, \overline{inp}, \overline{rdp}\}$ defined by the CLP clauses of Table 4. A *store constraint* is a conjunction of store atoms and arithmetic formulas.

The effect of an action like $out(a_i)$ executed at time x can then be described via the store atom $out(a_i, x)$. Its definition describes the addition of a new occurrence of message a_i to the store. More in general, store constraints will allow us to express in a logical way the composition operators of our process algebra. In order to formally relate store constraints to the denotational semantics of a process, we need to refine the notion of solutions as follows.

Definition 8 (Solutions). Given a store constraint φ with variables in V,

Table 4. Definition of store atoms: a_i is a message in Msg

$$out(a_i, x) \Leftarrow s_i(x+1) = s_i(x) + 1 \wedge \bigwedge_{j=1, j \neq i}^{k} s_j(x+1) = s_j(x)$$
$$in(a_i, x) \Leftarrow s_i(x) \geq 1 \wedge s_i(x+1) = s_i(x) - 1 \wedge \bigwedge_{j=1, j \neq i}^{k} s_j(x+1) = s_j(x)$$
$$inp(a_i, x) \Leftarrow s_i(x) \geq 1 \wedge s_i(x+1) = s_i(x) - 1 \wedge \bigwedge_{j=1, j \neq i}^{k} s_j(x+1) = s_j(x)$$
$$\overline{inp}(a_i, x) \Leftarrow s_i(x) = 0 \wedge \bigwedge_{j=1}^{k} s_j(x+1) = s_j(x)$$
$$rdp(a_i, x) \Leftarrow s_i(x) \geq 1 \wedge \bigwedge_{j=1}^{k} s_j(x+1) = s_j(x)$$
$$\overline{rdp}(a_i, x) \Leftarrow s_i(x) = 0 \wedge \bigwedge_{j=1}^{k} s_j(x+1) = s_j(x)$$

- a *partial solution* ν of φ is a map from V to $\mathbb{Z}$ such that φ evaluates to true under some $\mathcal{T}_{E+Z}$-interpretation which extends ν in the natural way;
- a *solution* extends a partial solution with maps from the function symbols s_i to the functions $\overline{s_i} : \mathbb{Z} \to \mathbb{Z}$ for $i : 1, \ldots, k$;
- a (partial) solution ν is *injective* for $U \subseteq V$ if $\nu(x) \neq \nu(y)$ for every $x, y \in U$ such that $x \neq y$;
- given a subset $U \subseteq V$ with cardinality m, a (partial) solution ν is *closed* for U if $\nu(x) \in [1, m]$ for any $x \in U$.

A C_{E+Z} constraint is satisfiable if it has a solution. The *store function* of a solution ν is defined as $s(t) = \langle \overline{s_1}(t), \ldots, \overline{s_k}(t) \rangle$. Notice that $s(t)$ represents the content of the store at time t, i.e, the multiset in which a_i has $s_i(t)$ occurrences for $i : 1, \ldots, k$. An injective solution associates distinct values to updates represented in a store constraint. A closed solution associates values from a closed interval. With a closed injective solution each variable in $U = \{x_1, \ldots, x_k\}$ is assigned a distinct value in $1, \ldots, k$. This notion will be used to consider sequences of a closed system in which events occur at time instants with no gaps between them.

Remark 1. It is important to notice that, given a store constraint φ and a subset U of its variables, the injective (closed) solutions of φ for U coincide with the solutions of the constraint $\varphi \wedge \bigwedge_{x,y \in U} x \neq y$ ($\varphi \wedge \bigwedge_{x \in U} 1 \leq x \leq |U|$).

4.1 Formal Definition of the Translation

The translation of a process term P with definitions in D to a CLP(C_{E+Z}) program is inductively defined via the functions $\mathcal{T}_p$, $\mathcal{T}_d$ and $\mathcal{T}$. The function $\mathcal{T}_p$ takes as input the process term P and a variable x, representing a time instant, and returns a CLP program $Prog$ and a goal G. The CLP program contains the clauses defining some new process constants that we introduce in order to represent the choice operator and the conditionals. The goal G is the translation of P where we interpret actions in terms of store constraints, parallel in terms of conjunction and process constants as predicate symbols taken from Π (and therefore not in Σ). The function $\mathcal{T}_d$ translates process definitions in D into CLP(C_{E+Z}) clauses. Finally, the function $\mathcal{T}$, defined on top of $\mathcal{T}_d$ and $\mathcal{T}_p$, is used to encode the closed process $\langle P, D \rangle$.

Definition 9 (Translation). Let $\mathcal{V}$ be a denumerable set of variables, then the translation functions $\mathcal{T}_p$, $\mathcal{T}_d$ and $\mathcal{T}$ are formally defined as follows.

Halt: The process halt does not produce any constraint on the store. Thus, for any $x \in \mathcal{V}$ we define $\mathcal{T}_p(halt, x) = \langle \emptyset, true \rangle$.

In/Out: The precondition and effect of the action $out(a)$ performed at time x can be described via the store atom $out(a, x)$ of Table 4. Since the effect of $out(a)$ is visible at time $x + 1$, if y is the time starting from which actions in P may occur, we can enforce the sequentiality between the execution of $out(a)$ and of the actions in P by requiring that $x < y$. Thus, if $\mathcal{T}_p(P, y) = \langle Prog, G \rangle$ for some variable $y \in \mathcal{V}$, then we define

$$\mathcal{T}_p(out(a).P, x) = \langle Prog, out(a, x) \wedge x < y' \wedge G' \rangle$$

for any variable $x \in \mathcal{V}$, and renamings y', G' of y, G such that $x \notin Var(G') \cup \{y'\}$. The treatment of the term $in(a).P$ is analogous (predicate out is replaced by in).

Conditional: The preconditions and effects of the "if" branch of the conditional $rdp(a)?P_1 : P_2$ performed at time x can be described by using the store atom $rdp(a, x)$ in conjunction with the formula representing P_1. Similarly, the store atom $\overline{rdp}(a, x)$ can be used to model the "else" branch. In order to describe the whole conditional construct we then use a new predicate symbol which models the choice point in terms of two mutually exclusive clauses, corresponding to the two different cases. Formally, if $\mathcal{T}_p(P_1, y_1) = \langle Prog_1, G_1 \rangle$ and $\mathcal{T}_p(P_2, y_2) = \langle Prog_2, G_2 \rangle$ for two variables $y_1, y_2 \in \mathcal{V}$, then we define

$$\mathcal{T}_p(rdp(a)?P_1 : P_2, x) = \langle Prog \cup Prog_1 \cup Prog_2, p(x) \rangle$$
$$Prog = \{p(u) \Leftarrow rdp(a, u) \wedge u < y'_1 \wedge G'_1, \ p(u) \Leftarrow \overline{rdp}(a, u) \wedge u < y'_2 \wedge G'_2\}$$

for variables $x, u \in \mathcal{V}$, a new predicate symbol p, and renamings y'_1, y'_2, G'_1, G'_2 of y_1, y_2, G_1, G_2 such that $x, u \notin Var(G'_1) \cup Var(G'_2) \cup \{y'_1, y'_2\}$. The treatment for the term $inp(a)?P_1 : P_2$ is analogous.

Choice: To model internal choice we use the CLP non-determinism in the selection of the clause to apply to a given goal. Thus, also in this case we introduce a new predicate symbol and we define it using two different CLP clauses, which correspond to the two branches of the choice construct. Formally, if $\mathcal{T}_p(P_1, y_1) = \langle Prog_1, G_1 \rangle$ and $\mathcal{T}_p(P_2, y_2) = \langle Prog_2, G_2 \rangle$ for $y_1, y_2 \in \mathcal{V}$, then

$$\mathcal{T}_p(P_1 + P_2, x) = \langle Prog \cup Prog_1 \cup Prog_2, p(x) \rangle$$
$$Prog = \{p(y_1) \Leftarrow G_1, \ p(y_2) \Leftarrow G_2\}$$

for any $x \in \mathcal{V}$ and a new predicate symbol p. Notice that, differently from the case of the conditionals, here the two clauses in $Prog$ are not mutually exclusive.

Parallel: The translation of parallel composition is more subtle. The only constraint that we can put on the set of events occurring in two distinct processes running in parallel is that they will occur after the whole system started its execution. Now, suppose that y_1 represent the time at which process P_1 starts,

y_2 the time at which process P_2 starts and assume that x is the starting point of $P_1 \parallel P_2$. Then the constraint $x \leq y_1 \wedge x \leq y_2$ must hold. Notice that the encoding that we use does not forbid solutions that map time variables associated to distinct actions to the same value. This might lead to illegal traces, i.e., traces where two different basic actions are performed on the store at the same time instant, which is not acceptable since we use an interleaving model for the parallel operator. Illegal traces can be ruled out by considering solutions that are *injective* (see Def. 8) for the set of variables associated to actions, thus ensuring that different variables assume distinct values (see Theorem 3). Formally, if $\mathcal{T}_p(P_1, y_1) = \langle Prog_1, G_1 \rangle$ and $\mathcal{T}_p(P_2, y_2) = \langle Prog_2, G_2 \rangle$ for $y_1, y_2 \in \mathcal{V}$, then

$$\mathcal{T}_p(P_1 \parallel P_2, x) = \langle Prog_1 \cup Prog_2, (x \leq y'_1 \wedge x \leq y'_2) \wedge G'_1 \wedge G'_2 \rangle,$$

for any variable $x \in \mathcal{V}$, and renamings y'_1, y'_2, G'_1, G'_2 of y_1, y_2, G_1, G_2 such that $Var(G'_1) \cap Var(G'_2) = \emptyset$, and $x \notin Var(G'_1) \cup Var(G'_2) \cup \{y'_1, y'_2\}$.

Constant: A process constant p occurring in a term can be viewed as the invocation of a process definition. In our translation this maps naturally to a goal denoting a call to a clause defining p. Formally, for any $x \in \mathcal{V}$ we define

$$\mathcal{T}_p(p, x) = \langle \emptyset, p(x) \rangle.$$

Definition: Given $p =_{def} P$ and $\mathcal{T}_p(P, x) = \langle Prog, G \rangle$ for $x \in \mathcal{V}$, we define

$$\mathcal{T}_d(p =_{def} P) = Prog \cup \{p(x) \Leftarrow G\}.$$

Program: Now let $D = \{d_1, \ldots, d_n\}$ be process definitions, $y \in \mathcal{V}$, and P be a process term such that $\mathcal{T}_p(P, x) = \langle Prog, G \rangle$ for some $x \in \mathcal{V}$. Then we define

$$\mathcal{T}(\langle P, D \rangle) = \langle \{\mathcal{T}_d(d_1, y), \ldots, \mathcal{T}_d(d_n, y)\} \cup Prog, G \rangle.$$

Notice that the function $\mathcal{T}$ introduces an arbitrary initial time variable x.

4.2 Properties of the Encoding

The translation of a process term into a $CLP(\mathcal{C}_{E+Z})$ program allows us to reconstruct the denotational semantics of Def. 4 and the observables of Def. 2, by considering the answer constraint semantics of Def. 6. To make this claim more formal, it will be convenient to use the following terminology.

Given a (possibly instantiated) store constraint φ, we will use $A(\varphi)$ to denote the set of store atoms occurring in φ, and given a store atom $\alpha(t, x)$, we define $T(\alpha(a, t)) = t$ and $E(\alpha(a, t)) = \alpha(a)$. T and E are extended to sets, formulas, and sequences in the natural way. In particular, T applied to a store constraints φ returns the set of variables (time-stamps for an instantiated constraint) associated to the actions $E(\varphi)$. Furthermore, let ν be a partial solution of φ injective for $T(\varphi)$, and let $A(\nu(\varphi)) = \{A_1, \ldots, A_n\}$, then

$$S_\nu(\varphi) = \{A_{i_1} \cdot A_{i_2} \cdot \ldots \cdot A_{i_n} \mid T(A_{i_k}) \leq T(A_{i_{k+1}}) \text{ for } k : 1, \ldots, n-1\}$$

We are now ready to state the adequacy of our encoding. The following theorem shows that the answer constraints of the CLP(C_{E+Z}) program associated with a closed process is an implicit representation of its denotational semantics.

Theorem 2. For any closed process $S = \langle P, D \rangle$ such that $\mathcal{T}(S) = \langle Prog, G \rangle$, we have that $\mathcal{D}[\![S]\!] = \{E(s) \mid s \in S_\nu(\varphi), \varphi \in \mathcal{A}(Prog, G),$ and ν is a partial solution of φ injective for $T(\varphi)\}$.

Now, given a solution of φ with store function s (see Section 4) from φ we can extract the history of updates by evaluating s on the time-stamps $T(\nu(\varphi)) = \{t_1 \leq \ldots \leq t_k\}$ (recall that $s(t)$ is an alternative representation of a multiset):

$$O_\nu(\varphi) = \{\mathcal{M}_1 \cdot \ldots \cdot \mathcal{M}_k \mid \mathcal{M}_i = s(t_i),\ for\ i : 1, \ldots, k\}$$

The following theorem shows that in order to extract the observables of a *terminating* process we need to consider solutions that are both injective and closed (see Def. 8). Indeed a closed solution assigns to time variables values taken from a closed interval whose cardinality corresponds to the number of possible actions of the original process term.

Theorem 3. For any closed process $S = \langle P, D \rangle$ such that $\mathcal{T}(S) = \langle Prog, G \rangle$, we have that $\mathcal{O}_h(S) = \{s \mid s \in O_\nu(\varphi), \varphi \in \mathcal{A}(Prog, G),$ and ν is a solution of φ injective and closed for $T(\varphi)\}$.

5 Conclusions

In this paper we have shown how to use CLP to obtain a constraint-based symbolic representation of the set of finite computations of asynchronous processes communicating via a common store. Theorem 2 and 3 are at the basis of a possible compositional verification method for this computational model. Indeed CLP enjoys the *and-compositionality* property: an answer constraint for $G_1 \wedge G_2$ can be obtained by conjoining the answer constraints of G_1 and G_2 [10]. And-compositionality can be exploited for a compositional analysis of composed systems as follows. Suppose that the translation of processes P_1 and P_2 returns the goal G_1 and G_2 and a CLP program $Prog = Prog_1 \cup Prog_2$. Since the combination of P_1 and P_2 can be expressed as a constraint $G_1 \wedge G_2 \wedge \psi$ (ψ depends on the composition operator) we can use and-compositionality to separately analyze G_1 and G_2 and then join the results. Indeed, from Theorem 2, we know that answer constraints characterize open traces of processes that can be further combined with other traces. Furthermore, Theorem 3 can be exploited to combine CLP and a C_{E+Z} solver for checking properties of observables. To illustrate, let us consider a process P encoded via the CLP program $Prog$ and goal G and suppose we are interested in checking if a given store can be reached starting from an initial one. We first notice that we can encode a configuration $\mathcal{M}$ of the store at time t using the constraint $s_1(t) = n_1, \ldots, s_k(t) = n_k$ where n_i is the number of occurrences of message a_i in $\mathcal{M}$ for $i : 1, \ldots, k$. Reachability can be reduced then constraint computing and solving as follows. We first exploit the CLP component to compute an answer constraint φ of G (by exploiting and-compositionality this can be done separately for the subcomponents). Suppose now that the time-stamps of actions in φ are $T(\varphi) = \{x_1, \ldots, x_m\}$, where x_1 is the initial point of the evaluation of G. Furthermore, let ψ_0 be a constraint expressing initial condition on the store at time x_1, and ψ_1 be a constraint expressing conditions on the final store (i.e. at a time greater or equal than $x_1, \ldots, x_m$. By Theorem 3, the existence of an injective and closed solution for $\psi_0 \wedge \varphi \wedge \psi_1$ can be used as sufficient condition for the original reachability question. (Notice that to obtain a complete test we need to generate all answer constraints.) From remark 1 and from the encoding of store atoms in C_{E+Z} of Fig. 4 the latter problem can be reduced to a satisfiability problem of a C_{E+Z} formula. This kind of reasoning can be viewed as an extension of the bounded model checking paradigm [6] in which the encoding of a bounded execution of a system is constructed in a compositional way. We are currently working on a prototype implementing this verification method for our process algebra.

References

1. W. Ackermann. Solvable cases of the decision procedure. North-holland, 1954.
2. F. S. de Boer, M. Gabbrielli, and M. C. Meo. A Denotational Semantics for a Timed Linda Language. PPDP 2001: 28-36.
3. F. S. de Boer, M. Gabbrielli, and M. C. Meo. A Timed Concurrent Constraint Language. *Information and Computation*, 161(1): 45-83, 2000.
4. J.P Banatre and D. Le Metayer. Programming by Multiset Transformation. *Communication of the ACM*, 36(1): 98-111, 1993.
5. C. Barrett and S. Berezin. CVC Lite: A New Implementation of the Cooperating Validity Checker. CAV 2004: 515-518.
6. A. Biere, A. Cimatti, E. Clarke, Y. Zhu Symbolic Model Checking without BDDs TACAS 1999: 193-207.
7. M.Bozzano, R.Bruttomesso, A.Cimatti, T.Junttila, P.v.Rossum, S.Schulz, and R.Sebastiani. An incremental and Layered Procedure for the Satisfiability of Linear Arithmetic Logic. TACAS 2005: 317-333.
8. S. Brookes. A Fully Abstract Semantics of a Shared Variable Parallel Language. LICS 1993: 98-109.
9. G. Delzanno and A. Podelski. Model Checking in CLP. TACAS 1999: 223–239.
10. M. Gabbrielli and G. Levi. Modeling Answer Constraints in Constraint Logic Programs. ICLP 1991: 238-252.
11. D. Gelernter. Generative Communication in Linda. *TOPLAS*, 70(1): 80-112, 1985.
12. S. Ginsburg and E. H. Spanier. Mappings of languages by two-tape devices *JACM*, 12(3): 423-434, 1965.
13. J. Jaffar and M. J. Maher. Constraint Logic Programming: A Survey. *JLP*, 19-20:503-582, 1994.
14. B. Jonsson. A Model and a Proof System for Asynchronous Processes. PODC 1985: 49-58.
15. G. Nelson and D. C. Oppen. Fast Decision Procedures based on Congruence Closure. JACM, 27(2):356-364, 1980.
16. V.A. Saraswat and M. Rinard. Concurrent Constraint Programming. POPL 1990: 232–245.

Optimal Spaced Seeds for Faster Approximate String Matching

Martin Farach-Colton[1], Gad M. Landau[2], S. Cenk Sahinalp[3], and Dekel Tsur[4]

[1] Dept. of Computer Science and DIMACS, Rutgers University
[2] Dept. of Computer Science, University of Haifa
[3] School of Computing Science, Simon Fraser University
[4] Dept. of Computer Science and Engineering, University of California, San Diego

Abstract. *Filtering* is a standard technique for fast approximate string matching in practice. In filtering, a quick first step is used to rule out almost all positions of a text as possible starting positions for a pattern. Typically this step consists of finding the exact matches of small parts of the pattern. In the followup step, a slow method is used to verify or eliminate each remaining position. The running time of such a method depends largely on the quality of the filtering step, as measured by its false positives rate. The quality of such a method depends on the number of true matches that it misses, that is, on its false negative rate.

A *spaced seed* is a recently introduced type of filter pattern that allows gaps (i.e. don't cares) in the small sub-pattern to be searched for. Spaced seeds promise to yield a much lower false positives rate, and thus have been extensively studied, though heretofore only heuristically or statistically.

In this paper, we show how to optimally design spaced seeds that yield no false negatives.

1 Introduction

Given a pattern string P of length m, a text string T of length ℓ, and an integer k, the *approximate pattern matching problem* is to find all substrings of T whose edit distance or Hamming distance to P is at most k.

The basic idea employed in many approximate pattern matching algorithms [17, 8] and commonly used software tools such as BLAST [1] is *filtering* based on the use of the pigeonhole principle: Let P and S be two strings with edit distance or Hamming distance at most k. Then P and S must have identical substrings (contiguous blocks) whose sizes are at least $(m-k)/(k+1)$. This simple observation can be used to perform efficient approximate pattern matching through the following approach. (i) *Anchor finding:* consider each substring of P of size $b \leq (m-k)/(k+1)$ and find all of its exact occurrences in T. (ii) *Anchor verification:* verify whether each initial *exact match* extends to a complete *approximate match*, through the use of (a localized) dynamic program or any other appropriate method.

When T is available off-line, the anchor finding step above can be implemented very efficiently: (i) build a compact trie of all substrings in T of size b and (ii) search each substring in P of size b on the compact trie. By the use of *suffix links*, the compact trie can be built in $O(\ell)$ time and the pattern processing can be completed in $O(m)$ time, both independent of the size of b.

The running time of the anchor verification step depends on the specific method for extending an initial exact match and the value of b: As b increases, the number of *false positives* is expected to decrease, but if $b > (m-k)/(k+1)$ some actual occurrences of the pattern may be missed, yielding *false negatives*.

In the remainder of this discussion we will focus on filtering processes with no false negatives, except as noted. Under this constraint, much of the literature on pattern matching via filtering focuses on improving the specific method for verifying anchors. The fastest approximate pattern matching algorithms based on filtering have a running time of $O(\ell(1+\frac{poly\ k \cdot polylog\ \ell}{m}))$ and thus are especially powerful for "small" values of k [17, 8, 2]. In general, pattern matching under edit distance can be solved in time $O(\ell k)$ [13], whereas pattern matching under Hamming distance can be solved in time $O(\ell\sqrt{k \log k})$ [2].

1.1 The Performance of the Filtering Approach

Although the filtering approach does not always speed up pattern matching, it is usually quite efficient on high-entropy texts, such as those in which each character is drawn uniformly at random from the input alphabet (of size σ). Given a pattern P, suppose that the text string T is a concatenation of (1) *actual matches of P:* substrings of size m whose Hamming distance to P is at most k; (2) *high entropy text:* long stretches of characters, determined uniform i.i.d. from the input alphabet. On this T and P we can estimate the performance of the anchor verification step and thus the filtering approach in general as follows. Let the number of actual matches be #*occ*. each such match will be identified as an anchor due to the pigeonhole principle. The expected number of false positives, which in this case are substrings from the high entropy text that will be identified as anchors, can be calculated as follows. The probability that a substring $T[i : i+m-1]$ is identified as an anchor, i.e. has a block of size b which exactly matches its corresponding block in P, is $\leq m\sigma^{-b}$. The expected number of anchors from the high entropy text is thus $\leq \ell \cdot m \cdot \sigma^{-b}$. This implies that the running time of the filtering approach is proportional to $\#occ + \ell m \sigma^{-b}$ as well as the time required to verify a given anchor.

The above estimate of the performance of the filtering approach is determined mostly by problem specific parameters, $\#occ, \ell, m$ or the time for verifying a given anchor, none of which can be changed. There is only one variable, b, that can be determined by the filter designer. Unfortunately, in order to avoid false negatives b must be $\leq (m-k)/(k+1)$.

It is possible to relax the above constraint on b by performing filtering through the use of non-contiguous blocks, namely substrings with a number of *gaps* (i.e. *don't care* symbols). To understand why searching with blocks having don't care symbols can help, consider the case when $k = 1$, that is, we allow one mismatch.

When contiguous blocks are used, the maximum value of b is $(m-1)/2$. Now consider using a block of size $b+1$ with one don't care symbol in the center position. How large can b be while guaranteeing that each substring of T with a single mismatching character with P will be found? No matter where the mismatch occurs, we are guaranteed that such a substring can be found even when $b = \lfloor (2m-1)/3 \rfloor$. This is a substantial improvement over ungapped search, where $b \leq (m-1)/2$, reducing the time spent on false positives by a factor of $\approx \sigma^{m/6}$.

1.2 Previous Work

The idea of using gapped filters (which are called *spaced seeds*) was introduced in [7] and the problem of designing the *best* possible seed was first posed in [6] without providing any theoretical bound on the quality of the seed design. Available literature on designing "optimal" spaced seeds relies on experimental/statistical techniques [16, 6, 15, 4, 5, 3, 12], or assumes that the input is generated by a random data model [10]. More recent work on the use of multiple spaced seeds are also experimental in nature [11, 19, 18].

Some of the above approaches for designing spaced seeds allow false negatives. However, it is desirable to construct seeds that guarantee no false negatives, i.e., that finds *all* matches of a any pattern P of length m under Hamming distance k. Alternatively, one can want to find all matches of any pattern P' of length $m' \geq m$ within error rate k/m. Unfortunately this *combinatorial seed design* problem has remained largely open (see, for example [6]) since the introduction of spaced seeds in early nineties [7]. In fact, it was proven in [14] that for a specific pattern and a set of substrings that are no more than Hamming distance k from the pattern, the combinatorial seed design problem is NP-hard.

1.3 Our Contributions

In this paper we show how to solve the *general* combinatorial seed design problem optimally. That is, we give worst-case bounds for the combinatorial seed design problem for all possible pattern/text pairs, whereas a particular pattern and text may admit a "better" seed.

Our specific results are as follows. The combinatorial seed design problem has four parameters: minimum pattern length m, the number of "solid" symbols in the seed b, the number of "don't care" symbols in the seed g, and the maximum number of allowed errors between the pattern and its match k. We denote by n the seed length, namely $n = g + b$. Our method can optimize any one of the parameters, given the values of the other three parameters. In this paper we focus on two variants of this optimization problem; solutions to the other variants can be derived from these without much difficulty.

1. Given m, n, g, we show how to maximize the number of errors k; i.e. we show how to compute the *maximum possible number of errors k* and the associated spaced seed (with length n and g don't cares) which guarantees that *all* matches of a pattern of length m within Hamming distance k are

found. This spaced seed also maximizes the error rate k'/m' for the problem of finding all matches of any given pattern P' whose length is $m' \geq m$.
2. More interestingly, given the number of errors k and minimum pattern length m, we show how to compute the largest possible b such that $b + g = n \leq m$, and the associated seed that guarantees no false negatives. Clearly this seed minimizes the time spent on false positives and thus maximizes the performance of the filtering approach for any given pattern of size m with k errors. This spaced seed also provides the maximum possible b for the problem of finding all matches of any pattern P' of length $m' \geq m$ within error rate k/m.

Our final result is on the design of multiple seeds: for any fixed pattern length m and number of errors k (alternatively minimum pattern length $m \leq m'$ and error rate k/m), we show that by the use of $s \geq m^{1/k}$ seeds one can guarantee to improve on the maximum size of b achievable by a single seed.

2 Preliminaries

For the rest of the paper, the letters A and B will be used to denote strings over the alphabet $\{0, 1\}$. For a string A and an integer l, A^l denotes the concatenation of A l times. Let ONES(A) be the number ones in the string A. $A[i : j]$ denotes the substring of A that starts at the i-th character of A and ends at the j-th character.

For two strings A and B of equal lengths, we write $A \geq B$ if $A[i] \geq B[i]$ for $i = 1, \ldots, |A|$. Note that this differs from lexicographic ordering. We say that a string A covers a string B if there is a substring B' of B of length $|A|$ such that $A \geq B'$. In words, A covers B if we can align A against B such that every character 1 in the aligned region of B, is aligned against a character 1 in A. We will say that such an alignment covers B.

The connection between the above definitions and the seed design problem is as follows: a seed of length n will be represented by a string A of length n such that $A[i] = 1$ if the i-th symbol of the seed is a don't care, and $A[i] = 0$ otherwise. Given a pattern string P and a substring T' of some text T of length m, create a string B of length m, where $B[i] = 0$ if $P[i] = T'[i]$, and $B[i] = 1$ otherwise. Then, the filtering algorithm using seed A will find a match between P and T' if and only if A covers B.

We define $k(n, g, m)$ to be the maximum k such that there is a string A of length n containing g ones that covers every string B of length m with at most k ones. In other words, for a seed length of n with g don't cares, $k(n, g, m)$ is the maximum possible number of errors between any P and any substring T' of length m that is guaranteed to be detected by the best possible seed. Also, $b(k, m)$ is the maximum b such that there is a string A with b zeros that covers every string B of length m with at most k ones. In other words, given the maximum number of errors k, $b(k, m)$ is the maximum number of solid symbols one can have in a seed so that a match between any P and T' with k errors could be detected by the best possible seed.

In the next sections we will give upper and lower bounds on $k(n,g,m)$ and $b(k,m)$ for various values of parameters, effectively solving the combinatorial seed design problem.

3 Maximizing k

We first present our results on maximizing the number of errors k when the other parameters n, g and m are fixed. Our results also extend to the problem of maximizing the error rate k'/m' for any pattern P' of length $m' \geq m$ with fixed n and g.

3.1 Constant g

Theorem 1. *For every fixed g, $k(n,g,m) = (2 - \frac{1}{g+1}) \cdot \frac{m}{n} \pm O(\max(1, \frac{m}{n^2}))$.*

Proof. We first show a lower bound on $k(n,g,m)$. The main idea of the proof is that if B contains "few" 1 characters, then using some form of the pigeonhole principle, there will be an "isolated" 1 in B. Thus, we can align A over B such that the isolated one in B is aligned against some one in A, and there are no more ones in the aligned region of B.

Suppose that n is divisible by $2g+1$. Let A be the string of length n that contains ones in positions $\frac{n}{2g+1}, 2 \cdot \frac{n}{2g+1}, \ldots, g \cdot \frac{n}{2g+1}$, and zeros elsewhere. We will show that A covers any string B of length m with at most $(2 - \frac{1}{g+1}) \cdot \frac{m}{n} - 2$ ones. Let B such a string, and denote by $y_1, \ldots, y_f$ the indices in which the character 1 appears in B.

If $y_1 > n$, then $A \geq B[1:n] = 0^n$, and we are done. In the following, we assume that $y_1 \leq n$. Let $L = \frac{g+1}{2g+1}n$, $x_1 = y_1 - L$, $x_i = y_i - y_{i-1}$ for $i = 2, \ldots, f$, and $x_{f+1} = m + 1 - y_f$. We note that the reason x_1 is defined as above rather than $x_1 = y_1$ is technical, and will be revealed in the proof of Lemma 1. The average of the numbers $x_1, \ldots, x_{f+1}$ is

$$\frac{\sum_{i=1}^{f+1} x_i}{f+1} = \frac{m - \frac{g+1}{2g+1}n + 1}{f+1} > \frac{m - \frac{g+1}{2g+1}n}{\frac{2g+1}{g+1} \cdot \frac{m}{n} - 1} \geq \frac{g+1}{2g+1}n = L.$$

Lemma 1. *There is an index $i > 1$ such that $x_i > L$ and $x_i + x_{i-1} > 2L$.*

Proof. Suppose conversely that for every index $j > 1$ such that $x_j > L$, we have that $x_{j-1} + x_j \leq 2L$. We say in this case that x_{j-1} and x_j are *matched*. We have that $x_1 \leq L$ because otherwise, $y_1 = x_1 + L > 2L > n$, contradicting the assumption that $y_1 \leq n$. Suppose that there are r pairs of matched x_j-s. From the above, we have that the sum of the matched x_j-s is at most $r \cdot 2L$, and the sum of the unmatched x_j-s is at most $(f+1-2r) \cdot L$. Therefore, $\sum_{j=1}^{f+1} x_j \leq (f+1)L$, a contradiction. □

By Lemma 1, the string $B' = B[y_i - 2L : y_i - 1]$ has at most one occurrence of the character one, which may appear at position at most L. There are $2L - n + 1 =$

$\frac{1}{2g+1}n+1$ ways to align A against B', and at least one of these alignments cover B'. More precisely, if B' does not contain the character 1, then every alignment of A and B' covers B'. Otherwise, let j be the index such that $B'[j] = 1$, and let $s \leq g$ be the integer for which $\frac{s}{2g+1}n < j \leq \frac{s+1}{2g+1}n$. For $s = 0$, the alignment of A and B' in which $A[1]$ is aligned with $B'[j+1]$ covers B'. For $s > 1$, the alignment of A' and B in which the s-th one in A is aligned against $B'[j]$ covers B'

For every n which is not divisible by $2g+1$, let A' be the string constructed above for the length $n' = (2g+1)\lceil n/(2g+1) \rceil$, and let A be the prefix of length n of A' (if A contains less than g ones, arbitrarily change some zeros into ones). A covers every string that is covered by A'. Therefore, A covers any string B of length m with at most $\left(2 - \frac{1}{g+1}\right) \cdot \frac{m}{n'} - 2$ ones. Thus,

$$k(n,g,m) \geq \left(2 - \frac{1}{g+1}\right) \cdot \frac{m}{n'} - 2 \geq \left(2 - \frac{1}{g+1}\right) \cdot \frac{m}{n+2g} - 2$$
$$= \left(2 - \frac{1}{g+1}\right) \cdot \frac{m}{n} - \frac{(2 - \frac{1}{2g+1})2g \cdot m}{n(n+2g)} - 2.$$

Upper bound. We now give an upper bound on $k(n,g,m)$. Let A be some string of length n with g ones. We will construct strings $B_0, \ldots, B_g$ that are not covered by A, such that at least one of these strings has at most $(2 - \frac{1}{g+1}) \cdot \frac{m}{n} + O(\max(1, \frac{m}{n^2}))$ ones.

Let $y_1, \ldots, y_g$ be the indices of the ones in A. Let $Y = \{y_j - y_i : 1 \leq i < j \leq g\}$, and let Z be the multi-set $\{\max(y_i - 1, n - y_i) : i = 1, \ldots, g\}$. Denote the elements of Z in a non-decreasing order by $z_1, \ldots, z_g$, and denote $z_{g+1} = n$. Define $d_i = \max(\{0, \ldots, z_{i+1}\} \setminus Y)$ for $i = 0, \ldots, g$, and $d'_i = \max(\{0, \ldots, n - 1 - z_i\} \setminus Y)$ for $i = 1, \ldots, g$.

The strings $B_0, \ldots, B_g$ are constructed as follows: Let B_0 be the prefix of length m of the string $(0^{d_0-1}1)^m$. The string B_i is the prefix of length m of $(0^{d_i-1}10^{d'_i-1}1)^m$. If either $d_i = 0$ or $d'_i = 0$ for some i, we say that B_i is undefined.

We now show that A does not cover the defined strings in $B_0, \ldots, B_g$. To see that A does not cover B_0 (if it is defined), suppose conversely that there is an alignment of A with B_0 that covers B_0. Since $d_0 < n$, the aligned region of B_0 contains one in some position j, and this one must be aligned with a one in A. Suppose that $B_0[j]$ is aligned with $A[y_i]$. We can break the string A into two parts: the characters to the left of y_i, and the characters to the right of y_i, whose lengths are $y_i - 1$ and $n - y_i$, respectively. By the definition of z_1, it follows that the size of the larger part is at least z_1. W.l.o.g. assume that the larger part is the part of the characters to the right of y_i. Since $d_0 \leq z_1$, the aligned region of B_0 contains another one at position $j + d_0$. From the definitions of d_0 and Y, this position must be aligned with a zero in A, contradicting the assumption that the alignment covers B_0.

Now, suppose that there is an alignment of A with B_l that covers B_l. The aligned region of B_l must contain one in some position j, which is aligned with $A[y_i]$ for some i. We again break the string A into two parts, and we have that either the larger part is of size at least z_{l+1}, or the smaller part is of size at most $n - 1 - z_l$. From the definition of d_l, d'_l, and Y, it follows that there is a one in the aligned region of B_l that is aligned against a zero in A, a contradiction.

As A doesn't cover the strings $B_0, \ldots, B_g$, we obtain that $k(n, g, m) < \min\{\text{ONES}(B_i) : i = 0, \ldots, g\}$. If $z_1 \geq \frac{g+1}{2g+1}n$, then $d_0 \geq \frac{g+1}{2g+1}n - |Y| \geq \frac{g+1}{2g+1}n - \frac{g^2}{2}$. For large enough n, $d_0 > 0$, so B_0 is defined and

$$\text{ONES}(B_0) \leq \frac{m}{d_0} \leq \frac{m}{\frac{g+1}{2g+1}n - \frac{g^2}{2}} = \frac{2g+1}{g+1} \cdot \frac{m}{n} + O(\tfrac{m}{n^2}).$$

Otherwise $z_{g+1} - z_1 > \frac{g}{2g+1}n$, so there is an index $i \geq 1$ such that $z_{i+1} - z_i > \frac{1}{2g+1}n$. Thus, $d'_i \geq n - 1 - z_i - |Y| \geq z_{i+1} - z_i - \frac{g^2}{2} > \frac{1}{2g+1}n - \frac{g^2}{2}$, so for large n, $d'_i > 0$ (and also $d_i > 0$). Moreover, $d_i + d'_i \geq n - 1 + z_{i+1} - z_i - 2|Y| \geq n + \frac{1}{2g+1}n - g^2$, and

$$\text{ONES}(B_i) \leq 2\left[\frac{m}{d_i + d'_i}\right] \leq \frac{2m}{n + \frac{1}{2g+1}n - g^2} + 2 = \frac{2g+1}{g+1} \cdot \frac{m}{n} + O(\max(1, \tfrac{m}{n^2})). \quad \square$$

3.2 Non-constant g

Theorem 2. *For every fixed integers $r \geq 2$ and $l \geq r$, $k(n, (1+o(1))l \cdot n^{1-1/r}, m) \geq (r + 1 - \frac{r}{l-r+2}) \cdot \frac{m}{n} - O(\max(1, \frac{m}{n^{1+1/r}}))$.*

Proof. Recall that in the construction of Theorem 1, we were able to find an alignment of A and B such that the aligned region of B contained the character 1 at most once, and this character was aligned against one of the ones in A. Here, we will find an alignment that contains at most r ones in the aligned region of B, which will be aligned against ones in A.

We first prove the theorem for the case $r = 2$. Suppose that $\sqrt{n}$ is integer, and that $\sqrt{n}$ is divisible by $3l - 2$. Let A be a string that consists of $\sqrt{n}$ blocks of size $\sqrt{n}$. The blocks numbered $\frac{2i}{3l-2}\sqrt{n}$ for $i = 1, \ldots, l-1$ contain only ones. The other blocks contain $\sqrt{n} - 1$ zeros, and a one at the end.

We will show that A covers any string B of length m with

$$\text{ONES}(B) \leq (3 - \tfrac{2}{l}) \cdot \frac{m}{n} - \frac{m\sqrt{n}}{\frac{l}{3l-2}n(\frac{l}{3l-2}n + \sqrt{n})} - 5.$$

Let B such a string, and denote by $y_1, \ldots, y_f$ the indices of the ones in B.

If $y_2 \geq n + \sqrt{n}$, then the string $B[1 : n + \sqrt{n} - 1]$ contains at most one occurrence of the character 1, at position y_1. If $y_1 > n$ we are done. Otherwise, consider the alignment of A against B in which $A[1]$ is aligned against $B[1]$. Let d be the distance from position y_1 to the nearest occurrence of 1 in A to the left of y_1 ($d = 0$ if $A[y_1] = 1$), and $d = y_1$ if there is no such occurrence. From the construction of A, $d < \sqrt{n}$. Therefore, if we move A d positions to the right along B, the 1 at position y_1 will either be aligned with a 1 of B (if $d \neq y_1$), or will be outside the aligned region of B (if $d = y_1$). Moreover, the aligned region of B will contain no more ones (as $y_2 \geq n + \sqrt{n}$). Therefore, this alignment covers B. For the rest of the proof we assume that $y_2 < n + \sqrt{n}$.

Let $L = \frac{l}{3l-2}n$, $x_1 = y_1 - 2L$, $x_2 = y_2 - y_1 - 2L$, $x_i = y_i - y_{i-1}$ for $i = 3, \ldots, f$, and $x_{f+1} = m + 1 - y_f$. We claim that there is an index $i > 2$ such that $x_i > L + \sqrt{n}$ and $x_i + x_{i-1} + x_{i-2} > 3L + 3\sqrt{n}$. The proof of this claim is similar to the proof of Lemma 1.

If $x_i > n$, then A covers B. Otherwise, let $B' = B[y_i - 3L - 3\sqrt{n} : y_i - 1]$. B' contains at most two ones, where the rightmost one is at position at most $2L + 2\sqrt{n}$. W.l.o.g. assume that B' contains two ones, at positions j and j', with $j' < j$. Suppose that $j > \frac{2}{3l-2}n + 2\sqrt{n}$ (the case $j \leq \frac{2}{3l-2}n + 2\sqrt{n}$ is similar, and we omit its proof). Let $s \leq l - 1$ be the integer such that $\frac{2s}{3l-2}n + 2\sqrt{n} < j \leq \frac{2(s+1)}{3l-2}n + 2\sqrt{n}$.

Consider an alignment of A and B' in which $B'[j]$ is aligned against $A[\frac{2s}{3l-2}n]$. Note that $A[\frac{2s}{3l-2}n]$ is the last character of block number $\frac{2s}{3l-2}\sqrt{n}$ in A. If $B'[j']$ is not in the aligned region, or it is aligned against a 1 in A then we are done. Otherwise, denote by $i = j' + \frac{2s}{3l-2}n - j$ the position in A which is aligned against $B'[j']$. Let d be the distance from position i to the nearest occurrence of 1 in A to the left of i, and $d = i$ if there is no such occurrence. Since $d < \sqrt{n}$, by moving A d positions to the right, we obtain an alignment that covers B'.

The case when $\sqrt{n}$ is not integer, or when $\sqrt{n}$ is not divisible by $3l - 2$, is handled in the same way as in Theorem 1: We build a string A' of length n' as described above, where n' is the minimal integer greater than n such that $\sqrt{n'}$ is an integer divisible by $3l - 2$, and we take A to be the prefix of length n of A'.

We now deal with the case of $r > 2$. If $n^{1/r}$ is an integer divisible by $(r+1)(l-r+1)+1$, then we build the string A as follows: We begin by taking $A = 0^n$. We then partition A into blocks of different levels. Level i ($i = 0, \ldots, r-1$) consists of $n^{1-i/r}$ blocks of size $n^{i/r}$ each. For every $i = 0, \ldots, r-2$, and every j which is divisible by $n^{1/r}$, we change block number j in level i to consists of all ones. Furthermore, for $j = 1, \ldots, l - r + 1$, we change block number $j \cdot \frac{l-r+2}{(r+1)(l-r+1)+1} \cdot n^{1/r}$ in level $r - 1$ to consists of all ones. The blocks that were changed are called the *ones blocks*.

For every string B of length m with

$$\text{ONES}(B) \leq (r + 1 - \frac{r}{l-r+2}) \cdot \frac{m}{n} - O(\max(1, \frac{m}{n^{1+1/r}})),$$

we have that either there is a substring B' of B of length $(1 + \frac{r}{(r+1)(l-r+1)+1})n + o(n)$ that contains at most r ones, or the prefix of B of length $n + O(n^{1-1/r})$ contains at most $r - 1$ ones. We describe below the proof for the former case (the proof for the latter case is similar). Suppose w.l.o.g. that B' contains exactly r ones. We create an alignment of A and B that covers B as follows: First, we align the rightmost one in B' with the rightmost character of the appropriate ones block of level $r - 1$. Then, for $i = 2, \ldots, r$, we move A to the right, until the i-th one from the right in B' is either aligned against the rightmost character of some ones block of level $r - i$, or it is outside the aligned region. By our construction, the movement of A is no more than $n^{1-(i-1)/r} - 1$ positions. Moreover, during the movements of A the following invariant is kept: For every $j \leq i-1$, after the i-th movement, the j-th one from the right in B' is aligned against a character of some ones block of level j', where $r - i - 1 \leq j' \leq r - j$. In particular, at the end, all the ones in B' are aligned against ones in A, and therefore A covers B.

The case when $n^{1/r}$ is not an integer, or when $n^{1/r}$ is not divisible by $(r+1)(l-r+1)+1$, is handled the same as before. □

The following theorem gives an upper bound that matches the lower bound in Theorem 2 (we omit the proof due to lack of space).

Theorem 3. *For every integer r, if $g \leq n^{1-1/r}$ then $k(n,g,m) \leq r \cdot \frac{m}{n} + O(1)$.*

4 Maximizing b

We now focus on the problem of maximizing b, the number of solid symbols in a seed, given the number of errors k and the pattern length m. This result also provides the maximum b for the problem of finding all matches of any pattern P' of size $m' \geq m$ within error rate $k'/m' = k/m$.

Theorem 4. *For every $k < \frac{1}{2} \log m$,*

$$b(k,m) \geq \begin{cases} m - O(km^{1-1/(k+1)}) & \text{if } k < \log \log m \\ m - O(m^{1-1/(k+1)}) & \text{if } k \geq \log \log m \end{cases}$$

and $b(k,m) \leq m - \Omega(m^{1-1/(k+1)})$.

Proof. We begin with showing the lower bound on $b(k,m)$. Let $s = \lfloor m^{1/(k+1)} \rfloor$. We construct a string A of length $n = m - 2\sum_{i=1}^{k} s^i$ by dividing it into blocks in k levels, similarly to Theorem 2: The i-th level of blocks ($i = 0, \ldots, k-1$) consists of blocks of size s^i each (the last $m \bmod s^i$ characters of A do not belong to a level i block). For every $i \leq k-1$ and every j which is divisible by s, we make block number j in level i a ones block.

We need to show that A covers every string B of length m with k ones. Let B be such string, and let $y_1, \ldots, y_k$ be the indices in which the character 1 appears in B. Let $y_{k+1} = m + 1$. If $y_{i+1} - y_i \leq 2s^i$ for all $i \leq k$ then we have that $y_1 \geq m + 1 - \sum_{i=1}^{k} 2s^i = n + 1$. Therefore $B[1:n] = 0^n$, and A covers B. Otherwise, let j be the maximum index such that $y_{j+1} - y_j > 2s^j$. Note that from the maximality of j, $y_{j+1} = m + 1 - \sum_{i=j+1}^{k}(y_{i+1} - y_i) \geq m + 1 - \sum_{i=j+1}^{k} 2s^i$, so $y_{j+1} - n > 2s^j$.

We align A over B such that $A[n]$ is aligned with $B[\max(n, y_j)]$. Then, for $i = j, j-1, \ldots, 1$, we move A to the right until $B[y_i]$ is against the rightmost character of some level $i-1$ ones block in A. The total movement of A is at most $\sum_{i=1}^{j} s_j \leq 2s^j < y_{j+1} - \max(n, y_j)$, and therefore at the end of this process the alignment covers B.

We therefore have that $b(k,m) \geq n - g$, where g is the number of ones in B. The lower bound for $k < \log \log m$ follows from the fact that $g = O(km^{1-1/(k+1)})$.

Now, suppose that $k \geq \log \log m$. We randomly construct a string A of length $n = m - \lfloor m^{1-1/(k+1)} \rfloor$: Each character of A is 1 with probability $p = 100/m^{1/(k+1)}$, and 0 otherwise, where all the choices are are independent. By Markov's inequality, with probability at least $1/2$, the number of ones in A is at most $2pn = O(m^{1-1/(k+1)})$. We will show that with probability at least $2/3$, A covers every string of length m with k ones. Therefore, there exists a string of length n with $O(m^{1-1/(k+1)})$ ones that covers every string of length m with k ones, and the lower bound follows.

Let B be some string of length m with k ones at positions $y_1, \ldots, y_k$. There is a set $X \subseteq \{0, \ldots, m-n\}$ of size at least $(m-n)/k^2$ such that for every $x, x' \in X$, there are no indices i and j such that $y_i + x = y_j + x'$ (Such a set X can be built by starting from $X = \phi$ and $X' = \{0, \ldots, m-n\}$ and moving elements from X' to X. Each element added to X rules out at most $k(k-1)$ elements in X', and therefore $|X| \geq (m-n+1)/(k(k-1)+1) > (m-n)/k^2$).

For every $x \in X$, the probability that the alignment that aligns $A[1]$ with $B[1+x]$ does not cover B is at most $1 - p^k$ (the probability is less than $1 - p^k$ if some of the ones of B are outside the aligned region). From the definition of X, the events above are independent, so the probability that A does not cover B is at most $(1-p^k)^{|X|}$. Using the union bound we obtain that the probability that there is a string B that is not covered by A is at most

$$\binom{m}{k}(1-p^k)^{(m-n)/k^2} \leq m^k e^{-p^k m^{1-1/(k+1)}/k^2} < \frac{1}{3}.$$

Upper bound. Denote $M = \lceil \frac{1}{2} m^{1-1/(k+1)} \rceil$. Let A be a string with b zeros and g ones, and suppose that $b \geq m - M + 1$. We will show that there is a string B of length m with k ones that is not covered by A. Clearly, $g \leq m - b \leq M - 1$.

Define $I_i = \{(i-1)M + 1, \ldots, iM\}$ for $i = 1, \ldots, 2m^{1/(k+1)} - 1$. Let Y be the set of all k-tuples $(j, i_1, \ldots, i_{k-1})$ such that (1) $j \leq 2m^{1/(k+1)} - 1$ and $M + 1 \leq i_1 < \cdots < i_{k-1} \leq jM$, and (2) there is an index $x \in I_j$ such that $A[x] = A[x - i_1] = \cdots = A[x - i_{k-1}] = 1$. We have that

$$|Y| \leq \binom{g}{k} < \binom{M}{k} < \frac{\binom{m-2M}{k}}{M}.$$

Since the number of k-tuples that satisfy (1) above is at least $\binom{m-2M}{k}/M$, we have that there is a k-tuple $(j, i_1, \ldots, i_{k-1})$ that satisfies (1) but does not satisfy (2). We now construct a string B of length m that contains ones in positions $jM, jM - i_1, \ldots, jM - i_{k-1}$ and zeros elsewhere. If A covers B, then consider some alignment of A and B that covers B, and suppose that $A[1]$ is aligned with $B[y]$. Since the length of A is at least $b \geq m - M + 1$, we have that $y \leq M$. Therefore, $B[jM]$ is aligned with $A[x]$ for some $x \in I_j$. It follows that $(j, i_1, \ldots, i_{k-1}) \in Y$, a contradiction. Therefore, A does not cover B, and the upper bound follows. □

Theorem 5. *For every $k \geq \frac{1}{2}\log m$, $b(k,m) = \Omega(\frac{m}{k}\log\frac{m}{k})$ and $b(k,m) = O(\frac{m}{k}\log m)$.*

Proof. The lower bound is trivial if $k = \Omega(m)$, so assume that $k < \frac{m}{10}$. By Theorem 4, there is a string A of length $n = \frac{1}{2} \cdot \frac{m}{k}\log\frac{m}{k}$ that contains $b = \Theta(n)$ ones and covers every string of length $2n$ with at most $\log n$ ones. If B is a string of length m with at most k ones, then by the pigeon-hole principle, there is a substring B' of B of length $2n$ that has at most $k \cdot \frac{2n}{m} = \log\frac{m}{k} \leq \log n$ ones, and therefore A covers B'. Thus, A covers B.

Upper bound. Suppose that A is a string of length n with $b \geq \frac{m}{k}\log m$ zeros. Let $z_1,\ldots,z_b$ be the positions of the zeros in A. Construct a collection of sets $S_1,\ldots,S_m$, where $S_x = \{y \in \{0,\ldots,m-n\} : \exists j \text{ s.t. } z_j + y = x\}$. For every $y = 0,\ldots,m-n$, there are b sets among $S_1,\ldots,S_m$ that contain y. Therefore, there is a set cover of $\{0,\ldots,m-n\}$ using $l \leq \log(m-n+1)/\log(1/(1-b/m))$ sets from $S_1,\ldots,S_m$ [9], namely, there are indices $x_1,\ldots,x_l$ such that $\bigcup_{i=1}^{l} S_{x_i} = \{0,\ldots,m-n\}$. Now, let B be a string of length m that contains ones in positions $x_1,\ldots,x_l$ and zeros elsewhere. If we align A and B, where $A[1]$ is aligned with $B[1+y]$ for some $0 \leq y \leq m-n$, then there is an index x_i such that $y \in S_{x_i}$, that is, there is an index j such that $z_j + y = x_i$. This implies that the character $A[z_j] = 0$ is aligned against $B[x_i] = 1$. Therefore, A doesn't cover B. The number of ones in B is at most

$$\frac{\log(m-n+1)}{\log\frac{1}{1-b/m}} \leq \frac{m}{b} \cdot \log m \leq k,$$

and the upper bound follows. □

4.1 Multiple Seeds

To model multiple seeds, we define that a set of strings $\{A_1,\ldots,A_s\}$ covers a string B if at least one string A_i from the set covers B. Let $b(k,m,s)$ be the maximum b such that there is a set of strings $\{A_1,\ldots,A_s\}$ that covers every string B of length m with at most k ones, and each string A_i contains at least b zeros. The following theorem shows that using multiple seeds can give better results than one seed, namely, $b(k,m,m^{1/k})$ is slightly larger than $b(k,m)$ (see Theorem 4).

Theorem 6. *For every $k < \log\log m$, $b(k,m,m^{1/k}) \geq m - O(km^{1-1/k})$.*

Proof. We take $s = \lfloor m^{1/k} \rfloor$, and build a string A of length $n = m - \sum_{l=0}^{k-1} s^l$, that has $k-1$ levels of blocks as in the proof of Theorem 4. Then, we build strings $A_1,\ldots,A_{s-1}$, where A_i is obtained by taking the string A and adding ones at positions $js - i$ for $j \geq s$. It is easy to verify that $\{A_1,\ldots,A_{s-1}\}$ covers every string of length m with at most k ones. □

References

1. S. Altschul, W. Gisch, W. Miller, E. Myers, and D. Lipman. Basic local alignment search tool. *J. of Molecular Biology*, 215(3):403–410, 1990.
2. A. Amir, M. Lewenstein, and E. Porat. Faster algorithms for string matching with k-mismatches. In *Proc. ACM-SIAM SODA*, pp 794–803, 2000.
3. B. Brejova, D. G. Brown, and T. Vinar. Vector seeds: an extension to spaced seeds allows substantial improvements in sensitivity and specificity. In *Proc. WABI*, pp 39–54, 2003.
4. J. Buhler. Provably sensitive indexing strategies for biosequence similarity search. In *Proc. ACM RECOMB*, pp 90–99, 2002.
5. J. Buhler, U. Keich, and Y. Sun. Designing seeds for similarity search in genomic DNA. In *Proc. ACM RECOMB*, pp 67–75, 2003.
6. S. Burkhardt and J. Karkkainen. Better filtering with gapped q-grams. *Fundamenta Informaticae*, 56:51–70, 2003.
7. A. Califano and I. Rigoutsos. Flash: a fast look-up algorithm for string homology. In *Proc. ISMB*, pp 56–64, 1993.
8. R. Cole and R. Hariharan. Approximate string matching, a simpler, faster algorithm. In *Proc. ACM-SIAM SODA*, pp 463–472, 1997.
9. M. Karpinski and A. Zelikovsky. Approximating dense cases of covering. *Electronic Colloquium on Computational Complexity*, 4(4), 1997.
10. U. Keich, M. Li, B. Ma, and J. Tromp. On spaced seeds for similarity search. *Discrete Applied Mathematics*, 138(3):253–263, 2004.
11. G. Kucherov, L. Noé, and M. Roytberg. Multi-seed lossless filtration. In *Proc. CPM*, pp 297–310, 2004.
12. G. Kucherov, L. Noé, and Y. Ponty. Estimating seed sensitivity on homogeneous alignments. In *Proc. IEEE BIBE*, pp 387–394, 2004.
13. G. M. Landau and U. Vishkin. Fast parallel and serial approximate string matching. *Journal of Algorithms*, 10(2):157–169, 1989.
14. M. Li, B. Ma, D. Kisman, and J. Tromp. Patternhunter II: Highly sensitive fast homology search. *J. of Bioinformatics and Computational Biology*, 2(3):417–439, 2004.
15. B. Ma, J. Tromp, and M. Li. PatternHunter: Faster and more sensitive homology search. *Bioinformatics*, 18:440–445, 2002.
16. P. Pevzner and M. Waterman. Multiple filtration and approximate pattern matching. *Algorithmica*, 13:135–154, 1995.
17. S. C. Sahinalp and U. Vishkin. Efficient approximate and dynamic matching of patterns using a labeling paradigm. In *Proc. IEEE FOCS*, pp 320–328, 1996.
18. Y. Sun and J. Buhler. Designing multiple simultaneous seeds for DNA similarity search. In *Proc. ACM RECOMB*, pp 76–84, 2004.
19. J. Xu, D. Brown, M. Li, and B. Ma. Optimizing multiple spaced seeds for homology search. In *Proc. CPM*, pp 47–58, 2004.

Fast Neighbor Joining

Isaac Elias and Jens Lagergren

Dept. of Numerical Analysis and Computer Science,
Royal Institute of Technology, Stockholm, Sweden
{isaac, jensl}@nada.kth.se

Abstract. Reconstructing the evolutionary history of a set of species is a fundamental problem in biology and methods for solving this problem are gaged based on two characteristics: accuracy and efficiency. Neighbor Joining (NJ) is a so-called distance-based method that, thanks to its good accuracy and speed, has been embraced by the phylogeny community. It takes the distances between n taxa and produces in $\Theta(n^3)$ time a phylogenetic tree, i.e., a tree which aims to describe the evolutionary history of the taxa. In addition to performing well in practice, the NJ algorithm has optimal reconstruction radius.

The contribution of this paper is twofold: (1) we present an algorithm called Fast Neighbor Joining (FNJ) with optimal reconstruction radius and optimal run time complexity $O(n^2)$ and (2) we present a greatly simplified proof for the correctness of NJ. Initial experiments show that FNJ in practice has almost the same accuracy as NJ, indicating that the property of optimal reconstruction radius has great importance to their good performance. Moreover, we show how improved running time can be achieved for computing the so-called correction formulas.

1 Introduction

The evolutionary history of a set of species is a central concept in biology that is commonly described by a phylogenetic tree. Frequently it is the case that the phylogenetic tree is unknown and the only information available are the genetic sequences from the extant species, i.e., currently living species. It is therefore a fundamental problem to reconstruct the phylogenetic tree given genetic sequences. Several reconstruction methods have been suggested, and it is natural to compare these based on how accurate they are in reconstructing the correct phylogeny. Unfortunately though, of these methods the more accurate are much too slow to be used in studies that involve reconstructing large or many phylogenies. The focus of this paper is to build an algorithm that is accurate and has quadratic running time in the number of species.

As more genetic information is collected it becomes possible to answer more complex questions. An obvious question that involves reconstructing a large phylogeny is to relate all living species in the tree of life. Another very central question is to relate large sets of genes and from such phylogenies draw conclusion about their function and origin. However, reconstruction of large phylogenies is not the only case in which efficient reconstruction is necessary. There are other cases that involve many reconstructions, e.g., studies where phylogenies are built for each gene shared by a set of species. The common technique of bootstrapping also requires many reconstructions in order to obtain significance values for a single phylogeny.

Throughout the paper, phylogenetic trees are leaf-labeled binary trees with edge lengths. Thus each phylogenetic tree T naturally induces an additive leaf-to-leaf distance function D_T. The reconstruction methods for which most complexity results have been shown are the so-called distance methods. These algorithms take as input an estimated distance function D (normally computed from the genomic sequences) and construct a phylogeny whose additive distance function is close to D. The problem of finding the closest additive distance function under the infinity norm is known to be NP-hard [1].

The Neighbor Joining (NJ) algorithm is a distance method introduced by Saitou and Nei in [15]. As shown in [8], when NJ is given an *additive* distance function D_T, it reconstructs the unique tree T. However, as Atteson [2] proved NJ reconstructs the closest tree for even more cases. A distance function D is *nearly additive* if there is an additive distance function D_T such that

$$|D - D_T|_\infty < \mu(T)/2, \qquad (1)$$

where $\mu(T)$ is the minimum edge length in T. All the additive distance functions for which Equation 1 holds have the same topology, i.e., disregarding the edge lengths, T is the unique tree for which the equation holds. The NJ algorithm has *optimal reconstruction radius* in the sense that: (a) given a nearly additive distance function it reconstructs the unique tree T and (b) there can be more than one tree for which $|D - D_T| < \delta$ holds if $\delta \geq \mu(T)/2$. In practice most distances are far from being nearly additive. Thus, although important, optimal reconstruction radius is not sufficient for an algorithm to be useful in practice.

The estimated distances that are given as input to distance methods are normally deduced from genomic sequences and a probabilistic model. There are various Markov models of sequence evolution which describe how sites evolve independently and identically from the root down toward the leafs. Many of these models have an associated closed correction formula for inverting the model and giving an estimated evolutionary distance for a pair of sequences. These formulas are consistent in the sense that the estimated distance approaches the underlying additive distance as the sequence length approaches infinity. As a result, the NJ algorithm is a consistent method for recovering the correct phylogeny, i.e., NJ reconstructs the correct phylogeny given infinitely long sequences.

An interesting line of research is to design fast-converging algorithms, i.e., algorithms that reconstruct the correct phylogeny from sequences whose length is polynomial in the number of sequences [7, 9, 12, 5]. However, except from the Disc-Covering Method (DCM) [9, 12] these algorithms have had little or no practical impact. The only variation of DCM that is fast-converging and of practical interest uses NJ to construct small sub-phylogenies that are later patched together into one larger phylogeny, i.e., NJ is used as a subroutine.

Although the NJ algorithm is not fast-converging, it has in experimental studies been shown to perform very well [13]. Moreover, with $O(n^3)$ as the worst case running time it has become the reconstruction algorithm that is most frequently used in practice. Heuristic implementations of NJ have been given which, without leading to better worst case analysis of the time complexity, in practice show improved running time [3, 17].

There are two major contributions in this paper: (1) we present an algorithm called Fast Neighbor Joining (FNJ) with optimal reconstruction radius and optimal run time complexity $O(n^2)$ and (2) we present a greatly simplified proof for the correctness of the NJ algorithm. Initial experiments show that the FNJ algorithm in practice has almost the same accuracy as the NJ algorithm; this indicates that it is the optimal reconstruction radius and other similarities with NJ that give FNJ its good performance. We also describe how a better running time for computing the correction formulas can be achieved, in theory, through matrix multiplication and, in practice, through table lookups.

The FNJ algorithm is useful in its own right. But it is also important to note that FNJ together with the proof of optimal reconstruction radius presents a good foundation for building reconstruction algorithms that are both practically useful and fast-converging. For example the running time of DCM can be improved by a factor $O(n)$ by simply replacing NJ with FNJ. It will be interesting to see how the running time of extensions of NJ, such as Weighbor and BioNJ, can be improved using our ideas.

The paper is organized as follows. The next section contains some basic definitions and a description of the NJ algorithm. In Section 3, the FNJ algorithm is introduced. Subsequently we give the proof of the FNJ algorithm and also a more economical and intuitively appealing proof of Atteson's theorem. Finally, in Section 7, we approach the practical problem of computing the correction formulas and also show that the FNJ algorithm in practice performs almost exactly as good as the NJ algorithm. Except for Lemma 1 below, which in [2] (Lemma 12) is proved by straightforward algebraic verification, the present paper is self-contained.

2 Definitions and the Neighbor Joining Algorithm

A $n \times n$ *distance function* D, for a set of taxa $\mathcal{N}(D)$, is a function $\mathcal{N}(D)^2 \to R^+$, where $|\mathcal{N}(D)| = n$, which is symmetric and satisfies $D(x,x) = 0$ for every $x \in \mathcal{N}(D)$. For two distance functions D_1 and D_2 such that $\mathcal{N}(D_1) = \mathcal{N}(D_2) = \mathcal{N}$, their distance is defined as $\max_{x,y \in \mathcal{N}} |D_1(x,y) - D_2(x,y)|$ and denoted $|D_1 - D_2|_\infty$. By a *phylogenetic tree* we mean a tree T given together with an edge length function $l_T : E(T) \to R^+$. For a phylogenetic tree T, $\mu(T)$ denotes the minimum edge length of T, i.e., $\min_{e \in E(T)} l(e)$. The unique path in a tree T between two of its vertices u and v is denoted $P_T(u,v)$. Every phylogenetic tree T induces a distance function for the leafs in the tree, i.e., $D_T : L(T)^2 \to R^+$ where $D_T(a,b) \triangleq \sum_{e \in P_T(a,b)} l(e)$.

A distance function D is *additive* if there is a phylogenetic tree T such that $D = D_T$; the tree is said to *realize* D, it is unique, and it is denoted $T(D)$. A distance function D is *nearly additive* if there is a phylogenetic tree T such that $|D - D_T|_\infty < \mu(T)/2$; again, the tree is said to *realize* D, it is unique, and it is denoted $T(D)$ [2]. The *parent* of a leaf a in a tree T is the unique neighbor of a in T. A pair of leaves of a tree T are *siblings* if they have the same parent in T (note only leaf-siblings).

The NJ algorithm builds a tree by iteratively combining pairs of taxa. It takes as input a distance function D for n taxa and attempts to identify two siblings by selecting the pair of taxa (a, b) that minimizes the *NJ function*, defined by

$$S_D(x,y) \triangleq (|\mathcal{N}(D)| - 2) \cdot D(x,y) - \sum_{z \in \mathcal{N}(D)} (D(z,x) + D(z,y)). \quad (2)$$

Thereafter the pair (a, b) is *reduced* to a a new node c, representing the parent, which gives a new distance function D' with $\mathcal{N}(D') = (\mathcal{N}(D) \setminus \{a,b\}) \cup \{c\}$ defined by

$$D'(x,y) \triangleq \begin{cases} D(x,y), & \text{if } c \notin \{x,y\} \\ \frac{D(z,a)+D(z,b)}{2}, & \text{otherwise } z \in \{x,y\} \setminus \{c\}. \end{cases} \quad (3)$$

Finally the algorithm is applied iteratively on the new distance function D'. A formal description of the NJ algorithm is given below.

Algorithm NJ(D_1)

1. For each $i \leftarrow 1$ to $n - 3$ do
   (a) $(a_i, b_i) \leftarrow \operatorname{argmin}_{x \neq y \in \mathcal{N}(D_i)} S_{D_i}(x,y)$
   (b) Reduce a_i and b_i to a new node c_i and let D_{i+1} be the new distance function given by the reduction in Equation 3.
   (c) Connect a_i and b_i to c_i by adding edges (a_i, c_i) and (b_i, c_i).
2. Connect the three nodes of $\mathcal{N}(D_{n-3})$ in a star and return the resulting tree.

Theorem 1 (Atteson's Theorem). *Given a nearly additive distance function D NJ outputs $T(D)$. Moreover, in each iteration i, D_i is nearly additive and $T(D_i) = T(D_{i-1}) \setminus \{a_{i-1}, b_{i-1}\}$.*

In Section 3, we will show the analogous theorem for the FNJ algorithm, by showing that for nearly additive distance functions it gives exactly the same output as NJ. In Section 5, we give a proof of Atteson's theorem above.

3 The Fast Neighbor Joining Algorithm

In Step 1a of the NJ algorithm and for $i \leq n/2$, the minimum is taken over $\Omega(n^2)$ pairs which implies a running time of $\Omega(n^3)$. In the FNJ algorithm an $O(n^2)$ running time is obtained by using two ideas. First, the minimum is taken over a set, called the *visible set*, of cardinality $O(n)$. Second, using the auxiliary function R, introduced below, the updated NJ function can be computed in constant time. It should be noted that the resulting trees of NJ and FNJ are only guaranteed to be the same if the input is nearly additive.

A pair (a, b) is *visible from* a w.r.t. a distance function D if

$$b = \mathrm{argmin}_{x \in \mathcal{N}(D) \setminus \{a\}} S_D(a, x).$$

A pair (a, b) is *visible* w.r.t. D if it is visible from either a or b. Hence the number of visible pairs is $O(n)$. In the next section it is shown that for each nearly additive distance function D, any sibling pair in $T(D)$ is visible w.r.t. D.

To enable an overall $O(n^2)$ running time, the NJ function is computed using an auxiliary function R defined by $R_D(a) \triangleq \sum_{x \in \mathcal{N}(D)} D(a, x)$, i.e., R is the row sums. It is straightforward to verify that for a D' defined as in Equation 3,

$$R_{D'}(x) = R_D(x) - \frac{D(x, a) + D(x, b)}{2}. \tag{4}$$

Hence, given R_D it is possible to compute the updated row sums $R_{D'}$ in time $O(n)$. Moreover, since $S_D(x, y) = (|\mathcal{N}(D)| - 2) \cdot D(x, y) - R_D(x) - R_D(y)$, the NJ function can be computed in constant time, for any given pair (x, y).

It should be clear, from the formal description below that the FNJ algorithm runs in time $O(n^2)$. Note that the input actually has size $\Omega(n^2)$.

Algorithm FNJ(D_1)

1. The first visible set $\mathcal{V}_1$ is initialized to the set of pairs visible w.r.t. D_1.
2. For each $a \in \mathcal{N}(D)$, $R_{D_1}(a)$ is initialized to $\sum_{x \in \mathcal{N}(D_1)} D_1(a, x)$.
3. For each $i \leftarrow 1$ to $n - 3$ do
   (a) $(a_i, b_i) \leftarrow \mathrm{argmin}_{(x,y) \in \mathcal{V}_i} S_{D_i}(x, y)$
   (b) Reduce a_i and b_i to a new node c_i and let D_{i+1} be the new distance function given by the reduction in Equation 3.
   (c) Connect a_i and b_i to c_i by adding edges (a_i, c_i) and (b_i, c_i).
   (d) Compute $R_{D_{i+1}}$.
   (e) $\mathcal{V}_{i+1} \leftarrow (\mathcal{V}_i \setminus \{(x, y) : x = a_i \text{ or } x = b_i\}) \cup \{(c_i, d)\}$ where (c_i, d) is the pair visible from c_i w.r.t. D_{i+1}.
4. Connect the three nodes of $\mathcal{N}(D_{n-3})$ in a star and return the resulting tree.

---

4 Correctness of FNJ

According to Theorem 1, given a nearly additive distance function D, NJ outputs $T(D)$, i.e., it outputs the unique tree that is close to D. Here we prove that FNJ has the same property. Since NJ constructs the correct tree, we know that in each iteration the minimum pair over the NJ function is a sibling pair in $T(D_i)$.

Hence, to prove the correctness of FNJ, it suffices to show that in each iteration the minimum pair is in the visible set, $\mathcal{V}_i$. The proof is in two steps; first the Visibility lemma is presented. According to this, if a has a sibling b in $T(D)$, then (a,b) is in the visible set. Second, in Theorem 2, the Visibility lemma together with the correctness of NJ is used to prove the correctness of FNJ.

Before we proceed to prove the Visibility lemma, we state an observation and a lemma which in Atteson [2] are proved through straightforward algebraic verification. For any tree T, edge e of T, and leaf a of T, let $\mathcal{L}_T(a,e)$ denote the set of leaves of T belonging to the same connected component of $T \setminus \{e\}$ as a.

Observation 1 (Atteson). *If D_T is an additive distance function, then,*

$$S_{D_T}(a,b) = \sum_{e \in E(T)} w_e(a,b)\, l(e), \text{ where}$$

$$w_e(a,b) = \begin{cases} -2 & \text{if } e \in E(P_T(a,b)) \\ -2|L(T) \setminus \mathcal{L}_T(a,e)| & \text{otherwise.} \end{cases}$$

Lemma 1 (Atteson, Lemma 12). *Let D_T and D be two n-domain distance functions such that D_T is additive and D is nearly additive w.r.t. D_T. For any $a,b,x,y \in \mathcal{N}(D)$, the value of $S_D(a,b) - S_{D_T}(a,b) + S_{D_T}(x,y) - S_D(x,y)$ is*

$$> \begin{cases} -3(n-4)\mu(T) & \text{if } \{a,b\} \cap \{x,y\} = \emptyset \\ -2(n-3)\mu(T) & \text{if } |\{a,b\} \cap \{x,y\}| = 1. \end{cases}$$

Lemma 2 (The Visibility Lemma). *Let D_T and D be two n-domain distance functions such that D_T is additive and D is nearly additive w.r.t. D_T. If a has a sibling b in T, then (a,b) is visible from a w.r.t. D, i.e.,*

$$b = \mathrm{argmin}_{x \in \mathcal{N}(D) \setminus \{a\}} S_D(a,x).$$

Proof. As in Figure 1, let $c \in \mathcal{N}(D) \setminus \{a,b\}$ and let e_a, e_b, and e_c be the edges of T incident with a, b, and c, respectively. Moreover, let e be the edge incident with the parent of a and b which is not incident with either a or b. Consider D_T, by definition of the weights in Observation 1 the following is true

(i) $w_f(a,b) = -2 = w_f(a,c)$ for any $f \in \{e_a, e_b, e_c\}$,
(ii) $w_f(a,b) \leq -3 < w_f(a,c)$ for any $f \in E(P_T(a,c)) \setminus \{e_a, e_c\}$,
(iii) $w_f(a,b) = w_f(a,c)$ for any $f \in E(T) \setminus E(P_T(a,c))$.

Moreover, since $w_e(a,b) = -2(n-2)$ and $w_e(a,c) = -2$, it follows that $S_{D_T}(a,c) - S_{D_T}(a,b) \geq 2(n-3)\mu(T)$. Finally, by Lemma 1,

$$S_D(a,c) - S_D(a,b) =$$

$$\underbrace{S_D(a,c) - S_{D_T}(a,c) + S_{D_T}(a,b) - S_D(a,b)}_{> -2(n-3)\mu(T)} + \underbrace{S_{D_T}(a,c) - S_{D_T}(a,b)}_{\geq 2(n-3)\mu(T)} > 0. \quad \square$$

We are now ready to prove that given a nearly additive distance function D, FNJ in each iteration selects the same sibling pair as NJ, i.e., FNJ outputs $T(D)$. By Atteson's theorem NJ outputs $T(D)$ by in each iteration reducing a pair of siblings such that $T(D_i) = T(D_{i-1}) \setminus \{a_{i-1}, b_{i-1}\}$. Since FNJ uses the same reduction as NJ it is sufficient to show that all sibling pairs are in the visible set. In the next section, we give a short and intuitively appealing proof of Atteson's theorem, which together with the Visibility lemma gives a direct proof of the theorem below.

Theorem 2. *Given a nearly additive distance function D, FNJ outputs $T(D)$.*

Proof. We prove by induction that, for each $i = 1, \ldots, n-3$, $\mathcal{V}_i$ contains all sibling pairs of $T(D_i)$ (here $D_1 = D$). By the Visibility lemma it is clear that the statement is true for $i = 1$. Assume that the statement holds for each $i = 1, \ldots, j$.

By the correctness of NJ, if (a_j, b_j) is the minimum over the NJ function, then (a_j, b_j) is a sibling pair $T(D_j)$. Therefore, by the induction assumption, (a_j, b_j) is in $\mathcal{V}_j$. Consequently, since the minimum over the NJ function is a sibling pair, FNJ and NJ select the same sibling pair in iteration j.

After reducing (a_j, b_j) to c_j, by the Visibility lemma, if c_j has a sibling d in $T(D_{j+1})$, then in Step 3e (c_j, d) is added to $\mathcal{V}_{j+1}$. Moreover, by the assumption, all other sibling pairs of $T(D_{j+1})$ are in $\mathcal{V}_j$ and therefore also in $\mathcal{V}_{j+1}$. Hence, by induction and the correctness of NJ, FNJ outputs $T(D)$. □

5 Atteson's Theorem - Correctness of NJ

The proof of Atteson's theorem is in two steps. The first step consists of the key technical lemma below, of which we give a much more concise and direct proof. The central idea in this proof, is to show that for any additive distance function the difference is large between the value of NJ function applied to a sibling pair, and applied to a pair of leaves which are not siblings. In fact, the difference is so large that even when the distance function is nearly additive the NJ function is minimized by a sibling pair. The final step in proving Atteson's theorem consists of showing that the distance function, after a reduction, remains nearly additive.

Lemma 3. *If D is a nearly additive distance function, $a, b \in \mathcal{N}(D)$, and $S_D(a, b) = \min_{x \neq y \in \mathcal{N}(D)} S_D(x, y)$, then (a, b) is a sibling pair in $T = T(D)$.*

Proof. According to the Visibility lemma, if a has a sibling b then $S_D(a, b) < S_D(a, x)$ for any $x \neq b$. Hence, the lemma follows if for any two leaves, x and y, of which none has a sibling in T, there exists a sibling pair (a, b), such that $S_D(x, y) - S_D(a, b) > 0$. Let D_T be an additive distance function such that $|D - D_T| < \mu(T)/2$. Notice that

$$S_D(x, y) - S_D(a, b)$$
$$= S_D(x, y) - S_{D_T}(x, y) + S_{D_T}(a, b) - S_D(a, b) + S_{D_T}(x, y) - S_{D_T}(a, b)$$

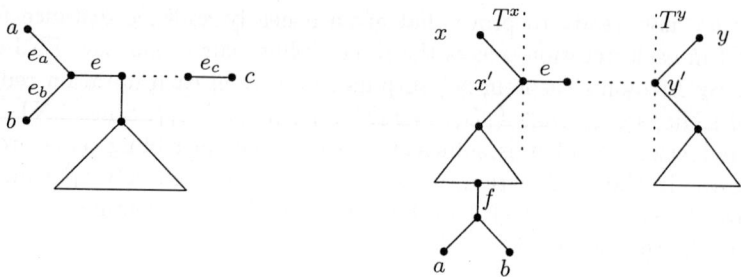

Fig. 1. To the left the figure for the Visibility lemma. To the right the figure for Lemma 3

$$> -3(n-4)\mu(T) + S_{D_T}(x,y) - S_{D_T}(a,b),$$

where the inequality follows by Lemma 1. We proceed by showing that $S_{D_T}(x,y) - S_{D_T}(a,b) > 3(n-4)\mu(T)$.

In T let x' and y' be the unique neighbors of x and y, respectively (see Figure 1). Further, let T^x and T^y be the subtrees of $T \setminus P_T(x', y')$ containing x and y, respectively. W.l.o.g., assume that $|L(T^x)| \le |L(T^y)|$, and hence that $|L(T^x)| \le n/2$. Let e be the edge of $P_T(x', y')$ incident to x'. Since neither x nor y has a sibling, both T^x and T^y contain a sibling pair of T. Let a and b be siblings in T^x, and let f be the edge incident with their parent but not a and not b.

First note that $w_g(a,b) \le w_g(x,y)$ for any $g \in E(T) \setminus \{e, f\}$. The only edges for which the latter inequality is non-trivial are those of $P_T(a, x')$; for those the inequality follows from the assumption that $|L(T^x)| \le n/2$. Using the definition of weights, it is straightforward to verify that $w_e(a,b) = -2|L(T) \setminus L(T^x)| \ge -n$ while $w_e(x,y) = -2$, and that $w_f(a,b) = -2(n-2)$ while $w_f(x,y) = -4$. It follows that

$$\begin{aligned} S_{D_T}(x,y) - S_{D_T}(a,b) &\ge \bigl(-2 - 4 + n + 2(n-2)\bigr)\mu(T) \\ &= (3n - 10)\mu(T) \\ &> 3(n-4)\mu(T). \end{aligned}$$
□

(Proof Theorem 1). The proof is by induction. First note that the theorem holds when $|\mathcal{N}(D)| = 3$. Assume that the theorem holds when $|\mathcal{N}(D)| = n - 1$. We now prove that it holds for $|\mathcal{N}(D)| = n$.

Since D is nearly additive, by the lemma above, NJ in the first iteration reduces a pair (a, b) that are siblings in $T = T(D)$ to a new node c, representing their parent. Denote the distance function after the reduction by D'. We need to prove that D' is nearly additive and that $T(D') = T \setminus \{a, b\}$.

Let S be the tree $T \setminus \{a, b\}$ with the edge length function defined as follows:

$$l_S(u, v) \triangleq l_T(u, v)$$

for all $u, v \in V(S) \setminus \{c\}$, and

$$l_S(c, c') \triangleq l_T(c, c') + \frac{l_T(c, a) + l_T(c, b)}{2}$$

for the unique neighbor c' of c in S. It should be clear that $\mu(T) \leq \mu(S)$.
We now show that $|D' - D_S| < \mu(S)/2$, i.e., that $T(D') = S = T \setminus \{a, b\}$.
From this, the theorem follows immediately. For $u, v \in L(S) \setminus \{c\}$,

$$|D'(u, v) - D_S(u, v)| = |D(u, v) - D_T(u, v))| < \frac{\mu(T)}{2} \leq \frac{\mu(S)}{2}$$

For all $u \in L(S)$,

$$\begin{aligned}
&|D'(u, c) - D_S(u, c)| \\
&= \left| \frac{D(u, a) + D(u, b)}{2} - D_S(u, c') - l_S(c', c) \right| \\
&= \left| \frac{D(u, a) + D(u, b)}{2} - D_T(u, c') - l_T(c', c) - \frac{l_T(c, a) + l_T(c, b)}{2} \right| \\
&\leq \left| \frac{D(u, a) - D_T(u, a)}{2} + \frac{D(u, b) - D_T(u, b)}{2} \right| \\
&\leq \left| \frac{D(u, a) - D_T(u, a)}{2} \right| + \left| \frac{D(u, b) - D_T(u, b)}{2} \right| \\
&< \frac{\mu(T)}{4} + \frac{\mu(T)}{4} = \frac{\mu(T)}{2} \leq \frac{\mu(S)}{2}.
\end{aligned}$$

□

6 Improved Computations of Correction Formulas

As was mentioned in the introduction, the real input to a reconstruction problem is usually n sequences of length l. The assumption is that these sequences have evolved from an original ancestor sequence down the branches of the phylogeny, according to a model of sequence evolution. The distance method approach, to the reconstruction problem, is to first use the sequences to estimate the actual distances between every pair of leaves, and thereafter find a phylogeny that fits the estimated distances. That is, from the n sequences of length l, an $n \times n$ distance function is computed through a correction formula. This formula is dependent on the model assumed to have generated the sequences; the most common models are Jukes-Cantor (JC) [10] and Kimura 2 parameter (K2P) [11]. Most correction formulas are in a sense functions of the hamming distance, e.g., the JC correction formula is given by

$$JC(s_1, s_2) \triangleq -\frac{3}{4} \cdot \log \left(1 - \frac{4 \cdot H(s_1, s_2)}{3l}\right),$$

where H is the hamming distance. Clearly, the straightforward way of computing this function takes $O(l)$ time, and as a result the overall running time of computing all estimated distances is $O(ln^2)$. Since l typically is larger than n, the computation of the correction formula is the bottleneck in fast reconstruction algorithms.

Computing all n^2 pairwise hamming distances for n strings is a special case of matrix multiplication, and can therefore be done in $O(ln^{1.376})$ time [4]. The reduction for strings from the alphabet $\{A, C, G, T\}$, is by representing each string by a row in the matrix M, and code each symbol by the unary code, e.g., by letting $A = 1000$. Thereby, the elements in the matrix MM^T are $l - H(s_i, s_j)$. It should be noted that the general belief is that matrix multiplication can be done in $O(ln)$ time, which would imply that the correction formulas can be computed in optimal time. Unfortunately, all existing matrix multiplication algorithms are slow in practice.

Below we present an algorithm that improved the computations of the correction formula by more than a factor of 3, compared to the straightforward approach. The idea is to first represent each symbol by 2 bits, and then use a precomputed table with 2^{2k} entries to look up the distance for k symbols at a time. In our tests, $k = 7$ resulted in the best running time.

1. Code the symbols of the sequences as follows: $A = 00$, $C = 01$, $G = 10$, $T = 11$.
2. For each pair of compacted strings c_i and c_j
   (a) Compute the xor $X_{ij} = c_i \bigoplus c_j$.
   (b) Read $2k$ bits of X_{ij} at a time and use the table to look up the distance for the associated k symbols.

7 Experiments

In this paper it has been shown that both NJ and FNJ have optimal reconstruction radius. However, there are many distance matrices that are not nearly additive and for which both algorithms reconstruct the closest tree. And for yet more matrices the algorithms fail to reconstruct the tree, but they do not fail by much. Therefore, it is of major interest to know how well the two algorithms perform in practice.

Several studies have been made on the accuracy of different reconstruction algorithms, the most notable work being that by Nakhleh et al. [13]. In that paper, four different methods are examined: NJ, DCM-NJ+MP, Weighbor, and Greedy Parsimony. And it is noted that the NJ algorithm, because of its speed, is the method of choice when the input data are accurate, i.e., when the sequence length is large and the corrected distances are close to additive. In this section, we replicate some of the experiments and show that although the NJ algorithm perform slightly better than the FNJ algorithm, when the input data are accurate the performance is in fact close to the same.

The test data were produced in the same way as in [13]. First, the model trees were generated through a random birth-death process using the r8s[16] software package. These trees where then made non-ultrametric, i.e., root to leaf paths where made to vary in length, by multiplying the edge lengths with a

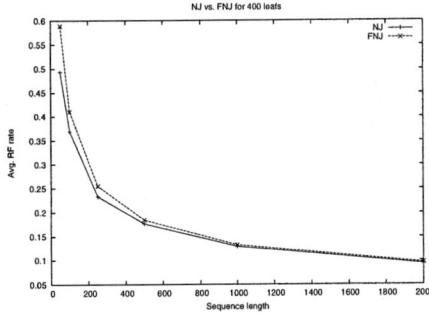

Algo.	Time (min)	Avg. RF (%)
FNJ	4	10.5
NJ	52	10.1
GME	26	14.1

Fig. 2. Left: Trees with 400 taxa. Right: Comparing the running time for 10 trees of 4000 taxa

random number in different intervals[1]. Subsequently, sequence data was generated according to the JC model using the Seq-Gen[14] program. The JC correction formula was then applied to get the distances, and for saturated data a fix factor of 1 was used.

To measure the accuracy we used the normalized Robinson-Foulds (RF) distance between the model tree and the tree given by the method. To get statistically robust results we performed 20 runs on each test size, and computed the average RF rate and standard deviation. In Figure 2, to the left, we plot the average RF rate as a function of the sequence length for trees with 400 taxa. Notice that both methods converge to the true tree as the sequence length increases, and that for accurate data the methods perform almost the same. For these experiments the standard deviation varied between 1-4% except for sequences of length 50. Many more experiments have been performed and the same pattern emerges there too but due to space limitations these data have been omitted.

7.1 Comparison with GME

In Desper et al. [6], an $O(n^2)$ algorithm called GME is introduced that, although it does not have optimal reconstruction radius, in practice it has acceptable accuracy. However, as is clearly shown in Figure 2, for 10 trees of 4000 taxa each, FNJ outperforms both GME and NJ. When accuracy is concerned the best algorithm is NJ, tightly followed by FNJ. In addition to GME, Desper et al. present a clever nearest neighbor interchange (NNI) algorithm, that in many cases improves the accuracy of reconstruction algorithms. It is therefore reasonable to believe that FNJ in conjunction with NNI would be a very fast and accurate combination.

---

[1] Following [13] we used ultrametric deviation 4 and generated sequences with diameter factors 0.05, 0.10, 0.25, and 0.5. E.g. diameter factor 0.25 yields the interval $[1/16, 1]$

Acknowledgments

We would like to thank Luay Nakhleh and Tandy Warnow for discussions on experimental studies of the NJ algorithm and for helping us replicate their experiments. We are also grateful to Johan Håstad for valuable comments and ideas.

References

1. R. Agarwala, V. Bafna, M. Farach, M. Paterson, and M. Thorup. On the approximability of numerical taxonomy (fitting distances by tree metrics). *SICOMP*, 28(3):1073–1085, 1999.
2. K. Atteson. The performance of neighbor-joining methods of phylogenetic reconstruction. *Algorithmica*, 25, 1999.
3. G.S. Brodal, R. Fagerberg, T. Mailund, C.N. Pedersen, and D. Phillips. Speeding up neighbour-joining tree construction. Technical Report ALCOMFT-TR-03-102, 2003.
4. D. Coppersmith and S. Winograd. Matrix multiplication via arithmetic progressions. In *STOC '87*, pages 1–6, 1987.
5. M. Csűrös. Fast recovery of evolutionary trees with thousands of nodes. In *RECOMB-01*, pages 104–113, 2001.
6. R. Desper and O. Gascuel. Fast and accurate phylogeny reconstruction algorithms based on the minimum-evolution principle. *Journal of Computational Biology*, 19(5):687–705, 2002.
7. P.L. Erdös, M.A. Steel, L.A. Szekely, and T.J. Warnow. A few logs suffice to build (almost) all trees (I). *RSA: Random Structures & Algorithms*, 14:153–184, 1999.
8. O. Gascuel. Concerning the NJ algorithm and its unweighted version, UNJ. *American Mathematical Society*, pages 149–170, 1997.
9. D.H. Huson, S. Nettles, and T. Warnow. Disk-covering, a fast-converging method for phylogenetic tree reconstruction. *Journal of Computational Biology*, 6(3/4):369–386, 1999.
10. T.H. Jukes and C.R. Cantor. Evolution of protein molecules. *Mammalian Protein Metabolism*, pages 21–132, 1969.
11. M. Kimura. A simple model for estimating evolutionary rates of base substitutions through comparative studies of nucleotide sequences. *Journal of Molecular Evolution*, 16:111–120, 1980.
12. J. Lagergren. Combining polynomial running time and fast convergence for the disk-covering method. *JCSS: Journal of Computer and System Sciences*, 65, 2002.
13. L. Nakhleh, B.M.E. Moret, K. St John, J. Sun, U. Roshan, and T. Warnow. The accuracy of fast phylogenetic methods for large datasets. *PSB-02*, pages 211–222, 2002.
14. A. Rambaut and N.C. Grassly. Seq-gen: An application for the monte carlo simulation of dna sequence evolution along phylogenetic trees. *Comp. Appl. Biosci.*, 13:235–238, 1997.
15. N. Saitou and M. Nei. The neighbor-joining method: a new method for reconstructing phylogenetic trees. *Mol. Biol. Evol.*, 4:406–425, 1987.
16. M. Sanderson. r8s software package. http://ginger.ucdavis.edu/r8s/.
17. CN. Pedersen T. Mailund. Quickjoin–fast neighbour-joining tree reconstruction. *Bioinformatics*, 2004.

Randomized Fast Design of Short DNA Words*

Ming-Yang Kao, Manan Sanghi, and Robert Schweller

Department of Computer Science,
Northwestern University,
Evanston, IL 60201, USA
{kao, manan, schwellerr}@cs.northwestern.edu

Abstract. We consider the problem of efficiently designing sets (codes) of equal-length DNA strings (words) that satisfy certain combinatorial constraints. This problem has numerous motivations including DNA computing and DNA self-assembly. Previous work has extended results from coding theory to obtain bounds on code size for new biologically motivated constraints and has applied heuristic local search and genetic algorithm techniques for code design. This paper proposes a natural optimization formulation of the DNA code design problem in which the goal is to design n strings that satisfy a given set of constraints while minimizing the length of the strings. For multiple sets of constraints, we provide high-probability algorithms that run in time polynomial in n and any given constraint parameters, and output strings of length within a constant factor of the optimal. To the best of our knowledge, this work is the first to consider this type of optimization problem in the context of DNA code design.

1 Introduction

In this paper we study the problem of efficiently designing sets (codes) of DNA strings (words) of near optimal length that fulfill certain combinatorial constraints. Many applications have emerged in recent years that depend on the scalable design of such words. One such problem is in DNA computing where inputs to computational problems are encoded into DNA strands for the purpose of computing via DNA complementary binding [1]. Another application involves implementing Wang tile self-assembly systems by encoding glues of Wang tiles into strands of DNA [17]. DNA words can also be used to store information at the molecular level [4], act as molecular bar codes for identifying molecules in complex libraries [5,4,13], or implement DNA arrays [3].

For a set of DNA words to be effective for the above applications, they must fulfill a number of combinatorial constraints. Of particular importance is the need for specific hybridization between a given word and its unique Watson-Crick complement. That is, we need to make sure that hybridization does not

---

* Supported in part by NSF Grant EIA-0112934.

Table 1. This table summarizes our results regarding the efficient design of DNA words. Here n is the number of words; k denotes the maximum of the constraint parameters for constraints 1 through 6 (see Section 2); and $\ell = \Theta(k + \log n)$ denotes the optimal achievable word length for the listed word design problems (see Theorems 1, 3, 4 and 6)

	Word Length and Time Complexity for DNA Word Design			
	\multicolumn{2}{c}{Word Length}	\multicolumn{2}{c}{Time Complexity}
	Lower Bound	Upper Bound	Lower Bound	Upper Bound
$DWD_{1,2,3,4,5,6,7}$	$\Theta(\ell)$ (Thm. 1)	(Thm. 3)	$\Omega(n\ell)$ (Thm. 1)	$\Theta(n\ell)$ (Thm. 3)
$DWD_{1,2,3,7,8}$	$\Theta(\ell)$ (Thm. 1)	(Thm. 4)	(Thm. 1)	$\Theta(n\ell)$ (Thm. 4)
$DWD_{1,2,3,4,5,6,9}$	$\Theta(\ell)$ (Thm. 1)	(Thm. 6)	$\Omega(n\ell)$ (Thm. 1)	$O\left(\min\left\{\begin{array}{l}\ell^{1.5}\log^{0.5}\ell + n\ell,\\ n\ell\log\ell\end{array}\right\}\right)$ (Thm. 6)

occur among a word and the complement of a different word in the set, or even of any word with any other word in the set. For this requirement Marathe et al. [12] have proposed the *basic Hamming constraint*, *reverse complement Hamming constraint*, and *self-complementary constraint*. We further consider the more restricting *shifting Hamming constraint* which requires a large Hamming distance between all alignments of any pair of words [6].

We also consider three constraints not related to Hamming distance. The *consecutive base constraint* limits the length of any run of identical bases in any given word. Long runs of identical bases are considered to cause hybridization errors [14, 6]. The *GC content constraint* requires that a large percentage of the bases in any given word are either G or C. This constraint is meant to give each string similar thermodynamic properties [14, 16, 15]. The *free energy constraint* requires that the difference in free energy of any two words is bounded by a small constant. This helps ensure that each word in the set has a similar melting temperature [6, 12].

In addition to the above constraints, it is desirable for the length ℓ of each word to be as small as possible. The motivation for minimizing ℓ is evident from the fact that it is more difficult to synthesize longer strands. Similarly, longer DNA strands require more DNA to be used for the respective application.

There has been much previous work in the design of DNA words [6, 12, 4, 9, 10, 11, 13, 16, 15]. In particular, Marathe et al. [12] have extended results from coding theory to obtain bounds on code size for various biologically motivated constraints. However, most work in this area has been based on heuristics, genetic algorithms, and stochastic local searches that do not provide provably good words provably fast.

In this work we provide algorithms with analytical guarantees for combinatorial structures and time complexity. In particular, we formulate an optimization problem that takes as input a desired number of strings n and produces n length-ℓ strings that satisfy a specified set of constraints, while at the same time minimizing the length ℓ. We restrict our solution to this problem in two ways. First, we require that our algorithms run in time only polynomial in the number of strings n as well as any given constraint parameters. Second, we require that our algorithms produce sets of words that achieve word length ℓ that is within a constant multiple of the optimal achievable word length, while at the same time fulfilling the respective constraints with high probability. For various subsets of the constraints we propose, we provide algorithms that do this. We thus provide fast algorithms for the creation of sets of short words.

Paper Layout: In Section 2, we describe the different biologically motivated combinatorial constraints we use. In Section 3 we solve the design problem with subsets of constraints including the Hamming constraints, the consecutive bases constraint, and the GC content constraint. In Section 4 we extend our algorithms to deal with the free energy constraint.

2   Preliminaries

2.1   Notations

Let $X = x_1 x_2 \ldots x_\ell$ be a word where x_i belongs to some alphabet Π. In this paper we deal with two alphabets, namely, the binary alphabet $\Pi_B = \{0, 1\}$ and the DNA alphabet $\Pi_D = \{A,C,G,T\}$. The elements of an alphabet are called *characters*. We will use capital letters for words and small letters for characters. Our goal is to design DNA words but some of our algorithms generate binary words in intermediate steps.

The *reverse* of X, denoted by X^R, is the word $x_\ell x_{\ell-1} \ldots x_1$. The *complement* of a character x is denoted by x^c. The complements for the binary alphabet are given by $0^c = 1$, $1^c = 0$, and for the DNA alphabet we have $A^c = T$, $C^c = G$, $G^c = C$, $T^c = A$.

The *complement* of a word is obtained by taking the complement of each of the characters in the word, i.e., $X^C = x_1^c x_2^c \ldots x_\ell^c$. The *reverse complement* of X is the complement of X^R, $X^{RC} = x_\ell^c x_{\ell-1}^c \ldots x_1^c$. The *Hamming distance* $H(X,Y)$ between two words X and Y is the number of positions where X differs from Y.

We are interested in designing a set $\mathcal{W}$ of n words over Π_D each of length ℓ which satisfy the constraints defined in Section 2.2 below.

2.2   Constraints

The constraints we consider can be classified into two categories: *non-interaction* constraints and *stability* constraints. Non-interaction constraints ensure that unwanted hybridizations between two DNA strands are avoided, and stability constraints ensure that the DNA strands are stable in a solution. The first six constraints below are non-interaction constraints while the remaining three are stability constraints.

$C_1(k_1)$: **Basic Hamming Constraint** (k_1) = for any words $Y, X \in \mathcal{W}$, $H(Y, X) \geq k_1$.

This constraint limits non-specific hybridizations between the Watson-Crick complement of some word Y with a distinct word X.

$C_2(k_2)$: **Reverse Complementary Constraint** (k_2) = for any words $Y, X \in \mathcal{W}$, $H(Y, X^{RC}) \geq k_2$.

This constraint is intended to limit hybridization between a word and the reverse of another word.

$C_3(k_3)$: **Self Complementary Constraint** (k_3) = for any word Y, $H(Y, Y^{RC}) \geq k_3$.

This constraint prevents a word from hybridizing with itself.

$C_4(k_4)$: **Shifting Hamming Constraint** (k_4) = for any two words $Y, X \in \mathcal{W}$,

$$H(Y[1..i], X[(\ell - i + 1)..\ell]) \geq k_4 - (\ell - i) \text{ for all } i.$$

This is a stronger version of the Basic Hamming Constraint.

$C_5(k_5)$: **Shifting Reverse Complementary Constraint** (k_5) = for any two words $Y, X \in \mathcal{W}$,

$$H(Y[1..i], X[1..i]^{RC}) \geq k_5 - (\ell - i) \text{ for all } i; \text{ and}$$

$$H(Y[(\ell - i + 1)..\ell], X[(\ell - i + 1)..\ell]^{RC}) \geq k_5 - (\ell - i) \text{ for all } i.$$

This is a stronger version of the Reverse Complementary Constraint.

$C_6(k_6)$: **Shifting Self Complementary Constraint** (k_6) = for any word $Y \in \mathcal{W}$,

$$H(Y[1..i], Y[1..i]^{RC}) \geq k_6 - (\ell - i) \text{ for all } i; \text{ and}$$

$$H(Y[(\ell - i + 1)..\ell], Y[(\ell - i + 1)..\ell]^{RC}) \geq k_6 - (\ell - i) \text{ for all } i.$$

This is a stronger version of the Self Complementary Constraint.

$C_7(\gamma)$: **GC Content Constraint** (γ) = γ percentage of bases in any word $Y \in \mathcal{W}$ are either G or C.

The GC content affects the thermodynamic properties of a word [14, 16, 15]. Therefore, having the same ratio of GC content for all the words will assure similar thermodynamic characteristics.

$C_8(d)$: **Consecutive Base Constraint** (d) = no word has more than d consecutive bases for $d \geq 2$.

In some applications, consecutive occurrences (also known as runs) of the same base increase the number of annealing errors.

$C_9(\sigma)$: **Free Energy Constraint** (σ) = for any two words $Y, X \in \mathcal{W}$, $FE(Y) - FE(X) \leq \sigma$ where $FE(W)$ denotes the free energy of a word defined in Section 4.

This constraint ensures that all the words in the set have similar melting temperatures which allows hybridization of multiple DNA strands to proceed simultaneously [13].

For each of the given constraints above we assign a shorthand boolean function $C_i(t)$ to denote whether or not a given set of words $\mathcal{W}$ fulfills constraint C_i with respect to parameter t. For a given integer n, the goal of DNA word design is to efficiently create a set of n length-ℓ words such that a given subset of the above constraints are satisfied, while trying to minimize ℓ. That is, for a given subset of constraints $\{C_{\pi_1}, C_{\pi_2}, \ldots, C_{\pi_r}\} \subseteq \{C_1, C_2, \ldots, C_9\}$, the corresponding DNA word design (DWD) optimization problem is as follows.

Problem 1 $(\text{DWD}_{\pi_1, \pi_2, \ldots, \pi_r})$.
INPUT: Integers $n, t_1, t_2, \ldots, t_r$.
OUTPUT: A set $\mathcal{W}$ of n DNA strings each of the minimum length such that for all $1 \leq i \leq r$ the constraint $C_{\pi_i}(t_i)$ is satisfied over set $\mathcal{W}$.

For this problem we have the following trivial lower bounds for time complexity and the word size ℓ when any one of the first six constraints is applied.

Theorem 1. *Consider a set $\mathcal{W}$ of n DNA words each of length ℓ.*

1. *If $\mathcal{W}$ fulfills any one of the constraints $C_1(k), C_2(k), C_3(k), C_4(k), C_5(k)$, and $C_6(k)$, then $\ell = \Omega(k + \log n)$.*
2. *The time complexity of producing a set $\mathcal{W}$ that fulfills any one of the constraints $C_1(k), C_2(k), C_3(k), C_4(k), C_5(k)$, and $C_6(k)$ is $\Omega(nk + n \log n)$.*

The goal of DNA word design is to simultaneously satisfy as many of the above nine constraints as possible while achieving words within a constant factor of the optimal length ℓ for the given set of constraints. In Section 3 we show how to accomplish this goal for various subsets of the constraints.

3 Algorithms for DNA Word Design

In this section we develop randomized algorithms to generate sets of length-ℓ DNA words that satisfy certain sets of constraints while keeping ℓ within a constant of the optimal value. In particular, we first show how simply generating a set of n words at a specific length $\ell = O(k + \log n)$ uniformly at random is sufficient to fulfill constraints 1, 2, 3, 4, 5, and 6 simultaneously with high probability. We then propose three extensions to this algorithm to fulfill different subsets of constraints within a constant factor of the optimal word length. The first extension yields an algorithm for fulfilling the GC content constraint while the second yields one for the consecutive base and GC content constraints at the cost of the shifting constraints. Finally, we extend the basic randomized algorithm to fulfill the free energy constraint. The first is thus an algorithm for simultaneously fulfilling constraints 1, 2, 3, 4, 5, 6, and 7, the second simultaneously fulfills constraints 1, 2, 3, 7, and 8, and the last one fulfills constraints 1, 2, 3, 4, 5, 6 and 9.

Algorithm. $FastDWD_{1,2,3,4,5,6}(n, k_1, k_2, k_3, k_4, k_5, k_6)$

1. Let $k = \max\{k_1, k_2, k_3, k_4, k_5, k_6\}$.
2. Generate a set $\mathcal{W}$ of n words over Π_D of length $\ell = 9 \cdot \max\{k, \lceil \log_4 n \rceil\}$ uniformly at random.
3. Output $\mathcal{W}$.

Fig. 1. A randomized algorithm for generating n DNA strings satisfying constraints $C_1(k_1)$, $C_2(k_2)$, $C_3(k_3)$, $C_4(k_4)$, $C_5(k_5)$, and $C_6(k_6)$

3.1 A Simple Randomized Algorithm

Problem 2 ($DWD_{1,2,3,4,5,6}$).
INPUT: Integers n, k_1, k_2, k_3, k_4, k_5, k_6.
OUTPUT: A set $\mathcal{W}$ of n DNA strings each of the minimum length such that the constraints $C_1(k_1)$, $C_2(k_2)$, $C_3(k_3)$, $C_4(k_4)$, $C_5(k_5)$, $C_6(k_6)$ hold.

The next theorem shows that Algorithm $FastDWD_{1,2,3,4,5,6}$ $(n, k_1, k_2, k_3, k_4, k_5, k_6)$ in Figure 1 yields a polynomial-time solution to the $DWD_{1,2,3,4,5,6}$ problem with high probability. We omit the proof in the interest of space.

Theorem 2. *Algorithm* $FastDWD_{1,2,3,4,5,6}$ *produces a set* $\mathcal{W}$ *of n DNA words of optimal length* $\Theta(k + \log n)$ *in optimal time* $\Theta(n \cdot k + n \cdot \log n)$ *satisfying constraints* $C_1(k_1)$, $C_2(k_2)$, $C_3(k_3)$, $C_4(k_4)$, $C_5(k_5)$ *and* $C_6(k_6)$ *with probability of failure* $o(1/(n + 4^k))$, *where* $k = \max\{k_1, k_2, k_3, k_4, k_5, k_6\}$.

Proof (Sketch). The probability that two random words violate any of the constraints $C_1(k_1)$, $C_2(k_2)$, $C_4(k_4)$, and $C_5(k_5)$, can be bounded using Chernoff type bounds. Similarly, we can bound the probability of a random word violating any of the constraints $C_3(k_3)$ and $C_6(k_6)$.

We can then apply the Boole-Bonferroni Inequalties to yield a bound on the probability that any pair of words in a set of n random words violates constraints $C_1(k_1)$, $C_2(k_2)$, $C_4(k_4)$, or $C_5(k_5)$; or that any single word violates constraints $C_3(k_3)$ or $C_6(k_6)$. □

3.2 Incorporating the GC Content Constraint into $FastDWD_{1,2,3,4,5,6}$

Now we show how to modify Algorithm $FastDWD_{1,2,3,4,5,6}$ so that it produces a set of words that also satisfies the GC content constraint. That is, we will show how to solve the following problem.

Problem 3 ($DWD_{1,2,3,4,5,6,7}$).
INPUT: Integers n, k_1, k_2, k_3, k_4, k_5, k_6, γ.
OUTPUT: A set $\mathcal{W}$ of n DNA strings each of the minimum length such that the constraints $C_1(k_1)$, $C_2(k_2)$, $C_3(k_3)$, $C_4(k_4)$, $C_5(k_5)$, $C_6(k_6)$, $C_7(\gamma)$ hold.

We modify Algorithm $FastDWD_{1,2,3,4,5,6}$ to get Algorithm $FastDWD_{1,2,3,4,5,6,7}$ shown in Figure 2. The next theorem shows that $FastDWD_{1,2,3,4,5,6,7}$ yields a

Algorithm. $FastDWD_{1,2,3,4,5,6,7}(n, k_1, k_2, k_3, k_4, k_5, k_6, \gamma)$

1. Let $k = \max\{k_1, k_2, k_3, k_4, k_5, k_6\}$.
2. Generate a set $\mathcal{W}$ of n words over the binary alphabet Π_B of length $\ell = 10 \cdot \max\{k, \lceil \log_2 n \rceil\}$ uniformly at random.
3. For each word $W \in \mathcal{W}$, for any $\lceil \gamma \cdot \ell \rceil$ characters in W, replace 0 by G and 1 by C. For the remaining characters replace 0 by A and 1 by T to get W'. Let $\mathcal{W}'$ be the set of all words W'.
4. Output $\mathcal{W}'$.

Fig. 2. A randomized algorithm for generating n DNA strings satisfying constraints $C_1(k_1)$, $C_2(k_2)$, $C_3(k_3)$, $C_4(k_4)$, $C_5(k_5)$, $C_6(k_6)$, and $C_7(\gamma)$

polynomial-time solution to $DWD_{1,2,3,4,5,6,7}$ with high probability. We omit the proof in the interest of space.

Theorem 3. *Algorithm* $\text{FastDWD}_{1,2,3,4,5,6,7}$ *produces a set $\mathcal{W}$ of n DNA words of optimal length $\Theta(k + \log n)$ in optimal time $\Theta(n \cdot k + n \cdot \log n)$ satisfying constraints $C_1(k_1)$, $C_2(k_2)$, $C_3(k_3)$, $C_4(k_4)$, $C_5(k_5)$, $C_6(k_6)$, and $C_7(\gamma)$ with probability of failure $o(1/(n + 2^k))$, where $k = \max\{k_1, k_2, k_3, k_4, k_5, k_6\}$.*

3.3 Incorporating the Consecutive Bases Constraint into FastDWD$_{1,2,3,4,5,6,7}$

Now we modify Algorithm FastDWD$_{1,2,3,4,5,6,7}$ so that it produces a set that satisfies both the GC content constraint and the consecutive base constraint at the cost of the shifting constraints. That is, we will show how to solve the following problem.

Problem 4 (DWD$_{1,2,3,7,8}$).
INPUT: Integers n, k_1, k_2, k_3, γ, d.
OUTPUT: A set $\mathcal{W}$ of n DNA strings each of the minimum length such that the constraints $C_1(k_1)$, $C_2(k_2)$, $C_3(k_3)$, $C_7(\gamma)$, $C_8(d)$ hold.

We use Algorithm BreakRuns shown in Figure 3 to break long runs for a binary word so that it satisfies the consecutive bases constraint with parameter d. Intuitively what this algorithm does is for a given word X, it outputs X' by inserting characters at intervals of $d - 1$ from the left and the right in a manner such that there are no consecutive runs of length greater than d. We need to add characters from both ends to ensure that $H(X, Y^{RC}) \leq H(X', Y'^{RC})$ where X' and Y' are the respective outputs for X and Y from BreakRuns.

We modify Algorithm FastDWD$_{1,2,3,4,5,6,7}$ to get Algorithm FastDWD$_{1,2,3,7,8}$ shown in Figure 3. The next theorem shows that FastDWD$_{1,2,3,7,8}$ yields a polynomial-time solution to DWD$_{1,2,3,7,8}$ with high probability. We omit the proof in the interest of space.

Theorem 4. *Algorithm* $\text{FastDWD}_{1,2,3,7,8}$ *produces a set $\mathcal{W}$ of n DNA words of optimal length $\Theta(k + \log n)$ in optimal time $\Theta(n \cdot k + n \cdot \log n)$ satisfying constraints*

Algorithm. $BreakRuns(X, d)$

1. Let $X = x_1 x_2 \ldots x_\ell$. For $0 < i \le \lceil \frac{\ell}{2(d-1)} \rceil - 1$, let $x'_{\ell_i} = x^c_{i(d-1)}$ and $x'_{r_i} = x^c_{\ell - i(d-1)}$. Let $x'_{\text{mid}} = x^c_{\lfloor \ell/2 \rfloor}$.
2. Output $X' = x_1 \ldots x_{d-1} x'_{\ell_1} x_d \ldots x_{\lfloor \ell/2 \rfloor} x'_{\text{mid}} x_{\lfloor \ell/2 \rfloor + 1} \ldots x_{\ell-(d-1)-1} x'_{r_1} x_{\ell-(d-1)} \ldots x_\ell$.

Algorithm. $\text{FastDWD}_{1,2,3,7,8}(n, k_1, k_2, k_3, \gamma, d)$

1. Let $k = \max\{k_1, k_2, k_3\}$.
2. Generate a set $\mathcal{W}$ of n words over the binary alphabet Π_B of length $\ell = 10 \cdot \max\{k, \lceil \log_2 n \rceil\}$ uniformly at random.
3. For each word $W \in \mathcal{W}$, let $W' = \text{BreakRuns}(W, d)$. Let $\mathcal{W}'$ be the set of all words W'.
4. For each word $W' \in \mathcal{W}'$, for any $\lceil \gamma \cdot \ell \rceil$ characters in W', replace 0 by G and 1 by C. For the remaining characters replace 0 by A and 1 by T to get W''. Let $\mathcal{W}''$ be the set of all words W''.
5. Output $\mathcal{W}''$.

Fig. 3. Algorithms for generating n DNA strings satisfying constraints $C_1(k_1)$, $C_2(k_2)$, $C_3(k_3)$, $C_7(\gamma)$, and $C_8(d)$

$C_1(k_1)$, $C_2(k_2)$, $C_3(k_3)$, $C_7(\gamma)$, and $C_8(d)$ with probability of failure $o(1/(n+2^k))$, where $k = \max\{k_1, k_2, k_3\}$.

4 Incorporating the Free Energy Constraint into $\text{FastDWD}_{1,2,3,4,5,6}$

Now we give an alternate modification of Algorithm $\text{FastDWD}_{1,2,3,4,5,6}$ such that the free energy constraint is satisfied. The free-energy $\text{FE}(X)$ of a DNA word $X = x_1 x_2 \ldots x_\ell$ is approximated by $\text{FE}(X) = \text{correction factor} + \sum_{i=1}^{\ell-1} \Gamma_{x_i, x_{i+1}}$, where $\Gamma_{x,y}$ is the pairwise free energy between base x and base y [7]. For simplicity, we denote the free energy as simply the sum $\sum_{i=1}^{\ell-1} \Gamma_{x_i, x_{i+1}}$ with respect to a given pairwise energy function Γ. Let $\Gamma_{\max}$ and $\Gamma_{\min}$ be the maximum and the minimum entries in Γ respectively. Let $D = \Gamma_{\max} - \Gamma_{\min}$.

We now show how to satisfy the free energy constraint $C_9(\sigma)$ for a constant $\sigma = 4D + \Gamma_{\max}$, while simultaneously satisfying constraints $1, 2, 3, 4, 5$, and 6. That is, we show how to solve the following problem.

Problem 5 ($\text{DWD}_{1,2,3,4,5,6,9}$).
INPUT: Integers n, k_1, k_2, k_3, k_4, k_5, k_6.
OUTPUT: A set $\mathcal{W}$ of n DNA strings each of the minimum length such that the constraints $C_1(k_1)$, $C_2(k_2)$, $C_3(k_3)$, $C_4(k_4)$, $C_5(k_5)$, $C_6(k_6)$, $C_9(4D + \Gamma_{\max})$ hold.

We modify Algorithm $\text{FastDWD}_{1,2,3,4,5,6}$ to get Algorithm $\text{FastDWD}_{1,2,3,4,5,6,9}$ shown in Figure 4 for solving $DWD_{1,2,3,4,5,6,9}$. The following lemmas identify the properties of symbols $\Delta, \mathcal{W}, \mathcal{W}_{\max}, \mathcal{W}_{\min}, S_i, \hat{S}_j, \alpha, \beta,$ and W'_i defined in Figure 4 and are used for proving the correctness of Algorithm $\text{FastDWD}_{1,2,3,4,5,6,9}$.

Algorithm. $FastDWD_{1,2,3,4,5,6,9}(n, k_1, k_2, k_3, k_4, k_5, k_6)$

Let $\hat{S}^1, \hat{S}^2, \ldots, \hat{S}^{4^m}$ be all possible sequences of length $m = 2\ell$ where ℓ is as defined in Step 2 below such that $\text{FE}(\hat{S}^1) \leq \text{FE}(\hat{S}^2) \leq \cdots \leq \text{FE}(\hat{S}^{4^m})$. For two strings X and Y of respective lengths ℓ_X and ℓ_Y where ℓ_Y is even, let $X \otimes Y$ be the string $Y[1..(\ell_Y/2)] \, X[1..\ell_X] \, Y[(\ell_Y/2+1)..\ell_Y]$. Let $\Delta = \max_i \{\text{FE}(\hat{S}^{i+1}) - \text{FE}(\hat{S}^i)\}$.

1. Let $k = \max\{k_1, k_2, k_3, k_4, k_5, k_6\}$.
2. Generate a set $\mathcal{W}$ of n DNA words of length $\ell = 9 \cdot \max\{k, \lceil \log_4 n \rceil\}$ uniformly at random.
3. Let $\mathcal{W}_{\max} = \max_{X \in \mathcal{W}}\{\text{FE}(X)\}$ and $\mathcal{W}_{\min} = \min_{X \in \mathcal{W}}\{\text{FE}(X)\}$.
   if $\mathcal{W}_{\max} - \mathcal{W}_{\min} \leq 3D$, then output $\mathcal{W}$.
   else
4. Let $\alpha = \mathcal{W}_{\max} + \hat{S}^1$ and $\beta = \alpha + \Delta$. For each $S_i \in \mathcal{W}$, find $\hat{S}_j$ such that $\alpha \leq \text{FE}(S_i) + \text{FE}(\hat{S}_j) \leq \beta$. Let $W'_i = S_i \otimes \hat{S}_j$.
5. output $\mathcal{W}' = \{W'_1, \ldots, W'_n\}$.

Fig. 4. A randomized algorithm for generating n DNA strings satisfying constraints $C_1(k_1), C_2(k_2), C_3(k_3), C_4(k_4), C_5(k_5), C_6(k_6)$, and $C_9(4D + \Gamma_{\max})$

Lemma 1. $\Delta < 2D$.

Lemma 2. *If* $\mathcal{W}_{\max} - \mathcal{W}_{\min} > 3D$, *then* $\mathcal{W}_{\max} - \mathcal{W}_{\min} + 2D \leq \text{FE}(\hat{S}^{4^m}) - \text{FE}(\hat{S}^1)$.

Lemma 3. *For each* $S_i \in \mathcal{W}$, *there exists* $\hat{S}_j$ *such that* $\alpha \leq \text{FE}(S_i) + \text{FE}(\hat{S}_j) \leq \beta$.

Lemma 4. *For all* i, $\alpha - D \leq \text{FE}(W'_i) \leq \beta + D + \Gamma_{\max}$.

Section 4.1 discusses the details for Step 4 of the algorithm. Finally, Section 4.2 establishes its correctness and time complexity.

4.1 Computing Strings with Bounded Energies

In Step 4 of Algorithm $FastDWD_{1,2,3,4,5,6,9}$ we need to produce a set of n DNA strings $\hat{S}_1, \hat{S}_2, \ldots \hat{S}_n$, each of a given length $L = m$, such that $A_i \leq \text{FE}(\hat{S}_i) \leq B_i$ for some A_i, B_i such that $B_i - A_i \leq \Delta$. That is, we need to solve the following problem.

Problem 6 (Bounded-Energy Strand Generation).
INPUT:

1. Integers A_i and B_i for $i = 1$ to n such that
   (a) $A_i \geq \mathcal{W}_{\min}$;
   (b) $B_i \leq \mathcal{W}_{\max}$;
   (c) $B_i - A_i \leq \Delta$.
2. Length L.

OUTPUT: Strings $\hat{S}_1, \hat{S}_2, \ldots \hat{S}_n$ each of length L and respective energy E_i such that $A_i \leq E_i \leq B_i$.

Our solution to this problem involves transforming the blunt of the computational task into the problem of polynomial multiplication. Consider the following polynomial.

Definition 1. *For any integer $\ell \geq 1$, let $f_{\ell,a,b}(x)$ be the polynomial $\sum_{z=0}^{\ell \cdot m} \zeta_z x^z$ where coefficient ζ_z is the number of length-ℓ strings whose first character is a, last character is b, and free energy is z.*

For $f_\ell(x) = \sum_{\forall a,b \in \Pi} f_{\ell,a,b}(x)$ the coefficient of x^i denotes the number of strings of length ℓ and free energy i. As a first step towards our solution, we use a subroutine BUILD(L) which computes Φ, the polynomials $f_{L,a,b}(x), f_{\lfloor L/2 \rfloor, a, b}(x), \ldots, f_{1,a,b}(x)$, for all $a, b \in \Pi$ in $O(L \log L)$ time. The efficient computation of these polynomials relies on the following recursive property.

Lemma 5. *For any integers $\ell_1, \ell_2 \geq 1$,*

$$f_{\ell_1+\ell_2,a,b}(x) = \sum_{d_1,d_2 \in \Pi} f_{\ell_1,a,d_1}(x) \cdot f_{\ell_2,d_2,b}(x) \cdot x^{\Gamma_{d_1,d_2}}.$$

The problem of determining the number of strings of length L and free energy E is considered in [12] and a dynamic programming based $O(L^2)$-time algorithm is provided. However, exploiting the recursive property of Lemma 5 and Fast Fourier Transforms [8] for polynomial multiplication the subroutine BUILD solves this problem in faster $O(L \log L)$ time and may be of independent interest.

Our algorithm for Problem 6 has two phases, the *build* phase and the *extract* phase. The build phase constructs a data structure that permits the extract phase to be executed quickly. In the extract phase, an extraction routine is run n times to output $\hat{S}_i$ for each $i \in [1, n]$. Since the extraction routine is executed n times and the build routine only once, the phase that constitutes the bottleneck for our algorithm for Problem 6 depends on the values of n and L. We thus provide two forks for the algorithm to take, one with a fast build routine and a modestly fast extract routine, and the other with a slower build routine but an optimally fast extract routine. In particular, if n is sufficiently larger than L, our algorithm for Problem 6 calls a routine SlowBuild(L) which improves the runtime of Extract. Otherwise, only a faster BUILD function is called in the first phase, leading to a slower Extract routine. The algorithm for Problem 6 is given in Figure 5.

Algorithm ConstructStrings makes use of three subroutines – Build, SlowBuild and Extract. The procedure Build(L) computes Φ, a data structure containing for all $a, b \in \Pi$ and a given L, the polynomials $f_{\ell,a,b}(x)$ for $\ell = L, \lfloor \frac{L}{2} \rfloor, \lfloor \frac{L}{4} \rfloor, \lfloor \frac{L}{8} \rfloor, \ldots, 1$. This

---

Algorithm. $ConstructStrings(\{A_i\}, \{B_i\}, L)$

1. Let $\Phi \leftarrow \text{Build}(L)$.
2. if $n \geq \sqrt{\frac{L}{\log L}}$, then $\Psi \leftarrow \text{SlowBuild}(L)$, else $\Psi \leftarrow \text{NULL}$.
3. For each $i = 1$ to n, find a nonzero coefficient ζ_{E_i} of X^{E_i} in some polynomial $f_L^{a,b}(x) \in \Phi$ such that $A_i \leq E_i \leq B_i$.
4. For $i = 1$ to n, set $\hat{S}_i = \text{Extract}(E_i, \Phi, \Psi)$.

Fig. 5. This algorithm solves the Bounded Energy Strand Generation Problem (Problem 6)

permits Extract(E, Φ, Ψ) to obtain a length L string of energy E in time $O(L \log L)$. A call to SlowBuild(L) of time complexity $O(L^{1.5} \log^{0.5} L)$ improves the complexity of Extract(E, Φ, Ψ) to $O(L)$ by computing Ψ, a data structure containing for every non-zero term x^i in $f_{\lfloor \frac{L}{2^a} \rfloor, a, b}$ a corresponding pair of non-zero terms x^j and $x^{i-j-\Gamma_{d_1, d_2}}$ in $f_{\lfloor \frac{L}{2^{a+1}} \rfloor, a, d_1}$ and $f_{\lfloor \frac{L}{2^{a+1}} \rfloor, d_2, b}$ respectively. This yields the following theorem.

Theorem 5. *Algorithm* ConstructStrings$(\{A_i\}, \{B_i\}, L)$ *solves Problem 6 in time* $O(\min\{nL \log L, L^{1.5} \log^{0.5} L + nL\})$.

4.2 Putting It All Together for DWD$_{1,2,3,4,5,6,9}$

Theorem 6. *Algorithm* FastDWD$_{1,2,3,4,5,6,9}$ *produces a set of n DNA words of optimal length $\Theta(k + \log n)$ in time $O(\min\{n\ell \log \ell, \ell^{1.5} \log^{0.5} \ell + n\ell\})$ satisfying the constraints $C_1(k_1)$, $C_2(k_2)$, $C_3(k_3)$, $C_4(k_4)$, $C_5(k_5)$, $C_6(k_6)$, and $C_9(4D + \Gamma_{\max})$ with probability of failure $o(1/(n + 4^k))$, where $k = \max\{k_1, k_2, k_3, k_4, k_5, k_6\}$.*

Proof. From Theorem 2 we know that $\mathcal{W}$ satisfies constraints $C_1(k_1)$, $C_2(k_2)$, $C_3(k_3)$, $C_4(k_4)$, $C_5(k_5)$, and $C_6(k_6)$ with probability of failure $o(1/(n+4^k))$. If $\mathcal{W}_{\max} - \mathcal{W}_{\min} \leq 3D$, then FastDWD$_{1,2,3,4,5,6,9}$ outputs $\mathcal{W}$ which satisfies $C_9(3D)$ and hence also satisfies $C_9(4D+\Gamma_{\max})$. Otherwise, it is easy to verify that since $\mathcal{W}$ satisfies these six constraints, so does $\mathcal{W}'$. From Lemma 3 we know that there always exists a string $\hat{S}_j$ as required in Step 4 of FastDWD$_{1,2,3,4,5,6,9}$. Further, Lemma 4 shows that $\mathcal{W}'$ satisfies $C_9(\Delta + 2D + \Gamma_{\max})$. Therefore, $\mathcal{W}'$ satisfies constraints $C_1(k_1)$, $C_2(k_2)$, $C_3(k_3)$, $C_4(k_4)$, $C_5(k_5)$, $C_6(k_6)$, and $C_9(4D + \Gamma_{\max})$ with the stated failure probability.

The length of any word $W' \in \mathcal{W}'$ is at most 3ℓ where $\ell = \Theta(k + \log n)$, which is optimal from Theorem 1.

Generating $\mathcal{W}$ takes $O(n \cdot k + n \cdot \log n)$ time. The bulk of the time complexity for the algorithm comes from Step 4, which is analyzed in Section 4.1 to get $O(\min\{nL \log L, L^{1.5} \log^{0.5} L + nL\})$ (see Theorem 5) where $L = O(\ell)$. □

5 Future Work

A number of problems related to this work remain open. It is still unknown how to generate words of optimal length that simultaneously satisfy the free energy constraint and the consecutive bases constraint. We also have not provided a method for combining the consecutive bases constraint with any of the shifting constraints.

Another open research area is the verification problem of testing whether or not a set of words satisfy a given set of constraints. This problem is important because our algorithms only provide a high-probability assurance of success. While verification can clearly be done in polynomial time for all of our constraints, the naive method of verification has a longer runtime than our algorithms for constructing the sets. Finding faster, non-trivial verification algorithms is an open problem.

A third direction for future work involves considering a generalized form of the basic Hamming constraint. There are applications in which it is desirable to design sets of words such that some distinct pairs bind with one another, while others do not [2, 14]. In this scenario, we can formulate a word design problem that takes as input a matrix of pairwise requirements for Hamming distances. Determining when such a problem is solvable and how to solve it optimally when it is are open problems.

References

[1] L. M. ADLEMAN, *Molecular Computation of Solutions to Combinatorial Problems*, Science, 266 (1994), pp. 1021–1024.

[2] G. AGGARWAL, M. H. GOLDWASSER, M.-Y. KAO, AND R. T. SCHWELLER, *Complexities for Generalized Models of Self-Assembly*, in Proceedings of the 15th Annual ACM-SIAM Symposium on Discrete Algorithms, 2004, pp. 880–889.

[3] A. BEN-DOR, R. KARP, B. SCHIWKOWSKI, AND Z. YAKHINI, *Universal DNA Tag Systems: A Combinatorial Design Scheme*, in Proceedings of the 4th Annual International Conference on Computational Molecular Biology, 2000, pp. 65–75.

[4] S. BRENNER, *Methods for Sorting Polynucleotides using Oligonucleotide Tags*. US Patent Number 5,604,097, February 1997.

[5] S. BRENNER AND R. A. LERNER, *Encoded Combinatorial Chemistry*, in Proceedings of Natianal Academy of Science, vol. 89, June 1992, pp. 5381–5383.

[6] A. BRENNERMAN AND A. E. CONDON, *Strand Design for Bio-Molecular Computation*, Theoretical Computer Science, 287 (2001), pp. 39–58.

[7] K. J. BRESLAUER, R. FRANK, H. BLOCKER, AND L. A. MARKY, *Predicting DNA Duplex Stability from the Base Sequence*, in Proceedings of the National Academy of Sciences, vol. 83, 1986, pp. 3746–3750.

[8] T. H. CORMEN, C. L. LEISERSON, R. L. RIVEST, AND C. STEIN, *Introduction to Algorithms*, MIT Press, Cambridge, MA, 2nd ed., 2001.

[9] R. DEATON, M. GARZON, R. MURPHY, D. FRANCESCHETTI, AND S. STEVENS, *Genetic Search of Reliable Encodings for DNA Based Computation*, in Proceedings of the 1st Annual Conference on Genetic Programming, 1996, pp. 9–15.

[10] A. G. FRUTOS, Q. LIU, A. J. THIEL, A. M. W. SANNER, A. E. CONDON, L. M. SMITH, AND R. M. CORN, *Demonstration of a Word Design Strategy for DNA Computing on Surfaces*, Nucleic Acids Research, 25 (1997), pp. 4748–4757.

[11] M. GARZON, R. DEATON, P. NEATHERY, D. FRANCESCHETTI, AND R. MURPHY, *A New Metric for DNA Computing*, in Proceedings of the 2nd Genetic Programming Conference, 1997, pp. 472–278.

[12] A. MARATHE, A. CONDON, AND R. M. CORN, *On Combinatorial DNA Word Design*, Journal of Computational Biology, 8 (2001), pp. 201–219.

[13] D. D. SHOEMAKER, D. A. LASHKARI, D. MORRIS, M. MITTMANN, AND R. W. DAVIS, *Quantitative Phenotypic Analysis of Yeast Deletion Mutants Using a Highly Parallel Molecular Bar-coding Strategy*, Nature, 16 (1996), pp. 450–456.

[14] S. A. TSAFTARIS, *DNA Computing from a Signal Processing Viewpoint*, IEEE Signal Processing Magazine, 21 (2004), pp. 100–106.

[15] D. C. TULPAN AND H. H. HOOS, *Hybrid Randomised Neighbourhoods Improve Stochastic Local Search for DNA Code Design*, in Lecture Notes in Computer Science 2671: Proceedings of the 16th Conference of the Canadian Society for Computational Studies of Intelligence, Y. Xiang and B. Chaib-draa, eds., Springer-Verlag, New York, NY, 2003, pp. 418–433.

[16] D. C. TULPAN, H. H. HOOS, AND A. CONDON, *Stochastic Local Search Algorithms for DNA Word Design*, in Lecture Notes in Computer Science 2568: Proceedings of the 8th International Workshop on DNA-Based Computers, M. Hagiya and A. Ohuchi, eds., Springer-Verlag, New York, NY, 2003, pp. 229–241.

[17] E. WINFREE, F. LIU, L. WENZLER, AND N. SEEMAN, *Design and Self-Assembly of Two-Dimensional DNA Crystals*, Nature, 394 (1998), pp. 539–544.

A Quantum Lower Bound for the Query Complexity of Simon's Problem

Pascal Koiran, Vincent Nesme, and Natacha Portier

Laboratoire de l'Informatique du Parallélisme,
Ecole Normale Supérieure de Lyon, 46, allée d'Italie,
69364 Lyon Cedex 07, France
{Pascal.Koiran, Vincent.Nesme, Natacha.Portier}@ens-lyon.fr

Abstract. Simon in his FOCS'94 paper was the first to show an exponential gap between classical and quantum computation. The problem he dealt with is now part of a well-studied class of problems, the hidden subgroup problems. We study Simon's problem from the point of view of quantum query complexity and give here a first nontrivial lower bound on the query complexity of a hidden subgroup problem, namely Simon's problem. More generally, we give a lower bound which is optimal up to a constant factor for any Abelian group.

1 Introduction

Given an Abelian group G and a subgroup $H \leq G$, a function $f : G \to X$ is said to be hiding H if f can be defined in a one-to-one way on G/H. More precisely, f hides H if and only if

$$\forall g, g' \in G \ (f(g) = f(g') \iff \exists h \in H \ g = g' + h)$$

Suppose G is a fixed group and f is computed by an oracle: a quantum black-box. We are interested here in algorithms that find the hidden subgroup H. A large amount of documentation about the hidden subgroup problem can be found in the book of Nielsen and Chuang [13][1]. Among all work already done about such algorithms one can cite Shor's famous factoring algorithm [17]: it uses a period-finding algorithm, which is a special case of a hidden subgroup problem. In recent years, attention has shifted to non-Abelian hidden subgroup problems but we will restrict our attention here to Abelian groups.

In general, two kinds of complexity measures for black-box problems can be distinguished: query complexity, i.e., the number of times the function f is evaluated using the black-box, and computational or time complexity, i.e., the number of elementary operations needed to solve the problem. Typically, a hidden subgroup algorithm is considered efficient if its complexity (in query or in

---

[1] History of the problem on page 246 and expression of many problems (order-finding, dicrete logarithm...) in terms of hidden subgroup problems on page 241.

time, depending on the interest) is polynomial in the logarithm of the cardinality of G. For example, Kuperberg's algorithm [10] for the (non-Abelian) dihedral hidden subgroup problem is subexponential (but superpolynomial) in both time and query complexities.

Our main result is that the query complexity of finding a subgroup hidden in G is of order $r(G)$ for any Abelian group G, where $r(G)$ denotes the *rank* of G, that is, the minimal cardinality of a generating set of G (for instance, $r((\mathbb{Z}/p\mathbb{Z})^n) = n$ if $p \geq 2$ is an arbitrary integer). The proof of this result is naturally divided into an upper bound and a lower bound proof. The upper bound is achieved through a tight analysis of the standard Fourier sampling algorithm. It is a folklore theorem in quantum computation that this algorithm solves the hidden subgroup problem in Abelian groups with polynomial query complexity (see for instance [7], [5], [2] or [8]), but strangely enough no precise analysis seems to be available in the litterature.

The greatest part of this paper is devoted to the lower bound proof. Here all the important ideas already appear in the analysis of Simon's problem, to which our preprint [9] is devoted. It is therefore fitting to recall the history of this problem, which is defined as follows. We are given a function f from $G = (\mathbb{Z}/2\mathbb{Z})^n$ to a known set X of size 2^n, and we are guaranteed that the function fulfills Simon's promises, that is either:

(1) f is one-to-one, or
(2) $\exists s \neq 0 \; \forall w, w' \; f(w) = f(w') \iff (w = w' \lor w = w' + s)$.

The problem is to decide whether (1) or (2) holds. Note that (1) is equivalent to "f hides the trivial subgroup $H = \{(0,\ldots,0)\}$" and (2) is equivalent to "f hides a subgroup $H = \{(0,\ldots,0), s\}$ of order 2". The original problem [18] was to compute s and the problem considered here is the associated decision problem. Of course, any lower bound on this problem will imply the same one on Simon's original problem. In his article, Simon shows that his problem can be solved by a quantum algorithm which makes $O(n)$ queries in the worst case and has a bounded probability of error. The time complexity of his algorithm is linear in the time required to solve an $n \times n$ system of linear equations over $(\mathbb{Z}/2\mathbb{Z})^n$. He also shows that any classical (probabilistic) algorithm for his problem must have exponential query complexity. As a side remark, note that Simon also gives a Las Vegas version of his algorithm with expected query complexity $O(n)$. Even better, Brassard and Høyer [4] have given an "exact polynomial time" quantum algorithm for Simon's problem (i.e., their algorithm has a polynomial worst case running time and zero probability of error).

The two main methods for proving query complexity lower bounds in quantum computing are the adversary method of Ambainis and the polynomial method (for an excellent review of these methods in French, read [16]). We shall use the polynomial method, which was introduced in quantum complexity theory in [3]. There are recent interesting applications of this method to the collision and element distinctness problem [1, 12]. All previous applications of the polynomial method ultimately rely on approximation theory lemmas of Paturi [15] or Nisan and Szegedy [14]. Besides the application to a new type of problems (namely, the hidden subgroup problems) we also contribute to the development of the method by applying it in a situation where these lemmas are not applicable. Instead, we use an apparently new (and elementary) approximation theory result: Lemma 3 from section 3.

The remainder of this paper is organized follows. After some preliminaries in section 2 we give in section 3 an $\Omega(n)$ lower bound for groups of the form $(\mathbb{Z}/p\mathbb{Z})^n$, where p is a prime number. The general case of arbitrary Abelian groups (lower and upper bound) is treated in section 4. The longer version of this paper will also contain lower bounds for other query models than the standard model presented in section 2. Obtaining tight bounds for non-Abelian groups is of course a natural open problem.

2   Preliminaries

From now on, p denotes a prime number and the problem of distinguishing the trivial subgroup from a group of order p in $(\mathbb{Z}/p\mathbb{Z})^n$ will be called "Simon's problem in $(\mathbb{Z}/p\mathbb{Z})^n$" (or sometimes just "Simon's problem"). More precisely, we are given a function f from $G = (\mathbb{Z}/p\mathbb{Z})^n$ to a known set X of size p^n, and we are guaranteed that the function fulfills Simon's promises, that is, either:

(1) f is one-to-one, or
(2) $\exists s \neq 0 \ \forall w, w' \ [f(w) = f(w') \iff w - w' \in \langle s \rangle]$, where $\langle s \rangle$ is the group generated by s.

Again, the problem is to decide whether (1) or (2) holds. As pointed out in the introduction, Simon considered only the case $p = 2$.

We assume here that the reader is familiar with the basic notions of quantum computing [13, 6] and we now present the polynomial method. Let A be a quantum algorithm solving Simon's decision problem. Without loss of generality, we can suppose that for every n the algorithm A acts like a succession of operations

$$U_0, O, U_1, O, \ldots, O, U_{T(n)}, M$$

on a m-qubit, for some $m \geq 2n$, starting from state $|0\rangle^{\otimes m}$. The U_i are unitary operations independent of f and O is the call to the black-box function: if x and y are elements of $\{0,1\}^n$ then $O|x, y, z\rangle = |x, y \oplus f(x), z\rangle$. The operation M is the measure of the last qubit. There are some states of $(m-1)$-qubits $|\phi_0(f, n)\rangle$ and $|\phi_1(f, n)\rangle$ (of norm possibly less than 1) such that

$$U_{T(n)} O U_{T(n)-1} O \ldots O U_0 |0\rangle^{\otimes m} = |\phi_0(n, f)\rangle \otimes |0\rangle + |\phi_1(n, f)\rangle \otimes |1\rangle.$$

After the measure M, the result is 0 (reject) with probability $\|\phi_0(n, f)\|^2$ and 1 (accept) with probability $\|\phi_1(n, f)\|^2$. The algorithm A is said to solve Simon's problem with bounded error probability ϵ if it accepts any bijection with probability at least $1 - \epsilon$ and rejects every other function fullfilling Simon's promise with probability at least $1 - \epsilon$. By definition, the query complexity of A is the function T. In section 3 we will prove the following lower bound.

Theorem 1. *If A is an algorithm which solves Simon's problem in $(\mathbb{Z}/p\mathbb{Z})^n$ with bounded error probability ϵ and query complexity T, then for every large enough integer n we have:*

$$T(n) \geq \frac{\log_2\left((2-4\epsilon)\frac{p^{n+3}}{p-1}\right) - 1}{2\log_2\left(\frac{p^3}{p-1}\right) + 2}.$$

Altghough it might not be self-evident that $T(n) = \Omega(n)$, this bound is indeed in the expected range. Indeed, it can be checked easily that the right-hand side is equivalent, for large values of n, to $A(p).n$, where $A(p)$ is positive and $\lim_{p \to +\infty} A(p) = \frac{1}{4}$. For $p = 2$ we obtain the result presented in our preprint [9]: $T(n) \geq \frac{n+2+\log_2(2-4\epsilon)}{8}$.

As explained in the introduction, our proof of this theorem is based on the polynomial method. Lemma 1 below is the key observation on which this method relies. We state it using the formalism of [1]: if s is a partial function from $(\mathbb{Z}/p\mathbb{Z})^n$ to X and f a function from $(\mathbb{Z}/p\mathbb{Z})^n$ to X, $|\text{dom}(s)|$ denotes the size of the domain of s. Moreover, we define:

$$I_s(f) = \begin{cases} 1 \text{ if } f \text{ extends } s \\ 0 \text{ otherwise.} \end{cases}$$

Lemma 1. *[3] If A is an algorithm of query complexity T, there is a set S of partial functions from $(\mathbb{Z}/p\mathbb{Z})^n \to E$ such that for all functions $f : (\mathbb{Z}/p\mathbb{Z})^n \to E$, A accepts f with probability*

$$P_n(f) = \sum_{s \in S} \alpha_s I_s(f)$$

where for every $s \in S$ we have $|\text{dom}(s)| \leq 2T(n)$ and α_s is a real number.

The goal is now to transform $P_n(f)$ into a low-degree polynomial of a single real variable. This is achieved in Proposition 1. We can then prove and apply our lower bound result on real polynomials (Lemma 3).

3 Lower Bound Proof

An algorithm for Simon's problem is only supposed to distinguish between the trivial subgroup and a hidden subgroup of cardinality p (we recall that p is a prime number). To establish our lower bound, we will nonetheless need to examine its behavior on a black-box hiding a subgroup of arbitrary order (a similar trick is used in [1] and [12]). Note that this "generalized Simon problem" (finding an arbitrary hidden subgroup of $(\mathbb{Z}/p\mathbb{Z})^n$) can still be solved in $O(n)$ queries and bounded probability of error by essentially the same algorithm, see for instance [6].

From now on we suppose that A is an algorithm solving Simon's problem with probability of error bounded by $\epsilon < \frac{1}{2}$ and query complexity T. Moreover, $P_n(f) = \sum_{s \in S} \alpha_s I_s(f)$ as given by lemma 1.

For $0 \leq d \leq n$ and $D = p^d$, let $Q_n(D)$ be the probability that A accepts f when f is chosen uniformly at random among the functions from $(\mathbb{Z}/p\mathbb{Z})^n$ to X hiding a subgroup of $(\mathbb{Z}/p\mathbb{Z})^n$ of order D. Of course, $Q_n(D)$ is only defined for some integer values of D and it can be extended in many different ways. By abuse of language we will say that Q_n is a polynomial of degree δ if it can be interpolated by a polynomial of degree δ.

The point of this definition is that we have a bound on some values of Q_n, and a gap between two of them. Namely, we have:

1. for any integer $d \in [0;n]$, $0 \leq Q_n(p^d) \leq 1$ (this number is a probability), and
2. $Q_n(1) \geq 1-\epsilon$ and $Q_n(p) \leq \epsilon$, hence $|Q'_n(x_0)| \geq \frac{1-2\epsilon}{p-1} > 0$ for some $x_0 \in [1;2]$.

If we denote by X_D the set of functions hiding a subgroup of order D, by Lemma 1 we have $Q_n(D) = \sum_{s \in S} \left(\frac{\alpha_s}{|X_D|} \sum_{f \in X_D} I_s(f) \right)$. Hence

$$Q_n(D) = \sum_{s \in S} \alpha_s Q_n^s(D), \tag{1}$$

where $Q_n^s(D)$ is the probability that a random function f hiding a subgroup of order D extends s. We now prove that Q_n is a low-degree polynomial. By (1), it suffices to bound the degree of Q_n^s. Let us start by counting subgroups:

Lemma 2. *Let n and k be nonnegative integers. The group $(\mathbb{Z}/p\mathbb{Z})^n$ has exactly $\beta_p(n,k) = \prod_{0 \leq i < k} \frac{p^{n-i}-1}{p^{k-i}-1}$ distinct subgroups of order p^k.*

Proof. We look at $(\mathbb{Z}/p\mathbb{Z})^n$ as a vector space over the field $\mathbb{Z}/p\mathbb{Z}$: from this point of view the subgroups are the subspaces. We start by counting the number of free k-tuples of vectors. For the first v_0, we can choose anything but 0, so there are $p^n - 1$ choices. For the second vector v_1 we can choose any element not in the subspace generated by v_0; $p^n - p$ possibilities remain. For the third vector, any linear combinaison of v_0 and v_1 is forbidden: there are p^2 of them. In general, the number of free k-tuples of vectors is $\alpha_p(n,k) = \prod_{0 \leq i < k} (p^n - p^i)$. Each subspace of dimension k can be generated by $\alpha_p(k,k)$ different k-tuples, so the total number of subspaces of dimension k is $\frac{\alpha_p(n,k)}{\alpha_p(k,k)} = \prod_{0 \leq i < k} \frac{p^{n-i}-1}{p^{k-i}-1}$. Note that this formula is correct even if $k > n$, in which case $\alpha_p(n,k) = 0$.

Proposition 1. *The polynomial Q_n is of degree at most $2T(n)$.*

Proof. By (1), it suffices to show that for all partial functions $s : (\mathbb{Z}/p\mathbb{Z})^n \to E$ such that $|\text{dom}(s)| \leq 2T(n)$, the probability $Q_n^s(D)$ that a random function f hiding a subgroup of order D extends s is a polynomial in D of degree at most $2T(n)$. So, let s be such a partial function. We will proceed in three steps: we first examine the case where s is a constant function, then the case where s is injective and finally the general case.

Let us therefore suppose that s is constant and note $\text{dom}(s) = \{a_i / i = 1 \ldots k\}$, with $k \leq 2T(n)$, the a_i's being of course all different. A function f hiding a subgroup H extends s if and only if $\{a_i - a_1 / i = 1 \ldots k\} \subseteq H$ and $f(a_1) = s(a_1)$. So $Q_n^s(D) = Q_n^{s'}(D)$ where $s'(x) = s(x - a_1)$. We will thus suppose without loss of generality that $a_1 = 0$. Since E, the possible range for f, is of size p^n, we have $Q_n^s(D) = \frac{\lambda}{p^n}$, where λ is the proportion, among the subgroups of order D, of those containing $\text{dom}(s)$. Let H' be the subgroup generated by $\text{dom}(s)$, and $D' = p^{d'}$ its order, d' being the dimension of H' as a vector space. The number of subgroups of order D containing H' is equal to the number of subgroups of order $\frac{D}{D'}$ of $(\mathbb{Z}/p\mathbb{Z})^n / H'$, which is isomorphic to $(\mathbb{Z}/p\mathbb{Z})^{n-d'}$; so there are $\beta(n - d', d - d')$ of them. We then have $Q_n^s(D) = \frac{1}{p^n} \frac{\beta(n-d',d-d')}{\beta(n,d)} = \frac{1}{p^n} \prod_{0 \leq i < d'} \frac{p^{d-i}-1}{p^{n-i}-1}$, which is a polynomial in D of degree $d' < |\text{dom}(s)| \leq 2T(n)$.

Let us now suppose that s is injective. We still note in the same way $\text{dom}(s) = \{a_i / i = 1 \ldots k\}$. A function f hiding a subgroup H extends s if and only if the a_i's lie in distinct cosets of H and f takes appropriate values on these cosets; so $Q_n^s(D) = \nu\lambda$, where λ is the probability for a subgroup H of order D to contain none of the $a_i - a_j (i \neq j)$ and ν is the probability to extend s for a function h hiding a subgroup H of order D that does not contain any of the $a_i - a_j (i \neq j)$. First we compute ν. For each subgroup H of order D that does not contain any of the $a_i - a_j (i \neq j)$ there are $(p^n)(p^n - 1) \ldots (p^n - p^n/D + 1)$ possible functions f: choose a different value for each coset of H. Among these functions, the number of them extending s is $(p^n - k)(p^n - k - 1) \ldots (p^n - p^n/D + 1)$: choose a value for each coset not containing any a_i. So $\nu = \frac{(p^n - k)!}{(p^n)!}$. The probability λ is equal to $1 - \mu$, where μ is the probability for a subgroup H of order D to contain some $a_i - a_j$ for some $i \neq j$.

By the inclusion-exclusion formula, we can expand λ as follows:

$$\lambda = 1 - \left[\begin{array}{l} \sum_{i \neq j} \Pr(a_i - a_j \in H) \\ - \sum_{\substack{i_1 \neq j_1 \\ i_2 \neq j_2 \\ \{i_1; j_1\} \neq \{i_2; j_2\}}} \Pr(a_{i_1} - a_{j_1} \in H \land a_{i_2} - a_{j_2} \in H) \\ + \cdots \\ - \cdots \\ \vdots \\ + \Pr(\forall i \neq j \ a_i - a_j \in H) \end{array} \right]$$

Our study of the first case above shows that each term in this sum is a polynomial in D of degree less than d', where the order of the subgroup generated by the a_i-a_j's is $p^{d'}$. Since a_i-a_j is always in the subgroup generated by dom(s), $d' \leq |\text{dom}(s)| \leq 2T(n)$.

Finally, in the general case the partial function s is defined by conditions of the form
$$\begin{cases} s(a_1^1) = s(a_2^1) = \cdots = s(a_{k_1}^1) = b_1 \\ s(a_1^2) = s(a_2^2) = \cdots = s(a_{k_2}^2) = b_2 \\ \quad \vdots \\ s(a_1^l) = s(a_2^l) = \cdots = s(a_{k_l}^l) = b_l \end{cases}$$

with $b_1, \ldots, b_l$ all different. In the same way as before, we will suppose without loss of generality that $a_1^1 = 0$. Furthermore, since $f(a_i^j) = f(a_1^j)$ is equivalent to $f(a_i^j - a_1^j) = f(0)$ (i.e. a_i^j and a_1^j are in the same coset of H) we can remove each a_i^j, for $i,j > 1$ from dom(s) and replace them by adding the point $a_i^j - a_1^j$ to dom(s) associated to the value b_1. The size of dom(s) does not increase. It may happen that s was already defined on one of these entries and that our new definition is contradictory. In that case there is simply no subgroup-hiding function f extending s, so Q_n^s is simply the null polynomial and we are done. We will therefore consider only conditions of the form:
$$\begin{cases} s(0) = s(a_2^1) = \cdots = s(a_{k_1}^1) = b_1 \\ s(a^2) = b_2 \\ \quad \vdots \\ s(a^l) = b_l \end{cases}$$

The probability $Q_n^s(D)$ that a function f hiding a subgroup of order D extends s is the probability Q_1 that f satisfies $f(0) = f(a_2^1) = \cdots = f(a_{k_1}^1) = b_1$ times the probability Q_2 that f extends s given that $f(0) = f(a_2^1) = \cdots = f(a_{k_1}^1) = b_1$. We have already computed the first probability: this is the case where s is constant. Let H' be the subgroup generated by the a_i^1's and $D' = p^{d'}$ its order; then $Q_1 = \frac{1}{p^n} \prod_{0 \leq i < d'} \frac{p^{d-i}-1}{p^{n-i}-1}$. Let us define s' on G/H' as the quotient of s if it exists (if not, this means again that Q_n^s is the null polynomial, and we are done). If f satisfies $f(0) = f(a_2^1) = \cdots = f(a_{k_1}^1) = b_1$ then we can define f' on G/H' as the quotient of f; the condition "f extends s and hides a subgroup of order D" is equivalent to "f' extends s' and hides a subgroup of order D/D'". Since s' is defined by the condition $s'(H') = b_1, s'(a^2 + H') = b_2, \ldots, s'(a^l + H') = b_l$ and is injective, our study of the second case shows that $Q_2 = Q_n^{s'}(D/D')$ is a polynomial in D of degree less than $|\text{dom}(s')|$. Hence, $Q_n^s(D)$ is a polynomial in D of degree at most $d' + |\text{dom}(s')| \leq |\text{dom}(s)| \leq 2T$.

Now that we have an upper bound on the degree of Q, let us find a lower bound. The following analogue of the lemmas of Paturi [15] and Nisan-Szegedy [14] will help.

Lemma 3. *Let $c > 0$ and $\xi > 1$ be constants and P a polynomial with the following properties:*

1. *For any integer $0 \leq i \leq n$ we have $|P(\xi^i)| \leq 1$.*
2. *For some real number $1 \leq x_0 \leq \xi$ we have $|P'(x_0)| \geq c$.*

Then $\deg(P) = \Omega(n)$, and more precisely: $\deg(P) \geq \min\left(\frac{n}{2}, \frac{\log_2(\xi^{n+3}c) - 1}{\log_2\left(\frac{\xi^3}{\xi-1}\right) + 1}\right)$.

Proof. Let d be the degree of P, and let us write $P'(X) = \lambda \prod_{i=1}^{d-1}(X - \alpha_i)$, where the α_i's are real or complex numbers. The polynomials P' and P'' are respectively of degree $d-1$ and $d-2$, so there exists an integer $a \in [n - 2d + 2; n - 1]$ such that P'' has no real root in $(\xi^a; \xi^{a+1})$, and P' has no root whose real part is in this same interval. If $d \geq n/2$ there is nothing to prove, so we may and we will assume that $d \leq \frac{n}{2}$. This implies in particular that $\xi^a \geq \xi^2$.

The polynomial P' is monotone on $(\xi^a; \xi^{a+1})$, for P'' has no root in it. This means that P is either convex or concave on this interval, so that the graph of P is either over or under its tangent at the middle point of the interval, which is equal to $\frac{\xi^a + \xi^{a+1}}{2} = \frac{1+\xi}{2}\xi^a$. Suppose that $P'\left(\frac{1+\xi}{2}\xi^a\right)$ is nonnegative (the case when it is negative is similar). Then P is increasing on $(\xi^a; \xi^{a+1})$, since P' has no root in this interval. Let $y = t(x)$ be the equation of the tangent of P at $\frac{1+\xi}{2}\xi^a$. If $t(\xi^{a+1}) > 1$, then $P(\xi^{a+1}) < t(\xi^{a+1})$, so P is concave on $(\xi^a; \xi^{a+1})$, hence $-1 \leq P(\xi^a) \leq t(\xi^a)$. But, since P is monotone on $(\xi^a; \xi^{a+1})$, $t\left(\frac{1+\xi}{2}\xi^a\right) = P\left(\frac{1+\xi}{2}\xi^a\right) \leq 1$. Since $t(\xi^{a+1}) - t\left(\frac{1+\xi}{2}\xi^a\right) = t\left(\frac{1+\xi}{2}\xi^a\right) - t(\xi^a)$, it follows that $t(\xi^{a+1}) \leq 3$ and $t(\xi^{a+1}) - t(\xi^a) \leq 4$. The same inequality can also be derived if we assume $t(\xi^a) < -1$, and it is of course still true if $t(\xi^a) \geq -1$ and $t(\xi^{a+1}) \leq 1$. We conclude that the inequality $t(\xi^{a+1}) - t(\xi^a) \leq 4$ always holds, which implies that $0 \leq P'\left(\frac{1+\xi}{2}\xi^a\right) \leq \frac{4}{\xi^a(\xi-1)}$. If we now include the case where P' is negative, we obtain the inequality

$$\left|P'\left(\frac{1+\xi}{2}\xi^a\right)\right| \leq \frac{4}{\xi^a(\xi-1)}.$$

We therefore have

$$\left|\frac{P'\left(\frac{1+\xi}{2}\xi^a\right)}{P'(x_0)}\right| \leq \frac{4}{c\xi^a(\xi-1)} \leq \frac{4}{c\xi^{n-2d+2}(\xi-1)}. \tag{2}$$

To conclude we need to state a simple geometric fact. Let MBC be a triangle, M' the orthogonal projection of M onto (BC), and (d) the perpendicular bisector of $[BC]$. Let us suppose that M is "at the right of (d)", i.e. $MC \leq MB$.

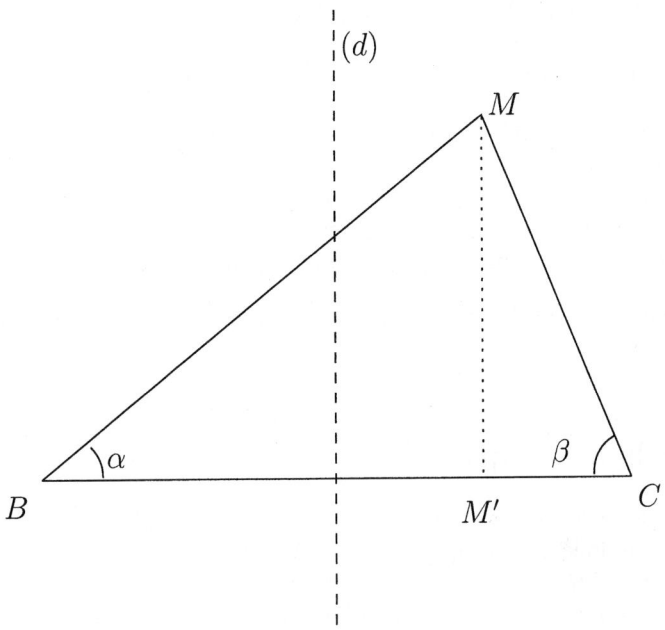

Since C is closer to the line (MM') than B, $\tan\alpha = MM'/BM' \leq \tan\beta = MM'/CM'$. Hence $\alpha \leq \beta$, and $\cos\alpha \geq \cos\beta$, i.e.:

$$\frac{MC}{MB} \geq \frac{M'C}{M'B}. \tag{3}$$

Let $f : \begin{pmatrix} \mathbb{R} \setminus \{x_0\} \to \mathbb{R} \\ x \mapsto \left|\frac{\frac{1+\xi}{2}\xi^a - x}{x_0 - x}\right| \end{pmatrix}$. Since $x_0 < \xi^a < \frac{1+\xi}{2}\xi^a < \xi^{a+1}$, a quick study of this function shows that for all $x \in \mathbb{R} \setminus (\{x_0\} \cup (\xi^a; \xi^{a+1}))$, $f(x) \geq \min(1, f(\xi^a), f(\xi^{a+1})) \geq \frac{\xi-1}{2\xi}$.

We will distinguish two cases for each $i \in \{1; \ldots; d-1\}$.

1. If $\Re(\alpha_i) \leq \frac{1}{2}\left(\frac{1+\xi}{2}\xi^a + x_0\right)$, then $\left|\frac{\frac{1+\xi}{2}\xi^a - \alpha_i}{x_0 - \alpha_i}\right| \geq 1$.

2. If $\Re(\alpha_i) > \frac{1}{2}\left(\frac{1+\xi}{2}\xi^a + x_0\right)$, let us apply (3) to the points $M = \alpha_i$, $M' = \Re(\alpha_i)$, $B = x_0$ and $C = \frac{1+\xi}{2}\xi^a$. We obtain the inequality

$$\left|\frac{\frac{1+\xi}{2}\xi^a - \alpha_i}{x_0 - \alpha_i}\right| \geq \left|\frac{\frac{1+\xi}{2}\xi^a - \Re(\alpha_i)}{x_0 - \Re(\alpha_i)}\right|.$$

Remember though that no root of P' has its real part in $(\xi^a; \xi^{a+1})$, so that $\left|\frac{\frac{1+\xi}{2}\xi^a - \alpha_i}{x_0 - \alpha_i}\right| \geq \frac{\xi-1}{2\xi}$.

We conclude that $\left|\frac{\frac{1+\xi}{2}\xi^a - \alpha_i}{x_0 - \alpha_i}\right| \geq \frac{\xi-1}{2\xi}$ in both cases. Taking (2) into account, we finally obtain the inequality $\left(\frac{\xi-1}{2\xi}\right)^{d-1} \leq \frac{4}{c\xi^{n-2d+2}(\xi-1)}$, hence $d \geq \frac{\log_2(\xi^{n+3}c)-1}{\log_2\left(\frac{\xi^3}{\xi-1}\right)+1}$.

We can now complete the proof of Theorem 1. Let A be our algorithm solving Simon's problem with bounded error probability ϵ and query complexity T. As pointed out before Lemma 2, the associated polynomial Q_n satisfies $|Q'_n(x_0)| \geq 1-2\epsilon$ for some $x_0 \in [1,\xi]$ and $Q_n(\xi^i) \in [0,1]$ for any $i \in \{0,1,\ldots,n\}$. An application of Lemma 3 to the polynomial $P = 2Q_n - 1$ therefore yields the inequality $\deg(Q_n) \geq \min\left(\frac{n}{2}, \frac{\log_2\left((2-4\epsilon)\frac{p^{n+3}}{p-1}\right)-1}{\log_2\left(\frac{p^3}{p-1}\right)+1}\right)$. Theorem 1 follows since $\deg(Q_n) \leq 2T(n)$ by Proposition 1.

4 Abelian Groups

In this section we give lower and upper bounds for the quantum query complexity of Abelian hidden subgroup problems. As explained in the introduction, our bounds are optimal up to constant factors.

Let G be a finite Abelian group, $\hat{G}$ its dual group, i.e. the group of its characters (see for example [6]). For each subgroup H of G, we note $H^\perp$ the orthogonal of H, which is a subgroup of $\hat{G}$ consisted of those characters χ such that $\chi(h) = 1$ for all $h \in H$. According to basic representation theory, $\hat{G}$ is isomorphic to G and, for all subgroup $H \leq G$, the index of $H^\perp$ in $\hat{G}$ is equal to the order of H.

The well-established method of Fourier sampling allows one, with one query to the black-box function, to pick a uniformly random element of the orthogonal of the hidden subgroup. In order to solve the hidden subgroup problem for G, this routine is run k times so as to generate k random elements $x_1,\ldots,x_k \in H^\perp$. The algorithm outputs the orthogonal of the group generated by $x_1,\ldots,x_k$. This output is correct if $x_1,\ldots,x_k$ generate all of $H^\perp$.

We will now show that this algorithm is optimal if we know when to stop, i.e., how many random elements should be picked in $H^\perp$. The following lemma states that the query complexity of the cyclic subgroup problem is constant. The proof is ommited from this extended abstract (note that this fact is already pointed out in [17]).

Lemma 4. *For any integer $M \geq 1$, two random elements chosen uniformly and independently in $\mathbb{Z}/M\mathbb{Z}$ generate all of this group with probability at least $\frac{1}{2}$.*

We recall that (following for instance [11]) the rank $r(G)$ of a group G is the minimal cardinality of a generating set of G. According to the fundamental theorem of finite Abelian groups, G is isomorphic to $\mathbb{Z}/m_1\mathbb{Z}\times\mathbb{Z}/m_2\mathbb{Z}\times\cdots\times\mathbb{Z}/m_{r(G)}\mathbb{Z}$ where m_i divides m_{i-1} for every $i \in \{2,\ldots,r(G)\}$, and this decomposition is unique.

Proposition 2. *For any $\epsilon > 0$ there exists an integer k such that for any finite Abelian group G, $k.r(G)$ random elements chosen uniformly and independently in G generate all of this group with probability at least $1 - \epsilon$.*

Proof. Let us denote by $\mathcal{E}_n$ the supremum of the expectations of the number of random elements of G needed to generate G, taken over the groups G such that $r(G) \leq n$. We can assume that $G = \mathbb{Z}/m_1\mathbb{Z} \times \cdots \times \mathbb{Z}/m_{r(G)}\mathbb{Z}$, where $m_{r(G)}|\ldots|m_1$. To generate G we can proceed with the two following steps.

First we pick enough random elements $\left(x_1^1,\ldots,x_1^{r(G)}\right),\ldots,\left(x_k^1,\ldots,x_k^{r(G)}\right)$ in G so that $x_1^1,\ldots,x_k^1$ generate $\mathbb{Z}/m_1\mathbb{Z}$; the expectation of k is at most $\mathcal{E}_1$. By Lemma 4, $\mathcal{E}_1$ is finite; we can very roughly bound it in the following way.

First pick two random elements in $\mathbb{Z}/m_1\mathbb{Z}$. With probability $p_{\leq 2}$ they generate $\mathbb{Z}/m_1\mathbb{Z}$ and with probability $p_{>2}$ they do not; when they fail to generate, just forget about them and renew the experiment with two new random elements. In the first case the expectation of the number of elements is 2, in the second case it is at most $2 + \mathcal{E}_1$, so we have $\mathcal{E}_1 \leq 2p_{\leq 2} + (2+\mathcal{E}_1)p_{>2}$. Clearly $p_{\leq 2} + p_{>2} = 1$ and according to Lemma 4 we have $p_{\leq 2} \geq \frac{1}{2}$. This shows that $\mathcal{E}_1 \leq 4$.

Then the subgroup generated by these elements contains some element $y = (y^1,\ldots,y^{r(G)})$ such that the of order of y^1 is m_1. The rank of $G/\langle y \rangle$ is equal to $r(G) - 1$ since $G/\langle y \rangle$ is isomorphic to $\mathbb{Z}/m_2\mathbb{Z} \times \cdots \times \mathbb{Z}/m_{r(G)}\mathbb{Z}$. This isomorphism follows from the fact the classes of $e_2,\ldots,e_{r(G)}$ generate $G/\langle y \rangle$, where e_i denotes the element of G whose i^{th} coordinate is equal to 1 and all other coordinates equal to 0. We now pick enough random elements $x_{k+1},\ldots,x_{k+l} \in G$ so that their images in $G/\langle y \rangle$ generate all of it; the expectation of l is of course at most $\mathcal{E}_{r(G)-1}$. Putting it together, we get $\mathcal{E}_{n+1} \leq \mathcal{E}_1 + \mathcal{E}_n$, so $\mathcal{E}_n \leq 4n$. By Markov's inequality, if we choose $\lfloor \frac{4}{\epsilon} \rfloor r(G)$ random elements in a group G, we generate all of this group with probability at least $1 - \epsilon$.

We can now prove our main result.

Theorem 2. *The quantum query complexity of the hidden subgroup problem in a finite Abelian group G is $\Theta(r(G))$.*

Proof. The upper bound is achieved with the standard method: one just applies Proposition 2 to the orthogonal of the hidden subgroup, which is isomorphic to a subgroup of G, using the fact that r is an nondecreasing function on finite Abelian groups.

The lower bound of course comes from Theorem 1. Since for every finite Abelian group G there is some prime p such that $(\mathbb{Z}/p\mathbb{Z})^{r(G)}$ is isomorphic to some subgroup of G, we need only to state that the hidden subgroup problem for a subgroup of G reduces correctly to the hidden subgroup problem for G. Indeed, let H be a subgroup of G and let $H + t_0, \ldots, H + t_k$ be the cosets of H in G, where $t_0 = 0$. If $\gamma : H \to X$ hides a subgroup of H, we can define a function $\gamma' : G \to X \times \{t_i / 0 \leq i \leq k\}$ which hides the same subgroup. Namely, we define $\gamma'(x + t_i) = (\gamma(x), t_i)$ for $x \in H$. Moreover, a call to γ' uses just one call to γ, so we are done.

Acknowledgments

Many thanks to Xavier Caruso, Yves de Cornulier and Joël Riou for useful help. Thanks also go to Frédéric Magniez and to the anonymous referees for bibliographical hints.

References

1. Scott Aaronson and Yaoyun Shi. Quantum lower bounds for the collision and the element distinctness problems. *Journal of the ACM*, 51(4):595–605, July 2004.
2. Robert Beals. Quantum computation of Fourier transforms over symmetric groups. In *Proceedings of the 29th Annual ACM Symposium on the Theory of Computation (STOC)*, pages 48–53. ACM Press, 1997.
3. Robert Beals, Harry Buhrman, Richard Cleve, Michele Mosca, and Ronald de Wolf. Quantum lower bounds by polynomials. *J. ACM*, 48(4):778–797, 2001.
4. Gilles Brassard and Peter Høyer. An exact quantum polynomial-time algorithm for Simon's problem. In *Israel Symposium on Theory of Computing Systems*, pages 12–23, 1997.
5. Lisa R. Hales. *The Quantum Fourier Transform and Extensions of the Abelian Hidden Subgroup Problem*. PhD thesis, UC Berkeley, 2002.
6. Mika Hirvensalo. *Quantum Computing (Natural Computing Series)*. SpringerVerlag, 2001.
7. Peter Høyer. Conjugated operators in quantum algorithms. *Phys. Rev. A*, 59:3280–3289, may 1999.
8. R. Jozsa. Quantum algorithms and the Fourier transform. *Proc. R. Soc. of London A*, 454, 1998.
9. Pascal Koiran, Vincent Nesme, and Natacha Portier. A quantum lower bound for the query complexity of Simon's problem. http://www.arxiv.org/pdf/quant-ph/0501060.
10. Greg Kuperberg. A subexponential-time quantum algorithm for the dihedral hidden subgroup problem. Quantum Physics e-Print Archive, 2003.
11. Hans Kurzweil and Bernd Stellmacher. *The Theory of Finite Groups, An Introduction*. Universitext. Springer, 2004.
12. Samuel Kutin. Quantum lower bound for the collision problem. *quant-ph/0304162*, 2003.
13. Michael A. Nielsen and Isaac L. Chuang. *Quantum computation and quantum information*. Cambridge University Press, 2000.
14. Noam Nisan and Mario Szegedy. On the degree of boolean functions as real polynomials. *Comput. Complex.*, 4(4):301–313, 1994.
15. Ramamohan Paturi. On the degree of polynomials that approximate symmetric boolean functions (preliminary version). In *STOC '92: Proceedings of the twenty-fourth annual ACM symposium on Theory of computing*, pages 468–474, 1992.
16. Pierre Philipps. Bornes inférieures en calcul quantique : Méthode par adversaire vs. méthode des polynômes. Rapport de stage de DEA, effectué au LRI sous la direction de Frédéric Magniez, http://www.lri.fr/~magniez/stages-dea.html, 2003.
17. Peter W. Shor. Polynomial-time algorithms for prime factorization and discrete logarithms on a quantum computer. *SIAM J. Comput.*, 26(5):1484–1509, 1997.
18. David R. Simon. On the power of quantum computation. *SIAM Journal on Computing*, 26(5):1474–1483, 1997.

All Quantum Adversary Methods Are Equivalent

Robert Špalek[1,*] and Mario Szegedy[2,**]

[1] CWI, Amsterdam
sr@cwi.nl
[2] Rutgers University
szegedy@cs.rutgers.edu

Abstract. The quantum adversary method is one of the most versatile lower-bound methods for quantum algorithms. We show that all known variants of this method are equal: spectral adversary [1], weighted adversary [2], strong weighted adversary [3], and the Kolmogorov complexity adversary [4]. We also present a few new equivalent formulations of the method. This shows that there is essentially *one* quantum adversary method. From our approach, all known limitations of all versions of the quantum adversary method easily follow.

1 Introduction

1.1 Lower-Bound Methods for Quantum Query Complexity

In the query complexity model, the input is accessed using oracle queries and the query complexity of the algorithm is the number of calls to the oracle. The query complexity model is helpful in obtaining time complexity lower bounds, and often this is the only way to obtain time bounds in the random access model.

The first lower-bound method was the hybrid method of Bennett, Bernstein, Brassard, and Vazirani [5] to show an $\Omega(\sqrt{n})$ lower bound on the quantum database search. Their proof is based on the following simple observation: If the value of function f differs on two inputs x, y, then the output quantum states of any bounded-error algorithm for f on x and y must be almost orthogonal. On the other hand, the inner product is 1 at the beginning, because the computation starts in a fixed state. By upper-bounding the change of the inner product after one query, we lower bound the number of queries that need to be made.

The second lower-bound method is the polynomial method of Beals, Buhrman, Cleve, Mosca, and de Wolf [6]. It is based on the observation that the measure-

---

* Supported in part by the EU fifth framework project RESQ, IST-2001-37559.
** Supported by NSF grant 0105692, and in part by the National Security Agency (NSA) and Advanced Research and Development Activity (ARDA) under Army Research Office (ARO), contract number DAAD19-01-1-0506.

ment probabilities can be described by low-degree polynomials in the input bits. If t queries have been made, then the degree is at most $2t$. Since the measurement probabilities are always inside $[0, 1]$, one can apply some degree lower bounds for polynomials and thus obtain good lower bounds for quantum query complexity.

The third lower-bound method is the quantum adversary method of Ambainis [7]. It extends the hybrid method. Instead of examining a fixed input pair, Ambainis takes an average over many pairs of inputs. In this paper, we study different variants of the quantum adversary method.

The fourth lower-bound method is the semidefinite programming method of Barnum, Saks, and Szegedy [1]. It exactly characterizes quantum query complexity by a semidefinite program. The dual of this program gives a lower bound that encompasses the quantum adversary bound.

1.2 The Variants of the Quantum Adversary Method

The original quantum adversary method, let us call it unweighted, was invented by Ambainis [7]. It was successfully used to obtain the following tight lower bounds: $\Omega(\sqrt{n})$ for Grover search [8], $\Omega(\sqrt{n})$ for two-level And-Or trees (see [9] for a matching upper bound), and $\Omega(\sqrt{n})$ for inverting a permutation.

Some functions, such as sorting or ordered search, could not be well lower-bounded by the unweighted method. Høyer, Neerbek, and Shi used a weighting argument [10] to obtain tight bounds for these problems. Barnum, Saks, and Szegedy proposed a general method [1] that gives necessary and sufficient conditions for the existence of a quantum query algorithm. They also described a special case, so-called *spectral method*, which gives a lower bound in terms of spectral norms of an adversary matrix. Ambainis also published a *weighted version of his adversary method* [2]. He applied it to get a lower bound for several iterated functions. Zhang observed that Ambainis had generalized his oldest method [7] in two independent ways, so he unified them, and published a *strong weighted adversary method* [3]. Finally, Laplante and Magniez used Kolmogorov complexity in an unusual way and described a *Kolmogorov complexity method* [4].

A few relations between the methods are known. The strong weighted adversary is clearly at least as good as the weighted adversary. Laplante and Magniez showed [4] that the Kolmogorov complexity method is at least as strong as all the following methods: the Ambainis unweighted and weighted method, the strong weighted method, and the spectral method. The method of Høyer et al. [10] is a special case of the weighted adversary method.

In addition it was known that there were some limitations for lower bounds obtained by the adversary method. Szegedy observed [11] that the weighted adversary method is limited by $\min(\sqrt{C_0 n}, \sqrt{C_1 n})$, where C_0 is the zero-certificate complexity of f and C_1 is the one-certificate complexity of f. Laplante and Magniez proved the same limitation for the Kolmogorov complexity method [4], which implies that all other methods are also bounded. Finally, this bound was improved to $\sqrt{C_0 C_1}$ for total f by Zhang [3] and independently by us.

1.3 Our Results

In this paper, we clean up the forest of adversary methods. We show that there is essentially only one quantum adversary method and that all the former methods [1, 2, 3, 4, 10] are just different formulations of the same method. This means that the quantum adversary method is a very robust concept. Furthermore, we also present a new simple proof of the $\min(\sqrt{C_0 n}, \sqrt{C_1 n})$ limitation of the quantum adversary method for partial f, resp. $\sqrt{C_0 C_1}$ for total f.

This paper is an extended abstract with some proofs omitted. The full version can be downloaded from http://arxiv.org/abs/quant-ph/0409116.

1.4 Separation Between the Polynomial and Adversary Method

The polynomial method and the adversary method are incomparable. There are examples when the polynomial method gives better bounds and vice versa.

The polynomial method was successfully applied to obtain tight lower bound $\Omega(n^{1/3})$ for the collision problem and $\Omega(n^{2/3})$ for the element distinctness problem [12] (see [13] for a matching upper bound). The adversary method is incapable of proving such lower bounds due to the small certificate complexity of the function. Furthermore, the polynomial method often gives tight lower bounds for the exact and zero-error quantum complexity, such as n for the Or function [6]. The adversary method can only provide bounded-error lower bounds.

On the other hand, Ambainis exhibited some iterated functions [2], for which the adversary method gives better lower bounds than the polynomial method. The biggest proved gap between the two methods is $n^{1.321}$. Furthermore, the polynomial method did not yet succeed in proving several lower bounds that are very simple to prove by the adversary method. A famous example is the two-level And-Or tree. The adversary method gives a tight lower bound $\Omega(\sqrt{n})$ [7], whereas the best bound obtained by the polynomial method is $\Omega(n^{1/3})$ and it follows from the element distinctness lower bound [12].

There are functions for which none of the methods is known to give a tight bound. A long-standing open problem is the binary And-Or tree. The best known quantum algorithm just implements the classical zero-error algorithm by Saks and Wigderson [14] running in expected time $O(n^{0.753})$. Adversary lower bounds are limited by $\sqrt{C_0 C_1} = \sqrt{n}$. Recently, Laplante, Lee, and Szegedy showed [15] that this limitation $\sqrt{n}$ holds for every read-once $\{\wedge, \vee\}$ formula. The best known lower bound obtained by the polynomial method is also $\Omega(\sqrt{n})$ and it follows from embedding the parity function. The polynomial method might prove a stronger lower bound. Another example is triangle finding. The best upper bound is $O(n^{1.3})$ [16] and the best lower bound is $\Omega(n)$. Again, the adversary method cannot give better than a linear bound, but the polynomial method might.

The semidefinite programming method [1] gives an exact characterization of quantum query complexity. However, it is too general to be applied directly. It is an interesting open problem to find a lower bound that cannot be proved by the adversary or polynomial method.

2 Preliminaries

2.1 Quantum Query Algorithms

We assume familiarity with quantum computing [17] and sketch the model of quantum query complexity, referring to [18] for more details, also on the relation between query complexity and certificate complexity. Suppose we want to compute some function f. For input $x \in \{0,1\}^N$, a *query* gives us access to the input bits. It corresponds to the unitary transformation, which depends on input x in the following way: $O_x : |i,b,z\rangle \mapsto |i, b \oplus x_i, z\rangle$. Here $i \in [N] = \{1, \ldots, N\}$ and $b \in \{0,1\}$; the z-part corresponds to the workspace, which is not affected by the query. We assume the input can be accessed only via such queries. A T-query quantum algorithm has the form $A = U_T O_x U_{T-1} \cdots O_x U_1 O_x U_0$, where the U_k are fixed unitary transformations, independent of x. This A depends on x via the T applications of O_x. The algorithm starts in initial S-qubit state $|0\rangle$. For a Boolean function f, the output of A is obtained by observing the leftmost qubit of the final superposition $A|0\rangle$, and its acceptance probability on input x is its probability of outputting 1.

2.2 Kolmogorov Complexity

An excellent book about Kolmogorov complexity is the book [19] by Li and Vitányi. A deep knowledge of Kolmogorov complexity is not necessary to understand this paper. Some results on the relation between various classical forms of the quantum adversary method and the Kolmogorov complexity method are taken from Laplante and Magniez [4], and the others just use basic techniques.

A set is called *prefix-free* if none of its members is a prefix of another member. Fix a universal Turing machine M and a prefix-free set S. The *prefix-free Kolmogorov complexity* of x given y, denoted by $K(x|y)$, is the length of the shortest program from S that prints x if it gets y on the input. Formally, $K(x|y) = \min\{|P| : P \in S, M(P, y) = x\}$.

2.3 Notation

Let $[n] = \{1, 2, \ldots, n\}$. Let Σ^* denote the set of all finite strings over alphabet Σ. All logarithms are binary. Let I denote the *identity* matrix. Let A^T denote the *transpose* of A. Let diag (A) denote the column vector containing the *main diagonal* of A. Let tr (A) be the *trace* of A and let $A \cdot B$ be the scalar product of A and B. For a column vector x, let $|x|$ denote the ℓ_2-norm of x. Let $\lambda(A)$ denote the *spectral norm* of A, formally $\lambda(A) = \max_{x:|x|\neq 0} |Ax|/|x|$. We say that a matrix is *Boolean*, if it contains only zeroes and ones. Let AB denote the usual *matrix product* and let $A \circ B$ denote the *Hadamard (point-wise) product* [20]. Let $A \geq B$ denote the *point-wise comparison* and let $C \succeq D$ denote that $C - D$ is *positive semidefinite*. Let $r_x(M)$ denote the ℓ_2-*norm* of the x-th row of M and let $c_y(M)$ denote the ℓ_2-*norm* of the y-th column of M. Let $r(M) = \max_x r_x(M)$ and $c(M) = \max_y c_y(M)$.

We call a function $f : S \to \{0,1\}$ *total*, if $S = \{0,1\}^n$, otherwise it is called *partial*. Let f be a partial function. A *certificate* for an input $x \in S$ is a subset $I \subseteq [n]$ such that fixing the input bits $i \in I$ to x_i determines the function value. Formally, $\forall y \in S : y|_I = x|_I \Rightarrow f(y) = f(x)$, where $x|_I$ denotes the substring of x indexed by I. A certificate I for x is called minimal, if $|I| \leq |J|$ for every certificate J for x. Let $\mathcal{C}_f(x)$ denote the lexicographically smallest *minimal certificate* for x. Let $C_0(f) = \max_{x:f(x)=0} |\mathcal{C}_f(x)|$ be the *zero-certificate complexity* of f and let $C_1(f) = \max_{x:f(x)=1} |\mathcal{C}_f(x)|$ be the *one-certificate complexity* of f.

3 Main Result

In this section, we present several equivalent quantum adversary methods and a new simple proof of the limitations of these methods. We can categorize these methods into two groups. Some of them solve conditions on the primal of the quantum system [1]: these are the spectral, weighted, strong weighted, and generalized spectral adversary; and some of them solve conditions on the dual: these are the Kolmogorov complexity bound, minimax, and the semidefinite version of minimax. Primal methods are mostly used to give lower bounds on the query complexity, while we can use the duals to give limitations of the method.

Theorem 1. *Let $n \geq 1$ be an integer, let $S \subseteq \{0,1\}^n$, and let $f : S \to \{0,1\}$ be a partial Boolean function. Let $Q_\varepsilon(f)$ be the ε-error quantum query complexity of f. Then $\frac{Q_\varepsilon(f)}{1 - 2\sqrt{\varepsilon(1-\varepsilon)}} \geq \mathrm{SA}(f) = \mathrm{WA}(f) = \mathrm{SWA}(f) = \mathrm{MM}(f) = \mathrm{SMM}(f) = \mathrm{GSA}(f) = \Theta(\mathrm{KA}(f))$, where SA, WA, SWA, MM, SMM, GSA, and KA are lower bounds given by the following methods:*

- **Spectral adversary [1].** *Let D_i, F be $|S| \times |S|$ Boolean matrices that satisfy $D_i[x,y] = 1$ iff $x_i \neq y_i$ for $i \in [n]$, and $F[x,y] = 1$ iff $f(x) \neq f(y)$. Let Γ denote an $|S| \times |S|$ non-negative symmetric matrix with $\Gamma \circ F = \Gamma$. Then*

$$\mathrm{SA}(f) = \max_{\Gamma} \frac{\lambda(\Gamma)}{\max_i \lambda(\Gamma \circ D_i)}. \tag{1}$$

- **Weighted adversary [2].**[1] *Let w, w' denote a weight scheme as follows:*
  - *Every pair $(x,y) \in S^2$ is assigned a non-negative weight $w(x,y) = w(y,x)$ that satisfies $w(x,y) = 0$ whenever $f(x) = f(y)$.*
  - *Every triple $(x,y,i) \in S^2 \times [n]$ is assigned a non-negative weight $w'(x,y,i)$ that satisfies $w'(x,y,i) = 0$ whenever $x_i = y_i$ or $f(x) = f(y)$, and $w'(x,y,i)w'(y,x,i) \geq w^2(x,y)$ for all x,y,i such that $x_i \neq y_i$.*

---
[1] We use a different formulation [4] than in the original Ambainis papers [7,2]. In particular, we omit the relation $R \subseteq A \times B$ on which the weights are required to be nonzero, and instead allow zero weights.

For all x, i, let $wt(x) = \sum_y w(x,y)$ and $v(x,i) = \sum_y w'(x,y,i)$. Then

$$\mathrm{WA}(f) = \max_{w,w'} \min_{\substack{x,i \\ f(x)=0 \\ v(x,i)>0}} \sqrt{\frac{wt(x)}{v(x,i)}} \cdot \min_{\substack{y,j \\ f(y)=1 \\ v(y,j)>0}} \sqrt{\frac{wt(y)}{v(y,j)}}. \qquad (2)$$

- **Strong weighted adversary** [3]. *Let w, w' denote a weight scheme as above. Then*

$$\mathrm{SWA}(f) = \max_{w,w'} \min_{\substack{x,y,i \\ w(x,y)>0 \\ x_i \neq y_i}} \sqrt{\frac{wt(x)wt(y)}{v(x,i)v(y,i)}}. \qquad (3)$$

- **Kolmogorov complexity** [4].[2] *Let $\sigma \in \{0,1\}^*$ denote a finite string. Then*

$$\mathrm{KA}(f) = \min_\sigma \max_{f(x) \neq f(y)} \frac{1}{\sum_{i: x_i \neq y_i} \sqrt{2^{-K(i|x,\sigma) - K(i|y,\sigma)}}}. \qquad (4)$$

- **Minimax over probability distributions** [4]. *Let $p: S \times [n] \to \mathbb{R}$ denote a set of probability distributions, that is $p_x(i) \geq 0$ and $\sum_i p_x(i) = 1$ for every x. Then*

$$\mathrm{MM}(f) = \min_p \max_{\substack{x,y \\ f(x) \neq f(y)}} \frac{1}{\sum_{i: x_i \neq y_i} \sqrt{p_x(i)\, p_y(i)}} \qquad (5)$$

$$= 1 \Big/ \max_p \min_{\substack{x,y \\ f(x) \neq f(y)}} \sum_{i: x_i \neq y_i} \sqrt{p_x(i)\, p_y(i)}. \qquad (6)$$

- **Semidefinite version of minimax.** *Let D_i, F be Boolean matrices as above. Then $\mathrm{SMM}(f) = 1/\mu_{\max}$, where $\mu_{\max}$ is the maximal solution of the following semidefinite program:*

$$\begin{aligned}
& \text{maximize } \mu \\
& \text{subject to } \quad \forall i: \ R_i \succeq 0, \\
& \qquad\qquad\quad \sum_i R_i \circ I = I, \\
& \qquad\qquad\quad \sum_i R_i \circ D_i \geq \mu F.
\end{aligned} \qquad (7)$$

- **Generalized spectral adversary.** *Let D_i, F be Boolean matrices as above. Then $\mathrm{GSA}(f) = 1/\mu_{\min}$, where $\mu_{\min}$ is the minimal solution of the following semidefinite program:*

$$\begin{aligned}
& \text{minimize } \mu = \operatorname{tr} \Delta \\
& \text{subject to } \qquad \Delta \text{ is diagonal} \\
& \qquad\qquad\qquad Z \geq 0 \\
& \qquad\qquad\qquad Z \cdot F = 1 \\
& \qquad\quad \forall i: \ \Delta - Z \circ D_i \succeq 0.
\end{aligned} \qquad (8)$$

---

[2] We use a different formulation than Laplante and Magniez [4]. They minimize over all algorithms A computing f and substitute σ = source code of A, whereas we minimize over all finite strings σ.

Before we prove the main theorem in the next sections, let us draw some consequences. We show that there are limits that none of these quantum adversary methods can go beyond.

Theorem 2. *Let $S \subseteq \{0,1\}^n$ and let $f : S \to \{0,1\}$ be a partial Boolean function. The max-min bound (6) is upper-bounded by $\mathrm{MM}(f) \leq \min(\sqrt{C_0(f)n}, \sqrt{C_1(f)n})$. If f is total, then $\mathrm{MM}(f) \leq \sqrt{C_0(f)C_1(f)}$.*

Proof. The following simple argument is due to Ronald de Wolf. We exhibit a set of probability distributions p such that

$$m(p) = \min_{\substack{x,y \\ f(x) \neq f(y)}} \sum_{i:x_i \neq y_i} \sqrt{p_x(i)\,p_y(i)} \geq \frac{1}{\sqrt{C_0 n}}, \text{ resp. } \frac{1}{\sqrt{C_0 C_1}}.$$

The max-min bound is $\mathrm{MM}(f) = 1/\max_p m(p)$ and the statement follows.

Let f be partial. For every $x \in f^{-1}(0)$, distribute the probability uniformly over any minimal certificate $C_f(x)$, and for every $y \in f^{-1}(1)$, distribute the probability uniformly over all input bits. Formally, $p_x(i) = 1/|C_f(x)|$ iff $i \in C_f(x)$, $p_x(i) = 0$ for $i \notin C_f(x)$, and $p_y(i) = 1/n$. Take any x,y such that $f(x) = 0$ and $f(y) = 1$, and the zero-certificate $I = C_f(x)$. Since $y|_I \neq x|_I$, there is a $j \in I$ such that $x_j \neq y_j$. Now we lower-bound the sum of (6):

$$\sum_{i:x_i \neq y_i} \sqrt{p_x(i)\,p_y(i)} \geq \sqrt{p_x(j)\,p_y(j)} = \sqrt{\frac{1}{|C_f(x)|} \cdot \frac{1}{n}} \geq \frac{1}{\sqrt{C_0 n}}.$$

If f is total, then we can do even better. For every $x \in \{0,1\}^n$, distribute the probability uniformly over any minimal certificate $C_f(x)$. Formally, $p_x(i) = 1/|C_f(x)|$ iff $i \in C_f(x)$, and $p_x(i) = 0$ otherwise. Take any x,y such that $f(x) \neq f(y)$, and let $I = C_f(x) \cap C_f(y)$. There must exist a $j \in I$ such that $x_j \neq y_j$, otherwise we could find an input z that is consistent with both certificates. (That would be a contradiction, because f is total and hence $f(z)$ has to be defined and be equal to both 0 and 1.) After we have found a j, we lower-bound the sum of (6) in the same way as above. □

Some parts of the following statement have already been observed for individual methods by Szegedy [11], Laplante and Magniez [4], and Zhang [3]. This corollary rules out all adversary attempts to prove good lower bounds for problems with small certificate complexity, such as element distinctness [12], binary And-Or trees [14, 21, 9], or triangle finding [16].

Corollary 1. *All quantum adversary lower-bounds are at most $\min(\sqrt{C_0(f)n}, \sqrt{C_1(f)n})$ for partial functions and $\sqrt{C_0(f)C_1(f)}$ for total functions.*

4 Equivalence of Spectral and Strong Weighted Adversary

In this section, we give a linear-algebraic proof that the spectral bound [1] and the strong weighted bound [3] are equal. The proof has three steps. First, we show that the weighted bound [2] is at least as good as the spectral bound. Second, using a small combinatorial lemma, we show that the spectral bound is at least as good as the strong weighted bound. The third step is trivial, since the strong weighted bound is always at least as good as the weighted bound. The generalization of the weighted adversary method thus does not make the bound stronger, however its formulation is easier to use.

First, let us state two useful statements upper-bounding the spectral norm of a Hadamard product of two non-negative matrices. The first one is due to Mathias [20]. The second one is our generalization and its proof is omitted.

Lemma 1. [20] *Let S be a non-negative symmetric matrix and let M and N be non-negative matrices such that $S \leq M \circ N$. Then $\lambda(S) \leq r(M)c(N) = \max_{x,y} r_x(M)c_y(N)$. Moreover, for every S there exist M, N such that $S = M \circ N$ and $\lambda(S) = r(M)c(N)$.*

Lemma 2. *Let S be a non-negative symmetric matrix and let M and N be non-negative matrices such that $S \leq M \circ N$. Let $B(M, N) = \max_{\substack{x,y \\ S[x,y]>0}} r_x(M)c_y(N)$. Then $\lambda(S) \leq B(M, N)$.*

Now we reduce the spectral adversary to the weighted adversary.

Theorem 3. $SA(f) \leq WA(f)$.

Proof. Let Γ be any non-negative symmetric matrix with $\Gamma \circ F = \Gamma$ as in equation (1). Assume without loss of generality that $\lambda(\Gamma) = 1$. Let δ be the principal eigenvector of Γ, that is $\Gamma\delta = \delta$. Define the following weight scheme: $w(x,y) = w(y,x) = \Gamma[x,y] \cdot \delta[x]\delta[y]$. Furthermore, for every i, decompose every $\Gamma_i = \Gamma \circ D_i$ into a Hadamard product of two non-negative matrices $\Gamma_i = M_i \circ N_i$ such that $\lambda(\Gamma_i) = r(M_i)c(N_i)$. This is always possible by Lemma 1. We ensure that $r(M_i) = c(N_i) = \sqrt{\lambda(\Gamma_i)}$ by multiplying M_i and dividing N_i by the same constant. Define w':

$$w'(x,y,i) = \begin{cases} (M_i[x,y]\,\delta[x])^2 & \text{iff } f(x) = 0,\ f(y) = 1,\ \text{and } x_i \neq y_i, \\ (N_i[y,x]\,\delta[y])^2 & \text{iff } f(x) = 1,\ f(y) = 0,\ \text{and } x_i \neq y_i, \\ 0 & \text{otherwise.} \end{cases}$$

Let us verify that w, w' is a weight scheme. From the definition, $w(x,y) = w'(x,y,i) = 0$ if $f(x) = f(y)$, and also $w'(x,y,i) = 0$ if $x_i = y_i$. Furthermore, if $f(x) = 0$, $f(y) = 1$, and $x_i \neq y_i$, then $w'(x,y,i)w'(y,x,i) = (M_i[x,y]\,\delta[x])^2 (N_i[x,y]\,\delta[y])^2 = (\Gamma_i[x,y]\,\delta[x]\delta[y])^2 = w(x,y)^2$. Finally, let us compute the bound (2) given by the weight scheme. Let $v_b = \max_{\substack{x,i \\ f(x)=b}} \frac{v(x,i)}{wt(x)}$. Then

$$wt(x) = \sum_y w(x,y) = \delta[x] \sum_y \Gamma[x,y]\delta[y] = \delta[x]\,(\Gamma\delta)[x] = \delta[x]^2,$$

$$v_0 = \max_{\substack{x,i \\ f(x)=0}} \frac{\sum_y w'(x,y,i)}{wt(x)} \leq \max_{\substack{x,i \\ f(x)=0}} \frac{\sum_y (M_i[x,y])^2\,\delta[x]^2}{\delta[x]^2} \leq \max_i (r(M_i))^2,$$

and, analogously, $v_1 \leq \max_i (c(N_i))^2$. Since $r(M_i) = c(N_i) = \sqrt{\lambda(\Gamma_i)}$, both $v_0, v_1 \leq \max_i \lambda(\Gamma_i)$. Hence $1/\sqrt{v_0 v_1} \geq 1/\max_i \lambda(\Gamma_i)$, and the weight scheme w, w' gives at least as good bound as the matrix Γ. □

Now we reduce the strong weighted adversary to the spectral adversary.

Theorem 4. $\mathrm{SWA}(f) \leq \mathrm{SA}(f)$.

Proof. Let w, w' be any weight scheme as in equation (2). Define the following symmetric matrix Γ on $S \times S$: $\Gamma[x, y] = \frac{w(x,y)}{\sqrt{wt(x)wt(y)}}$. We also define column vector δ on S such that $\delta[x] = \sqrt{wt(x)}$. Let $W = \sum_x wt(x)$. Then $\lambda(\Gamma) \geq \delta^T \Gamma \delta / |\delta|^2 = W/W = 1$. Next, we show that, for every i, we have $\lambda(\Gamma_i) \leq \sqrt{u_i}$ for $u_i = \max_{w(x,y)>0, x_i \neq y_i} \frac{v(x,i)v(y,i)}{wt(x)wt(y)}$. Once we prove this, we are done, since the strong weighted bound (3) is $1/\max_i \sqrt{u_i}$. Let $A = f^{-1}(0)$ and $B = f^{-1}(1)$. Fix i and define the following rectangular matrices on the index set $A \times B$:

$$M_i[x, y] = \sqrt{\frac{w'(x, y, i)}{wt(x)}}, \qquad N_i[x, y] = \sqrt{\frac{w'(y, x, i)}{wt(y)}}.$$

Every weight scheme satisfies $w'(x, y, i) w'(y, x, i) \geq w^2(x, y)$ for all x, y, i such that $x_i \neq y_i$. It follows that if we reorder Γ_i to put A first and B last, then

$$\Gamma_i = \Gamma \circ D_i \leq \begin{pmatrix} 0 & M_i \circ N_i \\ M_i^T \circ N_i^T & 0 \end{pmatrix} = \begin{pmatrix} 0 & \overline{M}_i \\ \overline{N}_i^T & 0 \end{pmatrix} \circ \begin{pmatrix} 0 & \overline{N}_i \\ \overline{M}_i^T & 0 \end{pmatrix},$$

where $\overline{M}_i = \sqrt{\frac{c(N_i)}{r(M_i)}} M_i$ and $\overline{N}_i = \sqrt{\frac{r(M_i)}{c(N_i)}} N_i$. This is done for balancing the row norm of M_i and column norm of N_i: $r(\overline{M}_i) = c(\overline{N}_i) = \sqrt{r(M_i)c(N_i)}$. Evaluate

$$B(M_i, N_i^T) = \max_{\substack{x,y \\ \Gamma_i[x,y]>0}} \sqrt{\sum_k \frac{w'(x,k,i)}{wt(x)} \sum_\ell \frac{w'(y,\ell,i)}{wt(y)}} = \max_{\substack{x,y \\ w(x,y)>0 \\ x_i \neq y_i}} \sqrt{\frac{v(x,i)v(y,i)}{wt(x)wt(y)}}.$$

By Lemma 2, $\lambda(\Gamma_i) \leq B(\overline{M}_i, \overline{N}_i^T) = B(M_i, N_i^T) = \sqrt{u_i}$, as claimed. □

5 Equivalence of Minimax and Generalized Spectral Adversary

In this section, we prove that the minimax bound is equal to the generalized spectral bound. We first get rid of the reciprocal by taking the max-min bound. Second, we write this bound as a semidefinite program. An application of duality theory of semidefinite programming finishes the proof.

Theorem 5. $\mathrm{MM}(f) = \mathrm{SMM}(f)$.

Proof. Let p be a set of probability distributions as in equation (6). Define $R_i[x,y] = \sqrt{p_x(i)\, p_y(i)}$. Since p_x is a probability distribution, we get that $\sum_i R_i$ must have all ones on the diagonal. The condition $\min_{\substack{x,y \\ f(x)\neq f(y)}} \sum_{i:x_i \neq y_i} R_i[x,y] \geq \mu$ is rewritten into $\forall x,y : f(x) \neq f(y) \implies \sum_{i:x_i \neq y_i} R_i[x,y] \geq \mu$, which is $\sum_i R_i \circ D_i \geq \mu F$. However, the matrices R_i are rank-1 and they have nonnegative entries. We have replaced that condition by $R_i \succeq 0$ to get semidefinite program (7). Hence the program (7) is a relaxation of the condition of (6) and $\mathrm{SMM}(f) \leq \mathrm{MM}(f)$.

Let us show that every solution R_i of the semidefinite program can be changed to an at least as good rank-1 solution R'_i. Take a Cholesky decomposition $R_i = X_i X_i^T$. Define a column-vector $q_i[x] = \sqrt{\sum_j X_i[x,j]^2}$ and a rank-1 matrix $R'_i = q_i q_i^T$. It is not hard to show that all R'_i satisfy the same constraints as R_i. First, R'_i is positive semidefinite. Second, $R'_i[x,x] = \sum_j X_i[x,j]^2 = R_i[x,x]$, hence $\sum_i R'_i \circ I = I$. Third, by a Cauchy-Schwarz inequality,

$$R_i[x,y] = \sum_j X_i[x,j] X_i[y,j] \leq \sqrt{\sum_k X_i[x,k]^2} \sqrt{\sum_\ell X_i[y,\ell]^2} = R'_i[x,y],$$

hence $\sum_i R'_i \circ D_i \geq \sum_i R_i \circ D_i \geq \mu F$. We conclude that $\mathrm{MM}(f) \leq \mathrm{SMM}(f)$. □

Theorem 6. $\mathrm{SMM}(f) = \mathrm{GSA}(f)$.

Proof. Omitted; it only uses the duality theory of semidefinite programming.

6 Equivalence of Generalized Spectral and Spectral Adversary

In this section, we prove that the generalized spectral adversary bound is equal to the spectral adversary bound. The main difference between them is that the generalized method uses a positive diagonal matrix Δ as a new variable.

Theorem 7. $\mathrm{GSA}(f) = \mathrm{SA}(f)$.

Proof. Let Z, Δ be a solution of (8). First, let us prove that $\Delta \succ 0$. Since both $Z \geq 0$ and $D_i \geq 0$, it holds that $\mathrm{diag}\,(-Z \circ D_i) \leq 0$. We know that $\Delta - Z \circ D_i \succeq 0$, hence $\mathrm{diag}\,(\Delta - Z \circ D_i) \geq 0$, and $\mathrm{diag}\,(\Delta) \geq 0$ follows. Moreover, $\mathrm{diag}\,(\Delta) > 0$ unless Z contains an empty row, in which case we delete it (together with the corresponding column) and continue. Second, $\Delta - Z \circ D_i \succeq 0$ implies that $Z \circ D_i$ is symmetric for every i. It follows that Z must be also symmetric.

Take a column vector $a = \mathrm{diag}\,(\Delta^{-1/2})$ and a rank-1 matrix $A = aa^T \succ 0$. It is simple to prove that $A \circ X \succeq 0$ for every matrix $X \succeq 0$.

Since $\Delta - Z \circ D_i \succeq 0$, also $A \circ (\Delta - Z \circ D_i) = I - Z \circ D_i \circ A \succeq 0$ and hence $\lambda(Z \circ D_i \circ A) \leq 1$. Now, define the spectral adversary matrix $\Gamma = Z \circ F \circ A$. Since $0 \leq Z \circ F \leq Z$, it follows that

$$\lambda(\Gamma \circ D_i) = \lambda(Z \circ F \circ A \circ D_i) \leq \lambda(Z \circ D_i \circ A) \leq 1.$$

It remains to show that $\lambda(\Gamma) \geq 1/\operatorname{tr} \Delta$. Let $b = \operatorname{diag}(\sqrt{\Delta})$ and $B = bb^T$. Then

$$1 = Z \cdot F = \Gamma \cdot B = b^T \Gamma b \leq \lambda(\Gamma) \cdot |b|^2 = \lambda(\Gamma) \cdot \operatorname{tr} \Delta.$$

Γ is clearly symmetric, $\Gamma \geq 0$, and $\Gamma \circ F = \Gamma$. The bound (1) given by Γ is bigger than or equal to $1/\operatorname{tr} \Delta$, hence $\operatorname{SA}(f) \geq \operatorname{GSA}(f)$.

For the other direction, let Γ be a non-negative symmetric matrix satisfying $\Gamma \circ F = \Gamma$. Let δ be its principal eigenvector with $|\delta| = 1$. Assume without loss of generality that $\lambda(\Gamma) = 1$ and let $\mu = \max_i \lambda(\Gamma_i)$. Take $A = \delta \delta^T$, $Z = \Gamma \circ A$, and $\Delta = \mu I \circ A$. Then $Z \cdot F = \Gamma \cdot A = \delta^T \Gamma \delta = 1$ and $\operatorname{tr} \Delta = \mu$. For every i, $\lambda(\Gamma_i) \leq \mu$, hence $\mu I - \Gamma \circ D_i \succeq 0$. It follows that $0 \preceq A \circ (\mu I - \Gamma \circ D_i) = \Delta - Z \circ D_i$. The semidefinite program (8) is satisfied and hence its optimum is $\mu_{\min} \leq \mu$. We conclude that $\operatorname{GSA}(f) \geq \operatorname{SA}(f)$. □

7 Proof of the Main Theorem

In this section, we close the circle of reductions. We use the results of Laplante and Magniez, who recently proved [4] that the Kolmogorov complexity bound is asymptotically lower-bounded by the weighted adversary bound and upper-bounded by the minimax bound. The upper bound is implicit in their paper, because they did not state the minimax bound as a separate theorem.

Theorem 8. [4–Theorem 2] $\operatorname{KA}(f) = \Omega(\operatorname{WA}(f))$.

Theorem 9. $\operatorname{KA}(f) = O(\operatorname{MM}(f))$.

Proof. Take a set of probability distributions p as in equation (5). The query information lemma [4–Lemma 3] says that $K(i|x,p) \leq \log \frac{1}{p_x(i)} + O(1)$ for every x, i such that $p_x(i) > 0$. This is true, because any i of nonzero probability can be encoded in $\lceil \log \frac{1}{p_x(i)} \rceil$ bits using the Shannon-Fano code of distribution p_x, and the Shannon-Fano code is prefix-free. Rewrite the inequality as $p_x(i) = O(2^{-K(i|x,p)})$. The statement follows, because the set of all strings σ in (4) includes among others also the descriptions of all probability distributions p. □

Proof (of Theorem 1). We have to prove that $\frac{Q_\varepsilon(f)}{1-2\sqrt{\varepsilon(1-\varepsilon)}} \geq \operatorname{SA}(f) = \operatorname{WA}(f) = \operatorname{SWA}(f) = \operatorname{MM}(f) = \operatorname{SMM}(f) = \operatorname{GSA}(f) = \Theta(\operatorname{KA}(f))$, Put together all known equalities and inequalities:

- $SA(f) = WA(f) = SWA(f)$ by Theorem 3 and Theorem 4,
- $MM(f) = SMM(f)$ by Theorem 5,
- $SMM(f) = GSA(f)$ by Theorem 6,
- $GSA(f) = SA(f)$ by Theorem 7,
- $KA(f) = \Theta(WA(f))$ by Theorem 8 and Theorem 9.

Finally, one has to prove one of the lower bounds. For example, Ambainis proved [2] that $Q_2(f) \geq (1 - 2\sqrt{\varepsilon(1-\varepsilon)}) \, WA(f)$. □

Acknowledgments

We thank Ronald de Wolf for many fruitful discussions, for his suggestions concerning Theorem 2, and for proofreading, and Troy Lee for discussions.

References

1. Barnum, H., Saks, M., Szegedy, M.: Quantum decision trees and semidefinite programming. In: Proc. of the 18th IEEE Conference on Computational Complexity. (2003) 179–193
2. Ambainis, A.: Polynomial degree vs. quantum query complexity. In: Proc. of the 44th IEEE Symp. on Foundations of Computer Science. (2003) 230–239
3. Zhang, S.: On the power of Ambainis's lower bounds. In: Proceedings of 31st ICALP. (2004) 1238–1250 quant-ph/0311060.
4. Laplante, S., Magniez, F.: Lower bounds for randomized and quantum query complexity using Kolmogorov arguments. In: Proceedings of 19th IEEE Conference on Computational Complexity. (2004) 294–304 quant-ph/0311189.
5. Bennett, H., Bernstein, E., Brassard, G., Vazirani, U.: Strengths and weaknesses of quantum computing. SIAM Journal on Computing **26** (1997) 1510–1523 quant-ph/9701001.
6. Beals, R., Buhrman, H., Cleve, R., Mosca, M., de Wolf, R.: Quantum lower bounds by polynomials. Journal of the ACM **48** (2001) 778–797 Earlier version in FOCS'98. quant-ph/9802049.
7. Ambainis, A.: Quantum lower bounds by quantum arguments. Journal of Computer and System Sciences **64** (2002) 750–767 Earlier version in STOC'2000. quant-ph/0002066.
8. Grover, L.K.: A fast quantum mechanical algorithm for database search. In: Proceedings of 28th ACM STOC. (1996) 212–219 quant-ph/9605043.
9. Høyer, P., Mosca, M., de Wolf, R.: Quantum search on bounded-error inputs. In: Proceedings of 30th ICALP. (2003) 291–299 LNCS 2719, quant-ph/0304052.
10. Høyer, P., Neerbek, J., Shi, Y.: Quantum complexities of ordered searching, sorting, and element distinctness. Algorithmica **34** (2002) 429–448 Special issue on Quantum Computation and Cryptography. quant-ph/0102078.
11. Szegedy, M.: On the quantum query complexity of detecting triangles in graphs. quant-ph/0310107 (2003)
12. Aaronson, S., Shi, Y.: Quantum lower bounds for the collision problem. Journal of the ACM **51** (2004) 595–605 quant-ph/0111102.

13. Ambainis, A.: Quantum walk algorithm for element distinctness. In: Proceedings of 45th IEEE FOCS. (2004) 22–31 quant-ph/0311001.
14. Saks, M., Wigderson, A.: Probabilistic Boolean decision trees and the complexity of evaluating games trees. In: Proc. of the 27th Annual Symp. on FOCS. (1986) 29–38
15. Laplante, S., Lee, T., Szegedy, M.: The quantum adversary method and formula size lower bounds. In: Proceedings of 20th IEEE Conference on Computational Complexity. (2005) To appear. quant-ph/0501057.
16. Magniez, F., Santha, M., Szegedy, M.: Quantum algorithms for the triangle problem. In: Proceedings of 16th SODA. (2005) 1109–1117 quant-ph/0310134.
17. Nielsen, M.A., Chuang, I.L.: Quantum Computation and Quantum Information. Cambridge University Press (2000)
18. Buhrman, H., de Wolf, R.: Complexity measures and decision tree complexity: A survey. Theoretical Computer Science **288** (2002) 21–43
19. Li, M., Vitányi, P.M.B.: An Introduction to Kolmogorov Complexity and its Applications. Second edn. Springer, Berlin (1997)
20. Mathias, R.: The spectral norm of a nonnegative matrix. Linear Algebra and its Applications **139** (1990) 269–284
21. Barnum, H., Saks, M.: A lower bound on the quantum query complexity of read-once functions. Journal of Computer and Systems Sciences **69** (2004) 244–258 quant-ph/0201007.

Quantum Complexity of Testing Group Commutativity[*]

Frédéric Magniez[1] and Ashwin Nayak[2]

[1] CNRS–LRI, UMR 8623 Université Paris–Sud, France
[2] University of Waterloo and Perimeter Institute for Theoretical Physics, Canada

Abstract. We consider the problem of testing the commutativity of a black-box group specified by its k generators. The complexity (in terms of k) of this problem was first considered by Pak, who gave a randomized algorithm involving $O(k)$ group operations. We construct a quite optimal quantum algorithm for this problem whose complexity is in $\tilde{O}(k^{2/3})$. The algorithm uses and highlights the power of the quantization method of Szegedy. For the lower bound of $\Omega(k^{2/3})$, we introduce a new technique of reduction for quantum query complexity. Along the way, we prove the optimality of the algorithm of Pak for the randomized model.

1 Introduction

A direction of research in quantum computation pioneered by Grover [1] around search problems in unstructured, structured, or partially structured databases has recently seen an extraordinary expansion. In contrast to problems based on Hidden Subgroup Problem (HSP) (see for instance Ref. [2]), the speed up for these search problems is often only polynomial.

Usually in search problems, the access to the input is done via an oracle modeling access to the input. This leads to the notion of query complexity which measures the number of accesses to the oracle. While no significant lower bounds are known for quantum time complexity, the oracle constraint sometimes enables us to prove such bounds in the query model. For promise problems quantum query complexity indeed can be exponentially smaller than the randomized one. A prominent example is the HSP. On the other hand, for total functions, deterministic and quantum query complexities are polynomially related [3].

In the HSP, the group with its all structure is known to the algorithm designer, and the group operations are generally efficiently computable. In the event that the group is not explicitly known, or the group operations are not efficient to implement, it is appropriate to model the group operations by an oracle or a black-box. The notion of *black-box groups* was introduced by Babai and Szemerédi [4]. In this model, the elements of a group are encoded by words

---

[*] For their research support, F.M. thanks the EU 5th framework program RESQ and the French Research Ministry, and A.N. thanks Canada's NSERC and CIAR.

over a finite alphabet, and the group operations are performed by an oracle (the black-box). The groups are assumed to be specified by generators, and the encoding of group elements is not necessarily unique: different strings may encode the same group element. Mosca [2] showed that one can learn in quantum polynomial time the structure of any black-box abelian group. Such a task is known to be hard classically. Then Watrous [5] pioneered the study of black-box group properties in the quantum context.

In this context, we study the problem of testing commutativity of a black-box group (GROUP COMMUTATIVITY) given by its generators. The classical complexity of this problem was first considered by Pak [6]. The straightforward algorithm for the problem has complexity $O(k^2)$, where k is the number of generators, since it suffices to check if every pair of generators commute. Pak presented a surprising randomized algorithm whose complexity is linear in k, and also showed that the deterministic lower bound is quadratic. The linear upper bound on complexity may also be obtained by applying quantum search [1] to locate a pair of generators that do not commute. Using the quantization of random walks by Szegedy [7], we instead present a *sublinear* algorithm in $\tilde{O}(k^{2/3})$ (**Theorem 3**), where the $\tilde{O}$ notation means that logarithmic multiplicative factors are omitted.

GROUP COMMUTATIVITY bears a deceptive resemblance to ELEMENT DISTINCTNESS. The aim in the former is to detect the presence of a pair of generators which collide in the sense that they do not commute. However, since the group structure is unknown, whether or not a pair of generators collide can only be determined by invoking the group oracle. Moreover, the group oracle provides access to elements from the entire group spanned by the given generators, which may be used towards establishing commutativity.

These differences necessitate the use of ideas from Pak's algorithm, the theory of rapidly mixing Markov chains, and perhaps most remarkably, the Szegedy quantization of walks. GROUP COMMUTATIVITY appears to be the first natural problem for which the approach of Szegedy has no equivalent using other known techniques for constructing quantum algorithms—Grover search [1], or the type of quantum walk introduced by Ambainis [8]. A recent result of Buhrman and Spalek [9] on matrix product verification is in the same situation for its time complexity but not for the query complexity, since the approach of Ambainis gives an algorithm whose query complexity is the same in the worst case.

We also prove that our algorithm is almost optimal by giving an $\Omega(k^{2/3})$ lower bound for the quantum query complexity of GROUP COMMUTATIVITY (**Theorem 6**). Simultaneously, we give an $\Omega(k)$ lower bound for its randomized query complexity (**Theorem 5**). This lower bound shows that the algorithm of Pak [6] is optimal, and to our knowledge is new. We first state an easier lower bound using a simple reduction from the problem of detecting a unique collision pair of a function, which is a special case of ELEMENT DISTINCTNESS, when one allows non-unique encoding of the black-box group (**Theorem 4**). For the lower bound for uniquely encoded black-box groups, the proof gets more complex. The randomized case relies upon an adversary argument. The quantum case is subtle. We show the said lower bound for the number of accesses to the given generators. The lower bound also holds for the number of group operations in *generic* quantum algorithms (see Section 4 for a definition). This is shown using a new kind of reduction based on approximation degree of the problem.

2 Preliminaries

2.1 Black-Box Groups

We will suppose that the elements of the group G are encoded by binary strings of length n for some fixed integer n, which we call the *encoding length*. The groups will be given by generators, and therefore the *input size* of a group is the product of the encoding length and the number of generators. For simplicity, we also assume that the identity element of the group is given. Note that the encoding of group elements need not be unique, a single group element may be represented by several strings. If the encoding is not unique, one also needs an oracle for identity tests. Unless otherwise specified, we assume that the encoding is unique in this paper. All of our results apply when the encoding is not unique if one is given an oracle for identity tests.

Since we will deal with black-box groups we shall shortly describe them in the framework of quantum computing (see also [2] or [5]). For a general introduction to quantum computing the reader might consult [10, 11]. We will work in the quantum circuit model. For a group G of encoding length n, the black-box will be given by two oracles O_G and its inverse O_G^{-1}, both operating on $2n$ qubits. For any group elements $g, h \in G$, the effect of the oracles is the following: $O_G|g\rangle|h\rangle = |g\rangle|gh\rangle$, and $O_G^{-1}|g\rangle|h\rangle = |g\rangle|g^{-1}h\rangle$.

In this notation we implicitly use the encoding of a group element. We will do that everywhere in the paper when there is no ambiguity. Also, not every binary string of length n necessarily corresponds to a group element. In this case, the behavior of the black box can be arbitrary.

2.2 Query Model

The quantum query model was explicitly introduced by Beals, Buhrman, Cleve, Mosca, and de Wolf [3]. In this model, as in its classical counterpart, we pay for accessing the oracle, but unlike the classical case, the machine can use the power of quantum parallelism to make queries in superposition.

The state of the computation is represented by three registers, the query register g, the answer register h, and the work register z. The computation takes place in the vector space spanned by all basis states $|g, h, z\rangle$. In the *quantum model* the state of the computation is a complex combination of all basis states which has unit length in the ℓ_2 norm.

For a black-box group the query operator will be O_G together with its inverse O_G^{-1}. For oracle function $F : X \to Y$ the query operator is $O_F : |g\rangle|h\rangle \mapsto |g\rangle|h \oplus F(g)\rangle$, where $\oplus$ denotes the bitwise xor operation.

Non-query operations are independent of the oracle. A *k-query algorithm* is a sequence of $(k + 1)$ operations $(U_0, U_1, \ldots, U_k)$ where each U_i

is unitary. Initially the state of the computation is set to some fixed value $|\bar{0},\bar{0},\bar{0}\rangle$. In case of an oracle function, the sequence of operations $U_0, O_F, U_1, O_F, \ldots, U_{k-1}, O_F, U_k$ is applied. For black-box groups, the modified sequence of operations $U_0, O_G^{b_1}, U_1, O_G^{b_2}, \ldots, U_{k-1}, O_G^{b_k}, U_k$ is applied, where $b_i = \pm 1$. Finally, one or more qubits designated as output bits are measured to get the outcome of the computation. The quantum algorithms we consider might give an erroneous answer, but the probability of making an error is bounded by some fixed constant $\gamma < 1/2$.

In the query model of computation each query adds one to the *query complexity* of an algorithm, but all other computations are free. The *time complexity* of the algorithm is usually measured in terms of the total circuit size for the unitary operations U_i. We will however take a more coarse grained view of time complexity, and assume that operations such as accessing qubits containing group encodings or updating them, take unit time.

2.3 Quantum Walks

We state a simple version of the recent result of Szegedy [7]. Let P be an ergodic and symmetric Markov chain on a graph $G = (V, E)$ on N vertices. We denote by $P[u, v]$ the transition probability from u to v. Let M be a set of marked nodes of V. Assume, one is given a database D that associates some data $D(v)$ to every node $v \in V$. From $D(v)$ we would like to determine if $v \in M$. We expedite this using a quantum procedure Φ. When operating with D three types of cost are incurred. The cost might denote any measure of complexity such as query or time complexities.

Setup cost S: The cost to set up $D(v)$ for a $v \in V$.
Update cost U: The cost to update $D(v)$ for a $v \in V$, i.e. moving from $D(v)$ to $D(v')$, where the transition from v to v' is allowed by the Markov chain P.
Checking cost C: The query complexity of $\Phi(D(v))$ for a $v \in V$.

Concerning the quantization of the walk P, one needs to consider the quantum time complexity of its implementation in terms of the following parameters.
Initialization time I: The time complexity for constructing the superposition $\frac{1}{\sqrt{N}} \sum_{u,v} \sqrt{P[u,v]} |u, v\rangle$.
Transition time T: The time complexity of realizing the transformation $|u, v\rangle \mapsto 2\sqrt{P[u,v]} \sum_{v'} \sqrt{P[u,v']} |u, v'\rangle - |u, v\rangle$.

In the following theorem, the notation $O(\cdot)$ denotes the existence of a universal constant so that the expression is an upper bound. We now state the main result of [7].

Theorem 1 (Szegedy [7]). *Let δ be the eigenvalue gap of P, and let $\frac{|M|}{|V|} \geq \varepsilon > 0$ whenever M is non-empty. There exists a quantum algorithm that determines if M is non empty with cost $S + O((U + C)/\sqrt{\delta \varepsilon})$, and an additional time complexity of $I + O(T/\sqrt{\delta \varepsilon})$.*

Moreover, if P is state transitive then the cost of finding a marked element of M is the same as above.

Note that in this theorem, when the cost denotes the time complexity, we need to add to it the additional time complexity term given in the theorem.

2.4 The Problems

Here we define the problems we are dealing with. The focus of the paper is on

GROUP COMMUTATIVITY
Oracle: Group operations O_G and O_G^{-1} for an encoding in $\{0,1\}^n$
Input: The value of n and the encoding of generators $g_1, \ldots, g_k$ of G
Output: Yes if G is commutative, and No otherwise (if there are two indices i, j such that $g_i g_j \neq g_j g_i$)

The next problem is a special instance of a well-studied problem, ELEMENT DISTINCTNESS.

UNIQUE COLLISION
Oracle: A function F from $\{1, \ldots, k\}$ to $\{1, \ldots, k\}$
Input: The value of k
Output: Yes if there exists a unique collision pair $x \neq y \in \{1, \ldots, k\}$ such that $F(x) = F(y)$, and No if the function is a permutation

This is a promise problem (or a relation) since we do not require a definite output for certain valid oracle functions. We will also use a further specialization of the problem when k is even, UNIQUE SPLIT COLLISION, where one element of the colliding pair has to come from $\{1, \ldots, k/2\}$ and the other from $\{k/2+1, \ldots, k\}$. We call this a *split* collision. Note that in the positive instances of this problem, the restriction of the function to the two intervals $\{1, \ldots, k/2\}$ and $\{k/2+1, \ldots, k\}$ is injective.

2.5 Approximation Degree

We describe the notion of approximation degree for oracle decision problems. Let $\mathcal{S}$ be the set of functions from $\{1, \ldots, k\}$ to $\{1, \ldots, k\}$. An *oracle decision problem* is a boolean function on the set $\mathcal{S}$. For every function $F \in \mathcal{S}$, we define the variables x_{ij} which are 1 if $F(i) = j$ and 0 otherwise.

Definition 1 ([3, 12]). *Let $\Phi : \mathcal{S} \to \{0, 1\}$ be an oracle decision problem. Then the approximation degree of Φ is the lowest degree of real multivariate polynomials P in variables x_{ij}, such that $|P(x) - \Phi(F)| \leq 1/3$, for every $F \in \mathcal{S}$*

The following powerful result relates approximation degree to quantum query complexity.

Proposition 1 ([3, 12]). *If the quantum query complexity of Φ is T, then the approximation degree of Φ is at most $2T$.*

A beautiful application of the polynomial method gives us the optimal query complexity of UNIQUE COLLISION.

Theorem 2 ([12, 13, 14]). *The approximation degree of* UNIQUE COLLISION, *and hence its quantum query complexity, is* $\Omega(k^{2/3})$.

The original result of the works cited above refer to the more general problem ELEMENT DISTINCTNESS, which requires the detection of one or more colliding pairs. This was proven by a randomized reduction from the problem COLLISION which detects between a bijection and a two-to-one function. However, the reduction is still valid for the special case we consider. As noticed by Ambainis [14], this reduction also implies the lower bound on the approximation degree.

3 A Quantum Algorithm for Group Commutativity

We are given a black-box group G with generators $g_1, \ldots, g_k$. The problem is to decide if G is abelian. For technical reasons (see the proof of Lemma 1), and without loss of generality, we assume that g_1 is the identity element.

We denote by S_l the set of all l-tuples of distinct elements of $\{1, \ldots, k\}$. For any $u = (u_1, \ldots, u_l) \in S_l$, we denote by g_u the group element $g_{u_1} \cdots g_{u_l}$. Our algorithm is based on the quantization of a random walk on S_l^2. We will also adapt an approach due to [6]. For this we generalize Lemma 1.3 of [6] for random elements from S_l. Then we show how to walk on S_l^2 for finding a non commutative element in G, if there is any. We will conclude using Theorem 1.

In this section, we let $p = \frac{l(l-1)+(k-l)(k-l-1)}{k(k-1)}$. Observe that when $k = 2l$, then $p = \frac{l-1}{2l-1} \leq \frac{1}{2}$. Moreover, when $l = o(k)$, then $1 - p = \Theta(l/k)$.

Lemma 1. *Let $K \neq G$ be a subgroup of G. Then $\Pr_{u \in S_l}[g_u \notin K] \geq \frac{1-p}{2}$.*

Proof. First we fix a total order (equivalently, a permutation) σ of $\{1, \ldots, k\}$, and we denote by S_l^σ that subset of l-tuples in S_l which respect the total order σ. In other words, $u = (u_1, \ldots, u_l) \in S_l^\sigma$ iff $\sigma^{-1}(u_i) < \sigma^{-1}(u_{i+1})$ for all $1 \leq i < l$. Since $(S_l^\sigma)_\sigma$ is an equitable partition of S_l, picking a random element from S_l is the same as first picking a random permutation σ, and then picking a random element $u \in S_l^\sigma$. Therefore it is enough to prove the theorem for any fixed order σ. The reader may find it helpful to take σ to be the identity permutation to understand the idea behind the proof.

We denote $g'_j = g_{\sigma(j)}$ for every $j = 1, \ldots, k$. Let i be the smallest index such that $g'_i \notin K$. Such an i exists since $K \neq G$. Let j be such that $g'_j = g_1$, the identity element.

Fix an ordered l-tuple u such that $g'_i \notin u$ and $g'_j \in u$. We denote by v the ordered l-tuple where g'_j has been deleted from u, and g'_i has been inserted into it at the appropriate position (that respects the total order). Formally, if $u = (u_1, \ldots, u_m, u_{m+1}, \ldots, u_l)$ such that $\sigma^{-1}(u_m) < i < \sigma^{-1}(u_{m+1})$, then v is obtained by deleting g'_j from the $(l+1)$-tuple $(u_1, \ldots, u_m, \sigma(i), u_{m+1}, \ldots, u_l)$.

Let $a = g_{u_1} g_{u_2} \cdots g_{u_m}$, and $b = g_{u_{m+1}} \cdots g_{u_l}$. Then $g_u = ab$ and $g_v = ag_i'b$. Note that because of the choice of i, $a \in K$. If $g_u = ab \in K$ so that $b \in K$ as well, then $g_v \notin K$. Therefore $\Pr_{u \in S_l^\sigma}[g_u \in K | i \in u \text{ xor } j \in u] \leq \frac{1}{2}$.

Since $\Pr_{u \in S_l^\sigma}[i, j \in u \text{ or } i, j \notin u] = p$, we conclude that $\Pr_{u \in S_l^\sigma}[g_u \in K] \leq (1-p) \times \frac{1}{2} + p \times 1$. □

With the approach of [6] and from Lemma 1, we can generalize easily Lemma 1.1 of [6].

Lemma 2. *If G is non commutative then $\Pr_{u,v \in S_l}[g_u g_v \neq g_v g_u] \geq \frac{(1-p)^2}{4}$.*

Proof. If G is non-commutative, then the centre $C(G)$ of G is a proper subgroup. With probability at least $(1-p)/2$, g_u does not belong to $C(G)$ for a random $u \in S_l$ (Lemma 1). Conditioned upon this event, the probability that for a random $v \in S_l$, g_v does not belong to the centralizer of g_u is also at least $(1-p)/2$. □

Let t_u be the balanced binary tree with l leaves, whose leaves are from left to right the elements g_{u_i}, for $i = 1, \ldots, l$, and such that each internal node is the group product of its two successors. If l is not a power of 2, we put the deepest leaves to the left.

The random walk on S_l^2 that forms the basis of our quantum algorithm will consist of two independent simultaneous walks on S_l. For a pair (u, v) of l-tuples, we will maintain the binary trees t_u, t_v as described above as the data.

The random walk on S_l

Suppose the current state is $u \in S_l$.
With probability $1/2$ stay at u; with probability $1/2$, do the following:
- Pick a uniformly random position $i \in \{1, \ldots, l\}$, and a uniformly random index $j \in \{1, \ldots, k\}$.
- If $j = u_m$ for some m, then exchange u_i and u_m, else, set $u_i = j$.
- Update the tree t_u (using $O(\log l)$ group operations).

Lemma 3. *The spectral gap of the walk described above is at least $\frac{c}{l \log l}$, for a universal constant $c \geq \frac{1}{8e}$, provided $l \leq k/2$.*

Proof. First, we will show that the random walk mixes rapidly using a "coupling argument". Then, using a relation between mixing time and the second largest eigenvalue, we will get a bound on the spectral gap.

Note that the walk is ergodic and has the uniform distribution on S_l as its stationary distribution π. Thus $\pi(u) = \frac{(k-l)!}{k!}$ for all u. Moreover, because of the self-loops, all the eigenvalues of the walk are non-negative.

Let P_x^t be the probability distribution on S_l obtained by performing t steps of the walk starting at x. Let $\Delta(t)$ be the maximum over all starting states $x \in S_l$ of the total variation distance $\|P_x^t - \pi\|$. Let τ (the *mixing time*) be the smallest t such that $\Delta(t') \leq \frac{1}{2e}$ for all $t' \geq t$.

A *coupling* for a Markov chain is a stochastic process on pairs of states (U_t, V_t) such that U_t and V_t, viewed marginally, each evolve according to the Markov chain, and if $U_t = V_t$, then $U_{t+1} = V_{t+1}$. The *coupling time* T is the maximum expected time (over all pairs of initial states (u,v)) for the states U_t, V_t to coincide: $T = \max_{u,v} E[\mathrm{argmin}_t \{U_t = V_t, U_0 = u, V_0 = v\}]$.

We will use the following facts about mixing of Markov chains:

1. [15–Proposition 2.2, Chapter 2] For walks with only non-negative eigenvalues, $\lambda^t \leq \Delta(t) \cdot (\min_u \pi(u))^{-1}$, where λ is the second largest eigenvalue. This bounds the second largest eigenvalue in terms of the total variation distance.
2. (see e.g., Ref. [16]) $\Delta(t) \leq 2 \exp(-\lfloor \frac{t}{\tau} \rfloor)$. This relates the total variation distance at any time t to the mixing time τ.
3. [17] $\tau \leq 2eT$. This bounds the mixing time τ in terms of the coupling time T.

Combining all three relations, taking t-th roots, and letting $t \to \infty$, we see that $\lambda \leq \exp(-\frac{1}{2eT}) \leq 1 - \frac{1}{4eT}$. Thus, the spectral gap is $1 - \lambda \geq \frac{1}{4eT}$.

A coupling for which $T \leq l \log l$ is the obvious one: for any pair $u, v \in S_l$, follow one step of the random walk with the same choice of random position i and index j. This is clearly a valid coupling.

Let d be the hamming distance between the two tuples u, v. This distance never increases during the process described above. Moreover, in one step of the process, the distance goes down by 1 with probability at least $\frac{d}{2l}$. This is because with probability d/l, the position i is one where u and v are different, and with probability at least $(k-l)/k$, the index j is not one from the positions where u and v are the same. Since $l \leq k/2$, the net probability that the distance decreases by 1 is at least $d/2l$.

By a straightforward calculation, the expected time T for the distance to go to zero is at most $2l \log l$ (since $d \leq l$). Using the relation described above between λ and T, we get our bound on the spectral gap. □

Theorem 3. *There is a quantum algorithm that solves* GROUP COMMUTATIVITY *problem with $O(k^{2/3} \log k)$ queries and time complexity $O(k^{2/3} \log^2 k)$.*

Proof. The walk is the above described walk on S_l^2. The database associated with a tuple $u \in S_l$ is the binary tree t_u. Using Szegedy's theorem 1, we need only compute the eigenvalue gap of the random walk and the initial success probability (in the uniform distribution).

The stationary distribution for the walk is the uniform distribution on $S_l \times S_l$. So, from Lemma 2 above, the success probability is at least $(1-p)^2/4$. The spectral gap for the walk is the same as that on S_l, i.e. $c/(l \log l)$, from Lemma 3.

Since we start with a uniform distribution over $|u, t_u\rangle|v, t_v\rangle$, where $u, v \in S_l$. The setup cost is at most $2(l-1)$ and the updating cost of the walk is $O(\log l)$. We will choose $l = o(k)$ so that $1 - p = \Theta(l/k)$. The total query cost is then

$$2(l-1) + O\left(\frac{1}{1-p}\sqrt{l \log l} \cdot \log l\right) = 2(l-1) + O\left(\frac{k}{\sqrt{l}} \log^{3/2} l\right).$$

This expression is minimized when $l = k^{2/3} \log k$, and the cost is $O(k^{2/3} \log k)$.

The time complexity overhead comes from the initialization and transition times that are both essentially equal to the time complexity of performing a Grover diffusion operation. For the initialization, we use a diffusion over S_l^2, whose time complexity is $O(\log(|S_l|^2)) = O(l \log k)$. For the transition, we use a diffusion over a set of size 2 tensor product with a diffusion over a set of size kl, therefore the corresponding time complexity is $O(\log(kl)) = O(\log k)$. □

4 Reduction from Unique Split Collision

We begin our presentation of the lower bound by considering the complexity of UNIQUE SPLIT COLLISION. This problem is at least as hard as UNIQUE COLLISION in its query complexity since any bounded-error algorithm for the former can be used to detect an arbitrary collision. The proof is omitted due to the lack of space.

Proposition 2. *The approximation degree and the quantum query complexity of* UNIQUE SPLIT COLLISION *are both* $\Omega(k^{2/3})$. *The randomized query complexity of this problem is* $\Omega(k)$.

We conclude by proving the same lower bound for GROUP COMMUTATIVITY as well. We thus show that the algorithm described in the previous section is almost optimal.

The group involved in the proof of the lower bound will be a subgroup G of $U(2k)$, the group (under matrix multiplication) of $2k \times 2k$ unitary matrices. The generators of G will be block diagonal, each with k blocks of dimension 2×2. Each block will be one of the following three (Pauli) matrices:

$$I = \begin{pmatrix} 1 & 0 \\ 0 & 1 \end{pmatrix}, \quad X = \begin{pmatrix} 0 & 1 \\ 1 & 0 \end{pmatrix}, \quad Z = \begin{pmatrix} 1 & 0 \\ 0 & -1 \end{pmatrix}.$$

The group may also involve the remaining Pauli matrix $Y = XZ = \begin{pmatrix} 0 & -1 \\ 1 & 0 \end{pmatrix}$. No pair of matrices amongst X, Y and Z commute. An encoding of the group consists in words $\sigma_1 \ldots \sigma_k$ of length k over the alphabet $\{I, X, Y, Z\}$ together with a sign vector $s = (s_1, s_2, \ldots, s_k)$ in $\{+1, -1\}^k$. A tuple $(s, \sigma_1, \ldots \sigma_k)$ represents the matrix $\text{diag}(s_1 \sigma_1, \ldots, s_k \sigma_k)$. We will call this encoding the *explicit encoding*.

Let a_j and b_j be generators that have the identity matrix in all their blocks except for the j-th. The j-th block is Z in a_j and X in b_j.

We describe a connection between UNIQUE SPLIT COLLISION and GROUP COMMUTATIVITY. Suppose the oracle for the problem UNIQUE SPLIT COLLISION computes the function $F : \{1, \ldots, k\} \to \{1, \ldots, k\}$. We associate a generator g_i of the type described above with each element i in the domain. The generator g_i is $a_{F(i)}$ if $i \leq k/2$, and it is $b_{F(i)}$ if $i > k/2$. As long as the function F is injective on the two intervals $\{1, \ldots, k/2\}, \{k/2+1, \ldots, k\}$, the set of generators $\{g_i\}$ consists of k distinct elements. None of these generators is contained in the span of the remaining generators.

It is straightforward to check that there is a collision in F (with one point on either side of $k/2$) iff the group generated by $\{g_i\}$ is non-commutative. We use this connection for proving our lower bound. While the main result uses a non-standard method, we first prove a weaker result which explains the intuition behind the final proof to the reader.

Theorem 4. *If non-unique encoding of group elements is allowed, the randomized and the quantum query complexity of* GROUP COMMUTATIVITY *are respectively* $\Omega(k)$ *and* $\Omega(k^{2/3})$.

Proof. Suppose we allow non unique encoding of the group G. We show that any algorithm A solving GROUP COMMUTATIVITY may be adapted to solve UNIQUE SPLIT COLLISION, with at most four times the query complexity of A. We then conclude our theorem using Proposition 2.

We construct a black-box group which may invoke the oracle for UNIQUE SPLIT COLLISION to implement group operations. The encoding of the group elements will be either the explicit encoding defined above, or an element of $\{1,\ldots,k\}$. When the encoding is an integer $i \in \{1,\ldots,k\}$, it represents the generator g_i. When an integer i is involved in a group operation, we query the oracle for F at i, and construct g_i as defined above. One more query to F is required to erase the value of the function. Group operations can be performed without incurring any further calls to F. Operations on previously computed products also do not cost any queries. Therefore a group operation involves F at most four times, when both of the elements are encoded by integers. The oracle hides the group G with this non-unique encoding, and the input is the sequence of encodings $1, 2, \ldots, k$. □

In the case of unique encoding of group elements by the black-box, the reduction above is not guaranteed to work. The reason is that non-trivial products of generators may evaluate to the value of a generator. These products are represented in explicit form, and therefore our simulation possibly uses two different representations for the generators. We can nevertheless modify our simulation to work, while maintaining essential properties of the algorithm. In the classical model, our simulation preserves the number of oracle queries. In the quantum model, our simulation will produce a polynomial that approximates UNIQUE SPLIT COLLISION and has degree of the order of the number of queries made by the commutativity algorithm.

In our arguments, we assume that the algorithm never queries the black-box with encodings that did not result from previous queries to the oracle. This family of algorithm are usually called *generic algorithms*. This notion was introduced to cryptography by Nechaev [18] and Shoup [19]. By suitably randomizing the encoding such as in [20], we can ensure that the probability that the algorithm chances upon a valid pair of encoded group elements is $o(1)$, if this input does not result from previous queries. If we choose n, the encoding length, to be $\Omega(\log|G|)$, this probability would be exponentially small in n. We can therefore assume in the black-box group setting that a correct algorithm is always generic, and we make this assumption until the end of this section. These arguments do not generalize easily to the quantum setting; we leave a proof to a more complete version of this paper. We start with the classical simulation.

Theorem 5. *With unique encoding of group elements, the randomized query complexity of* GROUP COMMUTATIVITY *is* $\Omega(k)$.

Proof. Our reduction from UNIQUE SPLIT COLLISION works essentially because all the generators are distinct, and because no generator is contained in the span of the remaining $k-1$ generators. We modify the simulation in the proof of Theorem 4 so that we record the value of the function F at any point which is queried. We retain the explicit encoding for all group elements except the generators as the encoding for the black-box group. The generators are represented by integers $1, \ldots, k$. A non-trivial product may equal a generator g_i only if the product contains this generator. The value of F at point i would necessarily have been queried for the algorithm to have computed this product. The index i can therefore be located by examining the record of all queries made thus far in the algorithm, and used to encode the generator. □

For the modified simulation in the quantum case, we introduce another *implicit* encoding of elements of G as tuples $(s, x_1, x_2, \ldots, x_k)$, where $s \in \{+1, -1\}^k$, and $x_i \in \{0, 1\}$. A word in this implicit encoding represents the group element $\mathrm{diag}(s_1 I, s_2 I, \ldots, s_k I) \cdot g_1^{x_1} g_2^{x_2} \cdots g_k^{x_k}$. This is a unique encoding of elements in G.

As in the proof of the classical lower bound, we restrict ourselves to generic algorithms. In a generic quantum black-box group algorithm, along every computational path, the queries to the oracle involve either a generator, or a product that was previously computed along that path.

Theorem 6. *With unique encoding of group elements, any generic quantum algorithm for* GROUP COMMUTATIVITY *performs* $\Omega(k^{2/3})$ *group operations (queries to the group oracle).*

Proof. In using a generic algorithm for GROUP COMMUTATIVITY to solve UNIQUE SPLIT COLLISION we now use the implicit encoding of group elements. The generators are specified in this notation, and we will maintain this representation for all the intermediate products that are computed during the algorithm. The query cost of simulating a group operation is no longer $O(1)$. Indeed, we may need up to $O(k)$ queries to F to implement a multiplication of two elements. Nevertheless, we argue that the degree of the polynomial that results from this simulation is of the order of the query cost of the commutativity algorithm.

We refine the proof of Proposition 1 due to [3, 12] to claim that after t queries:

– The amplitude of any basis state is a polynomial of degree at most $O(t)$ in the variables x_{ij}.
– Fix a classical basis state. If a generator g_i occurs in the implicit encoding of group elements stored in the special registers, then a variable x_{ij} (for some j) is a factor of the polynomial that gives the amplitude of that state.

This ensures that in our simulation, the degree of the polynomial corresponding to a basis state does not increase as we query F to implement a group operation involving a previously computed product. In making this claim, we rely on the fact that $x_{ij} \in \{0, 1\}$ for inputs of interest to us, so $x_{ij}x_{ij'} = \delta_{jj'}x_{ij}$. (The degree goes up by $O(1)$ when at least one of the operands is a fresh generator.)

As a consequence, we derive a polynomial of degree of the order of the number of queries made by the commutativity algorithm, and this approximates UNIQUE SPLIT COLLISION. We can thus conclude the same lower bound as in Proposition 2 for testing commutativity as well. □

Note that the complications due to the unique encoding requirement do not arise if we are concerned with the number of accesses to the input generators. For arbitrary (possibly non-generic) quantum algorithms, the same reduction also directly gives a bound on this notion of complexity.

Proposition 3. *The lower bound of $\Omega(k^{2/3})$ above also holds for the query complexity of any quantum algorithm, if the generators to a possibly known group are specified by an input oracle.*

References

1. Grover, L.: A fast quantum mechanical algorithm for database search. In: Proc. of 28th ACM STOC. (1996) 212–219
2. Mosca, M.: Quantum Computer Algorithms. PhD thesis, Univ. of Oxford (1999)
3. Beals, R., Buhrman, H., Cleve, R., Mosca, M., Wolf, R.: Quantum lower bounds by polynomials. J. of the ACM **48** (2001) 778–797
4. Babai, L., Szemerédi, E.: On the complexity of matrix group problems I. In: Proc. of 25th IEEE FOCS. (1984) 229–240
5. Watrous, J.: Quantum algorithms for solvable groups. In: Proceedings of 33rd Symposium on Theory of Computing, ACM (2001) 60–67
6. Pak, I.: Testing commutativity of a group and the power of randomization. Electronic version at http://www-math.mit.edu/~pak/research.html (2000)
7. Szegedy, M.: Quantum speed-up of markov chain based algorithms. In: Proc. of 45th IEEE FOCS. (2004) 32–41 Also arXiv.org report quant-ph/0401053.
8. Ambainis, A.: Quantum walk algorithm for Element Distinctness. In: Proceedings of 45th IEEE FOCS. (2004) 22–31
9. Buhrman, H., Spalek, R.: Quantum verification of matrix products. Technical Report quant-ph/0409035, arXiv archive (2004)
10. Nielsen, M., Chuang, I.: Quantum Computation and Quantum Information. Cambridge University Press (2000)
11. Kitaev, A., Shen, A., Vyalyi, M.: Classical and Quantum Computation. Volume 47 of Graduate Studies in Mathematics. AMS (2002)
12. Aaronson, S., Shi, Y.: Quantum lower bound for the collision problem. J. of the ACM **51** (2004) 595–605
13. Kutin, S.: A quantum lower bound for the collision problem. Technical Report quant-ph/0304162, arXiv archive (2003)
14. Ambainis, A.: Quantum lower bounds for collision and element distinctness with small range. Technical Report quant-ph/0305179, arXiv archive (2003)
15. Sinclair, A.: Algorithms for Random Generation and Counting: A Markov Chain Approach. Progress in theoretical computer science. Birkhäuser, Boston (1993)

16. Aldous, D.: Random walks on finite groups and rapidly mixing Markov chains. In: Séminaire de Probabilités XVII. Volume 986 of Lecture Notes in Mathematics., Springer-Verlag (1981-82) 243-297
17. Griffeath, D.: Coupling methods for Markov processes. In Rota, G.C., ed.: Studies in Probability and Ergodic Theory. Academic Press (1978) 1-43
18. Nechaev, V.: Complexity of a determinate algorithm for the discrete logarithm. Mathematical Notes **55** (1994) 165-172
19. Shoup, V.: Lower bounds for discrete logarithms and related problems. In: Proc. of Eurocrypt. (1997) 255-266
20. Schnorr, C., Jakobsson, M.: Security of signed ElGamal encryption. In: Proc. of 6th Asiacrypt. (2000) 73-89

Semantic-Based Code Obfuscation by Abstract Interpretation

Mila Dalla Preda and Roberto Giacobazzi

Dipartimento di Informatica, Università di Verona,
Strada Le Grazie 15, 37134 Verona Italy
dallapre@sci.univr.it, roberto.giacobazzi@univr.it

Abstract. In this paper we introduce a semantic-based approach for code obfuscation. The aim of code obfuscation is to prevent malicious users to disclose properties of the original source program. This goal can be precisely modeled by abstract interpretation, where the hiding of properties corresponds to abstract the semantics. We derive a general theory based on abstract interpretation, where the potency of code obfuscation can be measured by comparing hidden properties in the lattice of abstract interpretations. Semantic-based code obfuscation is applied to show that well known program transformation methods, such as constant propagation, can be seen as code obfuscation.

Keywords: Code Obfuscation, Abstract Interpretation, Program Transformation, Semantics.

1 Introduction

Code obfuscation is a program transformation typically intended to prevent reverse engineering [2, 4, 5, 6]. A number of results are known in the literature providing obfuscation algorithms such as: *Layout* transformations, which remove source code formatting and scramble identifiers; *control* transformations, which affect the control flow of the program; and *data* transformations, which operate obfuscating the data structures used in the program [5]. The major negative result on code obfuscation is given in Barak *et al.* [1]. They prove that there is no obfuscation method that works for any program and it is able to transform them in such a way that the only properties which can be disclosed are those which can be derived from the input/output semantics. This result is not as bad as it might seen, in fact even though the "ideal" obfuscator of Barak *et al.* does not exist, software obfuscation would be still useful when employed for hiding specific code properties and working for specific classes of programs [14]. The classical notion of code obfuscation of Collberg *et al.* [4, 5, 6] defines an obfuscator as a potent transformation that preserves the *observable behavior* of programs. In this setting a transformation is *potent* when the obfuscated program is more complex than the original one. Clearly this definition of code obfuscator relies on a fixed metric for measuring program complexity, and finding such metrics is a major challenge in practical code obfuscation algorithms.

The Problem

The major drawback of most software obfuscation techniques is that they do not have a well found theoretical base, and thus it is unclear how effective they are. Even if semantic preservation is implied in code obfuscation [18], the lack of a complete formal setting where these program transformations can be studied defects any possibility of comparing them with respect to their ability to obfuscate program properties. The main problem here is to fix a measure for potency. Usually syntactic (textual) measures are considered, such as code length, nesting-levels, fan-in-out complexity, branching, *etc* [5]. Semantic-based measures are instead less common, even thought they may provide a deeper insight in the true potency of code obfuscation. In order to understand this point we need to model attackers, *i.e.*, code de-obfuscation techniques. Static program analysis is the standard method for making reverse-engineering. Recently dynamic attacks have also been considered in [3] for strengthening static ones for de-obfuscation. Both static and dynamic attacks strongly relies upon program semantics: the first corresponds precisely to a decidable semantic abstraction, while the last are based on the concrete semantics, *e.g.*, interpreters. In both cases syntactic measures can be misleading. More significant measures have to be derived from semantics and this, as far as we know, is an open problem.

Main Results

In this paper we consider the lattice of abstract interpretations as the domain for measuring potency. The goal of code obfuscation is to prevent reverse engineering (*e.g.*, by static or dynamic analysis) to grasp sensible properties on program's structure and semantics. It is well known that static analysis can be completely and fully specified as an abstract interpretation [10], *i.e.*, as an approximation of the concrete semantics of programs. Similarly, dynamic analysis can be seen as a possibly non decidable approximation of the concrete semantics. In this sense, code obfuscation can be seen as a way to prevent that some information about program behaviour is disclosed by an abstract interpretation of its semantics. In order to apply abstract interpretation as a model for attackers, we need to replace syntactic code obfuscators with corresponding semantic transformations. Recently, Cousot & Cousot in [12] have introduced a semantic-based formalization of program transformation based on abstract interpretation. In this construction the relation between syntactic and semantic transformations is specified by an abstract interpretation, where syntax is an abstraction of semantics. This allows us to mirror code obfuscators, viewed as syntactic transformations, in the semantics, by considering corresponding semantic obfuscators instead. Therefore the role of abstract interpretation in a semantic-based approach to code obfuscation is crucial: By fixing a formal relation between syntactic and semantic transformations, abstract interpretation provides the most general setting where attackers can be compared by comparing abstractions, leading us to derive a semantic-based metric for the potency of code obfuscation. Traditionally code obfuscation is intended to preserve input/output (denotational) semantics of programs. This is again an unreasonable restriction: Semantics at different levels of abstraction can be related by abstract interpretation in a hierarchy of semantics [9]. Therefore, any program transformation τ which preserves a given semantics in the hierarchy may act as a code obfuscation for those properties that are not preserved by the transformation. The idea is that the transformed program $\tau[\![P]\!]$ is more complex (obscure) then the original program P, *i.e.*, τ is potent, when there exists a semantic property, *i.e.*, an abstract semantics, which is not preserved by the transformed program $\tau[\![P]\!]$. Potency is therefore strictly related with the rate of abstraction of the most concrete preserved semantics of a code transformation. This corresponds to map code transformations to the lattice of abstract interpretations to measure their obfuscating potency. We introduce a constructive systematic method for deriving the most concrete preserved abstraction of a generic code transformation. Then we generalize Collberg *et al.* definition of code obfuscation [4,5] by considering as obfuscators those transformations that mask some abstractions in the lattice of abstract interpretations. This means that, in principle, *any program transformation may potentially act as a code obfuscator*. We show how a well known program transformation for solving the constant propagation problem, can be seen as a code obfuscator. In this case the transformation acts as a code obfuscator relatively to the command-line structure.

2 Preliminaries

Abstract Interpretation. The notation $\langle C, \leq, \vee, \wedge, \top, \bot \rangle$ denotes a complete lattice C, with ordering $\leq$, *lub* $\vee$, *glb* $\wedge$, greatest element $\top$, and least element $\bot$. $x \in C$ is *meet-irreducible* if $x = a \wedge b$ then $x \in \{a, b\}$. The set of meet-irriducibile elements in C is denoted $Mirr(C)$. The downward closure of $S \subseteq C$ is $\downarrow S \stackrel{\text{def}}{=} \{x \in C | \exists y \in S . x \leq y\}$, and for $x \in C$, $\downarrow x$ is a shorthand for $\downarrow \{x\}$, while the upward closure $\uparrow$ is dually defined. Abstract domains can be formulated either in terms of Galois connections or upper closures operators [11]. An *upper closure operator* on a ordered set C, $\mu : C \to C$, is a monotone, idempotent, and extensive ($\forall x \in C . x \leq \mu(x)$) function. With $uco(C)$ we denote the set of all upper closure operators of C. Each closure $\mu \in uco(C)$ is uniquely determined by the set of its fix-points $\mu(C)$ [17]. $X \subseteq C$ is a set of fix-points of an upper closure operator on C iff X is a *Moore family* of C, *i.e.*, $X = \mathcal{M}(X) \stackrel{\text{def}}{=} \{\wedge S | S \subseteq X\}$ - where $\wedge \emptyset = \top \in \mathcal{M}(X)$. If C is a complete lattice also $uco(C)$ is a complete lattice denoted by $\langle uco(C), \sqsubseteq, \sqcup, \sqcap, \lambda x.\top, \lambda x.x \rangle$ where, given two closures η and μ, $\eta \sqsubseteq \mu$ iff $\mu(C) \subseteq \eta(C)$. On the other side a concrete domain C and an abstract domain A form a *Galois Connection* (GC), denoted by (C, α, A, γ), when $\alpha : C \to A$ and $\gamma : A \to C$ define an adjunction, namely $\forall x \in C, y \in A : \alpha(x) \leq y \Leftrightarrow x \leq \gamma(y)$. Given a GC (C, α, A, γ) the closure corresponding to the abstract domain A is $\rho = \gamma \circ \alpha$ on C. If ρ is a closure in C and $\iota : \rho(C) \to A$ is an isomorphism of complete lattices (with inverse ι^{-1}), then $(C, \iota \circ \rho, A, \iota^{-1})$ is a GC. Given a function $f : C \to C$ and a closure $\alpha \in uco(C)$, then $f^\sharp$ is a *correct (sound)* approximation of f in α when $f^\sharp : \alpha(C) \to \alpha(C)$ and $\alpha \circ f \circ \alpha \leq f^\sharp$. $f^\alpha \stackrel{\text{def}}{=} \alpha \circ f \circ \alpha$ is the *best correct approximation* of f in α [11]. The abstraction α is *complete*

Table 1. Abstract syntax

Arithmetic expressions	$E \in \mathbb{E} \quad E ::= n \mid X \mid E_1 - E_2$
Boolean expressions	$B \in \mathbb{B} \quad B ::= E_1 < E_2 \mid B_1 \vee B_2 \mid \neg B_1 \mid \textit{true} \mid \textit{false}$
Program actions	$A \in \mathbb{A} \quad A ::= X := E \mid X := ? \mid B$
Commands	$C \in \mathbb{C} \quad C ::= L_1 : A \to L_2$ where $L_1, L_2 \in \mathbb{L} \cup \{\text{\textit{f}}\}$ and $\text{\textit{f}}$ is the undefined label

when $\alpha \circ f = f^\alpha$ [10, 16]. The point-wise ordering on $uco(C)$ corresponds to the standard ordering used to compare abstract domains with regard to their precision: Let $\rho_i \in uco(C)$ and $A_i = \rho_i(C)$, then A_1 is more precise then A_2 (*i.e.*, A_2 is an abstraction of A_1) iff $A_1 \sqsubseteq A_2$ in $uco(C)$. Let $\{A_i\}_{i \in I} \subseteq uco(C)$: $\sqcup_{i \in I} A_i$ is the most concrete among the domains in $uco(C)$ which is abstraction of all the A_i's, *i.e.*, $\sqcup_{i \in I} A_i$ is the *least* (w.r.t $\sqsubseteq$) *common abstraction* of all the A_i's, and $\sqcap_{i \in I} A_i$ is the well-known *reduced product* of all the A_i's, namely the most abstract among the domains in $uco(C)$ which is more concrete then every A_i. *Complementation* corresponds to the inverse of reduced product [7], namely an operator that, given two domains $C \sqsubseteq D$, gives as result the most abstract domain $C \ominus D$, whose reduces product with D is exactly C (*i.e.*, $(C \ominus D) \sqcap D = C$). Therefore we have that $C \ominus D \stackrel{\text{def}}{=} \sqcup \{E \in uco(C) \mid D \sqcap E = C\}$. It has been proved in [15] that $\mathcal{M}(Mirr(C) \setminus Mirr(D)) = C \ominus D$.

Semantics. Given a transition system $\langle \Sigma, \frown \rangle$, where Σ is a nonempty set of states and $\frown \subseteq \Sigma \times \Sigma$ is the transition relation over states, we denote by Σ^+ and $\Sigma^\omega \stackrel{\text{def}}{=} \mathbb{N} \to \Sigma$ respectively the finite nonempty and the infinite sequences of symbols in Σ. Let $\sigma \in \Sigma^\infty \stackrel{\text{def}}{=} \Sigma^+ \cup \Sigma^\omega$ be a generic (finite or infinite) sequence of states then $|\sigma| \in \mathbb{N} \cup \omega$ is its length, σ_i is its i-th element, and σ_f is the final state if $\sigma \in \Sigma^+$. A finite (infinite) sequence of states σ is a program *trace* when for all $i < |\sigma| : \langle \sigma_i, \sigma_{i+1} \rangle \in \frown$, denoted $\sigma = \sigma_1^\frown \sigma_2^\frown \ldots^\frown \sigma_i^\frown \ldots$. In the following $\sigma_i^\frown \sigma$ denotes the concatenation of a state σ_i with a trace σ. In this context μ is a *subtrace* of σ if there exists $i, j \in [0, |\sigma|)$, where $i < j$, and $\mu = \sigma_i^\frown \ldots^\frown \sigma_j$. The *maximal trace semantics* of a transition system is $\tau^\infty \stackrel{\text{def}}{=} \tau^+ \cup \tau^\omega$, where τ^+ is the set of finite traces and τ^ω the set of infinite traces [9] of τ. From now on the trace semantics τ^∞ of the program P is considered as the concrete semantics $S[\![P]\!] \subseteq \Sigma^\infty$. In [9] Cousot defines a hierarchy of semantics, where the big-step, termination and nontermination, Plotkin's natural, Smyth's demonic, Hoare's angelic relational and corresponding denotational, Dijkstra's predicate transformer weakest-precondition and weakest-liberal precondition and Hoare's partial and total axiomatic semantics, have all been derived by successive abstractions from the maximal trace semantics of a transition system τ^∞. In this setting $uco(\wp(\Sigma^\infty))$ is the lattice of abstract semantics, namely each closure in $uco(\wp(\Sigma^\infty))$ represents an abstraction of trace semantics. For example the (natural) denotational semantics $\mathcal{D} \in uco(\wp(\Sigma^\infty))$ can be formalized as an abstract interpretation of τ^∞: $\mathcal{D}(X) = \{\, \sigma \in \Sigma^+ \mid \exists \delta \in X^+. \sigma_0 = \delta_0 \wedge \sigma_f = \delta_f \,\} \cup \{\, \sigma \in \Sigma^\omega \mid \exists \delta \in X^\omega. \sigma_0 = \delta_0 \,\}$, where $X^+ \stackrel{\text{def}}{=} X \cap \Sigma^+$ and $X^\omega \stackrel{\text{def}}{=} X \cap \Sigma^\omega$. $\mathcal{D}$ ab-

Table 2. Semantics

Arithmetic expressions	$\mathbf{A}[\![n]\!]\rho \stackrel{\text{def}}{=} n$, $\mathbf{A}[\![X]\!]\rho \stackrel{\text{def}}{=} \rho(X)$
$\mathbf{A}: \mathcal{E}[\![\chi]\!] \to D_\Upsilon$ and $var[\![E]\!] \subseteq \chi$	$\mathbf{A}[\![E_1 - E_2]\!]\rho \stackrel{\text{def}}{=} \mathbf{A}[\![E_1]\!]\rho - \mathbf{A}[\![E_2]\!]\rho$
Boolean expressions	$\mathbf{B}[\![E_1 < E_2]\!]\rho \stackrel{\text{def}}{=} \mathbf{A}[\![E_1]\!]\rho < \mathbf{A}[\![E_2]\!]\rho$
$\mathbf{B}: \mathcal{E}[\![\chi]\!] \to \{true, false, \Lambda\}$ and $var[\![B]\!] \subseteq \chi$	$\mathbf{B}[\![B_1 \vee B_2]\!]\rho \stackrel{\text{def}}{=} \mathbf{B}[\![B_1]\!]\rho \vee \mathbf{B}[\![B_2]\!]\rho$
	$\mathbf{B}[\![\neg B]\!]\rho = \neg \mathbf{B}[\![B]\!]$, $\mathbf{B}[\![true/false]\!]\rho \stackrel{\text{def}}{=} true/false$
Program actions	$\mathbf{S}[\![B]\!]\rho \stackrel{\text{def}}{=} \{\,\rho' \mid \mathbf{B}[\![B]\!]\rho' = true \wedge \rho' = \rho\,\}$
$\mathbf{S}: \mathbf{A} \to (\mathcal{E}[\![\chi]\!] \to \wp(\mathcal{E}[\![\chi]\!]))$ and $var[\![A]\!] \subseteq dom(\rho)$	$\mathbf{S}[\![X :=?]\!]\rho \stackrel{\text{def}}{=} \{\,\rho' \mid \exists z \in \mathbb{Z}: \rho' = \rho[X := z]\,\}$
	$\mathbf{S}[\![X := E]\!]\rho \stackrel{\text{def}}{=} \{\rho[X := \mathbf{A}[\![E]\!]\rho]\}$
	$\mathbf{S}[\![true]\!]\rho \stackrel{\text{def}}{=} \mathbf{S}[\![skip]\!]\rho \stackrel{\text{def}}{=} \{\rho\}$

stracts away the history of the computation by observing the input/output relation of finite traces and the input of diverging computations only. In the following we consider the simple imperative language defined in [12] with abstract syntax in Table 1, together with the following basic functions: $lab[\![L_1 : A \to L_2]\!] \stackrel{\text{def}}{=} L_1$, $lab[\![P]\!] \stackrel{\text{def}}{=} \{lab[\![C]\!] | C \in P\}$, $var[\![L_1 : A \to L_2]\!] \stackrel{\text{def}}{=} var[\![A]\!]$, $var[\![P]\!] \stackrel{\text{def}}{=} \cup_{C \in P} var[\![C]\!]$, $succ[\![L_1 : A \to L_2]\!] \stackrel{\text{def}}{=} L_2$ and $act[\![L_1 : A \to L_2]\!] \stackrel{\text{def}}{=} A$. Each variable $X \in var[\![P]\!]$ has values in the semantic domain D, where the undefined value is denoted by Υ and $D_\Upsilon \stackrel{\text{def}}{=} D \cup \{\Upsilon\}$. We define an *environment* $\rho \in \mathcal{E}$ as a map from variables $X \in dom(\rho)$ to values $\rho(X) \in D_\Upsilon$. Let χ be a subset of variables, then $\rho|_\chi$ is the restriction of environment ρ to the domain $dom(\rho) \cap \chi$. The semantics is in Table 2. A *state* $\sigma_i \in \Sigma \stackrel{\text{def}}{=} \mathcal{E} \times \mathbb{C}$ is a pair $\langle \rho, C \rangle$, where C is the next command to be executed in the environment ρ. The *transition relation* between states specifies the set of states that can be reached from a given state: $R_P(\langle \rho, C \rangle) \stackrel{\text{def}}{=} \{\,(\rho', C') \mid \rho' \in \mathbf{S}[\![act(C)]\!]\rho,\ succ[\![C]\!] = lab[\![C']\!],\ \rho, \rho' \in \mathcal{E}[\![P]\!],\ C' \in P\,\}$. The *finite traces* of a program $P \in \mathbb{P}$ are obtained by the computation of the $lfp_\subseteq F[\![P]\!]$, while the *infinite traces* by the computation of the $gfp_\subseteq F[\![P]\!]$, where $F[\![P]\!](X) \stackrel{\text{def}}{=} X \cup \{\,\sigma_i \frown \sigma_i' \frown \sigma \mid \sigma_i' \in R_P(\sigma_i),\ \sigma_i' \frown \sigma \in X\,\}$. We denote with $S[\![P]\!]$ the set of finite and infinite traces of P.

Program Transformation. In [12] the authors define a language-independent methodology for systematically derive syntactic program transformations from semantic ones by mean of abstract interpretation (see Fig. 1). Given a program $P \in \mathbb{P}$ and a *syntactic transformation* $\tau : \mathbb{P} \to \mathbb{P}$ which returns the transformed program $\tau[\![P]\!] \in \mathbb{P}$, we have that the corresponding *semantic transformation* t takes the semantics $S[\![P]\!] \in \wp(\Sigma^\infty)$ of P, and returns the semantics $S[\![\tau[\![P]\!]]\!]$ of the transformed program. We consider programs as an abstraction of their semantics, as formalized by the Galois connection $(\langle \wp(\Sigma^\infty); \sqsubseteq \rangle, p, \langle \mathbb{P}; \leq \rangle, S)$, where $p[\mathcal{S}]$ is the simplest program whose semantics upper approximates $\mathcal{S} \in \wp(\Sigma^\infty)$ [12]. Therefore the *semantic transformation* $t : \wp(\Sigma^\infty) \to \wp(\Sigma^\infty)$ induced by the syntactic transformation τ is $t(S[\![P]\!]) = S[\![\tau[p(S[\![P]\!])]\!]]$, while the *syntactic transformation* τ induced by the semantic transformation t is $\tau[\![P]\!] = p(t(S[\![P]\!]))$. A program transformation is *correct* if at some level of abstraction it is meaning

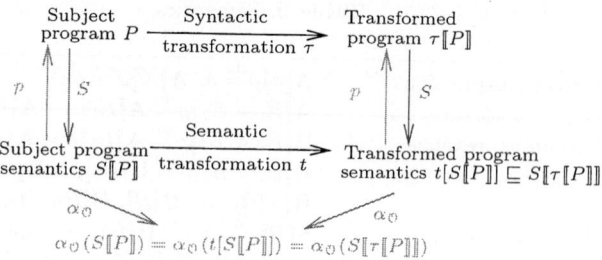

Fig. 1. Syntactic-Semantic Program Transformations

preserving, namely a syntactic transformation τ is correct w.r.t. an observational abstraction $\alpha_\mathcal{O}$ if $\alpha_\mathcal{O}(S[\![P]\!]) = \alpha_\mathcal{O}(S[\![\tau[\![P]\!]]\!])$ [12].

3 Semantic-Based Program Obfuscators

Traditionally a program transformation $\tau : \mathbb{P} \to \mathbb{P}$ is an obfuscation if: 1) it is a potent transformation and 2) P and $\tau[\![P]\!]$ have the same observational behavior, *i.e.,* if P fails to terminate or it terminates with an error condition then $\tau[\![P]\!]$ may or may not terminate; otherwise $\tau[\![P]\!]$ must terminate and produce the same output as P [4,5,6]. This definition requires that given an initial state σ_0, if $\sigma_0^\frown \sigma \in S[\![P]\!]^\omega$ or $\sigma_0^\frown \sigma^\frown \sigma_e \in S[\![P]\!]^+$, where $\sigma_e \in Err$ with Err being the set of error states, then $\sigma_0^\frown \sigma' \in S[\![\tau[\![P]\!]]\!]^\infty$; otherwise if $\sigma \in S[\![P]\!]^+$ then $\sigma \in \mathcal{D}(S[\![\tau[\![P]\!]]\!]^+)$. We denote by $\mathcal{C}$ the family of these code transformations.

3.1 Abstract Interpretation-Based Code Obfuscation

In this section we specify what is hidden by a syntactic transformation, by introducing a novel definition of code obfuscation based on semantics.

Definition 1. *Let $\tau : \mathbb{P} \to \mathbb{P}$ be a program transformation. τ is potent if there is a property $\alpha \in uco(\wp(\Sigma^\infty))$ such that: $\alpha(S[\![P]\!]) \neq \alpha(S[\![\tau[\![P]\!]]\!])$.*

Therefore a transformation τ is potent when there exists a property α and a program P such that the approximation of the concrete trace semantics of P is different from the same approximation for the concrete trace semantics of $\tau[\![P]\!]$.

Example 1. Let us consider a program P and its transformed version P':

```
P:                              P':
L1: X:= ?   → L2                L1: X:= ?     → L2
L2: Y:= ?   → L3                L2: Y:= ?     → L3
L3: out:= (X - Y)²  → ↯         L3: out:= X²  → L4
                                L4: out:= out - 2XY → L5
                                L5: out:= out - Y²  → ↯
```

Let us define $\alpha \in uco(\wp(\Sigma^\infty))$ as $\alpha(X) \stackrel{def}{=} \{\,\sigma \in \Sigma^+ \,|\, \exists \eta \in X : |\sigma| = |\eta|\,\}$. The property α observes the length of program traces. In this case we have $\alpha(S[\![P]\!]) = \{\,\sigma \in \Sigma^+ \,|\, |\sigma| = 3\,\}$, while $\alpha(S[\![\tau[\![P]\!]]\!]) = \{\,\sigma \in \Sigma^+ \,|\, |\sigma| = 5\,\}$. It is clear that $\alpha(S[\![P]\!]) \neq \alpha(S[\![\tau[\![P]\!]]\!])$, i.e., the transformation obfuscates α.

In order to factor the observational semantics into preserved and masked properties, we define the most concrete property preserved by a transformation $\tau : \mathbb{P} \to \mathbb{P}$, as follows $\delta_\tau \stackrel{def}{=} \sqcap \{\varphi \in uco(\wp(\Sigma^\infty)) | \varphi(S[\![P]\!]) = \varphi(S[\![\tau[\![P]\!]]\!])\}$.

Lemma 1. *Given a transformation* $\tau : \mathbb{P} \to \mathbb{P}$, $\delta_\tau(S[\![P]\!]) = \delta_\tau(S[\![\tau[\![P]\!]]\!])$.

Given a program transformation τ we want to characterize the set of properties that are not preserved, i.e., obfuscated, by τ. By considering the transformation τ and the property $\alpha \in uco(\wp(\Sigma^\infty))$ that the attacker wants to observe, we note that $\alpha \ominus (\delta_\tau \sqcup \alpha)$ is precisely what the transformation τ hides of the property α. In fact when $\alpha \ominus (\delta_\tau \sqcup \alpha) \neq \top$ some parts of the property α has been lost in the transformation. In this case we say that the property α is *obfuscated* by the transformation τ, because that property cannot be observed on the semantics of the transformed program.

Definition 2. $O_{\delta_\tau} = \{\,\alpha \in uco(\wp(\Sigma^\infty)) \,|\, \alpha \ominus (\delta_\tau \sqcup \alpha) \neq \top\,\}$ *is the set of properties obfuscated by* $\tau : \mathbb{P} \to \mathbb{P}$.

Given $\delta \in uco(\wp(\Sigma^\infty))$, we define a δ-obfuscator as any potent program transformation $\tau : \mathbb{P} \to \mathbb{P}$, such that every program is equivalent to its obfuscated version w.r.t. the particular observational semantics δ.

Definition 3. $\tau : \mathbb{P} \to \mathbb{P}$ *is a δ-obfuscator if* $\delta = \delta_\tau$ *and* $O_\delta \neq \varnothing$

Lemma 2. *Given a δ-obfuscator* $\tau : \mathbb{P} \to \mathbb{P}$, *then for each* $\alpha \in O_\delta$ *there exists a program P such that* $\alpha(S[\![P]\!]) \neq \alpha(S[\![\tau[\![P]\!]]\!])$.

The above definition of obfuscator is a generalization of the classical one by Collberg et al., introduced at the beginning of this Section.

Theorem 1. *If τ is a $\mathcal{D}$-obfuscator then $\tau \in \mathcal{C}$.*

In particular we can define a *partial order* between obfuscating transformations, by considering the set of properties that the transformations hide. Given two program transformations τ and τ', then τ' is more potent than τ, denoted $\tau \leq_P \tau'$, if $O_{\delta_\tau} \subseteq O_{\delta_{\tau'}}$. We can also compare the potency of τ and τ' w.r.t. a particular property α, namely by measuring the approximation in the knowledge of that property that can be obtained by observing semantics. Let $\alpha \in O_{\delta_\tau} \cap O_{\delta_{\tau'}}$, then τ' is more potent than τ w.r.t. α, denoted $\tau \leq_\alpha \tau'$, if $\alpha \ominus (\delta_{\tau'} \sqcup \alpha) \sqsubseteq \alpha \ominus (\delta_\tau \sqcup \alpha)$. In this case τ' obfuscates the property α more than what τ does. From the structure of the lattice of abstract interpretations $uco(\wp(\Sigma^\infty))$ we can derive the some basic properties of O_δ and therefore of the potency of code obfuscations.

Proposition 1. *Let* $\delta, \mu \in uco(\wp(\Sigma^\infty))$ *and* τ, τ' *be program transformations:*
1) $O_\delta = \{\,\alpha \in uco(\wp(\Sigma^\infty)) \,|\, \alpha \not\sqsubseteq \uparrow \delta\,\}$; *2) If* $\mu \sqsubset \delta$ *then* $O_\mu \subset O_\delta$, *i.e.,* $\tau_\mu \leq_P \tau_\delta$;
3) $O_{\delta \sqcup \mu} = O_\delta \cup O_\mu$; *4) If* $\tau \leq_\delta \tau'$, *then for each property* $\mu \sqsubseteq \delta$ *we have* $\tau \leq_\mu \tau'$;
5) If $\tau \leq_{\delta \sqcap \mu} \tau'$, *then* $\tau \leq_\delta \tau'$ *and* $\tau \leq_\mu \tau'$.

3.2 Constructive Characterization of Potency

In this section we define a method for deriving δ_τ for a given syntactic transformation $\tau : \mathbb{P} \to \mathbb{P}$. By Definition 2, this provides a characterization of the potency of a code transformation. Let $t : \wp(\Sigma^\infty) \to \wp(\Sigma^\infty)$ be a semantic transformation and $K_{P,t} : uco(\wp(\Sigma^\infty)) \to uco(\wp(\Sigma^\infty))$ be a domain transformer that, given a property $\mu \in uco(\wp(\Sigma^\infty))$, returns the closest abstraction preserved by t on $P \in \mathbb{P}$: $K_{P,t} \stackrel{def}{=} \lambda\mu. \sqcap \{ \varphi \in uco(\wp(\Sigma^\infty)) \,|\, \mu \sqsubseteq \varphi \wedge \varphi(S[\![P]\!]) = \varphi(t(S[\![P]\!])) \}$. For a given $\mu \in uco(\wp(\Sigma^\infty))$, $K_{P,t}(\mu)$ is a closure operator on sets of traces. In order to characterize the set of its fix-points, we have to specify the set of traces $X \subseteq \Sigma^\infty$ preserved by the transformation t on a particular program P. The predicate $Pres_{P,t}(X)$, where $X \in \wp(\Sigma^\infty)$, precisely captures this notion, in fact $Pres_{P,t}(X)$ evaluates to *true* if and only if $\forall Y \subseteq S[\![P]\!] : Y \subseteq X \Rightarrow t(Y) \subseteq X$.

Lemma 3. $\{ X \in \wp(\Sigma^\infty) \,|\, Pres_{P,t}(X) \}$ *is preserved by t on program P.*

Theorem 2. $K_{P,t}(id) = \{ X \in \wp(\Sigma^\infty) \,|\, Pres_{P,t}(X) \}$

$K_{P,t}(id)$ models the most concrete property preserved by the transformation t for the program P. The generalization of this notion for all programs follows straightforwardly.

Corollary 1. *Let $\tau : \mathbb{P} \to \mathbb{P}$, then* $\delta_\tau = \bigsqcup_{P \in \mathbb{P}} K_{P, p \circ \tau \circ S}(id)$.

4 An Example: Obfuscation by Program Specialization

In this section we consider constant propagation as code obfuscation. This proves that our semantic-based approach to code obfuscation is adequate both to include a wide range of program transformation techniques and to compare them by extracting what is actually masked by the transformation. We follow Cousot & Cousot [12] in the definition of an algorithm for constant propagation.

4.1 Cousot's Constant Propagation

The *residual* $R[\![E]\!]\rho$ of an arithmetic or boolean expression E in an environment ρ is the expression resulting by the specialization of E in such environment (see Table 3). An expression $E \in \mathbb{E} \cup \mathbb{B}$ is *static* in the environment ρ, denoted $static[\![E]\!]\rho$, if it can be fully evaluated in ρ, that is $var[\![E]\!] \subseteq dom(\rho)$; otherwise E is *dynamic*. With $R[\![A]\!]\rho$ we denote the specialization of action A in the environment ρ (see Table 3). We consider syntactic and semantic program transformations relative to the constant propagation problem as formalized in [12], where the arguments of the program transformation are the semantics $S[\![P]\!]$ and the result of a constant detection static analysis $S^C[\![P]\!]$.

Abstract semantics: It is defined as $S^C = \alpha^C \circ S$, where α^C is given by:

$$\alpha^C(S[\![P]\!]) = \lambda L.\lambda X. \bigsqcup \{ \rho(X) \,|\, \sigma \in S[\![P]\!], C \in \mathbb{C}, \sigma_i = \langle \rho, C \rangle, lab[\![C]\!] = L \}$$

Table 3. Expression and Action specialization

$$
\begin{aligned}
&\underline{\text{Expressions}} \quad R \in \mathbb{E} \times \mathcal{E} \to \mathbb{E} \\
&R[\![n]\!]\rho = n \\
&R[\![X]\!]\rho = \text{if } X \in dom(\rho) \text{ then } \rho(X) \text{ else } X \\
&R[\![E_1 - E_2]\!]\rho = \text{let } E_1^r = R[\![E_1]\!]\rho \text{ and } E_2^r = R[\![E_2]\!]\rho \\
&\qquad \text{if } E_1^r = \Lambda \text{ or } E_2^r = \Lambda \text{ then } \Lambda \\
&\qquad \text{else if } E_1^r = n_1 \text{ and } E_2^r = n_2 \text{ then } n = n_1 - n_2 \text{ else } E_1^r - E_2^r \\
&\qquad\qquad R \in \mathbb{B} \times \mathcal{E} \to \mathbb{B} \\
&R[\![E_1 < E_2]\!]\rho = \text{let } E_1^r = R[\![E_1]\!]\rho \text{ and } E_2^r = R[\![E_2]\!]\rho \\
&\qquad \text{if } E_1^r = \Lambda \text{ or } E_2^r = \Lambda \text{ then } \Lambda \\
&\qquad \text{else if } E_1^r = n_1 \text{ and } E_2^r = n_2 \text{ and } b = n_1 < n_2 \\
&\qquad \text{then } b \text{ else } E_1^r < E_2^r \\
&R[\![B_1 \vee B_2]\!]\rho = \text{let } B_1^r = R[\![B_1]\!]\rho \text{ and } B_2^r = R[\![B_2]\!]\rho \\
&\qquad \text{if } B_1^r = \Lambda \text{ or } B_2^r = \Lambda \text{ then } \Lambda \\
&\qquad \text{else if } B_1^r = true \text{ or } B_2^r = true \text{ then } true \\
&\qquad \text{else if } B_1^r = false \text{ then } B_2^r \\
&\qquad \text{else if } B_2^r = false \text{ then } B_1^r \\
&\qquad \text{else } B_1^r \vee B_2^r \\
&R[\![\neg B]\!]\rho = \text{let } B^r = R[\![B]\!]\rho \\
&\qquad \text{if } B^r = \Lambda \text{ then } \Lambda \\
&\qquad \text{else if } B^r = true \text{ then } false \\
&\qquad \text{else if } B^r = false \text{ then } true \\
&\qquad \text{else } \neg B^r \\
&R[\![true]\!]\rho = true \\
&R[\![false]\!]\rho = false \\
&\underline{\text{Actions}} \quad R \in \mathbb{A} \times \mathcal{E} \to \mathbb{A} \\
&R[\![B]\!]\rho = \langle \rho, R[\![B]\!]\rho \rangle \\
&R[\![X := ?]\!]\rho = \langle \rho \setminus X, X := ? \rangle \\
&R[\![X := E]\!]\rho = \text{if } \mathbf{static}[\![E]\!]\rho \text{ then } \langle \rho[X := R[\![E]\!]\rho], skip \rangle \\
&\qquad \text{else } \langle \rho \setminus X, X := R[\![E]\!]\rho \rangle
\end{aligned}
$$

This function, given a label L and a variable X, returns the least upper bound $\sqcup$ on the flat lattice of program values for X at the program point L.

Semantic transformation: The semantic transformation t^C is defined in [12] where: $t^C[S[\![P]\!], S^C[\![P]\!]] \stackrel{\text{def}}{=} \{t^C[\sigma, S^C[\![P]\!]] | \sigma \in S[\![P]\!]\}$ is the transformation of semantics, $t^C[\sigma, S^C[\![P]\!]] \stackrel{\text{def}}{=} \lambda i. t^C[\sigma_i, S^C[\![P]\!]]$ is the transformation of traces, and $t^C[\langle \rho, C \rangle, S^C[\![P]\!]] \stackrel{\text{def}}{=} \langle \rho, t^C[C, \tau^C(lab[\![C]\!])] \rangle$ is the transformation of states, where $t^C[L_1 : A \to L_2, \rho^C] \stackrel{\text{def}}{=} L_1 : t^C[A, \rho^C] \to L_2$ is command specialization and $t^C[A, \rho^C] = \text{let } \langle \rho_r, A_r \rangle \stackrel{\text{def}}{=} R[\![A]\!]\rho|_{\{X \in \mathbb{X} | \rho^C(X) \in D_T\}} \in A_r$ is action specialization. The syntactic transformation τ^C can be systematically derived as shown in [12].

Observational semantics: Given a partial trace σ, the observational semantics $\alpha_\emptyset^C$ returns the sequence of its environments: $\alpha_\emptyset^C(S[\![P]\!]) \stackrel{\text{def}}{=} \{\alpha_\emptyset^C(\sigma) | \sigma \in S[\![P]\!]\}$, $\alpha_\emptyset^C(\sigma) \stackrel{\text{def}}{=} \lambda i. \alpha_\emptyset^C(\sigma_i)$, and $\alpha_\emptyset^C(\langle \rho, C \rangle) \stackrel{\text{def}}{=} \rho$.

Table 4. A simple program from [8]

a:= 1; b:=2; c:=3; d:=3; e:=0;	L_1 : a:= 1; b:=2; c:=3; d:=3; e:=0; $\to L_2$
while B do	L_2 : B $\to L_3$
	L_2 : ¬B $\to L_5$
b:=2*a; d:=d+1; e:=e-a;	L_3 : b:=2*a; d:=d+1; e:=e-a; $\to L_4$
a:=b-a; c:=e+d;	L_4 : a:=b-a; c:=e+d; $\to L_2$
endw	L_5 : stop $\to \not{l}$

4.2 Code Obfuscation by Constant Propagation

In order to specify the properties obfuscated by constant propagation τ^C, we derive δ_{τ^C}. Let us define the property θ as follows: $\theta(S[\![P]\!]) \stackrel{def}{=} \{\theta(\sigma) | \sigma \in S[\![P]\!]\}$, $\theta(\sigma) \stackrel{def}{=} \lambda i.\theta(\sigma_i)$, and $\theta(\langle \rho, C \rangle) \stackrel{def}{=} \langle \rho, lab[\![C]\!], succ[\![C]\!]\rangle$.

Theorem 3. $\theta = \delta_{\tau^C}$ and $O_\theta \neq \varnothing$.

It is worth noting that θ is the most concrete property preserved by constant propagation τ^C. Let us consider the property $\vartheta \in uco(\wp(\Sigma^\infty))$, observing the environment, the labels, and the type of actions: $\vartheta(\tau^\infty) \stackrel{def}{=} \{\vartheta(\sigma) | \sigma \in \tau^\infty\}$, $\vartheta(\sigma) \stackrel{def}{=} \lambda i.\vartheta(\sigma_i)$, and $\vartheta(\langle \sigma, C \rangle) \stackrel{def}{=} \langle \rho, lab[\![C]\!], succ[\![C]\!], type(act[\![C]\!])\rangle$, where *type* maps actions into the following set of action types {*assign, skip, test*}. This property belongs to O_θ, meaning that $O_\theta \neq \varnothing$, as requested by Definition 3 and shown in Example 2 below. This proves that τ^C is actually a program obfuscator, hiding the type of the command actions. This is a consequence of the structure of the masked closure $\vartheta \ominus (\vartheta \sqcup \theta)$, which characterizes what is obfuscated from ϑ by τ^C. Consider the closure η which observes the *type* of actions defined as follows:

$$\eta = \lambda X. \left\{ \sigma \;\middle|\; \begin{array}{l} \sigma' \in X \text{ and } \forall i.\; \sigma_i = \langle \rho_i, C_i \rangle, \sigma'_i = \langle \rho'_i, C'_i \rangle \\ type(C_i) = type(C'_i) \end{array} \right\}$$

The following theorem specifies that η is masked by the constant propagation, i.e., τ^C obfuscates η.

Theorem 4. $\vartheta \ominus (\vartheta \sqcup \theta) \sqsubseteq \eta$.

Example 2. As observed above, ϑ is not preserved by t^C, namely it could happen that: $\vartheta(S[\![P]\!]) \neq \vartheta(t^C[S[\![P]\!], S^C[\![P]\!]])$. In the following we represent the environment as a tuple $(v_a, v_b, v_c, v_d, v_e)$ of values corresponding to the variables a, b, c, d, e in a certain execution point. Let us run the program in Table 4, and consider the states $\sigma_2 = \langle (1, 2, 3, 3, 0), L_3 : b := 2*a; d := d+1; e := e-a; \to L_4 \rangle$ and $\sigma_3 = \langle (1, 2, 3, 4, -1), L_4 : a := b-a; c := e+d \to L_2 \rangle$. Their transformed versions are: $t^C(\sigma_2) = \langle (1, 2, 3, 3, 0), L_3 : d := d+1; e := e-a; \to L_4 \rangle$ and $t^C(\sigma_3) = \langle (1, 2, 3, 4, -1), L_4 : skip \to L_2 \rangle$. In this case $\vartheta(\sigma_2) = \langle (1, 2, 3, 3, 0), L_3, L_4, assign \rangle$ and $\vartheta(\sigma_3) = \langle (1, 2, 3, 4, -1), L_4, L_2, assign \rangle$; while $\vartheta(t^C(\sigma_2)) = \langle (1, 2, 3, 3, 0), L_3, L_4, assign \rangle$ and $\vartheta(t^C(\sigma_3)) = \langle (1, 2, 3, 4, -1), L_4, L_2, skip \rangle$, showing that the property η is not preserved.

5 Discussion

In this paper we introduce a notion of code obfuscation providing a general enough definition including both most program transformation techniques as obfuscators and the standard definition of Collberg et al. [5] as special cases. Moreover, it provides advanced techniques for comparing obfuscating algorithms relatively to their potency in the lattice of abstract interpretations. This definition can be considered as the first step towards a semantic-based theory for code obfuscation. Note however that in Definition 1 we consider abstractions of concrete semantics and not arbitrary sound, possibly incomplete, abstract interpretations. This makes Definition 1 too strong for modeling attackers which can be any decidable sound approximation of semantics, *e.g.*, arbitrary static program analyzers. Consider the following example of program transformation:

P:
L_1: X:= ? → L_2
L_2: Y:= ? → L_3
L_3: out:= (X - Y)^2 → ↯

P':
L_1: X:= ? → L_2
L_2: Y:= ? → L_3
L_3: out:= X^2 - 2XY + Y^2 → ↯

Let's consider the property of the sign of a variable, *i.e.*, the abstract domain $Sign \stackrel{def}{=} \{\mathbb{Z}, \mathbb{Z}^+, \mathbb{Z}^-, \varnothing\}$. By considering the sign of the variable out, when X is positive and Y is negative, we have that: $Sign(S[\![P]\!]) = Sign(S[\![\tau[\![P]\!]]\!]) = \mathbb{Z}^+$, while $S^{Sign}[\![P]\!] = \mathbb{Z}^+$ and $S^{Sign}[\![\tau[\![P]\!]]\!] = \mathbb{Z}$. Therefore the previous transformation obfuscates program P for the sign analysis, because the static sign analysis of the transformed program is unable to get the sign of out. This is not captured by Definition 1, because $Sign$ is incomplete for integer addition [16]. Indeed, in Definition 1, the abstraction is applied to the concrete semantics, returning $\mathbb{Z}^+$ in both cases. It would be important to weaken our notion of potency by considering abstract semantics derived by sound (possibly incomplete) approximations of the concrete semantics. This is crucial in order to include in our model of attackers arbitrary sound program analyzers. Moreover the systematic design of program transformation by abstract interpretation [12] can be applied for the systematic design of code obfuscation algorithms, driven by the abstraction (semantic property) to be masked. We can observe that other program transformations, such as *abstract watermarking* [13], can be seen as a particular code obfuscation. The relation between our notion of code obfuscation and abstract watermarking deserves further investigation.

References

1. B. Barak, O. Goldreich, R. Impagliazzo, and S. Rudich. On the (Im)possibility of Obfuscating Programs. In *Advances in Cryptology, Proc. of Crypto'01*, 2001, volume 2139 of LNCS, pages 1-18. Springer-Verlag.
2. C. Collberg and C. Thomborson. Watermarking, Tamper-Proofing, and Obfuscation-Tools for Software Protection. In *IEEE Trans. Software Eng.*, pages 735-746, 2002.
3. S. Chandrasekharan and S. Debray. Deobfuscation: Improving Reverse Engineering of Obfuscated Code. Draft, 2005.

4. C. Collberg and C. Thomborson. Breaking Abstractions and Unstructural Data Structures. In *Proc. of the 1994 IEEE Internat. Conf. on Computer Languages (ICCL '98)*, pages 28-37, 1998.
5. C. Collberg, C. Thomborson, and D. Low. A Taxonomy of Obfuscating Transformations. Technical Report 148, Dept. of Computer Science, The Univ. of Auckland, 1997.
6. C. Collberg, C. Thomborson, and D. Low. Manufacturing Cheap, Resilient, and Stealthy Opaque Constructs. In *Proceedings of the 25th ACM SIGPLAN-SIGACT Symposium on Principles of programming languages (POPL '98)*, pages 184-196. ACM Press, 1998.
7. A. Cortesi, G. Filé, R. Giacobazzi, C. Palamidessi, and F. Ranzato. Complementation in abstract interpretation. In *ACM Trans. Program. Lang. Syst.*, 19(1):7-47, 1997.
8. P. Cousot. Méthodes itératives de construction et d'approximation de points fixes d'opérateurs monotones sur un treillis, analyse sémantique des programmes. PhD Thesis, Université Scientifique et Médicale de Grenoble, Grenoble, France. 1978.
9. P. Cousot. Constructive Design of a Hierarchy of Semantics of a Transition System by Abstract Interpretation. *Theoretical Computer Science*, 277(1-2):47-103, 2002.
10. P. Cousot and R. Cousot. Abstract interpretation: A unified lattice model for static analysis of programs by construction or approximation of fixpoints. In *Conference Record of the 4th ACM Symp. on Principles of Programming Languages (POPL '77)*, pages 238–252. ACM Press, New York, 1977.
11. P. Cousot and R. Cousot. Systematic design of program analysis frameworks. In *Conference Record of the 6th ACM Symp. on Principles of Programming Languages (POPL '79)*, pages 269–282. ACM Press, New York, 1979.
12. P. Cousot and R. Cousot. Systematic Design of Program Transformation Frameworks by Abstract Interpretation. In *Conference Record of the Twentyninth Annual ACM SIGPLAN-SIGACT Symposium on Principles of Programming Languages*, pages 178-190, New York, NY, 2002. ACM Press.
13. P: Cousot and R: Cousot. An Abstract Interpretation-Based Framework for Software Watermarking. *Conference Record of the Thirtyfirst Annual ACM SIGPLAN-SIGACT Symposium on Principles of Programming Languages*, pages 173-185. ACM Press, New York, NY, 2004.
14. L. D'Anna, B. Matt, A. Reisse, T. Van Vleck, S. Schwab, and P. LeBlanc. Self-Protecting Mobile Agents Obfuscation Report. Technical report, Network Associates Laboratory, 2003.
15. G. Filé and F. Ranzato. Complementation of abstract domains made easy. In *Proceedings of the 1996 Joint International Conference and Symposium on Logic Programming (JICSLP '96)*, pages 348-362. The MIT Press,Cambridge, Mass., 1996.
16. R. Giacobazzi, F. Ranzato, and F. Scozzari. Making abstract interpretations complete. *J. of the ACM.*, 47(2):361–416, 2000.
17. J. Morgado. Some results on the closure operators of partially ordered sets. *Portug. Math.*, 19(2):101-139, 1960.
18. R. Paige. Future directions in program transformations. In *ACM SIGPLAN Not.*, volume 32, pages 94-97, 1997.

About Hoare Logics for Higher-Order Store*

Bernhard Reus[1],[**] and Thomas Streicher[2]

[1] University of Sussex, Brighton BN1 9QH, UK
[2] TU Darmstadt, 64298 Darmstadt, Germany

Abstract. We present a Hoare logic for a simple imperative while-language with stored commands, ie. stored parameterless procedures. Stores that may contain procedures are called higher-order. Soundness of our logic is established by using denotational rather than operational semantics. The former is employed to elegantly account for an inherent difficulty of higher-order store, namely that assertions necessarily describe recursive predicates on a recursive domain. In order to obtain proof rules for mutually recursive procedures, assertions have to explicitly refer to the code of the procedures.

1 Introduction and Motivation

Hoare logic for imperative languages has been invented in the late 60es [7] and since then extended in many directions (for a survey see e.g. [4]). Procedures are a typical example. For a simple while language with parameterless recursive procedures it is common to apply the following rule (see [4]) for a procedure p declared with body C:

$$(proc) \quad \frac{\{P\}\, p\, \{Q\} \vdash \{P\}\, C\, \{Q\}}{\{P\}\, p\, \{Q\}}$$

In order to verify the effect of a procedure call, one has to show that the procedure body satisfies the very same effect under the assumption that the call already does so. Semantically, this corresponds to a form of fixpoint induction where admissibility of the semantical predicates is guaranteed automatically as store is modeled by a flat domain. Thus, in rule $(proc)$ the $\{P\}\, p\, \{Q\}$ in the conclusion refers to the fixpoint of the definition rec $p \Leftarrow C[p]$ whereas on the left hand side in the premise it refers to an arbitrary implementation of p.

The situation changes dramatically if one allows stored procedures, ie. if procedures are kept in the store – in the same way as basic data like numbers – and called by their (variable) name. For example, run x invokes the procedure

---

[*] Both authors have been partially supported by APPSEM II (Applied Semantics), a thematic network funded by the IST programme of the European Union, IST-2001-38957.
[**] The first author has been partially supported by the EPSRC under grant GR/R65190/01, "Programming Logics for Denotations of Recursive Objects".

stored in variable x. The semantics of programs, being state transformers, now becomes implicitly higher-order, as they depend on the (code in the) store which contains such transformers itself. For this reason such stores are sometimes called "higher-order" or even recursive.

Landin had already observed[1] that in such situations one is able to tie "knots through the store". Put differently, recursion through the store becomes available such that additional fixpoint operators are obsolete. Consider e.g. the example $x := C$; run x which first stores a command C in x and then runs this command. But C itself may contain a command run x, in which case we obtain a recursive procedure. In traditional semantics (see [9, 14, 6]) the semantics of a procedure is a fixpoint. This is fine as long as newly added code can only call the old procedures and not vice-versa. In object-oriented languages, however, new subclasses can change the semantics of the old classes (that is the whole point of object-orientation) and the traditional semantics cannot cope with that. For languages with higher-order store this is no problem, as recursion is through the store and not by fixpoint.

To the best of our knowledge, there is no Hoare-calculus for partial correctness of (even simple) imperative languages with higher-order store in the literature. However, in [8] a calculus for *total* correctness of programs with higher-order store has been presented recently where soundness is based on induction on a termination measure. The semantics does not make use of domain theory and does not seem to be easily extendible to partial correctness.

Several (fully-abstract) models using games (or abstract versions of games) have been developed but they focus on observational equivalence, e.g. [3, 10, 11].

A Hoare-like calculus for an object-based language, Abadi and Cardelli's imperative object calculus [1], has been suggested in [2]. In that language, simple field values and method closures are kept together in the same store. Hence, the store is higher-order. In [2], the program logic does not use Hoare-triples but specifications that refer to the state before and after method execution. Consequently, in [2], method specifications can only use static information about other methods and thus cannot cope with callbacks or dynamic loading where specifications may change at runtime. Method update had to be disallowed. Note that in this paper our stored procedures can be updated.

In [16,17], we have presented a denotational technique to understand and model such object logics. This has been extended to a complete analysis of the entire Abadi-Leino calculus in [15]. Separately, in [5], Calcagno and O'Hearn set out to put ideas of separation logic [12] into a program logics for objects in a traditional Hoare-triple style but had problems with the object introduction rule.

In this paper we present a simple imperative language (Sect. 2) with higher-order store but without objects, and an assertion language (Sect. 3). We present some new proof rules (Sect. 4), give examples (Sect. 5) and prove soundness

---

[1] As Peter O'Hearn pointed out to the first author.

Table 1. BNF syntax of $\mathcal{L}$

$x \in \mathsf{Var}$		variable
$e \in \mathsf{Exp}$ ::=	x	variable expression
	k	numbers and other constants
	e o_2 e	binary operators
	o_1 e	unary operators
	's'	quote (command as expression)
$s \in \mathsf{Com}$ ::=	nop	no op
	$x := e$	assignment
	$s;s$	sequential composition
	if e then s else s	conditional
	run x	unquote (run the command in x)

(Sect. 6). We finish with an outlook where the results of this paper may lead us and how related work could be helpful.

The language in use is arguably the simplest language that uses higher-order store. It is thus an ideal candidate to investigate the problems caused by higher-order store in isolation. Using denotational semantics we will discover where exactly the difficulties of higher-order store are rooted.

2 The Programming Language

Syntax First we define the programming language syntax of our language, called $\mathcal{L}$. Let Var be the set of (countable) program variables, Exp the side effect free expressions, and Com the statements (commands). A BNF grammar for $\mathcal{L}$ is presented in Table 1.

The simplest non-terminating loop can be written as $x :=$ 'run x'; run x.

Semantics The semantics is developed in a category of cpo-s and partial continuous maps (predomains). For any (pre-)domain there is, as usual, a partial order $\sqsubseteq$, and for a partial continuous function $f \in A \rightharpoonup B$ and $a \in A$, we write $f(a)\downarrow$ to state that the application is defined.

Let BVal be the set of basic first-order values like numbers or booleans ordered discretely. Values and stores are defined by the following system of (pre-)domain equations

$$\mathsf{Val} = \mathsf{BVal} + [\mathsf{St} \rightharpoonup \mathsf{St}] \qquad \mathsf{St} = \mathsf{Val}^{\mathsf{Var}}$$

Stores in St map variables into values in Val. The fact that state transformers can be values reflects the fact that the store is higher-order. Note that for a store σ, a variable x, and a value a we write $\sigma[x \mapsto a]$ for the map σ' defined as

$$\sigma'(y) = \begin{cases} a & \text{if } y \equiv x \\ \sigma(y) & \text{otherwise} \end{cases}$$

Table 2. Semantics of $\mathcal{L}$

$$[\![x]\!]^e\,\sigma = \sigma(x)$$
$$[\![\text{'}s\text{'}]\!]^e\,\sigma = [\![s]\!]$$
$$[\![k]\!]^e\,\sigma = k$$
$$[\![e_1\circ_2 e_2]\!]^e\,\sigma = \circ_2([\![e_1]\!]^e\,\sigma,[\![e_2]\!]^e\,\sigma)$$
$$[\![\circ_1\,e_1]\!]^e\,\sigma = \circ_1([\![e_1]\!]^e\,\sigma)$$

$$[\![\mathsf{nop}]\!]\,\sigma = \sigma$$
$$[\![x := e]\!]\,\sigma = \sigma[x \mapsto [\![e]\!]^e\,\sigma]$$
$$[\![s_1;s_2]\!]\,\sigma = [\![s_2]\!]([\![s_1]\!]\,\sigma)$$
$$[\![\text{if } e \text{ then } s_1 \text{ else } s_2]\!]\,\sigma = \begin{cases} [\![s_1]\!]\sigma & \text{if } [\![e]\!]^e\sigma = true \\ [\![s_2]\!]\sigma & \text{if } [\![e]\!]^e\sigma = false \\ \text{undefined} & \text{otherwise} \end{cases}$$
$$[\![\mathsf{run}\,x]\!]\,\sigma = \sigma(x)(\sigma)$$

The equations for higher-order store can thus be rewritten in one equation as follows:
$$\mathsf{St} = (\mathsf{BVal} + [\mathsf{St}\rightharpoonup\mathsf{St}])^{\mathsf{Var}}$$
or equivalently, by setting $\mathsf{Cl} = [\mathsf{St}\rightharpoonup\mathsf{St}]$,
$$\mathsf{Cl} = [(\mathsf{BVal}+\mathsf{Cl})^{\mathsf{Var}} \rightharpoonup (\mathsf{BVal}+\mathsf{Cl})^{\mathsf{Var}}]$$

The mixed-variant functor for which Cl is the solution is given by its object and morphism part below:
$$F(X,Y) = [(\mathsf{BVal}+X)^{\mathsf{Var}} \rightharpoonup (\mathsf{BVal}+Y)^{\mathsf{Var}}]$$

If $e : X \rightharpoonup Y$ then let $\hat{e} = (\mathsf{BVal} + e)^{\mathsf{Var}} : (\mathsf{BVal} + X)^{\mathsf{Var}} \rightharpoonup (\mathsf{BVal} + Y)^{\mathsf{Var}}$. More precisely, for a store $\sigma \in (\mathsf{BVal} + X)^{\mathsf{Var}}$, $\hat{e}(\sigma)$ is defined as follows:

$$\hat{e}(\sigma)(x) = \begin{cases} \sigma(x) & \text{if } \sigma(x) \in \mathsf{BVal} \\ e(\sigma(x)) & \text{if } \sigma(x) \in X \end{cases}$$

Now we can define the morphism part:
$$F(e,f) = \lambda h : F(X,Y).\,\hat{f} \circ h \circ \hat{e}.$$

For $e \in [\mathsf{Cl}\rightharpoonup\mathsf{Cl}]$ let $e^A \in [\mathsf{Cl}^A\rightharpoonup\mathsf{Cl}^A]$ be defined by $e^A(h)(a) = e(h(a))$. We can interpret $\mathcal{L}$ using an interpretation function for expressions $[\![\_]\!]^e$: Exp $\rightarrow$ [St$\rightharpoonup$Val] and commands $[\![\_]\!]$: Com $\rightarrow$ [St$\rightharpoonup$St] as presented in Table 2.

The last equation $[\![\mathsf{run}\,x]\!]\sigma = \sigma(x)(\sigma)$ is reminiscent of the self-application semantics of method call in OO-languages.

3 The Assertion Language

The assertion language is based on the assertions of the classic Hoare-calculus with the difference, though, that expressions can also refer to stored procedures.

Table 3. Syntax of Assertions

$n, p \in \mathsf{AuxVar}$		
$\tau \in \mathsf{Type}$	$::= \mathsf{bool} \mid \mathsf{int} \mid \mathsf{com}$	types
$e \in \mathsf{Exp}$	$::= n \mid p$	auxiliary variables in BVal and Cl
	$\mid x \mid k \mid e \circ_2 e \mid \circ_1 e \mid \text{'}s\text{'}$	
$P \in \mathsf{Asrt}$	$::= \mathit{false}$	falsity
	$\mid P \wedge P$	conjunction
	$\mid \neg P$	negation
	$\mid \forall n.\, P$	universal quantification
	$\mid \tau?\, e$	type check
	$\mid e \leq_\tau e$	comparison

Table 4. Semantics of Assertions

$$\begin{aligned}
(\!|\mathit{false}|\!)\,\eta &= \emptyset \\
(\!|P \wedge Q|\!)\,\eta &= (\!|P|\!)\,\eta \cap (\!|Q|\!)\,\eta \\
(\!|\forall n.\, P|\!)\,\eta &= \bigcap_{v \in \mathsf{Val}} (\!|P|\!)\,\eta[n \mapsto v] \\
(\!|\tau?\, e|\!)\,\eta &= \{\sigma \mid [\![e]\!]^e\,\eta\,\sigma \in [\![\tau]\!]\} \text{ where } [\![\mathsf{com}]\!] = \mathsf{Cl},\ [\![\mathsf{int}]\!] = \mathbb{Z},\ [\![\mathsf{bool}]\!] = \mathbb{B} \\
(\!|\neg P|\!)\,\eta &= \{\sigma \mid \sigma \notin (\!|P|\!)\,\eta\} \\
(\!|e_1 \leq_\tau e_2|\!)\,\eta &= \{\sigma \mid [\![e_1]\!]^e\,\eta\,\sigma \in [\![\tau]\!] \wedge [\![e_2]\!]^e\,\eta\,\sigma \in [\![\tau]\!] \wedge [\![e_1]\!]^e\,\eta\,\sigma \sqsubseteq [\![e_2]\!]^e\,\eta\,\sigma\}
\end{aligned}$$

Syntax The syntax of assertions is presented in Table 3. They may contain expressions of $\mathcal{L}$ which have no side effects. Note that $e = e$ and $\phi \vee \phi$ can be expressed using $\leq$, and $\wedge$ and $\neg$, respectively.

As already known from classic Hoare-calculus one needs "ghost variables" (also called *auxiliary variables*) to be able to refer to values in the pre-execution state. For example, in the Hoare-triple $\{x = n\}\, fac\, \{x = n!\}$ we have a program variable x, and an auxiliary variable n. The countable set of auxiliary ghost variables is called AuxVar. Throughout the paper we use x, y, z to denote instances of program variables in Var and n, p, q to denote instances of auxiliary variables in AuxVar where n is usually used for auxiliary variables of basic type and p and q for auxiliary variables for commands (procedures). It is important not to confuse those different types of variables.

Semantics The denotational semantics of assertions is standard but we have to take care of auxiliary variables. For those variables an additional environment is in use of type $\mathsf{Env} = \mathsf{Val}^{\mathsf{AuxVar}}$. Correspondingly, an interpretation function for assertions must have type $(\!|\_|\!) : \mathsf{Asrt} \to \mathsf{Env} \to \mathcal{P}(\mathsf{St})$ and its equations can be found in Table 4.

Observe that $[\![\_]\!]^e$ has to be extended to auxiliary variables. Therefore, we stipulate $[\![n]\!]^e\,\eta\,\sigma = \eta(n)$ and assume that the definitional equations for $[\![\_]\!]^e$ in Table 2 have been changed accordingly.

Table 5. Syntax of pure assertions without undesired comparisons of commands

$\phi, \in \mathsf{BAsrt} ::= \mathit{false} \mid \phi \wedge \phi \mid \neg \phi \mid \forall n.\, \phi \mid \tau?\, e \mid e \leq_{\mathsf{nat}} e \mid e \leq_{\mathsf{bool}} e$
$P \in \mathsf{DClAsrt} ::= \phi \mid \forall n.\, P \mid x \leq_{\mathsf{com}} n \mid x \leq_{\mathsf{com}} \text{`}s\text{'} \mid \text{`}s\text{'} \leq_{\mathsf{com}} \text{`}s\text{'} \mid P \wedge P \mid P \vee P$

A particular subset of downward-closed assertions Equality between procedures ($\leq_{\mathsf{com}}$) is problematic as it is not a downward-closed predicate which will be important for our semantics. We therefore identify a particular sublanguage of assertions, DClAsrt, which are more restrictive with respect to comparisons of commands, in particular they do not admit equality on procedures. To define DClAsrt, we introduce "basic assertions" BAsrt first, which do not use any comparison between expressions of type com at all. The exact definitions can be found in Table 5.

Assertions like $x =_{\mathsf{com}} \text{`}s\text{'}$, i.e. $x \leq_{\mathsf{com}} \text{`}s\text{'} \wedge \text{`}s\text{'} \leq_{\mathsf{com}} x$, are not in DClAsrt as $\text{`}s\text{'} \leq_{\mathsf{com}} x$ is *not* in DClAsrt. The assertions in DClAsrt all satisfy three conditions explained in Lemma 1 below. These properties will turn out to be crucial to obtain a semantics that validates the proof rules introduced in the next section.

Lemma 1. *For any assertion $P \in \mathsf{DClAsrt}$ its semantics $(\!|P|\!) \in \mathsf{Env} \to \mathcal{P}(\mathsf{St})$ has the following properties:*

1. $(\!|P|\!)\, \eta$ *is a downward closed predicate for all $\eta \in \mathsf{Env}$.*
2. $(\!|P|\!)$ *is monotonic in its (procedure environment) argument.*
3. $(\!|P|\!)\, \eta\, \sigma$ *implies* $(\!|P|\!)\, \hat{e}(\eta)\, \hat{e}(\sigma)$ *for all $e \in \mathsf{Cl} \to \mathsf{Cl}$, $\sigma \in \mathsf{St}$, and $\eta \in \mathsf{Env}$ where $\hat{e}(\eta)$ is defined analogously to $\hat{e}(\sigma)$ with the only difference that the variables used in* Env *are* AuxVar *whereas those used in* St *are* Var.

Proof. We only have to consider assertions on commands. Assertions on basic types trivially fulfill the requirements since BVal is ordered discretely. Since $\forall$, $\wedge$, and $\vee$ preserve the conditions above we only need to show each of them for the three comparisons on commands (and the type check assertion com?e which trivially fulfills all conditions). (1) and (2) are immediate by definition of the assertions. For (3) we show the interesting case: $[\![x \leq_{\mathsf{com}} n]\!]\, \eta\, \sigma$ iff $\sigma(x) \sqsubseteq \eta(n)$ (†). Now $[\![x \leq_{\mathsf{com}} n]\!]\, \hat{e}(\eta)\, \hat{e}(\sigma)$ iff $\hat{e}(\sigma)(x) = e(\sigma(x)) \sqsubseteq \hat{e}(\eta)(n) = e(\eta(n))$ which follows from (†) as e is monotonic.

4 Proof Rules

First of all, the standard rules for assignment (A), composition (S), conditional (I), weakening (W), and no operation (ϵ) are in use as presented in Fig. 1.

New rules are needed to deal with stored procedures as outlined in Fig. 2. The run-rule (R) is canonical for non-recursive procedure calls. Rule (H) is like (R) for cases where the code in a variables is not known but described by an auxiliary variable. Finally, the recursion rule (μ) is used for stored procedures that are (mutually) recursive. This is necessary as rule (R) is not able to get rid of the

$$(A) \; \frac{}{\{P[e/x]\} \, x := e \, \{P\}} \qquad (S) \; \frac{\{P\} \, C_1 \, \{R\} \quad \{R\} \, C_2 \, \{Q\}}{\{P\} \, C_1; C_2 \, \{Q\}}$$

$$(I) \; \frac{\{b \wedge P\} \, C_1 \, \{Q\} \quad \{\neg b \wedge P\} \, C_2 \, \{Q\}}{\{P\} \, \text{if } b \text{ then } C_1 \text{ else } C_2 \, \{Q\}} \qquad (W) \; \frac{P \Rightarrow P' \quad \{P'\} \, C \, \{Q'\} \quad Q' \Rightarrow Q}{\{P\} \, C \, \{Q\}}$$

$$(\epsilon) \; \frac{}{\{P\} \, \text{nop} \, \{P\}}$$

Fig. 1. Standard Rules

$$(R) \; \frac{\{P \wedge x \leq \text{`}C\text{'}\} \, C \, \{Q\}}{\{P \wedge x \leq \text{`}C\text{'}\} \, \text{run } x \, \{Q\}} \qquad Q \in \text{DClAsrt}$$

$$(H) \; \frac{}{\{P \wedge x \leq p\} \, p \, \{Q\} \vdash \{P \wedge x \leq p\} \, \text{run } x \, \{Q\}} \qquad Q \in \text{DClAsrt}$$

$$(\mu) \; \frac{\bigwedge_{1 \leq i \leq n} \{P_1\} \, p_1 \, \{Q_1\} \ldots \{P_n\} \, p_n \, \{Q_n\} \vdash \{P_i\} \, C_i \, \{Q_i\}}{\bigwedge_{1 \leq i \leq n} \{P_i[C/p]\} \, C_i \, \{Q_i[C/p]\}} \qquad \forall 1 \leq i \leq n. \, P_i, Q_i \in \text{DClAsrt}$$

Fig. 2. New Rules for Stored Procedures

circular reference to the procedure. Rule (μ) is able to do just that analogously to the standard procedure rule $(proc)$ mentioned in the introduction. In fact, by using first (R) and then (μ), one obtains the derived rule stated below (left). For comparison the standard recursive procedure rule is repeated next to it (right).

$$\frac{\{P \wedge x \leq p\} \, p \, \{Q\} \vdash \{P \wedge x \leq p\} \, C \, \{Q\}}{\{P \wedge x \leq \text{`}C\text{'}\} \, \text{run } x \, \{Q\}} \qquad (proc) \; \frac{\text{rec } p \Leftarrow C[p]}{\{P\} \, p \, \{Q\} \vdash \{P\} \, C \, \{Q\}}{\{P\} \, p \, \{Q\}}$$

Whereas for $(proc)$ the definition of the procedure p is separate, for stored procedure x one needs to use an auxiliary variable p to denote the content of x during execution of its body which may change x.

Note that throughout the rest of the paper we simply write $\leq$ instead of $\leq_{\text{com}}$ when one of its arguments is obviously of type com. The necessity of the side conditions for (μ), (R), and (H) will become clear when we discuss the soundness of these rules.

5 Sample Derivations

We present some sample derivations to demonstrate how the proof rules above are to be used.

Example 1. A derivation for a specification of our introductory example of a non-terminating loop, $\{true\} \, x := \text{`run } x\text{'} ; \text{run } x \, \{false\}$, is outlined in Figure 3.

Example 2. Because of recursion through the store we can simulate a while loop while B do C od as $z := \text{`if } B \text{ then } C; \text{run } z \text{ else nop'} ; \text{run } z$. Of course, program

$$\frac{\overline{\{\text{'run } x\text{'} \sqsubseteq \text{'run } x\text{'}\}\, x := \text{'run } x\text{'}\, \{x \sqsubseteq \text{'run } x\text{'}\}}\ (A)}{\{true\}\, x := \text{'run } x\text{'}\, \{x \sqsubseteq \text{'run } x\text{'}\}}\ (W) \qquad \frac{\overline{\{x \sqsubseteq q\}\, q\, \{false\} \vdash \{x \sqsubseteq q\}\, \text{run } x\, \{false\}}\ (H)}{\{x \sqsubseteq \text{'run } x\text{'}\}\, \text{run } x\, \{false\}}\ (\mu)$$

$$\overline{\{true\}\, x := \text{'run } x\text{'};\, \text{run } x\, \{false\}}\ (S)$$

Fig. 3. Derivation for Example 1

variable z is not supposed to occur in C. When doing the proof it becomes clear that it is enough that z is not altered by C. The standard rule for while and its derived equivalent for the encoding in $\mathcal{L}$ read as follows:

$$\frac{\{B \wedge I\}\, C\, \{I\}}{\{I\}\, \text{while } B \text{ do } C \text{ od } \{\neg B \wedge I\}} \qquad \frac{\{B \wedge I \wedge z \leq p\}\, C\, \{I \wedge z \leq p\}}{\{I\}\, \text{while } B \text{ do } C \text{ od } \{\neg B \wedge I\}}$$

The encoded form has to state that z is invariant, i.e. that the content of this cell is not changed by the body of the while statement. This is expressed using an auxiliary variable $p \in \mathsf{AuxVar}$.

For the proof assume that (∗) $\{B \wedge I \wedge z \leq p\}\, C\, \{I \wedge z \leq p\}$ and let IF abbreviate the expression (of command type) 'if B then C; run z else nop'. The first part of the derivation is straightforward (see Fig. 4). The remaining open goal $\{I \wedge z \leq IF\}\, \text{run } z\, \{\neg B \wedge I\}$ is then derived in (β) in the second prooftree of Fig. 4. The application of the recursion rule (μ) will introduce the hypothesis (†) $\{I \wedge z \leq p\}\, p\, \{\neg B \wedge I\}$ to be used at the top of subtree (β).

$$\frac{\overline{\{I \wedge IF \leq IF\}\, z := IF\, \{I \wedge z \leq IF\}}\ (A)}{\{I\}\, z := IF\, \{I \wedge z \leq IF\}}\ (W) \qquad \frac{(\beta)}{\{I \wedge z \leq IF\}\, \text{run } z\, \{\neg B \wedge I\}}\ (R)$$

$$\overline{\{I\}\, z := IF;\, \text{run } z\, \{\neg B \wedge I\}}\ (S)$$

where (β) is the following prooftree

$$\frac{(*)}{\{B \wedge I \wedge z \leq p\}\, C\, \{I \wedge z \leq p\}} \quad \frac{\overline{\{I \wedge z \leq p\}\, p\, \{\neg B \wedge I\}}\ (\dagger)}{\{I \wedge z \leq p\}\, \text{run } z\, \{\neg B \wedge I\}}\ (H) \qquad \frac{\overline{\{\neg B \wedge I\}\, \text{nop}\, \{\neg B \wedge I\}}\ (N)}{\{\neg B \wedge I \wedge z \leq p\}\, \text{nop}\, \{\neg B \wedge I\}}\ (W)$$

$$\frac{\{B \wedge I \wedge z \leq p\}\, C;\, \text{run } z\, \{\neg B \wedge I\}}{\{I \wedge z \leq p\}\, \text{if } B \text{ then } C;\, \text{run } z \text{ else nop}\, \{\neg B \wedge I\}}\ (I)$$

$$\frac{}{\{I \wedge z \leq IF\}\, iteBC;\, \text{run } z\, \text{nop}\, \{\neg B \wedge I\}}\ (\mu)$$

$$\overline{\{I \wedge z \leq IF\}\, \text{run } z\, \{\neg B \wedge I\}}\ (R)$$

Fig. 4. Derivation for Example 2

Table 6. Semantics of Triples

$(\eta,\sigma) \models \{P\} C \{Q\}$	$\Leftrightarrow \forall \sigma' \in \mathsf{St}. (\!(P)\!) \eta \sigma \wedge [\![C]\!] \sigma = \sigma' \Rightarrow (\!(Q)\!) \eta \sigma'$
$(\eta,\sigma) \models \{P\} p \{Q\}$	$\Leftrightarrow \forall \sigma' \in \mathsf{St}. (\!(P)\!) \eta \sigma \wedge \eta(p)(\sigma) = \sigma' \Rightarrow (\!(Q)\!) \eta \sigma'$
$\models \{P_1\} p_1 \{Q_1\}, \ldots, \{P_n\} p_n \{Q_n\}$	$\Leftrightarrow \forall \eta \in \mathsf{Env}. (\forall \sigma \in \mathsf{St}. \bigwedge_{1 \leq i \leq n} (\eta, \sigma) \models \{P_i\} p_i \{Q_i\})$
$\vdash \{P\} C \{Q\}$	$\Rightarrow \forall \sigma \in \mathsf{St}. (\eta, \sigma) \models \{P\} C \{Q\}$

Example 3. This example shows how procedures can modify themselves so that different invocations of x behave differently. Let

$$S \equiv \text{`if } z{=}0 \text{ then nop else } (y := y + z; z := z{-}1; \text{run } x)\text{'}$$

and consider the program $x := \text{`}y := y{+}1; x := S\text{'}; \text{run } x; \text{run } x$ for which we can derive the following annotations:

$\{y = n \wedge z = m\}$
$x := \text{`}y := y{+}1; x := S\text{'}$ \hfill (A),(W)
$\{y = n \wedge z = m \wedge x \leq \text{`}y := y{+}1; x := S\text{'}\}$
run x \hfill (R),(A),(S),(W)
$\{y = n{+}1 \wedge z = m \wedge x \leq S\}$
run x \hfill (R),(μ),(I),(S),(A),(W),(ϵ)
$\{y = n{+}1 + \sum_1^m i \wedge z = 0 \wedge x \leq S\}$

6 Soundness

Pre-condition P and post-condition Q of Hoare triples do not only depend on the store but also on the values for the auxiliary variables in environment Env. We write $(\eta, \sigma) \models \{P\} C \{Q\}$ meaning that $\{P\} C \{Q\}$ is valid in σ and η.

Definition 1. *The semantics of Hoare triples for commands, for closure variables, and for commands in context, resp., is given in Table 6. Note that triples express* partial *correctness.*

Now we are in a position to formally prove the soundness of the new rules.[2]

Theorem 1. *The (run-)rules (R) and (H) are sound.*

Proof. Let C be a command. Assume that (1) $\models \{P \wedge x \leq \text{`}C\text{'}\} C \{Q\}$. We have to show $\models \{P \wedge x \leq \text{`}C\text{'}\} \text{run } x \{Q\}$. Therefore, assume that $\eta \in \mathsf{Env}$, and $\sigma \in \mathsf{St}$ such that (2) $(\!(P \wedge x \leq \text{`}C\text{'})\!) \eta \sigma$ and that (3) $[\![\text{run } x]\!]\sigma = \sigma'$. It remains to show that $(\!(Q)\!) \eta \sigma'$. Define (4) $\sigma'' := [\![C]\!] \sigma$. From (1), (2) and (4) we obtain $(\!(Q)\!) \eta \sigma''$. Since $(\!(Q)\!)$ is downward-closed by Lemma 1(2) and $Q \in \mathsf{DC|Asrt}$, it suffices to prove that $\sigma' \sqsubseteq \sigma''$. But

$$\sigma' =_{(3)} [\![\text{run } x]\!]\sigma = \sigma(x)(\sigma) \sqsubseteq_{(2)} [\![C]\!]\sigma =_{(4)} \sigma''\ .$$

---

[2] Soundness of the old rules is standard.

The proof for (H) is carried out analogously.

In the next proof we will make use of a binary operation $+ : \text{Env} \times \text{Cl}^A \to \text{Env}$ that represents "overwriting of environments" where $A \subseteq \text{AuxVar}$. Accordingly, $(\eta+\delta)(n) = \eta(n)$ if $n \in A$ and $\delta(n)$ otherwise.

Theorem 2. *The rule (μ) is correct.*

Proof. Let $(\!|P_i|\!), (\!|Q_i|\!) \subseteq \text{Env} \times \text{St}$ for $1 \leq i \leq n$ be the denotations of the predicates in the assertions of (μ). We basically follow the ideas of [16] where similar arguments were used to prove the correctness of the object introduction rule of [2]. For arbitrary $\eta \in \text{Env}$ let $A_\eta \subseteq \text{Cl}^{\{p_1,\ldots,p_n\}} \times \text{Cl}^{\{p_1,\ldots,p_n\}}$ be defined as follows:

$$A_\eta(\psi, \phi) \equiv \forall 1 \leq j \leq n. \; \forall \sigma \in \text{St}.$$
$$(\!|P_j|\!) (\eta+\psi) \sigma \wedge \phi(p_j)(\sigma){\downarrow} \Rightarrow (\!|Q_j|\!) (\eta+\psi) (\phi(p_j)(\sigma))$$

Let $p_i \mapsto [\![C_i]\!]$ be the environment in $\text{Cl}^{\{p_1,\ldots,p_n\}}$ that assigns $[\![C_i]\!]$ to p_i for $1 \leq i \leq n$. Then verifying rule (μ) amounts to showing that for arbitrary $\eta \in \text{Env}$:

(†) $\quad \forall \phi \in \text{Cl}^{\{p_1,\ldots,p_n\}}. \; A_\eta(\phi, \phi) \Rightarrow A_\eta(\phi, p_i \mapsto [\![C_i]\!])$

implies

$$A_\eta(p_i \mapsto [\![C_i]\!], p_i \mapsto [\![C_i]\!])$$

since in general $\forall \phi \in \text{Env}. \; R \phi$ is equivalent to $\forall \phi \in \text{Env}. \forall \psi \in \text{Cl}^{\{p_1,\ldots,p_n\}}. \; R(\phi+\psi)$. Let S be a predicate on $\text{Cl}^{\{p_1,\ldots,p_n\}}$ such that for all $\phi \in \text{Cl}^{\{p_1,\ldots,p_n\}}$

(1) $\quad S(\phi) \iff \forall \psi \in \text{Cl}^{\{p_1,\ldots,p_n\}}. \; S(\psi) \Rightarrow A_\eta(\psi, \phi)$

from which it follows that for any ϕ

(2) $\quad S(\phi) \Rightarrow A_\eta(\phi, \phi).$

Now from (2) and assumption (†) it follows that

(3) $\quad S(\phi) \Rightarrow A_\eta(\phi, p_i \mapsto [\![C_i]\!])$

i.e. $S(p_i \mapsto [\![C_i]\!])$ due to (1) as ϕ was arbitrary such that by (1) again we obtain $A_\eta(p_i \mapsto [\![C_i]\!], p_i \mapsto [\![C_i]\!])$ as desired. The existence of an appropriate S in (1) is shown in Lemma 2.

Lemma 2. *There is a $S \subseteq \text{Env}$ such that $S(\phi) \iff \forall \rho \in \text{Env}. \; S(\rho) \Rightarrow A_\eta(\rho, \phi)$ for any $\eta \in \text{Env}$.*

Proof. Let $\mathcal{L}$ denote the admissible subsets of $\text{Cl}^{\{p_1,\ldots,p_n\}}$ ordered by $\subseteq$ and

$$\Phi : \mathcal{L}^{op} \to \mathcal{L} : S \mapsto \{\, \phi \in \text{Cl}^{\{p_1,\ldots,p_n\}} \mid \forall \rho \in \text{Cl}^{\{p_1,\ldots,p_n\}}. \; S(\rho) \Rightarrow A_\eta(\rho, \phi)\,\}$$

for which it is necessary that $A_\eta(\rho, -)$ specifies an admissible subset of $\text{Cl}^{\{p_1,\ldots,p_n\}}$ for all $\rho \in \text{Cl}^{\{p_1,\ldots,p_n\}}$ which follows from Lemma 1(1). For guaranteeing existence (and uniqueness) of such an S by Pitts' Theorem [13] we have to show that $e : S_1 \subseteq S_2$ implies $F(e, e) : \Phi(S_2) \subseteq \Phi(S_1)$ for all (idempotent) $e \sqsubseteq \text{id}_{\text{Cl}}$ where for $X, Y \subseteq \text{Cl}^{\{p_1,\ldots,p_n\}}$ one defines $e : X \subseteq Y$ iff $\forall x \in X. \; \hat{e}(x){\downarrow} \Rightarrow \hat{e}(x) \in Y$. Suppose $e \sqsubseteq \text{id}_{\text{Cl}}$ with $e : S_1 \subseteq S_2$ and $\Phi(S_2)(\phi)$. We have to show $\Phi(S_1)(\widehat{F(e,e)}(\phi))$. For this to be possible we need that $(\!|P_j|\!)$ and $(\!|Q_j|\!)$ satisfy all three properties of Lemma 1.

7 Loose Ends and Related Work

Our programming language $\mathcal{L}$ contains only constant command expressions. Self-modifying programs would also modify commands, in other words, use operations on commands.

Consider, for example, an operation make_com that concatenates two commands returning a new one, such that modifications like $x := \mathsf{make\_com}(x, y)$ become possible. To allow for such expressions one needs to axiomatise make_com algebraically, i.e. the underlying first-order logic on data types needs to be enriched by axioms like $\mathsf{make\_com}(`s_1\text{'}, `s_2\text{'}) = `s_1; s_2\text{'}$ and a monotonicity axiom for make_com w.r.t. $\leq_{\mathsf{com}}$.

Procedures with parameters have not been discussed here but are certainly an issue. We expect, however, that they are not more difficult to cope with than in ordinary Hoare-calculus with recursive procedures. In particular, if parameters are passed by value.

The logic presented is not modular as all code must be known in advance and must be carried around in assertions. In [5], a calculus for objects was suggested that uses nested Hoare triples in order to state properties of programs in the store rather than referring to their explicit code without a hint on a denotational semantics nor a soundness proof. The meaning of such a "specification" triple in our setting would be a recursively defined predicate:

$$(\sigma, \eta) \models \{P\} x \{Q\} \Leftrightarrow \forall \sigma', \sigma'' \in \mathsf{St}.$$
$$(\sigma', \eta) \models \{P\} x \{Q\} \wedge (\!|P|\!)\, \sigma' \wedge \sigma(x)(\sigma') = \sigma'' \Rightarrow (\!|Q|\!)\, \sigma''$$

Unfortunately, such a recursively defined semantics of triples, $\models$, does not allow for a separate specification for mutual recursive procedures (there is no "Bekic Lemma"). This suggests that it is highly unlikely that there is a modular logic for $\mathcal{L}$. On the other hand, in [2, 15] it has been shown that modularity can be achieved for Abadi and Leino's object logic in a setting where assignments to procedure variables are disallowed. It is unclear yet, to which extent this restriction can be relaxed. In [17] we have discussed an idea for a relaxation but it will have to be tested on a fully-fledged soundness proof where frame properties are to be shown.

Alternative models for higher-order store have been suggested by [3, 10, 11] using games or locally boolean domains. It might be worthwhile to consider the domain equation for Cl within such locally boolean domains (or equivalently the category of sequential algorithms). This would already provide us with a fully abstract model for higher-order store in the sense of Idealized Algol.

Another open question is relative completeness of our logic. It is no problem to express weakest liberal preconditions, at least when extending the assertion language such that it can express arbitrary inductively defined predicates. The hard problem is to show, by induction on S, that $\{\mathsf{wlp}(C, P)\} C \{P\}$ is derivable for all post-conditions P.

The results of this paper, however, are the first step to get a grip on higher-order store from a programming logic point of view.

Acknowledgements. We would like to thank Cristiano Calcagno, Peter O'Hearn, and Jan Schwinghammer for discussions and comments on this work. We are grateful to the anonymous referees for their suggestions to improve readability.

References

1. M. Abadi and L. Cardelli. *A Theory of Objects*. Springer Verlag, 1996.
2. M. Abadi and K. R. M. Leino. A logic of object-oriented programs. In Nachum Dershowitz, editor, *Verification: Theory and Practice*, pages 11–41. Springer, 2004.
3. S. Abramsky, K. Honda, and G. McCusker. A fully abstract game semantics for general references. In *LICS '98: Proceedings of the 13th Annual IEEE Symposium on Logic in Computer Science*, page 334. IEEE Computer Society, 1998.
4. K.R. Apt. Ten Years of Hoare's Logic: A Survey – Part I. *TOPLAS*, 3(4):431–483, 1981.
5. C. Calcagno and P.W. O'Hearn. A logic for objects. Slides of a Talk, 2001.
6. A.V. Hense. Wrapper semantics of an object-oriented programming language with state. In *Proceedings Theoretical Aspects of Computer Software*, volume 526 of *Lecture Notes in Computer Science*, pages 548–568, Berlin, 1991. Springer.
7. C.A.R. Hoare. An axiomatic basis for computer programming. *Comm. ACM*, 12:576–583, 1969.
8. K. Honda, N. Yoshida, and M. Berger. An observationally complete program logic for imperative higher-order functions. In *20th Symp. on Logics in Computer Science, LICS*. To appear in IEEE, 2005.
9. S.N. Kamin and U.S. Reddy. Two semantic models of object-oriented languages. In Carl A. Gunter and John C. Mitchell, editors, *Theoretical Aspects of Object-Oriented Programming: Types, Semantics, and Language Design*, pages 464–495. The MIT Press, 1994.
10. J. Laird. A categorical semantics of higher-order store. *Electronic notes in Theoretical Computer Science*, 69, 2002.
11. J. Laird. Locally boolean domains. 2004. Submitted.
12. P.W. O'Hearn, J.C. Reynolds, and H. Yang. Local reasoning about programs that alter data structures. In *CSL*, volume 2142 of *Logic in Computer Science*, pages 1–19, Berlin, 2001. Springer.
13. A. M. Pitts. Relational properties of domains. *Information and Computation*, 127:66–90, 1996.
14. B. Reus. Modular semantics and logics of classes. In *Computer Science Logic*, volume 2803 of *Lecture Notes in Computer Science*, pages 456–469, Berlin, 2003. Springer.
15. B. Reus and J. Schwinghammer. Denotational semantics for Abadi and Leino's logic of objects. In Mooly Sagiv, editor, *European Symposium on Programming*, volume 3444 of *Lecture Notes in Computer Science*, pages 263–278, Berlin, 2005. Springer.
16. B. Reus and Th. Streicher. Semantics and logics of objects. In *Proceedings of the 17th Symp. Logic in Computer Science*, pages 113–122, 2002.
17. B. Reus and Th. Streicher. Semantics and logic of object calculi. *TCS*, 316:191–213, 2004.

The Polyranking Principle*

Aaron R. Bradley, Zohar Manna, and Henny B. Sipma

Computer Science Department,
Stanford University,
Stanford, CA 94305-9045
{arbrad, zm, sipma}@theory.stanford.edu

Abstract. Although every terminating loop has a ranking function, not every loop has a ranking function of a restricted form, such as a lexicographic tuple of polynomials over program variables. The *polyranking principle* is proposed as a generalization of polynomial ranking for analyzing termination of loops. We define *lexicographic polyranking functions* in the context of loops with parallel transitions consisting of polynomial assertions, including inequalities, over primed and unprimed variables. Next, we address *synthesis* of these functions with a complete and automatic method for synthesizing *lexicographic linear polyranking functions* with supporting linear invariants over linear loops.

1 Introduction

Guaranteed termination of program loops is necessary in many settings, such as embedded systems and safety critical software. Additionally, proving general temporal properties of infinite state programs requires termination proofs, for which automatic methods are welcome [13, 9, 11]. We propose a termination analysis based on lexicographic polynomial *polyranking* functions, which subsume lexicographic polynomial *ranking* functions.

Although every terminating loop has a ranking function, not every terminating loop has a ranking function of a restricted form, such as a lexicographic tuple of polynomials over program variables. In [2], we present a method for identifying bounded expressions that are eventually negative over loops with parallel transitions expressed as simultaneous assignments of polynomial expressions to variables. For showing termination, a function that *eventually* ranks is as good as a function that always ranks. That transitions are only assignments allows an efficient analysis based on *finite difference expressions* with respect to transitions. In this paper, we generalize our approach to loops with assertional transition relations, rather than just assignments. Including assertional transition relations in a loop abstraction language makes it more widely applicable for

---

* This research was supported in part by NSF grants CCR-01-21403, CCR-02-20134, CCR-02-09237, CNS-0411363, and CCF-0430102, by ARO grant DAAD19-01-1-0723, and by NAVY/ONR contract N00014-03-1-0939. The first author was additionally supported by a Sang Samuel Wang Stanford Graduate Fellowship.

modeling infinite state systems. The main idea for analyzing these loops is to identify eventually decreasing expressions that upper bound differences.

In [1], we show how to synthesize lexicographic linear ranking functions with supporting linear invariants over loops with linear assertional transition relations. We extend the machinery in this paper to synthesize lexicographic linear polyranking functions with supporting linear invariants over linear loops. This extension requires solving complex structural constraints induced by the recursive definition of polyranking functions. The synthesis method solves efficiently a combination of structural and numerical constraints. It searches a space of *partial* structures, solving induced constraint systems to guide the search. These partial structures represent sets of complete structures so that an infeasible induced constraint system excludes entire sets. For our application, the induced constraint systems are numerical. We believe that the general strategy is applicable to other problem domains.

Termination of loops has received a fair amount of attention recently. Synthesis of linear ranking functions over linear loops with multiple paths and assertional transition relations is accomplished via polyhedral manipulation in [4,5]. In [10], Podelski and Rybalchenko specialize the technique to single-path linear loops without an initial condition, providing an efficient and complete synthesis method for linear ranking functions based on linear programming. We generalize these results in [1]. Cousot shows how polynomial ranking functions can be synthesized over nonlinear loops using nonlinear convex optimization [6]. Tiwari proves that the termination of a class of single-path loops with linear guards and assignments is decidable [14]. Related efforts on *verification diagrams* [9] and *transition invariants* [11] (see also [7,8,3]) use automatic termination analysis. These frameworks either abstract the state-space or the transition relation to isolate a finite number of different forms of infinite behavior. Termination analysis can then find the reason for termination of each isolated behavior.

The rest of the paper is organized as follows. Section 2 introduces our loop abstraction and basic concepts. Section 3 describes the polyranking principle. Sections 4 and 5 describe our synthesis technique. Section 6 concludes.

2 Preliminaries

We define our abstraction of loops, present several basic concepts, and recall Farkas's Lemma, a mathematical result that we employ in synthesis.

Definition 1 (Polynomial Assertion). A *real variable* is a variable x that ranges over the reals, $\mathbb{R}$. A *polynomial term* has the form $c \prod_i x_i^{d_i}$ for constant $c \in \mathbb{R}$, real variables x_i, and degrees $d_i \in \mathbb{N}$ ranging over the nonnegative integers. A *polynomial expression* P is the sum of polynomial terms. A *polynomial atom* is the comparison $P_1 \bowtie P_2$ of two polynomial expressions, for $\bowtie \in \{<, \leq, =, \geq, >\}$. A *polynomial assertion* is the conjunction of polynomial atoms.

Definition 2 (Polynomial Loop). A *polynomial loop* $L : \langle \mathcal{V}, \theta, \mathcal{T} \rangle$ consists of variables $\mathcal{V}$, *initial condition* θ, and set of *transitions* $\mathcal{T}$. $\theta(\mathcal{V})$ is a polynomial assertion over $\mathcal{V}$ expressing what is true before entering the loop. A transition $\tau(\mathcal{V}, \mathcal{V}')$ over variables $\mathcal{V} \cup \mathcal{V}'$ is a polynomial assertion over unprimed and primed variables, where a primed variable x' represents the next-state value of x.

Definition 3 (Linear Loop). A *linear loop* is a polynomial loop in which all assertions and expressions are affine.

When discussing affine expressions, assertions, and loops, we use the notation $\mathbf{c}^T \mathbf{x} \geq 0$ to mean a linear atom, where $\mathbf{x} = (x_1, x_2, \ldots, x_m, 1)^T$ is a homogenized vector corresponding to the variables $\mathcal{V}$. A linear assertion is then $A\mathbf{x} \geq 0$, for homogenized matrix A. By $\tau(\mathbf{xx}') \geq 0$, we mean the linear assertion corresponding to τ's transition relation, where $(\mathbf{xx}')$ stands for the homogenized vector corresponding to $\mathcal{V} \cup \mathcal{V}'$: $(\mathbf{xx}') = (x_1, x_2, \ldots, x_m, x_1', x_2', \ldots, x_m', 1)^T$.

Definition 4 (Loop Satisfaction). A loop $L : \langle \mathcal{V}, \theta, \mathcal{T} \rangle$ satisfies an assertion φ, written $L \models \varphi$, if φ holds in all reachable states of L.

Definition 5 (Infinitely Often). For infinite computation σ of loop $L : \langle \mathcal{V}, \theta, \mathcal{T} \rangle$, $io(\sigma) \subseteq \mathcal{T}$ is the set of transitions that occur *infinitely often* in the computation.

Definition 6 (Lexicographic Polynomial Ranking Function). A *lexicographic polynomial ranking function* for a loop $L : \langle \mathcal{V}, \theta, \mathcal{T} \rangle$ is an n-tuple of polynomial expressions $\langle r_1(\mathcal{V}), \ldots, r_n(\mathcal{V}) \rangle$ such that for some $\epsilon > 0$ and for each $\tau \in \mathcal{T}$, for some $i \in \{1, \ldots, n\}$,

(**Bounded**) $L \models \tau(\mathcal{V}, \mathcal{V}') \rightarrow r_i(\mathcal{V}) \geq 0$,
(**Ranking**) $L \models \tau(\mathcal{V}, \mathcal{V}') \rightarrow r_i(\mathcal{V}') - r_i(\mathcal{V}) \leq -\epsilon$;
(**Unaffecting**) for $j < i$, $L \models \tau(\mathcal{V}, \mathcal{V}') \rightarrow r_j(\mathcal{V}') - r_j(\mathcal{V}) \leq 0$.

If $n = 1$, then the result is simply a polynomial ranking function.

For some loops, invariants are necessary to show that a ranking function (and, more generally, a polyranking function) is bounded in the loop; we say such invariants are *supporting* invariants.

Definition 7 (Polynomial Inductive Invariant). A formula φ is an invariant of a loop $L : \langle \mathcal{V}, \theta, \mathcal{T} \rangle$ if it satisfies the following conditions:

(**Initiation**) $\theta(\mathcal{V}) \rightarrow \varphi(\mathcal{V})$, and
(**Consecution**) for every $\tau \in \mathcal{T}$, $\varphi(\mathcal{V}) \wedge \tau(\mathcal{V}, \mathcal{V}') \rightarrow \varphi(\mathcal{V}')$.

Example 1. Consider the loop SIMPLE in Figure 1. Initially, $x + y$ is nonnegative. Transitions τ_1 and τ_2 are both guarded by $x \leq N$. Termination is relatively straightforward: τ_2 at least increments x, while τ_1 increases x if $x + y > 0$ and always increases y. Discovering the invariant $x + y \geq 0$ shows that incrementing y makes $x + y > 0$. However, a linear ranking function over $\{x, y, N\}$ cannot prove termination, as, for example, τ_1 does not change x when $x = -y < 0$.

$$\theta : \{x + y \geq 0\}$$
$$\tau_1 : \{x \leq N\} \Rightarrow \{x' \geq 2x + y,\ y' \geq y + 1,\ N' = N\}$$
$$\tau_2 : \{x \leq N\} \Rightarrow \{x' \geq x + 1,\ y' = y,\ N' = N\}$$

Fig. 1. Program SIMPLE, written as a loop

$$\theta : \{p \geq 0,\ q \geq 1,\ x = 0,\ y = 0\}$$
$$\tau_1 : \{x + y \leq N\} \Rightarrow \left\{\begin{array}{c} x + e - q \leq x' \leq x + e + q, \\ y + n - q \leq y' \leq y + n + q, \\ n + e + 1 \leq n' + e' \leq n + e + p, \\ p' = p,\ q' = q,\ N' = N \end{array}\right\}$$
$$\tau_2 : \{x + y \leq N,\ n + e \geq 2q + 1\} \Rightarrow \left\{\begin{array}{c} x + e - q \leq x' \leq x + e + q, \\ y + n - q \leq y' \leq y + n + q, \\ n' = n,\ e' = e,\ p' = p,\ q' = q,\ N' = N \end{array}\right\}$$
$$\tau_3 : \{p \geq 0\} \Rightarrow \left\{\begin{array}{c} n' + e' \leq -(n + e), \\ p' = p - 1,\ q' = \frac{1}{2}q, \\ x' = x,\ y' = y,\ N' = N \end{array}\right\}$$

Fig. 2. Description of the erratic robot as the program ERRATIC

Example 2. Consider the program in Figure 2, written as a loop. It defines the behavior of an erratic robot on a partially-bounded plane. Its current position is given by (x, y); its tendency to move *north* is given by n (which may be negative), while its tendency to move *east* is given by e (which may also be negative). In τ_1 and τ_2, the robot's next position is determined by its current position and (e, n), along with an error parameter q. In τ_1, n and e change so that their new sum is at least one greater than previously, but in τ_2, n and e remain constant. Finally, τ_3 adjusts parameters p and q and makes $n + e$ potentially negative (countering τ_1), but it does not change the robot's position.

Does the robot's program eventually halt? Intuitively, τ_3 may only be taken a finite number of times, while τ_1 and τ_2 *eventually* make $x + y$ only increase.

Finally, for synthesis, Farkas's Lemma [12] will serve both as a device for generating constraint systems and as the foundation for completeness claims.

Theorem 1 (Farkas's Lemma). *Consider the following system of linear inequalities over real variables* $\mathcal{V} = \{x_1, \ldots, x_m\}$:

$$S : \begin{bmatrix} A_{1,1} x_1 + \cdots + A_{1,m} x_m + A_{1,m+1} \geq 0 \\ \vdots \qquad\qquad \vdots \qquad\qquad \vdots \\ A_{n,1} x_1 + \cdots + A_{n,m} x_m + A_{n,m+1} \geq 0 \end{bmatrix}$$

If S is satisfiable, it entails linear inequality $c_1 x_1 + \cdots + c_m x_m + c_{m+1} \geq 0$ iff there exist real numbers $\lambda_1, \ldots, \lambda_n \geq 0$ such that

$$c_1 = \sum_{i=1}^{n} \lambda_i A_{i,1} \quad \cdots \quad c_m = \sum_{i=1}^{n} \lambda_i A_{i,m} \quad c_{m+1} \geq \left(\sum_{i=1}^{n} \lambda_i A_{i,m+1}\right).$$

(Negative) $E(\mathcal{V})$ is bounded: $L \models E(\mathcal{V}) \leq -\epsilon$ for some $\epsilon > 0$
(Eventually Negative) For each $\tau \in \overline{\mathcal{T}}$, one of the following conditions holds:
  (Nonincreasing) $\tau \in \overline{\mathcal{T}} - \mathcal{A}$ and τ does not increase $E(\mathcal{V})$:
    1. $\tau \in \overline{\mathcal{T}} - \mathcal{A}$
    2. $L \models \tau(\mathcal{V}, \mathcal{V}') \rightarrow E(\mathcal{V}') - E(\mathcal{V}) \leq 0$
  (Eventually Decreasing) τ does not increase $E(\mathcal{V})$ by more than some eventually negative $F(\mathcal{V})$:
    1. $L \models \tau(\mathcal{V}, \mathcal{V}') \rightarrow E(\mathcal{V}') - E(\mathcal{V}) \leq F(\mathcal{V})$
    2. $F(\mathcal{V})$ is eventually negative by $\{\tau\} \subseteq \overline{\mathcal{T}}$

Fig. 3. $E(\mathcal{V})$ is eventually negative by $\mathcal{A} \subseteq \overline{\mathcal{T}}$ on loop $L : \langle \mathcal{V}, \theta, \mathcal{T} \rangle$

Furthermore, S is unsatisfiable iff S entails $-1 \geq 0$.

3 The Polyranking Principle

The polyranking principle is based on the following recursive definition of what it means for an expression $E(\mathcal{V})$ to be *eventually negative*.

Definition 8 (Eventually Negative by $\mathcal{A} \subseteq \overline{\mathcal{T}}$). Given loop $L : \langle \mathcal{V}, \theta, \mathcal{T} \rangle$ over $\mathcal{V}$, an expression $E(\mathcal{V})$ is *eventually negative by* $\mathcal{A} \subseteq \overline{\mathcal{T}} (\subseteq \mathcal{T})$, for $\mathcal{A} \neq \emptyset$, if either case **Negative** or case **Eventually Negative** holds in Figure 3. For case **Eventually Negative** to hold, the recursion must have finite depth.

Intuitively, transitions in $\mathcal{A}$ should make progress toward decreasing $E(\mathcal{V})$, while transitions in $\overline{\mathcal{T}}$ should not interfere counterproductively.

We represent an application of this definition as a tree of expressions, called an *EN-tree*. $E(\mathcal{V})$ is the root; the **Negative** case introduces the $-\epsilon$ expressions as leaves; the **Nonincreasing** case introduces 0-leaves; and the **Eventually Decreasing** case introduces $F(\mathcal{V})$ expressions as inner nodes. Branches are labeled by transitions.

Example 3. Figure 4 shows three examples of EN-trees. The tree in Figure 4(a) for SIMPLE represents the property that $N - x$ is eventually negative by $\{\tau_1, \tau_2\} \subseteq \{\tau_1, \tau_2\}$. The first difference of $N - x$ by τ_1 is at most $-x - y$, which is itself eventually negative. The supporting invariant $x + y \geq 0$ is required for proving that the first difference of $-x - y$ by τ_1 is at most -1.

Lemma 1. *If for loop $L : \langle \mathcal{V}, \theta, \mathcal{T} \rangle$, $E(\mathcal{V})$ is eventually negative by $\mathcal{A} \subseteq \overline{\mathcal{T}}$ ($\subseteq \mathcal{T}$), then on any nonterminating computation of L, γ, such that $io(\gamma) \subseteq \overline{\mathcal{T}}$ and $io(\gamma) \cap \mathcal{A} \neq \emptyset$, eventually henceforth $E(\mathcal{V}) \leq -\epsilon$ for some $\epsilon > 0$.*

Proof. We proceed by induction on the structure of Definition 8 and the depth of the recursion, which we know is finite. As the base case, consider when **Negative** applies to $E(\mathcal{V})$; then the conclusion is immediate. For the inductive case,

Eventually Negative applies. Consider $\tau \in \mathcal{T}$. If $\tau \notin io(\gamma)$, then we can, without loss of generality, assume that τ is never taken in γ — just skip a finite prefix — so that τ has no effect on $E(\mathcal{V})$. Otherwise, $\tau \in io(\gamma)$. There are two cases. If $\tau \notin \mathcal{A}$ and **Nonincreasing** applies to τ, then τ does not increase $E(\mathcal{V})$. Otherwise, **Eventually Decreasing** applies to τ, so that τ increases $E(\mathcal{V})$ by at most $F(\mathcal{V})$, while $F(\mathcal{V})$ is eventually negative by $\{\tau\} \subseteq \overline{\mathcal{T}}$. Since $io(\gamma) \cap \{\tau\} \neq \emptyset$, we have by induction that eventually henceforth $F(\mathcal{V}) \leq -\epsilon$ for some $\epsilon > 0$. By assumption, there is at least one $\tau \in \mathcal{A} \cap io(\gamma)$. Once each of these τ's $F(\mathcal{V})$ is henceforth negative, $E(\mathcal{V})$ is decreased by at least some $\epsilon > 0$ infinitely many times. Since no other transition can increase $E(\mathcal{V})$, eventually henceforth $E(\mathcal{V}) \leq -\epsilon$.

Example 4. SIMPLE always terminates because the expression $N - x$ is bounded from below by 0 in the loop and eventually negative, as shown in Figure 4(a).

Definition 9 (Lexicographic Polyranking Function). *The ℓ-tuple of functions $\langle r_1(\mathcal{V}), r_2(\mathcal{V}), \ldots, r_\ell(\mathcal{V}) \rangle$ is a lexicographic polyranking function of loop $L: \langle \mathcal{V}, \theta, \mathcal{T} \rangle$ if for some $\pi: \mathcal{T} \to \{1, \ldots, \ell\}$,*

(Bounded) *for $\tau \in \mathcal{T}$, $L \models \tau(\mathcal{V}, \mathcal{V}') \to r_{\pi(\tau)}(\mathcal{V}) \geq 0$;*
(Polyranking) *$r_i(\mathcal{V})$ is eventually negative by $\{\tau : \pi(\tau) = i\} \subseteq \{\tau : \pi(\tau) \geq i\}$ for $i \in \{1, \ldots, \ell\}$.*

Theorem 2. *If loop $L: \langle \mathcal{V}, \theta, \mathcal{T} \rangle$ has a lexicographic polyranking function, then it always terminates.*

Proof. Suppose L does not always terminate, yet $\langle r_1(\mathcal{V}), r_2(\mathcal{V}), \ldots, r_\ell(\mathcal{V}) \rangle$ with map $\pi: \mathcal{T} \to \{1, \ldots, \ell\}$ is a lexicographic polyranking function for L. Let γ be an infinite computation, and let $i \in \{1, \ldots, \ell\}$ be the lexicographic index such that $i = \min_{\tau \in io(\gamma)} \pi(\tau)$. By Definition 9, $r_i(\mathcal{V})$ is eventually negative by $\{\tau : \pi(\tau) = i\} \subseteq \{\tau : \pi(\tau) \geq i\}$. Moreover, by selection of i, $\{\tau : \pi(\tau) = i\} \cap io(\gamma) \neq \emptyset$; and by selection of i and Definition 9, $io(\gamma) \subseteq \{\tau : \pi(\tau) \geq i\}$. Then by Lemma 1, $r_i(\mathcal{V})$ eventually becomes negative and stays negative, disabling transitions in $\{\tau : \pi(\tau) = i\}$ and thus increasing i for the remaining computation. Repeating this argument at most ℓ times proves that all transitions are eventually disabled, a contradiction.

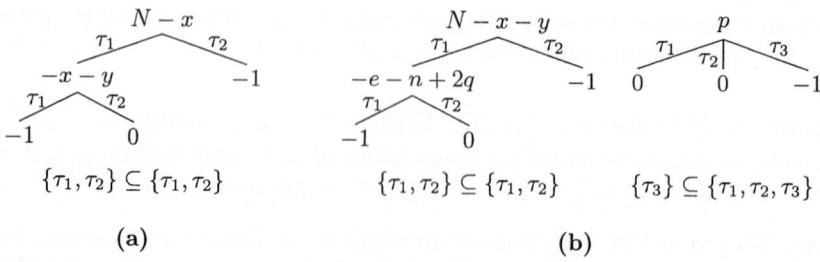

Fig. 4. EN-trees for **(a)** SIMPLE and **(b)** ERRATIC

$\mathbb{I}:$ $\begin{array}{cc}\Theta\mathbf{x} & \geq 0 \\ \hline \mathbf{Ix} & \geq 0\end{array}$	$\mathbb{D}_i:$ $\begin{array}{cc}\mathbf{Ix} & \geq 0 \\ \tau_i(\mathbf{xx'}) & \geq 0 \\ \hline -1 & \geq 0\end{array}$	$\mathbb{C}_i:$ $\begin{array}{cc}\mathbf{Ix} & \geq 0 \\ \tau_i(\mathbf{xx'}) & \geq 0 \\ \hline \mathbf{Ix'} & \geq 0\end{array}$
$\mathbb{B}_{ij}:$ $\begin{array}{cc}\mathbf{Ix} & \geq 0 \\ \tau_i(\mathbf{xx'}) & \geq 0 \\ \hline \mathbf{c_j}^T\mathbf{x} & \geq 0\end{array}$	$\mathbb{R}_{ij}:$ $\begin{array}{cc}\mathbf{Ix} & \geq 0 \\ \tau_i(\mathbf{xx'}) & \geq 0 \\ \hline \mathbf{c_j}^T\mathbf{x} - \mathbf{c_j}^T\mathbf{x'} - \epsilon & \geq 0\end{array}$	$\mathbb{U}_{ij}:$ $\begin{array}{cc}\mathbf{Ix} & \geq 0 \\ \tau_i(\mathbf{xx'}) & \geq 0 \\ \hline \mathbf{c_j}^T\mathbf{x} - \mathbf{c_j}^T\mathbf{x'} & \geq 0\end{array}$
$\mathbb{D}^{\star}_{ijk}:$ $\begin{array}{cc}\mathbf{Ix} & \geq 0 \\ \tau_i(\mathbf{xx'}) & \geq 0 \\ \hline \mathbf{c_k}^T\mathbf{x} + \mathbf{c_j}^T\mathbf{x} - \mathbf{c_j}^T\mathbf{x'} & \geq 0\end{array}$		

Fig. 5. Farkas's Lemma specializations

Example 5. ERRATIC has lexicographic linear polyranking function $\langle p, N-x-y \rangle$ where $\pi : \{\tau_1 \mapsto 2, \tau_2 \mapsto 2, \tau_3 \mapsto 1\}$. Clearly, $N - x - y$ is bounded from below by 0 by both τ_1 and τ_2, while p is bounded from below by 0 by τ_3. Figure 4(b) presents the EN-trees representing the proofs that $N - x - y$ is eventually negative by $\{\tau_1, \tau_2\} \subseteq \{\tau_1, \tau_2\}$ and p is eventually negative by $\{\tau_3\} \subseteq \{\tau_1, \tau_2, \tau_3\}$. No supporting invariants are required.

4 Constraint Generation

The remainder of this paper focuses on a complete method of synthesizing lexicographic linear polyranking functions with supporting linear invariants for linear loops. We adapt the machinery of [1] to this more general setting. The main idea is to invoke a set of Farkas's Lemma *specializations* on a loop and a set of *template* ranking functions and invariants.

Definition 10 (Template Expression). A *template expression* over $\mathcal{V}$ is a linear expression $\mathbf{c}^T\mathbf{x}$, with unknown coefficients $\mathbf{c}$. A *template assertion* is a linear assertion $\mathbf{Cx} \geq 0$ with matrix of unknown coefficients $\mathbf{C}$.

Farkas's Lemma takes a system of linear assertions and templates and returns a *dual* numeric constraint system over the λ-multipliers and the unknown template coefficients. Given a loop $L : \langle \mathcal{V}, \theta, \mathcal{T} \rangle$, the supporting invariant template $\mathbf{Ix} \geq 0$, and template expressions $\{\mathbf{c_1}^T\mathbf{x}, \ldots, \mathbf{c_n}^T\mathbf{x}\}$, the following Farkas's Lemma *specializations* are applied to encode the appropriate conditions.

Definition 11 (Farkas's Lemma Specializations). See Figure 5.

- $\mathbb{I}$ (*Initiation*): The supporting invariant includes the initial condition.
- $\mathbb{D}_i$ (*Disabled*): Transition $\tau_i \in \mathcal{T}$ and the invariant may contradict each other, indicating "dead code."
- $\mathbb{C}_i$ (*Consecution*): For transition $\tau_i \in \mathcal{T}$, if the invariant holds and the transition is taken, then the invariant holds in the next state.

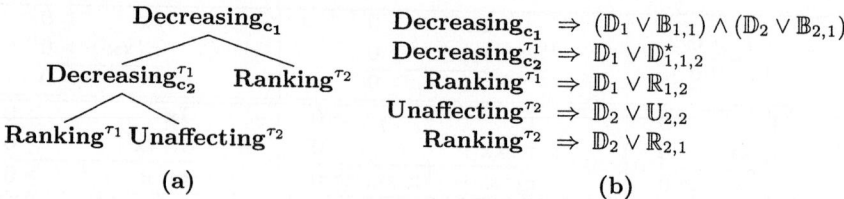

Fig. 6. (a) Template tree. (b) Generating tree constraints

- $\mathbb{B}_{ij}$ (*Bounded*): For transition $\tau_i \in \mathcal{T}$ and template expression $\mathbf{c_j}^T\mathbf{x}$, the invariant and transition imply the nonnegativity of the expression.
- $\mathbb{R}_{ij}$ (*Ranking*): For transition $\tau_i \in \mathcal{T}$ and template expression $\mathbf{c_j}^T\mathbf{x}$, taking the transition decreases the value of the expression by at least some positive amount ($\epsilon > 0$).
- $\mathbb{U}_{ij}$ (*Unaffecting*): For transition $\tau_i \in \mathcal{T}$ and template expression $\mathbf{c_j}^T\mathbf{x}$, taking the transition does not increase the value of the expression.
- $\mathbb{D}^\star_{ijk}$ (*Decreasing*): For transition $\tau_i \in \mathcal{T}$ and template expressions $\mathbf{c_j}^T\mathbf{x}$ and $\mathbf{c_k}^T\mathbf{x}$, the first difference of $\mathbf{c_j}^T\mathbf{x}$ over τ_i is upper bounded by $\mathbf{c_k}^T$.

All specializations except $\mathbb{D}^\star_{ijk}$ are used in [1] for synthesizing linear ranking functions, and thus explained in greater depth there. $\mathbb{D}^\star_{ijk}$ is the key specialization for polyranking function synthesis. It allows us to express the condition that the difference $\mathbf{c_j}\mathbf{x}' - \mathbf{c_j}\mathbf{x}$ is upper bounded by the expression $\mathbf{c_k}\mathbf{x}$, which in turn is constrained to be eventually negative.

Applying a set of specializations to a loop and a set of template expressions induces a numeric constraint system. However, we must define how to apply the specializations. For now, we assume that we are given a tuple of *template trees* and a map π mapping transitions to tuple components. The ith template gives the form of an EN-tree for the property that the root expression is eventually negative by $\{\tau : \pi(\tau) = i\} \subseteq \{\tau' : \pi(\tau') \geq i\}$.

Definition 12 (Template Tree). A *template tree* for loop $L : \langle V, \theta, \mathcal{T} \rangle$ and $\mathcal{A} \subseteq \overline{\mathcal{T}}$ gives the form of an EN-tree for the property that the root expression is eventually negative by $\mathcal{A} \subseteq \overline{\mathcal{T}}$. The root node and inner nodes are **Decreasing** nodes (corresponding to the root expression and **Eventually Decreasing** case of Definition 8, respectively) labeled with template expressions, while leaves are **Ranking** nodes (the **Negative** case) or **Unaffecting** nodes (the **Nonincreasing** case). Non-root nodes are also labeled with some $\tau \in \overline{\mathcal{T}}$. Each inner node has $|\overline{\mathcal{T}}|$ children.

Definition 8 indicates when a leaf node must be **Ranking**$^\tau$ or **Unaffecting**$^\tau$. Specifically, if the node is a child of the root and $\tau \in \mathcal{A}$, or if its parent is a node labeled by τ, then the node is **Ranking**$^\tau$; otherwise, it is **Unaffecting**$^\tau$.

Example 6. The template tree for $\{\tau_1, \tau_2\} \subseteq \{\tau_1, \tau_2\}$ of SIMPLE in Figure 6(a) includes the EN-tree in Figure 4(a).

Definition 13 (Tree Constraints). Given a loop $L : \langle \mathcal{V}, \theta, \mathcal{T} \rangle$ and a template tree for $\mathcal{A} \subseteq \overline{\mathcal{T}}$, the *tree constraints* are generated by applications of the Farkas's Lemma specializations as follows:

Parent is **Decreasing**$_{c_j}^*$ and node is ...
  ... **Ranking**$^{\tau_i}$ $\Rightarrow$ $\mathbb{D}_i \vee \mathbb{R}_{ij}$
  ... **Unaffecting**$^{\tau_i}$ $\Rightarrow$ $\mathbb{D}_i \vee \mathbb{U}_{ij}$
  ... **Decreasing**$_{c_k}^{\tau_i}$ $\Rightarrow$ $\mathbb{D}_i \vee \mathbb{D}_{ijk}^*$
Node is *root* and **Decreasing**$_{c_j}$ $\Rightarrow$ $\mathbb{D}_i \vee \mathbb{B}_{ij}$ for each $\tau_i \in \mathcal{A}$

Example 7. The tree constraints of the template tree in Figure 6(a) are generated as in Figure 6(b). The *disabled* cases are often ignored in practice, so the template tree induces numerical constraints corresponding to $\mathbb{B}_{1,1} \wedge \mathbb{B}_{2,1} \wedge \mathbb{D}_{1,1,2}^* \wedge \mathbb{R}_{1,2} \wedge \mathbb{U}_{2,2} \wedge \mathbb{R}_{2,1}$.

We use template trees and tree constraints to formalize the synthesis of lexicographic linear polyranking functions. This theorem mimics Definition 9 in the context of synthesis.

Theorem 3. *Loop $L : \langle \mathcal{V}, \theta, \mathcal{T} \rangle$ has an ℓ-lexicographic linear polyranking function supported by an n-conjunct linear inductive invariant iff there exists a mapping $\pi : \mathcal{T} \to \{1, \ldots, |\mathcal{T}|\}$ and an ℓ-tuple of template trees $\langle T_1, T_2, \ldots, T_\ell \rangle$ such that (1) the ith template tree is for $\{\tau : \pi(\tau) = i\} \subseteq \{\tau' : \pi(\tau') \geq i\}$, and (2) the ℓ numeric constraint systems induced by*

$$\mathbb{I} \wedge \bigwedge_{\tau_i \in \mathcal{T}} (\mathbb{D}_i \vee \mathbb{C}_i) \wedge \text{tree\_constraints}(T_j) \quad \text{for} \quad j \in \{1, \ldots, \ell\}$$

are satisfiable.

5 Synthesis

Theorem 3 applies to a given tuple of template trees. But in practice, we must find this tuple of templates. This section addresses this issue with an effective search, which seeks a tuple of trees such that the root expressions form a lexicographic linear polyranking function for a loop $L : \langle \mathcal{V}, \theta, \mathcal{T} \rangle$. Three different forms of constraints are handled by the full algorithm. The *lexicographic constraints* induce the transition sets $\mathcal{A}$ and $\overline{\mathcal{T}}$ for each component tree. These transition sets plus the *well-formedness constraints* from the definition of EN-trees induce a *numeric constraint system* via the tree constraints. Only when the two sets of structural constraints and the induced numeric constraints are satisfied is a solution found.

The first algorithm searches for EN-trees. Its input is a loop; an n-conjunct invariant template; the two sets of transitions, $\mathcal{A}$ and $\overline{\mathcal{T}}$; and a maximum tree height. It returns whether an EN-tree exists of at most the maximum height, possibly supported by an n-conjunct invariant. Its search strategy consists of

```
let exists_tree A T̄ =
  let initial_tree =
    Decreasing_{c_1} ( {ShouldRank^τ  : τ ∈ A} ∪
                       {ShouldUnaffect^τ : τ ∈ T̄ − A} )
  in
  let sat en = num_sat (inv_cts ∧ (tree_cts en)) in
  let next_tree en extend = ... in
  let rec search en =
        height en ≤ max_height
    and sat en
    and ( complete (en)
          or search (next_tree en false)
          or search (next_tree en true) )
  in
  search initial_tree
```

Fig. 7. ($exists\_tree$ $\mathcal{A}$ $\overline{\mathcal{T}}$) searches for an EN-tree proving that there is a bounded expression $E(\mathbf{x})$ that is eventually negative by $\mathcal{A} \subseteq \overline{\mathcal{T}}$.

incrementally building partial template trees. A partial template tree induces a numeric constraint system that is satisfiable if some completion of the tree induces a satisfiable constraint system. Thus, solving partial constraint systems guides the search. If a complete template tree is found that induces a satisfiable constraint system, the algorithm reports success.

Partial template trees are represented as in Section 4, except that two additional nodes are introduced. **ShouldRank** and **ShouldUnaffect** nodes are directions for making a partial template tree complete. In a complete template tree, all such nodes are replaced by one of the three original nodes. Indeed, complete trees must have a form obeying Definition 12. The definition of tree constraints is extended to the new nodes by simply ignoring them: **ShouldRank** and **ShouldUnaffect** nodes do not induce constraints. In this section, we represent the children of **Decreasing** nodes as a set labeling the node, rather than as figures like Figure 6(a).

Figure 7 describes the procedure as the function $exists\_tree$. It searches for an EN-tree of at most the specified maximum height, max_height, corresponding to the property that the root expression is eventually negative by $\mathcal{A} \subseteq \overline{\mathcal{T}}$. It defines the initial tree, $initial\_tree$, as a **Decreasing** node with **ShouldRank** $\mathcal{A}$ children and **ShouldUnaffect** $\overline{\mathcal{T}} - \mathcal{A}$ children. The function sat builds the numerical constraint system induced by the given tree en and returns its satisfiability. The constraints include those on the template invariant (inv_cts).

The search proceeds by exploring the space of partial template trees via $next\_tree$. For current tree en and boolean $extend$, ($next\_tree$ en $extend$) returns the next tree in the search space. The next tree is constructed by replacing some **ShouldRank**^{τ} (**ShouldUnaffect**^{τ}) node with a **Ranking**^{τ} (**Unaffecting**^{τ}) or a **Decreasing**$^τ_{c_k}$ node. The choice is controlled by $extend$. When $extend$ is **true**, the tree is "extended" by replacing the node with a **Decreasing**$^τ_{c_k}$

node, which has a fresh template expression c_k and a set of **ShouldRank**$^{\tau'}$ and **ShouldUnaffect**$^{\tau'}$ children, where a child is **ShouldRank**$^{\tau'}$ only if $\tau = \tau'$.

Example 8. Consider ERRATIC with a linear transition order $\tau_3 \prec \tau_1 \prec \tau_2$. The initial partial template trees for the lexicographic components of τ_1, τ_2, and τ_3 are the following:

τ_1 : **Decreasing**$_{c_1}$ {**ShouldRank**$^{\tau_1}$, **ShouldUnaffect**$^{\tau_2}$}
τ_2 : **Decreasing**$_{c_2}$ {**ShouldRank**$^{\tau_2}$}
τ_3 : **Decreasing**$_{c_3}$ {**ShouldUnaffect**$^{\tau_1}$, **ShouldUnaffect**$^{\tau_2}$, **ShouldRank**$^{\tau_3}$}

The next tree for τ_1, without extending the tree, could be

Decreasing$_{c_1}$ {**Ranking**$^{\tau_1}$, **ShouldUnaffect**$^{\tau_2}$};

however, the corresponding numeric constraint system is unsatisfiable. Therefore, with extension, the next tree is

$$\textbf{Decreasing}_{c_1} \left\{ \begin{array}{l} \textbf{Decreasing}_{c_4}^{\tau_1}\{\textbf{ShouldRank}^{\tau_1}, \textbf{ShouldUnaffect}^{\tau_2}\}, \\ \textbf{ShouldUnaffect}^{\tau_2} \end{array} \right\}.$$

Finally, the search runs as follows. It checks that the given template tree, en, has a height at most max_height and that the numeric constraint system induced by en is feasible. If the system is infeasible, then no extension of en can induce a feasible constraint system, so the search halts on the branch. If both checks are satisfied and the tree is *complete*, then a satisfying tree has been found, so the search returns **true**. If the tree is still incomplete, it recurses on the next tree, first without extension and then with extension. Each of the two calls to *next_tree* should modify the same node.

Lemma 2. *If num_sat is a decision procedure, (exists_tree $\mathcal{A}$ $\overline{\mathcal{T}}$) returns* **true** *iff there is an expression $E(\mathbf{x})$ and an n-conjunct supporting linear invariant $I\mathbf{x} \geq 0$ such that (1) $E(\mathbf{x})$ is lower bounded by 0 by each $\tau \in \mathcal{A}$ relative to $I\mathbf{x} \geq 0$, and (2) $E(\mathbf{x})$ is eventually negative by $\mathcal{A} \subseteq \overline{\mathcal{T}}$ relative to $I\mathbf{x} \geq 0$, and the associated EN-tree has height at most max_height.*

Example 9. After several more iterations for each template tree in Example 8, each **ShouldRank** (**ShouldUnaffect**) node is replaced by a **Ranking** (**Unaffecting**) node. One possible solution is discussed in Example 4, where the trees for τ_1 and τ_2 have been merged into one tree.

The second algorithm, *lex*, adapted from [1], works on top of the first to find a lexicographic function. It associates with each transition $\tau \in \mathcal{T}$ a template expression $T(\tau)$. It seeks a linear order among the transitions by incrementally searching over partial orders. A partial order $\prec$ over $\mathcal{T}$ induces a set of constraints: for each τ, $T(\tau)$ must be eventually negative by $\{\tau\} \subseteq \{\tau' : \tau' \prec \tau \vee \tau' = \tau\}$; and $T(\tau)$ must be lower bounded by 0 when τ is enabled. These constraints are checked by the first algorithm. If $\prec$ induces an unsatisfiable constraint system, then no linear extension can induce a satisfiable constraint system. Thus, solving partial constraint systems guides this search, as well. If a linear order is found that induces a satisfiable constraint system, the second algorithm has found a lexicographic polyranking function.

Proposition 1. *Consider loop $L : \langle \mathcal{V}, \theta, \mathcal{T} \rangle$, an n-conjunct template invariant, and maximum tree height max_height. (lex L) returns* true *iff L has a lexicographic linear polyranking function such that the associated EN-trees have height at most max_height and are each supported by an n-conjunct linear invariant.*

6 Conclusion

We implemented the lexicographic linear polyranking approach to proving termination of linear loops in our tool, linsys. The implementation found the polyranking function and supporting invariant described in Example 4 in about 2 seconds and the lexicographic polyranking function described in Example 5 in about 1 second. The primary challenge in scaling our synthesis method to assertional polynomial loops is solving the polynomial constraint systems. Approximation methods like in [6] may address scalability.

We believe that two ideas from this paper are useful in domains other than termination analysis. First, the recursive definition of an eventually negative expression may be of interest for hybrid system analysis and control theory. Second, the form of our synthesis method for solving structural and numeric constraints should be applicable to other constraint solving and synthesis tasks.

Acknowledgments. We thank the reviewers for their insightful comments.

References

1. BRADLEY, A. R., MANNA, Z., AND SIPMA, H. B. Linear ranking with reachability. In *CAV* (2005).
2. BRADLEY, A. R., MANNA, Z., AND SIPMA, H.B. Termination of polynomial programs. In *VMCAI* (2005), pp. 113–129.
3. CODISH, M., GENAIM, S., BRUYNOOGHE, M., GALLAGHER, J., AND VANHOOF, W. One lop at a time. In *WST* (2003).
4. COLÓN, M., AND SIPMA, H. B. Synthesis of linear ranking functions. In *TACAS* (2001), pp. 67–81.
5. COLÓN, M.A., AND SIPMA, H. B. Practical methods for proving program termination. In *CAV* (2002), pp. 442–454.
6. COUSOT, P. Proving program invariance and termination by parametric abstraction, lagrangian relaxation and semidefinite programming. In *VMCAI* (2005), pp. 1–24.
7. DERSHOWITZ, N., LINDENSTRAUSS, N., SAGIV, Y., AND SEREBRENIK, A. A general framework for automatic termination analysis of logic programs. *Applicable Algebra in Engineering, Communication and Computing 12* (2001), 117–156.

8. LEE, C.S., JONES, N. D., AND BEN-AMRAM, A.M. The size-change principle for program termination. In *POPL* (2001), pp. 81–92.
9. MANNA, Z., BROWNE, A., SIPMA, H. B., AND URIBE, T. E. Visual abstractions for temporal verification. In *Algebraic Methodology and Software Technology* (1998), pp. 28–41.
10. PODELSKI, A., AND RYBALCHENKO, A. A complete method for the synthesis of linear ranking functions. In *VMCAI* (2004), pp. 239–251.
11. PODELSKI, A., AND RYBALCHENKO, A. Transition invariants. In *LICS* (2004), pp. 32–41.
12. SCHRIJVER, A. *Theory of Linear and Integer Programming*. Wiley, 1986.
13. SIPMA, H. B., URIBE, T. E., AND MANNA, Z. Deductive model checking. In *CAV* (1996), pp. 209–219.
14. TIWARI, A. Termination of linear programs. In *CAV* (2004), pp. 70–82.

Approximate Guarding of Monotone and Rectilinear Polygons

Bengt J. Nilsson

Technology and Society, Malmö University College,
SE-205 06 Malmö, Sweden
Bengt.Nilsson@ts.mah.se

Abstract. We show a constant factor approximation algorithm for interior guarding of monotone polygons. Using this algorithm we obtain an approximation algorithm for interior guarding rectilinear polygons that has an approximation factor independent of the number of vertices of the polygon. If the size of the smallest interior guard cover is OPT for a rectilinear polygon, our algorithm produces a guard set of size $O(OPT^2)$.

1 Introduction

The art gallery problem is perhaps the best known problem in computational geometry. It asks for the minimum number of guards to guard a space having obstacles. Originally, the obstacles were considered to be walls mutually connected to form a closed Jordan curve, hence, a simple polygon. Tight bounds for the number of guards necessary and sufficient were found by Chvátal [7] and Fisk [14]. Subsequently, other obstacle spaces, both more general and more restricted than simple polygons have also been considered for guarding problems, most notably, polygons with holes and simple rectilinear polygons [16, 22].

Art gallery problems are motivated by applications such as line-of-sight transmission networks in polyhedral terrains, e.g., signal communications and broadcasting, cellular telephony, and other telecommunication technologies as well as placement of motion detectors and security cameras.

We distinguish between two types of guarding problems. Vertex guarding considers only guards positioned at vertices of the polygon, whereas interior guarding allows the guards to be placed anywhere in the interior of the polygon.

The computational complexity question of guarding simple polygons was settled by Aggarwal [1] and Lee and Lin [19] independently when they showed that the problem is NP-hard for both vertex guards and interior guards. Further results have shown that already for very restricted subclasses of polygons the problem is still NP-hard [3, 21]. Also, Chen et al. [5] claim that vertex guarding a monotone polygon is NP-hard, however the details of their proof are omitted and still to be verified.

The approximation complexity of guarding polygons has been studied by Eidenbenz and others. Eidenbenz [13] shows that polygons with holes cannot be efficiently guarded by fewer than $\Omega(\log n)$ times the optimal number of interior or vertex guards, unless P=NP, where n is the number of vertices of the polygon. Brodén et al. and Eidenbenz [3, 12] independently prove that interior guarding simple polygons is APX-hard, thus showing that, unless P=NP, no PTAS is possible for this problem.

Any polygon (with or without holes) can be efficiently vertex guarded with logarithmic approximation factor in the number of vertices of the polygon. The algorithm is a simple reduction to SET COVER and goes as follows [15]: compute the arrangement produced by the visibility polygons of the vertices. Next, let each vertex v correspond to a set in the set cover instance consisting of elements corresponding to the faces of the arrangement that lie in the visibility polygon of v. The greedy algorithm for SET COVER will then produce a guard cover having logarithmic approximation factor.

The above result can be improved for simple polygons using randomization, giving an algorithm with expected running time $O(nOPT^2 \log^4 n)$ that produces a vertex guard cover with approximation factor $O(\log OPT)$ with high probability, where n is the size of the polygon and OPT is the smallest vertex guard cover for the polygon [11].

Recently, interesting constant factor approximation algorithms have been developed for the special case of one-dimensional terrains [2, 8]. A terrain is a two-dimensional region above a monotone chain.

We prove polynomial time approximation algorithms for interior guarding of monotone and rectilinear polygons. As mentioned, vertex guarding of monotone polygons is believed to be NP-hard, and furthermore, it is known that covering rectilinear polygons with the minimum number of convex pieces (rectangles) is NP-hard [9]. This suggests that interior guarding these two classes of polygons is difficult and provides the basis for our interest in approximation algorithms for these problems.

The next section contains some useful definitions and an Sections 3 and 4 we describe the algorithms for monotone and rectilinear polygons respectively.

2 Definitions

A polygon **P** is *l-monotone* if there is a line of monotonicity l such that any line orthogonal to l has a simply connected intersection with **P**. When we talk about monotone polygons, we will henceforth assume that they are x-monotone, i.e., the x axis is the line of monotonicity for the polygons we consider.

The boundary of a monotone polygon **P** can be subdivided into two chains, the *upper chain* U and the *lower chain* D. Let s and t be the leftmost and rightmost vertices of **P** respectively. The chain U consists of the boundary path followed from s to t in clockwise direction, whereas D is the boundary path followed from s to t in counterclockwise direction.

A polygon **P** is *rectilinear* if the boundary of **P** consists of axis parallel line segments. Hence, at all vertices, the interior angle between segments are either 90 or 270 degrees; see Figure 1(a).

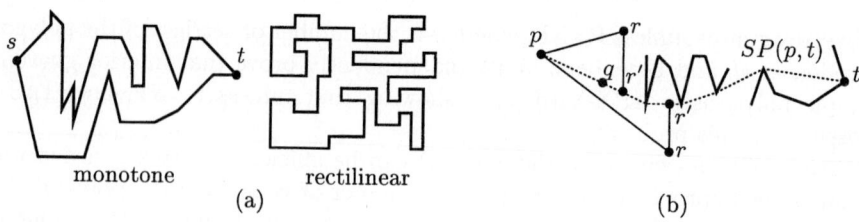

Fig. 1. Illustrating definitions and Lemma 1

Let $\mathbf{VP}(p)$ denote the visibility polygon of $\mathbf{P}$ from the point p, i.e, the set of points in $\mathbf{P}$ that can be connected with a line segment to p without intersecting the outside of $\mathbf{P}$.

Consider a partial set of guard points $g_1, \ldots, g_m$ in $\mathbf{P}$ and the union of their visibility polygons $\bigcup_{i=1}^{m} \mathbf{VP}(g_i)$, the set $\mathbf{P} \setminus \bigcup_{i=1}^{m} \mathbf{VP}(g_i)$ is the region of $\mathbf{P}$ not seen by the points $g_1, \ldots, g_m$. This region consists of a set of simply connected polygonal regions called *pockets* bounded by either the polygon boundary or the edges of the visibility polygons.

The following definitions are useful for monotone polygons. Let q be a point in $\mathbf{VP}(p)$ that lies to the right of p. We denote by $\mathbf{VP}_R(p, q)$ the part of $\mathbf{VP}(p)$ that lies to the right of q. Also, $\mathbf{VP}_R(p) = \mathbf{VP}_R(p, p)$.

A pocket in a monotone polygon $\mathbf{P}$ is an *upper pocket* if it is adjacent to the upper boundary U of $\mathbf{P}$, otherwise it is a *lower pocket*. Note that an upper pocket can be adjacent to D whereas a lower pocket is never adjacent to U.

Let $SP(p, q)$ denote the shortest (Euclidean) path between points p and q inside $\mathbf{P}$.

Lemma 1. *If q is a point on $SP(p, t)$ inside a monotone polygon $\mathbf{P}$, then $\mathbf{VP}_R(p, q) \subseteq \mathbf{VP}_R(q)$.*

Proof. Let r be a point to the right of q in $\mathbf{P}$ that is visible from p. To prove that r is seen from q consider the vertical line through r and its intersection point r' with $SP(p, t)$. The three points p, r, and r' define a polygon in $\mathbf{P}$ having three convex vertices and possibly some reflex vertices on the path $SP(p, r')$. Since r sees both p and r', r sees all of the path $SP(p, r')$ and hence also the point q; see Figure 1(b). □

3 Interior Guarding Monotone Polygons

Our algorithm for guarding a monotone polygon $\mathbf{P}$ will incrementally guard $\mathbf{P}$ starting from the left and moving right. Hence, we are interested in the structure of the pockets that occur when guarding is done in this way. We first define *kernel expansions* of the pockets given a partial guard cover $\mathcal{G}_p$, and then taking maximal nonempty intersection of these we produce the main region that we will be interested in. This region is called a *spear* and with this we can define a well behaved guard cover $\mathcal{G}^*$ that has small size. We finally prove that our incremental algorithm produces a guard cover at most a constant times larger than $\mathcal{G}^*$.

Assume that we have a partial guard cover $\mathcal{G}_p$ in **P** and that everything to the left of the rightmost guard is seen. Consider the upper pockets resulting from this guard cover and enumerate them $\mathbf{p}_1^U, \ldots$ from left to right. The lower pockets are enumerated from left to right $\mathbf{p}_1^D, \ldots$ in the same way; see Figure 2.

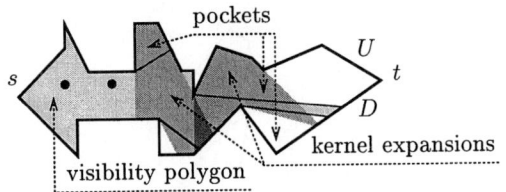

Fig. 2. Illustrating pockets and kernel expansions

Let $\mathbf{p}^U$ be an upper pocket. The *kernel expansion* $\mathbf{ke}(\mathbf{p}^U)$ consists of all the points in **P** that see everything in $\mathbf{p}^U$ to the left of themselves. For the lower pockets we define the kernel expansion symmetrically. The definition of kernel expansion is valid also when no guards have as yet been placed in the polygon. In this case, we take all of the polygon **P** to be an upper pocket.

Let k be the largest index so that $\bigcap_{i=1}^{k} \mathbf{ke}(\mathbf{p}_i^U)$ is nonempty. This nonempty intersection of kernel expansions is called the *upper spear*, also denoted $\mathbf{sp}^U$. We can in the same way define the *lower spear* $\mathbf{sp}^D$ as the maximal nonempty intersection of the kernel expansions for the lower pockets.

Given the partial guard cover $\mathcal{G}_p$ the upper spear $\mathbf{sp}^U$ can be computed in linear time as follows. Let r_1^U be the leftmost point of $\mathbf{p}_1^U$. An edge e of the pocket that is also part of U is defined to have the same direction as when it is traversed during a traversal of U from s to t. Following the boundary of the upper pockets starting at r_1^U, for each edge e of the pocket that is also an edge of U, we issue a half line from e having the same direction as e. When the traversal of a pocket $\mathbf{p}_i^U$ reaches the last edge we establish the last point q_i^U of the edge not seen by $\mathcal{G}_p$ and we extend the directed half line issuing from q_i^U toward the vertex v on U between q_i^U and r_{i+1}^U so that this half line has minimum interior angle and does not intersect the exterior of **P** above U. Using these half lines in the order they were computed we incrementally find their right half plane intersection in the same way as is done to compute the kernel of a polygon [17, 18]. This gives us the upper boundary of the spear.

To compute the lower boundary of the spear we follow the lower boundary D of **P** from the point having the same x-coordinate as r_1^U toward t and maintain the half lines issuing from r_1^U having maximal interior angle to U and such that they do not intersect the exterior of **P** below D; see Figure 3. The intersection point between the upper and lower boundary of the spear is called the upper *spear tip* and we denote it u^U.

In a similar manner we can compute the lower spear $\mathbf{sp}^D$ and its lower spear tip u^D.

To every spear $\mathbf{sp}$ we also associate a region called the *shadow* of the spear, denoted $\mathbf{shd}(\mathbf{sp})$. If the spear tip lies on the lower boundary D the shadow is

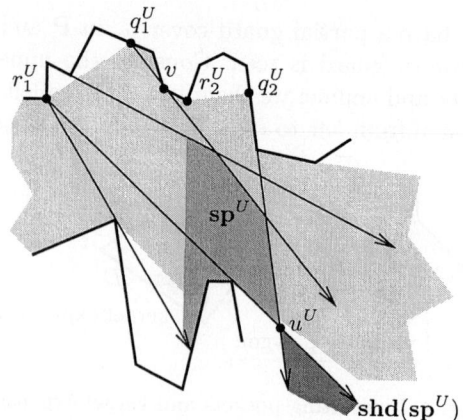

Fig. 3. Computing the upper spear and the shadow

the empty set. If the spear tip lies in the interior of **P**, the shadow of the spear is the region to the right of the spear tip between the two half lines bounding the spear that intersect at the spear tip; see Figure 3.

The upper and lower spears are dependent on the placement of the previously placed guards so we will henceforth refer to them as $\mathbf{sp}^U(\mathcal{G}_p)$ and $\mathbf{sp}^D(\mathcal{G}_p)$ given the partial guard set $\mathcal{G}_p$. For each spear, $\mathbf{sp}^U(\mathcal{G}_p)$ and $\mathbf{sp}^D(\mathcal{G}_p)$ we denote the upper spear tip $u^U(\mathcal{G}_p)$ and the lower spear tip $u^D(\mathcal{G}_p)$. If $\mathcal{G}_p = \emptyset$, the upper spear $\mathbf{sp}^U(\emptyset)$ and the upper spear tip $u^U(\emptyset)$ are well defined.

We prove the following two lemmas.

Lemma 2. *If $\mathcal{G}_p$ and $\mathcal{G}'_p$ are two partial guard covers of **P** such that $\mathbf{sp}^U(\mathcal{G}_p)$ and $\mathbf{sp}^U(\mathcal{G}'_p)$ do not intersect, then $\mathbf{shd}(\mathbf{sp}^U(\mathcal{G}_p)) \cap \mathbf{shd}(\mathbf{sp}^U(\mathcal{G}'_p)) = \emptyset$.*

Proof. We make a proof by contradiction and assume that the two shadows intersect. Assume that $\mathbf{sp}^U(\mathcal{G}_p)$ lies to the left of $\mathbf{sp}^U(\mathcal{G}'_p)$ and let p be a point in the intersection of the two shadows. We can connect p to $u^U(\mathcal{G}_p)$ with a line segment and then follow the line segment from $u^U(\mathcal{G}_p)$ back to its starting point r_1^U at the leftmost point of the first upper pocket associated to $\mathbf{sp}^U(\mathcal{G}_p)$. From r_1^U we follow the upper boundary of the pocket to the rightmost point q_k^U of the last upper pocket associated to $\mathbf{sp}^U(\mathcal{G}'_p)$, from this point on to $u^U(\mathcal{G}'_p)$, and then back to p. This traversal bounds a polygon interior to **P** that contains completely the lower boundary segment of $\mathbf{sp}^U(\mathcal{G}'_p)$. However, this is not possible because by construction this segment must intersect the lower boundary D of **P**, giving us a contradiction; see Figure 4. □

At this point, it is important to note that a single guard placed in a spear will guard all pockets associated to the spear. However, these pockets can also be guarded by placing one guard above and below the shadow and possibly one or more guards inside the shadow. Now, we are interested in bounding from below the number of guards needed to the right of a spear for the case that no guard is in the spear.

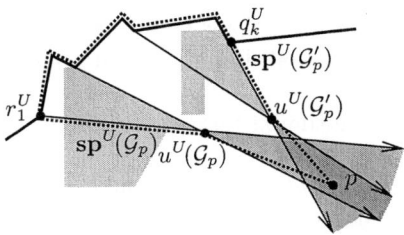

Fig. 4. Illustrating the proof of Lemma 2

Let $\mathcal{G}$ be any guard cover for **P** and let $\mathcal{G}_p \subset \mathcal{G}$ be a possibly empty partial guard cover. If $\mathcal{G}_p$ is nonempty assume that it has g as its rightmost guard and assume further that all of **P** to the left of g is guarded by $\mathcal{G}_p$. Let $\mathcal{G}_f = \mathcal{G} \setminus \mathcal{G}_p$ and define the following sets recursively.

$$\mathcal{G}_0 = \mathcal{G}_p$$
$$\mathcal{G}_i = \mathcal{G}_{i-1} \cup \{u^U(\mathcal{G}_{i-1})\} \qquad \text{for } i > 0$$

Lemma 3. *Let $\mathcal{G}_f$ and $\mathcal{G}_0, \ldots, \mathcal{G}_k$ be sets as defined above. If all guards of $\mathcal{G}_f$ lie to the right of $u^U(\mathcal{G}_k)$, then $\mathcal{G}_f$ contains at least $k+2$ guards.*

Proof. By the construction of the sets $\mathcal{G}_i$, we know that their corresponding spears do not intersect, and hence, from Lemma 4 their shadows do not intersect either. Let g be the rightmost guard of $\mathcal{G}_p = \mathcal{G}_0$.

If some shadow $\mathbf{shd}(\mathbf{sp}^U(\mathcal{G}_i))$ does not intersect the vertical line through $u^U(\mathcal{G}_k)$ then at least one guard is needed in the interval between g and $u^U(\mathcal{G}_k)$, contradicting that $\mathcal{G}$ is a guard cover for **P**. Hence, all shadows intersect this vertical line. Now, assume that $\mathcal{G}_f$ contains at most $k+1$ guards. By the pigeon-hole principle there are two consecutive shadows $\mathbf{shd}(\mathbf{sp}^U(\mathcal{G}_i))$ and $\mathbf{shd}(\mathbf{sp}^U(\mathcal{G}_{i+1}))$ such that at least one of them and the region between them does not contain any guard, i.e., all guards of $\mathcal{G}_f$ lie either above the upper boundary of $\mathbf{shd}(\mathbf{sp}^U(\mathcal{G}_i))$ and below the upper boundary of $\mathbf{shd}(\mathbf{sp}^U(\mathcal{G}_{i+1}))$ or above the lower boundary of $\mathbf{shd}(\mathbf{sp}^U(\mathcal{G}_i))$ and below the lower boundary of $\mathbf{shd}(\mathbf{sp}^U(\mathcal{G}_{i+1}))$. In both cases this means that there are points of the last pocket associated to $\mathbf{sp}^U(\mathcal{G}_i)$ that are not seen by $\mathcal{G}$, giving us a contradiction; see Figure 5. □

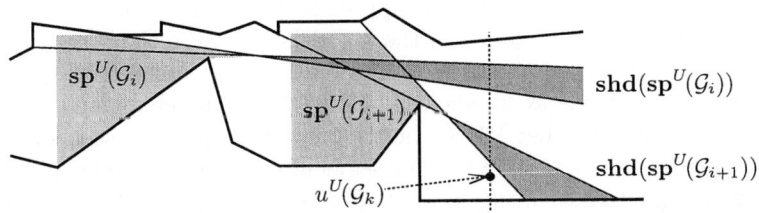

Fig. 5. Illustrating the proof of Lemma 3

We say that a guard *belongs* to a shadow $\mathbf{shd}(\mathbf{sp}^U(\mathcal{G}_i))$, if it lies in the interior of $\mathbf{shd}(\mathbf{sp}^U(\mathcal{G}_i))$ or it lies below $\mathbf{shd}(\mathbf{sp}^U(\mathcal{G}_i))$ but above all other shadows lying below $\mathbf{shd}(\mathbf{sp}^U(\mathcal{G}_i))$.

We can, of course, prove similar results as those in Lemmas 2 and 3 for the shadows of lower spears.

A guard cover is called *serial*, if the following invariant condition holds when we incrementally place the guards of the cover in **P** one by one in order from left to right.

If g_m is the mth guard placed in the order, then g_m either lies in the upper spear or the lower spear of the guards $g_1, \ldots, g_{m-1}$.

The next lemma shows that there is a serial guard cover of small size.

Lemma 4. *If $\mathcal{G}$ is a guard cover for the monotone polygon* **P**, *then there is a serial guard cover $\mathcal{G}^*$ for* **P** *such that $|\mathcal{G}^*| \leq 3|\mathcal{G}|$.*

Proof. Given a guard cover $\mathcal{G}$ we transform it to be serial as follows. Order the guards of $\mathcal{G} = \{g_1, \ldots, g_m\}$ from left to right. The transformation incrementally adds guards moving from left to right into two sets $\mathcal{G}^U$ and $\mathcal{G}^D$ ensuring that the next guard place lies in a spear. To make the constructed guard set serial we employ a plane sweep approach moving from left to right. As soon as the sweep line reaches a guard, the guard is attached to the sweep line and moves along it following the shortest path to t. By Lemma 1 this does not decrease visibility to the right. Now, as the sweep proceeds one of two things happen. Either a guard becomes the last guard to leave a spear (with respect to the previously released guards) and it is then released from the sweep line or the sweep reaches the spear tip without having released a guard. In this case, the spear has been completely empty of guards. In this latter case, we add a guard at the spear tip, placing it in $\mathcal{G}^U$ if the spear is an upper spear and in $\mathcal{G}^D$ if the spear is a lower spear. When the sweep line reaches t, those guards still attached to it are removed (except for possibly one) giving us the serial guard cover $\mathcal{G}^*$.

To count the number of extra guards placed by this process we can associate each new guard placed at a spear tip with one belonging to the shadow of the associated spear. From Lemma 3 we have that a guard belonging to an upper (lower) shadow can at worst belong also to a lower (upper) shadow. Hence, $|\mathcal{G}^U| \leq |\mathcal{G}|$ and $|\mathcal{G}^D| \leq |\mathcal{G}|$ giving us that $|\mathcal{G}^*| \leq |\mathcal{G}| + |\mathcal{G}^U| + |\mathcal{G}^D| \leq 3|\mathcal{G}|$. □

We can now give the details of the incremental algorithm, displayed in Figure 6, and prove its correctness, approximation factor and time complexity.

To prove the complexity of the algorithm we note that the loops of Steps 2 and 3 are performed $O(n)$ times. Computing the spear and its spear tip can be done in linear time as we showed before. Hence, it remains to show how to do Step 2.4 efficiently. Let $\mathbf{VP}(\mathcal{G})$ be the part of the polygon seen so far. We begin by placing g' at the top of the line segment l and compute the upper spear with g' in the guard set. Then, we slide g' along l continuously updating the point $u^U(\mathcal{G} \cup \{g'\})$ as we go along. The structural changes of $\mathbf{sp}^U(\mathcal{G} \cup \{g'\})$ occur at certain key points on l. These are

```
Algorithm    Guard-Monotone-Polygons
Input:       A monotone polygon P
Output:      A guard cover for P
1  Let G := ∅
2  while not all upper pockets are guarded do
2.1          Compute sp^U(G) and u^U(G)
2.2          Place a guard g at u^U(G); G := G ∪ {g}
2.3          Compute ⋃_{g∈G} VP(g), let p^U be the first upper pocket in P, and let l
             be the vertical line segment through the leftmost boundary point of p^U
2.4          Place a guard g' on l so that u^U(G ∪ {g'}) lies as far to the right as
             possible; G := G ∪ {g'}
   endwhile
3  Repeat Step 2 for the lower pockets to guard these
   return G
End  Guard-Monotone-Polygons
```

Fig. 6. The algorithm for monotone polygons

1. when the convex vertex of $\mathbf{VP}(\mathcal{G}) \cup \mathbf{VP}(g')$ on an edge adjacent to an upper pocket becomes incident to a vertex of the polygon boundary U.
2. when an edge of the boundary of $\mathbf{sp}^U(\mathcal{G} \cup \{g'\})$ becomes incident to two vertices of the upper boundary U.
3. when three consecutive half lines issuing from pockets intersect at the same point.

The key points occur at an at most cubic number of discrete points on l. (The maximum number of possible common intersection points between three lines among n lines.) Moving g' in between the key points will make $u^U(\mathcal{G} \cup \{g'\})$ move monotonically to the right or to the left. Hence, by computing the key points, which can be done incrementally in at most linear time, we can find the point on l where $u^U(\mathcal{G} \cup \{g'\})$ lies as far to the right as possible; see Figure 7.

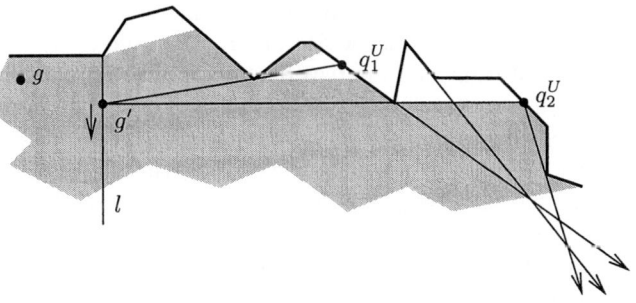

Fig. 7. Computing the rightmost spear tip

We have the following theorem.

Theorem 1. *The algorithm Guard-Monotone-Polygons computes a guard cover for a monotone polygon* $\mathbf{P}$ *of size at most* $12\,OPT$ *in polynomial time, where* OPT *is the size of the smallest guard cover for* $\mathbf{P}$.

Proof. To prove correctness, the algorithm incrementally guards everything to the left of the rightmost guard, hence, it will completely guard the polygon.

To prove the approximation factor, consider any serial guard cover $\mathcal{G}^* = \{g_1^*, \ldots, g_m^*\}$ ordered from left to right in the polygon. We prove by induction that after the ith iteration of the loop at Step 2, the last guard placed thus far lies further to the right than g_i^*. For the base case, after the first iteration of the loopi, note that the algorithm places a guard at the first spear tip, the rightmost point of the first spear, and hence, to the right of g_1^*. Assume now that after the ith iteration of the loop, the guard g lies to the right of g_i^*. Since we place g' so that the rightmost point $u^U(\mathcal{G} \cup \{g'\})$ of $\mathbf{sp}^U(\mathcal{G} \cup \{g'\})$ lies as far to the right as possible it has to lie at least as far to the right as g_{i+1}^*. Associating g and g' during the ith iteration to g_i^* we see that the loop will place at most $2|\mathcal{G}^*|$ guards in the polygon. Thus, a total of at most $4|\mathcal{G}^*|$ guards are placed. By Lemma 4 we can choose $\mathcal{G}^*$ as the smallest serial guard cover which in turn is bounded by $3\,OPT$.

The complexity of the algorithm follows from the previous discussion. □

4 Interior Guarding Rectilinear Polygons

The algorithm for computing a guard cover in a simple rectilinear polygon consists of two main steps. First, we find a subdivision of the polygon into monotone pieces, second, we use the previously given algorithm to compute a guard cover in each monotone piece.

Consider a simple rectilinear polygon $\mathbf{P}$. To every reflex vertex v we can associate two *extensions*, i.e., the two maximal line segments in $\mathbf{P}$ through v and collinear to the two edges adjacent to v. We associate a direction to an extension e collinear to an edge e_v by giving e the same direction as e_v gets when $\mathbf{P}$ is traversed in counterclockwise order. This allows us to refer to the regions to the left and right of an extension, meaning to the left or right of e if e is directed upward.

Given two extensions e and e', we say that e *dominates* e', if all points in $\mathbf{P}$ to the left of e are also to the left of e'. Using the algorithm of Chin and Ntafos [6] in conjunction with Chazelle's triangulation algorithm [4], we can in linear time compute the most dominant extensions that we call the *essential extensions*. Assume that this computation gives us k essential extensions. An essential extension e_i is collinear to an edge with one reflex and one convex vertex. Let v_i denote the convex vertex. This gives us k convex vertices and we choose one of them, say v_k, as root in a shortest rectilinear path tree T_R to each of the other vertices v_i, for $1 \leq i < k$, that can be computed in linear time [10]. The shortest rectilinear path tree consists of paths that are shortest in the L_1-metric connecting v_k to all the other vertices v_i, for $1 \leq i < k$.

To each rectilinear path $SP_R(v_i, v_k)$ connecting v_i with v_k we define a vertical and a horizontal *histogram expansion*. The horizontal histogram expansion $\mathbf{H}_i^H$ consists of those points in $\mathbf{P}$ that can be connected to the path $SP_R(v_i, v_k)$ with vertical line segments contained in $\mathbf{P}$. We define the vertical histogram expansion $\mathbf{H}_i^V$ in a similar manner. A histogram expansion can be computed in linear time using an algorithm by Levcopoulos [20]. Each horizontal histogram expansion consists of a number of x-monotone polygons with the property that no guard in one monotone polygon can see anything in any of the others. Similarly a vertical histogram expansion subdivides into y-monotone pieces with the same property; see Figure 8.

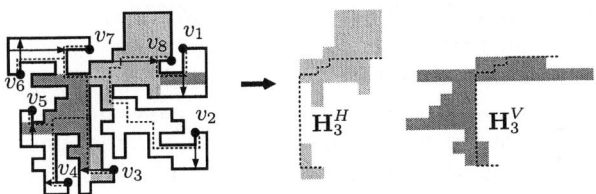

Fig. 8. Illustrating the algorithm

Lemma 5. *If $\mathbf{P}$ can be guarded with OPT guards, then a histogram expansion can also be guarded with at most OPT guards interior to the region.*

Proof. Let p be a point that sees into a monotone piece $\mathbf{R}$ of a histogram expansion. Assume that $\mathbf{R}$ is x-monotone and that p lies in a region adjacent to the lower boundary D of $\mathbf{R}$. Let l be the line segment that separates $\mathbf{R}$ from the piece containing p. Consider the intersection $\mathbf{VP}(p) \cap \mathbf{R}$. The intersection subdivides $\mathbf{R}$ into left pockets and right pockets. Traversing the boundary of $\mathbf{VP}(p)$ clockwise starting at a point outside $\mathbf{R}$ will first reach the edges that are incident to left pockets, then a boundary chain C of $\mathbf{R}$, and finally the edges that are incident to right pockets of $\mathbf{R}$. Take any point q of C and let p' be the intersection of the line segment between p and q with l. Any point in $\mathbf{R}$ seen by p will also be seen by p', which proves the lemma. □

We use the *Guard-Monotone-Polygon* algorithm of the previous section to guard each monotone piece with at most $12m$ guards, where m is the smallest guard cover for the monotone piece. From Lemma 5 we know that each histogram expansion can be guarded with OPT guards interior to the histogram expansion, and hence, our algorithm guards it with at most $12OPT$ guards.

Furthermore, one guard can see at most two of the vertices v_i, for $1 \leq i \leq k$, hence, $k/2 \leq OPT$. Since we construct a total of $2k$ horizontal and vertical histogram expansions, the union $\bigcup_{i=1}^{k-1} \mathbf{H}_i^H \cup \mathbf{H}_i^V$ can be guarded by at most $48OPT^2$ guards.

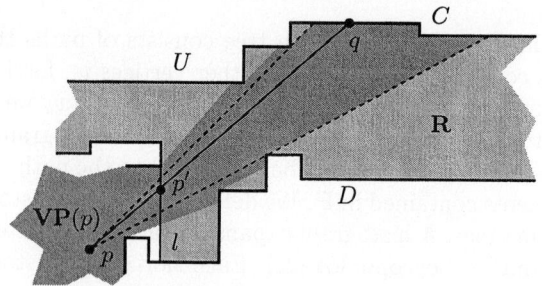

Fig. 9. Illustrating the proof of Lemma 5

The set $\mathbf{P} \setminus (\bigcup_{i=1}^{k-1} \mathbf{H}_i^H \cup \mathbf{H}_i^V)$ partitions into a number of connected regions. These *extra pieces*; see Figure 10; are monotone with respect to both the x- and the y-axis and we can guard each of them with one extra guard. An extra piece is also adjacent to one horizontal and one vertical histogram expansion and each monotone piece in a histogram expansion can be adjacent to at most two extra pieces. Hence, to count the number of extra pieces, i.e., the number of additional guards we place, we associate each extra piece with the horizontal or vertical monotone piece of a histogram expansion that is closer to v_k. Thus, the number of guards placed to see all of $\mathbf{P}$ has at most doubled.

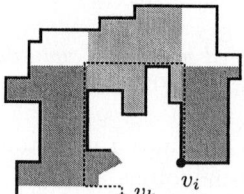

Fig. 10. Handling the extra pieces

We have proved the following theorem.

Theorem 2. *There is a polynomial time algorithm that computes a guard cover of size $96\,OPT^2$ in a rectilinear polygon $\mathbf{P}$, where OPT is the size of the smallest guard cover for $\mathbf{P}$.*

5 Conclusions

We have proved polynomial time algorithms for approximate interior guarding of monotone and rectilinear polygons. Our contribution is that the approximation factors for both algorithms is independent of the size of the polygon. Interesting open problems are to improve the approximation bounds for monotone and rectilinear polygons, to find approximation algorithms for other classes of polygons, and ultimately approximate guarding of the general class of simple polygons.

References

1. A. AGGARWAL. *The Art Gallery Theorem: Its Variations, Applications and Algorithmic Aspects*. PhD thesis, Johns Hopkins University, 1984.
2. B. BEN-MOSHE, M.J. KATZ, J.S.B. MITCHELL. A Constant-Factor Approximation Algorithm for Optimal Terrain Guarding. In *Proc. 16th ACM-SIAM Symposium on Discrete Algorithms, SODA'05*, 2005.
3. B. BRODÉN, M. HAMMAR, B.J. NILSSON. Guarding Lines and 2-Link Polygons is APX-hard. In *Proc. 13th Canadian Conference on Computational Geometry, CCCG'01*, pages 45–48, 2001.
4. B. CHAZELLE. Triangulating a Simple Polygon in Linear Time. In *Proc. 31st Symposium on Foundations of Computer Science*, pages 220–230, 1990.
5. D.Z. CHEN, V. ESTIVILL-CASTRO, J. URRUTIA. Optimal Guarding of Polygons and Monotone Chains. In *Proc. 7th Canadian Conference on Computational Geometry, CCCG'95*, pages 133–138, 1995.
6. W. CHIN, S. NTAFOS. Optimum Watchman Routes. *Information Processing Letters*, 28:39–44, 1988.
7. V. CHVÁTAL. A Combinatorial Theorem in Plane Geometry. *Journal of Combinatorial Theory B*, 13(6):395–398, 1975.
8. K.L. CLARKSON, K. VARADARAJAN. Improved Approximation Algorithms for Geometric Set Cover. In *Proc. 21st ACM Symposium on Computational Geometry*, 2005.
9. J.C. CULBERSON, R.A. RECKHOW. Covering Polygons is Hard. In *Proc. 29th Symposium on Foundations of Computer Science*, pages 601–611, 1988.
10. M. DE BERG. On Rectilinear Link Distance. *Computational Geometry: Theory and Applications*, 1(1):13–34, 1991.
11. A. EFRAT, S. HAR-PELED. Locating Guards in Art Galleries. In *Proc. 2nd IFIP International Conference on Theoretical Computer Science*, 2002.
12. S. EIDENBENZ. Inapproximability Results for Guarding Polygons without Holes. In *Proc. 9th Annual International Symposium on Algorithms and Computation*, pages 427–436, 1998.
13. S. EIDENBENZ. *Inapproximability of Visibility Problems on Polygons and Terrains*. PhD thesis, ETH, Zurich, 2000.
14. S. FISK. A Short Proof of Chvátal's Watchman Theorem. *Journal of Combinatorial Theory B*, 24:374, 1978.
15. S.K. GHOSH. Approximation Algorithms for Art Gallery Problems. In *Proceedings of the Canadian Information Processing Society Congress*, 1987.
16. F. HOFFMANN, M. KAUFMANN, K. KRIEGEL. The Art Gallery Theorem for Polygons with Holes. In *Proc. 32nd IEEE Symposium on the Foundations of Computer Science*, pages 39–48, 1991.
17. B. JOE, R.B. SIMPSON. Correction to Lee's Visibility Polygon Algorithm. *BIT*, 27:458–473, 1987.
18. D.T. LEE. Visibility of a Simple Polygon. *Computer Vision, Graphics, and Image Processing*, 22:207–221, 1983.
19. D.T. LEE, A.K. LIN. Computational Complexity of Art Gallery Problems. *IEEE Transactions on Information Theory*, IT-32:276–282, 1986.
20. C. LEVCOPOULOS. *Heuristics for Minimum Decompositions of Polygons*. PhD thesis, University of Linköping, Linköping, Sweden, 1987.
21. B.J. NILSSON. *Guarding Art Galleries — Methods for Mobile Guards*. PhD thesis, Lund University, 1995.
22. J. O'ROURKE. *Art Gallery Theorems and Algorithms*. Oxford University Press, 1987.

Linear Time Algorithms for Clustering Problems in Any Dimensions

Amit Kumar[1], Yogish Sabharwal[2], and Sandeep Sen[3]

[1] Dept of Comp Sc & Engg, Indian Institute of Technology,
New Delhi-110016, India
amitk@cse.iitd.ernet.in
[2] IBM India Research Lab, Block-I, IIT Delhi, Hauz Khas,
New Delhi-110016, India
ysabharwal@in.ibm.com
[3] Dept of Comp Sc & Engg, Indian Institute of Technology,
Kharagpur, India
ssen@cse.iitkgp.ernet.in

Abstract. We generalize the k-means algorithm presented by the authors [14] and show that the resulting algorithm can solve a larger class of clustering problems that satisfy certain properties (*existence of a random sampling procedure* and *tightness*). We prove these properties for the k-median and the discrete k-means clustering problems, resulting in $O(2^{(k/\varepsilon)^{(1)}} dn)$ time $(1+\varepsilon)$-approximation algorithms for these problems. These are the first algorithms for these problems linear in the size of the input (nd for n points in d dimensions), independent of dimensions in the exponent, assuming k and ε to be fixed. A key ingredient of the k-median result is a $(1+\varepsilon)$-approximation algorithm for the 1-median problem which has running time $O(2^{(1/\varepsilon)^{(1)}} d)$. The previous best known algorithm for this problem had linear running time.

1 Introduction

The problem of clustering a group of data items into similar groups is one of the most widely studied problems in computer science. Clustering has applications in a variety of areas, for example, data mining, information retrieval, image processing, and web search ([5,7,16,9]). Given the wide range of applications, many different definitions of clustering exist in the literature ([8,4]). Most of these definitions begin by defining a notion of distance (similarity) between two data items and then try to form clusters so that data items with small distance between them get clustered together.

Often, clustering problems arise in a geometric setting, i.e., the data items are points in a high dimensional Euclidean space. In such settings, it is natural to define the distance between two points as the Euclidean distance between them. Two of the most popular definitions of clustering are the k-*means clustering problem* and the k-*median clustering problem*. Given a set of points P, the k-means clustering problems seeks to find a set K of k centers, such that $\sum_{p\in P} d(p,K)^2$ is minimized, whereas the k-median clustering problems seeks to find a set K of k centers, such that $\sum_{p\in P} d(p,K)$ is minimized. Note that the points in K can be arbitrary points in the Euclidean space. Here $d(p,K)$ refers to the distance between p and the closest center in K. We can think of this as each point in P gets assigned to the closest center. The points that get assigned to the same center form a cluster. These problems are NP-hard for even $k=2$ (when dimension is not fixed). Interestingly, the center in the optimal solution to the 1-mean problem is the same as the center of mass of the points. Howvever, in the case of the 1-median problem, also known as the Fermat-Weber problem, no such closed form is known. We show that despite the lack of such a closed form, we can obtain an approximation to the optimal 1-median in $O(1)$ time (independent of the number of points). There exist variations to these clustering problems, for example, the discrete versions of these problems, where the centers that we seek are constrained to lie on the input set of points.

1.1 Related Work

A lot of research has been devoted to solving these problems exactly (see [11] and the references therein). Even the best known algorithms for the k-median and the k-means problem take at least $\Omega(n^d)$ time. Recently, some work has been devoted to finding $(1+\varepsilon)$-approximation algorithm for these problems, where ε can be an arbitrarily small constant. This has led to algorithms with much improved running times. Further, if we look at the applications of these problems, they often involve mapping subjective features to points in the Euclidean space. Since there is an error inherent in this mapping, finding a $(1+\varepsilon)$-approximate solution does not lead to a deterioration in the solution for the actual application.

The following table summarizes the recent results for the problems, in the context of $(1+\varepsilon)$-approximation algorithms. Some of these algorithms are randomized with the expected runing time holding good for any input.

Problem	Result	Reference
1-median	$O(n/\varepsilon^2)$	Indyk [12]
k-median	$O(n^{O(1/\varepsilon)+1})$ for $d=2$	Arora [1]
	$O(n + \varrho k^{O(1)} \log^{O(1)} n)$ (discrete also) where $\varrho = exp[O((1+\log 1/\varepsilon)/\varepsilon)^{d-1}]$	Har-Peled et. al. [10]
	$O(2^{(k/\varepsilon)^{(1)}} d^{O(1)} n \log^{O(k)} n)$	Badoiu et al. [3]
discrete k-median	$O(\varrho n \log n \log k)$	Kolliopoulos et al. [13]
k-means	$O(n/\varepsilon^d)$ for $k=2$	Inaba et al. [11]
	$O(n\varepsilon^{-2k^2 d} \log^k n)$	Matousek [15]
	$O(g(k,\varepsilon) n \log^k n)$ $g(k,\varepsilon) = exp[(k^3/\varepsilon^8)(\ln(k/\varepsilon)\ln k]$	de la Vega et al. [6]
	$O(n + k^{k+2}\varepsilon^{-(2d+1)k} \log^{k+1} n \log^k \frac{1}{\varepsilon})$	Har-Peled et al. [10]
	$O(2^{(k/\varepsilon)^{(1)}} dn)$	Kumar et al. [14]

1.2 Our Contributions

In this paper, we generalize the algorithm of authors [14] to a wide range of clustering problems. We define a general class of clustering problems and show that if certain conditions are satsified, we can get linear time $(1+\varepsilon)$-approximation algorithms for these problems. We then use our general framework to get the following results. Given a set of n points P in $\Re^d$, we present

1. a randomized algorithm that generates a candidate center set of size $O(2^{1/\varepsilon^{O(1)}})$, such that at least one of the points in this set is a $(1+\varepsilon)$-approximate 1-median of P with constant probability. The running time of the algorithm is $O(2^{1/\varepsilon^{O(1)}}d)$, assuming that the points are stored in a suitable data structure such that a point can be randomly sampled in constant time. This improves on the algorithm of Badoiu et al. [3] which generates a candidate center set of size $O(2^{1/\varepsilon^4}\log n)$ in time $O(d2^{1/\varepsilon^4}\log n)$.
2. a randomized $(1+\varepsilon)$-approximation algorithm for the 1-median problem which runs in time $O(2^{1/\varepsilon^{O(1)}}d)$, assuming that the points are stored in a suitable data structure such that a point can be randomly sampled in constant time.
3. a randomized $(1+\varepsilon)$-approximation algorithm for the k-median problem which runs in $O(2^{(k/\varepsilon)^{O(1)}}nd)$ time.
4. a randomized $(1+\varepsilon)$-approximation algorithm for the discrete k-means clustering which runs in $O(2^{(k/\varepsilon)^{O(1)}}nd)$ time.

All our algorithms yield the desired result with constant probability (which can be made as close to 1 as we wish by a constant number of repetitions). As mentioned earlier, we generalize the result of the authors in [14] to solve a larger class of clustering problems satisfying a set of conditions (c.f. section 2). We then show that the k-median problem and the discrete k-means problem fall in this class of clustering problems. One important condition that the clustering problems must satisfy is that there should be an algorithm to generate a candidate set of points of size independent of n, such that at least one of these points is a close approximation to the optimal center when we desire only one cluster. Armed with such a subroutine, we show how to approximate all the centers in the optimal solution in an iterative manner.

It is easy to see that our algorithms for the k-median and the discrete k-means problems have better running time than the previously known algorithms for these problems, specially when d is very large. In fact, these are the first algorithms for the k-median and the discrete k-means clustering that have running time linear in the size of the input for fixed k and ε.

For the 1-median problem, the candidate center set generation and the actual approximation algorithm have better running time than all previously known algorithms. The algorithms in this paper have the additional advantage of simplicity inherited from generalizing the approach of Kumar et al. [14].

The remaining paper is organized as follows. In Section 2, we describe a general approach for solving clustering problems efficiently. In the subsequent sections we give applications of the general method by showing that this class of problems includes the k-median, the k-means and the discrete k-means problems. In section 4.3, we also describe an efficient approximation algorithm for the 1-median problem.

2 Clustering Problems

In this section, we give a general definition of clustering problems. Our algorithms will work on any of these problems provided certain conditions are satisfied. We will state these conditions later in the section.

We shall define a clustering problem by two parameters – an integer k and a real-valued cost function $f(Q,x)$, where Q is a set of points, and x is a point in an Euclidean space. We shall denote this clustering problem as $\mathcal{C}(f,k)$. The input to $\mathcal{C}(f,k)$ is a set of points in a Euclidean space.

Given an instance P of n points, $\mathcal{C}(f,k)$ seeks to partition them into k sets, which we shall denote as *clusters*. Let these clusters be $C_1, \ldots, C_k$. A solution also finds k points, which we call *centers*, $c_1, \ldots, c_k$. We shall say that c_i is the center of cluster C_i (or the points in C_i are assigned to c_i). The objective of the problem is to minimize the quantity $\sum_{i=1}^{k} f(C_i, c_i)$.

This is a fairly general definition. Let us see some important special cases.

- k-median : $f(Q,x) = \sum_{q \in Q} d(q,x)$.
- k-means : $f(Q,x) = \sum_{q \in Q} d(q,x)^2$.

We can also encompass the discrete versions of these problems, i.e., cases where the centers have to be one of the points in P. In such problems, we can make $f(Q,x)$ unbounded if $x \notin Q$.

As stated earlier, we shall assume that we are given a constant $\varepsilon > 0$, and we are interested in finding $(1+\varepsilon)$-approximation algorithms for these clustering problems.

We now state the conditions the clustering problems should satisfy. We begin with some definitions first. Let us fix a clustering problem $\mathcal{C}(f,k)$. Although we should parameterize all our definitions by f, we avoid this because the clustering problem will be clear from the context.

Definition 1. *Given a point set P, let $\mathrm{OPT}_k(P)$ be the cost of the optimal solution to the clustering problem $\mathcal{C}(f,k)$ on input P.*

Definition 2. *Given a constant α, we say that a point set P is (k,α)-irreducible if $\mathrm{OPT}_{k-1}(P) \geq (1+150\alpha)\mathrm{OPT}_k(P)$. Otherwise we say that the point set is (k,α)-reducible.*

Reducibility captures the fact that if P is (k,α)-reducible for a small constant α, then the optimal solution for $\mathcal{C}(f,k-1)$ on P is close to that for $\mathcal{C}(f,k)$ on P. So if we are solving the latter problem, it is enough to solve the former one. In fact, when solving the problem $\mathcal{C}(f,k)$ on the point set P, we can assume that P is (k,α)-irreducible, where $\alpha = \varepsilon/1200k$. Indeed, suppose this is not the case. Let i be the highest integer such that P is (i,α)-irreducible. Then, $\text{OPT}_k(P) \leq (1 + 150k\alpha)^{k-i}\text{OPT}_i(P) \leq (1 + \varepsilon/4)\text{OPT}_i(P)$. Therefore, if we can get a $(1 + \varepsilon/4)$-approximation algorithm for $\mathcal{C}(f,i)$ on input P, then we have a $(1 + \varepsilon)$-approximation algorithm for $\mathcal{C}(f,k)$ on P. Thus it is enough to solve instances which are irreducible.

The first property that we want $\mathcal{C}(f,k)$ to satisfy is a fairly obvious one – it is always better to assign a point in P to the nearest center. We state this more formally as follows :

Closeness Property : Let Q and Q' be two disjoint set of points, and let $q \in Q$. Suppose x and x' are two points such that $d(q,x) > d(q,x')$. Then the cost function f satisfies the following property

$$f(Q,x) + f(Q',x') \geq f(Q - \{q\}, x) + f(Q' \cup \{q\}, x').$$

This is essentially saying that in order to find a solution, it is enough to find the set of k centers. Once we have found the centers, the actual partitioning of P is just the Voronoi partitioning with respect to these centers. It is easy to see that the k-means problem and the k-median problem (both the continuous and the discrete versions) satisfy this property.

Definition 3. *Given a set of points P and a set of k points C, let $\text{OPT}_k(P,C)$ be the cost of the optimal solution to $\mathcal{C}(f,k)$ on P when the set of centers is C.*

We desire two more properties from $\mathcal{C}(f,k)$. The first property says that if we are solving $\mathcal{C}(f,1)$, then there should be a simple random sampling algorithm. The second property says that suppose we have approximated the first i centers of the optimal solution closely. Then we should be able to easily extract a large number of points in P which get assigned to these centers. We describe these properties in more detail below :

- ***Random Sampling Procedure*** : There exists a procedure $\mathcal{A}$ that takes a set of points $Q \in \Re^d$ and a parameter α as input. $\mathcal{A}$ first randomly samples a set R of size $\left(\frac{1}{\alpha}\right)^{O(1)}$ points from Q. Starting from R, $\mathcal{A}$ produces a set of points, which we call $\text{core}(R)$, of size at most $2^{\left(\frac{1}{\alpha}\right)^{O(1)}}$. $\mathcal{A}$ satisfies the condition that with constant probability there is at least one point $c \in \text{core}(R)$ such that $\text{OPT}_1(Q, \{c\}) \leq (1 + \alpha)\text{OPT}_1(Q)$. Further the time taken by $\mathcal{A}$ to produce $\text{core}(R)$ from R is at most $O(2^{\left(\frac{1}{\alpha}\right)^{O(1)}} \cdot dn)$.

- ***Tightness Property*** : Let P be a set of points which is (k,α)-irreducible for some constant α. Consider an optimal solution to $\mathcal{C}(f,k)$ on P – let $C = \{c_1, \ldots, c_k\}$ be the centers in this solution. Suppose we have a set of i points $C'_i = \{c'_1, \ldots, c'_i\}$, such that $\text{OPT}_k(P, C') \leq (1 + \alpha/k)^i \text{OPT}_k(P)$, where $C' = \{c'_1, \ldots, c'_i, c_{i+1}, \ldots, c_k\}$. Let $P'_1, \ldots, P'_k$ be the partitioning of P if we choose C' as the set of centers (in other words this is the Voronoi partitioning of P with respect to C'). We assume w.l.o.g. that P'_{i+1} is the largest cluster amongst $P'_{i+1}, \ldots, P'_k$. Then there exists a set of points S such that the following conditions hold :

(a) S is contained in $P'_1 \cup \ldots \cup P'_i$.
(b) Let $x \in S, x' \in P - S$. Then, $d(x, \{c'_1, \ldots, c'_i\}) \leq d(x', \{c'_1, \ldots, c'_i\})$.
(c) $P - S$ contains at most $\frac{|P'_{i+1}|}{\alpha^{(1)}}$ points of $P'_1 \cup \ldots \cup P'_i$.

3 A General Algorithm for Clustering

We can show that if a clustering problem $\mathcal{C}(f, k)$ satisfies the conditions stated in the previous section, then there is an algorithm which with constant probability produces a solution within $(1+\varepsilon)$ factor of the optimal cost. Further the running time of this algorithm is $O(2^{(-)^{(1)}} \cdot dn)$. The techniques are very similar those in [14] and are omitted. We now give applications to various clustering problems. We show that these clustering problems satisfy the tightness property and admit a random sampling procedure as described in the previous section.

4 The k-Median Problem

As described earlier, the clustering problem $\mathcal{C}(f, k)$ is said to be the k-median problem if $f(Q, x) = \sum_{q \in Q} d(q, x)$. We now exhibit the two properties for this problem.

4.1 Random Sampling Procedure

Badoiu et al. [3] showed that a small random sample can be a used to get a close approximation to the optimal 1-median solution. Given a set of points P, let $\texttt{AvgMed}(P)$ denote $\frac{\texttt{OPT}_1(P)}{|P|}$, i.e., the average cost paid by a point towards the optimal 1-median solution.

Lemma 1. *[3] Let P be a set of points in $\Re^d$, and ε be a constant between 0 and 1. Let X be a random sample of $O(1/\varepsilon^3 \log 1/\varepsilon)$ points from P. Then with constant probability, the following two events happen: (i) The flat $\texttt{span}(X)$ contains a point x such that $\texttt{OPT}_1(P, \{x\}) \leq (1+\varepsilon) \texttt{OPT}_1(P)$. and (ii) X contains a point y at distance at most $2\texttt{AvgMed}(P)$ from x.*

We now show that if we can upper and lower bound $\texttt{AvgMed}(P)$ upto constant factors, then we can construct a small set of points such that at least one of these is a good approximation to the optimal center for the 1-median problem on P.

Lemma 2. *Let P be a set of points in $\Re^d$ and X be a random sample of size $O(1/\varepsilon^3 \log 1/\varepsilon)$ from P. Suppose we happen to know numbers a and b such that $a \leq \texttt{AvgMed}(P) \leq b$. Then, we can construct a set Y of $O(2^{(1/\varepsilon)^{(1)}} \log(b/\varepsilon a))$ points such that with constant probability there is at least one point $z \in X \cup Y$ satisfying $\texttt{OPT}_1(P, \{z\}) \leq (1+2\varepsilon) \texttt{OPT}_1(P)$. Further, the time taken to construct Y from X is $O(2^{(1/\varepsilon)^{(1)}} d)$.*

Proof. Our construction is similar to that of Badoiu et al. [3]. We can assume that the result stated in Lemma 1 holds (because this happens with constant probability). Let x and y be as in Lemma 1.

We will carefully construct candidate points around the points of X in $span(X)$ in an effort to get within close distance of x.

For each point $p \in X$, and each integer i in the range $[\lfloor \log \frac{\varepsilon}{4} a \rfloor, \lceil \log b \rceil]$ we do the following – let $t = 2^i$. Consider the grid $G_p(t)$ of side length $\varepsilon t/(4|X|) = O(t\varepsilon^4 \log(1/\varepsilon))$ in $span(X)$ centered at p. We add all the vertices of this grid lying within distance at most $2t$ from p to our candidate set Y. This completes the construction of Y. It is easy to see that the time taken to construct Y from X is $O(2^{(1/\varepsilon)^{O(1)}} d)$.

We now show the existence of the desired point $z \in X \cup Y$. Consider the following cases:

1. $d(y,x) \leq \varepsilon \text{AvgMed}(P)$: Using triangle inequality, we see that
$$f(P,y) \leq f(P,x) + |P|d(y,x) \leq (1 + 2\varepsilon)\text{OPT}_1(P).$$
Therefore y itself is the required point.

2. $d(y,x) > \varepsilon \text{AvgMed}(P)$: Consider the value of i such that $2^{i-1} \leq AvgMed(P,1) \leq 2^i$ – while constructing Y, we must have considered this value of i for all points in X. Let $t = 2^i$. Clearly, $t/2 \leq \text{AvgMed}(P) \leq t$.
Observe that $d(y,x) \leq 2\text{AvgMed}(P) \leq 2t$. Therefore, by the manner in which we have constructed $G_y(t)$, there must be a point $p \in G_y(t)$ fow which $d(p,x) \leq \varepsilon t/2 \leq \varepsilon \text{AvgMed}(P)$. This implies that
$$f(P,p) \leq f(P,x) + |P|d(x,p) \leq (1 + 2\varepsilon)\text{OPT}_1(P).$$
Therefore p is the required point.

This completes the proof of the lemma.

We now show the existence of the random sampling procedure.

Theorem 1. *Let P be a set of n points in $\Re^d$, and let ε be a constant, $0 < \varepsilon < 1/12$. There exists an algorithm which randomly samples a set R of $O((\frac{1}{\varepsilon})^{O(1)})$ points from P. Using this sample only, it constructs a set of points $\text{core}(R)$ such that with constant probability there is a point $x \in \text{core}(R)$ satisfying $f(P,x) \leq (1 + O(\varepsilon))\text{OPT}_1(P)$. Further, the time taken to construct $\text{core}(R)$ from R is $O(2^{(1/\varepsilon)^{O(1)}} d)$.*

Proof. Consider the optimal 1-median solution for P – let c be the center in this solution. Let T denote $\text{AvgMed}(P)$. Consider the ball B_1 of radius T/ε^2 around c. Let P' be the points of P contained in B_1. It is easy to see that $|P'| \geq (1 - \varepsilon^2)n$.

Sample a point p at random from P. With constant probability, it lies in P'. Randomly sample a set Q of $1/\varepsilon$ points from P. Again, with constant probability, these points lie in P'. So we assume that these two events happen. Let $v = \sum_{q \in Q} d(q,p)$. We want to show that v is actually close to $\text{AvgMed}(P)$.

Let B_2 denote the ball of radius εT centered at p. One of the following two cases must happen :

- There are at least $2\varepsilon|P'|$ points of P' outside B_2 : In this case, with constant probability, the sample Q contains a point outside B_2. Therefore, $v \geq \varepsilon T$. Also notice that any two points in B_1 are at distance at most $2T/\varepsilon^2$ from each other. So, $v \leq 2T|Q|/\varepsilon^2$. We choose $a = \frac{v\varepsilon^2}{2|Q|}$ and $b = v/\varepsilon$. Notice that b/a is $O(1/\varepsilon^{O(1)})$. We can now use the Lemma 2 to construct the desired core set.
- There are at most $2\varepsilon|P'|$ points of P' outside B_2 : Suppose $d(p,c) \leq 4\varepsilon T$. In this case $f(P,p) \leq (1+O(\varepsilon))\mathrm{OPT}_1(P)$ and we are done. So assume this is not the case. Note that the number of points outside B_2 is at most $|P - P'| + 2\varepsilon|P'| \leq \varepsilon^2 n + 2\varepsilon(1-\varepsilon^2)n \leq 3\varepsilon n$. Now suppose we assign all points of P from c to p. Let us see the change in cost. The distance the points in B_2 have to travel decreases by at least $d(c,p) - 2\varepsilon T$. The increase in the distance for points outside B_2 is at most $d(c,p)$. So the overall decrease in cost is at least

$$|B_2|(d(c,p) - 2\varepsilon T) - (n - |B_2|)d(c,p) > 0$$

if we use $|B_2| \geq n(1-3\varepsilon)$ and $d(c,p) \geq 4\varepsilon T$. This yields a contradiction because c is the optimal center. Thus we are done in this case as well.

4.2 Tightness Property

We now show the existence of tightness property. We will use the same notation as used while defining the tightness property in Section 2. We need to show the existence of the desired set S.

Consider the closest pair of centers between the sets $C' \setminus C'_i$ and C'_i – let these centers be c_l and c'_r respectively. Let $t = d(c_l, c'_r)$. Let S be the set of points $\mathcal{B}(c'_1, t/4) \cup \cdots \cup \mathcal{B}(c'_i, t/4)$, i.e., the points which are distant at most $t/4$ from $C'_i = \{c'_1, \ldots, c'_i\}$.

Clearly, S is contained in $P'_1 \cup \cdots \cup P'_i$. This shows (a). Also, for any $x \in S, x' \in P - S$, $d(x, \{c'_1, \ldots, c'_i\}) \leq d(x', \{c'_1, \ldots, c'_i\})$. This proves (b).

Suppose $P - S$ contains more than $|P_l|/\alpha$ points of $P'_1 \cup \cdots \cup P'_i$. In that case, these points are assigned to centers at distance at least $t/4$. It follows that $\mathrm{OPT}_k(P, C')$ is at least $\frac{t|P_l|}{4\alpha}$. This implies that $t|P_l| \leq 4\alpha\mathrm{OPT}_k(P, C')$. But then if we assign all the points in P_l to c'_r, the cost increases by at most

$$|P_l|t \leq 4\alpha\mathrm{OPT}_k(P, C') \leq 4\alpha(1+\alpha/k)^i \mathrm{OPT}_k(P) \leq 4\alpha(1+\alpha/k)^k \mathrm{OPT}_k(P) \leq 12\alpha\mathrm{OPT}_k(P).$$

But this contradicts the fact that P is (k, α)-irreducible.

4.3 Applications to the 1-Median Problem

In this section, we present an algorithm for the 1-median problem. Given a set of n points in $\Re^d$, the algorithm with constant probability produces a solution of cost at most $(1+\varepsilon)$ of the optimal cost for any constant $\varepsilon > 0$. The running time of the algorithm is $O(2^{1/\varepsilon^{O(1)}} d)$, assuming that it is possible to randomly sample a point in constant time.

Our algorithm is based on the following idea presented by Indyk [12].

Lemma 3. *[12] Let X be a set of n points in $\Re^d$. For a point $a \in \Re^d$ and a subset $Q \subseteq X$, define $S_Q(a) = \sum_{x \in Q} d(a, x)$ and $S(a) = S_X(a)$. Let ε be a constant, $0 \le \varepsilon \le 1$. Suppose a and b are two points such that $S(b) > (1+\varepsilon)S(a)$. Then,*

$$Pr\left(\sum_{x \in Q} d(a,x) \ge \sum_{x \in Q} d(b,x)\right) < e^{-\varepsilon^2 |Q|/64}.$$

We now show the existence of a fast algorithm for approximating the optimal 1-median solution.

Theorem 2. *Let P be a set of n points in $\Re^d$, and let ε be a constant, $0 < \varepsilon < 1$. There exists an algorithm which randomly samples a set R of $O((\frac{1}{\varepsilon})^{O(1)})$ points from P. Using this sample only, it finds a point p such that $f(P, x) \le (1 + O(\varepsilon))\text{OPT}_1(P)$ with constant probability (independent of ε). The time taken by the algorithm to find such a point p from R is $O(2^{(1/\varepsilon)^{(1)}} d)$.*

Proof. We first randomly sample a set R_1 of $O((\frac{1}{\varepsilon})^{O(1)})$ points from P and using Theorem 1, construct a set $\text{core}(R_1)$ of $O(2^{(1/\varepsilon)^{(1)}})$ points such that with constant probability, there is a point $x \in \text{core}(R_1)$ satisfying $f(P,x) \le (1 + O(\varepsilon))\text{OPT}_1(P)$.

Now we randomly sample a set R_2 of $O((1/\varepsilon)^{O(1)})$ points and find the point $p \in \text{core}(R_1)$ for which $S_{R_2}(p) = f(R_2, p)$ is minimum. By Lemma 3, p is with constant probability a $(1 + O(\varepsilon))$-approximate median of P.

Clearly, the time taken by the algorithm is $O(2^{(1/\varepsilon)^{(1)}} d)$.

Also note that we can boost the success probability to an arbitrarily small constant by selecting a large enough (yet constant) sample R.

5 k-Means Clustering

In this problem, $f(Q, x) = \sum_{q \in Q} d(q, x)^2$. The two properties for the k-means problem were shown by the authors in [14]. For a set of points T, let $c(T)$ denote their centroid. The random sampling property follows from the following fact showed by Inaba et al. [11].

Lemma 4. *[11] Let T be a set of m points obtained by independently sampling m points uniformly at random from a point set P. Then, for any $\delta > 0$,*

$$f(S, c(T)) < \left(1 + \frac{1}{\delta m}\right) \text{OPT}_1(P)$$

holds with probability at least $1 - \delta$.

The proof of tightness property is similar to that for the k-median problem.

6 Discrete k-Means Clustering

This is same as k-means problem with the extra constraint that the centers must be from the input point set only. We now show the two properties here.

6.1 Random Sampling Procedure

We first show that given a good approximation to the center of the optimal (continuous) 1-means problem, we can get a good approximation to the center of the optimal discrete 1-means problem. Let us have some notation first. Let P be a set of n points in $\Re^d$. Let c be the center of the optimal solution to the (continuous) 1-means problem on P.

Lemma 5. *Let α be a constant, $0 < \alpha < 1$, and c' be a point in $\Re^d$ such that $\sum_{p \in P} d(p, c')^2 \leq (1 + \alpha) \sum_{p \in P} d(p, c)^2$. Let x' be the point of P closest to c'. Then $\mathsf{OPT}_1(P, \{x'\}) \leq (1 + O(\sqrt{\alpha}))\mathsf{OPT}_1(P)$.*

Proof. Let x be the center of the optimal discrete 1-means solution, i.e., $\mathsf{OPT}_1(P, \{x\}) = \mathsf{OPT}_1(P)$. Let T be the average cost paid by the points of P in the optimal 1-means solution, i.e., $T = \frac{\sum_{p \in P} d(p,c)^2}{|P|}$.

Then $\mathsf{OPT}_1(P) = |P|(T + d(c, x)^2)$ and $\mathsf{OPT}_1(P, \{x'\}) = |P|(T + d(c, x')^2)$. From the definition of c', we know that $d(c, c')^2 \leq \alpha T$. Notice that

$$d(c, x') \leq d(c, c') + d(c', x') \leq d(c, c') + d(c', x) \leq 2d(c, c') + d(c, x).$$

We know that $f(P, x) = |P|(T + d(c, x)^2)$ and $f(P, x') = |P|(T + d(c, x')^2)$. So

$$f(P, x') - f(P, x) = |P|(d(c, x')^2 - d(c, x)^2) \leq |P| \left((2d(c, c') + d(c, x))^2 - d(c, x)^2\right)$$
$$\leq 4|P| \left(d(c, c')^2 + d(c, c')d(c, x)\right) \leq 4|P| \left(\alpha T + \sqrt{\alpha T} d(c, x)\right)$$
$$\leq 4|P| \left(\alpha T + \sqrt{\alpha}(T + d(c, x)^2)\right) \leq O(\sqrt{\alpha})\mathsf{OPT}_1(P).$$

We now show the existence of the random sampling procedure.

Theorem 3. *Let α be a constant, $0 < \alpha < 1$. There exists an algorithm which randomly samples a set R of $O\left(\frac{1}{\alpha}\right)$ points from P. Using this sample, it finds a singleton set $\mathsf{core}(R)$ such that with constant probability the point $x \in \mathsf{core}(R)$ satisfies $f(P, x) \leq (1 + O(\sqrt{\alpha}))\mathsf{OPT}_1(P)$. Further, the time taken to construct $\mathsf{core}(R)$ from R is $O((\frac{1}{\alpha} + n)d)$.*

Proof. Using Lemma 4, we can get a point c' such that $\sum_{p \in P} d(p, c')^2 \leq (1 + \alpha) \sum_{p \in P} d(p, c)^2$. As mentioned in the lemma, we do this by by taking the centroid of a random sample of $O(1/\alpha)$ points of P. This takes time $O(\frac{1}{\alpha} \cdot d)$.

The rest follows from the previous lemma.

6.2 Tightness Property

We now show the existence of tightness property. We will use the same notation as used while defining the tightness property in Section 2. We need to show the existence of the desired set S.

Consider the closest pair of centers between the sets $\{c'_1, \ldots, c'_i\}$ and $\{c_{i+1}, \ldots, c_k\}$ – let these centers be c'_r and c_l respectively. Let $t = d(c_l, c'_r)$. Let S be the set of points $\mathcal{B}(c'_1, t/4) \cup \cdots \cup \mathcal{B}(c'_i, t/4)$, i.e., the points which are distant at most $t/4$ from $C'_i = \{c'_1, \ldots, c'_i\}$.

Clearly, S is contained in $P'_1 \cup \cdots \cup P'_i$. This shows (a). Also, for any $x \in S, x' \in P - S$, $d(x, \{c'_1, \ldots, c'_i\}) \leq d(x', \{c'_1, \ldots, c'_i\})$. This proves (b).

Suppose $P - S$ contains more than $|P_l|/\alpha^2$ points of $P'_1 \cup \cdots \cup P'_i$. In that case, these points are assigned to centers at distance at least $t/4$. It follows that $\mathrm{OPT}_k(P, C')$ is at least $\frac{t^2|P_l|}{16\alpha^2}$. This implies that $t^2|P_l| \leq 16\alpha^2 \mathrm{OPT}_k(P, C')$.

Let m_l and m'_r be the centers of the optimal (continuous) 1-means solution of P_l and P'_r respectively. Let T_l and T'_r be the average cost paid by P_l and P'_r in this optimal solution respectively, i.e., $T_l = \frac{\sum_{\in} d(p,m)^2}{|P|}$ and $T'_r = \frac{\sum_{\in} d(p,m')^2}{|P'|}$. Observe that $f(P_l, c_l) = |P_l|(T_l + d(c_l, m_l)^2)$ and $f(P_l, c'_r) = |P_l|(T_l + d(c'_r, m_l)^2)$. Therefore, if we assign the points in P_l from c_l to c'_r, the increase in cost is

$$|P_l|\left(d(c'_r, m_l)^2 - d(c_l, m_l)^2\right) \leq |P_l|\left((d(c'_r, c_l) + d(c_l, m_l))^2 - d(c_l, m_l)^2\right)$$
$$\leq |P_l|\left(t^2 + 2t d(c_l, m_l)\right)$$

We know that the first term above, i.e., $|P_l|t^2$ is at most $16\alpha^2 \mathrm{OPT}_k(P, C')$. We now need to bound the second term only. We consider two cases

- $t \leq \alpha d(c_l, c_m)$: In this case, $|P_l| \cdot 2td(c_l, m_l) \leq 2\alpha d(c_l, m_l)^2|P_l| \leq 2\alpha f(P_l, c_l) \leq 2\alpha \mathrm{OPT}_k(P, C')$.
- $t > \alpha d(c_l, c_m)$: In this case, $|P_l| \cdot 2td(c_l, m_l) \leq \frac{2t^2|P|}{\alpha} \leq 32\alpha \mathrm{OPT}_k(P, C')$.

Thus, in either case, the cost increases by at most

$$48\alpha \mathrm{OPT}_k(P, C') \leq 48\alpha(1+\alpha/k)^i \mathrm{OPT}_k(P) \leq 48\alpha(1+\alpha/k)^k \mathrm{OPT}_k(P) \leq 144\alpha \mathrm{OPT}_k(P).$$

But this contradicts the fact that P is (k, α)-irreducible.

7 Concluding Remarks

The framework in this paper can be extended to handle the situation when each of the points has an associated (integral) weight. The solution to the above clustering problems for the weighted version is the same as the solution to the unweighted version where a point p with weight w is replaced by w points of unit weight. It can be verified that for handling the weighted case: the closeness property remains unchanged; in condition (c) for the tightness property, the size of the set gets replaced by the weight of the set; The random sampling procedure requires time at most linear in n (number of remaining distinct points) in order to perform the required weighted sampling.

The running time thus obtained for the algorithm in the weighted case is $O(2^{(k/\varepsilon)^{(1)}} n \cdot d \log^k W)$ where W is the sum of all the weights.

References

1. Arora, S.: Polynomial time approximation schemes for Euclidean TSP and other geometric problems. Proceedings of the 37th Annual Symposium on Foundations of Computer Science (1996) 2–11
2. Arora, S., Raghavan, P., Rao, S.: Approximation schemes for Euclidean k-medians and related problems. Proceedings of the thirtieth annual ACM symposium on Theory of computing (1998) 106–113
3. Badoiu, M., Har-Peled, S., Indyk, P.: Approximate clustering via core-sets. Proceedings of the thiry-fourth annual ACM symposium on Theory of computing (2002) 250–257
4. Bern, M., Eppstein, D.: Approximation algorithms for geometric problems. Approximating algorithms for NP-Hard problems. PWS Publishing Company (1997) 296–345
5. Broder, A., Glassman, S., Manasse, M., Zweig, G.: Syntactic clustering of the Web. Proc. of 6th International World Wide Web Conference (1997) 391–404
6. de la Vega, W. F., Karpinski, M., Kenyon, C., Rabani, Y.: Approximation schemes for clustering problems. Proceedings of the thirty-fifth annual ACM symposium on Theory of computing (2003) 50–58
7. Deerwester, S., Dumais, S. T., Landauer, T. K., Furnas, G. W., Harshman, R. A.: Indexing by latent semantic analysis. Journal of the American Society for Information Science **41(6)** (1990) 391–407
8. Duda, R. O., Hart, P. E., Stork, D. G.: Pattern Classification. Wiley-Interscience, New York, 2nd edition (2001)
9. Faloutsos, C., Barber, R., Flickner, M., Hafner, J., Niblack, W., Petkovic, D., Equitz, W.: Efficient and effective querying by image content. Journal of Intelligent Information Systems **3(3)** (1994) 231–262
10. Har-Peled, S., Mazumdar, S.: On coresets for k-means and k-median clustering. Proceedings of the thirty-sixth annual ACM symposium on Theory of computing (2004) 291–300
11. Inaba, M., Katoh, N., Imai, H.: Applications of weighted Voronoi diagrams and randomization to variance-based k-clustering. Proceedings of the tenth annual symposium on Computational Geometry (1994) 332–339
12. Indyk, P.: High Dimensional Computational Geometry. Ph.D. Thesis. Department of Computer Science, Stanford University (2004)
13. Kolliopoulos, S., Rao, S.: A nearly linear time approximation scheme for the Euclidean k-medians problem. Proceedings of the 7th European Symposium on Algorithms (1999) 362-371
14. Kumar, A., Sabharwal, Y., Sen, S.: A simple linear time $(1+\varepsilon)$-approximation algorithm for k-means clustering in any dimensions. Proceedings of the 45th Annual Symposium on Foundations of Computer Science (2004) 454–462
15. Matousek, J.: On approximate geometric k-clustering Discrete and Computational Geometry **24** (2000) 61–84
16. Swain, M. J., Ballard, D. H.: Color indexing. International Journal of Computer Vision (1991) 11–32

Dynamic Diffusion Load Balancing

Petra Berenbrink[1], Tom Friedetzky[2], and Russell Martin[3,*]

[1] School of Computing Science, Simon Fraser University,
Burnaby, B.C., Canada
[2] Department of Computer Science, University of Durham,
Durham, UK
[3] Department of Computer Science, University of Warwick,
Coventry, UK

Abstract. We consider the problem of dynamic load balancing in arbitrary (connected) networks on n nodes. Our load generation model is such that during each round, n tasks are generated on arbitrary nodes, and then (possibly after some balancing) one task is deleted from every non-empty node. Notice that this model *fully saturates* the resources of the network in the sense that we generate just as many new tasks per round as the network is able to delete. We show that even in this situation the system is *stable*, in that the total load remains bounded (as a function of n alone) over time. Our proof only requires that the underlying "communication" graph be connected. (It of course also works if we generate less than n new tasks per round, but the major contribution of this paper is the fully saturated case.) We further show that the upper bound we obtain is asymptotically tight (up to a moderate multiplicative constant) by demonstrating a corresponding lower bound on the system load for the particular example of a linear array (or path). We also show some simple negative results (i.e., instability) for work-stealing based diffusion-type algorithms in this setting.

1 Introduction

The use of parallel and distributed computing is established in many areas of science, technology, and business. One of the most crucial parameters of parallel machines is the efficient utilization of resources. Of greatest importance here is an even distribution of the workload among the processors. In particular applications exposing some kind of "irregularity" require the use of a load balancing mechanism.

A well known and much studied load balancing approach is the so-called *diffusion load balancing*, first introduced by Cybenko and Boillat ([11], [10]). The algorithm works in synchronized rounds. The basic idea is that in every

---

* A portion of this work was performed during a visit to the School of Computing Science at Simon Fraser University. Supported in part by the EPSRC grant "Discontinuous Behaviour in the Complexity of Randomized Algorithms".

round, every processor p balances load with all its neighbors (independently, i.e., pair-wise). Let ℓ_p be the load of p, ℓ_q the load of some of p's neighbor q, and let Δ denote the maximum degree of the underlying graph. In the discrete setting, p transfers $\max\{0, \lfloor(\ell_p - \ell_q)/(\Delta + 1)\rfloor\}$ tasks to q. Some of many advantages of diffusion-type algorithms are the *locality* (no global knowledge regarding the overall load situation, or, in fact, anything except the strict neighborhood of any vertex is needed), its *simplicity*, and its *neighborhood preservation* (tasks tend to stay close to the processors where they are generated, which may help to maintain small communication overhead).

The diffusion load balancing algorithm has been thoroughly analyzed for *static* scenarios, where each processor has some initial number of tasks, and the objective is to distribute *this* load evenly among the processors as quickly as possible. Much work has been done under the assumption that every edge is only allowed to forward one task per round [16, 17, 19] or when a constant number of tasks can be passed by each processor [15]. We refer to these scenarios as *token distribution problems*. In addition [12, 13, 14] have studied the diffusion algorithm where tasks can be split arbitrarily.

In contrast to the static case of load balancing and token distribution, in the dynamic setting during each time step new tasks are generated (in some manner) on the set of processors, load is balanced amongst neighbors, then tasks are deleted from non-empty processors.

Much of the past work has studied the dynamic token distribution problem. Muthukrishnan and Rajaraman [18] studied a dynamic version where processors can forward a single task in each round. They assume an adversarial load generation model. The adversary is allowed to generate and to delete tokens from the network in every round. The simple and elegant algorithm they consider is due to [2]: A node sends a task to its neighbor if the load difference between them is at least $2\Delta + 1$. They show that the system is stable if the load change in every subset S of the nodes minus $a|S|$ is at most $(1 - \epsilon)e(S)$ for $\epsilon > 0$. Here $e(S)$ is the number of outgoing edges of S and a is the change in the average load. Their system is said to be stable if the deviation of the load of any processor from the average load can be bounded. Muthukrishnan and Rajaraman left open the question whether the system is also stable for $\epsilon = 0$.

Anshelevich, Kempe, and Kleinberg [4] gave a positive result for token distribution when $\epsilon = 0$. They showed that under the above load generation model no processor has more than average load $+(2\Delta + 1) \cdot n$. Anshelevich, et al. also showed how their result can be generalized for edges that can forward c tokens per time step. A node sends $\min\{c, \rho\}$ tasks to its neighbor if the load difference is at least $2\Delta c + \rho$. In this setting no processor has more than average load $\pm(2\Delta + 1)c \cdot n$ as long as the load change in every subset S of the nodes minus $a|S|$ is at most $c \cdot e(S)$. Additionally, they showed that a generalization of the algorithm is stable for two distinct types of jobs, and they extended their results to related flow problems.

In [6, 7] Awerbuch and Leighton use a variant of the token distribution model under the assumption that tokens can be split into arbitrarily sized parts. They use a "balancing" algorithm to approximate the multi-commodity flow problem with capacitated edges. Their method is an iterative approach where flow is queued at the vertices of the graph. In each step, the commodity which has the largest excess is shipped from one vertex to another, and then new flow is injected into the system. In this balancing process, edge capacities must always be respected. These edge capacities are analogous to the restrictions on the number of tasks that can be passed over any single edge in the token distribution problems. Furthermore, their model does not actually allow full use of those edge capacities, which is similar to the case in [18] where $\epsilon > 0$ was required to ensure stability. The work in [1] and [5] expands the results of Awerbuch and Leighton for packet routing, but again in these cases only a constant number of tasks can be moved across any edge in a single time step.

Clearly the condition that processors can forward only a single task (or a constant number) per edge in each round significantly restricts the number and distribution of tasks that can be generated on (or deleted from) processors in each round and still obtain a stability result. Thus, in the results of [18] and [4] some dependence on the quantity $e(S)$ (or some measure of the "edge expansion") is to be expected.

Anagnostopoulos et al. [3] consider the setting where there are no restrictions on the number of tasks balanced between processors in a time step, and they allow a broad range of injection models. Their protocol is similar to that studied in [15] for a static setting, but is not the typical diffusion load balancing procedure. In their setting, in each step nodes are matched randomly with adjacent neighbors and matched nodes equalize their load. Hence, every processor is only involved in a single load balancing action. They show that the system is stable as long as at most $wn\lambda$ tasks (in expectation) are generated in a time interval of length w, where $\lambda < 1$. Their proof method unfortunately cannot be generalized to the case of *full saturation* when $\lambda = 1$, which is the main focus of this paper.

1.1 Our Results

In this paper we present the first analysis of the simple diffusion scheme for the dynamic load balancing problem that allows full saturation of the resources. We assume that n new tasks are generated per round and, after load balancing, every non-empty processor deletes one task each round. (With small modifications our proofs will carry through to the case when we generate *at most n* tasks per round.) In contrast to [4] and [18], the newly generated tasks may be arbitrarily distributed among the nodes of the network, regardless of any "edge expansion" type of condition as in those models. For example, the tasks may always be generated on the same processor, or all tasks may be generated on one processor but the processor can change from round to round, or alternatively, the tasks may be allocated at random each round. Note that, obviously, without load balancing the total number of tasks in the system may grow unboundedly with time (in the worst case, we generate n new tasks per step but delete only one).

We show that the system of processors is *stable* under the diffusion load balancing scheme and the generation model described above. By stable, we mean that the total load in the system does not grow with time. In particular, we show that the total system load can be upper-bounded by $O(\Delta n^3)$, where Δ denotes the maximum degree of the network. Furthermore, we present a simple, asymptotically matching lower bound when the network is a path.

Our technique also captures a different scenario, similar to that in [4, 18], where stability is defined in terms of deviation of any processor's load from the average. In this scenario we have two separate phases, the first where tasks are generated on and/or deleted from nodes, and the second where tasks are then balanced amongst nodes. Let $\bar{L}^t(S)$ denote the total load of the nodes in the set S after the task generation/deletion phase, and $L^t(S)$ denote the total load of S after the balancing step at time t. Assume that the generation/deletion phase satisfies the following condition:

$$(L(S)^t - \bar{L}(S)^{t-1}) \leq (\text{avg}(t) - \text{avg}(t-1)) \cdot |S| + \rho,$$

where $\text{avg}(t)$ denotes the average system load in step t. Then the total load of S can be bounded by $|S| \cdot \text{avg}(t) + 5\Delta n\rho$.

For both proofs of our results we use a potential function. Although the potential function we use looks similar to the one used in [4], the proof technique is very different. The proof method in [4] very much relies upon the restriction of their generation/deletion model, where the number of tasks inserted into/deleted from a set S is bounded by a function of $e(S)$, the number of edges that join the set S to its complement $\bar{S}$. This, together with the bounded capacities on the edges of the graph, allows for a direct analysis of how the loads of sets might change in a single step of their process. The arbitrary distribution of tasks in our generation model and the unrestricted capacity of the edges in our network (i.e. unknown bounds on load transferred into a set S in a single step) does not allow us to directly obtain similar results, so we need a different proof to show stability under our model.

Another approach to show our results would be to demonstrate an upper bound on the number of tasks that can be moved over a single edge during any time step of our algorithm. If this is possible, the results in [4] could then be used to prove stability under either model that we have described. However, the authors of this paper feel that showing this result is not easier than the proof method we used, especially in the second model where the number of tasks inserted into the system at any time can be unbounded.

In the final part of our paper we discuss a different method of load balancing, one which is commonly referred to as *work stealing*. In this framework, processors that are empty after task generation will balance with processors that are not empty, but no other balancing actions are permitted.

We show that for this work-stealing protocol there are graphs for which the system cannot be stable for a significant class of generation parameters. These results show that restricting balancing actions to empty processors is not sufficient in general.

In contrast, Berenbrink, Friedetzky, and Goldberg [8] showed stability of a work stealing algorithm under a load generation model that is similar to many of those already mentioned. They consider a flexible distribution of n generators among the nodes of the network, where each generator is allowed to generate a task with probability *strictly smaller* than one. In this setting a very simple, parameterized work-stealing algorithm achieves stability (in our sense) for a wide range of parameters. The important point to note is that their model applies only when the set of processors (and their communication linkages) forms a complete graph, and their results only hold for the case where strictly less than n tasks (in expectation) are generated during any time step.

Our model is defined in the next section, and the formal definition of the diffusion approach to load balancing is given following that.

1.2 Our Model

Our parallel system is modeled by a connected graph $G = (V, E)$. The nodes V of the graph model our processors $\mathcal{P} = \{P_1, \ldots, P_n\}$, and the edges E model the underlying communication structure. If two nodes are connected with each other, this means that the processors modeled by the nodes can communicate directly. For us, this means that they are allowed to exchange tasks. Nodes not connected by an edge have to communicate via message passing. Furthermore, let Δ be the maximum degree of the graph. We assume that each processor maintains a queue in which yet-to-be-processed tasks are stored. One round looks as follows:

1. n generators are arbitrarily distributed over the processors, and each generator generates one task at the beginning of every time round. For $1 \leq i \leq n$, let $k_i^t = j$ if generator i is allocated to processor P_j in round t, and $k_i^t = 0$ if the generator is not allocated to any processor in that round.
2. Every processor balances its load with some or all its neighbors in the network (according to a well-defined scheme for doing this operation).
3. Every non-empty processor deletes one task.

Let $\hat{\ell}_i^t$ be the load of P_i directly after the load deletion phase in round t. A system is called *stable* if the number of tasks $\hat{L}^t(\mathcal{P}) = \sum_{i=1}^n \hat{\ell}_i^t$ that are in the system at the end of round t does not grow with time, i.e. the total load $\hat{L}^t(\mathcal{P})$ is bounded by a number that might depend on n, but not on the time t.

We will mainly focus on one load balancing method called the *diffusion approach*. Every processor is allowed to balance its load with all its neighbors. As mentioned previously, we briefly consider a second approach in Section 4 where only empty processors are allowed to take load from their non-empty neighbors. We call this second method the *work stealing approach*.

Diffusion approach. We begin with a detailed description of the first approach, an integral variant of the *First Order Diffusion scheme*. Let $\bar{\ell}_i^t$ be the load of processor P_i directly before the load balancing phase, and ℓ_i^t the load directly after the load balancing phase. Let $\alpha_{i,j}^t$ be the load that is to be sent from P_i to P_j in round t for $(i, j) \in E$ ($\alpha_{i,j}^t = 0$ otherwise). Then $\alpha_{i,j}$ and ℓ_i are calculated as follows:

$$\alpha_{i,j}^t := \max\left\{0, \lfloor(\bar{\ell}_i^t - \bar{\ell}_j^t)/2\Delta\rfloor\right\} \qquad \ell_i^t := \bar{\ell}_i^t - \sum_{(i,j)\in E} \alpha_{i,j}^t + \sum_{(j,i)\in E} \alpha_{j,i}^t.$$

To compute $\hat{\ell}_i^t$, the load of processor P_i after load deletion, it remains to subtract one if $\ell_i^t > 0$, thus $\hat{\ell}_i^t := \max\{0, \ell_i^t - 1\}$. Note that the "standard" diffusion approach divides $\bar{\ell}_i^t - \bar{\ell}_j^t$ by $\Delta + 1$ instead of 2Δ. We need the 2Δ for our analysis.

Our main contributions are as follows. In Section 2, we prove Theorem 1, which states that we can upper-bound the total system load by $3\Delta n^3$. This generalizes the results of [4] to the case of unbounded edge capacities and, hence, analyzes the standard diffusion approach. Theorem 8 in Section 3 provides an asymptotically matching lower bound, showing that our upper bound is tight, up to a multiplicative constant. In Section 4 we discuss the problem of combining the diffusion-approach with the work-stealing approach and show that certain assumptions necessarily lead to instability.

2 Analysis of the Dynamic Diffusion Algorithm

In this section we will show that the diffusion approach yields a stable system. Moreover, we are able to upper bound the maximum load that will be in the system by $O(\Delta n^3)$. Throughout, we assume that $n \geq 2$ and $\Delta \geq 2$. Apart from what has been defined above, we will also use notation like $\bar{L}^t(S) = \sum_{i:P_i\in S} \bar{\ell}_i^t$ for a subset $S \subseteq \mathcal{P}$, with similar definitions for $L^t(S)$ and $\hat{L}^t(S)$. Our main result about the diffusion approach to load balancing is

Theorem 1. *Let $n \geq 2$ denote the number of processors in the system, and an upper bound on the number of tasks that are generated during each time round. Let $\Delta \geq 2$ denote the maximum degree of the graph G that specifies the communication linkages in the network. Then, starting with an empty system, for all $t \geq 1$ we have $\hat{L}^t(\mathcal{P}) = \sum_{i=1}^n \hat{\ell}_i^t \leq 3\Delta n^3$.*

We will prove this theorem by first giving a series of preliminary results. The proof of Theorem 1 uses a similar potential function as the one that was used in [4] (though what follows is very different). This idea is to prove an invariant that for all $t \geq 1$, every subset $S \subseteq \mathcal{P}$ satisfies the following inequality:

$$\hat{L}^t(S) \leq \sum_{i=n-|S|+1}^{n} i \cdot (4\Delta) \cdot n. \qquad (1)$$

Then, Inequality (1) will immediately imply Theorem 1 (by taking $S = \mathcal{P}$). We will often have occasion to refer to the right hand side of Inequality (1) for many sets, so to make our proofs that follow easier to read, we define the following function $f : \{1, \ldots, n\} \to \mathbb{N}$ in this way

$$f(k) = \sum_{i=n-k+1}^{n} i \cdot (4\Delta) \cdot n. \qquad (2)$$

Definition 2. *In what follows, we will refer to sets as being* bad *after load generation in round t, or after the load balancing phase of round t, etc., meaning that the load of the set at that particular time violates Inequality (1). For example, if we say that a set S is bad after load generation in round t, we mean that* $\bar{L}^t(S) > f(|S|)$.

Conversely, we will also refer to a set as being good *(after load generation, or load balancing, etc.) if it satisfies Inequality (1) (at the time in question).*

The first lemma states that if we consider any (non-empty) set S at the end of round t, there must have existed a set S' so that the load of S' before load balancing was at least as large as the load of S after load balancing, i.e. $\bar{L}^t(S') \geq L^t(S) \geq \hat{L}^t(S)$. The fact that might not be obvious is that we can assert that the two sets contain the same number of processors. This is the statement of the following lemma.

Lemma 3. *Let* $\emptyset \neq S \subseteq \mathcal{P}$ *denote an arbitrary subset of processors. There exists a set* $|S'|$ *such that*

1. $|S'| = |S|$, *and*
2. $\bar{L}^t(S') \geq L^t(S)$.

Proof. The claim is clear if $S = \mathcal{P}$, since in this case we have $L^t(\mathcal{P}) \geq \hat{L}^t(\mathcal{P})$ and $\bar{L}^t(\mathcal{P}) = L^t(\mathcal{P})$. Taking $S' = \mathcal{P}$ then satisfies the conclusions of the theorem.

So we suppose that S is not the entire set of processors. Then let $S_{in} = \{v : v \in S \text{ and } \exists w \notin S \text{ with } \alpha^t_{wv} > 0)\}$. In other words, S_{in} is the subset of S consisting of processors that received tasks from *outside* of S during load balancing.

Case 1: $S_{in} = \emptyset$. This case is essentially the same as when $S = \mathcal{P}$. Since no processors in S received load from outside of S, the elements of S can only exchange load among themselves or send load to processors outside of S. Then it is clear that $\bar{L}^t(S) \geq L^t(S)$, so taking $S' = S$ again satisfies the desired conclusions.

Case 2: $S_{in} \neq \emptyset$. Let $R = \{w : w \notin S \text{ and } \exists v \in S_{in} \text{ with } \alpha^t_{wv} > 0\}$. In other words, R is the set of nodes *not* in S that pushed tasks into S during load balancing. The main idea of what follows is that we are going to swap some elements of R for elements of S_{in} on a one-for-one basis to find the set S' we desire. More formally, let $L_{in} = \sum_{w \in R, v \in S_{in}} \alpha^t_{wv}$ denote the total flow from R to S during load balancing. We aim to find sets $R_1 \subseteq R$ and $S_1 \subseteq S_{in}$ with (i) $|R_1| = |S_1|$, and (ii) $\bar{L}^t(R_1) \geq L^t(S_1) + L_{in} +$ (flow from S_1 to $S \setminus S_1$). Then we will take $S' = S \setminus S_1 \cup R_1$. Our choice of the set R_1 guarantees that S' will satisfy $\bar{L}^t(S') \geq L^t(S)$, since the elements of R_1 account for all flow that enters S during load balancing, plus all flow that passes from elements in S_1 to elements in $S \setminus S_1$ as well.

To do this, let $E_1 = \{(w, v) : w \in R, v \in S_{in}, \alpha^t_{vw} > 0\}$. Consider an edge $e_1 = (w_1, v_1) \in E_0$ where $\alpha^t_{e_1}$ is largest. Then, from the definition of α^t_{wv}, we see that $\bar{\ell}^t_{w_1} \geq 2\Delta \alpha^t_{w_1 v_1} + \bar{\ell}^t_{v_1}$. The key observation is that by choosing the largest edge, the expression $\bar{\ell}^t_{w_1}$ accounts for all possible load that v_1 could have received during load balancing, *and* all tasks that w_1 pushes into the set S too (and any tasks that v_1 might happen to pass to other elements in S, since this is counted in the term $\bar{\ell}^t_{v_1}$). We set $R_1 := \{w_1\}$ and $S_1 := \{v_1\}$, and $E_2 = E_1 \setminus (\{(w_1, v') : v' \in S_{in}\} \cup \{(w', v_1) : w' \in R\})$.

Then, we iteratively apply this argument, i.e., take a largest edge $e_2 = (w_2, v_2) \in E_2$. (Note that $w_2 \neq w_1$ and $v_2 \neq v_1$.) The choice of largest edge then allows us to swap w_2 for v_2, again accounting for all tasks that w_2 pushes into S during load balancing, all tasks that v_2 receives, and any tasks that v_2 passes to other elements in S. Then, we add w_2 to R_1, i.e. $R_1 := R_1 \cup \{w_2\}$, add v_2 to S_1, so $S_1 := S_1 \cup \{v_2\}$, and delete the appropriate set of edges from E_1. Thus, $E_2 = E_1 \setminus (\{(w_2, v') : v' \in S_{in}\} \cup \{(w', v_2) : w' \in R\})$.

We continue to iterate this procedure, selecting an edge with largest α^t_{wv} value, and performing an exchange as before, until we finish step k with a set $E_k = \emptyset$. It is possible that this procedure terminates at a step when $R_1 = R$ or $S_1 = S_{in}$ (or both), or with one or both of R_1, S_1 being proper subsets of their respective sets. In any case, we have constructed sets R_1 and S_1 (each with $k \leq \min\{|S_{in}|, |R|\}$ elements), so that by taking $S' = (S \setminus S_1) \cup R_1$, this set S' satisfies the two conditions of the theorem.

From the previous lemma, we see that we have proven an inequality about the load of the sets of highest loaded processors, before and after load balancing (which, of course, need not be equal to each other). Thus we can conclude the following result:

Corollary 4. *For $i \in [n]$, let $\bar{M}^t_i$ denote a set of i largest loaded processors before load balancing (in round t). Also let M^t_i denote a corresponding set of i largest loaded processors after load balancing. Then $\bar{L}^t(\bar{M}^t_i) \geq L^t(M^t_i)$.*

We also conclude another result from Lemma 3.

Corollary 5. *Fix $i \in \{1, \ldots, n\}$. Suppose that every subset with i processors is good after the load generation phase of round t. Then, after the load balancing phase (and thus after the task deletion phase), every subset with i processors is still good. (Of course, provided that $\bar{M}^t_i$ is good after load generation, we actually get the same conclusion from Corollary 4.)*

Our next result tells us that if a set is made bad by load generation, then the load balancing and deletion phases are sufficient to make that set good again. The proof is omitted due to space limitations.

Lemma 6. *Suppose that at the end of round t, every set $S \subseteq \mathcal{P}$ satisfies (1). Further, suppose that after the load generation phase in round $t+1$, there is some set $S \subseteq \mathcal{P}$ such that $\bar{L}^{t+1}(S) > f(|S|)$. Then, at the end of round $t+1$, S again satisfies Inequality (1).*

Lemma 6 tells us that if a set is made bad by the load generation phase, then the load balancing and deletion phases are sufficient to make this set good. The essential task that remains to be shown is that load balancing cannot, in some way, change a good set into a bad one. Corollary 5 tells us half the story. We need a little more to cover all possible sets. The proof of the following lemma is omitted due to space limitations.

Lemma 7. *Suppose that all sets are good at the end of round t, but that after load generation in round $t+1$, there exists a bad set S with $|S| = i$. Then after load balancing and deletion, there exists no bad set with i processors.*

Now we are prepared to prove our main result.

Proof. [**Theorem 1**] We prove this theorem by induction on t. Inequality (1) holds when $t = 1$, for however we inject the first n tasks into the system, all sets are good at the end of the first round. So assume that at the end of round t, all sets are good. Fix $i \in \{1, \ldots, n\}$. If all sets of i processors are good after the load generation phase, then from Corollary 5 they are all good at the end of round $t+1$. If there is some bad set of i processors after load generation, then Lemmas 6 and 7 show that all sets of size i are still good at the end of round $t+1$. Finally, it is not possible that during load balancing a (good or bad) set of i processors will lead to the creation of a bad set of $j(\neq i)$ processors. For suppose there is some bad set of $j(\neq i)$ processors at the end of round $t + 1$. Lemma 3 tells us that there must exist a set of j processors that was bad before the load balancing phase, but then Lemmas 6 and 7 again tell us that there is no bad set of j processors at the end of round $t+1$, a contradiction to our assumption that there was a bad set of j processors at the end of the round.

On the first glance it might look as if the our proof strategy is overly complicated and that there is a much simpler proof. In the course of proving our result, we show that there is a gap of $4n\Delta$ tasks between a processor in the bad set S and a processor outside of the bad set before balancing whenever S is bad after balancing. Hence, at least n tasks were sent away from S in this step and the invariant could not have been violated by S. But unfortunately it is possible to create a different bad set of processors during load balancing (possibly with a different number of processors), and we have to discount this case too. Hence, we have to show that if we can find a bad set after load balancing, then there was another bad set S' before load balancing, which leads us to a contradiction through our series of lemmas above.

3 A Matching Lower Bound

In this section we provide a simple example that asymptotically matches the upper bound from Section 2. Consider the linear array $G = (V, E)$ with $V = \{P_0, \ldots, P_{n-1}\}$ and $E = \{(P_i, P_{i+1}) | 0 \leq i < n - 1\}$. Furthermore, suppose that during every time step, n new tasks are generated on processor P_{n-1}.

Theorem 8 below implies that the preceding analysis of our algorithm is tight up to a multiplicative constant, because the line graph has maximum degree $\Delta = 2$, and thus we have an upper bound of $O(n^3)$ on the system load. The proof is omitted due to space constraints.

Theorem 8. *The system described above on the linear array is stable with a total steady-state system load of $\Theta(n^3)$.*

4 Some Instability Results for Work Stealing

In this section we will consider a variation of our load balancing process where we may transfer tasks to empty processors only. This method is usually referred to as *work stealing*. It is similar to the diffusion approach, only the computation of the $\alpha_{i,j}^t$ is different. The value of $\alpha_{i,j}^t$, the load that is sent from P_i to P_j, is larger than zero iff P_j is empty (and P_i non-empty). A frequently used transfer function is $\alpha_{i,j}^t = \lfloor \frac{\bar{\ell}_i^t}{\Delta+1} \rfloor$ and $\alpha_{i,j}^t = 0$ otherwise. Note that the result below also hold when we divide by 2Δ instead of $\Delta+1$. We use the above definition as worst case assumption. In [8] the authors showed that simple work stealing yields a stable system. They assumed that there are at most $(1-\epsilon)n$ new tasks generated per round, for some $\epsilon \in (0,1]$. The important point to note is that in [8], the processor communication links correspond to a complete graph on n vertices. Here we will see that the work stealing method can fail (in the sense that the total load is unbounded over time) if the graph is no longer the complete graph. We consider the line network, i.e., nodes $P_1, \ldots, P_n$ and edges between nodes P_i and P_{i+1} for $1 \leq i \leq n-1$. Hence, the maximum degree is 2. Due to space limitations, we state the following observation without proof.

Observation 9. *Assume we have n processors connected as a line and n generators are all on processor 1. Then the diffusion work stealing system is not stable.*

In a similar manner, under adversarial injection schemes, it is easy to show that the work stealing protocol will not be stable for many classes of graphs, even under a probabilistic injection pattern. For example, we can simply define the process in a way such that the expected load of a processor increases between two load balancing actions.

5 A Different Model for Task Generation/Deletion

In this section we define a load generation model similar to [18] and [4]. Rather than bounding the total number of tasks that are generated per round, we bound the *load change* in any subset of the processors. During each round, tasks can be added or deleted from processors, subject to the restriction in Inequality (3) below. The processors then balance load amongst themselves as before.

In the following, ℓ_i^t (respectively, $L^t(S)$) denotes the load of processor P_i (resp. the total load of all processors in set the S) after we have generated and deleted tasks, and ℓ_i^t (resp. $L^t(S)$) is the load of processor P_i (resp. the total load of all processors in the set S) immediately after the load balancing phase. Let $\mathrm{avg}(t)$ be the average load of the processors in round t after load generation and deletion, i.e. $\text{avg}(t) = \frac{1}{n} \cdot \sum_{i=1}^{n} \bar{\ell}_i^t$. Again, $L^t(\mathcal{P})$ denotes the total system load at the end of step t. One round looks now as follows:

1. Tasks are generated and deleted according to the following *generation restriction*:

$$\bar{L}^t(S) - L^{t-1}(S) \leq |S| \cdot (\text{avg}(t) - \text{avg}(t-1)) + n. \tag{3}$$

2. Every processor balances its load with some or all its neighbors in the network using the diffusion operation defined in Section 1.2.

We can show the following result, the proof of which is not included in this extended abstract due to space constraints.

Theorem 10. *Let $n \geq 2$ denote the number of processors in the system. Let $\Delta \geq 2$ denote the maximum degree of the graph G that specifies the communication linkages in the network. Assume the load generation and deletion fulfills the generation restriction in (3). Then, starting with an empty system, for all $t \geq 1$ and all $S \subseteq \mathcal{P}$ we have $L^t(S) \leq |S| \cdot \text{avg}(t) + 5\Delta n^3$. Furthermore, the maximum number of tasks per processor is $\text{avg}(t) + 5\Delta n^2$.*

5.1 Further Extensions

We can easily generalize our results to other load generation processes, and the proofs of the following results are much like those of Theorem 10 and are omitted due to space limitations.

Theorem 11. *Let $n \geq 2$ denote the number of processors in the system. Let $\Delta \geq 2$ denote the maximum degree of the graph G that specifies the communication linkages in the network. Assume the load generation and deletion fulfills the generation restriction $\bar{L}^t(S) - L^{t-1}(S) \leq |S| \cdot (\text{avg}(t) - \text{avg}(t-1)) + K$. Then, starting with an empty system, for all $t \geq 1$ and all $S \subseteq \mathcal{P}$ we have $L^t(S) \leq |S| \cdot \text{avg}(t) + 5\Delta n K$. Furthermore, the maximum number of tasks per processor is $\text{avg}(t) + 5\Delta n K$.*

Furthermore, we can improve our results to a load generation model where the imbalance that we allow to be generated in any set depends on the number of outgoing edges.

Theorem 12. *Let $n \geq 2$ denote the number of processors in the system. Let $\Delta \geq 2$ denote the maximum degree of the graph G that specifies the communication linkages in the network. Let $e(S)$ be the number of outgoing edges of the set S. Assume the load generation and deletion fulfills the generation restriction $\bar{L}^t(S) - L^{t-1}(S) \leq |S| \cdot (\text{avg}(t) - \text{avg}(t-1)) + K \cdot e(S)$. Then, starting with an empty system, for all $t \geq 1$ and all $S \subseteq \mathcal{P}$ we have $L^t(S) \leq |S| \cdot \text{avg}(t) + 5\Delta n K$. Furthermore, the maximum number of tasks per processor is $\text{avg}(t) + 5\Delta n K$.*

References

1. W. Aiello, E. Kushilevitz, R. Ostrovsky, and A. Rosen. Adaptive packet routing for bursty adversarial traffic. *J. Computer and Systems Sciences* **60** (2000), pp. 482–509.
2. W. Aiello, B. Awerbuch, B. Maggs, and S. Rao. Approximate load balancing on dynamic and asynchronous networks. *Proceedings of the 25th Annual ACM Symposium on Theory of Computing (STOC 1993)*, pp. 632–641.
3. A. Anagnostopoulos, A. Kirsch, and E. Upfal. Stability and efficiency of a random local load balancing protocol. *Proceedings of the 44th Annual IEEE Symposium on Foundations of Computer Science (FOCS 2003)*.
4. E. Anshelevich, D. Kempe, and J. Kleinberg. Stability of load balancing algorithms in dynamic adversarial systems. *Proceedings of the 34th Annual ACM Symposium on Theory of Computing (STOC 2002)*, pp. 399–406.
5. B. Awerbuch, P. Berenbrink, A. Brinkmann, and C. Scheideler. Simple routing strategies for adversarial systems. *Proceedings of the 32nd Annual ACM Symposium on Theory of Computing (STOC 2001)*, pp. 158–167.
6. B. Awerbuch and T. Leighton. A simple local control algorithm for multicommodity flow. *Proceedings of the 34th IEEE Symposium on Foundations of Computer Science (FOCS 1993)*, pp. 459–468.
7. B. Awerbuch and T. Leighton. Improved approximation algorithms for the multicommodity flow problem and local competitive routing in dynamic networks. *Proceedings of the 26th Annual ACM Symposium on Theory of Computing (STOC 1994)*, pp. 487–496.
8. P. Berenbrink, T. Friedetzky, and L.A. Goldberg. The natural work-stealing algorithm is stable. *SIAM Journal of Computing, SICOMP* **32** (2003), pp. 1260–1279.
9. P. Berenbrink, T. Friedetzky, and E. W. Mayr. Parallel continuous randomized load balancing. *Proceedings of the 10th Annual ACM Symposium on Parallel Algorithms and Architectures (SPAA'98)*, 1998, pp.192-201.
10. J.E. Boillat. Load balancing and Poisson equation in a graph. *Concurrency: Practice and Experiences* **2** (1990), pp. 289–313.
11. G. Cybenko. Load balancing for distributed memory multiprocessors. *J. Parallel and Distributed Computing* **7** (1989), pp. 279–301.
12. R. Diekmann, A. Frommer, and B. Monien. Efficient schemes for nearest neighbor load balancing. *J. Parallel Computing* **25** (1999), pp. 789–812.
13. R. Elsässer and B. Monien. Load balancing of unit size tokens and expansion properties of graphs. *Proceedings of the 15th Annual ACM Symposium on Parallel Algorithms and Architectures (SPAA 2003)*, pp. 266–273.
14. R. Elsässer, B. Monien, and R. Preis. Diffusion schemes for load balancing on heterogeneous networks. *Theory of Computing Systems* **35** (2002), pp. 305–320.
15. B. Gosh and S. Muthukrishnan. Dynamic load balancing by random matchings. *J. Computer and Systems Science, 53 (1996), pp. 357-370*.
16. B. Ghosh, F.T. Leighton, B.M. Maggs, S. Muthukrishnan, C.G. Plaxton, R. Rajaraman, A.W. Richa, R.E. Tarjan, and D. Zuckerman. Tight analyses of two local load balancing algorithms. *Proceedings of the 27th Annual ACM Symposium on Theory of Computing (STOC 1995)*, pp. 548–558.
17. F.M. auf der Heide, B. Oesterdiekhoff, and R. Wanka. Strongly adaptive token distribution. *Algorithmica* **15** (1996), pp. 413–427.

18. S. Muthukrishnan and R. Rajaraman. An adversarial model for distributed load balancing. *Proceedings of the 10th Annual ACM Symposium on Parallel Algorithms and Architectures (SPAA 1998)*, pp. 47–54.
19. D. Peleg and E. Upfal. The token distribution problem. *SIAM J. Computing* **18** (1989), pp. 229–243.
20. Y. Rabani, A. Sinclair, and R. Wanka. Local divergence of Markov chains and the analysis of iterative load-balancing schemes. *Proceedings of the 39th IEEE Symposium on Foundations of Computer Science (FOCS 1998)*, pp. 694–703.

On the Power of Random Bases in Fourier Sampling: Hidden Subgroup Problem in the Heisenberg Group

Jaikumar Radhakrishnan[1], Martin Rötteler[2], and Pranab Sen[2]

[1] School of Technology and Computer Science,
Tata Institute of Fundamental Research, Mumbai, India and
Toyota Technological Institute, Chicago, USA
jaikumar@tifr.res.in
[2] NEC Laboratories America, Inc.
4 Independence Way, Princeton, NJ 08540, USA
{mroetteler, pranab}@nec-labs.com

Abstract. The hidden subgroup problem (HSP) offers a unified framework to study problems of group-theoretical nature in quantum computing such as order finding and the discrete logarithm problem. While it is known that Fourier sampling provides an efficient solution in the abelian case, not much is known for general non-abelian groups. Recently, some authors raised the question as to whether post-processing the Fourier spectrum by measuring in a random orthonormal basis helps for solving the HSP. Several negative results on the shortcomings of this *random strong* method are known. In this paper however, we show that the random strong method can be quite powerful under certain conditions on the group G. We define a parameter $r(G)$ and show that $O((\log |G|/r(G))^2)$ iterations of the random strong method give enough classical information to solve the HSP. We illustrate the power of the random strong method via a concrete example of the HSP over finite Heisenberg groups. We show that $r(G) = \Omega(1)$ for these groups; hence the HSP can be solved using polynomially many random strong Fourier samplings followed by a possibly exponential classical post-processing without further queries. The quantum part of our algorithm consists of a polynomial computation followed by measuring in a random orthonormal basis. As an interesting by-product of our work, we get an algorithm for solving the *state identification problem* for a set of nearly orthogonal pure quantum states.

1 Introduction

The hidden subgroup problem (HSP) is defined as follows: We are given a function $f : G \to S$ from a group G to a set S with the promise that there exists a subgroup $H \leq G$ such that f is constant on the left cosets of H and takes distinct values on distinct cosets. Here, f is given via a black box. The task is to find a set of generators for H while making as few queries to f as possible. The abelian HSP (i. e. G is abelian) encompasses several interesting problems such as finding the order of an element in a group and the discrete logarithm problem. Factoring an integer n can be reduced to order finding in the group $\mathbb{Z}_n^*$. The problems of graph isomorphism and graph automorphism can be cast as hidden subgroup problems over the non-abelian group S_n.

Given the success of Fourier sampling in solving the abelian HSP, one can similarly ask whether Fourier sampling over the non-abelian group G helps in solving the HSP over G. The Fourier transform over a (in general, non-abelian) group G gives us a superposition over (ρ, i, j) where ρ is an irreducible unitary representation of G and i, j are the row and column indices of the matrix ρ. The choice of basis for ρ gives us a degree of freedom in defining the Fourier transform over G. This is in contrast to the abelian case, where all representations are one-dimensional and hence only their names ρ matter. Exploiting the symmetries in the above quantum state, one can show that (see e.g. [Kup03], [Ip03], [MRS05]) the optimal measurement to recover the hidden subgroup from a coset state consists of applying the Fourier transform to the first register, measuring the name of an irreducible representation ρ, followed by a POVM on the column space of the resulting state. By *strong Fourier sampling*, we usually mean that we measure the column space using some orthonormal basis instead of a general POVM. In *weak Fourier sampling*, one measures the names ρ of the representations only, and ignores the row and column indices (i, j).

Hallgren, Russell and Ta-Shma [HRTS03] showed that polynomially many iterations of weak Fourier sampling give enough information to reconstruct normal hidden subgroups. More generally, they show that the normal core of the hidden subgroup can be reconstructed via the weak method. Grigni, Schulman, Vazirani and Vazirani [GSVV04] and Gavinsky [Gav04] extend the weak method to some more cases of group/subgroup pairs. The main shortcoming of the weak method is that it gives exactly the same probability distribution if the hidden subgroup is H or a conjugate gHg^{-1} of H. This leads us to consider the *strong* method where we measure both representation names as well as column indices (ρ, j). The amount of additional information about the hidden subgroup H that can be extracted by measuring j depends in general on the basis used for ρ. In a recent paper, Moore, Russell and Schulman [MRS05] showed that for the symmetric group S_n, for *any* choice of bases for the representations, there are order two subgroups that require exponential number of Fourier samplings in order to distinguish them from the identity subgroup. Grigni, Schulman, Vazirani and Vazirani [GSVV04] study the *random strong* method where a random basis is used for each representation ρ. They define a group-theoretic parameter α depending on G and H and show that if α is exponentially large, the additional advantage of the random strong method over the weak method is exponentially small. In particular, this is case when $G = S_n$ and $H \leq S_n, |H| = 2^{O(n \log n)}$.

Our contributions. In this paper, we analyze the power of the random strong method and show, for the first time, that under certain (different) general conditions on G polynomially many iterations of the random strong method do give enough classical information to identify H. We illustrate the power of the random strong method via a concrete example of the HSP over finite Heisenberg groups $\mathcal{H}_p$ of order p^3, p prime. $\mathcal{H}_p$ is defined as the following set of upper triangular matrices:

$$\mathcal{H}_p := \left\{ \begin{pmatrix} 1 & x & z \\ 0 & 1 & y \\ 0 & 0 & 1 \end{pmatrix} : x, y, z \in \mathbb{F}_p \right\}. \tag{1}$$

A convenient encoding for the elements of $\mathcal{H}_p$ is to write (x, y, z), where $x, y, z \in \mathbb{F}_p$ match the components in equation (1). It is easy to see that the classical randomised query complexity of the HSP on $\mathcal{H}_p$ is $\Theta(p)$. The generic quantum algorithm of Ettinger, Høyer and Knill [EHK04] achieves $O(\log p)$ query complexity, but at the expense of $p^{O(\log p)}$ quantum operations. An algorithm with $2^{\Theta(\sqrt{\log p})}$ quantum operations can be obtained by using the orbit coset reduction of [FIM+03] together with the algorithm of [Kup03] to solve orbit coset over the group $\mathbb{Z}_p$. However, the query complexity of this algorithm is also $2^{\Theta(\sqrt{\log p})}$. It seems non-trivial to design a quantum algorithm with $(\log p)^{O(1)}$ query complexity and total running time $p^{O(1)}$. Various existing methods for non-abelian HSP fail to achieve this goal. We show how the random strong method attains this goal, illustrating the power of random bases in Fourier sampling.

It can be shown that $\mathcal{H}_p$ is a semidirect product of the form $\mathbb{Z}_p \ltimes (\mathbb{Z}_p \times \mathbb{Z}_p)$, where the normal subgroup is given by $N_\infty := \{(0, y, z) : y, z \in \mathbb{F}_p\}$ and the complement by $A_{0,0} := \{(x, 0, 0) : x \in \mathbb{F}_p\}$. The commutator subgroup of $\mathcal{H}_p$ is given by $[\mathcal{H}_p, \mathcal{H}_p] = \{(0, 0, z) : z \in \mathbb{F}_p\}$, which is also the centre $\zeta(\mathcal{H}_p)$. The commutator subgroup is isomorphic to $\mathbb{Z}_p$; hence it is abelian but not *smoothly abelian* [FIM+03]. The *Baer subgroup* [GSVV04] turns out to be $\zeta(\mathcal{H}_p)$. Thus, the methods of [Gav04, IMS03, FIM+03] are not applicable in order to solve the HSP for $\mathcal{H}_p$ efficiently. For more details about the Heisenberg group, see Section 2.

The chief obstacle to finding hidden subgroups in $\mathcal{H}_p$ arises from the order p subgroups of $\mathcal{H}_p$ other than its centre. There are $(p^2 + p)$ such order p subgroups; we shall call them $A_{i,j}$, $i \in \mathbb{F}_p \cup \{\infty\}$, $j \in \mathbb{F}_p$. The *forgetful abelian method* (i.e. Fourier sampling over the abelian group $\mathbb{Z}_p \times (\mathbb{Z}_p \times \mathbb{Z}_p)$ instead of the non-abelian group $\mathcal{H}_p \cong \mathbb{Z}_p \ltimes (\mathbb{Z}_p \times \mathbb{Z}_p)$), weak Fourier sampling, strong Fourier sampling in the natural representation basis of $\mathcal{H}_p$ (i.e. the representation basis adapted to the distinguished subgroup tower $\{1\} \triangleleft N_\infty \triangleleft \mathcal{H}_p$) as well as strong Fourier sampling in the $\mathbb{Z}_p$-Fourier transform of the natural representation basis give exponentially small information about the index i of $A_{i,j}$. The details are left for the full version of the paper. For now, we give an intuitive description of the main difficulty posed by these subgroups. Suppose the hidden subgroup is $A_{i,j}$ for some $i \in \mathbb{F}_p \cup \{\infty\}$, $j \in \mathbb{F}_p$. With exponentially high probability, Fourier sampling over $\mathcal{H}_p$ gives us a representation uniformly at random from one of the $(p-1)$ irreducible representations ρ_k of degree p for $k = 1, \ldots, p-1$ of $\mathcal{H}_p$. Suppose one such representation ρ_k shows up. The state essentially collapses to a vector $|\psi_{k,i,j}\rangle \in \mathbb{C}^p$, i.e., $(\mathcal{H}_p, A_{i,j})$ is a Gelfand pair for all i, j (see also [MR05] for Gelfand pairs in the context of the HSP). The vectors $|\psi_{k,i,j}\rangle$ have the property that

$$|\langle \psi_{k,i,j} | \psi_{k,i',j'} \rangle| = \begin{cases} \frac{1}{\sqrt{p}} & : i \neq i', \text{ for all } j, j', \\ \delta_{j,j'} & : i = i', \end{cases}$$

i.e., they form a set of $(p+1)$ mutually unbiased bases of $\mathbb{C}^p$ [WF89]. The main difficulty is that it is not clear a priori that there is any orthonormal basis that can pairwise distinguish between these $(p^2 + p)$ vectors with inverse polynomial probability. Note that the so-called *hidden conjugate problem* [MRRS04] is easy to solve information-theoretically for $\mathcal{H}_p$; the conjugacy classes of the order p groups are defined by i and the above property says that $\{|\psi_{k,i,j}\rangle\}_j$ is an orthonormal basis of $\mathbb{C}^p$, so given the conjugacy class i one can measure in this orthonormal basis to determine the actual hidden subgroup $A_{i,j}$. In view of this, the main challenge in solving the HSP for $\mathcal{H}_p$ is to identify the conjugacy class i.

In this paper however, we show that a random representation basis for ρ_k does in fact pairwise distinguish between $|\psi_{k,i,j}\rangle$ with constant probability. In fact, we refine the method of random measurement bases to distinguish between families of nearly orthogonal subspaces. We combine the geometric ideas of random measurement bases together with representation-theoretic techniques and exhibit a parameter $r(G)$ of a group G and show that it is a lower bound on the total variation distance between the distributions on pairs (ρ, j) of representation names and column indices obtained by the random strong method for possible candidate hidden subgroups H. The parameter $r(G)$ is defined in terms of the ranks and overlaps of the projectors obtained by averaging representations ρ over various subgroups H. We show that $O\left(\frac{\log s(G)}{r^2(G)}\right)$ iterations of the random strong method give sufficient classical information to identify the hidden subgroup H, where $s(G)$ denotes the number of distinct subgroups of G. Note that $s(G) \leq 2^{\log^2 |G|}$ for any group G.

We will see later in Section 2 that $s(\mathcal{H}_p) = O(p^2)$. In Section 4, we show that $r(\mathcal{H}_p) = \Omega(1)$, implying that $O(\log p)$ iterations of the random strong method give sufficient information to extract the hidden subgroup in $\mathcal{H}_p$. This gives us an algorithm solving the HSP over $\mathcal{H}_p$ with $O(\log p)$ query complexity, $O(\log^3 p)$ quantum operations for implementing the non-abelian Fourier transforms, $\tilde{O}(p^2)$ quantum operations to measure in a random basis, and $\tilde{O}(p^4)$ classical post-processing operations. This gives the first example of a group where random representation bases do help in solving the HSP and for which no explicit representation bases are known that solve the problem with $(\log p)^{O(1)}$ Fourier samplings.

As an interesting by-product of our work, we get an algorithm for solving the following *quantum state identification* problem: Consider a set of pure quantum states $\{|\psi_1\rangle, \ldots, |\psi_m\rangle\} \in \mathbb{C}^n$ with the property that $|\langle \psi_i | \psi_j \rangle| \leq \delta$ for all $i \neq j$, where δ is a sufficiently small constant (and typically $m \gg n$). We are given t independent copies of $|\psi_i\rangle$. The task is to identify the index i. We show that $t = O(\log m)$ independent random complete von Neumann measurements in $\mathbb{C}^n$ suffice to identify i with high probability.

Relation to other work. Moore, Rockmore, Russell and Schulman [MRRS04] use non-abelian strong Fourier sampling to give an efficient algorithm for the HSP over the q-hedral group $\mathbb{Z}_q \ltimes \mathbb{Z}_p$ when p, q are prime, $q \mid (p-1)$ and $(p-1)/q = (\log p)^{O(1)}$. Our techniques show that for p, q prime, $q \mid (p-1)$, $q = \Omega(\sqrt{p})$, $r(\mathbb{Z}_q \ltimes \mathbb{Z}_p) = \Omega(1)$, which proves that polynomially many random strong Fourier samplings suffice to find an arbitrary hidden subgroup of $\mathbb{Z}_q \ltimes \mathbb{Z}_p$ in this case. For prime p, $q \mid (p-1)$, $q = \Omega(p^{3/4})$, subgroups H_1, H_2 conjugate to $\mathbb{Z}_q \leq \mathbb{Z}_{p-1}$, our techniques show that $r(\mathbb{Z}_{p-1} \ltimes \mathbb{Z}_p; H_1, H_2) = \Omega\left(\sqrt{\frac{q}{p}}\right)$. Moore et al. [MRRS04] prove a nearly matching upper bound of $r(\mathbb{Z}_{p-1} \ltimes \mathbb{Z}_p; H_1, H_2) = O\left(\sqrt{\frac{q}{p} \log p}\right)$. Thus, a polynomial amount of random strong Fourier sampling can solve the hidden conjugate problem for subgroup $\mathbb{Z}_q \leq \mathbb{Z}_{p-1}$ of the affine group $\mathbb{Z}_{p-1} \ltimes \mathbb{Z}_p$ if and only if $p/q = (\log p)^{O(1)}$.

In this paper, we confine ourselves to random strong Fourier sampling. Our quantum operations always factor into a tensor product over the coset states obtained by querying the function oracle. This distinguishes the Heisenberg group from the symmetric group for which Moore, Russell and Schulman [MRS05] show that single register Fourier sampling is not sufficient to solve the HSP. The quantum part of our algorithm consists of a polynomial computation followed by measuring in a random orthonormal basis. In fact, if a suitable kind of pseudo-random unitary transformation can be generated and implemented efficiently, then the quantum part of the algorithm can be made fully polynomial. Various notions of pseudo-random unitary transformations have been studied (see e.g. [EWS+03], [Eme04]), but it has to be investigated whether they are sufficient for our purposes.

Very recently, Bacon, Childs, and van Dam [BCvD05] have given a fully polynomial time quantum algorithm for the HSP in the Heisenberg group. Their algorithm is based on the square root measurement (also known as "pretty good measurement") on a pair of registers as opposed to the single register experiments considered in this paper.

2 Heisenberg Groups over Z_p

Subgroup Lattice. Since the order of $\mathcal{H}_p$ is p^3 we can expect to find subgroups of order p and p^2 besides the trivial subgroup $\{1\}$ and $\mathcal{H}_p$. The centre of $\mathcal{H}_p$ is given by $\zeta(\mathcal{H}_p) = \langle (0,0,1) \rangle = \{(0,0,z) : z \in \mathbb{F}_p\}$. Note that $|\zeta(\mathcal{H}_p)| = p$. There are $p+1$ subgroups N_i of order p^2, where $i \in \mathbb{F}_p \cup \{\infty\}$. They are given by $N_i := \langle (1,i,0), (0,0,1) \rangle = \{(x, xi, z) : x, z \in \mathbb{F}_p\}, \forall i \in \mathbb{F}_p$. The group N_∞ is given by $N_\infty := \langle (0,1,0), (0,0,1) \rangle = \{(0,y,z) : y, z \in \mathbb{F}_p\}$. It is easy to see that for all $i \in \mathbb{F}_p \cup \{\infty\}$, $\zeta(\mathcal{H}_p) \triangleleft N_i$. Furthermore the N_i are normal subgroups, $N_i \triangleleft \mathcal{H}_p$ and $N_i \cong \mathbb{Z}_p \times \mathbb{Z}_p$. For each $i \in \mathbb{F}_p \cup \{\infty\}$, we have that N_i contains p subgroups $A_{i,j}$ for $j \in \mathbb{F}_p$. The subgroups $A_{i,j}$ satisfy $|A_{i,j}| = p$, whence $A_{i,j} \cong \mathbb{Z}_p$. For $i, j \in \mathbb{F}_p$ we have the following explicit desciption of the elements of $A_{i,j}$: $A_{i,j} := \langle (1,i,j) \rangle = \{(\mu, \mu i, \binom{\mu}{2}i + \mu j) : \mu \in \mathbb{F}_p\}$. For $i = \infty, j \in \mathbb{F}_p$ we obtain $A_{\infty,j} := \langle (0,1,j) \rangle = \{(0, \mu, \mu j) : \mu \in \mathbb{F}_p\}$. It is easy to see that $A_{i,j} \not\leq N_{i'}$ if $i \neq i'$, and their normalizer in $\mathcal{H}_p$ is given by $N_{\mathcal{H}_p}(A_{i,j}) = N_i$. The above groups form a complete list of distinct subgroups of $\mathcal{H}_p$.

The Irreducible Representations of $\mathcal{H}_p$. Since we want to perform Fourier analysis on the groups $\mathcal{H}_p$ we have to determine the irreducible representations of $\mathcal{H}_p$. The reader not familiar with the standard notations of representation theory is referred to standard references like [CR62] or [Ser77]. Observe that $\mathcal{H}_p = A_{0,0} \ltimes N_\infty \cong \mathbb{Z}_p \ltimes (\mathbb{Z}_p \times \mathbb{Z}_p)$. This semidirect product structure can be used to construct the irreducible representations of $\mathcal{H}_p$. First, there are p^2 one-dimensional representations $\chi_{a,b}$ for $a, b \in \mathbb{F}_p$ which come from the factor group $\mathcal{H}_p/\zeta(\mathcal{H}_p) \cong \mathbb{Z}_p^2$. In the following, let ω denote a fixed pth root of unity in the complex numbers. Then the one-dimensional irreducible representations of $\mathcal{H}_p$ are given by $\chi_{a,b}((x,y,z)) := \omega^{ax+by}$ $a, b \in \mathbb{F}_p$.

Let $\mathbb{F}_p^*$ denote the group of non-zero elements of $\mathbb{F}_p$ under multiplication. There are $p-1$ irreducible representations ρ_k, $k \in \mathbb{F}_p^*$ of degree p. They are obtained in the following way: Take a nontrivial character of the centre $\zeta(\mathcal{H}_p)$, extend it to the abelian group N_∞, and induce it to $\mathcal{H}_p$. Explicitly, we obtain the following representations: For each $k \in \mathbb{F}_p^*$, we have a nontrivial character ϕ_k of $\zeta(\mathcal{H}_p)$ given by $\phi_k((0,0,z)) := \omega^{kz}$. Since $\zeta(\mathcal{H}_p) \triangleleft N_\infty$ and N_∞ is abelian, we can extend ϕ_k to a character $\overline{\phi}_k$ of N_∞ by simply defining $\overline{\phi}_k((0,y,0)) := 1$. We choose the elements of $A_{0,0}$ as transversals for N_∞ in $\mathcal{H}_p$. Then ρ_k is defined to be the induction $\rho_k := \overline{\phi}_k \uparrow_{A_{0,0}} \mathcal{H}_p$. On the generators of $\mathcal{H}_p$, we find that ρ_k takes the following values: $\rho_k((1,0,0)) = \sum_{a \in \mathbb{F}_p} |a\rangle\langle a+1|$, $\rho_k((0,1,0)) = \sum_{a \in \mathbb{F}_p} \omega^{ka}|a\rangle\langle a|$ and $\rho_k((0,0,1)) = \omega^k \mathbf{1}_p$, where $\mathbf{1}_p$ denotes the identity operator in $\mathbb{C}^p$. Since $(x,y,z) = (0,0,z)(0,y,0)(x,0,0)$ for all $x,y,z \in \mathbb{F}_p$, we obtain that $\rho_k((x,y,z)) = \omega^{kz} \sum_{a \in \mathbb{F}_p} \omega^{kya}|a\rangle\langle a+x|$. It can be readily checked that the $\chi_{a,b}$, for $a,b \in \mathbb{F}_p$ and ρ_k, for $k \in \mathbb{F}_p^*$ form a complete set of inequivalent irreducible representations of $\mathcal{H}_p$.

Calculating Rank and Overlap of the Projectors. Define the projector $P_{k;i,j} := \frac{1}{p} \sum_{a \in A_{i,j}} \rho_k(a)$. In order to calculate the parameter $r(\mathcal{H}_p)$ (see Section 4 for the details of the calculation) we have to compute the ranks of $P_{k;i,j}$ and pairwise *overlaps* $\|P_{k;i,j} P_{k;i',j'}\|$ (the reason for the nomenclature of *overlap* will become clear later). For $i,j \in \mathbb{F}_p$, we obtain that $P_{k;i,j} = \frac{1}{p} \sum_{\mu,\nu \in \mathbb{F}_p} \omega_p^{k(\binom{\mu}{2}i + \mu j - \binom{\nu}{2}i - \nu j)} |\nu\rangle\langle\mu|$. Hence, $P_{k;i,j} = |\psi_{k;i,j}\rangle\langle\psi_{k;i,j}|$, where $|\psi_{k;i,j}\rangle = \frac{1}{\sqrt{p}} \sum_{\mu \in \mathbb{F}_p} \omega^{-k(\binom{\mu}{2}i + \mu j)} |\mu\rangle$, $i, j \in \mathbb{F}_p$, $k \in \mathbb{F}_p^*$. In the case $i = \infty$, $j \in \mathbb{F}_p$, we get $P_{k;\infty,j} = |\psi_{k;\infty,j}\rangle\langle\psi_{k;\infty,j}|$, where $|\psi_{k;\infty,j}\rangle = |-j\rangle$ $j \in \mathbb{F}_p$, $k \in \mathbb{F}_p^*$. Thus for all $k \in \mathbb{F}_p^*$, $i \in \mathbb{F}_p \cup \{\infty\}$, $j \in \mathbb{F}_p$, $\text{rank}(P_{k;i,j}) = 1$ and $P_{k;i,j}$ is an orthogonal projection onto $|\psi_{k;i,j}\rangle$. For $j, j' \in \mathbb{F}_p$, $\|P_{k;\infty,j} P_{k;\infty,j'}\| = \delta_{j,j'}$. For $i, i', j' \in \mathbb{F}_p$, we get $\|P_{k;i,j} P_{k;\infty,j'}\| = \frac{1}{\sqrt{p}}$. For $i, i', j, j' \in \mathbb{F}_p$, we get $\|P_{k;i,j} P_{k;i',j'}\| = |\langle\psi_{k;i,j}|\psi_{k;i',j'}\rangle| = \frac{1}{p} \sum_{\mu \in \mathbb{F}_p} \omega^{k(\binom{\mu}{2}(i-i') + \mu(j-j'))}$. To evaluate the last term above, we need the following fact about quadratic Weil sums in $\mathbb{F}_p$.

Fact 1 ([LN94–Theorem 5.37]). *Let $h(X) \in \mathbb{F}_p[X]$ be a degree two polynomial. Then,* $\left|\sum_{x \in \mathbb{F}_p} \omega^{h(x)}\right| = \sqrt{p}$.

By Fact 1, if $i \neq i'$, $|\langle\psi_{k;i,j}|\psi_{k;i',j'}\rangle| = \frac{1}{\sqrt{p}}$ irrespective of j and j'. If $i = i'$, it is easy to see that $|\langle\psi_{k;i,j}|\psi_{k;i',j'}\rangle| = \delta_{j,j'}$. To summarise, we have shown the following result:

Lemma 1. *Suppose p is an odd prime. Let $i, i' \in \mathbb{F}_p \cup \{\infty\}$, $j, j' \in \mathbb{F}_p$ and $A_{i,j}, A_{i',j'}$ be two order p subgroups of $\mathcal{H}_p$ other than the centre $\zeta(\mathcal{H}_p)$. Let ρ_k, where $k \in \mathbb{F}_p^*$, be an irreducible representation of $\mathcal{H}_p$ of degree p. Let $P_{k;i,j}$ be defined by $P_{k;i,j} := \frac{1}{p} \sum_{a \in A_{i,j}} \rho_k(a)$ and let $P_{k;i',j'}$ be defined similarly. Then $P_{k;i,j}, P_{k;i',j'}$ are rank one orthogonal projections, and their overlap is given by*

$$\|P_{k;i,j} P_{k;i',j'}\| = \begin{cases} \frac{1}{\sqrt{p}} & : i \neq i', \text{ for all } j, j', \\ \delta_{j,j'} & : i = i'. \end{cases}$$

Thus, for any $k \in \mathbb{F}_p^*$, the vectors $|\psi_{k;i',j'}\rangle$ form a set of $(p+1)$ mutually unbiased bases for $\mathbb{C}^p$.

3 Random Bases and Fourier Sampling

Nearly orthogonal vectors. In this subsection, we state some results about sets of nearly orthogonal unit vectors in a Hilbert space. We use $\|\cdot\|$ to denote the ℓ_2-norm of vectors as well as the ℓ_2-induced operator norm of matrices. We use $\|v\|_1$ to denote the ℓ_1-norm of a vector v. We let $\|M\|_{\mathrm{tr}} = \mathrm{Tr}\sqrt{M^\dagger M}$ denote the trace norm of a matrix M. For subspaces V_1, V_2 having trivial intersection, their *overlap* is defined as $\mathrm{ovlap}(V_1, V_2) = \max_{v_1, v_2} |\langle v_1 | v_2 \rangle|$, where v_i range over unit vectors in V_i. Let Π_i denote the orthogonal projection operator onto V_i. It is easy to see that $\mathrm{ovlap}(V_1, V_2) = \|\Pi_1 \Pi_2\|$.

Proposition 1. *Let V_1, V_2 be subspaces of a Hilbert space having trivial intersection. Let σ_2 denote the totally mixed state in V_2. Let V_2' denote the orthogonal complement of V_1 in $V_1 + V_2$ and σ_2' denote the totally mixed state in V_2'. Let $\delta = \mathrm{ovlap}(V_1, V_2)$. Then, $\|\sigma_2 - \sigma_2'\|_{\mathrm{tr}} \leq 2\delta^{1/2}(1 - \delta^2)^{-1/4}$.*

Proposition 2. *Let $v_1', \ldots, v_n'$ be unit vectors in a Hilbert space. Let $0 \leq \delta < \frac{1}{2n}$. Suppose for all $i, j, i \neq j$, $|\langle v_i' | v_j' \rangle| \leq \delta$. Let $v_1, \ldots, v_n$ be unit vectors obtained by Gram-Schmidt orthonormalising $v_1', \ldots, v_n'$. Then $\| |v_i\rangle\langle v_i| - |v_i'\rangle\langle v_i'| \|_{\mathrm{tr}} < 2\sqrt{6} \cdot \delta \sqrt{n}$, for any i, $1 \leq i \leq n$,*

The proofs of the above propositions are left to the full version of the paper.

Random orthonormal vectors. In this subsection, we state some facts about random orthonormal sets of vectors in $\mathbb{C}^d$. One way of generating a random unit vector in $\mathbb{C}^d$ is as follows: Consider $(y_1, \ldots, y_{2d}) \in \mathbb{R}^{2d}$, where each y_i is independently chosen according to the one dimensional Gaussian distribution with mean 0 and variance 1 (i.e. y_i is a real valued random variable with probability density function $\frac{1}{\sqrt{2\pi}} \exp(-y^2/2)$). Normalise to get the unit vector $(x_1, \ldots, x_{2d})$, where $x_i = \frac{y_i}{\sqrt{y_1^2 + \cdots + y_{2d}^2}}$ (note that any $y_i = 0$ with zero probability). We thus get a random unit vector in $\mathbb{R}^{2d}$. Identifying a pair of real numbers with a single complex number, we get a random unit vector $(z_1, \ldots, z_d)$ in $\mathbb{C}^d$. To generate a random orthonormal ordered set $\{v_1, \ldots, v_m\}$ of vectors in $\mathbb{C}^d$, we can first sample m unit vectors $\{v_1', \ldots, v_m'\}$ in $\mathbb{C}^d$ and then do Gram-Schmidt orthonormalisation on them to get $\{v_1, \ldots, v_m\}$ (note that with probability 1, $\{v_1', \ldots, v_m'\}$ are linearly independent).

The following fact follows by combining Theorem 14.3.2 and Proposition 14.3.3 of [Mat02–Chapter 14] and using the concavity of the square-root function.

Fact 2. *Let $t > 0$, and $|v\rangle, |w\rangle$ independent random unit vectors in $\mathbb{C}^d$. Then,*

$$\Pr\left[|\langle v|w\rangle| > t + \frac{10}{\sqrt{d}}\right] \leq 2\exp(-t^2 d).$$

We will require the following upper and lower bounds on the tails of the chi-square distribution (the chi-square distribution with d degrees of freedom is the sum of squares of d independent Gaussians with mean 0 and variance 1).

Fact 3. Let $(X_1, \ldots, X_d)$ be independent random variables such that X_i is Gaussian with mean 0 and variance 1. Let $X^2 = X_1^2 + \cdots + X_d^2$. Let $0 \leq \epsilon < 1/2$. There exists a universal constant $\gamma > 0$ such that

1. $\Pr[|X^2 - d| > d\epsilon] < 2\exp(-d\epsilon^2/6)$,
2. $\Pr[X^2 > d + \sqrt{d}] > \gamma$, $\Pr[X^2 < d - \sqrt{d}] > \gamma$.

The following result follows easily from Fact 3. A similar result appears as Lemma 5.1 in [MRRS04].

Fact 4. Let $V = \{a^1, \ldots, a^p\}$ be a random orthonormal set of p vectors in $\mathbb{C}^d$. Let a_j^i denote the jth coordinate of vector a^i. Define the d-dimensional probability vector S as follows: $S_j = \frac{1}{p}\sum_{i=1}^{p} |a_j^i|^2$. Let $0 \leq \epsilon < 1/2$. Suppose $p = \Omega(\epsilon^{-2}\log d)$. Let U denote the uniform probability distribution on $\{1, \ldots, d\}$. Then, with probability at least $1 - \exp(-\Omega(\epsilon^2 p))$ over the choice of V, $\|S - U\|_1 \leq \epsilon$.

We will also need the following Chernoff upper bounds on the tail of the sum of d independent identically distributed binary random variables.

Fact 5 ([AS00–Cor. A.7, Theorem A.13]). Suppose $(X_1, \ldots, X_d)$ are independent binary random variables such that $\Pr[X_i = 1] = p$. Let $X = X_1 + \cdots + X_d$. Let $0 \leq \epsilon < 1/2$. Then,

1. $\Pr\left[\left|\frac{X}{d} - p\right| > \epsilon\right] < 2\exp(-2\epsilon^2 d)$,
2. $\Pr[X < \frac{dp}{2}] < \exp(-dp/8)$.

Hidden subgroup problem and Fourier sampling. In this subsection, we recall the standard approach to solving the hidden subgroup problem based on Fourier sampling. The quantum Fourier transform over G, QFT_G, is the $\mathbb{C}$-linear map defined as follows:

$$|g\rangle \mapsto \sum_\rho \sqrt{\frac{d_\rho}{|G|}} \sum_{i,j=1}^{d_\rho} \rho_{ij}(g)|\rho, i, j\rangle,$$

where ρ runs over inequivalent irreducible unitary representations of G, d_ρ denotes the dimension of ρ and i, j run over the row and column indices of ρ. For a subset $T \subseteq G$, define $|T\rangle = \frac{1}{\sqrt{|T|}}\sum_{t \in T} |t\rangle$ to be the uniform superposition over elements of T. For a representation ρ, define the matrix $\rho(T) = \frac{1}{\sqrt{|T|}}\sum_{t \in T} \rho(t)$. If $H \leq G$, it can be shown (see e.g. [HRTS03]) that $\frac{1}{\sqrt{|H|}}\rho(H)$ is an orthogonal projection onto the subspace V_H^ρ of the representation space of ρ consisting of all vectors $|v\rangle$ such that $\rho(h)|v\rangle = |v\rangle$ for all $h \in H$. Thus, $\mathrm{rank}(\rho(H)) = \dim V_H^\rho$.

In the standard Fourier sampling method for the hidden subgroup problem, we begin by forming the uniform superposition $\frac{1}{\sqrt{|G|}}\sum_{g \in G}|g\rangle|0\rangle$ and then query f to get the superposition $\frac{1}{\sqrt{|G|}}\sum_{g \in G}|g\rangle|f(g)\rangle$. We then measure the second register to get a uniform mixture over vectors $|gH\rangle$ in the first register. Assuming the first register is in state $|gH\rangle$ for some particular $g \in G$, its state after the application of QFT_G

becomes $\frac{1}{\sqrt{|G||H|}}\sum_{\rho,i,j}\sqrt{d_\rho}\sum_{h\in H}\rho_{ij}(gh)|\rho,i,j\rangle$. If we now measure the representation name and column index, we sample (ρ,j) with probability

$$P_H^G(\rho,j) = \frac{d_\rho}{|G|}\sum_i |\rho_{ij}(gH)|^2 = \frac{d_\rho}{|G|}\|\rho(gH)|j\rangle\|^2 = \frac{d_\rho}{|G|}\|\rho(H)|j\rangle\|^2.$$

The third equality above follows from the fact that $\|\rho(gH)|j\rangle\| = \|\rho(g)\rho(H)|j\rangle\| = \|\rho(H)|j\rangle\|$, since $\rho(g)$ is unitary. Thus, as long as we measure just the representation name and column index (ρ,j), the probabilities are independent of the actual coset gH that we find ourselves in. This fact can be viewed as the non-abelian generalisation of the fact that in abelian Fourier sampling the probability distribution on the characters is independent of the actual coset that we land up in. Also, it can be shown that (see [GSVV04])

$$P_H^G(\rho) = \sum_{j=1}^{d_\rho} \frac{d_\rho}{|G|}\|\rho(H)|j\rangle\|^2 = \frac{d_\rho|H|}{|G|}\mathrm{rank}(\rho(H)) = \frac{d_\rho|H|}{|G|}\dim V_H^\rho.$$

In *weak Fourier sampling*, we only measure the names ρ of the representations and ignore the column indices j. It can be shown (see e.g. [HRTS03]) that for normal hidden subgroups H, no more information about H is contained in the column space of the resulting state after the measurement of ρ. Thus, weak Fourier sampling is the optimal measurement to recover a normal hidden subgroup starting from the uniform mixture of coset states.

Define a distance measure $w(G;H_1,H_2) = \sum_\rho |P_{H_1}^G(\rho) - P_{H_2}^G(\rho)|$ between subgroups $H_1, H_2 \leq G$. Hence, $w(G;H_1,H_2)$ is the total variation distance between the probability distributions, when the hidden subgroup is H_1 or H_2, on the names of the representations obtained via weak Fourier sampling. Hallgren, Russell and Ta-Shma [HRTS03] showed that $O(\log|G|)$ weak Fourier samplings suffice to reconstruct the *normal core* $c(H)$ of the hidden subgroup H, where $c(H)$ is the largest normal subgroup of G contained in H. Adapting their arguments, we prove the following result.

Proposition 3. *Let $H_1, H_2 \leq G$, $c(H_1) \neq c(H_2)$. Then, $w(G;H_1,H_2) \geq 1/2$.*

The proof is left to the full version of the paper. For a normal subgroup $N \triangleleft G$, define the *normal core family* of N, $\mathrm{ncf}(N) = \{H : H \leq G, c(H) = N\}$. In view of Proposition 3, the remaining challenge is to distinguish between subgroups H_1, H_2 from the same normal core family.

The success of strong Fourier sampling depends on how much statistical information about H is present in the probability distribution $P_H^G(\rho,j)$. The amount of information, in general, depends on the choice of basis for each representation ρ, i. e., on the choice of basis for j; see [MRRS04] for more details. Grigni et al. [GSVV04] show that under certain conditions on G and H, the *random strong* Fourier sampling method, where a random choice of basis is made for each representation, gives exponentially small information about distinguishing H from the identity subgroup. In the next section, we prove a complementary result viz. under different conditions on G, $(\log|G|)^{O(1)}$ random strong Fourier samplings do give enough information to reconstruct the hidden subgroup H with high probability.

4 Power of the Random Strong Method

We now define a parameter $r(G)$ for a group G which, if at least $(\log|G|)^{-O(1)}$, suffices for the random strong method to identify the hidden subgroup with $(\log|G|)^{O(1)}$ Fourier samplings. Let $H_1, H_2 \leq G$. We first define a distance measure $r(G; H_1, H_2)$ between H_1, H_2. In what follows, we use the notation of Section 3.

Definition 1 ($r(G; H_1, H_2; \rho)$). *Suppose ρ is an irreducible d_ρ-dimensional unitary representation of G. Let Π_i denote the orthogonal projection onto $V^\rho_{H_i}$, i.e. $\Pi_i = \frac{1}{|H_i|}\sum_{h \in H_i} \rho(h)$. Let $\Pi_{1,2}$ denote the orthogonal projection onto $V^\rho_{H_1} \cap V^\rho_{H_2}$. It is easy to check that $V^\rho_{H_1} \cap V^\rho_{H_2} = V^\rho_{\langle H_1, H_2\rangle}$, where $\langle H_1, H_2\rangle$ denotes the subgroup of G generated by H_1 and H_2. Thus, $\Pi_{1,2} = \frac{1}{|\langle H_1, H_2\rangle|}\sum_{h \in \langle H_1, H_2\rangle} \rho(h)$. Define $\Pi'_i = \Pi_i - \Pi_{1,2}$. Π'_i is the orthogonal projection onto the subspace V'_i defined as the orthogonal complement of $V^\rho_{H_1} \cap V^\rho_{H_2}$ in $V^\rho_{H_i}$. V'_1 and V'_2 have trivial intersection. Define $r_i = \mathrm{rank}(\Pi_i)$ and $r'_i = \mathrm{rank}(\Pi'_i)$. Define $\hat{h} = \max\{|H_1|r_1, |H_2|r_2\}$, $\tilde{h} = |(|H_1|r_1 - |H_2|r_2)|$ and $\delta = \|\Pi'_1 \Pi'_2\|$. Recall that $\delta = \mathrm{ovlap}(V'_1, V'_2)$. Consider the following three cases:*

1. *When $\frac{\sqrt{d_\rho}}{\log|G|} = \Omega((r_1+r_2)^{3/2})$. Loosely speaking, r_1, r_2 are both small. In this case, define*

$$r(G; H_1, H_2; \rho) = \max\left\{\frac{\hat{h}}{2}\left(\Omega\left(\frac{\sqrt{r'_1}}{r_1} + \frac{\sqrt{r'_2}}{r_2}\right) - 2\delta^{1/2}(1-\delta^2)^{-1/4}\right), \tilde{h}\right\}.$$

2. *When $\frac{\sqrt{d_\rho}}{\log|G|} = \Omega(r_1)$ and $\frac{r_2}{r_1} = \Omega(\log^2|G|)$. Loosely speaking, r_1 is small and r_2 is relatively large with respect to r_1. In this case, define*

$$r(G; H_1, H_2; \rho) = \max\left\{\frac{\hat{h}}{2} \cdot \Omega\left(\frac{1}{\sqrt{r_1}}\right), \tilde{h}\right\}.$$

3. *Otherwise, define $r(G; H_1, H_2; \rho) = \tilde{h}$.*

Definition 2 ($r(G; H_1, H_2), r(G)$). *Let $H_1, H_2 \leq G$. Define $r(G; H_1, H_2) = \sum_\rho \frac{d_\rho}{|G|} \cdot r(G; H_1, H_2; \rho)$ and $r(G) = \min_{H_1, H_2} r(G; H_1, H_2)$.*

Definition 3 ($P^{G,\mathcal{B}}_{H,\mathcal{B}}$). *Let $\mathcal{B}$ be a set of orthonormal bases for the irreducible unitary representations of G. Suppose $H \leq G$. $P^{G,\mathcal{B}}_H$ denotes the probability distribution on the representation names and column indices (ρ, j) got by strong Fourier sampling the state $|H\rangle$ according to $\mathcal{B}$.*

The significance of $r(G; H_1, H_2)$ arises from the following theorem.

Theorem 1. *With probability at least $1 - \exp(-\Omega(\log^2|G|))$ over the choice of random representation bases $\mathcal{B}$ for Fourier sampling, $\|P^{G,\mathcal{B}}_{H_1} - P^{G,\mathcal{B}}_{H_2}\|_{\mathrm{tr}} \geq r(G; H_1, H_2)$.*

Using this theorem, we can apply a 'minimum-finding-like' algorithm to identify the hidden subgroup.

Corollary 1. *Let $s(G)$ denote the number of distinct subgroups of G. With probability at least $2/3$ over the choice of random bases for representations of G, $O\left(\frac{\log s(G)}{r^2(G)}\right)$ random strong Fourier samplings give enough classical information to identify a hidden subgroup in G. In particular, $O\left(\left(\frac{\log |G|}{r(G)}\right)^2\right)$ random strong Fourier samplings suffice.*

The proof is left to the full version of the paper. The rest of the section is devoted to proving Theorem 1. We first state some necessary technical lemmas.

Lemma 2. *Let $W = \{a^1, \ldots, a^p\} \cup \{b^1, \ldots, b^q\} \cup \{c^1, \ldots, c^r\}$ be a random orthonormal set of $p+q+r$ vectors in $\mathbb{C}^d$. Let a_j^i denote the jth coordinate of vector a^i; similar notations will be used for the vectors b^i, c^i too. Define two d-dimensional probability vectors S, T as follows:*

$$S_j = \frac{1}{p+r}\left(\sum_{i=1}^{p}|a_j^i|^2 + \sum_{i=1}^{r}|c_j^i|^2\right), T_j = \frac{1}{q+r}\left(\sum_{i=1}^{q}|b_j^i|^2 + \sum_{i=1}^{r}|c_j^i|^2\right).$$

Then there exists $\delta = \Theta((p+q+r)^{-3/2})$ such that the following holds: Define $\alpha = d\delta^2 - 2\log(p+q+r)$. Suppose $\alpha = \Omega(1)$. Then, with probability at least $1 - \exp(-\Omega(\alpha))$ over the choice of W, $\|S - T\|_1 = \Omega\left(\frac{\sqrt{p}}{p+r} + \frac{\sqrt{q}}{q+r}\right)$.

Lemma 3. *Let $W = \{a^1, \ldots, a^p\}$ be a random orthonormal set of p vectors in $\mathbb{C}^d$. Let a_j^i denote the jth coordinate of vector a^i. Define the d-dimensional probability vector S as follows: $S_j = \frac{1}{p}\sum_{i=1}^{p}|a_j^i|^2$. Then there exists $\delta = \Theta(p^{-1})$ such that the following holds: Define $\alpha = d\delta^2 - 2\log p$. Suppose $\alpha = \Omega(1)$. Let U denote the uniform probability distribution on $\{1, \ldots, d\}$. Then, with probability at least $1 - \exp(-\Omega(\alpha))$ over the choice of V, $\|S - U\|_1 = \Omega(p^{-1/2})$.*

We are now in a position to finally prove Theorem 1.

Proof. **(of Theorem 1, sketch)** Note that $\tilde{h}$ describes the efficiency of the weak standard method, and so the total variation distance between the probability distributions $P_{H_1}^G(\rho, j)$ and $P_{H_2}^G(\rho, j)$ is always lower bounded by $w(G; H_1, H_2)$. When case 1 of Definition 1 applies, we use Lemma 2, Proposition 1 and symmetry arguments to prove the lower bound on the total variation distance. When case 2 of Definition 1 applies, we use Lemma 3, Fact 4 and symmetry arguments. For each representation ρ, the confidence bound in applying the above random basis arguments is at least $1 - \exp(-\Omega(\log^2 |G|))$. Since there are at most $|G|$ representations, the total confidence bound is at least $1 - \exp(-\Omega(\log^2 |G|))$.

We now have all the tools to prove that $r(\mathcal{H}_p) = \Omega(1)$. In fact, we can now prove the following theorem. The proof is left for the full version of the paper.

Theorem 2. *The random strong method is sufficient to solve the hidden subgroup problem in the Heisenberg group $\mathcal{H}_p$. The query complexity of this algorithm is $O(\log p)$. The quantum part of the algorithm consists of a circuit of size $O(\log^4 p)$ followed by a circuit of size $\tilde{O}(p^2)$ for implementing the measurement in a random orthonormal basis. The classical post-processing does not make any queries and has a running time of $\tilde{O}(p^4)$.*

Acknowledgements

We thank Frédéric Magniez, Leonard Schulman, Cris Moore, Alex Russell, and Avery Miller for useful discussions. This work was carried out while M. R. and P. S. were with the Institute for Quantum Computing, University of Waterloo.

References

[AS00]   N. Alon and J. Spencer. *The probabilistic method*. John Wiley and Sons, 2000.

[BCvD05]   D. Bacon, A. Childs, and W. van Dam. From optimal measurement to efficient quantum algorithms for the hidden subgroup problem over semidirect product groups. ArXiv preprint quant–ph/0504083, 2005.

[CR62]   W. C. Curtis and I. Reiner. *Representation Theory of Finite Groups and Algebras*. Wiley and Sons, 1962.

[EHK04]   M. Ettinger, P. Høyer, and E. Knill. The quantum query complexity of the hidden subgroup problem is polynomial. *Information Processing Letters*, 91(1):43–48, 2004. See also ArXiv preprint quant–ph/0401083.

[Eme04]   J. Emerson. Random quantum circuits and pseudo-random operators: theory and applications. ArXiv preprint quant–ph/0410087, 2004.

[EWS+03]   J. Emerson, Y. Weinstein, M. Saraceno, S. Lloyd, and D. Cory. Pseudo-Random unitary operators for quantum information processing. *Science*, 302:2098–2100, 2003.

[FIM+03]   K. Friedl, G. Ivanyos, F. Magniez, M. Santha, and P. Sen. Hidden translation and orbit coset in quantum computing. In *Proceedings of the Symposium on Theory of Computing (STOC)*, pages 1–9, 2003.

[Gav04]   D. Gavinsky. Quantum solution to the hidden subgroup problem for poly-near-Hamiltonian groups. *Quantum Information and Computation*, 4(3):229–235, 2004.

[GSVV04]   M. Grigni, L. Schulman, M. Vazirani, and U. Vazirani. Quantum mechanical algorithms for the nonabelian hidden subgroup problem. *Combinatorica*, pages 137–154, 2004.

[HRTS03]   S. Hallgren, A. Russell, and A. Ta-Shma. The Hidden Subgroup Problem and Quantum Computation Using Group Representations. *SIAM Journal on Computing*, 32(4):916–934, 2003.

[IMS03]   G. Ivanyos, F. Magniez, and M. Santha. Efficient quantum algorithms for some instances of the non-abelian hidden subgroup problem. *International Journal of Foundations of Computer Science*, pages 723–740, 2003.

[Ip03]   L. Ip. Shor's algorithm is optimal. Unpublished manuscript, 2003.

[Kup03]   G. Kuperberg. A subexponential-time quantum algorithm for the dihedral hidden subgroup problem. ArXiv preprint quant–ph/0302112, 2003.

[LN94]   R. Lidl and H. Niederreiter. *Introduction to finite fields and their applications*. Cambridge University Press, 2nd edition, 1994.

[Mat02]   J. Matoušek. *Lectures on Discrete Geometry*. Graduate Texts in Mathematics. Springer-Verlag, 2002.

[MR05]   C. Moore and A. Russell. For distinguishing conjugate hidden subgroups, the pretty good measurement is as good as it gets. ArXiv preprint quant–ph/0501177, 2005.

[MRRS04]   C. Moore, D. Rockmore, A. Russell, and L. Schulman. The power of basis selection in Fourier sampling: hidden subgroup problems in affine groups. In *Proceedings of the Fifteenth Annual ACM-SIAM Symposium on Discrete Algorithms (SODA'04)*, pages 1113–1122, 2004. ArXiv preprint quant–ph/0503095.

[MRS05]　C. Moore, A. Russell, and L. Schulman. The symmetric group defies strong Fourier sampling: Part I. ArXiv preprint quant–ph/0501056, 2005.
[Ser77]　J. P. Serre. *Linear Representations of Finite Groups*. Springer, 1977.
[WF89]　W. Wootters and B. Fields. Optimal state-determination by mutually unbiased measurements. *Ann. Physics*, 191(2):363–381, 1989.

On the Hardness of Embeddings Between Two Finite Metrics

Matthew Cary, Atri Rudra, and Ashish Sabharwal

Computer Science and Engineering,
University of Washington, Box 352350,
Seattle, WA 98195-2350, U.S.A.
{cary, atri, ashish}@cs.washington.edu

Abstract. We improve hardness results for the problem of embedding one finite metric into another with minimum distortion. This problem is equivalent to optimally embedding one weighted graph into another under the shortest path metric. We show that unless P = NP, the minimum distortion of embedding one such graph into another cannot be efficiently approximated within a factor less than 9/4 even when the two graphs are unweighted trees. For weighted trees with the ratio of maximum edge weight to the minimum edge weight of α^2 ($\alpha \geq 1$) and all but one node of constant degree, we improve this factor to $1+\alpha$. We also obtain similar hardness results for extremely simple line graphs (weighted). This improves and complements recent results of Kenyon et al. [13] and Papadimitriou and Safra [18].

1 Introduction

For two n-point metric spaces (X,ρ) and (Y,σ), the *expansion* of a bijection $\varphi : X \to Y$ is defined as $exp(\varphi) = \max_{a \neq b \in X} \frac{\sigma(\varphi(a),\varphi(b))}{\rho(a,b)}$. The *distortion* of φ, denoted $dist(\varphi)$, is the product of $exp(\varphi)$ and $exp(\varphi^{-1})$. The expansion of φ^{-1} is also referred to as the *contraction* of φ and denoted $con(\varphi)$. The *distortion between X and Y*, denoted $dist(X,Y)$, is the minimum distortion over all such bijections and may be thought of as a difference measure between these metric spaces. This paper addresses the computational hardness of the problem of embedding one finite metric space into another with minimum distortion.

The notion of distortion was originally studied for infinite metrics [12] in the analysis of Banach spaces. More recently the embedding of finite metrics into Euclidean and other L_p metrics has been very successful for applications in theoretical computer science, including approximation, learning, on-line algorithms, high-dimensional geometry, and others [6, 17, 16, 11]. This notion has been extended in such directions as embedding a finite metric into a distribution of metrics which has again found great success in approximation algorithms [1, 8]. This continues to be an active area of research [2, 15].

We point out that the problems addressed in the works mentioned above are combinatorial in nature– that is, they are concerned with embedding a finite metric into another *class* of metrics and the focus is on providing bounds for the distortion itself. However, we are interested in the algorithmic problem of embedding a *specific* metric into another *specific* metric– i.e. we are interested in the worst case ratio of the distortion obtained by the algorithm under consideration and the best possible distortion. This problem was introduced by Kenyon et al. [13]. The recent work of Bădoiu et. al. [3] considers the algorithmic question of finding embeddings of a specific metric into a class of metrics.

In addition to the fact that the problem of finding low-distortion embeddings between two finite metrics is a very natural question that by itself merits investigation, the problem is also likely to have much wider use than theoretical computer science. To mention three examples, theorem proving and symbolic computation [20], database problems such as queries over heterogeneous structured databases [21], and matching gels from electrophoresis [10] can all be expressed as tree embedding problems. The problem has several other applications as well [13].

We note a basic fact that any n-point metric may be realized as the shortest path metric of a weighted undirected graph over n nodes, for example by making a complete graph whose adjacency matrix is the matrix of metric distances. Due to this correspondence, we will exclusively focus on the problem of optimally embedding one graph into another. We will implicitly identify a graph with the metric given by shortest paths on that graph. For a set of weighted graphs, their *weight ratio* is the ratio of the maximum to the minimum weights of edges in the graphs.

1.1 Previous Results

The only upper bounds on this problem known to us are by Kenyon et al. [13]. Given two point sets on the real line with the L_1 distance metric that have distortion less than $3 + 2\sqrt{2}$, there is a polynomial time algorithm to find an embedding with the minimum distortion. Their second result finds the minimum distortion between an arbitrary graph and a tree, in polynomial time if the degree of the tree and the distortion are constant. Their algorithm is exponential in the degree of the tree and doubly-exponential in the distortion. Both algorithms are based on dynamic programming; the latter is similar to those based on tree decompositions of graphs.

The situation for hardness results is a little more clear. Determining if there is an isometry—a distortion 1 embedding—between two graphs is the graph isomorphism problem, which is not known to be in P but which is probably not NP-hard either. Kenyon et al. [13] show the problem is NP-hard to approximate within a factor of 2 for general graphs and a factor of 4/3 in the case where one of the graphs is an unweighted tree and the other is a weighted graph with weights 1/2 or 1. Papadimitriou and Safra [18] show that it is NP-hard to approximate within a factor of 3 the distortion between any two finite metrics realized as point sets in $\mathbb{R}^3$ where the distance metric is the L_2 norm.

1.2 Our Results

Unweighted Trees (Section 3.3) The problem is NP-hard to approximate within a factor less than 9/4 for unweighted trees. As far as we know, this is the first hardness result for embedding an unweighted graph into another. It also improves the factor of 2 result for general graphs [13] even when the graphs are unweighted.

Weighted Trees (Section 3.2) The problem is NP-hard to approximate within a factor less than $1 + \alpha$ for any $\alpha \geq 1$ and tree graphs with weight ratio $\Omega(\alpha^2)$. This is the first hardness result for embedding trees into trees and improves the bound of 2 for general graphs [13] at the expense of a larger weight ratio. Our result also holds when all but one node of the underlying graphs have degree ≤ 4; the problem is known to be easy in the unweighted case when all nodes have constant degree and the distortion is small [13]. This result also improves the bound of 3 by Papadimitriou and Safra [18].

Weighted Line Graphs (Section 4) The problem is NP-hard to approximate within a factor of α for any $\alpha > 1$ and line graphs with weight ratio $\Omega(\alpha^2 n^4)$, where n is the number of nodes in the two graphs. This is the only bound known for graphs with constant degrees and large weights.

2 Preliminaries

We begin with some basic properties of the distortion resulting from embedding a weighted undirected graph G into another such graph H. Let $[m]$ denote the set of integers from 1 to m. Let d_G and d_H denote the shortest path distances in G and H, respectively. Fix a bijection $\varphi : G \to H$. We state the following results for $exp(\varphi)$. Analogous results hold for $con(\varphi)$ which is nothing but $exp(\varphi^{-1})$.

Lemma 1 ([13]). φ achieves its maximum expansion at some edge in G, i.e., $exp(\varphi) = \max_{\{a,b\} \in E(G)} \frac{d_H(\varphi(a), \varphi(b))}{d_G(a,b)}$.

Corollary 1. If G and H are unweighted then $exp(\varphi)$ is an integer.

Lemma 2. If G and H are unweighted and H has no edge-subgraph that is isomorphic to G then $exp(\varphi) \geq 2$.

Proof. Let u and v be nodes of G such that $(u,v) \in E(G)$ but $(\varphi(u), \varphi(v)) \notin E(H)$. Such nodes must exist because H has no edge-subgraph isomorphic to G. $d_G(u,v) = 1$ and $d_H(\varphi(u), \varphi(v)) \geq 2$, implying an expansion of at least 2. □

We now state the problem we use in the reductions for our NP-hardness proofs. It is a generalization of the Hamiltonian cycle problem [9]. Let $\mathcal{G} = (V, E)$ be a directed graph over n vertices. $\mathcal{G}$ has a *disjoint cycle cover* if there is a collection of vertex-disjoint cycles in $\mathcal{G}$ that contain every node in V, i.e., there exists a permutation $\sigma : [n] \to [n]$ such that for all $i \in [n]$, $(v_i, v_{\sigma(i)}) \in E$. $\mathcal{G}$

has a *loose* disjoint cycle cover if it has a disjoint cycle cover after adding two arbitrarily chosen edges to E.

The *loose directed disjoint cycle cover testing* problem is a property testing problem defined as follows. Given a directed graph $\mathcal{G}$, output 1 if $\mathcal{G}$ has a disjoint cycle cover and 0 if $\mathcal{G}$ does not even have a loose disjoint cycle cover. Note that in the remaining scenario, one is allowed to output anything.

Lemma 3. *The loose directed disjoint cycle cover testing problem is NP-hard for graphs with indegree ≤ 4 and outdegree $= 3$.*

Proof. This can be shown by an extension of the ideas used in the NP-completeness proof of the directed disjoint cycle cover problem in an earlier paper by the authors [5] using in addition the fact that the Vertex Cover problem is hard to approximate [7]. We omit the details. □

Finally, we mention a combinatorial result about sum-free sequences that is used in one of our constructions. A sequence of n integers is *k-way sum-free* if all n^k sums of k integers (not necessarily distinct) in it are distinct. Khanna et al. [14] suggest a greedy algorithm to construct 3-way sum-free sequences. Their result can be generalized to the following.

Lemma 4. *There exists a strictly increasing sequence of size n in $[n^{2k-1}]$ that is k-way sum-free and is computable in time $O(n^{2k-1})$.*

3 Hardness of Embeddings Between Tree Graphs

Consider the problem of finding a minimum distortion embedding between two given undirected tree graphs. We give reductions from the loose directed disjoint cycle cover testing problem to the decision version of this embedding problem on weighted as well as unweighted trees. The result for the weighted case holds even for graphs with all but one node of degree at most 4. We begin with a general construction that will be used in both reductions.

Given a directed graph $\mathcal{G}$ with outdegree $= 3$ and indegree ≤ 4, we will construct a source tree $\mathcal{S}$ and a destination tree $\mathcal{D}$ with the property that there exist $0 < a < b$ such that

1. if $\mathcal{G}$ has a disjoint cycle cover then $dist(\mathcal{S},\mathcal{D}) \leq a$, and
2. if $\mathcal{G}$ has no *loose* disjoint cycle cover then $dist(\mathcal{S},\mathcal{D}) \geq b$.

It follows from Lemma 3 that it is NP-hard to approximate $dist(\mathcal{S},\mathcal{D})$ within a factor less than b/a.

3.1 The Construction

We describe in this section the construction of $\mathcal{S}$ and $\mathcal{D}$ from $\mathcal{G}$. Let $\mathbb{Z}^+$ denote the set of positive integers and $s : \mathbb{Z}^+ \to \mathbb{Z}^+$ be a strictly increasing monotonic function. Let $v_1, \ldots, v_n$ be the vertices of $\mathcal{G}$.

We will need two types of gadgets, a *center gadget* and for each $i \in [n]$, a *size gadget* T_i. The center gadget is a rooted tree consisting of n leaves, all at depth 1. All edges in this gadget have weight $y \in \mathbb{Z}^+$. Its root is denoted by c_r and leaves by c_ℓ. The size gadget T_i is a rooted tree consisting of $s(i)$ leaves, all at depth 1. All edges in T_i have weight 1. The root of T_i is denoted by g_r and the leaves by g_ℓ.

The source tree $\mathcal{S}$ is constructed as follows (see Fig. 1). Start with a copy of the center gadget and associate with each c_ℓ node of it a distinct vertex v_i of $\mathcal{G}$. For any $i \in [n]$, let the successors of v_i in $\mathcal{G}$ be the vertices v_{i_1}, v_{i_2}, and v_{i_3}. Attach to the c_ℓ node corresponding to v_i copies of the three size gadgets T_{i_1}, T_{i_2}, and T_{i_3} by adding edges with weight $x \in \mathbb{Z}^+$ to the g_r nodes of these gadgets. Copies of any size gadget T_i in $\mathcal{S}$ will henceforth be denoted by S_i.

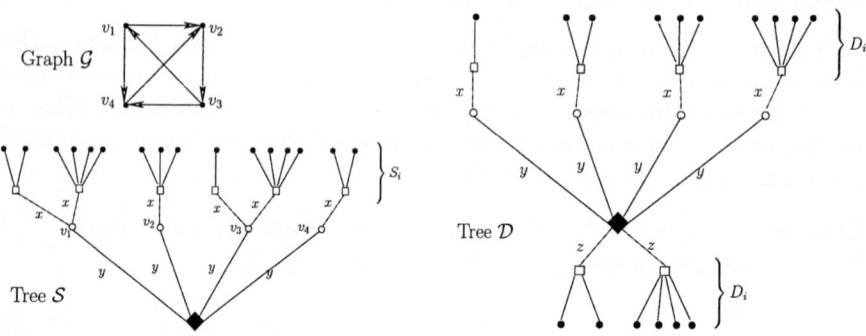

Fig. 1. A directed graph and the source and destination trees corresponding to it. For simplicity of depiction, $s(i) = i$. Unmarked edges have a weight of 1

The destination tree $\mathcal{D}$ is constructed similarly. As before, start with a copy of the center gadget. Fix an arbitrary ordering of its c_ℓ nodes. For all $i \in [n]$, attach to the ith c_ℓ node a copy of the size gadget T_i by adding an edge of weight x to its g_r node. These n size gadgets are called *non-spare* size gadgets. Now let $\mathcal{P}$ be the multi-set $\{i \mid \text{gadget } T_i \text{ is used in } \mathcal{S}\}$. We may assume that $\mathcal{P} \supseteq [n]$, otherwise a disjoint cycle cover cannot exist. For each $i \in \mathcal{P} \setminus [n]$, attach a copy of the size gadget T_i directly to the c_r node by adding edges of weight $z \in \mathbb{Z}^+$ to their g_r node. These are called *spare* size gadgets. Copies of any size gadget T_i in $\mathcal{D}$ will henceforth be denoted by D_i.

Note that both $\mathcal{S}$ and $\mathcal{D}$ have the same number of nodes and for every $i \in [n]$, the same number of copies of the size gadget T_i. Further, $\mathcal{S}$ and $\mathcal{D}$ each have exactly one c_r node, n c_ℓ nodes, and $3n$ g_ℓ nodes (recall the outdegree of every vertex of $\mathcal{G}$ is 3). Consider a mapping φ from $\mathcal{S}$ to $\mathcal{D}$. Let A and B be sets of nodes in $\mathcal{S}$ and $\mathcal{D}$, respectively. φ *fully* maps A to B if $\{\varphi(u) \mid u \in A\} = B$. φ maps A *exactly* to B if A and B are size gadgets with g_r nodes a and b, respectively, φ fully maps A to B, and $\varphi(a) = b$.

The basic idea of the construction is that $\mathcal{S}$ encodes the input graph $\mathcal{G}$ while $\mathcal{D}$ is setup so that the relationships between the c_ℓ nodes and the non-spare size gadgets induce (via a low distortion embedding) a permutation on the vertices of $\mathcal{G}$. This construction balances two conflicting desires. On one hand, it must be possible to match unused size gadgets to the spare gadgets with small distortion when a disjoint cycle cover exists. Thus, the spare gadgets cannot be too far from the successor-selection part $\mathcal{D}$. On the other hand, a node corresponding to a vertex in $\mathcal{G}$ must be far enough from size gadgets not corresponding to its own successors so that choosing a predecessor incorrectly gives large distortion.

Lemma 5. *If $\mathcal{G}$ has a disjoint cycle cover then $dist(\mathcal{S}, \mathcal{D}) \leq (y+z)(x+y)/(xz)$.*

Proof. As $\mathcal{G}$ has a disjoint cycle cover, there is a permutation $\sigma : [n] \to [n]$ such that for all $i \in [n]$, $(i, \sigma(i))$ is an edge in $\mathcal{G}$. We construct a small distortion embedding φ of $\mathcal{S}$ into $\mathcal{D}$. Consider any $i \in [n]$. By the definition of σ, an S_i gadget A is attached to the c_ℓ node u corresponding to $v_{\sigma(i)}$ in $\mathcal{S}$. Let φ map A exactly to the non-spare D_i gadget B of $\mathcal{D}$ and u to the c_ℓ node attached to B. This leaves $2n$ size gadgets of $\mathcal{S}$ not yet mapped. Map each of these exactly to spare size gadgets of $\mathcal{D}$. Finally, let φ map the c_r node of $\mathcal{S}$ to the c_r node of $\mathcal{D}$.

We claim that $exp(\varphi) = (y+z)/x$. By Lemma 1, we only need to consider the expansion of the edges of $\mathcal{S}$. The (g_r, g_ℓ) and (c_r, c_ℓ) edges in $\mathcal{S}$ have an expansion of 1. A (g_r, c_ℓ) edge in $\mathcal{S}$ has an expansion of 1 if the corresponding S_i gadget is mapped to a non-spare D_i gadget and $(y+z)/x$ otherwise. This proves the claim. We further claim that $exp(\varphi^{-1}) = (x+y)/x$. Again using Lemma 1, the only edges in $\mathcal{D}$ that have expansion different from 1 are the (c_r, g_r) edges in $\mathcal{D}$ that give an expansion of $(x+y)/z$. This completes the proof. $\square$

Let φ be any embedding of $\mathcal{S}$ into $\mathcal{G}$. Since both $\mathcal{S}$ and $\mathcal{D}$ contain edges of weight 1 and all edge weights are in $\mathbb{Z}^+$, we have the following.

Proposition 1. $exp(\varphi) \geq 1$ *and* $con(\varphi) \geq 1$.

Lemma 6. *If $\mathcal{G}$ has no disjoint cycle cover and φ fully maps every non-spare D_i gadget from an S_i gadget and c_ℓ nodes from c_ℓ nodes, then both $exp(\varphi)$ and $con(\varphi)$ are at least $1 + 2y/x$.*

Proof. For $i \in [n]$, consider the S_i gadget A_i that maps to the non-spare D_i gadget B_i of $\mathcal{D}$. Let A_i be attached to the c_ℓ node u_j of $\mathcal{S}$ corresponding to vertex v_j of $\mathcal{G}$. Let B_i be attached to the c_ℓ node w_i of $\mathcal{D}$. If u_j maps to w_i and $(v_j, v_i) \in E(\mathcal{G})$, think of vertex v_i being chosen as the successor of vertex v_j in $\mathcal{G}$. Since $\mathcal{G}$ does not have a disjoint cycle cover, there must exist $i \in [n]$ such that u_j, as defined above, does not map to w_i. Fix such i and j. Let $\varphi(u_j) = w_{k_1}$ and $\varphi(u_{k_2}) = w_i$, where $k_1 \neq i$ and $k_2 \neq j$. Let r be the g_r node of A_i and r' be that of B_i. The edge (u_j, r) in $\mathcal{S}$ gives an expansion of at least $(x+2y)/x = 1 + 2y/x$ because φ maps u_j to w_{k_1} and r to a node within B_i. Similarly, the edge (w_i, r') in $\mathcal{D}$ gives a contraction of at least $1 + 2y/x$ because φ^{-1} maps w_i to u_{k_2} and r' to a node within A_i. $\square$

Lemma 7. *If $\mathcal{G}$ has no loose disjoint cycle cover and φ fully maps every S_i gadgets to a D_i gadget, then both $exp(\varphi)$ and $con(\varphi)$ are at least $1 + 2y/x$.*

Proof. Since every S_i gadget fully maps to a D_i gadget, the center gadget of S fully maps to the center gadget of D. We first consider the case when φ maps the c_r node of S to the c_r node in D. Every c_ℓ node of S must then map to a c_ℓ node of D and Lemma 6 completes the proof.

Now suppose that φ maps the c_r node of S to a c_ℓ node w_i of D. As all gadgets are fully mapped, there is a c_ℓ node u_j of S corresponding to vertex v_j of $\mathcal{G}$ be mapped to the c_r node of D. Let B_i be the D_i gadget attached to w_i. From the arguments we made above, it follows that if we want at least one of $exp(\varphi)$ and $con(\varphi)$ to be strictly less than $1 + 2y/x$, then only one of two things can happen. First, a size gadget A_i in S that does not correspond to a successor of v_j is mapped to B_i and every other size gadget maps correctly w.r.t. the successor relationship in $\mathcal{G}$. In this case, $exp(\varphi) \geq 1 + 2y/x$ while $con(\varphi)$ may be at most $1 + y/x$. However, if A_i corresponds to vertex v_i, by adding the edge (v_j, v_i), we have a disjoint cycle cover, contradicting the absence of a *loose* cycle cover. Second, B_i and at most two other non-spare size gadgets B_k and B_ℓ in D are mapped from size gadgets in S that correspond to successors v_i, v_k and v_ℓ of v_j, and every other size gadget maps correctly w.r.t. the successor relationship in $\mathcal{G}$. In this case, $con(\varphi) \geq 1 + 2y/x$ while $exp(\varphi)$ may be at most $1 + y/x$. The successor of v_j is well-defined in this case, but v_k and v_ℓ may not be successors of the c_ℓ nodes in S mapped to the c_ℓ nodes of B_k and B_ℓ. If those nodes are v_s and v_t, by adding edges (v_s, v_k) and (v_t, v_ℓ), we have a disjoint cycle cover, again contradicting the absence of a loose cover. □

3.2 Hardness for Weighted Trees

We first consider general weighted trees with unbounded degree and then modify the reduction so that exactly one node in both S and D has non-constant degree. Let φ be an embedding of S into D. We begin by showing that for suitably weighted S and D, the distortion is large if φ does not map size gadgets correctly.

Lemma 8. *If $s(1) > n$ and φ does not fully map every S_i gadget to a D_i gadget, then $dist(\varphi) \geq x \cdot \min\{x, z\}$.*

Proof. Suppose $exp(\varphi) < \min\{x, z\}$. For $i \in [n]$, $s(i) \geq s(1) > n$. Since the center gadgets have only $n + 1$ nodes, every size gadget in S must have at least one node that φ maps to a size gadget in D. Recall that all edges within size gadgets in S have weight 1 while every edge going out of size gadgets in D has weight $\min\{x, z\}$. To keep $exp(\varphi) < \min\{x, z\}$, every node of any size gadget in S must map within a single size gadget in D. Since for all $i \in [n]$, S and D have the same number of S_i and D_i gadgets, respectively, this can happen only if every S_i gadget fully maps to a D_i gadget. A similar argument shows that $exp(\varphi^{-1}) < x$ only if every D_i gadget fully maps to an S_i gadget. □

Theorem 1. *For $\alpha \geq 1$, it is NP-hard to approximate the distortion between two trees with weight ratio $O(\alpha^2)$ within a factor less than $1 + \alpha$.*

Proof. Let $\mathcal{G}$, $\mathcal{S}$, and $\mathcal{D}$ be as in Section 3.1 with $x = \alpha + 1$, $y = \alpha(\alpha+1)/2$, $z = x + y = (\alpha+1)(\alpha+2)/2$, and $s(i) = i + n$ for $i \in [n]$. The weight ratio of $\{\mathcal{S}, \mathcal{D}\}$ is $(\alpha+1)(\alpha+2)/2$. If $\mathcal{G}$ has a disjoint cycle cover then by Lemma 5 $dist(\mathcal{S}, \mathcal{D}) \leq 1 + 2y/x = 1 + \alpha$. If $\mathcal{G}$ does not have a loose disjoint cycle cover then by Lemmas 7 and 8, $dist(\mathcal{S}, \mathcal{D}) \geq \min\{x \cdot \min\{x, z\}, (1 + 2y/x)^2\}$, which is $(1+\alpha)^2$. The result follows from Lemma 3. □

Let N be the number of nodes in $\mathcal{S}$ (and $\mathcal{D}$). In the above construction, $N = \Theta(n^2)$. The c_r nodes of $\mathcal{S}$ and $\mathcal{D}$ have degrees n and $3n$, respectively, which is $\Theta(\sqrt{N})$. The c_ℓ nodes have degrees 4 and 2, respectively. The g_r nodes have degrees between $n = \Theta(\sqrt{N})$ and $2n$, while the g_ℓ nodes have degree 1. By replacing each S_i and D_i gadget with a line graph, we can show the following theorem. The complete proof may be found in the full version of the paper [4].

Theorem 2. *For $0 < \epsilon \leq 1/2$ and $\alpha \geq 1$, it is NP-Hard to approximate the distortion between two trees with N nodes, weight ratio $\Omega(\alpha^2)$, exactly one node of degree $\Theta(N^\epsilon)$, and all other nodes of degree ≤ 4 within a factor less than $1 + \alpha$.*

3.3 Hardness for Unweighted Trees

The construction from Section 3.1 needs slight modification in order to obtain hardness results for the unweighted case. Let $\mathcal{G}$, $\mathcal{S}$, and $\mathcal{D}$ be as in Section 3.1 with $x = y = z = 1$ and $s(i) = 2c \cdot (f(i) + 2n^5)$, where $c = 4n + 2$ and f is a strictly increasing 3-way sum-free sequence of size n in $[n^5]$ guaranteed by Lemma 4. These parameters imply six useful properties of s, namely, $s(\cdot)$ is even, $s(\cdot)$ is a multiple of c, $2s(1) \gg s(n)$, $2s(n) < 3s(1)$, $|s(i) - s(j)|$ is large for $i \neq j$, and $s(1), s(2), \ldots, s(n)$ is a strictly increasing 3-way sum-free sequence. Furthermore, we have that $c > |\text{Edges}(\mathcal{G})| = 3n$. We will repeatedly use the fact that $\mathcal{S}$ and $\mathcal{D}$ each have $n + 1$ center gadget nodes and $3n$ g_r nodes.

The only change to the construction is to modify the non-spare size gadgets in $\mathcal{D}$. Instead of being depth one trees with $s(i)$ leaves, they are now depth two trees with $s(i)/2$ nodes at depth one, each of which has a single depth two leaf. The root and depth one nodes are denoted by g_r and g_ℓ as before, the depth two leaves are denoted by g'_ℓ, and the depth one and two nodes are together denoted by $\overline{g_\ell}$. All other notation is unchanged. Like the original construction, both $\mathcal{S}$ and $\mathcal{D}$ have the same number of nodes and for each S_i gadget there is a corresponding D_i gadget with the same number of nodes.

Let φ be any embedding of $\mathcal{S}$ into $\mathcal{D}$. We will prove the following lemmas in the rest of this section using Propositions 2 and 3, respectively.

Lemma 9. *If $\mathcal{G}$ has no disjoint cycle cover and φ does not fully map every S_i gadget to a D_i gadget, then $exp(\varphi) \geq 3$.*

Lemma 10. *If $\mathcal{G}$ has no disjoint cycle cover and φ does not fully map every S_i gadget to a D_i gadget, then either $con(\varphi) \geq 3$ or $exp(\varphi) \geq 5$.*

Theorem 3. *It is* NP*-Hard to approximate the distortion between two unweighted trees within a factor less than* 9/4.

Proof. If $\mathcal{G}$ has a disjoint cycle cover then by an argument similar to Lemma 5, $dist(\mathcal{S}, \mathcal{D}) \leq 4$. Assume that $\mathcal{G}$ does not have a loose disjoint cycle cover (and hence no disjoint cycle cover either). If φ fully maps every S_i gadget to a D_i gadget then by an argument similar to Lemma 7, $dist(\varphi) \geq 9$. If it does not then Lemmas 2, 9, and 10 imply $dist(\varphi) \geq 9$. The result follows from Lemma 3. □

Proposition 2. *If any of the following fail,* $exp(\varphi) \geq 3$.

1. No size gadget in $\mathcal{S}$ maps to the $\overline{g_\ell}$ nodes of two distinct size gadgets in $\mathcal{D}$.
2. Nodes of no two size gadgets in $\mathcal{S}$ are both mapped to the $\overline{g_\ell}$ nodes of a single size gadget in $\mathcal{D}$.
3. No node of an S_i gadget maps to a $\overline{g_\ell}$ node of a D_j gadget for $j \neq i$.
4. The g_r node of any S_i gadget A maps within the unique D_i gadget B whose $\overline{g_\ell}$ nodes A maps to, or possibly to the c_r node of $\mathcal{D}$ if B is a spare gadget.
5. The c_r node of $\mathcal{S}$ is not mapped to a non-spare gadget or the g_ℓ nodes of a spare gadget in $\mathcal{D}$.
6. No c_ℓ node in $\mathcal{S}$ is mapped to a non-spare gadget or the g_ℓ nodes of a spare gadget in $\mathcal{D}$.

Proof. Suppose (1) fails and a size gadget A in $\mathcal{S}$ maps to the $\overline{g_\ell}$ nodes of two distinct size gadgets B and C in $\mathcal{D}$. Any $\overline{g_\ell}$ node of B is at least distance 5 away from any $\overline{g_\ell}$ node of C, while all nodes in A are within distance 2 of each other. Hence, $exp(\varphi) \geq 5/2$. By Corollary 1, $exp(\varphi) \geq 3$.

Suppose (2) fails with an S_i gadget A and an S_k gadget C mapping to the $\overline{g_\ell}$ nodes of a single D_j gadget B. A and C together have at least $s(i) + s(k) - s(j) \geq 2s(1) - s(n) \geq 2c(2 + n^5)$ nodes mapped outside B. Since there are only $4n + 1$ non-$\overline{g_\ell}$ nodes in $\mathcal{D}$ ($n + 1$ in the center gadget and $3n$ g_r nodes), a node of A or C must map to a $\overline{g_\ell}$ node of a size gadget in $\mathcal{D}$ other than C. This violates (1).

The proofs of the remaining cases are similar and may be found in [4]. □

Proof of Lemma 9. By Proposition 2 (1), (4), (5), and (6), no node other than that of a unique S_i gadget can be mapped to any non-spare D_i gadget or the g_ℓ nodes of a spare D_i gadget. It follows that all non-spare gadgets are fully mapped. We further claim that all c_ℓ nodes of $\mathcal{S}$ are mapped to c_ℓ nodes of $\mathcal{D}$, in which case the proof is complete by Lemma 6. The claim holds because of the following. Observe that since all non-spare gadgets are fully mapped, all c_ℓ nodes of $\mathcal{S}$ must map within the center gadget of $\mathcal{D}$ to ensure $exp(\varphi) \leq 2$. Further, by the assumption in the lemma, at least one spare gadget B is partially mapped from a gadget A in $\mathcal{S}$. By (4), the g_r node of A must map to the c_r node of $\mathcal{D}$, making the latter unavailable for the c_ℓ nodes of $\mathcal{S}$. □

We begin the contraction argument by stating a straightforward but crucial property of the g'_ℓ nodes of the size gadgets in $\mathcal{D}$.

Observation 1 *If a g'_ℓ node of a size gadget B in $\mathcal{D}$ does not have as its image under φ^{-1} in $\mathcal{S}$ a node with neighbors only those nodes that are images of nodes of B, the c_ℓ node attached to B, or the c_r node of $\mathcal{D}$, then $exp(\varphi) \geq 5$.*

Define the *successor cluster* X corresponding to a vertex v of $\mathcal{G}$ to be the c_ℓ node u of $\mathcal{S}$ corresponding to v and the three size gadgets A_{ix}, A_{jx}, and A_{kx} attached to it. Let $Q_X^\varphi \subseteq \{1,\ldots,n\}$ be the *multi*-set defined by $Q_X^\varphi = \{r \mid$ some non-g'_ℓ node of a D_r gadget maps under φ^{-1} to a non-c_ℓ node of $X\}$. The multiplicity of r in Q_X^φ is the number of D_r's that map in this way to X. Since the number of center gadget nodes in $\mathcal{D}$ is only $n+1$, $s_X^\varphi = \sum_{r \in Q_X^\varphi} s(r)$ can be less than $s_X = s(i_X) + s(j_X) + s(k_X)$ by at most $n+1$. However, since s_X^φ and s_X are both multiples of $c > n+1$, $s_X^\varphi \geq s_X$.

Proposition 3. *If any of the following fail, $con(\varphi) \geq 3$ or $exp(\varphi) \geq 5$.*

1. $Q_X^\varphi = \{i_X, j_X, k_X\}$.
2. The g_r node of any D_i gadget B is mapped within the unique successor cluster X to which B's non-g'_ℓ nodes map.
3. The c_r node of $\mathcal{D}$ maps to the c_r node of $\mathcal{S}$.
4. The g_ℓ nodes of $\mathcal{S}$ are occupied only by the size gadget nodes of $\mathcal{D}$.
5. If a c_ℓ node of $\mathcal{D}$ is mapped to a node of a successor cluster X, then nodes from exactly three size gadgets of $\mathcal{D}$ map into X. (X may have other c_ℓ nodes of $\mathcal{D}$ mapped into it as well.)
6. If a c_ℓ node of $\mathcal{D}$ is mapped to the c_ℓ node of a successor cluster X, then three size gadgets of $\mathcal{D}$ fully map to the non-c_ℓ nodes of X.
7. If no c_ℓ node of $\mathcal{D}$ is mapped to a node of a successor cluster X, then nodes from exactly three size gadgets of $\mathcal{D}$ map into X and the c_ℓ node of X is occupied by a node from a fourth size gadget of $\mathcal{D}$.
8. Every successor cluster in $\mathcal{S}$ is fully mapped from exactly one c_ℓ node and three size gadgets of $\mathcal{D}$.
9. If a c_ℓ node v in $\mathcal{D}$ is mapped to a successor cluster X, then the root r of the size gadget B attached to v is mapped to X.

Proof. Unless mentioned otherwise the mapping under consideration in this proof is φ^{-1}. We refer the reader to the full version [4] for most cases of the proof.

If (5) fails, let v be a c_ℓ node of $\mathcal{D}$ that is mapped to a node u of X. By (4), u is either a c_ℓ node or a g_r node. Suppose first that it is a c_ℓ node. Since (5) fails, there is a size gadget B in $\mathcal{D}$ that has a node mapping to X and another adjacent node mapping outside X. Then B contains two nodes that are mapped at least distance 3 apart because they cannot map to the c_ℓ node of X or to the c_r node of $\mathcal{S}$.

Suppose on the other hand that u is the g_r node of size gadget A in X. Consider the set $\mathcal{Z}$ of size gadgets in $\mathcal{D}$ that have a node mapping to a g_ℓ node of A. Since the size of each gadget in $\mathcal{Z}$ is 1 mod c, the number of g_ℓ nodes of A is 0 mod c, and $|\mathcal{Z}| \leq |\text{Edges}(\mathcal{G})| = 3n < c$, there exists a size gadget $B \in \mathcal{Z}$ that also maps outside A. In particular, B must have a node mapped to the c_ℓ node of X which is the only node of $\mathcal{S}$ outside A within distance 2 of the g_ℓ nodes of $\mathcal{A}$. Since by (4) the c_r node of $\mathcal{S}$ is already occupied by the c_r node of $\mathcal{D}$, no size gadget (other than possibly B) mapping to a node outside X can also map within X without causing $exp(\varphi^{-1}) \geq 3$. (5) now follows from (1). □

Proof of Lemma 10. From Proposition 3 (8) and (9), any non-spare size gadget in $\mathcal{D}$ and the c_ℓ node of $\mathcal{D}$ it is attached to must map within the same successor cluster in $\mathcal{S}$ under φ^{-1}. Consequently, Proposition 3 (8) can be strengthened to say that every successor cluster in $\mathcal{S}$ corresponding to a node v is fully mapped from exactly one c_ℓ node in $\mathcal{D}$, the size gadget B attached to it, and two spare size gadgets. As $s(\cdot)$ is sum-free, B must correspond to a successor of v. Since there are exactly n c_ℓ nodes in $\mathcal{D}$, this assigns a unique successor to each node v, establishing a disjoint cycle and the contradiction which proves the lemma. □

4 Hardness of Embeddings Between Line Graphs with Large Weights

A *line graph* is an acyclic connected graph of maximum degree two, that is, a line of vertices. We have the following result whose proof appears in the full version of the paper [4].

Theorem 4. *Given two line graphs with n nodes and weight ratio $\Omega(b^2)$, for any $k > 1$ and b with $b = \Omega(kn^2)$, it is NP-hard to determine if the distortion between them is less than b/k or at least b.*

Corollary 2. *For $\alpha > 0$, it is NP-hard to approximate the distortion between two line graphs with n nodes and weight ratio $\Omega(\alpha^2 n^4)$ within a factor of α.*

5 Conclusion

We have shown that the problem of finding a minimum distortion embedding between two metrics is hard to approximate within constant factors on even extremely simple graphs, such as weighted lines or unweighted trees. While our constants improve previous results, we believe they are still far from the true story: it seems likely that even approximating distortion in unweighted graphs is much harder than what we know.

One natural relaxation to the graph embedding problem is to find the distortion of embedding a constant fraction of one graph to another. While this quantity will in general be far from the true distortion, it may provide a good enough measure of graph difference for certain applications. Other notions of distortion may also be useful. Rabinovich [19] has used average distortion to study the MinCut-MaxFlow gap in uniform-demand mulitcommodity flow. Other possibly interesting measures are max-distortion, which is the maximum of expansion and contraction rather than the product, and Gromov-Hausdorff distance, which has applications in analysis. The problem remains open in all these scenarios.

References

1. Y. Bartal. Probabilistic approximation of metric spaces and its algorithmic applications. In *FOCS*, pages 184–193, Burlington, VT, 1996.
2. M. Bădoiu, E. Demaine, M. Farach-Colton, M. Hajiaghayi, and A. Sidiropoulos. Ordinal embeddings of minimum relaxation: General properties, trees, and ultrametrics. In *SODA*, pages 650–659, Vancouver, BC, Jan. 2005.
3. M. Bădoiu, K. Dhamdhere, A. Gupta, Y. Rabinovich, H. Raecke, R. Ravi, and A. Sidiropoulos. Approximation algorithms for low-distortion embeddings into low-dimensional spaces. In *SODA*, pages 119–128, Vancouver, BC, Jan. 2005.
4. M. Cary, A. Rudra, and A. Sabharwal. On the hardness of embeddings between two finite metrics. Technical Report UW-CSE-TR-2005-04-02, Univ. of Washington, 2005.
5. M. Cary, A. Rudra, A. Sabharwal, and E. Vee. Floodlight illumination of inifinite wedges. In *14th Annual Fall Workshop on Comp. Geom.*, Boston, MA, Nov. 2004. Technical Report UW-CSE-2004-10-4, University of Washington.
6. M. Deza and M. Laurent. *Geometry of Cuts and Metrics*, volume 15 of *Algorithms and Combinatorics*. Springer, 1997.
7. I. Dinur and S. Safra. The importance of being biased. In *STOC*, pages 33–42, Montreal, Canada, May 2002.
8. J. Fakcharoenphol, S. Rao, and K. Talwar. A tight bound on approximating arbitrary metrics by tree metrics. In *STOC*, pages 448–455, San Diego, CA, June 2003.
9. M. R. Garey and D. S. Johnson. *Computers and Intractability: A Guide to the Theory of NP-Completeness*. W. H. Freeman and Company, NY, 1979.
10. F. Hoffmann, K. Kriegel, and C. Wenk. Matching 2d patterns of protein spots. In *Proc. 14th Ann. Symp. on Comp. Geom.*, pages 231–239, 1998.
11. P. Indyk. Algorithmic aspects of geometric embeddings, Oct. 2001. FOCS '01 tutorial. Available at http://thoery.lcs.mit.edu/~indyk/tut.html.
12. W. B. Johnson and H. Lindenstrauss, editors. *Handbook of the Geometry of Banach Spaces*. North-Holland, 2003.
13. C. Kenyon, Y. Rabani, and A. Sinclair. Low distortion maps between point sets. In *STOC*, pages 272–280, Chicago, IL, June 2004.
14. S. Khanna, N. Linial, and S. Safra. On the hardness of approximating the chromatic number. *Combinatorica*, 20(3):393–415, 2000.
15. J. Lee. On distance scales, embeddings, and efficient relaxations of the cut cone. In *SODA*, pages 92–101, Vancouver, BC, Jan. 2005.
16. N. Linial. Finite metric spaces – combinatorics, geometry and algorithms. In *Proc. Int. Congress of Math. III*, pages 573–586, Beijing, China, 2002.
17. J. Matoušek. *Lectures on Discrete Geometry*, volume 212 of *Graduate Texts in Mathematics*. Springer, 2002.
18. C. Papadimitriou and S. Safra. The copmlexity of low-distortion embeddings between point sets. In *SODA*, pages 112–118, Vancouver, BC, Jan. 2005.
19. Y. Rabinovich. On average distortion of embedding metrics into the line and into l1. In *STOC*, pages 456–462, San Diego, CA, June 2003.
20. R. Ramesh and I. V. Ramakrishnan. Nonlinear pattern matching in trees. *Journal of the ACM*, 39(2):295–316, 1992.
21. T. Schlieder and F. Naumann. Approximate tree embedding for querying XML data. In *ACM-SIGIR Work., XML and Info. Retrieval*, Athens, Greece, July 2000.

Improved Lower Bounds for Locally Decodable Codes and Private Information Retrieval

Stephanie Wehner[*] and Ronald de Wolf[*]

CWI, Kruislaan 413, 1098 SJ, Amsterdam, the Netherlands
{wehner, rdewolf}@cwi.nl

Abstract. We prove new lower bounds for *locally decodable codes* and *private information retrieval*. We show that a 2-query LDC encoding n-bit strings over an ℓ-bit alphabet, where the decoder only uses b bits of each queried position, needs code length $m = \exp\left(\Omega\left(\frac{n}{2^b \sum_{i=0}^{b} \binom{\ell}{i}}\right)\right)$. Similarly, a 2-server PIR scheme with an n-bit database and t-bit queries, where the user only needs b bits from each of the two ℓ-bit answers, unknown to the servers, satisfies $t = \Omega\left(\frac{n}{2^b \sum_{i=0}^{b} \binom{\ell}{i}}\right)$. This implies that several known PIR schemes are close to optimal. Our results generalize those of Goldreich et al. [8], who proved roughly the same bounds for *linear* LDCs and PIRs. Like earlier work by Kerenidis and de Wolf [12], our classical bounds are proved using quantum computational techniques. In particular, we give a tight analysis of how well a 2-input function can be computed from a quantum superposition of both inputs.

1 Introduction

1.1 Locally Decodable Codes

Error correcting codes allow reliable transmission and storage of information in noisy environments. Such codes often have the disadvantage that one has to read almost the entire codeword, even if one is only interested in a small part of the encoded information. A *locally decodable* code $C : \{0,1\}^n \to \Sigma^m$ over alphabet Σ is an error-correcting code that allows efficient decoding of individual bits of the encoded information: given any string y that is sufficiently close to the real codeword $C(x)$, we can probabilistically recover any bit x_i of the original input x, while only looking at k positions of y. The code length m measures the cost of the encoding, while k measures the efficiency of decoding individual bits. Such codes have had a number of applications in recent computer science research, including PCPs and worst-case to average-case reductions. One can also think of applications encoding a large chunk of data in order to protect it from noise, where we are only interested in extracting small pieces at a time. Imagine for

---

[*] Supported by EU project RESQ IST-2001-37559 and NWO Vici grant 2004-2009.

example an encoding of all books in a library, where we would like to retrieve only the first paragraph of this paper.

The main complexity question of interest is the tradeoff between m and k. With $k = \text{polylog}(n)$ queries, the code length can be made polynomially small, even over the binary alphabet $\Sigma = \{0,1\}$ [3]. However, for fixed k, the best upper bounds are superpolynomial. Except for the $k = 2$ case with small alphabet Σ, no good lower bounds are known. Katz and Trevisan [10] showed superlinear but at most quadratic lower bounds for constant k. Goldreich et al. [8] showed an exponential lower bound for *linear* codes with $k = 2$ queries and constant alphabet, and Kerenidis and de Wolf [12] extended this to *all* codes, using techniques from quantum computing. For $\Sigma = \{0,1\}^\ell$ they prove $m = 2^{\Omega(n/2^{5\ell})}$. They also slightly improved the polynomial bounds of [10] for $k > 2$.

Clearly the above lower bound becomes trivial if each position of the codeword has $\ell \geq \log(n)/5$ bits. In this paper we analyze the case where ℓ can be much larger, but the decoder uses only b bits out of the ℓ bits of a query answer. The b positions that he uses may depend on the index i he is interested in and on his randomness. This setting is interesting because many existing constructions are of this form, for quite small b. Goldreich et al. [8] also analyzed this situation, and showed the following lower bound for *linear* codes: $m = 2^{\Omega(n/\sum_{i=0}^{b}\binom{\ell}{i}))}$. Here we prove a slightly weaker lower bound for all codes: $m = 2^{\Omega(n/2^b \sum_{i=0}^{b}\binom{\ell}{i}))}$. In particular, if $b = \ell$ (so the decoder can use all bits from the query answers) we improve the bound from [12] to $m = 2^{\Omega(n/2^{2\ell})}$. We lose a factor of 2^b compared to Goldreich et al. This factor can be dispensed with if the decoder outputs the parity of a subset of the bits he receives. All known LDCs are of this type.

Our proofs are completely different from the combinatorial approach of Goldreich et al. Following [12], we proceed in two steps: (1) we reduce the two classical queries to one *quantum* query and (2) show a lower bound for the induced one-quantum-query-decodable code by deriving a *random access code* from it. The main novelty is a tight analysis of the following problem. Suppose we want to compute a Boolean function $f(a_0, a_1)$ on $2b$ bits, given a quantum superposition $\frac{1}{\sqrt{2}}(|0, a_0\rangle + |1, a_1\rangle)$ of both halves of the input. We show that *any* Boolean f can be computed with advantage $1/2^{b+1}$ from this superposition, and that this is best-achievable for the parity function. This may be of independent interest. In fact, Kerenidis [11] recently used it to exhibit an exponential quantum-classical separation in multiparty communication complexity, and in an interesting new approach to improve depth lower bounds for *classical* circuits.

1.2 Private Information Retrieval

There is a very close connection between LDCs and the setting of *private information retrieval*. In PIR, the user wants to retrieve some item from a database without letting the database learn anything about what item he asked for. In the general model, the user retrieves the ith bit from an n-bit database $x = x_1 \ldots x_n$ that is replicated over $k \geq 1$ non-communicating servers. He communicates with each server without revealing any information about i to individual servers, and at the end of the day learns x_i. This is a natural cryptographic problem that has applications in systems where privacy of the user is important, for example databases providing medical information. Much research has gone into optimizing the communication complexity of one-round PIR schemes. Here the user sends a t-bit message ("query") to each server, who responds with an ℓ-bit message ("answer"), from which the user infers x_i. A number of non-trivial upper bounds have been found [7, 1, 4, 6], but, as in the LDC case, the optimality of such schemes is wide open. In fact, the best known constructions of LDCs with constant k come from PIR schemes with k servers. Roughly speaking, concatenating the servers' answers to all possible queries gives a codeword $C(x)$ of length $m = k2^t$ over the alphabet $\Sigma = \{0,1\}^\ell$ that is decodable with k queries. The privacy of the PIR scheme translates into the error-correcting property of the LDC: since many different sets of k queries have to work for recovering x_i, we can afford some corrupted positions. Conversely, we can turn a k-query LDC into a k-server PIR scheme by asking one query to each server (so $t = \log m$). The privacy of the resulting PIR scheme follows from the fact that an LDC can be made to have a "smoothness" property, meaning that most positions are about equally likely to be queried, independent of i.

Here we restrict attention to 2 servers, which is probably the most interesting case. The paper by Chor et al. [7] that introduced PIR, gave a PIR scheme where both the queries to the servers and the answers from the servers have length $\Theta(n^{1/3})$ bits. Later constructions gave alternative ways of achieving the same complexity, but have not given asymptotic improvements for the 2-server case (in contrast to the case of 3 or more servers [6] and the case of 2 *quantum* servers [12]). Though general lower bounds for 2-server PIRs still elude us, reasonably good lower bounds can be proved for schemes that only use a small number b of bits from each possibly much longer answer string. This b is sometimes called the *probe complexity* of the scheme. As stated in [5], small probe complexity is a desirable property of a PIR scheme for a number of reasons: the user needs less space; the schemes can be more easily applied recursively as in [6]; and such PIR schemes induce locally decodable codes where the codelength m is relatively small while the codeword entries are allowed to have many bits each, but the decoder needs only few bits from each codeword entry it read.

As was implicitly stated by Katz and Trevisan [10] and formalized by Goldreich et al. [7], it is possible to translate 2-server PIRs to 2-query LDCs, where the property of only using b bits from each ℓ-bit string carries over. Combining this lemma with our LDC lower bounds gives the following bound for 2-server PIRs with t-bit queries, ℓ-bit answers, and probe complexity b: $t = \Omega(n/2^b \sum_{i=0}^{b} \binom{\ell}{i})$. In particular, for fixed b the overall communication is $C = 2(t+\ell) = \Omega(n^{1/(b+1)})$. This is tight for $b=1$ (we describe an $O(\sqrt{n})$ scheme in Section 2) and close to optimal for $b=3$, since a small variation of the Chor et al. scheme achieves $C = O(n^{1/3})$ using only 3 bits from each answer [1], while our bound is $\Omega(n^{1/4})$. Similar results were established for *linear* PIR schemes by Goldreich et al., but our results apply to *all* PIR schemes. They imply that in improved 2-server PIR schemes, the user needs to use more bits from the servers' answers. For general schemes, where $b = \ell$, we obtain $t = \Omega(n/2^{2\ell})$. This improves the $n/2^{5\ell}$ bound from [12]. It implies a lower bound of $5 \log n$ on the total communication $C = 2(t + \ell)$. This is incredibly weak, but without any assumptions on how the user handles the answers, and still improves what was known [13, 12].

2 Preliminaries

We use $a_{|S}$ to denote the string a restricted to a set of bits $S \subseteq [n] = \{1,\ldots,n\}$, e.g., $11001_{|\{1,4,5\}} = 101$. We identify a set $S \subseteq [n]$ with n-bit string $S = S_1 \ldots S_n$, where $i \in S$ if and only if the ith bit $S_i = 1$. We use e_i for the n-bit string corresponding to the singleton set $S = \{i\}$. If $y \in \Sigma^m$ where $\Sigma = \{0,1\}^\ell$, then $y_j \in \Sigma$ denotes its jth entry, and $y_{j,i}$ with $i \in [\ell]$ is the ith bit of y_j. We assume general familiarity with the quantum model [15]. Our proofs depend heavily on the notion of a quantum query. We consider queries with ℓ-bit answers, where $\ell \geq 1$. For $\Sigma = \{0,1\}^\ell$, a quantum query to a string $y \in \Sigma^m$ is the unitary map $|j\rangle|z\rangle \mapsto |j\rangle|z \oplus y_j\rangle$, where $j \in [m]$, $z \in \{0,1\}^\ell$ is called the target register, and $z \oplus y_j$ is the string resulting from the xor of the individual bits of z and y_j, i.e. $z \oplus y_j = (z_1 \oplus y_{j,1})\ldots(z_\ell \oplus y_{j,\ell})$. It is convenient to get the query result in the phase of the quantum state. To this end, define $|z_T\rangle = \frac{1}{\sqrt{2^\ell}} \bigotimes_{i=1}^{\ell} (|0\rangle + (-1)^{T_i}|1\rangle)$ where T_i is the ith bit of the ℓ-bit string T. Since $|0 \oplus y_{j,i}\rangle + (-1)^{T_i}|1 \oplus y_{j,i}\rangle = (-1)^{T_i \cdot y_{j,i}}(|0\rangle + (-1)^{T_i}|1\rangle)$, a query maps $|j\rangle|z_T\rangle \mapsto |j\rangle(-1)^{T \cdot y_j}|z_T\rangle$.

A locally decodable code is an error-correcting code that allows efficient decoding of individual bits.

Definition 1. $C : \{0,1\}^n \to \Sigma^m$ *is a (k, δ, ε)-locally decodable code (LDC), if there exists a classical randomized decoding algorithm A with input $i \in [n]$ and oracle access to a string $y \in \Sigma^m$ such that*

1. *A makes k distinct queries $j_1, \ldots, j_k$ to y, non-adaptively, gets query answers $a_1 = y_{j_1}, \ldots, a_k = y_{j_k}$ and outputs a bit $f(a_1, \ldots, a_k)$, where f depends on i and A's randomness.*
2. *For every $x \in \{0,1\}^n$, $i \in [n]$ and $y \in \Sigma^m$ with Hamming distance $d(y, C(x)) \leq \delta m$ we have $\Pr[f(a_1, \ldots, a_k) = x_i] \geq 1/2 + \varepsilon$.*

Here probabilities are taken over A's internal randomness. For $\Sigma = \{0,1\}^\ell$, we say the LDC uses b bits, if A only uses b predetermined bits of each query answer: it outputs $f(a_{1|S_1}, \ldots, a_{k|S_k})$ where the sets $S_1, \ldots, S_k$ are of size b each and are determined by i and A's randomness.

---

[1] A polynomial-based $O(n^{1/3})$-scheme from [4] does not have this "small b"-property.

In our arguments we will use *smooth* codes. These are codes where the decoding algorithm spreads its queries "smoothly" across the codeword, meaning it queries no code location too frequently.

Definition 2. $C : \{0,1\}^n \to \Sigma^m$ is a (k,c,ε)-smooth code (SC) if there is a randomized algorithm A with input $i \in [n]$ and oracle access to $C(x)$ s.t.

1. A makes k distinct queries $j_1, \ldots, j_k$ to $C(x)$, non-adaptively, gets query answers $a_1 = C(x)_{j_1}, \ldots, a_k = C(x)_{j_k}$ and outputs a bit $f(a_1, \ldots, a_k)$, where f depends on i and A's randomness.
2. For every $x \in \{0,1\}^n$ and $i \in [n]$ we have $\Pr[f(a_1, \ldots, a_k) = x_i] \geq 1/2 + \varepsilon$.
3. For every $x \in \{0,1\}^n$, $i \in [n]$ and $j \in [m]$, $\Pr[A \text{ queries } j] \leq c/m$.

The smooth code uses b bits, if A only uses b predetermined bits of each answer.

Note that the decoder of smooth codes deals only with valid codewords $C(x)$. The decoding algorithm of an LDC on the other hand can deal with corrupted codewords y that are still sufficiently close to the original. Katz and Trevisan [10-Theorem 1] showed that LDCs and smooth codes are closely related:

Theorem 1 (Katz & Trevisan). *If $C : \{0,1\}^n \to \Sigma^m$ is a (k,δ,ε)-LDC, then C is also a $(k, k/\delta, \varepsilon)$-smooth code (the property of using b bits carries over).*

The following definition of a one-query *quantum* smooth code is rather ad hoc and not the most general possible, but sufficient for our purposes.

Definition 3. $C : \{0,1\}^n \to \Sigma^m$ is a $(1,c,\varepsilon)$-quantum smooth code (QSC), if there is a quantum algorithm A with input $i \in [n]$ and oracle access to $C(x)$ s.t.

1. A probabilistically picks a string r, makes a query of the form

$$|Q_{ir}\rangle = \frac{1}{\sqrt{2}}\left(|j_{1r}\rangle \frac{1}{\sqrt{2^b}}\sum_{T \subseteq S_{1r}}|z_T\rangle + |j_{2r}\rangle \frac{1}{\sqrt{2^b}}\sum_{T \subseteq S_{2r}}|z_T\rangle\right)$$

and returns the outcome of some measurement on the resulting state.
2. For every $x \in \{0,1\}^n$ and $i \in [n]$ we have $\Pr[A \text{ outputs } x_i] \geq 1/2 + \varepsilon$.
3. For every x,i,j, $\Pr[A \text{ queries } j \text{ with non-zero amplitude}] \leq c/m$.

The QSC uses b bits, if the sets S_{1r}, S_{2r} have size b.

PIR allows a user to obtain the ith bit from an n-bit database x, replicated over $k \geq 1$ servers, without revealing anything about i to individual servers.

Definition 4. *A one-round, $(1-\eta)$-secure, k-server private information retrieval (PIR) scheme for a database $x \in \{0,1\}^n$ with recovery probability $1/2+\varepsilon$, query size t, and answer size ℓ, consists of a randomized algorithm (user) and k deterministic algorithms $S_1, \ldots, S_k$ (servers), such that*

1. *On input $i \in [n]$, the user produces k t-bit queries $q_1, \ldots, q_k$ and sends these to the respective servers. The jth server returns ℓ-bit string $a_j = S_j(x, q_j)$. The user outputs a bit $f(a_1, \ldots, a_k)$ (f depends on i and his randomness).*

2. For every $x \in \{0,1\}^n$ and $i \in [n]$ we have $\Pr[f(a_1,\ldots,a_k) = x_i] \geq 1/2 + \varepsilon$.
3. For all $x \in \{0,1\}^n$, $j \in [k]$, and any two indices $i_1, i_2 \in [n]$, the two distributions on q_j (over the user's randomness) induced by i_1 and i_2 are η-close in total variation distance.

The scheme uses b bits if the user only uses b predetermined bits from each a_i.

If $\eta = 0$, then the server gets no information at all about i. All known non-trivial PIR schemes have $\eta = 0$, perfect recovery ($\varepsilon = 1/2$), and one round of communication. We give two well-known 2-server examples from [7].

Square scheme. Arrange $x = x_1 \ldots x_n$ in a $\sqrt{n} \times \sqrt{n}$ square, then index i is given by two coordinates (i_1, i_2). The user picks a random string $A \in \{0,1\}^{\sqrt{n}}$, and sends $\sqrt{n}$-bit queries $q_1 = A$ and $q_2 = A \oplus e_{i_1}$ to the servers. The first returns $\sqrt{n}$-bit answer $a_1 = q_1 \cdot C_1, \ldots, q_1 \cdot C_{\sqrt{n}}$, where $q_1 \cdot C_c$ denotes the inner product mod 2 of q_1 with the cth column of x. The second server sends a_2 analogously. The user selects the bit $q_1 \cdot C_{i_2}$ from a_1 and $q_2 \cdot C_{i_2}$ from a_2 and computes $(A \cdot C_{i_2}) \oplus ((A \oplus e_{i_1}) \cdot C_{i_2}) = e_{i_1} \cdot C_{i_2} = x_i$. Here $t = \ell = \sqrt{n}$ and $b = 1$.

Cube scheme. A more efficient scheme arranges x in a cube, so $i = (i_1, i_2, i_3)$. The user picks 3 random strings T_1, T_2, T_3 of $n^{1/3}$ bits each, and sends queries $q_1 = T_1, T_2, T_3$ and $q_2 = (T_1 \oplus e_{i_1}), (T_2 \oplus e_{i_2}), (T_3 \oplus e_{i_3})$. The first server computes the bit $a = b_{T_1 T_2 T_3} = \bigoplus_{j_1 \in T_1, j_2 \in T_2, j_3 \in T_3} x_{j_1, j_2, j_3}$. Its answer a_1 is the $n^{1/3}$ bits $b_{T_1' T_2 T_3} \oplus a$ for all T_1' differing from T_1 in exactly one place, and similarly all $b_{T_1 T_2' T_3} \oplus b$ and $b_{T_1 T_2 T_3'} \oplus a$. The second server does the same with its query q_2. The user now selects those 3 bits of each answer that correspond to $T_1' = T_1 \oplus e_{i_1}$, $T_2' = T_2 \oplus e_{i_2}$, $T_3' = T_3 \oplus e_{i_3}$ respectively, and xors those 6 bits. Since every other x_{j_1, j_2, j_3} occurs exactly twice in that sum, what is left is $x_{i_1, i_2, i_3} = x_i$. Here $t, \ell = O(n^{1/3})$ and $b = 3$.

3 Computing $f(a_0, a_1)$ from Superposed Input

To prove the lower bound on LDCs and PIRs, we first construct the following quantum tool. Consider a state $|\Psi_{a_0 a_1}\rangle = \frac{1}{\sqrt{2}}(|0, a_0\rangle + |1, a_1\rangle)$ with a_0, a_1 both b-bit strings. We show that we can compute any Boolean function $f(a_0, a_1)$ with bias $1/2^{b+1}$ given one copy of this state. After that we show that bias is optimal if f is the $2b$-bit parity function. The key to the algorithm is the following:

Lemma 1. *For every $f : \{0,1\}^{2b} \to \{0,1\}$ there exist non-normalized states $|\varphi_a\rangle$ such that $U : |a\rangle|0\rangle \to \frac{1}{2^b} \sum_{w \in \{0,1\}^b} (-1)^{f(w,a)} |w\rangle|0\rangle + |\varphi_a\rangle|1\rangle$ is unitary.*

Proof. Let $|\psi_a\rangle = (1/2^b) \sum_{w \in \{0,1\}^b} (-1)^{f(w,a)} |w\rangle|0\rangle + |\varphi_a\rangle|1\rangle$. It is easy to see that U can be extended to be unitary if and only if $\langle \psi_a | \psi_{a'} \rangle = \delta_{aa'}$ for all a, a'. We will choose $|\varphi_a\rangle$ to achieve this. First, since $\langle w | w' \rangle = \delta_{ww'}$ and $\langle w, 0 | \varphi_a, 1 \rangle = 0$:

$$\langle \psi_a | \psi_{a'} \rangle = \frac{1}{2^{2b}} \sum_{w \in \{0,1\}^b} (-1)^{f(w,a) + f(w,a')} + \langle \varphi_a | \varphi_{a'} \rangle.$$

Let C be the $2^b \times 2^b$ matrix with entries $C_{aa'} = (1/2^{2b}) \sum_{w \in \{0,1\}^b} (-1)^{f(w,a)+f(w,a')}$ where the indices a and a' are b-bit strings. From the definition of $C_{aa'}$ we have $|C_{aa'}| \leq 1/2^b$. By [9–Corollary 6.1.5], the largest eigenvalue is

$$\lambda_{max}(C) \leq \min \left\{ \max_a \sum_{a' \in \{0,1\}^b} |C_{aa'}|, \max_{a'} \sum_{a \in \{0,1\}^b} |C_{aa'}| \right\} \leq \sum_{a \in \{0,1\}^b} \frac{1}{2^b} = 1.$$

However, $\lambda_{max}(C) \leq 1$ implies that $I - C$ is positive semidefinite and hence, by [9–Corollary 7.2.11], $I - C = A^\dagger A$ for some matrix A. Now define $|\varphi_a\rangle$ to be the ath column of A. Since the matrix $C + A^\dagger A = I$ is composed of all inner products $\langle \psi_a | \psi_{a'} \rangle$, we have $\langle \psi_a | \psi_{a'} \rangle = \delta_{aa'}$ and it follows that U is unitary. □

Theorem 2. *Suppose $f : \{0,1\}^{2b} \to \{0,1\}$ is a Boolean function. There exists a quantum algorithm to compute $f(a_0, a_1)$ with success probability exactly $1/2 + 1/2^{b+1}$ using one copy of $|\Psi_{a_0 a_1}\rangle = \frac{1}{\sqrt{2}}(|0, a_0\rangle + |1, a_1\rangle)$, with $a_0, a_1 \in \{0,1\}^b$.*

Proof. First we extend the state $|\Psi_{a_0 a_1}\rangle$ by a $|0\rangle$-qubit. Let U be as in Lemma 1. Applying the unitary transform $|0\rangle\langle 0| \otimes I^{\otimes b+1} + |1\rangle\langle 1| \otimes U$ to $|\Psi_{a_0 a_1}\rangle |0\rangle$ gives

$$\frac{1}{\sqrt{2}} \left(|0\rangle|a_0\rangle|0\rangle + |1\rangle \left(\frac{1}{2^b} \sum_{w \in \{0,1\}^b} (-1)^{f(w, a_1)} |w\rangle|0\rangle + |\varphi_{a_1}\rangle|1\rangle \right) \right).$$

Define $|\Gamma\rangle = |a_0\rangle|0\rangle$ and $|\Lambda\rangle = \frac{1}{2^b} \sum_w (-1)^{f(w, a_1)} |w\rangle|0\rangle + |\varphi_{a_1}\rangle|1\rangle$. Then $\langle \Gamma | \Lambda \rangle = \frac{1}{2^b}(-1)^{f(a_0, a_1)}$ and the above state is $\frac{1}{\sqrt{2}}(|0\rangle|\Gamma\rangle + |1\rangle|\Lambda\rangle)$. We apply a Hadamard transform to the first qubit to get $\frac{1}{2}(|0\rangle(|\Gamma\rangle + |\Lambda\rangle) + |1\rangle(|\Gamma\rangle - |\Lambda\rangle))$. The probability that a measurement of the first qubit yields a 0 is $\frac{1}{4}\langle \Gamma + \Lambda | \Gamma + \Lambda \rangle = \frac{1}{2} + \frac{1}{2}\langle \Gamma | \Lambda \rangle = \frac{1}{2} + \frac{(-1)^{f(a_0, a_1)}}{2^{b+1}}$. Thus by measuring the first qubit we obtain $f(a_0, a_1)$ with bias $1/2^{b+1}$. □

To prove that this algorithm is optimal for the parity function, we need to consider how well we can distinguish two density matrices ρ_0 and ρ_1, i.e., given an unknown state determine whether it is ρ_0 or ρ_1. Let $\| A \|_{tr}$ denote the trace norm of matrix A, which equals the sum of its singular values.

Lemma 2. *Two density matrices ρ_0 and ρ_1 cannot be distinguished with probability better than $1/2 + \| \rho_0 - \rho_1 \|_{tr}/4$.*

Proof. The most general way of distinguishing ρ_0 and ρ_1 is a POVM [15] with two operators E_0 and E_1, such that $p_0 = tr(\rho_0 E_0) \geq 1/2 + \varepsilon$ and $q_0 = tr(\rho_1 E_0) \leq 1/2 - \varepsilon$. Then $|p_0 - q_0| \geq 2\varepsilon$ and likewise, $|p_1 - q_1| \geq 2\varepsilon$, for similarly defined p_1 and q_1. By [15–Theorem 9.1], $\| \rho_0 - \rho_1 \|_{tr} = \max_{\{E_0, E_1\}}(|p_0 - q_0| + |p_1 - q_1|)$ and thus $\| \rho_0 - \rho_1 \|_{tr} \geq 4\varepsilon$. Hence $\varepsilon \leq \| \rho_0 - \rho_1 \|_{tr}/4$. □

Theorem 3. *Suppose that f is the parity of $a_0 a_1$. Then any quantum algorithm for computing f from one copy of $|\Psi_{a_0 a_1}\rangle$ has success probability $\leq 1/2 + 1/2^{b+1}$.*

Proof. Define ρ_0 and ρ_1 by $\rho_c = \frac{1}{2^{2b}-1} \sum_{a_0 a_1 \in f^{-1}(c)} |\Psi_{a_0 a_1}\rangle\langle\Psi_{a_0 a_1}|$, with $c \in \{0,1\}$. A quantum algorithm that computes parity of $a_0 a_1$ with probability $1/2 + \varepsilon$ can be used to distinguish ρ_0 and ρ_1. Hence by Lemma 2: $\varepsilon \leq \|\rho_0 - \rho_1\|_{tr}/4$. Let $A = \rho_0 - \rho_1$. It is easy to see that the $|0, a_0\rangle\langle 0, a_0|$-entries are the same in ρ_0 and in ρ_1, so these entries are 0 in A. Similarly, the $|1, a_1\rangle\langle 1, a_1|$-entries in A are 0. In the off-diagonal blocks, the $|0, a_0\rangle\langle 1, a_1|$-entry of A is $(-1)^{|a_0|+|a_1|}/2^{2b}$. For $|\phi\rangle = \frac{1}{\sqrt{2^b}} \sum_{w \in \{0,1\}^b} (-1)^{|w|} |w\rangle$ we have $|\phi\rangle\langle\phi| = \frac{1}{2^b} \sum_{a_0, a_1} (-1)^{|a_0|+|a_1|} |a_0\rangle\langle a_1|$ and $A = \frac{1}{2^b}(|0, \phi\rangle\langle 1, \phi| + |1, \phi\rangle\langle 0, \phi|)$. Let U and V be unitary transforms such that $U|0, \phi\rangle = |0, 0^b\rangle$, $U|1, \phi\rangle = |1, 0^b\rangle$ and $V|0, \phi\rangle = |1, 0^b\rangle$, $V|1, \phi\rangle = |0, 0^b\rangle$, then $UAV^\dagger = \frac{1}{2^b}(U|0, \phi\rangle\langle 1, \phi|V^\dagger + U|1, \phi\rangle\langle 0, \phi|V^\dagger) = \frac{1}{2^b}(|0, 0^b\rangle\langle 0, 0^b| + |1, 0^b\rangle\langle 1, 0^b|)$. The two nonzero singular values of $UAV^\dagger$ are both $1/2^b$, hence $\|\rho_0 - \rho_1\|_{tr} = \|A\|_{tr} = \|UAV^\dagger\|_{tr} = 2/2^b$. Therefore $\varepsilon \leq \|\rho_0 - \rho_1\|_{tr}/4 = 1/2^{b+1}$. □

4 Lower Bounds for LDCs That Use Few Bits

We now make use of the technique developed above to prove new lower bounds for 2-query LDCs over non-binary alphabets. First we construct a 1-query quantum smooth code (QSC) from a 2-query smooth code (SC), and then prove lower bounds for QSCs. In the sequel, we will index the two queries by 0 and 1 instead of 1 and 2, to conform to the two basis states $|0\rangle$ and $|1\rangle$ of a qubit.

Theorem 4. *If $C : \{0,1\}^n \to (\{0,1\}^\ell)^m$ is a $(2, c, \varepsilon)$-smooth code that uses b bits, then C is a $(1, c, \varepsilon/2^b)$-quantum smooth code that uses b bits.*

Proof. Fix index $i \in [n]$ and encoding $y = C(x)$. The 1-query quantum decoder will pick a random string r with the same probability as the 2-query classical decoder. This r determines two indices $j_0, j_1 \in [m]$, two b-element sets $S_0, S_1 \subseteq [\ell]$, and a function $f : \{0,1\}^{2b} \to \{0,1\}$ such that $\Pr[f(y_{j_0}|_{S_0}, y_{j_1}|_{S_1}) = x_i] = p \geq \frac{1}{2} + \varepsilon$, where the probability is taken over the decoder's randomness. Assume for simplicity that $j_0 = 0$ and $j_1 = 1$, and define $a_0 = y_{j_0}|_{S_0}$ and $a_1 = y_{j_1}|_{S_1}$. We now construct a 1-query quantum decoder that outputs $f(a_0, a_1)$ with probability $1/2 + 1/2^{b+1}$, as follows. The result of a quantum query to j_0 and j_1 is

$$\frac{1}{\sqrt{2}} \left(\underbrace{|0\rangle}_{j_0} \frac{1}{\sqrt{2^b}} \sum_{T \subseteq S_0} (-1)^{a_0 \cdot T} |z_T\rangle + \underbrace{|1\rangle}_{j_1} \frac{1}{\sqrt{2^b}} \sum_{T \subseteq S_1} (-1)^{a_1 \cdot T} |z_T\rangle \right).$$

Note that we write $a_0 \cdot T$ instead of $y_{j_0} \cdot T$, since $T \subseteq S_0$ and therefore the inner product will be the same. We can unitarily transform this to $\frac{1}{\sqrt{2}}(|0\rangle|a_0\rangle + |1\rangle|a_1\rangle)$. By Theorem 2, we can compute an output bit o from this such that $\Pr[o = f(a_0, a_1)] = 1/2 + 1/2^{b+1}$. The probability of success is then given by $\Pr[o = x_i] = \Pr[o = f(a_0, a_1)]\Pr[x_i = f(a_0, a_1)] + \Pr[o \neq f(a_0, a_1)]\Pr[x_i \neq f(a_0, a_1)] = (1/2 + 1/2^{b+1})p + (1/2 - 1/2^{b+1})(1 - p) \geq 1/2 + \varepsilon/2^b$. Since no j

is queried with probability more than c/m by the classical decoder, the same is true for the quantum decoder. □

Our lower bound for 2-query LDCs uses the following notion, due to [2].

Definition 5. *A* quantum random access code *is a mapping $x \mapsto \rho_x$ of the n-bit strings x into m-qubit states ρ_x, such that any bit x_i can be recovered with some probability $p \geq 1/2 + \varepsilon$ from ρ_x*

Note that we need not be able to recover *all* x_i's simultaneously from ρ_x, just any one x_i of our choice. Nayak [14] proved a tight bound on m:

Theorem 5 (Nayak). *Every quantum random access code has $m \geq (1 - H(p))n$.*

The main idea of our proof is to show how the following state $|U(x)\rangle$ induces a quantum random access code. For $u = \sum_{i=0}^{b} \binom{\ell}{i}$ define the pure states

$$|U(x)_j\rangle = \frac{1}{\sqrt{u}} \sum_{|T| \leq b} (-1)^{T \cdot C(x)_j} |z_T\rangle \text{ and } |U(x)\rangle = \frac{1}{\sqrt{m}} \sum_{j=1}^{m} |j\rangle |U(x)_j\rangle.$$

Lemma 3. *Suppose $C : \{0,1\}^n \to (\{0,1\}^\ell)^m$ is a $(1, c, \varepsilon)$-quantum smooth code that uses b bits. Then given one copy of $|U(x)\rangle$, there is a quantum algorithm that outputs 'fail' with probability $1 - 2^{b+1}/(cu)$ with $u = \sum_{i=0}^{b} \binom{\ell}{i}$, but if it succeeds it outputs x_i with probability at least $1/2 + \varepsilon$.*

Proof. Let us fix $i \in [n]$. Suppose the quantum decoder of C makes query $|Q_{ir}\rangle$ to indices j_{0r} and j_{1r} with probability p_r. Consider the following state

$$|V_i(x)\rangle = \sum_r \sqrt{p_r} |r\rangle \frac{1}{\sqrt{2}} \left(|j_{0r}\rangle |U(x)_{j_{0r}}\rangle + |j_{1r}\rangle |U(x)_{j_{1r}}\rangle \right).$$

We first show how to obtain $|V_i(x)\rangle$ from $|U(x)\rangle$ with some probability. Rewrite $|V_i(x)\rangle = \sum_{j=1}^{m} \alpha_j |\phi_j\rangle |j\rangle |U(x)_j\rangle$, where the α_j are nonnegative reals, and $\alpha_j^2 \leq c/(2m)$ because C is a QSC (the $1/2$ comes from the amplitude $1/\sqrt{2}$). Using the unitary map $|0\rangle |j\rangle \mapsto |\phi_j\rangle |j\rangle$, we can obtain $|V_i(x)\rangle$ from the state $|V_i'(x)\rangle = \sum_{j=1}^{m} \alpha_j |j\rangle |U(x)_j\rangle$. We thus have to show that we can obtain $|V_i'(x)\rangle$ from $|U(x)\rangle$. Define operator $M = \sqrt{2m/c} \sum_{j=1}^{m} \alpha_j |j\rangle\langle j| \otimes I$ and consider a POVM with operators $M^\dagger M$ and $I - M^\dagger M$. These operators are positive because $\alpha_j^2 \leq c/2m$. Up to normalization, $M|U(x)\rangle = |V_i'(x)\rangle$. The probability that the measurement succeeds (takes us from $|U(x)\rangle$ to $|V_i'(x)\rangle$) is $\langle U(x)|M^\dagger M|U(x)\rangle = \frac{2m}{c}\langle U(x)| \left(\sum_j \alpha_j^2 |j\rangle\langle j| \otimes I \right) |U(x)\rangle = \frac{2}{c} \sum_j \alpha_j^2 = \frac{2}{c}$. Now given $|V_i(x)\rangle$ we can measure r, and then project the last register onto the sets S_{0r} and S_{1r} that we need for $|Q_{ir}\rangle$, by means of the measurement operator $|j_{0r}\rangle\langle j_{0r}| \otimes \sum_{T \subseteq S_{0r}} |T\rangle\langle T| + |j_{1r}\rangle\langle j_{1r}| \otimes \sum_{T \subseteq S_{1r}} |T\rangle\langle T|$. This measurement succeeds with probability $2^b/u$, but if it succeeds we have the state corresponding to the answer to query $|Q_{ir}\rangle$, from which we can predict x_i. Thus, we succeed with probability $(2^b/u) \cdot (2/c)$, and if we succeed, we output x_i with probability $1/2 + \varepsilon$. □

We can avoid failures by taking many copies of $|U(x)\rangle$:

Lemma 4. *If $C : \{0,1\}^n \to (\{0,1\}^\ell)^m$ is a $(1, c, \varepsilon)$-quantum smooth code, then $|W(x)\rangle = |U(x)\rangle^{\otimes cu/2^{b+1}}$ is a $cu(\log(m) + \log(u))/2^{b+1}$-qubit random access code for x with recovery probability $1/2 + \varepsilon/2$ where $u = \sum_{i=0}^{b} \binom{\ell}{i}$.*

Proof. We do the experiment of the previous lemma on each copy of $|U(x)\rangle$ independently. The probability that all experiments fail simultaneously is $(1 - 2^{b+1}/(cu))^{cu/2^{b+1}} \le 1/2$. In that case we output a fair coin flip. If at least one experiment succeeds, we can predict x_i with probability $1/2 + \varepsilon$. This gives overall success probability at least $1/2(1/2 + \varepsilon) + (1/2)^2 = 1/2 + \varepsilon/2$. □

The lower bound for 2-query SCs and LDCs over non-binary alphabets is then:

Theorem 6. *If $C : \{0,1\}^n \to \Sigma^m = (\{0,1\}^\ell)^m$ is a $(2, c, \varepsilon)$-smooth code where the decoder uses only b bits of each answer, then $m \ge 2^{dn - \log(u)}$ for $d = (1 - H(1/2 + \varepsilon/2^{b+1}))2^{b+1}/(cu) = \Theta(\varepsilon^2/(2^b cu))$ and $u = \sum_{i=0}^{b} \binom{\ell}{i}$. Hence $m = 2^{\Omega(\varepsilon^2 n/(2^{2\ell} c))}$ if $b = \ell$.*

Proof. Theorem 4 implies that C is a $(1, c, \varepsilon/2^b)$-quantum smooth code. Lemma 4 gives us a random access code of $cu(\log(m) + \log(u))/2^{b+1}$ qubits with recovery probability $p = 1/2 + \varepsilon/2^{b+1}$. Finally, the random access code lower bound, Theorem 5, implies $cu(\log(m) + \log(u))/2^{b+1} \ge (1 - H(p))n$. Rearranging and using that $1 - H(1/2 + \eta) = \Theta(\eta^2)$ gives the result. □

Since a $(2, \delta, \varepsilon)$-LDC is a $(2, 2/\delta, \varepsilon)$-smooth code (Theorem 1), we obtain:

Corollary 1. *If $C : \{0,1\}^n \to \Sigma^m = (\{0,1\}^\ell)^m$ is a $(2, \delta, \varepsilon)$-locally decodable code, then $m \ge 2^{dn - \log(u)}$ for $d = (1 - H(1/2 + \varepsilon/2^{b+1}))\delta 2^b/u = \Theta(\delta \varepsilon^2/(2^b u))$ and $u = \sum_{i=0}^{b} \binom{\ell}{i}$. Hence $m = 2^{\Omega(\delta \varepsilon^2 n / 2^{2\ell})}$ if $b = \ell$.*

In all known non-trivial constructions of LDCs and SCs, the decoder outputs the parity of the bits that he is interested in. Then, we can prove:

Theorem 7. *If $C : \{0,1\}^n \to \Sigma^m = (\{0,1\}^\ell)^m$ is a $(2, c, \varepsilon)$-smooth code where the decoder outputs $f(g(a_{0|S_0}), g(a_{1|S_1}))$, with $f, g : \{0,1\}^2 \to \{0,1\}$ fixed functions, then $m \ge 2^{dn - \log(\ell')}$ for $d = \Omega(\varepsilon^2/(c\ell'))$ and $\ell' = \binom{\ell}{b}$.*

Proof. Transform C into a smooth code $C' : \{0,1\}^n \to (\{0,1\}^{\ell'})^m$ with $\ell' = \binom{\ell}{b}$ by defining $C'(x)_j$ to be the value of g on all $\binom{\ell}{b}$ possible b-subsets of the original ℓ bits of $C(x)_j$. We need only 1 bit of each $C'(x)_j$, and can apply Theorem 6. □

5 Lower Bounds for Private Information Retrieval

Here we derive improved lower bounds for 2-server PIRs from our LDC bounds. We use the following [8–Lemma 7.1] to translate PIR schemes to smooth codes:

Lemma 5 (GKST). *Suppose there is a one-round, $(1-\eta)$-secure PIR scheme with two servers, database size n, query size t, answer size ℓ, and recovery probability at least $1/2 + \varepsilon$. Then there is a $(2, 3, \varepsilon - \eta)$-smooth code $C : \{0,1\}^n \to (\{0,1\}^\ell)^m$, where $m \leq 6 \cdot 2^t$. If the PIR scheme uses only b bits of each server answer, then the resulting smooth code uses only b bits of each query answer.*

We now combine this with Theorem 6 to slightly improve the lower bound given in [12] and to extend it to the case where we only use b bits of each server reply.

Theorem 8. *A classical 2-server $(1 - \eta)$-secure PIR scheme with t-bit queries, ℓ-bit answers that uses b bits and has recovery probability $1/2 + \varepsilon$ satisfies $t = \Omega\left(\frac{n(\varepsilon - \eta)^2}{2^b u}\right)$ with $u = \sum_{i=0}^{b} \binom{\ell}{i}$. In particular, if $b = \ell$, then $t = \Omega(n(\varepsilon - \eta)^2 / 2^{2\ell})$.*

Proof. Using Lemma 5 we turn the PIR scheme into a $(2, 3, \varepsilon - \eta)$-smooth code $C : \{0,1\}^n \to (\{0,1\}^\ell)^m$ that uses b bits of ℓ where $m \leq 6 \cdot 2^t$. From Theorem 6 we have $m \geq 2^{dn - \log(u)}$ with $d = \Theta((\varepsilon - \eta)^2 / (2^b u))$. □

If b is fixed, $\varepsilon = 1/2$ and $\eta = 0$, this bound simplifies to $t = \Omega(n/\ell^b)$, hence

Corollary 2. *A 2-server PIR scheme with t-bit queries and ℓ-bit answers has communication $C = 2(t + \ell) = \Omega\left(n^{1/(b+1)}\right)$.*

For $b = 1$ this gives $C = \Omega(\sqrt{n})$, which is achieved by the square scheme of Section 2. For $b = 3$ we get $C = \Omega(n^{1/4})$, which is close to the $C = O(n^{1/3})$ of the cube scheme. As in Theorem 7, we can get the better bound $t = \Omega(n(\varepsilon - \eta)^2 / \binom{\ell}{b})$ for PIR schemes where the user just outputs the parity of b bits from each answer. All known non-trivial PIR schemes have this property.

The previous lower bounds on the query length of 2-server PIR schemes were significant only for protocols that use few bits from each answer. Here we slightly improve the best known bound of $4.4 \log n$ [12] on the overall communication complexity of 2-server PIR schemes, by combining our Theorem 8 and Theorem 6 of Katz and Trevisan [10]. We restate their theorem for the PIR setting, assuming for simplicity that $\varepsilon = 1/2$ and $\eta = 0$.

Theorem 9 (Katz & Trevisan). *Every 2-server PIR with t-bit queries and ℓ-bit answers has $t \geq 2 \log(n/\ell) - O(1)$.*

We now prove the following lower bound on the total communication $C = 2(t + \ell)$ of any 2-server PIR scheme with t-bit queries and ℓ-bit answers:

Theorem 10. *Every 2-server PIR scheme has $C \geq (5 - o(1)) \log n$.*

Proof. We distinguish three cases, depending on the answer length. Let $\delta = \log \log n / \log n$.

case 1: $\ell \leq (0.5 - \delta) \log n$. Theorem 8 implies $C \geq t = \Omega(n^{2\delta}) = \Omega((\log n)^2)$.
case 2: $(0.5 - \delta) \log n < \ell < 2.5 \log n$. Then from Theorem 9 we have
$C = 2(t+\ell) > 2\left(2\log(n/(2.5\log n)) - O(1) + (0.5 - \delta)\log n\right) = (5 - o(1))\log n$.
case 3: $\ell \geq 2.5 \log n$. Then $C = 2(t+\ell) \geq 5 \log n$. □

6 Conclusion and Future Work

Here we improved the best known lower bounds on the length of 2-query locally decodable codes and the communication complexity of 2-server private information retrieval schemes. Our bounds are significant whenever the decoder uses only few bits from the two query answers, even if the alphabet (LDC case) or answer length (PIR case) is large. This contrasts with the earlier results of Kerenidis and de Wolf [12], which become trivial for logarithmic alphabet or answer length, and those of Goldreich et al. [8], which only apply to *linear* schemes.

Still, general lower bounds without constraints on alphabet or answer size completely elude us. Clearly, this is one of the main open questions in this area. Barring that, we could at least improve the dependence on b of our current bounds. For example, a PIR lower bound like $t = \Omega(n/\ell^{\lceil b/2 \rceil})$ might be feasible using some additional quantum tricks. Such a bound for instance implies that the total communication is $\Omega(n^{1/3})$ for $b = 3$, which would show that the cube scheme of [7] is optimal among all schemes of probe complexity 3. Another question is to obtain strong lower bounds for the case of $k \geq 3$ queries or servers. For this case, no superpolynomial lower bounds are known even if the alphabet or answer size is only one bit.

References

1. A. Ambainis. Upper bound on communication complexity of private information retrieval. In *Proceedings of the 24th ICALP*, volume 1256 of *Lecture Notes in Computer Science*, pages 401–407, 1997.
2. A. Ambainis, A. Nayak, A. Ta-Shma, and U. Vazirani. Dense quantum coding and a lower bound for 1-way quantum automata. In *Proceedings of 31st ACM STOC*, pages 697–704, 1999.
3. L. Babai, L. Fortnow, L. Levin, and M. Szegedy. Checking computations in polylogarithmic time. In *Proceedings of 23rd ACM STOC*, pages 21–31, 1991.
4. A. Beimel and Y. Ishai. Information-theoretic private information retrieval: A unified construction. In *Proceedings of 28th ICALP*, pages 912–926, 2001.
5. A. Beimel, Y. Ishai, and E. Kushilevitz. General constructions for information-theoretical Private Information Retrieval. Manuscript, available on Amos Beimel's homepage. Includes [4], 2004.

6. A. Beimel, Y. Ishai, E. Kushilevitz, and J. Raymond. Breaking the $O(n^{1/(2k-1)})$ barrier for information-theoretic Private Information Retrieval. In *Proceedings of 43rd IEEE FOCS*, pages 261–270, 2002.
7. B. Chor, O. Goldreich, E. Kushilevitz, and M. Sudan. Private information retrieval. *Journal of the ACM*, 45(6):965–981, 1998. Earlier version in FOCS'95.
8. O. Goldreich, H. Karloff, L. Schulman, and L. Trevisan. Lower bounds for linear locally decodable codes and private information retrieval. In *Proceedings of 17th IEEE Conference on Computational Complexity*, pages 175–183, 2002.
9. R. A. Horn and C. R. Johnson. *Matrix Analysis*. Cambridge Univ. Press, 1985.
10. J. Katz and L. Trevisan. On the efficiency of local decoding procedures for error-correcting codes. In *Proceedings of 32nd ACM STOC*, pages 80–86, 2000.
11. I. Kerenidis. Quantum multiparty communication complexity and circuit lower bounds. Apr 12, 2005. quant-ph/0504087.
12. I. Kerenidis and R. de Wolf. Exponential lower bound for 2-query locally decodable codes via a quantum argument. *Journal of Computer and Systems Sciences*, 69(3):395–420, 2004. Earlier version in STOC'03. quant-ph/0208062.
13. E. Mann. Private access to distributed information. Master's thesis, Technion - Israel Institute of Technology, Haifa, 1998.
14. A. Nayak. Optimal lower bounds for quantum automata and random access codes. In *Proceedings of 40th IEEE FOCS*, pages 369–376, 1999. quant-ph/9904093.
15. M. A. Nielsen and I. L. Chuang. *Quantum Computation and Quantum Information*. Cambridge University Press, 2000.

Preservation Under Extensions on Well-Behaved Finite Structures

Albert Atserias[1,*], Anuj Dawar[2], and Martin Grohe[3]

[1] Universitat Politècnica de Catalunya, Barcelona, Spain
[2] University of Cambridge, Cambridge, UK
[3] Humboldt Universitat zu Berlin, Berlin, Germany

Abstract. A class of relational structures is said to have the extension preservation property if every first-order sentence that is preserved under extensions on the class is equivalent to an existential sentence. The class of all finite structures does not have the extension preservation property. We study the property on classes of finite structures that are better behaved. We show that the property holds of classes of acyclic structures, structures of bounded degree and more generally structures that are *wide* in a sense we make precise. We also show that the preservation property holds for the class of structures of treewidth at most k, for any k. In contrast, we show that the property fails for the class of planar graphs.

1 Introduction

The subject of model theory is concerned with the relationship between syntactic and semantic properties of logic. Among classical results in the subject are preservation theorems which relate syntactic restrictions on first-order logic with structural properties of the classes of structures defined. A key example is the Łoś-Tarski Theorem which asserts that a first-order formula is preserved under extensions on all structures if, and only if, it is logically equivalent to an existential formula (see [7]). One direction of this result is easy, namely that any formula that is purely existential is preserved under extensions, and this holds on any class of structures. The other direction, going from the semantic restriction to the syntactic restriction makes key use of the compactness of first-order logic and hence of infinite structures.

In the early development of finite-model theory, when it was realized that finite structures are the ones that are interesting from the point of view of studying computation, it was observed that most classical preservation theorems from model theory fail when only finite structures are allowed. In particular, the Łoś-Tarski theorem fails on finite structures [9, 6]. These results suggest that the class of finite structures is not well-behaved from the point of view of model theory. However, when one considers the computational structures that arise in practice and are used as interpretations for logical languages (for instance, program models interpreting specifications or databases interpreting queries), in many cases they are not only finite but satisfy other structural

---

[*] Supported in part by CICYT TIN2004-04343 and by the European Commission through the RTN COMBSTRU HPRN-CT2002-00278.

restrictions as well. This motivates the study, not just of the class of finite structures, but of well-behaved subclasses of this class.

There are certain structural restrictions that have proved especially useful from an algorithmic point of view. For instance, many intractable computational problems become tractable when restricted to planar graphs or structures of bounded treewidth [2]. This is also the case in relation to evaluation of logical formulas [5]. A common generalization of classes of bounded treewidth and planar graphs are classes of structures that exclude a minor which have also been extensively studied.

A study of preservation properties for such restricted classes of finite structures was initiated in [1]. There, the focus was on the homomorphism preservation theorem, whose status on the class of finite structures was open. It was shown that this preservation property holds on any class of structures of bounded degree, bounded treewidth or that excludes some minor (and has certain other closure properties). In the present paper, we investigate the Łoś-Tarski extension preservation property on these classes of finite structures. Note that the failure of the property on the class of all finite structures does not imply its failure on subclasses. If one considers the non-trivial direction of the preservation theorem on a class $\mathcal{C}$, it says that any sentence φ that is preserved under extensions *on* $\mathcal{C}$ is equivalent *on* $\mathcal{C}$ to an existential sentence. Thus, restricting to a subclass $\mathcal{C}'$ of $\mathcal{C}$ weakens both the hypothesis and the consequent of the statement.

We show that the extension preservation theorem holds on any class of finite structures closed under substructures and disjoint unions that is also *wide* in the sense that any sufficiently large structure in the class contains a large number of elements that are far apart. This includes, for instance, any class of structures of bounded degree. While classes of structures of bounded treewidth are not wide, they are nearly so in that they can be made wide by removing a small number of elements. We use this property and show that it implies the extension preservation theorem for the class $\mathcal{T}_k$—the class of structures of treewidth k or less (note this is not as general as saying that the property holds for all classes of bounded treewidth). Finally, we show that the construction does not extend to classes defined by excluded minors. Indeed, the extension preservation theorem does not hold for the class of planar graphs, as we show through a counterexample. This contrasts with the results obtained for the homomorphism preservation property in [1] as this property was shown to hold on all classes excluding a graph minor and closed under substructures and disjoint unions.

The main methodology in establishing the preservation property for a class of structures $\mathcal{C}$ is to show an upper bound on the size of a minimal model of a first-order sentence φ that is preserved under extensions on $\mathcal{C}$. The way we do this is to show that for any sufficiently large model $\mathbf{A}$ of φ, there is a proper substructure of $\mathbf{A}$ and an extension of $\mathbf{A}$ that cannot be distinguished by φ. In Section 3 we establish this for the relatively simple case of acyclic structures by means of a Hanf locality argument. Section 4 contains the main combinatorial argument for wide structures which uses Gaifman locality and an iterated construction of the substructure of $\mathbf{A}$. In Section 5, the combinatorial argument is adapted to the classes $\mathcal{T}_k$. Finally, in Section 6 we conclude and discuss the existence of a counterexample in the case of planar graphs. We begin in Section 2 with some background and definitions.

2 Preliminaries

Our notation and terminology is standard (see [3]).

Relational structures. A *relational vocabulary* σ is a finite set of *relation symbols*, each with a specified *arity*. A σ-*structure* $\mathbf{A}$ consists of a *universe* A, or *domain*, and an *interpretation* which associates to each relation symbol $R \in \sigma$ of some arity r, a relation $R^\mathbf{A} \subseteq A^r$. A *graph* is a structure $\mathbf{G} = (V, E)$, where E is a binary relation that is symmetric and anti-reflexive. Thus, our graphs are undirected, loopless, and without parallel edges.

A σ-structure $\mathbf{B}$ is called a *substructure* of $\mathbf{A}$ if $B \subseteq A$ and $R^\mathbf{B} \subseteq R^\mathbf{A}$ for every $R \in \sigma$. It is called an *induced substructure* if $R^\mathbf{B} = R^\mathbf{A} \cap B^r$ for every $R \in \sigma$ of arity r. Notice the analogy with the graph-theoretical concept of *subgraph* and *induced subgraph*. A substructure $\mathbf{B}$ of $\mathbf{A}$ is proper if $\mathbf{A} \neq \mathbf{B}$. If $\mathbf{A}$ is an induced substructure of $\mathbf{B}$, we say that $\mathbf{B}$ is an *extension* of $\mathbf{A}$. If $\mathbf{A}$ is a proper induced substructure, then $\mathbf{B}$ is a proper extension. If $\mathbf{B}$ is the disjoint union of $\mathbf{A}$ with another σ-structure, we say that $\mathbf{B}$ is a *disjoint extension* of $\mathbf{A}$. If $S \subseteq A$ is a subset of the universe of $\mathbf{A}$, then $\mathbf{A} \cap S$ denotes the *induced substructure generated by* S; in other words, the universe of $\mathbf{A} \cap S$ is S, and the interpretation in $\mathbf{A} \cap S$ of the r-ary relation symbol R is $R^\mathbf{A} \cap S^r$.

The *Gaifman graph* of a σ-structure $\mathbf{A}$, denoted by $\mathcal{G}(\mathbf{A})$, is the (undirected) graph whose set of nodes is the universe of $\mathbf{A}$, and whose set of edges consists of all pairs (a, a') of elements of A such that a and a' appear together in some tuple of a relation in $\mathbf{A}$. The *degree* of a structure is the degree of its Gaifman graph, that is, the maximum number of neighbors of nodes of the Gaifman graph.

Neighborhoods and treewidth. Let $\mathbf{G} = (V, E)$ be a graph. Moreover, let $u \in V$ be a node and let $d \geq 0$ be an integer. The d-*neighborhood* of u in $\mathbf{G}$, denoted by $N_d^\mathbf{G}(u)$, is defined inductively as follows:

1. $N_0^\mathbf{G}(u) = \{u\}$;
2. $N_{d+1}^\mathbf{G}(u) = N_d^\mathbf{G}(u) \cup \{v \in V : (v, w) \in E \text{ for some } w \in N_d^\mathbf{G}(u)\}$.

If $\mathbf{A}$ is a σ-structure, a is a point in $\mathbf{A}$, and $\mathbf{G}$ is the Gaifman graph of $\mathbf{A}$, we let $N_d^\mathbf{A}(a)$ denote the d-neighborhood of a in $\mathbf{G}$. Where it causes no confusion, we also write $N_d^\mathbf{A}(a)$ for the substructure of $\mathbf{A}$ generated by this set.

A *tree* is an acyclic connected graph. A *tree-decomposition* of $\mathbf{G}$ is a labeled tree $\mathbf{T}$ such that

1. each node of $\mathbf{T}$ is labeled by a non-empty subset of V;
2. for every edge $\{u, v\} \in E$, there is a node of $\mathbf{T}$ whose label contains $\{u, v\}$;
3. for every $u \in V$, the set X of nodes of $\mathbf{T}$ whose labels include u forms a connected subtree of $\mathbf{T}$.

The *width* of a tree-decomposition is the maximum cardinality of a label in $\mathbf{T}$ minus one. The *treewidth* of $\mathbf{G}$ is the smallest k for which $\mathbf{G}$ has a tree-decomposition of width k. The *treewidth* of a σ-structure is the treewidth of its Gaifman graph. Note that trees have treewidth one.

First-order logic, monadic second-order logic, and types. Let σ be a relational vocabulary. The *atomic formulas* of σ are those of the form $R(x_1,\ldots,x_r)$, where $R \in \sigma$ is a relation symbol of arity r, and $x_1,\ldots,x_r$ are first-order variables that are not necessarily distinct. Formulas of the form $x = y$ are also atomic.

The collection of *first-order formulas* is obtained by closing the atomic formulas under negation, conjunction, disjunction, universal and existential first-order quantification. The collection of *existential* first-order formulas is obtained by closing the atomic formulas and the negated atomic formulas under conjunction, disjunction, and existential quantification. The semantics of first-order logic is standard.

The collection of *monadic second-order formulas* is obtained by closing the atomic formulas under negation, conjunction, disjunction, universal and existential first-order quantification, and universal and existential second-order quantification over sets. The semantics of monadic second-order logic is also standard.

The quantifier rank of a formula, be it first-order or monadic second-order, is the maximum nesting of quantifiers of its subformulas.

Let $\mathbf{A}$ be a σ-structure, and let $a_1,\ldots,a_n$ be points in $\mathbf{A}$. If $\varphi(x_1,\ldots,x_n)$ is a formula with free variables $x_1,\ldots,x_n$, we use the notation $\mathbf{A} \models \varphi(a_1,\ldots,a_n)$ to denote the fact that φ is true in $\mathbf{A}$ when x_i is interpreted by a_i. If m is an integer, the first-order m-type of $a_1,\ldots,a_n$ in $\mathbf{A}$ is the collection of all first-order formulas $\varphi(x_1,\ldots,x_n)$ of quantifier rank at most m, up to logical equivalence, for which $\mathbf{A} \models \varphi(a_1,\ldots,a_n)$. The monadic second-order m-type of $a_1,\ldots,a_n$ in $\mathbf{A}$ is defined analogously.

Preservation under extensions and minimal models. Let $\mathcal{C}$ be a class of finite σ-structures that is closed under induced substructures. Let φ be a first-order sentence. We say that φ is *preserved under extensions* on $\mathcal{C}$ if whenever $\mathbf{A}$ and $\mathbf{B}$ are structures in $\mathcal{C}$ such that $\mathbf{B}$ is an extension of $\mathbf{A}$, then $\mathbf{A} \models \varphi$ implies $\mathbf{B} \models \varphi$. We say that $\mathbf{A}$ is a *minimal model* of φ if $\mathbf{A} \models \varphi$ and every proper induced substructure $\mathbf{A}'$ of $\mathbf{A}$ is such that $\mathbf{A}' \not\models \varphi$. The following Lemma states that the existential sentences are precisely those that have finitely many minimal models. Its proof is part of the folklore:

Lemma 1. *Let $\mathcal{C}$ be a class of finite σ-structures that is closed under induced substructures. Let φ be a first-order sentence. Then, the following are equivalent:*

1. *φ is equivalent on $\mathcal{C}$ to an existential sentence,*
2. *φ has finitely many minimal models in $\mathcal{C}$.*

In the rest of the paper, we use several times the implication from 2. to 1. Just for completeness, this is proved by taking the disjunction of the existential closure of the atomic types of the finitely many minimal models.

3 Acyclic Structures

We begin with the simple case of acyclic structures, by which we mean structures whose Gaifman graph is acyclic. We show that any class of such structures satisfying certain closure properties admits the extension preservation property. Note, for structures whose Gaifman graphs are acyclic, there is no loss of generality in assuming that the vocabulary σ consists of unary and binary relations only.

The proof is based on Hanf locality and will be presented in detail in the journal version of the paper. Here we only present an overview. We begin by considering structures whose Gaifman graph consists of a simple path. In a large enough structure $\mathbf{A}$ of this type, we can find a segment $\mathbf{C}$ of the path such that the structure $\mathbf{A}'$ obtained by removing $\mathbf{C}$ and the structure $\mathbf{B} = \mathbf{A} \oplus \mathbf{C}$ are equivalent under $\equiv^m$.

Lemma 2. *For every vocabulary σ and every $m > 0$ there is a p such that if $\mathbf{A}$ is a σ-structure whose Gaifman graph is connected, acyclic and of degree at most 2 and $|A| > p$, then there is a disjoint extension $\mathbf{B}$ of $\mathbf{A}$ and a proper substructure $\mathbf{A}'$ of $\mathbf{A}$ such that $\mathbf{A}' \equiv^m \mathbf{B}$.*

A reduction extends this result to acyclic structures without a degree bound.

Lemma 3. *For every vocabulary σ and every $m > 0$ there is a p such that if $\mathbf{A}$ is a structure whose Gaifman graph is connected and acyclic and which contains a path with more than p elements, then there is a disjoint extension $\mathbf{B}$ of $\mathbf{A}$ and a proper substructure $\mathbf{A}'$ of $\mathbf{A}$ such that $\mathbf{A}' \equiv^m \mathbf{B}$.*

This lemma covers the key case in the proof of the main theorem of this section.

Theorem 4. *Let $\mathcal{C}$ be a class of acyclic finite structures, closed under substructures and disjoint unions. Then, on $\mathcal{C}$, every first-order sentence that is preserved under extensions is equivalent to an existential sentence.*

4 Wide Structures

This section will focus on structures that are *wide*, meaning that they contain many points that are pairwise far apart from each other. The goal is to show that the preservation under extensions property holds for classes of such structures. The precise definition will be introduced later on in this section. At this point let us just mention that an interesting particular case is the class of all structures of degree bounded by a constant.

Unfortunately, the techniques and arguments of Section 3 based on Hanf locality will not be enough for our current purpose. Instead, we will have to resort to Gaifman locality, for which we provide the necessary background first.

For every integer $r \geq 0$, let $\delta(x, y) \leq r$ denote the first-order formula expressing that the distance between x and y in the Gaifman graph is at most r. Let $\delta(x, y) > r$ denote the negation of this formula. A *basic local sentence* is a sentence of the form

$$(\exists x_1) \cdots (\exists x_n) \left(\bigwedge_{i \neq j} \delta(x_i, x_j) > 2r \wedge \bigwedge_i \psi^{N_r(x_i)}(x_i) \right), \qquad (1)$$

where ψ is a first-order formula with one free variable. Here, $\psi^{N_r(x_i)}(x_i)$ stands for the relativization of ψ to $N_r(x_i)$; that is, the subformulas of ψ of the form $(\exists x)(\theta)$ are replaced by $(\exists x)(\delta(x, x_i) \leq r \wedge \theta)$, and the subformulas of the form $(\forall x)(\theta)$ are replaced by $(\forall x)(\delta(x, x_i) \leq r \rightarrow \theta)$. The *locality radius* of a basic local sentence is r. Its *width* is n. Its *local quantifier rank* is the quantifier rank of ψ. We will use the fact that basic local sentences are preserved under disjoint extensions. Note, however, that they may not be preserved under plain extensions since in that case the neighborhoods can grow.

The main result about basic local sentences is that they form a building block for first-order logic. This is known as Gaifman's Theorem (for a proof, see, for example, [3–Theo. 2.5.1]):

Theorem 5 (Gaifman Locality). *Every first-order sentence is equivalent to a Boolean combination of basic local sentences.*

We reach now the main technical part of the paper. Let us say that a set of points B in a σ-structure $\mathbf{A}$ is d-scattered if for every pair of distinct $a, b \in B$ we have $N_d^{\mathbf{A}}(a) \cap N_d^{\mathbf{A}}(b) = \emptyset$. We say that a class of finite σ-structures $\mathcal{C}$ is *wide* if for every d and m there exists an N such that every structure in $\mathcal{C}$ of size at least N contains a d-scattered set of size m.

Theorem 6. *Let $\mathcal{C}$ be a class of finite σ-structures that is wide and closed under substructures and disjoint unions. Then, on $\mathcal{C}$, every first-order sentence that is preserved under extensions is equivalent to an existential sentence.*

Proof. Let φ be a first-order sentence that is preserved under extensions on $\mathcal{C}$. By Gaifman's Theorem we may assume that $\varphi = \bigvee_{i \in I} \tau_i$, with

$$\tau_i = \bigwedge_{j \in J_i} \theta_j^i \wedge \bigwedge_{k \in K_i} \neg \theta_k^i, \tag{2}$$

where each θ_h^i is a basic local sentence. Now we define a list of parameters that we need in the proof (the reader may skip this list now and use it to look up the values when they are needed):

- r is the maximum of the locality radii of all θ_h^i,
- s is the sum of all widths of all θ_h^i,
- m is the maximum of the local quantifier ranks of all θ_h^i,
- ℓ is the number of disjuncts in φ, so $\ell = |I|$.
- $n = (\ell + 2)s$,
- $M = m + 4r + 1$,
- $d = 2(r + 1)(\ell + 1)s + 6r + 2$,
- q is the number of monadic second-order M-types with one free variable,
- N is such that every structure in $\mathcal{C}$ of size at least N contains a $(4dq + 2r + 1)$-scattered set of size $(n - 1)q + s + \ell s + 1$.

Our goal is to show that the minimal models of φ have size less than N. Suppose on the contrary that $\mathbf{A}$ is a minimal model of φ of size at least N. The *type* of a point $a \in A$ is the monadic second-order M-type of a in $\mathbf{A} \cap N_d(a)$. We say that a *realizes* its type. The reason we consider monadic second-order types, instead of first-order types, will become clear later in the proof. Let $t_1, \ldots, t_q$ be all possible types. We need a couple of definitions. Let C be a subset of A and t a type. We say that t is *covered by* C if for all realizations a of t we have $N_d(a) \subseteq C$. We say that t is *free over* C if there are at least n realizations $a_1, \ldots, a_n$ of t such that $N_d(a_i)$ and $N_d(a_j)$ are pairwise disjoint and do not intersect C.

Claim 1. *There exists a radius $e \leq 2dq$ and a set D of at most $(n-1)q$ points in A such that all types are either covered by $N_e(D)$ or free over $N_e(D)$.*

Proof. We define D and e inductively. Let $D_0 = \emptyset$ and $e_0 = 0$. Suppose now that D_i and e_i are already defined. Let $C = N_{e_i}(D_i)$. If all types are either covered by C or free over C, then let $D = D_i$ and $e = e_i$. Otherwise, let j be minimal such that type t_j is neither covered by C nor free over C. We define a set E inductively as follows. Let $E_0 = \emptyset$. Suppose now that E_t is already defined. If there is no realization of t_j outside $N_{2d}(C \cup E_t)$, then let $E = E_t$ and we are done with the construction of E. Otherwise, let a_{t+1} be a realization of t_j outside $N_{2d}(C \cup E_t)$ and let $E_{t+1} = E_t \cup \{a_{t+1}\}$. Note that this iteration cannot continue beyond $n-1$ steps since otherwise t_j would be free over C. This means that the iteration stops, and when it does $|E| \leq n-1$ and t_j is covered by any set that contains $N_{2d}(C \cup E)$, and in particular by $N_{e_i+2d}(D_i \cup E)$. Let $D_{i+1} = D_i \cup E$ and $e_{i+1} = e_i + 2d$. The construction stops after at most q steps because at each step one new type is covered and remains covered for the rest of the construction. This shows that $|D| \leq (n-1)q$ and $e \leq 2dq$, which proves the claim. □

In the following, we fix e and D according to Claim 1. We say that a type t is *frequent* if it is not covered by $N_e(D)$. Otherwise we say that t is *rare*.

We shall build a finite sequence of sets $S_0 \subseteq S_1 \subseteq \ldots \subseteq S_p \subseteq A$, with $p \leq \ell$, so that the last set S_p in the sequence will be such that the substructure of $\mathbf{A}$ induced by S_p is a proper substructure of $\mathbf{A}$ that satisfies φ. This will contradict the minimality of $\mathbf{A}$ and will prove the theorem. The sequence S_i is constructed inductively together with a second sequence of sets $C_0 \subseteq C_1 \subseteq \cdots \subseteq C_p \subseteq A$ called the *centers*, and a sequence of sets of indices $I_0 \subseteq I_1 \subseteq \cdots \subseteq I_p \subseteq I$ (recall that φ is the disjunction of the formulas τ_i from (2) for $i \in I$). Moreover, the following conditions will be preserved by the inductive construction for every $i < p$.

(a) The type of each $a \in C_i$ is frequent.
(b) $S_i \subseteq N_r(C_i)$.
(c) $|C_i| \leq is$.
(d) No disjoint extension of $\mathbf{A} \cap S_i$ satisfies $\bigvee_{j \in I_i} \tau_j$.
(e) $N_e(D)$ and $N_d(C_i)$ are disjoint.
(f) $|I_i| = i$.

Let $S_0 = C_0 = I_0 = \emptyset$, and let us assume that S_i, C_i and I_i have already been defined with the properties above. We construct S_{i+1}, C_{i+1}, and I_{i+1}. Let $\mathbf{B}$ be the disjoint union of $\mathbf{A}$ with a copy of $\mathbf{A} \cap S_i$.

$$\text{Since } \mathbf{B} \text{ is an extension of } \mathbf{A}, \text{ it satisfies } \varphi. \tag{3}$$

Therefore, there exists an $i' \in I$ such that $\mathbf{B}$ satisfies $\tau_{i'}$. By (d), since the extension is disjoint, we know that $i' \notin I_i$. Let $I_{i+1} = I_i \cup \{i'\}$. For the rest of the proof, the index i' will be fixed so we drop any reference to it. For example, we will write τ instead of $\tau_{i'}$ and θ_h instead of $\theta_h^{i'}$. Recall that

$$\tau = \bigwedge_{j \in J} \theta_j \wedge \bigwedge_{k \in K} \neg \theta_k.$$

Since **B** satisfies τ, in particular it satisfies the positive requirements: $\mathbf{B} \models \bigwedge_{j \in J} \theta_j$. Let W_j be a minimal set of witnesses in **B** for the outermost existential quantifiers in θ_j, and let $W = \bigcup_{j \in J} W_j$. We have $|W| \leq s$. Some of these witnesses may be in **A** and some may be in the disjoint copy of $\mathbf{A} \cap S_i$ in **B**. Let $W_A \cup W_B = W$ be such a partition, with W_A being the witnesses in A. The following claim shows that W_A can be chosen far from C_i. This will be needed later.

Claim 2. *There is a set W of witnesses such that $N_{r+1}(C_i) \cap N_r(W_A) = \emptyset$.*

Proof. Fix a set W of witnesses so that the number of points b in W_A for which $N_{r+1}(C_i)$ and $N_r(b)$ are not disjoint is minimal. Suppose that this number is not zero, and let $b \in W_A$ with $N_{r+1}(C_i) \cap N_r(b) \neq \emptyset$. Let $a \in C_i$ be such that $N_{r+1}(a) \cap N_r(b) \neq \emptyset$. Then $N_r(b) \subseteq N_{3r+1}(a) \subseteq N_d(a)$. By property (a), the type t of a is frequent. So let a' be a realization of t such that $N_{r+1}(W \cup C_i)$ and $N_{3r+1}(a')$ are disjoint. Such an a' exists because t is frequent and thus, by Claim 1, is free over $N_e(D)$ and thus has

$$n > (\ell + 1)s = |W \cup C_i|$$

realizations whose d-neighborhoods are pairwise disjoint and disjoint from $N_e(D)$.

Now, since $N_d(a)$ and $N_d(a')$ have the same monadic second-order M-type and $N_r(b) \subseteq N_{3r+1}(a) \subseteq N_d(a)$, there must exist a b' such that $N_r(b') \subseteq N_{3r+1}(a') \subseteq N_d(a')$ such that b and b' have the same first-order m-type on $\mathbf{A} \cap N_r(b)$ and $\mathbf{A} \cap N_r(b')$ respectively. Here we use $M = m + 4r + 1$ and the fact that we considered second-order types; this has the effect that every subset of $N_d(a)$ has an *equivalent* subset in $N_d(a')$. But then b' can replace b as a witness in W_A, and since $N_{r+1}(W \cup C_i)$ and $N_{3r+1}(a')$ are disjoint, so are $N_{r+1}(C_i)$ and $N_r(b')$. This contradicts the minimality of W. □

In the following, we fix a set W of witnesses such that $N_{r+1}(C_i) \cap N_r(W_A) = \emptyset$. We let **C** be the substructure of **A** induced by $N_e(D) \cup N_r(W_A) \cup S_i$. We claim that **C** satisfies the positive requirements of τ:

Claim 3. *$\mathbf{C}$ is a substructure of $\mathbf{A}$ such that $\mathbf{C} \models \bigwedge_{j \in J} \theta_j$.*

Proof. It is obvious that **C** is a substructure of **A**. The point, however, is that **C** is in fact the disjoint union of the substructure induced by $N_e(D) \cup N_r(W_A)$ and the substructure induced by S_i. This is because $S_i \subseteq N_r(C_i)$ and $N_{r+1}(C_i)$ is disjoint from $N_e(D)$ by property (e) and also disjoint from $N_r(W_A)$ by Claim 2. It follows that the witnesses from **B** in W_B can also be found in **C**. Obviously, also the witnesses from **B** in W_A can be found in **C**. This proves that **C** satisfies the positive requirements of τ. □

Consider φ on **C**. If **C** is a model of φ, let $S_p = N_e(D) \cup N_r(W_A) \cup S_i$ and we are done. Notice that **C** is a proper substructure of **A** because **A** contains $(n-1)q + s + \ell s + 1$ points that are $(4dq + 2r + 1)$-scattered, but $S_p \subseteq N_{2dq+r}(D \cup W_A \cup C_i)$ and

$$|D \cup W_A \cup C_i| \leq (n-1)q + s + \ell s.$$

If $\mathbf{C}$ is not a model of φ it cannot satisfy τ. However, by Claim 3, $\mathbf{C}$ satisfies the positive requirements $\bigwedge_{j \in J} \theta_j$. Therefore, $\mathbf{C}$ does not satisfy $\bigwedge_{k \in K} \neg \theta_k$. Let $k \in K$ such that $\mathbf{C} \models \theta_k$. In the next claim we find a substructure of $\mathbf{A}$ that extends $\mathbf{A} \cap S_i$ and forces all its disjoint extensions to satisfy θ_k.

Claim 4. *There exist $C_{i+1} \supseteq C_i$ and $S_{i+1} \supseteq S_i$ as required by conditions (a)–(e).*

Proof. We have $\mathbf{C} \models \theta_k$. Let V be a minimal set of witnesses in $\mathbf{C}$ for the outermost existential quantifiers in θ_k. We have $|V| \leq s$.

Necessarily, the type t of some $a \in V$ is frequent. Otherwise $N_r(V) \subseteq N_e(D) \subseteq A$, so $\mathbf{A} \models \theta_k$, and thus $\mathbf{B} \models \theta_k$, because $\mathbf{B}$ is a disjoint extension of $\mathbf{A}$. However, this is impossible because $\mathbf{B} \not\models \tau$.

So let $a \in V$ have frequent type t. Let Z be a set of s realizations of t such that

(i) $N_d(b) \cap N_d(b') = \emptyset$ for every pair of distinct $b, b' \in Z$,
(ii) $N_e(D) \cap N_d(Z) = \emptyset$,
(iii) $N_{r+1}(C_i) \cap N_r(Z) = \emptyset$.

Such a set Z exists because t is frequent, $n = (\ell + 2)s$, and $|C_i| \leq \ell s$ by property (c).

Now, let $F = N_r(a) \cap C$ be the subset of $N_r(a)$ that actually appears in $\mathbf{C}$. Note that $a \in F$. Since each $b \in Z$ has the same monadic second-order M-type as a, for every $N_r(b)$ there exists a set $F_b \subseteq N_r(b)$, equivalent to F in $N_r(a)$, so that a and b have the same first-order m-type in $\mathbf{A} \cap F$ and $\mathbf{A} \cap F_b$ respectively. This is the other place we use monadic second-order types. Define $C_{i+1} = C_i \cup Z$ and

$$S_{i+1} = S_i \cup \bigcup_{b \in Z} F_b.$$

Let us prove that C_{i+1} and S_{i+1} satisfy the properties (a), (b), (c), (d), and (e). Property (a) is clear since the type of each $b \in Z$ is t, and t is frequent. Property (b) is also clear since $F_b \subseteq N_r(b)$. For property (c) we have $|C_{i+1}| = |C_i| + s = (i+1)s$. Property (e) is satisfied by (ii) in our choice of Z.

Finally, for property (d) we argue as follows. First note that $\mathbf{A} \cap S_{i+1}$ is a disjoint extension of $\mathbf{A} \cap S_i$ because $N_{r+1}(C_i) \cap N_r(Z) = \emptyset$ by (iii) and $S_i \subseteq N_r(C_i)$ by (c). Therefore, no disjoint extension of $\mathbf{A} \cap S_{i+1}$ satisfies τ_j for any $j \in I_i$. It remains to show that no disjoint extension of $\mathbf{A} \cap S_{i+1}$ satisfies τ. However, this is clear from the construction because every disjoint extension of $\mathbf{A} \cap S_{i+1}$ will contain witnesses for the outermost existential quantifiers in θ_k; namely any subset of Z of size $|V|$. □

Note that I_{i+1} is constructed to satisfy property (f) as well. This completes the definition of the inductive construction. All it remains to see is that the construction stops in at most ℓ steps. Because suppose for contradiction that we have constructed S_ℓ, C_ℓ, and I_ℓ satisfying (a)–(f). Then $I_\ell = I$ by (f), and by (d), no disjoint extension of $\mathbf{A} \cap S_\ell$ satisfies $\varphi = \bigvee_{i \in I} \tau_i$. However,

the disjoint union $\mathbf{B}$ of $\mathbf{A} \cap S_\ell$ with $\mathbf{A}$ is an extension of $\mathbf{A}$ and hence (4) does satisfy φ.

This is a contradiction. □

As a direct application of Theorem 6, let us consider the class $\mathcal{D}_r$ of all finite σ-structures of degree bounded by r. Clearly, this class is both wide and closed under substructures and disjoint unions.

Theorem 7. *Let r be an integer. Then, on $\mathcal{D}_r$, every first-order sentence that is preserved under extensions is equivalent to an existential sentence.*

In the following section we show how the argument of Theorem 6 can be extended, in some cases, to classes of structures that are *almost wide*.

5 Bounded Treewidth Structures

The goal of this section is to show that the preservation under extensions property holds for the class $\mathcal{T}_k$ of all finite σ-structures of treewidth less than k. In other words, we want to prove the following result:

Theorem 8. *Let k be an integer. Then, on $\mathcal{T}_k$, every first-order sentence that is preserved under extensions is equivalent to an existential sentence.*

The proof of this result requires three ingredients. The first ingredient is a generalization of the disjoint union operation on structures by allowing some non-empty intersection. Let $\mathbf{A}$ and $\mathbf{B}$ be σ-structures, and let $C \subseteq A \cap B$ be such that $\mathbf{A} \cap C = \mathbf{B} \cap C$. The *union of $\mathbf{A}$ and $\mathbf{B}$ through C*, denoted by $\mathbf{A} \oplus_C \mathbf{B}$ is a new σ-structure defined as follows. The universe of $\mathbf{D} = \mathbf{A} \oplus_C \mathbf{B}$ is $A' \cup B' \cup C$, where A' is a disjoint copy of $A - C$ and B' is a disjoint copy of $B - C$. The relations of $\mathbf{D}$ are defined in the obvious way: If $a_1,\ldots,a_r$ are points in A and $a'_1,\ldots,a'_r$ are the corresponding points in $A' \cup C$, then $(a'_1,\ldots,a'_r) \in R^{\mathbf{D}}$ if and only if $(a_1,\ldots,a_r) \in R^{\mathbf{A}}$. Similarly, if $b_1,\ldots,b_r$ are points in B and $b'_1,\ldots,b'_r$ are the corresponding points in $B' \cup C$, then $(b'_1,\ldots,b'_r) \in R^{\mathbf{D}}$ if and only if $(b_1,\ldots,b_r) \in R^{\mathbf{B}}$. Observe that this construction is precisely the disjoint union of $\mathbf{A}$ and $\mathbf{B}$ when $C = \emptyset$.

The next Lemma is a straightforward generalization of the obvious fact that $\mathcal{T}_k$ is closed under disjoint unions. The Lemma states, roughly, that $\mathcal{T}_k$ is closed under unions through *subsets of bags of tree-decompositions*. More precisely:

Lemma 9. *Let k be an integer. Let $\mathbf{A}$ and $\mathbf{B}$ be two σ-structures, let $C \subseteq A \cap B$ be such that $\mathbf{A} \cap C = \mathbf{B} \cap C$, and let (T, L) and (T', L') be tree-decompositions of width k of $\mathbf{A}$ and $\mathbf{B}$, respectively. Then, if there exists nodes $u \in T$ and $u' \in T'$ such that $C \subseteq L(u) \cap L'(u')$, then the union of $\mathbf{A}$ and $\mathbf{B}$ through C has treewidth at most k.*

Proof: The tree-decomposition of the union is $(T'', L \cup L')$, where $T'' = T \cup T'$ with a new tree edge joining u and u'. □

The second ingredient is the fact that the class of structures of treewidth less than k is *almost wide*, in the sense that there exists a small set of vertices whose removal produces a large scattered set. Such a set is henceforth called a *bottleneck*. This was proved in [1] but here we state the stronger claim that the bottleneck can be found in the bag of a tree-decomposition. The proof is the same as in [1].

Lemma 10. *For every k, and for every d and m, there exists an N such that if $\mathbf{A}$ is a σ-structure of size at least N and (T, L) is a tree-decomposition of $\mathbf{A}$ of width k, then there exist $u \in T$ and $B \subseteq L(u)$ such that $\mathbf{A} - B$ contains a d-scattered set of size m.*

The third ingredient in the proof is a first-order interpretation of a wide structure inside an almost-wide structure. From now on we focus on graphs; the construction extends easily to the general case. Let $P_1, \ldots, P_k, Q_1, \ldots, Q_k$ be unary relation symbols and $\sigma = \{E, P_1, \ldots, P_k, Q_1, \ldots, Q_k\}$. For every graph $\mathbf{G} = (V, E^{\mathbf{G}})$ and every tuple $\mathbf{a} = (a_1, \ldots, a_k) \in V^k$ we define a σ-structure $\mathbf{A} = \mathbf{A}(\mathbf{G}, \mathbf{a})$ as follows:

1. $A = V$,
2. $E^{\mathbf{A}} = E^{\mathbf{G}} - \{(a, b) \in A^2 : \{a, b\} \cap \{a_1, \ldots, a_k\} \neq \emptyset\}$,
3. $P_i^{\mathbf{A}} = \{a_i\}$,
4. $Q_i^{\mathbf{A}} = \{b \in A : (a_i, b) \in E^{\mathbf{A}}\}$.

Let us call a σ-structure $\mathbf{A}$ *derived* if $E^{\mathbf{A}}$ is a symmetric and anti-reflexive binary relation, and there are elements $a_1, \ldots, a_k \in A$ such that $P_i^{\mathbf{A}} = \{a_i\}$ for $1 \leq i \leq k$ and a_i is isolated in the graph underlying $\mathbf{A}$; that is, for $1 \leq i \leq k$ there is no b such that $(a_i, b) \in E^{\mathbf{A}}$. Note that for every derived structure $\mathbf{A}$ there is a unique graph $\mathbf{G}(\mathbf{A})$ and a unique k-tuple $\mathbf{a}(\mathbf{A})$ of vertices of $\mathbf{G}(\mathbf{A})$ such that

$$\mathbf{A} = \mathbf{A}(\mathbf{G}(\mathbf{A}), \mathbf{a}(\mathbf{A})).$$

The point behind the construction of $\mathbf{A} = \mathbf{A}(\mathbf{G}, \mathbf{a})$ is that if $B = \{a_1, \ldots, a_k\}$ is a bottleneck of $\mathbf{G}$ in the sense that $\mathbf{G} - B$ contains a large scattered set, then $\mathbf{A}$ itself has a large scattered set and maintains all the information to reconstruct $\mathbf{G}$. Indeed, $\mathbf{G}(\mathbf{A})$ is first-order interpretable in $\mathbf{A}$, and thus we get the following lemma:

Lemma 11. *For every first-order sentence φ of vocabulary $\{E\}$ there is a sentence $\tilde{\varphi}$ of vocabulary σ such that for all σ-structures $\mathbf{A}$ we have:*

1. *If $\mathbf{A} \models \tilde{\varphi}$ then $\mathbf{A}$ is derived.*
2. *If $\mathbf{A}$ is derived, then $\mathbf{A} \models \tilde{\varphi}$ if and only if $\mathbf{G}(\mathbf{A}) \models \varphi$.*

This follows at once from a standard result on syntactical interpretations (cf., for example, Theorem VIII.2.2 of [4]).

Equipped with these three ingredients we are ready for the main argument.

Proof (of Theorem 8). Let φ be a first-order sentence that is preserved under extensions in $\mathcal{T}_k$. It suffices to show that φ has finitely many minimal models. Let $\mathbf{G} = (V, E^{\mathbf{G}})$ be a graph in $\mathcal{T}_k$ that is a minimal model of φ. Suppose for contradiction that $\mathbf{G}$ is very large. Let (T, L) be a tree-decomposition of width k of $\mathbf{G}$, and let $B = \{b_1, \ldots, b_k\} \subseteq V$ be a bottleneck; that is, a set such that $\mathbf{G} - B$ contains a large scattered set. By Lemma 10 we may assume that $B \subseteq L(u)$ for some $u \in T$. Let $\mathbf{A} = \mathbf{A}(\mathbf{G}, \mathbf{b})$, where $\mathbf{b} = (b_1, \ldots, b_k)$. The idea is to work with $\mathbf{A}$ and $\tilde{\varphi}$ instead of $\mathbf{G}$ and φ and proceed as in the proof of Section 4. The difference is that $\tilde{\varphi}$ is *not* preserved under extensions. However, preservation under extensions is used only twice in the proof of

Section 4 (in (3) and (4)), both times to prove that the disjoint union **B** of the structure **A** with $\mathbf{A} \cap S_i$ is a model of φ. Claim 5 shows that in both cases, **B** is a model of $\tilde{\varphi}$.

Claim 5. *Let $C \subseteq A$ such that the type of each $a \in C$ is frequent. Let $S \subseteq N_r(C)$ and let **B** be the disjoint union of **A** with a disjoint copy of $\mathbf{A} \cap S$. Then **B** is derived, **G** is an induced subgraph of $\mathbf{G}(\mathbf{B})$, and $\mathbf{G}(\mathbf{B})$ belongs to $\mathcal{T}_k$.*

Proof. Note that **B** is derived because the bottleneck points are not in S since their type is not frequent. Let $\mathbf{H} = \mathbf{G}(\mathbf{B})$. Clearly, **G** is an induced subgraph of **H**. Thus all we have to prove is that **H** belongs to $\mathcal{T}_k$. Let $\mathbf{A}' = \mathbf{A} \cap (S \cup B)$, where B is the bottleneck of **G**. Again, $\mathbf{A}'$ is derived. Let $\mathbf{G}' = \mathbf{G}(\mathbf{A}')$. Clearly, $\mathbf{G}'$ is an induced subgraph of **G**. In particular, $\mathbf{G}'$ is in $\mathcal{T}_k$ so it has a tree-decomposition of width k. More importantly, since $B \subseteq L(u)$, we can assume as well that B is a subset of some bag of the tree-decomposition of $\mathbf{G}'$. These two facts together imply that the union of **G** and $\mathbf{G}'$ through B, which is precisely **H**, is in $\mathcal{T}_k$ by Lemma 9. □

This shows then that the **B** in (3) and (4) is a model of $\tilde{\varphi}$. The proof proceeds until we construct a structure **C** that satisfies $\tilde{\varphi}$ and is a proper substructure of **A**. We claim that **C** is derived. This is because all bottleneck points have rare type, so they belong to D. Let $\mathbf{H} = \mathbf{G}(\mathbf{C})$. Note now that **H** is the union of two subgraphs $\mathbf{G}_1$ and $\mathbf{G}_2$ of **G** through the bottleneck B. Again B is a subset of a bag of the tree-decompositions of $\mathbf{G}_1$ and $\mathbf{G}_2$, so **H** belongs to $\mathcal{T}_k$ by Lemma 9. Moreover **H** is a proper induced subgraph of **G** and $\mathbf{H} \models \varphi$ by Lemma 11. This contradicts the minimality of **G**, which concludes the proof. □

6 Conclusions

The proof that the extension preservation theorem holds on the classes $\mathcal{T}_k$ is based on somewhat different properties to the proof in [1] that the homomorphism preservation theorem holds on all classes of bounded treewidth. Indeed, our proof requires stronger closure conditions. A natural question is whether the result could extend, as in the case of the homomorphism preservation theorem to all classes that are defined by excluding a graph minor. It turns out this is not the case. We are able to construct a counterexample that shows that the extension preservation does not hold on the class of planar graphs. This construction will be presented in the journal version of the paper.

References

1. A. Atserias, A. Dawar, and Ph. G. Kolaitis. On preservation under homomorphisms and unions of conjunctive queries. In *Proc. 23rd ACM Symp. on Principles of Database Systems*, pages 319–329, 2004.
2. R.G. Downey and M.R. Fellows. *Parametrized Complexity*. Springer-Verlag, 1999.
3. H-D. Ebbinghaus and J. Flum. *Finite Model Theory*. Springer, 2 edition, 1999.
4. H-D. Ebbinghaus, J. Flum, and W. Thomas. *Mathematical Logic*. Springer, 2nd edition, 1994.

5. M. Grohe, J. Flum, and M Frick. Query evaluation via tree-decompositions. *Journal of the ACM*, 49:716–752, 2002.
6. Y. Gurevich. Toward logic tailored for computational complexity. In M. Richter et al., editors, *Computation and Proof Theory*, pages 175–216. Springer Lecture Notes in Mathematics, 1984.
7. W. Hodges. *Model Theory*. Cambridge University Press, 1993.
8. L. Libkin. *Elements of Finite Model Theory*. Springer, 2004.
9. W. W. Tait. A counterexample to a conjecture of Scott and Suppes. *Journal of Symbolic Logic*, 24:15–16, 1959.

Unsafe Grammars and Panic Automata

Teodor Knapik[1], Damian Niwiński[2,*],
Paweł Urzyczyn[2,**], and Igor Walukiewicz[3,***]

[1] Université de la Nouvelle Calédonie
knapik@univ-nc.nc
[2] Institute of Informatics, Warsaw University
{niwinski, urzy}@mimuw.edu.pl
[3] CNRS LaBRI, Université Bordeaux-1
igw@labri.fr

Abstract. We show that the problem of checking if an infinite tree generated by a higher-order grammar of level 2 (hyperalgebraic) satisfies a given μ-calculus formula (or, equivalently, if it is accepted by an alternating parity automaton) is decidable, actually 2-EXPTIME-complete. Consequently, the monadic second-order theory of any hyperalgebraic tree is decidable, so that the safety restriction can be removed from our previous decidability result. The last result has been independently obtained by Aehlig, de Miranda and Ong. Our proof goes *via* a characterization of possibly unsafe second-order grammars by a new variant of higher-order pushdown automata, which we call *panic automata*. In addition to the standard pop_1 and pop_2 operations, these automata have an option of a destructive move called *panic*. The model-checking problem is then reduced to the problem of deciding the winner in a parity game over a suitable 2nd order pushdown system.

1 Introduction

Context-free tree grammars constitute the basic level in an infinite hierarchy of higher-order grammars introduced by W. Damm [8] (built on the earlier ideas of [11]). Courcelle [6] proved decidability of the monadic second-order (MSO) theory of any tree generated by an *algebraic* (context-free) tree grammar. Later Knapik *et al* [13, 14] attempted to extend this decidability result to all levels of the Damm hierarchy. This has been achieved partially, namely with an additional syntactic restriction imposed on the grammars, called *safety*: the MSO theory of any tree generated by a safe grammar of level n is decidable.

Higher-order grammars can be seen as program schemes, where functions can take higher-order arguments. The tree generated by such a grammar describes completely the semantics of the program scheme. Thus decidability of

---

* Partly supported by KBN Grant 4 T11C 042 25.
** Partly supported by KBN Grant 3 T11C 002 27.
*** The 2nd and the 4th author were also supported by the EC Research Training Network *Games*.

the MSO theory of such a tree implies decidability for a large class of properties of behaviours of higher-order program schemes.

The safety requirement, roughly speaking, prevents the use of individual parameters in the functional arguments of higher-order functions. The concept of safe (tree) grammars has been further justified by a characterization in terms of higher-order pushdown automata originally introduced by Maslov [16]: the trees generated by safe higher-order grammars of level n coincide with the trees recognized (in suitable sense) by pushdown automata of level n [14]. See [5, 7] for another characterization of this hierarchy.

Monadic second-order logic is extremely succinct, which yields the non-elementary complexity of the MSO theories, even for regular trees. However, over trees, the MSO logic has the same expressive power as the *µ-calculus*, which is much better tractable algorithmically. Cachat [3] showed that the model-checking problem for the µ-calculus and the trees recognized by the pushdown automata of level n is in n-EXPTIME, actually n-EXPTIME-complete [4].

In this paper we show that for the level 2 (hyperalgebraic), the model checking problem remains decidable, actually 2-EXPTIME-complete, also for grammars *without* safety restriction.

To this end, we first find an automata-theoretic counterpart of unsafe hyperalgebraic grammars, which is an extension of second-order pushdown automata by a new destructive operation that we call *panic*. This operation allows us to simulate the change of environment needed to evaluate parameters of unsafe productions. We further introduce 2nd order pushdown systems with panic (equipped with alternation and ranks), which can be viewed as a generalization of pushdown systems studied in [20, 3]. The model-checking problem then reduces to the problem of deciding the winner in a parity game over such a system. The key step is a reduction of this game to a game over a 2nd order pushdown system *without* panic, for which a 2-EXPTIMEprocedure is already known [3].

An immediate consequence of our result is that the model checking problem for the hyperalgebraic grammars is decidable also for the monadic second-order logic, consequently the MSO theory of any hyperalgebraic tree is decidable. This result has been recently independently obtained by Aehlig, de Miranda and Ong [2], by a different proof, based on transformations of infinite lambda terms. Compared to [2], the present paper has two new elements: (1) it gives a characterization of unsafe grammars by panic automata, (2) the actual proof yields an optimal decision procedure for the µ-calculus model checking.

At present we do not know if the safety requirement really restricts the generating power of the tree grammars. Recently, Aehlig, de Miranda and Ong have studied the safety restriction [1]. They have shown that it is inessential for the *word* grammars of level 2, where a grammar, as usual, can generate a set of words. To simulate a non-safe grammar by a safe one, they use nondeterminism in an essential way, hence their result is not directly applicable for trees.

Due to space limitations, many arguments are omitted or sketchy; they will appear in the full paper (see [15] for a preliminary version).

2   Trees

Types, terms, and trees. We fix the set of simple types τ constructed from a unique *basic* type $\mathbf{0}$, by the rules $\tau ::= \mathbf{0} \mid (\tau_1 \to \tau_2)$. The level of a type is defined by $\ell(\mathbf{0}) = 0$, and $\ell(\tau_1 \to \tau_2) = \max(1 + \ell(\tau_1), \ell(\tau_2))$. Thus $\mathbf{0}$ is the only type of level 0 and each type of level 1 is of the form $\mathbf{0}^n \to \mathbf{0}$ for some $n > 0$ (which abbreviates $(\mathbf{0} \to (\mathbf{0} \to (\cdots(\mathbf{0} \to \mathbf{0})\ldots)))$, with $n+1$ occurrences of $\mathbf{0}$).

A *typed alphabet* is a set Γ of symbols, each γ in Γ given with its type, $\gamma : \tau$. We inductively extend Γ to the set $T(\Gamma)$ of *(applicative) terms* over Γ; if $t : \tau_1 \to \tau_2$ and $s : \tau_1$ then $(ts) : \tau_2$. As usual we abbreviate $(\cdots((t_0 t_1) t_2) \cdots) t_n$ by $t_0 t_1 \ldots t_n$. For terms $t, t_1, \ldots, t_m$, and symbols $z_1, \ldots, z_m$, of appropriate types, term $t[z_1 := t_1, \ldots, z_k := t_k]$ results from simultaneous replacement in t of z_i by t_i.

The set of natural numbers is denoted by ω. A *tree* (over a set X) is any nonempty prefix-closed subset T of the free monoid X^*, with ε as the *root*. If $u \in T$, $x \in X$, and $ux \in T$, then ux is a *successor* of u in T.

Now let Σ be a *signature*, i.e., a typed alphabet of symbols of level ≤ 1. A Σ-*tree* is a mapping $t \colon \mathrm{dom}\, t \to \Sigma$, where $\mathrm{dom}\, t \subseteq \omega^*$ is a tree, and if $t(w)$ is a symbol of type $\mathbf{0}^k \to \mathbf{0}$ then w has exactly k successors, $w1, \ldots, wk$ (hence w is a leaf if $t(w) : \mathbf{0}$). The set of Σ-trees is written $T^\infty(\Sigma)$.

A *limit* of a sequence $t_0, t_1, \ldots$ of Σ-trees is defined, provided that for any k, there is $m = m(k)$, such that $t_n \restriction k$ and $t_{n'} \restriction k$ coincide, for all $n, n' \geq m(k)$ (where, in general, $t \restriction k$ is restriction of t to the set $\{w \in \mathrm{dom}\, t : |w| \leq k\}$). Then $\lim t_n \in T^\infty(\Sigma)$ is just the set-theoretical union of the functions $t_n \restriction m(n)$.

Grammars. We fix an infinite typed alphabet of *variables* (or *parameters*), $\mathcal{X}$. A *grammar* is a tuple $\mathcal{G} = (\Sigma, N, \mathcal{S}, E)$, where Σ is a *signature*, N is a finite typed alphabet of *nonterminals*[1], $\mathcal{S} \in N$ is a *start symbol* of type $\mathbf{0}$, and E is a set of productions of the form

$$\mathcal{F} z_1 \ldots z_m \Rightarrow w$$

where $\mathcal{F} : \tau_1 \to \tau_2 \cdots \to \tau_m \to \mathbf{0}$ is a nonterminal, z_i a variable of type τ_i, and w an applicative term in $T(\Sigma \cup N \cup \{z_1 \ldots z_m\})$ of type $\mathbf{0}$. The *level* of a grammar is the highest level of its nonterminals. Since we are interested in grammars as generators of (single) Σ-trees, we assume that for each $\mathcal{F}$, there is exactly one production with $\mathcal{F}$ on the left-hand side.

The single-step reduction relation $\to_\mathcal{G}$ between terms over $\Sigma \cup N$ is defined inductively by the following clauses.

1. $\mathcal{F} t_1 \ldots t_k \to_\mathcal{G} t[z_1 := t_1, \ldots, z_k := t_k]$ if there is a production $\mathcal{F} z_1 \ldots z_k \Rightarrow t$ with $z_i : \rho_i$, and $t_i \in T(\Sigma \cup N)$, where $t_i : \rho_i$, for $i = 1, \ldots, k$.
2. If $t \to_\mathcal{G} t'$ then $(st) \to_\mathcal{G} (st')$ and $(tq) \to_\mathcal{G} (t'q)$, whenever the expressions in question are applicative terms.

---
[1] Without loss of generality, we assume that the types of nonterminals are *homogeneous*, i.e., $\mathbf{0}$ or $\tau_1 \to \cdots \to \tau_n \to \mathbf{0}$, where if each τ_i is homogeneous and $\ell(\tau_1) \geq \ell(\tau_2) \geq \ldots \geq \ell(\tau_n)$.

That is, $t \to_{\mathcal{G}} t'$ whenever t' is obtained from t by replacing some occurrence of a nonterminal F by the right-hand side of the appropriate production in which all parameters are in turn replaced by the actual arguments of F.

In order to define the result of an infinite derivation, we extend Σ to $\Sigma^{\perp} = \Sigma \cup \{\perp\}$, with $\perp : \mathbf{0}$. With any term t over $\Sigma \cup N$, we inductively associate an expression $t^{\perp}$ over $\Sigma^{\perp}$, by setting $f^{\perp} = f$ (for $f \in \Sigma$), $X^{\perp} = \perp$ (for $X \in N$), and $(st)^{\perp} = (s^{\perp}r^{\perp})$ whenever $s^{\perp} \neq \perp$, otherwise $(st)^{\perp} = \perp$. Then, we define relation $t \to_{\mathcal{G}}^{\infty} t'$, where t is a term in $T(\Sigma \cup N)$ and t' a tree in $T^{\infty}(\Sigma^{\perp})$, by

- t' is finite, and there is reduction sequence $t = t_0 \to_{\mathcal{G}} \ldots \to_{\mathcal{G}} t_n = t'$, or
- t' is infinite, and there is an infinite reduction sequence $t = t_0 \to_{\mathcal{G}} t_1 \to_{\mathcal{G}} \ldots$ such that $t' = \lim t_n^{\perp}$.

To define a unique tree produced by the grammar, we recall a standard *approximation ordering* on $T^{\infty}(\Sigma^{\perp})$: $t' \sqsubseteq t$ if $dom\ t' \subseteq dom\ t$ and, for each $w \in dom\ t'$, $t'(w) = t(w)$ or $t'(w) = \perp$. Then the tree generated by $\mathcal{G}$ is defined by

$$[\![\mathcal{G}]\!] = \sup\{t \in T^{\infty}(\Sigma^{\perp}) : S \to_{\mathcal{G}}^{\infty} t\}$$

It is easy to see that, by the Church-Rosser property of our grammar, the above set is directed, and hence $[\![\mathcal{G}]\!]$ is well defined since $T^{\infty}(\Sigma^{\perp})$ with the approximation ordering is a cpo.

In this paper we only study grammars of level 2, which we call *hyperalgebraic*, as they constitute the next level above the algebraic (context-free) grammars.

Parity games. A *parity game* is a perfect information game of possibly infinite duration played by two players, say Eve and Adam. We present it as a tuple $(V_{\exists}, V_{\forall}, E, p_1, \Omega)$, where $V_{\exists}$ and $V_{\forall}$ are (disjoint) sets of positions of Eve and Adam, respectively, $E \subseteq V \times V$ is the relation of possible moves, with $V = V_{\exists} \cup V_{\forall}$, $p_1 \in V$ is a designated initial position, and $\Omega : V \to \omega$ is the ranking function.

The players start a play in the position p_1 and then move the token according to relation E (always to a successor of the current position), thus forming a path in the graph (V, E). The move is selected by Eve or Adam, depending on who is the owner of the current position. If a player cannot move, she/he looses. Otherwise, the result of the play is an infinite path in the graph, $v_0, v_1, v_2, \ldots$ Eve wins the play if $\limsup_{n \to \infty} \Omega(v_n)$, is even, otherwise Adam wins.

Parity games, introduced by Emerson and Jutla [9, 17], have been recognized as a combinatorial essence of many model checking problems. A crucial property is the *positional determinacy*: any position is winning for one of the players, and moreover a winning strategy of player θ can be made *positional*, i.e., represented by a (partial) function $\sigma : V_\theta \to V$.[2] We say simply that Eve *wins* the game $\mathcal{G}$ if

---

[2] Positional strategy σ is *winning* for θ if every play $p_1 = q_1, q_2, \ldots$ is won by θ, provided that $q_\ell \in V_{\exists}$ follows $q_{\ell+1} = \sigma(q_\ell)$.

she has a winning strategy, the similar for Adam. We refer the reader to [19, 12] for an introduction to parity games.

Model checking. The model checking problem in our consideration is to verify if the tree generated by a given grammar $\mathcal{G}$ satisfies a property φ expressed in some logical language, in symbols $[\![\mathcal{G}]\!] \models \varphi$. The most expressive logics considered in literature are the *monadic second-order logic* (MSO) or the *μ-calculus*; both have the same expressive power over Σ-trees [18]. In this paper we avoid a logical machinery by using an equivalent formalism of *alternating parity automata* (defined below). A polynomial-time translation from the μ-calculus to alternating automata, $\varphi \mapsto A_\varphi$, is known [9], such that, for any tree t, $t \models \varphi$ iff t is recognized by A_φ.

An alternating parity tree automaton over signature Σ can be presented as a tuple

$$\mathcal{B} = \langle \Sigma, Q_\exists, Q_\forall, q_1, \delta, \Omega \rangle$$

where $Q_\exists \cup Q_\forall = Q$ is a finite set of states with the initial state q_1, $\Omega : Q \to \omega$ is a ranking function, and δ is a set of transitions of the form $q \to f(q_1, \ldots, q_k)$, where $q, q_1, \ldots, q_k \in Q$ and $f \in \Sigma$ with type $f : \mathbf{0}^k \to \mathbf{0}$.

The acceptance of a tree $t \in T^\infty(\Sigma)$ by the automaton $\mathcal{B}$ can be presented by a suitable parity game. We first define a the computation tree, $r : dom\, r \to Q$ with $dom\, r \subseteq (\omega \times \omega)^*$, such that, for any $u \in dom\, r$, the projection $u\!\downarrow_1$ on the first component is a node in $dom\, t$. We let $r(\epsilon) = q_1$, and whenever $r(u) = q$ and $t(u\!\downarrow_1) = f$, then, for any transition of the form $q \to f(q_1, \ldots, q_k)$, the node u has a successor $u(i,j)$ with $r(u(i,j)) = q_i$, for some j, and $i = 1, \ldots, k$. Now consider a parity game $Game(\mathcal{B}, t)$ with $V = dom\, r$ partitioned such that $u \in V_\exists$ iff $r(u) \in Q_\exists$, the initial configuration ε, and $\Omega(u) = \Omega(r(u))$. We let $\mathcal{B}$ accept the tree t iff Eva wins this game.

The model checking problem addressed in this paper will be the following.

Problem 1. Given a 2nd order grammar $\mathcal{G}$ and an alternating parity tree automaton $\mathcal{B}$. Does $\mathcal{B}$ accept $[\![\mathcal{G}]\!]$?

3 Panic Automata

Classically the content of a pushdown store is just a word over the pushdown alphabet. For our purpose, it is convenient to consider pushdown symbols with "time stamps" (sort of). We let a *level 1 pushdown store* (or a *1-pds* or 1-stack) over an alphabet Γ be a non-empty word $a_1 \ldots a_k$ over $\Gamma \times \omega$. A *level 2 pds* (2-pds, 2-stack) is a non-empty sequence $s_1 \ldots s_l$ of 1-pds's, which may also be written as $[s_1][s_2] \ldots [s_l]$ or as $s'[s_l]$, where s' stands for $[s_1][s_2] \ldots [s_{l-1}]$. The 1-stack s_i is called the *i-th row* of s. We assume that push-down stores grows to the *right*, so that, for instance, the item (a, m) is on top of the 2-pds $s'[w(a,m)]$. By $top(s)$ we denote the topmost Γ-symbol of s, i.e., $top(s'[w(a,m)]) = a$.

The following operations are possible on level 2 push-down stores.

$$push_1\langle a\rangle([s_1][s_2]\ldots[s_l][w]) = [s_1][s_2]\ldots[s_l][w(a,l)]$$
$$pop_1(\alpha[w\xi]) = \alpha[w]$$
$$push_2(\alpha[w]) = \alpha[w][w]$$
$$pop_2(\alpha[v][w]) = \alpha[v]$$
$$panic([s_1][s_2]\ldots[s_m]\ldots[s_l][w(a,m)]) = [s_1][s_2]\ldots[s_m]$$
$$skip(s) = s$$

The operation pop_2 (resp. pop_1) is undefined on a 2-stack s if it contains only one row (resp. the top row of s has only one element).

Let $\bot$ be a symbol in Γ. It is easy to see that if a 2-pds $[s_1]\ldots[s_m]$ is generated from $[(\bot,0)]$ by the above operations, and $s_i = (a_{i,1}, m_{i,1})\ldots(a_{i,k_i}, m_{i,k_i})$ then $m_{i,j} \leq i-1$, and $j \leq j'$ follows $m_{i,j} \leq m_{i,j'}$. Intuitively, whenever a new symbol $a \in \Gamma$ is placed on the top of the stack, the second component registers the number of the stack row which is directly below the current top row. Later the symbol can be duplicated several times by subsequent executions of $push_2$, but the second component keeps record of the level when it has first appeared in the stack. The *panic* operation returns to the 2-stack previous to the first appearance of the actual top symbol.

Now let Σ be a signature, and let $\Sigma_r \subseteq \Sigma$ be the set of symbols of type $0^r \to 0$. A *panic automaton* is defined as a tuple

$$\mathcal{A} = \langle \Sigma, Q, \Gamma, q_1, \delta, \bot \rangle,$$

where Q is a finite set of *states*, with an *initial state* q_1, Γ is a stack alphabet, with a distinguished *bottom symbol* $\bot$, and $\delta: Q \times \Gamma \to \mathcal{I}$ is a *transition function*, where $\mathcal{I}$ is the set of possible instructions,

$$\mathcal{I} \subseteq Op^{\Gamma} \times Q \cup \bigcup_r \Sigma_r \times Q^r,$$

where $Op^{\Gamma} = \{push_1\langle a\rangle : a \in \Gamma\} \cup \{pop_1, push_2, pop_2, panic, skip\}$.

A *configuration* of an automaton $\mathcal{A}$ as above is a pair (q, s), where $q \in Q$, and s is a 2-stack (over Γ). The *initial configuration* is $(q_1, [(\bot, 0)])$. We define the relation $\to_{\mathcal{A}}$ on configurations as follows. Let $s = [s_1]\ldots[s_l][w(a,m)]$.

1. If $\delta(q, a) = \langle \alpha, q'\rangle$ with $\alpha \in Op^{\Gamma}$ then $\langle q, s\rangle \to_{\mathcal{A}} \langle q', \alpha(s)\rangle$.
2. If $\delta(q, a) = \langle f, p_1, \ldots, p_r\rangle$ then $\langle q, s\rangle \to_{\mathcal{A}} \langle p_i, s\rangle$, for all $i = 1, \ldots, r$.

In the first case we write $\to_{\mathcal{A}}^{\circ}$ instead of $\to_{\mathcal{A}}$. The symbol $\twoheadrightarrow_{\mathcal{A}}^{\circ}$ stands for the reflexive and transitive closure of $\to_{\mathcal{A}}^{\circ}$.

Let $t: dom\, t \to \Sigma$ be a Σ-tree. A *partial run* of $\mathcal{A}$ on t is a partial function ϱ from an initial segment of $dom\, t$ to the set of all configurations, such that if $\varrho(w) = \langle q, s\rangle$, for some $w \in dom\, t$, then $\delta(q, a) = \langle f, p_1, \ldots, p_r\rangle$, where $f = t(w)$ and $a = top(s)$. In addition $\langle p_i, s\rangle \twoheadrightarrow_{\mathcal{A}}^{\circ} \varrho(wi)$, for each $i = 1, \ldots, r$ when $\varrho(wi)$ is defined.

If a partial run is a total function over $dom\,t$, we call it a *run*. As our automaton is deterministic, there can be at most one tree over which $\mathcal{A}$ has a run. This is the *tree recognized by* $\mathcal{A}$.

4 Automata vs Grammars

Theorem 4.1. *For any panic automaton $\mathcal{A}$, one can construct a hyperalgebraic grammar $\mathcal{G}_\mathcal{A}$, such that if the automaton recognizes a tree t in $T^\infty(\Sigma)$ then $t = [\![\mathcal{G}_\mathcal{A}]\!]$. Conversely, for a hyperalgebraic grammar $\mathcal{G}$, one can construct a panic automaton $\mathcal{A}_\mathcal{G}$ such that if the grammars generates a tree in $T^\infty(\Sigma)$, this is the (unique) tree recognized by $\mathcal{A}_\mathcal{G}$. Both constructions can be realized in polynomial time.*

Here we only sketch the direction $\mathcal{G} \mapsto \mathcal{A}_\mathcal{G}$, which is essential for the upper bound result.

Let $\circ_1, \circ_2, \ldots$ be fresh identifiers of type $\mathbf{0}$, which we call *holes*. The pushdown alphabet Γ of $\mathcal{A}_\mathcal{G}$ consists of subterms of the right-hand sides of the productions of $\mathcal{G}$, possibly applied to some holes, so that the result is of type $\mathbf{0}$. More precisely, let $u = Ft_1 \ldots t_d$ be such a subterm, where F is an operator (variable or nonterminal) of type $\tau_1 \to \cdots \to \tau_d \to \mathbf{0}^k \to \mathbf{0}$. Then $u \circ_1 \ldots \circ_k \in \Gamma$. In particular, if $u : \mathbf{0}$ then simply $u \in \Gamma$.

The holes $\circ_1, \ldots, \circ_k$ represent "missing arguments" of the operator F. Since holes are new identifiers, one can safely identify $Ft_1 \ldots t_d \circ_1 \ldots \circ_k$ with $Ft_1 \ldots t_d$ if this is convenient.

The idea of our simulation is that the top of pds (an expression u) represents a variable-free expression u' occurring in a derivation of $\mathcal{G}$. Since the pds alphabet must be finite, the term u can only be an "approximation" of u'. This approximation is "evaluated" to yield an approximate representation of the next step of reduction. The contents of the pds represents an environment in which the evaluation takes place. The environment is searched if one needs to find the meaning of a variable, or to find a missing argument.

The bottom pds symbol is $\mathcal{S}$, the initial nonterminal. This is our approximation in the first step of reduction. The automaton then works in phases, each phase beginning and ending in the distinguished state q_1. We define automaton informally, by describing the possible behaviour in a phase, beginning with a configuration (q_1, s).

T1 Let $top(s) = \mathcal{F}u_1 \ldots u_n$, where $\mathcal{F}$ is a nonterminal, and let the corresponding production be $\mathcal{F}x_1 \ldots x_n \Rightarrow u$. The automaton executes the instruction $\delta(q_1, \mathcal{F}u_1 \ldots u_n) = (push_1\langle u \rangle, q_1)$, so that our next approximation is u.

T2 If $top(s) = ft_1 \ldots t_r$ where f is a terminal, then the automaton executes the instruction $\delta(q_1, ft_1 \ldots t_r) = (f, p_1, \ldots, p_r)$, followed (at the i-th branch of the run) by a pop_1 and $push_1\langle t_i \rangle$. Thus the next approximation at the i-th branch is t_i. (If f is a constant, there is no further step.)

T3 Let $top(s) = x$, where x is an (ordinary) variable of type $\mathbf{0}$. To "evaluate" x, the automaton restores the environment where x was defined. It executes

pop_1 and inspects the new top symbol. It should be of the form $\mathcal{F}t_1\ldots t_e$, where $\mathcal{F}$ is a nonterminal. In addition, the variable x should be one of the formal parameters of $\mathcal{F}$, say, the j-th one. The next approximation is t_j, and it should be evaluated in the present environment (that of the caller). Another pop_1 is now executed, followed by a $push_1\langle t_j \rangle$, and the machine returns to state q_1.

T4 Let $top(s) = \varphi u_1 \ldots u_h$ where φ is a variable of type $\mathbf{0}^h \to \mathbf{0}$. If we now executed a pop_1 as above then the information about the actual parameters $u_1,\ldots,u_h$ would be lost. Instead, a $push_2$ is executed, followed by a pop_1. As in the previous case, the top of the stack should be of the form $\mathcal{F}t_1\ldots t_e$ and φ should be one of the formal parameters of $\mathcal{F}$, say, the j-th one. However now the new approximant t_j is an expression of a functional type and we actually place $t_j \circ_1 \ldots \circ_h$ on the pds rather than t_j.

T5 The last case is when $top(s)$ is a hole, say $\circ_i$. The automaton gets now into a panic. After the panic move the top of the pds should be $\psi v_1 \ldots v_l$ for some variable ψ. The new approximation is v_i. We execute in order pop_1 and $push_1\langle v_i \rangle$ and return to state q_1.

To explain the last case let us first observe that holes (missing arguments) are created when we attempt to evaluate a function variable (the fourth case). Holes correspond to the arguments (actual parameters) of this function variable that were "left behind" for a while. Later, a hole is found at top of the pds when we need to evaluate such a missing argument of an operator. In order to do so, we must restore the situation from the time of the call. The crucial thing is that the "time stamp" of the topmost item points out to exactly the moment in the past we need, namely to the stage when the hole was created.

5 Pushdown Systems

By now, we have considered panic automata as tree acceptors, with the aim to characterize hyperalgebraic grammars. For the model-checking applications, it is convenient to use a related concept of pushdown systems which do not take inputs, but are equipped with ranks and alternation (c.f. [3, 20]).

A *2nd order pushdown system with panic* (or 2-PPDS, for short) is a tuple $\mathcal{C} = (P, P_\exists, P_\forall, \Gamma, p_1, \Delta, \Omega)$ where P is a finite set of control locations partitioned into $P_\exists$ and $P_\forall$, Γ is a finite store alphabet, $p_1 \in P$ is the initial location, $\Omega : P \to \omega$ is the rank function, and $\Delta \subseteq P \times \Gamma \times P \times Op^\Gamma$ is the finite set of (unlabeled) transition rules (where Op^Γ is as in Sect. 3). A transition $(p, a, p', \alpha) \in \Delta$ is often written by $p, a \to_\Delta p', \alpha$.

A *configuration* of a 2-PPDS $\mathcal{C}$ is a pair (p, s) where $p \in P$ and s is a 2-stack over Γ. A 2-PPDS $\mathcal{C}$ induces a parity game $Game(\mathcal{C}) = (V_\exists, V_\forall, E, (p_1, [\bot, 0]), \Omega)$, where

- $V_\exists$ is the set of configurations of $\mathcal{C}$ with a location from $P_\exists$, similarly for $V_\forall$;
- $(p, s)E(p', s')$ iff $\exists (p, a, p', \alpha) \in \Delta$, $top(s) = a$ and $s' = \alpha(s)$;
- $\Omega(p, s) = \Omega(p)$.

Problem 2. Given a 2-PPDS $\mathcal{C}$, decide if Eva wins the game $Game(\mathcal{C})$.

Proposition 5.2. *Problem 1 and Problem 2 are polynomial-time equivalent.*

Proof: *(Sketch)* By Theorem 4.1, we reduce Problem 1 to the question if a tree recognized by a panic automaton $\mathcal{A}$ is accepted by an alternating tree automaton $\mathcal{B}$. Then a suitable 2-PPDS $\mathcal{C}$ is obtained by a standard product construction, such that Eve wins $Game(\mathcal{C})$ if and only if she wins $Game(\mathcal{B}, t_\mathcal{A})$

The converse transformation of Problem 2 to Problem 1 is also easy, after introduction of a suitable functional signature Σ (depending on given $\mathcal{C}$). □

Before proceeding to the proof of our main result, we show a useful transformation of 2-PPDS's, which will allow us to control the dynamics of ranks by the topmost symbols on the pushdown store.

Let $Tr(\mathcal{C})$ be the tree obtained by unfolding the graph of $Game(\mathcal{C})$. That is, the root of $Tr(\mathcal{C})$ is labeled by the initial configuration, and a node labeled by (p, s) has a successor labeled (p', s'), whenever $(p, s)E(p', s')$. Note that a node of the tree determines a computation up to this node.

Definition 5.3 *Let a node v of $Tr(\mathcal{C})$ be labeled with $\langle q, s \rangle$ such that $panic(s)$ is defined. Let v' be the closest to v ancestor of v labeled with $\langle q', panic(s) \rangle$, for some q'. We call v' the* panic ancestor *of v. The* panic rank *of v is the maximal rank of a state occurring between the panic ancestor of v and v, including the rank of v and excluding the rank of the panic ancestor of v.*

A 2-PPDS $\mathcal{C}$ is rank-aware *iff there exists a function $Rank : \Gamma \to \mathrm{Rg}(\Omega)$ such that the panic rank of every node v of $Tr(\mathcal{C})$ labeled (q, s) is equal to $Rank(top(s))$. That is, the panic rank is determined by the top of the stack.*

Lemma 5.4. *For every 2-PPDS $\mathcal{C}$, one can construct in polynomial time a rank-aware 2-PPDS $\mathcal{C}'$, such that Eve wins $Game(\mathcal{C}')$ iff she wins in $Game(\mathcal{C})$.*

Proof: *(Idea)* The stack alphabet of $\mathcal{C}'$ is defined as $\Gamma' = \Gamma \times \{0, \ldots, d\}^2$, where d is the highest rank in $\mathrm{Rg}(\Omega)$. The construction ensures that if (a, m_p, m_l) is currently on top of the stack, then m_p is the panic rank of the node, and m_l is the highest rank of a state seen since the creation of the current top row.

The construction of transitions of $\mathcal{C}'$ goes naturally case by case. The correctness proof is straightforward, but tedious. □

6 Deciding the Winner in $Game(\mathcal{C})$

A *2nd order pushdown system* (or 2-PDS, for short) is like a 2-PPDS defined in section 5, but *without* panic, that is, $\Delta \subseteq P \times \Gamma \times P \times (Op^\Gamma - \{panic\})$. Then a level 1 pushdown store can be viewed just as a word over Γ (not $\Gamma \times \omega$). The concepts of *2-pds*, configuration, and $Game(\mathcal{C})$ are simplified accordingly.

Games over such pushdown systems have been considered by Engelfriet [10], Cachat [3], and others.

Lemma 6.5 ([10, 3, 4]). *It is decidable if Eve wins $Game(\mathcal{C})$ for a given 2nd-order pushdown system $\mathcal{C} = (P, P_\exists, P_\forall, \Gamma, p_1, \Delta, \Omega)$. There is an algorithm working in time[3] $2^{|\Gamma|2^{O(|P|)}}$, moreover the problem is 2-EXPTIME-complete.*

We now transform a rank-aware 2nd order pushdown system with panic $\mathcal{C} = (P, P_\exists, P_\forall, \Gamma, p_1, \Delta, \Omega)$ into a 2nd order pushdown system without panic $\mathcal{C}' = (P', P'_\exists, P'_\forall, \Gamma', p'_1, \Delta', \Omega')$, such that Eve wins $Game(\mathcal{C})$ iff she wins $Game(\mathcal{C}')$.

For technical convenience, we assume that a transition $p, a \to_\Delta p', \alpha$ with $\alpha \neq skip$ is possible only for $p \in P_\exists$. Clearly, we can always achieve this property by duplicating locations and using $skip$ operation.

Let $Rank : \Gamma \to \text{Rg}(\Omega) = \{0, \ldots, d\}$ be the function of Definition 5.3. We introduce an ordering on ranks: $m \preccurlyeq n$ iff $(-1)^m \cdot m \leq (-1)^n \cdot n$ (e.g., for $d = 6$, this yields an ordering $5, 3, 1, 0, 2, 4, 6$).

We define the set of *returns* as

$$Ret = P \dashrightarrow \{0, \ldots, d\} \ .$$

The intention is that a partial function $R \in Ret$ assigns to a location p the worst, in $\preccurlyeq$-order, panic rank still acceptable for the panic moves *ending* in position p. We let

$P'_\exists = P_\exists \cup \{\bot\},$
$P'_\forall = P_\forall \cup \{p_{push}, p_{ver}, p_{rank=i} : p \in P, i \in \{0, \ldots, d\}\} \cup \{\top\} \cup \{ver\} \cup Aux$
$\Gamma' = Ret \cup (\Gamma \times Ret).$

The rank function Ω' is 0 for all the locations except for

$$\Omega'(p) = \Omega(p), \quad \Omega'(p_{rank=i}) = i; \quad \text{for all } p \in P.$$

The transition rules Δ' are defined by the following clauses.

- If $p, a \to_\Delta p', skip$ then $p, (a, R) \to_{\Delta'} p', skip$, for all $R \in Ret$.
- If $p, a \to_\Delta p', push_1\langle a \rangle$ then $p, (a, R) \to_{\Delta'} p'_{ver}, push_1\langle (a', R') \rangle$, for all $R, R' \in Ret$. Additionally we have $p'_{ver}, (a', R') \to_{\Delta'} ver, skip$, and $p'_{ver}, (a', R') \to_{\Delta'} p', skip$.
- From the location ver the moves are defined in such a way that Eve wins iff R in the top letter of the top-most 1-stack is the same as R at the top of 1-stack just below the top-most 1-stack.
- If $p, a \to_\Delta p', push_2$ then $p, (a, R) \to_{\Delta'} p'_{push}, push_1\langle R' \rangle$, for all $R, R' \in Ret$. Additionally we put in Δ' the rules[4] $p'_{push}, R' \to_{\Delta'} p', push_2; pop_1$, as well as

---

[3] The single exponential dependence on $|\Gamma|$ is not made explicit in these references, but it follows from the analysis of level 1 pushdown systems in [20].
[4] Notation like $p, a \to_{\Delta'} p', \alpha_1; \alpha_2$ clearly abbreviates two rules.

$p'_{push}, R' \to_{\Delta'} p''_{rank=R'(p'')}, pop_1$, and $p''_{rank=R'(p'')}, (a, R) \to_{\Delta'} p'', skip$, for all p'' such that $R'(p'')$ is defined.
- If $p, a \to_\Delta p', pop_1$ then $p, (a, R) \to_{\Delta'} p', pop_1$, for all $R \in Ret$.
- If $p, a \to_\Delta p', pop_2$ then $p, (a, R) \to_{\Delta'} p', pop_2; pop_1$, for all $R \in Ret$.
- If $p, a \to_\Delta p', panic$ then $p, (a, R) \to_{\Delta'} \top, skip$ if $Rank(a) \succcurlyeq R(p')$, and we have $p, (a, R) \to_{\Delta'} \bot, skip$ otherwise, i.e., when $Rank(a) \prec R(p')$ or $R(p')$ is undefined.
- There are no transitions from states $\top$ and $\bot$. This implies that Eve wins in $\top$ and loses in $\bot$.

Proposition 6.6. *Eve wins in $Game(\mathcal{C}')$ iff she wins in $Game(\mathcal{C})$.*

Proof: *(Idea)* The construction guarantees that a 2-pds in a reachable configuration of $\mathcal{C}'$ has a form $v = [w_1 R_1] \ldots [w_k R_k][w_{k+1}]$, where $w_i \in (\Gamma \times Ret)^*$ and $R_i \in Ret$. We say that v *represents* a 2-pds $s = [s_1] \ldots [s_{k+1}]$ of $\mathcal{C}$ if, for all i, $s_i \downarrow_1 = w_i \downarrow_1$, and if $s_i \downarrow_2 = \ell$ then $w_i \downarrow_2 = R_\ell$. Note that a possibly infinite domain of "time stamps" (in $\mathcal{C}$) is represented here by the finite set Ret. Both players can ensure that the 2-stacks reachable in $Game(\mathcal{C}')$ represent the 2-stacks of $\mathcal{C}$. Together with the mechanism of returns, this allows them to transfer the strategies from $Game(\mathcal{C})$ to $Game(\mathcal{C}')$. □

We are now ready to state our main result.

Theorem 6.7. *The problem of checking if Eva wins $Game(\mathcal{C})$, for a given 2nd order pushdown system with panic $\mathcal{C}$, is 2-EXPTIME-complete (Problem 2). Consequently, so is the problem of checking if an alternating parity tree automaton $\mathcal{B}$ accepts the tree $[\![\mathcal{G}]\!]$ generated by an hyperalgebraic grammar $\mathcal{G}$ (Problem 1).*

Proof: The 2-EXPTIME hardness result follows from Lemma 6.5, and Proposition 5.2. To show the upper bound, let $\mathcal{C}$ be a 2-PPDS. By Lemma 5.4 we can assume that $\mathcal{C}$ is rank-aware. To solve $Game(\mathcal{C})$ we construct a 2-PDS $\mathcal{C}'$ (without panic), as in Proposition 6.6. The system $\mathcal{C}'$ has the number of stack letters which is exponential in the size of $\mathcal{C}$, but the number of locations of $\mathcal{C}'$ is linear in the number of locations of $\mathcal{C}$. By the first part of Lemma 6.5, the winner in the game $Game(\mathcal{C}')$ can be decided in time doubly exponential in the number of states and singly exponential in the number of stack symbols, which gives us the desired complexity. □

By the relation between automata, μ-calculus, and monadic logic (cf. Section 2), we obtain the following.

Corollary 6.8. *It is 2-EXPTIME-complete to check if $[\![\mathcal{G}]\!] \models \varphi$, for a given 2nd order grammar $\mathcal{G}$, and a given μ-calculus formula φ. Consequently, the analogous problem for the monadic second-order logic (MSO) is decidable, in particular the MSO theory of any hyperalgebraic tree is decidable.*

References

[1] Aehlig, K., de Miranda, J.G., and Ong, L., Safety is not a restriction at level 2 for string languages. In: Proc. FOSSACS '05, Springer LNCS 3441 (2005), 490–504.

[2] Aehlig, K., de Miranda J.G., and Ong, L., The monadic second order theory of trees given by arbitrary level-two recursion schemes is decidable. In: Proc. TLCA '05, Springer LNCS 3461 (2005), 39–54.

[3] Cachat, T., Higher Order Pushdown Automata, the Caucal Hierarchy of Graphs and Parity Games. In: Proc. ICALP 2003, Springer LNCS 2719 (2003), 556–569.

[4] Cachat, T., Walukiewicz, I., The complexity of games on higher order pushdown automata, manuscript, 2004.

[5] Caucal, D., On infinite terms having a decidable monadic second-order theory. In: Proc. MFCS 2002, Springer LNCS 2420 (2002), 65–176.

[6] Courcelle, B., The monadic second-order theory of graphs IX: Machines and their behaviours. *Theoretical Comput. Sci.*, 151:125–162, 1995.

[7] Courcelle, B., Knapik, T., The evaluation of if first-order substitution is monadic second-order compatible *Theoretical Comput. Sci.*, 281(1-2):177–206, 2002.

[8] Damm, W., The IO- and OI-hierarchies. *Theoretical Comput. Sci.*, 20(2):95–208, 1982.

[9] Emerson, E. A., Jutla, C. S., Tree automata, mu-calculus and determinacy. In: *Proceedings 32th Annual IEEE Symp. on Foundations of Comput. Sci.*, IEEE Computer Society Press, 1991, pp. 368–377.

[10] Engelfriet, J., Iterated push-down automata and complexity classes. In: *Proc. 15th STOC*, 1983, pp. 365–373.

[11] Engelfriet, J., Schmidt, E.M., IO and OI, *J. Comput. System Sci.* **15**, 3, 1977, pp. 328–353, and **16**, 1, 1978, pp. 67–99.

[12] Grädel, E., Thomas, W., and Wilke, T., Editors, *Automata, Logics, and Infinite Games. A Guide to Current Research*, LNCS 1500, Springer-Verlag, 2002.

[13] Knapik, T., Niwiński, D., and Urzyczyn, P., Deciding monadic theories of hyperalgebraic trees. In: *Typed Lambda Calculi and Applications, 5th International Conference*, Springer LNCS 2044 (2001), 253–267.

[14] Knapik, T., Niwiński, D., Urzyczyn, P., Higher-order pushdown trees are easy. In: *Proc. FoSSaCS'02*, Springer LNCS 2303 (2002), 205–222.

[15] Knapik, T., Niwiński, D., Urzyczyn, P., Walukiewicz, I., Unsafe grammars and panic automata, draft, http://www.mimuw.edu.pl/~ niwinski/prace.html.

[16] Maslov, A.N., The hierarchy of indexed languages of an arbitrary level, *Soviet Math. Dokl.*, **15**, pp. 1170–1174, 1974.

[17] A. W. Mostowski. Games with forbidden positions. Technical Report 78, Instytut Matematyki, University of Gdansk, 1991.

[18] Niwiński, D., Fixed points characterization of infinite behaviour of finite state systems. *Theoret. Comput. Sci.*, 189:1–69, 1997.

[19] Thomas, W., Languages, automata, and logic. In G. Rozenberg and A. Salomaa, editors, *Handbook of Formal Languages*, volume 3, Springer-Verlag, 1997, pp. 389–455.

[20] Walukiewicz, I., Pushdown processes: Games and model checking. *Information and Computation*, 164(2):234–263, 2001.

Signaling P Systems and Verification Problems*

Cheng Li[1], Zhe Dang[1,**], Oscar H. Ibarra[2], and Hsu-Chun Yen[3]

[1] School of Electrical Engineering and Computer Science,
Washington State University, Pullman, WA 99164, USA
zdang@eecs.wsu.edu
[2] Department of Computer Science,
University of California, Santa Barbara, CA 93106, USA
[3] Department of Electrical Engineering,
National Taiwan University, Taipei, Taiwan 106, R.O.C.

Abstract. We introduce a new model of membrane computing system (or P system), called signaling P system. It turns out that signaling systems are a form of P systems with promoters that have been studied earlier in the literature. However, unlike non-cooperative P systems with promoters, which are known to be universal, non-cooperative signaling systems have decidable reachability properties. Our focus in this paper is on verification problems of signaling systems; i.e., algorithmic solutions to a verification query on whether a given signaling system satisfies some desired behavioral property. Such solutions not only help us understand the power of "maximal parallelism" in P systems but also would provide a way to validate a (signaling) P system in vitro through digital computers when the P system is intended to simulate living cells. We present decidable and undecidable properties of the model of non-cooperative signaling systems using proof techniques that we believe are new in the P system area. For the positive results, we use a form of "upper-closed sets" to serve as a symbolic representation for configuration sets of the system, and prove decidable symbolic model-checking properties about them using backward reachability analysis. For the negative results, we use a reduction via the undecidability of Hilbert's Tenth Problem. This is in contrast to previous proofs of universality in P systems where almost always the reduction is via matrix grammar with appearance checking or through Minsky's two-counter machines. Here, we employ a new tool using Diophantine equations, which facilitates elegant proofs of the undecidable results. With multiplication being easily implemented under maximal parallelism, we feel that our new technique is of interest in its own right and might find additional applications in P systems.

1 Introduction

P systems [18, 19] are abstracted from the way the living cells process chemical compounds in their compartmental structure. A P system consists of a finite number of

---

* The work by Cheng Li and Zhe Dang was supported in part by NSF Grant CCF-0430531. The work by Oscar H. Ibarra was supported in part by NSF Grants CCR-0208595 and CCF-0430945.
** Corresponding author.

membranes, each of which contains a multiset of objects (symbols). The membranes are organized as a Venn diagram or a tree structure where a membrane may contain other membranes. The dynamics of the P system is governed by a set of rules associated with each membrane. Each rule specifies how objects evolve and move into neighboring membranes. In particular, a key feature of the model of P systems is that rules are applied in a nondeterministic and maximally parallel manner. Due to the key feature inherent in the model, P systems have a great potential for implementing massively concurrent systems in an efficient way that would allow us to solve currently intractable problems (in much the same way as the promise of quantum and DNA computing). It turns out that P systems are a powerful model: even with only one membrane (i.e., 1-region P systems) and without priority rules, P systems are already universal [18, 22]. In such a one-membrane P system, rules are in the form of $u \rightarrow v$, which, in a maximally parallel manner, replaces multiset u (in current configuration which is a multiset of symbol objects) with multiset v.

Signals are a key to initiate biochemical reactions between and inside living cells. Many examples can be found in a standard cell biology textbook [3]. For instance, in signal transduction, it is known that guanine-nucleotide binding proteins (G proteins) play a key role. A large heterotrimeric G protein, one of the two classes of G proteins, is a complex consisting of three subunits: G_α, G_β, and G_γ. When a ligand binds to a G protein-linked receptor, it serves as a signal to activate the G protein. More precisely, the GDP, a guanine nucleotide, bound to the G_α subunit in the unactivated G protein is now displaced with GTP. In particular, the G protein becomes activated by being dissociated into a G_β-G_γ complex and a G_α-GTP complex. Again, the latter complex also serves as a signal by binding itself to the enzyme adenylyl cyclase. With this signal, the enzyme becomes active and converts ATP to cyclic AMP. As another example, apoptosis (i.e., suicide committed by cells, which is different from necrosis, which is the result from injury) is also controlled by death signals such as a CD95/Fas ligand. The signal activates caspase-8 that initiates the apoptosis. Within the scope of Natural Computing (which explores new models, ideas, paradigms from the way nature computes), motivated by these biological facts, it is a natural idea to study P systems, a molecular computing model, augmented with a signaling mechanism.

In this paper, we investigate one-membrane signaling P systems (signaling systems in short) where the rules are further equipped with signals. More precisely, in a signaling system M, we have two types of symbols: object symbols and signals. Each configuration is a pair consisting of a set S of signals and a multiset α of objects. Each rule in M is in the form of $s, u \rightarrow s', v$ or $s, u \rightarrow \Lambda$, where s, s' are signals and u, v are multisets of objects. The rule is enabled in the current configuration (S, α) if s is present in the signal set S and u is a sub-multiset of the multiset α. All the rules are fired in maximally parallel manner. In particular, in the configuration as a result of the maximally parallel move, the new signal set is formed by collecting the set of signals s' that are emitted from all the rules actually fired during the move (and every signal in the old signal set disappears). Hence, a signal may trigger an unbounded number of rule instances in a maximally parallel move.

We focus on verification problems of signaling systems; i.e., algorithmic solutions to a verification query on whether a given signaling system does satisfy some desired behavioral property. Such solutions not only help us understand the power of the maximally parallelism that is pervasive in P systems but also would provide a way to validate a (signaling) P system in vitro through digital computers when the P system is intended to simulate living cells. However, since one-membrane P systems are Turing-complete, so are signaling systems. Therefore, to study the verification problems, we have to look at restricted signaling systems. A signaling system is non-cooperative if each rule is in the form of $s, a \to \Lambda$ or in the form of $s, a \to s', bc$, where a, b, c are object symbols. All the results can be generalized to non-cooperative signaling systems augmented with rules $s, a \to s', v$. We study various reachability queries for non-cooperative signaling systems M; i.e., given two formulas $Init$ and $Goal$ that define two sets of configurations, are there configurations C_{init} in $Init$ and C_{goal} in $Goal$ such that C_{init} can reach C_{goal} in zero or more maximally parallel moves in M? We show that, when $Init$ is a Presburger formula (roughly, in which one can compare integer linear constraints over multiplicities of symbols against constants) and $Goal$ is a region formula (roughly, in which one can compare multiplicities of symbols against constants), the reachability query is decidable. Notice that, in this case, common reachability queries like halting and configuration reachability are expressible. We also show that introducing signals into P systems indeed increases its computing power; e.g., non-cooperative signaling systems are strictly stronger than non-cooperative P systems (without signals). On the other hand, when $Goal$ is a Presburger formula, the query becomes undecidable. Our results generalize to queries expressible in a subclass of a CTL temporal logic and to non-cooperative signaling systems with rules $S, a \to S', v$ (i.e., the rule is triggered with a set of signals in S). We also study the case when a signal has bounded strength and, in this case, non-cooperative signaling systems become universal.

Non-cooperative signaling systems are also interesting for theoretical investigation, since the signaling rules are context-sensitive and the systems are still nonuniversal as we show. In contrast to this, rules $a \to v$ in a non-cooperative P system are essentially context-free. It is difficult to identify a form of restricted context-sensitive rules that are still nonuniversal. For instance, a communicating P system (CPS) with only one membrane [21] is already universal, where rules are in the form of $ab \to a_x b_y$ or $ab \to a_x b_y c_{come}$ in which a, b, c are objects (the c *comes* from the membrane's external environment), x, y (which indicate the directions of movements of a and b) can only be *here* or *out*. Also one membrane catalytic systems with rules like $Ca \to Cv$ (where C is a catalyst) are also universal. More examples including non-cooperative signaling systems with promoters, which will be discussed further in this section, are also universal. Our non-cooperative signaling systems use rules in the form of $s, a \to s', v$, which are in a form of context-sensitive rules, since the signals constitute part of the triggering condition as well as the outcome of the rules.

At the heart of our decidability proof, we use a form of upper-closed sets to serve as a symbolic representation for configuration sets and prove that the symbolic representation is invariant under the backward reachability relation of a non-cooperative signaling system. From the studies in symbolic model-checking [7] for classic transition systems, our symbolic representation also demonstrates a symbolic model-checking procedure at least for reachability. In our undecidability proofs, we use the well-known result on the Hilbert's Tenth Problem: any r.e. set (of integer tuples) is also Diophantine. We note that, for P systems that deal with symbol objects, proofs for universality almost always use the theoretical tool through matrix grammar with appearance checking [16] or through Minsky's two-counter machines. Here, we employ a new tool using Diophantine equations, which facilitates elegant proofs of the undecidable results. With multiplication being easily implemented under maximal parallelism, we feel that our new technique is of interest in its own right and might find additional applications in P systems.

Signaling mechanisms have also been noticed earlier in P system studies. For instance, in a one-membrane P system with promoters [4], a rule is in the form of $u \to v|p$ where p is a multiset called a promoter. The rule fires as usual in a maximally parallel manner but only when objects in the promoter all appear in the current configuration. Notice that, since p may not be even contained in u, a promoter, just as a signal, may trigger an unbounded number of rule instances. Indeed, one can show that a signaling system can be directly simulated by a one-membrane P system with promoters. However, since one-membrane non-cooperative P systems with promoters are known to be universal [4], our decidability results on non-cooperative signaling systems have a nice implication: our signals are strictly weaker than promoters (and hence have more decidable properties). The decidability results also imply that, as shown in the paper, non-cooperative signaling systems and vector addition systems (i.e., Petri nets) have incomparable computing power, though both models have a decidable configuration-to-configuration reachability. This latter implication indicates that the maximal parallelism in P systems and the "true concurrency" in Petri nets are different parallel mechanisms. Other signaling mechanisms such as in [2] are also promoter-based.

2 Preliminaries

We use $\mathbf{N}$ to denote the set of natural numbers (including 0) and use $\mathbf{Z}$ to denote the set of integers. Let $\Sigma = \{a_1, \cdots, a_k\}$ be an alphabet, for some k, and α be a (finite) multiset over the alphabet. In this paper, we do not distinguish between different representations of the multiset. That is, α can be treated as a vector in $\mathbf{N}^k$ (the components are the multiplicities of the symbols in Σ); α can be treated as a word on Σ where we only care about the counts of symbols (i.e., its Parikh map). For a $\sigma \subseteq \Sigma$, we use σ^* to denote the set of all multisets on σ.

A set $S \subseteq \mathbf{N}^k$ is a *linear set* if there exist vectors $v_0, v_1, \ldots, v_t$ in $\mathbf{N}^k$ such that $S = \{v \mid v = v_0 + a_1 v_1 + \cdots + a_t v_t, \ a_i \in \mathbf{N}\}$. A set $S \subseteq \mathbf{N}^k$ is *semilinear* if it is a finite union of linear sets. Let $x_1, \cdots, x_k$ be variables on $\mathbf{N}$. A *Presburger formula* is a Boolean combination of linear constraints in the following form: $\sum_{1 \leq i \leq k} t_i \cdot x_i \sim n$, where the t_i's and n are integers in $\mathbf{Z}$, and $\sim \in \{>, <, =, \geq, \leq, \equiv_m\}$ with $0 \neq m \in \mathbf{N}$. It is known that a set of multisets (treated as vectors) is semilinear iff the set is definable by a Presburger formula. Also, Presburger formulas are closed under quantification.

A signaling system is simply a P system [18] augmented with signals. Formally, a (1-membrane) *signaling system* M is specified by a tuple $\langle \Sigma, Sig, R \rangle$, where $\Sigma = \{a_1, \cdots, a_k\}$ is the alphabet, Sig is a nonempty finite set of *signals*, and R is a finite set of *rules*. Each rule is in the form of $s, u \to s', v$, where $s, s' \in Sig$ and u and v are multisets over alphabet Σ. (Notice that a rule like $s, u \to v$ (without emitting signal) can be treated as a short hand of $s, u \to s_{\text{garbage}}, v$ where s_{garbage} is a "garbage" signal that won't trigger any rules.) A *configuration* C is a pair consisting of a set S of signals and a multiset α on Σ. As with the standard semantics of P systems [18, 19, 20], each evolution step, called a *maximally parallel move*, is a result of applying all the rules in M in a maximally parallel manner. More precisely, let $s_i, u_i \to s'_i, v_i$, $1 \le i \le m$, be all the rules in M. We use $\mathbf{R} = (r_1, \cdots, r_m) \in \mathbf{N}^m$ to denote a multiset of rules, where there are r_i instances of rule $s_i, u_i \to s'_i, v_i$, for each $1 \le i \le m$. Rule $s_i, u_i \to s'_i, v_i$ is *actually fired* in $\mathbf{R}$ if $r_i \ge 1$ (there is at least one instance of the rule in $\mathbf{R}$). Let $C = (S, \alpha)$ and $C' = (S', \alpha')$ be two configurations. The rule multiset $\mathbf{R}$ is *enabled* under configuration C if

- multiset α contains multiset $\cup_{1 \le i \le m} r_i \cdot u_i$ (i.e., the latter multiset is the multiset union of r_i copies of multiset u_i, for all $1 \le i \le m$), and
- set $S \supseteq \{s_i : r_i > 0, 1 \le i \le m\}$ (i.e., for every rule actually fired in $\mathbf{R}$, the signal s_i that triggers the rule must appear in the set S of the configuration C).

(We say that a rule is enabled under configuration C if the rule multiset that contains exactly one instance of the rule is enabled under the configuration.) The result $C' = (S', \alpha')$ of applying $\mathbf{R}$ over $C = (S, \alpha)$ is as follows: set S' is obtained by replacing the entire S by the new signal set formed by collecting all the signals s'_i emitted from the rules that are actually fired in $\mathbf{R}$, and, multiset α' is obtained by replacing, in parallel, each of the r_i copies of u_i in α with v_i. The rule multiset $\mathbf{R}$ is *maximally enabled* under configuration C if it is enabled under C and, for any other rule multiset $\mathbf{R}'$ that properly contains $\mathbf{R}$, $\mathbf{R}'$ is not enabled under the configuration. Notice that, for the same C, a maximally enabled rule multiset may not be unique (i.e., M is in general nondeterministic). C can reach C' through a maximally parallel move, written $C \to_M C'$, if there is a maximally enabled rule multiset $\mathbf{R}$ such that C' is the result of applying $\mathbf{R}$ over C. We use $C \leadsto_M C'$ to denote the fact that C' is reachable from C; i.e., for some n and $C_0, \cdots, C_n$, we have $C = C_0 \to_M \cdots \to_M C_n = C'$. We simply say that C is reachable if the initial configuration C' is understood. We say that configuration C is *halting* if there is no rule enabled in C.

When the signals are ignored in a signaling system, we obtain a 1-membrane P system. Clearly, signaling systems are universal, since, as we have mentioned earlier, 1-membrane P systems are known to be universal. A non-cooperative signaling system is a signaling system where each rule is either a *split-rule* in the form of $s, a \to s', bc$ or a *die-rule* in the form of $s, a \to \Lambda$, where $s, s' \in Sig$ and symbols $a, b, c \in \Sigma$. The two rules are called a-rules (since a appears at the LHS). Intuitively, the split-rule, when receiving signal s, makes an a-object split into a b-object and a c-object with signal s' emitted. On the other hand, the die-rule, when receiving signal s, makes an a-object die (i.e., becomes null). In particular, for a configuration C, an a-object is *enabled* in C if there is an enabled a-rule in C; in this case, we also call a to be *an enabled symbol in C*. In the rest of the paper, we will focus on various reachability queries for non-cooperative signaling systems.

3 Configuration Reachability

We first investigate the *configuration-reachability* problem that decides whether one configuration can reach another.

Given: a non-cooperative signaling system M and two configurations C_{init} and C_{goal},
Question: Can C_{init} reach C_{goal} in M?

In this section, we are going to show that the problem is decidable. The proof performs backward reachability analysis. That is, we first effectively compute (a symbolic representation of) the set of all configurations C' such that $C' \leadsto_M C_{\text{goal}}$. Then, we decide whether the initial configuration C_{init} is in the set.

Before proceeding further, we first introduce the symbolic representation. Let $\mathcal{C}$ be a set of configurations. We say that $\mathcal{C}$ is *upper-closed* if $\mathcal{C} = \{(S, \alpha) : \alpha$ is the multiset union of β and some multiset in $\sigma^*\}$, for some $S \subseteq Sig$, multiset β and some symbol-set $\sigma \subseteq \Sigma$. In this case, we use $[S, \beta, \sigma^*]$ to denote the set $\mathcal{C}$. We say that $\mathcal{C}$ is *m-bounded* if $|\beta| \leq m$. Let $\mathcal{C}$ be a finite union of upper-closed sets of configurations. The *pre-image* of $\mathcal{C}$ is defined as $Pre_M(\mathcal{C}) = \{C' : C' \to_M C \in \mathcal{C}\}$. We use $Pre_M^*(\mathcal{C})$ to denote the set of all configurations C' such that $C' \leadsto C$ for some $C \in \mathcal{C}$. The main result of this section is as follows.

Theorem 1. *Let $\mathcal{C}$ be a finite union of upper-closed sets of configurations in M. Then, $Pre_M^*(\mathcal{C})$ can also be effectively represented as a finite union of upper-closed sets of configurations in M.*

The complex proof of Theorem 1 constructs an intermediate signaling system $\hat{M}$ whose $Pre_{\hat{M}}^*$ is easier to compute. The theorem can be established after we prove that Pre_M^*-computation can be realized by $Pre_{\hat{M}}^*$-computation and that $Pre_{\hat{M}}^*(\mathcal{C})$ can be effectively represented as a finite union of upper-closed sets.

Now, we can show that the configuration-reachability problem for non-cooperative signaling systems is decidable. This result implies that non-cooperative signaling systems are not universal (the set of reachable configurations is recursive). Notice that $\mathcal{C} = \{C_{\text{goal}}\}$ is an upper-closed set. Since, from Theorem 1, $Pre_M^*(\mathcal{C})$ is effectively a finite union of upper-closed sets, one can also effectively answer the reachability at the beginning of this Section by checking whether C_{init} is an element in one of the upper-closed sets. Hence,

Theorem 2. *The configuration reachability problem for non-cooperative signaling systems is decidable.*

Reachability considered so far is only one form of important verification queries. In the rest of this section, we will focus on more general queries that are specified in the computation tree logic (CTL) [6] interpreted on an infinite state transition system [5]. To proceed further, more definitions are needed.

Let M be a non-cooperative signaling system with symbols Σ and signals Sig. We use variables $\#(a), a \in \Sigma$, to indicate the number of a-objects in a configuration and use variable S over 2^{Sig} to indicate the signal set in the configuration. A *region formula* F (the word "region" is borrowed from [1]) is a Boolean combination of formulas in the following forms: $\#(a) > n$, $\#(a) = n$, $\#(a) < n$, $S = sig$, where $a \in \Sigma, n \in \mathbf{N}$, and $sig \subseteq Sig$. Region-CTL formulas f are defined using the following grammar:

$f ::= F \mid f \wedge f \mid f \vee f \mid \neg f \mid \exists \circ f \mid \forall \circ f \mid f \, \exists \mathcal{U} \, f \mid f \, \forall \mathcal{U} \, f$, where F is a region formula. In particular, the eventuality operator $\exists \diamond \, f$ is the shorthand of $true \, \exists \mathcal{U} \, f$, and, its dual $\forall \square \, f$ is simply $\neg \exists \diamond \, \neg f$. We use Region-CTL$^\circ$ to denote a subset of the Region-CTL, where formulas are defined with: $f ::= F \mid f \wedge f \mid f \vee f \mid \neg f \mid \exists \circ f \mid \forall \circ f \mid \exists \diamond \, f \mid \forall \square \, f$, where F is a region formula. Each f is interpreted as a set $[f]$ of configurations that satisfy f, as follows:

- $[F]$ is the set of configurations that satisfy the region formula F;
- $[f_1 \wedge f_2]$ is $[f_1] \cap [f_2]$; $[f_1 \vee f_2]$ is $[f_1] \cup [f_2]$; $[\neg f_1]$ is the complement of $[f_1]$;
- $[\exists \circ f_1]$ is the set of configurations C_1 such that, for some execution $C_1 \to_M C_2 \to_M \cdots$, we have $C_2 \in [f_1]$;
- $[\forall \circ f_1]$ is the set of configurations C_1 such that, for any execution $C_1 \to_M C_2 \to_M \cdots$, we have $C_2 \in [f_1]$;
- $[f_1 \, \exists \mathcal{U} \, f_2]$ is the set of configurations C_1 such that, for some execution $C_1 \to_M C_2 \to_M \cdots$, we have $C_1, \cdots, C_n$ are all in $[f_1]$ and C_{n+1} is in $[f_2]$, for some n;
- $[f_1 \, \forall \mathcal{U} \, f_2]$ is the set of configurations C_1 such that, for any execution $C_1 \to_M C_2 \to_M \cdots$, we have $C_1, \cdots, C_n$ are all in $[f_1]$ and C_{n+1} is in $[f_2]$, for some n.

Below, we use P to denote a Boolean combination of Presburger formulas over the $\#(a)$'s and formulas in the form of $S = sig$, where $sig \subseteq Sig$. The Region-CTL model-checking problem for non-cooperative signaling systems is to answer the following question:

Given: a non-cooperative signaling system M, a Region-CTL formula f, and a Presburger formula P,

Question: Does every configuration satisfying P also satisfy f?

It is known that the Region-CTL model-checking problem for non-cooperative P systems with rules $a \to b$ is undecidable [8]. From this result, one can show that the Region-CTL model-checking problem for non-cooperative signaling systems is undecidable as well.

Theorem 3. *The Region-CTL model-checking problem for non-cooperative signaling systems is undecidable.*

In contrast to Theorem 3, the subset, Region-CTL$^\circ$, of Region-CTL is decidable for non-cooperative signaling systems:

Theorem 4. *The Region-CTL$^\circ$ model-checking problem for non-cooperative signaling systems is decidable.*

Using Theorem 4, the following example property can be automatically verified for a non-cooperative signaling system M:

"From every configuration satisfying $\#_a - \#_b < 6$, M has some execution that first reaches a configuration with $\#_a > 15$ and then reaches a halting configuration containing the signal s_1 and with $\#_b < 16$."

Notice that, above, "halting configurations" (i.e., none of the objects is enabled) form a finite union of upper-closed sets.

4 Presburger Reachability

Let M be a non-cooperative signaling system and C_{init} be a given initial configuration. In this section, we are going to investigate a stronger form of reachability problems. As we have mentioned earlier, a multiset α (over alphabet Σ with k symbols) of objects can be represented as a vector in $\mathbf{N}^k$. Let $P(x_1, \cdots, x_k)$ be a Presburger formula over k nonnegative integer variables $x_1, \cdots, x_k$. The multiset α *satisfies* P if $P(\alpha)$ holds. A configuration (S, α) of the non-cooperative signaling system M satisfies P if α satisfies P. An *equality* is a Presburger formula in the form of $x_i = x_j$, for some $1 \leq i, j \leq k$. An *equality formula*, which is a special form of Presburger formulas, is a conjunction of a number of equalities. The *Presburger-reachability* problem is to decide whether a non-cooperative signaling system has a reachable configuration satisfying a given Presburger formula:

Given: a non-cooperative signaling system M, an initial configuration C_{init}, and a Presburger formula P,
Question: is there a reachable configuration satisfying P?

In contrast to Theorem 2, we can show that the Presburger-reachability problem is undecidable. The undecidability holds even when M has only one signal (i.e., $|Sig| = 1$) and P is an equality formula (i.e., the *equality-reachability problem*). In fact, what we will show is a more general result that characterizes the set of reachable configurations in M satisfying P exactly as r.e. sets. Notice that, for P systems that deal with symbol objects, proofs for universality almost always use the theoretical tool through matrix grammar with appearance checking [16]. Here, we employ a new tool using Diophantine equations. Before we proceed further, we recall some known results on Diophantine equations (the Hilbert's Tenth Problem).

Let $m \in \mathbf{N}$, $Q \subseteq \mathbf{N}^m$ be a set of natural number tuples, and $E(z_1, \cdots, z_m, y_1, \cdots, y_n)$ be a Diophantine equation system. The set Q is *definable by* E if Q is exactly the solution set of $\exists y_1, \cdots, y_n . E(z_1, \cdots, z_m, y_1, \cdots, y_n)$; i.e., $Q = \{(z_1, \cdots, z_m) : E(z_1, \cdots, z_m, y_1, \cdots, y_n) \text{ holds for some } y_1, \cdots, y_n\}$. An *atomic* Diophantine equation is in one of the following three forms: $z = xy + \frac{1}{2}x(x+1)$, $z = x + y$, $z = 1$, where x, y, z are three distinct variables over $\mathbf{N}$. A conjunction of these atomic equations is called a Diophantine equation system of atomic Diophantine equations. It is well known that Q is r.e. iff Q is definable by some Diophantine equation system [17]. From here, it is not hard to show the following:

Lemma 1. *For any set $Q \subseteq \mathbf{N}^m$, Q is r.e. iff Q is definable by a Diophantine equation system of atomic Diophantine equations.*

We now build a relationship between Diophantine equations and non-cooperative signaling systems. Recall that Q is a subset of $\mathbf{N}^m$. We say that Q is (M, C_{init}, P)-*definable* if there are designated symbols $Z_1, \cdots, Z_m$ in M such that, for any numbers $\#(Z_1), \cdots, \#(Z_m)$,

> $(\#(Z_1), \cdots, \#(Z_m))$ is in Q iff there is a reachable configuration from C_{init} in M satisfying P and, for each i, the number of Z_i-objects in the configuration is $\#(Z_i)$.

When P is $true$ and C_{init} is understood, we simply say that Q is definable by M. The non-cooperative signaling system M is *lazy* if, for any reachable configuration and any number n, if the configuration is reachable from C_{init} in n maximally parallel moves, then it is reachable in t maximally parallel moves for any $t \geq n$. We first show that solutions to each atomic Diophantine equation can be defined with a lazy non-cooperative signaling system M with only one signal.

Lemma 2. *The solution set to each atomic Diophantine equation is definable by some lazy non-cooperative signaling system M (starting from some C_{init}) with only one signal.*

Now, we can show the following characterization.

Theorem 5. *For any set $Q \subseteq \mathbf{N}^m$, Q is r.e. iff Q is (M, C_{init}, P)-definable for some non-cooperative signaling system M with one signal, some configuration C_{init}, and some equality formula P.*

From Theorem 5, we immediately have

Theorem 6. *The equality-reachability problem for non-cooperative signaling systems with only one signal is undecidable. Therefore, the Presburger-reachability problem for non-cooperative signaling systems is undecidable as well.*

All the decidable/undecidable results presented so far can be generalized to the case when non-cooperative signaling systems are augmented with rules in the following forms: $s, a \rightarrow s', v$, where v is a multiset. From now on, we let non-cooperative signaling systems contain these rules by default.

The results in Theorem 5 and Theorem 6 can be used to obtain a new result on non-cooperative P systems $\hat{M}$ where $\hat{M}$ has only one membrane and each rule is in the form of $a \rightarrow v$, where v is a multiset. Notice that $\hat{M}$ is very similar to a non-cooperative signaling system M with only one signal. Indeed, one can easily show that they are effectively equivalent in the following sense:

Lemma 3. *For any set $Q \subseteq \mathbf{N}^m$, Q is definable by some non-cooperative P system $\hat{M}$ iff Q is definable by some non-cooperative signaling system M with only one signal.*

It is known that $\hat{M}$ is not a universal P system model; multisets generated from $\hat{M}$ form the Parikh map of an ET0L language [15]. We now augment $\hat{M}$ with a *Presburger tester* that, nondeterministically at some maximally parallel move during a run of $\hat{M}$, tests (*for only once*) whether the current multiset satisfies a given Presburger formula P. When P is an equality formula, the tester is called an *equality tester*. If yes, the tester outputs the multiset and $\hat{M}$ shuts down. Otherwise, $\hat{M}$ crashes (with no output). Let $X_1, \cdots, X_m$ be designated symbols in $\hat{M}$. We say that $Q \subseteq \mathbf{N}^m$ is *output-definable* by $\hat{M}$ if Q is exactly the set of tuples $(\#(X_1), \cdots, \#(X_m))$ in the output multisets. Directly from Lemma 3 and Theorem 5, one can show that non-cooperative P systems (as well as non-cooperative signaling systems with only one signal) with an equality tester are universal:

Theorem 7. *For any set $Q \subseteq \mathbf{N}^m$, Q is r.e. iff Q is the output-definable by a non-cooperative P system (as well as a non-cooperative signaling system with only one signal) with an equality (and hence Presburger) tester.*

Hence,

Corollary 1. *The equality-reachability problem for non-cooperative P systems is undecidable. Therefore, the Presburger-reachability problem is undecidable as well.*

With the current technology, it might be difficult to implement the equality tester device to achieve the universality, which requires, e.g., external multiset evaluation during an almost instantaneous chemical reaction process. As we already know, a more natural way to perform the evaluation is to wait until the system *halts*; i.e., none of the objects in the current configuration is enabled. In this way, one can similarly formulate the halting-definability and the Presburger/equality-halting-reachability problems for non-cooperative signaling systems as well as for non-cooperative P systems, which concern halting and reachable configurations (instead of reachable configurations). We first show that non-cooperative signaling systems with only one signal has semilinear halting-definable reachability sets. This result essentially tells us that the number of signals matters, as far as halting configurations are considered: non-cooperative signaling systems with multiple signals are strictly stronger than non-cooperative signaling systems with only one signal (as well as non-cooperative P systems). This is because a non-semilinear set like $\{(n, 2^n) : n \geq 0\}$ can be easily halting-definable by a non-cooperative signaling system.

Theorem 8. *For any $Q \subseteq \mathbf{N}^m$, Q is a semilinear set iff Q is halting-definable by a non-cooperative signaling system with only one signal (as well as by a non-cooperative P system).*

One can similarly augment $\hat{M}$ as well as $\check{M}$ with a Presburger tester but only test and output when a halting configuration is reached; i.e., a *Presburger halting tester*. The following result shows that non-cooperative signaling systems with only one signal and with a Presburger halting tester are not universal, while non-cooperative signaling systems with two signals and with an equality halting tester are universal. That is, again, the number of signals matters.

Theorem 9. *For any $Q \subseteq \mathbf{N}^m$, (1). Q is a semilinear set iff Q is output-definable by a non-cooperative signaling system with only one signal (as well as a non-cooperative P system) and with a Presburger halting tester. (2). Q is r.e. iff Q is output-definable by a non-cooperative signaling system with two signals and with an equality (and hence Presburger) halting tester.*

From Theorem 9, we immediately have:

Theorem 10. *(1). The halting Presburger reachability problem for non-cooperative signaling systems with two signals is undecidable. (2). The halting Presburger reachability problem for non-cooperative signaling systems with only one signal is decidable.*

5 Discussions and Future Work

In our set-up, a signal in a non-cooperative signaling system M has unbounded strength; i.e., it can trigger an unbounded number of instances of an enabled rule. If we restrict the strength of each signal in M to be B (where B is a constant), the resulting M is called a B-bounded non-cooperative signaling system. A move in such M is still maximally parallel. However, each signal can fire at most B instances of rules. From Theorem 2, we know that (unbounded) non-cooperative signaling systems are not universal. In contrast to this fact, we can show that bounded non-cooperative signaling systems are universal. The universality holds even when $B = 2$. The case for $B = 1$ is open.

There is an intimate relationship between some classes of P systems and VAS (vector addition systems, or, equivalently, Petri nets)[13, 14]. Though non-cooperative signaling systems as well as VAS are not universal, they are incomparable in terms of the computing power. This is because, the Presburger-reachability problem of VAS is decidable [9] while, as we have shown, the same problem for non-cooperative signaling systems is undecidable. On the other hand, the Pre^*-image of a non-cooperative signaling system is always upper-closed while this is not true for VAS.

In the definition of a non-cooperative signaling system, a rule is in the form of $s, a \rightarrow s', v$, where s and s' are signals. Now, we generalize the definition by allowing rules in the form of $S, a \rightarrow S', v$ where S and S' are sets of signals (instead of signals). The maximally parallel semantics of the rules can be defined similarly. The differences are that the rule is enabled when every signal in S is in the current configuration and, after the rule is fired, every signal in S' is emitted. Hence, the rule now is triggered by exactly all of the signals in S. Such a rule is called a *multi-signal rule*. Let M be such a non-cooperative signaling system with multi-signal rules. The proof of Theorem 1 can be adapted easily for such an M. Therefore, Theorem 2 and Theorem 4 still hold for non-cooperative signaling system with multi-signal rules. In fact, the results can be further generalized as follows.

Our study of non-cooperative signaling system was restricted to one membrane. We can generalize the model to work on multiple membranes (as in the P system), where each membrane has a set of rules, and in each rule $S, a \rightarrow S', v$ (we are using multi-signal rules) we specify the "target" membranes where each object in v as well as each signal in S' are transported to. Notice that we do not use priority rules nor membrane dissolving rules. We call this generalized model as a multimembrane non-cooperative signaling system with multi-signal rules. Observe that multimembranes can be equivalently collapsed into one membrane through properly renaming (signal and object) symbols in a membrane. That is, each membrane is associated with a distinguished set of symbols. Of course, in doing so, the number of distinct symbols and signals in the reduced one-membrane system will increase as a function of the number of membranes in the original system. Therefore, Theorem 2 and Theorem 4 can be further generalized to multimembranes non-cooperative signaling systems with multi-signal rules.

It is known that there are nonuniversal P systems where the number of membranes induces an infinite hierarchy in terms of computing power [12]. However, the above generalization says that the hierarchy collapses for non-cooperative signaling systems. Is there a hierarchy in terms of the number of membranes for a restricted and nonuniversal form of signaling systems (which is stronger than non-cooperative signaling systems)? We might also ask whether for one-membrane signaling systems, there is a hierarchy in terms of the numbers of symbols and signals used (since the conversion described above from multimembrane to one membrane increases the number of symbols and signals). As defined, a non-cooperative signaling system is a "generator" of multisets. For a given configuration C, there may be many configurations C' that satisfy $C \rightarrow_M C'$. Hence, a (maximally parallel) move is nondeterministic. Can we define an appropriate model of non-cooperative signaling system e.g., an "acceptor" of multisets (rather than a generator) such that the next move is unique (i.e., deterministic)? Deterministic P systems have been found to have some nice properties [11].

References

1. R. Alur and D. L. Dill. A theory of timed automata. *TCS*, 126(2):183–235, April 1994.
2. Ioan I. Ardelean, Matteo Cavaliere, and Dragos Sburlan. Computing using signals: From cells to P Systems. In *Second Brainstorming Week on Membrane Computing, Sevilla, Spain, February 2-7 2004*, pages 60–73, Sevilla, Spain, February 2-7 2004.
3. W. M. Becker, L. J. Kleinsmith, and J. Hardin. *The World of the Cell (5th Edition)*. Benjamin Cummings (San Francisco), 2003.
4. P. Bottoni, C. Martin-Vide, Gh. Paun, and G. Rozenberg. Membrane systems with promoters/inhibitors. *Acta Informatica*, 38(10):695–720, 2002.
5. T. Bultan, R. Gerber, and W. Pugh. Model-checking concurrent systems with unbounded integer variables: symbolic representations, approximations, and experimental results. *TOPLAS*, 21(4):747–789, 1999.
6. E. M. Clarke, E. A. Emerson, and A. P. Sistla. Automatic verification of finite-state concurrent systems using temporal logic specifications. *TOPLAS*, 8(2):244–263, April 1986.
7. E. M. Clarke, O. Grumberg, and D. A. Peled. *Model Checking*. MIT Press, 1999.
8. Z. Dang, O. H. Ibarra, C. Li, and G. Xie. On model-checking of P systems. 2005 (submitted).
9. Javier Esparza. Decidability and complexity of Petri net problems - an introduction. In *Petri Nets*, pages 374–428, 1996.
10. R. Freund, L. Kari, M. Oswald, and P. Sosik. Computationally universal P systems without priorities: two catalysts are sufficient. Available at *http://psystems.disco.unimib.it*, 2003.
11. R. Freund and Gh. Paun. On deterministic P systems. Available at http://psystems.disco.unimib.it. 2003.
12. O. H. Ibarra. The number of membranes matters. In *WMC'03*, volume 2933 of Lecture Notes in Computer Science, pages 218–231, Springer, 2004.
13. O. H. Ibarra, Z. Dang, and O. Egecioglu. Catalytic P systems, semilinear sets, and vector addition systems. *Theoretical Computer Science*, 312(2-3):379–399, 2004.
14. O. H. Ibarra, H. Yen, and Z. Dang. The power of maximal parallelism in P systems. In *DLT'04*, volume 3340 of Lecture Notes in Computer Science, pages 212–224, Springer, 2004.
15. M. Ito, C. Martin-Vide, and Gh. Paun. A characterization of Parikh sets of ET0L languages in terms of P systems. In *Words, Semigroups, and Transducers*, pages 239–254. World Scientific.
16. C. Martin-Vide and Gh. Paun. Computing with membranes (P systems): universality results. In *MCU'01*, volume 2055 of *LNCS*, pages 82–101. Springer, 2001.
17. Y. V. Matiyasevich. *Hilbert's Tenth Problem*. MIT Press, 1993.
18. Gh. Paun. Computing with membranes. *JCSS*, 61(1):108–143, 2000.
19. Gh. Paun. *Membrane Computing: An Introduction*. Springer, 2002.
20. Gh. Paun and G. Rozenberg. A guide to membrane computing. *TCS*, 287(1):73–100, 2002.
21. P. Sosik. P systems versus register machines: two universality proofs. In *WMC'02*, volume 2597 of *LNCS*, pages 371–382, Springer, 2003.
22. P. Sosik and R. Freund. P systems without priorities are computationally universal. In *WMC'02*, volume 2597 of *LNCS*, pages 400–409. Springer, 2003.

Author Index

Abadi, Martín 664
Abdulla, Parosh Aziz 1089
Alstrup, Stephen 78
Alur, Rajeev 1102
Ambühl, Christoph 1139
Asarin, Eugene 1031
Atserias, Albert 1437
Avin, Chen 677

Bădoiu, Mihai 866
Badouel, Eric 742
Baldamus, Michael 1202
Bao, Feng 459
Baudet, Mathieu 652
Béal, Marie-Pierre 397
Beame, Paul 1176
Beaudry, Martin 421
Berenbrink, Petra 1386
Bille, Philip 66
Borissov, Yuri 324
Borodin, Allan 943
Bradley, Aaron R. 1349
Braeken, An 324
Brodal, Gerth Stølting 576

Cary, Matthew 1412
Cashman, David 943
Castagna, Giuseppe 30
Catalano, Dario 298
Chan, Wun-Tat 614
Chatterjee, Krishnendu 878
Chattopadhyay, Arkadev 994
Chenou, Jules 742
Chevalier, Yannick 639
Chiniforooshan, Ehsan 179
Chlebus, Bogdan S. 347
Cohen, Reuven 335
Collins, Pieter 1031
Conforti, Giovanni 766
Cortier, Véronique 652
Czeizler, Eugen 410
Czumaj, Artur 866

Dalla Preda, Mila 1325
Dang, Zhe 1462

Datta, Anupam 16
Dawar, Anuj 1437
de Alfaro, Luca 878
De Nicola, Rocco 1226
de Wolf, Ronald 1424
Delzanno, Giorgio 1239
Deneux, Johann 1089
Derek, Ante 16
Di Crescenzo, Giovanni 216, 816
Diehl, Scott 982
Dietzfelbinger, Martin 166
Dimitrov, Nedialko B. 702
Donato, Debora 717
Droste, Manfred 513

Efthymiou, Charilaos 690
Eisenbrand, Friedrich 1151
Elbassioni, Khaled 1115
Elias, Isaac 1263
Englert, Matthias 627
Epstein, Leah 602
Ercal, Gunes 677
Etessami, Kousha 891

Fagerberg, Rolf 576
Farach-Colton, Martin 1251
Farzan, Arash 179
Fiala, Jiří, 360
Fischlin, Marc 779
Fishkin, Aleksei V. 1115
Fokkink, Wan 755
Fomin, Fedor V. 191
Ford, Jeff 1163
Fraigniaud, Pierre 335
Franceschini, Gianni 90
Friedetzky, Tom 1386
Frisch, Alain 30

Gabbrielli, Maurizio 1239
Gairing, Martin 51, 828
Gál, Anna 1163
Galindo, David 791
Gastin, Paul 513
Gentry, Craig 803
Giacobazzi, Roberto 1325

Gąsieniec, Leszek 347
Golovach, Petr A. 360
Gorla, Daniele 1226
Gørtz, Inge Li 66, 78
Grandoni, Fabrizio 191, 1151
Grohe, Martin 1076, 1437
Grossi, Roberto 90
Gu, Qian-Ping 373
Guillou, Goulven 742
Gupta, Anupam 1051

Hajiaghayi, MohammadTaghi 853
Haneda, Mitsuhiro 539
Hansen, Kristoffer Arnsfelt 994
Hast, Gustav 956
Henzinger, Thomas A. 878
Hermida, Claudio 1018
Hopper, Nicholas 311
Horvitz, Omer 128
Hromkovič, Juraj 385

Ibarra, Oscar H. 1462
Ilcinkas, David 335
Indyk, Piotr 866

Jain, Kamal 853
Jampala, Hema 563

Kaligosi, Kanela 103
Kao, Ming-Yang 1275
Karakostas, George 1043
Kari, Jarkko 410
Katz, Jonathan 128
Kavitha, Telikepalli 273
Kawazoe, Mitsuru 539
Kayal, Neeraj 551
Kempe, David 1127
Kiayias, Aggelos 216
Kiltz, Eike 434
Kleinberg, Jon 1127
Knapik, Teodor 1450
Koch, Christoph 1076
Koiran, Pascal 1287
Könemann, Jochen 930
Korman, Amos 335
Kovács, Annamária 840
Kowalski, Dariusz R. 347
Kowaluk, Miroslaw 241
Kratsch, Dieter 191
Kratochvíl, Jan 360
Kremer, Steve 652

Kumar, Amit 1374
Kumar, Viraj 1102
Kursawe, Klaus 204

Lagergren, Jens 1263
Laird, James 904
Lam, Tak-Wah 614
Landau, Gad M. 1251
Lemieux, François 421
Leonardi, Stefano 717, 930
Levy, Meital 602
Li, Cheng 1462
Libkin, Leonid 35
Lin, Henry 497
Lingas, Andrzej 241
Lipmaa, Helger 459
Lombardy, Sylvain 397
Lücking, Thomas 51

Macedonio, Damiano 766
Madhusudan, P. 1102
Magen, Avner 943
Magniez, Frédéric 1312
Makkai, Michael 1018
Manna, Zohar 1349
Martin, Russell 1386
Maurer, Ueli 472
Mehlhorn, Kurt 103
Mirzazadeh, Mehdi 179
Mislove, Michael W. 1006
Mitchell, John C. 16
Mityagin, Anton 434
Monien, Burkhard 51, 828
Moran, Tal 285
Moruz, Gabriel 576
Mousavi, Mohammad Reza 1214
Munro, J. Ian 103
Murawski, Andrezej 917
Mustafa, Nabil H. 1115

Nain, Sumit 755
Naor, Moni 285
Nayak, Ashwin 1312
Nesme, Vincent 1287
Nikova, Svetla 324
Nilsson, Bengt J. 1362
Niwiński, Damian 1450

Ong, Chin-Hao Luke 917
Oriolo, Gianpaolo 1151
Ouaknine, Joël 1089

Author Index

Pál, Martin 1051
Pătraşcu, Corina E. 969
Pătraşcu, Mihai 969
Panjwani, Saurabh 434
Parrow, Joachim 1202
Peleg, David 335
Pemmaraju, Sriram V. 1064
Persiano, Giuseppe 228
Pitassi, Toniann 1176
Plaxton, C. Greg 702
Portier, Natacha 1287
Pous, Damien 730
Preneel, Bart 324
Pugliese, Rosario 1226

Radhakrishnan, Jaikumar 1399
Radzik, Tomasz 347
Raghavan, Barath 434
Raman, Rajiv 1064
Ramzan, Zulfikar 803
Rauhe, Theis 78
Reniers, Michel A. 1214
Reus, Bernhard 1337
Roditty, Liam 249, 261
Rötteler, Martin 1399
Roughgarden, Tim 497
Rudra, Atri 1412
Rusinowitch, Michaël 639

Sabharwal, Ashish 1412
Sabharwal, Yogish 1374
Sahinalp, S. Cenk 1251
Sakarovitch, Jacques 397
Sanders, Peter 103
Sanghi, Manan 1275
Sassone, Vladimiro 766
Schäfer, Guido 930
Schnitger, Georg 385
Schweikardt, Nicole 1076
Schweller, Robert 1275
Segerlind, Nathan 1176
Sen, Pranab 1399
Sen, Sandeep 1374
Shmatikov, Vitaly 16
Shoup, Victor 204
Sipma Henny B. 1349
Sitters, René, 1115
Sjödin, Johan 472
Skutella, Martin 1151
Sohler, Christian 866

Špalek, Robert 1299
Spirakis, Paul G. 690
Streicher, Thomas 1337
Szegedy, Mario 1299

Takahashi, Tetsuya 539
Talwar, Kunal 853
Tamaki, Hisao 373
Tamassia, Roberto 153
Tardos, Éva 497, 1127
Tesson, Pascal 526
Thorup, Mikkel 78, 261
Thérien, Denis 421, 526
Tiemann, Karsten 51
Triandopoulos, Nikos 153
Trolin, Mårten 446
Tsaparas, Panayiotis 717
Tsur, Dekel 1251
Turuani, Mathieu 16

Urzyczyn, Paweł 1450

Valiant, Leslie G. 1
van Breugel, Franck 1018
van Melkebeek, Dieter 982
van Zwam, Stefan 930
Victor, Björn 1202
Visconti, Ivan 228, 298, 816
Viswanathan, Mahesh 1102

Walkover, Asher 497
Walukiewicz, Igor 917, 1450
Wang, Guilin 459
Warinschi, Bogdan 664
Wee, Hoeteck 140
Wegener, Ingo 589
Wehner, Stephanie 1424
Weidling, Christoph 166
Westermann, Matthias 627
Wikström, Douglas 446, 1180
Woclaw, Andreas 828
Wong, Prudence W.H. 614
Worrell, James 1018, 1089

Yannakakis, Mihalis 891
Yen, Hsu-Chun 1462

Zeh, Norbert 563
Zhang, Li 485
Zimand, Marius 115
Zwick, Uri 78, 249, 261

Lecture Notes in Computer Science

For information about Vols. 1–3476

please contact your bookseller or Springer

Vol. 3580: L. Caires, G.F. Italiano, L. Monteiro, C. Palamidessi, M. Yung (Eds.), Automata, Languages and Programming. XXV, 1477 pages. 2005.

Vol. 3578: M. Gallagher, J. Hogan, F. Maire (Eds.), Intelligent Data Engineering and Automated Learning - IDEAL 2005. XVI, 599 pages. 2005.

Vol. 3576: K. Etessami, S.K. Rajamani (Eds.), Computer Aided Verification. XV, 564 pages. 2005.

Vol. 3574: C. Boyd, J.M. González Nieto (Eds.), Information Security and Privacy. XIII, 586 pages. 2005.

Vol. 3573: S. Etalle (Ed.), Logic Based Program Synthesis and Transformation. VIII, 279 pages. 2005.

Vol. 3572: C. De Felice, A. Restivo (Eds.), Developments in Language Theory. XI, 409 pages. 2005.

Vol. 3570: A. S. Patrick, M. Yung (Eds.), Financial Cryptography and Data Security. XII, 376 pages. 2005.

Vol. 3569: F. Bacchus, T. Walsh (Eds.), Theory and Applications of Satisfiability Testing. XII, 492 pages. 2005.

Vol. 3567: M. Jackson, D. Nelson, S. Stirk (Eds.), Database: Enterprise, Skills and Innovation. XII, 185 pages. 2005.

Vol. 3565: G.E. Christensen, M. Sonka (Eds.), Information Processing in Medical Imaging. XXI, 777 pages. 2005.

Vol. 3562: J. Mira, J.R. Álvarez (Eds.), Artificial Intelligence and Knowledge Engineering Applications: A Bioinspired Approach, Part II. XXIV, 636 pages. 2005.

Vol. 3561: J. Mira, J.R. Álvarez (Eds.), Mechanisms, Symbols, and Models Underlying Cognition, Part I. XXIV, 532 pages. 2005.

Vol. 3560: V.K. Prasanna, S. Iyengar, P.G. Spirakis, M. Welsh (Eds.), Distributed Computing in Sensor Systems. XV, 423 pages. 2005.

Vol. 3559: P. Auer, R. Meir (Eds.), Learning Theory. XI, 692 pages. 2005. (Subseries LNAI).

Vol. 3557: H. Gilbert, H. Handschuh (Eds.), Fast Software Encryption. XI, 443 pages. 2005.

Vol. 3556: H. Baumeister, M. Marchesi, M. Holcombe (Eds.), Extreme Programming and Agile Processes in Software Engineering. XIV, 332 pages. 2005.

Vol. 3555: T. Vardanega, A. Wellings (Eds.), Reliable Software Technology – Ada-Europe 2005. XV, 273 pages. 2005.

Vol. 3554: A. Dey, B. Kokinov, D. Leake, R. Turner (Eds.), Modeling and Using Context. XIV, 572 pages. 2005. (Subseries LNAI).

Vol. 3553: T.D. Hämäläinen, A.D. Pimentel, J. Takala, S. Vassiliadis (Eds.), Embedded Computer Systems: Architectures, Modeling, and Simulation. XV, 476 pages. 2005.

Vol. 3552: H. de Meer, N. Bhatti (Eds.), Quality of Service – IWQoS 2005. XV, 400 pages. 2005.

Vol. 3551: T. Härder, W. Lehner (Eds.), Data Management in a Connected World. XIX, 371 pages. 2005.

Vol. 3548: K. Julisch, C. Kruegel (Eds.), Intrusion and Malware Detection and Vulnerability Assessment. X, 241 pages. 2005.

Vol. 3547: F. Bomarius, S. Komi-Sirviö (Eds.), Product Focused Software Process Improvement. XIII, 588 pages. 2005.

Vol. 3543: L. Kutvonen, N. Alonistioti (Eds.), Distributed Applications and Interoperable Systems. XI, 235 pages. 2005.

Vol. 3541: N.C. Oza, R. Polikar, J. Kittler, F. Roli (Eds.), Multiple Classifier Systems. XII, 430 pages. 2005.

Vol. 3540: H. Kalviainen, J. Parkkinen, A. Kaarna (Eds.), Image Analysis. XXII, 1270 pages. 2005.

Vol. 3537: A. Apostolico, M. Crochemore, K. Park (Eds.), Combinatorial Pattern Matching. XI, 444 pages. 2005.

Vol. 3536: G. Ciardo, P. Darondeau (Eds.), Applications and Theory of Petri Nets 2005. XI, 470 pages. 2005.

Vol. 3535: M. Steffen, G. Zavattaro (Eds.), Formal Methods for Open Object-Based Distributed Systems. X, 323 pages. 2005.

Vol. 3533: M. Ali, F. Esposito (Eds.), Innovations in Applied Artificial Intelligence. XX, 858 pages. 2005. (Subseries LNAI).

Vol. 3532: A. Gómez-Pérez, J. Euzenat (Eds.), The Semantic Web: Research and Applications. XV, 728 pages. 2005.

Vol. 3531: J. Ioannidis, A. Keromytis, M. Yung (Eds.), Applied Cryptography and Network Security. XI, 530 pages. 2005.

Vol. 3530: A. Prinz, R. Reed, J. Reed (Eds.), SDL 2005: Model Driven. XI, 361 pages. 2005.

Vol. 3528: P.S. Szczepaniak, J. Kacprzyk, A. Niewiadomski (Eds.), Advances in Web Intelligence. XVII, 513 pages. 2005. (Subseries LNAI).

Vol. 3527: R. Morrison, F. Oquendo (Eds.), Software Architecture. XII, 263 pages. 2005.

Vol. 3526: S.B. Cooper, B. Löwe, L. Torenvliet (Eds.), New Computational Paradigms. XVII, 574 pages. 2005.

Vol. 3525: A.E. Abdallah, C.B. Jones, J.W. Sanders (Eds.), Communicating Sequential Processes. XIV, 321 pages. 2005.

Vol. 3524: R. Barták, M. Milano (Eds.), Integration of AI and OR Techniques in Constraint Programming for Combinatorial Optimization Problems. XI, 320 pages. 2005.

Vol. 3523: J.S. Marques, N. Pérez de la Blanca, P. Pina (Eds.), Pattern Recognition and Image Analysis, Part II. XXVI, 733 pages. 2005.

Vol. 3522: J.S. Marques, N. Pérez de la Blanca, P. Pina (Eds.), Pattern Recognition and Image Analysis, Part I. XXVI, 703 pages. 2005.

Vol. 3521: N. Megiddo, Y. Xu, B. Zhu (Eds.), Algorithmic Applications in Management. XIII, 484 pages. 2005.

Vol. 3520: O. Pastor, J. Falcão e Cunha (Eds.), Advanced Information Systems Engineering. XVI, 584 pages. 2005.

Vol. 3519: H. Li, P. J. Olver, G. Sommer (Eds.), Computer Algebra and Geometric Algebra with Applications. IX, 449 pages. 2005.

Vol. 3518: T.B. Ho, D. Cheung, H. Liu (Eds.), Advances in Knowledge Discovery and Data Mining. XXI, 864 pages. 2005. (Subseries LNAI).

Vol. 3517: H.S. Baird, D.P. Lopresti (Eds.), Human Interactive Proofs. IX, 143 pages. 2005.

Vol. 3516: V.S. Sunderam, G.D.v. Albada, P.M.A. Sloot, J.J. Dongarra (Eds.), Computational Science – ICCS 2005, Part III. LXIII, 1143 pages. 2005.

Vol. 3515: V.S. Sunderam, G.D.v. Albada, P.M.A. Sloot, J.J. Dongarra (Eds.), Computational Science – ICCS 2005, Part II. LXIII, 1101 pages. 2005.

Vol. 3514: V.S. Sunderam, G.D.v. Albada, P.M.A. Sloot, J.J. Dongarra (Eds.), Computational Science – ICCS 2005, Part I. LXIII, 1089 pages. 2005.

Vol. 3513: A. Montoyo, R. Muñoz, E. Métais (Eds.), Natural Language Processing and Information Systems. XII, 408 pages. 2005.

Vol. 3512: J. Cabestany, A. Prieto, F. Sandoval (Eds.), Computational Intelligence and Bioinspired Systems. XXV, 1260 pages. 2005.

Vol. 3510: T. Braun, G. Carle, Y. Koucheryavy, V. Tsaoussidis (Eds.), Wired/Wireless Internet Communications. XIV, 366 pages. 2005.

Vol. 3509: M. Jünger, V. Kaibel (Eds.), Integer Programming and Combinatorial Optimization. XI, 484 pages. 2005.

Vol. 3508: P. Bresciani, P. Giorgini, B. Henderson-Sellers, G. Low, M. Winikoff (Eds.), Agent-Oriented Information Systems II. X, 227 pages. 2005. (Subseries LNAI).

Vol. 3507: F. Crestani, I. Ruthven (Eds.), Information Context: Nature, Impact, and Role. XIII, 253 pages. 2005.

Vol. 3506: C. Park, S. Chee (Eds.), Information Security and Cryptology – ICISC 2004. XIV, 490 pages. 2005.

Vol. 3505: V. Gorodetsky, J. Liu, V. A. Skormin (Eds.), Autonomous Intelligent Systems: Agents and Data Mining. XIII, 303 pages. 2005. (Subseries LNAI).

Vol. 3504: A.F. Frangi, P.I. Radeva, A. Santos, M. Hernandez (Eds.), Functional Imaging and Modeling of the Heart. XV, 489 pages. 2005.

Vol. 3503: S.E. Nikoletseas (Ed.), Experimental and Efficient Algorithms. XV, 624 pages. 2005.

Vol. 3502: F. Khendek, R. Dssouli (Eds.), Testing of Communicating Systems. X, 381 pages. 2005.

Vol. 3501: B. Kégl, G. Lapalme (Eds.), Advances in Artificial Intelligence. XV, 458 pages. 2005. (Subseries LNAI).

Vol. 3500: S. Miyano, J. Mesirov, S. Kasif, S. Istrail, P. Pevzner, M. Waterman (Eds.), Research in Computational Molecular Biology. XVII, 632 pages. 2005. (Subseries LNBI).

Vol. 3499: A. Pelc, M. Raynal (Eds.), Structural Information and Communication Complexity. X, 323 pages. 2005.

Vol. 3498: J. Wang, X. Liao, Z. Yi (Eds.), Advances in Neural Networks – ISNN 2005, Part III. XLIX, 1077 pages. 2005.

Vol. 3497: J. Wang, X. Liao, Z. Yi (Eds.), Advances in Neural Networks – ISNN 2005, Part II. XLIX, 947 pages. 2005.

Vol. 3496: J. Wang, X. Liao, Z. Yi (Eds.), Advances in Neural Networks – ISNN 2005, Part II. L, 1055 pages. 2005.

Vol. 3495: P. Kantor, G. Muresan, F. Roberts, D.D. Zeng, F.-Y. Wang, H. Chen, R.C. Merkle (Eds.), Intelligence and Security Informatics. XVIII, 674 pages. 2005.

Vol. 3494: R. Cramer (Ed.), Advances in Cryptology – EUROCRYPT 2005. XIV, 576 pages. 2005.

Vol. 3493: N. Fuhr, M. Lalmas, S. Malik, Z. Szlávik (Eds.), Advances in XML Information Retrieval. XI, 438 pages. 2005.

Vol. 3492: P. Blache, E. Stabler, J. Busquets, R. Moot (Eds.), Logical Aspects of Computational Linguistics. X, 363 pages. 2005. (Subseries LNAI).

Vol. 3489: G.T. Heineman, I. Crnkovic, H.W. Schmidt, J.A. Stafford, C. Szyperski, K. Wallnau (Eds.), Component-Based Software Engineering. XI, 358 pages. 2005.

Vol. 3488: M.-S. Hacid, N.V. Murray, Z.W. Raś, S. Tsumoto (Eds.), Foundations of Intelligent Systems. XIII, 700 pages. 2005. (Subseries LNAI).

Vol. 3486: T. Helleseth, D. Sarwate, H.-Y. Song, K. Yang (Eds.), Sequences and Their Applications - SETA 2004. XII, 451 pages. 2005.

Vol. 3483: O. Gervasi, M.L. Gavrilova, V. Kumar, A. Laganà, H.P. Lee, Y. Mun, D. Taniar, C.J.K. Tan (Eds.), Computational Science and Its Applications – ICCSA 2005, Part IV. LXV, 1362 pages. 2005.

Vol. 3482: O. Gervasi, M.L. Gavrilova, V. Kumar, A. Laganà, H.P. Lee, Y. Mun, D. Taniar, C.J.K. Tan (Eds.), Computational Science and Its Applications – ICCSA 2005, Part III. LXV, 1340 pages. 2005.

Vol. 3481: O. Gervasi, M.L. Gavrilova, V. Kumar, A. Laganà, H.P. Lee, Y. Mun, D. Taniar, C.J.K. Tan (Eds.), Computational Science and Its Applications – ICCSA 2005, Part II. LXV, 1316 pages. 2005.

Vol. 3480: O. Gervasi, M.L. Gavrilova, V. Kumar, A. Laganà, H.P. Lee, Y. Mun, D. Taniar, C.J.K. Tan (Eds.), Computational Science and Its Applications – ICCSA 2005, Part I. LXV, 1234 pages. 2005.

Vol. 3479: T. Strang, C. Linnhoff-Popien (Eds.), Location- and Context-Awareness. XII, 378 pages. 2005.

Vol. 3478: C. Jermann, A. Neumaier, D. Sam (Eds.), Global Optimization and Constraint Satisfaction. XIII, 193 pages. 2005.

Vol. 3477: P. Herrmann, V. Issarny, S. Shiu (Eds.), Trust Management. XII, 426 pages. 2005.